Springer
Proceedings in Physics

87

Springer
Berlin
Heidelberg
New York
Barcelona
Hong Kong
London
Milan
Paris
Singapore
Tokyo

Springer Proceedings in Physics

48 *Many-Atom Interactions in Solids*
Editors: R. M. Nieminen, M. J. Puska,
and M. J. Manninen

49 *Ultrafast Phenomena in Spectroscopy*
Editors: E. Klose and B. Wilhelmi

50 *Magnetic Properties of Low-Dimensional
Systems II: New Developments*
Editors: L. M. Falicov, F. Mejía-Lira, and
J. L. Morán-López

51 *The Physics and Chemistry of Organic Superconductors*
Editors: G. Saito and S. Kagoshima

52 *Dynamics and Patterns in Complex Fluids:
New Aspects of the Physics–Chemistry Interface*
Editors: A. Onuki and K. Kawasaki

53 *Computer Simulation Studies in Condensed-
Matter Physics III*
Editors: D. P. Landau, K. K. Mon,
and H.-B. Schüttler

54 *Polycrystalline Semiconductors II*
Editors: J. H. Werner and H. P. Strunk

55 *Nonlinear Dynamics and Quantum Phenomena
in Optical Systems*
Editors: R. Vilaseca and R. Corbalán

56 *Amorphous and Crystalline Silicon Carbide III,
and Other Group IV-IV Materials*
Editors: G. L. Harris, M. G. Spencer,
and C. Y. Yang

57 *Evolutionary Trends in the Physical Sciences*
Editors: M. Suzuki and R. Kubo

58 *New Trends in Nuclear Collective Dynamics*
Editors: Y. Abe, H. Horiuchi,
and K. Matsuyanagi

59 *Exotic Atoms in Condensed Matter*
Editors: G. Benedek and H. Schneuwly

60 *The Physics and Chemistry of Oxide Superconductors*
Editors: Y. Iye and H. Yasuoka

61 *Surface X-Ray and Neutron Scattering*
Editors: H. Zabel and I. K. Robinson

62 *Surface Science: Lectures on Basic Concepts
and Applications*
Editors: F. A. Ponce and M. Cardona

63 *Coherent Raman Spectroscopy:
Recent Advances*
Editors: G. Marowsky and V. V. Smirnov

64 *Superconducting Devices and Their Applications*
Editors: H. Koch and H. Lübbig

65 *Present and Future of High-Energy Physics*
Editors: K.-I. Aoki and M. Kobayashi

66 *The Structure and Conformation of Amphiphilic
Membranes*
Editors: R. Lipowsky, D. Richter, and K. Kremer

67 *Nonlinearity with Disorder*
Editors: F. Abdullaev, A. R. Bishop,
and S. Pnevmatikos

68 *Time-Resolved Vibrational Spectroscopy V*
Editor: H. Takahashi

69 *Evolution of Dynamical Structures in Complex
Systems*
Editors: R. Friedrich and A. Wunderlin

70 *Computational Approaches in Condensed-Matter
Physics*
Editors: S. Miyashita, M. Imada, and H. Takayama

71 *Amorphous and Crystalline Silicon Carbide IV*
Editors: C. Y. Yang, M. M. Rahman,
and G. L. Harris

72 *Computer Simulation Studies in Condensed-Matter
Physics IV*
Editors: D. P. Landau, K. K. Mon,
and H.-B. Schüttler

73 *Surface Science: Principles and Applications*
Editors: R. F. Howe, R. N. Lamb, and K. Wandelt

74 *Time-Resolved Vibrational Spectroscopy VI*
Editors: A. Lau, F. Siebert, and W. Werncke

75 *Computer Simulation Studies
in Condensed-Matter Physics V*
Editors: D. P. Landau, K. K. Mon,
and H.-B. Schüttler

76 *Computer Simulation Studies
in Condensed-Matter Physics VI*
Editors: D. P. Landau, K. K. Mon,
and H.-B. Schüttler

77 *Quantum Optics VI*
Editors: D. F. Walls and J. D. Harvey

78 *Computer Simulation Studies
in Condensed-Matter Physics VII*
Editors: D. P. Landau, K. K. Mon,
and H.-B. Schüttler

79 *Nonlinear Dynamics and Pattern Formation
in Semiconductors and Devices*
Editor: F.-J. Niedernostheide

80 *Computer Simulation Studies
in Condensed-Matter Physics VIII*
Editors: D. P. Landau, K. K. Mon,
and H.-B. Schüttler

81 *Materials and Measurements in Molecular
Electronics*
Editors: K. Kajimura and S. Kuroda

82 *Computer Simulation Studies
in Condensed-Matter Physics IX*
Editors: D. P. Landau, K. K. Mon,
and H.-B. Schüttler

83 *Computer Simulation Studies
in Condensed-Matter Physics X*
Editors: D. P. Landau, K. K. Mon,
and H.-B. Schüttler

84 *Computer Simulation Studies
in Condensed-Matter Physics XI*
Editors: D. P. Landau and
H.-B. Schüttler

85 *Computer Simulation Studies
in Condensed-Matter Physics XII*
Editors: D. P. Landau, S. P. Lewis and
H.-B. Schüttler

86 *Computer Simulation Studies
in Condensed-Matter Physics XIII*
Editors: D. P. Landau, S. P. Lewis and
H.-B. Schüttler

87 *Proceedings of the 25th International
Conference on the Physics of Semiconductors*
Editors: N. Miura, T. Ando

88 *Starburst Galaxies: Near and Far*
Editors: D. Lutz, L. Tacconi

Volumes 1–47 are listed at the end of the book

N. Miura T. Ando (Eds.)

Proceedings of the 25th International Conference on the Physics of Semiconductors

Part I

Osaka, Japan,
September 17–22, 2000

 Springer

Professor Dr. Noboru Miura
Professor Tsuneya Ando

Institute for Solid State Physics
Roppongi, Minato-ku
106-8666 Tokyo, Japan

Cataloging-in-Publication Data applied for.

Die Deutsche Bibliothek – CIP-Einheitsaufnahme

International Conference on the Physics of Semiconductors <25, 2000, Osaka>:
Proceedings of the 25th International Conference on the Physics of Semiconductors: Osaka, Japan, September 17–22, 2000 /
N. Miura; T. Ando (ed.). –
Berlin; Heidelberg; New York; Barcelona; Hong Kong; London; Milan; Paris; Singapore; Tokyo: Springer
(Springer proceedings in physics; ...)
Pt. 1.– (2001)
(Springer proceedings in physics; 87)

ISBN-13: 978-3-642-63993-7   e-ISBN-13: 978-3-642-59484-7
DOI: 10.1007/978-3-642-59484-7

Springer-Verlag Berlin Heidelberg New York
a member of BertelsmannSpringer Science+Business Media GmbH

http://www.springer.de

Typesetting: Camera ready copy from the authors/editors
Printed on acid-free paper   SPIN: 10765254   57/3141/XO - 5 4 3 2 1 0

Preface

The 25th International Conference on the Physics of Semiconductors (ICPS25) was held at the Osaka International Convention Center and the Rihga Royal Hotel, Osaka, Japan on 17–22 September, 2000. The ICPS has a long tradition that began in 1950 with the conference at Reading, counting 110 participants. The ICPS25 conference was attended by 1,048 participants from 36 countries, and, altogether, 1,038 papers were presented, including 11 plenary papers, 32 invited papers and 995 contributed papers (168 oral and 827 poster presentations). These proceedings consist of two volumes that contain most of the contributions presented at the conference.

This conference was the last ICPS in the 20th century; at this special occasion a Plenary Symposium was organized on the subject of the interface between fundamental research and device applications. One of the invited speakers at this special symposium was Prof. Z.I. Alferov of the Ioffe Physico-Technical Institute at St. Petersburg (Russia). It is an auspicious sign of continuing achievements in semiconductor physics that he was awarded the Nobel Prize for Physics soon after he gave his excellent lecture at the conference.

The proceedings contains the plenary presentations including those given in the special symposium (Sect. 1), and sections for the following subject areas:

2. Bulk and Dynamics
    2.1 Electronic Structure
    2.2 Phonons
    2.3 Optical Properties and Excitons
    2.4 Transport and Localization
    2.5 Carrier Relaxation and Ultrafast Phenomena
    2.6 Phase Transition and Ordering
    2.7 Magnetic and Semimagnetic Semiconductors
3. Growth, Surfaces, and Interfaces
    3.1 Atomic Structure
    3.2 Electronic Structure
    3.3 Adsorption and Surface Reactions
    3.4 Bulk Growth and Thin Film Epitaxy
    3.5 Growth of Self-Assembled Dots
    3.6 Semiconductor–Nonsemiconductor Interfaces
    3.7 STM, AFM, and Near-Field Studies of Surfaces and Interfaces
4. Heterostructures and Superlattices: Optical
    4.1 Structural Properties
    4.2 Electronic Structures
    4.3 Magnetic Superlattices
    4.4 Interband Transitions and Excitons
    4.5 Nonlinear Optical Studies
    4.6 Carrier Dynamics and Relaxation
    4.7 Microcavities
    4.8 Near-Field Studies
    4.9 Intersubband Transitions
5. Heterostructures and Superlattices: Transport
    5.1 Transport, Magneto-Transport, and Coherent Effects
    5.2 Magnetic Superlattices
    5.3 Tunneling and Resonant Tunneling
    5.4 Hopping, Localization
    5.5 Phonons, Plasmons
    5.6 High-Frequency Transport, Microwaves, Noise, Short Pulses
6. 2D Systems with High Magnetic Fields
    6.1 Magneto-Transport, Coherent Effects
    6.2 High-Frequency Transport and Noise
    6.3 Quantum Hall Effect
    6.4 Skymions
    6.5 Correlation Effect
    6.6 Phonons, Plasmons
7. One- and Zero-Dimensional Systems
    7.1 Transport in Wires and Dots
    7.2 Optical Studies of Wires and Dots
    7.3 Mesoscopic Systems
    7.4 Surface Probe Studies
8. Defects and Impurities
    8.1 Shallow Impurities
    8.2 Deep Impurities and Defects
    8.3 Magnetic Impurities
    8.4 Structural Defects
    8.5 Amorphous Silicon
9. Wide Band-Gap Materials
    9.1 Growth
    9.2 Optical Studies
    9.3 Transport Studies
    9.4 Laser Diodes
10. New Materials and New Concepts
    10.1 Novel Semiconductors
    10.2 Novel Semiconductor Devices
    10.3 Clusters
    10.4 Nanotechnology and Micromachining
11. Technology and Devices
    11.1 Silicon Devices
    11.2 III–V Devices
    11.3 Optical Devices
    11.4 Mesoscopic Devices

The invited papers are marked by the word "invited" in the table of contents.

The conference was greatly honored by the presence of Their Imperial Highnesses Prince and Princess Takamado who attended the oral sessions on the first day of the conference. At the opening ceremony of the conference, Prince Takamado gave a gracious speech; we are honored and pleased to include this in the proceedings.

Osaka, November, 2000

Noboru Miura
Tsuneya Ando

Sponsors and Supporters

Sponsors

The International Union of Pure and Applied Physics
The Science Council of Japan
The Physical Society of Japan
The Japan Society of Applied Physics

Patronage

Osaka Prefecture
Osaka City

Institutional Supporters

Commemorative Association
    for the Japan World Exposition (Osaka, 1970)
Hoso-Bunka Foundation
Inoue Foundation for Science
The Kao Foundation for Arts and Science
The Murata Science Foundation
Nippon Sheet Glass Foundation for Materials Science
    and Engineering
The Ogasawara Foundation for the Promotion
    of Science and Engineering
Osaka Convention Bureau
Research Foundation for Opto-Science and Technology

Industrial Supporters

Advantest Corporation
Asahi Breweries, Ltd.
Canon Inc.
Chubu Electric Power Co., Inc.
Chugoku Electric Power Co., Inc.
Denso Corporation
Fujistu Limited
Fujitsu AMD Semiconductor, Ltd.
Hitachi Cable, Ltd.
Hitachi Chemical Co., Ltd.
Hitachi Metals, Ltd.
Hitachi, Ltd.
Hokkaido Electric Power Co., Inc.
Hokuriku Electric Power Co., Inc.
IBM Japan, Ltd.
Kansai Electric Power Co., Inc.
KDD Corporation
Kirin Brewery Co., Ltd.
Kyushu Electric Power Co., Inc.
Kyushu Matsushita Electric
Matsushita Electric Industrial Co., Ltd.
Mitsubishi Electric Corporation
NEC Corporation
Nikon Corporation
Nippon Life Insurance Company
Nissin Ion Equipment Co., Ltd.
NTT Corporation
NTT Electronics Corporation
Oki Electric Industry Co., Ltd.
Ozeki Corporation
Rorze Corporation
Sanyo Electric Co., Ltd.
Semiconductor Leading Edge Technologies, Inc.
Sharp Corporation
Shikoku Electric Power Co., Inc.
Shinko Electric Industries Co., Ltd.
Sony Corporation
Springer-Verlag
Sumitomo Electric Industries, Ltd.
Suntory Limited
Tohoku Electric Power Co., Inc.
Tokyo Electric Power Co., Inc.
Toppan Printing
Toshiba Corporation
Toyota Motor Corporation

Cooperation

Japan National Tourist Organization (JNTO)
Rihga Royal Hotel

Committees

Organizing Committee

H. Kamimura, Conference Chair
L. Esaki, Honorary Chair
K. Murase, Vice-Chair & Local Committee Chair
H. Watanabe, Vice-Chair
T. Ando, Program Chair
Y. Shiraki, Finance Chair
N. Miura, Publication Chair
H. Sakaki, Exhibits and Poster Chair
M. Nakayama, Excursion Chair
C. Hamaguchi, Conference Secretary
N. Mori, Associate Secretary

IUPAP C8 Commission

M. Cardona, Germany (Chair)
R. Woltjer, The Netherlands (Vice-Chair)
E. Molinari, Italy (Secretary)
M. Caldas, Brazil
E. Gornik, Austria
M. Grynberg, Poland
M. Heiblum, Israel
K. Kash, USA
J.-Y. Marzin, France
N. Miura, Japan
M.S. Skolnick, UK
L. Viña, Spain
J.F. Young, Canada

International Advisory Committee

G. Abstreiter, Germany
I. Akasaki, Japan
S.J. Allen, Jr., USA
G. Bauer, Austria
L.L. Chang, China
E. Cohen, Israel
E.A. Davies, UK
T. Dietl, Poland
M.S. Dresselhaus, USA
A. Frova, Italy

M. Heiblum, Israel
M. Hirose, Japan
S. Kawaji, Japan
G. Landwehr, Germany
J.R. Leite, Brazil
G. Martinez, France
B. Monemar, Sweden
J.E. Mooij, The Netherlands
L.J. Sham, USA
H.L. Stormer, USA
R.A. Stradling, UK
M.L.W. Thewalt, Canada
K. von Klitzing, Germany
F. Yndurain, Spain

Program Committee

T. Ando, Japan (Chair)
H. Suzuura, Japan (Secretary)
Y. Arakawa, Japan
Y. Horikoshi, Japan
E. Kapon, Switzerland
S. Katsumoto, Japan
M.J. Kelly, UK
S. Komiyama, Japan
J.P. Kotthaus, Germany
I.V. Kukushkin, Russia
S.G. Louie, USA
J.C. Maan, The Netherlands
Y. Masumoto, Japan
J.-Y. Marzin, France
Y. Mochizuki, Japan
R.J. Nicholas, UK
Y. Oka, Japan
A. Oshiyama, Japan
A. Pinczuk, USA
N. Sawaki, Japan
M. Scheffler, Germany
S. Tarucha, Japan
R.M. Tromp, USA

Opening Address

Hiroshi Kamimura

Chairman, Organizing Committee of ICPS25

Your Imperial Highnesses, Distinguished Guests, Ladies and Gentlemen:

On behalf of the Organizing Committee, I declare the 25th International Conference on the Physics of Semiconductors open, and I am very happy to welcome you to this meeting in the new International Convention Center in Osaka. In particular, it is a great honour for us to welcome Their Imperial Highnesses Prince and Princess Takamado.

This conference sponsored by the IUPAP has been organized by the Science Council of Japan, the Physical Society of Japan, and the Japan Society for Applied Physics. I would like to express my sincere gratitude to their contributions and also to the IUPAP Semiconductor Commission, as well as to all of you, 45 Japanese companies and 9 scientific foundations who have contributed to this conference. Today there are 1048 participants and 68 accompanying persons from 36 countries attending. The conference program was structured by the International Program Committee chaired by Tsuneya Ando, with valuable suggestions and comments by the members of the International Advisory Committee and IUPAP Semiconductor Commission. In this conference we will have 5 plenary papers, 6 plenary symposium papers, and 32 invited papers, 995 contributed papers taken from about 1300 submitted abstracts.

Now I would like to explain our ICPS25 logo. Please look at this symbol inside the stone wall of Osaka Castle. Some Japanese and Chinese participants might read this as the first letter of the Chinese character of Osaka. This is the transistor invented in 1947.

After this discovery, the first ICPS conference was held in Reading, UK in 1950. In the proceedings of the first conference the preface said that during recent years physicists in many countries had made rapid and important advances in the field of solid-state physics. Semiconducting materials, in particular, had become a subject of great interest by reason of their numerous applications. This is still true, even after 50 years. This was the opening of the modern semiconductor physics era.

In the 1970s we entered a new era called the "Tailormade Material Age", stimulated by the proposal of semiconductor superlattices by Leo Esaki and Raphael Tsu in 1969. Indeed, this new material age is characterized by our ability to create materials unknown to nature, ranging widely from intercalation compounds, two-dimensional superlattices and hetero-structures, to one-

dimensional quantum wires, and even zero-dimensional quantum dots.

With these new tailor-made materials, an important doorway to new technologies was opened in the decade starting in 1980. The benefit to fundamental physics, too, has been measureless. The development of two-dimensional systems has provided the essential background for the discovery of the integer and the fractional quantum Hall effects. The latter, in particular, has led to a new theoretical discovery of the novel ground state of the two-dimensional collectivity of electrons, whose elementary excitations are quasiparticles of fractional charges. It was astonishing to learn from this discovery that these quasiparticles of fractional charges were neither fermions nor bosons, but anyons. Our logo showing the fractional quantum Hall effect commemorates the winning of the Nobel Prize in Physics by Robert Laughlin, Horst Stormer and Daniel Tsui in 1998.

In the 1980–90s, the ability to fabricate ultra-small systems drove the exploration of the quantum world down to the nanometer scale, allowing the observation of fundamental quantum confinement and interference phenomena, and single-electron phenomena, as well as important changes in the behavior of a quantum system going through a microscopic to mesoscopic transition. I call this era the "Age of Reduced Dimensionality".

In this age semiconductor nanotechnology also developed remarkably by strong interactions with nanoscale basic science. For example, the number of transistors per chip has increased substantially. In this way semiconductor physics and technology have made great progress by strong interactions between both fields in the last decade of the 20th century, and established the basis for the present and future society of Information Technology (IT) and DNA science. Thus, I call the last decade of this century the "Age of Interactions and Development" In this respect I should say that Japanese industries have noticed an important role of the ICPS25 in the transition period from the 20th to the 21st century and helped us enormously in organizing this conference.

In this context the Organizing Committee organized the plenary symposium on the interface between the fundamentals and device applications in the morning of 20 September, aimed at predicting future prospects for the development of semiconductor physics, semiconductor materials and device applications in the 21st century. The heterostructure laser, quantum cascade laser, quantum computing, gallium–nitride transistors and carbon nanotubes are the subjects of this symposium.

In concluding my address, please look again at our logo. The Osaka Castle was originally built about 400 years ago. This castle, called "Tenshukaku" in Japanese, is 55 meters high. From the top of Tenshukaku you can see great distances. I hope that during the conference you will succeed in discovering something new and the feeling of seeing greater distances from the top of the castle. If you succeed, and many of you say at the end that it was worthwhile coming to this conference and to Osaka, then it would certainly be the nicest compliment to the organizers.

Thank you for your kind attention.

Word of Welcome

Manuel Cardona

Chairman, Commission C8: Semiconductors, International
Union of Pure and Applied Physics (IUPAP)

Your Imperial Highnesses, Authorities, Colleagues, and
Friends:

On behalf of the International Union of Pure and
Applied Physics (IUPAP) I welcome you to the 25th International Conference on the Physics of Semiconductors
(ICPS). The youngest among you may wonder what is
the IUPAP. A detailed description can be seen in my
opening address to the 24th ICPS, held in Jerusalem two
years ago. It is printed in the proceedings of the conference and therefore I shall only say a few words about it.
IUPAP is a private organization, with about 50 member
countries, which coordinates international activities in
the field of physics, among them the large international
conferences. One of these conferences is the ICPS. It
falls under the supervision of Commission C8, which I
have the honor to chair.

Japan is one of four countries that have hosted the
ICPS three times since the inception of the biannual
conferences in 1950, shortly after the invention of the
transistor. The other countries that have hosted three
ICPSs are the United Kingdom, the United States and
Germany. Japan hosted its previous two conferences in
Kyoto (in 1966 and 1980).

In my Jerusalem address I mentioned the names of
three Nobel Prize Laureates for Peace who had made
the hosting of the ICPS by Israel possible: Arafat, Peres
and Rabin. Japan has hosted two conferences and is
now hosting a third one. So, I tried to look for possible individuals who were largely responsible for this
unusual fact. First among these persons I would like to
mention Emperor Meiji (Meiji Tenno), the great grand-
father of Takamadonomiya Denka, who, together with
Takamadonomiya Hidenka, honors us today with his
presence. At the age of 16, Meiji Tenno forged Japan
from isolation and feudal rule into a modern, Western-
like administration, open to international influence and
eager to absorb the tenets of Western science. In his
Charter Oath (Gokajou no goseimon), pronounced on 6
April 1868, Meiji Tenno said "Chisiki wo sekai ni mo-
tome, ooini kouki wo shinki subeshi" (Knowledge will be
sought all over the world, so as to strengthen the Nation). One can hardly think of a better justification for
a conference like ours: to bring here knowledge from all
around the world.

Without Meiji's Restoration (in the West we would
say revolution) we would not have been able to meet here
in Osaka. Foreigners were only allowed to be on a small
artificial island, Dejima, off Nagasaki's coast, and moreover, they had to be Dutch. One of the most important
tasks of the IUPAP is to make sure that all nationals,
not only the Dutch, are allowed to participate in international physics conferences.

Meiji Tenno was 16 years old when he took the Charter Oath. Obviously, his complex and sweeping revolution would not have been possible without the participation of a great many advisers, scholars, patriots and
other public figures. As an example, I single out Sakuma
Shozan (1811–1864), a thinker and military strategist
who expressed his philosophy as "Toyo dotoku, seiyo
geijutsu"[1] [Eastern ethics, Western technology (freely
translated)]. Sakuma was assassinated in 1864 by reac-
tionary forces, two years before the Charter Oath took
place.

Last, but not least, I would like to mention a for-
eigner, a US Naval Officer, who in the year of Mutsuhito's birth (later Meiji Tenno) triggered a series of
events that led, in a rather tortuous way, to the Meiji
Restoration. I am speaking of Commodore Matthew C.
Perry, USN born and raised in the New England state of
Rhode Island, the smallest but one of the most beautiful
states of the US Union. I spent the best years of my
life in that state of which I still am a registered voter.
Our children were taught in Rhode Island schools that

---

[1] 東洋道徳 西洋芸術

"Commodore Perry opened Japan to the West." They did not, of course, quite understand the meaning.

Perry, with a flotilla of 4 men-of-war, sailed up the Tokyo Bay (then Edo Wan) on 8 July 1853, a feat that had not been accomplished by any foreign boats since the Bakufu (Shogunate) of the Tokugawa family had closed Japan to all external influence in 1638. The purpose of Perry's unannounced visit was to deliver a letter from President Pierce to either the Tenno or the Shogun (I am not sure that Perry knew the difference at that point) demanding a treaty that would open Japan to international intercourse and trade. Within four days he managed to deliver the letter and promised to return in the following Spring. This he did on 8 March 1854, with an impressive flotilla of 3 steamships and 7 sailing ships, 2000 men, and 200 guns which he never used with live ammunition. Negotiations started soon, with the Rector of Edo University, Hayashi Noboru Daigaku no kami, a wise and distinguished scholar, as head of the Japanese team. A treaty acceptable to both parties was signed on 31 March 1854. After this event political affairs moved fast in Japan, and the influence of the young Mutsuhito (later to become Meiji Tenno) grew even faster. The outcome has already been mentioned.

Before closing I want to share with you some of Perry's prophetical thoughts about Japan and its prospects once isolation ended (written in 1856, two years before his death, as part of a report to the US Government on his mission in Japan): "I have not met in any part of the world, even in Europe, with a people of more unaffected grace and dignity ... In the practical and mechanical arts, the Japanese show great dexterity, the perfection of their manual skill appears marvelous, especially when one considers the rudeness of their tools ... Once possessed of the acquisitions of the past and present in the Western world, the Japanese will enter as powerful competitors the race for mechanical success ... (semiconductors were not yet known at that time, otherwise he would have mentioned them)."

Many believe that the secret of Japanese success lies indeed in the unique amalgamation of Eastern ethics and Western science. Some voices even warn against upsetting this equilibrium in favor of the latter, among them the great movie director Kurosawa Akira in Dream Number Eight: "Shizenatteno ningen nanoni, sono shizenwo ranbouni ijikurimawashi, oretachi wa motto iimonowo tsukureru to omotteiru. Tokuni gakusha niwa, atama wa iinokamo shirenaiga, shizen no fukai kokoro ga sappari wakaranai mono ga ooinode komaru."[2] (Although people can survive only within nature, they treat it rudely for they believe that they can create better things. It is particularly annoying that some scientists are very clever but do not know at all the profound heart of nature.)

Let us pray that this conference, under Imperial auspices, brings us closer to nature's heart. Thank you and domo arigato.

---

[2] 自然あっての人間なのに、その自然を乱暴にいじくりまわし、おれたちはもっといいものを作れると思っている。特に学者には頭はいいのかも知れないが、自然の深い心がさっぱり分からない者が多いので困る。

Welcome Address

Hiroyuki Yoshikawa

President, Science Council of Japan

Your Imperial Highnesses Prince and Princess Takamado, Mr. Chairman, distinguished guests of the International Conference on the Physics of Semiconductors, ladies and gentlemen:

Today it is our greatest honor and privilege to have Their Imperial Highnesses Prince and Princess Takamado at this opening ceremony of the Conference. On behalf of the Science Council of Japan, it is my great pleasure to express our sincere welcome to all of you, who represent 50 nations from all over the world.

The Science Council of Japan was established in 1949 as a governmental organization to represent the qualified Japanese scientists in all fields of cultural, social and natural sciences. The aim of the Council is to promote scientific research and to enhance its influence on social activities, industry and civil life.

Since the foundation of the Council, we have been working to contribute to the progress of science in cooperation with academic organizations throughout the world, by sponsoring various international congresses here in Japan, and by sending Japanese delegates to international congresses held overseas. We do this because we believe that the promotion of international scientific exchange is one of our most important missions.

We have opened today the 25th International Conference jointly with the Physical Society of Japan and the Japan Society of Applied Physics. It is my great pleasure to meet with distinguished physicists working at the frontiers of semiconductor physics from all over the world, and to have an opportunity to attend fascinating lectures and presentations.

I heard that this is the third conference of this series held in Japan. We regard it as very delightful and meaningful that the 25th International Conference on the Physics of Semiconductors is being held in Japan and that approximately 1000 participants have gathered from 50 countries.

At this conference, it is expected that a number of new research results will be reported across various research fields, and that intensive discussions will be held on these new results, the future direction of research, international collaboration and so forth. Particularly, as this is the concluding conference of this series in the 20th century, it would be a good opportunity to review the

progress of semiconductor physics in this century and to consider the future prospects in the next century. In addition, I think that stimulating discussion will be made on the interface between the fundamentals and applications as we can see in the title of the plenary special symposium.

Semiconductor physics is one of the most important research areas of basic physics, and many important discoveries or the creation of new concepts in this area have greatly influenced the other research fields of basic science. On the other hand, the physics of semiconductors provides a vast background for applications in materials and devices, forming a firm basis for electronics, which is indispensable for our modernized world. Such communication between the different sciences and also between science and society is becoming a very important factor allowing science to develop. The organizers of the World Conference of Science which was held in June–July of 1999 in Budapest, the ICSU (International Council of Science) and UNESCO, published a statement that emphasized the increasing importance of scientific contributions to society. The ICSU is now trying to understand the general process through which science brings about benefits to society, while minimizing the possible threats caused by the wrong application of scientific knowledge.

The history of the development of Semiconductors, (in theory and practice) is a typical welcome example of such a process. Other sciences may be able to learn relevant processes from it in order to develop their fields further in harmony with social developments.

I believe that Semiconductor researchers will continuously generate important basic knowledge in physics in the coming years. Such basic knowledge has already promoted, and will steadily continue to promote the information society. I heartily hope that semiconductor researchers will play a role in not only developing the physical science itself, but also in informing researchers in other scientific fields how basic scientific knowledge can contribute to social developments benefiting human beings.

In conclusion, I sincerely wish great success for the 25th International Conference on the Physics of Semiconductors. For those participants from foreign countries, I hope that your stay in Japan will be enjoyable and that you will form friendships with Japanese scientists, making this congress a very memorable one for you.

Thank you for your kind attention.

Gracious Speech

Prince Takamado

Ladies and Gentlemen:

It is my great pleasure to be able to attend the opening of the 25th International Conference on the Physics of Semiconductors. I would like to welcome all the participants, especially those from abroad. We are very happy to have this conference for the third time in Japan in the last year of the 20th century in which one of the most remarkable discoveries was semiconductors.

The purpose of the physics of semiconductors is to study the basic properties of semiconductors, the essential materials for electronics that is the basis of the modern science and technology. Since transistors were invented in the United States in 1948, the application of semiconductors to electronic or optoelectronic devices has been developed incessantly, and, at present, semiconductors greatly contribute to the welfare of our modern society as indispensable advanced materials for our highly developed information-oriented community.

In this field, Japan is one of the countries, I believe, that has been very active in research, development and manufacturing, and Osaka and its surrounding Kansai area is one of the major centres for the research of semiconductor physics and its industry. There are also a variety of advanced facilities and institutions in the area and modern science and technology are growing at a remarkable pace. This International Conference on the Physics of Semiconductors should play an important role in encouraging not only the physics and industrial communities in Japan but also those in the other parts of Asia where the physics of semiconductors and its application are rapidly increasing their importance.

Last night, I had the pleasure of meeting some of the members taking part in this conference, and was pleasantly surprised. The subject in which you specialise is a difficult one for the average layman to understand. We suspect that your everyday conversation consists solely of semiconductors, transistors, carbon nanotubes, etc. But from the nice conversations that I had with some of you over a reception and dinner last night, I found that, though you research in a very specialised field, there is some artistic element in pursuing such a world. I myself am an avid music lover and an amateur musician, and I thought, maybe there is not much difference between us. Just a tiny difference, whether dealing with a conductor or a semiconductor. In this age of rapid globalisation, it is comforting to think that we speak a common language ...I sincerely hope that through free and frank discussions among the participants from all over the world, you will have a productive conference. I am convinced that it will serve to show the correct direction for the further development of the physics and the industrial application of semiconductors, and that it will greatly benefit mankind in the upcoming 21st century.

Thank you.

Telegram Message

Yoshiro Mori

Prime Minister

Ladies and Gentlemen:

I am pleased to extend a hearty welcome to all the participants from all over the world on this occasion of the Opening Ceremony of the 25th International Conference on the Physics of Semiconductors in Osaka, in the presence of Their Imperial Highnesses Prince and Princess Takamado, under the joint sponsorship of the Science Council of Japan, the Physical Society of Japan and the Japan Society of Applied Physics.

I wish great success for this international conference in the advancement of the field of the physics of semiconductors.

Contents

Part I

Preface   v

Sponsors and Supporters   vii

Committees   ix

Opening Address
Hiroshi Kamimura   xi

Word of Welcome
Manuel Cardona   xiii

Welcome Address
Hiroyuki Yoshikawa   xv

Gracious Speech
Prince Takamado   xvii

Telegram Message
Yoshiro Mori   xix

1. Plenary Papers

Twenty years of quantum Hall effect
K. von Klitzing   1

Optical manipulation of electron and nuclear spins in semiconductors
D.D. Awschalom   2

Controlling the conductivity of wide-band-gap semiconductors
C.G. Van de Walle and J. Neugebauer   3

High performance quantum cascade lasers for the mid-to far-infrared : from Band structure engineering to commercialization
F. Capasso   9

Heterostructure lasers: development of new physics and new technology
Zh.I Alferov   14

Quantum computation
Y. Yamamoto and F. Yamaguchi   20

Hybrid structures of fullerenes and single-wall carbon nanotubes
S. Iijima   24

The science, technology and impact of gallium nitride-based transistors
U.K. Mishra   25

Correlation, decoherence, dephasing and relaxation in semiconductors
D.S. Chemla   26

Experimental signatures of broken symmetries in the quantum Hall regime
J.P. Eisenstein   34

2. Bulk and Dynamics

2.1 Electronic Structure

New effective-mass theory for degenerate bands in semiconductors
B.A. Foreman
41

Antiferromagnetic coupling between the conduction electrons and the $4f$ electrons in $Eu@C_{60}$
S. Suzuki, M. Kushida, S. Amamiya, S. Okada, and K. Nakao
43

First-principles approach to the oxygen vacancies in $SrTiO_3$
T. Schimizu and T. Kawakubo
45

Hartree-Fock study on the phase diagrams of alkali-metal-doped C_{60}
J. Hirosawa, S. Suzuki, and K. Nakao
47

Dynamical-mean-field study on the photoemission spectra of alkali-metal-doped C_{60}
T. Chida, S. Suzuki, and K. Nakao
49

Spin effects in HgSe in megagauss fields
M. von Ortenberg, I. Stolpe, O. Portugall, N. Puhlmann, H.-U. Mueller, M. von Truchsess, C.R. Becker,
A. Pfeuffer-Jeschke, and G. Landwehr
51

Temperature dependence of effective band edge mass in n-InSb by magnetophonon resonance up to $T = 320$ K and influence of nonparabolicity corrections
D. Schneider, S. Krull, C. Brink, G. Irmer, and P. Verma
53

Self-energy correction to the mass of the wide gap semiconductors
M. Oshikiri and F. Aryasetiawan
55

First-principles study of electronic structure of $Si_{1-x}Ge_x$ and $Si_{1-x-y}Ge_xC_y$ disordered alloys
M. Ohfuti, Y. Sugiyama, Y. Awano, and N. Yokoyama
57

Electronic structures of the C_{82} and $La@C_{82}$ crystals by the relativistic LCAO method
S. Amamiya, S. Okada, S. Suzuki, and K. Nakao
59

New determination of the camel's back in AlAs
L.E. Bremme, H. Im, H. Choi, P.C. Klipstein, R. Grey, and G. Hill
61

Theory of the electronic structure of $Ga_{1-y}In_yN_xAs_{1-x}$
A. Lindsay and E.P. O'Reilly
63

Influence of Ag on the energy spectrum of $(Bi_{1-x}Sb_x)_2Te_3$
V.A. Kulbachinskii, A.Yu. Kaminsky, P. Lostak, and C. Drasar
65

2.2 Phonons

Raman study of isotopically tailored CuBr
J. Serrano, F. Widulle, T. Ruf, C.T. Lin, and M. Cardona
67

Isotope disorder effects in the Raman spectrum of Silicon
F. Widulle, T. Ruf, M. Konuma, I. Silier, W. Kriegseis, V.I. Ozhogin, and M. Cardona
69

Observation of extraordinary phonon polaritons in ZnO
T. Azuhata, K. Torii, M. Ono, and T. Sota
71

Disorder-activated resonant Raman scattering in GaNAs/GaAs structures
I.A. Buyanova, W.M. Chen, H.P. Xin, and C.W. Tu
73

Collective dynamics in liquid Ge obtained by an inelastic x-ray scattering experiment
S. Hosokawa, Y. Kawakita, W.-C. Pilgrim, F. Hensel, H. Sinn, and M. Krisch
75

2.3 Optical Properties and Excitons

Non-perturbative terahertz sideband generation from bulk GaAs
M.A. Zudov, J. Kono, A.P. Mitchell, A.H. Chin, and K. Johnsen
77

Contents

Ab initio study of optical absorption spectra of semiconductors and conjugated polymers
M.L. Tiago, E.K. Chang, M. Rohlfing, and S.G. Louie 79

Optical response of incoherent excitons in a dense electron-hole plasma
T. Schmielau, G. Manzke, D. Tamme, and K. Henneberger 81

Effect of reabsorption on electroluminescence of excitonic origin in Cu_2O
Y. Nakamura, N. Naka, and N. Nagasawa 83

Null Fresnel drag in a resonant moving medium
F. Bassani, M. Artoni, G.C. La Rocca, and I. Carusotto 85

Coupled plasmon-LO phonon modes at high magnetic fields
A. Wysmolek, M. Potemski, and T. Slupinski 87

Excitation induced flips of the phase of transmitted pulses
G. Manzke, K. Henneberger, J.S. Nägerl, B. Stabenau, G. Böhne, and R.G. Ulbrich ... 89

Below bottleneck polaritonic radiation in ultra high quality AlGaAs alloys
T.S. Shamirzaev, A.I. Toropov, A.K. Bakarov, K.S. Zhuravlev, A.Yu. Kobitsky, H.P. Wagner, and
D.R.T. Zahn 91

Coherent control of speckle: Resonant Rayleigh scattering from a conjugated polymer
S.P. Kennedy, N. Garro, and R.T. Phillips 93

Fine structure of excitons in diamond: A cathodoluminescence study
N. Teofilov, R. Sauer, K. Thonke, T.R. Anthony, and H. Kanda 95

Transient reflection from anisotropic semiconductors studied with coherent control method
Y. Mitsumori, M. Sato, H. Tobioka, and F. Minami 97

Exciton-phonon droplets with Bose-Einstein condensate: transport and optical properties
D. Roubtsov, Y. Lépine, and I. Loutsenko 99

Role of the excitonic continuum in the polariton problem
E.A. Muljarov and R. Zimmermann 101

Far IR radiation from p-Ge under crossed directions of electric field and uniaxial pressure
A.A. Abramov, V.I. Akimov, A.T. Dalakyan, V.N. Tulupenko, D.A. Firsov, V.M. Bondar, V.N. Poroshin,
V.I. Gavrilenko, and M.S. Kagan 103

Anomalous transport of excitons in Cu_2O
S.G. Tikhodeev and N.A. Gippius 105

Pump-probe spectroscopy by infrared probe light in highly excited semiconductors: Beyond the BCS-like
mean-field theory
T.J. Inagaki and M. Aihara 107

Strain-and field-induced optical anisotropies of GaAs measured by RDS
N. Kumagai, T. Yasuda, T. Hanada, and T. Yao 109

Electromagnetically induced transparency for intrinsic exciton lines in bulk semiconductors
M. Artoni, G.C. La Rocca and F. Bassani 111

Reversible photoinduced structural changes in $GeSe_2$ glass at low-temperature
T. Nakaoka, Y. Wang, O. Matsuda, K. Inoue, and K. Murase 113

Photoluminescence due to the scattering process from heavy-hole to light-hole excitons in CuI thin films
I. Tanaka and M. Nakayama 115

Identification of fine structures of magneto-oscillatory spectra in cuprous oxide as quantum manifestation
of classical non-integrability
K. Hammura, K. Sakai, and M. Seyama 117

Optical studies of alloy semiconductors $Ge_{1-x}C_x$ $(x < 0.05)$ grown on Si substrates by combined low-energy
ion beam and molecular beam epitaxy
M. Sakai, Y. Kitayama, S. Takaku, S. Kimura, and H. Shibata 119

Optical properties of Ga_2Se_3
H. Nishimura, M. Takumi, K. Nagata, and Y. Miyamoto   *121*

Exciton sublattice of incommensurate phase in low-dimensional $TlInS_2$
N. Mamedov, T. Aoki-Matsumoto, B. Gadjiev, H. Uchiki, N. Yamamoto, and S. Iida   *123*

2.4 Transport and Localization

Isotopic mass and lattice constants of Si and Ge: X-ray standing wave measurements (invited)
J. Zegenhagen, A. Kazimirov, L.X. Cao, M. Konuma, E. Sozontov, D. Plachke, H.D. Carstanjen, G. Bilger, E. Haller, V. Kohn, and M. Cardona   *125*

Metal-insulator transition in doped semiconductors (invited)
K.M. Itoh   *128*

Isotope effect on the thermal conductivity of silicon
T. Ruf, R.W. Henn, M. Asen-Palmer, E. Gmelin, M. Cardona, H.-J. Pohl, G.G. Devyatych, P.G. Sennikov, and J. Bollmann   *132*

Axisymmetric Gunn effect
L.L. Bonilla, R. Escobedo, and F.J. Higuera   *134*

DC field response of hot carriers under circular polarized intense microwave fields in semiconductors
N. Ishida   *136*

Transient photoconductivity of pure CuO
H. Yamaguchi, T. Ito, and T. Masumi   *138*

A microscopic simulation of the evolution of local perturbations
P. Gaubert, L. Varani, J.C. Vaissière, J.P. Nougier, E. Starikov, P. Shiktorov, and V. Gruzhinskis   *140*

Dispersive energy transport, relaxation current and Einstein relationship in band tails
O. Bleibaum, H. Böttger, V.V. Bryksin, and A.N. Samukhin   *142*

Angle-resolved tunneling spectroscopy of Si conduction band using bonded Si(111) wafer pair
A. Kazama, H. Bando, Y. Miyahara, H. Enomoto, and H. Ozaki   *144*

Identification of physical mechanisms of tensoeffects in highly strained Si and Ge
S.I. Budzulyak, V.M. Ermakov, Yu.P. Dotsenko, V.V. Kolomoets, Yu.M. Shwarts, E.F. Venger, M. Fukuzawa, P. Verma, M. Yamada, E. Liarokapis, and D.P. Tunstall   *146*

Metal-insulator transition in anisotropic systems
F. Milde, R.A. Römer, and M. Schreiber   *148*

Hot electron degradation of the extraordinary magneto-resistance in inhomogenous InSb
D. Poplavskyy, A.C.H. Rowe, R.A. Stradling, and S.A. Solin   *150*

Critical exponents for the metal-insulator transition of ^{70}Ge:Ga in magnetic fields
M. Watanabe, K.M. Itoh, M. Morishita, Y. Ootuka, and E.E. Haller   *152*

Boron in diamond at high concentrations
M.J.R. Hoch, J.E. Lowther, and T. Tshepe   *154*

Impurity scattering in metallic carbon nanotubes with superconducting pairs
K. Harigaya   *156*

Self-organization of current density filaments in n-GaAs Corbino discs
K. Aoki   *158*

Electrical observation of impurity levels in boron doped homoepitaxial diamond
T. Inushima, R.F. Mamin, T. Matsushita, S. Ohoya, and H. Shiomi   *160*

Monte Carlo simulation of temperature-and field-dependent impact ionization for GaAs
H.K. Jung, S.W. Ko, C.K. You, and K. Taniguchi   *162*

A study on temperature-and field-dependent impact ionization coefficient for silicon using Monte Carlo simulation
H.K. Jung, C.K. You, S.W. Ko, and K. Taniguchi   *164*

Critical behavior of the thermoelectric transport properties in amorphous systems near the metal-insulator transition
C. Villagonzalo, R.A. Römer, M. Schreiber, and A. MacKinnon   166

Photo-quenching in small-polaronic conduction in $LiMn_2O_4$
K. Kushida and K. Kuriyama   168

Low-frequency oscillations and chaos in semi-insulating GaAs subjected to a high magnetic field
A. Neumann, A.G.M. Jansen, P. Wyder, and R. Deltour   170

The influence of the temperature in the scattering mechanisms in pure GaSb
L.G.O. Messias and E. Marega Jr.   172

2.5 Carrier Relaxation and Ultrafast Phenomena

Reversible quantum dynamics of impurity-bound electrons in GaAs (invited)
B.E. Cole, J.B. Williams, M.S. Sherwin, and C.R. Stanley   174

Magnetic field enhanced terahertz emission from semiconductor surfaces
R. McLaughlin, A. Corchia, M.B. Johnston, C.M. Ciesla, D.D. Arnone, G.A.C. Jones, E.H. Linfield,
A.G. Davies, and M. Pepper   178

Dephasing of coherent LO phonon-plasmon coupled modes in polar semiconductors
S. Katayama, M. Hase, M. Iida, and S. Nakashima   180

Femtosecond spectroscopy of large-momentum excitons in GaAs
M. Betz, G. Göger, A. Leitenstorfer, M. Bichler, W. Wegscheider, and G. Abstreiter   182

A novel biexcitonic, non-radiative electron-hole recombination mechanizim and its application in hydrogenated silicon semiconductors
S.B. Zhang and H.M. Branz   184

Annihilation of coherent LO phonon-plasmon coupled modes by lattice defects in n-GaAs
M. Hase, K. Ishioka, K. Ushida, and M. Kitajima   186

Ultrafast electro-absorption at the transition between classical and quantum response
J. Kono and A.H. Chin   188

Deviations from diffusive transport in 100 nm - 100 fs spatio-temporal pump-probe experiments on GaAs
J. Hetzler, A. Brunner, M. Wegener, S. Leu, S. Nau, and W. Stolz   190

Femtosecond optical excitation of acoustic phonons in bulk GaAs
O.B. Wright, B. Perrin, O. Matsuda, and V.E. Gusev   192

Theory of hot luminescence in pulse-excited semiconductors
K. Hannewald, S. Glutsch, and F. Bechstedt   194

Dynamics of Bloch oscillations under the influence of scattering
F. Löser, Yu.A. Kosevich, K. Köhler, and K. Leo   196

Nonlinear response due to propagating excitonic polaritons in GaAs heterostructure
K. Akiyama, N. Tomita, Y. Nishimura, Y. Nomura, and T. Isu   198

Dynamic inter-sideband Fano interference of excitons in ac-driven superlattices
R.-B. Liu and B.-F. Zhu   200

Ultrafast inter-conduction band carrier dynamics in SiC
T. Tomita, S. Saito, T. Suemoto, H. Harima, and S. Nakashima   202

Time-resolved stimulated photon echoes in layered semiconductors
H. Tobioka, Y. Mitsumori, and F. Minami   204

Two-color time-resolved electronic Raman measurement in semiconductors
S. Saito and T. Suemoto   206

Fermi edge singularities in out-of-equilibrium semiconductors
M. Combescot, B. Roulet, and C. Tanguy   208

Theory of coherent oscillation of phonon-polaritons
C.S. Kim, A.M. Satanin, G.D. Sanders, and C.J. Stanton   210

2.6 Phase Transition and Ordering

Self-organized electron density patterns in n-GaAs induced by microwaves
V. Novák, V.V. Bel'kov, D.M. Mathúna, S.D. Ganichev, and W. Prettl   212

Novel observation of magnetic-field induced metal-insulator transition in n-GaAs through far-infrared photoconductivity measurements
H. Kobori, M. Inoue, and T. Ohyama   214

Magnetization of single MnTe (sub)monolayers embedded in nonmagnetic quantum wells
G. Prechtl, W. Heiss, S. Mackowski, A. Bonanni, E. Janik, H. Sitter, and W. Jantsch   216

Energy relaxation by phonon scattering in long-range ordered $(Al_{0.5}Ga_{0.5})_{0.5}In_{0.5}P$
T. Kita, K. Yamashita, T. Nishino, Y. Wang, and K. Murase   218

Low-dimensional magnetism in polymorphic compounds, $VOMoO_4$ and $(V_{0.56}, Mo_{0.44})_2O_5$
I. Shiozaki, S. Takada, S. Wada, and T. Toyoda   220

Highly efficient luminescence in partially ordered $GaInP_2$
X.H. Zhang, S.J. Chua, and J.R. Dong   222

Kinetics of the amorphous-to-crystalline phase transformation in Ge:Sb:Te alloys used for the phase-change optical memory applications
E. García-García, E. Morales-Sánchez, J. González-Hernández, E. Prokhorov, Yu. Vorobiev, and S. Kostylev   224

The intensity and wavelength dependence for photo-induced crystallization process in amorphous $GeSe_2$
K. Sakai, K. Maeda, H. Yokoyama, and T. Ikari   226

Stiffness transition and connectivity in amorphous chalcogenide semiconductors
Y. Wang, T. Nakaoka, M. Nakamura, O. Matsuda, and K. Murase   228

Effect of pressure on electrical transport in electron doped perovskite manganite $Sr_{0.9}Ce_{0.1}MnO_3$
T. Eto, F. Honda, G. Oomi, and A. Sundaresan   230

2.7 Magnetic and Semimagnetic Semiconductors

Ferromagnetism in diluted magnetic semiconductors
J. König, H.-H. Lin, and A.H. MacDonald   232

Ferromagnetic interactions in p- and n-type II-VI diluted magnetic semiconductors
T. Andrearczyk, J. Jaroszyński, M. Sawicki, L. van Khoi, T. Dietl, D. Ferrand, C. Bourgognon, J. Cibert, S. Tatarenko, T. Fukumura, Z. Jin, H. Koinuma, and M. Kawasaki   234

Theoretical predictions for transparent ferromagnets with transition metal atom doped ZnO
K. Sato and H. Katayama-Yoshida   236

sp-d exchange interaction in GaMnAs investigated by resonant Kerr effect under high magnetic field
F. Michelini, N. Nègre, G. Fishman, M. Goiran, J. Sadowski, E. Vanelle, and S. Askénasy   238

Giant tunability of excitonic photoluminescence transitions in antiferromagnetic EuTe epilayers
W. Heiss, G. Prechtl, and G. Springholz   240

Hybridization-induced exchange interaction between the conduction band electrons and Mn ions in diluted magnetic semiconductors
A.K. Bhattacharjee and J. Pérez-Conde   242

Cross-sectional scanning tunneling microscope (STM) study of Mn-doped GaAs layers
T. Tsuruoka, R. Tanimoto, N. Tachikawa, S. Ushioda, F. Matsukura, and H. Ohno   244

Anomalous enhancement of the cyclotron mass of electrons in $Cd_{1-x}Mn_xTe$ observed at very high magnetic fields
Y.H. Matsuda, T. Ikaida, T. Yasuhira, N. Miura, S. Kuroda, F. Takano, and K. Takita   246

Electronic structure of superlattices of II–VI/III–V diluted magnetic semiconductors

T. Kamatani and H. Akai   *248*

Tunneling into amorphous Gd_xSi_{1-x} at the metal-insulator transition and its independence of magnetic impurities in the barrier

W. Teizer, F. Hellman, and R.C. Dynes   *250*

Spin-lattice relaxation in semimagnetic quantum wells with a 2DEG

D.R. Yakovlev, A.V. Scherbakov, B. König, W. Ossau, A.V. Akimov, T. Wojtowicz, G. Karczewski, and J. Kossut   *252*

Magnetic-field-driven metal-insulator transition in magnetic semiconductor (Ga,Mn)As

T. Hayashi, Y. Hashimoto, S. Yoshida, S. Katsumoto, and Y. Iye   *254*

The effect of stress on magnetic properties of ferromagnetic EuS-PbS and EuS-PbSe semiconductor structures

T. Story, M. Arciszewska, M. Chernyshova, W. Dobrowolski, W. Mac, A. Twardowski, R. Świrkowicz, and A.Yu. Sipatov   *256*

Fundamental properties of Fe-based III-V magnetic alloy semiconductor (Ga, Fe)As

R. Moriya, T. Kondo, Y. Katsumata, H. Munekata and S. Haneda   *258*

The waves of spin and charge densities in diluted magnetic semiconductors near the paramagnetic-ferromagnetic phase transition

Yu.G. Semenov and V.N. Sokolov   *260*

Electronic structure of $Ga_{1-x}Mn_xAs$ studied by photoemission spectroscopy

J. Okabayashi, A. Kimura, O. Rader, T. Mizokawa, A. Fujimori, T. Hayashi, and M. Tanaka   *262*

Electronic structure and magnetic properties of (Zn,Mn)Se with carriers doping studied by FLAPW method

K. Matsushita, H. Harima, A. Yanase, and H. Katayama-Yoshida   *264*

Cyclotron resonance of spin polaron in CdMnTe/CdMgTe 2D electron systems

Y. Imanaka, T. Takamasu, G. Kido, G. Karczewski, T. Wojtowicz, and J. Kossut   *266*

Far infrared absorption spectra in ferromagnetic $Ga_{1-x}Mn_xAs$

Y. Nagai, T. Kunimoto, K. Nagasaka, H. Nojiri, M. Motokawa, F. Matsukura, and H. Ohno   *268*

Dynamical aspects of exciton magnetic polaron formation in diluted magnetic semiconductor quantum wells

T. Stirner and W.E. Hagston   *270*

Cryobaric exciton magnetophotoluminescence in $Cd_{1-x}Mn_xSe$

N. Kuroda and Y.H. Matsuda   *272*

Line shape of excitation spectra of photoluminescence in CdMnTe and its quantum wells

K. Takamura, S. Yamamoto, and J. Nakahara   *274*

3. Growth, Surfaces, and Interfaces

3.1 Atomic Structure

Possibility of the quantum fluctuation of the Si(001) surface at low temperature

Y. Yoshimoto and M. Tsukada   *277*

(2×4) and (4×2) reconstructions of GaAs(001): The surface phase diagram re-examined

W.G. Schmidt, S. Mirbt, and F. Bechstedt   *279*

Theory of $Al_2O_3(0001)$ surfaces and their employment as a substrate for nitride growth

R. Di Felice and J.E. Northrup   *281*

Influence of strain and diffusion on the growth of V-groove InGaAs/GaAs quantum wires

K. Leifer, F. Lelarge, A. Rudra, S. Stauss, and E. Kapon   *283*

Atomic structure at a Si (001) oxidation front

N. Ikarashi, K. Watanabe, and Y. Miyamoto   *285*

Binding and migration paths of Au adatoms on the GaAs (001)-$\beta 2(2\times 4)$ surface
A.A. Bonapasta and F. Buda
287

Surface topography of Si(111)-7×7 reconstruction: first-principles investigations
S.H. Ke, T. Uda, and K. Terakura
289

A structural model of Si(111)$(2\sqrt{3}\times 2\sqrt{3})$ R30°-Sn
T. Ichikawa, K. Cho, T. Onodera, A. Mizoguchi, and T. Ohkoshi
291

Synchrotron radiation photoemission studies of surface reconstruction on GaAs (001)
K. Ono, T. Mano, K. Nakamura, M. Mizuguchi, S. Nakazono, K. Horiba, T. Kihara, H. Kiwata, I. Waki, M. Oshima, N. Koguchi, and A. Kakizaki
293

Structural ordering on Si(111)$\sqrt{3} \times \sqrt{3}$-Ag surface : Monte Carlo simulation based on first-principles calculations
Y. Nakamura, Y. Kondo, J. Nakamura, and S. Watanabe
295

STM and RHEED study of Ge(110) reconstructions
T. Ichikawa, H. Fujii, and A. Sugimoto
297

3.2 Electronic Structure

Terrace and step contributions to the surface optical anisotropy of Si(001)
W.G. Schmidt, F. Bechstedt, and J. Bernholc
299

Nano-scale ferromagnets on semiconductors: Ga adsorbates on Si (100) surfaces
S. Okada and A. Oshiyama
301

AlGaN/GaN lateral polarity heterostructures
A.P. Lima, C. Miskys, O. Ambacher, M. Stutzmann, R. Dimitrov, V. Tilak, M.J. Murphy, and L.F. Eastman
303

Modification of the Si surface electronic properties by Ge nanostructures: Surface photovoltage studies
K. Nauka and T.I. Kamins
305

Ab initio study of SiC/Ti polar interfaces
S. Tanaka and M. Kohyama
307

3.3 Adsorption and Surface Reactions

Microscopic mechanism of Si oxidation (invited)
K. Shiraishi, H. Kageshima, and M. Uematsu
309

Electron paramagnetic resonance of a single-crystal surface: the Si(111)-7×7 surface and its oxidation process
T. Umeda, M. Nishizawa, T. Yasuda, J. Isoya, S. Yamasaki, and K. Tanaka
313

Atomic layer oxidation of H terminated Si(100) surface : Domino reaction via oxidation and H migration
K. Kato, H. Kajiyama, S. Heike, T. Hashizume, and T. Uda
315

Time-dependent density-functional simulations of desorption dynamics of H and Br terminated Si surfaces induced by electronic excitations
Y. Miyamoto and O. Sugino
317

Mechanism of migration and dissociation for As_2 molecule on GaAs(001) surfaces during the MBE growth: A computational study
A. Ishii, K. Seino, and T. Aisaka
319

Photoemission spectroscopy of Si(001) surfaces oxidized by hyperthermal O_2 molecular beams
Y. Teraoka and A. Yoshigoe
321

Formation of first InSb molecular layer on Si(111) substrate: Role of In(4×1) reconstruction
B.V. Rao, D. Gruznev, T. Tambo, and C. Tatsuyama
323

Laser-induced electronic instability on semiconductor surfaces of Si (111)-(7x7) and InP (110)-(1x1)
K. Tanimura, J. Kanasaki, and K. Ishikawa
325

Surface states of (100)n-GaAs with adsorbed oxygens and their dependence on chemical treatment
Y. Kasai, T. Tsuzuku, Y. Ohta, T. Inokuma, K. Iiyama, and S. Takamiya 327

Real-time observation of initial oxidation on highly B-doped Si (100)-2×1 surfaces using scanning tunneling microscopy
K. Ohmori, M. Tsukakoshi, H. Ikeda, A. Sakai, S. Zaima, and Y. Yasuda 329

Fabrication of boron delta-doped structures in Si by solid phase epitaxy
T. Ishikawa, H. Nagai, K. Ishii, and S. Matsumoto 331

Effect of gas adsorption on the interfacial states at n-SnO$_2$/p-Si heterojunction
R.B. Vasiliev, A.M. Gaskov, M.N. Rumyanseva, and L.I. Ryabova 333

Structure of a SiN layer on Si(111) surface
T. Yamasaki, C. Kaneta, and T. Uda 335

3.4 Bulk Growth and Thin Film Epitaxy

Distinct morphological evolution of Si$_{1-x}$Ge$_x$ films on Si(100) during gas-source MBE and photo-CVD
H. Akazawa 337

Toward predictive growth simulations: MBE on GaAs(001)
P. Kratzer and M. Scheffler 339

Dual Au-Si bonding character and Au surfactant effect on Au/Si(111) surfaces
M. Murayama, T. Nakayama, and A. Natori 341

Theoretical study on epitaxial growth of lattice-mismatched semiconductor systems
N. Miyagishima, K. Okajima, K. Takeda, N. Oyama, T. Ohno, K. Shiraishi, and T. Ito 343

Control of electronic structure and thermodynamic stability of SrTiO$_3$ with Ru or Nb doping from first-principles calculation
T. Schimizu and T. Kawakubo 345

Modeling and simulation of FBAR devices fabricated on Si
D.H. Kim, G. Yoon, and H.D. Park 347

Nucleation mechanism of stacking fault during Si epitaxial growth by chemical vapor deposition with dichlorosilane
Y. Takakuwa, M.K. Mazumder, and N. Miyamoto 349

Self-limiting etching prior to self-limiting growth in ultra-high vacuum for obtaining clean interface
N. Otsuka, J. Nishizawa, Y. Oyama, H. Kikuchi, and K. Suto 351

Reflectance-difference spectroscopy of (001) InAs surfaces in ultrahigh vacuum
T. Kita, H. Tango, K. Tachikawa, K. Yamashita, and T. Nishino 353

3.5 Growth of Self-Assembled Dots

Controlling of lateral and vertical order in self-organized PbSe quantum dot superlattices (invited)
G. Springholz, M. Pinczolits, V. Holy, P. Mayer, G. Bauer, H. H. Kang, and L. Salamanca-Riba 355

Shape analysis of single and stacked InAs quantum dots at the atomic level by cross-sectional STM
D.M. Bruls, J.W.A.M. Vugs, P.M. Koenraad, M.S. Skolnick, M. Hopkinson, and J.H. Wolter 359

Formation procees of InAs dots including Mn atoms and their physical properties
S. Okumura, H. Asahi, Y.K. Zhou, J. Asakura, M. Kanamura, K. Asami, H. Kubo, C. Hamaguchi, and S. Gonda 361

Shape and size of buried SiGe islands
J. Stangl, V. Holý, A. Daniel, T. Roch, G. Bauer, T.H. Metzger, J. Zhu, K. Brunner, and G. Abstreiter 363

Optical anisotropy of Stranski-Krastanov growth surface of InAs on GaAs (001)
T. Kita, H. Tango, K. Tachikawa, K. Yamashita, T. Nishino, T. Nakayama, and M. Murayama 365

Lasers based on self-assembled InAs/GaAs and InP/InGaP quantum dots
O.G. Schmidt, M.O. Lipinski, Y.M. Manz, H. Heidemeyer, W. Winter, and K. Eberl 367

Effect of vertical size uniformity on diffraction contrast images of stacked InGaAs/GaAs quantum dots
M. De Giorgi, A. Passaseo, R. Cingolani, A. Taurino, and M. Catalano   *369*

Lateral distribution of buried self-assembled InAs quantum dots on GaAs
K. Zhang, J. Falta, Ch. Heyn, Th. Schmidt, and W. Hansen   *371*

Nucleation site control in self-assembling Si quantum dots on ultrathin SiO_2/c-Si
S. Miyazaki, M. Ikeda, E. Yoshida, N. Shimizu, and M. Hirose   *373*

Reversibility of the island shape, volume and density in Stranski-Krastanow growth
N.N. Ledentsov, V.A. Shchukin, R. Heitz, D. Bimberg, V.M. Ustinov, N.A.Cherkashin, A.R. Kovsh, Yu.G. Musikhin, B.V. Volovik, A.E. Zhukov, G.E.Cirlin, and Zh.I. Alferov   *375*

Effect of Si diffusion on growth of GeSi self-assembled islands
A.V. Novikov, N.V. Vostokov, S.A. Gusev, Yu.N. Drozdov, Z.F. Krasil'nik, D.N. Lobanov, L.D. Moldavskaya, M. Miura, V.V. Postnikov, M.V. Stepikhova, Y. Shiraki, and N. Usami   *377*

Observation of a universal behavior in the growth of InAs self-assembled quantum dots on patterned substrates
S.W. Hwang, M.H. Son, B.H. Choi, D. Ahn, C.K. Hyon, S.-H. Song, Y.J. Park, and E.K. Kim   *379*

Monte Carlo simulation of the self-organised growth of quantum dots with anisotropic surface diffusion
M. Meixner, R. Kunert, S. Bose, E. Schöll, V.A. Shchukin, D. Bimberg, E. Penev, and P. Kratzer   *381*

Growth of InAs quantum dots on (110) -oriented cleaved GaAs surfaces
M. Gerling, S. Jeppesen, A. Gustafsson, and L. Samuelson   *383*

Fabrication of compound-semiconductor quantum dots on a Si(111) substrate terminated by bilayer-GaSe
K. Ueno, K. Saiki, and A. Koma   *385*

Optimized growth procedure for self-organized InAs quantum dots
E. Steimetz, T. Wehnert, P. Kratzer, L.G. Wang, Q.K.K. Liu, H. Kirmse, J.-T. Zettler, W. Neumann, M. Scheffler, and W. Richter   *387*

Effect of inter-island interaction on the growth of self-assembled quantum dots
S. Vannarat, S.T. Chui, K. Esfarjani, and Y. Kawazoe   *389*

Properties of CdSe/ZnSe based quantum heterostructures with and without lateral confinement potentials
E. Kurtz, M. Schmidt, B. Dal Don, S. Wachter, D. Litvinov, D. Gerthsen, H. Kalt, and C. Klingshirn   *391*

Growth, morphology and optical properties of metal clusters on semiconductor surface (Au/GaAs)
N. Dmitruk, T. Barlas, I. Dmitruk, T. Mikhailik, and V. Romaniuk   *393*

Formation of GaAsN nanoinsertions in a GaN matrix
A.F. Tsatsul'nikov, I.L. Krestnikov, W.V. Lundin, A.V. Sakharov, D.A. Bedarev, A.S. Usikov, B.Ya. Ber, V.V. Tret'yakov, Zh.I. Alferov, N.N. Ledentsov, A. Hoffmann, D. Bimberg, T. Riemann, J. Christen, Yu.G. Musikhin, I.P. Soshnikov, D. Litvinov, A. Rosenauer, D. Gerthsen, and A. Plaut   *395*

Extremely uniform InAs/GaAs quantum dots emitting at 1.46 mkm at room temperature grown by MOCVD with Bi doping
B.N. Zvonkov, I.A. Karpovich, N.V. Baidus, D.O. Filatov, Yu.Yu. Gushina, S.V. Morozov, and S.B. Levichev   *397*

Effects of InAs coverage on the Ga diffusion into InAs self-assembled quantum dots on GaAs(100)
N. Matsumura, T. Haga, S. Muto, Y. Nakata, and N. Yokoyama   *399*

Photoluminescence and atomic force microscopy studies of InAs/InSb nanostructures grown by MBE
Ya.V. Terent'ev, A.A. Toropov, V.A. Solov'ev, B.Ya. Mel'tser, M.M. Moiseeva, S.V. Ivanov, B. Magnusson, B. Monemar, and P.S. Kop'ev   *401*

Three dimensional self-organization of InAs quantum-dot multilayers
J.C. González, F.M. Matinaga, W.N. Rodrigues, M.V.B. Moreira, A.G. de Oliveira, M.I.N. da Silva, J.M.C. Vilela, M.S. Andrade, D. Ugarte, and P.C. Silva   *403*

Investigation of the properties of molecular beam epitaxy grown self-organised ZnSe quantum dots embedded in ZnS
C.L. Yang, L.W.Lu, W.K. Ge, Z.H. Ma, I.K. Sou, and J.N. Wang   *405*

Exciton localization in alloy/alloy interfaces of InGaAs/GaAs(001) stepped quantum wells
A. D'Andrea, F. Fernández-Alonso, M. Righini, D. Schiumarini, S. Selci, and N. Tomassini   *407*

Characterization of Ge nanocrystal embedded in SiO₂ films
H. Fukuda, S. Sakuma, T. Yamada, S. Nomura, M. Nishino, T. Higuchi, and S. Ohshima   *409*

3.6 Semiconductor–Nonsemiconductor Interfaces

Control of quantum confinement in metal-clad InAs quantum wells
S. Tsujino, S.J. Allen, M. Rüfenacht, M. Thomas, J.P. Zhang, J. Speck, T. Eckhause, and B. Gwinn   *411*

Unsaturated cyclic-hydrocarbon molecules on a Si (001) surface: A first-principles approach
K. Akagi and S. Tsuneyuki   *413*

Organic modified GaAs(100) Schottky contacts
Th. Lindner, T.U. Kampen, S. Park, and D.R.T. Zahn   *415*

Interface effects on exciton states in a CdTe/(Cd, Mn)Te quantum well
*H. Yokoi, Y. Kakudate, Yu.G. Semenov, S. Takeyama, S.W. Tozer, Y. Kim, T. Wojtowicz, G. Kar-
czewski, and J. Kossut*   *417*

Reaction of atomic hydrogen with the Si(100)/SiO₂ interface defects
C. Kaneta, T. Yamasaki, and T. Uda   *419*

Temperature dependence of low frequency noise mechanisms in Schottky barrier structure
J.I. Lee, I.K. Han, J. Brini, A. Chovet, and C.A. Dimitriadis   *421*

Interfacial transition regions of gate dielectrics in advanced silicon devices
G. Lucovsky, J.C. Phillips, and M.F. Thorpe   *423*

ESR and magnetic measurement of Ni/GaAs composite system
Y. Seino, S.A. Haque, A. Matsuo, Y. Yamamoto, S. Yamada, and H. Hori   *425*

3.7 STM, AFM, and Near-Field Studies of Surfaces and Interfaces

Imaging of Friedel oscillations at epitaxially grown InAs(111)A surfaces using scanning tunneling microscopy
(invited)
K. Kanisawa, M.J. Butcher, H. Yamaguchi, and Y. Hirayama   *427*

Probing of subsurface dopants buried in silicon by scanning tunneling microscopy
Y. Suwa, T. Hitosugi, S. Matsuura, S. Heike, S. Watanabe, T. Onogi, and T. Hashizume   *431*

Effect of tip morphology on AFM images: *Ab initio* simulations on GaAs(110) surface
S.H. Ke, T. Uda, I. Štich, and K. Terakura   *433*

Novel investigation technique for interior III/V-semiconductor interfaces
S. Nau, G. Bernatz, and W. Stolz   *435*

Observation of dopant-atom dimers on hydrogen-terminated Si(100)-2×1 surface by scanning tunneling
microscopy
S. Matsuura, M. Fujimori, S. Heike, Y. Suwa, T. Onogi, H. Kajiyama, K. Kitazawa, and T. Hashizume   *437*

Reconstruction on Si(100) surface induced by the type-*A* defects near T_c
M. Okamoto, T. Yokoyama, and K. Takayanagi   *439*

Space and energy distribution of surface gap states on MBE-grown and silicon-covered (001) GaAs surfaces
studied by scanning tunneling spectroscopy
N. Negoro, S. Kasai, and H. Hasegawa   *441*

Atomically resolved imaging of semiconductor surfaces using noncontact atomic force microscopy
S. Morita and Y. Sugawara   *443*

Atomic structure of GaP (001) and InP (001) reconstructions: Scanning tunneling microscopy and ab initio
theory
K. Lüdge, P. Vogt, O. Pulci, N. Esser, F. Bechstedt, and W. Richter   *445*

Bias voltage dependence of scanning tunneling microscopy images of a Si (100)2x3-Ba surface
K. Ojima, M. Yoshimura, and K. Ueda   447

Removal of particles from surface of silicon using clean solutions with surfactants and chelates
C. Baocheng, Y. Xinhao, M. Honglei, M. Jin, and L. Zhengli   449

4. Heterostructures and Superlattices: Optical

4.1 Structural Properties

Phase difference between coherent GaSb-like and AlSb-like LO phonons in GaSb/AlSb superlattices
H. Takeuchi, K. Mizoguchi, M. Nakayama, K. Kuroyanagi, and T. Aida   451

Plasmon-phonon coupling at $Ga_{0.5}In_{0.5}P$/GaAs heterointerfaces induced by CuPt-type ordering
K. Yamashita, T. Kita, T. Nishino, Y. Wang, K. Murase, C. Geng, F. Scholz, and H. Schweizer   453

Optical properties and band alignment of ZnSe/ZnMgBeSe heterostructures
K. Godo, M.W. Cho, J.H. Chang, S.K. Hong, H. Makino, T. Yao, M.Y. Shen, and T. Goto   455

Analysis of Raman spectra of quasiperiodic GaAs-AlAs heterostructures
E. Matsushita and T. Furuyama   457

4.2 Electronic Structures

Controlling Fermi-edge singularities by a periodic external potential
S. Nomura, T. Nakanishi, and Y. Aoyagi   459

Optical anisotropy change of buried SiO_2/Si interfaces during layer-by-layer oxidation
T. Nakayama and M. Murayama   461

Negatively charged excitons in semiconductor quantum wells: effects of longitudinal electric and magnetic fields
L.C.O. Dacal, M.J.S.P. Brasil, and J.A. Brum   463

The influence of intraband transitions on the resonant and extended electronic states in GaAs/AlGaAs asymmetric quantum wells as a function of their confinement
M. Levy, R. Beserman, R. Kapon, A. Sa'ar, V. Thierry-Mieg, and R. Planel   465

Zeeman mapping of probability densities in square quantum wells using ultranarrow probes
G. Prechtl, W. Heiss, A. Bonanni, E. Janik, S. Mackowski, G. Karczewski, and W. Jantsch   467

Tight-binding description of GaAs/AlAs quantum wells and superlattices: Gap transitions and intervalley couplings
R.B. Capaz, J.G. Menchero, T.G. Dargam, and B. Koiller   469

Large quantum confinement effect of conduction electrons in ZnSe/BeTe type II heterostructures
R. Akimoto, Y. Kinpara, and K. Akita   471

A study of the GaAs/partially ordered GaInP interface
T. Kobayashi, K. Inoue, A.D. Prins, K. Uchida, and J. Nakahara   473

Analytical and transfer matrix solutions of the $\mathbf{k} \cdot \mathbf{p}$ Hamiltonian for the X-band in AlAs confined systems
L.E. Bremme and P.C. Klipstein   475

Relation between the in-plane polarization anisotropy of the optical properties and the microscopic atomic configuration in $(Ga_{0.5}In_{0.5}As)$/(InP) superlattices
R. Magri and S. Ossicini   477

Comparison of empirical pseudopotential and $\mathbf{k} \cdot \mathbf{p}$ calculations in p-doped strained layer SiGe quantum wells
Z. Ikonić, R.W. Kelsall, and P. Harrison   479

Plasmons in laterally density modulated 2D electron gas in shallow etched single-heterostructures
T. Tagawa and S. Katayama   481

Electronic properties of quantum wire superlattices elaborated by the 'Atomic Saw' method
F. Michelini, L. Ressier, G. Fishman, E. Vanelle, F. Laruelle, and J.P. Peyrade   483

Cyclotron resonance of holes and shallow acceptor photoconductivity in strained MQW Ge/GeSi heterostructures in strong magnetic fields
V.Ya. Aleshkin, I.V. Erofeeva, V.I. Gavrilenko, O.A. Kuznetsov, M.D. Moldavskaya, V.L. Vaks, and
D.B. Veksler   485

Breakdown of rotational symmetry at semiconductor interfaces
O. Krebs, S. Cortez, and P. Voisin   487

4.3 Magnetic Superlattices

Binding energy and internal magnetic field of exciton magnetic polarons in a single semimagnetic quantum dot
G. Bacher, A.A. Maksimov, A. McDonald, M.K. Welsch, H. Schömig, V.D. Kulakovskii, A. Forchel,
Ch. Becker, G. Landwehr, and L.W. Molenkamp   489

Spin transport of excitons in asymmetric double quantum wells of $Cd_{1-x}Mn_xTe$
K. Kayanuma, E. Shirado, M.C. Debnath, I. Souma, S. Permogorov, and Y. Oka   491

Raman scattering in MnTe/ZnTe and ZnSe/ZnMnSe multilayer semiconductor superlattices
K. Ozawa, Y. Tanaka, K. Iio, N. Nakajima, Y. Nabetani, and T. Matsumoto   493

4.4 Interband Transitions and Excitons

Exciton-trion coupling in modulation doped quantum well structures
V.P. Kochereshko, G.V. Astakhov, R.A. Suris, D.R. Yakovlev, W. Ossau, J. Nurnberger, W. Faschinger,
and G. Landwehr   495

Rapid recombination process of free trions
D. Sanvitto, R.A. Hogg, A.J. Shields, D.M. Whittaker, M.Y. Simmons, D.A. Ritchie, and M. Pepper   497

Spin dependent exciton-exciton interaction in hot and cold 2D exciton gases controlled by an electric field
G. Aichmayr, L. Viña, S.P. Kennedy, R.T. Phillips, and E.E. Mendez   499

From excitons to Fermi edge singularity
G. Yusa, H. Shtrikman, and I. Bar-Joseph   501

Magnetic traps for excitons in $GaAs/Al_xGa_{1-x}As$ quantum wells
J.A.K. Freire, F.M. Peeters, A. Matulis, V.N. Freire, and G.A. Farias   503

Excess carrier effects upon the excitonic absorption thresholds of remotely doped GaAs/AlGaAs quantum wells
R. Kaur, A.J. Shields, R.A. Hogg, J.L. Osborne, M.Y. Simmons, D.A. Ritchie, and M. Pepper   505

Observation of heavy-hole minibands in ultra-short period superlattices
M. Eckardt, W. Geisselbrecht, S. Malzer, G.H. Döhler, K. Maranowski, and A.C. Gossard   507

CdTe quantum wells as ideal systems for the study of negatively charged excitons: A spectral and temporal analysis
V. Ciulin, P. Kossacki, M. Kutrowski, A. Esser, S. Haacke, J.-D. Ganière, T. Wojtowicz, and B.
Deveaud   509

Collapse of the excitonic states at $r_s=8$ in high quality GaAs/AlGaAs single quantum wells
S.I. Gubarev, I.V. Kukushkin, S.V. Tovstonog, M.Yu. Akimov, L.V. Kulik, J. Smet, K. von Klitzing,
and W. Wegscheider   511

Phase sensitive detection of transmitted femtosecond pulses in GaAs quantum wells
R. Kawahara, T. Kuroda, Y. Mitsumori, and F. Minami   513

Oscillatory behavior of the Γ-X coupling with AlAs thickness in type II GaAs/AlAs heterostructures
C. Gourdon, D. Martins, V. Voliotis, P. Lavallard, and E.L. Ivchenko   515

Blue Stark shift in composite quantum wells
A.Yu. Silov, B. Aneeshkumar, M.R. Leys, H. Vonk, and J.H. Wolter   517

Spectroscopic studies of the interaction of an electron gas with photoexcited electron-hole pairs in modulation-doped GaAs/AlGaAs quantum wells
T. Yeo, H.A. Nickel, G. Comanescu, H.D. Cheong, M. Furis, A. Petrou, and B.D. McCombe ... 519

Resonant Rayleigh scattering of exciton-polaritons in multiple quantum well structures
A.V. Kavokin, G. Malpuech, W. Langbein, and J.M. Hvam ... 521

Speckle-correlation spectroscopy on localized spin-split exciton states in quantum wells
R. Zimmermann, W. Langbein, E. Runge, and J.M. Hvam ... 523

Optical properties of trions in semiconductor nanostructures
A. Esser, E. Runge, and R. Zimmermann ... 525

Internal transitions of charged magneto excitons in II-VI quantum well heterostructures
C.J. Meining, M. Furis, H.A. Nickel, D.R. Yakovlev, W. Ossau, A. Petrou, and B.D. McCombe ... 527

Many-body corrections in n-type 2D telluride structures
V. Huard, R.T. Cox, K. Saminadayar, C. Bourgognon, S. Tatarenko, and L. Besombes ... 529

Possibility of excitonic polymers in type-II superlattices
T. Tsuchiya ... 531

Localised exciton transitions in high-quality GaAs/AlGaAs quantum wells
A.G. Steffan, A. García-Cristóbal, R.T. Phillips, and D.A. Ritchie ... 533

Field-induced optical anisotropy in semiconductor superlattices: the Wannier-Pockels effect
S. Cortez, O. Krebs, J.C. Harmand, J.L. Gentner, and P. Voisin ... 535

Origin of optical-phonon Raman spectra in multiple quantum wells
B.-F. Zhu and S.F. Ren ... 537

Temperature dependence of the band gap in (InGa)(AsN)/GaAs single quantum wells
A. Polimeni, M. Capizzi, M. Geddo, M. Fischer, M. Reinhardt, and A. Forchel ... 539

Charge transfer of carriers by interband photoexcitation in asymmetric GaAs/AlGaAs coupled quantum wells
M. Levy, Yu.L. Khait, R. Beserman, A. Sa'ar, V. Thierry-Mieg, and R. Planel ... 541

Optical properties of InGaAsN/GaAs QWs for long-wavelength lasers on GaAs
H. Riechert, A.Yu. Egorov, Gh. Dumitras, and B. Borchert ... 543

Effective mass theory for a magneto-exciton in the type II superlattices or confinement structures and the origin of anomalous photoluminescence
K. Kouzu, M. Nishimura, and H. Kamimura ... 545

Modulation spectroscopy of critical points in excitonic energy structure of thick GaAs quantum wells
V.G. Davydov, Yu.K. Dolgikh, Yu.P. Efimov, S.A. Eliseev, A.V. Fedorov, I.Ya. Gerlovin, I.V. Ignatiev, I.E. Kozin, V.V. Petrov, V.V. Ovsyankin, I.A. Yugova, H.-W. Ren, S. Sugou, and Y. Masumoto ... 547

Exciton-phonon coupling in wida bandgap II-VI quantum wells
B. Urbaszek, A. Balocchi, C. Morhain, C. Bradford, X. Tang, C.M. Townsley, C.B. O'Donnell, S.A. Telfer, K.A. Prior, B.C. Cavenett, and R.J. Nicholas ... 549

Magnetooptical evidence of many-body effects in spin-polarized 2D electron gas
C. Testelin, A. Lemaître, C. Rigaux, T. Wojtowicz, G. Karczewski, and F. Teran ... 551

Forbidden interband transitions and the indirect valence band in strained GaInAs/InP quantum wells
A. Dörnen, J. Shao, V. Härle, and F. Scholz ... 553

Magnetic field and pressure effects of photoluminescence in a GaP/AlP short-period superlattice
K. Uchida, N. Miura, F. Issiki, and Y. Shiraki ... 555

Optical properties of interacting excitons in quantum wells
S. de-Leon and B. Laikhtman ... 557

Recombination processes of GaNAs/GaAs structures : Effect of rapid thermal annealing
I.A. Buyanova, W.M. Chen, G. Pozina, P.N. Hai, N.Q. Thinh, H.P. Xin, and C.W. Tu ... 559

Dressed excitons and shallow impurities in low-dimensional semiconductor systems within a renormalized effective-mass approach
H.S. Brandi, A. Latgé, and L.E. Oliveira   561

Superlinear photoluminescence in GaAs/(GaAl)As heterojunctions
J.X. Shen, R. Pittini, and Y. Oka   563

The exciton dead layer revisited
M. Combescot, R. Combescot, and B. Roulet   565

Picosecond dynamics of polarized resonance photoluminescence in the GaAs/AlGaAs- superlattices
I.Ya. Gerlovin, Yu.K. Dolgikh, S.A. Eliseev, V.V. Ovsyankin, Yu.P. Efimov, I.V. Ignatiev, I.E. Kozin, V.V. Petrov, and Y. Masumoto   567

Magneto-photoreflectance of the above barrier state transitions in $GaAs/Al_{0.3}Ga_{0.7}As$ double quantum wells
G. Sek, M. Nowaczyk, L. Bryja, K. Ryczko, J. Misiewicz, M. Bayer, J. Koeth, and A. Forchel   569

Indirect transitions between barrier (X) electrons and two-dimensional hole gas in mixed type I - type II quantum wells
R. Guliamov, E. Lifshitz, E. Cohen, A. Ron, and L.N. Pfeiffer   571

Electronic states of interface Al-2p core excitons in GaAs/AlAs/GaAs heterostructures
K. Inoue, Y. Ishiwata, and S. Shin   573

Enhancement of Fermi energy optical emission induced by the band structure in strained-layer InGaAs/InP quantum wells
H.A.P. Tudury, F. Iikawa, E. Ribeiro, J.A. Brum, W. Carvalho Jr., A.A. Bernussi, and A.L. Gobbi   575

Band offset determination and excitons in SiGe/Si (001) quantum wells
H.H. Cheng, S.T. Yen, and R.J. Nicholas   577

4.5 Nonlinear Optical Studies

Few-cycle THz spectroscopy of semiconductor quantum structures (invited)
K. Unterrainer, R. Bratschitsch, T. Müller, R. Kersting, J.N. Heyman, and G. Strasser   579

Extreme mid-infrared nonlinear optics in semiconductors
J. Kono, A.H. Chin, and O.G. Calderson   583

Fine structure of the amplified spontaneous emission of ZnSe laser structures
R. Heinecke, U. Neukirch, P. Michler, and J. Gutowski   585

Biexcitonic signatures in femtosecond pulse propagation
J. Meinertz, I. Gösling, U. Neukirch, and J. Gutowski   587

Femtosecond response times of large optical nonlinearity in low-temperature grown GaAs/AlAs multiple quantum wells
T. Okuno, Y. Masumoto, Y. Sakuma, M. Ito, and H. Okamoto   589

Asymmetric short-period GaAs/AlAs superlattices for observation of low-threshold laser effect
V.G. Litovchenko, D.V. Korbutyak, S.G. Krylyuk, A.I. Bercha, H.T. Grahn, and K.H. Ploog   591

Nonlocality induced size-enhancement of excitonic nonlinear response in high quality samples
H. Ishihara, K. Cho, K. Akiyama, N. Tomita, Y. Nomura, and T. Isu   593

Recombination mechanism of anti-Stokes photoluminescence in partially ordered GaInP-GaAs heterostructure
S.J. Xu, Q. Li, H. Wang, M.H. Xie, S.Y. Tong, and J.R. Dong   595

Spontaneous Raman scattering in GaP-AlGaP heterostructure waveguides
T. Saito, K. Suto, M. Kawasaki, T. Kimura, A. Watanabe, and J. Nishizawa   597

Generation mechanisms of coherent phonons in quantum wells revealed
K.J. Yee, Y.S. Lim, and D.S. Kim   599

4.6 Carrier Dynamics and Relaxation

Probing and controlling spin-relaxation in GaAs quantum wells (invited)
Y. Ohno, R. Terauchi, T. Adachi, F. Matsukura, and H. Ohno
601

Speckle-averaged resonant Rayleigh scattering from quantum well excitons
G.R. Hayes, B. Deveaud, V. Savona and S. Haacke
605

Exciton and spin transport by surface acoustic waves in GaAs quantum wells
T. Sogawa, P.V. Santos, S.K. Zhang, S. Eshlaghi, A.D. Wieck, and K.H. Ploog
607

Coherent vs. incoherent emission in quantum wells studied by polarisation- and time-resolved spectroscopy
G. Aichmayr, L. Viña, S.P. Kennedy, R.T. Phillips, and K. Ploog
609

Ultrafast dynamics of holes in $Si_{1-x}Ge_x/Si$ multiple quantum wells
R.A. Kaindl, M. Wurm, K. Reimann, M. Woerner, T. Elsaesser, C. Miesner, K. Brunner, and G. Abstreiter
611

Mechanism of positively charged exciton spin relaxation in CdTe and CdMnTe quantum wells
E. Vanelle, P. Kossacki, J. Cibert, T. Amand, P. Renucci, X. Marie, and S. Tatarenko
613

Microscopic simulation of hot-carrier intersubband relaxation in quantum-cascade lasers
R.C. Iotti and F. Rossi
615

Reduction of Coulomb scattering in a GaAs/AlGaAs mesoscopic 2DEG disk
N. Suzumura, M. Yamaguchi, and N. Sawaki
617

Collective plasma response of interacting electrons localized in disordered GaAs/AlGaAs superlattices
Yu.A. Pusep, W. Fortunato, P.P. González-Borrero, A.I. Toropov, and J.C. Galzerani
619

Ultrafast relaxation dynamics of photoexcited carriers in an $In_{0.53}Ga_{0.47}As/InP$ multiple-quantum well
Y. Hamanaka, A. Nakamura, K. Tanase, R. Ohga, Y. Nonogaki, Y. Fujiwara, and Y. Takeda
621

Spin relaxation of negatively charged excitons in CdTe-based quantum wells
P. Kossacki, V. Ciulin, M. Kutrowski, J.-D. Ganière, T. Wojtowicz, and B. Deveaud
623

Kinetics of a low density system of indirect excitons in double quantum wells
V.I. Yudson
625

Photoexcitation-energy-dependent capture dynamics of excitons in electronically isolated GaAs quantum wells
K. Fujiwara, H.T. Grahn, L. Schrottke, and K.H. Ploog
627

Magnetic polaron dynamics at high excitation densities in $Cd_{1-x}Mn_xTe/ZnTe$ multiple quantum wells
R. Pittini, J.X. Shen, M. Takahashi, and Y. Oka
629

Excitonic vs free-carrier spin-relaxation in III-V quantum wells
A. Malinowski, P.A. Marsden, R.S. Britton, K. Puech, A.C. Tropper, and R.T. Harley
631

1S-2S-continuum coupling and dephasing in 5 nm GaAs quantum wells
A.A. Busch, J.M. Watson, P. Paddon, Z. Wasilewski, and J.F. Young
633

Electric-field-dependent carrier dynamics in an (Al,Ga)As/GaAs double-quantum-well superlattice
L. Schrottke, R. Hey, and H.T. Grahn
635

Direct and indirect radiative recombination in strongly excited ZnSe/BeTe superlattices
A.A. Maksimov, S.V. Zaitsev, I.I. Tartakovskii, V.D. Kulakovskii, N.A. Gippius, D.R. Yakovlev, W. Ossau, G. Reuscher, A. Waag, and G. Landwehr
637

Radiative recombination and spin relaxation of excitons in $Cd_{1-x}Mn_xTe$ quantum wells
M.C. Debnath, J.X. Shen, I. Souma, R. Pittini, and Y. Oka
639

Nonequilibrium electrons in double quantum well structures: Boltzmann equation approach
S. Khan-ngern and I.A. Larkin
641

Direct study of spin relaxation processes in 2D ZnCdSe/ZnSe systems through time resolved measurements of polarized exciton emission
S. Permogorov, Y. Oka, R. Pittini, J.X. Shen, K. Kayanuma, A. Reznitsky, L. Tenishev, and S. Verbin
643

4.7 Microcavities

Mixed neutral and negatively charged microcavity polaritons (invited)
R. Rapaport, E. Cohen, A. Ron, E. Linder, R. de Picciotto, R. Harel, and L.N. Pfeiffer   645

Confined optical modes in photonic molecules and crystals (invited)
M. Bayer, A. Forchel, T.L. Reinecke, and P.A. Knipp   649

Non-linear spin polarization dynamics in semiconductor microcavities
P. Renucci, X. Marie, T. Amand, M. Paillard, P. Senellart, and J. Bloch   653

Optically pumped quantum dot lasers using high-Q microdisk cavities
P. Michler, A. Kiraz, C. Becher, Lidong Zhang, E. Hu, A. Imamoglu, W.V. Schoenfeld, P.M. Petroff   655

Incoherent amplification phenomena in semiconductor microcavities
G. Dasbach, T. Baars, M. Bayer, A. Larionov, and A. Forchel   657

Selectively in situ probing of self-assembled InGaAs quantum dots in a planar GaAs microcavity by angle-resolved detection of photoluminescence spectrum
J.H. Chen, J.H. Zhao, F.H. Yang, P.H. Tan, J.D. Zhang, Y.P. Zeng, J.Q. You, and H.Z. Zheng   659

Spin quantum beats of exciton-polariton in semiconductor microcavities
P. Renucci, X. Marie, T. Amand, M. Paillard, and E. Vanelle   661

Observation of enhanced spontaneous emission coupling factor in blue InGaN microcavities
S. Kako, T. Someya, and Y. Arakawa   663

Ballistic transport of exciton-polaritons in a graded quantum microcavity
B. Sermage, G. Malpuech, A. Kavokin, and V. Thierry-Mieg   665

Ultrafast polarization switching in a CdTe microcavity
M.D. Martín, H. Davies, L. Viña, and R. André   667

Cavity QED of quantum dots embedded in dielectric microspheres
H. Wang, X. Fan, and M. Lonergan   669

Polarization of magnetopolaritons in a semiconductor microcavity
M.D. Martín, S. Burgas, M. Alonso, L. Viña, F.J. Terán, M. Potemski, and E.E. Mendez   671

Rabi splitting enhancement in planar semiconductor microcavities
A. D'Andrea and L. Pilozzi   673

Luminescence properties of CdS quantum dots embedded in monolithic II-VI microcavity
T. Tawara, H. Yoshida, H. Kumano, S. Tanaka, and I. Suemune   675

Strongly detuned IV-VI microcavity and microdisk resonances: mode splitting and lasing
T. Schwarzl, W. Heiss, G. Springholz, S. Gianordoli, G. Strasser, M. Aigle, and H. Pascher   677

Bottleneck and resonant enhancement in free electron – cavity polaritons scattering in GaAs/AlGaAs microcavities
R. Rapaport, A. Qarry, E. Cohen, A. Ron, E. Linder, and L.N. Pfeiffer   679

Biexcitons or bipolaritons in a semiconductor microcavity?
P. Borri, W. Langbein, U. Woggon, J.R. Jensen, and J.M. Hvam   681

Theory of microcavity resonant Rayleigh scattering
D.M. Whittaker   683

Spatially extended cavity polaritons arising from weakly confined excitons
H. Ishihara and J. Kishimoto   685

Stimulation of polariton emission in a homogeneously broadened semiconductor microcavity
V. Mizeikis, J. Erland, J.R. Jensen, N.A. Mortensen, and J.M. Hvam   687

Bloch oscillations of light in laterally confined Bragg mirrors and multiple coupled microcavities
G. Malpuech, A. Kavokin, A. Di Carlo, P. Lugli, and G. Panzarini   689

Enhanced light transmission through nano-structured surfaces
S. Meinlschmidt, R. Windisch, A. Knobloch, P. Kiesel, P. Heremans, and G.H. Döhler   691

Optically pumped lasing at 1.3 μm of GaInNAs-based VCSEL structures
M. Hetterich, M.D. Dawson, A.Yu. Egorov, and H. Riechert
693

The interaction between exciton states near the quantum well bandgap and confined photons in GaAs/AlAs microcavities
A. Qarry, R. Rapaport, E. Cohen, A. Ron, E. Linder, and L.N. Pfeiffer
695

Resonant Raman scattering in an InAs/GaAs monolayer structure
J. Maultzsch, S. Reich, A.R. Goñi, and C. Thomsen
697

Light-exciton coupling in semiconductor microcavities of cylindrical and spherical symmetry
R.A. Abram, S. Brand, M.A. Kaliteevski, V.V. Nikolaev, M.V. Maximov, N.N. Ledentsov, C.M. Sotomayor Torres, and A.V. Kavokin
699

4.8 Near-Field Studies

Near-field spectroscopy of delocalized excitons in single quantum wires
F. Intonti, V. Emiliani, Ch. Lienau, T. Elsaesser, R. Nötzel, and K.H. Ploog
701

Scanning near-field optical spectroscopy of buried semiconductor heterostructures
M. Hauert, R. Roshan, A.C. Maciel, J. Kim, J.F. Ryan, A. Schwarz, A. Kaluza, Th. Schäpers, and H. Lüth
703

Hints for a Non-thermal distribution of excitons in CdSe/ZnSe quantum islands
G. von Freymann, E. Kurtz, C. Klingshirn, Th. Schimmel, and M. Wegener
705

4.9 Intersubband Transitions

Quantum optics and the observation of electromagnetically induced transparency with QW subbands (invited)
C.C. Phillips, E. Paspalakis, G.B. Serapiglia, C. Sirtori, and K.L. Vodopyanov
707

Towards quantum well hot hole lasers
P. Kinsler and W.Th. Wenckebach
711

Character of electronic excitations in GaAs-AlGaAs quantum structures
E. Ulrichs, C. Steinebach, C. Schüller, Ch. Heyn, W. Hansen, and D. Heitmann
713

Correlation of vertical transport and infrared absorption in GaAs/AlGaAs superlattices
M. Helm and G. Strasser
715

Influence of in-plane magnetic field on cyclotron resonance in double-layer two dimensional electron system
H. Aikawa, S. Takaoka, A. Kuriyama, K. Oto, K. Murase, S. Shimomura, S. Hiyamizu, T. Jungwirth, and L. Smrčka
717

Circular photogalvanic effect in p-GaAs/AlGaAs MQW
S.D. Ganichev, E.L. Ivchenko, H. Ketterl, L.E. Vorobjev, M. Bichler, W. Wegscheider, and W. Prettl
719

Cascading effect in type-II InAs/GaSb/AlSb intersubband light emitter
K. Ohtani, H. Sakuma, and H. Ohno
721

Magneto-optical transitions involving a 2DEG confined in Cd(Mn)Te/CdMgTe quantum wells
F.J. Teran, M.L. Sadowski, M. Potemski, P. Kossacki, P. Hawrylak, and G. Karczewski
723

Comparison of intersubband relaxation times in GaN/AlGaN and in InGaAs/AlGaAs quantum wells
T. Asano, S. Yoshizawa, S. Noda, N. Iizuka, K. Kaneko, N. Suzuki, and O. Wada
725

Electronic inelastic light scattering in a periodic δ-doping GaAs multiple quantum well structure
C. Kristukat, A.R. Goñi, S. Rutzinger, W. Wegscheider, G. Abstreiter, and C. Thomsen
727

Intersubband electroluminescence using X-Γ carrier injection in a GaAs/AlAs double-quantum-well superlattice
C. Domoto, N. Ohtani, K. Kuroyanagi, P.O. Vaccaro, T. Nishimura, H. Takeuchi and M. Nakayama
729

Hot electron optical phenomena in GaAs/AlAs MQW structures in strong lateral electric field
L.E. Vorobjev, S.N. Danilov, I.E. Titkov, D.A. Firsov, V.A. Shalygin, A.E. Zhukov, A.R. Kovsh, V.M. Ustinov, V.Ya. Aleshkin, A.A. Andronov, E.V. Demidov, and Z.F. Krasilnik
731

Novel mid-infrared laser designs based on intraband and interband carrier transitions in quantum wells
L.E. Vorobjev, D.A. Firsov, G.G. Zegrya, V.L. Zerova, E. Towe, J.W. Cockburn, and Z.F. Krasil'nik   733

5. Heterostructures and Superlattices: Transport

5.1 Transport, Magneto-Transport, and Coherent Effects

Is there a true metallic state in two dimensions? (invited)
M.Y. Simmons, A.R. Hamilton, M. Pepper, E.H. Linfield, P.D. Rose, and D.A. Ritchie   735

Self-induced Shapiro effect in semiconductor superlattices
F. Löser, M.M. Dignam, Yu.A. Kosevich, K. Köhler, and K. Leo   739

Effects of a parallel magnetic field on the novel metallic behavior in two dimensions
K. Eng, X.G. Feng, D. Popović, and S. Washburn   741

Optics with ballistic electrons: Anti-reflection coatings for GaAs-AlGaAs superlattices
C. Pacher, G. Strasser, E. Gornik, F. Elsholz, A. Wacker, and E. Schöll   743

Chaotic quantum transport in superlattices
T.M. Fromhold, A.A. Krokhin, A.E. Belyaev, C.R. Tench, S. Bujkiewicz, P.B. Wilkinson, F.W. Sheard, L. Eaves, and M. Henini   745

The magnetoresistance of a two-dimensional electron gas in the presence of a spatially random magnetic field
A.W. Rushforth, B.L. Gallagher, P.C. Main, C.H. Marrows, B.J. Hickey, E.D. Dahlberg, A.C. Neumann, and M. Henini   747

Giant negative magnetoresistance in two-dimensional antidot arrays: ballistic orbital effect and ballistic weak localization
T. Osada, H. Nakamura, and Y. Shiraki   749

Correlation of optical and transport properties of (AlGa)As/GaAs heterostructures
L. Gottwaldt, F.J. Ahlers, E.O. Göbel, G. Hein, S. Nau, K. Pierz, and W. Stolz   751

A method of determining potential barrier heights at semiconductor heterointerfaces
G.-H. Kim, H.-S. Sim, M.Y. Simmons, D.A. Ritchie, C.-T. Liang, A.C. Churchill, and W.S. Han   753

Conductance quantization in an array of ballistic constrictions
S. de Haan, A. Lorke, J.P. Kotthaus, W. Wegscheider, and M. Bichler   755

Transport in asymmetric two-dimensional lateral surface superlattices
S. Chowdhury, A.R. Long, J.H. Davies, D.E. Grant, E. Skuras, and C.J. Emeleus   757

Suppression of miniband transport in magnetic field: role of injection process
A.A. Krokhin, A.E. Belyaev, T.M. Fromhold, C.R. Tench, H.M. Murphy, S. Bujkiewicz, P.B. Wilkinson, F.W. Sheard, L. Eaves, and M. Henini   759

Quantitative evaluation of electron-electron scattering rate in two-dimensional electron gas by magnetic lateral superlattice
M. Kato, A. Endo, M. Sakairi, M. Hara, S. Katsumoto, and Y. Iye   761

Shubnikov-de Haas effect with several filled subbands
U. Ekenberg   763

On the origin of beat patterns in the quantum magneto-resistance of gated InAs/GaSb and InAs/AlSb quantum wells
A.C.H. Rowe, R.S. Ferguson, and R.A. Stradling   765

Rashba spin splitting in 2D electron and hole systems: Implications for the metal-insulator transition
R. Winkler   767

Weak localization in periodically modulated magnetic field
Y. Blum, A. Tsukernik, A. Palevski, T.A. Shutenko, B.L. Altshuler, I.L. Aleiner, A. Rudra, and E. Kapon   769

Intrinsic mobility limits in polarization induced two-dimensional electron gases
D. Jena, Y.P. Smorchkova, C.R. Elsass, A.C. Gossard, and U.K. Mishra   771

Combined S- and Z-shaped current bistability induced by charging of quantum dots
A.E. Belyaev, L. Eaves, S.A. Vitusevich, P.C. Main, M. Henini, A. Förster, N. Klein, and S.V. Danylyuk   773

Interaction of Mu with spin current in GaAs/GaAsP/Si
E. Torikai, Y. Ikedo, A. Ihori, K. Shimomura, K. Nagamine, T. Saka, and T. Kato   775

Metallic behaviour and temperature dependent screening in p-SiGe
V. Senz, T. Ihn, T. Heinzel, K. Ensslin, G. Dehlinger, U. Gennser, and D. Gruetzmacher   777

Anisotropic magneto-transport properties of 70 nm-period lateral surface superlattices in high magnetic
fields
M. Akabori, J. Motohisa, and T. Fukui   779

Thermopower of a two-dimensional antidot lattice
A.G. Pogosov, M.V. Budantsev, A.E. Plotnikov, A.K. Bakarov, and A.I. Toropov   781

Magnetotransport in two-dimensional lateral superlattices in strongly coupled electron-hole gases
B. Kardynal, R.J. Nicholas, J. Rehman, K. Takashina, and N.J. Mason   783

Semiclassical origin of the 2D metallic state in high mobility Si-MOS and Si/SiGe structures
G. Brunthaler, A. Prinz, G. Pillwein, G. Bauer, K. Brunner, G. Abstreiter, T. Dietl, and V.M. Pudalov   785

Zero-bias conductance anomaly in GaAs/AlGaAs modulation doped field-effect transistors
S. Skaberna, U. Kunze, D. Reuter, and A.D. Wieck   787

Magnetotransport and capacitance investigations of strongly coupled but spatially separated 2D electron
and hole systems
M. Pohlt, M. Lynass, W. Dietsche, K. von Klitzing, K. Eberl, and R. Mühle   789

First principles study of spin-electronics: Zero-field spin-splitting in superlattices
J.A. Majewski, P. Vogl, and P. Lugli   791

A comparative study of 'metallic' and 'insulating' behaviour of the two-dimensional electron gas on (100)
and vicinal surfaces of Si MOSFET
S.H. Roshko, S.S. Safonov, A.K. Savchenko, A.G. Pogosov, and Z.D. Kvon   793

The role of surface-localized states in the in-plane transport properties of superlattices
A.B. Henriques   795

High-mobility heterostructure as a new kind of chaotic billiards
L.D. Shvartsman   797

Gate-controlled very large spontaneous spin-splittings in normal $In_{0.75}Ga_{0.25}As/In_{0.75}Al_{0.25}As$ heterojunctions grown on GaAs substrates
S. Yamada, Y. Sato, S. Gozu, and T. Kikutani   799

Field domains in semiconductor superlattices: Dynamic scenarios of multistable switching
A. Amann, A. Wacker, L.L. Bonilla, and E. Schöll   801

Angular dependent magnetoresistance oscillations in semiconductor superlattice with incoherent interlayer
coupling
M. Kuraguchi, E. Ohmichi, T. Osada, and Y. Shiraki   803

Optical and transport properties of modulation doped InAs/GaAs superlattices
V.A. Kulbachinskii, R.A. Lunin, V.G. Kytin, A.V. Golikov, V.A. Rogozin, V.G. Mokerov, Yu.V. Fedorov, and A.V. Hook   805

Weak localization effects in a wide parabolic quantum well
N.M. Sotomayor, G.M. Gusev, J.R. Leite, N.T. Moshegov, and A.I. Toropov   807

Magnetotransport properties of multisubband semiconductor structures
N.S. Averkiev, L.E. Golub, S.A. Tarasenko, and M. Willander   809

Effect of buffer layer thickness on improvement of modulation doped CdTe/CdMgTe heterostructures grown on GaAs substrate
D. Wasik, M. Baj, L. Dmowski, J. Siwiec-Matuszyk, J. Przybytek, E. Janik, T. Wojtowicz, and G. Karczewski 811

Domain formation in a one-dimensional superlattice with phonon scattering
L.G. Mourokh, A.Yu. Smirnov, N.J.M. Horing, and V.I. Gavrilenko 813

Asymmetric carrier diffusion and phonon-wind-driven transport in an InGaAs-InP quantum well
A.F.G. Monte, S.W. da Silva, P.C. Morais, J.M.R. Cruz, and A.S. Chaves 815

5.2 Magnetic Superlattices

Nonzero Hall resistance in a spatially fluctuating magnetic field with zero mean
A.A. Bykov, G.M. Gusev, J.R. Leite, A.K. Bakarov, N.T. Moshegov, D.K. Maude, M. Casse, and J.C. Portal 817

5.3 Tunneling and Resonant Tunneling

Optical detection of ballistically injected electrons in III/V heterostructures
M. Kemerink, K. Sauthoff, P.M. Koenraad, J.W. Gerritsen, H. van Kempen, and J.H. Wolter 819

Electron tunneling time between double quantum dot
A. Tackeuchi, Y. Nakata, T. Kuroda, K. Mase, and N. Yokoyama 821

Direct measurement of the AlAs X-band Fermi surface
H.-S. Im, L.E. Bremme, P.C. Klipstein, A.V. Kornilov, R. Grey, and G. Hill 823

Miniband transport and Stark-cyclotron-resonance in InAs/GaSb superlattices
V.J. Hales, R.J. Nicholas, and N.J. Mason 825

Effective mass anisotropy of Γ-electrons in GaAs/AlGaAs quantum wells due to interface band mixing
T. Reker, H. Im, H. Choi, L.E. Bremme, Y.C. Chung, R. Grey, G. Hill, and P.C. Klipstein 827

Vertical transport and interband luminescence in type II InAs/GaSb/InAs heterostructures
M. Roberts, N.J. Mason, S.G. Lyapin, Y.C. Chung, and P.C. Klipstein 829

Resonant tunneling of holes under in-plane uniaxial stress
Y.C. Chung, T. Reker, L.E. Bremme, R. Grey, and P.C. Klipstein 831

Resonant tunneling through zero-dimensional impurity states: Effects of a finite temperature
P. König, U. Zeitler, J. Könemann, T. Schmidt, and R.J. Haug 833

Temperature dependence of resonant tunneling characteristics in a p-type GaAs/AlAs double-barrier structure
M. Ono, N. Nishioka, M. Morifuji, and C. Hamaguchi 835

Negative differential resistance and current self-oscillation in doped GaAs/AlAs superlattices
J.N. Wang, C.Y. Li, X.R. Wang, B.Q. Sun, Y.Q. Wang, W.K. Ge, D.S. Jiang, and Y.P. Zeng 837

Photocurrent self-oscillations in weakly coupled, type-II GaAs/AlAs superlattices embedded in p-i-n and n-i-n diodes
N. Ohtani, M. Rogozia, C. Domoto, T. Nishimura, and H.T. Grahn 839

Zener-phonon resonances in the quantum transport of multiband semiconductor superlattices
P. Kleinert 841

Correlation between a remote electron and a two-dimensional electron gas in resonant tunneling devices
H. Kato and F.M. Peeters 843

Spin effects in InAs quantum dots: Tunneling experiments in tilted magnetic fields
J.M. Meyer, I. Hapke-Wurst, U. Zeitler, R.J. Haug, H. Frahm, A.G.M. Jansen, and K. Pierz 845

Carrier transport affected by hole-subband resonances in a strained GaAs/InAlAs superlattice
M. Hosoda, K. Kuroyanagi, N. Ohtani, and T. Aida 847

Disorder-enhanced tunneling transport through doping barriers
R. Elpelt, O. Wolst, H. Willenberg, S. Malzer, and G.H. Döhler 849

Response time of the double-barrier heterostructures with resonant tunneling
M.N. Feiginov
                                                                                                                851

Current self-oscillations with discrete frequencies in weakly coupled semiconductor superlattices
M. Rogozia, H.T. Grahn, and R. Hey
                                                                                                                853

Observation of the scattered electrons in the resonant tunneling regime using a three-terminal quantum-well heterostructure
G.-G. Kim, D.-W. Roh, S.-W. Paek, K.-M. Koh, K.E. Pyun, and C.-H. Kim
                                                                                                                855

5.4 Hopping, Localization

Nature of the localized phase in a two-dimensional electron system
I.E. Itskevich, R.J.A. Hill, S.T. Stoddart, H.M. Murphy, A.S.G. Thornton, P.C. Main, L. Eaves, M. Henini, D.K. Maude, and J.C. Portal
                                                                                                                857

Evidence for screening breakdown near the metal-to-insulator transition in two dimensions
W. Jantsch, Z. Wilamowski, N. Sandersfeld, and F. Schäffer
                                                                                                                859

Coulomb interaction and density of states in amorphous Si_xGe_{1-x} films
K. Nakada, K. Nara, N. Aoki, and Y. Ochiai
                                                                                                                861

Studies of localization in an interacting two-electron system
J. Talamantes, M. Pollak, and I. Varga
                                                                                                                863

5.5 Phonons, Plasmons

Bridging the gap with cleaved edge overgrowth superlattices : on minigaps, magnetic breakdown and quantum interference in artificial bandstructures (invited)
R.A. Deutschmann, W. Wegscheider, C. Albrecht, J.H. Smet, M. Rother, M. Bichler, and G. Abstreiter
                                                                                                                865

Electron wires driven by a surface acoustic wave and nonlinear acoustoelectric interactions in quantum wells
A.O. Govorov, A.V. Kalameitsev, V.M. Kovalev, H.-J. Kutschera, M. Streibl, M. Rotter, and A. Wixforth
                                                                                                                869

Generation and detection of picosecond acoustic phonon pulses in a double quantum well structure
I. Ishii, O. Matsuda, T. Fukui, J.J. Baumberg, and O.B. Wright
                                                                                                                871

Amplification and generation of high-frequency coherent acoustic phonons under the drift of 2D-electrons
M.A. Stroscio, S.M. Komirenko, K.W. Kim, A.A. Demidenko, and V.A. Kochelap
                                                                                                                873

Observation of a two-dimensional plasmon in a metallic monolayer on silicon surface
T. Nagao, T. Hildebrandt, M. Henzler, and S. Hasegawa
                                                                                                                875

First-and second-order Raman spectroscopy of $^{70}Ge_n/^{76}Ge_n$ isotope superlattices
K. Morita, K.M. Itoh, M. Nakajima, H. Harima, K. Mizoguchi, Y. Shiraki, and E.E. Haller
                                                                                                                877

Raman scattering in Ge quantum dot superlattices
A. Milekhin, N. Stepina, A. Yakimov, A. Nikiforov, S. Schulze, T. Kampen, and D.R.T. Zahn
                                                                                                                879

Plasmon-phonon coupled mode excitation in a hot 2DEG observed by a phonon pulse technique
J.K. Wigmore, H.A. Al Jawhari, A.G. Kozorezov, and M. Sahraoui-Tahar
                                                                                                                881

Peculiarities of phonons in strained short-period GaN/AlN superlattices: A *first-principles* study
J.-M. Wagner and F. Bechstedt
                                                                                                                883

5.6 High-Frequency Transport, Microwaves, Noise, Short Pulses

Study of ambipolar diffusion and drift of spatially separated charge carriers
M. Beck, M. Vitzethum, D. Streb, P. Kiesel, C. Metzner, S. Malzer, and G.H. Döhler
                                                                                                                885

Localization of carriers and Bloch oscillations in quantum dot superlattices in dc electric field
R.A. Suris and I.A. Dmitriev
                                                                                                                887

A novel quantum transport mechanism in biased quantum-box superlattices under terahertz irradiation
P. Kleinert and V.V. Bryksin
                                                                                                                889

On the spectra of field and current oscillations due to laser irradiation in superlattices
Yu.A. Romanov and Ju.Yu. Romanova
                                                                                                                891

1 Plenary Papers

Proc. 25th Int. Conf. Phys. Semicond., Osaka 2000 (Eds. N. Miura and T. Ando)

Twenty years of quantum Hall effect

K. von Klitzing

Max-Planck-Institut für Festkörperforschung
Heisenbergstr. 1, 70569 Stuttgart, Germany

Twenty years ago, at the 15th ICPS in Kyoto, a talk about the Quantum Hall Effect (QHE) could not be included since its discovery was after the deadline for the submission of paper. In addition, the question arose whether an effect which seems to be material independent has something to do with semiconductor physics. Today the QHE is a synonym for electrons in strong magnetic fields with connections to other research areas like astrophysics (edge states in gravity and black hole physics) or high energy physics (quantum Hall quarks). Experimentally, the GaAs heterostructure is the best characterized material used for QHE research and most of the fundamental discoveries in this field are made with this semiconductor. Highlights are the discovery of the fractional QHE with excitations described as fractional charges, the introduction of the composite fermion picture, the spin- and pseudospin phonomena including skyrmions, coupled quantum wells and electron spin-nuclear spin interactions, the discussion of the QHE on the basis of the Landauer formalism and edge channels (Luttinger liquid) or the speculations about striped phases and charge density waves. The talk covers both, the historical aspect of the QHE, including its application in metrology and a very incomplete presentation of new experimental results, which demonstrates the variety of phenomena and open questions in this modern research field.

Optical manipulation of electron and nuclear spins in semiconductors

D. D. Awschalom

Department of physics, University of California
Santa Barbara, CA 93106 USA

While conventional electronic devices rely on charge for the transport of information, electron spin has recently demonstrated potential as a storage medium for classical and quantum information within semiconductors [1]. This may eventually form the basis for new paradigms of device operation with improved speed and fundamentally different functionality. Femtosecond-resolved optical experiments reveal a remarkable resistance of quantum spin states to environmental decoherence in a variety of semiconductors and nanostructures, proving hope for the use of these systems as a foundation for the emerging field of 'spintronics'. Optical pulses are used to create a superposition of the basis spin states defined by an applied magnetic field, and to follow the phase, amplitude, and location of the resulting electronic spin precession. The data reveal that spin lifetimes can exceed 100 nanoseconds and that spin packets can be transported 100's of microns. Spatial imaging and dynamical magnetometry of local fields monitors decoherence and dephasing of itinerant spin information as it flows not only through semiconductors and across dissimilar material interfaces [2], but also into systems of localized moments such as nuclei. Periodic excitation of the electronic spin system can be used to resonantly operate on the nuclear spin system of the semiconductor host and to detect associated changes in the nuclear magnetization, thereby demonstrating all-optical NMR [3]. In an effort to electrically drive these dynamics, we discuss the fabrication of all-semiconductor light-emitting spintronic devices, where electrical spin injection occurs in zero magnetic field [4]. These results provide exciting opportunities for the transport and storage of quantum information in the solid state.

References

1. D.D. Awschalm and J.M. Kikkawa, Physics Today 52, 33 (1999)
2. I. Malajovich et al., Phys. Rev. Lett. 84, 1015 (2000)
3. J.M. Kikkawa and D.D. Awschalom, Science 287, 473 (2000)
4. Y. Ohno et al., Nature 402, 790 (1999)

Controlling the conductivity of wide-band-gap semiconductors

Chris G. Van de Walle[1], J. Neugebauer[2]

[1] Xerox Palo Alto Research Center, 3333 Coyote Hill Road, Palo Alto, CA 94304, USA e-mail: `vandewalle@parc.xerox.com`
[2] Fritz-Haber-Institut der Max-Planck-Gesellschaft, Faradayweg 4-6, D-14 195 Berlin-Dahlem, Germany

Abstract Wide-band-gap semiconductors often exhibit limitations in the ability to control n-type or p-type doping. A theoretical framework is presented for studying doping of semiconductors, with key parameters derived from first-principles calculations. The formalism is illustrated with examples for GaN and ZnO, where increased understanding can lead to improved control of doping through *defect and impurity engineering*.

1 Introduction

The ability to control n-type and p-type conductivity is essential for design and fabrication of electronic and optoelectronic devices. Such conductivity control has traditionally been very difficult in wide-band-gap semiconductors, and native point defects have often been invoked to explain these problems. We will describe the formalism that we have developed to address the energetics of defect and impurity incorporation, and the effects of these species on the electronic properties. While building on notions developed in the 1950s, the current formulation in terms of formation energies and chemical potentials allows more direct insight into the mechanisms that govern incorporation of intrinsic and extrinsic defects.

Progress in computational physics now allows explicit evaluation of formation energies and the resulting defect and impurity concentrations entirely from first principles, i.e., without any input from experiment. Such state-of-the-art calculations are based on density functional theory and pseudopotentials, and carried out in a supercell geometry [1]. These fundamental concepts will be illustrated with examples from our recent work on III-V nitride and II-VI oxide semiconductors. We will review some key results for GaN, showing that the "conventional wisdom" regarding the role of native defects in doping and compensation is not always correct. For many years, nitrogen vacancies were thought to be the major source of the n-type conductivity that is commonly observed in as-grown GaN; our computational results showed, however, that formation of nitrogen vacancies is energetically unfavorable in n-type GaN [2]. We will also discuss the challenge of p-type doping in III-nitrides, and various strategies for impurity engineering that are aimed at overcoming the doping limitations [3].

We will also present some recent results for ZnO, a semiconductor with a great variety of applications. Recently, ZnO has come to the forefront as a promising candidate for new optoelectronic and electronic devices, but this potential can only be fulfilled if the conductivity of the material can be brought under control. Our calculations elucidate the role of native defects – showing, for instance, that the oxygen vacancy is a deep, rather than a shallow donor. As in GaN, native point defects are unlikely to be the source of the widely observed unintentional n-type conductivity. As an alternative source of n-type doping, we will discuss the incorporation of hydrogen impurities [4]. Hydrogen acts as a donor in ZnO – in contrast to its amphoteric behavior in other semiconductors. These results provide important guidelines for optimizing dopant impurity incorporation and activation.

The theoretical approach described here is quite general in nature, and can be applied to any semiconductor (as well as other materials) in which impurities and defects play a role. The current applications to wide-band-gap semiconductors constitute a nice illustration of the power of the approach.

2 Methodology

2.1 First-principles calculations

First-principles computational theory has significantly contributed to the fundamental understanding of defects and impurities in semiconductors. Such calculations yield the following types of information:

- microscopic structure, including atomic relaxations of the host atoms, derived by minimizing the total energy with respect to the atomic coordinates
- charge densities, which provide insight into the chemical bonding
- wave functions, which can be used for calculating hyperfine parameters, allowing direct comparison with experimental results obtained, e.g., with electron paramagnetic resonance
- band structures, which contain information about the optical and electronic properties
- frequencies of local vibrational modes (derived by evaluating the energy change or forces when atoms are displaced from their equilibrium positions)
- total energies, which allow evaluation of the relative stability of various configurations.

In addition, we have developed the methodology for using the total energies to provide a direct measure of the abundance of a defect or impurity in the crystal, as described in Sect. 2.2.

The first-principles calculations used in our studies are founded on density-functional theory, using a supercell geometry and *ab initio* pseudopotentials [5]. It is widely known that the density-functional approach underestimates the band gap; this may affect the results for point defects that have occupied electronic states in the gap. Significant improvements in the band gaps of wide-band-gap semiconductors have been obtained by introducing self-interaction corrections. In particular, we have used an approach based on self-interaction and relaxation-corrected pseudopotentials, as described in Refs. [6] and [7]. Further details and references for the computational approach can be found in Refs. [2] and [4].

2.2 Formalism for calculating defect and impurity concentrations

The equilibrium concentration of an impurity or point defect is given by

$$c = N_{\text{sites}} \exp^{-E^f/k_B T} \qquad (1)$$

where E^f is the *formation energy*, N_{sites} is the number of sites (per unit volume) on which the defect or impurity can be incorporated, k_B is the Boltzmann constant, and T the temperature. Equation (1) shows that defects with a *high* formation energy will occur in *low* concentrations.

The formation energy is not a constant but depends on the growth conditions. Let us illustrate this with the example of a nitrogen vacancy in GaN. The formation energy of this defect is determined by the relative abundance of Ga and N atoms, as expressed by the chemical potentials μ_{Ga} and μ_{N}. If the nitrogen vacancy (V_{N}) is charged (as is expected when it is electrically active), the formation energy depends further on the Fermi level (E_F), which acts as a reservoir for electrons. Forming a nitrogen vacancy requires the removal of one N atom, which is placed in the thermodynamic reservoir for N atoms, with energy μ_{N}; the formation energy is therefore:

$$E^f(V_{\text{N}}^+) = E_{\text{tot}}(V_{\text{N}}^+) - E_{\text{tot}}(\text{GaN bulk}) + \mu_{\text{N}} + E_F. \quad (2)$$

First-principles calculations allow explicit derivation of $E_{\text{tot}}(V_{\text{N}}^+)$, the total energy of a system containing a nitrogen vacancy. Similar expressions apply to other impurities and to the various native point defects.

Equation 1 relies on an assumption of thermodynamic equilibrium, which can be expected to be satisfied at the high temperatures at which bulk growth or metal-organic chemical vapor deposition (MOCVD) of wide-band-gap semiconductors is carried out. At lower temperatures, such as those used in molecular-beam epitaxy (MBE), deviations from equilibrium may occur.

2.3 Chemical potentials

We emphasized above that the chemical potentials are free parameters, reflecting the relative abundance of various sources in the growth system. However, the chemical potentials are subject to specific bounds. For instance, in the case of GaN the upper limit on the Ga chemical potential is given by $\mu_{\text{Ga}} = \mu_{\text{Ga[bulk]}}$; indeed, pushing the Ga chemical potential to higher values would result in precipitation of bulk Ga rather than growth of GaN. This

upper limit on μ_{Ga} places a lower limit on μ_{N}, because in equilibrium: $\mu_{\text{Ga}} + \mu_{\text{N}} = E_{\text{tot}}[\text{GaN}]$ where $E_{\text{tot}}[\text{GaN}]$ is the total energy of a two-atom unit of bulk GaN, calculated for the structurally optimized wurtzite structure. For the N-rich case, the upper limit on μ_{N} is given by $\mu_{\text{N}} = \mu_{\text{N}[\text{N}_2]}$, i.e., the energy of N in an N_2 molecule at $T = 0$; this yields a lower limit on μ_{Ga}.

For ease of presentation, we will choose specific values for the atomic chemical potentials when plotting our results; however, a general case can always be addressed by referring back to Eq. (2).

The Fermi level E_F is not an independent parameter, but is always determined by the condition of charge neutrality. In principle, equations such as (2) can be formulated for every native defect and impurity in the material; the complete problem (including free-carrier concentrations in valence and conduction bands) can then be solved self-consistently, imposing charge neutrality. However, it is instructive to plot formation energies as a function of E_F in order to examine the behavior of defects and impurities when the doping level changes.

The formalism described here is in principle equivalent to writing down mass-action relations for all defect reactions, as discussed, e.g., in Ref. [8]. However, the use of formation energies and chemical potentials (rather than partial pressures) renders the process amenable to general solution. In addition, the recent advances in computational physics now allow derivation of the key parameters from first principles. Indeed, the state-of-the-art calculations used to derive E_{tot} in Eq. (2) do not require any adjustable parameters or any input from experiment.

3 Results for GaN

3.1 Native defects and n-type doping

Figure 1 shows calculated formation energies for all native point defects in GaN, as a function of Fermi level, for Ga-rich conditions. The zero of E_F is located at the top of the valence band. For each defect, only the line segment is shown that corresponds to the charge state that gives rise to the lowest energy at a particular value of E_F. The change in slope of the lines therefore represents a change in the charge state of the defect [see Eq. (2)], and the Fermi-level position at which this change occurs corresponds to a transition level that can be experimentally measured.

Two major conclusions can immediately be drawn:

1. In thermodynamic equilibrium self-interstitials and antisites are too high in energy to form in significant concentrations. Only vacancies have sufficiently low energies to play a role in the electrical properties of the material.

2. Nitrogen vacancies have a high formation energy in n-type GaN; the corresponding concentration is too low to affect the electrical conductivity. Nitrogen vacancies can thus not be responsible for the commonly observed unintentional n-type conductivity in GaN.

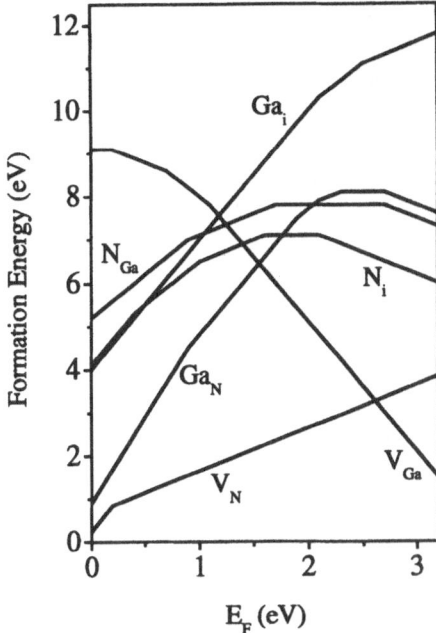

Fig. 1 Formation energies as a function of Fermi level for
native point efects in GaN under Ga-rich conditions. $E_F = 0$
corresponds to the top of the valence band.

We have, instead, attributed this conductivity to the
unintentional incorporation of donor impurities, such as
silicon or oxygen [9]. These insights allow better control
of the n-type doping of GaN, providing a good starting
point for the achievement of p-type doping.

3.2 Hydrogen and p-type doping

Magnesium is currently the main p-type dopant used
for nitrides. Unfortunately, its large ionization energy
(210 meV, see Ref. [10]) poses severe limitations. About
10^{20} cm^{-3} Mg acceptors can be incorporated in GaN,
but at room temperature, the hole concentration is only
on the order of 10^{18} cm^{-3}. Increasing the Mg concentration does not improve the conductivity; in fact, it leads
to lower hole concentrations [11]. We have performed
extensive investigations in order to understand this behavior.

Hydrogen is known to play an important role in p-
type doping of GaN [12–14]; indeed, many of the common growth techniques for GaN introduce large quantities of hydrogen into the growth environment. We have
therefore devoted particular attention to the properties
of interstitial hydrogen. The calculated formation energies for H in GaN are shown in Fig. 2.

Hydrogen prefers the positive charge state in p-type
material, acting as a donor; it prefers the negative charge
state in n-type material, acting as an acceptor. Hydrogen
is thus an amphoteric impurity and always counteracts
the prevailing conductivity of the semiconductor. This
is similar to the behavior of hydrogen in other semiconductors such as silicon [15] and GaAs [16].

Let us now focus on the behavior in p-type GaN.
Figure 3 displays formation energies of defects and im-

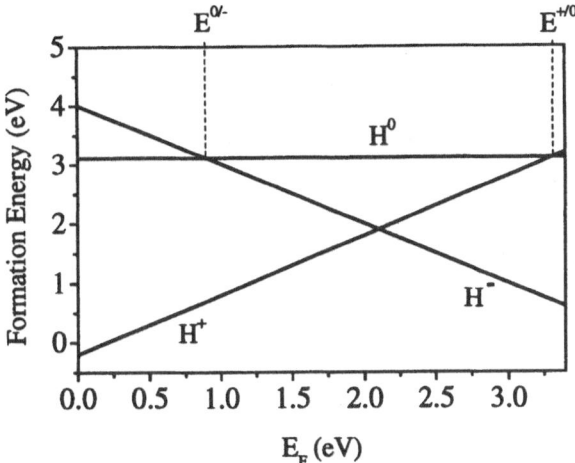

Fig. 2 Formation energies of various charge states of hydrogen in GaN as a function of Fermi level. $E_F = 0$ corresponds
to the top of the valence band. The chemical potential of
hydrogen was chosen to correspond to H$_2$ molecules at $T=0$.

purities relevant for p-type doping. The formation energy of Mg$_{Ga}$ (the acceptor species) was determined by
assuming equilibrium with Mg$_3$N$_2$; this corresponds to
the solubility limit, i.e., the maximum impurity concentration that can be obtained in thermodynamic equilibrium. Trying to push the Mg concentration in GaN
beyond this limit may result in precipitation of Mg$_3$N$_2$,
which may be responsible for a decrease in crystal quality
and the observed reduction in hole concentrations [11].
Incorporation of Mg on interstitial sites (Mg$_i$) or antisites (Mg$_N$), which had been suggested as an explanation
for the limited effectiveness of Mg doping, is unlikely to
be a problem, given the high formation energies of these
species.

We note that compensation by nitrogen vacancies
may be a problem in p-type GaN. Nitrogen vacancies
have high formation energies in n-type GaN, but their
donor character leads to a significant reduction of the
formation energy in p-type material. In the presence of
hydrogen, however, compensation by nitrogen vacancies
will be suppressed. Indeed, the formation energy of hydrogen is lower than that of V_N, and hence hydrogen is
more likely to incorporate as the compensating donor.
Another beneficial effect due to hydrogen is an increase
in the solubility of the acceptor. Imposing the condition of charge neutrality leads to the condition that the
concentrations (and hence the formation energies) of acceptor and donor species must be equal at the growth
temperature (at least when E_F is far enough from the
band edges to allow neglect of free carriers). We see that
the presence of hydrogen leads to a Fermi-level position
higher in the band gap than in the absence of hydrogen;
this leads to a decrease in the formation energy of Mg$_{Ga}$,
and hence a higher solubility.

After the growth, the material contains equal concentrations of Mg and H, and is thus electrically inactive;
however, the binding energy of H to Mg, and the migra-

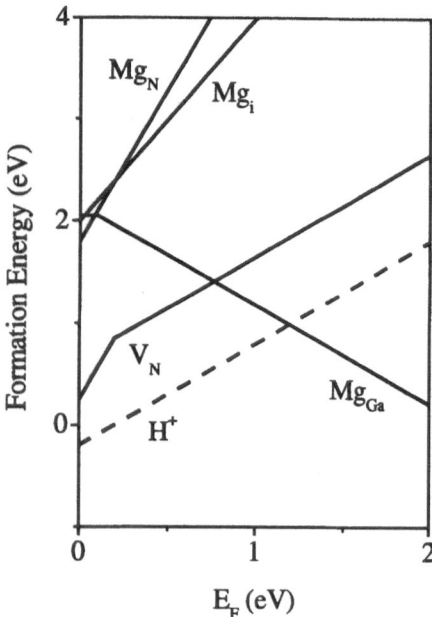

Fig. 3 Formation energies as a function of Fermi level for defects and impurities relevant for p-type GaN, under Ga-rich conditions. $E_F = 0$ corresponds to the top of the valence band.

tion barrier for H diffusion are low enough [14] to allow hydrogen to be removed from the vicinity of the acceptors by a thermal annealing procedure. Such an activation anneal is indeed known to render the Mg acceptors electrically active [12].

The incorporation of hydrogen along with Mg during the growth of p-type GaN can be regarded as a successful example of *impurity engineering*: the introduction of this additional impurity indeed enhances incorporation of the desired impurity, and leads to improved quality due to reduced compensation. This application of *codoping* is successful thanks to the relative ease with which hydrogen can be removed from the acceptor-doped layer after the growth [14]. Co-doping with donors that can *not* be removed from the acceptor-doped layer is unlikely to provide any benefits.

Finally we note that we have carried out extensive theoretical investigations [17] of alternative acceptors in GaN. Only Be has emerged as a potential improvement over Mg, exhibiting both a higher solubility and lower ionization energy. Unfortunately, Be suffers from potential self-compensation due to the incorporation of Be interstitials, which act as donors. We are currently engaged in investigations to determine whether this problem can be solved.

4 Results for ZnO

4.1 n-type doping, hydrogen

We have recently witnessed a resurgence of interest in ZnO as an optoelectronic and electronic material, triggered by advances in ZnO bulk crystal growth [18]. ZnO is an extremely versatile material, with important applications in varistors, piezoelectric transducers, and phos-

phors. The recent improvements in crystal quality have raised hopes for applications in light emitters; however, such device applications require improved control over conductivity.

Zinc oxide almost always exhibits strong n-type conductivity. Despite years of investigations, the source of this conductivity has remained controversial. Because of its prevalence, the n-type conductivity has traditionally been attributed to native defects. Recent first-principles investigations [19,20], however, have revealed that none of the native defects exhibits characteristics consistent with a high-concentration shallow donor. Only the vacancies have sufficiently low energies to form during synthesis of the material; zinc vacancies behave as deep acceptors, and oxygen vacancies as deep donors. The prevailing n-type conductivity can therefore not be attributed to native defects; it must thus be caused by impurities that are unintentionally incorporated.

The obvious candidates for donors in ZnO (Al or Ga on the Zn site, F or Cl on the O site) are unlikely to systematically occur as contaminants in the variety of growth environments that are used for ZnO. Hydrogen, however, has turned out to be a likely candidate [4]. This may seem surprising, since (as highlighted in Sect. 3.2) hydrogen behaves as an amphoteric impurity in the semiconductors in which it has previously been investigated. That means it always *compensates* the prevailing conductivity of the material, and cannot act as a *source* of conductivity. Hydrogen in ZnO, however, behaves differently, as can be seen from the calculated formation energy in Fig. 4: only the *positive* charge state is stable, causing hydrogen to behave exclusively as a donor. Stabilization of the neutral and negative charge states would require moving the Fermi level far above the conduction-band minimum. The hydrogen-induced level in the band structure is thus a resonance in the conduction band. Electrons placed in this resonance of course relax to the conduction-band minimum, where they can be bound to the donor in a hydrogenic state. Hydrogen thus behaves as a shallow donor in ZnO.

The reason for the high stability of H^+ in ZnO is the strong bond that can be formed between hydrogen and oxygen; indeed, interstitial H^+ sits at a distance of about 1 Å from an oxygen atom, in either a bond-center or anti-bonding configuration, with the bond between oxygen and one of its neighboring Zn atoms effectively broken. For the bond-center (BC) configuration the Zn atom moves outward over a distance equal to 40% of the bond length (0.8 Å), to a position slightly beyond the plane of its nearest neighbors.

We have calculated the vibrational frequencies of the O-H stretching and wagging modes (in the harmonic approximation); for H^+ at BC we find 3680 cm^{-1} (stretch) and 450 cm^{-1} (wag). Not surprisingly, the value for the stretch modes is close to that for H_2O molecules.

It is also interesting to consider complex formation between hydrogen and native defects. We pointed out

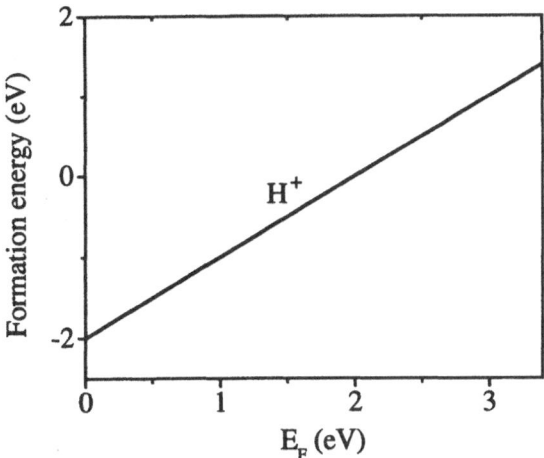

Fig. 4 Formation energy of the positive charge state of hydrogen in ZnO as a function of Fermi level. $E_F = 0$ corresponds to the top of the valence band. The chemical potential of hydrogen was chosen to correspond to H_2 molecules at $T=0$. Charge states other than the positive are not stable.

before that the oxygen vacancy is a *deep donor*; however, complex formation with hydrogen turns it into a *shallow donor*. The calculated binding energy, expressed with respect to H^+ and V_O^0, is 0.8 eV. The hydrogen atom is located close to the center of the vacancy (to within 0.05 Å); this configuration can thus also be regarded as a substitutional hydrogen impurity located on an oxygen lattice site.

4.2 Comparison with experiment

Experimental indications for hydrogen's behavior as a donor in ZnO were actually reported as early as the 1950s [21–23], ZnO being the first semiconductor in which the properties of hydrogen were systematically studied. Those results, however, went largely unnoticed during the upsurge in research activity on hydrogen in semiconductors that started about 30 years later. Mollwo [21] observed an increase in the conductivity of ZnO crystals exposed to hydrogen at temperatures above 200°C. The increase in the conductivity was demonstrated to be due to in-diffusion of hydrogen. An increase in conductivity upon exposure to H_2 has also been observed by Baik *et al.* [24], and by Kohiki *et al.* [25] who introduced hydrogen by proton implantation followed by annealing at 200°C. All of these experiments indicate that introducing hydrogen into ZnO does *not* result in a reduction of the conductivity, which is the expected behavior for hydrogen in other semiconductors. Instead, hydrogen shows strong behavior as a donor, consistent with our theoretical predictions.

Hydrogen is a likely candidate for an unintentional donor, since it is present in many of the growth environments commonly used for ZnO, including vapor-phase transport [18], hydrothermal growth [26], and chemical vapor deposition (MOCVD) [27]. When growth is carried out in air, water vapor can act as a source of hydrogen, and techniques such as laser ablation [28] or sputtering

[29] are sometimes intentionally carried out in a hydrogen atmosphere. In addition, H_2 or H_2O may always be present as a residual gas in any high vacuum system, serving as a source of hydrogen in techniques such as molecular beam epitaxy.

Ion channeling experiments [30] may provide further evidence for the presence of interstitial H in ZnO. Ohta *et al.* [30] used ion channeling to investigate the lattice positions of the host atoms in ZnO. They found that Zn atoms have a much greater tendency to displace from the atomic rows along the $< 0001 >$ direction than oxygen atoms. This is in agreement with our result that for the lowest-energy configuration of H^+, at the bond-center site, large displacements of the Zn atoms occur. The concentration of the point defect giving rise to the displacement was estimated to be on the order of 10^{20} cm^{-3} [30]. These results are consistent with the presence of hydrogen as an unintentional donor.

Finally, we comment on the possibility of achieving *p*-type conductivity in ZnO. Given its size and chemical similarity to oxygen, nitrogen is probably the most promising candidate acceptor. Our preliminary investigations indicate that nitrogen would be a reasonably shallow acceptor, but with a low solubility. In addition, compensation by oxygen vacancies may occur. Co-doping with hydrogen could potentially be beneficial, as it was in the case of *p*-type doping of GaN. The success of this approach will depend on the ability to remove hydrogen from the *p*-type layer after the growth.

5 Summary

We have described a theoretical approach for studying doping of semiconductors, taking the effects of (intentional and unintentional) impurities as well as point defects into account. The framework is generally applicable, but has been applied here to identify and overcome bottlenecks in doping of wide-band-gap semiconductors. Inspection of formation energies for the relevant impurities and defects provides immediate insight in their electronic properties, as well as their expected concentrations in the material.

In GaN, we have shown that nitrogen vacancies are unlikely to be responsible for unintentional *n*-type conductivity; rather, impurities such as silicon or oxygen act as unintentional dopants. *p*-type GaN benefits from the incorporation of hydrogen during growth, which suppresses compensation and enhances the solubility of the acceptor. This is an example of *defect and impurity engineering*.

For ZnO, native defects are again unlikely as the cause of unintentionally *n*-type conductivity. In particular, the widely discussed oxygen vacancy is a *deep*, rather than a *shallow* donor. We have proposed hydrogen as a candidate for the unintentional donor. In contrast to other semiconductors, hydrogen is not amphoteric in ZnO, but occurs exclusively as a donor. Controlling the conductivity of ZnO thus requires careful control of hydrogen exposure during and after growth. We suggest that the behavior observed here for hydrogen in ZnO is likely to occur in other oxide materials as well. The fundamental insights provided by the present work should prove useful for studying and controlling hydrogen in other systems.

Acknowledgements   CVdW would like to acknowledge fruitful collaborations with C. Stampfl, S. Limpijumnong, J. McCaldin, S. Pantelides, A. Kohan, G. Ceder, D. Vogel, P. Krüger, J. Pollmann, and M. Fuchs, and express his gratitude to the Fritz-Haber-Institut and Paul-Drude-Institut, Berlin, Germany, for their hospitality, and to the *Alexander von Humboldt Foundation* for a *US Senior Scientist Award*. The work described here was supported in part by the Office of Naval Research (Contract No. N00014-99-C-0161), the Air Force Office of Scientific Research (Contract No. F49620-00-C-0019).

References

1. C. G. Van de Walle, D. B. Laks, G. F. Neumark, and S. T. Pantelides, Phys. Rev. B **47** (1993) 9425.

2. J. Neugebauer and C. G. Van de Walle, Phys. Rev. B **50** (1994) 8067.

3. J. Neugebauer and C. G. Van de Walle, Appl. Phys. Lett. **68** (1996) 1829.

4. C. G. Van de Walle, Phys. Rev. Lett. **85** (2000) 1012.

5. M. Bockstedte, A. Kley, J. Neugebauer, M. Scheffler, *Comp. Phys. Commun.* **107** (1997) 187.

6. D. Vogel, P. Krüger, and J. Pollmann, Phys. Rev. B **54** (1996) 5495; Phys. Rev. B **55** (1997) 12836.

7. C. Stampfl, C. G. Van de Walle, D. Vogel, P. Krüger, and J. Pollmann, Phys. Rev. B **61** (2000) 7846.

8. F. A. Kröger, *The Chemistry of Imperfect Crystals* (North-Holland, Amsterdam 1964).

9. J. Neugebauer and C. G. Van de Walle, in *Proceedings of the 22th International Conference on the Physics of Semiconductors*, edited by D. J. Lockwood (World Scientific Publishing Co Pte Ltd., Singapore 1994), p. 2327.

10. W. Götz, R. S. Kern, C. H. Chen, H. Liu, D. A. Steigerwald, and R. M. Fletcher, Mater. Sci. Engin. B **59** (1999) 211.

11. D. P. Bour, H. F. Chung, W. Götz, L. Romano, B. S. Krusor, D. Hofstetter, S. Rudaz, C. P. Kuo, F. A Ponce, N. M. Johnson, M. G. Craford, and R. D. Bringans, Mater. Res. Soc. Symp. Proc. **449** (1997) 509.

12. S. Nakamura, N. Iwasa, M. Senoh, and T. Mukai, Jpn. J. Appl. Phys. **31** (1992) 1258.

13. J. Neugebauer and C. G. Van de Walle, Phys. Rev. Lett. **75** (1995) 4452.

14. J. Neugebauer and C. G. Van de Walle, Appl. Phys. Lett. **68** (1996) 1829.

15. N. M. Johnson and Chris G. Van de Walle, in *Hydrogen in Semiconductors II*, edited by N. H. Nickel, *Semiconductors and Semimetals*, Vol. 61, Treatise Editors R. K. Willardson and E. R. Weber (Academic Press, Boston 1999), p. 13.

16. L. Pavesi and P. Gianozzi, Phys. Rev. B **46** (1992) 4621.

17. J. Neugebauer and Chris G. Van de Walle, J. Appl. Phys. **85** (1999) 3003.

18. D. C. Look, D. C. Reynolds, J. R. Sizelove, R. L. Jones, C. W. Litton, G. Cantwell, and W. C. Harsch, Solid State Commun. **105** (1998) 399.

19. A. F. Kohan, G. Ceder, D. Morgan, and Chris G. Van de Walle, Phys. Rev. B **61** (2000) 15019.

20. C. G. Van de Walle *et al.* (unpublished).

21. E. Mollwo, Z. Physik **138** (1954) 478.

22. D. G. Thomas and J. J. Lander, J. Chem. Phys. **25** (1956) 1136.

23. J. J. Lander, J. Phys. Chem. Solids **3** (1957) 87.

24. S. J. Baik, J. H. Jang, C. H. Lee, W. Y. Cho, K. S. Lim, *Appl. Phys. Lett.* **70** (1997) 3516.

25. S. Kohiki, M. Nishitani, T. Wada, T. Hirao, *Appl. Phys. Lett.* **64** (1994) 2876.

26. M. Suscavage, M. Harris, D. Bliss, P. Yip, S.-Q. Wang, D. Schwall, L. Bouthilette, J. Bailey, M. Callahan, D. C. Look, D. C. Reynolds, R. L. Jones, and C. W. Litton, MRS Internet J. Nitride Semicond. Res. **4S1** (1999) G3.40.

27. S. Y. Myong, S. J. Baik, C. H. Lee, W. Y. Cho, and K. S. Lim, Jpn. J. Appl. Phys. **36** (1997) L1078.

28. H. Kordi Ardakani, Thin Solid Films **287** (1996) 280.

29. A. Valentini, F. Quaranta, M. Penze, and F. R. Rizzi, J. Appl. Phys. **73** (1993) 1143.

30. Y. Ohta, T. Haga, Y. Abe, *Jpn. J. Appl. Phys.* **36** (1997) L1040.

High Performance Quantum Cascade Lasers for the Mid-to Far-Infrared

from Band-structure engineering to Commercialization

F. Capasso

Bell Laboratories, Lucent Technologies
Murray Hill, NJ 07974

Abstract The state of the art of quantum cascade (QC) lasers is briefly reviewed. These unipolar semiconductor lasers have unprecedented power, because of the cascade effect, and their wavelength has been tailored, using the same combination of materials, from 3.4 to 19.4 μm, thus covering most of the mid-infrared range. Recently the first far-infrared QC laser has been reported at a wavelength of 24 μm. Distributed feedback (DFB) QC lasers have been demonstrated, which have displayed a large single mode tuning range by temperature tuning of up to 0.15 μm, in pulsed mode. Linewidths as small as 200 kHz in cw, free running mode, and of 20 kHz, stabilized by side locking to a molecular transition, have been demonstrated at liquid nitrogen temperature. QC DFB lasers have been used for trace gas analysis detecting chemicals down to parts per billion in volume with various spectroscopic techniques. QC laser technology has recently become commercially available through licensing by Lucent Technologies.

1 Quantum Cascade Lasers: Band-structure Engineering, Applications and Commercialization

Semiconductor lasers are widely established as light sources in laser pointers, laser printers, CD players, as well as in optical communications. In the latter application (typically around 1.3 and 1.55μm) they have reached a very high standard of sophistication, and are mostly built from InP based semiconductor heterostructure materials. They represent, however, only a small fraction of the wide variety of semiconductor lasers that are available. While optical data-storage evolves towards blue lasers, and optical communications around near-infrared wavelengths, another wavelength range - the mid-infrared, ranging from $\sim$3.5 to 12μm - is attracting a lot of attention. Most trace gases of importance, from byproducts of burning fossil fuel to constituents of the human breath, have telltale absorption features in this wavelength range - known as the molecular "finger-print" region of the spectrum - as a result of molecular rotational-vibrational transitions. Narrow linewidth, tunable semiconductor lasers in this wavelength range are used to spectrally map out and qualitatively/quantitatively detect these trace gases, by a measurement technique called "tunable infrared laser diode absorption spectroscopy (TILDAS)" [1]. The advantages of TILDAS are its high sensitivity and specificity in addition to its non-invasive and real-time nature typical of optical methods. To these add the compactness and robustness of a packaged semiconductor laser device.

The fundamentals of conventional semiconductor lasers dictate that the material band-gap determines to a large degree the emission wavelength. Therefore, in order to reach ever-longer wavelengths, different materials have to be used. For the wavelength range around 3 - 4μm, antimony based materials are the materials of choice [2,3], while for longer wavelengths up to 20μm lead-salt combinations are indicated [4]. However, under many circumstances and particular for longer wavelengths, these materials are less developed and less reliable than the well-known and mature InP or GaAs-based semiconductor hetero-structure materials of the near infrared. Therefore, semiconductor lasers have - albeit their promises - only been used sparsely in trace gas sensing applications, and predictions on their market dominance were less optimistic.

This situation changed with the invention of the quantum cascade (QC) laser in 1994 [5–7] following about fifteen years of research on band-structure engineering and Molecular Beam Epitaxy (MBE) of new artificially structured semiconductors and related devices [8–11]. Using band-structure engineering the energy diagram can be designed and spatially tailored in an almost arbitrary way by a suitable combination of building blocks such as quantum wells and superlattices, compositionally graded materials, delta doping, etc. In particular energy levels, wavefunctions, optical matrix elements and the scattering rates of carriers can be tailored at will. This approach is the basis for designing and modifying in unprecedented ways the electronic, transport and optical properties, which has led often to altogether new materials and to new devices which are having or are poised to have a major impact on electronics and photonics. Among the latter one should recall quantum well lasers and electro-absorption modulator lasers, quantum well infrared detectors, modulation doped field effect transistors, heterojunction bipolar transistors and graded base bipolar transistors.

Several important milestones contributed to the invention of the QC laser. Among these one should recall: the proposal by Kazarinov and Suris of light amplification based on intersubband transitions in quantum wells electrically pumped by resonant tunneling [12], and the direct observation of sequential resonant tunneling through many quantum wells [13]. The concept of an energy staircase created by applying an electric field to a compositionally graded sawtooth heterostructure is also central to the QC laser [14].

QC lasers do not involve the band-gap of the material for the generation of light. Therefore, InP- and GaAs-based III-V semiconductor materials can now be used for the generation of long-wavelength mid-infrared light. These materials are furthermore straightforward to process and pattern. This is essential for the more sophisticated device geometries such as distributed feedback lasers (DFB), which require a periodic modulation (i.e. a grating) of a material parameter (e.g. the refractive index or loss coefficient) to be integrated into the device.

DFB lasers provide a very elegant and reliable method to achieve a well-defined single-wavelength emission (called single-mode operation) as opposed to the usually multiple mode emission of free-running Fabry-Perot resonators. Repeated scattering from the grating favors one wavelength (the Bragg wavelength) and it is the grating period rather than the peak position of the emission spectrum, which determines the single-mode emission wavelength. QC-DFB lasers have first been demonstrated in 1996 [15,16].

What primarily distinguishes QC lasers from conventional diode lasers is the light generation scheme (intersubband/interminiband emission) that allows to achieve very high power by recycling many times (equal to the number N of cascaded stages) the electron making the optical transition. As a consequence a single electron injected above threshold generates N laser photons leading to a high output power. Record output peak powers in excess of 0.5 Watts have been achieved at room temperature at 5 and 8 μm wavelengths; similar power levels have been achieved in cw mode.

Intersubband /interminiband emission also allows one to tailor the wavelength over an extremely wide range by controlling the thickness of the nanometer wide quantum wells. In this way almost the entire mid-infrared spectrum (3.5 μm - 19 μm) has been covered using the AlInAs/GaInAs material system. Very recently we have demonstrated the first far infrared QC laser, which operated at a wavelength of 24 μm.

QC lasers can now be purchased from a company, Applied Optoelectronics Inc., based in Sugarland, Texas (..) www.ao-inc.com, to which Lucent Technologies has transferred and licensed its technology.

2 Single mode continuously tunable QC lasers.

The characteristic QC-laser dielectric waveguide described in section II provides several possibilities to produce a grating modulation strong enough to provide the necessary wavelength selective feedback. In the first type the grating is etched into the surface of the waveguide. If the highly doped top-most waveguide layer is made thin or entirely removed and then directly overlaid with

metal patterned as a grating three phenomena appear. First, the waveguide loss in the position of the grating grooves is higher than in the position of the grating ridges (the unetched portion of the grating), leading to a loss-modulation of the waveguide. Second, the metal layer in the grating grooves – now closer to the low-doped section of the waveguide – pulls the mode towards itself, giving it some characteristics of a surface plasmon, as discussed before. As the mode is pulled into the top-cladding layer, its effective refractive index (roughly defined as the average refractive index of the various waveguide layers, weighted by their overlap with the optical mode) changes. Thirdly, this displacement of the mode also results in a modulation of the overlap of the guided mode with the active material, inducing a modulation of the modal gain, which has the same character as a loss modulation and usually is designed to increase the effect of the latter. Depending on the actual etch depth compared to the thickness of the cladding layers, one component of waveguide modulation may win over the other ones (the modulation of the gain via the confinement factor is usually the weakest component). All our gratings were fabricated by optical contact lithography (the grating periods vary from 800 nm for $\lambda \sim$ 5 μm emission wavelength to 1.25 μm for $\lambda \sim 8$ μm) and wet chemical etching. Since a duty-cycle of the grating (i.e. the ratio of groove to ridge width) close to unity is preferred in order to obtain a strong first Fourier component of the grating shape function, the obtainable etch depth is for practical reasons limited (e.g. to $\sim$ 350nm for a grating period of $\sim$ 1 μm).

The first QC-DFB lasers were "complex-coupled" structures [15,16]; i.e. they had both, a modulation of the loss and the refractive index, with both being comparable in strength. The plasmon-layer thickness was $> 0.6\mu$m and was only partly etched by the grating. Single-mode emission was achieved in pulsed operation for two different QC-DFB lasers around 5.3 and 7.8 μ m [15]. The best lasers operated single-mode in a temperature range from $\sim$ 100 K up to room temperature (300 K) and displayed single-mode tuning ranges of 70 and 150 nm, respectively. As the temperature is increased both, the Bragg resonance via the temperature dependence of the refractive index and the gain spectrum via the temperature dependence of the intersubband structure, are shifted to longer wavelength. The red-shift of the peak gain is approximately twice as strong as the shift of the Bragg resonance. The interplay of this temperature induced detuning with the strength of the Bragg-grating is ultimately responsible for the extent of the single-mode tuning range (an initial blue-detuning of gain peak and Bragg resonance can be favorably used to increase the tuning range and single-mode yield). The grating "strength" can be roughly approximated by the coupling coefficient $\kappa = \pi \cdot \Delta n_{eff}/(2\lambda_0) + i \cdot (\Delta\alpha + \Delta\Gamma \cdot g_{th})/4$ [17]. The emission wavelength is denoted by λ_0; g_{th} is the gain coefficient at laser threshold; Δn_{eff}, $\Delta\alpha$, and $\Delta\Gamma$ are the differences between the effective refractive indices (n_{eff}), waveguide attenuation coefficients (α), and confinement factors (Γ) for the waveguides at the locations of the grating grooves and plateaus, respectively. For our early devices we estimated a coupling coefficient $|\kappa| \sim$ 2 - 3 cm^{-1}.

Recent QC-DFB laser designs [18,19] improved the top-grating approach by using a thinner highly doped, top-most waveguide layer, which could be entirely removed in the grating grooves by etching. This resulted in a greater grating strength with a coupling coefficient $|\kappa| \geq$ 15 cm^{-1}, which is now clearly dominated by the modulation of the effective refractive index (by a factor of $\sim$4). The lasers were designed for $\sim$ 4.6 μm wavelength [20] and also for the wavelength range of 9.5 - 10.5 μm [19]. The single-mode tuning ranges with heat sink temperature and the lasers operated in pulsed mode were 65 and 150 nm, respectively.

Although the top grating approach yields quite good results and provides for a straightforward time-saving processing technology, it suffers the fundamental drawback that the Bragg-grating is located only in the exponentially decaying wing of the waveguide mode. This ultimately limits the maximum achievable strength of the grating.

Therefore, early on we took another, parallel, approach to QC-DFB lasers, which positions the grating close to the active waveguide core, where the mode intensity is high [16]. In a first growth cycle by MBE, the active waveguide core - a several 100 nm thick InGaAs layer followed by the stack of many ($\sim$ 30) periods of alternated active regions and injectors and capped by another $\sim$ 500 nm thick InGaAs layer is grown. The wafer is then removed from the growth chamber, and the Bragg-grating is fabricated into the upper InGaAs layer using the conventional technique. The wafer is then transferred back into another growth chamber, where an InP top cladding is grown on top of the Bragg-grating using solid-source MBE. The refractive index contrast between InGaAs ($\sim$ 3.48) and InP ($\sim$ 3.1) and the strong overlap of the grating with the mode provide for a strong modulation of the effective refractive index of the waveguide and a large coupling coefficient $|\kappa| \sim$30-80cm^{-1}. Although the procedure of two growth cycles increases the fabrication efforts, QC-DFB lasers with buried gratings made for the best single-mode laser devices. We have worked extensively in the 5 and 8 μm wavelength range with this type of devices [16,20,?]. Very high single-mode output power $\sim$150mW at liquid Nitrogen temperature is achieved for these lasers.

A third method of incorporating a Bragg-grating into a QC-laser with surface plasmon waveguide has recently been discussed and demonstrated by us [22]. It is fabricated from alternating stripes of two different metals and is a based on the periodic modulation of the skin depth in the metal. It can have sufficient strength for single-mode operation at the very long wavelengths. We

demonstrated a pulsed single-mode QC-DFB laser based on this principle using a Bragg-grating of alternating titanium and gold (resulting in $\kappa \sim$ 7cm^{-1}) with a periodicity of 2.0μm. The lasers emitted at 16.2 μm wavelength with 50 nm tuning range (between 10 and 100K heat sink temperature - which was also the temperature range of laser action for this device).

R. Williams et al. [23] measured the intrinsic linewidth of several of our QC-DFB lasers around 8 μm wavelength. They observed fluctuations of the collected optical intensity when the laser light was passing through a sample cell containing a gas (N_2O) with well-know absorption features and the laser was being tuned to the side of one such absorption line. This resulted in a linewidth of $\sim$ 1 MHz over $\sim$ 1 ms integration time. The laser could furthermore be electronically stabilized (and locked to the wavelength of the N_2O absorption feature) by using the detector signal as feedback to the drive current. These stabilized QC lasers had linewidths < 20 kHz, which is very narrow for any as-cleaved semiconductor-DFB laser.

3 Trace gas-sensing applications.

Two large regions of good transparency in the atmosphere are found around 5 μm and from 7.5 to 13 μm wavelength, the two so-called "atmospheric windows". Furthermore, most trace gases of importance have characteristic absorption features in the mid-infrared, between 3 and 17 μm wavelength. As discussed before, QC-DFB lasers can be designed to emit at any wavelength in this spectral range. Therefore, it was straightforward to examine the lasers' potential in demanding gas-sensing applications. To that aim, many fruitful collaborations with expert spectroscopists were established; some of the results are presented in the following.

E. A. Whittaker et al. [24] at Stevens Institute of Technology, NJ, used pulsed QC-DFB lasers with a top-grating at near room temperature conditions to measure mid-IR ($\lambda \sim$ 7.8 μm) absorption spectra of N_2O and CH_4 diluted in N_2 (prepared in a 10 cm long single-pass gas cell). They employed a measurement technique known as wavelength-modulation spectroscopy which measures the derivative of the spectrum. The noise equivalent sensitivity limit of the measurement was 50 ppm.

D. Sonnenfroh and coworkers of Physical Sciences Inc. have used low duty cycle (< 1%) pulsed room temperature 5.4 μm QC-DFB lasers and balanced ratiometric detection in a direct absorption experiment of trace gases [25]. They demonstrated sensitivities for N_2O of 10 ppm-m and for NO of 55 ppb-m. These results are very promising for the realization of sensitive, very compact and portable sensors for real world applications.

Pulsed QC-DFB lasers show a wavelength chirp caused by heating during the current pulse which results in an integrated linewidth of several 100 MHz. This is sufficiently narrow to measure the pressure-broadened absorption features of trace gases dispersed in standard atmosphere with confidence. It is, however, too broad for gases at low pressure, when the absorption width (few 10 MHz) is predominantly determined by the motion of the molecules (Doppler-broadening). To achieve the necessary narrow linewidth, the lasers have then to be operated in cw. Therefore, the following measurements all used cw operated buried-grating QC-DFB devices at LN2 temperature. The tuning of the single-mode output over the absorption feature(s) of the target gas is accomplished by varying the current through the device. A rough glimpse of such an experiment can be seen in Fig. 8. The dips in the light output versus current characteristics (indicated by arrows) are due to absorption of the laser light by water vapor in the few cm of room air between the laser cryostat and the detector. In a real gas sensing experiment the emission wavelength as a function of the heat sink temperature and laser drive current would be calibrated, resulting in the position of the water absorption line, and the depth of the dip would correspond to the amount of water in the air. (The feature shown in Fig. 8 is actually a doublet of absorption lines, pressure-broadened into one broader feature.)

S. W. Sharpe, et al. [26] at the Pacific Northwest National Laboratory (PNNL), WA, conducted high-resolution, Doppler-limited, direct absorption measurements of NO and NH_3 using QC-DFB lasers at 5.2 and 8.5 μm, respectively. The laser drive current was a saw-tooth ratchet with 6 - 11 kHz repetition rate; with the rising current the laser would tune $\sim 2.5 cm^{-1}$, covering e.g. 11 absorption features of NH_3. The noise equivalent sensitivity limit in these measurements was 3×10^{-6} absorbance.

B. Paldus, et al. [27] at Stanford University, CA, together with J. Oomens, et al. at the University of Nijmegen, The Netherlands, reported photoacoustic spectroscopy on NH_3 and H_2O diluted in N_2 using a cw QC-DFB laser emitting at 8.5 μm wavelength. The noise-limited minimum detectable concentration of NH_3 was 100 ppbv for 1 s integration time. B. Paldus and coworkers at Informed Diagnostics have also demonstrated sub-ppbv (NH_3) sensitivity measurements using cavity ring down spectroscopy [28].

A. A. Kosterev, et al. [29,30] at Rice University, TX, reported on measurements of the concentration of $^{12}CH_4$, its natural isotopes $^{13}CH_4$ and $^{12}CH_3D$, H_2O, N_2O, and C_2H_5OH diluted in standard air using a direct absorption technique around 7.95 μm wavelength. With a 100m multi-pass gas cell and employing a new background subtraction method they were able to demonstrate a sensitivity limit in the ppbv concentration range, e.g. 125 ppbv of C_2H_5OH.

J. F. Remillard and coworkers at the Ford Research Laboratories have used a quantum cascade distributed feedback laser operating at 5.2 μm to obtain sub-Doppler resolution limited saturation features in a Lamb-dip experiment on the $R(13.5)_{1/2}$ and $R(13.5)_{3/2}$ transitions of NO [31]. The laser is operated CW in a LN_2 Dewar. Lamb dips appear as transmission spikes with full widths

of ~ 4.3 MHz. At this resolution the 73 MHz lambda doubling of the $R(13.5)_{3/2}$ line, which is normally obscured by the 130 MHz Doppler broadening, is easily resolved.

C. Webster and coworkers [32] at JPL, California Institute of Technology, conducted measurements of the concentration of CH_4 and N_2O in the earth's atmosphere from ground level to the stratosphere ($\sim 70,000$ ft) using a cw operated 7.95 μm QC-laser and a wavelength modulation technique. The laser in a LN2 dewar was on board a high-altitude air-plane; the surrounding air would be sucked into a multi-pass gas cell aboard the plane, as the plane makes 8 hr long flights to map the atmosphere for the trace gases. The noise-equivalent sensitivity limit was ~ 2 ppbv.

4 Acknowledgements

Collaborations with C. Gmachl, A. Y. Cho, D. L. Sivco, J. Faist, C. Sirtori, A. Tredicucci, G. Scamarcio, A. L. Hutchinson are gratefully acknowledged.

References

1. H. I. Schiff, G. I. Mackay, and J. Bechara in "Air monitoring by spectroscopic techniques", ed. M. W. Sigrist, Wiley Interscience, New York, 1994.

2. D. Garbuzov, M. Maiorov, H. Lee, V. Khalfin, R. Martinelli, J. Connolly, Appl. Phys. Lett., vol. 74, pp. 2990-2992, 1999; W. W. Bewley, C. L. Felix, I. Vurgaftman, D. W. Stokes, E. H. Aifer, L. J. Olafsen, J. R. Meyer, M. J. Yang, B. V. Shanabrook, H. Lee, R. U. Martinelli, A. R. Sugg, Appl. Phys. Lett., vol. 74, pp. 1075 - 1077, 1999.

3. H. K. Choi, G. W. Turner, J. N. Walpole, M. J. Manfra, M. K. Connors, L. J. Missaggia, SPIE Proceedings, vol. 3284, pp. 268 - 273, 1998; B. Lane, Z. Wu, A. Stein, J. Diaz, M. Razeghi, Appl. Phys. Lett., vol. 74, pp. 3438-3440, 1999.

4. Maurus Tacke, " New developments and applications of tunable IR lead salt lasers", Infrared Phys. Technol., vol. 36, 447 - 463, 1995; M. Hodges, U. W. Schiessl, "Lead salt tunable diode lasers: key devices for high sensitivity gas analysis", SPIE Proceedings, vol. 3628, pp. 113-121, 1999.

5. J. Faist, F. Capasso, D. L. Sivco, C. Sirtori, A. L. Hutchinson, and A. Y. Cho, "Quantum Cascade Laser", Science, vol. 264, pp. 553-556, 1994.

6. F. Capasso, C. Gmachl, D. L. Sivco, and A. Y. Cho, "Quantum cascade lasers" Physics World, vol. 12 pp. 27-33, June 1999

7. F. Capasso, C. Gmachl, A. Tredicucci, A. L. Hutchinson, D. L. Sivco, and A. Y. Cho, "High Performance Quantum Cascade Lasers", Optics and Photonics News, vol. 10, pp. 31-37, 1999.

8. F. Capasso, "Band-gap engineering: From Physics and Materials to New Semiconductor Devices", Science, vol. 235, pp. 172-176, 1987

9. F. Capasso and A. Y. Cho, "Band-gap engineering of semiconductor heterostructures by Molecular Beam Epitaxy: Physics and Device Applications", Surface Sci. vol. 299/300, pp. 878-891, 1994

10. F. Capasso, J. Faist and C. Sirtori, "Mesoscopic Phenomena in Semiconductor Naostructures by Quantum Design", *J. Math. Phys.* vol. **37**, pp. 4775-4792, 1996

11. A. Y. Cho, Ed. "Molecular Beam Epitaxy", New York: AIP Press, 1994

12. R. Kazarinov and R. A. Suris: "Amplification of Electromagnetic Waves in a Semconductor Superlattice", *Sov. Phys. Semicond.* vol. **5**, pp. 707-709, 1971

13. F. Capasso, K. Mohammed, and A. Y. Cho, "Sequential Resonant Tunneling through a Multiquantumwell Superlattice", *Appl. Phys. Lett.* vol. **48**, pp.478-48, 1986

14. F. Capasso, W. T. Tsang and G. F. Williams, "Staircase solid-state Phtomultipliers and Avalanche Photodiodes with Enhanced Ionization Rates Ratio", *IEEE Trans. Electron Dev.* vol. **ED-30**, pp. 381-390, 1983.

15. J. Faist, C. Gmachl, F. Capasso, C. Sirtori, D. L. Sivco, J. N. Baillargeon, A. L. Hutchinson, and A. Y. Cho, "Distributed feedback quantum cascade lasers", *Appl. Phys. Lett.*, vol. **70**, pp. 2670 - 2672, 1997.

16. C. Gmachl, J. Faist, J. N. Baillargeon, F. Capasso, C. Sirtori, D. L. Sivco, S. N. G. Chu, and A. Y. Cho, "Complex-Coupled Quantum Cascade Distributed-Feedback Laser", *IEEE Photon. Technol. Lett.*, vol. **9**, pp. 1090 - 1092, 1997.

17. G. Morthier, P. Vankwinkelberge, *Handbook of distributed feedback laser diodes*, Boston, MA, Artech House Inc., 1997.

18. R. K · ler, C. Gmachl, F. Capasso, A. Tredicucci, D. L. Sivco, S. N. G. Chu, and A. Y. Cho, "Single-mode tunable, pulsed and continuous wave quantum-cascade distributed feedback lasers at $\lambda \cong 4.6$ - $4.7\mu m$", *Appl. Phys. Lett.*, vol. **76**, pp. 1092-1094, 2000.

19. R. K · ler, C. Gmachl, F. Capasso, A. Tredicucci, D. L. Sivco, and A. Y. Cho, "Single-Mode, Tunable Quantum Cascade Lasers in the Spectral Range of the CO_2 Laser at $\lambda=9.5$-$10.5\mu m$", *Photon. Techn. Lett.* vol. **12**, pp. 474-476, 2000.

20. C. Gmachl, F. Capasso, J. Faist, A. L. Hutchinson, A. Tredicucci, D. L. Sivco, J. N. Baillargeon, S. N. G. Chu, and A. Y. Cho, "Continuous-wave and high-power pulsed operation of index-coupled distributed feedback quantum cascade laser at $\lambda \approx 8.5$ μm", *Appl. Phys. Lett.*, vol. **72**, pp. 1430-1432, 1998.

21. C. Gmachl, F. Capasso, A. Tredicucci, D. L. Sivco, J. N. Baillargeon, A. L. Hutchinson, and A. Y. Cho, "High power, continuous wave, current tunable, single-mode quantum cascade distributed feedback lasers at $\lambda \cong 5.2$ and $7.95\mu m$", *Opt. Lett.*, vol. **25**, pp. 230-232, 2000.

22. A. Tredicucci, C. Gmachl, F. Capasso, A. L. Hutchinson, D. L. Sivco, and A. Y. Cho, "Single-mode surface plasmon laser", *Appl. Phys. Lett.*, vol. **76**, pp. 2164-2166.

23. R. M. Williams, J. F. Kelly, J. S. Hartman, S. W. Sharpe, M. S. Taubman, J. L. Hall, F. Capasso, C. Gmachl, D. L. Sivco, J. N. Baillargeon, and A. Y. Cho, "Kilo-Hertz Linewidth from Frequency Stabilized Mid-Infrared Quantum Cascade Lasers", Opt. Lett., vol. **24**, pp. 1844-1846, 1999.

24. K. Namjou, S. Cai, and E. A. Whittaker, J. Faist, C. Gmachl, F. Capasso, D. L. Sivco, and A. Y. Cho, "Sensitive absorption spectroscopy with a room-temperature distributed-feedback quantum-cascade laser", *Opt. Lett.*, vol. **23**, pp. 219-221, 1998.

25. D. M. Sonnenfroh, W. Terry Rawlins, M. G. Allen, C. Gmachl, F. Capasso, A. L. Hutchinson, D. L. Sivco, J. M. Baillargeon and A. Y. Cho, "Application of Balanced Detection to Absorption Measurements of Trace Gases with Room Temperature, Quasi-CW QC lasers ", Applied Optics, in press.

26. S. W. Sharpe, J. F. Kelly, J. S. Hartman, C. Gmachl, F. Capasso, D. L. Sivco, J. N. Baillargeon, and A. Y. Cho, "High-resolution (Doppler-limited) spectroscopy using quantum-cascade distributed feedback lasers", *Opt. Lett.*, vol. **23**, pp. 1396-1398, 1998.

27. B. A. Paldus, T. G. Spence, R. N. Zare; J. Oomens, F. M. J. Harren, D. H. Parker; C. Gmachl, F. Capasso, D. L. Sivco, J. N. Baillargeon, A. L. Hutchinson, and A. Y. Cho, "Photoacoustic spectroscopy using quantum-cascade lasers", *Opt. Lett.*, vol. **24**, pp. 178-180, 1999.

28. B. A. Paldus, C. C. Harb, T. G. Spence, R. N. Zare, C. Gmachl, F. Capasso, D. L. Sivco, J. N. Baillargeon, A. L. Hutchinson, and A. Y. Cho, "Cavity Ring-down Spectroscopy using Mid-Infrared Quantum Cascade Lasers", *Opt. Lett.* vol. **25**, pp. 666-668, 2000.

29. A. Kosterev, R. F. Curl, F. K. Tittel; C. Gmachl, F. Capasso, D. L. Sivco, J. N. Baillargeon, A. L. Hutchinson, and A. Y. Cho, "Methane concentration and isotopic composition measurements with a mid-infrared quantum cascade laser", Opt. Lett., vol. **24**, pp. 1762-1764, 1999.

30. A. A. Kosterev, R. F. Curl, F. K. Tittel, C. Gmachl, F. Capasso, D. L. Sivco, J. N. Baillargeon, A. L. Hutchinson, and A. Y. Cho, "Detection of simple and complex molecules in ambient air with a quantum cascade distributed feedback laser", *Appl. Opt.*, in press.

31. J. T. Remillard, D. Uy, W. H. Weber, F. Capasso, C. Gmachl, A. L. Hutchinson, D. L. Sivco, J. N. Baillargeon, and A. Y. Cho, "Sub-Doppler resolution limited Lamb-dip spectroscopy of NO with a quantum cascade distributed feedback laser ", Opt. Lett. to be published.

32. C. R. Webster, G. J. Flesch, D. C. Scott, J. Swanson, R. D. May, and W. S. Woodward, C. Gmachl, F. Capasso, D. L. Sivco, J. N. Baillargeon, A. L. Hutchinson, and A. Y. Cho, "Quantum cascade laser measurements of stratospheric methane (CH_4) and nitrous oxide (N_2O)", *Appl. Opt.*, in press

Heterostructure lasers: development of new physics and new technology

Zhores Alferov

Ioffe Physico-Technical Institute, Russian Academy of Sciences, St Petersburg, Russia

Abstract　Short historical review of the physics and technology of heterostructure lasers based on double heterostructures — bulk and quantum wells is described. Recent progress in quantum dots laser structures and future trends in developement of physics and technology of these new types of heterostructures are discussed.

1 Introduction

It would be very difficult today to imagine solid state physics without semiconductor heterostructures. Semiconductor heterostructures and especially double heterostructures, including quantum wells, quantum wires, and quantum dots, currently comprise the object of investigation of two thirds of all research groups in the physics of semiconductors.

While the feasibility of controlling the type of conductivity of a semiconductor by doping it with various impurities and the concept of nonequilibrium carrier injection are the seeds from which semiconductor electronics has sprung, heterostructures provide the potential means for solving the far more general problem of controlling fundamental parameters in semiconductor crystals and devices, such as the width of the bandgap, the effective masses and mobilities of charge carriers, the refractive index, and the electron energy spectrum.

The development of the physics and technology of semiconductor heterostructures has brought about tremendous changes in our everyday lives. Heterostructure-based electron devices are widely used in many areas of human activity. Life without telecommunication systems utilizing double-heterostructure (DH) lasers, without heterostructure light-emitting diodes (LEDs) and bipolar transistors, or without the low-noise, high-electron-mobility transistors (HEMTs) used in high-frequency devices, including satellite television systems, is scarcely conceivable. The DH laser is now found in virtually every home as part of the compact-disk (CD) player. Solar cells incorporating heterostructures are used extensively in both space and terrestrial programs; for fifteen years the "Mir" space station has been utilizing solar cells based on AlGaAs heterostructures.

2 The DHs-concept and its application for semiconductor lasers

The idea of using heterostructures in semiconductor electronics emerged at the very dawn of electronics. Already in the first patent associated with p-n junction transistors W. Shockley [1] proposed the application of a wide-gap emitter to achieve one-way injection. Some of the most important theoretical explorations in this early stage of heterostructure research were carried out by H. Kroemer, who introduced the concept of quasielectric and quasi-magnetic fields in a graded heterojunction and hypothesized that heterojunctions could possess extremely high injection efficiencies in comparison with homojunctions [2].

The next important step was taken several years later, when we and Kroemer [3] independently formulated the concept of DH-based lasers. In our patent we noted the feasibility of attaining a high density of injected-carriers and population inversion by "double" injection. We specifically mentioned that homojunction lasers "do not provide continuous lasing at elevated temperatures," and to demonstrate an added benefit of DH lasers, we explored the possibility of "increasing the emitting surface and utilizing new materials to achieve emission in different regions of the spectrum".

At the beginning theoretical research significantly outpaced its experimental implementation. In 1966, we predicted [4] that the injected-carried density could well be several orders of magnitude greater than the carrier density in a wide-gap emitter (the "superinjection" phenomenon). At the same year, in a paper [5], I generalized our conception of the principal advantages of DHs for various devices, particularly for lasers and high-power rectifiers:

"The regions of recombination, light emission, and population inversion coincide and are concentrated entirely in the middle layer. Owing to potential barriers at the boundary of semiconductors with different bandgap widths, even for large displacements in the direction of transmission, there is absolutely no indirect passage of electron and hole currents, and the emitters have zero recombination (in contrast with $p-i-n$, $n-p-p^+$, where recombination plays a decisive role).

"Population inversion to generate stimulated emission can be achieved by pure injection means (double injection) and does not require a high doping level of the middle region and especially does not require degeneracy. Because of the appreciable difference in the dielectric constants, light is concentrated entirely in the middle layer, which functions as a high-Q waveguide, and optical losses in the passive regions (emitters) are therefore nonexistent."

The following are the most important physical phenomena predicted in heterostructures: (i) superinjection of carriers; (ii) optical confinement; (iii) electron confinement (Fig. 1).

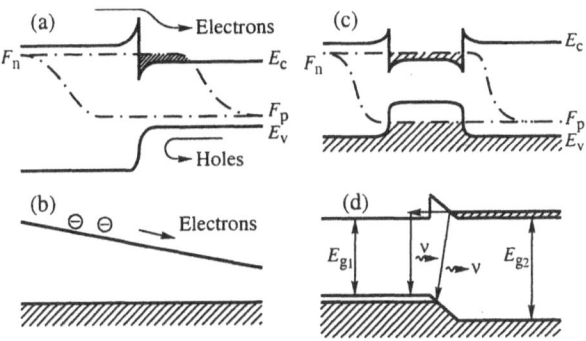

Fig. 1 Main physical phenomena in classical heterostructures. (a) One-side injection and superinjection; (b) Deffusion in an imbedded quasielectric field; (c) Electron and optical confinement; (d) Diagonal tunneling through a heterojunction.

At that time there was widespread skepticism regarding the feasibility of fabricating an "ideal" heterojunction with a defect-free boundary and especially one that exhibited the theoretically predicted injection properties. The actual construction of efficient, wide-gap emitters was regarded as a sheer impossibility, and many viewed the patent for a DH laser as a "paper" patent.

The discovery of the first "ideal" AlGaAs heterostructures [6] and demonstration of the first lasers operating at room temperature [7] experimentally confirmed predicted earlier physical phenomena and became the basis for modern optoelectronics.

Particularly important was continuous wave (CW) operation at room temperature [8] (Fig. 2). The latter event represented a principal point for semiconductor lasers: fiber-optical communication systems were born due to this achievement.

It was soon clear in this early stage of development of the physics and technology of heterostructures that we needed to look for new lattice-matched heterostructures in order to cover broad area of energy spectrum. The first important step was taken in 1970, in the paper [9] was reported that various lattice-matched heterojunctions based on quaternary III–V solid solutions were possible, which permitted independent variation between lattice constant and band gap.

The basic concepts of the DFB semiconductor lasers we formulated in 1971 [10]: (i) Fabry–Perot cavity re-

placed by diffraction grating resonator; (ii) The diffraction grating is formed not in the bulk, but on the surface of the wave guide layer. (iii) The interaction of wave guide modes with the surface diffraction grating produces not only distributed feedback, but also well-collimated radiation at the output.

Now these principles widely used in telecommunication diode lasers.

The creation of DHS lasers led not only to new light-emitting device concept and new physics, but also to important technological peculiarities: (i) fundamental need for structures with well-matched lattice-parameters; (ii) the use of multi component solid solutions to match the lattice parameters; (iii) fundamental need for epitaxial growth technologies.

Because of electron confinement in double heterostructures, DH lasers have essentially become the direct precursors to quantum-well structures, which have a narrow-gap middle layer with a thickness of a few hundred angstroms, an element that has the effect of splitting the electron levels as a result of quantum-size effects. However, high-quality DHs with ultrathin layers could not be attained until new methods were developed for the growth of heterostructures. Two principal modern-day epitaxial growth techniques with precision monitoring of thickness, planarity, composition, etc., were developed in the seventies. Today molecular-beam epitaxy (MBE) has grown into one of the most important technologies for the growth of heterostructures using III–V compounds, primarily through the pioneering work of A. Cho [11]. The basic concepts of metal-organic vapor-phase epitaxy (MOVPE) were set forth in the early work of H. Manasevit [12] and have enjoyed widespread application for the growth of heterostructures from III–V compounds, particularly in the wake of a paper by R. Dupuis and P. Dapkus reporting the successful use of this technique to create a room-temperature injection DH laser in the system AlGaAs [13].

The distinct manifestation of quantum-well effects in the optical spectra of GaAs-AlGaAs semiconductor heterostructures with an ultrathin GaAs layer (quantum well) was demonstrated by R. Dingle *et al.* in 1974 [14]. The authors observed a characteristic step structure in the absorption spectra and a systematic shift of the characteristic energies as the thickness of the quantum well was decreased.

Lasing by means of quantum wells was first accomplished by J.P. van der Ziel in 1975 but the lasing parameters fell short of average DH lasers. It was 1978 before R. Dupuis and P. Dapkus in collaboration with N. Holonyak reported the first construction of a quantum-well (QW) laser with parameters to match those of standard DH lasers [15]. The term "quantum well" first surfaced in this paper. The real advantage of QW lasers was demonstrated much later by W.T. Tsang of Bell Telephone Laboratories. Through a major improvement

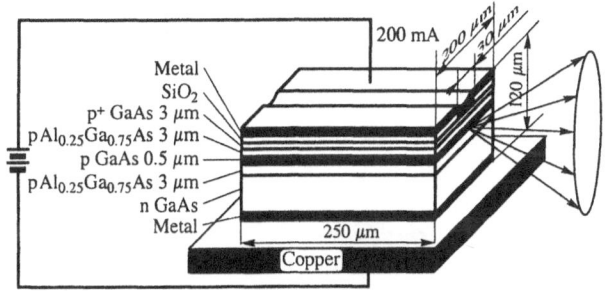

Fig. 2 Schematic view of the structure of the first injection DH laser operating in the CW regime at room temperature.

in MBE growth technology it was possible to lower the threshold current density to 160 A/cm^2 [16].

The most complex QW laser structure, consolidating a single quantum well and short-period superlattices (SPSs), was grown in our laboratory in 1988 [17]. We obtained threshold current densities $J = 52$ A/cm^2 and, after a certain optimization, $J = 40$ A/cm^2, which was the world record for semiconductor injection lasers untill the late 1990s and affords a good demonstration of the effective use of quantum wells and superlattices in electron devices.

The concept of stimulated emission in superlattices, set forth by R. Kazarinov and R. Suris [18], was made a reality by F. Capasso et al. [19] almost a quarter-century later. The previously proposed structure was substantially optimized, and the cascade laser developed by Capasso gave birth to a new generation of unipolar lasers operating in the mid-IR range.

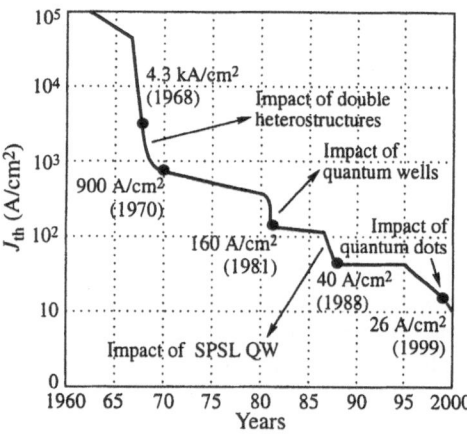

Fig. 3 Evolution of the threshold current of semiconductor lasers.

From a certain standpoint, the history of semiconductor lasers is the history of the campaign to lower the threshold current, as graphically illustrated in Fig. 3. The most significant changes in this endeavor did not take place until the concept of DH lasers had been introduced. The application of SPS quantum wells actually brought us to the theoretical limit of this most important parameter. Subsequent possibilities associated with the use of new structures utilizing quantum wires and quantum dots will be discussed in the next section of the article.

Applications of the quantum-well and superlattice heterostructures in semiconductor lasers permitted to get:

- shorter emission wavelengths, lower threshold current, higher differential gain, and weaker temperature dependence of the threshold current in semiconductor lasers;
- infrared quantum cascade lasers;
- lasers with quantum wells bounded by short-period superlattices;

- optimization of electron and optical confinement and of the waveguide characteristics in semiconductor lasers.

There were important technological consequences:

- no need to carefully match lattice parameters;
- fundamental need to use slow-growth technologies (MBE and MOVPE);
- submonolayer growth method;
- suppression of the propagation of mismatch dislocations during epitaxial growth;
- radical diversification of materials available for heterostructure components.

3 Quantum-dot heterostructure lasers

The principal advantage application of quantum-size heterostructure for lasers originates from the noticeable increasing of the density of states with reducing of the dimensionalities for electron gas (Fig. 4)

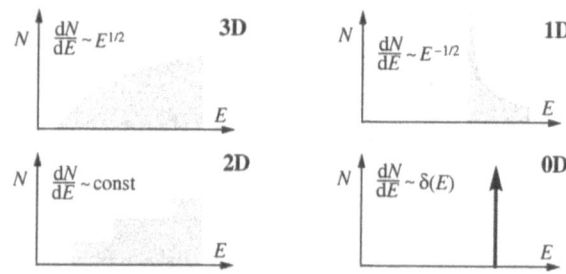

Fig. 4 Density of states for charge carriers in structures with different dimensionalities.

During the 1980s, progress in 2D-quantum well heterostructures physics and its applications attracted many scientists to studying systems of far less dimensionality — quantum wires and quantum dots. In contrast to quantum "wells" where carriers are localized in the direction perpendicular to the layers but move freely in the layer plane, in quantum "wires" carriers are localized in two directions and move freely along the wire axis. And being confined in all three directions quantum "dots" — "artificial atoms" with a totally discrete energy spectrum are created (Fig. 5).

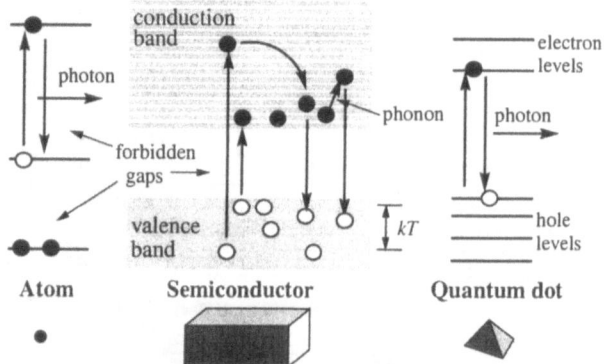

Fig. 5 Schematic representation of energy diagrams in case of a single atom (left), a bulk crystal (center), and a quantum dot (right).

Experimental work on fabrication and investigation of quantum wire structures began more than 15 years ago. Theoretical consideration of problems concerning one of the most interesting applications — quantum wire lasers was done at the same time [20]. The authors [20] pointed to the possibility to weaken the threshold current dependence on temperature for QWR lasers and full temperature stability for QD lasers. By now there is a significant number of both theoretical and experimental papers in this field.

The first semiconductor dots based on II–VI microcrystals in glass matrix were proposed and demonstrated by A.I. Ekimov and A.A. Onushchenko [21]. However, since the semiconductor quantum dots were introduced in an insulating glass matrix and the quality of the interface between glass and semiconductor dot was not high, both fundamental studies and device applications were limited. Much more exciting possibilities appeared since three dimensional coherent quantum dots had been fabricated in semiconductor matrix [22].

Several methods were proposed for the fabrication of these structures. Indirect methods, such as the post-growth lateral patterning of 2D quantum well suffer often from insufficient lateral resolution and interface damage caused by the patterning procedure. A more promising way is the fabrication by direct methods, i. e., growth in V-grooves and on corrugated surfaces which may result in formation of quantum wires and dots. The groups of the Ioffe Institute and Berlin Technical University — last years we carried out this research in close cooperation — contributed significantly to the last direction.

Finally we came to the conclusion that the most exciting method of the formation of ordered arrays of quantum wires and dots is the self-organization phenomena on crystal surfaces. Strain relaxation on step or facet edges may result in formation of ordered arrays of quantum wires and dots both for lattice-matched and lattice-mismatched growth.

The first very uniform arrays of three dimensional quantum dots exhibiting also lateral ordering were realized in the system InAs–GaAs both by MBE and MOCVD growth methods [23, 24].

Elastic strain relaxation on facet edges and island interaction via the strained substrate are driving forces for self-organization of ordered arrays of uniform, coherently strained islands on crystal surfaces [25].

In lattice-matched heteroepitaxial systems the growth mode is determined solely by the relation between the energies of two surfaces and the interface energy. If the sum of the surface energy of epitaxial layer γ_2 and energy of interface γ_{12} is lower than the substrate surface energy, $\gamma_2 + \gamma_{12} < \gamma_1$, i. e., if the material 2 being deposited wets the substrate, then we have the Frank–van der Merve growth. Changing the $\gamma_2 + \gamma_{12}$ value may result in a transition from the Frank–van der Merve mode to on Volmer–Weber one where 3D islands are formed an a bare substrate.

In a heteroepitaxial system with lattice mismatch between the material being deposited and the substrate the growth may initially proceed in a layer-by-layer mode.

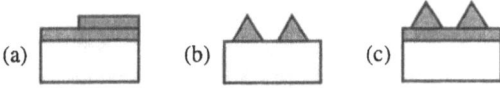

Fig. 6 (a) Frank–van der Merve, (b) Volmer–Weber, (c) Stranski–Krastanow growth modes.

However, a thicker layer has a higher elastic energy, and the elastic energy tends to be reduced via formation of isolated islands. In these islands the elastic strains relax and, correspondingly, the elastic energy decreases. This results in a Stranski–Krastanow growth mode (Fig. 6). The characteristic size of islands is determined by the minimum in the energy of an array of 3D coherently strained islands per unit surface area as a function of the island size (Fig. 7) [25]. Interaction between islands via elastically strained substrate would results in lateral island ordering typical of the square lattice.

Experiments show in most cases rather narrow size distribution of the islands, and on top of that coherent islands of InAs form under certain conditions a quasi-periodic square lattice (Fig. 8) Shape of quantum dots can be significantly modified during regrowth or post-growth annealing, or by applying complex growth sequences. Short period alternating deposition of strained materials leads to a splitting of QDs and to formation of vertically coupled quantum dot superlattice structures (Fig. 8) [26]. Ground state QD emission, absorption and lasing energies are found to coincide [23]. Observation of ultranarrow (<0.15 meV) luminescence lines from single quantum dots [23], which do not exhibit broadening with temperature, is the proof of the formation of an electronic quantum dot.

Quantum dot lasers are expected to have superior properties with respect to conventional QW lasers. High differential gain, ultralow threshold current density and

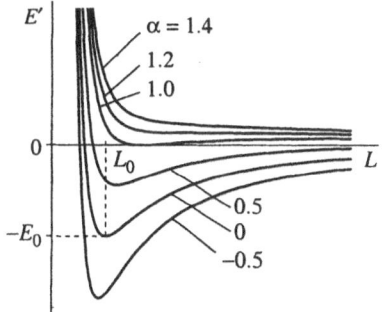

Fig. 7 Energy of a sparse array of 3D coherently strained islands per unit surface area as a function of island size. The parameter α is the ratio between the change in the surface energy upon island formation and the contribution from island edges to the elastic relaxation energy. When $\alpha > 1$, the system tends thermodynamically toward island coalescence. When $\alpha < 1$, there exists an optimal island size and the system of islands is stable against coalescence.

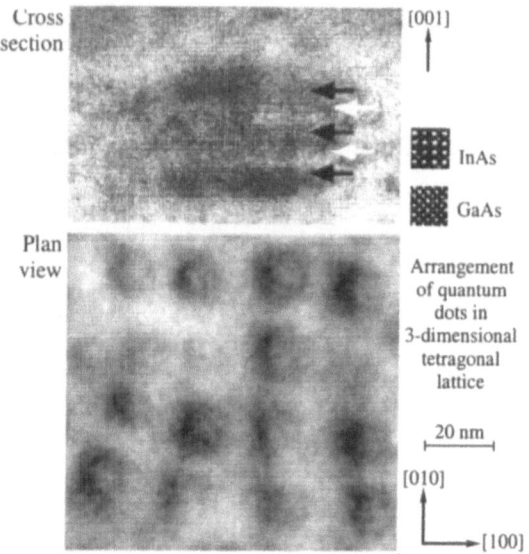

Fig. 8 Vertical and transverse ordering of coupled QDs in the system InAs-GaAs.

high temperature stability of threshold current density are expected to occur simultaneously. Additionally ordered arrays of scatterers formed in an optical waveguide region may result in distributed feedback and (or) in stabilization of single-mode lasing. Intrinsically buried quantum dot structures spatially localize carriers and prevent them from recombining nonradiatively at resonator faces. Overheating of facets being one of the most important problems for high-power and high efficiency operation of AlGaAs-GaAs and AlGaAs-InGaAs lasers, may thus be avoidable.

Since the first realization of QD lasers [27], it has become clear that the QD size uniformity was sufficient to achieve good device performance. But even at that time, it was recognized that the main obstacle for QDHS laser operation at room and elevated temperatures was connected to temperature-induced evaporation of carriers from QD's. Different methods were developed to improve the laser performance: (i) the increase of the

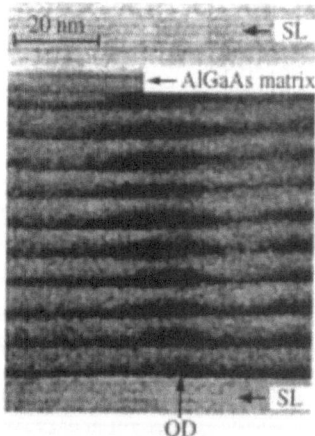

Fig. 9 Transmission electron microscopy image of the active region of high-power QDHS laser.

density of QD's by stacking of QD's (Fig. 9); (ii) the insertion of QD's into a QW sheet; (iii) the use of a matrix material with a higher bandgap energy. As a result, we got many parameters of QDHS lasers better than ones for QWHS lasers based on the same materials. As an example, the world-record threshold-current density of 26 A/cm^2 has been recently achieved [28]. Further, the cw-output power up to 3.5–4.0 W (CW) for a 100-μm strip width, the quantum efficiency of 95% and the wall-plug efficiency of 50% were obtained [29].

Significant activities in theoretical understanding of QD lasers with realistic parameters have been performed. For a QD size dispersion of about 10% and other practical structure parameters, the theory [30] predicts typical threshold-current densities of 5 A/cm^2 at room temperature. The value of 10 A/cm^2 at 77 K [31] and even 5 A/cm^2 at 4 K [32] have been experimentally observed.

In view of advanced device applications of QD's, the incorporation of QD's in vertical-cavity surface-emitting lasers (VCSEL's) seems to be very important. QD VCSEL's with parameters, which fit to the best values for QW devices of the similar geometry, have been demonstrated [33]. Recently, very promising results for 1.3-μm QD VCSEL's on a GaAs substrate to use in fiber optical communications have been obtained [34].

In a free-standing 3D island formed on a lattice-mismatched substrate, the strains can relax elastically, without the formation of dislocations. Thus, sufficiently large volume of a coherent narrow-gap QD material can be realized. This makes possible to cover a spectral range of 1.3–1.5 μm using a GaAs substrate and to develop wavelength-multiplexing systems on the base of QD VCSEL's in the future.

4 Future trends

Recently very impressive results for short wave-length light sources have been achieved on the base of II–VI selenides and III–V nitrides. The success in this research was mostly determined by application of heterostructure concepts and methods of growth which had been developed for III–V quantum wells and superlattices. The natural and most predictable trend is the application of the heterostructure concepts as well as technological methods and peculiarities to new materials. Different III–V, II–VI and IV–VI heterostructures, developed in recent time, are good examples of this statement.

But from a general and more deep point of view, heterostructures (it concerns all of them: the classical, QWs and SLs, QWRs and QDs) are the way of creation of new types of materials — Heterosemiconductors. By using Leo Esaki words — instead of "God made crystals" we create by ourselves — "Man made crystals".

The classical heterostructures, quantum wells and superlattices are quite mature and we exploit many of their unique properties. Quantum wires and dots structures are still very young: exciting discoveries and new unexpected applications are awaiting us on this way. Even now we can say that ordered equilibrium arrays of quantum dots may be used in many devices: lasers, light modulators, far-infrared detectors and emitters, etc. Resonant tunneling via semiconductor atoms introduced in larger bandgap layers may lead to significant improvement in device characteristics. More generally speaking, QD structures will be developed both in "width" and in "depth". In "width" means new material systems to cover new energy spectrum. The life-time problems of the green and blue semiconductor lasers and even more general problems of the creation defect-free structures based on wide-gap II–VI and III–V (nitrides) would be solved by using QDs structures in these systems.

In "depth" it's necessary to mention that degree of ordering depends on very complicated growth conditions, materials constants, concrete values of the surface free energy. The way to resonant tunneling and "single" electron devices including optical one is a deep detailed investigation and evaluation of these parameters in order to achieve the maximal possible degree of ordering. In general, it's necessary to find out more strong self-organization mechanisms for ordered arrays of QDs creation.

It is hardly possible to describe even the main directions of the modern physics and technology of semiconductor heterostructures. There are much more than it was mentioned. Many scientists contributed to this tremendous progress which not only defines to a great extent the future prospects of condensed matter physics and semiconductor laser and communication technology but, in a sense also, the future of the human society. I would like also to emphasize the impact of scientists of previous generations who prepared our way. I am very happy that I had a chance to work in this field from the very beginning. I am even more happy that we can continue to contribute to the progress in this area now.

References

1. W. Shockley, U.S. Patent No 2,569,347 (Sept. 25, 1951).
2. H. Kroemer, Proc. IRE **45**, (1957) 1535; RCA Rev. **28**, (1957) 332.
3. Zh.I. Alferov and R.F. Kazarinov, Inventor's Certificate No. 181737 [in Russian], Appl. No. 950840, priority as of March 30, 1963; H. Kroemer, US Patent No 3.309,553 (1967) (Filed Aug. 16, 1963).
4. Zh.I. Alferov, V.B. Khalfin, and R.F. Kazarinov, Sov. Phys. Solid State **8**, (1966) 2480.
5. Zh.I. Alferov, Sov. Phys. Semicond. **1**, (1967) 358.
6. Zh.I. Alferov, V.M. Andreev, V.I. Korol'kov, D.N. Tret'yakov, and V.M. Tuchkevich, Sov. Phys. Semicond. **1**, (1967) 1313; H.S. Rupprecht, J.M. Woodall, and G.D. Pettit, Appl. Phys. Lett. **11**, (1967) 81.
7. Zh.I. Alferov, V.M. Andreev, E.L. Portnoy, and M.K. Trukan, Sov. Phys. Semicond. **3**, (1969) 1107.
8. Zh.I. Alferov, V.M. Andreev, D.Z. Garbuzov, Yu.V. Zhilyaev, E.P. Morozov, E.L. Portnoi, and V.G. Trofim, Sov. Phys. Semicond. **4**, (1970) 1573; I. Hayashi, M.B. Panish, P.W. Foy, and S. Sumski, Appl. Phys. Lett. **17**, (1970) 109.
9. Zh.I. Alferov, V.M. Andreev, S.G. Konnikov, V.G. Nikitin, and D.N. Tret'yakov, Proc. of Int. Conf. Phys. Chem. Semicond. Heterojunctions and Layer Structures, Budapest, October, 1970, Vol. 1, edited by G. Szigeti (Academiai Kiado, Budapest, 1971), p. 93.
10. Zh.I. Alferov, V.M. Andreev, R.F. Kazarinov, E.L. Portnoi, and R.A. Suris, Inventor's Certificate No. 392875 [in Russian], Appl. No. 1677436, priority as of July 19, 1971.
11. A.Y. Cho, J. Vac. Sci. Technol. **8**, (1971) 31.
12. H.M. Manasevit, Appl. Phys. Lett. **12**, (1968) 156.
13. R.D. Dupuis and P.D. Dapkus, Appl. Phys. Lett. **31**, (1977) 466.
14. R. Dingle, W. Wiegmann, and C.H. Henry, Phys. Rev. Lett. **33**, (1974) 827.
15. R.D. Dupuis, P.D. Dapkus, N. Holonyak, Yr., E.A. Rezek and R. Chin, Appl. Phys. Lett. **32**, (1978) 295.
16. W.T. Tsang, Appl. Phys. Lett. **40**, (1982) 217.
17. Zh.I. Alferov, A.I. Vasil'ev, S.V. Ivanov, P.S. Kop'ev, N.N. Ledentsov, M.E. Lutsenko, B.Ya. Mel'tser, and V.M. Ustinov, Sov. Tech. Phys. Lett. **14**, (1988) 782.
18. R.F. Kazarinov and R.A. Suris, Sov. Phys. Semicond. **5**, (1971) 619; Sov. Phys. Semicond. **6**, (1972) 96; Sov. Phys. Semicond. **7**, (1973) 246.
19. J. Faist et al., Science **264**, 553, (1994); Electron. Lett. **30**, (1994) 865.
20. Y. Arakawa and H. Sakaki, Appl. Phys. Lett. **40**, (1982) 939.
21. A.I. Ekimov and A.A. Onushchenko, JETP Lett. **34**, (1981) 345.
22. L. Goldstein, F.Glas, J.Y. Marzin, M.N. Charasse and G. Le Roux, Appl. Phys. Lett. **47**, (1985) 1099.
23. N.N. Ledentsov et al., Proc. 22nd Int. Conf. Phys. Semicond., (Vancouver, Canada, 1994). World Scientific, Singapore, 1995.
24. Zh.I. Alferov et al. Semiconductors **30**, (1996) 197.
25. V.A. Shchukin, N.N. Ledentsov, P.S. Kop'ev and D. Bimberg, Phys. Rev. Lett. **75**, (1995) 2968.
26. Zh.I. Alferov et al. Semiconductors **30**, (1996) 194.
27. N. Kirstaedter et al., Electron. Lett. **30**, (1994) 1416.
28. G.T. Liu, H. Li, K.J. Malloy, and L.F. Lester, Electron. Lett. **35**, (1999) 1163.
29. A.E. Zhukov, A.R. Kovsh, S.S. Mikhrin, N.A. Maleev, V.M. Ustinov, D.A. Lifshits, I.S. Tarasov, D.A. Bedarev, M.V. Maximov, A.F. Tsatsul'nikov, I.P. Soshnikov, P.S. Kop'ev, Zh.I. Alferov, N.N. Ledentsov, and D. Bimberg, Electron. Lett. **35**, (1999) 1845.
30. L.V. Asryan and R.A. Suris, Semicond. Sci. Tehnol. **11**, (1996) 554.
31. A.E. Zhukov, V.M. Ustinov, A.Yu. Egorov, A.R. Kovsh, A.F. Tsatsul'nikov, N.N. Ledentsov, S.V. Zaitsev, N.Yu. Gordeev, P.S. Kop'ev, and Zh.I. Alferov, Jpn. J. Appl. Phys., pt. 1, **36**, (1997) 4216.
32. G. Park, O.B. Shchekin, S. Csutak, D.L. Huffaker, and D. Deppe, Appl. Phys. Lett., **75**, (1999) 3267.
33. J.A. Lott, N.N. Ledentsov, V.M. Ustinov, A.Yu. Egorov, A.E. Zhukov, P.S. Kop'ev, Zh.I. Alferov, and D. Bimberg, Electron. Lett. **33**, (1997) 1150.
34. J.A. Lott, N.N. Ledentsov, V.M. Ustinov, N.A. Maleev, A.E. Zhukov, A.R. Kovsh, M.V. Maximov, B.V. Volovik, Zh.I. Alferov, and D. Bimberg, Electron. Lett., (2000) in press.

Quantum Computation

Y. Yamamoto[1], F. Yamaguchi[2]

[1] Quantum Entanglement Project, ICORP, JST, Edward L. Ginzton Laboratory, Stanford University, and NTT Basic Research Laboratories
[2] Quantum Entanglement Project, ICORP, JST, Edward L. Ginzton Laboratory, Stanford University

Abstract Quantum computation is an interdisciplinary subject and vast field of intensive research in recent years. In this paper we will overview the three important research areas of quantum computation: quantum algorithms, oracle implementation and quantum hardwares.

1 Basic concepts in quantum interference

If we must choose one word to describe the principle of a quantum computer, it should be an interferometer, not classical but quantum interferometer. There are four basic concepts in quantum interference: entanglement, which-path measurement, quantum erasure and non-locality (or non-separability).

Figure 1 shows a two-photon interferometer [1] A photon-pair (signal photon and idler photon) is generated by a parametric down-converter, in which a second-order nonlinear crystal is excited by a strong pump beam. The energy and momentum conservation laws apply for the absorbed pump photon and generated signal and idler photons: $\omega_p = \omega_s + \omega_i$ and $k_p = k_s + k_i$. A Young's double slit interferometer is formed for signal and idler photons independently. Two photon coincidence counts with two detectors D_s and D_i are measured as well as single count either by D_s or D_i.

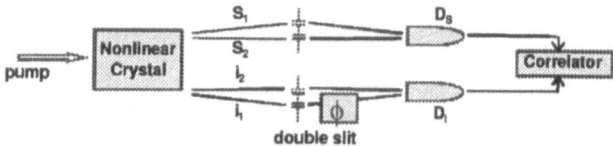

Fig. 1 A two-photon interferometer with a parametric down-converter.

Entanglement

The signal photon reaching the detector D_s simultaneously passes through the two slits, which can be mathematically described by a linear superposition of two orthogonal states $|s_1\rangle$ and $|s_2\rangle$. The same is true for the idler photon. The idler photon is also described by a linear superposition of two orthogonal states $|i_1\rangle$ and $|i_2\rangle$. Moreover, these two photons are correlated. If the signal photon passed through the slit $s_1(s_2)$, the idler photon must have passed through the slit $i_1(i_2)$. This correlation is a consequence of the above mentioned momentum conservation, $k_p = k_s + k_i$. Consequently, we can write the two photon state in front of the two detectors:

$$|\chi\rangle_{si} = \frac{1}{\sqrt{2}} \left(|s_1\rangle|i_1\rangle + |s_2\rangle|i_2\rangle \right) \tag{1}$$

The quantum correlated superposition state like (1) is called an entangled state.

Which-path measurement

When the detector D_s is scanned along the vertical direction with recording the single count rate, there is no interference pattern observed even though a Young's double slit interferometer is constructed for the detected signal photons. The reason is that we throw away the idler photon which carries the which-path information (if $|s_1\rangle$ is true or $|s_2\rangle$ istrue). Mathematically, the dissipation of the idler photon into reservoirs is represented by the partial trace,

$$\rho_s = Tr_i \left[|\chi\rangle_{si\,si}\langle\chi| \right] = \frac{1}{2} \left[|s_1\rangle\langle s_1| + |s_2\rangle\langle s_2| \right] \quad , \tag{2}$$

and the resulting state is no more superposition state but a statistical mixture. An actual measurement for the idler photon state ($|i_1\rangle$ or $|i_2\rangle$) is not necessary. The possibility of performing such a which-path measurement is enough to eliminate an interference effect.

Quantum erasure

If two photon coincidence count rates are recorded with scanning the position of the detector D_s, while the position of the detector D_i is fixed, the interference pattern can be observed. In this case, the detection of the idler photon at a certain position (of D_i) erases the which-path information and projects the signal photon state onto

$$|\chi\rangle_s = \frac{1}{\sqrt{2}}(|s_1\rangle + |s_2\rangle) \quad . \tag{3}$$

If the detector (D_i) position is shifted along the vertical direction by a half wavelength, the same idler photon detection by D_i projects the signal photon state onto $|\chi\rangle_s = \frac{1}{\sqrt{2}}(|s_1\rangle - |s_2\rangle)$. Therefore, the above two cases produce the two-photon interference pattern shifted by a half period. In a sence, no interference pattern in the single count rate mentioned above is the consequence of the sum of two interference patterns shifted by a half period.

Non-locality or non-separability

If one of two arms for the idler interferometer is phase modulated, i.e. $|i_1\rangle \rightarrow e^{i\phi}|i_1\rangle$, in the two photon coincidence count measurement, the observed interference pattern is shifted accordingly. This is because the idler photon count at a fixed detector (D_i) position projects the signal photon state onto

$$|\chi\rangle_s = \frac{1}{\sqrt{2}} \left(e^{i\phi}|s_1\rangle + |s_2\rangle \right) \quad , \tag{4}$$

instead of Eq. (3). The phase modulation imposed on the idler photon is not a local effect but is shared by the signal photon which is propagating in a different spatial path. This is due to the non-local correlation, Eq. (1), between the two photons. The entangled two photon state preserves the correlation and shares the same perturbation that is physically introduced into one constituent even though the constituents are separated spatially.

2 Quantum algorithms

Let us consider how a quantum algorithm works using a Deutsch-Jozsa (DJ) algorithm as an example. The DJ problem is a following game [2]. Alice sends one integer number $x \epsilon [0, 2^n - 1]$ to Bob. Bob calculates a funciton $f(x)$ for a given x and responds to Alice. Bob's function takes only either 0 or 1 and there are two types: $f(x)$ is all zero or all one (type I) or $f(x)$ is half-zero and half-one (type II). How fast can Alice tell the type of Bob's function? The best classical algorithm requires the $2^{n-1} + 1$ inquiries in the worst case. For instance, in the $n = 3$ bit DJ problem, if the first two answers are 0 and 0, Alice still cannot determine the type of Bob's function. Only after the third answer is 0(1), she can determine Bob's function is type I (type II). The solution requires an exponentially increasing computation time. For $n = 1$ bit DJ problem, she needs only 2 inquiries, but for $n = 10$ and 20 bits DJ problems, she needs 513 inquiries and 524.289 inquiries, respectively.

On the other hand, a quantum algorithm shown in Fig. 2 needs only one computational time irrespective of the problem size, so quantum computation promises an exponential speed-up over classical computation [2]. For that purpose, we first prepare n-qubit control registers and one-qubit work register all in their ground states:

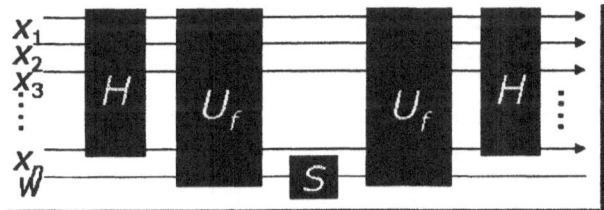

Fig. 2 A Duetsch-Jozsa quantum algorithm.

$$|o\rangle_c |0\rangle_w = (|0\rangle_1 |0\rangle_2 \cdots |0\rangle_n) |0\rangle_w \quad . \tag{5}$$

Next we perform the Walsh-Hadamard transform H for all control registers,

$$|0\rangle_c |0\rangle_w \xrightarrow{H} \frac{1}{\sqrt{2^n}} \sum_{x=0}^{2^n - 1} |x\rangle_c |0\rangle_w \quad . \tag{6}$$

Note that n-qubit control registers are in a linear superposition state and simultaneously represent all 2^n different x values. This means that simultaneous calculation of $f(x)$ for 2^n different inputs x is now possible. This property is often referred to as quantum parallelism.

A unitary evolution U calculates $f(x)$ and places this result in the work resister for all x values in a following

way:

$$\frac{1}{\sqrt{2^n}} \sum_{x=0}^{2^n-1} |x\rangle_c |0\rangle_w \xrightarrow{U} \frac{1}{\sqrt{2^n}} \sum_{x=0}^{2^n-1} |x\rangle_c |f(x)\rangle_w \quad . \tag{7}$$

The control and work resisters are now in an entangled state. Even though we have all the necessary information to determine the type of Bob's-function, we cannot perform a simple quantum measurement on the work register. If we did, we would get only one bit of information (either 0 or 1) and collapses the wavefunctions of both control and work registers. We need to use a quantum interference effect to extract a desired information. At the heart of DJ algorithm is a non-local phase modulation directly incurred by $f(x)$ stored in the work register,

$$\frac{1}{\sqrt{2^n}} |x\rangle_c |f(x)\rangle_w$$
$$\xrightarrow{S} \frac{1}{\sqrt{2^n}} \sum_{x=0}^{2^n-1} (-1)^{f(x)} |x\rangle_c |f(x)\rangle_w \quad . \tag{8}$$

Note that the information of $f(x)$ is imprinted on the phase of each ket vectors $|x\rangle_c$ of the control register and so the constant (type I) or balanced (type II) can be determined by using an interference effect. In order to realize such quantum interference between different $|x\rangle$ vectors, we must eliminate first which-path information stored in the work register $|f(x)\rangle_w$. For that purpose, we perform the same unitary evolution as Eq. (7):

$$\frac{1}{\sqrt{2^n}} \sum_{x=0}^{2^n-1} (-1)^{f(x)} |x\rangle_c |f(x)\rangle_w$$
$$\xrightarrow{U} \frac{1}{\sqrt{2^n}} \sum_{x=0}^{2^n-1} (-1)^{f(x)} |x\rangle_c |0\rangle_w \quad , \tag{9}$$

where we use the $x \oplus x = 0$ (exclusive or) for $x = 0$ and $x = 1$.

Now we are ready to perform the second Walsh-Hadamard transform for all control registers,

$$\frac{1}{\sqrt{2^n}} \sum_{x=0}^{2^n-1} (-1)^{f(x)} |x\rangle_c |0\rangle_w$$
$$\xrightarrow{H} \frac{1}{2^n} \sum_{x,y} (-1)^{f(x)+x,y} |y\rangle_c |0\rangle_w \quad . \tag{10}$$

The final state is in general, a linear superposition state of all $|y\rangle_c$ states, but the probability amplitude of $|y = 0\rangle_c$ state has the special character,

$$\frac{1}{\sqrt{2^n}} \sum_{x=0}^{2^n-1} (-1)^{f(x)} = \begin{cases} \pm 1 : \text{type I} \\ 0 : \text{type II} \end{cases} \quad . \tag{11}$$

The above result indicates that if we measure the n-qubit control registers and obtain $|0\rangle_c = |0\rangle_1 |0\rangle_2 \cdots |0\rangle_n$, we can conclude Bob's function is type I. If we obtain other results, Bob's function is type II.

This DJ algorithm requires the two time evaluations of $f(x)$ by a unitary logic circuit, but if we prepare the work register initially in $\frac{1}{\sqrt{2}}(|0\rangle - |1\rangle)_w$ instead of $|0\rangle_w$, we can eliminate the second unitary evolution. In this way, the DJ algorithm allows an exponentially fast computation over the classical method ($2^{n-1} + 1$ vs. 1).

The discovery of DJ algorithm is followed by Shor's factoring and discrete logarithm algorithms [3] and Grover's data search algorithm [4], which are practically useful and also more efficient algorithms as will be discussed next. However, the essential structures of those new algorithms are the same as one used in the DJ.

3 Oracle implementation

The simultaneous computation of $f(x)$ for all possible x values can be realized by the two different ways. One way is to decompose the oracle ($f(x)$ for all x values) into two elementary gates, one-bit arbitrary rotation and two-bit controlled-NOT, in cascade [5]. These two gates form a universal gate, that is, can perform an arbitrary unitary evolution. The advantage of this method is both one-bit gate and two-bit gate can be relatively easily constructed in many physical systems. The disadvantage of this method is that many gates must be cascaded to implement a certain oracle. For instance, the number of gates required to implement DJ algorithm, Shor's algorithm and Grover's algorithm scale as $O(n2^n)$, $O(n^3)$ and $O(n)$, respectively. Since a qubit system has a finite decoherence time, such sequential implementation of many gates places a serious limitation on the feasibility of quantum computation.

The other way to implement the oracle is to prepare n-qubit system with 2-body, 3-body, $\cdots$ and n-body interactions [6]:

$$\mathcal{H} = \sum_i \hbar u_i \sigma_z^i + \sum_{ij} \hbar u_{ij} \, \sigma_z^i \sigma_z^j + \cdots$$
$$+ \hbar u_{12 \cdots n} \sigma_z^1 \sigma_z^2 \cdots \sigma_z^n \quad . \qquad (12)$$

If we can adjust all the coupling constants $u_i, u_{ij}, \cdots u_{12 \cdots n}$ arbitrarily, any function $f(x)$ in Eq. (9) can be implemented by only one time evolution. The advantage of this method is the simultaneous and collective interactions among all qubits, which leads to only one unitary time evolution for the implementation of the oracle. This property would lift off a stringent requirement on the decoherence time of a quantum system. The disadvantage of this method is that it is not easy to find such a highly nonlinear system with adjustable coupling strengths.

4 Quantum hardwares

What kind of physical devices or systems can be used as a building block of a quantum computer? This is a hard question to answer, because a well isolated quantum system is difficult to couple and to measure, whereas the introduction of necessary couplings and probes leads to the devastating effects of decoherence. This is precisely a reason why no one ever observed a Schrödinger's cat.

Table I summarizes the proposed quantum hardwares for sequential implementation of the oracle. Presence or absence of a photon, as well as two orthogonal polarization of a photon, can be used as a basis for a qubit. A

standard construction of a two-bit gate requires a nonlinear switch (or π-phase shift) by a single photon [7]. Such a single photon Fredkin gate is a building block for an optical quantum computer [7]. Instead of using such a nonlinear switch, we can use a single photon source or entangled photon source combined with a linear optical system [8]. Optical implementation of quantum information processing is highly desirable in connection with its application to the emerging fields of a quantum key distribution and quantum repeater systems.

Table 1 Quantum hardwares

Schemes	References
Nonlinear optics	[7]
Linear optics with nonclassical source	[8]
Ion trap	[9]
Solution NMR	[10] [11]
Nuclear spin in Si	[12]
Electron spin in Si/Ge	[13]
Josephson junction	
Charge state	[14]
Flux state	[15]
Quantum dots	
Coulomb interaction	[16]
Cavity QED	[17]
Electrons on liquid the surface	[18]
Crystal lattice	[19] [20]

The discrete (metastable) energy levels of an ion trapped in a vacuum can also be used as a qubit [9]. A two-bit gate can be constructed by a collective vibrating motion (phonon) of such an ion system. This scheme is a most actively investigated experimental system at present time.

The nuclear spins of a molecule in a solution offers a very stable qubit system [10],[11]. A well established nuclear magnetic resonance (NMR) pulse sequence, together with a nuclear dipole coupling, can easily implement one-bit and two-bit gates. This scheme is a most advanced experimental system at present time.

An implanted donor impurity (such as ^{31}p) in a Si substrate can be used as a qubit nuclear spin [12]. A two-bit gate is constructed via the nuclear spin-spin coupling mediated by a bound electron to the donor atom. Alternatively, the electron spin bound to the donor can be used as a qubit [13]. This interesting idea requires the development of a single nuclear/electron spin detection technique, which is quite challenging.

The excess charge [14] or quantized flux [15] trapped in a Josephson junction circuit can be used as a qubit. These proposals have a certain advantage in that the device fabrication and integration technique and the qubit detection techniques based on a single electron transistor or SQUID are relatively well established techniques. This is also an active research field at present time.

The electron spin in a semiconductor quantum dot also attracts great attention as a qubit system. A two-bit gate can be constructed by using a direct Coulomb interaction [16] or stimulated Raman scattering with a laser pulse [17]. The electrons trapped on liquid He surface by the electrostatic potential offers interesting possibility as a qubit system, too [18].

Finally, the nuclear spins of a constituent atom in a solid crystal can be used as a qubit system [19]. By applying a strong magnetic field gradient, nuclear spins in each atom layer have different Zeeman frequencies, so each layer can be addressed individually by a RF field. On the other hand, all nuclear spins in the same atomic layer serves as a copy of the same qubit. In a sense, the ensemble of a microscopic one-dimensional quantum computer is realized in a natural bulk crystal. For that purpose, a quasi-one dimensional crystal, such as a Fluorapatite for ^{19}F nuclear spin, is used with a rather sophisticated decoupling/selective recouping NMR pulse sequence [20].

5 Conclusion

A quantum information theory has rapidly developed for the last ten years and many fundamental theorems are well founded by now. These include Bell's theorem [21], no measurement [22] and no cloning [23] principles, von Neumann entropy and Holove information [24], and quantum error correction codes [25]. Theoretical understanding of quantum computation is also well established.

On the other hand, an experimental progress on a hardware side has been rather slow. We are still not sure if a quantum computer is ever feasible. This is the first time that a scientist needs an almost perfectly isolated quantum system and must compete with an environment in the measurement of a delicate quantum system. Deeper understanding of decoherence and clever engineering of reservoirs are decisive factors for answering this question [26].

References

1. E. Wolf and L. Mandel, *Quantum Optics* (Cambridge Univ. Press, Cambridge, 1995) p. 1074.
2. D. Deutsch and R. Jozsa, Proc. of R. Soc. London, **A439**, (1992) 553.
3. P.W. Shor, SIAM J. Comput. **26**, (1997) 1484.
4. L.K. Grover, Phys. Rev. Lett. **79**, (1997) 325.
5. A. Barenco et al., Phys. Rev. **A52**, (1995) 3457.
6. F. Yamaguchi, S. Master, and Y. Yamamoto, quant-ph/0005128 (2000).
7. I.L. Chuang and Y. Yamamoto, Phys. Rev. **A52**, (1995) 3489.
8. E. Knill et al., quant-ph/0006088 (2000).
9. J.I. Cime and P. Zoller, Phys. Rev. Lett. **74**, (1995) 4091.
10. D.G. Cory et al., Physica **D120**, (1998) 82.
11. N.A. Gershenfeld and I.L. Chuang, Science **275**, (1997) 350.
12. B.E. Kane, Nature **313**, (1998) 133.
13. R. Vrijen et al., quant-pa/9905096 (1999).
14. Y.Makhlin et al., Nature **398** (1999) 395.
15. J.E. Mooij et al., Science **285**, (1999) 1036.
16. D. Loss and D.P. DiVincenze, Phys. Rev. **B59**, (1999) 2070.
17. A. Imamoglu et al., Phys. Rev. Lett. **83**, (1999) 4204.
18. P.M. Platzman et al., Science **284**, (1999) 1967.
19. F. Yamaguchi and Y. Yamamoto, Appl. Phys. **A68**, (1999) 1.
20. T. Ladd et al., quant-ph/0009122 (2000).
21. J.S. Bell, *Speakable and Unspeakable in Quantum Mechanics* (Cambridge Univ. Press, Cambridge, 1987).
22. O. Alter and Y. Yamamoto, Phys. Rev. Lett. **74**, (1995) 4106.
23. W.K. Wootters and W.H. Zurek, Nature **299**, (1982) 802.
24. A.S. Holevo, *Probabilistic and Statistial Aspects of Quantum Theory* (North-Holland, Amsterdam, 1982).
25. P.Shor, quant-ph/9605011 (1996).
26. S. Haroche, Physica Scripta **T76** (1998), 159.

Hybrid structures of fullerenes and single-wall carbon nanotubes

S. Iijima

Meijo University, NEC and JST-ICORP

Carbon nanotubes and fullerenes have been providing unique opportunities to nano-scale science and technology. In the last decade carbon nanotubes and fullerenes were studied independently and there was no cross-talking between them in spite of strong structural similarity. The situation seems to be over now. In this talk I will present hybrid structures of fullerenes and carbon nanotubes.

In a rare occasion a SWNT could trap fullerenes in its central hollow during laser ablation of graphite (1), which is so called "peapod" (2). The observation motivated us to make SWNTs encapsulating fullerenes in a simple vapor phase reaction by doping with C60 after an appropriate treatment of SWNTs to open their caps. The encapsulation of fullerenes succeeded equally with various higher fullerenes and metal-endhedral fullerenes. The latter contains metal atoms such as Gd, La, Sm, Sc. TEM and EELS reveald clearly individual metal atoms inside carbon cages which are suspended inside SWNTs (3). These trapped fullerenes aligned so regularly inside the central hollows of SWNTs that they can be considered as one-dimensional fullerene crystals. Such SWNTs will behave differently in their physical and chemical properties from conventional ones, providing a new series of hybrid structures of fullerene and nanotube. Some of their properties will be presented. We propose the use of nano-scale space of SWNTs for building nano-structures.

Other aspects of SWNTs encapsulating fullerenes are concerned with structural evolution of encapsulated fullerene molecules within the central hollows of SWNTs. We observed that encapsulated fullerene can be fused together at an elevated temperature to form smaller diameter SWNTs inside the host SWNTs, leading to double-walled carbon nanotubes (DWNTs) (4). In our ex-situ heat treatment the DWNTs were formed in a large quantity so that their Raman signals could show clearly a characteristic co-relation among radial breathing mode peaks. The formation of DWNTs inside existing SWNTs does not involve any catalytic metal and thus their growth mechanism is simpler, providing useful information for the growth of SWNTs.

References

1. Y. Zhang, S. Iijima, Z. Shi and Z.Gu, Phil.Mag. Lett. 79, 473 (1999)
2. B.W. Smith, M. Monthioux and D.E.Luzzi, Nature 396, 323 (1993)
3. K. Hirahara, et al. Physical Revies Letters, submitted.
4. J. Sloan, et al. Chem. Phys. Lett., 316, 191 (2000)

Proc. 25th Int. Conf. Phys. Semicond., Osaka 2000 (Eds. N. Miura and T. Ando)

25

The science, technology and impact of gallium nitride-based transistors

U.K. Mishra

ECE Department UCSB, Santa Barbara, CA. 93108 USA

Gallium Nitride (GaN) is potentially one of the most important emerging compound semiconductors because of the exstremely wide scope of its impact. Along with its family of compounds based on Al and In, (the Al, Ga, In)N family, it has revolutionized the world of light emitters via the commercialization of the blue, green and white LEDs and the blue laser. However, a potentially equally exiting impact is soon to come in the field of GaN electronics. Here, the wide bandgap of GaN is used to achieve a large breakdown voltage and the high electron mobility and velocity toproduce large currents, the product enabling an order of magnitude higher power than current GaAs-based technology. This changes the rules of system design for RADARS and Narrow and Broadband Communications. The two devices that are currently being pursued are the AlGaN/GaN High Electron Mobility Transistor (HEMT) and the Heterojunction Bipolar Transistor (HBT). The former has already achieved a power density of approximately 10W/mm at 10GHz (compared to 1W/mm from GaAs). This enormous potential extends to frequencies over 30GHz, perhaps being the enabler for low-cost LMDS, a last-mile solution. The AlGaN/GaN HBT is relatively immature being more sensitive to dislocations and surface and bulk point defects.

Lastly, the properties of the AlGaN/GaN heterostructures are fundamentally determined by the large polarization fields in the structures. These impact the materials properties, device physics and device design in a wholly new manner. Several new device concepts are being proposed because of these unique properties.

In this talk and attempt will be made to link aspects of this field from science through to applications to both show how much has been achieved but also how vast the research is that needs to be done.

Correlation, Decoherence, Dephasing and Relaxation in Semiconductors

D.S. Chemla

Department of Physics, University of California and
Materials Sciences Division, Lawrence Berkeley National Laboratory
1 Cyclotron Rd., MS 2-346, Berkeley, CA 947200

Abstract Ultimate semiconductor devices which are now considered involve only a few electrons and photons and exploit all attributes of quantum mechanics. Yet, very often: (i) coherence and relaxation are considered independently and are opposed, (ii) many-body processes are considered taboo. In this conference I will emphasize that coherence and relaxation both stem from Coulomb Correlation, a fundamentally many-body phenomenon which prevails in almost all situations, including extremely low densities. Furthermore, these interactions produce the nonlinearities that are prerequisite for performing any interesting function.

1 Introduction

The most striking difference between classical and quantum systems is that the latter are capable of being in entangled states, superpositions of states with well defined relative phases. Experiments on simple atomic systems have demonstrated fascinating properties of these correlated states, but it remains a challenge to make similar observations in condensed matter. In that case, because of the extreme electron density, $\approx 10^{23} cm^{-3}$, the Coulomb force induces both correlation and decoherence. Indeed, many dephasing processes occur in solids, which turn coherent excitations into incoherent occupations and lead to qualitative changes in the behavior of initially coherent systems.

Electrons in condensed matter are dressed by their interaction with the other electrons and the nuclei. This dressing gives rise to the concept of quasi–particles, introduced by Landau. Therefore, when we try to develop a description of such a system we immediately face the problem of Coulomb correlation with its two factets: (i) correlations drive coherences which result in the formation of the quasi–particles, (ii) (Coulomb) interactions between quasi–particles destroy their coherence and limits their life time. Over the last decade a wealth of new information on Coulomb correlation and electronic dynamics has been obtained in regimes where traditional assumptions fail. In this context, semiconductors play a very special role: (i) because of their technological importance, very advanced fabrication and processing techniques have been developed so that almost perfect samples and artificial quantum structures are now available, (ii) a lot is known about their fundamental properties, and sound theoretical techniques, based on well established approximations, are at hand for describing their ground state and linear response, (iii) the large dielectric constant and small effective mass in semicon-ductors allows the description of their electronic states in terms of envelope wavefunctions whose energy, time and length scales are mesoscopic, half way between those of atoms and molecules and those of macroscopic systems. For all these reasons it is possible to investigate in semiconductors, with table–top experiments, quantum mechanical, many–body and quantum kinetic phenomena that would be nearly impossible to observe in other systems. Semiconductors and their heterostructures are thus ideal laboratories for investigating such phenomena and also serve as models for other, more complex, condensed matter systems. Interestingly, this also has a crucial importance for technology because many modern electronic and optoelectronic devices operate precisely in the regime where fundamental many-body effects are dominant. Furthermore, attempts for reaching fundamental limits with existing devices and proposals for new device concepts call for manipulation and engineering of these many–body process. In that respect, the case of semiconductors is again particular since modern synthesis techniques now provide us with the ability to manipulate the electronic envelope wavefunctions and the dielectric function profile in heterostructures, thus opening many new opportunities. It is worth noting to end this introduction, that although the linear properties of semiconductors are similar to those of two-level systems (excitons and exciton–complexes), (i) this linear regime is only reached at extremely low densities almost never encountered in practice, (ii) all interesting devices are based on nonlinearities which, as we shall see, are dominated by Coulomb interaction effects.

In this conference I will give a comprehensive and balanced account of both experimental and theoretical advances of the last decade. I shall try to focus on the most important physics and, as much as possible, give an intuitive picture of the new phenomena that have been observed and explained.

2 Nonlinear Optical Spectroscopy

Interactions between quasi–particles obviously lead to nonlinearities in the optical response. Therefore, nonlinear optical spectroscopic techniques are well adapted to the study of these processes. With the availability of very reliable ultrashort pulsed lasers experiments such as pump/probe (P/p) and coherent wave mixing (WM) measurements are now routine. They offer tremendous flexibility for driving coherent excitations or non equilibrium processes and probing them [7,8].

The principle of these experiments is very simple. In the simplest configuration a short pulse propagating along direction $\mathbf{k}_1$ excites the sample at time $t = 0$. A second pulse, derived from the same laser and propagating along $\mathbf{k}_2$ is delayed and excites the sample at $t = \Delta t$. A P/p experiment measures the changes in transmission of a weak probe–pulse, along $\mathbf{k}_1$, induced by a strong pump–pulse, along $\mathbf{k}_2$. If this signal is spectrally resolved it essentially gives the changes of absorption spectrum, $\alpha(\omega)$, induced by the pump as function of the delay, $\Delta\alpha(\omega, \Delta t) = [\alpha_{probe}(\text{pump on}) - \alpha_{probe}(\text{pump off})]$. In a WM measurement, the pulses interfere in the sample and generate nonlinear polarization waves which emit photons in momentum conserving and background free directions such as (a) $\mathbf{k}_{s1} = 2\mathbf{k}_2 - \mathbf{k}_1$ for a four–WM (FWM) measurement or (b) $\mathbf{k}_{s2} = 3\mathbf{k}_2 - 2\mathbf{k}_1$ in a six–WM (SWM) measurement. These signals can be spectrally resolved (SR), $S_{SR}(\omega, \Delta t)$, or temporally resolved (TR), $S_{TR}(t, \Delta t)$, at fixed time delays. One can also use a slow detector and time–integrate (TI) the signals as the delay Δt is varied, $S_{TI}(\Delta t)$. In the case of non–interacting two–level systems whose nonlinearity just reflects the Pauli blocking, the spectral response $S_{SR}(\omega, \Delta t)$ is a Lorentzian and the corresponding the temporal response $S_{TR}(t, \Delta t)$ is a damped oscillation starting right after the second pulse. Its decay measures half the dephasing time, T_2, of the nonlinear polarization and corresponds to the inverse linewidth of the Lorentzian. Because of the very simple response in this case the TI–signal $S_{TI}(\Delta t)$ has as a function of Δt the same behavior as the TR–signal as a function of the real time t (the integral of an exponential is an exponential.) It thus contains the same information: it is zero for $\Delta t < 0$, peaks at $\Delta t = 0$ and decays exponentially with a time constant $T_2/2$. Any departure from this behavior is a signature of quasi–particle interactions and carries information relevant to their nature and symmetry. It is important to note that the TI signal on one hand, and the SR and TR signals on the other give information on two type of variables that are *not* quantum mechanical conjugates, Δt and ω or t. In analogy with NMR, coherent–WM measurements involving more complicated pulse sequences are easily implemented and can provide more refined information.

3 High Order Correlations

Two seminal experiments have shown that the response of semiconductors is dramatically different from that of non–interacting two–level systems. In the first [1,2], the time-integrated FWM response of a high quality GaAs quantum wells (QW) was measured with a time resolution $\approx 300fs$. A strong $S_{TI}(\Delta t)$ signal was observed for $\Delta t < 0$ extending more than 50 times beyond the time resolution of the experiment with a rise time constant approximately half that of the decay. This result, in

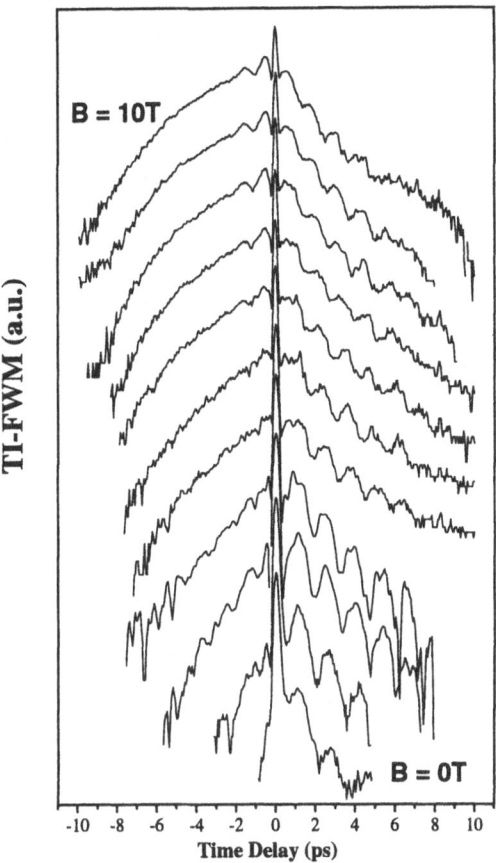

Fig. 1 Time integrated four wave mixing vs time delay in GaAs for manetic field $B = 0 \to 10T$.

contradiction with the case of non–interacting two–level systems, provided the first evidence of correlation effects in the nonlinear optical response of semiconductors. In the second experiment [3–5] TR-FWM measurments on GaAs QW showed that the TR-signal, $S_{ST}(t, \Delta t)$, does not peak at the second pulse as expected in atomic-like systems but can be considerably delayed. These observations triggered a flurry of theoretical and experimental activities.

The only material parameter that enters into the Maxwell Equations and describes all optics is the polarization. It is related to the expectation value of a product of electronic creation and destruction operators. The Coulomb interaction between electrons, holes or electrons and holes is described by products of four creation and destruction operators. Therefore, accounting for that interaction in the equations of motion of operator pairs results in an infinite hierarchy of coupled differential equations for products of 2, 4, 6 ... operators that cannot be solved in general. The simplest approach for taking into account the lowest order correlation is a mean-field theory, based on the random phase approximation (RPA), that factors the expectation value of the 4-particle operators into the products of expectation values of 2-particle operators. This formalism, called the semiconductor Bloch equations (SBE) [6], explains well the aforementioned experiments and many others of the same type [7]. In the SBE approach the Rabi frequency, describing the light-matter coupling, and the energy of the electrons are renormalized. This introduces two types of nonlinearities, one that translates the Pauli Blocking, i.e., the saturable electron-photon coupling, and a second that is a Coulomb mediated self interaction of the polarization. For non-interacting two-level systems, the FWM signal results from the Pauli Blocking only, and appears as an interaction between the field of one pulse by the populations created by the other one. Thus it does not exists when they do not overlap, e.g., for $\Delta t < 0$. The Coulomb mediated nonlinearities produce interactions between polarization waves created by the two pulses which overlap whatever the pulse sequence. This creates nonlinear polarization waves both for $\Delta t > 0$ and $\Delta t < 0$ which, in addition, needs time to build up and have a delayed emission. Thus the Coulomb nonlinearities are at the origin of the unexpected results mentioned above. It is important to note that, although they contribute to the signal for all Δt, the $\Delta t < 0$ signal is *entirely due to Coulomb correlation*. Therefore, by a proper choice of the pulse sequence on can isolate the Coulomb mediated effects and investigate them as shown by the examples given below.

The traditional theoretical description of semiconductors assumes a quasi–stationary limit in which the mean-field RPA is valid. At low density and on timescales short compared to the time between collisions one quasi–particle does not interact with enough of its neighbors to feel a mean-field. Thus it becomes possible to observe deviations from mean-field theory, in a regime where high order correlations become dominant. Recent comparative experimental and theoretical studies have revealed features related to genuine 4-particle and 6-particle correlations. Effects of 4-particle correlations in the continuum of exciton (X) scattering states are evidenced in Fig. 1 [9,10] which shows the TI-FWM signal measured in a GaAs sample, with two exciton resonances, in a magnetic field B. For $\Delta t > 0$, they have the expected profile: beats between the two excitons superimposed on a dephasing exponential. This profile changes only slightly with B. In contrast, the interaction induced signal for $\Delta t < 0$ changes drastically as B increases. Its magnitude increases and its profile becomes highly non-exponential and extends as far as 100 times the pulse duration. This effect is due to a quadrupole interaction between excitons distorted by the magnetic field. It creates a coherent 4-particle correlation between the eletron–hole pairs generated by the $\mathbf{k_2}$-pulse. This two–photon coherence cannot produce a

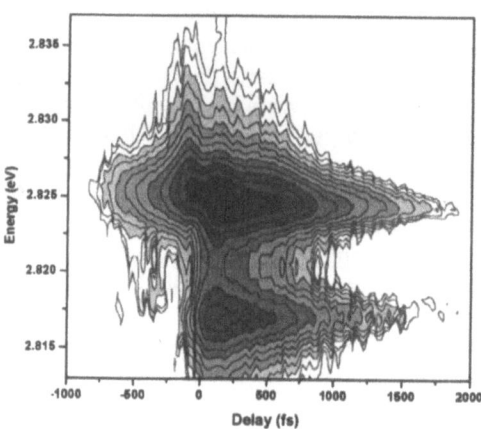

Fig. 2 Measured six-wave-mixing emission shown over 3 orders of magnitude in intensity in the $\{x, y, y\}$ polarization configuration.

one–photon emission by itself. Thus, it is stored in the medium until the $\mathbf{k_1}$-pulse triggers the emission, i.e., for $\Delta t < 0$. It is interesting to note that high order correlation functions appear naturally in the theory, but are rarely accessible to direct measurement. Through the type of experiments described above, not only can they be observed, but their properties can be investigated. For example the dephasing of 4-particle correlation functions by phonon scattering and by collision with excitons and free carriers has been studied recently for the first time [11].

To better isolate higher order correlation one has to turn to WM involving more photons. Indeed, recent SWM studies have revealed 6-particle correlations in a single ZnSe quantum well (QW) [12]. Fig. 2 shows a contour map of the SR-SWM, $S_{SR}(\omega, \Delta t)$, generated by laser pulses polarized along the sample's $\mathbf{x}$ and $\mathbf{y}$ axes, and a measured signal polarized along $\mathbf{y}$. Two contributions are seen, one at the exciton energy $\Omega_X = 2.825 eV$ and the other at the difference between the biexciton and the exciton energies, $\Delta \Omega_{X_2 - X} = \Omega_{X_2} - \Omega_X \approx 2.818 eV$. Both features decay smoothly without temporal or spectral beats. Fig. 3 shows the result of theoretical modeling of the experiments at three levels of sophistication. The upper graph accounts only for the 2-particle and 4-particle correlation involved in the coherent limit, i.e., 1-photon and 2-photon coherences. The center graph in addition accounts for the 4-particle correlation that describes incoherent exciton densities, and the bottom one also accounts for the 6-point correlation that describe transitions from incoherent exciton densities to two-pair states. The three theoretical plots are obtained by solving numerically the coupled equations of motion of the 2-, 4- and 6-particle correlations and thus the results include contributions of infinite order in the laser field. Both the upper and middle graphs show major disagreements with experiment. But adding the 6-point corre-

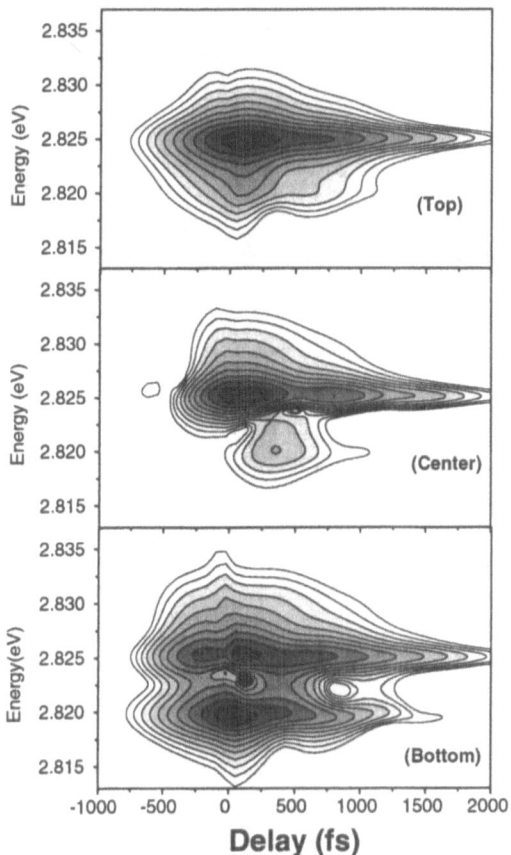

Fig. 3 Theoretical six-wave-mixing emission shown over 3 orders of magnitude in intensity in the $\{x, y, y\}$ polarization configuration. Top: One and Two Photon Coherences. Center: One and Two Photon Coherences, plus Incoherent Exciton Populations. Bottom: One and Two Photon Coherences, Incoherent Exciton Populations, plus Six Particle Correlations.

lation brings the theory it in excellent agreement with experiment, showing that this correlation function is absolutely necessary to describe the SWM experiments.

The optical response is often described in terms of a "few-level" model where the exciton and biexciton resonances are represented by discrete levels. Therefore, it is interesting to study the relative influence of the bound biexciton states, X_2, and of the two–pair scattering continuum, X-X. This is most easily done in numerical computation [13]. The results are shown in Fig. 4. The top panel shows the contribution of X_2 alone whereas the bottom panel shows that of X-X alone. In that case the magnitude of the emission at the energy $\Delta\Omega_{X_2-X}$ dramatically drops down whereas it is of course stronger in upper graph. However, in contrast with the experiment and the full calculation, this emission is still much weaker than the emission at Ω_X. Another remarkable

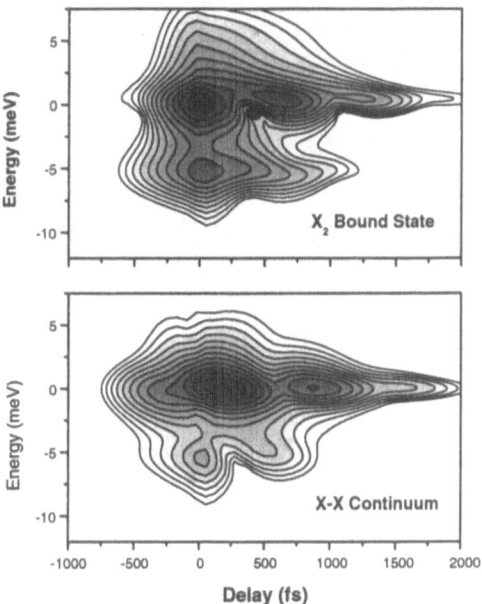

Fig. 4 Theoretical six-wave-mixing emission shown over 3 orders of magnitude in intensity in the $\{x, y, y\}$ polarization configuration. Top: Bound biexciton states only. Bottom: Two–exciton unbound states only.

aspect of Fig. 4 is the appearance of beats at Ω_X and at $\Delta\Omega_{X_2-X}$ which are not seen experimentally nor in the full theory. The suppression of these beats thus implies a subtle interference and a delicate balance between the discrete and continuum contributions of the 6-particle correlation. In these experiments evidence of effects 7th order and higher in the applied field were observed down to the lowest excitaton explored [12].

To conclude this section, it is interesting to see if the effects we are considering are relevant in practice. Among modern optoelectronics devices, semiconductor microcavities (SMC) have attracted much attention because they allow independent control of the two components of the light-matter interaction i.e., the modes of the electromagnetic field and the electronic states of the active medium [14]. This could have profound implications on information technology by opening the way for cavity quantum electrodynamics in the solid-state. SMC nonlinearities appear as soon as several excitons are generated in the active medium. Indeed recent P/p experiments on a SMC containing a single 5nm ZnSe QW have clearly revealed spectrally resolved coherent features due to polariton–biexciton transitions as shown in Fig. 5 [15]. The nonlinear P/p spectra are much more complex than intuitively anticipated. In Fig. 5 the upper spectrum is obtained with co–circular, σ^+/σ^+, polarized pump and probe a configuration where Pauli Blocking is active, but where the biexciton is not. The lower spectrum is obtained with counter–circular, σ^-/σ^+, po-

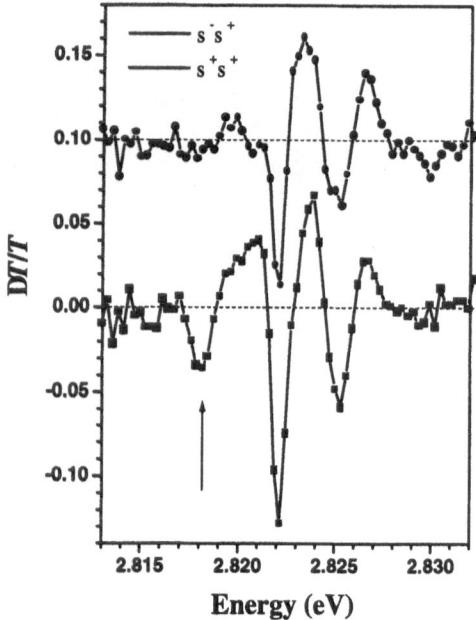

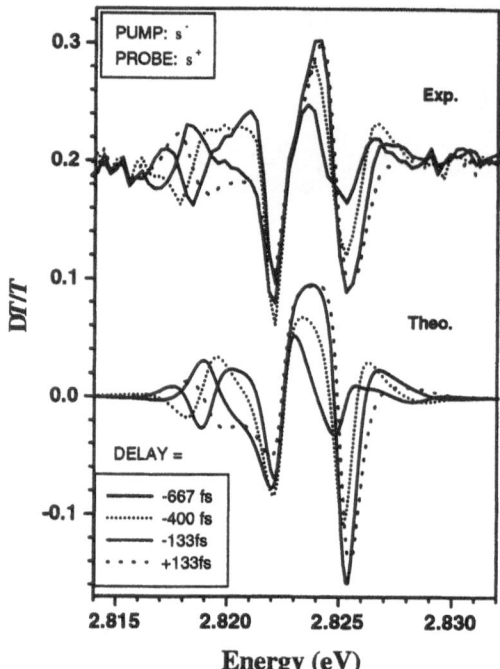

Fig. 5 Differential transmission of a single ZnSe quantum well microcavity for a detuning of −2.4 meV and co-circularly polarized (upper spectrum) and counter-circularly polarized pump and probe beams at a time delay of −533 fs.

Fig. 6 Differential transmission of the microcavity for a detuning of −2.4 meV for counter-circularly polarized beams for several delays. Top: Experimental data, Bottom: Theory with four particle correlation only.
Time delays: -667fs, -400fs, -133fs, +133fs.

larizations where the biexciton is active and the Pauli Blocking is not. It exhibits a new spectral feature around $\Delta\Omega_{X_2-Up} = 2.818\,eV$, the energy of the transition from the SMC upper-polariton to the biexciton. The dependence of the P/p spectra on polarization and Δt agrees very well with a theoretical model that includes 4-particle correlations. Importantly, as shown in Fig. 6 [15], in the coherent regime the σ^-/σ^+ SMC response is dominated by antiparallel-spin exciton–exciton correlations beyond the RPA. Two nonlinear sources beyond RPA contribute, the exciton screening and the biexciton-exciton interaction. It is found that the σ^-/σ^+ SMC response is almost entirely explained by the biexciton-exciton interaction. Although a slight improvement is obtained by adding a small exciton screening contribution, the biexciton-exciton interaction accounts for all the features seen experimentally including details such as the complex Δt-dependence of the P/p oscillations around $\Delta\Omega_{X_2-Up}$.

In summary, nonlinear optical effects in semiconductors and the devices built up from their heterostructures appear at extremely low densities and are dominated by Coulomb correlation.

4 Correlation, dephasing and decoherence

A regime where traditional assumptions fail is that of correlated dynamics. In that case it is no longer justified to treat the interactions between quasi–particles as scat-

tering events local in space and time, a key assumption of the Boltzmann theory or Fermi's golden rule. These approaches describe relaxation and dephasing processes in terms of instantaneous collisions between classical point-like particles. Thus the only time scale that appears is the mean free time between collisions. This picture is in contradiction with the quantum mechanical wave-like nature of the collective excitations in solids which interact through interferences with a finite duration. A crystal, as any quantum mechanical system, is described by equations that are local in time. Therefore, the knowledge of the Hamiltonian and the state at one time is enough to determine the future. In practice, however, because of the complexity of the system, it is necessary to restrict the quantum mechanical treatment to a subsystem consisting of only a few degrees of freedom, electrons or lattice vibrations for example, and treat all the other degrees of freedom as a thermal reservoir. Then the equations of motion of the sub-system becomes nonlocal in time and exhibits memory effects. If the time resolution of an experient is shorter than the period of the interacting quasi–particle waves memory effects become observable and, in some case, controllable.

In the case of semiconductors the main processes that result in dissipation are the electron-electron interaction and the electron-phonon interaction. The electron-phonon interaction is conceptually somewhat easier to consider since the degrees of freedom of one sub-system (the electrons) and the reservoir (the lattice) are distinct. Let us consider first that case, i.e., the scattering between electrons and LO-phonons in the polar semiconductor GaAs. The electrons bands are strongly dispersive, but the LO-phonons are rather dispersionless, $\Omega(k) = \Omega_{LO}$. They form approximately a "single frequency reservoir". Fig. 7, shows the results of FWM experiments performed with linearly polarized light, extremely short pulses of duration $\tau_\ell \approx 15$fs and where the pulse propagating in the direction $\mathbf{k}_1$ is replaced by a pair of phase-locked pulses separated by a time delay $\Delta t_{11'}$ defined to better than ± 40 attoseconds [16–18]. The figure displays a series of TI-signal traces, $S_{TI}(\Delta t)$, at selected values of $\Delta t_{11'}$ covering approximately *one optical cycle*, plotted vs. $\Delta t = t_{21'} = t_2 - t_{1'}$. For certain values of $\Delta t_{11'}$ clear oscillations with a period of about $95 - 100$fs are seen as Δt is varied, wehereas the oscillations are completely absent for other values of $\Delta t_{11'}$. The oscillation period, close but not equal to that of the GaAs LO-phonons $T_{LO}(\text{GaAs}) = 2\pi/\omega_{LO} = 115$fs, corresponds to the beat period between interband transitions whose conduction band states are separated by one LO-phonon, $\Omega_{osc} = \Omega_{LO}(1 + m_e/m_h)$. An intuitive interpretation of these results is the following. An electron with wavevector $\mathbf{k}$ is first excited by a $\mathbf{k}_1$-photon from the valence band, $E_{VB}(\mathbf{k})$, to the conduction band, $E_{CB}(\mathbf{k})$, and then through interaction with an LO-phonon is scattered within the conduction band to a state at $\mathbf{k}'$ and energy $E_{CB}(\mathbf{k}') = E_{CB}(\mathbf{k}) \pm \hbar\Omega_{LO}$. In the "long time" regime this scattering would have the time to be completed, the electrons would stabilize in the state at $\mathbf{k}'$, and thus a dissipative process would have occurred. In the Quantum Kinetics regime, however, the scattering takes some time to be completed. During the interference between the electronic-wave and the lattice-wave, the electron oscillates back and forth between the $\mathbf{k}'$ and the $\mathbf{k}$ states. This oscillation modulates the interband transition amplitudes involving these states. Since the interband transition is responsible for the diffraction of the $\mathbf{k}_2$-photons in the $\mathbf{k}_{s1}$ direction, this results in an oscillation of the TI-FWM signal as the delay Δt varies. Note that in the latter process, no phonon is involved and thus the spectrum of the FWM signal does not show "phonon-sidebands" in agreement with experiment. Most interesting is the $\Delta t_{11'}$ dependence. Indeed, for specific values of $\Delta t_{11'}$ there is destructive (constructive) interferences between the electronic-wave and the lattice-wave that exactly reverse (reinforce) the electron-phonon scattering so that the modulation of the coherent emission is attenuated (enhanced). This process can only occur between coherent waves and therefore requires the pair of $\mathbf{k}_1$-pulses to be phase-locked. This intuitive argument indeed captures the essential physics and is sub-

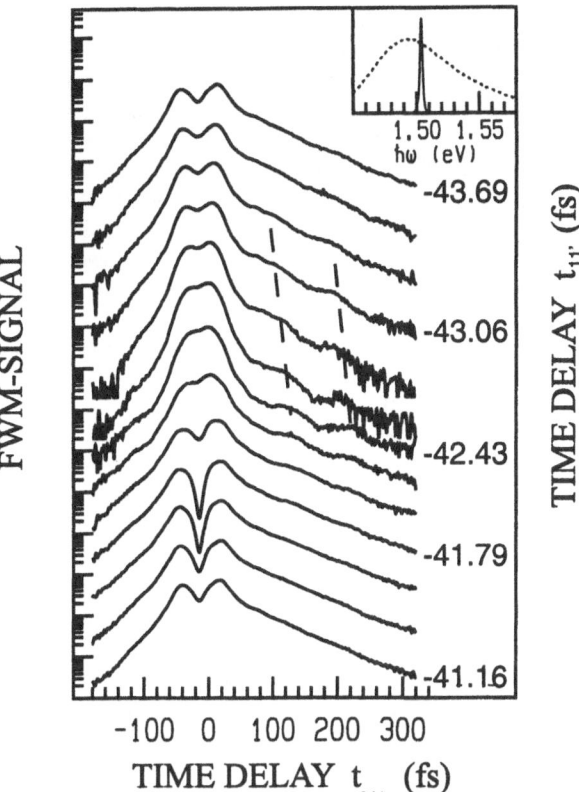

Fig. 7 Four wave-mixing signal measured with a pair of phase-locked pulses in the direction $\mathbf{k}_1$, versus the time delay $\Delta t_{21'}$ for selected values of time delay between the phase-locked pulses $\Delta t_{11'}$ at $T = 77$ K. Low carrier density regime and resonant excitation of the band edge.

stantiated by the complete Quantum Kinetics theory [19, 20].

In the case of electron-electron scattering the sub-system coincides with the reservoir, thus requiring a self consistent treatment much more complex to describe. Quantum Kinetics effects in electron-electron scattering have been observed recently [21–24]. Electron-electron scattering becomes non–trivial in a two dimensional electron gas (2DEG) in the presence of a magnetic field. There the kinetic energy is quenched and correlation is enhanced. When the 2DEG occupies only the lowest Landau level (LL), there are no interactions between photoexcited pairs, unless there is an asymmetry between electron and hole wavefunctions [25]. When the lowest LL is partially filled, the dephasing originates mainly from the scattering of the photoexcited carriers with the intra-LL collective excitations of the strongly correlated 2DE-liquid [26–28].

FWM experiments on modulation–doped QW's using laser pulses resonantly exciting only one LL, have recently revealed very strong variation of the interband dephasing time as a function of the filling factor, ν, as well as direct evidence of memory effects [29]. The TI-FWM signal was measured on samples with doped carrier density under illumination $n = 2.5 \rightarrow 5 \times 10^{11}\text{cm}^{-2}$ with

σ^+-polarized $\tau_\ell = 300$fs laser pulses and $B = 0 \rightarrow 12$T. The excitation intensity was kept low enough for the density of photogenerated e-h pairs, n_{eh}, to remain small compared to the doping density of electrons, $n_{eh} < n/10$, and the laser was tuned to excite electrons only into the highest partially occupied LL, which contains the Fermi energy, E_F. For large B the $S_{TI}(\Delta t)$ profile is complicated and shows non-exponential behavior for short time delays. The decay time increases as B is lowered and the $S_{TI}(\Delta t)$ profile becomes a single exponential with an unusually long decay time. However, as E_F moves to the next LL the $S_{TI}(\Delta t)$ profile suddenly exhibits a very fast decay and again becomes non exponential. Extracting an overall decay time gives an estimate of the interband polarization dephasing time T_2. The results are displayed by the full squares in the upper and lower panels of Fig. 8 for two samples with different doping. It is striking to note the very large jumps each time the system passes through even filling factors and in particular at $\nu = 2$. Since these features are reproducible as a function of ν for samples with different densities, one can assert that this is an effect of the cold 2DEG. The non-exponential behavior at high field is characterized by a change in slope that occurs in the sample of the upper pannel at $\Delta t \approx 4.2 \rightarrow 2.5$ ps as $B \approx 7.5 \rightarrow 11.5$ T. This indicates memory effects in the polarization dynamics. These are also seen in the frequency domain where the $S_{SR}(\omega, \Delta t)$ profile changes from a Lorentzian lineshape with a constant width, $\Gamma \propto T_2^{-1}$, to an asymmetric frequency dependent width, $\Gamma(\omega)$. Such a profile corresponds to a polarization relaxation term $\propto \Gamma(\omega)P(\omega)$ or in the time domain a dephasing with memory structure, $(\partial P/\partial t)_{scatt} = \int_{-\infty}^{t} dt' \Gamma(t - t') P(t')$. The memory kernel within the lowest LL can be expressed as $\Gamma(t - t') = (2\nu^{-1} - 1)\kappa(t - t')$. The factor $(2\nu^{-1} - 1)$, expected on general physical grounds, is proportional to N_{empty}, the number of empty states available for scattering within the LL containing E_F, and $\kappa(t)$ is a smooth function. In Fig. 8 the full circles give N_{empty}^{-1} normalized so that the maximum height coincides with that of the T_2 curve. For low fields the agreement is striking. However, there are significant differences in the B-dependence of T_2 for strong field. This transition from Markovian to non-Markovian behavior is interpreted as being due to the suppression of the inter-LL scattering relative to the dynamical response of the collective excitations of the 2DE-liquid, i.e., magnetorotons [29]. The magnetoroton dephasing mechanism is somewhat similar to that of acoustic phonon scattering with a resonance in $\Gamma(\omega)$ near the magnetoroton energy that leads to non-Markovian behavior with a characteristic response time of approximately the inverse of this energy. The latter is estimated from the gap at the magnetoroton dispersion minimum, $\Delta \sim 0.1(e^2/\epsilon l)$ which for $B = 10$ T is ≈ 1.5meV, in good agreement with the experimental data that implies a reaction time $T_r \approx 2.5$ps $\rightarrow 4$ps for the 2DE-liquid collective excitations.

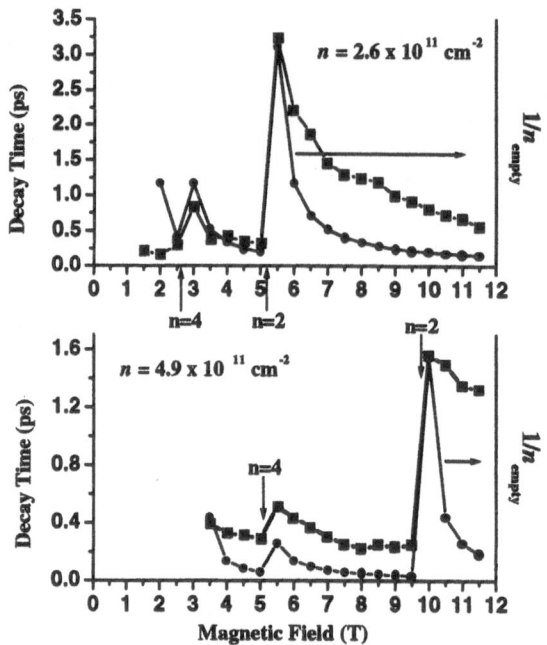

Fig. 8 TI-FWM decay times versus magnetic field for two samples. The squares are the experimental data points, and the circles correspond to $N_{empty}^{-1} = \nu/(2(N + 1) - \nu)$, where N is the Landau level number and ν the filling factor. The data was taken using σ^+ polarized light.

5 Conclusion

The never ending demand on technology for increased speed and capacity continuously calls for more performing devices and systems. Ultimately devices will involve only a few electrons and photons and will exploit all attributes of quantum mechanics. Yet, for many in R&D device optimization, coherence and relaxation are often opposed, and many-body processes are considered taboo. In this conference I have tried to emphasize that coherence and relaxation both stem from Coulomb Correlation, a fundamentally many-body phenomenon. Not only these processes are intimately related, but Coulomb mediated many-body effects prevail in almost all situations, including extremely low densities. Furthermore, it is the Coulomb interactions that produce the very nonlinearities which are prerequisite for performing any interesting function. Also, as shown by the two examples given above, aspects of dissipation and dephasing can be manipulated and even controlled. Thus the fundamental limits can be approached and eventually reached only by understanding and taming the physics behind coherence and relaxation and make it works for us. I believe that Coulomb correlation in well designed semiconductor quantum structures can provide us with many unexploited opportunities for developing nonlinearities and novel functionalities.

6 Acknowledgments

It is a pleasure to thanks my collaborators, in particular, Martin Axt, Sarah Bolton, Neil Fromer, Peter Kner, Reinhold Lövenich, Uli Neukirch, Ilias Perakis, Tigran Shahbazyan, Lu Sham, Wilfried Schäfer, Christian Schüller, Jerome Tignon, Martin Wegener and Markus Wehner.

This work was supported by the Director, Office of Energy Research, Office of Basic Energy Sciences, Division of Material Sciences of the U.S. Department of Energy, under Contract No. DE-AC03-76SF00098.

References

1. K. Leo, M. Wegener, J. Shah, D.S. Chemla, E.O. Göbel, T.C. Damen, S. Schmitt-Rink, and W. Schäfer, Phys. Rev. Lett. **65**, 1340 (1990).

2. M. Wegener, D.S. Chemla, S. Schmitt-Rink, and W. Schäfer, Phys. Rev. A **42**, 5675 (1990).

3. M.-A. Mycek, S. Weiss, J.-Y. Bigot, S. Schmitt-Rink, D.S. Chemla, Appl. Phys. Lett. **60**, 2666 (1992).

4. S. Weiss, M.-A. Mycek, J.-Y. Bigot, S. Schmitt-Rink, and D.S. Chemla, Phys. Rev. Lett. **69**, 2685 (1992).

5. D.S. Kim, J. Shah, T.C. Damen, W. Schäfer, F. Jahnke, S. Schmitt-Rink, and K. Köhler, Phys. Rev. Lett. **69** 2725 (1992).

6. H. Haug and S. W. Koch, *Quantum theory of the optical and electronic properties of semiconductors*, 2nd edition, World Scientific, Singapore, (1993).

7. D. S. Chemla, *Ultrafast Transient Nonlinear Optical Processes in Semiconductors* in *Nonlinear Optics in Semiconductors* Willardson and Beers Eds. Academic Press, NY (1999)

8. J. Shah, *Ultrafast Spectroscopy of Semiconductors and Semiconductor Nanostructures*, Springer Solid State Sciences **115**, 2nd edition Springer-Verlag Berlin, (1999).

9. P. Kner, S. Bar-Ad, M.V. Marquezini, W. Schäfer, and D.S. Chemla. Phys. Rev. Lett. **78**, 1319 (1997)

10. P. Kner, S. Bar-Ad, M. V. Marquezini, D. S. Chemla, R. Lövenich, W. Schäfer. Phys. Rev. **B 60**, 4731 (1999)

11. P. Kner, W. Schäfer, R. Lövenich, and D. S. Chemla. Phys. Rev. Lett. **81**, 5386 (1998)

12. S.R. Bolton, U. Neukirch, L.J. Sham, D.S. Chemla and V. M. Axt, Phys. Rev. Lett. **85**, 2002, (2000)

13. V. M. Axt, S.R. Bolton, U. Neukirch, L.J. Sham and D.S. Chemla, submitted to Phys. Rev. B

14. G. Khitrova, H.M. Gibbs, F. Jahnke, M. Kira, S.W. Koch, Rev. Mod. Phys. **71**, 1591 (1999).

15. U. Neukirch, S. Bolton, N. A. Fromer, L. J. Sham, and D. S. Chemla Phys. Rev. Lett. **84**, 2215 (2000)

16. M.U. Wehner, M.H. Ulm, D.S. Chemla, and M. Wegener. Phys. Rev. Lett. **80**, 1992 (1998)

17. M. U. Wehner, D. S. Chemla, and M. Wegener, Phys. Rev. **B 58** 3590 (1998)

18. M. Wegener and D. S. Chemla. Chem. Phys. **251**, 269 (2000)

19. L. Bányai, D.B. Tran Thoai, E.Reistamer, H. Haug, D. Steinbach, M. U. Wehner, M. Wegener, T. Marschner, W. Stolz, Phys. Rev. Lett. **75**, 2188 (1995)

20. D. Steinbach et al., J. Optical Soc. Am. **B 15**, 1231 (1998)

21. S. Bar-Ad, P. Kner, M.V. Marquezini, D.S. Chemla, K. El Sayed, Phys. Rev. Lett. **77**, 3177 (1996)

22. F. X. Camescase, A. Alexandrou, D. Hulin, L. Bányai, D.B. Tran Thoai, H. Haug, Phys. Rev. Lett. **77**, 5429 (1996)

23. S. Bar-Ad, D.S.Chemla, Materials Science and Engineering **B48**, 83 (1997)

24. W. A. Hügel, M. F. Heinrich, M. Wegener, Q. T. Vu L. Bányai, H. Haug, Phys. Rev. Lett. **83**, 3313 (1999)

25. I. V. Lerner and Yu. E. Lozovik, Zh. Exp. Teor. Fiz. **80**, 1488 (1981) [Sov. Phys.–JETP **53**, 763 (1981)].

26. S. Girvin, A. H. MacDonald, P. M. Platzman Phys. Rev. B **33**, 2481 (1986)

27. R. Haussmann, Phys. Rev. B **53**, 7357 (1996).

28. A. Pinczuk B.S. Dennis, L. N. Pfeiffer, K. West, Phys. Rev. Lett. **70**, 3983 (1993)

29. N. Fromer, C. Schüller, D. S. Chemla, T. V. Shahbazyan, I. E. Perakis, K. Maranowski, A. C. Gossard, Phys. Rev. Lett. **83**, 4646 (1999)

Experimental Signatures of Broken Symmetries in the Quantum Hall Regime

J.P. Eisenstein

California Institute of Technology, Pasadena, California 91125, USA

Abstract Experiments on high mobility single and double layer two-dimensional electron systems have revealed evidence for broken symmetry ground states in the quantum Hall regime. In single layer systems, transport anomalies have been observed at very low temperature when three or more orbital Landau levels are occupied. These anomalies include giant anisotropies of the longitudinal resistance near half filling of the valence Landau level and intriguing re-entrant integer quantized Hall states near $\frac{1}{4}$ and $\frac{3}{4}$ filling. These features are widely believed to reflect the spontaneous development of charge density wave order with broken rotation and/or translational symmetry. A more exotic form of broken symmetry is exhibited by double layer 2D systems. If the two layers are close enough together, interlayer Coulomb interactions can produce a ferromagnetic ground state. Recent tunneling experiments on such systems have shown a spectacular resonant enhancement of the zero-bias tunneling when the ferromagnetic phase is entered. This feature represents the direct detection of the Goldstone mode of the broken symmetry ground state.

1 Introduction

Research on two-dimensional electron systems (2DES) exposed to high magnetic fields has highlighted the role of broken symmetries. The most familiar example is the single layer 2DES at Landau level filling fraction $\nu = 1$. At this filling the density of electrons N_s is equal to the degeneracy eB/h of a single spin-resolved Landau band. It has been long known that the strength of the integer quantized Hall effect observed at this filling is dominated by electron-electron interaction effects, not by the single particle Zeeman energy. In fact, even in the absence of any Zeeman energy, the exchange energy of the system leads to spontaneous ferromagnetic alignment of the electron spins. Since the Coulomb interaction is spin-independent, the orientation of the macroscopic polarization is arbitrary. In reality, of course, the small Zeeman energy orients the moment along the external magnetic field direction. This particular *quantum Hall ferromagnet* has been extensively studied both theoretically and experimentally.

In this contribution I will discuss experiments on two different kinds of broken symmetry states. First, transport data which suggest the existence of new collective states in ultra-high mobility 2D electron systems at high Landau level occupancy will be described. These new states, which are very different from the familiar fractional quantized Hall states, appear to exhibit spontaneously broken rotational, and possibly translational, symmetry. Second, I will briefly describe very recent ex-

periments on a remarkable ferromagnetic state occuring in double layer 2D systems. This system displays an exotic broken symmetry known as *spontaneous interlayer phase coherence* in which each electron is in a single linear superposition of individual layer eigenstates. Using the technique of tunneling spectroscopy, the direct detection of the elusive Goldstone mode of this system has recently been achieved.

2 New Phases of 2D Electrons in High Landau Levels

In a sufficiently strong perpendicular magnetic field, all electrons in a 2D system are in the lowest (N=0) Landau level (LL). In this "extreme quantum limit" the kinetic energy of the electrons is quenched and Coulomb interactions dominate the physics[1]. The interplay of these interactions with the external magnetic field leads to the condensation of numerous exotic many body states. At many odd-denominator fractional fillings of the lowest LL (e.g. ν=1/3, 2/5, 3/7, etc.) quantized Hall states appear while at prominent even-denominator fillings (e.g. ν=1/2 and 1/4) composite fermion metallic states are found. In general, for each such state in the lower spin branch of the N=0 LL (i.e. $0 < \nu < 1$) an analogous state appears at $2 - \nu$ in the upper spin branch of the same LL ($1 < \nu < 2$).

In comparison to the N=0 Landau level, the nature of collective phenomena in higher Landau levels is much less well understood. Already in the first excited LL (N=1, $2 < \nu < 4$) things are quite different than in the lowest LL. Very few fractional quantum Hall states are seen, even in the very best samples, and two of these (at $\nu = 5/2$ and $7/2$) violate the famous odd-denominator rule valid in the lowest LL. These even-denominator states (the only known examples in a single layer 2DES) may result from a BCS-like pairing of composite fermions[2], but the problem remains open.

In the N=2 and higher LLs the situation changes again. Magneto-transport data of Lilly, et al.[4] and Du, et al.[5] have revealed a drastic change in the nature of electrical conduction occurs once the third (N=2) LL is entered. These and subsequent related data have generated great excitement as they suggest the existence of a whole new family of many-electron states.

2.1 Anisotropy

Figure 1 displays one of the two main findings from transport studies on high mobility GaAs/AlGaAs heterostructures at high Landau level occupancy. The figure

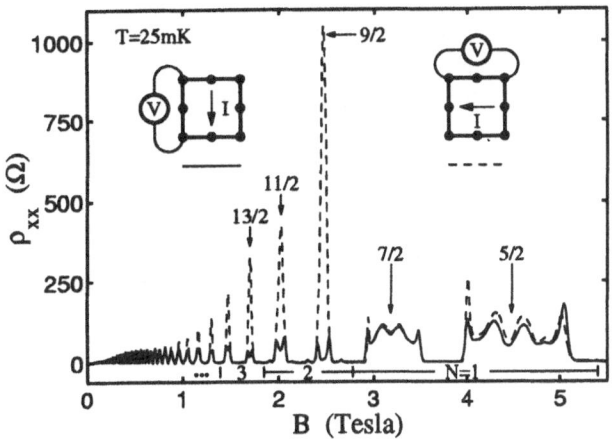

Fig. 1 Anisotropy of longitudinal resistance at T=25mK. Dashed curve: average current along $\langle 1\bar{1}0 \rangle$; solid curve: average current along $\langle 110 \rangle$. Insets denote contact configurations.

shows the longitudinal resistance measured at T=25mK in a sample having a mobility of about $9 \times 10^6 cm^2/Vs$ and a sheet density of $2.67 \times 10^{11} cm^{-2}$. Two traces are shown; the only difference being the average direction of current flow through the sample and the positions of the voltage probes. The two configurations are depicted in the insets to the figure. It is quite clear that a large change occurs in the nature of transport in the sample at about B=2.75T, corresponding to $\nu = 4$. As the magnetic field falls below this value, the third and higher Landau levels become occupied.

The data in Fig. 1 show that a strong anisotropy in the longitudinal resistance exists at low temperature near half filling of the N$\geq$2 Landau levels, i.e. at $\nu = 9/2$, 11/2, 13/2, etc. The resistance measured with the average current flow along the $\langle 1\bar{1}0 \rangle$ crystal direction shows a strong peak at these filling factors, while in the orthogonal $\langle 110 \rangle$ direction the resistance shows a pronounced minimum. At $\nu = 9/2$ resistance anisotropy ratios exceeding 3000 have been seen in some samples.

This effect, which develops very rapidly as the temperature is reduced below about 100mK, is quite robust. It has been seen in dozens of high mobility GaAs/AlGaAs heterostructures, including both conventional single heterointerfaces and square quantum wells. While most transport measurements are done after brief illumination of the sample at low temperature, the anisotropy has also been seen in samples cooled in the dark. Provided the magnetic field is perpendicular to the 2D plane, all samples behave qualitatively as in Figure 1: Little, if any, systematic anisotropy is observed at B=0 or in the semi-classical regime at very low magnetic fields (B$\ll$1T). Anisotropy gradually becomes evident at high half-odd

integers fillings and is strongest at $\nu = 9/2$. Remarkably, the effect abruptly disappears as the magnetic field is raised above $\nu = 4$ and the Fermi level falls into the N=1 and then N=0 LL. In all cases we have examined, the "hard" and "easy" transport directions at $\nu = 9/2, 11/2$, etc. are $\langle 1\bar{1}0 \rangle$ and $\langle 110 \rangle$ respectively. When a magnetic field component parallel to the 2D plane is added[6, 7], fascinating additional effects are seen. These are described briefly below.

These findings strongly suggest the development of some new kind of anisotropic ordered phases at low temperatures in half filled high Landau levels. The intricate filling factor dependence of the anisotropy, plus its sudden development at very low temperature, points to an intrinsic mechanism. Although extrinsic effects (e.g. anisotropy in the disorder potential) undoubtedly play a role (determining, presumably, the preferred orientation of the anisotropy axes), it is clear from the data that electrons in high LLs behave quite differently than they do in the ground and first excited Landau levels.

2.2 Re-entrant Integer Quantized Hall States

Anisotropy is not the only remarkable phenomenon exhibited by high mobility 2D electrons in high Landau levels. Away from half filling of the N$\geq$2 LLs, the transport becomes isotropic once again. Furthermore, near both $\frac{1}{4}$ and $\frac{3}{4}$ filling of each of these LLs (e.g. near $\nu = 4 + \frac{1}{4}$) deep minima are observed in the resistance measured in both directions. At low temperatures these minima become "zeroes" in the quantum Hall sense and suggest the existence of new *fractional* quantized Hall states. Figure 2 exhibits this behavior for the case of $5 > \nu > 4$. As the figure proves, these isotropic minima in the flanks of the LL are separated from the broad zeroes associated with the integer QHE states at $\nu = 4$ and 5 by small peaks in the resistance.

Examination of the Hall resistance, however, reveals a startling fact. While the Hall resistance *is* quantized over the field ranges where the suggestive zeroes are seen, it is at the value of the neighboring *integer* quantized Hall states. As with the longitudinal resistances, these *re- entrant integer quantized Hall states* (RIQHE) are separated from the main integer QHE plateaus by a narrow field ranges in which quantization is lost. RIQHE states are also seen near $\nu = 4 + \frac{3}{4}$ and at the analogous locations in higher LLs.

In common with main integer QHE states, the phenomena of zero longitudinal resistance and integer quantized Hall resistance still implies that the quasiparticles in the valence LL are in an insulating state. On the other hand, the re-entrant character of these new states suggests that the conventional picture of single particle localization does not apply. In the usual picture quasiparticles delocalize as their density increases. The RIQHE shows that these quasiparticles can localize again at some higher density. As with the anisotropy at half filling, this phenomenon points to the formation

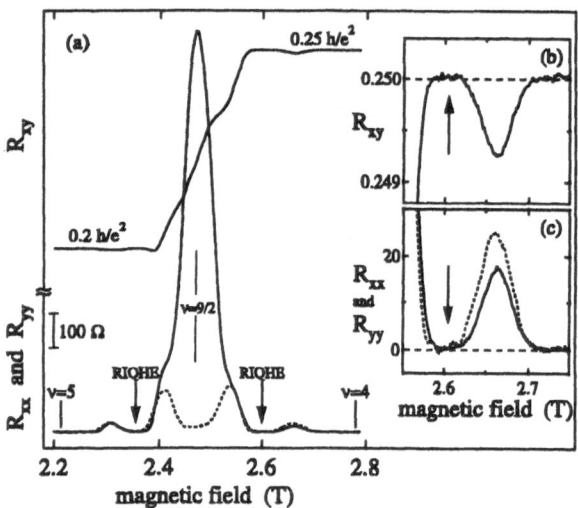

Fig. 2 (a) Longitudinal and Hall resistances between 5 > ν > 4 at T=50mK. Arrows point to RIQHE features. Insets (b) and (c) magnify region near $\nu = 4 + \frac{1}{4}$.

of a new collective state. Since both the RIQHE and the anisotropy effects occur under similar circumstances (high mobility, low temperature and high Landau level index), it seems plausible that a common explanation exists.

2.3 The CDW Interpretation

These observations have generated a great deal of interest in the theory of 2D electron systems. Even prior to the experiments however, there were predictions that unusual new phases might appear in high Landau levels. Koulakov, Fogler, and Shklovskii (KFS)[8] and Moessner and Chalker (MC)[9] applied Hartree-Fock theory to conclude that the ground state of 2D electrons in high LLs would in fact be a charge density wave. Of course, a similar prediction made long ago[10] for the ground Landau level was upset by the discovery of the fractional quantum Hall effect. Unlike the N=0 lowest LL, the higher LLs all have *nodes* in their orbital wavefunctions. In effect, these nodes soften the short-range part of the direct Coulomb interaction and allow the attractive exchange energy to promote phase separation. Indeed, both KFS and MC predicted that near half filling of high Landau levels the ground state of the system would be a unidirectional charge density wave, or stripe phase. The length scale for this phase separation is expected to be a few times the classical cyclotron radius R_c. For example, at $\nu = 9/2$ the 2DES breaks up into stripes of $\nu = 4$ and $\nu = 5$ with a period of approximately $\lambda \approx 3R_c$. While MC concentrated on the large N limit, KFS predicted that the stripe phases would defeat Laughlin-like alternatives only for N≥2.

On moving away from half-filling one set of stripes broadens while the other narrows. There is a critical point at which the narrower stripes break up and a triangular "bubble" phase forms. Each bubble is expected to contain an integer number of electrons. In principle, a hierarchy of bubble phases is expected, each phase characterized by the number of electrons per bubble. Deep in the flanks of the LL this hierarchy terminates in a Wigner crystal with one electron per bubble. The nature of the transitions between different bubble phases is not yet understood.

While transport theories of such charge density wave (CDW) phases have begun to appear only recently[11, 12], it is certainly plausible that the stripe phases near half filling would exhibit highly anisotropic transport in agreement with experiment. In fact, preliminary analyses of the transport coefficients[13] based on a classical model of the current distribution[14] have shown encouraging agreement with these theories. On the other hand, at very low temperature the stripes might themselves become modulated[15,11]. Such a "stripe crystal" would be pinned and thus an insulator. Whether the crystallization occurs at an inaccessibly low temperature[11] or is prevented by quantum fluctuations[15] remains an interesting question.

The RIQHE may also find a natural explanation in terms of the predicted bubble phases. As triangular CDWs these states are likely isotropic pinned insulators and the Hall resistance would be quantized to the nearest integer value as in experiment. Support for this scenario has come from recent experiments on RIQHE transport in the non-linear regime. Cooper,*et al*[16] have observed discontinuous, hysteretic, and noisy transitions between the insulating RIQHE phase and a conducting configuration. Examples of such transitions are shown in Figure 3. Transitions of this sort have been seen in the RIQHE states near $\nu \approx 4 + \frac{1}{4}$, $4+\frac{3}{4}$, $5+\frac{1}{4}$ and $5+\frac{3}{4}$.

2.4 Open Questions

The early Hartree Fock theories of stripe and bubble phases have received considerable support from recent numerical few-particle exact diagonalization studies[17]. Looking beyond Hartree-Fock, there has been has also been keen interest in the effects of quantum and thermal fluctuations on the CDW states. An intriguing connection with nematic liquid crystals was advanced by Fradkin and Kivelson[18] and continues to be intensely studied[19].

There remain, however, significant unanswered questions. The most obvious of these concerns the reproducible orientation of the anisotropy relative to the underlying crystal axes. As mentioned above, the hard transport direction at $\nu = 9/2$, 11/2, etc. is always $\langle 1\bar{1}0 \rangle$, provided that the magnetic field is perpendicular to the 2D plane. There must therefore exist a native anisotropy field within the GaAs/AlGaAs heterostructure which effectively orients the stripe phases. Although there have

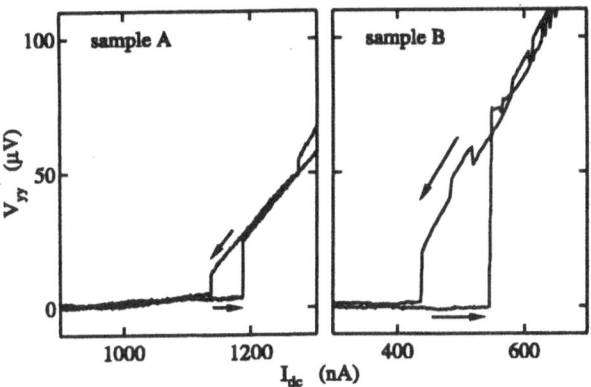

Fig. 3 Typical current-voltage characteristics observed in the RIQHE near $\nu = 4 + \frac{1}{4}$.

been various suggestions[20], a conclusive answer remains elusive.

Fortunately, there is relatively easy way to estimate the strength of the native anisotropy field. Remarkably, a small ($\sim 0.5T$) magnetic field component lying in the 2D plane can reorient the anisotropy axes[6,7]. This effect has been examined by Jungwirth, et al.[21] who have performed detailed calculations of the effect of the in-plane field on putative stripes. Their calculations reveal a complex interplay of various effects of the in-plane field (subband mixing leading to anisotropic coulomb interactions, etc.) but nonetheless explain the experimental observation that the hard transport direction prefers to be parallel to the in-plane field, at least in typical samples. A simple estimate of the native anisotropy energy is obtained by evaluating the calculated field orientation energy at the experimentally measured "switching" field. In typical samples, anisotropy energies of only about $\sim 1mK$ per electron are found. This is quite small compared to the temperature ($\sim 100mK$) at which the anisotropy develops and supports a picture, analogous to ferromagnetism, of a collective broken symmetry state weakly oriented by an external field.

3 Goldstone Mode in a Double Layer Quantum Hall Ferromagnet

Double layer 2D electron systems exhibit broken symmetry states associated with the layer index degree of freedom[22]. One of the simplest cases occurs when each 2D system, viewed independently, is at Landau filling factor $\nu = \frac{1}{2}$. If the layers are far apart the system behaves as two independent 2DES's, each a spin polarized gapless composite fermion metal. If the layers are close together, interlayer couplings drive a transition to an in-

trinsically bilayer state which exhibits a quantized Hall effect. The Hall resistance of this state is characterized by the *total* filling factor $\nu_T = 1$.

If the only significant coupling between the layers is tunneling, then the $\nu_T = 1$ QHE state may be viewed as a filled Landau level of electrons each of which is in the symmetric combination of the individual layer eigenstates. In this case the energy gap for the QHE will be just the symmetric-antisymmetric tunnel splitting, Δ_{SAS}[23]. If, on the other hand, tunneling is insignificant, interlayer Coulomb interactions can stabilize a bilayer QHE phase if the layer spacing is small enough. The situation is analogous to a single layer 2DES at $\nu = 1$ where both the single particle Zeeman energy and the Coulomb exchange interaction act together to produce the QHE.

The analogy with spins in a single 2D layer can be made clearer by introducing a *pseudospin* observable σ in the double layer problem. An electron definitely in one layer is in an eigenstate $|\uparrow\rangle$ of σ_z with eigenvalue $+1$ while one definitely in the other layer ($|\downarrow\rangle$) has eigenvalue -1. In this language, a symmetric double layer state is in an eigenstate $(|\uparrow\rangle + |\downarrow\rangle)/\sqrt{2}$ of σ_x with eigenvalue $+1$. Thus, a full Landau level of symmetric double layer states is a fully pseudospin polarized state, just as the single layer system at $\nu = 1$ is a fully spin polarized state. In both cases exchange interactions favor complete spin or pseudospin polarization and a single particle energy (Zeeman or tunneling) serves to orient the moment. Both systems are examples of quantum Hall ferromagnetism.

There is, however, a fundamental difference between the single and double layer cases. In the single layer case the Coulomb interaction is spin independent while in the double layer case the non-zero separation of the layers means the interaction is pseudospin dependent. This enriches the double layer case enormously. Indeed, the collapse of the $\nu_T = 1$ QHE when the layer separation exceeds a critical value is a phase transition having no analog in the single layer $\nu = 1$ problem. This difference in symmetry has another qualitative consequence. While the ferromagnetism of the single layer is isotropic, the pseudomagnetism of the bilayer case is easy-plane. The reason for this is simple: Owing to the finite layer separation, there is a capacitive energy penalty if the pseudospin moment moves out of the x-y plane. In the ordered phase each electron exists in the same linear combination of the individual layer states: $(|\uparrow\rangle + e^{i\phi}|\downarrow\rangle)/\sqrt{2}$. In the absence of tunneling the phase ϕ is arbitrary. This collective state, which exhibits "spontaneous interlayer phase coherence", is anticipated to possess numerous intriguing properties, including counterflow superfluidity, textural and finite temperature Kosterlitz-Thouless phase transitions, exotic vortex-like charged excitations known as "merons", etc.[22]. Of particular interest here is the expectation that the long wavelength collective modes will have an acoustic linear dispersion.

These pseudospin waves are the Goldstone modes of the broken symmetry state.

3.1 Tunneling Spectroscopy

The bilayer QHE state at $\nu_T = 1$ has been experimentally studied primarily via transport[22]. As an alternative, we have recently begun to examine this state via the method of tunneling spectroscopy[24]. Tunneling measurements on double layer 2D systems have been performed regularly since the development of a robust technique for making separate electrical contact to the individual layers[25]. Until now, however, the samples used always had too large a layer separation to support any intrinsically bilayer QHE states. At high magnetic field such samples show a deep suppression of the zero bias tunneling conductance (dI/dV at V=0) that is largely independent of the filling factor in each layer[26]. This suppression stems from the energetic penalty associated with the rapid injection (or extraction) of an electron into a highly correlated single layer 2D system. The net tunneling characteristic is a convolution of the spectral functions of two independent layers.

Samples with strong interlayer Coulomb correlations generally have relatively thin barrier layers separating the two quantum wells. For tunneling experiments, however, it is essential to keep the tunneling resistance much higher than the resistances of the 2D layers themselves. At high magnetic fields this becomes a serious problem and increased barrier thicknesses are inevitable. To simultaneously satisfy these two requirements we have employed samples with nearly pure AlAs barriers and low 2D densities in each of the GaAs quantum wells. The high Al fraction gives a large barrier height[27] and thus minimizes the tunneling matrix elements. The low electron density helps since it is the ratio of the layer separation d to the mean electron separation (which, at a given filling factor, is proportional to the magnetic length $\ell = (\hbar/eB)^{1/2}$) that determines the importance of interlayer correlations. A typical sample consists of two 18nm GaAs wells separated by a 10nm $Al_{0.9}Ga_{0.1}As$ barrier. As grown, each well contains roughly $5.5\times10^{10}cm^{-2}$ carriers and has a mobility of about $10^6 cm^2/Vs$. The estimated tunnel splitting for this sample is only $\Delta_{SAS} \sim 90\mu K$. At $\nu_T = 1$ $d/\ell \approx 2.3$ while the phase boundary separating the QHE and non-QHE phases has been found to be near $d/\ell \approx 2$ in similar samples. Front and back gates allow for in situ adjustment of the layer densities and thereby continuous control of the crucial d/ℓ ratio.

Figure 4 shows the key result of our recent tunneling experiments[24]. Both traces are taken at $\nu_T = 1$ and T=40mK but one is at a high density ($N_1 = N_2 = 5 \times 10^{10}cm^{-2}$; $d/\ell = 2.22$) and one at a low density ($N_1 = N_2 = 3 \times 10^{10}cm^{-2}$; $d/\ell = 1.72$). The high density trace shows the characteristic suppression of the tunneling conductance at zero bias that is seen in all samples in which the interlayer correlations are weak. In

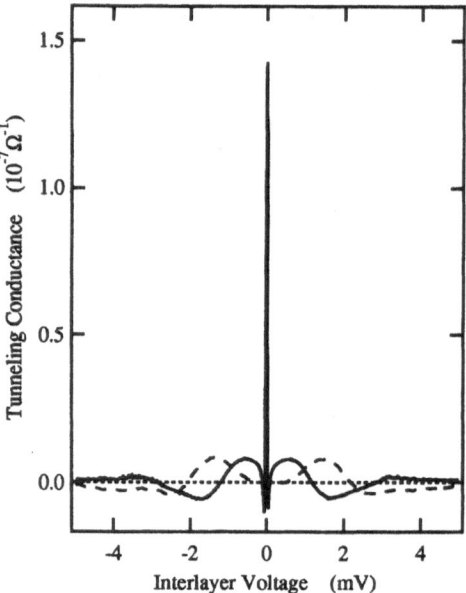

Fig. 4 Tunneling spectra at $\nu_T = 1$ and T=40mK at two different densities. Solid curve: $N_1 = N_2 = 3 \times 10^{10}cm^{-2}$; dashed curve: $N_1 = N_2 = 5 \times 10^{10}cm^{-2}$. Low density data is from within the QHE phase, high density data is from the non-QHE compressible phase.

contrast, the low density trace shows a spectacular enhancement of the zero bias conductance. This peak first appears as a small zero bias feature at about $d/\ell = 1.85$ but grows quickly as the density is reduced further. As a function of temperature, we have found this peak to grow rapidly as the temperature is reduced. The occurence of the peak seems to be coincident with the development of the QHE in transport. This conclusion requires further study, however, since the QHE is weak in these samples, especially near the phase boundary. Finally, we remark that the zero bias peak shown in Fig. 4 is quite narrow, its full width at half maximum is only $40\mu V$. To our knowledge, this is the narrowest 2D-2D tunnel resonance ever reported. Interestingly, it is even narrower (by a factor of 2) than the resonance observed at $B = 0$ in the same sample.

This sharp tunnel resonance has been interpreted[24, 28, 29] as a direct manifestation of the Goldstone collective mode of the broken symmetry ground state at $\nu_T = 1$. In the presence of tunneling (which is extremely weak in these samples) the collective mode becomes gapped at long wavelength and can efficiently transfer charge between the layers. The gap is expected to scale as $\sqrt{\Delta_{SAS}}$ but is estimated[30] to be only about 70mK in this sample. This is much smaller than the resonance width. Tunneling (in a perpendicular magnetic field) samples the collective mode spectrum at $q = 0$ and therefore the resonance appears at zero bias.

In a different interpretation[31], these tunneling results at $\nu_T = 1$ have been viewed in terms of the Josephson effect[32]. Indeed, the actual current-voltage charac-

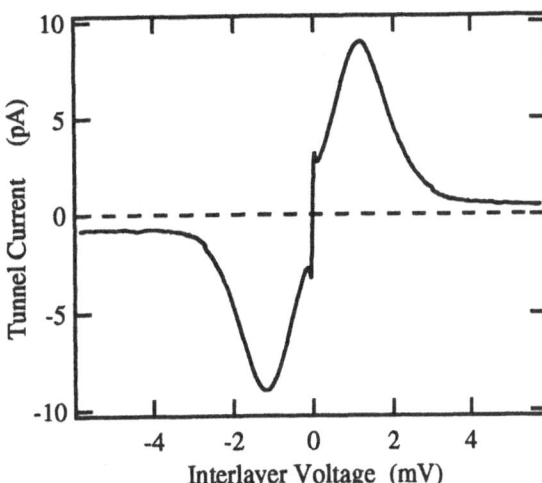

Fig. 5 Tunneling current-voltage characteristic at $\nu_T = 1$ and T=40mK. Density is $N_1 = N_2 = 3 \times 10^{10} cm^{-2}$, placing the data within the QHE phase.

teristic for our devices do resemble those of a current-biased Josephson junction. Figure 5 displays the I-V curve which was recorded simultaneously with the dI/dV data shown in Fig. 4. As expected, there is sharp step at zero bias. The slope of this step is of course not infinite. It will be very interesting to see how this finite slope (i.e. dI/dV) depends on the precise conditions of the experiment. Already there is evidence that that resonance width is readily increased by electrical interference and therefore may ultimate become even narrower. The prospect of detecting a Josephson effect in a non-superconducting tunnel junction is tantalizing indeed.

4 Conclusion

There are several examples of broken symmetry states of 2D electron systems in high magnetic fields. In this paper, I have discussed experimental signatures of two of these: the apparent charge density wave (or liquid crystal) phases in high Landau levels and a ferromagnetic state in double layer 2D systems at unit filling factor. Transport measurements are so far the only experimental tool to be applied to the relatively new high Landau level problem. While double layer QHE phenomena have been studied for several years, I have here discussed new results obtained via the method of tunneling spectroscopy. Neither of these problems can yet be regarded as solved.

5 Acknowledgements

The results described here represent the efforts of a collaboration with M.P. Lilly, K.B. Cooper, and I.B. Spielman at Caltech and with L.N. Pfeiffer and K.W. West at Bell Laboratories, Lucent Technologies. This work is supported by the National Science Foundation under Grant DMR-0070890 and the Department of Energy under Grant DE-FG03-99ER45766.

References

1. For a recent review of the fractional quantum Hall effect, composite fermions, and other interaction phenomena in 2D electron systems, see *Perspectives in Quantum Hall Effects*, edited by S. Das Sarma and A. Pinczuk (John Wiley, New York, 1997).
2. G. Moore and N. Read, Nucl. Phys. B **360**, 362 (1991).
3. E.H. Rezayi, F.D.M. Haldane and Kun Yang, Phys. Rev. Lett. **84**, 4685 (2000).
4. M.P. Lilly, et al., Phys. Rev. Lett. **82**, 394 (1999).
5. R.R. Du, et al., Solid State Commun. **109**, 389 (1999).
6. M.P. Lilly, et al., Phys. Rev. Lett. **83**, 824 (1999).
7. W. Pan, et al., Phys. Rev. Lett. **83**, 820 (1999).
8. A.A. Koulakov, M.M. Fogler and B.I. Shklovskii, Phys. Rev. Lett. **76**, 499 (1996); Phys. Rev. B **54**, 1853 (1996); M.M. Fogler and A.A. Koulakov, Phys. Rev. B **55**, 9326 (1997).
9. R. Moessner and J.T. Chalker, Phys. Rev. B **54**, 5006 (1996).
10. H. Fukuyama, P.M. Platzman, and P.W. Anderson, Phys. Rev. B **19**, 5211 (1979).
11. A.H. MacDonald and M.P.A. Fisher, Phys. Rev. B **61**, 5724 (2000).
12. F. von Oppen, B.I. Halperin and A. Stern, Phys. Rev. Lett. **84**, 2937 (2000).
13. J.P. Eisenstein, et al., Physica E, to be published.
14. S. Simon, Phys. Rev. Lett. **83**, 4223 (1999).
15. H. Fertig. Phys. Rev. Lett. **82**, 3693 (1999).
16. K.B. Cooper, et al., Phys. Rev. B **60**, R11285 (1999).
17. E.H. Rezayi, F.D.M. Haldane and Kun Yang, Phys. Rev. Lett. **83**, 1219 (1999) and cond-mat/0001394.
18. E. Fradkin and S. Kivelson, Phys. Rev. B **59**, 8065 (1999) and E. Fradkin, et al., Phys. Rev. Lett. **84**, 1982 (2000).
19. For a recent contribution, see C. Wexler and A. Dorsey, cond-mat/0009096.
20. See, for example, D.V. Fil, Low Temp. Phys. **26**, 581 (2000).
21. T. Jungwirth, et al. Phys. Rev. B **60**, 15574 (1999).
22. For a review of theoretical and experimental aspects of bilayer quantum Hall systems, see the chapters by S.M. Girvin and A.H. MacDonald and by J.P. Eisenstein, respectively, in reference 1.
23. Provided that the tunneling splitting is smaller than the spin splitting. We shall assume that the true spins are fully polarized, and thus irrelevant, throughout our discussion of double layer systems.
24. I.B. Spielman, J.P. Eisenstein, L.N. Pfeiffer and K.W. West, Phys. Rev. Lett. **84**, 5808 (2000).
25. J.P. Eisenstein, L.N. Pfeiffer and K.W. West, Applied Phys. Lett. **57**, 2324 (1990).
26. J.P. Eisenstein, L.N. Pfeiffer and K.W. West, Phys. Rev. Lett. **69**, 3804 (1992).
27. It appears that the tunneling is via the Γ point, even in these high Al concentration barriers.
28. A. Stern, et al., cond-mat/0006457.
29. L. Radzihovsky and L. Balents, cond-mat/0006450.
30. K. Moon, et al., Phys. Rev. B **51**, 5138 (1995).
31. M. Fogler and F. Wilczek, cond-mat/0007403.
32. X.G. Wen and A. Zee, Phys. Rev. Lett, **69**, 1811 (1992) and Phys. Rev. B **47**, 2265 (1993).

2 Bulk and Dynamics

New effective-mass theory for degenerate bands in semiconductors

Bradley A. Foreman

Department of Physics, Hong Kong University of Science and Technology, Clear Water Bay, Kowloon, Hong Kong, China

Abstract A new effective-mass theory is developed by introducing a mixing parameter of arbitrary strength into the Luttinger-Kohn basis states for degenerate bands. The extra degree of freedom provides valuable insight into the physics of degenerate bands, and sheds new light on a recent controversy over the correct form of the Hamiltonian for the valence bands of zinc-blende semiconductors in an electric field.

1 Introduction

One of the most important advances in 20th-century semiconductor physics was the development of an effective-mass theory for degenerate energy bands by Luttinger and Kohn [1,2] and Kittel *et al.* [3,4]. This gave physicists the ability to study in detail the behavior of electrons and holes in real semiconductors. The theory has since been adapted for narrow-gap semiconductors [5], strained materials [6], and semiconductor heterostructures [7], and it remains in widespread use. However, the fundamental structure of this theory is the same today as it was in 1955 [1]. No alternative to the original set of basis functions has ever been proposed.

This paper presents such an alternative, based on a qualitative change in the Luttinger-Kohn basis functions. The new basis allows one, within certain limits, to *design* an effective-mass Hamiltonian for the needs of the problem at hand. The resulting flexibility offers advantages in terms of both physical understanding and computational efficiency.

This basis is used here to reexamine a recent controversy [8–10] over the correct form of the Hamiltonian for degenerate bands in an electric field. The debate centers on a proposal by Khurgin and Voisin [8] that the Luttinger-Kohn valence-band Hamiltonian should contain a mixing term proportional to the scalar product of a dipole matrix element and the electric field. The author has shown that a correct application of the Luttinger-Kohn formalism yields no evidence for such a term [9, 10], and that the proposed Hamiltonian is invalid because the dipole moment of a unit cell is ill defined [10]. However, as demonstrated below, one can construct a valid Hamiltonian that is similar to the Khurgin-Voisin model simply by introducing an adjustable parameter into the Luttinger-Kohn basis states.

2 Theory

For simplicity, I shall neglect spin-orbit coupling and use symmetry relations appropriate for the Γ point of a zinc-blende crystal. The analysis for the general case [10] follows the same lines, but is more cumbersome. The initial stages of the present theory are identical to those of Ref.

[1]. The Luttinger-Kohn basis functions are $\langle \mathbf{x}|n\mathbf{k}\rangle = e^{i\mathbf{k}\cdot\mathbf{x}}u_{n0}(\mathbf{x})$, where $\mathbf{k}$ is a wave vector in the first Brillouin zone and $u_{n0}(\mathbf{x})$ is a zone-center Bloch function. Every operator considered below has matrix elements of the form $\langle n\mathbf{k}|A|n'\mathbf{k}'\rangle = \langle n|A|n'\rangle\delta(\mathbf{k}-\mathbf{k}')$, where $\langle n|A|n'\rangle$ is a $\mathbf{k}$-dependent operator. Thus, for brevity I shall work with $\langle n|A|n'\rangle$.

The Hamiltonian is $H = H_0 + U$, where $H_0 = H^{(0)} + H^{(1)}$ is periodic, with matrix elements [1]

$$\langle n|H^{(0)}|n'\rangle = (E_n + \hbar^2 k^2/2m)\delta_{nn'}, \tag{1}$$

$$\langle n|H^{(1)}|n'\rangle = (\hbar/m)k_\alpha p_{nn'}^\alpha. \tag{2}$$

Here E_n is the energy of state u_{n0}, m is the mass of a free electron, and $p_{nn'}^\alpha$ is the α component of the momentum matrix element between u_{n0} and $u_{n'0}$ (a sum on α is implied). In a zinc-blende crystal, $p_{nn'}^\alpha = 0$ whenever $E_n = E_{n'}$ [11], so $H^{(1)}$ does not couple states of the same energy. For an electron of charge $-e$ in a uniform electric field $\mathcal{E}$, the potential energy U has the form $U(\mathbf{x}) = e\mathcal{E}\cdot\mathbf{x}$, so its matrix elements are given by [1,10]

$$\langle n|U|n'\rangle = ie\mathcal{E}\cdot\nabla_\mathbf{k}\delta_{nn'}. \tag{3}$$

The next step in the theory of Luttinger and Kohn is to eliminate the first-order $\mathbf{k}\cdot\mathbf{p}$ coupling (2). This is done by introducing new basis kets

$$|\overline{n\mathbf{k}}\rangle = e^S|n\mathbf{k}\rangle, \tag{4}$$

in which the anti-Hermitian operator S is defined below. The matrix elements of H in the new basis $\{|\overline{n\mathbf{k}}\rangle\}$ are the same as the matrix elements of

$$\bar{H} = e^{-S}He^S = H + [H,S] + \tfrac{1}{2!}[[H,S],S] + \cdots \tag{5}$$

in the old basis $\{|n\mathbf{k}\rangle\}$. Substituting $H = H^{(0)} + H^{(1)} + U$ into (5), one finds that $H^{(1)}$ can be eliminated by choosing S such that

$$H^{(1)} + [H^{(0)}, S] = 0. \tag{6}$$

Luttinger and Kohn define S to be the simplest operator that satisfies Eq. (6):

$$\langle n|S|n'\rangle = -ik_\alpha \xi_{nn'}^\alpha, \tag{7}$$

in which

$$\xi_{nn'}^\alpha = \begin{cases} \dfrac{p_{nn'}^\alpha}{im\omega_{nn'}} & \text{if } E_n \neq E_{n'} \\ 0 & \text{if } E_n = E_{n'} \end{cases} \tag{8}$$

is the "crystal coordinate" matrix [12], and $\hbar\omega_{nn'} = E_n - E_{n'}$. Note that to first order in $\mathbf{k}$, the transformation (7) does not mix states of the same energy.

In this paper, the basis kets (4) are defined by the more general relation

$$\langle n|S|n'\rangle = -ik_\alpha X_{nn'}^\alpha, \tag{9}$$

in which

$$X_{nn'}^{\alpha} = \xi_{nn'}^{\alpha} + C_{nn'}^{\alpha}, \tag{10}$$

where $C_{nn'}^{\alpha}$ is a Hermitian matrix. Equation (6) is satisfied by any operator of the form (9), provided $C_{nn'}^{\alpha}$ mixes degenerate states only (i.e., $\omega_{nn'}C_{nn'}^{\alpha} = 0$). In addition, $C_{nn'}^{\alpha}$ is chosen to have the same symmetry properties as $\xi_{nn'}^{\alpha}$. Otherwise, however, the value of $C_{nn'}^{\alpha}$ is arbitrary. A transformation similar to (9) was used by Kohn [13] to fix the phase of nondegenerate Bloch functions. The consequences of (9) for degenerate states have not previously been considered.

With the operator S defined by (9), the leading terms in the Hamiltonian $\bar{H}$ are [10]

$$\bar{H} = H^{(0)} + \tfrac{1}{2}[H^{(1)}, S] + \tfrac{1}{3}[[H^{(1)}, S], S] + U + [U, S] \tag{11}$$

The interband terms in the bulk part of $\bar{H}$ are of order k^2 and higher, so Eq. (11) is accurate to order k^3. The term $\tfrac{1}{2}[H^{(1)}, S]$ is just the usual effective-mass coupling

$$\tfrac{1}{2}\langle n|[H^{(1)}, S]|n'\rangle = \frac{\hbar k_{\alpha} k_{\beta}}{2m^2} \sum_i p_{ni}^{\alpha} p_{in'}^{\beta} (\omega_{ni}^{-1} + \omega_{n'i}^{-1}), \tag{12}$$

where $C_{nn'}^{\alpha}$ has been discarded because it yields only interband terms of order k^2. The k^3 terms are given by

$$\tfrac{1}{3}\langle n|[[H^{(1)}, S], S]|n'\rangle = -\frac{\hbar k_{\alpha} k_{\beta} k_{\gamma}}{3m}$$
$$\times \sum_{i,j}(p_{ni}^{\alpha} X_{ij}^{\beta} X_{jn'}^{\gamma} - 2X_{ni}^{\alpha} p_{ij}^{\beta} X_{jn'}^{\gamma} + X_{ni}^{\alpha} X_{ij}^{\beta} p_{jn'}^{\gamma}), \tag{13}$$

while $[U, S]$ is a dipole-like coupling:

$$\langle n|[U, S]|n'\rangle = e\mathcal{E}_{\alpha} X_{nn'}^{\alpha}. \tag{14}$$

Hence, the parameters $C_{nn'}^{\alpha}$ govern directly the field-induced mixing of degenerate states.

The states of interest are the Γ_1 conduction ($n = S$) and Γ_{15} valence ($n = X, Y,$ or Z) states of a zinc-blende crystal. Since $C_{nn'}^{\alpha}$ is chosen to have the same symmetry as $\xi_{nn'}^{\alpha}$, its value for the conduction band is $C_{SS}^{\alpha} = 0$. For the valence band, there is only one independent parameter:

$$C_{XY}^z = C_{YZ}^x = C_{ZX}^y \equiv C = C^*, \tag{15}$$

with all other matrix elements zero. Thus, a representative valence-band mixing term would be

$$\langle X|[U, S]|Y\rangle = e\mathcal{E}_z C. \tag{16}$$

If the crystal is free of stress, the converse piezoelectric effect gives rise to a mixing of the same form [9]:

$$\langle X|H_{\text{piezo}}|Y\rangle = \frac{\sqrt{3}}{2} d d_{14} \mathcal{E}_z, \tag{17}$$

where d is a deformation potential [6] and d_{14} is the piezoelectric coefficient. Finally, the k^3 coupling (13) is

$$\tfrac{1}{3}\langle X|[[H^{(1)}, S], S]|Y\rangle$$
$$= i[G - (L - M + N)C](k_x^2 - k_y^2)k_z, \tag{18}$$

in which L, M, and N are the valence-band effective-mass parameters defined in Refs. [4–6], and

$$G = \frac{\hbar}{im^3}\sum_{i,j} \frac{[(p_{Xi}^x + p_{Zi}^z)p_{ij}^z + p_{Xi}^x p_{ij}^x]p_{jY}^x}{\omega_{Xi}\omega_{Xj}}. \tag{19}$$

To calculate optical transition rates, one also needs to know the momentum operator, which is given by $\bar{\mathbf{p}} = e^{-S}\mathbf{p}e^S = \mathbf{p} + [\mathbf{p}, S] + \cdots$. To first order in k, one has (for $n \neq n'$)

$$\langle n|\bar{p}^{\alpha}|n'\rangle = p_{nn'}^{\alpha} - ik_{\beta}\sum_i(p_{ni}^{\alpha}X_{in'}^{\beta} - X_{ni}^{\beta}p_{in'}^{\alpha}). \tag{20}$$

This yields the following result for interband transitions:

$$\frac{\hbar}{im}\langle S|\bar{\mathbf{p}}|X\rangle = P\hat{\mathbf{x}} - i(B + PC)(k_z\hat{\mathbf{y}} + k_y\hat{\mathbf{z}}), \tag{21}$$

where $P = (\hbar/im)p_{SX}^x$ and

$$B = \frac{\hbar}{m^2}\sum_i^{\Gamma_{15}} p_{Si}^y p_{iX}^z(\omega_{Xi}^{-1} + \omega_{Si}^{-1}). \tag{22}$$

In the diamond structure, $C = d_{14} = G = B = 0$ due to inversion symmetry.

3 Conclusions

Since the mixing parameter C is arbitrary, one is free to choose whatever value of C is most convenient. There are three obvious possibilities: (i) $C = -\sqrt{3}dd_{14}/2e$, which eliminates all linear-field terms from the Hamiltonian; (ii) $C = G/(L - M + N)$, which eliminates the k^3 terms (18); or (iii) $C = -B/P$, which eliminates the first-order $\mathbf{k}$ dependence of the interband momentum matrix (21). The fact that all three phenomena are controlled by one parameter shows clearly that they are interrelated and to some degree equivalent.

The Khurgin-Voisin model [8] achieves even greater simplicity by omitting both the k^3 terms and the $\mathbf{k}$ dependence of the momentum matrix. However, this is not consistent with the present theory because $G/(L - M + N) \neq -B/P$. Also, the Khurgin-Voisin dipole moment is supposed to be an objective property of the valence states, which C clearly is not. Therefore, despite some obvious similarities, the present theory is not equivalent to that of Khurgin and Voisin.

Acknowledgements This work was supported by Hong Kong RGC Grants No. DAG99/00.SC27 and HKUST6139/00P.

References

1. J. M. Luttinger and W. Kohn, Phys. Rev. **97**, (1955) 869.
2. J. M. Luttinger, Phys. Rev. **102**, (1956) 1030.
3. C. Kittel and A. H. Mitchell, Phys. Rev. **96**, (1954) 1488.
4. G. Dresselhaus, A. F. Kip, and C. Kittel, Phys. Rev. **98**, (1955) 368.
5. E. O. Kane, J. Phys. Chem. Solids **1**, (1957) 249.
6. G. L. Bir and G. E. Pikus, *Symmetry and Strain-Induced Effects in Semiconductors* (Wiley, New York 1974).
7. M. G. Burt, J. Phys. Condens. Matter **4**, (1992) 6651.
8. J. B. Khurgin and P. Voisin, Semicond. Sci. Technol. **12**, (1997) 1378; Phys. Rev. Lett. **81**, (1998) 3777.
9. B. A. Foreman, Phys. Rev. Lett. **84**, (2000) 2505.
10. B. A. Foreman, J. Phys. Condens. Matter **12**, (2000) R435.
11. G. Dresselhaus, Phys. Rev. **100**, (1955) 580.
12. E. N. Adams, J. Chem. Phys. **21**, (1953) 2013.
13. W. Kohn, Phys. Rev. **115**, (1959) 1460.

Antiferromagnetic coupling between the conduction electrons and the $4f$ electrons in Eu@C$_{60}$

S. Suzuki, M. Kushida, S. Amamiya, S. Okada, K. Nakao

Institute of Materials Science, University of Tsukuba, Tsukuba 305-8573, Japan e-mail: `shugo@ims.tsukuba.ac.jp`

Abstract We study the electronic structure of solid Eu@C$_{60}$ by using the scalar relativistic full-potential LCAO method based on the density functional theory. We find that the coupling between the conduction electrons and the $4f$ electrons in Eu is antiferromagnetic. Furthermore, we study the coupling between the Eu spins mediated by the conduction electrons. The results show that the Eu-Eu coupling is very weak.

1 Introduction

So far, a number of compounds based on C$_{60}$ have been synthesized and their properties have been studied extensively[1]. The superconductivity, for example, is one of the most attractive topics in this field[2–4]. The extensive studies have revealed that some outstanding features of these compounds are characteristic of C$_{60}$-based materials but not observed in other fullerene-based materials. However, among the studies on C$_{60}$-based materials, that on the solid state properties of metal endohedral C$_{60}$ is still in early stage because of the difficulty of its synthesis despite that this class of material is expected to show some remarkable properties. Recently, a successful synthesis of Eu@C$_{60}$ has been reported and the study on the solid state properties of this materials has been started[5–7].

One of the most interesting properties expected on this materials is the magnetism. Since Eu has the open $4f$ shell, exotic magnetic behaviors can be observed. Furthermore, since the ionization energy of Eu is very small, the charge transfer from Eu to C$_{60}$ is also expected. It is then natural to consider what type of interaction arises between the conduction electrons and the $4f$ electrons. Moreover, it is also important to study the magnetic coupling between the Eu spins mediated by the conduction electrons, i.e., the RKKY interaction. In the present study, we investigate the magnetic coupling between the conduction electrons and the $4f$ electrons and the RKKY interaction between the Eu spins by using the scalar relativistic full-potential linear-combination-of-atomic-orbitals (LCAO) method.

2 Method

Since Eu is a heavy element, it is important to include relativistic effects appropriately. In the present study, we have used the scalar relativistic version of our full-potential LCAO method based on the density functional theory with in the local density approximation[8–10]. This method deals with the relativistic effects due to the Darwin term and the mass-velocity term. The basis set employed in the calculations is a split valence class; we use a single basis function for a core orbital while we use double basis functions for a valence orbital. The atomic orbitals employed for the C atoms are the $1s$, $2s$, and $2p$ orbitals of a neutral C atom and the $2s$, $2p$ orbitals of a C^{2+} atom. Also, the atomic orbitals employed for the Eu atom are the $1s$, $2s$, $2p$, $3s$, $3p$, $3d$, $4s$, $4p$, $4d$, $4f$, $5s$, $5p$, $5d$, and $6s$ orbitals of a neutral Eu atom and the $6p$ orbitals of a Eu$^+$ atom and the $5d$ and $6s$ orbitals of a Eu^{2+} atom and the $6p$ orbitals of a Eu^{3+} atom.

We first optimize the geometry of molecular Eu@C$_{60}$ by the conjugate gradient optimization scheme without any restriction on the geometry such as an impose of a symmetry on the geometry. We next calculate the electronic structure of solid Eu@C$_{60}$ by placing the molecules so that they form the fcc lattice, assuming the lattice constant being the same as that in the pristine C$_{60}$ crystal.

3 Results and discussion

We begin with the results of the calculations on molecular Eu@C$_{60}$. In the optimized geometry, the Eu atom is placed at about 1.2 Å away from the center of the C$_{60}$ cage. Two $6s$ electrons are transfered to the C$_{60}$ molecule as valence electrons while seven $4f$ electrons remain on the Eu atom. The Eu atom in molecular Eu@C$_{60}$ thus exists as Eu^{2+}. Furthermore, the spin direction of the valence electrons is antiparallel to that of the $4f$ electrons. That is, the coupling between the valence electrons and the $4f$ electrons is antiferromagnetic. The strength of this coupling can be estimated from the energy difference between the up- and down-spin energy levels of the valence electrons and is found to be about 0.1 eV. This means that the antiferromagnetic coupling is strong enough against thermal fluctuation as well as magnetic field. It is thus expected that the magnetic susceptibility measurement observes Curie-like behavior with the spin of 5/2.

We next show the results of the calculations on the electronic structure of solid Eu@C$_{60}$. The Eu atom is assumed to be placed along the $\langle 111 \rangle$ direction. The band structure is shown in Fig. 1(a) for the up-spin electrons and in Fig. 1(b) for the down-spin electrons. In the figure, the Fermi level is shown by the dotted line. We find that the bands on which the Fermi level lies originate in the lowest unoccupied molecular orbitals (LUMO) of C$_{60}$. Also, the five bands at the bottom

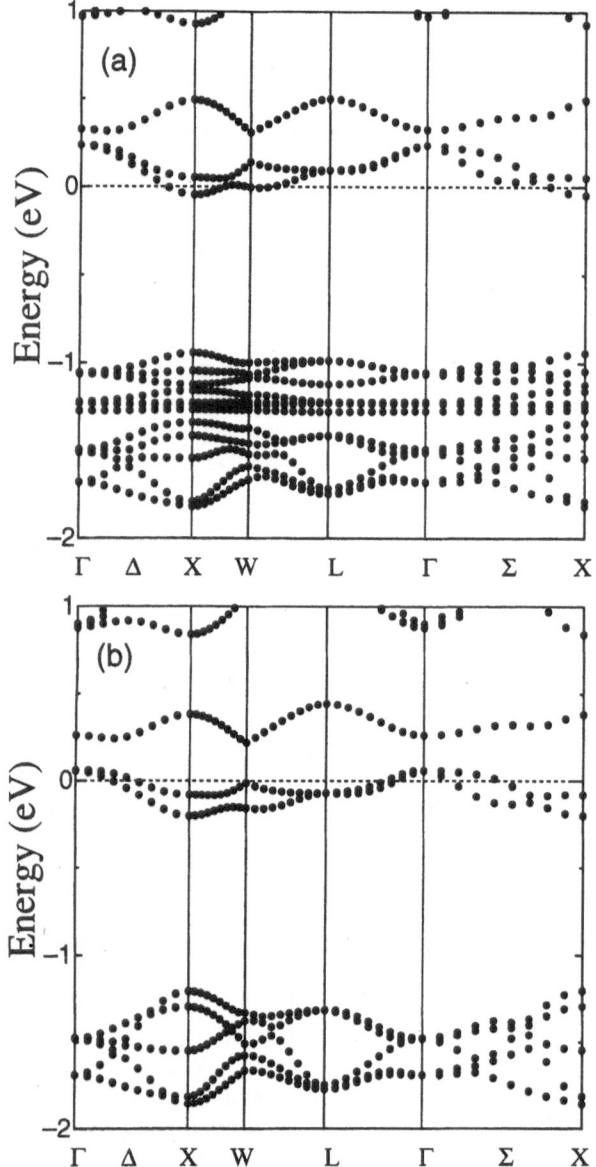

Fig. 1 Band structure of Eu@C$_{60}$ for (a) up-spin electrons and (b) down-spin electrons. The dotted line shows the Fermi level.

in each figure originate in the highest occupied molecular orbitals (HOMO) of C$_{60}$. Furthermore, there exist seven dispersion-less bands just above the HOMO-derived bands in Fig. 1(a) while not in Fig. 1(b). These up-spin bands originate in the $4f$ atomic orbitals of Eu and corresponding down-spin bands exist at about 3 eV above the Fermi level. Also, the bands derived from the $6s$ atomic orbitals of Eu exists at about 2 eV above the Fermi level and are not shown in the figure.

These results show the following points. Firstly, seven $4f$ electrons still remain in Eu atom also in solid Eu@C$_{60}$ and two $6s$ electrons in Eu are transfered to the LUMO-derived bands. That is, the Eu atom in solid Eu@C$_{60}$ exists also as Eu^{2+}. Secondly, the spin direction of the conduction electrons is antiparallel to that of the $4f$ elec-

trons in Eu. That is, the coupling between the conduction electrons and the $4f$ electrons in Eu is antiferromagnetic. It is noted, however, that the spin polarization is not perfect in contrast to the case of molecular Eu@C$_{60}$; there exist up-spin electrons in the LUMO-derived bands although the majority-spin electrons are still down-spin electrons.

Finally, we investigate the coupling between the Eu spins mediated by the conduction electrons, i.e., the RKKY interaction. To this end, we have calculated the total energies of two configurations of the Eu spins by using a unit cell consisting of two Eu@C$_{60}$ molecules; one configuration is ferromagnetic one while the other is antiferromagnetic one in which we assume alternate stacking of up- and down-spin Eu@C$_{60}$ layers along the ⟨111⟩ direction. The results show that the ferromagnetic spin configuration is more stable than the antiferromagnetic one by about 0.001 eV per Eu atom. This result, however, is beyond the accuracy of our calculations and accordingly we should regard the result as the indication that the RKKY interaction between the Eu spins is very weak.

References

1. H. Ehrenreich, F. Spaepen(Eds.), *Solid State Physics*, Vol. 48, (Academic Press, New York, 1994) and references therein.
2. A. F. Hebard, M. J. Rosseinsky, R. C. Haddon, D. W. Murphy, S. H. Glarum, T. T. M. Palstra, A. P. Ramirez, and A. R. Kortan, Nature **350**, (1991) 600.
3. M. J. Rosseinsky, A. P. Ramirez, S. H. Glarum, D. W. Murphy, R. C. Haddon, A. F. Hebard, T. T. M. Palstra, A. R. Kortan, S. M. Zahurak, and A. V. Makhija, Phys. Rev. Lett. **66**, (1991) 2830.
4. K. Tanigaki, T. W. Ebbesen, S. Saito, J. Mizuki, J. S. Tsai, Y. Kubo, and S. Kuroshima, Nature **352**, (1991) 222.
5. Y. Kubozono, K. Hiraoka, Y. Takabayashi, T. Nakai, T. Ohta, H. Maeda, H. Ishida, S. Kashino, S. Emura, S. Ukita, and T. Sogabe, Chem. Lett., (1996) 1061.
6. Y. Takabayashi, Y. Kubozono, K. Hiraoka, T. Inoue, K. Mimura, H. Maeda, and S. Kashino, Chem. Lett., (1997) 1019.
7. T. Inoue, Y. Kubozono, S. Kashino, Y. Takabayashi, K. Fujitaka, M. Hida, M. Inoue, T. Kanbara, S. Emura, and T. Uruga, Chem. Phys. Lett. **316**, (2000) 381.
8. S. Suzuki and K. Nakao, J. Phys. Soc. Jpn. **69**, (2000) 532.
9. P. Hohenberg and W. Kohn, Phys. Rev. **136**, (1964) B864.
10. W. Kohn and L. J. Sham, Phys. Rev. **140**, (1965) A1133.

Proc. 25th Int. Conf. Phys. Semicond., Osaka 2000 (Eds. N. Miura and T. Ando)

45

First-principles approach to the oxygen vacancies in SrTiO$_3$

Tatsuo Schimizu, Takashi Kawakubo

Advanced LSI Technology Laboratory, Corporate Research & Development Center, Toshiba Corp., Saiwai-ku, Kawasaki 212-8582, Japan. E-mail : schimizu@mdl.rdc.toshiba.co.jp

Abstract The authors report on a first-principles calculation of not only the electronic and geometric structure of a single oxygen vacancy in SrTiO3, but also vacancy-vacancy interaction. The unexpected result obtained is that the energy level of the oxygen vacancy is found to appear inside the conduction band. The single-oxygen-vacancy state supplies the conduction band with electrons. Therefore, only a small amount of electron doping via oxygen vacancies makes SrTiO3 metallic. The second novel result is that total energy is predicted to be stabilized in the case that two oxygen vacancies are in a line. This interaction between two oxygen vacancies creates a brand-new scheme of the oxygen vacancies. This could also explain the reduction of the carrier density relative to the density of electrons introduced by oxygen vacancies.

1. Introduction

There has been engineering and scientific interest in the electronic properties of oxygen vacancies (VO) in SrTiO3 thin films respecting electric conduction [1-5]. Oxygen vacancies dope electrons into the system, and such electron doping transforms insulating SrTiO3 into a metallic state with an extremely small extent of doping[3]. SrTiO3-δ is expected to have 2δ number of electron. Hall data suggest, however, that a large number of doped electrons do not contribute to the electric conduction [3]. In ultraviolet-photoemission spectroscopy (UPS) experiments of oxygen-vacancy-rich SrTiO3 (001) surface prepared by argon-ion bombardment and annealing in ultrahigh vacuum (UHV) [4,5], an inside-gap state appears at about 1.3eV below the Fermi level. The inside-gap state has not been explained.

The objective of this study is not only to investigate the electronic and geometric structures of an oxygen vacancy in SrTiO3 bulk crystal, but also to investigate interactions between oxygen vacancies in order to interpret experimentally observed important aspects of oxygen vacancies in SrTiO3 for which clarification has long been sought: the metallization with an extremely small extent of oxygen vacancies [1], the existence of a inside-gap state[5], and the reduction of the carrier density relative to the density of electrons introduced by oxygen vacancies[3].

2. Method

The first-principles calculations used in this study are based on the density functional theory with the local spin density approximation. The exchange-correlation term used here is of the Ceperley-Alder [6] form parameterized by Perdew and Wang[7]. Throughout this work, the paramagnetic solutions are obtained except for oxygen molecule. For each ion, an ultrasoft pseudopotential as proposed by Vanderbilt [8] is generated. The cutoff energy, 30.25 Ryd., is used throughout the present work. A total of 64 k points were used for integration over the Brillouin zone. From these potentials, the lattice constant of a0= 3.88 A is obtained for cubic perovskite SrTiO3. This is consistent with the experimental value (3.90 A [9]) within an error of -0.6%.

3. Calculations and Results

To characterize the effect of electron density delocalization of the oxygen vacancy state in SrTiO3, we calculate the formation energy, E(Form), of the oxygen vacancy and an oxygen molecule relative to the defect-free material. Supercell size is defined by the notation (NML). Here N $\times$ M and L are the size of the supercell in the SrO layer (ab-plane) and in the direction of the Ti-VO axis (c-axis), respectively, in the unit of the calculated lattice constant a0

We need to consider the relaxation energy using (nnm) supercells. Without relaxtion, the calculated formation energy converges to 6.26 eV by using more than (223) supercell. Fig.1 shows the supercell size dependence of the calculated formation energy of the oxygen vacancy with (open circles) and without (filled circles) structural relaxation. The deviations of the calculated formation energy of (225) and ($\sqrt{5}$ $\sqrt{5}$ 4) supercell from that of (224) supercell are vary small rerative to the relaxation energy (about 1.05 eV). The range of the effect of the oxygen vacancy is estimated to be about (224) times unit cell. The oxygen vacancy formation energy from SrTiO3 bulk crystal is found to be about 5.21 eV per oygen vacancy with structural relaxation.

In order to investigate the vacancy state, the band structure of (224) supercell is calculated. Fig.2 shows the band structure with structural relaxation. The narrow vacancy state labeled by the filled circles, which has large charge density at the oxygen vacancy site, is found to be inside the conduction band. The single-oxygen-vacancy state has been thought to be an origin of the inside-gap state, but it is proved not to be so. According to the above band structure (Fig.2), the single-oxygen-vacancy state supplies the conduction band with electrons. Therefore, only a small amount of electron doping via oxygen vacancies makes SrTiO3 metallic [1].

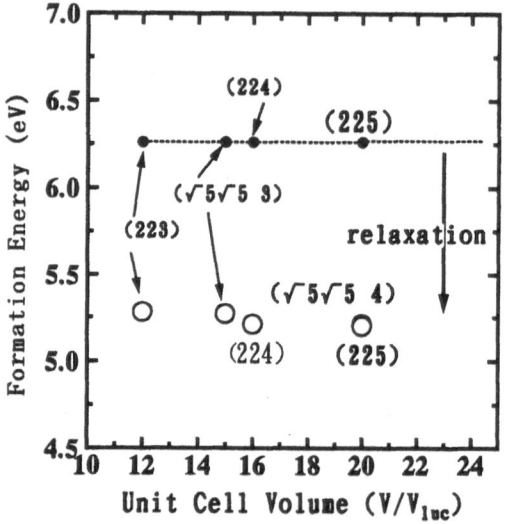

FIG.1. The unit cell volume dependence of the formation energy of the oxygen vacancy per oxygen atom. Supercell size (nnm) is shown as a parameter.

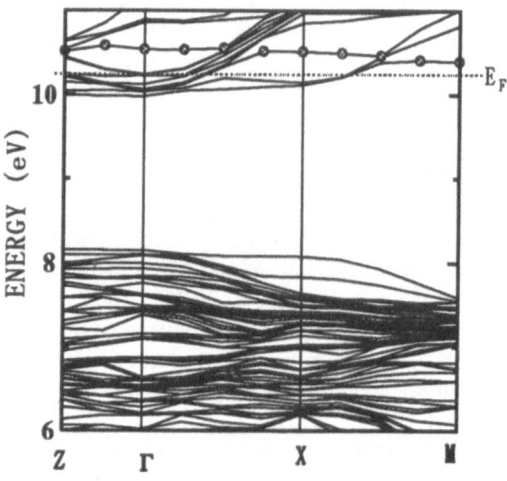

FIG.2. Band structure of (224) supercell. The band with filled circles is the vacancy state.

The interactions between oxygen vacancies are investigated in order to discuss clustering of oxygen vacancies in the range of the second nearest oxygen vacancies. There are three different configurations (A-C) arising from the distribution of two oxygen vacancies in the range. In the (A) and (B) configurations, two oxygen vacancies are in the same oxygen octahedron cage. The angles of the two directions from center atom Ti to vacancies are 180 and 90 degrees, respectively. On the other hand, in the configuration (C), there is no Ti ion between two oxygen vacancies.

To estimate the interaction between oxygen vacancies, we calculate the formation energy of two vacancies per oxygen vacancy, E(2V, Form). If the stabilization energy per vacancy, Est = E(2V, Form)-E(Form), is negative, the interaction between oxygen vacancies is attractive. Est's are -0.09eV, +0.30 eV and +0.41eV for the configurations (A), (B) and (C), respectively. Therefore, the configuration (A) is found to be the only preferable configuration. The interaction between oxygen vacancies in the line, -O-Ti-VO-Ti-VO-Ti-O-, is found to be attractive. Moreover, the vacancy states inside the conduction band interact with each other, and the hybridized dispersion-less inside-gap state appears. The inside-gap state is about 0.7eV below the Fermi level. Because the inside-gap state is a dispersion-less filled state, the electrons in the state do not contribute to the electric conduction. Therefore the effective carrier density is half of the expected one. The inside-gap state is found to be attributable to isolated Ti ion between two oxygen vacancies owing to the outward movement of the edge Ti ions in the Ti-VO-Ti-VO-Ti chain. The density of oxygen vacancies increases, three or more oxygen vacancies can be in a line. In that case, the number of the isolated Ti ions increases, and the effective carrier density declines as shown in Hall measurements[3].

References

1. P. Calvani, et al., Phys. Rev. B47 8917 (1993).
2. H. P. R. Frederikse, W. R. Thurber, and W. R. Hosler, Phys. Rev. 134 A442 (1964).
3. W. Gong, et al., J. Solid state Chem. 90 320 (1991).
4. B. Cord and R. Courths, App. Surf. Sci.162 34 (1985).
5. V. E. Henrich, G. Dresselhaus, and H. J. Zeiger, Phy. Rev. 17 4908 (1978).
6. D. M. Ceperley and B. J. Alder, Phys. Rev. Lett. 45 566 (1980).
7. J. Perdew and   Y. Wang: Phys. Rev. B45 (1992) 13244.
8. D. Vanderbilt, Phys. Rev. B41 7892 (1990).
9. Landolt and Bornstein, Numerical Data and Functional Relations in Science and Technology – Crystal and Solid State Physics (Springer, Berlin) (1982).

Proc. 25th Int. Conf. Phys. Semicond., Osaka 2000 (Eds. N. Miura and T. Ando)

47

Hartree-Fock study on the phase diagrams of alkali-metal-doped C_{60}

J. Hirosawa, S. Suzuki, and K. Nakao

Institute of Materials Science, University of Tsukuba, Tsukuba 305-8573, Japan e-mail: `jin@ims.tsukuba.ac.jp`

Abstract   We investigate the phase diagrams of alkali-metal-doped C_{60}, A_3C_{60} and A_4C_{60}, with respect to the strength of interactions, employing a model that takes account of both the electron-electron and electron-phonon interactions. We find that A_3C_{60} has four phases: paramagnetic metals, superconductors, orbital-polarized paramagnetic metals, and spin- and orbital-polarized insulators. On the other hand, we find that A_4C_{60} has three phases: paramagnetic metals, superconductors, and orbital-polarized insulators.

1 Introduction

In the last decade, the electronic properties of alkali-metal-doped C_{60} have been studied extensively [1]. In particular, the superconductivety with the transition temperatures beyond 30 K found in A_3C_{60} (A =K,Rb, etc) has attracted many researchers in this field [2,3]. Furthermore, A_4C_{60}, which is well known material in addition to A_3C_{60}, has been investigated as well [4-9]. The extensive studies have shown that there is anomalous behavior of these materials as described below and the elucidation of its origin is desired strongly.

Since the conduction bands of C_{60} consist of threefold degenerate molecular orbitals which are the lowest unoccupied molecular orbitals of C_{60}, the t_{1u} orbitals, A_xC_{60} for $0 < x < 6$ are predicted to be metallic. Also, the theoretical study employing the density functional method performed by Saito and Oshiyama has shown that the band width is about 0.4 eV [10]. One of the great successes of the band calculations is the explanation of the relation between the superconducting transition temperature and the lattice constant in A_3C_{60} [1].

However, many experiments have shown that the rigid band model cannot be applied to the electronic states of A_4C_{60}. For example, it has been found that A_4C_{60} are nonmagnetic insulators according to the magnetic susceptibility measurement employing NMR and ESR [4-6]. Also, the infrared reflection experiment has shown that the spectrum of A_3C_{60} shows the Drude behavior but that of A_4C_{60} does not [7]. In the photoemission and inverse photoemission study, it has been observed that there exists an extraordinary change of spectrum near the Fermi level as the valence is varied [8]. This cannot be explained by the rigid band model. Furthermore, the μSR experiment has shown that the lowest electronic excitation in K_4C_{60} occurs at 0.33 eV [9]. More recently, it has been revealed that $(NH_3)A_3C_{60}$ is the Mott-Hubbard insulator with antiferromagnetic spin order [11]. These experiments reveal that alkali-metal-doped C_{60} has not only metallic phases but also both magnetic and nonmagnetic insulating phases.

In the present study, we investigate the phase diagrams of A_3C_{60} and A_4C_{60} with respect to the strength of interactions, employing a model in which both the electron-electron and electron-phonon interactions are taken into account. As the above experiments show, the rigid band model cannot be applied to the electronic state of A_4C_{60}. By considering both the electron-electron and electron-phonon interactions, we can elucidate possible phases including not only metallic or superconducting phases but also magnetic or nonmagnetic insulating phases.

2 Results and discussion

We study possible phases in A_3C_{60} and A_4C_{60} as follows. We adopt the effective Hamiltonian in which the electron-phonon interaction is included into the electron-electron interaction in the second order perturbation method [12,13], and employ the Hartree-Fock (HF) approximation using the transfer integrals that reproduce the band structure given in Ref. 10 [12]. Also, for A_4C_{60}, we take account of the crystal field splitting of 0.2 eV [12]. In the phase diagrams, the horizontal axis is the intraorbital repulsion V_{intra} and the vertical axis is the pair transfer interaction K with changing its sign. Also, we use the condition for the superconductivity given in Ref. 13 and draw a line in the phase diagrams accordingly.

As a result, it is found that A_3C_{60} have four phases. The phase diagram is shown in Fig. 1(a). The obtained phases are paramagnetic metal (PM), superconductor (SC), orbital-polarized paramagnetic metal (OPM), and spin- and orbital-polarized insulator (SOI). Also, we find that OPM and SOI exist only for A_3C_{60}. It is worthwhile to note that SOI is the Mott-Hubbard insulator. As shown in Fig. 1(a), this phase exists in the region of large V_{intra}. On the other hand, metallic phases, PM, SC, and OPM, exist in the region of small V_{intra}. Furthermore, OPM exists in the region of large K. In the experiments, it has been observed that A_3C_{60} is a superconductor and $(NH_3)A_3C_{60}$ is a magnetic insulator. Therefore, A_3C_{60} belongs to SC and $(NH_3)A_3C_{60}$ belongs to SOI. Nevertheless, since estimated values of V_{intra} and K are about 0.2 eV and about -0.1 eV, respectively, A_3C_{60} belongs to OPM in this phase diagram in conflict with the experimental result that A_3C_{60} is a superconductor.

Next, we show that A_4C_{60} have three phases. The phase diagram is shown in Fig. 1(b). In addition to PM

(a)

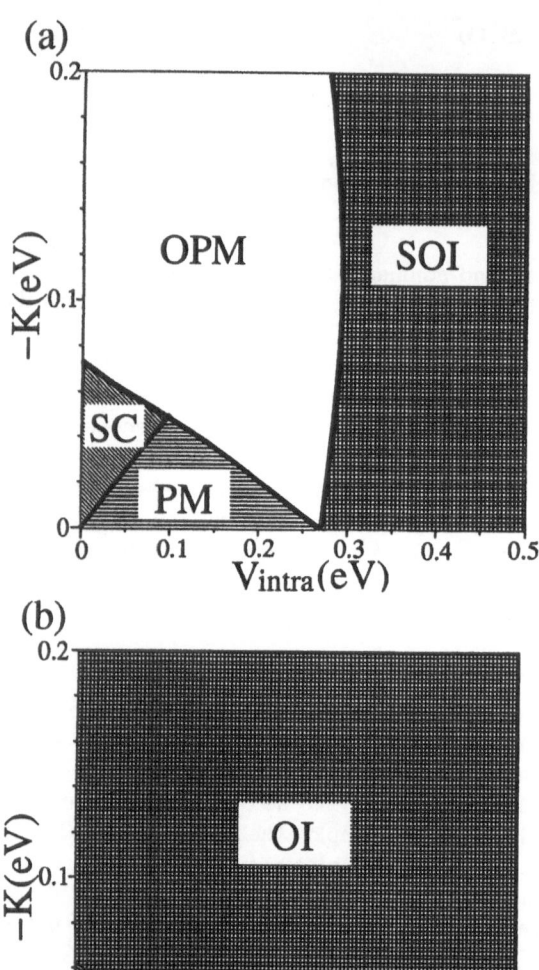

(b)

Fig. 1 Phase diagrams of (a) A_3C_{60} and (b) A_4C_{60}. The horizontal axis is the intraorbital repulsion V_{intra} and the vertical axis is the pair transfer interaction K with changing its sign: PM shows a paramagnetic metal. SC shows a superconductor. OPM shows an orbital-polarized paramagnetic metal. SOI shows a spin- and orbital-polarized insulator. OI shows an orbital-polarized insulator.

and SC, there is a phase of orbital-polarized insulators (OI). This insulator is nonmagnetic and exists only for A_4C_{60}. As in A_3C_{60}, PM and SC exists in the region of small V_{intra} and K. In the experiments employing NMR, ESR, and μSR, it has been observed that all known A_4C_{60} are nonmagnetic insulators [4–6]. Then, all of A_4C_{60} belong to OI. Since the estimated values of V_{intra} and K are about 0.2 eV and about -0.1 eV, respectively, A_4C_{60} belongs to OI in this phase diagram in agreement with the experimental results. Also, it is possible that some of them can be even metallic or superconducting, for example, under high pressures because

there exsit PM and SC in the region of small V_{intra} and K in our phase diagram.

We finally discuss the reason why our results fail to explain experimental observation for A_3C_{60} while succeed for A_4C_{60}. A most serious deficiency in the HF approximation is the ignorance of fluctuations. In paticular, A_3C_{60} possesses a number of low energy solutions of the HF equations with almost the same energy. Accordingly, the HF approximation is not suitable for A_3C_{60} because the fluctuations are expected to play important roles when we study beyond the HF approximation. On the other hand, the HF approximation is suitable for A_4C_{60} because it possesses only one HF ground state much lower in enegy well separated from the exited states. In the future study, it is thus necessary to consider many body effects for obtaining the reliable phase diagram of A_3C_{60}.

References

1. H. Ehrenreich, F. Spaepen(eds.), Solid State Physics, Vol. 48, (Academic Press, New York, 1994) and references therein.
2. M. J. Rosseinsky, A. P. Ramirez, S. H. Glarum, D. W. Murphy, R. C. Haddon, A. F. Hebard, T. T. M. Palstra, A. R. Kortan, S. M. Zahurak, and A. V. Makhija, Phys. Rev. Lett. **66**, (1991) 2830.
3. K. Tanigaki, T. W. Ebbesen, S. Saito, J. Mizuki, J. S. Tsai, Y. Kubo, and S. Kuroshima, Nature **352**, (1991) 222.
4. R. Tycko, G. Dabbagh, M. J. Rosseinsky, D. W. Murphy, A. P. Ramirez, and R. M. Fleming, Phys. Rev. Lett. **68**, (1992) 1912.
5. M. Kosaka, K. Tanigaki, I. Hirosawa, Y. Shimakawa, S. Kuroshima, T. W. Ebbesen, J. Mizuki, and Y. Kubo, Chem. Phys. Lett. **203**, (1993) 429.
6. I. Lukyanchuk, N. Kirova, F. Rachdi, C. Goze, P. Molinie, and M. Mehring, Phys. Rev. B **51**, (1995) 3978.
7. Y. Iwasa and T. Kaneyasu, Phys. Rev. B **51**, (1995) 3678.
8. J. H. Weaver, J. Phys. Chem. Solids **53**, (1992) 1433.
9. R. F. Kiefl, T. L. Duty, J. W. Schneider, A. MacFarlane, K. Chow, J. W. Elzey, P. Mendels, G. D. Morris, J. H. Brewer, E. J. Ansaldo, C. Niedermayer, D. R. Noakes, C. E. Stronach, B. Hitti, and J. E. Fischer, Phys. Rev. Lett. **69**, (1992) 2005.
10. S. Saito and A. Oshiyama, Phys. Rev. Lett. **66**, (1991) 2637.
11. H. Tou, Y. Maniwa, Y. Iwasa, H. Shimoda, and T. Mitani, Phys. Rev. B **62**, (2000) 775.
12. S. Suzuki and K. Nakao, Phys. Rev. B **52**, (1995) 14206.
13. S. Suzuki, S. Okada, and K. Nakao, J. Phys. Soc. Jpn. **69**, (2000) 2615.

Dynamical-mean-field study on the photoemission spectra of alkali-metal-doped C$_{60}$

T. Chida, S. Suzuki, K. Nakao

Institute of Materials Science, University of Tsukuba, 1-1-1 Tennoudai, Tsukuba, Ibaraki 305-8573, Japan

Abstract We study the spectral density for A_3C_{60} taking account of the electron-electron and electron-phonon interactions. The calculated spectral density successfully explain the experimental photoemission spectra. The broadening of the lowest-unoccupied-molecular-orbital derived band is originated in the multiplet splitting in an isolated C_{60}^{3-} molecule.

1 Introduction

Many experiments have shown that the electron-electron and electron-phonon interactions play important roles in alkali-metal-doped C$_{60}$, A_xC_{60} where A is an alkali metal. A_2C_{60} and A_4C_{60} are nonmagnetic insulators [1] in contradiction to the results of band calculations that these materials are metals. The theoretical studies predict that these materials favor the nonmagnetic insulator due to the the electron-electron and electron-phonon interactions [2–4].

Furthermore, there exists experimental evidence for A_3C_{60} to be an anomalous metal although the metallic behavior is in agreement with the results of band calculations [5]. In particular, the photoemission spectra for A_3C_{60} have some anomalies [6–9]. Firstly, the lowest-unoccupied-molecular-orbital (LUMO) derived band is much broader than predicted by band calculations. Also, a shoulder at about -1.5 eV below the Fermi level appears on the highest-occupied-molecular-orbital (HOMO) derived band. As well as in A_2C_{60} and A_4C_{60}, it is expected that the electron-electron and electron-phonon interactions play important roles in the electronic structures of A_3C_{60}.

In this paper, we show the calculated spectral density for A_3C_{60} and compare our results with the experimental photoemission spectra. To understand the result for A_3C_{60}, we discuss the result for an isolated C_{60}^{3-} molecule in detail.

2 Method

In this section, we explain the method of calculations of the spectral density. In the present study, we employ a model which takes account of both the electron-electron and electron-phonon interactions within antiadiabatic approximation [2,10]. We also consider only t_{1u} orbitals, x, y and z, which are three-fold degenerate. We calculate the spectral densities for A_xC_{60} based on the dynamical mean-field theory (DMFT) by using the exact diagonalization method, which is a powerful method to investigate many-body effects [11,12]. In the limit of an isolated molecule, our model can be diagonalized and the spectral density can be calculated analytically. The spectral density for N-electron system is given by

$$A(\omega) = \sum_{a,\sigma} A_{a\sigma}(\omega) \qquad (1)$$

where

$$A_{a\sigma}(\omega) = \sum_i | < \psi_i^{N-1}|a_\sigma|\psi_{gs}^N > |^2 \delta(\omega + E_i^{N-1} - E_{gs}^N)$$
$$+ \sum_i | < \psi_i^{N+1}|a_\sigma^\dagger|\psi_{gs}^N > |^2 \delta(\omega - E_i^{N+1} + E_{gs}^N).$$

$$(2)$$

E_{gs}^N is the ground-state energy for N-electron system and $E_i^{N\pm1}$ is the ith eigen value for $N \pm 1$-electron system. Also, a_σ ($a_\sigma^\dagger$) represents the annihilation (creation) operator of the t_{1u} electron with spin σ in the orbital a.

3 Results and discussion

We find six peaks in the spectral density for an isolated C_{60}^{3-} molecule as shown in Fig. 1. The multiplet splitting occurs due to the interactions between t_{1u} electrons. In this paper, we denote the six peaks by the T_{1g}, H_g ,A_g, $\bar{A}_g$, $\bar{H}_g$ and $\bar{T}_{1g}$ peak as shown in Fig. 1. Here, we consider the photoemission processes for understanding the three peaks for the occupied states. The initial state in the photoemission processes is the ground state of an isolated C_{60}^{3-} molecule. On the other hand, there are three final states in the photoemission processes because the eigen states for an isolated C_{60}^{2-} molecule split into three multiplets; $t_{1u} \times t_{1u}$ is reduced to $A_g + H_g + T_{1g}$. Firstly, the ground state for an isolated C_{60}^{2-} molecule is of the A_g symmetry and spin-singlet. Thus, the A_g peak at about -0.03 eV in Fig. 1 is produced by the process from the initial state to this state. Secondly, the first excited state of an isolated C_{60}^{2-} molecule is of the H_g symmetry and spin-singlet. The process to this state produces the H_g peak at about -0.3 eV. Finally, the highest excited state of an isolated C_{60}^{2-} molecule is of the T_{1g} symmetry and spin-triplet. The process to this state produces the T_{1g} peak at about -0.55 eV. Similarly, we can understand three peaks for the unoccupied states by considering the inverse photoemission processes. In Fig. 1, we also find that the energy gap between the occupied and unoccupied states is very small. The reason for this small energy gap is that the electron-electron and electron-phonon interactions compete in an isolated C_{60}^{3-} molecule and we discuss in detail elsewhere [13].

We compare the spectral density for A_3C_{60} with that for an isolated C_{60}^{3-} molecule. The spectral density for A_3C_{60} consists of two parts. One is the continuous features around the Fermi level ranged from -1 eV to $+1$

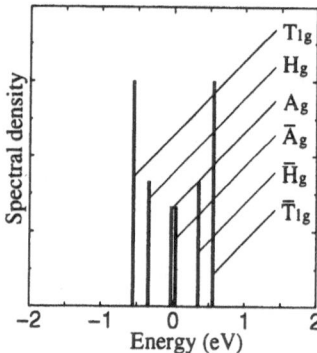

Fig. 1 Spectral density for an isolated C_{60}^{3-} molecule.

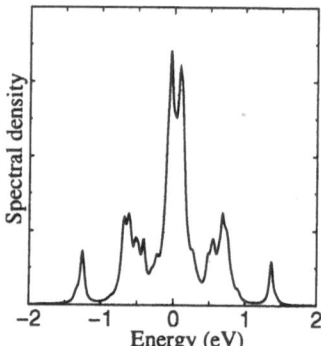

Fig. 2 Spectral density for A_3C_{60}.

eV. The other is the satellites at about $\pm$ 1.3 eV. These continuous features around the Fermi level are directly derived from the peaks for the isolated C_{60}^{3-} molecule in Fig. 1; each peak is broadened by the band formation. However, the existence of satellites at ±1.3 eV cannot be explained only by considering the spectral density for an isolated C_{60}^{3-} molecule. Recently, we have successfully explained that the satellites are derived from the charge fluctuation and we discuss this point elsewhere [13]. The existence of satellites indicate clearly the fact that A_3C_{60} is an anomalous metal.

Our results can explain the experimental photoemisson spectra for K_3C_{60} and Rb_3C_{60}. In the experiments, the LUMO-derived band is much broader than predicted by band calculations with the width of about 1.0 eV [6–8]. Our results reproduce this LUMO-derived band as the continuous features below the Fermi level ranged from -1 eV to 0 eV. Also, in the experimental photoemission spectra, two features are observed below the Fermi edge; one feature exists at about -0.3 eV and the other feature exists at about -0.6 eV [6–8]. These features may originate in the H_g peak and the T_{1g} peak in Fig. 1, respectively. However, it is difficult to resolve these features in Fig. 2. This may be due to the finite size effect in our calculations. Furthermore, the shoulder observed at about -1.5 eV on the HOMO-derived band [6,7] corresponds to the satellite at about -1.3 eV in our result.

Finally, we propose the interpretation of the optical conductivity spectra from our results. Iwasa *et al.* report that there are not only the Drude component but also the midinfrared absorption at 0.4−0.5 eV in the optical conductivity spectra of K_3C_{60} and Rb_3C_{60} [1]. Furthermore, Degiorgi *et al.* observe the midinfrared absorption at 0.1 eV in Rb_3C_{60} [14]. We suggest that this midinfrared absorption is produced by the intermolecular charge transfer transition; it should be noted that the intramolecular transition is forbidden due to the selection rule. In Fig. 1, the energy difference between one peak for unoccupied states and that for occupied states is the energy needed to produce one C_{60}^{2-} and one C_{60}^{4-} molecules from two C_{60}^{3-} molecules. In other words, this energy difference corresponds to the energy needed for the intermolecular charge transfer. We find that there are several transitions, which need the energy of about 0.4−0.5 eV. For example, the energy difference between the $\bar{T}_{1g}$ and A_g peak is about 0.5 eV. The transition associated with these peaks can produce the midinfrared absorption at 0.4−0.5 eV in the the optical conductivity spectra. On the other hand, the energy difference between the $\bar{A}_g$ and A_g peak is about 0.1 eV. Thus, the transition associated with these peaks can produce the midinfrared absorption at about 0.1 eV.

References

1. Y. Iwasa and T. Kaneyasu, Phys. Rev. B **51** (1995) 3678.
2. S. Suzuki and K. Nakao, Phys. Rev. B **52** (1995) 14206.
3. M. Fabrizio and E. Tosatti, Phys. Rev. B **55** (1997) 13465.
4. J. E. Han, E. Koch and O. Gunnarsson, Phys. Rev. Lett. **84** (2000) 1276.
5. S. Saito and A. Oshiyama, Phys. Rev. B **44** (1991) 11536.
6. M. Knupfer, M. Merkel, M. S. Golden, J. Fink, O. Gunnarsson and V. P. Antropov, Phys. Rev. B **47** (1993) 13944.
7. P. J. Benning, F. Stepniak and J. H. Weaver, Phys. Rev. B **48** (1993) 9086.
8. A. Goldni, S. L. Friedmann, Z.-X. Shen, and F. Pamigiani, Phys. Rev. B **58** (1998) 11023.
9. T. Chida, S. Suzuki and K. Nakao, J. Phys. Soc. Jpn. **69** (2000) 1249.
10. S. Suzuki, S. Okada and K. Nakao, J. Phys. Soc. Jpn. **69** (2000) 2615.
11. M. Caffarel and W. Krauth, Phys. Rev. Lett. **72** (1994) 1545.
12. A. Georges, G. Kotliar, W. Krauth and M. J. Rozenberg, Rev. Mod. Phys. **68** (1996) 13.
13. T. Chida, S. Suzuki and K. Nakao, submitted to J. Phys. Soc. Jpn.
14. L. Degiorgi, G. Grüner, P. Wachter, S.-M. Huang, J. Wiley, R. L. Whetten, R. B. Kaner, K. Holczer and F. Diederich, Phys. Rev. B **46** (1992) 11250.

Spin effects in HgSe in Megagauss fields

M. von Ortenberg[1], I. Stolpe[1], O. Portugall[1*], N. Puhlmann[1], H.-U. Mueller[1],
M. von Truchsess[2], C.R. Becker[2], A. Pfeuffer-Jeschke[2], G. Landwehr[2]

[1] Institute of Physics, Chair for Magnetotransport, Humboldt-University at Berlin, Invalidenstrasse 110, D-10115 Berlin, Germany
[2] Institute of Physics of the University of Wuerzburg, Chair EP III, Am Hubland, D-97974 Wuerzburg, Germany

Abstract   We demonstrate for HgSe epitaxial layers that transient magnetic fields in the megagauss regime prove as a new tool to investigate detailed information on spin-lattice relaxation using the delayed population change of the spin levels in the presence of rapidly varying magnetic fields up to 150 T.

1 Introduction

Mercuryselenide belongs to the II-VI-family of zero-gap semiconductors crystallizing in the zincblende structure [1]. Due to the high mobility of the electrons this material is an ideal matrix to study additional quantization effects in low dimensional structures and due to semimagnetic properties. The theoretical concept of the k^*p-model has been confirmed and the application also extended to the megagauss regime by a new parameter set emphasizing the far-off-band contributions [2]. Most magnetic field applications so far could be treated in the quasi-static approximation, since all internal population relaxations in the presence of a magnetic field variation were considered to be "immediate". This condition is, however, not fulfilled in pulsed magnetic megagauss fields having a pulse length of only some microseconds, so that the response of the carrier system in HgSe for these transient strong magnetic fields shows pronounced hysteresis effects.

2 Experiment

Using molecular beam epitaxy high purity HgSe-layers of 1 μm thickness were grown on (100) GaAs substrates having a 5 μm thick buffer layer of ZnTe. Different characterization methods revealed an electron concentration at T=4.2 K between some 10^{16} cm^{-3} (Hall effect and SdH) and $3*10^{15}$ cm^{-3} (cyclotron resonance). The Hall-mobility was $\mu=1.5*10^4$ cm^2/Vsec. Different samples of size 2x2 mm^2 of the same wafer were investigated by IR-megagauss transmission experiments in Faraday configuration in a temperature range between T=6 K and room temperature using the sophisticated equipment of the *Humboldt High Magnetic Field Center* [3].The total pulse length for up- and down-sweep of the magnetic field generated by the single-turn-coil

*   *Present address:* Laboratoire National des Champs Magnétiques Pulsés, F-31432 Toulouse, France

technique was 6 μsec. Due to the excellent signal/noise ratio even small transmission changes could be unambiguously detected. In Fig. 1 we have reproduced the data of the relative transmission for 117.2 meV-laser radiation as a function of the magnetic field B for the temperatures T=6, 207, 300 K. Up- and down-sweep are indicated by solid and broken curves, respectively. All spectra exhibit pronounced resonance lines. Whereas in the low-temperature measurements hysteresis effects in the intensity are rather negligible, at higher temperatures they dominate the results.

3 Results and Discussion

Due to the freeze out the spectrum for T=6 K in Fig. 1 is dominated by interband transitions "I" and the fundamental cyclotron resonance "CR1" at about B=32 T is essentially suppressed. All observed transitions can be modeled within an 8x8 k^*p-matrix with the parameter set of table 1 emphasizing the far-off-bands contribution by a large value for F [2]. The resultant Landau-levels for $k_z=0$ split into two spin states "a" and "b" and are plotted in Fig. 2.

With increasing temperature there is an increased population of quasi-free carrier states in both the conduction and valence bands. Due to the large hole

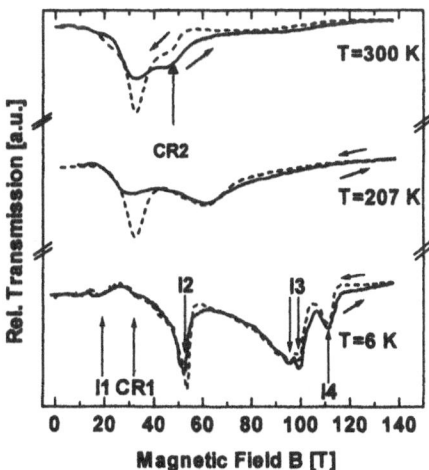

Fig. 1 Experimental data of magneto transmission on HgSe for 117.2 meV radiation.

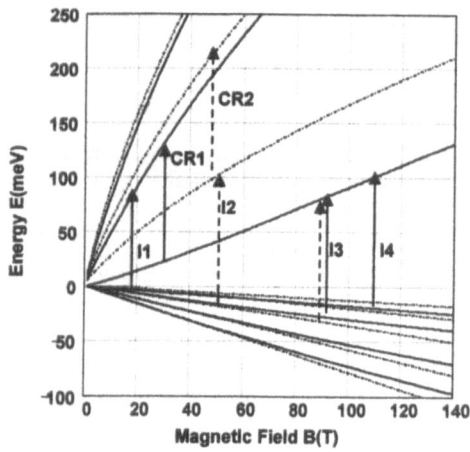

Fig. 2 the scheme of Landau levels of HgSe
The arrows indicate the observed transitions.
___ a-set   _._._ b-set

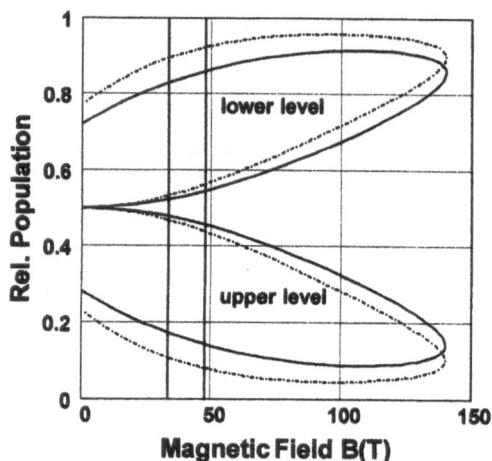

Fig. 3 The non-equilibrium distribution for both
initial spin-states of the cyclotron transitions
CR1 and CR2 indicated by the vertical bars at
32 T and 47.45 T, respectively.
____ T=300 K   _._._ T=207 K

masses the corresponding cyclotron resonance transitions are beyond the magnetic field range of the present experiments. The electron cyclotron resonance becomes dominant and obscures the absorption intensity contributed from the interband transitions. All resonance lines contributed to cyclotron resonance transitions exhibit pronounced hysteresis effects in the line intensity as can be seen in the spectra for up- and down-sweeps. The interband transition still observed for T=207 K at about 59 T in Fig. 1 is practically not affected. Similar hysteresis effects were observed by Arimoto et al. for single InAs/GaAs quantum wells and tentatively explained by a delayed population adjustment of the levels involved due to the finite spin-lattice relaxation time in pulsed transient magnetic fields [4]. We have confirmed this concept in our experiments and can definitely exclude any equipment related artifacts since we observe in one and the same spectrum both, intraband and interband transitions with pronounced and without any hysteresis, respectively. From the physical point of view this means that the momentum relaxation tin the carrier system is "immediate" compared to the pulse length of the magnetic field in contrast to the spin relaxation, so that for the same magnetic field value in up- and down-sweep quite different populations of the initial state are experienced. Applying a two-level system for spin states of the same energy equally populated at zero magnetic field, we are able to simulate the relative of the magnetic field pulse B(t) in connection with the theoretical results of the k^*p-model. The finite spin-

population as shown in Fig. 3. In the simulation we take into account of the actual experimental time dependence lattice relaxation time τ=1.2 μsec produces the hysteresis in the resulting non-equilibrium distribution functions by a corresponding convolution process as shown in Fig. 3 for T=300 K and T=207 K by the solid and broken curves, respectively. The vertical lines indicate the resonance fields for the two cyclotron-resonance positions. We like to emphasize that only the intraband transitions are sensitive to the finite spin-lattice relaxation due to the change in population of the initial state. For interband transitions the electron population of the valence band states as well as the "emptiness" of the final states are practically not affected, so that the corresponding transitions do not experience any hysteresis effect.

References

[1] M. von Truchsess, A. Pfeuffer-Jeschke, S. Einfeldt, C.R. Becker, E. Batke, Proc. ICPS24, Jerusalem 1998, ed. D. Gershoni, World Scientific (1999)

[2] I. Stolpe, O. Portugall, N. Puhlmann, H.-U. Mueller, M. von Ortenberg, M. von Truchsess, C.R. Becker, A. Pfeuffer-Jeschke, G. Landwehr, Proc. RHMF2000, Porto 2000, Physica B, in print

[3] O. Portugall, N. Puhlmann, H.-U. Mueller, M. Barczewski, I. Stolpe, M. von Ortenberg, J. Phys. D 32 (1999) 2354

[4] H. Arimoto, N. Miura, R.A. Stradling, Proc. "9th Intnl. Conf. on Narrow Gap Semiconductors", Berlin 1999, ed. N. Puhlmann, H.-U. Mueller, M. von Ortenberg, Magnetotransport in Solids, Humboldt-University Berlin (2000) p. 10

E_g meV	Δ meV	Δ_{hl} meV	γ_1	$-\gamma_2$	γ_3	P meV*m	$-\kappa$	F
273	387	3.5	1.5	0.5	0.4	$7.50*10^5$	1	1

Table 1 k^*p-parameters for HgSe in megagauss fields

Temperature dependence of effective band edge mass in n-InSb by magnetophonon resonance up to T=320 K and influence of nonparabolicity corrections

D. Schneider[1], S. Krull[1], C. Brink[1], G. Irmer[2], P. Verma[2]

[1] Institut für Technische Physik und Hochmagnetfeldanlage, Technische Universität Braunschweig, Mendelssohnstr. 2, D-38106 Braunschweig, Germany, e-mail: `d.schneider@tu-bs.de`

[2] Institut für Theoretische Physik, TU Bergakademie Freiberg, Bernhard-von-Cotta-Str. 4, D-09596 Freiberg, Germany

Abstract   We measured the magnetophonon (MP) resonance in the range from $T = 40$ to 320 K, by the transverse magnetoresistance, for the first time above 260 K. Raman scattering was used to determine precisely the frequencies of LO-phonons in that range. We extracted the fundamental resonance field B_0 by a full-curve-fit of the Barker formula to the experimental MP oscillations for each temperature. Proving different approximations for the nonparabolicity (np) we evaluate the band edge mass $m_0^*(T)$ in the range from $T = 40$ to 320 K. Its temperature change is in accordance with the expectation from $E_g(T)$, if we use the np-correction by Stradling and Wood (1970) and take into account the resonance scattering of electrons up to sufficiently high Landau levels. Additionally our result agrees with a theoretical curve, based on the three band model by Kane (1957).

1 Introduction

The compound semiconductor InSb is characterized by a small energy gap ($E_g = 0.24$ eV) and both a relatively large spin-orbit-interaction ($\Delta_0 = 0.8$ eV) and g-factor ($g^* = 50$). It has been used as a model system for studying the interaction of electrons (including spin), optical phonons, photons and magnetic fields [1]. However, still to be clarified is the correct relation for the strong nonparabolicity of the conduction band, responsible e.g. for the temperature dependence of effective electron mass $m^*(T)$. The band edge mass $m_0^*(T)$, determined by magneto phonon resonance (MPR) including nonparabolicity (np) corrections, shows only half of the change with temperature, as concluded from the optical energy gap [2], [3]. We study the temperature dependence of $m_0^*(T)$ by MPR, similar to the investigation in [4], [5]. We measured both magnetoresistance and Raman scattering and apply suitable relations from [2], [6], [7] and [8] for extracting and discussing the relevant physical parameter.

2 Results and Discussion

Our samples were prepared from an one side polished, (111)-oriented InSb wafer No. 36, cut with 0.5 mm in thickness from an undoped single crystal (No. ISD 005/un). It was obtained as a gift from WAFER TECHNOLOGY, Ltd, UK. Two samples, No. 36 M and 36 R, have been cut from the wafer with 7x2 mm^2 and 5x5 mm^2 in surface area, for magnetoresistance and Raman experiments, respectively. At 77 K an electron concentration of $n = 2.2 \cdot 10^{14}$ cm^{-3} and a mobility of $\mu = 4.0 \cdot 10^5$ cm^2/Vs

were determined for sample 36 M by Hall effect. The electrical current was directed parallel to the long side of the sample, oriented in [11-2] direction and the magnetic field was applied perpendicular to that axis. The sample M 36 was contacted by five 70 μm thick Pt wires, as described earlier [9], [10]. For measurements below 300 K we used a superconducting magnet system (SPECTROMAG 2000, Oxford Instruments) and above that temperature a water cooled Bitter magnet (see [10]).

The dependence of the longitudinal optical (LO) phonon frequencies on temperature between 77 and 400 K for sample 36 R was obtained by Raman scattering experiments. They were performed by employing the 514.5 nm wavelength of an argon-ion laser, the Jobin Yvon triple stage monochromator T64000, a CCD detector and usual electronics. The probing laser power of 4 mW and focal spot of about 100 μm was chosen in order to prevent laser induced heating of the sample. A good spectral resolution was obtained by reducing the slit width to 50 μm, the spectra were corrected for the slit function using a mathematical deconvolution. The calibration of the phonon frequencies was performed with plasma lines of the laser tube. Due to the back-scattering geometry used, photons with wavevectors $|\mathbf{q}_{ph}| = 2n$ $|\mathbf{q}_{laser}| \approx 10^6$ cm^{-1} near the centre of the first Brillouin zone were measured, n being the refractive index of InSb. The spectra for each temperature were recorded a few times. Fig. 1 gives the result, a fitted curve from theory and values used by [2], [3] for a linear approximation.

We measured the magnetoresistance $\rho(B)$ up to $B = 4$ T at 35 different temperatures between 49 and 321 K. The observed ratio $\Delta\rho/\rho_0 \equiv [\rho(B) - \rho(B=0)]/\rho(B=0)$ increases at that field range about a factor of 50 at 49 K and of 5 at 321 K. The normal part of the magnetoresistance $\rho_n(B)$ is superposed by MP oscillations $\rho_{osc}(B)$, i.e. $\rho = \rho_n + \rho_{osc}$. Those detectable oscillations are related to quantum numbers $N = 1$ to 5. The ratio $\Delta\rho_{osc}/\rho_0$ extends between the range of about 10^{-4} at 321 K and 0.2 at 110 K.

MP oscillations are periodic in $1/B$. The effective electron mass m^* is related to the observed period $1/B_0$ by $1/B_0 = e/m^* \omega_{LO}$. B_0 is the so called fundamental field [1]. It is extracted by a full curve fit of the first two terms of the Fourier expansion by Barker [6] to ρ_{osc}

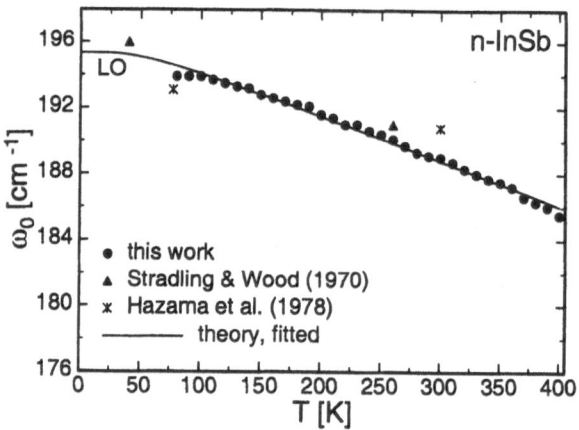

Fig. 1 Experimental values of LO phonon frequencies ω_{LO}.

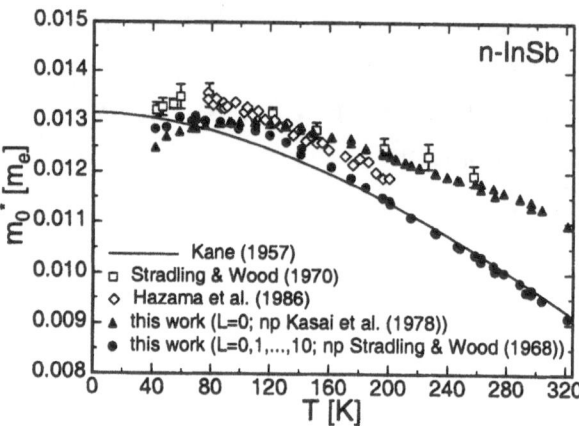

Fig. 2 Temperature dependence of effective electron mass m_0^* extracted from MPR experiments and Kane theory.

[4]. To obtain the value of the mass m_0^*, we applied non-parabolicity and polaron corrections.

We proceeded similar to [5] : **1.** Separation of ρ_{osc} directly by subtraction of ρ_n, which has been determined by a polynomial fit corresponding to the case of Shubnikov-de Haas oscillations [10]. **2.** Rescaling the magnetic field $B_{rec}(B, T_i, B_0) = f_{np}(B, T_i, B_0) \cdot B$ for each $T_i = \text{const.}$, starting with a first approximation for B_0 and iteration for f_{np} (cf. [11]). f_{np}^N is known as np-correction, valid for each discrete maximum $N = 1, 2, 3, \ldots$. We introduced a transition to a continuous correction function $f_{np}^N \to f_{np}(B)$ valid for the whole course of the curve $\rho_{osc}(B)$. **3.** Fit of Barker's formula [6] for MPR to the rescaled curve $\rho_{osc}(B_{resc})$ and determination of B_0 in second approximation. **4.** Iterative procedure of steps (2) and (3) up to convergence of f_{np}. Fig. 1 in [5] gives an example for that procedure. **5.** Applying of the polaron correction, taken from [12]. **6.** For determination of m_0^* the LO phonon frequency from Fig. 1 has been applied.

For comparison, we used two different np-corrections. The first one has been introduced by [12] (equ.(13)). We considered the occupation of the Landau levels L by the Boltzmann expression and took into account the scattering from $L = 0, 1, 2$ up to $L = 10$ to higher levels. The second one is based on [7] (equ.A4). In that approximation only the scattering from the lowest Landau level $L = 0$ is included.

Our findings are displayed in Fig. 2 together with earlier results [2], [3]. They deviate partly by the amount and the slope. The reasons may be : **1.** The experimental method of MPR detection as discussed in [11]. We observed the magnetoresistance directly and determined B_0 for the first time by a selfconsistent full curve fit of theory including nonparabolicity, deviating from the method of derivatives and restriction to a few extrema [2], [3]. **2.** We applied the measured $\omega_0(T)$ dependence of our sample, contrary to deviating approximations (cf. Fig. 1). **3.** We took into account the transition N of electrons starting from level $L = 0$ up to 10, compared to the only one ($L = 0$) in [7]. In [3]

L is not specified. Fig. 2 shows clearly that the np correction restricted to small numbers of L (minimum $L = 0$), just as used by [7], flats the slope of $m_0^*(T)$ above 100 K and makes it steeper to lower temperatures. Our result is in accordance with the calculated change of 22 % in m_0^* for $T = 40$ to 260 K from the optical band gap by [2]. Additionally it agrees well with a calculated curve, based on the three band model by Kane [8] if we use the parameter $\Delta_0 = 0.9$ eV, $P^2 = 23.95$ eV and $E_g(T) = 0.2352 - [0.6T^2/(T + 500)]$ eV.

More details of our investigation will be presented in a forthcoming paper. The gift of the sample by WAFER TECHNOLOGY, Ltd, UK is gratefully acknowledged.

References

1. Yu. A. Firsov, V. L. Gurevich, and R. V. Parfeniev, in G. Landwehr, E. I. Rashba (Eds.), *Landau Level Spectroscopy* (Elsevier, Amsterdam 1991) pp. 1182-1302.

2. R. A. Stradling and R. A. Wood : J. Phys. C: Solid State Phys. **3**, (1970) L 94.

3. H. Hazama, T. Sugimase, T. Imachi, and Ch. Hamaguchi, J. Phys. Soc. Japan **55**, (1986) 1282.

4. D. Schneider, K. Fricke, J. Schulz, G. Irmer, and M. Wenzel, *Proc. 23rd Int. 'l Conf. Phys. Semicond.*, ed. M. Scheffler and R. Zimmermann (World Scientific, Singapore 1996) p. 221.

5. D. Schneider, Ch. Keilhack, S. Krull, and O. P. Hansen, *Proc. 24th Int. 'l Conf. Phys. Semicond.*, ed. D. Gershoni (World Scientific, Singapore 1999) ISBN: 981-02-4030-9(CD).

6. J. R. Barker, J. Phys. **C5**, (1972) 1657.

7. K. Kasai, T. Shirakawa, and Ch. Hamaguchi, J. Phys. Soc. Japan **44** (1978) 216.

8. E. O. Kane, J. Phys. Chem. Sol. **1**, (1957) 249.

9. D. Schneider, C. Brink, G. Irmer, and P. Verma, Physica B **256** (1998) 625.

10. D. Schneider, B. Himstedt, A. Schlachetzki, and G.-P. Tang, J. Appl. Phys. **85**, (1999) 6542.

11. D. Schneider, D. Rürup, B. Schönfelder, and A. Schlachetzki, Z. Phys. B **100** (1996) 33.

12. R. A. Stradling and R. A. Wood J. Phys. C: Solid State Phys. **1**, (1968) 1711.

Self-energy correction to the mass of the wide gap semiconductors

Mitsutake Oshikiri[1], Ferdi Aryasetiawan[2]

[1] National Research Institute for Metals, 3-13 Sakura, Tsukuba, Ibaraki 305-0003, Japan
[2] Joint Research Center for Atom Technology, 1-1-4 Higashi, Tsukuba, Ibaraki 305-0046, Japan

Abstract The effective mass of wide gap semiconductors has been calculated within the GW approximation using the linear muffin tin orbital basis. The quasiparticle effective mass is compared with the effective mass obtained by the conventional local density approximation and with experiment. The mass obtained by the conventional local density approximation is corrected by the energy and k-dependence of the self-energy.

1 Introduction

Most ab initio calculations on band structure so far have been based on the density functional theory (DFT) [1, 2] within the local density approximation (LDA) [1,2]. They are intended to describe the ground state and cannot describe the excited states in principle except for the highest occupied state. In the conventional DFT-LDA, the important properties of non-locality and energy dependence in the exchange-correlation potential are neglected and the self-interaction problem for the occupied states exists. These defects bring the smaller band gap than experiment in semiconductors and sometimes give wrong effective mass. The mass obtained by the DFT-LDA is not in general the mass of the quasiparticle (QP) formed by the dynamical many-body interaction.

This work is intended to correct the LDA mass of the conduction band by the GW approximation (GWA) [3, 4] methods in wide gap semiconductors (AlN, GaN and ZnO) which are on the focus of the recent technological interest.

The GWA is known as a relatively simple way to remove these defects in the conventional DFT-LDA since the self-energy (SE) correction is taken into account. It is originally derived from a many-body perturbation theory based on Green's function and can take into account both the non-locality and dynamic (i.e. energy dependent) feature of correlation in many-body system within random phase approximation (RPA) [5,6]. The self-interaction problem is also removed. But electron-phonon coupling is not included.

2 Theory and Computational Procedure

The theoretical framework is as follows. The standard linear muffin tin orbital (LMTO) basis [7,8] with the atomic sphere approximation (ASA) [7,8] is used to obtain the LDA eigen values and wavefunctions by solving the equation $[E_j^{lda} - H_0]\varphi_j - V_{xc}^{lda}\varphi_j = 0$, which are employed to make an initial Green function for the GWA scheme. The H_0 includes the kinetic energy operator, potential due to the ions and the Hartree potential

of the electrons. V_{xc}^{lda} is the exchange correlation potential within the LDA. j denotes the band index. The 3d orbitals of Ga and Zn are treated as valence states explicitly in this study.

The polarization function P is computed with the Green function which is constructed from the LDA eigen energies and functions, $P = -iGG$, where $G = G_0(E_j^{lda}, \varphi_j^{lda})$. The dynamically screened potential is calculated within the RPA using the dielectric function.

$$W(r, r'; \omega) = \int d^3r'' \epsilon^{-1}(r, r''; \omega) v(|r'' - r'|), \quad (1)$$

where $\epsilon^{-1}(r, r''; \omega)$ is an inverse dielectric matrix obtained from $\epsilon = 1 - vP$ and $v(|r'' - r'|)$ is a bare Coulomb potential $1/(|r'' - r'|)$. The SE is obtained by the convolution of the Green function and the screened potential.

$$\Sigma(r, r'; \omega) = i \int \frac{d\omega'}{2\pi} G(r, r'; \omega + \omega') W(r, r'; \omega') e^{i\delta\omega'}, \quad (2)$$

where $G(r, r'; \omega + \omega')$ is the full Green function and $\delta = 0^+$. The many-body exchange and relatively long-range correlation corrections (i.e. screening) are taken into account by this nonlocal and energy-dependent SE operator. The LDA eigen energy is corrected by the obtained SE and the QP energy structure is obtained with arbitrary k point as follows;

$$E_j^{qp} = E_j^{lda} + \langle j|\Sigma - V_{xc}^{lda}|j\rangle \quad (3)$$

Finally, the next equation should be solved (see Ref. [9] in English and Ref. [10] in Japanese).

$$[E_j^{qp} - H_0]\varphi_j(r; \omega) - \int dr' \Sigma(r, r'; \omega)\varphi_j(r'; \omega) = 0 \quad (4)$$

The procedure of the effective mass correction is as follows. The effective mass of the quasiparticle is defined as the next equation [11].

$$\frac{dE_j^{qp}}{dE_j^{lda}} = \frac{m_j^{lda}}{m_j^{qp}}, \quad (5)$$

where $E_j^{qp} = E_j^{lda} + \Delta \, \mathrm{Re} \, \Sigma(\vec{k}, E_j^{qp}) \quad (6)$

After differentiating the eq. (5), the effective mass at around Γ point of the QP in the k_α direction is obtained as follows;

$$\frac{m^{lda}_{j,\vec{k_\alpha}}}{m^{qp}_{j,\vec{k_\alpha}}} = \lim_{|\vec{k_\alpha}|\to 0} \frac{1 + \frac{m^{lda}_j}{\hbar^2 |\vec{k_\alpha}|}\frac{\partial\Delta\,\mathrm{Re}\,\Sigma(\vec{k},\omega)}{\partial \vec{k}}|_{\vec{k}=\vec{k_\alpha}}}{1 - \frac{\partial\Delta\,\mathrm{Re}\,\Sigma(\vec{k},\omega)}{\partial\omega}|_{\omega=E^{qp}_j}} \qquad (7)$$

Therefore, the LDA effective mass is corrected by both the energy dependence (the denominator of eq.(7)) and the k-dependence (the numerator of eq.(7)) of the SE. The convergence of the energy dependence of the SE is much better than that of the k-dependence. We define the part of the energy dependence as $Zfac$;

$$Zfac = \left[1 - \frac{\partial\Delta\,\mathrm{Re}\,\Sigma(\vec{k},\omega)}{\partial\omega}|_{\omega=E^{qp}_j}\right]^{-1} \qquad (8)$$

The experimental lattice constants (Table. 1 [12]) are used for more meaningful comparison with experimental data and theoretical results. In this work, the spin-orbit interaction is not included and the spin direction is not distinguished.

Table 1 Experimental lattice constant [12] (in Å) of the compound semiconductors in the wurtzite structure.

Compound	a axis	c axis	u parameter
AlN	3.110	4.980	0.382
GaN	3.190	5.189	0.377
ZnO	3.24961	5.20653	0.345

3 Results

For simplification, we ignore the cross term of the effective mass such as $m_{k_x k_y}$. In this paper, the k_x is defined as the direction from Γ to M point in the k space of the 1st. Brillouin zone. The k_z is the direction from Γ to A point. The k_x, k_y and k_z form the Cartesian coordinate but the k_y is identical to the direction from Γ to K point due to the wurtzite symmetry. Since it is not easy to obtain the k-dependence of the SE with high accuracy due to the very slow SE convergence, we limit the k-dependence calculation to the AlN system which is the simplest system among the three.

3.1 Mass enhancement by the energy dependence of SE

The $Zfac$ which leads to the mass enhancement by the energy dependence of SE and the corrected mass are summarized in the Table.2. The LDA mass of the conduction band of AlN,GaN and ZnO are enhanced by 16,19 and 17 %, respectively. We do not mention the detail of the valence band here since it is sometimes sensitive to the choice of the empty sphere in the LMTO-ASA calculation and the spin orbit interaction is not taken into account at present, however, we wish to mention that there is a tendency that the effective mass of the highest valence band is enhanced by the energy dependence of the SE by larger amount than that of the conduction band by about a few or several %.

3.2 Mass enhancement by the k dependence of SE

The enhancement factors of the AlN system are about 6 %,6 % and < 21 % in the k_x,k_y and k_z directions, re-

Table 2 Conduction band effective mass corrected by the energy dependence of the self-energy. ( ) means the LDA mass without correction. The unit is the electron rest mass m_0

Compound	$Zfac$	m_{kx}	m_{ky}	m_{kz}
AlN	0.859	0.37(0.32)	0.37(0.32)	0.38(0.32)
GaN	0.843	0.21(0.17)	0.21(0.17)	0.19(0.16)
ZnO	0.855	0.24(0.21)	0.24(0.21)	0.21(0.18)

spectively. Consequently, the mass of the quasiparticle of the conduction band of AlN has been obtained to be $m^{qp}_{k_x} = 0.39$, $m^{qp}_{k_y} = 0.39$ and $m^{qp}_{k_z} < 0.46$. (The unit is the electron rest mass m_0.) In the k_z direction, we have not yet found the complete convergence even with the 432 k points for the SE calculation. The experimental data for AlN could not be found. On the other hand, the experimental effective masses of GaN and ZnO are reported to be about 0.20 [13] and 0.23 [13] (, which correpond to m_{kx} or m_{ky},) after subtracting the electron-phonon coupling, respectively.

4 Conclusion

The conduction band effective masses of AlN, GaN and ZnO have been calculated by correcting the LDA mass with self-energy obtained by GWA. The LDA mass has the tendency that the mass is often smaller than experiment, however, we have found the self-energy correction of the energy dependence by GWA can reduce the discrepancy from the experiment.

References

1. P. Hohenberg and W. Kohn, Phys. Rev. **136,** (1964) B864.
2. W. Kohn and L. J. Sham, Phys. Rev. **140,** (1965) A1133.
3. L. Hedin and S. Lundqvist, Solid State Physics **23,** (1969) 1.
4. M. S. Hybertsen and S. G. Louie, Phys. Rev. B **34,** (1986) 5390
5. R. W. Godby, M. Schlüter and L. J. Sham: Phys. Rev. B **37,** (1988) 10159.
6. F. Aryasetiawan, Phys. Rev. B **46,** (1992) 13051.
7. O. K. Andersen, Phys. Rev. B **12,** (1975) 3060.
8. O. K. Andersen and O. Jepsen, Phys. Rev. Lett. **53,** (1984) 2571.
9. F. Aryasetiawan and O. Gunnarsson, Rep. Prog. Phys. **61,** (1998) 271., M. Oshikiri and F. Aryasetiawan, J. Phys. Soc. Jpn. **69,** (2000) 2113.
10. M. Oshikiri and F. Aryasetiawan, BUTSURI **55,** (2000) 117.
11. G. D. Mahan, *Many-particle physics (2nd ed.)* (Plenum Press, 1993) p.157.
12. (AlN/GaN):H. Shulz and K. H. Thiemann, Solid State Com. **23,** (1977) 815.,(ZnO):F. S. Galasso, *Structure and properties of inorganic solids* (Pergamon Press, 1970) p.123.
13. (GaN):M. Drechsler, D. M. Hofmann, B. K. Meyer, T. Detchprohm, H. Amano, I. Akasaki: Jpn. J. Appl. Phys. 2, Lett. (Japan), **34,** (1995) L1178-9., (ZnO):M. Oshikiri, Y. Imanaka, F. Aryasetiawan and G. Kido, Physica B to be published.

First-principles study of electronic structure of $Si_{1-x}Ge_x$ and $Si_{1-x-y}Ge_xC_y$ disordered alloys

M. Ohfuti, Y. Sugiyama, Y. Awano, and N. Yokoyama

Fujitsu Laboratories Ltd., 10-1 Morinosato-Wakamiya, Atsugi 243-0197, Japan

Abstract We first develop a scheme for dealing with disordered alloys from first principles. We make a unit cell large enough that we can find arrangements that do not have extra symmetry other than primitive translations and then examine every such arrangement for the given unit cell. The scheme is shown to reproduce the band gap measured experimentally in relaxed $Si_{1-x}Ge_x$. The large reductions from the band gap derived by linear interpolation arise not at Δ but at X or L. When we use the scheme to predict the electronic structure of Si-lattice-matched $Si_{1-x-y}Ge_xC_y$, we find that the band gap for the alloy with 4.7% C is 0.6 eV and that the state at the bottom of the conduction bands has the s-orbital components of C atoms that lead to strong optical transition.

1 Introduction

$Si_{1-x-y}Ge_xC_y$ has attracted much attention, mainly because band engineering is possible with lattice matching to Si. Experiments with Si-lattice-matched $Si_{1-x-y}Ge_xC_y$ having C concentrations up to 1% have shown that the band gap can be reduced by as much as $7y$ eV [1]. We have reported that the reduction derived by calculations from first principles for 3.1% C agrees with that derived from experimental data [2]. The reduction calculated for 6.3% C, is much larger than that calculated for 3.1% C and different arrangements of C produce very different band gaps [3]. These results were obtained for ordered alloys, but the arrangements of $Si_{1-x-y}Ge_xC_y$ alloys can be disordered. We therefore need to know the electronic properties of a disordered alloy. We supposed that this problem is similar to one with regard to $Si_{1-x}Ge_x$ alloys, which are believed to have disordered arrangements. The band gap in relaxed $Si_{1-x}Ge_x$ has been evaluated experimentally by measuring photoluminescence (PL) [4], and the band gap reduction from the result of linear interpolation between the band gaps of pure Si and pure Ge is very large. The reduction calculated using the coherent potential approximation [5], however, is much smaller than that obtained experimentally and the bottom of the conduction bands remains at Δ for Ge concentrations up to 80% . Before dealing with $Si_{1-x-y}Ge_xC_y$, we need to be able to reproduce the experimental band gap of $Si_{1-x}Ge_x$.

2 First-principles method

Our first-principles method uses a typical local density functional scheme with norm-conserving pseudopotentials. The cutoff energy for the plane-wave expansion is 540 eV (40 Ry) for the convergence of the calculated results. The total energy is minimized with respect to both the energy-band structure and the atomic positions. The

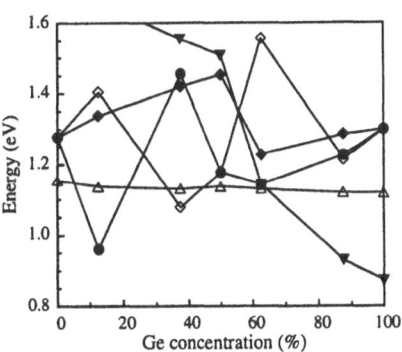

Fig. 1 Conduction-band energies for relaxed $Si_{1-x}Ge_x$ with an 8-atom unit cell and the point-group T_d. They are relative to top of the valence bands. Open and closed triangles indicate the energies at Δ and L, respectively. Both circles and squares indicate the energies at X.

density functional scheme is well-known to provide a band gap smaller than would be obtained experimentally, so a correction factor is needed when calculated results are compared with experimental results. The correction factor for alloys is defined as the weighted average of the correction factors for pure crystals of Si, Ge, and 3C-SiC.

3 Results for relaxed $Si_{1-x}Ge_x$

Conduction-band energies at symmetry points calculated for relaxed $Si_{1-x}Ge_x$ by using an 8-atom unit cell with point-group T_d are shown in Fig. 1. We refer to the symmetry points by the corresponding symbols of pure Si and Ge crystals. For pure Si and Ge, the energies at X are six-fold degenerate, but for other concentrations are split into a wide energy range and cause large reductions of the band gap. For 12.5% Ge, the Ge atoms are found to have only the s-orbital components at the bottom of the conduction bands. This is caused by the high symmetry of the crystal. The s-orbital level of the Ge atom is much lower than in the Si atom, and this biased s-orbital component seems to lead to the large reduction of the band gap.

Figure 2 shows the band gaps calculated for every independent arrangement of a 16-atom unit cell. They are scattered over a wide range of energies. Some of them lie lower than the experimental data. When we examine the number of symmetry operations of arrangements, the band gap distribution becomes narrower as the number is decreased. This shows extra symmetry causes a greater reduction of the band gap with biased s-orbital components. Because an actual disordered alloy

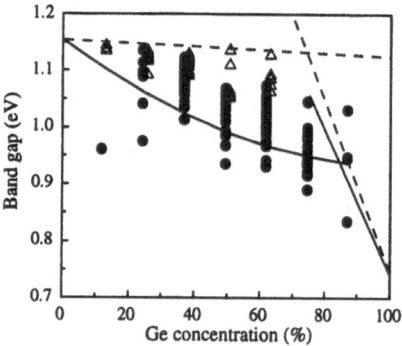

Fig. 2 Band gaps calculated using a 16-atom unit cell. Triangles and circles indicate that the bottom of the conduction bands lies at Δ and X, respectively. Broken lines represent the data for Δ and L interpolated between Si and Ge. Solid curve is for experimental data.

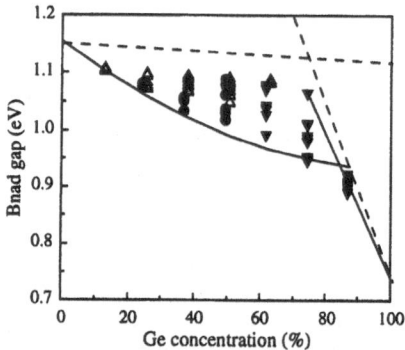

Fig. 3 Band gaps for arrangements without extra symmetry calculated by using a 32-atom unit cell. Open and closed triangles indicate the bottom of the conduction bands at Δ and L, respectively. Circles indicate that at X.

has no symmetry, we remove the data for crystals with extra symmetry other than the primitive translations. The distribution of the band gaps narrow enough that the band gaps lie between the values obtained by linear interpolation and the values obtained from experimental data. For low and high concentrations of Ge, all the crystals have extra symmetry. This is because the unit cell is too small. In our scheme, we therefore make a unit cell large enough that we can find arrangements that do not have extra symmetry. Then we examine every such arrangement for the given unit cell.

We can find arrangements without extra symmetry for each x when we use a 32-atom unit cell. Every such arrangement was examined and the band gaps are shown in Fig. 3. The smallest band gap agrees with the experimental PL data for the whole range of concentrations. For Ge concentration higher than 25%, the smallest band gap lies at X or L. Point L appears at Ge concentration lower than 80% and gives the largest reduction. The band gaps at Δ are larger and distributed across a narrower range, and the maximum width of the distribution is at 50%.

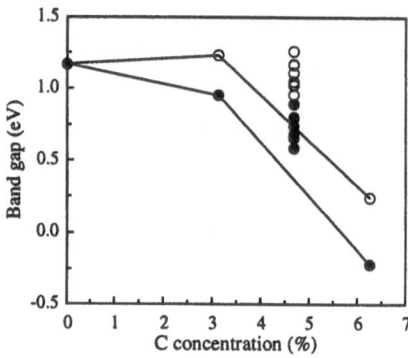

Fig. 4 Band gaps in $Si_{1-x-y}Ge_xC_y$ (closed circles) and $Si_{1-y}C_y$ (open circles). Data points connected by line segments were obtained using the smallest possible unit cell.

4 Results for $Si_{1-x-y}Ge_xC_y$

We use the scheme to predict the electronic structure of $Si_{1-x-y}Ge_xC_y$ with 4.7% C. We examined every independent arrangement of $Si_{1-y}C_y$ that does not have extra symmetry by using a unit cell of 64 atoms. The calculated band gaps (Fig. 4) are more widely distributed than those for $Si_{1-x}Ge_x$. This is because the s-orbital level of C is lower than the s-orbital level of Ge. The smallest band gap of $Si_{1-y}C_y$ with 4.7% C is 0.2 eV above the result of linear interpolation between the band gaps for 3% C and 6% C obtained by using the smallest possible unit cell. The band gap in $Si_{1-x-y}Ge_xC_y$ is predicted to be 0.6 eV because we have found that the effect of C on the band gap of $Si_{1-x-y}Ge_xC_y$ can be separated from that of Ge. The state at the bottom of the conduction bands having the s-orbital components of C leads to strong optical transition. All of the crystals for 4.7% C have the bottom of the conduction bands including the C s-orbital components. Therefore the disordered alloy can have the same optical properties. This is very important with regard to practical applications.

5 Summary

We have developed a scheme for analyzing disordered alloys from first principles and have shown that it reproduces the band gap measured experimentally in relaxed $Si_{1-x}Ge_x$. We have also used it to predict the electronic structure of Si-lattice-matched $Si_{1-x-y}Ge_xC_y$ and have calculated that the band gap for 4.7% C is 0.6 eV; the state at the bottom of the conduction bands has the s-orbital components of C. These results show that $Si_{1-x-y}Ge_xC_y$ is potentially very useful for electrical and optical applications.

References

1. O. G. Schmidt and K. Eberl, Phys. Rev. Lett. **80**, (1998) 3396.
2. M. Ohfuti et al., Phys. Rev. B **60**, (1999) 15515.
3. M. Ohfuti et al., Phys. Rev. B **60**, (1999) 13547.
4. J. Weber and M. I. Alonso, Phys. Rev. B **40**, (1989) 5683.
5. S. Krishnamurthy et al., Phys. Rev. B **33**, (1986) 1026.

Electronic structures of the C_{82} and La@C_{82} crystals by the relativistic LCAO method

S. Amamiya, S. Okada, S. Suzuki, K. Nakao

Institute of Materials Science, University of Tsukuba, Tsukuba 305-8573, Japan e-mail: **ama@ims.tsukuba.ac.jp**

Abstract We report on electronic structures of C_{82} and La@C_{82} crystals by using the scalar relativistic full-potential linear-combination-of-atomic-orbitals method based on the density functional theory. For the La@C_{82} crystal, we only find a metallic solution with half-filled energy band at the Fermi level. The band width around the Fermi level in La@C_{82} crystal is about 0.3 eV which is comparable to that of the fcc C_{60}. The energy bands of empty C_{82} cages are significantly modulated by encapsulating the La atom into its cage. This modulation is mainly due to the hybridization of La 5d and C π states.

1 Introduction

Encapsulation of metal atoms inside fullerene cages, endohedral metallofullerenes, is one of the most exciting topics in the field of fullerene science. It has been attracted special interest due to the possibility of designing the new materials with novel properties which are not expected in hollow fullerenes, such as superconductivity, ferroelectrics, and nonlinear optical response. In particular, La@C_{82} has been extensively studied as a typical monometallofullerene ever since the first success in extraction [1]. Electron spin resonance studies have revealed that there is two dominant signals corresponding to two La@C_{82} isomers [2]. Furthermore, the recent x-ray diffraction experiments on La@C_{82} powder, i.e., combination of the Rietveld analysis and the maximum entropy method, have revealed that one of the extracted La@C_{82} has C_{2v} geometry [3]. In addition to the isolated La@C_{82} molecule, a solvent-free single crystal of La@C_{82} has been also synthesized [4]. The x-ray diffraction experiments have revealed that the crystalline La@C_{82} has the face-centered cubic lattice structure at room temperature. However, the details of the electronic structures are still unknown.

In the present study, we investigate the electronic structures of C_{82} and La@C_{82} crystals possessing C_{2v} cage symmetry by the scalar relativistic density-functional theory.

2 Calculation methods

To study the geometry and the electronic structure of La@C_{82}, we employ the scalar relativistic full-potential linear-combination-of-atomic-orbitals method which is based on the density functional theory with the local-density approximation [5]. We use not only the atomic orbital of neutral atoms but also those of charged atoms to consider the variational flexibility. The atomic orbitals used for C atoms are 1s, 2s, and 2p atomic orbitals of neutral C atoms and 2s and 2p atomic orbitals of C^{2+}

atoms. Also, the atomic orbitals used for a La atom are 1s, 2s, 2p, 3s, 3p, 3d, 4s, 4p, 4d, 4f, 5s, 5p, 5d, and 6s atomic orbitals of a neutral La atom, 5d and 6s atomic orbitals of a La^{2+} atom, and 6p atomic orbital of La^+ and La^{3+} atoms. Exchange-correlation energy of interacting electrons is considered with functional form fitted to the Ceperley-Alder results [6]. We use 2064 and 4128 points per C and La atoms, respectively, to perform the three-dimensional numerical integration in the real space. Also, one k-point, i.e., Γ-point, is used for the integration in first Brillouin zone. The geometry of La@C_{82} molecule is optimized by the conjugate-gradient minimization scheme. In the optimized geometries, forces acting on the atoms are less than 0.5 eV/Å.

3 Results and Discussion

We optimize the geometry of the La@C_{82} molecule without any restriction. We show the optimized structure and the total valence charge density of the La@C_{82} molecule in Fig. 1 (a) and (b), respectively. We find that the La atom is below the center of the hexagon and on the C_2-axis of the C_{82} cage. The distance between the La atom and the nearest neighboring C atom on the hexagon is 2.58 Å. This distance is too large to generate an interatomic covalent bond. In fact, by analyzing the charge distribution of the La@C_{82} molecule, the charge between the La and C atoms is considerably smaller than that on the C-C bonds (Fig. 1). Hence, the strong covalent bond between La and C is unlikely to take place and the La

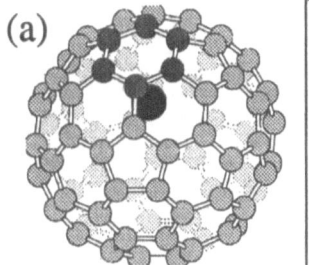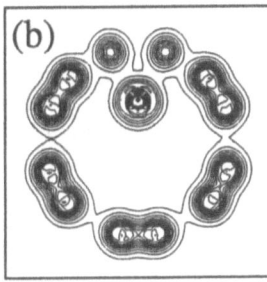

Fig. 1 (a) The structure of the La@C_{82} molecule with C_{2v} cage symmetry. The C and La atoms are denoted by light and dark shaded circles, respectively. The middle shaded circles denote the nearest neighbor C atoms on the hexagon. (b) Contour plot of the total valence charge density of the La@C_{82} molecule. The contour lines are drawn from 0.0 eÅ^{-3} with 0.16 eÅ^{-3} intervals.

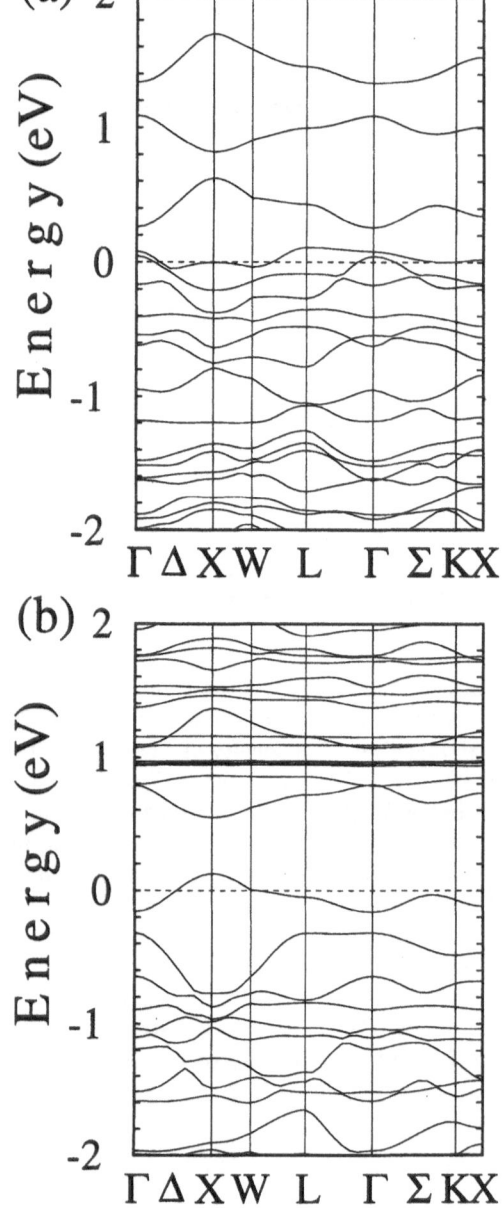

Fig. 2 Band structure of (a) C_{82} and (b) La@C_{82} crystals. Energies measured from the Fermi level energy, so that the dotted line denotes the Fermi level.

atom may freely move in the cage at room temperature. The calculated reaction energy for La@C_{82} is -8.4 eV. The reaction energy is defined by the difference between the total energy of La@C_{82} and the sum of the total energies of an isolated La atom and a C_{82} molecule. This result shows that the reaction La + C_{82} → La@C_{82} is exothermic. The electric dipole moment of the La@C_{82} molecule obtained is 2.2 D per molecule which is almost the same as that of H_2O (1.9 D). It is most likely that the electric dipole moment plays an important role to determine the arrangement of La@C_{82} molecules in crystalline

form. The La@C_{82} molecule is found to have the open shell electronic structure and the highest occupied state is singly occupied. The sum of spin density of the C_{82} cage is about 1.0 per molecule. The polarized electrons are mostly distributed on the carbon cages and extended over the whole network of the cage.

We show the electronic structure of C_{82} and La@C_{82} crystals in Fig. 2. In this calculation, we assume that the C_{82} and La@C_{82} molecules have the fcc arrangement with experimentally observed lattice parameters for the La@C_{82} crystal [4], in which the C_2 symmetry axis of the molecules is parallel to the ⟨111⟩ direction. In the La@C_{82} crystal, the electron spin is not polarized : we only find a metallic solution with a half-filled band at the Fermi level. The band width around the Fermi level is about 0.3 eV which is comparable to that of the fcc C_{60} [7]. The energy bands corresponding to the $4f$, $5d$, and $6s$ orbitals of the La atom are located at about 1 eV above the Fermi level. These energy bands exhibit extremely small dispersion in the whole Brillouin zone. The energy bands of empty C_{82} cages are significantly modulated by encapsulating the La atom into its cage. This modulation is mainly due to the hybridization of La $5d$ and C π states. Especially, the bands related to the C atoms surrounding La atom are considerably modulated by the hybridization.

Acknowledgement

Part of this study is supported by the Grant-in-Aid for Scientific Research from the Ministry of Education, Science, Sports and Culture of Japan and the grant from Research and Development Applying Computational Science and Technology of Japan Science and Technology Corporation (ACT-JST).

References

1. Y. Chai, T. Guo, C. Jin, R. E. Haufler, L. P. F. Chibante, J. Fure, L. Wang, J. M. Alford, R. E. Smalley, J. Phys. Chem. **95** (1991) 7564.

2. S. Suzuki, S. Kawata, H. Shiromaru, K. Yamauchi, K. Kikuchi, T. Kato, Y. Achiba, J. Phys. Chem. **96** (1992) 7159.

3. E. Nishibori, M. Takata, M. Sakata, M. Hasegawa, H. Shinohara, Meeting Abst. Phys. Soc. Jpn. **53** (1998) 319.

4. T. Watanuki, A. Fujiwara, K. Ishii, T. Shibata, H. Suematsu, H. Nakao, Y. Fujii, H. Kawada, Y. Murakami, K. Kikuchi, Y. Achiba, Y. Maniwa, Photon Factory Activity Report **14** (1996) 403.

5. S. Suzuki, K. Nakao, J. Phys. Soc. Jpn. **69** (2000) 532.

6. D. M. Ceperley, B. J. Alder, Phys. Rev. Lett. **45** (1980) 566.

7. S. Saito, A. Oshiyama, Phys. Rev. Lett. **66** (1991) 2637.

New determination of the camel's back in AlAs

Laura E. Bremme[1], H. Im[1], H. Choi[1], P. C. Klipstein[1], R. Grey[2], G. Hill[2]

[1] Clarendon Laboratory, Department of Physics, University of Oxford, Parks Road, Oxford, OX1 3PU, U.K.
Fax: +44-1865-272400; Email: l.bremme1@physics.ox.ac.uk
[2] III-V Facility, Department of Electrical Engineering, University of Sheffield, Mappin Street, Sheffield, S1 3JD, U.K.

Abstract We present the first direct determination of the X-point *k.p*-parameters and camel's back in AlAs. Using a new transfer matrix approach, we calculate the energy spectra of electrically or magnetically confined AlAs systems, as a function of the *k.p*-parameters. Fitting the parameters to a range of experimental data, we find a unique combination of $R = 3.39\ eV\mathring{A}$ and $m'_z = 0.16m_0$, which explains all experimental results. We have strong evidence for the existence of a camel's back in AlAs and present its new shape, which has a depth of ~8.5meV and a minimum 4.3% of the Brillouin zone away from the X-point.

1 Introduction

Over more than 20 years a considerable amount of work was published on the dispersion of the AlAs conduction band edge, near the X-point of the Brillouin zone [1,2]. The *k.p*-interaction between neighbouring X_1 and X_3 bands leads to a camel's back. Different values have been reported for the effective mass at the base of the camel's back, ranging between $1.1m_0$ and $10.7m_0$ [3,4]. To date, no general agreement has been found, and the exact strength of the interaction leading to this camel's back has never been determined properly. To the best of our knowledge, Bimberg [2] presented the only empirical determination of the *k.p*-interaction parameter, $R{\sim}1eV\mathring{A}$, and the effective mass of the unmixed X_1 and X_3 bands, $m'_z =1.56m_0$. Im *et al.* [5] adjusted these two *k.p*-parameters - which fully determine the camel's back dispersion - to fit their tunneling results on X_X-X_Y mixing. They claimed: $R{\sim}2.5eV\mathring{A}$ and $m'_z{\sim}0.3m_0$, already realising that $R=1eV\mathring{A}$ greatly underestimates the true value.

To properly deal with the camel's back in an AlAs *heterostructure* or a *quantizing magnetic field*, R and m'_z must be determined reliably, for which there is clearly a great need. Also, a treatment of quantum confinement, incorporating the highly non-parabolic dispersion, has to replace the present effective mass approach, almost universally used in heterostructures, where confinement energies are usually comparable with the camel's back depth. A parabolic effective mass treatment cannot be valid.

2 Results

We present a proper determination of the AlAs bulk *k.p*-parameters and an accurate determination of the camel's back, using AlAs heterostructures. We employ a two-band Hamiltonian to calculate the energy spectra in confined heterostructure systems and in magnetic fields, using a new transfer matrix approach [6]. We take the *k.p*-interaction into account that has often been neglected in the past, but which proves to be crucial. For a whole set of experimental data, we find R as a function of m'_z that will give a good fit to the observed results with the same degree of certainty.

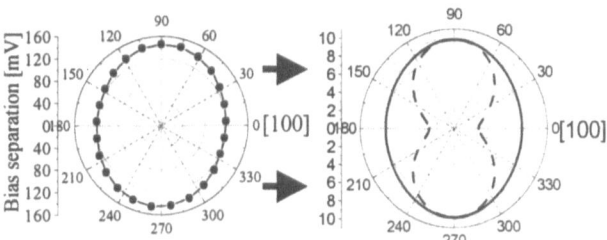

Fig. 1: Mapping of $X_{XY}(1)$-dispersion in AlAs QW, using resonant magneto-tunneling at 9 *kbar*: data and best fit

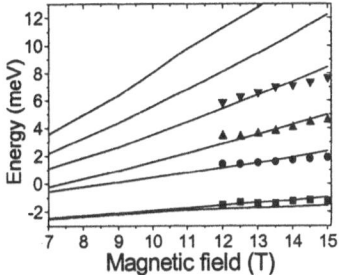

Fig. 2: Fan of $X_{X,Y}$ Landau levels in GaAs/AlAs double barrier structure: experimental data (symbols) and best fit

2.1 Direct Measurement of AlAs Fermi surface

We determine the R- m'_z -curve for a mapping of the $X_{XY}(1)$-dispersion in a 60 or 70Å AlAs QW, measured by magneto-tunneling at ~10kbar [7]. We reproduce the correct shape of the Fermi surface and the ratio of its principal axes with our model. Fig. 1 shows both the experimental results and the best fit. If we try to fit the data of this direct Fermi surface measurement *in the absence* of a 'camel's back', we are unable to obtain a satisfactory fit for any combination of R and m'_z. Our results therefore strongly confirm the existence of the camel's back in AlAs.

2.2 Tunneling between $X_{X,Y}$ Landau levels in AlAs

We also fit resonant tunneling data between $X_{X,Y}$ Landau levels, measured at ~10 *kbar* in GaAs/AlAs double-barrier structures, to our model [8]. Fig. 2 shows the experimental results as well as the modelled Landau level fan, including the *k.p*-interaction. This yields a *different* function, $R(m'_z)$.

The intersection of these two curves gives a unique pair of values of R and m'_z at ~10 *kbar*, see fig. 3. If we further assume that R is essentially independent of pressure (since its magnitude is determined by the symmetry of the bulk unit cell which is unperturbed by hydrostatic pressure) and that m'_z exhibits a similar pressure dependence to that recently measured for $m_{X,Y}$ [9], we can estimate the values of R and m'_z at ambient pressure. We find $R = 3.39\ eV\mathring{A}$ and $m'_z = 0.16\ m_0$.

2.3 Photoluminescence (PL) data

We next calculate the bandgaps in several type II superlattices with different AlAs thicknesses between 22 Å and 43 Å and find the R - m'_z -curve that gives the best fit to photoluminescence data, see fig. 3. At first sight, this R-m'_z-curve does not seem to agree well with the newly determined values of R and m'_z, derived from magneto-

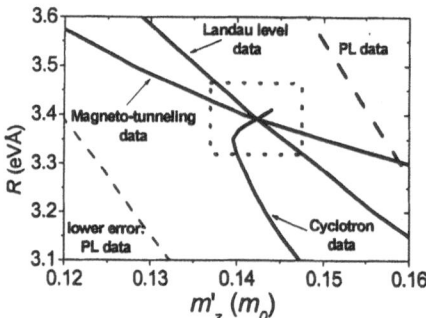

Fig. 3: Plot of R vs. m'_z. The crossing of all curves gives the unique combination, $R = 3.39\ eV\mathring{A}$ and $m'_z = 0.14m_0$ at ~10kbar

tunneling experiments. But as opposed to the latter, the evaluation of the PL data depends sensitively on the actual AlAs and GaAs layer widths, which can be estimated to be accurate within half a monolayer of the nominal values. That represents quite a large proportion of the total width, leading to a large error in the R-m'_z-curve. We find that our new **k.p**-parameters lie well within this error range.

2.4 Cyclotron resonance data

In addition to PL, we also compare our R and m'_z values with cyclotron resonance data, obtained by Miura et al. in

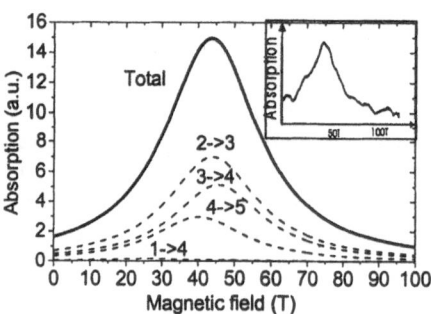

Fig. 4: Cyclotron resonance measurements: experiments performed by Miura et al. (inset) and best fit

1972 [1], see inset in fig. 4. Taking into account the **k.p**-interaction, we calculate the absorption curves for all possible optical transitions, using an empirical Lorentzian broadening parameter. There should be a strong transition between the first two Landau levels with a cyclotron resonance peak around 80T, plus substantial contributions from the higher Landau levels, which occur at lower fields. However, no significant feature was observed above ~70T. A likely explanation for this is the occurrence of magnetic freeze-out of carriers at high magnetic fields, consistent with the observed suppression of all signals below ~120K. If we only consider transitions between Landau levels

LL(i)→LL(i+1), where i = 2,3 and 4 (the next-strongest transitions), we indeed find excellent agreement between the experimental results and our modelled curves, using the new set of parameters (fig. 4).

Our results show good agreement with the whole set of four different experiments. Fig. 3 shows that a unique combination of R and m'_z can explain all the different experimental results. In addition, we have made possible a correct interpretation of results on X_X-X_Y mixing [5,7], for which the actual strength of the **k.p**-interaction is needed to

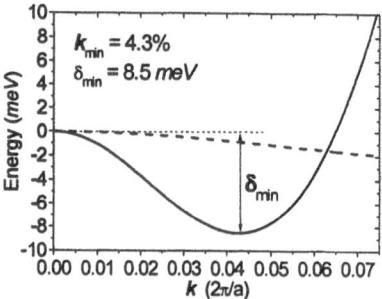

Fig. 5: The new camel's back in AlAs and the old one in comparison (dotted line)

estimate values for the mixing potentials.

With the new combination of **k.p**-parameters, we re-calculate the bulk camel's back in AlAs, shown in fig. 5 in comparison with the old camel's back. It is immediately clear that the dispersion has been substantially modified: The position of the minimum from the X-point moves from 10% down to 4.3% of the Brillouin zone, and the depth of the camel's back increases from 2.2meV up to 8.5meV.

3 Conclusions

We present the first direct determination of the **k.p**-parameters in AlAs, by fitting them to a whole range of experimental data: magneto-tunneling in GaAs/AlAs heterostructures, cyclotron resonance experiments and PL data. We have strong confirmation for the existence of a camel's back in AlAs and present its new shape.

1. N. Miura, G. Kido, M. Suekane, and S. Chikazumi, *J. Phys. Soc. Japan* **52**, (1983) 2838

2. D. Bimberg, *Solid State Communications* **37**, (1981) 987

3. B. Rheinländer, H. Neumann, P. Fischer, G. Kühn, *phys. stat. sol.(b)* **49**, (1972) K167

4. A. A. Kopylov, *Solid State Communications* **56**, (1985) 1

5. H. Im, P. C. Klipstein, R. Grey and G. Hill, *Phys. Rev. Lett.* **18**, (1999) 3693

6. L. E. Bremme and P. C. Klipstein, *proc. 25th Int. Conf. Phys. Semicond.*, Osaka, September 2000

7. H. Im, L. E. Bremme, P. C. Klipstein, R. Grey, G. Hill, *proc. 25th Int. Conf. Phys. Semicond.*, Osaka, September 2000

8. J. M. Smith, P. C. Klipstein, R. Grey and G. Hill, *Phys. Rev. B* **57**, (1998) 1746

9. H. Im, P. C. Klipstein, R. Grey, G. Hill, *Phys. Rev. B* in press (2000)

Theory of the electronic structure of Ga$_{1-y}$In$_y$N$_x$As$_{1-x}$

A. Lindsay and E.P. O'Reilly

Department of Physics, University of Surrey, Guildford, Surrey GU2 7XH, U.K. email: e.oreilly@surrey.ac.uk

Abstract We show using an sp^{3}s* tight-binding Hamiltonian that replacing a single As atom by N introduces a resonant "impurity" level above the conduction band edge in GaInAs, and demonstrate that the interaction of this level with the conduction band edge accounts for the strong band gap bowing observed in GaInN$_x$As$_{1-x}$. The energy of the resonant level and the magnitude of its interaction with the conduction band edge varies approximately linearly with the number of In neighbours but, for low N composition, the overall N interaction with the conduction band edge can still be described using a suitably chosen effective 2-level model, with the magnitude of the interaction scaling with nitrogen fraction, x, as $x^{1/2}$.

1 Introduction

When a small fraction of arsenic atoms in GaAs are replaced by nitrogen, the band gap initially decreases rapidly, at about 0.1 eV per % of N, for $x < 0.03$ [1]. In addition, the electron effective mass increases as the energy gap decreases, contrary to other alloy systems [2]. Shan *et al.* [3] used hydrostatic pressure measurements to propose that the reduction in energy gap is due to an interaction between the conduction band edge and a higher-lying band of localised N resonant states. This model is confirmed by detailed investigation of the conduction band structure carried out using an sp^{3}s* tight-binding Hamiltonian we have developed. Outline details of the tight-binding method used are presented elsewhere [4,5].

We use the tight-binding Hamiltonian here to investigate a number of features of the resonant state. We investigate how the nature of the isolated resonant state changes with local environment in GaInN$_x$As$_{1-x}$. We explicitly identify and track the resonant level in supercell calculations even to nitrogen compositions as high as $x = 0.25$. The conduction band edge energy can be described analytically assuming independent resonant states up to $x \sim 0.03$, but interactions between neighbouring resonances must be included for larger x. The energy of an isolated resonant state varies approximately linearly with the number of nearest neighbour indium atoms, and is calculated to inccrease from E$_N$ = 1.52 eV with four Ga neighbours to E$_N$ = 1.75 eV with four In neighbours in an otherwise random Ga$_{.75}$In$_{.25}$As alloy.

2 Theoretical model and results

To explain the observed pressure dependence of the conduction band edge energy of Ga$_{1-y}$In$_y$N$_x$As$_{1-x}$, Shan *et al.* [3] introduced a simple model of two interacting levels, one at energy E_c associated with the extended conduction band edge state ψ_{c0} of the GaInAs matrix, and the other at energy E_N associated with the localised N impurity states, ψ_N, with the two states linked by a matrix element V_{Nc} due to the interaction between them. The energy of the conduction band edge in GaInN$_x$As$_{1-x}$ can then be determined by finding the lower eigenvalue, E_0, of the 2 × 2 determinant

$$\begin{vmatrix} E_N & V_{Nc} \\ V_{Nc} & E_c \end{vmatrix} \qquad (1)$$

To investigate the resonant state and its behaviour, we first calculate the electronic structure of GaN$_x$As$_{1-x}$ supercells. We write the Hamiltonian matrix, H_1, for such a supercell as $H_1 = H_0 + \Delta H$, where H_0 is the Hamiltonian matrix for GaAs and ΔH the change in the Hamiltonian matrix due to the introduction of N. We calculate separately the wavefunctions ψ_{c0} and ψ_{c1} for H_0 and H_1. By comparing the calculated eigenvectors and eigenvalues, we can derive a nitrogen resonant level wavefunction, ψ_N, proportional to $\psi_{c1} - <\psi_{c0}|\psi_{c1}>\psi_{c0}$. Using ψ_N and ΔH we can then deduce the magnitude of V_{Nc}, E_N and E_c in any supercell structure.

We use a large (1728-atom) supercell containing a single N atom to estimate ψ_{N0}, the resonant wavefunction associated with an isolated N atom. This is highly localised, with 5.5% probability density on the N site, and 12.6% on each of the neighbouring Ga atoms, thereby giving over 50% probability density on these 5 sites. Because the N resonant level is so localised, we expect we can associate a similar localised resonant state with each N site. To test if this is so, we compare the calculated resonant wavefunctions, $\psi_{N,j}$ for a series of increasingly smaller unit cells with the predicted resonant state, $\psi_{N,j0}$, obtained by forming a normalised linear combination of resonant wavefunctions associated with isolated N atoms. Figure 1(a) shows $|<\psi_{N,j}|\psi_{N,j0}>|^2$, the modulus squared of the overlap between the calculated and predicted resonant states for a series of simple cubic (sc,•) and face-centred cubic (fcc,♦) arrays of N atoms. The calculated overlap remains over 94% for the sc structures, even up to $x = 0.25$ (a Ga$_4$NAs$_3$ sc unit cell), and even remains above 70% in a Ga$_4$NAs$_3$ 2 × 2 × 1 fcc structure, where 25% of the Ga atoms have 2 N nearest neighbours. Fig. 1(a) therefore provides striking evidence of the validity of assuming localised N resonant states even up to alloy compositions as large as $x \sim 0.25$. The filled data points in fig. 1(b) show the calculated resonant state energy, E_N, for each structure, found by evaluating $<\psi_{N,j}|H_1|\psi_{N,j}>$ in each case, while the filled data points in fig. 1(c) show how the conduction band edge is pushed down in energy with increasing N density. The open data points in fig. 1(b) show the resonant state energy calculated for each structure by directly evaluating the interaction between isolated resonant states, $E_{N,j0} = <\psi_{N,j0}|H_1|\psi_{N,j0}>$, while the open data points in fig. 1(c) are found by substituting $E_{N,j0}$, $V_{Nc} = <\psi_{N,j0}|H_1|\psi_{c0}>$, and $E_c = <\psi_{c0}|H_1|\psi_{c0}>$ into eq.(1) and then diagonalising to find E_0. These values are in excellent agreement with the values obtained from the full calculation, further confirming the validity of the localised resonance model. Finally the full solid line in fig. 1(c) shows the calculated variation of the conduction band edge assuming independent N states, with E_N = 1.63 eV, V_{Nc} = 1.95 $x^{1/2}$ eV and E_c varying as 1.42 − 1.28x eV in eq. (1). This analytical 2-band model is in excellent agreement with the data up to $x \sim 0.03$, but deviates from the full calculation at larger x, where E_N varies strongly with composition.

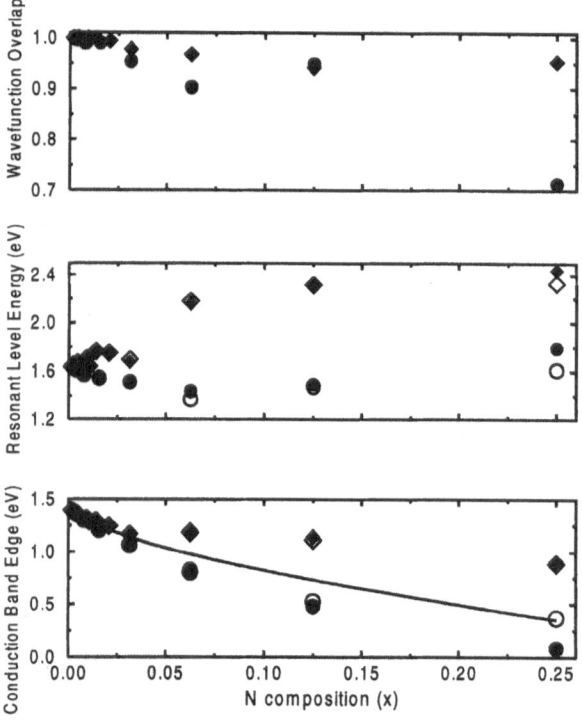

Fig. 1 (a) Overlap between the calculated and predicted resonant state wavefunctions as a function of N composition in simple cubic (•) and face-centred cubic (♦) supercells. (b) Resonant state energy, E_N, as a function of composition, based on a full calculation (solid data points) and a simplified model (open data points). (c) Conduction band edge energy, E_0 of GaN_xAs_{1-x}, calculated using the full Hamiltonian (solid data points), a two-band model (open data points), and a simplified analytical expression (solid line).

Having established that we can use a 2-band model to describe the variation of the conduction band edge energy with N composition and with ordering in the ternary alloy, GaN_xAs_{1-x}, we turn now to investigate the influence of an isolated N atom on the electronic structure of $Ga_{1-y}In_yAs$. We have taken a number of 216-atom supercell structures in which we constrain the central group V site to have a given number, m, of indium nearest neighbours ($m = 0$ to 4). We place In atoms at random on the remaining sites to give an overall indium fraction of $y = 0.25$, and then use the GULP molecular relaxation package [6] to calculate the equilibrium positions of all the atoms in each cluster, including appropriate bond-stretching and bond-bending force constants about each atom. We then calculate the conduction and valence band edge energies, E_c and E_v, and the conduction band edge wavefunction, ψ_{c0}, for each structure. The sp^3s^* tight-binding model used is fitted separately to the band structure of GaAs and of InAs, with the bond-length dependence of the Ga-As and In-As parameters chosen to fit a range of deformation potentials in both materials. We find that $E_c = 1.15 \pm 0.003$ eV and $E_v = 0.05 \pm 0.003$ eV for the five structures considered, where the zero of energy is at the valence band edge of bulk GaAs.

We now replace the central As atom in each structure by a single N atom, allow the structure to relax again using the GULP procedure (keeping the supercell volume constant in this model calculation) and we then calculate values of the new conduction

band edge energies. We find for the structures considered that E_N has a close to linear variation from $E_N = 1.52$ eV for 4 Ga neighbours to $E_N = 1.75$ eV for 4 In neighbours. The value of E_N does depend weakly on the atomic arrangement in the second shell of group III neighbours. Calculations for a range of supercell structures indicate that the atomic arrangement in the second shell only shifts E_N by $\sim \pm 0.02$ eV, down by an order of magnitude on the effects of variations in the nearest neighbour environment. The matrix element linking the N resonant state and the conduction band edge is calculated to vary between $2.00\ x^{1/2}$ eV for a structure where all N have four Ga neighbours to $1.35\ x^{1/2}$ eV for a structure where all N have four In neighbours. The conduction band edge energy, E_c, in the 2×2 matrix of eq. (1) then varies between 1.10 eV and 1.11 eV for the 216-atom supercells considered, and the band gap varies from 0.95 eV to 1.04 eV, as the number of In nearest neighbours m increases from 0 to 4. The average interaction in a larger unit cell containing a mixture of nearest neighbour environments is then given by an appropriately weighted average of the above parameters, again allowing the construction of an effective 2-band model to describe the variation of conduction band edge energy and energy gap in quaternary $GaInN_xAs_{1-x}$ alloys for low x values ($x < \sim 0.03$).

3 Conclusions

In summary, we have presented a tight-binding model which accounts for the strong band gap bowing observed in $GaInN_xAs_{1-x}$. The model introduced is not unique to this material system but can instead be used to describe the electronic structure and band gap bowing of all tetrahedrally bonded semiconductors. We conclude therefore that GaInNAs is not just an interesting material in its own right but will also provide insight into trends in the electronic structure of all tetrahedrally bonded semiconductor alloys.

Acknowledgements We thank the Engineering and Physical Sciences Research Council (UK) for financial support and Agilent Technologies for providing a CASE studentship for AL. We thank A.R. Adams, A.D. Andreev, P.J. Klar and B.A. Weinstein for useful discussions.

References

1. M. Weyers, M. Sato and H. Ando, Jpn. J. Apl. Phys. **31**, (1992) L983.

2. C. Skierbiszewski, P.Perlin, P. Wisniewski, W. Knap, T. Suski, W. Walukiewicz, W. Shan, K.M. Yu, J.W. Ager III, E.E. Haller, J.F. Geisz and J.M. Olson, Appl. Phys. Lett. **76**, (2000) 2409

3. W. Shan, W. Walukiewicz, J.W. Ager III, E.E. Haller, J.F. Geisz, D.J. Friedman, J.M. Olson and S.R. Kurtz, Phys. Rev. Lett. **82**, (1999) 1221

4. A. Lindsay and E.P. O'Reilly, Solid State Commun. **112**, (1999) 443

5. E.P. O'Reilly and A. Lindsay, phys. stat. sol. (b) **216**, (1999) 131

6. J. Gale, J. Chem. Soc. Faraday Trans. **93**, (1997) 629.

Influence of Ag on the energy spectrum of $(Bi_{1-x}Sb_x)_2Te_3$

V.A. Kulbachinskii[1], A.Yu. Kaminsky[1], P.Lostak[2], C.Drasar[2]

[1] Low Temperature Physics Department, Moscow State University, 119899, Moscow, Russia e-mai: kulb@mig.phys.msu.su
[2] Pardubice University, Pardubice, Czech Republic

Abstract The temperature dependence of resistivity, Hall effect, and the concentration and angular dependences of Shubnikov-de Haas effect in p-$(Bi_{1-x}Sb_x)_2Te_3Ag_y$ ($0 \leq x \leq 1.0$; $0 \leq y \leq 0.0152$) single crystals have been investigated in magnetic fields up to 40 T. Overstoichiometric doping of $(Bi_{1-x}Sb_x)_2Te_3$ crystals with silver showed that Ag exhibits acceptor properties.

1 Introduction

Semiconductors of the Sb_2Te_3 type are layered crystals having a rhombohedral structure. Crystals of Sb_2Te_3 have a very high concentration of holes because of the presence of a large number of charged points defects preferentially antistructural type that is an Sb atom in a Te position. Since defects are negatively charged, Sb_2Te_3 grown under stoichiometric conditions always possesses p-type conductivity. The weak polarity of the Sb-Te bonds is conducive to the formation of antistructural defects. A change in the polarity of the bonds by doping leads to a change in the concentration of the antistructural defects and therefore the carrier concentration. Consequently doping with elements in particular group of the periodic table may have a donor or acceptor effect not depending on the number of the group but as a result of the influence of the incorporated element on the bond polarity. As an example we can quote the Group III element indium, which has a donor effect in Sb_2Te_3 [1].

Mixed crystals, i.e., crystals of the type $(Bi_{1-x}Sb_x)_2Te_3$ or $Sb_2Te_{3-y}Se_y$ are of particular interest because the highest thermoefficiencies Z are observed in these. Very little data are currently available on the influence of doping with Group I elements on the energy spectrum of mixed $(Bi_{1-x}Sb_x)_2Te_3$ crystals. In the present study we investigated concentration and angular dependence of the SdH effect in magnetic field up to 40 Tesla. We also studied the influence of silver doping on galvanomagnetic properties of mixed $(Bi_{1-x}Sb_x)_2Te_3$ single crystals.

2 Samples

P-type single crystals of silver-doped Sb_2Te_3 and mixed crystals $(Bi_{1-x}Sb_x)_2Te_3$ were grown by Bridgman method.

We added silver to the stoichiometric polycrystal for doping so that the sample compositions will be subsequently given in the form $Sb_2Te_3Ag_y$ and $Bi_{0.5}Sb_{1.5}Te_3Ag_y$. After preparing a polycrystal of the required composition we grew the single crystals. The Ag content in the samples was determined by atomic absorption spectroscopy (AAS) for the every specific sample used for electrophysical measurements and listed in table 1.

Table 1 Ag content in the samples

Samples		y (AAS)	C (10^{19}atoms/cm^3)
$Sb_2Te_3Ag_y$	1	0.0035	2.2
	2	0.0093	5.9
	3	0.0100	6.2
	4	0.0152	9.6
$Bi_{0.5}Sb_{1.5}Te_3Ag_y$	1	0.0014	0.8
	2	0.0030	1.9
	3	0.0048	3.0

3 Results and discussion

For all samples the resistivity ρ decreases with decreasing temperature and saturates at low temperatures. The Hall coefficients R are positive for all samples and decrease monotonically with increasing Ag content, which indicates an increase in the hole concentration. The Hall coefficients do not depend on magnetic field. The Hall mobility μ in the silver-doped samples decreases with increasing silver content, the decrease of μ being greater for $Bi_{0.5}Sb_{1.5}Te_3Ag_y$ than for $Sb_2Te_3Ag_y$ as is illustrated in Fig. 1.

It can be seen from the results that the average efficiency of silver in $Bi_{0.5}Sb_{1.5}Te_3Ag_y$ crystals is lower than that in $Sb_2Te_3Ag_y$ crystals, that is the number of additional holes per silver atoms is lower in $Bi_{0.5}Sb_{1.5}Te_3Ag_y$. This result may be explained by the fact that in $Sb_2Te_3Ag_y$ silver mainly enters the Sb sublattice and forms negatively charge substitutional point defects whereas in Bi_2Te_3 silver forms interstitial atoms. This is because the Bi-Te bonds in Bi_2Te_3 crystals are more ionic than the Sb-Te bonds in Sb_2Te_3 crystals. Thus very few point defects where Bi is substituted by Ag are formed in the Bi_2Te_3 lattice. At the same time, the incorporation of Ag suppresses the formation of antistructural negatively charged Bi_{Te} defects. All this gives rise to a different type of defects and ultimately to different behavior of Ag: in Sb_2Te_3 silver behaves as an acceptor whereas in Bi_2Te_3 silver doping induces a donor effect [2]. It is then easy to understand why the efficiency of silver as an acceptor is lower in mixed $Bi_{0.5}Sb_{1.5}Te_3Ag_y$ crystals.

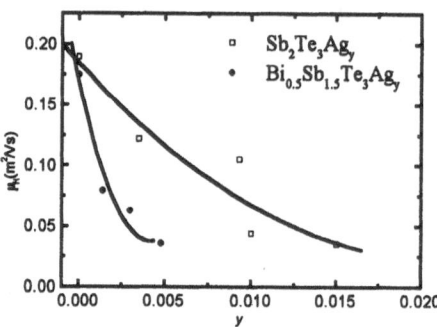

Fig. 1 Dependences of the Hall mobility μ at T=4.2 K on the Ag content y in $Sb_2Te_3Ag_y$ and $Bi_{0.5}Sb_{1.5}Te_3Ag_y$.

SdH effect was investigated at T=4.2 K with the magnetic field vector directed along the C_3 axis. For this orientation of the vector **B** the cross-sections of six ellipsoids of the upper valence band of the Fermi surface are the same. A single oscillation frequency F is observed in $Sb_2Te_3Ag_y$. In $Bi_{0.5}Sb_{1.5}Te_3$ two frequencies are observed. The second frequency is ascribed to the lower valence band. Under silver doping the second frequency disappeared.

Table 2 Frequency F of SdH oscillations, Fermi-energy E_F, and SdH hole concentration p (in 10^{19} cm^{-3})

	Samples	F (T)	E_F (meV)	P
	$Sb_2Te_3Ag_y$			
	0	52.0	98	2.20
	0.0035	76.5	144	5.64
y=	0.0093	111.0	210	9.87
	0.0100	115.0	217	10.4
	0.0152	120.8	228	11.2
	$Bi_{0.5}Sb_{1.5}Te_3Ag_y$			
	0	56.8	100	3.28
	0.0014	64.6	114	3.97
y=	0.0030	82.6	146	5.76
	0.0048	92.0	163	6.74

Using simple ellipsoidal nonparabolic model [3,4] we calculated the hole concentrations in the upper valence band and the Fermi energy. The results are given in the table 2.

For $Bi_{0.5}Sb_{1.5}Te_3Ag_{0.003}$ the SdH effect was also

studied when the magnetic filed vector **B** was rotated in the plane C_3C_1 (see Fig. 2). Fig. 3 gives the theoretical (curves) and experimental (symbols) dependencies of the extreme cross-sections of the Fermi-surface ellipsoids on the inclination of **B** relative to C_3 axis. Experimental data are in a good agreement with anisotropy of the Fermi-surface $\eta=S_{max}/S_{min}=3.8$.

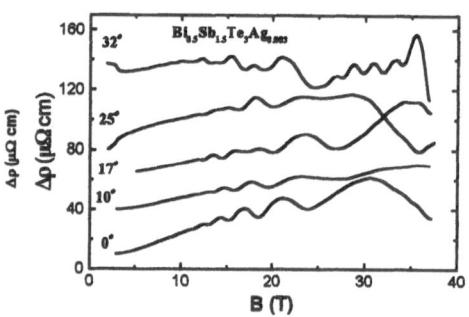

Fig. 2 Angular dependences of SdH oscillations in $Bi_{0.5}Sb_{1.5}Te_3Ag_{0.003}$

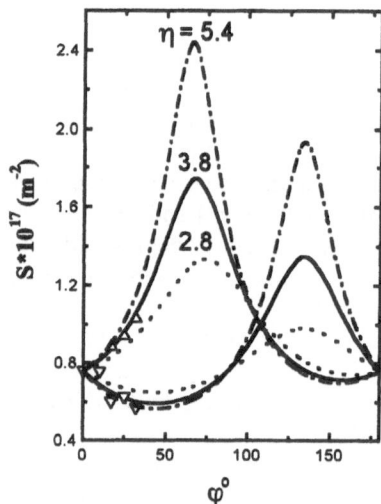

Fig. 3 Theoretical (curves) and experimental (symbols) dependencies of the extreme surface cross-sections S_H on the angle φ between the C_3 axis and vector **B**. The theoretical dependencies are plotted for various values of anisotropy $\eta=S_{max}/S_{min}$ of the Fermi-surface (indicated in figure).

References
1. V.A. Kulbachinskii, Z.M. Dashevskii, M. Inoue, *et al.*, Phys. Rev. B **52**, (1995) 10915.
2. J. Navratil, I. Klichova, S. Karamazov, *et al.*, Sol. State Chem. **140**, (1998) 29.
3. V.A. Kulbachinskii, N. Miura, H. Nakagawa, *et al.*, J.Phys.: Cond. Matter **11**, (1999) 5273.
4. V.A. Kulbachinskii, A.Yu. Kaminsky, V.G. Kytin, *et al.*, ZETF. **117**, (2000) 1242 (JETP, **90** (2000) 1081)).

Raman study of isotopically tailored CuBr

J. Serrano*, F. Widulle, T. Ruf, C.T. Lin, and M. Cardona.

Max-Planck-Institut für Festkörperforschung, Heisenbergstr. 1, D-70569 Stuttgart, Germany.

Abstract Raman measurements are reported for isotopically tailored CuBr at low temperature. The transverse optic (TO) mode shows the expected $\sim \mu^{-1/2}$ behavior with the reduced mass, with a small deviation that can be easily explained in terms of anharmonicity. However, for the longitudinal optic (LO) mode we observe a broad structure which can be resolved in three features of total width around 17 cm^{-1}. The nature of this structure is assessed by means of the Fermi resonance model, i.e., a two-phonon combination couples with the LO mode. The behavior of this structure with different Cu and Br isotopes is analyzed in detail.

Copper halides exhibit very interesting properties such as superionic conduction, anomalous thermal expansion and strong anharmonicity. Despite many studies, some basic properties related to their lattice dynamics remain still controversial, e.g., in CuCl two different theories have been proposed to explain the anomalous Raman spectrum of the *transverse optic* phonon. A recent study using isotopically tailored CuCl [1] succeeded in explaining this phonon lineshape in terms of anharmonic interaction between the TO-mode and the two-phonon density of states (DOS) as a so-called Fermi resonance (FR) [2,3]. However, earlier and even recent *ab-initio* studies [4–6] claim that this feature could be caused by off-center displacements of Cu cations.

In order to clarify this issue we have investigated the Raman lineshapes of isotopically modified CuBr at 2 K using the 647.1 nm Kr$^+$ line for excitation. The TO-mode has an almost perfect Lorentzian lineshape and its frequency evolves $\simeq \mu^{-1/2}$ with isotope substitution, where μ is the reduced mass. This behavior is displayed in Fig. 1. It shows also a small deviation in the TO frequency with respect to the $\mu^{-1/2}$ behavior, due to anharmonicity. This shift can be roughly estimated in the virtual-crystal approximation by the equation:

$$\omega = A\mu^{-1/2} - B\mu^{-1} \qquad (1)$$

Using Eq. 1 to fit the phonon energies of our isotopic samples we obtain $B\mu^{-1} = 7.6$ cm^{-1} for natural CuBr. The anharmonic contribution can be also obtained from the difference between the measured phonon frequency at $T = 0$ K, and the extrapolation to this temperature of the linear regime of the temperature dependence of the phonon frequency [7]. From the experimental data from Ref. [8], we obtain an anharmonic shift of ~6 cm^{-1}, in good agreement with our experimental findings.

In the LO case, however, a pronounced broad structure with a width of about 17 cm^{-1} is observed, rather similar to that observed for the TO-mode of CuCl [1,

---

* Corresp. author: jserrano@kmr.mpi-stuttgart.mpg.de

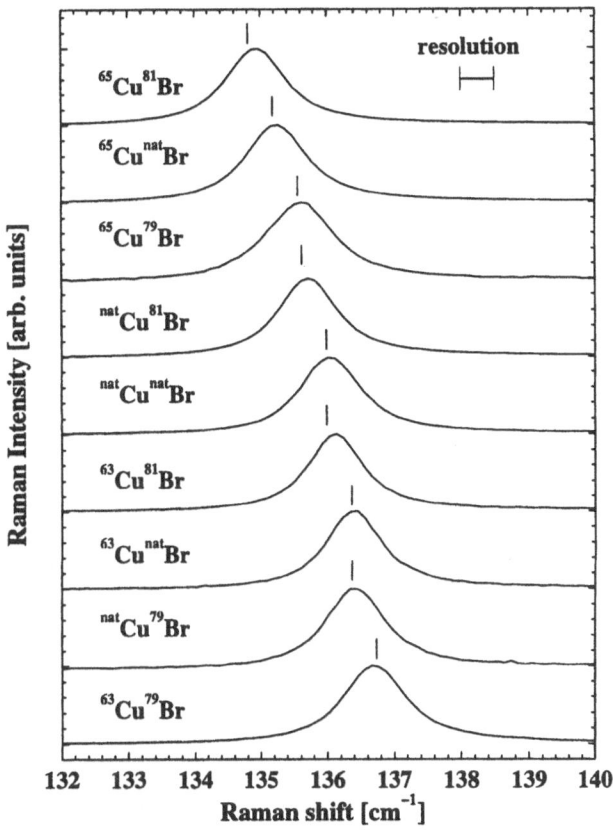

Fig. 1 TO phonon Raman spectra of CuBr for different isotopic compositions. The bars near the peaks represent the $\mu^{-1/2}$ behavior; the sample ^{63}Cu^{79}Br was used as reference.

2]. The Raman spectra of the LO phonon are shown in Fig. 2 for the different isotope combinations. The lineshapes consist of two broad peaks (A,B) and a small peak on the high energy side (C). We find that peak B changes its position with isotope substitution as $\mu^{-1/2}$, while the frequencies of peaks A and C are proportional mainly to the copper mass.

The model used to explain this Raman lineshape and its change with isotope substitution is similar to that employed for CuCl in Refs. [1,2]. It is based also on a Fermi resonance. The complex frequency-dependent phonon self-energy $\Sigma(\omega) = \Delta(\omega) - i\,\Gamma(\omega)$, where $\Gamma(\omega) = |V_3|^2 \rho_2(\omega)$ is proportional to the two-phonon DOS, $\rho_2(\omega)$, and V_3 is an effective cubic anharmonicity coefficient in a third-order perturbation approximation of the lattice potential. The Raman intensity is then given by

$$I_S(0,j;\omega) \propto \frac{\Gamma(0,j;\omega)}{[\omega - \omega(0,j) - \Delta(0,j;\omega)]^2 + \Gamma(0,j;\omega)^2} \qquad (2)$$

assuming a lorentzian phonon lineshape [7].

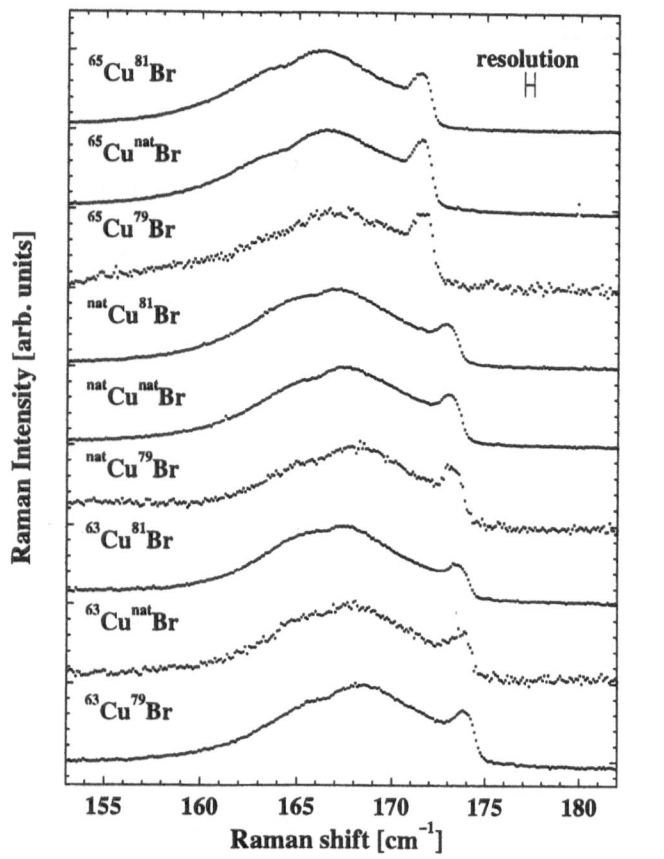

Fig. 2 Raman spectra of the LO mode of CuBr for different isotopic compositions.

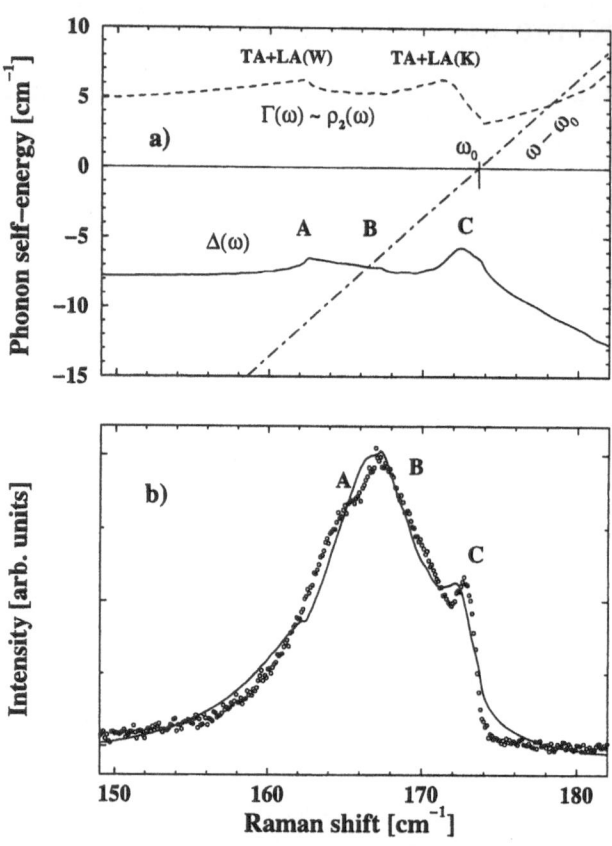

Fig. 3 (a) Imaginary part, $\Gamma(\omega)$, (dashed line) and real part, $\Delta(\omega)$, (solid line) of the phonon self energy in natural CuBr calculated with the latticel-dynamical model from Ref. [9] (see text). The dashed-dotted line represents $(\omega - \omega_0)$, where ω_0 is the harmonic unrenormalized LO frequency. (b) Raman spectrum of the LO phonon of natural CuBr at 2 K (open circles). The three features (A,B,C) described in the text are shown together with a fit (solid line) using the FR model and the phonon self-energy of (a).

We have calculated $\rho_2(\omega)$ with a 14-parameter shell model using the parameters of Set I from Ref. [9]. $\Delta(\omega)$ was then obtained from $\Gamma(\omega_{LO})$ by Kramers-Kronig transformation. The self-energy in the region of the LO is shown in Fig. 3. This DOS exhibits two peaks related to TA+LA phonon combinations near the W and K points. As can be seen from Eq. 2 and the dashed-dotted line indicating $\omega - \omega_0$ (ω_0 is the harmonic LO frequency) in Fig. 3(a), this structure in $\Sigma(\omega)$ is responsible for the peaks A and C. The peak B occurs at the solution of the equation $\omega = \omega_0 + \Delta(\omega)$. Figure 3(b) shows the excellent agreement between the calculation and the experimental phonon lineshape for natural CuBr. A value of $|V_3|^2 \simeq 45$ cm^{-1} was used in the calculation.

The flatness of $\rho_2(\omega)$ explains the evolution of B $\sim \mu^{-1/2}$ with isotope substitution. However, the experimentally observed dependence of peaks A and C on the copper mass indicates that these ions predominate in the acoustic-phonon eigenvectors which cause the lineshape renormalization. This is in contrast to the shell model [9] which predicts that the bromine ions are most important for the acoustic modes. Hence, a better lattice-dynamical model is needed in order to understand these unusual isotope effects. Whereas the phonon *dispersions* are rather well described by these lattice models, this is not the case for the *eigenvectors*, which are usually not taken into acount when fitting the parameters. Therefore, it is

not unusual to find wrong eigenvectors in such models. This might be prevented by future *ab-initio* calculations.

Similar to a recent investigation for GaP [10], these results widen the base of evidence for anharmonicity-related phonon lineshape renormalizations that can be well explained by the Fermi-resonance scenario.

References

1. A. Göbel *et al.*, Phys. Rev. B **56**, (1997) 210.
2. M. Krauzman *et al.*, Phys. Rev. Lett. **33**, (1974) 528;
3. G. Kanellis, W. Kress and H. Bilz, Phys. Rev. Lett. **56**, (1986) 938.
4. C. H. Park and D. J. Chadi, Phys. Rev. Lett. **76**, (1996) 2314;
5. S. R. Bickham *et al.*, Phys. Rev. Lett. **83**, (1999) 568.
6. M. Cardona, K. Syassen and T. Ruf, Phys. Rev. Lett. **84**, (2000) 4511.
7. J. Menéndez and M. Cardona, Phys. Rev. B **29**, (1984) 2051.
8. J. E. Potts, R. C. Hanson and C. T. Walker, Solid State Commun. **13**, (1973) 379.
9. S. Hoshino *et al.*, J. Phys. Soc. Jpn. **41**, (1976) 965.
10. F. Widulle *et al.*, Phys. Rev. Lett. **82**, (1999) 5281.

Proc. 25th Int. Conf. Phys. Semicond., Osaka 2000 (Eds. N. Miura and T. Ando)

69

Isotope disorder effects in the Raman spectrum of Silicon

F. Widulle[1*]**, T. Ruf**[1]**, M. Konuma**[1]**, I. Silier**[1]**, W. Kriegseis**[2]**, V. I. Ozhogin**[3]**, M. Cardona**[1]

[1] Max-Planck-Institut für Festkörperforschung, Heisenbergstraße 1, D-70569 Stuttgart, Germany
[2] I. Physikalisches Institut, Universität Giessen, Heinrich-Buff-Ring 16, D-35392 Giessen, Germany
[3] Institute of Molecular Physics, RRC "Kurchatov Institute", 123182 Moscow, Russia

Abstract We have measured disorder-induced effects of the phonon self-energy on the low-temperature Raman spectrum of isotopically tailored silicon crystals. The frequency and the linewidth of the Raman peak show isotope disorder-induced shifts and broadenings resulting in a nonlinear dependence on composition ("bowing"). This broadening exhibits an asymmetric variation with respect to the intermediate mass $M = 29$ amu. This dependence on composition is similar to, but much smaller than, the one seen in diamond and is attributed to higher-order perturbations in the mass disorder. Weak disorder-induced excitations are observed from $30 \, \text{cm}^{-1}$ to $60 \, \text{cm}^{-1}$ below the main Raman peak. Their spectral positions for various isotopic compositions do not simply follow the harmonic scaling law $\omega \propto \overline{M}^{-1/2}$ expected for a virtual crystal with average mass $\overline{M}$. In addition, their intensities are not proportional to the second moment g_2 of the mass-fluctuation parameter.

1 Introduction

Many details of the phonons of semiconductors can be investigated by varying their isotopic composition. In this way not only the average mass of the system is controllable but also the mass disorder. The effects of the latter on the Raman lineshape can often be evaluated by second-order perturbation theory [1,2]. For a monoatomic crystal the strength of the disorder-induced effects on the self-energy is then proportional to the second moment (g_2) of the mass-fluctuation parameter

$$g_n = \sum_{i=1}^{N} c_i \left(1 - \frac{M_i}{\overline{M}}\right)^n \qquad (1)$$

for a crystal where N is the number of isotopes with masses M_i and concentration c_i. However, this simple approach does not hold for some experimental observations.

In diamond the linewidth clearly shows an asymmetrically bowed mass dependence [3,4], i.e. the largest self-energy effect is not seen in the most disordered crystal, as measured by g_2. While second-order perturbation theory fails in explaining these experimental findings, the coherent potential approximation (CPA) has been successfully applied to the problem [4]. The CPA considers higher-order contributions to the self-energy which are proportional to g_n with $n > 2$.

The one-phonon density of states (DOS) $\rho(\omega_0)$ at the Raman frequency ω_0 is finite in diamond due to the overbending of the longitudinal optic branch along $\Gamma - X$

whereas it is zero for Si, Ge, and α-Sn within the harmonic approximation. Thus, in the latter three crystals isotope effects which are directly proportional to $\rho(\omega_0)$, e.g. the broadening, are very small and difficult to measure for Γ-point phonons. This situation requires a qualitatively different starting point for the theoretical treatment, for which anharmonicity should be included as it induces a finite DOS at ω_0.

Another feature of the Raman spectrum which originates from elastic scattering of phonons on mass-fluctuations are the weak excitations on the low-energy tail of the Raman peak. They have been observed in diamond [5], Ge [6], and α-Sn [7] and reflect the large DOS for the transverse optic (TO) branches at critical points of the Brillouin zone. Here also, the CPA qualitatively reproduces their spectral peculiarities.

Regarding the two effects described above, experimental self-energy data exist on all the $\rho(\omega_0) = 0$ crystals Ge [8], α-Sn [7], and Si [9], but they do not particularly cover the composition range which is necessary for the interpretation of higher-order contributions. Here we present the measured linewidths and weak disorder-induced excitations of bulk silicon crystals whose various isotopic abundances enable us to analyze higher-order perturbative terms experimentally and theoretically.

2 Experimental results

The experiments were performed at low temperature ($T = 6$ K) with the 647 nm krypton laser line. In this work we extend an earlier study [9] by including two additional samples with the nominal abundances $^{28}\text{Si}_{0.75}$ $^{30}\text{Si}_{0.25}$ and $^{28}\text{Si}_{0.25}$ $^{30}\text{Si}_{0.75}$. Their actual isotopic compositions were determined by SIMS and TGMS. Figure 1 shows the experimentally obtained intrinsic linewidths of the Raman phonon. Their values are obtained by fitting a Voigt profile to the measured lineshapes. We use the mean value between ^{28}Si and $^{\text{nat}}\text{Si}$ as the reference point and indicate the mass dependence of a purely anharmonic broadening with $\Gamma_{\text{anh}} \propto \overline{M}^{-1}$ (straight line). The maximum of the disorder-induced broadening does not occur for $^{28}\text{Si}_{0.5}$ $^{30}\text{Si}_{0.5}$, for which g_2 is maximum, but rather close to $^{28}\text{Si}_{0.25}$ $^{30}\text{Si}_{0.75}$. This suggests the influence of g_3 which contributes asymmetrically with respect to the intermediate mass $M = 29$ amu. The bowed solid curve in Fig. 1 is the fit of the data to the expected anharmonic behavior Γ_{anh} plus approximate expressions

* *Correspondence to* widulle@kmr.mpi-stuttgart.mpg.de

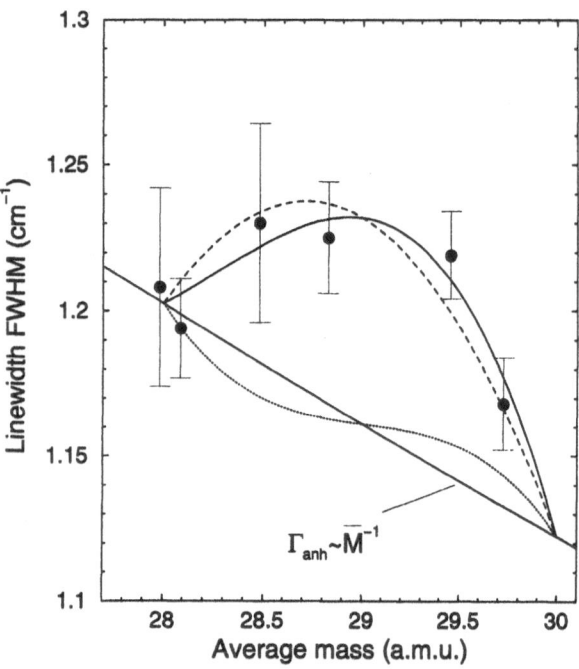

Fig. 1 Measured linewidths of the Raman peak in Si for various isotopic compositions. The straight line indicates a purely anharmonic line broadening $\Gamma_{\rm anh} \propto \overline{M}^{-1}$. The bowed solid curve through the data points represents a fit of the additional disorder-induced broadening taking into account g_2 (dashed line) and g_3 (dotted line).

for g_2 and g_3

$$\Gamma = \Gamma_{\rm anh}(\overline{M}) + c_{30}\,(1-c_{30})\,(A_2 + A_3\,2\,(c_{30}-0.5)) \quad (2)$$

with $c_{30} = (\overline{M}-28)/2$. For the fit parameters we find $A_2 = 0.28\,{\rm cm}^{-1}$ and $A_3 = 0.12\,{\rm cm}^{-1}$. The individual contributions of g_2 and g_3 to the additional broadening are displayed as dashed and dotted lines in Fig. 1, respectively.

Figure 2 displays the measured weak excitations on the low-energy tail of the Raman peak together with Voigt lines which simulate the lineshapes of merely anharmonically broadened spectra, without isotopic disorder. The spectra have been normalized to equal main peak intensity, shifted horizontally to match the Raman shift of ^{28}Si and also vertically for clarity. The broad structure around $440\,{\rm cm}^{-1}$ is due to second-order Raman scattering by 2TA(W) phonons [10]. The disorder-induced excitations arise in the spectral region between $465\,{\rm cm}^{-1}$ and the sharp increase of the main Raman peak. The two peaks observed are attributed to isotope disorder-induced scattering from TO(X,W) and TO(L) states [11]. The evolution of the spectra with varying average mass demonstrates that the second-order part of the spectrum, which essentially reflects the two-phonon DOS, remains unaltered whereas the disorder-induced part depends on the mass-fluctuations of the system. Concerning the latter, the TO(L) excitation approaches the main peak with increasing average mass. The TO(L) intensity is maximum for the sample with $c_{30} = 0.75$

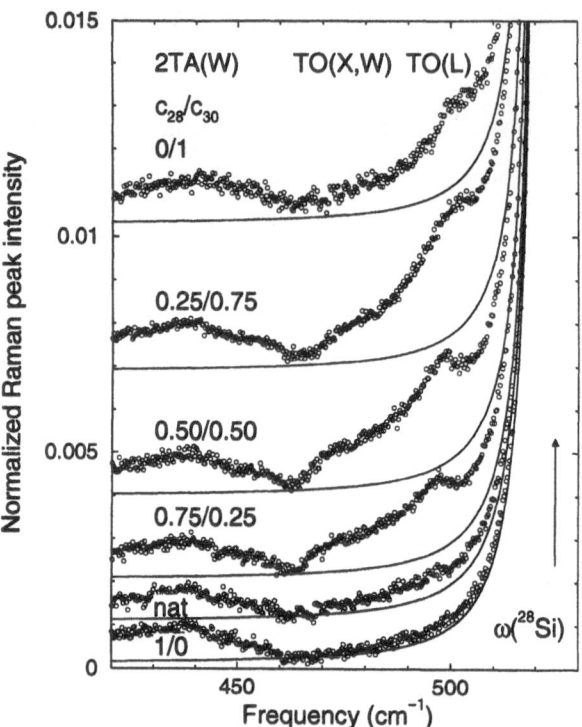

Fig. 2 Measured weak disorder-induced excitations arising on the low-energy tail of the Raman peak in Si. The numbers give the nominal isotope concentrations. The maximum intensity of the excitations is $\approx 0.4\%$ of the main peak which is normalized to one for all spectra. For a better comparison the individual spectra have been shifted vertically and also spectrally in order to coincide with the Raman shift of ^{28}Si. The solid lines indicate purely anharmonically broadened Raman lines in the absence of disorder.

and clearly observed even for the nominal ^{30}Si specimen ($c_{30} = 0.87$ from SIMS) while the TO(X,W) feature is smeared out.

A systematic theoretical study using the CPA for various isotopic compositions has been performed in order to provide a consistent picture of all disorder-induced effects on the Raman spectrum of silicon. These results will be published elsewhere.

References

1. S. Tamura, Phys. Rev. B **27**, (1983) 858.
2. J. Menéndez *et al.*, Phil. Mag. B **70**, (1994) 651.
3. R.M. Chrenko, J. Appl. Phys. **63**, (1988) 5873.
4. K.C. Hass *et al.*, Phys. Rev. B **44**, (1991) 12046; K.C. Hass *et al.*, Phys. Rev. B **45**, (1992) 7171.
5. J. Spitzer *et al.*, Solid State Commun. **88**, (1993) 509.
6. P. Etchegoin *et al.*, *Proceedings of the 22nd International Conference on the Physics of Semiconductors* (World Scientific, Singapore, 1994), p. 277.
7. D.T. Wang *et al.*, Phys. Rev. B **56**, (1997) 13167.
8. J. M. Zhang *et al.*, Phys. Rev. B **57**, (1998) 1348.
9. F. Widulle *et al.*, Physica B **263-264**, (1999) 381.
10. P. A. Temple and C. E. Hathaway, Phys. Rev. B **7**, (1973) 3685.
11. B. A. Weinstein and G. J. Piermarini, Phys. Rev. B **12**, (1975) 1172.

Proc. 25th Int. Conf. Phys. Semicond., Osaka 2000 (Eds. N. Miura and T. Ando)

71

Observation of extraordinary phonon polaritons in ZnO

T. Azuhata[1], K. Torii[2], M. Ono[2], and T. Sota[2]

[1] Department of Materials Science and Technology, Hirosaki University, Hirosaki, Aomori 036-8561, Japan e-mail: `azuhata@cc.hirosaki-u.ac.jp`

[2] Department of Electrical, Electronics, and Computer Engineering, Waseda University, Shinjuku, Tokyo 169-8555, Japan

Abstract   Raman scattering by phonon polaritons in ZnO was observed in forward scattering geometry with annular apertures. The signals were assigned to extraordinary phonon polaritons by calculating Raman scattering efficiencies for ordinary and extraordinary phonon polaritons. The assignment was confirmed by the fact that the scattering angle dependence of their frequencies shows good agreement with the theoretical dispersion curve.

1 Introduction

ZnO crystallizes in the uniaxial wurtzite structure with C_{6v} symmetry. There exist two types of transverse optical (TO) phonons. One is pure E_1 TO phonons and the other is quasi-TO phonons, which is a mixture of E_1 TO and A_1 TO phonons. These types of phonons are called ordinary and extraordinary phonons, respectively. As the result, there exist two types of bulk phonon polaritons, i.e., ordinary and extraordinary phonon polaritons. Ordinary phonon polaritons were observed by Porto et al. [1]. To the best of our knowledge, however, there is no report on extraordinary phonon polaritons. In this paper, we report the observation of extraordinary phonon polaritons in ZnO by polarized Raman spectroscopy.

2 Experimental

The sample with (0001) surface grown by a hydrothermal method was purchased from GOODWILL, Russia. According to the data sheet from GOODWILL, the sample is annealed and its resistivity is as high as $10^{11}\,\Omega$ cm.

Stokes Raman spectra were measured at room temperature using the second harmonics of a Nd^{3+}:YAG laser, a triple spectrometer including a subtractive filter with a focal length of 1 m, and an image-intensified multi-channel detector. The experimental configuration used for forward Raman scattering measurements is illustrated schematically in Fig. 1. To select the value of q, which is an absolute value of wave vectors of phonon polaritons observed, five annular apertures with different radii were used.

3 Results and Discussions

First, we measured Raman spectra of bulk phonons in backscattering and right angle scattering geometries. From the spectra, all frequencies of the optically-active phonons were obtained as shown in Table 1. These frequencies are consistent with the results by Damen et al. [2].

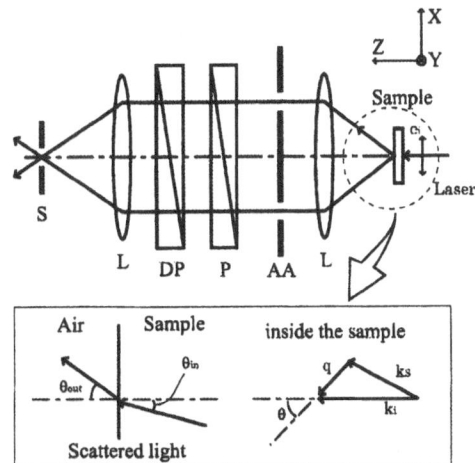

Fig. 1 Schematic illustration of the experimental configuration used for forward Raman scattering measurements. S, L, DP, P, and AA represent, respectively, spectrometer entrance slit, lens, de-polarizer, polarizer, and annular aperture. θ_{out} (θ_{in}) is the angle between the incident light and the scattered light outside (inside) the sample. k_i, k_s, and q are, respectively, wave vectors of incident light, scattered light, and phonon polaritons inside the sample.

Table 1 Optical phonon frequencies of ZnO in units of cm^{-1}.

	E_2		E_1		A_1	
	low	high	TO	LO	TO	LO
Damen et al. [2]	101	437	407	583	380	574
This work	101	438	409	586	375	575

Next, forward Raman scattering measurements with annular apertures were performed to investigate phonon polaritons. The incident light was parallel to the c-axis of the sample. Changing the radius of annular apertures makes it possible to obtain polariton dispersion. As shown in Fig. 2, observed were Raman signals which showed remarkable frequency lowering as the radius of the aperture was decreased. The signals are attributed to the lower branch of phonon polaritons. The peaks around 330 cm^{-1}, frequency of which is independent of θ_{out}, are due to multi-phonon scattering processes [2]. A(BC)D means that the incident light with polarization parallel to the B-axis propagates along the A-axis and the scattered light with polarization parallel to the C-axis propagates along the D-axis, for example.

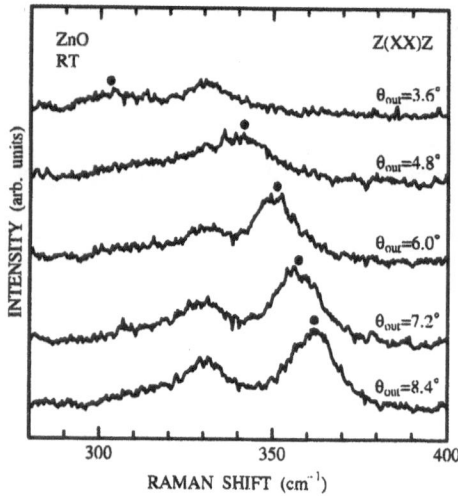

Fig. 2 Raman spectra of ZnO for Z(XX)Z. The signals marked with dots are due to phonon polaritons. The peaks around 330 cm^{-1} are attributed to multi-phonon scattering processes [2].

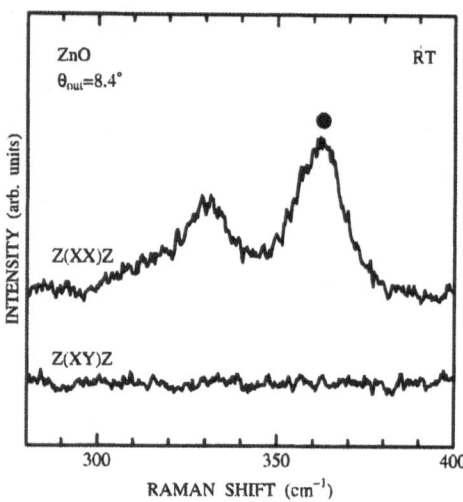

Fig. 3 Polarized Raman spectra of phonon polaritons (•) in ZnO.

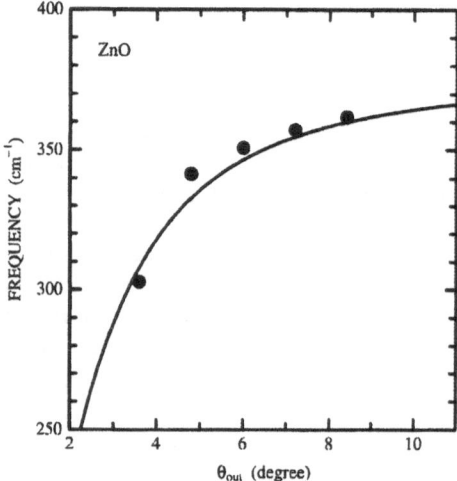

Fig. 4 Frequency variation of extraordinary phonon polaritons in ZnO as a function of θ_{out}. The dots are experimental data and the solid line is theoretical results calculated using Eq. (1).

Polarized Raman spectra of phonon polaritons for $\theta_{out}=8.4°$ are shown in Fig. 3. The peak due to phonon polaritons is labeled a dot. The peak appears in Z(XX)Z and disappears in Z(XY)Z. In order to assign the polariton signals, we calculated scattering efficiencies of both ordinary and extraordinary phonon polaritons for the configuration shown in Fig. 1. Although the lower branch of extraordinary phonon polaritons is not pure transverse and has certain amount of longitudinal components, its deviation from pure transverse nature is small enough in the case under consideration. Therefore, we assumed it is pure transverse as a good approximation. The results show that Raman intensity for extraordinary (ordinary) phonon polaritons should be higher in Z(XX)Z (Z(XY)Z) than in Z(XY)Z (Z(XX)Z). Thus, the polariton signals observed are assigned to the lower branch of extraordinary phonon polaritons.

Theoretical dispersion curve for extraordinary phonon polaritons can be easily obtained from Maxwell's equations as

$$\frac{q^2 c^2}{\omega^2} = \frac{\varepsilon_\perp(\omega)\varepsilon_{||}(\omega)}{\varepsilon_{||}(\omega)\cos^2\theta + \varepsilon_\perp(\omega)\sin^2\theta}, \quad (1)$$

where $\varepsilon_{||}(\varepsilon_\perp)$ is the dielectric function parallel (perpendicular) to the Z-axis, and θ is the angle between $\mathbf{q}$ and Z-axis as shown in Fig. 1. ω and c are, respectively, the angular frequency of phonon polaritons and the velocity of light. The dispersion curve calculated using phonon frequencies shown in Table 1 and values of 4.0 [1] for high frequency dielectric constants and 2.0 [3] for refractive indices around the incident and scattered light wavelengths is shown in Fig. 4 by a solid line. Note that we assumed that the anisotropy of high frequency dielectric constants and refractive indices is small enough to be neglected. Experimentally obtained frequencies, which are plotted with dots in Fig. 4, are in good agreement with

the theoretical curve of extraordinary phonon polaritons. The fact confirms that the signals originate from extraordinary phonon polaritons.

4 Conclusion

We have observed extraordinary phonon polaritons in ZnO by forward Raman scattering spectroscopy with annular apertures. The assignment was confirmed by Raman scattering efficiencies and theoretically obtained dispersion curves.

References

1. S. P. S. Porto, B. Tell, and T. C. Damen, Phys. Rev. Lett. **16**, (1966) 450.
2. T. C. Damen, S. P. S. Porto, and B. Tell, Phys. Rev. **142**, (1966) 570.
3. K. Torii, M. Ono, and T. Sota (unpublished).

Disorder-Activated Resonant Raman Scattering in GaNAs/GaAs Structures

I. A. Buyanova[1], W. M. Chen[1], H. P. Xin[2], and C. W. Tu[2]

[1]Department of Physics and Measurement Technology, Linköping University, S-581 83 Linköping, SWEDEN
[2] Department of Electrical and Computer Engineering, University of California, La Jolla, CA 92093-0407, USA

Abstract Resonant photoluminescence and resonant Raman (RR) spectroscopies are utilized to study effect of disorder on the optical and vibronic properties of GaNAs-based structures. New Raman modes located at 10.2, and 22 are observed under resonant excitation close to the bandgap of the GaNAs alloy and are tentatively attributed to disorder activated Raman scattering involving acoustic phonons which exhibit resonance with localized states. From temperature dependent RR measurements, the dephasing of the localized states at low temperatures is shown to be determined by phonon-assisted tunneling.

1 Introduction

Considerable differences in sizes and electronegativity of a N atom and other column V anions which are replaced lead to a huge bowing in band gap energy, which is the most distinguished property of III-V-N alloys. On the other hand, the large size mismatch and the existence of the large miscibility gap can also cause nonuniformity in the local strain field of the alloy and severe compositional disorder. Localization and disorder are known to introduce changes in electronic and vibronic properties of semiconductors. However very little is so far known regarding their effect on the N-containing III-V alloys.

In this work we employ resonant Raman scattering (RRS) and selectively excited photoluminescence (PL) to study effects of localization and disorder on optical properties of GaNAs/GaAs quantum wells and epilayer structures with N composition of about 1 %.

2 Samples and Methods

All the investigated structures were undoped and were grown by gas source MBE. Two types of sample structures were studied, i. e. 7 periods GaAs/GaN$_x$As$_{1-x}$ (70Å/ 200Å) multiple quantum well (MQW) structures and 1100 Å thick GaN$_x$As$_{1-x}$ epilayers. N composition x in the structures was about 1%.

PL and Raman scattering were excited either by the 514 nm line of an Ar$^+$ ion laser or by a tunable titanium-sapphire solid state laser. Spectra were measured by a Ge detector and a grating monochromator.

3 Experimental Results and Discussion

Under above band gap non-resonant excitation the PL spectra of GaNAs are dominated by the commonly observed [1] featureless localized exciton (LE) emission – see Fig.1. In contrast, when excitation energy is tuned close to the GaNAs band edge a series of additional narrow lines can be detected in the spectra – Fig.1. The spectral position of the sharp lines changes continuously while maintaining the same energy separations from the laser energy. Thus they can be attributed to Raman scattering.

Three of the Raman lines, i.e. the 10.2, 22 and 45 meV modes demonstrate rather unusual properties, which are the subject of this report (the 33.5 and 36.5 meV lines are due to the non-resonant Raman scattering involving host optical phonons in GaNAs [2]). First of all, they can be excited only within a narrow spectral range, i.e. exhibiting strongly resonant behavior – Fig.2. The resonant profile has a line width of about 20 meV and a maximum position at the high energy edge of the LE PL emission well below the

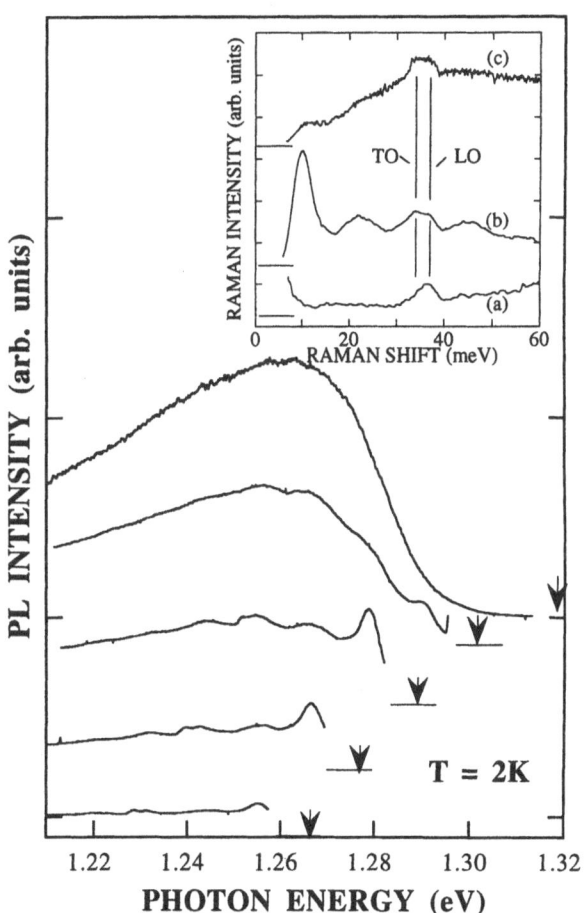

Fig.1. PL and Raman spectra of the GaN$_{0.012}$As$_{0.988}$/GaAs MQW structure with various excitation wavelengths (indicated by arrows). Spectra are shifted in vertical direction for clarity. The insert shows Raman spectra detected from the GaN$_{0.012}$As$_{0.988}$/GaAs MQW structure under non-resonant (a) and resonant excitation (b), i.e. with laser energy of 1.530 eV and 1.278 eV, respectively. For comparison, also shown in the insert is a Raman spectrum from the reference GaN$_{0.012}$As$_{0.988}$ epilayer sample (c) measured under the resonant conditions.

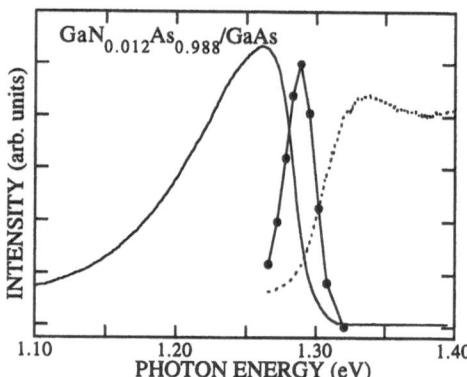

Fig. 2. Comparison of the PL (solid lines), PLE (dotted lines) and resonant Raman profile (dots) of the dominant 10.2 meV Raman mode.

maximum of the PL excitation (PLE) spectrum. This suggests that the resonance might occur via localized rather than free states. The spectral position of the RR profile depends on N composition providing evidence that this scattering is related to the GaNAs alloy. Unlike the normal Raman scattering, the intensity of the RR lines decreases almost exponentially with increasing temperature starting from 1.9K - Fig.3a. This is in contrast to the PL thermal dependence where a decrease of the PL intensity occurs only at temperatures higher than 20K- see Fig.3a.

Let us now address the possible origin of the RR modes. Considering their energy positions, possible mechanisms for the 10.2 meV and 22 meV lines include electronic Raman scattering (ERS) by shallow dopants, or RR scattering involving zone edge acoustic phonons. However the electronic structure of shallow impurities is known to change with quantum confinement [3] which is not observed in the experiment – see insert in Fig.1. Thus ERS by the shallow defects can be ruled out as the origin of the observed RR modes.

As an alternative, the RR lines has been proposed to represent disorder activated transverse acoustic (DATA) and longitudinal acoustic (DALA) modes involving zone edge phonons which exhibit strong resonance with the localized states. The 45 meV line can then be related to LA overtone scattering. The observed strong temperature dependence of the RRS can be taken as an additional proof that the intermediate electronic states involved in scattering process are the localized states [4]. Indeed, the intensity (I_R) of the RR scattering by the LE has been shown [4] to be inversely proportional to the homogeneous excitonic linewidth Γ. The drastic thermal quenching of the RRS in this case reflects a decrease of the lifetime of the intermediate electronic state involved in the scattering. It can be related to activation of excitons from the localized states with a long lifetime (small Γ) to the delocalized states or to the thermally-enhanced diffusion of the LE between different localized states. The former process should prevail in the same temperature range as the PL thermal quenching, i.e. above 20K in our experiments – Fig.3a. Below 20 K the dephasing of the intermediate states should be determined by thermally enhanced transfer

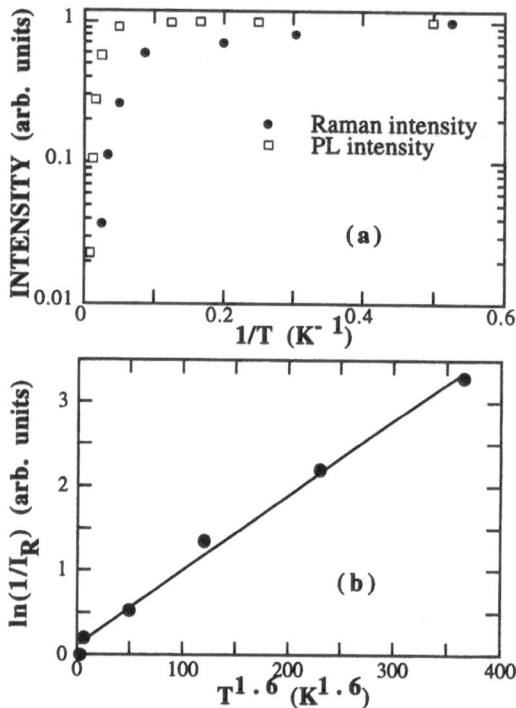

Fig.3. (a) Arrhenius plots of the integrated PL intensity (open squares) and the intensity of the dominant 10.2 meV Raman mode (filled dots) from the GaN$_{0.012}$As$_{0.988}$/GaAs MQW. (b) Dependence of the homogeneous linewidth (inversely proportional to the Raman intensity I_R) as a function of $T^{1.6}$

of the LEs. Taking into account that $I_R(T) \sim 1/\Gamma(T)$, we can determine the mechanism of linewidth broadening. The experimental data are well described by the dependence $\ln(1/I_R) \sim T^{1.6}$ – Fig. 3b. Thus the low-temperature dephasing of the localized states in the GaNAs is determined by the phonon-assisted tunneling [5].

4 Summary

In conclusion, we have found that the compositional disorder and localization in the GaNAs material lead to the appearance of the unusual Raman modes located at 10.2, and 22 meV. The RR lines have been proposed to be due to the disorder activated Raman scattering involving acoustic phonons which exhibit strong resonance with the localized states. From temperature dependent RR measurements, the dephasing of the localized states at low temperatures is shown to be determined by the phonon-assisted tunneling.

References
1. Buyanova et al, Appl. Phys. Lett. **75**, (1999) 501.
2. A. M. Mintrainov et al, Phys. Rev. B. **56**, (1997) 15 836.
3. C. Mailhiot, Y.-C. Chung, and T. C. McGill, Phys. Rev. B. **26**, (1982) 4449.
4. J. E. Zucker et al, Phys. Rev. B. **35**, (1987) 2892.
5. T. Takagahara, Phys. Rev. B. **32**, (1985) 7013.

Collective dynamics in liquid Ge obtained by an inelastic x-ray scattering experiment

S. Hosokawa[1], Y. Kawakita[1]*, W.-C. Pilgrim[1], F. Hensel[1],
H. Sinn[2], M. Krisch[3]

[1] Institut für Physikalische-, Kern-, und Makromolekulare Chemie, Philipps Universität Marburg, Hans-Meerwein-Str.
D-35032 Marburg, Germany e-mail: hosokawa@mailer.uni-marburg.de
[2] SRI-CAT, APS, Argonne National Laboratory, Argonne, IL 60439, USA
[3] ESRF, BP220, F-38043 Grenoble cedex 9, France

Abstract   We present the first results for the dynamic structure factor $S(Q,\omega)$ of liquid Ge at 980 °C obtained from inelastic x-ray scattering experiments. Distinct exitations resulting from propagating modes can be identified. The phonon dispersion obtained in the present experiments matches the hydrodynamic sound velocity in the low Q range, i.e., no *positive* dispersion can be detected. This result is, however, inconsistent with generalized hydrodynamics beyond $Q = 5$ nm^{-1}.

1 Introduction

Liquid (*l*-) Ge shows many unusual properties, which in the past have stimulated intensive experimental and theoretical investigations. Upon melting it undergoes a semiconductor-metal transition in which the density *increases* by about 4.7 % accompanied by significant structural changes. The coordination number grows from four in the solid state to about 6.5 in the liquid [1]. Despite its metallic nature, the structure of *l*-Ge is more complicated than that of typical liquid metals like liquid alkalis. Besides the undercoordination, the structure factor $S(Q)$ of *l*-Ge has a shoulder on the high-Q side of the first peak, a feature that cannot be reproduced using a simple hard-sphere model. These features were interpreted as indications that covalent structures persist in the liquid state [1]. In this paper we report the first results of the dynamic structure factor $S(Q,\omega)$ for *l*-Ge obtained from a high-resolution inelasic scattering experiment using intense x-rays from a third generation synchrotron facility.

2 Experimental procedure

Inelastic x-ray scattering is a modern technique which allows to study the Q dependence of excitations in the meV range. Hence, this method is suitable to investigate, e.g., liquid dynamics which has so far been a domain of inelastic neutron scattering. The

Present address: Department of Physics, Graduate School of Science, Kyushu University, Ropponmatsu, Chuo-ku, Fukuoka 810-8560, Japan

present experiments were carried out at the beamline ID28 of the European Synchrotron Radiation Facility (ESRF) in Grenoble, France, using a newly developed horizontal diffractometer. Details of this technique are given elsewhere [2].

The sample was located in a single-crystal sapphire cell [3], placed in a vessel equipped with Be windows capable to cover scattering angles of 0°-55° [4]. The vessel was filled with 2 bar of high-purity grade He gas. The temperature of 980 °C, well above the melting point, was achieved using a W resistance heater and measured using two W-5%Re:W-26%Re thermocouples.

The experiments were performed for nine Q values between 2-28 nm^{-1}. Since the energy resolution of the spectrometer, 2.48 meV (FWHM), was in the same range as the width of the peaks in $S(Q,\omega)$, a resolution correction needed to be performed by fitting the convolution between a model function and the experimental resolution to the experimental data. As the model function, a Lorenzian was chosen for the central line and a damped harmonic oscillator [5] for the inelastic peaks.

3 Results and discussion

Fig.1 shows the results of $S(Q,\omega)$ normalized to $S(Q)$ at $Q = 2$-12 nm^{-1}, where inelastic peaks can clearly be identified. Full circles represent the experimental data with error bars and the solid line is the best fit of the convolution integral to the data. Also given is the experimental resolution function by a dashed line at the bottom of the figure. At $Q = 2$ nm^{-1}, the sharp phonon peaks located at $\omega = \pm 3.8$ meV superimpose the central quasielastic line. With increasing Q, the width of both the elastic and inelastic peaks broaden, and especially the phonon peaks get to be highly damped. At $Q = 12$ nm^{-1}, which is about a half the Q value of the first maximum in $S(Q)$ (25 nm^{-1} [1]), the phonon peaks are located at ± 17.2 meV. The spectra in Fig. 1 demonstrate that the dynamics of *l*-Ge is dominated by longitudinal propagating modes,

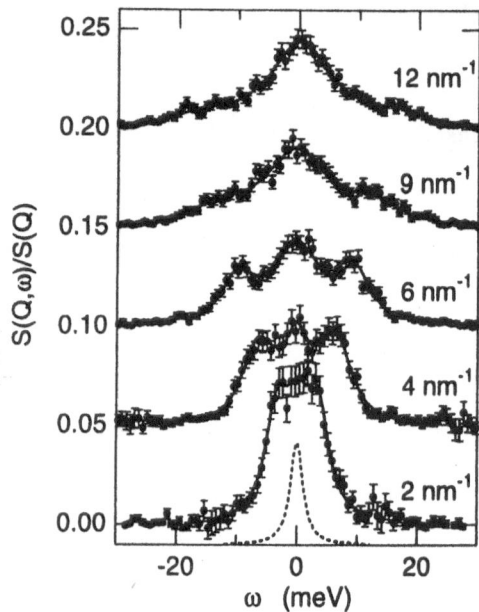

Fig. 1 $S(Q,\omega)$ normalized to $S(Q)$ at $Q = 2\text{-}12\ \text{nm}^{-1}$.

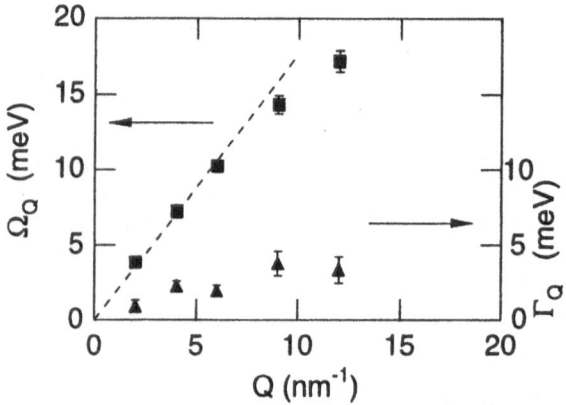

Fig. 2 The phonon excitation energy Ω_Q (squares) and the peak width Γ_Q (triangles). The dashed line represents the dispersion of the hydrodynamic sound.

analogous to liquid alkali metals where phonon excitations are well known [6].

The excitation energy Ω_Q and the peak width Γ_Q as a function of Q are depicted in Fig. 2 by squares and triangles, respectively. The dashed line represents the dispersion of hydrodynamic sound obtained from recent measurements of the adiabatic sound velocity (2682 m/s). Of particular interest in this dispersion relation is the observation that within the experimental error, the positions of the phonon peaks lie on the hydrodynamic line in the low Q range up to 6 nm^{-1}. Namely, the so-called *positive* dispersion relation observed in liquid alkali metals [6] is not

found in *l*-Ge.

The most fundamental approach to a theoretical understanding of the collective dynamics in simple monatomic fluids was obtained through the application of generalized hydrodynamics [7], which has successfully been applied to many liquid metal systems. In the hydrodynamic limit $(\omega\tau)^2 \ll 1$, where τ stands for the viscoelastic relaxation time introduced by Maxwell, the slope of the phonon dispersion is given by the adiabatic sound velocity. For $(\omega\tau)^2 \gg 1$, the viscoelastic damping starts to contribute to the relaxation process in the liquid, causing a shift of Ω_Q to higher ω, i.e., the *positive* dispersion. In this high-frequency limit, the dispersion relation has the limit $\Omega_Q = c_\infty Q$, where c_∞ is no longer determined by the compressibility of the system, but ruled by elastic moduli as in a solid.

Although *l*-Ge does not belong to these simple liquids, an attempt to check the applicability of this theory may be instructive. Using a potential given by Arnold *et al.* [8], c_∞ in the low Q limit could be calculate to be about twice c_s, hence a big *positive* dispersion should be expected. However, τ could also be calculated to give 8.5×10^{-2} ps. This value is considerably smaller than in liquid alkali metals, which might be due to the covalent nature of *l*-Ge. Since the crossover appears around $\omega\tau \approx 1$, a *positive* dispersion should be observed beyond 7.8 meV, which corresponds a Q of 5 nm^{-1}. Accordingly it is not reasonable that the present result does not show a *positive* deviation from the hydrodynamic line beyond $Q = 5\ \text{nm}^{-1}$. This inconsistency may be due to the fact that *l*-Ge does not belong to the class of liquids that can be approximated as simple monatomic fluids, i.e., three-body correlation or the covalent nature dominates the collective dynamics in *l*-Ge.

References

1. Y. Waseda and K. Suzuki, Z. Phys. B **20**, (1975) 339.
2. G. Ruocco and F. Sette, J. Phys.: Condens. Matter **11**, (1999) R259.
3. K. Tamura, M. Inui, and S. Hosokawa, Rev. Sci. Instrum. **70**, (1999) 144.
4. S. Hosokawa and W.-C. Pilgrim, Rev. Sci. Instrum., submitted.
5. B. Fak and B. Dorner, ILL Report No. 92FA008S, 1992.
6. W.-C. Pilgrim *et al.*, J. Non-Cryst. Solids **250-252**, (1999) 96, and references therein.
7. J. P. Boon and S. Yip, *Molecular Hydrodynamics* (McGraw-Hill. New York 1980).
8. A. Arnold, N. Mauser, and J. Hafner, J. Phys.: Condens. Matter **1**, (1989) 965.

Non-Perturbative Terahertz Sideband Generation from Bulk GaAs

M. A. Zudov[1], J. Kono[1], A. P. Mitchell[2], A. H. Chin[2], and K. Johnsen[3]

[1] Department of Electrical and Computer Engineering, Rice University, Houston, Texas 77005, U.S.A.
[2] W.W. Hansen Experimental Physics Laboratory, Stanford University, Stanford, California 94305, U.S.A.
[3] NORDITA, Blegdamsvej 17, DK-2100 Copenhagen, Denmark.

Abstract Using intense picosecond pulses of coherent terahertz (THz) radiation, we have investigated time-resolved non-resonant THz sideband generation from bulk GaAs. The THz power dependence clearly reveals a non-perturbative strong-field regime. In addition to the expected ω_{-2} sideband, we detected the ω_{-1} (or odd) sideband which has previously been observed only in a quantum well system where the inversion symmetry was intentionally broken.

1 Introduction

Semiconductors subject to an intense electric field exhibit phenomena that cannot be understood by treating the field as a small perturbation. Intense fields can strongly modify the lattice periodic potential, exerting significant influences on optical processes. Exploring these electro-optical effects in high frequency ac fields, especially in the THz range, is not only important technologically, but also allows us to explore new multi-photon, strong-field phenomena that cannot be probed by other traditional methods.

Here, we study THz sideband generation [1]. In this multi-photon process, a weak optical beam passes through a THz-driven semiconductor, acquiring sideband frequencies separated by integer multiples of the THz frequency:

$$\omega_n = \omega_{NIR} + n\omega_{THz}; \quad n = \pm 1, \pm 2, \pm 3.... \qquad (1)$$

These lines completely dominate the near-bandedge emission properties of THz-driven semiconductors [1]. Recent theoretical studies predicted new phenomena for THz sidebands such as dynamical symmetry breaking [2], non-monotonic power dependence [3], chaotic behavior [4], sideband disappearance under cyclotron resonance [5], and THz-induced subband hybridization [6]. However, none of these has been observed.

Previous observation of THz sidebands relied on *resonant enhancement* using magnetoexcitons [1] or quantum well subbands [7], and the main results were successfully explained by a *perturbation* theory [8]. In [7] the inversion symmetry of the system was intentionally broken, allowing the detection of first-order ($n = \pm 1$) sidebands, whereas in [1] only even ($n = \pm 2, \pm 4$) sidebands were observed. In either case, there was no evidence of bulk $\chi^{(2)}$ contribution. Finally, these previous studies were essentially CW, and thus did not provide any information on the time evolution of the sidebands. In what follows, we present results on *time-resolved* spectroscopy of *non-perturbative, off-resonance* THz sideband generation from *bulk* GaAs.

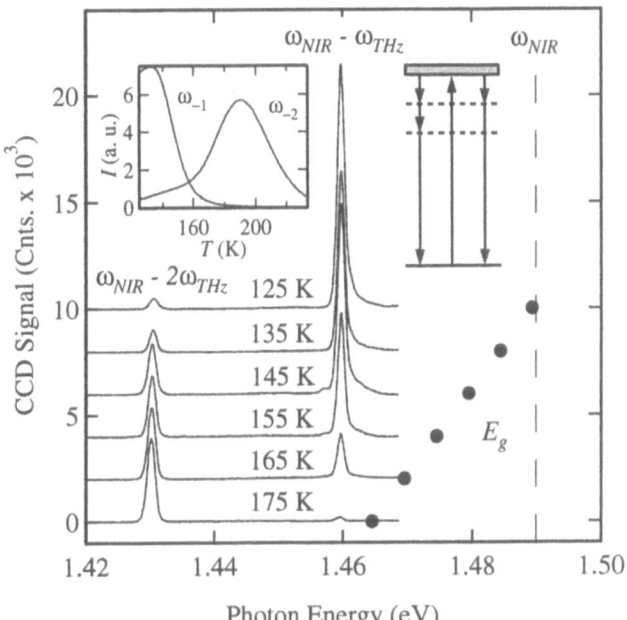

Fig. 1 Solid lines are offset for clarity sideband spectra at different T; vertical dashed line denotes the $\hbar\omega_{NIR}$; circles represent E_g position as measured by FTIR for each T; inset shows the calculated [Section 3] T-dependence of the sidebands.

2 Experimental Results and Discussion

We utilized the wide range (5 to 100 THz or 3 to 70 μm), short pulse duration (≥ 0.6 ps), and high peak powers (≤ 2 MW) of the Stanford Free Electron Laser. We combined the THz pulses with the near-infrared (NIR) pulses from a mode-locked Ti:Sapphire laser using a Pelicle plate and then focused the collinear beams onto the sample placed in a variable temperature (20 – 300 K) cryostat. The transmitted and emitted NIR radiation was then sent into a spectrometer/CCD imaging system. The temporal overlap between the two beams was adjusted via a delay stage. Most measurements were made with the sample oriented at 45° with respect to the incident beams. While similar results were obtained for a variety of THz and NIR wavelengths, here we present the data only for $\lambda_{NIR} = 832$ nm (1.49 eV) and $\lambda_{THz} = 42$ μm (0.03 eV).

In Fig. 1 we show typical sideband spectra acquired at different temperatures, T. The vertical dashed line represents the fundamental NIR frequency, ω_{NIR} and the circles show the bandedge of GaAs, E_g for each T,

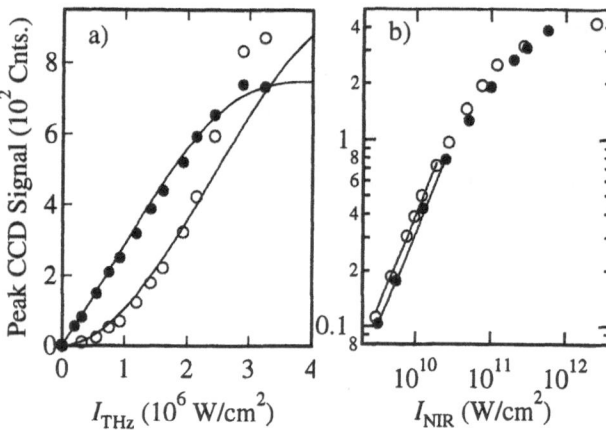

Fig. 2 Filled (open) circles: ω_{-1} (ω_{-2}) sideband intensity; solid curves in a) are calculated according to the model (3); lines in b) are linear fits to the data

obtained by FTIR interband absorption spectroscopy. We observe two strikingly strong lines which appear precisely at one and two ω_{THz} below the ω_{NIR}, identifying themselves as ω_{-1} and ω_{-2} sidebands, respectively [Eq. (1)]. The energy level diagram in Fig. 1 illustrates the experimental situation: dashed lines are the intermediate virtual levels and the shaded area represents the continuum of real states above the bandedge. With our available THz intensities, I_{THz} and detector sensitivity, we were not able to detect third or higher order sidebands.

As seen from Fig. 1 the intensity of the sidebands is strongly T-dependent. As T is raised from 125 K to 175 K, the ω_{-2} sideband increases while the ω_{-1} sideband decreases and finally goes away at $T \approx 180$ K. At higher T (not shown) ω_{-2} sideband weakens and vanishes at $T \approx 230$ K. This behavior can be qualitatively explained by the interplay between the absorption of the ω_{NIR} and that of the sidebands. While a larger density-of-states at the ω_{NIR} favors more intense sidebands, no density-of-states is desirable at the sideband frequencies. The inset in Fig. 1 represents the results of our calculations [Section 3], in qualitative agreement with experiment.

There are three new features that are worth pointing out. First, unlike the previous studies [1,7], the sidebands appear at energies *below* E_g, where there are no real states and therefore are not resonantly enhanced. Second, we observe an odd sideband (i.e., the ω_{-1}) without intentional breaking of the inversion symmetry. We found that the ω_{-1} sideband depends sensitively on the crystal orientation with respect to the laser beams, indicating that its origin is the lack of inversion symmetry of the GaAs crystal. In particular, it disappeared at normal incidence. Third, since we used picosecond THz pulses, we were able to monitor the evolution of the sidebands directly in the time domain and thus measure the temporal profile of the THz pulse using a Si CCD detector.

Finally, we studied power dependence presented in Fig. 2(a) and 2(b), where the sideband intensities are plotted vs. I_{THz} and I_{NIR}, respectively. At low I_{THz}

the ω_{-1} (ω_{-2}) sideband follows linear (quadratic) dependence, indicating a perturbative $\chi^{(2)}$ ($\chi^{(3)}$) process involving one (two) THz photons. At higher I_{THz}, we observe a deviation from the perturbative behavior which indicates the entrance into the strong THz field regime. Initially lienar with I_{NIR}, both sidebands eventually saturate, probably due to the free carrier absorption of the THz radiation since the $\hbar\omega_{NIR} \approx E_g$.

3 Theoretical Model

The experimental observation of the odd sideband implies the lack of inversion symmetry. While the observed angle dependence on the incident beams can be explained by a simple bulk $\chi^{(2)}$ consideration, it is also possible that the surface is playing a role via band-bending due to Fermi-level pinning and the presence of residual carriers. In order to model this effect we assume that a static electric field is present in addition to the THz and NIR fields. Once the static field is introduced, the first sideband can be explained as a $\chi^{(3)}$ process involving one NIR, one THz plus one zero-frequency photon of the static field. Using non-equilibrium Green function techniques developed in [9] and assuming a simple two band model, we calculate the two time susceptibility, including the THz and static fields nonperturbatively, which in frequency domain describes the sideband generation [2]. As seen from the inset of Fig. 1 and Fig. 2(a), this theory qualitatively accounts for both the T-dependence and the I_{THz}-dependence, including the departure from the perturbative regime. We stress, however, that further experiments are called for in order to certainly pinpoint the physical process leading to the odd sideband generation.

4 Summary

We have performed THz optical sideband generation from bulk GaAs and observed sidebands below the band gap where there are no states. We demonstrated a new scheme to measure the temporal profile of the THz pulses with a NIR detector. The THz power dependence clearly revealed a non-perturbative strong field regime. Finally, we detected an odd sideband, which has previously been observed only in an asymmetric quantum well system where the inversion symmetry was intentionally broken.

References

1. J. Kono *et al.*, Phys. Rev. Lett. **79**, 1758 (1997).
2. K. Johnsen, cond-matt/0001155 (1999).
3. T. Inoshita and H. Sakaki, in the Proc. of the 24th Int. Conf. on the Phys. of Semicond., Jerusalem, Israel, 1998, ed. D. Gershoni (World Scientific, Singapore, 1999).
4. K. Johnsen, unpublished.
5. T. Inoshita, Phys. Rev. B **61**, 15610 (2000).
6. D.S. Citrin, Phys. Rev. B **60**, 13695 (1999).
7. C. Phillips *et al.*, Appl. Phys. Lett. **75**, 2728 (1999).
8. T. Inoshita *et al.*, Phys. Rev. B **57**, 4604 (1998).
9. A.-P. Jauho and K. Johnsen, Phys. Rev. Lett. **76**, 4576 (1996); K. Johnsen and A.-P. Jauho, Phys. Rev. B **57**, 8860 (1998)

Proc. 25th Int. Conf. Phys. Semicond., Osaka 2000 (Eds. N. Miura and T. Ando)

79

Ab initio study of optical absorption spectra of semiconductors and conjugated polymers

M. L. Tiago[1], Eric K. Chang[1], Michael Rohlfing[2], and Steven G. Louie[1]

[1] Department of Physics, University of California at Berkeley, Berkeley, CA, 94720, USA and Materials Science Division , Lawrence Berkeley National Laboratory, Berkeley, CA, 94720

[2] Institut für Theoretische Physik II - Festkörperphysik, Universität Münster, Wilhelm-Klemm-Straße 10, 48149 Münster, Germany

Abstract The effects of electron-hole interaction on the optical properties of a variety of materials have been calculated using an *ab initio* method based on solving the Bethe-Salpeter equation. Results on selected semiconductors, insulators, and semiconducting polymers are presented. In the cases of alpha-quartz (SiO_2) and poly-phenylene-vinylene, resonant excitonic states qualitatively alter the absorption spectra.

1 Introduction

The role of electron-hole interactions in the optical properties of semiconductors and insulators has been long recognized (see *e.g.* [1]). It is possible now to get an accurate description of such properties for a large class of materials from first principles calculations [2–4]. The theoretical framework is based on solving a Bethe-Salpeter equation for the electron-hole amplitude of the 2-particle Green's function[5]. Our approach [2] takes into account electron self energy effects and the interaction between an optically excited electron and the hole left behind. The underlying one-particle (quasiparticle) description is taken within the GW approximation for the electron self-energy [6].

2 Theory

In moderately correlated electron systems, the photo-excited states can be written, to a good approximation, as a linear combination of free electron-hole configurations,

$$|S\rangle = \sum_{cvk} A_{cvk}^S |c\mathbf{k}; v\mathbf{k}\rangle \qquad (1)$$

where $|c\mathbf{k}; v\mathbf{k}\rangle$ represents a configuration in which a quasi-electron is promoted from the valence band v to the conduction band c. Only vertical transitions are taken into account. From the Bethe-Salpeter equation for the two-particle Green's function, the coefficients A_{cvk}^S satisfy an eigenvalue equation,

$$(E_{c,\mathbf{k}} - E_{v,\mathbf{k}}) A_{cvk}^S + \sum_{cvk,c'v'k'} \mathcal{K}_{cvk}^{c'v'k'} A_{c'v'k'}^S = \Omega^S A_{cvk}^S$$

$$(2)$$

where $\mathcal{K}_{cvk}^{c'v'k'}$ describes the interaction between the excited electron and hole and Ω^S is the energy of state $|S\rangle$. In general, $\mathcal{K}$ is a complicated function that describes the scattering of an electron-hole pair in configuration $|c\mathbf{k}; v\mathbf{k}\rangle$ to $|c'\mathbf{k}'; v'\mathbf{k}'\rangle$. This can be simplified to a screened attractive interaction plus a bare repulsive ex-

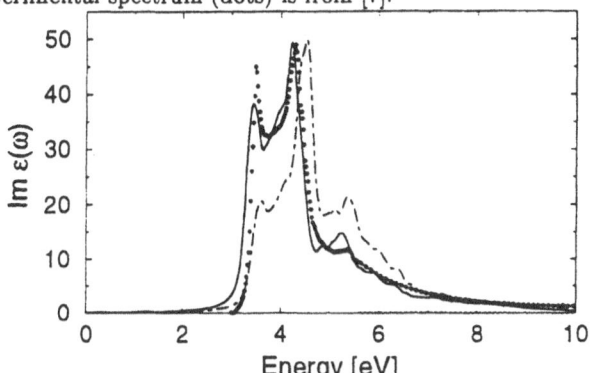

Fig. 1 Absorption spectrum of bulk silicon, with (full line) and without (dotted-dashed line) excitonic effects. The experimental spectrum (dots) is from [7].

change interaction [2]. Dynamical interactions can also be neglected if the energies Ω^S are close to the energy of non-interacting pairs.

The optical absorption is given by

$$\epsilon_2(\omega) = \frac{4\pi^2 e^2}{\omega^2} \frac{1}{V_c} \sum_S |\langle G|\lambda \cdot \mathbf{v}|S\rangle|^2 \delta(\Omega_S - \hbar\omega) \qquad (3)$$

where $|G\rangle$ is the ground state and λ and $\mathbf{v}$ are the polarization vector and the velocity operator, respectively.

The one-particle energies in Eq. (2) are taken from a GWA calculation, as described by Hybertsen and Louie [6].

3 Applications

The measured absorption spectrum of silicon, plotted in Figure 1, shows two well-pronounced peaks at energies about 3.5 eV and 4.2 eV [7]. The calculated results using Eq. (3) reproduce the experimental data very accurately. The minor discrepancies at the energy range 5.5-6.0 eV may be due to the finite sampling of points in the Brillouin zone. The large enhancement in the amplitude of the first peak is due to excitonic effects.

The optical properties of alpha quartz (SiO_2) have been a subject of debate since the 1960's [8]. The presence of a series of four sharp peaks in its absorption spectrum has been recently shown to be due to the formation of resonant excitons [3].

In Figure 2, we show the optical absorption of SiO_2 for polarization perpendicular to the hexagonal plane. The agreement between the experimental spectrum (with peaks at energies 10.3, 11.7, 14.0 and 17.3 eV) and the

Fig. 2 Absorption spectrum of alpha quartz (SiO_2) with excitonic effects (solid line) as compared to the interband transition theory (dotted-dashed line). Experimental data are given by the dashed line [8].

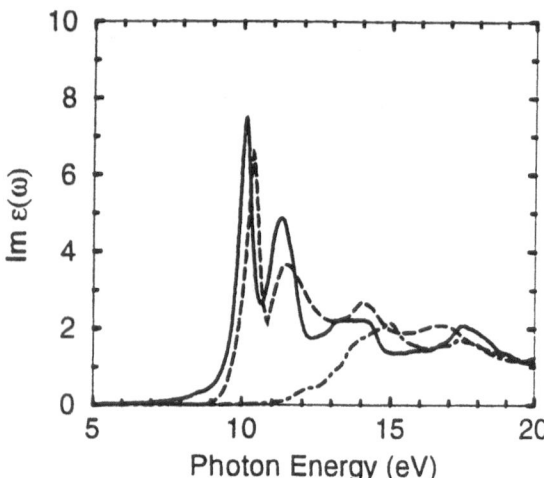

Table 1 Calculated binding energies of excitons in GaAs, compared with data from optical absorption [9] and from two-photon absorption [10].

	this work [meV]	Exp. [meV]
E_{1s}	4.0	4.2 [a]
E_{2s}	0.9	1.09 [a]
E_{1p}	0.2-0.7	~ 0.1 [b]

[a] Reference [9]
[b] Reference [10]

calculated spectrum, with excitonic effects included, is again excellent. The fact that an interband transition description gives an almost featureless absorption spectrum shows that the formation of optically active resonant excitonic states dominates the spectrum up to energies about 18 eV.

It is also possible to obtain the binding energy of bound excitons very accurately from Eq. (2), even when this binding energy is only a small fraction of eV. In this case, a careful sampling of the Brillouin zone is required, so that the electronic energy bands are described with the necessary accuracy. Table 1 shows the binding energy of the lowest energy excitonic states in GaAs [2]. In this calculation, we made use of a set of 1000 special k-points near the position of the band extrema.

As an example of a lower-dimensional system, we present results for the conjugated polymer, poly-phenylene-vinylene (PPV), in Figure 3. The difference between the results with and without excitonic effects is striking. The peak at 2.4 eV results from coherent coupled transitions between the highest occupied band and lowest unoccupied band. The resulting exciton has a large binding energy of 0.9 eV. The measured absorption peaks are located at energies 2.5, 3.7, 4.8 and 6.0 eV, which are in quite good agreement with the calculated peak posi-

Fig. 3 Calculated optical absorption spectrum of PPV, with (solid line) and without (dashed line) excitonic effects. An artificial broadening of 0.05 eV is included, and the polarization vector is along the chain.

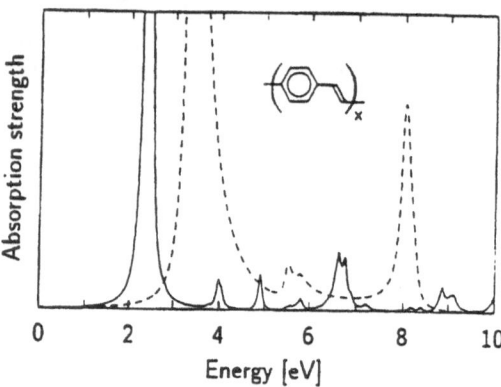

tions. As in the case of SiO_2, the spectrum of PPV is dominated by resonant excitons.

4 Conclusion

In summary, we have shown that an *ab initio* method, based on solving the Bethe-Salpeter equation, is capable of describing successfully the optical properties of a variety of semiconductors and insulators. The agreement with experimental data is remarkably good. The same analysis can be and have been applied to other systems including surfaces and clusters[2].

5 Acknowledgments

This work was supported by NSF Grant No. DMR-9520554, U.S. Department of Energy under Contract No. DE-AC03-76SF00098, and by the Deustche Forschungsgemeinschaft (Bonn, Germany) under Grant No. Ro-1318/1-1. Computational resources were provided by NERSC and by the NSF.

References

1. F. Bassani and G. Pastori Parravicini, *Electronic States and Optical Transitions in Solids* (Pergamon Press. Oxford 1975)
2. M. Rohlfing and S. G. Louie. Phys. Rev. Lett. **81**, 2312 (1998); Phys. Rev. B, **62**, 4927 (2000)
3. E. K. Chang, M. Rohlfing and S. G. Louie, to appear in Phys. Rev. Lett.
4. L. X. Benedict, E. L. Shirley and R. B. Bohn, Phys. Rev. Lett. **80**, 4514 (1998) ; Phys. Rev. B **59**, 5441 (1999) ; S. Albrecht, L. Reining, R. Del Sole and G. Onida, Phys. Rev. Lett. **80**, 4510 (1998)
5. G. Strinati, Phys. Rev. B **29**, 5718 (1984)
6. M. S. Hybertsen and S. G. Louie, Phys. Rev. Let. **55**, 1418 (1985); Phys. Rev. B **34**, 5390 (1986)
7. D. E. Aspnes and A. A. Studna, Phys. Rev B **27**, 985 (1983)
8. H. R. Philipp, Solid State Commun., **4**, 73 (1966)
9. D. D. Sell, Phys. Rev. B **6**, 3750 (1972)
10. J. S. Michaelis, K. Unterrainer, E. Gornik, and E. Bauser, Phys. Rev B **54**, 7917 (1996)
11. S. Mukamel *et al.*, Science **277**, 781 (1997)

Optical response of incoherent excitons in a dense electron-hole plasma

T. Schmielau, G. Manzke, D. Tamme, K. Henneberger

Department of Physics, University of Rostock, D-18051 Rostock, Germany
e-mail: tim@physik3.uni-rostock.de

Abstract The influence of a fraction of relaxed excitons in a dense electron-hole plasma onto single particle properties and optical response is studied. Using Green's function techniques the optical susceptibility is expressed in terms of the T-matrix taking into account selfenergies of the carriers depending both on wave-number and energy. Numerical calculation of the T-matrix allows a detailed study of the Mott transition of the excitons. For temperatures above $T \gtrsim 40\,\mathrm{K}$ we find, that the Mott transition is already finished when the chemical potential crosses the exciton, preventing excitonic gain discussed for II-VI semiconductors.

Semiconductor Bloch equations are a well established method for the description of the optical response of highly excited semiconductors. Taking into account as relevant many-body effects (1) the renormalization of the band gap, (2) screening of the Coulomb interaction between carriers, and (3) the diagonal as well as off-diagonal dephasing, the absorption can be described in a range from low excitation, where excitonic features dominate (see [1] and papers cited therein), up to higher excitations, where optical gain appears (see [2] and papers cited therein). However, up to now only situations were considered, where all carriers are unbound, e.g. where no population of incoherent (relaxed) excitons exists. The existence of an excitonic population becomes relevant if the Mott transition and the onset of gain are discussed and the question arises, whether an exciton gas or an electron-hole plasma dominates the properties of the gain in II-VI semiconductors [3].

To incorporate the effect of relaxed excitons, we use Green's function techniques and base our approach on the T-Matrix [4,5], which describes the two-particle states of the system. The retarded T-matrix obeys a Bethe-Salpeter equation (BSE),

$$T^{\mathrm{r}}_{\mathrm{eh},k,k'',q}(\omega) = -W^{\mathrm{eff}}_{k-k''}(\omega)$$
$$- \sum_{k'} W^{\mathrm{eff}}_{k-k'}(\omega)\, \chi^0_{k',q}(\omega)\, T^{\mathrm{r}}_{\mathrm{eh},k',k'',q}(\omega) \quad (1)$$

where W^{eff} is the effective screened interaction [1,5], and

$$\mathrm{Im}\,\chi^0_{k,q}(\omega) = \int \frac{d\omega'}{2\pi} \left\{ 1 - f^{\mathrm{e}}(\omega-\omega') - f^{\mathrm{h}}(\omega') \right\}$$
$$\times\, \hat{G}_{\mathrm{e},k}(\omega-\omega')\, \hat{G}_{\mathrm{h},k-q}(\omega') \quad (2)$$

is related to the spectral functions $\hat{G}_a$ of electrons and holes ($a = \mathrm{e, h}$), and the Fermi function $f^a(\omega) = 1/\{1 + \exp[(\omega-\mu_a)/kT]\}$ with the chemical potential μ_a and temperature T. The corresponding real part of $\chi^0_{k,q}(\omega)$

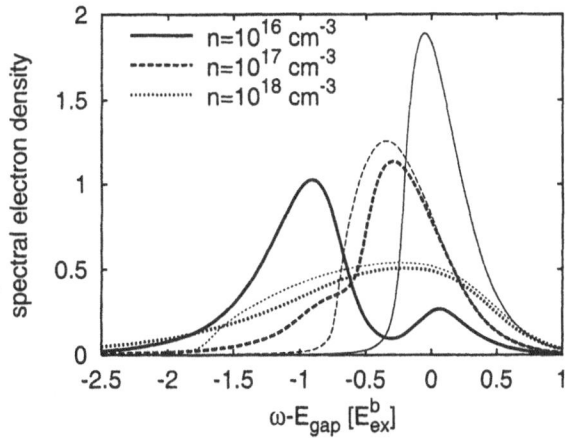

Fig. 1 Spectral carrier densities $n_{\mathrm{e}}(\omega)$ in units of the exciton binding energy $E^{\mathrm{b}}_{\mathrm{ex}}$ in dependence on the detuning with respect to the band gap for ZnSe at $T = 40\,\mathrm{K}$ with (thick lines) and without (thin lines) T-matrix contribution (6)

follows by Kramers-Kronig transformation, and the susceptibility is given by $\chi_k(\omega) = \sum_q \chi_{q,k}(\omega)$.

Utilizing the T-matrix, the susceptibility is given by

$$\chi_q(\omega) = \chi^0_q(\omega) + \sum_{k,k'} \chi^0_{k,q}(\omega) \cdot T^{\mathrm{r}}_{\mathrm{eh},k,k',q}(\omega) \cdot \chi^0_{k',q}(\omega). \quad (3)$$

This corresponds to the level of approximation of [1], if the QPA is used to simplify eq. (2) into

$$\chi^0_{k,q}(\omega) = \frac{1 - f^{\mathrm{e}}_k - f^{\mathrm{h}}_k}{\omega - \varepsilon^{\mathrm{e}}_k - \varepsilon^{\mathrm{h}}_k - \Sigma^{\mathrm{r}}_{\mathrm{e},k}(\omega-\varepsilon^{\mathrm{h}}_k) - \Sigma^{\mathrm{r}}_{\mathrm{h},k}(\omega-\varepsilon^{\mathrm{e}}_k)}, \quad (4)$$

where ε^a_k denote the HF-quasiparticle energies, $\Sigma^{\mathrm{r}}_{a,k}(\omega)$ are the retarded selfenergies and $f^a_k = f^a(\varepsilon^a_k)$. However, as this approximation cannot guarantee the correct crossover between absorption and gain at the chemical potential [6], it is avoided in this paper. Instead, (2) is directly calculated from the one-particle spectral functions

$$\hat{G}_{a,k}(\omega) = -2\,\mathrm{Im}\,[\omega - \varepsilon^a_k - \Sigma^{\mathrm{r}}_{a,k}(\omega)]^{-1}. \quad (5)$$

In order to include the influence of bound e-h pairs additionally to the well-known $G \cdot W$-contribution, describing collisions between single carriers, the T-matrix contribution of the carrier selfenergies has to be considered

$$\mathrm{Im}\,\Sigma^{\mathrm{r,T}}_{\mathrm{e},k}(\omega) = 2\sum_{k'} \int \frac{d\omega'}{2\pi} \left\{ b(\omega+\omega') + f^{\mathrm{h}}(\omega') \right\}$$
$$\times\, \mathrm{Im}\,T''^{\mathrm{r}}_{\mathrm{eh},\hat{k},\hat{k},k-k'}(\omega+\omega')\, \hat{G}_{\mathrm{h},k'}(\omega'), \quad (6)$$

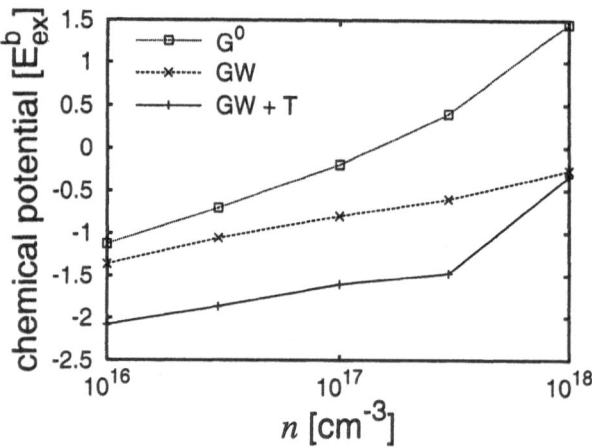

Fig. 2 Chemical potential of excitons $\mu - E_{\mathrm{gap}}$ in ZnSe at $T = 40\,\mathrm{K}$ in dependence on the carrier density with different approximations for the carrier selfenergy as indicated in the figure

where $b(\omega) = 1/\{1 - \exp[(\omega-\mu)/kT]\}$ is the Bose function with $\mu = \mu_{\mathrm{e}} + \mu_{\mathrm{h}}$ and the abbreviation $\hat{k} = (m_{\mathrm{e}}\,k + m_{\mathrm{h}}\,k')/m_{\mathrm{eh}}$ (m_{eh} - reduced mass). In this way the solution of the BSE for the T-matrix (1) and the determination of the carrier selfenergy (6) are coupled.

The influence of the the T-matrix selfenergy (6) is most obvious in the spectral carrier density

$$n_a(\omega) = \sum_k f^a(\omega)\,\hat{G}_{a,k}(\omega), \qquad (7)$$

representing the density of actually occupied states of energy ω. It is shown in fig. 1. Consideration of the T-matrix selfenergy results in two distinct peaks, corresponding to bound and free electrons, at low densities, while without the T-matrix selfenergy only unbound electrons are considered. With increasing density the spectrum of unbound electrons are shifted to lower energy by many-particle effects (band-gap shrinkage), while the bound electrons nearly remain unshifted. Finally both contributions join together (Mott transition). The bound states lead to a considerable reduction of the average electron energy at low densities, as does the band gap shrinkage at high densities (the band edge without many-particle effects is at $\omega = 0$). For higher temperatures the spectral contribution of excitons decreases rapidly, at $T = 77\,\mathrm{K}$ the fraction is below 10%.

This is also reflected in the chemical potential, which has to be calculated for a given temperature T and carrier density $n = n_e = n_h$ by

$$n = \sum_k f_k^a = \sum_k \int \frac{d\omega}{2\pi} f^a(\omega)\,\hat{G}_{aa,k}(\omega) \qquad (8)$$

self-consistently with eqs. (1),(2),(5),(6). It is plotted for different levels of approximation in fig. 2. For high densities the band-gap shrinkage obtained from the $G \cdot W$ selfenergy reduces the chemical potential compared to the case of free carriers (G^0). The T-matrix contribution leads to a further reduction at densities below the Mott

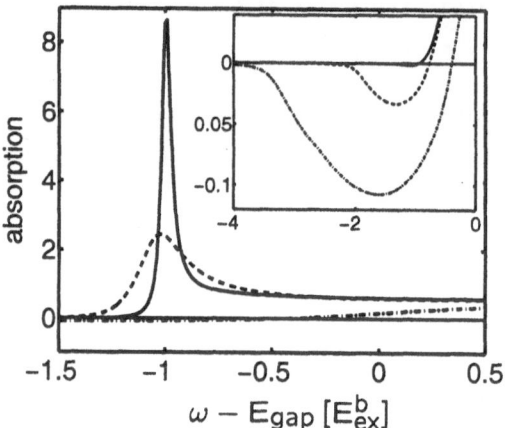

Fig. 3 Imaginary part of the susceptibility (absorption) $\mathrm{Im}\sum_q \chi_q(\omega)$ in dependence on the detuning with respect to the band gap at $T = 40\,\mathrm{K}$ and different carrier densities $n = 1\cdot 10^{17}cm^{-3}$ (solid line), $n = 3\cdot 10^{17}cm^{-3}$ (dashed), $n = 1\cdot 10^{18}cm^{-3}$ (dash-dotted) including the T-matrix selfenergy and without (inset)

transition, preventing a crossover of chemical potential and exciton position as long as excitons exist.

In Fig. 3 the imaginary part of the susceptibility is shown for three different carrier densities including the T-matrix contribution (6) to the selfenergy and without it (see inset). The influence of an excitonic fraction on the chemical potential is reflected here, too: Considering the T-matrix contribution the Mott transition and optical gain in the excitonic region is observed at clearly higher densities. Further, the Mott transition is completed before the chemical potential crosses the exciton position. This is important for the treatment of excitonic gain and Bose condensation of excitons. In II-VI semiconductors different gain mechanisms were discussed in the literature (see [2]), particularly excitonic gain was expected [3], since it was observed just around the exciton position. We find, that for the considered temperatures above $T \gtrsim 40\,\mathrm{K}$ there is no exciton population at densities where gain appears in the excitonic region. Instead, the gain mechanism is dominated by many-body effects between strongly Coulomb correlated carriers.

References

1. G. Manzke, Q.Y. Peng, K. Henneberger, U. Neukirch, K. Hauke, K. Wundke, J. Gutowski, and D. Hommel, Phys. Rev. Lett. **80**, 4943 (1998).
2. K. Henneberger, H. Güldner, G. Manzke, Q.Y. Peng and M.F. Pereira, Jr., Advances in Solid State Physics, **38**, 61 (1998). (Vieweg, Wiesbaden, 1998).
3. J. Ding, M. Hagerott, T. Ishihara, H. Jeon, and A.V. Nurmikko, Phys. Rev. B **47** (1993) 10528.
4. R. Zimmermann, Many - Particle Theory of Highly Excited Semiconductors, Teubner, Leipzig 1988.
5. T. Schmielau, G. Manzke, D. Tamme, and K. Henneberger, phys. stat. sol. (b) **221**, 215 (2000).
6. Q.Y. Peng, T. Schmielau, G. Manzke, and K. Henneberger, J. Crystal Growth **214-215**, 810 (2000).

Effect of reabsorption on electroluminescence of excitonic origin in Cu_2O

Y. Nakamura, N. Naka and N. Nagasawa

Department of Physics, Graduate School of Science, University of Tokyo, 7-3-1 Hongo, Bunkyo-ku, Tokyo 113-0033, Japan
e-mail: ynakam@exciton.phys.s.u-tokyo.ac.jp

Abstract   Spatial and temporal characteristics of electroluminescence (EL) have been studied in naturally grown single crystals of Cu_2O with Au-Al electrodes. Remarkable narrowing of the spectral width of the 1LO-phonon (Γ_3^-) assisted emission of 1s ortho-excitons is observed away from the electrodes at 77K. The spectral change has been completely reproduced by taking account of the reabsorption of the emission light that originates from the excitons generated by the injected carriers at the electrodes and traverses in the crystal. The interpretation is consistent with the temporal change of the EL. The reabsorbance is found to be three times larger than the usual absorbance of the relevant phonon-assisted transitions. At low temperatures as 2K, the effect is negligible and the line narrowing observed implies cooling of the exciton system.

1 Introduction

Electroluminescence (EL) of Cu_2O has been studied in 1960's as one of device-oriented researches [1]. Since the emission spectrum of excitonic origin is similar to relevant photoluminescence in band-to-band excitation regime [2], we became aware that the current injection can be a useful means to generate incoherent excitons of high-density without any excitation light sources. In the present paper, the space-characteristics of the EL is examined in view of current interests on the quantum statistical properties of the excitons as a quasi-bose gas system [3].

2 Experiments

Samples were cut from a natural single crystal of cuprite. The dimensions were $1.2 \times 2.5 \times 9.5$ and $0.8 \times 1.5 \times 6.0$ mm^3. We chose aluminum and gold evaporation layers as anode and cathode, respectively, to enhance the EL intensity [4]. The sample was immersed in liquid nitrogen or superfluid helium. The measurement of space-resolved EL spectra were performed by DC current injection with application of voltage of the order of 100V to the electrodes. We also tried to measure the temporal change by pulsed injection with an appropriate DC bias. An imaging spectrograph composed of a spectrometer (Jasco, CT-25T) and an ICCD camera system (LaVision, PicoStar HR12) was used for the detection.

3 Results and Discussions

Fig.1 shows an example of DC spectrograms of the EL in visible region at 77K. The emission due to oxygen vacancy, Vo (a), and 1LO-phonon assisted emission of 1s ortho-excitons, Xo-Γ_3^- (b), are seen. The corresponding photoluminescence spectrum is shown in the inset. The distance, d, was measured from the edge of the electrode.

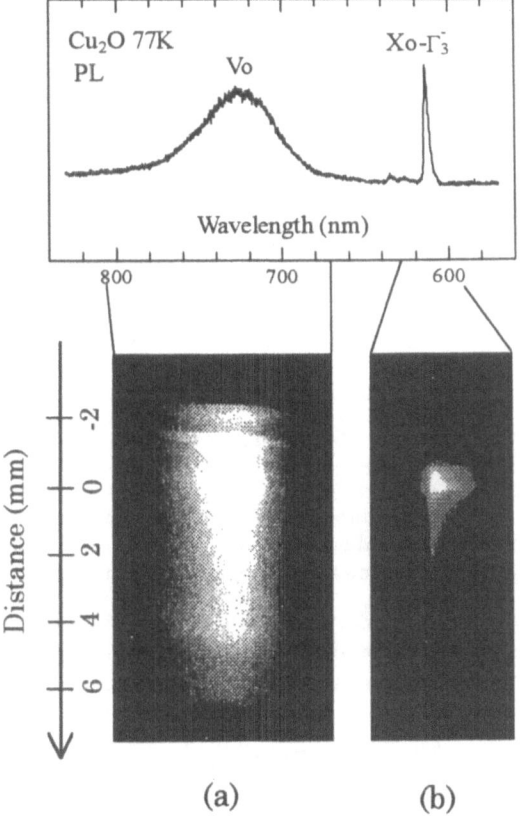

Fig. 1 Spectrogram of EL in Cu_2O at 77K in visible region. Inset shows a corresponding photoluminescence spectrum.

In addition to the EL strongly observed between the electrodes as expected, emission light spreading over the sample could also be seen. However, in contrast to Vo, the exciton emission shows strong narrowing with the increase of d. In particular, high-energy tail disappears quickly.

The solid curves in Fig. 2 are EL spectra of Xo-Γ_3^- at various d. The FWHM is plotted as a function of d in Fig. 3(a), where remarkable narrowing is clearly seen. As is well known, the spectral shape of the phonon-assisted emission reflects the distribution of excitons in momentum space, or, kinetic energy and effective temperature of the system. If we consider that the narrowing is due to the cooling of the excitons migrating through the sample, the temperature reaches down to 40K. Therefore, this interpretation is unrealistic for the present case. O'Hara et al. [5] have noticed that the reabsorption effect on the phonon-assisted emission modulates the spectral shape

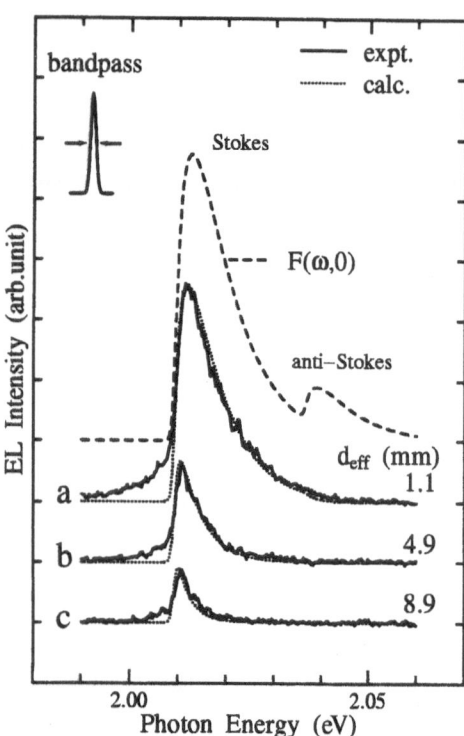

Fig. 2 Spectral shape of Xo-Γ_3^- at different spots on the sample. Corresponding d is 0 (a), 1.2 (b), and 2.4mm (c). The dashed line shows $F(\omega,0)$ explained in the text.

remarkably at high temperatures. We analyzed our results along this context. Dotted curves in Fig. 2 show the reproduction of the spectra observed, assuming that the modulated spectral shape is expressed by

$$F(\omega,d) = F(\omega,0) \exp[-\alpha(\omega) d_{eff}], \qquad (1)$$

where $\alpha(\omega)$ is the absorption coefficient at photon energy, $\hbar\omega$, and

$$F(\omega,0) = \sqrt{\omega_-} \exp[-\beta\hbar\omega_-] + \sqrt{\omega_+} \exp[-\beta\hbar(\omega_+ + \omega_{\Gamma_3^-})]$$

is the emission spectrum with no reabsorption effect. Here β represents $1/k_BT$, $\hbar\omega_\mp$ is the energy measured from the edge of Stokes and anti-Stokes component, and $\hbar\omega_{\Gamma_3^-}$ is the energy of the phonon involved. In the fitting, the effective length, d_{eff}, in (1) was introduced as a fitting parameter. The spectral shapes were completely reproduced except for the low energy tail due to indirect transition involving Γ_4^- phonons, which is not included in the calculation. An interesting fact is that d_{eff} obtained is about three times larger than d for each fitting (Fig. 3(b)). Similar discrepancy has been obtained in the off-axis transmission spectroscopy and the photo-voltaic spectroscopy in the same exciton system. The effect of 2-phonon assisted Rayleigh scattering is considered as a candidate to explain the deviation in problem.

The present analysis is based on the propagation of the relevant emission light in the crystal. This was also confirmed by the temporal characteristics of the emission. Time-resolved images of the EL after application

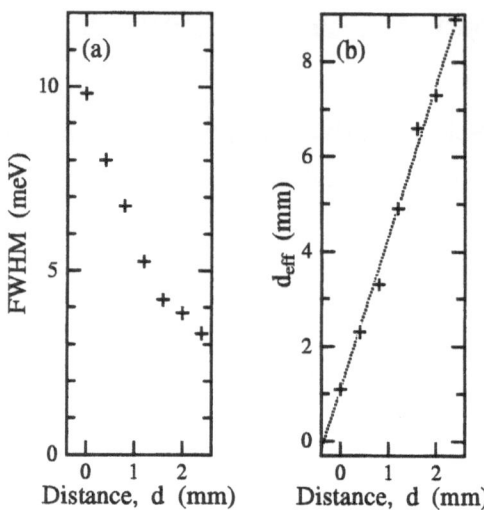

Fig. 3 Spectral width (a) and d_{eff} (b) vs. d.

of 13ns pulsed voltage were taken using the same ICCD camera in a fast-gated mode. From the rising edge of the EL signal at different points on the sample (i.e., by the time of flight measurement), we estimated the speed of the EL propagating to be faster than 5×10^7cm/s. This indicates that EL light is propagating in the sample as photons.

Finally, a few comments will be given on the spectral change at 2K. In this case, the exciton temperature between electrodes raised to 14K, probably by Joule heat. The reabsorption effect is expected to be negligible at this temperature. Nevertheless, slight narrowing was still seen in Xo-Γ_3^-. We consider that the narrowing might be due to the cooling of the excitons to about 10K as they propagate away from the electrodes. On the other hand, the direct emission, Xo, showed strong reabsorption, suggesting the effect of the scattering of the polaritons [6].

Acknowledgements We are most grateful to T Ueda and T. Nishi from Marubun Corporation for their courtesy of allowing us to use the ICCD camera system. This work was partially supported by The Mitsubishi Foundation for Scientific Researches, and the grant-in-aid for scientific research from The Ministry of Education, Science and Culture, Japan.

References

1. R. Frerichs and R. Handy, Phys. Rev., **113**, (1959) 1191.
2. F. I. Kreingol'd and V. M. Fokin, Soviet Phys.– Semiconductors, **4**, (1971) 1888.
3. *Bose-Einstein Condensation*, edited by A. Griffin, D. W. Snoke, and S. Stringari (Cambridge University Press, Cambridge, 1995).
4. A. G. Gol'dman and L. V. Toropkova, Soviet Phys.– Doklady, **7**, (1963) 1111.
5. J. T. Warren, K. E. O'Hara, and J. P. Wolfe, Phys. Rev. B, **61**, (2000) 8215.
6. N. Naka and N. Nagasawa, Solid State Commun., **110**, (1999) 153.

Null Fresnel drag in a resonant moving medium

F. Bassani[1], **M. Artoni**[2], **G.C. La Rocca**[3,1], **I. Carusotto**[1]

[1] INFM, Scuola Normale Superiore, Piazza dei Cavalieri, I-56126 Pisa, Italy
[2] INFM and LENS, Largo E. Fermi 2, I-50125, Florence, Italy
[3] Dipartimento di Fisica, Università di Salerno, I-84081 Baronissi (Sa), Italy

Abstract We discuss how the phase of a light plane–wave propagating in a resonant medium under electromagnetically induced transparency (EIT) is affected by the uniform motion of the medium. In a specific pump–probe EIT configuration the resonant probe beam experiences a phase-shift (Fresnel-Fizeau effect) that undergoes large variations, acquiring even vanishing values, due to the combined effects of the strong frequency dispersion and anisotropy induced by the pump beam. Detailed numerical simulations are presented based on the "yellow exciton" level scheme of Cu_2O.

The phase velocity of light depends on whether light is propagating in a moving or in a stationary medium. This effect, which gives rise to the familiar Fresnel light-drag [1], has been observed for the first time in Fizeau's flowing water experiment [2] and had a profound influence on the change of our perception of the nature of space and time at the turn of the century. Several other observations of light–drags have followed in which different dragging media and different interferometric measurement techniques have been employed. There still remains, however, a formidable challenge to perform high-precision measurements of light–drags as they have not yet reached the level of accuracy of other tests of special relativity [3].

In order to perform a high–precision measurement of the Fresnel-Fizeau effect one needs rather high sensitivity to velocity induced phase shifts which means large sample speeds. At the same time, in order to preserve high contrast of the interference fringe pattern, vibrations transmitted from the sample movement to the interfereometer itself have to be minimized. To this extent we here anticipate that precision can be improved through EIT [4,5] when this is devised to occur in a slab of cuprous oxide (Cu_2O) used as a dragging medium. Improvements consist in quite large positive and negative light–drags that enable one to achieve high interferometric sensitivities even for low drag velocities. Large drags are associated with the steep dispersion occurring within a transparency window centered at the $2P$ exciton resonance and arising from the quantum interference in a specific pump–probe Λ configuration which involves the ground, the $1S$ and the $2P$ exciton levels [6].

Not only can one increase precision, but in a sample rendered anisotropic by a suitable choice of the pump polarization one can also make the phase shift corresponding to the light-drag to vanish. This means that in a typical interferometric experiment no fringe shift would be observed for light propagating through the moving medium with respect to light propagating through the medium at rest. This somewhat surprising conclusion holds for all velocities of experimental interest while tuning of the light–drag to zero may turn out to be quite favorable for investigating the electromagnetic and mechanical balance of the momentum associated with light in EIT media.

We then examine Cu_2O under EIT [6] as dragging medium and look at the Fresnel drag experienced by a weak probe beam of frequency ω tuned about the $2P$ "yellow" exciton line of resonant frequency ω_{2P}, while a strong beam of Rabi frequency Ω_c is resonant with the $1S$ to $2P$ exciton transition. A Λ-type model hamiltonian can be developed to describe such a system and to calculate the effective dielectric tensor describing the optical response of the medium to the weak probe beam [6]. The quadrupole allowed threefold degenerate Γ_5^+ $1S$ exciton state has a small linewidth ($\hbar\gamma_{1S} \simeq 0.1$ meV) compared to that of the second class dipole allowed threefold degenerate Γ_4^- $2P$ exciton state ($\hbar\gamma_{2P} \simeq 1$ meV). For a pump Rabi frequency $\Omega_c \approx \sqrt{\gamma_{2P}\gamma_{1S}}$ EIT takes place whereby a narrow transparency window with a rather steep dispersion appears around the $2P$ exciton line [6].

The effective dielectric tensor seen by the probe in the presence of the pump is in general anisotropic, depending on the pump polarization and on the detailed structure of the exciton levels involved. For the sake of simplicity, we here assume that the Γ_4^- states are well separated from all other $2P$ states and the pump polarization is along the cubic axis $\hat{x}'$ (See Fig.1). The resulting dielectric tensor is uniaxial with the optical c-axis along $\hat{x}'$, i.e., $\epsilon_{x'x'} = \epsilon_\parallel$, $\epsilon_{y'y'} = \epsilon_{z'z'} = \epsilon_\perp$ and $\epsilon_{j'\neq k'} = 0$ and for a resonant pump beam one has

$$\epsilon_\perp = \epsilon_{EIT} = \epsilon_\infty + \frac{A\,\gamma_{2P}\,(\delta_p - i\gamma_{1S})}{(\delta_p - i\gamma_{2P})(\delta_p - i\gamma_{1S}) - \Omega_c^2/4},$$

while $\epsilon_\parallel$ is obtained by setting $\Omega_c \to 0$ in the above equation. Here $A \simeq 0.02$ a numerical constant proportional to the $2P$ exciton oscillator strength [6], $\epsilon_\infty \simeq 6.5 + i\, 2 \cdot 10^{-3}$ is the background dielectric constant whereas $\delta_p = \omega_{2P} - \omega$ is the probe detuning.

In a typical configuration [7,3] a slab of thickness L moves in one arm of a ring–cavity interferometer with constant velocity $v = \beta c$ with respect to an observer in the laboratory frame S as sketched in Fig.1. The quantity of experimental relevance is the velocity dependent phase-shift $\Delta\phi$ experienced by the probe upon single–

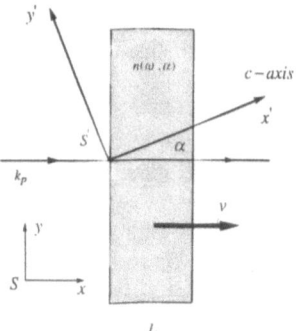

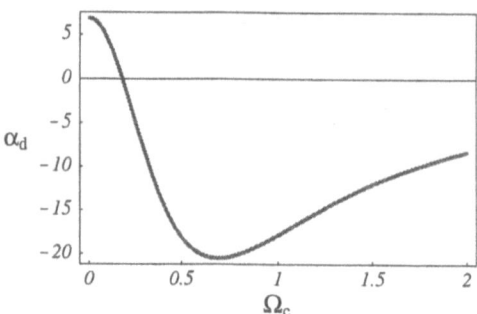

Fig. 1 Probe beam path across a Cu_2O slab moving with velocity v in the laboratory frame S. The probe wavevector k_p, the slab velocity and its surface normal are all in the same direction $\hat{x}$. The pump is directed along $\hat{z} = \hat{z}'$ and its polarization is parallel to the optical axis $\hat{x}'$ which makes a fixed angle α with $\hat{x}$.

pass transmission. We do not account here for small tilt angles between the surface normal and the cavity axis $\hat{x}$ employed in practice [3] to avoid multiple reflections. We give the phase–shift for a probe single–pass at normal incidence as it is sufficient here to illustrate the effect of frequency and angular dispersion associated with EIT on the light–drag [8].

The shift $\Delta\phi$ is a relativistic invariant and can be conveniently evaluated in the slab rest frame S'. Under EIT the dispersion equation for a probe polarized in the $x'y'$ plane is that of an extraordinary ray [9] whose complex refractive index is given by

$$n^2(\omega', \alpha) = \frac{\epsilon_\parallel(\omega')}{1 + \epsilon_r(\omega')\cos^2\alpha}; \quad \epsilon_r(\omega') \equiv \frac{\epsilon_\parallel(\omega')}{\epsilon_\perp(\omega')} - 1,$$

with n' and n'' respectively its real and imaginary parts. In the laboratory frame the single-pass phase shift, including contributions to first order in β, can be written as [8]

$$\Delta\phi \simeq \Delta\phi_0 - \beta\frac{\omega L}{c}\left(1 - n'(\omega, -\alpha) - \omega\frac{\partial n'(\omega, -\alpha)}{\partial\omega}\right)$$

$$\equiv \Delta\phi_0 - \beta\frac{\omega L}{c}\alpha_d,$$

where $\Delta\phi_0$ corresponds to the shift induced by a stationary slab and α_d represents the effective phase shift coefficient in terms of the probe laboratory frame frequency ω. The usual definition [10] of the drag coefficient for the phase velocity of light can be related to the measurable quantity α_d [8]. Owing to EIT the absorptive part n'' only appears as a higher–order correction to $\Delta\phi$ and can therefore be neglected.

The magnitude and sign of the coefficient α_d can be controlled through the cleavage angle α and the coupling beam Rabi frequency Ω_c. In the following, we discuss numerical results for a resonant pump. Fig.2 shows the variation of α_d as a function of Ω_c for a resonant probe. Notice that α_d vanishes at $\hbar\Omega_c \simeq 0.18$ meV but can be as large as 20, that is over one order of magnitude larger than the α_d's of dragging glasses [3] typically used for

Fig. 2 Coefficient α_d vs. Ω_c in units of γ_{2P} for a resonant probe crossing a Cu_2O slab. Here the cleavage angle is $\alpha \simeq 5°$ and $\gamma_{2P} = 1$ meV.

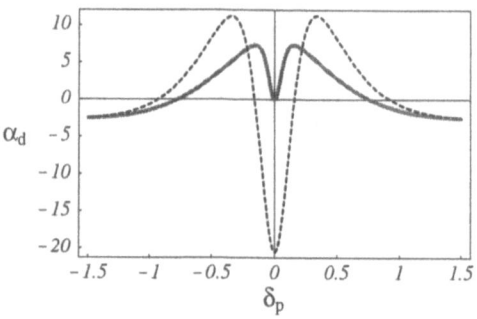

Fig. 3 Coefficient α_d vs. δ_p in units of γ_{2P} for a Rabi frequency of $\Omega_c/\gamma_{2P} = 0.18$ (solid) and 0.7 (dash). The other parameters are the same as in Fig.2.

high–precision measurements of the Fresnel–drag. This novel and important feature is due to the anisotropy and to the strong dispersion both characteristic of the EIT regime. Fig.3 shows instead the variation of α_d as a function of the probe detuning for fixed pump intensities. Because of EIT, transmission is quite large in the range $|\delta_p| < 0.5\,\gamma_{2P}$; for a slab thickness $L = 25\ \mu m$ this varies between $10 \div 30$ % despite the $2P$ exciton resonance[6]. Apart from the background absorption, the $1S$ exciton linewidth γ_{1S} is the material parameter that mostly limits the possibility of controlling the magnitude of α_d through EIT.

References

1. A. Fresnel, Ann. Chim. Phys. **9**, 57 (1818).
2. H. Fizeau, C. R. Acad. Sci. **B 33**, 349 (1851).
3. G.A. Sanders *et al.*, J. Opt. Soc. Am. B, **5**, 674, (1988).
4. S. Harris, Physics Today **50**, 36, (1997); E. Arimondo, *Progress in Optics*, Ed. E. Wolf (Elsevier Science, 1996) page 257; and references therein.
5. G. Alzetta, A. Gozzini, L. Moi, G. Orriols, Nuovo Cimento **B36**, 5 (1976);
6. M. Artoni, G.C. La Rocca and F. Bassani, EuroPhys. Lett.s **49**, 445, (2000).
7. W. Macek *et al.*, J. Appl. Phys. **35**, 2556 (1964).
8. A complete description including sample tilting will be presented elsewhere (M. Artoni et al., to be published).
9. M. Born and E. Wolf, *Principles of Optics*, (Pergamon Press, Oxford 1993)
10. L.D. Landau and E.M. Lifschitz, *Electrodynamics of continuous media*, (Nauka, Moscow 1982)

Coupled plasmon-LO phonon modes at high magnetic fields

A. Wysmolek[1,2], **M. Potemski**[1], and **T. Slupinski**[2]

[1] Grenoble High Magnetic Field Laboratory, MPI/FKF-CNRS, BP 166X, F-38042 Grenoble Cedex 9, France
    e-mail: wysmolek@polycnrs-gre.fr
[2] Institute of Experimental Physics, Warsaw University, Hoza 69, PL-00-681 Warszawa, Poland

Abstract We report on the results of high-magnetic-field (up to 28T) inelastic light scattering measurements of coupled plasmon-phonon modes in metallic GaAs samples with electron concentrations ranging from 10^{17} to $2 \cdot 10^{18}$ cm^{-3}. The zero-field Raman spectra of the investigated samples show peaks related to the transverse optical phonon (TO) as well as the coupled plasmon-phonon ω_- (below TO frequency) and ω_+ (above LO frequency) modes. In the backscattering Voigt geometry ($k \perp B$) in which the modes propagating across the field direction are probed, the magnetic field does not influence the TO resonance but leads to a splitting of both ω_+ and ω_- branches. For highly doped samples, the interaction between coupled plasmon-phonon modes with the Berstein mode is observed. The experimental data are approximated using the dielectric function formalism.

1 Introduction

Wave propagation in solids has been intensively studied in the sixties with significant interest focused on doped polar semiconductors in which the nowadays text-book effect of coupling between plasmons and optical phonon modes can be clearly observed as demonstrated for example, by the pioneering work of Mooradian and Wright [1] on light scattering in n-type GaAs samples. The dielectric function theory applied to coupled electron-phonon excitations has interesting consequences in the limit of high magnetic fields when the cyclotron frequency is tuned over all other undressed excitations present in the crystal [2].

We report, to our knowledge for the first time, on the results of high-magnetic-field (up to 28T) inelastic light scattering measurements of coupled plasmon-phonon modes in metallic GaAs samples.

2 Results and discussion

The Raman scattering experiments have been performed on bulk n-type Czochralski grown GaAs samples with room temperature concentration in the range from 10^{17} up to $2 \cdot 10^{18}$ cm^{-3}. A tunable Ti: sapphire laser operating in the range between 820-860 nm, typically with 100 mW power has been used for below band-gap excitation. A specially designed optical fibre system has been employed for the measurements in the magnetic field up to 28 T, supplied by a resistive magnet. The Raman spectra have been measured at liquid helium temperatures, in four different configuration of the exciting and scattered light wavevectors with respect to the magnetic field direction.

The zero-field Raman spectra of the investigated samples are shown in Fig. 1. As can be seen from Fig. 1 none of the spectrum displays signals related directly to the longitudinal optical phonon mode (LO) of undoped GaAs. It

is replaced be the coupled plasmon-phonon modes: ω_- observed below the TO frequency and ω_+ observed above the LO frequency. As expected, the ω_- and ω_+ frequencies follow the well known dependence [1, 3] on electron concentration.

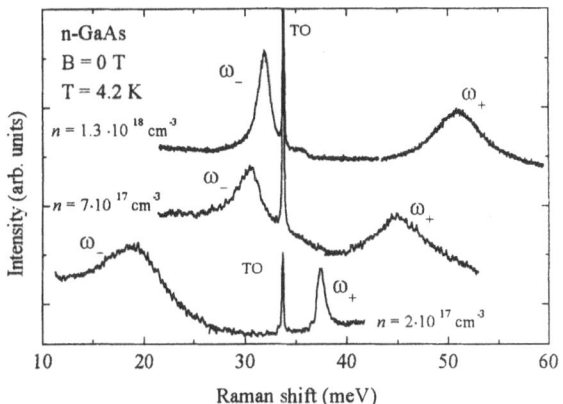

Fig. 1 Raman scattering spectra of n-GaAs samples with different electron concentrations, measured at T= 4.2 K.

The application of a magnetic field modifies the measured Raman response, although the effect is remarkably dependent on the configuration of the incident and scattered wavevectors (correspondingly, k_i and k_s) with respect to the magnetic field (B) direction. As expected, a magnetic field applied in the Faraday configuration hardly modifies the measured Raman spectra. Here, we concentrate on the results obtained in the backscattering Voigt geometry ($k \perp B$) in which the modes propagating across the field direction are probed. The Raman scattering spectra measured for two samples, with electron concentrations of $1.3 \cdot 10^{18}$cm^{-3} (sample (a)) and $2 \cdot 10^{17}$cm^{-3} (sample (b)) in the range of magnetic field up to 28T are shown in Fig 2. The corresponding diagrams of peak positions versus the magnetic field are presented in Fig. 3. Roughly speaking, the magnetic field behaviour of Raman scattering spectra for the case of sample (a) with high electron concentration can be understood in terms of the observation of hybrid cyclotron-plasmon-phonon modes shown with solid lines in Fig. 3a according to the following formula [4, 5]:

$$2\omega_{hyb}^{\pm} = \Omega_c^2 + \omega_l^2 \pm \left[\left(\Omega_c^2 + \omega_l^2 \right)^2 - 4\left(\omega_c^2 \omega_l^2 + \omega_p^2 \omega_t^2 \right) \right]^{1/2},$$

in which, $\Omega_c^2 = \omega_c^2 + \omega_p^2$, ω_c is the cyclotron frequency, ω_p the plasma frequency, and ω_l and ω_t are, respectively, the LO and TO phonon frequencies of an undoped material.

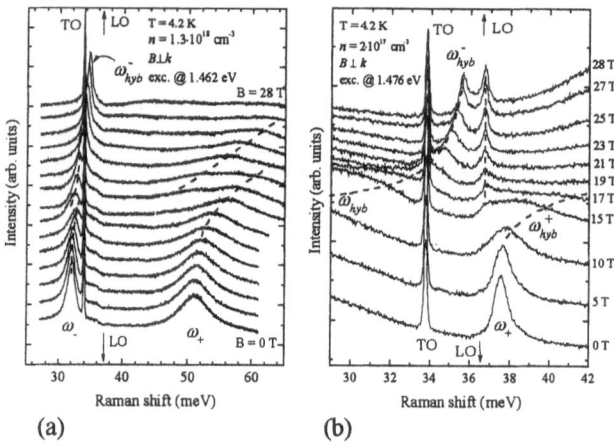

(a)   (b)

Fig. 2 The evolution of Raman scattering spectra measured for the backscattering Voigt configuration in magnetic fields up to 28 T for samples with (a) $n = 1.3 \cdot 10^{18}$ cm^{-3} and (b) $n = 2 \cdot 10^{17}$ cm^{-3}. Spectra shown in left panel were measured with 2 Tesla field step.

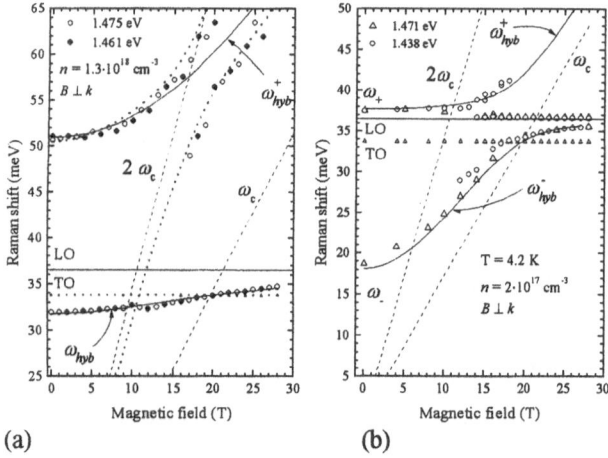

(a)   (b)

Fig. 3 Magnetic field dependence of the plasmon –LO phonon modes for the sample with (a) $n = 1.3 \cdot 10^{18}$ cm^{-3} and (b) $2 \cdot 10^{17}$ cm^{-3}. Different symbols denote the data obtained using different excitation energies. Dashed lines show single- and double-cyclotron energy. Solid and dotted lines represents the evolution of collective modes (see text). The LO phonon energy of undoped material is shown with solid horizontal lines.

As seen in Figs. 2 and 3, ω^-_{hyb} and ω^+_{hyb} shift towards higher energies with increasing magnetic field. ω^-_{hyb} crosses the TO mode and eventually approaches the energy of the undressed LO phonon of undoped material.

More subtle effects observed for sample (a) are the field dependent broadening and/or splitting of each ω^-_{hyb} and ω^+_{hyb} mode (Fig. 2a). The ω^-_{hyb} peak is relatively narrow but show appreciable broadening around 10-12T. One may even presume that at these fields, ω^-_{hyb} shows a rather nonmonotonic field dependence indicating some anticrossing effect. The anticrossing behaviour is more clearly pronounced for the ω^+_{hyb} mode, but occurs at higher magnetic fields, around ~16T. As illustrated in Fig. 3a, we believe that this anticrossing is a result of interaction between collective hybrid modes and collective cyclotron

excitations (Bernstein modes) with frequency of $2\omega_C$.[6, 7] Since ω^+_{hyb} mode shows more pronounced anticrossings than the ω^-_{hyb}, we deduce that the Bernstein modes predominantly couple with plasmon-like and not with phonon-like modes. Thus, a simple theoretical model [8] (dotted lines in Fig. 3a) well accounts for the anticrossing behaviour of the plasmon-like branch.

The consideration of only hybrid cyclotron-plasmon-phonon modes is not sufficient to interpret the magnetic field evolution of the Raman scattering spectra measured for sample (b) with a relatively small electron concentration. As can be seen in the raw data presented in Fig.2b, the upper branch of the plasmon-LO phonon peak splits, in this sample, into two components at fields around 15T. The remaining component of this upper branch, which persist at high magnetic fields, is a sharp peak. In sufficiently high magnetic fields, it is observed at the energy which is very close to the LO phonon energy of undoped material (Fig. 2). This observation seems to indicate that our analysis of the dielectric function is not sufficiently elaborated to explain the high magnetic field data.

3 Conclusions

Raman scattering experiments on heavily doped GaAs samples show the development of hybrid cyclotron-plasmon-phonon modes as well as certain other collective excitations. In the high field limit, the collective modes which resemble longitudinal optical phonon excitations propagating across the field direction have been observed. The interaction of coupled plasmon-phonon modes with Berstein modes have also been observed. This interaction is stronger if it involves collective modes of plasmon-like character.

Acknowledgements

A. Wysmolek gratefully acknowledges financial support from the Alexander von Humboldt Foundation.

References

1. A. Mooradian and G. B. Wright, Phys. Rev. Lett. **16**, 999 (1966)
2. B. Lax, *Proceedings of the International Conference on the Physics of Semiconductors*, Kyoto, 1966, Journal of the Physical Society of Japan **21**, 165 (1966)
3. B. B. Varga, Phys. Rev. **137**, A1896 (1965)
4. I. Yokota, *Proceedings of the International Conference on the Physics of Semiconductors*, Kyoto, 1966, Journal of the Physical Society of Japan, **21**, 738 (1966)
5. R. Kaplan, E. D. Palik, R. F. Wallis, S. Iwasa, E. Burstein, and Y. Sawada, Phys. Rev. Lett. **18**, 159 (1967)
6. C. K. Patel and R. E. Slusher, Phys. Rev. Lett. **21**, 1563 (1968)
7. V. Lopez, F. Comas, C. Trallero-Giner, T. Ruf, an M. Cardona, Phys. Rev. B **54**, 10502 (1996)
8. P. M. Platzman, P. A. Wolff, *Waves and interactions in Solid State Plasmas* (Academic Press, New York, 1973)

Excitation induced flips of the phase of transmitted pulses

G. Manzke[1], K. Henneberger[1], J. S. Nägerl[2], B. Stabenau[2], G. Böhne[2], and R. G. Ulbrich[2]

[1] University Rostock, Department of Physics, D-18051 Rostock, Germany, e-mail: manzke@physik3.uni-rostock.de
[2] University Göttingen, IV. Physical Institute, D-37073 Göttingen, Germany

Abstract   We investigate the phase and amplitude of $40fs$-pulses propagating through a 3.8μm thick GaAs platelet. Phase jumps of the transmitted light field appearing at the single beat nodes of the amplitude due to interference of polaritons change their sign with increasing excitation. While a simple treatment with a Lorentzian-broadened dielectric function fails, these changes are explained in terms of a many-body approach.

1 Introduction

The transmission of an optical pulse through a semiconductor slab at resonance with the exciton is strongly influenced by propagation effects of polaritons. Pronounced propagation beats were found for the first time in [1] in the vicinity of the quadrupole polariton resonance of Cu_2O. We report on interferometric cross-correlation measurements of both amplitude and phase of transmitted $40fs$-pulses through a 3.8μm thick GaAs platelet. At low excitation phase shifts occur at the single beats. With increasing excitation we find significant changes of these phase shifts. First the phase shift for later times turns from $+\pi$ to $-\pi$, then those for earlier times follow.

We demonstrate that many-body effects, such as diagonal and off-diagonal dephasing, renormalization of the interband energy and of the Coulomb interaction [5] have to be taken into account, depending both on wave number and the energy in the considered region around and below the band edge.

2 Experiment

The experiments were performed on a 3.8μm thick, high purity Gallium-arsenide platelet ($|N_d - N_a| < 2 \cdot 10^{14} cm^{-3}$) grown by gas-phase epitaxy, held freely between glass plates and immersed in liquid He at $2K$. The linewidth (fwhm) of the 1s resonance was $\hbar\gamma = 2 \cdot 44\mu eV$. For a detailed description of the used interferometric cross-correlation measurement technique see [2].

The behavior of amplitude and phase of the transmitted pulses with increasing laser fluence is shown in Fig. 1. The excitation level $0.4W/cm^2$ represents the linear case: further reducing the laser pulse fluence did not change the measured spectra any more. The incident pulse is split into an aperiodic train of smooth maxima and minima with increasing beat period. The beat nodes of the intensity are accompanied by phase shifts close to π and have negative sign for early $\tau = 0.7ps$ and late $7.8ps$ delay time and point upwards (positive sign) for $\tau = 2.4ps$ and $4.3ps$. Generally such a behavior of the phase can be understood as an interference effect between wave pack-

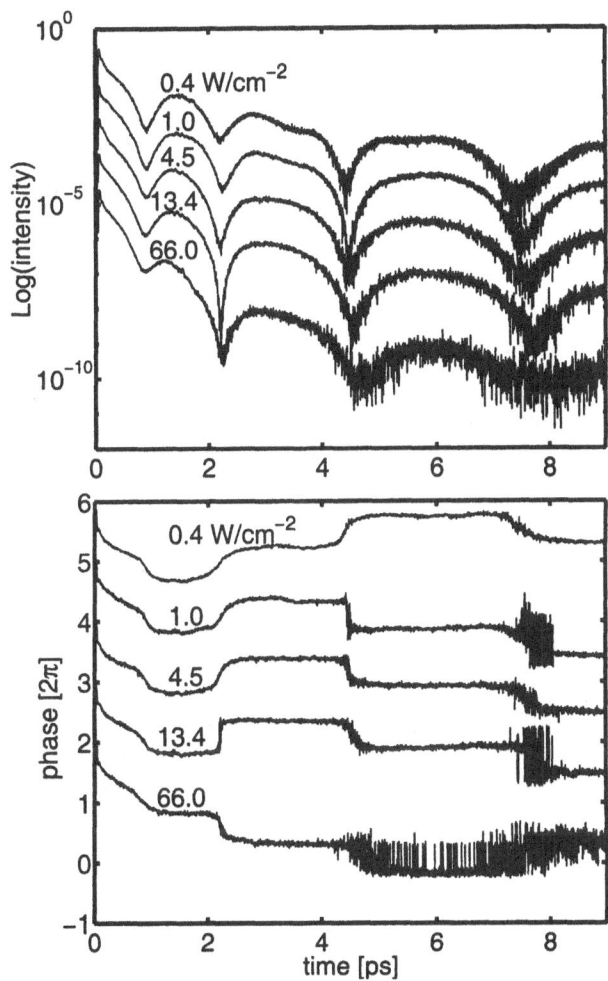

Fig. 1 Measured intensity and phase of the transmitted electric field for different excitations (noise for higher fluences and at later times is caused by the rapidly vanishing amplitude in this case).

ets of the upper and lower branch. Depending on their amplitudes and on the time, in which they reach the rear end of the sample, the phase shifts may be either positive or negative [2]. The smooth modulation of the amplitude signal around $3ps$ and $6ps$ is caused by the interference of 1s- and 2s-exciton emission signals. With increasing excitation the intensity shows the expected behavior: interferences between 1s- and 2s-excitons disappear, since the 2s-exciton is damped out and merges with the shrinking band edge. Furthermore the decay times become shorter corresponding to an increase of the excitonic linewidth. Pronounced changes are found for the phases (see Fig.1). With increasing excitation the signs of the observed phase shifts change. For laser power densities approaching $1W/cm^2$ the phase shift at $4.3ps$ first becomes steeper, then changes sign and finally becomes smoother again. At a laser fluence of $13.6W/cm^2$ the phase shift at $2.3ps$ flips.

This characteristic behavior cannot be modeled with an additional Lorentzian broadening. Symmetric Lorentzian lineshapes always lead to positive jumps of the phase. Even if the damping of the exciton resonances is increased in such a way that an apparent asymmetry of the 1s-resonance is introduced by broadening of higher lying exciton states, the sequential order of the phase flips cannot be reproduced. In [2] it was shown, that an asymmetric exciton lineshape is responsible for the flips of the phase shifts (for a detailed discussion see [3]). In the next section we present the results of a many-body approach [6] for the dielectric function, which gives an explanation for these asymmetries.

3 Theory

As a starting point to describe the correct behavior of phase and amplitude of the transmitted pulses (see Fig. 1) for low excitation, we model the damping as an asymmetric function of the frequency.

$$\gamma_0(\omega) = c_1 + c_2 \tanh(c_3 \omega + c_4). \qquad (1)$$

The parameters are chosen as follows: damping of the 1s-state is taken to be $\gamma_0(E_{gap} - E_{ex}^b) = 50\mu eV$ (E_{ex}^b - excitonic Rydberg), the damping of the 2s-state $\gamma_0(E_{gap} - E_{ex}^b/4) = 250\mu eV$, as the limit high in the band $\gamma_0(\omega \gg E_{gap}) = 300\mu eV$ (E_{gap} - gap energy) is set, and the asymptotic $\gamma_0(\omega \ll E_{gap} - E_{ex}^b) = 28\mu eV$ is fitted to get best agreement with the experiment. The overall shape of $\gamma_0(\omega)$ is shown in Fig. 2. In the case of low excita-

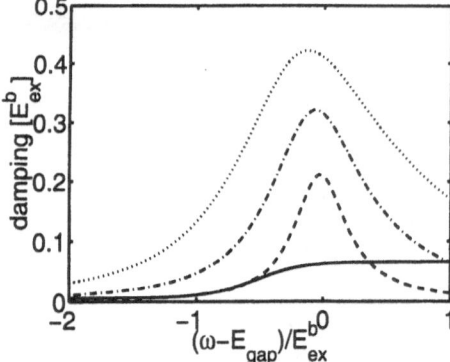

Fig. 2 Theoretical model for the damping of the non-excited sample (solid line) in dependence on the energy and carrier induced dephasing at $k = 0$ (k - wave number) for different carrier densities: $n = 1 \cdot 10^{13} cm^{-3}$ (dashed), $n = 1 \cdot 10^{14} cm^{-3}$ (dash-dotted), $n = 5 \cdot 10^{14} cm^{-3}$ (dotted).

tion (with the background damping $\gamma_0(\omega)$) our solution of the semiconductor Bloch equations agrees with the that of a broadened Elliot formula [4] and the calculated phase and intensity give a best fit to the experiments (see upper curves in Fig. 1 and Fig. 3).

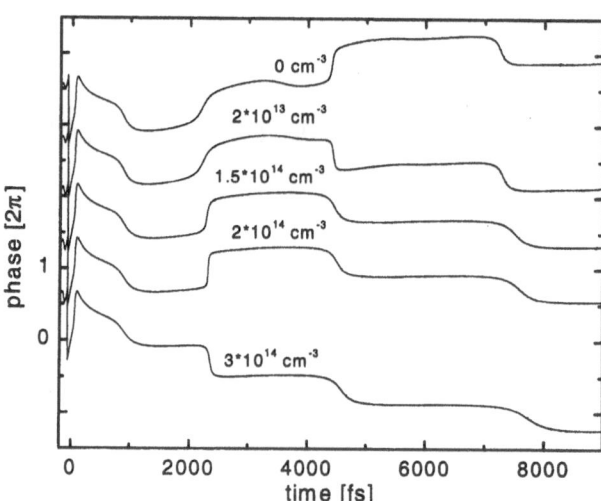

Fig. 3 Phase of transmitted pulses calculated for different carrier densities.

With increasing excitation many-body effects become important. The carrier-induced dephasing is presented in Fig. 2 for different carrier densities and a temperature of $T = 30K$. The dephasing becomes maximum at the renormalized band edge a strongly decreases towards to the exciton. Here the dephasing depends both on wave number and energy [6]. The latter was neglected in earlier treatments and the dephasing was only considered as a function of the wavenumber being independent of the energy (quasi-particle approximation).

We determine the optical susceptibility by solving the semiconductor Bloch equations including diagonal and off-diagonal dephasing, renormalization of the interband energy and of the Coulomb interaction [5]. All these effects have to be taken into account, depending both on wave number and the energy [6]. They result in an asymmetry of the exciton line on the high-energy tail, which is growing up with increasing excitation. Finally, this asymmetry influences the dispersion, group velocity and the damping of polaritons, particularly those of the upper branch. The order, in which the polaritons of the upper and lower branch reach the rear end of the sample changes and the phase shifts change their sign. For comparison the calculated phase of transmitted pulses is given in Fig. 3 for different carrier densities, where Fresnell's formula were used without spatial dispersion. The agreement with the experiment Fig. 1 is obvious.

References

1. D. Fröhlich et al., Phys. Rev. Lett. **67**, 2343 (1991).
2. J. S. Nägerl, et al., phys. stat. sol. **A 178**, 559 (2000).
3. J. S. Nägerl, et al., submitted to Phys. Rev. **B**.
4. C. Tanguy, Phys. Rev. Lett., **75**, 4090 (1995).
5. G. Manzke et al. Phys. Rev. Lett. **80**, 4943 (1998).
6. G. Manzke, and K. Henneberger, Progress in Nonequilibrium Green's functions, edited by M. Bonitz (World Scientific, Singapore 2000) p. 238.

Proc. 25th Int. Conf. Phys. Semicond., Osaka 2000 (Eds. N. Miura and T. Ando)

91

Below bottleneck polaritonic radiation in ultra high quality AlGaAs alloys

T.S. Shamirzaev[1], A.I. Toropov[1], A.K. Bakarov[1], K.S. Zhuravlev[1], A.Yu. Kobitski[1,2], H.P. Wagner[2], D.R.T. Zahn[2]

[1] Institute of Semiconductor Physics, Novosibirsk, Lavrentieva 13, 630090, Russia. e-mail: tim@isp.nsc.ru
[2] Institute für Physik, TU Chemnitz, D-09126, Chemnitz, Germany

Abstract We have studied the stationary and time-resolved polariton radiation in ultra high quality AlGaAs layers. It has been found that elastic exciton-exciton collisions lead to the appearance of a novel low-energy line of the polariton radiation. We show that the rate of the exciton to polariton transitions caused by elastic exciton-exciton collisions is determined not only by the density of the excitonic gas but also by its temperature in accordance with the existing theoretical prediction.

1 Introduction

It is well known that the spectrum of excitonic polariton radiation is determined by their spatial and energy distributions [1]. By now, experimental and theoretical studies of the formation and evolution of the energy distribution of excitonic polaritons shown that: (i) at low excitation levels, the energy distribution of excitonic polaritons is determined by the collisions (both elastic and inelastic) of excitons with impurities and phonons [2]; the polariton concentration is very low due to a small rate of the exciton to polariton transitions (Here and in the following, we refer to the polaritons as the particles with momentum $k < k_0$, (where k_0 is the momentum value at the crossover between excitons and photons on the dispersion curve) and to the excitons as the particles with $k > k_0$); (ii) at high excitation levels, the energy distribution of the excitonic polaritons is determined by the inelastic exciton-exciton scattering which leads to a high rate of the exciton-to-polariton transitions [3]. At the same time, the case of an intermediate level of optical excitation, when the energy distribution of the excitonic polaritons (and therefore, the spectrum of the polaritonic radiation) is determined by the elastic exciton-exciton collisions [4], has not been experimentally studied yet. It has been theoretically predicted that an elastic collision of two excitons results in a relaxation of one of them into a polariton and a transition of the second one into the high-momentum region of the lower polaritonic branch (LPB) [4]. This should lead to the appearance of an additional below bottleneck polariton luminescence line. Up to now such a line has not been observed experimentally. The main obstacle for the experimental observation of this line is its masking by the impurity- bound exciton lines, which are usually present in the near-bandgap luminescence spectra of semiconductor crystals.

Recently we reported on the growth of ultra high quality Al_xGa_{1-x} As layers by molecular beam epitaxy

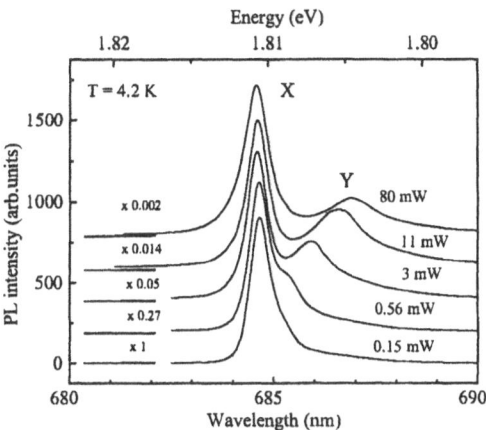

Fig. 1 Low-temperature photoluminescence spectra of a pure $Al_xGa_{1-x}As$ layer with an AlAs fraction $x=0.21$ measured at different excitation powers.

technique [5]. The X line of the polaritonic radiation dominates in the photoluminescence PL spectra of these layers, whereas the bound exciton lines are absent. In this paper we report on the experimental observation of a new polaritonic luminescence line which is due to elastic exciton-exciton collisions in ultra-high quality AlGaAs layers.

2 Results and discussion

The details of the layer growth and equipment used for the recording of PL were described elsewhere [5]. Figure 1 shows the 4.2 K PL spectra of the $Al_xGa_{1-x}As$ layer with an AlAs fraction of $x =0.21$ measured at different excitation levels. The line of the radiation of LPB labeled X dominates in all the spectra. A novel line labeled Y appears at the low-energy tail of the X line as the excitation power exceeds 0.5 mW. The Y line shifts towards the red with increasing excitation power by as much as 8 meV at the maximum excitation power used. The relative intensity of the Y line remains nearly constant when the excitation power is increased from 0.5 to 10 mW, whereas it decreases with further increase of the excitation power. The dependence of the Y line position on the laser power is described by a logarithmic function predicted by Bisti [4] for the additional line of the polaritonic radiation, which appears due to exciton-exciton collisions. The decrease of the Y line intensity with the excitation power is due to the decrease of the exciton-polariton transition probability with a rise of the

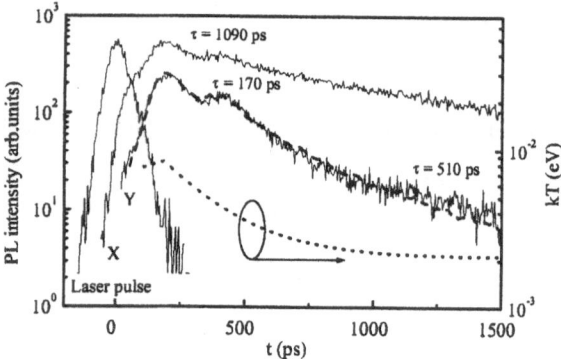

Fig. 2 The kinetics of intensity of the X and Y lines after a pulsed excitation (pulse energy 2 nJ). The thick dashed line is a calculated curve for the additional line of the polaritonic radiation appearing due to elastic exciton-exciton collisions. The time dependence of the temperature of the excitonic gas is shown by the dotted line.

exciton gas temperature (T_{ex}) [4]. In order to determine the value of T_{ex}, we fitted the high-energy tail of the X line with a temperature-dependent exponential function $f(\hbar\omega) = a \cdot exp(-\hbar\omega/kT_{ex})$, where a is a constant, $\hbar\omega$ is the photon energy, and k is the Boltzmann's constant. It was found that the relative intensity of the Y line on excitation power is proportional to $T_{ex}^{-5/4}$ in accordance with the theoretical prediction [4].

Figure 2 shows the PL kinetics of the X and Y lines after a pulsed laser excitation. The intensity of the Y line increases more rapidly than that of the X line, while the decay of both lines starts simultaneously. The X line decay curve is described by an exponential function with the decay time equal to 1090 ps. The decay of the Y line shows a biexponential behavior. The fast initial dynamics has a decay time of 170 ps followed by the decay with a time constant of 510 ps. A decrease of the excitation density leads to a change in the kinetics of the Y line. At low excitation densities the decay law of this line is a single exponential with a decay time of 510 ps. This value is half that of the X line, that is typical for a PL line appeared due to exciton-exciton collisions [6].

The modification of the Y line kinetics with the excitation power is also due to the dependence of the exciton to polariton transition probability on T_{ex}. Let us show that the biexponential behavior at high excitation powers is due to the cooling of the exciton gas with time after the excitation pulse. If the exciton-to- polariton transitions are due to collisions of excitons, then the kinetics of the concentration of polaritons is described by

$$dn_p/dt = -n_p/\tau_p + B(t) \qquad (1)$$

where n_p and τ_p is the concentration and the radiative lifetime of polaritons, respectively, and $B(t)$ is the rate of the exciton to polariton transitions due to exciton- exciton collisions. Since the excitonic gas has a Maxwellian

energy distribution, which is evidenced by the exponential shape of the high-energy tail of the X line in both the cw and time-resolved PL spectra, the rate of the exciton to polariton transitions due to elastic exciton-exciton collisions may be written as [4]: $B(t) \sim (n_{ex}^2(t)/kT_{ex}(t)) \cdot J(E/kT_{ex}(t)) \cdot exp(-E/kT_{ex}(t))$, where $n_{ex}(t)$ is the concentration of excitons, E is distance between exciton and polariton energy, and J(x) is a slowly-varying function of x calculated in Ref.4. The variation of the T_{ex}, (in units of kT) with delay time after the laser pulse is presented in Fig.2. It is seen that after the formation of the excitonic gas its temperature exceeds the lattice temperature. The temperature of excitons decreases with time probably due to the exciton-phonon scattering and the reduction of the gas density. Three distinct intervals are observed in the decay: (i) at delays smaller than 200 ps Tex increases with the density of the excitonic gas [7]; (ii) between 200 ps and 700 ps T_{ex} decreases,and (iii) at delays larger than 700 ps T_{ex} is close to the lattice temperature and does not change with time. The values of $n_{ex}(t)$ and E which are also required for the calculation of the $B(t)$ function have been derived from the experimental data. The concentration $n_{ex}(t)$ is proportional to the intensity of the X line, and E is equal to the value of the red-shift of the Y line. The kinetics of the intensity of the new polaritonic line calculated using expression (1) is shown in Fig 2. One can see that the calculated curve well describes the kinetics of the Y line during both the rise and the decay stages. The fast initial decay stage of the Y line at the high excitation level coincides with the cooling of excitonic gas.

3 Summary

In conclusion, we have observed a novel line of polaritonic radiation in the photoluminescence spectra of ultra high quality AlGaAs. We have shown that this line appears as a result of exciton-to-polariton transition caused by elastic exciton-exciton collisions. The rate of the elastic exciton to polariton scattering is determined not only by the density of the excitonic gas but by its temperature. We acknowledge the support of this work by the Russian Foundation for Basic Research and by the Graduiertenkolleg "Thin films and non-crystalline materials" of the Technische Universität Chemnitz.

References

1. S.I. Pekar, Sov. Phys. JETP **6**,(1958) 785.
2. E. Koteles, J. Lee, J.P. Salerno, and M.O. Vassell, Phys. Rev. Lett **55**, (1985) 867.
3. C. Klingshirn, and H. Haug, Physics Reports **70**, (1981) 315.
4. V.E. Bisti, Sov. Phys .Solid State **18**, (1976) 603.
5. K.S. Zhuravlev, A.I. Toropov, T.S. Shamirzaev, and A.K. Bakarov, Appl.Phys.Lett. **76**,(2000) 1131.
6. See for example, R. Huang, F. Tassone, and Y. Yamamoto, Phys. Rev. B **61**,(2000) R7854.
7. D. Snoke, and J.P. Wolfe, Phys. Rev. B **39**,(1989) 4030.

Coherent Control of Speckle: Resonant Rayleigh scattering from a conjugated polymer

S.P. Kennedy[1], N. Garro[1], R.T. Phillips[1]

Cavendish Laboratory, University of Cambridge, Cambridge CB3 0HE, U.K. e-mail: spk1000@cam.ac.uk

Abstract We show that the resonant Rayleigh scattering dominates the emission from the conjugated polymer poly(p-phenylene vinylene) excited with photons at energies below the threshold at which exciton migration is reduced. The persistence of a coherent polarisation is confirmed by performing coherent control by exciting with a phase-locked pair of pulses.

1 Introduction

Resonant optical excitation of excitons can lead to resonant emission of two types - coherent resonant Rayleigh scattering (RRS) and incoherent photoluminescence (PL). Distinguishing between these two components has been the focus of much recent work, using techniques such as spectral interferometry[1] and coherent control[2]. The coherent component shows angular variations in intensity known as speckle. Time-resolved analysis of this speckle has also been useful in probing the coherence of the emission[3]. Previous experiments have been performed mainly on GaAs quantum wells with inhomogeneous linewidths of the order of 1meV or less. In contrast, the conjugated polymer poly(p-phenylene vinylene) (PPV), with which LEDs and lasers have been produced, has an emission linewidth of 50meV.

Previous work on the dynamics of emission from PPV has revealed extremely fast energy relaxation when the exciting photons are at the peak of the absorption[4, 5]. This efficient relaxation is due to strong coupling between vibrational and electronic modes and excitonic migration to states at lower energies, and rules out the observation of the coherent effects seen in quantum wells. Here we show that by reducing the excitation energy the exciton relaxation rate in PPV is reduced, allowing coherent emission to dominate.

2 Enhancement of resonant emission

Time-integrated spectra for four different excitation energies are shown in Figure 1. The resonant emission, indicated in each case by an arrow, is much more intense for lower excitation energies. Similar results from GaAs quantum wells were attributed to resonant Rayleigh scattering by Hegarty et al[6] in 1982.

3 Dynamics of resonant emission

The results of two-colour upconversion with 230fs time-resolution are shown in Figure 2. The decay of the resonant emission changes dramatically as the excitation energy is changed from 2.317 to 2.385eV. For the lowest excitation energy the resonant emission persists well beyond the duration of the incident pulse, which has

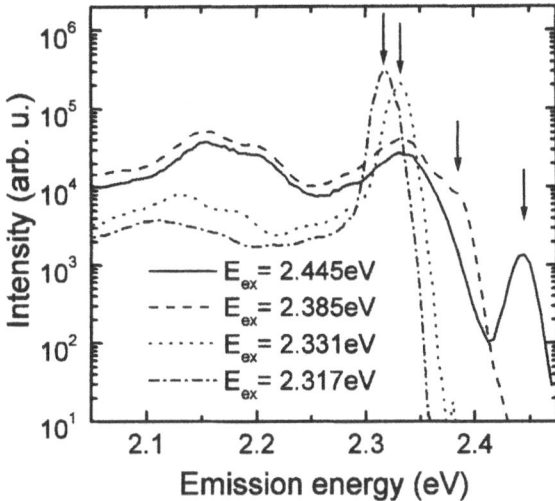

Fig. 1 Time-integrated spectra for four different excitation energies.

a cross-correlation indicated by the dashed curve. The results of exponential decay fits to the time-resolved resonant emission data are shown in the inset. The increase in the intensity and the longer lifetime of the RRS both correspond to longer phase coherence of the excited state. Since time-integrated studies of PPV have shown that exciting below ca. 2.37eV reduces excitonic migration, the dependence on excitation energy that we observe indicates that migration dominates dephasing of the excitons created in PPV just above this energy. At yet higher excitation energies, phase-breaking is probably dominated by relaxation of excited vibrational states.

4 Evidence of coherence: speckle and coherent control

Constructive and destructive interference between spatially distributed coherent oscillators produces speckle - variations in the far-field emission intensity at different angles, only observed for reduced excitation energies.

Given this demonstration of coherent resonant secondary emission, it seems appropriate to probe this coherence with the method of coherent control used in several experiments on inorganic semiconductors[2]. The sample is excited with a pair of phase-locked pulses, with the inter-pulse delay controlled to 50 attoseconds (corresponding to 1/40th of the period of the light field).

Figure 3(a) is a density map of the intensity of the time-integrated emission from PPV excited at 2.317eV

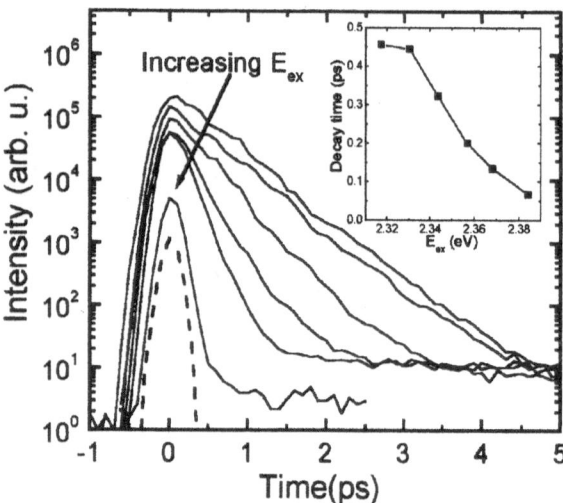

Fig. 2 The dynamics of the resonant emission for a range of excitation energies, from 2.317 to 2.385eV.

as a function of inter-pulse delay and emission angle. The vertical bands correspond to speckles. The bright and dark horizontal bands at early times reveal interference between the two pulses. The interference persists beyond the time for which the pulses overlap, and so can only be mediated by the material polarisation. This is direct evidence of the control of the quantum phase in the excitations of a conjugated polymer. In contrast, Figure 3(b) shows that for the slightly higher excitation energy of 2.368eV there is only purely optical interference during pulse overlap, which decays to an incoherent PL background for delays longer than the pulse duration.

The exact inter-pulse delay at which the emission is maximised or minimised is not identical for different speckles when the pulses do not overlap. For an inter-pulse delay of 302.5fs the intensity of one speckle is nearly a minimum while the intensity of another is at a local maximum. As a result, the angle-integrated intensity shows little variation with inter-pulse delay beyond 200fs.

Inhomogeneous broadening complicates coherent control: excitons with different energies have phases which evolve at different rates during the time between pulses. At longer delays the second pulse interferes constructively with some excitons but destructively with others. Angle-resolved detection reduces the averaging of the excited ensemble, permitting the observation of the coherent manipulation of the quantum states generated in PPV.

5 Conclusion

As the excitation energy is reduced, the resonant emission from PPV dramatically increases in intensity and persists several hundred femtoseconds after excitation. The dominance of coherent resonant Rayleigh scattering is revealed by speckle in the time-integrated emission. Manipulation with a second phase-locked pulse confirms

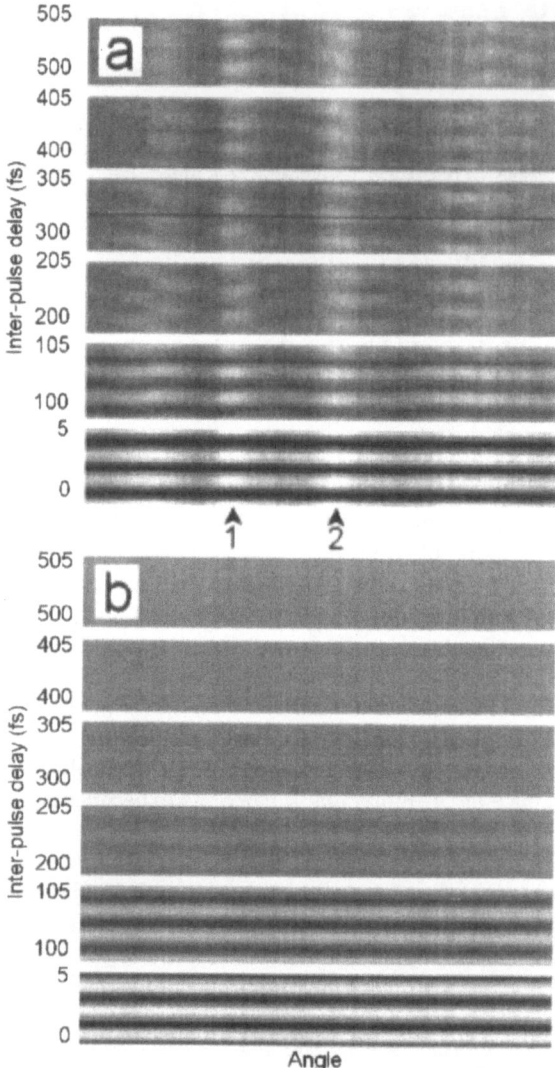

Fig. 3 The angle-resolved intensity variation as a function of inter-pulse delay for excitation with two phase-locked pulses at 2.317eV (a) and 2.368eV (b).

that a macroscopic polarisation remains in the sample even after the decay of the excitation pulse. With its large inhomogeneous broadening, PPV is a good material with which to demonstrate the effect of a range of oscillator energies on coherent control.

We are grateful to Neil Greenham and Richard Gymer for assistance in sample preparation. This work has been supported by the EPSRC, and by Renishaw plc.

References

1. D. Birkedal and J. Shah, Phys. Rev. Lett. **81**, (1998) 2372.
2. N. Garro et al, Phys. Rev. B **60**, (1999) 4497.
3. W. Langbein et al, Phys. Rev. Lett. **82**, (1999) 1040.
4. R. Kersting et al, Phys. Rev. Lett. **70**, (1993) 3820.
5. G.R. Hayes et al, Phys. Rev. B. **52**, (1995) 11569.
6. J. Hegarty et al, Phys. Rev. Lett. **49**, (1982) 930.

Proc. 25th Int. Conf. Phys. Semicond., Osaka 2000 (Eds. N. Miura and T. Ando)

95

Fine structure of excitons in diamond: A cathodoluminescence study

N. Teofilov[1], R. Sauer[1], K. Thonke[1], T.R. Anthony[2], H. Kanda[3]

[1] Abteilung Halbleiterphysik, Universität Ulm, D-89069 Ulm, Germany e-mail: rolf.sauer@physik.uni-ulm.de
[2] General Electric Corporate Research and Development, Schenectady, New York 12309, USA
[3] NIRIM National Institute for Research in Inorganic Materials, Tsukuba, Ibaraki 305, Japan

Abstract A complex fine structure of the free exciton luminescence in diamond is reported verifying recent data, and studied at temperatures from 9 K to $\approx$ 100 K.

1 Introduction

The intrinsic excitonic gap of diamond has been studied intensely in previous years in absorption [1] and in emission [2]. In the latter work, Dean et al. used 60 keV electrons to excite the excitonic luminescence while the specimens were kept at temperatures of (80 ... 100) K. Under these conditions, they found that the free exciton emission is a doublet where the higher-energy component, split apart by $\approx$ 7 meV, thermalizes with the lower-energy component. It was natural to associate these two lines with the recombination of excitons incorporating either a Γ_8^+ hole from the degenerate light and heavy hole bands, or a Γ_7^+ hole from the spin-orbit split-off band. (The spin-orbit interaction amounts to $\Delta_0 \approx 6$ meV following cyclotron resonance measurements with the excitation of holes from boron acceptor levels by tunable light in the range of 370 meV [3].) Recent work demonstrated that there is much more structure in the free exciton (FE) spectra visible at lower temperatures [4,5]. In particular, Sauer et al. [5] employing a special derivative technique to their cathodoluminescence (CL) spectra detected two similar groups of lines spaced by 10.3 meV in the TO-phonon replica as well as in the TA-phonon replica. The present CL work aims at verifying this complex structure and to study the thermalization features. This is accomplished by taking temperature-dependent high signal-to-noise CL spectra of the TO- and TA-phonon region adding rich information from the weak LO-phonon replica. A discussion on the physical nature of these multiple states is postponed in the present investigation since this is difficult owing to the possible incorporation of the split-off band in the excitonic hole states of diamond [5].

2 Experimental

The samples are HPHT synthetic diamonds with little or essentially no boron-bound exciton luminescence. All spectral details reported below for ^{12}C samples were observed identically in isotopically almost pure ^{13}C specimens, with energy shifts of + 17.5 meV (TA), + 19.2 meV (TO), and + 20.5 meV (LO) for the various wave vector conserving phonon replicas. The samples were mounted on the Cu cold finger of a helium flow cryo-stat and were excited by electrons from a conventional RHEED electron gun at an energy of 6 keV. The luminescence signals were dispersed by a monochromator (1 m focal length, grating with 1200 lines/mm blazed at $\lambda = 250$ nm) and detected by a LN$_2$-cooled UV-optimized CCD-camera. All sample temperatures indicated below were determined from the Boltzmann tails of the free exciton luminescence lines.

3 Results

Fig. 1 shows the intrinsic FE spectrum at $\approx$ 9 K consisting of the two lines 1 and 2 in the TA-, TO-, and LO-phonon replicas. The line notation follows that in Ref. [5]. An additional line (2' in the previous notation) at 5.277 eV belongs to the TO-replica.

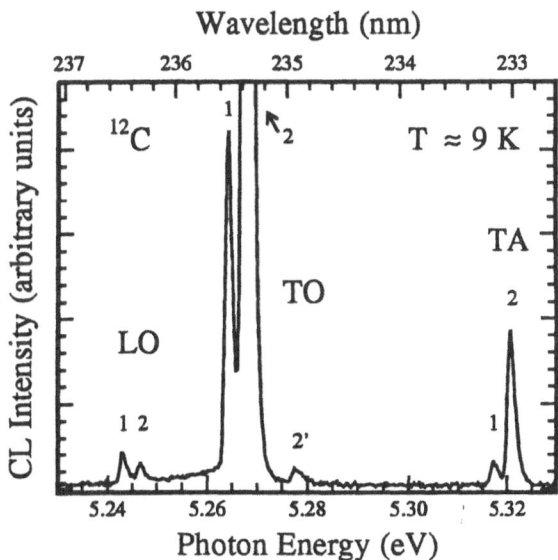

Fig. 1 9 K CL spectrum of free excitons (FE) with wave-vector conserving TA-, TO-, and LO-phonons. Line notation as in Ref. [5]. Original spectrum as recorded.

The spectrum implies 3 characteristic features: (a) The intensity ratio of line 2 to line 1 is around 5 for FE^{TA} and FE^{TO} but much smaller (around 0.5) for FE^{LO}. Quite generally, the relative line intensities are always very similar for FE^{TO} and FE^{TA}. (b) In all three cases, these intensity ratios are very much higher than would correspond to a Boltzmann factor $exp(-\Delta_{12}/kT)$ with $\Delta_{12} \approx 3$ meV being the line spacing. This indicates line 2 is not thermally activated with Δ_{12} at increasing temperatures. Experimentally, the intensity ratios

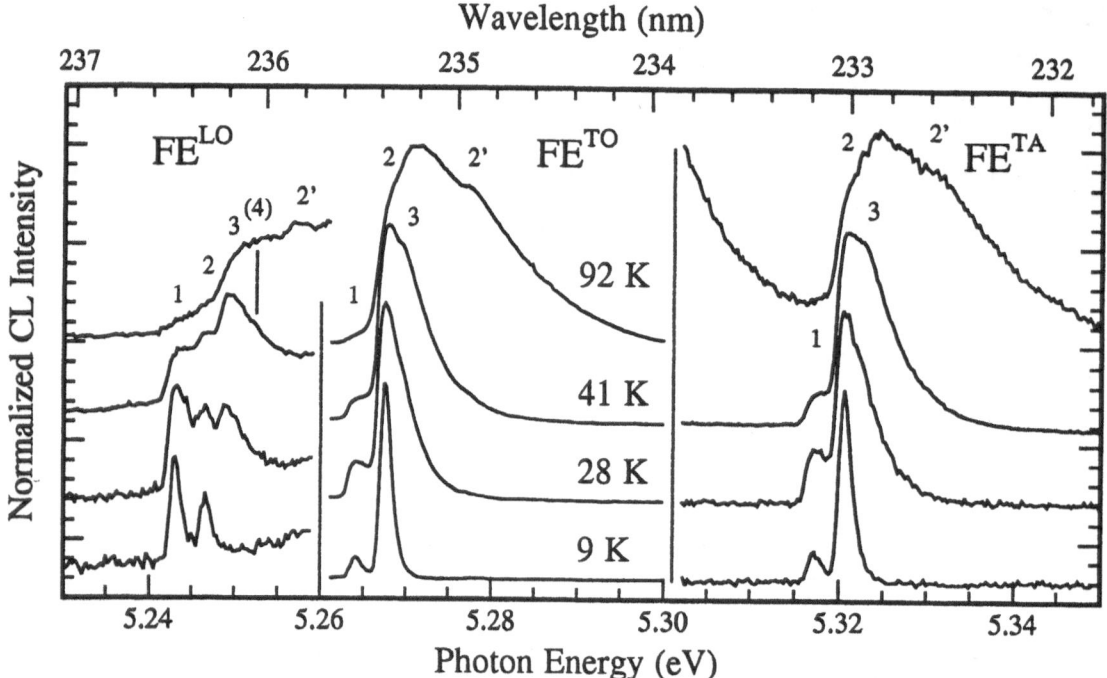

Fig. 2 Original free exciton (FE) CL-spectra at different temperatures. Vertical scales are expanded for FE^{LO} and FE^{TA} by factors of ≈ 25 and ≈ 20, respectively.

indeed remain roughly constant. (c) Also the line 2'(TO) at 5.277 eV is not in quasi-thermal equilibrium with the lines 1(TO) and 2(TO). It is important to note that all lines are free exciton components (as contrasted by excitons localized at very shallow impurities or potential fluctuations of the diamond lattice) since they develop Boltzmann tails at higher temperatures.

The evolution of the spectra with increasing temperatures is plotted in Fig. 2. In all three phonon replicas of the FE higher-energy lines come up. Line 3 is readily observable on all replicas. A further line 4 is detected weakly in some spectra in FE^{LO} at $\approx (30 ... 70)$ K. It corresponds to line 4 (TO) reported with some uncertainty in the derivative CL spectra in Ref. [5]. In the latter work, a second group of excited lines 1' ... 3' similar to 1 ... 3 but up-shifted by ≈ 10.3 meV as a whole was found for the TO- and TA-phonon replicas. Out of this group only 2' (TO, TA) is easily observed in CL probably because it alone has the same high oscillator strength as the corresponding line 2 (TO, TA). At temperatures above ≈ 70 K line 2' is also seen in the LO-phonon replica. The appearance of line 2' together with the overlapping lines 1, 2, and 3 leads to a spectral structure for FE^{TO} and FE^{TA} around 80 K and higher temperatures which looks like an exciton doublet when the spectral resolution is lower than in the present work. These were the experimental conditions under which Dean et al. [2] reported the excitonic doublet with a ≈ 7 meV splitting. Actually, this splitting is essentially the spacing between lines 3 and 2' which has no special physical relevance.

Our data contribute to figuring out the puzzling spectral structure of the exciton gap of diamond. We are far from understanding the physical nature of the electron-hole states involved. In particular, questions to be answered are the giant oscillator strength of line 2 for FE^{TO} and FE^{TA} and the non-thermalization of the basic low-temperature lines 1 and 2 in all phonon replicas.

Table 1 Photon energies of the observed line maxima in eV corrected for temperature shifts so as to correspond to T $\approx$ 9 K

line	hν (LO)	hν (TO)	hν (TA)
1	5.242$^+$	5.263$^+$	5.317
2	5.246	5.267	5.320
3	5.249	5.270	5.324$^+$
2'	5.256	5.277	5.330

References

1. P.J. Dean and J.C. Male, J. Phys. Chem. Solids **25** (1964) 311
2. P.J. Dean, E.C. Lightowlers, and D.R. Wight, Phys. Rev. **140** (1965) A 352
3. C.J. Rauch, Proc. Internat. Conf. Semicond. Phys., Exeter 1962 (Institute of Physics and the Physical Society, London1963), p. 276
4. S.J. Sharp and A.T. Collins, Mater. Res. Soc. Symp. Proc. **416** (1996) 125
5. R. Sauer, H. Sternschulte, S. Wahl, K. Thonke, and T.R. Anthony, Phys. Rev. Lett. **84** (2000) 4172

Transient reflection from anisotropic semiconductors studied with coherent control method

Yasuyoshi Mitsumori, Mariko Sato, Hideaki Tobioka, and Fujio Minami

Department of Physics, Tokyo Institute of Technology
Meguro-ku, Tokyo 152-8511, Japan

Abstract The transient reflection of subpicosecond optical pulse has been investigated under the Brewster angle of the incidence by using the coherent control technique with a phase-locked pulse pair to excite resonantly excitons in anisotropic crystal GaSe. The interferogram of the reflected pulse has a long tail corresponding to the free induction decay signal. We also observed the phase difference between the reflected light from the Γ_4 and Γ_6 excitonic polarizations.

1 Introduction

Coherent transients in semiconductors have provided us a detailed knowledge about the nature of the coherent interaction between the light and the matter, such as the pulse distortion, coherent control, and four wave mixing[1-4]. However, a few study of the transient reflectivity, which is one of the most basic interactions between light and matters, has been reported. Here, we report the transient reflectivity at the resonance with the excitons in anisotropic crystal GaSe under the Brewster angle incidence. Since the Brewster effect eliminates the nonresonant contribution to the reflection amplitude, coherent transients involving the excitons can be observed directly. We have applied the coherent control technique to analyze the temporal evolution of the reflected pulse. The resultant interferogram has a long tail corresponding to the coherent emission form the excitons. Moreover, by changing the polarization angle between the first and second pulses, the phase difference between the on-resonant and off-resonant polarizations is detected. A phase shifted beating structure is also observed in the decay of the signal. This beating behavior arises from the anisotropy of the crystal.

2 Theory

The reflected pulse shape $E_R(t)$ is given by

$$E_R(t) = \int d\omega\, E_I(\omega)\, \gamma(\omega, \theta)\, e^{i\omega t}, \quad (1)$$

where $E_I(t)$ represents the Fourier transformation of the incident electric field and $\gamma(\omega, \theta)$ is the frequency-dependent Fresnel reflection coefficient. When the dielectric function of semiconductor is modeled by the simple Lorenz oscillator, the excitonic nonresonant term ε_∞ is very large in comparison with the excitonic resonant term. Because of this large nonresonant term, the reflected pulse shape usually becomes almost identical to the incidence. However, under the Brewster angle of the p-polarized incidence $\theta_B = \sqrt{\varepsilon_\infty}$, the nonresonant contribution to reflection can be eliminated. In the weak resonant limit, the reflection coefficient $\gamma(\omega, \theta_B)$ is described by a simple Lorezian function. The temporal shape of the reflected pulse is expressed by convolution of the incident pulse with the exponentially damping of resonance. Consequently, the reflected pulse under the Brewster angle of the incidence shows only the temporal response of the excitonic polarization[5, 6].

3 Experiment

The samples we used in this work were the layered semiconductors GaSe grown by the Bridgman method and cleaved on the face normal to the c-axis. The 1S exciton state splits into two states: the Γ_4 and Γ_6 states. The Γ_6 exciton is optically active for the polarization $E \perp c$ and the Γ_4 exciton allowed for the polarization $E // c$. The Γ_4 exciton is situated at about 2meV higher than the Γ_6 exciton[7]. These excitons can be excited by the p-polarized beam at oblique incidence. The excitation source was an Optical Parametric Oscillator (OPO) system based on a cw mode-locked Ti-Sapphire laser. The temporal duration of the generating pulse is ~300fs. The excited wavelength was tuned to cover both of the exciton resonance. The experiments were performed as follows. Pulses from the OPO system were injected into a Michelson interferometer. One arm of the interferometer was varied in length. In order to analyze the temporal shape of the reflected pulse, we placed a $\lambda/2$ plate in the interferometer. The two beams were recombined to incident collinearly to the sample. A part of the recombined beam was used for a phase-locked loop to adjust an arm length of the interferometer. The incident angle was tuned to the Brewster angle. After the sample, a linear polarizer was placed in order to measure the interference intensity between the cross-polarized beams. The reflected intensity was detected as a function of the delay time. All the measurements were carried out at 5K.

4 Result and Discussion.

Fig. 1 shows the interferogram of the reflected pulses under the Brewster angle incidence, when the first and second pulses are p-polarized. The interference intensities are affected by the phase-locked angle and time delay. The interferogram has a long tail even when the first pulse and second pulses are temporally separated. Because the Brewster effect eliminates the contribution of the excitonic noresonant term to the reflection, the interferogram only reflects the interference between the polarizations created by the phase-locked pulse pair. The difference in the interference intensity in case of phase-locked angle 0 (constructive interference) and case of π (destructive interference) corresponds to the coherence of the excitonic polarization, i.e. free induction decay. A beating structure can be also observed in the decay of the signal. This beating structure can be explained by a beating polarization of the coherent superposition of the Γ_4 and Γ_6 excitons. The beating period is found to be ~ 2.2ps, which is calculated to be ~2meV in the energy domain. This value is in good agreement with the energy separation between the Γ_4 and Γ_6 excitons.

When the polarization of the second pulse changes from the p-polarization to the s-polarization, the shape of the interferogram dramatically changes, as shown in Fig. 2. In contrast to the case of the parallel-polarization, the shape of the interferogram becomes asymmetric with respect to the zero delay time, because the interferogram reflects the correlation between the different pulse shapes. We also changed the phase-locked angle from 0 (π) to π/2 (3π/2) in order to measure the envelope function of the interferogrm. This change of the phase-locked angle reveals the phase difference between the on-resonant and off-resonant polarizations. For the s-polarization, the reflected pulse is comprised of the emission from the excitonic off-resonant polarization, because the reflection coefficient is governed by the excitonic nonresonant term ε_∞. On the other hand, under the Brewster angle of the p-polarized incidence, the reflected pulse consists of the coherent emission form the exciton. Generally, the phase of the on-resonant polarization is *90°* out of phase in comparison with that of the off-resonant polarization, because the on-resonant polarization contains a damping factor. As the result of this phase difference, the phase-locked angle should be changed. A beating structure can be observed in the decay of the signal. This beating structure can be explained by the same reason as the case of the parallel-polarization. The phase of the beating structure is found to be *180°* out of phase in comparison with that of the parallel-polarization. This phase-shifted beating structure indicates that the phase of the light emitted form the Γ_4 exciton in the direction of the reflection is different from that of the Γ_6 excitons. From the selection rules of the two 1S excitons, the Γ_4 excitonic polarization oscillates parallel to the *c*-axis (*z*-axis) and the Γ_6 excitonic polarization normal to that (*x*-axis). In the dipole approximation, the Γ_4 and Γ_6 polarization can be described as $<\Gamma_4|z|g>$ and $<\Gamma_6|x|g>$, respectively. When these polarizations emit the coherent light in the direction of the reflection, the emitted optical field can be written by

$$E_R(t) \propto e_R \cdot \left(\langle \Gamma_4|z|g \rangle e^{i\omega_1 t} + \langle \Gamma_6|x|g \rangle e^{i\omega_2 t} \right)$$
$$\propto \cos\theta_B \langle \Gamma_4|z|g \rangle e^{i\omega_1 t} - \sin\theta_B \langle \Gamma_6|x|g \rangle e^{i\omega_2 t}, \quad (2)$$

where, e_R represents the unit polarization vector of the reflection under the p-polarization, ω_1 and ω_2 are the resonant frequencies of the Γ_4 and Γ_6 excitons, respectively and we neglect the damping factor of the oscillators. The sign of the first and second term in Eq. (2) becomes opposite, which leading to the π phase shifted beating structure, as shown in Fig. 2.

5 Conclusion

We have investigated the transient reflection of subpicosecond optical pulse under the Brewster angle of the incidence by using the coherent control technique in anisotropic crystal GaSe. The interferogram of the reflected pulse has a long tail corresponding to the free induction decay signal. When changing the polarization of the second pulse of the phase-locked pulse pair, a phase-shifted beating structure appears in the interferogram. This beating structure can be well explained by the reflection mechanism in the anisotropic crystal.

Acknowledgment

This work was supported by a Grand-in-Aid for Scientific Research.

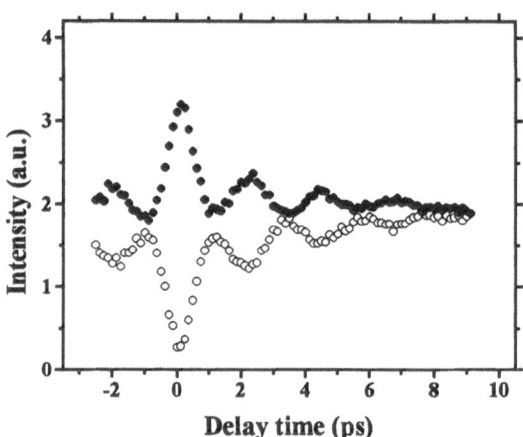

Fig. 1. Interferogram of the reflected pulses under the Brewster angle incidence for the p-polarization, when the first and second pulses are polarized along the same direction. The filled and open circles represent data for the phase-locked angle 0 and π, respectively.

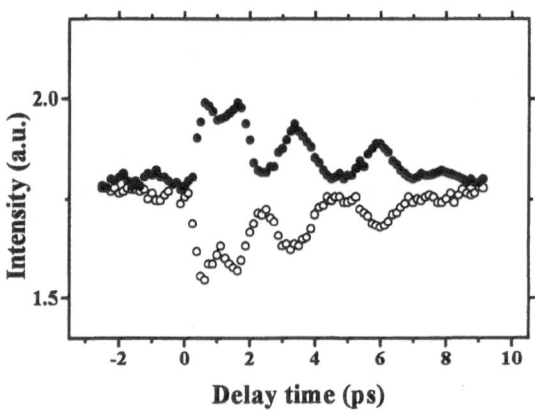

Fig. 2. Interferogram of the reflected pulses, when the first pulse is p-polarized, the second pulse is s-polarized. The filled and open circles represent data for the phase-locked angle π/2 and 3π/2, respectively.

References
1. D. K. Kim *et. al.*, Phys. Rev. B **50** (1994) 18240.
2. A. P. Heberle *et. al.*, Phys. Rev. Lett. **75** (1995) 2598.
3. Y. Mitsumori *et. al.*, J. Lumin. **87-89** (2000) 914.
4. L. Schultheis *et. al.*, Phys. Rev. B **34** (1996) 9027.
5. J. Aaviksoo *et. al.*, IEEE J. Quant. Electron. QE **25** (1989) 2523.
6. A. Hasegawa *et. al.*, J. Lumin. **58** (1994) 234.
7. E. Mooser *et. al.*, Nuovo Cimento B **18** (1973) 164.

Exciton-phonon droplets with Bose-Einstein condensate: transport and optical properties

D. Roubtsov[1], Y. Lépine[1], I. Loutsenko[2]

[1] GCM et Département de physique, Université de Montréal, C.P. 6128, succ. Centre-ville, Montreal, P.Q., H3C 3J7, Canada
[2] Lockheed Martin Canada et CRM, Université de Montréal, C.P. 6128, succ. Centre-ville, Montreal, P.Q., H3C 3J7, Canada

Abstract We discuss the possibility for a moving droplet of excitons and phonons to form a coherent state inside the packet. We describe such an inhomogeneous state in terms of Bose-Einstein condensation prescribing it a macroscopic wave function. Existence and, thus, coherency of the Bose-core inside the droplet can be checked experimentally by letting two moving packets to interact.

1 Introduction

Nowadays, there is a lot of experimental evidence that excitons in semiconducting crystals and heterostructures can form a strongly correlated state, which can be assigned to the excitonic Bose-Einstein condensate (BEC) [1]. In the most cases, however, the excitons are prepared in the ground state with $\langle \hbar k \rangle \simeq 0$ (or they are in a quasi-equilibrium state with some $T^*(t)$ cooling down toward the ground state). The conclusion about the presence of Bose-Einstein correlations among them is based on unusual properties of the direct PL signal from the excitonic cloud [1].

This article is motivated by experimental data on the transport properties of excitons in 3D crystals, such as Cu_2O [2], and 2D sheets in BiI_3 [3]. Surprisingly, a relatively dense cloud of excitons can be prepared in a moving state, and such a packet can move ballistically through the whole crystal at $T < T_c$. If $T > T_c$, however, the excitonic packet exhibits the standard diffusive behavior [2]. Thus, the 3D droplets that could contain the excitonic Bose-Einstein condensate are found in a spatially inhomogeneous state with the well-defined characteristic width L_{ch} in the direction of motion. Moreover, the registered ballistic velocities of such excitonic packets turn out to be always less, but relatively close to the longitudinal sound speed of the crystal, $v < c_s$. Note that the paraexcitons in pure Cu_2O crystals have an extremely large lifetime, $\tau \simeq 13 \, \mu s$, and a moving exciton with $\hbar k_x \sim m_x c_s$ cannot be converted into a photon directly. Then, one can even exclude the photons from simple models describing the transport of a *single* packet of excitons in a medium.

To understand the physics of anomalous excitonic transport, we assume that the macroscopic wave function $\Psi_0 \simeq \phi_0 e^{i\varphi_c}$ can be associated with the coherent part of the excitonic packet at $T < T_c$. Here, φ_c is the coherent phase of the condensate. Indeed, the experimental results [2],[3] suggest the following decomposition of the density of excitons in the packet,

$$n(\mathbf{x}, t) = n_{coh}(\mathbf{x}, t) + \Delta n(\mathbf{x}, t), \qquad (1)$$

where $n_{coh}(\mathbf{x}, t) \approx n_{core}(x - vt)$ is the ballistic (superfluid?) part of the packet,

$$n_{core}(x - vt) \simeq |\Psi_0|^2(x - vt), \qquad (2)$$

and $\Delta n(\mathbf{x}, t)$ is the noncondensed part of it. Therefore, the challenging problem is how to describe the spatially inhomogeneous state of the droplet with the excitonic BEC in terms of $\Psi_0(\mathbf{x}, t)$ and $\delta\hat{\psi}(\mathbf{x}, t)$, where $\delta\hat{\psi}$ is the "fluctuating" part of the exciton Bose-field. For example, $n_{core}(x/L_0)$ and $\Delta n(x) \simeq \delta n_{o,loc}(x/L_{ch}) + \delta n_{tail}$ can have different characteristic lengthes and coherence properties. Note that if the excitonic packet moves in a crystal (or another semiconductor structure), it interacts with thermal phonons, noncondensed excitons, impurities and other imperfections of the lattice, etc.. Then, the coherent core of the packet can be found in a quasi-stable state, so that the fluctuations of $\phi_0(x - vt)$ and, especially, $\varphi_c(x, t)$ can be of a great importance for possible experimental verifications of their existence.

2 Exciton-Phonon Condensate

To obtain the necessary density of excitons n_x in the excitonic cloud and, thus, meet the BEC conditions, the crystals are irradiated by laser pulses with $\hbar\omega_L \gg E_{gap}$, and $T \simeq 1 \sim 5 \, K$. If the cross-section area S of an excitation spot on the surface of the crystal can be made large enough, such as $S \approx S_{surf}$, the hot droplet of excitons can acquire an average momentum during its thermolization process $(T^*(t) \to T)$. Indeed, the phonon wind, or the flow of nonequilibrium phonons, blows unidirectionally from the surface into the bulk [4] and transfers the nonzero momentum to the excitonic cloud, see Fig. 1. As a result, the packet of moving excitons and nonequilibrium phonons of the phonon wind $(N_{ph} \sim N_x)$ is actually the system that undergoes the transition toward developing the Bose-Einstein correlations at $T^* < T_c$.

Let us assume that the condensate has been already formed inside the moving excitonic droplet, and the following representation of the exciton Bose-field holds: $\hat{\psi} = \Psi_0 + \delta\hat{\psi}$. For the displacement field of the crystal $\hat{\mathbf{u}}$, we introduce a nontrivial coherent part too, i.e., $\hat{\mathbf{u}} = \mathbf{u}_0 + \delta\hat{\mathbf{u}}$ and $\mathbf{u}_0 \neq 0$. In these terms, a moving packet contains the (mascoscopically occupied) exciton-phonon condensate, or the Bose-core $\Psi_0(\mathbf{x}, t) \cdot \mathbf{u}_0(\mathbf{x}, t)$, and out-of-condensate excitons and phonons as well. The macroscopic wave function of excitons $\Psi_0(x, t)$ is normalized as $\int |\Psi_0|^2(x, t) \, dx = S \int \phi_0^2(x, t) \, dx = N_o$, where N_o is the macroscopic number of condensed excitons, and, generally, $N_0(T) < N_x$. Within the stationary approximation, we have $N_0(T) = $ const.

To model the ballistic motion of a single packet, we use the following ansatz for the Bose-core of the packet:

$$\Psi_0(x,t) = e^{-i(\tilde{E}_g + m_x v^2/2 - |\mu|)t} e^{i(\varphi_c + k_0 x)} \phi_0(x - vt), \quad (3)$$

$$u_{0j}(x,t) = u_0(x - vt)\,\delta_{1j}, \quad (4)$$

where $\tilde{E}_g = E_{\text{gap}} - E_x$, E_x is the exciton Rydberg, $\varphi_c = $ const is the macroscopic phase, $\hbar k_0 = m_x v$, and $\mu = \mu(N_0) < 0$ is the effective chemical potential of the condensate. At $T \ll T_c$, one can disregard the interaction between the Bose-core and the out-of-condensate cloud and write down the following equations on the envelope functions $\phi_0(x/L_0)$ and $u_0(x/L_0)$ [5]:

$$-|\mu|\,\phi_0(x) = -(\hbar^2/2m_x)\partial_x^2 \phi_0(x) - |\tilde{\nu}_0|\,\phi_0^3(x) + \tilde{\nu}_1\,\phi_0^5(x),$$

$$\partial_x u_0(x) \approx -\text{const}_0\,\phi_0^2(x) + \text{const}_1\,\phi_0^4(x). \quad (5)$$

At $T \neq 0$, $T < T_c$, we choose the quasi-stationary approximation to write the decomposition of the exciton and phonon fields of the moving droplet,

$$\hat{\psi}_0(\mathbf{x}, t) = \exp\big(i\varphi_c(x,t)\big)\big\{\phi_0(x - vt) + \delta\hat{\psi}_0(x - vt, \mathbf{x}_\perp, t)\big\},$$

$$\hat{u}_{0,j}(\mathbf{x},t) = u_0(x - vt)\,\delta_{1,j} + \delta\hat{u}_{0,j}(x - vt, \mathbf{x}_\perp, t). \quad (6)$$

Then, the following correlation functions have to be included into an analog of Eq. (5): the 'anomalous' ones, such as $\tilde{m}(x) = \langle \delta\hat{\psi}_0 \delta\hat{\psi}_0 \rangle$, the exciton-phonon correlators, such as $\tilde{q}_j = \langle \partial_j \delta\hat{u}_{0,j}\,\delta\hat{\psi}_0(x, \mathbf{x}_\perp, t) \rangle$, and the out-of-condensate density of the excitons and phonons,

$$\delta n_0(x) = \langle \delta\hat{\psi}_0^\dagger \delta\hat{\psi}_0(x, \mathbf{x}_\perp, t) \rangle \text{ and } Q_{xx}(x) = \langle (\partial_x \delta\hat{u}_{0,x})^2 \rangle.$$

It is possible to generalize Eq. (5) to the case of $T \neq 0$. Here, we write out it in the following (tractable) form:

$$-(|\mu| + \delta\mu)\,\phi_0(x) = -(\hbar^2/2m^*)\partial_x^2\phi_0(x) + (\tilde{\nu}_0 + \delta\nu_0)\,\phi_0^3(x)$$

$$+ (\tilde{\nu}_1 + \delta\nu_1)\,\phi_0^5(x), \quad (7)$$

$$\partial_x u_0(x) \approx -\text{const}_0'\,\phi_0^2(x) + \text{const}_1'\,\phi_0^4(x) - |\text{const}_{\text{tail}}|,$$

We disregard all the dissipation terms and assume $T^* \simeq T$. Then, the corrections to $|\mu|$, the x-x and x-ph interaction vertices, $\tilde{\nu}_j$ and const_j, depend on temperature. This means Eq. (7) has to be considered together with the equations on the out-of-condensate excitons and phonons, $\delta\hat{\psi}_0$, $\delta\hat{\psi}_0^\dagger$, and $\delta\hat{u}_{0,j}$. Some qualitative results obtained from this approach are presented on Fig. 2.

References

1. J. L. Lin, J. P. Wolfe, Phys. Rev. Lett. **71**, (1993) 122;
   L.V. Butov, A. I. Filin, Phys. Rev. B **58**, (1998) 1980;
   T. Goto, M. Y. Shen et al., Phys. Rev. B **55**, (1997) 7609.
2. E. Fortin, S. Farad, and A. Mysyrowicz, Phys. Rev. Lett. **70**, (1993) 3951;
   A. Mysyrowicz, E. Benson, and E. Fortin, Phys. Rev. Lett. **77**, (1996) 896;
   E. Benson, E. Fortin et al., Europhys. Lett. **40**, (1997) 311.
3. H. Kondo, H. Mino et al., Phys. Rev. B **58**, (1998) 13835.
4. G. A. Kopelevich, S. G. Tikhodeev, and N. A. Gippius, JETP **82**, (1996) 1180.
5. D. Roubtsov, Y. Lépine, Phys. Rev. B **61**, (2000) 5237; see also in cond–mat archive at http://xxx.lanl.gov/.

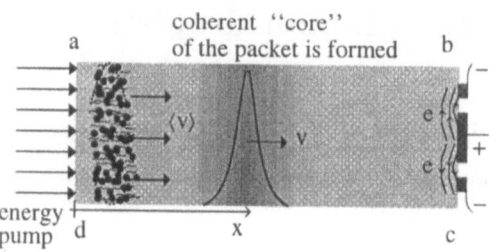

Fig. 1 A medium, in which the exciton-phonon droplet can propagate, is presented in the form of the channel 'abcd' on this Figure. After some amount of energy has been pumped into the medium during a short time interval and absorbed near a boundary, a localized excited state is formed near the face 'ad'. If there is a mechanism of the momentum transfer to the excited state, the droplet begins to move toward the opposite face 'bc' with the velocity $\langle v \rangle$. Then, such conditions can favor the appearance of an inhomogeneous coherent state inside the droplet if the average density of the excitons $n_x > n_c(T)$. The profile of the excitonic part of it, $n_{\text{core}}(x,t) \simeq |\Psi_0(x,t)|^2$, is shown by the bold line and the intensity of the elastic (phonon) part, $\partial_x u_{0,x}(x,t)$, is represented by changements of the intensity of the background color. When the packet reaches the face 'bc', the total density of excitons $n(x)$ is converted into an electric current, $i(t)$.

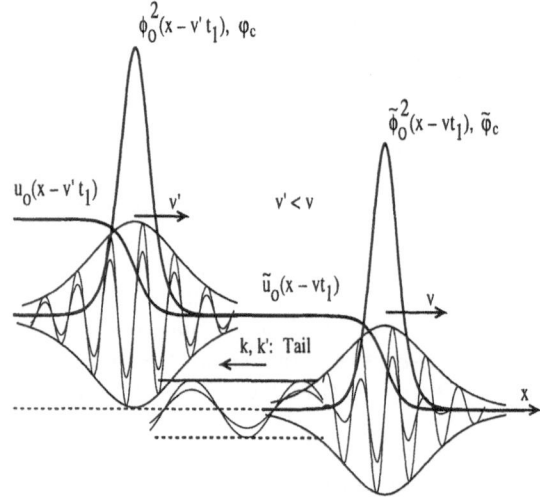

Fig. 2 Two ballistic packets with two exciton-phonon condensates ($e^{i\varphi_c(x,t)}\phi_0(x - vt) \cdot u_0(x - vt)\,\delta_{1j}$) inside were created with the same concentration of excitons and the same ballistic velocity, v, in a crystal. The time delay between them is a free parameter. Then, two different interaction regimes are possible. The first one corresponds to the case in which the Bose-cores of the packets overlap. This is a strong interaction case, and the packets can merge into one droplet. The second regime, in which the Bose-cores do not overlap, is the case of weak interaction between the packets. It is depicted on this Figure. However, the second moving packet (the left one) can "feel" the first packet (the right one) through the interaction with the exciton-phonon tail of the first one. As a result, the second packet slows down, $v' < v$, and becomes more broad.

Role of the excitonic continuum in the polariton problem

E. A. Muljarov[1] and R. Zimmermann[2]

[1] General Physics Institute, Russian Academy of Sciences, Vavilova 38, Moscow 117942, Russia; e-mail: muljarov@gpi.ac.ru

[2] Institut für Physik der Humboldt-Universität zu Berlin, Hausvogteiplatz 5-7, D-10117 Berlin, Germany

Abstract Applying approximate center-of-mass boundary conditions to the microscopic exciton polarization, the polariton problem is solved analytically in the cases of half space and slab geometries. This analytical solution depends explicitly on the bulk exciton dielectric function which allows a direct inclusion of all bound exciton states and the Sommerfeld-enhanced continuum (scattering states). The inclusion of the excitonic continuum allows us to explain quantitatively the experimental data on differential transmission of 500 nm GaAs slab. As a side effect, we have re-considered the problem of additional boundary conditions (ABC's) which traditionally arises in the polariton problem.

1 Introduction

In semiconductors, within dipole approximation the propagating electro-magnetic field exchanges momentum with a Coulomb coupled electron-hole pair, leading to a creation of a new type of excitations called polaritons [1, 2]. Whereas in bulk semiconductors the center-of-mass and relative motions of the electron-hole pair are decoupled, in bounded semiconductors formerly independent Hydrogen-like states of an exciton are strongly mixed, due to the breaking of translational symmetry at the surfaces, thus leading to the electro-magnetic field interacts not only with the exciton ground state, but also with all higher bound and scattering states. To provide the correct microscopic boundary conditions (BC's) on a semiconductor surface, namely, the vanishing of the coherent electron-hole amplitude [3], the contribution of bulk exciton states other than $1S$ to the full excitonic polarization is very important even in the energy region near the fundamental resonance.

The purpose of this the paper is twofold. Firstly, we present for the first time the results of calculations of the full optical spectra far in the continuum, in a close comparison with experimental results [4]. Secondly, we undertake an attempt to review critically the problem of ABC's in the context of recent debates [5,6].

2 Model

Within the microscopic model [3] the exciton polariton problem in the case of normal incidence of light is described by coupled material and Maxwell's equations

$$\left[-\frac{\hbar^2}{2M}\frac{\partial^2}{\partial Z^2} + \hat{H}_{\mathrm{ex}}(\mathbf{r}) + E_g - \hbar\omega - i\gamma \right] Y(\mathbf{r}, Z)$$
$$= \mathcal{M}\delta(\mathbf{r})E(Z), \qquad (1)$$

$$\frac{\partial^2}{\partial Z^2}E(Z) + q_0^2 \varepsilon_b E(Z) + q_0^2 \mathcal{M}Y(0, Z) = 0, \qquad (2)$$

where $Y(\mathbf{r}, Z)$ is the coherent electron-hole amplitude, $\mathcal{M}\delta(\mathbf{r})$ is the transition dipole density, $\hat{H}_{\mathrm{ex}}(\mathbf{r})$ is the Hamiltonian of the bulk exciton relative motion; $\mathbf{r}$ and Z are, respectively, the relative and center-of-mass coordinates, M is the exciton effective mass along Z; $q_0 = \omega/c$, where ω is the photon frequency, γ is a phenomenological damping, E_g is the semiconductor energy gap, ε_b is the background dielectric constant.

The electric field $E(Z)$ and electron-hole amplitude $Y(\mathbf{r}, Z)$ should satisfy certain BC's. According to the Maxwell's BC's both $E(Z)$ and $E'(Z)$ should be continuous everywhere. The BC for the polarization should be $Y = 0$, if the electron or hole coordinate approaches a semiconductor surface. Instead of this microscopically correct BC, we assume center-of-mass BC's

$$Y_Z'(\mathbf{r}, 0) = \alpha Y(\mathbf{r}, 0), \quad Y_Z'(\mathbf{r}, d) = -\alpha Y(\mathbf{r}, d), \qquad (3)$$

in the case of a semiconductor slab occupying the area $0 \leq Z \leq d$. Note that as soon as only the contribution of $1S$ exciton state to Y is taken into account, we immediately result in Pekar ABC's [1] for $\alpha = \infty$ and Ting-Frankel-Birman ABC's [7] for $\alpha = 0$, otherwise the case of arbitrary α is referred to as mixed ABC's. In other words, the traditional ABC's are generalized in Eq. (3), where the fact is taken into account that the real polariton is formed from the whole exciton spectrum.

We solve Eqs. (1) and (2) together with BC's Eq. (3) expanding the function $Y(\mathbf{r}, Z)$ into slab eigenmodes,

$$u_n(Z) = \frac{(\alpha + ik_n)e^{ik_n Z} - (\alpha - ik_n)e^{-ik_n Z}}{\sqrt{2[2\alpha + d(\alpha^2 + k_n^2)]}}, \qquad (4)$$

which in fact are a complete set of orthonormal functions and satisfy the same BC's, Eq. (3), provided that the eigenvalue k_n is a solution of the dispersion equation

$$e^{ik_n d} = \pm\frac{\alpha - ik_n}{\alpha + ik_n}. \qquad (5)$$

The electric field $E(Z)$ can be also expanded into $u_n(Z)$ with the only exception that a homogeneous part of the solution of Eq. (2) should be added to provide the necessary BC's for $E(Z)$, different from that for Y. This homogeneous part is fixed by Maxwell's BC's and then all the coefficients in the expansion are determined. Finally, in the case of 'Pekar' BC's ($\alpha = \infty$) the electric field in a slab takes the form

$$E(Z) = \frac{\pi i}{d^2}\sum_{n=-\infty}^{\infty} n\frac{E(0) - (-1)^n E(d)}{q_0^2 \varepsilon(\frac{\pi n}{d}, \omega) - (\frac{\pi n}{d})^2}e^{i\pi n Z/d}, \qquad (6)$$

where $\varepsilon(k, \omega)$ is the full dielectric function of the bulk exciton, which includes the contribution of all bound and

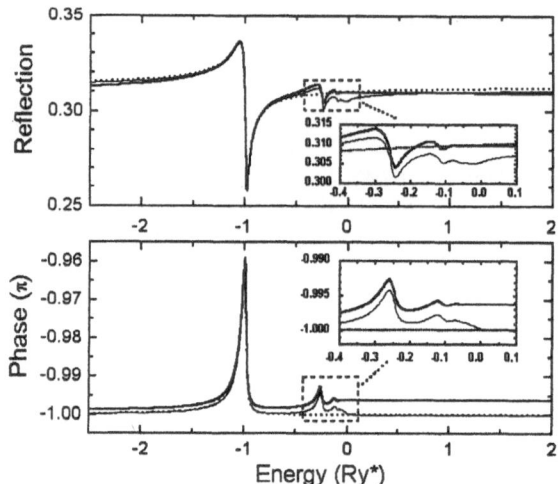

Fig. 1 GaAs reflectivity (top) and phase (bottom) calculated with account for both bound and scattering exciton states (thick lines), for bound states only (thin lines), and for the ground state only (dotted lined).

scattering states. Treating the half space situation as a limiting case of an extremely wide slab $d \to \infty$ we find

$$E(Z) = E(0) \frac{i}{\pi} \int_{-\infty}^{\infty} \frac{k \, dk \, e^{ikZ}}{q_0^2 \varepsilon(k, \omega) - k^2}. \qquad (7)$$

3 Results and Discussions

To demonstrate the role of the excitonic continuum in the reflectivity of semi-infinite crystals we plot in Fig. 1 reflection spectra of GaAs, calculated for $\alpha = \infty$, with the account for all states (bound and scattering), for only bound states, and for the ground state only. The difference in the spectra becomes visible even near the fundamental resonance and reaches the maximum closer to and above the gap energy. The phase of the reflected wave (bottom panel) has a break at $\hbar\omega = E_g$ (thin curve) which is removed due to the contribution of scattering states (thick curve). It is noticeable that above the band gap the phase calculated with the full exciton spectrum reaches the values larger than its usual quantity $(-\pi)$ far from an optical resonance.

A spectrum of the differential transmission (DT) of a sample containing GaAs slab 500 nm wide covered with an antirefractive coating is calculated for $\alpha = \infty$ and shown in Fig. 2 together with the experimental spectrum of Ref. [4]. The account for bound states only leads to a very large slope in the DT, whereas the DT spectrum calculated with the account for the excitonic continuum has almost no slope. The participation of scattering states in the transmission/reflection processes does not lead to new features in the spectra: there is a finite density of states in the continuum, which can produce a background rather than peculiarities. At finite but sufficiently small γ the peculiarities of $E(Z; \omega)$ and $E'(Z; \omega)$ manifest in maxima and minima of DT, which are known as spectral oscillations due to the effect

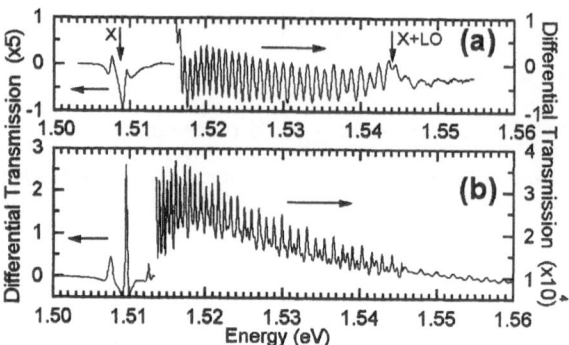

Fig. 2 Differential transmission of 500 nm GaAs slab: (a) experiment taken from Ref. [4]; (b) our calculation with $M_{hh} = 0.46 \, m_0$ and $M_{lh} = 0.232 \, m_0$.

of the exciton center-of-mass quantization in a slab [8]. These resonances appear at the frequencies close to $\hbar\omega = E_\nu + \hbar^2\pi^2n^2/2M$ (E_ν is the energy of the given exciton state), when the denominator in Eq. (6) has nodes.

We confirm the result of a number of previous works that Pekar ABC and its extended version of the present work (the case of $\alpha = \infty$) gives the best agreement with experimental data and is also microscopically justified, whereas the opposite case of $\alpha = 0$ is very close to the case when the spatial dispersion is neglected, and no oscillations are observed in the DT of a slab.

As in many early works, in a recent paper [5] it was undertaken an attempt to solve the polariton problem without going to any microscopic theory. However, it can be shown that the result of Ref. [5] is nothing else than the case of $\alpha = 0$ of the present work, which corresponds to the situation when almost no spatial dispersion comes into play. We conclude that any attempt to solve the polariton problem on purely macroscopic background always leads to a specific choice of ABC's and, at the same time, no one can deduce correct ABC's from the microscopic theory.

Acknowledgements. This work has been funded by Deutscher Akademischer Austauschdienst, NATO fellowship 325-A/99/02854. The authors are thankful to A. Stahl, G. Göger, and K. Henneberger for useful discussions.

References

1. S. I. Pekar, Zh. Eksp. Teor. Fiz. **33** (1957) 1022 [Soviet Phys. JETP **6** (1958) 785].
2. J. J. Hopfield, Phys. Rev. **112** (1958) 1555.
3. A. Stahl, Phys. Status Solidi B **94** (1979) 221.
4. G. Göger, M. Betz, A. Leitenstorfer, M. Bichler, W. Wegscheider, and G. Abstreiter, Phys. Rev. Lett. **84** (2000) 5812.
5. K. Henneberger, Phys. Rev. Lett. **80** (1998) 2889.
6. D. F. Nelson and B. Chen, Phys. Rev. Lett. **83** (1999) 1263; R. Zeyher, *ibid*, **83** (1999) 1264; K. Henneberger, *ibid*, **83** (1999) 1265.
7. C. S. Ting, M. J. Frankel, and J. L. Birman, Solid State Commun. **17** (1975) 1285.
8. A. Tredicucci, Y. Chen, and F. Bassani, Phys. Rev. B **47** (1993) 10348.

Far IR radiation from p-Ge under crossed directions of electric field and uniaxial pressure

A. A. Abramov[1], V. I. Akimov[1], A. T. Dalakyan[1], V. N. Tulupenko[1], D. A. Firsov[2], V. M. Bondar[3],

V.N.Poroshin[3], V.I. Gavrilenko[4], M.S. Kagan[5]

[1]Department of Physics, Donbass State Engineering Academy Shkadinova St. 72, Kramatorsk 84313, Ukraine, E-mail: tvn@laser.donetsk.ua
[2]St.Petersburg State Technical University Politechnicheskaya St. 29, St.Petersburg 195251, Russia
[3]Institute of Physics, National Academy of Science, Kiev 252650, Ukraine
[4]Microstructure Physics Institute of Russian Academy of Science, Nizhny Novgorod, 603600 Russia
[5]Institute of Radioengineering and Electronics of Russian Academy of Science, Moscow 103907, Russia

Abstract. The results of the first experiments investigating far-IR emission for crossed electric field and pressure are reported. The pressure dependences of the radiation intensity in various electric fields are explained by the different hole occupation of the states of an impurity center which are split by uniaxial pressure.

1 Introduction

The physical reason for the stimulated emission observed in [1] from p-Ge sample at parallel electric field **E** and uniaxial pressure (UP) **X** is inversion of the hole distribution on split by UP impurity states. One of these states at the certain X gets to the continuous spectrum of valence band and becomes a resonant one. The results of an investigation of the effect of resonant states on electric current flow at **E⊥X** for Ge samples cut in various crystallographic directions so that **X∥[111]** and **X∥[100]** were presented in [2]. In the present work we report the results of far IR radiation study for the same configurations.

2 Experiment

Investigations of radiation in the range $\lambda \approx 100$ μm (the maximum of the spectral sensitivity of a Ge:Ga photodetector with quartz and black polyethylene filters) for similar rectangular samples with orthogonal directions of the electric field and UP at T=4.2 K was performed. The optical signal was recorded with recorder with pressure scanning and electric field pulses with a constant amplitude. The pulse repetition frequency was about 70 Hz, and the pulse duration was 0.4 μs.

3 Discussion

Fig. 1 shows the computed dependences of the positions of the split by UP valence subbands ($\varepsilon_{\cdot p}$, ε_{-p}) of Ge [3] and the levels of the ground (ε_+, ε_-) and excited states of a shallow Ga impurity (concerning the construction of these curves see below) on the

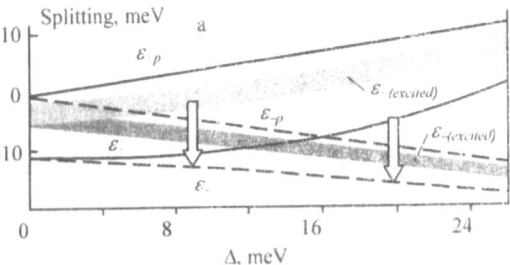

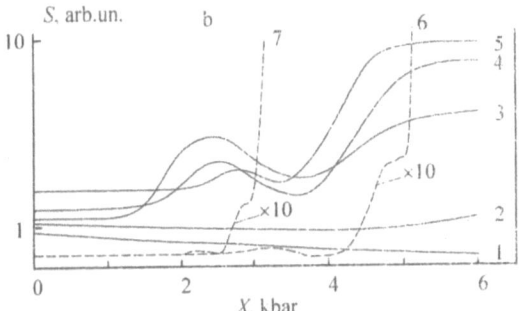

Fig. 1: a) Splitting of valence subbands and impurity levels against pressure. Transitions corresponding to the maximum of spectral sensibility of Ge(Ga) photodetector ($\lambda \approx 100$ μm) are shown with wide arrows. b) Radiation intensity versus the uniaxial pressure: T=4.2K; N_A-N_D, 10^{14}cm^{-3}: 1..6-6.8; 7-2.2. 1..6 - **X∥[111]**, 7 - **X∥[100]**. E, kV/cm: 1-0.5; 2-1.0; 3-1.4; 4-1.7; 5-1.8; 6-2.0; 7-2.2.

splitting Δ between the subbands and the experimentally determined pressure dependences of the spontaneous radiation intensity. We note the correspondence between the abscissas for **X∥[111]**: Δ=αX, α≅4 meV/kbar.

The nonmonotonic behavior of the radiation intensity is of interest. We attribute the increase in the optical signal near 2.5 kbar to optical transitions from a group of the first excited states $\varepsilon_{+\text{excited}}$ (which become resonant for X>1 kbar) to the ground state level ε_-. The subsequent decrease of the signal with increasing pressure seems to be due to broadening of these states and a decrease of their depth below the bottom of the upper subband ε_{-p}, which results in their being emptied in a strong electric field. The maxima near 5 kbar could be due to transitions between the ground states ε_+ and ε_-. We note that in this case the inhibiting parity of the states is lifted by the presence of a strong electric field. Similar transitions have been observed for parallel orientations of electric field and UP [1]. In Fig. 1 the wide arrows show the respective optical transitions with the same photon energy, corresponding to the sensitivity peak of the photodetector.

The pressure dependences of the position of the level ε_- were obtained by passing a smooth curve through three points. The first point is the depth of the Ga level in undeformed Ge (11.3 meV). The second one corresponds to the appearance of a resonant level (approximately 16 meV), and the third one, for the case of high pressures $\Delta \geq 40$ meV (this point falls outside the limits of the figure), was taken from [4]. The first three excited states [4], with their broadening taken into account, are shown by a single wide band $\varepsilon_{-excited}$ and were constructed in a similar manner. To ensure a unified approach, the dependences ε_- and $\varepsilon_{-excited}$ were found similarly, even though the positions of these levels as a function of pressure is well known from experiments [5] and calculations [6], and they agree well with the values obtained by the indicated method.

Qualitatively, we explain the relative position of the curves 1-5 at the first maximum (Fig. 1b) as follows. For appreciable spontaneous emission, the population of the lower states should not exceed that of the upper states. For a relatively weak electric field (curves 1 and 2), the level ε_- remains partially populated. As the electric field increases, the hole population of the level ε_- decreases, and appreciable spontaneous emission (curve 3) arises. As the electric field increases further, the excited states $\varepsilon_{+excited}$ are also emptied, apparently because of breakdown and transition of holes into the hole subband ε_{+p}. As a result, the spontaneous emission intensity decreases (curves 4 and 5).

The conclusion concerning the transition of holes into the hole subband is confirmed by the fact that at the second maximum the alteration of curves with increasing electric field does not occur. In this case, emptying of the ε_+ level (ground state) does not occur because of its large depth below the bottom of its subband. At the same time, since level ε_- goes closer to the bottom of ε_{-p} with increasing pressure, in identical electric fields the population of ε_- will be smaller for 5 kbar then 2.5 kbar. This is reflected the fact that the increment to the radiation intensity and the values of the maxima themselves are greater for 5 kbar then 2.5 kbar.

Results similar to those presented in Fig. 1 were also obtained for $X\|[100]$ ($\alpha \approx 6$ meV/kbar). If a radiation intensity is plotted as a function of the splitting Δ rather then pressure X, then the positions of the maxima on the abscissa will be practically the same for both directions of the pressure.

On this basis, it is not surprising that for some samples a jump in the optical signal by two orders of magnitude was observed at the maximum of the spontaneous emission both for $X\|[111]$ and for $X\|[100]$ (see Fig.1b, curves 6 and 7 correspondingly). We note first that the pressures corresponding to the threshold increase in the signal make it possible to attribute the signal to the appearance of resonant states ε_+. Then it can be concluded, by analogy to [1], that in strong electric fields lasing on total internal reflection modes arises due to intracenter inversion of the hole distribution.

This work was supported by UFBR, grant 2.4/970, Ukrainian-Russian grant 2M/88-2000.

References

1. I.V. Altukhov, M.S. Kagan and V.N. Sinis, JETP Lett. **47** (1988) 164.
2. V.M. Bondar, A.T. Dalakyan, V.N. Tulupenko and D.A. Firsov, JETP Lett. **35** (1982) 440.
3. G.L. Bir and G.E. Pikus, *Symmetry and Strain-Induced Effects in Semiconductors* (Wiley, NY 1974).
4. M.A. Odnobliudov, A.A. Pakhomov, I.N. Yassievich et al., Fiz. Tekh. Poluprovodn. **31** (1997) 1180.
5. E.I. Voevodin, E.M. Gershenzon, G.N. Gol'tsman et al., Fiz. Tekh. Poluprovodn. **8** (1986) 1356.
6. J. Broeckx and J. Vennik, Phys. Rev. B **35** (1987) 6165.

Proc. 25th Int. Conf. Phys. Semicond., Osaka 2000 (Eds. N. Miura and T. Ando)

105

Anomalous transport of excitons in Cu_2O

S. G. Tikhodeev[1] and N. A. Gippius[1,2]

[1] General Physics Institute, Russian Academy of Sciences, Vavilova 38, Moscow 117942, Russia; e-mail: tikh@gpi.ru
[2] Technische Physik, Würzburg Universität, Am Hubland, D-97074 Würzburg, Germany

Abstract Quite a number of experimental and theoretical work has been devoted last decade to the investigation of sonic nondiffusive transport of excitons in Cu_2O. However, the Bose-Einstein condensation interpretation of this transport is controversial. In this paper we summarize the interpretation of the anomalous exciton transport in Cu_2O via nonequilibrium phonon-assisted drag of excitons (phonon wind). We explain the intensity and temperature dependencies of the exciton transport and the nonlinearities under two-pulse and pulse & cw excitation and propose several experiments to check the predictions of the phonon wind model.

1 Introduction

It has been demonstrated experimentally [1–6] that the low temperature exciton transport in Cu_2O shows a great many of anomalous features: transition from a diffusive to a sonic ballistic regime with the increase of intensity of excitation and decrease of temperature, well pronounced nonlinearities under two-pulse and pulse& cw excitation. These effects have been attributed by the authors of the experiments to Bose Einstein condensation (BEC) in Cu_2O exciton system and its superfluidity. However, the BEC interpretation of this transport is controversial. The first theoretical attempts [7,8] failed to explain, e.g., why the velocity of the anomalous exciton transport is stabilized at the velocity of sound. Recently a model of sonic exciton solitons [9,10] has been proposed, but it predicts [11] several orders of magnitude shorter exciton pulses than measured experimentally.

We have shown in Refs. [12–17] that anomalous transport phenomena in Cu_2O can be quantitatively explained within the nonequilibrium phonon-assisted (phonon-wind) model. In this paper we review briefly the phonon wind model, and discuss several important questions concerning this model.

2 Phonon Wind Model

The nonequilibrium-phonon assisted drag of excitons (phonon wind) was well established in the physics of electron hole droplets in Ge and Si (see Refs. [18] and references therein). However, to the best of our knowledge, there was no direct experimental investigations of phonon wind phenomena in Cu_2O.

The phonon-wind model [12] was proposed firstly to explain the initial stage of the anomalous transport in Cu_2O [1], near the excitation surface. After experimental discovery [2] of the long-range (up to 1 cm) sonic nondiffusive transport of excitons accross the whole sample, and of striking nonlinearities under 2-pulse excitation [3,4], we have shown [13,14] that the equations of

the phonon wind model from [12] describe the long-range transport and its nonlinearities as well.

Next important experimental result was the discovery [5] of strong amplification of the ballistic exciton signal from pulse excitation, by simultaneous cw excitation. These experiments allowed [15] to estimate roughly the exciton density in the arriving exciton pulse. The latter appeared to be about 100 times smaller than the critical BEC density at the temperature of the experiment. Note that this estimate is still valid (see in [15]) if the paraexciton lifetime in the bulk is longer than $10\mu s$ - in this case the real exciton lifetime is governed by diffusion to the surface and surface recombination. Thus, the BEC interpretation of the anomalous transport becomes even more doubtful. As for the phonon wind model, it quantitatively explains the effect of cw-driven amplification [15,16].

In [6] the 2-pulse experiment was modified in order to excite the excitons in spatially separated surface spots. A detailed explanation of these experiments within the phonon wind model was given in [17].

3 Discussion

The most direct check of the phonon wind model will be just measuring of the ballistic phonon signal. But a significant change of the experimental setup is needed in order to register phonons, not excitons. However, several experiments can be proposed within the existent setup, which may help to distinguish between superfluid and phonon mechanisms of the ballistic exciton transport. We will focus here on several important predictions of the phonon wind model which can be checked experimentally.

1. Let us first discuss the motion of excitons, produced by a short surface excitation pulse, and which is due to the ballistic phonons, generated by the same excitation pulse. Within the phonon wind model, if the intensity of the excitation pulse is high enough and the sample is semi-infinite, the motion of the excitons separates into two phases. On the first phase, a cloud of excitons forms, which propagates into the bulk with sonic velocity, together with the front of the ballistic phonon pulse. On the second phase the phonons go ahead, leaving the exciton cloud at some distance $L(I)$ from the excitation surface. The latter is approximately proportional to the excitation intensity, $L(I) \propto I$. During this second phase the cloud expands diffusively.

The sample depth l is finite in the experiments, and the detector is collecting the arriving excitons on the back surface. If the excitation intensity is low, so that $L(I) < l$, the excitons reach the detector diffusively. But with the increase of excitation intensity above the "threshold" intensity, which can be found from the condition $L(I_{th}) = l$, the detector begins to measure the ballistic signal. Hence, within the phonon wind model the threshold excitation intensity has to increase with the sample depth, and this prediction can be checked on the experiment.

2. Within the phonon-wind model, the phonon pulse is calculated assuming that some part ε_{pw} of each absorbed light quantum goes into incoherent ballistic phonons born at the same spatial and temporal points where the absorption (and generation of the e-h pairs) takes place. From the comparison of our model with the experiments we found [12,13,17] that this energy is very small, $\varepsilon_{pw} \approx 1-5$ meV at 2 K, although the total excess energy of the excitation light quantum, which goes into nonequilibrium phonons is as large as 400 meV in case of surface excitation. But this energy goes mostly into optical and high-frequency acoustic phonons which are not ballistic and form so called hot spot near the excitation surface. Also it was established [18] that, in Ge and Si at least, ε_{pw} does not change with the change of the excitation light frequency. Hence, the phonon wind model predicts that the threshold between diffusive and ballistic exciton transport will not decrease with the increase of the excitation light frequency, although the number of nonequilibrium phonons per excitation quantum grows up. Moreover, the threshold even grows up, because with the increase of the excitation light frequency the absorption length is decreased, and excitons become more confined near the excitation surface.

3. Within the phonon wind model, the nonlinear (amplification) effects under two pulse and pulse & cw excitation are explained because the ballistic phonons from the second pulse can push to the detector also the excitons which were left in the sample after the first pulse or were generated by the cw excitation. And the "amplification" of the second signal is stronger if more excitons were near the detector at the arrival time of the second phonon pulse. Hence, the amplification can be tuned, e.g., by means of changing the delay between pulses, their relative intensities or spatial positions of excitation spots. For example, we have shown [17] that the evolution of the detector signal with time delay is qualitatively different for spatially coincident and separated excitation spots. This also can be checked experimentally.

Acknowledgements. This work has been funded in part by Russian Foundation for Basic Research, and Russian Ministry of Science program "Nanostructures".

References

1. D. W. Snoke, J. P. Wolfe, and A. Mysyrowicz, Phys. Rev. B **41**, (1990) 11171.

2. E. Fortin, S. Fafard, and A. Mysyrowicz, Phys. Rev. Lett. **70**, (1993) 3951.

3. A. Mysyrowicz, E. Fortin, E. Benson, S. Fafard, and Ei-ichi Hanamura, Solid State Commun. **92**, (1994) 957.

4. E. Benson, E. Fortin, and A. Mysyrowicz, Phys. Status Solidi (b), **191**, (1995) 345.

5. A. Mysyrowicz, E. Benson, and E. Fortin, Phys. Rev. Lett. **77**, (1996) 896.

6. E. Benson, E. Fortin, B. Prade and A. Mysyrowicz, Europhys. Lett. **40**, (1997) 311.

7. B. Link and G. Baym, Phys. Rev. Lett. **69**, (1992) 2959.

8. E. Hanamura, Solid State Commun. **91**, (1994) 889.

9. I. Loutsenko and D. Roubtsov, Phys. Rev. Lett. **78**, (1997) 3011.

10. D. Roubtsov, Y. Lepine, and I. Loutsenko, Physics Letters A **265**, (2000) 145.

11. S. G. Tikhodeev, Phys. Rev. Lett. **84**, (2000) 3502.

12. A. E. Bulatov and S. G. Tikhodeev, Phys. Rev. **B46**, (1992) 15058.

13. G. A. Kopelevich, N. A. Gippius, and S. G. Tikhodeev, Solid State Commun. **99**, (1996) 93.

14. G. A. Kopelevich, S. G. Tikhodeev, and N. A. Gippius, JETP **82**, (1996) 1180.

15. S. G. Tikhodeev, Phys. Rev. Lett. **78**, (1997) 3225.

16. S. G. Tikhodeev, G. A. Kopelevich, and N. A. Gippius, Phys. Status Solidi (b) **206**, (1998) 45.

17. S. G. Tikhodeev, N. A. Gippius, and G. A. Kopelevich, Phys. Stat. Sol. (a) **178**, (2000) 63.

18. J. P. Wolfe and C. D. Jeffries, in Modern Problems in Condensed Matter Science, eds. C. D. Jeffries and L. V. Keldysh (North-Holland, Amsterdam, 1983) p. 431; L. V. Keldysh and N. N. Sibeldin, in Nonequilibrium Phonons in Nonmetallic Crystals, eds. W. Eisenmenger and A. A. Kaplyanskii (Elsevier Science Publ. B. V. 1986) p.455 .

Pump-probe spectroscopy by infrared probe light in highly excited semiconductors:

Beyond the BCS-like mean-field theory

T. J. Inagaki *, M. Aihara

Graduate School of Materials Science, Nara Institute of Science and Technology, Ikoma, Nara 630-0101, Japan.

Abstract   We analyze the pump-probe spectroscopy with infrared or THz probe light as a direct measurement of the macroscopic quantum state of the optically excited electron-hole systems. The effect of the collective phase fluctuation associated with the center-of-mass motion of the electron-hole pairs is taken into account by solving the quasi-static Eliashberg equation for electron-hole systems, which makes it possible to calculate the renormalized band energy, BCS-like gap, and wave-function renormalization factor in the wide range of particle densities. The calculated spectra clearly show the BCS-like gap formation at quasi-Fermi level, and the strong photoexcitation leads to the extraordinary large BCS-like gap.

1 Introduction

The high-density electron-hole (e-h) system induced by an intense laser field has long been studied. This system is a promising candidate for observing the macroscopic quantum phenomenon because of the Bosonic nature of the bound e-h pairs (excitons) [1]. Recent developments in the experimental technique enable us to observe remarkable phenomena suggesting the macroscopic quantum phenomena in the optically generated nonequilibrium state. In particular, fast and coherent propagation of excitons are found in the highly excited Cu_2O [2] and BiI_3 [3]. These phenomena suggest the generation of the macroscopic quantum coherence in semiconductors because fast exciton transport is distinctly enhanced with increasing the exciton density. However, the current understanding of these phenomena still remains tentative because of the complicated experimental situation such as phonon effect, spatial inhomogeneity of the exciton density, finite life-time of excitons, and so forth.

When the particle density is so increased that the excitons are deeply overlapped with each other, the Fermion nature of electrons and holes, such as state-filling effect and band-gap renormalization effect, plays the crucial role. In this situation, the macroscopic quantum state is characterized by the BCS-like energy gap, and not by the exciton condensate; the bound e-h pairs are similar to the Cooper pairs in superconductors, and the exciton concept is no longer appropriate.

In this work, we first propose the pump-probe spectroscopy with the infrared or THz probe light as a direct measurement of the BCS-like energy gap at the quasi-Fermi level generated by an intense visible pump light.

* *Electronic address:* inagaki@ms.aist-nara.ac.jp

We incorporate the strong quantum fluctuation effect in the e-h BCS state by solving the quasi-static Eliashberg equation for e-h systems. We then calculate the intraband dielectric function by the generalized random-phase approximation (RPA) where the collective phase fluctuation associated with the center-of-mass motion of e-h pairs is considered.

2 Formulation

2.1 Model Hamiltonian

We consider an e-h system in a direct-gap semiconductor, which consists of the isotropic nondegenerate parabolic conduction and valence bands with identical electron and hole masses. The system is supposed to be in a stationary state driven by an intense monochromatic pump light with frequency ω_L; the pump-light is treated as a classical field. We analyze the following time-independent Hamiltonian for pump-light driven e-h system, which is obtained by the unitary transformation proposed by Galitskii *et al.* [4],

$$H = \sum_{\mathbf{k}} \left\{ \varepsilon_{\mathbf{k}}^e c_{\mathbf{k}}^\dagger c_{\mathbf{k}} + \varepsilon_{\mathbf{k}}^h d_{-\mathbf{k}}^\dagger d_{-\mathbf{k}} + \lambda c_{\mathbf{k}}^\dagger d_{-\mathbf{k}}^\dagger + \lambda d_{-\mathbf{k}} c_{\mathbf{k}} \right\}$$
$$+ \frac{1}{2} \sum_{\mathbf{k},\mathbf{p},\mathbf{q}} V_{\mathbf{k}} \left\{ c_{\mathbf{k}+\mathbf{q}}^\dagger c_{\mathbf{p}-\mathbf{q}}^\dagger c_{\mathbf{p}} c_{\mathbf{k}} + d_{\mathbf{k}+\mathbf{q}}^\dagger d_{\mathbf{p}-\mathbf{q}}^\dagger d_{\mathbf{p}} d_{\mathbf{k}} \right.$$
$$\left. -2 c_{\mathbf{k}+\mathbf{q}}^\dagger c_{\mathbf{k}} d_{\mathbf{p}-\mathbf{q}}^\dagger d_{\mathbf{p}} \right\}, \quad (1)$$

where λ is the interaction energy between pump-light and carriers, and $V_{\mathbf{k}} = 4\pi e^2/(\epsilon_0 q^2)$ is the Coulomb interaction with the background dielectric constant ϵ_0. Here $c_{\mathbf{k}}$ and $d_{\mathbf{k}}$ are the annihilation operators for electrons and holes, respectively; $\varepsilon_{\mathbf{k}}^e = k^2/(2m) + E_g - \omega_L/2$ ($\varepsilon_{\mathbf{k}}^h = k^2/(2m) - \omega_L/2$) is the single-particle excitation energy of electrons (holes) in this pump-light driven system.

2.2 Quasi-static Eliashberg equation

The quasi-static Eliashberg equation for optically excited e-h systems is obtained by a variational calculation with respect to the Bogoliubov parameters: $\delta\langle H\rangle/\delta u_{\mathbf{k}} = 0$. Here $\langle H\rangle$ is the expectation value of Eq. (1) with respect to the generalized RPA ground state. The quasi-static Eliashberg equation for optically driven e-h systems is a set of coupled integral equations for renormalized band $\zeta_{\mathbf{k}}$, the BCS-like gap $\Delta_{\mathbf{k}}$, and the wave-function renormalization factor $Z_{\mathbf{k}}$, and is expressed as follows,

$$\zeta_{\mathbf{k}} = \frac{1}{2}(\varepsilon_{\mathbf{k}}^e + \varepsilon_{\mathbf{k}}^h) - \frac{1}{2}\sum_{\mathbf{p}} V_{\mathbf{p}}$$

$$- \frac{1}{2}Z_{\mathbf{k}}\sum_{\mathbf{p}} V_{\mathbf{k}-\mathbf{p}}C_{\mathbf{p},\mathbf{p}}^{(2)}[1 + 2\chi_{\mathbf{k}-\mathbf{p}}(-E_{\mathbf{k}} - E_{\mathbf{p}})],$$

$$\Delta_{\mathbf{k}} = \lambda + Z_{\mathbf{k}}\sum_{\mathbf{p}} V_{\mathbf{k}-\mathbf{p}}C_{\mathbf{p},\mathbf{p}}^{(1)}[1 + 2\chi_{\mathbf{k}-\mathbf{p}}(-E_{\mathbf{k}} - E_{\mathbf{p}})],$$

$$Z_{\mathbf{k}}^{-1} = 1 + \sum_{\mathbf{p}} V_{\mathbf{k}-\mathbf{p}}C_{\mathbf{p},\mathbf{p}}^{(3)2}\left[\frac{\partial\chi_{\mathbf{k}-\mathbf{p}}(\omega)}{\partial\omega}\right]_{\omega = -E_{\mathbf{k}}-E_{\mathbf{p}}}, \quad (2)$$

where $E_{\mathbf{k}} = \sqrt{\zeta_{\mathbf{k}}^2 + \Delta_{\mathbf{k}}^2}$ is the single particle energy of Bogoliubov quasiparticles; the coherent factors are defined by $C_{\mathbf{k},\mathbf{p}}^{(1)} = u_{\mathbf{k}}v_{\mathbf{p}} + v_{\mathbf{k}}u_{\mathbf{p}}$, $C_{\mathbf{k},\mathbf{p}}^{(2)} = u_{\mathbf{k}}v_{\mathbf{p}} - v_{\mathbf{k}}u_{\mathbf{p}}$, and $C_{\mathbf{k},\mathbf{p}}^{(3)} = u_{\mathbf{k}}u_{\mathbf{p}} - v_{\mathbf{k}}v_{\mathbf{p}}$, where $u_{\mathbf{k}}$ and $v_{\mathbf{k}}$ are the Bogoliubov parameters given by $u_{\mathbf{k}}^2 = (1 - \zeta_{\mathbf{k}}/E_{\mathbf{k}})/2$ and $v_{\mathbf{k}}^2 = (1 + \zeta_{\mathbf{k}}/E_{\mathbf{k}})/2$. In Eq. (2), we introduced the partial screening function [5], $\chi_{\mathbf{k}}(\omega) = (1/\pi)\int_0^\infty dz\,\mathrm{Im}\epsilon_{\mathbf{k}}^{-1}(z)/(z-\omega)$; the dielectric function $\epsilon_{\mathbf{k}}(\omega)$ is evaluated by the generalized RPA, and the result is written as follows,

$$\epsilon_{\mathbf{q}}(\omega) = 1 - V_{\mathbf{q}}\sum_{\mathbf{k}} \frac{2(E_{\mathbf{k}+\mathbf{q}} + E_{\mathbf{k}})}{\omega^2 - (E_{\mathbf{k}+\mathbf{q}} + E_{\mathbf{k}})^2}C_{\mathbf{k}+\mathbf{q},\mathbf{k}}^{(3)2} \quad (3)$$

3 Numerical Calculations

For the sake of simplicity, we employ, in the numerical calculation, the partial screening function which is calculated by the quasi-static single-plasmon-pole approximation [5,6]. In the following discussion, we use the units where the exciton binding energy and the exciton Bohr radius being unity.

Fig. 1 shows the calculated mean interparticle distance $r_s = (4\pi n/3)^{-1/3}$, as a function of the pump-light frequency ω_L, where n is the e-h pair density. In the weak pump excitation (small λ), the resonatorless optical bistability [7] is also found to arise because of the instability of optically driven semiconductors due to the band-gap renormalization effect (the self-energy correction). It is however noted that the present analysis shows that the collective phase fluctuation makes the considerable reduction of the bistability region.

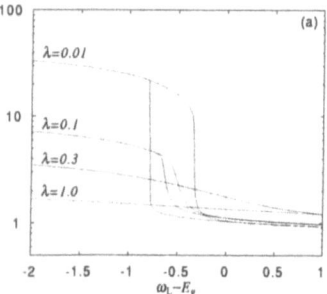

Fig. 1 The mean interparticle distance as a function of the pump-light frequency $\omega_L - E_g$.

Figures 2–3 show the reflectance and absorption spectra for various λ which are obtained by the numerical

calculation of $\epsilon(\omega) = \lim_{\mathbf{q}\to 0}\epsilon_{\mathbf{q}}(\omega)$. Here the pump-light frequency ω_L is set at E_g, and the exciton decay constant is chosen as $\gamma = 0.03$. When the coupling between the pump-light and carriers is relatively weak ($\lambda = 0.1$), we find the strong reflection and absorption below the plasma frequency ($\omega_{pl} \simeq 6$); this behavior is similar to that of metals. However, we find structures in relfectance and absorption spectra near $\omega \simeq 0$ which originates from the BCS-like gap formation at the quasi-Fermi level. With increasing λ, the magnitude of the BCS-like gap extraordinary grows and the transparent region is found near $\omega \simeq 0$ for $\lambda \geq 0.3$. This result indicates that the BCS-like gap is enhanced by the intense pump-light as is investigated by [8], and is expected to be directly verified by the pump-probe experiment with infrared probe-light.

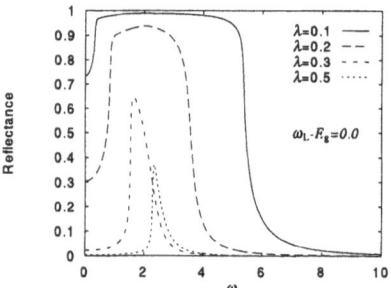

Fig. 2 The reflectance spectra for $\lambda = 0.1, 0.2, 0.3, 0.5$, where the pump-light frequency is chosen as $\omega_L - E_g = 0.0$.

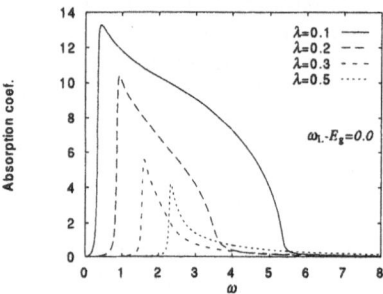

Fig. 3 The absorption spectra for $\lambda = 0.1, 0.2, 0.3, 0.5$, where the pump-light frequency is chosen as $\omega_L - E_g = 0.0$.

Acknowledgment

This work is partially supported by a Grant-in-Aid for Scientific Research from the Ministry of Education, Science, Culture and Sports in Japan.

References

1. S. A. Moskalenko and D. W. Snoke, *Bose-Einstein Condensation of Excitons and Biexcitons* (Cambridge University Press, Cambridge, 2000).
2. E. Fortin *et al.*, Phys. Rev. Lett. **70**, (1993) 3951.
3. H. Kondo *et al.*, Phys. Rev. B **58**, (1998) 13835.
4. V. M. Galitskii *et al.*, Sov. Phys. JETP **30**, (1979) 117.
5. R. Zimmermann, Phys. Status Solidi B **76**, (1976) 191.
6. H. Haug *et al.*, Prog. Quantum Electron. **9**, (1988) 3.
7. Th. Habrich and G. Mahler, Phys, Status Solidi B **117**, (1983) 635; T. Iida *et al.*, Phys. Rev. B **47**, (1993) 9328.
8. T. J. Inagaki *et al.*, to appear in Phys. Rev. B (2000).

Strain- and field-induced optical anisotropies of GaAs measured by RDS

N. Kumagai[1], T. Yasuda[2], T. Hanada[1], T. Yao[1]

[1] Institute for Materials Research, Tohoku University, 2-1-1 Katahitra, Aoba-ku, Sendai 980-8577, Japan, e-mail: kumagai@imr.tohoku.ac.jp
[2] Joint Research Center of Atom Technology, NAIR, 1-1-4 Higashi, Tsukuba, 305-0046, Japan

Abstract We have measured bulk optical anisotropies of GaAs which is induced by piezo-optic and/or linear electro-optic effects. Reflectance difference spectroscopy (RDS) was employed to measure these optical anisotropies. We evaluated the spectrum of the P_{44} component from RDS results. And we could find the possibility to determine the spectrum of the r_{41} component above the band gap.

1 Introduction

While the study on the piezo-optic and electro-optic effects of semiconductors has a long history, spectral dependence of the relevant tensor components have been reported for a limited number of materials. Especially, the linear electro-optic (LEO) coefficient above the band gap has not been measured yet. Reflectance difference spectroscopy (RDS), which has been mainly used to measure surface-induced anisotropy, is an ideal technique to study the strain- and field-induced bulk optical anisotropies with a high sensitivity ($\Delta\epsilon\sim0.001$). Making the best use of this advantage of RDS, Ronnow et al. recently reported the piezo-optic tensors of ZnSe and InP using bulk samples[1,2]. In this presentation, we first show the RD spectra for an GaAs(001) wafer under an uniaxial stress which was applied simply by bending the wafer. The P_{44} spectrum has been obtained from the measured RD spectra. Next, we report the RD spectra under an application of an electric fields from which we can determine the LEO tensor. Determination of these tensors is important for us to utilize in-situ RDS for real-time control of the doping and strain in device fabrication[3–5].

2 Experiment

2.1 Application of stress

Small tensile stress ($\sim$10 MPa) to undoped GaAs(001) substrate by bending along [110] or [1$\bar{1}$0] direction. The curvature and uniformity of the bending was check by letting a He-Ne laser beam reflected by the sample surface and observing the displacement of the reflected beam which is proportional to the curvature. The RD signal is definded as:

$$\frac{\Delta\tilde{r}}{\tilde{r}} = \frac{\tilde{r}_{1\bar{1}0} - \tilde{r}_{110}}{(\tilde{r}_{1\bar{1}0} + \tilde{r}_{110})/2}. \quad (1)$$

The obtained RD spectra were converted to the P_{44} spectra assuimg that the anisotropy is uniform in the depth direction within the penetration depth of light[6].

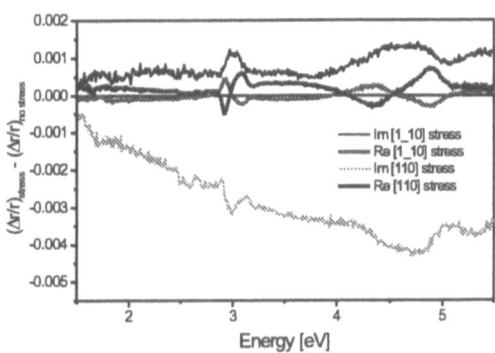

Fig. 1 Spectral changes of RD signalinduced by stress,$(\Delta r/r)_{stress}$-$(\Delta r/r)_{no\ stress}$. While real part shows good baseline, Imaginary part shows offset form baseline.

2.2 Application of electric field

An undoped GaAs layer of about 300 nm thickness was grown on p+-GaAs substrate by MBE, and a gold film of 10 nm thickness was evaporated on the sample surface to form a transparent electronode. Indium was pasted on the backside of the substrate. RD spectra were measured under an application of a DC voltage between the gold and indium electrodes.

3 Results and Discussion

3.1 RD spectra depended on stress

Figure 1 shows the spectral changes induced by the applied stress. As expected, the sign of the changes is reversed upon switching the bending direction. The amplitude of the change was proportional to the curvature of the bending. The P_{44} spectra can be calculated from the stress-induced spectral change, $\Delta r/r = (\Delta r/r)_{stress}$-$(\Delta r/r)_{no\ stress}$. The equations used for this transformation are:

$$Re(\tilde{P}_{44}) = \frac{T_1^2 - T_2^2}{c_{44}e(T_1^2 + T_2^2)}\{T_2(Im(\frac{\Delta\tilde{r}}{\tilde{r}}) - T_1Re(\frac{\Delta\tilde{r}}{\tilde{r}})\}, (2)$$

$$Im(\tilde{P}_{44}) = \frac{T_2^2 - T_1^2}{c_{44}e(T_1^2 + T_2^2)}\{T_2Re(\frac{\Delta\tilde{r}}{\tilde{r}}) + T_1Im(\frac{\Delta\tilde{r}}{\tilde{r}})\}, (3)$$

$$T_1 = n\epsilon_1 - \kappa\epsilon_2 - n, \quad T_2 = \kappa\epsilon_1 + n\epsilon_2 - \kappa$$

where c_{44} is the elastic stifness coefficient of GaAs, and e is the strain which we estimated to be 2.7 x 10^{-4}

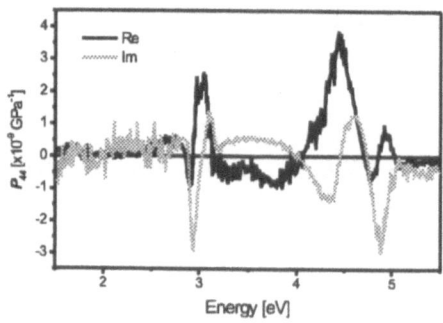

Fig. 2 P_{44} spectrum which is transformed from RDS results.

from the laser-beam displacement for the experiment in Fig.1. The P_{44} spectra thus obtained are shown in Fig. 2. In this transformation, we used the results for the [1$\bar{1}$0] stress because of the smaller offset in the imaginary part than for the [110] data set. The P_{44} spectra show clear resonance at the bulk critical-point energies. The overall spectral lineshape agrees with the previous report by Etchegoin et al[7]. in which a bulk GaAs sample under a large stress ($\sim$500 MPa) was measured by spectroscopic ellipsometry. However, the agreement is not sufficient near the E_2 energy. The cause for this disagreement is not clear at present.

3.2 RD spectra depended on electric field

The MBE-grown undoped GaAs layer was slightly n-type, as judged from the LEO feature near the E_1 energy observed under no DC field application [8,9]. This implies the p^+-n^- juction is formed between the undoped layer and the substrate. Therefore, polarity dependence was expected in the RD measurement under the DC field application. In Fig. 3 the field-induced changes of the RD spectra are shown. The changes are indeed dependent on the direction of DC field. Under the reverse bias, the field-induced RD signal is very small, because most of the potential drop occurs in the depletion layer. On the other hand, the LEO feature around the E_1 energy is clearly observed under the forward bias. Plotted in Fig.4 is the change of the RD signal at 3.1 eV as a function of the DC. It is seen that under the forward bias the RD change is proportional to the voltage, indicating that we can estimate the electric field in the undoped layer by dividing the DC voltage by it thickness. Transformation of the data in Fig. 3 to the r_{41} spectra needs an optical model which properly represents the optical response of the gold/ZnSe/GaAs structure, and efforts are under way to complete this transformation.

4 Summary

We have obtained the P_{44} spectra of GaAs by RDS measurements on a bent wafer sample. This technique is applicable to the other materials such as binary and ternary compound semiconductors for which preparation of a bulk sample is usually difficult. We have also shown

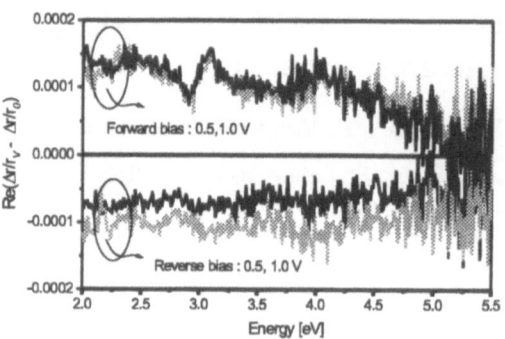

Fig. 3 Spectral changes of RD signalinduced by DC field,$(\Delta r/r)_V$-$(\Delta r/r)_0$. The black line denotes RD changes with ± 1.0 [V], the gray one denotes with ± 0.5 [V], respectively

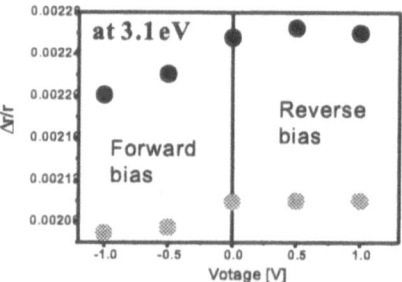

Fig. 4 RD changes at 3.1 eV depended on bias voltage.

the possibility to determine the r_{41} spectra from the RD measurements under the DC bias application.

Acknowledgement

This study was supported in part by a Grant-in-Aid from the Ministry of Science, Culture, Sports and Education (Japan Society for Promotion of Science) and the Visiting Researcher's Program of the Institute for Materials Research , Tohoku university.

References

1. D. Ronnow, L. F.Lastras-Martinez, M. Cardona, and P. V.Santos, J.Opt.Soc.Am.A,**16**,(1999)568.
2. D. Ronnow, M. Cardona, and L. F.Lastras-Martinez, Phys. Rev.,**B59**,(1999)5581.
3. T. Hanada, T. Yasuda, A. Ohtake, K. Hingerl, S. Miwa, K. Arai, and T. Yao, Phys. Rev,**B60**,(1999)8909.
4. K. Hingerl, T. Yasuda, T. Hanada, S. Miwa, K. Kimura, A. Ohtake and T. Yao, J.Vac.Sci.Technol.,**B16**,(1998)2342.
5. N. Kumagai, T. Yasuda, T. Hanada, and T. Yao, J.Cryst.Growth,**214/215**,(2000)547.
6. D. E.Aspnes and A. A.Studna, Phys. Rev. Lett.,**54**,(1985)1956.
7. P. Etchegoin, J. Kircher, M. Cardona, C. Grein and E. Bustarret, Phys. Rev.,**B46**,(1991)15139.
8. H. H.Farell, M. C.Tamargo, T. J.Gmitter, A. L.Weaverand and D. E.Aspnes, J.Appl.Phys.,**70**,(1991)1033
9. H. Tanaka, E. Colas, I. Kamiya, D. E.Aspnes and R. Bhat, Appl.Phys.Lett.**59**,(1991)3443.

Electromagnetically induced transparency for intrinsic exciton lines in bulk semiconductors

M. Artoni[1], G.C. La Rocca[2,3], F. Bassani[3]

[1] INFM and LENS, Largo E. Fermi 2, I-50125, Florence, Italy
[2] Dipartimento di Fisica, Università di Salerno, I-84081 Baronissi (Sa), Italy
[3] INFM, Scuola Normale Superiore, Piazza dei Cavalieri, I-56126 Pisa, Italy

Abstract Electromagnetically induced transparency (EIT) is a well established phenomenon in dilute atomic vapors. We predict here a strong EIT effect in an undoped bulk semiconductor where a quenching of the absorption nearly as large as that observed in atoms is expected to occur for sharp free-exciton lines corresponding to intrinsic delocalized electronic states. Numerical estimates are presented for the specific case of the second class $2P$ "yellow" exciton of Cu_2O, in which case all spectroscopic parameters are well known.

1 Introduction

Many intriguing effects due to coherence and interference have been demonstrated in atomic physics. The extension of laser cooling techniques to the micro-kelvin regime, spontaneous emission quenching and lasing without inversion are just few instances. The investigation of fundamental coherent optical effects in semiconductors is also an important field of research where similar phenomena could show up. We are here concerned with electromagnetically induced transparency (EIT), so far demonstrated mostly for dilute atomic vapors [1], whereby the lower doublet of a three-level system may be pumped into a coherent superposition state where the population is trapped (population trapping state) due to destructive interference between two distinct paths of absorption to the third level [2].

Condensed matter exhibits quite a variety of three-level systems where EIT could also be realized. Yet dephasings, which can easily break the coherence of the population trapping state, makes it difficult to observe a large EIT effect in solids. For microwaves, EIT has been observed in ruby [3] using a strong external magnetic field, while in the infrared EIT has been observed in intersubband transitions in a quantum well [4]. For optical frequencies, EIT in solids has only been observed in an inhomogeneously broadened hole-burning Pr3+-doped Y2SiO5 crystal [5].

We predict here a remarkable enhancement of electromagnetically induced transparency in an undoped bulk semiconductor exhibiting sharp free-exciton lines that correspond to intrinsic delocalized electronic states. We specifically consider the "yellow exciton" series in Cu_2O for which all the relevant spectroscopic parameters are available [6,7].

2 Model and Results

In Cu_2O both conduction and valence band extrema occur at the Γ point (O_h symmetry) and have, respec-

tively, Γ_6^+ and Γ_7^+ symmetry, each twofold degenerate spin included. The four states of the $1S$ exciton manifold split into an upper quadrupole allowed threefold degenerate Γ_5^+ level and a lower forbidden nondegenerate Γ_2^+ level. The twelve states of the $2P$ exciton manifold are classified according to their symmetries as follows: $\Gamma_2^- \oplus \Gamma_3^- \oplus \Gamma_4^- \oplus 2\Gamma_5^-$, among which only the threefold degenerate Γ_4^- states are dipole active and give rise to the second class $2P$ "yellow exciton" line [8]. In the following, we assume that the Γ_4^- optically active $2P$ states are sufficiently separated from all others so that we can restrict ourselves only to the Γ_5^+ $1S$ and Γ_4^- $2P$ exciton states. The longitudinal-transverse splitting of the second class $2P$ exciton is negligible. The inclusion of the complete $2P$ exciton manifold [9] would not change significantly the main results here discussed.

We are interested in the optical responce experienced by a weak *probe* beam of frequency ω_p tuned around the $2P$ "yellow" exciton line of resonant frequency ω_{2P}, while a strong *coupling* beam of frequency ω_c and Rabi frequency Ω_c is nearly resonant with the $1S$ to $2P$ exciton transition of resonant frequency $\omega_{2P-1S} = \omega_{2P} - \omega_{1S}$. All relevant parameters in the effective hamiltonian describing this Λ three exciton level configuration can be determined from the usual envelope function picture of Wannier-Mott exciton states [10]. The dipole forbidden $1S$ exciton state has a small linewidth ($\hbar\gamma_{1S} \simeq 0.1$ meV) compared to that of the second class allowed $2P$ exciton state ($\hbar\gamma_{2P} \simeq 1$ meV) and a well developed EIT can be established in the presence of the coupling beam. A transparency window about the $2P$ exciton line appears, while for very high pump intensities, or at least for Ω_c larger than about $2\gamma_{2P}$, the Autler-Townes regime is eventually reached [1,7,10]. The optical susceptibility experienced by a probe polarized along x in the presence of a pump polarized along y is given by

$$\chi_p = \frac{A\,\gamma_{2P}\,(\delta_p - \delta_c - i\gamma_{1S})}{(\delta_p - i\gamma_{2P})(\delta_p - \delta_c - i\gamma_{1S}) - \Omega_c^2/4} \;;$$

with $A \simeq 0.02$ is a numerical constant proportional to the $2P$ exciton oscillator strength[10] whereas $\delta_p = \omega_{2P} - \omega_p$ and $\delta_c = \omega_{2P-1S} - \omega_c$ are the two relevant detunings.

Fig.1 shows the imaginary part of the susceptibility for pump intensities characterizing both EIT and Autler-Townes regimes for a probe tuned to the $2P$ exciton line. The actual transmission through a slab, including the

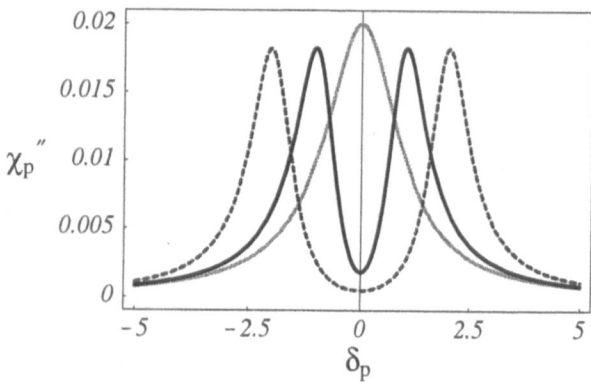

Fig. 1 Imaginary part of the susceptibility vs. the probe frequency detuning in units of γ_{2P} for a resonant coupling beam. The coupling Rabi frequencies are $\Omega_c/\gamma_{2P} = 2$ (solid) and 4 (grey dash) while in the absence of the coupling beam ($\Omega_c = 0$) the susceptibility is described by the solid grey curve.

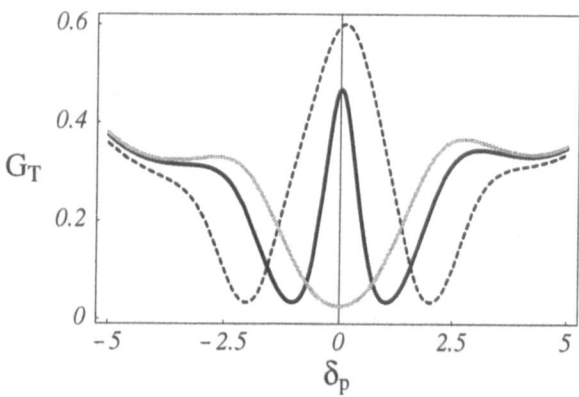

Fig. 2 Transmission coefficient G_T vs. the probe frequency detuning in units of γ_{2P}; the background dielectric constant is $\epsilon_\infty = 6.5 + i\,2 \times 10^{-3}$, the slab thickness $d=35\mu$m while the other parameters are the same as in Fig. 1.

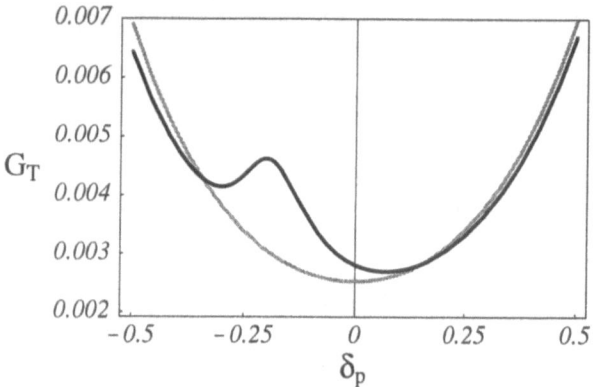

Fig. 3 Same as in Fig.2 except for a slab thickness $d=60\mu$m, a coupling Rabi frequencies $\Omega_c/\gamma_{2P} = 0.2$, and a pump detuning of $\delta_c/\gamma_{2P} = -0.2$. Notice the change of scale.

effect of the complex background dielectric constant, is shown instead in Fig.2 anticipating a nearly 50% transmission due to EIT. Apart from the background absorption, the $1S$ exciton linewidth γ_{1S} is the material parameter that mostly limits the possibility of achieving even larger EIT effects in this system. Finally, Fig.3 shows the transmission for parameters appropriate to the experiments by Fröhlich and coworkers [7] where a very tiny EIT effect could indeed be recognized: this figure reproduces well the differential transmission measured in [7] for a nearly resonant pump.

The observation of a fully developed EIT regime for intrinsic exciton lines in semiconductors would open the way to many opportunities involving either device applications of EIT or fundamental issues such as, e.g., the role of polaritonic effects and of ponderomotive forces in the EIT context.

It is a pleasure to thank prof. D. Fröhlich for very useful correspondence.

References

1. S. Harris, Physics Today **50**, (1997) 36; E. Arimondo, *Progress in Optics XXXV*, Ed. E. Wolf (Elsevier Science, Amsterdam 1996) 257; and references therein.
2. G. Alzetta, A. Gozzini, L. Moi, G. Orriols, Nuovo Cimento **B36**, (1976) 5; E. Arimondo, G. Orriols, Lett. Nuovo Cimento **17**, (1976) 333.
3. Y. Zhao, C. Wu, B. S. Ham, M. K. Kim, and E. Awad, Phys. Rev. Lett. **79**, (1997) 641.
4. G.B. Serapiglia, E. Parpalakis, C. Sirtori, K.L. Vodopyanov, C.G. Phillips, Phys. Rev. Lett. 84, (2000) 1019.
5. B. S. Ham, M. S. Shahriar, and R.P. Hemmer, Opt. Lett.s **22**, (1997) 1138; K. Ichimura, K. Yamamoto, and N. Gemma, Phys Rev. A **58**, (1998) 4116.
6. S. Nikitine, J.B. Grun, M. Sieskind, J. Phys. Chem. Solids **17**, (1961) 292; Landolt-Börnstein, series III, **17** (Semiconductors), (Springer, Berlin, 1984).
7. D. Fröhlich, A. Nothe, and K. Reimann, Phys. Rev. Lett. **55**, (1985) 1335; D. Fröhlich, Ch. Neumann, B. Uebbing and R. Wille, phys. stat. sol. (b) **159**, (1990) 297.
8. F. Bassani and G. Pastori Parravicini, *Electronic States and Optical Transitions in Solids* (Pergamon Press, Oxford 1975); P. Yu and M. Cardona, *Fundamentals of Semiconductors* (Springer, Berlin 1996).
9. M. Artoni et al., to be published.
10. M. Artoni, G.C. La Rocca and F. Bassani, EuroPhys. Lett.s **49**, (2000) 445.

Reversible photoinduced structural changes in GeSe$_2$ glass at low-temperature

T. Nakaoka[1], Y. Wang[1], O. Matsuda[2], K. Inoue[3], and K. Murase[1]

[1] Department of Physics, Graduate School of Science, Osaka University, 1–1 Machikaneyama, Toyonaka 560–0043, Japan
e-mail: nakaoka@mmm.phys.sci.osaka-u.ac.jp
[2] Division of Applied Physics, Graduate School of Engineering, Hokkaido University, Sapporo 060–8628, Japan
[3] Institute of Scientific and Industrial Research, Osaka University, 8–1 Mihogaoka, Ibaraki 567–0047, Japan

Abstract   We have observed resonant Raman enhancement of stretching vibration of Se–Se bonds relative to breathing vibration of GeSe$_{4/2}$ tetrahedra at room temperature and 15 K in GeSe$_2$ chalcogenide semiconducting glass. At 15 K, structural changes by illumination, which cause photodarkening, induce excess resonant enhancement. After the structural changes are saturated, we can observe a reversible change with the excitation power in resonant Raman spectra. It is found that the origin of the the reversible power dependence is the same as that of the excess resonant enhancement. Both of them are attributed to three-fold coordinated Se atoms forming Se–Se bonds, which cause the photodarkening.

1 Introduction

Various kinds of photoinduced structural changes in chalcogenide glasses are the subjects of considerable interest from both fundamental and practical points of view. The photoinduced structural changes result from the atomic rearrangements which are caused by the energy relaxation of the photoexcited electrons through strong electron-lattice interactions. Despite a great number of the studies [1], the microscopic mechanism responsible for the photoinduced structural changes is still not well understood. In this paper, we present a resonant Raman study of the photoinduced changes at 15 K in GeSe$_2$ glass, a typical chalcogenide semiconductor [2].

Glassy GeSe$_2$ was prepared by quenching the melts into ice water. Raman scattering was measured in the back scattering configuration. Light sources in an energy range of 1.81– 2.71 eV were a DCM dye laser or an Ar$^+$ ion laser. The incident light was focused onto a rectangular region of about 5×0.1 mm^2 (line-focusing) to reduce heating effects.

2 Results and discussion

Raman spectra measured at 15 K, excited at 1.81 eV and 2.41 eV in GeSe$_2$ are presented in Fig. 1. With increasing excitation energy, the stretching vibration of Se–Se mode around 275 cm^{-1} [3] is enhanced relative to the A_1 mode of GeSe$_{4/2}$ tetrahedra. The inset of Fig. 1 shows the excitation energy dependence of the intensity ratio of the Se–Se mode to the A_1 mode, I(Se–Se)/I(A_1). We attribute the enhancement of the I(Se–Se)/I(A_1) ratio around the optical gap of 2.4 eV to the resonance effect with the electronic transition between the Se lone-pair states and the antibonding states of the Ge–Se bonds because the former forms the top of the valence band and

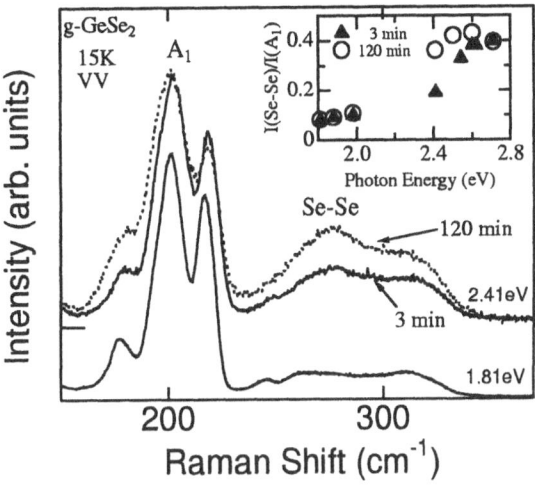

Fig. 1 Polarized resonant Raman spectra of GeSe$_2$ glass at 15 K, excited by the light with photon energies of 1.81 eV and 2.41 eV when illuminated for 3 min and 120 min. The power densities are 3 W/cm^2. The spectra are normalized to the intensity of the A_1 mode of GeSe$_{4/2}$ tetrahedra around 200 cm^{-1}. The inset shows the excitation energy dependence of the intensity ratio of the Se–Se mode to the A_1 mode for illuminated 3 min and 120 min at each excitation energy.

the latter forms bottom of the conduction band [4–6]. Although the I(Se–Se)/I(A_1) ratios excited by lights below 2.0 eV are independent of exposure time, that excited at 2.41 eV gradually increases with exposure time and finally saturates after the illumination for about 120 minutes. Simultaneously, we monitor the transmitted intensity. The transmitted light darkens with exposure time and saturates after about 120 minutes as well. The decrease of the transmitted light intensity is attributed to the narrowing of the optical gap, which is termed photodarkening. Obviously, the changes of the resonant Raman spectra with time are closely related to photodarkening. As shown in the inset of Fig. 1, structural changes following photodarkening should form the *excess* resonant electronic states around 2.4 eV, but does not form them at 2.71 eV. At room temperature, the resonant enhancement of the Se–Se mode is also observed, but the enhancement hardly changes with exposure time.

Excitation power density dependence is studied at 15 K after the 120 min illumination of light at 2.41 eV. With increasing excitation power, the resonantly enhanced intensity ratio I(Se–Se)/I(A_1) decreases, as shown in Fig.

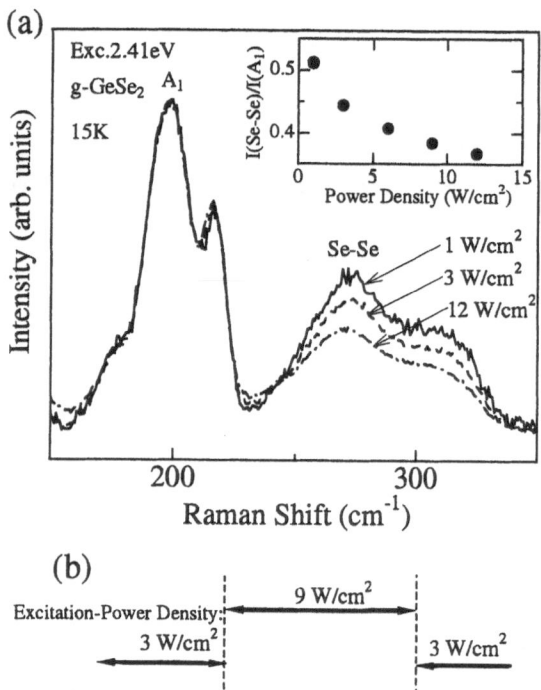

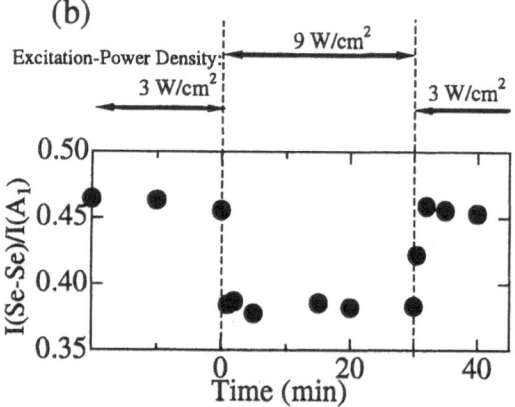

Fig. 2 (a) Excitation-power density dependence of the resonant Raman spectra at 15 K after 120 min exposure of the light of 2.41 eV. The spectra are normalized to the intensity of the A_1 mode. The inset shows the variation of the intensity ratio I(Se–Se)/I(A_1) with power density. With increasing power density, the intensity ratio rapidly decreases. (b) Response of the intensity ratio to the power density. The excitation power is switched from 3 W/cm^2 to 9 W/cm^2 at 0 min (corresponding to the time after illuminated for 120 min), and reversed at 30 min. The variation with power is fully reversible.

2 (a). The decrease of the intensity ratio is not due to heating effects because it is estimated that the increase of temperature under the 3 W/cm^2 illumination of 2.41 eV does not exceeds 5 K. It should be noted that the shape of the resonant Raman spectra is nearly independent of temperature from 10 K to 45 K.

In Fig. 2 (b), we show the response of the intensity ratio I(Se–Se)/I(A_1) to a rapid change of the excitation power density. The response time of the intensity ratio to the power is within a minute. The change with power density is fully reversible. Similar power dependent behavior is observed for the excitations of 2.4–2.6 eV lights. However, no power dependence is observed at the 2.71 eV excitation where no excess resonant enhancement is observed. Thus, a higher excitation-power

density does not reduce the *initial* resonant enhancement, occurring from the beginning, but reduces the *excess* resonant enhancement caused by illumination. We attribute the origin of the reversible power dependence to the photocreated three-fold Se atoms forming Se–Se bonds (>Se–Se⌐) which form the excess resonant electronic states. It was suggested from EXAFS (extended x-ray absorption fine structure) and Raman measurements [7] that photodarkening was caused by creation of under- and/or over-coordinated defects by illumination at low-temperatures. This supports our model for the excess resonant enhancement following photodarkening. We assume that the excess resonant electronic states can be transformed to nonresonant states, and they are relaxed to the ground states of the three-fold Se atoms. Trapping of photoexcited carriers seems to play a key role on the transformation to the nonresonant states because the excess resonant enhancement is also decreased by adding a weak sub-gap light (0.5 W/cm^2, 1.91 eV), which rarely causes structural changes such as photodarkening, to the prove light (3 W/cm^2, 2.41 eV). If the photocreation rate of the nonresonant states is much faster than that of the relaxation time, the population of excess resonant states rapidly decreases nearly to zero under the high-power illumination. Thus, the competition between the photoexcitation to nonresonant states and the relaxation into the structure of the ground state causes the power dependence of the resonant spectra. The newly observed photoinduced changes in resonant Raman spectra will be advantageous to understanding the dynamics of the photoexcited states in chalcogenide semiconductors.

Acknowledgements

This work was supported by Grants-in-Aid for Scientific Research (B) (No. 09440117), for Encouragement of Young Scientists (No. 11740173), and for Scientific Research on Priority Area 'Cooperative Phenomena in Complex Liquids', from the Ministry of Education, Science and Culture (Japan). One of us (Y. W.) acknowledges support from the Inamori Foundation.

References

1. For a review, see, K. Shimakawa, A. V. Kolobov, and S. R. Elliott, Adv. Phys. **44** (1995) 475.
2. For a recent review, see, P. Boolchand, *Insulating and Semiconducting Glasses*, edited by P. Boolchand (World Scientific, 2000), Chap. 6b, p. 369; K. Murase, ibid., Chap. 6c, p. 415.
3. R. J. Nemanich, G. A. N. Connell, T. M. Hayes, and R. A. Street, Phys. Rev. B **18** (1978) 6900.
4. K. Inoue, T. Katayama, K. Kawamoto, and K. Murase, Phys. Rev. B **35** (1987) 7496.
5. S. Hosokawa, K. Nishihara, Y. Hari, and M. Taniguchi, Phys. Rev. B **47** 3546 (1993).
6. M. Cobb, D. A. Drabold, and R. J. Cappelletti, Phys. Rev. B **54** (1996) 12 162.
7. A. V. Kolobov, H. Oyanagi, A. Roy, and K. Tanaka, J. Non-Cryst. Solids, **232-234** (1998) 80.

Photoluminescence due to the scattering process from heavy-hole to light-hole excitons in CuI thin films

I. Tanaka and M. Nakayama

Department of Applied Physics, Faculty of Engineering, Osaka City University 3-3-138 Sugimoto, Sumiyoshi-ku, Osaka 558-8585, Japan e-mail: `itanaka@a-phys.eng.osaka-cu.ac.jp`

Abstract We have investigated photoluminescence properties of CuI thin films with the thickness of 100 nm, which were grown on (001) NaCl substrates by vacuum deposition, under intense excitation conditions. The absorption spectra at low temperatures exhibit the thermal-strain-induced splitting of the heavy-hole and light-hole exciton energies, where the heavy-hole exciton energy is lower than the light-hole one. In the resonant excitation of the heavy-hole exciton, we have observed a new photoluminescence band whose intensity has a superlinear (almost quadratic) dependence on the excitation power. The energy spacing between this band and the heavy-hole exciton is almost equal to the splitting energy of the heavy-hole and light-hole excitons. These results indicate that the observed new photoluminescence band originates from the inelastic exciton-exciton scattering process from the heavy-hole exciton to the light-hole one.

1 Introduction

It is known in II-VI semiconductors such as CdS and ZnO that inelastic scattering processes of excitons lead to stimulated emission, so-called P emission [1]. In the scattering process of two $n = 1$ excitons, one is scattered into a higher excited state with $n \geq 2$, while the other is scattered onto a photon branch. Cuprous halides have been a model material for the investigation of excitonic properties because of the large exciton binding energies: 190 meV for CuCl and 58 meV for CuI [2]. Recently, we reported that the energies of heavy-hole (HH) and light-hole (LH) excitons at Γ point, which are degenerate in bulk crystals, are split by thermal-strain effects in CuI thin films grown on various substrates [3]. In the present work, we have investigated photoluminescence (PL) properties of CuI thin films under intense excitation conditions from the viewpoint of effects of the HH-LH splitting on the exciton-exciton scattering process.

CuI thin films were grown on (001) NaCl substrates by vacuum deposition. The absorption spectra at low temperatures clearly exhibit the splitting of the HH- and LH-exciton energies: The HH exciton is lower in energy than the LH exciton, similar to a quantum-well system. In the resonant excitation of the HH exciton, a new PL band (N band) appears on the low-energy side of the HH exciton: The energy spacing from the HH exciton is almost equal to the HH-LH splitting energy which is about a half of that between the $n = 1$ and $n = 2$ HH excitons. The intensity of the N band exhibits a superlinear (almost quadratic) dependence on the excitation power. From these results, we conclude that the N band

originates from the inelastic scattering of two excitons associated with the scattering process from the HH exciton to the LH one.

2 Experimental

CuI thin films with the thickness of 100 nm were grown on (001) NaCl substrates at 170 °C by a vacuum deposition method in high vacuum ($\sim 1 \times 10^{-6}$ Torr). All the optical measurements were performed at 10 K using a closed cycle helium-gas cryostat. In the PL measurement, we used a N_2 laser with a pulse width of 5 ns and a dye laser pumped by the N_2 laser for the excitation light. The PL spectra were measured by the combination of a 32-cm single monochromator and a multi-channel CCD detector with a resolution of 0.2 nm.

3 Results and discussion

Figure 1 (a) shows the absorption spectrum of the CuI thin film measured at 10 K. The observed two peaks are attributed to the HH and LH excitons, where the

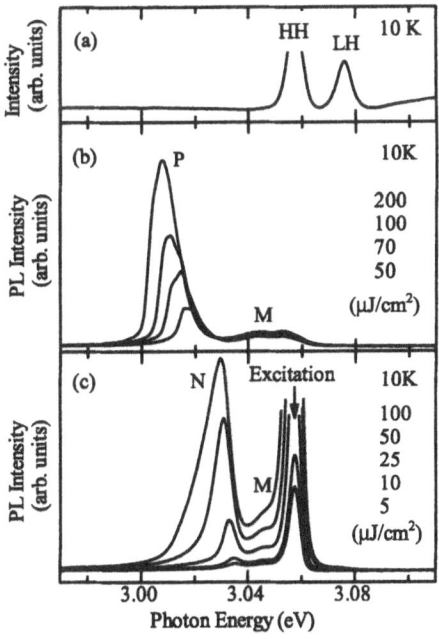

Fig. 1 (a) Absorption spectrum of the CuI thin film at 10K. (b) and (c) PL spectra under various excitation conditions. The excitation energy in (b) is much higher than the band gap energy, while that in (c) is equal to the HH-exciton energy. The excitation power for each PL spectrum is indicated on the right side of the figures.

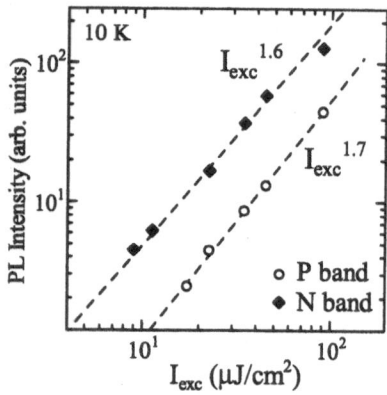

Fig. 2 PL intensity of the P and N bands as a function of the excitation power (I_{exc}). The broken lines show the fitting results.

peak region for the HH-exciton is eliminated because the absorption intensity is over the measurement range. The HH- and LH-exciton energies, which are degenerate in bulk crystals, are split by the thermal strain effect due to the difference of the thermal expansion coefficients of CuI and the substrate material [3].

Figure 1 (b) shows the PL spectra excited by the N_2 laser under various excitation conditions, where the excitation energy is much higher than the band gap energy. The free HH-exciton PL band and other two PL bands labeled M and P were observed. There has been no report on the excitonic molecule in CuI, however, its binding energy is theoretically estimated to be 6 meV [4]. Since the energy spacing between the HH exciton and the high-energy edge of the M band is about 6 meV, it is considered that the M band is due to the excitonic molecule. The P band appears on the low-energy side of the M band. The PL peak energy exhibits a low-energy shift with the increase of the excitation power, and the intensity increases superlinearly as shown in figure 2: $I_P \propto I_{exc}^{1.7}$. The superlinear (almost quadratic) dependence indicates the occurence of exciton-exciton collisions. At the lowest excitation power of 50 μJ/cm², the energy spacing between the P band and the HH exciton is almost equal to that between the $n = 1$ and $n = 2$ HH excitons [2]. Thus, we consider that the P band originates from the inelastic scattering of two excitons, the so-called P emission well known in II-VI semiconductors [1]. In this scattering process, one $n = 1$ HH exciton is scattered into the $n \geq 2$ HH exciton state, while the other is scattered onto the photon blanch with the enegy of

$$\hbar\omega_P = E_{HH}(n = 1) \tag{1}$$
$$- \{E_{HH}(n \geq 2) - E_{HH}(n = 1)\} - 3\delta k_B T_{eff},$$

where δ is a positive constant smaller than 1, and T_{eff} is an effective temperature [5]. It is considerd that the increase of the effective temperature with the increase of the excitation power causes the low-energy shift of the PL band. The P emission has been observed also in CuI bulk crystals [6].

Figure 1 (c) shows the PL spectra under various excitation conditions, where the excitation energy is tuned to the HH-exciton energy. In this case, a new PL band labeled N appears on the low-energy side of the M band. The PL peak energy exhibits a low-energy shift with the increase of the excitation power, and the intensity increases superlinearly as shown in figure 2: $I_N \propto I_{exc}^{1.6}$, which are similar to the P band. However, the PL-peak energy is remarkably different from that of the P band. We note that the energy spacing between the N band and the HH exciton at the lowest excitation power of 5 μJ/cm² is almost equal to the HH-LH splitting energy. From the above results, we consider that the N band originates from the following scattering process of two $n = 1$ HH excitons: One is scattered into the $n = 1$ LH exciton state, while the other is scattered onto the photon branch with the energy of

$$\hbar\omega_N = E_{HH}(n = 1) \tag{2}$$
$$- \{E_{LH}(n = 1) - E_{HH}(n = 1)\} - 3\delta k_B T_{eff}.$$

This PL process due to the inelastic scattering from the HH exciton to the LH one is the novel finding in the present work.

Finally, we briefly describe the optical gain in the energy range of the N band. The exciton-exciton scattering process generally leads to stimulated emission [1]. We confirmed that the N band has the optical gain about 600 cm^{-1} by using a variable stripe length method [7]. This indicates the stimulated emission characteristic of the N band. The details will be reported elsewhre.

In summary, we have observed a new PL band (N band) under the resonant excitation of the HH exciton in the 100-nm CuI thin film. The energy spacing from the HH exciton is almost equal to the HH-LH splitting energy, which is different from the P emission, and its intensity shows an almost quadratic dependence on the excitation power. These results clearly demonstrate that the N band originates from the inelastic scattering of two HH excitons associated with the scattering process from the HH exciton to the LH one. In addition, we have confirmed that the N band leads to stimulated emission.

References

1. C. Klingshirn, Phys. Rep. **70**, (1981) 315.
2. *Physics of II-VI and I-VII Compounds, Semimagnetic Semiconductors*, edited by O. Madelung, Landolt-Börnstein, New Series, Group 3, Vol. 17, Part b, p.258 and 270 (Springer, Berlin 1982).
3. D. Kim, M. Nakayama, O. Kojima, I. Tanaka, H. Ichida, T. Nakanishi, and H. Nishimura, Phys. Rev. B**60**, (1999) 13879.
4. W.T. Hung, phys. stat. sol. (b) **60**, (1973) 309.
5. C. Klingshirn, phys. stat. sol. (b) **71**, (1975) 547.
6. C.I. Yu, T. Goto, and M. Ueta, J. Phys. Soc. Japan **32**, (1972) 1671.
7. K.L. Shaklee, R.F. Leheny, and R.F. Nahory, Phys. Rev. Lett. **26**, (1971) 153.

Identification of fine structures of magneto-oscillatory spectra in cuprous oxide as quantum manifestation of classical non-integrability

K. Hammura[1], K. Sakai[2], M. Seyama[3],

[1] RIKEN (The Institute of Physical and Chemical Research), 2-1 Hirosawa, Wako-shi, Saitama 351-0198, Japan e-mail: `hammura@postman.riken.go.jp`

[2] School of Law, Meiji University, 1-9-1 Eifuku, Suginami-ku, Tokyo 168-0064, Japan

[3] Department of Physics, Faculty of Science, Science University of Tokyo, 1-3 Kagurazaka, Shinjuku-ku, Tokyo 162-8601, Japan

Abstract The magneto-oscillatory spectra in Cu_2O are measured precisely, and the multi-oscillatory structure thereof is successfully interpreted as a consequence of classical non-integrability of the Hamiltonian of an electron-hole (e-h) pair generated in the sample in magnetic fields, i.e., a magnetized exciton.

1 Introduction

In the field of solid state spectroscopy, the magneto-oscillatory spectra—excitonic absorption spectra in intermediate magnetic fields in the photon energy region above the excitonic threshold energy—have received a growing interest for identification[1]. It has been found[2, 3] that the energy distances between adjacent peaks at a fixed field are almost equal and energy positions increase almost linearly with increasing fields, but the spectra have not interpreted satisfactorily by now.

In this paper, we consider the Coulomb attractive force between an electron and a hole as essential role to the magneto-oscillatory spectra. This leads to consider the dynamics near the origin (in relative coordinate from a hole) to be important, and to concentrate the *transient* dynamics of an electron on the morrow of the birth of the electron-hole pair because the e-h pair is born at the origin.

Under the coexistence the system is classically non-integrable, where an electron relative to a hole does not have the unique trajectory. In stead of this, many profiles of trajectories are available. Therefore, we cannot use the traditional formula of absorption cross section $\bar{\Sigma}(\omega)$, which connect $\bar{\Sigma}(\omega)$ with stationary state, because the state corresponds to the classical unique trajectory. For our case of the non-existence of quantum stationary state, we introduce the alternative expression for $\bar{\Sigma}(\omega)$ as follows[4],

$$\bar{\Sigma}(\omega) = 8\pi\alpha\, a_0^2\, \omega \int_{-\infty}^{+\infty} d\tau\, \exp\left(i\{\omega + \frac{E_i}{\hbar}\}\tau\right)\cdot\langle\phi|\phi(\tau)\rangle,$$
(1)

where $|\phi\rangle \equiv \hat{\mu}|\chi_i\rangle$ denotes the photon absorbed wavefucntion (PAWF) and $|\phi(\tau)\rangle = \exp(-i\hat{H}\tau/\hbar)|\phi\rangle$ time developed state of $|\phi\rangle$ by the Hamiltonian $\hat{H}$.

We treat the dynamics of the PAWF in a semi-classical manner: the PAWF is regarded as the photon-absorbed wavepacket (PAWP), and the center of mass of the PAWP

moves as a classical orbit. This is confirmed by fact that the life time which is estimated from the width of the spectral peak is comparable to the *recurrence* time of the trajectory calculated here.

Procedure of this paper is as follows: in section 2, experiment for obtaining the magneto-oscillatory spectra. In section 3, analysis: (i) inverse fourier transform (IFFT) of the observed spectra to acquire the characteristic time until the classical PAWP *recur*. (ii) Numerically calculate the *recurrence* time of the classical trajectory of an electron in a relative reference from a hole of the Hamiltonian. (iii) Compare the characteristic time with the *recurrence* time.

2 Experiment

With magnetic field up to 4.5 T generated by a superconducting magnet, excitonic absorption spectra in thin Cu_2O are observed at liquid helium temperature above and below the threshold energy of a yellow series exciton in Cu_2O, $E_\infty = 17523\,cm^{-1}$ (Fig 1). Owing to the incident light being left circularly polarized, the σ^+-spectra are observed. The magneto-oscillatory spectra are obtained above E_∞, and the features of the spectra at $B = 3.6$ T are summarized as following two points:(1) spectra are composed of superposition of three approximately equally-spaced *quasi*-Landau levels. The triplet structure is found for the first time by this work.(2) All of the three spacings of the peaks of the oscillations are about 1.2 times the cyclotron energy.

3 Analysis

To obtain the characteristic time embedded in the spectra, IFFT of the reduced absorption coefficient $\alpha_{red.}$ is performed (Fig 2). The square of the IFFT is equivalent to $|\langle\phi|\phi(\tau)\rangle|^2$, which we see have three peaks at $\tau_A = 0.79 \pm 0.04$, $\tau_B = 1.71 \pm 0.17$, and $\tau_C = 2.70 \pm 0.4$.

To assign these three times, the *recurrence* time of the PAWP is numerically calculated with classical mechanics, where the model Hamiltonian is that of a magnetized hydrogen atom,

$$H = \frac{1}{2}p_\rho^2 + \frac{1}{2\rho^2}p_\varphi^2 + \frac{1}{2}p_z^2 + \frac{\gamma}{2}p_\varphi + \frac{\gamma^2}{8}\rho^2 - \frac{1}{\sqrt{\rho^2 + z^2}},$$
(2)

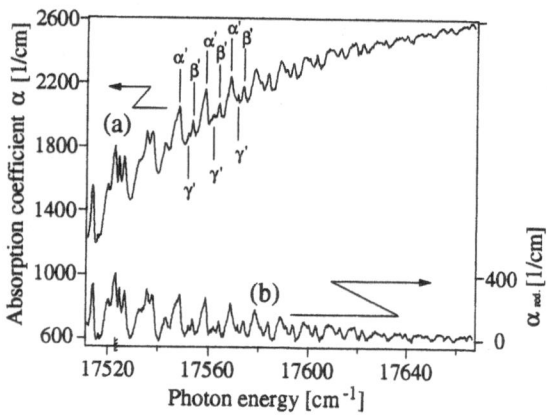

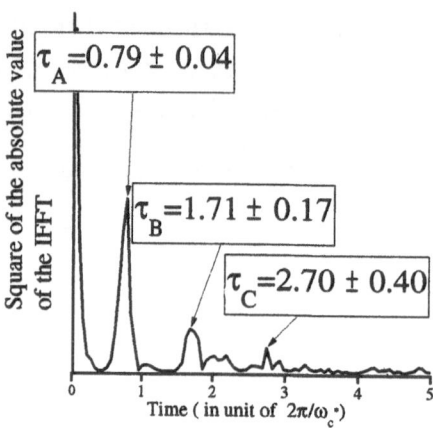

Fig. 1 (a) Absorption coefficient α versus photon energy at 4.2 K at $B = 3.6$ T. σ^+-spectra,i.e.,$m = +\hbar$, for left circularly polarized light. The spectra are composed of three sets of peaks denoted by α', β' and γ' in the figure. Resolving power is 0.5 cm^{-1}. (b) Reduced absorption coefficient $\alpha_{\rm red.}$ obtained by eliminating the spectra at $B = 0$ T from the spectra (a). The limiting energy of the exciton series in the yellow series exciton in Cu$_2$O is denoted by oblique lines on the horizontal axis near 17520 cm^{-1}.

Fig. 2 Square of IFFT of the observed absorption coefficient $\alpha_{\rm red.}(\omega)$ in Figure 1(b). This is equivalent to $|\langle\phi|\phi(\tau)\rangle|^2$. Time is scaled in units of the effective cyclotron period with effective reduced mass of the exciton. Three peaks are found and named $\tau_{\rm A} = 0.79 \pm 0.04$, $\tau_{\rm B} = 1.71 \pm 0.17$, $\tau_{\rm C} = 2.70 \pm 0.4$

as a non-dimensional expression with ordinary cylindrical coordinate. Under initial conditions; $\rho_0 = 0.5, \varphi_0 = 0, p_{\varphi 0} = 1, z_0 = 0, 0 < p_{z0} < 1.5 \times 10^{-4}$, with total energy $E = 2.368758 \times 10^{-2}$ and $\gamma = 5.647 \times 10^{-3}$ which means 17523 cm^{-1} and 3.6 T, respectively. $p_{\rho 0}$ is set so as to make the total energy E be constant. The nonzero values of p_{z0} reflect the thermal fluctuation from the temperature of 4.2 K.

By the calculations, the following three are found: (1) every *recurrence* occurs near the origin. (2) 1072 *recurrences*—about 20% of 5001 trajectories—are obtained. (3) Histogram of the *recurrence* time has 4 sharp peaks in time (Fig 3).

The shorter three times of the peak values in the histogram coincide with the observed $\tau_{\rm A}$, $\tau_{\rm B}$, and $\tau_{\rm C}$. From these results, the multi-oscillatory structure of the observed spectra is found to be caused by the classical non-integrability of the system, which does not yield an unique *recurrence* time but several ones.

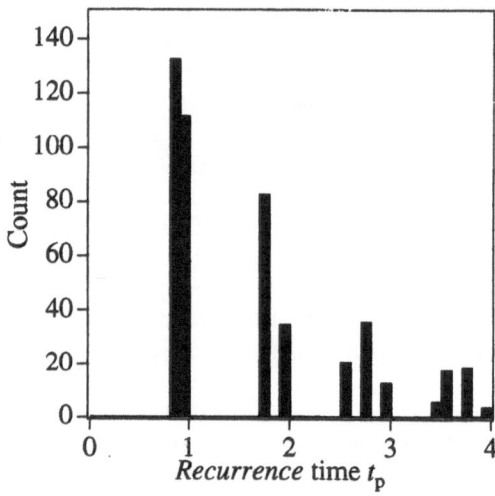

Fig. 3 A histogram of the calculated *recurrence* time $t_{\rm p}$. The unit of time (abscissa) is the cyclotron period at 3.6 T. The distribution of $t_{\rm p}$ is localized at four values 0.8, 1.7, 2.7, and 3.7

References

1. V. T. Agekyan, phys. stat. sol. (a) **43**, 11 (1977).
2. V. T. Agekyan, Yu. A. Stephanov, and I. P. Shiryapov, Sov. Phys. Solid State, **15**, 1698 (1973).
3. H. Sasaki and G. Kuwabara, J. Phys. Soc. Jpn. **34**, 95 (1973).
4. E. J. Heller, J. Chem. Phys. **68**, 2066 (1978).

Optical studies of alloy semiconductors $Ge_{1-x}C_x$ ($x < 0.05$) grown on Si substrates by combined low-energy ion beam and molecular beam epitaxy

M. Sakai[1], Y. Kitayama[1], S. Takaku[1], S. Kimura[2], H. Shibata[2]

[1] Department of Functional Materials Science, Saitama University, 255 Shimo-okubo, Urawa 338-8570, Japan e-mail: `sakai@fms.saitama-u.ac.jp`
[2] Electrotechnical Laboratory, 1-1-4 Umezono, Tsukuba 305-8568, Japan

Abstract We have investigated the optical properties of $Ge_{1-x}C_x$ ($x \leq 0.047$) thin films grown on Si substrates for photon energies ranged from 0.8 to 1.4 eV. Possibility of transformation into direct gap material is discussed based on the present study as well as a previous study by Orner *et al.* (J. Electron. Mater. **25**, (1996) 297.)

1 Introduction

Alloys of group IV elements such as $Si_{1-x}Ge_x$ and $Si_{1-x}C_x$ have been extensively investigated for use in heterostructure devices compatible with Si technology. In addition to these materials, recently an alloy semiconductor, $Ge_{1-x}C_x$, is gathering interest because of the following reasons: i) if sufficient C can be incorporated, the $Ge_{1-x}C_x$ alloys could be lattice matched to Si, ii) using known values of Ge and diamond as endpoints a direct bandgap material is predicted for $0.06 \leq x \leq 0.1$. Despite the predictions, few properties of $Ge_{1-x}C_x$ is known so far due to difficulty in synthesizing, i.e., the extremely low solubility of C in Ge at thermodynamic equilibrium; only the absorption characteristics associated with the indirect transition have been investigated up to x=0.03 at room temperature [1]. By one of authors metastable $Ge_{1-x}C_x$ solid solutions were fabricated up to $x=0.047$ that is the maximum value to date achieved by a combined ion beam and molecular beam epitaxy(CIBMBE) [2]. We report here on the optical properties of the $Ge_{1-x}C_x$ alloys ($x \leq 0.047$) for photon energies ranged from 0.8 to 1.4 eV over a temperature range of 67 to 300 K. Based on the present experiment we discuss influence of carbon incorporation on the direct transition energy in $Ge_{1-x}C_x$.

2 Experimental

The samples studied were grown on Si(001) substrates by the CIBMBE and hence consist of 350 nm thick $Ge_{1-x}C_x$ layer, 50 nm thick Ge buffer layer and 250 μm thick Si substrate. X-ray diffraction suggests that the deposited films are single crystals grown epitaxially on the substrate with twins on {111} planes: the C-1s chemical shift observed by X-ray photoelectron spectroscopy in the sample of $x = 0.047$ reveals that about 72 % of the C atoms are incorporated in substitutional and interstitial lattice, while about 28 % is in the form of precipitate [2]. Further details on sample preparation and the structural characteristics were given in ref. [2].

Reflectance(R) and transmittance(T) of the as-grown samples are measured at approximately normal incidence. The intensity of the lights reflected by and transmitted through the sample were detected by a cooled germanium-photodiode with a standard phase-sensitive-detection system. The temperature of the sample was controlled by a closed-cycle refrigerator with a temperature controller.

3 Results and discussion

Surface morphology measurement by an atomic force microscopy shows roughness of order of about 10 nm in as-grown samples except x=0. Poor surface morphology similar to our result is also reported in MBE-grown samples of ref. [1]. The presence of surface roughness should reduces R and hence T owing to scattering; it is shown from a simple calculation that the reduction factors in both R and T are the same and given by $1 - I_s/I_0$, where I_0 and I_s are the intensities of incident and scattered lights, respectively. Unfortunately the scattered component, I_s, were not detected in the present study. Thus we need another quantity instead of $1 - I_s/I_0$. Since the sample having $x=0$ has a fairly smooth surface, I_s for $x=0$ sample is substantially small compared to $x \neq 0$ samples. Supposed that R and T for $x \neq 0$ samples with, if possible, smooth surfaces have the same magnitude as those for $x=0$ (R_0 and T_0) at photon energies below the direct gap (D_1) of Ge, the quantities R/R_0 and T/T_0 observed below D_1 may give reduction factors associated with the surface-roughness scattering. We believe that this assumption is justified in only discussing energy dispersion of optical constants. In fact observed values of R/R_0 are almost equal to T/T_0 at photon energies in the range of 0.78 to 0.8 eV. In the following analysis, therefore, R/R_0 and T/T_0 will be considered as reduction factors.

Shown in Fig. 1 are spectra of R and T obtained at 67 K for various x-values by taking into account influence of the surface-roughness scattering. There exist x-variation in photon-energy dependence for R and T. This x-variation is likely to be caused by x-variation of the refractive index(n) and extinction coefficient(κ) of $Ge_{1-x}C_x$ because properties of another parts of sam-

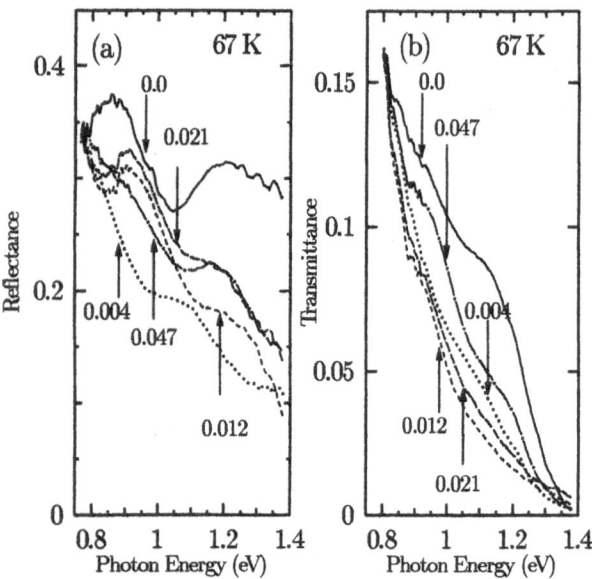

Fig. 1 Reflectance(a) and transmittance(b) spectra observed at 67 K are shown for various x-values in $Ge_{1-x}C_x$. The spectra are modified by considering influence of the scattering of probe light due to surface roughness.

ples, i.e., Si substrate and Ge buffer layer should remain unchanged when x-value is changed. We have also investigated temperature dependence of R and T to identify absorption characteristics found in spectra of κ.

In order to investigate quantitatively the energy dispersion of the optical constants, the formula of R and T including influence of the Si substrate as well as Ge buffer layer is first derived analytically by letting n and κ of $Ge_{1-x}C_x$ be unknown quantities and is secondly solved numerically using observed spectra of R and T to obtain n and κ. Results of κ thus obtained are shown in Fig. 2 for x=0.047 at various temperatures. It is shown that an absorption edge probably due to D_1 transition shifts toward higher energy with decreasing temperature. Magnitude of the energy shift found to be slightly large compared to that of D_1 gap in Ge [3].

To investigate x-variation of D_1 gap, spectra of κ obtained at 67 K are shown in Fig. 3 for several x-values. The absorption edge attributed to D_1 transition does not show significant energy shift when x-value is increased up to x=0.047. This result is in contrast to that of the indirect gap(E_g) showing a blue shift of 70 meV/at. % [1]. Since energy of E_g is about 170 meV lower than that of D_1 in Ge [4], the present result on x-variation of D_1 transition suggests that energy band structure of $Ge_{1-x}C_x$ is transformed into direct transition type at about x=0.024. This value is small compared with that predicted from a linear interpolation using known values of Ge and diamond as endpoints[1].

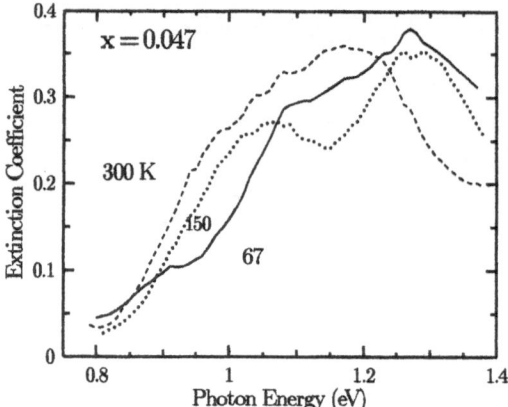

Fig. 2 Spectra of extinction coefficients obtained from observed values of reflectivity and transmittance are indicated for x=0.047 at several temperatures.

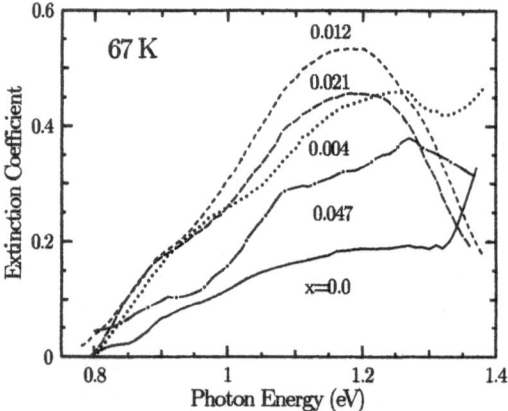

Fig. 3 Spectra of extinction coefficients obtained from low-temperature(67K) reflectivity and transmittance are indicated for several x-values in $Ge_{1-x}C_x$.

4 Conclusions

We have investigated influence of carbon atomic fraction on the infrared absorption characteristics associated with direct transition as well as refractive index in $Ge_{1-x}C_x$ thin films. Although possibility of transformation into direct gap material has been discussed quantitatively, further studies treating the surface-roughness scattering more rigorously are necessary to establish influence of carbon incorporation on the energy band structure.

References

1. B.A. Orner, A. Khan, D. Hits, F. Chen, K. Roe, J. Pickett, X. Shao, R.G. Wilson, P.R. Berger, and J. Kolodzey, J. Electron. Mater. **25**, (1996) 297.
2. H. Shibata, S. Kimura, P. Fons, A. Yamada, A. Obara, and N. Kobayashi, Jpn. J. Appl. Phys. **38**, (1999) 3459.
3. M. V. Hobden, J. Phys. Chem. Solids. **23**, (1962) 821.
4. G. G. Macfarlane, T. P. McLean, J. E. Quarrington, and V. Roberts, Phys. Rev. **108**, (1957) 1377.

Optical properties of Ga$_2$Se$_3$

H. Nishimura, M. Takumi, K. Nagata and Y. Miyamoto

Department of Applied Physics, Fukuoka University, 8-19-1 Nanakuma, Jonan-ku, Fukuoka 814-0180, Japan e-mail: nisimura@cis.fukuoka-u.ac.jp

Abstract Optical properties have been investigated for both α- and β-Ga$_2$Se$_3$, whose vacant sites are disordered and ordered, respectively, by optical absorption, photoacoustic, and photoluminescence measurements. In the β-Ga$_2$Se$_3$, an absorption peak due to free exciton state and an emission peak are observed at room temperature. From these peaks and the photoacoustic spectrum, it is suggested that the optical transition in β-Ga$_2$Se$_3$ is direct one and that the absorption edge is 2.3 eV. The temperature coefficient of the band gap energy is estimated to be -5×10^{-4} eV/K from the temperature dependences of the exciton and photoluminescence peak energies. On the other hand, in α-Ga$_2$Se$_3$, neither exciton nor emission peaks is observed. From this fact and the absorption and photoacoustic spectra, it is suggested that the optical transition in α-Ga$_2$Se$_3$ is indirect one and that the absorption edge is 2.1 eV.

1 Introduction

Gallium sesquiselenide, Ga$_2$Se$_3$, has a defect zinc-blende structure in which one third of the cation sites are vacant. Up to now, four kinds of crystalline allotropes, α- [1], β- [2], γ- [3], and orthorhombic [4] modifications are known to exist in Ga$_2$Se$_3$. In α-Ga$_2$Se$_3$, the vacancies are randomly distributed at the cation sites, while the vacancies in β- and orthorhombic Ga$_2$Se$_3$ are arranged in order in a zigzag and a straight-line manners, respectively. The Ga$_2$Se$_3$ has many potential applications for new type optoelectronic devices because of its interesting properties such as slow relaxation in photoconductivity, residual photoconductivity [5] and optical anisotropy [6]. The origin of these properties is expected to be attributed to a large number of vacancies. However, there are only a few reports on the effect of vacancy ordering on the optical property and the band structure of Ga$_2$Se$_3$ [7, 8]. In this paper, we report the optical properties investigated for both α- and β-Ga$_2$Se$_3$ by optical absorption, photoacoustic (PA), and photoluminescence (PL) measurements.

2 Experimental

α-Ga$_2$Se$_3$ was synthesized from commercial 99.999% pure gallium and selenium by Bridgman method. β-Ga$_2$Se$_3$ was obtained by annealing the α-Ga$_2$Se$_3$ at 580 °C for a week [9]. These crystals were checked to be pure by X-ray diffraction and Raman spectroscopic methods. The PA measurement was carried out by using microphone method at room temperature. The chopping frequency was 13.7 Hz. The intensity of the PA signal was normalized by dividing the intensity of signal of the sample by that of a carbon black. The optical absorption measurement was made in the energy range between 1.5 eV and

3.0 eV. The absorption spectra of α- and β-Ga$_2$Se$_3$ were measured at only room temperature and in temperature range from 148 K to 294 K, respectively. The PL measurements were carried out at room temperature and 77 K using the 488.0 nm argon ion laser as an excitation source.

3 Results and discussion

Figure 1 shows PA, optical absorption, and PL spectra of α- and β-Ga$_2$Se$_3$ at room temperature. The fundamental absorption edges of α- and β-Ga$_2$Se$_3$ are estimated to be 2.1 and 2.3 eV, respectively from the positions of knees in the PA spectra. The saturated PA signal of α-Ga$_2$Se$_3$ is larger than that of β-Ga$_2$Se$_3$. This shows that the recombination process due to non-radiative transition occurs more frequently in α-Ga$_2$Se$_3$ than that in β-Ga$_2$Se$_3$.

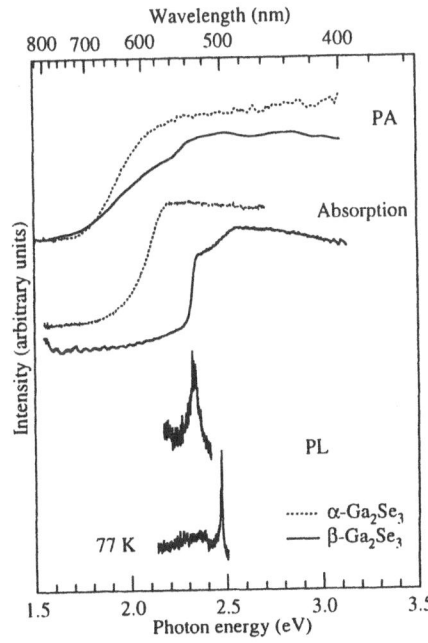

Fig. 1 Optical absorption and photoacoustic spectra of α- and β-Ga$_2$Se$_3$ at room temperature. Photoluminescence spectra of β-Ga$_2$Se$_3$ at room temperature and 77 K are also shown. Dotted and solid lines denote the spectra of α- and β-Ga$_2$Se$_3$, respectively.

In the optical absorption spectrum of β-Ga$_2$Se$_3$, the abrupt rise near 2.3 eV and a following gradual increase between 2.3 eV and 2.6 eV are observed. The abrupt rise near 2.3 eV in the absorption spectrum is attributed to an exciton absorption. Because, the free exciton (n=1) peak is observed near 2.3 eV at low temperature as shown in Fig. 2, which shows the absorption spectrum of β-Ga$_2$Se$_3$ at various temperatures. Since the exciton

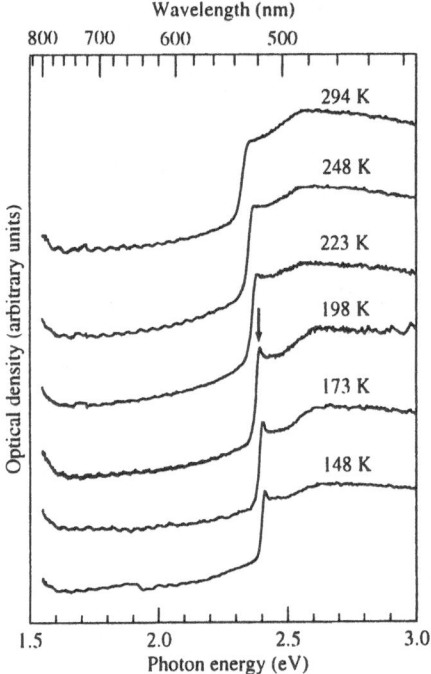

Fig. 2 Absorption spectra of β-Ga$_2$Se$_3$ at various temperatures. An exciton peak denoted by arrow appears at low temperature clearly.

peaks of higher levels (n$\geq$2) are not observed, the binding energy of the exciton can not be calculated. However, the excitons in most of semiconductors are Wannier excitons whose binding energy is in the range of 1 meV to 60 meV, and consequently, the exciton peak energy is almost equal to the band gap energy. Actually, the exciton peak energy of β-Ga$_2$Se$_3$ agrees well with the absorption edge estimated from the PA spectrum. On the other hand, in the optical absorption spectrum of α-Ga$_2$Se$_3$, an abrupt rise is observed near 2.0 eV and exciton peak is not observed. Though the energy of the abrupt rise in absorption spectrum of β-Ga$_2$Se$_3$ agrees with the absorption edge estimated from the PA spectrum, the energy of the abrupt rise in absorption spectrum of α-Ga$_2$Se$_3$ is 0.1 eV lower than the absorption edge estimated from the PA spectrum. It is difficult to determine the band gap energy of such vacancy disordered α-Ga$_2$Se$_3$ exactly. Because, the numerous defects due to vacancy disordering gives rise to many localized states near the edges of both conduction and valence bands and consequently, the absorption edge become unclear. However, attributing the absorption at energy from 2.0 eV to 2.1 eV to the optical transitions arising from these localized states, the band gap energy of α-Ga$_2$Se$_3$ is estimated to be about 2.1 eV.

In the PL spectrum of β-Ga$_2$Se$_3$ at room temperature, a sharp emission peak is observed near the fundamental absorption edge, 2.3 eV, while any signal of emission is not observed in the PL spectrum of α-Ga$_2$Se$_3$. Since the energy of emission peak of β-Ga$_2$Se$_3$ agrees

with that of exciton peak, the emission peak is ascribed to the radiative annihilation of the free exciton. From the above mentioned results, it is concluded that the optical transitions in α- and β-Ga$_2$Se$_3$ are indirect and direct ones, respectively, and that the band gaps of α- and β-Ga$_2$Se$_3$ are 2.1 and 2.3 eV, respectively. Thus, one can see that the ordering of vacancies in Ga$_2$Se$_3$ crystals affects their electronic band structures distinctly.

With decreasing temperature, the exciton peak of β-Ga$_2$Se$_3$ increases in intensity and shifts to higher energy region. Similarly, the PL peak at 77 K is sharp and located at higher energy region in comparison with that at room temperature. Figure 3 shows the temperature dependence of exciton and PL peak energies. The temperature coefficient of the bang gap energy is estimated to be about -5×10^{-4} eV/K from an average of the temperature coefficients of exciton and PL peak energies.

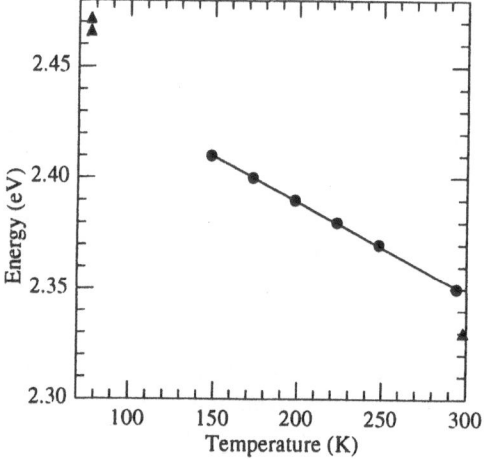

Fig. 3 Temperature dependences of exciton and PL peak energies in β-Ga$_2$Se$_3$. Filled circles and triangles denote exciton and PL peak energies, respectively.

References

1. H. Hahn and W. Klingler, Z. Anorg. Allg. Chem. **259**, (1949) 135.
2. D. Lübbers and V. Leute, J. Solid State Chem. **43**, (1982) 339.
3. L. S. Palatnik and E. K. Belova, Inorg. Mater. **1**, (1965) 1703.
4. A. Yamada, N. Kojima, K. Takahashi, T. Okamoto and M. Konagai, Jpn. J. Appl. Phys. **31**, (1992) L186.
5. G. B. Abdullaev, B. G. Tagiev, G. M. Niftiev, S. I. Aliev, Sov. Phys. Semicond. **16**, (1982) 1048.
6. T. Okamoto, N. Kojima, A. Yamada, M. Konagai, K. Takahashi, Y. Nakamura and O. Nittono, Jpn. J. Appl. Phys. **31**, (1992) L143.
7. T. Nakayama and M. Ishikawa, J. Phys. Soc. Jpn. **66**, (1997) 3887.
8. T. Hanada, *Ph. D. Thesis* (Faculty of Engineering, Tokyo Institute of Technology, Tokyo 1997).
9. E. Finkman, J. Tauc, R. Kershaw and A. Wold, Phys. Rev. B **11**, (1975) 3785.

Exciton Sublattice of Incommensurate Phase in Low-Dimensional TlInS$_2$

N. Mamedov[1], Tamao Aoki-Matsumoto[2], B. Gadjiev[3], H. Uchiki[4], N. Yamamoto[1] and S. Iida[4]

[1] Department of Physics and Electronics, Osaka Prefecture University, Gakuen-cho 1-1, Sakai, Osaka 599-8531, Japan e-mail: mamedov@pe.osakafu-u.ac.jp
[2] Department of Physics, Konan University, Okamoto 8-9-1, Higashinada-ku, Kobe 658-8501, Japan
[3] Department of Theory of Semiconductors, Institute of Physics, 33 H. Javid prospect, Baku, 370143, Azerbaijan
[4] Department of Electrical Engineering, Nagaoka University of Technology, Kamitomioka 1603-1, Nagaoka 940-2188, Japan

Abstract Basing on phenomenological considerations, a concept of excitons in incommensurate phase is proposed. The obtained data on optical absorption data in TlInS2 favors the proposed concept. A more accurate testing is suggested using low temperature spectroscopic ellipsometry. The room temperature dielectric function spectra given for the first time prove that all previously reported data on edge excitons in TlInS2 were obtained in partly forbidden configuration.

1. Introduction

Quasi-two-dimensional TlInS$_2$ is a semiconductor-ferroelectric with well-defined low-temperature exciton spectra [1-3] and interesting photo-induced memory effects [4]. In this material the phase transition from paraelectric phase (P) to ferroelectric (commensurate) phase (C) takes place through an intermediate (incommensurate) phase (I) and is accompanied by the anomalies of dielectric susceptibility [5], heat capacity [6], and linear expansion coefficients [7]. In the rather comprehensive work Kalomiros et al. [2] tried to characterize the temperature behavior of excitons in TlInS$_2$ using a Bose-Einstein statistical factor only. Although such treatment is correct in far away from the anomaly region, it may not be quite adequate in the extended temperature range of P-I-C phase transition. The obtained low values of the parameters related to the Debye temperature in TlInS$_2$ [2] are then indicative of the possible underestimation of the effect of the irregular behavior of the dielectric susceptibility. It also explains a controversy between the previous value of the exciton binding energy 13 meV [2] and that of 30meV or higher reported later [4].

In this work on TlInS$_2$ we report our results on a semi-phenomenological approach to the exciton behavior at P-I-C phase transition together with relevant experimental observations on optical absorption and dielectric function in the region of interband optical transitions.

2. Results

The soft mode in TlInS$_2$ semiconductor-ferroelectric with space group C_{2h}^6 [8] at room temperature is an optic vibration in the point $q=(0.012,0,0.25)$ [9] of the BZ. In the framework of this symmetry the consequent solution of the kinetic equations leads to the following expression

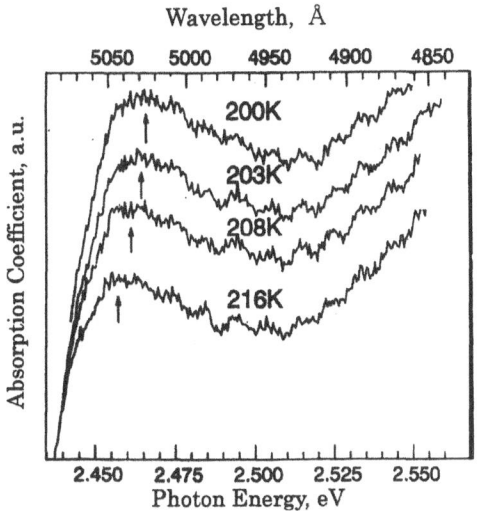

Wavelength, Å

Fig.1 Absorption spectra of TlInS2 in I-phase for partly forbidden configuration. The arrows indicate the exciton structures.

for the local susceptibility:

$$\chi(\omega = 0, T, x) = \chi_p(1 + c\frac{k^2}{1-k^2}\cos^2 amqx),$$

$$1 - k^2 \sim \frac{(T-T_c)}{(T_i-T_c)} \quad (1)$$

Here amu is the elliptic Jacobi function. The χ_p equals χ (ω=0, T, x) in P-phase. It follows from the Eq. (1) that I-phase represents an alternation of the commensuration and discommensuration regions with $\chi=\chi_p$ and $\chi\sim(T-T_c)^{-1}$, respectively. The binding energy of excitons in these regions scales down as

$$E_{exc} \sim \frac{1}{(1+4\pi\chi_{eff})^2} \quad (2)$$

and will be remarkably different. One can say that excitons in commensuration regions will form a sublattice of I phase. In the Eq. (2) the χ_{eff} is either

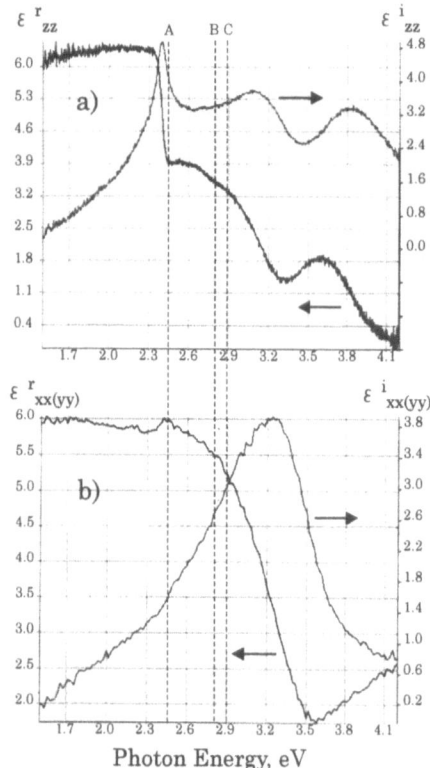

Fig. 2 Real and imaginary parts of dielectric function spectra of TlInS₂ at room temperature for allowed (a) and partly-allowed (b) configurations. The letters A, B and C indicate critical point locations determined previously [2].

dynamic or static dielectric susceptibility, depending on the condition between the value of the binding energy and the averaged phonon frequencies. If electron and hole belong to the different regions the χ_{eff} in the Eq. (2) is the mean susceptibility of the two. The static susceptibility is infinite at temperature T_C of the I-C phase transition. The dynamic susceptibility takes on though large but finite values under the same condition.

As shown in our previous work [3], the exciton absorption rather rapidly decreases from ~10^4 cm⁻¹ to ~10^3 cm⁻¹, with increasing the temperature up to T_C (~200K) in P-phase. Some increase of the exciton absorption is observed above the temperature T_i (~220K) of the P-I phase transition [3]. The absorption spectra in I-phase are given in Fig.1. The observed structures are definitely excitonic and evidence for the possible formation of the exciton sublattice. However, a more accurate model testing is highly desirable.

The room temperature dielectric function spectra, derived in uniaxial approximation from our ellipsometric data for the surface normal to the layer plane are shown in Fig. 2. In E∥Z orientation of the electrical vector (Fig. 2a) the line at 2.40 eV associates with direct allowed

exciton transitions at the fundamental gap. Apparently, in E⊥Z geometry (Fig. 2b) these transitions are partly forbidden. Structures A, B, and C obtained previously [2] are thought to correspond to the latter case. In the edge region the dielectric function spectra obtained from the mirror-like surface parallel to the layer plane exhibit a better resolved exciton line A and are in good agreement with those in Fig. 2b. The exciton absorption is at higher level than that at ~200K.

3. Summary

We have verified that at least in the region of P-I-C phase transition the exciton absorption in TlInS₂ is greatly influenced by the irregular behavior of the dielectric susceptibility. The data can be plausibly explained in terms of the suggested concept of the exciton sublattice of incommensurate phase.

Our firstly obtained dielectric function spectra drive into conclusion that all previous experiments on edge excitons in TlInS₂ were confined to the partly forbidden configuration and therefore need cross-checking in allowed geometry. Low temperature spectroscopic ellipsometry in allowed configuration is expected to provide a more accurate test on the exciton sublattice of I-phase. The results will be given in a later publication.

The authors are grateful to Mr. Y. Shim for his helpful assistance.

References

1. G.I. Abutalibov, S.G. Abdullaeva, and N.M. Zeinalov, Fiz Tekh. Poluprovodn. **16** (1982) 2086.
2. J.A. Kalomiros and A.N. Anagnostopoulos, Phys. Rev. B **50** (1994) 7488.
3. N. Mamedov, S. Iida, Tamao Matsumoto, H. Uchiki, Yo. Tanaka, *Proc. of 11ᵗʰ Int'l. Conf. Ternary and Multinary Compounds*, edited by R.D. Tomlinson, A.E. Hill and R.D. Pilkington (Institute of Physics Conference Series No. 152 Salford 1998) p.899.
4. H. Uchiki, D. Kanazawa, N. Mamedov, S. Iida, J. Luminescence **87-89** (2000) 664.
5. K.R. Allakhverdiev, N. Turetken, F.M. Salaev, F.A. Mikailov, Sol. Stat. Com. **96** (1995) 827.
6. B.R. Gadjiev, N.T. Mamedov, and F.B. Godjaev, Sov. Low Temp. Phys. **20** (1994) 7448.
7. N.A. Abdullaev, K.R. Allakhverdiev, G.L. Belenkii, T.G. Mamedov, R.A. Suleimanov, Ya.N. Sharifov, Sol. Stat. Comm. **53** (1985) 601.
8. W. Henkel, H.D. Hochheimer, C. Carlone, A. Werner, S. Ves, H.G. von Schnering, Phys. Rev. B **26** (1982) 3211.
9. S.B. Vakhruzhev V.V. Zhadanova, B.E. Kvyatkovskii, N.M. Okunev, K.R. Allakhverdiev, R.A. Aliev, R.M. Sardarl, Sov. Phys. JETP Lett. **39** (1984) 291.

Proc. 25th Int. Conf. Phys. Semicond., Osaka 2000 (Eds. N. Miura and T. Ando)

125

Isotopic Mass and Lattice Constants of Si and Ge: X-ray Standing Wave Measurements

J. Zegenhagen[12], A. Kazimirov[3], L.X. Cao[12], M. Konuma[1], E. Sozontov[1], D. Plachke[4], H.D. Carstanjen[4], G. Bilger[5], E. Haller[6], V. Kohn[7], M. Cardona[1]

[1] Max-Planck-Institut für Festkörperforschung, Heisenbergstraße 1, D-70569 Stuttgart, Germany
[2] European Synchrotron Radiation Facility, BP 220, F-38043 Grenoble Cedex, France
[3] Materials Science Department, Northwestern University, Evanston, IL 60208, USA
[4] Max-Planck-Institut für Metallforschung, Heisenbergstraße 1, D-70569 Stuttgart, Germany
[5] Institut für Physikalische Elektronik, Universität Stuttgart, Pfaffenwaldring 47, D-70569 Stuttgart, Germany
[6] University of Berkeley, 553 Evans Hall, CA 94720, USA
[7] Russian Science Center Kurchatov Institute, 123 182 Moscow, Russia

Abstract The molecular volume of materials depends on their isotopic mass. This effect is purely quantum-mechanically in origin, since it arises from the combined effect of the zero-point motion of the atoms and the anharmonicity of the potential. Associated differences are thus small. We measured the isotopic mass dependence of the lattice constants of Si and Ge with high accuracy. Scaled to a mass difference of $\Delta M = 1$ amu we find values of $(\Delta a/a)$ of -0.36 $\times 10^{-5}$ and -0.88 $\times 10^{-5}$ for Ge at $300K$ and $30K$, respectively. For Si, scaled to $\Delta M = 1$, we find -1.8 $\times 10^{-5}$ and -3.0 $\times 10^{-5}$ at $300K$ and $30K$, respectively. Our results represent a stringent test for theoretical calculations. The agreement is good with published results for Ge but less good with the two existing calculations for Si.

1 Introduction

The influence of the isotopic composition on the properties of materials is attracting increasing interest lately [1]. The reason for this is twofold. On the one hand, one obvious reason is the increased availability of isotopic pure elements and substances since after the end of the cold war isotope separators can now be used for peaceful purposes. On the other hand, materials with tailored isotopic composition are attracting growing interest because of their - sometimes exotic - features. Primarily influenced by the specific isotopic composition are the vibrational properties. Diamond made of isotopic pure ^{13}C, for example, should thus be harder than natural diamond and was suggested to be the hardest substance known to man [2]. The thermal conductivity of isotopic pure crystals can considerably exceed the values for the corresponding natural substances [3]. Thus, isotopically pure materials may have interesting applications were extreme stress tolerance and heat dissipation is required and isotopically pure crystals have, e.g., successfully been used as x-ray monochromators for third generation synchrotron sources [4]. Likewise are the electronic properties affected by the isotopic composition, allthough these influences are less dramatic. Compared to the result of the isotopic mass on the vibrational properties of crystals, the effect on the structure via changes in the molecular volume are subtle. For heavier elements only of the order of $\Delta d/d = 10^{-6}$ is expected at room

temperature for a 1% change of mass. However, even such subtle differences must be taken into account in some cases. E.g., present attempts to improve the accuracy in the determination of Avogadro's number in order to replace the archival Pt/Ir kilogram, kept at Sèvres near Paris, by an atomic standard are resting upon the precise determination of the silicon lattice constant and this in turn is eventually influenced by the isotopic composition [5].

The dependence of the lattice constant of crystals on their isotopic mass is purely a quantum-mechanical effect. It originates from the zero-point motion of the atoms in association with the anharmonicity of the potential which is also responsible for the temperature dependence of the molar volume. Following the early work of London [6], a few theoretical papers have been published recently on the issue of lattice constant versus isotopic mass [7–9]. Since the associated changes are very small, there are hardly any experimental data available. Measurements are additionally hampered by the fact that the lattice constant difference is largest at $0K$ and almost vanishes at temperatures above the Debye temperature of the particular crystal. Thus, until recently, the effect had been determined reliably only for diamond because (a) diamond has a very high Debye temperature and thus the lattice constant difference is almost at its maximum even at room temperature, (b) the relative mass difference is high, and (c) highly perfect crystals of diamond can be produced, allowing fairly accurate lattice constant measurements with standard diffraction techniques. However, diffraction techniques are bound to fail to provide precise lattice constant values for crystals with smaller relative mass differences and with low Debye temperatures. Thus, the results of an earlier study for the lattice constant difference between natural germanium and an enriched ^{74}Ge crystal using a modified diffraction technique [10] were later found to be in considerable disagreement with theoretical calculations [7].

We demonstrated lately [11] that the x-ray standing wave technique [12,13] can be used to determine lattice constant differences with very high accuracy. For a thorough description of the principle of the method we refer the reader to our recently published work [11] and restrict ourselfs here to a brief outline. By Bragg diffraction, an interference field is generated in a substrate of a certain isotopic composition on which an homoepitaxial film with a different isotopic composition is grown. The planes of the interference field are accurately in registry with the lattice planes of the substrate and thus, this is essence of the method, serve as a yardstick for the position of the *surface planes* of the overlayer. These are displaced by N times the lattice mismatch from the position which they would adopt if the epilayer would have had the same isotope composition as the substrate. Here, N is the number of lattice planes of the overlayer. Furthermore, the mismatch of the lattice planes normal to the surface is increased for a pseudomorphic, epitaxial layer compared to the lattice mismatch for an unstrained system by a factor ϵ since the film is in registry at the interface. Thus, the difference in lattice constant normal to the surface is enhanced as determined by the Poisson ratio.

By scanning a suitable Bragg reflection of the substrate crystal in angle or in energy, as in the present case, the planes, i.e. maxima and minima, of the x-ray standing wave move inward, antiparallel to the **Q**-vector by half the spacing of the diffracting planes [13]. The position of the surface planes of the isotopically modified epilayer in reference to the extrapolated bulk planes is assessed by recording the emitted photoelectrons while scanning the chosen Bragg reflection of the substrate with the diffraction vector (**Q**-vector) normal to the growth surface.

2 Experimental

Epitaxial films were grown on Ge(111) and Si(111) by molecular beam epitaxy (MBE) [11] and liquid phase epitaxy (LPE), respectively. Isotopically enriched ^{76}Ge films (86 % ^{76}Ge, 14 % ^{74}Ge), with a thickness of 1.36 $\pm$ 0.09 μm and 0.56 $\pm$ 0.01 μm, were grown on the (111) surfaces of a natural and an isotopically highly enriched (96% ^{70}Ge, 4% ^{72}Ge) single crystal [14], respectively. The isotopically enriched Si film (60 % ^{30}Si, 40 % ^{28}Si) with a thickness of 0.92 μm was grown by liquid phase epitaxy on a perfect single crystal Si(111) substrate with natural isotopic composition (91% ^{28}Si, 4% ^{29}Si, 5% ^{30}Si). Thickness and isotopic composition of the films had been determined with Rutherford backscattering spectrometry (RBS) and secondary ion mass spectroscopy, respectively [15]. The x-ray standing wave measurements [16] were carried out using Cu-K$_\alpha$ radiation from a stationary x-ray tube and at the Hamburg Synchrotron Radiation Laboratory (HASYLAB), using synchrotron radiation from the DORIS storage ring which was monochromatized by a Si(333)/Si(511) crystal combination with a bandpass of 0.3 eV at the energies at which the measurements were performed (around 7.59 keV for Ge and 8.00 keV for Si). The samples were mounted strain free on the cold finger of a He flow-

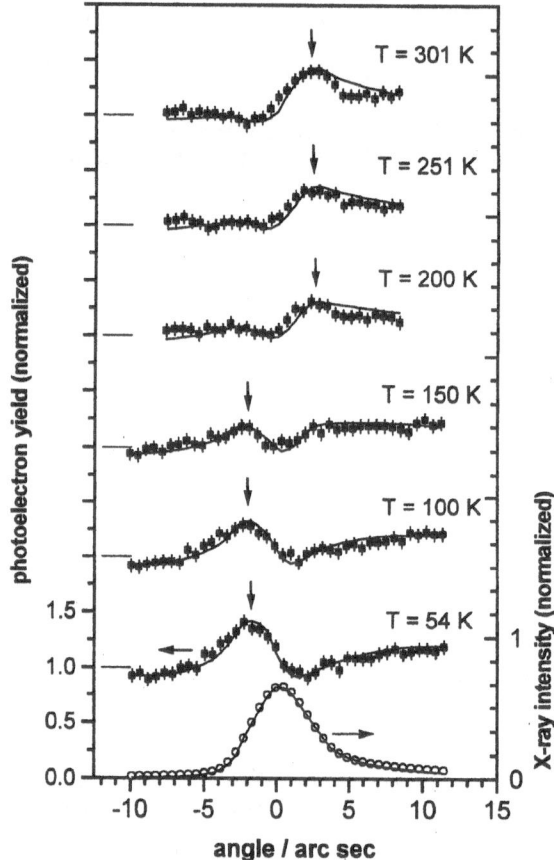

Fig. 1 Ge(333) reflectivity and photoelectron yield from a 1.36 μm epitaxial ^{76}Ge layer on $^{nat.}$Ge(111) as a function of glancing angle Θ for different temperatures (from ref. [11]). The solid lines are fits to the experimental data (symbols). The fit to the reflectivity determines the angular scale and fits to the photoelectron signal yield the surface phase shifts from which the d-spacing is calculated.

through cryostat with a diode temperature sensor placed close by. The total yield of the emitted photoelectrons was detected with a channeltron [17]. For the standing wave measurements, the monochromator was scanned in energy, passing the Bragg reflections of Si and Ge. For Si we chose the (333) reflection, whereas for Ge we chose the (444) reflection at a Bragg angle close to 90°. The ^{70}Ge(111) substrate crystal showed a mosaic spread of 0.3°. However, this did not compromise the accuracy of our measurement since the acceptance of the crystal is > 0.3° close to normal incidence.

3 Results and Discussion

The results of several XSW scans as a function of temperature are shown for ^{76}Ge on natural Ge(111) in Fig. 1. The change in the shape of the photoelectron yield curves with changing temperature can clearly be distinguished and is a signature for the change in the lattice constant difference between the substrate and the epi-

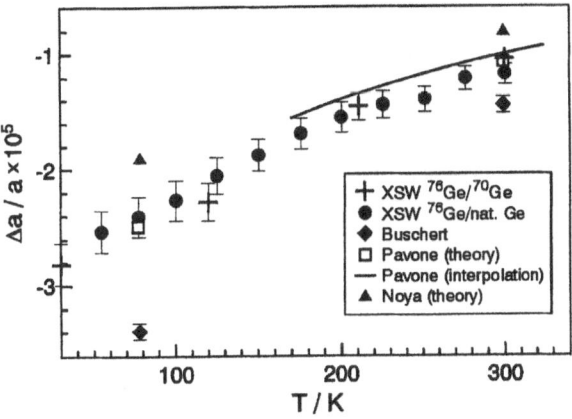

Fig. 2 The lattice constant difference of Ge with isotopic mass as a function of temperature. All data are scaled to a mass difference of $\Delta M = 3.05$ amu. Shown are the results of our recently published work [11] (solid dots) and the present XSW measurements (crosses) in comparison with existing calculations [7,9] and the results of earlier diffraction measurements by Bushert et al [10].

Table 1 Lattice constant difference for Si isotopes of different mass scaled to $\Delta M = 1.06$

T (K)	XSW results $\Delta a/a$ $\times 10^{-5}$	Herrero [20] $\Delta a/a$ $\times 10^{-5}$	Biernacki&Scheffler [21] $\Delta a/a$ $\times 10^{-5}$
0	-	-2.7	-4.1
30	-3.2	-	-4.2
90	-3.1	-	-4.2
100	-	-2.2	
150	-2.9	-2.5	-3.9
180	-2.6	-	-3.5
200	-	-2.3	
210	-2.5	-	-3.3
240	-2.3	-	-3.0
300	-2.0	-1.9	-2.5
	±6%	± ≈ 12%	

layer. Shown are also fits to the experimental data from which the value of the lattice constant difference for the surface layer of the thin film are deduced. To calculate the change in lattice constant, the elastic constants of Ge must be considered and the measured lattice mismatch has to be divided by a factor of 1.37 [19]. The experimental results obtained for the dependence of the lattice constant difference on the Ge isotope mass as a function of temperature are shown in Fig. 2. Data and calculations are scaled to $\Delta M = 3.05$. The outcomes of both sets of XSW measurements agree very well with each other and support the calculations of Pavone and Baroni [7].

The corresponding results for silicon, also corrected with the help of the elastic constants [19] (a correction factor of 1.44 for Si), are shown in table 1. For comparison, we are also listing values calculated by Herrero [20] and Biernacki and Scheffler [21]. We deduced them from the plots shown in those publications and compile them here for convenience. All lattice constant differences are scaled to $\Delta M = 1.06$ for silicon. The calculations of Herrero agree within the limits of error with our data whereas the calculations of Biernacki and Scheffler differ by about 30%. However, it is worth noting that the error of the calculations published by Herrero is about twice as large as our experimental error, too large to show the anomaly in the lattice expansion of Si below 150K which is clearly reflected in our experimental data and qualitatively shown in Biernacki and Schefflers calculation.

References

1. see e.g.: M. Cardona, in Advances in Solid State Physics, R. Helbig, Ed. (Vieweg, Braunschweig/Wiesbaden, Germany, 1994), vol. 34, pp. 35–50.

2. A.K. Ramdas, S. Rodrigues, M. Grimsditch, T.R. Anthony, W.F. Banholzer, Phys. Rev. Lett. **71**, (1993) 189.

3. see e.g. T. Ruf, R.W. Henn, M. Asen-Palmer, E. Gmelin, M. Cardona, H.-J. Pohl, G.G. Devyatych, P.G. Sennikov, Solid State Comm. **115**, (2000) 243, and references therein.

4. L.E. Berman, J.B. Hastings, D.P. Siddons, M. Koike, V. Stojanoff, S. Sharma, Synchr. Rad. News **6**, No.3 (1993) 23.

5. see e.g.: P. Becker, H. Bettin, L. Koenders, J. Martin, A. Nicolaus, S. Röttger, PTB-Mitteilungen **106**, (1996) 321.

6. H. London, Z. Physik Chem. **16**, (1958) 302.

7. P. Pavone and S. Baroni, Solid State Commun. **90**, (1994) 295.

8. A. Debernardi and M. Cardona, Phys. Rev. B **54**, (1996) 11305.

9. J.C. Noya, C.P. Herrero, R. Ramírez, Phys. Rev. B **56**, (1997) 237.

10. C. Buschert, A.E. Merlini, S. Pace S. Rodriguez, M.H. Grimsditch, Phys. Rev. B **38**, (1988) 5219.

11. A. Kazimirov, J. Zegenhagen, M. Cardona, Science **282**, (1998) 930.

12. B.W. Batterman, Phys. Rev. **133**, (1964) A759.

13. see e.g., J. Zegenhagen, Surf. Sci. Rep. **18**, (1993) 199.

14. The crystal had been produced by the Bridgman technique.

15. In the homoepitaxial Si layer, RBS had shown thickness fluctuations of 10 %.

16. The experimental set up is described in reference [11].

17. In this way the recorded electron signal was originating from a certain depth which was taken into account in the analysis [11,18].

18. M.V.Kovalchuk, V.G. Kohn, E.F. Lobanovich, Sov. Phys. Solid State **27**, (1985) 2034.

19. J. Honstra and W.J. Bartels, J. Cryst. Growth **44**, (1978) 513.

20. C.P. Herrero, Solid State Comm. **110**, (1999) 243.

21. S, Biernacki and M. Scheffler, J. Phys. Cond. Matter **6**, (1994) 4879.

Metal-Insulator Transition in Doped Semiconductors

Kohei M. Itoh*

Dept. Applied Physics and Physico-Informatics, Keio University, 3-14-1, Hiyoshi, Kohoku-ku, Yokohama 223-8522, Japan
e-mail: kitoh@appi.keio.ac.jp

Abstract   The values for conductivity critical exponent μ for doping induced metal-insulator transition in semiconductors have been debated for more than two decades. The present work shows convincingly that $\mu \approx 0.5$ for completely uncompensated semiconductors. $\mu \approx 1$ previously found for a wide variety of nominally uncompensated semiconductors is an artifact due most likely to the effect of compensation, and does not represent the intrinsic value of μ for uncompensated semiconductors.

1 Introduction

The metal-insulator (MI) transition in the presence of both disorder and electron-electron interaction turns out to be one of the most challenging subjects in condensed-matter physics. Despite many decades of theoretical [1–5] and experimental efforts [6], researchers are yet to agree upon an unified description of the phenomena. [7]

The doping-induced MI transition in single crystalline semiconductors is the best example of disorder and interaction induced transition that has been studied extensively via measurements of physical quantities such as electrical conductivity, dielectric constant, and heat capacity. In particular the critical behavior of the electrical conductivity at zero temperature $\sigma(0)$ has been evaluated as a function of a parameter t that describes the degree of the disorder and interaction;

$$\sigma(0) \propto |t/t_c - 1|^{\mu} \qquad (1)$$

where μ is the conductivity critical exponent and t_c is the critical value of t that separates the insulating and metallic phases. The MI transition in semiconductors has been investigated as a function of the doping concentration (N), externally applied magnetic field (B), and externally applied uniaxial stress (S), i.e., $t \equiv N, B, S$, and $t_c \equiv N_c, B_c, S_c$, respectively, in eq. (1). In this paper we consider the MI transition in the uncompensated system Ge:Ga when N is taken as a variable.

Since the classic experiment by Rosenbaum et al. that showed $\mu \approx 0.5$ for stress(S)-tuned Si:P [8], a wide variety of experiments has been performed on nominally uncompensated semiconductors. $\mu \approx 0.5$ has been found for doping(N)-tuned Si:As [9] and Ge:Ga [10,11] while $\mu \approx 0.65$ for N-tuned Si:B. [12] $\mu \approx 0.5$ is puzzling from a theoretical point of view since it violates Chayes et al.'s inequality $\mu > 2/3$[13] assuming $\mu = \nu$ [14] where ν is the critical exponent for the localization length ξ. [15] More recently, Löhneysen and co-workers questioned the use of a relatively wide range of N and S that can be

* Member of PRESTO-JST

fitted with $\mu \approx 0.5$ (typically up to $t = 1.2t_c$ or more) and obtained $\mu \approx 1.3$ on N-tuned Si:P by limiting the critical region to $N_c < N < 1.07N_c$. [16] Rosenbaum et al. immediately argued that $\mu \approx 1.3$ region analyzed in Ref. [16] was an artifact due to an inhomogeneous dopant distribution [17]. The question raised "How wide is the critical region?" has, therefore, become one of the most important issues along with the homogeneity of the dopant distribution in the samples.

In all of the above mentioned experiments, the zero temperature conductivity $\sigma(0)$ has been obtained from extrapolation of the finite temperature conductivity $\sigma(T)$ to $T = 0$ assuming a particular temperature dependence of $\sigma(T)$; typically $\propto T^{1/2}$ or $\propto T^{1/3}$. More recently researchers questioned the validity of extrapolation and started to employ so called finite temperature scaling of the form; [5]

$$\sigma(t, T) \propto T^x f(|t/t_c - 1|/T^y), \qquad (2)$$

where $y = 1/z\nu$ where z is the dynamical scaling exponent. The critical exponent is given by $\mu = x/y$. Eq. (2) has two advantages over the conventional analysis involving eq. (1). Firstly, eq. (2) allows us to use values of $\sigma(t, T)$ taken at finite temperatures, i.e., the conventional extrapolation to $T = 0$ can be avoided. Secondly, eq. (2) allows us to evaluate $\sigma(t, T)$ taken on the both sides of the transition ($t < t_c$ and $t_c < t$). The application of eq. (1), on the other hand, has been limited to the analysis of $\sigma(t, T)$ on the metallic side only. Using eq. (2), $\mu = x/y \approx 1.0$ and 1.6 have been presented for nominally uncompensated stress-tuned Si:P [18] and Si:B [19], respectively, for a very narrow range of S around S_c. It was demonstrated graphically that plots of $\sigma(S, T)/T^x$ vs. $|S/S_c - 1|/T^y$ collapse to form a single scaling curve on each side of the transition. [19] Unfortunately, an exact form of the mathematical expression $f(|S/S_c - 1|/T^y)$ in eq. (2) was not obtained for Si:P or Si:B, i.e.,it was not clear whether the same $f(|S/S_c - 1|/T^y)$ works on both sides of the transition. Nevertheless, the fact is that the values of $\mu = x/y \approx 1.0$ and 1.6 for Si:P and Si:B, respectively, are significantly larger than the typical values of μ previously obtained by the conventional extrapolation analysis on the metallic samples only, e.g., $\mu \approx 0.5$ for the S-tuned Si:P [8] and $\mu \approx 0.65$ for the N-tuned Si:B. [12]

The point is very clear. For nominally uncompensated semiconductors, the finite temperature scaling [eq. (2)] works with $\mu = x/y \approx 1.0$ for only a small region around

t_c. On the other hand, the finite temperature scaling does not work for a wide region of t to which the zero temperature scaling [eq. (1) has been applied successfully with $\mu \approx 0.5$. This observation has lead many to believe that the intrinsic value of μ is ≈ 1 for uncompensated semiconductors.

The present work demonstrates clearly that the intrinsic value of μ is ≈ 0.5 for completely uncompensated semiconductors. We argue $\mu \approx 1$ obtained for nominally uncompensated semiconductors is due actually to the effect of doping compensation, i.e., co-presence of donors and acceptors that happens unavoidably in every doped semiconductor. In our experiment, $\mu = 1.2 \pm 0.2$ has been obtained by the finite temperature scaling of the critical conductivity tuned by impurity concentration (N) in a nominally uncompensated Ge:Ga for the range limited to $N = N_c \pm 0.004 N_c$.[20] The exact form of the function $f(|t/t_c - 1|/T^y)$ has been obtained in terms of a third-order non-linear equation, and it describes very well the critical behavior of $\sigma(t, T)$ on both sides of the MI transition. In order to show that doping compensation is responsible for the $\mu \approx 1.2$ region we obtained near N_c, the localization length ξ and the impurity dielectric susceptibility χ_{imp} have been determined as a function of the Ga acceptor concentration, i.e., we find critical exponents ν for ξ and ζ for χ_{imp} using relations $\xi \propto (1 - N/N_c)^{-\nu}$ and $\chi_{imp} \propto (1 - N/N_c)^{-\zeta}$, respectively. As the results, $\nu = 1.2 \pm 0.3$ and $\zeta = 2.3 \pm 0.6$ have been obtained for the samples having $0.99 N_c < N < N_c$ to which finite temperature scaling leading to $\mu = 1.2 \pm 0.2$ was applied successfully.[21]For this region, Wegner's scaling law [14], i.e., $\mu \approx \nu \approx 0.5\zeta$ holds. Values of μ, ν, and ζ found for this region is very similar to those obtained for heavily compensated Ge:Ga.[22] A difference is that the width of N around N_c to which $\mu \approx 1$, $\nu \approx 1$, and $\zeta \approx 2$ apply is much larger for compensated Ge:Ga than that for our nominally uncompensated Ge:Ga. In other words, the finite- temperature scalable region broadens with the amount of doping compensation. Therefore, the values of $\mu \approx 1$, $\nu \approx 0.5$, and $\zeta \approx 2$, which have been found for the finite- temperature scalable region in nominally uncompensated Ge:Ga and for a wide region of N in compensated Ge:Ga, represent intrinsic values of μ, ν, and ζ for compensated semiconductors. Outside of the finite-temperature scalable region ($N < 0.99 N_c$), the values of the exponents drop to $\nu = 0.33 \pm 0.03$ and $\zeta = 0.62 \pm 0.05$.[21] $\mu = 0.50 \pm 0.04$ was found for the metallic samples having $N < 1.4 N_c$.[11] These values of μ, ν, and ζ should be regarded as the intrinsic values for completely uncompensated semiconductors.

The success of this work is due mostly to the high quality of the samples prepared for this study. Our sample fabrication technique, neutron-transmutation-doping (NTD) of isotopically enriched ^{70}Ge single crystals, leads to a completely random impurity distribution down to the atomic-level. [10,11] The situation is very different in the melt-doped samples that have been employed in most of the previous studies, [8,9,12,16,19,18] in which the spatial fluctuation of N due to dopant striations and segregation can easily be on the order of 1% across a typical sample for four-point conductivity measurements.[23] The present study differs from the previous ones since it analyzes $\sigma(N, T)$ taken from the $|N/N_c - 1| < 1\%$ region using truly homogeneous Ge:Ga samples.

2 Experimental

Chemically pure, isotopically enriched ^{70}Ge single crystals of isotopic composition ^{70}Ge(96.2 at%) and ^{72}Ge(3.8 at%) were grown and irradiated with thermal neutrons at University of Missouri Research Reactor. Upon capturing a thermal neutron ^{70}Ge becomes a ^{71}Ga acceptor via electron capture, while ^{72}Ge becomes ^{73}Ge which is stable. Therefore the crystal is doped exclusively with Ga with a compensation ratio less than 0.001. The concentration N of Ga after NTD is proportional to the neutron fluence n with the relation,

$$N(\text{cm}^{-3}) = 0.1155 \times n(\text{cm}^{-2}) \qquad (3)$$

for our irradiation condition. The following irradiation sequence was conducted in order to minimize the absolute error of N in each sample. All samples were doped together in a single irradiation run up to $N = 0.850 N_c$. At this point one sample was taken out of the reactor, then the irradiation continued with the remaining samples. Another sample was taken out after a short duration of the neutron irradiation corresponding to $\delta N \approx 0.001 N_c$, then the irradiation continued with the remaining samples. This cycle was repeated until the last sample was irradiated up to $N = 1.4 N_c$. The relative error in N between samples near N_c is estimated to be less than 0.005%. All the samples were annealed at 650°C for 10 sec in order to remove the structural defects that may have been introduced by the unavoidable flux of fast neutrons. The electrical conductivity measurements were performed in the temperature range 20mK to 1K using a dilution refrigerator.

3 Results and Discussions

Fig.1 (a) shows the finite temperature scaling plot of $\sigma(N, T)$ using eq. (2) with $x = 1/3$, $y = 2/3$, and $N_c = 1.860 \times 10^{17} \text{cm}^{-3}$. [20] $\sigma(N, T)$ taken between $T = 20$ and 750mK was used for the analysis in Fig.1. Fairly good scaling was obtained on the metallic side as expected, while scaling on the insulating side is clearly unsatisfactory. In order to find a better set of x, y, and N_c, a numerical fitting has been performed using the following non-linear equation;

$$\sigma(N, T) = T^x [a_0 + a_1 \frac{(N/N_c - 1)}{T^y} + a_2 \left(\frac{(N/N_c - 1)}{T^y} \right)^2$$
$$+ a_3 \left(\frac{(N/N_c - 1)}{T^y} \right)^3]. \qquad (4)$$

Fig.1(b) shows the result of the fitting analysis when the critical region was limited to $N = N_c \pm 0.004 N_c$, i.e., only

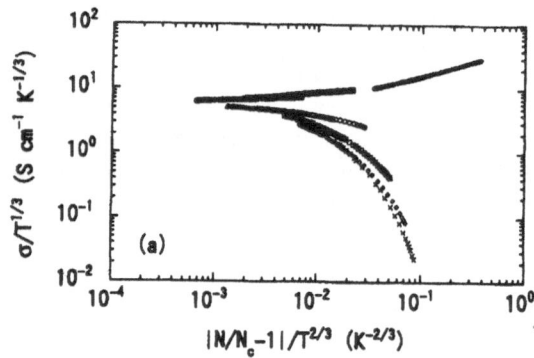

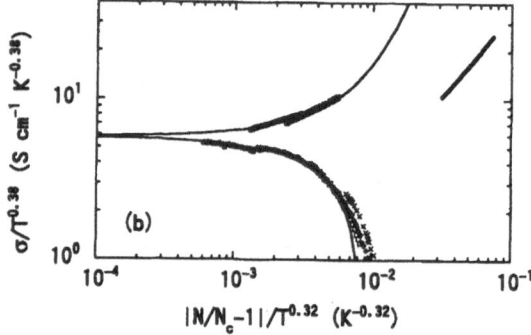

Fig. 1 Finite temperature scaling analysis of $\sigma(N, T)$ using eq. (2) with $x = 1/3$, $y = 2/3$, and $N_c = 1.860 \times 10^{17} \mathrm{cm}^{-3}$, and (b) $x = 0.38$, $y = 0.32$, and $N_c = 1.8590 \times 10^{17} \mathrm{cm}^{-3}$. The solid curve in (b) is the best fit to the data.

the data from three insulating and two metallic samples closest to the transition are fitted. Unfortunately, we could not achieve any reasonable fit when data from all the samples were included. The solid curve in Fig.1(b) is the fit with $x = 0.38$, $y = 0.32$, $N_c = 1.8590 \times 10^{17} \mathrm{cm}^{-3}$, $a_0 = 5.75$, $a_1 = 580$, $a_2 = 1.97 \times 10^4$, and $a_3 = 3.15 \times 10^6$ in eq. (4). A similar fit with a fourth-order equation lead to practically the same set of parameters with the absolute value of the fourth-order term being much smaller that that of the lower-order terms, i.e., eq. (4) of the third-order is sufficient for a good fit. The major consequence of this analysis is $\mu = x/y = 1.2 \pm 0.2$. Although the range of N we chose for scaling may appear to be too small, we emphasize again that the range is the only part that can be scaled with eq. (2).

The resistivity of insulating samples that have been included in the finite temperature analysis has been analyzed based on the variable-range-hopping theory of Efros and Shklovskii:[24]

$$\rho(T) \propto T^{-r} \exp(T_0/T)^{1/2}, \qquad (5)$$

where r and T_0 are the constants. $r = 1/3$ has been determined for our nominally uncompensated Ge:Ga.[25]

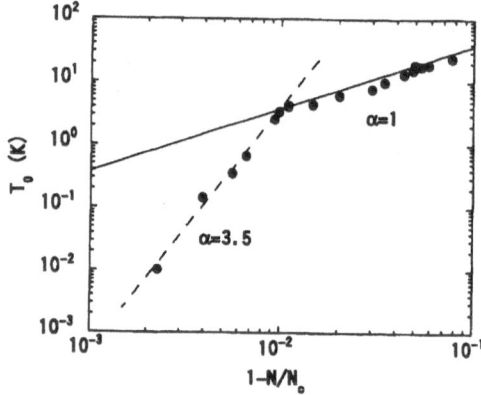

Fig. 2 T_0 as a function of $1 - N/N_c$

T_0 collapses as N approaches N_c from the insulating side:

$$k_B T_0 \approx \frac{1}{4\pi\epsilon_0} \frac{2.8e^2}{\epsilon(N)\,\xi(N)} \qquad (6)$$

in SI units, where $\epsilon(N)$ and $\xi(N)$ are the dielectric constant and the localization length, respectively, given by; $\epsilon(N) = \epsilon_0(1 - N/N_c)^{-\zeta}$ and $\xi(N) = \xi_0(1 - N/N_c)^{-\nu}$ as N approaches N_c from the insulating side. With these relations T_0 becomes

$$k_B T_0 = \frac{2.8e^2}{4\pi\epsilon_0\kappa_0\xi_0} (1 - N/N_c)^{\alpha}, \qquad (7)$$

where $\alpha = \zeta + \nu$ is the critical exponent for T_0. Fig.2 shows the values of α that have determined experimentally in our recent study.[25] Very interestingly, the value changes from 1 to 3.5 around $1 - N/N_c \approx 0.99 N_c$.

In order to obtain ζ and ν independently, we have performed magneto-transport measurements in the critical regime.[21] Fig. 3 shows the results. We find, that both ξ and χ_{imp} do not show simple dependencies on N in the range shown in Fig. 3, and that there is a sharp change of both dependencies at $N \approx 0.99 N_c$. On both sides of the change in slope, the concentration dependence of ξ and χ_{imp} are expressed very well by the scaling formula indicated by solid curves in Fig. 3.

We believe doping compensation is origin for the change in slope. Although our samples are nominally uncompensated, doping compensation of less than 0.1% may present due to residual isotopes that become n-type impurities after NTD. In addition to the doping compensation, the effect known as "self compensation" may play an important role near N_c. It is empirically known that the doping compensation affects the value of the critical exponents. Rentzsch et al. studied VRH conduction of n-type NTD Ge in the concentration range of $0.2 < N/N_c < 0.91$, and showed that T_0 vanishes as $T_0 \propto (1 - N/N_c)^{\alpha}$ with $\alpha \approx 3$ for $K = 38\%$ and 54% [22]. Since $\alpha \approx \nu + \zeta$ [Eq. (6)], we find for our NTD ^{70}Ge:Ga samples $\alpha = 3.5 \pm 0.8$ for $0.99 < N/N_c < 1$ and $\alpha = 0.95 \pm 0.08$ for $0.9 < N/N_c < 0.99$. Interestingly, $\alpha = 3.5 \pm 0.8$ agrees with $\alpha \approx 3$ found for compensated

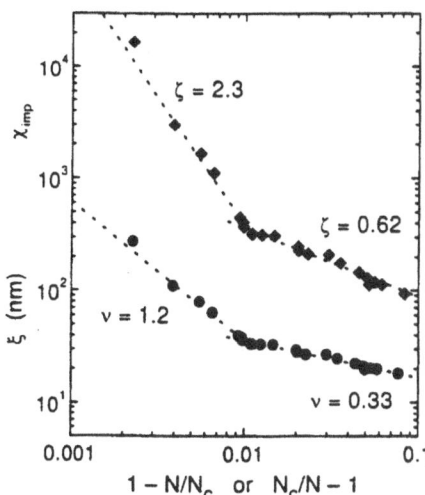

Fig. 3 Localization length ξ vs $1-N/N_c$ (lower data set) and the dielectric susceptibility χ_{imp} arising from the impurities vs $N_c/N - 1$ (upper data set).

samples. Moreover, we have seen that $\mu \approx 1$ in the same ^{70}Ge:Ga only within the very vicinity of N_c. An exponent of $\mu = 0.50 \pm 0.04$, on the other hand, holds for a wide region of N up to $1.4 N_c$ [10,11]. Again, $\mu \approx 1$ near N_c may be viewed as the effect of compensation. Therefore, it is likely that the region of N around N_c where $\nu \approx 1$ and $\mu \approx 1$ changes its width as a function of the doping compensation. In the limit of zero compensation, the part that is characterized by $\nu \approx 1$ and $\mu \approx 1$ vanishes, i.e., we propose $\nu = 0.33 \pm 0.03$, $\zeta = 0.62 \pm 0.05$, and $\mu = 0.50 \pm 0.04$ for truly uncompensated systems and that Wegner's scaling law of $\nu = \mu$ is not satisfied. In compensated systems, on the other hand, Wegner's law holds. The experiment on compensated AlGaAs:Si that showed $\zeta \approx 2$ and $\mu \approx 1$ [26] is also consistent with our interpretation.

4 Summary

Using homogeneously doped NTD Ge:Ga, we have determined directly the critical exponents for the electrical conductivity, localization length, and dielectric susceptibility for doping induced metal-insulator transition in semiconductors. We have shown that the intrinsic value of μ for uncompensated semiconductors is ≈ 0.5. The small amount of doping compensation that is unavoidably present in all nominally uncompensated samples has been responsible for $\mu \approx 1$ reported previously as an intrinsic value for uncompensated semiconductors.

Acknowledgements This work has been performed in collaboration with M. Watanabe, Y. Ootuka, and E. E. Haller. We acknowledge T. Ohtsuki, S. Katsumoto, K. Slevin, R. N. Bhatt, T. G. Castner, M. P. Sarachik, J. C. Phillips, and T. Kawarabayashi for valuable discussions. The work was supported in part by the Grant-in-Aid from the Ministry of Education, Science, Sports, and Culture.

References

1. N. F. Mott, *Metal-Insulator Transitions*, 2nd ed. (Taylor&Francis, London, 1990).
2. E. Abrahams, P. W. Anderson, D. C. Licciardello, and T. V. Ramakrishnan, Phys. Rev. Lett. **42** (1979) 673.
3. P. A. Lee and T. V. Ramakrishnan, Rev. Mod. Phys. **57** (1985) 287.
4. A. Kawabata, Prog. Theor. Phys. Suppl. **84** (1985) 16.
5. D. Belitz and T. R. Kirkpatrick, Rev. Mod. Phys. **66** (1994) 261.
6. M. P. Sarachik, in *Metal-Insulator Transitions Revisited*, edited by P. P. Edwards and C. N. Rao (Taylor&Francis, London, 1995).
7. E. Abrahams and G. Kotliar, Science **274** (1996) 1853.
8. T. F. Rosenbaum, K. Andres, G. A. Thomas, and R. N. Bhatt, Phys. Rev. Lett. **45** (1980) 1723.
9. P. F. Newman and D. F. Holocomb, Phys. Rev. B **28** (1983) 638; W. N. Shfarman, D. W. Koon, and T. G. Castner, Phys. Rev. B **40** (1989) 1216.
10. K. M. Itoh, E. E. Haller, J. W. Beeman, W. L. Hansen, J. Emes, L. A. Reichertz, E. Kreysa, T. Shutt, A. Cummings, W. Stockwell, B. Sadoulet, J. Muto, J. W. Farmer, and V. I. Ozhogin, Phys. Rev. Lett. **77** (1996) 4058.
11. M. Watanabe, Y. Ootuka, K. M. Itoh, and E. E. Haller Phys. Rev. B. **58** (1998) 9851.
12. P. Dai, Y. Zhang, and M. P. Sarachik, Phys. Rev. Lett. **66** (1991) 1914.
13. J. Chayes, L. Chayes, D. S. Fisher, and T. Spencer, Phys. Rev. Lett. **57** (1986) 2999.
14. F. Wegner, Z. Phys. B **25** (1976) 327.
15. There are theory that have attempted to explain $\mu \approx 0.5$ in uncompensated semiconductors. Example is: J. C. Phillips, J. Phys. Soc. Jpn. **67** (1999) 3346.
16. H. Stupp, M. Hornung, M. Lakner, O. Madel, and H. v. Löhneysen, Phys. Rev. Lett. **71** (1993) 2634.
17. T. F. Rosenbaum, G. A. Thomas, and M. A. Paalanen: Phys. Rev. Lett. **72** (1994) 2121.
18. S. Waffenschmidt, C. Pfleiderer, and H. v. Löhneysen, Phys. Rev. Lett.**83** (1999) 3005.
19. S. Bogdanovich, M. P. Sarachik, and R. N. Bhatt: Phys. Rev. Lett. **82** (1999) 137.
20. K. M. Itoh, M. Watanabe, Y. Ootuka, x and E. E. Haller Ann. Phys. (Leipzig) **58** (1998) 9851.
21. M. Watanabe, Y. Ootuka, K. M. Itoh, and E. E. Haller Phys. Rev. B **58** (1998) 9851.
22. R. Rentzsch, A. N. Ionov, Ch. Reich, M. Müller, B. Sandow, P. Fozooni, M. J. Lea, V. Ginodman, and I Shlimak, Phys. Stat. Sol. (b) **205** (1998) 269.
23. See, for example, F. Shimura, *Semiconductor Silicon Crystal Technology* (Academic Press, San Diego, 1988), p. 159-161.
24. B. I. Shklovskii and A. L. Efros, *Electronic Properties of Doped Semiconductors* (Springer-Verlag, Berlin, 1984).
25. K. M. Itoh, Phys. Stat. Sol. (b) **218** (2000) 211.
26. S. Katsumoto, in *Localization and Confinement of Electrons in Semiconductors*, edited by F. Kuchar, H. Heinrich, and G. Bauer (Springer-Verlag, Berlin, 1990), p. 117.

Isotope effect on the thermal conductivity of silicon

T. Ruf[1]*, R. W. Henn[1], M. Asen-Palmer[1], E. Gmelin[1], M. Cardona[1], H.-J. Pohl[2], G. G. Devyatych[3], P. G. Sennikov[3], J. Bollmann[4]

[1] Max-Planck-Institut für Festkörperforschung, Heisenbergstr. 1, 70569 Stuttgart, Germany
[2] VITCON Projectconsult GmbH, Otto-Schott-Str. 13, 07745 Jena, Germany
[3] Inst. of Chemistry of Highly-Pure Substances, Russ. Acad. of Sciences, Tropin Str. 49, 603600 Nizhny Novgorod, Russia
[4] Institut für Tieftemperaturphysik, Technische Universität Dresden, 01062 Dresden, Germany

Abstract   As an extension of our recent work, we have measured the thermal conductivity of another bulk crystal of highly enriched (99.896%) ^{28}Si for temperatures in the 80-320 K range, using a steady-state heat-flow technique. The data show same significant thermal-conductivity enhancement compared to natural silicon that we observed before. We document the quality of our material by transport data. This demonstrates that considerable problems in the materials preparation have been solved. Large quantities of high-quality isotopic silicon are now available for a wide range of applications, e.g., in microelectronics or X-ray optics.

1 Introduction

Isotope scattering is the most important mechanism that determines the maximum thermal conductivity κ of high-quality, i.e., chemically pure and dislocation-free, semiconductor crystals. The maximum of κ occurs at a temperature T_{max} separating the well-known regimes where the thermal conductivity is limited by boundary scattering ($\kappa \sim T^3$, $T \ll T_{max}$) and the umklapp region where anharmonic scattering processes dominate ($\kappa \sim T^{-1}$, $T \gg T_{max}$).

Thermal conductivity enhancements around T_{max} of about an order of magnitude have been observed in highly isotopically enriched diamond and germanium compared to crystals with the natural isotope content [1,2]. In order to study common trends and the scaling of parameters for the series of group-IV semiconductors diamond, silicon and germanium, a complete set of data is highly desirable.

Recently, large high-purity single crystals of ^{28}Si (several grams) have become available [3]. Some of them were grown by the crucible-free float-zone method. We have used such bulk material to measure isotope effects on the thermal conductivity with the conventional steady-state heat-flow technique over a wide temperature range and reported a maximum κ of about 30000 Wm^{-1}K^{-1} around T_{max} = 20 K [4]. Here we complement these studies by data from another crystal and provide additional evidence for the high chemical and electronic quality of our isotopically enriched bulk ^{28}Si.

2 Thermal conductivity

The filled circles in Fig. 1 are thermal conductivity data of a p-type, (100)-oriented ^{28}Si single crystal (labelled SI283 in Ref. [4], isotope enrichment: 99.896% ^{28}Si). The

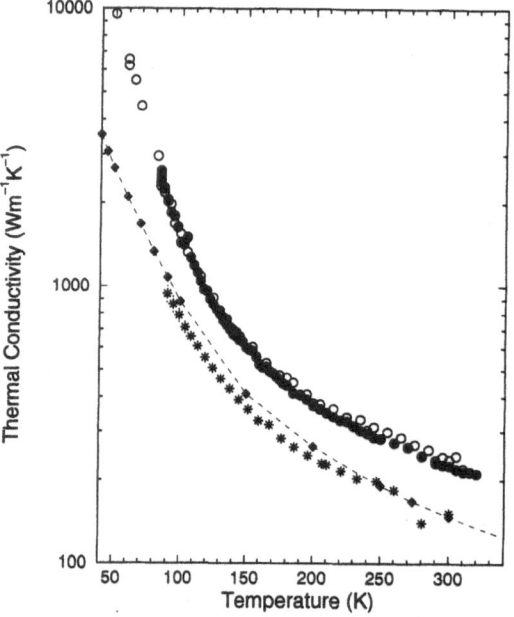

Fig. 1 Thermal conductivity data for two bulk ^{28}Si samples: filled circles: SI283, open circles: SI284 (from [4]). The asterisks (sample SIN1, from [4]) and the points connected by the dashed line (literature values, from [5]) are reference data for natural silicon.

measurements were performed on a $2.6 \times 1.4 \times 60$ mm^3 rectangular rod with polished (100)-oriented side faces. This sample was cut from the SI283 crystal shown "as grown" in Fig. 1 of Ref. [4]. The steady-state heat-flow technique was applied as described in Ref. [4], using a separation of 40.65 mm between the (inner) thermocouples across which the temperature gradient for different heating powers Q was measured. At each temperature, κ was determined from the slope of a linear regression of measurements for 5-10 different Q values. We verified in each case, that offsets due to parasitic thermovoltages in the experimental setup can be neglected. Measurements below 80 K were not possible for this large sample. They will be performed in the future on a rod with a smaller cross section.

The data for SI283 are basically identical to those for SI284, the n-type (111)-oriented crystal (isotope enrichment: 99.8588% ^{28}Si) studied in Ref. [4]. In both cases κ is significantly enhanced compared to natural Si. At high enough temperatures, the thermal conductivity does not depend on the isotope composition and the enhancement should vanish. This regime needs to be investigated in the future.

3 Transport data

Figure 2 shows Hall measurements for both SI283 (filled circles) and SI284 (open circles). At 300 K, the values for SI284 are $n = 9.8 \times 10^{14}$ cm^{-3}, $\rho = 8\,\Omega$cm, and $\mu = 825$ cm^{2}V^{-1}s^{-1}. From Fig. 2(a), we determine a donor activation energy of 34.0 meV (taking the slope around 0.02-0.03 K^{-1}) and 27.7 meV in the range above 0.035 K^{-1}. The chemical origin of these donors is presently unclear. The higher energy is close to the value for lithium. More work is required to verify this conjecture.

Figure 3(a) shows depth profiles of the free carrier density in SI283 and SI284 obtained from C-V-measurements at 80 K. We observe almost constant values of n(SI284) $= 5.0 \times 10^{14}$ cm^{-3} and p(SI283) $= 3.2 \times 10^{14}$ cm^{-3}. The difference to the results of Fig. 2(a) can be attributed to the fact that C-V-measurements are very reliable for bulk samples, while Hall data suffer from geometrical effects. DLTS spectra of both samples are shown in Fig. 3(b). No significant contribution from deep level centers within one half of the band gap is observed, thus limiting their concentration to 10^{10} cm^{-3} at most.

4 Applications

With single-isotope silicon (SISSI) ^{28}Si available in large quantities, various applications seem now worth to be

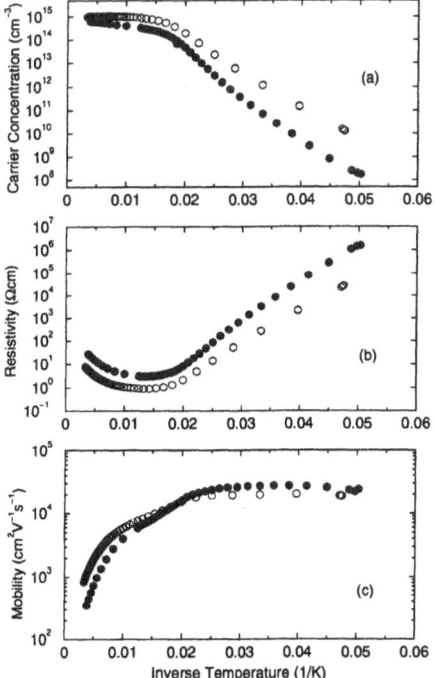

Fig. 2 (a) Carrier concentration, (b) specific resistivity, and (c) mobility vs. inverse temperature for ^{28}Si samples SI284 (open circles, n-type) and SI283 (filled circles, p-type, from [4]).

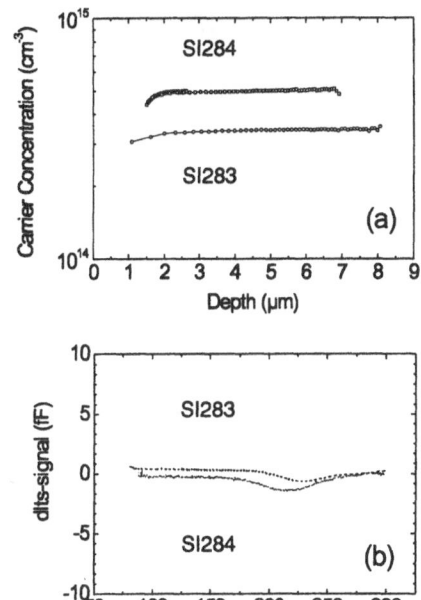

Fig. 3 (a) Carrier concentration from C-V-profiling (5MHz, 80K) and (b) DLTS spectra (rate window 500 s^{-1}, reverse bias 20 V, filling pulse 200 μs/0 V) for both ^{28}Si samples as indicated.

investigated further. In *microelectronics*, an additional SISSI layer on chips might help to dissipate heat more effectively from hot spots in the circuitry, thus allowing for faster operating speed or integration density. However, we do not expect any direct benefits of SISSI on transport properties, such as improved mobilities or scattering rates. Another area where SISSI might be very useful is high heat-load *X-ray optics*. In addition to improved energy removal at all temperatures up to more than 320 K, it should be noted that the maximum of κ in silicon around 20 K nearly coincides with a zero crossing of the thermal expansion coefficient α (around 18 K [6]). Thus the thermal slope error ($\sim \kappa/\alpha$) [7], an important figure of merit in X-ray optics, will be very small in devices like Bragg mirrors operated around this temperature.

References

[*] Corresponding author:
    ruf@kmr.mpi-stuttgart.mpg.de (T. Ruf)
1. L. Wei et al., Phys. Rev. Lett. **70**, (1993) 3764.
2. M. Asen-Palmer et al., Phys. Rev. B **56**, (1997) 9431 and references therein.
3. A. D. Bulanov et al., Crystal Research and Technology **35**, (2000) 1023.
4. T. Ruf et al., Solid State Commun. **115**, (2000) 243.
5. Y. S. Touloukian, R. W. Powell, C. Y. Ho, and P. G. Klemens, *Thermal Conductivity – Metallic Elements and Alloys*, in *Thermophysical Properties of Materials*, Vol. 1, (IFI/Plenum, New York – Washington, 1970).
6. K. G. Lyon et al., J. Appl. Physics **48**, (1977) 865.
7. A. K. Freund, J.-A. Gillet, and L. Zhang, SPIE (Proceedings: *Conference on Crystal and Multilayer Optics*) Vol. 3448, p. 362 (1998).

Axisymmetric Gunn effect

L. L. Bonilla[1], R. Escobedo[1]*, F. J. Higuera[2]

[1] Escuela Politécnica Superior, Universidad Carlos III de Madrid, Av. de la Universidad 30, 28911 Leganés, Spain.
[2] E.T.S. Ingenieros Aeronáuticos, Universidad Politécnica de Madrid, Pza. Cardenal Cisneros 3, 28040 Madrid, Spain.

Abstract   We have studied by numerical and asymptotic methods the solutions of a widely used drift-diffusion model of the Gunn effect in Corbino geometry (a circular disk with one contact on its circumference and a point contact at its center). The result is that axisymmetric pulses of the electric field are periodically shed by an inner circular cathode for a *dc* voltage bias above a certain onset ϕ_α. These waves decay during their journey to the outer anode, which they may not reach. Meanwhile the current continuously increases and then abruptly decreases when a new wave is shed, in agreement with existing experimental results of Willing and Maan[1]. Depending on the bias, more complex patterns with multiple shedding of pulses at the cathode are possible.

1 Introduction

Current self-oscillations in semiconductors with a region of negative differential resistivity in their current-field characteristic are known since J. B. Gunn's experiments on n-GaAs samples in 1963. Most studied are Gunn self-oscillations in one-dimensional spatial configurations which appear when planar contacts are placed in bulk semiconductor samples: during each period of the current oscillation, a charge dipole wave is triggered at the injecting contact, moves and is annihilated at the receiving contact. Dynamics of planar dipole waves can be surprisingly rich for systems with one-dimensional geometry: besides periodic self-oscillations, under *dc* voltage bias there may appear period doubling, frequency blocking and intermittency routes to chaos [5,6]. We present the Kroemer's model in the case of Corbino geometry and we sumarize our numerical and asymptotic results.

2 The Kroemer's model

The model is based on the idea that the electrons could transfer at higher electric fields from a high- to a low-mobility valley in the conduction band. Together with the assumption of very fast momentum relaxation, this leads to a N-shaped velocity-field relationship: $v(E) = (E + v_s E^4)/(1 + E^4)$. The electric field is radial in the axisymmetric case and it obeys the Ampère's equation:

$$\frac{\partial E}{\partial t} + v(E)\left[1 + \frac{1}{r}\frac{\partial(rE)}{\partial r}\right] - \delta\frac{\partial}{\partial r}\left[\frac{1}{r}\frac{\partial(rE)}{\partial r}\right] = \frac{J(t)}{r} \quad (1)$$

where $2\pi J(t)$ is the total current density through the cylindrical semiconductor. The problem must be solved

with the following bias and boundary conditions:

$$\frac{1}{L}\int_{r_c}^{r_a} E(r,t)\,dr = \phi, \quad (2)$$

$$E(r,t) = \rho\left(\frac{J(t)}{r} - \frac{\partial E(r,t)}{\partial t}\right) \quad \text{at} \quad r = r_c, r_a, \quad (3)$$

where $\Phi = \phi L$ is the applied voltage, $r_{c,a}$ are the radii of the inner and outer contacts respectively and $L \equiv r_a - r_c$.

3 Numerical simulations

We have solved (1)-(3) numerically with appropriate initial condition for $E(r,0)$ and $\rho = 2$, $\delta = 0.013$, $r_c = 10$ and $L = 40$ and 80, using ϕ as a control parameter. The main result is the J-ϕ characteristic, which presents the following typical features: a regime of sublinearity $(0, \phi_\alpha \approx 0.17)$, a short regime of small amplitude oscillations $(\phi_\alpha, 0.2)$ and a large regime where intervals of periodic oscillations and complicated patterns alternate $(0.2, \phi_\omega)$, where $\phi_\omega = \infty$ if $v_s = 0$ and $\phi_\omega < \infty$ if $v_s > 0$. For $\phi > \phi_\omega$ the stationary solutions are again stable.

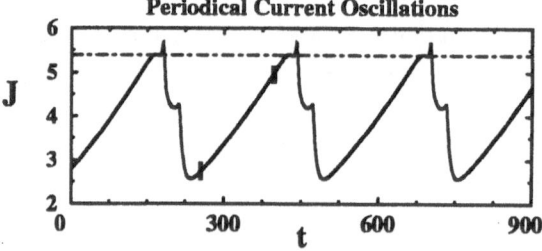

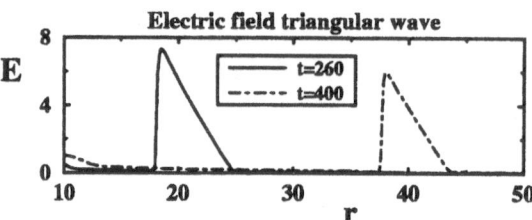

Fig. 1 During periodic oscillation the current continuously increases and then abruptly decreases when a new wave is shed; the electric field is a triangular wave which decreases during its trip towards the anode, which it may not reach.

4 Stationary states

The stationary solution of (1)-(3) is studied by solving

$$\frac{dE}{dr} = \frac{J/r - v(E)(1 + E/r)}{v(E)} \quad \text{with} \quad E(r_c) = \frac{\rho J}{r_c} \quad (4)$$

for a fixed value of J; the corresponding value of ϕ is given by (2). This study gives the critical current J_{c2} at

* *e-mail:* escobedo@math.uc3m.es

which $E(r)$ changes its shape abruptly and dipole waves may be generated. The stationary solution becomes unstable at $\phi_\alpha \approx (1/L) J_{c2} \ln |r_a/r_c|$. The J-ϕ characteristic may be evaluated asymptotically giving good agreement with numerical and experimental results. Similarly, approximations to ϕ_α and ϕ_ω can be obtained for large samples; see the neutral stability curve of Fig. 2.

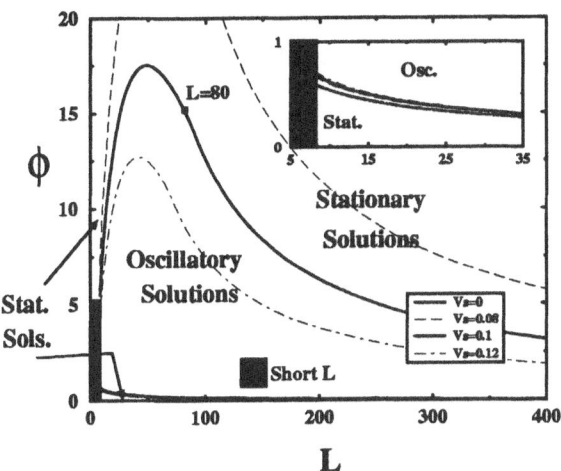

Fig. 2 Neutral stability curve: inside, the steady state is unstable, and stable outside. The lower and upper branches correspond respectively to $\phi_\alpha(L)$ and $\phi_\omega(L)$, onset and end of the oscillatory regime for a given value of L.

5 Asymptotics

Now we shall assume that $r_a - r_c = L \gg 1$ and denote $\epsilon = 1/L \ll 1$. We suppose moreover that $r_c \gg 1$ and $\phi > \phi_\alpha$; then a Gunn instability is to be expected.

5.1 Kinematics of the wave

A dipole wave is a straight triangle of height $E_+ - E_-$ and base $R_l - R_b$ which has a small size in comparison with the distance it travels: $r_w \gg r_w - R_{b,l} \gg 1$. Outside

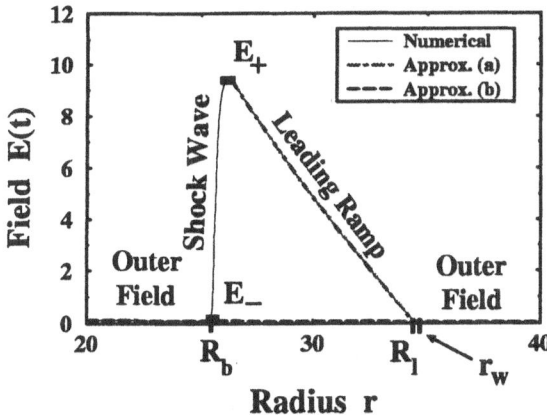

Fig. 3 Kinematics of the wave: $R_b(t)$: position of shock wave; $R_l(t)$: position of the intersection between leading front and outside field; $E_+(t)$: height of the wave; $E_-(t)$: height of outside field at $r = R_b(t)$. The parameter $r_w(t)$ is the intersection between the prolongation of the leading front and the r axis.

the wave, the electric field solves Eq.(1) with negligible space and time derivatives, so $v(E_-) = J/r$ when $J \ll r$ (see Approx. (a) in Fig. 3). The leading ramp is a region depleted of carriers: Poisson's equation gives $E = r_w - r$, so $E_+ = r_w - R_b$ (with $E_+ \gg E_-$; see (b) in Fig. 3).

5.2 Explicit expressions

Three equations determine the evolution of the three parameters of the problem, $J(t)$, $E_+(t)$ and $R_b(t)$: the equal area rule, the conservation of the triangular shape of the wave and the conservation of the area under $E(r,t)$:

$$\frac{dR_b(t)}{dt} = \frac{\pi}{4E_+(t)}, \tag{5}$$

$$\frac{dr_w(t)}{dt} = \frac{J(t)}{r_w(t)}, \tag{6}$$

and $\quad \phi = \phi_{in}(t) + \phi_{out}(t), \tag{7}$

$$= \epsilon \frac{1}{2} E_+^2(t) + \epsilon J(t) \ln(r_a/r_c).$$

This system can be studied by means of a phase plane analysis, and simplified when $|\phi/(\epsilon r_c)| > 1$, obtaining the following explicit expression for the current density:

$$J(t) \sim J_T \left(1 - \sqrt{\left(1 - \frac{J_{min}}{J_T}\right)^2 - \frac{t - t_0}{t_T}} \right), \tag{8}$$

where

$$J_T = \frac{2\phi}{3\epsilon \ln(r_a/r_c)}, \quad t_T = \frac{32\phi^2}{3\epsilon^2 \pi^2 \ln(r_a/r_c)} \tag{9}$$

and J_{min} is the initial value of $J(t)$ at $t = t_0$.

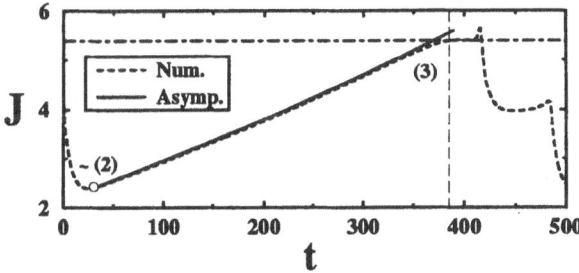

Fig. 4 Explicit expression of $J(t)$ obtained from asymptotic analysis, in good agreement with numerical simulations and qualitatively with experimental curve of Willing.

Explicit expressions for E_+ and R_b are now easy to find.

6 Conclusion

We have proposed an extension of the 1D Kroemer's model to the Corbino geometry ($1\frac{1}{2}$D), and we have found explicit expressions which are in good agreement with our numerical simulations and the existing experimental results.

References

1. B. Willing and J. C. Maan, Phys. Rev. B **49** (1994) 13995.
2. F. J. Higuera and L. L. Bonilla, Physica D **57** (1992) 161.
3. L. L. Bonilla et al., SIAM J. Appl. Math. **54** (1994).
4. L. L. Bonilla et al., SIAM J. Appl. Math. **55** (1995) n.6.
5. E. Gwinn et al., Phys. Rev. Lett. **57** (1986); **59** (1987).
6. J. Peinke et al., Phys. Lett. A **108** (1985) 407.

DC field response of hot carriers under circular polarized intense microwave fields in semiconductors

N. Ishida

Department of Physics, Faculty of Science and Engineering, Chuo University, 1-13-27 Kasuga, Bunkyo-ku, Tokyo 112-8551, Japan   e-mail: ishida@phys.chuo-u.ac.jp

Abstract   Hot carrier dynamics under intense microwave fields is investigated theoretically for the case that the dominant scattering process is optical phonon emission. When the microwave amplitude is appropriately large, an accumulated distribution of carriers in momentum space appears, and this is found to cause various peculiar DC fields response, e.g. strong nonlinearity, negative differential conductivity, and often negative response, under realistic physical conditions[1]. In the proper strength of microwave and DC electric fields, especially in the case of circular polarized microwave fields, the carrier motions are converged to some trajectories in momentum space, then, a new type of accumulated distribution of carriers in momentum space appears. This is one reason for the negative differential conductivity appeared in the drift velocity vs. DC field relation not found in the case of linear polarized microwave fields. Another structures in DC response, e.g. negative response, which is caused by the geometric property in carrier motions are also found only under circular polarized microwave fields.

1 Introduction

In many semiconductors, the carrier strongly interacts with the optical phonon. In rather pure crystals at low temperatures, the collision time of the carrier is long when the energy is less than the optical phonon energy $\hbar\omega_{op}$, while it is very short for the energy larger than $\hbar\omega_{op}$, because of the spontaneous optical phonon emission. It has long been known that a system of such carriers exhibits various characteristic behaviors under intense electric fields[2].

In previous paper [1], a possibility of an accumulated distribution of such carriers in momentum space under appropriately intense microwave electric fields, which would show various peculiar responses to DC fields applied in parallel to the microwave fields. Although the carrier accumulation in momentum space is intrinsic for such peculiarities in DC response, but in general, an accumulated distribution is broadened or dismissed as DC field strength is increased, because the carrier motion is complicated when the microwave and DC fields are applied simultaneously to a system.

As described in the following sections, a new type of carrier accumulation appears in a proper strength of circular polarized microwave electric field and DC Field, not found in the case of linear polarized microwave field.

2 Carrier Accumulation in Momentum Space

Under microwave fields, carriers make sinusoidal motion in momentum space. When the microwave amplitude is appropriately large but smaller than p_{op} (p_{op} denotes the momentum corresponding to $\hbar\omega_{op}$), carriers on most trajectories are accelerated above $\hbar\omega_{op}$ and scattered into low energy region within a period of the microwave field, while some carriers belonging to a region in momentum space do not arrive at $\hbar\omega_{op}$ and continue the free motion for their collision time. As a consequence, the carriers are accumulated in this region and move periodically in momentum space. This accumulated distribution is a dynamic population inversion and causes various peculiar effects.

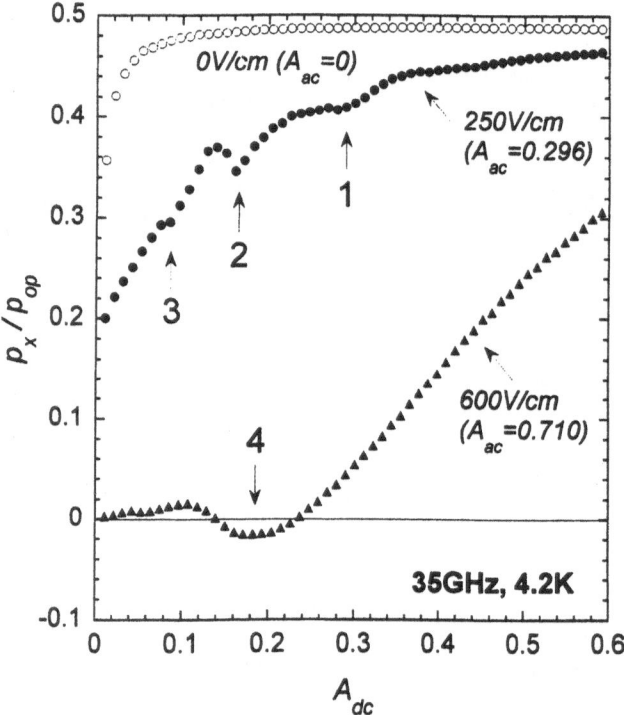

Fig. 1 Drift velocity vs. DC electric field relation calculated for heavy hole in germanium with spherical band and collision enhanced model, where p_x is drift momentum parallel to DC fields, A_{dc} and A_{ac} are DC field and microwave field strength respectively. Curves are two variations of microwave amplitude except a curve denoted as open circles. Here the collision rate for the optical phonon is larger, and for the acoustic phonon is smaller than these in real p-Ge. Structures pointed 1 and 4 in this figure are not found in the case of linear polarized microwave fields.

3 DC Responses

The response to DC electric fields (E_{dc}) in the same plane to the circular polarized microwave field exhibits many features. Figure 1 shows the computational result of the response (drift momentum p_x) to DC, where A_{dc} is expressed in dimensionless forms instead of E_{dc}($A_{dc} = e\,E_{dc}\,/\omega\,p_{op}$, where ω is a microwave frequency).

Some of the significant features are dip-like structures in the p_x–E_{dc} relation often accompanying negative differential conductivity (NDC) and negative response. Causes are classified roughly four reasons:

1. Carrier accumulation under low DC fields.
2. Discontinuity of collision time as increased in DC fields.
3. Synchronization with microwave frequency.
4. Trajectory convergence in momentum space.

Reason 1-3 are commonly predicted in the case of linear polarized microwave fields, but the last one is the only advent under circular polarized microwave fields. Figure 2, A and B are the scheme of this carrier motion in momentum space, Figure 3, A and B are the corresponding diagrams. Each two figures are the same amplitude(A_{ac}= $e\,E_0\,/\omega\,p_{op}$, where E_0 is a p-p value of the microwave field, i.e. E_{ac}=E_0sinωt) but A_{dc} are slightly different. So conditions of the convergence are severe.

Other remarks of DC response is pointed No.2 in Figure 1. This causes mainly by the reason 2 and 3 above. Carrier synchronization occurs in the condition: A_{dc}=1/(2$n\pi$), n=1,2,3,... In A_{ac}<0.5(exactly this value is larger than 0.5), discontinuity of the carrier's collision time(reason 2) is also appeared in the same conditions. So the dip-like structures in DC response are much clearer under the small value of A_{ac}. Same situations are found in the left side of point 2 in figure 1 (dip at A_{dc} =0.08=1/(4π), point 3).

4 Negative Differential Mobility and Negative Mobility

Figure 4, A and B show some trajectories of carriers in momentum space, where the conditions are corresponding to point 4 in figure 1. As A_{dc} is increased, carriers lose the opportunities to emit the optical phonons to travel long time staying in negative region. This negative contribution depends on the ratio of carriers move to negative region, and also the carrier convergence. Discontinuity points of collision time give an opportunity to create negative mobility so that the response in low DC field has a sharp dip. Only the carrier convergence is not all reason to be negative.

References

1. N. Ishida and T. Kurosawa, J. Phys. Soc. Jpn 64,(1995) 2996
2. For Example, S. Komiyama, T. Kurosawa, and T. Masumi, Chapter 6, in *Hot Electron Transport in Semiconductors.*, edited by L. Reggiani (Springer-Verlag, Berlin 1985)

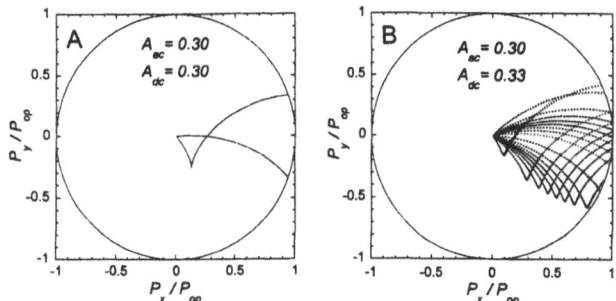

Fig. 2 Carrier motions in momentum space. In each figure microwave amplitudes are same, but A_{dc} is slightly different. A large circle in these figures denotes the equivalent energy surface in momentum space corresponding to $\hbar\omega_{op}$. Here is no scattering except the optical phonon emissions with complete inelastic collision.

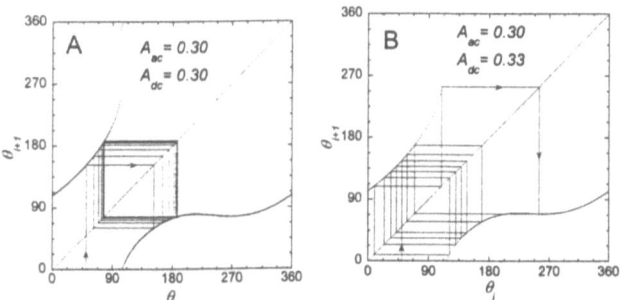

Fig. 3 Diagrams of carrier convergence corresponding to Fig. 2. Symbol θ=ωt is phase angle of microwave, and suffix i and i+1 denote the occurrence of collision, that is, the collision cycle becomes a function of phase angle. Figures are started with θ=50 degrees for example. In A, carrier converges on the angle 77.0 and 187.5 degrees, but in B, trajectories are spread.

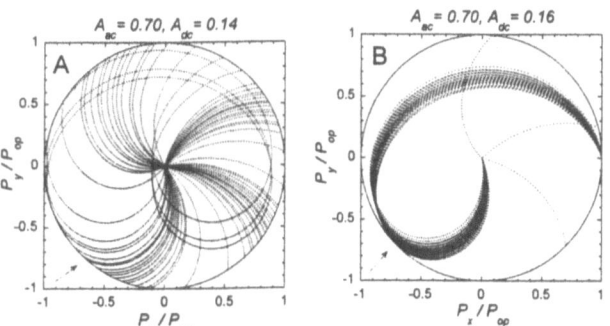

Fig. 4 Carrier motion in momentum space. From A to B, A_{dc} is slightly increased. An arrow pointed in A shows the opportunities to emit the optical phonons, in B, these are missing, but many trajectories stay in the negative region (p_x<0) instead.

Transient photoconductivity of pure CuO

H. Yamaguchi[1,2], **T. Ito**[2], **T. Masumi**[3,2]

[1] Department of Electronics and Information Systems, Akita Prefectural University, 84-4 Tsuchiya-Ebinokuchi, Honjo, Akita 015-0055, Japan e-mail: yamaguchi@akita-pu.ac.jp

[2] Department of Electronic Engineering, Gunma University, 1-5-1 Tenjincho, Kiryu, Gunma 376-8515, Japan

[3] National Research Institute for Metals, 1-2-1 Sengen, Tsukuba, Ibaraki 305-0047, Japan

Abstract We have studied transient photoconductivity, transient Hall current, magnetoresistance effect and cyclotron resonance absorption in CuO at low temperatures. In this report, we provide direct evidence of the existence of photoinduced large polarons in highly purified CuO. We also present a value of the effective mass of large polarons in CuO.

1 Introduction

The crystal structure of CuO is monoclinic but a rather simple with Cu-O network. Therefore, CuO has been regarded as a basic host insulator of high-T_C superconductors. Recent experimental and theoretical researches have shown that chemically doped carriers in CuO are self-localized small polarons [1] like as carriers doped in high-T_C superconductive compounds. [2] However, for transition metal oxides, it is difficult to conclude what kinds of mechanism are dominant for carrier transport phenomena. We have examined motion of photocarriers in purified CuO at low temperatures by measuring the Hall mobility and the magnetoresistance mobility. Furthermore, microwave absorption in cyclotron resonance of photocarriers at 35 GHz and at 4.2 K was also successfully observed. Here, we provide direct evidence of photoinduced large polaron conduction in pure CuO.

2 Experimental

CuO crystals were prepared by the FZ method. [3] Crystal surfaces were polished mechanically and etched chemically. Typical grain size of crystals was 500 μm to be free from grain boundary effect.

Pulsed photoconductivity, Hall current, magnetoresistance effect and cyclotron resonance absorption measurement was carried out by a fast pulse technique. For detailed experimental procedures, see reference. [4]

3 Results and Discussion

Figure 1 displays temperature dependence of photoconductivity $Q_x(T)$ at various excitation energy E_{ex}. While $Q_x(T)$ at E_{ex}=2.19eV increases with decreasing T, $Q_x(T)$ at E_{ex}=2.01eV decreases with decreasing T below 100K. At E_{ex}=1.53eV, temperature dependence of $Q_x(T)$ has a dip structure

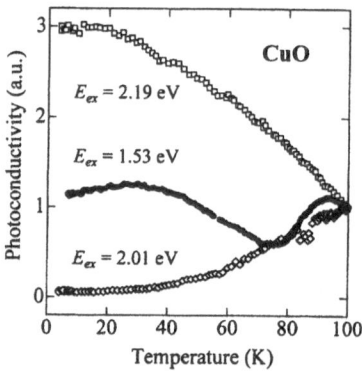

Fig. 1 Typical temperature dependence of transient photoconductivity $Q_x(T)$ in CuO. Values normalized at T=100K.

near 80K. In previous work, we reported temperature dependence of optical absorption spectra $\kappa(T)$ of CuO. [5] It is difficult to explain temperature dependence of $Q_x(T)$ by temperature dependence of $\kappa(T)$. We may suppose that lifetime $\tau_t(T)$ depends weakly on T on the basis of photocarrier number $n(T)$ experimentally observed. It is plausible that transient photoconductivity $Q_x(T)$ observed in present work reflects initial states at photoexcitation in the Cu(3d)O(2p) hybridized valence band. Namely, the electric conduction, measured in dark, due to thermal hopping conduction via in-gap trapping levels is negligible

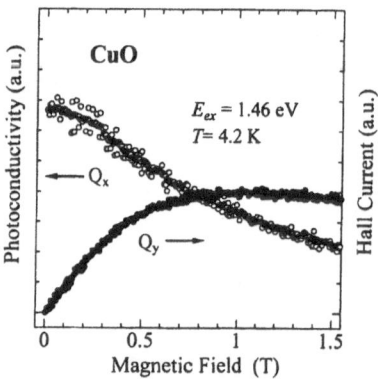

Fig. 2 Magnetic field dependence of Hall current $Q_y(B)$ and photoconductivity $Q_x(B)$ at T=4.2K and E_{ex}=1.46eV.

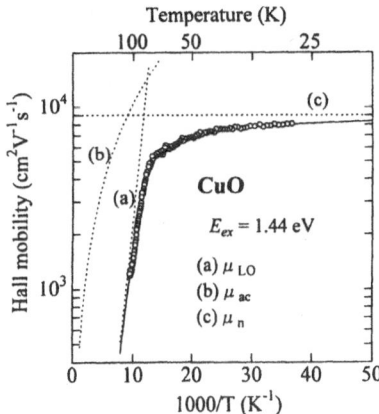

Fig. 3 Temperature dependence of Hall mobility of photoinduced holes in CuO. [6]

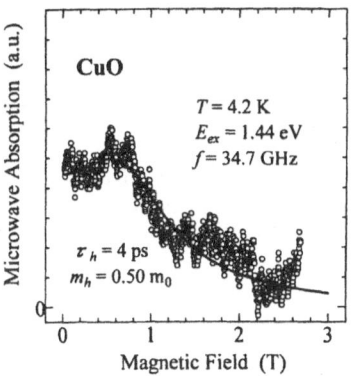

Fig. 4 Microwave absorption in cyclotron resonance of photocarriers in CuO at T=4.2K, E_{ex}=1.44eV and f=35GHz. Solid line is the best fitting curve of eq.(1) to the data.

here. We speculate that $Q_x(T)$ at E_{ex}=2.01eV is attributed to localized d levels in the Cu(3d)O(2p) hybridized valence band. On the other hand, $Q_x(T)$ at E_{ex}=2.19eV and at E_{ex}=1.53eV below 70K can be attributed to the p state. Figure 2 shows magnetic field dependence of Hall current $Q_y(B)$ and photoconductivity $Q_x(B)$ at T=4.2K at E_{ex}=1.46eV. Since the sign of Q_y / Q_x is positive, dominant carriers turn out to be positive holes. From these results of the positive magnetoresistance effect and transient Hall effect, both values of the magnetoresistance mobility μ_M and of the Hall mobility μ_H of photo-induced carriers are estimated to be 1×10^4cm^{2}V^{-1}s^{-1}at 4.2K. Figure 3 denotes temperature dependence of $\mu_H(T)$ at E_{ex}=1.44eV. The value of $\mu_H(T)$ increases with decreasing T below 100K and reaches 8×10^3cm^{2}V^{-1}s^{-1} at 25K. Temperature dependence of $\mu_H(T)$ is different from that for Cu$_2$O. Neither thermal hopping conduction model nor tunneling conduction model will account for this experimental result. $\mu_H(T)$ can be explained by the large polaron model. The polaron mass and the Fröhlich coupling constant are estimated to be 0.52m$_0$ and 1.3, respectively. The relaxation time τ due to scattering is estimated to be 2.4ps at 25K. [6] Figure 4 shows microwave absorption in cyclotron resonance of photocarriers in CuO at T=4.2K, E_{ex}=1.44eV and f=35GHz. In the standard case that the band mass of carrier is isotropic and uniform over its energy surface, the microwave absorption is represented by

$$P = \frac{1}{2}\sigma_0 E_\omega^2 \frac{1+(\omega_c^2 + \omega^2)\tau^2}{[1+(\omega_c^2 - \omega^2)\tau^2]^2 + 4\omega^2\tau^2}, \quad (1)$$

where E_ω, σ_0 and ω_0 denote the amplitude of microwave field, the DC conductivity and the cyclotron frequency, respectively. By fitting the line shape of the cyclotron resonance absorption in Fig.4 to Eq. 1, the polaron mass and scattering time at

4.2K are estimated to be 0.50m$_0$ and 4ps, respectively. Here, we should note that there exists no Cu$_2$O in the sample. [3] Mobility of photocarriers is calculated to be 1.4×10^4cm^{2}V^{-1}s^{-1}. These experimental results measured by various methods are consistent with each other. This indicates that it is possible to create the photoinduced large polarons in CuO.

4 Conclusion

Previously, small polaron conduction in CuO in dark has been reported. On the other hand, we provide direct evidence of photoinduced large polaron conduction in purified CuO at low temperatures. By investigating Hall current, magnetoresistance effect and cyclotron resonance absorption, the values of the polaron mass and the mobility of photocarriers at 4.2K are estimated to be about 0.5m$_0$ and 1×10^4cm^{2}V^{-1}s^{-1}, respectively. Here, besides small polaron conduction in dark, we propose a possibility of the large polaron conduction with photoinduced holes in CuO.

References

1. F. P. Koffyberg and F. A. Benko, J. Appl. Phys. **53**, (1982) 1173.
2. A. S. Alexandrov and Sir Nevill Mott, *POLARONS AND BIPOLARONS* (World Scientific, 1995)
3. T. Ito, H. Yamaguchi, K. Okabe, and T. Masumi, J. Mater. Sci. **33**, (1998) 3555.
4. T. Masumi, Phase Transit. **51**, (1994) 127.
5. T. Masumi, H. Yamaguchi, T. Ito, and H. Shimoyama, J. Phys. Soc. Jpn. **67**, (1998) 67.
6. H. Yamaguchi, T. Ito, and T. Masumi, J. Phys. Soc. Jpn. **67**, (1998) 1102.

A microscopic simulation of the evolution of local perturbations

P. Gaubert[1*], L. Varani[1], J.C. Vaissière[1], J.P. Nougier[1], E. Starikov[2], P. Shiktorov[2], and V. Gruzhinskis[2]

[1] Centre d'Electronique et de Micro-Optoélectronique de Montpelleir (CNRS UMR 5507), Université Montpellier II, Place E. Bataillon, 34095 Montpellier Cedex 5, France
[2] Semiconductor Physics Institute, Goshtauto 11, 2600 Vilnius, Lithuania

Abstract We report the results of simulations of electronic noise in submicronic p^+pp^+ Si structures using a new approach with a microscopic simulator developed in our laboratory. With respect to the standard impedance field method, this new approach takes into account microscopic effects such as the coupling between the noise sources which are important in deep submicron devices. We will see that to compute electronic noise, we need the knowledge of a new noise source and of a new propagator called the generalized impedance field. To compute the propagator we must study the response of the device in time and space when it is submitted to a local perturbation of velocity or energy.

1 Introduction

The classical impedance field method, developed first par Shockley [1], is one of the most powerful techniques for calculating electronic noise in devices. This method requires the knowledge of the propagator: the so-called impedance field which describes the evolution of a perturbation of electric field in point **r**, produced by a current perturbation introduced at point **r'**. It also requires the knowledge of the local noise source. The application of this method is particularly difficult in deep submicron devices, since space correlations of the carriers velocities [2] have a significant effect, although they can be neglected in longer devices. To overcome this difficulty, we have developed a new technique, called the generalized impedance field method (GIFM) [3], according to which diffusion noise is originated by the fluctuations of carriers acceleration due to scattering events. The GIFM needs the knowledge of two propagators describing the effect at point **r** respectively of a velocity perturbation and of an energy perturbation introduced at point **r'**. Then, we need also new noises sources. Due to the fact that the fluctuations of the carriers accelerations are local, instantaneous and not correlated, the noise sources are frequency independent and without space correlations which are taken into account only in the generalized impedance fields (GIF).

In this paper, we present a new microscopic method for determining the evolution of these perturbations, using the scattered packet method (SPM)[4].

The SPM follows the spatio-temporal evolution of the carrier population in different cells of the phase space us-

* *Present address:* Department of Electronic Engineering, Osaka University Suita City, Osaka 565-0871, Japan

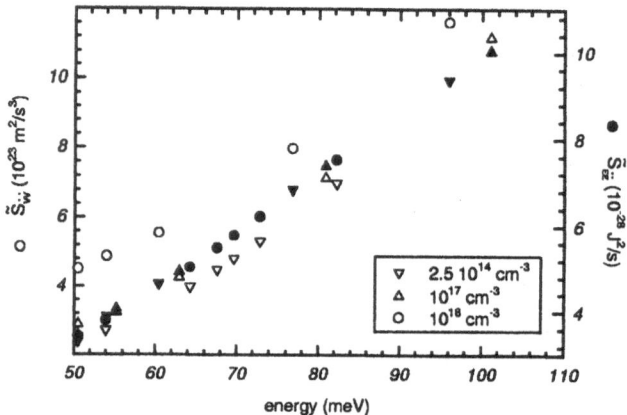

Fig. 1 Local noise sources $\tilde{S}_{\dot{v}\dot{v}}$ (empty symbols) associated with velocity and $\tilde{S}_{\dot{\varepsilon}\dot{\varepsilon}}$ (full symbols) associated with energy versus energy in p-Si bulk for three doping levels: $2.5 \times 10^{14} cm^{-3}$, $10^{17} cm^{-3}$ and $10^{18} cm^{-3}$.

ing an evolution operator. This operator (in particular, its collision part) can be employed to obtain the local noises sources corresponding to the acceleration fluctuations of velocity and energy[5]. To calculate the generalized impedance or admittance fields, it is necessary to follow the spatio-temporal evolution of a local perturbation of velocity or energy introduced initially without modifying the others quantities. Starting from a stationary state and using the constant voltage or current operation mode, the perturbation is introduced modifying the local carrier distribution function. As a consequence local quantities like velocity, energy, electric field, current density and carrier concentration are perturbated and evolve until coming back to the stationary state.

2 Results

The noises sources used with the GIF are independent of the frequency and depend only of the carrier mean energy in the bulk material [6]. Then, they can be tabulated versus the mean energy of carriers and the doping concentration. Figure 1 shows the behaviour of the two main noises sources in the p-Si bulk material. The source associated with the velocity of carriers $\tilde{S}_{\dot{v}\dot{v}}$ increases with the energy. For a given energy, this source is greater for high doping levels. This can be explained by the role of the collisions with impurities. The noise source associated with energy $\tilde{S}_{\dot{\varepsilon}\dot{\varepsilon}}$ increases with the energy and has a quasi-linear behaviour independently of the doping.

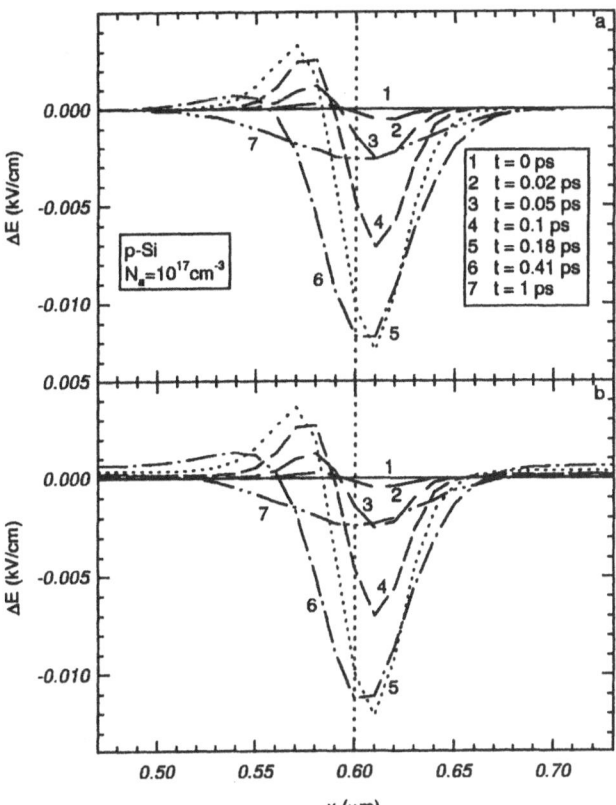

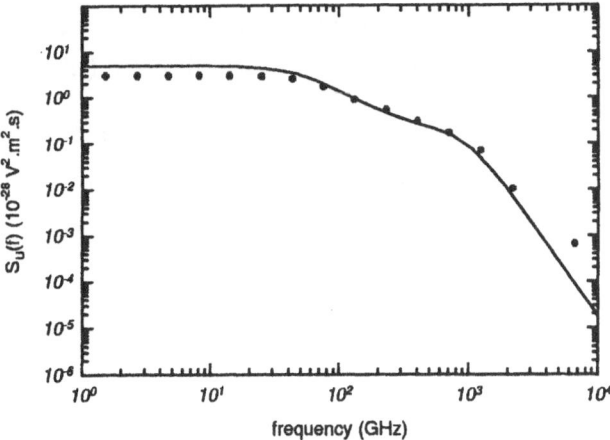

Fig. 3 Spectral density in tension in a Si p^+pp^+ diode 10^{17}-10^{16}-10^{17} cm^{-3} and 0.3-0.4-0.3 μm with a polarisation of 0.6 V. Full line: present method; symbols: results of the stochastic SPM simulation.

age spectral density of a p^+pp^+ Si diode calculated by the new method. Results are compared with those obtained using the GIFM and a hydrodynamic simulator and those obtained using the stochastic SPM developed by Aboubacar [7]

3 Conclusion

Using the SPM we studied the spatio-temporal evolution of different quantities such as the energy, velocity, electric field, current density, carrier concentration and distribution function in the frame of the GIFM. We have also calculated the noises sources associated with this new approach and the electronic noise in small devices.

Acknowledgement This work has been performed with the support of the high-level DRB4/MDL/no99-30 of the French Ministère de l'Education Nationale, de la Recherche et de l'Industrie and the French-Lithuanian bilateral Cooperation no 5380 of CNRS.

References

1. W. Shockley, J.A. Copeland and R.P. James, (Ed. P.O. Lowdin Academic, New York, 1966)
2. J.P. Nougier, J.C. Vaissière and C. Gontrand, Phys. Rev. Letters **51**, (1983) 513.
3. P. Shiktorov, E. Starikov, V. Gruzhinskis, L. Reggiani, T. Gonzalez, J. Mateos, D. Pardo and L. Varani, Phys. Rev. B **57**, (1998) 11866.
4. J.P. Nougier, L. Hlou, P. Houlet, J.C. Vaissière and L. Varani, *3rd Workshop on Computational Electronics* (Ed. M. Goodnick, Oregon State, 1994) 15.
5. P. Gaubert, PhD Dissertation (University of Montpellier II, 1999), available upon request.
6. P. Gaubert, L. Varani, J.C. Vaissière, J.P. Nougier, E. Starikov, P. Shiktorov, V. Gruzhinskis, Proceedings of the 7th International Workshop on Computational Electronics (Glasgow, may 2000), to be published.
7. M. Aboubacar, P. Houlet, J. P. Nougier, L. Varani et J. C. Vaissière, Lith. J. Phys. **36**, 583 (1996).

Fig. 2 Response of the electric field for a perturbation of energy applied at $t=0$ ps in a constant current operation mode (a) and in a constant voltage mode (b) for different times in p-Si.

The cross-correlated noise source terms have a negligeable contribution to the total noise and, in general, they can be neglected.

2.1 Local perturbations

We present on the Fig.2(a) the spatio-temporal behaviour of the electric field when an initial perturbation of the energy is applied in a resistance p-Si in a constant current operation mode. The response of this local quantity follows a complicated behaviour. We can verify that the electric field is not perturbated at $t=0$ ps. It becomes positive or negative depending of the observation point, reaches a maximum and comes back to the stationary state. We also applied the same perturbation in constant voltage operation mode. The evolution of the electric field is shown for different times in the device in Fig.2(b). The behaviour of the electric field around the position where we applied the initial perturbation is not exactly the same as in Fig. 2(a). As a matter of fact, in Fig. 2(b), we have a nonlocal distribution of the electric field in the device compount to have the integral of the electric field independant in time.

2.2 Electronic noise

The Fourier transform of the response function of the electric field gives the generalized impedance field and then the electronic noise. Figure 3 presents the volt-

Dispersive energy transport, relaxation current and Einstein relationship in band tails

O. Bleibaum[1], H. Böttger[1], V. V. Bryksin[2], A. N. Samukhin[2]

[1] Institut für Theoretische Physik, Otto-von-Guericke Universität Magdeburg, PF 4120, 39016 Magdeburg, Germany e-mail: olaf.bleibaum@physik.uni-magdeburg.de

[2] A. F. Ioffe Physico-Technical Institute, Politekhnicheskaya 26, 194021 St. Petersburg, Russia

Abstract　We study relaxation of photoexcited charge carriers hopping between localized tail states at zero temperature. We argue that due to disorder both particle and energy transport are dispersive, and we discuss the impact of the dispersion on the energy distribution function and the relaxation current for model densities of states (DOS).

1 Introduction

In recent years much attention has been devoted to the study of relaxation processes of non-equilibrium charge carriers in strongly localized systems, where transport proceeds via phonon-assisted hopping, as with photoexcited charge carriers in band tails (see e.g. [1]) and Anderson insulators (see e.g. [2]). In such systems particularly long relaxation rates are observed.

The theoretical investigation of such processes is difficult, since the system is both strongly disordered and in a transient state. Since furthermore, one is interested in time scales which are large compared to the time needed for a single hop, one has to take into account that the transport coefficients depend strongly on frequency, even for very low frequencies. In most problems both spatial and energetic disorder exist. Already due to spatial disorder particle diffusion is dispersive. Due to energetic disorder every transition is inelastic. At zero temperature every jump is accompanied by a transfer of energy from the electron to the phonon system. This energy flow is also affected by disorder, so that energy transport is also dispersive.

To get some analytical insight into the problem of dispersive energy transport we investigate the relaxation of photoexcited charge carriers via phonon-assisted hopping between localized states in model DOS at zero temperature. We assume that the number of charge carriers excited is small and that they are far apart from the Fermi energy. In this case the problem simplifies considerably, since Fermi-correlation is negligible. Transport can be modeled by the simple rate equation

$$\frac{d\rho_m}{dt} = \sum_n (\rho_n W_{nm} - \rho_m W_{mn}), \qquad (1)$$

where ρ_m is the number of particles at site m, and W_{nm} is the transition probability for a hop from site n to site m. Since the temperature is zero only hops from sites of higher energy to sites of lower energy are allowed. Furthermore, we restrict the consideration to systems with weak electron-phonon coupling. For such systems

the transition probability is given by

$$W_{mn} = \theta(V_m - V_n)\theta(\omega - V_m + V_n)\nu \exp(-2\alpha|R_{mn}|). (2)$$

Here ω is the upper bound for the energy transfered in one hop, ν is the attempt to escape frequency, α^{-1} is the localization length, $\theta(x)$ is the step function, R_m is the position vector of site m, and $R_{mn} = R_m - R_n$. The energy $V_m = \epsilon_m - e(ER_m)$ of an electron at site m depends on the postion of the site in the constant electric field E, and on ϵ_m, the site energy in the absence of the field. Both R_m and ϵ_m are random quantities in the problem of interest.

Note that, the second step function in Eq.(2) takes into account that for weak electron-phonon coupling only a finite amount of energy can be transfered to the phonon system in one hop. Since the electron-phonon coupling constant tends to zero for phonons with wave vector $q > 2\alpha$ high energy phonons are ineffective. Consequently, ω is smaller as the Debye energy by a factor $2\alpha a$, where a is the lattice constant of the host material. If furthermore, the localized state is an impurity state, the Debye screening length is large compared to the localization length, since otherwise the localized state would be unstable. In this case the electron-phonon coupling constant already rapidly tends to zero for phonons with wave vector $q > \lambda$, where λ^{-1} is the Debye screening length, so that we expect that in many materials ω can be considered as the smallest energy scale in the problem.

2 The diffusion propagator

Suppose that the initial particle distribution is given by $n_0(R, \epsilon) = \delta(R)\delta(\epsilon - \epsilon_0)$. The evolution of the particle distribution is determined by the diffusion propagator F, according to

$$n(R, \epsilon|t) = \int dR' d\epsilon' n_0(R', \epsilon') F(R' - R; \epsilon', \epsilon|t). \qquad (3)$$

To calculate F we use our effective-medium formalism [3]. In this formalism Eq.(1) is first transformed into the continuous representation. Thereafter, the configuration average is calculated diagrammatically. To this end, we assume that the sites are distributed homogeneously in space. The distribution function of site energies is assumed to be given by an arbitrary distribution function. The application of the averaging procedure results in a coupled system of integral equations, which is solved using the effective-medium approximation [4], and simplified further using the quasi-elastic approximation. The introduction of the latter approximation relies on the notion that ω is the smallest energy scale in the problem. Using this procedure we obtain the following equation for the diffusion propagator in the hydrodynamic limit:

$$sF(\mathbf{k}|\epsilon',\epsilon) = \delta(\epsilon'-\epsilon) - D(\epsilon,s)k^2 F(\mathbf{k}|\epsilon',\epsilon)$$

$$-i(\mathbf{Ek})u(\epsilon,s)F(\mathbf{k}|\epsilon',\epsilon)+\frac{\partial}{\partial\epsilon}(F(\mathbf{k}|\epsilon',\epsilon)v(\epsilon,s)). \quad (4)$$

Here s is the variable which corresponds to a Laplace transformation with respect to time. The transport coefficients $D(\epsilon,s)$, $u(\epsilon,s)$ and $v(\epsilon,s)$ in Eq.(4) are to be identified with the spectral diffusion coefficient, the spectral mobility and the energy relaxation rate, respectively. Explicit expressions for the diffusion coefficient and the energy relaxation rate can be found in Ref.[4], where also the range of applicability of Eq.(4) is discussed in detail.

3 Dispersive energy transport

In order to discuss the motion of a particle packet in energy space we focus on the energy distribution function $F(0|\epsilon_0,\epsilon;s)$. Since temperature is zero there is no energy diffusion, which is reflected in the absence of the second derivate with respect to energy in Eq.(4). Consequently, in the absence of dispersion the energy distribution function is a delta function, centered at the instantaneous position of the particle in energy space. It is the strong dependence of v on s, even for small s, that causes the spreading of the distribution function. We have investigated this point further for two DOS. We find that the motion of the packet in energy space is affected by two competing tendencies, if the DOS not decreases with increasing energy. On the one one hand, there is a packet spreading due to dispersive transport. On the other hand, there is a packet narrowing, if the DOS increases with increasing energy. It is the interplay between these two tendencies that determines the overall evolution. If the DOS is constant the second mechanism is absent. Then the distribution evolves into a Gaussian packet, moving in energy space with constant velocity and mean squared deviation

$$\sigma^2(t) =<(\epsilon-<\epsilon>(t))^2 >\propto t. \quad (5)$$

Here $<\epsilon>(t)$ is the mean energy of the distribution. If the DOS increases exponentially, according to $N(\epsilon) = N_0\exp(3\epsilon/\Delta)$, the competition between distribution narrowing and distribution widening results in an interesting balance. The packet becomes well localized in energy space. Its velocity

$$\frac{d<\epsilon>(t)}{dt} = -\frac{\Delta^2}{\omega}\frac{1}{\ln(\omega\nu t/\Delta)}\frac{1}{\nu t} \quad (6)$$

strongly slows down with time. Its dispersion

$$\sigma^2(t) = \Delta\omega \quad (7)$$

becomes time independent. Consequently, the motion of the packet in energy space is soliton-like. These results hold provided the variation of the relaxation rate $\delta v(\epsilon,0)$ over the distribution is small compared to the relaxation rate itself. However, since the particles lower their velocity when sinking down, this criterion is violated at a

certain instant of time. The variation of the relaxation rate over the distribution, being small compared to the relaxation rate itself in the first stage of the evolution, is getting large. When this happens the packet undergoes a restructuring to another stable form. The packet, being of Gaussian form before, is getting non-Gaussian with a width lower than before by a purely numerical factor. Then all moments agree with those of the distribution function $p(x) = -d/dx \exp(-x^2/2)$, after proper scaling. The latter result remains valid until the conditions for the applicability of the quasi-elastic approximation break down.

4 Einstein relationship and relaxation current

In order to derive the Einstein relationship we have restricted the consideration to small electric fields, which satisfy $|eER_c(\epsilon)/\omega| \ll 1$, where $R_c(\epsilon)$ is the critical hopping length at energy ϵ [4]. Using this restriction we obtain

$$u(\epsilon,s) = 2eD(\epsilon,s)\frac{d\ln N(\epsilon)}{d\epsilon}. \quad (8)$$

Note that, the mobility depends on the first derivative of the DOS. Consequently, for a constant DOS the relaxation current is zero. The relaxation current is positive if the DOS increases with increasing energy, and negative if the DOS decreases with increasing energy. This result, which on the first glance looks rather strange, is immediately obvious if the character of the transitions is taken into account: the charge carrier jumps into that direction where most sites are. If the DOS is constant no prefered direction is produced.

If we use Eq.(4) and the Einstein relationship (8) to calculate the relaxation current for an exponential DOS in saddle-point approximation we obtain

$$\mathbf{j}(t) \propto \frac{e^2 n\nu}{(2\alpha)^2\Delta}\ln(\frac{\omega\nu t}{\Delta})\frac{\Delta}{\omega\nu t}\mathbf{E}, \quad (9)$$

where n is the charge carrier density. The saddle-point approximation holds provided $\nu t \ll \sqrt{\Delta/\omega}\exp(\sqrt{\Delta/\omega})$. Note, that the quasi-elastic approximation applied holds only provided $\omega/\Delta \ll 1$. The result (9) can also be obtained if dispersion is ignored, and the $s \to 0$ limit of the transport coefficients is used. In this way the same result was also obtained in Ref.[1] for a model in which jumps are effectively of depth Δ. In the opposite limit, that is for $\nu t \gg \sqrt{\Delta/\omega}\exp(\sqrt{\Delta/\omega})$, dispersion becomes extremely important. Unfortunately, no analytical approximations could be obtained for this range so far.

References

1. D. Monroe, *Hopping Transport in Solids*, edited by M. Pollak and B.I. Shklovskii (North Holand, Amsterdam 1991) p. 49.
2. Z. Ovadyahu and M. Pollak, Phys. Rev. Lett. **79**, (1997) 459.
3. O. Bleibaum, H. Böttger, V. V. Bryksin, Phys. Rev. B **54**, (1996) 5444.
4. O. Bleibaum, H. Böttger, V. V. Bryksin, A. N. Samukhin, cond-mat/003408.

Angle-resolved tunneling spectroscopy of Si conduction band using bonded Si(111) wafer pair

A. Kazama[1], H. Bando[2], Y. Miyahara[3], H. Enomoto[4], H. Ozaki[1]

[1] Department of Electrical, Electronic and Computer Engineering, Waseda University, 3-4-1 Okubo, Shinjuku-ku, Tokyo 169-8555, Japan e-mail: ozaki@ozaki.elec.waseda.ac.jp
[2] Academic Frontier Promotion Center, Osaka Electro-Communication University, Neyagawa City, Osaka 572-8530, Japan
[3] Department de Microtechnique, Ecole Polytechnique Federale de Lausanne, CH-1015, Lausanne, Switzerland
[4] Department of Solid-State Electronics, Osaka Electro-Communication University, Neyagawa City, Osaka 572-8530, Japan

Abstract The angle-resolved tunneling spectroscopy has been performed on the conduction band of Si. Heavily doped n-Si(111) wafer pairs were bonded to form the SIS tunnnel junctions with various in-plane crystal-axis angles between the two crystals. The angle dependence of tunneling originates from the condition of the conservation of transverse component of momentum, together with the energy conservation, of the tunneling Bloch electrons. From the angle dependence of the tunneling characteristics, the conduction band structure of Si was derived along the orbit in the k-space, whose projection on the (111) plane forms a circle centered at $k=0$ and passes the bottom of the conduction band.

1 Introduction

Throughout the extensive applications of electron tunneling to device functions and spectroscopies of electronic states, the energy conservation through the tunneling has been exclusively taken into consideration as the selection rule of the tunneling. The electronic structure of a crystal has a symmetry which yields a degeneracy and an anisotropy in the electronic states. In order to investigate such an anisotropic nature of electronic states by utilizing tunneling, another selection rule is needed to be taken into consideration.

In the electron tunneling between the Bloch states, the transverse component of the wave vector is conserved, and thus, a specific anisotropy is expected to be observed in the tunneling characteristics [1,2]. Previously, we observed the anisotropy of the CDW energy gap in 2D materials 1T-TaS$_2$ [3], by the Angle-Resolved Tunneling Spectroscopy (ARTS).

In this study, we employ the ARTS and investigate a portion of conduction-band structure of Si.

2 Angle-Resolved Tunneling Spectroscopy

We consider the electron tunneling between the (111) surfaces of heavily doped n-Si. The angle of the corresponding in-plane crystal-axes between the two Si is θ, and it is variable. Fig.1(a) shows the Fermi surfaces of heavily doped n-Si (small six ellipsoids), projected on the (111) plane parallel to the surface of the tunnel barrier. The size of the Fermi surface in the figure approximately corresponds to that with carrier concentration 1×10^{19}cm^{-3}. As the size of the Fermi surface is much

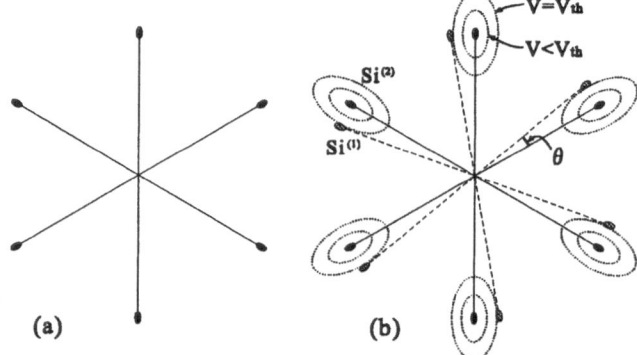

Fig. 1 (a)The Fermi surfaces of heavily doped n-Si projected on the (111) plane. (b)The Fermi surfaces of the two Si crystals, Si$^{(1)}$ and Si$^{(2)}$, projected on the (111) plane, with angle θ between the corresponding in-plane crystal-axes.

smaller than the spacing among them, we neglect the size of the Fermi surface in the following discussion.

When $\theta=0$, the projected Fermi surfaces of both Si on the (111) plane overlap completely. Thus, the tunneling can begins to occur by an infinitesimal bias voltage between the two Si, because both the energy and the transverse components of wave vectors (the projected positions on the (111) plane) can be conserved through the tunneling.

When $\theta>0$, as shown in Fig.1(b), the transverse components of the wave vectors of Fermi surfaces of Si$^{(1)}$ and Si$^{(2)}$ do not coincide each other. In this case, the energy of the empty state in Si$^{(2)}$, which has the same transverse components of wave vectors with those of the Fermi surfaces of Si$^{(1)}$, is different from the Fermi energy of Si$^{(1)}$. So, the direct tunneling does not occur for a small bias voltage.

When Si$^{(2)}$ is positively biased relative to Si$^{(1)}$ with $V=V_{th}$, such that the energy surface of Si$^{(2)}$ with energy V_{th} measured from its Fermi surface contacts the Fermi surface of Si$^{(1)}$ (as shown in Fig.1(b) with $V=V_{th}$), the electrons on the Fermi surface of Si$^{(1)}$ find the empty states in Si$^{(2)}$ with both energy and transverse component of the wave vector equal to those of the Fermi surface of Si$^{(1)}$, and thus, the direct tunneling occur. For $V>V_{th}$, the above condition is still fulfilled and the tunneling continues to occur. Thus, the bias voltage V_{th}

manifests the threshold voltage for the occurrence of the direct tunneling. That is, V_{th} represents the energy of the conduction band of Si[2], measured from its Fermi level, at k whose projected position on the (111) plane is marked by that of the Fermi surface of Si[1]. Thus, when θ is changed, V_{th} draws the conduction band structure along the orbit in the k-space whose projection on the (111) plane forms a circle passing the projections of six Fermi surfaces (Fig.1(b)), where the bottom of the conduction band stays at $\theta=0$.

3 Experiments

The samples for the ARTS were prepared by bonding the two n-Si(111) wafers whose surfaces were pre-oxidized to form the tunnel barrier. The carrier concentrations of the Sb-doped $400\mu m$ thick Si wafers were $1 \sim 5 \times 10^{19} cm^{-3}$. The surfaces of the wafers were firstly hydrogen terminated [4,5] and then oxidized in the $1.0\% O_2/Ar$ atmosphere for 5 min. at 900°C. The oxidized surfaces were then annealed [6] and bonded [5] with various in-plane angles θ between the two wafers. The bonded angle θ was determined by XRD.

The tunneling measurements were carried out by the modulation method using 1kHz, 1mV modulation voltage, and the current I and dI/dV were obtained as a function of dc bias voltage V. The d^2I/dV^2-V curves were obtained numerically from dI/dV data, in order to determine V_{th} in detail.

4 Results and Disscussion

Fig.2 shows the observed $I-V$ and $dI/dV-V$ tunneling characteristics, and d^2I/dV^2-V numerically calculated from dI/dV, for $\theta=3.8°$, 13.6° and 27.2° at $T=10K$. The V_{th} are shown by arrows at the d^2I/dV^2-V curves. The V_{th}, thus obtained, are plotted as a function of θ in Fig.3. The plots with ◯ are for $0° \leq \theta \leq 30°$, and △ for $30° \leq \theta \leq 60°$. The abscissa for $30° \leq \theta \leq 60°$ is folded up on that for $0° \leq \theta \leq 30°$. The figure shows that the plots ◯ and △ behave similarly, reflecting the symmetry of the energy band with respect to the ARTS.

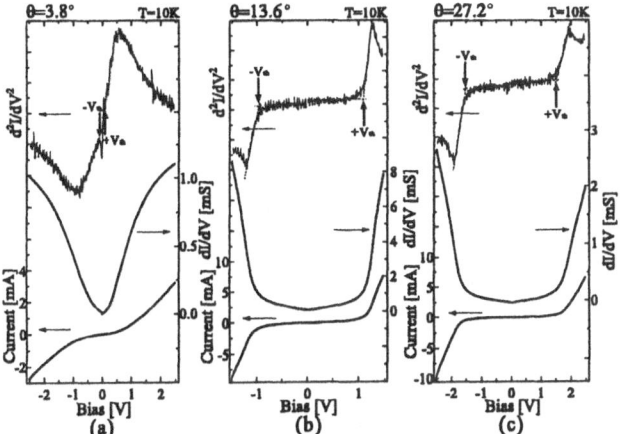

Fig. 2 The tunneling $I-V$, $dI/dV-V$ and d^2I/dV^2-V characteristics for $\theta=$(a)3.8°, (b)13.6° and (c) 27.2° at 10K.

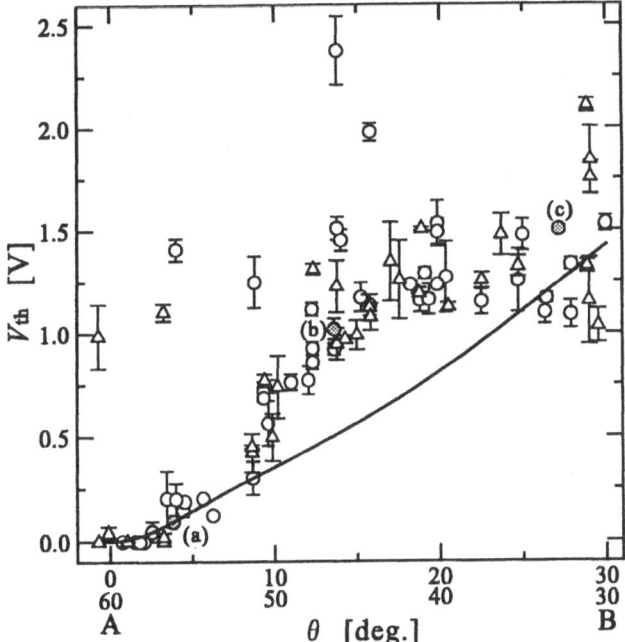

Fig. 3 The θ dependence of V_{th} derived from the tunneling characteristics. (a), (b) and (c) show the plots from Fig.2. A and B show the points in Fig.4(b).

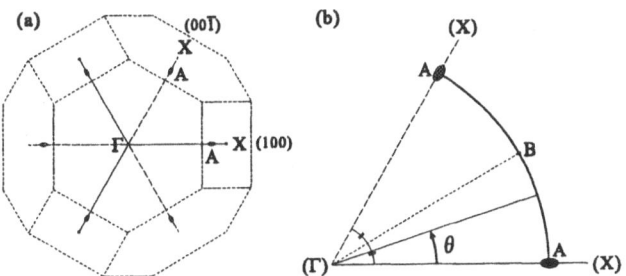

Fig. 4 (a)The Fermi surfaces in the Brillouin zone viewed from [111] direction. (b)The projection of Fermi surfaces on the (111) plane. The direct tunneling begins at $V=V_{th}$ along the circular arc A–B–A when θ is changed from 0° to 60°.

The solid curve depicts the energy band calculated by the pseudo-potential method of Chelikowsky and Cohen [7], along the circular arc A–B on the k-space projected on the (111) plane as shown in Fig.4. The result obtained by ARTS well reproduces the calculated band structure near the bottom of the conduction band ($\theta=0$). However, a remarkable discrepancy is noted between the ARTS result and the band calculation.

References

1. C.B. Duke, *Tunneling Phenomena in Solids* (edited by E. Burstein and S. Lundgvist, Plenum, NewYork 1969).
2. T. Ezaki, *et al.*, J. Vac. Sci. Technol. A**8**, (1990) 182.
3. H. Enomoto, *et al.*, J. Vac. Sci. Technol. B**9**, (1991) 1022.
4. T. Yasaka, *et al.*, J. J. Appl. Phys. **30**, (1991) 3567.
5. Q.-Y. Tong and U. Gösele, *Semiconductor Wafer Bonding* (John Wiley & Sons, NewYork 1999).
6. A. Ogura, J. Electrochem. Soc. **138**, (1991) 807.
7. J.R. Chelikowsky and M.L. Cohen, Phys. Rev. B**14** (1976) 556.

Identification of physical mechanisms of tensoeffects in highly strained Si and Ge

S. I. Budzulyak[1], V. M. Ermakov[1], Yu. P. Dotsenko[1], V. V. Kolomoets[1], Yu. M. Shwarts[1], E. F. Venger[1], M. Fukuzawa[2], P. Verma[2], M. Yamada[2], E. Liarokapis[3], D. P. Tunstall[4]

[1] Institute of Semiconductor Physics, National Academy of Sciences of Ukraine, Prospect Nauki 45, 03028 Kiev, Ukraine
e-mail: kolomoe@class.semicond.kiev.ua

[2] Department of Electronics and Information Science, Kyoto Institute of Technology, Matsugasaki, Sakyo-ku, Kyoto 606, Japan

[3] National Technical University of Athens, Department of Physics, Zografou Campus, GR 15773 Athens, Greece

[4] University of St. Andrews School of Physics and Astronomy, North Haugh, St. Andrews, Fife KY169SS, Scotland

Abstract   Transport phenomena and strain-induced transformation of band structure in highly strained n-Si, p-Si, n-Ge, p-Ge, and p-Ge/GaAs structures are analyzed to identify the main mechanisms of tensoresistive effects in a wide region of uniaxial pressure, impurity concentration and temperature changes. The developed methods of identification of the principal physical reasons of a non-trivial phenomena in extremely strained crystals are discussed.

1 Introduction

Since the first report by Smith [1], many investigations of tensoresistive effect (TRE) in semiconductors were carried out in the region of weak deformation, i.e., at low uniaxial pressure. The results of these investigations, however, did not allow to realize significant possibilities of this method. The TRE measurements in highly strained semiconductors are able to solve such problems as unambiguous determination of the TRE mechanisms and determination of the fundamental semiconductors parameters with a high accuracy.

An original equipment developed for investigation of TR effects at high deformation [2] allows a substantial increase of the higher limit of uniaxial pressure which does not destroy samples (3.5 GPa for Ge and 6.5 GPa for Si). Thus, the criteria of high uniaxial pressure $\delta\varepsilon \gg kT$ (where $\delta\varepsilon$ is the pressure-induced shift of the band minima or energy levels) is valid for equilibrium conditions up to room temperature.

In this work, we present the results of identification of dominant mechanisms of TRE for highly strained semiconductors and structures.

2 Results and discussion

The mechanism of TRE observed in [1] for n-Si and n-Ge was explained by electrons redistribution between equivalent minima of the conduction band [1,3]. Investigations of the longitudinal TRE in n-Ge and n-Si crystals with appreciable contribution of scattering on either acoustical phonons or ionized impurities showed substantially different regularities of the change in the dependencies $\rho_X/\rho_0 = f(X)$ in the high pressure region at different temperatures (Fig. 1). At the same time, there was no qualitative difference for these dependencies in the region of low pressures ($X < 0.05$ GPa). Thus, analyzing

the both regularities of the change of the dependencies $\rho_X/\rho_0 = f(X)$ in the high pressure region and temperature dependencies of resistivity or Hall-mobility it is possible to determine a pressure-induces change of the scattering mechanisms and, hence, to identify the TRE mechanisms. Moreover, the presence of the dominant contribution of f-transitions to the inter-valley scattering in n-Si was determined because in the high pressure region f-transitions can be completely eliminated at $X \| [001]$.

In order to identify the TRE mechanisms in many-valleys semiconductors, method of determination of deformation potential constant Ξ_u was adopted [4]. Since the energy gap between the equivalent valleys is proportional to the uniaxial pressure ($\delta\varepsilon \sim X$) and the electrons concentration ratio n_2/n_1, (where n_2 and n_1 is electron concentration in "upper" and "lower" valleys, respectively) for nondegenerative distribution of the electrons

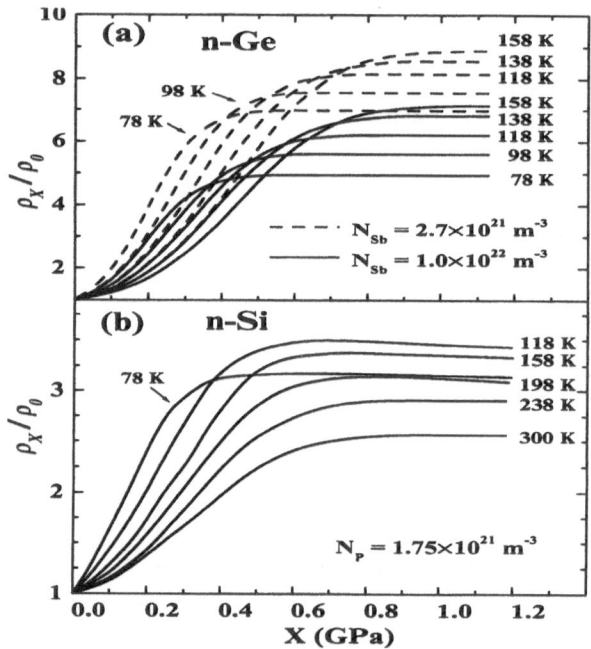

Fig. 1 Stress dependencies of the normalized resistivity of (a) n-Ge(Sb) and (b) n-Si(P) at different impurity concentrations and temperatures as indicated in the figure.

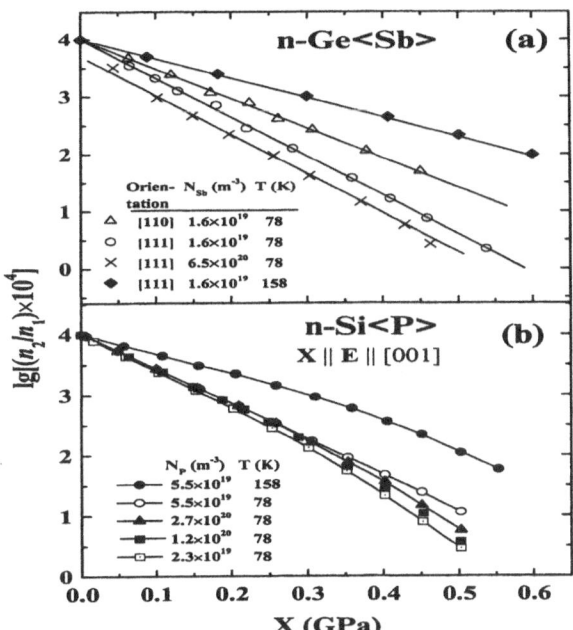

Fig. 2 Stress dependencies of $\lg[n_2/n_1 \times 10^4]$ for (a) n-Ge(Sb) and (b) n-Si(P).

can be presented as $n_2/n_1 = \exp(-\delta\varepsilon/kT)$, a linear dependence $\lg(n_2/n_1) = f(X)$ is obtained for the mechanism of intervalley redistribution. The data of longitudinal TRE measurements in a wide pressure region allow to determine the concentration ratio n_2/n_1. Since the relationships of $\lg(n_2/n_1) = f(X)$ for both n-Si and n-Ge can be obtained taking into account only the intervalley redistribution mechanism, any deviation from linearity in these dependencies indicate incompatibility of the measured effect to the Smith-Herring mechanism.

The analysis of the longitudinal TRE measurements in n-Ge shows that the mechanism of electron redistribution between the equivalent L_1-valleys (Fig. 2(a)) is the dominant mechanism in the wide region of pressure, concentration and temperature. In highly strained n-Si, the TRE mechanisms is determined by the electron redistribution between Δ_1-valleys only in a narrow temperature region (78-100 K) and shallow donor concentration $\approx (10^{19} - 10^{20})$ m^{-3} (Fig. 2(b)). At higher temperature, the dependence of f-scattering probability on pressure leads to a deviation from linearity for $\lg(n_2/n_1)$ (curve 1 in Fig. 2(b)). Additional ionization of phosphorous impurity in n-Si at $T = 78$ K also leads to a deviation from linearity for concentration $N_P \geq 10^{20}$ m^{-3} (curve 3 in Fig. 2(b)). The dependence of ionization energy of the thermo-donors on the pressure in neutron transmutation doped (NTD) n-Si(P) determines the non-linearity of the curves 4, 5 in Fig. 2(b) for concentrations of P $1.2 \cdot 10^{20}$ m^{-3} and $2.3 \cdot 10^{19}$ m^{-3}, respectively. Obtained conclusions were verified by the tenso-Hall effect measurement in (NTD) and γ-irradiated n-Si(P) crystals. Substantially non-linear and high TR effects are observed in n-Si in the absence of intervalley redistribution of electrons at $X \parallel [111]$. Investigations of TRE

at $X \parallel [111]$ up to 6 - 6.5 GPa confirmed the mechanism of increase of effective mass due to strain-induced non-parabolicity of Δ_1-valleys [5]. An increase of electron effective mass in degenerately doped crystals n-Ge at $X \parallel [001]$ and n-Si at $X \parallel [111]$ leads to some effects which is determined by transition from metallic type conductivity to the activation type one [6]. A substantial anisotropy of the TRE is observed in highly strained crystals p-Ge, p-Si and p-Ge/GaAs structures. An increase and a decrease of the hole effective mass in different valence subbands, appearance of a high anisotropy of the effective mass, strain-induced non-parabolicity of light- and heavy-hole bands determine the mechanisms of TRE in highly strained crystals and structures. Substantial difference of the TRE in p-Ge and p-Si is connected with an appreciable influence of spin-split-off V_3-subband in silicon and insignificant influence this band in germanium. In pure crystals n-Ge and p-Ge, the mechanisms of impact ionization of shallow impurities is determined by ballistic heating of carriers and streaming motion of electrons and holes [7].

3 Conclusions

On the basis of the transport phenomena analysis, the dominant mechanisms of the TRE in highly strained n-Si, p-Si, n-Ge, p-Ge and p-Ge/GaAs were determined in a wide region of uniaxial pressure, dopant concentration and temperature. The method of TRE mechanism discrimination is described for many-valleys semiconductors. Determination of peculiarities of band structure transformation in highly strained crystals allows to realize non-trivial mechanisms of strain-induced metal-insulator transition for n-Si and n-Ge. For both cases a remarkable increase of electron effective mass leads to the decrease of the Bohr radius and electron localization on impurity centers. Data of TRE investigation allow to determine with high accuracy the band structure parameters and parameters which characterize the scattering mechanisms in investigated crystals. Obtained data demonstrate the main merits of the method high uniaxial pressure opposite to the low deformation method in terms of identification of the TRE mechanisms and for development of the methods of determination of semiconductor parameters.

References

1. C. S. Smith, Phys. Rev **94**, (1954) 42.
2. V. N. Ermakov, V. V. Kolomoets, B. A. Suss, and V.E. Rodionov, Patent of Russia. 2040785 (1995).
3. C. Herring., Bell. Syst. Techn. J. **34**, (1995) 237.
4. P. I. Baranskii, I. S. Buda, I. V. Dachovskii, and V. V. Kolomoets. *Electrical and galvanomagnetical phenomena in anizotropic semiconductors* (Naukova Dumka, Kiev, 1977).
5. P. I. Baranskii, V. V. Kolomoets, and S. S. Korolyuk, Phys. Stat. Sol (b) **116**, (1983) K199.
6. S.I. Budzulyak et al., Sov. Phys. — Semiconductors, **34**, (2000) 1063.
7. A. E. Gorin, V. N. Ermakov, and V. V. Kolomoets, Sov. Phys. — Semiconductors, **29** (1995) 615.

Metal-insulator transition in anisotropic systems

F. Milde, R. A. Römer, M. Schreiber

Institut für Physik, Technische Universität, 09107 Chemnitz, Germany

Abstract We study the three-dimensional Anderson model of localization with anisotropic hopping, *i.e.*, weakly coupled chains and weakly coupled planes. In our extensive numerical study we identify and characterize the metal-insulator transition by means of the transfer-matrix method and energy level statistics. Using high accuracy data for large system sizes we estimate the critical exponent as $\nu = 1.6 \pm 0.3$. This is in agreement with its value in the isotropic case and in other models of the orthogonal universality class.

Previous studies of Anderson localization [1] in three-dimensional (3D) disordered systems with anisotropic hopping using the transfer-matrix method (TMM) [2–4], multifractal analysis (MFA) [5] and energy-level statistics (ELS) [6] show that an MIT exists even for very strong anisotropy. In Refs. [7,8], we studied critical properties of this second-order phase transition with high accuracy. Here we shall demonstrate the significance of irrelevant scaling exponents for an accurate determination of the critical disorder W_c and the critical exponent ν. Previous highly accurate TMM studies for isotropic systems of the orthogonal universality class reported $\nu = 1.54 \pm 0.08$ [9], $\nu = 1.58 \pm 0.06$ [10], $\nu = 1.61 \pm 0.07$, and $\nu = 1.54 \pm 0.03$ [11], whereas for anisotropic systems of weakly coupled planes $\nu = 1.3 \pm 0.1$ and $\nu = 1.3 \pm 0.3$ was found [3]. We emphasize that this variation in theoretical values has its counterpart in the experiments where a large variation of ν has been reported with values ranging from 0.5 [12] over 1.0 [13], 1.3 [14], up to 1.6 [15]. Possibly this experimental "exponent puzzle" [14] is due to other effects such as electron-electron interaction [15] or sample inhomogeneities [14,16,17].

A further important aspect of anisotropic hopping besides the question of universality is the connection to experiments which use uniaxial stress, tuning disordered Si:P or Si:B systems across the MIT [12–15]. Applying stress reduces the distance between the atomic orbitals, the electronic motion becomes alleviated, and the system changes from insulating to metallic. Thus, although the explicit dependence of hopping strength on stress is material specific and in general not known, it is reasonable to relate uniaxial stress in a disordered system to an anisotropic Anderson model with increased hopping between neighboring planes.

We use the standard Anderson Hamiltonian [1]

$$\mathbf{H} = \sum_{i \neq j} t_{ij} |i\rangle\langle j| + \sum_i \epsilon_i |i\rangle\langle i| \tag{1}$$

with orthonormal states $|i\rangle$ corresponding to electrons located at sites $i = (x, y, z)$ of a regular cubic lattice with periodic boundary conditions. The potential energies ϵ_i are independent random numbers drawn uniformly from $[-W/2, W/2]$. The disorder strength W specifies the amplitude of the fluctuations of the potential energy. The hopping integrals t_{ij} are non-zero only for nearest neighbors and depend on the spatial directions, thus t_{ij} can either be t_x, t_y or t_z. We study (i) *weakly coupled planes* with $t_x = t_y = 1$, $t_z = 1 - \gamma$ and (ii) *weakly coupled chains* with $t_x = t_y = 1 - \gamma$, $t_z = 1$ with hopping anisotropy $\gamma \in [0, 1]$. For $\gamma = 0$ we recover the isotropic case, $\gamma = 1$ corresponds to independent planes or chains. We note that uniaxial stress would be modeled by weakly coupled chains after renormalization of the hopping strengths such that the largest t is set to 1.

The MIT in the Anderson model of localization is expected to be a second-order phase transition [18,19]. It is characterized by a divergent correlation length $\xi_\infty(W) \propto |W - W_c|^{-\nu}$ [20]. To construct the correlation length of the *infinite* system ξ_∞ from finite size data ξ_M [3, 20–22], the one-parameter scaling hypothesis [23] $\xi_M = f(M/\xi_\infty)$ is employed. One might determine ν from fitting to ξ_∞ obtained by a FSS procedure [22]. Better accuracy can be achieved by fitting directly to the ξ_M data [9–11]. We use fit functions [10] which include two kinds of corrections to scaling: (i) nonlinearities of the disorder dependence of the scaling variable and (ii) an irrelevant scaling variable with exponent $-y$ (cp. Fig. 1). For the nonlinear fit, we use the Levenberg-Marquardt method [7,10]. The input data ξ_M for the FSS procedure are either (a) reduced localization lengths Λ_M obtained by TMM with 0.07% accuracy and system widths up to 17×17 for, *e.g.*, the case of weakly coupled planes with $\gamma = 0.9$ [8]; or (b) integrated Δ_3 statistics obtained from highly accurate ELS data (0.2% to 0.4%) and system sizes up to 50^3 [7].

When applying the TMM to our anisotropic systems, one has to consider two non-equivalent orientations of the axis of the quasi-1D bar: parallel and perpendicular to the planes or chains. The localization lengths in the perpendicular direction are smaller than in the parallel direction by a factor of about $1 - \gamma$ for coupled planes and $(1 - \gamma)^2$ for chains [3]. The critical disorder W_c should not depend on the orientation of the bar [3]. For strong anisotropies $\gamma \geq 0.9$ this is difficult to verify numerically due to strong finite size effects as shown in Fig. 1. By computing data for very large system sizes up to $M^2 = 22^2$ (46^2) for the case of weakly coupled planes with $\gamma = 0.9$ (0.96) we can show that this finite size effect can be sucessfully modelled (cp. Fig. 1) by an

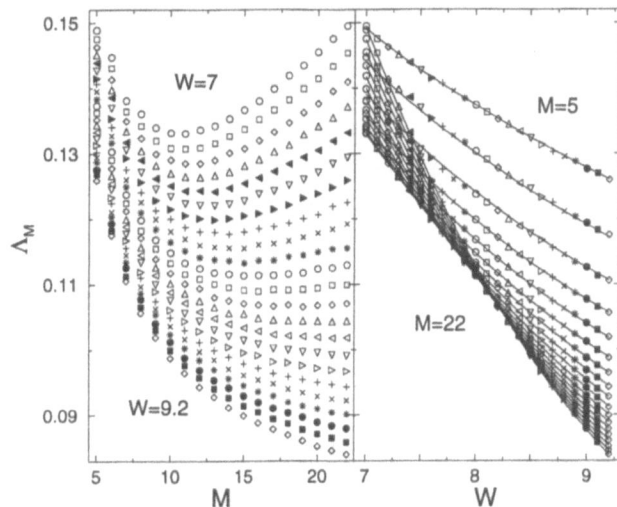

Fig. 1 Λ_M for coupled planes with $\gamma = 0.9$ (perpendicular orientation) with relative error 0.1%, $W = 7, 7.1, 7.2, \cdots, 9.2$ and $M^2 = 5^2, 6^2, 7^2, \cdots, 22^2$. The solid lines in the right part are fits to the data with $y = 2.05 \pm 0.08$.

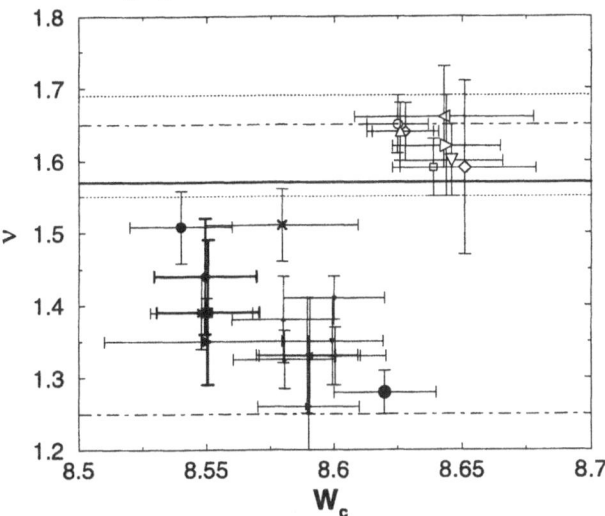

Fig. 2 Results for W_c and ν, for coupled planes with $\gamma = 0.9$, obtained from FSS of (parallel-direction) TMM data (open symbols) and ELS data. The error bars show the 95% confidence intervals. The dotted (dashed) lines represent the error bounds for $\nu = 1.62 \pm 0.07$ (1.45 ± 0.2) of TMM (ELS). The solid line marks the result of [10]. The goodness of a fit is reflected in the size of the symbol. The 2 thick error bars mark high quality ELS fits for large system sizes.

irrelevant scaling exponent and W_c is indeed the same for both orientations.

In Fig. 2, we show fitted values obtained by FSS of TMM data for different choices of expansion coefficients in the nonlinear fit procedure. We conclude $\nu = 1.62 \pm 0.07$ and $W_c = 8.63 \pm 0.02$. In Fig. 2, we also show the results for FSS of highly accurate ELS data (0.2% to 0.4%) and system sizes up to $N^3 = 50^3$. The error estimate is larger and the values of W_c and ν are much more scattered than before. Comparing the spreading of the W_c and ν values with their confidence intervals, the

error estimates appear to be too small. *E.g.*, the 95% confidence intervals of the smallest and largest W_c value do not overlap. We therefore estimate $\nu = 1.45 \pm 0.2$ and $W_c = 8.58 \pm 0.06$ [7].

In conclusion, our results confirm the existence of an MIT for anisotropy $\gamma < 1$ for weakly coupled planes found previsouly in studies using TMM [3], MFA [5], and recently by ELS [6]. We have shown that large system sizes, high accuracies [7,8] and irrelevant scaling exponents are necessary to determine the critical behavior reliably. Our results are in good agreement with other high accuracy TMM studies for the orthogonal universality class [9–11,24]. These numerical estimates seem to converge towards $\nu \approx 1.6$.

Acknowledgements We are grateful for the support of the DFG through Sonderforschungsbereich 393.

References

1. P. W. Anderson, Phys. Rev. **109**, 1492 (1958).
2. Q. Li, *et al.*, Phys. Rev. B **40**, 2825 (1989).
3. I. Zambetaki, *et al.*, Phys. Rev. Lett. **76**, 3614 (1996).
4. N. A. Panagiotides, S. N. Evangelou, and G. Theodorou, Phys. Rev. B **49**, 14122 (1994).
5. F. Milde, R. A. Römer, and M. Schreiber, Phys. Rev. B **55**, 9463 (1997).
6. F. Milde and R. A. Römer, Ann. Phys. (Leipzig) **7**, 452 (1998).
7. F. Milde, R. A. Römer, and M. Schreiber, Phys. Rev. B **61**, 6028 (2000), cond-mat/9909210.
8. F. Milde, R. A. Römer, M. Schreiber, and V. Uski, Eur. Phys. J. B **15**, 685 (2000), cond-mat/9911029.
9. A. MacKinnon, J. Phys.: Condens. Matter **6**, 2511 (1994).
10. K. Slevin and T. Ohtsuki, Phys. Rev. Lett. **82**, 382 (1999), cond-mat/9812065.
11. P. Cain, R. A. Römer, and M. Schreiber, Ann. Phys. (Leipzig) **8**, SI33 (1999), cond-mat/9908255.
12. M. A. Paalanen and G. A. Thomas, Helv. Phys. Acta **56**, 27 (1983).
13. S. Waffenschmidt, C. Pfleiderer, and H. v. Löhneysen, Phys. Rev. Lett. **83**, 3005 (1999), cond-mat/9905297.
14. H. Stupp *et al.*, Phys. Rev. Lett. **71**, 2634 (1993).
15. S. Bogdanovich, M. P. Sarachik, and R. N. Bhatt, Phys. Rev. Lett. **82**, 137 (1999).
16. T. F. Rosenbaum, G. A. Thomas, and M. A. Paalanen, Phys. Rev. Lett. **72**, 2121 (1994).
17. H. Stupp *et al.*, Phys. Rev. Lett. **72**, 2122 (1994).
18. D. Belitz and T. R. Kirkpatrick, Rev. Mod. Phys. **66**, 261 (1994).
19. E. Abrahams, *et al.*, Phys. Rev. Lett. **42**, 673 (1979).
20. B. Kramer and A. MacKinnon, Rep. Prog. Phys. **56**, 1469 (1993).
21. J.-L. Pichard and G. Sarma, J. Phys. C: **14**, L127 (1981).
22. A. MacKinnon and B. Kramer, Phys. Rev. Lett. **47**, 1546 (1981); Z. Phys. B **53**, 1 (1983).
23. D. J. Thouless, Phys. Rep. **13**, 93 (1974).
24. K. Slevin and T. Ohtsuki, Phys. Rev. Lett. **78**, 4083 (1997), cond-mat/9704192.

Hot electron degradation of the extraordinary magneto-resistance in inhomogenous InSb

D. Poplavskyy[1], A.C.H. Rowe[1], R.A. Stradling[1], S.A. Solin[2]

[1] Experimental Solid State Physics Group, Imperial College of Science, Technology and Medicine, Prince Consort Road, London SW7 2BZ, United Kingdom e-mail: a.c.rowe@ic.ac.uk
[2] NEC Research Institute Inc., 4 Independence Way, Princeton NJ 08540, USA

Abstract Pulsed high current resistivity and Hall effect measurements on metallically included InSb van der Pauw (vdP) discs and on InSb epilayer Hall bars are reported. It is shown that measurement of the Hall voltage in the inhomogeneous samples yields quantitative information on the extent of the current exclusion from the metallic inhomogeneity. This current exclusion is responsible for the observed extraordinary magneto-resistance (EMR). A degradation in the EMR of approximately 10 % at drive currents of 200 mA is attributed to a combination of a reduction in the mobility and an increase in the carrier concentration of the InSb due to hot electron effects.

1 Introduction

Semiconductor vdP discs with a single concentric metallic inclusion were first studied in detail by Wolfe *et al.* [1]. For this specific geometry they deduced formal expressions for the apparent resistivity at zero magnetic field, $\rho_{app}(0,\alpha)$, and the variation of the Hall coefficient with magnetic field, $R_H(B,\alpha)$, where α is the inclusion size defined as the ratio of the radius of the metallic inclusion to the disc radius. However, they did not explicitly deduce a form for $\rho_{app}(B,\alpha)$, the variation of the apparent resistivity with applied field, although this is a logical extension of their treatment. As a consequence perhaps, the potential for extraordinarily large magneto-resistance (dubbed EMR) of such devices was not realised until very recently [2]. Experimentally observed EMR at 300K up to 100 % at B=50 mT was reported in InSb based structures with a Ti/Pt/Au inclusion. As noted in Ref. 2, InSb with its high room temperature mobility (μ) is the material of choice since the MR is expected to increase roughly as the square of μ. However because of the high mobility, hot electron effects are to be expected at relatively low electric fields. According to an expression for $\rho_{app}(B,\alpha)$ derived from Ref. 1 [3], a reduction in the MR is expected with a decrease μ and an increase in the concentration (n).

To investigate this possibility, pulsed high field measurements on InSb epilayers fabricated into Hall bars and of $\rho_{app}(B,\alpha)$ on similar samples to those used in Ref. 2 are performed. Low electric field measurements of the Hall effect in the inhomogeneous samples provide a quantitative description of the current focussing caused by the metallic inclusion.

2 Results and Discussion

The Hall bars and the samples with inclusions were fabricated from 1.2 μm thick Te-doped InSb grown on GaAs substrates. A surface doping layer and the InSb-GaAs interface layer do not contribute significantly to the conduction at 300 K [2].

Consider first a circular vdP disc with four leads in a four-fold symmetric configuration. According to the method first outlined by van der Pauw [4] for the Hall geometry, the Hall voltage in a thin layer with contacts at the circumference is $V_H = I\rho\mu B/d$, where ρ is the sample resistivity, I is the current through the layer, B is the applied field and d is the sample thickness. In a sample with a metallic inclusion a portion I_s of the total current I, flows around the inclusion. The remaining portion I_m flows through the inclusion. Extending van der Pauw's approach to this situation, $V_H = V_{Hs} + V_{Hm} = I_s\rho_s\mu_s B/d + I_m\rho_m\mu_m B/d$, where the subscripts 's' and 'm' denote semiconductor and metal respectively. Equivalently, $V_{Hs}/V_{Hm} = I_s\rho_s\mu_s/I_m\rho_m\mu_m$ and since $\rho_s\mu_s/\rho_m\mu_m \sim 10^8$ then $V_{Hs} >> V_{Hm}$ and we may write

$$V_H(B,\alpha)/V_H(B,0) = I_s(B,\alpha)/I_s(B,0) \qquad (1)$$

where $I_s(B,0) = I$ and $V_H \sim V_{Hs}$. The two terms on the left hand side of Eq. (1) are measured and ratio-ed so that the total current flowing around the inclusion (I_s) can be estimated for different values of α and B.

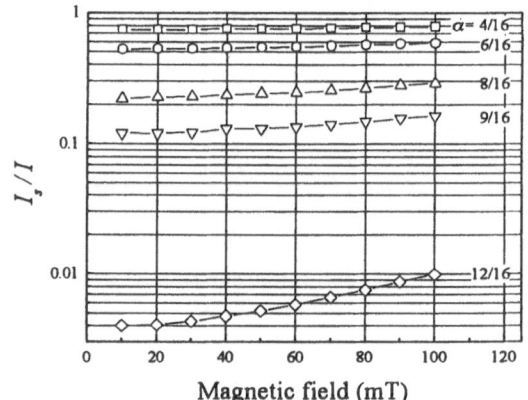

Fig. 1 The fraction of current flowing around the inclusion versus magnetic field for different inclusion sizes.

The current ratios obtained from low electric field DC Hall voltages are shown in Fig. 1. For $\alpha = 8/16$ (1/4 of the total disc area) at $B \to 0$, only 20 % of the current flows through the semiconductor. Upon application of a magnetic field, $B = 100$ mT, this increases to 30 %. In contrast, when the inclusion occupies roughly half the area ($\alpha = 12/16$), 0.4 % of the total current flows *only* through the semiconductor, but this increases considerably to 1 % at $B = 100$ mT. For $0 < B < 100$ mT this is the largest change in I_s

with applied field for any inclusion size, and is associated with the largest MR.

In high electric field measurements of $\rho_{app}(B,\alpha)$, a degradation in the EMR of ~ 10 % is observed when I is increased from 15 mA to 200 mA (see Fig. 2). Note that the current value is quoted instead of electric field because the latter is non-uniform in the vdP sample geometry. There are two hot-electron effects at work which may be responsible for this degradation; a decrease in the mobility and an increase in the carrier concentration.

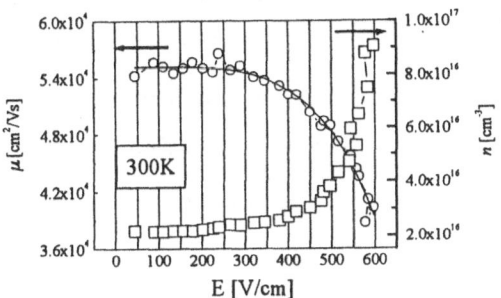

Fig 3 High electric field measurements on an epitaxial InSb Hall bar of the same material used to fabricate the inhomogeneous vdP discs. At 400 V/cm there is a 30% rise in n and a 5 % reduction in μ. The solid line is the catchment model fit to μ (Ref. 5).

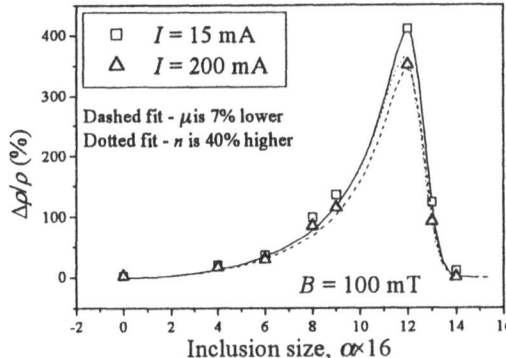

Fig 2 Variation in EMR as a function of inclusion size at B = 100 mT and two different currents. The experimental data is well fitted using the expression for $\rho_{app}(B,\alpha)$ derived from Ref. 1. The solid line is the fit using the low field values of n and μ. The dashed line is the same fit assuming a 7 % reduction in μ and the dotted line a 40% increase in n, both of which are the result of hot electron effects.

In order to quantify the hot electron effects in the InSb epilayer material, separate pulsed high-electric field measurements were also undertaken with Hall bar samples of the heteroepitaxial Te-doped InSb used to fabricate the EMR sensors [5].

The high field behaviour shown in Fig. 3 is quite different from that expected for pure bulk InSb where hot electron effects are significant at fields of the order of 200 V/cm. For thin heteroepitaxial InSb one expects a reduction in the low field mobility matched with an increase in the carrier concentration [6]. The lower mobility delays the onset of hot electron effects, and the higher concentration results in energy loss due to plasmon and phonon-plasmon coupled modes at high fields. The resulting quasi-Ohmic conduction up to fields greater than 300 V/cm is well described by the catchment model [5] which is based on these energy loss mechanisms (solid line, Fig. 3).

The expression derived for the EMR on the basis of the model in Ref. 1 and used to fit the data shown in Fig. 2 is dependent on both n and μ. Calculations show that the degradation in the EMR at high currents can be accounted for using the changes in n and μ observed at high electric fields in the Hall bar sample.

The dashed and dotted curves in Fig. 2 correspond to a decrease in μ by 7% and an increase in n by 40% respectively. These changes are similar to those measured at ~ 400 V/cm in the Hall bar (see Fig. 3). Thus we assign the reduction in the EMR of the inhomogeneous sample observed at high currents to hot electron changes in the InSb mobility and concentration.

It should be noted that the average estimated electric field across the InSb annulus of the inhomogeneous device is much lower than 400 V/cm (accounting for the contact resistance, the applied voltage across the 1 mm diameter disc is < 10 V). The design of the contacts in these experimental devices is such that the electric field near the contacts is very large. Thus the overall electrical response at high drive currents is a combination of Ohmic conduction in the bulk of the InSb annulus, and hot electron conduction near the contacts. A uniformly high electric field over the majority of the InSb annulus could therefore reduce the EMR even further. This will be the subject of a future study.

D.P. acknowledges support of a Shell Centenary Scholarship. A.C.H.R. acknowledges support of a Commonwealth Scholarship. We thank M. Pelczynski of EMCORE for sample preparation and S. Schwed (EMCORE) and D.R. Hines (NECI) for sample fabrication.

References

1. C.M. Wolfe, G.E. Stillman and J.A. Rossi, J. Electrochem. Soc.: Solid-State Science and Technology **119**, (1972) 250.
2. S.A. Solin, T. Thio, D.R. Hines and J.J. Heremans, *Science* **289**, (2000) 1530. Also presented at this conference as an invited talk [G11] by S.A. Solin.
3. D. Poplavskyy, *M.Sc Thesis, Imperial College 2000* (unpublished)
4. L.J. van der Pauw, Philips Res. Rept. **13**, (1958) 1.
5. A.C.H. Rowe, C. Gatzke, R.A. Stradling and S.A. Solin, Appl. Phys. Lett. **76**(14), (2000) 1902.
6. H.H. Weider, J. Vac. Sci Techn. **9**, (1972) 1193

Critical exponents for the metal-insulator transition of ^{70}Ge:Ga in magnetic fields

Michio Watanabe[1]*, Kohei M. Itoh[1]**, Masashi Morishita[2], Youiti Ootuka[2], Eugene E. Haller[3]

[1] Dept. Applied Physics and Physico-Informatics, Keio University, 3-14-1 Hiyoshi, Kohoku-ku, Yokohama 223-8522, Japan
[2] Institute of Physics, University of Tsukuba, 1-1-1 Tennodai, Tsukuba, Ibaraki 305-8571, Japan
[3] Lawrence Berkeley National Laboratory and University of California at Berkeley, Berkeley, California 94720, USA

Abstract We have measured the electrical conductivity of nominally uncompensated ^{70}Ge:Ga samples in magnetic fields up to $B = 8$ T at low temperatures ($T = 0.05 - 0.5$ K) in order to investigate the metal-insulator transition in magnetic fields. The values of the critical exponents in magnetic fields are consistent with the scaling theories.

1 Introduction

Doped crystalline semiconductors are ideal solids to probe the effects of both disorder and electron-electron interaction on the metal-insulator transition (MIT) in disordered electronic systems [1]. Important information about the MIT is provided by the critical exponent μ for the zero-temperature conductivity $\sigma(0)$ defined by

$$\sigma(0) \propto (N - N_c)^{\mu}, \tag{1}$$

where N is the dopant concentration and N_c is the critical concentration for the MIT, and $\mu \approx 0.5$ has been found in a number of nominally uncompensated semiconductors including our ^{70}Ge:Ga [2,3].

According to the theories of the MIT, the value of the critical exponents does not depend on the details of the system, but depends only on the universality class to which the system belongs. In this sense, the application of a magnetic field is important because the motion of carriers loses its time-reversal symmetry in magnetic fields, and the universality class changes. In our earlier work, we reported that a different exponent $\mu = \mu' = 1.1 \pm 0.1$ is obtained in magnetic fields [4]. Here, μ' characterizes the magnetic-field-induced MIT:

$$\sigma(0) \propto (B_c - B)^{\mu'}. \tag{2}$$

Since this result is based solely on the metallic samples, in this work we perform a finite-temperature scaling analysis [1] which uses the data on the both sides of the transition. Moreover, we investigate the temperature dependence of the conductivity on the insulating side in the context of variable-range-hopping (VRH) conduction.

2 Experiment

All the samples were prepared by neutron-transmutation-doping (NTD) of isotopically enriched ^{70}Ge single crystals. The NTD method assures a homogeneous Ga acceptor distribution which is a crucial condition for experimental studies of the MIT [2,3]. The electrical conduc-

---

* (post-doctoral) JSPS Research Fellow
** also at PRESTO-JST

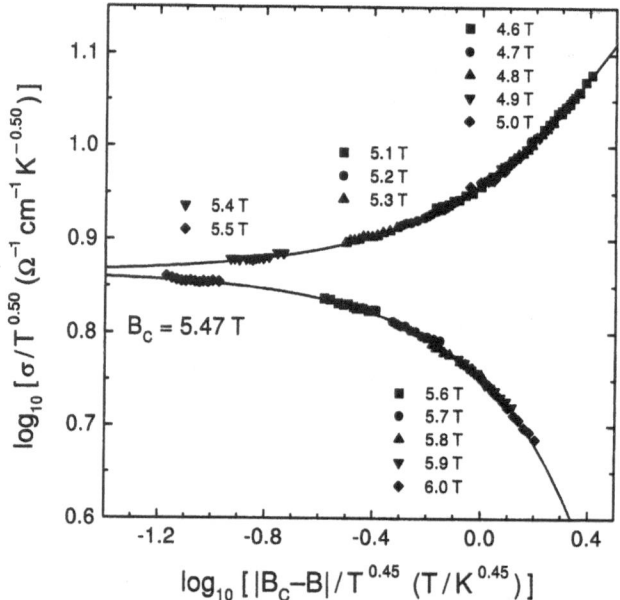

Fig. 1 Finite-temperature scaling plot for the magnetic-field-induced metal-insulator transition in the sample having $N = 2.004 \times 10^{17}$ cm^{-3}.

tivity was measured at low temperatures between 0.05 and 0.5 K.

3 Results and discussion

We show in Fig. 1 that $\mu' = 1.1$ (Ref. 4) yields an excellent finite-temperature scaling plot

$$\frac{\sigma(B,T)}{T^x} \propto f\left(\frac{|B_c - B|}{T^y}\right), \tag{3}$$

where x/y is equivalent to μ'. Here we employ B_c obtained by fitting Eq. (2). The temperature variation of the conductivity is proportional to $T^{1/2}$ even around the critical point in magnetic fields [4], leading to $x = 1/2$. Note that $y = x/\mu' = 0.45$, i.e., none of these parameters are used as a fitting parameter. Hence, Fig. 1 strongly supports $\mu' = 1.1$.

The temperature dependence of the conductivity on the insulating side of the MIT is shown in Fig. 2. We already reported that VRH conductivity at $B = 0$ obeys Efros and Shklovskii's (ES) law [6]

$$\sigma(N,0,T) = \sigma_0(N,0)\exp[-(T_0/T)^{1/2}] \tag{4}$$

even in the immediate vicinity of N_c ($0.99 N_c < N < N_c$) when an appropriate temperature dependence of the

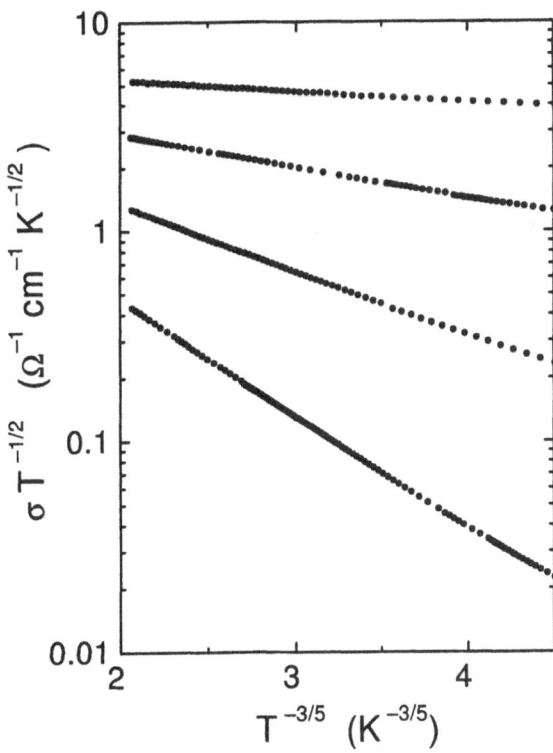

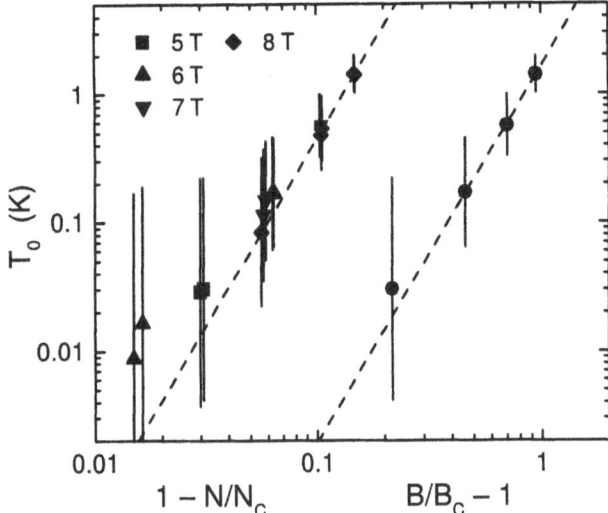

Fig. 2 Conductivity multiplied by $T^{-1/2}$ as a function of $T^{-3/5}$ for the sample having $N = 1.912 \times 10^{17}$ cm^{-3} in magnetic fields. From top to bottom in units of tesla, the magnetic induction is 5, 6, 7, and 8, respectively.

Fig. 3 T_0 determined by $\sigma(T) \propto T^{1/2} \exp[-(T_0/T)^{3/5}]$ as a function of $1 - N/N_c(B)$ in constant magnetic fields of $B = 5$, 6, 7, and 8 T (left data set), and as a function of $B/B_c - 1$ for the sample having $N = 1.912 \times 10^{17}$ cm^{-3} (right data set). The dashed lines represent the best fits to the data satisfying $T_0 > 0.05$ K.

4 Conclusion

The critical exponent of the zero-temperature conductivity for the metal-insulator transition of doped semiconductors in magnetic fields is confirmed to be close to unity. Other exponents in magnetic fields are also explained within the context of the scaling theories.

Acknowledgments

We are thankful to T. Ohtsuki for valuable discussions. Conductivity measurements were carried out at the Cryogenic Center, the University of Tokyo, Japan. This work was supported by a Grant-in-Aid for Scientific Research from the Ministry of Education, Science, Sports, and Culture, Japan, the Director, Office of Energy Research, Office of Basic Energy Science, Materials Sciences Division of the U. S. Department of Energy under Contract No. DE-AC03-76SF00098, and U. S. NSF Grant No. DMR-97 32707.

prefactor $\sigma_0 \propto T^r$ is taken into account [5]. Based on this finding, we analyze in this work the data at $B \geq 5$ T in the context of ES VRH conduction

$$\sigma(N, B, T) = \sigma_0(N, B) \exp[-(T_0/T)^{3/5}] \qquad (5)$$

in a strong field ($\sqrt{\hbar/eB} \ll \xi$, where ξ is the localization length) [6]. By assuming $\sigma_0 \propto T^{1/2}$, which is consistent with the above finite-temperature scaling analysis, the conductivity in magnetic fields is described well by Eq. (5) as seen in Fig. 2. The values of T_0 in Eq. (5) satisfy the relations $T_0 \propto (N_c - N)^\alpha$ and $T_0 \propto (B - B_c)^{\alpha'}$, and $\alpha = \alpha' \approx 2.9$ is obtained from the data satisfying $T_0 > 0.05$ K. (See Fig. 3.) Note that the condition $T < T_0$ is required for the ES theory to be valid, i.e., T_0 has to be evaluated only from the data obtained at temperatures low enough to satisfy this condition. Since $T_0 \propto (\xi\epsilon)^{-1}$ and both ξ and the dielectric constant ϵ diverge at N_c, $\alpha = \nu + \zeta$. Here, ν and ζ are given by $\xi \propto (N_c - N)^{-\nu}$ and $\epsilon \propto (N_c - N)^{-\zeta}$, respectively. The relation $2\nu \approx \zeta$, which was predicted theoretically [7], has been obtained at $B = 0$ for the present system by measuring magnetoresistance in weak fields ($\sqrt{\hbar/eB} \gg \xi$) [5]. Assuming the relation $2\nu = \zeta$, $\nu \approx 1$ is obtained for $B \neq 0$, and hence, the Wegner relation $\mu = \nu$ [8] holds in magnetic fields.

References

1. D. Belitz and T. R. Kirkpatrick, Rev. Mod. Phys. **66**, (1994) 261.
2. K. M. Itoh *et. al.*, Phys. Rev. Lett. **77**, (1996) 4958.
3. M. Watanabe *et. al.*, Phys. Rev. B **58**, (1998) 9851.
4. M. Watanabe *et. al.*, Phys. Rev. B **60**, (1999) 15817.
5. M. Watanabe *et. al.*, Phys. Rev. B **62**, (2000) R2255.
6. B. I. Shklovskii and A. L. Efros, *Electronic Properties of Doped Semiconductors* (Springer-Verlag, Berlin, 1984).
7. A. Kawabata, J. Phys. Soc. Jpn. **53**, (1984) 318.
8. F. J. Wegner, Z. Phys. B **25**, (1976) 327; *ibid.* **35**, (1979) 207.

Boron in diamond at high concentrations

M.J.R. Hoch, J.E. Lowther, T. Tshepe*

Department of Physics and Materials Physics Institute, University of the Witwatersrand, Johannesburg, PO Wits 2050, South Africa.

Abstract The critical boron concentration at the metal-insulator transition in diamond has previously been found to be $n_c \sim 4 \times 10^{21}$ cm^{-3}. Together with the Mott relation this suggests that the boron ions are at a separation comparable to the lattice spacing at the transition. Ab-initio electronic structure calculations for substitutional single boron ions and boron dimers in diamond provide support for the experimental conclusion that an impurity band evolves at high boron concentrations.

1 Introduction

In previous work[1, 2] it has been shown that boron concentrations in excess of 10^{21} cm^{-3} may be achieved in diamond using multiple ion implantation methods carried out at low temperature with high temperature annealing between successive implantations. It appears that the majority of the boron ions are activated and contribute to electrical conduction. The critical boron concentration for the metal-insulator (MI) transition in diamond is found to be $n_c \sim 4 \times 10^{21}$ cm^{-3}. There is some uncertainty in this value because of possible compensation effects due to vacancy centres produced in the ion implantation and annealing processes. Hall effect measurements are in progress in an effort to establish the carrier concentration in samples close to the transition.

Ab initio electronic structure calculations have been undertaken for pairs of boron ions in a diamond lattice in an attempt to gain insight into the nature of the MI transition in diamond. In particular, information on the effective radius of boron impurities is obtained.

2 Theoretical calculations

A 64-atom supercell calculation has been carried out for single substitutional boron atoms and for pairs of substitutional boron atoms in diamond. For the pairs, the boron atoms were located firstly at next nearest-neighbour sites and secondly at nearest-neighbour sites. A plane wave pseudopotential approach[3] in the local density approximation has been used with all atoms in the supercell allowed to relax.

The calculations lead to energy band structures which appear rather similar for a single substitutional B and for the B-dimer on next nearest-neighbour lattice sites. Energy levels near the origin in supercell k-space extend slightly above the Fermi level. However, for B atoms located on nearest-neighbour sites, the band structure shows important changes, with additional levels in the vicinity of the valence band edge for a number of points

---

* *Present address: Eskom Reactor Physics Group, Pretoria, South Africa.*

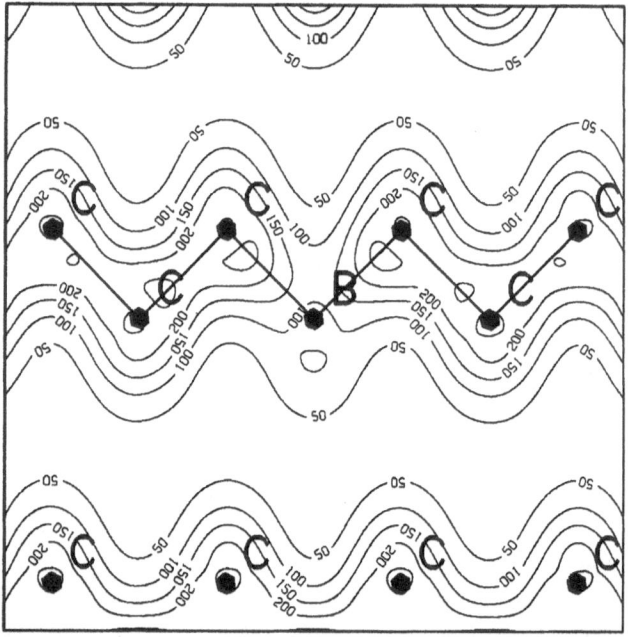

Fig. 1 Charge density distribution for a single substitutional boron atom in a diamond lattice. The contour spacing is 50 millibohr/au^3.

in supercell k-space. These levels extend well above the Fermi level. Details are presented elsewhere[4].

Binding energies for the B-dimers have been estimated using the relation,

$$\Delta E_{B:B} = E(B_2 C_{62}) + E(C_{64}) - 2E(BC_{63}) , \qquad (1)$$

where, for example, $E(BC_{63})$ is the total energy with a single boron atom in the supercell. While the local density approximation includes atomic energies only in an approximate way, these contributions cancel in the binding energy calculation. Values obtained for $\Delta E_{B:B}$ for boron ions on nearest-neighbour and next nearest-neighbour sites are +1 meV and +2 meV respectively. These values are very small and below the limits of accuracy of the present calculations. The binding energy results suggest that boron dimer formation can occur with the small positive binding energy taken up by lattice relaxation.

It is of interest to examine the electron density in the vicinity of the boron atoms for the cases considered above. Figures 1, 2, and 3 show the density distribution for the three cases considered. It can be seen that, for the nearest-neighbour pair, the region between the B atoms is depleted of charge compared to the bulk lattice. For the next nearest-neighbour pair, the charge distribution is similar to that for the single B impurity.

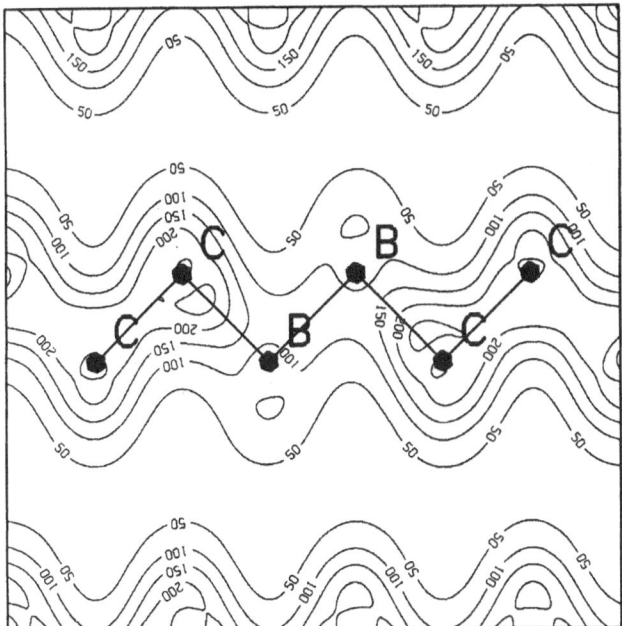

Fig. 2 Charge density distribution for a nearest-neighbour boron dimer in diamond.

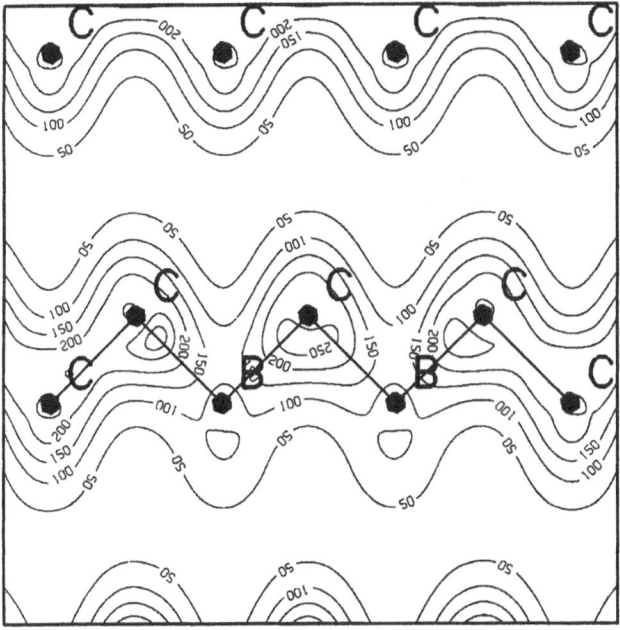

Fig. 3 Charge density distribution for a next-nearest-neighbour boron dimer in diamond.

3 Discussion

For B ions on nearest-neighbour sites, significant changes in the electronic structure are predicted compared to the single substitutional B in the diamond case. This is not the case for B-dimers in next nearest-neighbour positions. The results are consistent with the formation of an impurity band in B-doped diamond when sufficiently high concentrations are reached.

Using the available experimental value for n_c and the *Mott relation* $n_c^{1/3} a_H^* = 0.25$ *yields for the effective Bohr radius* $a_H^* \sim 0.16$ nm. This is comparable to the C–C bond length in diamond of 0.154 nm. The experimental results and the present theoretical predictions suggest that, for the formation of an impurity band in B-doped diamond, the B ions should be spaced on average at distances not much greater than the lattice spacing.

The critical exponent in B-doped diamond has been estimated as $\mu \approx 1.7$ [1]. This high value, compared to exponents found for other heavily doped semiconductors[5], may be regarded as evidence for quantum percolation. However, the site occupation probability for B acceptors at $n = n_c$ is approximately 0.02, which is much lower than the accepted site occupation probability for percolation to be important.

4 Conclusion

Theoretical calculations carried out using a plane wave pseudopotential approach in the local density approximation suggest that the MI transition in B-doped diamond occurs when the B concentration is sufficiently high that the B ions are on average separated by a distance not much greater than the lattice spacing. This finding supports the available experimental value for the critical concentration in this system.

5 Acknowledgements

Financial support from the National Research Foundation, the University of the Witwatersrand, and the De Beers Diamond Research Laboratories is gratefully acknowledged.

References

1. T. Tshepe, J.F. Prins and M.J.R. Hoch, Diamond and Related Materials **8**, (1999) 1508.
2. M.J.R. Hoch, T. Tshepe and J.F. Prins, Ann. Phys. **8**, (1999) SI-93.
3. M. Bockstede, A. Kley, J. Neugebauer and M. Scheffler, Comp. Phys. Comm. **107**, (1997) 187.
4. J.E. Lowther and M.J.R. Hoch, Appl. Phys. Lett. (submitted).
5. P. P. Edwards, R. L. Johnston, E. Hensel, C. N. R. Rao and D. P. Tunstall, in Solid State Physics Vol **52**, edited by H. Ehrenreich and F. Spaepen (Academic Press, New York 1999).

Impurity scattering in metallic carbon nanotubes with superconducting pairs

Kikuo Harigaya[*]

Electrotechnical Laboratory, Tsukuba 305-8568, Japan

Abstract   Effects of the superconducting pair potential on the impurity scattering processes in metallic carbon nanotubes are studied theoretically. The backward scattering of electrons vanishes in the normal state. In the presence of the superconducting pair correlations, the backward scatterings of electron- and hole-like quasiparticles vanish, too. The impurity gives rise to backward scatterings of holes for incident electrons, and it also induces backward scatterings of electrons for incident holes. Negative and positive currents induced by such the scatterings between electrons and holes cancel each other. Therefore, the nonmagnetic impurity does not hinder the supercurrent in the regions where the superconducting proximity effects occur, and the carbon nanotube is a good conductor for Cooper pairs. Relations with experiments are discussed.

1 Introduction

Recent investigations [1,2] show that the superconducting proximity effect occurs when the carbon nanotubes contact with conventional superconducting metals and wires. The superconducting energy gap appears in the tunneling density of states below the critical temperature T_c. On the other hand, the recent theories discuss the nature of the exceptionally ballistic conduction [3] and the absence of backward scattering [4] in metallic carbon nanotubes with impurity potentials at the normal states.

In this report, we study the effects of the superconducting pair potential on the impurity scattering processes in metallic carbon nanotubes, using the continuum $k \cdot p$ model for the electronic states. We find the absence of backward scatterings of electron- and hole-like quasiparticles in the presence of superconducting proximity effects, and the nonmagnetic impurity *does not hinder the supercurrent* in the regions where superconducting proximity effects occur. Therefore, the carbon nanotube is a good conductor for Cooper pairs as well as in the normal state. This finding is interesting in view of the recent experimental progress of the superconducting proximity effects of carbon nanotubes [1,2]. We note that the details of this short report have been published elsewhere [5].

2 Impurity scattering in normal nanotubes

We will study the metallic carbon nanotubes with the superconducting pair potential. The model is as follows:

$$H = H_{\text{tube}} + H_{\text{pair}}, \tag{1}$$

* *Present address:* Agency of Industrial Science and Technology (AIST), Tsukuba 305, Japan

H_{tube} is the electronic states of the carbon nanotubes, and the model based on the $k \cdot p$ approximation [4] represents electronic systems on the continuum medium. The second term H_{pair} is the pair potential term owing to the proximity effect.

The hamiltonian of a graphite plane by the $k \cdot p$ approximation [4] in the secondly quantized representation has the following form:

$$H_{\text{tube}} = \sum_{k,\sigma} \Psi_{k,\sigma}^{\dagger} E_k \Psi_{k,\sigma}, \tag{2}$$

where E_k is an energy matrix:

$$E_k = \begin{pmatrix} 0 & -i\gamma k_y & 0 & 0 \\ i\gamma k_y & 0 & 0 & 0 \\ 0 & 0 & 0 & i\gamma k_y \\ 0 & 0 & -i\gamma k_y & 0 \end{pmatrix}, \tag{3}$$

$k = (0, k_y)$, and $\Psi_{k,\sigma}$ is an annihilation operator with four components: $\Psi_{k,\sigma}^{\dagger} = (\psi_{k,\sigma}^{(1)\dagger}, \psi_{k,\sigma}^{(2)\dagger}, \psi_{k,\sigma}^{(3)\dagger}, \psi_{k,\sigma}^{(4)\dagger})$. Here, the fist and second elements indicate an electron at the A and B sublattice points around the Fermi point K of the graphite, respectively. The third and fourth elements are an electron at the A and B sublattices around the Fermi point K'. The quantity γ is defined as $\gamma \equiv (\sqrt{3}/2)a\gamma_0$, where a is the bond length of the graphite plane and γ_0 ($\simeq 2.7$ eV) is the resonance integral between neighboring carbon atoms. When the above matrix is diagonalized, we obtain the dispersion relation $E_{\pm} = \pm\gamma\sqrt{\kappa_{\nu\phi}^2(n) + k_y^2}$, where k_y is parallel with the axis of the nanotube, $\kappa_{\nu\phi}(n) = (2\pi/L)(n + \phi - \nu/3)$, L is the circumference length of the nanotube, n ($= 0, \pm 1, \pm 2, ...$) is the index of bands, ϕ is the magnetic flux in units of the flux quantum, and ν ($= 0, 1,$ or 2) specifies the boundary condition in the y-direction. The metallic and semiconducting nanotubes are characterized by $\nu = 0$ and $\nu = 1$ (or 2), respectively. Hereafter, we consider the case $\phi = 0$ and the metallic nanotubes $\nu = 0$.

The second term in eq. (1) is the pair potential:

$$H_{\text{pair}} = \Delta \sum_k (\psi_{k,\uparrow}^{(1)\dagger}\psi_{-k,\downarrow}^{(1)\dagger} + \psi_{k,\uparrow}^{(2)\dagger}\psi_{-k,\downarrow}^{(2)\dagger}$$
$$+ \psi_{k,\uparrow}^{(3)\dagger}\psi_{-k,\downarrow}^{(3)\dagger} + \psi_{k,\uparrow}^{(4)\dagger}\psi_{-k,\downarrow}^{(4)\dagger} + \text{h.c.}) \tag{4}$$

where Δ is the strength of the superconducting pair correlation of an s-wave pairing. We assume that the spatial extent of the regions where the proximity effect occurs is as long as the superconducting coherence length.

Now, we consider the impurity scattering in the normal metallic nanotubes. We take into account of the single impurity potential located at the point r_0:

$$H_{imp} = I \sum_{k,p,\sigma} e^{i(k-p)\cdot r_0} \Psi_{k,\sigma}^\dagger \Psi_{p,\sigma}, \tag{5}$$

where I is the impurity strength.

The scattering t-matrix at the K point is

$$t_K = I[1 - I\frac{2}{N_s}\sum_k G_K(k,\omega)]^{-1}, \tag{6}$$

where G_K is a propagator of a π-electron around the Fermi point K. The discussion about the t-matrix at the K' point is qualitatively the same, so we only look at the t-matrix at the K point. The sum for $k = (0, k_y)$, which takes account of the band index $n = 0$ only, is replaced with an integral:

$$\frac{2}{N_s}\sum_k G_K(k,\omega) = \rho \int d\varepsilon \frac{1}{\omega^2 - \varepsilon^2}\begin{pmatrix} \omega & -i\varepsilon \\ i\varepsilon & \omega \end{pmatrix}$$

$$\simeq -\rho\pi i \operatorname{sgn}\omega \begin{pmatrix} 1 & 0 \\ 0 & 1 \end{pmatrix}, \tag{7}$$

where $\rho = a/2\pi L\gamma_0$ is the density of states at the Fermi energy. Therefore, we obtain

$$t_K = \frac{I}{1 + I\rho\pi i \operatorname{sgn}\omega}\begin{pmatrix} 1 & 0 \\ 0 & 1 \end{pmatrix}. \tag{8}$$

The transformation into the energy-diagonal representation where the branches with $E = \pm\gamma|k_y|$ are diagonal has the same form of t_K.

The scattering matrix t_K in the energy-diagonal representation is diagonal, and the off-diagonal matrix elements vanish. This means that only the scattering processes from k_y to k_y and from $-k_y$ to $-k_y$ are effective. The scatterings from k_y to $-k_y$ and from $-k_y$ to k_y are cancelled. Such the absence of the backward scattering has been discussed recently [4].

3 Effects of superconducting pair potential

We consider the single impurity scattering when the superconducting pair potential is present. In the Nambu representation, the scattering t-matrix at the K point is

$$\tilde{t}_K = \tilde{I}[1 - \frac{2}{N_s}\sum_k \tilde{G}_K(k,\omega)\tilde{I}]^{-1}, \tag{9}$$

where $\tilde{G}_K$ is the Nambu representation of G_K and

$$\tilde{I} = I\begin{pmatrix} 1 & 0 & 0 & 0 \\ 0 & 1 & 0 & 0 \\ 0 & 0 & -1 & 0 \\ 0 & 0 & 0 & -1 \end{pmatrix}. \tag{10}$$

The sign of the scattering potential for holes is reversed from that for electrons, so the minus sign appears at the third and fourth diagonal matrix elements.

The sum over k is performed as in the previous section, and we obtain the scattering t-matrix (with the same form in the energy-diagonal representation):

$$\tilde{t}_K = \frac{I}{1 + (I\rho\pi)^2}$$

$$\times \begin{pmatrix} 1+\alpha\omega & 0 & -\alpha\Delta & 0 \\ 0 & 1+\alpha\omega & 0 & -\alpha\Delta \\ -\alpha\Delta & 0 & -1+\alpha\omega & 0 \\ 0 & -\alpha\Delta & 0 & -1+\alpha\omega \end{pmatrix} \tag{11}$$

where $\alpha = I\rho\pi i/\sqrt{\omega^2 - \Delta^2}$.

Hence, we find that the off-diagonal matrix elements become zero in the diagonal 2×2 submatrix. This implies that the backward scatterings of electron-line and hole-like quasiparticles vanish in the presence of the proximity effects, too. Off-diagonal 2×2 submatrix has the diagonal matrix elements whose magnitudes are proportional to Δ. The finite correlation gives rise to backward scatterings of the hole of the wavenumber $-k_y$ when the electron with k_y is incident. The back scatterings of the electrons with the wavenumber $-k_y$ occur for the incident holes with k_y, too. Negative and positive currents induced by such the two scattering processes cancel each other. Therefore, the nonmagnetic impurity *does not hinder the supercurrent* in the regions where the superconducting proximity effects occur. This effect is interesting in view of the recent experimental progress of the superconducting proximity effects [1,2].

Finally, we discuss the effects of more complex pairings. We consider possible pairings between electrons with the dispersion $E = \gamma|k_y|$ and electrons with $E = -\gamma|k_y|$. This type of pairing correlation with the strength Γ will be smaller than the simple correlation with Δ. However, the effects of Γ will be interesting, because the above electronic states are degenerate at the Fermi energy. As discussed in [5], we find that the scattering t-matrix has the same form where all the off-diagonal matrix elements of the submatrices become zero. Therefore, the additional pairing potential Γ does not alter the conclusion of this report.

4 Summary

We have investigated the effects of the superconducting pair potential on the impurity scattering processes in metallic carbon nanotubes. The backward scattering of electrons vanishes in the normal state. In the presence of the superconducting pair correlations, negative and positive currents induced by the scatterings between electrons and holes cancel each other. Therefore, the carbon nanotube is a good conductor for the Cooper pairs coming from the proximity effects.

References

1. A. Y. Kasumov et al, Science **284**, (1999) 1508.
2. A. F. Morpurgo, J. Kong, C. M. Marcus, and H. Dai, Science **286**, (1999) 263.
3. C. T. White and T. N. Todorov, Nature **393**, (1998) 240.
4. T. Ando and T. Nakanishi, J. Phys. Soc. Jpn. **67**, (1998) 1704.
5. K. Harigaya, J. Phys. Soc. Jpn. **69**, (2000) 1958.

Self-organization of current density filaments in n-GaAs Corbino discs

Kazunori Aoki

Department of Electrical and Electronics Engineering, Faculty of Engineering, Kobe University, Rokkodai, Nada, Kobe 657-8501, Japan

Abstract Self-organization of the curent density filaments has been investigated in the n-GaAs Corbino discs, by the measurements of the PL spatial patterns at 4.2 K. Under the pulsed sample voltage V_s with the width of 14.8 μs and the repetition frequency of 8.5 kHz, we ovserve a novel phenomenon of a long transient of the PL patterns as a function of the time τ. It is also a new finding that, at the initial stage of the pattern formation, the nascent filaments as well as the boundaries of the evolved filaments show the anomalously bright patterns.

1. Introduction

At low temperatures a current density filament can be formed in n-GaAs between the Ohmic contacts by applying an electric field of a few V/cm due to the impact ionization avalanche of neutral shallow donors. [1]   Recently, nucleation process of the current density filaments formed by impact ionization avalanche of the shallow donors has been studied numerically in a n-GaAs Corbino disc, [2] predicting a novel phenomenon of the **self-organization** and the symmetry-breaking process. In this paper, motivated by the numerical simulation, [2] we investigate the initial pattern formation of the filaments in n-GaAs Corbino discs. By the measurements of the photoluminescence (PL) spatial patterns, [1,3] we presents new findings on the long transient of the pre-filament states.

During the impact ionization avalanche the shallow donors are completely ionized inside the filamentary channel followed by the complete quenching of the PL intensity, whereby the pattern images of the current density filament can be well reconstructed for a lightly doped sample with the donor concentration of $N_D \approx 10^{15}$/cm³. For a high-purity sample with the donor concentration of $\sim 10^{14}$/cm³, the clear pattern image of the filament can be also reconstructed under the steady state condition, but the initial pattern formation during the self-organization process exhibits more complicated situation, as will be discussed in this paper.

2. Experimental

We used a high-purity n-GaAs sample grown on semi-insulating (SI) substrate with a 20.8 μm-thick epilayer, where carrier density and mobility were n=6.23×10^{13}/cm³, μ=1.23×10^5 cm²/Vs at 77 K.[3] Diameter of the outer Corbino contact was 5.0 mm and diameter of the inner electrode was 1.6 mm. All measurements were done at 4.2 K. Spatiotemporal images of the current filaments were measured by the PL patterns, illuminating a 20 mW He-Ne laser light over the whole sample surface by using a spatial filter and strongly attenuated by using three neutral density filters and detected by an image intensifier coupled with a

12-bit CCD camera. A pulsed sample voltage V_s with the width of 14.8 μs and the repetition frequency of 8.5 kHz was applied to the inner point contacts of the Corbino disc (Fig.1(a)). The current-voltage characteristic shows typically an S-shaped curve followed by the hysteretic behaviors, where the breakdown voltage for the current filamentation was V_b=0.63 V. In order to take snapshots of the PL pattern images, a trigger pulse of the width of 463 ns was applied to the image intensifier, adjusting the time delay τ which was measured from the onset of V_s.

3. Results and discussions

By applying the pulse voltage V_s slightly above the breakdown voltage (V_s=0.78 V), we can observe the initial pattern formation of the current density filaments as a function of the time delay τ.   At the onset of the sample pulsed voltage with τ~0 ns, several pre-filament states can be seen. In Fig.1(a) with τ~0 ns, we observe two filaments labeled by 1 and 2 (the PL quenched-patterns) and nascent filaments as labeled by 3~5, where the nascent filaments 3 and 4 appear as bright PL filaments. The nascent filaments 4 and 5 decay gradually and disappear with τ >1μs. Finally the filaments 1,2 and 5 survive, as observed by the PL quench-patterns with τ=14.0 μs (Fig.1(f)).   One of our new findings is the anomalous enhancement of the PL intensities for the nascent filaments 4 and 5 and also for the bright filament boundaries of 1 and 2.   For the time delay τ ≥1.7 μs, more precise evolution of the PL pattern is shown in Fig.2.   The bright filament boundaries for τ=1.7 μs in Fig.2(a) changes into the dark line structure for τ=2.4 μs (see arrows in Fig.2(b)). The line structure is similar to the formation of the Turing pattern observed recently in the high-purity n-GaAs.[4,5]   We consider

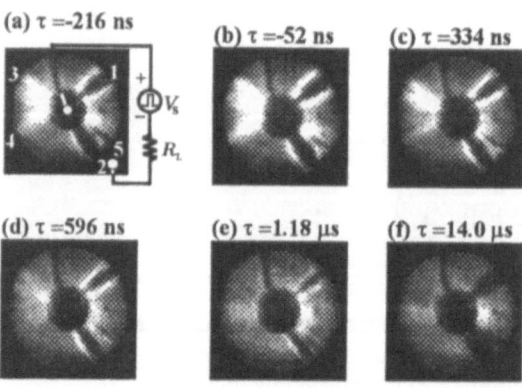

(a) τ =-216 ns   (b) τ =-52 ns   (c) τ =334 ns

(d) τ =596 ns   (e) τ =1.18 μs   (f) τ =14.0 μs

Fig. 1. Snapshots of the PL pattern images as a function of the time delay τ.

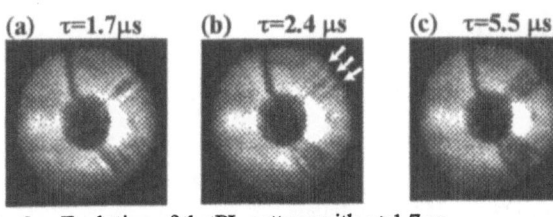

Fig. 2.   Evolution of the PL pattern with $\tau \geq 1.7 \ \mu$s.

that the long transient of the PL pattern in Fig.2 is caused by the transient behavior of the Turing bifurcation. By applying the pulsed sample voltage greater than the breakdown voltage, the load line passes through the Turing bifurcation point which locates close to the breakdown point in the negative differential resistance region of the S-shaped I-V curve [4,5], resulting in the transient appearance of the Turing pattern.

Fixing the time delay τ at a value of 364 ns, a series of the PL pattern images can be observed as a function of the pulsed sample voltage, in Fig.3. By increasing V_s, the nacent filaments labeled by 3 and 4 in Fig.3(a), (b) become the evolved filaments for $V_s \geq 1.18$ V in Fig.3(c), (d). The current density filament labeled by 2 appears clearly for V_s =0.78 V in Fig.3(b), but it disappears in Fig.3(c) for V_s =1.18 V when the rudimentary filaments 3 and 4 in Fig.3(a) becomes the evolved filament. This behavior seems to be related to the character of the self-organization process. [2] In Fig.3(d), the filament 2 appears again as well as the appearance of the filament 5.

Besides the long transient behaviors followed by the transient appearance of the Turing pattern in Fig.2, the anomalous brightening of the PL intensity for the nascent filament and also for the filament boundaries in Figs.1 and 3 is another point of the question. Under the open circuitry condition, for which only the inner contact of the Corbino disc was biased by the pulsed sample voltage with the outer contact being kept free from the applied pulse voltage, we can observe the significan brightening of the PL pattern, as shown in the top figure of Fig.4. The PL brightening was found to be caused by the initial charging current and the discharging current during the rise-time and the decay-time of the pulse voltage, respectively. The charging current and the discharging current flow in the epi-layer of the sample via the stray capacitance which originates from the sample geometry and the surround equipments in the intractable way.

The PL spectra observed under the open circutry condition are shown in Fig.4, where the photoexcitation density was $J_p \approx 20$ mW/cm². It is noted that the brightening of the PL intensity was less effective in Fig.4 for the relatively

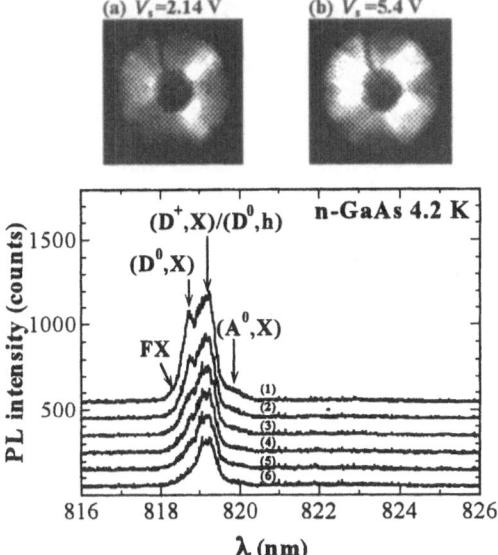

Fig. 4.   Top: the PL patterns observed under the open circuitry condition; (a) V_s =2.14 V, and (b) V_s =5.4 V. Bottom: the PL spectra as a function of the pulsed voltage V_s; (1) V_s=0 V, (2) V_s =1.98 V, (3) V_s=3.48 V, (4) V_s=5.44 V, (5) V_s=7.0 V, and (6) V_s=11.0 V. The photoexcitation density was $J_p \approx 400 \ \mu$W/cm² for the top figure, and $J_p \approx 20$ mW/cm² for the bottom.

high density of the photoexcitation $J_p \approx 20$ mW/cm². By applying the pulsed voltage, the PL peak intensity due to the exciton bound to the (D^0,X) line decreases appreciably, while the PL peak intensity due to the (D^0,h) and/or (D^+,X) emission line decreases only by a small amout. From the result, we conclude that the anomalous enhancement of the PL intensity is caused by the impact ionization process of the exciton bound to the neutral donor; $e^-+(D^0,X) \rightarrow 2e^-+(D^+,X)$ or $e^-+(D^0,X)\rightarrow e^-+D^0+X$, where e^- denotes the free electron, (D^0,X) the exciton bound to the neutral donor, (D^+,X) the exciton bound to the ionized donor, and X the free exciton, respectively.

4. Coclusion

In summary, we have found the long transient of the PL pattern during the self-organization of the current density filament, followed by the transient appearance of the Turing pattern. Anomalous enhancement of the PL intensity for the nascent filament and for the filament boundaries can be explained by the impact ionbization of the exciton bound to the neutral donor via the process; $e^-+(D^0,X)\rightarrow 2e^-+(D^+,X)$ or $e^-+(D^0,X)\rightarrow e^-+D^0+X$.

References

1. W. Eberle, J.b Hirschinger, U. Margull, W. Prettl, N. Novák and H. Kostial, Appl. Phys. Lett. 68 (1996) 3329.
2. G. Schwarz, C. Lehman and E. Schöll, Phys. Rev. **B61** (2000) 10194.
3. K. Aoki, *Nonlinear Dynamics and Chaos in Semiconductors*, (IOP, Bristol, 2000) 568 pages, in press.
4. K. Aoki, J. Phys. Soc. Jpn. **69** (2000) 324.
5. K. Aoki, Solid State Commun. **114** (2000) 203.

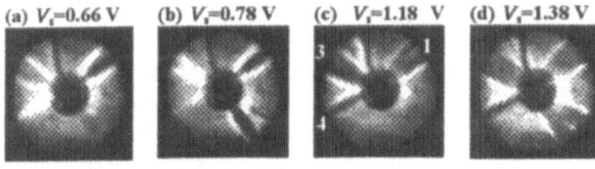

Fig. 3.   The PL pattern at the onset of the pused sample voltage as a function of V_s, where τ=364 ns.

Electrical observation of impurity levels in boron doped homoepitaxial diamond

T. Inushima[1], R.F. Mamin[1*], T. Matsushita[1], S. Ohoya[2], H. Shiomi[3]

[1] Department of Communications, Tokai University, Hiratsuka, 259-1292 Japan e-mail: `inushima@keyaki.cc.u-tokai.ac.jp`
[2] Kanagawa Industrial Technology Research Center, Shimo-Imaizumi, Ebina, 243-0435 Japan
[3] Itami Res. Lab., Sumitomo Electric Industries, Ltd., Koyakita, Itami, 664-0016 Japan

Abstract　Boron-doped p-type diamond has at least three activation energies. The first one is 0.30~0.35 eV, which is the energy from the ground states of boron to the valence band. The second one is 0.05~0.07 eV, which is the energy from the impurity band to the valence band which becomes observable when the impurity concentration increases. The third one is 0.003 eV or less, which is the energy of the variable range hopping.

1 Introduction

Impurity levels of wide band gap semiconductor play a very important role in the device application, but few precise reports have been made on them. Recent development of homoepitaxial crystal growth technique of diamond provides us with smooth and clean surfaces and we can fabricate thin diamond layers with accurate impurity concentration and with few trapping states at the interface[1]. The clean surface provides us with highly accurate and unique data of the electronic properties of impurity levels in diamond, and we have succeeded in observing electrically the impurity band position formed in the heavily doped diamond together with the multiplicity of charge states, and the emission rates of carriers[2].

In this report we present the temperature dependence of ac and dc conductance $\sigma(\omega)$ of p-type diamond by the use of a $p^+/p^-/p^+$ mesa structure, where p^+ is heavily-doped diamond with impurity band conduction and p^- is less-doped with valence band conduction. By the use of an equivalent circuit model, we analized the frequency dependence of the ac conductivity of the mesa structure. From the analysis we understood that there was a barrier height of 0.07 eV at the p^+/p^- interface, which indicates that the impurity band formed in the p^+ layer is located 0.07 eV above the valence band.

2 Experiment

The homoepitaxial films used in this experiment were grown on off-orientated substrates (the (001) substrate with a 7° off-angle toward ⟨110⟩) of polished Ib synthesized diamond by the microwave-enhanced chemical vapor deposition method. The use of these substrates enhanced the step-flow growth of the diamond[3].

The schematic drawing of the mesa structure is shown in Fig 1. To improve the quality of diamond, the reaction pressure was increased to 100 Torr and the CH_4 to H_2 ra-

* *Present address:* Kazan Physico-Technical Institute, Russian Academy of Science, Kazan, Russia, 420029, Russia

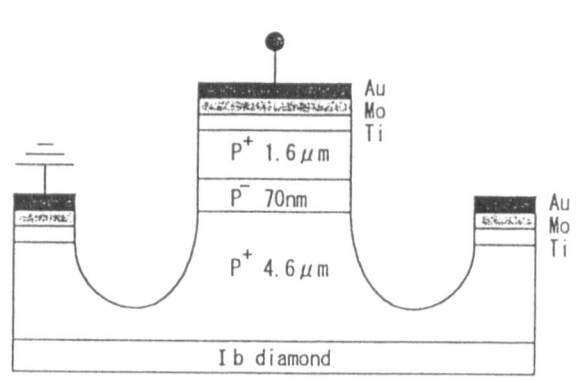

Fig. 1 Schematic drawing of the $p^+/p^-/p^+$ mesa structure.

tio was reduced to 2 %. B_2H_6 was used as the doping gas and the impurity concentration was controlled by varying the ratio of B_2H_6/CH_4. In this experiment we used heavily doped diamond (2500 ppm) as the electrodes and the conduction was monitored using a 250 ppm boron-doped sample with the thickness of 70 nm. The actual percentage of boron in the films was determined by SIMS and the impurity involvement and the gas phase mixture were almost same. The size of the substrates was $4 \times 4 \times 0.3$ mm^3 and the deposited film thickness was 4.6, 0.07 and 1.6 μm for p^+, p^- and p^+ layers, respectively. For the contact, Ti(50nm)/Mo(200nm)/Au(200nm) were fablicated. Oxygen plasma etching was conducted to form the mesa structure having the hight of 2.4 μm. DC and AC measurements were carried out using the combination of a cooled-helium circulation cryostat and HY 4140B, HP4184A and HP4185A.

3 Results and discussion

The temperature dependence of $\sigma(\omega)$ is shown in Fig. 2. At higher temperatures, $\sigma(\omega)$ approaches to σ_{DC}, and as the temperature decreases, it deviates from σ_{DC} and becomes nearly constant against the temperature change. In this flat region $\sigma(\omega) \sim \omega^{1.0}$ is satisfied. After that $\sigma(\omega)$ converges to the straight line having an activation energy similar to that of σ_{DC}. The Arrhenius plot of σ_{DC} between 150 and 300 K gives the activation energy of 0.06 eV, which indicates that there is a trap state in the p-layer 0.06 eV above the valence band.

When we plot the data given in Fig. 2 as a function of frequency, $\sigma(\omega)$ depends on ω in four steps. The first part is observed in the low frequency region where $\sigma(\omega)$ is constant against the frequency change. The second part is observed where $\sigma(\omega)$ is proportional to $\omega^{1.0}$. The third part is the second plateau, where $\sigma(\omega)$ becomes almost constant against the frequency change. The fourth is the part which shows $\sigma(\omega) \sim \omega^{1.8}$, which is observed up to the highest frequency region.

The relation of $\sigma(\omega) \sim \omega^{1.0}$, which holds over the three decades of ω, is discussed as the result of the nearest neighbor hopping mechanism between the occupied and unoccupied states in the boron doped Si[5].

The temperature and frequency dependences of $\sigma(\omega)$ given in Fig. 2 are well understood by an equivalent circuit model. In this model, we used the equivalent circuit which is composed of two capacitors C_1 and C_2 connected in series with the conductance σ_1 and σ_2, respectively. The first capacitor is due to the p^- layer and the second one is due to the barrier at the p^+/p^- interface. These two capacitors are connected with an electrode having the conductivity σ_3 in series. Fot the calculation, the thickness of the p^- layer ($d_1 = 70$ nm), the thickness of the p^+ layers ($d_2 = 1.6$ μm and $d_3 = 4.6$ μm), the electrode area $S = 0.73$ mm^2, the dielectric constant $\epsilon = 5.6$ are taken into account. The addmittance $Y(\omega, T)$ of the equivalent circuit is given as follows,

$$Y(\omega, T) = [\frac{1}{\sigma_1(T) + i\omega C_1} + \frac{1}{\sigma_2(T) + i\omega C_2} + \frac{1}{\sigma_3(T)}]^{-1}$$
$$\sigma_1(T) = \sigma_{10} exp(-\epsilon_{10}/kT) + \sigma_{11} exp(-\epsilon_{11}/kT)$$
$$\sigma_2(T) = \sigma_{20} exp(-\epsilon_{20}/kT)$$
$$\sigma_3(T) = \sigma_{30} exp(-\epsilon_{30}/kT). \tag{1}$$

For the fitting, C_1 is fixed at 550 pF because it is the capacitor of the p^- layer and is obtained by $C_1 = \epsilon S/d_1$. C_2 is fixed at 6 pF because it is the capacitor associated with the p^+ layer and $C_2 = \epsilon S/(d_2 + d_3)$. The capacitance formed at the p^+/p^- interface is involved in C_1, because we cannot distinguish it at zero bias condition.

By the use of this equivalent circuit, we can reproduce the temperature and frequency dependencies of $\sigma(\omega)$. Following parameters are obtained from the fitting; $\sigma_1(T)$ consists of two parts having different activation energies ϵ_{10} and ϵ_{11}. ϵ_{10} is 0.07 eV, which is similar to the energy observed by the carriers injected from p^+ layer[2]. ϵ_{11} is 0.003 eV, which should be the energy observed at low temperatures when the injected carrier is transferred by the hopping mechanism in the p^- layer. ϵ_{20} is 0.07 eV and it is regarded as the activation energy associated with the p^+/p^- interface. This value is close to what was observed by the thermionic current measurements at low temperatures[4] and is the energy difference between the impurity band formed at the p^+ layer and the valence band at the p^- layer. $\sigma_3(T)$ is the conductance of the bulk part of the p^+ layer, which has the length of 1μm and the area of about 0.0003 cm^2

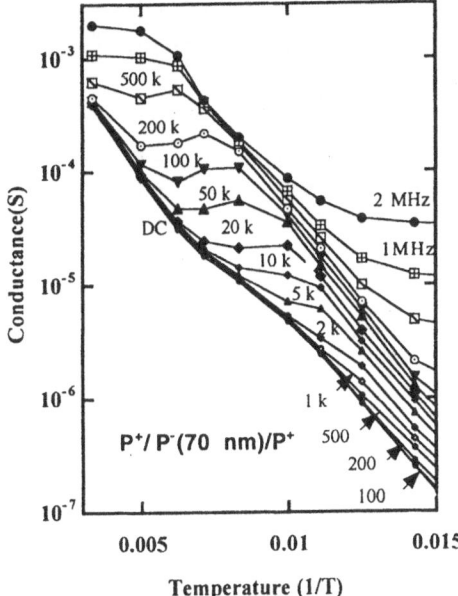

Fig. 2 Inverse temperature dependence of $\sigma(\omega)$ as a function of frequency. The measuring signal amplitude is 0.1 V.

as is seen in Fig. 1. This part shows a weak temperature dependence ($\epsilon_{30} = 3$meV), and is the most conductive among the three components ($\sigma_3(T) \rangle \sigma_1(T), \sigma_2(T)$).

In conclusion, p-type diamond has at least three activation energies. One is 0.30~0.35 eV for ϵ_1. The second is 0.05~0.07 eV for ϵ_2, which is the energy difference between the impurity band and the valence band. The third is $\epsilon_3 \sim 0.003$ eV, which is the activation energy of the variable range hopping.

-Acknowledgment-The authors are grateful to the Ministry of Education for the Grant-in-Aid for Scientific Research (C) 11650332 and to the Matsumae foundation.

References
1. H. Shiomi et al., IEEE ED Letters **16**, (1995) 36 .
2. T. Inushima et al., Appl. Phys. Lett. **77**, (2000) 1173.
3. H. Shiomi and Y. Kumazawa, Diamond Films and Technology **7**, (1998) 874.
4. T. Inushima, et al., Diamond and Related Mater. **9**, (2000) 1066.
5. M. Pollak and T. H. Geballe, Phys. Rev. **122**, (1961) 1742.

Proc. 25th Int. Conf. Phys. Semicond., Osaka 2000 (Eds. N. Miura and T. Ando)

Monte Carlo Simulation of Temperature- and Field-dependent Impact Ionization for GaAs

H.K. Jung[1]*, S.W. Ko[1], C.K. You[1], K. Taniguchi[2]

[1] School of Electronic and Information Engineering Kunsan National University, 68 Miryong-dong, Kunsan, Chonbuk 573-701, Korea e-mail: hkjung@ks.kunsan.ac.kr

[2] Department of Electronics and Information systems, Osaka University, Suita City, Osaka 565-0871, Japan

Abstract   To investigate temperature- and field-dependent characteristics of impact ionization(I.I.) for GaAs, we calculate I.I. coefficients by full band Monte Carlo simulator. We know that energy is increasing along increasing the field, while energy is decreasing along increasing the temperature since the phonon scattering rates for emission mode are very high at high temperature. The logarithmic fitting function of I.I. coefficients is described as a second orders function for temperature and field. We know, therefore, logarithm of I.I. coefficients have quadratic dependence on temperature and field, and we can save time of calculating the temperature- and field-dependent I.I. coefficients.

1. Introduction

As semiconductor device dimensions are fastly scaled down, an I.I. events are very important to analyze hot carrier transport in high energy region. The exact model of I.I. is, therefore, demanded on device simulation.

The interest for the full band model is rising, but these studies are yet insufficient since vast computer system and calculation time are demanded to analyze full band scattering rate.

In this study, we present full band model to analyze electron transport of GaAs at each temperature(4.2K, 77K, 300K, 400K, 500K) and calculate phonon scattering and I.I. rate using this model. We calculate I.I. coefficients using the full band Monte Carlo simulator and investigate dependence on temperature and field of GaAs. The Keldysh formula derived from parabolic energy band structure has been extensively used to describe I.I. though its physical meaning with the power exponent of 2 is lost at high field. We modify the Keldysh formula for full band I.I. rate. We use a wavevector- and frequency-dependent dielectric function to calculate the I.I. rate.

2. Full band model and scattering rates

A realistic energy band structure is necessary to investigate electron transport in a high electric field. We used a empirical pseudopotential method [1] to calculate realistic energy band structure. Here, we calculate a full band structure by using the eigenvalues of Hemiltonian using the reciprocal vector a set of $113\,G$. Since their effects are very small, nonlocal correction and spin-orbit splitting are not included. Prior to the derivation of the

---

* *Present address:* School of Electronic and Information Engineering Kunsan National University, 68 Miryong-dong, Kunsan, Chonbuk 573-701, Korea

phonon scattering rate, we calculate the phonon dispersion relation using a shell model [2]. The shell model has been mainly used for ionic compound semiconductors such as GaAs since the model well describes lattice dynamics of ionic crystals. In the model, a shell is assumed to be bound to its atomic core. The phonon scattering rates used in the Monte Carlo simulators have been adjusted to fit experimental impact ionization results, but this approach is not adequate to investigate especially hot electron transport which requires accurate phonon scattering rates in the high energy range. The empirical phonon scattering rate P_{ph} for the full band Monte Carlo simulation is given by deriving linear function for $P_{ph}/D(E)$ since the phonon scattering rate is linearly depended on density of states $D(E)$. We derive a linear function $aE + b$ to reduce computational time, and the empirical phonon scattering rate for the full band Monte Carlo simulation is given by;

$$P_{ph}(E) = (aE + b)D(E) \tag{1}$$

where a and b are constants, $D(E)$ is the density of states. Brillouin zone is divided into cubic mesh for calculating impact ionization rate, and the size of cubic mesh is $1/16(2\pi/a)$, where a is the lattice constant of GaAs. We use the tetrahedron method [3], in which each cubic is divided into six equal-volume-tetrahedron and each tetrahedron is integrated as the basic volume. The impact ionization is a kind of an electron-electron interaction process occured in semiconductor under high electric field. If a high energy electron in the conduction band collides with an electron in the valence band, electron-hole pair is generated, leaving two electrons in the conduction band and a hole in the valence band. The impact ionization rate was calculated from the Fermi's golden rules. Accurate calculation of impact ionization rate requires the use of a wavevector- and frequency-dependent dielectric function $\epsilon(q, \omega)$ [4]. To calculate the dielectric function, we use electron energies and wave functions of the lower eleven conduction bands and four valence bands obtained from the empirical pseudopotential method. For our full band Monte Carlo simulation, we use a modified Keldysh formula[5]. The calculated impact ionization rates are divided into two energy regions to reduce fitting errors.

3. Monte Carlo simulation

To calculate the impact ionization coefficients, we developed a full band Monte Carlo simulator which in-

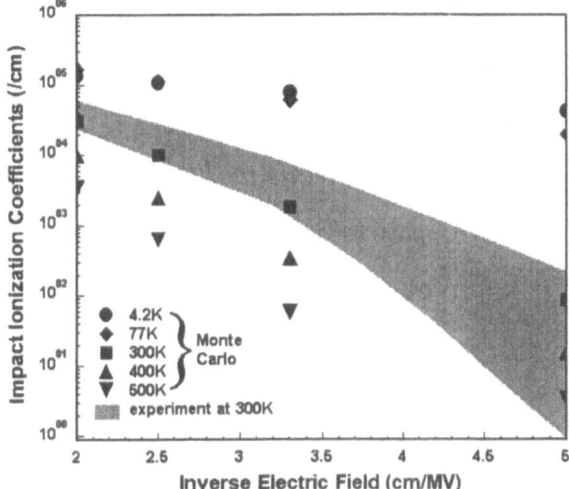

Fig. 1 Impact ionization coefficients as a function of inverse electric field calculated from the Monte Carlo simulation

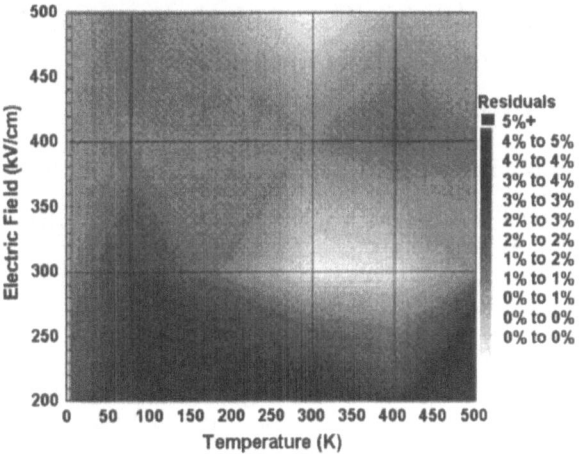

Fig. 2 Residuals of the logarithm fitting function for the impact ionization coefficients.

cludes the energy band structure derived from the empirical pseudopotential method, the phonon dispersion relation from the shell model, the empirical phonon scattering rates, the impact ionization rates derived from the modified Keldysh formula. We have generated a grid composed of 4,481 points in the first Brillouin zone, equally spaced with $1/10(2\pi/a)$. We calculated impact ionization coefficients at 4ps after electric fields are applied along the different directions in order to investigate transport characteristics for GaAs. As shown in Fig. 1, the calculated results are good agreement with a experimental data at 300K [6]-[8]. The impact ionization coefficients are decreasing along increasing temperature due to a high emission mode of phonon scattering. In this study, we fitted the impact ionization coefficients obtained from the Monte Carlo simulation to a second orders of logarithm as follow;

$$log[I.I.(T,F)] = \sum_{i=0}^{2}\sum_{j=0}^{2}a_{ij}b_{ij} \qquad (2)$$

where T is the temperature(K) and F is the electric field(kV/cm). The parameter a_{ij} and b_{ij} are presented in table 1.

Table 1 a_{ij} and b_{ij} of the logarithm equation fitted to a second orders on the temperature and field

parameter values		
$a_{00} = 4.34$	$a_{10} = -0.0276$	$a_{20} = 2.66\times10^{-5}$
$a_{01} = 0.0029$	$a_{11} = 0.000111$	$a_{21} = -1.57\times10^{-7}$
$a_{02} = -2.37\times10^{-6}$	$a_{12} = -1.15\times10^{-7}$	$a_{22} = 1.88\times10^{-10}$
$b_{00} = 1$	$b_{10} = T$	$b_{20} = T^2$
$b_{01} = F$	$b_{11} = TF$	$b_{21} = T^2F$
$b_{12} = F^2$	$b_{12} = TF^2$	$b_{22} = T^2F^2$

The values obtained by equation (2) are good agreement with those derived from Monte Carlo simulation within 5% error bound as shown in Fig. 2.

4. Conclusions

To investigate the temperature- and field-dependent impact ionization for GaAs, we calculated a full band structure using empirical psuedopotential method. We obtained the impact ionization coefficients by the Monte Carlo simulation using the phonon dispersion relation, the density of state and phonon scattering in the full band. The mean energies of secondary electrons are low at the high temperature due to the high emission mode of phonon scattering. As increasing the electric field, the impact ionization have been frequently occurred. The impact ionization for GaAs, therefore, is dependent on the temperature and the electric field.

We fitted the impact ionization coefficients obtained from the Monte Carlo simulation to a second order logarithm function for the temperature and the electric field. The residuals were within 5%. Using the presented logarithmic formula, we can save the calculation times for the impact ionization coefficients.

5. Acknowledgments

This paper is partially supported by Brain Korea 21 project.

References

1. M. L. Cohen and T. K. Bergstresser, Phys. Rev. **141**, (1966) 789.
2. K. Kunc and O. H. Nielsen, Phys. Comm. **17**, (1979) 413.
3. O. Jepson and O. K. Anderson, Solid State Comm. **9** (1971) B601
4. J. P. Walter and M. L.Cohen, Phys. Rev. **5**, (1972) 3101.
5. H. K. Jung, K. Taniguchi, C. Hamaguchi, J. Appl. Phys. **79** (1996) 2473.
6. G. E. Stillman, V. M. Robbins, K. Hess, Physica B **134** (1985) 241.
7. S. N. Shade, C. Yeh, J. Appl. Phys. **41** (1970) 4743.
8. H. D. Law, C. A. Lee, Solid State Electron, **21** (1978) 331.

A Study on Temperature- and Field-dependent Impact Ionization Coefficient for Silicon using Monte Carlo Simulation

H. K. Jung[1]*, C. K. You[1], S. W. Ko[1], K. Taniguchi[2]

[1] School of Electronic and Information Engineering, Kunsan National University 68 Miryong-dong, Kunsan, Chonbuk 573-701, Korea e-mail: hkjung@ks.kunsan.ac.kr

[2] Department of Electronics and Information Systems, Osaka University, Suita City, Osaka 565-0871, Japan

Abstract The impact ionization(I.I.) is necessary to analyze carrier transport properties under the influence of high electric field. The full band E-k relation and Fermi's golden rule are used for the calculation of impact ionization rate. We have investigated the temperature- and field-dependent impact ionization coefficient for silicon using full band Monte Carlo simulation. The impact ionization coefficients calculated by our impact ionization model are agreed with experimental data at 300K. We know that impact ionization coefficients and electron energies are decreasing along increasing temperature due to increase of phonon scattering, especially by emission. The logarithm of impact ionization coefficients are fitted to linear function for temperature and field.

1. Introduction

Recently high electric field effects such as I.I. play a very important role in hot carriers transport as device dimensions are fastly scaled down.

The Keldysh formula represented by the physical parameters E_{th}(threshold energy for I.I. event) and P(a measure to describe the magnitude of I.I. rate) has been extensively used to describe impact ionization : the E_{th} and P have been adjusted in such a way that calculated results of I.I. coefficient agree well with experimental results. And phonon scattering rates also have been adjusted without verifying the reasonableness [1].

But the physical meaning of the Keldysh formula with the power exponent of 2 is lost at high energy range because it was originally derived from parabolic band structures for both the conduction and valence bands.

Therefore the full band E-k relation and Fermi's golden rule are used for the calculation of I.I. rate [2], and the modified Keldysh formula that denotes the relation of the I.I. rate and energy is used for full band Monte Carlo simulation. We have investigated the temperature- and field-dependent impact ionization coefficient for silicon using full band Monte Carlo simulation.

2. Energy band structure and scattering rate

A realistic band structure is necessary to investigate hot electron transport in a high electric field because analytical band structures largely differ from the realistic band structure especially in the high energy range, in which I.I. effect is very strong. In the present study, the energy band structure of Si is calculated by the empirical pseudopotential method, in which the periodic part of the Bloch wave functions is expanded with a basis set of reciprocal lattice vectors G. We use a set of $113 G$ vectors for the expansion [3]. The local form factors are determined by using the steepest descent method [4]. Nonlocal corrections and spin-orbit splitting are not included since their effects are very small.

Table 1 Local empirical pseudopotential form factors [Ry]

	V_3^s	V_8^s	V_{11}^s
77K	-0.207480	0.049510	0.089600
200K	-0.207430	0.049580	0.088896
300K	-0.225800	0.056980	0.070709
400K	-0.207290	0.049780	0.087050
500K	-0.206910	0.049710	0.086270
600K	-0.206740	0.049730	0.085370
700K	-0.206590	0.049810	0.084370
800K	-0.206470	0.049910	0.083360

In order to calculate the density of states $D(E)$, we used the tetrahedron method, in which Brillouin zone is divided into $1/16(2\pi/a)$ [5]. The phonon scattering rate $P_{ph}(E)$ is given by deriving from linear function for P_{ph} (E) versus $D(E)$ since $P_{ph}(E)$ is linearly depended on $D(E)$ [6]. I.I. rate $P_{ii}(E)$ is calculated from the first principles theory. To calculate the wavevector- and frequency-dependent dielectric function, we use electron energies and wave functions of the lower eleven conduction bands and four valence bands obtained from the empirical pseudopotential method. Our impact ionization rate does not follow the simple Keldysh formula. For our full band Monte Carlo simulation, we use the modified Keldysh formula given by

$$
\begin{aligned}
P_{ii}(E) &= 2.05 \times 10^{11}(E - 1.20)^{4.60} & \text{at } 77K \\
&= 1.16 \times 10^{11}(E - 1.10)^{5.10} & \text{at } 200K \\
&= 9.74 \times 10^{10}(E - 1.10)^{5.19} & \text{at } 300K \\
&= 1.60 \times 10^{11}(E - 1.10)^{4.81} & \text{at } 400K \\
&= 1.97 \times 10^{11}(E - 1.10)^{4.63} & \text{at } 500K \\
&= 2.21 \times 10^{11}(E - 1.10)^{4.56} & \text{at } 600K \\
&= 2.35 \times 10^{11}(E - 1.07)^{4.34} & \text{at } 700K \\
&= 3.01 \times 10^{11}(E - 1.07)^{4.11} & \text{at } 800K \quad (1)
\end{aligned}
$$

* *Present address:* School of Electronic and Information Engineering, Kunsan National University 68 Miryong-dong, Kunsan, Chonbuk 573-701, Korea

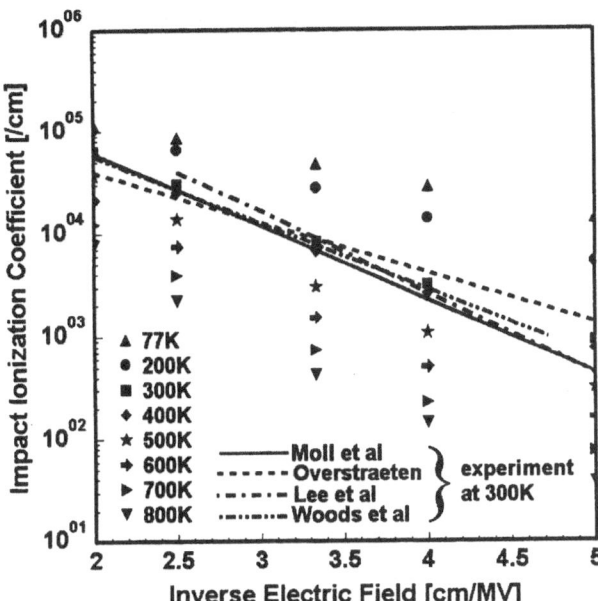

Fig. 1 I.I. coefficients as a function of inverse electric field. I.I. coefficients are decreasing along increasing temperature due to increase of phonon scattering.

3. Monte Carlo simulation

In the present study, we used the full band Monte Carlo simulator to investigate I.I. process in detail [7]. The full band Monte Carlo simulator includes the energy band structure derived from the empirical pseudopotential method, the phonon dispersion derived from an adiabatic bond-charge model [8], electron-phonon scattering rates derived from a rigid pseudo-ion model [9], the modified Keldysh formula. The phonon scattering rate is stored in a look-up table with the energy spacing of 5 meV. The polar scattering is negligible in high energy range where I.I. usually occurred. Brillouin zone is divided into $(k_x, k_y, k_z) = 1/10(2\pi/a)$ and the total number of grid points is 4,481. At these points, the electron energy, its gradients and second derivatives of five conduction bands are calculated and then stored in the look-up tables. The general full band Monte Carlo algorithm is used, and a kind of scattering, energy state, free flight time etc. are determined by the random numbers. We calculated I.I. coefficients at 4ps after electric fields are applied along the <100> direction. Fig. 1 shows the calculated I.I coefficient together with the experimental data at 300K. The results show good agreement with the experimental data at 300K. Therefore the phonon scattering rate and I.I. rate presented in this study is reasonable. The logarithm of I.I. coefficients is fitted to linear function for temperature and field as follows

$$\log[I.I(T, F)] = \sum_{i=0}^{1} \sum_{j=0}^{1} a_{ij} b_{ij} \quad (2)$$

where T is temperature[K], F is electric field[kV/cm] and parameter values of a_{ij} are a_{00}=3.82128, a_{01}=3.23145×10^{-3} [cm/kV], a_{10}=-4.77045×10^{-3} [/K] and a_{11}= 6.21162

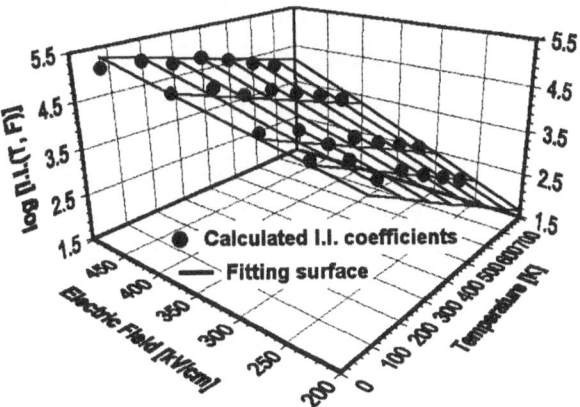

Fig. 2 Fitting surface of the logarithm for I.I. coefficients. The plane is fitting surface.

×10^{-6} [cm/K·kV], respectively. The b_{ij} are presented as b_{00} =1, b_{01}=F, b_{10}=T and b_{11}=FT. We know logarithmic I.I. coefficient is linearly dependent on temperature and field within the error bound of 5%.

4. Conclusions

The full band E-k relation is calculated by the empirical pseudopotential method, the density of states is calculated by the use of the tetrahedron method. $P_{ph}(E)$ is given by deriving from linear function for $P_{ph}(E)$ versus $D(E)$, the full band E-k relation and Fermi's golden rule are used for the calculation of I.I. rate. We used the full band Monte Carlo simulator to investigate impact ionization process in detail. I.I. coefficients are decreasing along increasing temperature due to increase of phonon scattering. The logarithm of I.I. coefficients is fitted to linear function for temperature and field. Therefore we know logarithmic I.I. coefficient is linearly dependent on temperature and field. The linear function which is presented in this study, we think, can be used in decreasing the calculating time of impact ionization coefficients.

5. Acknowledgments

This paper is partially supported by Brain Korea 21 project.

References

1. L. V. Keldysh, Sov. Phys. JETP **21**, (1965) 1113.
2. H. K. Jung, H. Nakano, K. Taniguchi, Physica B **272**, (1999) 244.
3. M. L. Cohen, T. K. Bergstresser, Phys. Rev. **144**, (1966) 789.
4. H. H. Rosenbrock, Comput. J. **3**, (1960) 175.
5. O. Jepson, O. K. Anderson, Solid State Comm. **9**, (1971) 1763.
6. H. Mizuno, K. Taniguchi, C. Hamaguchi, Phys. Rev. B **48**, (1993) 1512.
7. Y. Kamakura, K. Taniguchi et al., J. Appl. Phys. **75**, (1994) 3500.
8. W. Weber, Phys. Rev. B **15**, (1977) 4789.
9. S. Zollner, S. Gopalan, M. Cardona, J. Appl. Phys. **68**, (1990) 1682.

Critical Behavior of the Thermoelectric Transport Properties in Amorphous Systems near the Metal-Insulator Transition

C. Villagonzalo[1] *, R. A. Römer[1], M. Schreiber[1], A. MacKinnon[2]

[1] Institut für Physik, Technische Universität, 09107 Chemnitz, Germany
[2] Blackett Laboratory, Imperial College, Prince Consort Rd., London SW7 2BZ, U.K.

Abstract The scaling behavior of the thermoelectric transport properties in disordered systems is studied in the energy region near the metal-insulator transition. Using an energy-dependent conductivity σ obtained experimentally, we extend our linear-response-based transport calculations in the three-dimensional Anderson model of localization. Taking a dynamical scaling exponent z in agreement with predictions from scaling theories, we show that the temperature-dependent σ, the thermoelectric power S, the thermal conductivity K and the Lorenz number L_0 obey scaling.

The scaling description [1] of disordered systems, e.g. the Anderson model of localization, has cultivated our understanding of transport properties in such systems [2,3]. According to the scaling hypothesis, the behavior of the d.c. conductivity σ near the metal-insulator transition (MIT) in the Anderson model can be described by only a single scaling variable. As a result of the scaling theory, the dynamical conductivity in the three-dimensional (3D) Anderson model behaves as [4,5]

$$\frac{\sigma(t,T)}{T^{1/z}} = \mathcal{F}\left(\frac{t}{T^{1/\nu z}}\right) , \qquad (1)$$

where T is the temperature and t is the dimensionless distance from the critical point. For example, $t = |1 - E_F/E_c|$ where E_F and E_c are the Fermi energy and the mobility edge, respectively. The parameter ν is the correlation length exponent, which in 3D is equivalent to the conductivity exponent, $\sigma \propto t^\nu$, and z is the dynamical exponent, $\sigma \propto T^{1/z}$. It was further demonstrated that not only $\sigma(t,T)$ obeys scaling in the 3D Anderson model but also the thermoelectric power $S(t,T)$ [6,7], the thermal conductivity $K(t,T)$ and the Lorenz number $L_0(t,T)$ [7]. However, despite the quality of the scaling of σ, we obtained an unphysical value for z [7]. Scaling arguments for noninteracting systems predict $z = d$ in d dimensions [4,5]. But we found [7] $z = 1/\nu \ll 3$. In addition, values of $S(T)$ [8,9] are at least an order of magnitude larger than in measurements of doped semiconductors [10] and amorphous alloys [11,12].

In what follows, we show that we obtain the right order of magnitude [13] and good scaling for these thermoelectric transport properties by using a "modified" critical behavior of σ in the linear-response formulation for the Anderson model based on experimental data.

In the linear-response formulation, the thermoelectric transport properties can be determined from the kinetic coefficients L_{ij} [9], i.e.,

$$\sigma = L_{11}, \qquad K = \frac{L_{22}L_{11} - L_{21}L_{12}}{e^2 T L_{11}} ,$$

$$S = \frac{L_{12}}{|e| T L_{11}} , \quad \text{and} \quad L_0 = \frac{L_{22}L_{11} - L_{21}L_{12}}{(k_B T L_{11})^2} . \qquad (2)$$

The L_{ij} relate the induced charge and heat current densities to their sources such as a temperature gradient [9]. In the absence of interactions and inelastic scattering processes, the L_{ij} are expressed as [14–16]

$$L_{ij} = (-1)^{i+j} \int_{-\infty}^{\infty} A(E) \left[E - \mu(T)\right]^{i+j-2}$$
$$\cdot \left[-\frac{\partial f(E,\mu,T)}{\partial E}\right] dE , \qquad (3)$$

for $i,j = 1,2$, where μ is the chemical potential of the system, $f(E,\mu,T)$ is the Fermi distribution function, and $A(E)$ describes the system dependent features. In the Anderson model, one sets $A(E)$ to be equal to the critical behavior of $\sigma(E) \propto |1 - E/E_c|^\nu$ [2]. Note, however, that this behavior near the MIT does not contain a T dependence. Hence, the T dependence of the L_{ij} in Eq. (3) is merely due to the broadening of f and the T dependence of μ. The latter stems from the structure of the density of states, variations in which yield only negligible changes in L_{ij} [13]. Thus, in order to model a correct T dependence of the L_{ij} and, consequently, the transport properties, we need to reconsider what $A(E)$ should be.

A suitable $\sigma(E)$ may be obtained from appropriate experimental data. The recent measurements of σ by Waffenschmidt *et al.* [17] in Si:P at the MIT under uniaxial stress show that $\sigma(t,T)$ obeys scaling with $\nu = 1\pm0.1$ and $z = 2.94 \pm 0.3$. Obtaining (i) $z \approx d = 3$ in good agreement with scaling arguments [4,5] and (ii) ν which also agrees reasonably well with the numerical results for noninteracting systems [18–20] makes the experimental data in Ref. [17] an excellent empirical model for $A(E)$. In those experiments, t in $\sigma(t,T)$ is given in terms of the uniaxial stress and its critical value near the MIT. Here, we derive a functional form of $\sigma(E)$ by constructing a polynomial fit or a spline curve of the experimental data and setting $t = |1 - E/E_c|$. When using this data as

---

* *Permanent address: National Institute of Physics, University of the Philippines, Diliman, 1101 Q. C., Philippines*

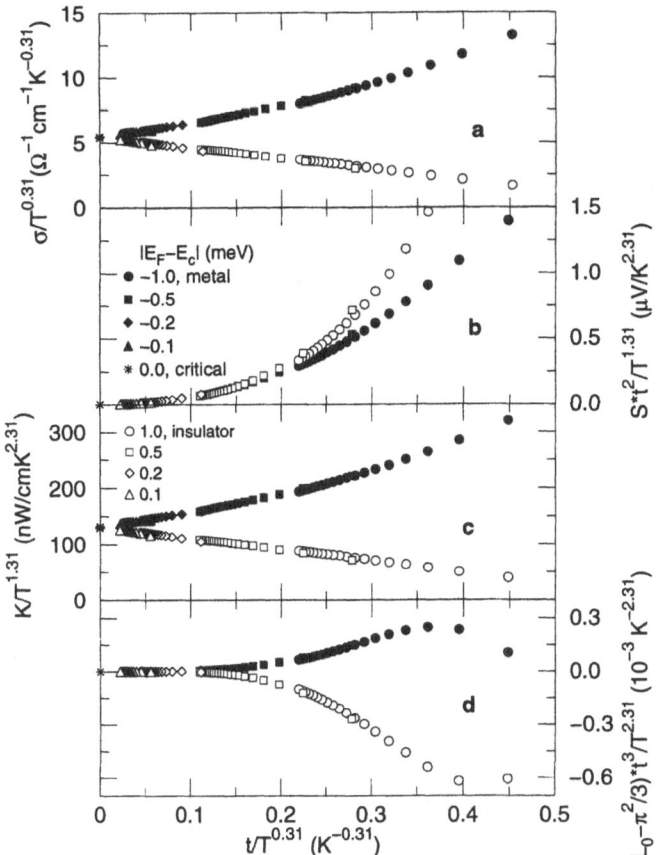

Fig. 1 Scaling of thermoelectric transport properties where $t = |1 - E_F/E_c|$. The different symbols denote the relative positions of various values of the Fermi energy E_F with respect to the mobility edge E_c.

input for Eq. (3) the difference in the order of magnitude in S as compared to experiments is removed [13]. Thus, following the approach of Ref. [13], we now study whether the T dependence of $\sigma(E,T)$ can be scaled as in Eq. (1) and whether S, K and L_0 also obey scaling.

In Fig. 1 we show that σ, S, K and L_0 data for different t and T parameters collapse onto scaling curves when plotted as a function of $|1 - E_F/E_c|/T^{0.31}$. For each figure, we clearly obtain two branches, one for the metallic regime and another for the insulating regime. As depicted in Fig. 1a, σ satisfies Eq. (1). With $\nu = 1$ in accordance with the experiment in Ref. [17], $1/\nu z = 0.31$ gives $z = 3.2$. This is in good agreement with the prediction $z = 3$ for 3D noninteracting systems. Furthermore, it indicates that the Harris criterion [21], $\nu z > 1$, is satisfied which in turn implies a sharp MIT.

The prefactor in K in Fig. 1c verifies that σ and K/T behave similarly as the MIT is approached. This confirms earlier results in the 3D Anderson model from various methods [9,22]. Meanwhile, the prefactors in both S and L_0 in Figs. 1b and 1d have not been observed in the respective scaling curves in the 3D Anderson model [7]. The prefactors serve as corrections to the scaling of

S and L_0 when an appropriate $\sigma(E,T)$ is used as input in Eq. (3).

As shown in Fig. 1d in accordance also with the results in Ref. [13], $L_0 \to \pi^2/3$, as the MIT is approached from the metallic or the insulating regime. This is the expected value in the Sommerfeld free electron theory. It is different from the result for the unmodified 3D Anderson model [9] for which the magnitude of L_0 depends on ν [9].

In conclusion, we find that by modifying $\sigma(E,T)$ in our calculations, the thermoelectric transport properties near the MIT obey scaling.

Acknowledgements The authors are grateful for the support of the DFG through Sonderforschungsbereich 393, the DAAD, the British Council and the SMWK.

References

1. E. Abrahams, P. W. Anderson, D. C. Licciardello, and T. V. Ramakrishnan, Phys. Rev. Lett. **42**, (1979) 673.
2. B. Kramer and A. MacKinnon, Rep. Prog. Phys. **56**, (1993) 1469.
3. P. A. Lee and T. V. Ramakrishnan, Rev. Mod. Phys. **57**, (1985) 287.
4. F. Wegner, Z. Phys. B **25**, (1976) 327.
5. D. Belitz and T. R. Kirkpatrick, Rev. Mod. Phys. **66**, (1994) 261.
6. U. Sivan and Y. Imry, Phys. Rev. B **33**, 551 (1986).
7. C. Villagonzalo, R. A. Römer, and M. Schreiber, Ann. Phys. (Leipzig) **8**, (1999) SI-269.
8. J. E. Enderby and A. C. Barnes, Phys. Rev. B **49**, (1994) 5062.
9. C. Villagonzalo, R. A. Römer, and M. Schreiber, Eur. Phys. J. B **12**, (1999) 179.
10. M. Lakner and H. v. Löhneysen, Phys. Rev. Lett. **70**, (1993) 3475.
11. C. Lauinger and F. Baumann, J. Phys.: Condens. Matter **7**, (1995) 1305.
12. G. Sherwood, M. A. Howson, and G. J. Morgan, J. Phys.: Condens. Matter **3**, (1991) 9395.
13. C. Villagonzalo, R. A. Römer, M. Schreiber and A. MacKinnon, preprint (2000), cond-mat/0006083.
14. G. V. Chester and A. Thellung, Proc. Phys. Soc. **77**, (1961) 1005.
15. R. Kubo, J. Phys. Soc. Japan **12**, (1957) 570.
16. D. A. Greenwood, Proc. Phys. Soc. **71**, (1958) 585.
17. S. Waffenschmidt, C. Pfleiderer, and H. v. Löhneysen, Phys. Rev. Lett. **83**, (1999) 3005.
18. K. Slevin and T. Ohtsuki, Phys. Rev. Lett. **82**, (1999) 382; **82**, (1999) 669.
19. F. Milde, R. A. Römer, and M. Schreiber, Phys. Rev. B **61**, (2000) 6028.
20. F. Milde, R. A. Römer, M. Schreiber, and V. Uski, Eur. Phys. J. B, (2000), 685.
21. A. B. Harris, J. Phys. C **7**, (1974), 1671.
22. G. Strinati and C. Castellani, Phys. Rev. B **36**, (1987) 2270.

Photo-quenching in small-polaronic conduction in LiMn$_2$O$_4$

K. Kushida and K. Kuriyama

College of Engineering and Research Center of Ion Beam Technology, Hosei University, Koganei, Tokyo 184-8584, Japan

Abstract   Electrical transport properties of spinel-LiMn$_2$O$_4$ films, synthesized on (100)-Si substrates by a sol-gel spin-coating method, were investigated. In dark d.c. conductance, the small-polaronic hopping conduction with an activation energy of ~307 meV was observed above T$_0$~280 K, arising from the localized electron hopping near the Fermi level in the Mn-t$_{2g}$ band. The dark conductance showed a transition from hopping to tunneling at around 220 K. The hopping conduction was quenched and showed the thermal recovery at around T$_0$ under the illumination with a wavelength of λ = 660 nm, although this phenomenon showed poor reproducibility in dependence on the quality of samples. The electronic structure of spinel-LiMn$_2$O$_4$ is also discussed using an optical absorption data.

1 Introduction

Recently, much interest has been attracted to various electrochemical and physical properties of spinel lithium-manganese-oxide (LiMn$_2$O$_4$) [1-5]. The authors performed the surface analysis of the films on Si with atomic-resolution in the previous study [6]. Most of the physical properties would be affected by its electronic structure. According to the ligand-field theory, Mn-3d bands of spinel-LiMn$_2$O$_4$, playing an important role in various electronic phenomena, consist of the Mn-t$_{2g}$ and Mn-e$_g$ bands. Since a proposal for the electron-configuration in spinel-LiMn$_2$O$_4$ by Goodenough et al.[1], it has been believed that equal amounts of Mn^{3+} (the electron-configuration: t$_{2g}^3$e$_g^1$) and Mn^{4+} (t$_{2g}^3$e$_g^0$) ions are distributed in the 16d sites of spinel-LiMn$_2$O$_4$ in consideration of its formal charge and charge neutrality. From this point of view, the electrical conduction has been interpreted to originate from a small-poralonic hopping between the e$_g$ levels. However, a recent calculation of the electronic structure by Liu et al. showed that the Fermi level (E$_f$) lied in the t$_{2g}$ band, not in e$_g$ bands [7], and a recent report about electrical properties suggested to be the Mn^{3+}/Mn^{4+} ratio <1 [3]. Moreover, there is no electrical conductivity data below 200 K to our knowledge. In the present study, we report electrical transport properties of spinel-LiMn$_2$O$_4$ films in dark and under the illumination in the temperature range of 15 < T <300 K. The electronic structure of spinel-LiMn$_2$O$_4$ is also discussed using an optical absorption data.

2 Experimental

LiMn$_2$O$_4$ films were synthesized on (100)-Si substrates (6×6×0.5 mm^3 in size) using a sol-gel method [6]. The spin-coated substrates were baked at 800 ℃ for 30 min under O$_2$ flow. The Ohmic contacts with a gap of 1mm were fabricated by evaporating Au on the films. d.c. conductance measurements were performed using an electrometer (KEITHLEY 617) with an input impedance of 100 tera-Ω. Photoconductance measurements were carried out under the illumination of a light emitting diode with a wavelength of λ = 660 nm. The optical absorption of the films on SiO$_2$-glass substrates was measured in the wavelength range of 400 – 1500 nm [8].

3 Results and discussion

Figure 1 shows the d.c. conductance (G) of the spinel-LiMn$_2$O$_4$ films (solid line) together with that of a neutron-irradiated GaAs (broken line) as a reference. The inset shows G vs. T$^{-1/4}$ plots for the both specimens. The conductance measurements for the neutron-irradiated GaAs, showing the Mott-type variable range hopping conduction [9], support the accuracy of the measuring system used here.

Figure 2 shows Arrhenius plots of G×T in dark and under the illumination. According to the calculated density of states of spinel-LiMn$_2$O$_4$ (see the inset in Fig.3)[7], E$_f$ lies in the Mn-t$_{2g}$ band, and the Mn-e$_g$ bands split into two bands with an energy separation of 900 meV. Therefore, the electrical transport would occur between the localized levels near the E$_f$ in the Mn-t$_{2g}$ band, not in the Mn-e$_g$ bands. Furthermore, small-polaron would originate from the interaction between electrons in t$_{2g}$ orbitals and O atoms

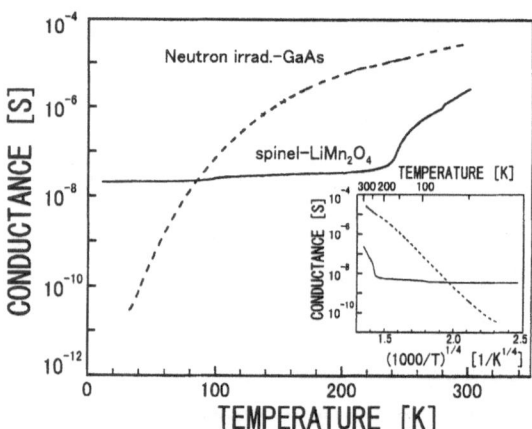

Fig. 1   The d.c. conductance G of spinel-LiMn$_2$O$_4$ films (solid line) and neutron-irradiated GaAs (broken line) as a reference. The inset shows the G vs T$^{-1/4}$ plots.

residing at the corners of the Mn octahedral matrix. Indeed, the small distortion of the Mn matrix has been observed using an x-ray absorption method [5]. Small-polaronic conduction has been predicted by the theoretical studies of polaron motion [10] as follows: Hopping conduction is represented by $G \propto T^{-1}\exp(W/kT)$ for $T > \theta_D/2$ and the tunneling conduction for $T < \theta_D/2$, where θ_D is the Debye temperature, W an activation energy, and k the Boltzman constant. θ_D of spinel-LiMn₂O₄ is unknown to our knowledge. As shown in Fig. 2, the dark conductance showed a small-poralonic hopping with the activation energy W of ~307 meV in the spinel-region (280 < T < 300 K), showing a rapid decrease of W from ~10^2 to ~10^{-1} meV via a transition region in the temperature range of 220 < T < 280 K. This indicates a transition from the small-polaronic hopping to the tunneling conduction as predicted by the above theory [10]. The discontinuous change was observed at around T_0 ~280 K, arising from the structural phase transition from spinel to tetragonal structure (c/a=1.011)[4], although some samples showed the obscure phase transition. The hopping conduction was quenched and showed the thermal recovery with an abrupt change at around T_0 under the illumination for the samples showing the phase transition clearly. The possible origin of the quenching phenomenon may be attributed to the existence of electrically inactive metastable states of Mn ion due to the lattice-relaxation induced by the illumination.

Figure 3 shows the optical absorption spectrum of spinel-LiMn₂O₄. The inset shows the schematic view of the electronic structure of spinel-LiMn₂O₄ calculated by Liu et al.[7]. The absorption peaks Ⅰ and Ⅱ were observed at around 760 nm (1.63 eV) and 620 nm (2.00 eV), respectively [8], which are assigned to the transitions from E_f in the Mn-t_{2g} band to the lower and upper Mn-e_g bands, respectively. The large absorption Ⅲ was also observed at around 400 nm (3.10 eV), which is associated with a transition from the O-2p valence band to E_f in the Mn-t_{2g} bands. These results suggest that the electrically inactive metastable state, probably Mn-e_g band, is induced under the illumination. However, this phenomenon was dependent on

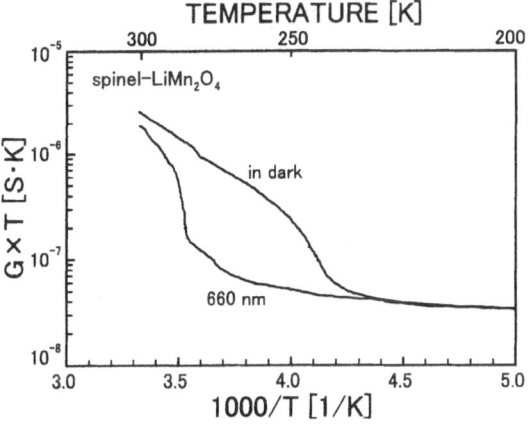

Fig. 2 Arrhenius plots of G×T in dark and under the illumination with λ = 660 nm taken in the temperature range of 200 - 300 K.

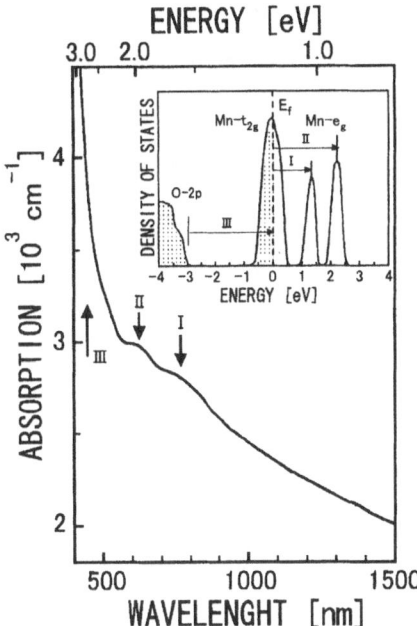

Fig. 3 The optical absorption spectrum of spinel-LiMn₂O₄ films taken at room temperature in the wavelength range of 400 - 1500 nm. The inset shows the schematic view of the electronic structure of spinel-LiMn₂O₄ calculated by Liu et al. (Ref.7).

the quality of samples.

4 Conclusion

Small polaronic hopping with an activation energy of ~307 meV was observed near the Fermi level in the Mn-t_{2g} band above 280 K. The conduction showed a transition from hopping to tunneling conduction below 220 K. We suggest that the electrical transport occurs the localized states in Mn-t_{2g} band. The hopping conduction was quenched and showed the thermal recovery with an abrupt change at around T_0 under the illumination with λ = 660 nm, although this phenomenon showed poor reproducibility. Further studies for the photoconductance will be required.

References

[1] J.B. Goodenough, A. Mnthiram and A. C. W. P. James, Mat. Res. Soc. Symp. Proc. **135,** (1989) 391.

[2] F. K. Shokoohi, J. M. Tarascon and B. J. Wilkens, Appl. Phys. Lett. **59,** (1991) 1262.

[3] E. Iguchi, N. Nakamura and A. Aoki, Phil. Mag. B, **78,** (1998) 65.

[4] A. Yamada and M. Tanaka, Mat. Res. Bull. **30,** (1995) 715.

[5] H. Yamaguchi, A. Yamada and H. Uwe, Phys. Rev. B. **58,** (1998) 8.

[6] K. Kushida and K. Kuriyama, Appl. Phys. Lett. **76,** (2000) 2238.

[7] Y. Liu, T. Fujiwara, H. Yukawa and M. Morinaga, Solid State Ionics, **126,** (1999) 209.

[8] K. Kushida and K. Kuriyama (to be submitted).

[9] M. Satoh and K. Kuriyama, Phys. Rev. B. **40,** (1989) 3473.

[10] T. Holstein, Ann. Phys. **8,** (1959) 343.

Low-frequency oscillations and chaos in semi-insulating GaAs subjected to a high magnetic field

A. Neumann[1,2,3]*, **A. G. M. Jansen**[2], **P. Wyder**[2], **R. Deltour**[3]

[1] School of Physics and Astronomy, University of Nottingham, Nottingham NG7 2RD, UK
[2] Grenoble High Magnetic Field Laboratory, Max-Planck-Institut für Festkörperforschung and Centre National de la Recherche Scientifique, BP 166, F-38042 Grenoble Cedex 9, France
[3] Université Libre de Bruxelles, Service de Physique des Solides, CP 233, Boulevard du Triomphe, B-1050 Brussels, Belgium

Abstract   We report on the influence of a strong transverse magnetic field on the current oscillations in semi-insulating GaAs to which a high DC voltage is applied. The oscillations are caused by travelling high-electric-field domains and are periodic at zero field. With increasing field they become quasiperiodic, then periodic again, and finally chaotic. This is demonstrated by means of their power spectrum, phase portrait, and the dimension of their attractor. Optical experiments reveal that the lifetime of the high-field domains decreases with increasing magnetic field, which allows us to explain the observed results.

1 Introduction

Semi-insulating (SI) GaAs exhibits spontaneous current oscillations with frequencies as low as several Hertz when a sufficiently high DC bias is applied to the sample. These oscillations are caused by travelling high-electric-field domains that are the result of a negative differential conductivity caused by electric-field-enhanced trapping of free electrons by the native EL2 defect. In this paper we describe how a strong magnetic field **B** turns initially periodic oscillations into quasiperiodic and chaotic ones. This work is an addition to previous studies of the influence of the applied voltage and illumination, see for example [1, 2].

2 Experiment

We had several samples of unintentionally doped bulk SI GaAs (Sumitomo and Wacker) grown by the liquid-encapsulated Czochralski technique, with a free-carrier concentration of $\simeq 10^{13}$ m^{-3} at zero field. For the transport measurements we used (100)-oriented samples of width 5 mm and thickness 0.45 mm as well as (110)-oriented samples of width 15 mm and thickness 0.60 mm. We did not observe any influence of the orientation on the transport properties of the samples. For the optical experiments we only used samples of the latter type. The electrical contacts consisted of parallel stripes of an evaporated Au/Ge/Ni alloy that covered the whole width of the sample. The contact spacing ranged between 5 mm and 12 mm. We used a low-noise DC power supply to apply voltages up to 1600 V and monitored the current by means of a 200 kΩ load resistor in series with the sample. An AND AD-3525 FFT Analyzer allowed us to

display simultaneously the current I as a function of time and its power spectrum from which we could determine the oscillation frequency. The experiments were carried out at or above room temperature (up to 65°C). The magnetic field was oriented perpendicular to the sample surface and thus perpendicular to the electric field inside the sample. For the transport measurements the samples were shielded against illumination.

For the optical experiments we illuminated the whole sample with an enlarged parallel laser beam of wavelength 904 nm. The sample was placed between two crossed linear polarisers at 45° with respect to the electric field inside the sample. By turning the second polariser slightly (but no more than a few degrees), we could detect in transmission, by means of a CCD camera, dark regions that moved across the sample. The shape and the velocity of the regions could accurately be linked to the temporal evolution and the frequency of the current oscillations, so that we identify the dark regions as the high-electric-field domains. The incident light can be absorbed by trapped electrons. Since the number of trapped electrons is highest inside the domains, the stronger absorption in the domains leads to dark regions.

3 Discussion

Our samples showed qualitatively similar results. In the following we will concentrate on one sample with contact spacing 7 mm to which a bias of 1200 V was applied at 42.6 ± 0.2°C. At $B = 0$ T we observed periodic oscillations with a main frequency $\nu_1 = 122.5$ Hz that are shown in Fig. 1 together with their power spectrum and their phase portrait. The power spectrum reveals the main frequency and its harmonics. The phase portrait has been obtained by a Takens-Crutchfield construction [3] by means of the current at time t and the current at time $(t + \Delta)$ where Δ can be an almost [4] arbitrarily chosen constant and in this case is 0.00039 s. The phase portrait represents the attractor of our system. In this case the attractor is circle-like in 2D, which means that the current oscillations are periodic.

Fig. 2 shows the situation at $B = 4.6$ T. One can notice that the main oscillation frequency ν_1 has increased to 155 Hz, which can be explained by the domains moving faster and decaying inside the sample before reach-

* e-mail: andreas.neumann@nottingham.ac.uk

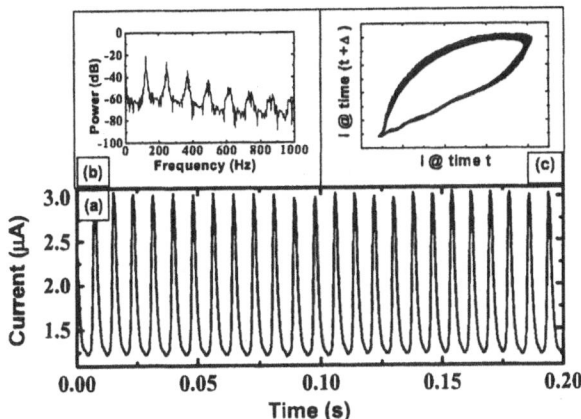

Fig. 1 (a) Periodic oscillations at $B = 0$ T; (b) power spectrum; (c) phase portrait.

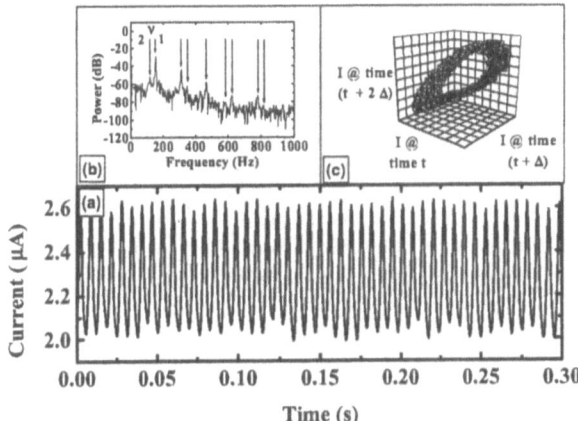

Fig. 2 (a) Quasiperiodic oscillations at $B = 4.6$ T; (b) power spectrum with arrows indicating ν_1 and ν_2 and their harmonics; (c) phase portrait.

ing the anode, as revealed by our optical experiments. In addition to this, the amplitude of the current oscillations varies in time, which leads to the existence of a second frequency $\nu_2 = 117.5$ Hz that can be seen in the power spectrum. The oscillation amplitude changes because the formation of every domain is influenced by the decay of the preceding one. Since at high magnetic fields the domains decay inside the sample and no longer at the anode, the conditions during the formation may change for different domains. The 3D phase portrait is torus-shaped with an elliptic Poincaré cross-section. Since the two frequencies ν_1 and ν_2 are incommensurate, the above results prove that the current oscillations are quasiperiodic.

At $B = 5.2$ T the oscillations are periodic again with a frequency $\nu_1 = 133.75$ Hz. This result could be confirmed by a power spectrum and a phase portrait similar to the ones displayed in Fig. 1. This evolution of the sample (periodicity - quasiperiodicity - periodicity) corresponds to the so-called frequency-locking (or mode-locking) scenario in the context of the quasiperiodic-breakdown route to chaos [5].

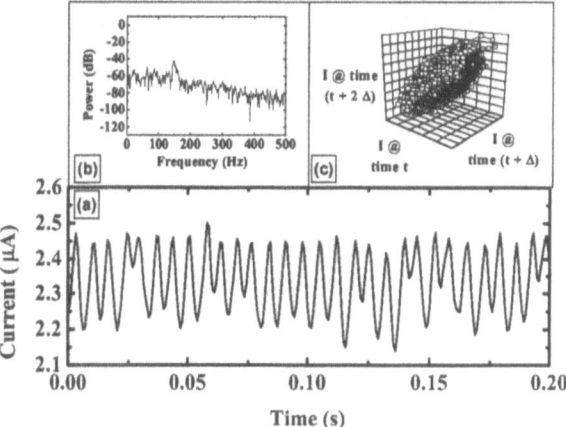

Fig. 3 (a) Chaotic oscillations at $B = 5.8$ T; (b) power spectrum; (c) phase portrait.

At even higher magnetic fields the domains do not enter the bulk of the sample anymore but appear and decay at the cathode. The corresponding current oscillations look "noisy", there is no dominant frequency anymore, and the 3D attractor is no longer torus-shaped, as can be seen in Fig. 3. From this we can conclude that the oscillations are neither periodic nor quasiperiodic. In addition we calculated the correlation dimension [6]

$$d = \lim_{\zeta \to 0} \left[\frac{1}{\log \zeta} \lim_{N \to \infty} \frac{1}{N^2} \sum_{a,b=1}^{N} \Theta(\zeta - |x_a - x_b|) \right] \quad (1)$$

where ζ is the sidelength of a volume element on the attractor, N is the number of points on the attractor, Θ is the Heaviside function, and x_a and x_b the coordinates of points a and b on the attractor. Choosing $\zeta = 0.001$, we obtained $d = 0.97$ with the original $N = 1024$ data points (scaled to lie between 0 and 1) and $d = 0.62$ for $N = 32768$ points produced by a piecewise-linear fit of the original data. These values indicate that the dimension of the attractor is not an integer and that the oscillations are chaotic.

4 Conclusion

A strong transverse magnetic field forces the current oscillations in SI GaAs under a high DC bias onto the quasiperiodic-breakdown route to chaos. This evolution is linked to a decreasing lifetime of the high-field domains that are at the origin of the current oscillations.

References

1. J. Pożela et al., Solid-State Electron. **31**, (1988) 805.
2. K. Karpińska and J. Łusakowski, Acta Phys. Pol. A **80**, (1991) 425.
3. J. Peinke et al., Encounter with Chaos (Springer, Berlin 1992).
4. J.-C. Roux et al., Physica D **8**, (1983) 257.
5. M. P. Shaw et al., The Physics of Instabilities in Solid State Electron Devices (Plenum, New York 1992).
6. P. Grassberger and I. Procaccia, Phys. Rev. Lett. **50**, (1983) 346.

The influence of the temperature in the scattering mechanisms in pure GaSb

L. G. O. Messias, E. Marega Jr.

Instituto de Física de Sao Carlos, Universidade de Sao Paulo,CP 369 Sao Carlos-SP CEP 13560-970, Brasil, e-mail:euclydes@if.sc.usp.br

Abstract   Theoretical mobility of slight n-doped GaSb as a function of temperature has been calculated by Monte Carlo method in the temperature range of 77-300K. The effect of the compensation on electron mobilities are investigated. Our results of the drift velocity at 300K with carrier concentration of 6.8×10^{16} cm^{-3} are in good agreement with available experimental results [1].

1 Introduction

In despite of a lot of research in GaSb growth mechanisms, there are few reports on bulk electronic transport properties in comparison with others III-V compounds. In a recent work Damayanthi et. al. [2] using a many-valley, anisotropic Monte Carlo model, shown some theoretical results for the electron mobility in bulk GaSb. However, their results are in disagreement with the experimental results of the W. Jantsch and H. Heinrich [1]. The Damayanthi work, predict that the drift velocity has Negative Differential Conductivity (NDC) at T = 300K, in contracdition with the results obtained by W. Jantsch and H. Heinrich for n-type GaSb with 6.8×10^{16} cm^{-3}. In another report the electronic transport in In$_{1-x}$Ga$_x$Sb [3], has been studied using a two valley model. No NDC was observed, but the results are in disagreement with W. Jantsch and H. Heinrich. In this case for high fields the contribution of the X valley should be considered [1].

2 Monte Carlo Method

In this work, we present calculations of drift velocity and mobility for bulk n-type GaSb based on Ensemble Monte Carlo method. The transport in n-type GaSb take in to account the symmetry points Γ, L and X [5] [6], so we has been used a three-valley model to represent the band structure. The dispersion relations were represented by nonparabolic analytical expression [7]. In our Monte Carlo model we included the following scattering mechanisms: acoustic modes via inelastic process (deformation potential), polar optical phonon (deformation potential), impurity scattering (Brooks Herring approach), equivalent and non-equivalent intervalley scatterings [7]. The acoustic phonon scattering has been performed as an inelastic process, because the condition $(k_b T \ll \hbar \omega)$ isn't more satisfied. In the calculations of ionized impurity scattering, we have take the aproach of Basinsky et. al.[8]. In this work, the total ionized impurity concentration used in the ionized impurity scattering rate expression is $N_i = n + 6N_a^{-2}$. We used the data of Arimoto et. al. [4] for the effective mass for L and Γ, and the well known relation to calculate effective mass

L [7]. In our analysis the influence of the temperature in the gap and in the differences of energy $\Delta \varepsilon_{\Gamma L}$ and $\Delta \varepsilon_{\Gamma X}$, were included like the usual way[6]. The parameters used in our simulation can be found in the work of Lee et. al. [6]. Parameters like the deformation potential was obtained simulating the experimental results of Heinrich [1]. The parameters used in our model are given in Table 1.

Table 1 Parameters used in Monte Carlo simulation

Γ-valley effective mass (m_Γ)	: 0.039
L-valley effective mass (m_L)	: 0.136
X-valley effective mass (m_X)	: 0.3
Acoustical deformation potential (eV)	: 3.0
Γ-L intervalley deformation potential (eV/cm)	: 1.7×10^9
Γ-X intervalley deformation potential (eV/cm)	: 0.1×10^9
L-L intervalley deformation potential (eV/cm)	: 0.6×10^9
L-X intervalley deformation potential (eV/cm)	: 0.08×10^9
X-X intervalley deformation potential (eV/cm)	: 0.08×10^9
Γ-L valley separation (meV)	: 65.0
Γ-X valley separation (meV)	: 89.0

3 Results

Figure 1 shows the temperature dependence of drift-velocity curves from 77 up to 300K as a function of the external field obtained from the Monte Carlo simulation. For temperatures in the range 150-300K no NDC was observed. The NDC will be obtained when energy separation $\Delta_{\Gamma L}$ is less or comparable to $4k_b T$. This is fulfilled for GaSb at 300K, and for 150K the energy separation begin to be comparable. However, when the temperature decrase below 100K, $\Delta_{\Gamma L}$ is greater than $4k_b T$ and the NDC start to be observed.

The effect of ionized-impurity scattering is shown in Figure 2, for three different values, $N_i = 5 \times 10^{15}$, $N_i = 1 \times 10^{16}$ and $N_i = 5 \times 10^{16}$ cm^{-3} to uncompensated material ($N_a = 0$). Clearly, for the velocity-field curve at $N_i = 5 \times 10^{16}$ cm^{-3} the NDC start to disappears, it's due to increase in the ionized-impurity scattering. Similar behavior was found by Ikoma et. al.[3] in Ga$_{0.9}$In$_{0.1}$Sb. This effect could explain why no NDC was observed at T = 100K.

Simulations results of the mobility as a function of the temperature with acceptors concentrations covering the range $N_a = 0$ to $N_a = 1.5 \times 10^{15}$ cm^{-3} are shown in Figure 3. At $N_a = 0$ was expected that the mobility decrease with the temperature [9]. When the acceptor concentration increase the mobility decrease in the range

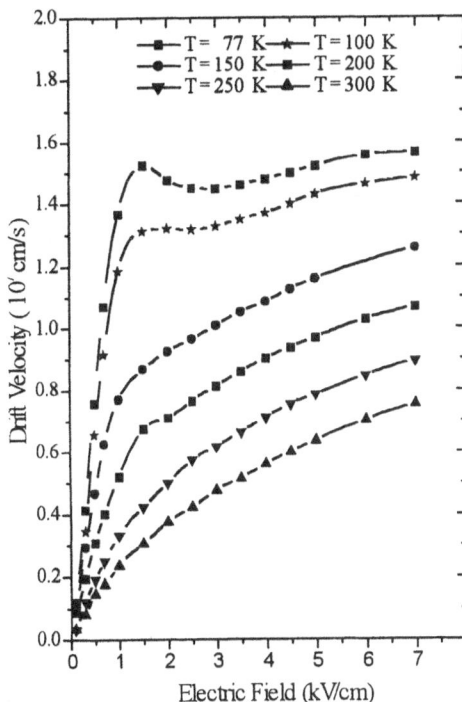

Fig. 1 Temperature dependence of the electron drift-velocity as a function of the external field for $N_i = 1 \times 10^{16}$ cm^{-3}. No compensated acceptors are considered.

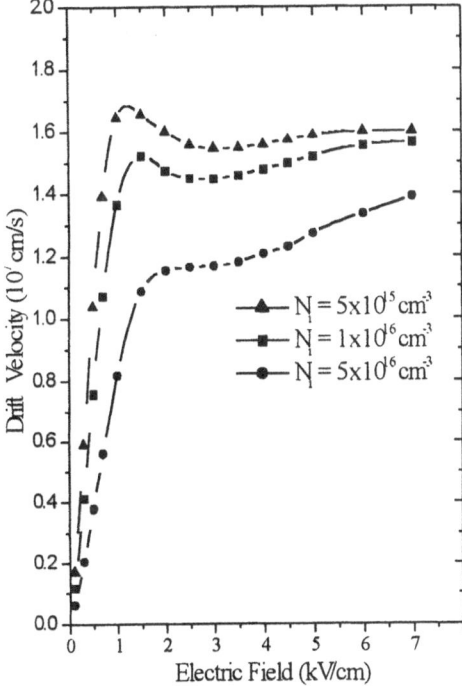

Fig. 2 Calculated electron drift velocity as a function of electric field for three value of the impurity concentration at T = 77K. No compensated acceptors are considered

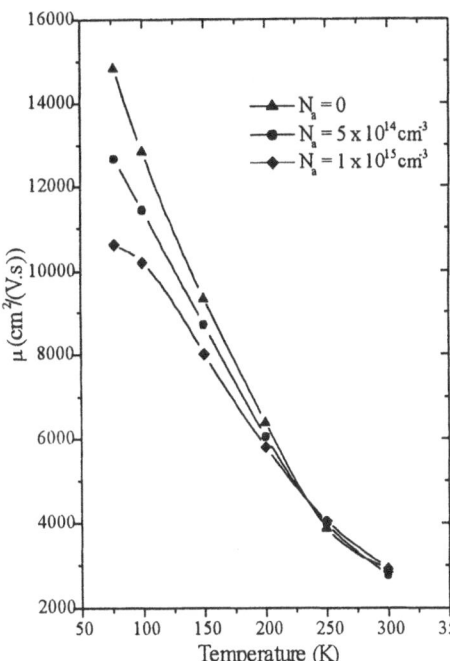

Fig. 3 Monte Carlo results of the electron mobility in GaSb as a function of temperature. We studied the influence of compensation on the mobility at $N_i = 1 \times 10^{16}$ cm^{-3}.

T = 77K to T = 200K. This behavior could be explained by the increase in the acceptor concentration to the same level of the impurity concentration, and consequently the increase in the ionized impurity scattering. In the region above 200K, the phonon scattering is predominating and the influence of the impurity scattering is small.

4 Conclusion

In conclusion, we have studied the electron transport in uncompensated and compensated slight n-doped GaSb. The NDC regime occurs at low temperature but can be suppressed by a small increase in the impurity concentration. The compensation of the carriers affect the mobility in low temperature, and in the region close the room temperature the phonon scattering masked the influence of the impurity scattering.

References

1. W. Jantsch, and H. Heinrich, Phys. Rev. B **3**, (1971) 420
2. P. Damayanthi and R. P. Joshi, J. Appl. Phys. Lett **86**, (1999) 5060
3. T. Ikoma, K. Sakai, Y. Adachi and H. Yanai Jap. J. Appl. Phys. **16**, (1977) 1379
4. H. Arimoto, N. Miura, R. J. Nicholas, N. J. Mason and P. J. Walker, Phys. Rev. B **47**, (1998) 4560
5. P. S. Dutta and H. L. Bhat, J. Appl. Phys. Lett, **81** , (1997) 5821
6. H. J. Lee and J. C. Wooley, Can. J. Phys. **59** (1981) 1844
7. K. Tomizawa, *Numerical Simulation of Submicron Semiconductor Devices*, (Arthec House, Boston-London 1993)
8. J. Basinsky, C. C. Y. Kwan, and J. C. Wooley, Can. J. Phys. **50**, (1972) 1068
9. D. L. Rode, Phys. Rev. B **2**, (1970) 1012

Reversible Quantum Dynamics of Impurity-Bound Electrons in GaAs

B. E. Cole[1*], **J. B. Williams**[1], **M. S. Sherwin**[1], **C. R. Stanley**[2]

[1] Quantum Institute, University of California at Santa Barbara, CA 93106, USA

[2] Department of Electronics and Electrical Engineering, University of Glasgow, Glasgow G12 8QQ, United Kingdom

Abstract   Here we describe the first observation known to us of Rabi oscillation in the time-domain of electronic motional states in a doped semiconductor. In particular, we report the dynamic response of impurity-bound electrons to intense terahertz (THz) radiation in lightly-doped GaAs. Photoconductivity is used to probe the state of the system at the end of each excitation pulse and the evolution of the 1s bound state population due to the THz driving field is measured as a function of time by varying the THz pulse length from a few ps to 50ps. When the THz frequency is resonant with the 1s-2p$^+$ impurity transition, the photocurrent executes damped oscillations whose frequency increases with THz field strength and detuning from resonance. The observed Rabi-oscillation frequency is in good agreement with that expected from calculations using the THz field strangth and the 1s-2p$^+$ transition matrix element of a hydrogenic system.

1 Introduction

The intraband dynamics of a semiconductor quantum system in the presence of an intense periodic driving field have been studied previously in the steady state, with the observation of many novel phenomena including photon-assisted tunnelling [1,2], AC-Stark and AC-Franz-Keldysh effect [3], sideband generation [4] and electromagnetically induced transparency [5]. Experimental investigation into the dynamic response due to an intense transient AC field have, to date, be confined to measurements of the relaxation characteristics of the system after the excitation field is removed. For example intersubband population relaxation rates [6], dephasing rates [7] and carrier cooling rates [8] have all been obtained. Here we report measurements of the temporal evolution of a donor-bound electron after the transient 'switch-on' of an intense THz driving field. Damped oscillation of the electron between the 1s and $2p^{m_j=+1}$ (or 2p$^+$) bound states (the so-called 'Rabi' oscillation [9]) is observed; this represents the first time the transient motional-dynamics of periodically-driven electrons in a semiconductor system have been investigated in the time domain. The oscillations indicate the reversible nature of the excitation process; electrons initially excited to the 2p$^+$ state can be coherently de-excited back to the ground 1s state by a driving pulse of appropriate length and phase.

Shallow donor states in GaAs are hydrogenic in character with a Bohr radius ~10 nm. The 1s-2p$^+$ transition lies in the THz frequency range and may be tuned from 1 THz to over 5 THz by the application of a magnetic field of just a few Tesla. Photoconductivity has been extensively used for CW spectroscopy of the energy level structure of the donor systems [10,11]. In common with quantum-dot states, donor state levels in GaAs have significantly longer lifetimes than 1-or-higher dimensional systems due to the low probability for acoustic phonon relaxation. Photoconductivity in donor systems provides us with a particularly convenient means of probing the system. Electrons left in upper bound states of the donor 'atom' after a THz excitation pulse do not return to the ground state directly but are preferentially ionized, by acoustic phonon absorption or emission, to the conduction band where the electron may contribute to the macroscopic conductivity of the sample, an easily accessed experimental quantity. The sequence of processes characterizing the photoconductive response of the system is summarized as follows: 1) THz photoexcitation out of the ground state 2) Ionization from the excited state 3) transport under an applied DC field 4) recapture 5) cascade to the donor ground state. In our experiment, the evolution of the donor-bound electron population out of the 1s ground state is determined by measurements of the transient photocurrent (step 3 above) as a function of the THz driving pulse *length*.

Previous studies[12] have determined the lifetime of the 2p$^+$ level to be limited to not more than 300ps, under weak THz driving conditions. Transition linewidths indicate a dephasing rate somewhat shorter than this, due to inhomogeneous broadening. Planken et al. [7] have observed the free-induction decay of the 1s-2p$^+$ n-GaAs donor transition. In this work, the dynamics of the system (the decay of the polarization) is probed after a THz pulse and the donors are not driven far from equilibrium. In the present work, intense THz pulses coherently empty and then repopulate the ground state of the donors.

2 Experimental Details

The sample was fabricated from a 15 μm thick, unintentionally doped GaAs layer grown by molecular beam epitaxy on a semi-insulating GaAs substrate[13]. The residual electron density was 2.8×10^{14} cm^{-3} with mobility of 140,000 V-1s-1cm2 as determined from Hall measurements at 77K [14]. Quasi-CW photoconductivity spectra

---

* *Present address:* Toshiba Research Europe Ltd., 260 Cambridge Science Park, Milton Road, Cambridge CB4 0WE, United Kingdom. email: bryan.cole@crl.toshiba.co.uk

indicate that a single donor species (sulphur) dominates in the sample. The sample had a 20 nm Si-doped (10^{18} cm^{-3}) cap layer to facilitate the formation of ohmic contacts (AuNiGe). This cap layer was etched off the remaining sample area after alloying the contacts. Care was taken to ensure a THz electric field sufficiently homogeneous over the active region of the sample to avoid blurring of field-dependent dynamics. Wet etching was used to define the strip of GaAs between the contacts. The 100 μm-by-100 μm active region of the sample, as defined by the ohmic contacts and the width of the mesa, was much smaller than the 500 μm FWHM size of the THz focus. To minimise distortions of the THz field by the ohmic contacts, these were made in thin lines perpendicular to the polarisation of the THz electric field. The back side of the sample had an impedance-matched anti-reflection coating to eliminate internal reflections [15]. The sample was placed in the bore of a superconducting solenoid and immersed in pumped liquid helium. The sample temperature was 2 K for all measurements. The design of the sample is illustrated in Figure 1a.

The sample was driven by suitably intense, tunable THz radiation provided by the UCSB free electron lasers [16]. These provide radiation from 0.2 THz to 4 THz at a kW power level in the form of μs-long pulses. A fixed frequency of 2.5 THz was chosen for this experiment, which coincides with a window in the atmospheric absorption. The 1s-2p$^+$ transition could be tuned to this frequency with a magnetic field of 3.51 T. In order to generate shorter THz pulses, of variable length, a 'pulse-slicing' facility has been implemented for use with the UCSB FELs . This system provides approximately 'flat-top' THz pulses, with the unique facility to vary the pulse duration continuously from a few ps up to 4 ns at any wavelength within the FEL tuning range. Details of the pulse-slicer design and performance are given elsewhere [17,18]. The pulse-slicer was optimised to obtain a 'contrast ratio'[1] of 5×10^6; at this level, the pre-pulse power was insignificant.

While pulses down to under one cycle in length may be switched at the output of the pulse-slicer, the filtering effect of atmospheric absorption between the pulse-slicer and sample cryostat limits the minimum pulse length at the sample to about 8 ps (4 ps risetime). This is consistent with an THz atmospheric transmission window about 80 GHz at a center frequency of 2.5 THz.

The sample was connected directly to a 50Ω coaxial cable which fed through a 'bias-tee' into a broadband pulse amplifier and a 750 MHz sampling oscilloscope. The sample was biased with an electric field of 10V/cm, below the threshold for impact ionization. The photocurrent transient is integrated over several ns after the end of the THz pulse. The temporal profile of the photocur-

---

[1] Defined as the ratio of main 'sliced' pulse power to any residual THz power breaking through the slicer-system before the main pulse.

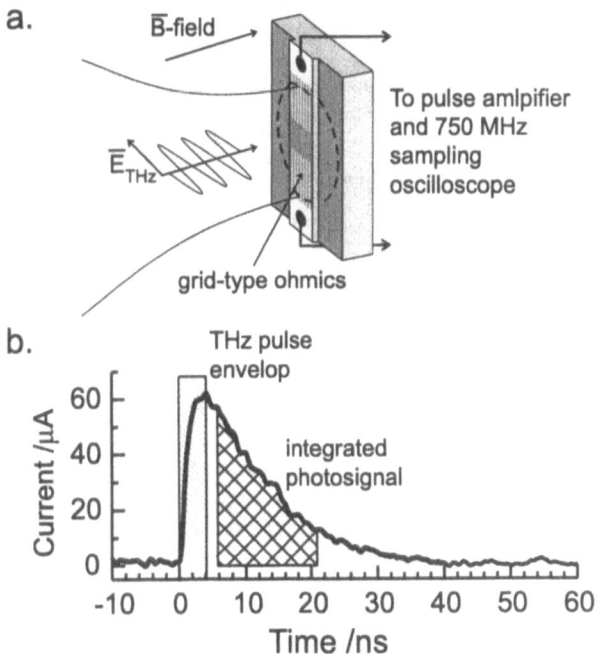

Fig. 1 a) The sample geometry is illustrated; grid-type ohmic contacts prevent near-field distortion of the THz field by the contacts. b) The trace shows a typical photocurrent transient, as measured directly on a sampling oscilloscope. The integrated photocurrent (cross-hatched) is measured vs. pulse length. The duration of the THz pulse (hatched) is shown.

rent transient is shown in Fig. 1b; the exponential decay of the photocurrent is determined by the electron recapture rate, a constant sample parameter for the purposes of this experiment. Excited electrons are known to ionize within the first ns after the end of the THz pulse [12]. The integrated photocurrent is thus proportional to the total fraction of electrons excited out of the 1s ground state at the end of the THz pulse. By measuring the integrated photocurrent as a function of pulse duration we obtain the excited electron population as a function of time. The repetition rate of the THz pulses was under 10 Hz ensuring that the system had ample time to return to equilibrium after each transient.

3 Experimental Results

The integrated photocurrent is plotted vs. pulse length in Fig. 2a for a range of THz field strengths. Each point represents a single THz pulse; no averaging was performed. The magnetic field was set to 3.56 T, corresponding to the on-resonance condition at the highest THz fields. One to two cycles of Rabi-oscillation are clearly visible and the oscillation frequency is seen to increase with increasing THz power, as expected [9]. Unexpectedly, the number of cycles of oscillation remains constant with increasing THz field, indicating that the dephasing rate increases with the THz field (see discussion below). Fig. 2b shows the integrated photocurrent vs. pulse length at a fixed THz field and three dif-

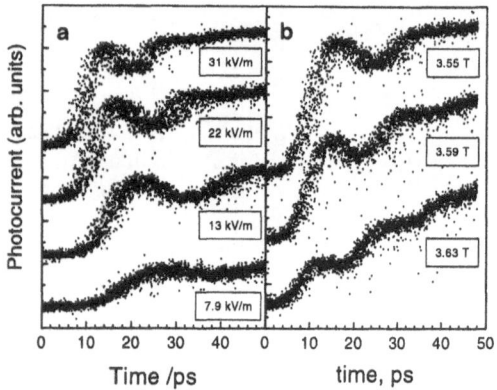

Fig. 2 a) The integrated photocurrent is plotted vs. THz pulse length for four values of THz field-strength. Each datapoint indicates one THz pulse. The magnetic field was set to 3.56 T. Traces are offset for clarity. b) The integrated photocurrent vs. pulse length for three different values of magnetic field (i.e. detuning). The THz field strength was 2.1×10^4 V/m.

ferent values of B-field detuning. As the system is detuned from resonance, the frequency of the oscillations increases while the depth of modulation decreases.

The magnetic field for which the photocurrent rise-time is a minimum corresponds to the 'on-resonance' condition (in the 2-level approximation). spectra plotted in Fig. 3a shows the 1s-2p$^+$ resonance in the weak-driving limit at 3.51 T. When the integrated photocurrent at a fixed pulse length is plotted as a function of magnetic field, the peak value, corresponding to the resonance, is found to be shifted up in magnetic field. The photocurrent for a series of pulse lengths at a THz field strengths of 2.1×10^4 V/m and 4.9×10^4 V/m is plotted in Figs. 3b and 3c respectively. The resonant field is indicated by an arrow in each case.

4 Discussion

The theoretical treatment of Rabi-oscillations is usually given in the context of a closed two-level 'atomic' system [19]. The oscillation frequency, Ω is given by $\Omega = \sqrt{\Omega_R^2 + \Delta^2}$, with the bare Rabi-frequency, $\Omega_R = \frac{eE_{THz}z_{12}}{\hbar}$ and the detuning, $\Delta = \omega_{THz} - \omega_{12}$, z_{12} denotes the transition dipole matrix element and ω_{12} denotes the bare transition energy. On resonance, where $\Delta = 0$, the oscillation frequency equals the bare Rabi frequency. For a THz field strength of 3.1×10^4 V/m and assuming a matrix element of 10 nm, a Rabi frequency of 4.7×10^{11} radians/s is predicted; this is well within the experimenetal error of the value found experimentally.

The Rabi oscillation frequency, as determined from the period of the 1st cycle of oscillation, is plotted against THz field strength in Fig. 4, at a magnetic field of 3.56 T. This field was chosen to correspond to the 'on-resonance' condition at the highest THz field strengths. A significant zero-field intercept for the Rabi frequency is apparent, indicating the de-tuning from resonance in the

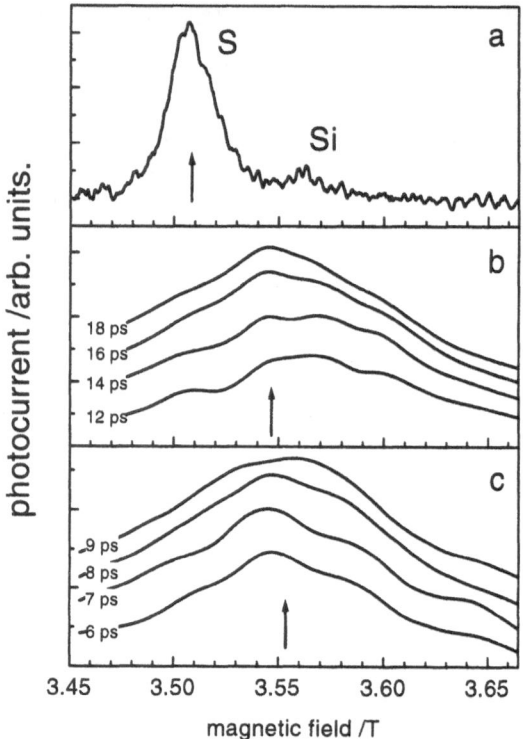

Fig. 3 a) Linear photoconductivity vs. magnetic field. b) The integrated photocurrent measured at four different THz pulse lengths (indicated) vs. THz pulse length and magnetic field at a THz field strength of 2.1×10^4 V/m. c) Integrated photocurrent vs. magnetic field for four pulse lengths at a THz field strength of 4.9×10^4 V/m. Arrows indicate the 'on-resonance' field for all three plots.

absence of THz. The data is fitted with a $\sqrt{\Omega_R^2 + \Delta^2}$ function (solid line in Fig. 4) and the detuning, Δ, found to be 22 GHz. This agrees well with the expected detuning, as obtained from the linear photocurrent spectra.

At the lowest excitation fields, The observed damping rate is close to that obtained from the linear spectra, 63×10^9 rad/s. In this regime, inhomogeneous broadening dominates the dephasing and originates from a background disorder potential due to the random distribution of impurities [20]. In addition to dephasing in the conventional sense, loss of electrons from the system (i.e. out of the 2p$^+$ state) may contribute to damping of the Rabi oscillations. Electrons that are ionized from the 2p$^+$ state during the THz pulse will contribute to the photocurrent transient in a similar way to those ionized after the pulse. Hence, the photocurrent tends toward its maximum value in the long-pulse limit (corresponding to the case where all donors are ionized). This contrasts with the ideal 2-level case for a fully dephased system where are carrier population is divided equally between the two levels. A third parameter, the ionisation rate, is

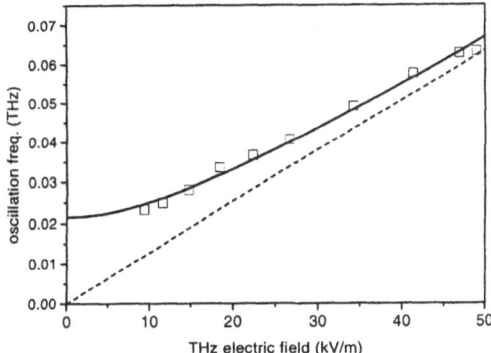

Fig. 4 The photocurrent oscillation frequency, obtained by measurement of the period of the first 'Rabi' oscillation is plotted against THz field strength (datapoints). The fit to the data is shown (solid) - see text. The expected THz field dependence for the zero-detuning case is also shown (dotted).

required to fully characterise the data. The experimental results indicate that as the THz field is increased the dephasing rate increases to approximately four times the low-THz-field value. Thus, we conclude that the THz field itself gives rise to rapid (photon-driven) ionisation from the $2p^+$ level.

The apparent shift in the resonant field to higher B as the THz field strength is increased provides further distinction from the 2-level case. This appears be an AC-Stark-like effect due to dipole coupling of the $2p^+$ level to another state(s) approximately 1 THz photon energy above. We can discount any perturbation of the 1s state through coupling to a state other than the $2p^+$; dipole coupling to another level would be readily observed in the linear absorption or photoconductivity spectra and, experimentally, is none is found sufficiently close to the $2p^+$ energy to explain our results. While the $2p^+$ state is degenerate with the N=0 Landau-level continuum at this magnetic field, dipole transitions between the $1s_{(m=0)}$ state and this $m = 0$ continuum are forbidden. Perturbation to the $2p^+$ energy by coupling to the $3d^{+2}$ might be expected; however, this level is known[21] to lie significantly less than the $1s$-$2p^+$ transition photon energy above the $2p^+$ and hence would tend to shift the $2p^+$ up in energy, in contrast to what we observe here. Thus, it seems a more detailed picture of the donor energy-level structure including the many possible $m_j = 2$ is required before further progress in understanding the THz-induced shift in $1s$-$2p^+$ energy can be made.

In conclusion, we have reported the first measurements of the time-domain response of donor-bound electrons to intense THz transients. Rabi oscillations between the 1s and $2p^+$ states have been observed and the dephasing rate of the system is found to increase with THz driving strength, consistent with photon-driven excitation to states outside the $1s$-$2p^+$ system. The $1s$-$2p^+$ level separation is found to shift to lower energies as the

THz drive strength is increase. Taken together with the field-dependent dephasing rate, dipole coupling of the $2p^+$ state to another unidentified state close to one THz photon energy above is suggested. In this work, the number of Rabi-oscillations through which the system can be driven is limited, possibly by strong optical coupling to higher donor levels and ultimately by phonon-emission, both these processes should be substantially reduced in a state several THz photons below the conduction band edge such as the $2p^-$ state. Future work aims to probe the intrinsic dephasing processes of this state with the aim to acheiving much longer coherence times.

The authors gratefully acknowledge financial support from the ARO, the ONR/Medical Free-Electron Laser Program, and the NSF. We thank C. Sean Roy for assistance with the experiments.

References

1. H. Drexler *et al.*, Appl. Phys. Lett. **67**, 2816 (1995).
2. L. Kouwenhoven *et al.*, Phys. Rev. Lett. **73**, 3443 (1994).
3. K. Nordstrom *et al.*, Phys. Rev. Lett. **81**, 457 (1998).
4. C. Phillips *et al.*, Appl. Phys. Lett. **75**, 2728 (1999).
5. C. Phillips, proceeding of this conference **N11**, (2000).
6. J. Heyman *et al.*, Appl. Phys. Lett. **68**, 3019 (1996).
7. P. C. M. Planken *et al.*, Phys. Rev. B **51**, 9643 (1995).
8. B. Murdin *et al.*, Phys. Rev. B **55**, 5171 (1997).
9. I. Rabi, Phys. Rev. **51**, 652 (1937).
10. T. O. Klaassen, J. L. Dunn, and C. A. Bates, in *Atoms and Molecules in strong external fields*, edited by P. Schmelcher and W. Schweizer (Plenum Press, New York, 1998), pp. 291–300.
11. G. E. Stillman, C. M. Wolfe, and J. O. Dimmock, Solid State Commun. **7**, 921 (1969).
12. J. Burghoorn, T. O. Klaassen, and W. T. Wenckebach, Semicond. Sci. Technol. **9**, 30 (1994).
13. C. R. Stanley *et al.*, Appl. Phys. Lett. **58**, 478 (1991).
14. C. R. Stanley *et al.*, in *Inst. Phys. Conf. Series*, edited by K. E. Singer (IOP Publishing Ltd., London, 1990), Vol. 112, pp. 67–72.
15. S. W. McKnight, K. P. Stewart, H. D. Drew, and K. Moorjani, Infrared Physics **27**, 327 (1987).
16. G. Ramian, Nuclear Instrumentation and Methods in Physics Research **A318**, 225 (1992).
17. F. A. Hegmann *et al.*, Appl. Phys. Lett. **76**, 262 (2000).
18. F. Hegmann and M. Sherwin, Proc. SPIE **2842**, 90 (1996).
19. R. W. Boyd, *Nonlinear Optics* (Academic Press, London, 1992).
20. U. Bockelmann, Phys. Rev. B **50**, 17271 (1994).
21. P. Planken *et al.*, Optics Communications **124**, 258 (1996).

Magnetic field enhanced terahertz emission from semiconductor surfaces

R. McLaughlin[1,2], A. Corchia[1,2], M. B. Johnston[1], C. M. Ciesla[2], D. D. Arnone[2], G. A. C. Jones[1], E. H. Linfield[1], A. G. Davies[1], M. Pepper[1,2]

[1] Semiconductor Physics Group, Cavendish Laboratory, University of Cambridge, Madingley Road, Cambridge. CB3 0HE. UK.
[2] Toshiba Research Europe Limited, Cambridge Research Laboratory, 260 Cambridge Science Park, Milton Road, Cambridge. CB4 0WE. UK.

Abstract   We present an experimental study of enhanced THz emission via surface field generation under the influence of a magnetic field for GaAs, InAs and InSb . Furthermore, we reveal the mechanism responsible for this behaviour using a Monte Carlo model of photo-generated carriers within the surface field depletion region.

1 Introduction

There is currently a great deal of interest in the production of bright, coherent Terahertz (THz=10^{12}Hz) sources to span the so-called "terahertz gap" in the electromagnetic spectrum from 100GHz to 10THz. Applications in this frequency range include spectroscopy [1] and more recently medical imaging [2]. There are currently no compact solid state sources of coherent THz radiation within this frequency range, although recent advances in ultrafast pulsed lasers operating in the near-infrared and visible have allowed coherent, broadband THz pulses to be produced from semiconductors.

We report the magnetic field enhancement of coherent THz emission due to the surface field photo-current. The high magnetic field enhancement from InAs has recently been reported [3,4] with the general features in [4] explained using a simple Lorentz force model. In this report InAs, GaAs and InSb results are presented and the mechanism for the enhancement is discussed using a *many carrier* Monte Carlo simulation.

2 Experimental Set-up

A mode-locked Ti:Sapphire laser producing 100fs pulses at a wavelength of 790nm and a repetition rate of 82MHz, yielding 0.8W average power was used to illuminate the samples within a cryostat with optical access. The illumination intensity could be varied using a $\frac{\lambda}{2}$ plate and polariser arrangement such that optical fluence-dependent measurements were possible. The optical cryostat, with a split coil super conducting magnetic, was capable of producing magnetic fields (**B**) from **B**=-8T to **B**=8T. The sample surface normal was at 45° to both the applied magnetic field and also the wavevector $\mathbf{k}_{pump}$ of the incident pump beam. Details of the experimental arrangement have been published previously in [4]. The sample characteristics are listed in Table 1.

Table 1 Sample characteristics at room temperature.

Characteristic	InAs	GaAs	InSb
n (cm^{-3})	3×10^{16}	1.2×10^{15}	3×10^{14}
μ_e (cm^{2}V^{-1}s^{-1})	2.5×10^4	9.2×10^3	4.9×10^4
m$_e^*$	0.023	0.067	0.014
Crystal Surface	(100)	(100)	(100)

2.1 Bolometer Power Measurements

The absolute power of the emitted THz radiation was measured using a calibrated Bolometer detector. A THz polariser allowed selection of the transverse magnetic (TM) or the transverse electric (TE) components of the THz radiation. Detection of absolute powers is possible, including both coherent and incoherent emission from the sample.

2.2 Electro-Optic Sampling (EOS)

This technique detects only coherent radiation and allows the direct measurement of the THz electric field as a function of time delay, using the ac Pockels effect within a 1mm (110) ZnTe crystal [5].The Fourier transform gives the power spectrum of the emitted THz within the detectable bandwidth of the 1mm ZnTe crystal [6].

3 Results

3.1 Magnetic field enhanced THz

Figure 1 shows the typical enhancement behaviour for GaAs, with both the power from EOS measurements and the absolute bolometer power. The excellent agreement between the two techniques strongly suggests that the majority of the THz radiation is coherent and thus useful for potential applications. The agreement between the two detection techniques was true for all the materials studied. As expected, only the TM component was observed at B=0T [7] and then the emergence of the TE component became evident as **B** was increased. This phenomena can be explained using the simple Lorentz force model with the carriers describing circular orbits once the magnetic field is applied. Reversal of the magnetic field direction confirms this simple model, as EOS reveals a 180° phase change in the TE component of the emitted THz, whilst the TM remained unchanged. Table 2 shows the enhancement observed by the application of the magnetic field for each material system. The

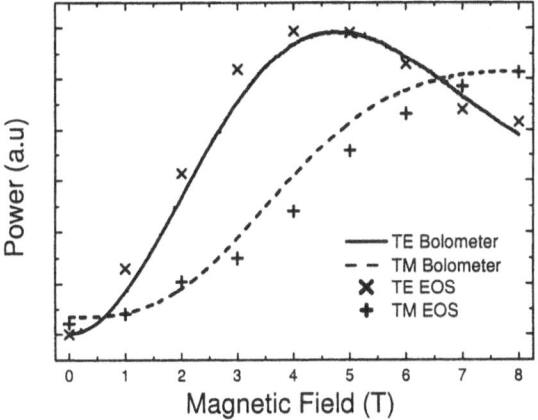

Fig. 1 Typical traces for GaAs emission with comparison of EOS and bolometer.

Table 2 Enhancement of Total THz Power from B=0T to B=8T

Enhancement	InAs	GaAs	InSb
$Power_{(B_{8T})}/Power_{(B_{0T})}$	7.6	25	30
THz $Power_{(B=8T)}$ (μW)	12	1	0.85

largest overall enhancement was from InSb, although the absolute power was less than from InAs. The origin of the large power available from InAs is thought to be the strong surface band bending found in this material and resultant high acceleration in the large surface electric field.

3.2 Optical fluence dependence

Although the Lorentz model explains the main features observed with the application of the magnetic field, it does not take into account the effects due to such large numbers of carriers being created within the surface region. With the optical excitation of many carriers, screening of the surface field and scattering is inevitable and is therefore included in our *many carrier* simulation. The screening is derived using an iterative process, constantly re-calculating the surface field as carriers are created and accelerated. As the results of the simulation show (Figure 2), the field is rapidly screened after illumination by the exciting pulse. With the application of the magnetic field the simulation reveals that the surface field is screened to a lesser extent, illustrating that the magnetic field suppresses screening and enhances THz emission in addition to accelerating photo-excited charge. Varying the optical fluence incident on the samples confirms the accuracy of the simulation. Figure 3 shows the results for InAs TE polarisation where the reduction in screening is clearly evident as the magnetic field is increased.

4 Conclusions

We have extensively studied the surface field generation of THz radiation in the presence of a magnetic field

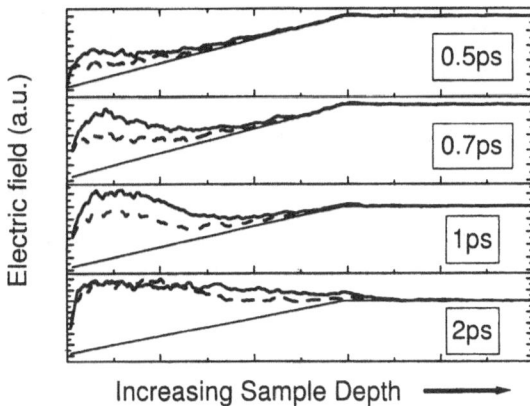

Fig. 2 Simulations of the surface field profile as photo-excited carriers are introduced. Solid line is **B=0T** and dashed line is **B=8T** compared with the surface field before the incident pulse.

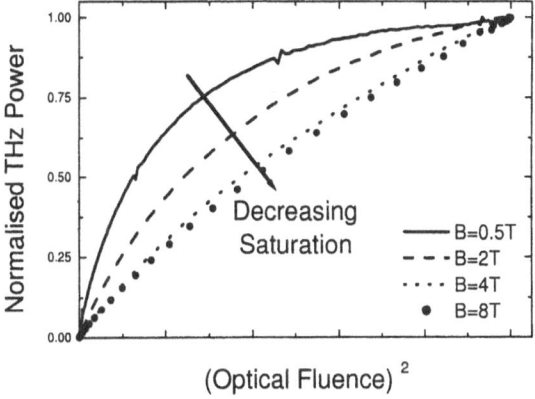

Fig. 3 Optical fluence results from InAs confirm the reduction of surface field screening with magnetic field.

for GaAs, InAs and InSb. All three material systems show an enhancement with the application of a magnetic field, with largest powers being observed from InAs. We explain the enhancement in terms of both the simple Lorentzian acceleration of carriers, and the reduction of screening as the magnetic field is increased, confirmed by both simulation and experiment.

References

1. M. C. Nuss et al., Phys. Rev. Lett. **58**, (1987) 2355.
2. D. D. Arnone et al., SPIE Proceedings **3828**, (1999) 209.
3. H. Ohtake et al., Appl. Phys. Lett. **76**, (2000) 1398.
4. R. McLaughlin et al., Appl. Phys. Lett. **76**, (2000) 2038.
5. Q. Wu et al., Appl. Phys. Lett. **68**, (1996) 2924.
6. P. Y. Han et al., Appl. Phys. Lett. **73**, (1998) 3045.
7. X. -C. Zhang et al., Appl. Phys. Lett. **71**, (1992) 326.

Dephasing of coherent LO phonon-plasmon coupled modes in polar semiconductors

S. Katayama[1], M. Hase[2], M. Iida[1], S. Nakashima[3]

[1] School of Materials Science, Japan Advanced Institute of Science and Technology, 1-1 Asahidai, Tatsunokuchi, 923-1292, Japan e-mail: s-kata@jaist.ac.jp

[2] National Research Institute for Metals, 1-2-1 Sengen, Tsukuba, 305-0047, Japan

[3] Department of Electrical and Electronic Engineering, Miyazaki University, 1-1 Gakuen-kiban-dai, Miyazaki, 889-2192, Japan

Abstract   We present a simple theoretical analysis of the observed dephasing time for the coherent coupled LO phonon-plasmon modes in n-type GaAs. The variation in the behavior of the upper (L_+) and lower (L_-) branch with photoexcited carrier density is interpreted in terms of interchange between the roles of the plasmon and phonon decays.

1 Introduction

Recently the coherent oscillation of the longitudinal optical (LO) phonon-plasmon coupled (LOPC) modes in n-type and p-type GaAs has been studied in experimental [1-4] and theoretical [5] works. The pump-probe experiments using laser pulses with a few ten fs duration by Cho et al.[1] and two of us and collaborators [2,3] have revealed that the dephasing of the coherent LOPC modes depends both on the density of photoexcited carriers (n_{ex}) and the density of doped carriers (n_d). The later work [3] has determined the dephasing time for the L_+ and L_- branches of the coherent LOPC modes as a function of n_{ex} in n-type GaAs with different n_d . Vallée et al. [6] have derived a decay formula for coherent LOPC modes to analyze their observed coherent Anti-Stokes Raman scattering data, but it is limited to the L_+ mode in the low carrier density regime. More detailed theoretical studies are required to the determination of the precise dephasing time over entire carrier density. The energy dependence of the dephasing time has also been observed by using a spectrally resolved pump-probe measurements.[4] The present paper presents a simple theoretical analysis of the observed dephasing time for the coherent LOPC modes.

2 Equations of Motion and Coherent LOPC Modes

We begin the equations of motion for the LO phonon coordinate q(t) and relative displacement field x(t) = $x_e - x_h$ of electrons and holes with effective mass m_e^* and m_h^* under an external field E^{ex}:

$$\frac{\partial^2 q}{\partial t^2} + \frac{2}{T_2^0}\frac{\partial q}{\partial t} + \omega_{LO}^2 q = \frac{e^*}{\mu}[E^{ex} + \frac{4\pi N_e e x}{\epsilon_\infty}], \quad (1)$$

$$\frac{\partial^2 x}{\partial t^2} + \frac{1}{\langle\tau\rangle}\frac{\partial x}{\partial t} + \Omega_p^2 x = -\frac{e}{m_r^*}[E^{ex} - \frac{4\pi N e^* q}{\epsilon_\infty}], \quad (2)$$

with

$$\Omega_p^2 = \frac{4\pi e^2}{\epsilon_\infty}(\frac{n_e}{m_e^*} + \frac{n_h}{m_h^*}), \frac{1}{\langle\tau\rangle_\infty} = \frac{m_h^*}{M}\frac{1}{\langle\tau_e\rangle} + \frac{m_e^*}{M}\frac{1}{\langle\tau_h\rangle}. (3)$$

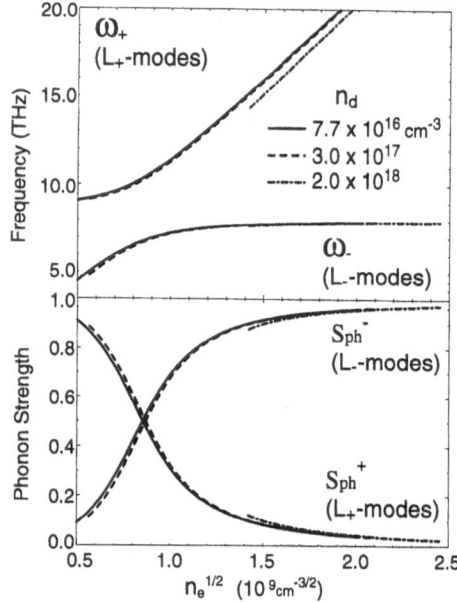

Fig. 1 Upper panel:Coupled modes frequencies as a function of $n_e = n_d + n_{ex}$ for n-type GaAs. the lower panel: the corresponding phonon strength of the coupled modes in the upper panel as a function of n_e.

In Eq.(1), ω_{LO} and $1/T_2^0$ are the frequency and the decay rate of the bare LO phonon, respectively, μ is the reduced mass of two ions per unit cell, and $N_e = m_r^*$ $[(n_e/m_e^*) + (n_h/m_h^*)]$, $n_e = n_d + n_{ex}$ and $n_h = n_{ex}$ being density of electrons and holes in n-type doping materials. In Eq.(2), Ω_p and $\langle\tau\rangle_\infty$ are the plasmon frequency of two-component plasma and the average carrier momentum scattering time in the high frequency limit, $1/m_r^*$ = $1/m_e^* + 1/m_h^*$ and $M = m_e^* + m_h^*$. By transforming from equations of motion for $q(t) = eQ(t)exp(-i\omega t)$ and $x(t) = eX(t)exp(-i\omega t)$ to those of the normal coordinates $Q^{(\pm)}$ for the coherent LOPC modes, we find the eigen frequencies $\omega_\pm$ as,

$$\omega_\pm^2 = \frac{1}{2}[(\omega_{LO}^2 + \Omega_p^2) \pm ((\omega_{LO}^2 + \Omega_p^2)^2 - 4\omega_{TO}^2\Omega_p^2)^{1/2}],(4)$$

where the signs + and − correspond to the L_+ and L_- mode, respectively.

Figure 1 shows the frequencies and the corresponding phonon strength $(S_{ph}^\pm)$ calculated using Eqs.(4) and (7) as a function of electron density. As is evident there appears a clear coupling between the LO phonon and plas-

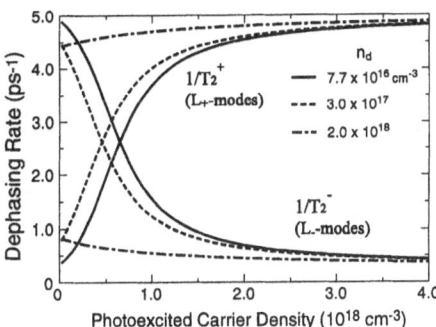

Fig. 2 Dependence of dephasing rate on photoexcited carrier density n_{ex} for three different n_d

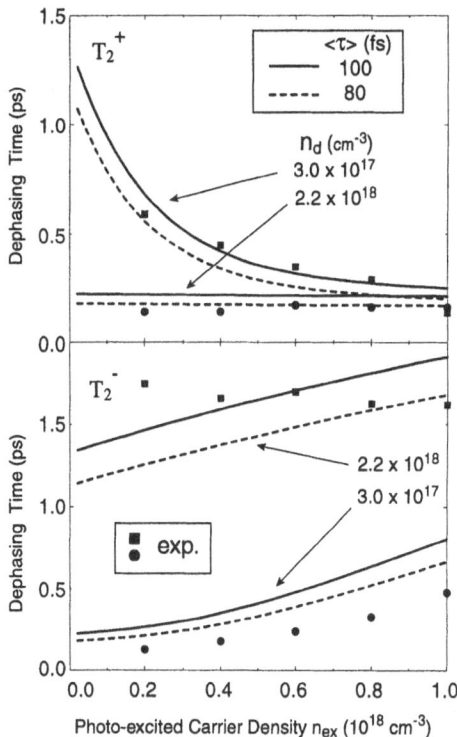

Fig. 3 The full and dashed curves show the calculated dephasing time $T_2^{\pm}$ as a function of n_{ex} for $\langle\tau\rangle_{\infty}$ =100 and 80 fs, and T_2^0 =4 ps, respectively, assuming n_d =3.0 × 10^{17} and 2.2 × 10^{18} cm^{-3} in n-type GaAs. The closed squares and closed circles denote the experimental results for n_d =3.0 × 10^{17} and 2.2 × 10^{18} cm^{-3}.

mon. Thus the equations of the coherent LOPC mode amplitudes are

$$\frac{\partial Q^{\pm}}{\partial t} + \frac{1}{T_2^{\pm}}Q^{\pm} + [\frac{1}{T_2^0} - \frac{1}{2\langle\tau\rangle_{\infty}}]S_{ph}^{\pm}Q^{\mp} = 0, \qquad (5)$$

where the dephasing rates for the L_+ and L_- mode are given by

$$\frac{1}{T_2^{\pm}} = \frac{S_{ph}^{\pm}}{T_2^0} + \frac{1 - S_{ph}^{\pm}}{2\langle\tau\rangle_{\infty}}, \qquad (6)$$

with

$$S_{ph}^{\pm} = \frac{\pm\omega_{\pm}^2 \mp \Omega_p^2}{\omega_+^2 - \omega_-^2}. \qquad (7)$$

The dephasing rate $1/T_2^{\pm}$ is composed of two terms associated with the LO phonon dephasing rate $1/T_2^0$ and the carrier momentum-relaxation rate $1/2\langle\tau\rangle_{\infty}$. The roles of the two terms in $1/T_2^+$ interchange with each other through the changes of phonon strength S_{ph}^+ from unity (phonon-like) to nearly zero (plasmon-like) as n_{ex} is increased, and vice visa in $1/T_2^-$.

In Fig. 2, we plot the depahsing rates as a function of photoexcited carrier density assuming n_d=0.08, 0.3 and 2.0×10^{18}cm^{-3}. Our choosed parameters are T_2^0=2 ps and $\langle\tau\rangle_{\infty}$=200 fs. Other material parameters are the same with ones in Fig.1.

In Fig. 3, we plot the calculated and experimental dephasing time as a function of the photoexcited carrier density. The observed remarkable dependence of T_2^+ on n_{ex} is reproduced well using $T_2^0 = 4$ ps and $\langle\tau\rangle_{\infty}$ =100 fs for $n_d = 3 \times 10^{17}$cm^{-3}, whereas there remain some discrepancies between calculation of T_2^- and experiments for n_d =2.2×10^{18}cm^{-3}. Our analysis clarifies the existence of the two competing dephasing mechanisms associated with the phonon damping and the carrier momentum-scattering (i.e. plqsmon decays). This suggests that the direct probing into the ultrafast carrier dynamics is possible by the time domain spectroscopy of the coherent LOPC modes.

3 Conclusions

It is demonstrated that the ultrashort dephasing times are dominated by the competing plasmon and phonon decays and that two decays interchange their contributions to the dephasing through the change of $S_{ph}^{\pm}$ with the photoexcited carrier density. Though the observed n_{ex}-dependence of $T_2^{\pm}$ in n_d= 3 × 10^{17}cm^{-3} was described well by our formula, there remain some discrepancies between calculation of T_2^- and masurements for higher doped density. These discrepancies would be originated from an inhomogeneous density distribution of photoexcited carriers due to a pump-beam profile. This effect lead to experimental errors in estimation of n_{ex}.[5]

References

1. G. C. Cho, T.Dekorsy,H.J.Barker,R.Hövel, and H.Kurz Phys. Rev. Lett.**77**, (1996) 4062.

2. M. Hase, K. Mizoguchi, H. Harima, F. Miyamaru, S. Nakashima, R. Fukazawa, M. Tani, and K. Sakai, J.lumin. **76, 77**, (1997) 68.

3. M. Hase, S. Nakashima, K. Mizoguchi,H. Harima, and K. Sakai, Phys. Rev.B **60**, (1999) 16526.

4. K. Mizoguchi, F. Miyamaru, M. Nakajima, M. Hase, and S. Nakashima, Physica B**272**, (1999) 367.

5. A.V.Kuznetsov and C.J.Stanton, Phys. Rev. B**51**, (1995) 7555.

6. F. Vallée, F. Ganikhanov,and F. Bogani, Phys. Rev.B **56**, (1997) 13141.

Femtosecond spectroscopy of large-momentum excitons in GaAs

M. Betz[1], G. Göger[1], A. Leitenstorfer[1], M. Bichler[2], W. Wegscheider[3], G. Abstreiter[2]

[1] Physik-Department E11, Technische Universität München, D-85748 Garching, Germany e-mail: `mbetz@ph.tum.de`
[2] Walter-Schottky-Institut, Technische Universität München, D-85748 Garching, Germany
[3] Institut für Angewandte und Experimentelle Physik, Universität Regensburg, D-93040 Regensburg, Germany

Abstract Highly energetic excitons with wave vectors much larger than that of an absorbed photon are excited in thin GaAs layers. The dispersion relations of these coherent excitations are measured employing femtosecond transmission spectroscopy. Ultrafast exciton damping via scattering with nonequilibrium carriers and with phonons is investigated.

1 Introduction

It is usually assumed that optical excitation and probing of free excitons in direct gap semiconductors is only possible near the center of the Brillouin zone since the photon momentum is negligible as compared to the reciprocal lattice vector. Especially, absorption of photons with energies corresponding to the absorption continuum is believed to result exclusively in the creation of unbound electron-hole pairs with total momentum close to zero. However, these restrictions do not hold in general due to the break of translational symmetry at any semiconductor interface. As a result, excitons with large wave vectors may be excited in addition to free electron-hole pairs. In time and energy resolved femtosecond transmission experiments we find spectral beats between a photonlike mode propagating with the velocity of light divided by the refractive index and weak coherent excitations having a phase velocity given by the excess energy and the effective mass of excitons [1]. Such exciton polariton phenomena have until now only been studied in experiments focusing on the frequency regime close to the fundamental resonance.

2 Experiment

We perform highly sensitive pump-probe experiments using a two-color Ti:sapphire laser with two independently tunable synchronized pulse trains. The high purity MBE grown samples consist of 500, 200 and 50 nm thick GaAs layers oriented along ⟨100⟩ followed by an AlGaAs cladding layer. The films are antireflection coated on both sides, contacted to transparent substrates and kept at a low lattice temperature inside a cryostat. The time delayed 17 fs probe pulse is spectrally dispersed in a double monochromator after having passed through the sample (spectral resolution: 0.6 meV). Transmission changes due to the excitation pulse are detected in the energy range from the 1s exciton up to excess energies of 300 meV.

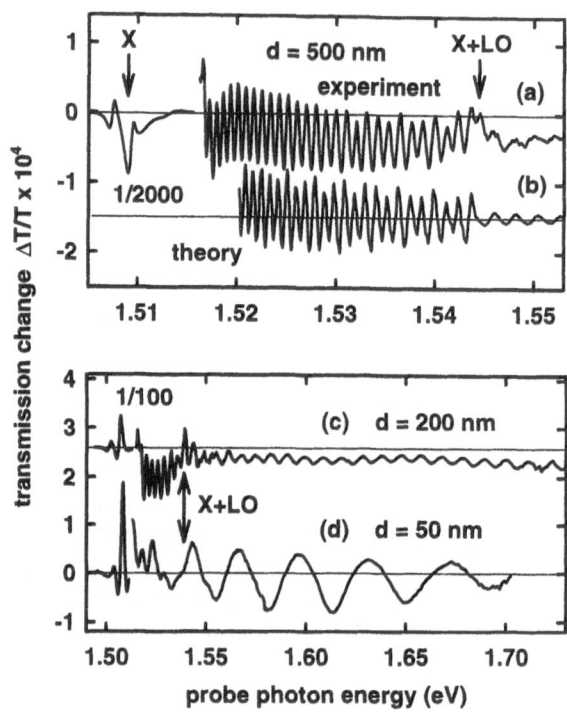

Fig. 1 Spectrally resolved transmission changes measured 10 ps after hot carrier injection in bulk GaAs samples of a thickness of 500 nm (a), 200 nm (c) and 50 nm (d) for a lattice temperature $T_L = 20$ K. Transmission changes as calculated for the 500 nm specimen are shown in Fig. 1(b).

3 Results for different film thicknesses

In Fig. 1(a) the pump induced transmission changes in the 500 nm specimen are shown, measured 10 ps after excitation of an electron-hole density of 3×10^{15} cm^{-3} by a pump pulse ($t_p = 80$ fs, E = 1.7 eV). At probe photon energies in the absorption continuum, nonequidistant oscillations appear in the differential spectra. This phenomenon is explained as follows: when probe photons are incident on the sample surface, excitonlike polaritons with wave vectors much larger than that of the electromagnetic wave are excited in addition to the predominant photonlike mode. For excess energies of a few meV above the fundamental resonance, these heavy hole (hh) and light hole (lh) excitonlike polaritons recover their parabolic center-of-mass dispersion. In the limit of weak dephasing, the polarization waves travel almost unattenuated to the other sample surface and are reconverted into photons. Spectral interference beats between the three modes reflect the different phase velocities of the polariton branches. The optical pump pulse creates free electron-hole pairs before or during the propagation of polaritons excited by the probe pulse. The efficient damping of excitonlike polarizations via scattering with these free carriers modifies the propagation beats in the probe beam. Therefore our photomodulation technique is very sensitive to polariton effects and avoids the background problem which would be faced in linear transmission. In this way, both stationary aspects like exciton dispersions as well as dynamical properties such as exciton-carrier scattering rates become accessible.

The periods of the oscillations are given by the mass and the excess energy of the excited excitons and the sample thickness. A sharp discontinuity of the oscillation amplitude and period is observed at a probing position corresponding to the 1s exciton plus one LO phonon energy (X+LO). We find that heavy hole excitons are strongly damped above this threshold energy and the spectral beating is exclusively determined by the light hole component.

As depicted in Fig. 1(b), the experimental data are quantitatively reproduced by a model based on Pekars's additional boundary conditions [2] assuming purely polar optical scattering of 1s excitons with LO phonons [3]. Surprisingly, the LO phonon emission by lh excitons turns out to be one order of magnitude smaller than that of hh excitons. This effect is ascribed to a destructive interference in the scattering matrix element for the case of similar electron and hole masses.

From the data in Fig. 1(a) and (c) we determine a hh effective mass of 0.52 m_0 in $\langle 100 \rangle$ direction. For the lh exciton mass a value of 0.19 m_0 is found, also reproducing the hh-lh beating structure near X+LO. In the 200 nm sample [see Fig. 1(c)] the high energy region of the lh exciton dispersion may be examined. It turns out that for excess energies in the order of 200 meV the oscillation period increases slower than expected in a parabolic bandstructure. This can be utilized to investigate the nonparabolicity of the conduction and the lh band. In the polariton interference spectrum of the 50 nm sample [see Fig. 1(d)] again a discontinuity is found at X+LO. Even for propagation times of a few hundred femtoseconds LO phonon emission apparently leads to an effective damping of the hh polariton wave packet. Since the propagation distance approaches the excitonic Bohr radius, a quantitative simulation of the spectrum would have to take into account the microscopic properties of the excitonic polarizability near the semiconductor interface.

4 Acoustic phonon scattering

Polariton interference spectra have also been studied as a function of temperature, see Fig. 2: The results obtained for the 500 nm sample at a delay time of 10 ps show an

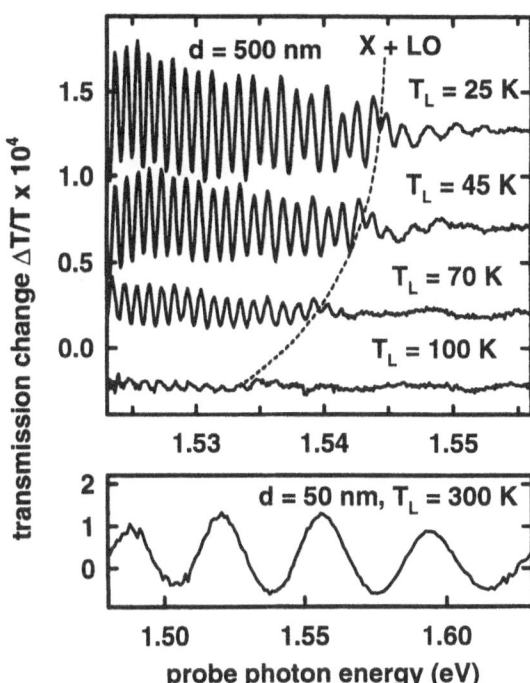

Fig. 2 Upper part: Spectrally resolved transmission changes for the 500 nm thick GaAs sample, measured 10 ps after hot carrier injection for various lattice temperatures. Lower part: transmission changes in the 50 nm thick GaAs specimen measured 2 ps after hot carrier injection at room temperature.

increasing red shift of the X+LO threshold with temperature due to band gap renormalization (see dashed line as a guide for the eye). Simultaneously, the oscillation amplitude decreases. At temperatures above 100 K the propagation beats vanish due to the efficient exciton dephasing via acoustic phonon scattering. From an analysis of the oscillation amplitude, we obtain an estimate for the linear coefficient σ of the increase of the homogeneous exciton linewidth with temperature. We deduce $\sigma = 20 \, \mu\text{eV/K}$, in surprising agreement with the value of $\sigma = 17 \, \mu\text{eV/K}$ found for the 1s exciton with negligible center-of-mass motion [4].

The lower part of Fig. 2 depicts a polariton interference spectrum of the 50 nm GaAs layer at room temperature $T_L = 300$ K. We still observe pronounced oscillations associated with the lh exciton polariton. This demonstrates that coherent propagation of highly energetic excitons is still observable for elevated temperatures where excitonic effects are usually believed to be of minor importance in bulk GaAs.

References

1. G. Göger, M. Betz, A. Leitenstorfer, M. Bichler, W. Wegscheider, G. Abstreiter, Phys. Rev. Lett. **84**, 5812 (2000)
2. S. I. Pekar, Sov. Phys. JETP **34**, 813 (1958)
3. Y. Toyozawa, Prog. Theor. Phys. **12**, 111 (1959)
4. L. Schultheis, A. Honold, J. Kuhl, K. Köhler, and C. W. Tu: Phys. Rev. B **34**, 9027 (1986)

A novel biexcitonic, non-radiative electron-hole recombination mechanism and its application in hydrogenated silicon semiconductors

S. B. Zhang, Howard M. Branz

National Renewable Energy Laboratory, Golden, Colorado 80401, USA

Abstract   A biexciton model is proposed for non-radiative electron hole (e-h) recombination in semiconductors based on first-principles calculations [Phys. Rev. Lett. **84**, 967 (2000)]. In hydrogenated Si, the trapping of photoexcited e-h pairs drives large structural reconfiguration, accompanied by the crossing of an electron and a hole level near midgap. This level crossing is the key to a fast and non-radiative recombination, without the slow multi-phonon emission processes.

Non-radiative recombination (NRR) of trapped electron hole (e-h) pairs in semiconductors is a field of broad interest. NRR is of particular interest in the technologically important silicon semiconductors. In particular, a-Si:H exhibits weak luminescence and rapid non-radiative recombination under conditions at which defect recombination is negligible. Because the probability of multi-phonon emission falls exponentially with the number of emitted phonons, band-to-band NRR by multi-phonon emission is not a significant process. Instead, photocarrier-induced excitation of H from Si-H bonds is a key step in light-induced creation of metastable dangling-bond (DB) defects. Such NRR mediated large structural reconfigurations are also known to exist in various solids, including alkali-halide salts, chalcogenide glasses, SiO_2, and GaAs, but the energy transfer mechanism from carriers to the metastable configuration remains unknown.

Previous theories postulating H emission driven by a single e-h recombination have failed to provide a convincing mechanism for low-barrier H dissociation from the Si-H bond. Here, we explore the trapping of biexcitons at a Si-H bond in hydrogenated silicon or at a Si-Si bond in c-Si. We find, using first-principles total energy calculations, that recombination of a biexciton in hydrogenated Si can stimulate low-energy-barrier H emission from a Si-H bond into a metastable configuration. There are several key features of this process: (i) Asymmetric distortion of a Si-Si bond creates localized levels that trap a biexciton. (ii) Subsequent H rotation toward a metastable bond center site (BC) on a neighboring Si-Si bond causes the trapped biexciton to recombine non-radiatively. This NRR occurs when the high-lying electron-occupied level falls and the low-lying hole-occupied level rises until they cross in the gap. (iii) A significant fraction of the biexciton energy is absorbed in promoting H to the metastable configuration, denoted as (H-BC, DB). (iv) From (H-BC, DB), the H may either escape to infinity as mobile H or be retrapped to the DB. This provides an explanation for carrier-induced

emission of H from Si-H bonds, as well as a mechanism for the defect-independent NRR in a-Si:H. In silicon, H participation is essential to the low-barrier pathway; in c-Si without H, the asymmetric distortion of the Si-Si bond cannot be stabilized nor can the electron-hole level crossing occur.

The simple H-rotation pathway: In the initial-hop, the H atom is rotated from a Si-H bond to a nearby bond center site to form (H-BC, DB). The energy barrier to this rotation is 1.4 eV. The energy barrier for the reverse process is 0.3 eV, which keeps the H atom from falling back freely. The final energy difference between (H-BC, DB) and the ground state Si-H configuration is 1.1 eV. Hopping further away to distant BC sites in a-Si:H requires another 1.1 eV. In this simple pathway, the lowest-unoccupied molecular orbital (LUMO) never drops significantly below the CBM, while the highest-occupied molecular orbital (HOMO) remains below the VBM until the H atom is rotated by about 50°. Hence, photogenerated e-h pairs can have no effect on the H dissociation through exciton localization until the H is displaced significantly over an energy-barrier greater than 0.8 eV.

The biexciton pathway: The Fig. 1 inset illustrates three coordinates that describe the pathway, d_{Si}, d_H, and Θ; $D_{Si(1)}$ is the position of Si(1). Along the entire path illustrated in Fig. 1, the Si-H bond length, d_H is nearly a constant. We describe the changes in $(d_{Si}, \Theta, D_{Si(1)})$ by $(\delta_{Si}, \Theta, \delta D)$ where δ_{Si} is the change in d_{Si}, and δD is the change in $D_{Si(1)}$. Figure 1a shows the total energy of the new low-energy path, for biexciton occupation. There are three steps in the process: (i) We displace Si(2) away from the stationary Si(1) by δ_{Si} while Θ remains roughly constant. This asymmetric bond stretching increases the total energy by only 0.26 eV at $\delta_{Si} \sim 0.4$ Å but immediately brings levels into the gap that trap the biexciton. (ii) We rotate the H into the BC site (from $\Theta \sim 110°$ to $\Theta \sim 0°$), while δ_{Si} remains roughly constant and find a maximum additional barrier of 0.15 eV. (iii) We relax Si(1) and Si(2) through δD without significant change of either δ_{Si} or Θ, which reduces the total energy by an additional 1.5 eV. We have not made an exhaustive search of the energy surface in the 3D-$(\delta_{Si}, \Theta, \delta D)$ configuration space; an optimal path with a much lower energy barrier could undoubtedly be found. However, there are several key features illustrated by the pathway shown in Fig. 1: (a) An asym-

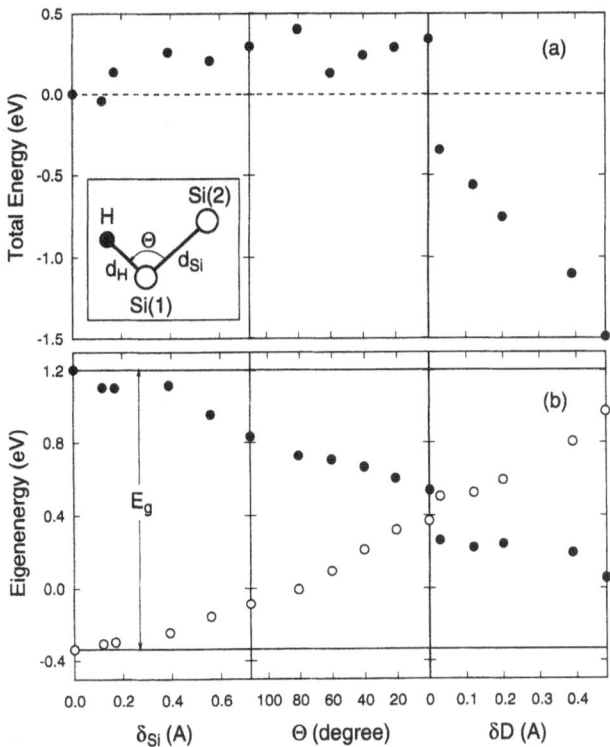

Fig. 1 (a) Total energies and (b) corresponding HOMO (•) and LUMO (○) eigenenergies along the biexcitonic pathway involving an embedded H-Si(1)-Si(2) configuration.

metric dilation brings localized states into the mobility gap of a-Si:H with negligible barrier. This permits biexciton trapping that leads to the recombination. (b) The HOMO-LUMO gap closes to zero upon H rotation. This is the key to reaching the low-energy final state over a low barrier but with neither photon nor phonon emission; the biexciton-occupied levels actually cross in the gap. (c) The level crossing is the non-radiative recombination of the biexciton. Fig. 1b shows how the two photogenerated electrons (which are initially in the LUMO-state near the CBM) evolve to new positions near the VBM, while the two photogenerated holes (which are initially in the HOMO-state near the CBM) evolve to new positions near the CBM. The biexciton is thus annihilated and a portion of the photoexcitation energy is absorbed into the H reconfiguration. This NRR by H emission proceeds without multiphonon emission and can therefore be an efficient process.

To better understand the low-energy path and the conditions for NRR by level crossing, we also studied the asymmetric [111] displacement of a Si (δ_{Si}) in c-Si, without H. Figure 2 shows (a) the total energies with zero-, one- and bi-exciton occupation of the localized states and (b) the change in the zero-exciton HOMO and LUMO eigen-energies. In order to compare energies, two e-h pairs are always included; the free e-h pairs are placed at the c-Si band edge. The eigenenergies are nearly independent of exciton occupation. With no trapped exciton, Fig. 2a shows the total energy increases

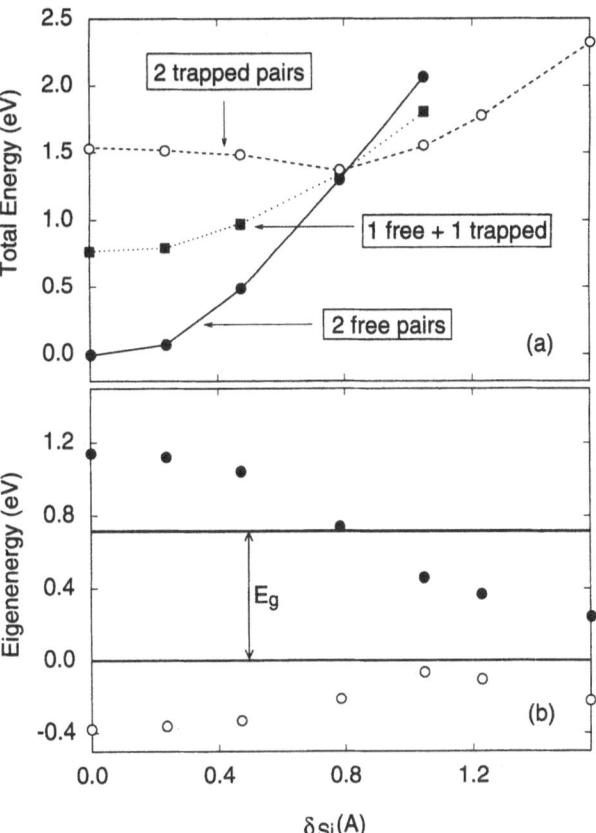

Fig. 2 Same as Fig. 1, for asymmetric dilation of a Si-Si bond in c-Si. Total energy curves are for zero-, one- and biexcitons trapped on the levels in (b).

quickly with δ_{Si} as expected, and Fig. 2b shows that the HOMO and LUMO states move into the band gap due to the Si displacement. Each additional trapped exciton raises the total energy because the trapping levels are well inside the c-Si band edges (Fig. 2b). However, the total-energy slope is reduced by the trapped excitons and above about $\delta_{Si} \sim 0.8$ Å, exciton trapping actually reduces the total energy. In the case of a trapped biexciton in c-Si, Fig. 2a shows that the total energy decreases slightly with Si displacement up to $\delta_{Si} = 0.8$ Å. The biexciton greatly "weakens" the Si-Si bond, primarily because of the double-occupancy of the LUMO level. However, Fig. 2b shows that the HOMO-LUMO gap does not close to zero. The minimum gap is 0.34 eV at $\delta_{Si} = 1.0$ Å. Non-radiative recombination is thus a slow-multiphonon emission process, in contrast to the hydrogenated Si path described earlier. Also, in contrast to a-Si:H and c-Si:H, there is no barrier in c-Si (Fig. 2a) that prevents reemission of the trapped biexciton and a return to $\delta_{Si} = 0$. Thus, excess carriers do not cause metastable structural change in c-Si as they do in a-Si:H.

Our biexciton model resolves two long-standing experimental puzzles in a-Si:H: 1) the non-radiative decay of, and 2) H emission from Si-H bonds induced by, light-induced carriers. The U.S. DOE supported this research under Contract DE-AC36-98G010337.

Annihilation of coherent LO phonon-plasmon coupled modes by lattice defects in n-GaAs

Muneaki Hase[1], Kunie Ishioka[1], Kiminori Ushida[2], Masahiro Kitajima[1]

[1] National Research Institute for Metals, 1-2-1 Sengen, Tsukuba, Ibaraki 305-0047, Japan e-mail: `hasedon@nrim.go.jp`
[2] RIKEN (The Institute of Physical and Chemical Research), 2-1 Hirosawa, Wako, Saitama 351-0198, Japan

Abstract   We have studied the effect of point defects on dephasing of coherent optical phonons in ion-implanted GaAs. Ultrafast dynamics of coherent phonons and photo-generated carriers in the femtosecond time-domain have been precisely measured by means of a reflection-type pump-probe technique. The coherent LO phonon-plasmon coupled (LOPC) modes disappear with increasing ion dose, and only the coherent LO phonon is observed at high ion doses. The relaxation time of the photo-generated carriers is decreased by ion irradiation, and this effect is due to trapping of carriers with deep defects. Our results suggest that point defects involve deep levels below the conduction band, and as a results, carrier trapping dominates annihilation of the coherent coupled modes.

1 Introduction

Defects in semiconductors have been widely investigated for device applications. GaAs is a key material for ultrafast switching and THz radiation because the carrier lifetime of the low-temperature (LT) grown GaAs is sub-picosecond, which is due to excess As in the form of interstitials, antisite defects, or gallium vacancies.[1] The dynamics of photo-excited carriers and coherent phonons have been investigated in as-grown and annealed LT-GaAs.[1–3] On the other hands, doping a semiconductor with foreign atoms can alter the electrical conductivity by many order of magnitude. GaAs doped with Si is n-type in which the plasmon and the LO phonon form coupled modes through Coulomb interactions when the doping level is around 10^{18} cm^{-3}. The frequencies of the LO phonon-plasmon coupled (LOPC) modes depend on the carrier density.[4] Ultrafast relaxation of the coherent LOPC modes in n-doped GaAs has been studied using pump-probe techniques, and indicated that electron-hole scattering play a dominant role in dephasing of the LOPC modes.[5,6] For highly n-doped GaAs, relaxation time of the LOPC modes did not depend on photo-excited carrier density and the dephasing of the coupled modes was thought to be caused by impurity or defect scattering when doping level is higher than photo excitation.[6] Here we report the effect of lattice defects on generation and relaxation of the coherent LOPC modes in He$^+$ irradiated n-GaAs for various ion doses.

2 Experiment

The samples used were Si doped n-GaAs with the doped carrier density of $n_{dop}=1.4\times10^{18}$cm^{-3}. In order to examine the effect of point defects, 5 keV He$^+$ ions were irradiated to n-GaAs samples at doses from 9.4×10^{12}

He$^+$/cm^2 to 5.0×10^{14} He$^+$/cm^2 in a UHV chamber with base pressure of 3×10^{-9} Torr.[7] The femtosecond pump-probe measurements were carried out at room temperature using a mode-locked Ti:sapphire laser whose pulse width was 25 fs. The pump- and probe-beam powers were fixed to 60 mW and 6 mW, respectively. Anisotropic reflectivity changes ($\Delta R_{eo}/R$) were measured using an EO sampling technique to obtain coherent lattice vibrations, and isotropic reflectivity changes ($\Delta R/R$) were measured to obtain relaxation dynamics of nonequilibrium carriers as a function of the time delay by scanning the optical path length of the probe beam.

3 Results and Discussions

3.1 Coherent LOPC Modes

Figure 1 shows the time derivatives of $\Delta R_{eo}/R$ signal obtained for n-GaAs with various ion doses. The beating

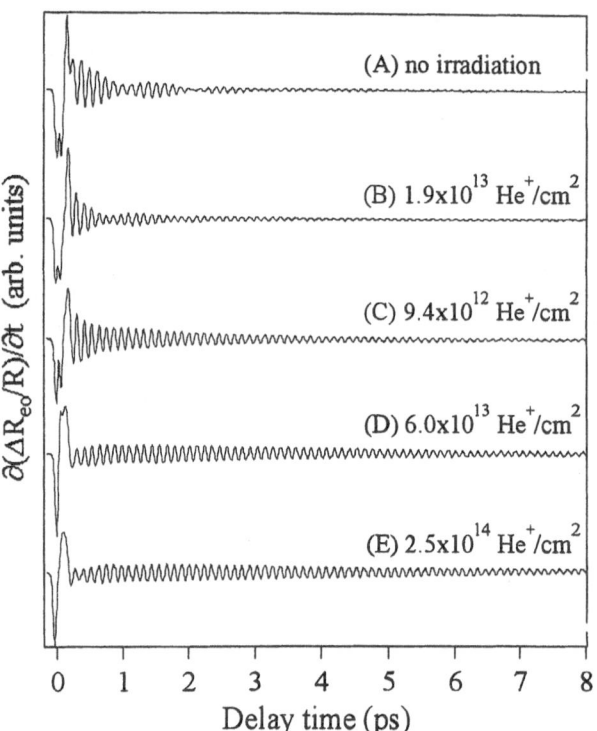

Fig. 1 The time derivatives of the transient reflectivity changes for various He$^+$ doses obtained by electro-optic detection.

pattern of the oscillation observed before ion irradiation is due to co-existence of the strong LO mode and the lower branch of the coupled modes (L$_-$) as shown in its

Fourier transformed spectrum (Fig. 2(A)).[5,6] Disappearance of the mode-beating pattern at high ion dose suggests that the pump pulse does not generate the L_- mode or the L_- mode loses its coherence in a few hundred femtoseconds. The decay time of the L_- mode estimated from fitting the time-domain data with damped harmonic oscillations decreases from 970 ± 40 to 340 ± 40 fs with increasing ion dose. The decay time of the coherent LO phonon increases from 2.1 to 7.9 ps and then decreases to 2.2 ps as increasing the ion dose. A possible explanation for these results is decrease in effective carrier density due to point defects introduced by He^+ irradiation. As a result, electron-phonon interaction becomes less efficient, which would lead to a longer lifetime of the LO phonon, and the L_- mode changes from a phonon-like mode with a long dephasing time to a plasmon-like mode with a short dephasing time.[5,6]

The FT spectrum for the non-irradiated n-GaAs reveals distinctly the bare LO phonon (8.75 THz), the L_- mode (7.98 THz), and the upper branch of the coupled modes (L_+ at 14.9 THz) as shown in Fig. 2(A). The frequency of the L_- mode shifts from 7.98 down to 7.70 THz, and that of the L_+ mode shifts from 14.9 down to 13.6 THz after ion irradiation. These frequency-shifts of the coupled modes are thought to be resultant of the change in the effective carrier density because the frequency of the LOPC modes ($\omega_\pm$) depend on the carrier density N,[4]

$$\omega_\pm = \frac{1}{\sqrt{2}}\left[\omega_{LO}^2 + \omega_p^2 \pm \sqrt{(\omega_{LO}^2 + \omega_p^2)^2 - 4\omega_p^2\omega_{TO}^2}\right]^{\frac{1}{2}}, \quad (1)$$

where $\omega_p = \sqrt{4\pi Ne^2/m^*\varepsilon_\infty}$ is the plasma frequency, ω_{LO} and ω_{TO} are the frequency of the LO and TO mode, respectively. The decay time of the L_+ mode obtained by a time partitioning Fourier transform (TPFT)[6] increases from 100 ± 40 fs (for the non-irradiated sample) to 150 ± 40 fs (for 9.4×10^{12} He^+/cm^2). The increase of the dephasing time can be explained by decrease in effective carrier density because the L_+ mode changes from a plasmon-like mode with a short dephasing time to a phonon-like mode with a long dephasing time as decreasing the carrier density.[6]

3.2 Carrier Dynamics

The isotropic reflectivity change $\Delta R/R$ reveals an initial fast relaxation (subpicosecond time scale) and a following slow relaxation. The fast one corresponds to carrier-carrier and carrier-phonon scattering, and the slow one corresponds to recombination of carriers. The relaxation times of the fast and slow components is deduced by fitting the time-domain data to a double exponential decay. At low ion doses the obtained fast scattering time does not depend on ion irradiation, being 220 ± 40 fs for all the samples. However, the slow relaxation time decreased from 15.0 ps to 2.4 ps as increasing ion dose. These results suggest that point defects involve deep levels (or deep centers) below the conduction bands, and as

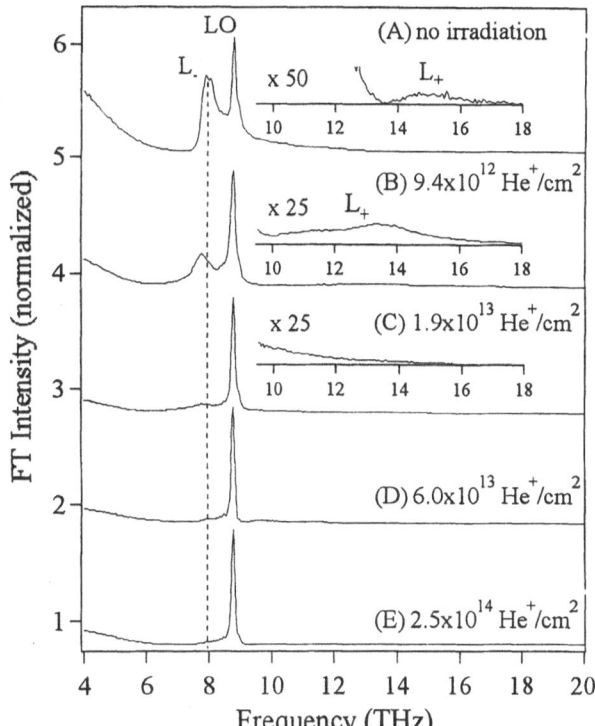

Fig. 2 Fourier transformed spectra obtained from time domain oscillations in Fig.1. The LO mode, the L_- mode and the L_+ mode are observed before ion irradiation. The dotted line shows the peak frequency of the L_- mode for the non-irradiated n-GaAs.

a result, photo-excited carriers decay into the deep levels during and after carrier-carrier and carrier-phonon scattering. Defect-induced deep centers would trap doped carriers in n-GaAs, resulting annihilation of the coherent LOPC modes at high ion doses.

4 Conclusions

The effect of point defects on the ultrafast dynamics of the coherent LO phonon-plasmon coupled modes in ion-irradiated n-GaAs has been studied by using a femtosecond pump-probe technique. The annihilation of the coherent LOPC modes is observed with He-ion irradiation, which is explained by trapping of majority carriers at deep levels below the conduction bands.

References

1. S. Gupta, J. F. Whitaker, and G. A. Mourou, IEEE J. Quantum Electron. **28** (1992) 2464.

2. T. Dekorsy, H. Kurz, X. Q. Zhou, and K. Ploog, Appl. Phys. Lett. **63** (1993) 2899.

3. S. D. Benjamin, H. S. Loka, A. Othonos, and P. W. E. Smith, Appl. Phys. Lett. **68** (1996) 2544.

4. A. Mooradian and A. L. MaWhorter, Phys. Rev. Lett. **19** (1967) 849.

5. G. C. Cho, T. Dekorsy, H. J. Bakker, R. Hövel, and H. Kurz, Phys. Rev. Lett. **77** (1996) 4062.

6. M. Hase, S. Nakashima, K. Mizoguchi, H. Harima, and K. Sakai, Phys. Rev. B **60** (1999) 16526.

7. M. Hase, K. Ishioka, M. Kitajima, K. Ushida, and S. Hishita, Appl. Phys. Lett. **76** (2000) 1258.

Ultrafast Electro-Absorption at the Transition between Classical and Quantum Response

J. Kono[1] and A. H. Chin[2]*

[1] Department of Electrical and Computer Engineering, Rice University, Houston, Texas 77005, U.S.A.
[2] W.W. Hansen Experimental Physics Laboratory, Stanford University, Stanford, California 94305, U.S.A.

Abstract   We report the first observation of unusually large induced absorption in semiconductors strongly driven by intense ultrashort mid-infrared laser fields. This ultrafast electro-absorption has the largest extent below the band edge ($\sim$1 eV) ever observed, which we interpret as being a manifestation of the dynamical Franz-Keldysh effect. The electro-absorption is observable when the ponderomotive potential is comparable to the photon energy of the applied field, i.e., when the system response is at the transition between the classical and quantum regimes.

1 Introduction

We have observed ultrafast induced absorption below the band edge in semiconductors strongly driven by intense mid-infrared (MIR) pulses, which we attribute to the dynamical Franz-Keldysh effect (DFKE) [1–3]. This ultrafast induced absorption extends $\sim$1 eV below the band edge and we believe that this represents the strongest electro-absorption ever observed. Phenomena of this type occur in matter when the applied AC field has a ponderomotive potential, $U_p = (2\pi e^2/mc)(I/\omega^2)$, comparable to the photon energy, $\hbar\omega$ [1–6]. Under such conditions, the behavior of matter is at the transition between classical and quantum response.

Ponderomotive potential effects in condensed matter (semiconductors in particular) have been considered ever since the invention of the laser [1,4]. However, strong electronic absorption (the typical semiconductor band-gap energy, $E_{gap} \sim 1$ eV) and competing high-intensity damage mechanisms have made such observations impossible using traditional laser sources ($E_{photon} \sim 1$ eV, $\lambda \sim 1~\mu$m). Using longer wavelength light avoids the electronic absorption problem and avoids high-intensity damage mechanisms because the required intensity to observe ponderomotive effects becomes smaller. In addition, the smaller effective masses (m^*) of carriers in semiconductors relative to the free electron mass (m_e) makes the ponderomotive potential larger for a given laser intensity than in the atomic case. Excitonic DFKE was recently observed in a quantum well semiconductor system using 1 THz (4 meV) FIR radiation [7]. The resonant coupling to exciton states led to the simultaneous observation of the AC Stark effect and the ponderomotive potential shifting present in the DFKE. However, the virtual nature of the process was not demonstrated,

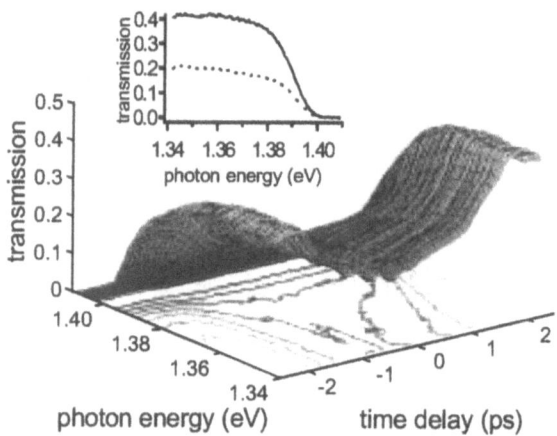

Fig. 1 Transmission of a near-infrared broadband probe pulse through GaAs (350 μm thick) as a function of time delay between the arrival of the intense 3.5 μm MIR driving pulse and the near-infrared probe pulse (nagative time delays correspond to MIR pulses arriving first). A contour plot is shown in the plane of the time delay and photon energy axes.

and no induced absorption below the band edge was observed in the previous work.

2 Experimental Details

In order to observe ultrafast induced absorption below the band edge, we used an optical parametric amplifier (OPA) system with difference-frequency mixing (DFM) as a source of tunable intense MIR pulses. The OPA uses a β-Barium Borate (BBO) crystal as a nonlinear medium that is pumped by the output of a Ti:Sapphire regenerative amplifier system (800 nm, 1 mJ, 1 kHz, $\sim$1 ps) to parametrically amplify the tunable signal (1.1 μm to 1.6 μm) and idler (1.6 μm to 2.9 μm) outputs. The signal and idler then undergo DFM in a AgGaS2 crystal to produce MIR pulses tunable from 3 μm to 10 μm.

We measured the transmission near the band edge of bulk semiconductors using broadband light as a probe. The broadband light was produced by continuum generation in a sapphire plate, using the residual pump pulse (800 nm) after the OPA. The broadband probe was temporally overlapped with the MIR pump by changing the time delay of the MIR pump relative to the probe. Spectra were obtained with the pump at various time delays with respect to the probe, and are subsequently normal-

* Present address: Lawrence Livermore National Laboratory, Information Science and Technology Division, Livermore, California 94550, U.S.A.

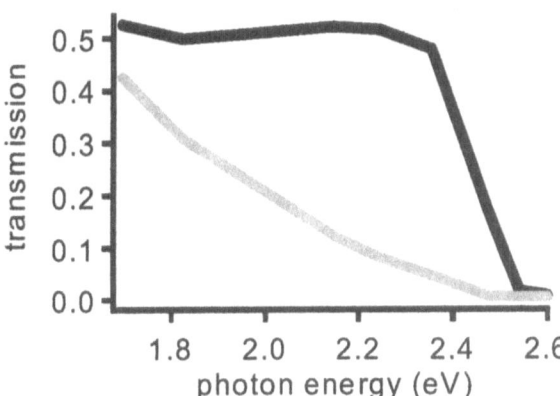

Fig. 2 Transmission below the band edge of ZnSe (3 mm thick) with (gray line) and without (black line) an intense 3.5 μm MIR driving field with $U_p \sim \hbar\omega$.

ized to the probe spectrum without the sample to obtain the absolute transmission.

3 Results and Discussion

Shown in Fig. 1 is transmission data taken using 3.5 μm, $\sim$1 ps MIR pulses with a peak intensity of $\sim$2 × 10^{10} W/cm^2 ($U_p \sim \hbar\omega = 0.35$ eV). Under these conditions, we observed a dramatic decrease in transmission that extends past 0.2 eV below the band edge ($E_{gap} = 1.4$ eV) of the GaAs sample (350 mm thick). This decreased transmission, due to induced absorption, occurs only during the presence of the intense MIR pulse. This clearly demonstrates the virtual nature of the effect, i.e., no MIR excitation of carriers across the band gap and/or lattice-heating effects are involved.

The effect is not specific to GaAs, as we also observed induced absorption in polycrystalline ZnSe ($E_{gap} = 2.7$ eV) and crystalline ZnTe ($E_{gap} = 2.3$ eV). The induced absorption in ZnSe (3 mm thick) with and without the presence of a 3.5 mm MIR driving field is shown in Fig. 2. Our measurements clearly demonstrate not only the magnitude of the effect, but also the extent ($\sim$1 eV below the band edge) of the absorption induced by MIR pulses, which far exceeds any previously observed induced absorption in semiconductors.

We attribute the observed induced absorption to the dynamical Franz-Keldysh effect. Since the dc Franz-Keldysh effect ($U_p >> \hbar\omega$) can be thought of as a tunneling assisted transition [8], the DFKE ($U_p \sim \hbar\omega$) may be intuitively thought of as the point where the tunneling time is comparable to the period of the applied electric field. Because the conduction band states are analogous to free electrons, an analogy can be made between the energy gap in a semiconductor and the ionization potential in a gas. Consequently, a quantitative description of the DFKE may be constructed in terms of effective mass Bloch-Volkov wavefunctions [9] that are analogous to Volkov wavefunctions that represent free electrons in an electromagnetic field. Using this perspec-

tive, the modified conduction band can be considered to consist of replicas of the original conduction band, each separated by the driving field photon energy. This modification is the manifestation of the combined classical and quantum character of the response in this strongly driven regime, i.e., where $U_p \sim \hbar\omega$. These states similar to dressed states in the AC Stark effect. Transitions from the valence band to "dressed states" in this modified conduction band are weighted by a function of the driving field.

The existing models predict an induced absorption with an absorption coefficient that is linear with U_p and with an extent that is on the order of U_p (or $\hbar\omega$, since $U_p \sim \hbar\omega$.) [1–3]. Both of these predictions are consistent with our data. Furthermore, using a model based on the Bloch-Volkov wavefunctions and the theory of x-ray absorption in gases in the presence of a strong laser field [10], we simulated the data in GaAs driven by a 3.5 μm MIR driving field [11]. There is good agreement between our model and observations.

4 Summary

We have made the first observation of ultrafast electro-absorption below the band edge in semiconductors due to the dynamical Franz-Keldysh effect. The observed ultrafast absorption induced by intense MIR pulses has an extent below the band gap that is the largest ever observed. The DFKE occurs at the transition between classical and quantum behavior, i.e., when the ponderomotive potential is comparable to the photon energy of the strong driving field.

5 Acknowledgements

We gratefully acknowledge support from NSF Grant No. DMR-9970962, ONR Grant No. N00014-94-1-1024, the Japan Science and Technology Corporation PRESTO Program, and the NEDO International Joint Research Grant Program.

References

1. Y. Yacoby, Phys. Rev. **169**, 610 (1968).
2. K. Johnsen and A.-P. Jauho, Phys. Rev. B **57**, 8860 (1998)
3. S. Hughes and D.S. Citrin, Optics Lett. **25**, 493 (2000).
4. L.V. Keldysh, Sov. Phys. JETP **20**, 1307 (1965).
5. Y I. Balkarei and E.M. Epshtein, Sov. Phys. Solid State **15**, 641 (1973)
6. L.C.M. Miranda, Solid State Commun. **45**, 783 (1983).
7. K. B. Nordstrom et al., Phys. Rev. Lett. **81**, 457 (1998).
8. W. Franz, Z. Naturforsch. **13A**, 484 (1958); L.V. Keldysh, Sov. Phys. JETP **34**, 788 (1958).
9. H.D. Jones and H.R. Reiss, Phys. Rev. B **16**, 2466 (1977).
10. M. Jain and N. Tzoar, Phys. Rev. A **15**, 147 (1977).
11. A.H. Chin, J.M. Bakker, and J. Kono, Phys. Rev. Lett. **85**, 3293 (2000).

Deviations from Diffusive Transport in 100 nm – 100 fs Spatio-Temporal Pump-Probe Experiments on GaAs

J. Hetzler[1], A. Brunner[1], M. Wegener[1], S. Leu[2], S. Nau[2], W. Stolz[2]

[1] Institut für Angewandte Physik, Universität Karlsruhe (TH), 76128 Karlsruhe, Germany
    e-mail: jochen.hetzler@physik.uni-karlsruhe.de
[2] Wiss. Zentrum für Materialwissenschaften, Philipps-Universität Marburg, 35032 Marburg, Germany

Abstract   We study the decay of an optically excited cloud of carriers as a function of its initial diameter d. Values as low as $d = 100$ nm are possible via apertures in a metal film, on a 100 nm thin film of bulk GaAs. We find that the decay time τ versus d scales as $\tau \propto d$ for submicron values of d, whereas one expects $\tau \propto d^2$ for diffusive transport. Furthermore, for submicron values of d, the transport dynamics depends on the initial carrier kinetic energy in sharp contrast to the diffusive regime.

1 Introduction

Consider a small cloud of carriers with transverse diameter d is impulsively excited by a short optical pulse in a thin film of a semiconductor. This cloud will spread out by carrier transport in the plane, thus the density in the initially excited area will decrease. This decay is not necessarily exponential, yet, for the sake of simplicity, we want to characterize it by a time constant τ. How does τ depend on d? For large values of d, one obviously expects diffusive transport. In this case it is simple to show that τ is given by $\tau = \frac{1}{8\ln 2}\frac{1}{D}d^2 \propto d^2$, with the diffusion constant D. Is this picture still correct on a scale of $d = 100$ nm? For ballistic transport[1] the carriers would obviously have some characteristic velocity v and one would naively expect the dependence $\tau = d/v \propto d$. In this article, we experimentally study the dependence $\tau(d)$ on a thin film of bulk GaAs for different temperatures and different excitation conditions with 100 fs pulses. This extends and complements other recent experiments [2–7] as well as theoretical work[8] on this issue.

2 Experiment

We employ a 100 nm thin film of bulk GaAs grown by metal-organic vapor phase epitaxy, so we can safely assume homogeneous excitation into the depth of the GaAs film. To avoid surface effects, the GaAs is clad between thin $Al_{0.3}Ga_{0.7}As$ barriers. A 100 nm thick aluminum film is evaporated on top(Fig. 1). Its measured intensity transmission coefficient is 10^{-5}. To minimize interaction effects between the metal and the semiconductor, we introduce a 10 nm layer of insulating MgF_2. Sets of well defined apertures in a metal film [9] are fabricated with different diameter by standard electron-beam lithography with negative resist and subsequent lift-off. Fig. 1 shows a typical set of apertures under an optical micro-

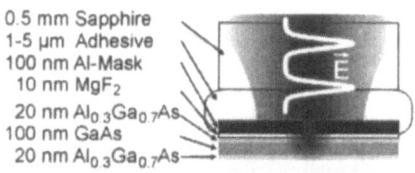

0.5 mm Sapphire
1-5 μm Adhesive
100 nm Al-Mask
10 nm MgF_2
20 nm $Al_{0.3}Ga_{0.7}As$
100 nm GaAs
20 nm $Al_{0.3}Ga_{0.7}As$

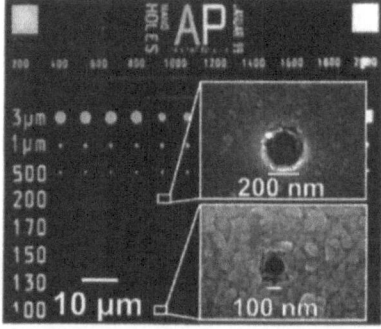

Fig. 1 Layer sequence of the sample, microscope image with an array of apertures and two selected examples of electron micrographs of individual apertures in the aluminum film.

scope and electron micrographs which reveal the quality of this process. Obviously, the GaAs film is within the optical near-field of these apertures. Thus, we do not expect the excited electron-hole cloud to be much broader than the physical size of the hole. Also, from what is known about surface plasmon effects at the holes in the metal film[10], we do not expect a significant broadening of the initial electron-hole cloud. Using a microscope lens (NA=0.4), we obtain a diffraction limited spot size on the sample of $1.2\,\mu m$. The pump intensity is ten times the probe intensity. Pump and probe propagate coaxially and have orthogonal circular polarization (to keep cylindrical symmetry). The carrier densities quoted in the following refer to that density which would be excited by the laser spot on a plane GaAs film (no aperture), assuming the unsaturated linear absorption coefficient.

We have performed pump-probe experiments at different excitation conditions through apertures from 100 nm to $1.5\,\mu m$ diameter. To complement these experiments, we have also performed standard pump-probe experiments with focused laser spots with diameters between $1.2\,\mu m$ and $8.8\,\mu m$ on the identical sample. The decay of the pump-probe traces shows a strong dependence on the aperture or spot diameter and on the excitation conditions. Even though the decays are usually non-exponential, we have fitted an exponential with time

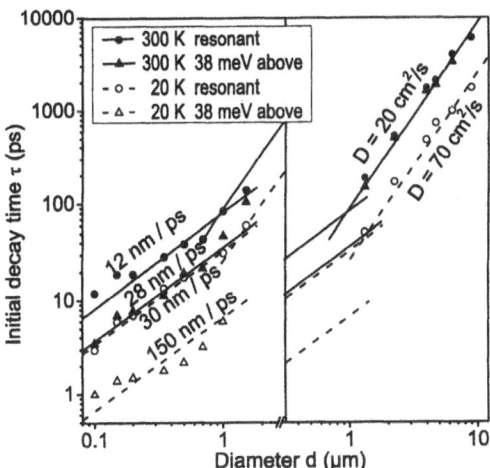

Fig. 2 Summary of decay times τ versus d on a double-logarithmic scale, left side: experiments with apertures, right side: experiments with different focus sizes.

constant τ to the initial decay. The dependence of this initial decay time τ on the diameter d for carrier densities around 10^{17}cm^{-3} for resonant and above band edge excitation at T=20 K and T=300 K is shown in Fig. 2.

To ensure that the observed decays are a result of the expansion of the small cloud of carriers, we also perform pump-probe experiments from the reverse side. The focused laser excites the GaAs film in front of the Al-mask. The probe beam is detected through the aperture. Ideally, one would expect no influence of d in this geometry. The small effects we have observed might be due to small modifications of the potential landscape in the GaAs film due to the presence of the metal aperture (strain or electrostatic). Further experiments show a shift of the photoluminescence energy under the holes in comparison to the plain GaAs film. This modification reduces for the higher excitation densities around 10^{17}cm^{-3}, possibly due to screening by the excited carriers.

3 Discussion

All measured decay times of the pump-probe signal are summarized in Fig. 2. Both for T=20 K and T=300 K we find a scaling according to $\tau \propto d$ for submicron values of d. For large d we find a scaling according to $\tau \propto d^2$ (see Fig. 2), as expected for diffusive behavior [11]. It is important to note that the observed decay times τ for submicron values of d are slower (and not faster) than one would expect from an extrapolation of the data at large d. As the intensity distribution under the near-field apertures is unlikely to be Gaussian but is rather expected to be constant under the aperture and zero elsewhere, one might expect faster decays due to large spatial gradients – in contrast to our observation. From this discussion we conclude that transport for submicron values of d is not consistent with a simple diffusive picture. Also, for diffusive transport, one expects no dependence on the initial kinetic energy of the carriers because the diffusion equa-

tion implies that the carriers have thermalized to the lattice temperature. This is indeed observed for large d. For submicron values of d, on the other hand, we find that τ does depend on the excess energy (Fig. 2). The decays become significantly faster for above-resonant excitation as compared to excitation resonant with the band edge. This difference strongly suggests that the carriers have not yet (fully) thermalized on the timescale of transport under the aperture. This is in agreement with the observed deviation from diffusive transport. If the transport at small d was ballistic for T=20 K (dashed line) and T=300 K, (solid line) one would obviously expect no sample-temperature dependence. Comparison of the solid and dashed lines shows that this is not the case. This indicates, that – at least for T=300 K – transport is not simply ballistic. Yet, we do find a dependence $\tau = d/v$ for small d even at T=300 K. This suggests that our results at small d cannot be explained by ballistic transport alone. Transport driven by the so-called Fermi-pressure[12] also leads to an expansion with a characteristic velocity v. Thus, one would also expect a dependence according to $\tau = d/v$. In fact, the values for v in the literature [13,14] are in the same range as the ones that we determine from Fig. 2. An expansion driven by Fermi-pressure should be density-dependent. In the range between 10^{17} cm^{-3} and 10^{18} cm^{-3} we have not found any significant influence of n_{eh} on the scenario shown in Fig. 2. However, these densities might still be too large to get below the critical density. The underlying difficulty to measure at lower densities becomes obvious when one calculates the total number of carriers below one 100 nm aperture created by the pump. For $n_{\mathrm{eh}} = 10^{17}$ cm^{-3} one finds a total number of only 78 electron-hole pairs.

4 Acknowledgements

This research has been supported by the DFG-SFB 195, the DFG-GK 284 and the Leibniz-Preis 2000, and is performed within the Institut für Nanotechnologie der Universität Karlsruhe (TH). We thank H. v. Löhneysen and his group for help with the electron-beam-lithography. We acknowledge stimulating discussions with Th. Schimmel, S. W. Koch, and A. Knorr.

References

1. A. C. Schaefer et al., Phys. Rev. B **54**, R 11046 (1996)
2. S. Grosse et al., phys. stat. sol. (b) **204**, 147 (1997)
3. S. Smith et al., ultramicroscopy **71**, 213 (1998)
4. A. Richter et al., Appl. Phys. Lett. **73**, 2176 (1998)
5. B. A. Nechay et al., Appl. Phys. Lett. **61**, 61 (1999)
6. M. Achermann et al., Phys. Rev. B **60**, 2101 (1999)
7. M. Vollmer et al., Appl. Phys. Lett. **74**, 1791 (1999)
8. B. Hanewinkel et al., Phys. Rev. B **60**, 8975 (1999)
9. D. Gammon et al., Phys. Rev. Lett. **76**, 3005 (1996)
10. C. Sönnichsen et al., Appl. Phys. Lett. **76**, 140 (2000)
11. H. Hillmer et al. Appl. Phys. Lett. **53**, 1937 (1988)
12. R. Ziebold et al, phys. stat. sol. (b), in press (2000)
13. K. M. Romanek et al., J. of Luminesc. **24/25**, 585 (1981)
14. M. Combescot, Phys. Lett. **85A**, 308 (1981)

Femtosecond optical excitation of acoustic phonons in bulk GaAs

O. B. Wright[1,2], B. Perrin[3], O. Matsuda[1], V. E. Gusev[4]

[1] Department of Applied Physics, Faculty of Engineering, Hokkaido University, Sapporo 060-8628, Japan
[2] PRESTO, Japan Science and Technology Corporation, Kawaguchi 332, Japan
[3] Laboratoire des Milieux Désordonnés et Hétérogènes, Université Pierre et Marie Curie, UMR 7603 CNRS, 4 Place Jussieu, 75252 Paris, France
[4] Laboratoire de Physique de l'Etat Condensé, UPRESA-CNRS 6087, Faculté des Sciences, Université du Maine, Av. O. Messiaen, 72085 Le Mans, France

Abstract Femtosecond spectroscopy of semiconductors has been extensively studied because it provides a window for the observation of ultrafast carrier distributions. Here we report an experimental investigation of the ultrafast coherent acoustic phonon generation mechanisms in bulk GaAs with ultrashort optical pulses. Comparison with a simple theory for ambipolar diffusion indicates that carrier diffusion has a significant effect on the spectrum of the phonon pulses generated.

Ultrafast carrier diffusion in semiconductors has been studied by a variety of experimental techniques (see, for example, Ref. 1). Laser picosecond ultrasonics is one such method that can be used to estimate the penetration of diffusing carriers perpendicular to a surface when excited with a picosecond or sub-picosecond optical pulse[2,3]. The carrier density is strongly coupled to strain through the deformation potential in semiconductors, and the acoustic pulse shape, typically with frequency components up to $\sim$100 GHz, plays the role of a depth profiler for the nonequilibrium carrier penetration. In order to obtain a high spatial resolution $\sim$10 nm, it is necessary to work with optical absorption depths of the same order (implying the use of relatively high photon excitation energies $\gtrsim$ 3 eV) and ultrashort optical pulses. Here we present the first such study of carrier diffusion in crystalline GaAs using sub-picosecond optical pulses.

Frequency doubled blue pump optical pulses of photon energy 3.3 eV (wavelength 375 nm), duration $\sim$200 fs and fluence $\sim$0.015 mJcm^{-2} were used to excite the surface of an approximately 2.6 μm thick slab of (100) n-doped GaAs (carrier concentration 1.3×10^{18} cm^{-3}), polished both sides and attached by ultraviolet-cured adhesive (several microns in thickness) to a glass substrate. The pump optical pulses launch coherent longitudinal acoustic phonon pulses into the solid. Infrared probe optical pulses of photon energy 1.65 eV of similar optical pulse duration derived from the same ultrashort pulse laser are used to detect transient changes in phase and reflectance in a Mach-Zehnder interferometer arrangement [4], and this allows the coherent phonon pulses returning to the surface to be monitored as a function of the delay time between the pump and probe pulses. Both pump and probe beams were focused to the same $\sim$35 μm diameter spot at low ($\sim$10°) angles of incidence. The laser repetition rate was 82 MHz.

Acoustic echoes due to the longitudinal acoustic phonon pulses returning to the surface after single or multiple reflection from the bottom surface of the GaAs could be detected. The shape of the first echoes in reflectance and phase are respectively shown by the solid curves in Figs. 1 and 2. The oscillatory reflectance changes arise because of the photoelastic coupling between the probe light electric field and the acoustic strain.[5] The overall duration of the acoustic pulses is determined by the probe light penetration ($\sim$600 nm). The phase changes arise from a combination of this photoelastic effect and a contribution proportional to the displacement of the free surface.[4,6] The surface motion occurs only while the phonon pulse is being reflected from the surface—a period corresponding to the raised portion of the signal in the centre of the echo in Fig. 2—and the results thus give a clear measure of the phonon pulse duration $\sim$25 ps (with a broad frequency distribution centred in the $\sim$15 GHz range). Using a simple model accounting for ambipolar carrier diffusion in GaAs and assuming that the acoustic strain generation results from the deformation potential mechanism[3,7], we have fitted the experimental results using the dimensionless parameters $D/v_l\zeta$, S/v_l, $dn/d\eta$ and $d\kappa/d\eta$ as variable

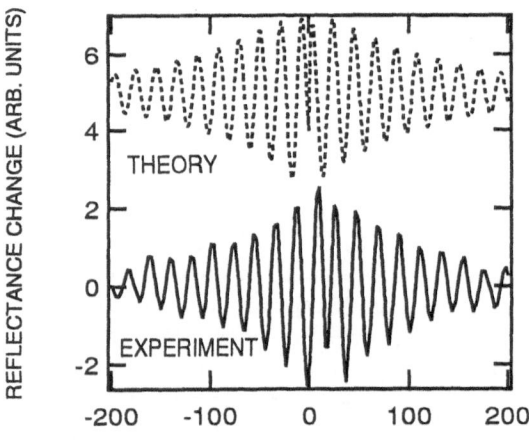

Fig. 1 Reflectance change corresponding to the first acoustic echo. The centre of the echo corresponds to a time delay of 1110 ps. The oscillations are caused by the photoelastic effect.

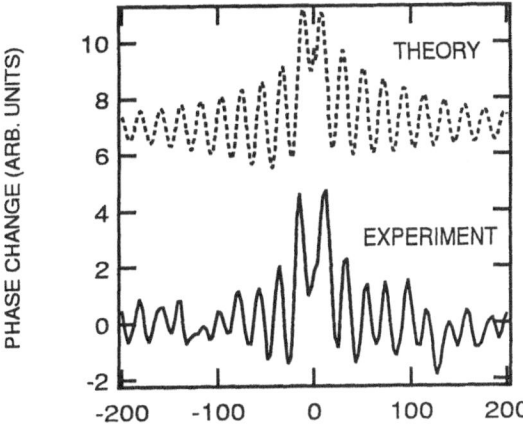

Fig. 2 Phase change corresponding to the first acoustic echo. The central, raised portion of the echo reveals the duration of the coherent phonon pulses to be ~5 ps.

data shown here. We also believe the effects of pulse broadening due to frequency dependent acoustic attenuation to be small[9]. The experimental conditions of our set-up are thus well suited to measure the diffusion of low-density electron-hole plasmas on sub-picosecond and picosecond timescales and length scales ~10 nm. In future we intend to probe the role of the transient populations of different bands in GaAs with measurements at different wavelengths. Possible practical application of this work may be in the field of GHz-THz acousto-optic modulation in ultrahigh speed semiconductor devices.

References

1. R. Ziebold et al., Phys. Rev. B **61**, (2000) 16610.
2. O. B. Wright, Phys. Rev. B **49**, (1994) 9985.
3. N. V. Chigarev et al., Phys. Rev. B **61**, (2000) 15837.
4. B. Perrin et al., Progress in Natural Science (China) **S6**, (1996) 444.
5. C. Thomsen et al., Phys. Rev. B **34**, (1986) 4129.
6. D. H. Hurley and O. B. Wright, Opt. Lett. **24**, (1999) 1305.
7. S. A. Akmanov and V. E. Gusev, Sov. Phys. Usp. **35**, (1992) 153.
8. T. Dekorsy et al., Phys. Rev. B **47**, (1993) 3842.
9. Chen et al., Phil. Mag. B **70**, (1994) 687.

quantities, where D is the ambipolar diffusivity, S is the surface recombination velocity, v_l is the longitudinal sound velocity, (fixed at 4730 ms^{-1}), ζ is the pump optical absorption depth (fixed at 13.8 nm) and $n + i\kappa$ is the complex refractive index at the probe wavelength (η the longitudinal strain). The dashed curves in Figs. 1 and 2, corresponding to $D = 2.6$ cm^{2}s^{-1}, $S/v_l = 0.2$, $dn/d\eta = -0.5$ and $d\kappa/d\eta = 2.7$, give a reasonable fit to the results, in particular to the central portion of the phase data. (Varying D and S by 25% and 50% respectively still produced acceptable fits.) Carrier diffusion broadens the acoustic pulse duration significantly (by a factor of $D/v_l\zeta \simeq 4$ from the value, $2\zeta/v_l \simeq 6$ ps, it would have in the absence of diffusion). This duration can be associated with the time taken for the electron-hole plasma to decelerate to the sound velocity. During this time the carriers penetrate to a typical depth of ~60 nm.

The value of D required to fit the results is significantly less than the standard value $D = 12$ cm^{2}s^{-1} determined by other methods for photoexcited carrier densities $\lesssim 10^{18}$ cm^{-3}. The low value of D derived may be related to the relatively high pump photon energy used.[1] However, the model we have used does not account for the effects of near-surface electric fields[8] (that will depend on the sample doping), strain generation by the thermoelastic effect (expected to give a ~30% contribution at the present pump photon energy[7]) and the complex carrier dynamics owing to intervalley scattering. Taking some of these factors into account would allow a more quantitative description of the acoustic pulse shape.

At the photoexcited carrier densities ($\sim 3 \times 10^{18}$ cm^{-3}) typically involved here we do not expect a significant carrier-density dependence of the diffusivity[1], and indeed the observed acoustic pulse shape for a pump intensity reduced by a factor of two was the same as the

Theory of hot luminescence in pulse-excited semiconductors

K. Hannewald, S. Glutsch, F. Bechstedt

Friedrich-Schiller-Universität Jena, Institut für Festkörpertheorie und Theoretische Optik, Max-Wien-Platz 1, 07743 Jena, Germany. e-mail: hannewd@ifto.physik.uni-jena.de

Abstract We present a theory of time-and-energy resolved photoluminescence (PL) from pulse-excited semiconductors. Our approach combines quantum kinetics of hot-carrier relaxation and quantum theory of spontaneous emission under consistent inclusion of Coulomb interaction. Model calculations show the transition from PL at the pump frequency via subsequent phonon replicas until the buildup of excitonic PL. We predict hot PL to be a sensitive measure of electron–LO-phonon quantum kinetics and bottleneck effects.

In recent years, tremendous progress in femtosecond laser spectroscopy has enabled time-and-energy-resolved studies of the interaction between laser-induced electron-hole pairs and lattice vibrations in semiconductors. Violation of classical energy conservation in electron–LO-phonon scattering was first observed in four-wave mixing experiments on bulk GaAs [1]. Another milestone in detecting quantum-kinetic effects was achieved by Leitenstorfer *et al.* [2,3] using pump-probe techniques, where the step-by-step relaxation of conduction electrons after pumping high above the band gap was investigated. Theoretical studies on pump-probe signals have confirmed the non-Markovian nature of the electron-phonon interaction on a sub-100-fs timescale [4,5]. Due to experimental progress it is now also timely to do similar investigations for the hot-luminescence signal after fs-pulse excitation.

In this paper, we present theoretical studies of time-resolved PL spectra from coherently excited semiconductors. These studies show that hot luminescence can serve as a very sensitive measure of the electron-lattice interaction on a sub-100-fs timescale. They indicate also how ultrafast PL spectroscopy can give insight into the conversion of secondary emission from hot luminescence after a few collisions into PL from thermalized carriers.

The numerical calculations are performed in two steps. First, we determine the time evolution of the distribution functions for electrons and holes after pulse excitation. The theoretical description of hot-carrier generation and relaxation is based upon electron–LO-phonon quantum-kinetic equations using phonon-assisted density matrices. The calculations are performed in second Born approximation, with Coulomb interaction consistently incorporated both in the coherent and in the scattering term. Recently, this method has been applied successfully to the calculation of FWM signals [1] and pump-probe signals [5]. Details of the numerical implementation including stability and convergency studies can be found in Ref. [5].

Subsequently, we convert the distribution functions into the PL signal by applying the recently developed theory of PL in semiconductors [6]. This approach relates the PL signal to the two-time two-wavevector polarization function which is then calculated from the Bethe-Salpeter equation in ladder approximation. This theory overcomes some shortcomings and/or limitations of previous approaches [7–9]. Most importantly, it is guaranteed that the PL signal is non-negative, and the theory does not rely on the explicit use of the Kubo-Martin-Schwinger relation, which allows us to treat arbitrary non-thermal situations, as exemplified in Ref. [6].

As a result, the time-dependent PL spectra $I(\omega, t)$ can be conveniently expressed as

$$I(\omega, t) \propto \frac{1}{\Omega} \sum_{\mathbf{k}_1 \mathbf{k}_2 \mathbf{k}} [H - \hbar(\omega + i\epsilon)]^{-1}_{\mathbf{k}_1 \mathbf{k}} \, 2\hbar\epsilon$$
$$\times (n_{\mathbf{k}})_{cc} [1 - (n_{\mathbf{k}})_{vv}] [H - \hbar(\omega + i\epsilon)]^{-1\dagger}_{\mathbf{k}\mathbf{k}_2}, \quad (1)$$

with an effective two-particle Hamiltonian

$$H_{\mathbf{k}_1 \mathbf{k}_2} = [E_c(\mathbf{k}_1) - E_v(\mathbf{k}_1)] \, \delta_{\mathbf{k}_1 \mathbf{k}_2}$$
$$- [(n_{\mathbf{k}_1})_{vv} - (n_{\mathbf{k}_1})_{cc}] \frac{1}{\Omega} V_{\mathbf{k}_1 - \mathbf{k}_2}. \quad (2)$$

The inclusion of Coulomb effects in Eqn. (2) allows for PL at the excitonic resonances and generalizes the text-book result of free-particle PL [10]. The band dispersions $E_j(\mathbf{k})$ include screening and band-gap renormalization.

For the explicit calculations we used bulk GaAs parameters, $E_g(300\,\text{K}) = 1.43\,\text{eV}$, $m_e = 0.067\,m_0$, $m_h = 0.442\,m_0$, $\hbar\epsilon = 0.94\,\text{meV}$, $\hbar\omega_{\text{LO}} = 36\,\text{meV}$, $\varepsilon_\infty = 11.1$, and $\varepsilon_s = 13.1$. The exciton binding energy is $1\,\text{Ry}^* = 4.7\,\text{meV}$. The pump-pulse parameters are fixed at a detuning of $\hbar\omega_p - E_g = 112.8\,\text{meV}$ ($24\,\text{Ry}^*$) and a maximum laser intensity of $I_p \approx 0.1\,\text{MW/cm}^2$ ($\mu E_p = 0.04\,\text{Ry}^*$) at $t = 0$. In order to optimize the spectral resolution, we choose a pump pulse length of 320 fs which is larger than the LO-phonon period $T_{\text{LO}} \approx 115\,\text{fs}$.

In Fig. 1, we present the time evolution of the energetic distributions of the electrons (a) and the corresponding PL spectra (b) for times of $t = -70\,\text{fs}$ to 2800 fs (from top to bottom). The lattice temperature is $T = 300\,\text{K}$.

From Fig. 1(a), it follows that the electrons start to relax already during the pumping process which is seen through a cascade-like buildup of phonon replicas of the initial distribution. These replicas start from an initially broad signal and narrow with increasing time. Such memory effects reflect the quantum-mechanical time-energy uncertainty principle and are a pronounced signature of quantum-kinetics. At later times, the electrons accumulate near the band minimum, and after 2.8 ps an almost thermal distribution is reached. The same is true for the holes (not shown).

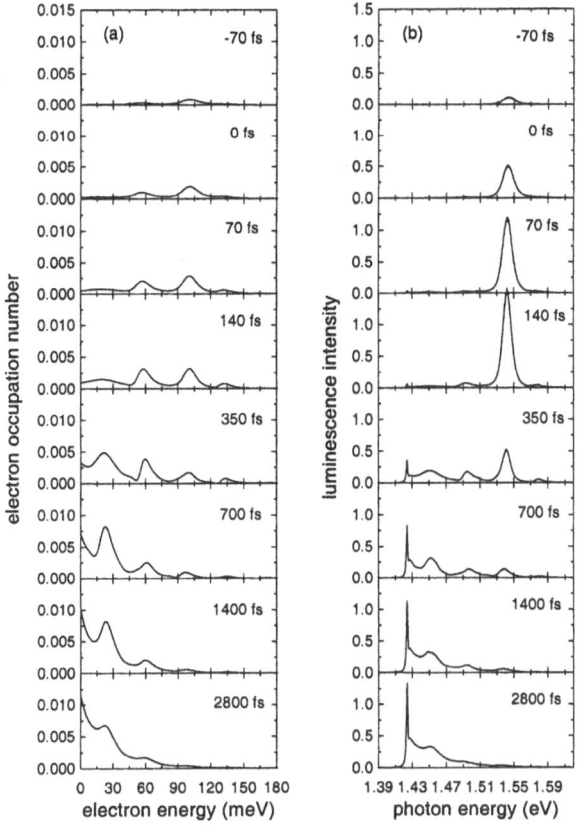

Fig. 1 (a) Electron distribution n_{cc} and (b) linear luminescence spectra $I(\omega, t)$ for times $t = -70, ..., 2800$ fs (from top to bottom). The pump parameters are $\hbar\omega_p - E_g = 112.8$ meV and $I_p \approx 0.1$ MW/cm^2, and the pulse length is 320 fs. The lattice temperature is $T = 300$ K.

In the PL spectra [Fig. 1(b)], the relaxation process is nicely resolved. After the strong initial PL at the pump frequency, subsequent phonon replicas emerge, both on the low- and high-energy side. Here, the non-Markovian broadening is especially pronounced at the second phonon replica ($\hbar\omega \approx 1.45$ eV). Furthermore, the relaxation process is accompanied by a gradual buildup of PL at the 1s exciton resonance, which becomes the dominant contribution after about 1 ps.

A logarithmic-scale plot also shows that for large times the PL intensity above the band gap evolves into an exponentially decreasing tail (see Ref. [11]). Assuming that electrons and holes are in quasi-equilibrium, an exponential fit of this tail may be used to define an effective temperature of the electron gas.

Because of bottleneck effects, the situation is more complicated for zero lattice temperature (not shown). Since upward scattering is not possible in this case, the conduction electrons do not completely thermalize within the first few ps. Instead, the electron distribution remains highly non-thermal even after 2.8 ps. Analogously, the holes do not significantly relax on this time scale. As a result, the PL signal at the pump frequency is still noticeable after 2.8 ps (see Ref. [11]).

In summary, we have presented the first quantum-kinetic studies of time-and-energy resolved photoluminescence from pulse-excited semiconductors. Our approach combines quantum kinetics of hot-electron relaxation and quantum theory of spontaneous emission under consistent inclusion of Coulomb interaction between the excited hot carriers. We predict time-resolved photoluminescence to be a sensitive measure of electron–LO-phonon quantum-kinetics and bottleneck effects. This is exemplified by model calculations that show how the luminescence intensity shifts from the initial signal at the pump frequency towards the excitonic resonance via emission of LO phonons. We feel that principally this scenario should be observable in sensitive time-resolved PL experiments. Effects beyond the present theory may be necessary for an improved description of the details of the relaxation process, but we believe they will not significantly alter the general picture presented here. Future refinements may include especially a density-dependent line broadening, e.g., by calculating the self-energy in T-matrix approximation, as recently done by Piermarocchi *et al.* within a simplified 1D model [8].

Financial support by the Carl Zeiss Jena GmbH and Thüringer Ministerium für Wissenschaft und Kultur is gratefully acknowledged.

References

1. L. Bányai *et al.*, Phys. Rev. Lett. **75**, (1995) 2188.
2. A. Leitenstorfer *et al.*, Phys. Rev. Lett. **76**, (1996) 1545.
3. C. Fürst *et al.*, Phys. Rev. Lett. **78**, (1997) 3733.
4. A. Schmenkel *et al.*, J. Lumin. **76&77**, (1998) 134.
5. K. Hannewald *et al.*, Phys. Rev. B **61**, (2000) 10 792.
6. K. Hannewald *et al.*, Phys. Rev. B **62**, (2000) 4519.
7. M. F. Pereira, Jr. and K. Henneberger, Phys. Rev. B **53**, (1996) 16 485; Phys. Rev. B **58**, (1998) 2064.
8. C. Piermarocchi *et al.*, Solid State Comm. **112**, (1999) 433.
9. M. Kira *et al.*, Phys. Rev. Lett. **81**, (1998) 3263.
10. See, e.g., S. L. Chuang, *Physics of Optoelectronic Devices* (Wiley, New York, 1995).
11. K. Hannewald *et al.*, submitted to Phys. Rev. Lett.

Dynamics of Bloch oscillations under the influence of scattering

F. Löser[1], Yu. A. Kosevich[2,3], K. Köhler[4], K. Leo[1]

[1] Institut für Angewandte Photophysik, Technische Universität Dresden, Germany, e-mail: leo@iapp.de
[2] Max-Planck-Institut für Physik komplexer Systeme, 01187 Dresden, Germany
[3] Moscow State University of Technology "STANKIN", 101472 Moscow, Russia
[4] Fraunhofer-Institut für Angewandte Festkörperphysik, 79108 Freiburg, Germany

Abstract The influence of scattering and plasmon coupling on temporal and spatial dynamics of Bloch oscillations in a semiconductor superlattice is investigated using optical techniques. The oscillations are strongly modified due to the scattering processes: The dependence of amplitude on k-space composition of the wave packet is modified due to a renormalization of the energy scale. For higher carrier densities, plasmon coupling causes anharmonic Bloch oscillations.

1 Introduction

The oscillatory motion of electrons in a periodic potential subject to an electric field (Bloch oscillations, BO) is a basic transport effect and has been recently demonstrated in a number of experiments in biased semiconductor superlattices (SL) [1]. The oscillations are characterized by the frequency $\omega_B = eF_{0z}d/\hbar$ and spatial amplitude $Z_B^{sc} = \Delta/(2eF_{0z})$ where F_{0z} is external field along the axis of the SL with period d and miniband width Δ. The frequency-space analog of BO is a ladder structure (the Wannier-Stark ladder, WSL) with resonances at $E_n = E_0 + neF_{0z}d$, $n = 0, \pm 1, \pm 2, ...$ with E_0 as the energy of a reference state, which has been observed in optical spectra [2,3]. Bloch oscillations were traced by time-resolved optical experiments [4–8].

We have recently investigated the dynamics of Bloch oscillations under the influence of scattering and coupling to coherent plasma oscillations [9]. The results confirm the theoretical prediction [10] that the spatial dynamics of the Bloch-oscillating carrier ensemble is clearly affected by these effects. The experiments are performed on a 67/17 Å GaAs/Al$_{0.3}$Ga$_{0.7}$As undoped superlattice with a reduced electron-hole miniband width $\Delta \approx 40\,meV$ held at $10\,K$. We use a a mode-locked Ti-Sapphire laser for optical excitation. The wave packet motion is detected by transient two-beam four-wave mixing [11]. We deduce the amplitude of the wave packet from the peak shift e_i: The oscillating electric field due to the Bloch oscillations modulates the WSL with shifts linked to the dipole moment of the carrier ensemble [7,12].

2 Influence of scattering on spatial dynamics

We investigate the spatial dynamics as a function of k-space composition of the Bloch wave packet to confirm the influence of scattering. Composition of the Bloch wave packet can be controlled by changing the spectral laser position $\hbar\omega_c$: Excitation well below or above the center of the WSL corresponds to wave packets close to

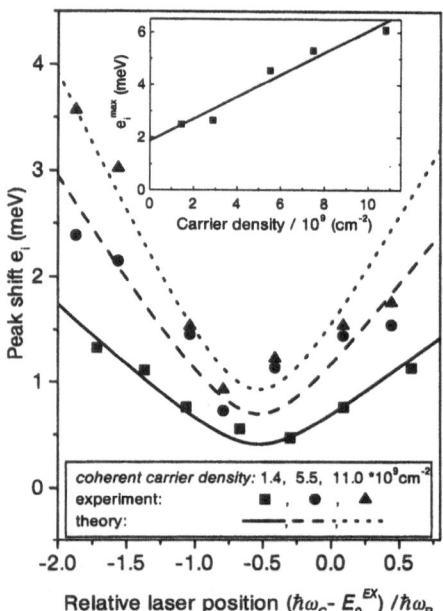

Fig. 1 Wannier-Stark ladder peak shift vs. laser spectral position for various excitation densities at $15\,kV\,cm^{-1}$ internal electric field. The inset shows the maximal value of the induced depolarization field energy as function of carrier density.

$k_z = 0$ or to the edges of the Brillouin zone, respectively, having large amplitudes. In contrast, excitation close to the WSL center corresponds to a wave packet spread over the Brillouin zone with little or no amplitude [8,13]. Without scattering, zero amplitude should be reached if the excitation is exactly at the center of the WSL, i.e., at E_0^{EX}. Figure 1 displays the experimentally observed dependence of the peak shift of the heavy-hole -1 transition (which is proportional to the BO amplitude [7]) for a broad range of laser excitation positions, given in units of the WSL splitting. The amplitudes are large for excitation above and below the center of the WSL; for exitation in between, a minimum is visible. The main observation is the downward shift of the minimum [8] compared to the simple model, which predicts it at zero relative energy. A recently developed theoretical model shows that this shift is caused mainly by the *influence of scattering on Bloch oscillations* (for a detailed description, refer to Ref. [9,10]). The model is based on a semi-

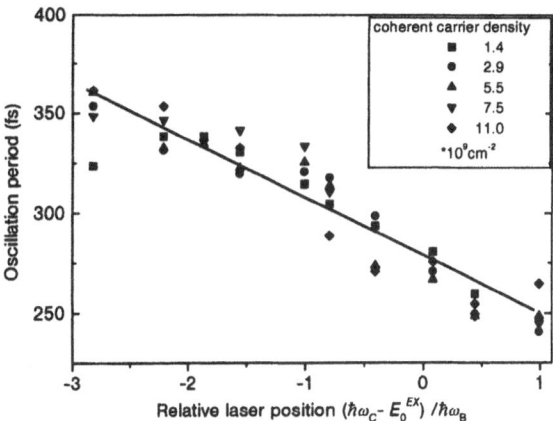

Fig. 2 The dynamic BO period as function of laser position. Changes of the period due to electron-hole interaction are clearly observed.

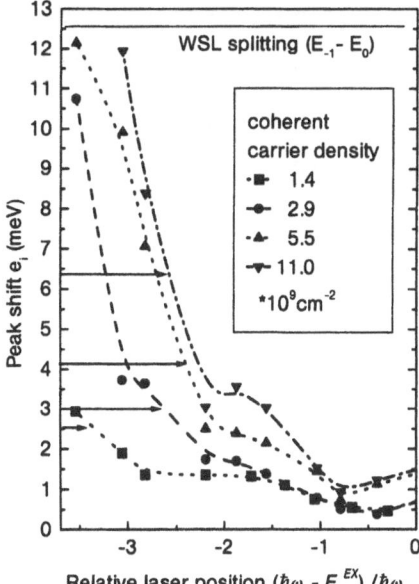

Fig. 3 Wannier-Stark ladder peak shift vs. laser spectral position for various excitation densities at $15\,kV\,cm^{-1}$ electric field. For laser detuning far from the WSL center, the peak shift strongly increases due to the anharmonic motion of the wave packet. The lines are guides for the eye. The arrows indicates the depolarization energy equivalent of the semiclassical amplitude.

classical description of the dynamics of the macroscopic-average kinetic energy and relative electron-hole velocity of an electronic wave packet (EWP) in a biased SL. The theory predicts the displacement $Z(t)$ of the harmonic BO when a finite phenomenological intraminiband scattering rate τ is included [10]. The spectral position and absolute value of the BO amplitude minimum are then determined by the relaxation rate $1/\tau$. In a collisionless SL ($1/\tau = 0$), the minimum is at the WSL center. For the finite scattering rate, the minimum of the BO amplitude *is prediced to be always below* the WSL center. The lines of Figure 1 display these theoretical predictions for the BO amplitude for different densities, in excellent agreement with the experimental data marked by symbols.

3 Influence of plasmon coupling

The interaction of BO with *coherent plasmons* [14] and the effect of electron-hole Coulomb coupling is included in the theory [10] by taking the total electric field F_z as a sum of the external, F_{0z}, and self-induced, $F_{iz} = -\frac{e}{\epsilon}Z(4\pi N + \beta/a_B^3)$, fields, where N is the electron density, $\epsilon = 12.8$ is a (static) background dielectric constant in $GaAs$, $a_B = \hbar^2\epsilon/me^2 = 10\,nm$ is the Bohr radius, the coefficient β is related to the exciton binding energy. The interplay of both fields leads to a prediction of a dynamic Bloch frequency ω_B^* [9,10], which is compared with experiment in Fig. 2 as a function of the spectral laser pulse position. The dependence of the BO period on density is removed by compensating the field screening by the bias voltage adjustment. The experimentally observed dependence is well described by the theory.

We also have investigated BO spatial dynamics for very high excitation densities, which is of prime importance for applications since useful THz emitter devices will need high carrier densities to give reasonable power levels. Figure 3 displays the amplitude for high various excitation densities. For larger detuning from the WSL center, the amplitude starts to increase faster with de-

tuning than close to the center, and peak shifts even *exceed the semiclassical value* (arrows in Fig. 3) when the induced field by the BO exceeds the bias field. It can be inferred from our results that operation of the emitter devices is limited to carrier densities on the order of several times 10^{10} per cm^2 per well.

We thank the DFG and VW-Stiftung for support.

References

1. L. Esaki and R. Tsu, IBM J. Res. Dev. **61**, (1970) 61.
2. E. E. Mendez, F. Agullo-Rueda, and J. M. Hong, Phys. Rev. Lett. **60**, (1988) 2426 .
3. P. Voisin et al., Phys. Rev. Lett. **61**, (1988) 1639.
4. J. Feldmann et al., Phys. Rev. B **46**, (1992) 7252.
5. K. Leo et al., Solid State Comm. **84**, (1992) 943.
6. C. Waschke et al., Phys. Rev. Lett. **70**, (1993) 3319.
7. V. G. Lyssenko et al., Phys. Rev. Lett. **79**, (1997) 301.
8. M. Sudzius et al., Phys. Rev. B **57**, (1998) R12693.
9. F. Löser et al. Phys. Rev. B **61**, (2000) R13373.
10. Yu. A. Kosevich, Ann. Phys. (Leipzig) **8**, (1999) 145.
11. T. Yajima and Y. Taira, J. Phys. Soc. Jpn. **47**, (1979) 1620.
12. K. Leo, Semicond. Sci. Technol. **13**, (1998) 249.
13. M. Dignam, J.E. Sipe, and J. Shah, Phys. Rev. B **49**, (1994) 10502.
14. A. W. Ghosh, L. Jönsson, and J. W. Wilkins, Phys. Rev. Lett. **85**, (2000) 1084.

Nonlinear response due to propagating excitonic polaritons in GaAs heterostructure

K. Akiyama, N. Tomita, T. Nishimura, Y. Nomura, T. Isu

Advanced Technology R&D Center, Mitsubishi Electric Corporation, 8-1-1 Tsukaguchi-honmachi, Amagasaki, Hyogo 661-8661, Japan e-mail: `akiyama@bio.crl.melco.co.jp`

Abstract   We report experiments on three beam degenerate four-wave mixing in a semiconductor heterostrucure including high quality GaAs layers. The temporal profiles of the signal at the exciton resonance show the energy dependence of the exciting picosecond pulses. This behavior is explained by propagation effects of polaritons that shows strong dispersion of the group velocity.

1 Introduction

Nonlinear optical properties of excitons in semiconductors have been a subject of intense experimental and theoretical investigations because of their particular importance for fundamental and practical aspects. Especially many studies have clarified the dynamics of excitonic processes by using progressing ultrafast spectroscopy [1]. These studies mainly have dealt with quantum wells, where spatial dispersion of the excitonic polarization can be neglected because of the short interaction length with the exciting electromagnetic (EM) field. However, when the sample thickness become larger than the Bohr diameter of excitons and the coherence of the exciton is maintained over the layers along the EM wave propagation direction, the excitonic polarization nonlocally couples with the EM field. In this case, the polariton nature can be considered at the exciton resonance. Such polariton effects in thin layers induce a strong modification of ultrafast coherent dynamics of excitons because propagating ultrashort pulses are severely distorted according to the dispersion of polaritons as observed in linear response of high quality semiconductors [2,3]. This feature can also affect nonlinear optical phenomena of excitons. We have observed a remarkable size dependence of degenerate four-wave mixing (DFWM) in high quality GaAs layers [4,5], which is explained by the nonlocal theory for thin layers. Recently, we have also demonstrated a high efficient and ultrafast all optical switching using a heterostrucure including size-controlled GaAs layers [6]. In this paper, we investigate coherent dynamics of excitons in such heterostrucure by three beam DFWM using picosecond pulses. The temporal behavior of the signals, that includes both coherent and incoherent process, is analyzed in terms of polariton propagation effects that show strong dispersion of the group velocity.

2 Experimental

The heterostructure sample is grown on a (100) GaAs substrate by molecular beam epitaxy. It consists of 3 pairs of very high quality 110 nm thick GaAs layers separated by 5 nm thick $Al_{0.3}Ga_{0.7}As$ layers and a dis-

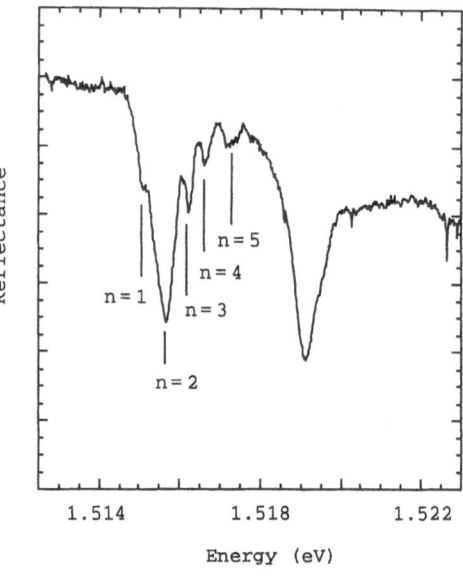

Fig. 1 Reflection spectrum of a heterostrucure including high quality GaAs layers obtained at T = 5 K.

tributed Bragg reflector (DBR). The DBR, which is composed of 24 pairs of n-doped GaAs (62 nm)/AlAs (74 nm), is used to collect the signal in a reflection geometry. The reflectance of the DBR is approximately 99% at the exciton resonance (1.5150 eV). In DFWM measurement, we use picosecond pulses (2.0 ps) from a mode-locked Ti:sapphire laser with a bandwidth of 0.9 meV. The two linearly polarized pump pulses simultaneously arrive at the sample and a grating of excitons is induced. Part of the probe pulse is diffracted by the grating and is detected as a function of the delay time between the two pump pulses and the probe pulse. The sample is cooled at 5 K during the measurements.

3 Results

The reflection spectrum of the sample exhibits several narrow dips at the higher energy side of the exciton resonance as shown in Fig. 1. These dips correspond to the quantized states of excitons due to confinement of center of mass motion along the growth direction. The damping of excitonic polarization estimated by the spectrum is 0.09 meV, which is less than the LT splitting energy 0.133 meV. This small damping verifies the polariton picture when the light propagates in the sample.

Figure 2-(a) shows the diffracted signal intensity obtained at the delay time T = 0. The signal is measured in a low excitation regime, where the power dependence of the signal shows third-order nonlinear response. The signal intensity has maximum value at 1.5151 eV which is slightly lower than the resonant energy of the n = 2 quantized state. This result is explained by the large absorption of this state. Figure 2-(b) shows the corresponding temporal profiles of the diffracted signals for several excitation pulse energies around the exciton resonance. The temporal profiles show two decay processes for higher excitation energies than 1.5150 eV. The fast temporal response at around T = 0 is due to the coherent excitons while the slow decay process with a time constant of several picoseconds originates from the incoherent excitons. The signal intensity from the coherent excitons is an order of magnitude larger than that of the incoherent excitons, which means that the dynamics of nonlinear optical process is dominated by the coherent excitons. The single exponential decay is observed for lower excitation energies than 1.5142 eV, which is ascribed to the biexcitons. The contribution of biexciton to nonlinear optical response of a thin film GaAs is observed for the excitation density dependence of four-wave mixing [8]. The rise and the decay time of the coherent signals strongly depend on the excitation pulse energy. For higher excitation pulse energies than 1.5150 eV, both the rise and the decay process are comparable to those of the excitation pulse. The decay becomes slow down when the pulse energies are tuned to the exciton resonance. Furthermore when the pulse is tuned to higher energies, the coherent process becomes fast and the incoherent process cannot be observed.

4 Discussions

The observed energy dependence of the coherent signals can be explained by a distortion of the excitation pulses due to spatial dispersion of excitonic polariton. The slow decay in the positive and negative delay region corresponds to slow propagation of the grating generated by the pump pulses and the probe pulse in the sample. The resonant polariton components constituting such pulses propagate with group velocity v_g which is reduced even to 10^4 m/s for bulk GaAs. The v_g of the upper polariton mode at the ω_T is several times smaller than that of the lower polariton mode because damping effects induce a modulation of the dispersion curve of the upper branch. By taking into account of the strong absorption of the excitation pulses, the signals are mainly emitted from the top GaAs layer. In the spectrum range for the experiment, the round trip time of the polariton pulses within the top GaAs layer is in the range from subpicoseconds to a few picosecons, which agrees with the decay rate of the experiment. The reason why the contribution of the upper polariton dominates the coherent response can be interfering effects including dumping for thin films. These results reveal that dynamics processes of nonlinear optical process of the coherent excitons are dominated by the polariton propagation.

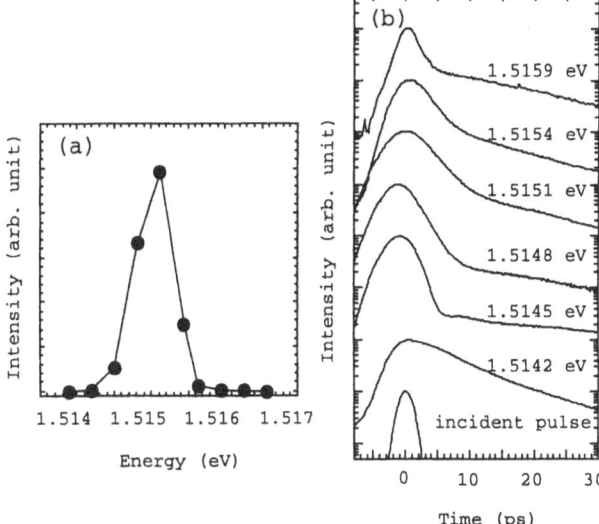

Fig. 2 Three beam degenerate four-wave mixing for different excitation pulse energies obtained at T = 0 and (b) corresponding temporal profiles. The density of the pump pulses is 2.5 kW/cm².

5 Conclusion

We have investigated the nonlinear optical response of a heterostrucure at exciton resonance by three beam degenerate four-wave mixing. The temporal profiles show the dependence of the excitation pulse energies, due to polariton dispersion. This result indicates that the grating is induced by the excitonic polariton and the decay of the response is determined by the polariton propagation in the layer.

6 Acknowledgment

A part of this work was performed under the management of a technological re search association, the Femtosecond Technology Research Association (FESTA), which is supported by New Energy and Industrial Development Organization (NEDO)

References

1. J. Shah, *Ulrtafast Spectroscopy of Semiconductors and Semiconductor Nanostructures* (Springer-Verlag, Berlin, 1996)
2. D. Fröhlich, A. Kulik, B. Uebbing, A. Mysyrowicz, V. Langer, H. Stolz, and W. von der Osten, Phys. Rev. Lett. **67**, (1991) 2343.
3. S. Nusse, P. H. Bolivar, K. Kurz, F. Levy, A. Chevy, and O. Lang, Phys. Rev. B**55**,(1997) 4620.
4. K. Akiyama, N. Tomita, Y. Nomura, and T. Isu, Appl. Phys. Lett. **75**,(1999) 475.
5. K. Akiyama, N. Tomita, Y. Nomura, and T. Isu, H. Ishihara, and K. Cho, Physica E**7**,(2000) 661.
6. N. Tomita, K. Akiyama, Y. Nomura, and T. Isu, IQEC 2000 (Nice)
7. H. Ishihara and K. Cho, Phys. Rev. B**53**, (1996)15823.
8. K. Akiyama, N. Tomita, Y. Nomura, and T. Isu, Physica B **272**,(1999) 505.

Dynamic inter-sideband Fano interference of excitons in ac-driven superlattices

Ren-Bao Liu, Bang-Fen Zhu

Center for Advanced Study, Tsinghua University, Beijing 100084, China
e-mail: bfzhu@castu.phys.tsinghua.edu.cn

Abstract Fano interference is predicted to occur between the sidebands of Floquet-state excitons in dc- and ac-driven superlattices, which leads to unique Fano shape in absorption spectra and to intrinsic decay in four-wave mixing traces.

1 Introduction

Quantum interference between discrete states and energetically degenerate continua [1], as named after Fano, have been intensively investigated in atomic and molecular systems. Recently, Fano interference (FI) between discrete excitons and the ionization continua of low-lying subbands was observed in semicondcutor systems [2–5], and transient four-wave mixing (FWM) experiments have demonstrated the FI as a fundamental type of dephasing mechanism [4,5].

In this contribution, we predict a novel type of FI between discrete Floquet-state excitons and sidebands of their ionization continua in dc- and THz-ac-driven superlattices. This inter-sideband FI is essentially a nonlinear optical process with respect to the THz driving field.

2 Model and Theory

The system considered is a superlattice (SL) subject to dc- and THz-ac-fields applied along the growth direction, which is modeled by a one-dimensional tight-binding lattice. This approximation can be justified by the fact that the ground Wannier-Stark (WS) exciton-state in dc-biased SL's usually possesses by far larger oscillator strength than its ionization continuum of in-plane motion.

By treating the Coulomb interaction in the Hartree-Fock approximation, Hamiltonian of the electron-hole (e-h) relative motion can be written in the quasimomentum representation $\{|k\rangle\}$ as

$$H(t) = \varepsilon_k |k\rangle\langle k| + i\left[\omega_{BO} + \omega_1 \cos(\omega t)\right] |k\rangle\partial_k\langle k| - V_{k,k'}|k\rangle\langle k'|, \qquad (1)$$

where $\varepsilon_k = E_0 - \frac{\Delta}{2}\cos k$ is the combined e-h miniband dispersion, ω_{BO} is the WS ladder spacing due to the dc field, ω_1 is the strength of the ac field with frequency ω (here the SL spatial period is set to be unit), and $V_{k,k'}$ is the Coulomb potential matrix element.

This time-periodic system has no stationary eigen state, but Floquet state and quasienergy can be defined as temporal analogue of Bloch state and quasimomentum in spatially periodic lattice [6]. The quasienergy ε_q is calculated by numerically diagonalizing the secular

equation $U(T+t,t)|q,t\rangle = |q,T+t\rangle = \exp(-i\varepsilon_q T)|q,t\rangle$, where $U(T+t,t) \equiv \hat{T}\exp[-i\int_t^{T+t} H(t)dt]$, $|q,t\rangle$ is the Floquet state, and $T \equiv 2\pi/\omega$. Obviously, for any integer number m, $|q,m\rangle \equiv |q\rangle\exp(im\omega t)$ is also a solution with quasienergy $\varepsilon_q + m\omega$, repeating in quasienergy Brillouin zones $\{[m\omega,(m+1)\omega]\}$ as sidebands induced by the THz ac field.

The interband optical spectra are obtained by integrating the semiconductor Bloch equations [7] for the exciton amplitude p_k and the e-h pair density f_k

$$\partial_t p_k = -i\varepsilon_k(t)p_k + [\omega_{BO} + \omega_1\cos(\omega t)]\partial_k p_k + i\chi^R(t)(1 - 2f_k) - \gamma_2 p_k, \qquad (2)$$

$$\partial_t f_k = [\omega_{BO} + \omega_1\cos(\omega t)]\partial_k f_k - 2\Im\left\{\chi^R(t)p_k^*\right\}, \qquad (3)$$

where $\varepsilon_k(t) = \varepsilon_k - 2\sum_{k'} V_{k,k'}f_{k'}$ is the renormalized miniband dispersion, $\chi^R(t) = \chi(t) + \sum_{k'} V_{k,k'}p_{k'}$ is the renormalized near-infrared optical excitation (with the dipole element assumed k-independent), and γ_2 is the phenomenologically introduced interband dephasing rate. In this time-dependent system, the linear absorption spectrum $\alpha(\Omega) \propto \Im[P^{(1)}(\Omega)/\chi(\Omega)]$ depends on the excitation profile $\chi(\Omega)$ [8], where $P^{(1)}(\Omega)$ is the linear part of the optical response spectrum $\sum_k p_k$. In this work, we consider the continuous wave absorption. The FWM signals are calculated following Ref. 9, and the two incoming pulses, delayed by a delay time τ, are assumed to have Gaussian shape with central frequency Ω_0 and spectrum width G.

Below we focus on a specific case: The ac field is resonant with the Bloch oscillation, and the Coulomb potential is of on-site type, i.e. $V_{k,k'} = V_0/N$ (N is the size

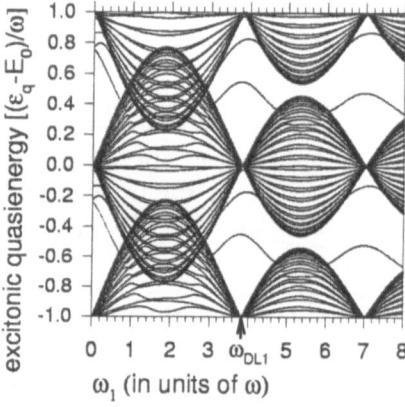

Fig. 1 Quasienergy sideband structure of excitons as functions of the ac field strength.

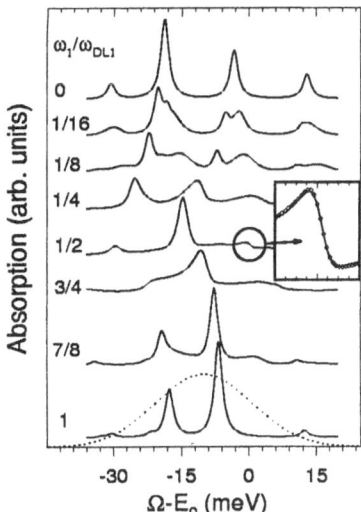

Fig. 2 Absorption spectra (solid lines) of the superlattice for various ac field strength as indicated by ω_1/ω_{DL1}. Inset is an enlarged example of the Fano resonance, where the squares are fitted with Eq. (4). The dotted line represents the excitation spectrum for calculating the FWM signals.

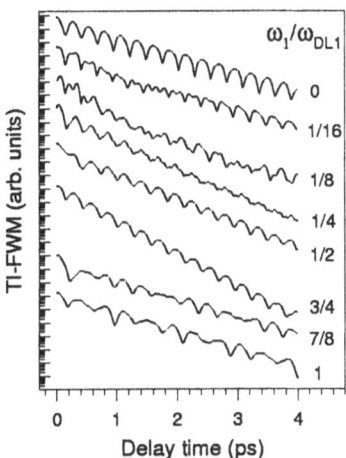

Fig. 3 TI-FWM signals as functions of delay time in the superlattice for various ac field strength indicated by ω_1/ω_{DL1}.

of the basis set $\{|k\rangle\}$). Parameters for our calculations are taken as follows. $V_0 = 10$ meV, $\Delta = 4V_0$, $\omega_{BO} = \omega = 1.5V_0$, $\gamma_2 = 0.1V_0$, $\Omega_0 = E_0 - V_0$, and $G = \omega_{BO}$. N is chosen to be 40 for calculating quasienergy, and 80 for optical spectra.

3 Results and Discussions

As shown in Fig. 1, the quasienergy spectrum consists of continuous minibands and a few well-separated discrete states below. From the discussions in Ref. 9, we know that the excitonic WS states are mixed through THz-photon-assisted hopping, forming the Floquet states, and the discrete states are constructed from the unequally spaced WS states with small e-h distance while the continua consist mainly of the almost equally spaced remote WS states. As the ac field strengthens, the miniband broadens and the discrete excited states (defined in a

quasienergy Brillouin zone) merge into the continuum one by one. Meanwhile, the discrete ground state is repelled by its continuum towards lower energy, and eventually dips from above into the ionization continuum of its neighbor sideband, causing a series of anticrossing in the spectrum. This coupling, however, is absent in the calculation neglecting multi-photon-assisted hopping between WS states [9], indicating the inter-sideband FI is a nonlinear optical process with regard to the Thz field. Further enhancement of the ac field may suppress the quasienergy miniband, then the discrete exciton states can be released from the miniband one by one. At certain strength, namely $\omega_1 = \omega_{DL1}$, the miniband collapses. After that the evolution described above will repeat with increasing the ac field.

Since only the WS states with e-h overlap possess oscillator strength, only those discrete quasienergy excitons are active in optical spectra, this is indeed observed in Fig. 2. The feature worth remarking is the broadened and asymmetric Fano resonance associated to the ground discrete state when it couples to the energetically degenerate continuum of its neighbor sideband ($\omega_1/\omega_{DL1} = 1/8$, 1/4, 1/2, and 3/4 in Fig. 2). As shown in the inset of Fig. 2, these resonances have perfect Fano lineshape as [1]

$$\alpha(\Omega) = \alpha_0 + \alpha_c(q\gamma + \Omega - E_x)^2 / [\gamma^2 + (\Omega - E_x)^2], \quad (4)$$

where α_0 is the background constant, α_c represents the continuum absorption without the inter-sideband coupling, q is the lineshape parameter, γ is related to the resonance broadening, and E_x denotes the position of the discrete state. When the discrete states emerge out of the continuum with increasing the ac field, the absorption spectrum evolves from Fano resonances to Lorentzian-shaped discrete lines.

The time-integrated (TI) FWM in Fig. 3 display exponential decay. The inter-sideband FI enhances significantly the decay rate which is otherwise determined by the dephasing rate and is roughly $2\gamma_2$, indicating that the dynamic FI plays a role as an intrinsic phase-breaking process as the usual static FI in semiconductors does [4,5].

References

1. U.Fano, Phys. Rev. **124**, (1961) 1866.
2. D. Y. Oberli, G. Böhm, G. Weimann, and J. A. Brum, Phys. Rev. B **49**, (1994) 5757.
3. S. Glutsch, U. Siegner, M.-A. Mycek, and D. S. Chemla, Phys. Rev. B **50**, (1994) 17 009.
4. U. Siegner, M.-A. Mycek, S. Glutsch, and D. S. Chemla, Phys. Rev. Lett. **74**, (1995) 470.
5. C. P. Holfeld et al., Phys. Rev. Lett. **81**, (1998) 874.
6. For general Floquet theory on time-periodic systems, see, e.g., J. H. Shirley, Phys. Rev. **138**, (1965) B979.
7. T. Meier, G. von Plessen, P. Thomas, and S. W. Koch, Phys. Rev. Lett. **73**, (1994) 902.
8. K. Johnsen and A.-P. Jauho, Phys. Rev. Lett. **83**, (1999) 1207.
9. R.-B. Liu and B.-F. Zhu, Phys. Rev. B **59**, (1999) 5759.

Ultrafast inter-conduction band carrier dynamics in SiC

T. Tomita[1], S. Saito[1], T. Suemoto[1], H. Harima[2], S. Nakashima[3]

[1] Institute for Solid State Physics, University of Tokyo, 5-1-5 Kashiwanoha, Kashiwa, Chiba, 277-8581, Japan e-mail: `tomita@issp.u-tokyo.ac.jp`

[2] Department of Applied Physics, Osaka University, 2-1 Yamadaoka, Suita, Osaka 565-0871, Japan

[3] Department of Electrical and Electronic Engineering, Miyazaki University, 1-1 Gakuenkibanadai-nishi Miyazaki 889-2192, Japan

Abstract The ultrafast inter-conduction band carrier dynamics in SiC was observed for the first time by using pump and probe transient absorption technique. The sum of the inter-conduction band electron-phonon scattering time and the electron cooling time which was determined from the experiment is 900 fs.

1 Introduction

Recently, silicon carbide (SiC) has been receiving more attention as a wide band gap semiconductor. One of its possible application is the high-speed and high-power devices which can exceed today's silicon devices. As far as the authors know, however, the ultrafast carrier dynamics in SiC has not systematically been investigated.

One of the specific features of SiC is the existence of many polytypes with various stacking orders of the SiC bilayers. These structures can be regarded as natural superlattices (homo-superlattices), and the electronic band structure of SiC depends strongly on the kind of polytypes. This fact allows us to study the carrier dynamics for various electronic structures in a series of materials which have the same composition.

The difficulties in investigating the ultrafast carrier dynamics in wide band gap materials lies in the preparation of ultrashort pulses in ultraviolet region. Although the valence to conduction band gap lies in ultraviolet region in SiC, the second conduction band (C.B.2) is located about 1-2 eV above the lowest conduction band (C.B.1) [1,2]. This band arrangement enables us to investigate the ultrafast inter-conduction band electron dynamics decoupled from the hole dynamics. In our paper, we use the term "C.B.1" as the general term for c_1 and c_2 bands in the Ref [2], "C.B.2" for c_3 and c_4 (see the inset of Fig. 2). In the following, we present the results of time-resolved transmission measurements for inter-conduction band transitions.

2 Experiment

2.1 Experimental setup

The light source used in this experiment is a 1 kHz regenerative amplifier (Spectra Physics, Spitfire) whose time duration is about 120 fs and wavelength is 800nm. A part of the beam (1.55 eV) was chopped by an optical chopper, and used as a pump beam (the excitation density was about 10 mJ/cm^2). The other part was focused into circulated water to produce the white probe

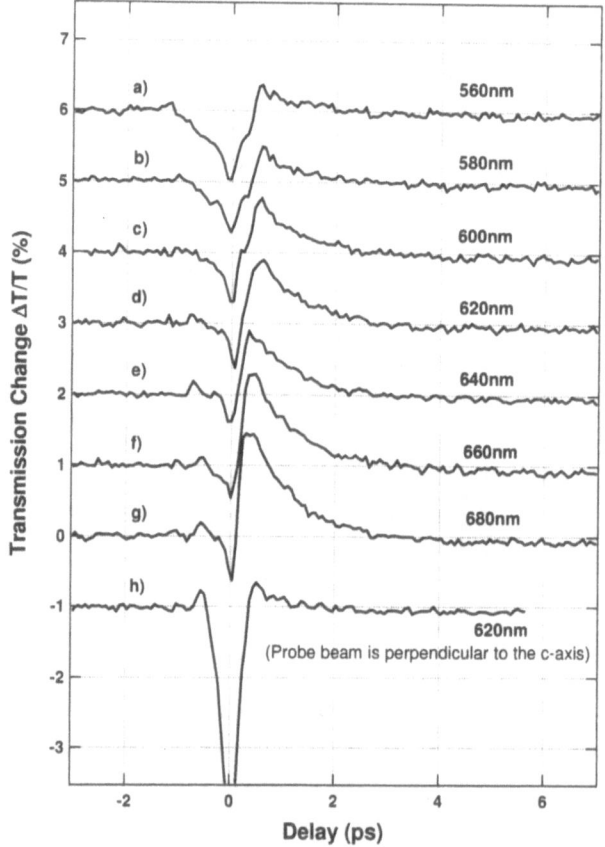

Fig. 1 Measured transmission changes in 6H-SiC for various probe wavelengths. The pump beam is perpendicular to the c-axis and the probe beam is parallel in a~g (perpendicular in h). The spectra are shifted by 1.0 along the ordinate for a convenient display.

beam. These two beams were directed into the sample, and transmitted probe beam was dispersed by a single monochromator (Nikon G250) and detected by a photomultiplier (Hamamatu R928) with a lock-in amplifier. The time resolution of this system is below 300 fs (for 680nm probe), and the detection limit is less than 10^{-3}.

The sample used in this experiment is a 6H-SiC crystal with an electron doping level of about 1.2×10^{18} cm^{-3}. The sample face is parallel to the c-axis.

2.2 Results

Measured transmission changes for various probe wavelengths are shown in Fig. 1. The polarization of pump beam was perpendicular and that of probe beam was parallel to the c-axis for a~g (perpendicular for h). We fitted the observed positive transmission changes (bleaching) by single exponential curves and determined the intensities and decay times of bleaching. The decay time of bleaching was about 900 fs at all the observed wavelengths. The negative transmission peak around 0 ps delay is due to the spectral artifacts which do not reflect carrier dynamics, and will not be discussed further [3].

The obtained intensities of bleaching for various probe energies were plotted in Fig. 2 by filled symbols. The experimentally obtained steady-state absorption, which is consistent with previously reported data was also plotted in Fig. 2 by a solid curve [4]. In this report, they assigned the absorption band peaked at 1.8 eV to the inter-conduction band transition from the C.B.1 to c_3 state. The subpicosecond absorption bleaching agrees well with the steady-state absorption in this polarization configuration. It is well known that the absorption profiles of SiC depend strongly on the polarization of the incident light. When the probe beam was perpendicular to the c-axis, the intensity profile of the bleaching was several times smaller and the peak appeared around 2.1 eV. These features are consistent with the steady-state absorption corresponding to C.B.1 to c_4 transition for this configuration. Moreover, the obtained decay time in this polarization is close to 900 fs. These facts indicate that this bleaching is due to the decreasing of the population of electrons in the C.B.1 state.

3 discussion

The schematic conduction band structure of 6H-SiC and the decay process of excited electrons are shown in the inset of Fig. 2 [5]. The pump beam (1.55 eV) is supposed to excite the electrons in both c_1 and c_2 band (C.B.1 state) to both c_3 and c_4 bands (C.B.2 state, from 1.2 eV to 2.0 eV above C.B.1), because of the breakdown of the symmetry and selection rule due to the impurity effect [2].

As for the subpicosecond relaxation process from C.B.2 to C.B.1 state, the electron transition by emitting large momentum phonon seems to be the first step, because the C.B.1 - C.B.2 separation is far larger than the typical phonon energies and the radiative decay should be very slow. The excited electrons in the C.B.2 state will be scattered to c_1 and c_2 bands (C.B.1) in the Γ-M direction by emitting phonons (inter-band scattering). The scattered electrons in the high energy part of C.B.1 state will then relax to the U-point in the c_1 band by emitting optical phonons (intra-band cooling). The observed recovery of bleaching with a time constant of 900 fs is considered to reflect the sum of these processes.

This inter-conduction band scattering process is similar to the inter-valley scattering process in the sense

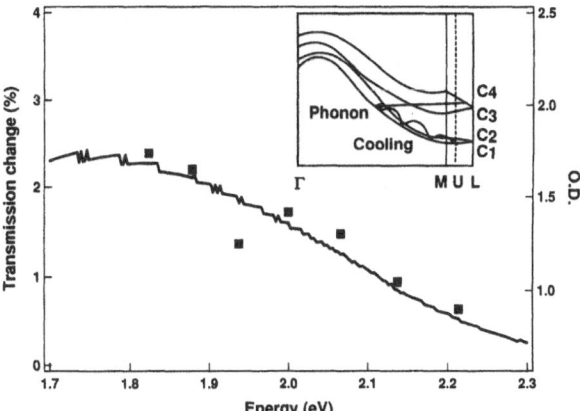

Fig. 2 The comparison between the intensity of transient transmission change (filled symbols) and the steady-state absorption band (solid line). Inset: Schematic band structure of 6H-SiC. "C.B.1" stands for c_1 and c_2 bands, "C.B.2" for c_3 and c_4. The decay process of electrons is shown.

that both processes are related to the transition of electrons by emitting large momentum phonons. The reported scattering time from X_7 to Γ valley was 0.60 ps for GaAs [6] and that for X_7 to X_6 valley scattering via Γ valley was 2 ps for GaP [7]. The reported non-equilibrium electron cooling time was 500 fs for GaN [8]. Comparing with these time constants, the observed 900 fs recovery time from C.B.2 to C.B.1 state seems reasonable.

4 Conclusion

Ultrafast inter-conduction band carrier dynamics in SiC was observed by using a femtosecond transient absorption spectroscopy. The polarization dependence was investigated for various probe wavelengths. The 900 fs recovery time of bleaching was understood to reflect the inter-band electron-phonon scattering process and the intra-band electron cooling process. The sum of the time constant for these processes in 6H-SiC which was determined from the experiment is 900 fs.

References

1. C. Persson and U. Linderfelt, Mater. Sci. Forum. **264-268** (1998) 275.

2. Sukit Limpijumnong, Walter R. L. Lambrecht, Sergey N. Rashkeev, and Benjamin Segal, Phys. Rev. B **59** (1999) 12890.

3. J. Mørk, A. Mecozzi, and Hultgren, Appl. Phys. Lett. **68** (1996) 449.

4. E. Biedermann, Solid State Commun **3** (1965) 343.

5. G. Wellenhofer and U. Rössler, Phys. Status. Solidi B, **202** (1997) 107.

6. M. A. Cavicchia and R. R. Alfano, Phys. Rev. B **48** (1993) 5696.

7. M. A. Cavicchia and R. R. Alfano, Phys. Rev. B **51** (1995) 9629.

8. C.-K. Sun, Y.-L. Huang, S. Keller, U. K. Mishra, and S. P. DenBaars, Phys. Rev. B **59** (1999) 13535.

Time-resolved stimulated photon echoes in layered semiconductors

Hideaki Tobioka, Yasuyosi Mitsumori, and Fujio Minami

Department of Physics, Tokyo Institute of Technology
Meguro-ku, Tokyo 152-8551, Japan

Abstract   Time-resolved stimulated photon echoes at the 1S exciton resonance in layered semiconductor GaSe is investigated by using up-conversion technique. For the short delay time between the second and third pulses, the temporal evolution of the echo signal shows an ideal photon-echo-like (gaussian-like) shape. On the other hand, for longer delay times, the temporal shape of the echo signal changes dramatically and has an extra peak at the arrival time of the third pulse. This behavior can be explained by spectral diffusion in the stimulated photon echo process.

1. Introduction

Recently, the development of the laser sources generating ultrashort pulses enables one to investigate ultrafast dynamical processes of carriers involving excitons in semiconductors. [1] One of the most powerful technique studying them is the photon echo spectroscopy. Most experiments have been performed by using the two-pulse photon echo method. [2] The two-pulse photon echo experiment, however, provides information only on dephasing processes. As the result, through this method, we cannot observe relaxation processes involving the energy relaxation which is important after dephasing. Here, by using the time-resolved stimulated photon echo (SPE) spectroscopy in which three pulses are utilized, we investigate the exciton dynamics occurring after the dephasing time in GaSe. In comparison with the usual two-pulse photon echoes, the SPE utilizes an extra pulse. In both cases, the relation of the echo signal to the separation time between the first and second pulses, τ_1, provides us with information on the dephasing process of the exciton polarization. In the SPE, however, the dephasing of the echo signal on the time interval between the second and third pulses, τ_2, gives us additional information, i.e., spectral diffusion. We measure the dependence of the time evolution of the SPE signals on the delay time τ_2. The temporal shape of the SPE shows strongly dependent on the delay time τ_2. For the short delay time τ_2, the temporal evolution of the echo signal is found to be an ideal photon-echo-like (gaussian-like) shape peaking at τ_1 after the third pulse arrives at the sample. On the other hand, for the longer delay time τ_2, the temporal shape of the echo signal changes dramatically and has an extra peak just after the third pulse. This dependence of the time evolution of the echo signal indicates that the spectral diffusion must occur in this system.

2. Experiment

The samples we used in this work were single crystals layered semiconductor GaSe($\sim$5 μ m) grown by the Bridgman method. At a temperature of 5K, the 1S exciton line had a peak at $\sim$2.11eV and a linewidth of $\sim$5meV. There exists inhomogeneous broadening due to stacking disorder in the sample. [3] The excitation source was an Optical Parametric Oscillator (OPO) system based on a

mode-locked Ti-Sapphire laser. The temporal duration of the pulse is $\sim$300fs. The wavelength was tuned to the 1S exciton resonance. All three pulses are polarized linearly in the same direction. This polarization can excite the lowest Γ_6 1S exciton state in GaSe. The temporal evolutions of the SPE signals were analyzed by measuring their cross-correlation function with the reference pulse. All the measurements were carried out at 5K.

3. Result and Discussion.

The dependence of the temporal evolutions of the SPE signals on the separation time τ_1 between the first and second pulses for $\tau_2 =0$ (the case of two-pulse photon echoes) are shown in Fig. 1. The time zero is set at an incident time of the third pulse. The signals show an ideal photon-echo-like (gaussian-like) and a symmetric shape with respect to the separation time τ_1. This temporal behavior arises from inhomogeneous broadening due to stacking disorder in the crystal. In Fig. 2, we show the temporal evolutions of the SPE signals with the different values of τ_2 and fixed τ_1 (=1.5 ps). The temporal shapes of the SPE signals show strongly dependent on the delay time τ_2. The arrow points in Fig. 2 indicate the peaking time of the ideal photon-echo intensity. When increasing the delay time τ_2, the temporal shape of the echo signal is becoming asymmetric with respect to the arrow point and an extra peak is found to be growing at zero delay, as shown in Fig. 2. Generally, in the SPE process in the case of inhomogeneous broadening, the first pulse excites a coherent superposition of many microscopic polarizations. Each microscopic polarization oscillates in time according to its own transition energy. Consequently, each microscopic polarization becomes rapidly out of phase, which leads the fading out of the macroscopic polarization. When the second pulse arriving within the dephasing time interacts with the microscopic polarizations in the system, a grating in the spectral domain as well as the spatial domain is formed by the interference of the two input pulses. The oscillation period of the modulation in the spectral domain is proportional to the reciprocal of the time separation τ_1 of the two input pulses. The third pulse arriving at the system interacts with the spectral modulation of the system and reverses the temporal evolution of the microscopic polarization. Therefore, each microscopic polarization is back in phase. Consequently, the macroscopic polarization recovers and emits an echo signal at the time interval τ_1 of the first and second pulses after the third pulse, [4,5] as shown in Fig. 1. If there exists a strong correlation between the microscopic polarizations and a strong coupling with a heat bath, these interactions cause the strong modification of the profile of the spectral modulation, i.e., spectral diffusion. [6] When the spectral diffusion begins, the temporal shapes of the echo signals should change. The spectral diffusion causes the washing out of the depth of the modulation in the spectral domain. The third pulse arriving in this situation interacts with the reduced modulation of the spectra. For the modulated part, the third pulse reverses the temporal evolution, which leads to the ideal photon-echo-like signal peaking at the delay time τ_1. For the washed out part, the third pulse need not reverse the temporal evolution because the microscopic polarizations have already been in phase. In other words, the washed out spectrum should be the same as that at the zero delay time between the first and second pulses. Therefore, the macroscopic polarization recovers immediately and emits the echo signal just after the third pulse, leading to the growing peaks at zero delay, as shown in Fig. 2. In the present experiments, the spectral diffusion is originated from the exciton migration between inter-layers. Since there are random stacking disorders in this system, stacking sequences seen by the excitons are different in space, and then the exciton resonance energies are broadened inhomogeneously. When the exciton confined in a certain stacking sequence migrates to another sequence due to the exciton-exciton and/or exciton-phonon scatterings, the washing out in the spectral domain occurs. By investigating the relation of the temporal echo shapes to the excitation densities, we confirmed that this migration process is dominated mainly by exciton-exciton scatterings. We will publish the details elsewhere.

4. Conclusion

The temporal evolutions of the SPE signals in layered semiconductors GaSe were observed. The temporal shape of the echo signal shows a strong dependence on the separation time between the second and third pulses. For the longer delay time between the second and third pulses, the temporal shape of the echo signal has an extra peak at the zero delay time. This behavior can be explained well by the spectral diffusion process in the SPE.

Acknowledgement

This work was supported by a Grant-in-Aid for Scientific Research.

References

1. L. Schultheis et al., Phys. Rev. B **34**, (1986) 9027.
2. K. Leo et al., Phys. Rev. B **44**, (1991) 5726.
3. E. Mooser and M. Shülter, Nuovo Ciment B**18**, (1973) 164.
4. Optical resonance and two-level atoms, edited by L.Allen and J. H. Eberly (Wiley, New York, 1975).
5. T.Yajima and Y.Taira, J. Phys. Soc. Jpn. **47**, (1979) 1620
6. Y.S. Bai and M.D. Fayer, Phys. Rev. B **39**, (1989) 11066.

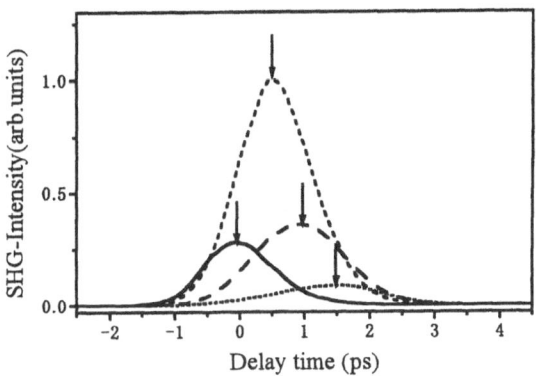

Fig. 1. The dependence of the temporal evolutions of the SPE signals on the separation time τ_1 for $\tau_2=0$ps. The solid line represents the data for $\tau_1=0$ps, the dashed line for $\tau_1=0.5$ps, the semi-dashed for $\tau_1=1.0$ps, the dotted line for $\tau_1=1.5$ps. The time zero is set at an incident time of the third pulse. The arrows indicate the peaking time of the ideal photon-echo intensity.

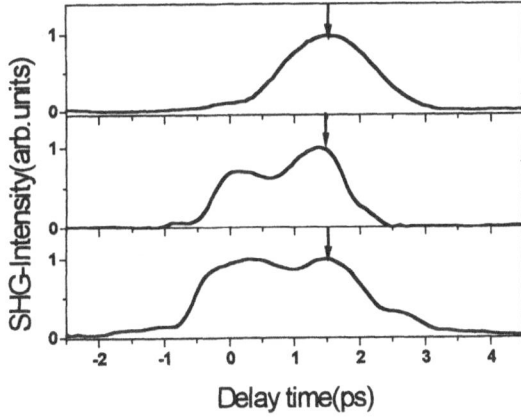

Fig. 2. The temporal evolutions of the SPE signals with the different values of τ_2 and fixed $\tau_1=1.5$ps: (a) $\tau_2=2.5$ps (b) $\tau_2=5$ps (c) $\tau_2=10$ps The time zero is set at an incident time of the third pulse. The arrows indicate the peaking time of the ideal photon-echo intensity.

Two-color time-resolved electronic Raman measurement in semiconductors

Shingo Saito and Tohru Suemoto

Institute for Solid State Physics, University of Tokyo, 5-1-5 Kashiwanoha, Kashiwa, Chiba 277-8581, Japan

Abstract   Time-resolved electronic Raman scattering of germanium was measured by pump-probe method under various pump energies. The time-resolved Raman intensities corresponding to the transition from heavy hole band to light hole band showed different features depending on the pump energy. The Raman intensities showed maximum at a few picosecond after excitation. These time developments are explained as the carrier cooling and diffusion. Furthermore, time development of Raman intensity in indium arsenide was measured and explained in the same way.

1 Introduction

The ultrafast carrier dynamics in semiconductors continues to be one of the most interesting and important subjects from a viewpoint of application and fundamental research. The thermalization and cooling processes of the photo-generated hot carriers are dominated by electron-electron, electron-hole and hole-hole collisions and also by carrier-phonon interactions.

A broad Raman band of Ge appears around 1500cm^{-1} under the 1.58eV excitation and is ascribed to the hole excitation from heavy hole to light hole band[1]. In the resonant electronic Raman scattering, the probe beam is resonant to the transition between the light hole band and conduction band at the wave vector k and the heavy hole is excited to the light hole band at the same k.

Time-resolved electronic Raman measurement of Ge was reported[2]. The rise of the intensity was interpreted in terms of the carrier population defined by the carrier temperature and the decay in terms of the spatial diffusion. In those works, the evolution of inter-valence band Raman scattering (IVRS) was observed by one color time-resolved Raman scattering measurement.

In order to study the behavior of the cooling and the diffusion systematically, we performed two-color time-resolved Raman scattering measurements using various pumping energies and discuss the cooling and the diffusion of carriers with various excess energies.

2 Experiment

We used the fundamental light (1.58eV) from 100kHz TiS regenerative amplifier as a probe beam. In all the measurements, the probe beam is passed through a spectrum trimmer consisting of a grating and a slit in order to remove the undesirable spectrum components due to the fluorescence from the TiS laser rod. The optical parametric amplifier pumped by regenerative amplifier and the second harmonics of TiS regenerative amplifier are used as a pump source. Polarization of the pump and probe beams are perpendicular each other. The spectral

resolution and the time resolution are about 300cm^{-1} and 400fsec, respectively. The probe beam is resonant to the band-to-band transition near the Γ point in Ge[3]. The diameters of the pump and probe beams on the sample were about 100μm. The initial distribution of carriers in the sample has a disk-like shape, because the penetration depth[4] of the pumping light is less than 0.2μm. The crystal had (001) plane as the face and scattering geometry was z(x, y)z'. All the measurements were performed at room temperature.

3 Results and Discussion

Data processing of time-resolved Raman measurement is mentioned in ref. [2]. We obtain Raman spectra at various time-delay by changing interval between the pump and the probe pulses.

The intensity of electronic Raman scattering, I, is the products of the hole distribution in the real space, N, and that in the momentum space, F.

Firstly, we consider a 1-dimensional diffusion model to treat the spatial distribution of the holes. The diffusion of particles which are placed at the surface as a Gaussian distribution at time $t=0$ is well described by a Fokker-Planck equation. Then the spatial population $N(z,t)$ is given by,

$$N(z,t) = [2D\pi(t+t_1)]^{-1/2} \cdot exp[-\frac{z^2}{4D(t+t_1)}], \quad (1)$$

where D is the diffusion constant of the holes. The parameter t_1 is introduced to describe the initial distribution of the holes: $2\sqrt{Dt_1}$ corresponds to the thickness of the distribution at $t=0$. To evaluate the Raman intensity, we have to consider the penetration depth of the probe beam and the re-absorption of the scattered Raman signal light and to perform a weighted integration. The total number of the heavy holes probed by Raman scattering, $N(t)$ is given by,

$$N(t) = \int_0^\infty exp(-2\alpha z) \cdot N(z,t)dz, \quad (2)$$

where α is the absorption coefficient of Ge for probe beam wavelength.

Next, we discuss the distribution in the momentum space. The cooling of the hole temperature, $T(t)$, is assumed to be,

$$T(t) = (T_i - T_c) \cdot exp(-t/\tau) + T_c, \quad (3)$$

where T_i is the initial temperature of photo-excited hole and T_c is the final temperature of photo-excited hole, room temperature. τ is fitting parameter. Furthermore, we use Boltzmann distribution , $f(\epsilon, T)$ and the density

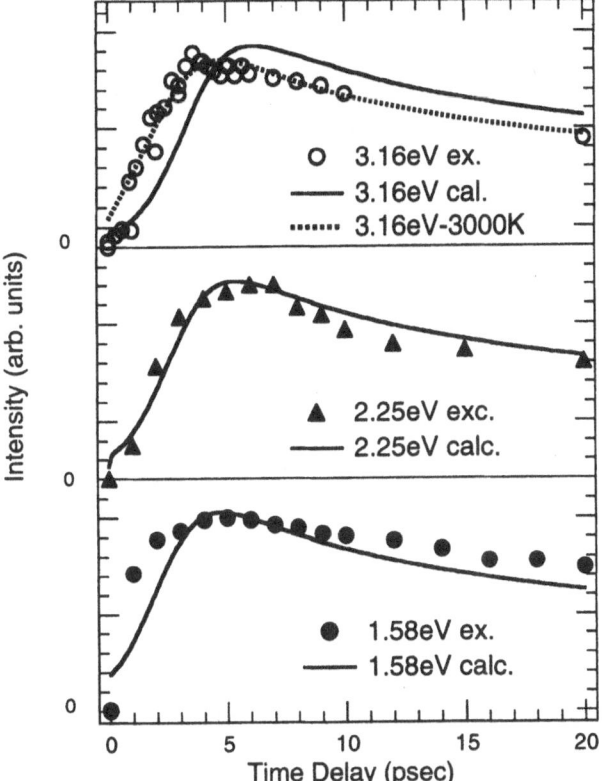

Fig. 1 The time evolution of Raman intensity of Ge from 0psec to 20psec probed by 1.58eV photon energy under the various pump energies. Marks show the peak heights (1500cm^{-1}) of the time-resolved Raman bands. Solid curve and broken line show the results calculated by the cooling and the diffusion model discussed in the text.

of state for the parabolic band, $\rho(\epsilon)$ for the calculation. As a result, the time development of Raman intensity is expressed as,

$$I(t) = f(\epsilon, T(t)) \cdot \rho(\epsilon) \cdot N(t)$$
$$= exp(-\epsilon/k_B T) \cdot \sqrt{\epsilon} \cdot N(t). \qquad (4)$$

Figures 1 shows the time evolution of Raman band in Ge and our model calculation from 0ps to 20ps under the various pumping energies. Closed circles, triangles, and open circles show the peak heights (1500cm^{-1}) of the time-resolved Raman spectra under the 1.58eV, 2.25eV and 3.16eV excitation conditions, respectively. Solid lines are the results from the cooling and diffusion model using constants explained below. The thickness[4] of the initial distribution of carriers for 1.58eV, 2.25eV, and 3.16eV excitation corresponds to t_1=0.5ps, 0.005ps, and 0.0025ps, respectively. Under the assumption that the excess energy of carrier corresponds to the initial hole temperature, the initial hole temperature should be about 3000K in case of the 1.58eV excitation, 4800K in the case of 2.25eV excitation, and 8000K in the case of 3.16eV excitation, respectively. Diffusion constant[5] of holes at RT, 46cm^2 s^{-1} is used in Eq. (1). The lines in Fig. 1 are the results of these calculations. The fitting parameters, τ is determined to be 1.3ps for both excitations.

The agreement with the experimental data in the time that shows the peak intensity is good in the case of 1.58eV and 2.25eV excitation. Though the calculation results in the case of 3.16eV excitation show the peak about 6psec delay time, the experimental data show the peak before 4psec delay time. Broken line is the results of the calculation under the assumption, $T_i = 3000K$. This result show good agreement with the experimental data and means that the initial temperature of hole is 3000K which is lower than that of assumed. It might be caused by following reasons. From the band structure[3] of germanium, the pump pulse of 3.16eV photon energy can excite the carrier not only between the highest valence band and the lowest conduction band far from the Γ point in the k-space but also between the highest valence band and the second lowest conduction band close to the Γ point. The holes created at the top of the valence band will have far smaller excess energy than the photo-generated carrier far from Γ point and these "cold carriers" might considerably lower the average temperature of the hole system.

In indium arsenide, a broad Raman band appears around 2000cm^{-1} under the 1.58eV probe beam condition. From the band calculation[3], the Raman shift shows agreement to the transition energy from the heavy hole band and the light hole band under this probing condition. We conclude this band is caused by IVRS. The time development of this Raman band intensity was measured under the 3.16eV and 1.58eV pumping conditions. The intensity under the 3.16eV excitation condition shows a maximum at a few picosecond after pumping. But it decreased monotonously under the 1.58eV excitation. These time developments are explained by the cooling and diffusion[6] of photo-excited hole.

In conclusion, from the time evolution of the Raman intensity under various excitation photon energies in Ge and InAs, we discussed the cooling process and the spatial diffusion of photo-excited holes. It has been shown that the two-color time-resolved Raman scattering measurement is a useful tool for investigating the dynamics of the energetic carriers.

Acknowledgements

This work was partially supported by a Grant-in-Aid for Scientific Research from the Ministry of Education, Science and Culture of Japan.

References

1. K. Tanaka, H. Ohtake, and T. Suemoto, Phys. Rev. **B, 50**, (1994) 10694.
2. K. Tanaka, H. Ohtake, H. Nansei, and T. Suemoto, Phys. Rev. **B, 52**, (1995) 10709.
3. J. R. Chelikowski and M. L. Cohen, Phys. Rev. **B, 14**, (1976) 556.
4. D. E. Aspnes and A. A. Studna, Phys. Rev. **B, 27**, (1983) 985.
5. F. J. Morin, Phys. Rev. **93**, (1954) 62.
6. O. G. Folberth, O. Madelung, and H. Weiss, Z. Naturforsch. **9a**, (1954) 954.

Fermi Edge Singularities in Out-of-equilibrium Semiconductors

M. Combescot[1], B. Roulet[1], C. Tanguy[2]

[1] Groupe de Physique des Solides des Universités Paris VI et Paris VII, CNRS, Tour 23, 2 place Jussieu, 75251 Paris Cedex 05, France
[2] France Telecom R&D DTD/CDP, 196 avenue Henri Ravera, BP 107, 92225 Bagneux Cedex, France

Abstract   We consider two types of out-of-equilibrium Fermi seas — extending between (μ_1,μ_2) or between (ε_1,μ_1) and (ε_2,μ_2) — as produced in bulk or quantum-well semiconductors by femtosecond laser pulses. We show that the lower threshold μ_1 of the case (μ_1,μ_2) induces a "convergent" singularity, whereas the possible excitation of Fermi sea electron-hole pairs with zero energy in the (ε_1,μ_1)-(ε_2,μ_2) case generates a lifetime which kills the singular power-law divergence. This last problem is of theoretical importance since the famous Nozières-De Dominicis method provides in this case a result which disagrees with both perturbative expansion and parquet diagram summation.

Fermi-edge singularities have been studied since many decades in the case of X-ray absorption by metals [1–5]: an electron is extracted from a deep level and joins the N electrons of the Fermi sea. These $N+1$ electrons react to the *sudden* appearance of the deep-hole potential by exciting Fermi sea electron-hole (e-h) pairs — which all have a positive energy. Various physical processes can be distinguished: the first one corresponds to the photocreated electron flying above the Fermi sea (see Fig. 1a); the others, which contain e-h pairs excitations, either involve the photoelectron as in Fig. 1b or merely "dress" the deep hole as in Fig. 1c.

The summation of processes like that of Fig. 1a, which corresponds to ladder diagrams, lead to the incorrect — though still widely quoted — concept of Mahan's exciton. Indeed, while such a process is the only existing one at first order in the electron-deep hole interaction, the second-order process of Fig. 1b is as large, making the restriction to ladder diagrams inconsistent. All the processes of Fig. 1 are in fact logarithmically singular close to the absorption threshold, making their summation necessary. Noting that the most singular terms come from parquet diagrams, Roulet, Gavoret, and Nozières [3] were the first to sum them up by using a standard but heavy parquet procedure. A few years later, Nozières and De Dominicis [4] unveiled the physical structure of the solution. Working with the time variable instead of the frequency variable, they showed that the time response factorizes into *(i)* a contribution coming from the photocreated electron dressed by its interaction with the Fermi sea, and *(ii)* a contribution coming from the dressed deep hole. The electron part of the resulting absorption has a frequency dependence $1/(\omega - \omega_T)^{2\delta/\pi - 1}$, whereas the deep-hole part behaves as $1/(\omega - \omega_T)^{1 - \delta^2/\pi^2}$,

making therefore the total absorption proportional to $1/(\omega - \omega_T)^{2\delta/\pi - \delta^2/\pi^2}$, where δ is the phase shift induced by the deep hole potential at the threshold ω_T. Let us stress that the power-law divergence at ω_T comes from the dressed-electron contribution through the exponent $2\delta/\pi$, while the dressed-hole contribution reduces this divergence through the exponent $-\delta^2/\pi^2$.

Novel out-of-equilibrium Fermi seas can now be produced in semiconductors by femtosecond pump pulses, and their observation is possible by using femtosecond probe beams. A few years ago, Fermi-edge singularities were observed in an out-of-equilibrium Fermi sea [6] similar to that of Fig. 2b. Such a Fermi sea has an interesting low-energy threshold μ_1 close to which it is possible to excite e-h pairs with *negative* energy. The excitation of these pairs can provide to a below 0 or below μ_2 photon the energy which is missing to complete absorption. Similarly, positive energy pairs allows the absorption of above μ_1 photons, so that absorption in the (μ_1,μ_2) transparency region is indeed possible (see Fig. 3a). In addition (as easily seen by calculating the first-order process of Fig. 1a), one can show that while the lower boundary μ_2 for the photocreated electron flights induces a *divergent* absorption similar to the one of the equilibrium $(0,\mu)$ Fermi sea, the upper boundary μ_1 for such flights induces a power-law *convergent* absorption (see Fig. 3a).

More recently, we have also studied the possible Fermi-edge singularities for out-of-equilibrium Fermi seas shown in Fig. 2c, which could be created in excited quantum wells. The new feature of this configuration [7] lies in the possibility to create e-h pairs with *zero* energy — in addition to those with positive or negative energies. These zero-energy pairs generate a lifetime which kills the singularities: no divergent power-law behavior exists anymore. This problem is also of importance on a theoretical point of view: The famous Nozières and De Dominicis method fails in this case since it provides a result which disagrees [8] with the one obtained by both the perturbative expansion in the electron-deep hole interaction — performed up to sixth order — *and* the parquet-diagram summation technique.

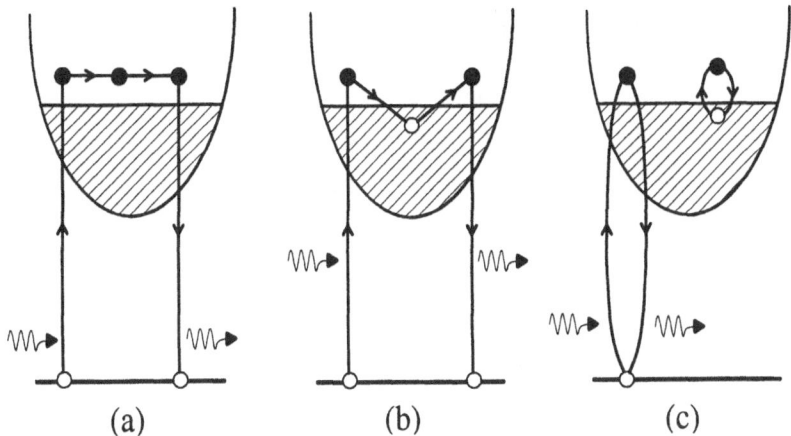

Fig. 1 Excitation processes for the $N + 1$ electrons of the Fermi sea, induced by the sudden appearance of a deep hole.

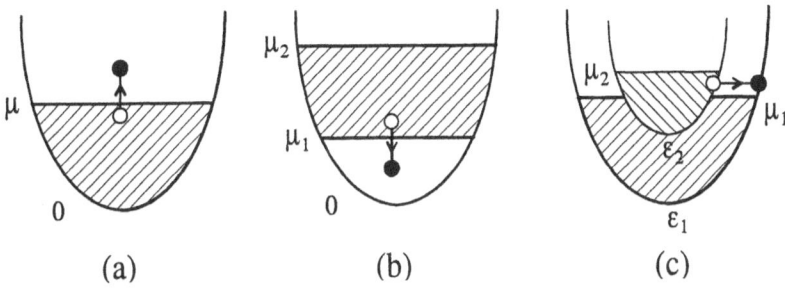

Fig. 2 The Fermi seas considered in this work, with the possible excitations of positive, negative, and zero energy e-h pairs.

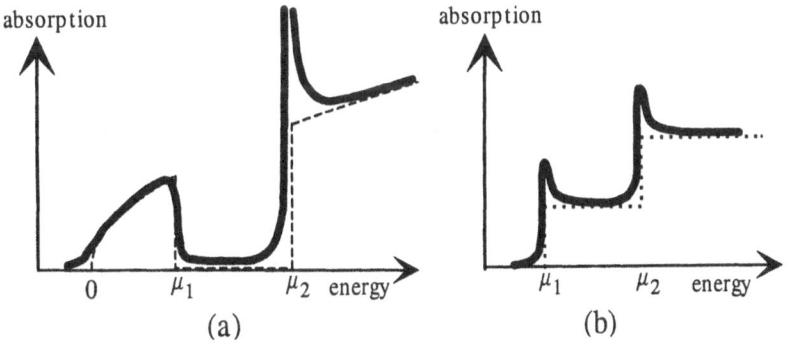

Fig. 3 Absorption edge singularities for the Fermi seas of Figs. 2b and 2c, respectively.

References

1. For a review, see K. Ohtaka and Y. Tanabe, Rev. Mod. Phys. **62**, (1990) 929.

2. G. D. Mahan, Phys. Rev. **163**, (1967) 612.

3. B. Roulet, J. Gavoret, and P. Nozières, Phys. Rev. **178**, (1969) 1078.

4. P. Nozières and C. T. De Dominicis, Phys. Rev. **178**, (1969) 1997.

5. M. Combescot and P. Nozières, J. Physique (Paris) **32**, (1971) 913.

6. J.-P. Foing, D. Hulin, M. Joffre, M. K. Jackson, J.-L. Oudar, C. Tanguy, and M. Combescot, Phys. Rev. Lett. **68**, (1992) 110; C. Tanguy and M. Combescot, Phys. Rev. Lett. **68**, (1992) 1935.

7. M. Combescot and B. Roulet, Phys. Rev. B **61**, (2000) 7609.

8. T. K. Ng, Phys. Rev. B **54**, (1996) 5814.

Theory of coherent oscillation of phonon-polaritons

C. S. Kim[1], A. M. Satanin[1], G. D. Sanders[2], and C. J. Stanton[2]

[1] Department of Physics, Chonnam National University, Kwangju 500-757, Korea
[2] Department of Physics, University of Florida, Gainesville, Florida 32611, U. S. A.

Abstract   We formulate a theory to describe the coherent dynamics of phonon-polaritons in nonlinear ionic crystals. By quantizing the Born-Huang theory of the coupled phonon-photon system we first construct our model Hamiltonian. Then, applying a nonequilibrium statistical mechanics for the open quantum system, we derive the macroscopic equations for the transverse, coherent amplitudes. The resulting nonlinear equations with damping are solved for the coupled harmonic oscillators numerically. Our theory predicts the nontrivial, nonlinear beating between two coherent polariton modes with the characteristic decay.

Optical pump and probe spectroscopy techniques with femtosecond time-resolution have allowed researchers to investigate the coherent dynamics of LO phonons in polar and nonpolar semiconductors [1]. Since the pump pulse creates photoexcited nonequilibrium carriers that modulate the reflectivity of Ge and a plasmon-phonon oscillation in GaAs, such a coherent phonon excitation is essentially electronically based [2].

In this paper, we consider the situation where coherent TO phonons are excited without electronic excitation. The coupling of ionic motion to the transverse radiation field results in a hybridization of the phonon and photon spectra, giving rise to a new quasiparticle, the *polariton*, in ionic crystals. The classical theory of the polariton was formulated by Born and Huang [3] in the early years and a subsequent quantum mechanical treatment was given later by Hopfield [4]. The polariton state is a coherent superposition of the phonon and photon modes. In recent years the coherent phonon-polariton excitations have been observed in ferroelectric crystals such as $LiTaO_3$ and $LiNbO_3$ [5]. These experiments provide some relevance to our theory.

We consider an ionic crystal having two basis ions per unit cell and adopt the idea of Born and Huang that: i) The wavelength of radiation is bigger than the lattice constant; ii) An ion in an elementary cell moves under the self-consistent field of the others; iii) The coupled mechanical and electromagnetic equations are used to handle the interplay between the ions and electromagnetic field [3]. After the standard canonical quantization of the classical Lagrangian, we obtain the Hamiltonian for the coupled phonon-photon system given by

$$\mathcal{H} = \mathcal{H}_0 + \mathcal{H}' + \mathcal{H}_{ex}. \tag{1}$$

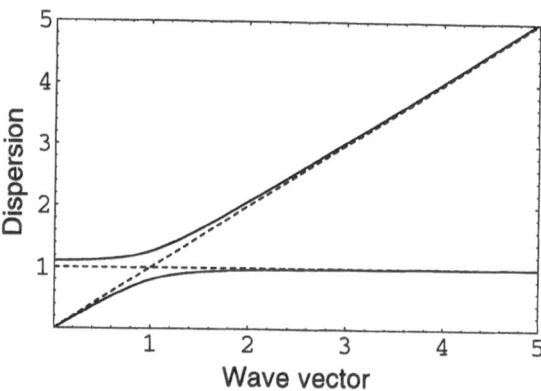

Fig. 1 The dispersion relations of two polariton branches are depicted as the solid curves. The dashed lines show the bare dispersion of TO phonon and photon; where the frequency and the wave vectors are in units of ω_{TO} and ω_{TO}/c, respectively.

The first term describes two linearly coupled harmonic oscillators:

$$\mathcal{H}_0 = \sum_k \hbar\Omega_k \hat{a}_k^\dagger \hat{a}_k + \sum_k \hbar\omega_k \hat{b}_k^\dagger \hat{b}_k$$
$$+ i \sum_k \hbar g_k (\hat{a}_{-k}^\dagger - \hat{a}_k)(\hat{b}_k + \hat{b}_{-k}^\dagger), \tag{2}$$

where $\hat{a}_k^\dagger$ ($\hat{b}_k^\dagger$) and a_k (b_k) are the creation and annihilation operators for photons with angular frequency Ω_k (phonons with frequency ω_k), both at momentum k, and g_k is the coupling constant between photon and phonon. The second term in Eq. (1) specifies the phonon-phonon interaction:

$$\mathcal{H}' = \sum_k (\hat{b}_k \hat{\Gamma}_{-k}^\dagger + \hat{\Gamma}_k \hat{b}_{-k}^\dagger) + \sum_k v_k (\hat{b}_k + \hat{b}_{-k}^\dagger)^3. \tag{3}$$

In deriving Eq. (3) we have treated the acoustic branches as a heat bath interacting with the TO mode; where $\hat{\Gamma}_k$ indicates the reservoir operator arising from this procedure. The first term in Eq. (3) describes the well known decay process of an optical mode into two acoustic modes (the nonlinear damping term is neglected in this work) [6], and the second term describes the self-interaction of the TO mode within the resonant approximation, with v_k being the associated coupling constant. The external Hamiltonian represents the direct interaction of the TO phonon with a short laser pulse,

$$\mathcal{H}_{ex} = -\bar{\alpha} \sum_k \sqrt{\frac{2\pi\hbar\epsilon_\infty}{\omega_k}} (\hat{b}_k + \hat{b}_{-k}^\dagger) E_k(t), \tag{4}$$

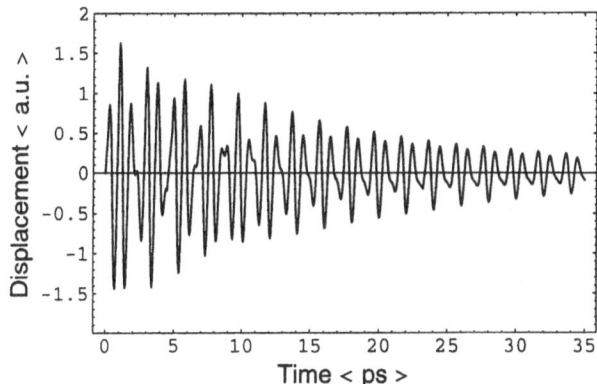

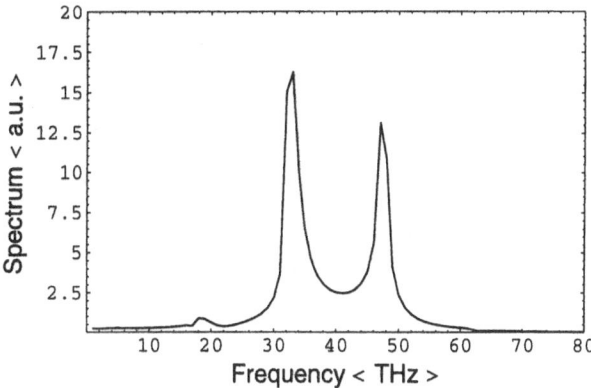

Fig. 2 Solution to two coupled nonlinear polariton oscillators excited by a short laser pulse in the form $E_k(t) = E_k \exp\{-t^2/2\tau^2\}\cos\omega_0 t$, with width $\tau = 2$ ps and central frequency $\omega_0 = 2.5\,\omega_{TO}$.

where $\bar{\alpha}^2 = ((\epsilon_0 - \epsilon_\infty)/\epsilon_0)\omega^2_{k=0}$, and $E_k(t)$ is the Fourier-component of the external field.

Next, we define the operators for the coherent amplitudes of phonon and photon field as

$$\hat{D}_k = \sqrt{\frac{\hbar}{2\omega_k}}(\hat{b}_k + \hat{b}^\dagger_{-k}), \quad \hat{A}_k = \sqrt{\frac{\hbar}{2\Omega_k}}(\hat{a}_k + \hat{a}^\dagger_{-k}),$$

and we introduce the Liouville - von Neumann equation to describe time-evolution of the density operator of the system, $\hat{\rho}(t)$. Finally, the classical equations of motion are obtained by employing a nonequilibrium statistical mechanics [7]. The results are

$$\frac{\partial^2 D_k}{\partial t^2} + (\omega_k^2 - \bar{\alpha}^2)D_k + \bar{\alpha}\frac{\partial A_k}{\partial t}$$

$$= -\gamma_k \frac{\partial D_k}{\partial t} - \lambda_k D_k^2 + F_k(t), \quad (5)$$

$$\frac{\partial^2 A_k}{\partial t^2} + \Omega_k^2 A_k - \bar{\alpha}\frac{\partial D_k}{\partial t} = 0. \quad (6)$$

where $D_k(t) = \mathrm{Tr}(\hat{\rho}(t)\hat{D}_k) \equiv \langle\hat{D}_k\rangle$ and $A_k(t) = \langle\hat{A}_k\rangle$, and $F_k(t)$ is the driving force proportial to $E_k(t)$. The damping coefficient γ_k is connected with a correlator of the reservoir operators and the anharmonic constant λ_k equals $3v_k(2\omega_k/\hbar)^{3/2}$.

Equations (5) and (6) constitute the basis of our theoretical analyses. When we retain terms related to the internal dynamics only in Eq. (5), we obtain two ideal polariton modes specified by the dispersion law:

$$\left(\omega_k^{(u,d)}\right)^2 = \frac{1}{2}\left(\Delta_k^2 \pm \sqrt{\Delta_k^4 - 4\Omega_k^2(\omega_k^2 - \bar{\alpha}^2)}\right) \quad (7)$$

where $\Delta_k^2 = \Omega_k^2 + \omega_k^2$. These polariton branches are drawn in Fig. 1 where the upper (lower) curve corresponds to $\omega_k^{(u)}$ ($\omega_k^{(d)}$); where $\omega_{TO} = 42.3$ THz and $\bar{\alpha} = 0.47\,\omega_{TO}$ were used. At the anti-crossing point, $k = \omega_{TO}/c$, the two bare polariton frequencies are given by $\omega_k^{(u)} = 47.2$ THz and $\omega_k^{(d)} = 33.8$ THz.

When the exciting field is weak, the nonlinear term $-\lambda_k D_k^2$ is not influential. Accordingly, only linear beating between two bare polariton oscillations is expected.

Fig. 3 Fourier spectrum of the lattice displacement given in Fig. 2: $|D(\omega)| = |\int dt\,e^{-i\omega t} D_k(t)|$. Two peaks are seen at the renormalized frequencies $\omega = 32.5, 46.7$ THz and an additional peak of a harmonic excitation at $\omega = 19.2$ THz.

We have solved the full set of Eqs. (5) and (6) numerically under strong excitation conditions, using the phenomenological material parameters: $\gamma_k = 0.1\,\omega_{TO}$ and $\lambda_k = 4\,\omega_{TO}^2/c\tau E_k$ at the fixed wavevector of $k = \omega_{TO}/c$. In Fig. 2 we draw the displacement of the coherent-phonon, $D_k(t)$. The damping term in Eq. (5) causes the decay of the amplitude, and the anharmonic term gives rise to the shift of the bare polariton frequencies and the *nonlinear* beating between the two branches. After performing the Fourier transform of the lattice displacement, we clearly observe two renormalized polariton modes and an additional harmonic in Fig. 3.

In conclusion, we have provided a theoretical framework for studying the excitation and subsequent dynamics of coherent TO phonons in an ionic crystal. Our preliminary results richly illustrate information on the polaritonic spectrum, and the damping and nonlinear beating between two polariton modes. We have found a qualitatively similar feature to the measured signal intensity of phonon-polaritons in LiTaO$_3$ [5].

Acknowledgements This work was supported by Chonnam National University through a Grant in the year of 1999. AMS is grateful to KISTEP for supporting his stay in Korea through the Brain Pool Program, Project No. 991-1-25.

References

1. For a recent review, see T. Dekorsy, G.C. Cho, and H. Kurz: In *Light Scattering in Solids VIII: Fullerenes, Semiconductor Surfaces, and Coherent Phonons* Topics Appl. Phys. **76**, ed. by M. Cardona and G. Güntherodt (Springer, Berlin, Heidelberg, 2000) pp. 169-209.
2. A.V. Kuznetsov, C.J. Stanton, Phys. Rev. Lett. **73**, (1994) 3243; Phys. Rev. B **51**, (1995) 7555.
3. M. Born and K. Huang, *Dynamical Theory of Crystal Lattices* (Clarendon, Oxford 1988).
4. J.J. Hopfield, Phys. Rev. **112**, (1958) 1555.
5. See, for instance, H.J. Bakker, S. Hunsche, and H. Kurz, Rev. Mod. Phys. **70**, (1998) 523.
6. P.G. Klemens, Phys. Rev. **148**, (1966) 845.
7. H. Carmichael, *An Open System Approach to Quantum Optics* (Springer-Verlag, Berlin 1991).

Self-organized electron density patterns in n-GaAs induced by microwaves

V. Novák[1,2], V.V. Bel'kov[3], D. Mac Mathúna[4], S.D. Ganichev[2,3], W. Prettl[2]

[1] Institute of Physics AS CR, Cukrovarnická 10, 162 53 Praha, Czech Republic e-mail: novakvit@fzu.cz
[2] Institut für Experimentelle und Angewandte Physik, Universität Regensburg, D93040 Regensburg, Germany
[3] A.F. Ioffe Physico-Technical Institute of the Russian Academy of Sciences, St. Petersburg 194021, Russia
[4] Physics Department, Trinity College, Dublin 2, Ireland

Abstract   Self-organized spatial patterns of enhanced electron density have been observed in a cooled n-GaAs layer under homogeneous microwave irradiation. The common origin of these patterns and the current filaments have been demonstrated. A global coupling between the microwave-field distribution and the power dissipation in the sample has been concluded as the pattern forming mechanism.

1 Introduction

If biased above a certain critical voltage, a moderately doped sample of n-GaAs cooled to 4.2 K exhibits an electric breakdown. The essential mechanism of this nonequilibrium phase transition is the autocatalytic generation of free carriers by the impact ionization of shallow impurities. Under the global constraint of the current supplying circuit the carrier multiplication results in the formation of distinct current channels. The geometry of these filaments inevitably reflects the geometry of the galvanic electrodes [1].

In the microwave field of sufficiently low frequency the carriers gain energy analogously to the case of static electric field, and thus analogous nonlinear effects can be expected to occur [2,3]. However, in contrast to case of the current filaments no external constraints like, e.g., the sample contacts a priori enforce a certain geometry of the possible pattern formation.

2 Experiment

The sample was cut from MBE-grown layer of n-type GaAs. The technological thickness of the layer was $4.3\mu m$, doping was $N_D = 3.1 \times 10^{15} cm^{-3}$ and $N_A = 5.1 \times 10^{14} cm^{-3}$, and the electron mobility at 77 K was $\mu = 40800 cm^2/Vs$. Dimensions of the sample were 3×7 mm, substrate thickness 0.4 mm. The sample was fixed inside a rectangular waveguide for the Ka-band of microwave frequencies, behind a fine-mesh metallic window in the waveguide wall. The waveguide was terminated by an absorber and immersed in the Helium bath. A klystron and a YIG-oscillator were used as microwave sources for frequencies from 26 to 36 GHz.

In order to detect the changes of the electron density in the epitaxial layer the sample was homogeneously illuminated by an interband light and its photoluminescence (PL) was photographed by an infrared-sensitive camera. Using a proper interference filter only the PL-line of excitonic recombination was selected, known to be quenched by the excess free electrons [4].

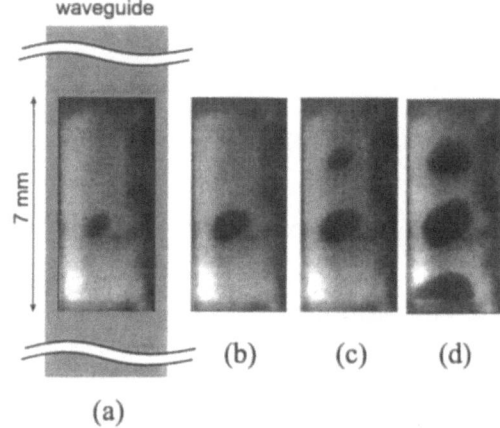

Fig. 1 Photoluminescence images of the sample in the waveguide. In (a) the positioning of the sample behind the window in the waveguide is schematically shown. The microwave power increases from (a) ($P = P_{th} = 15$ mW) to (d) ($P = 28$ mW).

The images in Fig. 1 show the changes of the photoluminescence when increasing the power of the incident microwaves. In the originally homogeneous luminescence a distinct, sharp edged dark spot of a roughly circular shape and a finite diameter abruptly emerges, if a certain critical power (15 mW for the reported sample) is exceeded, Fig. 1(a). Its size grows continuously as the power is further increased, Fig. 1(b), till another critical field strength is reached and a second spot is formed, Fig. 1(c). The scenario is repeated once again at a higher microwave power, Fig. 1(d). At the highest available power of about 50 mW the spots merge together, covering almost the whole area of the sample.

The critical microwave power necessary for the ignition of a particular spot is always higher then that of the spot extinction. Thus, hysteretic loops and intervals of bistability exist. It should be noted, however, that except for the close vicinity of the critical points the dark spot configuration is absolutely stable. The form of the spots does not change on varying the illumination intensity, the current spot configuration persists even after the interruption of the illumination. Any significant role of the light-generated charge carriers and excitons during the spot formation can thus be excluded, similarly as in the case of the current filaments.

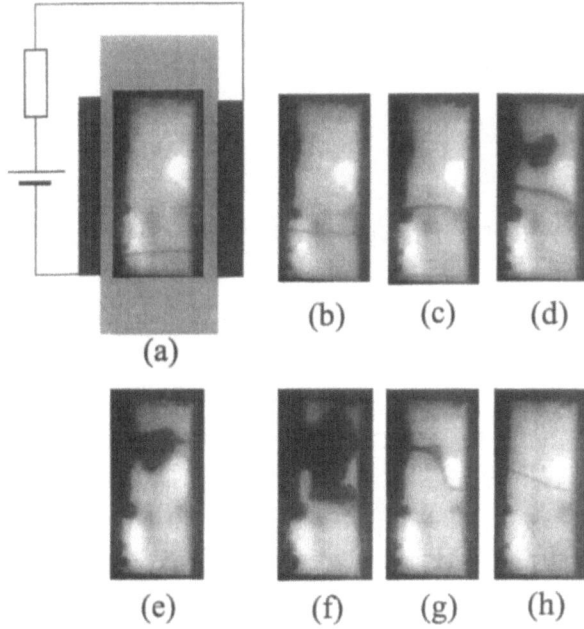

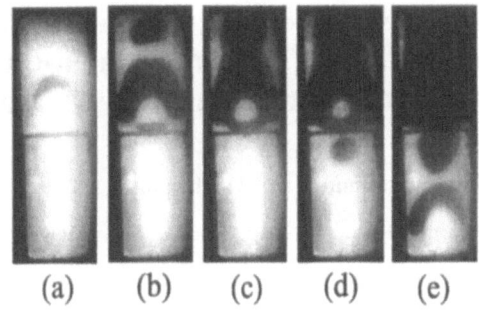

Fig. 3 Complex patterns of quenched PL in a galvanically divided sample. In the half-height of the images a horizontal gap is seen which separates the two parts of the sample. The microwave power increases from (a) to (e).

Fig. 2 Coalescence of the microwave-induced spots and the current filaments. In (a) the DC-contacts to the sample are schematically shown, leading through the openings in the side walls of the waveguide. The DC-current is between $110\mu A$ in (a) and (h), and $130\mu A$ in (f). The microwave power is zero in (a), increases to its maximum of 30 mW in (f), and then decreases again to zero in (h).

A common nature of the microwave-induced dark spots and the current filaments is demonstrated by combining the two effects. Additionally to the microwave irradiation, a static voltage bias is applied to the sample via a pair of stripe-like ohmic contacts, Fig. 2(a). In the solely static field, a thin current filament emerges on some preferred position in the sample, Fig. 2(a). On increasing the microwave power, the filament is apparently attracted to the place in the sample, where the microwave induced spot appears, Fig. 2(d), till the two structures finally coalesce, Fig. 2(e). Upon decreasing the microwave power (Fig. 2(f) to (h)) the DC-current path obviously prefers to cross the dark spots (Fig. 2(g)) till they extinct in a subcritical microwave field, Fig. 2(h).

The role of conductivity in the forming of the microwave induced structures has been studied on a sample with an insulating gap in the middle, Fig. 3. An obvious difference in the behaviour of the two parts of the sample is seen: the upper part (the one closer to the microwave generator) seems to absorb the most of the incident power. Only after this part is saturated with the high electron density, the dark spots emerge also in the bottom of the sample.

The shape and the configuration of the dark structures are influenced by several factors, as e.g. the quality of the sample, its exact positioning in the waveguide and the microwave frequency used. Thus, an equal spacing of the circular spots in Fig. 1 can be shown to coincide approximately with the half-wavelength of a standing wave,

which arises in the sample acting as a dielectric obstacle inside the waveguide. On the other hand, no such coincidence can be found in the case of more complex patterns observed in the galvanically split sample in Fig. 3.

3 Conclusions

The microwave induced structures correspond to regions of substantially enhanced density of free electrons due to the impact ionization of shallow donors in the microwave field. There exists a relation between the observed spatial form of the structures and the standing wave pattern induced in the waveguide by the presence of the dielectric obstacle. This relation is rather straightforward if the conductivity of the dark spots is low. If, however, a significant power is dissipated in the high-conducting areas, a non-linear coupling between the standing wave pattern and the conductivity distribution sets on, leading to the formation of complex spatial patterns.

Acknowledgement. The work was done under financial support of the Deutsche Forschungsgemeinschaft, the grant Nr. A1010011 of the GAAV, and the NATO linkage program. Support of V.N. by the Alexander von Humboldt Foundation is gratefully acknowledged.

References

1. J. Hirschinger, F.-J. Niedernostheide, W. Prettl, and V. Novák, Phys. Rev. **61**, (2000) 1952.
2. V.V. Bel'kov, J. Hirschinger, V. Novák, F.-J. Niedernostheide, S.D. Ganichev, and W. Prettl, Nature **397**, (1999) 398.
3. V.V. Bel'kov, J. Hirschinger, D. Schowalter, F.-J. Niedernostheide, S.D. Ganichev, W. Prettl, D. Mac Mathúna, and V. Novák, Phys. Rev. **61**, (2000) 13698.
4. W. Eberle, J. Hirschinger, U. Margull, W. Prettl, and V. Novák, Appl. Phys. Lett. **68**, (1996) 3329.

Novel Observation of Magnetic-Field induced Metal-Insulator Transition in n-GaAs through Far-Infrared Photoconductivity Measurements

H. Kobori [1], M. Inoue [2] and T. Ohyama [1]

[1] Department of Physics, Graduate School of Science, Osaka University, Machikaneyama 1-16, Toyonaka, Osaka 560-0043, Japan

[2] R&D Headquarters, 3D Project Display Group, Sanyo Electric Co., Ltd.
Dainichi-higashimachi 1-1, Moriguchi, Osaka 570-0016, Japan

Abstract

We propose a novel technique to observe the magnetic-field induced metal-insulator transition (MFMIT) in GaAs through the far-infrared photoconductivity (FIRPC) measurements. Below the critical concentrations of the MIT, features of the FIRPC spectra with different donor concentrations are basically the same each other except that linewidths of the resonance spectra increase with increasing the donor concentration. Above the critical concentration of the MIT, drastic change in the magnetic field dependence of the FIRPC spectra has been observed for donors in GaAs. In this case, features of the FIRPC spectra show entirely new aspects: 1) The FIRPC spectra show a single resonance-like ones (SRS) and complicated structures of the FIRPC spectra observed in the insulator phase completely disappear. 2) We have observed anomalous shifts of the magnetic field, which gives the maximum of the absorption, with increasing the donor concentration and with rising the temperature. 3) The linewidths of the SRS for the FIRPC broaden as the donor concentration increases. We have discussed these phenomena taking into consideration the capture process by ionized donors for electrons excited by the FIR light.

1 Introduction

Through the impurity cyclotron resonance (ICR) measurements with the far-infrared (FIR) laser, we can directly monitor energy states of impurity bands in the presence of a magnetic field, and furthermore observe the magnetic field induced metal-insulator transition (MFMIT). We have carried out experimental studies on the MFMIT by use of the ICR measurements and reported experimental results in ICPS 23 [1, 2]. In that study, we have observed that the ICR spectra suddenly change from the resonant type to the nonresonant one as the donor concentration increases, i.e., the MIT caused by the overlap between wavefunctions of adjacent donors. In addition, we have observed that the ICR spectra abruptly varied from the nonresonant type to resonant one as the magnetic field increases, i.e., the MFMIT. Following those researches, in this paper, we have studied the MIT through the FIR photoconductivity (FIRPC) measurements, especially aiming at studying the MFMIT in details. Although almost all studies on the MIT are related with electrical measurements, the FIRPC includes both physical properties obtained by the optical and electrical measurements and mediates their results. We have discussed these phenomena taking into consideration the capture process by ionized donors for electrons excited by the FIR light.

2 Experimental Procedures

We have employed six GaAs samples with different donor concentrations for this experiment. All samples are n-typed and were grown by the MBE method except for one, n-GaAs-1, which was grown by the LPE method. The typical size of these samples is 5mm x 5mm x 5 μm on semi-insulating GaAs substrate (whose thickness is 0.5mm). Characteristics of samples are listed in Table I. Where N_D and N_A represent the donor and acceptor concentrations, respectively. A discharge-type of the far-infrared (FIR) gas (H_2O and D_2O) laser has been employed as the source of electromagnetic waves for the FIRPC measurements. The wavelengths of the FIR laser light are 220 μm and 119 μm for H_2O and 172 μm and 84 μm for D_2O gases, respectively. The FIR photoconductivity measurements have been performed by use of conventional methods. The temperature was varied between 2.0 and 4.2K by adjusting the vapor pressure of liquid helium. The magnetic field is produced up to 9T by the conventional vertical type superconducting solenoid.

3 results and Discussions

The critical concentration N_c for the MIT in the absence of a magnetic field for hydrogenic donors in semiconductors is given by $N_c^{1/3} a_D = 0.26 \pm 0.05$ (a_D: effective Bohr radii of donors). For GaAs, we obtain $N_c = 1.6 \times 10^{16}$ cm^{-3} with a_D =9.9 nm. We note that the donor concentration in GaAs-5 is just on the N_c. Figure 1(a)~1(d) show the FIRPC spectra for the ICR in n-GaAs-1, 4, 5 and 6 at various temperatures for the wavelength of 119 μm. Below the critical concentrations of the MIT, namely, the sample is in the insulator phase, features of the FIRPC spectra with different donor concentrations are basically the same each other except that linewidths of resonance spectra increase with increasing the donor concentration. As shown in Fig. 1(a), we note that the FIRPC spectra for n-GaAs-1 are composed of sharp ICR lines related to transitions from the 1s to $2p_{+1}$, $3p_{+1}$, $4p_{+1}$ states and from the excited state to other excited ones. In Fig. 1(b), two-three broad ICR lines for transitions from the 1s to $2p_{+1}$, $3p_{+1}$, $4p_{+1}$ states in n-GaAs-4 are observed. In this case, it is difficult to identify other lines, since those are too broadened. The FIRPC spectra for n-GaAs-2 and –3 are very similar to those of n-GaAs-1 and –4. Above the critical concentration of the MIT, namely, the sample is in the metallic phase, drastic change in magnetic field dependence of the FIRPC spectra has been observed. As seen in Fig. 1(c) and 1(d), the FIRPC spectra entirely change the figure. We have scrupulously investigated the FIRPC spectra for n-GaAs-1~6 between 2.0K and 4.2K with the wavelengths of 220, 172, 119 and 84 μm. In consequence, we have found the followings results; 1) The FIRPC spectra show a single resonance-like absorption (SRS) and complicated structures of the FIRPC spectra observed in the insulator phase vanish altogether. 2) We have observed the anomalous shift of the magnetic field, which gives the maximum of the absorption, with increasing the donor concentration and with rising the temperature. 3) The linewidths of the SRS broaden as the donor concetration increases. Almost all electrons should be in the ground state of donors under present experimental conditions. Part of electrons excited from the ground to excited states by the FIR light are relaxed to the conduction band or the D^- band [3]. We consider that electrons in the conduction band (CBE) and electrons in the D^- band (DMBE) contribute to the FIRPC. Provided that the mobility of the CBE are roughly equal to that of the DMBE, the FIRPC is dominated by $\Delta n_c + \Delta n_{D-}$, where Δn_c and Δn_{D-} are the net increment of electron densities in the conduction and D^- bands by the FIR irradiation, respectively. We put the absorption coefficient of the FIR light related to the ICR and the capture probability of electrons in D^- by D^+ band with α_{ICR} and P_+^-, respectively. Then we can approximately express that $\Delta n_c + \Delta n_{D-} \propto \alpha_{ICR}(1-P_+^-)$. When the MIT occurs from the metallic to insulator phase with increasing the magnetic field, the capture probability P_+^- is expected to become small abruptly. In addition, it is considered that the capture probability P_+^- increases with increasing the temperature. We consider, thus, that $\Delta n_c + \Delta n_{D-}$, or the FIRPC, shows behavior described in 2). We believe that the character of 1) is caused by the fact that the ICR related to the $1s \rightarrow 2p_{+1}$ transition occurs at the largest resonance magnetic field and the capture probability P_+^- decreases with increasing the magnetic field. We regard the experimental observation of 3) as a common feature on the FIRPC for the ICR.

4. Conclusions

As a novel technique to observe the metal-insulator transition (MIT), especially the magnetic-field induced MIT (MFMIT), we have carried out the far-infrared photoconductivity (FIRPC) measurements for the impurity cyclotron resonance (ICR) in GaAs with various donor concentrations between 2.0K and 4.2K for wavelengths of 220, 172, 119 and 84 μm. Above the critical concentrations of the MIT, we have observed drastic

change in the magnetic field dependence of the FIRPC spectra for the ICR: 1) The FIRPC spectra show a broad single resonance-like absorption (SRS) and complicated structures of the FIRPC spectra observed in the insulator phase completely disappear. 2) We have observed the anomalous shift of the magnetic field that the spectra show the maximum of the absorption, with increasing the donor concentration and with rising the temperature. 3) The linewidths of the SRS for the FIRPC broaden as the donor concentration increases.

References

1. H. Kobori, M. Inoue and T. Ohyama, The 23rd International Conference on Physics of Semiconductors, 1 149-153 (1996)
2. H. Kobori, M. Inoue and T. Ohyama, The 9th Int. Conf. on Shallow-Level Centers in Semicond. (SLCS9) MoP-13
3. N. F. Mott, Metal-Insulator Transitions, Taylor&Francis Ltd. (1990)

Table I   List of Samples

Sample	Type	N_D $\times 10^{15}$ cm^{-3}	N_A $\times 10^{15}$ cm^{-3}	Compensation Ratio	Growth Method
GaAs-1	n	1.5	1.0	0.67	VPE
GaAs-2	n	2.3	0.3	0.13	MBE
GaAs-3	n	4.8	0.8	0.17	MBE
GaAs-4	n	10	$N_D \gg N_A$	~0	MBE
GaAs-5	n	16	$N_D \gg N_A$	~0	MBE
GaAs-6	n	25	$N_D \gg N_A$	~0	MBE

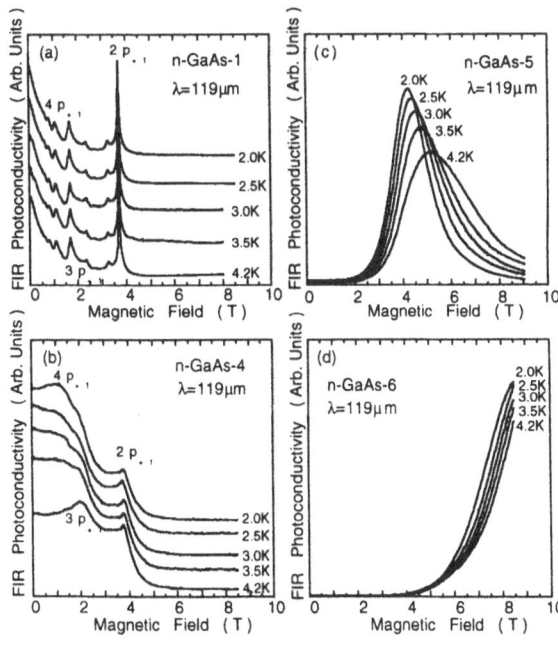

Fig. 1   FIRPC spectra for the ICR in n-GaAs-1,4,5 and 6 between 2.0 and 4.2K for the wavelength of 119 μm

Magnetization of single MnTe (sub)monolayers embedded in nonmagnetic quantum wells.

G. Prechtl[1], W. Heiss[1], S. Mackowski[2], A. Bonanni[1], E. Janik[2], H. Sitter[1], W. Jantsch[1]

[1] Institut für Halbleiter- und Festkörperphysik, Johannes Kepler Universität Linz, Altenbergerstraße 69, 4040 Linz, Austria
e-mail: gerhard.prechtl@jk.uni-linz.ac.at
[2] Institute of Physics, Polish Academy of Sciences, 02-668 Warsaw, Poland

Abstract: Single MnTe (sub)monolayer embedded in nonmagnetic quantum wells are investigated by magneto optical spectroscopy. In particular, the magnetization is probed as a function of temperature. This dependence shows a non Curie-Weiss behavior due to clusters with antiferromagnetic ordering. The experimental data agree with the magnetization predicted by Monte Carlo simulations performed for a diluted two dimensional arrangement of Mn ions.

We examine the magnetization of single MnTe layers with a thickness of one- or a fraction of one monolayer experimentally as well as theoretically. In particular, the single MnTe (sub)monolayers are embedded in nonmagnetic CdTe/CdMgTe quantum wells. This allows to probe the magnetization as function of temperature by measuring the exciton Zeeman splitting by polarization dependent photoluminescence excitation experiments. The dependence of the inverse Zeeman splitting is compared with Monte Carlo simulations on a diluted, single ML MnTe. This comparison allows also to determine the Mn diffusion profile in growth direction.

Since the MnTe (sub)monolayers are strained in the semiconductor quantum wells /1/ it is sufficient to take into account the antiferromagnetic Mn^{2+}-Mn^{2+} interactions by the Ising model. This leads to the Hamiltonian of a diluted antiferromagnet in a field (DAFF) which is given in units of the nearest neighbor coupling constant J by

$$H = \sum_{\langle i,j \rangle} \varepsilon_i \varepsilon_j \sigma_i \sigma_j - B \sum_i \varepsilon_i \sigma_i \qquad (1)$$

with the uniform field B>0 on all sites of the L × L lattice. Here σ_i denotes the spins of the Mn ions and x the fraction of lattice sites occupied with a spin is taken into account by setting $\varepsilon_i = 0,1$.

For MnTe it is sufficient to consider nearest-neighbor interactions. In the Monte Carlo simulation we used periodic boundary conditions and the well known heat-bath algorithm. The simulations for the DAFF are carried out for systems of sizes up to 150 × 150 lattice sites and the results are averaged over 100 different percolation configurations for the same x.

The magnetization as well as the magnetic susceptibility of a two-dimensional arrangement of Mn ions placed on a fcc lattice is a direct result of the Monte Carlo methods which we proved to be independent from the chosen lattice size. Without magnetic field and a 100% Mn coverage of the fcc lattice we find the antiferromagnetic-paramagnetic phase transition at a temperature of 80 K, which is significantly higher than the Néel temperature in cubic bulk MnTe, as indeed observed by neutron-scattering experiments [2]. The phase transition can be directly deduced from the computed magnetization and specific heat, as shown in Fig. 1(a) and (b), respectively, for a layer with x=1.

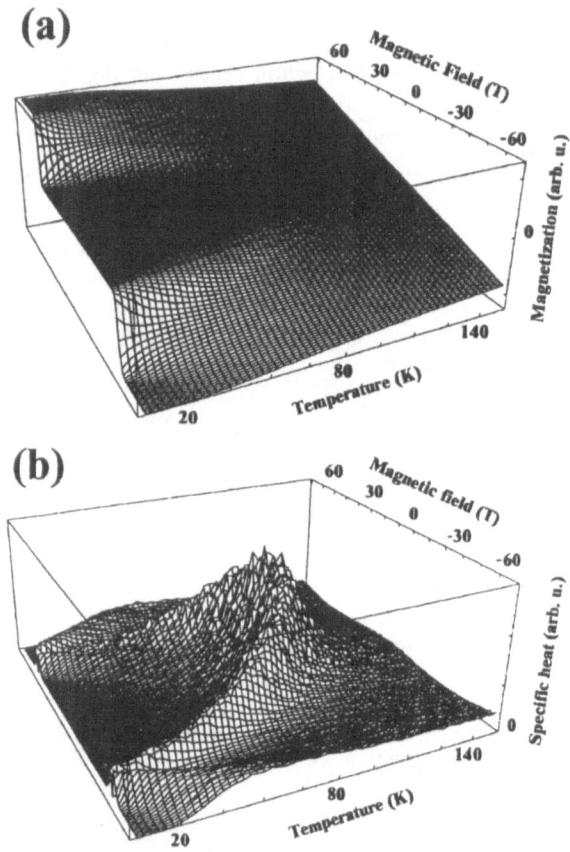

Fig. 1 Magnetization (a) and specific heat (b) of a single ML MnTe versus temperature and magnetic field.

At zero field, with decreasing coverage x, the Néel temperature decreases linearly down to 27 K at the percolation limit of $x_p \approx 60$ %. For lower coverages our simulations still shows a peak in the susceptibility corresponding to a transition from a spin glass to the paramagnetic phase. Below x_p, however, a smaller slope is observed in $T_{crit.}$ versus x than above the percolation limit. At finite fields any long range ordering breaks down, but still antiferrmagnetically ordered clusters exist causing a deviation from Curie-Weiss behavior at low temperatures.

Experimentally, we performed magneto-optical spectroscopy on single MnTe layers embedded in nonmagnetic $CdTe/Cd_{0.85}Mg_{0.15}Te$ quantum wells. We determine the temperature (T)- and magnetic field (B) dependence of the magnetization by measuring the e_1-hh_1 exciton spin splitting. The latter is strongly enhanced in our samples due to exchange interaction between the d electrons of the Mn^{2+} ions and the s-like conduction band and the p-like valence band electrons and holes, respectively. Therefore, the spin splitting induced by the insertion of a single MnTe submonolayer easily can be measured by polarization dependent photoluminescence excitation (PLE) experiments as well as by measuring the magneto-optical Kerr (MOKE) rotation.

For a sample containing a single, nominally one monolayer thick MnTe layer, we find a linear increase of the inverse spin splitting $(1/\Delta E)$ with increasing temperature up to 50 K. Above this temperature a smaller slope is observed in $1/\Delta E$ versus T, while at 90 K the slope again increases, as shown in Fig. 2. For samples containing fractional MnTe monolayers we find only one critical temperature where the slope of $1/\Delta E$ versus T is changing. In particular, for a ½ (1/4) monolayer we find a T_{crit} of 34 K (13 K).

In order to fit the measured temperature dependent inverse spin splitting with our calculations, the sp-d interaction has to be considered as well as the migration of Mn ions. In detail, the total magnetization is obtained by summing the magnetization of several layers with different x weighted by a factor which takes into account the sp-d exchange interaction. The weight-factor is obtained from the experimental dependence of the spin splitting on the Mn concentration as observed for bulk CdMnTe. This procedure gives a good fit of $1/\Delta E$ versus T for the sample containing one monolayer MnTe by assuming a central monolayer with a Mn coverage of 85 % and one monolayer with x=8 % on each side of this layer (see Fig. 2). This choice preserves the total Mn content present in the sample and gives a step like diffusion profile as an approximation for a Gaussian shaped Mn distribution. The used procedure is indeed very sensitive on the chosen Mn profile. This is demonstrated by the simulation performed for a center ML with x=75% (x=95%) with accompaning neighbor layers with x=13% (3%) strongly deviating from the experimental data, as shown in Fig. 2.

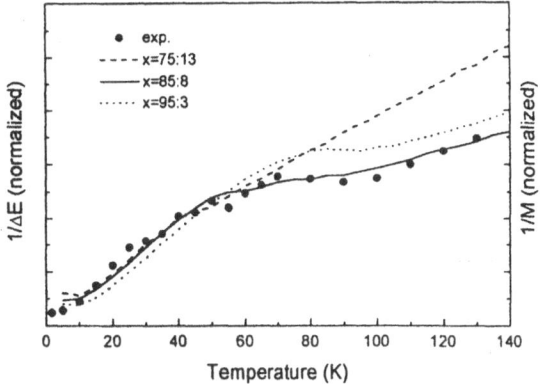

Fig. 2 Inverse Zeeman splitting of the e_1-hh_1 exciton transition measured at B=6 T as function of temperature. Full circles are experimental data and the dashed, solid and dotted lines are results from simulations performed for a $Cd_{1-x}Mn_xTe$ layer with x=75%, x=85% and x=95% with two neighbor layers with x=13%, x=8% and x=3%, respectively.

Therefore, our magneto-optical measurements allow not only to detect the magnetization of single, ultranarrow MnTe layers in nonmagnetic quantum wells but also to determine the migration of the Mn ions during growth and thus the Mn density profile in growth direction.

We thank W. Krepper for his enthusiastic interest and his excellent help at technical problems. This work has been supported by the "Fonds zur Förderung wissenschaftlicher Forschung (FWF-P12100 – PHY)", Austria, and in Poland by the grant PBZ28.11 from the State Committee for Scientific Research.

References

/1/ J.M. Yeomans, Statistical Mechanics of Phase Transitions, Clarendon Press Oxford, 1992
/2/ T.M.Giebultowicz, P. Klosowski, N.Samarth, H.Luo, J.K. Furdyna, J.J. Rhyne, Phys. Rev. B 48, 12817 (1993)

Energy relaxation by phonon scattering in long-range ordered $(Al_{0.5}Ga_{0.5})_{0.5}In_{0.5}P$

T. Kita[1], K. Yamashita[1], T. Nishino[1] *, Y. Wang[2], K. Murase[2]

[1] Department of Electrical and Electronics Engineering, Faculty of Engineering, Kobe University, Rokkodai 1-1, Nada, Kobe 657-8501, Japan

[2] Department of Physics, Graduate School of Science, Osaka University, 1-16 Machikaneyama, Toyonaka, Osaka 560-0043, Japan

Abstract Carrier relaxation and recombination in long-range ordered $(Al_{0.5}Ga_{0.5})_{0.5}In_{0.5}P$ have been studied by selectively excited PL spectroscopy. We found phonon assisted sharp resonant photoluminescence (PL) lines in long-range ordered (LRO) $(Al_{0.5}Ga_{0.5})_{0.5}In_{0.5}P$ by selectively excited PL spectroscopy. The observation of the resonant PL demonstrates energy relaxation from the excited state (X level) to the ground state (Γ level) by inelastic phonon scattering.

1 Introduction

Spontaneous CuPt-like ordering of $(Al_xGa_{1-x})_{0.5}In_{0.5}P$ has been widely observed in vapor-phase epitaxy on a lattice matched GaAs (001) substrate. [1] This ordering has a periodic stack of the cation sublattice along the [$\bar{1}11$] or [$1\bar{1}1$], the two CuPt$_B$ subvariants, which causes a reduction of the band gap and a splitting of the valence band maximum. The band-gap reduction and the valence-band splitting obey a square dependence on order parameter, which causes continuous changes according to degree of ordering.[2] Generally, ordered $(Al_xGa_{1-x})_{0.5}In_{0.5}P$ $(0 \le x \le 1)$ alloy comprises domains with different order parameter and different domain size. A statistical distribution of the order parameter in the epitaxial film causes a fluctuation of the band-gap energy in real space. In such band structure an anomalously large Stokes shift in PL has been observed at low temperature. In this study we focus our attention on energy relaxation process of photo-excited carriers in the ordered $(Al_{0.5}Ga_{0.5})_{0.5}In_{0.5}P$.

2 Experiments

An epitaxial film of $(Al_{0.5}Ga_{0.5})_{0.5}In_{0.5}P$ was grown on GaAs(001) substrate by organometallic vapor-phase epitaxy. Detailed sample growth conditions were reported in Refs. 3 and 4. A transmission-electron diffraction (TED) of the $(Al_{0.5}Ga_{0.5})_{0.5}In_{0.5}P$ on GaAs(001) shows super-reflection spots of the CuPt-type. $(Al_{0.5}Ga_{0.5})_{0.5}In_{0.5}P$-random alloy is an indirect gap material, while the optical transition in ordered $(Al_{0.5}Ga_{0.5})_{0.5}In_{0.5}P$ becomes a direct type, because the Γ conduction band energy is reduced and crosses the X valley level. [3]

3 Results and discussion

Figure 1 shows selectively excited PL spectra at 10 K for the ordered $(Al_{0.5}Ga_{0.5})_{0.5}In_{0.5}P$. The solid triangles

* *Present address*: Kobe City College of Technology Gakuen Higashimachi 8-3, Nishiku, Kobe 651-2194, Japan

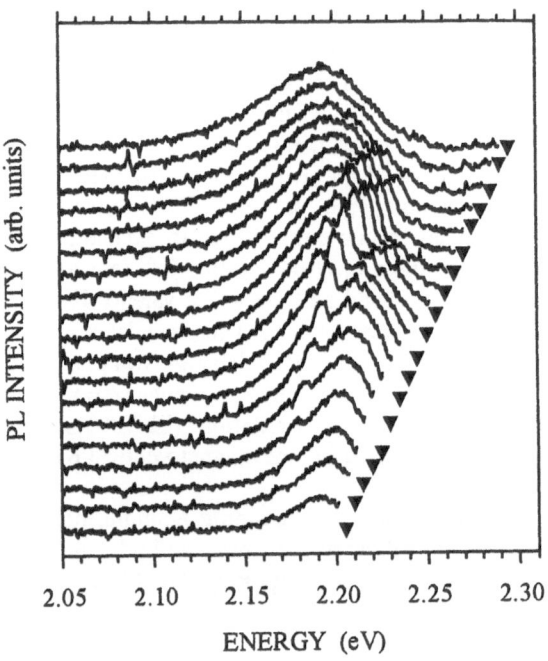

Fig. 1 Selectively excited PL spectra of a partially ordered sample. The triangles indicates the excitation energy.

indicate excitation energies for each spectrum. For excitation energies above 2.28 eV an inhomogeneous broad peak was observed indicating nonselective excitation. In the near-resonant excitation, on the other hand, we found a set of sharp PL peaks together with the inhomogeneous broad PL. With decreasing excitation energy the set of the sharp PL peaks shows a dramatic change. The resonance phenomenon was not observed in the weakly ordered $(Al_{0.5}Ga_{0.5})_{0.5}In_{0.5}P$ on the GaAs(115)A substrate and was not found in ordered $Ga_{0.5}In_{0.5}P$.

Figure 2 is a contour plot of the selectively excited PL as a function of the energy shift from the excitation energy. It is found that the energy shifts of these PL peaks do not depend on the excitation energy. The energy separations of the PL resonances lie in a range of typical phonon energies of the AlGaInP alloy. Dashed lines indicate Raman-scattering peaks. The sharp PL set well agrees with the Raman-scattering data of the LO phonons related to In-P, Ga-P, and Al-P bonds of the ordered sample. Also, we found new signals near 24 meV and 29 meV in the PL and the Raman scatter-

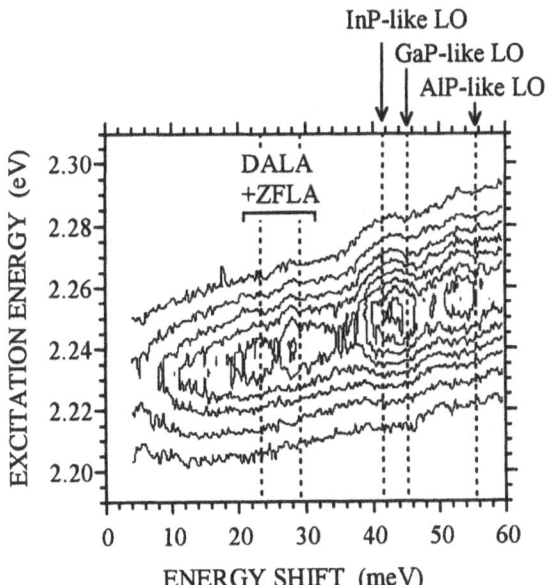

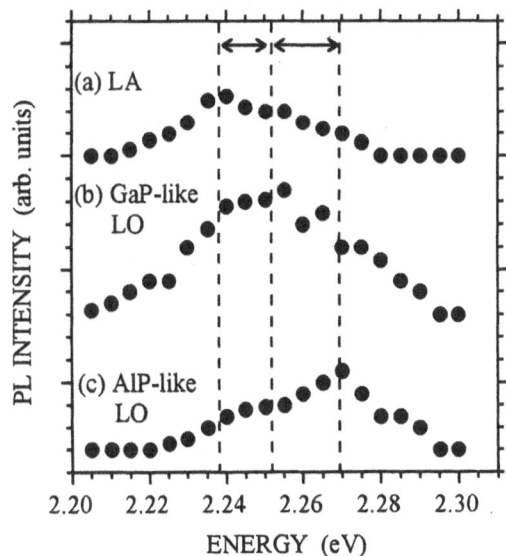

Fig. 2 Contour plot of the selectively excited PL. Dashed lines mark constant phonon resonances.

Fig. 3 Excitation energy dependence of the resonant PL intensities of LA, GaP-like LO and AlP-like LO resonances. The vertical lines represent resonant peaks.

ing of the ordered sample. These new structures can be attributed to a LA-phonon mode caused by zone-folding effects, i.e., zone-folded LA (ZFLA), in the long-range atomic ordering and a disorder activated LA phonon band (DALA), [5] while the low-frequency Raman scattering of the weakly ordered sample is weak.

The sharp PL intensity depends on the excitation energy, which reflects the density of domains with a matching ground-state transition energy. Figure 3 plots excitation energy dependence of the sharp PL intensities corresponding to the LA, GaP-like LO and AlP-like LO resonances. The excitation energy dependence shows different resonant profiles for the three resonances. The energy separation between the LA-resonant profile (a) and the GaP-like-LO resonant profile (b) nearly coincides with an energy difference between the corresponding phonon energies. The separation between the GaP-like-LO resonant profile (b) and the AlP-like-LO resonant profile (c) also coincides with an energy difference between the GaP- and AlP-lile LO phonons. These results deny a possibility of resonant Raman scattering. Therefore, we considered that the observed resonant phenomena in PL are caused by inelastic phonon scattering in energy relaxation process of excited carriers.

Under a high dense excitation above 6 mW/cm^2 the spectral shape saturates reveling an emission band from new *excited* states about 70 meV above the ground state transition. Let us consider two scenarios for the origin of the excited states; (1) low-dimensional system giving required discrete energy level and (2) a pseudodirect X states resulting from a Γ-X mixing. Generally, the domain size is not uniform and shows a statistical distribution in the epitaxial film. Then, sufficiently small

domains and sequence mutations [6], which may behave as low-dimensional system giving required discrete energy level, can contribute to such resonances. However, it was not confirmed in our TEM images that the dominant domain size is indeed sufficiently small to give the required discrete energy level structure. On the other hand, the scenario (2) originated from the X states is an important candidate. An energy position of the X exciton is close to the new emission band. [3,4] Furthermore, even in the weakly ordered sample we observed a strong no-phonon assisted PL band of the X exciton consisting of the X electron and Γ-heavy hole. Therefore, the observed excited state is considered to be a pseudodirect type transition of the X states with a finite oscillator strength resulting from a Γ-X mixing in the conduction band. Scattering effects at domain boundaries play an important role in the Γ-X mixing.

References

1. A. Zunger and S. Mahajan, in *Handbook of Semiconductors*, 2nd ed., edited by S. Mahajan (Elsevier, Amsterdam, 1994), Vol. 3, p. 1339, and references therein.
2. S. -H. Wei and A. Zunger, Phys. Rev. B **57**, 8983 (1998).
3. K. Yamashita, T. Kita, H. Nakayama, and T. Nishino, Phys. Rev. B **53**, 15713 (1996).
4. K. Yamashita, T. Kita, H. Nakayama, and T. Nishino, Phys. Rev. B **55**, 4411 (1997).
5. M. Cardona, Superlattice and Microstructures **5**, 27 (1989).
6. T. Mattila, Su-H. Wei, and A. Zunger, Phys. Rev. Lett, **83**, 2010 (1999).

Low-dimensional magnetism in polymorphic compounds, VOMoO$_4$ and (V$_{0.56}$, Mo$_{0.44}$)$_2$O$_5$

I.Shiozaki[1][*], S. Takata[2], S. Wada[2], and T.Toyoda[3]

[1] Department of Physics, Graduate School of Science, Tokyo Metropolitan University 1-1, Minami-Ohsawa, Hachiooji, Tokyo 192-03, Japan, e-mail: shiozaki@comp.metro-u.ac.jp

[2] Department of Physics, Faculty of Science, Kobe University, 1-1, Rokko-Dai, Nada, Kobe 657-8501, Japan

[3] Department of Applied Physics and Chemistry, The University of Electro-Communications, 1-5-1, Chofugaoka, Chofu, Tokyo 182-8585, Japan

Abstract Two kinds of low-dimensional magnetic oxides, VOMoO$_4$ and (V$_{0.56}$, Mo$_{0.44}$)$_2$O$_5$ were synthesized, where both have almost the same composition, V: Mo: O= 1: 1: 5. The former is 1D antiferromagnet, and the latter has 2D magnetic layers. Measurements of NMR, optical absorption by PAS, and electrical conductivity suggested the existence of two characteristic excitation energies in VOMoO$_4$. The lower one is about 70 meV. Another one was observed in PAS, and is 1.6 eV. Results were discussed referring to the difference of lattice dimensionality in two compounds.

1 Introduction

Crystal of VOMoO$_4$ contains 1D magnetic chains constructed by VO pairs along the tetragonal c-axis. Every chain is connected by MoO$_4$ tetrahedra with each other. Crystal of (V$_{0.56}$, Mo$_{0.44}$)$_2$O$_5$ is monoclinic, and has magnetic layered structure. Both are shown in Figs. 1(a) and (b). The magnetic susceptibility, χ, shows antiferromagnetic T dependence [1, 2] in both crystal. VOMoO$_4$ shows a Bonner-Fisher type broad maximum around 100 K, and goes to 3D ordered phase below 34 K. (V$_{0.56}$, Mo$_{0.44}$)$_2$O$_5$ also shows some kind of ordering behavior below 20 K, though the magnetic structure has not been determined, yet. In the former crystal Ref. 1 observed an anomalous T dependence of the lattice parameter, a, at 100 and 34 K.

In the present work we measured NMR and photo-acoustic (PA) spectroscopy. As both experiments suggested the energy gap formation in the crystals, electrical resistivity was also measured to confirm the gap value.

2 Experimental results

2.1 NMR

Temperature dependence of resonance shift, K, and nuclear spin-lattice relaxation times, T_1, were measured by NMR of ^{51}V in both compounds.

In case of VOMoO$_4$ the negative d-spin contribution to K showed a broad maximum around 100 K. The relaxation rate, T_1^{-1}, twice changed the gradient of the T_1^{-1} vs T curve at around 34 and 100 K. These behaviors

are corresponding to that of the static susceptibility, χ. Above 100 K T_1^{-1} increases as

$$T_1^{-1} = 0.01 + 10.4 \exp[-840/T] \quad (ms)^{-1}. \quad (1)$$

Signals in (V$_{0.56}$, Mo$_{0.44}$)$_2$O$_5$ show characteristic change around 15 K. Above 15 K Both of T_1^{-1} and K follow the equations

$$T_1^{-1} = 0.10 + 1.85 \exp[-200/T] \quad (ms)^{-1}, \quad (2)$$

$$K = 0.161 - 17.9/(T+79) \quad (\%). \quad (3)$$

Behaviors of T_1^{-1} are shown in Figs 2(a) and (b).

2.2 Photoacoustic spectroscopy

PA measurement of VOMoO$_4$ detected a strong absorption of UV and visible light, which suggested the existence of two different excitation energies. In case of direct allowed transition by optical absorption, the photon energy, $h\nu$, is proportional to $(Ph\nu)^2$, where P is PA signal intensity. Relation between them is shown in

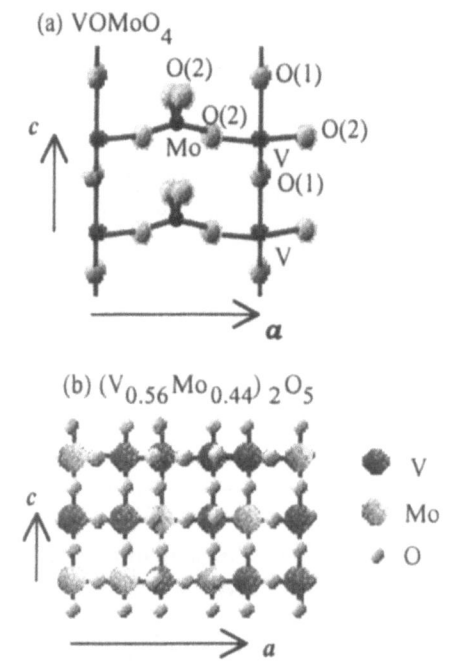

(a) VOMoO$_4$

(b) (V$_{0.56}$Mo$_{0.44}$)$_2$O$_5$

Fig. 1 Crystals of VOMoO$_4$ (a), and of (V$_{0.56}$Mo$_{0.44}$)$_2$O$_5$ (b).

Fig. 3, which clearly gives straight lines with different two gradients. That suggests there are two types of interband transition. The cross points of lines with the horizontal scale give the respective energy gaps, namely 0.9 and 1.6 eV.

2.3 Electrical resistivity

The electrical conduction in VOMoO$_4$ is thermally activated type. Arhenius plot of ρ, however, shows a curved line giving no characteristic carriers excitation energy as seen in Fig.4. Logarithmic plot of $\rho T^{-2/3}$ with the assumption of small polaron hopping is also failed. Carriers' excitation energy at 77 K was estimated from the gradient of tangential line of Arhenius curve, and was given as 60 meV.

3 Discussions

T_1^{-1} is attributed to 1/2 spin fluctuation accompanied by 1-magnon process. The exponential behavior in eq. (1) suggests the existence of an energy gap of 840K (=72 meV) in the energy band of VOMoO$_4$. The gap structure was also suggested from ρ (60 meV) and from PAS (0.9 eV), though there are some discrepancy in the numerical values. In Fig. 1(a) the bond length of V and O(1) alternatively changes along the c-axis. So that a slight splitting of t_{2g} orbits on V is expected. That is assigned to the origin of energy gap. From the T dependence of ρ the gap value decreases with increasing T. That should come from the increase of lattice distortion as discussed in [5 - 7].

In (V$_{0.56}$, Mo$_{0.44}$)$_2$O$_5$, K gives the Weiss temperature of -79 K. χ re-analyzed with this value gives the effective moments, 1.14 μ_B/formula unit, and suggests it belongs to 2D 1/2 spin AF system. Eq. 2 suggests the existence of energy gap coming from t_{2g} splitting, again.

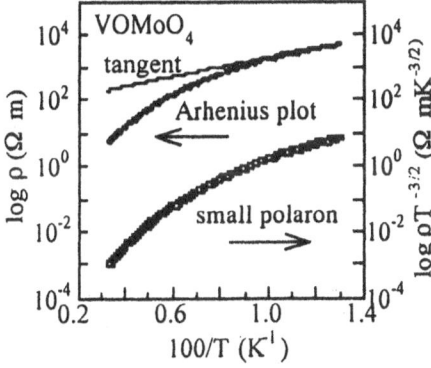

Fig. 4 Electrical resistivity as functions of T.
○ usual Arhenius plot, □ polaron hopping case.

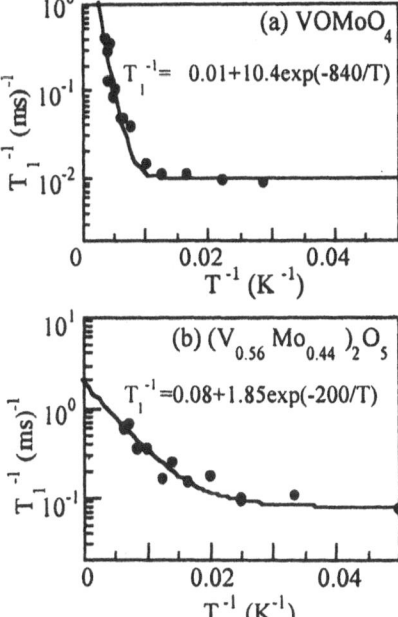

Fig. 2 Logarithmic plot of the spin-lattice relaxation rates against inverse T in two crystals.

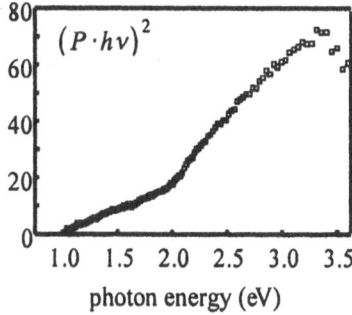

Fig. 3 Measured PAS in VOMoO$_4$. Square of PA intensity multiplied by $h\nu$ is plotted against $h\nu$.

Acknowledgement
The Authors owe Prof. K. Hiraga for the crystal analysis of (V$_{0.56}$ Mo$_{0.44}$)$_2$O$_5$.

References

1. I. Shiozaki, J. Phys. Condensed Matter **10** (1998) 9813.
2. I. Shiozaki, J. Magn. Magn. Mater. **181**(1998)261.
3. S. Takata, S. Wada, and I. Shiozaki, J. Phys. Condensed Matter, submitted.
4. T. Toyoda, and I. Shiozaki, Analytical Sci., submitted.
5. I. Shiozaki, M. Ohashi, and H. Kadowaki, J. Phys. Soc. JPN. **69**(2000), to be published.
6. I. Shiozaki, M. Ohashi, M. Koyano, and S. Katayama, Physica **B 284-288**(2000)1621.
7. M. Koyano, I. Shiozaki, D. J. Lockwood, and S. Katayama, Proc. ICPS25, August 2000, Osaka, submitted.

Highly efficient luminescence in partially ordered GaInP$_2$

X. H. Zhang[1], S. J. Chua[1,2], J. R. Dong[1]

[1] Institute of Materials Research and Engineering, 3 Research Link, Singapore 117602 e-mail: xh-zhang@imre.org.sg
[2] Department of Electrical Engineering, National University of Singapore, 10 Kent Ridge Crescent, Singapore 119260

Abstract   Continuous wave and time-resolved photoluminescence studies of partially ordered GaInP$_2$ grown by metal-organic chemical vapor deposition are presented. Analysis of the experimental results suggests the highly efficient excitonic recombination takes place in our ordered GaInP$_2$ sample and a type I alignment at the interface between ordered and disordered GaInP$_2$.

1 Introduction

Spontaneous long-range order has been observed in many III-V pseudobinary alloys under certain growth condition [1]. The most common is the CuPt type ordering that was first observed in GaInP$_2$ [2]. Certain samples, however, are not uniformly ordered but exhibit small domains of varying degree order in a disordered matrix. The optical properties of partially ordered GaInP$_2$ have shown several anomalies [3-5]. Based on these observations, it has been proposed that spatially indirect recombination takes place in this alloy [5] and the interface between ordered and disordered GaInP$_2$ has a type II alignment [6,7]. However, first-principles calculation has shown that the interface has a type I alignment [8]. The type II band alignment is difficult to reconcile with the theoretical calculation [8].

In this paper, we present the continuous wave (cw) and time-resolved photoluminescence (PL) studies on the partially ordered GaInP$_2$ samples grown by metal-organic chemical vapor deposition (MOCVD). Analysis of the experimental results suggests the highly efficient excitonic recombination in our ordered GaInP$_2$ sample and a type I alignment at the interface between ordered and disordered GaInP$_2$.

2 Experiment

The GaInP$_2$ samples used in this study were grown on GaAs substrate with a low pressure MOCVD system. Prior to the growth of 700 nm GaInP$_2$ layer, a 300 nm GaAs buffer layer was deposited. Details of growth conditions can be found elsewhere [9].

For the cw PL measurement, an Ar$^+$ laser operating at a wavelength λ = 488 nm was used for the excitation. The luminescence was dispersed in a 75 cm monochromator and detected by a cooled photomultiplier. The time-resolved PL spectra were measured using a streak camera under frequency-doubled femtosecond Ti:sapphire laser excitation. The samples were mounted on the cold finger of a temperature-variable closed-cycle He gas cryostat during the measurements.

3 Results and discussion

Figure 1 shows the anomalous behavior of PL peak energy as a function of temperature. With increasing temperature (6.4–30 K), an initial decrease in energy was observed, followed by an increase in energy (T from 30–60 K) and finally a decrease again: the so-called inverted-S shape.

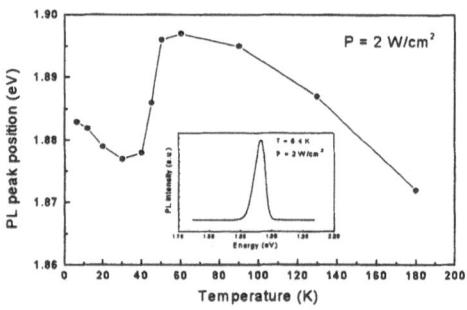

Fig. 1 The temperature dependence of the PL peak position. The inset depicts the PL spectrum recorded at 6.4 K.

The similar inverted-S shaped temperature dependence of the PL emission has previously been observed in quantum wells [10-12], superlattices [13-17] and Al$_x$In$_{1-x}$As epilayers [18]. For these quantum wells and superlattices, as well as the Al$_x$In$_{1-x}$As epilayers, strong carrier localization, due to a certain degree of disorder, was reported. Therefore, it appears the inverted-S shaped emission of PL emission is associated with carrier localization.

Our sample used in this study is not uniformly ordered but exhibit small domains of varying degree order in a disordered matrix [7]. Due to the fluctuations in the size of ordered domains and the varying degree order in the domains as well as the roughness of the order/disorder interface, there are a lot of localized band-tail states in the sample. Considering the smaller band-gap of the ordered domains, the ordered domains can be regarded as quantum wells with localized band-tail states. An explanation of the anomalous behavior of PL emission has been given in Ref. 10 in terms of localized states.

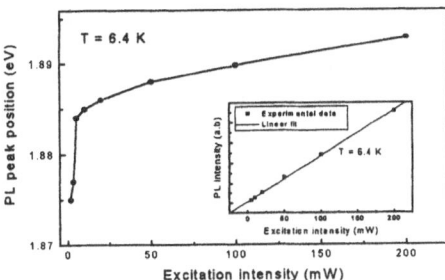

Fig. 2 The excitation intensity dependence of PL peak position at 6.4 K. The inset shows the PL intensity as a function of the laser excitation power.

Figure 2 shows the PL peak position as a function of excitation intensity measured at T = 6.4 K. The strong blue shift of the PL emission is another evidence for the presence of localized states. The inset to Fig. 2 depicts the integrated intensity of the luminescence as a function of excitation intensity at 6.4 K. A linear relation between the two intensities is indicative that the PL emission of our sample is due to excitonic recombination [19]. Our time-resolved PL measurements also support this view of excitonic recombination. Figure 3 shows typical temporal profile of the PL emission at 12 K. The measured carrier lifetime is 1.21 nanoseconds.

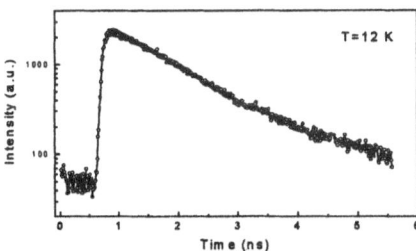

Fig. 3 Temporal profile of the PL emission at 12K.

For near-gap radiative recombination in direct gap semiconductors at low temperature, typical time constants are nanoseconds for excitonic recombination, tens of nanoseconds for band-acceptor, and tens to thousands of nanoseconds for donor-acceptor pair. The decay time of 1.21 nanoseconds we measured suggests that the direct excitonic recombination takes place in ours sample, which indicates that the order/disorder interface has a type I alignment and the ordered domains act like type I quantum wells. Using first-principles pseudopotential method, Froyen et al. [8] calculated a type I band alignment for the interface between ordered and disordered GaInP$_2$. Our results are consistent with their calculations.

4 Conclusions

In conclusion, we report the cw and time-resolved PL studies on the partially ordered GaInP$_2$ grown by MOCVD. Analysis of the results suggests that the direct excitonic recombination takes place in our partially ordered GaInP$_2$ sample and the interface between ordered and disordered GaInP2 has a type I alignment.

References

1. A. Zunger and S. Mahajan, *Handbook on Semiconductors*, edited by T. S. Moss (Elsevier, Amsterdam 1994), p. 1399.
2. A. Comyo, T. Suzuki, and S. Iijima, Phys. Rev. Lett. **60**, (1988) 2645.
3. M. Kondow, S. Minagawa, Y. Inoue, T. Nishino, and Y. Hamakawa, Appl. Phys. Lett. **54**, (1989) 1760.
4. M. Kondow and S. Minagawa, J. Appl. Phys. **64**, (1988) 793.
5. M. C. DeLong, W. D. Ohlsen, I. Viohl, P. C. Taylor, and J. M. Olson, J. Appl. phys. **70**, (1991) 2780.
6. R. A. J. Thomeer, F. A. J. M. Driessen, and L. J. Giling, Appl. Phys. Lett. **66**, (1995) 1960.
7. J.-R. Dong, Z.-G. Wang, X.-L. Liu, D.-C. Lu, and D. Wang, Appl. Phys. Lett. **67**, (1995) 1573.
8. S. Froyen, A. Zunger, and A. Mascarenhas, Appl. Phys. Lett. **68**, (1996) 2852.
9. Unpublished data.
10. M. S. Skolnick, P. R. Tapster, S. J. Bass, A. T. Pitt, and N. Apsley, Semicond. Sci. Technol. **1**, (1986) 29.
11. S. T. Davey, E. G. Scott, B. Wakefield, and G. J. Davies, Semicond. Sci. Technol. **3**, 365 (1988).
12. P. G. Eliseev, P. Perlin, J. Lee, and M. Osinski, Appl. Phys. Lett. **71**, (1997) 569.
13. A. Chomette, B. Deveaud, A. Regreny, and G. Bastard, Phys. Rev. Lett. **57**, (1986) 1464.
14. T. Yamamoto, M. Kasu, S. Noda, and Sasaki, J. Appl. Phys. **68**, (1990) 5318.
15. H. Nashiki, I. Suemune, H. Kumano, H. Suzuki, T. Obinata, K. Vesugi, and J. Nakahara, Appl. Phys. Lett. **70**, (1997) 2350.
16. S. Guha, Q. Cai, M. Chandrasekhar, H. R. Chandrasekhar, H. Kim, A. D. Alvarenga, R. Vogelgesang, A. K. Ramdas, and M. R. Melloch, Phys. Rev. B **58**, (1998) 7222.
17. L. Bergman, M. Dutta, M. A. Stroscio, S. M. Komirenko, R. J. Nemanich, C. J. Eiting, D. J. H. Lambert, H. K. Kwon, and R. D. Dupuis, Appl. Phys. Lett. **76**, (2000) 1969.
18. S. M. Olsthoorn, F. A. J. M. Driessen, A. P. A. M. Eijkelenboom, and L. J. Giling, J. Appl. Phys. **73**, (1993) 7798.
19. S. Jin, Y. Zheng, and A. Li, J. Appl. Phys. **82**, (1997) 3870.

Kinetics of the amorphous-to-crystalline phase transformation in Ge:Sb:Te alloys used for the phase-change optical memory applications

E. García-García[1], E. Morales-Sánchez[1,2], J. González-Hernández[2], E. Prokhorov[2], Yu. Vorobiev[2] and S. Kostylev[3]

[1] División de Estudios de Posgrado, Facultad de Ingeniería, UAQ, Querétaro, C.P. 76010, QRO., MÉXICO
[2] CINVESTAV del IPN, Unidad Querétaro, Juriquilla, Querétaro 76230, MÉXICO
[3] Energy Conversion Devices, 1675 West Maple Road, Troy, Michigan 48084, USA

Abstract. The kinetics of the amorphous-to-crystalline (fcc) phase transition in $Ge_2Sb_2Te_5$ was investigated under constant temperature and constant heating rate conditions. A procedure to determine the transformed volume fraction based on the electrical conductivity data is proposed. It is shown that the assumption widely accepted about the linear dependence between the conductivity variation and the transformed volume fraction is not correct. The use of this assumption could give an error in the determination of the crystalline volume fraction of up to one order of magnitude.

The amorphous to crystalline transformation kinetics observed agrees with the Johnson-Mehl-Avrami formalism. The transformation exhibits two distinct stages that are attributed to surface and bulk nucleation, and gives Avrami exponents close to 4 and 1, corresponding to the bulk and surface nucleation, respectively. The activation energy is around 5 eV; this value exceeds the previous estimations.

Rewritable phase-change optical and electrical memory devices based on the amorphous-to-crystalline phase transition in chalcogenide materials were originated by Ovshinsky [1] and extensively studied during the last thirty years [2-7 etc]. To monitor the transition, the electrical measurements are widely employed. Since the electric conductivity and dielectric constant of the two phases are essentially different, these measurements allow the *in situ* determination of the volume fraction of the transformed phase. However, details of the method of this determination were never analyzed; the majority of papers (like [5-7]) used the oversimplified assumptions, without justification. This may be one of the reasons for a controversy existing in the literature on the activation energy of transformation. Here we analyze the experimental data on the kinetics of the transition from amorphous to crystalline (fcc) phase in the alloys with the composition close to $Ge_2Sb_2Te_5$. The method of determination of the transformed volume fraction is discussed. The corrected value for the activation energy of the process is found.

The $Ge_{21}Sb_{26}Te_{53}$ films with the thickness of 50 nm were prepared by RF sputtering of the bulk alloy samples onto unheated corning glass substrate. The phase transformation was induced by sample heating, the temperature was controlled with a microprocessor-based controller programmed to produce a constant heating rate,

or to maintain a constant temperature. During the transformation, we monitored the electric resistivity and capacitance of the film. The morphology and structure of the samples after transformation were examined with a conventional transmission electron microscopy (TEM).

The resistance measurements during the sample isothermal annealing in the interval 120 - 140 °C, indicate the phase transformation: resistance decreases (i.e. conductivity grows) with an increase of the transformed volume fraction. In agreement with the previous publications [3-6], the transition time sharply decreases when the annealing temperature increases. The variation of capacitance has the character opposite to that of the resistance: when resistance decreases, capacitance grows, thus also reflecting the phase transition. The data obtained with the constant heating rate Z show that the transition temperature increases with an increase of Z, up to 200 °C for Z = 100 K/min.

The TEM and the electron diffraction data indicate that after the isothermal transformation at relatively low temperature, the large (around 1 μm) randomly orientated crystallites are formed, and a noticeable amount of amorphous phase is left. At higher annealing temperature, the crystallites are smaller (0.1 – 0.2 μm) and the amorphous phase is practically absent. In experiments with the constant heating rate, the lower rate corresponds to the larger crystallites formed, with the smaller amount of them.

All the papers using the electrical data for the transition monitoring, assume the linear dependence between the electrical conductivity σ and the volume fraction transformed x [6,7], namely:

$$x = (\sigma - \sigma_a)/(\sigma_c - \sigma_a)$$

where subindexes "c" and "a" denote the conductivity values for crystalline and amorphous phases, correspondingly. However, the applicability of this expression was never discussed; it obviously could be questioned. There are many models for the description of the electrical properties of a two-phase system (see [8]). The most simple and general, the Maxwell-Wagner model, assumes that one phase in the form of spheres is uniformly distributed in a continuous media of the other. If we take for our case, that at the initial stage of transformation the small spherical crystallites grow within the amorphous media, than, according to this model, the conductivity is equal

$$x = \frac{\sigma - \sigma_a}{\sigma_c - \sigma_a} \cdot \frac{2\sigma_a + \sigma_c}{2\sigma_a + \sigma}$$

where the first part of the product coincides with the empirical expression above, but the second one obviously exceeds 1; so, we see that the generally used empirical expression is not valid for any values of x, and its use will lead to large errors in estimation of the transformed volume fraction. In Fig. 1, the upper points shows the Maxwell-Wagner analysis, and the lower ones, the empirical law mentioned above. The great difference is obvious.

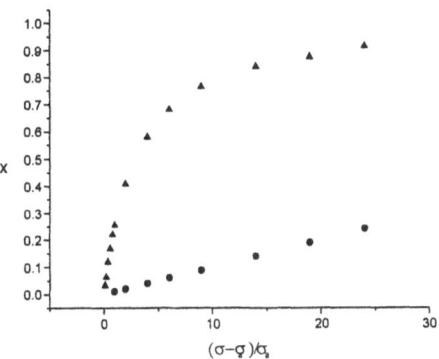

Fig. 1 Calculated dependence of the transformed volume fraction on the variation of conductivity (see text)

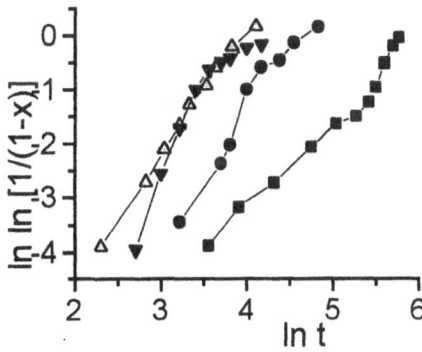

Fig. 2 The function used for determination of the Avrami exponent

According to the classic Johnson-Mehl-Avrami theory [9], the transformed volume fraction as a function of time and temperature is determined by the expression

$$x = 1 - \exp(-Kt^n) \tag{1}$$

where $K = K_o \exp(-E/kT)$, E being an effective activation energy, and n the Avrami exponent which value depends on the characteristics of nucleation and the dimensionality of growth. To find the activation energy of transformation E from the isothermal experiments, one has to find the temperature dependence of the time interval t_o necessary to obtain a certain degree of transformation (we took x = 0.2 corresponding to the ratio $\sigma/\sigma_a = 1.725$), and use the Arrhenius plot to find E from the slope of the adjusted straight line. From (1) we get

$$\ln t_o = Const + E/(nkT)$$

The corresponding experimental dependence gives E/n = 2.58 eV.

To find the Avrami exponent n, the plot of $\ln \ln [1/(1-x)]$ versus $\ln t$ is used. According to eq. (1), the slope of the line gives the value of n. The corresponding data for the isothermal transformation (Fig. 2, squares for 117, circles – 125, upsided triangles – 128.5, and downsided triangles for 130 °C) could be modeled with two straight lines; the values of n for each part are approximately 1 and 4. According to ref. [10], the initial high value of n corresponds to the transformation with a constant bulk nucleation rate. For the later stage of transformation, the nucleuses in the volume are exhausted, and new ones are formed on the grain boundaries, which gives n around 1.

In general, the process as a whole is characterized by the average value of n ≈ 2. The ratio E/n obtained from the Arrenius plot, is 2.58 eV; with the value of n above, it gives an activation energy E = 5.16 eV. We note that the value obtained is about two times larger than those given at the literature, with the exception of [11]; we may ascribe this to the inaccurate determination of x and, therefore, the errors in calculation of the Avrami exponent.

References
1. S.R. Ovshinski, Phys. Rev. Lett. **21** (1968) 1450
2. S.R. Ovshinski and D. Adler, J. Non-Cryst. Solids **90** (1987) 229
3. J. González-Hernández, B.S. Chao, D. Strand, S.R. Ovshinski, D. Pavlik, and P. Gasiorowski, Appl. Phys. Com. **11** (1992) 557
4. E. García-García, M. Yañez-Limón, Y. Vorobiev, F. Espinoza-Beltrán, and J. González-Hernández, Semicond. Phys., Quant. Electron. and Optoel. **1** (1998) 71
5. V.I. Odelevski, J. Techn. Phys. (USSR) **21** (1951) 673
6. T. Tien, G. Ottavian, K.N. Tu, J. Appl. Phys. **54** (1983) 7047
7. T.H. Geong, M.R. Kim, Hun Seo, Sang J. Kim, and Sang Y. Kim, J. Appl. Phys. **86** (1999) 774
8. J. Ross Macdonald, ed. *Impedance Spectroscopy*, John Wiley & Sons (1987)
9. W.A. Johnson, K.F. Mehl, Trans. Am. Inst. Mining Metall. Eng. **135** (1981) 315
10. J.W. Christian, *The Theory of Transformations in Metals and Alloys*, Pergamon Press, 1975
11. S.R. Ovshinski, *Proc. of 9th Symp. On Phase Change Record.* (Japan, November 1999)

The intensity and wavelength dependence for photo-induced crystallization process in amorphous GeSe$_2$

K. Sakai, K. Maeda, H. Yokoyama, T. Ikari

Department of Electrical and Electronic Engineering, Faculty of Engineering, Miyazaki University, 1-1 Gakuen Kibanadai-nishi, Miyazaki, 889-2192, Japan e-mail: sakai@pem.miyazaki-u.ac.jp

Abstract   The laser intensity and wavelength dependence for photo-induced crystallization process in amorphous GeSe$_2$ was studied by Raman and DSC measurements. We carried out the crystallization experiments with different wavelengths of laser irradiation. The photo-induced crystallization process is discussed in terms of the activation energy of the crystallization. We proposed a model of the decreasing of the activation energy for laser irradiated crystallization above the band-gap energy.

1 Introduction

It has been known that the amorphous chalcogenides easily change their structure by thermal and/or photo irradiation. The structural change occurs not only by a thermal effect but also by a photo-excitation due to a light absorption.

In this paper, we investigate the crystallization process with different wavelengths of laser irradiation in amorphous GeSe$_2$. And we discuss the difference between thermal and photo-induced crystallization processes in terms of the activation energy of the crystallization.

2 Experimental

The sample preparation of amorphous GeSe$_2$ has been shown in our previous paper [1]. The bulk samples, about $2\times2\times0.5$ mm^3, were annealed in an electric furnace with Ar gas flow between 380°C and 520°C for 0.2 to 500 h with laser irradiation. Light sources used were Ar-ion laser (457.9, 514.5 nm), diode-pumped YAG laser (532 nm) and He-Ne laser (632.8 nm). The total laser power of each laser was also adjusted almost the same which is about 20 mW. The full width at half maximum (FWHM) of each laser is about 0.4 mm except for He-Ne laser. The FWHM of He-Ne laser is about 0.2 mm. The laser beam vertically irradiated on the 2×2 mm^2 surface. The rise of the sample temperature by the laser irradiation was estimated to be less than 40°C from Raman shift.

The samples were quenched into water after annealing. Raman scattering measurements were performed at room temperature using He-Ne laser (3 mW) as a probing light to prevent further crystallization of the samples. we determined the form of the sample after annealing as either amorphous, LT or HT form, by the well-identified vibrational mode observed in the Raman spectra [2]. We can detect the crystallization area by our Raman measurement when a certain volume of the sample crystallizes by a laser irradiation.

The penetration depth of the irradiated laser light for a-GeSe$_2$ is 400 μm and 1 μm for 632.8 nm and 514.5 nm, respectively, calculated from the optical absorption spectra of amorphous and crystal GeSe$_2$ measured by Tronc et al [3]. Since the band-gap energy of the crystalline GeSe$_2$ (2.5eV) is higher than that of the amorphous form (2.2eV), the laser light penetrates deep into the sample after the crystallization [3].

3 Results

The time-temperature-transformation (T-T-T) diagram was plotted as annealing temperature vs. logarithmic annealing time. The T-T-T diagram of the crystallization process in dark condition was already reported in our previous paper [1].

In case of the laser irradiated crystallization, the boundaries of the phase change shift to lower temperature side about 20°C–30°C as compared with that in the dark condition. The slope of the boundary in photo-induced crystallization from amorphous state to LT-GeSe$_2$ became steeper than that in dark condition between 400°C and 450°C. When the laser power decreased to 8 mW, no effect of the laser irradiation was observed.

4 Discussion

The thermal and the photo-enhanced effect occur at the same time during the light absorption. The thermal effect of laser irradiated crystallization process causes a vertical shift of T-T-T diagram due to the sample temperature rises by light irradiation. When a sample was heated by the laser irradiation, the y-axis of T-T-T diagram should shift to higher temperature. But the shape of the T-T-T diagram did not change.

Isothermal annealing in the dark condition by the differential scanning calorimetry (DSC) was carried out to confirm the time of the crystallization. A boundary of amorphous and LT phases of T-T-T diagram in dark condition coincides with the exothermic peak by DSC measurements. The crystallization condition of the boundary in the T-T-T diagram corresponds to a certain volume fraction of crystal. Then, we can estimate the activation energy, E_a, supposing that the crystallization process is represented by the Arrhenius relation, that is, $t_c \sim \exp(-E_a/kT)$. Where k is Boltsman constant, and T is annealing temperature. t_c is the time when the

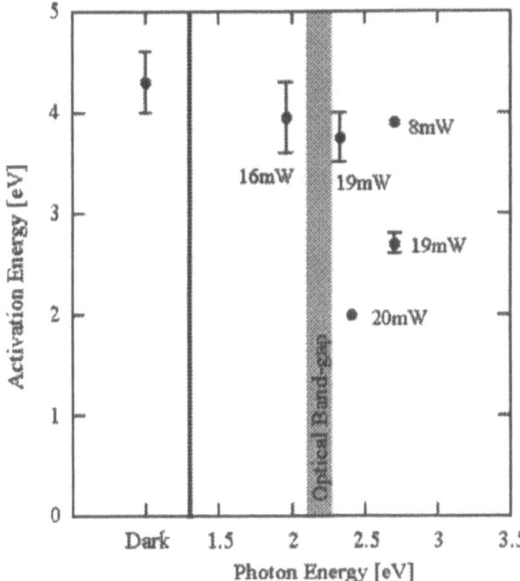

Fig. 1 Irradiated photon energy dependence of the activation energies for the laser crystallization.

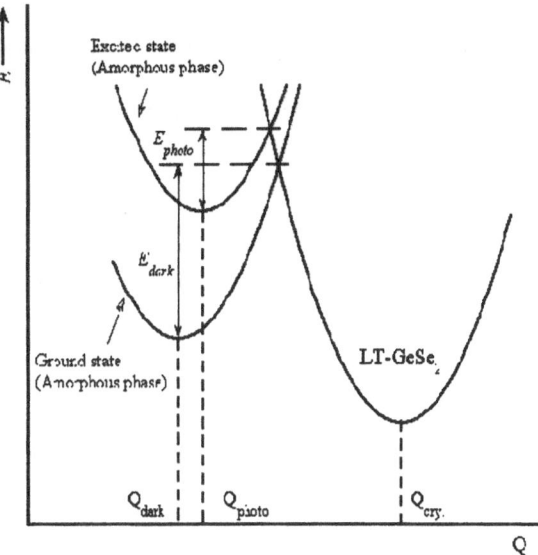

Fig. 2 The proposed configurational coordinate model for photo-induced crystallization process.

volume fraction of crystal becomes a certain value. The calculated activation energy of thermal crystallization was about 4.6 eV [1]. The activation energies under the laser illumination were also estimated in the same way.

The activation energies irradiated by different kinds of laser are shown in Fig. 1 as a function of the photon energy. It is known that the optical band-gap energy of a-GeSe₂ is 2.1–2.3 eV [3]. The activation energies of crystallization process under the irradiation below the band-gap energy were the same compared with that for the thermal crystallization process within the experimental error. For the laser irradiation above band-gap energy, the activation energy decreases. The activation process for different wavelength laser irradiation above band-gap energy may be the same as long as electrons are excited from valence to conduction band.

We use the well known configrational coordinate model to explain our experimental results as shown in Fig. 2. The vertical and horizontal axes show the total energy of the system and the configuration of the atoms in each phase, respectively. The amorphous state is excited by laser irradiation. Therefore, the thermal energy barrier to overcome to transform to the LT phase from this excited state (E_{photo}) becomes smaller than that from the ground state (E_{dark}).

The excited state is considered as follows. It is known that the lone-pair electrons occupy the top of the valence band. The lone-pair electrons will then be excited by laser irradiation just above band-gap energy for a-GeSe₂. The excited state of Se atom, that has a different atomic configuration from the stable state, is formed. When the photo-excited electrons relax their energy, they give the energy to phonons. This energy changes to the thermal energy. The bond which has lower bond energy is eas-

ily broken. The bond switching or the displacement of the atoms can be enhanced by this electron-phonon interaction. The crystallization process is accelerated in this way. Therefore, the crystallization progress at the lower activation process than that for thermal-induced crystallization.

5 Conclusion

We investigated the laser intensity and wavelength dependence for photo-induced crystallization process in amorphous GeSe₂. The differences for the growth condition of crystallization by different wavelengths of laser irradiation were evaluated in terms of the activation energy of crystallization. The photo-induced effect was observed as a decrease of the activation energy. The laser irradiation above optical band-gap energy enhances the crystallization process from amorphous state to LT form. We presented a model for the decreasing of the activation energy by laser irradiation.

Acknowledgements This work is partly supported by the Saneyoshi Scholarship Foundation.

References

1. K. Sakai, K. Yoshino, A. Fukuyama, H. Yokoyama, T. Ikari, and K. Maeda, Jpn. J. Appl. Phys. **39**, (2000) 1058.
2. K. Inoue, O. Matsuda, and K. Murase, Solid State Commun. **79**, (1991) 905.
3. P. Tronc, M. Bensoussan, A. Brenac, and C. Sebenne, Phys. Rev. B **8**, (1973) 5947.

Stiffness transition and connectivity in amorphous chalcogenide semiconductors

Y. Wang[1], **T. Nakaoka**[1], **M. Nakamura**[2], **O. Matsuda**[3], **K. Murase**[1]

[1] Department of Physics, Graduate School of Science, Osaka University, 1-1 Machikaneyama, Toyonaka 560-0043, Japan e-mail: `wang@phys.sci.osaka-u.ac.jp`
[2] KENS, High Energy Accelerator Research Organization (KEK), 1-1 Oho, Tsukuba, Ibaraki 305-0801, Japan
[3] Division of Applied Physics, Graduate School of Engineering, Hokkaido University, Sapporo 060-8628, Japan

Abstract Ge_xSe_{1-x} glasses have been investigated by Raman scattering. The ability of crystallization is obtained by studies of heating glasses from room temperature up to 1000 K. To understand the low-frequency dynamics of the network, we propose a fractal-structure view and a qualitative degree of the fragility. Transition with the connectivity for each studied property is universally related to the stiffness transition, derived from the concept of the constraint theory.

1 Introduction

Nobody doubts the importance of glassy materials for daily life. However, we do not have quantitative theory even for understanding the window (silicate) glass, which is the central to ceramics, although Kingery [1] has given an excellent qualitative guidance. Here, we demonstrate how the constraint theory of network glasses [2–4] succeeds in describing the critical changes of crystallization tendency, fractal behavior and fragility at stiffness transition in the typical covalent chalcogenide glasses, Ge_xSe_{1-x}, using Raman scattering technique. The mean coordination number $\langle r \rangle$ increases continuously with x to lead the stiffness transition at $\langle r \rangle = 2.4$ ($x=0.20$) where the floppy (underconstrained) glasses become rigid ones.

2 Experimental

Ge_xSe_{1-x} glasses are prepared by quenching the binary melts in ice water [5]. Raman spectra are measured in a back-scattering configuration. For avoiding light-induced events, a low power-density (less than 3 Wcm^{-2}) and long-wavelength (680–800 nm) probing light is applied. The scattered light is collected and analyzed with the triple grating polychromator (JOBIN YVON T64000) and CCD detector in polarized and depolarized configurations. Each spectrum is accumulated for 10 minutes at a fixed temperature. The temperature increase rate is among 3-5°C/minute between the successively measured temperatures.

3 Results and Discussions

3.1 Crystallization Tendency

From the spectral changes in Raman spectra with temperature [5], we obtain the glass-transition temperature, T_g ($\triangle$ in Fig. 1), which is in a good agreement with the result (squear) from DSC measurement. At crystallization temperature, T_c ($\bigcirc$), characteristic lines of crystalline (c-) Se or GeSe$_2$ appear in the spectra, and disappear at melting point, T_m ($\lozenge$). For $x<0.04$, as the small

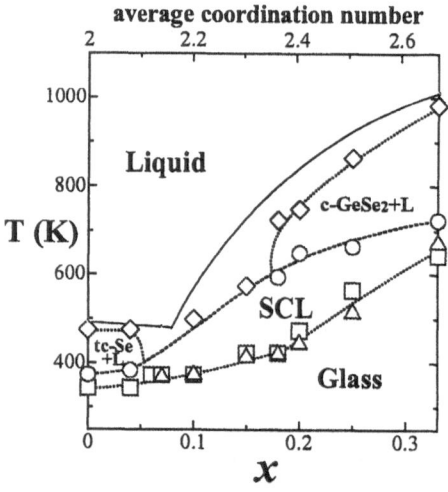

Fig. 1 Dynamic phase diagram for Ge-Se glass system.

amount of GeSe$_{4/2}$ tetrahedra fails to prevent the crystallization of Se, mixtures of crystalline c-Se and liquid Ge_xSe_{1-x} appear between the liquid and super-cooled liquid (SCL) states. On the other hand, for $x>0.18$, the large amount of GeSe$_{4/2}$ tetrahedra builds a medium-range structure topologically similar to the layered c-GeSe$_2$ to promote the crystallization of GeSe$_2$. Embryo of c-GeSe$_2$ constructs in SCL state for $x>0.10$, however, only those of $x>0.18$ evolve into nuclei and crystallize. Here, we define a crystallization ability by a ratio of crystallizable temperature range to SCL ones, $(T_m$-$T_g)/(T_c$-$T_g)$, as shown in Fig. 2. The crystallization tendency is derived from the ability of glass forming. Fig. 2 is very similar to the graph of glass forming difficulty summarized in Ref.[2]. The steep increase of the crystallization ability around x=0.20 should be treated as evidence of the proposed rigidity percolation.

3.2 Fractal behaviors

The reduced Raman spectra beyond Boson peak (BP) in 20-80 cm^{-1}, which qualitatively reflects the vibrational density of states $g(\omega)$, increase with ω described by $g(\omega) \propto \omega^{\tilde{d}-1}$ [6]. Spectral dimensionality, $\tilde{d}$, is shown in fig. 3 from fitting slopes of lines in the *log-log* plots of reduced intensity to ω. Here, we propose a fractal-structure picture to explain the fractal behavior qualitatively changing at the stiffness transition $\langle r \rangle = 2.4$. For the floppy glasses (x<0.20), the bending (#2) and stretch-

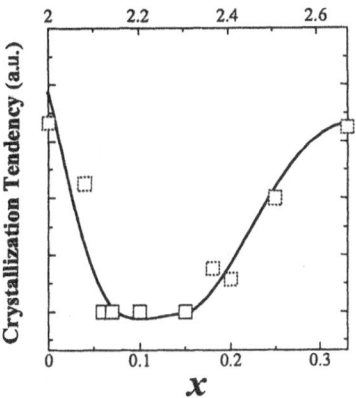

Fig. 2 Crystallization ability calculated form Fig. 1.

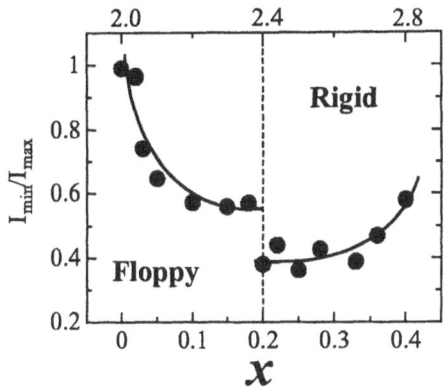

Fig. 4 $\langle r \rangle$ dependence of the intensity ratio regarded as a relative degree of fragility.

ing (#3) fracton coexist, however, only the stretching (#4) fracton is observed for the rigid glasses (x>0.20). The bending fracton vanishing in overconstrained glasses is assumed as that in the rigid region the elastic crossover length (l_c), which separates the bending and stretching behaviors, is longer than the structural crossover length, ξ. The vibrational excitation with wavelength longer than ξ belongs to extended modes as the network can be treated as a continuum. Thus, no bending fracton exists anymore. On the other hand, in the case of $\xi > l_c$, bending fracton contributes to the low frequency vibrational properties besides acoustic phonon and stretching fracton. Recently, Phillips suggests that [7]: as there is no longer any soft or "floppy" deformations available above $\langle r \rangle = 2.4$, which preserve both the bond-stretching and -bending constraints, it is no longer possible to separate and localize bending modes.

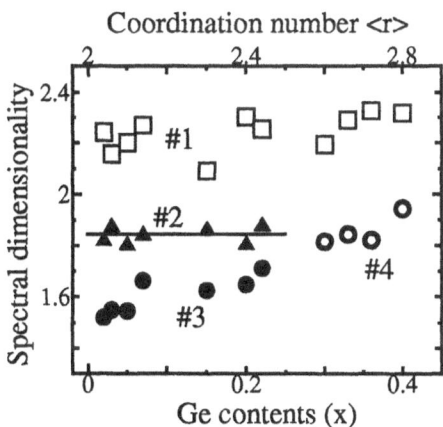

Fig. 3 The spectral dimensionalities obtained by fitting the reduced Raman spectra (from Ref.[6]).

3.3 Fragility

The low-energy Raman scattering comes from a quasielastic scattering, which is usually ascribed to some kind of relaxational motions, and the BP. Between the BP and relaxational contribution, a valley at 3–10 cm^{-1} is observed [8]. The intensity ratio of the valley to the BP, plotted in fig. 4, is qualitatively regarded as the degree

of the fragility. When the glass changes from floppy to rigid, the rapid drop of the intensity ratio points out a discontinuity of the fragility. The quasi-first-order transition suggests that the floppy-mode, related to the Se_n chains [9], exists in the underconstrained and disappears in the overconstrained glasses, since the fragility reflects the macroscopic properties of the network, mainly determined by the backbone structure. The slightly increase of fragility above $\langle r \rangle > 2.7$ is related to a nanophase separation of homopolar bonds that reduce the excessive constraints by breaking some chemical orders.

This work was supported by Grants-in-Aid for Scientific Research (B)(No. 09440117), Encouragement of Young Scientists (No. 11740173), and a grant for Scientific Research in the Priority Area "Cooperative Phenomena in Complex Liquids", from the Ministry of Education, Science and Culture (Japan). Y.W. acknowledges support from the Inamori Foundation.

References

1. W.D. Kingery, *Introduction to Ceramics* (Wiley, New York, 1976).
2. J.C. Phillips, J. Non-Cryst. Solids **34** (1979) 153 and M.F. Thorpe, J. Non-Cryst. Solids **57** (1983) 355.
3. P. Boolchand, X. Feng, D. Selvanathan and W.J. Bresser, *Rigidity Theory and Applications*, ed. by M.F. Thorpe and P.M. Duxbury, (Kluwer Academic, New York, 1999) p. 279.
4. K. Murase, *Insulating and Semiconducting Glasses*, ed. by P. Boolchand, (World Scientific, Singapore, 2000) p. 415.
5. Y. Wang, O. Matsuda, K. Inoue, O. Yamamuro, T. Matsuo, and K. Murase, J. Non-Cryst. Solids **232–234** (1998) 702.
6. M. Nakamura, O. Matsuda, and Murase. K. Phys. Rev. **B57** (1998) 10228 and M. Nakamura, O. Matsuda, Y. Wang, and K. Murase, Physica **B263&264** (1999) 330.
7. J.C. Phillips, the book listed in Ref. 3 p. 155.
8. Y. Wang, M. Nakamura, O. Matsuda, and K. Murase, Physica **B263&264** (199) 313 and Y. Wang, M. Nakamura, O. Matsuda, and K. Murase, J. of Non-Cryst. Solids **266–269** (2000) 872.
9. Y. Wang, M.K. Nakamura, T. Nakaoka, O. Matsuda, and K. Murase, to be published in J. of Non-Cryst. Solids.

Effect of pressure on electrical transport in electron doped perovskite manganite $Sr_{0.9}Ce_{0.1}MnO_3$

T. Eto[1], F. Honda[2], G. Oomi[2], A. Sundaresan[3]

[1] Research Center for Higher Education, Kyushu University, 4-2-1 Ropponmatsu, Fukuoka 810-8560, Japan

[2] Department of Physics, Kyushu University, 4-2-1 Ropponmatsu, Fukuoka 810-8560, Japan

[3] Electrotechnical Laboratory, 1-1-4 Umezono, Tsukuba, Ibaraki 305-8568, Japan

Abstract Temperature dependences of electrical resistivity and thermal expansion of $Sr_{0.9}Ce_{0.1}MnO_3$ have been studied at high pressure. It was found that the resistivity at low temperatures decreased and the band gap became narrow with increasing pressure. The Jahn-Teller transition temperature, T_{J-T}, largely decreased with an application of pressure at a rate of $dT_{J-T}/dp = -20(1)$ KGPa^{-1}.

1 Introduction

Investigation of perovskite manganites is one of the very topical subjects in not only in fundamental but also in applied magnetism due to the colossal magnetoresistance (CMR). Recently, it has been reported that an electron doping in $Sr_{1-x}Ce_xMnO_3$ by substituting Sr for Ce gives rise to decrease in the Jahn-Teller (J-T) distortion and antiferromagnetic (AF) interactions [1]. For $x=0.1$, it undergoes a first-order metal-insulator transition at 315 K which associated with a structural transition from cubic phase ($Pm3m$) to tetragonal phase ($I4/mcm$) due to the J-T effect, which stabilizes a chain like or C-type AF state with the spins aligned along the c-direction below $T_N \sim 290$ K. It was confirmed by resistivity measurement that the AF insulator state is independent of an applied magnetic field up to 7 T. This phenomenon is contrast to that of hole-doped manganites which shows metal-insulator transition and CMR effect [2]. Therefore, it is desired to make clear the magnetic interaction which are responsible for megnetoresitande in these materials.

In the present study, temperature dependence of electrical resistivity and thermal expansion in $Sr_{0.9}Ce_{0.1}MnO_3$ were measured as a function of pressure in order to obtain effect of pressure on these phase transitions.

2 Experimental

A polycrystalline sample $Sr_{0.9}Ce_{0.1}MnO_3$ was prepared by the solid reaction method from the stoichiometric mixtures of $SrCO_3$, CeO_2, and Mn_2O_3 at 1600 °C in air. The purity and the composition of the sample were checked by x-ray, neutron and electron diffraction measurements. The electrical resistivity and the thermal expansion, $\Delta L/L$, were measured by standard dc four-probe method and strain gauge method, respectively. Hydrostatic pressure up to 1.8 GPa was generated using Cu-Be piston-cylinder device and a 1:1 mixture of Fluorinert FC70 and FC77 as a pressure medium. The

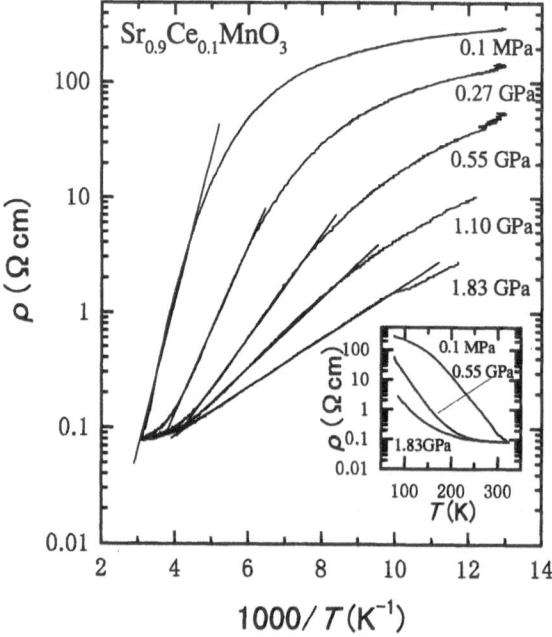

Fig. 1 ρ in a logarithmic scale as a function of $1000/T$ at various pressures. The inset shows whole part of $\log \rho$ vs. T.

measurement was performed in the temperature range of $77 \sim 320$ K.

3 Results and Discussion

The temperature dependence of the electrical resistivity $\rho(T)$ at various pressure up to 1.8 GPa is shown in Fig. 1. At ambient pressure above T_{J-T} (315 K), $\rho(T)$ behaves like metals due to occupation of degenerated e_g band. Below T_{J-T}, on the other hand, ρ increases logarithmically in the temperature range of about 220-300 K. This insulating nature in the tetragonal phase is consistent with the J-T distortion which results in localization of doping electrons [1]. With increasing pressure, ρ decreases in the whole temperature range investigated, the inclination of $\log \rho$ vs. $1000/T$ curves becomes small, and temperature range where the insulating behavior is observed spreads down to low temperatures. This indicates that the electronic state of $Sr_{0.9}Ce_{0.1}MnO_3$ becomes to be similar to that of the conventional semiconductor as pressure increases.

The activation energy for charge-carrier hopping (band

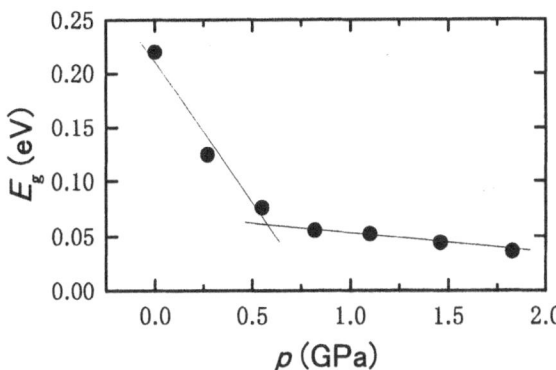

Fig. 2 Pressure dependence of E_g estimated from resistivity measurement. The lines are guides for the eye.

energy gap), E_g, was estimated by fitting the obtained data to the formula $\rho \sim \exp(E_g/2k_BT)$, where k_B is the Boltzmann constant. Fig. 2 indicates the pressure dependence of E_g. It is observed that E_g largely decreases with applying pressure at a rate of approximately $dE_g/dp \sim -200$ meV/GPa up to about 0.7 GPa. The effect of pressure on E_g would be resulted from an increase of kinetic energy, that is, an broadning of band width of the e_g electron. The decreasing of E_g by pressure is suppressed above 0.7 GPa.

Thermal expansion was also examined under high pressure in order to investigated the effect of pressure on T_{J-T} and T_N. The results are shown in Fig. 3; T_{J-T} is defined as an intermediate point between two extrapolated lines illustrated in the figure and T_N as a slightly inflection point of $\Delta L/L$-T curve. A hysteritic thermal behavior in $\Delta L/L$ was observed near T_{J-T} which indicates the J-T transition is a type of first order. With increasing pressure, T_{J-T} drastically decreases at a rate of $dT_{J-T}/dp = -20(1)$ KGPa^{-1}. Large negative dT_{J-T}/dp indicates an instability of the Jahn-Teller distortion of the MnO_6 octahedron under high pressure, which may lead to disappearance of the AF ground state below $T_N \sim 290$ K; actually, T_N cannot be detected within the experimental accuracy of the present study.

Large MR effect for the electron-doped manganites has been recently reported in the AF (G-type) insulator $CaMnO_3$ by substituting rare earths or Bi for Ca [3]. However, the magnitude of MR for the electron-doped system is smaller than that for hole-doped one. Above difference in the MR would be originated from the both electrical transport machanism. For the hole-doped manganites with Mn^{3+} and Mn^{4+} mixed system, the electristic nature is governed by double-exchange model [4] which describes the influence of magnetism on the electrical resistivity and the magnitude of the MR depends on the width of the e_g conduction band. In the electron-doped one, the C-type AF insulator caused by J-T ordering is almost insensitive to magnetic field, whereas the MR was found in A and CE-type insulator [5], though the datails were not made clear. The high pressure exper-

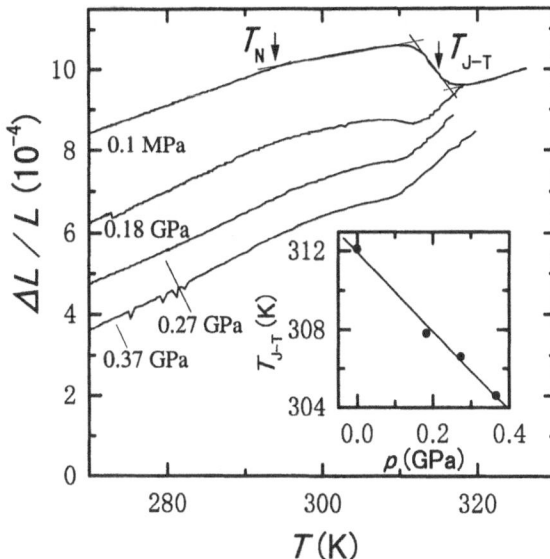

Fig. 3 Temperature dependence of $\Delta L/L$ under high pressure near T_{J-T} and T_N.

iment with magnetic field now planned would be useful to investigate the effect of conduction band width on the MR effect.

References

1. A. Sundaresan, J.L. Tholence, A. Maignan, C. Martin, M. Hervieu and B. Raveau, Eur. Phy. Journal **B14**, (2000) 430.

2. for example, A.P. Ramirez, J. Phys.:Condens. Matter. **9**, (1997) 8171.

3. for example, T. Murakami, D. Shindo, H. Chiba, M. Kikuchi, and Y. Syono, Phys. Rev. **B55**, (1997) 15043.

4. for examples, P.W. Anderson and H. Hasegawa, Phys. Rev. **100**, (1955) 675; P. -G. de Gennes, Phys. Rev. **118**, (1960) 141.

5. H. Kuwahara, Y. Tomioka, Y. Tokura, Science **270**, (1995) 961.

Ferromagnetism in Diluted Magnetic Semiconductors

Jürgen König[1,2]*, Hsiu-Hau Lin[3], Allan H. MacDonald[1,2]

[1] Department of Physics, Indiana University, Bloomington, IN 47405, USA e-mail: `koenig@gibbs.physics.indiana.edu`
[2] Department of Physics, The University of Texas at Austin, Austin, TX 78712, USA
[3] Department of Physics, National Tsing-Hua University, Hsinchu 300, Taiwan

Abstract We present a theory of carrier-induced ferromagnetism in diluted magnetic semiconductors (III,Mn)V which allows for arbitrary itinerant-carrier spin polarization and dynamic correlations. Both ingredients, which are absent in an RKKY description, are essential in identifying the system's elementary excitations and describing their properties. We find a Stoner continuum and two branches of collective modes (spin waves), and discuss their properties.

1 Introduction

The recent discovery of carrier-induced ferromagnetism [1,2] in doped diluted magnetic semiconductors (DMS) has generated intense interest, mainly because of the prospect of developing devices which combine information processing and storage functionalities in one material [3–9]. Critical temperatures T_c exceeding 100K have been realized [3] by using epitaxial growth at low temperature to introduce a high concentration N_{Mn} of randomly distributed Mn^{2+} ions in GaAs systems with a high hole density p. The tendency toward ferromagnetism and trends in the observed T_c's have been explained within a mean-field picture [10–13] in which uniform itinerant-carrier spin polarization mediates a long-range ferromagnetic interaction between the Mn^{2+} ions with spin $S = 5/2$.

We present here a theory [14] which accounts for dynamic correlations in the ordered state and derive the dispersion of the elementary spin excitations.

2 Independent spin-wave theory

The model we study provides an accurate description of undoped Mn based zincblende DMS's. Magnetic ions with spin $S = 5/2$ at positions $\mathbf{R}_I$ are antiferromagnetically coupled to valence-band carriers described by envelope functions,

$$H = H_0 + J_{\mathrm{pd}} \int d^3 r \, \mathbf{S}(\mathbf{r}) \cdot \mathbf{s}(\mathbf{r}), \qquad (1)$$

where $\mathbf{S}(\mathbf{r}) = \sum_I \mathbf{S}_I \delta(\mathbf{r} - \mathbf{R}_I)$ is the impurity-spin density. By coarse graining, this can be replaced by a smooth function. The itinerant-carrier spin density is expressed in terms of carrier field operators by the relation $\mathbf{s}(\mathbf{r}) = \frac{1}{2} \sum_{\sigma\sigma'} \Psi_\sigma^\dagger(\mathbf{r}) \boldsymbol{\tau}_{\sigma\sigma'} \Psi_{\sigma'}(\mathbf{r})$ where $\boldsymbol{\tau}$ is the vector of Pauli spin matrices. H_0 includes the valence-band envelope-function Hamiltonian. To simplify the present discussion we use a generic single-band model with quadratic dis-

persion and neglect interactions between the free carriers as well as the antiferromagnetic Mn-Mn interaction.

The model we use here is, thus, related to those for colossal magnetoresistance materials [15] and identical to those for dense Kondo systems, which simplify when the itinerant-carrier density p is much smaller than the magnetic ion density N_{Mn} [16]. The fact that $p/N_{\mathrm{Mn}} \ll 1$ in ferromagnetic-semiconductor materials is essential to their ferromagnetism. Similar models have been used for ferromagnetism induced by magnetic ions in nearly ferromagnetic metals such as palladium [17].

To derive the spin excitations we use the following procedure (for details see Ref. [14]). We represent the impurity spins in terms of bosonic degrees of freedom (Holstein-Primakoff bosons). Since the Hamiltonian is bilinear in fermionic fields, we can integrate out the itinerant carriers and arrive at an effective description for the localized spins only. The expansion of the effective action up to quadratic order in the boson fields yields as a kernel the inverse of the spin-wave propagator,

$$D^{-1}(\mathbf{k}, \nu_m) = -i\nu_m + J_{\mathrm{pd}} p \xi$$
$$+ \frac{N_{\mathrm{Mn}} J_{\mathrm{pd}}^2 S}{2\beta V} \sum_{n,\mathbf{q}} G_\uparrow^{MF}(\mathbf{q}, \omega_n) G_\downarrow^{MF}(\mathbf{q} + \mathbf{k}, \omega_n + \nu_m) \quad (2)$$

with the mean-field itinerant-carrier Green's function $G_\sigma^{MF}(\mathbf{q}, \omega_n) = -[i\omega_n - (\epsilon_{\mathbf{q}} + \sigma\Delta/2 - \mu)]^{-1}$. Here, $\Delta = N_{\mathrm{Mn}} J_{\mathrm{pd}} S$ is the zero-temperature spin-splitting gap, μ is the chemical potential, $\epsilon_{\mathbf{q}} = \hbar^2 q^2/(2m^*)$ with effective mass m^*, and $\xi = (p_\downarrow - p_\uparrow)/p$ is the fractional free-carrier spin polarization. A Debye cutoff $k_D = (6\pi^2 N_{\mathrm{Mn}})^{1/3}$ accounts for the correct number of modes, $|\mathbf{k}| \le k_D$.

If we neglect the spin polarization of the itinerant carriers in Eq. (2) and evaluate the static limit of the resulting spin-wave propagator, our results simplify to a description which assumes an RKKY interaction between magnetic ions. In the following we use as typical parameters $m^* = 0.5 m_e$, $J_{\mathrm{pd}} = 0.15\mathrm{eVnm}^3$, $N_{\mathrm{Mn}} = 1\mathrm{nm}^{-3}$, and $p = 0.1\mathrm{nm}^{-3}$, where m_e is the free-electron mass. For this set of parameters the mean-field itinerant-carrier system is fully polarized at $T = 0$, and the RKKY picture obviously breaks down.

3 Results

We obtain the spectral density of the spin-fluctuation propagator by analytical continuation, $i\nu_m \to \Omega + i0^+$ and $A(\mathbf{k}, \Omega) = \mathrm{Im}\, D(\mathbf{k}, \Omega)/\pi$. There are three different types of spin excitations.

* *Present address:* Department of Physics, RLM 5.208, The University of Texas at Austin, Austin, TX 78712, USA

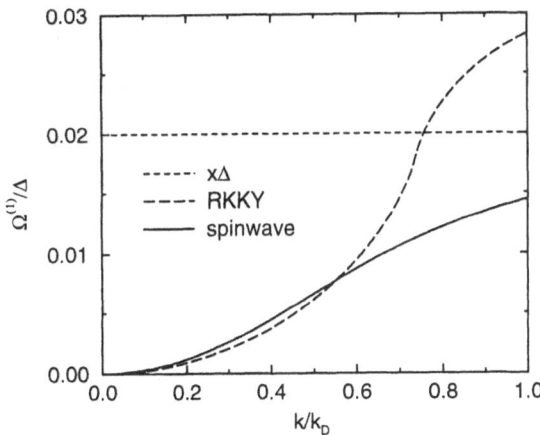

Fig. 1 Spin-wave dispersion. The short wavelength limit is the mean-field result $x\Delta$. For comparison, we show also the result obtained from an RKKY picture.

i) Our model has a gapless Goldstone-mode branch (see Fig. 1) reflecting the spontaneous breaking of rotational symmetry [18]. For $\Delta > \epsilon_F$ the $T = 0$ dispersion of the collective modes, where ϵ_F is the Fermi energy of the majority-spin band, is

$$\Omega_k^{(1)} = \frac{x}{1-x}\epsilon_k \left(1 - \frac{4\epsilon_F}{5\Delta}\right) + \mathcal{O}(k^4) \qquad (3)$$

and $\Omega_k^{(1)} = x\Delta$ for small and large momenta, respectively. At short wavelengths we obtain the mean-field result $x\Delta$, the spin splitting of a magnetic ion in the effective field produced by fully spin-polarized itinerant carriers. Note that the itinerant-carrier and magnetic-ion mean-field spin splittings differ by a factor of $x = p/(2N_{Mn}S) \ll 1$. At long wavelengths the magnon dispersion in an isotropic ferromagnet is proportional the spin stiffness ρ divided by the magnetization M. In the adiabatic limit [15,19], $\epsilon_F \ll \Delta$, our long-wavelength result reflects a spin stiffness due entirely to the increase in kinetic energy of a fully spin-polarized band when the orientation has a spatial dependence, $\rho = p\hbar^2/(4m^*)$, and a magnetization M which has opposing contributions from magnetic ions and itinerant carriers, $M = N_{Mn}S - p/2 = N_{Mn}S(1-x)$. In this limit, the mean-field critical temperature and the spin stiffness have opposite dependences on the itinerant-carrier mass.

ii) We find a continuum of Stoner spin-flip particle-hole excitations. They correspond to flipping a single spin in the itinerant-carrier system and, therefore, occur at much larger energies $\sim \Delta$ (see Fig. 2).

iii) We find additional collective modes associated primarily with the itinerant-carrier system at energies below the Stoner continuum (see Fig. 2). At $\Delta > \epsilon_F$ we get $-\Omega_{k=0}^{(2)} = \Delta(1-x)$.

In summary we presented a theory of ferromagnetism in DMS's. We determined the dispersion of the spin excitations in a regime where the RKKY picture breaks

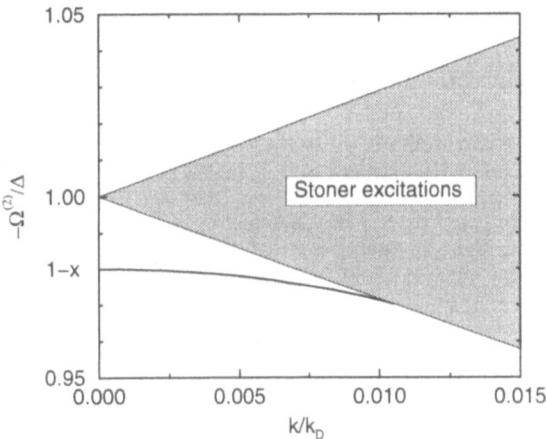

Fig. 2 Stoner excitations and collective modes in the free-carrier system. In an RKKY picture these modes are absent.

down. We predicted a collective mode in addition to the Goldstone branch and the Stoner continuum.

References

1. T. Story, R.R. Gałązka, R.B. Frankel, and P.A. Wolff, Phys. Rev. Lett. **56**, 777 (1986).
2. H. Ohno et al., Phys. Rev. Lett. **68**, 2664 (1992); H. Ohno et al., Appl. Phys. Lett. **69**, 363 (1996).
3. H. Ohno, J. of Magnetism and Magnetic Materials **200**, 110 (1999); Science **281**, 951 (1998).
4. G.A. Prinz, Science **282**, 1660 (1998).
5. T. Hayashi et al., J. Appl. Phys. **83**, 6551 (1998).
6. T.M. Pekarek, B.C. Crooker, I. Miotkowski, and A.K. Ramdas, J. Appl. Phys. **83**, 6557 (1998).
7. A. Van Esch et al., J. Phys. Condensed Matter **9**, L361 (1997); A. Van Esch et al., Phys. Rev. B **56**, 13103 (1997).
8. A. Oiwa et al., Phys. Rev. B **59**, 5826 (1999); F. Matsukura, H. Ohno, A. Shen, and Y. Sugawara, Phys. Rev. B **57**, R2037 (1998).
9. B. Beschoten et al., Phys. Rev. Lett. **83**, 3073 (1999).
10. T. Dietl, A. Haury, and Y.M. d'Aubigné, Phys. Rev. B **55**, R3347 (1997).
11. M. Takahashi, Phys. Rev. B **56**, 7389 (1997).
12. T. Jungwirth, W.A. Atkinson, B.H. Lee, and A.H. MacDonald, Phys. Rev. B **59**, 9818 (1999); B.H. Lee, T. Jungwirth, and A.H. MacDonald, Phys. Rev. B **61**, 15606 (2000); M. Abolfath, J. Brum, T. Jungwirth, and A.H. MacDonald, cond-mat/0006093.
13. T. Dietl et al., Science **287**, 1019 (2000); T. Dietl et al., cond-mat/0007190.
14. J. König, H. Lin, and A.H. MacDonald, Phys. Rev. Lett. **84**, 5628 (2000).
15. A.J. Millis, P.B. Littlewood, and B.I. Shraiman, Phys. Rev. Lett. **74**, 5144 (1995).
16. M. Sigrist, H.Tsunetsugu, and K. Ueda, Phys. Rev. Lett. **67**, 2211 (1991).
17. S. Doniach and E.P. Wohlfarth, Proc. R. Soc. London Ser. A **296**, 442 (1967).
18. Because of spin-orbit coupling in the valence bands, the spin-wave spectrum in realistic models will have a strain- and disorder-dependent gap.
19. E.L. Nagaev, Phys. Rev. B **58**, 827 (1998).

Ferromagnetic Interactions in p- and n-type II-VI Diluted Magnetic Semiconductors

T. Andrearczyk[1], J. Jaroszyński[1], M. Sawicki[1], Le Van Khoi[1], T. Dietl[1], D. Ferrand[2], C. Bourgognon[2], J. Cibert[2], S. Tatarenko[2], T. Fukumura[3], Zhengwu Jin[3], H. Koinuma[3], M. Kawasaki[3]

[1] Institute of Physics, Polish Academy of Sciences, al. Lotników 32/46, PL 02668 Warszawa, Poland; e-mail: dietl@ifpan.edu.pl

[2] Laboratoire de Spectrométrie Physique, Université Joseph Fourier Grenoble 1, CNRS (UMR 55 88) BP P87, 38402 Saint-Martin d'Heres Cedex, France

[3] Dept. of Innovative and Eng. Mater. Tokyo Inst. of Technology 4259 Nagatsuta, Midori-ku, Yokohama 226-8502, Japan

Abstract Measurements of magnetization, magnetoresistance and anomalous Hall effect indicate the ferromagnetic transition induced by the holes in doped p-$Zn_{1-x}Mn_x$Te. The experimental findings can be described in the framework of the mean-field model of the carrier induced ferromagnetism. The same model predicts that for n-type $Zn_{1-x}Mn_x$O:Al the temperature of ferromagnetic ordering is well below 1 K. Indeed, below 150 mK we observe hysteresis of magnetoresistance which could be a manifestation of carrier induced ferromagnetism.

1 Introduction

The possibility of controlling ferromagnetic interactions between the localized spins by the carriers [1–5], have recently renewed the interest in diluted magnetic semiconductors (DMS) [6]. Up to now, the carrier-induced ferromagnetism has been observed in lead-salt materials, $Pb_{1-x-y}Sn_yMn_x$Te [1], and in MBE-grown semiconductors with the zinc-blende structure: $In_{1-x}Mn_x$As [7] and $Ga_{1-x}Mn_x$As [2], as well as in p-doped $Cd_{1-x}Mn_x$Te quantum wells [4] and $Zn_{1-x}Mn_x$Te epilayers [5].

Here, in order to identify experimentally the dominant phenomena controlling the carrier-mediated ferromagnetic interactions in doped semiconductors, we have carried out millikelvin studies of II-VI diluted magnetic semiconductors by means of transport and magnetization measurements. In contrast to III-V compounds, the Mn ions in II-VI materials are electrically neutral, allowing variation of both the magnetic ions and the charge carriers concentrations independently. We extend our studies to n-type, heavily doped $Zn_{1-x}Mn_x$O:Al.

2 Samples

P-type $Zn_{1-x}Mn_x$Te:P samples were obtained by the Bridgman method in Warsaw while p-$Zn_{1-x}Mn_x$Te:N epilayers were deposited by MBE in Grenoble. Finally, n-type $Zn_{1-x}Mn_x$O:Al samples were grown in Tokyo by pulsed laser deposition technique. It should be stressed that in both materials record-high effective doping level above 10^{20} cm^{-3} was achieved.

3 Experimental results

Results of SQUID magnetometry of $Zn_{0.962}Mn_{0.038}$Te:N sample are shown in Fig. 1, which presents the temperature dependence of inverse magnetic susceptibility χ^{-1}, and examples of magnetization cycles taken at low tem-

peratures. The χ^{-1} vs. T plot lets us to determine the Curie-Weiss temperature T_{CW} and the effective concentration of localized spins x_{eff}. The value of x_{eff} is smaller than the total concentration of Mn ions x since the nearest-neighbor (n-n) Mn pairs are blocked antiparallel due to the intrinsic antiferromagnetic superexchange [6]. Antiferromagnetic interaction of distant pairs can be described by empirical parameter $T_{AF} > 0$. While purely antiferromagnetic behavior is observed on samples with low hole concentrations, the positive value of T_{CW} detected in heavily doped samples suggests that the Mn ions couple ferromagnetically when doping level exceeds $p \geq 10^{19}$cm^{-3}. As a further confirmation of this finding, characteristic ferromagnetic-like hysteresis are observed at temperatures below T_{CW}, as shown in Fig. 1b.

In addition to direct magnetization measurements, information on magnetic ordering has been obtained from magnetotransport studies, since spin effects are known to affect transport phenomena of doped semiconductors [8]. In particular, a large spin-orbit energy in the valence band leads to a sizable magnitude of the extraordinary Hall effect in p-type samples. Indeed, as it is seen in Fig. 2, below ferromagnetic transition temperature we observe clearly hysteresis on both Hall and longitudinal resistances.

It turned out that experimentally found ferromagnetic temperatures T_F can be reproduced without adjustable parameters with reasonable good accuracy. The

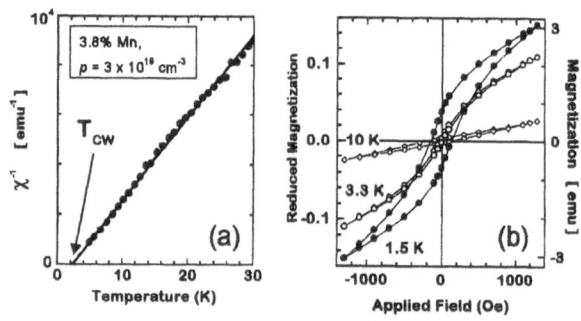

Fig. 1 (a) Inverse magnetic susceptibility for p-$Zn_{0.962}Mn_{0.038}$Te on the insulator side of the metal-to-insulator transition. Solid line shows the linear fit, which serves us to determine the effective Mn content x_{eff} and the Curie-Weiss temperature T_{CW}. (b) Magnetization cycles at various temperatures displayed also in the units of the saturation value.

theory is based on the RKKY model or its continuous-medium limit, the Zener model [10], with the effects of the competing antiferromagnetic interaction taken into account. According to that model $T_F \propto x_{\rm eff}\beta^2\tilde{\chi}_h$, where β stands for p-d exchange integral and $\tilde{\chi}_h$ denotes spin susceptibility of holes, which is proportional to their effective mass. In order to compare experimental and theoretical results for samples with various x and p, we introduce a normalized value of the ferromagnetic temperature, $\tilde{T}_F = (T_{CW} + T_{AF})/10^2 x_{\rm eff}$, where the corresponding values of T_{AF} are calculated from the experimentally determined equation: $T_{AF}[{\rm K}] = 58x - 150x^2$ [9]. According to the Zener model the normalized ferromagnetic temperature, as defined above, does not depend on the Mn content.

N-type materials have approximately four times weaker s-d exchange coupling between carriers and magnetic ions as well as lower effective mass. Accordingly the model predicts the ferromagnetic ordering to occur at much lower temperatures, providing that intrinsic antiferromagnetic superexchange does not suppress it at all. Expected T_{CW} for n-type $Zn_{0.97}Mn_{0.03}O:Al$ with the electron concentration $n = 1.4\times10^{20}$ cm^{-3} is well below 1 K. Indeed, as shown in Fig. 4, below $T = 160$ mK we observe hysteresis of the longitudinal resistance, as it is observed in case of p-type material below T_{CW}. This hysteresis could be a manifestation of carrier induced ferromagnetism. Low temperature SQUID measurements are under to verify this conjecture.

Acknowledgments

The work was supported under KBN Grant No. 2-PB03B-02417 and Foundation for Polish Science.

References

1. T. Story et al., Phys. Rev. Lett. **56**, (1986) 777; P. Lazarczyk et al., J. Magn. Magn. Mater. **169**, (1997) 151, and references cited therein.
2. F. Matskura, H. Ohno, A. Shen and Y. Sugawara, Phys. Rev. B 57, (1998) R2037; H. Ohno, Science **281**, (1998) 951, and references cited therein.
3. T. Dietl, A. Haury, and Y. Merle d'Aubigné, Phys. Rev. B 55, (1997) R3347; T. Dietl et al., Mater. Sci. Engin. B **63**, (1999) 103.
4. A. Haury et al., Phys. Rev. Lett. **79**, (1997) 511.
5. D. Ferrand et al., J. Crystal Growth, **214/215**, (2000) 387, see also http://arXiv.org/abs/cond-mat/0007502.
6. J.K. Furdyna, J. Appl. Phys. **64**, (1988) R29; T. Dietl, in: *Handbook on Semiconductors*, edited by T.S. Moss (North-Holland, Amsterdam, 1994) vol. 3b, p. 1251.
7. H. Ohno et al., Phys. Rev. Lett. **68**, (1992) 2664.
8. M. Sawicki et al., Phys. Rev. Lett. **56,** (1986) 508.
9. G. Barilero et al., Phys. Rev. B **32**, (1985) 5144; Y Shapira et al., Phys. Rev B **34**, (1986) 4187; J.P. Lascaray et al., Phys. Rev. B **35**, (1987) 6860; A. Twardowski et al., Solid State Commun. **59**, (1986) 199.
10. T. Dietl et al., Science **287**, (2000) 1019.

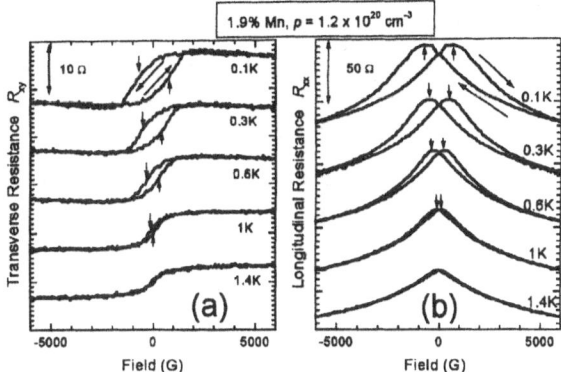

Fig. 2 Low temperature magnetoresistance (a) and Hall resistance (b) for metallic p-$Zn_{0.981}Mn_{0.019}Te$. The vertical arrows mark the width of the hysteresis loops.

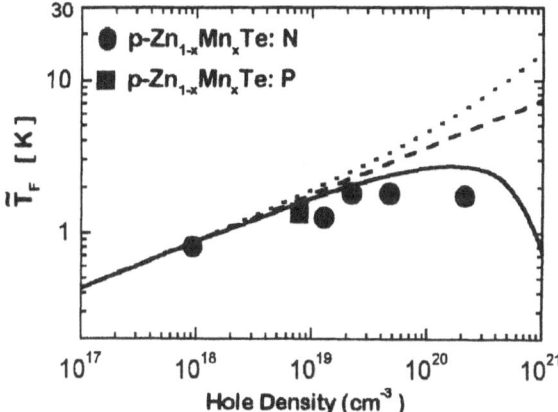

Fig. 3 Experimentally found (symbols) and calculated (lines) normalized ferromagnetic temperature normalized to the effective Mn content $x_{\rm eff}$: $\tilde{T}_F = (T_{CW} + T_{AF})/10^2 x_{\rm eff}$, versus the hole concentration. Dashed line: Zener model with the hole dispersion calculated from the 4 × 4 Luttinger spherical model for the Γ_8 band; dotted line: Zener model including the coupling between the Γ_8 and Γ_7 bands (6 × 6 Luttinger model); solid line: the RKKY and 6 × 6 Luttinger model for $x_{\rm eff} = 0.015$, taking into account the effect of the antiferromagnetic interactions on statistical distribution of unpaired Mn spins.

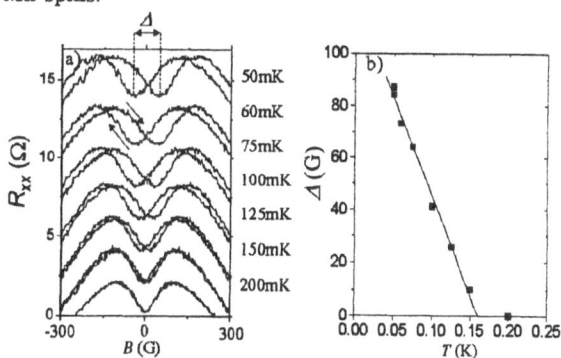

Fig. 4 (a) Diagonal resistance of n-$Zn_{0.97}Mn_{0.03}O:Al$ sample, with electron concentration $n = 1.4\times10^{20}$ cm^{-3} at selected temperatures, showing hysteresis of resistance. Traces were shifted vertically for clarity. (b) The width of hysteresis loops as a function of temperature. These data point to a transition in the vicinity of 150 mK.

Theoretical predictions for transparent ferromagnets with transition metal atom doped ZnO

K. Sato[1], H. Katayama-Yoshida[1]

The Institute of Scientific and Industrial Research, Osaka University, 8-1 Mihogaoka, Ibaraki, Osaka 567-0047, Japan. e-mail: ksato@cmp.sanken.osaka-u.ac.jp

Abstract   Ferromagnetism in a $3d$ transition metal atom doped ZnO was investigated by *ab initio* electronic structure calculations based on the local density approximation. It was shown that the ferromagnetic state was stable in V, Cr, Fe, Co or Ni, doped ZnO without any additional carrier dopants, and in Mn doped ZnO we found the carrier induced ferromagnetism by hole doping.

1 Introduction

Since the successful synthesize of new functional ferromagnetic materials by doping Mn atoms in the III-V semiconductor GaAs and InAs [1], diluted magnetic semiconductors (DMS) have been intensively studied with non-equilibrium crystal growth techniques. However, practical functional magnetic materials which have the Curie temperature (T_C) higher than room temperature and large magnetization has still been under exploring. In this paper, material design for a ZnO-based new DMS is proposed based on *ab initio* electronic structure calculations. Recently ZnO attracts much attention because of its cheapness, abundance and harmonisity with environment. Besides ZnO has wide band gap energy of 3.3 eV and large exiton binding energy of 60 meV, so it is one of the most promising substances for optoelectronic materials. Fukumura *et al.* [2] and Jin *et al.* [3] experimentally showed that $3d$ transition metal atoms were soluble up to several ten % in ZnO. Moreover, Joseph *et al.* successfully prepared p-type ZnO thin films [4] by using the codoping method. Therefore it is worth to investigate the carrier induced ferromagnetism in ZnO-based DMS.

2 Calculation Methods

Calculations were performed by the Korringa - Kohn - Rostoker (KKR) method based on the local density approximation [5] with the parameterization by Morruzi, Janak and Williams [6]. Independent ones of 64 k sampling points in the first Brillouin zone were calculated. The systems with impurities were treated by the super cell method. The super cell which consists four primitive unit cells of the wurzite structure was prepared and two of eight Zn atoms were substituted by two transition metal atoms. This substitution leads to fixed impurity concentration of 25%. Hole carriers were introduced into this system by substituting O atoms with N atoms, and electron carriers were introduced by putting Ga atoms instead of Zn atoms. The lattice constants of $a = 3.27$ Å, $c = 5.26$ Å and $u = 0.345$ were employed [2,7]. Muffin tin radii of 0.940 Å and 0.875 Å were adopted for cations and anions, respectively.

3 Results and Discussions

First, chemical trend of the magnetic state of (Zn, TM)O was investigated, where TM is one of the $3d$ transition metal atoms. Fig. 1 shows energy differences between anti-ferromagnetic state and ferromagnetic state for respective TM atom doped ZnO systems. Any additional carrier doping were not performed. As shown in the figure, the ferromagnetic state is stable for V, Cr, Fe, Co or Ni doped ZnO. The energy differences are as large as 15 mRy and almost saturated magnetic moments are expected for them. Jin *et al.* [3] showed that the $3d$ transition metal atom doped ZnO was transparent with visible ray. Therefore, they are the candidates for the transparent ferromagnet with high T_C and large magnetization.

When the super cell does not contain any carrier dopants, carrier concentration in (Zn, Mn)O is 0% because the substitution of Zn with Mn does not bring any carriers. At this point, the anti ferromagnetic configuration is stable than the ferromagnetic one. When the electron configuration of magnetic ions differs from d^5 as changing the valence number, the ferromagnetic state is stabilized. It seems that resulting mobile carriers participate to stabilization of the ferromagnetic state. This suggests the carrier induced ferromagnetism in (Zn, Mn)O. So, next we examined stability of the ferromagnetic state in (Zn, Mn)O as a function of the carrier concentration.

Fig. 2 shows the total energy differences in (Zn, Mn)O systems as a function of the carrier concentration. As shown in the figure, the carrier induced ferromagnetism was observed in (Zn, Mn)O. As increasing the hole concentration by the substitution of O with N, the ferromagnetic state is suddenly stabilized. In these calculations O atoms neighboring the Mn atoms were first substituted by N atoms. Roughly speaking the transition to the ferromagnetic state from the anti ferromagnetic one occurs at about 6% of hole carriers. The energy difference reaches to as large as 30 mRy at the maximum. Moreover, it was reported that N doped p-type ZnO was transparent with visible ray [4], so hole doped (Zn, Mn)O will also be a candidate for a high T_C transparent ferromagnet with large magnetization. If this material design is realized, it will have great impact on industrial application in magneto optical devices, because in this system its magnetic state will be controllable by several % change in the hole density.

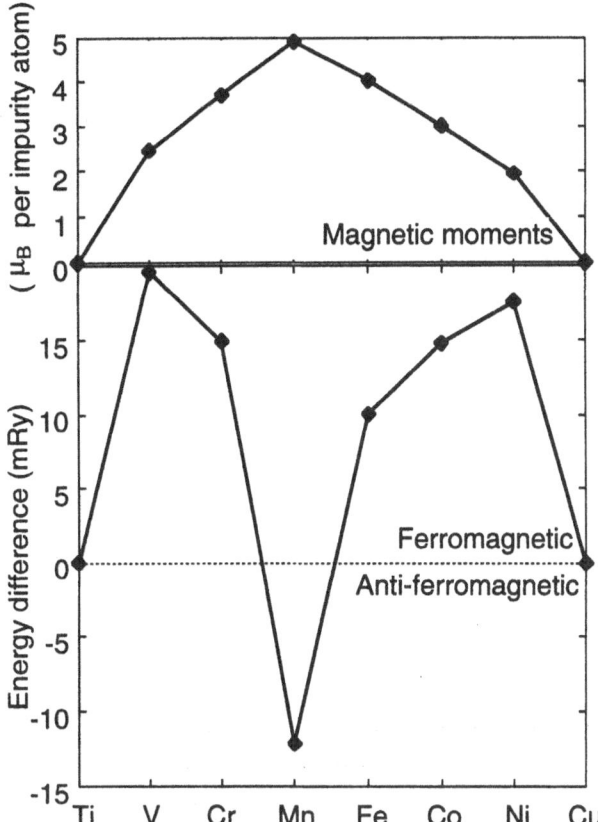

Fig. 1 Chemical trend of the magnetic states for 3*d* transition metal atom doped ZnO. The vertical axis shows the total energy difference per unit cell between the ferromagnetic and the anti-ferromagnetic state. Positive energy difference means that the ferromagnetic state is more stable than the anti-ferromagnetic one. Total magnetic moments per one transition metal atom are also shown.

Unlike the cases of hole doping, the ferromagnetic state did not become stable under the electron doping by the substitution of Zn with Ga. From the point of the double exchange mechanism, the ferromagnetic states might become stable under electron doping. Precise analysis of calculated DOS showed that the hole states mainly consisted of Mn 3*d* states, while the electron carriers were introduced into the host conduction band and did not occupy Mn 3*d* states. These observation suggests that the ferromagnetic state is stabilized by the itinerant *d* holes, and the stabilization of the ferromagnetic state is occurred due to the double exchange mechanism as was pointed out by Akai in the III-V based DMS [8].

Realistic material design for high T_C transparent ferromagnetic DMS was given in the present paper, however, we restricted our calculations to only one super cell and the arrangement of the dopants in the super cell was rather artificial. In real situation, TM atoms and dopant impurities are randomly distributed in ZnO. As is well known, the KKR method can be adapted to treat a disordered alloy within the framework of the coherent potential approximation (KKR-CPA) [8]. To confirm

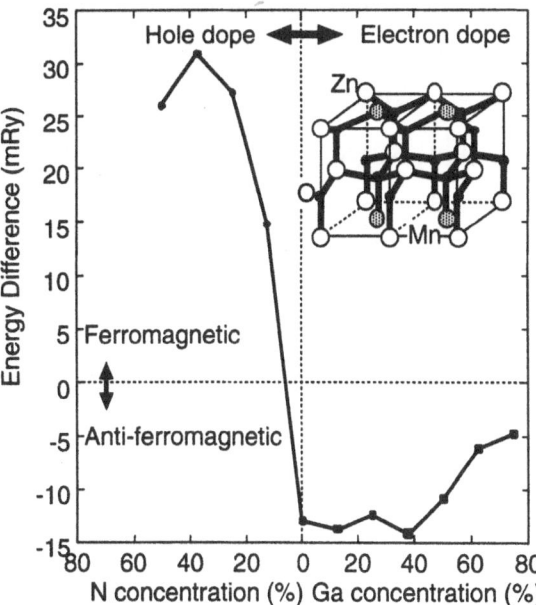

Fig. 2 Total energy difference between the ferromagnetic state and the anti-ferromagnetic state as a function of carrier concentration. The horizontal axis shows N dopant and Ga dopant concentration. Positive energy difference means that the ferromagnetic states is stable. The inset shows the simple orthorhombic super cell used in the present calculations.

the present material design the KKR-CPA calculation is strongly encouraged.

Acknowledgements This work was partially supported by a JSPS Research for the Future Program in the area of Atomic-Scale Surface and Interface Dynamics, JST-ACT, a Grant-In-Aid for Scientific Research on the Priority Area and Sanken-COE from the Ministry of Education, Science, Sports and Culture. We thank Prof. H. Akai for fruitful discussions and providing us with his KKR-CPA band structure calculation package.

References

1. H. Munekata, H. Ohno, S. von Molnar, Armin Segmüller, L. L. Chang and L. Esaki: Phys. Rev. Lett. **63** (1989) 1849.
2. T. Fukumura, Zhengwu Jin, A. Ohtomo, H. Koinuma and M. Kawasaki: Appl. Phys. Lett. **75** (1999) 3366.
3. Zhengwu Jin, M. Murakami, T. Fukumura, Y. Matsumoto, A. Ohtomo, M. Kawasaki and H. Koinuma, J. Cryst. Growth **214/215** (2000) 55.
4. M. Joseph, H. Tabata and T. Kawai: Jpn. J. Appl. Phys. **38** (1999) L1205.
5. H. Akai, M. Akai, S. Blügel, B. Drittler, H. Ebert, K. Terakura, R. Zeller and P. H. Dederichs: Prog. Theor. Phys. Supplement **101** (1990) 11.
6. V. L. Moruzzi, J. F. Janak and A. R. Williams: *Calculated Electronic Properties of Metals* (Pergamon, U.S.A, 1978)p. 11.
7. R. W. G. Wyckoff: *Crystal Structures* (Krieger, 1986) 2nd. ed., p. 112.
8. H. Akai: Phys. Rev. Lett. **81** (1998) 3002.

sp-d exchange interaction in GaMnAs investigated by resonant Kerr effect under high magnetic field

F. Michelini[1], N. Nègre[1], G. Fishman[2], M. Goiran[3], J. Sadowski[4], E. Vanelle[1], and S. Askénasy[2]

[1] Laboratoire de Physique de la Matière Condensée, UMR, I.N.S.A., 135 avenue de Rangueil, 31077 Toulouse Cedex, France
[2] Institut d'Electronique Fondamentale, bât. 220, Université Paris Sud, 91405 Orsay Cedex, France
[3] Laboratoire National des Champs Magnétiques Pulsés, CNRS, avenue de Rangueil, 31077 Toulouse Cedex, France
[4] MAX-lab, Lund University, P. O. Box 118, S-221 00 Lund, Sweden

Abstract   We report on the first observation of resonant Magneto-Optical Kerr Effect under magnetic fields up to 30 T, performed on p-type GaAs and $Ga_{1-x}Mn_xAs$ (x=0.03) thin layers on (001) GaAs substrates. The observed oscillations of the Kerr rotation angle correspond to resonant interband transitions between the valence bands (Γ_8 and Γ_7) and the conduction bands (Γ_6). We calculated the Landau level states and the allowed interband transition probabilities dealing with the axial approximation in an eight-band **k.p** model. In the case of GaAs, the numerical results fairly reproduce the spectrum oscillations. In the case of GaMnAs, the additional features observed on the spectra are related to the band structure modifications induced by the *s,p-d* exchange interaction. This analysis allows us to give an estimated value of $N_0\beta$, the exchange integral between the localized magnetic level and the valence bands.

1 Introduction

The magneto-optical experiments became relevant to investigate the strong *s,p-d* exchange interaction in the Diluted Magnetic Semiconductors (DMS) based on III-V compounds. In this paper, we present the first observations of Magneto-Optical Kerr Effect under high magnetic field on bulk GaAs and $Ga_{1-x}Mn_xAs$ epilayers.

Interband magnetospectroscopy calls for Landau level calculations. Parabolic band approximation can be assumed [1]. Platero and Altarelli took advantages of the Luttinger formalism within the axial or spherical approximation [2]. In the present case, the magnitude of the applied fields needs to consider an eight-band **k.p** model. We propose a general method of Landau level calculations formulating the axial approximation in the framework of the eight-band **k.p** theory. Therefore, the developed formalism allows one to perform the same kind of calculation in any case of applied field direction.

2 Experiments

(Ga,Mn)As epilayers have been grown by low temperature Molecular Beam Epitaxy. Their structural characterization and their magnetic properties are dealt with in Ref. [3]. Two samples of bulk GaAs and $Ga_{0.97}Mn_{0.03}As$/GaAs epilayer (400 nm) have been studied by Magneto-Optical Kerr Effect under high magnetic field at 5 K. The laser excitation energy is 1.960 eV. The $x = 0.03$-thin layer exhibits ferromagnetic behavior characterized by a Curie temperature of 45 K, and a saturation field of

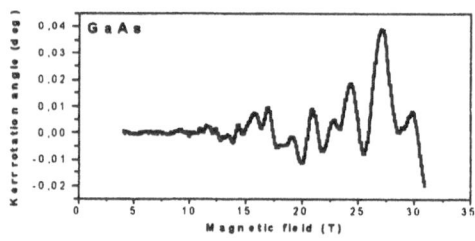

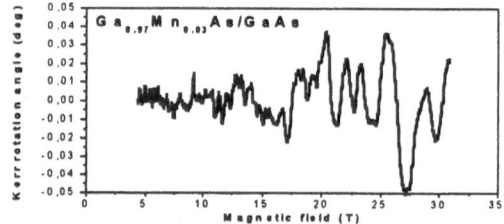

Fig. 1 Oscillatory behavior of the Kerr rotation angle in the case of bulk GaAs and $Ga_{0.97}Mn_{0.03}As$/GaAs epilayer (400 nm) samples.

15 T. Figure 1 displays the Kerr rotation angle θ_K upon the applied magnetic field. The oscillatory behavior of θ_K is associated to several resonance frequencies determined by Fourier transformation. In the case of GaAs, three frequencies are identified: 42 ± 2 T, 182 ± 4 T, and 276 ± 4 T. In the case of $Ga_{0.97}Mn_{0.03}As$/GaAs, an additionnal frequency of 141 ± 5 T is observed on the Fourier transform spectrum. The oscillatory behavior of θ_K has to be linked to the resonant direct transitions occuring between the Landau levels of the valence bands and the Landau levels of the conduction bands.

3 Model

We propose a general method of Landau level calculation using an eight-band **k.p** model within the axial approximation along the applied field direction. The tensor formalism introduced by Baldereschi, Lipari and Altarelli [4,5] in the case of the Γ_8, is generalized to the eight-band **k.p** model from the Γ_5 Luttinger Hamiltonian [6]. The details of these calculations will be published elsewhere. Assuming the axial approximation, the Landau level determination turns out possible in any case of applied field direction. In the case of (Ga,Mn)As, the *s,p-d* exchange interaction leads to additionnal terms in

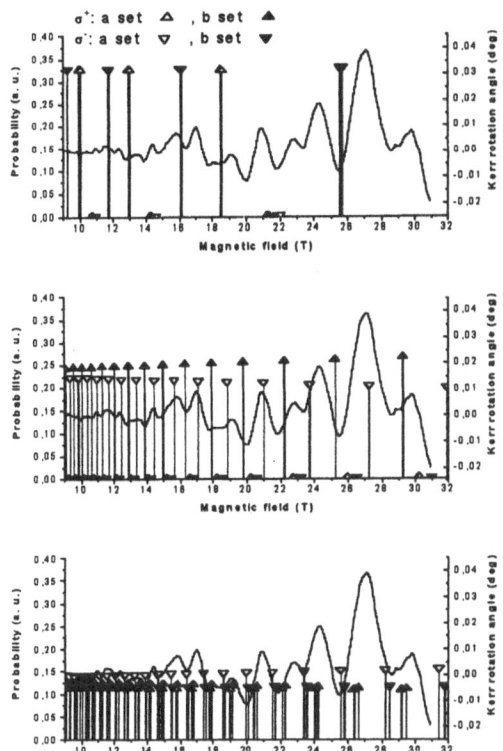

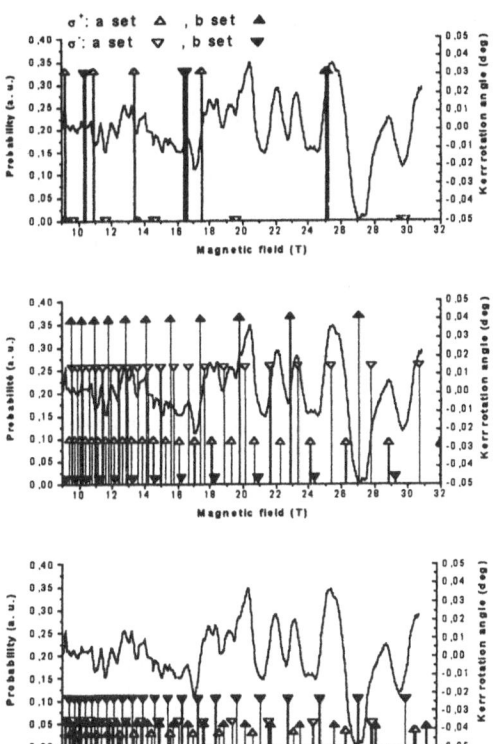

Fig. 2 Resonant allowed-transitions occuring between the conduction band and the split-off, light hole and heavy hole bands respectively. For the $k_z = 0$-heavy (-light) hole bands, the $\left|\pm\frac{1}{2}\right\rangle$ $\left(\left|\pm\frac{3}{2}\right\rangle\right)$ components remain greater than the $\left|\mp\frac{3}{2}\right\rangle$ $\left(\left|\mp\frac{1}{2}\right\rangle\right)$ ones.

Fig. 3 Resonant allowed-transitions occuring between the conduction band and the split-off, light hole and heavy hole bands respectively. $N_0\beta = 3$ eV and $N_0\alpha = 0$ eV is assumed. The resonance frequencies are 28 and 58 T concerned with the split-off bands, 146 and 291 T with the light hole bands, and 191 and 285 T with the heavy hole bands.

the Hamiltonian. According to the field magnitude, the calculations assume direct transitions in the center of the Brillouin zone. Therefore, the effective-mass Hamiltonian may be written in terms of two 4×4 matrices, related to the a and b sets of eigenstates [7].

4 Results and Discussions

We calculated the Landau level energies and the associated interband transition probabilities. In the case of GaAs, the resonant allowed-transitions between the Landau levels of the conduction bands and the split-off, light hole and heavy hole valence bands, are given in Fig. 2. The calculated peaks coincide with the different oscillations of the experimental spectrum. Indeed, the calculation of the resonance frequencies gives the following values: 43 T, 183 T, and 274 T respectively. Thus, an excellent agreement with the experimental data is obtained. In the case of $Ga_{0.97}Mn_{0.03}As/GaAs$, the exchange-induced energy splitting involves a frequency splitting and modifies the associated transition probabilities as shown in Fig. 3. $N_0\alpha$ $(N_0\beta)$ is the exchange integral between the localized magnetic level and the conduction (valence) band. Assuming $N_0\alpha = 0$ eV, the resonant transitions depend on the $N_0\beta$ value. Anyway, the highest probability is obtained for the resonant transitions occurring between the Landau levels of the con-

duction band and the light hole valence band. Therefore, the additional frequency of 141 T implies these interband transitions. The spectrum analysis allows us to estimate $N_0\beta$. We obtained $N_0\beta = -3.0 \pm 0.6$ eV.

5 Conclusion

We show that Magneto-Optical Kerr Effect under high magnetic field is a relevant technique to investigate the electronic properties of semiconductors. The estimated value of $N_0\beta$ is among the greatest ones obtained by magneto-optical measurements. We now intend to ensure the efficiency of this experimental method studying different alloys.

References

1. J. Szczytko *et al.*, Phys. Rev. B **59**, 12935 (1999).
2. G. Platero and M. Altarelli, Phys. Rev. B **39**, 3758 (1989).
3. J. Sadowski *et al.*, Acta Physica Polonica A **94**, 509 (1998).
4. A. Baldereschi and N. O. Lipari, Phys. Rev. B **8**, 2697 (1973).
5. N. O. Lipari and M. Altarelli, Phys. Rev. B **15**, 4883 (1977).
6. J. M. Luttinger, Phys. Rev. **102**, 1030 (1956).
7. C. R. Pidgeon and R. N. Brown, Phys. Rev. **146**, 575 (1966).

Giant tunability of excitonic photoluminescence transitions in antiferromagnetic EuTe epilayers

W. Heiss, G. Prechtl, G. Springholz,

Institut für Halbleiterphysik, Johannes Kepler Universität Linz, Altenbergerstr. 69, A-4040 Linz, Austria.

Abstract EuTe epilayers with high purity are investigated by magneto-optical spectroscopy. Low temperature photoluminescence experiments reveal narrow, exciton like emission peaks exhibiting a Stokes shift of 300 meV. These emission peaks can be tuned over a giant range of 160 meV by applying magnetic fields between zero and 5 T. This large tunability in magnetic fields is caused by the formation of magnetic polarons due d-f exchange interactions.

Diluted magnetic semiconductors (DMS) have attracted tremendous interest due to their unique physical properties, like the appearance of the huge exciton spin (Zeeman) splitting and the formation of magnetic polarons [1]. Both effects are related to exchange interactions between free carriers and localized magnetic moments. The large spin splitting leads to a shrinkage of the fundamental absorption with magnetic field and hence to a red shift of the exciton photoluminescence (PL) transitions. The polaron effect results in a Stokes shift of the PL lines in respect to the absorption edge which shrinks with increasing applied field in the case of DMS [2].

Anomalous large magnetic field induced red shifts of the PL transition energy as well as a large Stokes shift of the PL transitions [3] have also been observed in Eu-monochalcogenides (EuXc). These semiconductor-materials are classical Heisenberg magnets due to exchange interactions between the localized magnetic moments of the Eu^{2+} ions, resulting from their half filled 4f levels (spin 7/2). Because of the opposite sign of the nearest and next nearest neighbor exchange integral, EuTe is antiferromagnetic of type-II structure while the other EuXc exhibit ferromagnetic or meta magnetic behavior. The valence band is formed by the p-orbitals of the chalcogenes, while the conduction band is formed by the 5d and 6s orbitals of the Eu ions. Between the filled valence band and the empty conduction bands energy levels of the half filled 4f orbitals of the Eu^{2+} ions are located. Therefore, photoexcitation creates excitons consisting of strongly localized 4f-holes and d-electrons in the conduction band. Within the Bohr radius of the photoexcited electrons the local magnetic moments become ferromagnetically aligned due to d-f exchange interactions, thus MPs are formed.

In this work, we perform magneto-optical spectroscopy on novel antiferromagnetic EuTe layers grown by molecular beam epitaxy. We observe exceptional properties of excitonic PL transitions. In particular these transitions can be linearly tuned by an applied magnetic field over a giant spectral range.

The samples were grown by molecular beam epitaxy on BaF_2 (111) substrates using elemental effusion sources for Eu and Te. The epilayers show a very smooth surface, exhibiting only monolayer steps. Dislocations with a density of 10^8 cm^{-2} are formed due to the 6% lattice mismatch between the EuTe epilayer and the BaF_2 substrate. We obtained a rather good crystalline quality of the fully strain relaxed samples. This was proven by X-ray diffraction spectroscopy giving rocking curves of the (222) Bragg reflections with a peak width of 200 arcsec.

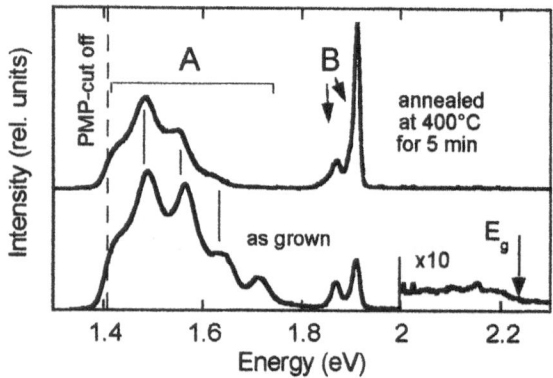

Fig. 1 Photoluminescence spectrum of the a 3 μm thick EuTe epilayer excited at 1.7 K by a 488 nm line of an Ar-ion laser. The arrow indicates the band gap energy of EuTe determined from transmission experiments.

Our EuTe samples exhibit unique luminescence properties: In contrast to all previous work [3] on Eu chalcogenides, at low temperatures we find two distinct luminescence bands, as shown in Fig. 1. One band (band A) consists of two narrow, exciton like photoluminescence peaks at transition energies around 1.9 eV with a full width at half maximum of about 10 meV. This emission band appears only in our epitaxial, high qualitiy samples whereas in previous investigations luminescence was only observed around 1.5 eV. There, a much broader, about 200 meV wide, emission band (band B) is observed. After annealing the sample at 400°C for 5 minutes the luminescence intensity of band B decreases substantially while the intensity of the excitonic luminescence increases about three times (see Fig. 1). This behavior indicates that the luminescence within band B is caused by self activated emission associated with deep centers. This assignment is in contrast to the previous interpretations, but it is confirmed by the fact that in epilayers heavily doped by BiTe only emission from band B is observed.

At 1.7 K, the photoluminescence of the excitonic peaks is 300 meV Stokes shifted in respect to the measured band gap at 2.25 eV. With increasing temperature up to the Néel temperature (T_N=9.6 K), the Stokes shift slightly increases while it decreases for temperatures above T_N.

The most striking feature of the excitonic emission lines is their huge red-shift in applied magnetic fields. Although EuTe is an antiferromagnetic semiconductor, we find that the excitonic luminescence lines in band A can be linearly tuned with a constant slope of the transition energy of − 34meV/T. This linear tuning results in a red shift of ,e. g., 160 meV at 5 T. This giant tuning range is to the best of our knowledge the largest ever observed in a semiconductor. The magnetic field dependence of the PL shift corresponds to an effective g factor of 1140, which is somewhat smaller as compared to the highest values in diluted magnetic semiconductors (DMS) of about g_{eff}=1500 at low fields. For EuTe, however, the g factor is essentially independent of applied magnetic field (H), whereas in DMS g_{eff} strongly decreases with increasing H. In contrast to the situation in DMS, the tunability of the photoluminescence transition is much larger than that of the absorption edge. The latter is due to the exciton Zeeman splitting and it causes a red shift of only 18 meV at 5 T.

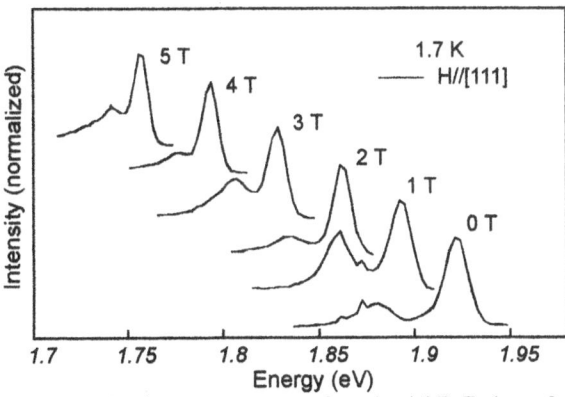

Fig. 2 Photoluminescence spectra of a epitaxial EuTe layer for various magnetic field values. The excitonic lines can be tuned to lower energies by more than 160 meV by applying fields between zero and 5 T.

The magnetic field dependence as well as the temperature dependence of the luminescence transition energies within band A are in good qualitative agreement with results of recent Monte Carlo simulations by M. Umehara [4]. In these numerical calculations the spontaneous magnetic ordering of the 4f spins and the MP formation induced by d-f exchange interactions is taken into account, as well as the

Coulomb interaction between photoexcited electron hole pairs and the electron (hole)-optical-phonon interactions. The calculations predict a linear tunability of the PL lines with magnetic field. Furthermore, the Stokes shift is calculated to be temperature independent up to the magnetic phase transition and to shrink with rising temperature above T_N. Thus, the expected dependencies agree with our experimental observations leading to the conclusion that the large Stokes shift and the sensitive dependence of the PL transition energy on magnetic field is dominated by the formation of magnetic polarons.

At zero field, two distinct excitonic PL-peaks are observed: A high peak at about 1.92 eV accompanied by a smaller one around 1.87 eV (see Fig. 2). The energy splitting between these peaks decreases with increasing magnetic field as well as with increasing temperature, whereas the line width of the satellite peak as function of temperature shows a maximum at the Neel temperature. A comparison of these experimental dependencies with that of luminescence transitons in antiferromagnetic insulators [5] indicates that the satellite peak at 1.87 eV represents a magnon side-band of the main PL peak.

We have demonstrated for the first time exciton like photoluminescence transitions in epitaxial EuTe samples with high purity. These excitonic transitions show a Stokes shift of 300 meV and can be tuned over a giant spectral range by an external magnetic field. The giant tunability of these lines provides interesting and new possibilities for applications in magneto-optical devices like widely magnetic field tunable lasers, detectors, and modulators. Furthermore, epitaxial EuTe could be used, alternatively to DMS, for the development of "spintronic" devices like spin filters or spin transistors as key elements for future solid state quantum computers.

References

1. for a review see, e. g.; J. K. Furdyna and J. Kossut, *Diluted Magnetic Semiconductors*, Semiconductor and Semimetals **25**, ed. R. K. Willardson and A. C. Beer, (Academic Press, 1988)
2. G. Mackh et al., Phys. Rev. B **50** (1994) 14069
3. R. Akimoto, M. Kobayashi, T. Suzuki, J. Phys. Soc. Jpn. **63** (1994) 4616 and references given in this article
4. M. Umehara, Phys. Rev. B **52** (1995) 8140
5. T. Tsuboi, W. Kleemann, Phys. Rev. B **27** (1983) 3762

Acknowledgements

This work is supported by the FWF, and the GME, in Vienna and by the Austrian Academy of Sciences.

Hybridization-induced exchange interaction between the conduction band electrons and Mn ions in diluted magnetic semiconductors

A. K. Bhattacharjee[1], J. Pérez-Conde[2]

[1] Laboratoire de Physique des Solides, CNRS - Université Paris-Sud, 91405 Orsay, France e-mail: `anadi@lps.u-psud.fr`
[2] Departamento de Física, Universidad Pública de Navarra, E-31006, Pamplona, Spain

Abstract   Our theoretical study based on the sp^3s^* tight-binding model for the band structure shows that the $\mathbf{k} - d$ hybridization between the conduction band states and the $3d$ orbitals of Mn, that vanishes at $k = 0$, increases with k and leads to a rather rapidly increasing antiferromagnetic contribution to the $sp - d$ exchange as the band energy approaches the electron affinity of the half-full $3d$ shell. A quantitative comparison indicates that the recently invoked "small gap" $\mathbf{k} \cdot \mathbf{p}$ model drastically overestimates the effect.

1 Introduction

The characteristic spin-related properties of the diluted magnetic semiconductors (DMS's) arise from the strong $sp - d$ exchange interactions between the band electrons and the transition-metal ions. In the most extensively studied group, Mn-based II-VI DMS's, the exchange parameter at the conduction band (cb) minimum $N_0\alpha$ is positive (ferromagnetic) and that at the valence band (vb) maximum $N_0\beta$ is negative (antiferromagnetic), the magnitude of the latter being 4 to 5 times that of the former. These experimental results were successfully interpreted years ago[1] as follows: α represents the Coulomb potential exchange integral and β is dominantly the effective exchange (Schrieffer-Wolff) arising from the $\mathbf{k} - d$ hybridization. This basic model of kinetic exchange for the valence band has been generalized in recent years to other transition-metal ions where orbital exchange also comes into play. However, until recently the kinetic exchange contribution to the cb was neglected, because the $\mathbf{k} - d$ hybridization vanishes at $k = 0$ by symmetry and is expected to be small even for finite k. Merkulov et al.[2] pointed out the possible importance of this contribution away from the zone center. In fact, they interpreted the strong confinement-induced reduction of both spin-flip Raman shift and exciton Zeeman splitting in CdMnTe/CdMgMnTe quantum wells by invoking a highly k-dependent kinetic exchange in the cb, calculated within the "small gap" $\mathbf{k.p}$ approximation. Here we present a theory based on the semiempirical sp^3s^* tight-binding (TB) model that accounts for the main features of the host semiconductor band structure. Both the potential and kinetic parts of the $sp - d$ exchange are calculated and the net value presented as a function of k for both cb and vb. Numerical results for $Cd_{1-x}Mn_xTe$ are compared with those from the $\mathbf{k} \cdot \mathbf{p}$ model.

2 Theory

In the case of the S-state ion Mn, the Kondo exchange Hamiltonian for the interaction between a band electron and a set of localized spins can be written in the Bloch function basis as

$$\langle \nu'\mathbf{k}'\sigma'|H|\nu\mathbf{k}\sigma\rangle = -\delta_{\nu'\nu}\sum_i e^{i(\mathbf{k}-\mathbf{k}')\cdot\mathbf{R}_i} J^{\nu}_{\mathbf{k}'\mathbf{k}}(\mathbf{s}\cdot\mathbf{S}_i)_{\sigma'\sigma} \quad (1)$$

Here ν, $\mathbf{k}$ and σ denote the band index, wavevector and the z-component of spin ($\mathbf{s}$), respectively. $\mathbf{R}_i$ and $\mathbf{S}_i$ represent the position and spin of the i-th Mn ion. The exchange parameter is a sum of the potential and kinetic parts.

$$J^{\nu(pot)}_{\mathbf{k}'\mathbf{k}} = \frac{2}{5}\sum_m \langle dm(1)\nu\mathbf{k}'(2)|v(|\mathbf{r}_{12}|)|dm(2)\nu\mathbf{k}(1)\rangle \quad (2)$$

where v is the sreened Coulomb potential between two electrons. The generalized Schrieffer-Wolff formula for the kinetic exchange is

$$J^{\nu(kin)}_{\mathbf{k}'\mathbf{k}} = -V^*_{\mathbf{k}d}V_{\mathbf{k}'d}\left[\frac{1}{E^+_{\mathbf{k}'}} + \frac{1}{E^+_{\mathbf{k}}} + \frac{1}{E^-_{\mathbf{k}'}} + \frac{1}{E^-_{\mathbf{k}}}\right] \quad (3)$$

where $E^-_{\mathbf{k}}$ ($E^+_{\mathbf{k}}$) is the excitation energy for virtual emission from (absorption onto) the d shell; in the usual one-electron language, $E^-_{\mathbf{k}} = E_\nu(\mathbf{k}) - \varepsilon_d$ and $E^+_{\mathbf{k}} = \varepsilon_d + U_{eff} - E_\nu(\mathbf{k})$. Also,

$$V^*_{\mathbf{k}d}V_{\mathbf{k}'d} \equiv \frac{1}{2S}\sum_m V^*_{\nu\mathbf{k}m}V_{\nu\mathbf{k}'m}. \quad (4)$$

$V_{\nu\mathbf{k}m} = \langle \nu\mathbf{k}|H_0|dm\rangle$ is a hybridization matrix element.

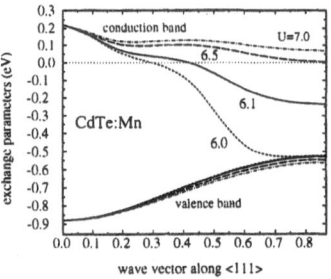

Fig. 1 Exchange parameters ($NJ^{c,\nu}_{\mathbf{k}\mathbf{k}}$) versus wave vector (in units of $2\pi/a$) along $< 111 >$ with $E_v - \varepsilon_d = 3.4$ eV and different values of U_{eff} (eV) for $Cd_{1-x}Mn_xTe$.

In the TB model the energy bands and the wave functions are obtained by diagonalizing the complex Hamiltonian matrix in the subspace of the Bloch sums:

$$|nb\mathbf{k}\rangle = \frac{1}{\sqrt{N}}\sum e^{i\mathbf{k}\cdot(\mathbf{R}+\mathbf{v}_b)}|nb\mathbf{R}\rangle. \quad (5)$$

Here $\mathbf{R}$ denotes the fcc lattice points, N being their total number in the crystal and $b = a(c)$ for anion (cation), while $n = s, p_x, p_y, p_z, s^*$ correspond to the atomic orbitals. The calculation of the $\mathbf{k} - d$ hybridization matrix elements is then straightforward. Consistently with our nearest-neighbor TB model, the $3d$ orbitals of Mn are assumed to couple only to the sp^3 orbitals of the four nearest-neighbor anions. The Slater-Koster interatomic matrix elements $sd\sigma$, $pd\pi$ and $pd\sigma$ are related through the Harrison ratios, leaving $pd\sigma$ as the only adjustable hybridization parameter. The potential exchange is approximated by the on-site integrals alone, so that $J_{\mathbf{kk}}^{\nu(pot)} = \frac{2}{N} \sum_n | < nck|\nu\mathbf{k} > |^2 J_{nd}$, where J_{nd} denotes the average intra-atomic exchange integrals J_{sd}, J_{pd} for the s and p orbitals, respectively.

In the 3-band $\mathbf{k} \cdot \mathbf{p}$ model the periodic part of the cb Bloch function for $\mathbf{k} \parallel \langle 001 \rangle$ is given by $|ck \uparrow\rangle = a(k)S| \uparrow\rangle + b(k)[-\frac{1}{\sqrt{2}}(X + iY)]| \downarrow\rangle + c(k)Z| \uparrow\rangle$ in the usual notations. Thus $|V_{kd}^c|^2 = |V_{0d}^v|^2|c(k)|^2$. Here the superscripts c and v denote cb and vb respectively. Note that there is no contribution $\propto |b(k)|^2$, in contradiction with the formula used in Ref.[2]. The k-dependence of the potential exchange is $J_{kk}^{c(pot)}/J_{00}^{c(pot)} = |a(k)|^2$.

3 Results and discussion

Numerical results for CdTe:Mn are presented in Fig. 1. The k-dependence of the net exchange $J_{\mathbf{kk}}^{\nu}$ with $\mathbf{k} \parallel < 111 >$ for the cb and vb are shown for different values of U_{eff} between 6.0 and 7.0 eV (see Ref.[4] for a comment on the values). The TB model parameters for CdTe are from Ref.[8]. And $E_v - \varepsilon_d = 3.4$ eV[4]. The Mn exchange integral J_{sd} was chosen to fit $N_0\alpha = 0.22$ eV. It turns out to be 0.13 eV, in good agreement with typical estimated values [5]; also, following Ref.[5], we assume $J_{pd} = \frac{4}{5}J_{sd}$. Note that, as expected, there is only a small potential exchange in the vb: ~ 0.05 eV. For each value of U_{eff}, in order to fit $N_0\beta = -0.88$ eV, we have a slightly different value of $V_{pd\sigma}$; they are however close to -1 eV, as usually assumed. Notice the drastic variation of the cb exchange for $U_{eff} < 6.5$ eV. The negative hybridization-induced (kinetic) contribution is rapidly enhanced as the cb energy approaches $\varepsilon_d + U_{eff}$, the virtual capture energy. However, one should be careful about the validity of the second-order perturbation formula (Eq.(3)) in this range of energy: the energy difference must be larger than the width of the resonance; in our case this is roughly satisfied if the magnitude of the resulting kinetic exchange is smaller than 1 eV.

The relatively small variation of the vb exchange found here contrasts with our earlier estimates based on a simple 4-orbital second-neighbor TB model [6]. However, the present 10-orbital nearest-neighbor model, certainly more realistic for the band structure, provides an alternative explanation for the reduced Zeeman splitting at the L point: the negative sign of the cb exchange. In fact, using Fig.1 we find that the ratio of the Zeeman

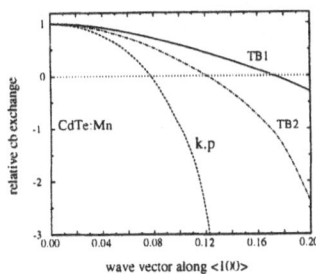

Fig. 2 The ratio of $NJ_{\mathbf{kk}}^c$ to its value at $k = 0$ versus wave vector (in units of $2\pi/a$) along $< 100 >$ with $E_v - \varepsilon_d = 3.5$ and $U_{eff} = 6.0$ eV. The curves labeled TB1 and TB2 correspond to the TB model with two different sets of parameters (see text); the third curve results from the $\mathbf{k}.\mathbf{p}$ model.

splitting at the L point to that at Γ varies from $\frac{1}{9}$ to almost zero as U_{eff} decreases from 6.1 to 6.0 eV. The experimental value of this ratio is $\sim \frac{1}{16}$. While the above U_{eff} values perfectly match that assumed in Ref.[2], they are much smaller than other estimates[7].

Fig. 2 focuses on the k-dependence of cb exchange in the neighborhood of the band minimum and compares our TB results with those of the $\mathbf{k}.\mathbf{p}$ model. Here we fixed $E_v - \varepsilon_d = 3.5$ eV and $U_{eff} = 6.0$ eV as in Ref.[2]. The curve labeled TB1 corresponds to the TB model with the same parameters as in Fig. 1. The curve TB2 is obtained with a modified parameter set [9] which yields a better fit to the cb effective mass. The third curve shows the results of the $\mathbf{k}.\mathbf{p}$ model, with the cb effective mass $m^* = 0.11$. The quantitative differences are indeed dramatic. The 3-band $\mathbf{k}.\mathbf{p}$ model drastically overestimates the variation of the cb exchange. We have checked that this results from an overestimated hybridization $|V_{kd}^c|^2$ that increases linearly with energy, much faster than in the TB model.

To conclude, the relative importance of the kinetic exchange in the cb is shown to depend crucially on the value of U_{eff}, calling for a precise experimental determination of the latter. However, the 3-band $\mathbf{k}.\mathbf{p}$ model used in Ref.[2] seems inadequate for a quantitative evaluation of the reduced spin splitting in confined structures.

References

1. A. K. Bhattacharjee et al., Physica **117B & 118B**, (1983) 449.
2. I. A. Merkulov et al., Phys. Rev. Lett. **83**, (1999) 1431.
3. P. Vogl et al., J. Phys. Chem. Solids **44**, (1983) 365.
4. B. E. Larson et al., Phys. Rev. B **37**, (1988) 4137.
5. C. Barreteau et al., Phys. Rev. B **61**, (2000) 7781.
6. A. K. Bhattacharjee, Phys. Rev. B **41**, (1990) 5696; *Proc. of 20th Int'l. Conf. Phys. Semicond.*, edited by E. M. Anastassakis and J. D. Joannopoulos (World Scientific, Singapore 1990) p. 763.
7. T. Mizokawa and A. Fujimori, Phys. Rev. B **48**, (1993) 14150; ibid. **56**, (1997) 6669.
8. A. Kobayashi et al., Phys. Rev. B **25**, (1982) 6367.
9. D. Bertho (unpublished), cited by V. Albe, Doctoral thesis, Université Montpellier II (1997).

Cross-sectional scanning tunneling microscope (STM) study of Mn-doped GaAs layers

T. Tsuruoka[1], R. Tanimoto[1], N. Tachikawa[1], S. Ushioda[1,2], F. Matsukura[2], H. Ohno[2]

[1] Research Institute of Electrical Communication, Tohoku University, and CREST-Japan Science and Technology (JST), Sendai 980-8577, Japan e-mail: tsuruoka@ushioda.riec.tohoku.ac.jp

[2] Laboratory for Electronic Intelligent Systems, Research Institute of Electrical Communication, Tohoku University, Sendai 980-8577, Japan

Abstract   Using cross-sectional scanning tunneling microscopy (XSTM), we have identified the dopant atoms and point defects in Mn-doped GaAs layers with different dopant concentrations. The electrically activated Mn acceptor concentration deduced from the STM images agrees well with the hole concentration determined by Hall measurements.

1 Introduction

The recent success in doping of III-V semiconductors with magnetic impurities, using low-temperature molecular-beam epitaxy (LT-MBE), has opened a new field in the development of diluted magnetic semiconductors (DMS's) [1]. In order to control the magnetic and magnetotransport properties of DMS's, it is important to know how the magnetic impurities are incorporated in host crystals under the low-temperature growth condition. However, there have been only a few reports on the local atomic structures surrounding magnetic impurities [2]. In this paper we present the results of direct observation of Mn-dopant atoms and defects in LT-MBE grown GaAs layers using cross-sectional scanning tunneling microscopy (XSTM).

2 Experiments

The samples used in this study were grown on p-type GaAs(100) substrates using LT-MBE. After growing a GaAs buffer layer and an AlGaAs layer (10 nm thick) at a substrate temperature of 560°C , a 5 μm thick Mn-doped GaAs layer was grown at 400°C. The AlGaAs layer was used as a marker layer in the XSTM measurements. We made three samples with hole concentrations of 1.3×10^{18} cm^{-3}, 3.7×10^{18} cm^{-3}, and 1.5×10^{19} cm^{-3}. The hole concentrations were determined by Hall measurements at room temperature. After growth the samples were thinned to 120 μm from the backside to enable reproducible cleavage along a scratch mark on the sample surface.

The XSTM measurements were performed at room temperature in an ultrahigh vacuum (UHV) system with a base pressure better than 5×10^{-9} Pa. The samples were cleaved in UHV to expose the atomically flat (110) surface. The STM was operated in a constant-tunneling-current mode using an electrochemically etched W tip.

3 Results and Discussion

Figure 1 shows a STM image ($20 \times 20 nm^2$) of the cleaved GaAs(110) surface with a hole concentration of $1.5 \times$

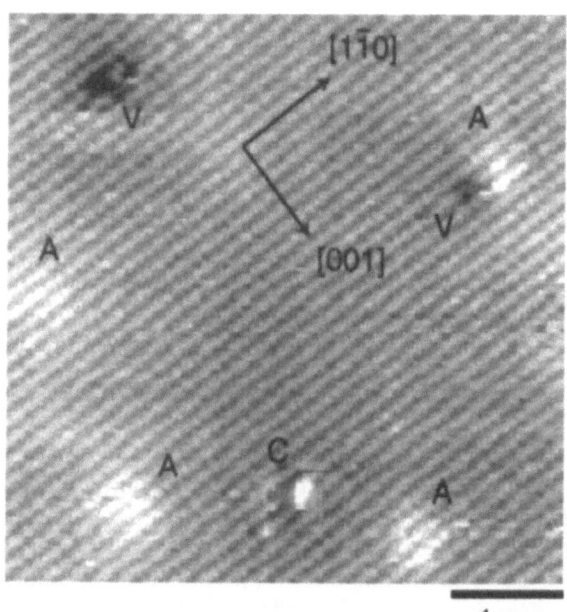

Fig. 1 STM image of Mn-doped GaAs(110) surface with a hole concentration of 1.5×10^{19} cm^{-3}.

10^{19} cm^{-3}. The image was taken with a sample bias voltage of -2.0 V and a tunneling current of 0.15 nA. The individual spots in the rows oriented along the [1$\bar{1}$0] direction correspond to the occupied dangling bonds of the As atoms [3]. Diffuse light areas (labeled "A") with a diameter of about 2 nm are observed superimposed on the background of the As atomic rows. The diffuse light area corresponds to a region where the tunneling barrier for the electrons is locally low. This local lowering of the tunneling barrier is caused by the Coulomb potential of the negatively charged Mn acceptor that substitutes a Ga atomic site [4].

We see that some of the As atoms are missing at sites labeled "V". These dark spots are surrounded by a diffuse dark area with a diameter of 1-2 nm, which is superimposed on the background of the As atomic rows. The diffuse dark area corresponds to a region where the tunneling barrier for the electrons is locally high, which is caused by the Coulomb potential of a positive charge. Hence we conclude that these dark spots correspond to the positively charged As vacancy.

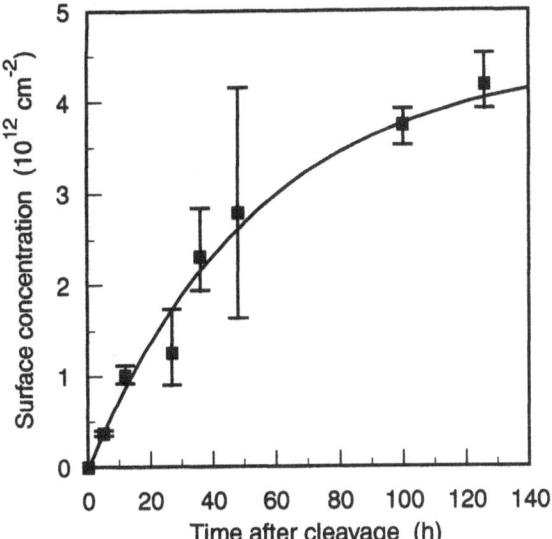

Fig. 2 Time evolution of the surface concentration of the As vacancy on the (110) surface of Zn-doped bulk GaAs. The solid line serves as a guide to the eye.

We found that the number of surface As vacancies of the Mn-doped GaAs sample increased with time. To understand the mechanism for the increase, we have observed the time evolution of the surface concentration of As vacancies of Zn-doped bulk GaAs with a hole concentration of 2×10^{19} cm^{-3}. The result is shown in Fig. 2. The As vacancy concentration just after cleavage was less than 2×10^{10} cm^{-2}. Then the concentration increased with time and reached 4×10^{12} cm^{-2} after 140 h. The vacancies are believed to be formed due to lowered formation energy for the surface anion vacancy [5], and also due to Langmuir desorption of As atoms from the surface in UHV [6]. If we assume that the As vacancies in the Mn-doped and Zn-doped samples behave similarly, the above result suggests that most of the observed As vacancies were formed on the surface after cleavage, and not present in the bulk before cleavage. Indeed after 24 h the surface As vacancy concentration of both samples reached a similar value.

As the surface As vacancy increased, another type of defect (labeled "C" in Fig. 1) appeared. This defect consists of very closely paired bright and dark spots. It is not surrounded by a diffuse area of nm scale. Thus this defect is not charged and does not induce local band bending. We conclude that this defect is an acceptor-vacancy complex in which the Mn acceptor is separeted by several atomic lattice distances from the surface As vacancy. Thus the charges of the complex are almost compensated, and they form an atomic-scale electric dipole.

From the surface and subsurface concentrations observable in the STM images, we have estimated the bulk concentration of the electrically activated Mn acceptors for the three samples. In this estimation Mn atoms compensated by the surface As vacancies were counted as the activated acceptors. The result is illusrated in Fig. 3

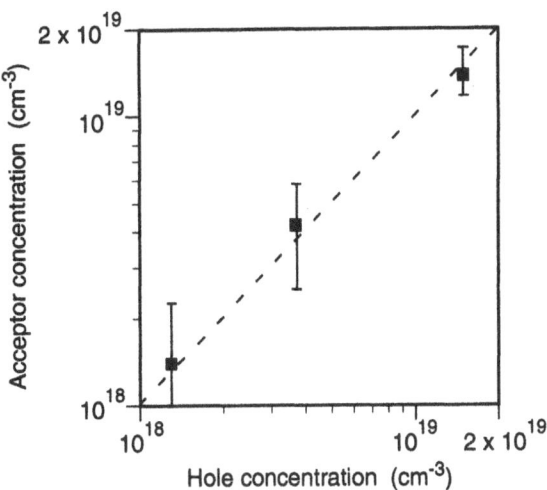

Fig. 3 Mn acceptor concentration estimated from the STM images as a function of the hole concentration determined by Hall meaurements.

as a function of the hole concentration. The Mn acceptor concentration deduced from the STM images agrees well with the hole concentration determined by Hall measurements. This agreement assures that our identification of the Mn atoms in the STM images is correct. No arsenic antisite defects and related complexes were found in the observed area (250×200 nm^2). Thus we conclude that almost all the Mn atoms are activated in the samples investigated here.

4 Conclusion

We have identified the dopant atoms and defects in the XSTM images of Mn-doped GaAs layers with hole concentartions ranging from 1.3×10^{18} cm^{-3} to 1.5×10^{19} cm^{-3}. The observed As vacancies were created on the surface after cleavage due to the desorption of surface As atoms. Some of the vacancies form the acceptor-vacancy complexes, resulting in electrical compensation of the Mn acceptors near the surface. When this compensation effect is taken into account, the electrically activated Mn acceptor concentration deduced from the STM images agrees well with the hole concentation determined by Hall measurements.

References

1. H. Ohno, A. Shen, F. Matsukura, A. Oiwa, A. Endo, S. Katsumoto, and Y. Iye, Appl. Phys. Lett. **69**, (1996) 363.

2. R. Shioda, K. Ando, T. Hayashi, and M. Tanaka, Phys. Rev. B **58**, (1998) 1100.

3. R. M. Feenstra, J. A. Stroscio, J. Tersoff, and A. P. Fein, Phys. Rev. Lett. **58**, (1987) 1192.

4. J. F. Zheng, X. Liu, N. Newman, E. R. Weber, D. F. Ogletree, and M. Salmeron, Phys. Rev. Lett. **72**, (1994) 1490.

5. S. B. Zhang and A. Zunger, Phys. Rev. Lett. **77**, (1996) 119.

6. M. Heinrich, Ph. Ebert, M. Simon, K. Urban, and M. G. Lagally, J. Vac. Sci. Technol. A **13**, (1995) 1714.

Anomalous enhancement of the cyclotron mass of electrons in $Cd_{1-x}Mn_xTe$ observed at very high magnetic fields

Y.H. Matsuda[1], T. Ikaida[1], T. Yasuhira[1], N. Miura[1], S. Kuroda[2], F. Takano[2], K. Takita[2]

[1] Institute for Solid State Physics, University of Tokyo 5-1-5 Kashiwano-ha, Kashiwa, Chiba 277-8581, Japan
e-mail: ymatsuda@issp.u-tokyo.ac.jp

[2] Institute of Materials Science, University of Tsukuba Ibaraki 305-8573, Japan

Abstract   Cyclotron resonance of the conduction electrons in $Cd_{1-x}Mn_xTe$ has been studied at very high magnetic fields exceeding 100 T. It has been found that the electron effective masses increase very rapidly with Mn concentration. The enhancement of the mass around 250 K can be well explained by not only the increase of the band gap but also the $sp - d$ hybridization effects. Even though the temperature dependence of the mass enhancement can be qualitatively understood by a modification of the spin splitting of the Landau levels due to the $s - d$ exchange interaction, it is not likely that the mean field exchange **k·p** Hamiltonian can fully explain the temperature dependence of the cyclotron masses.

Cyclotron resonance (CR) of II-VI and III-V diluted magnetic semiconductors (DMSs) has not been studied very actively so far, because the mobilities are generally not high enough to make CR observable. On the other hand, the effective mass of electrons (or holes) coupled strongly with localized magnetic moments is an intriguing physical problem. In strongly correlated metals, these electrons exhibit a variety of interesting physical phenomena, e.g., high temperature superconductivity or a very heavy effective mass. It can be expected that the effective masses of electrons and holes in DMSs are considerably influenced by the $sp - d$ exchange interaction, and the spin-dependent many-body effects like magnetic polarons are manifested in the CR spectra. In order to investigate this we have performed high field CR experiments in n−type $Cd_{1-x}Mn_xTe$, and we have measured CR in n−type $Cd_{1-x}Mg_xTe$ to compare the effective masses between the magnetic and the non-magnetic compound.

The CR measurement was done using very high magnetic fields over 100 T and several wavelengths from light sources in the infrared range. The single turn coil technique used in the present work can generate up to 200 T for experiments. CO_2 and H_2O gas lasers are used for a wavelength of 10.6 μm and for 16.9 and 28 μm, respectively. The single crystals of CdTe, $Cd_{1-x}Mn_xTe$ ($x=0.064, 0.097, 0.11$) and $Cd_{1-x}Mg_xTe$ ($x=0.041, 0.071, 0.098$) were grown by molecular beam epitaxy. The electron concentration in each sample is around 10^{17}-10^{18} cm^{-3}.

The cyclotron masses obtained at 10.6 μm in $Cd_{1-x}Mn_xTe$ are plotted as a function of temperature in Fig. 1. Representative CR spectra are shown in the inset. The CR absorption due to the conduction electrons is clearly

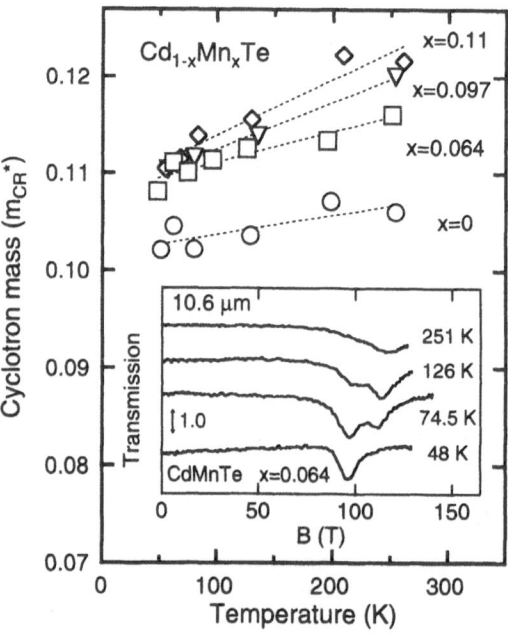

Fig. 1 Temperature dependence of the cyclotron masses in $Cd_{1-x}Mn_xTe$. Dotted lines are guides for eyes. The inset shows the magneto-transmission spectra in $Cd_{1-x}Mn_xTe$ ($x=0.064$).

observed as a broad single peak around 250 K. We can see two peaks in the spectra at low temperatures; the high field peak is due to the electron CR and the lower one is due to the impurity cyclotron resonance. The CR peak is diminished for decreasing temperature due to a carrier freeze-out effect; this makes it difficult to study CR at lower temperatures. Although a modulation doping technique can solve this problem, quantum structures make the analysis of CR more complicated due to a confinement effect. The present study is therefore focused on bulk samples for investigating the dependence of the electron masses on Mn concentration.

It is found that the cyclotron masses obtained around 250 K increase with the Mn concentration x. The relative increase of the cyclotron mass $\Delta m_{CR}^*(x)/m_{CR}^*(0)$ is approximately given by $1.4x$; it is significantly larger than the relative increase of the band gap $\Delta E_g(x)/E_g(0)$ $\sim 1.0x$. This experimental fact could be a manifestation of the influence of the $sp - d$ hybridization on the effective mass[1,2]. In the present study we carried out an 8-band **k·p** calculation based on the modified Pidgeon and Brown model [3] including the effective reduction of the momentum matrix element P due to the $sp - d$ hybridization effect [1,2]. For P^2/m_0 =8.54, 8.30, and 8.19 eV in CdTe, $Cd_{0.936}Mn_{0.064}Te$, and $Cd_{0.89}Mn_{0.11}Te$, respectively, the calculated CR energies are in good agreement with the magnetic field dependence of the CR energies given by the experimental results at 10.6 μm (117 meV), 16.9 μm (73.4 meV), and 28 μm (44 meV). From this analysis the x dependence of the band edge masses in $Cd_{1-x}Mn_xTe$ around 250 K is approximately given by the relation $m^* = (0.092 + 0.11x) m_0$. This gives a reasonable value of the band edge mass in CdTe (literature value $m^* \sim 0.090m_0$ in CdTe). Moreover, the value of $\Delta m^*(x)/m^*(0)$ seems to be very consistent with the theoretical results in Ref.[1]. We believe that the polaron effect is small since the energy range in the present study is far away from the LO-phonon energy in CdTe ($\sim$ 20 meV). The possible anti-crossing effect between the conduction band and the upper Hubbard level of the d-state may influence the cyclotron mass at very high fields [2]. The relative enhancement of m^*_{CR} around 100 T is slightly larger than that of m^*, however, we imagine that this kind of effect is not very large around 100 meV above the conduction band minimum.

We have found that the cyclotron masses in $Cd_{1-x}Mn_xTe$ decrease with decreasing temperature. Even though the electron masses in CdTe show the same tendency, we believe that the decrease of the masses is due to spin-related effects since the temperature coefficient changes systematically with the Mn concentration. If we assume that the temperature-dependent cyclotron mass in CdTe is due to the electron-phonon interaction and/or the disorder scattering effects [4] and that it is independent of x, we can see that the relative increase of the cyclotron mass $\Delta m^*_{CR}(x)/m^*_{CR}(0)$ decreases rapidly with decreasing temperature. We can expect that the temperature dependence of the spin splitting of the Landau levels explains this effect due to the $sp - d$ exchange interaction through the magnetization of the Mn ions.

It is found that the calculated temperature dependence of the cyclotron masses can explain the reduction of $\Delta m^*_{CR}(x)/m^*_{CR}(0)$ with decreasing temperature qualitatively. However, the degree of the calculated decrease of $\Delta m^*_{CR}(x)/m^*_{CR}(0)$ is smaller than the experimental results by a factor of 3; e.g. in the experiment we find $\Delta m^*_{CR}(x)/m^*_{CR}(0) \approx 1.4x$ and $0.8x$ at 250 K and 50 K, respectively, while in the calculation we find $\Delta m^*_{CR}(x)/m^*_{CR}(0) \approx 1.25x$ and $0.97x$ at 300 K and 50 K, respectively. This discrepancy may suggest that the effective mass cannot be fully understood by the **k·p** calculation within the framework of the mean field exchange theory. It is worth noting that in our recent study of InMnAs [5] the temperature dependence of the electron effective mass cannot be explained even qualitatively by the calculation in terms of the **k·p** method.

Figure 2 shows the relative increase of the cyclotron mass in $Cd_{1-x}Mn_xTe$ and $Cd_{1-x}Mg_xTe$ as a function

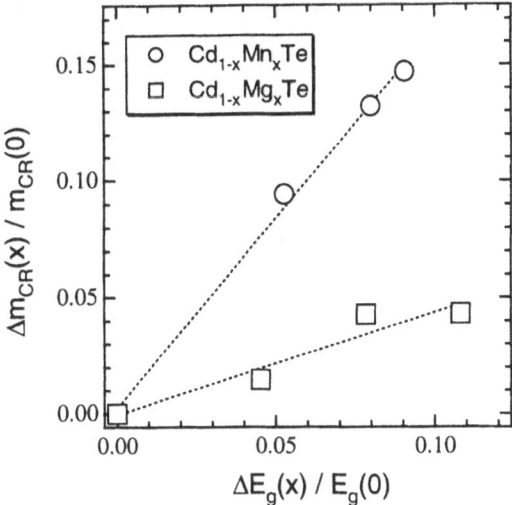

Fig. 2 Relative increase of the cyclotron masses at 117 meV and around 250 K are plotted as a function of relative increase of the band gap due to the Mn or Mg concentration. Dotted lines are guides for eyes.

of the relative increase of the band gap with Mn or Mg concentration. $Cd_{1-x}Mg_xTe$ is one of the most popular materials used for making the low dimensional electron gas in CdTe based quantum structures [6]. The band gap increases with increasing Mg concentration [7] at a slightly larger rate than that in $Cd_{1-x}Mn_xTe$ [8]. In Fig. 2 we find that the increase of the electron cyclotron masses of $Cd_{1-x}Mg_xTe$ is considerably smaller than that in $Cd_{1-x}Mn_xTe$. This experimental fact strongly suggests that the d-states of the Mn ions play an important role in making the electron mass anomalously large.

In summary, it has been found that the $sp - d$ hybridization makes the effective masses of conduction electrons much larger than expected from the increase of the band gap in $Cd_{1-x}Mn_xTe$. It is not likely that the temperature dependence of the cyclotron mass can be quantitatively understood by the **k·p** model in the framework of the mean field exchange theory.

References

1. P.M. Hui, H. Ehrenreich, and K.C. Hass, Phys. Rev. B **40**, (1989) 12346.
2. Y.H. Matsuda et al., Physica B, to be published.
3. J. Kossut, *Semiconductors and Senimetals vol. 25*, (Academic, Boston, MA,1988) 183.
4. J.T. Devreese et al., J. Cryst. Growth **214/215**, (2000) 465.
5. Y.H. Matsuda et al., submitted to *Proc. of Int'l. Conf. Phys. Appl. Spin-Related Phenomena Semicond.* (Sendai, 2000)
6. G. Karczewski et al., J. Crist. Growth **184/185**, (1998) 814.
7. S.G. Choi et al., Appl. Phys. Lett. **71**, (1997) 249.
8. J.K. Furdyna et al., J. Appl. Phys. **64**, (1988) R29.

Electronic structure of superlattices of II–VI/III–V diluted magnetic semiconductors

T. Kamatani and H. Akai

Department of Physics, Graduate School of Science, Osaka University, 1-1 Machikaneyama, Toyonaka, Osaka 560–0043, Japan
e-mail: kama@presto.phys.sci.osaka-u.ac.jp

Abstract   The possibility controlling the magnetism of III–V /II–VI superstructures of the diluted magnetic semiconductor AlAs/(Cd,Mn)Te is investigated by use of the KKR-CPA-LDA (Korringa–Kohn–Rostoker coherent–potential approximation and the local density approximation) first principles calculation. The superlattice whose unit cell is composed of a single AlAs layer and a single CdMnTe layer becomes ferromagnetic when the number of carrier holes are increased. In the case where the number of the CdMnTe layer in the unit cell is increased, the interlayer coupling between the CdMnTe layers becomes antiferromagnetic because the carrier holes reside only at the III–V /II–VI heterosurfaces.

1 Introduction

III–V diluted magnetic semiconductors (DMSs), which were fabricated by use of the low temperature molecular beam epitaxy (MBE)[1], exhibit the ferromagnetism. Since this ferromagnetism can be controlled by changing the carrier density, III–V DMSs are expected to open up a new possibility of devices using the magnetic and spin–dependent phenomena. However, the low solubility of the transitional elements in III–V compound semiconductors prevents us to obtain high T_c in III–V DMSs.

On the other hand, in II–VI compound semiconductors, Mn substitutes the group II element as much as 50%, forming II–VI DMSs[2]. The classical II–V DMSs such as CdMnTe and ZnMnSe, however, hardly show ferromagnetism, since it is rather difficult to increase the carrier density. Mn atoms in the DMSs show the ferromagnetic coupling through the double–exchange like mechanism[3], and in order for this mechanism to work, 3d carrier holes originated from Mn atoms are essential. It was reported, actually, that hole doped CdMnTe exhibited the ferromagnetism, although the Curie temperature was extremely low[4].

In this paper, we investigate the possibility that III–V /II–VI DMS superstructures, where the II–VI layer serves as a magnetic layer while the III–V layer provides carriers, exhibit controllable ferromagnetism by use of ab initio calculations based on the local density approximation.

2 method

Generally speaking, II–VI compound semiconductor has a wide gap. The band gap of CdTe is, however, narrower than that of AlAs and it is expected that the carriers move form III–V layer to II–VI layer. For this reason, we try AlAs/CdMnTe superlattices as a good candidate for III–V /II–VI magnetic superstructures.

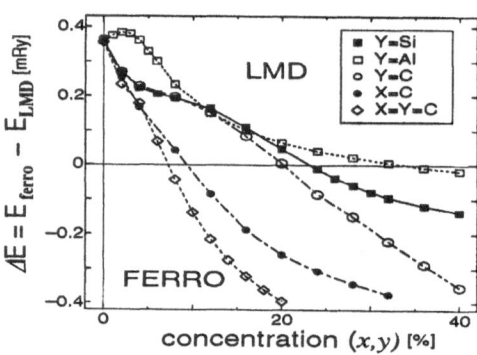

Fig. 1 The different $\triangle E = E_{ferro}\text{-}E_{LMD}$ in the total energy between the ferromagnetic and the LMD states as the function of the hole concentration x, y.

All results of band calculation in this paper were carried out by use of the the Korringa–Kohn–Rostoker (KKR) Green's function method and the coherent potential approximation (CPA) combined with the local density approximation (LDA). In order to see the relative stability between the ferromagnetic and spin–glass like local moment diorder (LMD) states, the total energy of these two states are compared. The potential form is assumed the maffin–tin type and $64k$ points in the first Brillouin zone were used. The Mn concentration in CdTe layer is 20% and the lattice constant is fixed to the avarage of AlAs and CdTe, 6.07Å.

3 Results and Discuss

$(Al_{1-x},X_x) (As_{1-y},Y_y) / Cd_{0.8}Mn_{0.2}Te$, whose unit cell is composed of a single AlAs layer and a single CdMnTe layer, is calculated, where X and Y are the impurities, such as Si, C, antisite As and antisite Al.

The results are summerized in Fig.1, where $E_{\mathbf{ferro}}$ and E_{LMD} are the total energies of ferromagnetic and LMD states, respectively, and $\triangle E = E_{\mathbf{ferro}} - E_{\mathrm{LMD}}$ is the difference in the total energies between these two states. When $\triangle E$ is negative, the ferromagnetic states is stabilized. Fig.1, which is the case of hole doping, shows that the ferromagnetic state is stabilized in the high concentration region of X and Y. On the other hand, in the case of electron doping, it is difficult to realize the ferromagnetism by increasing the number of carriers. These results are consistent with the experiments and the calculations on III–V DMSs, such as InMnAs and GaMnAs. The dependency of the ferromagnetism on the carrier

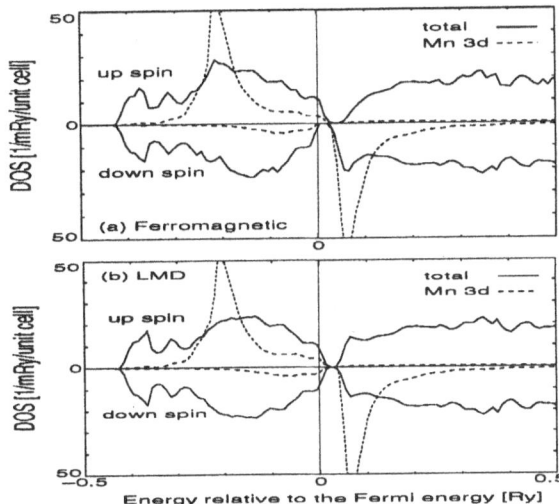

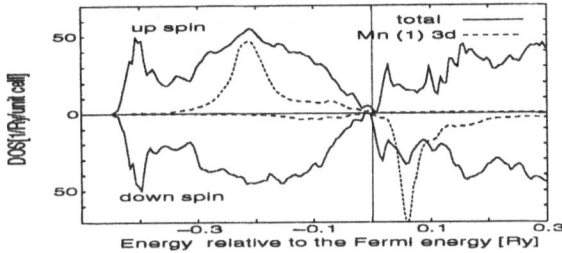

Fig. 4 The total DOS and the Mn 3d DOS in 1-site of (Al_{1-x},C_x) (As_{1-y},C_y) / $Cd_{0.8}Mn_{0.2}$Te for the $(\uparrow,\downarrow,\uparrow)$ state.

Fig. 2 The total DOS and the Mn 3d DOS of (Al_{1-x},C_x) (As_{1-x},C_x)/$(Cd_{0.8},Mn_{0.2})$Te for $x=0.32$, $y=0$. (a) is the ferromagnetic state and (b) is the LMD state.

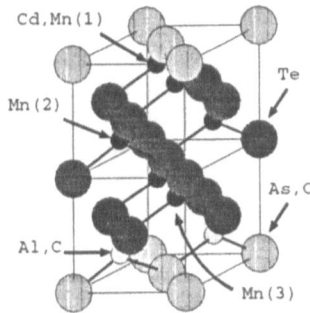

Fig. 3 The unit cell of calculation has single (Al,C)(As,C) layer and three CdMnTe layers. This unit cell has three defferent Mn sites.

density is also consistent with the argument about the energy gain due to the double–exchange, which is proportional to nt, where n is the number of carriers and t is the hopping integral.

The stabilization of ferromagnetism in II–VI layers depend on the species of the doping in III–V layers, and particularly, C doping is most efficient. C works as an acceptor in both Al and As sites since the impurity level of C is as deep as the valence band. From this arguments, (Al_{1-x},C_x) (As_{1-y},C_y) / $Cd_{0.8}Mn_{0.2}$Te seems to be most promising.

Fig.2 shows the total and Mn 3d DOS (density of states), where (a) is the case of the ferromagnetic state and (b) is for the LMD states at the 16% hole doping. The important fact seen in Fig.2 is that the carrier holes provided in III–V layer move to Mn 3d band in II–VI layer, and makes it possible for the double–exchang–like mechanism to work. Actually, Fig.2 shows that the width of Mn 3d band is widen slightly by the d-d hybridization in the majority spins, which is characteristic to the double–exchange.

The similar calculations for the larger unit cell, which is composed of two zincblend unit cells, were also perfromed. This unit cell has four layers, as is shown in Fig.3, where one layer is (Al,C)(As,C) and the other three layers are CdMnTe. In this case, there are 8 independent sites per unit cell (including Al, As, three different Cd(Mn) and Te sites), and the magnetic order of II–VI layer could be complex. Let us call the Mn site of the uppermost position in Fig.3 'site–1', the second one 2 and the lowest 3, Then the ferromagnetic order is defined as $(1,2,3)=(\uparrow,\uparrow,\uparrow)$. The magnetic structures of $(1,2,3)=(0,0,0)$, $(\uparrow,0,0)$, $(\uparrow,\uparrow,\uparrow)$ and $(\uparrow,\downarrow,\uparrow)$, whcih are the reasonable candidates for magnetic orders in this system, are examined. Here, 0 is the LMD state. The comparison of the total energies reveals that the ground state without carrier holes is $(\uparrow,\downarrow,\uparrow)$. The next stable state is $(0,0,0)$ and the energy relative to the ground state is about 0.2mRy.

To elucidate the mechanism that stabilizes $(\uparrow,\downarrow,\uparrow)$ state, DOS of $(\uparrow,\downarrow,\uparrow)$ state was shown in Fig.4. It shows that the double exchange coupling works only in Mn of site–1, since the 3d band of this Mn has some holes. The holes in site–1 originate from the Cd(Mn)–As bonds which provide acceptor levels. On the other hand, Te–Al bonds form donnor levels. When CdMnTe layer is thin, these holes and electrons conpensate each other. The interlayer coupling between the CdMnTe layers is antiferromagnetic since Mn's of site – 2 and 3 have no 3d holes. From such argument, it is concluded that the energy gain of $(\uparrow,\downarrow,\uparrow)$ state is larger than that of $(0,0,0)$ state. The ferromagnetic state becomes stable gradually by increasing the carrier holes in III–V layer. The ferromagnetic state is, however, hardly realized in CdMnTe layer even when 20% Al and As are substituted by C.

References

1. H. Munekata, H. Ohno, S. von Molnar, Arin Segümller, L. L. Chang, and L. Esaki, Phys. Rev. Lett. **63**, (1989) 1894.
2. J. K. Furdya, J. Appl. Phys, **64**(4), (1988) 15 R29.
3. H. Akai, Phys. Rev. Lett. **81**, (1998) 3002.
4. H.HUry, A.Wasiela, A.Arnoult, J.Cibert, S.Tatarenko, T.Dietl, YMerle d' Aubigne, Phys. Rev. Lett. **79** (1997) 511.

Tunneling into amorphous Gd_xSi_{1-x} at the Metal-Insulator Transition and its Independence of Magnetic Impurities in the Barrier

W. Teizer, F. Hellman, R. C. Dynes

Department of Physics, University of California, San Diego, La Jolla, CA 92093-0319, USA. teizer@physics.ucsd.edu

Abstract   We have determined the density of states of amorphous Gd_xSi_{1-x}, N(E), through the Metal-Insulator Transition by tunneling spectroscopy measurements. In a-Gd_xSi_{1-x}/oxide/Pb we found that $N(E) \propto dI/dV$ changes with magnetic field H and $N(0) \to 0$ at the Metal-Insulator Transition. Recent results on Al/Al_2O_3/a-Gd_xSi_{1-x} tunnel junctions confirm these results and indicate that they are independent of the nature of the tunnel barrier.

1 Introduction

Recently, it has become possible to simultaneously measure the transport conductivity, σ, and the tunneling conductance, dI/dV, of a material (a-Gd_xSi_{1-x}, hereafter referred to as GdSi) that can continuously be tuned through the three dimensional Metal-Insulator Transition (MIT) by application of a magnetic field H. [1] Since, under proper conditions, the density of states, N(E), is proportional to dI/dV, this development has cleared the way to unambiguously extract the behavior of the density of states at the MIT.

The details, how a disordered system with strong Coulomb interactions undergoes a MIT are poorly understood due to the complexity of the system under consideration. [2] However, it is expected that on the insulating side a soft Efros-Shklovskii gap ($N(E) \propto E^2$) opens. [3] On the metallic side a precursor behavior due to strong Coulomb interactions ($N(E)=N(0)+N_1E^{1/2}$) might be expected. [3]

Numerous studies have measured the transport conductivity, σ, in a continuously driven MIT [4] and one [5] has simultaneously measured σ and the tunneling conductance, dI/dV, in a doping driven MIT. While these studies have led to much insight into the MIT, the simultaneous measurement of σ and dI/dV in a continuously driven MIT allows a quantitative comparison of N(E) at various magnetic fields to the degree required to determine the scaling behavior of N(0) at the MIT. GdSi has been shown to have a large magnetoresistance [6] and, if properly doped, can be continuously and reversibly tuned through the metal insulator transition by the application of a magnetic field. [7] In recent work [1] we utilized GdSi/oxide/Pb tunnel junctions for measuring the tunnel conductance and as voltage probes for the transport measurement. The tunnel barrier in this type of junction is a native oxide of GdSi. It may be possible that magnetic impurities due to the oxides of Gd may be present in the tunnel barrier. It therefore remains to be shown that the observed change in

the tunneling conductance with magnetic field are not due to magnetic impurities in the tunneling barrier. To this end we present results from Al/Al_2O_3/GdSi tunnel junctions which confirm prior results [1] in a situation where magnetic impurities in the tunnel barrier are not present.

2 Experimental Issues

We compare the results from two types of tunnel junctions involving GdSi, GdSi/oxide/Pb [1] and Al/Al_2O_3/GdSi (Schematic tunnel junctions are shown in the insets of Fig. 2a,b). For the latter we thermally evaporated ~50nm of Al through a shadow mask to form three parallel stripes. After exposure to ambient conditions for ~30 min to produce a native oxide we electron beam co-evaporated Gd and Si through a shadow mask to form a stripe perpendicular to the Al stripes. We thus created three tunnel junctions with ~$10^{-7}m^2$ junction area. The tunnel junctions were also used as voltage probes for transport measurements. Observation of a superconducting gap of the counter electrode material in H=0 allowed for an unambiguous determination of single-step quantum tunneling across the tunnel barrier thus assuring $N(E) \propto dI/dV$.

3 Results

Figure 1a (b) shows the transport conductivity σ vs T of a GdSi film with Pb [1] (Al) electrodes at various magnetic fields H. The GdSi evaporation of the samples with both types of electrodes occurred simultaneously. The sample with Pb electrodes (Fig. 1a) is more conductive than the sample with Al electrodes (Fig. 1b) since its location during the deposition led to a slightly larger Gd content. Both samples are metallic at all magnetic fields, since the extrapolation of σ to T=0 is finite. The conductivity follows $\sigma(T,H)=\sigma_0(H)+\sigma_1(T)$ indicating that this data is acquired deep in the critical regime [7,1]. Figure 2a (b) shows the tunneling conductance dI/dV vs. V across a GdSi/oxide/Pb [1] (Al/Al_2O_3/GdSi) tunnel junction at T=100mK for various magnetic fields H. Data for Figs. 1a (b) and Figs. 2a (b) have been acquired on the same samples. In both parts of Figure 2 the H=0 data is offset for clarity. Focussing on Figure 2a, the superconducting Pb gap edge at V=+/-1.4mV and signatures of phonons at V=+/-6.1mV and +/-10.0mV arising from strong phonon coupling are observed. This observation indicates that the conductance across the tunnel junction is dominated by

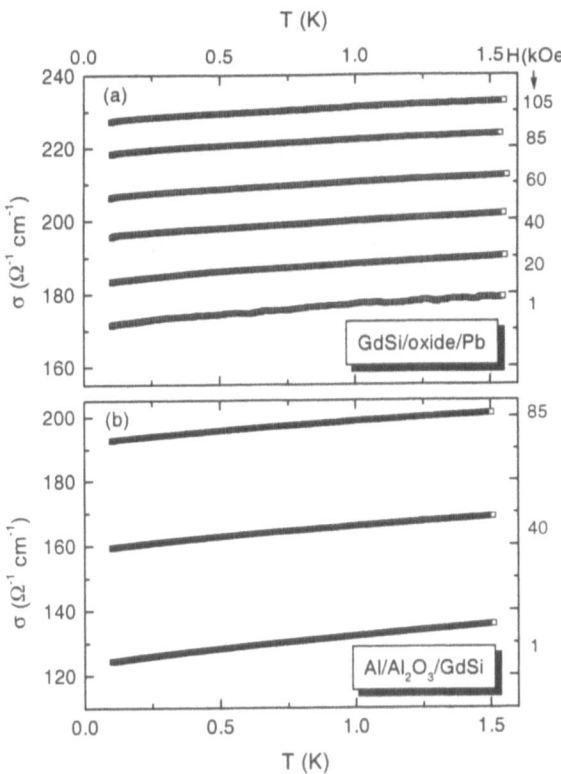

Fig. 1 Transport conductivity σ vs. T at various magnetic fields H for a GdSi sample with Pb electrodes (a) [1] and a GdSi sample with Al electrodes (b).

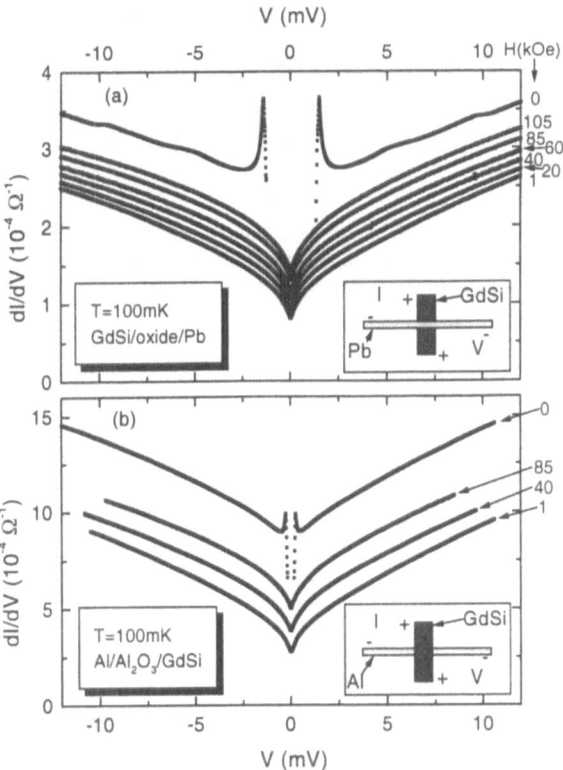

Fig. 2 Tunneling conductance dI/dV vs. V at T=100mK and various magnetic fields H for a GdSi/oxide/Pb (a) [1] and an Al/Al$_2$O$_3$/GdSi (b) tunnel junction.

single step quantum tunneling, thus assuring N(E)∝dI/dV(V). As a finite magnetic field is applied the Pb superconductivity is suppressed and these features disappear. The tunneling conductance for H≥1kOe follows (dI/dV)(V,H)=(dI/dV)$_0$(H)+(dI/dV)$_1$(V) and we can thus conclude: N(E,H)=N$_0$(H)+N$_1$(E). Figure 2b shows similar data for an Al/Al$_2$O$_3$/GdSi tunnel junction. The superconducting Al gap edge at V=+/-0.26mV is observed in the H=0 data. Again, for H≥1kOe, where superconducting features of the Al electrode are suppressed, (dI/dV)(V,H)=(dI/dV)$_0$(H)+(dI/dV)$_1$(V) is observed, indicating that this type of behavior is independent of the presence of magnetic impurities in the tunnel barrier. Measurements of the tunnel conductance of Al/Al$_2$O$_3$/GdSi tunnel junctions closer to the MIT (lower Gd content) have also confirmed prior results [1] on GdSi/oxide/Pb junctions. We conclude that, indeed, the change of the tunnel conductance with magnetic field is a signature of the change in the density of states at the Fermi energy of GdSi with magnetic field.

4 Conclusion

We have shown that the constant shift of the tunneling conductance dI/dV with H of GdSi/oxide/Pb tunnel junctions [1] is not due to the possible presence of magnetic impurities in the tunnel barrier. We have observed similar behavior in Al/Al$_2$O$_3$/GdSi tunnel junctions where no magnetic impurities are present in the

tunnel barrier. Work to determine the behavior of dI/dV at higher field and Hall measurements to determine the carrier density of GdSi are currently underway.

Acknowledgements

We acknowledge support by the NSF through DMR 97-05180 and DMR 97-053000 and by the AFOSR through F496209810264.

References

1. W. Teizer, F. Hellman, and R. C. Dynes, Phys. Rev. Lett. **85**, 848 (2000).
2. P. W. Anderson, Phys. Rev. **109**, 1492 (1958). For a recent review: B. Kramer and A. MacKinnon, Rep. Prog. Phys. **56**, 1469 (1993); P. A. Lee and T. V. Ramakrishnan, Rev. Mod. Phys. **57** 287 (1985); N. F. Mott, *Conduction in Non-Crystalline Materials* (Clarendon Press, Oxford, 1993).
3. A. L. Efros and B. I. Shklovskii, J. Phys. C **8**, L49 (1975).
4. Examples are: M. A. Paalanen et al., Phys. Rev. Lett. **48**, 1284 (1982). S. Waffenschmidt, C. Pfleiderer, and H. v. Löhneysen, Phys. Rev. Lett. **83**, 3005 (1999). S. v. Molnar et al., Phys. Rev. Lett. **51**, 706 (1983).
5. G. Hertel et al., Phys. Rev. Lett. **50**, 743 (1983).
6. F. Hellman et al., Phys. Rev. Lett. **77**, 4652 (1996). P. Xiong et al., Phys. Rev. B **59**, 3929 (1999).
7. W. Teizer, F. Hellman, and R. C. Dynes, Solid State Commun. **114**, 81 (2000).

Spin-lattice relaxation in semimagnetic quantum wells with a 2DEG

D.R.Yakovlev[1,2], **A.V.Scherbakov**[2], **B.König**[1], **W.Ossau**[1], **A.V.Akimov**[2], **T.Wojtowicz**[3], **G.Karczewski**[3], **J.Kossut**[3]

[1] Physikalisches Institut der Universität Würzburg, 97074 Würzburg, Germany, e-mail: yakovlev@physik.uni-wuerzburg.de
[2] A.F.Ioffe Physical-Technical Institute, Russian Academy of Sciences, 194021 St.Petersburg, Russia
[3] Institute of Physics, Polish Academy of Sciences, 02-668 Warsaw, Poland

Abstract The spin-lattice relaxation (SLR) dynamics of magnetic ions has been studied in semimagnetic $Cd_{0.99}Mn_{0.01}Te/Cd_{0.88}Mg_{0.12}Te$ quantum wells with a 2DEG by means of a method based on optical detection of injected nonequilibrium phonons. The relaxation rate is an increasing function of electron concentration and electron temperature.

It was demonstrated very recently that a layer of semimagnetic material could be included in a GaAs-based n-i-p light-emitting diode, where it serves as an effective injector of spin-polarized carriers [1]. A prerequisite for such structures is the presence of free carriers achieved by doping of the semimagnetic semiconductor (SS) layer. However, the experimental information on the influence of doping and of free carriers on the magnetic properties of wide-gap SS is very limited. One can expect that the energy- and spin-transfer between the system of carriers, magnetic ions and phonons (i.e. lattice) will be modified significantly in the presence of carrier gases.

We have reported recently that in $Cd_{0.99}Mn_{0.01}Te$-based QWs with a slow SLR rate the magnetic ion system can be substantially overheated by means of interaction with hot photocarriers, which is enhanced by the presence of a 2DEG [2].

Here we explore the question how magnetic properties of semimagnetic QWs, namely the spin-lattice relaxation rate of magnetic ions, can be modified by a 2DEG.

The investigated samples with an 80Å thick $Cd_{0.99}Mn_{0.01}Te/Cd_{0.76}Mg_{0.24}Te$ QW were grown by molecular beam epitaxy on (100) GaAs substrates. One is nominally undoped and contains a 2DEG with a density of 6×10^9 cm^{-2} provided by residual impurities. Other structures have a modulation doping of a barrier layer and contains up to 1.6×10^{11} cm^{-2} electrons. Details of growth and optical properties are published in Refs.[2-4]. The novel experimental technique we use combines injection of nonequilibrium phonons and optical detection of the induced changes via exciton photoluminescence

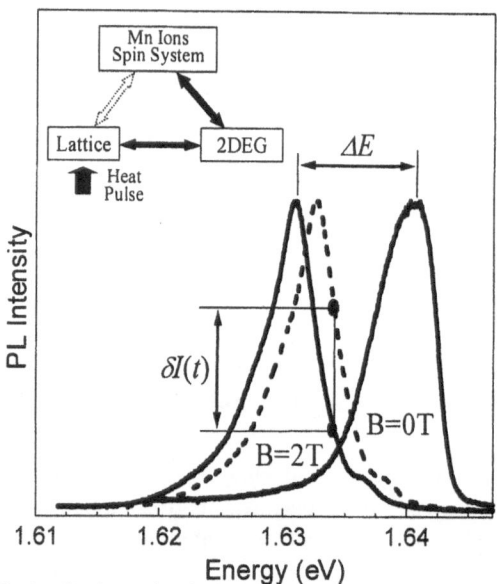

Fig.1 Exciton luminescence spectra of a doped $Cd_{0.99}Mn_{0.01}Te/Cd_{0.76}Mg_{0.24}Te$ QW measured in the absence (solid lines) and in the presence (dashed line) of nonequilibrium phonons. T=1.6K.

[5]. Nonequilibrium phonons were generated by a heat pulse technique. The phonon generator (a 10 nm constantan film) with an area 0.5x0.25 mm^2 was evaporated on the narrow side of the GaAs substrate and was heated by current pulses of duration 0.1 μs at a repetition rate of 5 kHz. Analysis with the time resolution up to 10 ns was achieved by means of a special interface board and a computer. The time-resolved detection of the Mn spin temperature is based on monitoring the dynamical shift of the exciton PL line shown in Fig.1 (for details see Refs. [5,6]). The sample was mounted inside a superconducting magnet in the Faraday configuration and was immersed in liquid helium. It is important to note here, that these measurements are not influenced by the phonon bottleneck effect observed in bulk materials.

The different energy reservoirs, characterized by temperatures and relaxation channels, in DMS QWs with a 2DEG are shown schematically in Fig.1. The heating pulse generates nonequilibrium phonons and heats the Mn system. The spin-lattice relaxation process (cooling of the Mn system), served by direct coupling of the Mn system and lattice (shown by an open arrow), is rather ineffective at low Mn concentrations [6]. However, the bypassing channel involving the energy reservoir of the 2DEG (filled arrows) should be very efficient, because a spin-flip scattering of electrons by the magnetic ions mediated by exchange interaction provides an efficient energy exchange between these two systems [2].

Fig.2a shows the normalized phonon-induced signal $\delta I(t) = I(t) - I_0$ (here $I(t)$ and I_0 are the PL intensities with and without phonon injection, respectively) measured at T=1.6K. We note here, that in the linear regime of the giant Zeeman splitting $\delta I(t)$ is proportional to the variation of the Mn spin temperature. The leading edges of $\delta I(t)$ have a width of about 2 μs, which is about the duration of the phonon pulse. The decay times τ, determined from the exponential fit of $\delta I(t)$, are the cooling times of Mn system and in our case, where the phonon bottleneck is absent, they are actually SLR times of the Mn ions.

It is clearly seen in Fig.2a that the cooling of the Mn sytem is faster in the sample with higher 2DEG density. Namely, the SLR time, determined from the exponential fit of $\delta I(t)$ decay, decreases from 83 μs in the undoped sample down to 20 μs in the sample with 1.5×10^{11} cm^{-2} electrons. SLR times vs electron concentration, measured at a magnetic fields of 2T, are presented in Fig.2b.

It was found that the SLR time depends strongly on the density of photoexcitation. It shortens significantly with the increase of laser power, as it is shown in Fig.2c for the structure with 1.5×10^{11} cm^{-2} electrons. We have checked that this effect is not caused by the lattice temperature increase, which is still very small for the used excitation densities. We believe that the elevated electron temperature is responsible for the reduction of the SLR time. This explanation is supported by calculations performed in the model developed in Ref.[2], which predict enhancement of the energy transfer between 2DEG and magnetic ions with an increase of electron temperature. This is valid for conditions when the thermal energy kT is smaller than the 2DEG Fermi energy. For CdTe-based QWs an electron density of 1.5×10^{11} cm^{-2} corresponds to a

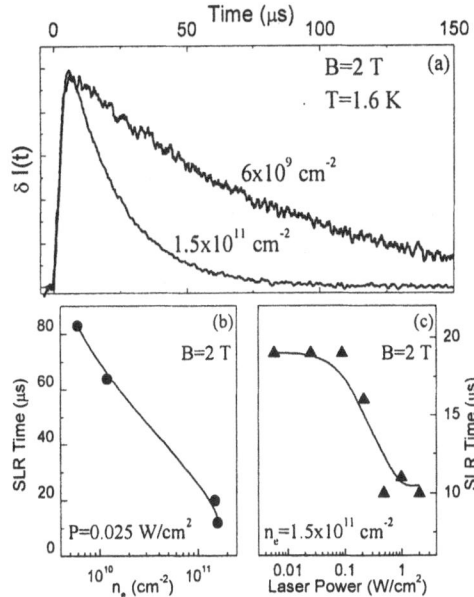

Fig.2 (a) Time evolution of the phonon-induced variation of the PL intensity (corresponding to variation of the Mn spin temperature) detected on the high energy side of PL line at 1.634 eV. SLR time vs electron concentration and laser power are presented in panels (b) and (c), respectively. T=1.6K.

Fermi energy of 3.5 meV.

To conclude, in modulation-doped quantum wells with a 2DEG of low density (Fermi level does not exceed 4 meV) a strong increase (by an order of magnitude) of SLR rates has been detected. It is evident for a new effective mechanism of SLR when the energy from Mn-ion system are firstly transferred to the 2DEG and then from the 2DEG further to the lattice.

We are thankful to I.A.Merkulov for stimulating discussions. This work is supported by the Deutsche Forschungsgemeinschaft (grants SFB 410 and 436 RUS 17/23/00), the NATO grant PST.CLG.976858, the Program for the young scientists of the Russian Academy of Sciences, and in Poland by grant PBZ 28.11/P8 from the State Committee for Scientific Research.

References

1. R.Fiederling et al., Nature **402**, (1999) 787.
2. B.König et al., Phys. Rev. B **61**,(2000) 16870.
3. T.Wojtowicz et al., Appl. Phys. Lett. **73**, (1998) 1379.
4. T.Wojtowicz et al., Phys. Rev. B **59**, (1999) R10437.
5. A.V.Scherbakov et al., Phys. Rev. B **60**, (1999) 5609.
6. A.V.Scherbakov et al., Phys. Rev B **62**, (2000) No.16.

Magnetic-Field-Driven Metal-Insulator Transition in Magnetic Semiconductor (Ga,Mn)As

T. Hayashi[1], Y. Hashimoto[1], S. Yoshida[1], S. Katsumoto[1,2] Y. Iye[1,2]

[1] Institute for Solid State Physics, University of Tokyo, 5-1-5 Kashiwanoha, Kashiwa, Chiba 277-8581, Japan
e-mail: thayashi@issp.u-tokyo.ac.jp
[2] CREST, Japan Science and Technology Corporation, 1-4-25, Mejiro, Tokyo 171-0031, Japan

Abstract Metal-insulator transition (MIT) in a diluted magnetic semiconductor (Ga,Mn)As is studied in detail. The material parameters were finely tuned by repeated low temperature annealing and finally the MIT was driven by external magnetic field. The temperature dependence in the vicinity of the transition was $T^{1/4}$ and the two-parameter scaling scheme was successfully applied to obtain the exponent 1.6± 0.2.

1 Introduction

For the metal-insulator transition (MIT) in disordered media, a kind of "scaling" has been believed to be applicable. For such scaling theories, universality class or symmetry of the system is essential[1]. Recently, also a kind of scaling is claimed to hold for MIT's in strongly correlated electron systems[2].

In magnetic materials, the time reversal symmetry is broken and the critical behavior is expected to be essentially different from that in non-magnetic ones. The MI transitions in semimagnetic semiconductors have been so far investigated mainly in II-VI based materials, in which magnetic polarons play an important role. Since the first synthesis of III-V based (Ga,Mn)As[3], it has captured researchers' attention by its exotic transport properties[4]. The MI transition in this material is also intriguing since the ferromagnetic interaction is mediated by current-carrying holes. In II-VI materials such as (Cd,Mn)Se[5], a new type of universality class, *i.e.*, "spin-polarized" symmetry has been claimed. In the case of (Ga,Mn)As, the local moments should be included into the system Hamiltonian and the universality class of the MIT would be changed even by the control parameter of the transition[2].

Here we report detailed measurement of the transport in the vicinity of critical point in (Ga,Mn)As, to which scaling analysis is successfully applied.

2 Experiment

We prepared a 200nm thick (Ga,Mn)As film with Mn concentration of 5% onto an (001) GaAs substrate by molecular beam epitaxy. As-grown film showed strongly insulating behavior. However, the quality of the film can be improved by annealing at comparatively low temperature as shown in Fig.1[6]. The hole concentration also increases by the annealing.

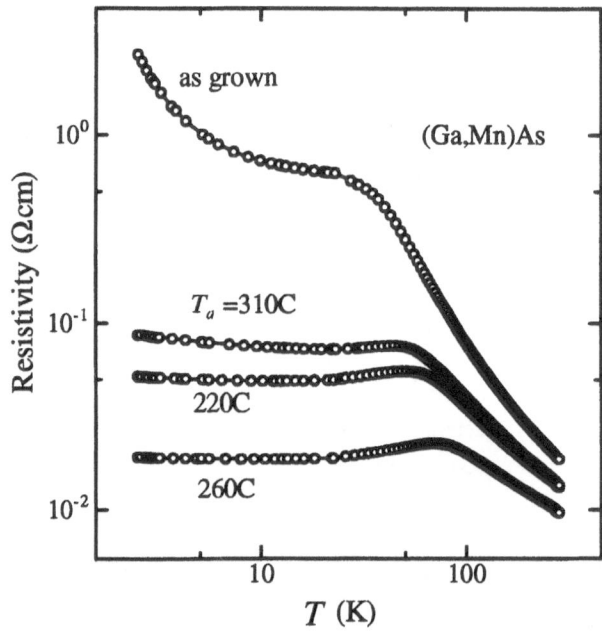

Fig. 1 Temperature dependence of the resistivity in (Ga,Mn)As, annealing temperature as a parameter. The annealing time was fixed to 10min.

The sample was etched into a Hall-bar and brought to marginally insulating state by repeated annealing at 230°C. It was directly immersed in a ^{3}He-^{4}He mixture in a mixing chamber of a dilution refrigerator together with a Ru$_2$O thermometer. Perpendicular magnetic field applied by a superconducting solenoid caused giant magnetoresistance[4] and resultant MIT.

3 Results and Discussion

A possible model for the ferromagnetism is the double exchange model[7], in which the energies of holes are also lowered by the mediation of the ferromagnetic interaction[8]. Disorder in the local spins thus causes localization of the holes. Such disorder in ferromagnetic phase appears in, *e.g.*, residual increase in the magnetization above the coercive force field[4]. The external magnetic field reduces the disorder and delocalize the holes (negative magnetoresistance). The time-reversal symmetry is already broken by the internal field and the external field can be ignored from this point of view. Hence we

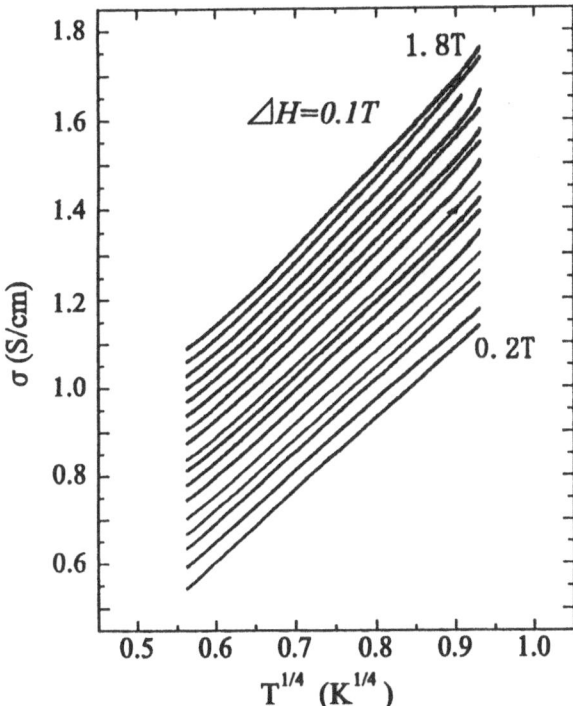

Fig. 2 Temperature dependence of the conductivity around the critical point. The temperature is scaled as $T^{1/4}$.

may say that purely the degree of disorder is varied as a control parameter.

In order to obtain the conductance extrapolated to absolute zero $\sigma(0)$, we assume simple temperature dependence $\Delta\sigma \propto T^\gamma$ We found that when $\sigma(0)$ falls around zero, $\Delta\sigma$ is best fitted by $\Delta\sigma(T) \propto T^{1/4}$ instead of $T^{1/2}$, which is often reported in other compensated semiconductor systems. Figure 2 shows the temperature dependence of the conductance as a function of $T^{1/4}$.

Although for quantitative understanding we should wait for development of a scaling theory, this steep temperature dependence can be understood as a result of co-operation of holes and local spins qualitatively.

Fortunately we can examine whether the conductance scales on H (magnetic field) and T consistently through two-parameter scaling form;

$$\sigma(H,T) \propto T^x f(|H/H_c - 1|/T^y), \qquad (1)$$

where f is a scaling function, x and y are parameters, giving the critical exponent $\nu = x/y$ regardless of the detailed contents of the scaling theory. We found that by adopting $f = 1.59024 + 0.38966X - 0.03027X^2 + 0.00623X^3$, the data around the critical point can be aligned along the scaling curve. The fitting gives the value of the critical exponent of the conductivity $\nu = 1.6 \pm 0.2$, which is defined by the critical behavior as

$$\sigma(H,0) \propto (H - H_c)^\nu. \qquad (2)$$

To the authors' knowledge, this is the first report of a successful application of the two-parameter scaling for magnetic field induced MI transition by adopting a *common* scaling function for both metallic and insulating sides.

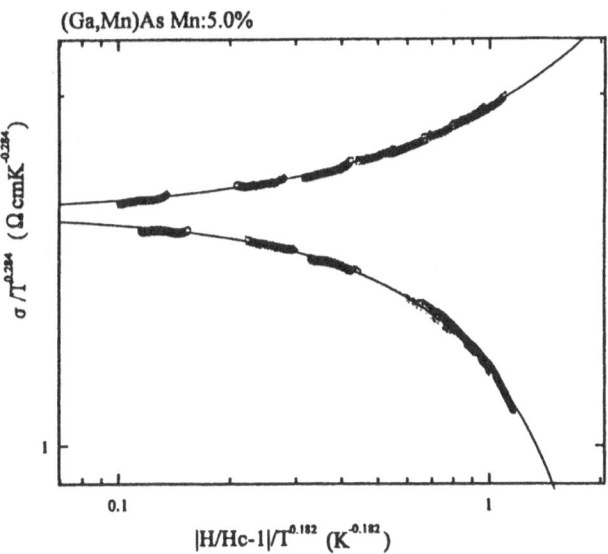

Fig. 3 Two parameter scaling plot of the data shown in Fig.2. The formula is shown in the text.

The exponents for temperature and the magnetic field are uncommon to those reported so far in MI transitions. This may be a manifestation that the present system belongs to a novel universality class. It should be noted that the value of ν is close to the results of numerical scaling in non-interacting systems[9]. It is hardly believed that the Coulomb repulsion is irrelevant in the present system. However, as noted above the external field purely controls the disorder and that would be the reason we got the value close to the result in non-interacting systems[10].

This work is partly supported by Grant-in-Aid for Scientific Research on the Priority Area "Spin Controlled Semiconductor Nanostructures" from the Ministry of Education, Science, Sports and Culture, Japan.

References

1. D. Belitz and T. R. Kirkpatrick, Rev. Mod. Phys. **66**, 261 (1994).
2. M. Imada, Fujimori and Y. Tokura, Rev. Mod. Phys. **70**, 1039 (1998).
3. H. Ohno et al., Appl. Phys. Lett. **69**, 363 (1996).
4. A. Oiwa et al., Solid State Commun. **103**, 209 (1997).
5. T. Dietl et al., in "Anderson Localization" p.58 (Springer, 1988).
6. T. Hayashi et al., submitted to Appl. Phys. Lett.
7. H. Akai, Phys. Rev. Lett. **81**, 3002 (1998).
8. S. Koshihara, A. Oiwa, M. Hirasawa, S. Katsumoto, Y. Iye, C. Urano, H. Takagi and H. Munekata, Phys. Rev. Lett. **78**, 4617 (1997).
9. T. Kawarabayashi, B. Kramer and T. Ohtsuki, Phys. Rev. **B57**, 11842 (1998).
10. We thank K. M. Ito for this comment.

The effect of stress on magnetic properties of ferromagnetic EuS-PbS and EuS-PbSe semiconductor structures

T. Story[1], M. Arciszewska[1], M. Chernyshova[1], W. Dobrowolski[1], W. Mac[2], A. Twardowski[2], R. Świrkowicz[3], A.Yu. Sipatov[4]

[1] Institute of Physics, Polish Academy of Sciences, Al. Lotników 32/46, 02-668 Warsaw, Poland
[2] Institute of Experimental Physics, Warsaw University, ul. Hoża 69, 00-681 Warsaw, Poland
[3] Faculty of Physics, Warsaw University of Technology, ul. Koszykowa 75, 00-662 Warsaw, Poland
[4] Kharkov State Polytechnical University, 21 Frunze Str., 310002 Kharkov, Ukraine

Abstract   Primary factors determining the Curie temperature of semiconductor ferromagnetic structures based on EuS are the thickness of magnetic layer and the stress present in the structure. We study these effects experimentally and theoretically for lattice-matched EuS-PbS and lattice-mismatched EuS-PbSe superlattices grown on KCl (100) substrate.

1 Introduction

EuS is a ferromagnetic member of the family of europium chalcogenides, the well known group of magnetic semiconductors [1]. In this work, we examine, in particular, the magnetic properties of EuS-PbS ferromagnetic-diamagnetic superlattices. The nonmagnetic layers in these structures are composed of PbS, the well known IV-VI semiconductor compound. Both EuS and PbS crystallize in rock salt crystal structure and their lattice parameters match very well, $\Delta a/a = 0.5\%$. Magnetically, EuS-PbS structures form a model ferromagnet-nonmagnet multilayer structure built entirely of non-metallic materials. Electronically, EuS-PbS structures form PbS multiple quantum well with ferromagnetic EuS barriers.

Recently, it was shown that the ferromagnetic Curie temperature T_C of EuS-PbS superlattices strongly depends on the thickness of EuS layer and the kind of substrate used: KCl (100) or BaF$_2$ (111) [2]. The latter effect was explained in terms of the thermal stress present in these structures due to the difference between thermal expansion coefficients (TEC) of the substrate and the structure. In this work, we further examine this effect in EuS-PbS/KCl (100) (an almost lattice-matched system) and EuS-PbSe/KCl (100) (lattice mismatch of 2.5%) superlattices. We also discuss the new approach to the theoretical analysis of the magnetic properties of stressed ultrathin EuS layers which is based on the analysis of the low energy magnetic excitations (spin waves).

2 Experimental results and discussion

We studied the set of 5x[EuS-PbS] and 5x[EuS-PbSe] multilayers grown epitaxially on KCl (100) substrates with EuS thickness varying in the range $d_{EuS} = 2\text{-}20$ML and with PbS (or PbSe) thickness being relatively large and approximately constant $d_{PbS} = d_{PbSe} = 170\text{-}200$Å. Fig. 1 shows the temperature dependence of magnetiza-

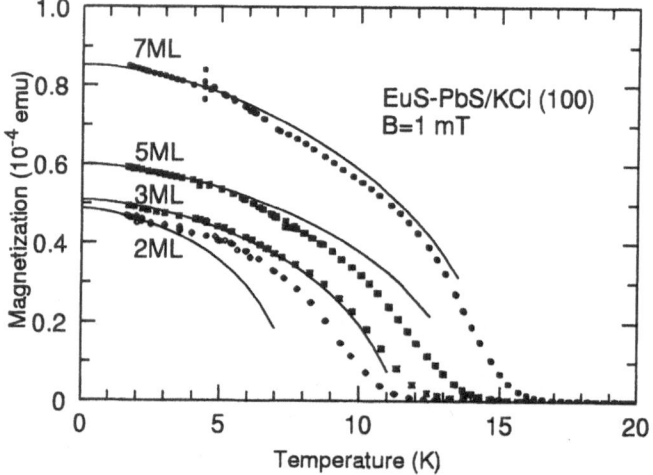

Fig. 1 The temperature dependence of magnetization for EuS-PbS superlattices on KCl (100) substrate. The thickness of EuS layer is given in the figure. The solid lines show the results of theoretical calculations.

tion for a series of EuS-PbS/KCl multilayers with varying EuS thickness. It clearly illustrates the effect of the reduction of the Curie temperature with decreasing EuS thickness. The analysis of these experimental data yields the thickness dependence of the Curie temperature in the form $T_C = T_C^b(1 - c/d_{EuS})$, where T_c^b is the Curie temperature of thick semi-bulk epitaxial layers of EuS on KCl and c is a numerical parameter [2]. For the equivalent set of EuS-PbSe/KCl multilayers our experimental data give $T_C(d_{EuS})$ dependence which is of similar form but differs from EuS-PbS in two respects: (1) for relatively thick EuS layers the Curie temperature of EuS-PbSe structures is about 4K lower than for similar EuS-PbS multilayers, (2) for thin EuS layers the $T_C(d_{EuS})$ dependence is much more pronounced for EuS-PbSe structures as compared to EuS-PbS.

The first effect we attribute to the action of stress which influences the interspin distances and, consequently, the exchange integrals and the ferromagnetic Curie temperature. An analysis of different mechanisms of stress which might operate in EuS-PbS multilayers suggests that the primary source of stress present in EuS-PbS is the thermal stress due to TEC variations while the lattice mismatch stress is assumed to be fully relaxed at growth temperature [2]. This assumption proved to be well justified in EuS-PbS structures which have negligible lattice mismatch. The case of EuS-PbSe systems is more complicated as the lattice mismatch is 5 times bigger and has different sign. For the relevant lattice parameters we have $a_{PbS} = 5.94\text{Å} < a_{EuS} = 5.97\text{Å} < a_{PbSe} = 6.12\text{Å}$. Therefore, under the action of lattice mismatch stress the EuS-PbS structures are expected to compress in the plane of the layer (leading to the increase of T_C), whereas the EuS-PbSe structures to expand in the plane of the layer (leading to the decrease of T_C). The effect of substrate induced stress can be demonstrated experimentally in a straightforward way by observing the differences in magnetic properties of EuS-PbS or EuS-PbSe multilayer grown on the substrate and the same structures without the substrate (free-standing layer) [2]. Fig. 2 shows the results of one of such experiments performed for 5x[EuS(80Å)-PbSe(200Å)] superlattice. The observed 2K decrease of ferromagnetic Curie temperature of free-standing layer as compared to the layer on the substrate is somewhat smaller for EuS-PbSe as compared to EuS-PbS structures where it amounts to 3K.

The second effect originates from the interdiffusion at the magnetic-nonmagnetic interfaces. In the mean field approach the ferromagnetic transition temperature scales with the average number of nearest magnetic neighbors z_{NN}. Since for Eu ions located close to the interface the number of magnetic neighbors is smaller than the bulk value, the Curie temperature becomes smaller for very thin layers, in which the interface region constitutes the large fraction of the total volume of the layer. This effect is, of course, observed even for ideal sharp interfaces but the presence of interdiffusion strongly enhances it. We suggest that the gradient of chemical composition present in EuS-PbSe anion sublattice provides the effective channel for an interdiffusion over a few monolayers. For EuS-PbS structures, which have the common anion, our data indicate interdiffusion of only ±1ML.

A more quantitative theoretical approach to the magnetic properties of stressed EuS-PbS multilayers can be obtained by analyzing the low energy magnetic excitations (spin waves) in stress-distorted rock salt crystal lattice of EuS. In particular, one can calculate the temperature dependence of magnetization for $T < T_C$. We performed such calculations applying Green function formalism [3]. The results are presented in Fig. 1 (solid lines). To account for different volumes and (possibly) domain structures each theoretical curve was adjusted to experimental data at the lowest temperature. No other fitting parameters were used and the whole set of curves was calculated for a single set of parameters (exchange

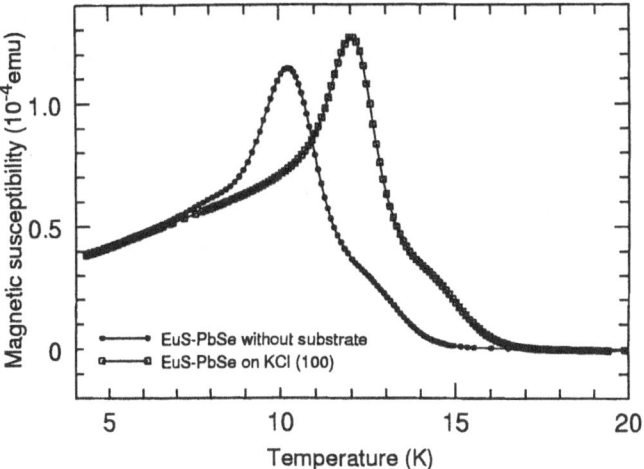

Fig. 2 The temperature dependence of magnetic susceptibility of EuS-PbS superlattices on KCl (100) substrate and without the substrate (free-standing structure).

integrals and volume and surface contributions to magnetic anisotropy).

3 Conclusions

We have studied experimentally and theoretically the ferromagnetic transition in EuS-PbS and EuS-PbSe superlattices grown on KCl (100) substrates. For relatively thick EuS layers (of the order of 100Å) the ferromagnetic Curie temperature T_C does not depend on the thickness of EuS but differs from the value T_C^b observed in bulk crystals: EuS-PbS structures show T_C higher than T_C^b whereas EuS-PbSe structures show T_C lower than T_C^b. We attribute these effects to the influence of stress present in these structures due both to the difference in thermal expansion coefficients of the layers and the substrate and, for EuS-PbSe, to the lattice mismatch. For very thin EuS layers the Curie temperature decreases with decreasing EuS thickness. The effect is particularly strong for EuS-PbSe what is related to the interdiffusion over the anion sublattice – the mechanism not present in EuS-PbS structures.

Acknowledgements This work was supported in part by KBN research project No 2 P03 B 154 18.

References

1. A. Mauger, C. Godart, Phys. Rep. **141**, (1986) 51.
2. A. Stachow-Wójcik, T. Story, W. Dobrowolski, M. Arciszewska, R.R. Gałązka, M.W. Kreijveld, C.H.W. Swuste, H.J.M. Swagten, W.J.M. de Jonge, A. Twardowski, A.Yu. Sipatov, Phys. Rev. B **60**, (1999) 15 220.
3. R. Świrkowicz, T. Story, J. Phys. C **12**, October 2000.

Fundamental properties of Fe-based III-V magnetic alloy semiconductor (Ga,Fe)As

R. Moriya, T. Kondo, Y. Katsumata, H. Munekata*, and S. Haneda

Imaging Science and Engineering Laboratory, Tokyo Institute of Technology
4259 Nagatsuda, Midori-ku, Yokohama 226-8503, Japan

Abstract   Preparation of (Ga,Fe)As has been studied systematically by molecular beam epitaxy for wide range of substrate temperature T_s. It has been found that the physical properties of epilayers can be classified into two different types depending on the T_s. Epitaxy at relatively low T_s (T_s = 260 °C) has yielded homogeneous, paramagnetic $Ga_{1-x}Fe_xAs$ epilayers, whereas epitaxy at higher temperatures (T_s = 350 – 580 °C) has resulted in GaAs epilayers with ferromagnetic Fe inclusions. The layers prepared at high T_s has exhibited magneto-optical effect with ferromagnetic characteristics.

1. Introduction

Up to know, studies on III-V-based magnetic alloy semiconductors (III-V-MAS) [1] have yielded carrier-induced magnetism in p-(In,Mn)As-based heterostructures [2] and metallic p-(Ga,Mn)As with Curie temperatures of 100K or higher [3]. While Mn appears to be an important magnetic element for III-V compound semiconductors, the III-V-MAS with other magnetic elements remains to be studied to further develop knowledge for spin-based cooperative effects and associated devices.

This paper describes the preparation and fundamental physical properties of (Ga,Fe)As. Referring the earlier works on Fe^{3+} deep centers in GaAs:Fe, we expect that incorporation of large amount of Fe would result in the realization of III-V-MAS with low background carrier concentration. This makes it possible to study the cooperative phenomena such as photo-carrier induced magnetism and excitonic magneto-optical effect at low carrier concentration regime.

2. Sample Preparation

Molecular beam epitaxy of (Ga,Fe)As has been studied for wide range of substrate temperatures T_s = 260-580 °C with Fe contents up to about 10 %. We found that reasonably homogeneous $Ga_{1-x}Fe_xAs$ epilayers, with zincblende structure and paramagnetic characteristics, can be prepared at relatively low substrate temperatures of T_s = 260 °C [4], whereas formation of ferromagnetic Fe nano-crystallites occurs with increasing the T_s. The sizes of the nano-crystallites, as evaluated by the cross-sectional TEM images, are 3 nm and 40 nm for samples prepared at T_s = 350 °C and 580 °C, respectively (Fig.1). The crystal quality of GaAs is fairly good even in the later case. The x-ray absorption fine

---

* Additional posts at PRESTO and KAST

structures (XAFS) show that incorporated Fe is surrounded by 3-5 As atoms in the homogeneous epilayers [5].

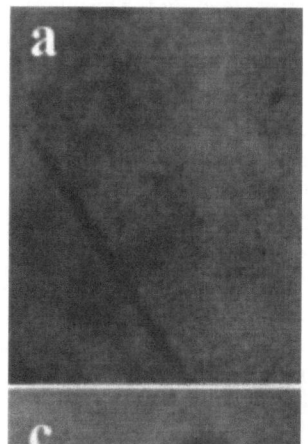
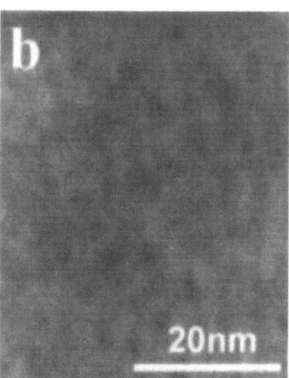
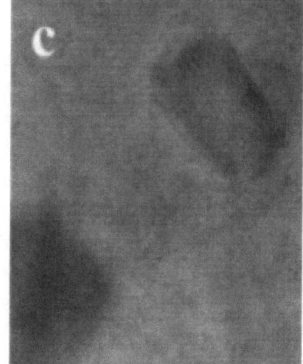

Fig. 1 : High resolution cross-sectional TEM images of (Ga,Fe)As layers grown at (a)T_s=260°C, (b)T_s=350°C and (c)T_s=580°C.

3. Physical Properties

Homogeneous samples are conductive (ρ = 0.1-1 Ω·cm) at room temperature, but become highly resistive (ρ > 10^3 Ω·cm) at low temperatures (Fig.2). Temperature dependence of ρ and Hall effect measurements suggest the occurrence of hopping conduction. Systematic studies for various samples with different Fe contents and annealing conditions suggest that the hopping may be attributed to both Fe and excess As.

Magnetic susceptibility χ is inversely-proportional to the temperature above 40K. Simply adopting the Curie-Weiss low to the plot of $1/\chi$ vs. T, we find the negative paramagnetic Curie temperature ($\Theta_p \approx$ -30 K for x = 0.04), together with the effective magnetic moment $p_{eff} \approx$ 5.9 ± 0.4. The former fact indicates the Fe-Fe spin exchange to be antiferromagnetic. The later value suggests that the Fe is

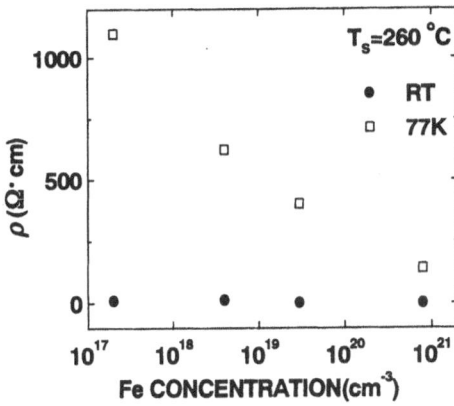

Fig. 2 Resistivity as a function of Fe concentration for (Ga,Fe)As layers grown at 260°C.

References
1. H. Munekata, H. Ohno, S. von Molnár, A. Segmuller, L.L. Chang, and L. Esaki, Phys. Rev. Lett. **63**, (1989) 1849.
2. H. Munekata, A. Zaslavski, P. Fumagalli, and R.J. Gambino, Appl. Phys. Lett. **63**, (1993) 2929.
3. F. Matsukura, H. Ohno, A. Shen, and Y. Sugawara, Phys. Rev. B **57**, (1998) R2037.
4. S, Haneda, M. Yamaura, Y. Takatani, K. Hara, S. Harigae, and H. Munekata, Jpn. J. Appl. Phys **39**, (2000) L9.
5. Y.L. Soo, G. Kioseoglou, S. Huang, S. Kim, Y.H. Kao, Y. Takatani, S. Haneda, and H. Munekata, to be appeared in Phys. Rev. B.
6. K. Ando, H. Hayashi, M. Tanaka, and A. Towardoski, J. Appl. Phys. B**83** (1998) 6548.
7. M. Abe, Phys. Rev. B**53** (1996) 7065.

incorporated in the form of Fe^{3+} ($3d^5$).

(Ga,Fe)As epilayers have showed magneto-optical (MO) effect at the near-infrared wavelength region (Fig.3). The MO effect for the T_s = 260 °C sample appears to be smaller than that reported for the (Ga,Mn)As epilayers [6]. The θ_K - H curve for the T_s = 580 °C sample shows ferromagnetic behavior reflecting the existence of Fe crystallites, and the magnitude of rotation is evidently larger than that for the host GaAs. The curve is accompanied by peaks at around ±0.5 T. The wavelength dependence on θ_K is different from that of the bulk Fe, which may suggest the modification of MO spectrum by the transparent dielectric GaAs host [7].

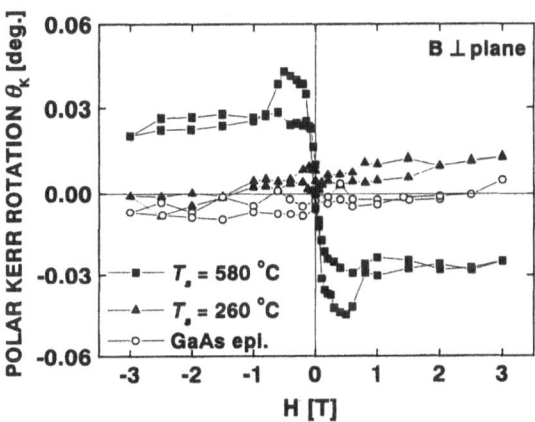

Fig.3 θ_K - H curves of the T_s = 260 °C sample and the T_s = 580 °C sample, together with a standard GaAs epilayer. Measurements were carried out at 4.5 K with probing wavelength of λ =900 nm.

The waves of spin and charge densities in diluted magnetic semiconductors near the paramagnetic-ferromagnetic phase transition

Yu. G. Semenov[1], V. N. Sokolov[1]

Institute of Semiconductor Physics, National Academy of Sciences of Ukraine, Prospect Nauki, 45, 03028, Kiev, Ukraine, e-mail: semenov@spin.kiev.ua

Abstract The carrier induced ferromagnetic (FM) phase ordering in the bulk diluted magnetic semiconductors (DMS) is considered with respect to the problem of non-homogeneous spatial structures formation. The physical reason for appearing such formations is as follows. The gain in the magnetic-ion exchange part of the free energy due to re-distribution of the free carriers into domains with local low concentration and domains with local high one but with local FM ordering can exceed the loss in the kinetic energy of the carriers and Coulomb interactions between the carriers into domains as well as inter-domain interaction. Thus, the FM phase ordering in DMS should take place through a relatively narrow temperature region corresponded to paramagnetic and ferromagnetic phases co-existence.

1 Introduction

Theoretical prediction and investigations of ferromagnetic to paramagnetic transition induced by free carriers in mean field approximation (MFA) for the DMS [1] originated long before the first experimental evidence of this phenomenon. The work [1] poses also the problem how does all of this fit into the scheme of indirect spin-spin interaction, which is known in metal physics as the RKKY interaction model.

More recently, the phenomenon of ferromagnetic ordering was experimentally found in the bulk DMS $Pb_{1-x-y}Sn_yMn_xTe$ (see, for example, Refs [2] and references therein). The RKKY interaction treated in the Curie–Weiss–field approximation (CWFA) was considered to be the single factor of importance for the induced ferromagnetism observed in these semiconductor systems. The reason for such an assumption is that the critical temperatures T_c calculated within the CWFA for the RKKY interaction and with the use of self-consistent procedure for the MFA just coincide. This peculiarity has also been deduced for low-dimensional systems at some additional assumptions (Ref.[3]) which can lead to a spurious conclusion about identical nature of the two aforementioned mechanisms of ferromagnetic ordering in the DMS. On the other hand, an analysis to this problem performed in the framework of exactly solvable model for infinite spin cluster (Ref.[4]) shows that the above mentioned mechanisms must simultaneously contribute to thermodynamic parameters of the system.

In this work, we present sequential calculations of the magnetic susceptibility χ_0 of the bulk DMS doped with free carriers. The proposed theory is based on the scrupulous analysis of different contributions from the carrier-ion exchange interaction to the free energy F of the magnetic system under consideration. Then, we present a proof of the theorem about instability of homogeneous free-carrier distributions in the DMS near the temperature of the FM phase transition.

2 Hamiltonian and free energy

In what follows we restrict ourselves by consideration of the carrier-ion exchange interaction in the model of the effective s(p)-d Hamiltonian. The total Hamiltonian of the system considered is

$$H = H_e + H_m + H_i\,. \tag{1}$$

This is composed of Hamiltonians for the free electrons (holes), H_e, the magnetic ion spin-spin interactions, H_m, and the carrier-ion exchange interactions, H_i, that have conventional form:

$$H_e = \sum_{p,\mathbf{k}}(\varepsilon_{p,\mathbf{k}} + \sigma\omega_e)a^\dagger_{p,\mathbf{k}}a_{p,\mathbf{k}}\,, \tag{2}$$

$$H_m = \frac{1}{2}\sum_{j,j'(j\neq j')}^{N_m} J_m(\mathbf{R}_{j,j'})\mathbf{S}^j\mathbf{S}^{j'} + \sum_{j=1}^{N_m}\omega_m S_z^j\,, \tag{3}$$

$$H_i = -\frac{1}{N_0}\sum_{p,p'}\sum_{\mathbf{k},\mathbf{k}'}\sum_{j=1}^{N_m}J_{b-d}(\mathbf{k},\mathbf{k}')e^{i(\mathbf{k}-\mathbf{k}')\mathbf{R}_j} \times$$
$$\times (p,\mathbf{k}\left|\mathbf{S}^j\mathbf{s}\right|p',\mathbf{k}')\,a^\dagger_{p,\mathbf{k}}a_{p',\mathbf{k}'}\,. \tag{4}$$

Here, $\varepsilon_{p,\mathbf{k}}$ is the electron (hole) energy in a band b with projection σ of the spin $\mathbf{s}$ on the z axis and wave vector $\mathbf{k}$, $p = \{b,\sigma\}$, $a^\dagger_{p,\mathbf{k}}$ and $a_{p,\mathbf{k}}$ are the creation and annihilation fermion operators, $\omega_e = g_e\mu_B B$ is Zeeman splitting of the band energy in a magnetic field $\mathbf{B}$ directed along the z axis; g_e is the electron g factor, μ_B is the Bohr magneton; $J_m(\mathbf{R}_{j,j'})$ is the exchange interaction constant for the magnetic ions with spins $\mathbf{S}^j$ and $\mathbf{S}^{j'}$ localized on the sites $\mathbf{R}_j$ and $\mathbf{R}_{j'}$, $\mathbf{R}_{j,j'} = \mathbf{R}_j - \mathbf{R}_{j'}$, N_m is the total number of magnetic ions, $\omega_m = g_m\mu_B B$ is Zeeman splitting for the magnetic ions with g factor g_m, S_z^j is z-component of the magnetic ion spin $\mathbf{S}^j$; N_0 is the number of primitive sells in the crystal; $J_{b-d}(\mathbf{k},\mathbf{k}')$ is the integral of carrier-ion exchange interaction that, in general, depends on the wave vectors $\mathbf{k}$ and $\mathbf{k}'$, with b denoting c or h for the conduction or valence band, respectively.

We will apply the carrier-ion exchange interaction Hamiltonian (4) to a single electron (hole) band. Under these conditions the interband transitions caused by the interaction H_i drop out from further considerations, so that index p relates to only spin states σ of this band. Moreover, we will neglect dependence on the wave vector in the exchange integral $J_{p-d}(\mathbf{k}, \mathbf{k}')$.

The Hamiltonian H_i includes both carrier-ion exchange interaction between the electron Bloch states without their modification ($\mathbf{k} = \mathbf{k}'$) and exchange scattering processes ($\mathbf{k} \neq \mathbf{k}'$) when each event of carrier-ion spin interaction is accompanied by the electron scattering. According to this property, the Hamiltonian H_i can be separated into two parts, the diagonal part H_{id} and its non-diagonal part H_{in}

One can see that the H_{id} has the form of Zeeman energy in the field $\mathbf{B}$

$$H_{id} = -J_{b-d}\Omega_0 n_e \frac{\xi}{2} \sum_{j=1}^{N_m} S_Z^j, \qquad (5)$$

where Ω_0 is the volume of primitive cell, the parameter ξ characterizing the electron spin polarization and the free carrier concentration n_e are expressed through occupation numbers $n_{\sigma,\mathbf{k}} = a_{\sigma,\mathbf{k}}^\dagger a_{\sigma,\mathbf{k}}$ as $\xi = \sum_{\mathbf{k}}(n_{+1/2,\mathbf{k}} - n_{-1/2,\mathbf{k}})/n_e$ and $n_e = \sum_{\mathbf{k}}(n_{+1/2,\mathbf{k}} + n_{-1/2,\mathbf{k}})$.

The non-diagonal part of the carrier-ion exchange interaction H_{in} in the second-order of perturbation theory with respect to electron quantum numbers $\mathbf{k}$ and $\mathbf{k}'$ takes the form of indirect spin-spin (RKKY) interaction. Because of this the Hamiltonian H_{in} is transformed to the form $H_{in}^{(2)} = \sum_{j,j'} J_{in}(\mathbf{R}_{j,j'})\mathbf{S}^j\mathbf{S}^{j'}$, where we have introduced an effective constant $J_{in}(\mathbf{R}_{j,j'})$ of the RKKY interaction.

The Hamiltonian $H_{in}^{(2)}$ and of spin-spin interactions in the H_m are of the same form. Thus, they should be added together forming an effective Hamiltonian of spin-spin interactions : $H_{SS}^{eff} = \sum_{j,j'} J_{j,j'}^{eff}\mathbf{S}^j\mathbf{S}^{j'}$, with $J_{j,j'}^{eff} \equiv J_m(\mathbf{R}_{j,j'}) + J_{in}(\mathbf{R}_{j,j'})$. Thereby, we have shown that the carrier-ion exchange interaction H_i can be reduced to magnetic one H_m. Their combination gives the effective spin Hamiltonian

$$H_m^{eff} = \frac{1}{2} \sum_{j,j'(j\neq j')}^{N_m} J_m^{eff}(\mathbf{R}_{j,j'})\mathbf{S}^j\mathbf{S}^{j'} +$$
$$+ \sum_{j=1}^{N_m} \left(\omega_m - J_{b-d}\Omega_0 n_e \frac{\xi}{2}\right) S_z^j. \qquad (6)$$

while initial Hamiltonian (1) can be approximated by $H = H_e + H_m^{eff}$. Because the H_e and H_m^{eff} are commuted, their contributions to the total free energy can be calculated independently as the sum of the electronic, Zeeman and magnetic parts: $F = F_e + F_Z + F_m$. The electronic part is calculated in a common manner. Calculation of the magnetic part can be performed in terms of phenomenological expression for the magnetization M_0 proposed in the work [5] with modified Brillouin function $M_0 = n_m g_m \mu_B S_0 B_S \left(\frac{g_m \mu_B BS}{T+T_0}\right)$. Two fitting parameters S_0 and T_0 just reflect spin-spin interaction. Taking

into account the relation $F_m = -\int_0^B M(B')dB'$ one can find

$$F(\xi) = \frac{9n_e^2}{40G(\varepsilon_F)}\left[(1+\xi)^{5/3} + (1-\xi)^{5/3}\right] + n_e g_e \mu_B B\frac{\xi}{2}$$
$$- n_m'(T + T_0 + T_{in}) \times$$
$$\times \ln Z_S\left(\frac{g_m \mu_B B - J_{b-d}\Omega_0 n_e \xi/2}{T + T_0 + T_{in}}\right). \qquad (7)$$

where T_0 and T_{in} correspond to two parts in the $J_m^{eff}(\mathbf{R}_{j,j'})$, $n_m' = n_m(S_0/S)$, $Z_S\left(\frac{g_m \mu_B B}{T}\right)$ being the partition function of a single spin S.

If we consider the $F(\xi)$ as a functional of ξ only, the procedure of free energy minimizing give rise to following expression for the critical temperature of FM ordering:

$$T_c = \frac{S(S+1)(J_{b-d}\Omega_0)^2 n_m' n_e^{1/3} m^*}{4 \cdot 3^{2/3}\pi^{4/3}\hbar^2} - T_{eff}. \qquad (8)$$

3 Homogeneous state stability

In order to obtain the thermodynamical conditions for spin and charge waves formation, one have to consider the F as the function of ξ and n_e. The chain of following conditions defines the stability of the system with respect to redistribution of free carriers for the domains with high and low local concentrations:

$$\frac{\partial F}{\partial \xi} = 0; \frac{\partial F}{\partial n_e} = 0; \qquad (9)$$

$$\frac{\partial^2 F}{\partial n_e^2} > 0; \frac{\partial^2 F}{\partial n_e^2}\frac{\partial^2 F}{\partial \xi^2} - \left(\frac{\partial^2 F}{\partial n_e \partial \xi}\right)^2 > 0. \qquad (10)$$

The analysis of these conditions carried out with Eq.(7) shows that the system loses stability for small carrier concentration fluctuations just at the temperature of FM ordering. The calculation of the spin/charge waves parameters have to incorporate to the expression of Eq.(6) the Coulomb carrier-carrier interaction. These results will be presented soon.

References

1. E.A.Pashitskij, S.M.Ryabchenko, Fiz. Tverd. Tela, **21**, 545 (1979).
2. T.Story, G.Karczewski, L.Swierkowski and R.R.Galazka, Phys.Rev. B **42**, 10477 (1990).
3. Y. Merle d'Aubigné, A. Arnoult, J. Cibert, T. Dietl, A.Haury, P.Kossacki, S. Tatarenko, A. Wasiela.Physica E, **3,**169 (1998).
4. Yu.G.Semenov, S.M.Ryabchenko. cond-mat/9908040.
5. J.A.Gaj, W.Grieshaber, C.Boden-Deshayes, J.Cibert, G.Feuillert, Y.Merle d'Aubigne and A.Wasiela. Phys.Rev. B **50**, 5512 (1994).

Electronic Structure of $Ga_{1-x}Mn_xAs$ Studied by Photoemission Spectroscopy

J. Okabayashi[1*], **A. Kimura**[2], **O. Rader**[3], **T. Mizokawa**[1], **A. Fujimori**[1], **T. Hayashi**[4], **M. Tanaka**[4]

[1] Department of Physics and Department of Complexity Science and Engineering, University of Tokyo, 7-3-1 Hongou, Bunkyou-ku, Tokyo 113-0033, Japan e-mail: jun@wyvern.phys.s.u-tokyo.ac.jp
[2] Department of Solid State Physics, Hiroshima-University, Higashi-Hiroshima 739-8526, Japan
[3] BESSY, Albert-Einstein-str. 15, D-12489 Berlin, Germany
[4] Department of Electronic Engineering, University of Tokyo, Bunkyo-ku, Tokyo 113-0033, Japan

Abstract We have obtained the Mn $3d$ partial density of states in $Ga_{1-x}Mn_xAs$ and energy-band dispersion by photoemission techniques to study the Mn doing effects in GaAs. We found that states near the Fermi level largely derived from As $4p$ states with some admixture of Mn $3d$ character. We also observed non-dispersive impurity-band like states near the Fermi level, which may give a key to understand the anomalous properties of $Ga_{1-x}Mn_xAs$.

The recent success of doping III-V semiconductors with magnetic impurities using the molecular beam epitaxy technique has opened up a new field in the research of diluted magnetic semiconductors (DMS) [1]. Because one can introduce both magnetic moments and charge carriers in the new DMS, they have the possibility of combining magnetics and electronics. Owing to the presence of conducting carriers, III-V based DMS show "carrier-induced ferromagnetism" in p-type samples. In a recent few years, a new III-V based DMS system $Ga_{1-x}Mn_xAs$ has attracted considerable interest because of its relatively high Curie temperature ($T_c \sim$ 110 K). In order to elucidate the magnetic and transport properties of the III-V based DMS, it is necessary to obtain information about their valence-band electronic structure, including the p-d exchange interaction between the Mn $3d$ electrons and the band electrons. The aim of this paper is to obtain the Mn $3d$ partial density of states (DOS) by resonant photoemission spectroscopy [2] and energy-band dispersion by angle-resolved photoemission spectroscopy (ARPES) [3], thereby revealing the electronic structure and obtaining a key to understand the anomalous properties of $Ga_{1-x}Mn_xAs$.

The photoemission experiments were performed at BL18A of the Photon Factory, High Energy Accelerator Research Organization. The total energy resolution was 100 meV because the measurements were done at room temperature. Photoemission measurements were performed in an ultra-high vacuum of 10^{-11} Torr. To remove oxidized surface layers and other contaminations, we made repeated Ar sputtering (1 kV) and annealing up to 240 °C in order to avoid the decomposition into GaAs and MnAs particles. After the annealing, we confirmed the ordered surfaces by observing 1×1 LEED patterns.

* *Present address:* Department of Physics and Department of Complexity Science and Engineering, University of Tokyo, 7-3-1 Hongou, Bunkyou-ku, Tokyo 113-0033, Japan

Figure 1 shows a series of valence-band spectra of $Ga_{0.931}Mn_{0.069}As$ for excitation photon energies ($h\nu$) in the Mn $3p$ - $3d$ core-excitation region. The intensities have been normalized to the photon flux. One can see the growth of a peak in going from $h\nu$=46 eV to $h\nu$=50 eV at a binding energy of 4.5 eV, which we attribute to the Mn $3d$ origin. The inset of Fig. 1 shows the Mn $3p$ core absorption spectra measured in the total electron yield mode, which shows a peak at $h\nu$=50 eV.

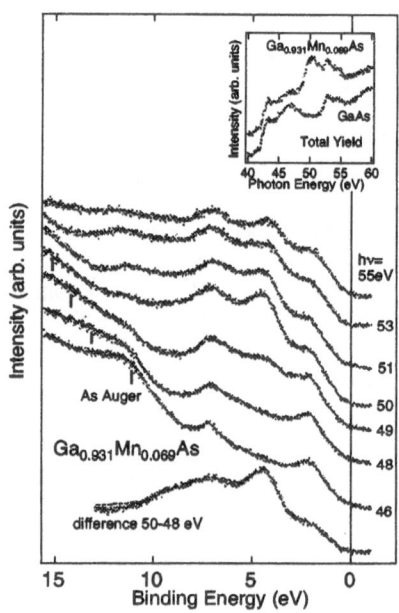

Fig. 1 Photoemission spectra of $Ga_{1-x}Mn_xAs$ for various photon energies near the Mn $3p \rightarrow 3d$ core excitation threshold. The hatched area is the As Auger contribution. The difference between the on-resonant ($h\nu = 50$ eV) and off-resonant (48 eV) spectra, which is a measure of the Mn $3d$ partial density of states, is shown at the bottom. The inset shows the absorption spectra of $Ga_{0.931}Mn_{0.069}As$ and GaAs recorded by the total electron yield method.

By subtracting the off-resonant ($h\nu$=48 eV) spectrum from the on-resonant (50 eV) one, we obtained the Mn $3d$ partial DOS as shown in the bottom panel of Fig. 1. The difference spectrum thus obtained corresponds to the Mn $3d$ partial DOS. This indicates the main peak at 4.5 eV and a strong satellite structure at 6-10 eV. The spectrum also shows a broad feature at a lower binding energy of ∼2 eV. The satellite at 6-10 eV binding energy cannot be reproduced by the band-structure calculation [4–6] and would therefore be attributed to a many-body effect. The line shape of the Mn 3d partial DOS is thus similar to that of Mn-doped II-VI compounds. There is little intensity at the Fermi level (E_F) in the Mn 3d spectrum, which is not reproduced by the local-density-approximation (LDA) band structure calculation [4,5], however, a calculation with Coulomb interaction (LDA+U) reproduces low Mn 3d spectral weight near E_F [6].

To know more details about the electronic structure of Ga$_{1-x}$Mn$_x$As, we performed ARPES experiments. Figure 2 shows the normal emission spectra. They show a dispersion along $\Gamma - \Delta - X$ direction in the Brillouin zone. Near E_F, new states are induced by Mn doping, which one can see by comparing the spectra of Ga$_{0.935}$Mn$_{0.035}$As and GaAs. These new states may be derived from the acceptor level of the neutral Mn impurity (A^0) in GaAs, which is located above the valence-band maximum by about 100 meV and may be viewed as Mn impurity states. Compared with GaAs, only the Δ_1 peak in Ga$_{1-x}$Mn$_x$As is shifted downward by 0.1-0.2 eV as a result of the hybridization with the Mn 3d states.

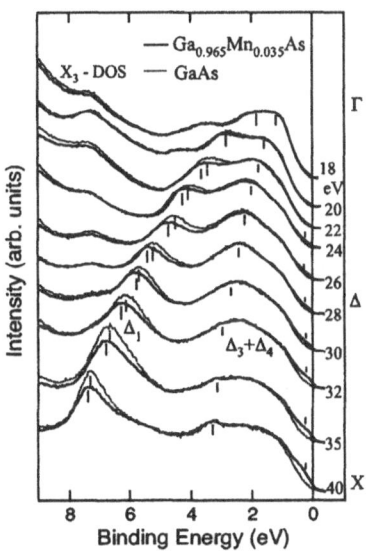

Fig. 2 Angle-resolved photoemission spectra of Ga$_{0.965}$Mn$_{0.035}$As (solid curves) and GaAs (dash curves) along the $\Gamma - \Delta - X$ ([001]) direction. Vertical bars show peak or shoulder positions.

Considering the resonant photoemission and ARPES results, we can discuss the origin of the impurity-band-like states. The Mn-induced states near E_F should be responsible for the metallic conduction in Ga$_{1-x}$Mn$_x$As and would be the origin of the carrier-induced ferromagnetism and the anomalous magneto-transport properties. If the new states induced by the Mn doping are derived from the acceptor level, which is originally located above E_F, large part of the impurity states would be located above E_F and ARPES sees only the bottom part of these states. From the previous cluster-model anal-

ysis [2,7], dominant components near E_F in the photoemission spectra come from a Mn 3d hole screened by charge transfer of a host valence electron ($d^5\underline{L}$ photoemission final state, where $\underline{L}$ denotes a hole in the As 4p valence band), because the original Mn 3d level is located far ($\sim$4.5 eV) below E_F. According to the Anderson-impurity model calculation, the neutral Mn in GaAs may induce a split-off state above the valence-band maximum through hybridization with Mn 3d states [8] and act as an acceptor. Therefore the wave function of the new states would have mainly As 4p character with hybridized Mn 3d character but has the same symmetry as the Mn 3d orbitals with respect to the Mn site. The Mn 3d component, although not dominant, would lead to the spin polarization of the impurity states. That is, the spin of the hole in the "impurity band" is coupled to the spin of the Mn 3d^5 configuration, and thus contribute to the magneto-transport phenomena in Ga$_{1-x}$Mn$_x$As. It should be noticed that the "impurity band" has a finite but low spectral weight at E_F and is spread over 0.5 - 1 eV from E_F. This behavior is similar to the doping-induced states in La$_{1-x}$Sr$_x$MnO$_3$ and implies a highly incoherent nature of the metallic states in Ga$_{1-x}$Mn$_x$As.

In conclusion, we have studied Ga$_{1-x}$Mn$_x$As by resonant photoemission and ARPES. We have observed the band dispersions for Ga$_{1-x}$Mn$_x$As in the $\Gamma - \Delta - X$ direction in the Brillouin zone. We have observed dispersionless "impurity band states" near E_F, which are also induced by the hybridization of the host valence-band states with the Mn 3d states.

References

1. H. Ohno, Science, **281** 951 (1998).
2. J. Okabayashi, A. Kimura, T. Mizokawa, A. Fujimori, T. Hayashi and M. Tanaka, Phys. Rev. B **59**, 2486 (1999).
3. J. Okabayashi, A. Kimura, O. Rader, T. Mizokawa, A. Fujimori, T. Hayashi and M. Tanaka, Phys. Rev. B, submitted.
4. H. Akai, Phys. Rev. Lett. **81**, 3002 (1998).
5. M. Shirai, T. Ogawa, I. Kitagawa, N. Suzuki, J. Magn. Magn. Mater., **177-181**, 1383 (1998).
6. J. H. Park, S.K. Kwon and B. I. Min, Physica B, **281, 282**, 703 (2000).
7. J. Okabayashi, A. Kimura, O. Rader, T. Mizokawa, A. Fujimori, T. Hayashi and M. Tanaka, Phys. Rev. B **58**, 4211 (1999).
8. T. Mizokawa, unpublished.

Electronic Structure and Magnetic Properties of (Zn,Mn)Se With Carriers Doping Studied by FLAPW Method

K. Matsushita[1], H. Harima[2], A. Yanase[2], H. Katayama-Yoshida[2]

[1] Department of Physics, Graduate School of Science, Osaka University 1-1 Machikaneyama, Toyonaka, Osaka, 560-0043, Japan, e-mail: kmatsu@presto.phys.sci.osaka-u.ac.jp

[2] Institute of Scientific and Industrial Research, Osaka University 8-1, Mihogaoka, Ibaraki, Osaka 567-0047, Japan

Abstract We have calcurated the electronic structures of $Zn_{0.5}Mn_{0.5}Se$ by using a FLAPW method. The anti-ferromagnetic states are stable upon electron and hole-doping up to 0.4 per $Zn_{0.5}Mn_{0.5}Se$ because of the anti-ferromagnetic super-exchange interaction in Mn-Se-Mn bonds with strong p-d hybridization. We propose three methods in order to suppress the anti-ferromagnetic interaction in Mn-Se-Mn bonds. We have found the effective two of them by using a FLAPW method.

1 Introduction

Diluted magnetic semiconductors (DMS) have been studied widely in order to fabricate new ferromagnetic materials since the discovery of the ferromagnetic III-V compound DMS with high enough density of carriers [1]. The ferromagnetic II-VI compound base DMS have been studied, because they have an advantage; i.e., the doping of transition atoms in II-VI compound semiconductor is easier than that in III-V compound semiconductor because of the higher solubility. The ferromagnetic II-VI compound DMS is discovered in modulation-doped $Cd_{1-x}Mn_xTe$ quantum well [2] and $Zn_{1-x}Mn_xTe$ with p type-doping system [3]. However, the ferromagnetic transition temperatures are very low (a few °K). Recently, the ferromagnetic DMS in $Zn_{0.75}Mn_{0.25}O$ system with hole-doping [4] has been proposed. We have calcurated the electronic structures of $Zn_{1-x}Mn_xSe$, in order to show the guidelines to fabricate new ferromagnetic materials with the transition temperature up to room temperatures.

2 The Method of Calculation

The electronic structure calculations have been carried out by using the full-potential linearized augmented plane wave method (FLAPW) within the local spin density approximation.

The super-cell lattice structures A in $Zn_{0.5}M_{0.5}Se$, (M = Cr, Mn and Fe) are shown in Fig. 1a). Zn layers of ZnSe in this super-lattice alternately are substituted by transition atoms layers. There are Mn-Se-Mn bonds due to the existence of Mn layers. The bond length d, the energy cut-off of FLAPW basis and a muffin-tin radius are set as 2.50 Å (as extrapolated from experimental values [5] for small x in $Zn_{1-x}Mn_xSe$), 10.77 Ry and 0.482 d, respectively. The effects of carrier-doping are investigated

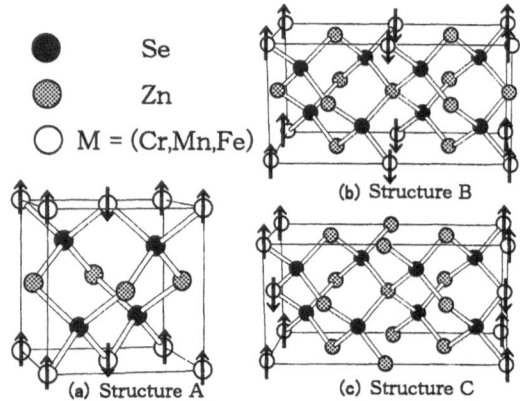

Fig. 1 (a) The super-cell of $Zn_{0.5}M_{0.5}Se$ (M = Cr, Mn, Fe). (b–c) The super-cells of $Zn_{0.75}Mn_{0.25}Se$. The arrows on M sites indicate spin directions of the anti-ferromagnetic states

in the band structure calculations with changing of the number of valence electrons and the nuclear charges of Se to keep charge neutrality.

The super-lattices in $Zn_{0.75}Mn_{0.25}Se$ are also shown in Fig. 1b–c. In the structures B, Mn atoms are uniformly distributed, while Mn layers only are formed in the structure C. Then the Mn-Se-Mn bonds are contained in the structure C. The energy cut-off of FLAPW basis and muffin-tin radius are chosen as 8.30 Ry and 0.489 d. The bond length is the same value of $Zn_{0.5}Mn_{0.5}Se$.

3 Result and Discussion

3.1 $Zn_{0.5}Mn_{0.5}Se$

In $Zn_{0.5}Mn_{0.5}Se$, the density of states (DOS) in the ferromagnetic state without carrier-doping are shown in Fig. 2a. The valence bands consist of Mn-$3d$ orbitals of majority spin states and Se-$4p$ orbitals. Mn-$3d$ bands in the valence bands are wide due to strong p–d hybridization. The conduction bands consist of Mn-$3d$ orbitals of minority spin states. In the ferromagnetic state, the valence bands and conduction bands are slightly overlapped. In the anti-ferromagnetic state, the composition of Mn-$3d$ orbitals and Se-$4p$ orbitals in valence bands and conduction bands are similar to that of the ferromagnetic state. There is a band gap in the anti-ferromagnetic state. Fig. 2b shows $\Delta E = E(Anti-Ferro) - E(Ferro)$, which is the difference of total energy

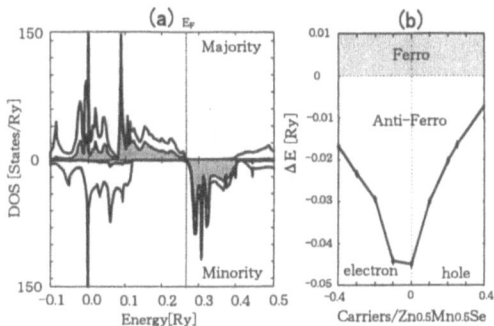

Fig. 2 (a) Dos of the ferromagnetic state without carrier-doping. (b) ΔE of $Zn_{0.5}Mn_{0.5}Se$. The gray zones of DOS are contribution of Mn-3d orbitals

between an anti-ferromagnetic state and a ferromagnetic state as a function of carrier concentration in $Zn_{0.5}Mn_{0.5}Se$. ΔE decreases with increasing the number of holes or electrons because of the ferromagnetic interaction is introduced with the carriers of Mn-3d band by gaining the band energy. However, the ferromagnetic state is unstable up to 0.4 holes and 0.4 electrons per $Zn_{0.5}Mn_{0.5}Se$. As shown Fig. 2a, Mn-3d orbitals and Se-4p orbitals hybridize strongly. Therefore, the instability of the ferromagnetic states are caused by the strong anti-ferromagnetic super-exchange interaction due to the strong p-d hybridization in the Mn-Se-Mn direct bonds.

Then, in order to stabilize the ferromagnetic state by suppressing the anti-ferromagnetic super-exchange interaction, we proposed the three methods as fellows: (i) The substitution other VI atoms for Se atoms in order to suppress the p-d hybridization. (ii) The substitution other transition atoms for Mn atoms in order to suppress the p-d hybridization. (iii) The extending of the Mn-Mn atoms in order to suppress the hopping of electrons which dominate the anti-ferromagnetic super-exchange interaction.

3.2 $Zn_{0.5}Cr_{0.5}Se$ and $Zn_{0.5}Fe_{0.5}Se$

In order to estimate the effectiveness of the method (ii), we have studied the magnetic properties of $Zn_{0.5}Cr_{0.5}Se$ and $Zn_{0.5}Fe_{0.5}Se$. Table 1 shows the results for ΔE . ΔE in the ferromagnetic $Zn_{0.5}Cr_{0.5}Se$ is the smallest in three materials. Then the p-d hybridization in $Zn_{0.5}Cr_{0.5}Se$ is the weaker than those of $Zn_{0.5}Mn_{0.5}Se$ and $Zn_{0.5}Fe_{0.5}Se$, because the energy levels of Cr-3d orbitals are situated above far those of Se-4p orbitals. The ferromagnetic states in $Zn_{0.5}Cr_{0.5}Se$ and $Zn_{0.5}Fe_{0.5}Se$ are more stable than That in $Zn_{0.5}Mn_{0.5}Se$ because of the carrier-induced ferromagnetic interaction.

We calcurated ΔE of $Zn_{0.5}Cr_{0.5}Se$ with 0.1 hole and 0.1 electron-doping, and we found that the ferromagnetic states are stabler than the anti-ferromagneticstates.

Table 1 ΔE in $Zn_{0.5}M_{0.5}Se(M = Cr, Mn$ and Fe) in $Zn_{0.75}Mn_{0.25}Se$

M : ΔE [mRy]/atom
Cr : -0.56
Mn : -2.81
Fe : -1.81

Table 2 ΔE in structure B and C

Structure : ΔE [mRy]/atom
B : -0.37
C : -1.24

3.3 $Zn_{0.75}Mn_{0.25}Se$

In order to estimate the effectiveness of the method (iii), we have studied the magnetic properties of structures B and C (see Fig. 1a–b) of $Zn_{0.75}Mn_{0.25}Se$ Table 1 shows the redults for ΔE . The ferromagnetic state of structure B is more stable than that of structure C because of the suppression of the electrons hopping which dominate the anti-ferromagnetic super-exchange interaction due to no Mn-Se-Mn bonds.

4 Summary

We have calcurated the electronic structures in order to study magnetic properties of $Zn_{0.5}Mn_{0.5}Se$ by using a FLAPW method with super-cells. The anti-ferromagnetic states are stable upon electron and hole-doping up to 0.4 per $Zn_{0.5}Mn_{0.5}Se$ because the anti-ferromagnetic super-exchange interaction in the Mn-Se-Mn bonds overcome the carrier-induced ferromagnetic interaction. In order to a ferromagnetic state is stabilized, We propose the three methods. We have studied the magnetic properties in $Zn_{0.5}Cr_{0.5}Se$, $Zn_{0.5}Fe_{0.5}Se$ and two kinds of Mn distribution in $Zn_{0.75}Mn_{0.25}Se$. we show the effectiveness of two methods.

References

1. H. Munekata, H. Ohno, S. von Molnar, A. Segmüller, L.L. Chang and L. Esaki, Phys. Rev. Lett. **63** (1989) 1849
2. A. Haury, A. Wasiela, A. Arnoult, J. Cibert, S. Tatarenko, T. Dietl, and Y. Merle, Aubigné, Phys. Rev. Lett. **79** (1997) 511
3. D. Ferrand, J. Cibert, C. Bourgognon, S. Tatarenko, A. Wasiela, G. Fishman, A. Bonanni, H. Sitter, S. Koleśnik, J. Jaroszyński, and T. Dietl, J. Cristal Growth. **214/215** (2000) 387–390
4. K. Sato, H. Katayama-Yoshida, Jpn. J. Appl. Phys. **39** (2000) L555
5. J.K. Furdyana J. Appl. Phys. **64(4)** 15 (1988), R29

Cyclotron resonance of spin polaron in CdMnTe/CdMgTe 2D electron systems

Y. Imanaka[1], T. Takamasu[1], G. Kido[1], G. Karczewski[2], T. Wojtowicz[2] and J. Kossut[2]

[1] National Research Institute for Metals 1-2-1 Sengen, Tsukuba, Ibaraki 305-0047, Japan e-mail: imanaka@nrim.go.jp
[2] Institute of Physics, Polish Academy of Science Al. Lotnikow 32/46.02-668 Warsaw, Poland

Abstract   Detailed far-infrared studies were performed in quantum Hall systems of diluted magnetic semiconductors, CdMnTe/CdMgTe single quantum well at magnetic fields up to 15T. The resonant width changed remarkably against the magnetic field and the drastic broadening of the resonant width with decreasing the magnetic field was observed at low magnetic fields. This rapid broadening is possible to be explained as the formation of the spin polaron.

1 Introduction

The cyclotron resonance is the most direct method to study the effective mass of semiconductors and information of various interactions is possible to be investigated through the magnetic field and the temperature dependence of the cyclotron resonance. In quantum Hall systems, many studies of cyclotron resonance were investigated so far for the extremely high mobility specimens. Anomalous cyclotron resonance originated to the electron-electron interaction or the electron-phonon interaction was observed in III-V or II-VI 2D systems [1, 2], whereas the exchange interaction between the carrier and the localized magnetic ion has not investigated by the cyclotron resonance yet. The exchange interaction causes various phenomena; the gigantic Zeeman effect, the carrier induced ferromagnetism and the spin polaron. Very few cyclotron resonance experiments were reported in past, however, the detailed study used to be impossible due to the low quality of specimens. Recent progress of the epitaxial growth technique enables us to study the quantum Hall effect in II-VI semiconductors and II-VI diluted magnetic semiconductors (DMS) [3,4]. Cyclotron resonance experiments are also possible even in II-VI DMS [5].

In this study, we performed the CR experiment in CdMnTe/CdMgTe single quantum well (CMT-SQW) at high magnetic fields up to 15T in the temperature range between 2K and 50K. The anomalous behavior of the cyclotron resonance in CMT-SQW was studied in details.

2 Experimental

Far-infrared transmission experiments were performed by the FT-IR (BOMEM DA8) with a superconducting magnet. An Hg lamp was used as a far-infrared source and a transmission signal was detected by a Si bolometer operated at 4.2K. The magnetic field up to 15T was applied perpendicular to a surface of the sample. CMT-SQW was grown by MBE on the (100) GaAs substrate. The iodine donor was doped in the center of the barrier layer. The mobility and the carrier concentration

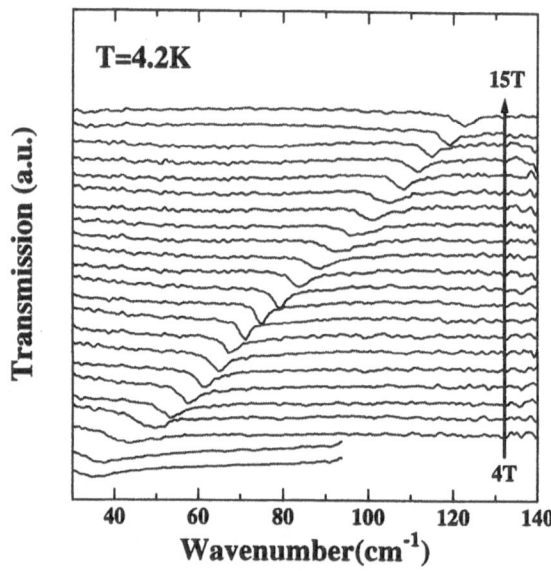

Fig. 1 The magnetic field dependence of the transmission spectra in CdMnTe/CdMgTe SQW between 4T and 15T

of CMT-SQW were estimated as $\mu=6\times10^4 \mathrm{cm}^2/\mathrm{V\cdot s}$ and $n=2\times10^{11}\mathrm{cm}^{-2}$ from the cyclotron resonance and the photoluminescence. The concentration of Mn and Mg was also obtained approximately as 3% and 17% respectively. Samples were wedged to avoid the interference effect by polishing.

3 Result and Discussions

Figure 1 shows the magnetic field dependence of the cyclotron resonance in CMT-SQW between 4T and 15T at every 0.5T. The cyclotron resonance was observed clearly in the whole range. The resonant width becomes narrower with increasing the magnetic fields up to 8.5T in Fig.1. The highest mobility was estimated from the resonant width as $\mu=6\times10^4\mathrm{cm}^2/\mathrm{V\cdot s}$.

The resonant width turns to be broad above 8.5T and it becomes sharp again at high magnetic fields. In comparison with 2D electron systems, CdTe/CdMgTe SQW, such a large variation of the width was not observed so far [6]. It is evident that anomalous behaviors in 2D electron systems of DMS attribute to the introduced Mn ion to the quantum well.

The spin of the carrier is polarized rapidly with applying the magnetic field due to the gigantic effective g-factor enhanced by the exchange interaction, whereas the Mn spin is not polarized up to moderate magnetic fields following the Brillouin function. The spin scattering between the carrier and the localized Mn ion becomes effective at low magnetic fields and this is one of possible explanations for the broadening of the cyclotron resonance width at low magnetic field. The localized behavior in CMT-SQW at low magnetic fields was observed in the energy dependence of the cyclotron resonance concerning with the disorder of the 2D system [5]. The disorder (randomness and spin fluctuation) also plays an important role for a localized behavior at low magnetic fields in our sample.

We found the asymmetrical shape of the cyclotron spectrum around 11T and the spectrum was fitted better by double Lorentzian curves than a single one in this region. This result suggests that the broadening around 11T is due to the overlap of two resonances. The magnetic field, at which the gigantic Zeeman energy becomes comparable with the cyclotron energy, is estimated around 10T from the simple calculation of the Brillouin function. The level crossing of different spin states of Landau levels may contributes to the asymmetrical shape of the cyclotron resonance in this region.

Figure 2a and 2b show the temperature dependence of the cyclotron resonance at 9T and 4T. The resonant width becomes narrow with decreasing the temperatures at 9T, whereas the one at 4T becomes broad below 20K. The scattering of the carrier is suppressed and the resonant width becomes narrower with decreasing temperatures in generally. If we employ the spin scattering model (between the carrier spin and the Mn spin) for explaining the temperature dependence, the CR width should become narrow with decreasing the temperature owing to the polarization of the Mn spin. Our result at 4T is in contradiction to this assumption.

The spin polaron (the magnetic polaron) is the very famous concepts for explaining the several anomalous behaviors in DMS and the spin polaron is more stable in low dimensional systems than bulk system. If the spin polaron model is employed, the temperature dependence of the resonant width at 4T is possible to explain as the stabilization of the spin polaron with lowering temperatures. The spin polaron can be considered as the appropriate candidate for explaining anomalous behaviors of the cyclotron resonance in DMS.

4 Summary

We performed the detailed cyclotron resonance experiment in the quantum Hall system of II-VI DMS, CdMnTe/CdMgTe SQW. The anomalous magnetic field dependence of the resonant width was observed. The magnetic polaron model is good candidate for explaining the rapid narrowing of the resonant width up to 8.5T rather than the spin scattering model. The origin of the width broadening around 11T is not clear at present. The level crossing of different spin states may cause the two cyclotron transitions and the overlap of these two resonances is the origin of the broadening at high magnetic

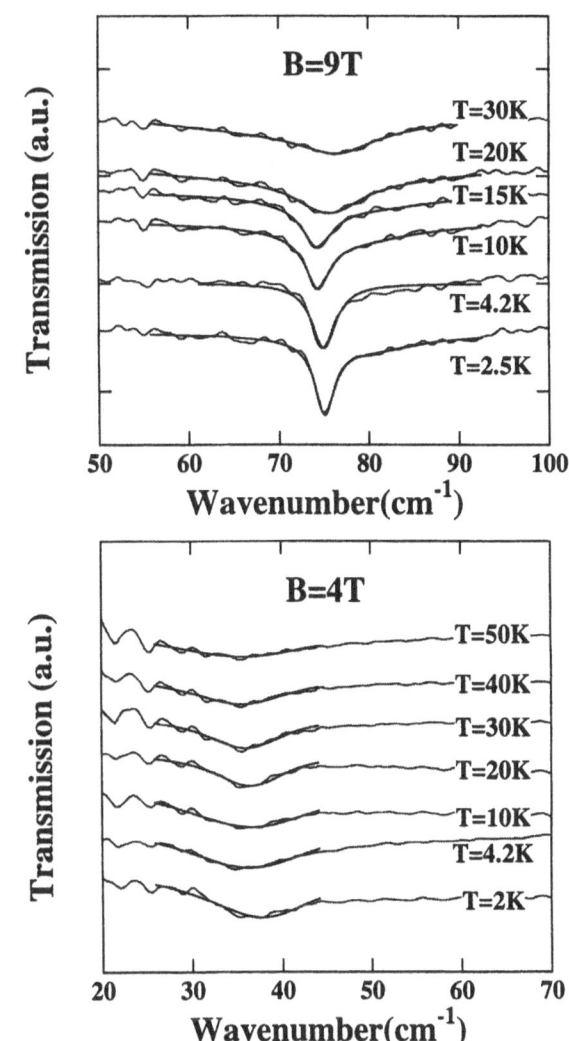

Fig. 2 The temperature dependence of the transmission spectra in CdMnTe/CdMgTe SQW at 4T and 9T.

fields region. Further experiments are needed to clarify the anomalous behavior of the quantum Hall system of II-VI DMS.

5 Acknowledgement

This work was supported in part by the grant 2P03B 119 14 from the Polish State Committee for Scientific Research.

References

1. J. G. Michels *et. al.*, Phys. Rev. B **54**, (1996) 13807.
2. Y.Imanaka *et. al.*, Physica B, **249-251**, (1998) 932.
3. G. Karczewski *et. al.*, J. Crystal Growth **184/185** (1998) 814.
4. J. Jaroszynski *et. al.*, Physica E **6** (2000) 790.
5. M. L. Sadowski *et. al.*, *Proc. of 24th Int. Conf. Phys. Semicond.*, edited by D Gershoni, (Singapore, 1999), Vol.4 B-4.
6. Y.Imanaka *et. al.*, Physica B, **256-258**, (1998) 457.

Far infrared absorption spectra in ferromagnetic Ga$_{1-x}$Mn$_x$As

Y. Nagai[1], **T. Kunimoto**[1*], **K. Nagasaka**[1], **H. Nojiri**[2], **M. Motokawa**[2], **F. Matsukura**[3] and **H. Ohno**[3]

[1] Department of Physics, Science University of Tokyo, 1-3 Kagurazaka, Shinjuku-ku, Tokyo 162-8601, Japan

[2] Institute for Materials Research, Tohoku University, 2-1-1 Katahira, Sendai 980-8577, Japan

[3] Laboratory for Electronic Intelligent Systems, Research Institute of Electrical Communication, Tohoku University, 2-1-1 Katahira, Sendai 980-8577, Japan

Abstract   Temperature and magnetic field dependence of far infrared absorption spectra have been measured on ferromagnetic Ga$_{1-x}$Mn$_x$As (x=0.034). The large enhancement of the absorption intensity was observed below the ferromagnetic ordering temperature T_C, which is considered as the evidence of the strong coupling between the conduction and the spin polarization in this material. The increase of the absorption intensity is also induced by the application of high magnetic fields, and a simple scaling relation between the magnetization and the absorption intensity is found.

1 Introduction

Diluted magnetic semiconductors exhibit many interesting properties caused by the interplay of spin and charge degrees of freedoms [1,2]. One of the most remarkable materials is a highly hole-doped Ga$_{1-x}$Mn$_x$As which shows ferromagnetism. It has been widely accepted that the ferromagnetic coupling between Mn ions is mediated by mobile carriers. However, the microscopic mechanism of the ferromagnetism has not been fully understood so far [3–5]. To study this problem experimentally, we have performed an infrared (IR) and far infrared (FIR) spectroscopy [6]. The following important features are found for x=0.034 sample with the Curie temperature T_C=43 K. (1) A broad absorption peak was found around 1600 cm^{-1}, which is very close to the Mn acceptor level in GaAs [7]. (2) The Drude weight is small and the FIR absorption spectrum is dominated by the hopping conduction, showing that the system is located near the metal-insulator (MI) transition point, probably in the insulating side. (3) The FIR absorption increases steeply below the T_C . The third point indicates that the hopping probability is strongly enhanced by the increase of the spin polarization caused by the ferromagnetic ordering. As is well known, a magnetic field is a useful mean to control the magnetization of the system. In the present work, we investigated the magnetic field dependence of FIR spectra to find out the relation between the FIR absorption and the magnetization.

2 Experimental

Ga$_{1-x}$Mn$_x$As (x=0.034) thin film with 2000 nm thickness was grown by molecular beam epitaxy technique on a semi-insulating GaAs (001) substrate. The Curie

temperature T_C was determined to be 43 K by the temperature dependence of residual magnetization. To obtain FIR transmission spectra, a Michelson interferometer was used with a Ga doped Ge bolometer. Magnetic field was applied in the Faraday configuration using a superconducting magnet up to 8 T.

3 Results and Discussion

FIR absorption spectra are shown in Fig. 1. The absorption peaks above 240 cm^{-1} are caused by the phonons. As mentioned before, the FIR spectra show that the hopping conduction is dominant. It is also found that the spectral shape does not change by the application of high magnetic fields and that the absorption is higher for 8 T. The field and temperature dependence of the integrated absorption coefficient α between 50 and 150 cm^{-1} is also shown in the inset. At zero field, the steep decrease of α by decreasing temperature indicates that "phonon-assisted hopping" is dominant for higher temperatures, while below T_C a steep upturn of the α is found. Since the phonon-assisted hopping is less effective

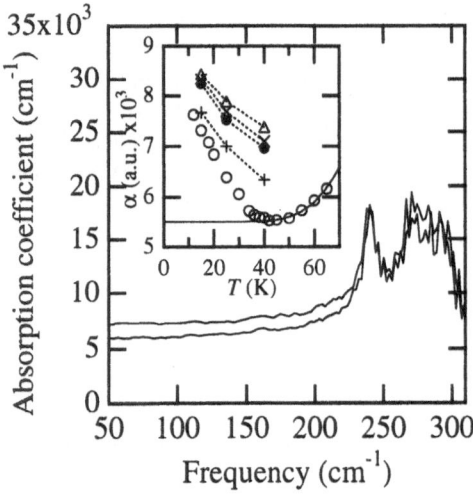

Fig. 1 FIR absorption spectra at 0 (lower trace) and 8 T (upper trace) at 15 K. The inset shows the field and temperature dependence of the integrated absorption coefficient α between 50 and 150 cm^{-1} at 0($\circ$), 2(+), 4($\bullet$), 6($\times$) and 8 T ($\triangle$). The solid curve shows the absorption due to the phonon-assisted hopping which is proportional to the phonon population number and the other absorption.

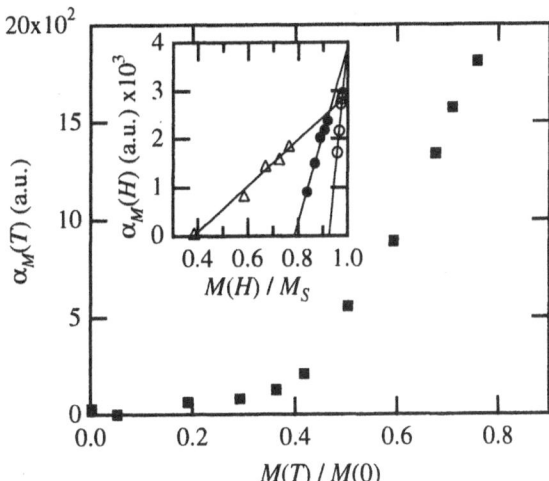

Fig. 2 The plot of α_M against the magnetization $M(T)/M(0)$ at zero field. Inset: The plot of α_M versus normalized magnetization $M(H)/M_S$ calculated by the molecular field theory at 15($\circ$), 25($\bullet$) and 40 K ($\triangle$).

at low temperatures, it is considered that the enhancement of the absorption is caused by the ferromagnetic spin alignment below T_C. If this interpretation is correct, at high fields, we can expect that the α is increased because of the increase of the magnetization. In fact, the increase of α at high fields are found as shown in the inset, which supports the above interpretation. It is also noticed that the field change of the α becomes smaller for lower temperature. This tendency can be explained considering the fact that the high field magnetization slope is less steep for lower temperatures.

To investigate the relation between the absorption and the magnetization, we plotted α_M as a function of $M(T)/M(0)$ for zero field in Fig. 2. The plot at high fields is also shown in the inset. In these plots, we defined α_M as follows: a component of "phonon-assisted hopping" absorption and the other absorption shown in the inset of Fig. 1 are subtracted by assuming that the "spin polarization dependent hopping" component is very small above T_C and at zero field, because of the zero magnetization. This assumption means that a short range spin correlation above T_C is neglected. As the magnetization at zero field, we employed the residual magnetization measured by a SQUID magnetometer. Since the magnetization curves below T_C are not available at present, we calculated the normalized magnetization M/M_S by using the conventional molecular field theory.

In Fig. 2, it is found that there is a strong correlation between the M and the α_M. The inflection point at $M(T)/M(0) \sim 0.4$ is found. It is noticed that the value of the magnetization is related to the 35 K, where a residual magnetization starts to deviate from the $M(T)$ curve calculated by the conventional molecular field theory. In fact, the temperature dependence of $M(T)$ is

much steeper than that of the theory at lower temperature , suggesting the existence of the two different types of magnetization components. Considering the fact that the present sample is located near the MI transition point, it is speculated that these two components are related to the major bulk ferromagnetic phase and the minor ferromagnetic cluster phase weakly coupled with the bulk phase. An alternative interpretation is the change of the conduction mechanism around 35 K, for example, the increase of the coherence of hopping conduction.

The relation between the $M(H)$ and the α_M shown in the inset of Fig. 2 is more clear. A simple scaling is found to be successful at each temperature. It is also found that the slope of the plot which is related to the coupling constant between the spin and the carrier becomes larger for lower temperature. This fact suggests that the temperature variation of the conduction mechanism or the coherence length of the hopping as suggested above is important for the present sample. However, it is unclear at present whether the former mechanism is also essential or not. To clarify this point, it would be useful to investigate the relation between M and α_M above T_C, which remains for future investigations. Despite the temperature dependence of the coupling constant, the experimentally found simple scaling relation between the M and the α_M convinced us that the "spin polarization dependent hopping mechanism" as the interpretation of the enhancement of α_M below T_C is plausible.

4 Summary

FIR absorption spectra have been measured on $Ga_{1-x}Mn_xAs$ ($x=0.034$) in the magnetic field below T_C. It is found that FIR absorption coefficient shows a simple scaling with the magnetization. It is also found that the coupling constant becomes larger below 35 K. This result suggests the change of the conduction mechanism or change of the coherence of hopping conduction at low temperatures.

Acknowledgements

This work was partially supported by Grant-in-Aid on Priority Area "Spin Controlled Semiconductor Nanostructures" (♯09244103) and "Development of Laser Terahertz Wave Technology" (♯11231203) from the Ministry of Education, Science, Sports and Culture of Japan.

References

1. H. Ohno, A. Shen, F. Matsukura, A. Oiwa, A. Endo, S. Katsumoto, and Y. Iye, Appl. Phys. Lett. **69**, (1996) 363.
2. H. Ohno, J. Magn. Magn. Mater. **200**, (1999) 110.
3. H. Ohno, H. Munekata, T. Penney, S. von Molnár, and L. L. Chang, Phys. Rev. Lett. **68**, (1992) 2664.
4. M. Shirai, T.Ogawa, I. Kitagawa, and N. Suzuki, J. Magn. Magn. Mater. **177-181**, (1998) 1383.
5. H. Akai, Phys. Rev. Lett. **81**, (1998) 3002.
6. Y. Nagai, T. Kunimoto, K. Nagasaka, H. Nojiri, M. Motokawa, F. Matsukura, and H. Ohno, preprint.
7. R. A. Chapman and W. G. Hutchinson, Phys. Rev. Lett. **18**, (1967) 443.

Dynamical aspects of exciton magnetic polaron formation in diluted magnetic semiconductor quantum wells

T. Stirner, W. E. Hagston

Department of Physics, University of Hull, Hull HU6 7RX, United Kingdom

Abstract   Calculations of the energies of magnetic polarons formed by excitons localized in magnetic $Cd_{1-x}Mn_xTe$ wells surrounded by $Cd_{1-y}Mn_yTe$ barriers with $y > x$ are presented. A comparison with recent photoluminescence experiments allows insight into the dynamical processes governing the formation of magnetic polarons. In particular it is argued that the time τ_f for the exciton to become localized in the plane of the quantum well is the rate determining step in the formation of the magnetic polaron and that, in the samples studied, τ_f is greater than τ_r, the exciton recombination time.

1 Introduction

Magnetic polaron complexes have recently received a great deal of attention [1–3]. However, very few investigations were concerned with polarons in magnetic quantum wells surrounded by magnetic barriers [4]. The original interest in these systems arose mainly from the large carrier-magnetic ion exchange interaction. However, as we will show, the observed values of the magnetic polaron energies in these systems permit an insight into the rate determining processes in the dynamical evolution of the magnetic polaron.

2 Theory

The approach we adopt to the exciton magnetic polaron (EMP) energies is an extension of that developed by Heiman, Wolff and Warnock [2] for single particle systems. The latter involves the minimization of the free energy

$$F = U - T S \qquad (1)$$

with respect to the orbital part of the carrier's wave function. Here U is the internal energy of the system, T is the lattice temperature and S the entropy. Introducing a function Ψ such that

$$\Psi = k N \ln(Z) \qquad (2)$$

where k is the Boltzmann constant, N is the number of particles involved and Z is the partition function, shows that concordance can be established between statistical and thermodynamical theory if we take (see Ref. [5])

$$S = \Psi + \frac{U}{T}. \qquad (3)$$

From eqs. (1) and (3) we see that $\Psi = -F/T$ and hence

$$F = -k T N \ln(Z), \qquad (4)$$

which is the starting point of Heiman, Wolff and Warnock.

Details of the theoretical formalism can be found in Ref. [3]. In short, the excitonic Bohr radius λ and the localization radius R of the centre-of-mass motion of the

Table 1   Structural details of the quantum wells. L_w denotes the well width, x is the Mn ion concentration in the well and y the Mn ion concentration in the barrier region.

sample	L_w (Å)	x (%)	y (%)
1	150	3.6	9.4
2	140	10.9	16.0
3	75	3.6	9.5

exciton are two variational parameters which have to be evaluated self-consistently by minimizing the total energy of the EMP complex. The EMP energies obtained by this optimization procedure represent the maximum values that are theoretically possible in that they are static calculations which give the EMP energy values under the assumption that the magnetic polaron forms fully. Being thermodynamic-type calculations they take no cognizance of the time required for the latter to occur.

3 Results and discussion

Three quantum well structures were investigated. These consisted of $Cd_{1-x}Mn_xTe$ well material surrounded by $Cd_{1-y}Mn_yTe$ barriers with $y > x$. Details of the structural data of the quantum wells are summarized in Table 1. The parameters chosen for the present theoretical study correspond to quantum well samples used in a recent experimental photoluminescence study [6]. The CdMnTe material parameters employed in the calculations were $m_e^* = 0.096\,m_0$, $m_h^* = 0.6\,m_0$ [7] and $N_0\alpha = 220$ meV, $N_0\beta = -880$ meV [8]. The exciton energy calculations [9] were performed with a value of the relative permittivity of $\epsilon_r = 10.6$.

Fig. 1 shows the calculated EMP energies for samples 1, 2 and 3 as a function of the localization radius of the centre-of-mass motion of the exciton. Also shown in Fig. 1 by the dashed lines are the experimentally determined values of the EMP energies taken from Ref. [6]. (The latter values are summarized, together with the photoluminescence and photoluminescence excitation/reflectivity peak positions, in Table 2.) There are several points to note from this figure. First, the EMP energy is a nonmonotonic function of the localization radius which decreases rapidly to zero at a small localization radius (as a result of the large kinetic energy associated with localization) as well as at a large localization radius (as a result of negligible polarization of the magnetic ions). Second, the observed EMP energies are much smaller than the maximum values permissible theoretically. This occurs

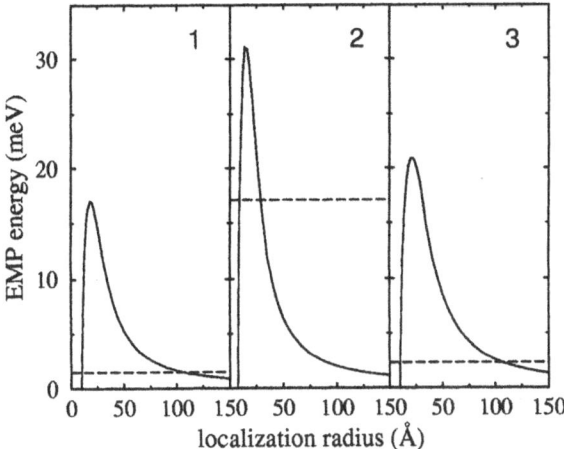

Fig. 1 EMP energy as a function of localization radius for samples 1, 2 and 3 ($T = 1.8$ K). The dashed lines show the experimentally observed EMP energies (see Ref. [6]).

Table 2 Experimental photoluminescence (PL), photoluminescence excitation (PLE) and reflectivity (R) peak positions for samples 1, 2 and 3 (see Ref. [6]). Also shown are the PL–PLE/R peak separations (ΔE), and the observed EMP energies ($T = 1.8$ K, $B = 0$ T). All energies are in meV.

sample	E(PL)	E(PLE/R)	ΔE	E_{EMP}
1	1658.7	1661.9 (PLE)	3.2	1.5
2	1760.4	1777.9 (R)	17.5	17.1
3	1670.3	1674.8 (PLE)	4.5	2.3

the scattering of the exciton into a localized orbit. The results also suggest that τ_{p} occurs on a much shorter time scale than τ_{f}, i.e. the polarization of the Mn ions, once the exciton is localized, is very rapid. Furthermore, the fact that none of the observed EMP energies reaches the maximum theoretical value is a consequence of the fact that τ_{f} exceeds τ_{r}, the excitonic recombination time.

4 Summary

EMP energies were calculated in three $Cd_{1-y}Mn_y$Te-$Cd_{1-x}Mn_x$Te-$Cd_{1-y}Mn_y$Te quantum wells with $y > x$. In agreement with physical intuition, the results show that an exciton needs to become localized in the plane of the quantum well in order that polaron formation can take place. In addition, the exciton will, in general, decrease its Bohr orbit during the polaron formation process. A comparison with recent photoluminescence experiments in these structures shows that the experimentally observed polaron energies are significantly smaller than the maximum theoretically possible values obtained from static polaron energy calculations. The exciton magnetic polaron has therefore to be viewed as a dynamically evolving complex for which, in the samples studied, the rate determining step is τ_{f}, the time required for the exciton to localize in the plane of the well. Furthermore, for these same samples τ_{f} is greater than τ_{r}, the exciton recombination time.

as a result of the dynamical processes associated with magnetic polaron formation which we will now address. There are two (interlinked) dynamical processes that determine the rate of magnetic polaron formation. The first is the time τ_{f} required for the exciton to form (or be scattered into) the optimum localization radius, whilst the second (and related effect) is the time τ_{p} for the localized carriers to polarize the magnetic ions within their orbit. Other things being equal, we might reasonably expect τ_{p} to increase with the number of magnetic ions that need to be polarized. Consequently we could anticipate that τ_{p} would increase along the sequence sample 1, sample 3 and sample 2. The reason is that sample 3 is identical with sample 1 apart from being much narrower which means that the carriers in sample 3 spend more time in the barriers where the Mn concentration is greater. Consequently the EMP in sample 3 would be expected to contain more magnetic ions than in sample 1. Similarly sample 2, since it has more magnetic ions in both the well and the barrier than either sample 1 or 3, would be expected to have the largest number of magnetic ions contained within the EMP. These considerations regarding the concentration of magnetic ions are also borne out by the maximum theoretical values of the EMP energies which follow the sequence 1, 3 and 2 (and in which the optimum localization radius is almost a constant thus justifying the "other things being equal" consideration). Given that polarization of the magnetic ions would be expected to be less rapid along the sample sequence 1, 3 and 2, it is significant that the largest observed EMP energies follow exactly the reverse sequence. This in turn suggests that it is not τ_{p} that is the rate determining step in these samples, but rather τ_{f}, and that the latter decreases along the sample sequence 1, 3 and 2. Since the exciton is surrounded by a greater concentration of magnetic ions along this sample sequence, we are led to the conclusion that the increasing lattice disorder that accompanies the increasing concentration of magnetic ions is probably the important physical factor that dominates

References

1. D. R. Yakovlev, Advances in Solid State Physics **32**, Ed. U. Rössler (Vieweg, Braunschweig 1992), p. 251.
2. D. Heiman, P. A. Wolff, and J. Warnock. Phys. Rev. B **27**, 4848 (1983).
3. W. E. Hagston, T. Stirner, J. P. Goodwin, and P. Harrison, Phys. Rev. B **50**, 5255 (1994).
4. G. Mackh, W. Ossau, D. R. Yakovlev, A. Waag, T. Litz, and G. Landwehr, Solid State Commun. **88**, 221 (1993).
5. D. F. Lawden, *Principles of Thermodynamics and Statistical Mechanics* (Wiley, Chichester, 1987), p. 65.
6. S. J. Weston, Ph.D. thesis, University of Hull, Hull, 1995; T. Stirner, M. R. Farrow, and W. E. Hagston, J. Phys.: Cond. Matter **12**, 701 (2000).
7. L. S. Dang *et al.*, Solid State Commun. **44**, 1187 (1982).
8. J. A. Gaj *et al.*, Solid State Commun. **29**, 435 (1979).
9. C. P. Hilton, J. Goodwin, P. Harrison, and W. E. Hagston, J. Phys. A **25**, 5365 (1992).

Cryobaric exciton magnetophotoluminescence in $Cd_{1-x}Mn_xSe$

N. Kuroda[1] and Y. H. Matsuda[2]

[1] Department of Mechanical Engineering and Materials Science, Faculty of Engineering, Kumamoto University, Kumamoto 860-8555, Japan   e-mail: **kuroda@msre.kumamoto-u.ac.jp**

[2] Institute for Solid State Physics, University of Tokyo, Kashiwa 277-8581, Japan

Abstract The mechanism of the distant-neighbor exchange interactions in a diluted magnetic semiconductor CdMnSe is studied on the basis of the pressure dependence of the energy shift of excitons under magnetic field.

1 Introduction

In diluted magnetic semiconductors of II-VI and III-V compounds the magnetic ions are scattered throughout the cation sublattice of the network of sp^3 covalent bonds. The majority of the magnetic ions are isolated and the rest form small clusters, i.e., pairs, triads, quartets and so on. The spin interactions among the magnetic ions depend strongly on the energy gap and carriers of the host semiconductor. For pairs in $Cd_{1-x}Mn_xSe$ the cryobaric study[1] of the exciton magnetophotoluminescence has shown that the interaction between the two Mn^{2+} spins constructing a pair can be described in terms of the kinetic exchange theory based on the three-level model of Larson et al.[2], the model being comprised of the upper and lower Hubbard states of localized d electrons and the valence band of extended anion p orbitals. At present, however, the mechanism of spin interactions among the small clusters themselves is to be studied yet.

In the present study we examine the properties of the nth-neighbor exchange energies J_ns of $n \geq 2$ in $Cd_{1-x}Mn_xSe$ of x=0.001, 0.05 and 0.10 by the cryobaric magnetophotoluminescence spectroscopy.

2 Experiment and results

We measure the pressure dependence of the near-gap magnetophotoluminescence spectrum at liquid helium temperatures using an optical system[3] consisting of a diamond anvil cell of clamp type and fiber optics. A hydrostatic environment is obtained using the condensed Ar as the pressure-transmitting medium. The steady magnetic field up to 23 T is generated with a hybrid magnet.

Figure 1 shows the magnetic-field-induced shift of the photoluminescence energy of excitons observed for the crystal of x=0.05 at 1.4 K under

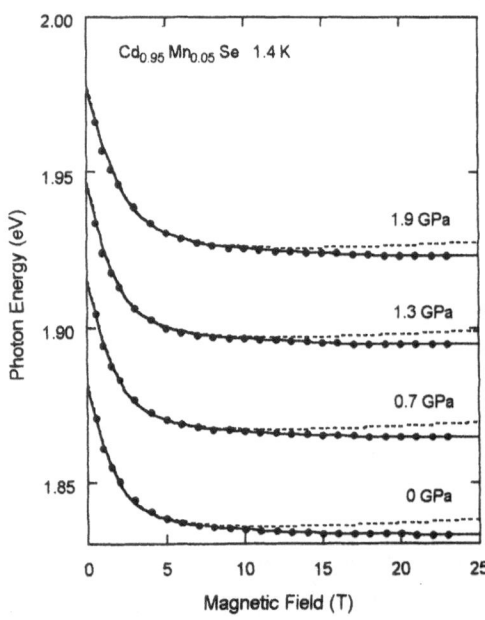

Fig. 1 Magnetic-field-induced energy shift of excitons in $Cd_{0.95}Mn_{0.05}Se$ at 1.4 K at several pressures. The magnetic field is applied parallel to the c-axis. The solid lines are the theoretical curves. The dotted lines are the curves calculated with the Mn^{2+}-pair contributions subtracted.

several pressures. The shift arises mainly from coupling of spins between excitons and Mn^{2+} ions[4]. The initial rapid shift reflects the magnetization of singles and triads of Mn^{2+} ions. The slope scales with the lattice temperature T and an effective temperature T_0 of Mn^{2+} spins as $(T+T_0)^{-1}$. If the magnetic field exceeds 10 T the influence of the staircasewise magnetization of pairs manifests itself as an additional shift. From the pair component we obtain $J_1/k = -7.4 \pm 0.4$ K at 1 atm and the pressure coefficient to be $d\ln|J_1|/dP$ =0.25±0.05 GPa^{-1}.

The data shown in Fig. 1 give T_0=2.0±0.2 K at 1 atm. Furthermore, T_0 remains positive under high pressures. The positive value of T_0 originates from antiferromagnetic internal field due to distant-neighbor spins. Consequently, T_0 is elevated prominently as the Mn content x increases, as shown in Fig. 2. We find that T_0 is elevated also by pressure as shown in Fig. 3.

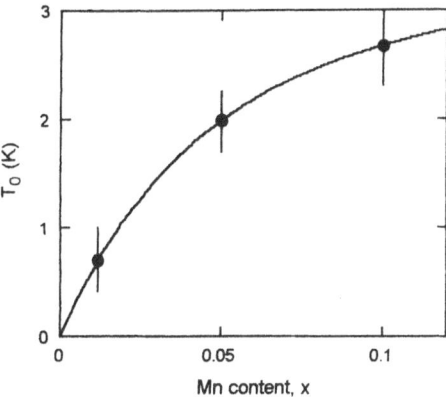

Fig. 2 The x dependence of T_0 in $Cd_{1-x}Mn_xSe$ at 1 atm. The solid line is the theoretical curve.

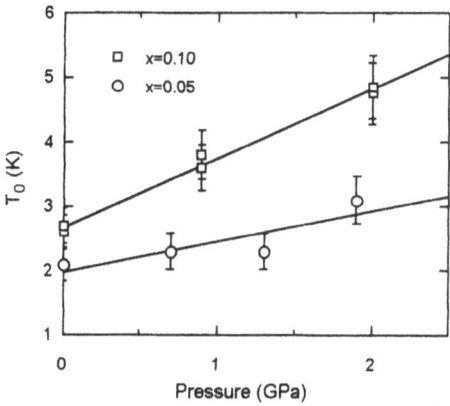

Fig. 3 Pressure dependence of T_0 in $Cd_{1-x}Mn_xSe$.

3 Discussion

In the mean-field approximation T_0 is expressed as[5]

$$kT_0 = -4p_1{}^*S(S+1)J^*, \qquad (1)$$

with an effective internal exchange constant J^* of

$$J^* = J_2 + \frac{10}{3}J_3 + 2J_4. \qquad (2)$$

The quantity $p_1{}^*$ in Eq.(1) is the probability that a Mn^{2+} ion behaves as singles. With the formula for $p_1{}^*$ given by Shapira[6], the experimental data shown in Fig.2 yield $J^*/k = -1.9\pm0.4$ K at 1 atm. From the pressure dependence of T_0 we obtain

$d\ln|J^*|/dP=0.24\pm0.1$ and 0.4 ± 0.1 GPa^{-1} for $x=0.05$ and 0.10, respectively.

The experimental values of $J_1/k = -7.4$ K and $J^*/k = -1.9$ K at 1 atm permit us to make a test of the validity of various proposals on the variation of J_n with n. For example, Larson's formula[2], $J_n=J_0\exp(-2.45r_n{}^2)$, claims $J^*/k = -0.85$ K, where r_n is the nth-neighbor distance normalized by the nearest-neighbor distance. Twardowski's[7] and Rusin's[8] power laws, $r_n{}^{-6.8}$ and $r_n{}^{-8.5}$, give $J^*/k = -1.47$ and -0.68 K, respectively, whereas Shen's independent-exchange-path model[9] gives $J^*/k = -4.1$ K if Shen's γ-parameter of 0.044 is adopted. It appears that Twardowski's power law is in accord with the case of diluted CdMnSe well.

We note from our experimental results that the pressure coefficient of J^* agrees with that of J_1 within the experimental errors. Although Larson's three-level model underestimates J^*, the above-mentioned exponential law predicts that the relative magnitudes of intersite spin interactions are independent of the crystal volume. In contrast the chemical bond picture such as the independent-exchange-path model presumes multiple super-exchanges along the chemical bonds connecting two spins, implying that the pressure coefficient of J^* is larger than twice that of J_1. In this sense our experimental results are in favor of Larson's picture of covalent spin interactions.

Acknowledgements

This work was supported in part by the Grant-in-Aid for Scientific Research on Priority Areas (B) from the Ministry of Education, Science, Sports and Culture.

References

1. N. Kuroda and Y. Matsuda, Phys. Rev. Lett. **77**, (1996) 1111.
2. B. E. Larson, K. C. Hass, H. Ehrenreich, and A. E. Carlsson, Phys. Rev. **B37**, (1988) 4137.
3. Y. Matsuda, N. Kuroda, and Y. Nishina, Rev. Sci. Instrum. **63**, (1992) 5764.
4. R. L. Aggarwal, S. N. Jasperson, P. Becla, and R. R. Galazka, Phys. Rev. **B 32**, (1985) 5132.
5. G. Barilero, C. Rigaux, N. H. Hau, J. C. Picoche, and W. Giriat, Solid State Commun. **62**, (1987) 345.
6. Y. Shapira, S. Foner, D. H. Ridgley, K. Dwight, and A. Wold, Phys. Rev. **B 30**, (1984) 4021.
7. A. Twardowski, H. J. Swagten, W. J. M. de Jonge, and M. Demianiuk, Phys. Rev. **B36**, (1987) 7013.
8. T. M. Rusin, Phys. Rev. **B53**, (1996) 12577.
9. Q. Shen, H. Luo, and J. K. Furdyna, Phys. Rev. Lett. **75**, (1995) 2590.

Line shape of excitation spectra of photoluminescence in CdMnTe and its quantum wells

K. Takamura*, S. Yamamoto and J. Nakahara

Division of Physics, Graduate School of Science, Hokkaido University, Sapporo 060-0810, Japan

Abstract Line shapes of photoluminescence (PL) excitation spectra are discussed in bulk, thin films and quantum wells of semimagnetic semiconductors CdMnTe(x=0.6). Considering saturation of direct excitation of Mn^{2+} ions and indirect one through band gap excitation, their observed spectra are reproduced qualitatively.

1 Introduction

Photoluminescence excitation spectra (PLE) in semiconductors have been investigated for long time. It is available to find out some levels related to PL. In addition, it is important to investigate excited states and relaxation processes. In semimagnetic semiconductor CdMnTe, different line shapes of excitation spectra of IRPL ($\sim$1 eV) or MnPL ($\sim$2 eV) are reported[1–3]. However there are only a few studies for their line shapes themselves.

In this paper, we observe excitation spectra of MnPL (MnPLE) in CdMnTe (x=0.6) bulk, its thin films and its quantum wells. In addition, model calculations of line shape of MnPLE considering the saturation of direct and indirect excitation of Mn^{2+} ions are performed. Calculated line shapes are qualitatively in good agreement with experimental results.

2 Model calculations, experimental results and discussions

Excitation spectra of MnPL in bulk, thin film (TF) of CdMnTe (x=0.6) and single quantum well CdTe/CdMnTe (x=0.6) (QW) at 2 K are shown in Fig.1. A bulk of CdMnTe (x=0.6) is prepared by Bridgman technique, TF and QW are grown on GaAs or sapphire substrates by MBE method. Thickness of TF is 100 nm. Well width of QW is about 13 A and its barrier widths are 500 A. The samples are excited by a dye laser with 5 ns pulse width pumped by YAG:Nd laser.

As shown in Fig.1, in bulk there is a peak at 2.2 eV and the intensity decreases gradually from 2.2 eV to band gap energy (E_g). In TF, however, there is no peak at 2.2 eV and the intensity increases forward E_g. Similar line shape is gotten in QW.

In order to discuss these line shapes of MnPLE, we perform model calculations at various excitation power density considering the saturation of excitation related to PL. This model, which is reported in previous paper[4], is assumed for excitation spectra of IRPL (IRPLE) in CdMnTe and can be adapted to MnPLE. Here,

* *Present address:R.I.E.C. Tohoku University, Sendai 980-8577, Japan e-mail: takamura@riec.tohoku.ac.jp*

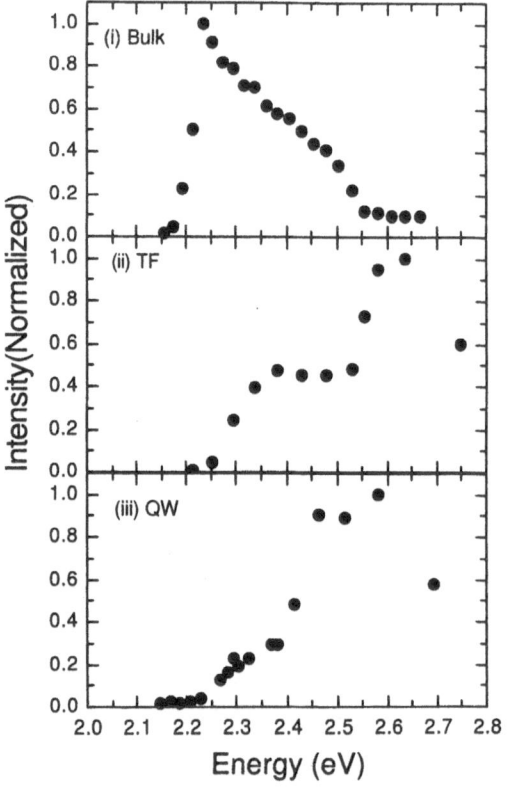

Fig. 1 Excitation spectra of MnPL in (i) $Cd_{0.4}Mn_{0.6}Te$ bulk, (ii) thin film with 100 nm thickness and (iii) CdTe(1.3 nm)/$Cd_{0.4}Mn_{0.6}Te$(50 nm) single quantum well at 2 K.

the saturation of direct excitation of Mn^{2+} ions and indirect one through band gap excitation for suggested model is considered. It is assumed that absorption profiles are shown in Fig.2(i). Band gap is at 2.54 eV where temperature of the sample is 2 K.

The typical calculated line shapes of MnPLE in TF whose thickness is 100 nm and bulk whose one is 1 mm are shown in Fig.2(ii) and Fig.2(iii), respectively. In bulk, a decrease of plateau from 2.2 to 2.3 eV as an increase of the excitation power density indicates the saturation of excitation related to MnPL. As an evidence of the saturation due to large excitation power density, we show experimental results of MnPLE excited by weak (20 W/cm^2) and strong (2 kW/cm^2) excitation power density at 140 K in Fig. 3. Indeed, as the excitation

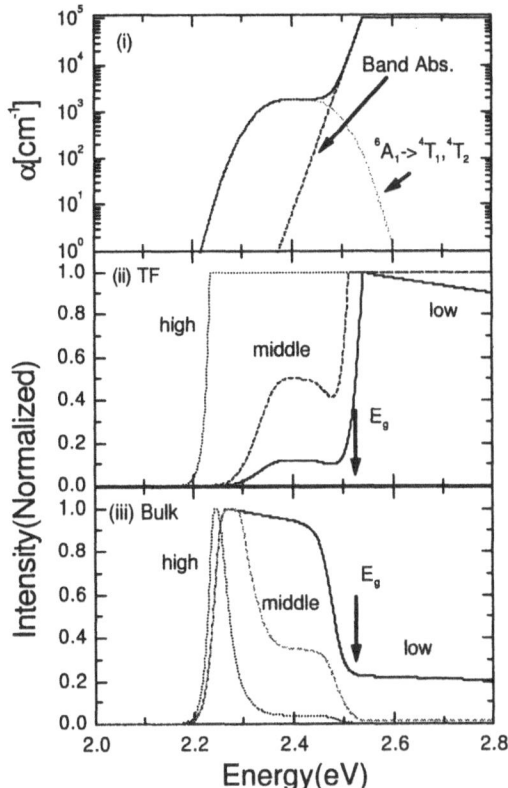

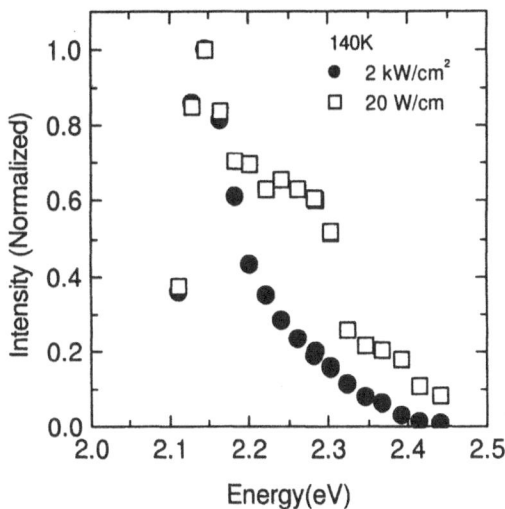

Fig. 3 Experimental excitation spectra of MnPL at 140 K in $Cd_{0.4}Mn_{0.6}Te$ bulk.

Fig. 2 (i) Assumed absorption coefficient at 2K in CdMnTe and calculated excitation spectra of MnPL in (ii) a thin film and (iii) a bulk at three different excitation power densities.

power density increases, it is observed that the intensity of the plateau from 2.2 to 2.3 eV decreases. In the case of weaker excitation power density like halogen lamp excitation, the clear plateau from 2.2 to E_g such as low case shown in Fig.2 (iii) is also reported[1,5]. In addition, calculated line shapes of IRPLE are in qualitatively good agreement with the experimental line shapes[4]. These results support our model calculations with saturation of energy transfer.

As shown in Fig 2.(ii), the intensity of plateau near 2.4 eV in TF is smaller than beyond E_g at low excitation power density. This is because that most of photo excitation lights near d-d absorption energy go through the sample without absorption. Then it is due to small d-d absorption coefficient and no saturation of excitation of Mn^{2+} ions occurs. If the excitation power density is enough large, intensity of its plateau becomes constant such as the high case in Fig.2(ii), because of the saturation of excitation of Mn^{2+} ions. The experimental result shown in Fig.1 (ii) is similar to the middle case in Fig.2 (ii).

In the case of the quantum well, a similar result to thin films has gotten. Spectrum shows practically MnPL of barrier. Then it can be thought that PLE spectrum

of QW is almost equivalent to that of 100 nm thin film which is the total thickness of barrier. Discussions of PL between quantum levels had been reported in previous report[6].

3 Summary

Excitation spectra of d-d transitions in CdMnTe (x=0.6) bulk, thin film and quantum wells are measured and compared with model calculation in various excitation powers. Calculated results reproduce well observed line shapes qualitatively. We find that the difference between their line shapes originates the saturation of excitation of Mn^{2+} ions because of high density excitation by pulsed laser lights .

References

1. M. M. Moriwaki, R. Y. Tao, R. R. Galazka, W. M. Becker, Physica **117b** & **118b**, (1983) 467.

2. G. Ambrazevicius, G. Babonas and Yu. V. Rud, Phys. Stat. Sol.(b), **125**, (1983) 759.

3. J. D. Park, S. Yamamoto, J. Watanabe, K. Takamura and J. Nakahara, J. Phys. Soc. Jpn. **66**, (1997) 3289.

4. K. Takamura, S. Yamamoto and J. Nakahara, in submission to Phys. Rev. B.

5. K. Takamura, S. Yamamoto, J. Watanabe and J. Nakahara, Inst. Phys. Conf. Ser. 152 (1998) 453.

6. J. Nakahara, K. Takamura, S. Yamamoto and K. Ando, Proceedings of 24th International Conference on the Physics of Semiconductors.

3 Growth, Surfaces, and Interfaces

Possibility of the quantum fluctuation of the Si(001) surface at low temperature

Yoshihide Yoshimoto[1*], **Masaru Tsukada**[1]

Department of Physics, Graduate School of Science, University of Tokyo, 7-3-1 Hongo, Bunkyo-ku, Tokyo 1130033, Japan

Abstract   Recently, several extremely low temperature STM observations suggested that Si(001) surface might have a novel phase of the reconstruction below 10 K. In these experiments, fluctuation between the $p(2 \times 2)$ and the $c(4 \times 2)$ asymmetric dimer structure, or the 2×1 symmetric dimer like STM image was reported. The present study proposes a theoretical explanation of this novel phase by a quantum fluctuation of the surface atomic structure.

1 Introduction

Recently, several studies by STM suggested that Si(001) surface might have yet another phase at extremely low temperature, below 10 K. One of these studies by Shigekawa et al.[1] claimed that the surface structure fluctuated between the $c(4 \times 2)$ structure and the $p(2 \times 2)$ structure and the phase boundary of the buckling order in a dimer row, the type-P defects, were moving during the experiment. Another two studies[2,3] reported that the surface turned into a $p(2 \times 1)$ symmetric structure from the $c(4 \times 2)$ structure when it was cooled below 10 K.

Meanwhile, the $p(2 \times 1)$ symmetric structure at room temperature is traditionally explained by thermal flip-flop of the asymmetric dimers. However, the traditional theory expects flip-flop rate to be astonishingly low at 6 K, 2×10^{-100} Hz.

Therefore the traditional understanding of this surface cannot explain this suggested phase. By the way, the Debye frequency of the silicon is rather high, 6×10^2 K, and the zero point energy of the dimer flip-flop motion is no less than 10^2 K. Therefore we studied the possibility of *quantum* fluctuation below 10 K. If it can be explained by the quantum fluctuation, it will mean that the tunneling of the degree of the freedom of the nucleus can be directly observed by STM.

The minimal fluctuation of this surface is the movement of the type-P defect which is caused by one dimer flip in the type-P defect. (Fig. 1) Probably, the type-P defects are permanently created by the frustration among the domains fixed by type-C defects arranged randomly. In this study, we treated the movement of the type-P defect.

2 Estimate the quantum fluctuation frequency

By the instanton (or the WKB) approximation, we tried to estimate the frequency of the quantum fluctuation.

* *Present address:* Tsuneyuki Group, Institute for Solid State Physics, University of Tokyo, 5-1-5 Kashiwa-no-ha, Kashiwa-shi, Chiba, 2778581, Japan

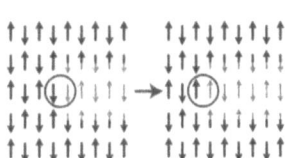

Fig. 1 Motion of a type-P defect enclosed in a circle. The directions of the dimer buckling is shown by the arrows. From the left figure to the right figure, the type-P defect moves by the unit distance per dimer.

First, we limit the degrees of the freedom within those of two dimer atoms only for the simplicity.

Secondly, we obtain the potential energy of the system V by first-principles calculations in a approximated form,

$$V \sim V_0(\theta) + \frac{1}{2}(\mathbf{w} - \hat{\mathbf{w}}(\theta))M(\theta)(\mathbf{w} - \hat{\mathbf{w}}(\theta))^t, \quad (1)$$

where $\mathbf{w}$ denotes the degrees of the freedom except for the buckling angle, θ. and $\hat{\mathbf{w}}(\theta)$ denotes the optimized structure for each θ. The vibrational frequency along the θ direction around the two minima of V which correspond to the two stable directions of the dimer is expressed by ω_0. The minimum of the potential energy $V_0(\theta)$ is taken as the origin of $V_0(\theta)$.

Finally, by the instanton approximation, the energy level split ΔE is given as $\Delta E \sim \hbar\omega_0 e^{-S_i/\hbar}$, where S_i is the action of the instanton which connects two stable buckling angles. The frequency ω_q of the quantum beat by this energy level split is $\hbar\omega_q = \Delta E$. We interpret ω_q as the frequency of the fluctuation of the dimer.

3 First-principles calculations

The potential V was obtained by the DFT calculation[4]. V_0 at each θ was determined by constrained structural optimization. $M(\theta)$ was determined by moving dimer atoms by 0.1 a.u. from $\hat{\mathbf{w}}(\theta)$ to each direction and obtaining the variation of the total energy. We performed the first-principles DFT calculations with various super cell sizes to obtain V. Each super cell contains a type-P defect and its buckling angle θ was altered to obtain V.

The condition of the calculations was as follows. The cutoff energy (E_c) was 16 Ry. The slab terminated by pseudo hydrogen atoms was five atomic layers thick and separeted by the four atomic layers thick vacuum. For the 7×4 and 9×4 super cell, the real space projection method was used and the cutoff radius of Si was 3.5 a.u. The maximal residual force of the relaxed structures was 1×10^{-5} and 3×10^{-4} Hartree/a.u. for 3×2 super cell and 7×4 or 9×4 super cell, respectively.

To check the convergence, the barrier height $V_0(\theta = 0°)$ is calculated altering the conditions.

3×2 super cell: each increase of E_c, the number of the sample k points (N_k), and the thickness of the vacuum and the slab, to 20.25 Ry, 6×6, and six and seven atomic layers caused the variation of $V_0(0)$ less than 2 meV.

7×4 super cell: Increasing E_c to 20.25 Ry changed $V_0(0)$ by 4 meV. $V_0(0)$ with and without the real space projection method were the same. $V_0(0)$ as a function of N_k in meV is 121, 119, 124, and 112 where N_k is 2×2, 2×3, 3×2, and 2×1, respectively.

The principal part of the values of the calculated $V_0(\theta)$ is summarized in Fig.2. $M(\theta)$ was also obtained at several sampled θ.

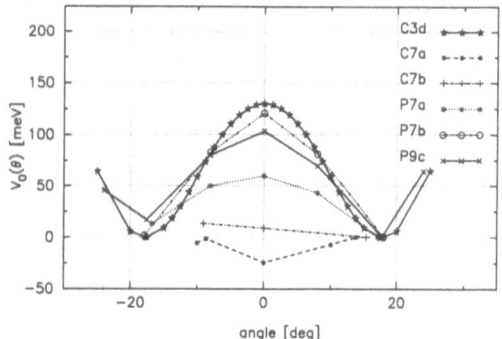

Fig. 2 $V_0(\theta)$ calculated with various conditions. The legends Xnx denotes the exchange correlation potential is X, the super cell size is n, and the number of the sample k points is x. The symbols X, n, and x take the following values.

X	C	Ceperley Alder	n	3	3×2	x	a	1×1
	P	PBE		7	7×4		b	2×2
				9	9×4		c	2×1
							d	4×4

4 Quantum fluctuation frequencies

Using the instanton method, we tried to estimate the quantum fluctuation frequencies, ω_q. Assumed value of ω_0 is 1.58×10^{13} Hz. This value of ω_0 is the zero point vibration frequency of θ in $V_0(\theta)$ obtained assuming that the dimer is a rigid rotor. Only the order of ω_0 is important, which is the order of the phonon frequency.

However, unfortunately, the calculated values of V_0 rather depend on the form of the exchange correlation potential, N_k along the dimer row, and the size of the super cell. On the dependence on the form of the exchange correlation potential, we think the result with the PBE type exchange correlation potential is more reliable, since it is conventionally believed that the GGA like the PBE type is more reliable to estimate the reaction barrier height than the LDA like the Ceperley Alder type. It should be also mentioned that the type-P defect is experimentally observed and this means $V_0(0) > 0$. The dependence on N_k along the dimer row perhaps means that the type-P defect is not completely isolated along the dimer row even in this 7×4 super cell. The 3×2 super cell have also problem that the focused dimer row is adjacent to itself and the density of the type-P defect in the super cell is too high, $2/3$. Judging from these dif-

fuculies, we can only say the barrier heigt, $V_0(0)$ might be from 60 meV to 120 meV.

Therefore we are oblidged to estimate only the possible range of ω_q by scaling the V_0 so that $V_0(0)$ become from 60 meV to 120 meV. The data by 3×2 super cell calculation was used as unscaled data and its $V_0(\theta)$ is scaled. ω_q as a function of $V_0(0)$ and the thermal flip-flop frequency are compared in Fig. 3. At least up to $T = 15K$, ω_q dominates than the thermal flip-flop rate from $V_0(0) = 60$ meV to $V_0(0) = 120$ meV.

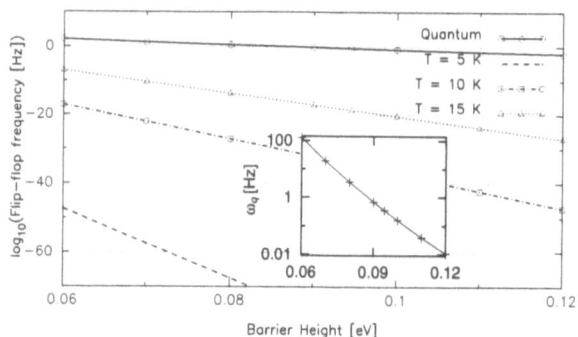

Fig. 3 The quantum flip-flop frequency, ω and thermal flip-flop frequencies as a function of $V_0(\theta = 0°)$, the barrier height. ω is magnified in the inset.

5 Discussion

This study suggests that the quantum fluctuation rate of the type-P defect might be significantly higher than the thermal fluctuation rate below 10 K. However the absolute rate itself is not clear enough to explain the experiment.

If the quantum fluctuation is possible, the experiment can be explained as follows. Above room temperature, the dimers thermally fluctuate and STM images become $p(2 \times 1)$ symmetric. At middle temperature, 100 K $\sim$ 200 K, the thermal fluctuation becomes small and the $c(4 \times 2)$ structure, the ground state, is observed. Below 10 K, the quantum fluctuation becomes strong and the dimers begin to fluctuate again. Therefore, STM image shows the $p(2 \times 1)$ symmetric structure, or the fluctuation between the $c(4 \times 2)$ structure and the $p(2 \times 2)$ structure is observed again when the rate of the fluctuation is not so high.

Acknowledgements This work was supported in part by a Grant-in-Aid from the Ministry of Education, Science, Sports and Culture of Japan.

References

1. H. Shigekawa *et al.*, Jpn. J. Appl. Phys. **35**, L1081 (1996).
2. T. Yokoyama and K. Takayanagi, Phys. Rev. B **61**, 5078 (2000).
3. Y. Kondo, T. Amausa, M. Iwatsuki, and H. Tokumoto, Surf. Sci. **453**, L318 (2000).
4. J. Yamauchi, M. Tsukada, S. Watanabe, and O. Sugino, Phys. Rev. B **54**, 5586 (1996).

(2×4) and (4×2) reconstructions of GaAs(001): The surface phase diagram re-examined

W.G. Schmidt[1], S. Mirbt[2], F. Bechstedt[1]

[1] Institut für Festkörpertheorie und Theoretische Optik, Friedrich-Schiller-Universität Jena, 07743 Jena, Max-Wien-Platz 1, Germany e-mail: W.G.Schmidt@ifto.physik.uni-jena.de

[2] Fysiska institutionen, Uppsala universitet, Box 530, 75121 Uppsala, Sweden

Abstract Total-energy calculations for a series of (2×4) and (4×2) reconstructed GaAs(001) surfaces not included in previous theoretical studies are presented. Their formation energy is compared with results for seemingly well established (2×4) and (4×2) geometries as well as c(4×4) and (2×6) reconstructions. A new $\alpha 2$(2×4) surface model containing single anion dimers in the first and third atomic layers is predicted for a balanced surface stoichiometry. It is more stable than the two-As-dimer α structure assumed previously, due to its lower electrostatic energy. For Ga-rich surfaces the ζ(4×2) geometry due to Lee *et al.* is found to be lower in energy than the $\beta 2$(4×2) surface suggested by Biegelsen and co-workers.

1 Introduction

The GaAs(001) surface exhibits a rich variety of ordered phases whose occurrence depends on the preparation conditions [1]. Farrel and Palmstrøm [2] correlated characteristic RHEED patterns with the surface stoichiometry and distinguished between three (2×4) phases, called α, β and γ. The α(2×4) and $\beta 2$(2×4) structures shown in Fig. 1 are generally accepted structural models for the α and β phase [3]. The γ phase occurring for more As-rich surfaces is assumed to be a mixture of the β phase and the c(4×4) surface, with the surface As coverage varying depending on the growth conditions [4]. The Ga-rich GaAs(001)(4×2) reconstruction is usually explained in the picture of the $\beta 2$(4×2) structure due to Biegelsen et al. [5].

While the GaAs(001) surface structures have long been considered model systems valid also for other III-V(001) surfaces, more recently a series of exceptions were found. For InP and GaP a single dimer (2×4) structure was found to be more stable than the α structure, irrespective of the surface chemical potentials [6]. This structure, which we call $\alpha 2$ in line with the nomenclature for GaAs, is a very plausible candidate geometry also for GaAs(001) surfaces. Furthermore, general considerations about the stability of III-V surfaces [7] show that under cation-rich conditions structures other than the GaAs(001)$\beta 2$(4×2) surface should be energetically favoured.

Our paper re-examines the phase diagram of GaAs(001) by means of *first principles* calculations.

2 Method

We use density-functional theory in the local-density approximation together with *ab initio* pseudopotentials to

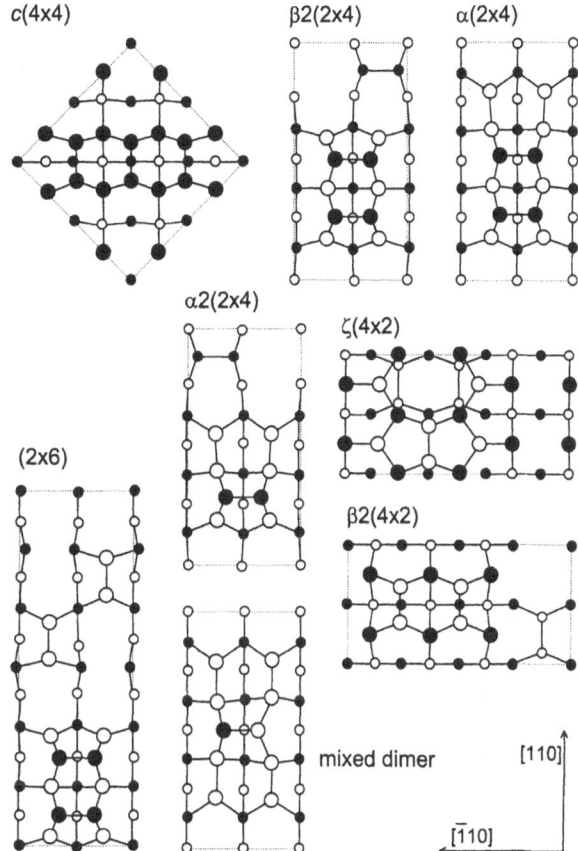

Fig. 1 Top view of relaxed GaAs(001) surface structures. Empty (filled) circles represent Ga (As) atoms. Positions in the uppermost two atomic layers are indicated by larger symbols.

determine the structurally relaxed ground states of the surface structures. A massively parallel, real-space finite-difference method [8] is used to deal efficiently with the large unit cells. A multigrid technique is employed for convergence acceleration. We model the surfaces by periodic supercells containing material slabs about 12 Å thick, separated by 12 Å of vacuum. The surface dangling bonds at the bottom layer are saturated with fractionally charged pseudohydrogens. Further details can be found in Ref. [9].

3 Results and Discussion

We probed a large number of plausible surface geometries for the GaAs(001) surface. In addition to the structures discussed in detail in Ref. [9] we included the three-As-dimer model for the $c(4\times4)$ reconstructed surface, the (2×6) surface structure proposed in Ref. [5], a modified version of that structure shown in Fig. 1, and the $\zeta(4\times2)$ structure suggested by Lee *et al.* [10]. The surface energies of the energetically favoured structures are plotted in Fig. 2.

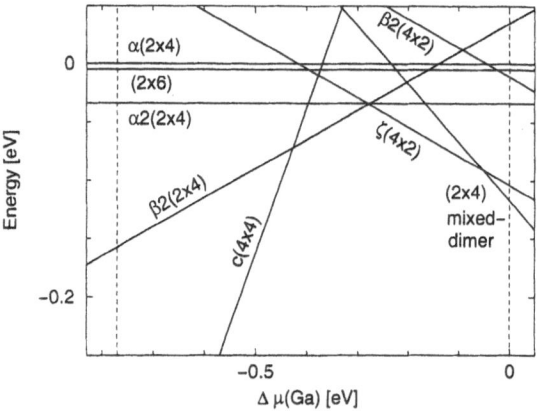

Fig. 2 Relative formation energy per (1×1) unit cell for the GaAs surface reconstructions shown in Fig. 1 vs the cation chemical potential. Dashed lines mark the approximate anion- and cation-rich limits of the thermodynamically allowed range of $\Delta\mu(Ga)$.

Our results for As-rich surfaces confirm previous findings: the stability of the three-As-dimer $c(4\times4)$ surface for extreme As-rich growth conditions, followed by the $\beta2(2\times4)$ structure for a more moderate As/Ga ratio.

For a balanced surface stoichiometry we consider the two-As-dimer structure known as α geometry, the single-dimer $\alpha2$ structure favoured for InP and GaP(001)(2×4) surfaces and two (2×6) reconstructions. The surface energy of the $\alpha2$ structure is 0.034 eV per (1×1) unit cell lower than that of the α model. As both structures have the same stoichiometry there is no dependence on the chemical potentials of the surface constituents. The α structure will be instable with respect to $\alpha2$ irrespective of the surface preparation conditions. That outcome is somewhat surprising, as the α model is seemingly well established [1]. In view of the electrostatic interactions at the surface, however, the lower energy of the $\alpha2$ structure seems plausible: Since the anion dimer bond accommodates 6 electrons [11] in addition to the 8 electrons forming the 4 bonds to the substrate, one expects a Coulomb repulsion between the negatively charged dimers. The surface may lower its electrostatic energy by distributing the dimers more uniformly, as it is the case for the $\alpha2$ structure. In order to estimate the Coulomb contribution to the energy difference between the two geometries we perform a Madelung summation for a periodic lattice of point charges,

$$S = \frac{1}{2}\sum_{i,j}\frac{q_iq_j}{|\mathbf{r}_i - \mathbf{r}_j|}, \tag{1}$$

where the vectors $\mathbf{r}_i$ are the positions of the surface atoms which have been assigned a charge q_i according to the electron counting rule. To obtain a quantitative estimate we approximate the screening by simply dividing S by the static dielectric constant of GaAs ($\epsilon \sim 13$). We thus obtain a difference in electrostatic energies between α and $\alpha2$ of 0.038 eV per (1×1) surface unit cell, very close to the energy difference obtained from *first principles*. The (2×6) structure proposed by Biegelsen and co-workers [5] is 0.006 eV higher in energy than the $\alpha(2\times4)$ surface. The (2×6) structure shown in Fig. 1 is slightly more favoured. It is 0.005 eV lower in energy than the $\alpha(2\times4)$ surface, but still unstable with respect to the $\alpha2$ model.

Our total-energy calculations favour for Ga-rich conditions the $\zeta(4\times2)$ structure proposed by Lee *et al.* over the $\beta2(4\times2)$ model assumed previously. The energy difference of 0.093 eV definitely excludes the occurrence of the $\beta2(4\times2)$ geometry. Again, the energy difference between $\beta2$ and ζ can be traced back to the more favourable electrostatic interaction between the surface atoms of the latter structure [10].

For extreme Ga-rich GaAs(001) surfaces the (2×4) mixed-dimer structure known from InP and GaP surfaces [6] has the lowest energy among the structures considered here. The appearance of (2×4) periodicities, however, is not observed for Ga-rich GaAs surfaces [1]. Either that extreme Ga-rich limit cannot be reached experimentally, or surface structures not included in the present study, such as (4×6) reconstructions, are even lower in energy.

Acknowledgments

We acknowledge financial support by the Göran-Gustafsson foundation and grants of computer time from the John von Neumann-Institut Jülich and the Höchstleistungsrechenzentrum Stuttgart.

References

1. Q.-K. Xue, T. Hashizume, and T. Sakurai, Prog. Surf. Sci. **56** (1997) 1.
2. H. H. Farrel and C. J. Palmstrøm, J. Vac. Sci. Technol. B **8** (1990) 903.
3. J. E. Northrup and S. Froyen, Phys. Rev. B **50** (1994) 2015.
4. T. Hashizume *et al.*, Phys. Rev. Lett. **73** (1995) 2208.
5. D. K. Biegelsen *et al.*, Phys. Rev. B **41** (1990) 5701.
6. N. Esser *et al.*, J. Vac. Sci. Technol. B **17** (1999) 1691.
7. S. Mirbt, *et al.*, Surf. Sci. **422** (1999) L177.
8. E. L. Briggs, D. J. Sullivan, and J. Bernholc, Phys. Rev. B **54** (1996) 14362.
9. W. G. Schmidt, S. Mirbt, and F. Bechstedt, Phys. Rev. B **62** (in press) .
10. S.-H. Lee, W. Moritz, and M. Scheffler, Phys. Rev. Lett. (submitted) .
11. W. G. Schmidt and F. Bechstedt, Phys. Rev. B **54** (1996) 16742.

Theory of Al$_2$O$_3$(0001) surfaces and their employment as a substrate for nitride growth

R. Di Felice[1], J. E. Northrup[2]

[1] INFM and Dipartimento di Fisica, Università di Modena e Reggio Emilia, Via Campi 213/A, 41100 Modena, Italy
[2] Xerox PARC, 3333 Coyote Hill Road, Palo Alto, CA 94304

Abstract We present the results of first-principle calculations for the equilibrium structures and the energetics of clean and hydrogenated Al$_2$O$_3$(0001) surfaces, as well as of AlN(0001) thin films on Al$_2$O$_3$(0001). The stable 1×1 clean surface is stoichiometric, terminated by 1/3 monolayer (ML) of Al atoms. A hydrogenated surface terminated by OH dimers may form exothermically and be exploited in the process of Atomic Layer Epitaxy (ALE) of Al$_2$O$_3$ films. Finally, we have investigated the suitability of c-plane sapphire as a substrate for nitride epitaxy, and present our predictions for the polarity of AlN films.

1 Introduction

Sapphire [1] is an allotropic form of aluminas, with many applications in catalytic reactions, in atmospheric treatments, and in the microelectronic industry. Because of its large availability at low cost, it has become a standard substrate for the growth of nitride films to be used at an industrial scale [2]. However, the GaN films deposited on Al$_2$O$_3$(0001) substrates exhibit a large number of threading dislocations originating at the interface, due to the high lattice mismatch (13%). Despite lateral epitaxial overgrowth has now been demostrated capable to reduce the concentration of such defects [3] and produce good-quality films, a microscopic understanding of the substrate/overlayer interface is still demanded.

To this purpose, we have investigated the structure of AlN/Al$_2$O$_3$(0001) thin films. This choice is due to the common employment of an AlN buffer layer (either through intentional deposition of Al and N, or through nitridation) at the interface with sapphire. Due to the importance of the substrate surface for establishing epitaxy, and because of the renewed interest in aluminum oxide for industrial applications [4], we have also studied the clean and hydrogenated c-plane surfaces of Al$_2$O$_3$. In particular, for the first time we have compared the energetics of different surface terminations as a function of the Al chemical potential, while the stoichiometric surface was usually assumed in the past without accurate demonstration [5].

2 Method

We have performed *ab initio* calculations in the framework of the density functional theory in the local density approximation, using pseudopotentials for the electron-ion interaction, and a cutoff of 50 Ry for the plane wave expansion of the electron wavefunctions [6]. Special **k** points were used for Brillouin Zone (BZ) sums. With our technique, we accurately reproduced the structure

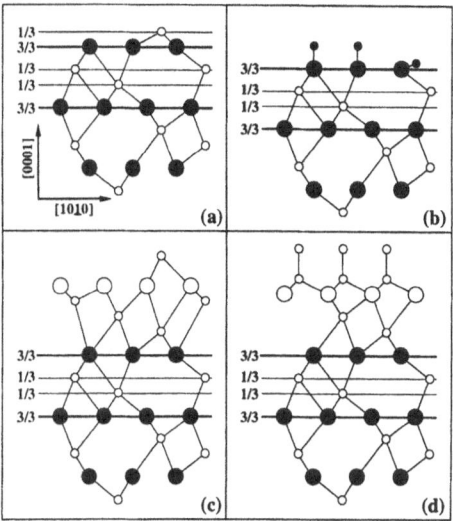

Fig. 1 Side views of (a) Al$_{1/3}$: the stable clean 1×1 Al$_2$O$_3$(0001) stoichiometric surface; (b) H: the stable hydrogenated 1×1 Al$_2$O$_3$(0001) surface; (c) AlN(000$\bar{1}$)/Al$_1$: the lowest-energy, (000$\bar{1}$)-polar, AlN thin film on c-plane sapphire. (d) AlN(0001)/Al$_{2/3}$: the lowest-energy, (0001)-polar, AlN thin film on c-plane sapphire. Large (small) dots represent O (H) atoms, and large (small) circles represent N (Al) atoms. Thick (thin) horizonatal lines represent O (Al) planes in sapphire. The fractional numbers indicate the stoichiometries of the planes (1/3 means 1 atom/layer).

of bulk Al$_2$O$_3$ [6] and bulk AlN [7]. The surfaces and the thin films have been simulated by repeated supercells of 1×1 lateral periodicity, containing 6 layers of O (3 atoms/layer), 2 layers of Al (1 atom/layer) between any two O planes, plus the additional Al atoms needed to accomplish the desired surface or thin film structures, 0 or 1 layer of N or of H (3 atoms/layer), and about 8 Å of vacuum. Models are shown in Fig. 1.

3 Results and Discussion

3.1 1×1 Al$_2$O$_3$(0001) surfaces.

Experiments demonstrated that the clean Al$_2$O$_3$(0001) surface undergoes an Al-rich $\sqrt{31} \times \sqrt{31}$ reconstruction upon high temperature annealing. The 1×1 phase is observed at lower temperatures. We have simulated 1×1 surfaces with Al coverage ranging between 1/3 ML and 1 ML. Additionally, we have investigated surfaces with Al coverage between 1 ML and 2 ML in order to inquire into the formation of the high-order reconstruction.

We have calculated the surface energies as a function of the Al chemical potential, with a method described elsewhere [6]. The stoichiometric surface, illustrated in Fig. 1(a), has a formation energy of 124 meV/Å^2 and is stable against surfaces with a higher or lower Al composition, independently of the Al chemical potential. The minimum energy cost to form other 1×1 structures containing additional Al atoms is 52 meV/Å^2: see the second line of Table 1. To gain insight into structures having more than 1 Al ML on the surface, we have studied an fcc Al(111) ML separated by the outermost O plane through an interface layer composed of 1/3, 2/3, or 1 ML of Al atoms in octahedral sites. We find that the presence of an incomplete interface layer with 2/3 ML Al composition, allows the addition of an Al(111) wetting layer without any energy cost in extremely Al-rich conditions: the two structures in the second and third lines of Table 1 have the same formation energy in Al-rich conditions. In contrast, the addition of such wetting layer, for the same Al bulk chemical potential, would cost 42 meV/Å^2 (105 meV/Å^2) above an interface layer composed of 1/3 (1) Al ML. Thus, our lowest energy Al-rich 1×1 surface has a total Al coverage of 5/3 ML. The experiments report an Al coverage of 1.69 ML for the $\sqrt{31} \times \sqrt{31}$. Hence, it is very likely that the extended reconstruction is similar to the 1×1 surface proposed by us, with slight interface adjustments in order to allow for charge neutrality and energy reduction.

For the 1 ML hydrogenated surface, the data in Table 1 show that it becomes energetically favorable with respect to the clean surface for Al-poor conditions, that is when the Al chemical potential is less than that of bulk Al by at least 1.5 eV (assuming the highest possible H chemical potential, as in H$_2$ molecules). H atoms bind to O atoms forming a mixture of planar and perpendicular OH dimers: This mixed dimer configuration is favorable by 0.18 eV/dimer with respect to an ordered array of planar dimers [4]. By analyzing the energetics of possible chemical reactions, we have found that hydrogenation of sapphire may occur exothermically upon H$_2$O adsorption on the clean surface Al$_{1/3}$. In this reaction, the clean surface is converted to the hydrogenated surface, and an additional layer of Al$_2$O$_3$ is added to one half of the surface [6]. Such chemical path is a viable self-limiting step for cyclic Atomic Layer Epitaxy of Al$_2$O$_3$ films, leading to a growth-rate of 1.08 Å/cycle, in good agreement with experiments [8].

3.2 $\sqrt{3} \times \sqrt{3}$ AlN/Al$_2$O$_3$ (0001) thin films.

The optimal commensurate epitaxial condition for AlN on sapphire is an AlN $\sqrt{3} \times \sqrt{3}$ cell (with side of 5.27 Å) on an Al$_2$O$_3$ 1×1 cell (with side of 4.69 Å), with a relative rotation of 30°, and a compressive strain of 11% for the AlN film. In such conditions, the AlN films are energetically unfavorable with respect to pure three-dimensional growth, independently of their surface geometry and of the substrate/overlayer interface. In fact, the lowest energy films reported in Table 1 have formation energies larger that that of the bare substrate sur-

Table 1 Formation energies of relevant surfaces and thin films. The structures are labeled according to the atomic composition of the layers above the outermost O plane. The subscripts indicate fractions of MLs. Equilibrium with bulk AlN and bulk Al$_2$O$_3$ is assumed to evaluate the data in this Table.

structure	ΔE(Al-rich) (meV/Å^2)	ΔE(Al-poor) (meV/Å^2)
Al$_{1/3}$	0	0
Al$_{2/3}$	+52	+236
Al(111)/Al$_{2/3}$	+52	>> +236
H	+102	-72
AlN(0001)/Al$_{2/3}$	+71	>> +134
*AlN(0001)/Al$_{2/3}$	+139	+139
AlN(000$\bar{1}$)/Al$_1$	+155	+155

face, in equilibrium with bulk AlN islands. However, it is possible to draw conclusions about the relative stability of different films, and we obtained results for the film polarity in agreement with experiments. Lowest-energy surface reconstructions have been chosen according to the film polarity [T4 Al ad-atom for AlN(000$\bar{1}$), Al adlayer for the AlN(0001)] [7]. As can be seen in Table 1, in Al-rich conditions the lowest energy film has the (0001) polarity (5th line). In equilibrium with bulk AlN, a film with the same polarity is favorable also in N-rich conditions (6th line). However, releasing the constraint of equilibrium, a film with the inverse (000$\bar{1}$) polarity, and the proper surface reconstruction [7,9], forms with a very small energy cost (7th line). In summary, films with (0001) polarity form under Al-rich conditions, which characterize Metal Organic Chemical Vapor Deposition of nitrides. However, after a nitridation process in Molecular Beam Epitaxy, which is thought as an out-of-equilibrium process, it is favorable to stabilize the (000$\bar{1}$) polarity.

References

1. Correctly, sapphire is a doped form of α-Al$_2$O$_3$. We use this term here for consistency with the nitride literature.
2. S. Nakamura, M. Senoh, S. Nagahama, N. Iwasa, T. Yamada, T. Matsuhita, H. Kiyoku, and Y. Sugimoto, Jpn. J. Appl. Phys. **35**(1B), (1996) L74-L76.
3. H. Marchand, X. H. Wu, J. P. Ibbetson, P. T. Fini, P. Kozodoy, S. Keller, J. S. Speck, S. P. DenBaars, and U. K. Mishra, Appl. Phys. Lett. **73**, (1998) 747-749.
4. K. C. Hass, W. F. Schneider, A. Curioni, and W. Andreoni, Science **282**, (1998) 265-268.
5. I. Manassidis and M. J. Gillan, J. Am. Ceram. Soc. **77**, (1994) 335-338; J. Guo, D. E. Ellis, and D. J. Lam, Phys. Rev. B **45**, (1992) 13647-13656.
6. R. Di Felice and J. E. Northrup, Phys. Rev. B **60**, (1999) R16287-R16290.
7. J. E. Northrup, R. Di Felice, and J. Neugebauer, Phys. Rev. B **55**, (1997) 13878-13883.
8. A. W. Ott, K. C. McCarley, J. W. Klaus, J. D. Way, and S. M. George, Appl. Surf. Sci. **107**, (1996) 128-136.
9. R. Di Felice and J. E. Northrup, Appl. Phys. Lett. **73**, (1998) 936-938.

Influence of strain and diffusion on the growth of V-groove InGaAs/GaAs quantum wires

K.Leifer, F. Lelarge, A. Rudra, S. Stauss and E. Kapon

Physics Department, Swiss Federal Institute of Technology (EPFL), 1015 Lausanne, Switzerland

Abstract We show that in V groove $In_yGa_{1-y}As$ quantum wires (QWR) grown by metal organic chemical vapor deposition (MOCVD) the incorporation of In in the QWRs and in the different facets depends sensitively on the growth conditions. Even at In concentrations of only $y=7\%$ growth on both bottom and sidewall facets differs strongly from growth of GaAs layers. In quantum wires with a nominal In concentration of 15% an actual concentration $y=30\%$ is observed. In concentration anisotropies between sidewall and ridge facet correlate with the pressure on these facets.

1 Introduction

The growth of InGaAs/GaAs V groove quantum wires (QWRs) with high In concentrations on GaAs substrates would open up the possibility for high quality diode laser structures emitting in the $1.3\mu m$ range. We show that in InGaAs QWRs grown by metal organic chemical vapor deposition (MOCVD) the incorporation of In in the QWRs and in the different facets depends sensitively on the growth conditions.

2 Experimental

To obtain the In concentration profile, two transmission electron microscope techniques were developed: electron energy loss spectroscopy (EELS) mapping and quantitative dark-field imaging [1] [2]. Using the first technique we obtain In concentration maps with a nanometer range resolution. To simulate dark-field image contrasts, the strain field the V-groove layer was calculated using the finite element method. From a subsequent Bloch wave calculation we obtain the dark field image contrasts. This enables us to separate strain and chemical contrasts in the experimental images and to obtain the chemical profile of the InGaAs layer with a resolution in the Angstrom range over distances of microns.

3 Facet and quantum wire growth

The InGaAs layers were deposited using MOCVD; different samples were prepared varying systematically growth temperature and In concentration. In the dark field images in figure 1 both the QWR geometry and sidewall inclination angle varies as a function of growth temperature and In concentration. Whereas at low temperatures the radius of curvature of the groove bottom is small, the QWR top facet becomes nearly horizontal for the high growth temperatures. The increase of the radius of curvature indicates an enhanced group III atom diffusion to the groove centre [3].

Whereas the $In_{0.07}Ga_{0.93}As$ layer in figure 1b follows the radius of curvature of the GaAs substrate, the top facet of the $In_{0.15}Ga_{0.85}As$ QWR grown at the same temperature (figure 1f) is nearly horizontal. Only when the growth temperature for specimens containing 7% In is increased to 650°C, the QWR top facet becomes horizontal indicating an enhanced group III adatom diffusion. This means that the radius of curvature in specimens with $y=0.07$ develops in a similar way as the one in specimens with $y=0.15$, but with an offset of about 50°C. Therefore, both higher temperature and higher In concentration increase the radius of curvature of the top facet of the InGaAs QWR leading to a nearly triangular shaped QWR.

4 In distribution in quantum wire, ridge and sidewall facets

To understand the contribution of Ga and In adatoms to surface diffusion along the growth front, the In concentration in the QWR and in the quantum well (QW) facets was analysed using EELS chemical mapping. First, in all specimens, growth was dominated by In diffusion leading to an increase of the overall In concentration in the quantum wire as compared to the surrounding (111)A sidewall facet (figure 2). This is expected in analogy to the V groove AlGaAs system, where capillarity is the driving force that concentrates the more mobile species (i.e. Ga in the AlGaAs system) in the groove centre [3]. Second, an In rich vertical well (VQW) running through the central 5-10nm of the QWR and containing up to 30% In (figure 2b,c) is formed. This In rich VQW is analogous to the Ga rich AlGaAs VQWs [4]. The In excess concentration is highest for a growth temperature of 600°C (figure 2c). This could be explained as follows: At a low growth temperature (550°C), the In adatom diffusion length is superior to the diffusion length of the Ga adatoms. When the temperature increases, the In adatom diffusion length increases strongly, whereas the Ga diffusion length still remains small (estimated Ga diffusion length at 550°C following the model in [3]: 30nm). Therefore, the In excess concentration increases at 600°C. At 650°C, the Ga adatom diffusion length increases to about 100nm, i.e. an increased Ga flux towards the groove centre reduces the In concentration in the InGaAs VQW.

Concentration differences appear also on the InGaAs QW facets (sidewall and ridge). The In concentration on the vicinal (111)A sidewall is lower than the nominal In concentration. This correlates with the elastic energy density obtained from finite element calculations (figure 3). The elastic energy is highest on the sidewall,

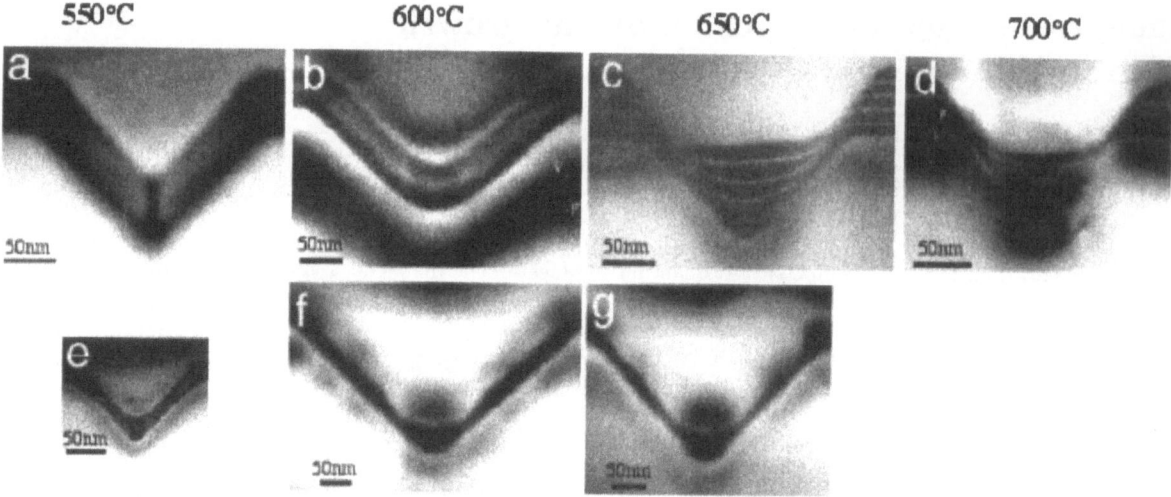

Fig. 1 InGaAs growth as a funciton of growth temperature (550-700°C) and In concentration (y=0.07, upper row, y=0.15, lower row). Specimens c and d contain AlGaAs markers.

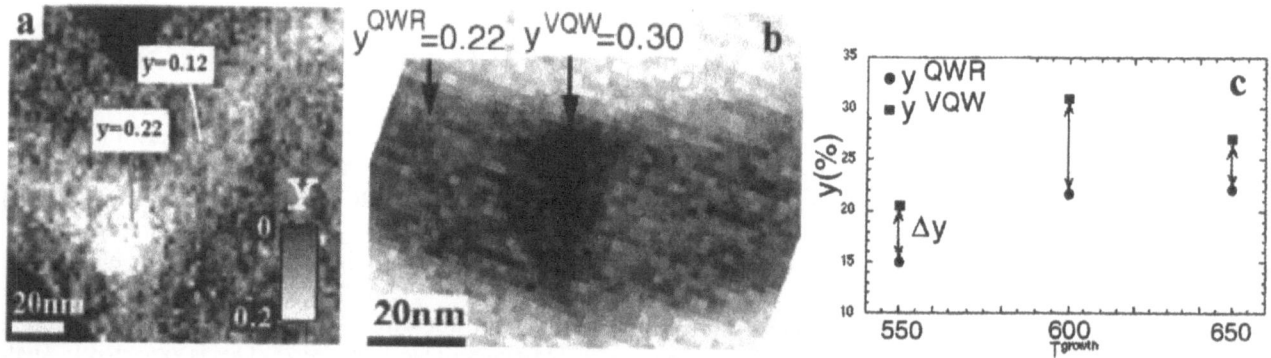

Fig. 2 In concentration maps at the bottom of the groove (a) and in the QWR (b). The specimens contain 15% In and were deposited at 550°C (a) and 600°C (b) respectively. c) In concentration in QWR as a function of growth temperature. The In excess concentration in the VQW is indicated by arrows.

which might contribute to In diffusion away from the sidewall facet.

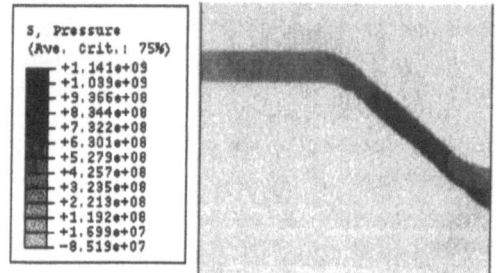

Fig. 3 Elastic energy density or pressure in 25nm thick V groove $In_{0.15}Ga_{0.85}As$ layer. The calculation was carried out using the ABAQUS finite element package.

In conclusion we observe an increase in the In concentration in the QWR. This makes V-groove InGaAs QWRs an interesting candidate for $1.3\mu m$ emitting optical devices deposited on GaAs.

References

1. K. Leifer, A. Rudra, G. Biasiol, H. Michler, E. Blank, P. Buffat, E. Kapon, Inst. Phys. Conf. Ser. **164**, (1999), 27.
2. A. M. Condo, K. Leifer, A. Rudra, J. Michler, E. Blank and E. Kapon, Inst. Phys. Conf. Ser. **164**, (1999), 185.
3. G. Biasiol, E. Kapon, Phys. Rev. Lett. **81**, (1998), 2962.
4. R. Bhat, E. Kapon, D.M. Hwang, M.A. Koza,C.P. Yun, J. of Cryst. Growth **93**, (1988) 850.

Atomic structure at a Si (001) oxidation front

Nobuyuki Ikarashi*, Koji Watanabe, and Yoshiyuki Miyamoto

System Devices and Fundamental Research, NEC Corporation, 34 Miyukigaoka, Tsukuba 305-8501, Japan
e-mail: n-ikarashi@cq.jp.nec.com

Abstract We used cross-sectional high-resolution transmission electron microscopy (HREM) to directly observe the atomic structure at a SiO_2/Si (001) interface. These observations provide the first direct evidence that cristobalite-like crystalline SiO_2 exists at the interface. We thus infer that the formation of crystalline SiO_2 at the oxidation front results in preservation of the initial atomic-scale flatness of the interface; that is, it results in layer-by-layer oxidation.

1 Introduction

To fabricate ultra-thin SiO_2 gate dielectrics in advanced metal-oxide-semiconductor transistors, the SiO_2/Si interfacial structure must be controlled on an atomic scale. However, the interfacial structure and the oxidation mechanism are not fully understood.

Accordingly, we used very thin HREM specimens (less than 5 nm thick) to enable direct observations of atomic structure at a SiO_2/Si(001) interface. In such thin specimens, dynamical diffraction effects, which hampered the direct observations of the interfacial structures in previous HREM experiments, are greatly reduced. Quantitative intensity analysis of the HREM images and first-principles calculations clearly show that cristobalite-like SiO_2 exists at the interface.

2 Experiments

A SiO_2 layer about 3 nm thick was grown on a Si (001) substrate by using radical oxygen (5×10^{-3} Torr partial pressure) at a substrate temperature of 750 °C [1]. Then HREM specimens were prepared by using CF_4 dry etching [2]. An imaging plate system was used to record HREM images quantitatively.

3 Results

A representative HREM image of the SiO_2/Si interface is shown in Fig. 1. Each black dot in the Si substrate corresponds to a closely spaced atomic column pair (an atomic dumbbell). The rectangular lattice shows the periodicity of the atomic dumbbells in bulk Si. It is evident that the interfacial black dots are often considerably displaced from their bulk positions toward the SiO_2 side.

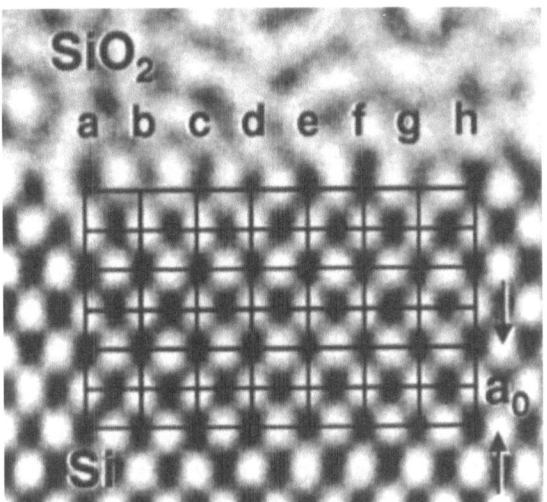

Fig. 1 <110> cross-sectional HREM image of SiO_2/Si interface. a_0=0.543 nm.(↑)

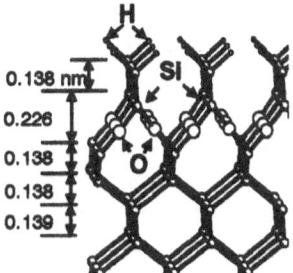

Fig. 2 Calculated Si-O geometry. (←)

The most plausible cause of this large displacement is back-bond oxidation. To examine the atomic structure at the back-bond-oxidated Si (001) interface, we calculated the stable atomic geometry of an oxidated Si surface model (Fig. 2) by using first-principles calculations within the framework of the local-density-approximation. We used a repeating-slab geometry in which all the back bonds of the second Si layers at the surfaces were oxidated. The surfaces of the slab were terminated by H. The calculated geometry shows that the oxidation increased the distance between the second and third Si (001) planes by about 0.09 nm.

Using the calculated Si-O geometry, we carried out HREM image simulation. Figure 3 shows projected potentials used in the simulations (left) and corresponding simulated images (center): (a) non-oxidated model, (b) first-layer-oxidated model, and (c) first-and-second-layer-oxidated model. The open

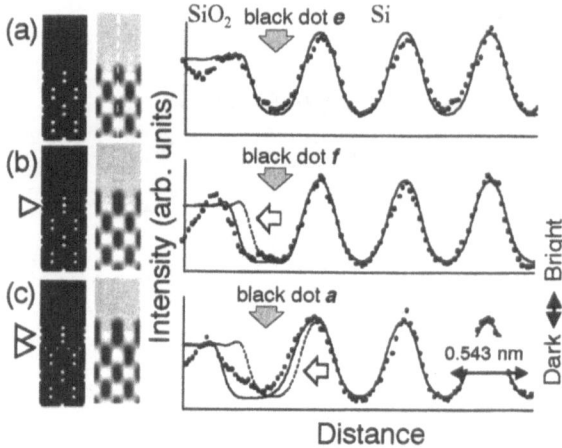

Fig. 3 Simulated HREM images of the interfaces. (a) non-oxidated model, (b), (c) oxidated model. Solid circles in the right panels show the measured intensities in Fig. 1.

triangles on the left indicate the O positions. In the right panels, solid curves show image intensity distributions in the simulated images along the [001] direction (along the dotted vertical line in (a), for example). Dotted curves in (b) and (c) show the image intensity distribution in (a). Figure 3(b) shows that the back-bond oxidation of the first Si layer elongates the interfacial black dot; that is the SiO$_2$-side slope of the interfacial black dots shifts toward the SiO$_2$ side. Figure 3(c) shows that the back-bond oxidation of the second Si layer shifts the interfacial black dot toward the SiO$_2$-side.

In the right panels of Fig. 3, solid circles show measured intensity distributions in Fig. 1 (black dots **e**, **f**, and **a**). The calculated intensity distributions agree well with the experimental ones. Hence the observed displacements in the interfacial black dots can be explained in terms of back-bond oxidation.

A schematic of the interfacial structure in Fig. 1 is shown in Fig. 4. Shaded areas show the back-bond oxidated Si. The back-bond oxidation occurred in at

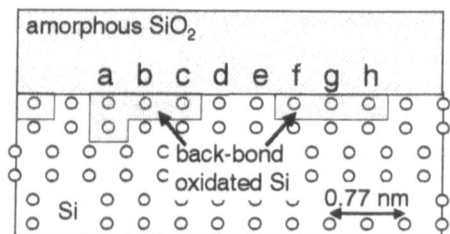

Fig. 4 Interfacial structure in Fig. 1.

least two interfacial Si layers, thus forming crystalline SiO$_2$ protrusions beneath the amorphous SiO$_2$. Since the Si-O-Si geometry in the crystalline SiO$_2$ is similar to that in cristobalite SiO$_2$, we conclude that epitaxial cristobalite-like SiO$_2$ exists at the oxidation front.

4 Oxidation mechanism

The above results leads us to infer that the formation of the crystalline SiO$_2$ at the oxidation front is an essential process for layer-by-layer oxidation [3, 4]. Our results show that oxidation occurs vertically beneath amorphous SiO$_2$, forming crystalline SiO$_2$ protrusions. Further vertical oxidation, however, is suppressed at the top of the crystalline SiO$_2$[5]. Consequently, the height of the interfacial roughness remains within one or two atomic-layers during the oxidation, resulting in layer-by-layer oxidation from a macroscopic viewpoint. When the area of the crystalline SiO$_2$ increases, a crystalline-amorphous transition is likely to occur because of the large elastic strain; the vertical oxidation will start again at the amorphous SiO$_2$/Si interface.

5 Conclusion

We have shown that Si back-bond oxidation occurs vertically at a Si(001) oxidation front, forming cristobalite-like protrusions beneath the amorphous SiO$_2$ layer. We thus infer that the suppression of further vertical oxidation at the top of the crystalline SiO$_2$ protrusions keeps the initial interfacial flatness on an atomic scale, resulting in layer-by-layer oxidation from a macroscopic viewpoint.

Acknowledgement

We thank Dr. H. Watanabe, Dr. H. Ono, and Dr. T. Tatsumi for their helpful discussions.

References

[1] K. Watanabe and T. Tatsumi, Jpn. J. Appl. Phys. **38**, L1055 (1999).

[2] N. Ikarashi and K. Watanabe, Jpn. J. Appl. Phys. **39**, 1278 (2000).

[3] J. M. Gibson and M. Y. Lanzerotti, Nature **340**, 128 (1989).

[4] H. Watanabe et al., Phys. Rev. Lett. **80**, 345 (1998).

[5] H. Kageshima and K. Shiraishi, Phys. Rev. Lett. **81**, 5936 (1998).

Binding and migration paths of Au adatoms on the GaAs (001)-β2(2x4) surface.

A. Amore Bonapasta[1], F. Buda[2]

[1]CNR - Istituto di Chimica dei Materiali (ICMAT), Via Salaria Km 29.5, CP 10, 00016 Monterotondo Stazione, Italy
[2]Department of Physical and Theoretical Chemistry, Vrije Universiteit, Amsterdam De Boelelaan 1083, NL-1081 HV Amsterdam, The Netherlands

Abstract The binding and the migration paths of an isolated Au adatom on the GaAs (001)-β2(2x4) reconstructed surface have been investigated by performing Car-Parrinello total-energy calculations. The calculated potential energy surface for the Au adatom shows that the most interesting Au binding sites are located at short-bridge sites next the As-As dimers. Difference charge-density maps show that the Au chemical binding does not give rise to the formation of covalent bonds, it is characterized instead by a strong polarization of the Au atomic electronic charge. These results suggest that Au adatoms do not modify the surface reconstruction. The Au diffusion results to be very anisotropic and faster in directions parallel to the dimer rows than perpendicular to the dimer rows.

The degradation of metal-semiconductor interfaces can be caused by migration of metal along the free semiconductor surface.[1] The quality of the metal contacts is also related to the morphology of the metal films which is fixed by the first steps of the metal deposition. Thus, several technological problems are related to the interaction of single metal adatoms with a semiconductor surface. The present study is focussed on the chemical binding and the migration paths of an Au adatom on the GaAs (001)-β2(2x4) surface. Gold is used in different metallization schemes and seems involved in degradation processes which originate from a lateral migration of metallic atoms.[1] The β2(2x4) reconstructed surface is quite stable. It is characterized by two As-As dimers and two missing dimers, see Fig. 1(a). The pairs of As-As dimers on the top (first) layer form dimer rows in the (110) directions which are separated by two missing rows of Ga atoms on the second layer. Further As-As dimers are formed on the third layer. The chemical binding and the migration paths of an Au adatom on the GaAs surface have been investigated by estimating the adiabatic potential experienced by the adatom.[2] An adatom is placed above the reconstructed surface in the *xy* plane. It is kept fixed at a given *(x,y)* position, while all substrate atoms and the *z* coordinate of the adatom are fully relaxed. The relaxation procedure is repeated for all *(x,y)* positions of a regular grid, thus providing a mapping of the potential energy surface (PES) *E(x,y)*. In some *(x,y)* positions, the relaxation procedure has been repeated by starting with atomic geometries where the As-As dimer close to the Au atom is broken in order to reveal deeper energy minima corresponding to broken dimer (BD) configurations.[2] In correspondance with the most interesting sites of the adatoms, the Au chemical binding has been investigated by analysing the local geometry and the distribution of the (valence) electron charge density. Difference density maps have been also calculated which permit to reveal even small

displacements of the electronic charge induced by bonding interactions between the interacting atoms. As an example, in the case of an As-As diatomic molecule, a charge-density sum D_{sum} is calculated by the addition of the charge densities of two *non-interacting* As atoms located at the same distance they have in the molecule, i.e., D_{sum} represents the electronic density of the two As atoms when their interaction is "switched off". Then, the charge density D for the As-As molecule is calculated. Difference densities D^+ and D^- are given by the positive and negative values of the difference $D-D_{sum}$, respectively. When the As-As interaction is "switched on" and the electronic charge density is rearranged the above difference has positive

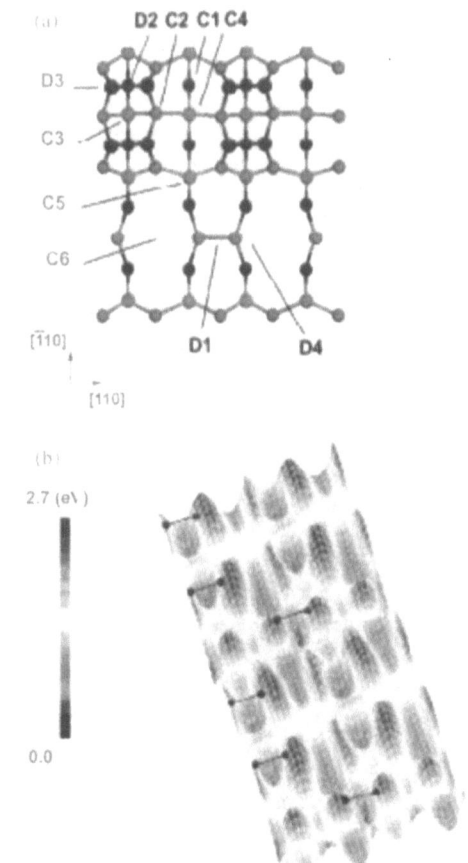

Fig. 1 (a) Top view of the GaAs (001)-β2(2x4)surface. The As and Ga atoms on the different layers are represented by different colors: red (As), cyan (Ga), green (As) and blue (Ga). D and C letters identify different Au sites. (b) potential energy surface (PES) for an Au adatom on the GaAs surface.

(negative) values where the electronic density increases (decreases), thus permitting a fine description of the atom-atom interaction.

The PES has been estimated by performing Car-Parrinello total-energy calculations in a supercell approach.[3] The interaction between the valence electrons and the frozen cores is described by soft first-principle pseudopotentials. A satisfactory convergence has been achieved by using a kinetic energy cutoff of 18 Ry. Only the Γ point is used for the k-space integration. The adatom-substrate system is modeled by a supercell geometry with a (4x4) periodicity parallel to the surface. The supercell has been tested to be sufficiently large to have negligible adatom-adatom interactions. The supercell also contains a vacuum of six layers of GaAs perpendicular to the (001) surface and an additional layer of H atoms which saturate the bonds of the lower surface.

The most interesting sites of the Au adatom are shown in Fig. 1(a). A map of the PES calculated for the isolated Au adatom is shown in Fig. 1(b). The absolute energy minimum corresponds to an Au short-bridge site, site D_1 in Fig. 1(a), where the Au adatom forms an As-Au-As complex with a dimer of the third layer. A local minimum 0.57 eV higher in energy has been found at the D_2 site, which corresponds to a short-bridge site for a dimer of the top layer. The highest energy barriers (~3 eV) correspond to the D_3 and D_4 sites. Other local minima and energy barriers are found along the channels around the dimers, e.g., the C_1 and C_2 sites are local minima and the C_3 and C_4 sites are saddle points. The Au adatom does not break the As-As dimers when located at the D_1 and D_2 binding sites. BD configurations have been also investigated at these sites because they play a key role in the case of the As-Ga-As complexes formed by a Ga adatom located at the X_1 and X_2 sites, which correspond to the D_1 and D_2 sites of Au, respectively.[2] It has been found that the BD configurations of the Au complexes are close in energy with the not BD ones and that high energy barriers (~2 eV) separate the corresponding energy minima. At variance with the case of Ga adatoms, the BD configurations can be therefore disregarded in the case of the Au adatoms. This also implies that the Au adatoms *do not affect the surface reconstruction*. The characteristics of the Au-dimer interaction have been compared with those of the Ga-dimer interaction in the case of dimers of the top layer, see Fig. 2. At the X_2 site, the As-Ga-As complex has a BD configuration characterized by an almost "on line" geometry, see Fig. 2(c), corresponding to a strong bonding Ga-As interaction. This interaction is clearly shown by the isosurfaces of D^- and D^+ charge densities given in Figs. 2(c) and (d), respectively, which show a displacement of electronic charge at positions midway between the Ga and As atoms. The Au adatom located at the D_2 site has not a BD configuration and is closer to one of the As atoms of the dimer. The corresponding D^- and D^+ isosurfaces, see Figs. 2(a) and 2(b), show that there is no displacement of the electronic charge toward the center of the Au-As bonds. There is instead a rearrangement of the Au electronic charge around the atom itself, i.e., a *polarization of the atomic charge*, when the adatom-dimer interaction is

"switched on". This polarization is characterized by a D^+ density with a shape similar to that of a p orbital pointing towards the As atom closer to the Au atom. The nature of the Au chemical binding is therefore quite different from that of the Ga adatom. This result accounts for the higher barriers that the Au adatom must overcome to break an As dimer with respect to the Ga adatom and for the different effects that the BD configurations have on the PES, negligible in the case of Au adatoms, significant in the case of Ga adatoms.[2]

The Au migration paths have been investigated here within the framework of the transition-state theory.[2] An analysis of the saddle points on the PES shows that the activation energies for the Au migration range from 0.6 eV to 3.6 eV and from 1.5 eV to 2.2 eV in the directions parallel and perpendicular to the dimer rows, respectively. These results, in particular the existence of an energy barrier of only 0.6 eV, support the existence of a strong anisotropy in the two directions. This is also immediately shown by the existence of "channels" of an almost homogeneous color in the map of Fig. 1(b) only in a direction parallel to the dimer rows. Present results suggest therefore a dominant motion of the Au atoms parallel to the dimer rows which should be revelead by STM experiments.

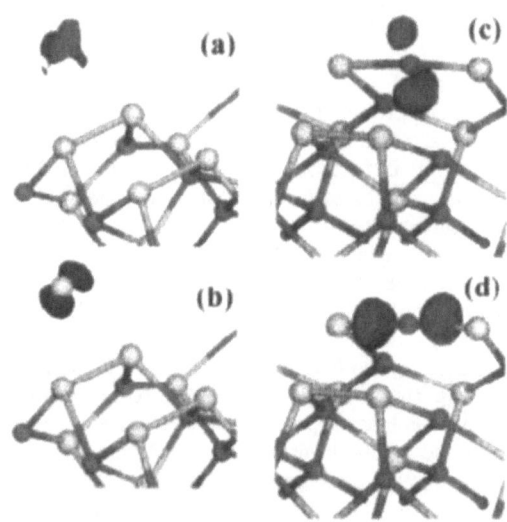

Fig. 2 Isosurfaces of difference densities: (a) and (b), D^- and D^+ densities for an Au adatom located at the site D_2, respectively; (c) and (d), D^- and D^+ densities for a Ga adatom located at the site X_2, respectively. The Au, As, Ga and H atoms are represented by the colors yellow, red, blue and green, respectively. The isosurfaces correspond to an electron density of 0.010 e/au^3.

References

1. K. Bock and H. L. Hartnagel, Semicond. Sci. Technol. **9**, (1994) **1005**.
2. A. Kley, P. Ruggerone, and M. Scheffler, Phys. Rev. Lett. **79**, (1997). **5278**.
3. Galli G.; Pasquarello, A. , *Computer Simulation in Chemical Physics*, edited by M. P. Allen and D.J. Tildsley (Kluwer, Amsterdam 1993).

Surface topography of Si(111)-7×7 reconstruction: first-principles investigations

S.H. Ke, T. Uda, K. Terakura

Joint Research Center for Atom Technology (JRCAT), 1-1-4 Higashi, Tsukuba, Ibaraki 305, Japan

Abstract The surface structure of Si(111)-7×7 reconstruction is calculated by using *ab initio* pseudopotential method adopting slab models of different geometries and different thicknesses. It is shown that the converged surface topography can be obtained when more than 5 Si layers are allowed to relax. The coverged surface topography turns out to be consistent with the experimental results from low-energy electron diffraction, scanning tunneling microscope, and non-contact atomic-force microscope. This result explains reasonably the discrepancy between the experimentally observed image and previous theoretical results.

1 Introduction

Si(111)-7×7 surface is one of the most complex surfaces found so far. Since its discovery through low-energy electron diffraction (LEED) about fourty years ago [1], an enormous amount of experimental effort and theoretical effort have been expended to understand the properties of this important surface. Now it is accepted that the geometry of this reconstruction is described by a dimer-adatom-stacking-fault (DAS) model [2]. This geometry has an extremely large area and vertical extent in its unit cell (see Fig.1). In spite of the extensive studies, some ambiguity still remains in the detailed feature of the surface topography of the 12 adatoms.

An elaborate LEED analysis [3] showed that the heights of the adatoms are in the following decreasing order: CoF > CeF > CoU > CeU, where the "Co" and "Ce" denote the corner and the center adatoms, respectively, and the "F" and "U" denote the faulted and the unfaulted halves, respectively (see Fig.1). The height difference for each step is 0.04Å. The images of scanning tunneling microscopy (STM) [4,5] showed the same height sequence but with different height differences: 0.28, 0.10, and 0.15 Å for the three steps, respectively. Recent years, the non-contact atomic-force microscope (nc-AFM) was developed as a novel technique for obtaining atomic-scale images with true atomic resolution. For a Si tip, a nc-AFM image [6] showed the same height sequence as obtained by the LEED and STM but with different height differeces: 0.10, 0.10, 0.05Å for the three steps, respectively. Another experiment [7] showed a slightly different height sequence: CoF > CeF > CoU ~ CeU. For a W tip [8], however, the image contrast is reversed: the center adatoms appear 0.13Å higher than the corner adatoms.

In the aspect of *ab initio* pseudopotential calculations which have been shown to be powerfull for determining structures of semiconductor surfaces, two elaborate calculations were reported by Brommer *et al.* [9] and Stich *et al.* [10]. Their results showed that the height

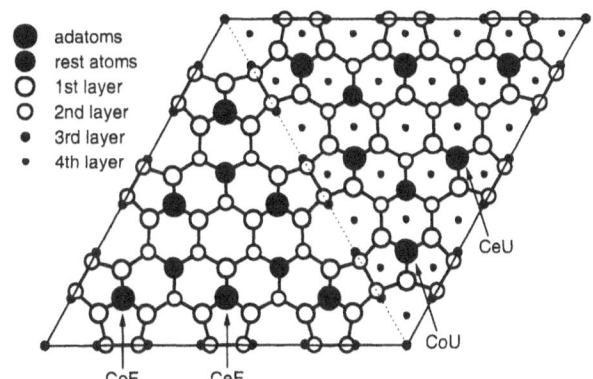

Fig. 1 Top view of the unit cell of the Si(111)-7×7 reconstruction. The faulted half and the unfaulted half are divided by a dotted line.

sequence of the adatoms is in the following decreasing order: CoF > CoU > CeF > CeU, being different from the experimental results. However, this theoretical result was recently regarded as the real surface topotgraphy and was used to analyse the nc-AFM image contrasts [6–8]. In these calculations only the adatom layer and the first three layers were allowed to relax ($n=3$), therefore, it is not certain that the obtained surface topography has been converged to the slab thickness. In this paper, we report careful *ab initio* pseudopotential calculations which show convincingly the converged surface topography of the Si(111)-7×7 reconstruction.

2 Computational details

we have considered three different geometries for the supercell of the slab model: (1) an inverse symmetry center is imposed to the supercell, as used in Ref.[10]; (2) the surface system is cut between two Si bilayers and the dangling bonds (one per atom) are saturated by 49 H atoms; (3) the surface system is cut inside a Si bilayer and the dangling bonds (three per atom) are saturated by 49 Al atoms. By using the three different geometries we can have a look at the effect from the boundary of the supercell and eventually exclude it in the conclusion of the subtle problem we want to investigate. For each supercell geometry we have considered several different slab thinknesses from $n = 3$ to 8. Between the slabs a vacuum region of 11Å is introduced. The supercells contain from 298 to 694 atoms, or eqivalently about 700 to 1100 bulk atoms in total. The calculation was performed by using plane-wave pseudopotential method with the Γ point for the k-sampling and 9Ry or 12Ry for the cut-

Table 1 Experimental and theoretical results of the height differences among the four adatoms along the long diagonal (in units of Å): ΔH_1 = height(CoF) – height(CeF), ΔH_2 = height(CeF) – height(CoU), ΔH_3 = height(CoU) – height(CeU). "Al–" and "H–" mean that the dangling bonds on the base layer of the Si slab are saturated by Al and H atoms, respectively, and "Si–" means that the Si slab has an inverse symmetry center.

				ΔH_1	ΔH_2	ΔH_3
LEED, Ref.12				0.04	0.04	0.04
STM, Ref.17				0.28	0.10	0.15
STM, Ref.18				+	+	+
nc-AFM, Ref.15				0.10	0.10	0.05
nc-AFM, Ref.16				+	+	$\sim$0
Cal.(Si–,LDA,8Ry,n=3),Ref.24				0.054	-0.007	0.038
Cal.(Si–,LDA,7Ry,n=3),Ref.21				0.039	-0.001	0.003
Si–	LDA	9Ry	n=3	0.054	-0.057	0.065
			n=6	0.040	0.054	0.015
Al–	LDA	9Ry	n=3	0.051	-0.018	0.034
			n=4	0.049	-0.005	0.034
			n=5	0.055	0.046	0.012
			n=6	0.056	0.049	0.011
			n=8	0.061	0.056	0.015
		12Ry	n=6	0.053	0.051	0.011
	GGA	9Ry	n=3	0.069	-0.034	0.049
			n=6	0.070	0.047	0.014
H–	LDA	12Ry	n=3	0.052	-0.020	0.041
			n=5	0.040	0.045	0.015

off energy which is significantly higher than those values used in the previous calculations [9,10]. Both the local density approximation (LDA) and the generalized gradient approximation (GGA) to the electron exchange-correlation were considered. The convergence criterion for the forces on the atoms is set to be 0.01 eV/Å which is much stricter than those adopted in the two previous calculations: 0.15 eV/Å in Ref.[9] and 0.1 eV/Å in Ref.[10].

3 Results and conclusions

The calculated height differences among the four adatoms along the long diagonal are listed in Table 1 together with the experimental results and the previous theoretical results. We should notice in Table 1 that for n = 3 the present LDA calculation for the inverse-symmetry system gives the same height sequence as obtained by the two previous LDA calculations: CoF > CoU > CeF > CeU, although the quantitative values are somewhat different among the three calculations.

Let us first have a look at the results of the LDA calculation for the Al-terminated system. For n = 3 the obtained height sequence is the same as that of the inverse-symmetry system. As the value of n is increased from 3 to 5 the height difference between CeF and CoU is actually reversed and the height difference between CoU and CeU decreases remarkably, indicating that the surface topography is not yet converged. However, from n = 5 to n = 8, these changes become very small ($\leq$ 0.01Å)

and the height sequence of the four adatoms keeps unchanged. These results show that the converged surface topography will be reached in the case that more than 5 Si layers on the surface are allowed to relax. However, we must keep in mind that what we are dealing with is a very subtle problem: the quantity is only several percentages of an angstrom. So we should check the result very carefully for the following items besides the slab thickness: (1) cutoff energy; (2) approximation to the electron exchange-correlation (LDA or GGA); (3) the boundary condition of the slab in the supercell approach.

For checking the convergence to the cutoff energy, we have carried out a further calculation for n = 6 adopting a higher cutoff energy of 12Ry for the Al-terminated system. It appears that the difference between the results from the 9Ry-calculation and the 12Ry-calculation is as small as negligible, indicating that the surface topography obtained has well converged with respect to the cutoff energy. For checking the effects of LDA and GGA to the electron exchange-correlation, we have further performed GGA calculations for the Al-terminated system (n = 3 and n = 6). The results show the same variation tendency with the increase of n as in the case of LDA, and the difference from the LDA result is minor. For checking the effect of the boundary condition, we can see in Table 1 that for n = 3 the results for the inverse-symmetry (9Ry, LDA), Al-terminated (9Ry, LDA or GGA), and H-terminated systems (12Ry, LDA) are in the same sequence, and the differences among the slab-thickness-converged results for the three boundary conditions are also just minor. This convincingly shows that the effect from the boundary condition on the height sequence of the adatoms is fairly small and does not affect the conclusion.

The slab-thickness-converged surface topography for all of the three systems turns out to be qualitatively consistent with the experimental results from the LEED [3], STM [4,5], and the nc-AFM [6,7]. Especially, the present result is in very good agreement with the LEED and nc-AFM results. This good agreement comfirms further the real topography of the adatoms on the Si(111)-7×7 reconstructed surface. It also indicates that the *ab initio* pseudopotential calculation is successful even for this very subtle problem.

References

1. R. E. Schlier, *et al.*, J. Chem. Phys. **30**, 917 (1959).
2. K. Takayanagi, *et al.*, J. Vac. Sci. Technol. A **3**, 1502 (1985); Surf. Sci. **164**, 367 (1985).
3. S. Y. Tong, *et al.*, J. Vac. Sci. Technol. A **6**, 615 (1988).
4. R. J. Hamers, *et al.*, Phys. Rev. Lett. **56**, 1972 (1986).
5. P. Avouris and R. Wolkow, IBM Research Report No. RC 13884, 1988 (unpublished).
6. T. Uchihashi, *et al.*, Phys. Rev. B **56**, 9834 (1997).
7. Nobuyuki Nakagiri, *et al.*, Surf. Sci. **373**, L329 (1997).
8. Ragnar Erlandsson, *et al.*, Phys. Rev. B **54**, R8309 (1996).
9. K.D. Brommer, *et al.*, Phys. Rev. Lett. **68**, 1355 (1992).
10. I. Štich, *et al.*, Phys. Rev. Lett. **68**, 1351 (1992).

A structural model of Si(111)($2\sqrt{3} \times 2\sqrt{3}$)R30°-Sn

T. Ichikawa, K. Cho, T. Onodera, A. Mizoguchi and T. Ohkoshi

Department of Physics, School of Science and Technology, Meiji University, Tama-ku, Kawasaki 214-8571, Japan

Abstract   High-temperature and room-temperature STM observations were performed of Si(111) ($2\sqrt{3} \times 2\sqrt{3}$)R30° structure, which revealed that a 1 x 1 structure above 430°C was an disordered one. A new ($2\sqrt{3} \times 2\sqrt{3}$)R30° structural model is proposed, being based on obtained results.

1 Introduction

When small amounts of Sn were deposited on clean Si(111) surfaces and then annealed at high temperatures (HT), 7 x 7, ($\sqrt{3} \times \sqrt{3}$)R30°,($2\sqrt{3} \times 2\sqrt{3}$)R30° and long-period superstructures form on t surfaces, depending on Sn amounts [1]. Hereafter the ($2\sqrt{3} \times 2\sqrt{3}$)R30° and ($\sqrt{3} \times \sqrt{3}$)R0° structures are abbreviated to $2\sqrt{3}$ and $\sqrt{3}$ ones. The $\sqrt{3}$ structure is accepted to consist of 1/3 monolayer of Sn adatoms adsorbed at T_4 sites of Si(111) surfaces [2], while no structural model is established for the $2\sqrt{3}$ structure.

The first $2\sqrt{3}$ model was proposed by Törnevik et al. [3], being based on their scanning tunneling microscopy (STM) observations at room temperature (RT) and accepted Sn coverage 1.1. The model consists of double layer of Sn adatoms where ten Sn adatoms in a unit mesh form a lower overlayer to reduce dangling bonds of surface Si atoms at an ideal Si(111) surface and another four Sn adatoms are located on the overlayer so as to explain four protrusions in their $2\sqrt{3}$ STM images. The Sn coverage of the model is 14/12, being nearly equal to 1.1, and the double-overlayer structure well explains height difference between the $2\sqrt{3}$ and the $\sqrt{3}$ one. A subsequent surface X-ray diffraction experiment by Levermann et al. [4], however, interposed a strong objection to the model, because intensity features of reflections and derived Patterson function were quite different from those calculated from the Törnevik model.

We have performed HT STM observations of the $2\sqrt{3}$ structure and have obtained several intriguing findings, leading to a new structural model which is not inconsistent with results the STM observations and the surface X-ray diffraction experiment. We present the findings and new structural model in the present paper.

2 Experimental

A 3 - 5 Ωcm B-doped mirror-polished Si(111) wafer, 0.3 mm thick, was cut into strips of dimensions 1 mm × 9 mm. After the strip was washed in acetone and rinsed in ultrapure water, it was mounted on a sample plate of an Omicron VT STM in a clean bench and then introduced into an analysis chamber. Surface cleaning of the strip was carried out by flashing at about 1200°C and small amounts of Sn were deposited on the surface. The strip was transferred into a STM chamber in an ultrahigh vacuum of 10^{-9} Pa and the surface was observed with STM.

3 Results and discussion

Figures 1 (a) and (b) are topographic STM images of the

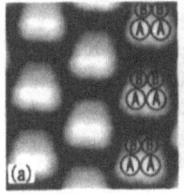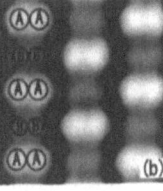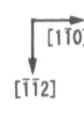

Fig. 1 Topographic STM images from $2\sqrt{3}$ structure at RT (a) and at 170°C (b).

$2\sqrt{3}$ structure taken at RT and at 170°C, in which two bright A and two less bright spots B are visible in every $2\sqrt{3}$ unit mesh. In fig. 1(a), the less bright spots stand close to lower bright spots, which is a noticeable feature already reported by Törnevik et al. [3]. On the other hand, in fig 1(b) the less bright spots shift their positions to the middle of the upper and lower bright spots. This is a new finding, which indicates that two kinds of $2\sqrt{3}$ structure, i.e. RT $2\sqrt{3}$ and HT $2\sqrt{3}$ phases, form on Sn-adsorbed Si(111) surfaces

At 190°C, the $2\sqrt{3}$ structure reversibly changes to a 1 × 1 one [1]. Figure 2 is a topographic STM image from the surface at 187°C just transforming from the $2\sqrt{3}$ to the 1 × 1 structure. The left portion of the image is surface region of the $2\sqrt{3}$ structure, while the right is $\sqrt{3}$ region with less Sn coverage of 1/3 monolayer. The phase boundary of the two structures are present between the two regions and the transformation has been initiated at the boundary. The 1 × 1 region proceeded toward the interior of the $2\sqrt{3}$ region as the surface temperature approached the transition temperature of 190°C. Any contrast with the periodicity of 1 x 1 is not observed in the 1 × 1 region, which reveals that the region is not the structure with the same periodicity as the Si(111)(1 × 1) substrate but is a disordered one. Törnevik et al. observed a large height

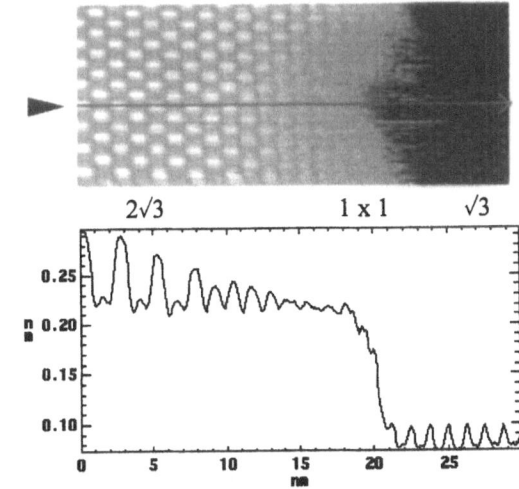

Fig. 2 (a) STM image of surface transforming from $2\sqrt{3}$ to 1 × 1 structure. (b) Height profile of $2\sqrt{3}$, 1 × 1 and $\sqrt{3}$ regions.

difference 2.1 Å between the $2\sqrt{3}$ and $\sqrt{3}$ structures and concluded that Sn adatoms in the $2\sqrt{3}$ structure should consist of a double-overlayer structure like α-Sn(111) atomic plane. The large height difference was confirmed also in the present work. Since the layer separation within an α-Sn(111) double layer is 1Å the interval between the lower and upper Sn overlayers can be estimated to be 1Å or so. Therefore, the Sn lower overlayer lies 2.54 Å (= 2.1 + 1.44 - 1 Å) above the Si surface, because the $\sqrt{3}$ atoms sit 1.44 Å above the Si surface [2]. Such a large separation means that Sn adatoms in the lower overlayer do not bind so strongly with Si surface atoms as in the $\sqrt{3}$ structure. It is noteworthy that the disordered structure lies at the same height as the lower adlayer, as shown in fig.2(b). Hence, we suppose that Sn adatoms form double-overlayer structure very similar to α-Sn(111) double layer, as suggested by Levermann et al.[4], notwithstanding they lie on the Si(111) substrates. Since unit mesh lengths of Si(111)1×1 and α-Sn(111)1×1 surfaces are 3.840 and 4.589 Å, respectively, six times the former almost equals five times the latter. Supposing that α-Sn(111)1×1 structure is 0.4 % expanded, atomic geometry in the composite structure of the two has a periodicity of 6×6. There is a close relationship between 6×6 and $2\sqrt{3}$ structures, because if every unit mesh of the 6 x 6 structure splits in three equivalent unit meshes, originating in some atomic rearrangement, the structure changes to a $2\sqrt{3}$ one. We modified coverage and positions of Sn adatoms in the 6 x 6 structure to yield a $2\sqrt{3}$ structure, taking account for the Sn coverage of 1.1 and the appearance of the four spots in the $2\sqrt{3}$ STM image. Figure 3 is a derived new $2\sqrt{3}$ structural model, where four and nine Sn adatoms form upper and lower overlayers, respectively. The present Sn coverage is $13/12 = 1.08$ and is a little smaller than that of the Törnevik model. Reflection 5/6 5/6 calculated from the present model has a strong intensity, which is one of major results obtained from the surface X-ray diffraction work [4]. Figure 4 is a Patterson function derived from the present model in which maxima are visible at positions where peaks are observed in the experimental Patterson function [4], though correspondence in height between the two functions is poor. This implies that the present model is too ordered and should be disturbed.

The height profile of fig. 2(b) suggests that Sn adatoms in the upper overlayer sink in the lower overlayer and the

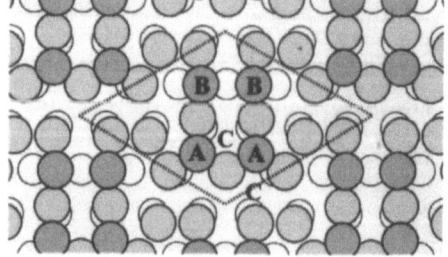

Fig. 3 Gray, light gray and open circles represent upper overlay er Sn adatoms, lower overlayer Sn adatoms and topmost Si substrate atoms, respectively. A $2\sqrt{3}$ unit mesh is marked with dotted lines for reference.

double-adlayer structure of the $2\sqrt{3}$ structure disappears on disordering at 190°C. The understanding of this disordering process, temperature change of the $2\sqrt{3}$ structure and the refinement of the present model are a future work to study.

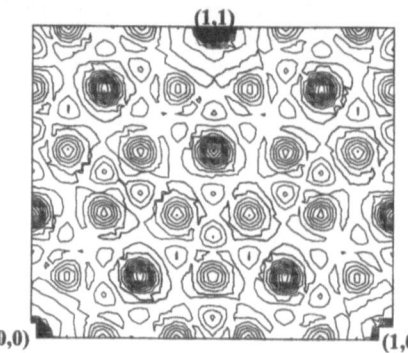

Fig. 4 Patterson function of the present $2\sqrt{3}$ model.

A $2\sqrt{3}$ structure having different image appeared at some places, as shown in a lower right region of fig. 5. This is a new type of $2\sqrt{3}$ structure probably stabilized by a local strain, because this was often observed on narrow terraces. It is noteworthy that if two adatoms A move to sites C in fig. 3, the new $2\sqrt{3}$ structure forms. Direction in which pairs of adatoms A stack is different from that of the normal $2\sqrt{3}$ structure. The difference in direction is explainable in terms of the two structural models.

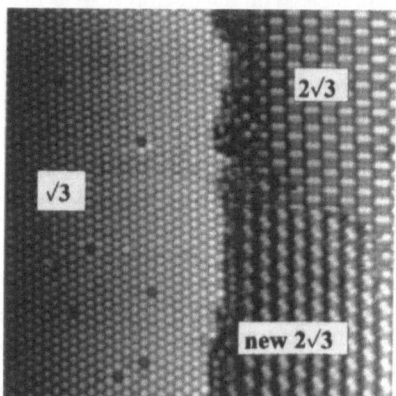

Fig. 5 New type of $2\sqrt{3}$ structure

4 Conclusion

STM observations of Si(111) $2\sqrt{3}$ surfaces at HT and RT clarified that the $2\sqrt{3}$ STM image at 170°C is different from at RT and the HT 1 x 1 structure is a disordered one. A $2\sqrt{3}$ structural model was newly proposed, based on obtained results, which explained the results by Levermann et al. fairly well.

References

1. T. Ichikawa : Surf. Sci. 140 (1984) 37.
2. K. M.Conway et al. : Surf. Sci. 215 (1989) 555.
3. C. Törnevik et al. : Phys. Rev. B44 (1991) 13144.
4. A.H. Levermann et al. : Applied Surf. Sci. 104/105 (1996) 124.

Synchrotron radiation photoemission studies of surface reconstruction on GaAs(001)

K. Ono[1], T. Mano[1], K. Nakamura[1], M. Mizuguchi[1], S. Nakazono[1], K. Horiba[1], T. Kihara[1], H. Kiwata[1], I. Waki[1], M. Oshima[1], N. Koguchi[2], A. Kakizaki[3]

[1] Graduate School of Engineering, University of Tokyo, 7-3-1 Hongo, Bunkyo-ku, Tokyo 113-8656, Japan
e-mail: ono@sr.t.u-tokyo.ac.jp
[2] National Research Institute for Metals, 1-2-1 Sengen, Tsukuba-shi, Ibaraki 305-0047, Japan
[3] Institute for Materials Structure Science, High Energy Accerelator Research Organization, 1-1 Oho, Tsukuba-shi, Ibaraki 305-0801, Japan

Abstract The surface core level shifts of (2×4), c(4×4), and (4×6) restructured GaAs(001) surfaces, which were prepared in optimized molecular beam epitaxy growth conditions, have been investigated by in situ synchrotron radiation high-resolution photoemission. The numbers of chemically inequivalent Ga and As surface sites were determined from the curve-fitting results of the high-resolution core level photoemission spectra.

1 Introduction

Surface atomic structures of GaAs(001) grown by molecular beam epitaxy (MBE) have been one of the most hot topics in semiconductor surface science in the last 20 years, since almost all GaAs-based devices are fabricated onto (001) surfaces [1–4]. The GaAs(001) surface exhibits various reconstructions such as c(4×4), (2×4), (2×6), (4×2), (4×6), and others depending on the surface coverage and the experimental conditions. However, the atomic structure of surface reconstructions on GaAs(001) which has been in long-standing controversies still remains unresolved although numerous models have been proposed. The necessity of in situ MBE-growth equipment makes it difficult for investigating GaAs(001) surfaces using state of the art surface science techniques.

The As-rich (2×4) and c(4×4) phases are the most intensively investigated, since the As-rich condition is most important in the crystal growth and device technology [5,6]. However, only a few successful characterizations have been reported on the Ga-rich (4×6) phase [7].

Surface core level shifts are one of the most powerful ways to determine the chemical structures of surface elements. In this study, we have performed the surface core level shift analysis for MBE-grown GaAs(001) surfaces and determined the numbers of chemically inequivalent Ga and As surface sites

2 Experimental

We have installed an MBE chamber to the high-resolution angle-resolved photoemission beamline BL-1C of the Photon Factory. MBE-grown samples can be transferred into a photoemission chamber without breaking ultrahigh vacuum (UHV). The MBE-photoemission system will be described in detail elsewhere.

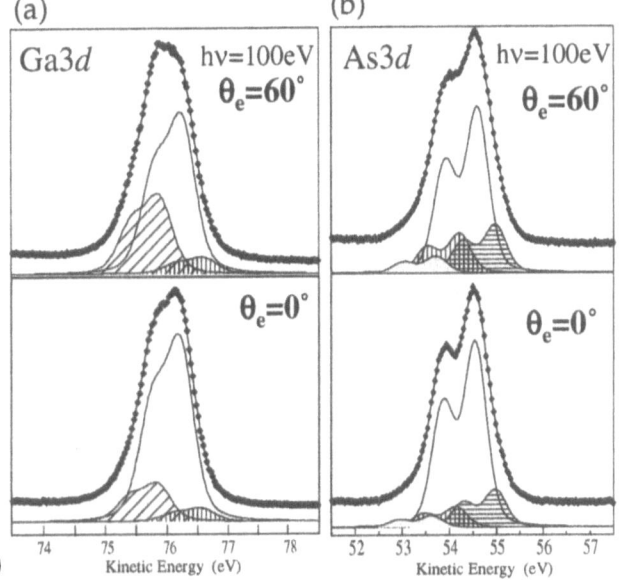

Fig. 1 Photoemission spectra of (a) Ga 3d and (b) As 3d of (2×4)-GaAs(001). The shaded components indicate the surface components.

An n^+-GaAs(001) epiready substrates were mounted on a molybdenum sample holder using indium solder without any additional wet process prior to the introduction to the MBE chamber. After annealing under an As flux at 580°C, a 3000Å buffer layer was grown in optimized conditions controlled by the reflection high-energy electron diffraction (RHEED) oscillations. The surface structures were monitored by RHEED during the growth. After the growth, the sample was transferred to the photoemission chamber without breaking UHV conditions. The photoemission measurements were carried out with the photon energy of 100 eV. The photoemission spectra of As 3d and Ga 3d core levels with 0 and 60 degree photoelectron emission angles were measured for each sample. The surface structures were checked by low energy electron diffraction (LEED) before and after each measurement.

3 Results and Discussion

Figure 1 shows the Ga 3d and As 3d photoemssion spectra of (2×4)-GaAs(001) taken at 0 and 60 degrees. In the Ga 3d spectra, two surface components as well as

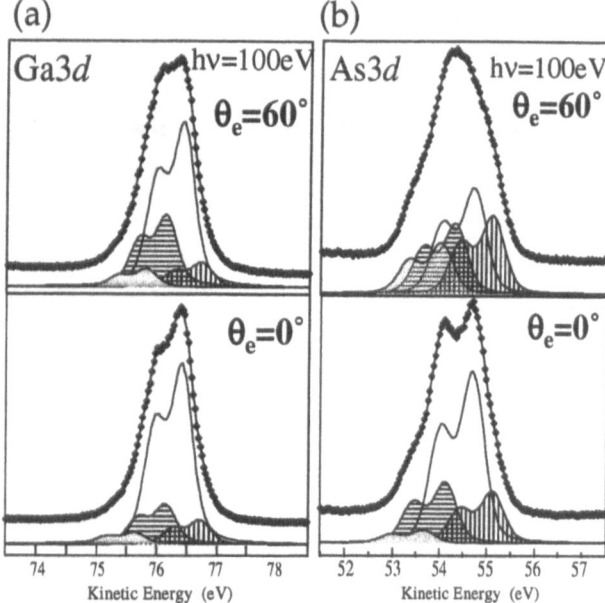

Fig. 2 Photoemission spectra of (a) Ga 3d and (b) As 3d of c(4×4)-GaAs(001). The shaded components indicate the surface components.

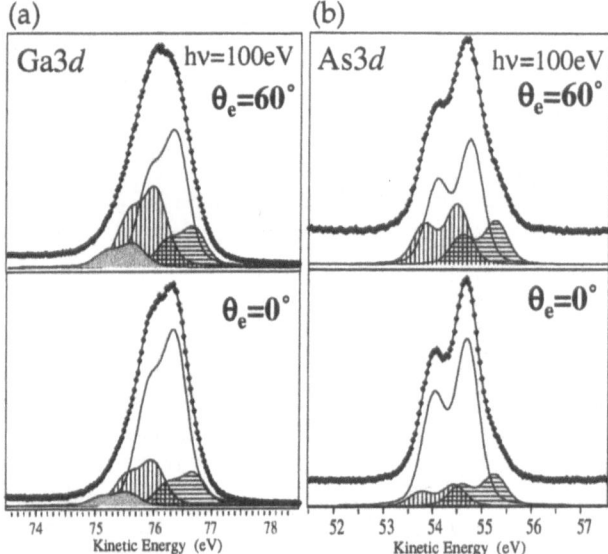

Fig. 3 Photoemission spectra of (a) Ga 3d and (b) As 3d of (4×6)-GaAs(001). The shaded components indicate the surface components.

a bulk component were observed. The intensity of the surface components increases at a take-off angle of 60 degrees which is surface sensitive condition compared with a take-off angle of 0 degree. This result suggests that there are two chemically inequivalent Ga-sites on the (2×4)-GaAs(001) surface, whereas in the case of As 3d spectra, three surface components were observed, indicating there are three chemically inequivalent As-sites.

Figure 2 shows the Ga 3d and As 3d photoemssion spectra of c(4×4)-GaAs(001). The spectral shapes are much different from the (2×4) phase. Curve-fitting results indicate that there are three surface sites for both Ga and As.

Figure 3 shows the Ga 3d and As 3d photoemssion spectra of (4×6)-GaAs(001). Curve-fitting results indicate that there are three surface Ga-sites and two surface As-sites.

The surface core level shifts of GaAs(001) reconstructured surfaces are summarized in Table 1. The above information is very useful to determine the correct model from numerous proposed models. However the one-to-one correspondence between surface core level shifts and atomic structures is not straightforward and theoretical investigation will be needed.

Table 1 Surface core level shifts of GaAs(001) reconstructured surfaces

(2×4)	Ga 3d	B	-
		S1	-0.27 eV
		S2	+0.34 eV
	As 3d	B	-
		S1	-0.40 eV
		S2	+0.37 eV
		S3	+0.86 eV
c(4×4)	Ga 3d	B	-
		S1	-0.30 eV
		S2	+0.27 eV
		S3	+0.76 eV
	As 3d	B	-
		S1	-0.42 eV
		S2	+0.49 eV
		S3	+0.79 eV
(4×6)	Ga 3d	B	-
		S1	-0.32 eV
		S2	+0.36 eV
		S3	+0.78 eV
	As 3d	B	-
		S1	-0.53 eV
		S2	+0.24 eV

4 Conclusion

In conclusion, we have determined the surface core level shifts of (2×4)-, c(4×4)-, and (4×6)-GaAs(001) surfaces, which were prepared in optimized molecular beam epitaxy growth conditions, by *in situ* synchrotron radiation high-resolution photoemission. The numbers of chemically inequivalent Ga and As surface sites were determined from the curve-fitting results of the high-resolution core level photoemission spectra.

References

1. R. Arthur, Surf. Sci. **43**, (1974) 449.
2. A. Y. Cho, J. Appl. Phys. **47**, (1976) 2841.
3. P. K. Larsen, J. H. Neave, and B. A. Joyce, J. Phys. **C14**, (1981) 167.
4. D. J. Chadi, J. Vac. Sci. Technol. **A5**, (1987) 834.
5. D. K. Biegelsen, R. D. Bringans, J. E. Northrup, and L.-E. Swartz, Phys. Rev. **B41**, (1990) 5701.
6. V. P. LaBella, H. Yang, D. W. Bullock, P. M. Thibado, P. Kratzer, and M. Scheffler, Phys. Rev. **B41**, (1990) 5701.
7. Q. Xue, T. Hashizume, J. M. Zhou, T. Sakata, T. Ohno, and T. Sakurai, Phys. Rev. Lett. **74**, (1995) 3177.

Structural ordering on Si(111)√3 × √3-Ag surface: Monte Carlo simulation based on first-principles calculations

Y. Nakamura[1,2], **Y. Kondo**[1,2], **J. Nakamura**[3], **S. Watanabe**[1,2]

[1] Department of Materials Science, The University of Tokyo, 7-3-1 Hongo, Bunkyo-ku, Tokyo 113-8656, Japan e-mail: naka@cello.mm.t.u-tokyo.ac.jp
[2] Core Research for Evolutional Science and Technology (CREST), Japan Science and Technology Corporation
[3] Surface and Interface Laboratory, RIKEN, 2-1 Hirosawa, Wako-shi, Saitama 351-0198, Japan

Abstract We study the arrangement of Ag atoms of the Si(111)√3× √3-Ag surface by Monte Carlo simulations (MCS) based on results of first-principles calculation (FPC). Introducing three-body term into the adiabatic potential for Ag atoms is very important to describe energies of possible configurations. Calculated STM images based on results of MCS clearly show that the observed STM images with a honeycomb pattern are well understood by the fluctuation of Ag atoms.

1 Introduction

After the years of controversy, the "Honeycomb-Chained Triangle (HCT)" model [1] has been widely accepted as the atomic structure of the Si(111)√3 × √3-Ag surface so far. Scanning tunneling microscopy (STM) images of the surface with a honeycomb pattern observed at room temperature [2,3] has been believed to correspond to the HCT structure [4]. Recently, however, a new model called the "InEquivalent Triangle (IET)" one [5] has been shown to be more stable by first-principles calculations (FPC) and supported by low-temperature STM experiments [5]. In this new model, Ag atom positions in the unit cell are a little rotated in the same direction from those of the HCT structure, and consequently two domains of opposite phase are allowed in accordance with the rotational direction.

This provides two possible different interpretations on the observed STM images at room temperature: Do the STM images correspond to the HCT structure or a time-average of fluctuation of Ag atoms? This also concerns the feature of the phase transition between the HCT and the IET structures. Though theoretical studies concerning the problem has already started [6], there remains disagreements with experimental results. In this study, we reexamine the arrangement of Ag atoms by Monte Carlo simulation (MCS) based on FPC results [7].

2 Model and calculations

We treat only thermal motion of Ag atoms in MCS explicitly on the following assumptions [6]: (i) the structure of a given configuration of Ag atoms is optimized with respect to the degrees of freedom for all the Si atoms; (ii) each Ag atom is allowed to move along the tangent touching the circle in the unit cell on which the three Ag atom positions of the HCT structure lie. We express total energy E_{tot} of the system as follows.

$$E_{tot} = \frac{1}{2}\sum_{i,j}\phi_2(r_{ij}) + \sum_i\sum_{\alpha=1,2}\phi_3(\theta_{i,\alpha}), \qquad (1)$$

$$\phi_2(r) = \sum_{k=1}^{3}a_k(r_k - r)^4 H(r_k - r), \qquad (2)$$

$$\phi_3(\theta) = k_1(\theta - \theta_0)^2 + k_2(\theta - \theta_0)^4, \qquad (3)$$

where $H(x) = 0$ for $x < 0$ and $H(x) > 1$ for $x > 0$. The ϕ_2 and ϕ_3 represent a two-body [8] and a three body interactions, respectively. We have confirmed that the three-body term, which is considered to come from the covalent-like bonding between the Ag atom and the underlying Si atom [9], is indispensable to describe energies of possible configurations. The r_{ij} and $\theta_{i,\alpha}$ are distances and angles among atoms forming chained triangle, respectively. The r_k is the knot point such that $r_1 > r_2 > r_3$ [8]. The k_1 and k_2 are taken positive. We assume that $\theta_0 = \pi/3$ [rad] and $r_3 = 14$ (in the unit of magnitude of displacement of Ag atom positions of the IET structure from those of the HCT structure). The values of parameters are given so as to reproduce energies of several configurations of Ag atoms calculated from first principles within the local density functional approach [7], which are summarized in Table 1.

With this potential, we perform MCS on the system consists of $N = 7,500$ Ag atoms under the periodic boundary conditions. The local-order parameter for the IET structure is defined as $P_i = \langle u_i \rangle$, where u_i and $\langle...\rangle$ represent the displacement of i-th Ag atom from the po-

Table 1 Obtained values of parameters in eqs. (2) and (3).

$a_1 = +0.13101 \times 10^0$ [meV]
$a_2 = -0.40768 \times 10^2$ [meV]
$a_3 = +0.45005 \times 10^2$ [meV]
$k_1 = +0.20283 \times 10^3$ [meV/rad^2]
$k_2 = +0.27950 \times 10^5$ [meV/rad^4]
$r_1 = 14.0$
$r_2 = 12.0$
$r_3 = 11.9$
$\theta_0 = \pi/3$ [rad]

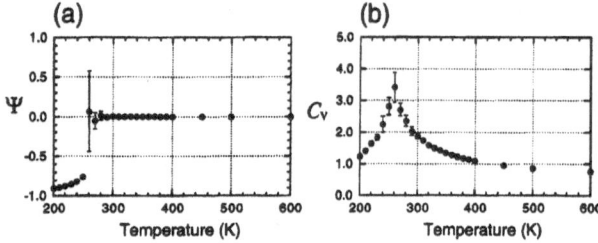

Fig. 1 Temperature dependence of (a) long-range order parameter Ψ and (b) specific heat C_V (in the unit of 10^{-5} meV/K).

sition of the HCT structure and statistical average, respectively. The IET structure is represented as $u_i = \pm 1$ for all i. The long-range order parameter Ψ is defined as the average of P_i over i. The specific heat is defined as $C_V = (\langle E^2 \rangle - \langle E \rangle^2)/k_B T^2$, where E, k_B, and T represent energy, the Boltzmann constant, and the temperature, respectively. STM images are approximated by a superposition of a two-dimensional Gaussian function centered at the centroid of each chained triangle [10].

3 Results and discussion

Figures 1(a) and 1(b) show the temperature dependence of the long-range order parameter Ψ and the specific heat C_V, respectively. As the temperature is lowered through 260-250 K, the absolute value of Ψ increases rapidly from nearly zero to almost unity (Fig. 1(a)). This result supports that of the previous MCS [6]. The system transforms into a single domain of the IET structure. In addition, the specific heat C_V has a sharp peak at 260 K (Fig. 1(b)), which means the system undergoes a second-order transition. From these results, the transition temperature T_C is estimated to be about 260K. This is consistent with experimental facts that the feature of the STM images observed at room temperature [2,3] differs completely from that at 62K [5].

Figure 2(a) shows the calculated STM image of a typical examples of snapshots at 300K. The surface is dominated by bright protrusions with hexagonal lattice patterns of opposite phase, which corresponds to the IET domains of opposite phase. The arrangement of the Ag atoms changes from snapshot to snapshot. The feature of STM images of these snapshots are quite different from that of the observed STM images at room temperature [2,3]. However, after we take a statistical average of the snapshots, STM image becomes a honeycomb pattern as shown in Fig.2(b). The present study has clearly shown that the observed STM images with a honeycomb pattern corresponds to a time-average of the fluctuation of each Ag atom. Recent FPC calculations for noncontact atomic force microscope (AFM) images of the Si(111)$\sqrt{3} \times \sqrt{3}$-Ag surface also support the possibility of such fluctuation [11].

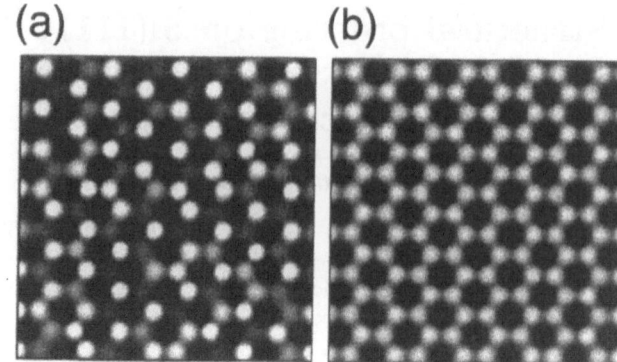

Fig. 2 Calculated STM images at 300K. (a) An example of snapshots. (b) Averaged images for the same area as (a).

4 Conclusions

We have studied the arrangement of Ag atoms of the Si(111)$\sqrt{3} \times \sqrt{3}$-Ag surface by MCS based on FPC. MCS using the three-body potential proposed in the present study showed that the surface undergo an order-disorder transition. The transition temperature estimated by the present MCS is consistent with results of the STM observations, which is not achieved by the previous MCS. Calculated STM images based on results of MCS clearly show that the observed STM images with a honeycomb pattern are well understood by the fluctuation of Ag atoms.

Acknowledgements The authors are grateful to Dr. N. Sasaki for valuable discussions.

References

1. T. Takahashi, S. Nakatani, N. Okamoto, T. Ichikawa, and S. Kikuta, Jpn. J. Appl. Phys. **27**, (1988) L753.
2. R.J. Wilson, and S. Chiang, Phys. Rev. Lett. **58**, (1987) 369.
3. E.J. van Loenen, J.E. Demuth, R.M. Tromp, and R.J. Hamers, Phys. Rev. Lett. **58**, (1987) 373.
4. S. Watanabe, M. Aono, and M. Tsukada, Phys. Rev. B **44**, (1991) 8330.
5. H. Aizawa, M. Tsukada, N. Sato, and S. Hasegawa, Surf. Sci. **429**, (1999) L509.
6. K. Kakikani, H. Kaji, Y. Yagi, and A. Yoshimori, Meeting Abst. Phys. Soc. Jpn. **55**, Iss.1, Part 4 (2000) 784.
7. S. Watanabe, Y. Kondo, Y. Nakamura, and J. Nakamura, submitted to Sci. Technol. Adv. Mater.
8. G.J. Ackland, G. Tichy, V. Vitek, M.W. Finnis, Phil. Mag. A **56**, (1987) 735.
9. H. Aizawa, M. Tsukada, N. Sato, and S. Hasegawa, Phys. Rev. B **59**, (1999) 10923.
10. Y. Nakamura, Y. Kondo, J. Nakamura, and S. Watanabe, submitted to Surf. Sci.
11. N. Sasaki, H. Aizawa, and M. Tsukada, submitted to Surf. Sci.

STM and RHEED study of Ge(110) reconstructions

T. Ichikawa[1], H. Fujii[2], and A. Sugimoto[1]

[1] Department of Physics, School of Science and Technology, Meiji University, Tama-ku, Kawasaki 214-8571, Japan
[2] Kanagawa Industrial Technology Research Institute, Shimoizumi, Ebina 243-0435, Japan

Abstract Some problems on reconstructed structures of Ge(110) surfaces are left open to argument. They were studied in the present study through in-situ STM and RHEED observations and very interesting results were obtained.

1 Introduction

When Ge(110) surfaces are cleaned and annealed at high temperature, surface reconstruction takes place and superstructures form on the surfaces. The first study of the surface reconstruction is a low energy electron diffraction work by Olshanesky et al. [1], in which they reported that a c(8 x 10) structure appeared by annealing at temperatures below 380°C and above 430°C and a (17 15 1) faceting occurred in the intermediate temperature range. This implies that Ge(110) surface reconstruction is anomalous, because on elevating temperature c(8 x 10) surface once changes to an undulated surface consisting of (17 15 1) facets and then is restored to the c(8 x 10) one. However, a subsequent reflection high energy electron diffraction (RHEED) work by Noro and Ichikawa [2] revealed that the intermediate structure was not a faceted one but a superstructure with the periodicity of 16 x 2.

Many superstructures associated with surface reconstruction change to disordered 1 x 1 structures on heating. A well-known example is structural change from 7 x 7 to 1 x 1 structure on Si(111) surface at 830°C [3]. A recent core-level and valence band photoemission spectroscopy work by Santoni et al. suggested that disordering took place on Ge(110) surface at 750 ± 50 K [4] as well. This temperature corresponds to the second transition temperature 430°C. Generally it is not so straightforward to derive structural information from superlattice reflections in diffraction patterns taken from heated surfaces because of a high background. Thus the aspect that the high-temperature phase is the c(8 x 10) structure is not so conclusive.

The purpose of the present study is to clarify unestablished results on Ge(110) reconstruction, e.g. whether the high-temperature phase is really the c(8 x 10) structure or disordered one and why the (17 15 1) facets were observed as the intermediate structure, carrying out scanning tunneling microscopy (STM) and RHEED observations.

2 Experimental

A 40 Ωcm intrinsic germanium wafer, 0.4 mm thick, with a mirror-polished (110) surface was cut into strips of dimensions 1 mm × 7 mm and 3 mm × 22 mm for STM and RHEED observations, respectively. The strip was washed in acetone and next in pure water and then introduced into ultrahigh vacuum of a JEOL JSTM-4500XT STM instrument and a home-built RHEED apparatus.

Cleaning of the Ge(110) surfaces was carried out by cycles of Ar-ion-etching at room temperature followed by annealing at about 720°C. Heating of the surface was performed by flowing DC current through the strip. The temperature of the strip was estimated from the reading of an ammeter, using a current-temperature characteristic calibrated by infrared thermometer.

3 Results and discussion

Figure 1 is a topographic STM image of Ge(110) surface just transforming at 430°C, in which the left region shows contrast of the 16 x 2 structure, while any periodic contrast is not observed in the right region. Since the interface of the two regions moves leftward by a slight temperature elevation, the right region is the high-temperature phase. The contrast of the right region shows that the high-temperature phase is not the c(8 x 10) and a 1 x 1 structure but a disordered one. An elaborate RHEED observation was undertaken to exclude an unlikely possibility that surface atoms are cooperatively moving about, holding the c(8 x 10) periodicity. This is due to the reason that snapshots of arrangement of surface atoms cannot be obtained in STM because of its sequential scanning. As was expected, no superlattice reflections of the c(8 x 10) structure were not observed in RHEED patterns taken from high-temperature-phase Ge(110) surfaces. In addition,

Fig. 1 STM image from surface just transforming from (16 x 2) to 1 x 1 structure.

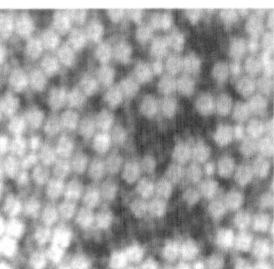

Fig. 2 STM image from quenched surface.

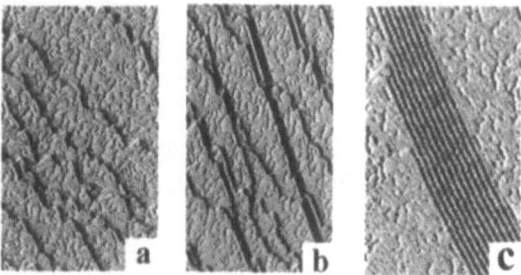

Fig. 3 STM images from quenched Ge(110) surfaces

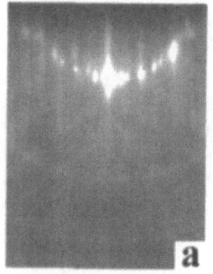

Fig. 5 RHEED patterns from annealed Ge(110) surfaces

the surfaces were quenched to room temperature. Figure 2 is a topographic STM image of the quenched surfaces. If the high-temperature phase has the c(8 x 10) or 1 x 1 periodicity, the quenched surface should also have the c(8 x 10) or 1 x 1 one. The image, however, seems to consist of a random assembly of five-membered rings [5] of Ge adatoms. These evidences reveal that the high-temperature phase is a disordered structure.

Next, we pursue a question why Olshanetsky et al. reported the (17 15 1) facet as the intermediate-temperature phase of reconstructed Ge(110) surfaces. Figures 3 (a), (b) and (c) are STM current images of surfaces quenched from 440, 430 and 415 °C. Individual steps disperse in fig. 3(a), while steps associate each other and form the (17 15 1) facet in fig. 3(c). In fig. 3(b) the association is about to initiate. These indicate that a reversible change of step distribution, i.e. association of steps to form facet and its dissociation, takes place at 430 °C. Suppose a rough Ge(110) surface. Such roughness is often introduced, originating from an inappropriate surface cleaning. Since the roughness is composed of a high density of steps, quenching from 430 °C or a little lower temperatures associates steps to form the (17 15 1) facet. We guess that the wrong conclusion [1] by Olshanetsky et al. may be due to a high density of steps of their Ge(110) sample surfaces.

When a high-temperature phase Ge(110) surface is quenched to temperatures a little lower than 380 °C and annealed at the temperatures for an hours or so, a well-developed c(8 x 10) structure forms on the surface, as reported in the previous works [1, 2]. Figures 4 (a) and (b) are models of the c(8 x 10) and 16 x 2 structures, respectively, derived from high-quality c(8 x 10) and 16 x 2 STM images. In the models five-membered rings of Ge adatoms are marked as principal building blocks common to the two structures, though they were described as

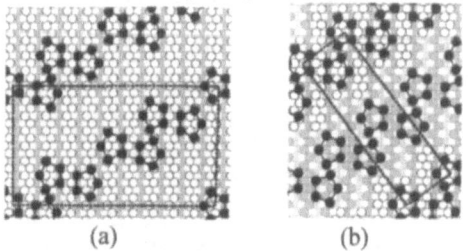

(a)　　　　　　　　　　　(b)

Fig. 4 Models of (a) c(8 x 10) and (b) 16 x 2 structures. Parallelograms represent the respective unit meshes.

clusters of adatoms in a previous study [5]. The same conclusion has been derived in a recent STM work [6]. The rings are depicted in the models with solid lines for convenience. It, hence, should be noticed that the lines do not necessarily represent bonding between Ge adatoms, because the separation between adatoms is much larger than bond length in Ge crystal. The rings arrange regularly on a flat surface in (a), while they are positioned on a periodic up-and-down sequence of terraces in (b). This indicates that building of the up-and-down terraces is required to form the 16 x 2 structure in addition to the regular arrangement of the rings. Therefore, an activation energy for the formation of the 16 x 2 structure is probably larger than that for the formation of the c(8 x 10) one. If so, the appearance of the c(8 x 10) structure by annealing at temperatures below 380° may originate in kinetics that the c(8 x 10) reconstruction proceeds faster than the 16 x 2 one. In order to establish this aspect, sufficient long annealing was carried out for the c(8 x 10) surfaces in the present study. An hour's annealing grew well-developed c(8 x 10) structure on the surface, as seen from RHEED pattern of fig. 5(a), while following one-night's annealing converted the structure to the 16 x 2 one, as seen from fig. 5(b). This RHEED observation shows that the 16 x 2 structure is the most stable one even in the temperature range below 430 °C and the misinterpretation of the c(8 x 10) structure as the low-temperature phase is due to an insufficient annealing.

4 Conclusion

In-situ STM and RHEED observations were performed in the present work and the following three intriguing results were obtained: (1) the high-temperature phase above 430 °C is a disordered structure, (2) the c(8 x 10) structure so far interpreted as the low-temperature phase is a transient one and (3) the conclusion by Olshanetsky et al. that the (17 15 1) facet is stable between 380 °C and 430 °C may be due to roughness of their sample surface.

References

1. B.X. Olshanetsky et al. : Surface Sci. 64 (1977) 224.
2. H. Noro and T. Ichikawa : Jpn. J. Appl. Phys. 24 (1985) 1288.
3. S. Ino : Jpn. J. Appl. Phys. 16 (1977) 891.
4. A. Santoni et al. : Surface Sci. 444 (2000) 156.
5. T. Ichikawa et al. : Solid State Commun. 93 (1995) 541.
6. Z. Gai et al. : Phys. Rev. B57(12) (1998) R6795.

Terrace and step contributions to the surface optical anisotropy of Si(001)

W.G. Schmidt[1], F. Bechstedt[1], J. Bernholc[2]

[1] Institut für Festkörpertheorie und Theoretische Optik, Friedrich-Schiller-Universität Jena, 07743 Jena, Max-Wien-Platz 1, Germany e-mail: `W.G.Schmidt@ifto.physik.uni-jena.de`

[2] Department of Physics, North Carolina State University, Raleigh, NC 27695-8202, USA

Abstract The contributions of flat terraces as well as of S_A, S_B and D_B steps to the optical anisotropy of Si(001) surfaces have been calculated using a real-space multigrid method. We find a distinct influence of the dimer arrangement on the optical spectra. The signal measured for atomically smooth terraces is well described by the calculated spectrum for the Si(001)$c(2\times4)$ surface. Surface steps are responsible for the significant optical anisotropy around 3 eV observed for vicinal surfaces. The influence of sub-surface atomic relaxations on the anisotropy signal is found to be surprisingly small.

1 Introduction

Reflectance anisotropy/difference spectroscopy (RAS/RDS) is an extremely versatile tool of surface analysis. Thanks to the development of both computational techniques and resources it is now possible to calculate anisotropy spectra from *first principles* in quantitative agreement with experiment [1]. Such calculations are very involved and have by now been restricted to ideal surfaces. The complete understanding of RAS spectra, however, requires to consider realistic surfaces, containing defects such as surface steps. We calculate the RAS of flat Si(001) surfaces forming (1×2), $p(2\times2)$ and $c(2\times4)$ reconstructions and of S_A, S_B and D_B surface steps from *first-principles*. Particular attention is paid to the origin of bulk-related features in the optical spectra.

2 Method

Our massively parallel, real-space finite-difference calculations [2] are based on density-functional theory in the local-density approximation and use a multigrid technique for convergence acceleration. The surface is modeled by periodic supercells containing asymmetric slabs of 11 atomic Si layers separated by 12 Å of vacuum. The dangling bonds at the bottom layer are saturated with hydrogen. S_A and S_B/D_B steps, respectively, are modeled by missing dimer structures in (2×6) and (2×7) surface unit cells. To calculate the dielectric function we use a sampling of about 500 – 1000 points in the full (1×1) surface Brillouin zone. The slab polarizability calculated in the independent-particle approximation is used to calculate the RAS spectra. Further details can be found in our study on stepped Si(111):H surfaces. [3]

3 Results and Discussion

3.1 Flat surface

The RAS spectra calculated for (1×2), $p(2\times2)$, and $c(2\times4)$ reconstructed Si(001) are shown in Fig. 1. The

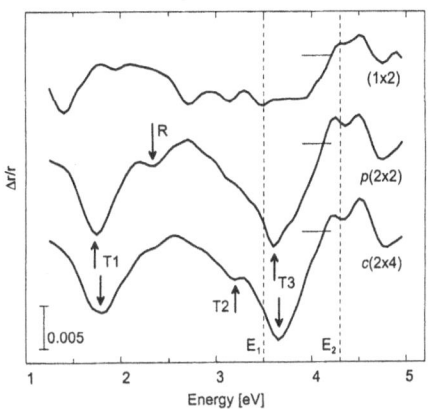

Fig. 1 RAS spectra $[Re\{(r_{[\bar{1}10]} - r_{[110]})/< r >\}]$ calculated for the Si(001)(1×2), $p(2\times2)$ and $c(2\times4)$ surfaces. Dashed lines mark the positions of bulk critical point energies.

spectrum for (1×2) shows a pronounced minimum at 1.4 eV, a rather broad negative anisotropy between 2.5 and 4.0 eV and a maximum at 4.5 eV, consistent with previous *ab initio* results [4,5]. The alternating arrangement of the asymmetric dimers in a $p(2\times2)$ rather than a (1×2) structure leads to pronounced spectral changes. Two distinct minima, $T1$ and $T3$, appear. The dimer-dimer interaction extends even beyond the dimer rows: differences in the RAS also appear in the spectra calculated for $p(2\times2)$ and $c(2\times4)$ surfaces. $T1$ shifts from 1.7 to 1.8 eV and the local minimum R evolves into a shoulder. A new feature, $T2$, appears at 3.2 eV.

Comparing the RAS calculated for (1×2), $p(2\times2)$ and $c(2\times4)$ reconstructed surfaces with experiment, we find that the calculated anisotropy for $c(2\times4)$ surfaces agrees well with the anisotropy measured for highly oriented samples [6,7].

To determine the origin of the spectral features we separate the contributions to the RAS from electronic transitions within the uppermost four atomic layers and from the bulk-like layers underneath. Fig. 2 shows that surface-state related transitions are mainly responsible for the optical anisotropies below the E_1 energy, $T1$ and $T2$. $T3$ and the two maxima below and above the E_2 en-

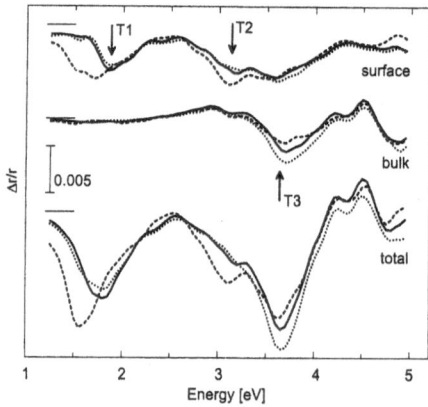

Fig. 2 RAS for the Si(001)$c(2\times4)$ surface, calculated for the fully relaxed configuration (solid lines) in comparison to calculations for a geometry where the relaxation was restricted to the uppermost two atomic layers (dashed lines). Spectra calculated for an anisotropically strained slab (see text) are shown by dotted lines. Shown are contributions to the RAS from electronic transitions within the uppermost four atomic layers, from the layers underneath, and the total signal.

ergy originate from bulk layers. They depend only little on the surface symmetry and appear for (1×2), $p(2\times2)$, and $c(2\times4)$ periodicities.

The appearance of optical anisotropies *not* related to transitions between surface states was suggested to be related to subsurface stress induced by the dimerization of the surface atoms [8]. We investigated this hypothesis by additional calculations. First a geometry was studied where only the uppermost two atomic layers were allowed to relax, while the remaining atoms occupied ideal bulk positions. Dashed lines in Fig. 2 show the resulting RAS. The surface-related features are distinctly changed. Only small modifications, in particular around the $T3$ feature, however, are observed for the RAS arising from bulk layers.

The dimerized surface is stretched along the dimer bonds and compressed across the dimer rows. As a second test we therefore relax the Si(001)$c(2\times4)$ surface in a unit cell whose lateral dimensions are shrunk/enlarged by 0.5 % along [110]/[$\bar{1}$10]. The spectra calculated for this configuration are shown by dotted lines in Fig. 2. Again, the optical anisotropies arising from subsurface transitions are only slightly affected. The effect is strongest for the $T3$ feature, which may thus be partially attributed to reconstruction-induced stress beneath the surface. The salient RAS features originating from bulk layers, however, are neither affected by stress nor relaxation. This indicates that they are caused by the anisotropic surface potential.

3.2 Surface steps

In the present study we focus on the three most commonly observed step configurations: S_A steps and rebonded steps of type S_B and D_B. The calculated step-generated RAS signal is compared with the experimental

findings [7] in Fig. 3. Given the limitations of our study, in particular with respect to the slab size, the comparison is gratifying. In agreement with experiment we find that surface steps on Si(001) give rise to a broad negative anisotropy below the E_1 energy, with a minimum at around 3 eV. Experimentally, a positive anisotropy between 3.5 and 4 eV is observed for miscut angles larger than 4°. A similar feature appears in the calculation for D_B steps. As biatomic steps form for larger miscuts only, the observed changes of the measured line shape with vicinality finds a natural explanation.

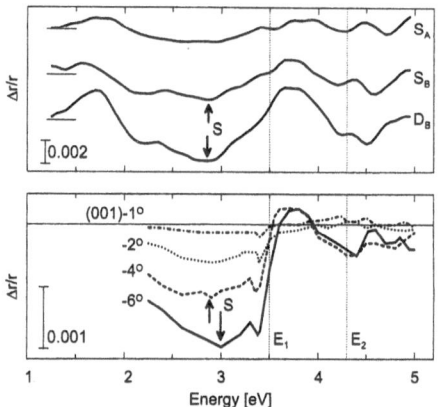

Fig. 3 Upper panel: Step-induced optical anisotropy calculated for S_A, S_B and D_B steps. Lower panel: Step-induced optical anisotropy measured for different miscut angles [7].

Combining our results for flat, $c(2\times4)$ reconstructed terraces and for the step related anisotropies, the peculiar RAS lineshape observed for vicinal samples [4,9] can be understood as superposition of terrace- and step-generated optical anisotropies.

Acknowledgments

We thank D.E. Aspnes, K. Hingerl and U. Rossow for suggesting this work and numerous useful discussions. The computer time was provided by the the DoD Challenge Program, the John von Neumann-Institut Jülich and the Höchstleistungsrechenzentrum Stuttgart. JB was supported in part by ONR and DoE.

References

1. W. G. Schmidt *et al.*, Phys. Rev. B. **61** (2000) R16335.
2. E. L. Briggs, D. J. Sullivan, and J. Bernholc, Phys. Rev. B **54** (1996) 14362.
3. W. G. Schmidt and J. Bernholc, Phys. Rev. B **61** (2000) 7604.
4. L. Kipp *et al.*, Phys. Rev. Lett. **76** (1996) 2810.
5. M. Palummo, G. Onida, R. Del Sole, and B. S. Mendoza, Phys. Rev. B **60** (1999) 2522.
6. R. Shioda and J. van der Weide, Phys. Rev. B **57** (1998) R6823.
7. S. G. Jaloviar *et al.*, Phys. Rev. Lett. **82** (1999) 791.
8. K. Hingerl *et al.*, Phys. Rev. B (accepted) .
9. T. Yasuda, L. Mantese, U. Rossow, and D. E. Aspnes, Phys. Rev. Lett. **74** (1995) 3431.

Nano-scale ferromagnets on semiconductors: Ga adsorbates on Si (100) surfaces

Susumu Okada[1], Atsushi Oshiyama[2]

[1] Institute of Material Science, University of Tsukuba, Tennodai, Tsukuba 305-8573, Japan
[2] Institute of Physics, University of Tsukuba, Tennodai, Tsukuba 305-8571, Japan

Abstract We report on total-energy electronic-structure calculations within the density functional theory performed for an atomic wires consisting of Ga atoms which are deposited on hydrogenated Si (100) surfaces. We find that there are three distinct stable atomic structures and that the ground state is ferromagnetic in one of the three structures. Stability of the ferromagnetic state against carrier doping, charge density wave generation, and anti-ferromagnetic ordering is discussed.

1 Introduction

Nanometer scale manipulation of atoms on surfaces has been partly realized by scanning tunneling microscope (STM) technique. An important example of such manipulations is the fabrication of a Ga wire on the hydrogenated Si (100). It has been demonstrated that H atoms on the hydrogenated Si(100) are removed along a Si dimer row [1,2], and then Ga atoms are placed on the dangling-bond array of Si atoms [2]. Electronic structure calculations by local density approximation (LDA) show that there are several metastable geometries for a Ga wire on the hydrogenated Si(100) and that energy bands near the Fermi level (E_F) are extremely flat for some of the geometries [3]. Further investigation using a tight-binding model suggests a possibility of spin polarization along the Ga wire [4]. The wire is then speculated to be an example of flat-band ferromagnetism in which subtle balance of electron transfers among atomic sites induces flat bands and carriers there are spin-polarized [4,5].

In this work, we study atomic and electronic structures of the Ga wires on the hydrogenated Si(100) surfaces by using the density functional theory (DFT) with the local spin density approximation (LSDA) and generalized gradient approximation (GGA) in which spin degrees of freedom are taken into account. Starting from three distinct initial structures which have been reported in the previous LDA calculations [3], we reach a new stable geometry for each structure and find that one of three structures exhibits certain magnetic ordering with non magnetic elements alone.

2 Methods

All calculations have been performed on the basis of the density functional theory [6,7]. Exchange-correlation interaction is essential to describe magnetism. Hence we use local spin density approximation and generalized gradient approximation and hereby extract common features irrespective of the approximation level. For the exchange-correlation energy, we adopt a functional form

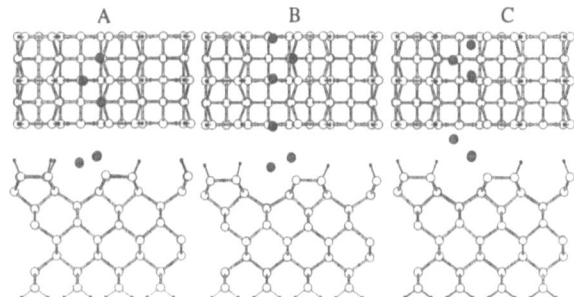

Fig. 1 Top and side views of optimized three atomic structures, A, B, and C, of a Ga wire on the hydrogenated Si(100) surface. White, shaded and black circles denote Si, Ga and H atoms, respectively.

by Perdew and Zunger [8,9] in LSDA and by Perdew and Wang [10] in GGA, respectively. Norm-conserving pseudopotentials generated by using the Troullier-Martins scheme are adopted to describe the electron-ion interaction [11,12]. We consider the partial core correction for the Ga atoms in the treatment of the exchange-correlation energy [13]. The valence wave functions are expanded in terms of plane-wave basis set with the cutoff energy of 10 Ry. The surface is simulated by a repeating slab model in which 8 Si atomic layers and the 5.6 Å-vacuum region are included. The bottom of the slab has a bulk-like structure with each Si saturated by two H atoms. On the top of the slab, which simulates the real surface, an H adlayer plus Ga atoms are considered. We adopt 4×2 lateral periodicity in which three Ga atoms are included [2,3]. Eight k-points are used for the integration over surface Brillouin zone. When we examine the stability of the 4×2 structure against the 4×4 modulations such as the charge density wave or the spin density wave, a 4×4 lateral cell with five atomic layers is adopted in the calculation. Geometry optimization has been performed for all atoms in the slab except for the bottom-most Si and H atoms. We use the conjugate-gradient minimization scheme both for the electronic structure calculation and for the geometry optimization [14]. In the optimized geometries the remaining forces acting on the atoms are less than 0.005 Ry/Å.

3 Results and discussion

Optimized structures of three stable geometries which are obtained by the GGA are shown in Fig. 1. The LSDA leads to qualitatively same optimized structures. Yet there are sizable differences in structure which cause the modification in energetics, as is shown below. In the

Table 1 Relative total energies E (eV) per unit cell obtained by LSDA and GGA. The energy is measured from the structure A. In the structure C, both paramagnetic and ferromagnetic spin configurations are obtained, whereas only paramagnetic configuration is obtained in A and B. The number of polarized electrons, $\Delta = n_{up} - n_{down}$, is a difference of the number of up spins (n_{up}) and down spins(n_{down}).

		A	B	C para	C ferro
E(eV)	LSDA	0	0.279	0.444	0.376
	GGA	0	-0.001	0.173	0.027
Δ	LSDA, GGA	0	0	0	1

three structures, adsorbed Ga atoms form trimers which are aligned in the direction parallel to the Si dimer row. The Ga atoms form covalent bonds with each other and with subsurface Si atoms. Table 1 shows calculated relative total energies for each structure. The structure A is the most stable among the three structures in the LSDA, whereas the structure B is the most stable in the GGA calculation. The structure C is not a global minimum in total energy. It is also found that there is an energy barrier of 0.4 eV along the reaction pathway from C to B. Therefore the structure C is certainly metastable and realized at room temperature.

In the structures A and B, electron spin is not polarized: We only find paramagnetic solutions. In the structure C, however, we find both paramagnetic and ferromagnetic states. The calculated number of polarized spin in the ferromagnetic state is one per unit cell. The ferromagnetic state is lower in energy than the paramagnetic state by 0.15 (0.07) eV in the GGA (LSDA) (Table 1). As a result, the ferromagnetic state in the structure C is higher in energy only by 28 meV in the GGA than the most stable paramagnetic state in the structure B.

The electronic energy bands, which are obtained by the GGA, of the three optimized structures are shown in Fig. 2. Since the structure C has ferromagnetic electronic structure, we show energy bands for both up and down spins. In all structures, the electron state α around E_F exhibits extremely small dispersion along the direction perpendicular to the wire. This means that the interaction between the adjacent wires is sufficiently small. Along the direction parallel to the wire, the state α shows substantial dispersion ($\sim$ 0.4 eV) in the structures A and B. In the structure C, however, the state α exhibits small dispersion even along the direction parallel to the wire: By analysing the spin density of the α state in the structure C, we found that the spin density is primarily distributed on Ga trimers. Yet it slightly penetrates to subsurface Si atoms and inter Ga trimer region. Hence, this localized nature causes small dispersion of the α band and thus ferromagnetic ground state.

Finally we examine the stability of the ferromagnetic state against the carrier doping, charge density wave generation(CDW), and the anti-ferromagnetic (AF) or-

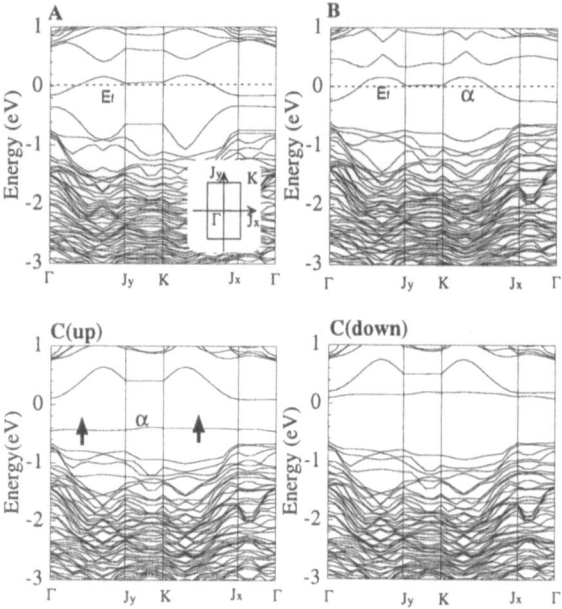

Fig. 2 Energy bands of a Ga wire on Si(100)/H with the atomic structures A, B, and C shown in Fig 1. In the cases of A and B structures, the energy bands for up and down spins are degenerate. In the structure C, the highest occupied band splits by $\sim$ 0.6 eV for up and down spins, and the band only for the majority spin is occupied.

dering. The present ferromagnetic state is found to be energetically favorable compared with the paramagnetic state for all number of carriers. Furthermore, the CDW is also unlikely to take place and the ferromagnetic state is robust against the lattice distortion. On the other hand, the calculated total energy of AF state is almost identical to that of the ferromagnetic state at the present accuracy.

Computations were done at ISSP, and IMS. This work was supported in part by JSPS (RFTF96P00203) and by Grant-in-Aid for Scientific Research (11740219).

References

1. T. C. Shen, et al,Science, **268**, (1995) 1590.
2. T. Hashizume, et al.,Jpn. J. Appl. Phys. **35**, (1996) L1085.
3. S. Watanabe, et al.,Jpn. J. Appl. Phys. **36**, (1997) L929.
4. M. Ichimura, et al.,Phys. Rev. B **58**, (1998) 9595.
5. For a review on the flat-band ferromagnetism, H. Tasaki, Prog. Theo. Phys. **99**, (1998) 489.
6. P. Hohenberg and W. Kohn, Phys. Rev. **136**, (1964) B864.
7. W. Kohn and L. J. Sham, Phys. Rev. **140**, (1965) A1133.
8. D. M. Ceperley and B. J. Alder, Phys. Rev. Lett. **45**, (1980) 566.
9. J. P. Perdew and A. Zunger, Phys. Rev. B **23**, (1981) 5048.
10. J. P. Perdew, et al.,Phys. Rev. B **23**, (1996) 16533.
11. N. Troullier and J. L. Martins, Phys. Rev. B **43**, (1991) 1993.
12. L. Kleinman and D. M. Bylander, Phys. Rev. Lett. **48**, (1982) 1425.
13. S. G. Louie, S. Froyen and M. L. Cohen: Phys. Rev. B **26**, (1982) 1738.
14. O. Sugino and A. Oshiyama, Phys. Rev. Lett. **68** (1992) 1858.

AlGaN/GaN lateral polarity heterostructures

A. P. Lima[1], C. Miskys[1], O. Ambacher[1], M. Stutzmann[1], R. Dimitrov[2], V. Tilak[2], M. J. Murphy[1,2], and L. F. Eastman[2]

[1]Walter Schottky Institut, TU München, Am Coulombwall, 85748 Garching, Germany
[2]School of Electrical Engineering, Cornell University, Ithaca, NY 14853, USA

Abstract   AlGaN/GaN heterostructures were grown by Plasma Induced Molecular Beam Epitaxy on patterned AlN thin nucleation layers deposited on c-plane sapphire substrates. It is shown, by the localization of the 2DEG, that the heterostructures grown in the region covered by the AlN are Ga-face and the heterostructures grown directly on sapphire are N-face. This provides the new possibility to obtain lateral polarity heterostructures.

1 Introduction

The strong internal electric fields in wurtzite AlGaN/GaN heterostructures resulting from fixed charges induced at surfaces or hetero-interfaces by an abrupt change of the spontaneous and piezoelectric polarization in III-nitrides have drastic effects on the band profiles, carrier distribution and electronic transport properties of these wide bandgap semiconductors. [1,2] Due to these bound charges at interfaces and their screening by free carriers, it is possible to achieve two dimensional hole and electrons gases with high sheet carrier concentrations without the conventional modulation doping known from the AlGaAs/GaAs heterosystem [3]. Furthermore, a significant reduction of the effective activation energy of the relatively deep acceptors in III-nitrides is observed in AlGaN/GaN superlattices. In these structures, the orientation of the piezoelectric and spontaneous polarization, the sign of interface bound sheet charges, and the exact location of 2DEGs and 2DHGs is determined by the respective polarity of the epitaxial film, which can be either Ga- or N-face.

It was shown recently that the polarity of GaN layers grown by Plasma Induced Molecular Beam Epitaxy (PIMBE) on (0001) sapphire substrate can be controlled to be either N- or Ga-face by use of suitable nucleation layers. GaN deposited directly on oxygen terminated sapphire substrates by PIMBE always exhibits N-face polarity. In contrast, GaN with Ga-face polarity is achieved by prior deposition of a thin ($\approx$ 5-10 nm) AlN nucleation layer completely covering the sapphire substrate. This allows the realization of lateral polarity heterostructures (LPH), with potential applications in novel electronic devices.

In the present work, we describe the preparation by molecular beam epitaxy and first electronic properties of such AlGaN/GaN LPHs.

2 PIMBE growth and patterning

The samples were grown in two steps, using a Tectra MBE chamber equipped with conventional effusion cells and an Oxford Applied Research CARS25 plasma source for the generation of nitrogen radicals. First we grew a 10 nm thick AlN nucleation layer on 2 inch c-plane sapphire wafers.   The samples were then patterned ex-situ by wet etching with a hot (100°C) 50% KOH solution. The pattern was defined by optical lithography using the combination of a metal/photoresist mask with stripes and squares having lateral dimensions between 1 and 100 μm. The etch procedure turned out to be the crucial step of the process. [4] Damage to the sapphire surface and its usual oxygen termination can lead to the formation of GaN with mixed polarity (uncontrolled growth of small domains with different polarity). In addition there are problems related with an efficient protection of the AlN layers by the mask against the etching. After optimized patterning of the nucleation layer, the substrate showed clean areas of undamaged sapphire adjacent to the non-etched AlN nucleation layer, as shown in the AFM image of figure 1.

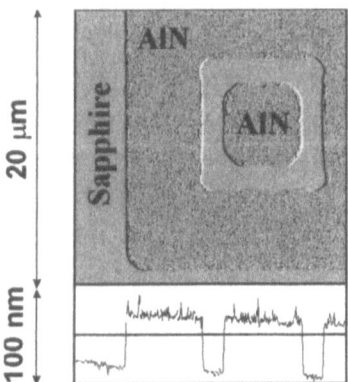

Fig. 1 AFM image and linear cross section of a patterned AlN nucleation layer on sapphire

The patterned substrate was overgrown by PIMBE with a heterostructure consisting of a 1000 nm thick GaN buffer layer, 30 nm $Al_{0.3}Ga_{0.7}N$ and 30 nm GaN. The alloy composition was determined by high resolution X-ray diffraction. Figure 2 shows optical microscopy images of two different areas of the sample surface after the overgrowth. The sharp borders between the areas grown directly on sapphire and on the AlN nucleation layers demonstrate the high quality of the etch and overgrowth procedure used. The optical contrast is due to a different surface roughness and slightly different growth rates for N- and Ga-face regions, respectively.

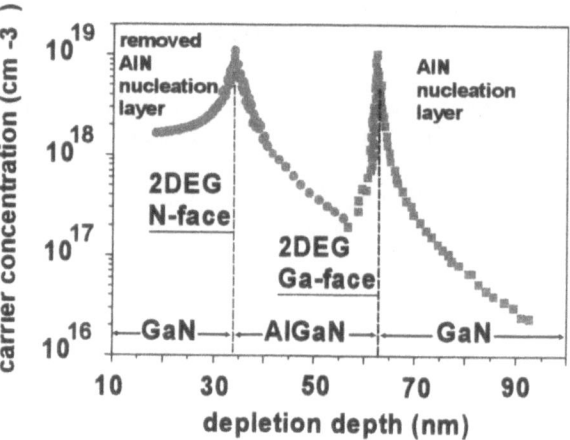

Fig. 3 Electron concentration profiles of GaN/AlGaN/GaN heterostructures with different polarities. Circles correspond to the area grown directly on the sapphire substrate, squares to the area grown on the AlN nucleation layer.

 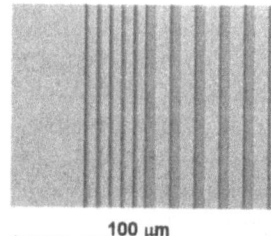

Fig. 2 Optical microscopy of the overgrown patterned substrate.

3 Polarity determination

The polarity of the overgrown GaN/AlGaN/GaN heterostructures was determined by means of C-V profiling measurements [2]. Figure 3 shows the result of these measurements performed on the area grown directly on the sapphire surface as well as on the area overgrown on the AlN nucleation layer.

In both cases, spontaneous formation of a two dimensional electron gas with similar sheet concentration occurs at the heterointerface. The 2DEG is located at the upper GaN/AlGaN interface for the heterostructure grown on the sapphire substrate without AlN nucleation layer, as expected for N-face polarity. On the other hand, the 2DEG is located at the lower AlGaN/GaN interface for the heterostructure grown on the AlN nucleation layer, which corresponds to a Ga-face polarity. [3]

4 Conclusion

We have demonstrated the possibility to grow AlGaN/GaN lateral polarity heterostructures with N-face polarity side by side with Ga-face polarity by means of a simple etching procedure prior to layer growth. Not only was it shown that the different polarities can be grown on the same substrate, but the overall crystalline quality is high enough that a 2DEG can be confined at each side of the heterointerface. This provides additional design freedom and interesting new possibilities for electronic devices.

References

1. V. Fiorentini, F. Bernardini, Phys. Rev. **B 60**, (1999) 8849.
2. R. Dimitrov, A. Mitchell, L. Wittmer, O. Ambacher, M. Stutzmann, J. Hilsenbeck, and W. Rieger, Jpn. J. Appl. Phys. **38**, (1999) 4962.
3. O. Ambacher, J. Smart, J. R. Shealy, N. G. Weimann, K. Chu, M. Murphy, W. J. Schaff, L. F. Eastman, R. Dimitrov, L. Wittmer, M. Stutzmann, J. Hilsenbeck, J. Appl. Phys. **85**, (1999) 3222.
4. R. Dimitrov, V. Tilak, M. Murphy, W. J. Schaff, L. F. Eastman, A. P. Lima, C. Miskys, O. Ambacher, and M. Stutzmann, Mat. Res. Soc. Symp. **622**, (2000) T4.6.1

Modification of the Si surface electronic properties by Ge nanostructures: Surface photovoltage studies

K. Nauka, T.I. Kamins

Hewlett-Packard Laboratories, Palo Alto, CA 94304, USA

Abstract The Surface Photovoltage technique was used to investigate surface barriers determined by Ge nanostructure islands on Si substrate. The initial surface barrier value depended uniquely on the size of the islands. Time evolution of the surface barrier was determined by oxidation of the nanostructures that consumed the small islands and reduced the size of the large nanostructures.

1. Introduction

Spontaneous formation of small Ge islands on Si substrates has recently attracted considerable attention as a potential path to manufacturing low dimensional, sub–100 nm Si-based devices [1,2]. The major advantage of Ge is its compatibility with mainstream Si integrated circuit technology. In fact, integrated circuit devices based on Si-Ge alloys are already commercially available [3]. Published reports demonstrated that Ge nanostructures with dimensions below 50 nm are easily formed on Si substrates by chemical vapor deposition equipment used in routine Si-device manufacturing [2,3].

Until now most of the investigation was focused on the formation mechanisms and structure of Ge islands fabricated on Si substrates. Little is known about their electronic properties and the relationship between the electronic nature of self-assembled Ge nanostructures and Si surfaces. In this work Surface Photovoltage is used to monitor changes of the Si substrate surface barrier caused by overlying Ge nanostructures.

2. Experimental

Ge islands were grown on (001)-oriented Si wafers in a lamp-heated, single wafer reactor with wafers resting on a support with moderate thermal mass. After loading, wafers were in-situ heat-treated at about 1150°C in a H_2 ambient to remove native oxides. Then, thin Ge layers were deposited at approximately 600°C using GeH_4 and H_2 carrier gas at pressures between 5 Torr and 600 Torr. In some cases thin Si epitaxial layers were grown first using the SiH_4 or SiH_2Cl_2, and then Ge was deposited. After deposition, Si wafers were cooled in an inert ambient and withdrawn from the reactor at room temperature. Some wafers were in-situ annealed in an inert ambient after Ge deposition, but before they were withdrawn from the reactor.

The amount of Ge deposited was measured with Rutherford Backscattering Spectroscopy (RBS), using 2 MeV He^{++} ions, and was expressed in terms of the equivalent number of Ge monolayers (eq-ML) even when Ge islands were present (1 eq-ML=$6.6*10^{14}$ cm^{-2}). Since RBS integrates Ge signal over a macroscopic region it offers a reliable way of quantifying the amounts of deposited Ge. Ge thickness was varied from less than a monolayer to a few tens of monolayers. The surface features were investigated with atomic force microscopy (AFM) [2] and complementary scanning tunneling microscopy studies [4] examined islands fabricated by physical vapor deposition.

The Surface Photovoltage technique (SPV) used to measures Si substrate surface barrier using surface-charge measurements [5]. The SPV signal is obtained by illuminating the front side of the wafer with monochromatic photon flux (λ = 800 nm), while keeping the back side in the dark. Non-equilibrium carriers generated by absorbed light lower the surface barrier and generate SPV signal. At high photon fluxes barrier disappears and the corresponding SPV value determines the equilibrium surface barrier height. In the present study the initial surface barrier was measured immediately after withdrawing the Si substrates from the inert reactor ambient and the surface barrier was subsequently monitored over an extended period of time during which wafers were exposed to the air. No contact is made to the sample surface during SPV measurement ensuring that the surface barrier is not disturbed [5].

3. Results and Discussion

Due to the lattice mismatch deposited Ge follows the Stranski – Krastanov growth mode, forming at first a two-dimensional wetting layer and then, after its thickness exceeds the critical value, forming three-dimensional islands. It has been shown [2] that deposition conditions used in this experiment form a wetting layer up to about 4 eq-ML. When the Ge thickness exceeds the critical value but remains below 8 eq-ML square based pyramids with a length/height ratio of 10 and height less than 8 nm form. Even higher Ge doses lead to formation of multifaceted "domes" with height about 15 nm and diameter of about 80 nm. Ge pyramids and domes are lattice matched to the Si substrate; even larger islands are frequently relaxed, with corresponding structural defects. The H_2 ambient used during Ge deposition provides a natural surface passivation that, at least partially, determines the surface barrier. The effect of H_2 on a Si surface had been previously measured [6] and was used in the present work to understand the influence of Ge nanostructures on the Si surface barrier.

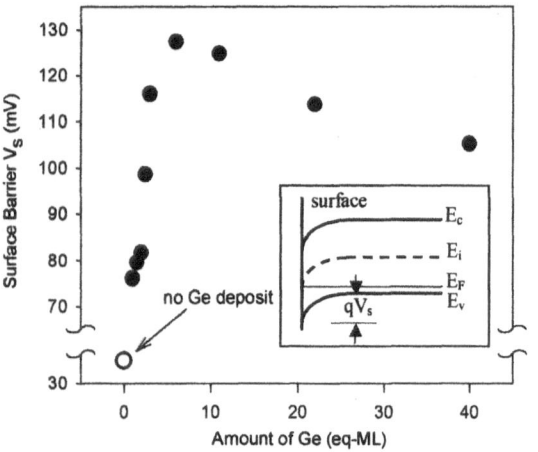

Fig.1 Initial surface barrier as a function of the Ge dose.

The initial surface barrier depended strongly on the amount of Ge per unit area (the "dose") (Figure 1). A rapid increase of the barrier at Ge doses below the onset of nanostructure formation corresponded to microscopic roughening of the wetting layer before pyramid-like nanostructures were formed [4]. The highest surface barriers corresponded to Ge doses at which strained pyramids and strained multifaceted domes were formed. A further increase of the nanostructure sizes led to their relaxation by defect formation and a corresponding decrease of the initial surface barrier. The measured barrier height represents a mean barrier value determined by local barriers beneath the individual nanostructures, regions between the nanostructures, and the Si substrate.

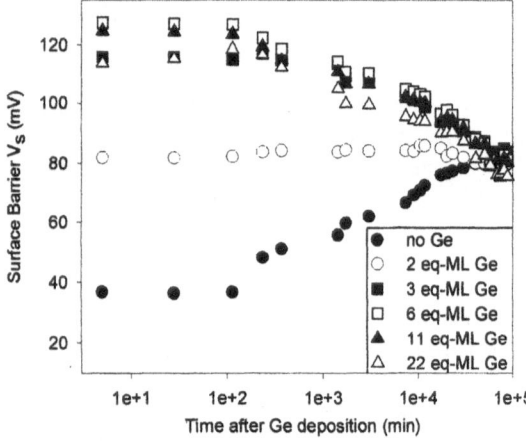

Fig.2 Evolution of the surface barrier due to oxidation of the Ge nanostructures.

Prolonged air exposure caused reduction of the surface barrier, as shown in Figure 2. Ge nanostructures with H_2 initially terminated surfaces gradually oxidized,

eliminating H_2 and replacing it with a thin sub-oxide SiO_x (x < 2) layer. The final barrier height of 78 meV was primarily determined by the barrier height at the Si / SiO_x interface [6]. Small Ge nanostructures were likely completely consumed by the oxidation while large structures were capped with an oxide layer. The detailed time dependence of the surface potential suggests that the nanostructures underwent a series of sequential oxidation processes that depended on the individual nanostructure sizes.

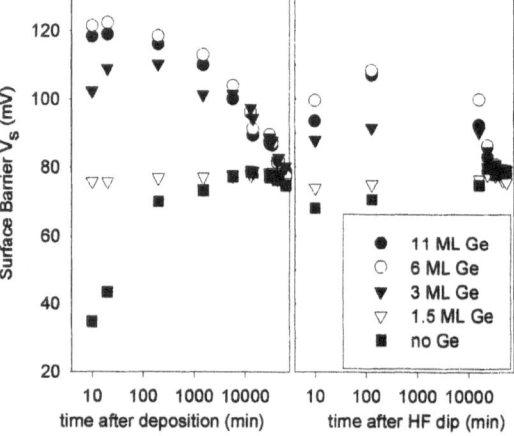

Fig.3 Change of the surface barrier after the oxide is removed.

Native oxide can be removed from Ge islands by dipping the wafers into the aqueous HF solution. Oxide removal and possible hydrogen termination increased the surface barrier immediately after the HF treatment (Figure 3). However, the surface barrier value for a given sample was below its original value and corresponded to the initial barrier value of a smaller island. This is likely due to the reduction in size of the Ge island as the surface oxide layer was removed.

Conclusions.

The Si substrate surface barrier is depends on the size and type of nanoscale Ge islands present on the surface. Large changes of the surface barrier are observed during the initial surface coarsening even before distinct Ge nanostructures are formed. In addition, the surface barrier can be used to monitor reduction of the island size by room temperature oxidation.

References

1. M.Krishnamurthy, J.S.Druker, J.A.Venables, J.Appl.Phys. **69** (1991), 6461.
2. T.I.Kamins, E.C.Carr, R.S.Williams, S.J.Rosner, J.Appl. Phys. **81** (1997), 211.
3. D.L.Harame et al., presented at the IEEE Internat. Solid-State Circuit Conference '95, San Francisco, CA (1995).
4. G.Medeiros-Ribeiro et al., Science **279** (1998), 353.
5. K.Nauka, Microelectron.Eng. **36** (1997), 351.
6. K.Nauka, T.I.Kamins, J.Electrochem.Soc. **146** (1999), 292.

Ab initio study of SiC/Ti polar interfaces

S. Tanaka (SWING), M. Kohyama

Osaka National Research Institute, 1-8-31 Midorigaoka, Ikeda, Osaka 563-8577, Japan e-mail: swing@onri.go.jp

Abstract *Ab initio* calculations of the polar interfaces between SiC substrate and deposited Ti layers have been performed. The Si-terminated and the C-terminated interfaces are different atomic and electronic features. The C-Ti bond at the C-terminated interfaces have covalent interactions with ionicity, while the Si-Ti bond at the Si-terminated interfaces have metallic behaviour. Adhesive energy values for the C-terminated interfaces are larger than Si-terminated ones. Calculated values of the *p*-type Schottky-barrier height show clearly different for each interface, in which it derives from the difference of interface dipole.

1 Introduction

SiC/metal interfaces are very important for the electronic and optoelectronic SiC devices as the structural ceramics. Among the various interfaces, the SiC/Ti interfaces have interested features derived from the strong reactivity. Recently, the estimation of the Schottky barrier height (SBH) have been done the epitaxial Ti thin films deposited on *n*-type Si-terminated 6H-SiC(0001) [1]. The estimated *n*-type SBHs from XPS and *I-V* and *C-V* measurements are different between as-deposited and annealed contacts. TEM observations of Ti thin films deposited on the flat (0001) face of the 6H-SiC have been performed by Sugawara et al [2]. They found that the structure of the deposited Ti is face-centered cubic, although structure of the bulk Ti is usually hexagonal close-packed at room temperature or body centered cubic at high temperature.

On the theoretical point of view, a first-principles calculations of the 3C-SiC(111)/TiC(111) interface were carried out using the full-potential linear-muffin-tin orbital method [3]. However, it is unlikely to lead to Ti-Si as pairs of interface plane since the bonding has a partially ionic character in both materials. In a such interface, it cannot be discussed the dependence of the atomic species.

In this paper, *ab initio* pseudopotential calculations of the 3C-SiC(111)/Ti polar interface are performed, and are compared with our previous calculations of the 3C-SiC(001)/Ti interface [4]. For both the Si-terminated and the C-terminated interfaces, stable atomic configurations at zero temperature, adhesive energies and SBHs are obtained . The bonding nature and adhesion of ceramic/metal interfaces are dominated by the following two factors. The first is the bonding nature of respective ceramics and metals and their combination. The second is the problem of polar interfaces, termination atoms, or interface stoichiometry. In the 3C-SiC(001)/Ti interfaces [4] and the 3C-SiC(001)/Al interface [5], the Si-terminated and the C-terminated interfaces have quite

different features including atomic configurations, adhesion and bonding nature.

The SBH which is an important factor for electronic devices with ceramic/metal interface is often discussed using the simplified models without the realistic atomic and electronic structures. It is shown that SBHs depend on the interface dipole.

2 Theoretical method

The details of the computational scheme are given in previous papers [5]. Total energies, stable configurations and electronic properties are given in the framework of the *ab initio* pseudopotential method based on the density functional theory within the local density approximation (DFT-LDA). Each supercell contains a slab of 14 3C-SiC(111) atomic layers with 2 stacking-faults atoms and two sets of 4 fcc-Ti(111) layers. Two free metal surfaces are separated by a vacuum region of about 18 a.u. in the supercell, which ensures stable interlayer distances without any constraint. All the configurations have special symmetry of the point group D_{3h}. The 16 special k-points per irreducible Brillouine zone are used.

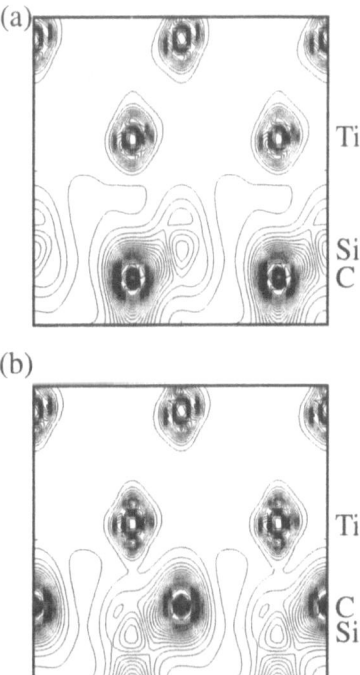

Fig. 1 Stable atomic configuration and valence charge distribution on the [$\bar{1}$10] cross-sections for the (a) Si- and (b) C-terminated interfaces.

3 Results and discussion

Figure 1 shows the stable atomic configuration and valence charge distribution on the $[\bar{1}10]$ cross-sections for the (111) (a) Si- and (b) C-terminated interfaces. We consider three geometric position; (1) Ti on top of the surface atoms: T_1 (2) Ti above the hollow site and (3) Ti atoms in the 4-fold coordinated position above the second-layer atoms of SiC substrate: T_4. In both interfaces, the most stable configuration is T_4 as shown in Fig. 1. The atomic configuration and electron distribution recover the bulk features quickly at the back Si-C and back Ti-Ti bonds. Concerning the charge distribution, in the C-terminated interface, the interface Ti atoms have special hybridization different from the other Ti atoms, although not stronger than (001) interface [4]. It can be said that the C-Ti bonds with p-d covalent interactions are formed at the interface. On the other hand, for the Si-terminated interface, it has a metallic character similar to the Si-terminated (001) interfaces [4].

Adhesive energies of the interfaces are obtained from the difference of the total energies between the relaxed supercell of interfaces and the relaxed SiC and Ti slabs with surfaces. Results are listed in Table I with those of the (001) interfaces. The adhesive energies of the Si- and the C-terminated interfaces are 4.673 and 5.213 Jm^{-2} per one interface, respectively. In comparison with (001) interfaces, the difference of the energies between the interfaces is small. It is due to that the number of the coordination of the interface Ti atoms and the number of the dangling bonds is different. Therefore, the adhesive energy of the C-terminated interface is larger that that of the Si-terminated one in SiC/Ti system so that the former is stronger than the latter.

Table 1 Calculated adhesive energy and Schottky-barrier height of the Si- and the C-terminated SiC/Ti interfaces.

		(111) int.	(001) int.
adhesive energy (Jm^{-2})	Si-terminated	4.67	2.52
	C-terminated	5.21	8.74
Schottky-barrier height (eV)	Si-terminated	0.81	0.50
	C-terminated	0.22	0.22

A p-type SBH can be obtained by supercell calculations as the difference between the Fermi level and the valence-band top (VBT) of the bulk SiC region. In this paper, the quantity of the SBH is determined by analyzing the local density of states (LDOS) for each interface. The LDOS is calculated for each region between successive (111) layers of the supercell. Figure 2 shows the LDOS in the bulk-SiC region. In the Si-terminated interface, the LDOS of the bulk-SiC region has a tiny peak near the VBT, so called metal-induced gap states (MIGS). On the other hand, in the C-terminated interface, MIGS is not clear. Calculated SBH values are listed in Table I with those of the (001) interface. The values of p-type SBH for the Si- and the C-terminated interface are 0.81eV and 0.22eV, respectively. Thus, the calculated p-type SBH of the C-terminated interface is smaller than the Si-terminated one in SiC/Ti system. As the previous calculations [4,5], the present calculations show that the values of SBHs between the different interface configurations in the same ceramic/metal system are different, although the problems of the *ab initio* calculations within the DFT-LDA are included. Conventional models with respect to SBH, not considered the realistic interface structures, cannot explain the series of the calculated results for the reactive interface. It is notice that the defect model cannot be discussed in the present supercell. Consequently, we consider that it is essential to calculate the interface structures, atomically and electronically, using the *ab initio* technique for SiC/Ti system.

Acknowledgements This work was supported by *"Promoted Research Projects for High Performance Computing"* in Tsukuba Advanced Computing Center (TACC) of the Agency of Industrial Science and Technology in Japan.

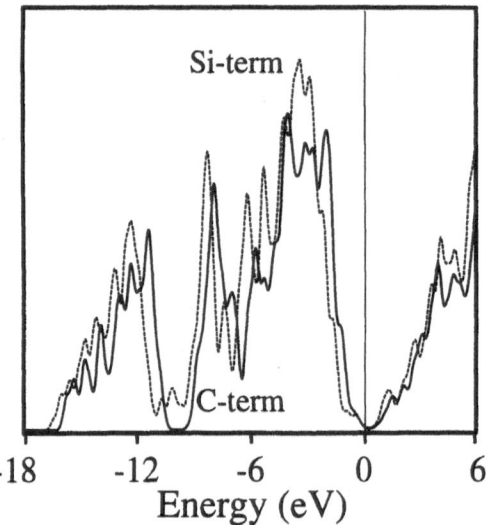

Fig. 2 Local density of states of the bulk SiC region in the 3C-SiC(111)/Ti supercell. The vertical line in figure is the Fermi level.

References

1. L. M. Porter, R. F. Davis, J. S. Bow, M. J. Kim, R. W. Carpenter and R. C. Glass, J. Mater. Res **10**, (1995) 668.
2. Y. Sugawara, N. Shibata, S. Hara, and Y. Ikuhara, J. Mater. Res **15**, (2000) in print.
3. S. N. Rashkeev, W. R. L. Lambrecht, and B. Segall, Phys. Rev. B **55**, (1997) 16472.
4. M. Kohyama, and J. Hoekstra, Phys. Rev. B **61**, (2000) 2672.
5. J. Hoekstra, and M. Kohyama, Phys. Rev. B **57**, (1998) 2334.

Microscopic mechanism of Si oxidation

Kenji Shiraishi, Hiroyuki Kageshima, Masashi Uematsu

NTT Basic Research Laboratories, 3-1 Morinosato-Wakamiya, Atsugi-shi, Kanagawa 243-0198, Japan
e-mail: siraisi@will.brl.ntt.co.jp

Abstract We do a first-principles calculation of the microscopic mechanism of Si oxidation. Our results indicate that oxidation preferentially proceeds laterally at the interface and that a large amount of the interfacial Si should be emitted as a means of releasing the accumulated oxidation-induced strain. We also reformulated the macroscopic diffusion equations by including terms based on the Si emission model. These equations can account for the initial enhanced oxidation phenomena over a wide range of oxidation temperatures (800—1200 °C).

1 Introduction

Si oxidation is one of the outstanding problems of semiconductor science and technology. Modern large-scale integration (LSI) technology is based on the characteristic behavior of Si oxidation phenomena. It is well known that the oxide thickness can be well reproduced by the macroscopic theory of Deal and Grove [1] when the oxide thickness is larger than 30 nm. In Deal-Grove theory, two main processes govern Si oxidation. One is the oxidant diffusion from the SiO_2 surface to the Si/SiO_2 interface. The other is the oxidation reaction at the Si/SiO_2 interface. Although Deal-Grove theory is a well-established theory, it cannot correctly describe oxidation processes for oxide thicknesses less than 30 nm. The experimentally obtained oxidation rate in this thin oxide region is much larger than that given by Deal-Grove theory. This is known as initial enhanced oxidation [1].

Recent Si devices have the fabrication of very thin gate oxides (thinner than 2 nm) and thus fall within the scale of initial enhanced oxidation. We believe that the discrepancy between Deal-Grove theory and experiments originates from a lack of understanding of the microscopic Si oxidation mechanism. Therefore, control of oxide thicknesses at nano-meter scales requires an atomistic understanding of the Si oxidation mechanism. To this end, many theoretical and experimental studies have been reported [2-9]. However, none of them has yielded a sufficient understanding of Si oxidation.

In this paper, we first examine the microscopic mechanism of Si oxidation on the basis of first principles calculations and identify the microscopic Si oxidation phenomena that should be added to the Deal-Grove theory. Next, we reformulate the diffusion equations, including terms for new processes predicted by the first-principles calculations. Finally, we show that our theory can reproduce the observed initial enhanced oxidation behavior over a wide range of

oxidation temperatures (800—1200 °C).

2. First principles calculations

2.1 Method

We used the ultra-soft pseudo potential method developed by Vanderbilt et al. [10-14]. The exchange-correlation potential was approximated by the Ceperley-Alder potential that was parameterized by Perdew and Zunger [15]. Plane waves up to 20.25 Ry were used for the bases. Eight k points in the (1x1) surface unit cell were used for the Brillouin zone integrals. We used a repeated slab model with a sufficiently thick vacuum region. In the optimization, the force on any atoms was smaller than 3×10^{-3} HR/a.u. Some of the first-principles calculations were done with TAPP (Tokyo *Ab-initio* Program Package), which was developed by our group [12-14].

In this paper, the insertion of oxygen atoms into Si-Si bonds is taken to oxidation. We investigate the interface oxidation by using the Si/SiO_2 (quartz) interface model. The models are shown in Fig. 1 [16].

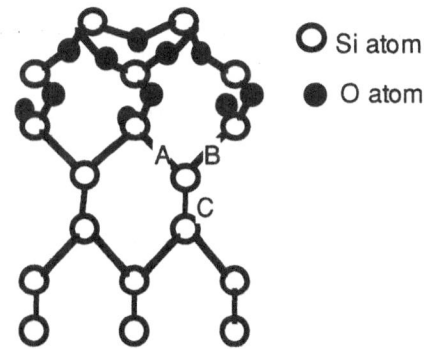

Fig. 1 The $Si(001)/SiO_2$ interface model used in the calculations. The SiO_2 layer has a quartz structure.

2.2 Oxide growth directions

To analyze the oxide at the Si/SiO_2 interface, we placed the first oxygen atom at the back bond site of the Si/SiO_2 (quartz) interface (Site A in Fig. 1) and looked for a favorable oxidation site next to it for the second oxygen atom. The calculation shows that the total energy of B site is lower than C site by about 0.29 eV. These results indicate that the oxide nucleas at the Si/SiO_2 interface preferentially grow laterally, *i.e.*, parallel to the interface. These results are in agreement with the recent experimental observation of layer-by-layer oxidation [6, 8].

We also investigated the preferential oxide growth direction on the Si(001) surface by using calculations similar to those described above. The results indicate that an oxide nucleus on the (001) surface preferentially grows vertically, *i.e.*, into the substrate [16].

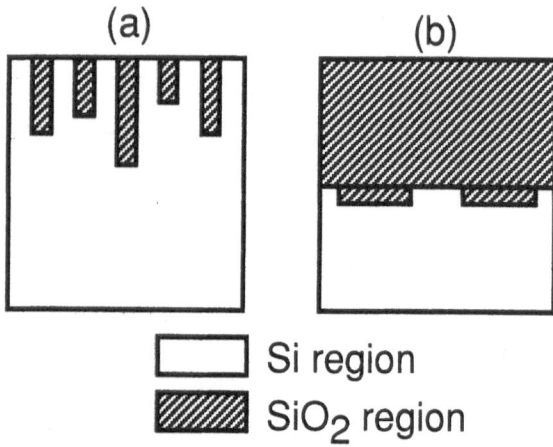

Fig. 2 Schematic views of surface (a) and interface (b) oxidation. Oxide regions grow vertically and laterally on the surface and at the interface, respectively.

Figure 2 shows schematic views of oxide growth obtained by our first-principles calculations. The preferential growth direction of an oxide nucleus on a surface is vertical, whereas direction at the interface is lateral. From the viewpoint of control of the interface characteristics in semiconductor process technology, lateral growth is favorable. Our calculated results imply that a uniform oxide layer can be obtained with any thickness as a result of thermal oxidation once a uniform surface oxide layer is formed. Therefore, the preparation of the initial surface oxide is crucial. X-ray photoelectron spectroscopy (XPS) and scanning tunneling microscopy (STM) measurements confirm that layer-by-layer oxidation occurs after preparation of uniform native oxide layers under the conditions of the lower oxide temperature and the lower oxygen pressure [6, 8].

This difference between surface and interface oxidation can be qualitatively explained as follows. The Si-O-Si bond formation on the surfaces, which causes expansion in the vertical direction, occurs easily. This is because the surface atoms can move upwards with almost complete freedom, while it is not easy for the bonds to expand laterally. Thus, oxide nucleas on the surface grow vertically in order to minimize stress. On the other hand, the vertical expansion of Si-O-Si bonds is constrained at the interface. Therefore, the energy gain due to stress release during vertical growth is quite

low. As a result, the initial oxide nucleas at the interface grow laterally in order to minimize the interface energy. We have confirmed these results by examining the strain distribution around the oxidation sites [17].

2.3 Interfacial Si emission during oxidation

As the oxidation proceeds, the Si/SiO$_2$ interface structures change as shown in Fig. 3(a)-(c). In Fig. 3(c), the oxidation-induced strain accumulates at the interface when three oxygen atoms are inserted into Si-Si bonds. The distance between Si and oxygen atoms indicated by the arrows in Fig. 3(c), becomes very small. This implies that bond reconstruction tends to occur because of the accumulated strain. To release the accumulated strain, we removed the shaded Si atom in Fig. 3(c) from the interface.

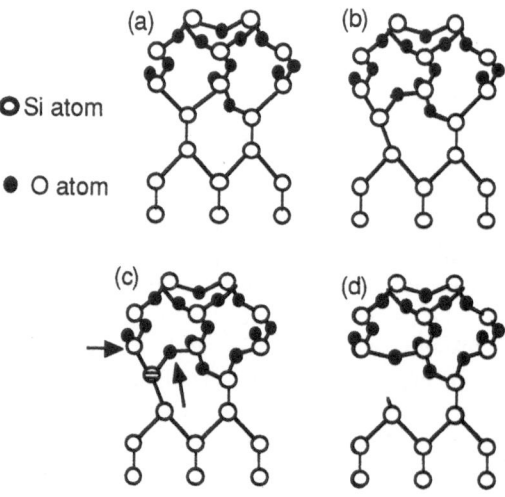

Fig. 3 Change in the interface structure as the oxidation proceeds. (a) Calculated interface structure after insertion of one O atom, (b) two O atoms, (c) three O atoms. (d) The calculated interface structure after one Si atom is removed from the structure (c).

After the removal of the interfacial Si atom, the Si/SiO$_2$ interface structure changes into that shown in Fig. 3(d). This interface structure preserves the Si/SiO$_2$ (quartz) structure of Fig. 1. Next, we examined the stability of the structure after Si emission by comparing the total energies of structures (c) and (d). The results show that the Si-removed structure is more stable than structure (c) by about 0.4 eV. In this estimation, we assumed that the emitted Si atoms are absorbed in a bulk Si crystal. This indicates that a large number of Si atoms must be emitted during oxidation to release the accumulated strain near the interface. This interfacial Si emission is consistent with the disappearance of Si during the oxidation of Si nano-structures [18].

The emitted Si atoms should play an important role in

the oxidation process. Small numbers of emitted Si atoms become the source of Si interstitial, which are known to be the origin of oxidation-induced stacking faults (OSF) [19, 20] and oxidation-enhanced diffusion (OED) [21, 22].

Next, we investigate where emitted Si atoms tend to go by estimating the reaction energy. The calculated reaction energy of "Si (emitted) + $O_2 \rightarrow SiO_2$" is much lower than that of "Si (emitted) $\rightarrow$ Si (bulk-interstitial)" by ~ 10 eV [23]. Therefore, emitted Si atoms preferentially back-flow into the oxide region. In fact, some researchers have already proposed this large Si emission into the SiO_2 region [24-27].

This leads to a modification of conventional macroscopic view of thermal oxidation, in that a large amount of the interfacial Si emission should be present. The conventional and modified oxidation processes are illustrated in Fig. 4.

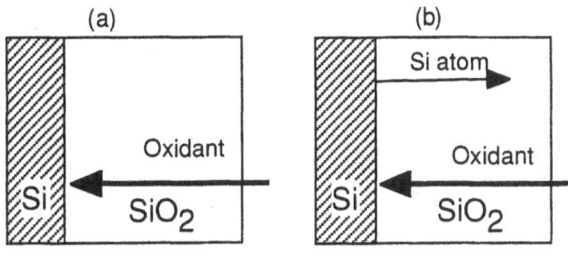

Fig. 4 (a) Conventional views of thermal oxidation. (b) Modified views of oxidation taking into account the interfacial Si emission.

Moreover, we should expect that emitted Si would seriously affect the macroscopic properties of oxidation. In the next section, we show that the initial enhanced oxidation, which has been a puzzling problem for over 30 years, can be accounted for by the interfacial Si emission model.

3. Modified macroscopic theory of Si oxidation

3,1 Modified equations taking account of Si emission

We reformulate the macroscopic diffusion equations by taking into account Si emission at the interface [23, 27]. We include four new atomic processes. (1) Interstitial Si atoms are created by oxidation at the interface. (2) The emitted Si ineterstitials diffuse into the oxide layer. (3) The diffused Si interstitials are re-oxidized in the SiO_2 region and on the SiO_2 surface. (4) The density of the Si interstitials at the interface reduces the oxidation rate at the interface as well as the emission rate of Si interstitials. Process (4) is described by the equation for the interfacial reaction-rate constant.

$$k = k_0(1 - C_{Si}^1/C_{Si}^0), \qquad (1)$$

where k is the reduced reaction rate at the interface, k_0 is the maximum reaction rate constant, C_{Si}^1 is the density of Si interstitials in the SiO_2 region near the interface, and C_{Si}^0 is the maximum density of Si interstitials in the SiO_2 region. We obtain a set of coupled partial differential equations:

$$\frac{\partial C_{Si}}{\partial t} = \frac{\partial}{\partial x}\left(D_{Si}\frac{\partial C_{Si}}{\partial x}\right) - R_1 - R_2 \qquad (2)$$

$$\frac{\partial C_O}{\partial t} = \frac{\partial}{\partial x}\left(D_O\frac{\partial C_O}{\partial x}\right) - R_1 - R_2 - R_3, \qquad (3)$$

where R_1, R_2, and R_3 are the reaction terms that represent the oxidation at the oxide surface, oxidation in the SiO_2 region, and oxidant transfer from the gas to the oxide surface, respectively, such that

$$R_1 = k' C_O^S C_{Si}^S \qquad (4.1)$$

$$R_2 = \kappa_1 C_O C_{Si} + \kappa_2 (C_O)^2 C_{Si} \qquad (4.2)$$

$$R_2 = h (C_O^S - C_O^*), \qquad (4.3)$$

where C_O and C_{Si} are the concentration of oxidant and Si interstitials, D_O and D_{Si} are the diffusion coefficient of oxidant and Si interstitials in the oxide, k' is the oxidation rate of Si atoms at the oxide surface, κ_1 and κ_2 are the oxidation of rates of the emitted Si atoms in the oxide, h is the gas-phase mass-transfer coefficient, and C_O^* is the solubility of the oxidant in the oxide. C_O^S and C_{Si}^S are the concentration of oxidant and Si atoms at the oxide surface, respectively. The boundary conditions for the Si interstitials and the oxidant at the interface ($x = 0$) are given by

$$D_{Si}\frac{\partial C_{Si}}{\partial x}\bigg|_{x=0} = -k\upsilon C_O^1 \quad \text{and} \quad D_O\frac{\partial C_O}{\partial x}\bigg|_{x=0} = kC_O^1 \quad (5)$$

where υ is the emission rate of Si atoms from the interface and C_O^1 is the concentration of the oxidant at the interface.

3,2 Simulated oxide thickness during oxidation

We performed a simulation based on the above equations [28]. The results are summarized in Fig. 5 for dry oxidation of Si(001) substrates under oxygen pressure of 1 atm at 800—1200°C. As shown in this figure, our theory works well over a wide range of oxidation temperature, including thicknesses that experience the initial enhanced oxidation that cannot be explained by conventional Deal-Grove theory. Note also that even the empirical model did not fit the thin film region for 800 and 900°C unless it included the second empirical term [29]. In addition, we would like to mention that our work could also reproduce the oxidation rate under high oxygen pressures (1-20 atm) [30].

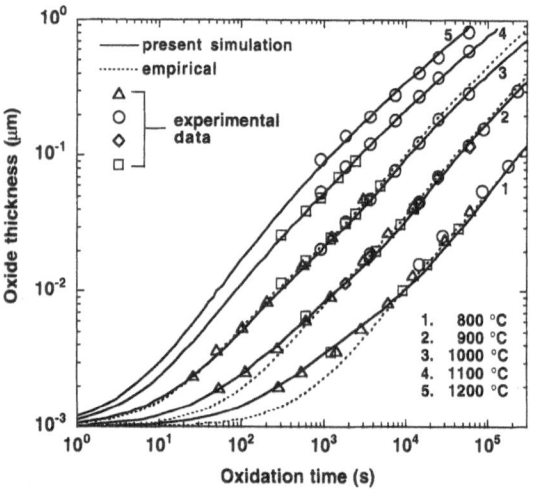

Fig. 5 Simulated (solid lines) and experimental oxide thickness for dry oxidation of Si(001) substrates under 1-atm oxygen pressure at 800-1200℃. Calculations using empirical models are also shown (dashed lines).

4. Conclusion

We investigated the microscopic mechanism of Si oxidation by using first-principles calculations. The results indicate that oxidation preferentially proceeds laterally at the interface and that a large amount of the interfacial Si should be emitted as a means of releasing the accumulated oxidation-induced strain. We also reformulated the macroscopic diffusion equations by including terms based on the Si emission model, thereby solving the long-standing problem of initial enhanced oxidation. Our results are valid for a wide range of oxidation temperatures (800—1200 °C).

Acknowledgements

We would like to thank Dr. H. Watanabe, Dr. M. Ichikawa, Prof. K. Taniguchi, Prof. Y. Takakuwa, Prof. T. Hattori, Dr. M. Nagase, Dr. K. Sumitomo, Dr. Y. Takahashi, Prof. A. Oshiyama and Prof. M. Umeno for their stimulating discussions. We also thank Dr. K. Murase, Dr. T. Mukai and Dr. T. Ogino for their continuous encouragement throughout this work. This work was partly supported by JSPS under Contract No. RFT96P00203.

References

1. B. E. Deal and A. S. Grove, J. Appl. Phys. **36**, (1965) 3770.
2. F. J. Grunthaner *et al.*, Phys. Rev. Lett. **43**, (1979) 1683.
3. Y. Miyamoto and A. Oshiyama, Phys. Rev. B **41**, (1990) 12680.
4. D. G. Cahill and Ph. Avouris, Appl. Phys. Lett. **60**, (1992) 326.
5. Y. Ono, M. Tabe and H. Kageshima, Phys. Rev. B **48**, (1993) 14291.
6. K. Ohishi and T. Hattori, Jpn. J. Appl. Phys. **33**, (1994) L675.
7. T. Uchiyama and M. Tsukada, Phys. Rev. B **53**, (1996) 7917.
8. H. Watanabe, K. Kato, T. Uda, K. Fujita, M. Ichikawa, T. Kawamura and K. Terakura, Phys. Rev. Lett. **80**, (1998) 345.
9. K. Kato, T. Uda and K. Terakura, Phys. Rev. Lett. **80**, (1998) 2000.
10. D. Vanderbilt, Phys. Rev. B **41**, (1990) 7892.
11. K. Laasonen, A. Pasquarello, R. Car, C. Lee and D. Vanderbilt, Phys. Rev. B **47**, (1993) 10142.
12. M. Tsukada, computer program package TAPP, University of Tokyo, Tokyo, Japan, 1983—2000
13. J. Yamauchi, M. Tsukada, S. Watanabe and O. Sugino, Phys. Rev. B, **54**, (1996) 5586
14. H. Kageshima and K. Shiraishi, Phys. Rev. B **56**, (1997) 14985.
15. J. P. Perdew and A. Zunger, Phys. Rev. B **23**, (1981) 5048.
16. H. Kageshima and K. Shiraishi, Phys. Rev. Lett. **81**, (1998) 5936.
17. H. Kageshima and K. Shiraishi, Surf. Sci. **438**, (1999) 102.
18. H. I. Liu, D. K. Biegelsen, N. M. Johnson, F. A. Ponce and R. F. W. Pease, J. Vac. Sci. Technol. B **11**, (1993) 2532.
19. D. J. Thomas, Phys. Status Solidi **3**, (1963) 2261.
20. S. M. Hu, Appl. Phys. Lett. **27**, (1975) 165.
21. S. Mizuo and H. Higuchi, Jpn. J. Appl. Phys. **20**, (1981) 735.
22. T. Y. Tan and U. Gosele, Appl. Phys. A **37**, (1985) 1.
23. H. Kageshima, K. Shiraishi and M. Uematsu, Jpn. J. Appl. Phys. **38**, (1999) L971.
24. S. T. Dunhum and J. D. Plummer, J. Appl. Phys. **59**, (1986) 2541.
25. K. Taniguchi, Y. Shibata and C. Hamaguchi, J. Appl. Phys. **65**, (1989) 2723.
26. T. Tamura, N. Tanaka, M. Tagawa, N. Ohmae and M. Umeno, Jpn. J. Appl. Phys. **32**, (1993) 12.
27. Y. Takakuwa, M. Nihei, and N. Miyamoto, Appl. Surf. Sci. **117/118**, (1997) 141.
28. M. Uematsu, H. Kageshima and K. Shiraishi, Jpn. J. Appl. Phys. **39**, (2000) L699.
29. H. Z. Massoud, J. D. Plummer and E. A. Irene, J. Electrochem. Soc. **132**, (1985) 2685.
30. M. Uematsu, H. Kageshima and K. Shiraishi, Jpn. J. Appl. Phys. **39**, (2000) L952.

Electron paramagnetic resonance of a single-crystal surface: the Si(111)-7×7 surface and its oxidation process

T. Umeda*, M. Nishizawa, T. Yasuda, J. Isoya, S. Yamasaki, K. Tanaka

Joint Research Center for Atom Technology (JRCAT) - National Institute for Interdisciplinary Research - Angstrom Technology Partnership, 1-1-4 Higashi, Tsukuba, 305-8562, Japan

Abstract We have carried out electron paramagnetic resonance (EPR) measurements on the Si(111)-7×7 surface and its oxidation process. This is the first report on EPR signals of a semiconductor single-crystal surface. Our results clearly show that there exist two sorts of new paramagnetic centers, a dangling-bond center and an oxygen-induced defect, on the oxygen-chemisorbed Si(111)-7×7 surface.

1 Introduction

EPR is widely used to study defect structures and chemical reactions that involve unpaired electrons. Although EPR has great potential as a surface probe, its application in surface studies has been limited to powder or porous samples [1]. In this paper, we report that the combination of a state-of-the-art EPR system and a carefully-designed ultrahigh-vacuum (UHV) sample cell has enabled us to detect paramagnetic centers on single-crystal surfaces, thus opening up a new approach in surface science analysis. The power of this UHV-EPR technique is demonstrated by revealing, for the first time, dangling-bond states that appear upon oxidation of the Si(111)-7×7 surface [2][3][4]. Since this well-studied semiconductor surface reconstruction shows nineteen surface dangling bonds (DBs) per unit cell, it is supposedly an ideal sample for us to demonstrate the capability of the UHV-EPR technique.

2 UHV-EPR setup

Our experimental setup consists of an X-band EPR spectrometer and a UHV chamber to which a sample cell (a silica glass tube of 1 cm in diameter) is attached. The UHV chamber (base pressure = 9×10^{-9} Pa) is arranged so that the sample cell is located within the microwave cavity of the EPR system. Si(111)-7×7 surfaces were prepared in the chamber by direct-current heating of a Si wafer above 1400 K. The clean surface was inserted into the UHV sample cell for EPR measurements. All measurements and surface oxidations were carried out at room temperature.

3 Results and Discussions

The EPR spectrum of the Si(111)-7×7 surface is shown in the top of Fig. 1. A reflection high energy electron diffraction (RHEED) measurement (inset of Fig. 1) en-

* *Present address:* System Devices and Fundamental Research, NEC corporation, e-mail: t-umeda@da.jp.nec.com, 34 Miyukigaoka, Tsukuba, 305-8501, Japan

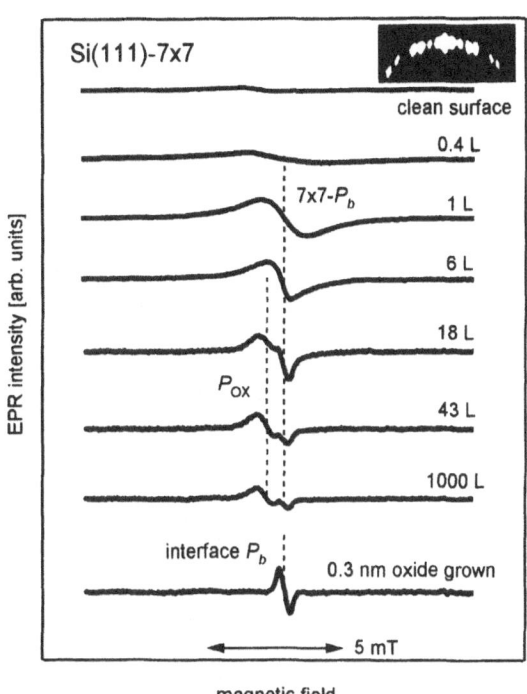

Fig. 1 EPR spectra of the Si(111)-7×7 surface as a function of O_2 dose. Each spectrum was accumulated for 2 to 6 hours with a microwave power of 1 to 200 mW and magnetic field parallel to the [111] axis.

sures well-ordered 7×7 reconstruction on our surface. Although high-density DBs exist on the surface, EPR is largely insensitive to the DBs at room temperature. A similar phenomenon has been observed for powder systems that contain a high density of surface paramagnetic ions [1]. In such systems the high degree of coupling between neighboring unpaired electrons broadens the EPR line, making the signal unobservable [1].

Chemisorption of oxygen on Si(111)-7×7 eliminates the DB states of the topmost Si adatoms, as reported in previous scanning tunneling microscope (STM) studies [2][3]. We have found that a new EPR signal appears in conjunction with such a reduction in the surface DBs. In Fig. 1, the spectra measured for different oxygen exposures are also shown. As is clear in the figure, the spectra taken at 1-L exposure (1 L = 10^{-4} Pa×sec) reveal a new resonance. We name this new paramagnetic center 7×7-P_b, since its g tensor resembles to that of

the well-known P_b center at the $SiO_2/Si(111)$ interface
[5]. The 7×7-P_b center shows an axial symmetry about
the [111] axis with principal g factors of $g_\parallel = 2.0015$
and $g_\perp = 2.0064$. These features are commonly observed
for the P_b center. It is well established that the P_b centers are [111]-oriented Si DBs with three Si-Si backbonds
[5]. Therefore we ascribe the 7×7-P_b center to the [111]-
oriented DB states of the unreacted Si adatoms (i.e.,
with three Si-Si backbonds).

As the O_2 exposure was increased, the 7×7-P_b signal
decayed and EPR spectra were increasingly dominated
by another new resonance (see Fig. 1). We tentatively
call it P_{OX}. This center appears to have an axial symmetry about the [111] axis with a small g anisotropy,
$g_\parallel = 2.0054$ and $g_\perp = 2.0043$. The g map of P_{OX} does
not agree with that for any known center in the SiO_2/Si
structure. Further, it dose not agree with that for any
of a variety of oxygen complexes and radicals in Si and
SiO_2. The origin of this signal is not clear at present.
We point out that the P_{OX} centers are a type of oxygen-
induced surface defect, because they exist only transiently during oxidation of the adatom and rest-atom
layers. Even at 1000-L exposure, the RHEED image re-
tained a faint 7×7 pattern. This observation indicates
that the above oxidation process primarily involves the
adatom and rest-atom layers, in consistency with previous results [4]. Once oxidation proceeds to the next
surface layers, the P_{OX} centers disappear completely, as
is shown in the bottom spectra in Fig. 1.

The bottom spectra in the figure were measured after
an oxide layer of approximately 0.3 nm in thickness was
formed on the surface by exposure of 1-atm O_2. The oxide thickness was estimated by Auger electron measurements. The 7×7 RHEED pattern was no longer detected
from the surface. We attributed the observed EPR signal to the P_b center at the oxide-Si(111) interface, since
the spectrum is identical to that reported for this center
[5].

Figure 2 summarizes the above EPR results. For the
clean 7×7 surface, the 7×7-P_b centers are not detectable
as discussed earlier. The signal line width narrows (Fig.
2a) as the density of unreacted adatom DBs is decreased
by the oxygen chemisorption. We found that the 7×7-P_b
signal becomes detectable at 0.4-L exposure. An additional exposure to approximately 10 L of O_2 reduces
both the line width and density of the 7×7-P_b center as
shown in Fig. 2. Reduction of the 7×7-P_b density corresponds to the fact that oxidation of the Si adatoms
proceeds with O_2 exposure. The observed change is consistent with previous observations by STM [2] and photoemission [4] that unreacted Si adatoms disappeared at
10 to 20 L. The formation of the P_{OX} center is induced
by oxygen exposure with 2 to 4 L, but its density (Fig.
2b) is much lower than that of the oxidized sites. As a
result, we conclude that most of the oxidized sites do not
have unpaired electrons.

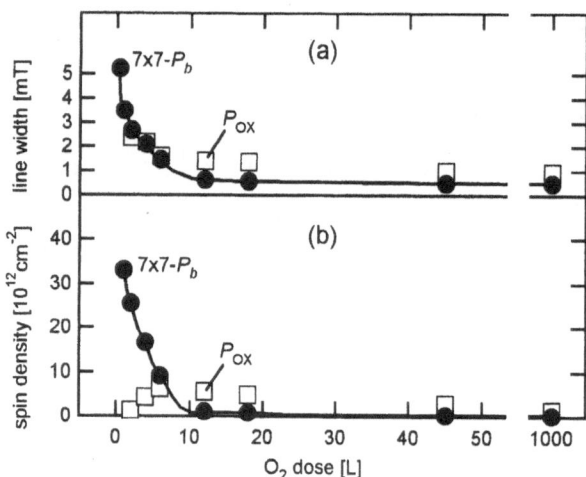

Fig. 2 Oxidation-induced changes of the 7×7-P_b and P_{OX}
signals as a function of O_2 dose. (a) Change in full-widths-
at-half-maximum of the integrated (i.e., absorption) signals.
(b) Change in the spin densities.

We note that the broadened 7×7-P_b signal remained
unsaturated even at a microwave power exceeding 200
mW and that its line shape was well fit by a Lorentzian
distribution function. These facts indicate that the lifetime broadening of the signal is caused by a very short
spin-lattice relaxation time (T_1) of unpaired electrons.
The origin of such a short T_1 may be attributed to metallic electronic states of the Si(111)-7×7 surface that originate from a high density of adatom DBs. Specifically, if
the unpaired electrons of the adatom DBs are delocalized at the surface, T_1 values as short as those for the
conduction electrons in metals and doped semiconductors can be obtained. Reduction of the signal broadening
effect by submonolayer adsorption of O_2 (Fig. 2a) is supposedly a consequence of loss of metallic surface states
on the Si(111)-7×7 surface.

4 Summary

Our experiments have demonstrated that surface defects
on Si single crystal can be resolved by the UHV-EPR
technique. The results obtained here are of technologically interest because oxidation of Si is an important
reaction in Si technology. This method is expectedly applicable to surface studies on a variety of clean surfaces
and their reaction processes.

References

1. R.F. Howe, *Chemistry and Physics of solid surfaces vol.5*,
   edited by R. Vanselow and R.F. Howe (Springer, Berlin
   1984), p.39.
2. F.M. Leibsle, A. Samsavar, T.-C. Chiang, Phys. Rev. B
   38 (1988) 5780.
3. I.-W. Lyo, Ph. Avouris, B. Schubert, R. Hoffmann, J.
   Chem. Phys. **94**, (1990) 4400; R. Martel, Ph. Avouris,
   I.-W. Lyo, Science **272**, (1996) 385.
4. P. Morgen, U. Höfer, W. Wurth, E. Umbach, Phys. Rev.
   B **39**, (1989) 3720.
5. K.L. Brower, Appl. Phys. Lett. **43**, (1983) 1111.

Atomic layer oxidation of H terminated Si(100) surface

Domino reaction via oxidation and H migration

K. Kato[1], H. Kajiyama[2], S. Heike[2], T. Hashizume[2], T. Uda[3]

[1] Advanced Materials and Devices Laboratory, Toshiba Corporate R & D Center, 1 Komukai Toshiba-cho, Saiwai-ku, Kawasaki 210-8582, Japan

[2] Advanced Research Laboratory, Hitachi, Ltd., Hatoyama, Saitama 350-0395, Japan

[3] Joint Research Center for Atom Technology, Angstrom Technology Partnership, 1-1-4 Higashi, Tsukuba, Ibaraki 305-0046, Japan

Abstract We found that oxidation occurs easily at an exposed DB and that the dissociated O atoms are chemisorbed at a dimer-bond and a back-bond, based on the first-principles calculations. It was found to induce the adjacent H atom to migrate onto the DB. As a consequence of the alternate oxidation and subsequent H atom migration processes, the atomic layer oxidation is actually found to occur along the H-terminated Si dimer-row at low temperatures without desorbing H atoms through our STM experiment.

1 Introduction

Under increasing demands for growing thinner gate insulating films for MOS LSI devices, layer-by-layer oxidation has been found to occur on a clean Si(100) surface [1,2]. However, clean Si surfaces are easily oxidized in the atmosphere even at room temperatures. Si wafers have to be processed in an oxygen-free chamber where pre-oxidized films are removed. When dangling-bonds (DBs) on Si surfaces are terminated with H atoms, the Si surfaces are hardly oxidized in the atmosphere even at 660 K. If we try to oxidize a H terminated Si surface by increasing the temperature, the layer-by-layer oxidation mode fails under those high-temperature treatments.

One possible measure to overcome those difficulties in oxidizing the H-terminated Si surface with the layer-by-layer mode is to begin the oxidation at a DB isolated on the H-terminated Si surface. In the case of oxidation at a single DB, only a single electron is available for a molecular dissociation process. The spin-state also remains in a doublet state during chemisorption. The oxidation process at a DB isolated in H-terminated Si surfaces must be fairly different from that on clean Si surfaces[1].

To understand the atomic processes of the Si oxidation at a single DB on a H-terminated Si surface, we have employed first-principles theoretical calculations. Then, we have elucidated the effect of the oxidation processes to H atom motions on the H-terminated Si(100) surface. Oxidation of H-terminated Si(100)-(2x1) surface have been also analyzed by a scanning tunneling microscope (STM) experiment.

2 Calculations

Our calculations are performed with density functional theory (DFT), including spin-polarization and generalized gradient approximation (GGA). Our major concern is oxidation processes at a DB isolated in the H-terminated Si(100). Figure 1 illustrates the energy variation for the molecular dissociation process at a DB, together with local geometries at representative positions. A spontaneous energy gain appears at first because of O_2 molecule chemisorption with the DB. The DB is hybridized with an antibonding $2p\pi_g^*$ state of the O_2 molecule. The bond length of the O_2 molecule is elongated 10 %, but the molecular dissociation does not occur at this stage possibly due to weak chemisorption with a single electron of the DB. The adiabatic energy barriers arising from further O atom migration into more stable sites are at most 0.39 eV, which is substantially lower than the energy barrier of more than 1 eV emerging from the molecular chemisorption process in the case of the clean Si(100) surfaces[1]. Energies needed for overcoming those small energy barriers can be easily supplied by the first energy gain of 1.63 eV, revealing that apparent barrier-less oxidation occurs and that the dissociated O atoms migrate into stable sites. When an O_2 molecule approachs the Si surface with a different orientation and from a different direction to the Si surface, the order of O atoms migrating into the back-bond and the dimer-bond will be exchanged. In this case, we found that the calculated energy variation shows a similar shape with lower activation barriers, leading also to apparent barrier-less oxidation. The indifference to orientation and direction of the O_2 molecule chemisorbed at a DB together with low energy barriers and no spin conversion will enhance the sticking probability of an O_2 molecule on the Si surface.

When electronic orbitals of O and Si atoms are hybridized, the total valence electronic charge of the O atom increases from 6 to 6.5 at the most. As the Si atom with a DB loses electronic charge slightly to the O atoms through this hybridization, the energy level of the DB will be subsequently lowered because the Si atom is positively charged, resulting in energy lowering of a H atom

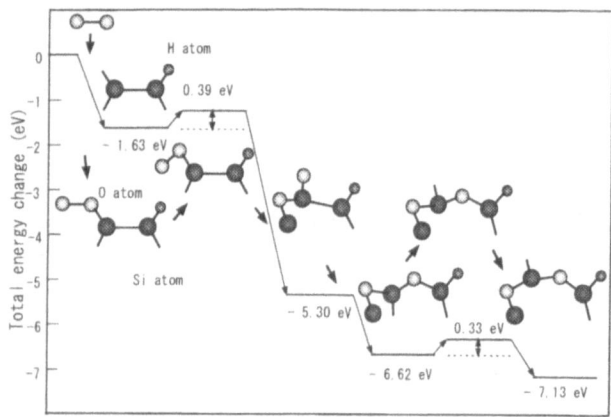

Fig. 1 Energy variation for an O_2 molecular dissociation process at a DB and its local geometries. The large, medium, and small circles represent Si, O, and H atoms, respectively.

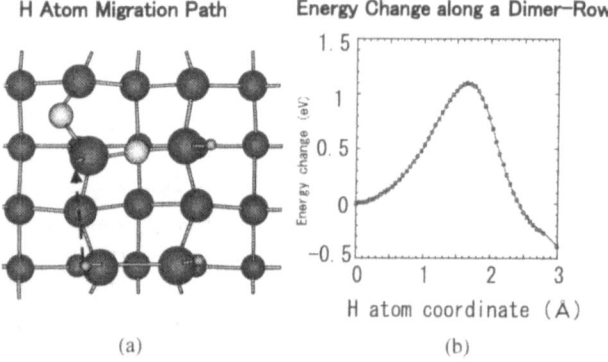

H Atom Migration Path

Energy Change along a Dimer-Row

(a)   (b)

Fig. 2 (a) H atom migration path along an inter-dimer and (b) its energy variation

chemisorbed structure. This favors adjacent H atom migration and adsorption onto the DB. In the case of a clean Si(100) surface, H atom migration occurs on the Si surface because of the small migration energy. Therefore, we can expect that a H atom migrates at moderately high temperatures without desorbing H atoms as a H molecule[3].

Figures 2(a) and (b) show a typical example of H atom migration along an inter-dimer from a dimer to the adjacent dimer with a DB and its energy variation where a single O_2 molecule is dissociatively chemisorbed via the DB. The energy of the structure after the H atom migration is -0.39 eV lower than that before the H atom migration, due to valence electronic charge transfer from the Si atom to the O atoms. The energy barrier for the H atom migration is 1.10 eV, which is substantially reduced from the energy needed for H atom migration on clean Si surfaces (1.68 eV)[4].

As one elemental process of the oxidation and subsequent H atom migration ends, a new DB emerges after the H atom migration, being a nucleus for another elemental process. The elemental processes of the oxida-

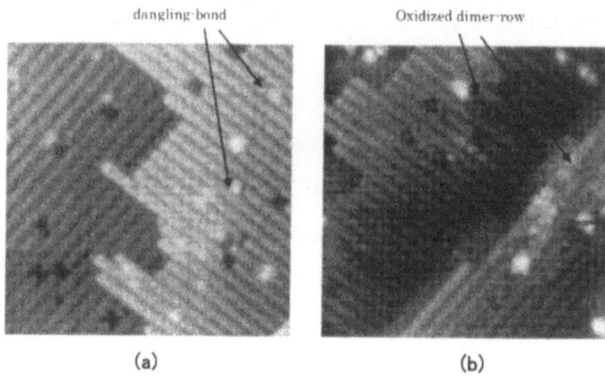

dangling-bond   Oxidized dimer-row

(a)   (b)

Fig. 3 A filled-state STM image of a H-terminated Si(100)-(2x1) (a) before and (b) after exposure to O_2 molecules.

tion and subsequent H atom migration will be lead to "domino-reaction" of the alternate oxidation and subsequent H atom migration. When we apply this process to a H-terminated Si(100)-(2x1) surface, we can expect the atomic layer oxidation along a H-terminated Si dimer-row.

3 Experiment

Traces of the domino-reaction at exposed DBs ware actually observed, by using an ultra-high vacuum STM. A small number of isolated DBs is naturally formed in the process of hydrogen adsorption for the preparation of a H-terminated Si(100)-(2x1) surface. Figures 3(a) and (b) show the STM images before and after exposure to 10 L of O_2 molecules at 530 K. The dimer rows including isolated DBs turned out to be oxidized atomic-wire structures. The newly-appeared irregular structure spreads up to 15 dimer units, where the spot brightness is almost the same as that of the H-terminated surface, indicating that the oxidized regions are still terminated with H atoms.

4 Summary

Oxidation at a DB isolated in a H-terminated Si(100) surface is found to occur far more easily than on clean Si(100) surfaces, inducing a H atom closer to the DB to migrate onto the DB. By means of the alternate oxidation and subsequent H migration process, atomic layer oxidation occurs along the dimer-row on the H-terminated Si(100) surface.

The work was partly supported by NEDO.

References

1. K. Kato, T. Uda, and K. Terakura, Phys. Rev. Lett. 80, (1998) 2000.
2. H. Watanabe, K. Kato, T. Uda, K. Fujita, M. Ichikawa, T. Kawamura, and K. Terakura, Phys. Rev. Lett. 80, (1998) 345.
3. E. Penev, P. Kratzer, and M. Scheffler, J. Chem. Phys. 22, (1999) 3986.
4. J. H. G. Owen, D. R. Bowler, C. M. Goringe, K. Miki, and G. A. D. Briggs, Phys. Rev. B 54, (1996) 14153.

Time-dependent density-functional simulations of desorption dynamics of H and Br terminated Si surfaces induced by electronic excitations

Yoshiyuki Miyamoto and Osamu Sugino

Fundamental Research Laboratories, System Devices and Fundamental Research, NEC Corporation, 34 Miyukigaoka, Tsukuba, Ibaraki 305-8501, Japan   e-mail: **y-miyamoto@ce.jp.nec.com**

Absract Hydrogen and halogen desorptions induced by electronic excitation have been examined by performing the first-principles molecular dynamics simulations. In order to take the finite lifetime of the excitation into account, real-time electron-ion dynamics has been dealt with the time-dependent Schrödinger equation of electron wave functions. Sharp contrast in computed lifetimes has been seen between the hydrogen- and halogen-cases, which originates from whether the excited states are in the resonance with bulk bands.

1 Introduction

Atomic scale manipulation on semiconductor surfaces has attracted much attention because of applications to nanotechnology. Electronic excitation with use of scanning tunneling microscopy tip can stimulate atomically controlled hydrogen desorption [1] and halogen desorption [2] from Si surfaces.

In the case of the hydrogen desorption, the yield of desorbed H atoms were in the order of 10^{-6} per injected electron. This low yield was ascribed to the short lifetime of the Si-H bonding to anti-bonding excitation. [1] Present simulation has confirmed that the lifetime should be less than 10 fs.

On the other hand, the halogen desorption was found to be induced by electron emission from halogen valence s bands. [2] The present simulation has shown that the excitation causes acceleration of Br-desorption being dependent on the Br adsorption sites with longer lifetime than 60 fs.

2 Computational methods

Density functional theory within the local density approximation or the generalized gradient approximation was applied to express exchange-correlation energy of valence electrons. Plane wave basis set and the pseudopotentials were used in treating the valence wave functions. For the real-time electron dynamics, the Suzuki-Trotter split operator method [3] was used to express time-evolution operator of the Kohn-Sham Hamiltonian (H_{KS}). When the simulation followed the adiabatic potential energy surfaces, the

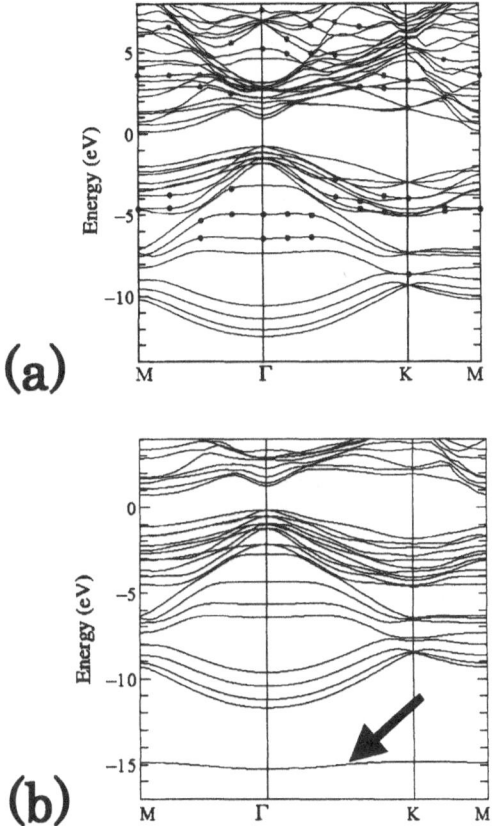

(a)

(b)

Fig. 1 Band structures of (a) H-terminated and (b) Br-terminated Si(111) surfaces. Dots in (a) denote the Si-H bonding (below 0 eV) and anti-bonding (above 0 eV) components. An arrow in (b) denotes the Br $4s$ band.

Hellmann-Feynman forces were used to describe ionic motion by solving the Newton's equation of motion. Deviation from the adiabatic surface occurred when the matrix of the H_{KS} with respect to the time-evolving wave functions became off diagonal. The detail of the present method (electron-ion dynamics) is described in Ref. [4].

3 Results

3.1 Band structures

The H-terminated Si(111) surface has Si-H bonding
and anti-bonding states in the resonance of the Si bulk
bands, as shown in Fig. 1 (a). On the other hand, in the
case of Br-terminated surface, the halogen valence s
band locates below the Si valence band minimum and
thus is a surface bound state, as shown in Fig. 1 (b).
These band structures are consistent with previous
calculations. [5] Such a difference in the energy band
structures between the H- and Br-terminated surface
gives rise to a sharp contrast in the lifetimes of the
excitations as shown below.

3.2 Desorption dynamics and decay processes

We first investigated the hydrogen motion on the
Si(111) surface induced by Si-H bonding to
anti-bonding excitations that were mimicked by
promoting electronic occupations. The hole state
initially consisted of the Si-H bonding state showed
rapid mixing with Si valence states. Such decay was
due to the location of the energy level of the Si-H
bonding state in the resonance of the Si valence bands.
Note that the former theoretical work [6] had assumed
the direct recombination from anti-bonding to bonding
to be the dominant decay process. On the contrary, the
mixing was found more rapid from our calculations,
demonstrating an importance of performing the
real-time electron-ion dynamics simulations.

Next, we investigated the Br dynamics triggered
by electron emission from Br $4s$ band. The emission
was mimicked by promoting single hole in the Br $4s$
band, which generated the driving force of 0.1
HR/(Bohr radius) to the Br atom in the direction to the
vacuum. The origin of this driving force was
coexistence of Br-Si bonding and anti-bonding states
below the Fermi level. Since the Br $4s$ band actually
consisted of Br $4s$ and Si $3s$ and $3p$ orbitals in the
Br-Si bonding phase, creating a hole in the Br $4s$ band
broke power balance of the system toward the
anti-bonding nature. The computed lifetime of the Br
$4s$ hole was substantially long: 66 fs when the Br atom
was above the surface Si atom, 90 fs when the Br atom
was above the surface Si adatom. Figure 2 shows
atomic motion and the Br $4s$ charge density in the
cases of (a) the Br atom above the surface Si atom and
(b) the Br atom above the surface Si adatom. An
interesting feature is faster motion above the surface
Si adatom since the adatom and its back bonding Si
atoms played a role of a catapult.

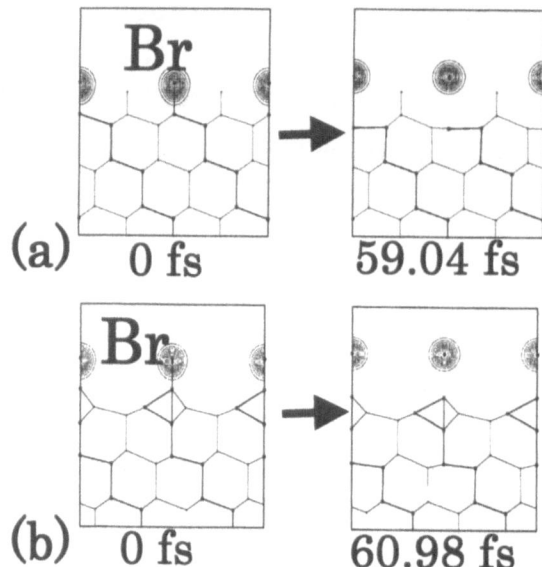

Fig. 2 Charge contour maps of Br $4s$ hole for the Br
desorptions (a) from a surface Si atom and (b) from a
surface Si adatom. The minimums of the contour lines are
set to be a common value. The simulation times (fs) are
also denoted.

4 Summary

The present simulation highlighted the sharp
difference in lifetimes of excitations between the H-
and Br-terminated Si(111) surfaces. The difference
originated form the contrast in electronic structures,
i.e., absence and presence of surface bound state
respectively on H- and Br-terminated surfaces. We
think that performing real-time electron-ion dynamics
instead of conventional molecular dynamics
simulation will be useful for intuitive understanding of
dynamics induced by electronic excitations on other
surfaces.

References

1. T. –C. Shen, et al., Science **268**, (1995) 1590.
2. K. Mochiji and M. Ichikawa, Phys. Rev. B**62**, (2000)
    2090.
3. M. Suzuki, J. Phys. Soc. Jpn. **61**, (1992) L3015.
4. O. Sugino and Y. Miyamoto, Phys. Rev. B**59**, (1999)
    2579.
5. M. Schlüter, and M. L. Cohen, Phys. Rev. B**17**, (1978)
    716.
6. Ph. Avouris et al., Chem. Phys. Lett. **257**, (1996) 148; T.
    Vondrak and X.-Y. Zhu, Phys. Rev. Lett. **82**, (1999)
    1967.

Proc. 25th Int. Conf. Phys. Semicond., Osaka 2000 (Eds. N. Miura and T. Ando)

319

Mechanism of migration and dissociation for As$_2$ molecule on GaAs(001) surfaces during the MBE growth: A computational study

Akira Ishii, Kaori Seino, Tsuyoshi Aisaka

Department of Applied Mathematics and Physics, Tottori University, Koyama, Tottori 680-8552, Japan

Abstract   The first-principle calculations for the processes of migration and dissociation for an As$_2$ molecule on GaAs(001) surfaces is performed. The calculation suggests the Ga atom clusters on the top layer will play an important role for the dissociation of As$_2$ molecules after they adsorb on the surface during MBE growth.

1 Introduction

The investigation on the molecular beam epitaxial(MBE) growth of GaAs(001) has been studied as a typical example of the growth of compound semiconductor. Using the recent two-component kinetic Monet Carlo simulation[1–4], it is found that GaAs(001) surface is not covered completely by As atoms and Ga terminated area exists. Thus, in order to reproduce the growth of GaAs(001) surface, we should know at least the migration of Ga atom on the As terminated surface area and that of As atom on the Ga terminated surface area.

During the real MBE growth of GaAs(001), the As atoms are supplied to the sample as As$_2$ or As$_4$ molecules. Since the dynamids of As$_4$ is considered to be very complex[5], we focus here our attention to the dynamics of As$_2$. In the case of As$_2$ beam supply, As$_2$ molecules are finally built into the GaAs cyrstal of zincblende structure. However, the timing of the dissociation of As$_2$ into two individual As atoms are still unknown. Morgan et al.[6] state that *the As$_2$ molecule does not dissociate in any of the adsorption sites* on the As terminated GaAs(001) surface. If the GaAs surface was covered completely with As atoms, As atoms would always incorporate into the GaAs crystal as As$_2$ molecule. On the other hand, according to Kratzer et al.[7], As$_2$ molecule can bind strongly on the four-Ga atom cluster on their "mountain" site where the four Ga atoms are on the three As dimers. It means that Ga atom on the top of the surface will play an important role to dissociate As$_2$ molecule into individual As atoms. Thus, in order to discuss the As atom migration on the GaAs(001) surface in detail, it is very important to discuss whether the dissociation of As$_2$ molecules occurs or not on Ga atoms. For simplicity of the computational task, we can consider here whether As atoms migrate on the "Ga-terminated surface" as single atoms or as As$_2$ molecules.

The migration for *single* As adatom on Ga-terminated GaAs(001) surface during GaAs molecular beam epitaxial growth has been studied previously using first-principle total-energy calculations[8,9]. From the calculation, it is confirmed that As atoms on the Ga-terminated

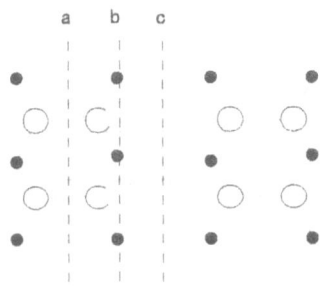

Fig. 1 Model of pair-migration of two As atoms on Ga-terminated GaAs(001)β2-(2×4) surface. Two As atom position is set to be normal to the Ga dimer. The position a, b and c correspond to 0, 2, and 4 Å, respectively. The dashed open circles are the As adatoms, the open circles are the substrate Ga dimers and the solid circles are the substrate As atoms of the second surface layer.

area migrate easier toward the dimer direction. Therefore, the most probable migration way for As$_2$ molecule is also the dimer direction of the Ga terminated area. We focus our attention to the migration of an As$_2$ molecule on the short-bridge sites of the Ga dimer toward the dimer direction.

2 Model

There are some different reconstructions reported on the Ga-terminated GaAs(001). In this paper, we focus on the β2(4 × 2) structure. The β2(4 × 2) surface that we use in this calculation is based on the reconstructed surface by first-principle calculation [10]. The unit cell on the surface has the 4× periodicity along the orientation parallel to the Ga dimer rows and the 2× periodicity perpendicular to the Ga dimer rows.

The pair of As atoms are positioned on the short-bridge site of the substrate two Ga dimers of the β2(4 × 2) surface. The positions of the two As atoms in the initial of the actual first-principle calculation is shown in Fig. 1 where the dashed open circles are As adatoms. Here we assume that the direction of the As ad-dimer is normal to the substrate Ga dimers. The x coordinate of the two As adatoms is kept to be fixed while y and z coordinate of them are fully relaxed during the calculation. In fig.1, "a" means that the two As adatoms are relaxed fully within the plane of $x = 0$. For "b" and "c", the As adatoms are relaxed within the plane of $x = 2$ and $x = 4$, respectively. In the actual calculation, we calculate several points of x between a and c.

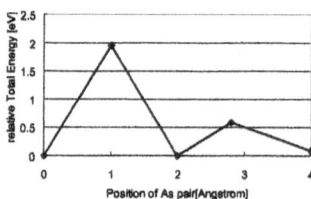

Fig. 2 Relative total energy as a function of the position of the As atom pair.

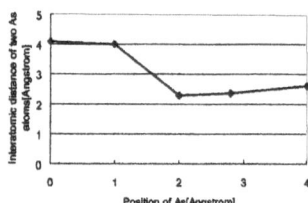

Fig. 3 Interatomic distance of the two As atoms as a function of the position of the As atom pair.

3 Results

In Fig. 2, we show the relative total energy as a function of the x coordinate of the adsorved As atom pair. We obtain three energy minimum; $x=0$ Å, 2 Åand 4 Å. The "a", "b" and "c" in fig.1 correspond to the three energy minimum. The position corresponds to $x = 0$ Å, "a", means that Each As adatoms positioned on the short-bridge sites of the two Ga dimers, breaking the dimer bonds. The position $x = 4$ Å, "c", corresponds that the As pair positioned at the long-bridge sites between the two dimer rows. The interesting result is the position $x = 2$ Å, "b", where the two As atoms positioned almost above the Ga atoms of the one-side of the dimers. For a single As migration, this position corresponds to the maximum of the energy.

In order to understand the "third" energy minimum at the position "b", we check the interatomic distance of the two As adatoms in the calculation as we show in Fig. 3. According to the figure, we find the two interatomic distances in the calculation; the distance corresponding to the distance of the two Ga dimers and the distance neary equal to the atomic distance of the isolated As_2 molecule.

In the figure, we find that at the position "b", the "third" energy minimum, the two As adatoms form As_2 molecules. The formation of the molecule bond decrease the total energy of the system.

4 discussion and conclusion

We obtain two new results in our present calculation. One is that the As adatom pair positions at the short-bridge sites of the Ga dimers of the Ga terminated GaAs (001)β2-(2×4) surface do *not* form ad-dimer. It is very different from the results of Morgan et al.[6] for the As-

terminated GaAs(001) β2-(2×4) surface. Thus, at least near the Ga atom clusters on the top-layer, As atoms can dissociate and migrate as an individual sigle adatoms. At the situation of the top mountain site of Kratzer et al.[7], it will be same that the two As adatoms on the mountain do *not* form ad-dimer.

The other is that the formation of molecule can make new energy stable site. The position "b" is energetically unstable for a single As adatom[9]. For the case of the two As atom pair, the total energy makes lower because of the binding energy of the As_2 molecule state.

The results show us that As atom pair can adsorb on the Ga terminated surface area as the individual adatoms or as the As_2 molecule. This results suggest us the speculation that the Ga dimers appeared on the GaAs(001) surface during the MBE growth play an important role to dissociate As_2 molecules. After dissociation of the As_2 molecule into the two individual As atoms, the As atoms can migrate on the surface as the single atoms because of the entropy.

In conclusion, we calculate the migration of the two As atom pair on the Ga terminated GaAs(001)β2-(2×4) surface. The two As atoms adsorb on the Ga terminated surface individually or adsorb as As_2 molecule depend on the position. It suggests that the Ga dimers will play dominant role for the dissociation of As_2 molecule during the MBE growth.

Acknowledgements This work was supported by the New Frontier Program Grand-in-Aid for Scientific Research (No.09 NP1201). The numerical computations were performed on SX4 system supported by the New Frontier Program Grand-in-Aid for Scientific Research, and on Supercomputing Centers at Institute for Solid State Physics, the University of Tokyo.

References

1. A. Ishii and T. Kawamura, Appl. Surf. Sci. **130-132** (1998) 403.
2. A. Ishii and T. Kawamura, Surf. Sci. **436** (1999) 38.
3. T. Kawamura and A. Ishii, Surf. Sci. **438** (1999) 155.
4. T. Kawamura and A. Ishii, Jpn. J. Appl. Phys. **39** (2000) 4376
5. G. R. Bell, M. Itoh, T. S. Jones, B. A. Joyce, Surf. Sci. **423** (1999) L280
6. C. G. Morgan, P. Kratzer and M. Scheffler, Phys. Rev. Lett. **82** (1999) 4886
7. P. Kratzer, C. G. Morgan, M. Scheffler, Phys. Rev. B **59** (1999) 15246.
8. K. Seino, A. Ishii and T. Aisaka, Surf. Sci. **438** (1999) 43.
9. K. Seino, A. Ishii and T. Kawamura, Jpn. J. Appl. Phys. **39** (2000) 4285
10. T. Ohno, Private communication.
11. C. T. Foxon and B. A. Joyce, Surf. Sci. **50** (1975) 434.
12. C. T. Foxon and B. A. Joyce, Surf. Sci. **64** (1977) 293.
13. E. S. Tok, J. H. Neave, J. Zhang, B. A. Joyce, and T. S. Jones, Surf. Sci. **374** (1997) 397.
14. J. Sudijono, M. D. Johnson, C. W. Snyder, M. B. Elowitz, and B. G. Orr, Phys. Rev. Lett. **69**, 2811 (1992).

Photoemission spectroscopy of Si(001) surfaces oxidized by hyperthermal O_2 molecular beams

Y. Teraoka and A. Yoshigoe

Synchrotron Radiation Research Center, Japan Atomic Energy Research Institute, 1-1-1 Kouto, Mikazuki, Sayo, Hyogo 679-5148, Japan

Abstract An experimental apparatus for reaction analysis on semiconductor surfaces has been constructed at the soft x-ray beamline, BL23SU, in the SPring-8. The effects of translational energy of incident O_2 molecules for initial oxidation of Si(001) surfaces have been investigated by using a supersonic molecular beam technique and photoemission spectroscopy. The Si sub-oxide structure in the Si-2p photoemission spectra showed the increase of oxidation number of Si atoms with increasing O_2 translational energy. It was also found that the sub-oxide structure was affected by the translational energy even at surface temperature of 600 ℃.

1. Introduction

A role of translational energy (E_t) of incident atoms and molecules for chemisorption on surfaces is worth studying to clarify surface reaction dynamics. As a model reaction system, Si oxidation is an interesting research subject. Thermal oxidation of Si surfaces has been widely studied theoretically and experimentally [1]. Dissociative chemisorption of O_2 molecules takes place on Si(001) surfaces at room temperature without no E_t barrier [2]. Recently it has been pointed out from the first-principles calculation [2] that backbond oxidation of top dimer Si atoms proceeds with E_t larger than 0.8 eV. Although E_t of incident O_2 molecules has been known as an effective factor for the chemisorption [1], the known data concerning the influence of E_t are restricted below about 1 eV. The E_t dependence of dissociative chemisorption of O_2 molecules on Si(001) surfaces, therefore, has been investigated by using O_2 supersonic molecular beams (SSMB) and x-ray photoemission spectroscopy (XPS) of Si-2p core level with a wide E_t range larger than 1 eV.

2. Experimental

A brief description is given here concerning the experimental procedure. The end-station is primarily composed of six ultra-high-vacuum chambers [3]. The synchrotron radiation (SR) beam is introduced into the surface reaction analysis chamber through the SR beam monitor chamber. The n^+-Si(001) samples, previously processed by Shiraki method [4], are firstly flashed up to about 1000 ℃ several times in the cleaning chamber. The 2x1 low energy electron diffraction pattern was confirmed after the flashing. The sample is heated again up to 1000 ℃ in the surface reaction analysis chamber prior to SSMB irradiation. The SSMB is continuously generated by adiabatic expansion of a mixture of O_2, He and Ar. The nozzle tube can be heated up to about 1150 ℃ so that E_t of O_2 molecules is obtained up to about 3 eV (calculated value). The beam flux was estimated to be typically about 2×10^{14} molecules/cm^2/sec on a sample surface. A hemispherical electron energy analyzer is usable simultaneously with SSMB as the pressure is kept to be less than 6×10^{-5} Pa even during SSMB operation.

3. Results and discussion

The Si(001) surface was exposed to SSMB until achieving oxygen saturation. The saturated oxygen amount on surfaces measured as a function of E_t [5]. Two E_t thresholds were found at 1.0 eV and 2.6 eV. These values are very close to the predicted values (0.8 eV and 2.4 eV) from the first-principles calculation [2]. These observed threshold energies were, therefore, assigned to backbond oxidation of top dimer Si atoms and subsurface Si oxidation, respectively [5]. An E_t region, therefore, can be divided into three regions: below 1.0 eV (region I), 1.0 eV-2.6 eV (region II) and above 2.6 eV (region III).

The surface oxidation states were analyzed by Si-2p XPS using the monochromated SR beams with elliptic polarization and 409.4 eV photon energy. Representative Si-2p XPS spectra are indicated in Fig. 1. The electron detection angle was 70 degree in these cases with respect to surface normal so that the electron escape depth was estimated to be about 0.24 nm. The Si-2p core level splits into $2p_{3/2}$ and $2p_{1/2}$ with statistical weight ratio of 2:1 due to spin-orbit interaction. We referred the Si-2p core level shifts due to oxidation from the recent report [6]: 1.0 eV for Si^{1+}, 1.81eV for Si^{2+}, 2.63 eV for Si^{3+} and 3.6 eV for Si^{4+} (SiO$_2$). According to the literature [7], a shoulder, appeared beside the Si-2p$_{3/2}$ main peak in Fig. 1(a), corresponds to up-atoms of top Si dimers. The shoulder disappeared by oxidation. The Si-2p XPS spectrum for the oxidized surface by the residual O_2 molecules with E_t of 0.04 eV is shown in Fig. 1(b) as representative in the region I. The slight contribution from Si^{2+} must be recognized in the spectrum (b). This fact means that an oxygen atom inserted into the backbond of top dimer even below the threshold energy of 1.0 eV. Representative Si-2p XPS spectra in the region II are shown in Fig. 1(c) and (d) for 1.5 eV and 2.0 eV, respectively. The spectrum (c) indicates that the oxidation proceeds to the extent of Si^{3+} formation. The Si^{4+} formation must be involved in the spectrum (d). These features are achieved by direct backbond oxidation. The O_2 molecules with E_t of 3.0 eV can adsorb dissociatively between the second and the third Si layer according to the prediction of the theory [2]. The subsurface (the second layer) Si atoms, therefore, can be surrounded by oxygen atoms as well as top dimer atoms in the region III so that the fraction of Si^{3+} and Si^{4+} may increase comparing to the region II. The Si-2p XPS spectrum for 3.0 eV indicates that Si^{3+} and Si^{4+} atoms contribute mainly in the satellite peak structure as shown in Fig. 1(e).

In order to investigate the influence of E_t at high surface temperature, the molecular beams were exposed to clean surfaces at 600 ℃ with about 5.5 L ($1L \equiv 1.33 \times 10^{-4}$ Pa · sec). The results are shown in Fig. 2. The detection angle was 9 degree in these cases with respect to surface normal. On the contrary to the room temperature cases, satellite peaks for Si^{4+} component are involved even in the Si-2p XPS spectrum for E_t of 0.6 eV. This fact must be due to thermal migration of oxygen atoms from the top layer into more deeper positions. Surprisingly, it seems that E_t affects to sub-

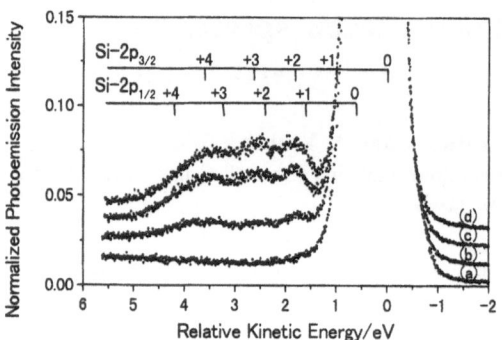

Fig. 2 Si-2p XPS spectra depending on translational energy of O_2 molecules: (a) clean surface, (b) 0.6 eV, (c) 2.0 eV and (d) 2.9 eV, respectively. The oxidation was performed at 600 ℃.

oxide structures even at 600 ℃. That is, relative intensity of satellite peak for Si^{3+} was increased with increasing E_t. As Si^{3+} and Si^{4+} oxidation states are not formed with direct dissociation of O_2 molecules with 0.6 eV, Si^{3+} and Si^{4+} components must be attributed to the thermal migration of oxygen atoms. As shown in Fig. 2(c), Si^{3+} component becomes to be larger than Si^{4+} component. This feature may be caused by oxygen insertion into the two backbonds of a top dimer Si atom due to enough E_t for direct backbond oxidation. The Si^{3+} component grows slightly as shown in Fig. 2(d) comparing to the spectrum (c). This may be caused by direct subsurface backbonds oxidation by E_t of 2.9 eV.

References

1. T. Engel, Surf. Sci. Rep. **18**, (1993) 91 and references therein.

2. T. Uda and K. Kato, J. Surf. Sci. Soc. Jpn. **19**, (1999) 173 (Japanese); K. Kato, T. Uda and K. Terakura, Phys. Rev. Lett. **80**, (1998) 2000.

3. Y. Teraoka and A. Yoshigoe, Jpn. J. Appl. Phys. **38**, Suppl. 38-1, (1999) 642.

4. A. Ishizuka and Y. Shiraki, J. Electrochem. Soc. **133**, (1986) 666.

5. A. Yoshigoe and Y. Teraoka, submitted to Jpn. J. Appl. Phys.

6. H. W. Yeom, H. Hamamatsu, T. Ohta and R. I. G. Uhrberg, Phys. Rev. B, **59**, (1999) R10413.

7. T.-W. Pi, C.-P. Cheng and I.-H. Hong, Surf. Sci. **418**, (1998) 113.

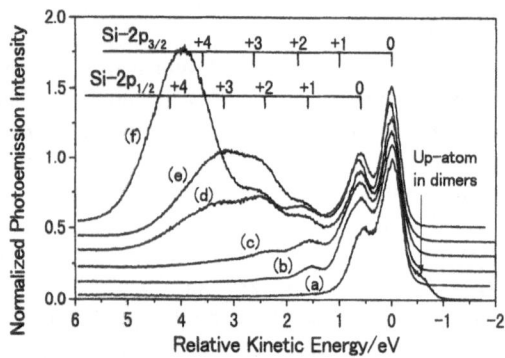

Fig. 1 Si-2p XPS spectra depending on translational energy of incident O_2 molecules: (a) clean surface, (b) 0.04 eV, (c) 1.5 eV, (d) 2.0 eV (e) 3.0 eV and (f) SiO_2 thin film. The oxidation was performed at room temperature.

Formation of first InSb molecular layer on Si(111) substrate: Role of In(4×1) reconstruction

B.V. Rao[1], D. Gruznev, T. Tambo, C. Tatsuyama
Faculty of Engineering, Toyama University, Toyama, Japan
[1]e-mail: bvrao1@hotmail.com

Abstract   Present report details the scanning tunneling microscopy (STM), reflection high-energy electron diffraction (RHEED) and Auger electron spectroscopy (AES) study of InSb molecular layer formation on the Si(111) substrate. The InSb layer is prepared by depositing 1-monolayer (ML) of Sb on the Si(111)-In(4×1) reconstruction. As the lattice mismatch between Si and InSb is over 19%, a molecular layer sustaining such a large strain is very difficult to envisage. This opens up various reaction paths and we will detail them in terms of replacement of In-layer by the Sb atoms. This study is also effective to understand the structure of top Si layer and its role in controlling the subsequent deposition processes. Also, this study answers some outstanding questions about the structure of the In(4×1) reconstruction.

1. Introduction

InSb/Si is among the most difficult systems for the direct heteroepitaxial growth due to large lattice mismatch (> 19%). Theoretical predictions suggest that above 10-12% mismatch, growth can takes place only in the Volmer-Weber mode [1], i.e., three dimensional island growth. However, recent reports suggested that (111) crystal orientation is effective in inhibiting the 3D island formation [2] and we could grow high quality InSb films on Si(111) substrate without buffer layer [3]. This inspired us to study the formation of first InSb molecular layer on Si(111) substrate. Both In and Sb form various well ordered reconstructions on Si(111) substrate at low and high temperatures, respectively [4] and it will be interesting to see how these surface phases react with each other to form the InSb layer on Si substrate with 19% strain. As InSb growth proceeds by the alternate stacking of In and Sb layers (in principle), we can start the InSb growth by either In or Sb layer. However, Sb does not form any ordered reconstructions on Si(111) substrate at low temperatures and we avoid the high temperature growth due to 3D island formation. Therefore, it is ideal to start the growth with low-temperature deposited In-reconstruction as the starting layer.

Si(111)-In(4×1) reconstruction can be prepared at a nominal In coverage of about 1 ML and therefore we have chosen this reconstruction as starting surface. In this article, we detail the structure and composition of various surface phases resulting from the Sb adsorption on Si(111) substrate covered with In(4×1) reconstruction at various substrate temperatures.

All the depositions were performed using MBE and the resulting surface phases were characterized in an attached surface analysis chamber equipped with STM/XPS/AES measurement systems. In and Sb were evaporated from different K-cells onto the Si(111) substrate to prepare the surface reconstructions.

2. Results and Discussions

Replacement of In-surface phase by Sb atoms is intriguing and is interesting because the replacement kinetics are very sensitive to the deposition conditions. STM images of the room-temperature (RT) Sb deposition on the Si(111)-In(4×1) reconstruction [denoted as In(4×1) phase] clearly showed that the 4×1 phase is partially damaged by the Sb atoms. Fig.1 shows the STM images of (a) In(4×1) reconstruction and (b) 0.1 ML Sb deposition on In(4×1) surface at RT. From the Fig.1(a) one may observe that the long stripes that form the 4×1 reconstruction are almost defect free (substrate orientation is different for Figs. 1(a) and 1(b)). After the Sb deposition, however, defects start cropping these stripes due to the replacement of In-atoms forming the surface phase (Fig. 1(b)). The replaced In atoms (and possibly some of the deposited Sb), form island structures on these stripes. From the large-scale STM images, we estimated that the area covered by these islands is nearly equal to the defected 4×1 areas. Also, it must be noted that most of the defects break the stripes at a "strange angle" of about 70°. This suggests that the In atomic chains forming the 4×1 phase make 70° with the row axis and supports the assumption made by Saranin et al., in proposing the atomic model for the In(4×1) reconstruction [5]. However, their model proposes that the central row In-atoms are loosely bound to the edge In-atoms by metallic bonds, are NOT bonded to the substrate. Such loosely bonded atoms are not stable when the edge In-atoms are replaced by the Sb. However, the STM image in Fig. 1(b) shows stable middle row In-atoms even when Sb replaces edge In-atoms. This aspect needs further attention to solve the structure of In(4×1) phase.

Continues Sb deposition causes the In-surface phase to become more disordered and this processes is highly sensitive to the deposition temperature. RHEED patterns of the 1-ML Sb deposition on In(4×1) surface at 210°C showed a twinned InSb molecular layer with the epitaxial relationship Si(111)‖InSb(111) and Si(110)‖InSb(112).

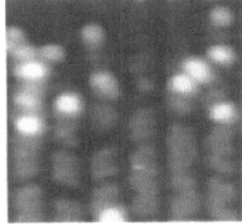

Fig. 1. (a)Si(111)-In(4x1), (b)0.1 ML Sb on (a) at RT

RHEED streak spacing suggested the twinned InSb lattice spacing (not shown here), and we could not observe any well-ordered structure in the STM images. This may be due to the partial replacement of In-layer by Sb and the replaced In-atoms might have formed 2D islands on the Sb-terminated Si surface. That is an In-terminated InSb surface is formed. The nature of In-Sb reaction kinetics seems complicated and we cannot completely rule out this possibility from the present study alone.

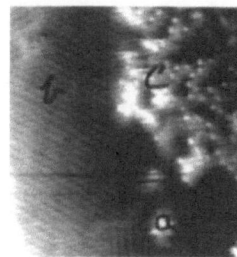

Fig. 2. In(4×1)/Sb 0.5 ML at 300 °C

Sb deposition on In(4×1) reconstruction above 250°C causes almost complete replacement of In-layer. Subsequently, Sb atoms form various well-ordered reconstructions on the Si surface. The structure and the composition of these surface phases vary with the deposition conditions. STM image of the 0.5-ML Sb deposited on In(4×1) surface at about 300°C is shown in Fig. 2. Auger spectra of this surface suggested a nearly stoichiometric composition. We believe that under these deposition conditions replaced In atoms rearrange themselves on the Sb-terminated Si surface. This surface is composed of three different structures. First, the area marked (a) shows rows of Sb atoms with a_0 spacing along the row (a_0 = Si lattice spacing = 3.84Å). These rows are separated by $\sqrt{3}a_0$. This is a (2×1) reconstruction with long axis along <110> direction. It is interesting to note that this structure is completely different from the 2×1 reconstruction formed by the direct Sb adsorption on Si(111)-7×7 surface. Formation of this new surface phase must have been necessitated from the re-reconstructed the top layer Si atoms beneath the In(4×1) phase. Exact structure of the top layer Si atoms beneath In(4×1) phase is still under debate and the present study may possibly offer some clues to arrive at a solution.

Area marked (b) in Fig. 2 consists of elongated dots separated by a_0 in straight rows. Spacing between the rows is again $\sqrt{3}a_0$ suggesting 2×1 reconstruction. However, the intensity profile of these spots is different from the area (a), and we estimate that these spots consist of two atoms arranged in a buckled dimmer structure. As the covalent radius of In and Sb are nearly same (In = 1.44Å and Sb = 1.4 Å), it is difficult to distinguish these two atoms in the STM images. However, high-resolution STM images, suggested minute difference in the intensity profile, and we estimated that the top dimmer atom is of In and bottom one

is of Sb. Area (c) in Fig.2 is a partially ordered In($\sqrt{3}\times\sqrt{3}$) surface, which is formed on the top of the area (b).

The replaced In-atoms may form either partially disordered surface phases on the Sb-terminated Si surface or together with Sb they form epitaxial InSb layers on the Si surface. The process of InSb molecular layer formation is very complex, as the lattice mismatch between Si and InSb is >19%, and it is very difficult for a molecular layer to sustain such a large strain. Besides, the replaced In-atoms may form different reconstruction on the above two surfaces. Our AES results showed that the coalescence of In-layer depends on the amount of Sb deposition and temperature of deposition. This means higher Sb coverage completely coalesces the In layer into the nanometer sized islands. Also, high Sb coverage seems to lead stronger bonding between Sb atoms forming the row structure. STM image of 1-ML Sb deposition on In(4×1) phase at 350°C is shown in Fig. 3, where straight line structures formed by Sb, and randomly distributed nanometer sized In islands are seen. We could not identify the atomic positions in the Sb rows, possibly due to different bonding scheme than the one observed in

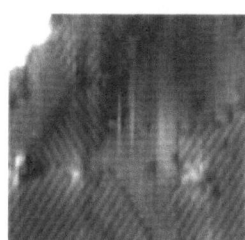

Fig. 3. In(4×1)/Sb 1-ML at 350 °C

Fig. 2. A more detailed report of this reaction path will be published elsewhere.

3. Conclusions

We explained the processes involved in the adsorption of Sb on the In(4×1) reconstruction at various temperatures and detailed the formation of InSb molecular layer on Si(111) substrate. Sb deposition below 250°C resulted in a twinned InSb molecular layer, whereas above 250°C Sb formed various complex surface phases together with In.

Acknowledgments

The present work has been supported in-part by a Grant-in-Aid for Scientific Research (10450005) from the Ministry of Education, Science, Sports and Culture.

References

1. I. Daruka, et al., Phys. Rev. Lett., 79 (1997) 3708.
2. K. Kanizawa, et al., Appl. Phys. Lett., 76(2000)589.
3. B.V. Rao, et al., Appl. Surf. Sci.,159-160 (2000) 335.
4. V.G. Lifshits, et al., (Eds.), Surface Phases on Silicon, (John Wiley, England, 1994), p. 247, p 362.
5. A.A. Saranin et al., Phys Rev. B 60 (1999) 14372.

Laser-induced electronic instability on semiconductor surfaces of Si (111)-(7x7) and InP (110)-(1x1)

K. Tanimura, J. Kanasaki*, K. Ishikawa

Department of Physics, Nagoya University, Furo-cho, Chikusaku, Nagoya 464-8602, Japan

Abstract Structural changes on clean surfaces of Si(111)-(7x7) and InP(110)-(1x1) induced by ns- and fs-laser irradiation for fluences well below thresholds of surface melting and ablation have been studied. Atomic structures of irradiated surfaces were imaged by scanning tunneling microscopy (STM) and neutral atoms desorbed by bond breaking were investigated by means of high sensitive laser-ionization spectroscopy. STM observation of irradiated Si(111)-(7x7) reveals the electronic instability to break the bond of adatoms at intrinsic surface sites, resulting desorption of Si atoms in the electronic ground state, with a strong site-dependent and wavelength dependent efficiency. On the other hand, STM images of the irradiated InP(110)-(1x1) surfaces have revealed the preferential removal of the outermost P atoms, with significant formation yields of vacancy strings consisting of several adjacent vacancies on the quasi-one dimensional P rows. The electronic bond breaking for both surfaces is shown to originate from non-linear localization of excited species in surface electronic states.

1. Introduction

Surfaces of tetrahedrally bound semiconductors show drastic reconstruction and/or relaxation of atomic structures. Consequently, the surface electronic states characteristic of a surface-specific structure are formed, which show quasi-two-dimensional nature. Thus, several properties of surfaces are significantly different from those in bulk. Structural response under photoexcitation is a typical example. In this paper, we show that structural modifications are induced via electronic mechanism on typical covalent semiconductor surfaces, like Si(111)-(7x7), InP(110)-(1x1).

2 Experimental

Clean surfaces of several semiconductors, prepared in an UHV chamber with basic pressure of 1×10^{-10} torr, were excited with ns- and fs-laser pulses for fluences well below thresholds of surface melting and ablation. Atomic structures of irradiated surfaces were imaged by scanning tunneling microscopy (STM), and neutral atoms desorbed by bond breaking were investigated by means of high sensitive laser-ionization spectroscopy with high sensitivity of 10^{-7} monolayer per pulse.

* *Present address:* Department of Intelligent Materials Engineering, Osaka-City university, Sugimoto 3-3-138, Sumiyoshi, Osaka 558-8585, Japan

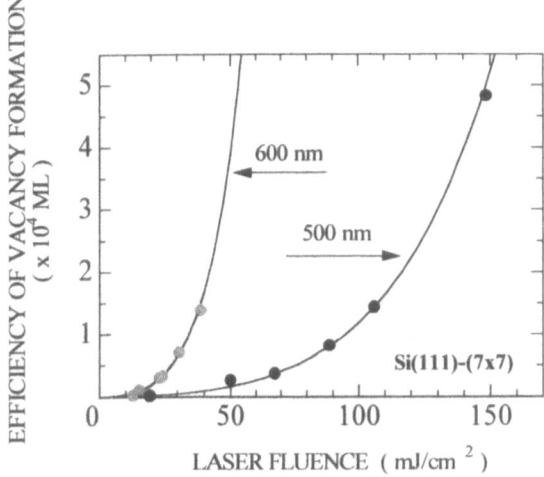

Fig.1   The bond breaking efficiencies as a function of excitation intensity for Si(111)-(7x7).

3 Results and discussions

STM observation of irradiated Si(111)-(7x7) reveals the electronic instability to break the bond of adatoms, and Si atoms in the electronic ground state are desorbed with a peak translational energy of 0.06 eV, as a direct consequence of the bond breaking.[1,2] Center adatoms are removed by three times more efficiently than corner adatoms, and adatom vacancies are formed mostly at individual adatom site; no preferential clustering of vacancies are observed. The efficiency of the electronic bond breaking depends strongly on the excitation wavelength; the efficiency shows the highest value for the photon energy of 2.0 eV. Also, most significant feature is the super-linear dependence of the efficiency on the excitation intensity. In Fig.1(a), we show the bond breaking efficiency as a function of excitation intensity for two different wavelengths of 500 and 600 nm. Detailed spectroscopic studies in the spectral range from 800 to 250 nm indicate that the super-linear dependence of the efficiency is nothing to do with multi-photon processes. The super-linearlity can be explained well by the two-hole localization of the surface holes; the solid curves in the figure are the fit of the model to the data; it described almost perfectly for the whole fluence range studied.

Examination of STM images of the irradiated n-type InP (110)-(1x1) at wavelength corresponding to the surface-

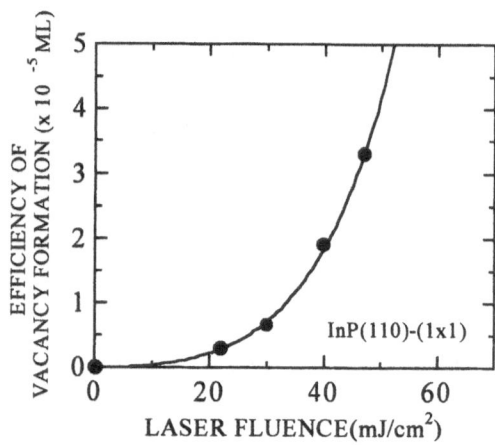

Fig. 2 The efficiencies of bond breaking at the intrinsic P-atom sites as a function of excitation intensity for InP(110)-(1x1).

References

[1] J. Kanasaki, T. Ishida, K. Ishikawa, and K. Tanimura., Phys. Rev. Lett. **80**, 4080(1998).

[2] J. Kanasaki, K. Iwata, and K. tanimura, Phys. Rev. Lett. **82**, 644(1999).

[3] K. Tanimura and J. Kanasaki, Proceeding of Laser Applications in Microelectronic and Optoelectronic Manufacturing IV **3618**, 26(1999)

specific transition of 460 nm show preferential removal of the top-most P atoms, with significantly enhanced formation yields of vacancy strings consisting of several adjacent vacancies on the quasi-one dimensional P rows. [3] The isolated In vacancies are also formed, but with a much smaller yield; showing specimen-dependent efficiency of electronic bond breaking. Similarly to the case of Si(111)-(7x7), the bond breaking efficiency on this surface also depends super-linearly on the excitation intensity. Figure 2 shows the results of the bond breaking efficiency of P atoms on this surface as a function of fluence of excitation laser pulses. The solid curve in the figure is again the fit of two-hole localization model to the data.

4 Conclusion

The surface-structural changes induced by laser-induced electronic instability are intrinsic in the sense that breaking is not related to pre-existing surface defects, and show characteristic features of strongly site dependent, specimen sensitive, and wavelength dependent. Also, for both surfaces, the efficiency of bond breaking shows an interesting common feature of super-linear dependence on the excitation intensity. Quantitative analysis suggests strongly that the instability originates from non-linear localization of surface-excited species; the mechanism of two-hole localization describes well the interesting feature of super-linear dependence of the bond-breaking efficiency on the excitation intensity.

Surface States of (100)n-GaAs with Adsorbed Oxygens and Their Dependence on Chemical Treatment

Y. Kasai, T.Tsuzuku, Y. Ohta, T. Inokuma, K. Iiyama, S.Takamiya

Graduate School of Natural Science and Technology, Kanazawa University, 2-40-20 Kodatsuno, Kanazawa, Ishikawa 920-8667, Japan e-mail: kasai@karies.ec.t.kanazawa-u.ac.jp

Abstract   Treatment of (100)n-GaAs wafers with HCl (pH =1) and NH_4OH (pH=13) resulted in different densities of surface oxygen and different barrier heights of Ni/n-GaAs Schottky junctions which were formed on them. In order to investigate the effect of the surface oxygen, the authors studied electronic status of GaAs(100)-(1×1) with adsorbed oxygens by DV-$X\alpha$ method.

1 Introduction

It is widely known that a surface potential of a semiconductor depends upon a status of the surface, and a Schottky barrier height is influenced by a chemical treatment prior to forming the barrier[1]-[3]. The authors investigated the effect of pretreatment by pH-controlled chemicals on Schottky barrier heights of Ni/n-GaAs junctions. An X-ray photoelectron spectroscopy(XPS) analysis of the treated surface suggested that difference in surface oxygen densities is the cause of the different electrical characteristics. In order to clarify it's mechanism, electronic status of GaAs(100) surfaces with and without adsorbed oxygen atoms were studied using DV-$X\alpha$ molecular orbital calculation method [5].

2 Experimental

2.1 Pretreatment with pH-controlled chemicals

The (100)n-GaAs wafers were ultrasonic-cleaned in acetone to remove organic contamination, and etched-off their native oxide by immersing them into buffered hydrofluoric acid, then rinsed by deionized water. After that, they were pretreated for 20 minutes by pH controlled chemicals; a) dilute HCl (pH=1, room temp.), b) dilute NH_4OH (pH=13, 50°C.). They are *ex-situ* processed in the room air. After rinse-and-dried, Ni and AuGe were selectively evaporated, to form Schottky (diameter of 320 μm) and ohmic contacts, both on the top surfaces of a set of the wafers. Let us denote the above 2 liquids as a) and b) chemicals, and the wafers and Schottky diodes treated by the respective liquid will be denoted in the following by a_1 and b_1.

Surface and insides of the a_1 and b_1 wafers of the other set were analyzed by XPS, before and after Ar ion etching. The results show that the density of adsorbed oxygens at the surface of the a_1 wafer is about two times higer than the b_1 wafer.

The idealiy factors in the forward currents and "effective" Schottky barrier heights of the samples, which were obtained from Richardson plots under reverse biased conditions, are summarized by unparenthesized numbers in Table.1. As a whole, the sample b_1 has 0.2 to 0.3eV lower effective barrier heights than those of the a_1.

Table 1 Barrier heights ϕ [eV] and ideality factors n of Ni/n-GaAs diodes. Unparenthesized numbers are different wafers. Numbers in ( ) or [ ] corresponds to the Case I or the Case II.

	a_1	a_2	a_3	b_1	b_2	b_3
n	1.2	(1.1)	[1.1]	1.2	(1.1)	[1.2]
$\phi_{0.1}$*	0.79	(0.73)	[0.76]	0.60	(0.63)	[0.48]
$\phi_{0.3}$	0.78	(0.71)	[0.74]	0.54	(0.54)	[0.41]
$\phi_{0.5}$	0.77	(0.67)	[0.71]	0.47	(0.44)	[0.33]
$\phi_{0.7}$	0.73	(0.59)	[0.68]	0.43	(0.39)	[0.27]

*0.1 to 0.7 are reverse voltages.

2.2 Two levels of Ni/n-GaAs Schottky barriers on a wafer

In order to realize two levels of Ni/n-GaAs Schottky barrier heights on a wafer, the authors tried two process sequences as illustrated in Figure.1. After cleaning

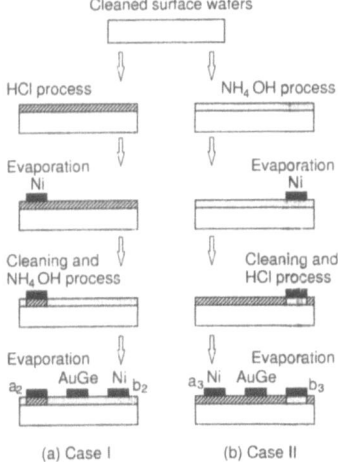

Fig. 1 Process sequence to form two levels of Ni/n-GaAs Schottky barrier heights.

by acetone and buffered hydrofluoric acid, a wafer was first pretreated by the dilute HCl (pH=1, room temperature), followed by Ni evaporation and the dilute NH_4OH (pH=13, 50°C) pretreatment, as shown by Case I of Figure.1, Another wafer was processed by the reverse sequence (Case II), in order to investigate whether the later cleaning-and-treatment process completely erases the memories of the first pretreatments and generates a status which is completely controlled by the later treatment. The Ni/n-GaAs Schottky diodes formed by the Case I process on the HCl-treated portion is denoted as a_2, and that on the NH_4OH-treated portion b_2, and those for the Case II are denoted by a_3 and b_3, respectively (Figure.1).

Measured ideality factors and effective barrier heights at reverse bias of 0.1 to 0.7 V are shown in Table 1 by ( ) for the Case I, and by [ ] for the Case II. The both cases reproduced the differences which were described in §2.1, and successfully realized the two different Schottky barrier heights by Ni/n-GaAs junctions on a wafer.

3 Theoretical study

The authors calculated the electronic status of GaAs surface with adsorbed oxygen atoms in order to clarify what influence do oxygen atoms give to the GaAs surfaces. We used cluster models of GaAs(100)-(1×1) surface, with chalcogen atoms(O, S, Se, Te) adsorbed at the bridge site [4], to do the molecular orbital calculation by DV-$X\alpha$ method [5].

By adsorption of chalcogen atoms on the GaAs surface, electrons shift from GaAs surface to the chalcogen atoms due to difference of electronegativities of the elements. Figure 2 shows net charges of the adsorbed chalcogen atoms and Ga(As surface) and As(Ga surface) atoms on the GaAs(100)-(1×1) surface. More electrons shift from the Ga-terminated surface than the As-termianted surface, due to smaller electoronegativity of Ga than that of As.

Density of states (DOS) of Ga terminated GaAs(100)-(1×1) surface is shown in Figure.3 (a), and that of the oxygen adsorbed surface is shown in Figure.3 (b). Without oxygens, no energy gap is seen at the GaAs surface, because dangling bond states of the top Ga layer occupy the GaAs energy gap. On the other hand, the oxygen adsorbed surface shows that the surface state density in the energy gap is remarkably reduced. It is because that oxygen-Ga bonding state lies in the GaAs valence band, and oxygen-Ga anti-bonding state lies in the conduc-

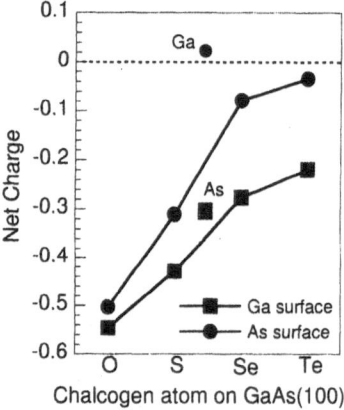

Fig. 2 Net charge of chalcogen and Ga and As atoms on GaAs(100) surface.

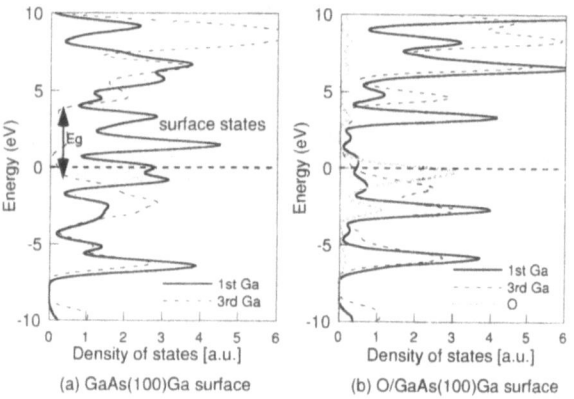

Fig. 3 Density of states of Ga terminated GaAs(100)-(1×1) surface (a) and oxygen adsorbed surface (b).

tion band. However on a As terminated surface, oxygen adsorption does not reduce the surface states, because oxygen-As anti-bonding state is located in the GaAs energy gap.

4 Discussion

XPS analysis suggests that the different densities of adsorbed oxygens on GaAs surface is the cause of the difference of Schottky barrier heights between HCl-treated and NH_4OH-treated samples. As the calculated net charge shows, oxygen adsorption cause strong shift of electrons from GaAs surface. Thus, polarization occurs at the oxygen adsorbed GaAs surface. It has a significant effect on characteristics of a Schottky junction formed on a GaAs. In case of a n-type GaAs surface with adsorbed oxygen atoms, oxygens act as minus ions and bend the energy band upward so that the Schottky barrier is kept high.

5 Conclusion

Effects of GaAs surface pretreatement by pH-controlled chemicals were experimentally invesitgated. Oxygens on the surface affect the characteristics of Ni/n-GaAs(100) Schottky diode which are formed after the treatment. We calculated the electronic status of GaAs(100)-(1×1) surface with adsorbed oxygens and other chalcogen atoms by DV-$X\alpha$ method. The calculation results indicated that oxygens cause a strong shift of electron, and oxygen-Ga bond remarkably decrease surface states in the forbidden band. These results seem to qualitatively explain the results of the experiments.

References

1. H. Hasegawa : Solid-State Electron. **41**, (1997) 1441.
2. O. Sugino, and B. D. Yu : Oyo Buturi. **68**, (1999) 1397 [in Japanese].
3. K. Hirose : Trans. IEICE. **C-II 82**, (1999) 1 [in Japanese].
4. T. Ohno : Surf. Sci. **255**, (1991) 229.
5. H. Adachi, Y. Owada, I. Tanaka, H. Nakamatsu, M. Masataka : *Hajimete no densi jyoutai keisan* (SANKYO SHUPPAN, Tokyo 1998) [in Japanese].

Real-time observation of initial oxidation on highly B-doped Si(100)-2x1 surfaces using scanning tunneling microscopy

K. Ohmori[1,2], M. Tsukakoshi[1], H. Ikeda[1], A. Sakai[1], S. Zaima[3], Y. Yasuda[1]

1   Graduate School of Engineering, Nagoya University, Furo-cho, Chikusa-ku, Nagoya 464-8603, Japan
    e-mail: ohmori@alice.xtal.nagoya-u.ac.jp
2   Venture Business Laboratory, Nagoya University, Furo-cho, Chikusa-ku, Nagoya 464-8603, Japan
3   Center for Cooperative Research in Advanced Science & Technology, Nagoya University, Furo-cho, Chikusa-ku, Nagoya 464-8603, Japan

Abstract   We have studied the initial oxidation on highly B-doped Si(100) surfaces by real-time scanning tunneling microscopy observation.   No change of the segregated B atoms on the surface are observed during O_2 exposure.   However, it is found that stabilized oxygen adsorption sites and etching sites are different from those on intrinsic Si(100) surfaces.   This means that the segregated B atoms influence the oxidation process on a long-range scale of the structure.

1. Introduction

It has been reported that boron (B) penetration is one of critical problems for dielectric breakdown of metal-oxide-semiconductor (MOS) devices [1]. Therefore, understanding the impurity influences on local bonding structures and electronic states of the gate oxide film has become essential for further miniaturization of ultra-large scale integrated circuits (USLIs).   In the present paper, we have investigated the initial oxidation processes on highly B-doped Si(100) surfaces by real-time observation using scanning tunneling microscopy (STM) for clarifying the impurity influences on the surface reaction of oxygen.

2. Experimental

All experiments were performed in an ultrahigh vacuum system equipped with variable-temperature STM and x-ray photoelectron spectroscopy (XPS) apparatuses, whose base pressure was about 2×10^{-8} Pa.   Highly B-doped Si(100) wafers with a resistivity of 0.007-0.015 Ωcm (B concentration: $\sim 5 \times 10^{18}$ cm^{-3}) were used.   After a sample was outgassed at a temperature of 600°C for about 10 hours, a clean Si(100)-2x1 surface was prepared by direct current heating at 1200°C in a pressure less than 5×10^{-8} Pa. The density of B-related structures on the clean surface increases with increasing the heating time at

1200°C due to the surface segregation of B atoms [2]. We used samples with a number density of 4×10^{12} cm^{-2} of B-related structures by heating for 15 min. In order to observe the changes in the surface structures during the oxidation directly, real-time STM observation was carried out under the condition that the clean surface was exposed to O_2 molecules with a pressure of 1.3×10^{-6} Pa at room temperature.

3. Results and Discussion

Figures 1(a)-(d) are STM images of a clean surface and surfaces exposed to oxygen of 0.4, 0.8 and 1.2 Langmuir (L) at room temperature, respectively, showing exactly the same area.   In Fig. 1(a), B-related structures are observed on the clean surface as indicated by black rectangles.   According to a model of the B-related structure proposed by Komeda et al. [3], one B-related structure include two B atoms, and each of them is bonded to two neighboring Si-Si dimers at their bridging sites on a dimer row.   By O_2 exposure, bright spots and dark regions appear and buckling of Si dimers is induced partially. These morphological changes indicate that oxidation and etching occur simultaneously [4].   The bright spots in white circles of Fig. 1(b) change into the dark region in Fig. 1(c) by increasing the oxygen exposure.   On the other hand, the bright spots in white squares in Fig. 1(c) remain in Fig. 1(d).   These results indicate that there are two kinds of oxygen adsorption sites: the site where the oxygen atom remains stably and the site where the oxygen atom moves into the back bond.   Although these changes are observed on a terrace randomly, B-related structures are found not to be modified by the O_2 exposure.

Occupied-state and unoccupied-state STM images of a surface exposed to 7 L oxygen are shown in Figs. 2(a) and 2(b), respectively.   We focus on three bright spots indicated by R, S and T in Fig. 2(a). The bright spot S is located between dimer rows, while the bright spots R and T, that are frequently observed on this surface, are located just above dimer rows. Figure. 2(c) is line profiles corresponding to M-M' and N-N' shown in STM images of Fig. 2(a) and 2(b), respectively. It should be noted that the bright spots R and T have a larger diameter in the unoccupied-state STM image (N-N') than that in the occupied-state STM image (M-M'), and the bright spot S vice versa. This characteristic of the bright spots R and T is in good agreement with that of the theoretical STM image of an oxygen atom adsorbing just above a Si-Si dimer bond (named an OD site) [5]. Thus, the bright spots R and T correspond to oxygen atoms adsorbing on the OD site. These spots were stable during STM observation. This is a marked contrast to the case of the intrinsic Si surface where the bright spots just above the dimer rows are unstable and turn to a dark region in the subsequent STM scanning [6]. It has been reported that the segregated B atoms induce comb-shaped step structures due to the anisotropic surface stress [3]. Considering this result, we deduce that long-range strain induced by the B atoms on the surface affects the stability in the oxygen adsorption site, hence leading to the stable oxygen atom on the OD site.

4. Conclusions

Initial oxidation processes on highly B-doped Si(100)-2x1 surfaces have been investigated on an atomic scale. The B-related structures remain during the initial oxidation. On the other hand, surface reactions with oxygen are influenced by highly-doped surface B atoms that give rise to long-range effects on the surface such as strain.

References

1 D. Wristers, H. H. Wang, L. K. Han, C. Lin, T. S. Chen, D. L. Kwong and J. Fulfoed, *The Physics and Chemistry of SiO₂ and the SiO₂-Si Interfaces-3* (Electrochemical Society, Pennington, 1996) p. 733.

2 V. V. Korobtsov, V. G. Lifshits, and A. V. Zotov, Surf. Sci. **195**, (1988) 466.

3 T. Komeda and Y. Nishioka, Surf. Sci. **405**, (1998) 38.

4 H. Ikegami, K. Ohmori, H. Ikeda, H. Iwano, S. Zaima and Y. Yasuda, Jpn. J. Appl. Phys. **35**, (1996) 1593.

5 K. Kato and T. Uda, Phys. Rev. Lett. **80**, (1998) 2000.

6 Ph. Avouris and D. Cahill, Ultramicroscopy **42-44**, (1992) 838.

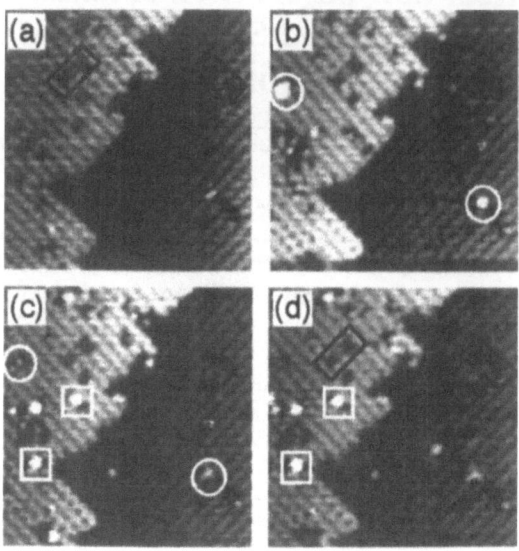

Fig. 1 STM images of (a) a clean Si(100)-2x1 surface and surfaces exposed to (b) 0.4, (c) 0.8 and (d) 1.2 L oxygen at room temperature, showing exactly the same region. 1 L is 1.3×10^{-4} Pa·s. The STM images were obtained at a sample bias of −1.5 V.

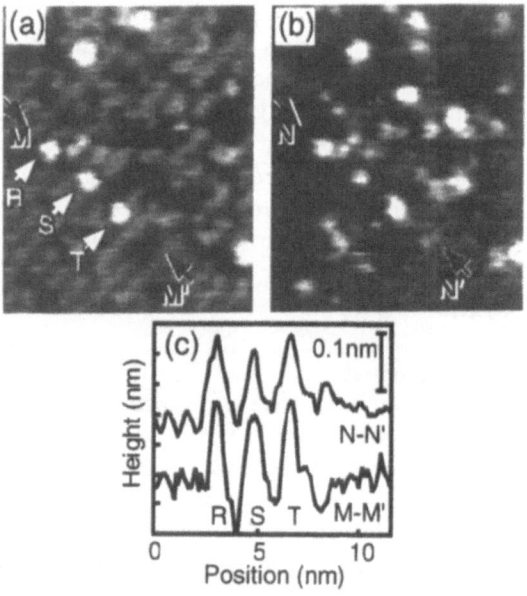

Fig. 2 (a) Occupied-state (-2.0 V) and (b) unoccupied-state (+2.0 V) STM images of the surface exposed to 7 L oxygen at room temperature. (c) Line profiles at M-M' and N-N' in Fig. 2(a) and 2(b), respectively.

Fabrication of boron delta-doped structures in Si by solid phase epitaxy

T. Ishikawa, H. Nagai, K. Ishii, and S. Matsumoto

Department of Electrical Engineering, Keio University

3-14-1, Hiyoshi, Yokohama 223-8522, Japan e-mail: matumoto@elec.keio.ac.jp

Abstract   Almost perfect $\sqrt{3}$x $\sqrt{3}$ R30° -B reconstructed surface was prepared by evaporating elemental boron on Si(111)7x7 surface, followed by annealing at 700 °C .   An ordered delta-doping structure was fabricated by growing thin Si layers on the $\sqrt{3}$x $\sqrt{3}$ R30° -B surface by solid pahse epitaxy method, keeping the periodic array of boron. In these samples, a mobility of $\sim$110 cm^2/Vs was obtained

1   Introduction

Ordered delta-doping embedded in Si consisting of a periodic array of dopant atoms in mono-layer plane has been of much interest for the possibility of new devices with lower-dimensional structures. One can also expect the reduction in ionized impurity scattering by forming a periodic array of dopant atoms.

Among many dopant atoms in Si, adsorbed boron is known to form a stable reconstructed surface with a $\sqrt{3}$x $\sqrt{3}$ R30° structure on Si (111)7x7 surface. It is very interesting to embed such an ordered doping structure in Si without breaking the order.

It has been reported that samples containing a boron delta-doping layer in Si have mobility of $\sim$20 cm^2/V s [1]. However, in this experiment, the periodic array of dopant atoms was not confirmed in an atomic level.

In this work, we investigated the almost complete formation of a $\sqrt{3}$x $\sqrt{3}$ R30° -B reconstructed structure in an atomic level on Si(111)7x7 surface and obtained   mobility of $\sim$ 100 cm^2/V s in samples embedded with   the ordered   boron monolayer in Si.

2   Experimental

N-type Si(111) wafers with resistivity of 0.7-1.3 Ω cm   were   cleaned   by   1250 °C flash-anneal cycles in UHV chamber(below 1x10^{-10} Torr) to form a clean 7x7 Si(111) surface.

Elemental boron was deposited on a clean Si (111)7x7 surface in UHV chamber at room temperature (RT), followed by annealing at 600$\sim$ 900°C. Thin Si capping layers were grown by solid phase epitaxy (SPE) method on the $\sqrt{3}$x $\sqrt{3}$ R30° B surface at annealing temperature of 500°C$\sim$700°C in UHV chamber successively.

Scanning tunneling microscopy was used to observe   the   surface   morphology.   Hall measurement was performed for electrical characteristics of the ordered delta-doped layer.

3   Results and Discussion

In the case of annealing for 3 min at 700°C immediately after B deposition, $\sqrt{3}$x $\sqrt{3}$ R30° -B surface was not formed and a lot of clusters were present on 7x7 Si(111) surface. On the other hand, by keeping the substrate at RT for 30 min after B deposition, almost complete $\sqrt{3}$x $\sqrt{3}$ R30° structures were formed at annealing temperature of 700C. Figure 1 shows the STM image of $\sqrt{3}$x $\sqrt{3}$ R30° B surface. In comparison with the previous MBE method of B deposition at an elevated temperature (900°C) [2], the solid phase epitaxy (SPE) method enables to lower the formation temperature. The present result also indicates that phase transition from 7x7 to 1x1surface, which occurs at around 830°C, is not necessarily required for the formation of $\sqrt{3}$x $\sqrt{3}$ R30° -B surface.

When a thin Si cap layer was deposited on the $\sqrt{3}$x $\sqrt{3}$ R30° -B surface, the $\sqrt{3}$x $\sqrt{3}$ R30° -B surface was observed on a top Si layer at annealing temperature of 600°C and 700°C. It indicates that segregation of B to the top layer occurred during Si capping layer growth at these annealing temperatures. On the other hand, at annealing temperature of of 500℃, it was confirmed with STM that such B segregation was not observed after the growth of Si capping layer. This means that the $\sqrt{3}x\ \sqrt{3}$ R30° -B surface was maintained at the interface between Si capping layer and the substrate.

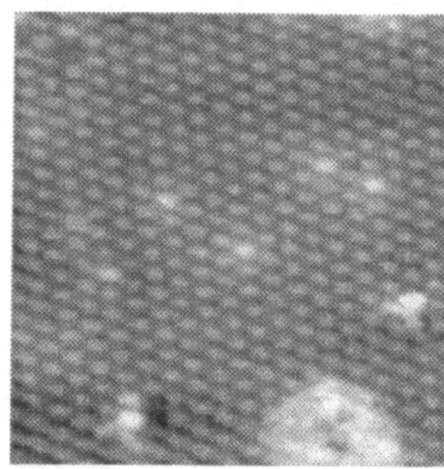

Fig.1 STM image of $\sqrt{3}x\ \sqrt{3}$ R30° -B surface annealed at 700℃ (12nmx12nm).

Figure 2 shows Hall mobility of samples with a structure of Si /$\sqrt{3}x\ \sqrt{3}$ R30° -B /Si substrate for different annealing temperatures (500, 600 and 700℃) during Si capping layer growth. The sample annealed at 500℃ has larger mobility values than those of 600 and 700℃. This is due to the fact that the $\sqrt{3}x\ \sqrt{3}$ R30° -B surface was maintained at the interface for a sample annealed at 500℃.

Figure 3 shows Hall mobility of a sample with a structure of Si capping layer/$\sqrt{3}x\ \sqrt{3}$ R30° -B /Si substrate. In Fig. 3, data of Hall mobility obtained from a sample with a structure of Si capping layer/B/Si substrate were also shown. In this sample, B was evaporated on Si(111)7x7 at RT and Si capping layer was grown sequentially. Then, $\sqrt{3}x\ \sqrt{3}$ R30° -B surface was not

formed in this case. It is found that a higher mobility of ∼110 cm²/V s is obtained for a sample with the ordered B delta-doped structure as compared with the sample of normal B delta-doped layer.

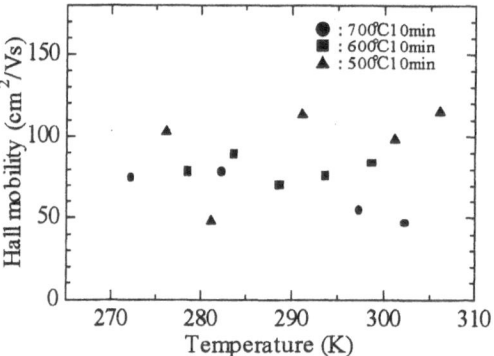

Fig.2 Annealing temperature dependence

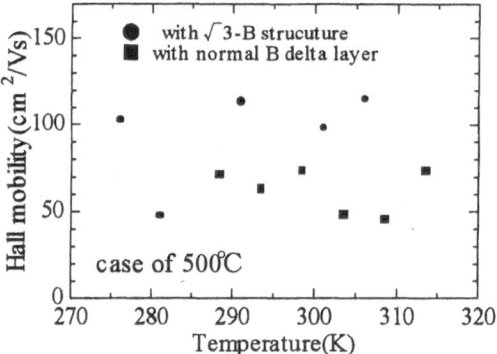

Fig.3 Hall mobitily of delta-doped layer

Considering the above experimental results, it is concluded that the ordered B dlta-doped structure can enhance the mobility by reducing the ionized impurity scattering.

References

1. B.E.Weir, L.C. Feldman, D. Monroe, H.-J. Gossmann, R.L.Headrick, and T.R. Hart, Appl. Phys. Lett. **65**, (1994)737.
2. T. Yamamoto, K. Ezoe, H. Kuriyama, K. Ishi, and S. Matsumoto, Appl. Surf. Sci., **130-132**, (1998)1.

Proc. 25th Int. Conf. Phys. Semicond., Osaka 2000 (Eds. N. Miura and T. Ando)

333

Effect of gas adsorption on the interfacial states at n-SnO$_2$/p-Si heterojunction

R.B.Vasiliev, A.M.Gaskov, M.N.Rumyanseva, and L.I.Ryabova

Moscow State University, Chemistry Department, Moscow 119899, Russia e-mail: **mila@mig.phys.msu.su**

Abstract n-SnO$_2$/p-Si and n-SnO$_2$(Me)/p-Si (Me: Cu, Pd, Ni) heterostructures were prepared by an aerosol pyrolysis of SnO$_2$ layers on single crystalline p-Si. To explain the current-voltage (*I-V*) and capacitance-voltage (*C-V*) curves behaviour a complicated band modulation at the interface region and the interface state (IS) contribution were regarded. The IS density is significantly modified by gas adsorption. NO$_2$ adsorption leads to the reduction of the IS contribution to electric parameters while C$_2$H$_5$OH adsorption results in the IS density increase. Sensitivities of heterostructures to adsorbed gas molecules are found to be bias dependent.

1 Introduction Electrical properties of semiconductor heterojunctions depend significantly on the IS character. Depletion layer depth, excessive currents in tunnel diodes, dipole layers formation, frequency dependent capacitance are among the factors affected by the IS. Therefore the IS existence leads to modification of *I-V* and *C-V* characteristics of tunnel and Schottky diodes. Metal oxide film being one of the elements of the heterojunction at the same time should be a porous membrane allowing gas molecules to reach the interface region and modify the IS density.

2 Investigated samples and experimental technique Nanocrystalline SnO$_2$ and SnO$_2$(Me) films of 1 μm thickness were grown by pyrolysis of an aerosol on a <100> single crystal p-Si substrate [1]. Microstructure of the films was controlled by scanning electron microscopy and X-Ray diffraction. The grain size of SnO$_2$ layer was about 10 nm. Au contact with the size less than the SnO$_2$ film area was deposited on the films surface and Al contact was created at the back side of the substrate. The ohmic behavior of the contacts was proved [2]. Static *I-V* and *C-V* characteristics at the frequencies of the reference signal *f* from 500 Hz up to 20 kHz were measured at the bias variation from − 3V to 3V, in the temperature range 180-300 K.

3 Experimental results and discussion Main features of the *I-V* curves for p-SnO$_2$/n-Si (see Fig. 1) and n-SnO$_2$(Me)/p-Si heterostructures in air may be summarised as follows. The reverse branches of *I-V* curves are characterised by a prolonged saturation

region. Temperature dependence of the reverse current at a fixed bias corresponds to a thermally activation process with an energy of about 0.3 eV. Forward

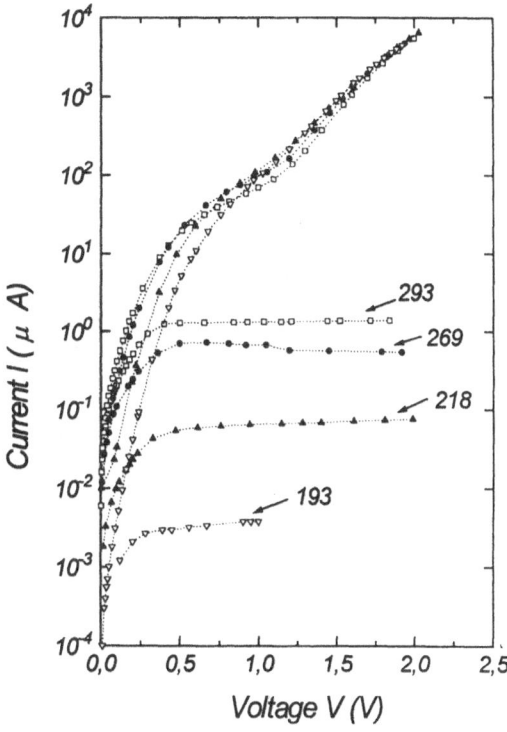

Fig. 1 *I-V* curves of SnO$_2$/Si heterostructure. Figures at the curves show temperature *T* (K). Arrows point at the reverse branches of the curves.

branches are non-linear in semi-logarithmic plot revealing a narrow saturation region at *V* ~ 1 V followed by the exponential growth of current value at *V*>1 V. At high forward bias current is nearly thermally independent. To explain the experimental data we suggest that the barrier at SnO$_2$/Si interface is composed of a high narrow barrier partially transparent for tunnel processes and a wide but lower barrier responsible for current saturation in the reverse bias region. Therefore current increase at reverse bias is limited by the Schottky barrier while the increase of the forward bias is limited by tunnel current. Since no tunnel current is observed at the reverse bias region the idea of interface states dominating role in the tunnel process seems preferable compared to tunnelling induced by local levels in the energy gap of semiconductors.

The character of *C-V* curves taken at different frequencies of the reference signal *f* supports this suggestion (see Fig.2a). The contribution of the IS to the capacitance value reduces with *f* increase since the characteristic time of the IS recharge becomes lower than $1/2\pi f$.

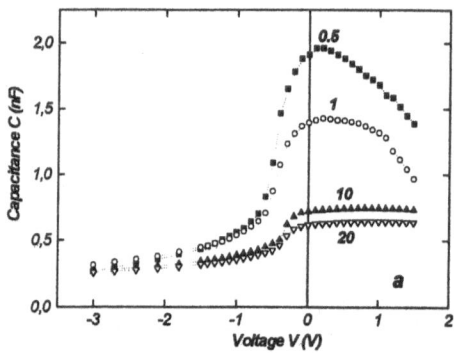

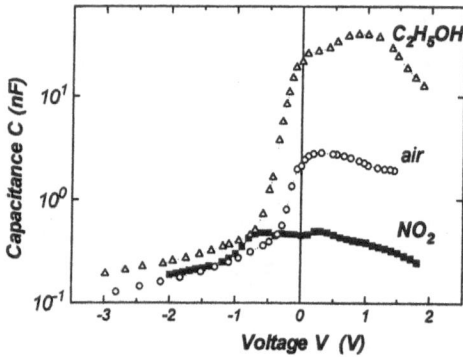

Fig. 2 *C-V* curves for SnO₂(Ni)/Si heterostructure; **a** -taken at different frequencies *f* of the reference signal. Figures at the curves - *f*, kHz. **b** - taken at *f*=1 kHz in air and gas mixtures N₂+10⁴ ppm C₂H₅OH and N₂+10³ ppm NO₂. T=300 K.

Modification of *I-V* and *C-V* curves induced by NO₂ and C₂H₅OH molecules adsorption is similar in main features for all investigated samples, but the amplitude of the response is maximal for Ni doped structure. The effect C₂H₅OH adsorption results in the reduction of the reverse current value and the rise of forward current. The effect of NO₂ is quite an opposite. So it may be concluded that adsorption

affects both the Schottky barrier height and probability of tunnelling. For *C-V* curves the mostly pronounced effect of gas adsorption is observed at a forward bias region. NO₂ adsorption leads to the reduction of the capacitance at the forward bias while C₂H₅OH induces the increase of *C* in the whole bias range (Fig. 2b). The behaviour of *I-V* and *C-V* curves are in good accordance. Reduction (NO₂) or increase (C₂H₅OH) of the tunnel current is accompanied by respective decrease and rise of the capacitance values proving the interfacial states contribution.

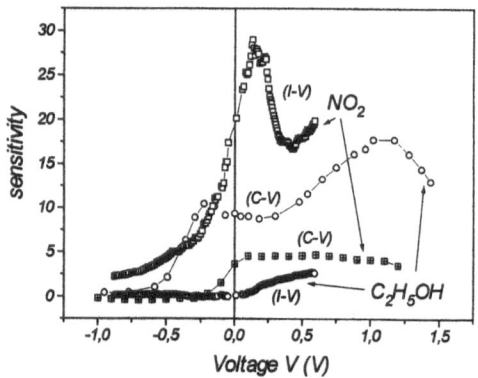

Fig. 3 Bias dependence of sensitivities *S* to ethanol and NO₂ of SnO₂(Ni)/Si heterostructure calculated from *I-V* and *C-V* data. T=300 K.

Bias dependence of sensitivity *S* (the ratio of the signal change under gas adsorption to signal value) is characterised by pronounced peaks for *I-V* data obtained under NO₂ adsorption and for *C-V* data under C₂H₅OH adsorption (Fig.3). The peak of *I-V* sensitivity to NO₂ corresponds to the reduction of tunnel current, while the peak of *C-V* sensitivity to C₂H₅OH corresponds to maximal contribution of interfacial states to the capacitance value. The bias dependence of sensitivity allows to choose the mostly preferable regimes for operation of the heterostructure as a gas sensor.

Acknowledgment This work was done under a partial financial support of RFBR under Projects N 00-03-32083a and N 98-03-32843a.

References

1. M.N.Rumyantseva, A.M.Gaskov, L.I.Ryabova, J.P.Senateur, B.Chenevier, and M.Labeau, Mater.Sci.Eng.B **41**, (1996) 333.
2. B.A.Akimov B A, A.M.Gaskov, S.E.Podguzova, M.N.Rumyantseva, L.I.Ryabova, M.Labeau and A.Tadeev, Semiconductors **33**, (1999) 175.

Structure of a SiN layer on Si(111) surface

Takahiro Yamasaki[1], Chioko Kaneta[1], and Tsuyoshi Uda[2]

[1] Fujitsu Laboratories Ltd. 10-1 Morinosato-Wakamiya, Atsugi 243-0197, Japan
[2] Joint Research Center for Atom Technology, Angstrom Technology Partnership 1-1-4 Higashi, Tsukuba 305-0046, Japan

Abstract   Structure of a silicon-nitride monolayer on the Si(111) surface was investigated by using a first principles molecular dynamics method based on the density functional theory. Most of the silicon atoms in the silicon-nitride layer displace vertically downward or upward and have sp^3 bonding nature. The majority of the downward protruding silicon atoms have covalent bonds with underlying silicon atoms of the Si(111). The other downward and all upward silicon atoms have dangling bonds. A small number of silicon atoms possibly have sp^2 bonding nature according to the relative position of the silicon-nitride layer to the underling Si(111). Almost all of the nitrogen atoms have planer sp^2 bonding nature, and their local density of states are similar regardless of the relative positions to the underlying Si(111). It can be concluded that $8/3 \times 8/3$ and $3\sqrt{3}/4 \times 3\sqrt{3}/4$ superstructures observed by scanning tunneling microscopy are caused by the upward protruding silicon atoms.

1 Introduction

Nitridation of the Si surface is one of the most important issues in silicon device technology. Initial reaction layers on the silicon surfaces has been studied experimentally with various techniques. The SiN layer formed by the nitridation of the Si(111) surface shows two kinds of phases: '8×8' and '3/4×3/4', when it is observed by low-energy electron diffraction (LEED) or electron energy-loss spectroscopy (EELS) [1–5]. The '8×8' phase has an $8/11 \times 8/11$ triangular lattice (a = 2.79 A) with respect to the Si(111)-1×1 lattice, and the '3/4 × 3/4' phase has a 3/4 × 3/4 triangular lattice with a rotation of $\pm 5°$ and $\pm 10°$ with respect to the substrate Si(111)-1×1 lattice. And '8×8' and '3/4×3/4' phases show super-structures of 8/3 × 8/3 and $3\sqrt{3}/4 \times 3\sqrt{3}/4$, respectively, when they are observed by scanning tunneling microscopy (STM) [6–8]. Models have been proposed for the two phases. Nishijima et al. proposed a Si_3N triangular pyramid model[3,5] and a planner quadruplet SiN model[3–5] for '8×8' and '3/4×3/4', respectively. Morita et al. proposed a two domain (in-phase and out-of-phase) model[7] and a hexagonal pyramid model[8] for these two phases, respectively. However, they are not definitive yet. For the purpose of evaluation of those models, we have first optimized structure of the SiN layer on Si(111) surface by using a first principles molecular dynamics method.

2 Methodology and Models

The present calculation is based on the density functional theory and employs pseudopotentials (PP's) (a norm-conserving PP for Si and ultrasoft Vanderbilt PP's [9] for N and H) and plane wave basis. We add generalized gradient correction [10] to the local density approximation for the exchange-correlation potential. Spin polarization is taken into account.

Our calculation model is SiN-4×4 on Si(111)-3×3, where the initial structure of the SiN layer is just what was proposed by Nishijima et al.[3,5] for the '8×8' phase and the SiN lattice size is expanded by +3.5% from the experimental value of 2.79 Å of the '8×8' phase. The substrate Si(111) is modeled with a slab of 6 Si layers and hydrogen atoms terminating the backsurface Si dangling bonds. The unit cell contains 95 atoms in all. In the configuration optimization process, the hydrogen atoms and bottom two Si layers are fixed, and other atomic positions are relaxed. The SiN layer have in-phase and out-of-phase regions, where in the in-phase region topmost Si atoms of the underlying Si(111) can make covalent bonds with Si atoms of the SiN layer without position shifts from their initial ones. The pattern of the in-phase and out-of-phase regions changes according to the lateral shift of the SiN layer to the substrate Si(111) surface. We have prepared two initial structures, where the difference is lateral positions of the SiN layers relative to the substrate: (case-1) one Si atom of the SiN layer is just above a topmost Si atom of the substrate; (case-2) the lateral position of the SiN layer is adjusted as the number of covalant bonds between the Si atoms of the SiN layer and those of the substrate would become to be maximum.

3 Results and Discussion

3.1 Optimized Structures

After optimization, all dangling bonds of the substrate topmost Si atoms disappeared and they were replaced by covalent bonds with the Si atoms of the SiN layer (a few of them are weaker). Even in the out-of-phase domain, dangling bonds of the underlying substrate Si atoms also disappeared by making weak covalent bonds with the Si atoms in the SiN layer accompanied with upward shift of the substrate Si atoms. The optimized structure of the case-2 is more stable than that of the case-1 by 2.3 eV per unit cell. The SiN framework is maintained and the lateral positions of the atoms do not change so much. However, vertical positions of the Si atoms were shifted remarkably from their initial ones. Fig. 1 shows topviews of the optimized structures, where upward and planer Si atoms are labeled with 'U' and 'P', respectively.

Graphs in Fig. 2 show distributions of protrusion of the atoms in the SiN layer for the two cases. The majority of the Si atoms protrude downward. Most of

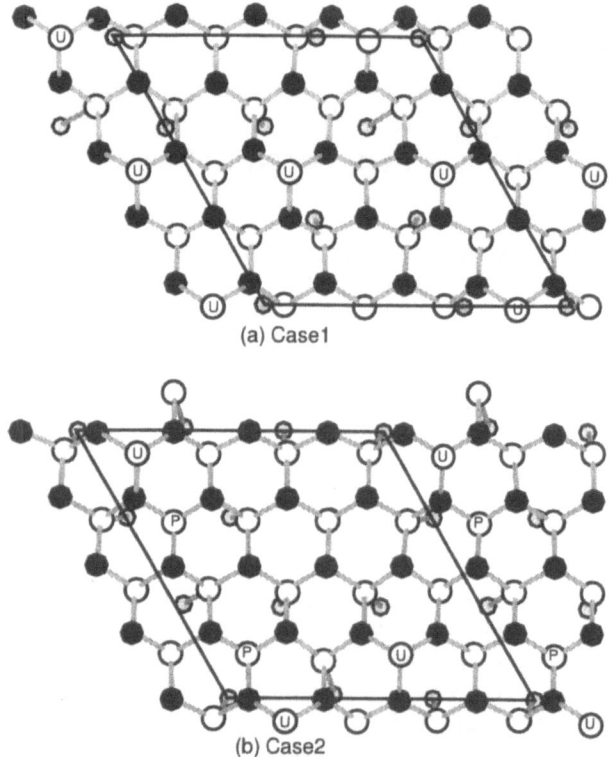

Fig. 1 Topviews of the optimized structures for (a) case-1 and (b) case-2. Small shaded, large open, and large closed circles denote the substrate topmost Si, the Si in the SiN layer, and the N atoms in the SiN layer, respectively. Unit cells of the Si(111)-3×3 are also shown on the topviews.

them make covalent bonds with the substrate topmost Si atoms. The other downward and all upward Si atoms in the SiN layer have dangling bonds. They all have sp^3 bonding nature. A few Si atoms in the case-2 have sp^2 bonding nature. The N atoms show planer sp^2 bonding nature, except one in the case-1. They have similar local density of states regardless of the lateral positions relative to the underlying Si(111).

3.2 Comparison with Proposed Models

Most of the proposed models so far assume that vertical positions of the Si atoms in the SiN layer are the same, except the hexagonal pyramid model[8] proposed to explain the $3\sqrt{3}/4 \times 3\sqrt{3}/4$ superstructure. However, our result show that the Si vertical shifts depend on their lateral positions relative to the substrate Si(111). Some of the Si atoms protrude upward with having dangling bonds, and will be detectable to STM experiments. Morita's model of hexagonal pyramid is a reasonable one. On the other hands, models proposed for '8×8' need to be revised.

4 Conclusion

In the SiN layer on the Si(111) substrate, most of the Si atoms have sp^3 bonding nature and protrude not only downward but also upward. The superstructures like 8/3×8/3 and $3\sqrt{3}/4\times3\sqrt{3}/4$ observed by STM ex-

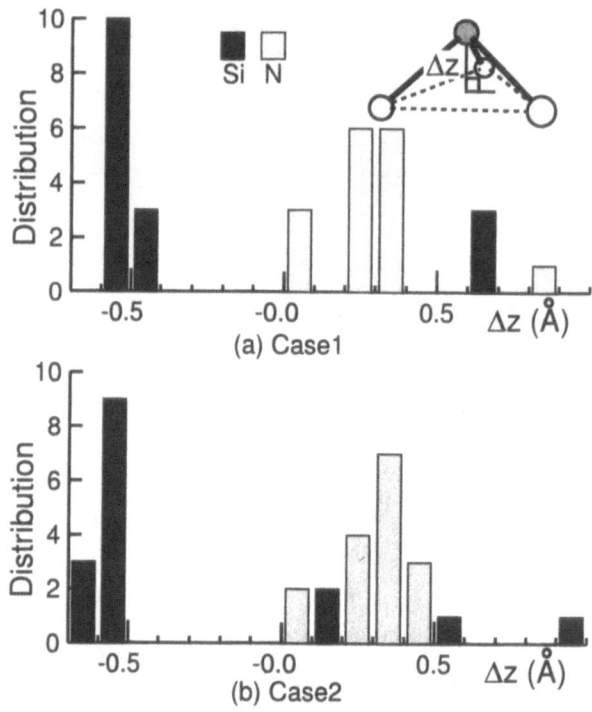

Fig. 2 Distributions of protrusion of the Si and the N atoms in the SiN layer for the case-1 and the case-2. The protrusion is defined by the distance between an object Si (N) atom and a plane made with the surrounding three N (Si) atoms. Positive (negative) value means protrusion upward (downward).

periments are thought to be reflections of those upward protruding Si atoms. On the other hand, most of N atoms in the SiN layer have planer sp^2 bonding nature and their local electronic states are similar each other. Dangling bonds of the substrate topmost Si atoms, even in the out-of-phase domain, disappear and are replaced by covalent bonds with the downward protruding Si atoms of the SiN layer.

Acknowledgement

This work was partly supported by the New Energy and Industrial Technology Development Organization (NEDO).

References

1. A. G. Schrott, J. S. C. Fain, Surf. Sci. **111** (1981) 39.
2. A. G. Schrott, J. S. C. Fain, Surf. Sci. **123** (1982) 204.
3. M. Nishijima, H. Kobayashi, K. Edomoto, M. Onchi, Surf. Sci. **137** (1984) 43.
4. M. Nishijima, K. Edamoto, Y. Kubota, H. Kobayashi, M. Onchi, Surf. Sci. **158** (1985) 422.
5. K. Edamoto, S. Tanaka, M. Onchi, M. Nishijima, Surf. Sci. **167** (1986) 285.
6. E. Bauer, Y. Wei, T. Muller, A. Pavlovska, I. S. T. Tsong, Phys. Rev. B **51** (1995) 17891
7. Y. Morita, H. Tokumoto, Surf. Sci. **443** (1999) L1037.
8. Y. Morita, H. Tokumoto, Surf. Sci. accepted.
9. D. Vanderbilt, Phys. Rev. **B41**, 7892 (1990).
10. J. P. Perdew, in *Electronic Structure of Solids '91*, edited by P. Ziesche and H. Eschrig (Akademie Verlag, Berlin, 1991)

Distinct morphological evolution of $Si_{1-x}Ge_x$ films on Si(100) during gas-source MBE and photo-CVD

H. Akazawa

NTT Telecommunications Energy Laboratories, 3-1 Morinosato Wakamiya, Atsugi, Kanagawa 243-0198, Japan e-mail:
akazawa@aecl.ntt.co.jp

Abstract When $Si_{1-x}Ge_x$ films are grown on Si(100) surfaces by gas-source molecular beam epitaxy, the transition from two-dimensional to Stranski-Krastanov mode proceeds through gradually undulating the surface and the morphology of dislocation-free $Si_{1-x}Ge_x$ films settles at a long-wavelength sinusoidal-like ripple structure. At higher Ge content, the ripple structure is converted to cusp-shape islands with introducing dislocations. When $Si_{1-x}Ge_x$ films are grown at low temperatures by photochemical vapor deposition, uniform films with numerous dislocations are obtained. Above 400℃, however, the increased mobility of Ge adatoms and Si monohydrides on H-terminated surfaces results in significant initial islanding. The grain boundaries between islands that form as a result of Volmer-Weber growth completely relax the strains, leading to early coalescence.

1 Introduction

When $Si_{1-x}Ge_x$ films are grown by gas-source molecular beam epitaxy (GSMBE) or by solid-source MBE, the growth mode can, depending on the conditions, be either two-dimensional (2D) or Stranski-Krastanov (SK). Although various types of undulation and island structures have been reported [1-4], there have been very few studies on real-time monitoring the phase transition between specific structures [5]. To clarify the stepwise structural transition from 2D to islands through undulation during GSMBE, we used cross-sectional transmission electron microscopy (XTEM) and in-situ spectroscopic ellipsometry (SE), which are sensitive to the evolution of the surface morphology. And during photochemical vapor deposition (photo-CVD) we observed a distinct Volmer-Weber (VW) growth mode, which is kinetically controlled by the presence of H atoms.

2 Experimental

$Si_{1-x}Ge_x$ films were grown in a UHV-CVD chamber connected to the compact electron storage ring "Super-ALIS" [6]. The energy of photons exciting the gas and substrate in photo-CVD was between 10 and 1500 eV, and the maximum photon flux was at 100 eV. After a Si(100) wafer was pretreated in a 2.5% HF solution, a Si buffer layer was grown on it at 700℃ by Si_2H_6-GSMBE. A $Si_{1-x}Ge_x$ overlayer was then grown, either by GSMBE or photo-CVD, at various temperatures and various rates of GeH_4 and Si_2H_6 flow. The ellipsometry measures the Fresnel reflection coefficient ratio ρ between light p-polarized and s-polarized with respect to the solid surface, and the ellipsometric angles ψ and Δ are defined by $\rho = R_p/R_s = \tan\psi \exp(i\Delta)$.

3 2D-SK transition during GSMBE of $Si_{1-x}Ge_x$ films

$\psi - \Delta$ trajectories observed at 3.4 eV during GSMBE of $Si_{1-x}Ge_x$ films are plotted in Fig. 1. The rapidly converging one-turn spiral O → A observed at 600℃ (open circles) can be simulated by assuming deposition of a 2D film with a Ge content of 0.35. The trajectory at lower Si_2H_6 flow rate (solid circles) is connected smoothly to an another small branch before converging at the destination point of the 2D film (trace A → B), which indicates the onset of three-dimensional (3D) nucleation. The convergence in the second branch indicates that an equilibrium surface morphology is reached while the thickness of the 2D base layer increases. The XTEM image in Fig. 2(a) shows that the surface of the $Si_{0.60}Ge_{0.40}$ film is slightly undulated, exhibiting a long-wavelength sinusoidal-like ripple [1] without dislocations. The 2D-SK transition is therefore the second kind. Misfit strains are relieved by strain modulation at the cost of surface energy.

Traces O → P → C and O → Q → D at 700℃ correspond to the genuine SK growth. The trace O → Q deviates from the 2D spiral at around point K, indicating the prevailing rippling at an early stage. Clear inflection points P and Q appear when the ripple is converted to cusp-shape islands consisting of high-index facet planes at trenches between islands [2] (see Fig. 2(b)). The second branch from the critical point extends to smaller ψ and Δ regions and 3D morphology is sustained. Because strains are relieved only partially by engraving trenches associated with introducing dislocations [3], coalescence occurs later. At higher Ge content, dome-shape islands were observed.

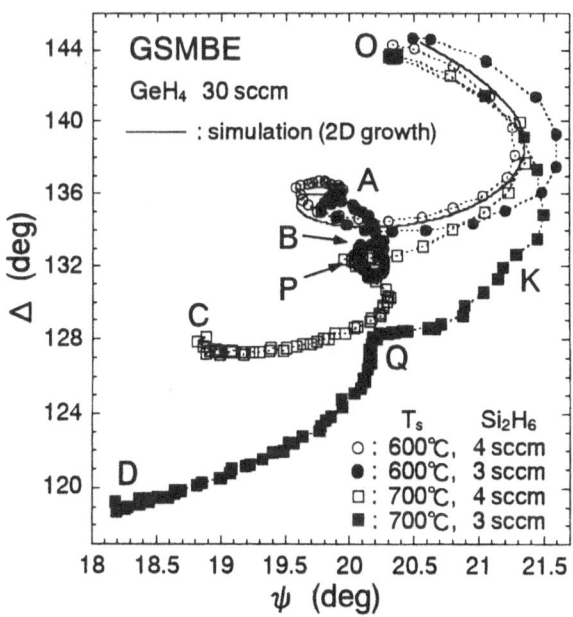

Fig. 1 $\psi - \Delta$ trajectories at 3.4 eV during GSMBE.

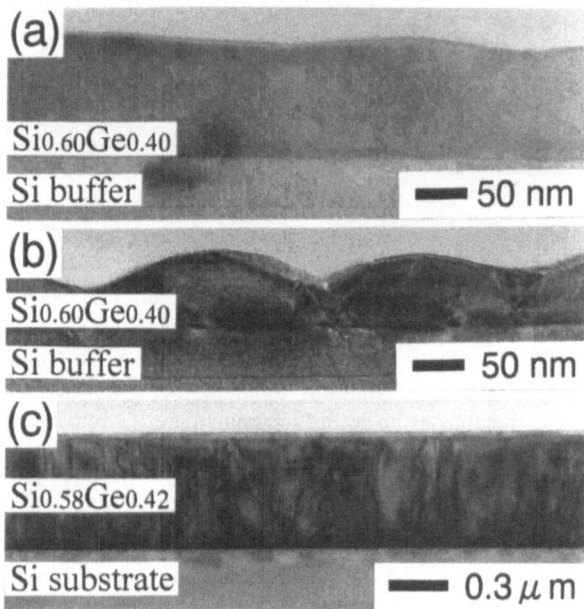

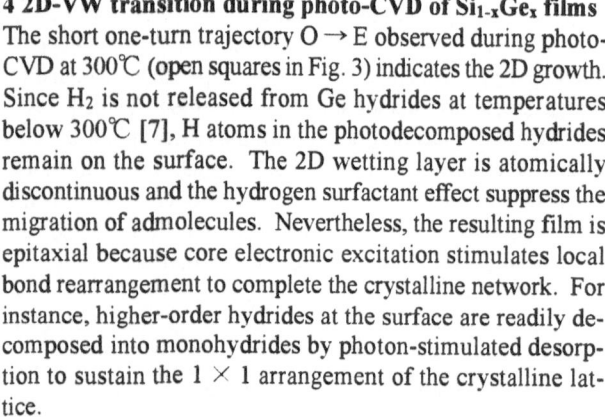

Fig. 2 XTEM images of (a) rippled and (b) cusp-shape islanded Si$_{0.60}$Ge$_{0.40}$ films grown by GSMBE at 600℃ and 700℃, respectively, and (c) a Si$_{0.58}$Ge$_{0.42}$ film grown by photo-CVD at 450℃.

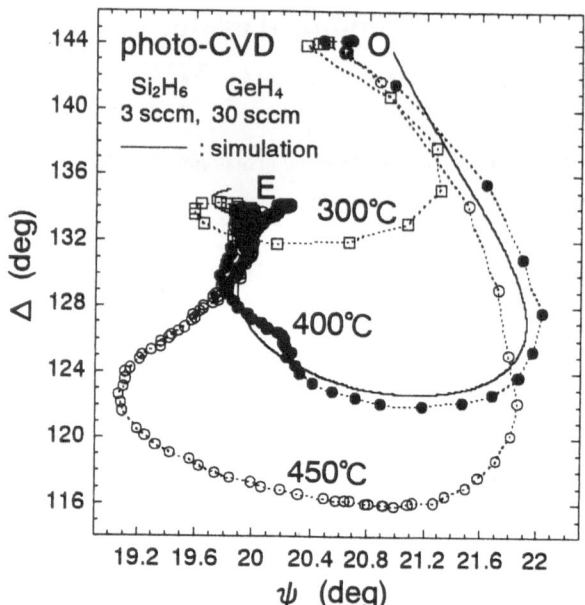

Fig. 3 ψ - Δ trajectories at 3.4 eV during photo-CVD.

4 2D-VW transition during photo-CVD of Si$_{1-x}$Ge$_x$ films

The short one-turn trajectory O → E observed during photo-CVD at 300℃ (open squares in Fig. 3) indicates the 2D growth. Since H$_2$ is not released from Ge hydrides at temperatures below 300℃ [7], H atoms in the photodecomposed hydrides remain on the surface. The 2D wetting layer is atomically discontinuous and the hydrogen surfactant effect suppress the migration of admolecules. Nevertheless, the resulting film is epitaxial because core electronic excitation stimulates local bond rearrangement to complete the crystalline network. For instance, higher-order hydrides at the surface are readily decomposed into monohydrides by photon-stimulated desorption to sustain the 1 × 1 arrangement of the crystalline lattice.

When the films were grown at 400℃ or 450℃, the initial trajectories extended rapidly to the region with smaller Δ angles. At these temperatures most Si adatoms are in the form of monohydrides and Ge adatoms are free of H atoms. As the number of H atoms decreases, the mobility of admolecules is enhanced, but the nucleation sites are restricted due to hydrogen termination. The reduction in the Δ angle reflects the development of Si$_{1-x}$Ge$_x$ islands at dangling-bond-terminated sites on Si(100); that is, it reflects the VW growth mode. The height of the islands depends on the adatom mobility, which is controlled by the number of H atoms. After reaching the lowest Δ region, the trajectories directed upward and the end point coincided with that of the 2D trajectory. When the misfit strains are fully relieved by the formation of grain boundaries between islands, migration on faceted surface is energetically disadvantageous and coalescence recovers a flat surface as seen in Fig. 2(c).

Because Si$_{0.50}$Ge$_{0.50}$ film grown at 400℃ is optically thick

at 3.4 eV, the (ψ , Δ) angles primarily reflect the dielectric property of the near-surface layer. The dielectric response change can be modeled by deposition of a void-containing graded film, in which the dielectric function of each slab layer is given by mixing the component dielectric function of Si$_{0.50}$Ge$_{0.50}$ and voids under Bruggeman effective medium approximation. The void volume fraction f$_v$ is assumed to change with time as follows: f$_v$=f$_{v0}$-R$_v$t ($0 \leqq t \leqq$ f$_{v0}$/R$_v$) and f$_v$=0 (t > f$_{v0}$/R$_v$), where f$_{v0}$ is the initial void volume fraction, and the film thickness increases lineraly with time. As shown by the solid line in Fig. 3, the experimental ψ - Δ trajectory at 400℃ is close to the simulation with f$_{v0}$=0.5. That only 50% of the Si(100) surface is initially covered by the wetting layer confirms the VW growth mode.

References

1. A.G. Cullis, D.J. Robbins, S.J. Barnett, and A.J. Pidduck, J. Vac. Sci. Technol. **A12**, (1994) 1924.
2. D.E. Jesson, S.J. Pennycook, J.-M. Baribeau, and D.C. Houghton, Phys. Rev. Lett. **71**, (1993) 1744.
3. J. Tersoff and F.K. Legoues, Phys. Rev. Lett. **72**, (1994) 3570.
4. N.-E. Lee, D.G. Cahill, and J.E. Greene, J. Appl. Phys. **80**, (1996) 2199.
5. C. Pickering, R.T. Carline, D.J. Robbins, W.Y. Leong, D.E. Gray, and R. Greef, Thin Solid Films **233**, (1993) 126.
6. H. Akazawa, Phys. Rev. B **59**, (1999) 3184.
7. H. Akazawa, J. Appl. Phys. **79**, (1996) 9396.
8. M. Copel and R.M. Tromp, Appl. Phys. Lett. **58**, (1991) 2648.

Toward Predictive Growth Simulations: MBE on GaAs(001)

P. Kratzer[1], M. Scheffler[1]

Fritz-Haber-Institut der Max-Planck-Gesellschaft, Faradayweg 4-6, D-14195 Berlin, Germany

Abstract   Using the results of accurate density-functional theory (DFT) calculations for molecular processes on the $\beta 2$-reconstructed GaAs(001) surface, including adsorption, desorption, surface diffusion and nucleation, we show how the MBE growth of GaAs can be modeled in atomistic detail. We perform kinetic Monte Carlo (kMC) simulations which take into account the surface reconstruction and include the rates of more than 30 molecular processes calculated from DFT. We find that film growth starts with an unusually high nucleation density, and proceeds by coalescence of small, two-dimensional islands elongated along the $[\bar{1}10]$ direction.

1 Introduction

When modeling epitaxial growth, the properties of interest (e.g. the growth morphology) develop over time scales of the order of seconds and involve several thousands of atoms, while the ruling microscopic processes operate in the length and time domains of 0.1 - 1 nm, and femto- to pico-seconds. Hence, we are challenged by the need to bridge this gap in the length and time scales by many orders of magnitude. Kinetic Monte Carlo (kMC) simulations offer an efficient and accurate way to cope with this difficulty. Until recently, kMC simulations usually employed empirically derived parameters, and often only one effective species. For this reason, simulations of this kind, though useful, lack a microscopic understanding, and are not predictive. In this situation it appears appealing to derive the rates entering in a kMC simulation directly from density-functional theory (DFT) calculations that are parameter-free and have predictive power.

2 The physical chemistry of the GaAs(001) surface

In an approach that combines the strengths of both the DFT and kMC methodology, we model the molecular beam epitaxy of GaAs, a system which is experimentally well studied. It is our goal to simulate the atomistic processes of island growth including all relevant microscopic details, such as the surface reconstruction, different kinetic properties of the species involved, etc., based on the understanding we obtain from first-principles calculations of the energetics. Specifically, we model diffusion, island nucleation and growth on the well-known $\beta 2(2 \times 4)$ reconstruction of GaAs(001) that prevails under frequently used growth conditions. This As-rich surface reconstruction is terminated by pairs of As dimers, which alternate with "trenches" (two As dimers missing) running in the $[\bar{1}10]$ direction.

Both experimental knowledge and previous theoretical studies agree that arsenic and gallium behave quite differently on this surface. While Ga atoms adsorb with unit sticking probability, As_2 molecules only stick to the surface after Ga has been deposited [1,2]. Previous calculations using DFT have shown that the As_2 molecule does not break its strong molecular bond easily, but instead prefers to get incorporated without full dissociation [3]. On the (001)-$\beta 2$ surface, dissociation is indeed not necessary for an adsorption pathway where a gas-phase As_2 attaches to two Ga adatoms in the trench and adds itself to the surface As dimers of the $\beta 2$ reconstruction. Consequently, we concentrate in the present model on this low-energy pathway to arsenic incorporation. As another consequence of the calculations, it appears that the mobility of arsenic on the surface is not mainly due to diffusion of As adatoms, but originates from As_2 in a mobile molecular precursor state [3]. This conclusion is corroborated by earlier empirical modeling of experimental results where such a precursor had been postulated [4]. Furthermore, the calculated adsorption energies of As_2 [3] show that strong chemisorption occurs at specific Ga-rich sites, corroborating the experimental findings that pre-adsorption of Ga is required before incorporation of arsenic can occur. Since there is only a limited number of sites available for As_2 chemisorption, the molecular precursor is also essential for having sufficient As incorporation during growth by enhancing the local concentration of As close to the surface. We account for this effect of the precursor in our model by working with an effective As_2 flux that is a factor of 100 higher than the external flux from the As_2 source.

For the Ga species, on the other hand, the mobility is completely determined by surface diffusion. The potential energy surface for Ga diffusion [5,6] was been studied previously using first principles calculations. Most importantly, the strongest adsorption sites for Ga were found within As surface dimers [6]. We find that Ga mobility is essential for growth because it is necessary to have three or four Ga atoms in neighboring sites before sufficiently strong binding sites for As_2 adsorption in the new layer become available. The role of the Ga atoms in growth is to break up existing As surface dimers, and the inserted Ga atoms, but also small Ga clusters, subsequently act as selective sites for further adsorption of As_2. For instance, bound pairs of Ga adatoms may originate either from dimerization of Ga atoms or from an indirect interaction of Ga atoms in the trenches that is mediated by the substrate reconstruction [7].

3 Kinetic Monte Carlo simulations

In order to understand how the morphology of a growing film is governed by atomistic processes, it is important to understand the interplay of the various processes and their relative importance, which in turn may dependent on the growth conditions. To this end, we have performed kinetic Monte Carlo simulations. This technique yields stochastic solutions of the master equations that lie on the basis of a microscopic description within the framework of non-equilibrium thermodynamics. In our simulations, time evolution proceeds by discontinuous changes of the occupation of discrete lattice sites. These events may be either adsorption or desorption of atoms and molecules, or the hopping of an atom from one site to another. Each event is characterized by a rate, which we determine in a parameter-free approach using the energetics obtained from our DFT calculations. Within transition-state theory, the rate law has the well-known Arrhenius form, with the activation energy in the exponent given by the energy difference between the initial state and the transition state. The prefactors, though in principle also accessible by DFT calculations, are all set to a common value of $10^{13}\,\mathrm{s}^{-1}$. This value is justified by empirical knowledge, and by the limited variation in prefactors reported earlier for systems this high activation barriers [6].

In this paper, we concentrate on the morphology of two-dimensional islands forming after sub-monolayer deposition, where we can compare to detailed experimental studies using scanning tunneling microscopy. The simulations reveal that the island growth is governed by the interplay of processes that occur on a hierarchy of time scales. In the following, we illustrate the situation for deposition at a surface temperature of 800 K. The shortest times scales ($10^{-12}\,\mathrm{s} - 10^{-9}\,\mathrm{s}$) are set by the hopping diffusion of Ga adatoms. Since hopping in [$\bar{1}10$] direction along the trenches is faster than hopping from one trench to another in [110] direction, Ga diffusion on the $\beta2(2 \times 4)$GaAs(001) surface is anisotropic. The terminating As dimers of the $\beta2$ reconstruction, in particular those in the trenches, can act as traps for diffusing Ga atoms, where they are immobilized for 10^{-8} s. At these sites, where a Ga atom has inserted itself into an As dimer in the trench, gas-phase As_2 molecules can adsorb readily, thereby estabilshing three bonds to surface Ga atoms. At temperatures below 800 K, the complexes of one Ga adatom plus one As ad-dimer are stable for a sufficiently long time (0.1 s) so that they can react with another diffusing Ga adatom, which will further increase their stability. Analysing the balance of adsorbing and desorbing As_2 fluxes shows that the transiently formed $Ga-As_2$ species is important for growth since it acts as a "doorway" state for the incorporation of As_2. As a side effect, the sticking probability of As_2 molecules is first order in the Ga coverage, a finding that is in agreement with the experimentally observed reaction order [1]. Eventually, the desorption of As_2 from groups of three As surface dimers (β reconstruction) sets the largest timescale in the problem (~ 100s).

4 Conclusions

The kMC simulations enable us to aquire an understanding of island growth on the level of atomistic processes, i.e. to analyse the relative importance of competing processes and to study their effect on the growth morphology, for instance on the saturated island density during stationary growth.

The competition of processes can be illustrated by the following example: The DFT calculations show that the strongest interaction between Ga adatoms is reached when they form a Ga dimer, oriented perpendicularly to the As dimers in the top layer [7]. Two such Ga dimers in adjacent sites can act as a strong adsorption site for As_2. Thus there is a competition between two possible growth mechanisms, where the first As_2 incorporation either occurs in the incomplete surface layer of the $\beta2$ structure, or alternatively in a new added layer. Our simulations show that at 800 K it is the first process, adsorption of an As_2 onto a Ga adatom in the trench, that initiates island growth. However, growth quickly extends into the new layer in those surface regions where material has already been incorporated in the trenches.

As another result, the simulations show that a steady-state growth regime with a constant island density is reached after deposition of 0.1–0.15 ML of GaAs. For growth at 800 K, we find a saturated island density of about 8000 μm^{-2}, in satisfactory agreement with experimental results ($3400 - 6600\ \mu m^{-2}$) obtained by analysing STM images of samples that were quenched to room temperature immediately after deposition [8,9]. We note that this nucleation density is an order of magnitude higher than what would be expected from standard nucleation theory. In its simplest version, the island density would scale with $(F/D)^{1/3}$, where F is the deposited flux of Ga atoms, and D is the diffusion constant of the Ga adatoms. In contrast to this prediction, but in agreement with our simulations, a homoepitaxial film of GaAs evolves by coalescence of an unusually large number of small islands elongated along the [$\bar{1}10$] direction.

References

1. C. T. Foxon and B. A. Joyce, Surf. Sci. **64**, 293 (1977).
2. E. S. Tok, J. H. Neave, J. Zhang, B. A. Joyce, and T. S. Jones, Surf. Sci. **374**, 397 (1997).
3. C. G. Morgan, P. Kratzer, and M. Scheffler, Phys. Rev. Lett. **82**, 4886 (1999).
4. M. Itoh et al., Phys. Rev. Lett. **81**, 633 (1998).
5. K. Shiraishi, Thin Solids Films **272**, 345 (1996).
6. A. Kley, P. Ruggerone, and M. Scheffler, Phys. Rev. Lett. **79**, 5278 (1997).
7. P. Kratzer, C. G. Morgan, and M. Scheffler, Phys. Rev. B **59**, 15246 (1999).
8. G. R. Bell, M. Itoh, T. S. Jones, and B. A. Joyce, Surf. Sci. **423**, L280 (1999).
9. B. Joyce et al., J. Cryst. Growth **201/202**, 106 (1999).

Dual Au-Si bonding character and Au surfactant effect on Au/Si(111) surfaces

M. Murayama[1], T. Nakayama[2], A. Natori[1]

[1] Department of Electronics Engineering, The University of Electro-Communications, 1 − 5 − 1 Chofugaoka, Chofu, Tokyo 182-8585, Japan

[2] Department of Physics, Faculty of Science, Chiba University, 1 − 33 Yayoi, Inage, Chiba 263-8522, Japan

Abstract   Electronic structures of Au/Si(111) surfaces have been calculated by the *ab initio* pseudopotential method in a local density approximation. We found that there are strong electronic-state couplings between Au trimers and surface Si, which induce the charge transfer from Si to Au-Si covalent bonds and the charge redistribution from d to s orbitals around Au. It is shown that Au overlayer acts as a surfactant through Au-Si rebonding during Si growth.

1 Introduction

The Au/Si contact systems are one of the most important fundamentals in semiconductor technology. Restricted to Au-overlayered Si(111) surfaces, however, to our knowledge, the detailed analysis of electronic structures for these surfaces has never been examined. The aim of this work is to investigate the electronic structures of Au/Si(111) systems. Especially, in this study, we investigate the role of Au overlayer as the surfactant, which effect was observed by Wilk *et al.*[1] during high-quality Si(111) film homoepitaxy, and the band structures of the $(\sqrt{3} \times \sqrt{3})R30°$-Au/Si(111) surface, which is the most stable surface with 1 monolayer Au and called the conjugate honeycomb-chained-trimers (CHCT) model[2]. From the former study, we found the Au-Si rebonding and the charge transfer from Si to Au, while from the latter analysis we found the charge redistribution around Au caused by a strong electronic-state hybridization between Au and surface Si and the increases of work function and core-level binding energies.

2 Method

The Au/Si(111) surface structures have been realized on isolated slabs, whose bottoms are terminated with hydrogens. These slabs are periodically arranged along the [111] direction, and their electronic structures have been calculated in a local density approximation (LDA) with the *ab initio* pseudopotentials method.[3] The wave functions were expanded by plane waves with a 12.96 Ry cutoff energy to obtain the optimized atom configurations, while a 20.25 Ry cutoff energy was used to evaluate the well-converged total energies and the band structures. Details of this were reported in our previous works.[4,5]

3 Calculational results

3.1 Surfactant effect of Au

In order to study the role of Au overlayer as a surfactant, we consider the atom movement during the Si [111] epitaxial growth using the 1 × 1 unit. In the case of layer-by-layer growth, two kinds of surfaces, the 1 × 1-Au/Si(111)-α and -β, should appear alternatively.[5] The 1 × 1-Au/Si(111)-β is realized by deposition of 1ML Si on the T_1 site of the 1 × 1-Au/Si(111)-α, followed by the movement of Au overlayer onto the surface. The calculated adiabatic potential profile is shown in Fig. 1 as a function of the Au atom position, where the energies are measured relative to that at the position e, *i.e.*, 1 × 1-Au/Si(111)-β, which is the most stable state.

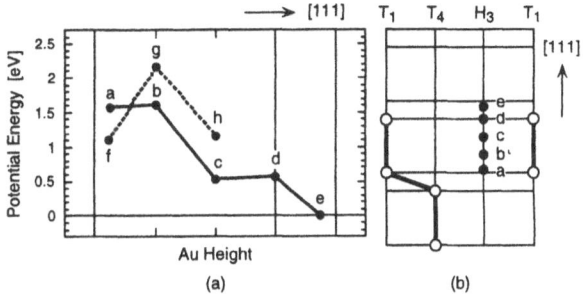

Fig. 1 (a) Adiabatic potential energy as a function of the Au position. The Au atom positions are shown in (b), where the open and solid circles, respectively, denote Si and Au atoms in a (110) plane.

In Fig. 1, one can notice that there is a little barrier for Au to float up onto the Si surface. The movement of Au atom from the position a to e is accompanied by the covalent rebonding with neighboring Si atoms. When the Au atom is located at the position a, there exist the bonds between Au and Si atoms in the second- and third-top layers. Once the Au moves to the position b, however, such bonds are broken. This is the reason why the position b has the highest energy. After the Au atom passes the position b, the rebonding begins between Au and surface-top Si. These new bonds terminate three dangling bonds of surface Si and remarkably stabilize the surface compared to the position a. In particular, these bonds continue to increase the strength as the Au atom floats up to reach the stable position e, where Au overlayer is located at the positions higher than the Si top layer and the charge transfer occurs from Si to Au-Si bonds. This higher Au position and the charge transfer are indispensable to explain the observed larger work function of $\sqrt{3} \times \sqrt{3}$-Au/Si(111) surface, which will be discussed later.

3.2 Electronic structures

Using the calculated surface energies, we found that any Au/Si(111) surfaces with 1 monolayer Au are more stable than various reconstructed clean Si(111) surfaces. In addition, the CHCT model has the lowest surface energy of 0.44 [J/m²] among these surfaces. This $\sqrt{3} \times \sqrt{3}$-Au/Si(111) surface is realized by the Au-trimer formation of 1×1-Au/Si(111)-β, whose surface energy is 1.01 [J/m²]. In the following, we study the electronic structure of this surface.

In Fig. 2, we show the band structure of the $\sqrt{3} \times \sqrt{3}$-Au/Si(111) system. One can first notice that this surface is semi-metallic; the top two valence bands, which is denoted as a in Fig. 2, are partially empty and the one-fold conduction band, which is denoted as b in Fig. 2, is a little occupied, both around Γ point. By analyzing the wavefunction probability density, we found that the former top two valence bands, a, originate from the top valence-band states of bulk Si. While, the latter one-fold conduction band, b, is produced by the strong antibonding hybridization of a Au-trimer state made of d orbitals and a dangling-bond state of surface-top Si. In a Au monolayer system, this Au-trimer state is fully occupied. However, once this Au monolayer is put on Si (111) substrate, the state b in Fig. 2 becomes almost empty. Namely, due to the hybridization of Si and Au orbitals, the definite charge depletion occurs from the d-orbital components of Au. By analyzing the other bands, we found the Au charge redistribution; the depletion charge flows into Au-Si bonding-like bands made of Au s orbitals.

At last, the calculated work function of $\sqrt{3} \times \sqrt{3}$-Au/Si(111) surface is 4.69 eV, which is 0.25 eV larger than 4.44 eV of the clean relaxed 1×1-Si(111) surface. This increase of work function is caused by the higher Au position than the Si top layers and the charge transfer from Si to Au-Si bonds. While such charge transfer and the above-mentioned charge redistribution around Au become origins of observed higher binding-energy shifts of Si $2p$ and Au $4f$ core levels.[6]

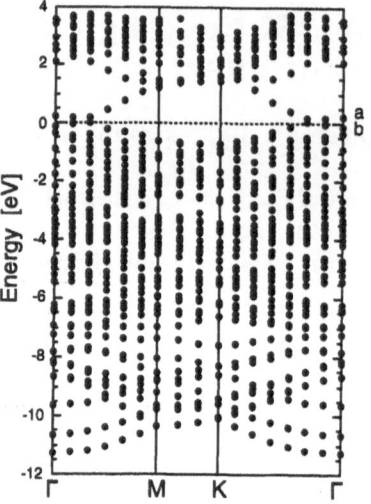

Fig. 2 Calculated band structure of $\sqrt{3} \times \sqrt{3}$-Au/Si(111) reconstructed surface structure. Fermi's energy is chosen as the energy origin.

4 Conclusions

Electronic structures of 1×1-Au/Si(111) and $(\sqrt{3} \times \sqrt{3})R30°$-Au/Si(111) systems have been calculated by the *ab initio* pseudopotential method in a local density approximation. We found that Au overlayer is located higher than the Si topmost layer to lower the dangling bond-energies of surface Si and the charge transfer occurs from Si to Au-Si covalent bonds, which both induce the surfactant functionality of Au through Au-Si rebonding. While Au atoms produce Au trimers on Si substrate, the charge redistribution occurs from d to s orbitals around Au due to a strong electronic-state hybridization between Au trimers and surface-top Si. All these cause the increases of both the work function and the core-level binding energies of Au and Si by Au adsorption observed in experiments.

References

1. G. D. Wilk, R. E. Martinez, J. F. Chervinsky, F. Spaepen, and J. A. Golovchenko, Appl. Phys. Lett. **65** (1994) 866.
2. Y. G. Ding, C. T. Chan, and K. M. Ho, Surf. Sci. Lett. **275** (1992) 691.
3. G. B. Bachelet, D. R. Hamann, and M. Schlüter, Phys. Rev. B **26** (1982) 4199.
4. M. Murayama, T. Nakayama, and A. Natori, Appl. Surf. Sci. **159/160** (2000) 45.
5. M. Murayama, T. Nakayama, and A. Natori, to be published in Jpn. J. Appl. Phys.
6. T. Okuda, H. Daimon, S. Suga, Y. Tezuka, and S. Ino, Appl. Surf. Sci. **121/122** (1997) 89.

Theoretical study on epitaxial growth of lattice-mismatched semiconductor systems

Nori MIYAGISHIMA[1*], Ko OKAJIMA[1], Kyozaburo TAKEDA[1], Norihisa OYAMA[2], Takahisa OHNO[2], Kenji SHIRAISHI[3], and Tomonori ITO[4]

[1] School of Science and Engineering, Waseda University, 3-4-1 Okubo, Shinjuku, Tokyo 169-8555, Japan,
[2] National Research Institute for Metals, Tsukuba 305-0047, Japan,
[3] NTT Basic Research Laboratories, Kanagawa 243-0198, Japan,
[4] Mie University, Tsu, 514-8507, Japan

Abstract   Focusing on the formations of misfit-dislocations (MDs) and islands, the energetics in the heteroepitaxial system was theoretically studied. We obtain several phenomenological parameters such as an effective interfacial elastic constant ($\tilde{M}$) and the formation energy of MDs (E_d) numerically from the microscopic total energy calculations, and determine the suitable growth process by tracing the trajectory of the minimized free energy. The characteristic growth behavior in GaSb/GaAs(100) e. g., is well demonstrated.

1 Introduction

The heteroepitaxial growth in the semiconductor system includes an inherent strain energy because of its lattice-mismatching. Various phenomena such as misfit-dislocations (MDs) creation, three dimensional (3D) islands formation and the intermixing of the constituting elements are induced to relax this misfit strain energy. Consequently, numbers of characteristic growth processes appear.

Much experimental effort has caused a success in classifying those growing processes into the following types; FM (Frank-van der Merwe) type of a simple two dimensionally (2D) wetting growth, VW (Volmer-Wever) type of the formation of 3D islands directly on the substrate, and SK (stranski-Krastanov) type of the islands formation on the 2D wetting substrate. Characteristics in the individual growth modes have been theoretically investigated by several pioneering works. Here, with including the formation of MDs and islands, we integrate and reform those previous results, and give a systematic understanding on the heteroepitaxial growth.

2 Free energy of the system

Based on the continuum medium treatment, we first describe the free energy of the heteroepitaxial system. The key to understand the growing behavior is to describe how the lattice strain should be relaxed. We consider the two typical relaxation processes of the MDs and island formations. The intermixing is excluded in the present consideration, because several heteroepitaxial systems produce an extremely steep interface of the constituting elements.

The system's free energy (per unit area of the substrate) can be described by the sum of the surface part

* E-mail: miya@qms.cache.waseda.ac.jp

(E^{surf}), the lattice-mismatched strain part (E^{strain}), and the MDs creation part (E^{MDs}) as follows:

$$E(\bar{h}) = E^{surf} + E^{strain} + E^{MDs}. \tag{1}$$

Here, the symbol $\bar{h}$ is an averaged layer thickness.

By using the surface energies of the epi-layer (γ_e) and substrate (γ_s), the first term E^{surf} is described as

$$E^{surf} = \gamma_e(1+\beta)\xi + \gamma_s(1-\xi), \tag{2}$$

where the symbol β denotes an effective increase in the epi-layer surface due to the islands formation, and ξ is a covering ratio of the epi-layer by those islands.

The lattice-mismatching causes a strain energy, which is, however, relaxed partly by the 3D island formation. We describe this relaxation by using a reduction parameter α. Thus, the relaxed strain energy is given as,

$$E^{strain} = \frac{1}{2}(1-\alpha)\tilde{M}\epsilon_0^2\bar{h}, \tag{3}$$

Here, ϵ_0 is an inherent strain generated by the difference in the lattice constants.

The formation of MDs also relax the above strain energy partly. According to the continuum medium approach, the inherent strain ϵ_0 should be changed into the relaxed one of $\epsilon = \epsilon_0(1 - \frac{L_0}{L})$. Here, the MDs are created with an interval L, and L_0 is the MDs' minimum interval at which lattice strain is completely relaxed.

Moreover, one should notice the opposite role of MDs. The creation of MDs destabilizes the system's energy by E_d (per one MD). When the MDs are created under the VW islands having the covering ratio ξ, the corresponding energy (per unit area of the substrate) is given as,

$$E^{MDs} = \frac{E_d}{L}\xi. \tag{4}$$

Thus, three variables of the averaged layer thickness ($\bar{h}$), the MDs' interval (L) and the covering ratio (ξ) determine the free energy as follows:

$$E(\bar{h}, L, \xi) = \gamma_e(1+\beta)\xi + \gamma_s(1-\xi)$$
$$+ \frac{1}{2}(1-\alpha)\tilde{M}\epsilon^2\bar{h}\frac{Ed}{L}\xi. \tag{5}$$

3 Microscopic features of MDs and islands

Equation (5) reveals the following microscopic features of the heteroepitaxial growing system. Let us first discuss an interval of MDs. Since the MDs creation functions oppositely to the system's energy, those MDs should be created with its optimal interval of MDs (L^{opt}). This L^{opt} is obtained by solving $\frac{\partial E}{\partial L} = 0$ as follows:

$$L^{opt} = L_0(1 - \frac{E_d\xi}{(1-\alpha)\tilde{M}\epsilon_0^2 L_0\bar{h}})^{-1}. \qquad (6)$$

Thus, the growing processes having a wetting epi-layer (SK and FM types) let ξ be unity, and the resulting free energy is depend only to $\bar{h}$ because the MDs are introduced with the above L^{opt}. However, when VW islands are created on the substrate, three variables $\bar{h}$, L and ξ should be included to determine the free energy. This is because the covering ratio ξ should be depend on the shape of the created islands and varied during the epitaxial growth.

Since several systems, GaSb/GaAs(001) e. g., produce flat-topped VW islands, we assume a simple form of those rectangular solids, and characterize their shapes by the aspect ratio σ. Thus, ξ and β are expressed as a function of σ as $\xi = (\frac{\bar{h}\sqrt{\sigma}}{\sigma})^{\frac{2}{3}}$ and $\beta = 4\sigma$, respectively.

The optimal aspect ratio σ^{opt} is also obtained at the given layer thickness by solving $\frac{\partial E}{\partial \sigma} = 0$. In the coherently growing process (L=∞), the resulting σ_{co}^{opt} is given as,

$$\sigma_{co}^{opt} = \frac{\gamma_e - \gamma_s}{2\gamma_e}. \qquad (7)$$

The characteristic point is that this value is not dependent on the averaged layer thickness $\bar{h}$ but determined only by the two surface energies γ_e and γ_s. The reason is that the VW islands should be created with giving their minimum surfaces. Equation (7) also reveals that the value σ_{co}^{opt} should be equal to zero when γ_e is nearly equal to γ_s. Thus, more flat-shaped islands are expected to be created when two layers have similar surface energies. On the contrary, all those flat-topped VW islands become a unique form having their aspect ratios $\sigma_{co}^{opt} \sim \frac{1}{2}$, when γ_e is extremely larger than γ_s.

When the MDs are introduced in the system, how does the optimal aspect ratio change with the averaged layer thickness? We show the calculated $\sigma_{MDs}^{opt}(\bar{h})$ in Fig. 1. The creation of the MDs increase the value of $\sigma_{MDs}^{opt}(\bar{h})$ from σ_{co}^{opt}, and lets it approach to $\sigma_{co}^{opt} + \frac{E_d}{2\gamma_e L_0}$. This limited value corresponds to the maximum aspect ratio when MDs are arranged with their minimum interval L_0. The energetical destabilization due to the MDs formation should be relaxed by increasing the volume (height) of the VW islands.

Giving the L^{opt} and setting the material parameters ξ, E_d, and $\tilde{M}$ numerically, we can show the change in the free energy with varying $\bar{h}$ (Fig. 2). The trace of the trajectory in the lower free energies well demonstrates the difference in the three characteristic growing processes.

4 Growth in GaSb/GaAs(001) system

Finally, we show the change in the free energies in GaSb/GaAs(001) system in Fig. 3. Those material parameters are determined by the total energy calculations due to the first-principles and/or empirical force-filed methods.

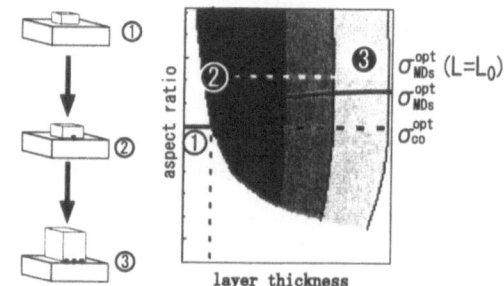

Fig. 1 Change in the aspect ratio of the VW islands.

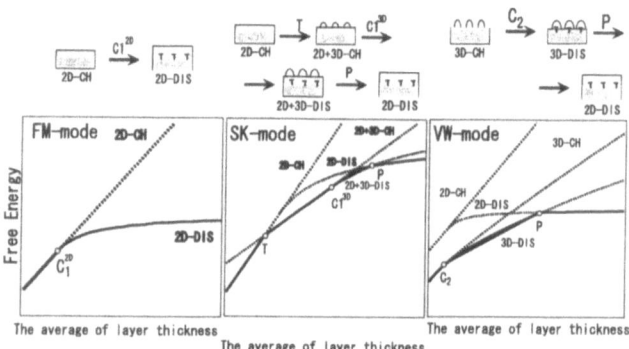

Fig. 2 Change in the calculated free energies for FM, SK and VW modes

Details in the parameters determination will be reported elsewhere.

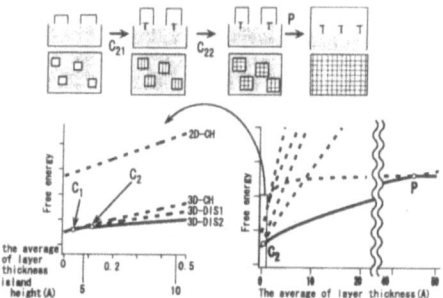

Fig. 3 Change in the calculated free energies of GaAs/GaSb(001) system.

In the initial stage of the growing, the VW-type islands having no MDs are created directly on the substrate (C_1). With growing those islands conserving σ_{co}^{opt}, the MDs are induced periodically but one-directionally (C_2). The continuous growth should introduce the other MDs perpendicular to the former MDs (P), and these VW islands merge to be a 2D wetting layer having the cross-linked MDs.

References

1. K. Okajima, K Takeda, N. Oyama, E. Ohta, K. Shiraishi and T. Ohno, Jpn. J. Appl. Phys., inpress.

Control of electronic structure and thermodynamic stability of SrTiO3 with Ru or Nb doping from first-principles calculation

Tatsuo Schimizu ,Takashi Kawakubo

Advanced LSI Technology Laboratory, Corporate Research & Development Center, Toshiba Corp. Komukai Toshiba-cho, Saiwai-ku, Kawasaki, Kanagawa 212-8582, Japan .   e-mail: schimizu@mdl.rdc.toshiba.co.jp

Abstract   Not only the electronic structure but also the thermodynamic stability of Sr(Ti,Ru)O3 and Sr(Ti,Nb)O3 is investigated using the first-principles method with ultrasoft pseudopotentials. The calculations lead us for the first time to propose Ru-doped or Nb-doped SrTiO3 electrode with much higher stability than SrRuO3 electrode by artificially controlling doping concentrations. Ru-doped SrTiO3 with a Ru concentration of about 50 mol % and Nb-doped SrTiO3 with a small amount of Nb concentration can be used as the promising candidates for stable electrodes.

1 Introduction

Conductive perovskite oxide materials are attracting much attention for application, instead of Pt or Ir, as electrodes of ultralarge-scale integration memory-cell dielectric capacitors, because of their congeniality with dielectric perovskite oxide materials, such as BaTiO3 or Pb(Zr,Ti)O3. In the last few years, SrRuO3 thin film is thought to be one of the most attractive electrodes [1,2]. SrRuO3 electrode, however, often oxidizes underlying barrier layer, and peels off.

The objective of this study is to investigate the thermodynamic stability, as well as the electronic structure, of conductive perovskite oxide materials, Sr(Ti,Ru)O3 and Sr(Ti,Nb)O3 ,using the first-principles calculations in order to obtain electrodes with much higher stability than SrRuO3 electrode. As a criterion respecting thermodynamic stability, we investigate whether neighboring (Ti1-x,Alx)N metal layer, which is often used as a barrier metal in memory-cell capacitors, is oxidized or not.

2 Total Energy Calculations

The present first-principles calculations are based on the density functional theory with the local spin density approximation. The exchange-correlation term used here is of the Ceperley-Alder [2] form parameterized by Perdew and Wang [3]. The paramagnetic solutions are obtained, with the exception of that for SrRuO3 bulk and electronic structure and oxygen molecule. Therefore, we do not refer to the magnetic properties. For each ion, an ultrasoft pseudopotential as proposed by Vanderbilt [4] is generated. We are able to get sufficient convergence of the total energy of cubic-perovskite

SrTiO3, SrRuO3, and SrNbO3 at the cutoff energy Ecut=30.25 Ry. A total of 64 k points are used for integration over the Brillouin zone. From these potentials, the lattice constants of a0(SrTiO3)= 3.88 A , a0(SRO3)= 3.91 A, and a0(SNO3)= 3.994 A are obtained for cubic perovskite SrTiO3, SrRuO3, and SrNbO3, respectively. This is consistent with the experimental values (3.90 A, 3.93 A, and 4.025 A, respectively [5]).

3 Calculations and Results

3.1 Thermodynamic Stability

As a criterion respecting thermodynamic stability, we investigate whether neighboring (Ti,Al)N metal layer, which is often used as a barrier metal in memory-cell capacitors, is oxidized or not. The energy loss by dissolution of cubic SrRuO3, SrNbO3 and SrTiO3 per oxygen molecule can be estimated by means of the first-principles total energy calculations. The dissolution process which occurs with the least energy loss per oxygen molecule must be considered as follows.
SrRuO3 = SrO(bulk) + Ru(bulk) +   O2 -   5.0 eV.
SrRuO3 = SrO(bulk) + Nb(bulk) +   O2 -   10.0 eV.
SrTiO3 = SrO(bulk) + Ti(bulk) +   O2 -   11.0 eV.

Next, the energy loss by dissolution of Sr(Ti1-x,Rux)O3 and Sr(Ti1-x,Nbx)O3 mixed crystal system (for x=1/8, 4/8, 7/8) per oxygen molecule is also calculated. We use $2 \times 2 \times 2$ supercell. The total energy calculations are performed with relaxations of the lattice constants and internal atoms' positions.

Although only homogeneous models of a mixed crsytal of SrTiO3 and SrRuO3 are adopted for simplicity, we obtain stabilized mixed crystal state relative to the separate SrTiO3 and SrRuO3 in all three cases. As for Sr(Ti,Nb)O3, we obtain stabilized mixed crystal state relative to the separate SrTiO3 and SrNbO3 in the case of x=1/8, 4/8.

Fig.1 shows the enthalpy of Sr(Ti1-x,Rux)O3 and Sr(Ti1-x,Nbx)O3 as a function of its composition, x. SrRuO3 is much less stable than underlying (Ti,Al)N layer. On the other hand, SrTiO3 is much more stable than (Ti,Al)N layer, and could grow on (Ti,Al)N directly. The thermodynamic stability of Sr(Ti,Ru)O3 is enhanced drastically relative to that of SrRuO3, as a Ru concentration decreases. Sr(Ti,Ru)O3 with a Ru concentration of less than 60 mol % is thought to be a candidate for a new electrode with higher themodynamic stability. Sr(Ti,Nb)O3 is always much stable relative to SrRuO3, but there is a concern that Sr(Ti,Nb)O3 is unstable with respect to oxidization of itself with a large amount of Nb ions, just as SrNbO3 is oxidized into insulating Sr2Nb2O7. Sr(Ti,Nb)O3 with a small amount of Nb concentration can be used as another stable electrode.

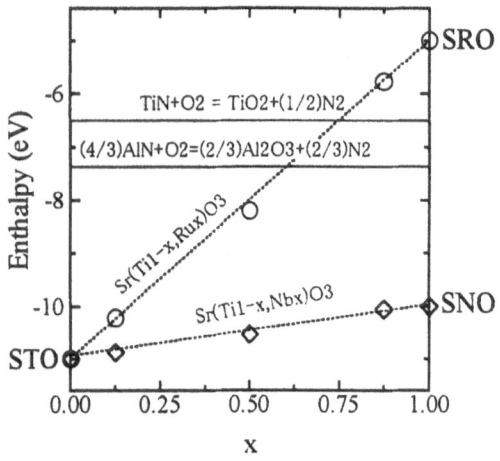

Fig.1   Enthalpy   of   Sr(Ti1-x,   Rux)O3   and Sr(Ti1-x,Nbx)O3   as a function of its composition, x.

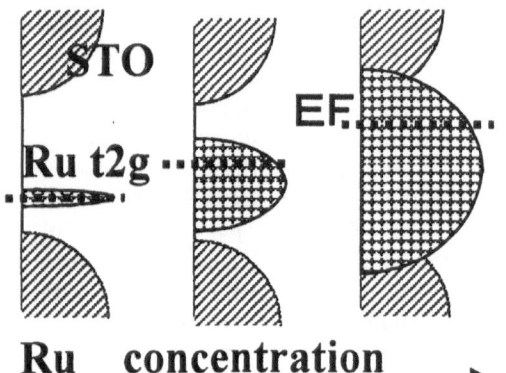

Fig.2 Schematic illustration for the change of the band structure owing to the increase of the Ru concentration.

3.2 Electronic Structure of Sr(Ti,Ru)O3 and Sr(Ti,Nb)O3

Fig.2 shows the schematic illustration for the change of the band structure of Sr(Ti,Ru)O3 owing to the increase of the Ru concentration. The electronic structure of the Sr(Ti, Ru)O3 with small Ru concentration, the deep level states, which mainly consist of the Ru 4d t2g states, appear below the middle of the band gap of SrTiO3. As

a Ru concentration increases to more than 50 mol %, the inside-gap states become delocalized, and the sufficiently conductive Sr(Ti, Ru)O3 thin film is obtained. The band structure with a Ru concentration of 87.5 % shows that the band gap of STO is filled.

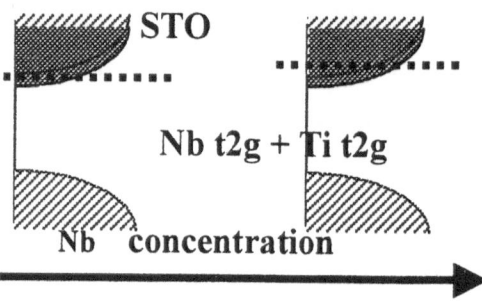

Fig.3 Schematic illustration for the change of the band structure owing to the increase of the Nb concentration.

Fig.3 shows the schematic illustration for the change of the band structure of Sr(Ti,Nb)O3 owing to the increase of the Nb concentration. Nb t2g states appear just below the SrTiO3 conduction band. These Nb states strongly hybridize with Ti states. As a Nb concentration increases, the Fermi energy shifts into the conduction band. In other words, electron doping can be described within a rigid-band model. Only a small amount of Nb doping is found to make SrTiO3 metallic, unlike Ru doping. The larger amount of Nb concentration, the better conduction can be expected.

4. Summary

According to the calculated thermodynamic data and electronic structures, Ru-doped SrTiO3 with a Ru concentration of about 50 mol % and Nb-doped SrTiO3 with a small amount of Nb concentration can be used as the promising candidates for stable electrodes.

References

1. M. Izuha,   Appl. Phys. Lett. 70 1405 (1997),   K. Abe, IECE TRNS. ELECTRON E81-C 505 (1998).
2. D. M. Ceperley and B. J. Alder: Phys. Rev. Lett. 45 (1980) 566.
3. J. Perdew and Y. Wang: Phys. Rev. B45 (1992) 13244.
4. D. Vanderbilt: Phys. Rev. B41 (1990)   7892.
5. Landolt and Bornstein: *Numerical Data and Functional Relations in Science and Technology- Crystal and Solid State Physics* (Springer, Berlin   1982).
6. K. Kubaschewski , *Metallurgical thermochemistry* , (Springer, Belrin   1951).

Modeling and simulation of FBAR devices Fabricated on Si

D. H. Kim[1*], Giwan Yoon[1], and H. D. Park[2]

[1] Information & Communications University, 58-4 Hwaam-dong, Yusong-gu, Taejon 305-348, KOREA
  Phone: +82-42-866-6131, Fax: +82-42-866-6227, E-mail: gwyoon@icu.ac.kr

[2] Research Center, Sang Shin Electric Co. Ltd., 614-1 Wolhari, Suhmyon, Yunkikun, Choongnam, KOREA

Abstract   Film bulk acoustic resonators for radio frequency wireless applications are presented. Various simulations and modeling were carried out. The impedance of a five-layered FBAR showed the same trend of the wideband characteristics as that of an ideal FBAR, but the characteristics of higher modes are much more suppressed. Also, the wideband impedance desreased with increasing device size. The resonance characteristics depend on the physical dimensions. At least five-layered reflector is required to obtain an efficient resonance response.

1. Introduction

Recently, film bulk acoustic resonator (FBAR) has become one of the most promising components mainly due to its small size and high device performance and its strong potential for MMICs applications [1]. FBAR is composed of two important parts: piezoelectric film sandwiched between conductors (for resonance) and multiple reflection layers (for acoustic isolation), affecting the overall resonance characteristic [2]. In spite of considerable efforts on piezoelectric ZnO films [3], few studies have been reported on the effects of reflector and thin film layers. In this paper, we present theoretical analysis and fabrication of FBAR.

2. Simulation & Device Fabrication

Various simulations were carried out using a derived input impedance equation on the FBAR with 0.13 μm Al for top and bottom electrodes as well as FBAR with 0.13um Al for top electrode and 0.13 μm Au for bottom electrode. For the simulation, the ZnO thickness was 1.4 μm, and both W and SiO_2 films were λ/4. Two-port FBAR devices were fabricated and measured for S-parameter extraction.

3. Results and Discussions

The resonance characteristics were simulated to understand the acoustic mass loading effects, resonance area effects, and reflector layers number as well as thickness variation effects. Fig. 1 (a) & (b) show a schematic structure and SEM micrograph of the FBAR with the five-layered reflector, respectively. Fig. 2 (a) & (b) show reflection loss (S11) and insertion loss (S21) plots, respectively where Q was 500 when calculated from the S21. Fig. 3 shows a shift in the resonance frequency due to acoustic mass loading effect where Au and Al were used as bottom electrodes. Fig. 4 shows the wideband response of the FBAR with

various resonance areas where the impedance has the same wideband capacitive characteristics as that of ideal FBAR. In addition, the wideband impedance level decreases with increasing device size, and the optimum size was 150 x 150μm² with 50Ω impedance. Fig. 5 (a) shows the simulated narrowband response of FBAR with SiO_2 thickness variation where the thinning of SiO_2 film from λ/4 results in higher resonance frequency. Also, W thickness variation showed the same trend, but much less shift than SiO_2 film as shown in Fig. 5 (b).

4. Conclusion

We have demonstrated the resonance dependence of the FBAR device on mass loading, resonance area, reflector layer number and variation. High-Q of 500 was obtained at 2 GHz mainly due to an optimized design and process. This FBAR technology will be very promising for RF filter applications.

References
1. S. V. Krishnaswamy et al., IEEE Ultrasonics Symp. (1990) 529
2. K. M. Lakin, et al., IEEE Trans. on Microwave Theory and Techniques **43** (1995) 2933
3. F. C. M. Van De Pol. et al., Thin Solid films **204** (1991) 349

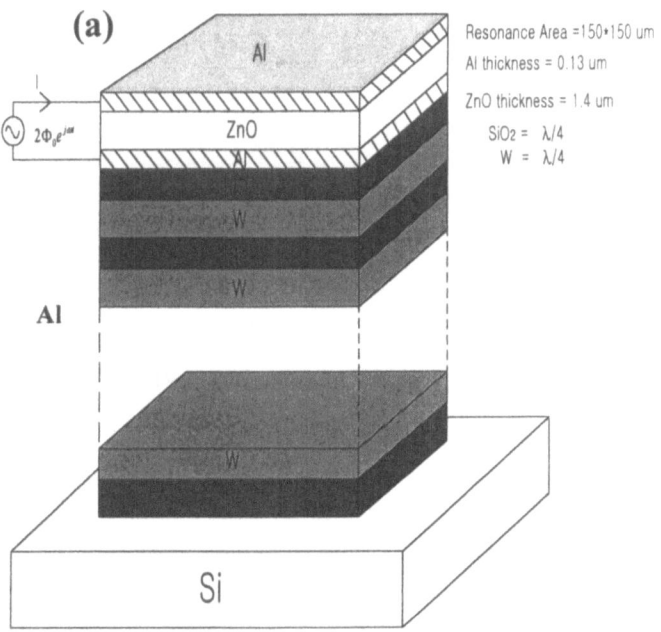

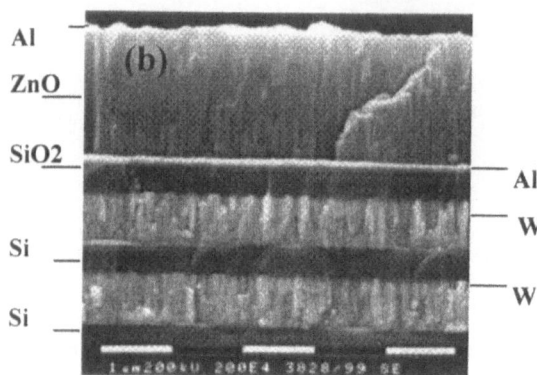

Fig. 1 shows (a) a schematic structure of the FBAR and (b) the SEM micrograph of the fabricated FBAR with five-layer reflector, respectively

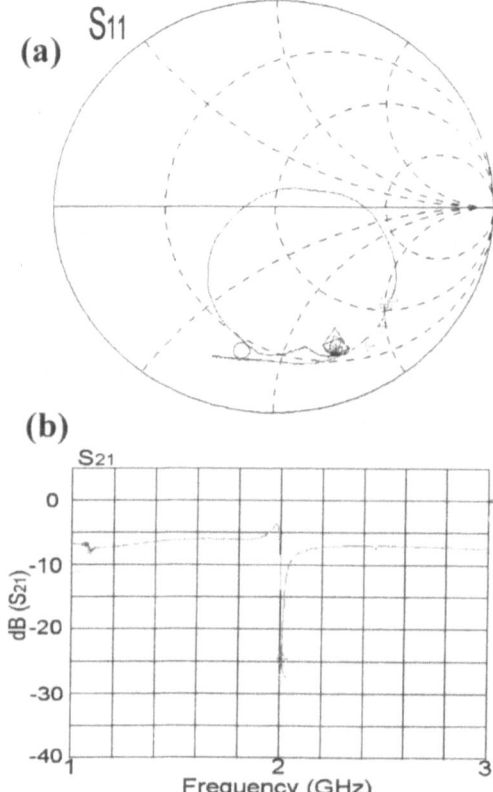

Fig. 2 (a) & (b) Reflection loss (S11) and insertion loss (S21) for the FBAR with five-layered reflector, respectively.

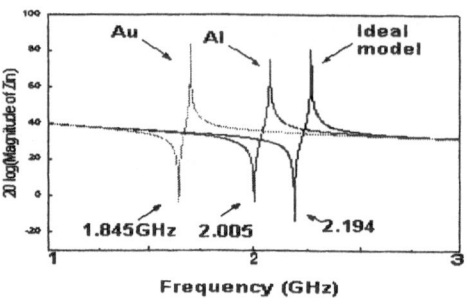

Fig. 3 Simulated resonance frequency shifts of FBAR devices due to acoustic mass loading effect of bottom electrodes

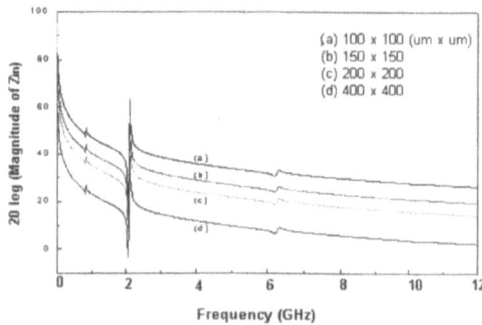

Fig. 4 Simulated wideband responses of FBAR devices with various resonance areas

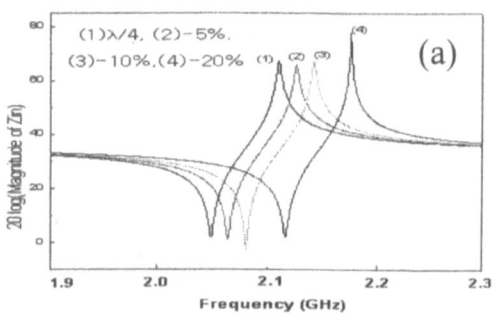

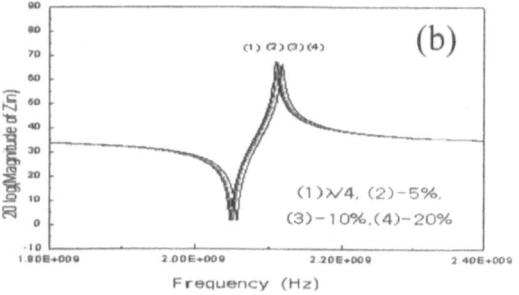

Fig. 5 shows simulated narrowband response of FBAR with thickness variation (a) in SiO_2 and (b) in W.

Nucleation mechanism of stacking fault during Si epitaxial growth by chemical vapor deposition with dichlorosilane

Y. Takakuwa[1], M.K. Mazumder[2] and N. Miyamoto[3]

[1]Research Institute for Scientific Measurements, Tohoku University, 2-1-1 Katahira, Aoba-ku, Sendai 980-8577, Japan   e-mail: takakuwa@rism.tohoku.ac.jp

[2]ULSI Laboratory, Mitsubishi Electric Co., 4-1 Mizuhara, Itami 664-0005, Japan

[3]Faculty of engineering, Tohoku Gakuin University, 1-13-1 Cyuo, Tagajyo 985-0873, Japan

Abstract   The nucleation kinetics of stacking fault during Si epitaxial growth was investigated for chemical vapor deposition on Si(111) with dichlorosilane (SiH_2Cl_2). It was observed that the nucleation of stacking fault is a second-order reaction of the surface monochloride coverage and rate-limited by $SiCl_2$ desorption. The nucleation mechanism of stacking fault is discussed in terms of the steric hindrance of a Cl atom.

1 Introduction

Chemical vapor deposition (CVD) of a Si thin layer with high crystallographic quality is a key process for fabricating Si epitaxial wafers, which are needed for an advanced metal-oxide-semiconductor field effect transistor and a bipolar transistor. Origins of defects such as stacking fault and dislocation in Si epitaxial layers are generally ascribed to contaminants, scratches and defects on the Si substrate [1]. In the previous report [2], we have observed that the nucleation of stacking fault occurs successively during CVD. This suggests that the stacking fault is nucleated not only by the above extrinsic origins but also by the origin intrinsic to Si growth in CVD. In this study, the nucleation kinetics of stacking fault was investigated for Si epitaxial growth by CVD on Si(111) with dichlorosilane (SiH_2Cl_2). The stacking fault density measured as functions of the growth temperature and line velocity is interpreted using a reaction model of the surface adsorbate.

2 Experimental

The CVD experiments of Si growth were performed using a horizontal reactor with rf induction heating at atmospheric pressure [2]. The carrier and source gases were H_2 of 99.99999% purity and H_2-diluted 2% SiH_2Cl_2, respectively. The sample was a Si(111) wafer with a miscut angle of 0.5°. The temperature of the sample was measured by an optical pyrometer. The line velocity (cm/s) is defined by dividing the total gas flow (l/min) by the net cross sectional area of the reaction tube at the sample position, where the gases can pass through (6.3 cm²). It was changed by increasing H_2 flow rate from 1 l/min to 8 l/min at a fixed SiH_2Cl_2 flow rate of 2 l/min.

The surface morphology of an as-grown surface was observed with a differential interference microscope. The stacking fault density d_{SF} was evaluated from the number of flat-top pyramidal hillocks without any chemical etching, because it was previously confirmed that the stacking fault always leads to the growth of a flat top hillock on the

Si(111)-0.5°off surface [3]

3 Results and discussion

The temperature dependence of d_{SF} is shown as an Arrhenius plot in Fig. 1, compared with that of the growth rate. The growth rate is divided into two temperature regions at 982°C. In a temperature region above and below 982°C, the Si growth is rate-limited by the gas diffusion and surface reaction, respectively [4]. In the surface-reaction rate-limiting region, adsorption of SiH_2Cl_2 on the surface is inhibited by the surface adsorbate such as chloride and hydride, resulting in a decrease of the growth rate.

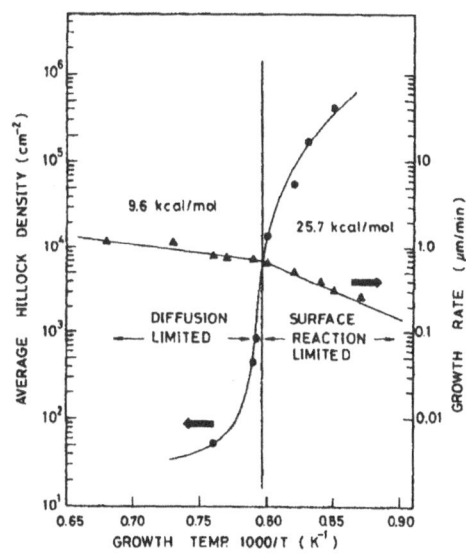

Fig.1 Arrhenius plot of d_{SF} and growth rate.

On the other hand, the d_{SF} increases drastically just in the surface-reaction rate-limiting region of the growth rate. From the analogy of the growth rate, this means that the nucleation of stacking fault is associated with the surface adsorbate. Yarmoff et al. [5] reported that the surface adsorbate on the SiH_2Cl_2-exposed Si(111) surface is not dichloride (Si-Cl_2) but monochloride (Si-Cl). Therefore, it is suggested that the Si-Cl is concerned with a nucleation of stacking fault.

In Fig. 2, is shown the line velocity dependence of d_{SF} and growth rate obtained at 970°C. The growth rate decreases gradually from 0.38 μm/min at 6 cm/s to 0.19 μm/min at 22 cm/s. This is due to a decrease of the SiH_2Cl_2

volume concentration $c_{SiH2CL2}$, because SiH_2Cl_2 is further diluted by H_2 at faster line velocity. By contrast, the d_{SF} decreases rapidly and reaches zero at 19 cm/s. Thus the nucleation of stacking fault can be suppressed completely at fast line velocity, while a Si epitaxial layer still grows with a decrease of ~40% in growth rate. No nucleation of stacking fault above 19 cm/s seems to originate from the same reason as in the gas-diffusion rate-limiting region of Fig. 1. Therefore, the line velocity dependence of d_{SF} is considered in terms of the surface adsorbate in the following.

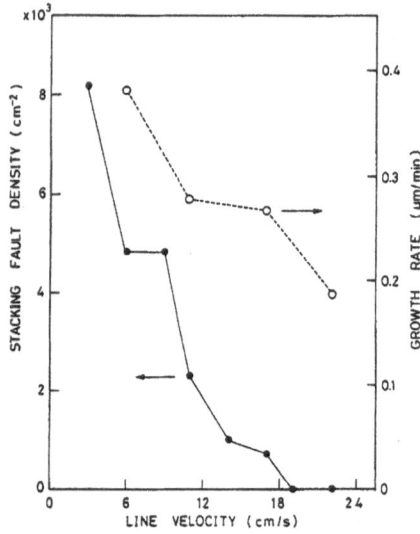

Fig. 2 Line velocity dependence of d_{SF} and growth rate.

Near the transition temperature from gas diffusion to surface reaction, the Si-Cl coverage θ_{Si-Cl} is approximately represented to be proportional to the $c_{SiH2CL2}$, because the θ_{Si-Cl} is much smaller than the unity. When the line velocity dependence of d_{SF} in Fig. 2 is plotted against the square of $c_{SiH2CL2}$, a good linear correlation is obtained as shown in Fig. 3. This indicates that the nucleation of stacking fault is a second-order reaction of θ_{Si-Cl}. Furthermore, there is an incubation concentration at $c_{SiH2CL2}{}^2$ of 2.8×10^{-3} (%²). The incubation in the nucleation of stacking fault is interpreted by that the rate-limiting reaction of Si growth changes from surface reaction to gas diffusion at the critical $c_{SiH2CL2}$. Namely, no Si-Cl remains on the surface in the incubation region of $c_{SiH2CL2}$.

The second-order reaction of θ_{Si-Cl} means a formation of Si-Cl_2, which may give rise to a nucleation of stacking fault. Therefore, if $SiCl_2$ desorption occurs from the surface, no stacking fault would appear in the Si epitaxial layer. The migration length of Si-Cl_2 λ_{Si-Cl2} until the nucleation of stacking fault is estimated from an average distance between stacking faults, which is approximately given by the root of an inverse of d_{SF}. The λ_{Si-Cl2} shows a good linear correlation in an Arrhenius plot, giving the activation energy of 64.9 kcal/mol. This is very close to the value of $SiCl_2$ desorption, 67 kcal/mol [6]. The agreement means that the nucleation of stacking fault is rate-limited by $SiCl_2$ desorption.

The nucleation of stacking fault by Si-Cl_2 is explained

using a steric hindrance of Cl atom. Due to a large ionic radius of Cl atom, a Si adatom, which has one back bond and three dangling bonds on Si(111) and is adjacent to the Cl atom, may be easily rotated 60°. In the case of Si-Cl_2, such a steric hindrance takes place for two Si adatoms, resulting in an appearance of rotated dangling bonds adjacent to each other. Their bond directions could be fixed by adsorption of a Si atom on them, giving rise to a nucleation of stacking fault. For Si-Cl, a rotated Si adatom is isolated, so that it would give no influence on the ordering of other Si adatom.

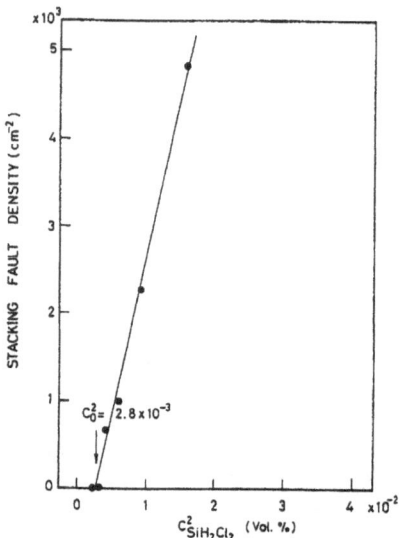

Fig.3 d_{SF} versus square of SiH_2Cl_2 concentration.

4 Conclusions

The nucleation kinetics of stacking fault during Si epitaxial growth by CVD on Si(111) with SiH_2Cl_2 was investigated. It was observed that the stacking fault density increases drastically in the surface-reaction rate-limiting region of the growth rate and is a parabolic function of the SiH_2Cl_2 concentration. It is concluded that the nucleation of stacking fault is a second-order reaction of Si-Cl and rate-limited by $SiCl_2$ desorption.

References

1. C.W. Pearce, *VLSI Technology* 2nd edition, edited by S.M. Sze (McGraw-Hill, New York, 1988) p. 55.
2. Y. Takakuwa, M.K. Mazumder, and N. Miyamoto, J. Electrochem. Soc. **141**, (1994) 2567.
3. M.K. Mazumder, Y. Mashiko, M.H. Koyama, Y. Takakuwa, and N. Miyamoto, J. Cryst. Growth **155**, (1995) 183.
4. J. Bloem, and W.A.P. Claassen, Philips Tech. Rev. **41**, (1983/84) 60.
5. J.A. Yarmoff, D.K. Shuh, T.D. Durbin, C.W. Lo, D.A. Lapiano-Smith, F.R. McFeely, and F.J. Himpsel, J. Vac. Sci. Technol. A10, (1992) 2303.
6. P. Gupta, P.A. Coon, B.G. Koehler, and S.M. George, J. Chem. Phys. **93**, (1990) 282.

Self-limiting etching prior to self-limiting growth in ultra-high vacuum for obtaining clean interface

N. Otsuka[1*], J. Nishizawa[1,2], Y. Oyama[1,2,3], H. Kikuchi[1,2], K. Suto[1,2,3]

[1] Telecommunications Advancement Organization of Japan, SENDAI Research Center, 19 Koeji, Nagamachi Aoba, Sendai 980-0868, Japan   e-mail: otsuka@crl.mei.co.jp
[2] Semiconductor Research Institute of Semiconductor Research Foundation, Kawauchi Aoba, Sendai 980-0862, Japan
[3] Dep. Materials Science and Engineering, Tohoku University, Aramaki Aoba, Sendai 980-8579, Japan

Abstract Intermittent injections of tertiarybutyl-phosphine (TBP) have been applied for molecular self-limiting etching (MSE) of InP substrates in order to obtain clean epilayer/substrate interface prior to growth of InP by molecular layer epitaxy (MLE). The concentrations of impurities, such as C, O, and Si at the interface using MSE decreased compared to those using pre-heating procedure with continuous TBP supply. The MSE mechanism was studied on (111)B oriented InP substrate. It was assumed that C and Si on InP surface and the In atoms on (111)B surface are removed by molecules of TBP or TBP related reactant because of their small binding energies in MSE.

1. Introduction

Atomic layer epitaxy (ALE) [1] and molecular layer epitaxy (MLE) [2] are promising technologies for the fabrication of electrical and optical devices with very thin layered structures of atomic accuracy. Although the presence of interface states at an etched/grown or epilayer/substrate interface is harmful for device performance, one cannot expect a clean interface by ex-situ processes due to possible contamination from the atmosphere. Thus, the combination of in-situ etching and epitaxial growth in identical apparatus has emerged as a promising technique for surface cleaning and fabrication of nanoscale structures.

We have previously reported on molecular self-limiting etching (MSE) [3] of InP using intermittent injections of tertiarybutylphosphine (TBP) at etching temperatures (T_e) below 360°C in ultrahigh vacuum (UHV). The etching depth is controlled by injection cycles and independent of injection time and pressure of TBP in a self-limiting fashion. Recently, we reported on self-limiting growth of InP as low as 320°C using alternate injections of triethylindium (TEI) and TBP by MLE in UHV [4]. In order to realize clean epilayer/substrate interface at low temperature, the combination of MSE and MLE in identical apparatus has been expected.

In this study, MSE have been applied for cleaning of the InP surface prior to growth of InP layer by MLE in UHV. Impurity concentrations are compared between MSE and conventional cleaning procedure. MLE mechanism is studied on (111)B InP surface.

* *Present address:* Matsushita Electric Industrial Co., Ltd., 3-1-1 Yagumonakamachi, Moriguchi, Osaka, 570-8501 Japan

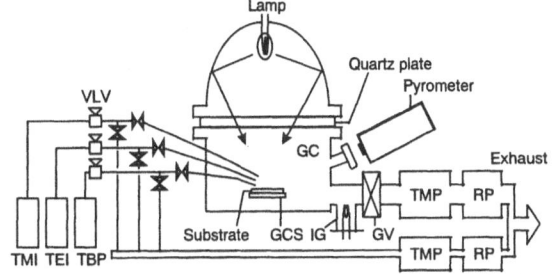

Fig. 1. Main part of experimental setup. GC; growth chamber, TMP; turbomolecular pump, RP; rotary pump, VLV; variable leak valve, GCS; glassy carbon sheet, IG; ion gauge, GV; gate valve.

2. Experiment

The experimental set up is shown in Fig. 1. The MSE and MLE conditions are described in Refs. 3 and 4. The concentration of impurities, such as carbon (^{12}C), oxygen (^{16}O), and silicon (^{28}Si) in the InP layer and substrate is measured by secondary ion mass spectroscopy (SIMS).

3. Results and discussion

The impurity concentrations using MSE and conventional pre-heating procedure (PH) with continuous TBP supply are shown in Fig. 2. The process temperature and the impurity concentration at the interface decrease with the introduction of MSE procedure.

With increasing the injection cycles (N_c) of MSE, the sheet impurity concentrations of C, O, and Si at the

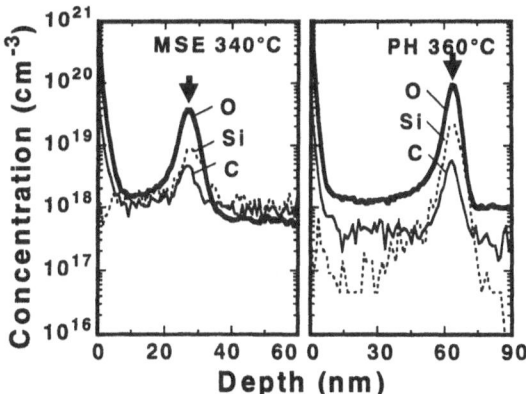

Fig. 2. SIMS profiles at the interface obtained using MSE and pre-heating (PH). Injection cycles of MSE is 100. Arrows show interface position.

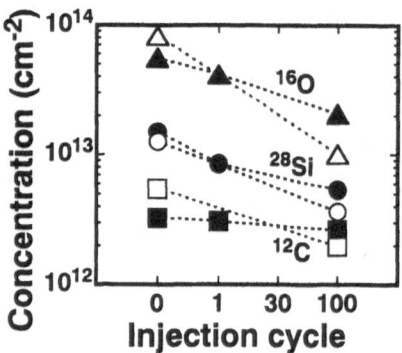

Fig. 3. Dependence of sheet impurity concentrations on injection cycle of MSE. Fe-doped (001) InP substrate grown by liquid-encapsulated Czochralski method (Fe-LEC, closed symbols), and S-doped (001) InP substrates grown by vapor pressure controlled Czochralski method (S-VCZ, open symbols). T_e=360°C.

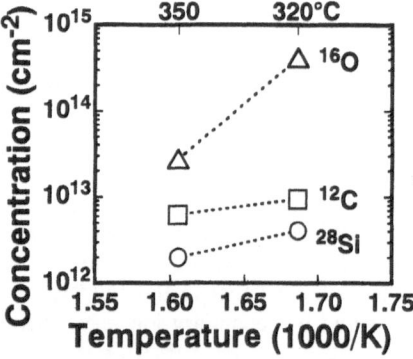

Fig. 4. Dependence of sheet impurity concentrations on MSE temperature as a parameter of impurities. Fe-VCZ (001) InP substrates. N_e= 30.

interface decrease to 2.0×10^{12}, 1.0×10^{13}, and 3.7×10^{12} cm^{-2}, respectively, as shown in Fig. 3. The reduction ratios of the sheet impurity concentrations are 36, 12, and 29% compared to those obtained in a conventional pre-heating procedure (N_e=0) using continuous TBP supply. Reported C, O, and Si concentrations at the GaAs epilayer/substrate interface using the normal pre-growth oxide desorption with ramping the substrate temperature from 500 to 650°C in molecular beam epitaxy (MBE) are $2{\sim}50 \times 10^{12}$, $0.5{\sim}1.0 \times 10^{13}$, and $3{\sim}7 \times 10^{12}$ cm^{-2} [5]. Decrease in the process temperature around 290°C is achieved in MSE realizing the almost identical impurity concentrations in MBE.

MSE temperature dependence is measured to study surface cleaning mechanism in MSE, as shown in Fig. 4. The activation energies (E_a) of C, O, and Si concentrations are 10, 67, and 17 kcal/mol. In order to study the desorption energy of impurities, the activation energy of etch rate on (111)B InP surface is studied, as shown in Fig. 5. The specular surface is achieved, thus the acti-

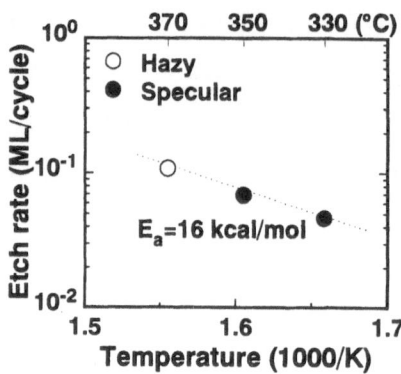

Fig. 5. Dependence of etch rate (monolayer per cycle) on etching temperature for Fe-LEC (111)B InP substrate. The surface morphology observed by Nomarski-interference microscopy is summarized as hazy and specular.

vation energy for desorption of In atoms is assumed to be 16 kcal/mol. The activation energies of C and Si correspond to that for etch rate on (111)B surface as well as that of 18 kcal/mol on (001) surface [3]. The activation energy of O corresponds to that of 61 kcal/mol for thermal decomposition of InP [3]. Thus, it is assumed that binding energies of C and Si on InP surface decrease in MSE owing to the reaction with TBP or TBP related reactants. This mechanism is assumed to be identical to In desorption mechanism in MSE.

4. Conclusions

The combination with MSE and MLE was studied to obtain clean epilayer/substrate interface. The concentrations of impurities, such as C, O, and Si, at the interface using MSE decreased to 36,12, and 29% compared to those using continuous TBP supply. It was assumed that the indium atoms on (111)B surface are easily removed by molecules of TBP or TBP related reactant in MSE with the activation energy of 16 kcal/mol. C and Si concentrations exhibited small activation energies of 10 and 17 kcal/mol, thus it was assumed that binding energies of C and Si on InP surface decrease in MSE owing to the reaction with TBP or TBP related reactants.

References

1. T. Suntola and J. Antson, Finnish Patent No. 52395 (1974) and US Patent No. 4058430 (1977).
2. J. Nishizawa and Y. Kokubun, in Ext. Abstr. 16th Conference on Solid State Device and Materials (1984), p. 1.
3. N. Otsuka, J. Nishizawa, Y. Oyama, H. Kikuchi, and K. Suto, Jpn. J. Appl. Phys. 38 (1999) 2529.
4. N. Otsuka, J. Nishizawa, H. Kikuchi, and Y. Oyama, J. Vac. Sci. Technol. A17 (1999) 3008.
5. A. J. SpringThorpe, W. T. Moore, A. Majeed and R. W. Steater, J. Vac. Sci. Technol. B11 (1993) 127.

Reflectance-Difference Spectroscopy of (001) InAs Surfaces in Ultrahigh Vacuum

T. Kita, H. Tango, K. Tachikawa, K. Yamashita, T. Nishino*

Department of Electrical and Erectronics Engineering, Faculty of Engineering, Kobe University, 1-1 Rokkodai, Nada ,Kobe 657-8501, Japan e-mail: kita@eedept.kobe-u.ac.jp

Abstract Reflectance-difference spectroscopy (RDS) is employed to study *in situ* the (4×2), (2×4), and (4×4) reconstructions of (001) InAs surfaces prepared in ultrahigh vacuum by molecular-beam epitaxy and simultaneously characterized by RHEED. The (001) InAs surface displays a similar phase diagram to the case of the (001) GaAs surfaces; an In-rich (4×2) structure and As-rich structures of (2×4) and (4×4). However, characteristic spectral features in terms of electronic excitations involving surface dimers of In and As are different from those of the (001) GaAs surfaces.

1 Introduction

The (001) InAs surface has attracted much attention during the past few decades for both technological and scientific reasons. As shown by scanning tunneling microscopy (STM) and diffraction probes such as reflection high-energy electron diffraction (RHEED) and x-ray diffraction (XRD), (001) InAs exhibits a variety of reconstructions depending on substrate temperature and surface stoichiometry.[1] The InAs surface shows a similar evolution of the surface reconstructions to that of GaAs. However, the detail is different; for example, the (2×4) of InAs shows the $\alpha2$ structure in contrast to the $\alpha1$ structure of GaAs, and the InAs surface shows a discontinuous first-order phase transition with hysteresis.

In this paper we present a systematic study of the various (001) InAs surfaces by reflectance-difference spectroscopy (RDS) for the first time. For optically isotropic materials, surface optical absorption contributes to the RD spectroscopy, which can extract information about dynamic processes at the surface. RD couples to local electronic structure rather than to long-range order, RD spectra not only determine surface reconstruction but also provide details not accessible by RHEED. [2,3]

2 Experiment

Our RD spectromator was mounted on a solid-source MBE system. The schematic diagram of the optical set up is shown in Fig.1. RD was measured between the complex near-normal-incidence reflectances of light linearly polarized along two principle axes [$\bar{1}$10] and [110], which can be expressed in terms of the surface dielectric anisotropy

$$\frac{\Delta \tilde{r}}{\tilde{r}} = \frac{4\pi i d}{\lambda} \frac{\epsilon_{\bar{1}10} - \epsilon_{110}}{\epsilon_s} \tag{1}$$

---

* *Present address:*Kobe City College of Technology Gakuen Higashimachi 8-3, Nishiku, Kobe 651-2194, Japan

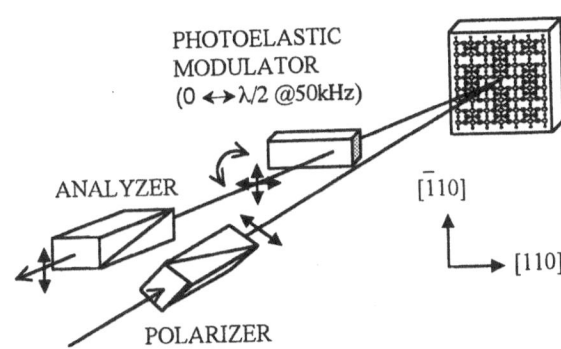

Fig. 1 Schematic diagram of RDS optics.

where ϵ_s is the bulk dielectric function of InAs, $\epsilon_{\bar{1}10}d$ and $\epsilon_{110}d$ are the surface dielectric response along [$\bar{1}$10] and [110] axis, respectively, and λ is the wavelength of the light. The MBE system is equipped with standard RHEED optics.

A 120-nm-buffer layer of InAs was grown at 540°C. RD spectra were measured as substrate temperature was lowered from 520°C to 100°C. RDS and RHEED measurements were carried out simultaneously under As$_4$ flux of about 5×10^{-5} Torr beam equivalent pressure. A one-monolayer deposition time was 60 s.

3 Results and Discussion

3.1 (4×2) and (2×4) reconstructions

Figure 2 summarizes RD spectra obtained as substrate temperature was lowered from 520°C to 100°C in an As$_4$ flux of 5×10^{-5} Torr. The observed RHEED patterns are listed. Above 500°C, the (001) InAs suface shows an In-stabilized (4×2) structure. The corresponding RD spectra indicate a negative signal near 1.5 eV due to the In-In dimers. With decreased temperature, the change in surface structure between In-stabilized (4×2) and As-stabilized (2×4) structures shows a discontinuous first-phase transition with hysteresis.[1]

A RD spectrum of the As-stabilized (2×4) at 450°C shows a positive signal near 2.2 eV corresponding to the optical transition occurring at the top-layer As-dimer atoms. This signal shifts to the higher energy side with decreased temperature in other words with increased As coverage, which results from a change of the As-dimer length. The same phenomena were expected in GaAs. Theoretical calculation by the *ab initio* pseudopotential

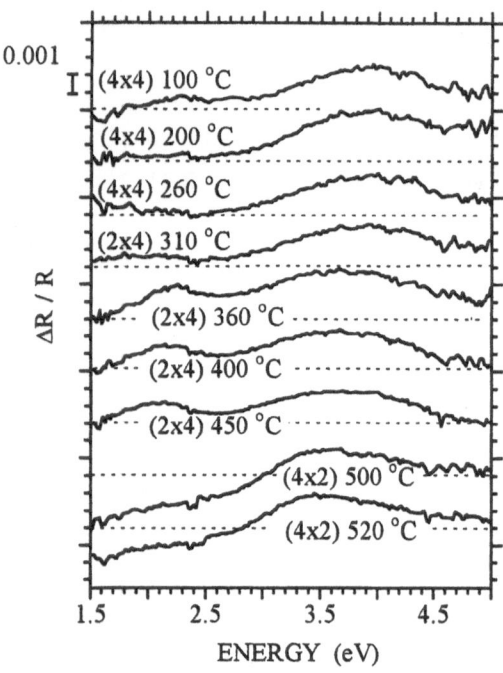

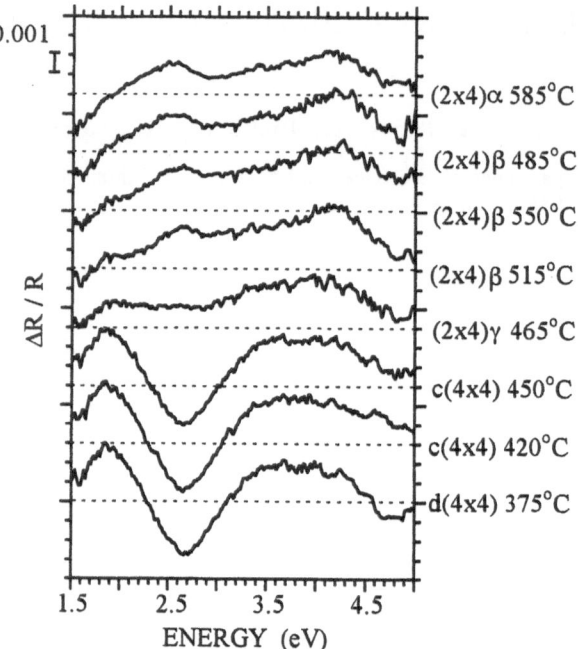

Fig. 2 RD spectra of a (001) InAs surface, obtained as substrate temperature was lowered from 520°C to 100°C in an As₄ flux of 5×10^{-5} Torr. The observed RHEED patterns are listed.

Fig. 3 RD spectra of a (001) GaAs surface, obtained as substrate temperature was lowered from 585°C to 375°C in an As₄ flux of 5×10^{-5} Torr. The observed RHEED patterns are listed.

method in local-density approximation predicted that the As-dimer length of the $(2\times4)\alpha1$ is longer than that of the $(2\times4)\beta2$.[4] In contrast to the GaAs case, the surface of InAs shows a transition from the $(2\times4)\beta2$ to the $(2\times4)\alpha2$.[1] Thus, the energy shift in the RD spectra indicates that the As-dimer length of the $(2\times4)\alpha2$ is longer than that of the $(2\times4)\beta2$.

3.2 c(4×4)reconstruction

At 310°C the RD signal due to the As-dimers becomes isotropic and its RHEED pattern was very similar to that of $(2\times4)\gamma$ GaAs. This isotropic feature comes from a coexistence of the [110]- and [1̄10]-As dimers. However, the further decrease of temperature does not cause an anisotropy near the As-dimer related signal, although the RHEED pattern shows an apparent c(4×4) surface as observed in GaAs. Figure 3 plots RD spectra of GaAs for various surface structures obtained as substrate temperature was lowered from 520°C to 100°C under the same As₄ flux. The results well agree with the previous report by Kamiya et al.[2] The RD spectra show a sign inversion near 2.5 eV corresponding to a rotation of top As-dimer direction. In the case of the c(4×4) GaAs, excess As atoms dimerize along the [110] and cause the strong negative RD signal near 2.5 eV. On the other hand, the anisotropy for the As dimers disappears on the c(4×4) InAs surface. Since in As-rich surfaces, it is important to note that the largest RD spectra feature

can be interpreted in term of the optical transition occurring at the top-layer As-dimer atoms, the lack of the strong negative signal in InAs indicates a isotropy of the top layer. This may caused by a *super* adsorption of As on this top-layer.

4 Summary

We investigated various (001) InAs surfaces in ultrahigh vacuum by RDS. RHEED patterns of the (001) InAs surfaces show the similar phase diagram to that of the (001) GaAs, while the RD spectra are quite different. We found that (1) the As dimer length of the $(2\times4)\alpha2$ is longer than that of the $(2\times4)\beta2$, and that (2) the surface dielectric fearture caused by the As dimers and bonds is isotropic in the c(4×4) surface.

References

1. H. Yamaguchi and Y. Horikoshi, Phys. Rev. **B51**, 9836 (1995).
2. I. Kamiya, D.E. Aspnes, L. T. Florez, and J. P. Harbison, Phys. Rev. **B46**, 15894 (1992).
3. R. Ares, J. Hu, P. Yeo, S. P. Watkins, J. Cryst. Growth. **195**, 234 (1998).
4. M.Maruyama and T.Nakayama, Jpn. J. Apply. Phys. **37**, 4109 (1998)

Proc. 25th Int. Conf. Phys. Semicond., Osaka 2000 (Eds. N. Miura and T. Ando)

355

Controlling of lateral and vertical order in self-organized PbSe quantum dot superlattices

G. Springholz,[1] M. Pinczolits,[1] V. Holy,[2] P. Mayer,[1] G. Bauer,[1] H. H. Kang,[3] L. Salamanca-Riba[3]

[1] Institut für Halbleiter- und Festkörperphysik, Johannes Kepler Universität Linz, A-4040 Linz, Austria
   G.Springholz@hlphys.uni-linz.ac.at
[2] Department of Solid State Physics, Masaryk University, Brno 61137, Czech Republic
[3] Department of Materials and Nuclear Engineering, University of Maryland at College Park, MD 20742-2115, USA

Abstract Self-organized lateral and vertical ordering in PbSe/Pb$_{1-x}$Eu$_x$Te quantum dot superlattices is investigated as a function of spacer thickness. From transmission electron and atomic force microscopy we find the occurrence of three different ordered dot structure: (a) for small spacer thicknesses, the dots are vertically aligned with weak hexagonal ordering tendency, (b) for intermediate spacer thicknesses, a well defined trigonal dot lattice with *fcc*-stacking and nearly perfect lateral ordering is formed, and for thick spacers (c), uncorrelated and laterally disordered superlattices are obtained. By finite element calculations, the formation of these different dot correlations is explained by the strong changes of the surface strain fields as a function of spacer thickness. Apart from the marked differences in the ordering tendency, also a qualitatively different scaling behavior of the lateral dot distances versus spacer thickness is observed in the different regimes.

1 Introduction

The spontaneous formation of three dimensional (3D) islands in the Stranski-Krastanow growth mode of highly lattice-mismatched heteroepitaxial layers has recently evolved as a novel technique for the fabrication of self-assembled semiconductor quantum dots [1,2]. It is based on the natural tendency of strained layers to form 3D islands on a wetting layer surface due to the high degree of strain relaxation allowed by the lateral expansion or compression of the islands. Under appropriate conditions, these islands exhibit sizes in the nanometer regime and are fully coherent (*i.e.*, dislocation free [3,4]) to the substrate. When embedded in a higher band gap matrix material, quantum boxes with atomic-like optical and electronic properties are formed. Due to the statistical nature of growth, however, these dots are usually not very uniform in size, shape and spacing, which poses severe limitations for device applications. Multilayering is one possible route for improving the uniformity of self-assembled dots [5-8]. Although in such structures the dots are separated by spacer layers, due to their interaction via elastic strain fields [5,8-10], long-range *vertical* correlations are formed, which may lead to a *lateral* ordering as well [5-9]. For various materials systems, different types of correlations have been observed, ranging from vertically aligned dot columns for InAs/GaAs [6,9] or SiGe/Si [11] superlattices, to trigonal dot lattices with *fcc* stacking for IV-VI superlattices [8]. This is due to the differences in the elastic anisotropy of the materials systems [10].

In the present work, we have investigated the self-organized vertical and lateral ordering in PbSe/Pb$_{1-x}$Eu$_x$Te quantum dot superlattices. It is shown that different types of cor-

relations are formed just by changes in the spacer thickness. Whereas for small spacer thicknesses, the dots are vertically aligned, an *ABCABC...* stacking sequence is formed for intermediate spacer thicknesses, and no correlations are found for thick spacer layers. The occurrence of these different structure is explained on the basis of finite element calculations of the elastic strain fields induced by the buried PbSe dots, showing a characteristic change when the spacer thickness becomes comparable to the buried dot size.

2 Experimental

A series of superlattice samples was grown by molecular beam epitaxy on PbTe buffer layers predeposited on (111) oriented BaF$_2$ substrates. The superlattice stacks consisted of 30 periods of 5 monolayers (ML) PbSe alternating with Pb$_{1-x}$Eu$_x$Te spacer layers of thicknesses ranging from 200 to 1000 Å. Due to the -5.4% lattice-mismatch with respect to PbTe, PbSe grows in the Stranski-Krastanow mode with the formation of 3D islands once the critical coverage of 1.5 ML is exceeded [12]. During the overgrowth of these islands, a rapid replanarisation takes place such that after 200 Å a smooth 2D surface is regained [7]. Identical growth conditions were employed for all samples with a substrate temperature of 360° C and PbSe Pb$_{1-x}$Eu$_x$Te growth rates of 0.08 ML/sec and 3.5 Å/sec, respectively. Thus, the only difference in the superlattices was the thickness and composition of the spacer layers, where the latter was adjusted in order to provide a complete strain symmetrization of the SL stack with respect to the PbTe buffer layer [7]. This prevents misfit dislocation formation and ensures identical strain conditions for the dot layers in all samples.

3 Results

For three representative samples, the atomic force microscopy (AFM) images of the last dot layer and the cross sectional transmission electron microscopy (TEM) images are shown in Figs. 1 to 3 for Pb$_{1-x}$Eu$_x$Te spacer thicknesses of d_S = 320, 465 and 660 Å, respectively. For small spacer thicknesses $d_S < 370$ Å (Fig. 1), we find that as the number of superlattice periods increases larger and larger PbSe dots are formed on the surface, with a dot density decreasing from the initial value of 550 μm^{-2} for the single layers to about 70 μm^{-2} after 30 SL periods. Correspondingly, the average dot height of 180 Å and base width of 700 Å is much larger as compared to that of single dot layers grown under identical growth conditions. The cross sectional TEM image of this sample (Fig. 1 b)) reveals that the dots are vertically aligned in *columns*, is similar as in InAs/GaAs [3,9] or SiGe/Si [11] dot superlattices.

Increasing the spacer thickness to 370 - 540 Å leads to a completely different dot arrangement. As shown in Fig. 2a), a *nearly perfect* hexagonal 2D lattice of dots is formed on the surface after 30 SL periods, with a substantial narrowing of the size dispersion [7]. In addition, the dot density of about 250 μm^{-2} is about four times larger as compared to the samples with thinner spacer layers, with a corresponding decrease of the dot height and width to 120 and 300 Å, respectively. As demonstrated by the cross sectional TEM of Fig. 2b), the PbSe dots are now aligned in directions *inclined* by 39° with respect to the growth direction (dashed lines in Fig. 2b)), and a well ordered trigonal 3D *lattice* of dots with a vertical *ABCABC...* stacking sequence is formed, in perfect agreement with our previous results from x-ray diffraction studies [8]. For spacer thicknesses larger than 550 Å (Fig. 3), another striking change occurs where the 2D hexagonal dot lattice is replaced by a highly disordered dot arrangement with an even higher dot density of

550 μm^{-2}. The cross-sectional TEM image of this sample (Fig. 3b) shows that the PbSe dots in each layer are randomly positioned relative to those in the adjacent layers. Thus, the whole SL structure just represents an uncorrelated repetition of disordered 2D single dot layers.

The different lateral ordering tendency is clearly reflected by the Fourier transformation (FFT) power spectra of the AFM images shown as insets in Figs. 1 -3. For the sample with intermediate spacer thicknesses, many well defined satellite peaks appear in the FFT image (see Fig. 2a)). The clear six-fold symmetry and the large number of higher order peaks indicates the formation of large perfectly ordered 2D hexagonal dot domains. The exceedingly high efficiency of the lateral ordering process is also corroborated by the corresponding cross-sectional TEM image, revealing that the lateral ordering sets in already within the first superlattice layers. For spacer thicknesses larger than 550 Å (Fig. 3a)), only a diffuse ring is observed in the FFT image. This indicates the lack of any lateral ordering tendency and a rather broad dispersion of dot spacings. For samples with small spacer thicknesses $d_s \le 370$ Å, although the dots seem to nucleate at random positions, six broad satellite side peaks are still visible in the FFT image (arrows in the

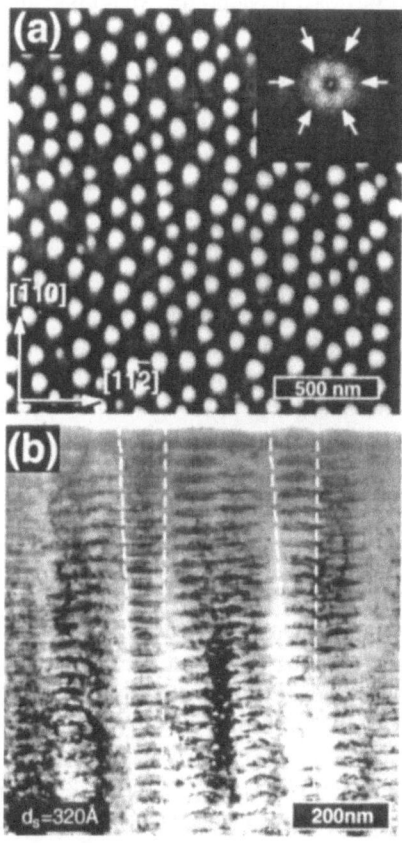

Fig 1 (a) AFM surface image and (b) cross-sectional TEM micrograph of a vertically aligned self-assembled PbSe dot superlattice with 320 Å Pb$_{1-x}$Eu$_x$Te spacer layer thickness and 30 SL periods. Insert: 2D FFT power spectrum of the AFM image.

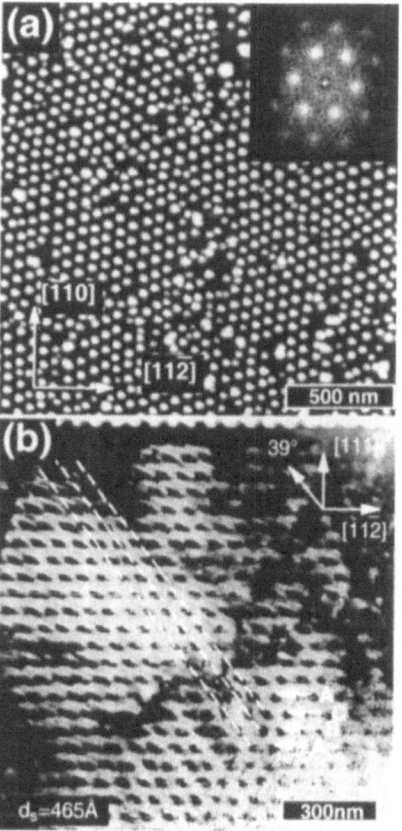

Fig 2 (a) AFM surface image and (b) cross-sectional TEM micrograph of a *fcc*-stacked self-assembled PbSe dot superlattice with 465 Å Pb$_{1-x}$Eu$_x$Te spacer layer thickness and 30 SL periods. Insert: 2D FFT power spectrum of the AFM image

inset of Fig. 1a)). This indicates a weak hexagonal ordering tendency with a rather large preferred lateral dot spacing. However, the absence of any higher order FFT satellites shows that only a short range order exists in this case.

A strikingly different behavior is also found for the scaling of the preferred in-plane dot separation as a function of spacer thickness. For small SL periods $D \leq 380$ Å, the lateral dot separation L is around 1300 Å, varying only slowly as $L = (0.6\ D + L_0)$ with $L_0 = 1080$ Å. At a critical period of $D_1^c = 390$ Å, L drops abruptly by a factor of three to 550 Å, and a nearly perfect 2D dot lattice is formed. In this regime, L scales *exactly* linearly with D as $L = \sqrt{3} \times D \tan \alpha$, where α is the layer-to-layer correlation angle of 39° observed by TEM. For SL periods exceeding a critical value of $D_2^c = 570$ Å, the mean lateral dot distance again drops abruptly to a value of 450 Å, and remains constant for all larger D. The corresponding dot density then equal to that of single dot layers grown under identical growth conditions. Both structural transitions are not completely abrupt, *i.e.*, for D around 390 Å actually mixed structures with coexisting trigonal and columnar dot regions are observed by TEM. Similarly, the second transition is manifested by a rapid but continuos drop of the layer-to-layer correlation probability within the trigonal dot arrangement.

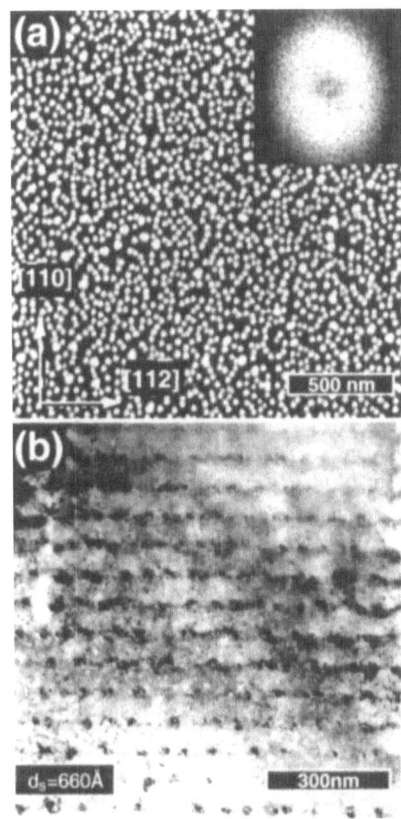

Fig 3 (a) AFM surface image and (b) cross-sectional TEM micrograph of an uncorrelated self-assembled PbSe dot superlattice with 660 Å $Pb_{1-x}Eu_xTe$ spacer layer thickness and 30 SL periods. Insert: 2D FFT power spectrum of the AFM image.

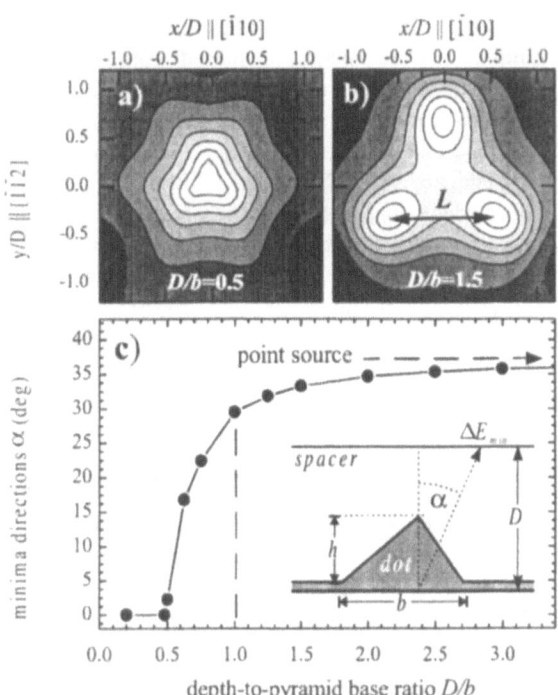

Fig 4 Surface strain energy distriubtions above a pyramidal PbSe quantum dot in a PbEuTe matrix derived from finite element calculations and plotted as a function of dimensionless surface coordinates x/D for two different dot depths of (a) $D = 0.5 \times b$ and (b) $D = 1.5 \times b$, where b is the dot base width. Brighter areas in the contour plots correspond to lower strain energies. (c) Directions α of the strain energy minima on the surface relative to the surface normal plotted as a function of the D/b ratio. The dot configuration for the calculations is shown in (c) as insert.

Another interesting feature is the very different evolution of dot sizes and shapes during superlattice growth. As evident from Fig. 1b), for samples with vertical dot alignment a pronounced successive increase of the island size and a broadening of the island shape occurs as the number of SL periods increases, similar as in vertically aligned InAs or SiGe dot superlattices. From AFM measurements this is found to be due to a gradual transition from initially sharp PbSe pyramids to truncated pyramids with a progressive increase of the top plateau width. For samples with *fcc* stacking, not only is a highly regular dot arrangement formed already in the first SL layers, but even more, the lateral dot spacing as well as the dot sizes remain *constant* throughout the whole SL growth. Because of this unique property and the narrowing of the size dispersion [7] extraordinarily homogenous 3D dot lattices are obtained.

The formation of different vertical correlations are explained as follows. Due to the lattice distortions around the buried dots a non uniform strain distribution is imposed on the epitaxial surface. This leads to a preferred island nucleation at the minima of the strain energy on the surface. As shown by our previous work [8,10], due to the very high elastic anisotropy of the IV-VI compounds and the chosen (111) growth orientation, the strain energy distribution caused by each buried PbSe dots exhibits three pronounced side minima that are laterally displaced from the surface normal direction. The preferred dot nucleation at these strain minima leads to an $ABCABC$... vertical dot stacking sequence [8] as well as to a very efficient hexagonal lateral ordering tendency. This is observed for the superlattices with intermediate spacer thicknesses. To explain the other types of dot correlations, the finite thickness of the spacer layer as well as the finite size of the buried quantum dots has to be taken into account. For this purpose we have performed a series of finite element calculations of the stress distribution around dots located at various depths D below the surface. According to our previous AFM studies [12], the PbSe dots were modelled as triangular pyramids with (100) side facets. This yields an aspect ratio of $b/h = 2.45$, where b is the dot base width and h the dot height. Introducing the dimensionless surface coordinates $x' = x/D$ along [-110] and $y' = y/D$ along [-1-12], it turns out that the shape of the energy distribution is only determined by the ratio of D/b, i.e., on the ratio of the depth to the island base width. The results of these calculations are shown in Fig. 4.

For spacer thicknesses larger than the island base, i.e., $D/b > 1$, the strain energy distributions $E(x', y')$ closely resemble that obtained from our previous point source model [8,9], with three well separated minima along the $<-1-12>$ directions. This is exemplified in Fig. 4 b) for $D/b = 1.5$. The directions of these minima are inclined by about $\alpha = 35°$ with respect to the surface normal, in close agreement to the correlation angle observed by TEM. However, when D/b decreases below 1, α rapidly decreases to zero (see Fig. 4 c)). This means that the three minima are replaced by one central minimum located exactly above the buried dot. This is illustrated by the calculated strain energy distribution shown in Fig. 4 a) for the case of $D/b = 0.5$. As a consequence, preferential nucleation of new dots then occurs directly above each buried dot, which corresponds to a crossover from the fcc dot stacking to a vertical alignment of the dots, as is observed by the experiments. A more detailed analysis shows, that this cross-over takes place when the lateral separation L of the strain minima becomes smaller than the dot base width. Then, the island size becomes incompatible with the trigonal dot arrangement. Application of this condition to the actual parameters of our experiments leads to an expected critical spacer thickness of $D_1^c = 310$ Å, for this transition, in close agreement with the experiments. On the other hand, as the spacer thickness increases, the depth of these strain minima ΔE_{min} rapidly decreases as $1/D^3$ [13]. As a result, the layer-to-layer dot correlation probability drops to zero at a certain spacer thickness, which is experimentally the case for $D_2^c = 560$ Å, the upper boundary for the trigonal dot structure.

Due to the changes in the PbSe dot size as a function of the growth conditions, the positions of the critical superlattice periods D^c also change upon by changes in the growth conditions. In particular, at lower substrate temperatures, which result in smaller dot sizes in the first SL layer, the phase boundaries are shifted to smaller values, which means that fcc-stacked dot superlattices can be also obtained for SL periods smaller than 380 Å. This effect can therefore be utilized in order to extend the tunability range of the trigonally ordered dot superlattice structure. At the same time, however, the corresponding smaller dot sizes also result in more shallow strain energy minima on the surface. Thus, the interlayer dot correlations are then lost already for SL periods below 550 Å.

4 Conclusions

In conclusion, different types of vertical correlations are observed in PbSe quantum dot superlattices as a function of the $Pb_{1-x}Eu_xTe$ spacer thickness. The different vertical correlations have dramatic effects on the lateral ordering tendency as well as on the scaling of the lateral dot spacing as a function of spacer thickness. In addition, a different evolution of dot sizes and shapes occurs. The transition between the differently correlated structures are caused by the dependence of the elastic interaction between the dots on the thickness of the spacer layer, as well as by the finite lateral dot sizes. Similar trends are be expected also for other material systems.

Acknowledgements

This work was supported by the Fond zur Förderung der Wissenschaftlichen Forschung, the Austrian Academy of Science and the Gesellschaft für Mikroelektronik, Austria.

References

1. D. Leonard, M. Krishnamurty, C. M. Reaves, S. P. Denbaar, and P. Petroff, Appl. Phys. Lett. **63** (1993) 3203.
2. J.M. Moison, F. Houzay, F. Barthe, L. Leprince, E. Andre, and O. Vatel, Appl. Phys. Lett. **64** (1994) 196 .
3. S. Guha, A. Madhukar, and K.C. Rajkumar, Appl. Phys. Lett. **57** (1989) 2110.
4. D.J. Eaglesham and M. Cerullo, Phys. Rev. Lett. **64** (1990) 1943.
5. J. Tersoff, C. Teichert, and M.G. Lagally, Phys. Rev. Lett. **76** (1996) 1675.
6. G.S. Solomon, S. Komarov, J.S. Harris, and Y. Yamamoto, J. Cryst. Growth **175/176** (1997) 707.
7. M. Pinczolits, G. Springholz, and G. Bauer, Phys. Rev. B **60** (1999) 11524.
8. G. Springholz, V. Holy, M. Pinczolits, and G. Bauer, Science **282** (1998) 734.
9. Q. Xie, A. Madhukar, P. Chen, and N. Kobayashi, Phys. Rev. Lett. **75** (1995) 2542.
10. V. Holy, G. Springholz, M. Pinczolits, and G. Bauer, Phys. Rev. Lett. **83** (1999) 356.
11. E. Mateeva, P. Sutter, J.C. Bean, and M.G. Lagally, Appl. Phys. Lett. **71** (1997) 3233.
12. M. Pinczolits, G. Springholz, and G. Bauer, Appl. Phys. Lett. **73** (1998) 250.
13. G. Springholz, M. Pinczolits, P. Mayer, V. Holy, G. Bauer, H. Kang, and L. Slamanca-Riba, Phys. Rev. Lett. **84** (2000) 4669.

Shape analysis of single and stacked InAs quantum dots at the atomic level by cross-sectional STM

D.M. Bruls[1], J.W.A.M. Vugs[1], P.M. Koenraad[1], M.S. Skolnick[2], M. Hopkinson[2] and J.H. Wolter[1]

[1]COBRA Inter-University Research Institute, Eindhoven University of Technology, P.O. Box 513, 5600 MB Eindhoven, The Netherlands, e-mail: p.m.koenraad@tue.nl.
[2]Department of Physics, University of Sheffield, Hounsfield Road, Sheffield, S3 7RH, United Kingdom

Abstract. We present a study of InAs self-assembled quantum dots in GaAs by cross-sectional scanning tunneling microscopy (X-STM). Our results shows that the dots consist of an InGaAs alloy and that the indium content increases towards the top. The analysis of the height versus base length relation obtained from cross-sectional images of the dots show that the shape of the dots resembles that of a truncated pyramid with a square base that is oriented along the [100] and [010] directions. Our results about the shape and size agree with the analysis of previous photocurrent measurements on these samples.

1 Introduction

There is currently a strong interest in self-assembled quantum dots. Their unique electronic properties, mostly due to their delta-shaped density of states, make many novel applications and the improvement of existing structures possible. There is also a very strong interest in the fundamental physics of these zero-dimensional structures. InAs quantum dots in a GaAs matrix, grown by MBE in the Stranski-Krastanov growth mode, are of particular interest as they can be obtained most easy. Most applied and fundamental research is hampered by the fact that no proper knowledge exists about the shape, size and composition of these self-assembled InAs dots. It is possible to investigate these dots using atomic force microscopy (AFM), but after covering them with a capping layer, the dot shape can change significantly and accordingly its electronic properties [1]. Therefore cross-sectional techniques are highly needed to study the shape, size and composition of covered self-assembled quantum dots [2,3]. In this paper we present a X-STM study of the shape, size and composition of InAs self-assembled dots that were previously characterized by extensive photo-current measurements [4]. We will compare the results obtained by both techniques.

2 Experimental

The X-STM measurements were performed under UHV ($p<4*10^{-11}$ torr) conditions, using an Omicron STM1, TS-2 scanner. The tips were prepared as described in [5]. The measurements were performed on *in situ* cleaved (101) or (011) surfaces. All structures were grown by MBE at a growth temperature of 512 ^{0}C and contained five layers of low growth rate (0.01ML/s) InAs dots within a GaAs matrix. Coupling between the dots is unimportant as the

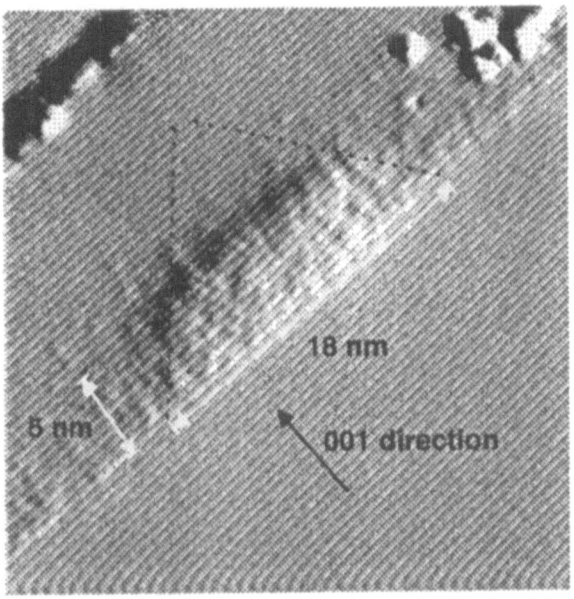

Fig. 1 X-STM current image (35 nm by 35 nm) of a self-assembled InAs quantum dot and the wetting layer.

spacing between the dot layers was 50 nm. The amount of InAs deposited during formation of the dots and the wetting layers was 2.4 ML.

3 Results and discussion

We investigated a series of self-assembled quantum dots by X-STM. A typical cross-sectional image of an investigated dot is presented in figure 1. The images were obtained in the constant current mode during which both the topography and the current image were recorded. As the height regulation is never exactly instantaneous the current image can be interpreted as the spatial derivative of the topographic image. In these current images the atomic corrugation and contrast are very pronounced. This enables us to make a clear distinction between the InAs self-assembled dot and the surrounding GaAs matrix, thus facilitating a better determination of the size and shape of the dots. Of this particular dot the height is 5 nm and the base length 18 nm and it has a trapezoidal shape. To the left and right of the dot, also a part of the wetting layer can be distinguished.

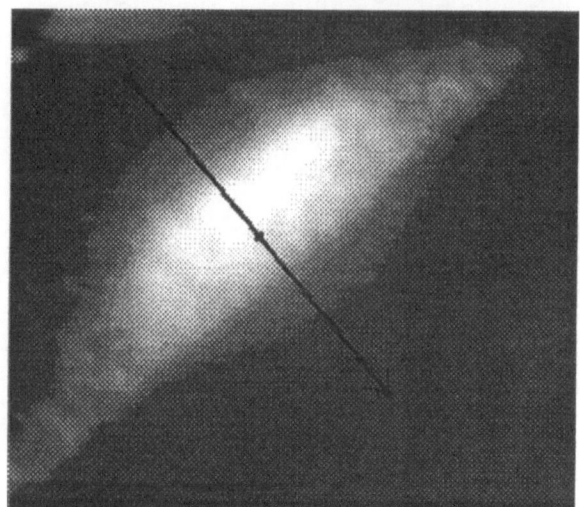

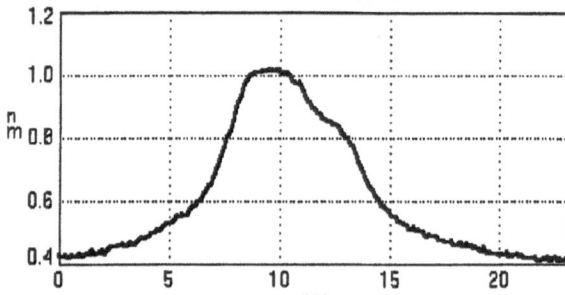

Fig. 2 Unfiltered topography image (15 nm by 17.5 nm) and line profile taken across the dot

The self-assembled dots do not consist of pure InAs. In the dots many small, short ranged height fluctuations are visible, which indicate that an InGaAs alloy is produced during the formation of the quantum dots. During growth, several growth processes like segregation [6] and strain related indium incorporation [7] could lead to an indium gradient in the self-assembled dot, resulting in a higher indium concentration in the top of the dot.

After cleavage the self-assembled dot relaxes outwards due to the strain resulting from the large lattice mismatch between InAs and GaAs (7%). This is clearly visible in a line-profile taken across the dot in figure 2. We assume that the magnitude of this outward relaxation is linked to the local indium concentration inside the quantum dot [6,7]. At the top of the dot the relaxation is the strongest indicating that the indium concentration increases towards the top of the dot.

Performing measurements on the natural (110) cleavage plane implies that the shape and size of the dot in the cross-sectional images depends on the position of the dots with respect to the cleavage plane and the actual dot-shape. To determine whether the quantum dots are shaped like a full or truncated pyramid, many dots were investigated and their cross-sectional dimensions measured. Various models for shape and orientation of the dots are proposed in the literature [2,8] and it has been shown that the shape

before capping is already strongly dependent on the growth parameters during MBE [9]. Therefore it is very important to determine the 3 dimensional shape, size and orientation of capped self-assembled dots. We considered 4 different dot geometries; namely full or truncated pyramids aligned either along the (100) direction or the (110) direction. The height and base length of 18 different dots found in our X-STM measurements were analyzed. Full pyramids aligned along the (100) direction would always give triangular cross-sections as the natural cleavage is along the (110) direction. This is not observed. If the dots would be aligned along the (110) direction we would observe cross-sections shaped like a trapezium having all the same base length but with a varying height. Again this is not the case. In agreement with the geometry of truncated pyramids aligned the (100) direction we observe that the cross-sections have the same height for a large range of base lengths.

For base lengths from 5 nm to 25 nm we observe the same height of about 5 nm. The maximum observed base length of 25 nm is the diagonal of the square base, so the actual base length of the pyramid is 18 nm. Exactly the same self-assembled dot structure was investigated using extensive photocurrent measurements [4]. In these measurements the dots were biased by an external electric field applied over the p-i-n structure. Fitting the transition energies as a function of the electric field over the dots showed that they could have the shape of a truncated pyramid with about the same dimensions as we found.

In literature, there are also models that propose lens shaped or cylindrical shaped dots. These models do not agree with our measurements, as these shapes would result in completely different cross-sectional images that are not observed.

References

1. N.N. Ledentsov *et al.*, Phys. Rev. B **54** , 8743 (1996)
2. O. Flebbe, H. Eisele, T. Kalka, F. Heinrichsdorff, A. Krost, D. Bimberg, J.Vac. Sci. Technol. B **17**, 1639 (1999)
3. Legrand, B. Grandidier, J.P. Nys, D. Stiévenard, Appl. Phys. Lett. **73**, 96 (1998)
4. P.W. Fry *et al.*, Phys. Rev. Lett. **84**, 733 (2000)
5. G.J. de Raad, P.M. Koenraad, J.H. Wolter, J. Vac. Technol. B **17**, 1946 (1999)
6. M. Pfister, M.B. Johnson, S.F. Alvarado, H.W.M. Salemink, Appl. Phys. Lett. **67**, 1459 (1995)
7. N. Liu, J.Tersoff, O. Baklenov, A.L. Holmes, C.K. Shih, Phys. Rev. Lett. **84**, 334 (2000)
8. V.A. Shchukin, D. Bimberg, Rev. Mod. Phys. **71**, 1123 (1999)
9. I. Daruka, J. Tersoff, A.L. Barabási, Phys. Rev. Lett. **82**, 2753 (1999)

Formation process of InAs dots including Mn atoms and their physical properties

S. Okumura, H. Asahi, Y.K. Zhou, J. Asakura, M.Kanamura, K. Asami, H. Kubo*, C. Hamaguchi* and S. Gonda

The Institute of Scientific and Industrial Research, Osaka University
8-1, Mihogaoka, Ibaraki, Osaka 567-0047, Japan
*Department of Electronic Engineering, Osaka University, Suita Osaka 565-0871, Japan

Abstract Mn-including InAs quantum dot structures are fabricated on (100) semi-insulating GaAs substrates by metal-organic molecular beam epitaxy (MOMBE) and their formation process is fully studied with atomic force microscopy. InAs grows along lateral direction around Mn nuclei on GaAs substrate at the early stage of growth and then grows along vertical direction, forming InAs dots including Mn atoms. Magneto-photoluminescence (M-PL) studies are performed up to 10 T in the Faraday configuration. Larger diamagnetic shift of ground state electron-hole transition of the M-PL spectra are compared with that of self-formed S-K mode grown InAs/GaAs quantum dots is observed.

1. Introduction

Recently III-V diluted magnetic semiconductors (DMS), such as (In,Mn)As and (Ga,Mn)As, have offered great opportunities to study interesting magneto-optical phenomena arising from the interaction of carries with local spins [1]. Because of the lower equilibrium solubility of Mn ions into the III-V semiconductor, it is difficult to fabricate low-defect nanostructures of III-V DMS. So it has been mainly studied in II-VI related materials rather than that of III-V related materials [2].

In order to overcome these difficulties, we have suggested the growth of the quantum dot (QD) structures including Mn atoms as a nucleus grown by the use of enhanced decomposition effect of metaloranic (MO) group-III precursors at mono-atomic steps. To fabricate Mn-including InAs dots trismethylindium (TMIn) and trisdimethylaminoarsenic (TDMAAs) are used as group III and group V sources, respectively, in the metalorganic molecular beam epitaxy (MOMBE) growth system [3].

In this paper we report the formation process of Mn-including InAs quantum dot structures and their M-PL studies.

2. Experiment

The growth of Mn-including InAs quantum dot structures was conducted in the MOMBE system [3]. After the growth of 300 nm-thick GaAs buffer layer on GaAs substrate, Mn atoms were first deposited for 3-10 seconds and MnGa or MnAs islands including several hundreds of Mn atoms in each island were formed.

Then TMIn molecules were supplied with a cycle of 1 second-supply and 9 seconds-interruption, while TDMAAs flux was supplied without interruption. Because of the enhanced decomposition effect of TMIn at mono-atomic steps of Mn islands, Mn-including InAs dots were formed. During growth, the sample surface was monitored with reflection high energy electron diffraction (RHEED). The formation process was studied by AFM measurements on the grown surfaces as a function of TMIn supply time.

M-PL measurements were performed on these Mn-including Mn-including dots samples (sample A) as well as on the self-formed Stranski-Krastanov (S-K) mode-grown InAs dots samples (sample B) in which Mn atoms are not included. Both samples are overgrown by 60 nm-thick GaAs cap layer.

In the M-PL measurements, the magnetic field was applied up to 10 T normal to the sample plane.

3. Results and discussion

3.1 Formation process of Mn-including InAs dots

To study the formation process of Mn-including InAs dots the sum of the TMIn supply time was varied from 5 to 9 sec. During the growth of these time periods, RHEED showed streak pattern. Fig.1 (a) and (b) show the histograms of the diameter (lateral size) and the height of Mn-including InAs dots, respectively.

The dashed curves show the fitting curves with a Gaussian function. With increasing TMIn supply time, the diameter of the dots steadily increased. On the other hand, the height does not increase until 7 sec. When the TMIn supply time is increased to 9sec, the height of the dot becomes large rapidly.

The formation process is considered as follows. At first, InAs dots grow along the lateral direction due to the preferential decomposition of TMIn at the atomic-step edges of the Mn-including islands. The growth of InAs along the lateral direction results in the formation of disk-like InAs dots, whose height and diameter are about 5-6 Å and about 200 Å, respectively. When the lateral size exceeds some critical size the growth along the vertical direction and/or change of the dot shape toward vertical direction occur to reduce the total energy of the system and the height of the Mn-including InAs dots rapidly increases.

When we stop the TMIn supply before the rapid increase in the height of the dots, we can obtain the disk-like dots including Mn atoms (sample A). The disk-like dots are expected to have lower defects compared with Mn-including InAs dots after the rapid growth along vertical direction.

3.2 Magneto-photoluminescence studies

M-PL was studied in the range of magnetic field 0-10 T in Faraday configuration. The evolutions of PL peak energies as a function magnetic field are shown in Fig.2 (a) disk-like Mn-including dots, (b) InAs wetting layer (WL) for sample A and Fig.3 (a) self-formed InAs dots, (b) InAs WL for sample B. All four peaks show diamagnetic shift which is related to s-like ground state electron-hole transition and can be written by following Eq.(1)

$$\Delta E = e^2 < r^2 > B^2/4\mu \qquad (1)$$

where e is electric constant, $\Delta r = 2 < r^2 >^{1/2}$ is a measure of the in-plane spatial extent of wave function perpendicular to the magnetic field, and μ is reduced mass of exciton [4]. The value of $e^2 < r^2 >/4\mu$ is diamagnetic constant whose unit is $\mu eV/T^2$. We adopt exciton reduced mass of both dots and WL for $\mu = 0.035 m_e$ from Ref. [5]. Solid curves in Fig.2 and Fig.3 are fitting curves of parabolic function. From coefficients of these parabolic functions, diamagnetic constants are estimated and in-plane spatial wave functions are calculated.

Diamagnetic constants of WL for sample A (58.53 $\mu eV/T^2$) and sample B (58.53 $\mu eV/T^2$) are almost same and larger than those of dots for sample A (48.67 $\mu eV/T^2$) and B (22.33 $\mu eV/T^2$). This result is reasonable from viewpoint of the spatial extent of the exciton wave function along lateral direction.

On the other hand, diamagnetic constant of Mn-including InAs dots is two times as large as that of S-K mode grown InAs dots. The calculated Δr of exciton wave function in the dots for sample A and B are 12.5 nm and 8.4 nm, respectively. From the histogram of Fig.1 the lateral size of Mn-including InAs dots is about 20 nm, which is almost the same size as that of S-K mode-grown InAs dots. So there are some discrepancies between calculated Δr and dot sizes.

The reason of this result is not clear at present. But it seems that the Mn d-like wave function affects something to expand the exciton s-like wave function in the Mn-including quantum dots.

Acknowledgment

This work was partly supported by a Giant-in-Aid for the Center of Excellence (COE) from the Ministry of Education, Science, Sports and Culture.

References

1. H. Ohno *et al*, Phys. Rev. Lett., **68**, 2664 (1992)
2. Y. Oka *et al*, Journal of Luminescence, **70** (1996) 35-47
3. J. Sato *et al*, J. Cryst. Growth. 188 (1998) 363
4. R.K. Hayden *et al*, Physica B 249-251 (1998) 262-266
5. I.E. Itskevich *et al*, Appl. Phys. Lett. **70** (4)

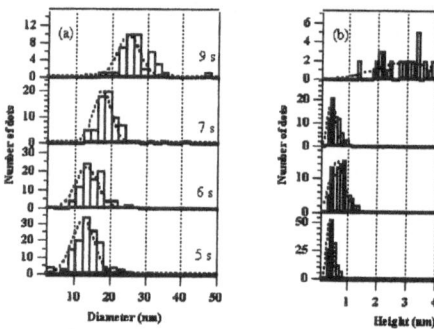

Fig.1 Histograms for (a) diameter and (b) height for the Mn-including InAs dots as a function of the TMIn supply times (5-9) seconds. AFM measurement was conducted on the area of 1x1 μm^2. The dashed curves show Gaussian fitting curves.

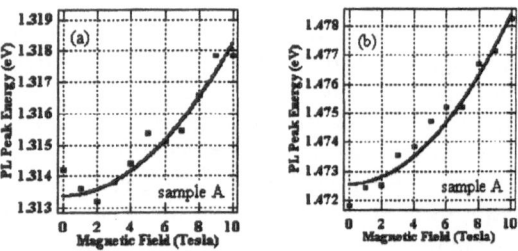

Fig.2 10 K PL peak energies as a function of magnetic field for (a) disk-like Mn-including dots on GaAs substrate (b) InAs WL. Magnetic field is perpendicular to the sample plane.

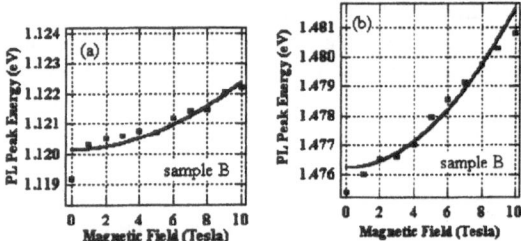

Fig.3 10 K PL peak energies as a function of magnetic field for (a) self-formed S-K mode grown InAs dots on GaAs substrate and (b) InAs WL.

Shape and size of buried SiGe islands

J. Stangl[1], V. Holý[2], A. Daniel[1], T. Roch[1], G. Bauer[1], T.H. Metzger[3], J. Zhu[4], K. Brunner[4], G. Abstreiter[4]

[1] Institut für Halbleiterphysik, Johannes Kepler Universität Linz, Austria e-mail: j.stangl@hlphys.uni-linz.ac.at
[2] Laboratory for Thin Films and Nanostructures, Masaryk University Brno, Czech Republic
[3] European Synchrotron Radiation Facility, Grenoble, France
[4] Walter Schottky Institut, Technische Universität München, Germany

Abstract We present a method to obtain the shape and the lateral correlation properties of *buried* islands from grazing incidence small angle x-ray scattering experiments. From reciprocal space maps recorded for various penetration depths, the island parameters are obtained by comparison with simulations based on distorted wave Born approximation. The method is demonstrated on self-organized SiGe islands in a Si/SiGe multilayer. It is possible to detect different shapes of the islands at the sample surface and those embedded in the multilayer.

1 Introduction

The shape of quantum dots influences considerably their electronic properties, and hence many studies on quantum dots addresses the their shape and its dependence on the growth conditions. For virtually any structure in a practical application, the dots have to be overgrown, which often is accompanied by a considerable change of their size and shape [1]. Hence methods are required, which allow for the characterization of *buried* quantum dots. Beside transmission electron microscopy (TEM), x-ray scattering techniques have successfully been applied to the study of free-standing and buried dots. In order to identify solely the *shape* of buried dots, the scattered intensity around the (000) Bragg reflection can be used [2, 3]. In this paper, we present an approach for the quantitative interpretation of grazing incidence small angle x-ray scattering (GISAXS) data. Describing the scattering process in the framework of the distorted-wave Born approximation (DWBA) [4], information on the shape and the positional correlation of buried *and* free-standing quantum dots is obtained. For this purpose, 2D reciprocal space maps (RSMs) are recorded in GISAXS geometry for various vertical momentum transfer values. From these maps, the autocorrelation spectra are calculated, which are simulated using a short-range order model of the dot positions and different models of the shape of the quantum dots [5].

2 Experiment

We have performed GISAXS measurements on SiGe islands embedded in a Si/SiGe superlattice on (001) oriented Si. The sample was grown by molecular beam epitaxy at a growth temperature of $550\,°C$ and consists of 20 bilayers of 2.5 nm thick $Si_{0.55}Ge_{0.45}$ and a 10 nm thick Si spacer layer [6]. On the top surface a dot layer with different parameters than that within the superlat-

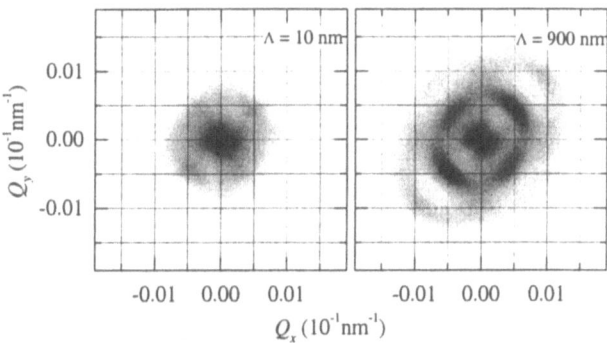

Fig. 1 In-plane intensity distributions measured in GISAXS geometry for two limiting information depths Λ.

tice was grown. X-ray diffraction and TEM show that Ge rich islands with about 120 nm spacing and 1.5 nm height form within the multilayer. From the GISAXS measurements, we can clearly distinguish between the islands at the sample surface and those within the superlattice, demonstrating the sensitivity of our method.

The experiments were carried out at TROÏKA II beamline (ESRF, Grenoble). Scans of the diffusely scattered intensity were measured as a function of the in-plane angle ϕ, for various incidence and exit angles $\alpha_{i,f}$, determining the penetration depth Λ and the vertical momentum transfer Q_z. From the scans measured at given Q_z we constructed the 2D intensity distribution in the Q_x-Q_y plane. The RSMs for maximum and minimum penetration depth are shown in Fig. 1. Performing the numerical 2D Fourier transformations of these maps, we obtained the functions $J(\mathbf{x}; Q_z)$ shown in Fig. 2. This function contains several contributions [5]:

$$J(\mathbf{x}; Q_z) = Sn_0 \left[\Phi(\mathbf{x}; q_z) + p(\mathbf{x}) \otimes \Phi(\mathbf{x}; q_z) \right], \quad (1)$$

where

$$\Phi(\mathbf{x}; q_z) = \frac{1}{4\pi^2} \int d^2\mathbf{Q}_\| \left| \Omega^{FT}(\mathbf{Q}_\|, q_z) \right|^2 e^{i\mathbf{Q}_\| \cdot \mathbf{x}} \quad (2)$$

depends on the shape of the dots described by $\Omega(\mathbf{x})$, $p(\mathbf{x})$ is the pair correlation function of the dot positions, describing the in-plane dot distribution, and $\otimes$ means the convolution. For small $|\mathbf{x}|$ the second term in Eq. (1) is constant, and the shape of the function J depends only on the *dot shape* [5]. For larger $|\mathbf{x}|$, the first term in Eq. (1) is zero. Assuming that the dot distance is much larger than the lateral dot size, the convolution in Eq. (1) can be replaced by a simple product, and $J(\mathbf{x})$ is proportional to the *dot correlation* function $p(\mathbf{x})$.

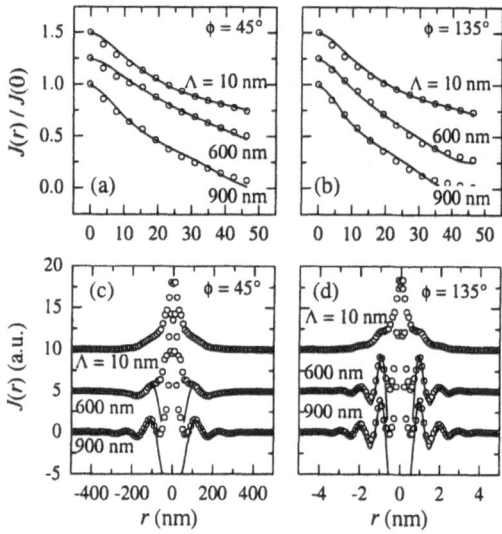

Fig. 2 (a,b) measured (points) values of J in the region for small $r = |\mathbf{x}|$ for two different azimuthal directions ϕ and three different information depths Λ and their fits using an ellipsoidal dot model (thin lines). (c,d) for large r the experimental data for J (circles) have been fitted using the SRO model for the dot arrangement (lines).

3 Results

For small $|\mathbf{x}|$ we have compared the the values of J with a simple model assuming that the dots have the shape of an upper half of an ellipsoid. From the fit of the function J to that model we determined the lateral half-axes of the ellipsoid a and b ($a > b$), the vertical half-axis c (the dot height) and the azimuthal angle α of the longer axis of the ellipsoid with respect to [110] direction. The quality of the fits can be judged from Fig. 2(a,b), where we have plotted radial cuts from the 2D distributions of J in the azimuthal directions $\phi = 45°$ and $135°$ with respect to [110] along with the fits.

As follows from Table 1, the values of a, b and c depend on Q_z. This fact can be ascribed to the Q_z dependence of the x-ray information depth $\Lambda = 1/\mathrm{Im}(q_z)$, ranging from 10 nm to about 1 μm. For the smallest Λ, only the topmost dot layer is probed by the x-rays, and consequently the obtained parameters are those of the islands of the surface. With increasing Λ we obtain parameters of the dot shape which represent an average over all dot layers weighted by the actual x-ray intensities inside the sample. For the measurement with the smallest Λ the values of a and b were found similar, so that the azimuthal angle α of the axis a could not be determined with sufficient accuracy.

The pair correlation function of the dot positions was obtained from the radial cuts of J for larger $|\mathbf{x}|$. We have compared these cuts with the 1D pair-correlation function assuming a short-range-order (SRO) model of the dot distribution [7] with Gamma distributed dot distances. From the fits to the experimental data shown in Fig. 2(c,d), we have determined the mean dot distance $\langle L \rangle$ and the order m of the Gamma distribution as func-

Table 1 Obtained parameters of the dot shapes and lateral ordering.

Λ (nm)	a (nm)	b (nm)	c (nm)	α (deg)
10	38 ± 6	36 ± 6	4.5 ± 0.4	–
600	35 ± 2	23 ± 2	2.6 ± 0.1	51 ± 6
900	25 ± 2	20 ± 2	2.3 ± 0.2	48 ± 20

Λ (nm)	ϕ (deg)	$\langle L \rangle$ (nm)	m
10	–	–	–
600	45	120 ± 5	5
600	135	101 ± 5	10
900	45	120 ± 5	9
900	135	101 ± 5	14

tions of the azimuthal angle ϕ (see Table 1). The function J obtained from the GISAXS scans with the smallest Λ exhibits no lateral maxima, so that the determination of $\langle L \rangle$ and m was not possible. Likely, the number of the irradiated interfaces, and consequently, the number islands is too small in this case, so that the wave scattered from the dots is blurred by, e.g., scattering from surface roughness with larger lateral correlation length.

The data in Table 1 show that the dots are arranged approximately in a square array along $\langle 100 \rangle$ directions, however, the dispersion of the dot distances in the [100] direction parallel to the miscut is smaller than in the other one due to the step bunching during growth of the multilayer [6]. The positions of the dots obey very well the SRO model.

In conclusion, we have demonstrated that GISAXS experiments on *buried and freestanding* islands can be quantitatively interpreted using a theoretical model based on the DWBA. Autocorrelation spectra are simulated using a SRO model of the dot positions and different models of the shape of the quantum dots. From the fits to the experimental data, the shape as well as the lateral correlation of the buried islands is obtained. Since the illuminated area is of the order of several mm^2 , the GISAXS results represent a statistical average over a large number of dots. Additionally, GISAXS scans for sufficiently small information depth give information solely on the top layer of unburied Ge islands.

References

1. E. Mateeva, P. Sutter, and M. G. Lagally, Appl. Phys. Lett. **74**, 567 (1999).
2. M. Schmidbauer, Th. Wiebach, H. Raidt, M. Hanke, R. Köhler, H. Wawra, Phys. Rev. B **58**, 10523 (1998).
3. M. Rauscher, R. Paniago, H. Metzger, Z. Kovats, J. Domke, J. Peisl, H.D. Pfannes, J. Schulze, I. Eisele, J. Appl. Phys. **86**, 6763 (1999).
4. S.K. Sinha, E.B. Sirota, S. Garoff, H.B. Stanley, Phys. Rev. B **38**, 2297 (1988).
5. J. Stangl, V. Holý, T. Roch, A. Daniel, G. Bauer, J. Zhu, K. Brunner, G. Abstreiter, Phys. Rev. B, in print.
6. J.H. Zhu, K. Brunner, G. Abstreiter, Appl. Phys. Lett. **72**, 424 (1998).
7. P.R. Pukite, C.S. Lent, P.I. Cohen, Surface Science **161**, 39 (1985).

Optical Anisotropy of Stranski-Krastanov Growth Surface of InAs on GaAs (001)

T. Kita[1] H. Tango[1], K. Tachikawa[1], K. Yamashita[1] T. Nishino[1]* T. Nakayama[2] M. Murayama[3]

[1] Department of Electrical and Erectronics Engineering, Faculty of Engineering, Kobe University, 1-1 Rokkodai, Nada ,Kobe 657-8501, Japan e-mail: kita@eedept.kobe-u.ac.jp

[2] Department of Physics, Faculty of Science, Chiba University, 1-33 Yayoi, Inage, Chiba 263-8522, Japan

[3] Department of Electronics Engineering, The University of Electro-Communication s,1-5-1 Chofugaoka, Chofu, Tokyo 182-8585, Japan

Abstract Reflectance-difference spectroscopy (RDS) is employed to study *in situ* the Stranski-Krastanov (S-K) growth surface of InAs on GaAs (001) during molecular-beam epitaxy and simultaneously characterized by RHEED. We found three-type time evolutions of (In,Ga)As wetting layer depending on substrate temperature. The surface structures of the wetting layer were studied by comparing experimentally observed RD spectra measured at various InAs coverage with calculations by the *ab initio* pseudopotential method in a local density approximation and the tight-binding surface-linear-response method.

1 Introduction

Recently, self-assembled quantum dots (QDs) in semiconductor heterostructures are of great interest [1], because of their discrete atomlike energy levels, good optical properties and promising device applications such as QD laser and infrared photo-detector. In lattice mismatched system such as InAs/GaAs, self-assembled QDs can be achieved by the Stranski-Krastanov (S-K) growth mode. To control the growth mode, it is important to understand a relation between the structure of the two dimensional wetting layer and the island formation. Reflectance difference (RD) spectroscopy has been emerging as a powerful tool to characterize *in situ* growth surfaces in various environments.[2,3] For optically isotropic materials, surface optical absorption contributes to the RD spectroscopy, which can extract information about dynamic processes at a growth front. We present a systematic study of initial stages of self-assembled growth of InAs dots on GaAs (001), starting from As-rich c(4×4) surface, by using a solid-source molecular-beam epitaxy equipped with RDS and RHEED optics.

2 Experiment

Our RD spectromator was mounted on a solid-source MBE system. RD was measured between the complex near-normal-incidence reflectances of light linearly polarized along two principle axes [$\bar{1}$10] and [110], i.e., $(r_{\bar{1}10}-r_{110})/(r_{\bar{1}10}+r_{110})$. The MBE system is equipped with RHEED. A 360-nm-buffer layer of GaAs was grown at 530°C before depositing InAs. A temperature range for the InAs deposition was from 300°C to 480°C. We confirmed the c(4×4) reconstruction of the As-rich GaAs(001)

* *Present address:* Kobe City College of Technology Gakuen Higashimachi 8-3, Nishiku, Kobe 651-2194, Japan

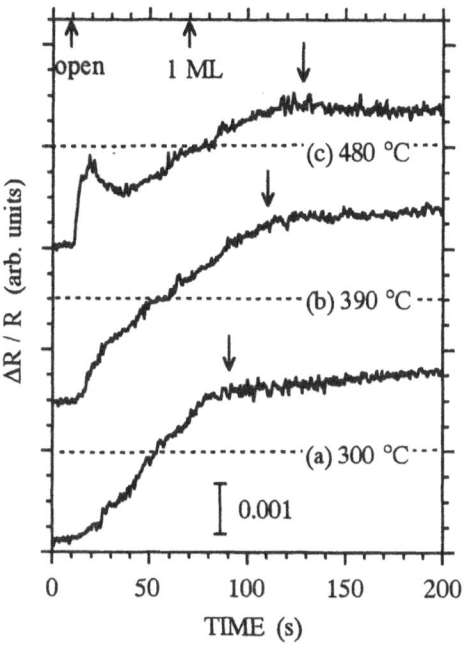

Fig. 1 Typical time resolved RD intensities at 2.5 eV during the InAs deposi tion as a function of substrate temperature. The arrows indicate the growth-mo de transition from the two-dimensional growth to the three-dimensional island growth, which was estimated from the ($\bar{1}$15)-RHEED spot intensity.

surface before the deposition. The As$_4$ beam equivalent pressure was about 5×10^{-5} Torr. A one-monolayer deposition time of InAs was 60 s.

3 Results and Discussion

3.1 Dynamic change in RD intensity

RD intensity was monitored at 2.5 eV, which corresponds to the optical transition occurring at the top-layer As-dimer atoms. Figure 1 plots the data. Three type evolutions of the wetting layer were found as a function of growth temperature. (a) At a growth temperature of 300°C, the negative RD signal due to the [110]-As dimersis gradually inverted after opening the In-cell shutter. This indicates a formation of As bonds along [$\bar{1}$10] according to the appearance of a new (1×3) surface of the wetting layer. The (1×3) surface is a (In,Ga)As ternary alloy having an excess As-As dimers [4], which appears

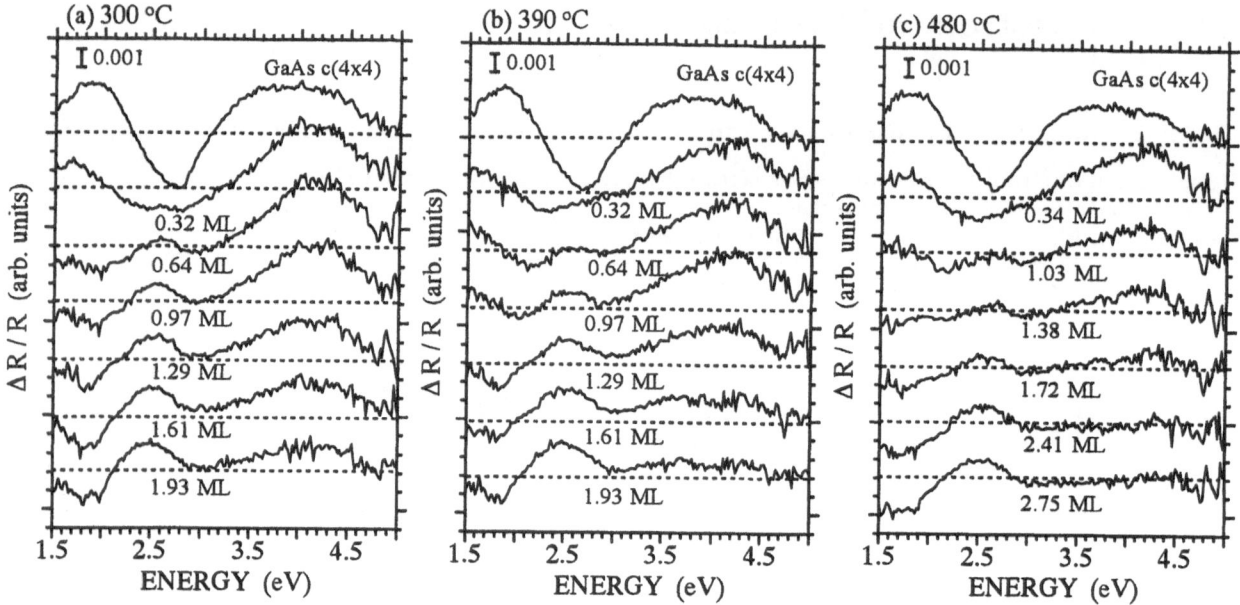

Fig. 2 RD spectra taken at various deposition of well defined amounts of InAs on GaAs (001).

under an As-rich condition. A (2×3) surface appears near the growth-mode transition from the two-dimensional growth to the three-dimensional island growth. The inverted RD signal shows saturation after the island growth near 1.5-ML deposition of InAs. (b) At 390°C, on the other hand, the growth-mode transition occurs near 1.7-ML deposition. Furthermore, this growth type shows a transition of the wetting surface from the (1×3) to the (2×4) via a (2×3) incommensurate phase near the growth-mode transition. (c) At 480°C, the commensurate and incommensurate phase transition was clearly observed just after opening the In-cell shutter. The rapid decrease of the anisotropy indicates that a small amount of deposition of In causes desorption of excess As atoms.

3.2 Evolution of RD spectra during InAs deposition

The surface structures of the wetting layer at various growth temperatures were confirmed by RD spectra measured at the initial and successive steps of wetting-layer formation. Figure 2 shows RD spectra measured at 300°C, 390°C, and 480°C. The As-dimer related signal near 2.5 eV shows the sign inversion and a red shift with increased deposition. This red shift indicates an expansion of the As-dimer lengh. We found that the experimentally observed spectra are well explained when the wetting surfaces changes from (1×3) to (2×4)α2 via (2×3)α2. These surfaces have both the In-In and As-As dimers. At 300°C, strongly localized states of the [110]-As dimer in the (1×3) cause the negative peak near 2 eV, while at high temperature this peak becomes weak and a signal due to the cation dimer along the [110] becomes dominant instead. The cation dimer related signal shits to the lower energy side with increased InAs deposition, becase of cation mixing. From our total energy calculation, the

(2×4)α2 structure is not stable under an As-rich condition. Then, the high growth temperatures of 390°C and 480°C causes the As-poor surface and the three-dimensional island growth occurs on this InAs-wetting surface.

4 Summary

We studied the wetting layer surface during the S-K growth of InAs on GaAs(001) by RDS. The surface structures of the wetting layer were obtained by comparing experimentally observed RD spectra measured at various InAs coverage with calculations by the *ab initio* pseudopotential method in a local density approximation and the tight-binding surface-linear-response method. At low substrate temperature, the growth transition occurs on the (2×3)α2 surface of (In,Ga)As, while at high temperature the transition occurs at the (2×4)α2 surface of InAs. Both In-In and As-As dimers are necessary to explain the observed RD spectra. RD spectra show expansion of the As-dimer length and the cation dimer length at the initial growth stage with increased InAs deposition.

References

1. D. Bimberg, M. Grundmann, and N. N. Ledentsov, Quantum Dot Heterostructures (J. Wiley, New York, 1998).
2. I. Kamiya, D.E. Aspnes, L. T. Florez, and J. P. Harbison, Phys. Rev. B**46**, 15894 (1992).
3. E. Steimentz, J. -T. Zetter, W. Richiter, D. I. Westwood, D. A. Woolf, and Z. Sobiesierski, J. Vac. Sci. Technol. B**14**, 3058 (1996).
4. M. Sauvage-Simkin, Y.Garreau, R.Pinchaux, M.B.Véron, J.P.Landesman, and J.Nageru, Phys. Rev. Lett. **75**, 3485 (1995).

Lasers based on self-assembled InAs/GaAs and InP/InGaP quantum dots

O. G. Schmidt, M. O. Lipinski, Y. M. Manz, H. Heidemeyer, W. Winter, K. Eberl

Max-Planck-Institut für Festkörperforschung, Heisenbergstraße 1, D-70569 Stuttgart, Germany

Abstract Room-temperature lasing is demonstrated for self-assembled InP/InGaP quantum dots. A 2 mm long laser diode exhibits a threshold current density of 2.3 kA/cm^2 in pulsed operation. Stimulated emission occurs via the ground state at $\lambda = 728$ nm. A maximum light output power of 250 mW is obtained without any saturation effects. 1.3 μm light emission is achieved with self-assembled InAs/GaAs quantum dots on GaAs (001) substrates using an extremely low deposition rate of only 0.01 monolayers/s at a growth temperature of 500 °C. The photoluminescence peak widths are as narrow as 23 meV. Lasing occors on the third excited state of an Al-free device. Internal quantum efficiencies and internal losses are determined at room temperature.

1 Introduction

Quantum dot (QD) lasers promise to have superior characteristics than quantum well lasers [1,2] and extremely low threshold current densities have been presented, recently [3,4]. In particular, the possibility to fabricate GaAs based quantum dot lasers, which emit at the important 1.3 μm wavelength [5], has made these devices extremely attractive. Here, we report room temperature lasing of red-light emitting InP/GaP quantum dots and infra-red lasing of InAs/GaAs quantum dots.

2 InP/InGaP quantum dot laser

We concentrate on one specific sample, which is grown with solid source molecular beam epitaxy (SSMBE). The optical active region of the laser structure consists of a 3-fold stack of nominally 3.3 monolayers (ML) InP quantum dots separated by 4 nm thick Ga$_{0.52}$In$_{0.48}$P spacer layers. The dots are vertically aligned and are electronically coupled as described earlier [6]. The quantum dot stack is inserted in the middle of a 145 nm thick Ga$_{0.52}$In$_{0.48}$P waveguide. A detailed description of the growth procedure is given in Ref. [7]. The sample is processed into 50 μm broad-stripe lasers with uncoated cleaved facets to form a 2 mm long cavity.

Fig. 1 shows the light output power versus current density of the described device. An optical power of 250 mW is measured at a current density of 4 kAcm^{-2}. There are no saturation effects and the slope efficiency η_s is 8.5 %. Extrapolation of the linear regime yields a threshold current density of $j_{th} = 2.3$ kA/cm^2.

It is interesting to compare the edge emission of the laser with the corresponding surface emission from the same device at a current density of $j = 1.2 \times j_{th}$, as presented in the inset of Fig. 1. The spontaneous surface emission reveals luminescence peaks originating from the QD ground and excited state as well as from the wetting

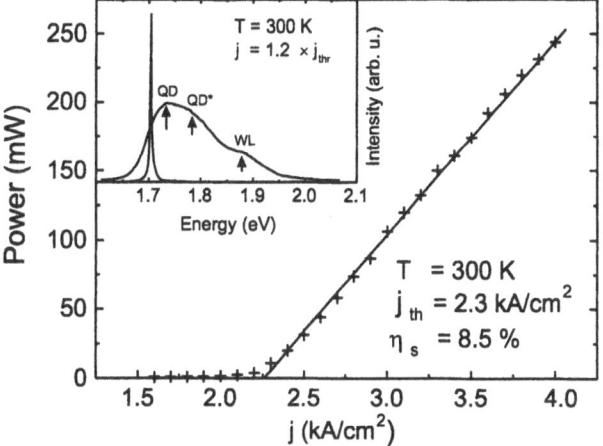

Fig. 1 Light output power versus current density of the InP/InGaP laser device. The threshold current density is 2.3 kAcm^{-2}. A maximum output power of 250 mW is measured without output power saturation. Inset: Comparison of edge and surface emission of same device at $j = 1.2 \times j_{th}$. Surface emission reveals ground, excited, and wetting layer states. Lasing occurs via the ground state.

layer. The lasing line appears on the low energy side of the QD ground state, thus prooving that lasing occurs on the ground state transition of the quantum dots.

3 InAs/GaAs quantum dot laser

The laser structures are fabricated by SSMBE on n$^+$-GaAs(001) substrates. The active layer consists of three layers of 2 ML InAs, grown at a very low In flux of 0.01 ML/s at a substrate temperature of $T_s = 500$ °C. Due to thermal In desorption a net InAs growth rate of approximately 0.007 ML/s is observed. The GaAs spacer between the QD layers was chosen thick enough as to prevent vertical alignment, which causes non-resonant emission of the different QD layers [8]. During the QD overgrowth the first 5 ML of the GaAs spacer are deposited at $T_s = 485$ °C. While growing the rest of the laser structure T_s is raised again. A detailed analysis on how growth rate and overgrowth temperature affect the emission wavelength has been described in Ref. [8] and Ref. [9]. The active layer is positioned between n-and p-doped Al-free Ga$_{0.52}$In$_{0.48}$P cladding layers.

The sample is processed into 50 μm broad-stripe lasers with uncoated cleaved facets to form 0.5,1,2, and 3 mm long cavities. Electroluminescence (EL) and lasing experiments are carried out with 50μm long pulses at a repitition rate of 20 kHz.

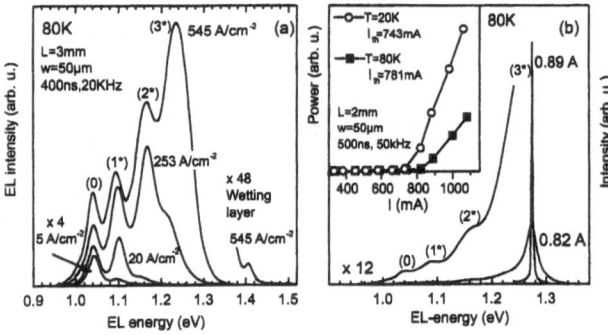

Fig. 2 (a) EL spectra at different current densities. Ground and excited state transitions are well-resolved. (b) Lasing spectrum of a 2 mm device. Note that lasing occurs on the third excited state (3*). Inset shows threshold at 20 K and 80 K.

Low-temperature (80 K) EL spectra are presented in Fig. 2 (a) for a 3 mm long device for current densities from 5Acm^{-2} to 545Acm^{-2}. Well-resolved peaks originating from the ground state (0) and excited state transitions (1*-3*) are observed (The structure is grown in such a way that ground state emission occurs exactly at 1.3 μm at room temperature). The peak linewidths are as narrow as 23 meV. At about 1.4 eV the wetting layer related energy transition is visible. It is noteworthy that with increasing current density excited states of higher order experience larger intensities than the ground state and excited states of lower order. We attribute this effect to larger degeneries and hence larger populations of excited states – very simiar to the increased degeneracy of outer shells in atoms. It is therefore not surprising that at even higher current densities the structure starts to oscillate via the third excited state (Fig. 2 (b)). There are simply more states available at higher energy, which can provide enough gain to compensate the mirror and internal losses of the device. The EL spectra are given for 0.82 A and 0.89 A. The inset shows the output power – current characteristic at two different temperatures for a 2 mm long laser. At 20 K the threshold current is 743 mA, whereas at 80 K it increases to 781 mA.

Lasing via excited states is also observed at room-temperature. Fig. 3 (a) shows the threshold current density as a function of the mirror losses, which are varied by changing the cavity length [10]. Fitting the data with a straight line yields a transparency current of $j_\infty = 614$ Acm^{-1}. There is only a minor wavelength shift with shorter cavities. This stability suggests that lasing occurs in a linear gain-current regime, where gain saturation plays practically no role. A detailed analysis of this effect was given in Ref. [10]. In Fig. 3 (b) the inverse external quantum efficiency η_d is plotted as a function of the cavity length for laser cavitiy widths $w = 50, 100,$ and 200 μm. From the slope an internal quantum efficiency $\eta_i = 0.25$ and internal losses $\alpha_i = 12$ cm^{-1} are determined [11].

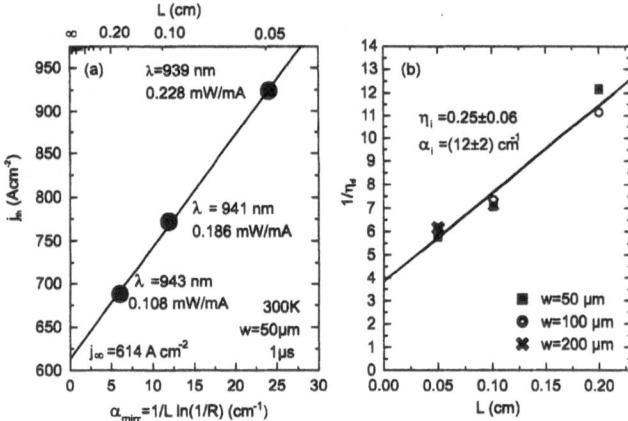

Fig. 3 (a) Threshold current density j_{th} as a function of mirror losses α_{mirr}, i.e. cavity length. Extrapolation yields threshold for a laser without mirror losses (j_∞). (b) shows the inverse external quantum efficiency η_d versus the cavity length for different stripe widths w. Internal quantum efficiency η_i and losses α_i are determined from (b).

4 Conclusion

We have presented room-temperature lasing of different lasers incorporating InP/InGaP as well as InAs/GaAs quantum dots. The InP/InGaP quantum dot laser oscillates via the ground state at 728 nm and has a 250 mW output power without saturation effects. 1.3 μm spontaneous ground-state emission is achieved for InAs/GaAs quantum dots. In an Al-free structure lasing occurs via excited states, which provide enough gain to outweigh the laser losses. Cavity length dependent measurements of threshold current densities and external quantum efficiencies suggest a linear gain-current regime for excited state lasing.

References

1. Y. Arakawa and H. Sakaki, Appl. Phys. Lett. **40**, (1982) 939-941.
2. M. Asada, Y. Miyamoto, and Y. Suematsu, IEEE J. Quantum Electron. **22**, (1986) 1915-1921.
3. G. Park, O. B. Shchekin, and D. L. Huffaker, and D. G. Deppe, IEEE Photonics Techn. Lett. **13**, (2000) 230-232.
4. X. Huang, A. Stintz, C. P. Hains, G. T. Liu, J. Cheng, and K. J. Malloy, IEEE Photonics Techn. Lett. **12**, (2000) 227-229.
5. D. L. Huffaker, G. Park, Z. Zou, O. B. Shchekin, and D. G. Deppe, Appl. Phys. Lett. **73**, (1998) 2564-2566.
6. M. K. Zundel, P. Specht, K. Eberl, N. Y. Jin-Phillipp, and F. Phillipp, Appl. Phys. Lett. **71**, (1997) 2972-2974.
7. Y. M. Manz, O. G. Schmidt, and K. Eberl, Appl. Phys. Lett. **76**, (2000) 3343-3345.
8. M. O. Lipinski, H. Schuler, O. G. Schmidt, K. Eberl, and N. Y. Jin-Phillipp, Appl. Phys. Lett. **77**, (2000) 1789-1791.
9. M. O. Lipinski, N. Y. Jin-Phillipp, O. G. Schmidt, and K. Eberl, Proc. 12th Internat. Conf. Indium Phosphide and Related Materials 2000, 215-218.
10. O. G. Schmidt et al., El. Lett. **32**, (1996) 1302-1304.
11. O. G. Schmidt et al., Proc. 8th Internat. Conf. Indium Phosphide and Related Materials 1996, 727-730.

Effect of vertical size uniformity on diffraction contrast images of stacked InGaAs/GaAs quantum dots

M. De Giorgi[1], A. Passaseo[1], R. Cingolani[1], A. Taurino[2] and M. Catalano[2]

[1] INFM, Dip. di Ingegneria dell'Innovazione, Università di Lecce, 73100 Lecce, Italy email: milena.degiorgi@unile.it
[2] Istituto CNR-IME, Consiglio Nazionale delle Ricerche, via Arnesano 73100 Lecce, Italy

Abstract A detailed TEM investigation was performed on capped single and vertically stacked $In_{0.5}Ga_{0.5}As$/GaAs quantum dots (QDs). The different diffraction contrast observed from the plan-view TEM images in the samples was correlated to the strain fields in the 3D islands.

1 Introduction

In recent years, quantum dots (QDs) have attracted much interest for their improved electronic and optical performances. In particular, QDs stacked in vertically organized columns are required to increase the optical density of the active medium and to change the emission wavelength. Their size is a critical parameter for controlling the optoelectronic properties of the devices. We present a criterion to evaluate the dot-size uniformity along the stacking direction by plan view TEM image. We show that in the vertically stacked dots, the strain fields associated to dots with different size coherently superpose inducing a modulation of the total strain field. This results in a modification of the electron diffraction conditions.

2 Experimental Setup

The samples consist of $In_{0.5}Ga_{0.5}As$/GaAs capped single and 6-fold stacked quantum dots grown in the same growth conditions by MOCVD on (001) exactly oriented GaAs substrates, by the well known Stranski-Krastanov method. TEM investigations were performed by using a Jeol 4000 EX microscope, operating at 400 kV, with an interpretable resolution limit of 0.16 nm.

3 Experimental Results

Fig.1 shows the on-zone plan-view bright field (B. F.) images of both single (a) and 6-stacked dot layers (b), obtained in the [001] zone axis. In the single dot, the contrast is characterized by an external dark region, of nearly circular-shape, with a bright spot at the center. The stacked dots show a completely different feature, i.e. an intensity modulation, resulting in a flower-like pattern. The contrast line-scan performed along the <010> directions is shown in Fig.1c-d. A single central maximum is observed for the single dot, whereas the stacked dots show three maxima of different intensity: the external ones being weaker and symmetric with respect to the central one.

4 Theoretical model

The white/black diffraction contrast in the TEM images is due to the inhomogeneous lattice strain associated to the 3D islands [1]. This induces a local variation of the lattice planes orientation, resulting in a local modification of the electron diffraction conditions. The images of the cross-section

image of a stack of quantum dots shows that the base diameter changes along the stacking direction.

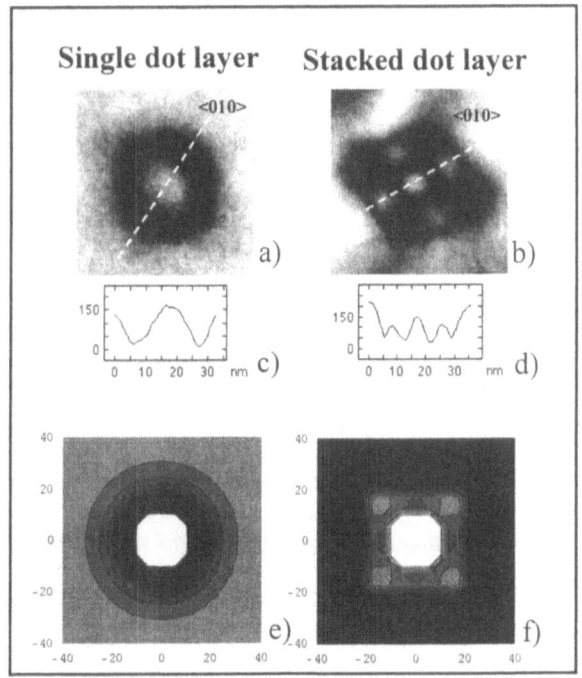

Fig. 1 [001] plan view-images obtained in the on-zone B.F. imaging conditions from a single (a) and a vertically stacked dot sample (b). The contrast line scans are also reported (c-d). Contour plot of the calculated strain field ($\varepsilon_{xx} + \varepsilon_{yy}$) for single (e) and six vertically stacked quantum dots (f) showing close analogy with the TEM diffraction contrast.

As a result, the observed differences in the contrast pattern in Fig. 1a-b are due to the different size of the dots in the 6-fold stacked samples. We believe that the strain field associated to each dot family overlaps inducing a modulation of the strain field along the stacking direction which results in a modulation of the transmitted electron beam.

In order to test this hypothesis, we have studied the effect of the linear combination of the strain fields associated to dots of different dimensions. The exact strain field in capped quantum dots is generally calculated with the finite element method. However, at first approximation we assumed it is equal to the strain field felt by quantum dots induced by stressors of simple parallelepiped shape and calculated the QD strain field by using the analytical method reported in ref.2. The obtained field has the same characteristics of one obtained with the finite element method.

We developed the functions describing the surface profile (the stressor pattern) in a Fourier series: the stress tensor σ is then calculated by superimposing the stress field associated to each cosine component. Since the amplitude of the relevant Fourier components are much smaller than the corresponding wavelength, we looked for a solution of the elasticity equations

$$(1+\nu)\, \nabla^2\sigma + \nabla^T\nabla\mathrm{Trace}(\sigma) = 0$$
$$\nabla\cdot\sigma = 0 \qquad\qquad\qquad (1)$$

having the form of a series expansion:

$$\sigma\,(x, y, z) = \sum_{\alpha=0}^{\infty} t^\alpha \sigma^\alpha\,(x, y, z) \qquad (2)$$

where ∇^T is the transpose of the gradient vector ∇ and ν is the Poisson's ratio. The boundary conditions are given by the requirement that no net force acts on the free surface. The strain tensor ε is then obtained by applying the Hooke's law and the displacement vector u is related to the ε by the well known equation

$$\varepsilon_{ij} = \frac{1}{2}\left(\frac{\partial u_i}{\partial x_j} + \frac{\partial u_j}{\partial x_i}\right) \qquad (3)$$

Only the strain components leading to a local variation of lattice planes which are parallel to the electron beam give rise to contrast in the TEM plan-view images, namely ε_{xx}, ε_{zz}, ε_{xz} and ε_{zx}. Since the off-diagonal components are smaller than the diagonal terms, in our discussion we consider only the term $\varepsilon_{xx} + \varepsilon_{zz}$.

In a simple model with only two stacked quantum dots of different size, we observe that the dilated region of the dot A (bigger dot) (Fig.2a) is larger than the dot B (smaller dot) (Fig.2b). As a consequence, the electrons transmitted at the center of the stacked dots (around $x = z = 0$) feel a strain dilatation in both quantum dots, whereas the electrons at the edges of the dots (at about 15 nm far from the center) feel a strain dilatation in the dot A and a strain compression in the dot B. The resulting strain field felt by the electrons is the coherent superposition of the strain fields associated to the two dots (Fig.2c). A main maximum occurs at the center of the structure, due to the combination of the strain field related to expanded regions (Fig.2a-b), whereas the secondary maxima are generated by the combination of regions where the structure is expanded (in the bigger dot A) and regions where the structure is compressed (in the smaller dot B). The induced modulation of the strain (Fig.2c) results in a correspondent modulation of the electron diffraction conditions. Our model is readily extended to the real case, in which the sample consists of six dot layers whose size changes continuously from the bottom to the top layers. In Fig.1e-f we display the calculated contour plot of the component $\varepsilon_{xx} + \varepsilon_{zz}$ for single and six stacked dots. The white and black zones correspond to the expanded and compressed regions, respectively, whereas the different gray zones display the intermediate conditions. As expected, in the single dot (Fig.1e), strain dilatation is observed into the island whereas strain concentration occurs at the edges. As a consequence, the transmitted electron beam is diffracted in a different way at the center and at the edges of the quantum dots resulting in the diffraction contrast of Fig.1a. In the six vertically stacked layers, the modulation of the strain field becomes evident and a flower-like pattern occurs. For dots

with equal size the contour plot is the same to that of a single dot.

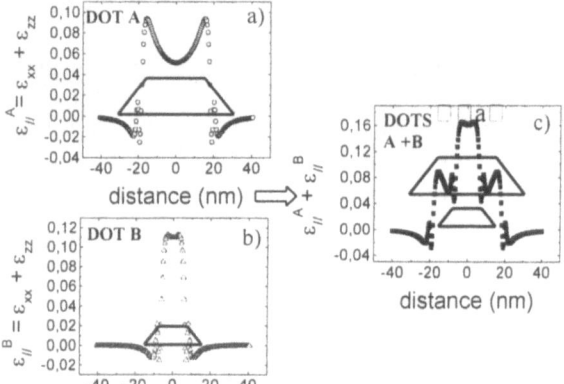

Fig. 2 Plots of the calculated strain components $\varepsilon_{xx} + \varepsilon_{zz}$ along the diagonal direction for two dots with different sizes. The circles (a) and triangles (b) display the strain field of big (dot A) and small dot (dot B), respectively. The squares (c) represent the coherent superposition of the strain field associated to dot A and dot B.

It is clear that the calculated strain patterns are not perfectly identical to the TEM contrast. The remote areas surrounding the dots are dark reflecting small strain whereas the corresponding areas are bright in the TEM images. This is surprising because in one case we look at a strain pattern and in the other case at a diffraction pattern. However, the goal of our model is to show that the coherent superposition of strain fields associated to quantum dots of different sizes induces a modulation of the total strain field that results necessary in a modulation of the electron diffraction contrast.

In conclusion, the result of our calculations are found to describe qualitatively the strain modulation for single and 6 stacked layer dots with a surprisingly good accuracy, shining light on the diffraction contrast observed in the plan view TEM images of our samples. Even though we are aware that a quantitative analysis of the contrast pattern and the strain field would need advanced simulations of the TEM images in which the modeled strain field is used as an input for the dynamical electron scattering [3-4], our experimental observations and theoretical results suggest that the different diffraction contrast can be due to a coherent superposition of the strain fields associated to dots of different sizes. This induces a modulation of the total strain field resulting in a modification of the electron diffraction conditions. Therefore from the plan-view diffraction contrast, we can get information about the uniformity of the dots in the stack and of the variations of their relative sizes.

References

1. J. Y. Yao, T. G. Andersson and G. L. Dunlop, *J. Appl. Phys*, **69**, 2224 (1991)
2. M. Mazzer, M. De Giorgi, R. Cingolani, G. Porello, F. Rossi and E. Molinari, *J. Appl. Phys*, **84**, 1 (1998)
3. X. Z. Liao, J. Zou, X. F. Duan, D. J. H. Cockayne, R.
4. T. Benabbas, P. Francois, Y. Androussi and A. Lefebvre, *J. Appl. Phys.*, **80**, 2763 (1996)

Lateral distribution of buried self-assembled InAs quantum dots on GaAs

K. Zhang[1], J. Falta[2], Ch. Heyn[1], Th. Schmidt[2], W. Hansen[1]

[1] Institute for Applied Physics, University of Hamburg, Jungiusstrasse 11, 20355 Hamburg, Germany

[2] Institute for Solid State Physics, University of Bremen, Postfach 330440, 28334 Bremen, Germany

Abstract   We have performed grazing incidence small angle x-ray scattering (GISAXS) experiments to characterize the lateral distribution of InAs quantum dots (QDs) in a single dot layer buried beneath a 10 nm GaAs cap layer grown with molecular beam epitaxy. The azimuthal distribution of satellite peak intensities caused by the diffuse scattering of InAs QDs are observed to be anisotropic. By analyzing the peak profiles, the mean dot-dot distance and correlation length of QD lateral distribution are derived. We find that the most pronounced ordering of QD distribution is in [110] direction.

1 Introduction

Semiconductor heterostructures exhibiting quantum confinement in three dimensions (3D) - socalled quantum dots (QDs) - are of high interest for fundamental research as well as technology [1]. For the preparation of QDs focus of interest has recently been devoted to self-assembling mechanisms occurring in III-V materials grown by molecular beam epitaxy (MBE) with highly strained heterolayers, e.g. InAs/GaAs. A few years ago the first experimental evidence for the existence of zero-dimensional electronic states in self-assembled InAs QDs was obtained with capacitance and Far Infrared Spectroscopy [2]. The structural organization of the dots is a crucial element in such systems, which may strongly influence their optoelectronic property [3-5]. Dots with coherent strain that are dislocation-free, and have a surprisingly narrow size distribution are highly desired. In the ideal case one would envisage the dots with high ordering in lateral distribution. For electronic device applications, the QDs in general must be capped/or buried by a so-called cap layer. This strongly obstructs direct investigations of the dot lateral distribution if performed by atomic force microscopy (AFM). Grazing incidence small angle x-ray scattering (GISAXS) generated by synchrotron radiation has been demonstrated to be also suited for the determination of statistical parameters describing the lateral distribution [6]. Here, we show that it is even possible to apply this technique on dot layers buried beneath a cap layer.

2 Experimental

The sample growth procedure is similar to that published previously [6]. The InAs deposit was chosen to 2.2 ML at the substrate temperature of 500^0C. As soon as the InAs deposition was finished, a 10 nm-thick GaAs layer was deposited at the same substrate temperature in order to cap the InAs dot layer. For the GaAs cap growth, we utilized 2 As-flux cells, each of which had pressure of 5×10^{-6} mbar. With the enhanced flux of two cells, we expect to reduce interdiffusion and intermixing between In and Ga atoms during capping. Prior to GISAXS experiments, we investigated the morphology of the sample surface by using AFM. No evidences for the existence of InAs QD morphology were found, implying that the InAs QDs are perfectly capped beneath the GaAs layer. The sample surface roughness is determined by AFM to be less than 1 nm. GISAXS measurements were performed at Hamburger synchrotron radiation laboratory beamline BW1 at a wavelength of 1.17 Å. The experimental setup is sketched in Fig. 1a. The incidence angle $\alpha_i = 0.26^0$ slightly exceeds the total external reflection angle of α_c. At this angle the x-ray penetrates into the sample by a depth of more than 10 nm. α_f is the reflection angle. The scattering intensities were recorded by a position sensitive detector (PSD) mounted parallel to the sample surface. Scattering vector $\vec{q_\parallel}$ is perpendicular to sample azimuthal orientation. The lateral resolution of the PSD is 2.1×10^{-3} Å^{-1}, with which mean dot-dot distances up to 300 nm can be determined. The reflection intensity from the surface, i.e. the specular beam intensity, is very strong as compared to uncapped dot samples [6]. In order to reduce the specular beam intensity, we move the PSD to a large exit angle of $\alpha_f = 0.8^0$. Ω is the sample azimuthal orientation angle. $\Omega=0^0$ corresponds to sample azimuth of [1$\bar{1}$0].

3 Results and discussions

The central, black and narrow line in Fig. 1b represents the specular beam intensity for all measured azimuths. Furthermore, non- specular diffuse scattering satellite peaks are found at symmetric positions on both sides of the specular beam for all measured azimuthal orientations of $0^0 - 180^0$, although in some azimuths the satellite peaks are quite weak. We attribute them to diffuse scattering of the buried InAs QDs. The most pronounced satellite peak intensity is found at azimuth orientation of 0^0 or 180^0 as indicated by white arrows, i.e. along [1$\bar{1}$0] or [$\bar{1}$10] azimuth, where the corresponding diffuse scattering vector $\vec{q_\parallel}$ is parallel to [110] or [$\bar{1}\bar{1}$0] direction. This significantly concentrated distribution of satellite peak intensity indicates a strong anisotropic ordering of the QD lateral distribution. Moreover, the largest distance between satellite peak and specular beam is also found around [110] direction. We quantitatively analyze the $q_\parallel$ dependent diffuse scattering spectra for all measured

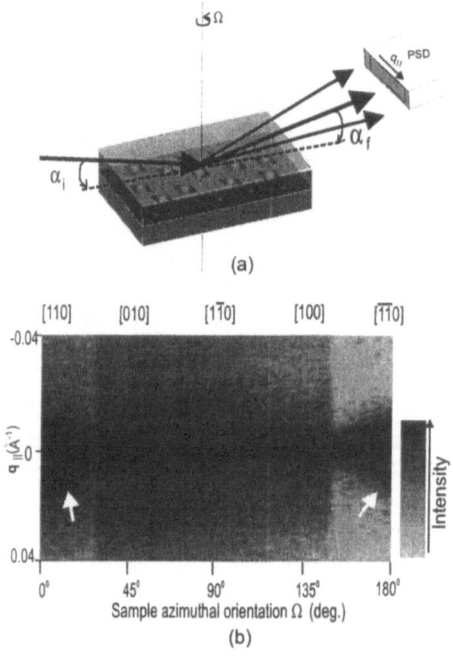

Fig. 1 (a) Schematic illustration of experimental setup of GISAXS. (b) Grey scale plot of the GISAXS intensity versus the scattering vector $\vec{q}_\parallel$ and sample azimuthal orientations. The sample azimuth of $[1\bar{1}0]$ is set to 0^0. On top of the figure (b), directions parallel to the scattering vector $\vec{q}_\parallel$ are labeled.

sample azimuths. As an example, we show the scattering spectra for $q_\parallel$ in [110], $[1\bar{1}0]$ and [100] directions in Fig. 2a and b. By fitting the specular beam intensity and the satellite peaks with Gaussian and Lorentzian functions, respectively, as shown in the inset of Fig. 2a and b, we derive the structure parameters of the QD lateral distribution. The mean dot-dot distances $\bar{d}$ of the InAs QD lateral distribution are determined to be anisotropic. The shortest one is around [110] $\pm15^0$ direction i. e. $\bar{d}=76 \pm 2$ nm, whereas, in other measured azimuths $\bar{d}$ values are approximately the same and equal to $\bar{d}=120\pm20$ nm, e.g. in $[1\bar{1}0]$ and [100] directions. From this, we infer that the lateral QD position distribution is isotropic in all directions except the [110] direction, where we find significantly smaller inter-dot distance. Moreover, we obtain the standard deviation of the dot-dot distance $\langle \sigma/\bar{d} \rangle$ for all measured azimuths. We get $38 \pm 2\%$ in [110] direction, and $60 \pm 5\%$ in other azimuths, e. g. in $[1\bar{1}0]$ and [100] directions. Assuming short-range order type of the correlation function [6,7], this corresponds to correlation lengths of 80 nm in [110], and 50 nm in other azimuths. The correlation length is inversely proportional to the standard deviation $\langle \sigma/\bar{d} \rangle$ value. We suggest that the QD lateral distribution is better ordered in [110] direction.

In summary, we determine the lateral ordering of InAs dot layers capped with GaAs by GISAXS. From the analysis of the structure parameters such as satellite peak intensity, mean dot-dot distance, standard de-

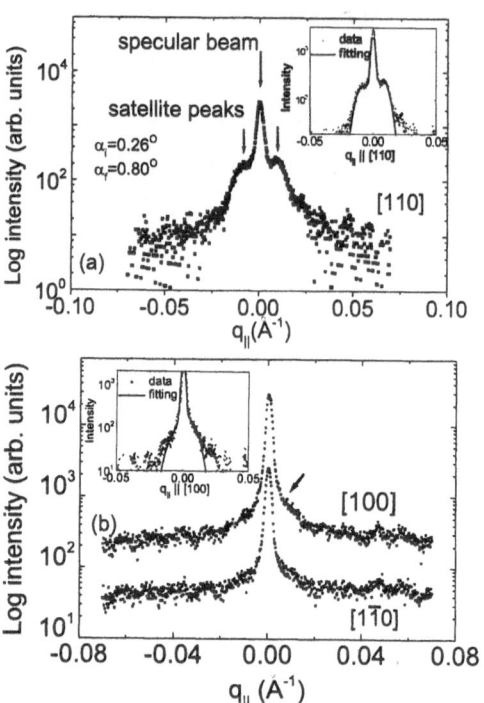

Fig. 2 Logarithmic GISAXS intensity curve versus $q_\parallel$ along (a): the [110] and (b): [100], $[1\bar{1}0]$ directions. The specular peak is located at $q_\parallel = 0\text{Å}^{-1}$. In (a) case, two satellite peaks are symmetrically distributed beside the specular peak. The inset in (a) represents a fitting to the data. In (b) case, only very tiny scattering intensity from QDs is observed as indicated by an arrow. The inset in (b) indicates a fitting to the measured data.

viation of the mean dot-dot distance, and correlation length, we conclude that the lateral distribution of the buried InAs QDs is strongly anisotropic and better ordered in [110] direction, these results are consistent with the distribution found in uncapped dot samples [6].

Acknowledgement : Financial funding of the Deutsche Forschungsgemeinschaft via SFB508 and Graduiertenkolleg "Physik Nanostrukturierter Festkörper" is gratefully acknowledged.

References

1. D. Heitmann and J. P. Kotthaus, Physics Today **46**, (1993) 56.
2. H. Drexler, D. Leonard, W. Hansen, J. P. Kotthaus and P. M. Petroff, Phys. Rev. Lett. **73**, (1994) 2252.
3. M. Grundmann, O. Stier, and D. Bimberg, Phys. Rev. B **52**, (1995) 11969.
4. K. Zhang, Ch. Heyn, W. Hansen, Th. Schmidt and J. Falta, Appl. Phys. Lett., in press for the issue on August 28, 2000.
5. K. Zhang, A. Foede, Th. Schmidt, P. Sonntag, Ch. Heyn, G. Materlik, W. Hansen and J. Falta, Phys. Stat. Soli. B **215**, (1999) 791.
6. K. Zhang, Ch. Heyn, W. Hansen, Th. Schmidt and J. Falta, Appl. Phys. Lett. **76**, (2000) 2229.
7. M. Schmidbauer, Th. Wiebach, H. Raidt, M. Hanke, R. Köhler and H. Wawra, Phys. Rev. B **58**, (1998) 10523.

Nucleation Site Control in Self-Assembling Si Quantum Dots on Ultrathin SiO$_2$/c-Si

S. Miyazaki, M. Ikeda, E. Yoshida, N. Shimizu and M. Hirose

Department of Electrical Engineering, Hiroshima University, 1-4-1 Kagamiyama, Higashi-Hiroshima 739-8527, Japan e-mail: miyazaki@sxsys.hiroshima-u.ac.jp

Abstract   We have investigated the nucleation and growth of silicon dots on ultrathin SiO$_2$ layers in low pressure chemical vapor deposition (LPCVD) of a monosilane gas and developed a fabrication technique for positioning the Si dots in conjunction with nanometer-scale modification of SiO$_2$ surfaces using a scanning tunneling microscope (STM) prior to LPCVD. We found that reactive sites for the Si dot growth are effectively created on the SiO$_2$ surface by electron beam irradiation from a Pt-Ir STM tip in H$_2$ ambient with a pressure of 1.3×10^{-5}Pa without causing dielectric degradation. This can be interpreted in terms that the field-emission of not only electrons but also adsorbed atomic hydrogen from the tip apex to the SiO$_2$ surface results in the formation of surface Si-H and/or Si-OH bonds which act as nucleation sites during LPCVD.

1 Introduction

Nanometer scale Si structures have attracted much attention in their potential applications in novel digital devices such as single-electron tunneling transistors [1-4] and quantum-dot floating gate memories [5-7]. For such nanoelectronic devices to operate reliably even at room temperature, the fabrication of Si nanostrucures with good size uniformity and low defect density is the first concern. The self-assembling formation of nanometer Si dots during low-pressure chemical vapor deposition (LPCVD) has received particular interest, which enables us to obtain the Si dots of high areal density on an ultrathin gate SiO$_2$ layer without process-induced damages for memory devices with the Si dots as a floating gate. Previous studies [8, 9] on the early stages of LPCVD using a SiH$_4$ gas have shown that Si nanocrystals are formed randomly on SiO$_2$ and the areal dot density is dramatically enhanced on SiO$_2$ surface treated by a dilute HF solution just before LPCVD, and also suggested that surface OH bonds can provide nucleation sites during LPCVD [9]. To further improve the size uniformity and the spatial distribution of the self-assembling Si dots, the control of the arrangement of Si dots remains a major challenge.

In this paper, we report a new approach to control of the nucleation sites, in which nanometer-scale modification of the initial SiO$_2$ surface before LPCVD has been made by using the tip of a scanning tunneling microscope (STM) in low pressure H$_2$ ambient, and discuss the growth mechanism of Si dots on such surface-modified SiO$_2$.

2 Experimental

A 1nm-thick SiO$_2$ layer was first grown on chemically-cleaned p-Si (100) at 750°C in 40Pa O$_2$ ambient. SiO$_2$/Si (100) so prepared was transferred in vacuum to another chamber for surface modification by STM. The STM chamber was filled with H$_2$ gas up to 1.3×10^{-5} Pa and then a Pt (20%Ir) STM tip with the apex of 15nm in radius was scanned on the as-grown SiO$_2$ surface with a constant tunnel current of 0.5nA to form reactive sites for nucleation of Si dots as schematically illustrated in Fig. 1. The total electron dose was changed from 0.13 to 1.3C/cm^2 at a substrate bias in the range of -10 to +10V. It was confirmed by tunneling spectroscopy that no electrical degradation of the SiO$_2$ layer is caused up to 100C/cm^2 electron does at a substrate bias of +10V as reported in Ref. 10. Subsequently, hemispherical, single-crystalline Si dots were formed from 550°C decomposition of pure SiH$_4$ at 0.2 Torr for 60s on the locally-modified SiO$_2$ surface. Under such LPCVD conditions, the areal density of Si dots on as-grown SiO$_2$ surface was 6.0×10^9cm^{-2}.

3 Results and Discussion

Topographic images measured by atomic force microscopy (AFM) after Si dot formation show that the Si dot density in the region raster-scanned by the Pt-STM tip is markedly increased compared with that in the region of as-grown oxide. As shown in Fig. 2, the Si dots with an areal density of 5.5×10^{10}cm^{-2} were formed in the region scanned by the Pt-STM tip at a substrate bias of +8V with an electron dose of 1.3C/cm^2 prior to LPCVD, whereas in the region of as-grown oxides, few dots are formed. When the Pt-STM tip was scanned at the same bias conditions in vacuum with a H$_2$ pressure below 4×10^{-8}Pa prior to LPCVD, no enhancement of the areal dot density was observed. Therefore, the remarkable increase in dot density by electron beam (EB) irradiation

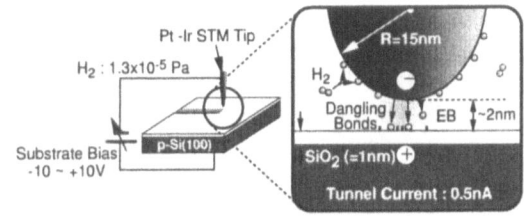

Fig. 1 Schematic illustration of nanometer-scale modification of SiO$_2$ surface by scanning a Pt-STM tip in H$_2$ ambient of 1.3×10^{-5}Pa at a substrate bias of -10 to +10V. Hydrogen atoms are adsorbed on the tip surface through dissociative adsorption and emitted by applying a positive or negative bias as much as 3.5V in absolute value between the tip and the sample[11, 12].

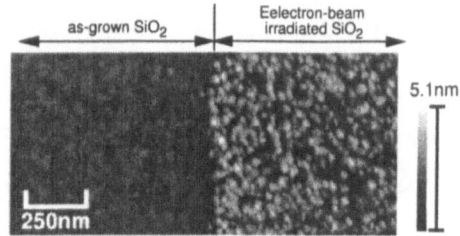

Fig. 2 AFM image obtained after Si dot formation at 550°C. In the right side from the center, the EB irradiation was made with an electron dose of 1.3C/cm² form the Pt-STM to the SiO₂ surface prior to LPCVD, while in the left, no EB irradiation was performed.

in H₂ ambient before LPCVD can not be attributed to surface damages, if any, induced by low energy EB such as the formation of surface dangling bonds, but to the role of hydrogen on the surface chemical modification under EB irradiation. It was reported that H atoms are adsorbed on the Pt-STM tip through dissociative adsorption of H₂ molecules at room temperature and deposited on Si(111)7×7 surface by field-enhanced emission at substrate biases of ±3.5V[11,12]. Accordingly, we suggest that Si-OH or Si-H bonds, which acts as reactive sites during LPCVD [9], are efficiently formed on the SiO₂ surface by field emissions of both H atoms and electrons. As represented in Fig. 3, the dot density is increased with electron dose and substrate bias. Note that, at a substrate bias of -10V, the dot density obtained even for 1.3C/cm² electron doses is 3.5×10^{10}cm^{-2} at most, which is only about 60% of the value obtained at +8V. This result indicates that the kinetic energy of electrons reaching the SiO₂ surface is a crucial factor to promote the reaction of atomic hydrogen with SiO₂. As demonstrated in Fig. 4, by line-scanning the Pt-STM tip with a scan speed of 0.3μm/s at a substrate bias of +10V prior to LPCVD, Si dots with an average dot height of 10nm are aligned densely. Notice that, in-between the lines of densely-aligned Si dots with a space below 150nm, the areal density of spontaneously-grown Si dots is appreciably decreased compared with that in the area being sufficiently away from the aligned Si dots. In addition, the average size of such intentionally-nucleated Si dots in the line is about twice as large as that of spontaneously nucleated dots on as-grown SiO₂. By employing a one-dimensional diffusion model of deposition precursors to explain the decreased dot density near the aligned dots, the mean surface-migration length of the precursors is estimated to be 50nm.

4 Conclusions

The effect of the substrate surface modification using a Pt-STM tip on the nucleation and growth of Si dots on SiO₂ during LPCVD has been investigated. We have found that EB irradiation to the SiO₂ surface under low pressure H₂ ambient makes nucleation sites for Si dot formation effectively without causing the dielectric degradation and leads us to control of positioning of the Si dots.

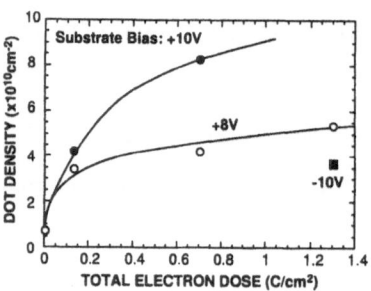

Fig. 3 The areal dot density as a function of total electron dose through the SiO₂ at different substrate biases.

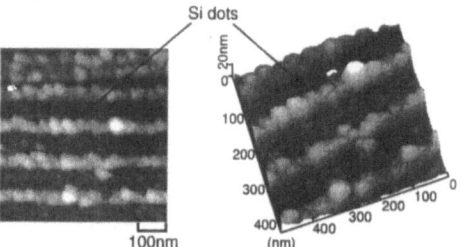

Fig. 4 AFM images obtained after Si dot formation at 550°C on the SiO₂ surface line-scanned by the STM tip. In the STM surface modification prior to LPCVD, the substrate bias and the scan speed were +10V and 0.3μm/s, respectively.

Acknowledgements

This work was supported in part by a Grant-in-Aid for Scientific Research from the Ministry of Education, Science, Sports and Culture of Japan, and the Core Research for Evolutional Science and Technology (CREST) of Japan Science and Technology Corporation (JST).

References

1. Y. Takahashi, M. Nagase, H. Namatsu, K. Kurihara K. Iwadate, Y. Nakajima, S. Horiguchi, K. Murase and M. Tabe, Electron. Lett. **31** (1995) 136.
2. E. Leobandung, L. Guo, Y. Wang and S. Y. Chou, Appl. Phys. Lett. **67** (1995) 938.
3. K. Yano, T. Ishii, T. Hashimoto, T. Kobayashi, F. Murai and K. Seki, IEEE Trans. Electron Devices **41** (1994) 1628.
4. K. Uchida, J. Koga, A. Ohata and A. Toriumi, Nanotechnol. **10** (1999) 198.
5. S. Tiwari, F. Rana, H. Hanafi, A. Hartstein, E. F. Crabb and K. Chan, Appl. Phys. Lett. **68** (1996) 1377.
6. A. Kohno, H. Murakami, M. Ikeda, H. Nishiyama, S. Miyazaki and M. Hirose, Ext. Abs. of Int. Conf. Solid State Devices and Materials (1998, Hiroshima) 174.
7. T. Maeda, E. Suzuki, I. Sakada, M. Yamanaka and K. Ishii, Nanotechnol. **10** (1999) 127.
8. K. Nakagawa, M. Fukuda, S. Miyazaki and M. Hirose, Mat. Res. Proc. Symp. **452** (1997) 243.
9. S. Miyazaki, Y. Hamamoto, E. Yoshida, M. Ikeda and M. Hirose, Thin Solid Films **369** (2000) 55.
10. H. Watanabe, T. Baba and M. Ichikawa, J. Appl. Phys **85** (1999) 6704.
11. H. Kuramochi, H. Uchida and M. Aono, Phys. Rev. Lett. **72** (1994) 932.
12. K. Oura, V. G. Lifshits, A. A. Saranin, A. V. Zotov and M. Katayama, Surf. Sci. Rept. **35** (1999) 1.

Reversibility of the Island Shape, Volume and Density in Stranski-Krastanow Growth

N.N. Ledentsov[1,2], V.A. Shchukin[1,2], R. Heitz[1], D. Bimberg[1], V.M. Ustinov[2], N.A. Cherkashin[2], A.R. Kovsh[2], Yu.G. Musikhin[2], B.V. Volovik[2], A.E. Zhukov[2], G.E. Cirlin[2], Zh.I. Alferov[2]

[1] Technische Universität Berlin, Institut für Festkörperphysik, Hardenbergstr. 36, 10623 Berlin, Germany
[2] A.F.Ioffe Physico-Technical Institute of the RAS, Politekhnicheskaya 26, 194021 St.Petersburg, Russia

Abstract We report on reversible phenomena in size-limited Stranski-Krastanow growth of strained InAs islands on GaAs(001) surface. Plan view and cross-sectional transmission electron microscopy and photoluminescence spectroscopy reveal that, with increasing the substrate temperature, the average volume of the island decreases, and the island density decreases as well, while the lateral size of the islands increases and islands strongly flatten. If, after the formation of islands, the substrate temperature is reduced, the average volume and the density of the islands increase, while the lateral size of the island shrinks. These observations confirm the equilibrium nature of the array of islands.

1 Introduction

Spontaneous formation of coherently strained three-dimensional (3D) nanometer-scale islands on top of a thin strained wetting layer (Stranski-Krastanow growth mode) is the basis of the modern technology of quantum dot (QD) fabrication [1]. A key property of arrays of 3D islands, observed in many heteroepitaxial systems, is the self-limited island growth (SLIG) providing a narrow size distribution around some optimum size and the stability of islands upon growth interruption. This, on the one hand, ensures applications of the QDs, and, on the other hand, is a challenge for theory as it disagrees with the conventional picture of Ostwald ripening.

To explain the SLIG, two classes of theoretical models were applied. In kinetic models, SLIG is assumed to occur under strongly non-equilibrium conditions. The absence of ripening is explained by strain-induced diffusion barriers for adatoms to attach to the island edge [2] or for a new atomic layer to nucleate on the facet [3]. The barriers are higher for larger islands, resulting in the SLIG. On the contrary, thermodynamic models describe an array of 3D islands ordered in shape, size and relative arrangement as an equilibrium surface structure [4], where, in a certain range of material parameters, e. g., surface energies and surface stresses, ripening is energetically unfavorable.

2 Experiment

We study experimentally the formation mechanism of InAs/GaAs(001) 3D strained islands and the influence of substrate temperature and tuning of the substrate temperature after the InAs deposition on the resulting shape, volume, and density of the islands formed in a SLIG mode. The samples were grown by a conventional molecular beam epitaxy (MBE) [5]. The islands were formed by the 3 ML InAs deposition followed by a 10 s growth interruption (GI) to ensure that the limiting size of the islands is completely reached. InAs islands are capped by a 10 nm thick GaAs at the same substrate temperature. For some of the samples, the substrate was tuned to lower temperatures after the InAs deposition, and the InAs islands were capped at the final temperature. Plan-view and cross-sectional transmission electron microscopy (TEM) and photoluminescence (PL) measurements of capped samples were carried out.

3 Results and discussion

Figure 1 shows the dependence of the photon energy of the QD-related optical transition, the average lateral size and the density of the islands revealed in plan-view TEM (see also Figs. 2a, 2b) on the substrate temperature. It follows from Figs. 1 and 2, that the lateral size and the height of the islands demonstrate opposite behavior: The increase of the substrate temperature leads to an increase of the lateral size and the decrease of the height. To obtain the dependence of the volume on temperature, we focus on the photon energy of the QD-related transition. Figure 1a demonstrates the blue shift with increasing temperature that basically corresponds to a decrease of the QD volume. The effect of the QD shape on the transition energy calculated in [6] shows that, for an InAs/GaAs QD of a fixed volume, the flattening of the QD shape leads to a red shift of about 30 meV. Thus, the observed behavior of the PL peak in Fig. 1a can only be explained by a decrease of the QD volume with the increasing temperature where the blue shift due to the volume decrease overcomes the red shift due to the flattening of the islands.

The decrease of the island volume with temperature disagrees with the kinetic models of the SLIG [2,3] that predict the increase of the island volume with temperature. On the contrary, the thermodynamic model

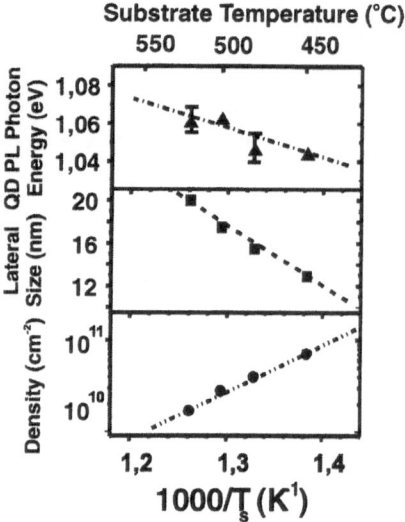

Fig.1 Effect of the substrate temperature on parameters of the array of 3D coherently strained InAs/GaAs islands. (a) The position of the PL peak (77 K). (b) The average lateral size of the islands. (c) The island density.

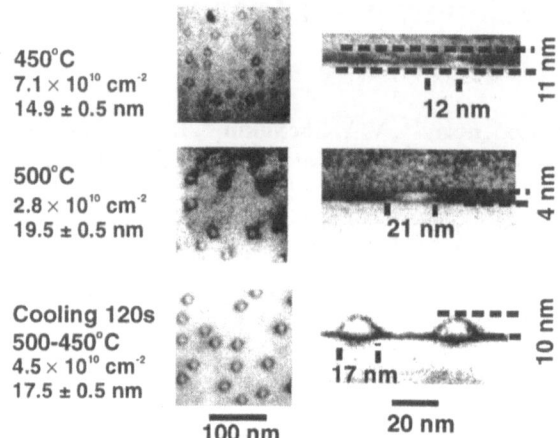

Fig.2. Effects of temperature and temperature ramping on the lateral size and the height of InAs/GaAs islands. (a), (b), (c): Plan view and. (d), (e), (f): cross-sectional TEM images.

developed in [7] for islands with a fixed height, predicts an entropy-driven decrease of the island volume with temperature. General trends are the same also for 3D islands, as the entropy effects favor islands of smaller volumes at higher temperatures. Thus, an observed decrease of the island volume with temperature emphasizes entropy effects and strongly supports the thermodynamic picture of the island formation.

We note that the studies of Indium segregation on the compositional profile for InGaAs/GaAs QDs [8] show that the extension of the In profile in samples grown at 500^0C is about 2 nm and exceeds the one in samples grown at 450^0C by about only 1 nm. Such a change can affect the volume, the shape, and the PL spectrum of the QDs only weakly.

Let us now focus on the effect of the lowering of the substrate temperature on the island density, volume, shape and PL spectra. The comparison of the array of islands after 120 s cooling (Figs. 2b, 2e) with the array formed and capped at 500^0C (Figs. 2c and 2f) indicates that cooling results in a decrease of the lateral size of the islands. Thus, the lateral size of the islands after cooling is an intermediate one between its value for the array deposited and capped at 450^0C (Figs. 2a, 2d), and the value for the array deposited and capped at 500^0C (Figs. 2b, 2e). The same is valid for the island density and the height, which are significantly increased after cooling (Figs. 2c, 2f). This indicates a partial reversibility of changes of the arrays of islands with temperature. The simultaneous increase of the island volume and density of the islands after cooling can only

occur due to the condensation of Indium adatoms from the wetting layer at the islands. Adatom condensation must be accompanied by the thinning of the wetting layer that is confirmed by the blue shift of the PL peak due to the wetting layer [9].

4 Conclusions

To conclude, plan-view and cross-section TEM and PL studies indicate the decrease of the island volume and the flattening of the islands with the increase of the formation temperature. Cooling of the array of InAs islands after deposition and before capping shows that the parameters of the array of islands are determined mostly by the final temperature and demonstrate reversible changes upon change of temperature. Thus, the array of SLIG InAs/GaAs(001) 3D coherent strained islands is governed, for used growth conditions, mostly by thermodynamics.

References

1. D. Bimberg, M. Grundmann, N.N. Ledentsov, *Quantum Dot Heterostructures* (Wiley, Chichester, 1998).
2. N. Kobayashi, *et al.* Appl. Phys. Lett. **68**, 3299 (1996).
3. D.E. Jesson, *et al.* Phys. Rev. Lett. **77**, 1330 (1996).
4. V.A. Shchukin and D. Bimberg, Rev. Mod. Phys. **71**, 1125 (1999), and references therein.
5. N.N. Ledentsov, In: *Molecular Beam Epitaxy*, Springer Tracts in Modern Physics **156**, Berlin (1999).
6. V.A. Shchukin, *et al.* Mat. Res. Soc. Symp. Proc. V. **618**, Pittsburgh, (2000).
7. V.A. Shchukin, *et al.* Mat. Res. Soc. Symp. Proc. V. **583**. Pittsburgh, (2000).
8. A. Rosenauer, *et al.* Phys. Rev. B **61**, 8276 (2000).
9. N.N. Ledentsov, *et al.*, to be published.

Effect of Si diffusion on growth of GeSi self-assembled islands

A.V.Novikov[1], N.V.Vostokov[1], S.A.Gusev[1], Yu.N.Drozdov[1], Z.F.Krasil'nik[1], D.N.Lobanov[1], L.D.Moldavskaya[1], M.Miura[2], V.V.Postnikov[1], M.V.Stepikhova[1], Y.Shiraki[2], N.Usami[3]

[1] Institute for Physics of Microstructures RAS, 603600, Nizhny Novgorod, GSP-105, Russia e-mail: anov@ipm.sci-nnov.ru
[2] RCAST, The University of Tokyo, 4-6-1 Komaba, Meguro-ku, Tokyo, 153-8904, Japan
[3] Institute for Materials Research, Tohoku University, Sendai 980-8577, Japan

Abstract The effect of alloying on growth and parameters of GeSi self-assembled islands grown at various temperatures is investigated by X-ray diffraction, atomic force and electron microscopy. The correlation between the sizes and composition of islands is obtained. The Ostwald ripening and alloying cause a significant spread of the size distribution of islands grown at 750°C. The correlation between composition and elastic strain of islands, obtained by X-ray analysis, and photoluminescence spectra is discussed.

1 Introduction

Heterostructures containing GeSi strain-induced self-assembled islands provide a promising material to realize Si-based optoelectronic devices operating at 1.2÷2 μm. This range is particularly important for telecommunication industry. High crystal quality of structures is critical to the success of such applications. In general, rather high growth temperatures (500°C÷800°C) are used for fabrication of GeSi structures with perfect crystal quality. The diffusion and alloying processes largely affect growth and parameters of structures at these temperatures. Additional flows of atoms are induced by the non-uniform elastic strain fields in case of islands growth [1].

In this paper, we present the results of investigation of self-assembled GeSi islands growth on Si (001) at various temperatures. The effect of alloying on the size and shape of GeSi self-assembled islands is investigated by X-ray diffraction, atomic force and electron microscopy. The information about sizes, composition and elastic strain of the islands is used for interpretation of photoluminescence (PL) spectra of structures with islands.

2 Results and discussion

The samples under investigations are grown by molecular beam epitaxy from a solid source (SSMBE) and a gas source (GSMBE) on Si (001) substrates at growth temperatures T_g=600°C, 700°C and 750°C. Further details of the growth process are presented elsewhere [2,3]. After substrate cleaning and growth of a Si buffer layer a Ge layer with the equivalent thickness from 3 to 11 ML is deposited. Structures for optical investigation have a Si cap layer grown at the same T_g as the islands. Surface morphology is determined by *ex situ* atomic force (AFM) and scanning electron microscopy (SEM). The X-ray diffraction measurements are performed with a "DRON-4" double

crystal X-ray diffractometer (at a wavelength of 1.54 Å). PL is detected by a liquid nitrogen cooled Ge detector using a standard lock-in technique. Ar+ laser is used as an excitation source. The samples are cooled to a temperature of 5 K.

The parameters (sizes, surface density, composition and elastic strain) of GeSi islands grown at 700°C from a gas source coincide with the parameters of the islands, grown from a solid source and investigated earlier [3]. The structures with a narrow size distribution of multifaceted dome-islands (dispersions of lateral sizes and height less than 10%) are fabricated at this growth temperature by both deposition techniques. The average composition and elastic strain of dome-islands are defined by X-ray analysis from reciprocal space maps around the (004) and (224) reciprocal lattice points. In the strained layer approximation the content of Si in dome-islands is about 50% regardless of a deposition technique. Such a high value of Si content is related to strain-driven surface and volume diffusion of Si from the region of maximum elastic strain near the islands bases [3,4].

A decrease of growth temperature to 600°C inhibits the surface and volume diffusion of Si and Ge. As a result of reduction of Ge adatom surface diffusion, the surface density of islands at T_g=600°C increases four times in comparison with that at T_g=700°C. Besides, the critical

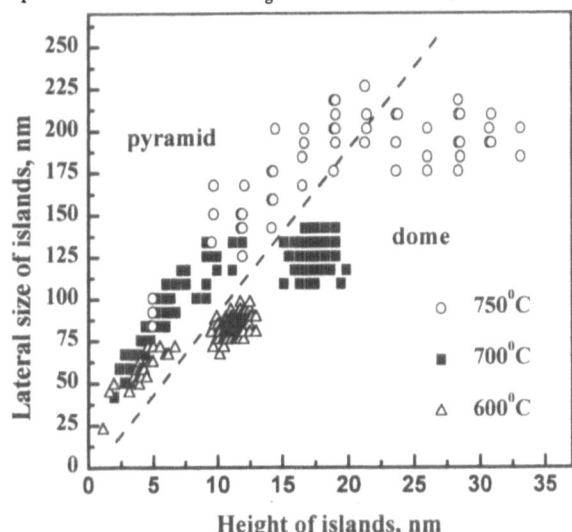

Fig. 1 The dependence of the island lateral size on its height obtained from AFM images. The dashed line separates the regions of different island shape.

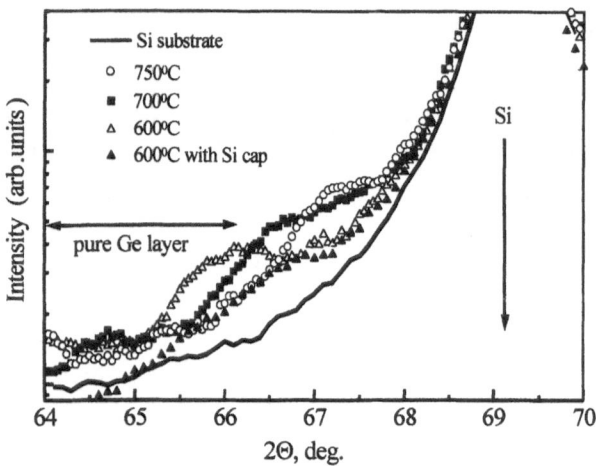

Fig.2 The ω/2Θ X-ray scans near (004) Si reflection for samples grown at different temperature. The arrows mark peak position from Si substrate and angle region of diffraction from pure Ge layers at different strain.

sizes of islands with pyramid-like shape ("pyramid"-islands), at which they transform to a dome shape, decrease about 2 times (Fig.1). One possible reason for this change of size is reduction of Si content in islands with a decrease of T_g. Figure 2 displays the ω/2Θ scans of samples grown at different temperatures near the symmetric (004) Si reflection. It is seen that the small peak from the islands shifts from the Si substrate peak to the region of diffraction from a pure Ge layer with a decrease of the growth temperature. The Ge concentration in islands obtained by analysis of reciprocal space maps changes from 50% at T_g=700°C to 75% at T_g=600°C. The increase of Ge content results in an increased elastic strain in islands and, according to the model of Ross [5], a decreased critical volume of pyramid-islands.

The effect of alloying on growth of GeSi islands is most pronounced at T_g=750°C. Besides the increase of pyramid and dome size (Fig.1), a high diffusion rate of Si into islands at this growth temperature initiates a reverse transformation of islands from dome to pyramid shape directly during Ge deposition. As a result, the islands of intermediate shape (with pyramid-like base and dome-like apex) are formed. Similar changes of the islands shape were observed during a post-growth annealing of structures grown at $T_g \leq$700°C and connected with a decrease of elastic strain in dome islands with an increase of Si content in the islands during annealing [3,6].

The Ostwald ripening mechanism also affects islands growth at T_g=750°C. During Ostwald ripening some islands build up at the expense of other islands that shrink and dissipate completely. This results in the spread of the islands sizes distribution. The surface of samples grown at T_g=750°C keeps traces of the dissociated islands as circumference-like trenches [3,6].

Energy band positions for the samples with dome-islands can be calculated using the data obtained by X-ray

analysis of composition and elastic strain. According to these calculations, the energy of the optical transition between the electrons in Si and the heavy holes in the islands is about 0.75 eV for samples grown at 700°C, and 0.67 eV, for samples grown at 600°C. The experimentally measured PL peak associated with the islands is in the region 0.78÷0.82 eV. The discrepancy between the calculated and the observed positions of the optical transition may be attributed to dissolution of the islands and decrease of Ge content in islands during Si overgrowth. The dissolution of islands is confirmed by the X-ray analysis of structures with a Si cap layer (Fig. 2).

3 Conclusion

In summary, we investigated the effect of alloying on growth and parameters of GeSi self-assembled islands. The change in maximum sizes of pyramid-islands with an increase of growth temperature is related to enhancement of alloying in GeSi islands. The Ostwald ripening and alloying broaden the islands sizes distribution at T_g=750°C. The discrepancy between the calculated and the experimental PL peaks from islands is associated with the changes of islands composition during Si overgrowth.

Acknowledgements

The work was supported by the RFBR grant # 99-02-16980, CRDF grant # RESC-02 (BRHE Program), and the State Programs on "Physics of solid-state nanostructures" (#99-2047) and "Technology and device prospects for micro- and nanoelectronics" (#02.04.1.1.16.E1). One of the authors (N.U.) would like to gratefully acknowledge the Core Research for Evolution Science and Technology (CREST) of JST and Grant-in-Aid for Scientific Research on Priority Area (Area No. 739) from the Ministry of Education, Science, Sports and Culture.

References

1. A.-L. Barabasi, Appl. Phys. Lett. **70**, (1997) 2565.
2. H.Sunamura, N.Usami, S.Fukatsu and Y.Shiraki, Appl. Phys. Lett., **66**, (1995) 3024.
3. Z.F.Krasil'nik, N.V.Vostokov, S.A.Gusev, I.V.Dolgov, Yu.N.Drozdov, D.N.Lobanov, L.D.Moldavskaya, A.V.Novikov, V.V.Postnikov, D.O.Filatov, Thin Solid Films, **367**, (2000) 171.
4. S.A.Chaparro, J.Druker, Y.Zhang, D.Chandrasekhar, M.R.McCartney, and D.J.Smith, Phys. Rev. Lett., **83**, (1999) 1199.
5. F.M.Ross, J.Tersoff and R.M.Tromp, Phys. Rev. Lett., **80**, (1998) 984.
6. T.I.Kamins, G.Medeiros-Ribero, D.A.A.Ohlberg and R.S.Williams, J. Appl. Phys., **85**, (1999) 1159.

Observation of a universal behavior in the growth of InAs self-assembled quantum dots on patterned substrates

S. W. Hwang[1,2], M. H. Son[1], B. H. Choi[1,2,3], D. Ahn[1], C. K. Hyon[2], S.-H. Song[2], Y. J. Park[3], and E. K. Kim[3]

[1] Institute of Quantum Information Processing and Systems, University of Seoul, Jeonnong, Dongdaemun, Seoul 130-743, Korea e-mail: swhwang@mail.korea.ac.kr
[2] School of Electrical Engineering, Korea University, Sungbuk, Anam, Seoul 136-075, Korea
[3] Semiconductor Materials Research Lab., Korea Institute of Science and Technology, P.O. Box 131, Cheongryang, Seoul 130-650, Korea

Abstract   In this paper, we have reported the observation of the depletion behavior in the patterned growth of SAQDs. It is difficult to find the relation between the width of the depletion region and the magnitude of the strain estimated from the simple formula. However, since all the detailed information related with the material preparation and the geometrical shape of the interface is necessary for the accurate determination of the strain during thermal cycling, more accurate universal law would require further works.

1 Introduction

Semiconductor quantum dots have been provided a lot of research interests both in the field of application and in condensed matter physics. Quantum dot memories and quantum dot transistors have become a promising candidate for a building block for future integrated circuits. The singularity in the zero-dimensional density of state endorses a possibility of realizing efficient lasers. Single electron transport through quantum dots reveals many interesting physics such as Kondo effect and spin blockade.

Most common method of fabricating semiconductor quantum dots has been the combination of nano-lithography and subsequent pattern transfer such as reactive ion etching or metal lift-off. In those quantum dots, there are inevitable irregularities from e-beam fluctuations and defects originated from the etch damage. On the other hand, self-assembled quantum dots (SAQDs) guarantees a one-step formation of defect-free, ultra-small zero-dimensional systems.

One major obstacle in the device application of SAQDs is natural randomness in the position of the dots. Recent researches therefore are mostly directed to the position control of SAQDs. There have been many different types of approaches so far and the examples are; (1) utilizing patterned mask material [1], (2) creating nucleation centers either by STM [2] or by AFM [3], (3) strain engineering by the patterning and the regrowth of the substrates [4]. We have also been trying various types of patterned SAQD growth, and in this paper, we report an interesting observation commonly occurring in most of our samples. The types of patterned substrates for the selective growth of SAQDs are; (1) Pt stripes on

GaAs, (2) W stripes on GaAs, (3) SiO$_2$ on Si, (4) AFM scratched grooves on GaAs. The essence of our observation is that the depletion of SAQDs occurs on the substrate near the masking materials. The magnitude of the SAQD depletion region is found to be the function of the masking material and the substrate. A simple model regarding the difference in the thermal expansion between the masking material and the substrate cannot explain our observation.

2 Experiments

All of our SAQDs are grown in an MOCVD chamber. The growth temperature for the SAQDs is $400 \sim 430$ °C and $2.1 \sim 2.5$ mono-layers of InAs are grown. The average density is 3×10^{11} cm^{-2} and the average dot diameter and the height are 30 and 8 nm, respectively. All of the growth results shown below utilize the similar conditions as the reference wafer.

Both Pt and W stripes on (100) GaAs wafers are prepared by an e-beam lithography, a sputtering, and lift-off processes. The thickness of the metals is maintained smaller than 50 nm. The SiO$_2$ masks are prepared by the e-beam patterning and subsequent reactive ion etching of the oxide covered on the (111) p-type Si wafer. The thickness of the SiO$_2$ layer is 100 nm and it is deposited by CVD. Finally, the nanometer size grooves are fabricated by the indentation with the air-operated AFM cantilevers [5].

Figure $1 \sim 4$ show either AFM or SEM images of SAQDs grown on those patterned substrates. While all the images demonstrate good selectivity against SAQDs, another interesting observation is that there always are SAQD depletion regions near the edges of the masks or grooves. In the case of Pt/GaAs (Fig. 1), the SAQDs are moved towards and piled at the edge of the Pt stripe, leaving approximately 1 μm of the depletion region. In the case of W/GaAs (Fig. 2) and SiO$_2$/Si (Fig. 3), there are clear depletion regions of the width 70 and 120 nm but it is difficult to identify any SAQDs at the edges. It suggests that the SAQDs are moved outward from the boundary between the masking pad and the substrate. In the case of AFM scratched groove (Fig. 4), many SAQDs

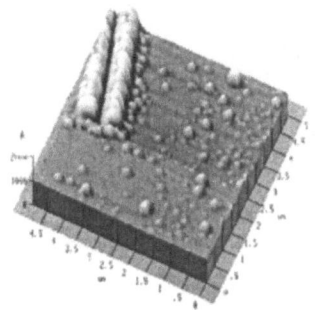

Fig. 1 AFM image of the patterned platinum on GaAs.

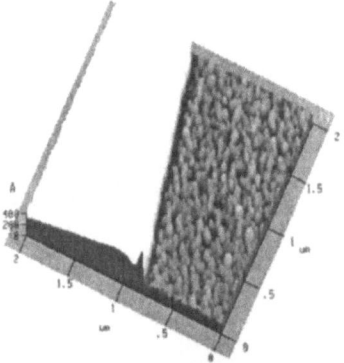

Fig. 2 AFM image of the patterned tungsten on GaAs.

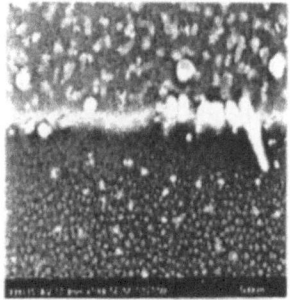

Fig. 3 SEM image of the patterned Si/SiO₂ on GaAs.

are deposited at the sidewall of the groove, causing the depletion near the groove.

A feasible model for our observation is the movement of the In adatoms due to the force originated from the strain near material interface. The simplest estimation for the direction and the amount of the strain between the masking material and the substrate during the growth can be expressed as the following equation,

$$P \propto \Delta T (C_{mask} - C_{sub})(a_{mask} - a_{sub}) \qquad (1)$$

where P, ΔT, C, and α are the total pressure at the interface, the temperature variation, Young's modulus, and the expansion coefficient, respectively. The polarity and the magnitude of the strain are calculated from the equation for the first three cases (except the case of AFM scratch). First of all, the direction of the strain determined from the polarity of the above equation is not consistent with the direction of the adatoms observed in Fig. 2 ∼ 4. Figure 5 summarizes the estimated strain

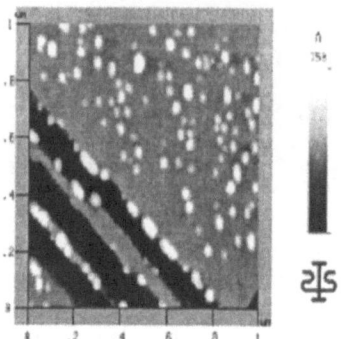

Fig. 4 AFM image of the AFM nano-carved GaAs.

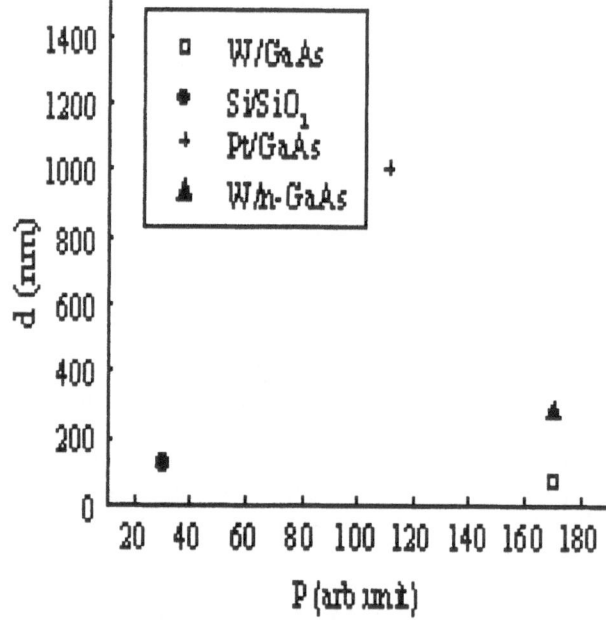

Fig. 5 Summary of the estimated strain and the depletion region.

and the depletion region width. It is also difficult to find the relation between the width of the depletion region and the magnitude of the strain.

3 Conclusions

In this paper, we have reported the observation of the depletion behavior in the patterned growth of SAQDs. It is difficult to find the relation between the width of the depletion region and the magnitude of the strain estimated from the simple formula. However, since all the detailed information related with the material preparation and the geometrical shape of the interface is necessary for the accurate determination of the strain during thermal cycling, more accurate universal law would require further works.

References

1. C. K. Han *et al.* Appl. Phys. Lett. **73** (1998), 2479.
2. S. Kohomoto *et. al.* Appl. Phys. Lett. **75** (1999), 3488.
3. C. K. Hyon *et. al.* to be published in APL (2000).
4. H. Lee *et. al.* QD2000 abstract (2000).
5. C. K. Hyon *et. al.* Appl. Phys. Lett. **75** (1999), 292.

Monte Carlo Simulation of the Self-Organized Growth of Quantum Dots with Anisotropic Surface Diffusion

M. Meixner[1], R. Kunert[1], S. Bose[1], E. Schöll[1], V. A. Shchukin[2,4], D. Bimberg[2], E. Penev[3], P. Kratzer[3]

[1] Institut für Theoretische Physik, Technische Universität Berlin, Hardenbergstr. 36, D-10623 Berlin, Germany
[2] Institut für Festkörperphysik, Technische Universität Berlin, Hardenbergstr. 36, D-10623 Berlin, Germany
[3] Fritz-Haber-Institut der Max-Planck-Gesellschaft, Faradayweg 4-6, D-14195 Berlin, Germany
[4] Ioffe Physicotechnical Institute, Politekhnicheskaya 26, St. Petersburg, 194021, Russia

Abstract By means of kinetic Monte Carlo techniques we investigate the self-organized growth of quantum dots in the Stranski-Krastanov growth mode. In our simulations we consider deposition from the vapor phase, anisotropic diffusion of atoms on the surface mediated by surface reconstruction, and self-consistent strain fields generated by the quantum dots, which are computed self-consistently from elastic theory. The size distribution of dots, their spatial arrangement and shape are discussed in dependence on relevant growth parameters like temperature, deposition rate, and coverage.

1 Introduction

The spontaneous formation of nanometer scale structures on semiconductor surfaces has aroused considerable interest in recent years [1].

In our theoretical work we consider the growth of quantum dots in the Stranski-Krastanov growth mode. Here we aim to explain the sharp size distribution of the quantum dots forming on top of the wetting layer as is observed in various experiments. In contrast to previous phenomenological work [2,3] we quantify the effects of surface strain which is responsible for spatially correlated island growth, and incorporate anisotropic diffusion. The spontaneous spatial ordering of quantum dot arrays is a technologically most interesting effect, since it gives way for the fabrication of close packed quantum dot devices.

However, it has to be noted that the validity of our simulations extends only towards systems in which the transition to three dimensional island growth is not yet reached. The proper formation of three dimensional islands is beyond the scope of our numerical routines.

2 Theory

As a tool for simulating the growth processes we make use of a continuous time Monte Carlo scheme. We take into account the deposition of material on the surface. Once an atom has reached the wetting layer it cannot evaporate again. We simulate the surface kinetics by means of activated diffusion. Here the hopping probability of an atom to an adjacent lattice site is proportional to an Arrhenius factor which contains the local binding energy

$$E = E_s + n \cdot E_b + E_{ad} - v\, E_{ae}(x,y),$$

where E_s is the binding energy to the wetting layer, E_b describes the bonds to the n nearest neighbors in the growth plane [4]. E_{ad} is an energy correction that mimics anisotropic diffusion caused by the surface reconstruction. Therefore it differs from zero only for hopping in a certain direction. E_{ae} is the spatially dependent elastic strain energy density in the wetting layer and v the volume of a lattice site. For compressive strain we assume the strain to reduce the binding energy, hence it appears as a negative contribution.

The evaluation of the strain field makes use of elastomechanics of continuous media. The island boundaries S act as the sources P of the strain. The acting forces then define the displacements u in the wetting layer via a Green's formalism:

$$\mathbf{u(r)} = -\oint_S d\mathbf{r}'\, \mathbf{G(r,r')}\mathbf{P(r')}$$

The strain tensor ε follows from the derivatives of the local displacements and finally the strain energy is given for cubic systems as

$$E_{ae} = \frac{c_{11}}{2}\left(\varepsilon_{xx}^2 + \varepsilon_{yy}^2\right) + c_{12}\varepsilon_{xx}\varepsilon_{yy} + 2c_{44}\varepsilon_{xy}^2$$

with elastic constants c_{11}, c_{12}, c_{44}.

The strain field plays a vital role in our simulations for it limits the island growth to a certain size and thus prevents Ostwald ripening. On the other hand it allows for spatial ordering by generating a path for island interactions.

3 Results

In the simulations we have varied experimentally relevant parameters such as growth temperature, flux to the surface, and the final coverage.

Generally, on short time scales an increase of temperature results in an increase of the average island size. Since the size distribution also broadens with increasing temperature one should aim at a proper tradeoff between average size and its deviations. The variation of the flux has very much the same effects as one can observe with different temperatures.

With increasing coverage we observe again an increase of the average island size. For coverages of about 30% we find good spatial ordering of the islands with a

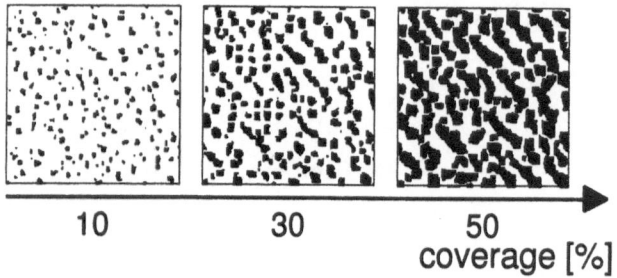

Fig. 1 Simulation results for different coverages. Growth temperature $T = 720$K, flux $F = 0.25$Ml/s. Strain and diffusion are assumed to be isotropic ($E_s = 1.2$eV, $E_b = 0.3$eV). Simulation time 10s.

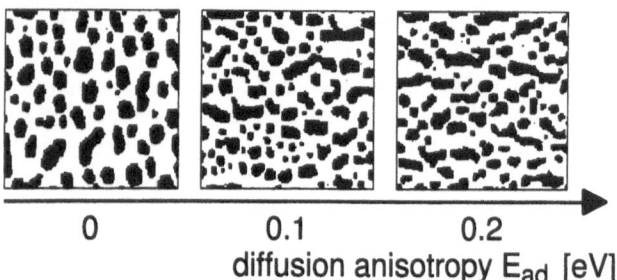

Fig. 2 Simulation results for anisotropic diffusion. E_{ad} denotes the additional diffusion barrier that is associated with diffusion in north-south direction. Simulation parameters are coverage $c = 30\%$, temperature $T = 720$K, flux $F = 0.1$Ml/s. Strain is assumed to be isotropic ($E_s = 1.2$eV, $E_b = 0.3$eV).

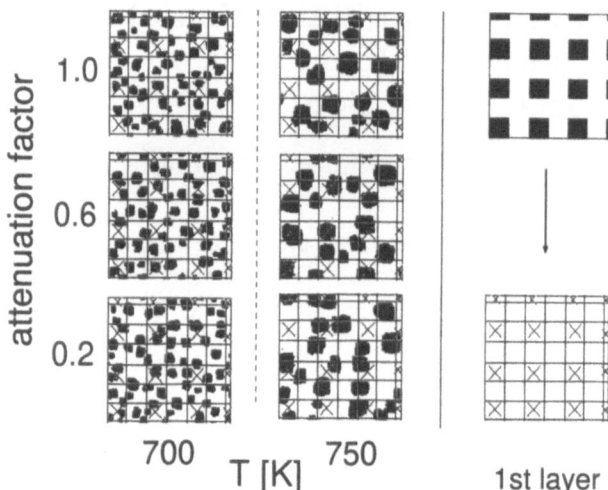

Fig. 3 Correlation and anticorrelation in quantum dot stacks in dependence of the buffer layer thickness corresponding to different attenuation factors and for two different temperatures. The buried quantum dots of the first layer are marked by crosses ($F = 0.01$Ml/s, $c = 30\%$).

sharp size distribution (Fig.1) if the other parameters are chosen appropriately. For much higher coverages there occurs coalescence of islands and clustering.

To improve the spatial ordering as well as the size distribution it has been found to be very helpful to introduce growth interruptions of a few 10 seconds. Then the self ordering effects of the strain field can act on the islands giving optimal growth results.

If the surface diffusion is assumed anisotropic , the islands become elongated in the direction of fast diffusion [5](Fig.2). By using anisotropic strain parameters for the material system SiGe/Si we were able to identify parameter ranges for which the formation of island chains in the (100) direction can be found. Since the strain does not extend far into the material in the elastically hard directions, nucleation of islands takes place favorably in the (100) direction where the atoms are more strongly bound to the surface. The same effect has been observed experimentally in samples grown by liquid phase epitaxy [6].

Furthermore, we present first results on the formation of stacks of quantum dots [7–9](Fig.3). For the first layer (buried quantum dots) we assume a regular array of quantum dots (upper right), whose strain field is superimposed onto the next layer with an attenuation factor. The growth simulation of the top layer is shown in the two columns on the left hand side, where the positions of the quantum dots in the first layer are marked by crosses. Thicker buffer layers correspond to larger at-

tenuation factors. Our simulations indicate that for thin buffer layers (attenuation 0.2) the quantum dots have a slight preference to form at vertically correlated positions, while for thicker buffer layers (1.0), especially at higher temperature, they are more likely to be anticorrelated.

4 Conclusion

We have simulated the growth of self-assembled quantum dots in the framework of a kinetic Monte Carlo scheme. The variation of parameters such as temperature, coverage, and flux yields results for the average island size and the spatial ordering that are in good agreement with the experiments. We have been able to explain effects of anisotropy, e.g. the formation of island chains, and have presented first results on the growth mechanism for stacks of quantum dots exhibiting (anti)correlation effects in dependence on the buffer layer thickness.

References

1. D. Bimberg, M. Grundmann, N.N. Ledentsov, *Quantum Dot Heterostructures* (Wiley, New York 1999).
2. E. Schöll and S. Bose, Sol. State El. **42**, (1998) 1587.
3. S. Bose and E. Schöll, *Proc. 24th Int'l Conf. Phys. Semicond.* ed. by D. Gershoni (World Scientific, 1999).
4. A. Kley, P. Ruggerone, and M. Scheffler, Phys. Rev. Lett. **79**, (1997) 5278.
5. C. Mottet, R. Ferrando, F. Hontinfinde, and A. C. Levi, Surf. Sci. **417**, (1998) 220.
6. M. Schmidtbauer, Th. Wiebach, H. Raidt, M. Hanke, and R. Köhler, Phys. Rev. B **58**, (1998) 10523.
7. Q. Xie, A. Madhukar, P. Chen, and N. P. Kobayashi, Phys. Ref. Lett. **75**, (1995) 2542.
8. O. Flebbe, H. Eisele, T. Kalka, F. Heinrichsdorff, A. Krost, D. Bimberg, and M. Dähne-Prietsch, Journal of Vacuum Science and Technology B **17**, (1999) 1639.
9. V. A. Shchukin and D. Bimberg, Rev. Mod. Phys. **71**, (1999) 1125.

Growth of InAs quantum dots on {110}-oriented cleaved GaAs surfaces

Maria Gerling, Søren Jeppesen, Anders Gustafsson and Lars Samuelson

Solid State Physics, Lund University, 221 00 Lund, Sweden

Abstract
We have investigated the growth of self-assembled InAs quantum dots on cleaved {110} edges of GaAs substrates. Low-temperature cathodoluminescence and atomic force microscopy demonstrate the formation of quantum dots on the cleaved edge of the GaAs substrates.

1 Introduction

Self-assembled InAs quantum dots (SAQDs) in a GaAs host result in strongly confined carriers, generating a discrete spectrum of electronic energy levels. SAQDs are therefore considered as possible building blocks for future nanoscale electronics. The SAQDs form spontaneously in a transition from a 2D to a Stranski-Krastanov growth mode. This transition minimizes the elastic energy, associated with the lattice mismatch strain between the InAs and the GaAs. It is important to control the position of the SAQDs and to be able to selectively place them at will [1]. We propose to overgrow SAQDs on a {110}-oriented cleaved edge of a (001)-strained layer superlattice for that purpose. Overgrowth on a strained layer superlattice was previously shown to successfully yield quantum confinement in two dimensions [2]. ZnCdSe dots have been found to form on the ZnSe layer on the cleaved edge of a ZnSe/GaAs substrate [3], and on a cleaved edge of GaAs [4].

The main difficulty in utilizing this method for generating ordered rows of SAQDs, lies in the epitaxial growth on the cleaved non-polar {110} surfaces in general, and growth of SAQDs in particular.

2 Experimental

The growth was performed in a chemical beam epitaxy, (CBE), system. For further details on the growth, see ref. [5]. The layers were grown on the cleaved {110} surface of a semi-insulating (001)-oriented GaAs wafer. The {110} surfaces were prepared by cleaving (001) GaAs substrates in air. A reference (001) substrate was always inserted along with the {110} cleaved edge. The surface reconstruction during growth was monitored on the (001) surface with the RHEED beam directed along one of the <110> directions. First, a GaAs buffer layer of about 100 nm was grown at 550 °C. The temperature was lowered to 490 °C prior to island growth. In order to find the optimized growth conditions on the {110} surface, we made three sets of samples with the TBAs control pressure at 0.6,

1.0 and 1.5 mbar. Each set contained several samples of different InAs deposition times. The TMIn control pressure was kept at 0.03 mbar and the temperature at 490 °C for all samples. The samples for optical studies had a GaAs cap layer of about 90 nm. A growth interruption of 30 sec was introduced before the GaAs capping layer was grown at 490 °C.

We know from previous studies when the dot formation starts at different TBAs pressures. We have used this knowledge and increased the growth time, since we expect a slower growth rate on the non-polar {110} surface. It has previously been reported that InAs growth on {110} GaAs substrates gives a 2D-growth mode and that Stranski-Krastanov growth is not possible though the lattice mismatch is almost 7 % [6].

The layers were characterized by atomic force microscopy, (AFM), used in contact mode and low-temperature cathodoluminescence, (CL).

3 Results and discussion

Fig. 1 shows AFM images of two samples grown with an InAs deposition time of 210 s and with a TBAs pressure of 1. 0 or 0.6 mbar, respectively. We observe dots on the surface, although they are larger than the dots we normally observe on (001) GaAs. An underlying saw-tooth structure is

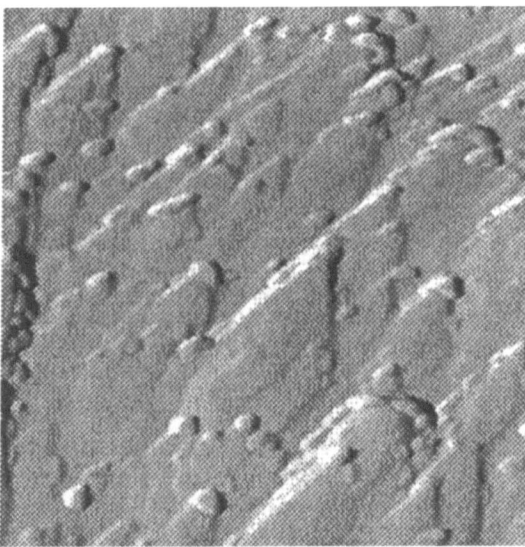

Fig. 1a 3μmx3μm AFM image of a {110} surface with InAs deposition time of 210 s and a TBAs control pressure of 1.0 mbar.

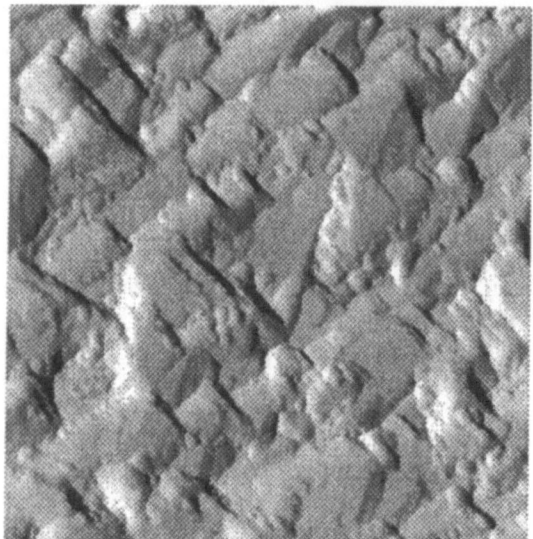

Fig. 1b 3μmx3μm AFM image of a {110} surface with InAs deposition time of 210 s and a TBAs control pressure of 0.6 mbar.

also observed, which probably originates from the GaAs layer. This has previously been observed in GaAs growth on {110} surfaces [7]. Fig. 2 and 3 shows the CL spectra of the same structure as Fig. 1a. The intensity variations in the CL images, Fig. 3, show complementary behavior. We conclude from the CL images, that the high energy peak originates from extended areas, (layer related). The low energy peak is of more local nature, (dot related). The InAs layer related peak has an energy position of 1.41-1.46 eV depending on how much material has been deposited.

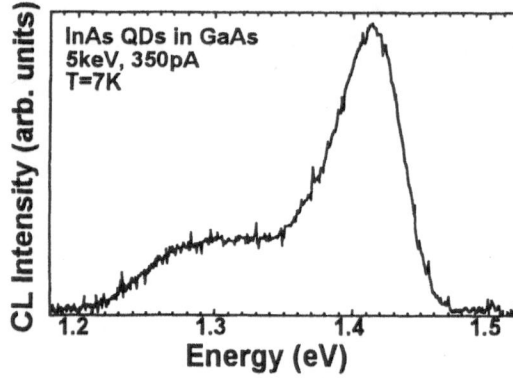

Fig. 2 Integrated CL spectra of a sample with InAs deposition time of 210 s and a TBAs control pressure of 1.0 mbar

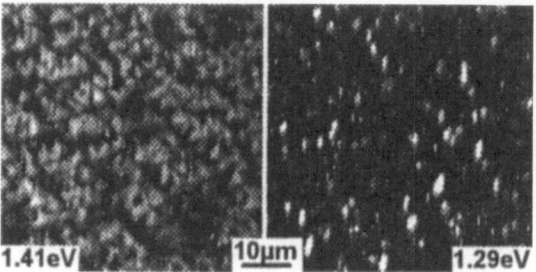

Fig. 3 CL images of the two CL peaks in Fig. 2.

The dot luminescence becomes more prominent for the sample with the longest InAs deposition time. The peak energy position of the dot luminescence is between 1.28-1.39, for all samples. In the spectra of the samples with shorter deposition time we observe that the dot luminescence has a long low-energy tail. We observe two peaks in the CL spectra, with the same spatial signature, for samples grown with a TBAs control pressure of 0.6 mbar and 210 s and 250 s of InAs deposition.

4 Summary

In summary we observe SAQDs on cleaved {110}-edges of GaAs substrates at different growth conditions. In the CL spectra two peaks are observed. From CL images we conclude that one peak is layer related whereas the other peak is dot related.

This work was supported by the Swedish National Science Research Council.

References

1. S. Jeppesen, M. S. Miller, D. Hessman, B. Kowalski, I. Maximov and L. Samuelson, Appl.Phys.Lett. **68**, 2228 (1996)
2. D.Gershoni, J.S. Wiener S.N.G. Chu et al. Phys. Rev. Lett. **65**, 1631 (1990)
3. B. P. Zhang, T. Yasuda, Y. Segawa. H. Yaguchi, K. Onabe, E. Edamatsu and T. Itoh, Appl. Phys. Lett. **70**, 2413 (1997)
4. B. P Zhang, W. X. Wang, T. Yasuda, Y. Segawa, K. Edamatsu and T. Itoh, Appl. Phys. Lett. **71**, 3370 (1997)
5. M. S. Miller, S. Jeppesen, B. Kowalski, I. Maximov, L. Samuelson, J. Cryst. Growth **164**, 345 (1996)
6. B. A. Joyce, J. L. Sudijono, J. G. Belk, H. Yamaguchi, X. M. Zhang, H. T. Dobbs, A. Zangwill, D. D. Vvedensky, T. S. Jones, Jpn. J. Appl. Phys. **36**, 4111 (1997)
7. P. Tejedor, P. Šmilauer, C. Roberts and B. A. Joyce, Phys. Rev. **B 59**, 2341 (1999)

Fabrication of compound-semiconductor quantum dots on a Si(111) substrate terminated by bilayer-GaSe

K. Ueno[1], K. Saiki[2], A. Koma[1]

[1] Department of Chemistry, Graduate School of Science, The University of Tokyo, 7-3-1, Hongo, Bunkyo-ku, Tokyo 113-0033, Japan e-mail: kei@chem.s.u-tokyo.ac.jp

[2] Department of Complexity Science and Engineering, Graduate School of Frontier Sciences, The University of Tokyo, 7-3-1, Hongo, Bunkyo-ku, Tokyo 113-0033, Japan

Abstract   We have developed a novel method to fabricate self-assembled quantum dots (QDs) of compound semiconductors on a Si(111) substrate, whose surface is terminated by bilayer-GaSe. This surface is formed by depositing 1 monolayer Ga on a clean Si(111)-7 × 7 surface and successive annealing in a Se flux at 520°C. Then the surface was irradiated with a Ga flux to form Ga droplets and annealed in an As flux at 250°C. It has been revealed that GaAs nano-crystals with the average size of 30 nm diameter and 10 nm height are formed on the substrate with density of $1.2 \times 10^{10} \mathrm{cm}^{-2}$. This method will open a new way to fabricate QDs of many kinds of compound semiconductors.

1 Introduction

Fabrication of semiconductor quantum dots (QDs) is of considerable interest because they have possibility for high-performance optical and electronic devices [1, 2]. Usually high-quality QDs of compound semiconductors are formed by the self-organization during Stranski-Krastanow (SK) mode growth [3,4]. In this case, however, the successful combination of the substrate and the grown materials has been limited to several cases, because the SK mode growth essentially requires an adequate amount of the lattice mismatch between them.

Our approach to fabricate QDs of compound semiconductor uses so-called 'droplet epitaxy' technique. If a Ga flux, for example, is irradiated under ultrahigh-vacuum (UHV) condition on an inactive surface at an appropriate substrate temperature, Ga atoms freely migrate on the surface and form a large number of uniform, nm-scale Ga droplets. Then they can be transformed into GaAs QDs by annealing in an As flux. Until now, QDs of III-V compound semiconductors have been fabricated on a S or Se-terminated GaAs(001) substrate by the droplet epitaxy [5–7]. However, this surface is not enough stable to form high-quality QDs, which requires high-temperature annealing in the As flux [8].

In order to fabricate high-quality QDs by the droplet epitaxy, we examined a new type of a substrate surface: a bilayer-GaSe terminated Si(111) (in the following abbreviate to 'BGS'). The structure of the BGS surface has been reported as shown in Fig. 1 [9,10]. In this surface, each Si atom in the top and the second layer of the ideal Si(111) surface is replaced by Se and Ga atoms, respectively. It has been reported by Eddrief et al. [9, 11,12] that an epitaxial film of GaSe, which has a two-

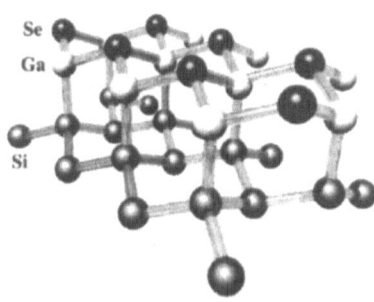

Fig. 1 Structure of a bilayer-GaSe terminated Si(111) surface

dimensional layered structure, can be grown on the BGS surface in the manner of the van der Waals epitaxy [13, 14] in spite of the 2.2% lattice mismatch and different crystal structures. Furthermore, the substrate temperature could be raised to 500°C during the growth of GaSe. Therefore we expected that the BGS surface has low surface energy without any active dangling bond that disturbs the formation of nanometer-scale Ga droplets, and is enough stable against the high-temperature annealing in the As flux.

2 Experimental

The BGS surface and GaAs QDs were fabricated in an UHV molecular beam epitaxy (MBE) chamber with a base pressure of 3×10^{-8} Pa. A clean Si(111)-7 × 7 surface was obtained by the direct current heating. After the substrate temperature was set to 520°C, 1 monolayer (ML) equivalent Ga atoms were evaporated from a Knudsen cell onto the clean Si(111)-7 × 7 surface. Successively the sample was annealed for several minutes at 520°C in a Se flux with the typical intensity of 3×10^{-4} Pa. After the substrate temperature was lowered to 250°C, Ga atoms were deposited onto the BGS surface with the rate of 0.25 ML/min. Then the sample was annealed at 250°C for 30 min in an As flux with the typical flux intensity of 3×10^{-4} Pa. During the fabrication of the BGS surface and GaAs QDs, surfaces were monitored by reflection high-energy electron diffraction (RHEED). After the growth of GaAs QDs the sample was taken out of the MBE chamber and observed by atomic force microscope (AFM) in air using SEIKO In-

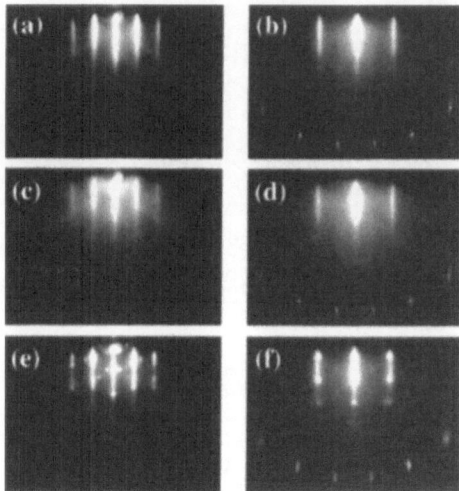

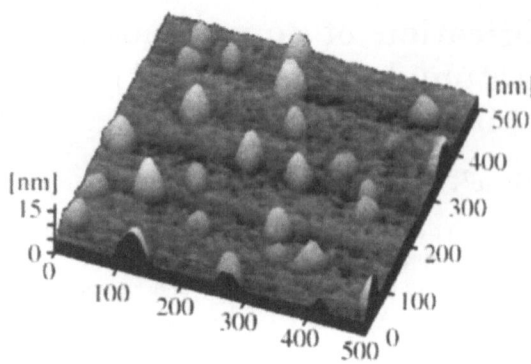

Fig. 3 An AFM image of GaAs QDs fabricated on a BGS surface.

Fig. 2 Evolution of RHEED images during the fabrication of a BGS surface and GaAs QDs. Images (a, c, e) and (b, d, f) were observed along the [10Ī] and [11Z̄] azimuthes of the Si substrate, respectively.

struments SPI-3800/SPA-300 system. AFM images were collected in the non-contact mode with a Si cantilever.

3 Results and Discussion

Figure 2 indicates the evolution of RHEED images during the QD fabrication process. RHEED images of the BGS surface (a, b) show clear streaks, indicating the flatness of the surface. After 1.5 ML equivalent Ga atoms were deposited onto the BGS surface, some rings appeared on the streak pattern (c, d), showing the formation of polycrystalline Ga droplets. Immediately after the start of annealing in the As flux, however, these rings vanished away, and many spots appeared over the streak pattern of the BGS surface (e, f). Some twin spots could be observed between streaks. No ring or halo pattern existed after the irradiation of the As flux, which means that Ga droplets were almost completely transformed into epitaxial nano-crystals of GaAs. No change was observed in the streak image coming from the BGS surface except the overlap of the spotty pattern originating from GaAs QDs. Then it is suggested that the bare BGS surface is exposed between QDs without being covered by the wetting GaAs layer.

Figure 3 indicates an AFM image of the same sample. Each QD does not seem to be spherical, but to be surrounded by some facets. QDs have 30–40 nm diameter and 10–15 nm height with an average dot density of $1.2 \times 10^{10} \mathrm{cm}^{-2}$. Rough calculation of the number of Ga atoms included in these GaAs QDs well agrees with the amount of deposited Ga atoms (1.5 ML equivalent).

We have already succeeded in reducing the density of GaAs QDs by increasing the substrate temperature, or by decreasing the intensity of the Ga flux when Ga atoms are deposited onto the BGS surface. Therefore it is suggested that Ga droplets are fabricated via the two-

dimensional nucleation without being trapped at active sites which may exist on the BGS surface.

Currently we are trying to fabricate QDs of other compound semiconductors on the BGS surface. Preliminary experiments have revealed that In droplets can be also formed on the BGS surface, and they can be transformed into InAs QDs. We think the inactiveness of the BGS surface will also enable us to form droplets of such III- or II- column metals as Al, In or Zn, Cd, Hg, and to transform them into QDs of III-V or II-VI semiconductors by annealing in fluxes of V- or VI-column elements such as N, P, As, Sb or O, S, Se, Te, respectively.

References

1. Y. Arakawa, and H. Sakaki, Appl. Phys. Lett. **40**, (1982) 939.
2. H. Sakaki, Jpn. J. Appl. Phys. **28**, (1989) L314.
3. B. A. Joyce, and D. D. Vvedensky: *Thin Films: Heteroepitaxial Systems*, eds. W. K. Liu and M. B. Santos (World Scientific Publishing, Singapore, 1999) Chap. 8, p. 368.
4. D. Leonard, M. Krishnamurthy, C. M. Reaves, S. P. Denbaars, and P. M. Petroff, Appl. Phys. Lett. **63**, (1993) 3203.
5. N. Koguchi, S. Takahashi, and T. Chikyow, J. Cryst. Growth **111**, (1991) 688.
6. N. Koguchi, and K. Ishige, Jpn. J. Appl. Phys. **32**, (1993) 2052.
7. N. Koguchi, K. Ishige and S. Takahashi, J. Vac. Sci. & Technol. B **11**, (1993) 787.
8. T. Mano, K. Watanabe, S. Tsukamoto, H. Fujioka, M. Oshima, and N. Koguchi, Jpn. J. Appl. Phys. **38**, (1999) L1009.
9. A. Koëbel, Y. Zheng, J. F. Pétroff, M. Eddrief, L. T. Vinh, and C. Sébenne, J. Cryst. Growth **154**, (1995) 269.
10. S. Meng, B. R. Schroeder, and M. A. Olmstead, Phys. Rev. **B61**, (2000) 7215.
11. N. Jedrecy, P. Pinchaux, and M. Eddrief, Phys. Rev. **B56**, (1997) 9583.
12. K. Amimer, M. Eddrief and C. A. Sébenne, J. Cryst. Growth **217**, (2000) 371.
13. A. Koma, K. Sunouchi and T. Miyajima, J. Vac. Sci. & Technol. B **3**, (1985) 724.
14. K. Ueno, M. Sakurai and A. Koma, J. Cryst. Growth **150**, (1995) 1180.

Proc. 25th Int. Conf. Phys. Semicond., Osaka 2000 (Eds. N. Miura and T. Ando)

387

Optimized growth procedure for self-organized InAs quantum dots

E. Steimetz [1], T. Wehnert [1], P. Kratzer [2], L.G. Wang [2], Q. K. K. Liu [3], H. Kirmse [4], J.-T. Zettler [1], W. Neumann [4], M. Scheffler [2], and W. Richter [1]

[1] Institut für Festkörperphysik, Technische Universität Berlin, Sekr. PN 6-1, Hardenbergstr. 36, D-10623 Berlin Germany
[2] Fritz-Haber-Institut der Max-Planck-Gesellschaft, Faradayweg 4-6, D-14195 Berlin, Germany
[3] Abteilung Theoretische Physik, Hahn-Meitner-Institut, Glienicker Str. 100, D-14109 Berlin, Germany
[4] Humboldt Universität Berlin, Inst. für Physik, Invalidenstr. 110, D-10115 Berlin, Germany

Abstract   A redistribution of InAs from partially covered islands to the GaAs cap layer was monitored in real-time by reflectance anisotropy spectroscopy (RAS) and ellipsometry. Introducing a growth interruption during cap layer growth, a narrower size distribution of the InAs quantum dot ensemble could be achieved. In order to explain this behavior we performed calculations in a hybrid approach combining *ab initio* surface energy calculations with continuum elasticity theory.

1 Introduction

Self-assembled islands formed in a Stranski-Krastanow growth mode have attracted much attention for making QD-structures due to their potential application in optoelectronic devices such as QD lasers [1]. However, problems such as nonuniform size distribution diminish the advantages predicted by theory. Especially the appearance of large relaxed clusters degrades the luminescence and often causes surface roughening. Therefore, smooth interfaces, crucial for device applications and stacking of QDs, can only be achieved with carefully optimized parameters for the cap layer growth.

In this paper we show that the QD-layer quality can be improved by a growth interruption during GaAs cap layer growth. This is due to a rearrangement of InAs from partially capped large islands into a new wetting layer on top of the GaAs cap. The driving force for this effect is demonstrated to be the energy reduction due to the InAs modified cap layer surface reconstruction. This causes a total energy minimum for the rearranged QD configuration.

2 Experimental

The experiments were performed in a low-pressure horizontal MOVPE reactor with three windows for in-situ RAS and ellipsometry measurements [2]. Arsine was used as group V precursor, while trimethylgallium (TMG) and trimethylindium (TMI) were used as group III precurcors. The InAs deposition was controlled by RAS, as this technique is capable of determining the 2D-3D transition in Stranski-Krastanow-growth mode [3]. The after growth surface morphology was studied by atomic force microscopy (AFM) in air. The QD-stack quality was controlled by cross-sectional TEM-micrographs and photoluminescence.

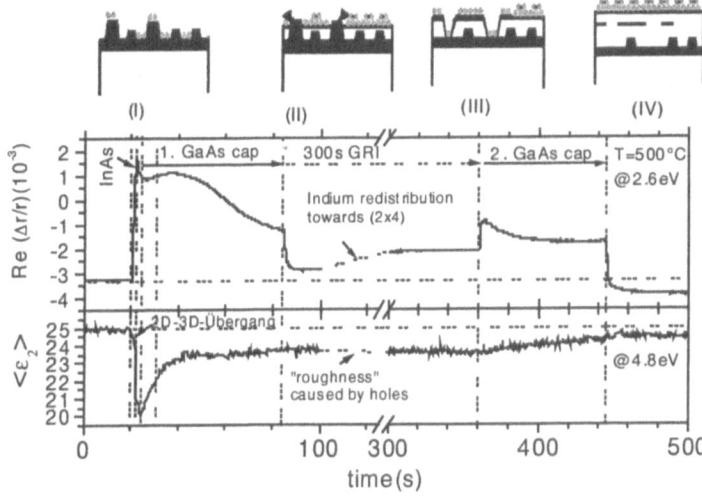

Fig. 1 Model of island redistribution during growth interruption as derived from in-situ studies. RAS-transients during InAs-QD growth show Indium enrichment of the surface during growth interruption (II to III). Ellipsometry indicates an increasing porosity of the layer.

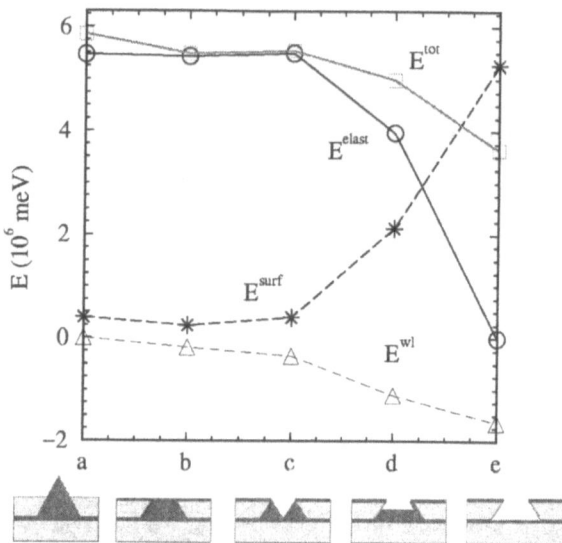

Fig.2 Total energy E^{tot} of the QD ensemble during the different stages of cap layer growth interruption: $E^{tot} = E^{surf}+E^{elast}+E^{wl}$. E^{wl} is the energy gain due to the spontaneous formation of a second wetting layer, E^{surf} is the surface energy of side facets and E^{elast} is the elastic energy of the structure's strain fields.

3 Results and Discussion

In order to study how the planarization of InAs QD islands proceeds during cap layer overgrowth, the InAs-QD samples were first capped with GaAs layers having nominal thicknesses ranging from 3.5 nm to 10.5 nm. Analyzing their surface stoichiometry (RAS) and morphology (AFM), indications of a material redistribution from partially covered islands were found. In AFM the surfaces are basically smooth, but holes with depths equal to the thickness of the deposited GaAs layer were found. Their size is approximately the same as that of uncovered InAs clusters. Thus, introducing an optimized growth interruption during GaAs cap layer growth can dissolve unwanted clusters.

On top of Fig. 1 the proposed model for an optimized growth process avoiding large clusters is given: (I) shows the island surface after InAs-growth. The 1 ML of As atoms on top symbolizes that the surface is (2x4)-reconstructed. The onset of dissolution for partially overgrown islands is sketched in (II) and (III) gives the surface after the growth interruption. The InAs from the islands is transformed into a second wetting-layer while holes remain in the cap layer. The expected final structure after a second GaAs-overgrowth is shown in (IV). The redistribution of material was monitored by changes in RAS- and ellipsometry transients during GRI shown in Fig.1. The redistribution of the InAs islands during the growth interruption should lead to the formation of a second wetting layer, since the temperature is too low to evaporate Indium atoms. The second fractional wetting layer was also evidenced by cross section TEM measurements on a 3-fold InAs-Quantum dot stack .

This concept is supported also by theoretical evidence: We perform calculations for a particular realization of the proposed scenario (Fig.2). The stability of various intermediate stages during the transformation process is analyzed using a hybrid approach [4]: The total energy is decomposed into a contribution from surface energies that are calculated using density-functional theory and contributions from elastic distortions that are determined within continuum elasticity theory. We find that the energy gained when the InAs forms a wetting layer on the GaAs capping layer is the main driving force for dissolving the large InAs clusters.

4 Summary

We demonstrated that it is possible to dissolve large clusters in InAs self-organized QD ensembles by introducing an interruption during the growth of the capping layer at a suitable moment, when these large clusters are only partially covered. The energy gained when the InAs forms a wetting layer on the GaAs capping layer is the main driving force for dissolving the large InAs clusters.

References

1. Y. Arakawa and H. Sakaki, Appl. Phys. Lett., **40**, (1982) 939; D. Bimberg, N. Ledentsov, N. Kitstaedter, O. Schmidt, M. Grundmann, V.M. Ustinov, A.Yu. Egorov, A.E. Zhukov, M.V. Maximov, P.S. Kop'ev , Zh. I. Alferov, S.S. Ruvimov, U. Gösele, J. Heydenreich, Jpn. J. Appl. Phys. **35**, (1996) 1311.

2. J.-T. Zettler, Progress in Crystal Growth and Characterization of Materials, **35**, (1998) 1.

3. E. Steimetz, J.-T. Zettler, F. Schienle, T. Trepk, T. Wethkamp, W. Richter, I. Sieber, Appl. Surf. Science **107**, (1996) 203.

4. L.G. Wang, P. Kratzer, M. Scheffler and N.Moll, Phys. Rev. Lett. **82**, 4042 (1999).

Proc. 25th Int. Conf. Phys. Semicond., Osaka 2000 (Eds. N. Miura and T. Ando)

389

Effect of inter-island interaction on the growth of self-assembled quantum dots

S. Vannarat[1,2], S. T. Chui[1,3], K. Esfarjani[1], Yoshiyuki Kawazoe[1]

[1] Institute for Materials Research, Tohoku University, 2-1-1 Katahira, Aobaku, Sendai, 980-8577 Japan
[2] National Electronic and Computer Technology Center, 73/1 Rama VI Rd., Rajdhevee, Bangkok 10400 Thailand
[3] Bartol Research Institute, University of Delaware Newark, DE 19716, USA

Abstract The elastic interaction between two coherent islands, derived from a simple model, is found to be a dipole-dipole liked repulsive interaction. Using the molecular dynamics simulation result of a two dimensional system of parallel dipoles, the phase diagram of a system of self-assembled quantum dots(SAQDs) is calculated as a function of the average island size and density. A system of islands should change from a liquid phase to a solid phase as the island size increase during the growth process. The island formation barrier is derived as a function of the average size and density of the pre-existing islands. The barrier increases rapidly at a critical value of the average size and density.

1 Introduction

The self-assembled quantum dots (SAQDs) refer to the coherent 3-dimensional islands spontaneously formed during the Stranski-Krastanov (SK) growth of highly lattice mismatched semiconductor heterostructures [1–8] under some growth conditions [9]. The island formation is known to be a strain relaxation mechanism. The strain field within the substrate also gives raise to the interaction between islands. In this paper, we show our results of deriving the interaction between two islands and the effects of the inter-island interaction on the island arrangement and the island formation.

2 Model and Assumptions

For simplicity we assume the case of isotropic materials and the island is assumed to be cylindrically symmetric around an axis perpendicular to the substrate. The islands are assumed to have such a shape that, when it is isolated, the strain at its interface with the substrate is uniform and can be described by two scalar values, ε_{so} on the substrate side and ε_{io} on the island side of the interface plane. The lattice coherency requires that the lattice mismatch

$$\varepsilon_o = \varepsilon_{io} - \varepsilon_{so}. \tag{1}$$

The strain energy per island volume is given by

$$\widetilde{E}_1 = \frac{1}{2}\lambda_s \varepsilon_{so}^2 + \frac{1}{2}\lambda_i \varepsilon_{io}^2, \tag{2}$$

where the constants λ_s and λ_i depend on the elastic constants of the island and the substrate, and on the shape of the island. By minimizing the strain energy with the constrain of the lattice coherency condition(eq.1), the value of ε_{io} and ε_{so} are determined as

$$\varepsilon_{io} = \frac{\lambda_s \varepsilon_o}{\lambda_i + \lambda_s} \text{ and } \varepsilon_{so} = \frac{-\lambda_i \varepsilon_o}{\lambda_i + \lambda_s}. \tag{3}$$

These values also correspond to the equilibrium of the forces between the island and the substrate

$$\lambda_i \varepsilon_{io} + \lambda_s \varepsilon_{so} = 0 \tag{4}$$

For a coherent heterostructure consisting of a uniform layer of the deposited material on a substrate, the strain occurs entirely in the deposited layer, and the strain energy per volume of the deposited material can be written as

$$\widetilde{E}_o = \frac{1}{2}\lambda \varepsilon_o^2,$$

where λ is the elastic constant of the deposited material. The strain energy of the island, $\widetilde{E}_1$, is less than the strain energy of the uniform deposited layer, and the relaxation factor, κ, can be defined as

$$\widetilde{E}_1 = \frac{1}{2}(1 - \kappa)\lambda \varepsilon_o^2.$$

The strain field in the substrate outside the island base can be calculated from the Green's tensor for the equations of equilibrium of a semi-infinite medium [10], and can be written in the form of

$$\varepsilon_s(\rho) = \varepsilon_{so} g(\rho), \tag{5}$$

where ρ is the reduced distance from the center of the island base. For an island with a lateral size L, $\rho = r/L$. The magnitude of the function $g(\rho)$ monotonically decreases toward zero with the distance away from the island.

3 Inter-island Interaction

When there are two islands, A and B on the same substrate, the strain field from one island also exists in the base area of the other. If the distance d between the two islands is large enough, the strain field from one island is almost constant within the base area of the other. The strain field in the substrate at the base of island A can, therefore, be written as,

$$\varepsilon_{sA} = \varepsilon_{sAo} + \varepsilon_{sBo} g(d/L_B), \tag{6}$$

where L_B is the size of island B. ε_{sAo} and ε_{sBo} are to be determined from the coherency condition and the equilibrium condition.

The coherency condition requires that the strain field on the island side at the base of island A, ε_{iAo}, satisfies

$$\varepsilon_o = \varepsilon_{iAo} - \varepsilon_{sA}. \tag{7}$$

The equilibrium condition is given by

$$\lambda_i \varepsilon_{iAo} + \lambda_s \varepsilon_{sAo} = 0. \tag{8}$$

Three equations of the same form as eq.6–8 can be written for the strain field at the base of island B. Solving these equations the values of ε_{iAo}, ε_{sAo}, ε_{iBo} and ε_{sBo} can be determined. The interaction energy between the two islands can be calculated from the their strain energy,

$$\widetilde{E}_o = \frac{1}{2}\lambda_s \left(\varepsilon_{sAo}^2 + \varepsilon_{sBo}^2 \right) + \frac{1}{2}\lambda_i \left(\varepsilon_{iAo}^2 + \varepsilon_{iBo}^2 \right). \tag{9}$$

It has been found that the function $g(\rho) \approx g/\rho^3$ for $\rho \gg 1$ with $g \cong 0.902$, and the interaction between two islands with sizes L_A and L_B, can be shown to be

$$E_{\mathrm{int}}(d) = \frac{2\xi(1-\kappa)g\alpha L_A^3 L_B^3 \lambda \varepsilon_o^2}{d^3}, \tag{10}$$

where $\xi = \lambda_i/(\lambda_s + \lambda_i)$ and α is the volume of island with the size $L = 1$. This inter-island interaction is a dipole-dipole liked interaction, i.e. the interaction energy is inversely proportional to the third power of the distance between the islands.

4 Phase of Island Arrangement

A molecular dynamics study [11] of systems with a dipole-dipole liked interaction asserts that such systems have two phases, namely the liquid phase and the solid phase. In the liquid phase, the particles are randomly distributed while in the solid phase, the particles are arranged in a 2 dimensional lattice. The transition temperature [11] between these two phases is given by

$$\Gamma = \frac{A}{a^3 k T_m} = 62 \mp 3, \tag{11}$$

where a is the average distance between two particles, T_m is the transition temperature, k is the Boltzmann's constant, and A is the dipole-dipole interaction coefficient defined by $E_{\mathrm{int}}(d) = A/d^3$.

From the coefficient of the inter-island interaction in eq.10, the critical island size, L_c, and density, n_c, at a temperature T can be calculated as

$$L_c^6 n_c^{3/2} = (62 \mp 3)\frac{kT}{2\alpha(1-\kappa)g\xi\lambda\varepsilon^2}. \tag{12}$$

When $L_c^6 n_c^{3/2}$ is smaller than the critical value, the system is in the liquid phase; as the growth progresses the average island size and density increase and the system changes into the solid phase.

5 Island Nucleation Energy Barrier

The energy of an island of size L as compared to the uniform layer of the same volume consists mainly of two parts the relaxation of the strain energy, which is proportional to the volume of the island, and the increasing surface energy. For a sufficiently small island the surface energy dominates and the island is energetically unfavorable, while for larger islands the relaxation of the strain energy becomes more important and the islands are more stable than the uniform layer. The maximum of the island energy is called the nucleation energy barrier, and the size of the island that has the maximum energy is called the critical nucleus size.

During the growth process, new islands may be formed when there have been a number islands already exist on the substrate. The nucleation barrier of the new island is increased by its interaction with the existing islands. This effect is studied by approximating the existing islands with a regular array of islands with a size l and a density n. The energy of an island of size L on a substrate with such an array of islands, as compared to the uniform layer of the same volume, is

$$\begin{aligned} E(L) = &-\kappa\lambda\varepsilon^2\alpha L^3 + \gamma\beta L^2 \\ &+ 2\xi(1-\kappa)l^3 gn^{3/2}\phi\alpha L^3\lambda\varepsilon^2. \end{aligned} \tag{13}$$

The first term is the strain energy relaxation; the second term is the increasing surface energy. The summation of the interaction between the new island with all the other islands gives the third term with a multiplication factor ϕ (for an equilateral triangle array, $\phi \cong 9.247$.) The nucleation barrier as a function of the existing island average size and density is

$$E_B(l,n) = \frac{4\gamma^3\beta^3}{27\alpha^2\lambda^2\varepsilon^4\left(\kappa - 2\xi(1-\kappa)l^3 gn^{3/2}\phi\right)^2}. \tag{14}$$

The nucleation barrier increase with $l^3 n^{3/2}$, and becomes infinite as $l^3 n^{3/2}$ approaches $\kappa/(1-\kappa)2\xi g\phi$, when the total interaction energy is the same as the strain relaxation.

6 Conclusion

The elastic interaction between two self-assembled islands is found to be a dipole-dipole liked repulsive interaction. The arrangement of the islands can be in two phases, namely, the liquid phase and the solid phase. The critical size for the transition between the two phases is derived from a molecular dynamics simulation result. The island nucleation barrier is derived as a function of the pre-existing island average size and density.

References

1. F. Houzay, C. Guille, J.M. Moison, P. Henoc and F. Barthe, J. Cryst. Growth **81**, (1987) 67.
2. S. Guha, A. Maduhkar and K.C. Rajkumar, Appl. Phys. Lett. **57**, (1990) 2110.
3. C.W. Snyder, B.G. Orr, D. Kessler and L.M. Sander, Phys. Rev. Lett. **66**, (1991) 3032.
4. D. Leonard, M. Krishnamurthy, S. Fafard, J. L. Merz and P. M. Petroff, J. Vac. Sci. B **12**, (1994) 1063.
5. K. Georgsson, N. Carlsson, L. Samuelson, W. Seifert and L. R. Wallenberg, Appl. Phys. Lett. **67**, (1995) 2981.
6. C. M. Reaves, R. I. Pelzel, G. C. Hsueh, W. H. Weinberg and S. P. DenBaars, Appl. Phys. Lett. **69**, (1996) 3878.
7. D. J. Eaglesham and M. Cerullo, Phys. Rev. Lett. **64**, (1990) 1943.
8. F. K. LeGoues, M. Copel and R. M. Tromp, Phys. Rev. B **42**, (1990) 11 690.
9. I. Daruka and A. L. Barabási, Phys. Rev. Lett. **79**, (1997) 3708.
10. L. D. Landau and E. M. Lifshitz, *Theory of Elasticity* (Butterworth-Heinemann Ltd, Oxford UK, 1986) 25.
11. V. M. Bedanov, G. V. Gadiyak, and Yu. E. Lozovik, Sov. Phys. Solid State **25**, (1983) 328.

Properties of CdSe/ZnSe based quantum heterostructures with and without lateral confinement potentials

E. Kurtz[1][*], M. Schmidt[1], B. Dal Don[1], S. Wachter[1], D. Litvinov[2], D. Gerthsen[2], H. Kalt[1], C. Klingshirn[1]

[1] Institut für Angewandte Physik, Universität Karlsruhe, Kaiserstr. 12, D-76128 Karlsruhe, Germany;
   e-mail: Elisabeth.Kurtz@phys.uni-karlsruhe.de
[2] Labor für Elektronenmikroskopie, Universität Karlsruhe, Kaiserstr. 12, D-76128 Karlsruhe

Abstract   Using CdS and Se, we have grown CdSe/ZnSe quantum structures with improved optical and structural properties exploiting an exchange reaction which leads to the substitution of sulfur by selenium. Depending on the growth mode it is possible to obtain quantum wells without strong lateral localization or as quantum island structures of good homogeneity. The differences in the exciton relaxation behaviour in these structures is demonstrated by photoluminescence measurements.

1 Introduction

Self-organized quantum islands have a great potential for future opto-electronic applications. While in the III-V system island formation is already well established [1], II-VI semiconductor based structurs suffer from a strong tendency of Cd segregation and CdSe/ZnSe intermixing, which prevents a classical Stranski-Krastanow type islanding [2]. In standard molecular beam epitaxy of such structures it seems impossible to suppress Cd concentration fluctuations which are often referred to as Type A [3] or natural quantum islands.

We will demonstrate, that using CdS as Cadmium source it is possible to grow CdSe/ZnSe based quantum wells with improved quality, where no type A islanding occurs and discuss the specific optical properties.

2 Experimental

The samples were grown by MBE on (001) GaAs wafers using elemental Zn and Se for ZnSe buffer growth at 280°C. CdSe based quantum wells and islands were obtained by inserting thin CdSe layers obtained by a substitutional reaction of CdS compound with elemental Se which leads to an almost complete replacement of the sulfur by selenium. Optical measurements indicate a sulfur contamination below 2%.

There are several advantages to this method. The rather high CdS oven temperature of 650°C improves the surface diffusion of adatoms. Furthermore, an effect similar to a self surfactant effect plays an important role which is caused by the exchange reaction between sulfur and Se. Quantum wells (QW) with outstanding optical quality like narrow photoluminescence (PL) line width and very weak lateral confinement were obtained em-

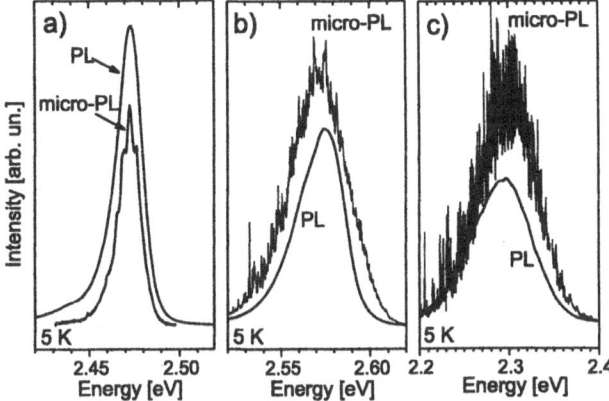

Fig. 1 PL and micro-PL spectra of three representative samples which will be discussed in the text. **a:** quantum well sample 1; **b:** quantum island sample 2; **c:** quantum island sample 3. The micro-PL had a spatial resolution of 1 μm

ploying a migration enhanced growth mode with periodic shutter cycles. Further detailes on the growth process can be found elsewhere [4,5]. The structural properties were obtained by compositional evaluation by lattice fringe analysis [6].

3 Results and Discussion

Fig. 1 illustrates clearly some of the differences between the island and quantum well samples. Sample 1, Fig. 1a is a QW with a nominal deposition of ~2.5-3 monolayers (ML) based on the observation of half a ML oscillation per deposition cycle by in-situ RHEED. However, while Cd segregation was suppressed in the growth mode employed, interdiffusion and intermixing still occured similar to observations in other CdSe based systems [2]. The quantum well is broadened to 13 MLs and contains an average Cd concentration of 32% while the interfaces are much better defined than in conventionally grown CdSe [5]. The very narrow line width of only 16 meV suggests already the absence of deep lateral localization sites. This becomes even clearer when measuring micro-PL. No sharp discrete single dot lines are observed but rather broad features. This indicates that fluctuations within the well have a lateral extension of the order of the diffusion length of an exciton.

Sample 2, Fig. 1b, contains a layer of nominally 2 ML CdSe which has been reformed into Cd enriched islands

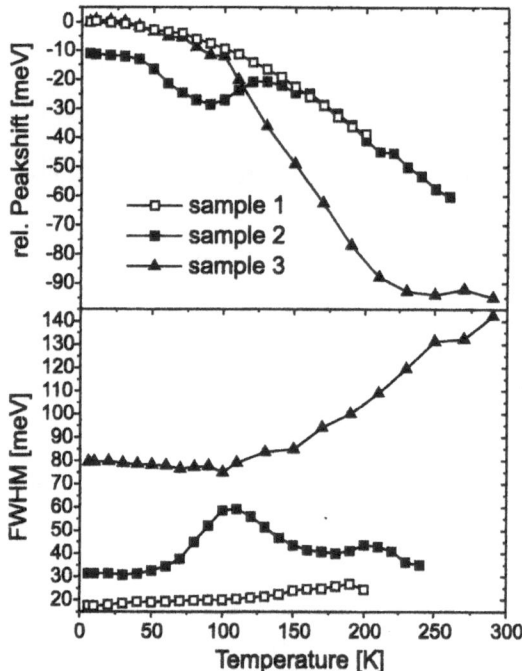

Fig. 2 *upper graph:* temperature dependent relative PL peak shifts for sample 1–3 as discussed in the text. *lower graph:* FWHM of the PL peaks

peratures. Below 40 K excitons created will be captured in the nearest potential fluctuations and will slowly relax to deeper neighbouring fluctuations, see [8]. At higher temperatures thermal excitation enhances this process, leading to a pronounced red shift until thermal activation becomes large enough to lead to delocalization at which point the shift follows the band gap shift. In case of sample 3 deep localization makes the shift follow the band edge initially. A much higher thermal activation is necessary to transfer excitons into deeper neighbouring fluctuations and the process sets in at 100 K. For this sample the fully delocalized state is never reached within the temperature range investigated. This type of deep confinement would be the most suitable for room temperature operated devices. However, transfer to nonradiative centers within the larger coalesced islands strongly quenches the PL efficiency at higher temperatures.

4 Conclusion

Substituting the conventional elemental Cd source with a CdS compound source and utilizing the S-Se exchange reaction opens new possibilities to grow extremely high quality CdSe based quantum wells and island structures. It was possible to suppress Cd segregation in the QW structures, which enabled the investigation of the localization behaviour in structures with varying lateral confinement potential depth.

The authors would like to thank the Deutsche Forschungsgemeinschaft and the Institute of Nanotechnology at University of Karlsruhe for support.

References

1. D. Bimberg, M. Grundmann, N.N. Ledentsov; *Quantum Dot Heterostructures* (John Wiley Sons, Chichester 1999)
2. D. Livinov, A. Rosenauer, D. Gerthsen, N.N. Ledentsov; Phys. Rev. B **61**, (2000) 16819
3. M. Strassburg, Th. Denizou, A. Hoffmann, R. Heitz, U.W. Pohl, D. Bimberg, D. Livtinov, A. Rosenauer, D. Gerthsen, S. Schwedhelm, K. Lischka und D. Schikora; Appl. Phys. Lett. **76** (2000) 685
4. E. Kurtz, M. Schmidt, M. Baldauf, S. Wachter, M. Grün, D. Litvinov, S.K. Hong, J.X. Shen, T. Yao, D. Gerthsen, H. Kalt, C. Klingshirn; J. Cryst. Growth **214-215** (2000) 712
5. E. Kurtz, M. Schmidt, B. Dal Don, S. Wachter, D. Litvinov, D. Gerthsen, H. Kalt, C. Klingshirn; 11th *Int. Conf. on MBE (MBE-XI), Peking China (2000)* to appear in J. Cryst. Growth (2001)
6. A. Rosenauer and D. Gerthsen; Advances in Imaging and Electron Physics, Vol. 107 (1999) 121-230
7. E. Kurtz, M. Schmidt, M. Baldauf, D. Litvinov, D. Gerthsen, H. Kalt, C. Klingshirn; *Int. Conf. on Quantum Dots QD2000, München, Germany (2000),* to appear in phys. stat. sol. (a) (2001)
8. E. Kurtz, T. Sekiguchi, Z. Zhu, T. Yao, J.X. Shen, Y. Oka, M.Y. Shen, T. Goto; Superlattices and Microstructures **25** (1999) 119

inserted in a broadened, intermixed wetting layer with a width of 10 MLs and average Cd concentration of ~15%. Two types of islands can be distinguished by Transmission Electron Microscopy (TEM) [7]: firstly, natural type A islands constitute of Cd concentration fluctuations (up to 30%) with a typical lateral extension of 3 nm and rather high densities close to $10^{12} cm^{-2}$. Secondly, self-organized islands of type B with a higher Cd concentration of up to 60%, diameters of typ. 5 nm and a density of $\sim 5 \cdot 10^{10} cm^{-2}$. Sample 3, Fig. 1c, has a similar deposition to sample 2, however, the island formation was influenced by an annealing step at 320°C for 5 min. Optically not active large coalesced islands coexist with better defined type B islands as well as natural type A fluctuations. The presence of these islands mirrors directly in the high number of sharp discrete lines observed in micro-PL. The shape and the distribution of lines suggests that in sample 2 type A quantum well fluctuations contribute mainly to the high energy side, while B-type islands make up the low energy tail due to their deeper confinement potentials. In sample 3 the PL observed is believed to mainly originate from the type B islands.

The various degrees of localization mirror directly in the temperature dependent PL. The relative shifts $E(T) - E(0K)$ are shown in Fig. 2. Sample 1 basically follows the shift of the band edge expected for a CdSe/ZnSe alloy of the given composition. The full width at half maximum (FWHM) of the PL line shows no peculiarities. In case of sample 2 the upper curve has been shifted to follow the shift of the basic energy gap at higher tem-

Growth, morphology and optical properties of metal clusters on semiconductor surface (Au/GaAs)

N. Dmitruk[1], T. Barlas[1], I. Dmitruk[2], T. Mikhailik[1], V. Romaniuk[1]

[1] Institute of Semiconductor Physics NAS of Ukraine, prospect Nauki 45, 03028 Kyiv, Ukraine
e-mail: nicola@dep39.semicond.kiev.ua

[2] Kyiv Taras Shevchenko University, prospect acad. Glushkova 6, 03127, Kyiv, Ukraine

Abstract Technology of local electrochemical deposition of noble metal nanoparticles (Au) on semiconductor substrate (GaAs) is elaborated. It allows to obtain as well isolated Au particles with the 2-50 nm diameter as some chain structures perhaps of fractal type. According to the TEM and AFM data the gold clusters have the form of a truncated spheres surrounded by uncovered microrelief areas of semiconductor. Optical reflectance spectra are interpreted by the modified Maxwell-Garnett theory taking into account the pair multipole interaction between the nanoparticles (at low filling factor) or by the symmetrical Bruggeman's approximation (at large gold covering).

1 Introduction

The electrochemical growth of discontinuous gold films on GaAs substrates from aqueous solution of $AuCl_3$ is presented. This technology is similar to electrodeposition of metals onto a semiconductor substrate, which is subject of fundamental and practical significance. Namely, such investigations are important for the understanding of the semiconductor/metal interface formation. The direct deposition of metal onto semiconductor is also of interest for Schottky diode behavior [1,2].

2 Technology and morphology of metallized semiconductor surface

Gold islets in size from 1 to 100 nm were deposited from an aqueous solution of $AuCl_3$ with the Au^{3+} ions concentration C of $(1-16) \times 10^{-5}$ gram-ion/l [1,2]. An illumination for the generation of free electron-hole pairs promoted the process of electrochemical deposition because after adsorption the Au^{3+} ions have been neutralized by the capture of valence or free electrons according to the reactions:

$$GaAs - 6e^- + 3H_2O \rightarrow Ga^{3+} + H_3AsO_3 + 3H^+,$$

$$Au^{3+} + 3e^- \rightarrow Au^0,$$

i.e. continuous dissolution of substrate around a gold islet for creation of microrelief takes place. Increasing the value of substrate exposition in the solution, Ct, $t = 1 \div 5$ min., the filling factor f_s (the metal fraction) may be changed in the range 0.01-0.7.

Analysis of electron micrographs (with shadowing) show, that the gold islets take the form of truncated spheres, separated by open areas of GaAs surface which have the form of wells (Fig.1). From the AFM image analysis it can be assumed that the growth mode of gold

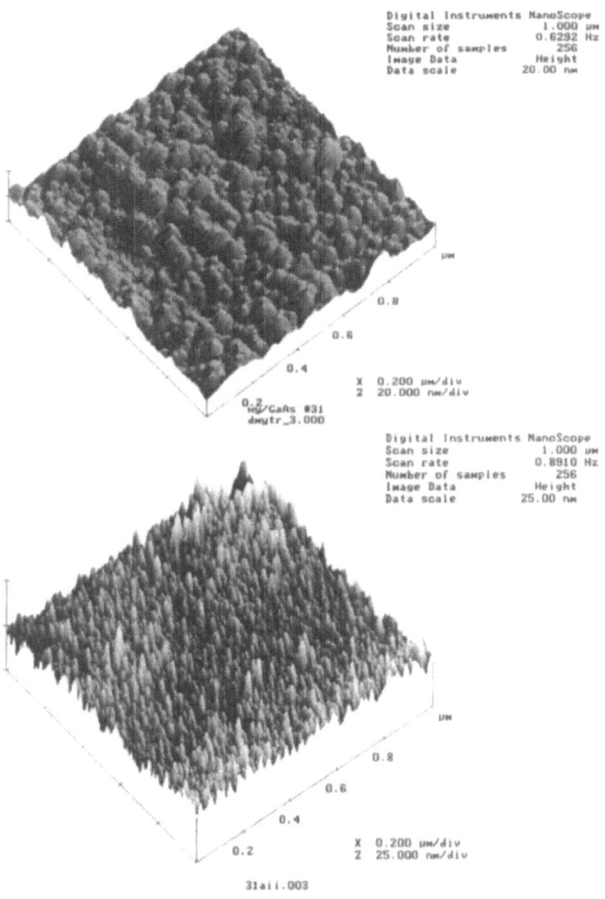

Fig. 1 AFM images of the Au/GaAs surface at $Ct = 1.6 \times 10^{-5}$ (a) and 1.6×10^{-4} g-ion/l (b), $t = 1$ min. GaAs (111) with $N_d = 2.5 \times 10^{18}$ cm^{-3}.

on GaAs is three dimensional. At increasing exposition of substrate in solution Ct, the concentration of islets decreases from $\sim 10^{12}$ cm^{-2} to 5×10^{10} cm^{-2}, during which the average diameter increases from $d_{av} \simeq 5$ nm to 50 nm. This probably indicates that the coalescence of islets takes place.

The surface coating of substrate by metal particles/ clusters, f_s have been calculated by a detailed statistical analysis of the TEM micrographs (without shadowing) and by statistical processing of the phase AFM images. Because the optical effects of metal islets are determined by the volume filling factor f_v of some surface overlayer with thickness d_{eff}, we have calculated the relation between f_s and f_v assuming monolayer coating model and truncated sphere form: $f_v/f_s = 1/3(3-x)/(2-x)$, where R is the radius and $x = h/R$, h is the height of spherical segment.

The statistical parameters of disordered (microrelief) surfaces as the rms roughness δ and the autocorrelation length σ have been determined by section analysis of AFM data as well: $\delta = 1 \div 2$ nm, and after electrochemical deposition of Au islets it increases to $\delta = 3 \div 6$ nm. Autocorrelation length is equal approximately $\sigma = 100 - 200$ nm in both cases.

According to the AFM-image of the metallized microrelief surface the Au particles were flattened in the direction orthogonal to the substrate with an axial ratio of $h/D \simeq 1/3$, i.e. $x = 2/3$ and $f_v/f_s \simeq 7/12$. This value is very important for the interpretation of optical results. It should be noted that similar results were obtained theoretically in [3] by thermodynamics of small systems and experimentally in [4] by electron microscopy.

3 Results and discussion

Typical reflectance spectra (Fig.2) show some minimum in the vicinity of the wavelength $\lambda \simeq 500$ nm. As is well known, this value corresponds to so-called Fröhlich frequency of surface (local) plasmon resonance excitation. There is also an additional channel for absorption/reflection of light induced by surface roughness.

The main idea of this investigations was the comparison of two samples with various concentration and size of gold particles: the first stage of deposition with small coating GaAs by gold, but large concentration of islets $N > 10^{12}$ cm^{-2}, and the final stage when at small concentration $N \sim 10^{10}$ cm^{-2} we have large gold coating.

For taking into account the plasmon and roughness effects, we used at the large coating the well known symmetrical Bruggeman's effective medium approximation (EMA) in the form:

$$\sum_i f_i \frac{\varepsilon_i - \varepsilon_{eff}}{\varepsilon_i + (D-1)\varepsilon_{eff}} = 0, \qquad (1)$$

where f_i and ε_i are the filling fraction and dielectric permittivity of i-th film component, ε_{eff} is an effective film permittivity, D is the lattice dimensionality connected with depolarization factor (Fig.2,a,b).

For small filling factor the metallized GaAs surface can be described as matrix disperse system, for which the well-known Maxwell-Garnett approximation (MGA) can be used. However for adequate description of the reflectance spectra observed we used the improved version of MGA for effective dielectric function $\varepsilon_{eff}(\omega)$ [6], which accounts exactly the pair multipole interaction between the particles belonging to matrix disperse system:

$$\frac{\varepsilon_{eff} + 2\varepsilon_0}{\varepsilon_{eff} - \varepsilon_0} = (f\frac{\varepsilon - \varepsilon_0}{\varepsilon + 2\varepsilon_0})^{-1} - \frac{2}{3}\ln\frac{3\varepsilon + 5\varepsilon_0}{2\varepsilon + 6\varepsilon_0}. \qquad (2)$$

Using Eq. (2) we can describe the reflectance spectra in detail (Fig.2,a). So, the consideration of interaction of

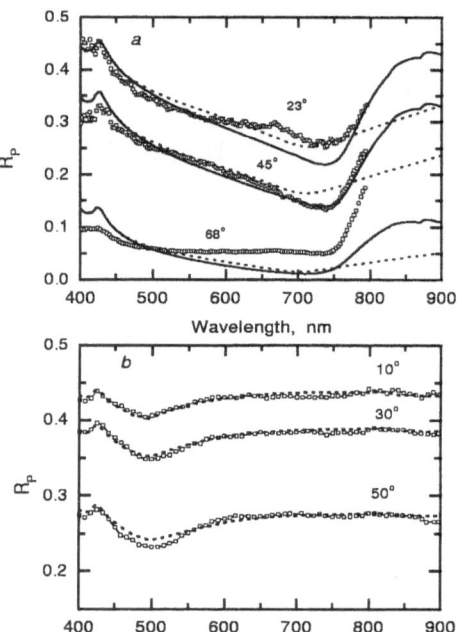

Fig. 2 Experimental (points) and calculated reflectance spectra of p-polarized light according to (1) -(dashed line) and (2) - (solid lines) at different angles of light incidence. $f = 0.029 - (a)$ and $f = 0.413 - (b)$. At fitting the optical parameters of gold were taken from [5].

small gold particles is necessary on the first stage of deposition when their concentration is very large, $N > 10^{12}$ cm^{-2}. On the next stage this concentration decreases approximately by an order of magnitude, MGA became inapplicable and we used the Bruggeman approximation (1) successfully (Fig.2b).

4 Conclusions

In conclusion it should be noted that the error in both the optical measurements and the following calculations is rather large, and now we can not confirm with confidence about the size dependence of $\varepsilon(\omega_p, \gamma)$ and the polarizability $\alpha = 3\frac{\varepsilon(\omega)-1}{\varepsilon(\omega)+2}$ of gold clusters with average diameter in the range of $10 \div 50$ nm [7].

References

1. N. Dmitruk, G. Kolbasov, O. Maeva, V. Poludin. Thin Solid Films. **75**, (1981) 341.
2. N. Dmitruk, O. Maeva, V. Poludin. Fiz. Tverd. Tela. **10**, (1976) 1925.
3. M. Alymov, M. Shorshov, *Proc. of 4th Int'l. Conf. Nanostruct. Materials.* (June 14-19, Stockholm, Sweden 1998) p. 367.
4. G.A.Niklasson, P.A.Bobbert, H.G.Craighead. *Proc. of 4th Int'l. Conf. Nanostruct. Materials.* (June 14-19, Stockholm, Sweden 1998) p. 93.
5. E. Palik. *Handbook of optical constants of solids.* (Academic Press, New York 1985).
6. L. Grechko, A. Blank, A. Pinchuk, L. Garanina. Radiophysics and Radioastronomy. **2**, (1997) 19.
7. N. Dmitruk, T. Lepeshkina, M. Pavlovska, L. Zabashta. Nanostructured Materials. **12**, (1999) 295.

Formation of GaAsN nanoinsertions in a GaN matrix

**A.F. Tsatsul'nikov[1],   I.L. Krestnikov[1],   W.V. Lundin[1],   A.V. Sakharov[1],   D.A. Bedarev[1],   A.S. Usikov[1],
B.Ya. Ber[1], V.V. Tret'yakov[1], Zh.I. Alferov[1], N.N. Ledentsov[2], A. Hoffmann[2], D. Bimberg[2], T. Riemann[3],
J. Christen[3], Yu.G. Musikhin[4], I.P. Soshnikov[4], D.Litvinov[4], A.Rosenauer[4], D. Gerthsen[4], A. Plaut[5]**

[1] A.F.Ioffe Physico-Technical Institute RAS, Politekhnicheskaya 26, 194021 St.Petersburg, Russia
[2] Technische Universität Berlin, Institut für Festkörperphysik, PN 5-2, Hardenbergstr. 36, 10623 Berlin, Germany
[3] Otto-von-Guericke-Universität Magdeburg, Universitätsplatz 2, 39106 Magdeburg, Germany
[4] Universität Karlsruhe, Kaiserstr.12, 76128 Karlsruhe, Germany
[5] Exeter University Stocker Road Exeter EX4 4QL UK

Abstract We report on the technique for incorporation of As in GaN layers using MOCVD. We formed GaAs deposits on GaN surfaces and converted then to GaAsN during GaN overgrowth. Stacked GaAsN insertions separated by the GaN spacers were also fabricated. Cathodoluminescence (CL) emission related to As having spectral position and relative intensity defined by deposition parameters is observed.

1 Introduction

Development of GaN technology motivates interest in wide gap mixed anion nitride alloys such as GaAsN and related quaternary semiconductor compounds (e.g. InGaAsN). These materials may bridge the emission energy range from below 1 eV to GaN. It was shown that the band gap energy of GaAsN alloy is characterised by a very large band gap bowing parameter and even very small additions of As may result in strong decrease in the band gap [1,2]. This makes this material very promising for devices operating in a broad wavelength range. However, direct attempts of GaAs incorporation in GaN [3-5] appeared to be not successful.

2 Experiment

Investigated structures were grown on c-plane sapphire using a horizontal type MOCVD reactor operating at low pressure. Ammonia and trimethylgallium were applied as component precursors. 10 % arsine diluted in (AsH_3) in hydrogen was used as source of As. All samples comprised of a 2.5μm-thick buffer GaN layer deposited at 1050°C at the total pressure of 200 mbar, active region and a 0.1 μm thick cap layer. Reference structures without GaAsN insertions as well as the samples with single and stacked thin GaAsN insertions were grown. In one of the samples after the GaN buffer layer growth the temperature was decreased to 730°C and a 75 nm thick GaN layer was deposited and annealed under arsine flow at 1050°C during 74". Active region of the sample with single GaAsN insertion

was formed by the deposition of 75 nm thick GaAs layer with the annealing/conversion and GaN overgrowth steps. An active region of the sample with stacked GaAsN insertions comprised of 10 GaAsN layers formed by conversion of 3 nm thick GaAs deposits at 810°C and separated by 7 nm thick GaN layers formed at 885°C. Cap layer in the all structures was grown at the same temperature as the buffer GaN layer.

3 Results and Discission

Formation and conversion of GaAs deposits to GaAsN was monitored *in situ* (Fig.1). The technique is based on

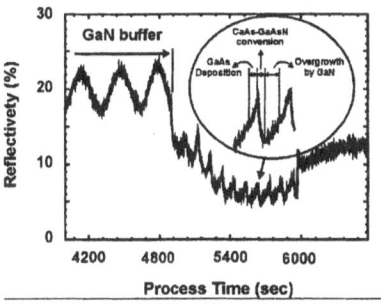

Fig.1 Reflectivity vs. time

the measurement of the intensity of reflected laser beam from the sample surface during the MOCVD epitaxial process. In Fig.1 one can clearly distinguish the modulation of the reflectivity during the formation of GaAs deposits and GaN overgrowth. On the initial stage of the overgrowth the reflectivity is decreased indicating increased roughness of the surface at the stage of the conversion of GaAs deposits to GaAsN. The increase in reflectivity during the overgrowth shows flattening of the surface during the GaN overgrowth.

Secondary ion mass spectroscopy (SIMS) of the structures (insert Fig.2) with stacked and single GaAsN insertions demonstrated As incorporation in the layers at concentrations close to 10^{20} cm^{-3}. Separate GaAsN insertions in the structure with multiple GaAsN insertions were not resolved due to surface morphology modulations comparable to the periodicity of the structure. "Blue-green" CL emission was revealed in the CL spectra of both investigated samples (Fig.2).

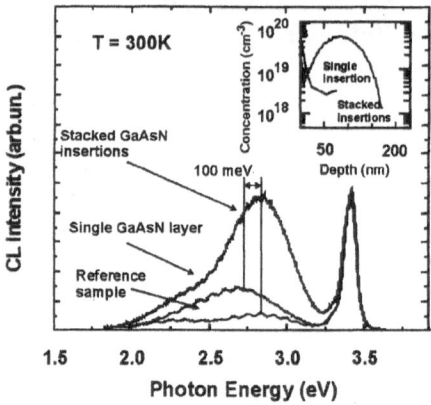

Fig.2. CL spectra of the investigated structures.

Similar emission was attributed to incorporation of As in GaN[6]. As follows from the spectrum of the annealed sample weak blue CL line is observed also in the annealing case. For the stacked insertions this line is dominant in spectrum. Scanning electron microscopy of this sample demonstrated flat surface. As opposite in the case of one single thick GaAs predeposit layer besides areas with flat surface, local formation of mesoscopic 3D islands on the surface occurs. We attribute these surface features to 3D GaAsN islands originating from the GaAs islands, initially formed due to large GaAs deposit thickness in this case and large lattice-mismatch between GaN and GaAs. CL mapping[7] shows that green component of the emission comes from these As-rich 3D islands on the surface (Fig.3a). These islands presumably have high As content as SIMS profile shows near surface.

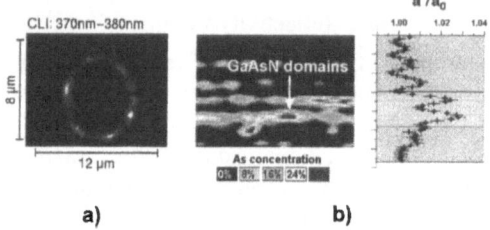

Fig.3 CL images of the samples with single GaAsN insertion (a), color-coded map of the local lattice parameter in the vertical direction (b).

TEM shows that areas between these islands are defect free. DALI [6,7] processing of the high resolution TEM (HRTEM) images reveals formation of coherent GaAsN nanodomains geometrically at the GaN buffer layer –

GaAs interface. These islands have lateral sizes of 3-4 nm, height about 1 nm and As content >30%.

In the case of stacked GaAsN insertions HRTEM shows local formation of staking faults (SF) in c-plane. These staking faults are a visible in the HRTEM image taken along the [11-20] direction (arrows in fig 4.

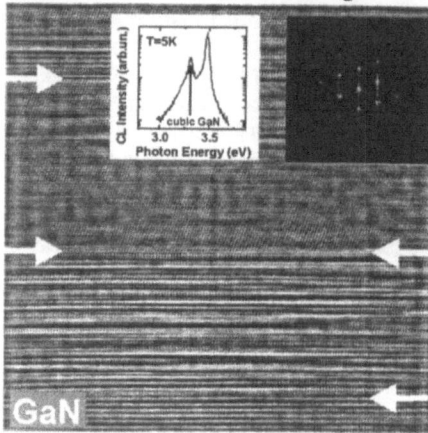

Fig.4. HRTEM image of the sample with stacked GaAsN insertions.

Cubic crystal structure is locally formed as confirmed by Fourier pattern of this area (insert in Fig.4). Height of the cubic phase domains is varied in the 10-50 nm range, and the lateral sizes are about 100 nm. The reason for the cubic phase may may be due to the cubic nature of the GaAs deposits initially formed on a GaN surface, and then converted to Ga(As)N. These regions of the cubic phase result in additional near gap CL line at ~3.3 eV . To avoid this phase the thickness of the GaAs deposits is to be further optimized

To conclude, we investigated optical and structural properties of GaAsN insertions in GaN. Formation of coherent GaAsN nanodomains is possible for very thin GaAs predeposits. Increase in GaAs thickness results in either formation of 3D GaAsN clusters on the surface or in the formation of the cubic phase.

References

1 L.Bellaiche, et. al. Appl.Phys.Lett. **70**, (1997) 3558.
2 W.G.Bi, et. al. Appl.Phys.Lett. **70**, (1997) 1608.
3 Z.Z. Bandić, et. al. Appl.Phys.Lett., **68**, (1996) 1510.
4 R.S. Goldman, et. al. Appl.Phys.Lett., **69**, (1996) 3698.
5 X. Li, et. al Appl.Phys.Lett., **72**, (1998) 1990.
6 A.F.Tsatsul'nikov, et. al Sem. Sc. and Techn., **15**, (2000) 766.
7 F. Bertram, et. al Appl.Phys.Lett., **74**, (1999) 359.

Extremely uniform InAs/GaAs quantum dots emitting at 1.46 mkm at room temperature grown by MOCVD with Bi doping

B. N. Zvonkov[1], I. A. Karpovich[2], N. V. Baidus[1], D. O. Filatov[3], Yu. Yu. Gushina[3], S. V. Morozov[2], S. B. Levichev[2]

[1] Physical-Technical Research Institute, University of Nizhny Novgorod, 23/3 Gagarin Ave, Nizhny Novgorod 603600 Russia
[2] Department of Physics of Semiconductors and Optpelectronics, University of Nizhny Novgorod, 23 Gagarin Ave, Nizhny Nogorod 603600 Russia
[3] Research and Educational Center for Scanning Probe Microscopy, University of Nizhny Novgorod, 23 Gagarin Ave, Nizhny Novgorod 603600 Russia e-mail: spm@phys.unn.runnet.ru

Abstract Doping of InAs quantum dots during growth by Metal Organic Chemical Vapor Deposition by Bi depresses coalescence of the nanoclusters and improves the uniformity of the quantum dot size distribution. By this method the quantum dot structures emitting at 1.46 μm at room temperature with the emission linewidth as narrow as 25 meV have been obtained.

1 Introduction

One of the serious problems in growing the self-assembled quantum dots (QDs) by MOCVD is to avoid coalescence of nanoclusters usually occurring at the growth temperatures, which are optimal for their optical properties [1]. Coalescence results in increasing of dispersion of the QDs in size, in decreasing their surface density and in formation of a considerable number of large dislocated clusters having a continuous energy spectrum. We have studied possibility of depressing the coalescence by doping of QDs during growth by Bi. It was suggested that presence of big and massive Bi atoms on the growth surface would result in limitation of diffusion mobility of In and As atoms and to depress the coalescence

2 Experiment

The InAs QDs were grown on (001) GaAs substrates misoriented by 3° towards [110] by atmospheric pressure MOCVD. The intentionally undoped ($n_0{\sim}5{\times}10^{15}$ cm^{-3}) GaAs buffer layer has been grown at 600°C. Then temperature was decreased down to 490°-530°C, and InAs QDs were grown in an alternating monolayer epitaxy mode. The nominal thickness of InAs was ~5 ML (~1.5 nm). Bi was deposited onto the InAs growing surface by sputtering of the Bi bulk target placed in the reactor by a Q-switched YAG pulsed laser. The estimated surface density of Bi was ~10^{14}-10^{15} cm^{-2}. A number of structures both with and without 15 nm GaAs cladding layer for optical and morphological investigations respectively were grown. The surface morphology of the samples was investigated by Atomic Force Microscopy (AFM) using TopoMetrix "Accurex" TMX-2100 AFM in contact mode. The optical properties of the QDs were investigated by photoluminescence (PL) spectroscopy at 77 and 300K. Also the photoelectric properties of the QDs at 300K were investigated by two techniques. The first was Capacitive Photo Voltage (CPV) spectroscopy (Bergmann's probe [2]). The details of this

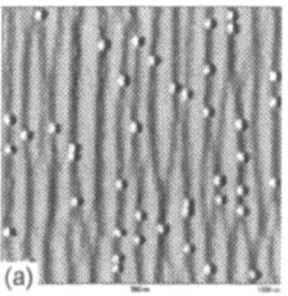

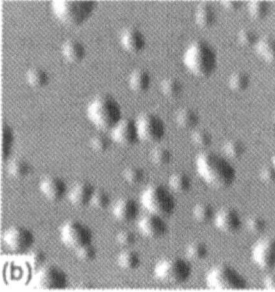

Fig. 1 AFM images of the QDs grown with (a) and without (b) Bi. Frame size 1×1 μm.

technique and its application to the QDs were described in [3]. The second was the surface photovoltage spectroscopy in a liquid electrolyte cell [4] (KCl solution in a mixture of glycerin with water). The main advantage of the latter is a possibility to apply an external bias and therefore to increase the photosensitivity that allowed us to resolve the QDs with surface density as low as 10^9 cm^{-2}.

3 Results and discussion

Doping of the QDs during growth by Bi was found to suppress coalescence of the QDs. Fig. 1 shows the surface topography of the QD structures without GaAs cap layer grown with (a) and without (b) Bi. The latter picture shows large dispersion of the QDs in size and height. Also there are some large dislocated clusters with the lateral size up to 200 nm and the height ~50 nm. The Bi-assisted grown dots are highly uniform in lateral size (41 ± 2 nm) as well as in height (5.8 ± 0.2 nm). Although there were almost no big clusters the surface density of QDs was rather low (4×10^9 cm^{-2}). Increasing the In flux increases the QD density, but

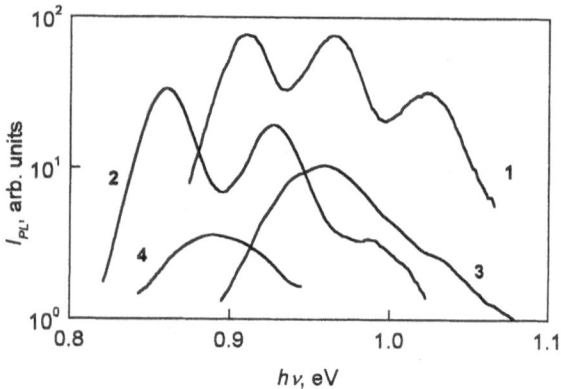

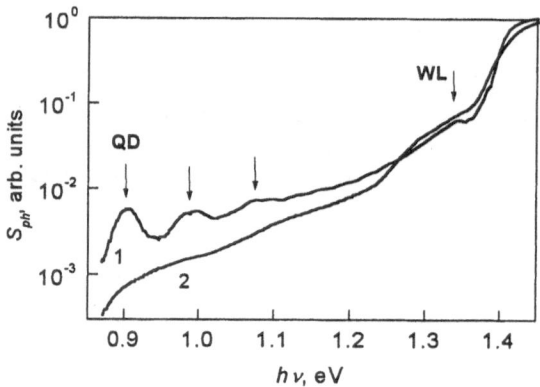

Fig. 2 The PL spectra of the QDs grown with (1, 2) and without Bi (3, 4) at 77 K (1, 3) and 300 K (2, 4).

Fig. 3 The CPV spectra (300K) of the QDs grown with (1) and without Bi (2).

efficiency of coalescence suppression becomes less. So optimization of In and Bi flows is necessary to increase the QD's density without surface morphology degradation.

The PL spectra of the undoped QDs have FWHM ~60 meV (Fig. 2, curves 3, 4). The Bi doped QDs have much narrower peaks (curves 1, 2). Dependence of the QDs PL spectra on the growth temperature of the InAs QDs T_D has been studied. Narrower peaks shifted to lower energies were observed at reduced growth temperatures, but their intensities were decreased probably due to increasing defect concentration. When $T_D > 550^{\circ}C$ the PL peaks broaden and shift to higher energies. The optimum temperatures for the Bi-assisted growth were found to be $510 \div 530^{\circ}C$. The lowest ground transition energy observed was 0.91 eV at 77K and 0.85 eV at 300K (emitting wavelength 1.46 μm). So it turns to be possible to cover the 1.3 μm band important for the optical communication by the devices based on InAs/GaAs system well-developed technologically. Recently 1.35 μm emission at 300K from the InAs/GaAs QDs grown by MBE has been reported [5]. However, this result has been achieved only when the QDs were covered by an external $In_xGa_{1-x}As$ quantum well (QW). The lowest FWHM observed was 25 meV. To our knowledge, this value is one of the lowest reported in the literature.

Besides the ground transition the exited state transitions were usually observed in the Bi-assisted grown QDs (Fig. 2). The energy gaps between the PL peaks ΔE was ~70-80 meV. The excited state transitions were observed also in the PL spectra measured at 300K (Fig 2, curve 2). This can be explained by rather high values of $\Delta E > 2E_{LO}$ where E_{LO} is the LO phonon energy in GaAs. It means that the single optical phonon-assisted relaxation of the excited carriers from the upper levels to the ground ones is difficult, so a finite probability of the radiative recombination through the excited states remains even at room temperature.

Formation of so "deep" QDs without misfit dislocations can be explained by formation of a transient InGaAs layer at the QDs' interface as a result of diffusion intermixing of In and Ga that results in partial relaxation of the elastic strain. An indirect evidence for this interface layer was derived out

from our experiments with combined QWs and QDs. Covering the InAs QDs by an $In_{0.2}Ga_{0.8}As$ QW resulted in considerably less red shift of the ground transition (by 20-40 meV only) compared to the one in the QW/QDs grown by MBE reported in [5]. It can be explained suggesting that an InGaAs transition layer already present at the InAs/GaAs interface plays a role of an external QW.

In the photosensitivity (PS) spectra of the QD structures grown without Bi the PS edge (≈0.9 eV) related to the ground transition in the QDs can be observed only (Fig.3, curve 2). The PS bands from the ground and the excited transitions are resolved poorly because of too high FWHM. In the Bi-assisted structures the ground transition and up to 2 excited transitions are well resolved (curve 1). The positions and the FWHMs of the peaks are in fair agreement with the ones in the PL spectra.

4 Conclusions

In conclusion, the Bi-assisted growth of the InAs/GaAs QDs is shown to be powerful tool for improvement of the morphology and the optical properties of the structures grown by MOVPE. The photoemitting and photosensitive structures for the waveband down to 1.46 μm were obtained.

Acknowledgements

This work was supported by the RFBR (98-02-16688 and 00-02-17598), by the Physics of Solid State Nanostructures Program (99-1141), by Russian Ministry of Education (97-7.1-204 and "Universities of Russia" Program 015.06.01.37), and by BRHE Program (REC-001).

References

1. R. Leon, *Proc. of the 10th Conf. Semiconducting and Insulating Materials SIMC-X* (Berkeley, California 1998) p. 193.
2. L. Bergmann, Phys. Zeitschrift **33**, (1932) 209.
3. B. N. Zvonkov, I. G. Malkina, E. R. Linkova, V. Ya. Aleshkin, I. A. Karpovich, and D. O. Filatov, Semicond. **31** (1997) 941.
4. X. He and M. Raseghi, Appl. Phys. Lett. **262**, (1993) 618.
5. H. Saito, K. Nishi, and S. Sugou, Appl. Phys. Lett. **74**, (1999) 1111.

Effects of InAs coverage on the Ga diffusion into InAs Self-Assembled Quantum Dots on GaAs(100)

N. Matsumura[1], T. Haga[1], S. Muto[1], Y. Nakata[2], N. Yokoyama[2]

[1] Department of Applied Physics, Hokkaido University, Sapporo 060-8628, Japan e-mail: nmatsu@eng.hokudai.ac.jp
[2] Fujitsu Laboratories Ltd., Morinosato-Wakamiya, Atsugi 243-0197, Japan

Abstract   InAs quantum dots (QDs) on GaAs(100) grown by molecular-beam epitaxy were structurally characterized by ion channeling. Lattice deformation of the InAs QDs and diffusion of Ga atoms into InAs QDs were clearly observed to depend strongly on the InAs coverage. It was revealed that the diffusion was significantly enhanced when the InAs coverage is changed from 1.53 to 1.71 monolayer (ML). During this change, lattice deformation was reduced while the average size (base diameter) of dots was decreased. These phenomena suggest that some growth process change occurred.

1 Introduction

Recently, a great interest in the growth of InAs Self-Assembled Quantum Dots (SAQDs) grown by Stranski-Krastanow growth mode has arisen [1]–[4]. In our previous report [5], by using Rutherford backscattering (RBS), diffusion of Ga atoms into InAs QDs was observed, and an accurate value of InAs coverage of uncapped samples was determined. In this study, we carefully determined the InAs coverage for the capped samples, and we investigated the dependence of the crystallographic structure of InAs QDs on the InAs coverage by using ion channeling observed by RBS and particle induced x-ray emission (PIXE).

2 Experiments

All samples contained a single InAs layer, and were grown by molecular-beam epitaxy (MBE). The InAs was grown at 510°C with growth rate of 0.1 ML/s. After InAs growth, the sample was annealed for 60 sec before cooling down or growth of GaAs cap layer. To evaluate lattice deformation, we used the normalized minimum yield, χ_{min}, which is defined as a ratio of aligned yields to random ones. Here, the lattice deformation is defined as the lateral displacement of atoms from the GaAs host lattice, since it is what is observed as χ_{min}.

We note that we successfully determined the accurate value of the InAs coverage of samples having a GaAs cap layer as well as uncapped ones. The accuracy is estimated to be ± 0.02ML.

3 Results and Discussion

In order to confirm the diffusion, we performed chemical etching of uncapped samples with concentrated HCl for 1 min to remove InAs dots and a wetting layer selectively. After the treatment, we carried out RBS experiments with 1.00 MeV He+ ions. If the diffusion of Ga atoms into InAs exists, In signals are expected to remain, since

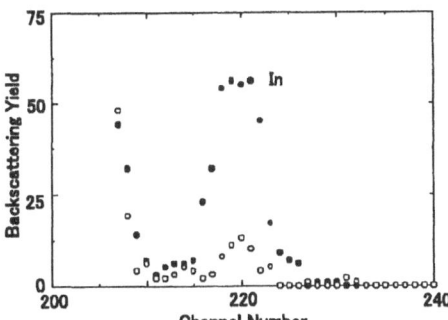

Fig. 1 Magnified RBS spectra of In for as-grown and etched InAs QD's with concentrated HCl for 1 min. InAs layer thickness: 2.23ML.

HCl cannot remove InGaAs [5]. Magnified RBS spectra around In signals of these samples are shown in Fig.1. In these figures, open (filled) circle indicates the spectrum for the etched (as-grown) sample. In Fig.1, we can see that the In signals even for the etched samples. This, together with its AFM observation, indicates the existence of the diffusion of Ga atoms into InAs QDs [5].

In order to observe lateral deformation of each element, PIXE/channeling experiments were carried out on capped samples with 1.00MeV H+ ions. Figure 2 shows χ_{min}'s of PIXE as a function of the InAs coverage. The magnitude of χ_{min} for In changed drastically with varying InAs coverage. The value of χ_{min} for In of sample having 1.02 ML was very close to that for RBS. This indicates that the InAs layer is pseudomorphically grown on the GaAs substrate, which means that InAs is grown two-dimensionally (2D). In fact, no dots were suggested to be formed by reflected high energy electron diffraction (RHEED) during the growth of this sample. The value of χ_{min} for In drastically increased, and showed the maximum when InAs coverage was 1.56 ML. This indicates that the InAs growth mode shifted from 2D to 3D and lattice deformation of InAs QDs was observed. The critical coverage for 2D-3D growth transition, θ_c, has been reported to be around 1.5 ML [2], [6], [7]. For 1.78 ML, the magnitude of χ_{min} for In suddenly decreased. This is consistent with the decrease of average dot diameter [3]. Since χ_{min} for In reflects the lattice deformation of InAs QDs, lattice deformation of the InAs QDs strongly depends on the InAs coverage. While χ_{min}'s for RBS, As, and In at 1.78 ML are smaller than those at 1.56 ML, slight increase of χ_{min} for Ga is observed at this

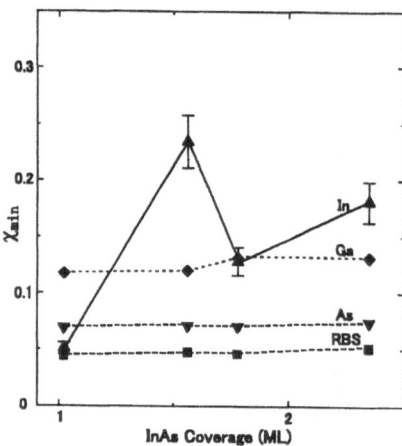

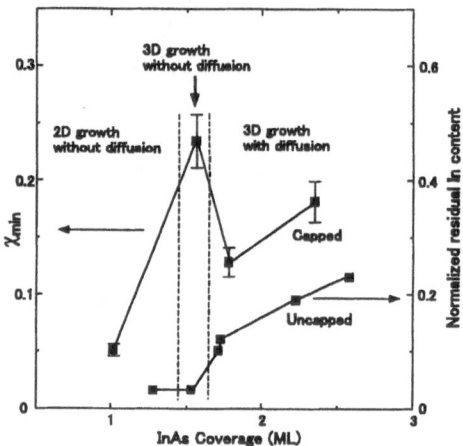

Fig. 2 The variation of the normalized minimum yield for each element and RBS as a function of the InAs coverage. All samples have 150nm GaAs capping layers.

Fig. 3 The residual In content of etched samples normalized by the In of as-grown samples as a function of the InAs coverage. For comparison, χ_{min}'s for In are also shown.

point. This indicates that Ga atoms diffused to form a part of dots.

Figure 3 shows residual In signals of etched samples normalized by those of the as-grown ones and χ_{min} for In as a function of the InAs coverage. When InAs coverage was 1.53 ML, the residual In signal was very small. However, When InAs coverage was 1.71 ML, about 12 % In signal was observed. The difference in the growth time for InAs is only a couple of seconds. Therefore, the simple diffusion theory does not seem to explain this difference. Rather, this result indicates that there are two 3D growth processes. At 1.53ML, InAs QDs grow without any Ga diffusion. On the other hand, InGaAs dots grow when InAs coverage reaches 1.71ML. There are three regions of growth suggested. These are 2D growth without Ga diffusion(I), 3D growth without Ga diffusion(II), and 3D growth with Ga diffusion(III) as we increase the InAs coverage. It should be noted that the three regions specified above are for the uncapped samples and are free from the correspondence between the capped and the uncapped.

The comparison of the capped sample to the uncapped is still a subtle issue. To be precise, we have to take the following features into account: 1) change in the deformation of dots when cap layer was overgrown, 2) the deformation of GaAs cap layer especially close to the dot, and 3) additional diffusion between the dots and cap layers. These are issues to be addressed in the future. However, a remarkable thing is that, in spite of these complexities, the strong correlation between deformation of capped samples to the three regions of uncapped samples is easily explained by the dot formation and the Ga diffusion from the GaAs substrate to the dot observed in uncapped samples.

4 Summary

In summary, by using ion channeling, we characterized lattice deformation of InAs self-assembled quantum dots grown by MBE. From the RBS and PIXE experiments, the accurate value of the InAs coverage was obtained even for the capped samples. A change in the amount of Ga diffusion into InAs QDs with varying InAs coverage was observed by the etching experiments of the uncapped samples. It was also traced in the Ga deformation of capped samples. No Ga diffusion was evidenced in InAs QDs when InAs coverage was 1.53ML. Thus, diffusion-free 3D growth region and diffusion-enhanced growth region are suggested to exist.

Acknowledgements The authors would like to thank Prof. Y. Abe for helpful discussions. The authors would also like to thank I. Toriumi and T. Ishigure for the experimental support. This work is partially supported by a Grant-in-Aid for Scientific Research from the Ministry of Education, Science, Sports and Culture, Japan.

References

1. D. Leonard, K. Pond, and P. M. Petroff, Phys. Rev. **B50**, (1994) 11687.
2. J. M. Moison, F. Houzay, F. Barthe, L. Leprince, E. Andre, and O. Vatel, Appl. Phys. Lett. **64**, (1994) 196.
3. N. P. Kobayashi, T. R. Ramachandran, P. Chen, and A. Madhakar, Appl. Phys. Lett. **68**, (1996) 3299.
4. P.B.Joyce, T.J.Krzyzewski, G.R.Bell,B.A.Joyce and T.S.Jones, Phys. Rev. **B58** (1998) 15981.
5. T. Haga, M. Kataoka, N. Matsumura, S. Muto, Y. Nakata, and N. Yokoyama, Jpn. J. Appl. Phys. **36**, (1997) L1113.
6. J. -Y. Marzin, J.-M. Gerard, A. Izrael, D. Barrier, and G. Basterd, Phys. Rev. Lett. **73**, (1994) 716
7. T. R. Ramachandran, R. Heitz, P. Chen, and A. Madhukar, Appl. Phys. Lett. **70**, (1997) 640.

Photoluminescence and atomic force microscopy studies of InAs/InSb nanostructures grown by MBE

Ya. V. Terent'ev[1], A. A. Toropov[1], V. A. Solov'ev[1], B. Ya. Mel'tser[1], M. M. Moiseeva[1], S. V. Ivanov[1], B. Magnusson[2], B. Monemar[2], and P. S. Kop'ev[1]

[1]Ioffe Physico-Technical Institute, RAS, 194021 St. Petersburg, Russia
[2]Department of Physics and Measurement Technology, University of Linköping, S - 581 83 Linköping, Sweden

Semiconductor heterostructures based on narrow gap III-V compounds and their alloys are under intensive study as highly promising materials for mid-IR optoelectronics and resonant tunneling devices. InSb nanostructures are of special interest due to the narrowest possible band gap and smallest carrier effective masses among these materials. However, until recently there have been no publications devoted to InSb/InAs nanostructures.

In this paper we report on the first MBE growth of InSb fractional monolayer (ML) insertions (in sub-monolayer and above 1 ML scale) in an InAs matrix, as well as photoluminescence (PL) and structural studies of these heterostructures. The latter involve the atomic force microscopy (AFM) and transmission electron microscopy (TEM) techniques.

The active region of the sample grown for PL studies consists of a 0.2 μm InAs layer centered with the InSb insertion. The nominal thickness of the insertion differs in the samples from 0.5 to 1.5 ML. The samples for AFM studies have additionally a 20 nm-AlSb confining layer atop followed by a 100 Å thick InAs layer covered with InSb of the same thickness and grown under the same conditions as the respective capped InSb insertion intended for PL studies. The buffer structure of all the samples, involving a 0.5 μm InAs layer followed by a 20 nm thick AlSb barrier, was grown at a substrate temperature T=480°C on an n-InAs substrate having electron concentration n~2×10^{16} cm^{-3}. The active region was grown at T=420°C. The existence of an optimum temperature (T~400°C) for MBE growth of GaInSb/GaSb strained quantum wells (QWs) has recently been proved by PL studies [1]. Degradation of optical properties was observed, either due to the formation of Sb clusters at lower temperatures, or due to structural defects arising from the enhanced group V molecule re-evaporation from the surface at higher temperatures [2]. One should stress especially that the main problem here is possible intermixing of group V elements at the interfaces. Since we used conventional solid sources for As and Sb, the growth rate of the InSb/InAs active region was chosen five times lower as compared to that of the InAs buffer layer, with respective reduction of the As flux intensity [3], which

allowed us to suppress the As incorporation into the InSb insertion.

The band alignment of the InAs/InSb heterostructure is predicted to be type-II broken gap [4]. Note also that due to 7% lattice mismatch between InSb and InAs either the formation of a QW or self-organized growth of QDs is expected.

PL measurements of the structure were performed in the temperature range from 2 to 102 K using for excitation an Ar$^+$ laser operating at 514.5 nm. The excitation power varied from 50 to 300 mW and the laser spot size was about 2 mm. A Bomem DA8 Fourier transform spectrometer equipped with an InSb detector was used for registration of PL spectra.

A relatively narrow PL peak (FWHM ~20 meV) is observed in the samples with 0.5 ML InSb insertion at

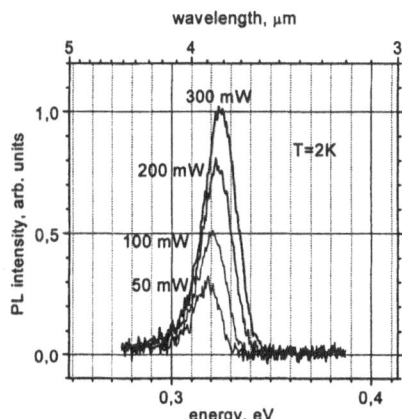

Fig. 1 PL spectra for different excitation densities in the sample with 0.5 ML InSb insertion.

an energy less than the InAs band gap (see Figs. 1,2). The PL maximum shifts towards higher energies with increasing excitation density (Fig. 1), whereas the temperature rise up to 102 K causes mainly a decrease of PL intensity (Fig. 2). The observed blue shift of the PL maximum with increasing excitation density is inherent for type-II structures [5]. The fact that the energy of the PL maximum does not follow the known temperature dependence of the InAs band gap also confirms that PL originates from radiative recombination between holes in the InSb insertion and electrons weakly confined by electrostatic interaction in the InAs barrier. The relatively small values of the PL peak width indicates, in our opinion, the formation of a QW in the structures with 0.5 ML InSb insertions since PL from QDs exhibits generally a significantly wider peak due to the inhomogeneous distribution of the dot sizes.

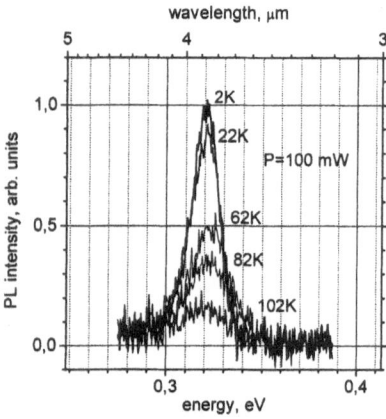

Fig. 2 Temperature dependence of PL spectra in the sample with 0.5 ML InSb insertion..

The PL from the samples with 0.7 and 1.5 ML insertions exhibits also narrow peaks in the wavelength region of 4.2-4.6 μm. Since the sample with the 1.5 ML InSb insertion reveals the formation of QDs (as discussed below), we suppose that the observed PL is due to the wetting layer because a QD emission must lay in a longer-wavelength range .

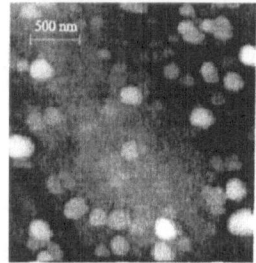 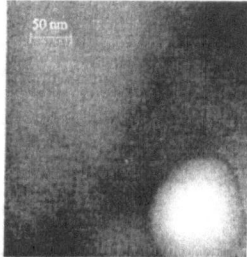

Fig. 3 AFM images of the surface taken in the sample with 1.5 ML InSb insertion.

The samples were studied at room temperature in ambient by AFM (NT-MDT) in a contact mode. No features related to QDs are observed in the sample with a 0.7 ML thick insertion. In contrast to that, the formation of dots is clearly revealed in the samples with

the InSb thickness ranging from 1.3 to 1.7 ML. For the 1.5 ML thick insertion the average dots density, lateral size and height are estimated as $2.5*10^9$ cm^{-2}, 70±30 nm and 10±5 nm, respectively (see Fig. 3).

Cross-section images taken by TEM confirm the formation of QDs in the samples with 1.5 ML insertion. Figure 4 shows two InSb QDs (dark contrast spots) and wetting layer (thin dark line crossing the QDs) surrounded by bulk InAs as well as 20 nm AlSb barriers (light stripes) confining the active region. A good correlation between AFM and TEM images is observed.

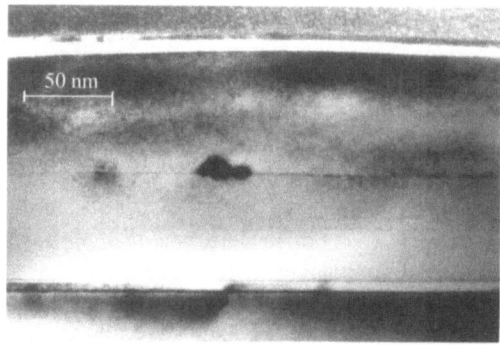

Fig. 4 TEM image of the active region of the sample with a 1.5 ML InSb insertion.

In summary, we have grown nanostructures containing an InSb fractional ML insertions in an InAs matrix. PL studies of the structures have demonstrated the type-II band alignment. Structural studies have shown that the critical thickness of the self-organized formation of InSb/InAs QDs lies between 0.7 and 1.3 ML. The structures display bright PL up to the wavelength of 4.6 μm, which makes them very attractive for mid-IR opto-electronic applications.

Acknowledgement - This work was supported in part by RFBR Grant #98-02-18211, the Program of the Ministry of Sciences of RF "Physics of Solid State Nanostructures", EOARD Contract # F61775-99-WE016, and CRDF Grant #6870.

References

1. N. Bertru, A. Baranov, Y. Cuminal, G. Almuneau, F. Genty, A. Joullie, O. Brandt, A. Mazuelas, K.H. Ploog, Semicond. Sci. Technol., **13**, (1998) 1.
2. P. V. Neklyudov, S. V. Ivanov, B. Ya. Mel'tser, P. S. Kop'ev, Semiconductors, **31**, (1997) 1067.
3. B. R. Bennett, B. V. Shanabrook, R. J. Wagner, J. L. Davis, J. R. Waterman, *Appl. Phys. Lett.* **63**, (1993) 949.
4. Su-Huai, A. Zunger, Phys. Rev.B **25**, (1995) 12039.
5. M. P. Mikhailova, A. N. Titkov, Semicond. Sci. Technol. **9**, (1994) 1279.

Three Dimensional Self-organization of InAs Quantum-dot Multilayers

J.C. González[1*], F.M. Matinaga[1], W.N. Rodrigues[1], M.V.B. Moreira[1], A.G. de Oliveira[1], M.I.N. da Silva[2*], J. M. C. Vilela[2], M. S. Andrade[2], D. Ugarte[3], and P.C. Silva[3]

[1] Departamento de Física, ICEx, Universidade Federal de Minas Gerais, C.P. 702, 30123-970, Belo Horizonte, Minas Gerais, Brazil. e-mail: wagner@fisica.ufmg.br

[2] Laboratório de Nanoscopia/CETEC, Av. José Cândido da Silveira 2000, 31170-000, Belo Horizonte, Minas Gerais, Brazil.

[3] Laboratório Nacional de Luz Síncrotron-LNLS, C.P. 6192, 13083-970, Campinas, São Paulo, Brazil.

Abstract   We report on experiments aiming to obtain three-dimensional self-organization in InAs quantum-dot multilayers embedded in GaAs. The InAs/GaAs quantum-dot multilayers have been grown by Molecular Beam Epitaxy. Employing Atomic Force Microscopy (AFM), Transmission Electron Microscopy (TEM) and Photoluminescence Spectroscopy (PL) we have studied samples with different number of periods of InAs/GaAs bilayers, as well diferent InAs coverages.

1   Introduction

InAs grows epitaxially on GaAs (100) substrates in "Stranski-Krastanov" mode [1]. The lattice misfit (~7%) between the InAs and the substrate causes coherency stresses and strains[1]. After a wetting layer (WL) of approximately 1.5 monolayers (MLs), the excess of InAs relaxes elastically by forming three-dimensional islands. Capping the InAs islands with a GaAs layer, and by iterating the deposition of InAs layers and GaAs spacers, one can produce multilayers of nanoscopic InAs islands buried in GaAs. Moreover, the coherency of the stresses between InAs and GaAs leads to self-assembled vertical alignment of the islands positions across the GaAs layers.

The above procedure has great potential for realizing high-density regular array of quantum-dots. However, in order to obtain buried islands with a narrow size distribution and a high degree of position alignment one needs to optimize different growth parameters: GaAs spacer thickness[2] d, InAs coverage[3] θ, growth interruption time[4] t, and the number of periods of the multilayer[1] n. In this paper, we present experimental results on multilayers of InAs buried in GaAs. We have analyzed the influence of the number of periods and InAs coverage on the morphological and optical properties of the samples by using Atomic Force Microscopy (AFM) and Photoluminescence (PL) techniques. The results show a clear tendency to spatially 3-D self-ordering of the quantum-dots.

2   Experimental

Four InAs quantum-dot multilayers with different number of periods (n = 0, 10, 50, 100) and five multilayers structures with different InAs coverages (Θ_{InAs}= 1.9,

2.0, 2.1, 2.3 and 2.5 ML) were grown by Molecular Beam Epitaxy (MBE) on (100) GaAs substrate. The samples with different number of periods were prepared as follows. After growing a 1 μm-thick GaAs buffer layer at 610 °C, 1.9 MLs of InAs were deposited at 510 °C, 5×10^{-5} Torr As$_4$ pressure, and at a growth rate of 0.16 ML/s. The islands were buried with a 30 ML-thick GaAs spacer grown at the same temperature of the islands, with 3.6×10^{-5} Torr As$_4$ pressure and a growth rate of 1.0 ML/s. This period of bilayers (InAs islands + GaAs spacer) was repeated n times. Finally, a new layer of islands was deposited on the last GaAs spacer layer. The samples with increasing InAs coverage were prepared in the same way, but with n = 50 and a 40 ML thick GaAs spacer.

AFM measurements were done using a Nanoscope III; Digital Instruments in tapping mode. High Resolution TEM (HR-TEM) measurements were done in a JEOL JEM 3010 instrument, operating at 300 kV. The PL measurements were done in a back scattering configuration. We used a Ti-sapphire laser for excitation at λ=750 nm with ~300 W/cm². The PL signal was dispersed in a Jobin Yvon T64000 monochromator with a 300 groves/mm grating, and collected on a charge coupled device (CCD camera). We estimated a penetration depth of ~1000 Å for this λ, and so we only have information from the last 15 periods of the sample.

3   Results

The AFM results, as shown in Fig.1 reveals a decrease and a tendency to saturation of the island density with the increase of the number of periods. This is interpreted as a 3-D self-organization characteristic of these samples [1, 3].

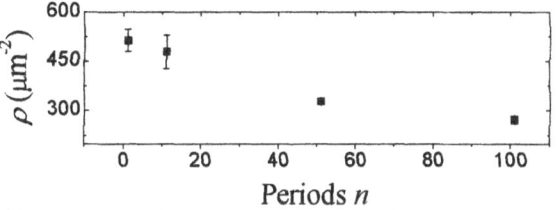

Fig.1 *Density of InAs islands on the surface of the samples, as a function of the number of periods in the multilayer.*

In our TEM images we found columns up to 25 vertical aligned islands. The PL spectra at T=77 K of our samples are shown in Fig. 2. A QD line, resulting from recombination of non-equilibrium carriers via the ground state of the QDs, can be seen in the spectra of

* *Present address: Analytical Instrumentation Facility, North-Carolina State University, 1010 Main Campus Drive, Raleigh, NC, 27606 – 1357, USA.*

samples containing 10, 50, and 100 periods. A second line from the recombination of the first excited state of the QDs is clearly present in the samples with $n = 10$, and 50 periods, but it is masked by the high intensity emission of the ground state line in the 100 period sample. The PL line associated to the WL, which is expected to be constant or to increase with n[5], was not observed. Both PL peaks have a shift toward the long-wavelength, indicating an increase in the mean volume of the QDs, as n increases. The full width at half maximum (Γ) of the QD peak shows an almost constant behavior with the increase of the number of periods. The behavior of PL agrees with the AFM data, which likewise show that increasing the number of periods results in an enlargement of the islands mean size. The relative constant behavior of the Γ denotes that the width of the islands size distribution is almost independent of the number of periods in the samples.

The second process leading to self-organization occurs near of the coherent-incoherent transition, that occurs for InAs coverages near of 2 ML. Around this critical coverage the islands surface density is almost constant[8] and the adatoms on the surfaces will attach to the existing islands. As the first incoherent islands appear, they "steal" material from the others. This phenomenon leads to a decrease of the size of the coherent islands and to a narrower island size distribution. For coverages below the critical value of this transition the island size distribution is broad, and for coverages above that value the number of relaxed islands increase quickly, inducing a broadening of the island size distribution. On the light of these arguments, there is an optimum value for the InAs coverage that will maximize the pairing probability and the narrowing of the island size distribution, simultaneously. We understand that these conditions were reached in the 2.1 ML sample.

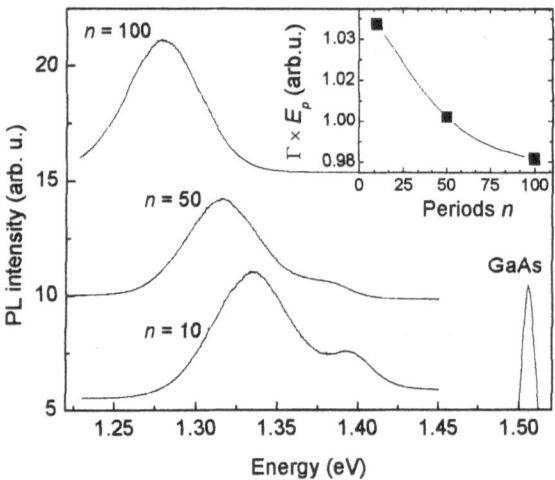

Fig. 2. *PL spectra of the multilayer samples. The exciton peak of bulk GaAs is shown for comparison. In the insert, the full width at half maximum Γ of the PL peak multiplied by its energetic position E_p is plotted as a function of the number of multilayer periods n.*

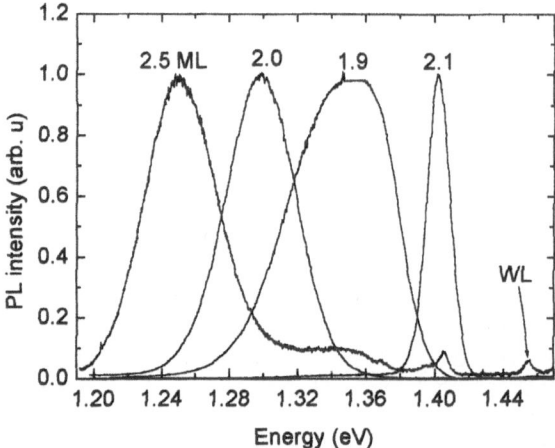

Fig.3 *Normalized PL spectra from 50 periods InAs QDs multilayer samples with different InAs coverages*

In Fig. 4 the PL spectra for increasing InAs coverages are shown. We can see that the sample with of 1.9 ML of InAs, shows a broad PL band of 71 meV FWHM at 1.35 eV. For the others samples with larger InAs coverages, with exception of the 2.1 ML sample, the PL peaks are shifted to lower energies indicating an increment of the mean size of the QDs. The FWHM oscillates between 63 and 43 meV, which are typical values for QDs samples with about ±10 % of size uniformity. However, for the sample with 2.1 ML of InAs, the PL peak occurs at a higher energy (1.40 eV), and the FWHM was reduced to the extremely low value of 13.6 meV. To our knowledge, this is the lowest value of PL FWHM until now reported for InAs QDs samples.

Two mechanisms seem to contribute to the narrowing of the QDs PL line of the 2.1 ML sample: First, the three dimensional self-organization of the QDs in multilayers samples[6], and second, the sharp changing of the QDs size at the coherent-incoherent transition[7].

The spectrum of the 2.5 ML sample in figure 1 shows clearly a residual population of QDs with similar properties to the 2.1 ML sample. This residual population of QDs appears in the spectrum of the 2.3 ML sample (not showed in Fig.1) too, but it is not present in the samples with InAs coverage below of 2.1 ML. This fact confirms that a family of smaller QDs is formed only in the moment of the coherent-incoherent transition. We understand that the formation of the family of smaller QDs explain the shift to higher energies of the PL line of the 2.1 ML sample.

Work supported by CLAF, CNPq, CAPES, FAPEMIG, FAPESP and MCT-PRONEX.

References

1. J. C. González et al., Appl. Phys. Lett. **76**, (2000) 3400, and references therein.
2. Q. Xie, et al., Phys. Rev. Lett. **75**, (1995) 2542.
3. J. Tersoff, Phys. Rev. Lett. **76**, (1996).1675.
4. E. Schöll and S. Bose, Solid-State Electron. **42**, (1998).1587
5. O. G. Schmidt, et al, Appl. Phys. Lett., **74**, (1999) 1272.
6. G. Springhoz, et al, Science **282**, (1998) 734.
7. D. Leonard, et al, Phys. Rev. **B 50**, (1994) 11687.
8. T. R. Ramachandran, et al. J. Vac. Sci. Technol. **B 16**, (1998) 1330.

Investigation of the properties of molecular beam epitaxy grown self-organized ZnSe quantum dots embedded in ZnS

C. L. Yang, L.W. Lu, W. K. Ge, Z. H. Ma, I. K. Sou, and J. N. Wang[*]

Department of Physics, Hong Kong University of Science and Technology, Clear Water Bay, Kowloon, Hong Kong

Abstract ZnSe quantum dots (QDs) were grown on ZnS buffer with 1nm or 100 nm ZnS capping layer. Their optical and electrical properties were investigated using photoluminescence, capacitance-voltage, and deep level transient Fourier spectroscopy. PL peak located at 3.19 eV was observed at low temperature (10K) but the intensity was quenched rapidly with increasing temperature and finally disappeared at about 100 K. Apparent carrier density profile obtained from the C-V measurement exhibited a maximum at the depth of about 100 nm below the surface, which was in good agreement with where the single QDs layer was located. The electronic ground state energy level of ZnSe QDs was determined as about 0.11 eV below the conduction band of ZnS by DLTFS.

1 Introduction

Low dimensional structures such as quantum wires or quantum dots (QDs) based on wide-gap II-VI and nitride semiconductor, which are expected to be short-wavelength light sources and detectors, have been actively studied. Up to now, studies of self-organized II-VI QDs have been focused on the relatively narrow gap CdSe [1] and CdTe [2]. Few researches have been carried out on the wide gap materials such as ZnSe [3-6]. In this article we describe the investigation of electronic structures of the ZnSe QDs embedded in ZnS grown by molecular beam epitaxy (MBE) using capacitance–voltage (C–V), photoluminescence (PL), and deep level transient Fourier spectroscopy (DLTFS) measurements. We observed a PL peak located at 3.19 eV at low temperature (10K) but its intensity was quenched rapidly with increasing temperature and finally disappeared at about 100 K. Obtained from the C-V measurement the apparent carrier density profile showed a maximum at the depth of about 100 nm below the surface, which was in good agreement with where the single QDs layer was located. The ground electronic energy level of ZnSe QDs was determined to be about 0.11 eV below the conduction band edge of ZnS by DLTFS.

2 Experiments

The QD structures studied in this work were grown using a VG V80H MBE system on GaP substrates. Details of the growth of QDs can be found elsewhere [7]. For optical measurements, a 3.2 ML ZnSe

was grown and followed by 3 minutes growth interruption on ZnS buffer layer. During the growth interruption a streaky to spotty transition was observed by reflected high-energy electron diffraction. About 1 nm ZnS was grown on the top as the capping layer. For the electrical measurements, the sample consists of a 3.2 ML ZnSe layer sandwiched between two 100 nm n-ZnS (Al: 1×10^{18}) barriers. The structure was grown on a 300 nm n$^+$-ZnS (Al:1×10^{19}) buffer layer used as the ohmic contact. Au spots of 0.5 mm^2 were evaporated on the surface of the structure to form the Schottky diodes.

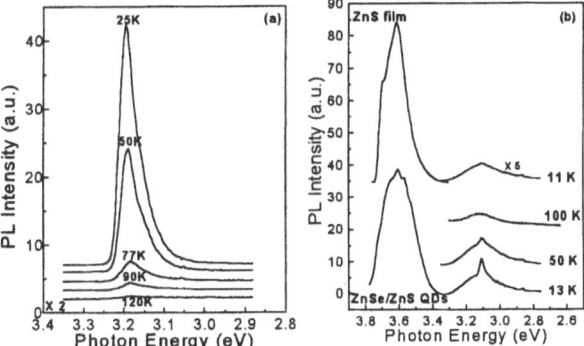

Fig. 1(a) PL spectra of ZnSe QDs with 1 nm ZnS as capping layer at various temperatures and (b) PL spectra of ZnS and ZnSe QDs with 100 nm ZnS:Al capping layer at three temperatures as indicated.

3 Results and discussion

Typical air atomic force microscope image shows that the self-assembled ZnSe QDs are ~20 nm in height and ~80 nm in width [7]. PL spectra of ZnSe QDs structure with 1nm ZnS capping are shown in Fig. 1(a). At low temperature, we can observe a strong PL peak at about 3.19 eV. But with increasing temperature, the PL intensity decreases rapidly. When the temperature is up to 120 K, the PL is almost unobservable. The PL spectra of ZnS thin film sample and ZnS/ZnSe QDs structure with 100nm ZnS capping for electrical characterization have also been measured, see Fig. 1(b). The ZnS PL spectrum shows the near band-edge emission at about 3.65 eV together with a broad and weak band centered at about 3.11 eV, which is believed due to some deep levels in ZnS. In comparison with ZnS PL spectrum, an extra sharp peak at about 3.116 eV is observed in PL spectrum of the ZnS/ZnSe QDs structure. The intensity of this sharp peak decreases

---

*Corresponding Author:,E-mail:phjwang@ust.hk

rapidly with increasing temperature and finally disappears at around 100K. We believe that the PL emissions at 3.19eV for ZnSe QDs sample and at 3.116eV for ZnS/ZnSe QDs sample are originated from ZnSe QDs. It has been shown[5] that the PL emission of the Se nano-clusters is in the range of 1.69 and 1.91 eV. In addition, the effect of the oxidation of ZnSe to form SeO2 clusters can also be ignored as our samples are capped by ZnS layers. We can therefore rule out the possibilities of forming Se cluster[9] or SeO2 cluster[10] on ZnSe layer surface during the growth. The weak intensity of ZnSe QDs related sharp peak from ZnS/ZnSe sample as compared to that of ZnS is likely caused by the 100 nm thick ZnS capping layer. It has been found that the QDs emission with a 100 nm ZnS capping has a 74 meV red shift compared to that with only 1 nm ZnS capping. This red shift may result from slight structure change and (or) confinement potential difference caused by a thick capping layer.

The apparent carrier concentration in the depth direction N_{CV} was obtained from the measured capacitance-voltage (C-V) method at sufficiently high frequencies[11] by using the expression

$$N_{CV}(W) = \frac{2}{A^2 e \varepsilon \varepsilon_0}[\frac{d}{dV}(\frac{1}{C^2})]^{-1} \quad (1)$$

where W is the thickness of the space charge layer, A is the contact area, e is the electron charge, and $\varepsilon \varepsilon_0$ is the dielectric constant. The depletion depth W was calculated from the depletion capacitance C using

$$W \quad (V) \quad = \quad \varepsilon \quad \varepsilon_0 \quad A/C(V) \quad (2)$$

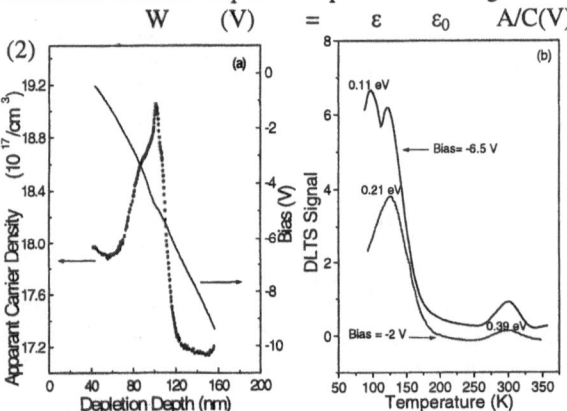

Fig. 2(a) Caculated apparent carrier density distribution in the growth direction (left) and the bias dependence of the depletion depth (right) based on the measured Capacitance-Voltage curve. (b) DLTFS spectra of the ZnSe QDs sample at the negative biases of 2 V and 6.5 V.

For the Schottky diode made from ZnS/ZnSe QDs structure, figure 2(a) shows the calculated depletion depth (solid line) against applied reverse bias and the apparent carriers density distribution (dotted line) along the depth deduced from the measured C-V curves. It is clearly shown that the carrier density is peaked at the location of ZnSe QDs layers, which is 100nm below the sample surface. According to Fig 2(b), the depletion region reaches the QDs layer at a reverse bias of about 4.5V. As a result, adjusting the reverse bias below and above this value allows the study of DLTFS without and with ZnSe QDs inclusion.

The typical DLTFS spectra for ZnS/ZnSe QDs structure Schottky diode are shown in Fig. 2(b) with reverse bias of 2 V and 6.5 V, respectively. For bias at 2V two electron-emission peaks are observed. They are attributed to two deep electron traps, subsequently referred to as $E1$ and $E2$. The activation energies of the traps evaluated from Arrhenius plots are 0.21 and 0.39 eV below the conduction band, respectively. Our previous work[12] has concluded that E1 and E2 are originated from Al dopants in ZnS. For bias at 6.5 V, an additional peak at about 90 K is observed. It corresponds to the activation energy of 0.11 eV. We attribute this additional trap to ZnSe QDs states as the QDs layer is included in the depletion region at this bias.

In conclusion, self-assembled ZnSe QDs on ZnS have been grown successfully by MBE. Detailed optical investigation show a peak at about 3.19 eV in the PL spectra at low temperature (< 100 K). Furthermore, techniques of CV and DLTFS were firstly employed on this system to study the electronic structure of the ZnSe QDs. The electronic state of the ZnSe QDs determined by DLTFS is about 0.11 eV below the conduction band of ZnS.

Acknowledgements The authors would like to thank the Research Grant Council of Hong Kong for the financial support of this work.

Reference
1. S. H. Xin, P. D. Wang, Aie Yin, C. Kim, M. Dobrowolska, J. L. Merz, and J. K. Furdyna, Appl. Phys. Lett.69, (1996) 3884.
2. Marsal, H. Mariette, Y. Samson, J. L. Rouviere, and E. Picard, Appl. Phys. Lett.73, (1998) 2974.
3. Yi-hong Wu, Kenta Arai, Takafumi Yao, Phys. Rev. B 53, (1996) R10485.
4. M. C. Harris Liao, Y. H. Chang, Y. F. Chen, J. W. Hsu, J. M. Lin and W. C. Chou, Appl. Phys. Lett.70, (1997) 2256.
5. C. A. Smith, H. W. H. Lee, V. J. Leppert and S. H. Risbud, Appl. Phys. Lett.75, (1999) 1688.
6. Takehiko Tawara, Satoru Tanaka, Hidekazu Kumano, and Ikuo Suemune, Appl. Phys. Lett.75, (1999) 235.
7. Z. H. Ma, W. D. Sun, I. K. Sou, and G. K. L. Wong, Appl. Phys. Lett.73, (1998) 1340.
9. X. B. Zhang and S. K. HarK, Appl. Phys. Lett, 74, (1999) 3857.
10. J. B. Smathers, E. Kneedler, B. R. Bennett, and B. T. Jonker, Appl. Phys. Lett.72, (1998) 1238.
11. P. Blood and J. W. Orton, The electrical characterization of Semiconductors: Majority Carriers and Electron States (Academic, London, 1992), PP. 220, 295.
12. Liwu Lu, Weikun Ge, I. K. Sou, Y. Wang, J. Wang, Z. H. Ma, W. S. Chen, J. Appl. Phys. 82, (1997) 4413.

Proc. 25th Int. Conf. Phys. Semicond., Osaka 2000 (Eds. N. Miura and T. Ando)

407

Exciton localization in alloy/alloy interfaces of InGaAs/GaAs(001) stepped quantum wells

A. D'Andrea[1], F. Fernández-Alonso[2], M. Righini[2], D. Schiumarini[3], S. Selci[2], N. Tomassini[1]

[1] Istituto di Metodologie Avanzate Inorganiche, MITER-CNR, CP 10 Monterotondo Stazione, 00016 Roma (Italy)
[2] Istituto di Struttura della Materia, CNR, via del Fosso del Cavaliere 100, 00133 Roma (Italy) e-mail: righini@ism.rm.cnr.it
[3] Istituto di Chimica dei Materiali, MITER-CNR, CP 10 Monterotondo Stazione, 00016 Roma (Italy)

Abstract We have performed reflectivity and photoluminescence measurements on a set of asymmetric InGaAs/GaAs quantum wells. Reflectivity spectra are in good agreement with our theoretical predictions. Luminescence measurements show a large discrepancy with respect to what it is observed in symmetric quantum wells of similar structural characteristics. We are led to conclude that such differences can be mainly ascribed to the role played by the In-alloy/In-alloy interface.

1 Introduction

InGaAs/GaAs asymmetric/stepped quantum wells (AQW) are of considerable importance for both fundamental and practical reasons. The large oscillator strengths of Wannier-exciton transitions for well thicknesses in the range of quasi two-dimensional confinement can be used to enhance their non-linear optical response (i.e., second-harmonic generation) [1].

From a fundamental viewpoint, little is known about the characteristics of epitaxially grown In-alloy/In-alloy interfaces, i.e., the role of strain and alloy fluctuations on electronic and optical properties. For such a purpose we have studied a set of InGaAs/GaAs AQW's where the fundamental excitonic state is mainly confined in the deepest well. This exciton serves as a local probe of the properties of the alloy/alloy interface.

2 Experimental methods and results

We have performed normal-incidence reflectivity (R) and photoluminescence (PL) measurements on $In_xGa_{1-x}As/GaAs(001)$ single quantum wells (SQW) and AQW's. PL has been investigated as a function of laser power (Ar$^+$ laser source) and temperature (10÷300 K). The SQW's serve as our reference system and are characterized by an indium mole fraction of $x = 0.09$, 0.185 and well widths ranging between 1.5 and 14.0 nm [2]. The AQW's consist of a double layer of $In_xGa_{1-x}As/In_yGa_{1-y}As$ alloys ($x = 0.064$, $y = 0.149$) of equal thickness corresponding to $L_x = L_y = 3.2$, 4.2, and 5.2 nm for samples SS22, SS23, and SS24 respectively. Two indium furnaces at different temperatures ($T_1 = 596$, $T_2 = 696$ °C) were used to grow the two ternary alloy layers without growth interruption in a molecular-beam-epitaxy (MBE) machine (Varian GENII). The samples were characterized in-situ by reflection high-energy electron diffraction (RHEED) and ex-situ by high-resolution X-Ray diffraction. For the sake of a meaningful comparison, both SQW's and AQW's have been grown in the same machine and under similar conditions.

Fig. 1 shows a comparison between the R and PL spectra of the smallest AQW (SS22) and those typically observed in a SQW. As shown in Fig. 1a, the SQW PL peak is very narrow (FWHM 4 meV) and well aligned with the fundamental-state transition observed in the R spectrum (Stokes shift < .2 meV). This finding is at variance with the results shown in Fig. 1b corresponding to AQW SS22. In this case, we observe a FWHM of 19 meV and an evident Stokes shift of 4.2 meV. Moreover, the PL lineshape appears to be more complex, with contributions from at least two distinct transitions. The other AQW samples behave in a similar fashion with Stokes shifts of 3.1 (SS23) and 1.4 meV (SS24), and FWHM of 13 and 10 meV respectively. The differences in energy values for different AQW's may be attributed to a change in well dimension, namely, by increasing the well width there is a monotonic decrease of both the PL FWHM and Stokes shift.

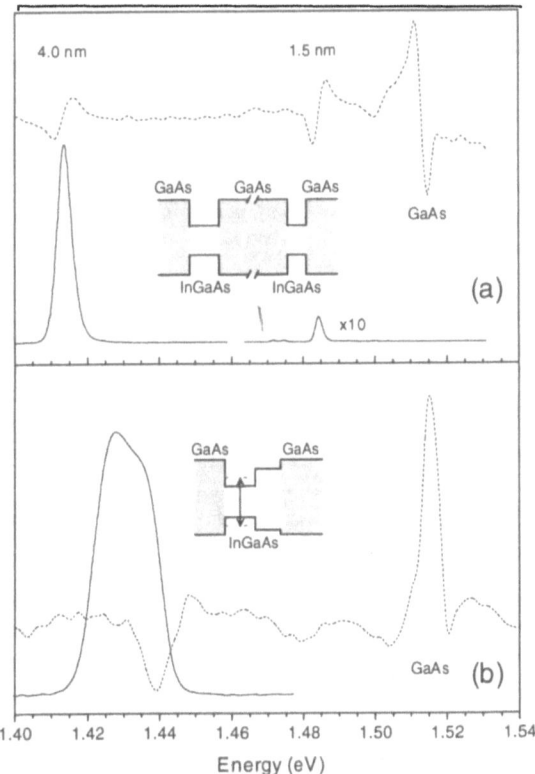

Fig. 1 Comparison of R (dashed lines) and PL (solid lines) spectra at 10 K. (a) Two isolated SQW's with x = 0.185, L = 4.0 and 1.5 nm, respectively. (b) Corresponding spectra for AQW SS22. The insets show a schematic diagram of the confining potentials.

The peculiar behavior of the AQW's is further corroborated by a power and temperature dependence of the PL lineshape. Fig. 2a shows the evolution of the PL with laser power for sample SS22. As the laser power is decreased, there is a linear decrease in PL intensity accompanied by a relative increase of the red tail of the peak. A similar trend is also observed when the sample temperature is decreased as shown in Fig. 2b. Under similar experimental conditions, the SQW's do not display these anomalous dependences.

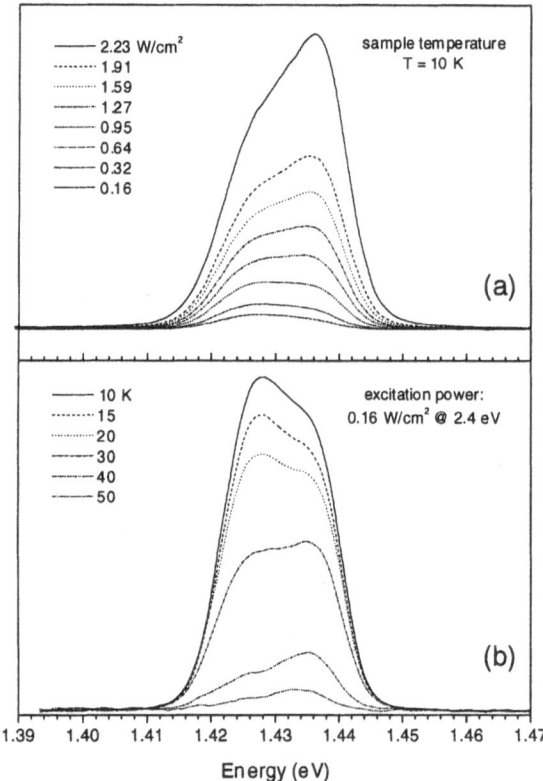

Fig. 2 AQW SS22 PL dependence on (a) laser power, and (b) sample temperature.

3 Discussion

The R spectra of all our samples have been previously compared with first-principles theoretical calculations using a parameter-free exciton envelope function model within the effective mass approximation [3]. The R spectra show resolved exciton peaks also at higher energies with respect to the lowest level responsible for the observed PL. In particular, our AQW R spectra show good agreement between experiment and theory in regard to energy positions, lineshapes, and peak relative intensities, confirming the presence of an abrupt potential discontinuity as expected for a stepped quantum well system [1]. Such findings are also in agreement with our X-ray diffraction measurements.

The small Stokes shifts and narrow PL peaks observed in the SQW reference system (Fig. 1a) suggest a high-quality $In_xGa_{1-x}As$ alloy material for indium mole fractions up to $x = 0.185$, as well as a standard behavior of the GaAs/InGaAs interfaces. We would like to stress the fact

that such an interface is common to both our SQW's and AQW's. This is the expected outcome of our MBE protocol which implemented growth interruption between barrier and well layers [4]. From these considerations we hypothesize that the major cause for the differences between our results for the SQW's and AQW's is related to the In-alloy/In-alloy interface.

Furthermore, the AQW PL features (Figures 1b and 2) are sufficient to invoke a composite peak with at least two major contributions. Their relative weight has a clear dependence on both laser power (density of photogenerated carriers) and temperature (non-radiative recombination channels or thermal spill-off). Several examples of a similar phenomenology have been reported in the literature. These include: monolayer fluctuations at AlGaAs/GaAs interfaces [5]; exciton trapping by In-rich islands at InGaAs/GaAs interfaces [4]; and excitons bound to photogenerated impurity states in AlGaAs/GaAs [6] as well as in InGaAs/GaAs QW's [7]. However, the PL peak widths reported in the literature are well below our values.

A simple-minded analysis of the PL lineshapes using only two gaussian components reveals supralinear ($\alpha \sim 1.3$) and sublinear ($\alpha \sim 0.8$) power dependences for the higher- and lower-energy peaks in all three AQW's. This saturation-like behaviour, also consistent with the observed temperature trend, can be explained by a lower density of states for the lowest energy transition. At present, studies of In-alloy/In-alloy interfaces are scarce in the literature [8] and do not allow for a clear-cut discrimination among the major mechanisms giving rise to our observations. Additional measurements such as the PL dependence on excitation energy in the presence and absence of an optical bias may provide further insight into this problem. These measurements are currently underway.

Acknowledgements

We thank "Progetto Finalizzato MADESS II (Materials and Devices for Solid State Electronics)" for partial financial support. S. Rinaldi, M. Luce, and M. Rinaldi are gratefully acknowledged for their technical assistance.

References

1. N. Tomassini, A. D'Andrea, M. Righini, S. Selci, L. Calcagnile, R. Cingolani, D. Schiumarini, and M. S. Simeone, Appl. Phys. Lett. **73**, (1998) 2245.
2. A. D'Andrea, N. Tomassini, L. Ferrari, M. Righini, S. Selci, M. R. Bruni, M. G. Simeoni, and N. Gambacorti, Phys. Rev. B **52**, (1995) 10713.
3. R. Atanasov, F. Bassani, A. D'Andrea, and N. Tomassini, Phys. Rev. B **50**, (1994) 14381.
4. H. Yu, C. Roberts, and R. Murray, Appl. Phys. Lett. **66**, (1995) 2253.
5. R. Köhrbrück, S. Munnix, D. Bimberg, D. E. Mars, and J. N. Miller, J. Vac. Sci. Technol. B **8**, (1990) 798.
6. V. Srinivas, Y. J. Chen, C. E. C. Wood, Solid State Commun. **89**, (1994) 611.
7. P. Borri, M. Gurioli, M. Colocci, F. Martelli, M. Capizzi, A. Patané, and A. Polimeni, J. Appl. Phys. **80**, (1996) 3011.
8. S. Marcinkevicius, U. Olin, and G. Treideris, J. Appl. Phys. **74**, (1993) 3587.

Characterization of Ge nanocrystal embedded in SiO$_2$ films

H. Fukuda[1], S. Sakuma[1], T. Yamada[1], S. Nomura[1], M. Nishino[2], T. Higuchi[3] and S. Ohshima[3]

[1] Department of Electrical and Electronic Engineering, Muroran Institute of Technology, 27-1 Mizumoto-cho, Muroran-shi, Hokkiado 050-8585, Japan   e-mail: fukuda@mmm.muroran-it.ac.jp

[2] Department of Information Technology, Hokkaido Polytechnic College, 3-190 Zenibako, Otaru-shi, Hokkaido 047-0292, Japan

[3] Fuji Research Institute Co., 2-3 Kandanishiki-cho, Chiyoda-ku, Tokyo 101-8443, Japan

Abstract Metal-insulator-semiconductor (MIS) structures with a Ge crystallite embedded in SiO$_2$ film were fabricated by Ge ion implantation and subsequent annealing. The data of Raman spectroscopy indicate the existence of crystallized Ge in the SiO$_2$ films. The Ge size is 3 to 5 nm on average and its density is 1×10^{12}/cm^2. Capacitance-voltage characteristics indicate the flat-band voltage shifts of 1.02 V after the electron injection. An anomalous leakage was also observed in the current-voltage characteristics. The precise simulation of quantum electron transport in the SiO$_2$ film indicates that the anomalous conduction is attributed to resonant tunneling in the SiO$_2$/Ge/SiO$_2$ multi-band structure.

1 Introduction

During the past few years, a considerable amount of attention has been focused on the strong photoluminescence (PL) in the visible wavelength range from nanostructures made of group IV elements, such as porous Si, nanocrystal Si, and Ge prepared by various methods [1,2]. For the Ge nanocrystals, most reports of visible PL from low-dimensional structures have involved Ge nanocrystals embedded in SiO$_2$. Their visible emission is considered to be due to three or two-dimensional quantum size effects. To our knowledge, although the optical properties of these nanostructures have been extensively investigated, there are few studies on their electrical properties.   In this study, we demonstrate a simple growth technique of Ge embedded in SiO$_2$ and evaluate the crystallinity and electrical properties of the SiO$_2$/Ge/SiO$_2$ structure.

2 Experimental

The starting materials were p-type (3-4 Ωcm), Si (100) wafers, which were then cut into 1.5×1.5 cm square chips. First, a SiO$_2$ film was grown on the Si in a dry oxygen ambient at 900℃ for 15 min. Then, ^{74}Ge$^+$ ion implantation into SiO$_2$ film was performed at an energy of 25 keV and a dose of 1×10^{14} ions/cm^2. After ion implantation, anneal of the films was also carried out in N$_2$ at 500 to 1100℃ for 1 hour. The total film thickness of the SiO$_2$ film containing Ge was 25 nm.

The Raman spectra was obtained using a Raman spectrometer with a 514.4 nm line of an Ar$^+$ laser (100mW) as the excitation source. All electrical measurements were conducted on films an a MIS configuration. Several aluminum (Al) dot electrodes of 0.5 mm diameter were evaporated over the area of the films. After Al deposition, the sintering was performed at 350 ℃ for 10 min in N$_2$ ambient. Capacitance-voltage (C-V) and current-voltage (I-V) measurements were carried out using C-V/I-V automatic electrical analyzer at room temperature.

3 Results and Discussion

Figure 1 shows the typical Raman spectra of samples for as-implanted and annealed at 500, 900 and 1000℃. As-implanted sample indicates no specific peak, resulting in amorphous state. The crystallization annealed above 500℃ is evidenced the sharp peak at around 300 cm^{-1} due to Γ_{25}' phonon mode. The peak position hardly changes but its full width at half-maximum (FWHM) decreases as annealing temperature increase.

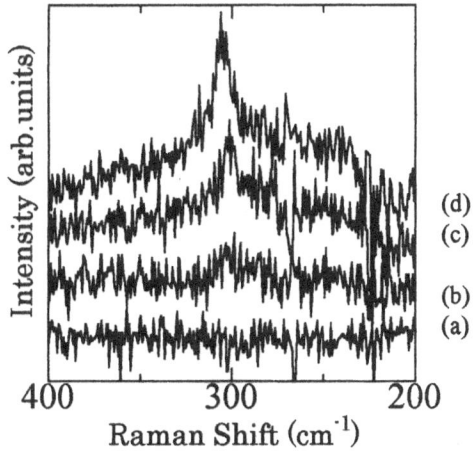

Fig. 1 Raman spectra for as-implanted (a),   annealed at 500℃(b), 900℃(c) and 1100℃(d).

We have confirmed that the Raman spectra of the Ge single crystal shows Ge-Ge peak at 300.5 cm^{-1} with a FWHM of 2.7 cm^{-1}. The peak intensity ratio of the unstrained Ge to nanocrystalline Ge in SiO$_2$ was typically 18:1. A qualitative relationship between the FWHM and average particle size was determined by use of one-phonon confinement model for Raman line shape of semiconductor nanocrystals. We can calculate the FWHM and mean crystallite size. The samples annealed at 900 and 1000°C indicate the FWHM of 25 and 19 cm^{-1}, respectively. These values correspond to average size of 3.3 and 4.7 nm, respectively. The size of the Ge crystals increases with increasing annealing temperature, in good agreement with the results reported elsewhere [3]

C-V measurements before and after electrical stress were used to investigate the charge storage in the Ge nanocrystals. As shown in Fig.2, a positive voltage shift in the C-V curve after the injection due to the presence of the interface-trapped charge is observed. The flat-band voltage (ΔV_{FB}) is not observed in the MIS capacitors without a Ge nanocrystal in SiO$_2$. The number of electrons ($N_{inj}=3.2\times10^{16}$/cm^2) that were injected into the oxide during stress was calculated by the integration of the injected gate current density ($J_{ox}=5.1$ mA/cm^2) at a constant current stress over the injection time of 1 s. After the electron injection, V_{FB} shifts to 1.02 V, in the positive direction. This means that the negative charge buildup in SiO$_2$ occurs due to electron tunneling in electron traps from the substrate. If the charge buildup is significant, the experimental C-V curve will be shifted from the ideal theoretical curve, in which the magnitude of ΔV_{FB}.

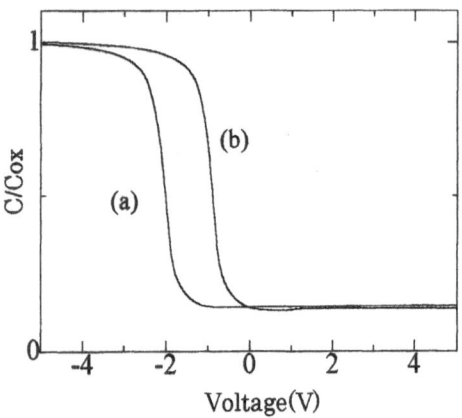

Fig. 2 C-V curves for the initial state (a) and after electron injection (b).

For the 5-nm-diameter Ge nanocrystals with a density of 1×10^{12} cm^{-2}, the voltage shift is calculated to be 1.0 V for one electron per nanocrystal. This value is in

good agreement with the ΔV_{FB} value determined from the C-V curve. The I-V characteristics of Ge-embedded MIS diodes are shown in Fig.3. An anomalous leakage was observed in the voltage range of 10 to 30 V in the I-V curves. In contrast, the I-V characteristics of the pure SiO$_2$ film showed typical Fowler-Nordheim tunneling characteristics. We performed a simulation of quantum transport using Esaki-Tsu one-dimensional integral approximation. The best fit is obtained for the case of mean Ge size of 3 nm with a dispersion of 0.5 nm. In the I-V characteristics, negative differential conduction could not be observed. This is due to the fact that the electrons do not tunnel into the resonant states in the SiO$_2$/Ge/SiO$_2$ potential barrier, but travel to bound states at a higher energy level in the well.

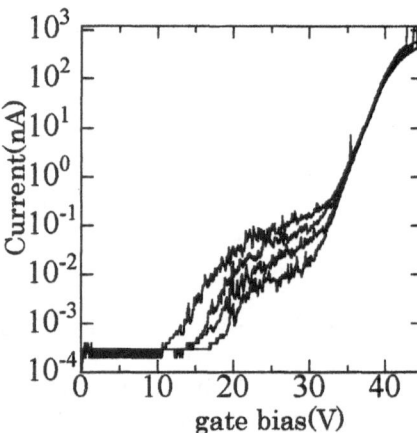

Fig. 3 Room temperature I-V curves of a SiO$_2$/Ge/SiO$_2$ MIS diodes.

4 Conclusion

We have fabricated a Ge-embedded SiO$_2$ films by Ge ion implantation. The Raman spectra showed the Ge crystallization process after annealing. The electrical properties indicate that electron capture and resonant tunneling via SiO$_2$/Ge/SiO$_2$ double-barrier structure.

Acknowledgement

This work was supported by a Grant-in-Aid for Scientific Research from The Ministry of Education, Science, Sports and Culture of Japan and Nippon Sheet Glass Foundation for Materials Science and Engineering.

References

1. Y. Maeda, N. Tsukamoto, Y. Yazawa, Y. Kanemitsu and Y. Matsumoto, Appl. Phys. Lett. **59**, (1991) 3168.
2. S. Y. Chou and A. E. Gordon, Appl. Phys. Lett. **60**, (1992) 1827.
3. H. Richter, Z. P. Wang and L. Ley, Solid State Commun. **39**, (1981) 625.

Control of quantum confinement in metal-clad InAs quantum wells

S. Tsujino[1], S.J. Allen[1], M. Rüfenacht[1], M. Thomas[2], J.P. Zhang[2], J. Speck[2], T. Eckhause[1], B. Gwinn[1]

[1]Institute of Quantum Science Engineering and Technology, University of California, Santa Barbara, CA 93106, U.S.A.
[2]Materials department, University of California, Santa Barbara, CA 93106, U.S.A.

Abstract We explore intersubband optical transitions in InAs quantum wells terminated by various metal-InAs interfaces and demonstrate that electrons can be quantum mechanically confined in InAs quantum well defined by a metal-semiconductor interface. We found that the transmission of electrons at the Pt-InAs interface is sensitively controlled by chemical reaction from highly transmissive (~ 70%) to highly reflective.

Quantum mechanical boundary condition at the metal-semiconductor interface is important for nanostructured semiconductor electronic and photonic devices, as well as for the spin-transport [1] and quasi-particle transport [2]. Although decades of studies have been conducted to understand the metal-semiconductor interface, there are few experiments that explore the boundary conditions. Nearly all aspects of the band structure between the metals and the semiconductors are different, so the transmission and the reflection at the interface are difficult to anticipate a priori.

In this work, we explore and demonstrate the quantum confinement of conduction electrons in InAs quantum wells (QWs) defined by a metal-semiconductor interface. Since the InAs-metal interface forms no Schottky barrier, it is usually assumed that the interface is transparent and quantum mechanical reflection can be ignored. The experiments in InAs QWs terminated with various metals showed this is not likely and that the quantum mechanical reflection at the metal-semiconductor interface leads to quantized electronic states within the InAs thin layer and *quantum confinement without walls* [3,4](Fig.1).

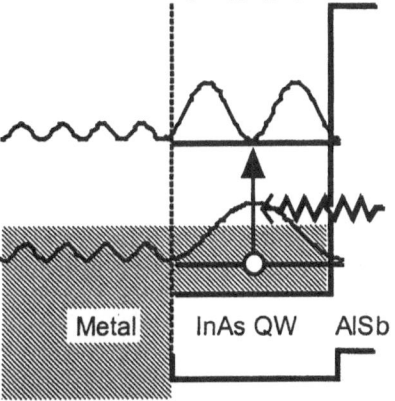

Fig,1 Confinement without walls in metal-clad InAs quantum well structure. The electron in the InAs quantum well is quantum mechanically reflected at the metal-InAs interface because of the band structure difference forming a quantized state, even though the metal-InAs interface is Ohmic.

We explore the electronic structure of the metal-clad InAs QWs using infrared optical absorption. With this method, the formation of quantized electronic states is detected by the intersubband optical transitions. Furthermore, intersubband transitions are sensitive to the amplitude and the phase of the reflection coefficient R of the conduction electrons at the interface.

In the experiment, we prepared metal-clad structures terminated by various metals, including Al, Nb, Sb, W, Pt, Ni, Ag, In, Ti, or Au. Previous study showed that Al-GaAs interface [5,6] and Al-InAs interface [3] are highly reflective. We also found that Nb-InAs interface is also highly reflective showing the transmission probability of the quasi-particles across the interface is low [4]. On the other hand, carefully prepared W-InAs interface can be highly transmissive and the intersubband absorption of W-clad InAs QWs exhibit auto-ionization of the quantized states of the InAs layer by the metallic states [3]. In the following, we focus on quantum mechanical reflection at the Pt-InAs interface controlled by chemical reaction.

The metal-clad InAs QW are produced by starting with an InAs-AlSb QW, removing the top barrier using selective chemical etching, and depositing metal over the InAs. The thickness of the InAs layer is 15 nm. We irradiated the InAs surface with Ar ions before the metal deposition in-situ, to remove a few ML of the material from the InAs surface including oxide.

The original InAs QWs are modulation doped in the surrounding AlSb layers and the electron concentration is nominally $7 \times 10^{12} \text{cm}^{-2}$. In this case, both the ground subband and the first excited subband are populated and the intersubband absorption of the QW shows two resonance's (Fig.2). The sharp absorption at frequency 174 meV with the full width at the half maximum (FWHM) of 7 meV corresponds to the intersubband transition from the first excited subband to the second excited subband. The weak and broader resonance at 110 meV corresponds to the transition from the ground subband to the first excited subband.

Figure 2 shows the absorption spectrum of the as deposited Pt-clad InAs sample produced from the same original QW sample shown in the Fig.2. A broad but well-defined absorption is observed at 170 meV. Comparing with the absorption spectrum of the original QWs, the absorption of the Pt-InAs QW is unambiguously ascribed to the intersubband absorption of electrons from the first excited subband to the second excited subband confined by the Pt-InAs interface. The spectrum exhibits a FWHM equal to 59 meV which is 52 meV broader than the original QW resonance. The long low frequency tail can be ascribed to the finite transmission through the Pt-InAs interface. When InAs is bounded by an infinite barrier, the reflection coefficient R at the interface is -1, forming a sharp quantized subband. But when the amplitude of the reflection $|R|$ is less than 1, the quantized states in the InAs layer are auto-ionized [7] by the coupling to the metallic states, which causes a broadening of the subbands. While processing damage can also cause a broadening by randomly thinning the InAs [8], the induced inhomogeneous broadening is only toward higher frequency. Therefore the low frequency tail cannot be accounted for by the inhomogeneous broadening.

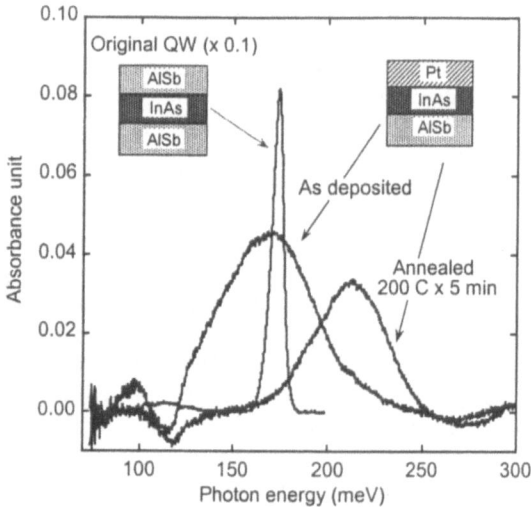

Fig,2 Intersubband absorption spectra of Platinum clad-InAs quantum well structures. The curves show the absorption spectra of the original InAs-AlSb quantum well, of the as-deposited Pt-InAs quantum well, and of the annealed Pt-InAs quantum well.

Figure 2 also shows the absorption spectrum of the Pt-InAs sample after annealed for 5 min at 200 °C. The observed spectrum at 212 meV with FWHM equal to 45 meV is narrowed and shifted toward higher frequency compared with the spectrum of the as-deposited sample. We note that the low frequency absorption below the original QW resonance is quenched by annealing. When the sample is annealed, we promote the creation of an interfacial layer, reducing the electron transmission. Therefore, the narrowing of the peak and the quenching of the low frequency absorption occur. The higher frequency shift is ascribed to the thinning of the InAs layer by the formation of the interface layer of about 2.6 nm. In fact, further annealing at 300 °C for 5 min quenched the spectrum.

The transmissive, non-annealed interface is characterized by a reflection coefficient $|R|^2 = 0.34$ from the increase of the FWHM of the as-deposited sample

from the original QW resonance. We assume the contribution of the inhomogeneous broadening is negligible in the as-deposited sample, since the low frequency absorption tail is of the same order as the increase of the half width.

Contrary to the as-deposited sample, annealed interface is highly reflective and the broadening of the annealed sample is solely ascribed to inhomogeneous broadening. When the broadening is mainly created by the inhomogeneous interface layer formation, the increase of FWHM in the annealed sample from the original QW should be of the same order as the peak shift, assuming that the chemical reaction at the interface progressed randomly. In the annealed Pt-InAs sample, we effectively observed that the amount of the increase of the FWHM and the higher frequency shift from the original QW are same (38 meV). It is interesting to note that similar features were observed in as-deposited Al-InAs QWs [3], suggesting that Al is highly reactive with InAs at room temperature.

In conclusion, we explored quantum confinement without walls in metal-clad InAs QW structures. We found that the transmission of electrons across the Pt-InAs interface is sensitively controlled by chemical reaction.

Acknowledgements One of the authors (S.T.) acknowledges to Prof. H. Sakaki and Prof. H. Kroemer for helpful discussions. This work is partially supported by National Science Foundation Science and Technology Center for Quantized Electronic Structures (QUEST), DMR 91-20007.

References

1. S. Datta and B. Das, Appl. Phys. Lett. **56**, 665 (1990).
2. A. Kastalsky, A. W. Kleinsasser, L. H. Greene, R. Bhat, F. P. Milliken, and J. P. Harbison, Phys. Rev.Lett. **67**, 3026 (1991).
3. S. Tsujino, S.J. Allen, M. Thomas, J.P. Zhang, J. Speck, T. Eckhause, E. Gwinn, M. Rüfenacht, H. Sakaki, Superlattice and Microstruc. **27**, 470 (2000).
4. T. Eckhause, S. Tsujino, K. Lehnert, E. Gwinn, S.J. Allen, M. Thomas, and H. Kroemer, Appl. Phys. Lett., **76**, 212 (1999).
5. M. V. Weckwerth, J. P. A. van der Wagt, and J. S. Harris, Jr., J. Vac. Sci. Technol. B12, 1303 (1994).
6. A.J. North, et.al., Phys. Rev. B **57**, 1847 (1998).
7. U. Fano, Phys. Rev. **124**, 1866 (1961).
8. P. H. Magnee, S. G. den Hartog, B. J. van Wees, T. M. Klapwijk, W. van de Graaf, and G. Borghs, Appl. Phys. Lett. **67**, 3569 (1995).

Unsaturated cyclic-hydrocarbon molecules on a Si(001) surface: A first-principles approach

Kazuto Akagi, Shinji Tsuneyuki

Institute for Solid State Physics, University of Tokyo

Abstract   Chemisorption of unsaturated cyclic-hydrocarbon molecules such as C_6H_{10}, C_6H_8 and C_5H_8 on Si(001) clean surface was investigated based on first-principles calculations, and the adsorption structures were discussed comparing with the experimental data. As for C_6H_{10}, two kinds of stable conformation were determined, which are consistent with two types of characteristic STM image and spectroscopic data. On the other hand, we also found two kinds of adsorption structures for C_6H_8 molecules, but the structure supported by STM and UPS observations is slightly unstable compared with the other one, and the situation is still mysterious.

1 Introduction

Various kinds of molecular species can be chemisorbed to Si(001) clean surface because of its dangling bonds. If we can arrange the orientation of chemsorption, various functional surfaces can be designed by controlling adsorption patterns and kinds of organic molecules. From such a point of view, we are investigating "unsaturated cyclic-hydrocarbon / Si(001)" systems collaborating with Yoshinobu group at ISSP[1].

The target molecules are cyclohexene(C_6H_{10}), 1,4-cyclohexadiene(C_6H_8) and cyclopentene(C_5H_8). By the experiments such as LEED, XPS and UPS, it is commonly found that the 2×1 structure from the silicon surface is kept even after the chemisorption, that silicon dangling bonds almost disappear and that Si-C bonds are formed. Especially, as for the C_6H_{10} system, two kinds of STM images (asymmetric and symmetric one, respectively) are reported. As for the C_6H_8 system, only one kind of STM image is reported, which is situated between two adjacent silicon dimers along a silicon dimer row and the π-bond like node appears depending on variation of the bias voltage.

2 Results

Hereafter, we focus the discussion mainly on the adsorption structure of the C_6H_{10} system. The calculation method used here is a conventional one based on the density functional theory (DFT) with GGA(PBE type)[3]. The adsorption structures were obtained by minimization of the total energy of the system based on the density functional method. Initial configurations were properly given for each system considering the ideal structures of adsorbate and substrate. We used the plane wave basis set, and explicitly treated only valence electrons by the pseudo potential method. The energy cutoff is basically 56.0[Ry], and we properly checked some results with 81.0[Ry]. The super-cell method was used, and the size is $1.0 \times 1.0 \times 2.2$, $2.0 \times 1.0 \times 2.2$ and $\sqrt{2} \times \sqrt{2} \times 2.2$, in

the unit of the lattice constant of Si(001) (2×2) surface. The slab model has five-layers thickness. The number of sample k-points was four. The criterion of force convergence was set to 5.0×10^{-3}[Ht/au].

Generally, the chemisorption of a unsaturated hydrocarbon molecules to Si(001) clean surface means a reaction between a surface silicon asymmetric dimer and a C=C π-bond. Here, the adsorption structures are expected to be classified by whether it is 'Cis-Type' or 'Trans-Type' as a first classification, and by its conformation such as 'Chair-Type' and 'Boat-Type' under a gas phase condition as a second classification. Fig.1 shows the C_6H_{10} cases. The adsorption energy at 0.5ML coverage is 1.28eV (cis1), 1.34eV (cis2) and 0.96eV (trans), respectively. The 'Trans-Type' chemisorption, however, will not be achieved on Si(001) clean surface because of the chemical restriction such as reported in the ethylene case[2].

Calculated local density of states (LDOS) for these two cis-type confirmations qualitatively explain the experimentally reported asymmetric (cis1) and symmetric (cis2) STM images. Furthermore, the energy barrier between these two conformations estimated by the first-principles calculations under some restrictions to degrees of freedom is turned to be 0.1eV or so. It is easy for C_6H_{10} molecules to gone over such a barrier at room temperature. By modifying the molecule, it will get possible to control the change of the conformation by STM and to keep the state for a long time.

Fig.2 shows the relation of the variation of free energy of the system depending on the chemical potential of the C_6H_{10} molecule[4]. It seems difficult to grow the chemisorption region along the silicon dimer rows.

Finally, we simply mention the C_6H_8 system which looks mysterious. As similar to the case of C_6H_{10} molecules, there are two kinds of adsorption structures, but the structure which is supported by UPS and STM experiments is slightly unstable compared with the other one. Taking account of the estimated low energy barrier (at most 0.1eV) between these two conformations, two kinds of STM images should be observed corresponding to both of these structures, but only one kind is reported actually. This system, which is also interesting from the viewpoint of electronic states, requires more investigation.

3 Acknowledgment

The authors would like to thank J.Yoshinobu, Y.Yamashita and K.Hamaguchi for fruitful discussions. The numerical

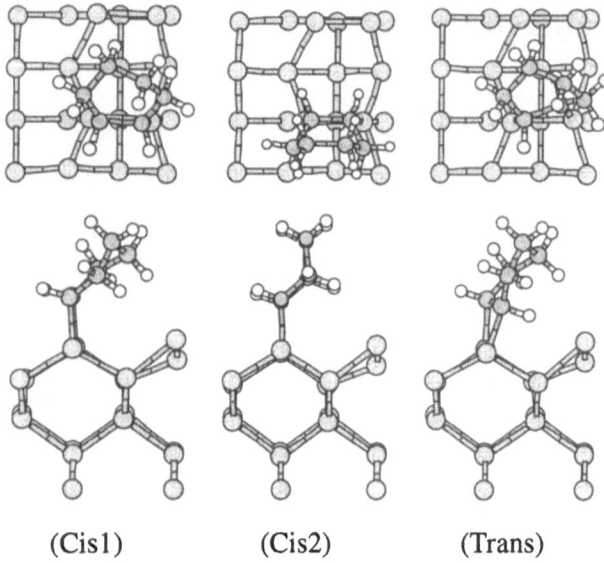

Fig. 1 Two kinds of the cis-type adsorption structure and one trans-type adsorption structure optimized by the first-principles calculations.

calculations were performed on NEC SX4 system supported by the New Program Project of the Ministry of Education and Culture and Hitachi SR8000 system at ISSP, Univ. of Tokyo.This work was supported in part by a Grant-in-Aid from the Ministry of Education, Science.

References

1. K.Hamaguchi *et al.*, J.Phys.Chem. B (2000) August.
2. J.Yoshinobu *et al.*, J.Chem.Phys. **87,** (1987) 7332.
3. The theoretical calculation was performed with TAPP(Tokyo Ab-initio Program Package), which has been developed in our group. (1)M. Tsukada *et al.*, computer program package TAPP, University of Tokyo, Tokyo, Japan, 1983–2000. (2)J.Yamauchi, M.Tsukada, S.Watanabe, and O.Sugino, Phys.Rev.B, **54,** (1996) 5586.
4. J.E.Northrup, Phys.Rev.B **44,** (1991) 1349.

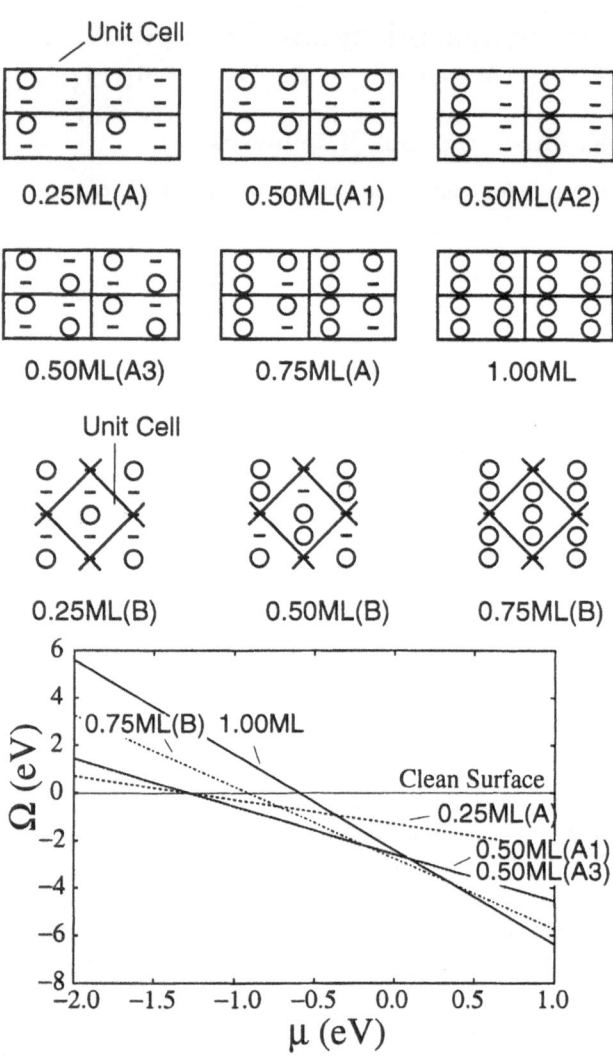

Fig. 2 Variation of free energy of the system depending on the chemical potential of the C_6H_{10} molecule. The origin of Ω and μ is set to a silicon clean surface and formation energy of a C_6H_{10} molecule from isolated carbon atoms and hydrogen atoms, respectively.

Organic modified GaAs(100) Schottky contacts

Th. Lindner, T. U. Kampen, S. Park, D. R. T. Zahn

Institut für Physik, Technische Universität Chemnitz, D-09107 Chemnitz, Germany;
e-mail: thomas.lindner@physik.tu-chemnitz.de

Abstract We investigated the current-voltage-characteristics of Ag/n-GaAs(100) diodes with organic interlayers of various film thickness. The organic semiconductor 3,4,9,10-perylenetetracarboxylic dianhydride (PTCDA) serves as the organic film. The results can be described in terms of a model including thermionic emission and space-charge-limited currents. At low current densities thermionic emission over the organic/inorganic contact barrier dominates, while at high current densities space-charge effects govern the charge transport. The effective barrier heights of Ag/PTCDA/n-GaAs(100) diodes decrease with increasing organic film thickness.

1 Introduction

The electronic properties of metal-semiconductor or Schottky contacts are characterized by their barrier heights. The barrier is the energy distance between the Fermi level and the respective majority-carrier band edge right at the interface which is the conduction band minimum for n-type semiconductors. The barrier of a given metal-semiconductor contact may be reproducibly changed by a controlled modification of the interface. Hydrogen at the interface of Pb/Si(111) Schottky contacts decreases and increases the barrier height on n-type and p-type doped samples, respectively [1]. The results can be explained by an additional hydrogen-induced interface dipole, which results in a charge of positive sign on the semiconductor side of the interface. The sulfur modification of In/GaAs(100) contacts results in the same qualitative change in barrier heights with respect to the unmodified contacts [2]. A further parameter for the controlled tuning of the electronic properties of metal-semiconductors, i.e. barrier height engineering, is the thickness of the interlayer. For this purpose we have used the organic prototype molecule perylene-3,4,9,10-tetracarboxylic dianhydride (PTCDA). The organic film thickness used ranged from 3 to 200 monolayers (ML). The change in the barrier height as a function of the thickness of the organic interlayer was determined from I/V-characteristics recorded *in situ*.

2 Experiment

Tellurium doped n-GaAs(100) with a doping concentration of $2 \cdot 10^{17} cm^{-3}$ (Freiberger Compound Materials) served as a substrate. The substrate was degreased in acetone, ethanol, and finally deionised water using an ultrasonic bath. To reduce the oxide on the surface the sample was etched in HCl for 30s. The samples were then attached to a copper holder with an In-Ga alloy. After transfer into the UHV chamber, which has a base pressure of $1 \cdot 10^{-10} mbar$, the samples were annealed

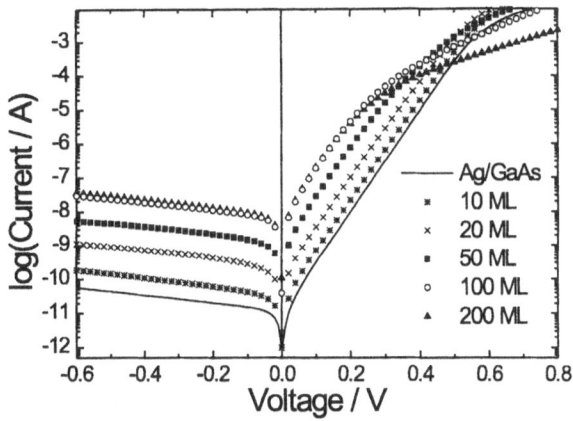

Fig. 1 I/V characteristics of samples with various PTCDA layer thickness. The contact area of $2.1 \cdot 10^{-7} m^2$ is constant for each characteristic.

for 30 min up to a temperature of $300^\circ C$. This leads to the formation of an InGaAs layer on the backside of the sample resulting in a low series resistance of less than 20Ω [3,4]. During the annealing of the sample a hydrogen plasma treatment was carried out for 5 min at an hydrogen pressure of $3 \cdot 10^{-3} mbar$. Atomic hydrogen reacts with arsenic- and gallium oxides to volatile compounds which leads to a further reduction of the oxide layer [5,6,7]. Organic molecular beam deposition was used for the growth of the PTCDA layers at room temperature. Before use the organic material was purified by vacuum sublimation. The deposition rates are between 0.3 and $0.8 nm/min$ and the thickness of the organic layer is controlled by an quartz crystal microbalance. One monolayer corresponds to $0.321 nm$, which is the distance between the molecular planes in a PTCDA crystal [8]. Silver contacts with an area of $2.1 \cdot 10^{-7} m^2$. were evaporated through a shadow mask with a deposition rate of $7 nm/min$. The I/V-characteristics were recorded *in situ* using an HP semiconductor test system.

3 Results and Discusion

Figure 1 shows the current-voltage characteristics of Ag/ PTCDA/n-GaAs(100) contacts as a function of the PTCDA film thickness. For the pure Ag/n-GaAs(100) Schottky contact we get a straight line in the $ln(I)$ vs. V plot at forward biases due to thermionic emission over the barrier. A least squares fit to this linear region gives a barrier height of $(0.81 \pm 0.01) eV$ and an ideality factor

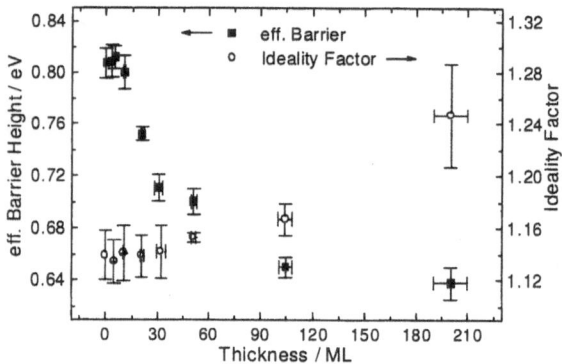

Fig. 2 Effective barrier height and ideality factor in dependence of the layer thickness. The values are calculated from the linear part of the forward I/V characteristics.

of 1.14 ± 0.02. A series resistance results in a deviation from this linear behaviour at forward voltages higher than 0.5 V. With increasing PTCDA layer thickness the current at reverse bias and lower forward increases. The increase of current saturates for organic layer thicknesses above 100 monolayers. The characteristics for the sample with 100 and 200 monolayers PTCDA are almost identical at reverse bias and forward bias lower than 0.2 V. With increasing forward bias the characteristics depart with the sample having a higher film thickness showing a lower current. At low forward biases we always observed a linear part in the $ln(I)$ $vs.$ V plot. This indicates that in this regime the current is exponentially dependent on voltage and therefore governed by thermionic emission. Figure 2 shows the effective barrier height and the ideality factor as a function of the PTCDA layer thickness. We suppose that the PTCDA/GaAs barrier is the relevant one. The values are mean values obtained from several characteristics measured on several dots of the sample. With increasing PTCDA layer thickness the effective barrier height and the ideality factor decrease and increase, respectively. It seems, however, that for very low coverages the barrier slightly increases before decreasing with the film thickness. For 200 monolayers PTCDA the effective barrier is reduced by 0.16 eV compared with the reference sample and the saturation current is nearly three orders of magnitude higher, despite of the low intrinsic free carrier concentration in PTCDA of approximately $5 \cdot 10^{14} cm^{-3}$ [9]. This behaviour is not understood at present. Photoemission experiments are planned in the future which should give some more detailed information on the interface formation. Space charge effects lead to a thickness dependence in the higher forward bias regime. In space charge limited transport, which is well known for PTCDA [9,10], there is a power law between current density j and the voltage U_o over the organic layer:

$$\frac{j}{d} \propto \left(\frac{U_o}{d^2}\right)^m \tag{1}$$

where m depends on the trap distribution in the material and d is the thickness of the layer [11]. In the trap free case $m = 2$ and we get the Mott-Gurney-law:

$$j \propto \frac{U_o^2}{d^3} \tag{2}$$

If the voltage drop over the organic layer is significant large compared with the voltage drop over the PTCDA/GaAs interface, the I/V-characteristics follow equation (2) and the currents decrease with increasing PTCDA film thickness. This is clearly observed for the two samples with the highest PTCDA coverage. For lower PTCDA coverages the voltage drops almost completely over the PTCDA/GaAs interface.

4 Summary

We investigated Ag/PTCDA/GaAs structures for various PTCDA interlayer thickness by current voltage measurement. The experimental data can be described by thermionic emission+space-charge-limited current model. A decreasing effective barrier height or increasing saturation current was observed for increasing thicknesses of the PTCDA interlayer. This decrease in barrier height saturates for coverages higher than 100 monolayer film thickness.

References

1. T. U. Kampen, and W. Mönch, Surf. Sci., **331-333**,(1995) 490.
2. St. Hohenecker, T .U. Kampen, T. Werninghaus, and D. R. T. Zahn, Appl. Surf. Sci., **142**,(1999) 28.
3. S. A. Ding, and C. C. Hsu, Appl. Phys., **A62**,(1996) 241.
4. A. A. Lakhani, J. Appl. Phys., **56**,(1984) 1888.
5. T. Akatsu, A. Plößl, H. Stenzel, and U. Gösele, J. Appl. Phys., **86**,(1999) 7146.
6. K. A. Elamrawi, and H. E. Elsayed, J. Vac. Sci. Technol., **A17**,(1999) 823.
7. M. Yamada, and Y. Die, Jpn. J. Appl. Phys., **33**,(1994) L671.
8. Y. Hirose, W. Chen, E. I. Haskal. S. R. Forrest, and A. Kahn, J. Vac. Sci. Technol. B, **12**,(1994) 2616.
9. S. R. Forrest, M. L. Kaplan, and P. H. Schmidt, J. Appl. Phys., **55**,(1984) 1492.
10. S. R. Forrest, M. L. Kaplan, and P. H. Schmidt, J. Appl. Phys., **56**,(1984) 543.
11. K. C. Kao, and W. Hwang, *Electrical transport in solids* (Oxford: Pergamon Press, 1981).

Proc. 25th Int. Conf. Phys. Semicond., Osaka 2000 (Eds. N. Miura and T. Ando)

Interface effects on exciton states in a CdTe/(Cd, Mn)Te quantum well

H. Yokoi[1], Y. Kakudate[1], Yu. G. Semenov[2], S. Takeyama[3], S. W. Tozer[4], Y. Kim[5]*, T. Wojtowicz[6], G. Karczewski[6], J. Kossut[6]

[1] National Institute of Materials and Chemical Research, 1-1 Higashi, Tsukuba, Ibaraki 305-8565, Japan, e-mail: yokoi@nimc.go.jp
[2] Institute of Semiconductor Physics, National Academy of Sciences of Ukraine, Prospect Nauki 45, Kiev 252650, Ukraine
[3] Department of Physics, Chiba University, 1-33, Yayoi-cho, Inage-ku, Chiba-shi, Chiba 263-8522, Japan
[4] National High Magnetic Field Laboratory, Florida State University, 1800 E. Paul Dirac Drive, Tallahassee, FL 32310, U.S.A.
[5] National High Magnetic Field Laboratory, Los Alamos National Laboratory, MS E536, Los Alamos, NM 87545, U.S.A.
[6] Institute of Physics, Polish Academy of Sciences, Al. Lotników 32/46, 02-668 Warsaw, Poland

Abstract Zeeman shifts of the band-edge exciton energies for a 1.9 nm thick CdTe quantum well and a $Cd_{1-x}Mn_xTe$ (x=0.24) barrier in a single quantum well structure were obtained through photoluminescence measurements at a temperature of 4 K, magnetic fields to 60 T and hydrostatic pressures to 2.45 GPa. Taking advantage of an additional red shift observed in the Zeeman shift for the quantum well, the interface effect on the exciton energy in the quantum well was described precisely in terms of the δ-localized exchange field. It is suggested that sp-d interaction in the interface region is enhanced by a pressure.

1 Introduction

In diluted magnetic semiconductors (DMSs), such as (Cd, Mn)Te, band electrons interact with local d-electrons via sp-d exchange interaction strongly. [1] This feature makes DMSs as a promising material for so called spintronics. In order to improve their spin functions, fabrication techniques for quantum structures, such as quantum wells (QWs) and quantum dots, have been elaborated.

In lower dimensional structures, existense of the interface as well as the dimensionality becomes more important. In a quantum well structure with a non-magnetic well and magnetic barriers, for example, magnetic properties of magnetic ions in the interface region would be different from those inside the barrier, due to the lack of magnetic ions to interact with in the QW side, and interdiffusion of the magnetic ions. This leads to the difference of the exchange interaction between the excitons in the QW and the magnetic ions in the interface (interface effect), from that between those excitons and the magnetic ions in the barrier (barrier contribution). However, overlap of the barrier contribution makes it difficult to extract the interface effect.

In previous studies on the interface effect [2,3], there was a disadvantage in that no distinct feature to distinguish one from the other was observed experimentally. We recently reported a sign observed in photoluminescence (PL) measurements for a non-magnetic quantum well with magnetic barriers, which is related to the bar-

rier contribution. [4] In the present paper, we will demonstrate the interface effect extracted from the experimental data in the framework of the δ-potential model [5], in which the Shrödinger equation can be solved analytically. Pressure effect on the interface effect will be also discussed.

2 Experiment and calculation

The sample that we used for this study is a 1.9 nm thick CdTe QWs separated by 48 nm thick $Cd_{1-x}Mn_xTe$ (x=0.24) barriers. Hybrid buffer layers between the quantum well and a (100) GaAs substrate were constructed.

External magnetic fields to 60 T were generated in a long pulsed magnet with typical pulse duration of 2 sec. The magnet was charged with a 1430 MVA/600 MJ inertial energy storage generator. A plastic diamond anvil cell (DAC) was employed in order to avoid the increase of sample temperature due to the eddy current during pulsed fields. [6]

Excitation light of 442 nm from a He-Cd laser was transferred to the DAC through a single optical fiber. PL signals were transferred back through the same fiber to a spectrometer equipped with a liquid-nitrogen-cooled back-illuminated charge-coupled-device (CCD) (Princeton Instruments, LN/CCD-100EB). Each spectrum was recorded at a rate of one per 2.1 msec. The two second long pulse of the magnet allowed us to take high resolution PL spectra every 0.05 T below 15 T and every 0.3 T between 30 and 60 T.

We have described the interface potential in terms of a δ-localized exchange field, as follows,

$$H_{if}^{(0)} = A_b\delta(z \pm L_w/2)\sigma, \tag{1}$$

with spin projection $\sigma = \pm 1/2$ and the quantum well width L_w. The factor A_b is expressed for an electron (b=e) and hole (b=h) as follows,

$$A_b = a_b L_w E_{0b} B_s, \tag{2}$$

where B_s is the Brillouin function for spin S, taking into account an antiferromagnetic temperature T_{if} phenomenologically, and $E_{0b} = \hbar/(2m_b^*L_w^2)$, m_b^* is an effective mass. The parameters a_b and T_{if} are determined through the fitting to the experimental results.

---

* *Present address:* Department of Physics, Quantum-functioning Semiconductor Research Center, Dongguk University, Seoul, South Korea

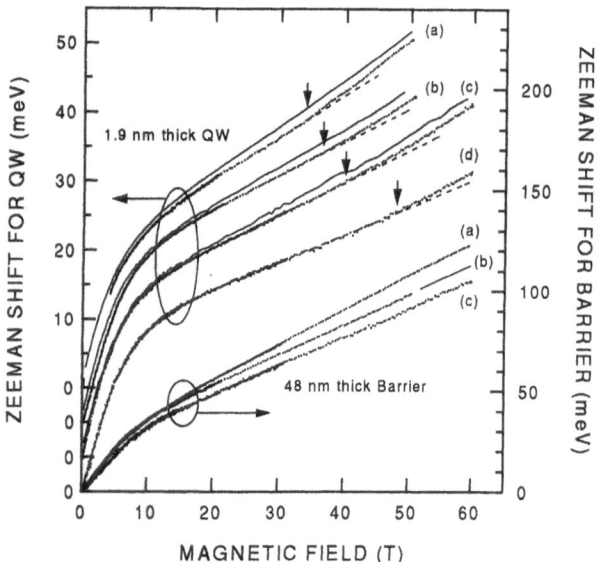

Fig. 1 Zeeman shifts of exciton energies for the 1.9 nm thick CdTe QW (left axis) and the $Cd_{1-x}Mn_xTe$ (x=0.24) barrier (right axis), obtained from the experiments (dots) and from the calculations (solid lines, shifted vertically by 1 meV) at 4 K and pressures of (a) 1.0×10^{-4}, (b) 1.11, (c) 1.46, and (d) 2.45 GPa, as a function of magnetic field. Broken lines show extrapolation of the experimental data taken between 25 T and 30 T. Arrows exhibit onset of the additional red shift.

3 Result and discussion

PL signals from band-edge excitons in the QW and the barrier were observed and Zeeman shifts of their peak energies were obtained as shown in Fig. 1. For the barrier, the Zeeman shift was linear as a function of magnetic field above 15 T, which reflects linear field dependence of the magnetization in (Cd, Mn)Te with large Mn concentration [7]. In contrast, an additional (super-linear) red shift was observed for the QW in the region of the linear magnetization. We have assigned the phenomenon to the accelerated penetration of the hole wavefunction into the barrier as the barrier height is decreased by a magnetic field.

The additional red sift was reproduced by use of the δ-potential model as shown in Fig. 1. In the fitting procedure, the exciton Zeeman shift for the barrier was utilized as magnetic field dependence of the barrier height, and the value of 0.37 was used for the band offset. We have noticed that the onset of the additional red shift was very sensitive to the value of the band offset. This would support our assignment for the origin of the additional red shift, and prove the accuracy of our estimate for the barrier contribution to the exciton energy.

The figure 2 exhibits Zeeman shift caused by the interface effect, extracted from the exciton Zeeman shift for the QW, at several pressures. The Zeeman shift saturated above 15 T, which would mean the spins in the interface region are isolated in contrast to spins in the barrier of $Cd_{1-x}Mn_xTe$ (x=0.24). The fitting parameters a_e and T_{if} have been determined as shown in Ta-

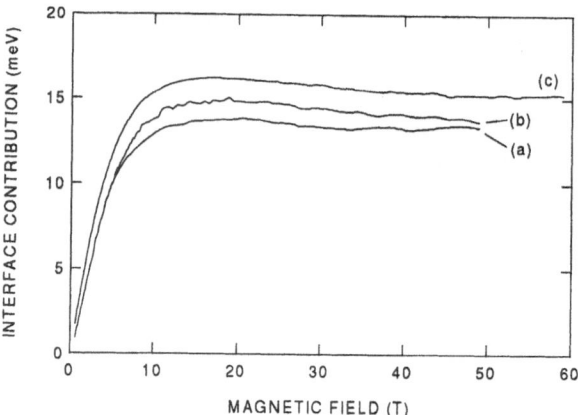

Fig. 2 Calculated interface effect on the exiton energy in a 1.9 nm thick $CdTe/Cd_{1-x}Mn_xTe$ (x=0.24) QW as a function of magnetic field. Notation of (a)-(c) is same as in Fig. 1.

ble 1. The monotonic increase of a_e as a function of pressure would reflect the enhancement of the sp-d interaction by pressure, which is in line with the case of bulk DMSs with low magnetic ion concentration. No change of T_{if} with pressure does not agree with the case of bulk DMS, where the corresponding parameter increases due to the enhancement of the antiferromagnetic coupling by pressure. Further study is required to clarify the origin of this discrepancy.

Table 1 Fitting parameters of the interface potential

Pressure(GPa)	a_e	T_{if}
1.0×10^{-4}	0.055	3.5
1.11	0.059	3.5
1.46	0.063	3.5

In conclusion, the exciton Zeeman shift caused by the interface effect in a 1.9 nm thick $CdTe/Cd_{1-x}Mn_xTe$ (x=0.24) QW has been described in terms of δ-potential model, taking advantage of the additional red shift, and is found to show a Brillouin function-like behavior saturating above 15 T. It is suggested that the sp-d interaction in the interface region is enhanced by pressure.

The work conducted at the National High Magnetic Field Laboratory was supported by NSF, the State of Florida, and DOE. This work is partially supported by a Grant-in-Aid for Scientific Research on the Priority Area "Spin Controlled Semiconductor Nanostructures" (No. 09244106) from the Ministry of Education, Science, Sports, and Culture. Work at Warsaw was partially supported by Polish Committee for Scientific Research grant PBZ 28.11/P8.

References

1. J.K. Furdyna, J. Appl. Phys. **64**, (1988) R29.
2. J.A. Gaj et al. Phys. Rev. B **50**, (1994) 5512.
3. W.E. Hagston et al. Phys. Rev. B **59**, (1999) 5784.
4. H. Yokoi et al. J. of Crys. Growth **214/215**, (2000) 428.
5. Yu.G. Semenov, Acta Phys. Polon. A, **94**, (1998) 526.
6. S.W. Tozer, unpublished.
7. D. Heiman et al. Phys. Rev. B **35**, (1987) 3307.

Reaction of atomic hydrogen with the Si(100)/SiO$_2$ interface defects

C. Kaneta[1], T. Yamasaki[1], T. Uda[2]

[1] Fujitsu Laboratories Limited, 10-1 Morinosato-Wakamiya, Atsugi 243-0197, Japan e-mail: kaneta@flab.fujitsu.co.jp

[2] JRCAT-ATP, 1-1-4 Higashi, Tsukuba 305-8562, Japan

Abstract The dissociation energy and the reaction path of a H atom terminating a silicon dangling bond (SDB) defect at the Si(100)/SiO$_2$ interface is investigated by employing the first-principles method based on the density functional theory. We found that more than 3.2 eV is required to remove the H atom from the Si-H bond to an interstitial site in SiO$_2$. No clear energy barrier is found for this reaction. The dissociation of the H atom into the bulk Si requires less energy than into SiO$_2$. The possibility of the carrier enhanced dissociation of H from the SDB defect at the Si(100)/SiO$_2$ is also shown.

1 Introduction

The Si dangling-bond (SDB) are the typical intrinsic defects at the Si/SiO$_2$ interface. They have been considered to form electrically active states at the interface which are harmful to the performance and the reliability of the MOS (metal-oxide-semiconductor) devices. The P_{b0} and P_{b1} centers [1] at the Si(100)/SiO$_2$ interface have been known as the SDB-type defects. In order to minimize the effect of the electrically active states, the defects are exposed to hydrogen, the idea being that the H atom will passivate these defects by forming a Si-H bond. Controlling the passivation and depassivation of the SDB-type defects by H atoms is crucial in the fabrication of high quality Si/SiO$_2$ interfaces. However, the dissociation mechanism of the Si-H bond at the Si/SiO$_2$ interface has been discussed only by the analogy of a H atom in the Si crystal and at the surface [2].

Here we present *ab initio* density functional calculations of the interactions of a H atom with the SDB-type defects at the Si(100)/SiO$_2$ interface.

2 Calculation Method

As a model of the Si(100)/SiO$_2$ interface, we employ the structure which consists of tridymite-type SiO$_2$ on Si(100) substrate [3]. The unit cell contains seven Si(100) 2 × 2 layers, two SiO$_2$ layers, H atoms terminating the dangling bonds at the surfaces of SiO$_2$ and Si, and a vacuum layer. This structure has no dangling bonds at the interface. A prototype of SDB defect •Si≡Si$_3$ is introduced into the interface by partly modifying the atomic configuration at the interface as shown in Fig.1. Here, the symbol • denotes a dangling bond. We call this defect SDB0. This includes a •Si-Si dimer and no O atom, which is basically the same structure as a recently proposed model for the P_{b1} center [4]. Then a H atom is introduced near the SDB0 to

investigate the interaction with it. To examine the difference of the behavior of the H atom around the SDB-type defect with a different structure, calculations are also done for the system including another type of SDB defect •Si≡Si$_2$O, which include a •Si-O bond instead of •Si-Si dimer and is labeled as SDB1.

The atomic configurations of the systems including a SDB defect and a H atom are optimized and their electronic structures are calculated with the first-principles method based on the density functional theory [5]. We use a norm-conserving pseudopotential for Si atoms, and ultrasoft pseudopotentials [6] for O and H atoms, respectively. The generalized gradient correction and spin polarization are also taken into account. In the optimization process, two bottom Si layers and the H atoms at the Si surface are fixed.

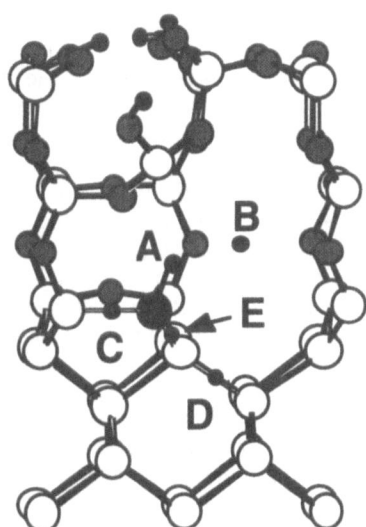

Fig. 1 Various positions of the H atom near the Si(100)/SiO$_2$ interface. Large white circles are Si atoms. Large black one is a defect Si atom. Intermediate gray and small black circles are O and H atoms, respectively. The position with Si-H bond, interstitial site in SiO$_2$, BC sites at the interface and in Si are labeled **A** - **E**.

3 Results and Discussion

Various positions of the H atom near the Si(100)/SiO$_2$ interface are shown in Fig.1. In the configuration with the H-Si≡Si$_3$, which we label SDB0-H, no defect states are found in the band gap region of Si. The length of the Si-H bond is 1.49 Å. It is almost the same as the Si-H bond in the SiH$_4$ molecule, where it is 1.48 Å.

The dissociation behavior of the H atom from the interface SDB defect is expected to be somewhat complicated because of the interaction with the SiO$_2$ network, interface, and Si lattice. Here, we compare the difference of the total energy between both sides of the reaction SDB0-H $\rightarrow$ SDB0 + H for various configurations of the H atom shown in Fig.1.

We find that more than 3.2 eV is required to remove the H atom from the Si-H bond to an interstitial position in SiO$_2$ (site **B**). The corresponding energy is 3.5 eV for the system with SDB1. No clear energy barriers are found in the adiabatic potentials for the Si-H distance. These values are rather high than the activation energy of the Si-H dissociation 2.56 eV which was obtained by the analysis of the experimental data on the Si(111)/SiO$_2$ interface [7].

When the H atom is removed to the bond-center (BC) site of a Si-Si in the bulk Si (site **D**), the increase of the total energy is 2.2 eV. The site **D** is a metastable position for the H atom. The Si-H distances between the bond-centered H atom and the adjacent Si atoms are 1.60 and 1.61 Å, respectively. The Si-H distances and the increase of the total energy are close to the values calculated for the corresponding phenomena in the crystalline Si [8, 9]. The corresponding energy is 2.5 eV for the system with SDB1. The difference between the total energy for the H in the site **B** and **D** is almost the same in both SDB0 and SDB1 systems, because the structural difference between them is only in the O atom adjacent to the •Si atom at the interface.

The position between the two dimered Si atoms at the Si(100)/SiO$_2$ interface (site **C**) is also a metastable one for the H atom. To move the H atom to the site **C** from the site **A**, the required energy is 1.2 eV, which is much smaller than in the above mentioned two cases. The distances from the H atom in the site **C** to the adjacent two Si atoms are 1.64 and 1.74 Å, respectively. The energy barrier of this reaction is 1.5 eV, and that of the reverse reaction is only 0.3 eV in the potential surface for the angle between the Si-H bond and the Si-Si dimer. Thus the reverse reaction can easily occur to form the Si-H configuration again.

The total energies for various configurations of the H atom in the SDB0 system are compared in Fig. 2.

No stable or metastable positions for the H atom are found near the BC site between the defect Si (•Si) and the Si atom in the Si substrate (site **E**) for the SDB0 system. On the other hand, a metastable position of 1.6 eV is found near the site **E** in the system with SDB1.

In the configuration with the H atom at the site **C**, new gap states are generated near the band edge of Si. The doubly occupied state near the valence band edge is related to the defect Si, while the unoccupied state near the conduction band edge is related to the Si-H-Si

structure. These levels can capture carriers. The carrier-enhanced dissociation is one of the possible mechanisms to generate an atomic H.

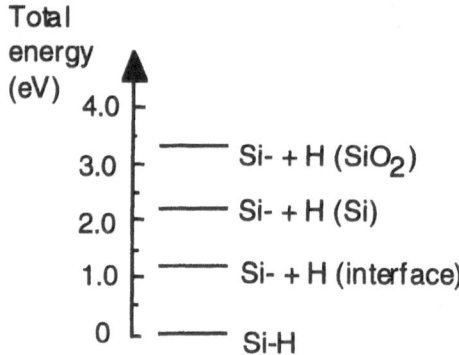

Fig. 2 Comparison of the total energies for various configurations of the H atom in the SDB0 system.

4 Conclusion

We compared the total energies for various configurations of the H atom near the Si(100)/SiO$_2$ interface, and investigated the reaction path for the dissociation of the H atom terminating a SDB defect at the interface, by employing the first-principles method based on the density functional theory. We found that more than 3.2 eV is required to remove the H atom from the Si-H bond to an interstitial position in SiO$_2$. No clear energy barrier was found for this reaction. On the other hand, the dissociation of the H atom into the Si or interface requires less energy than into the SiO$_2$. Our calculation has also shown the possibility of the carrier enhanced dissociation of H from the SDB defect at the Si(100)/SiO$_2$.

This work was supported by NEDO.

References

1. E. H. Poindexter, P. J. Caplan B. E. Deal, and R. R. Razouk, J. Appl. Phys., **52**, 879 (1981).
2. B. Tuttle and C. G. Van de Walle: Phys. Rev. **B59**, 12884 (1999).
3. C. Kaneta, T. Yamasaki, T. Uchiyama, T. Uda, and K. Terakura: Microelectronic Engineering **48**, 117 (1999).
4. A. Stesmans, B. Nowen, and V. V. Afanas'ev: Phys. Rev. **B58**, 15801 (1998).
5. R. Car and M. Parrinello, Phys. Rev. Lett., **55**, 2471 (1985).
6. D. Vanderbilt, Phys. Rev. **B41**, 7892 (1990).
7. K. L. Brower, Phys. Rev. **B42**, 3444 (1990).
8. C. G. Van de Walle and R. A. Street : Phys. Rev. **B49**, 14766 (1994).
9. C. G. Van de Walle, P. J. H. Denteneer, Y. Bar-Yam, and S. T. Pantelides : Phys. Rev. **B39**, 10791 (1989).

Temperature dependence of low frequency noise mechanisms in Schottky barrier structure

J. I. Lee[1,*], I.-K. Han[1], J. Brini[2], A. Chovet[2], C. A. Dimitriadis[3]

[1] Photonics research center, Korea Institute of Science and Technology, Seoul 130-650, Korea
[2] LPCS, UMR 5531 CNRS, ENSERG, Rue des Martyrs, BP 257, 38016 Grenoble, France
[3] Department of Physics, University of Thessaloniki, 54006 Thessaloniki, Greece

Abstract Temperature dependence of different low frequency noise generation mechanisms are compared quantitatively as a function of doping concentration, bulk and/or interface trap density for Schottky barrier structure. Non-uniform as well as uniform distribution of traps is considered.

1 Introduction

Schottky barrier structure is important as device structure and for noise spectroscopy. Schottky diode, one of the oldest semiconductor devices, is still extensively used in microwave circuits. Younger generation of semiconductor electronic devices such as metal-semiconductor field effect transistors (MESFETs), modulation-doped field effect transistors (MODFETs), poly-crystalline devices, and optical devices such as metal-semiconductor-metal (MSM) photodiodes, etc., also have Schottky barrier along and/or near the current path in the device structure.

Schottky barrier provides favorable circumstance for noise spectroscopy where band bending due to the barrier exposes the energetic and/or spatial distribution of the traps to the Fermi level. Those half-filled traps at around the Fermi level are important in generation of number fluctuation and the noise measurements can give useful information on the traps involved.

Low frequency noise is also important for the performance of the devices when they are used as mixer or oscillator where the low frequency noise is known to be upconverted to microwave frequency range and dominate the phase noise.

We scrutinize all the possible mechanisms for generation of low frequency excess noise, both number fluctuation and mobility fluctuation, and their temperature dependence, and show how they can be utilized as a spectroscopy tool to investigate the nature utilized as a spectroscopy tool to investigate the nature

2. Noise generation mechanisms

In general, noise generation mechanisms can be classified into two categories, number fluctuation and

mobility fluctuation. In number fluctuation theory for semiconductor devices, the necessary distribution of the time constants can be obtained in three ways, via thermal activation, tunneling [1], and random walk [2] of charge carriers. For Schottky barrier diode, Hsu first proposed thermal activation [3] and tunneling [4] models involving bulk traps. Random walk of electrons at the metal-semiconductor interface involving interface states has been considered rather recently [5,6].

2.1 Thermal activation and tunneling

In thermal activation and tunneling model the fluctuation in the occupancy of bulk traps in the depletion region modulate the barrier height to result in current noise intensity. Since the current depends on the barrier height exponentially, the current noise S_I is proportional to the barrier height fluctuation S_Φ such that [7],

$$S_I = (qI_F/kT)^2 S_\Phi, \tag{1}$$

where I_F is the forward current and kT is thermal energy.

The barrier height fluctuation is from the Schottky barrier lowering or image force-induced barrier lowering involving trapping and detrapping of bulk traps in the depletion layer.

Uniform distribution of bulk traps with linear approximation for the band bending provides the necessary distribution of time constants for 1/f noise. The barrier height fluctuation due to thermal activation, $S_{\Phi a}$ and due to tunneling, $S_{\Phi u}$ are given, respectively, as follows,

$$S_{\Phi a} = (1/f)(kT/4\varepsilon_s)^2 (qD_t/4\pi N_d A(V_D-V_F)), \tag{2}$$

and

$$S_{\Phi u} = (1/f)(q/4\pi\varepsilon_s)^2 (qD_t/3\pi\varepsilon_s\gamma^2 WA), \tag{3}$$

where V_D is the diffusion voltage corresponding to the barrier height, V_F the forward bias, kT the thermal energy, D_t the bulk trap density, N_d the doping concentration, A the area, $\gamma^2 = (4\pi q/3h)^2(2m^*N_dW/\varepsilon_s)$ the quantity related to the tunneling constant, and

---

* *Present address:* Basic S&T Program, KISTEP, 9F Dongwon Industry Bldg., 275 Yangjae-dong, Seocho-gu, Seoul 137-130, Korea e-mail: jil@kistep.re.kr

$W=(2\varepsilon_s(V_D-V_F)/qN_d)^{1/2}$ the depletion layer width [4].

The resulting current noise is independent of temperature for thermal activation, or proportional to $1/T^2$ for tunneling. The ratio is [8],

$$\Gamma_{ua}\equiv S_{Iu}/S_{Ia}=(q/kT)^2 N_d(3h/4\pi)^2/3 \; m^*\varepsilon_s, \qquad (4)$$

The ratio only depends on the temperature and the doping concentration. For tunneling mechanism to be dominant at room temperature we need doping concentration larger than 7.5×10^{18}/cm^3. For normal doping, the tunneling mechanism* can be important at low temperatures.

2.2 Random walk of electrons

Random walk of electrons at the interface also modulates the barrier height [6,9]. The barrier height fluctuation due to the random walk of electrons involving the interface states is given by,

$$S_{\phi r}=(G/f)(q/4\varepsilon_s)^2(kTD_{it}/ 4\pi N_d AW), \qquad (5)$$

where G is the distribution constant of the time constants for random walk and has a value of practically about 0.1.

The ratio of current noise intensities due to random walk and thermal activation can be calculated as [6],

$$\Gamma_{ra}\equiv S_{Ir}/S_{Ia}=4G(q/kT)(V_D-V_F)(D_{it}/D_tW). \qquad (6)$$

Note that random walk involving interface states can dominate if the interface states density D_{it} is comparable to or larger than the effective areal density of the bulk traps, D_tW, at room temperature and become more dominant at lower temperatures.

In a limiting case where the current is dominated by the surface current (ideality factor near 2) the noise intensity is directly related to the fluctuation of occupancy of interface states density [5,10].

2.3 Non-uniform distribution

Nonuniformity in energetic or spatial distribution of the bulk traps will result in the variation of the exponent of the frequency from unity [11-13]. Within thermal activation model, one can derive a simple relation between the energy distribution constant, E_0 and the exponent of the frequency, η such that, $\eta=1-(kT/E_0)$, and extract E_0 from the measured η [13].

2.4 Mobility and diffusivity fluctuation

Modbility and diffusivity fluctuation in Schottky barrier structure were first studied by Kleinpenning [14] and later corrected by Luo et al. [15] which resulted in

the current noise proportional to the current such that,

$$S_{Im}=(\alpha I_F/12\pi f)[kTm^*\varepsilon_s/\pi qN_d(V_D-V_F)]^{1/2}(q/\mu m^*)^2, \qquad (7)$$

where α is Hooge parameter. The critical current where the transition from linear to quadratic current dependence occurs will decrease with the temperature.

By examining the critical current one can get useful information about how certain process conditions affect bulk and/or interface properties in Schottky barrier structures [6,9,16].

References

1.   Van der Ziel, Proc. IEEE, **76**, (1988) 233.
2.   L. Bess, Phys. Rev., **91**, (1953) 1569.
3.   S. T. Hsu, IEEE Trans. Electron Devices, **ED-17**, (1970) 496.
4.   S. T. Hsu, IEEE Trans. Electron Devices, **ED-18**, (1971) 882.
5.   O. Jaentsch, IEEE Trans. Electron Devices, **ED-34**, (1987) 1100.
6.   J. I. Lee, J. Brini, A. Chovet, and C. A. Dimitriadis, Solid-St. Electron., **43**, (1999) 2185.
7.   C. Meva'a, X. Letartre, P. Rojo-Romeo, and P. Viktorovitch, Solid-St. Electron., **41**, (1997) 857.
8.   J. I. Lee et al., J. Kor. Phys. Soc., in press.
9.   J. I. Lee et al., Appl. Surf. Sci., **142**, (1999) 390.
10. F. V. Farmakis et al., Semicond. Sci. Technol., **13**, (1998) 1284.
11. J. I. Lee, J. Brini, A. Chovet, and C. A. Dimitriadis, Solid-St. Electron., **43**, (1999) 2181.
12. F. Z. Bathaei and J. C. Anderson, Philos. Mag. B, **55**, (1987) 87.
13. C. A. . Dimitriadis et al., J. Appl. Phys., **84** (1999) 3934.
14. T. G. M. Kleinpenning, Solid-St. Electron, **22**, (1979) 121.
15. M. Luo, G. Bosman, van der Ziel, and L. L. Hench, IEEE Trans. Electron Devices, **35**, (1988) 1351.
16. C, A. Dimitriadis et al., J. Appl. Phys., **84** (1999) 4238.

Interfacial transition regions of gate dielectrics in advanced silicon devices*

G. Lucovsky[1], J.C. Phillips[2], M.F. Thorpe[3]

[1] Department of Physics, North Carolina State University, Raleigh,
   NC 27695-8202, USAe-mail: gerry_lucovsky@ncsu.edu
[2] Lucent Bell Labs, Murray Hill, NJ 07974, USA
[3] Department of Physics and Astronomy, Michigan State Univ.,
   East Lansing, MI, 48824 ,USA

Abstract This paper addresses two different aspects of interfacial transition regions that apply to qualitatively different classes of replacement gate dielectrics that are under consideration for advanced silicon devices. The interfacial issues for SiO_2, Si_3N_4 and silicon oxynitride alloys are defects due to mechanical bond strain that derives from bond coordination constraints, whereas for interfaces with high-k dielectrics such as Al_2O_3 and Zr and Hf silicate alloys, the most important interfacial issues derive from their increased bond-ionity relative to SiO_2.

1 Introduction

The scaling of lateral device dimensions of Si field effect to less than 100 nm requires replacement of thermally-grown SiO_2 by deposited dielectrics with higher dielectric constants than SiO_2. This has stimulated much research on alternative thin film elemental and binary oxides, nitrides and oxynitride alloys with dielectrics higher than that of SiO_2. This paper presents analysis of the factors that are important in interfaces between these replacement dielectrics and either crystalline Si substrates or nitrided SiO_2 interfacial layers.

Since the replacement of thermally-grown SiO_2 with deposited dielectrics will proceed in two steps, first with deposited silicon nitride and oxynitride alloys, and then with metal oxides, and transition metal oxides, silicates, and aluminates, it is beneficial to place these materials into a prespective that derives from microscopic aspect of their bond chemistry.

2 Classification scheme for alternative dielectrics

An empirical classitication scheme based on the concept of bond ionicity of Pauling is introduced. If X(O) is the *Pauling electronegativity* of oxygen, 3.44, and X(Si) is the *Pauling electronegativity* of silicon, 1.90, then ΔX, the electronegativity difference of 1.54 is a convenient metric for characterizing bond-ionicity, I_b. Using the definition of bond-ionicity of Pauling, $I_b = 1- \exp(-0.25(\Delta X)^2)$, and is ~ 45 % for Si-O bonds. Dielectrics with $\Delta X < 1.6$, such as SiO_2, B_2O_3, As_2O_3, P_2O_5, $As_2S(Se)_3$, are generally good glass formers and have continuous random network, crn, bonding structures. In addition, they meet the Phillips criteria [1] for the average number of bonds/atom, N_{av}, and average number of constraints per atom, C_{av}, for continuous random networks with low densities of mechanical strain induced defects, e.g., C_{av} is less than or equal to 3. There is a second glass of good glass formers, the fluoride glasses that are considerably more ionic, including BeF_2 ($\Delta X = 2.41$) and ZnF_2 ($\Delta X = 2.33$). with bond ionicities greater than 70 %. Bonding coordination is essentially the same as SiO_2 with $N_{av} = 2.67$; however, since the bonding is not covalent, a valence force field model is not appropriate, and the constraint theory of Ref. 1 cannot be applied. A better way to describe these glasses is as random closed packed, rcp, *ionic structures* [1].

Another class of glasses with partial ionic character are the silicates i.e., binary alloys of SiO_2 and metal oxides such as Na_2O, CaO, Al_2O_3, ZrO_2, PbO, etc. In these silicates, the metal atoms are network modifying ions that are compensated by terminal, negatively oxygen atoms that are covalently bonded to Si-atoms the Si-O network [2]. The bonding in silicates is amphoteric with the Si-O network being the acidic negatively charged component, and the metal ion the basic positively charged component. Since metals are generally less electronegative than silicon, the *average* electronegativty difference will be greater than that of SiO_2, but less than that of the metal oxide. Since homogeneous as-deposited thin films can be prepared with compositions up to about 50 atomic % metal oxide, ΔX approach 2. As an example, ΔX for alloys of 70 atomic % SiO_2 and 30 atomic % or ZrO_2 and La_2O_3 are 1.74 and 1.81, respectively.

Based on i) large differences in ΔX and bond ionicity between the crn glasses, and rcp glasses, and ii) intermediate values of I_b in silicate glasses, non-crystalline oxides can be grouped into three classes according to ΔX or I_b: i) continuous random networks, crn's, with $\Delta X < 1.6$ and $I_b < 48$ %, ii) modified crn's or mcrn's, with ΔX between 1.6 and 2.0, and I_b between about 48 and 65 %, and iii) rcp ionic structures with $\Delta X > 2.0$ and $I_b > 65$ %. The boundaries between these structures are not *rigid*, but are representative of ionicities at which there are significant changes in bonding structure and microstructure. There are a small number of elemental oxides with ΔX between ~1.6 and ~2.0, that can also be characterized amphoteric and fall into the mrcn group.

3 Interfacial limitations for silicon based dielectrics

The limitations on the performance and reliability of field effect transistors with SiO_2, and Si nitride and oxynitride gate dielectrics with EOT < 2.5 nm have been shown to derive primarily from strain-induced bonding defects at the Si-dielectric interfaces. Mechanical bonding constraints at the interface have been characterized in terms of the average number of bonds per atom in the interfacial region [3]. The interfacial bonding structure is *defined* by 0.5 molecular layers of Si (0.5 atoms and two bonds), and

1.5 molecular layers of the dielectric film (SiO_2 or Si_3N_4). Interface nitridation has been taken into account by inserting one atomic layer of nitrogen between the Si substrate and SiO_2 layer. The Si-SiO_2 interface is used as a reference interface and is characterized by an average number of interfacial bonds/atom, $N_{av}^* = 2.86$. Using the model interface bonding model described above, N_{av} equals 3.47 for a Si-Si_3N_4 interface, and 2.89 for a monolayer nitrided Si-SiO_2 interface. The concentration of defects relative to the Si-SiO_2 interface scales as $(N_{av} - N_{av}^*)^2$. Based on this scaling the defect concentration at a Si-Si_3N_4 interface is anticipated to be more than 3 orders of magnitude higher than at a Si-SiO_2 interface. This is exactly what has been found in experimental studies of these interfaces. Introduction of ultra-thin SiO_2 interface layers between the Si and Si_3N_4 reduces the density of defects by about three orders of magnitude, but introduces strain induced defects in the form of fixed positive charge at the internal SiO_2-Si_3N_4 dielectric interface [4]. These defects shift threshold voltages of devices, but do not significantly degrade channel transport properties.

4 Interfacial limitations for metal oxides and silicates

In marked contrast, the average number of interfacial bonds/atom at a Si-dielectric interface of the second class, where the dielectric is an mcrn oxide or silicate is not sufficiently high to induce defects. However, a second mechanism comes into play that is directly related to the ionic character of the dielectric. Interfaces between Si and Ta_2O_5, Al_2O_3 and Zr silicate alloys were formed by remote plasma processing. In-line studies of interface chemistry indicated no detectable oxidation to a limit of ~ 0.5 nm during film deposition. Capacitance-voltage studies indicated fixed charge at these interfaces [5]. The sign of the charge is correlated with the charge on the metal atom of the dielectric. Each Si atom of the substrate of SiO_2 interface layer makes a covalent bond with O, however, the second neighbors to this Si atom are qualitatively different: i) the negatively charged network component of Al_2O_3, the positively charged network component of Ta_2O_5, and a combination of Zr^{4+} ions and neutral Si for the Zr silicate alloys. The Si and SiO_2 are covalently bonded, whereas the dielectric bonding at the interfaces is significantly more ionic, and analogous to the inherent heterovalency of Ge-GaAs interfaces on *polar* GaAs (100) or (111) faces [6]. The magnitude of fixed charge is significantly lower than the ion concentration at the interface. This can be explained in two ways, it either i) represents the effective charge of an interface dipole, or ii) is reduced by statistical fluctuations of oppositely charged species in the interface region. Since heterovalent bonding can lead to high densities of interfacial fixed charge, it is necessary to identify new high-k dielectrics that can provide isovalent interfaces with Si or with SiO_2. Our research suggests two possible approaches for achieving these interfaces that are based on the isovalent character of Ge-GaAs interfaces on *non-polar* GaAs (110) faces [6]. Based on the discussion presented above, three

aluminate alloy systems, which fall into the mcrn group have the potential for neutral bonding at Si-dielectric interfaces. The alloys are based on the AlO_2^{1-} negative network forming bonding group, which can be combined with a positive network forming group, such as TaO_2^{1+}, as in $TaAlO_4$, or with more electropositive group III and IV atoms such as Y and La, and Zr, and Hf, in stochiometric compositions, including $Y(La)(AlO_2)_3$ and $Zr(Hf)(AlO_2)_4$. Finally, the effects of statistical fluctuations in the neutral interface bonding of these aluminate alloys or compounds on channel transport must be determined.

The results obtained for GaN-dielectric interfaces lend support to the model [7]. The electron occupation of Ga dangling bonds on polar faces of GaN is *fractional* rather than *integral*, resulting in an inherent mismatch between electronic and nuclear charge for bonding to deposited dielectrics such as SiO_2. To form device-quality GaN MOS interfaces, a redistribution of electronic charge in the Ga atom dangling bonds that is *integral* is needed. This takes place during the plasma-assisted oxidation that forms a thin interfacial Ga_2O_3 layer. The interfacial Ga_2O_3 layer provides a substrate for deposition of SiO_2. Bonding at the internal dielectric interface is considerably more isovalent and low defect densities result.

5 Conclusions

Based on studies of i) Si_3N_4 and silicon oxynitride alloys, and ii) the metal oxides and silicates on Si and SiO_2 interface layers, it has been found that chemical bonding constraints are qualitatively different at these two types of Si-dielectric interfaces, but still lead to interface defects that will adversely effect device scaling based on substitution of deposited dielectrics other than SiO_2. In many cases insertion of an ultra-thin nitrided plasma-grown SiO_2 layer provides a marked reduction in defect density, typically about two orders of magnitude. However, interfacial layerss contribute about 0.35 nm of 'equivalent oxide thickness', or EOT to the dielectric. This limits ultimate device scaling by reducing the physical thickness of the higher-k oxide or silicate by a thickness in nm of the order of 0.9 k. This thickness reduction is equivalent to an increased direct tunneling current of more than three orders of magnitude.

*Supported by ONR, AFOSR and SEMATECH/SRC.

1. J.C. Phillips, J. Non-Cryst Solids 34 (1979) 153; 43 (1981) 37.
2. R. Zallen, *The Physics of Amorphous Solids* (Wiley-Interscience, New York, 1983).
3. G. Lucovsky, Y. Wu, H. Niimi, V. Misra and J.C. Phillips, Appl. Phys. Lett. 74 (1999) 2005.
4. G. Lucovsky et al., J. Vac. Sci. Tech. 18 (2000) 1745.
5. G. Lucovsky et al., AIP Conf. Proc. (in press).
6. W.A. Harrison, et al., Phys Rev B 18 (1978) 4402.
7. R. Therrien R G. Lucovsky and R.F. Davis, Physics Status Solid 176 (1999) 793.

ESR and magnetic measurement of Ni/GaAs composite system

Y. Seino, S. A. Haque, A. Matsuo, Y. Yamamoto, S. Yamada, H. Hori

School of Materials Science, Japan Advanced Institute of Science and Technology (JAIST), 1-1 Asahidai, Tatsunokuchi, Ishikawa 923-1292, Japan e-mail: seino@jaist.ac.jp

Abstract Ni films prepared by use of n-GaAs(001) substrate grown by molecular beam epitaxy(MBE) method in order to investigate their magnetic coupling and composite effect. Magnetization and ESR for these samples have been observed down to liquid He temperature. The Ni film made by MBE shows drastic change in characteristics on hysteresis and high field susceptibility. Besides usual ferromagnetic resonance(FMR) signals, weak but characteristic signals related with the interface and domain wall are observed.

1 Introduction

Hybrid ferromagnet-semiconductor system have recently stimulated much interest, both for purely scientific and application point of view. In the previous work[1], the anomalous magnetic domain effect of submicron-size Ni stripes on GaAs substrate was reported. To explain the structure, we proposed the model of quantum mechanical exchange interaction between Ni and GaAs substrate through the interface. Such an exchange coupling has been also theoretically proposed[2]. If such an exchange interaction is coupled with the electronic states in GaAs, it can be expected that some magnetic effects are likely to be introduced into the long coherent state in GaAs. To investigate the magnetic coupling and the composite effects of Ni films on epitaxial n-GaAs(001), ESR and magnetization were measured.

2 Sample preparation

Ni and GaAs films were grown in an ultrahigh vacuum chamber using a molecular beam epitaxy (MBE) deposition system in a base pressure of 2×10^{-10}Torr. The growth chamber was equipped with a reflection high-energy electron diffraction (RHEED) unit and electron-beam (EB) evaporator (for Ni) and effusion cells (for Ga, Ge and As$_4$ flux sources). Prior to insert the growth chamber, a GaAs wafer was prepared by sulfuric solution etch and rinsed deionized water. After oxide desorption, doped and nondoped GaAs layers were grown on a 1μm thick buffer layer. The process of epitaxial growth and the surface reconstruction were confirmed by taking the in-situ RHEED images. The doped layer had a carrier concentration of $\sim 10^{18}$cm^{-3}, which generated the electrons from pure Ge. After growing the cap GaAs layer on n-GaAs, ferromagnetic Ni film was deposited in the same growth chamber by EB evaporator in a pressure less than 10^{-9} Torr, to keep the GaAs surface clean. After removal of excess As atoms form the surface, Ni film was deposited. This method was different from the usual procedure and confirmed us to get a flat interface

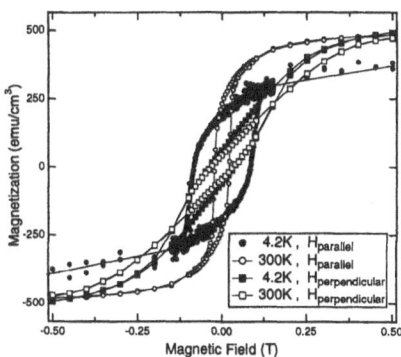

Fig. 1 Magnetization data of Ni film on epitaxial GaAs film made by MBE system.

between Ni and GaAs. It also confirmed contamination free interface.

3 Experimental Results and discussion

Figure 2 and figure 3 show overall profiles of ESR(X-band) in the fields parallel and perpendicular to the surface. Besides the strong FMR signals, quite weak spectra were also observed at the position of A and B. Such weak signals were not observed in other Ni film and bulk state of Ni.

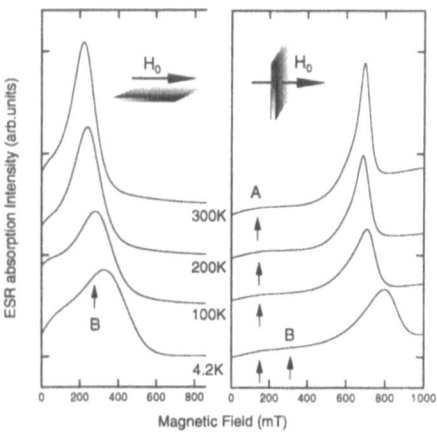

Fig. 2 Temperature dependence of ESR absorption spectra in both parallel and perpendicular field.

The profile of the signal in the parallel field looks like two peak spectra. The low field peak obtained by decomposition is at around 0.12 T. This lower peak position corresponds to the top of the hysteresis loop shown in Fig.1. Therefore, the peak is not intrinsic, although demagnetizing effect of domain structure makes absorption peak in low field. The peak in higher field is considered to be a typical FMR in single phase. This signal is in the high field susceptibility region in Fig.1. Assuming $g=2.1$ which is g-value of bulk Ni using demagnetization field estimated from experimental magnetization, the signals are consistent with the theoretical formulae of FMR.

The weak spectra have following characteristics:

1. the weak spectra A and B are observed at about 153 and 320 mT and they correspond to the position of $\Delta S_Z=2$, and 1 transitions, respectively.
2. A is forbidden transition with $\Delta S_Z=2$ and B is at the position of usual allowed transition with $\Delta S_Z=1$.
3. g-values of both A and B are approximately 2.1. Demagnetizing effect can be ignored in both spectra.
4. A is observed only under the perpendicular field, while B is observed in both field configurations.
5. A and B do not show temperature shift, although FMR signals show clear temperature shift because of the change in magnetization.
6. B is not observed in high temperature.

The intensities of A and B are in the order of 0.01% in comparison with the FMR signals. Such a small intensity ratio makes us infer that their origin is domain wall of Ni or the interface between Ni and GaAs surface. Because the ratio of constituent spin number is same order. In fact these signals are not observed in bulk Ni. These are not simple impurity resonance, because the width of about 20 mT on A and B is much broader than the width ~1 mT of usual impurity resonance. It is noticed that A and B are observed only on Ni film made by MBE and could not observed in the samples made by usual evaporation method. Because of their experimental results, the characteristics on these ESR signals are related with the magnetization process. According to the magnetization curve in Fig.1, the samples made by MBE have quite large hysteresis loop and wide range of high field susceptibility above the loop.

Because A is observed in hysteresis loop, origin of A is considered to be directly related to the magnetic domain. Most typical ESR in the ferromagnetic domain is soliton ESR. Especially, such an excitation does not have demagnetizing field[3], because the spin change in ESR is produced only by shift of the domain wall. Such a soliton ESR should be at the position of $\Delta S_Z=1$ with $g=2.1$, but the field is much higher than the hysteresis loop. However, if a domain consists of two domain walls, two soliton can be excited in both edges of the domain. Such double quantum transition with $\Delta S_Z=2$ can be produced in the hysteresis loop in perpendicular field. The ESR, however, should not be observed in parallel field, because the hysteresis curve is lower than the resonance field in the configuration. Thus, A should be assigned as double quantum ESR in the ferromagnetic domain wall or soliton ESR. Spectrum B is observed above the hysteresis curve and it is in high field susceptibility region at low temperature. However, in high

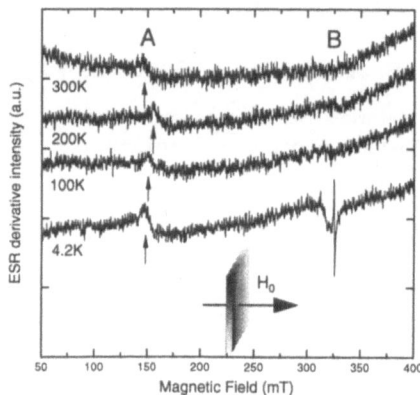

Fig. 3 ESR derivative intensities at the peaks of A and B in perpendicular field.

temperature, the high field susceptibility region becomes much smaller and B is not observed. Thus, B is strongly related with high field susceptibility. As is discussed in ref.[4], the high field susceptibility is induced by antiferromagnetic(AF) interaction between Ni and GaAs substrate.

High field susceptibility means magnetization reduction by AF interaction. Spin state of Ni at the interface is generally different from Ni states inside the Ni-film. The spins at the interface can be responsible for the ESR transition from high field susceptibility state to ferromagnetic saturation state. This mechanism is extended the higher order ESR theory. Detail model on B which satisfy the demagnetization effect, however, has not been given on clear experimental result. Following can be speculated as an example: transition from $S_Z=-1$ to $S_Z=+1$ on spin at the interface is excited together with change in $\Delta S_Z=-1$ of the spin inside the film. The demagnetization effect on the interface spin cancels the effect of the inside spin. This mechanism explains the reason of weak intensity and also condition of (6) as follows: region of high field susceptibility in low temperature includes the resonance fields in both field configurations. But in high temperature, the susceptibility region shrinks extremely and resonance field is out of the region. More detailed information by other experiment such as NMR make the problem clear.

This work was partly performed through the financial support by a Grant-in Aid for Scientific Research from the ministry of Education, Science, Sports and Culture of the Japanese Government.

References

1. M.Tona, Y.Seino *et al*, Proc. of the 24th ICPS, (Jerusalem, 1998), in CD-ROM.
2. V.L.Korenev, Phys. Solid State **38**, (1996) 502.
3. J.P.Boucher, L.P.Regnault *et al*, Sold Srare Comm. **33**, (1980) 171.
4. S.A.Haque, A.Matsuo *et al*, J. Magn. Magn. Mater., to be published.

Imaging of Friedel oscillations at epitaxially grown InAs(111)A surfaces using scanning tunneling microscopy

K. Kanisawa[1], M. J. Butcher[1,2], H. Yamaguchi[1], Y. Hirayama[1,3]

[1] NTT Basic Research Laboratories, 3-1 Wakamiya, Morinosato, Atsugi, Kanagawa, 243-0198, Japan
[2] School of Physics and Astronomy, University of Nottingham, Nottingham NG7 2RD, United Kingdom
[3] CREST-JST, 4-1-8 Honmachi, Kawaguchi, Saitama, 331-0012, Japan

Abstract The low-temperature scanning tunneling microscopy study of the local density of states (LDOS) was performed at the epitaxially grown InAs surface on the GaAs(111)A substrate. In the simultaneously mapped dI/dV images, LDOS oscillation patterns were clearly imaged and showed bias voltage dependence at surface defect sites. Clear oscillatory LDOS patterns made of concentric circles were observed due to Friedel oscillations at an isolated defect. It is found that the main wave feature is determined by semiconductor two-dimensional electron gas (2DEG) states in the surface accumulation layer. In the semiconductor nanostructures, symmetric and regular patterns are observed. Furthermore, zero-dimensional states were observed in the small triangular cages.

1 Introduction

Since the transport properties are dramatically affected by small defect concentrations in semiconductors, quantum mechanical understanding of electron scattering and interference is one of essentially important approaches in mesoscopic regime [1–4].

Recently, a scanning tunneling microscopy (STM) [5] and the application to spectroscopy in ultra-high vacuum (UHV) make it possible to explore nanometer-scale electron behavior directly in real space [6–20]. Especially on the noble metal surfaces, clear standing waves are observed at step edges [7–9], point defects [7,8] within metallic cages [10–12] and in quantum corrals [13,14]. Though there are several arguments to interpret such standing waves [13–16], it reaches a consensus that these waves can be attributed to the electron density oscillations due to long range screening phenomena [21] around perturbation potentials at surfaces.

On the semiconductor surfaces, the only clean surfaces to show interference effects have been three-dimensional electron gas (3DEG) at cleaved (110) surfaces [17,18] and one-dimensional electron gas (1DEG) at the Si (001) surface [19,20]. To date, there is no report on semiconductor two-dimensional electron gas (2DEG). Such semiconductor 2DEG is, however, especially important, because it is not only an ideal two-dimensional system but also a core of a field-effect transistor. In this paper, we report a low-temperature STM (LT-STM) study of 2DEG local density of states (LDOS) in the electron accumulation layer using an epitaxially-grown InAs(111)A surface. Furthermore, interference of 2DEG

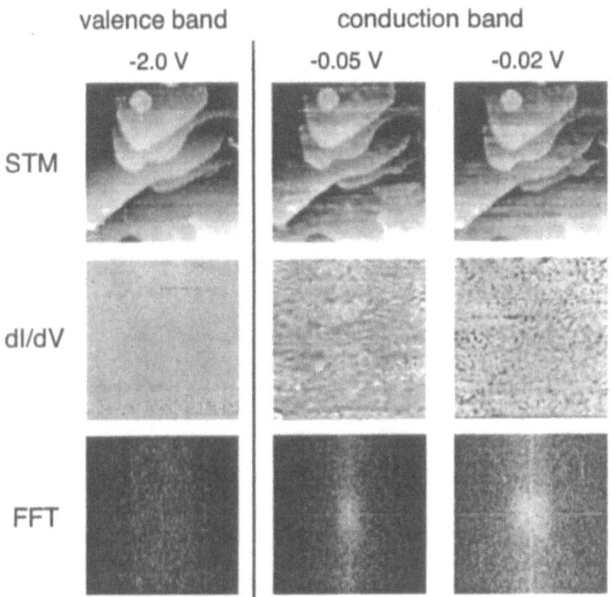

Fig. 1 LT-STM and dI/dV images of occupied states at bias voltages -2.0 V, -0.05 V and -0.02 V at 5.3 K. Each area size is 268 nm × 268 nm. Each FFT pattern was transformed from each dI/dV image. The slightly elliptical shape of FFT spectrum is due to a distortion of dI/dV image caused by insignificant drift during scanning. The vertical lines in FFT patterns are caused by horizontal line noise in the images.

states and the zero-dimensional quantum size effect in semiconductor nanostructures are demonstrated.

2 Experiments

A 100 - 200 nm thick InAs layer was grown on an n-GaAs(111)A substrate by molecular beam epitaxy (MBE). The (111)A surface is advantageous as it has an atomically smooth semiconducting (2×2) reconstruction and a surface electron accumulation layer. The accumulation layer is due to surface Fermi level (E_F) pinning above the bottom of the conduction band (E_c) [22–24]. This unusual electronic properties depend on the charge neutrality level above E_c [25] and surface E_F pinning to the level probably due to unidentified donor-type surface states [26]. This 2DEG formation at the InAs(111)A surface very much differs to that of the 3DEG dominant feature at E_F unpinned clean (110) InAs surface [18,27,

(a)

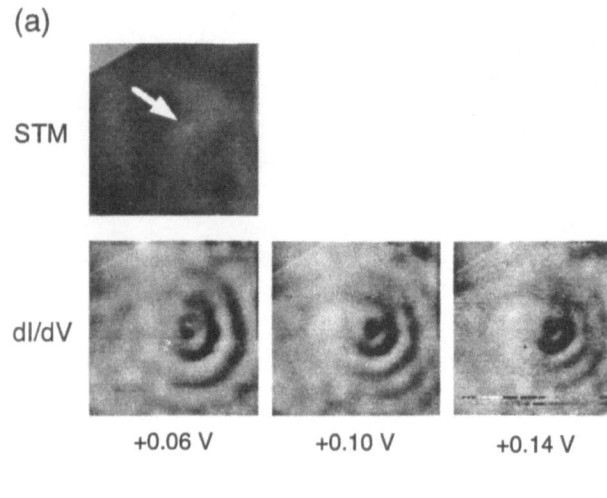

(b)

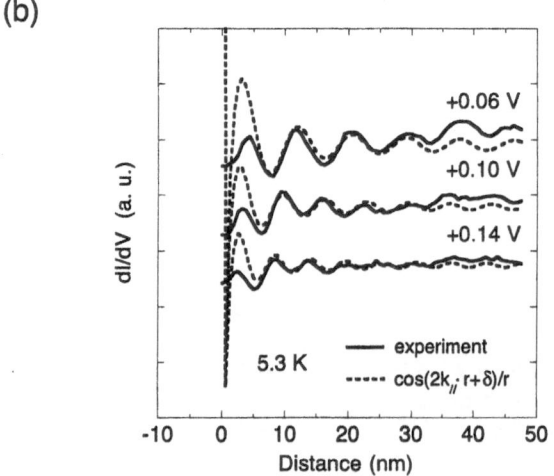

Fig. 2 (a) Constant current 67 nm x 67 nm STM image of the InAs(111)A surface and a simultaneously mapped dI/dV images of unoccupied states taken at 5.3 K. In the dI/dV images, circular LDOS standing waves around a point scatterer (shown by an arrow) due to Friedel oscillation show bias voltage dependence. (b) The measured LDOS oscillation amplitude profiles (solid lines) at a point scatterer in Fig. 2(a). Simulation curves (dashed lines) are fitted assuming 2DEG LDOS oscillations.

28]. The InAs growth proceeds in a layer-by-layer growth mode even on a 7.2 % lattice-mismatched GaAs(111)A substrate [29].

After the growth, the sample was transferred to the LT-STM stage and cooled. The LT-STM measurements were performed in ultra-high vacuum (UHV) below 10^{-10} mbar. Topographic images were obtained in constant current STM mode at 5.3 K. We applied a modulation (5 - 7 mV r.m.s. at 400 - 600 Hz) to the tunneling bias voltage during the topographic imaging and detected differential conductance, dI/dV, signal simultaneously using a lock-in technique. In case of this dI/dV measurement method, the dI/dV image is proportional to the sample surface LDOS map with very small deviations [10, 31–33]. The brighter regions in dI/dV images

are due to a higher LDOS. Through our experiments, the E_F pinning position is confirmed to be alway about 0.2 eV above E_c using scanning tunneling spectroscopy (STS). This value is comparable to a reported value of 0.21 ± 0.05 eV [23] on the MBE-grown InAs(111)A surface.

3 Results and Discussions

3.1 Imaging of occupied LDOS

Occupied state STM and dI/dV images taken at the negative bias voltages are shown in Fig. 1. In the STM images, an identical surface topography was always observed at each bias voltage. There was almost no feature in dI/dV image at -2.0 V (tunneling from the valence band). However, there were complex wavy corrugation patterns in the dI/dV images at -0.05 V and -0.02 V (tunneling from the conduction band). When we analyze the dI/dV images by taking fast Fourier transforms (FFT), it is found that larger wave number were contained in the image taken at higher bias voltage. The deferent appearance in the dI/dV images for the conduction band from that for the valence band is explained in terms of the different character of electronic states in each band. Electrons in the conduction band are delocalized and behave as free electron waves. However, electrons in energy levels much lower than the top of valence band (E_v) are concentrated in the bonds and overlap only with those of the nearest neighbors. The LDOS of such states do not show clear oscillatory patterns.

We numerically calculated Schrödinger's equation and Poisson's equation consistently and compared calculation results with experimental values. The existence of the two-dimensional subband at the position about 0.07 eV below E_F is found to be necessary to explain the experimental results. The two-dimensional surface electron accumulation layer is obtained to consist of more than 90 % 2DEG and less than 10 % 3DEG. This suggests

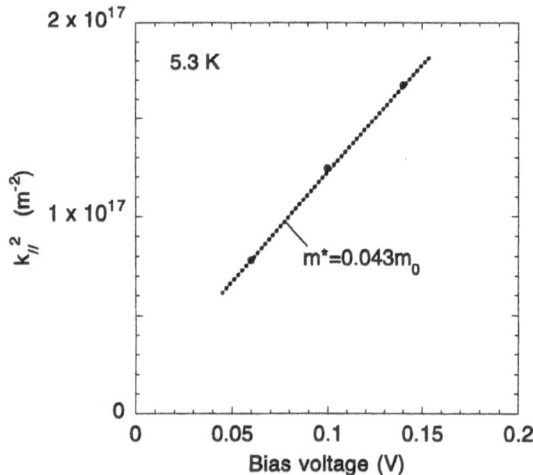

Fig. 3 Bias voltage dependence of lateral wave number square calculated from the oscillation periods in Fig. 2(b).

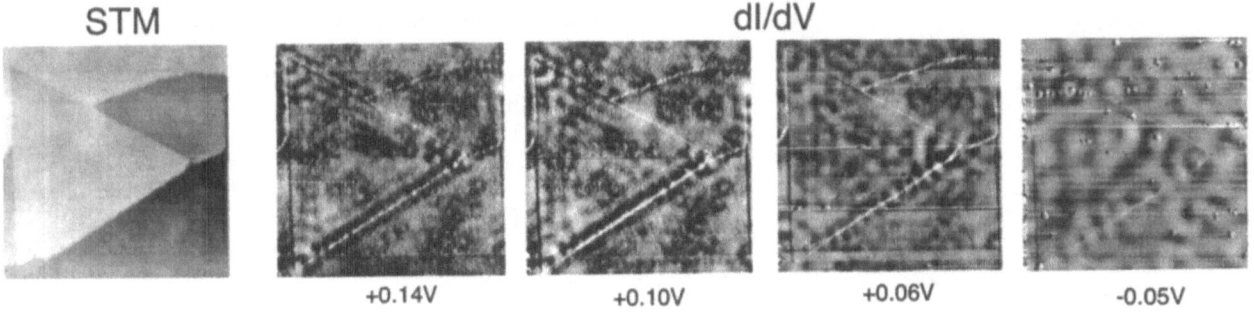

Fig. 4 Constant current STM image of the stacking fault tetrahedron observed in 134 nm × 134 nm area at 5.3 K. Simultaneously mapped dI/dV images at bias voltages between -0.05 V and +0.14 V are also shown. A lot of small adsorbates scattered on the surface at the bias voltage -0.05 V were dropped off from the tip during imaging.

that the 2DEG LDOS corrugation is mainly mapped in the dI/dV images. It should be noted that such occupied 2DEG LDOS were considerably perturbed at surface state sites and often showed complicated standing waves.

3.2 Imaging of unoccupied LDOS

In Fig. 2, a point defect, probably arises due to a threading dislocation, is shown. In each dI/dV images taken at positive bias voltages, concentric circular oscillation is observed at the defect. Oscillation period becomes shorter with increasing tunneling bias.

The amplitude of each LDOS oscillation in Fig. 2(a) is shown by solid line in Fig. 2(b). The dashed lines indicate calculation results assuming 2DEG LDOS oscillation $\Delta LDOS(E, \mathbf{r}) \propto \cos(2\mathbf{k}_{//} \cdot \mathbf{r} + \delta)/r$. Here, $\mathbf{k}_{//}$ is the wave vector parallel to the surface at an electron energy E, $\mathbf{r}$ is position vector from the scatterer, and δ is scattering phase shift, respectively. The experimental results are consistently explained. This comparison suggests that the measured dI/dV oscillation adopts a two-dimensional manner. We can therefore conclude that the observed dI/dV patterns are caused by interference of 2DEG LDOS at the surface accumulation layer in the conduction band.

Figure 3 shows bias voltage dependence of lateral wave number square calculated from oscillation periods. Electron effective mass m_* at about bias voltage +0.1 V was found $0.043m_0$. Though this value is larger than $0.024m_0$ of the intrinsic single InAs crystal, the obtained value is consistent with the value reported in carrier concentration dependence of m_* [30].

It should be noted that we found slight downward bowing tendency in the bias voltage dependence of wave number square, though the graph in Fig. 3 showed linear dependence. This bowing may be attributed to the non-parabolicity of $E - k$ dispersion relationship and further investigation will make this effect clear.

3.3 Interference of 2DEG LDOS in nanostructures

As shown in STM image in Fig. 4, equilateral triangular defect was often observed. Such equilateral tri-

angular region has 1/3 or 2/3 monolayer height difference and is a surface of stacking fault tetrahedron. This structure is formed during the sample preparation procedures. There are three {111}-stacking fault planes underneath and each intersecting summit of these planes and a surface is a Frank partial (or Lomer-Cottrell sessile) dislocation. The stacking fault is a phase boundary in InAs crystal lattice with polarization due to the different bond configuration there. The Frank partial dislocation is known as a donor-type defect array. These electronic features suggest that each side of the triangle can play as positively charged reflection wall.

Although the stacking fault tetrahedra are three-dimensional, they effectively act as two-dimensional triangular quantum cages for the 2DEG states. In dI/dV images, We observed standing wave patterns inside of the triangle as shown in Fig. 4. Regularly arranged LDOS peaks are observed inside and their distance becomes larger due to bias voltage decrease. Through the positive bias region to negative bias region, such regular patterns are mapped in the dI/dV images reflecting strong 2DEG LDOS scattering inside.

When the bias voltage condition becomes close to E_F, especially at negative bias condition, it is found that irregularly mapped pattern often appear and the regular distribution of LDOS peaks is disturbed. This is possibly due to the effect of scattering at surface state position, which has energy level close to E_F, and detailed analysis may clarify the origin of such states and their role in surface E_F pinning above E_c.

3.4 Zero-dimensionally confined states

Since the side length of such nanostructure is often less than 100 nm, we can consider the structure as a semiconductor nanostructure like a triangular wave resonator or a quantum dot. By using such nanostructure, we investigated electron interference manner and quantum size effect. In the small triangle shown in Fig. 5, three-fold symmetric patterns are observed. When we calculate LDOS taking zero-dimensional (0D) states into account, observed patterns are found to be equivalent to

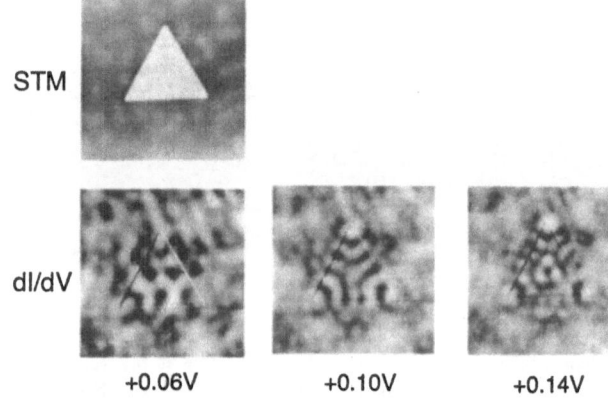

Fig. 5 STM and dI/dV images acquired at bias voltages between +0.06 V and +0.14 V at 5.3 K. Small stacking fault tetrahedron is observed and it has a side length of 48.7 nm.

the calculated 0D states. In each dI/dV image of Fig. 5, two or three 0D states are simultaneously mapped. This is because larger bias modulation amplitude than energy splitting between states. It is necessary to observe smaller structure for direct investigation of each discrete state, but the qualitative tendency of the bias-dependent pattern change is at least consistent with the calculation. Since the size of such nanostructure is comparable with quantum dots or artificial atoms [34], this may contribute to the investigation of 0D states in semiconductor nanostructures.

4 Summary

In summary, the characterization of 2DEG LDOS at the MBE-grown InAs(111)A surface was performed by using LT-STM in UHV. When LT-STM topography was performed at an energy regime of the conduction band, wave phenomena of energy-resolved Friedel oscillation were clearly observed at surface defect sites in simultaneously mapped dI/dV images. Such features were found to be oscillation of 2DEG LDOS of subband state in the surface accumulation layer. Especially, concentric circular patterns and three-fold regular patterns due to electron interference were consistently explained using the 2DEG LDOS. Zero-dimensional states were demonstrated in the small semiconductor triangular cages. The semiconductor nanoscale quantum laboratory is now available to explore the properties of low-dimensional electron systems.

Acknowledgment

Authors thank Y. Tokura for helpful discussions. This study was partly supported by the NEDO collaboration program (NTDP-98) and the Japan Society for the Promotion of Science ("Research for the Future" Program JSPS-RFTF96P00103). M. J. Butcher would like to thank EPSRC.

References

1. W. Shockley and J. Bardeen, Phys. Rev. **77**, (1950) 407.
2. W. Shockley, Phys. Rev. **91**, (1953) 228.
3. W. T. Read, Phil. Mag. **45**, (1954) 775.
4. V. Heine, Phys. Rev. **146**, (1966) 568.
5. G. Binnig, H. Rohrer, Ch. Gerber, and E. Weibel, Phys. Rev. Lett. **49**, (1982) 57.
6. H. A. Mizes and J. S. Foster, Science **244**, (1989) 559.
7. M. F. Crommie, C. P. Lutz, and D. M. Eigler, Nature **363**, (1993) 524.
8. Y. Hasegawa and Ph. Avouris, Phys. Rev. Lett. **71**, (1993) 1071.
9. N. Sato, S. Takeda, T. Nagao, and S. Hasegawa, Phys. Rev. B **59**, (1999) 2035.
10. J. Li, W.-D. Schneider, R. Berndt, and S. Crampin, Phys. Rev. Lett. **80**, (1998) 3332.
11. L. Bürgi, O. Jeandupeux, A. Hirstein, H. Brune, and K. Kern, Phys. Rev. Lett. **81**, (1998) 5370.
12. O. Jeandupeux, L. Bürgi, A. Hirstein, H. Brune, and K. Kern, Phys. Rev. B **59**, (1999) 15926.
13. M. F. Crommie, C. P. Lutz, and D. M. Eigler, Science **262**, (1993) 218.
14. E. J. Heller, M. F. Crommie, C. P. Lutz, and D. M. Eigler, Nature **369**, (1994) 464.
15. É. A. Pashitskiĭ and A. É. Pashitskiĭ, JETP Lett. **60**, (1994) 35.
16. H. K. Harbury and W. Porod, Phys. Rev. B **53**, (1996) 15455.
17. M. C. M. M. van der Wielen, A. J. A. van Roji, and H. van Kempen, Phys. Rev. Lett. **76**, (1996) 1075.
18. R. Dombrowski, Chr. Wittneven, M. Morgenstern, and R. Wiesendanger, Appl. Phys. A **66**, (1998) S203.
19. T. Yokoyama, M. Okamoto, and K. Takayanagi, Phys. Rev. Lett. **81**, (1998) 3423.
20. T. Yokoyama and K. Takayanagi, Phys. Rev. B **59**, (1999) 12232.
21. J. Friedel, Phil. Mag. **43**, (1952) 153.
22. L. Ö. Olsson, C. B. M. Andersson, M. C. Håkansson, J. Kanski, L. Ilver, and U. O. Karlsson, Phys. Rev. Lett. **76**, (1996) 3626.
23. L. Ö. Olsson, L. Ilver, J. Kanski, P. O. Nilsson, C. B. M. Andersson, U. O. Karlsson, and M. C. Håkansson, Phys. Rev. B **53**, (1996) 4734.
24. G. R. Bell, T. S. Jones, and C. F. McConville, Appl. Phys. Lett. **71**, (1997) 3688.
25. J. Tersoff, Phys. Rev. B **32**, (1985) 6968.
26. R. G. Egdell, S. D. Evans, R. A. Stradling, Y. B. Li, S. D. Parker, and R. H. Williams, Surf. Sci. **262**, (1992) 444.
27. H. U. Baier, L. Koenders, and W. Mönch, Solid State Commun. **58**, (1986) 327.
28. Chr. Wittneven, R. Dombrowski, M. Morgenstern, and R. Wiesendanger, Phys. Rev. Lett. **81**, (1998) 5616.
29. H. Yamaguchi, M. R. Fahy, and B. A. Joyce, Appl. Phys. Lett. **69**, (1996) 776.
30. W. Zawadzki, Adv. Phys. **23**, (1974) 435.
31. R. M. Feenstra, J. A. Stroscio, and A. P. Fein, Surf. Sci. **181**, (1987) 295.
32. G. Hörmandinger, Phys. Rev. B **49**, (1994) 13897.
33. J. Li, W.-D. Schneider, and R. Berndt, Phys. Rev. B **56**, (1997) 7656.
34. L. P. Kouwenhoven T. H. Oosterkamp, M. W. S. Danoesastro, M. Eto, D. G. Austing, T. Honda, and S. Tarucha, Science **278**, (1997) 1788.

Probing of subsurface dopants buried in silicon by scanning tunneling microscopy

Yuji Suwa[1], Taro Hitosugi[2], Shinobu Matsuura[2], Seiji Heike[1], Satoshi Watanabe[3], Toshiyuki Onogi[1], Tomihiro Hashizume[1]

[1] Advanced Research Laboratory, Hitachi, Ltd., Hatoyama, Saitama 350-0395, Japan
[2] Dept. of Superconductivity, Univ. of Tokyo, Bunkyo-ku, Tokyo 113-8656, Japan
[3] Dept. of Materials Science, Univ. of Tokyo, Bunkyo-ku, Tokyo 113-8656, Japan

Abstract We report on a novel imaging of surface adatom migration that allowed us to probe the impurity atoms buried in a semiconductor, by using scanning tunneling microscopy (STM). We adsorbed Ga atoms onto chemically inactive H-terminated Si(100) surface, and directly observed thermal Ga-atomic motion near 100 kelvin. By exploiting that the STM image reveals the thermodynamic distribution function of atomic trajectories, we made highly sensitive detection of a positional variation of the surface potential. The position of subsurface P dopants was thereby obtained, in combination with first principles calculations of electronic states.

1 Introduction

Creating atomically uniform surface and examining its homogeneity including its subsurface are becoming important due to the development of the atomic scale fabrication technology[1–3]. For making nano-structures, thin films, and clean interfaces, uniform surfaces are necessary. As a sensitive detector of surface inhomogeneities, recently we found an Scanning Tunneling Microscope (STM) images of a bar structure which we call Ga-bar[4, 5]. (See Fig. 1A.) That is an image of migrating Ga atom in a trough between the two Si-dimer rows. We have predicted that it can be a local probe of the surface energy variation because the image height reflects the energy variation. Here we show a new result of the detection of very large energy variation produced by subsurface impurities by using a kind of this "Ga-bar".

2 "Ga-bar" structures

Ga-bar structures (Fig. 1A) are found when the adsorbed Ga atoms on a H-terminated Si(100) surface are observed by STM at the temperature range 77K - 110K. A P-doped n-type Si(100) sample was used as the substrate. The average dopant concentration was $1\text{-}5\times10^{18}$ cm^{-3} (with resistivity of 7 to 18 mΩ·cm). Details of experimental conditions have been described in Ref. [4]. First principles calculations[4,5] revealed that it is an image of the linear migration of a Ga atom confined in a trough between two adjacent Si-dimer rows. The migrating Ga atom is confined in a finite area by two dihydride-Si-dimer structures, which exist randomly on the H-terminated Si(100) surface and act as stoppers[4, 5].

Because the STM is operating in a constant current mode, $I \cdot p$ is kept constant during the scan, where I is the peak tunneling current while the migrating Ga atom

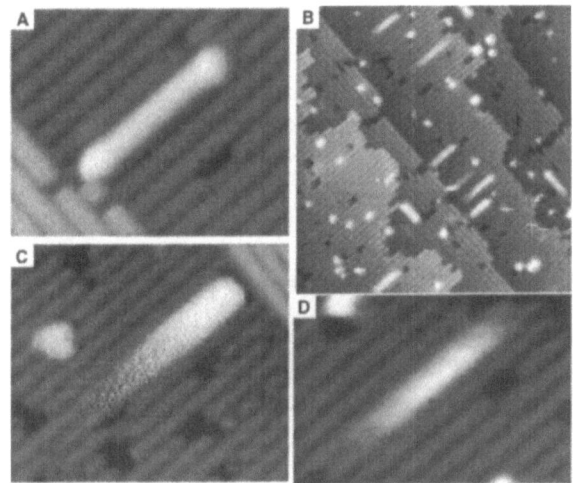

Fig. 1 STM images of (A) Ga-bar (9.3 × 7.3 nm), (B) Ga-bars and Ga-comets (36 × 36 nm), (C) Ga-comet (11.9 × 9.7 nm), and (D) Ga-comet with tails at both ends (11.2 × 8.7 nm).

occupies the site directly below the STM tip, and p is the site occupation probability, which is proportional to the Boltzmann factor, $e^{-E/kT}$. According to the WKB approximation[6] for I, the tip to Ga spacing z is determined by

$$I \cdot p = A \exp\left(-\frac{2\sqrt{2m\phi}}{\hbar}z\right) \cdot \exp\left(-\frac{E}{kT}\right) = \text{constant}. \tag{1}$$

Here, A is a constant, m is the electron mass, and $\hbar$ is the Plank constant divided by 2π. The work function at the Ga atom position was estimated to be $\phi =5.55$ eV, from our calculation. Thus, if E varies spatially, z accordingly varies:

$$\Delta z = -\frac{\hbar}{2kT\sqrt{2m\phi}}\Delta E. \tag{2}$$

This equation indicates that the relative height variation of the STM tip is proportional to the adsorption energy variation, with high sensitivity of $\Delta z/\Delta E = 4.8$ nm/eV at $T = 100$ K.

3 "Ga-comet" structures

Figure 1B shows a large-scale view of an image obtained by using the above technique at $T =100$ K. Therein, in addition to the Ga-bar structure, we find a qualitatively different structure, hereafter called a "Ga-comet". A typical example of a Ga-comet is shown in Fig. 1C. The tail structure at one end gradually fades, while the other remains sharply terminated by a dihydride. In the case shown, the length of the fading-out tail is 8 nm, and the relative image-height variation Δz is 0.21 nm, which corresponds to the adsorption energy variation (ΔE) of 43 meV. The average concentration of observed Ga-comets amounted to 0.001 nm^{-2}. The lengths of the tails were between 2.5 and 8 nm. In addition, we observed sub-specific comet structure with both-sided tails, as can be seen in Fig. 1D.

Let us consider the origin of such a long Ga-comet tail, which implies a long-range energy variation for the Ga atom. We propose here that the origin is a Coulomb interaction between the Ga adsorbate and a subsurface P dopant, because they have significant non-neutral electric charges, as discussed later. In fact, the concentration of the observed Ga-comets was comparable to the dopant concentration expected near the subsurface layers. This picture is also consistent with the length of the tail and the height of the Ga-comet. One might tempt to consider other possible origins, such as lattice distortion accompanied by dihydride defects or the frequently observed surface step edges (see Fig. 1B). However, these possibilities can be ruled out because of the variety of tail lengths and the existence of Ga-comets parallel to the step edges.

To reinforce the above *a priori* proposition quantitatively, we performed an *ab initio* electron-population analysis for subsurface P dopants and for the Ga adsorbate. Details of the calculation of electronic states are described in Ref. [4]. The population analysis was performed following the definition of the Mulliken charge[7], and its formulation using a plane-wave-basis set[8] was extended in order to apply it to the calculation with ultrasoft pseudopotentials.

As a result, we found that the P dopant in silicon has a negative electric charge of $Q_{\mathrm{P}} = -0.24e$ (e: an elementary electric charge). Because of the 2.1 electronegativity of the P atom and 1.8 of Si, it is reasonable that the P atom is negatively charged. One might argue that a P donor should have positive electric charges due to an electronic excitation providing one electron to the conduction band. However, the charge deviation of a P atom coming from such an excitation is estimated to be quite small at 100 K, i.e., $+0.015e$ at most. As for a Ga atom on the H-terminated Si surface, we found that it is positively charged with $Q_{\mathrm{Ga}}^{\mathrm{Surface}} = +0.30e$, thus yielding an attractive Coulomb interaction between the P dopant.

Let us estimate the order of Coulomb interaction energy E_c between the Ga atom and P atom taking into account the dielectric constant of the semi-infinite silicon substrate, $\varepsilon = (1 + \varepsilon_{\mathrm{Si}})/2 = 6.5$. We can find that E_c is 46 meV when the distance between the two point charges is 0.35 nm while it is 2 meV when the distance is 8.3 nm. From these values, we can say that the or-

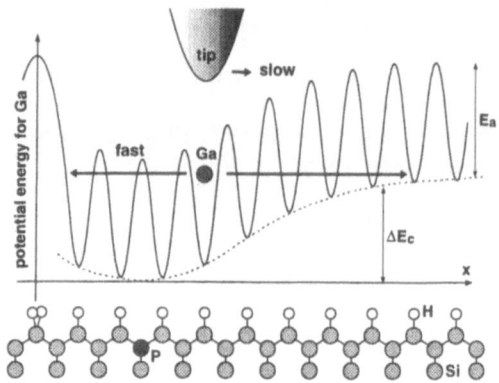

Fig. 2 Schematic picture of Ga-comet. Coulomb potential difference ΔE_c is produced by P atom. Ga atom migrates in a modulated potential well shown by a solid line.

der of magnitudes is reasonable to explain the origin of Ga-comets.

Figure 2 schematically illustrates how a Ga-comet structure is created. The solid line shows the spatial variation of the adsorption energy of a Ga atom. While the Ga atom spends most of the time in the local minimums of the adsorption energy, it thermally hops from site to adjacent site over the activation barrier, $E_a = 0.215 \pm 0.001$ eV, with an estimated hopping frequency of 240 ± 40 s^{-1} at $T = 100$ K[4]. Also, one side of the Ga migration is terminated by the potential barrier (0.56 eV) of the dihydride dimers. The dotted line shows the envelope of the minimum energies and it is modulated by Coulomb energy (typically $\Delta E_c \approx 43$ meV in Fig. 1C) due to dopant atom. As the consequence of Coulomb-force-modulated thermal hoppings of the Ga atom, we can observe the Ga-comet.

In the present observation, the surface position of a Ga-comet was determined by the random positions of naturally formed dihydride Si-dimer structures. However, we can control the position of Ga-comet if we use the atomic manipulation technique[9] to make dihydrides at desired positions. This technique offers a new opportunity for the study of atomic-scale mapping of dopants distributed inside a semiconductor.

References

1. D. M. Eigler, E. K. Schweizer, Nature **344**, (1990) 524.
2. T. C. Shen *et al.*, Science **268**, (1995) 1590.
3. T. Hashizume *et al.*, Jpn. J. Appl. Phys. **35**, (1996) L1085.
4. T. Hitosugi *et al.*, Phys. Rev. Lett. **83**, (1999) 4116.
5. Y. Suwa *et al.*, Transactions of the Materials Research Society of Japan **24**, (1999) 217.
6. J. G. Simmons, J. Appl. Phys. **35**, (1964) 2472.
7. R. S. Mulliken, J. Chem. Phys. **23**, (1955) 1833.
8. H. Aizawa, S. Tsuneyuki, Surface Science **399**, (1998) L364.
9. M. Sakurai, C. Thirstrup, T. Nakayama, M. Aono, Surface Science **386**, (1997) 154.

Effect of tip morphology on AFM images:
Ab initio simulations on GaAs(110) surface

S.H. Ke[1], T. Uda[1] I. Štich[2], K. Terakura[1]

[1] Joint Research Center for Atom Technology (JRCAT), 1-1-4 Higashi, Tsukuba, Ibaraki 305, Japan
[2] Department of Physics, Slovak Technical University (FEI STU), Ilkovičova 3, SK-812 19 Bratislava, Slovakia

Abstract The effect of tip morphology on the AFM image contrast for GaAs(110) surface is investigated theoretically by considering three different tip apexes of a Si tip: (1) Si apex with a half-filled dangling bond; (2) Ga apex with an empty dangling bond; and (3) As apex with a full-filled dangling bond. It is shown that the dangling bond state of the tip apex has significant effects on the image contrast: the Ga apex will image the As sublattice, but the As apex will image the Ga sublattice, and in the case of the Si apex, it is possible to image only the As sublattice or both the As and Ga sublattices, depending on the tip-sample separation.

1 Introduction and computation

Although true atomic resolution can be achieved by the atomic force microscope (AFM) in the non-contact (nc) mode, the physical origin of image contrasts has not been fully understood. It was thought that the short-range tip-sample interaction from dangling bonds plays an important role in the image formation on reactive semiconductor surfaces. Intuitively, this kind of interaction and therefore the resulting image contrast should be affected significantly by the tip morphology, especially the dangling bond state of the very end of the tip. This point has been corroborated by some experimental evidences. However, up to now, there is no quantitative investigation of this problem. In this paper, we report a theoretical investigation on the effect of tip morphology on AFM images by performing first-principles simulation for GaAs(110) surface with a Si tip.

We consider a supercell which contains a GaAs(110) slab and a Si tip as shown in Fig.1. In order to investigate the effect of tip morphology, we consider three different tip apexes: (1) Si apex with a half-filled dangling bond; (2) Ga apex with an empty dangling bond; (3) As apex with a full-filled dangling bond. Hereafter, the three different tips are denoted by Si/Si, Ga/Si, and As/Si tip, respectively. The operation of the AFM in the lateral scanning mode is simulated by making small movements of the tip parallel to the surface along A-A', B-B', C-C' and D-D' lines (see Fig.1b) at several constant tip-surface distances (d) which is defined as the unrelaxed vertical distance between the tip apex and the topmost surface atom. The calculation is performed by using the *ab initio* plane-wave pseudopotential method as described in Ref.[1]. At each step of the scans the atoms of the first three layers of the slab and the tip apex are allowed to relax to their equilibrium positions for the particular tip position.

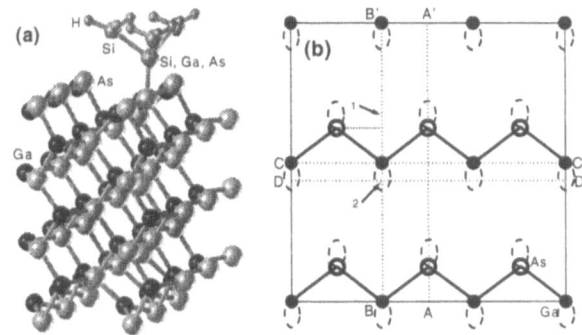

Fig. 1 Side view (a) and top view (b) of the supercell. In (b) the paths along which the tip performs lateral scans are denoted by the dotted lines, and the dangling bonds on the Ga and As atoms are denoted by ellipses of broken line.

2 Results and conclusions

As demonstrated in Ref.[1], in the nc-AFM under typical experimental conditions the quantity detected is approximately the geometric mean of the tip-sample potential energy and the normal force. Hence, in what follows (Figs. 2 - 4) we show the calculated potential energy and normal force variations in the lateral scans along the different lines. Here, the tip-sample potential energy is determined by $E = E_{tot}(\text{tip} + \text{slab}) - E_{tot}(\text{tip}) - E_{tot}(\text{slab})$. The results indicate that, except for some special cases such as those in Figs.2 (a)-(b) where a substantial tip-induced relaxation of the surface occurs, the variation of the normal force and potential energy follow the same trend. This general feature is very useful because we can "see" directly the qualitative image contrast from the result of the lateral scan.

From the maximum potential energy shown in Figs. 2-4 we can see that (1) the tip-surface interaction for the Ga/Si tip is slightly stronger than that for the Si/Si tip, and both of them are much stronger than that for the As/Si tip; (2) along A-A' and B-B' lines the maxima in the force curves shift slightly from the exact atomic position, which is consistent with the orientation of the dangling bonds on the As or Ga atoms.

For the Si/Si tip and relatively large tip-surface distance our previous work [1] showed that the Ga atoms will be "hidden" by the As atoms and only the As atoms will be visible in the image. This finding is consistent with experiments [2]. However, the present calculation shows (see Fig.2) that as the tip-surface distance is fur-

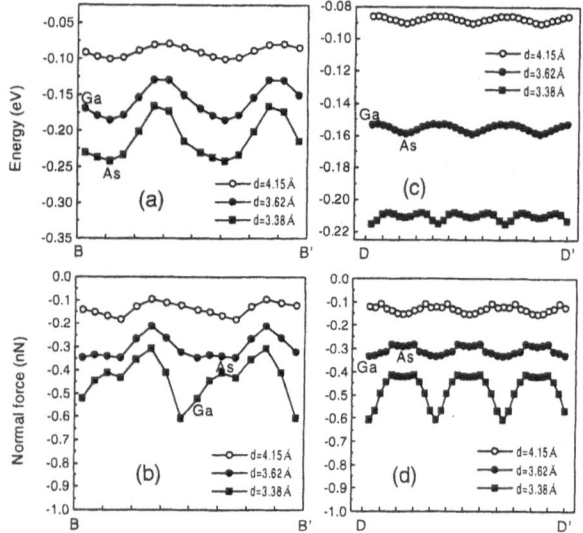

Fig. 2 Results of the lateral scans for the Si/Si tip.

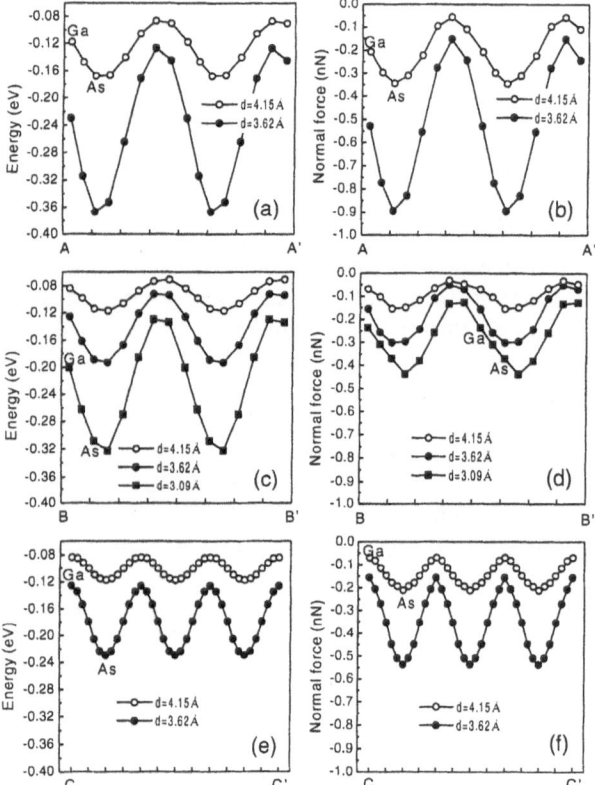

Fig. 3 Results of the lateral scans for the Ga/Si tip.

ther reduced, it is possible to image both As and Ga sublattices, and the pattern of the image contrast will appear as a structure of bright spots with small shoulders. This feature is related to the tip-induced relaxation of the Ga atoms on the surface. Our result may have some correspondences to a recent experimental report[3, 4] for a related system, InAs(110), which also showed that under certain experimental conditions the two sublattices on the surface can be resolved and that in some

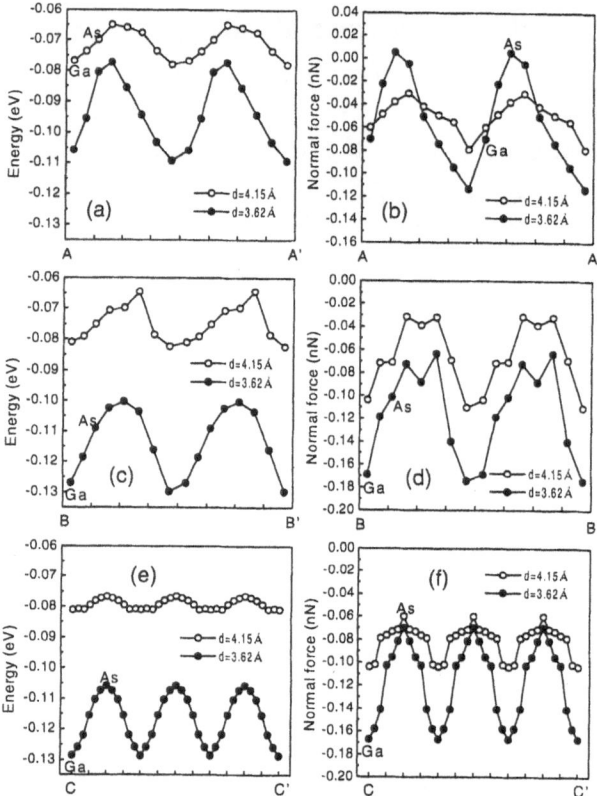

Fig. 4 Results of the lateral scans for the As/Si tip.

cases the image contrast has also the structure of bright spots with small shoulders.

For the Ga/Si and As/Si tips, we can find an interesting thing in Figs. 3-4: the change of tip apex atom from Ga to As will produces reversal in the image contrast. For the Ga/Si tip the As sublattice is imaged while for the As/Si tip the Ga sublattice. However, the topological pattern of the image contrasts are almost the same, the only difference being the strength of the image: the image contrast from the Ga/Si tip is much stronger than that from the As/Si tip. It is very interesting to notice that the As/Si tip will image the lower surface atoms (Ga) rather than the upper surface atoms (As). All of the above features can be explained reasonably by the dangling-bond type of interaction between the tip apex and the surface atoms.

Overall, the present investigation shows a significant effect of the tip morphology on the nc-AFM image formation. For this reason experimental images should be analysed carefully considering the tip morphology.

The present work was partly supported by NEDO.

References

1. S.H. Ke, *et al.*, Phys. Rev. B **60**, (1999) 11631.
2. Y. Sugawara, *et al.*, Appl. Surf. Sci. **140**, (1999) 371.
3. A. Schwarz, *et al.*, Appl. Surf. Sci. **140**, (1999) 293.
4. A. Schwarz, *et al.*, Phys. Rev. B, in press.

Novel investigation technique for interior III/V-semiconductor interfaces

S. Nau, G. Bernatz, and W. Stolz

Materials Science Center and Department of Physics, Philipps-University, D-35032 Marburg, Germany

Abstract   We have developed a novel investigation technique which allows for the quantitative evaluation of real interface structures in GaAs-based quantum well heterostructures. We apply this novel method combinig highly selective etching and subsequent AFM to study the (GaIn)As/AlAs and GaAs/AlAs interior interface morphology of samples grown by MOVPE on (100) exact substrates. A high density of mesoscopic islands is found superimposed on top of larger-scale monolayer island and terraces. The mesoscopic islands are unaffected by growth interruptions, but depend on the growth temperature. It can be shown that the mesoscopic interface structure to a large extent dominates the optical properties of quantum well samples.

1. Introduction

The structure of heterointerfaces is of key importance for the optical as well as transport properties of low-dimensional III/V-semiconductor carrier systems. Interior interfaces of semiconductor heterostructures are in fact overgrown and thus frozen epitaxial surfaces. Therefore, they are closely related to the surface processes determining the morphology of the epitaxial layer. Studying the morphology of interior interfaces thus leads to a better understanding of the epitaxial growth process.

In this study a novel investigation technique is applied to the structural evaluation of (GaIn)As/ GaAs quantum well heterostructures (QWH). This technique combines highly selective etching to prepare interior interfaces with the subsequent examination by atomic force microscopy (AFM). A novel mesoscopic island structure is determined for the first time. Its characteristics as a function of deposition temperature and growth interruption time is studied in detail. The correlation to the optical properties of QWH is examined.

2. Experimental

The (GaIn)As/GaAs-QWH have been deposited by metal organic vapour phase epitaxy (MOVPE). Here we concentrate on two series of samples, where first the growth temperature has been changed between 525°C and 675°C with a constant interruption time of 40 sec at both QWH interfaces and second the growth interruption time at the respective interface was varied up to 120 sec for a growth temperature of 525°C.

The overgrowth by a 75 nm thick AlAs-layer was applied to freeze-in the respective GaAs- or (GaIn)As-interface structure at growth temperature. After highly selective wet chemical removal of the AlAs-layer the uncovered interior interface is investigated by AFM. Further details of the interior interface preparation and examination technique have been published separately [1].

Standard optical spectroscopy techniques (PL/PLE) were applied to study QWH, which have been deposited under indentical growth conditions as the samples for structural characterization.

3. Results

On interior GaAs, (GaIn)As interfaces, respec-tively, deposited at high temperatures two different types of interface structures are clearly detectable by AFM (see Fig. 1a): The macroscopic interface structure consists of monolayer terraces and islands. An additional mesoscopic island structure of one monolayer height and a lateral extension of 35 nm at 675°C superimposes the macroscopic terraces.

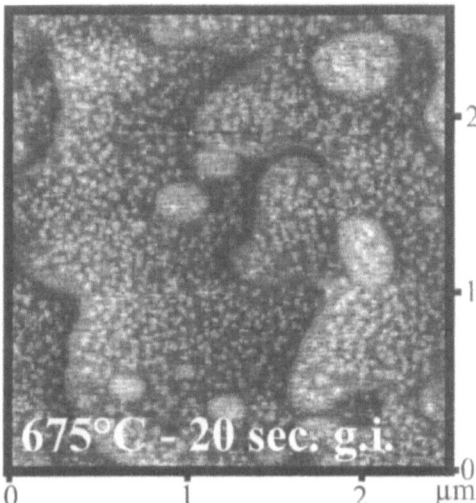

Fig. 1a AFM images of interior GaAs interfaces grown on exact GaAs (100) substrate under the indicated growth conditions. For all images the z scale is 0.7 nm from black to white.

The macroscopic structures smooth during growth interruptions. The number of islands decreases, the average size increases, terraces become less fragmented and terrace edges become smooth with reduced curvature. The mesoscopic island structure, however, is unaffected by the growth interruption time (see Fig. 1b/1c). In contrast, the mesoscopic island structure shows a specific temperature dependence. With a reduction in temperature the island size decreases and vanishes for deposition below 575°C for GaAs (see Fig. 1d).

Varying the growth interruption time has nearly no effect on the optical properties as can be seen by the constant linewidths and Stokes-shifts in PL/PLE experiments (Fig. 2) although we find a substantial macroscopic restructuring of the interface morphology.

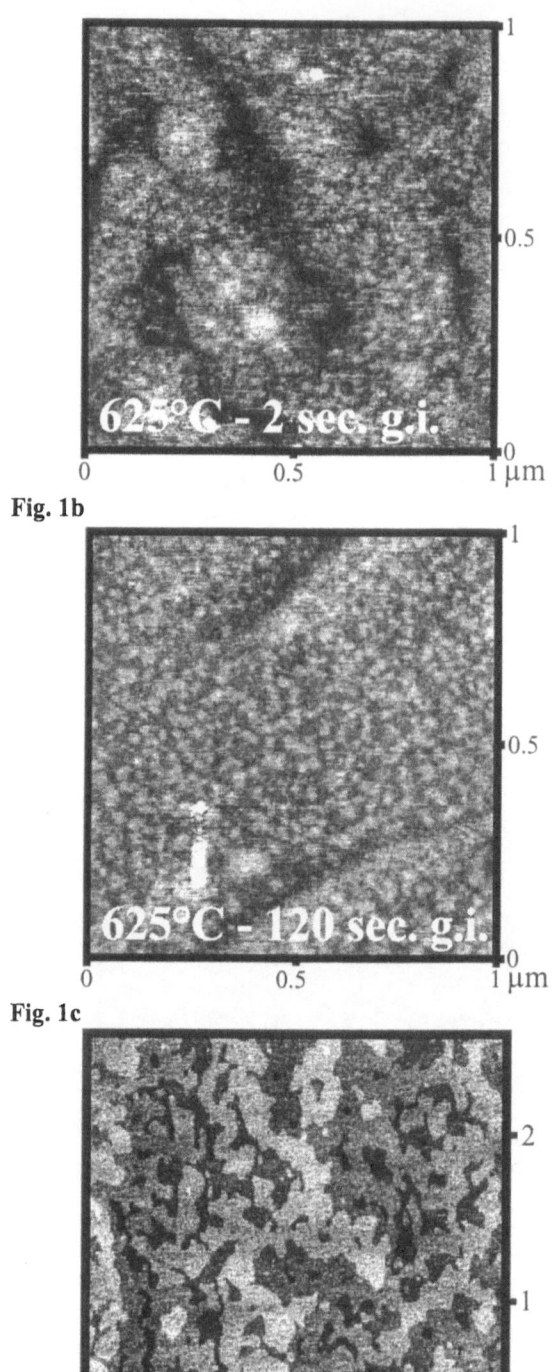

Fig. 1b

Fig. 1c

Fig.1d

Reducing the lateral size of the mesoscopic island structure, however, results in a significant reduction in PL/PLE linewidth and Stokes-shift (Fig. 3), yielding excellent low values of approximately 2 meV PL linewidth, 2.5 meV PLE-linewidth and 0.5 meV Stokes-shift for 3.8 nm thin ternary (GaIn)As/GaAs QWH. Thus, the novel detected mesoscopic island structure dominates the optical properties of the QWH.

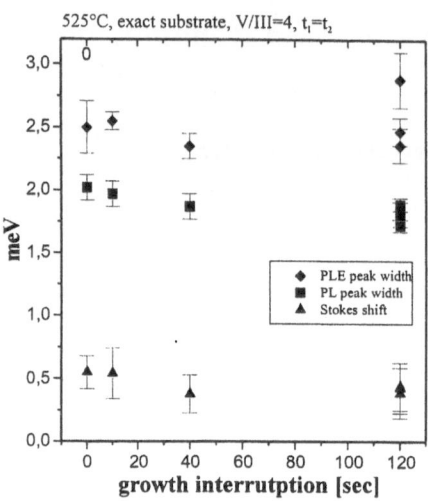

Fig. 2 PL/ PLE/ Stokes-shift of $(Ga_{0.89}In_{0.11})As/$ GaAs-QHW, d(QW)=3.8 nm, grown at 525°C

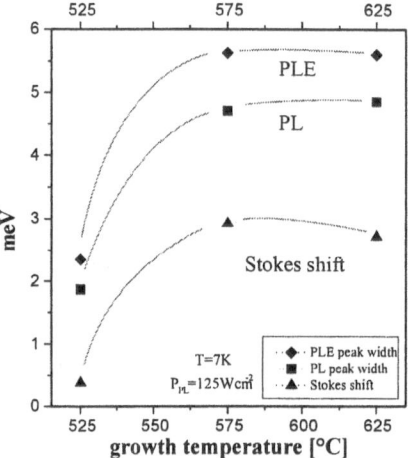

Fig. 3 PL/ PLE/ Stokes-shift of $(Ga_{0.89}In_{0.11})As/$ GaAs-QHW, d(QW)=3.8 nm, 40 sec. growth inter-ruption before and after the QW

4. Conclusion

The developed novel evaluation technique for interior interfaces reveals unambiguously a novel mesoscopic island structure at GaAs based QWH interfaces. The structural properties of the mesoscopic island structure as a function of the growth conditions are determined. Optical spectroscopic techniques underline the important characteristics of this mesoscopic island structure for high quality GaAs based QWHs.

Acknowledgement

Various parts of this work have been supported by the Volkswagen Stiftung and the Deutsche Forschungs-gemeinschaft (DFG) within the framework of the Sonder-forschungsbereich „Disorder in solids on mesoscopic scales" and the Graduiertenkolleg „Optoelectronics of mesoscopic semiconductors".

References

1.   G. Bernatz et al, J. Appl. Physics **86**, (1999), 6752

Observation of dopant-atom dimers on hydrogen-terminated Si(100)-2×1 surface by scanning tunneling microscopy

S. Matsuura[1], M. Fujimori[2], S. Heike[2], Y. Suwa[2], T. Onogi[2], H. Kajiyama[3], K. Kitazawa[1], T. Hashizume[2]

[1] Dept. of Superconductivity, Univ. of Tokyo, Hongo, Tokyo, 113-8656, Japan, e-mail: tt87218@mail.ecc.u-tokyo.ac.jp
[2] Advanced Research Laboratory, Hitachi, Ltd., Hatoyama, Saitama, 350-0395, Japan
[3] Hitachi Research Laboratory, Hitachi, Ltd., Omika, Ibaraki 319-1292, Japan

Abstract   A new dimer structure on a hydrogen-terminated n-type Si(100)-2×1-H surface was found by scanning tunneling microscopy (STM). The STM images of the structure are distinguishable from Si dimers only in the empty state. The results were obtained for arsenic-doped and phosphorus-doped samples, whereas no such structure was found in boron-doped samples. It was found that the surface density of the dimer structures was correlated with the dopant density, and the bias dependent STM images of the newly-found structure showed good agreement with first-principles calculations for dopant dimers on the Si(100)-2×1-H surface.

1 Introduction

The feature size of semiconductor devices continues to decrease and this has necessitated greater homogeneity in the distribution of dopant atoms. Asenov executed a three-dimensional atomistic simulation and showed that not only the number of dopants but also their individual arrangement has significant contribution to fluctuations in the threshold voltage of silicon devices [1]. Thus, precise control of dopant distribution should become more and more important. Scanning tunneling microscopy(STM) of dopant atoms has been reported for compound semiconductor surfaces [2]. In their studies, accurate atomic positions were not determined, since the observations were made by charge density oscillation or Friedel oscillation, extending over a few nm on the surface. In the case of Si surface, scanning tunneling spectroscopy was used for determining doping type and density, which affected I-V characteristic curves [3, 4]. There have been many studies on boron (B) atoms incorporated into Si(100) subsurface for B-doped samples or for B-deposited samples, where B-induced surface reconstruction has been observed [5,6]. There are, however, no reports on dopant atoms on Si surface top-layer, which is expected to directly affect electronic properties of Si surface. In this paper we report the findings of a new dimer structure on a hydrogen-terminated Si(100)-2×1 surface. We show that the surface density of the new structure is proportional to the dopant density. We conclude that the new structure is a dopant dimer incorporated into a Si surface top-layer.

2 Experimental

Surface preparation and measurements were carried out in an ultrahigh-vacuum (UHV) STM system. The scan-ning tips were made from ⟨111⟩-oriented single-crystal tungsten wires by electrochemical etching. The Si(100) samples used for this study were 1-3 mΩ·cm (arsenic (As)-doped), 1-6 mΩ·cm (As-doped), 7-18 mΩ·cm (phosphorous (P)-doped), 0.5-1 Ω·cm (P-doped) and 6.4-8.5 mΩ·cm (B-doped). To avoid metal contamination, non-metal tweezers were used to handle the samples and the sample holder was made of tantalum. The details of the preparation of clean and hydrogen-terminated surfaces have been described previously [7]. All STM images were taken in the constant-current (10 pA) mode at a temperature of approximately 80 K.

3 Results and Discussion

Figures 1 (a) and 1 (b) show filled- and empty-state images of the same hydrogen-terminated Si(100)-2×1 surface. Examples of dihydride defects and local 3×1 structures are indicated by an open circle and a black arrow, respectively. While regular Si dimers in rows are observed as shown in Fig. 1 (a), four striking structures indicated by white arrows are clearly distinguished in the empty state image (Fig. 1 (b)). Magnified STM images are shown in Figs. 1 (c) and 1 (d), corresponding to the white squares in Figs. 1 (a) and 1 (b), respectively. In Fig. 1 (c), only a slight deviation in electronic distribution from surrounding Si dimers was observed (indicated by the white arrows), whereas the new dimer structures (ghost-dimer structures) were notable in the empty state image (Fig. 1 (d)). The ghost-dimer structures were randomly distributed and their locations were independent of the characteristic surface structures, such as missing-dimer defects, step edges and dangling bonds. Such an empty state image has never been reported on a clean Si(100)-2×1 surface. Considering the symmetry of the ghost-dimer structure, the possible origins of the structure are (i) substitution of dimers with Si dimers, (ii) an interstitial atom just below the center of a Si dimer, (iii) an adatom on top of a dimer, (iv) adatoms on Si atoms in one dimer, and (v) missing Si-dimer defects. Since the filled state image of the ghost-dimer structure has only a slight difference from the surrounding Si dimer, adatoms and defects can be excluded from the candidates. We examined the surface density of the ghost-dimer structure for samples with different doping and found that the density was proportional to the dopant density of each

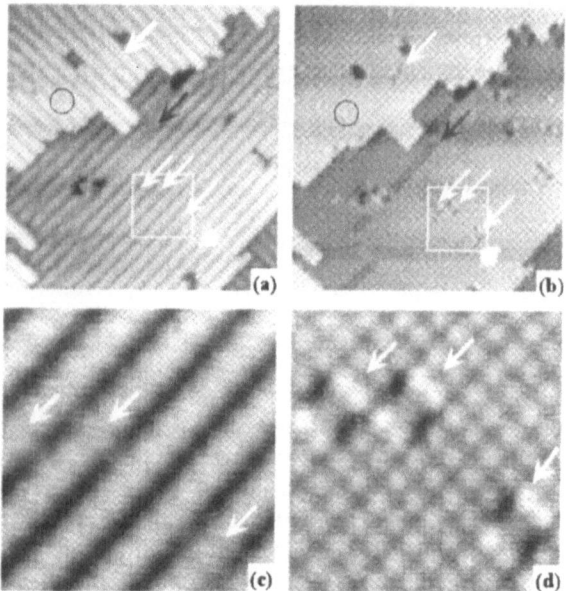

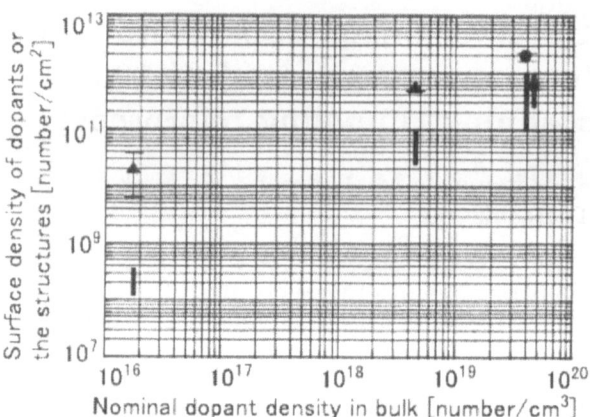

Fig. 2 Comparison of the surface density of the ghost-dimer structure with surface dopant density. Black circles and triangles with error bars show experimental data on As-doped and P-doped surfaces, respectively, and thick lines indicate the estimated surface density of dopants.

Fig. 1 STM images of Si(100)-2×1-H surface taken at a sample bias of -2.0 V (a, c) and 2.0 V (b, d). White squares in (a) and (b) (20 nm×20 nm) indicate the imaged area in (c) and (d) (4 nm×4 nm), respectively. White arrows show the positions of dimer structures. The open circle and black arrow show examples of dihydride structures and local 3×1 structures, respectively.

sample except for the B-doped sample. Figure 2 shows the surface density of the ghost-dimer structure and surface dopant density estimated from resistivity of each sample as functions of nominal dopant density in the bulk, assuming that the ghost-dimer structure is made of two atoms. The density of the B-doped sample is not shown because no such structure was observed. As shown in Fig. 2, the density of the ghost-dimer structures is approximately proportional to the dopant density, meaning that the structure can be caused by the dopant atoms.

Ramamoorthy *et al.* calculated the energies of adsorption and incorporation of dopants at a clean Si(100) surface and concluded that segregation of As and of P atoms to surface top-layer is favorable, while B atoms prefer to be incorporated into the second layer [10]. The absence of the ghost-dimer structure on the B-doped surface may be attributed to the lack of B dimers in the surface top-layer. Models of subsurface B atoms incorporation into the second layer [5] or into the third layer [11] have been proposed from previous STM observations on a clean Si(100) surface. Although identification of the most stable site for B atoms in a Si(100) surface is still a point of controversy, it can definitely be said that B atoms replace subsurface Si atoms rather than segregate in the surface layer. We believe that the same argument applied to the case of a hydrogen-terminated Si(100) surface because the B-Si bond formation is energetically more stable than the B-H bond formation. In addition, we examined the bias dependence of STM images by first-principles calculations for a dimer of As atoms and of P atoms each in a Si(100)-2×1-H surface and found

that good agreement with the images of the ghost-dimer structure, a result will be discussed elsewhere.

4 Conclusion

In conclusion, we found a new dimer structure on a hydrogen-terminated Si(100)-2×1 surface by STM. The origin of the ghost-dimer structure is a dopant dimer substituting for a Si dimer in the surface layer.

The authors (Matsuura and Kitazawa) would like to express their thanks to the Japan Science and Technology Corporation (JST) for the support of CREST (Core Research for Evolutional Science and Technology). One of the authors (Matsuura) is supported by JSPS Research Fellowships for Young Scientists.

References

1. A. Asenov, IEEE Trans. Eelectron Devices, **45,** (1998) 2505.
2. . B. Siemans, C. Domle, M. Heinrich, Ph. Ebert, and K. Urban, Phys. Rev. B, **59,** (1999) 2995, and references therein.
3. H. Fukutome, K. Tanaka, H. Yasuda, K. Maehashi, S. Hasegawa, and H. Nakashima, Appl. Surf. Sci., **130,** (1998) 346.
4. H.-A. Lin, R. Jaccodne, and M. S. Freund, Appl. Phys. Lett., **72,** (1998) 1993.
5. Y. Wang, R.J. Hamers, and E. Kaxiras, Phys. Rev. Lett., **74,** (1995) 403.
6. M.A. Kulakov, Z. Zhang, A.V. Zotov, B. Bullemer, and I. Eisele, Appl. Surf. Sci., **103,** (1996) 443.
7. T. Hashizume, S. Heike, M. I. Lutwyche, S. Watanabe, Y. Wada, K. Nakajima and T. Nishi, Jpn. J. Appl. Phys. **35,** (1996) L1085.
8. X.R. Qin and M. G. Lagally, Phys. Rev. B, **59,** (1999) 7293.
9. X.R. Qin, B.S. Swartzentruber, and M. G. Lagally, Phys. Rev. Lett., **84,** (2000) 4645.
10. M. Ramamoorthy, E.L. Briggs, and J. Bernholc, Phys. Rev. Lett., **81,** (1998) 1642.
11. Z. Zhang, M.A. Kulakov, B. Bullemer, I. Eisele, and A.V. Zotov, J. Vac. Sci. Technol. B, **14,** (1996) 2684.

Reconstruction on Si(100) surface induced by the type-A defects near T_c

Masakuni Okamoto[1], Takashi Yokoyama[2], Kunio Takayanagi[3]

[1] Joint Research Center for Atom Technology (JRCAT)-Angstrom Technology Partnership (ATP), 1-1-4 Higashi, Tsukuba, Ibaraki 305-0046, Japan e-mail: `momo@jrcat.or.jp`
[2] Super-Molecule Photonics Research Laboratory, NRIM, 2268-1 Shimo-Shidami, Moriyama-ku, Nagoya 463-0003, Japan
[3] Department of Material Science and Engineering, Tokyo Institute of Technology, 4259 Nagatsuta, Midori-ku, Yokohama 226-8502, Japan

Abstract   We studied the dynamical properties of the Si(100) surface by using a Monte Carlo technique on the model Ising system, regarding the buckled dimers as Ising spins. We determined the interaction parameters by the *ab initio* calculations within local density approximation. The simulated STM images were in good agreement with our observed images in the whole temperature range.

1 Introduction

The Si(100) surface is very important for silicon devices because of its smaller surface/interface trap state density [1] and higher mobility [2] than any other surface orientations. Extensive studies both theoretical [3] and experimental [4,5] have been made. They have revealed that the Si(100) surface consists of buckled dimers.

The Si(100) surface contains several defects. They are classified into three groups; type-A, type-B, and type-C defects [6]. The type-A and the type-B defects are well identified as a single-dimer vacancy [7], and a double-dimer vacancy, respectively. Very recently, we observed that the type-A defects induced characteristic buckling patterns near T_c^{exp} by using the detailed low-temperature scanning tunneling microscopy (STM) [8]; (i) When $T \gg T_c^{exp}$, the (2×1) reconstruction was observed in the whole area. (ii) When $T \approx T_c^{exp}$, the dimer row with a type-A defect was still symmetric, while the neighboring two dimer rows were buckled. The buckling was pinned such that the atoms adjacent to the type-A defects were protruded. (iii) Well below T_c^{exp}, the $c(4 \times 2)$ reconstruction was observed in the whole area. In the previous papers [8,9], we have shown that the above phenomena might be explained qualitatively by introducing the strain energy caused by the type-A defect.

In the present paper [10], we determine the strain parameter A by using reliable *ab initio* calculations and perform a detailed Monte Carlo simulation.

2 Strain energy

The dynamical properties of the Si(100) surface has been investigated successfully on the Ising model, where the buckled dimers are regarded as Ising spins [11]. The arrows on the buckled dimers in Fig. 1(a) correspond to spin directions ($\sigma = \pm 1$). The spins were located on the square lattice and the interactions between Ising spins V, H and D and are illustrated in Fig. 1(b). We also introduced in Fig. 1(b), the strain parameter A, which pins

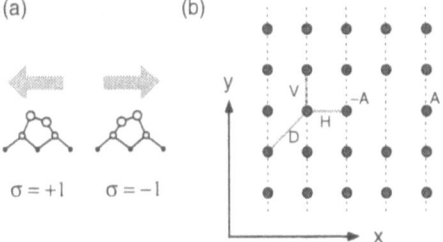

Fig. 1 (a) Correspondence between dimer and spin, and (b) definition of interaction parameters V, H, D and A for the Ising model.

the directions of two spins adjacent to a missing "spin". The Hamiltonian $\mathcal{H}$ of this Ising model is as follows;

$$
\begin{aligned}
\mathcal{H} = \ & -V \sum_{\langle i,j \rangle} \sigma_{i,j}\sigma_{i,j+1} - H \sum_{\langle i,j \rangle} \sigma_{i,j}\sigma_{i+1,j} \\
& - D \sum_{\langle i,j \rangle} \sigma_{i,j}(\sigma_{i+1,j+1} + \sigma_{i+1,j-1}) \\
& - A \sum_{\langle i_d,j_d \rangle} (\sigma_{i_d+1,j_d} - \sigma_{i_d-1,j_d}),
\end{aligned}
\tag{1}
$$

where positions of the type-A defects are denoted by (i_d, j_d), and $\langle i, j \rangle$ indicates the set of corresponding sites i and j.

In order to determine the strain parameter A, we have calculated total energies of the surface with the type-A defect by using *ab initio* calculation based on the local density functional approach. In the calculation, frozen core approximation, localized numerical basis functions of the 6-31G** level and exchange-correlation functional of Vosko, Wilk, and Nusair were used. Hydrogen atoms were fixed during structure optimization. We used clusters consisting of 4-layers and 8 dimers for the surface with a type-A defect. The dangling bonds of the subsurface layers are terminated by the hydrogen atoms.

We considered two geometries, DA_1 and DA_2. The optimized structures are shown in Fig. 2(a) and (b), respectively. The energy difference was calculated to be 25.35 meV, giving $A = 6.34$ meV. In other word, the type-A defect plays a role as a pinning center for spins 2 and 8. This effect gives rise to the frustration against the $c(4 \times 2)$ reconstruction. Thus the observed phenomena may be understood qualitatively by the interplay between these two factors.

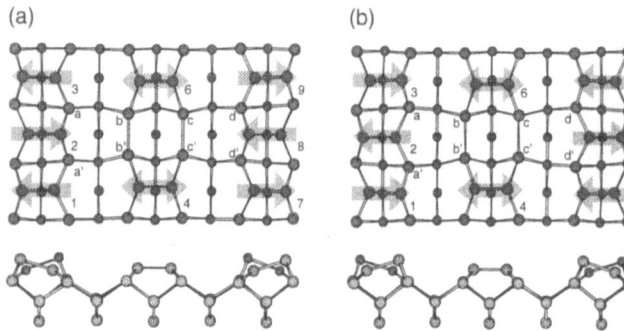

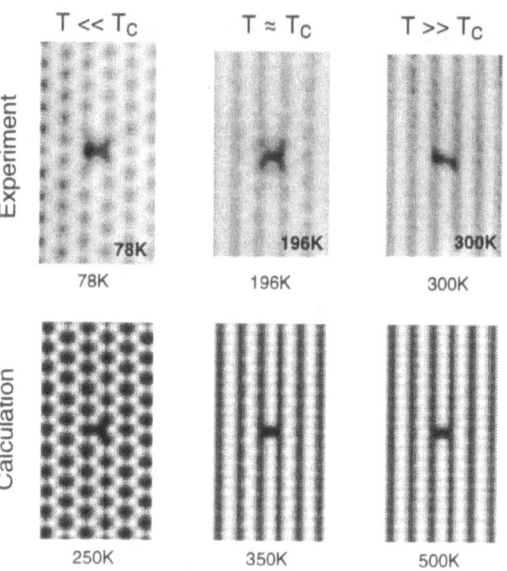

Fig. 2 Optimized structure of (a) DA_1 model and (b) DA_2 model. The cluster model consists of $Si_{75}H_{60}$. Hydrogen atoms are not shown. Arrows indicate buckled directions of dimers defined in Fig. 1(a). Dimers 4 and 6 were calculated to be symmetric while dimers 1, 2, 3, 7, 8 and 9 were buckled. DA_1 and DA_2 are the lowest and the highest energy arrangements of dimers, respectively.

3 Phase transition

We performed a Monte Carlo simulation based on the Ising spin model [9,12,13] with $V = -51.9$ meV, $H = 6.6$ meV, $D = 3.6$ meV, and $A = 6.34$ meV. We took the lattice size as 20×200. At each temperature, we performed 2×10^5 Monte Carlo steps (MCS) for annealing and then averaged next 10^5 MCS. Critical temperature T_c^{calc} of this Ising model was calculated to be $T_c^{calc} = 320$ K, which was consistent with the previous result [12].

When single type-A defect was introduced, STM images showed characteristic local structures around the type-A defect, which is shown in Fig. 3. Upper parts of Fig. 3 show our experimental STM images of the Si(100) surface at 63 K and 300 K, which were obtained using a constant-current mode of 100 pA at a negative sample voltage of -1.0 V. The experimental details have been described in Ref. [8]. Lower parts are calculated STM images following the method developed in Ref. [13]. When $T \gg T_c$, the STM images were rather symmetric. When $T \approx T_c^{exp}$, a local structure of nanometer size appeared around the type-A defect for both experiment and calculation. Below T_c, the $c(4 \times 2)$ structure appeared and the "Y"-shaped dark spot was observed in the STM images.

In case of two type-A defects on the Si(100) surface, we observed experimentally that the local structure was enhanced or suppressed depending on the relative location of two defects [8]. Our calculated STM images for two cases that the distance between two defects in the same row were *odd* and *even* clearly show the enhancement and suppression of the local structures.

4 Summary

We have calculated the strain energy caused by the type-A defect on the Si(100) surface using the *ab initio* cluster calculation. The obtained strain energy parameter was $A = 6.34$ meV. Then we performed a Monte Carlo simulation for the Ising model regarding the dimers as Ising spins using the interaction parameters obtained by

Fig. 3 STM image of the Si(100) surface with single type-A defect. Left, middle and right figures are at $T \ll T_c$, $T \approx T_c$ and $T \gg T_c$, respectively.

the *ab initio* calculation. A local structure of nanometer size appeared only slightly above T_c in the vicinity of the type-A defect. The local structure was enhanced or suppressed depending on the distance between two type-A defects above T_c. The calculated STM images reproduced well the experimental ones not only for the images around an isolated type-A defect but also for the correlation of the images between two defects.

This work was partly supported by New Energy and Industrial Technology Development Organization (NEDO).

References

1. M. H. White and J. R. Cricchi, IEEE Trans. Electron Devices **ED-19**, (1972) 1280.
2. T. Sato, Y. Takeishi, H. Hara, and Y. Okamoto, Phys. Rev. B **4**, (1971) 1950.
3. D. J. Chadi, Phys. Rev. Lett. **43**, (1979) 43.
4. R. A. Wolkow, Phys. Rev. Lett. **68**, (1992) 2636.
5. M. Kubota and Y. Murata, Phys. Rev. B **49**, (1994) 4810.
6. R. J. Hamers and U. K. Köhler, J. Vac. Sci. Technol. **A7**, (1989) 2854.
7. K. C. Pandey, *Proc. of the 17th International Conference on Physics of Semiconductors*, edited by D. J. Chadi and W. A. Harrison (Springer, New York 1985) p. 55.
8. T. Yokoyama and K. Takayanagi, Phys. Rev. B **56**, (1997) 10483.
9. M. Okamoto, T. Yokoyama, and K. Takayanagi, Surf. Sci. **402-404**, (1998) 851.
10. M. Okamoto, T. Yokoyama, T. Uda, and K. Takayanagi, Phys. Rev. B, (in press).
11. J. Ihm, D. H. Lee, J. D. Joannopoulos, and J. J. Xiong, Phys. Rev. Lett. **51**, (1983) 1872.
12. K. Inoue, Y. Morikawa, K. Terakura, and M. Nakayama, Phys. Rev. B **49**, (1994) R14774.
13. Y. Nakamura, H. Kawai, and M. Nakayama, Phys. Rev. B **52**, (1995) 8231.

Space and Energy Distribution of Surface Gap States on MBE-Grown and Silicon-Covered (001) GaAs Surfaces Studied by Scanning Tunneling Spectroscopy

N. Negoro, S. Kasai, H. Hasegawa

Research Center for Interface Quantum Electronics, and Graduate School of Electronics and Information Engineering, Hokkaido University, North-13, West-8, Kita-ku, Sapporo 060-8628, Japan e-mail: negoro@rciqe.hokudai.ac.jp

Abstract Microscopic topological and spectroscopic properties on MBE-grown (001) surfaces of III-V compound semiconductors with and without covering by ultrathin Si layers were investigated by the scanning tunneling microscope (STM) and scanning tunneling spectroscopy (STS) techniques. Spots with anomalously large STS conductance gaps existed on all the sample surfaces. The rate of finding such spots was strongly material-dependent, increasing in the order of InGaAs, GaAs and AlGaAs. The rate remarkably decreased after Si deposition. On the basis of a detailed computer simulation, anomaly was explained by a tip-induced local charging of surface states. Empty state STM images showed light and dark areas, and they exhibited strong correlation with the spatial distribution of normal and anomalous conductance gaps of the STS spectra, indicating presence of pinning areas with continuous space and energy distribution of surface gap states.

1 Introduction

III-V compound semiconductors are important materials for high speed electronic and photonic devices. However, as compared with silicon, they are known to possess high density of surface states which cause the so-called Fermi level pinning and various related unfavorable phenomena. To solve this problem, a Si interface control layer (Si-ICL)-based passivation method was proposed by Hasegawa et al.[1], and this has been found extremely effective for controlling III-V compound semiconductor surfaces and interfaces. However, the microscopic mechanism of the Fermi level pinning and its control by the Si ICL on III-V compound semiconductor surfaces is not well understood. Scanning Tunneling Spectroscopy (STS) is suitable for a microscopic study of the surfaces. However, only few studies has been done on the technologically most important (001) III-V compound semiconductor surfaces[2-5].

In this paper, we attempt to clarify the space and energy distribution of surface gap states responsible for pinning on MBE grown (001) surfaces of GaAs, $Al_{0.3}Ga_{0.7}As$ and $In_{0.5}Ga_{0.5}As$ by using the UHV STM/STS technique.

2 Experimental

All the growth and characterization experiments were carried out in an UHV-based multi-chamber system. In this system, a solid source MBE chamber, an XPS chamber, a UHV contactless capacitance-voltage (C-V) chamber and a UHV STM chamber were connected by a common UHV transfer chamber.

Si-doped GaAs, $Al_{0.3}Ga_{0.7}As$ lattice-matched to n^+-GaAs (001) substrates and $In_{0.5}Ga_{0.5}As$ lattice-matched to n^+-InP (001) substrates were grown by a conventional solid-source MBE. Then, substrate temperature was cooled down to 300°C and the Si-ICL was grown on GaAs (001) and AlGaAs (001) surfaces.

Then, the sample was transferred into the UHV STM/STS chamber, and STM images and STS spectra were taken using a JEOL JSTM-4600 microscope without breaking the UHV condition. The macroscopic surface Fermi level position and the surface state distribution were also characterized by XPS and contactless C-V techniques and they were compared with STM/STS results.

3 Result and Discussion

Examples of the STS spectra on surfaces of GaAs (001), $Al_{0.3}Ga_{0.7}As$ (001) and $In_{0.5}Ga_{0.5}As$ (001) are shown in Fig.1. As indicated by solid curves in Fig.1, normal STS spectra showing conductance gaps which equal to band gap of the material were obtained only on limited spots. On the other hand, anomalous spectra (thin solid curves in Fig.1) showing much wider conductance gaps than the band gap energy were also observed on all the sample surfaces where normal and anomalous spectra co-existed on the same sample surface. As compared with the normal spectra, anomalous spectra extended wider in both positive and negative bias directions. The rate of finding anomalous spectra on bare surfaces, however, was very much material dependent. On the InGaAs surface, it was smaller than on the GaAs surface. In the case of AlGaAs, we could not find spots with normal spectra at all. After Si deposition, the rates of finding spots with normal spectra remarkably increased both on GaAs and AlGaAs surfaces.

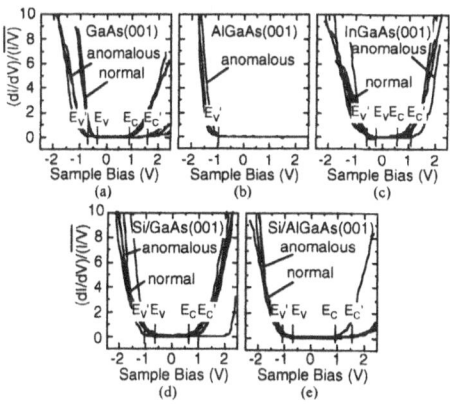

Fig.1 Examples of STS spectra on various surfaces.

We evaluated microscopic surface Fermi level positions (E_{FS}) from the normal spectra. In the case of GaAs (001) surface, E_{FS} was about 1.0eV from the conduction band edge (E_C), indicating strong pinning near the charge neutral level, E_{HO}[6]. The position of E_{FS} shifted to 0.7eV from E_C after Si deposition, indicating decrease of the surface state density. This large shift of the Fermi level position by 300meV is in agreement with the movement of the macroscopic E_{FS} position evaluated by XPS.

We have recently shown[5] that the origin of anomalous STS spectra with large conductance gaps is due to the surface band bending caused by local charging of surface states by the STM tip which takes place when the tip current is relatively large in nA range.

Based on this interpretation, the energy position of the tip with respect to the charge neutrality level, E_{HO}, was calculated for GaAs vs. sample bias, V_S, for different surface state distributions having discrete acceptor states at E_{HO}, discrete donor states at E_{HO} and a U-shaped continuous distribution including both donor and acceptor states below and above E_{HO}, respectively. In the case of discrete donor states, only downward movement of the tip potential was prevented at the discrete state level and

anomalous extension of the conductance gap took place under sample negative bias. On the other hand, by having discrete acceptor states, only upward movement of the tip potential was prevented under the positive bias, causing extension of the conductance gap toward tip positive bias direction. Obviously, the U-shaped continuous surface states extend the conductance gap both in negative and positive bias directions. The calculated STS spectra are shown in Fig.2. This shows that anomalous STS spectra were caused by the U-shaped continuous distributions of surface states. It was also quantitatively confirmed that the conductance gap width increased as the surface state density was increased.

In order to correlate spatial distributions of surface states with the STM image, empty state STM images were taken. As an example, Figure 3(a) shows an empty state STM image taken on GaAs (001) surface. It was found that the STM image had light and dark regions. The light region corresponded to the area where the normal spectra were obtained. On the other hand, dark regions corresponded to regions with anomalous STS spectra with large conductance gaps. We found that the dark regions extended spatially and the darkness changed gradually by position. We also found that the dark regions enlarged with decrease of V_S. From the results of the simulation of the conductance gap, these features indicate that surface states are not discrete states localized at particular defect spots, but they form a pinning area with a spatially and energetically continuous distribution of the surface states as schematically shown in Fig.3(b).

4 Conclusion

In order to clarify the space and energy distribution of surface gap states, a detailed UHV STM/STS study was made on MBE-grown (001) surfaces of GaAs, AlGaAs and InGaAs with and without Si ICL.

Spots with normal and anomalous STS spectra with normal and anomalous conductance gaps co-existed on the same surface in all the samples. Si ICL reduces the number of anomalous spots.

Regions with anomalous STS spectra showed strong correlation with STM images, and surface states are not localized to particular defects, but form pinning areas with continuous space and energy distributions.

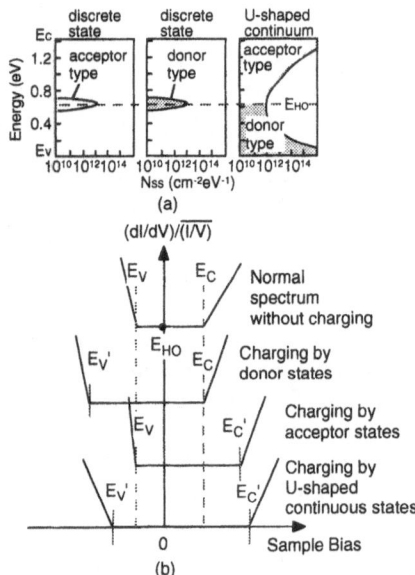

Fig.2 (a) Surface state distributions and (b) schematic STS spectra.

References
1. H. Hasegawa, M. Akazawa, H. Ishii and K. Matsuzaki, J. Vac. Sci. Technol. B7 (1989) 870.
2. M. D. Pashley, K. W. Haberern and P. M. Feenstra, J. Vac. Sci. Technol. B10 (1992) 1874.
3. V. Bressler-Hill, M. Wassermeier, K. Pond, R. Maboudian, G. A. D. Briggs, P. M. Petroff, and W. H. Weinberg, J. Vac. Sci. Technol. B10 (1992) 1881.
4. T. Takahashi and M. Yoshita, Appl. Phys. Lett. 70 (1997) 2162.
5. H. Hasegawa, N. Negoro, S. Kasai, Y. Ishikawa and H. Fujikura, J. Vac. Sci. Technol. B18 (2000) 2100.
6. H. Hasegawa and H. Ohno, J. Vac. Sci. Technol. B4 (1986) 1130.

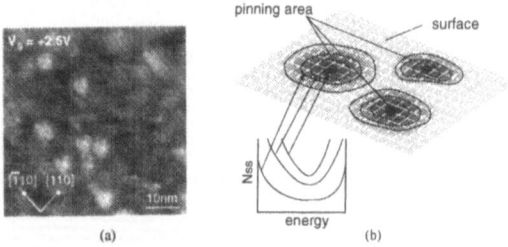

Fig.3 (a) An empty state STM image taken on the (001) GaAs surface and (b) schematic illustration of the spatial and energy distribution of surface states.

Atomically resolved imaging of semiconductor surfaces using noncontact atomic force microscopy

S.Morita, Y.Sugawara

Department of Electronic Engineering, Graduate School of Engineering, Osaka University, 2-1 Yamada-Oka, Suita, 565-0871, Japan e-mail: smorita@ele.eng.osaka-u.ac.jp

Abstract Using a noncontact atomic force microscope (NC-AFM), we measured lateral shifts of Si dimers adjacent to missing dimers on Si(100)2x1 surface. As a result, we found 0.08 nm shift toward the missing dimer in case of the nearest neighbor dimer and 0.04 nm shift in case of the second nearest neighbor dimer. Present result exactly proves that the interaction force between Si dimers on Si(100)2x1 surface is repulsive in contrast to the theoretical prediction.

1 Introduction

The noncontact atomic force microscope (NC-AFM) is a unique tool based on a mechanical method which has the following characteristics; (1) it has true atomic resolution [1], (2) it can measure atomic force (so-called atomic force spectroscopy) [2]-[3], (3) it can even observe insulators [4], and (4) it can measure mechanical responses. However, there seems no measurement on the mechanical response until now.

In general, surface reconstruction frequently induces surface stress. For example, the dimer bond of Si(100)2x1 surface exerts attractive force between pairing silicon atoms of the dimer which results positive surface stress and exerts tensile stress to the Si substrate. On the other hand, repulsive force works between adjacent hydrogen atoms on the Si(100):H dihydride surface which results negative surface stress and exerts compressive stress to the Si substrate.

There are many calculations on the surface stress of Si(100)2x1. Theoretically, surface stress g_{xx} along the dimer bond is positive (attractive force between pairing Si atoms) in every case such as before reconstruction [5], under 2x1 symmetric dimer [5]-[8] and even under 2x1 buckled dimer models [8]. On the other hand, surface stress g_{yy} along the dimer row is a delicate problem. Calculated results predict that it is positive (attractive force) before reconstruction [5], but negative (repulsive force) under 2x1 symmetric dimer [5]-[8], further again positive (attractive force) under 2x1 buckled dimer model [8].

Experimentally, macroscopic method can measure only asymmetry of surface stress, that is, the difference of g_{xx} and g_{yy}, because of alternating 2x1 and 1x2 structures. To our knowledge, no direct measurement of stress in specific directions has been reported [9].

In the present experiment, we precisely investigated position of dimer around missing dimer on Si(100)2x1 surface using the NC-AFM. As a result, we found lateral shift of dimer position toward the missing dimer along the dimer row. This result exactly proves that the NC-AFM can measure the mechanical response of the atom due to the surface stress and can directly measure stress effect in specific directions on an atomic scale.

2 Experiment

Using a home-made UHV-NC-AFM, we observed clean Si(100)2x1 and Si(100)2x1:H monohydride surfaces at RT under pressures lower than $2x10^{-10}$ Torr [10]. The frequency modulation detection method was used to measure the small change of the frequency-shift of the mechanical resonance of the cantilever due to interaction force between the tip and the sample surfaces. Spring constant and mechanical resonance frequency of the conductive silicon cantilever used were about 48 N/m and 175 kHz, respectively.

3 Experimental Results

Figure 1 shows the atomic resolution image of Si(100)-

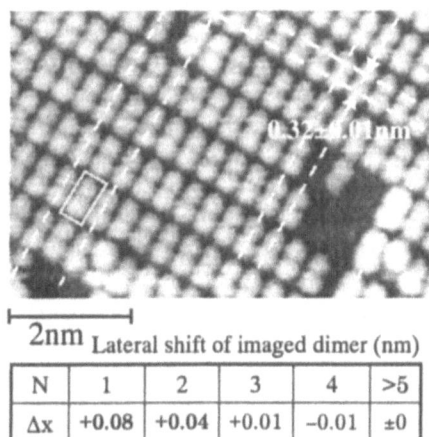

	Lateral shift of imaged dimer (nm)				
N	1	2	3	4	>5
Δx	+0.08	+0.04	+0.01	−0.01	±0

Fig.1 NC-AFM image of Si(100)2x1 surface and table of averaged lateral shift Δx (nm) of dimers toward missing dimers. N means the N-th neighbor dimer of missing dimers.

2x1 surface measured at the averaged frequency shift $\langle \Delta v \rangle$ = -30 Hz [10]. The paired bright spots (imaged dimer) constituting rows with a 2×1 symmetry was clearly observed. Further, the distance between paired bright spots is 0.32 ±0.01nm. The bright spots of Fig.1 don't image the silicon atom site, because the distance 0.32 ± 0.01nm between bright spots forming dimer structure of Fig.1 is lager than the distance between silicon atoms of dimer structure (maximum distance between alternating upper Si atoms on asymmetric dimer structure is 0.292nm). This result suggests that the chemical bonding interaction strongly works between the tilted dangling bond out of the silicon dimer and the dangling bond out of silicon tip apex. Therefore, we conjectured that the NC-AFM images the tilted dangling bonds on the Si(100)2x1 surface.

Further, we investigated position of imaged dimers adjacent to missing dimers to make clear the sign of the interaction force between dimers along the dimer row. As a result, we found lateral shift of the nearest neighbor (N=1) and the second nearest neighbor (N=2) dimers toward missing dimers as listed in the table of Fig.1. For N larger than 3, averaged lateral shift was less than the experimental error. This result exactly proves that the interaction force between dimers is repulsive along the dimer row. Thus, the NC-AFM can measure the mechanical response of the dimer due to the surface stress and can directly measure stress effect in specific directions.

Theoretically, surface stress g_{yy} along the dimer row is a delicate problem as shown in Table.1. Calculated results predict that it is positive (attractive force) before reconstruction [5], but negative (repulsive force) under 2x1 symmetric dimer [5]-[8], further again positive (attractive force) under 2x1 buckled dimer model [8]. It should be noted that present result (repulsive force) is in conflict with the calculated result.

Table.1 Calculated surface stress g_{yy} along the dimer row

	Along the dimer row $g_{yy}(eV/nm^2)$
Unreconstructed surface [5]	0.00058
2x1 symmetric dimer [5]	-0.00132
2x1 symmetric dimer [6]	-0.00035
2x1 symmetric dimer [7]	-0.00129
2x1 symmetric (semicond) [8]	-0.00054
2x1 symmetric (metallic) [8]	-0.00027
2x1 buckled dimer [8]	0.00020

We also investigated lateral shift of dimers on Si(100)2x1:H monohydride surface [10]. Figure.2 shows the atomic resolution image of hydrogen

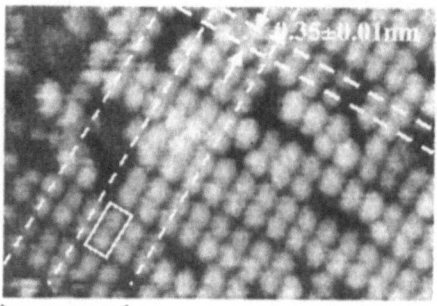

N	1	2	3	4	>5
Δx	±0	+0.01	+0.01	−0.01	±0

Fig.2 NC-AFM image of Si(100)2x1:H monohydride surface and table of averaged lateral shift Δx (nm) of dimers toward missing dimers. N means the N-th neighbor dimer of missing dimers.

terminated silicon [Si(100)2x1:H] surface measured at $\langle \Delta v \rangle$ = -11 Hz. The paired bright spots (imaged dimer) constituting rows with a 2×1 symmetry was observed. Further, the distance between paired bright spots is 0.35 ± 0.01nm, which approximately agrees with the distance between hydrogen atoms on monohydride surface, i.e., 0.352nm. Therefore, the NC-AFM images the individual hydrogen atoms on the Si(100)2x1:H monohydride surface.

In contrst to the result of Si(100)2x1 surface, we could not observe lateral shift of the dimers adjacent to the missing dimer on Si(100)2x1:H monohydride surface as listed in the table of Fig.2. This result means that the hydrogen termination removes even the surface stress between dimers along the dimer row.

References
1. Y.Sugawara, M.Ohta, H.Ueyama and S.Morita, Science, **270** (1995) 1646.
2. T.Minobe, T.Uchihashi, T.Tsukamoto, S.Orisaka, Y.Sugawara and S.Morita, Appl.Surf.Sci., **140** (1999) 298.
3. S.Morita, Y.Sugawara, K.Yokoyama and T.Uchihashi, Nanotechnology, **11** (2000) 120.
4. M.Bammerlin, R.Lüthi, E.Meyer, A.Baratoff, J.Lü, M.Guggisberg, Ch.Gerber, L.Howald and H.-J.Güntherodt, Probe Microscopy, **1** (1997) 3.
5. M.C.Payne, N.Roberts, R.J.Needs, M.Needels and J.D.Joannopoulos, Surf.Sci., **211/212** (1989) 1.
6. O.L.Alerhand, D.Vanderbilt, R.D.Meade and J.D.Joannopoulos, Phys.Rev.Lett., **61** (1988) 1973.
7. F.Liu and M.Lagally, Phys.Rev.Lett., **76** (1996) 3156.
8. A.Garcia and J.E.Northrup, Phys.Rev.B, **48** (1993) 17350.
9. F.Wu and M.G.Lagally, Phys.Rev.Lett., **75** (1995) 2534.
10. K.Yokoyama, T.Ochi, A.Yoshimoto, Y.Sugawara and S.Morita, Jpn.J.Appl.Phys., 39 (2000) L113.

Atomic Structure of GaP(001) and InP(001) Reconstructions: Scanning Tunneling Microscopy and ab initio Theory

K. Lüdge[1], P. Vogt[1], O. Pulci[2]*, N. Esser[1], F. Bechstedt[2], W. Richter[1]

[1] Institut für Festkörperphysik, Technische Universität Berlin, Hardenbergstr. 36, D-10623 Berlin, Germany e-mail: `luedge@physik.tu-berlin.de`

[2] Institut für Festkörperphysik und Theoretische Optik, Friedrich-Schiller-Universität, Max-Wien Platz 1, D-07743 Jena, Germany

Abstract We infer the structure of the Ga-rich GaP(001)-(2×4) and P-rich InP(001)(2×1) surface from a study of STM images obtained under UHV conditions on MOVPE-grown samples. STM images are compared with results of first–principles calculations for models energetically most favourable under Ga–rich and P–rich growth conditions, respectively. The comparison shows that the GaP(001)(2×4) surface unit cell consists of a mixed Ga-P dimer on top of a complete Gallium layer. For InP(001)(2×1) the structure cannot be explained completely.

1 Motivation

Because of their favorable properties for optoelectronics, the phosphorous-based zinc-blende III–V semiconductors InP and GaP have become technologically important during the last decade. Their surface structures, particularly the polar (001) faces, show differences to the group-III arsenides, typified by GaAs(001). As an example for the P-rich surfaces we will discuss the (2x1)/p(2x2) reconstruction of InP(001), that is up to now not well understood. The III-rich (2x4) reconstruction of InP(001) surface was already explained by the so called "mixed dimer" model [1]. In contrast just little is known about the surface structure of GaP(001), but a variety of measurement methods already indicates, that its Ga-rich surface structure is similar to that of InP(001)[2]. The following comparison of ab initio calculations and STM measurements will try to clarify the atomic arrangement of the P-rich InP(001)(2x1) surface and will infer the structure of the (2x4) reconstruction of GaP(001).

2 Experiment

Experiments were performed on MOVPE grown samples, which were in case of GaP capped by an amorphous As/P double layer in order to protect it before transport through air or in case of InP directly transferred into UHV using an UHV-transport chamber [4]. Under UHV conditions the cap of GaP(001) was thermally desorbed at a temperature of 400°C and subsequently annealed at 450°C. The InP(001)(2×1) surface was prepared by sequentially annealing the as grown sample at 380°C.

* *Present address:* Fritz-Haber Institut der MPG, Berlin

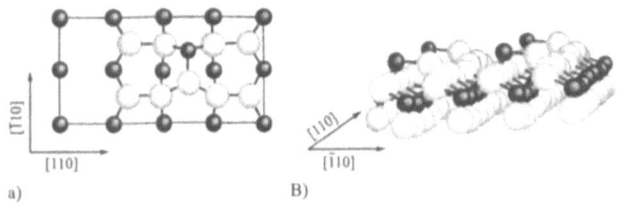

Fig. 1 a) Top view and b) Side view of the atomic arrangement within the first atomic layers of the "mixed–dimer" reconstruction. White (black) balls represent Ga (P).

3 Calculation

In order to interpret the STM images, ab initio calculations of the electron density were performed. The electron–wave–functions were calculated using the Density Functional Theory (DFT) within the Local Density Approximation (LDA)[7]. The electron–electron interaction was described using the Perdew–Zunger extrapolation. The electron–ion interaction was treated by using norm–conserving, fully separable pseudopotentials [8]. The single–particle wave functions were expanded into plane waves. They were restricted by a kinetic energy cutoff of 18 Ry. Once DFT–LDA electron–wavefunctions for the reconstructed surface were calculated, it was possible to simulate STM images using the Tersoff–Hamann approach [9].

4 Ga-rich GaP(001) surface

Experimental data of the GaP(001)(2×4) reconstructed surface show rows in the $[\bar{1}10]$ direction containing a triangular structure that repeats every two surface lattice constants. The rows are separated by 15.6 Å, corresponding to a fourfold surface lattice constant [3]. Fig.2 left shows a 3 nm long section of such a row. Total energy calculations for GaP(001) show that under Ga–rich conditions only those (2×4) reconstructions are stable which consist of either Ga-P (mixed–dimer) (Fig.1) or Ga–Ga (top Ga–dimer) dimers on top of a complete Ga layer. In Fig.2 the simulated STM images of both the mixed–dimer and the top-Ga-dimer reconstruction on GaP(100) are shown at an energy of 1.5 eV below the valence band maximum (VBM). For the mixed–dimer reconstruction a triangular structure similar to the experimental data appears, containing the more intensive circle like spot ("head") that is formed by the filled dan-

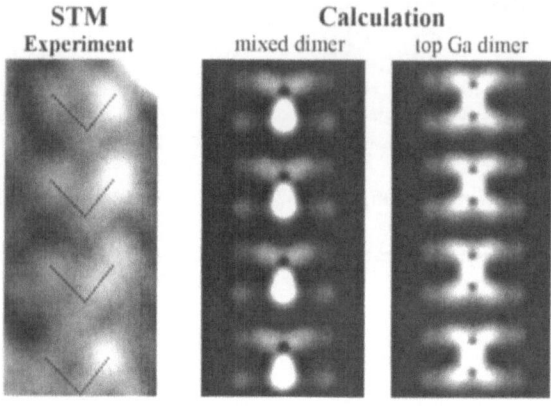

Fig. 2 Comparison between theoretical and experimental STM data for GaP(001)(2×4). Displayed area is always (3 nm x 1.5 nm) with axes along [$\bar{1}\bar{1}$0] x [110] .

gling bond at the P atom of the mixed dimer and two smaller spots ("ears") that originate from the filled backbonds between the Ga atom in the mixed dimer and neighbouring Ga atoms in the atomic layer beneath. In contrast, the simulated filled state STM images of the top–Ga–dimer reconstruction contains a symmetric X–pattern that is visible within the rows. The X–shape behaviour is a consequence of the backbonds between the Ga atoms in the top dimers and the Ga atoms underneath.

5 P-rich InP(001) surface

For the P-rich phase, STM observations show two different (2×1) reconstructions depending on the annealing time [4,5]. Here we concentrate on the most P–rich phase, since the zig–zag chains of the less P–rich phase were not yet reproduced by DFT-LDA calculations. For the as grown sample, a high number of defects characterizes the surface, that can locally be described in terms of P–dimer on top of a complete P–layer, giving rise to local (2×2), c(2×2) and c(4×2) structures as can be seen in Fig.3. Several different structures for this reconstruction

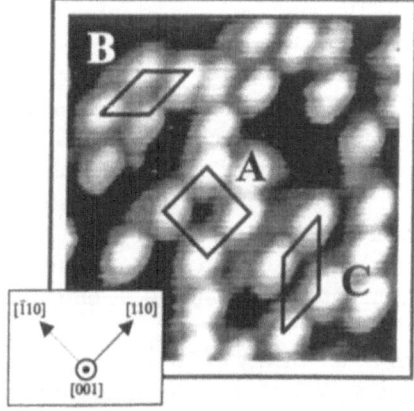

Fig. 3 STM image obtained for InP(001)(2×1) surface directly after transfer from MOVPE-reactor(U=-5.2 V, (6 nm x 6 nm)).

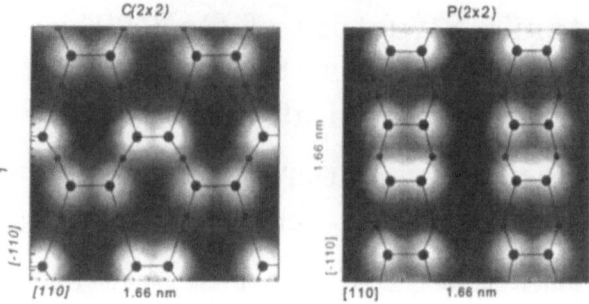

Fig. 4 Calculated STM images for the two energetically most stable structures on the InP(001) P-rich surface for occupied states at -1.5 eV below the VBM, i.e. U=-2.7 V

were characterized using total energy DFT–LDA calculations. The resulting phase diagram yields the c(2×2) and p(2×2) models to be most favourable [6]. Both structures consist of two P–P dimers on top of a complete P–layer. In Fig.4 the corresponding calculated STM images are shown. They contain a white oval spot at each position of the P–dimer of the topmost layer, and can therefore be correlated with the experimental structures B and A in Fig.3, respectively.

6 Conclusions

The comparison between STM data and ab initio theory shows, that for both InP(001) and GaP(001), under cation rich conditions the most favourable structure is the (2×4) "mixed-dimer" reconstruction. In the very phosphorous rich limit (2 ML P coverage) of InP(001) the two energetically most favorable structures of 12 proposed reconstructions are investigated. By the comparison of STM images for InP(001)(2×1) it has been shown, that the presence of local c(2x2) symmetries can be interpreted as P–dimers on top of a complete P–layer. However we would like to note, that adsorbed hydrogen might also play a role in surface formation.

References

1. W.G. Schmidt, F. Bechstedt, N. Esser, M. Pristovsek, Ch. Schultz and W. Richter, Phys. Rev. B **57**, 14596 (1998).
2. A.M. Frisch, W.G. Schmidt, J. Bernholc, M. Pristovsek, N. Esser, and W. Richter, Phys. Rev. B **60**, 2488 (1999).
3. K. Lüdge, P. Vogt, O. Pulci, N. Esser, F. Bechstedt, W. Richter, Phys. Rev. B**62**, (2000) (in press).
4. P. Vogt, Th. Hannappel, S. Visbeck, K. Knorr, N. Esser, and W. Richter, Phys. Rev. B**60**, R5117 (1999).
5. L. Li, Q. Fu, C.H. Li, B.-K. Han, and R.F. Hicks, Phys. Rev. B**61**, 10223 (2000).
6. O. Pulci, K. Lüdge, W.G. Schmidt, and F. Bechstedt Surf. Sci. , (in press).
7. P. Hohenberg and W. Kohn, Phys. Rev. **136**, B864 (1964); W. Kohn and L.J. Sham, Phys. Rev. **140**, A1133 (1965).
8. R. Stumpf and M. Scheffler, Comput. Phys. Commun. **79**, 447 (1994); M. Bockstedte, A. Kley, J. Neugebauer, and M. Scheffler, Comput. Phys. Commun. **107**, 187 (1997).
9. J. Tersoff and D.R. Hamann, Phys. Rev. B **31**, 805 (1985).

Bias voltage dependence of scanning tunneling microscopy images of a Si(100)2x3-Ba surface

K. Ojima, M. Yoshimura, K. Ueda

Toyota Technological Institute

2-12-1 Hisakata, Tempaku-ku, Nagoya 468-8511, Japan e-mail: ojima@toyota-ti.ac.jp

Abstract   We investigated the 2x3 structure on Ba-deposited Si(100) surface by means of scanning tunneling microscopy (STM). The 2x3 structure has a lattice vector with the length of $3a_0$ along the Si dimer row of original (100) plane. It is found that STM images of the 2x3 structure have sample bias voltage dependence, especially in empty-state images. At higher voltage than 1.5 V, only one protrusion is observed in a unit cell. On the other hand, at lower voltage than 1.5 V, an additional protrusion is observed. In comparison of empty-state images with filled-state images, it is found that the protrusion observed at high voltage is due to a Ba dimer and the additional one is due to Si dangling bonds.

1. Introduction

Barium (Ba) is one of important elements for electronic device material such as $BaTiO_3$, which is promising as a ferroelectricity in a random access memory. In order to make an electronic device, it is necessary to grow well-ordered thin film of a functional material on semiconductor substrates. For such purpose, the adsorption structure of Ba on a Si substrate, especially on a Si(100) surface, has been studied for a decade[1,2]. In the Ba-adsorbed Si(100) surface, some superstructures have been observed by low-energy electron diffraction (LEED) at the Ba coverage below 1 ML (1 ML = 7.88×10^{14} atom/cm^2). Among them, the 2x3 structure is intensively investigated by scanning tunneling microscopy (STM) very recently [3,4].

In this paper, the sample bias voltage dependence of the 2x3 structure is reported for the first time and its mechanism is discussed.

2. Experimental

Details of the experimental apparatus have been reported elsewhere [4]. The Si(100) substrate was cleaned by flashing at 1500K several times. Ba was deposited at room temperature by means of a commercial Ba getter (SAES getters). The temperature of the specimen was monitored by means of an infrared radiation thermometer.

3. Result and discussion

Figure 1 shows a high-resolution STM image of 0.15ML of Ba on Si(100) followed by annealing at 1160K for 2 min (sample bias voltage (Vs) = -1.53 V, tunneling current (It) = 0.13 nA). Three characteristic structures, a 2x3, a c(4x2) and a chain-shaped structure, are observed on the surface. Details for the c(4x2) and chain-shaped structures are described elsewhere[4]. A

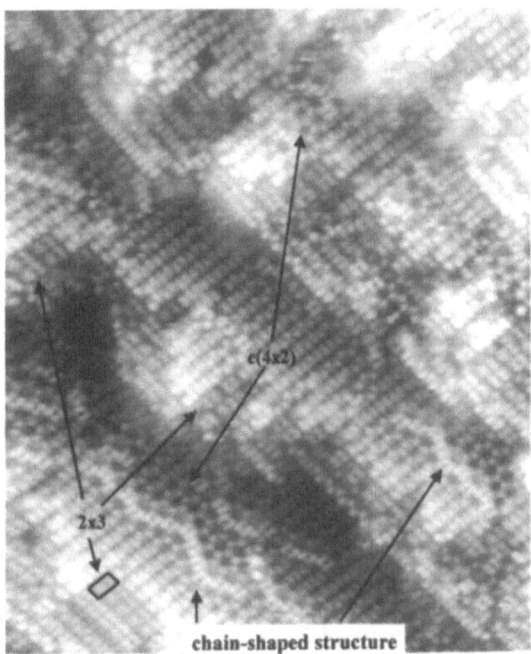

Fig. 1 A filled-state STM image of 0.15ML of Ba on Si(100) (Vs = -1.53 V, It = 0.13 nA). A unit cell of the 2x3 is indicated in a black rectangle.

unit cell of the 2x3 is illustrated as a rectangle in the figure. The length of the unit vector along the Si dimer row of original Si(100)2x1 surface is $3a_0$ (a_0, the distance of Si-Si on a bulk-terminated Si(100), 0.384 nm) and that perpendicular to the dimer row is $2a_0$.

Though the 2x3 structure is partially observed at the step edge, it also locates on the terrace in other images. The complete 2x3 phase results from spreading of partial 2x3 structure.

STM images of the 2x3 structure have a sample bias voltage dependence. This nature is more remarkable in empty-state images than in filled-state images. Figures 2 show empty-state STM images at (a) Vs = 1.3 V and (b) Vs = 2.0 V at the Ba coverage of ~1/3ML. A unit cell of the 2x3 is indicated in rectangle in both images where they show the identical position. At low sample bias voltage (LV), two protrusions are observed in a unit cell (Fig.2(a)), while at high sample bias voltage(HV), only one protrusion is observed (Fig. 2(b)). The STM measurements show that one protrusion is included in a unit cell above 1.5V and two protrusions below 1.5V. It is

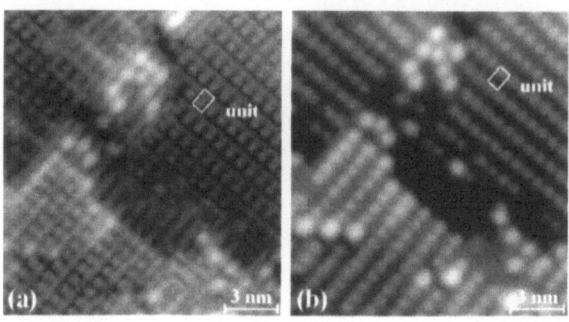

Fig.2 Empty-state STM images of the same area at (a) Vs = 1.3 V and (b) Vs = 2.0 V. A unit cell of the 2x3 is illustrated in both images.

also found from Figs.2 that a darker protrusion at LV corresponds to a bright protrusion at HV whereas a brighter protrusion at LV disappears at HV. Those results imply that the electronic density of states of the 2x3 structure is highly localized.

In order to examine the sample bias voltage dependence of STM images, we analyze the positional relation between protrusions in STM images of the 2x3 structure and atoms in a 2x3 model. Figure 3(a) shows positional relations between protrusions of STM images and the Si substrate. Figure 3(b) shows a model of the 2x3 structure [4]. In comparison of Fig. 3(a) with (b), two protrusions observed in empty-state images at LV locate on two sites;(1) the on-top site of a Ba dimer and (2) the site between two Si which do not bond with Ba atoms. As the electronegativity of Ba is 0.9 and that of Si 1.8, it is considered that valence electrons of Ba transfer to unoccupied state of Si. Then it is assumed that protrusions observed in empty-state images are due to Ba dimers. Since there is no Ba dimer at the site (2), the protrusion observed at the site (2) is due to two Si dangling bonds. In comparison of empty-state images with filled-state images (not shown here), it is found that the protrusion observed at HV in empty-state images locate on the site (1), wherefrom it is said that the other protrusion observed at LV locate on the site (2). It is assumed from this result that the unoccupied state of Ba dimer lies at higher energy than that of Si dangling bond of the 2x3 structure.

Though STM images are not shown here, filled-state images also have a sample bias voltage dependence. Circular and elliptical protrusions (Fig. 3(a)) have same brightness at HV, whereas elliptical one becomes brighter than circular one at LV.

In order to discuss sample bias voltage dependence of STM images in detail, further study such as scanning tunneling spectroscopy (STS) is needed.

4. Conclusion

The 2x3 structure on Ba-deposited Si(100) surfaces is investigated by STM. STM images of the 2x3 structure

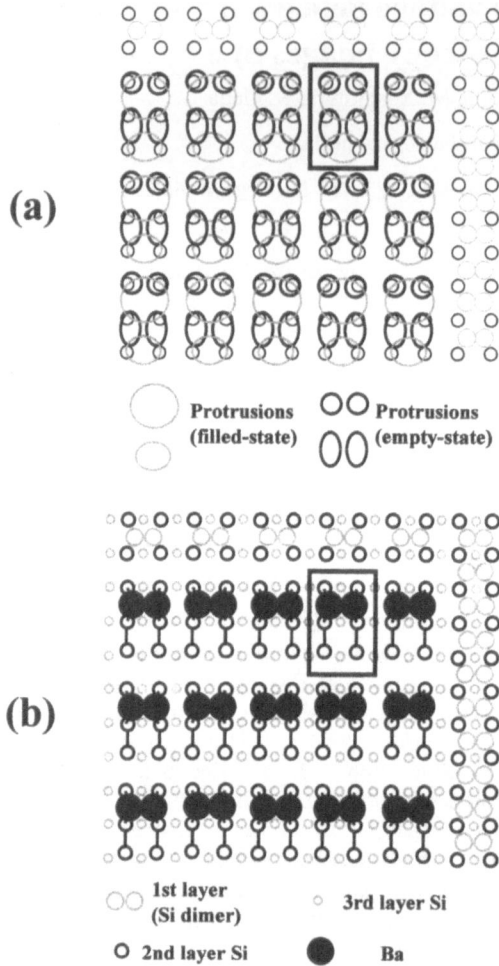

Fig. 3 (a) Positions of protrusions observed on Si(100) and (b) a model of the 2x3 structure.

have the sample bias voltage dependence, especially in empty-state images. Based on examination of both filled-state and empty-state images, it is suggested that the unoccupied state of Ba dimer lies at higher energy than that of Si dangling bond of the 2x3 structure.

Acknowledge

This work was supported by Grant-in-Aid for Scientist Research (Encouragement of Young Scientists #12750029).

References

1. W.C. Fan, and A. Ignatiev, Surf. Sci. **253**, (1991) 297.
2. T. Urano, K. Tamiya, K. Ojima, S. Hongo, and T. Kanaji, Surf. Sci. **357-358**, (1996) 459.
3. X. Hu, X. Yao, C.A. Peterson, D. Sarid, Z. Yu, J. Wang, D.S. Marshall, R. Droopad, J.A. Hallmark, and W.J. Ooms, Surf. Sci. **445**, (2000) 256
4. K. Ojima, M. Yoshimura, and K. Ueda, submitted to Surf. Sci.

Proc. 25th Int. Conf. Phys. Semicond., Osaka 2000 (Eds. N. Miura and T. Ando)

449

Removal of Particles from Surface of Silicon Using Clean Solutions with Surfactants and Chelates

Cao Baocheng[1*], Yu Xinhao[2], Ma Honglei[3], Ma Jin[4], Liu Zhengli[5]

[1-4] Institute of Optoelectronic Materials & Device, Shandong University, Jinan, 250100, P.R.CHINA

[5] Institute of Semiconduct, Chinese Academy of Sciences, Beijing CHINA

Abstract The most widely used system for silicon surface cleaning so far is the RCA standard clean developed by Kern in the fabrication of integrated circuits. This paper reports a new type of cleaning technique using clean solution with surfactants and chelates, and the results of cleaning using the two types of technique are compared. The results are in terms of the fixed charge density of the silicon dioxide layer obtained by high frequency C-V measurement, the surface metallic concentration obtained by the atomically absorption spectrometry, the silicon surface chemical composition obtained by X-ray photoelectron spectra, and the morphology obtained by atomic force microscope. It is demonstrated that the new technique is just as effective as the RCA standard clean technique, but as to the impact on the surface roughness of silicon wafer cleaned, the new technique is significantly better than the RCA standard system. In addition, the new technique has the advantages of being convenient in operation, free of toxicity, costing less, and health friendly.

1 Introduction

In the many years of studying the cleaning effect of the RCA standard technique, Gluck has demonstrated the RCA standard technique is very effective in removing metallic contaminants [1], Kern demonstrated that the RCA standard technique may leave a 1-1.5nm thick of silicon dioxide passivation film [2], and Okumura observed that the RCA technique has a fairly remarkable roughening effect [3]. This paper reports a new type of cleaning technique for semiconductor manufacture using clean solution with surfactants and chelates, and the results of cleaning using the two types of technique are compared.

2 Experiment

The new cleaning technique for semiconductor manufacture reported in this paper involves the use of new cleaning reagents designated as DGQ-1 and DGQ-2. Make a cleaning solution by mixing one part of DGQ-1 or DGQ-2 respectively with 19 parts of DI water, conduct ultrasonic cleaning at the temperature of 50-60°C, then rinse with DI water. Intentionally contaminate the polished surface of P type (1 1 1) 2 Ω-cm silicon wafer with finger prints and with wax at the back. Group them randomly into two. Clean the one group with the cleaning technique developed by us using DGQ-1 and DGQ-2 cleaning reagents, the other with RCA standard cleaning technique generally used on the CMOS line. Dry under infrared lamps before sealing them (all these are conducted in clean rooms of 100 order) and measuring them.

* Present address: Institute of Optoelectronic Materials & Devices, Shandong University, Shanda south road No:27, Jinan, 250100,CHINA.

3 Results and Discussion

3.1 High frequency C-V measurement

Nowadays high frequency C-V measurement is used in deciding and studying the cleaning effects. We separately conduct H O synthesized gate oxidation in silicon wafers cleaned with the new technique and those CMOS devices cleaned with the RCA standard technique. The oxidic layer of around 50 nm thick is obtained. After that, MOS capacitor is made and high frequency (1MHZ) C-V testing is conducted. Almost identical high frequency C-V curves of MOS capacity in both techniques are manifested, and fixed charge densities near SiO_2 side at the SiO_2-Si interface in both cases are about $1 \times 10^{11}/cm^2$.

3.2 capacity in cleaning alkali metals and heavy metal ions

In the manufacture of semiconductor devices, much attention is paid on the capacity of cleans in removing the alkali metal and heavy metal ions from the surface of silicon wafers. We conducted atomically absorption spectrometry in the silicon wafers cleaned using the RCA standard technique and those cleaned using the new technique. Results show that within instrumentally identifiable ranges, the two techniques have equal capacities for cleaning the above mentioned metallic vestiges. The alkali metallic vestige for Na and K is less then $3.1 Pg/cm^2$, fully meeting the cleaning requirements for MOS devices before gate oxidation.

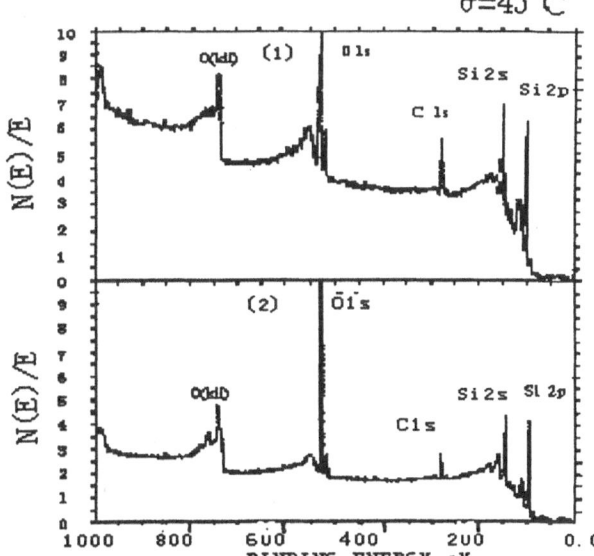

Figure 1: Full scanning of wafer surface by X-ray photoelectron spectra, (1) represents RCA standard technique, (2) represents the new technique.

3.3 Measurement with X-ray photoelectron spectra

There are two predominant ways to clean and passive silicon surfaces chemically. The first is to grow a thin layer of oxide in the act of cleaning. This is best accomplished using acidic or basic solutions mixed with hydrogen peroxide and is the basis of the RCA Standard clean developed by Kern in 1965. These clean leave1-1.5 nm of hydroxylated oxide on the silicon surface, which prevents recontamination of the Silicon[2]. Such surfaces are hydrophilic in nature and are easily wetted by aqueous solution. The second way to clean and passivate the surface is to dissolve the surface oxide completely in hydrofluoric acid. Then, what are the effects of the new type of cleaning technique using clean solution with surfactants and chelates? For this purpose, we separately used the RCA standard cleaning technique and the new type of clean with surfactants in processing silicon wafers and measured the results with X-ray photoelectron spectra. In Figure 1 that gives the full scan of the wafer, (1) represents RCA standard technique while (2) represents the new technique. It can be seen clearly in the figure that the wafers processed with the two types of technique yield similar surface electron structures. The wafer surface after cleaning is largely composed by Silicon, Oxygen and Carbon, in the concentration rates shown in Table 1, in which we see that the new type of cleaning technique yields less carbon contamination than the RCA standard type.

Table1:Elements on wafer surface and their concentration(%)

Elements	New technique Concentration(%)	RCA standard Concentration(%)
SI2p	45.45	39.52
C1s	16.16	20.85
O1s	38.39	39.63

It is known that the peaks in photoelectron energy spectra are correspondent to the related ion states. The peak with a binding energy of 284.7ev is a C1s. This peak comes from the contamination of hydrocarbons [4]. The peak with a binding energy of 532.84ev and 103.4 ev are the character peak of silicon dioxide. The peak with a binding energy of 99.45ev is the character peak of mono-crystal silicon. From the above measurement and analysis, we can say that the two techniques obtain similar wafer surfaces, with a thin silicon dioxide layer on the cleaned surfaces. This result complies with the conclusion made by Kern who reported that cleaning of semiconductor surface by the RCA standard technique produces a 1-1.5nm silicon dioxide passivation film. Moreover, the two techniques involve carbon contamination of equal quantities.

3.4 The microstructure on the wafer surface

It is known that the Microstructure State on the wafer surface has immediate effects on the features of future devices. Surface roughness causes lowered breakdown field intensity in the thin gate oxidation layer and reduces the channel mobility's[5-6]. We observed and photographed the wafer surfaces processed by both cleaning techniques with the atomic force microscope (See Figure 2). The silicon wafer surface cleaned with the new technique has less defect lattice

points and is smoother than that cleaned with the RCA technique. Results show: with respect to roughening effect, the new technique has advantages over the RCA standard technique.

4 Conclusion

Through high-frequency C-V measurement, measure-ments using atomically absorption spectrometry, X-ray photoelectron spectra, and atomic force microscope, we see that the RCA standard cleaning technique and the new type of cleaning technique using clean solution with surfactants and chelates have equal effects, both result in a thin film of silicon dioxide on the wafer surface, and both involve organic carbon contamination of equal quantities. With respect to the roughening effect, the new technique is better than the conventional one.

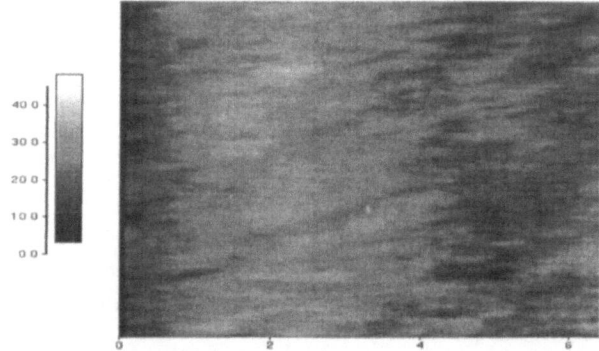

(1) Represents the new technique.

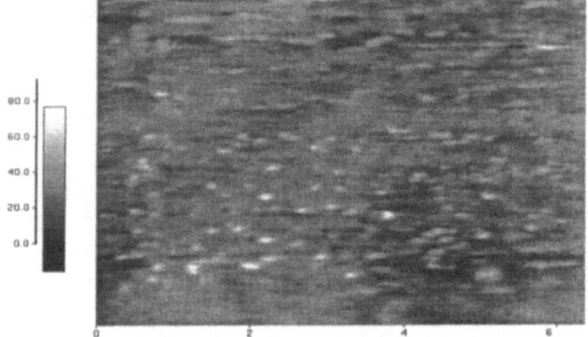

(2) Represents the RCA standard technique
Figure 2: AFM photograph of cleaned wafer.

References

1. R.M. Gluck, ECS Ext. Amstar. (1978) 78-2:640
2. W. Kern, "*Handbook of semiconductor wafer cleaning technology*" (Noyes Publication 1993).
3. H. Okumura, T. Akane, and Y. Tsubo, et al. Ibid. 144(11) (1997) 3765
4. J.Q. Wang, *Extended Discuss on XPS/XAES/UPS*, (Publishing House on National Defense Industry, 1992).
5. P.O.Hahn, and M.J.Henzler, Vac Sci. Technol. A2:574 (1984)
6. T.Ohmi, M.Miyashita, and T.Imaoka, *Proc. of the Microcontamination Meeting*, San Jose,CA,(Canon Communications 1991) p.491

4 Heterostructures and Superlattices: Optical

Phase Difference between Coherent GaSb-like and AlSb-like LO Phonons in GaSb/AlSb Superlattices

H. Takeuchi[1], K. Mizoguchi[1], M. Nakayama[1], K. Kuroyanagi[2], T. Aida[2]

[1] Department of Applied Physics, Faculty of Engineering, Osaka City University, 3-3-138 Sugimoto, Sumiyoshi-ku, Osaka 558-8585, Japan
e-mail: takeuchi@a-phys.eng.osaka-cu.ac.jp

[2] ATR Adaptive Communications Research Laboratories, 2-2 Hikaridai, Seika-cho, Soraku-gun, Kyoto 619-0288, Japan

Abstract The coherent longitudinal optical (LO) phonons in GaSb/AlSb superlattices are investigated by a pump-probe technique. We have simultaneously observed coherent GaSb-like and AlSb-like LO phonons which are confined in the quantum well and barrier layers, respectively. We have also observed the difference between the initial phase of the coherent GaSb-like and AlSb-like LO phonons, which indicates that the different generation mechanisms exist between the LO phonons confined in the well and barrier layers.

1 Introduction

Recently, coherent phonons in GaAs/AlAs superlattices (SLs) have been investigated extensively. Mizoguchi et al. discussed the generation and detection mechanism of coherent folded longitudinal acoustic (FLA) phonons observed by using a two-color pump-probe technique [1]. Yee et al. reported the effect of the interwell coupling of electrons on the generation of coherent GaAs-like longitudinal optical (LO) phonon-plasmon coupled mode [2]. However, there are no reports on the observation of the coherent LO phonons confined in the barrier layers of the SLs. Moreover, there are no studies on the phases of the coherent phonons in SLs although the time-domain spectroscopies with the femto-second pulse laser are sensitive to the phase of the oscillation [3,4]. In the present work, we report the first study on the coherent LO phonons in GaSb/AlSb SLs. We have detected the coherent AlSb-like LO phonons in the barrier layers in addition to the GaSb-like LO phonons in the well layers. By utilizing the time-domain spectroscopy, we have investigated the difference between the initial phase of the coherent GaSb-like and AlSb-like LO phonons. We discuss the generation mechanism of the coherent LO phonons in SLs.

2 Experimental Procedures

The samples used are the $(GaSb)_m/(AlSb)_m$ (m=8, 18, 33) SLs grown along the (100) direction by molecular-beam epitaxy, where the subscript m denotes the number of GaSb or AlSb monolayers (a thickness of one monolayer is 0.31 nm). Hereafter, we will call the $(GaSb)_m/(AlSb)_m$ superlattice "(m,m) SL". Reflection-type pump-probe experiments were performed at room temperature by using a mode-locked Ti:sapphire pulse laser with the duration time of 20 fs. The pump power was 150 mW. The photon energy of the pump and probe pulses was

1.55 eV, which is lower than the band gap energy of AlSb, while is higher than the first interband transition energy in the SLs. Thus, photo-excited carriers are generated only in the GaSb layers. The time derivative of the reflectivity change was recorded as a function of the time delay between the pump and probe pulses.

3 Results and Discussion

Figure 1(a) shows the oscillatory part of the time derivative of the reflectivity change in the (18,18) SL and bulk GaSb. The decay time of the oscillation in the (18,18) SL is about 5.5 ps which is longer than that in bulk GaSb (~1.1 ps). In bulk crystals, the relaxation processes of the coherent optical phonons are dominated by the energy relaxation into acoustic phonon branches [5]. In the SLs, the phonon dispersion is folded in the mini-Brillouin zone. This restricts the energy relaxation process of the LO phonon, which results in the longer decay time. The oscillation of the time-domain signal shows the beat in the (18,18) SL, which reflects that the several phonon modes are generated in the (18,18) SL. In order to analyze the oscillatory structures of these time-domain signals, we peformed the Fourier transform (FT) of the time-domain signals as depicted in Fig. 1(b). All spectra show the peak at 7.0 THz. This frequency agrees with that of the LO phonon of GaSb. The bandwidths of the coherent GaSb-like LO phonons in the SLs are narrower than that in bulk GaSb. This reflects the longer decay time in the SLs than that in bulk GaSb as shown in Fig.1 (a). The spectra of the SLs show another peak at 10.2 THz. This frequency agrees with that of the LO phonon of AlSb. Thus, both the coherent GaSb-like and AlSb-like LO phonons are simultaneously observed in the SLs. The full width at half maximum of the coherent AlSb-like LO phonon band is almost equal to that of the coherent GaSb-like LO phonon band (~0.1 THz): The relaxation time of the coherent AlSb-like LO phonon is the same as that of the coherent GaSb-like LO phonon. This indicates that the energy relaxation process of the coherent AlSb-like LO phonon is also restricted.

In order to investigate the initial phase difference between the coherent GaSb-like and AlSb-like LO phonons, we plotted the imaginary part of the FT components, $A(\omega)$, as a function of the real part of those shown in Fig. 2; the so-called Cole-Cole's plot. The $A(\omega)$ of the

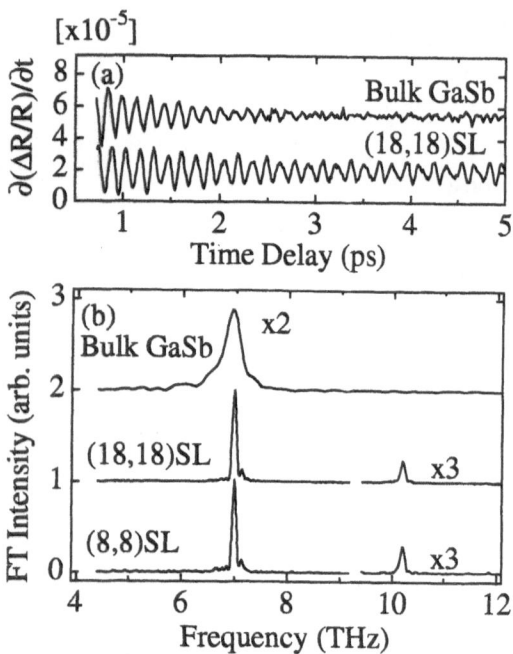

Fig. 1 (a) Oscillatory part of the time derivative of the reflectivity change as a function of the time delay at room temperature. (b) FT spectra in bulk GaSb, (18,18) SL and (8,8) SL.

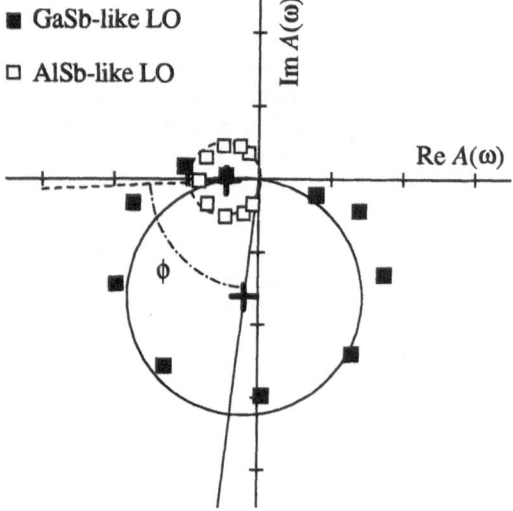

Fig. 2 Cole-Cole's plot of the FT components $A(\omega)$ of the coherent phonons in the (18,18) SL.

coherent GaSb-like LO phonon with the frequency of 7.0 THz and that of the coherent AlSb-like LO phonon with the frequency of 10.2 THz are indicated by closed and open squares, respectively. The fitted results of the data are represented by the two circles whose centers are depicted by the crosses. The angle ϕ between the solid and broken lines corresponds to the initial phase difference between the coherent GaSb-like and AlSb-like LO phonons. The initial phase difference in the (18,18) SL is about 90 degree. This value is almost the same in the

(8,8) and (33,33) SLs. When the coherent LO phonons confined in each constituent layer would be generated by the same mechanism, coherent LO phonons with the same symmetry oscillate almost in phase [3]. Our finding for the initial phase as shown in Fig.2 indicates the existence of the different generation mechanisms between the coherent LO phonons confined in the well and barrier layers. We consider the generation mechanism of the coherent GaSb-like and AlSb-like LO phonons as follows. At first, the pump pulse generates the carriers only in the GaSb layers. The photo-excited carriers generate the coherent GaSb-like LO phonons via the instantaneous electron-phonon interaction. Next, the polarization due to the coherent GaSb-like LO phonons induces the coherent AlSb-like LO phonons in the adjacent AlSb layers. This is responsible for the difference in the generation mechanism and the initial phase difference between the coherent GaSb-like and AlSb-like LO phonons. The details of the discussion will be reported elsewhere.

4 Summary

We have studied the coherent LO phonons in GaSb/AlSb SLs. The coherent GaSb-like and AlSb-like LO phonons have been simultaneously observed. The decay times of the coherent LO phonons in the SLs are longer than that in bulk GaSb, because the energy relaxation processes of the coherent LO phonons are restricted by folding the phonon dispersion in the SLs. The initial phase difference with 90 degree between the coherent GaSb-like and AlSb-like LO phonons will be due to the difference in the generation mechanism between the two coherent phonons.

5 Acknowledgement

This work is partially supported by a Grand-in-Aid for Scientific Research on Priority Areas, "Photo-induced Phase Transition and Their Dynamics", from the Ministry of Education, Science, Sports and Culture of Japan.

References

1. K. Mizoguchi, M. Hase, S. Nakashima and M. Nakayama, Phys. Rev. B**60**, (1999) 8262.
2. K. J. Yee, D. S. Yee, D. S. Kim, T. Dekorsy, G. C. Cho and Y. S. Lim, Phys. Rev. B**60**, (1999) 8513.
3. H. J. Zeiger, J. Vidal, T. K. Cheng, E. P. Ippen, G. Dresselhaus and M. S. Dresselhaus, Phys. Rev. B**45**, (1992) 768.
4. G. A. Garrett, T. F. Albrecht, J. F. Whitaker, and R. Merlin, Phys. Rev. Lett. **77**, (1996) 3661.
5. M. Hase, K. Mizoguchi, H. Harima, S. Nakashima and K. Sakai, Phys. Rev. B**58**, (1998) 5448.

Plasmon-phonon coupling at $Ga_{0.5}In_{0.5}P$/GaAs heterointerfaces induced by CuPt-type ordering

K. Yamashita[1], T. Kita[1], T. Nishino[1]*, Y. Wang[2], K. Murase[2], C. Geng[3], F. Scholz[3], H. Schweizer[3]

[1] Department of Electrical and Electronics Engineering, Faculty of Engineering, Kobe University, 1-1 Rokkodai, Nada, Kobe 657-8501, Japan e-mail: biwa@kobe-u.ac.jp

[2] Department of Physics, Graduate School of Science, Osaka University, 1-1 Machikaneyama, Toyonaka 560-0043, Japan

[3] Physikalisches Institut, Universität Stuttgart, Pfaffenwaldring 57, D-70550 Stuttgart, Germany

Abstract We found plasmon-phonon coupled modes and electric-field-induced mode in Raman-scattering spectra of long-range ordered $Ga_{0.5}In_{0.5}P$ and GaAs heterointerface. Raman shifts of the coupled modes enable us to estimate an exact carrier density accumulated at the interface. Photoluminescence excitation spectra of GaAs luminescence show Franz-Keldysh oscillations induced by internal electric field in the ordered $Ga_{0.5}In_{0.5}P$. These results suggest that the long-range ordering in $Ga_{0.5}In_{0.5}P$ is considered to play an important role in the carrier accumulation at the heterointerface.

1 Introduction

$Ga_{0.5}In_{0.5}P$ alloys, which are lattice-matched to a GaAs substrate, have been used for optical and electronic devices. Especially, to fabricate heterojunction-bipolar transistors (HBTs), a band alignment of the $Ga_{0.5}In_{0.5}P$ and GaAs heterointerface is very important. Interdiffusion [1] and carrier-accumulation [2] effects at the heterointerface have been suggested as origins of degradations in carrier-transport properties of the $Ga_{0.5}In_{0.5}P$/GaAs HBTs. Especially, it was suggested by C-V measurements that long-range ordering in $Ga_{0.5}In_{0.5}P$ leads a carrier accumulation at the heterointerface.[2] In this study, we investigated effects of the long-range ordering on the $Ga_{0.5}In_{0.5}P$/GaAs heterointerfaces by polarized Raman-scattering and photoluminescence excitation (PLE) measurements.

2 Experiments

$Ga_{0.5}In_{0.5}P$/GaAs samples measured in this study were grown on no-doped GaAs(001) substrates by metalorganic vapor-phase epitaxy. Order parameter η was controlled by varying the growth temperature. In Raman-scattering measurements, Ti: sapphire and DCM dye lasers were used for the sample excitation in order to excite the heterointerface selectively. We define that x, y, and z are unit vectors of the zincblende structure, and that X and Y are the $[\bar{1}10]$ and $[110]$ directions, respectively. In PLE measurements, a monochromized light of a tungsten lamp were used for an extremely weak excitation of ~ 50 $\mu W/cm^2$.

---

* *Present address*: Kobe City College of Technology, 8-3 Gakuen Higashimachi, Nishiku, Kobe 651-2194, Japan

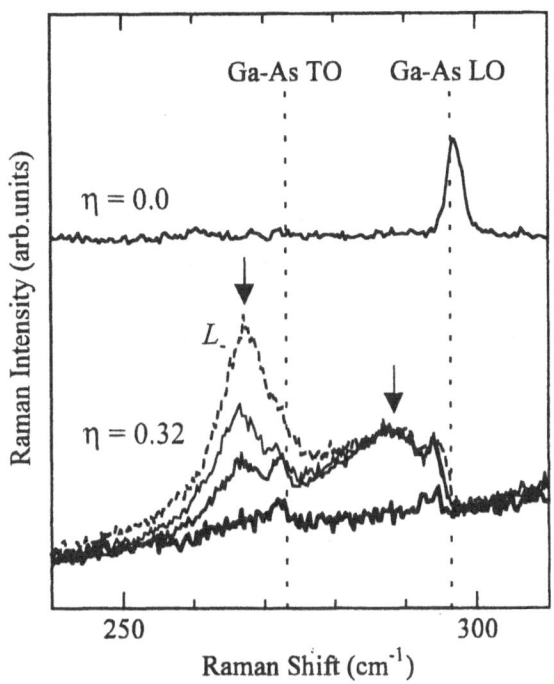

Fig. 1 Polarized Raman-scattering spectra of disordered ($\eta = 0.0$) and ordered ($\eta = 0.32$) $Ga_{0.5}In_{0.5}P$/GaAs heterointerface. Thick, solid, thin, and dashed lines plot the data at $\bar{z}(Y,Y)z$, $\bar{z}(x,x)z$, $\bar{z}(X,X)z$, and $\bar{z}(x,y)z$ configurations, respectively.

3 Results and Discussion

3.1 Plasmon-phonon coupling

Figure 1 shows (001) back scattering-Raman spectra of the $Ga_{0.5}In_{0.5}P$/GaAs samples at 10 K. Dashed lines are typical frequencies of transverse optical (TO) and longitudinal optical (LO) modes for a bulk GaAs. In contrast that the spectrum of the disordered sample ($\eta = 0.0$) shows only the Ga-As LO mode, the TO and LO modes and two broad features at ~ 266 cm^{-1} and ~ 287 cm^{-1} were observed in the ordered sample ($\eta = 0.32$). In polarization dependence for the ordered sample, the broad mode at ~ 287 cm^{-1} appears at parallel polarization. Therefore, the mode at ~ 287 cm^{-1} is considered to be an electric-field-induced GaAs-like LO mode of an interdiffused InGaAs layer. Since both TO and LO modes are slightly shifted to the lower frequency side from those of the bulk GaAs, the GaAs layer is expanded by the inter-

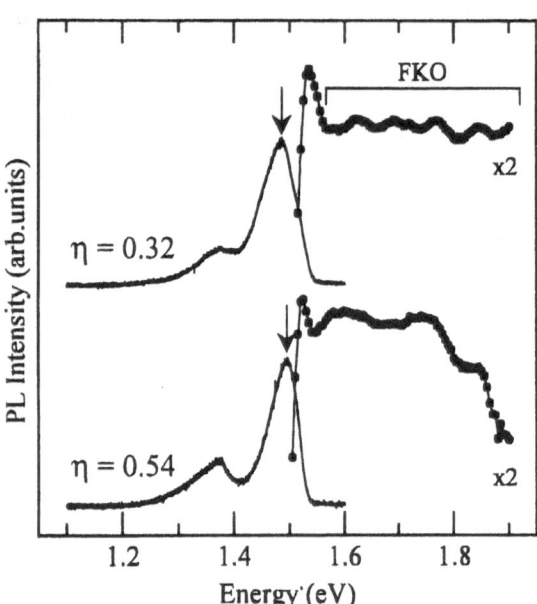

Fig. 2 PL (solid lines) and PLE (closed circles) spectra of ordered $Ga_{0.5}In_{0.5}P/GaAs$ samples with $\eta = 0.32$ and 0.54. The PLE spectra were detected at PL-peak energies indicated by arrows.

diffused InGaAs layer. On the other hand, the polarization dependence for the broad mode at ~266 cm^{-1} corresponds to that for lower branch of plasmon-phonon coupled mode (L_-).[3] In fact, we observed upper branch of the coupled mode (L_+) at ~530 cm^{-1}. Furthermore, we found that the frequency and the anisotropy of the L_+ and L_- modes depend on η. Generally, since p-type plasmon shows a large damping, the upper branch related to the hole plasmon is suppressed.[4] Therefore, accumulated carriers in this study are electrons, because of simultaneous appearance of the lower and upper branches. From the Raman shift of the upper branch of the coupled modes, the accumulated electron density is estimated to be ~3 x 10^{18} cm^{-3}.[3] The simultaneous observations of plasmon-phonon coupled mode and electric-field-induced mode suggest that internal electric field plays an important role for the electron accumulation.

3.2 Internal electric field

Figure 2 shows GaAs-PL spectra and their PLE spectra at 4.2 K for the $Ga_{0.5}In_{0.5}P/GaAs$ samples with η of 0.32 and 0.54. The PL spectra have two broad peaks at ~1.37 and ~1.49 eV. The PL peaks at ~1.49 eV are related to localized states of the GaAs, because their PLE spectra plotted by closed circles show absorption edges at ~1.52 eV. On the other hand, since PLE spectra of the PL peaks at ~1.37 eV revealed fundamental absorption edges at ~1.47 eV (which are not plotted in the figure). Therefore, the peaks at ~1.37 eV correspond to an emission from the interdiffused InGaAs layer. In the PLE spectra of the ~1.49-eV PL, we also found Franz-Keldysh oscillations (FKOs) caused by an internal electric field. Electric fields estimated from the FKOs

[5] were 128.7 and 238.7 kV/cm for the sample with η of 0.32 and 0.54, respectively. According to a previous paper, piezoelectric fields in ordered $Ga_{0.5}In_{0.5}P$ along [001] direction, which are calculated by the first principle pseudopotential method, are 230.9 and 923.8 kV/cm at $\eta = 0.5$ and 1.0, respectively.[6] The observed electric fields at the $Ga_{0.5}In_{0.5}P/GaAs$ heterointerfaces well agree with the calculated piezoelectric field when we assume that the piezoelectric field is proportional to η^2. From above results, it is considered that the piezoelectric field in the ordered $Ga_{0.5}In_{0.5}P$ causes the internal electric field near the heterointerface.

4 Summary

We found plasmon-phonon coupling in Raman-scattering measurements of long-range ordered $Ga_{0.5}In_{0.5}P/GaAs$ heterointerface. The accumulated carrier is electron of a density of ~3 x 10^{18} cm^{-3}. Furthermore, PLE spectra of the ordered samples reveal FKOs originated from piezoelectric field in the ordered $Ga_{0.5}In_{0.5}P$. Since electric-field-induced mode was observed together with the plasmon-phonon coupled modes in the Raman spectra, it is considered that the long-range ordering in $Ga_{0.5}In_{0.5}P$ plays an important role in the electron accumulation at the $Ga_{0.5}In_{0.5}P/GaAs$ heterointerface.

References

1. For example, C. Y. Tsai, M. Moser, C. Geng, V. Harle, T. Forner, P. Michler, A. Hangleiter, and F. Scholz, J. Cryst. Growth **145**, (1994) 786.
2. T. Kikkawa, K. Imanishi, K. Fukuzawa, T. Nishioka, M. Yokoyama, and H. Tanaka, Inst. Phys. Conf. Ser. **155**, (1996) 877.
3. A. Mooradian and A. L. McWhorter, Phys. Rev. Lett **19**, (1967) 849.
4. R. Fukasawa and S. Perkowitz, Phys. Rev. B **50**, (1994) 14119.
5. F. H. Pollak, in *Handbook of Semiconductor* (Elsevier, Amsterdam, 1994), Vol. 2, p. 527.
6. S. Froyen, A. Zunger. and A. Mascarenhas, Appl. Phys. Lett. **68**, (1996) 2852.

Optical properties and band alignment of ZnSe/ZnMgBeSe heterostructures

K.Godo[1], **M.W.Cho**[1]*, **J.H.Chang**[1], **S.K.Hong**[1], **H.Makino**[1], **T.Yao**[1] **M.Y.Shen**[2] and **T.Goto**[2]

[1] Institute for Material Research, Tohoku University, 2-1-1 Katahira, Sendai, 980-8577, Japan e-mail: `godo@imr.tohoku.ac.jp`
[2] Department of Physics, Graduate School of Science, Tohoku University,Sendai,980-8578, Japan

Abstract   Optical properties and band alignment of ZnSe/ZnMgBeSe heterostructures have been investigated. Two sets of ZnSe/ZnMgBeSe Multi quantum-wells that consists of 5 or 3 wells with different well thickness and 100 nm thick ZnMgBeSe barrier layers were grown on GaAs (100) substrates by molecular beam epitaxy (MBE). Low-temperature photoluminescence (PL) spectra show dominant sharp excitonic emission whose peak position systematically shifts to the higher energy side with decreasing the well thickness. These PL peaks show some stokes shifts which are attributed to excitons localized at hetero interfaces. Photoluminescence excitation (PLE) spectra showed optical transitions between excited quantum levels in addition to the ground levels. The standard analysis based on the effective-mass approximation gives the conduction band offset of $\Delta E_c = (0.6 \pm 0.1)\Delta E_g$. It is consistent with the semi-experimental calculation based on Harrison's LCAO Theory.

1 Introduction

Major developments in wide-band-gap II-VI semiconductor have recently occurred. Specially, Be chalcogenides have attracted considerable attention because of their strong covalent bonding that makes the materials robust against defect generation and propagation. This would result in improvement in laser performance. In fact, continuous operation of ZnCdSe/ZnMgBeSe QW LDs at 530 nm has been reported [1]. Most recently, the lasing wavelength as short as 455 nm has been demonstrated on ZnSe/ZnMgBeSe QW under optical pumping[2]. Although the optimization of the laser structure needs the precise information of the band alignment, there have been only few studies on the band alignment of ZnSe/ZnMgBeSe heterostructures.

In this report, we study the optical properties and the band offsets of ZnSe/ZnMgBeSe heterostructures by means of PL and PLE. In order to evaluate the band offsets, experimental results are compared with the excitonic transition energies calculated by the effective-mass approximation taking into account for the exciton binding energy and strain effects.

2 Experimental procedure

Two different ZnSe/ZnMgBeSe MQW structures were grown on GaAs (001) substrate by MBE. One MQW structure consists of 5 wells with different thickness of 0.62 , 1.2 , 2.5, 5.0, 10 nm and 100 nm thick ZnMgBeSe

---

* *Present address:* LG Electronics Institute of Technology MTS OE Team,Devices & Materials Lab, 16 Woomyeon-Dong, Seoul 137-724, Korea

cladding layers which lattice-match to GaAs substrate. The other consists of 3 wells with thickness of 6.3, 10, and 12 nm and ZnMgBeSe barriers with the same alloy composition(MgSe=15% , BeSe=10%). The well thickness was confirmed by transmission electron microscopy (TEM). PL were performed using a 32cm monochromator and a He-Cd laser. For PLE measurements, we used a Xenon lamp dispersed by a 32 cm monochromator as a tunable excitation source.

3 Results and discussion

XRD results of this structure shows that the (004) peak from the ZnMgBeSe cladding layer exhibits a line width as narrow as 80 arcsec reflecting high crystal quality. From the (115) diffraction peak, we found that ZnMgBeSe barrier lattice-matched to GaAs substrate, i.e., the ZnSe wells have a biaxial in-plane compressive strain from ZnMgBeSe barrier. This compressive strain remove the degeneracy of ZnSe valence bands and change the band gaps ($E_{g\ hh} = 2.824eV$, $E_{g\ lh} = 2.835eV$).

Dotted line in Fig1 shows PL spectra at 10K for a ZnSe/ZnMgBeSe MQW consists of 3 QWs. The emission peak 3.062eV is assigned to the ZnMgBeSe cladding layer and the three sharp peaks at around 2.82eV are assigned to excitonic emissions from each QWs as denoted in fig 1. They were compared with exciton absorption spectra measured by PLE or optical absorption. We noted that these PL peaks have stokes shifts of typically[19] meV. This indicate some possibility that the QW structures have interface fluctuations of a few monolayers in lateral size or fluctuations of ZnMgBeSe layer's composition. Such fluctuations are also reflected in the peak shape. In our case, the full-width at half maximum (FWHM) of the PL peak systematically change with the well thickness (not shown here). Effects of fluctuations for PL spectral shapes were reported by Singh and Bajaj[3]. If we calculate a well thickness dependence of the FWHM in compliance with the theory, it is in agreement with the tendency of experimental results. So the stokes shifts observed here are attributed to excitons localized at the hetero interface.

Fig 2 shows a plot of the peak positions of excitonic absorption against the well thickness. Inter-band transition energy between quantum states,(E_{c-v}), can be defined as

$$E_{c-v} = E_g \pm \Delta E_{strain} + E_c^{QW} + E_v^{QW} - E_{ex} \qquad (1)$$

where E_g is band gap energy of bulk ZnSe, and ΔE_{strain} is the change of band gap due to strain effects. E_c^{QW}

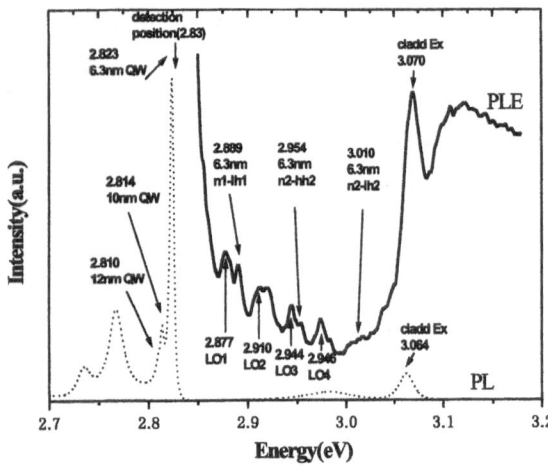

Fig. 1 PL(dotted line) and PLE(solid line) spectra at 10K from ZnSe / ZnMgBeSe MQW consists of 3 wells (6.3nm 10nm 12nm). The PLE detection energy was set to 2.830eV.

and E_v^{QW} are the quantum effect of conduction and valence band respectively calculated by the effective-mass approximation. E_{ex} is a quasi-two dimensional exciton binding energy calculated by H. Mathieu's method [4]. In this calculation, the heavy-hole, light-hole and electron effective masses of ZnMgBeSe are unknown parameters. Since there is not so much change in the results even though we vary the effective masses as estimated value by the $k \cdot p$ method, we employed the values of bulk ZnSe. The solid line indicates the calculated result for E_{1e-hh} as a function of the well thickness, assuming the conduction band offsets to be 60% of total band offsets($\Delta E_g = \Delta E_g^v + \Delta E_g^c$). We see a fairly agreement between the calculation and experimental results.

The solid line in Fig1 shows the PLE spectrum detected at 2.83eV which is a little higher energy than the PL peak of the 6.3 nm QW. It is reasonable to consider that the PLE spectra reveal optical transitions between quantum levels in 6.3nm QW. In addition to clear multi phonon relaxations, several transitions were carefully assigned as denoted in fig 1. This can be also compared with the above mentioned calculation. It gives the conduction band offsets $\Delta E_g^c = 0.6 \pm 0.1 \Delta E_g$.

Let us now turn to compare the estimated band offsets to some semi-empirical predictions. The most widely quoted model is the common cation anion rule. If a ZnSe/ZnMgBeSe hetero structure comply with the rule, the valence band offsets is hardly being. The evaluation here hardly agrees with the prediction. Next, we compared with another semi-experimental calculation based on Harrison's LCAO [5]. Here, we used recent reported value for valence band maximum (VBM) of MgSe [6] instead of the Harrison's one. It is well known that LCAO theory have a principle error in the omission of the cation d orbitals. In our case, the Zn atom has a fully occupied d shell, whereas the Mg atom in its ground state has

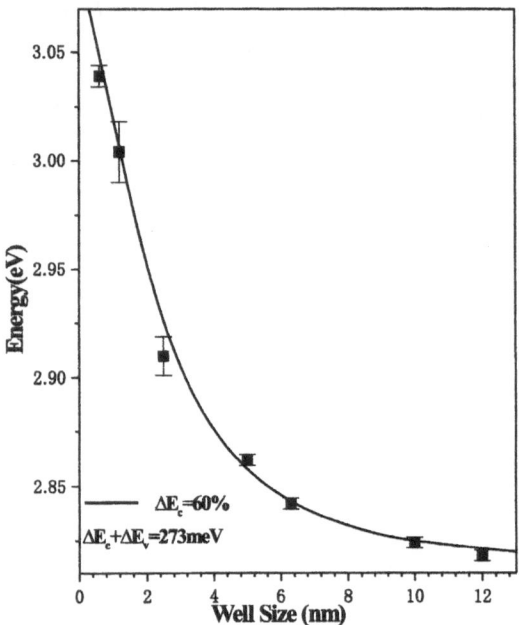

Fig. 2 A plot of the PLE and Absorption position against the well thickness.Solid line shows transition energy for E_{1e-hh} calculated by assuming the conduction band offsets to be 60% of total band offsets($\Delta E_g = \Delta E_g^v + \Delta E_g^c$)=273meV.

a partially occupied d shell, so Mg d orbitals affect the valence band. Actually, some experimental [7] and theoretical result[6] including the d orbital's effect showed that Mg-mixing the barrier layers of II-VI compound hetero structures increase the band offsets for both the heavy-hole and electron band. The obtained experimental result agrees fairly well with the predicted band offset of $\Delta E_c = 58\%$ by this caluclation.

4 conclusion

In conclusion, it is shown that the PL and PLE result indicate well confined exciton emission in ZnSe/ ZnMgBeSe MQW grown by MBE. The band offsets in this system are evaluated as $\Delta E_c = 0.6 \pm 0.1 \Delta E_g$ from the standard analysis based on the effective-mass approximation. The obtained result does not agree with the common cation rule but agrees fairly well with the estimated band offset of $\Delta E_c = 58\%$ by semi-experimental calculation based on Harrison's LCAO Theory.

References

1. A.Waag et al., J.Cryst.Growth ,**184/185,** (1998)1.
2. J.H. Chang et a., Appl. Phys. Lett,**75,**(1999)894.
3. Y.P.Varshni,Physica,**34,**(1967)149.
4. H.Mathieu et al.J.Appl. Phys,**72,**(1999)300.
5. W.A.Harrison,J.Vac.Technol.,**14,**(1977)1016
6. T.Nakayama,J.Phys.Appl.Phys,**33,**(1994)211
7. M.Wörz et al.,Phys.stat.sol.**202**(1997)805

Analysis of Raman spectra in quasiperiodic GaAs-AlAs heterostructures

E. Matsushita, T. Furuyama

Faculty of Engineering, Gifu University, Gifu 501-1193, Japan

Abstract   Theoretical analysis of Raman spectra of GaAs/AlAs heterostructures is studied to make clear essential properties of three types of one-dimensional quasi-periodic (Fibonacci, Thue-Morse and Dragon sequences) superlattices. The calculated Raman spectra are compared with the experimental data in good agreement, and are explained by identifying with the Fourier transformation of the quasi-periodic heterostructures composed of semiconducting GaAs and AlAs layers. The origin of characteristic properties to quasi-periodic structures such as the self-similarity, and the differences among Fibonacci, Thue-Morse and Dragon lattices are interpreted in terms of the correlation functions for their sequences and of X-ray diffraction patterns.

1 Introduction

In recent years, the basic research on semiconductor superlattices has been spurred to apply to the electronic and optical properties of quantum wells and the potential device with heterostructures. Fortunately, by the development of molecular beam epitaxy (MBE) technique, Raman scattering [1-3] and X-ray diffraction measurements [4,5] have been examined for new grown superlattices with one-dimensional (1D) GaAs/AlAs alternating layers.

In this paper, we present the theoretical analysis of Raman spectra in 1D GaAs/AlAs superlattices to investigate the effects of quasi-periodicity on phonon properties, and discuss the differences among three types of quasi-periodic superlattices in comparison with periodic and random lattices.

2 Model calculation

Here we adopt three types of 1D quasi-periodic sequences as

(a) Fibonacci: $S_1 = A$, $S_2 = AB$; $S_r = S_{r-1} + S_{r-2}$,

(b) Thue-Morse: $S_1 = A$, $S_2 = AB$; $S_r = S_{r-1} + S^*_{r-1}$,

(c) Dragon: $S_1 = A$, $S_2 = AAB$; $S_r = S_{r-1} + A + S'_{r-1}$,

where S^* is obtained by interchanging A and B in S, and S' by writing in backwards and interchanging A and B. The well-known order is (a) in which the ratio of the number of A and B is equal to the golden ratio $\tau = (1 + \sqrt{5})/2$.

Intensity of Raman scattering is described by the photoelastic model [6] as

$$I^\pm = \omega^\pm [n(\omega^\pm) + 1] |P|^2 , \qquad (1)$$

$$\omega^\pm = |k \pm q| v , \qquad (2)$$

where $n(\omega)$ is the Bose factor and $|P|$ is the Fourier

component of photoelastic coefficients which are taken as 0.48 for GaAs and 0.05 for AlAs with wave vector q. The sound velocities v are 4.72×10^5 and 5.71×10^5 [cm/sec] for GaAs and AlAs, respectively.   The superscript $\pm$ describes the phonon folding effect, because the doublet Raman peak frequencies $\omega^\pm$ of the folded acoustic phonon modes appear in quasi-periodic superlattices as a result of the change of Brillouin zone from $(-\pi/a \sim \pi/a)$ to $(-\pi/(m+n)a \sim \pi/(m+n)a)$ in $\{A\}_m/\{B\}_n$ heterostructure (the average lattice constant of GaAs and AlAs is estimated at $a \sim 5.66$Å). The calculated Raman spectra are shown in Figure 1 which is in good agreement with the results of Raman scattering measurements.

3 Results and discussion

Figure 1 reveals the calculated Raman intensity of quasi-periodic lattices of 14-th order of the sequence in the frequency region of acoustic phonons. As the sequence, $A=(GaAs)_{10}(AlAs)_{10}$ and $B=(GaAs)_{10}$ are assigned for (a)Fibonacci lattice (2048 layers), and $A=(GaAs)_{10}$, $B=(AlAs)_{10}$ are for (b)Thue-Morse (4096 layers) and for (c)Dragon (2048 layers) to make the same condition with the experimental data. [1], [3]   Zone-folded acoustic phonons in $\{A\}/\{B\}$ superlattices appear as doublet peaks of (m,n) modes in the Raman spectra. Thus Fourier spectra reflect the quasi-periodicity of the sequence.

Next to clarify the effect of quasi-periodicity, we indicate the intensity of the Fourier transformation of the superlattice chain in the 14-th order, $I(q)$, for monolayer stack of $\{A\}/\{B\}$ superlattices. In Figure 2, we can easily recognize that the characteristic properties to quasi-periodic structures such as the self-similarity are originated from the atomic sequence of GaAs/AlAs superlattices. Thus sharp diffraction peaks are shown at q/q_0 values in the superlattice with a strong correlation extending a great distance.

Here it is interesting to discuss the difference of the correlation functions $\Phi(n)$ in (a), (b) and (c), in comparison with periodic and random lattices. $\Phi(n)$ is corresponding to the structural factor of the quasi-periodic superlattice calculated from the Fourier transformation of the chain. Although superlattices have no periodicity, there exists strong correlation of lattice arrangements with the peculiar atomic sequence. The strength of the correlation is larger in the following order: periodic, Dragon, Fibonacci, Thue-Morse, random lattices.

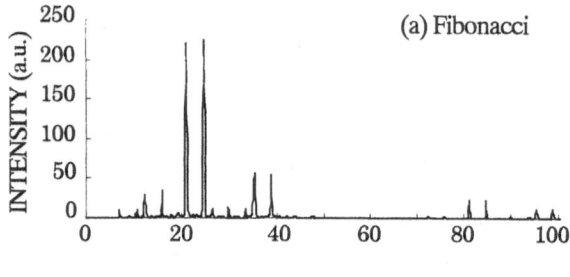

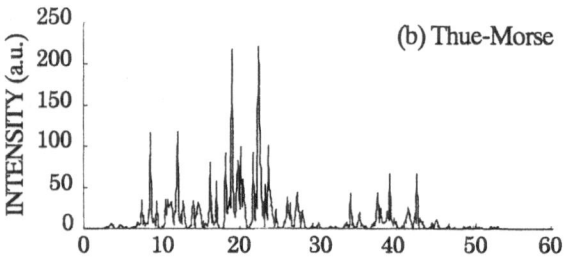

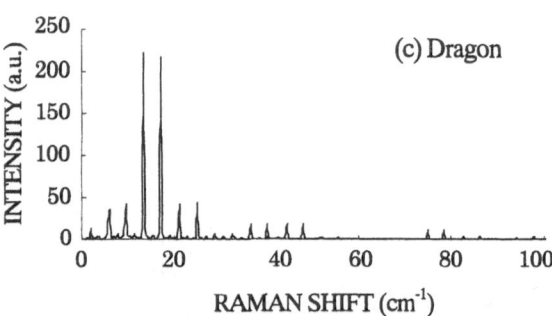

RAMAN SHIFT (cm⁻¹)

Fig. 1 Calculated Raman spectra of (a)Fibonacci, (b)Thue-Morse and (c)Dragon lattices for the generating sequence r=14. For (a), the superlattice becomes 2048 layers, and 4096 layers for (b) and 2048 layers for (c). Many sharp doublet peaks indicate the folded acoustic phonons.

The sharp satellite peaks which exhibit self-similarity of the structure were also observed in the X-ray diffraction pattern. Raman scattering has the potential to provide the structural information similar to satellite reflection which can be obtained by X-ray diffraction spectra.

In summary, essential properties of quasi-periodic sequences such as their self-similarities appear in the zone-folding doublet-peak frequency separation and come from only the atomic sequence in GaAs-AlAs superlattice. The characteristics of quasi-periodic heterostructures are interpreted in terms of the correlation functions in these superlattices which are quite different from them in periodic and random lattices. The Fibonacci lattice has most characteristic features to quasi-periodic superlattices within our theoretical analysis of Raman spectra of heterostructures. The detail will be discussed by combining with the calculated results for X-ray diffraction pattern in the next paper.

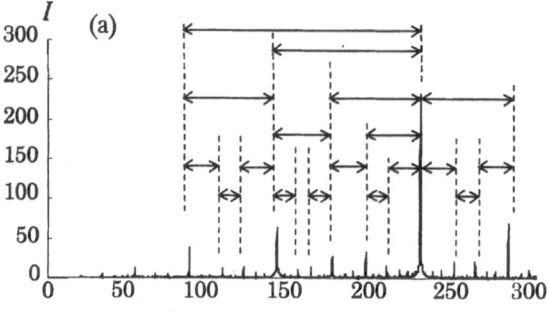

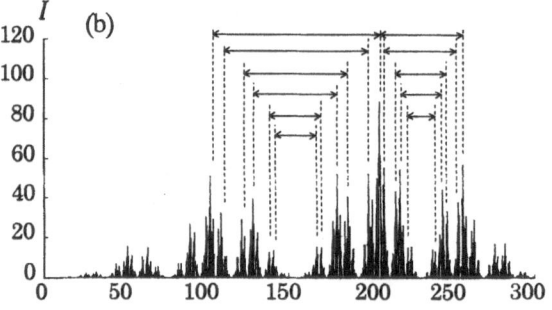

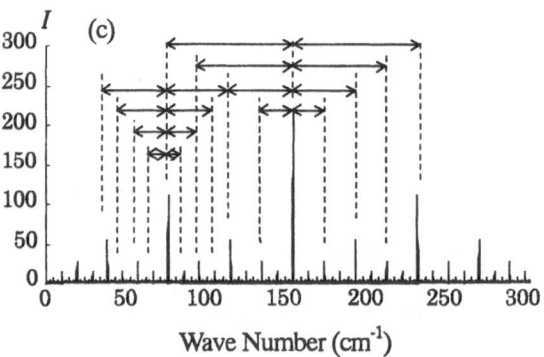

Wave Number (cm⁻¹)

Fig. 2 Intensity of the Fourier transformation of (a)Fibonacci, (b)Thue-Morse and (c)Dragon sequences. Self-similarity is shown explicitly as common feature to 1D quasi-periodic superlattices.

References

1. M. Nakayama, H. Kato and S. Nakashima, Phys. Rev. B **36**, (1987) 3472.
2. K. Mizoguchi, K. Matsutani, S. Nakashima, T. Dekorsy, H. Kurz and M. Nakayama, Phys. Rev. B **55**, (1997) 9336.
3. M. Ishida, K. Kamigaki, T. Morioka, H. Kato, N. Sano and H. Terauchi, J. Phys. Soc. Jpn. **61**, (1992) 149.
4. F. Axel and H. Terauchi, Phys. Rev. Lett. **66**, (1991) 2223.
5. H. Terauchi, M. Ishida, K. Kamigaki, H. Kato and N. Sano, J. Phys. Soc. Jpn. **61**, (1992) 1141.
6. C. Colvard, T.A. Gant, M.V. Klein, R. Merlin, R.Fischer, H. Morkoc and A.C. Gossard, Phys. Rev. B **31**, (1985) 2080.

Controlling Fermi-edge singularities
by a periodic external potential

Shintaro Nomura[1,2], Takeshi Nakanshi[2]*, and Yoshinobu Aoyagi[2]

[1] Institute of Physics, University of Tsukuba, 1-1-1 Tennoudai, Tsukuba, Ibaraki, Japan.
[2] The Institute of Physical and Chemical Research (RIKEN), 2-1 Hirosawa, Wako-shi, Saitama, Japan.

Abstract We demonstrate that the exponent of the Fermi-edge singularity can be controlled by a lateral periodic external potential in a n-type GaAs-GaAlAs modulation-doped quantum well structure with a free optically created hole. The asymmetric peak near the Fermi-level in the photoluminescence spectra shows the oscillation of the peak energy correlates with even filling factors of the Landau-levels in magnetic fields.

1 Introduction

The exponent of the power law divergence of the Fermi-edge singularity (FES) depends on the two competing effects, namely, the electrons-hole attractive Coulomb interaction and Anderson's orthogonality theorem [1]. The FES in systems with a finite-mass hole is critically depending on the dimensionality. In 3-dimensional (D) system, the FES is smeared out by the hole recoil effect, while the FES is present in 1-D regardless of the mass ratio between the electron and the hole [2]. This markedly contrasts with the original problem of the FES in systems with an infinite-mass or a localized hole considered by Mahan [1].

The balance between the electrons-hole attractive Coulomb interaction and Anderson's orthogonality theorem is critical in 2-D with a finite-mass hole. Theoretically disappearance of the FES was predicted at moderate electron densities [3–6]. On the other hand, the FES was observed in the absorption spectra [7], while the FES in the photoluminescence (PL) spectra was very weak in 2-D with a finite-mass hole [8]. It is thus possible to control the degree of the divergence by a weak lateral external potential in 2-D.

2 Experimental

The sample studied was based on molecular-beam epitaxy grown GaAs-AlGaAs n-type modulation-doped quantum well (MDQW) structures on a n-type GaAs substrate, which was used as a back contact [9]. The QW layer was embedded at 55 nm below the surface. The electron density was estimated to be 2.4×10^{11} cm^{-2} at 1.8 K by an optical Shubnikov-de Haas oscillations of the PL intensity at the Fermi energy and the lineshape of the PL spectra. Semi-transparent Schottky gate structure was fabricated on the surface with a square mesh of a period of 250 nm and a width of 25 nm by

* *Present address:* Department of Applied Physics and DIMES, Delft University of Technology, Lorentzweg 1, 2628 CJ Delft, The Netherlands

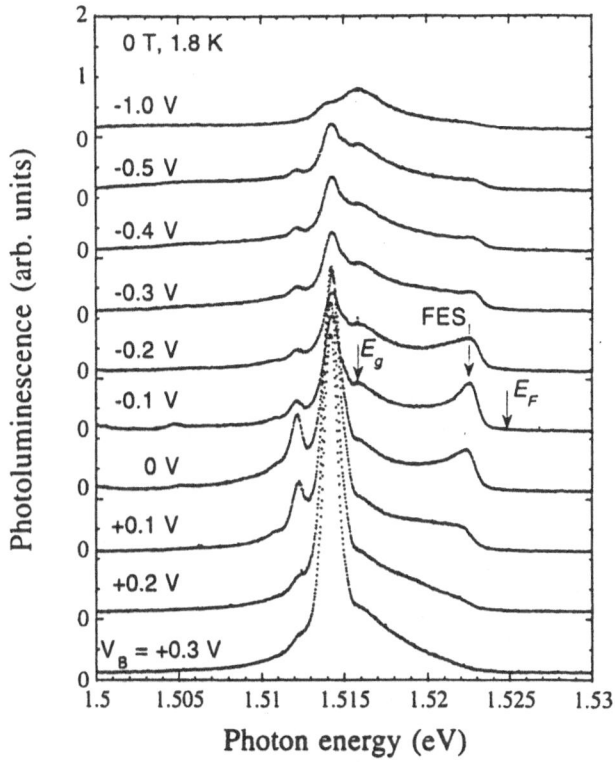

Fig. 1 Bias voltage dependent PL spectra at 1.8 K at 0 T. The band-gap (E_g), the asymmetric peak near the Fermi energy (FES) and the Fermi energy (E_F) are indicated by the arrows. Zero points of the vertical axis are shifted.

the electron beam lithography. A bias voltage V_B was applied between the surface mesh gate structure and a AuGe/Ni/Au ohmic back-contact. The strength of the lateral periodic potential is tuned by the bias voltage. The sample was excited by a 488-nm line of a continuous wave Ar ion laser at the incident power density of 1.6 mW/cm^2 or less through the metal mesh structure. Magnetic fields were applied perpendicular to the QW layer.

3 Results and discussion

In order to enhance divergence in the FES, the following schemes are considered: (i) Suppress hole recoil effect by creating geometrical restrictions for a hole; (ii) Enhance the electrons-hole Coulomb interaction by the confinement; (iii) For the FES in PL, relax the optical selection rule by mixing wave functions with different wave vectors; (iv) Change the electron density. A pronounced

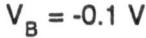

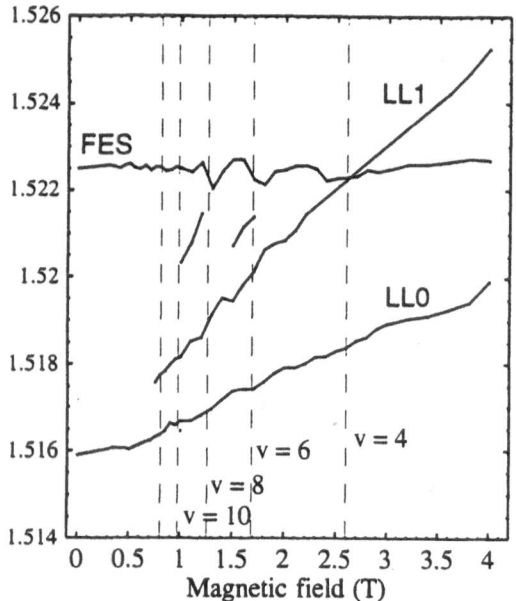

Fig. 2 PL peak energy as a function of magnetic field at $V_B = -0.1$ V. The energy of the peak due to the FES located near 1.5225 eV correlates with even filling factors.

FES was predicted by Bauer [6] in 2-D below a critical electron density. On the contrary, one of the methods to suppress the FES is to create a gap at the Fermi-level by suppressing low-energy electron-hole excitations around the Fermi-level.

By forming a lateral periodic potential to a MDQW it is considered that the degree of the divergence of the FES can be controlled externally.

Figure 1 shows bias voltage dependent PL spectra at 1.8 K at 0 T. A peculiar asymmetric peak is observed in the PL 2.5 meV below the Fermi energy under a weak lateral periodic potential at $V_B = -0.1$ V. The asymmetric peak vanishes at $V_B = -0.5$ V, where a gap opens due to the confinement and the Fermi surface disappears. The asymmetric peak shows strong temperature dependence, which is a characteristic feature for the FES.

The asymmetric peak in the PL at $V_B = -0.1$ V is shown to be due to the FES by magneto-PL measurements. The peak energy is found to oscillate as a function of magnetic field, as shown in Fig. 2. The oscillation of the peak energy correlates with even filling factors of the Landau-levels formed in magnetic fields. The PL intensity as a function of magnetic field also shows oscillations. The periods and the positions of the PL intensity of the asymmetric peak correlate with even filling factors. These results clearly show that the asymmetric peak is due to the recombination of the electrons at the Fermi level and a free optically created hole.

4 Conclusion

It is shown that the exponent of the singularity of the FES can be controlled externally by a lateral periodic potential. It has been found that the oscillations of the

energy positions of the asymmetric peak in PL due to the FES as a function of magnetic field correlate with even filling factors. Thus our sample is an ideal system to investigate the FES, a many-body effect between electrons and a hole.

Acknowledgement Part of this work is supported by the Sumitomo Foundation and by University of Tsukuba Research Projects.

References

1. G.D. Mahan, *Many-Particle Physics, 2nd ed.* (Plenum, 1990) 732.
2. T. Ogawa, A. Furusaki, and N. Nagaosa, Phys. Rev. Lett. **68**, (1992) 3638.
3. A.E. Ruckenstein and S. Schmitt-Rink, Phys. Rev. B **35**, (1987) 7551.
4. T. Uenoyama and L.J. Sham, Phys. Rev. Lett. **65**, (1990) 1048.
5. P. Hawrylak, Phys. Rev. B **44**, (1991) 3821.
6. G.E.W. Bauer, Phys. Rev. B **45**, (1992) 9153.
7. S.A. Brown et al., Phys. Rev. B **54**, (1996) R11082.
8. S.A. Brown et al., Phys. Rev. B **56**, (1997) 3937.
9. S. Nomura, T. Sugano, and Y. Aoyagi, Physica E **6**, (2000) 432.

Optical anisotropy change of buried SiO_2/Si interfaces during layer-by-layer oxidation

T. Nakayama[1], M. Murayama[2]

[1] Department of Physics, Chiba University, 1-33 Yayoi, Inage, Chiba 263-8522, Japan
[2] Department of Electronics, The University of Electro-Communications, 1-5-1 Chofugaoka, Chofu, Tokyo 182-8585, Japan

Abstract   Optical anisotropy spectra of SiO_2/Si (001) interfaces were theoretically investigated using the sp^3s^* tight-binding method. We found three types of optical transitions originating from the E_1 and E_2 transitions of bulk Si, the interface Si-Si bonds, and the dangling-bond states at the interface. It was shown that the sign of these transitions oscillates during the layer-by-layer oxidation, which indicates that by counting the oscillation one can determine the layer thickness of oxidized Si layers in an atomic scale.

1 Introduction

Oxidation of Si substrate is one of fundamental processes in the semiconductor technology and the atom-scale monitoring of oxidation is especially significant in these days, not only to clarify the growth mechanism of an oxide film but also to control its layer thickness for the microelectronics application. Recently Watanabe *et al.* showed by the scanning reflection electron microscopy (SREM) that the thermal oxidation of Si(001) surface proceeds layer-by-layer.[1] Since the SREM uses electrons as a probe, however, such observation is limited to the oxidation of ultrathin layers, *i.e.*, at most 1 nm.[1] On the other hand, the reflectance difference spectroscopy (RDS) is a useful *in situ* measurement to observe the electronic structures of surfaces/interfaces and their time evolution.[2] Since the light penetrates deep into the Si substrate, *i.e.*, an order of 10^3Å, the RDS can detect in principle the direction change of the inner interface bonds during the layer-by-layer oxidation, which has never been performed yet. Once such measurement is performed, one can monitor the oxidation in an atomic scale as the spectral oscillation, which is similar to the case of reflection high energy electron diffraction (RHEED) in the epitaxial growth. In the present work, we calculate the optical anisotropy spectra of SiO_2/Si (001) interfaces and their change during the layer-by-layer oxidation, and investigate the possibility of the RDS signal oscillation.

2 Calculation method

We assume that SiO_2 layers have quartz structure and adopt the interface model proposed by Kageshima and Shiraishi,[3] where, for example, the Si-Si and Si-O-Si layer thickness along [001] are 1.36 and 2.16 Å, respectively. The schematic pictures are shown in Figs. 1(a) and 1(b), which we call hereafter the interfaces, A and B, respectively. The interface, B, corresponds to the case when the monolayer oxidation is completed starting from the interface, A, and vice versa. Namely, these two interfaces alternatively appear during the layer-by-layer oxidation. For a convenience, the layers (bonds) are named as shown in the figures with the number, n. In the case of the interface, A, the Si-Si bonds in the odd and even-numbered layers direct to [-110] and [110], respectively.

Electronic structures are calculated using the empirical sp^3s^* tight-binding method[4,5] and the superlattice geometry made of twenty Si and six SiO_2 layers,[6] where the valence band offset of 3.75 eV is employed.[7] The optical anisotropy spectra, $\Delta\Lambda_2$, are calculated between [-110] and [110] directions based on the surface/ interface linear response theory developed in our previous works, which details are described elsewhere.[6,8,9]

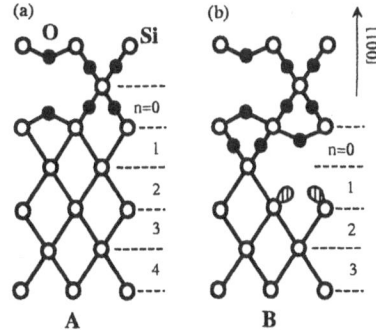

Fig. 1 Schematic picture of SiO_2/Si (001) interfaces, A and B. The layers (bonds) are named with the number, n, from the interface as shown in the right. The interface, B, has Si dangling bonds as denoted by tear drops.

3 Results and discussions

Figure 2 shows the calculated optical anisotropy spectra of SiO_2/Si interfaces, A and B, together with the imaginary part of bulk Si dielectric function, ε_2. To clarify the spectral origin, the contribution of respective layer is also shown with the layer number, n. We found that there exist the following three types of optical transitions in the calculated spectra below 8 eV; (i) the large and sharp peaks of energy-derivative-of-ε_2 shape are observed around 3.5 and 4.5 eV. As seen in the figure, apart from the sign, the contributions of inner Si layers with $n \geq 2$ are large at these energies and have a similar shape to ε_2. Since the optical anisotropy spectra are obtained by summing up all these contributions,[6] we can say that these peaks appear due to the abrupt termination of bulk-like Si-Si bonds at the second layer and originate from the Si E_1 and E_2 transitions modulated by the presence of interface. Especially, the contributions of

the $n=2$ and $n=3$ layers show small energy shifts from
those of the $n\geq4$ layers, which fact produces the energy-
derivative shapes.

(ii) The first ($n=1$) layer Si-Si bonds give the broad
positive/negative contribution in a higher energy region
from 5 to 7 eV for interfaces, A/B. This is because the
electronic states localized around such bonds have the
characters of Si atoms in both SiO_2 and Si layers and the
optical transitions originating from these states have the
energies respectively lower and higher than the band-gap
energies of SiO_2 and Si, 8 and 3 eV. On the other hand,
the SiO_2 layers have little contribution to the spectra
below 7 eV. (iii) In the case of the interface, B, other
optical transitions are seen below 3 eV. The interface, B,
has a number of dangling bonds around the $n=1$ layer
and the electronic states localized around these bonds
induce such transitions.

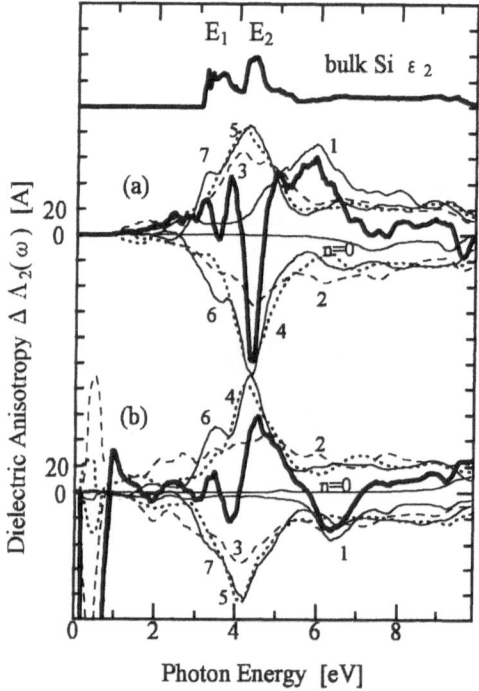

Fig. 2 Calculated optical anisotropy (RD) spectra of
SiO_2/Si (a) A and (b) B interfaces shown in Fig. 1. Bold
line denotes the spectra, while line with the number, n, is
the contribution of the n-th layer.

Roughly speaking, the spectra of the interfaces, A
and B, have the similar shapes and the opposite signs in
an energy region above 3 eV. This result indicates that
when the interfaces, A and B, appear alternatively during the layer-by-layer oxidation the RD signals oscillate
as shown in Fig. 3, which is similar to the case of RHEED
oscillation in the epitaxial layer-by-layer growth. Especially, such observation is apparently effective around
3.5/4.5 eV where the RD signal is large. In this case,
the period of oscillation corresponds to the bi-layer oxidation and one can know the number of oxidized layers
by counting the oscillation.

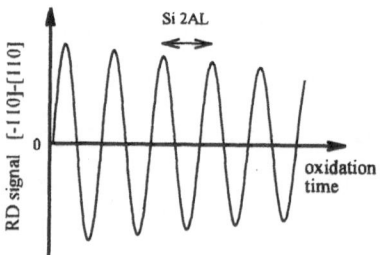

Fig. 3 Schematic time evolution of RD spectra of SiO_2/Si
interface during the layer-by-layer oxidation.

Then we consider how the spectra vary when Si layers receive the biaxial tensile strain due to oxidation or
some transition layers exist between SiO_2 and Si layers. Since the first type of transitions originate from the
Si bulk-like states, the tensile strain induces the energy
increase of these transitions. On the other hand, since
the second type transitions are responsible to the energy change of interface, the transition layers induce the
penetration of Si-like states into SiO_2 layers and decrease
the transition energies. These facts indicate that one can
know the strain and the presence of transition layers by
measuring the spectrum peak positions precisely.

Finally we comment on the RD spectra change of
(110) interface during the layer-by-layer oxidation. Since
at this interface the bonds along [001] exist in any stage
of oxidation, they always contribute to produce positive
peaks around the bulk E_1 and E_2 transitions. On the
other hand, the linear-chain bonds along [-110] induce
the positive E_1 and negative E_2-like transitions similar
to the (110) surfaces and such bonds appear and disappear during the layer-by-layer oxidation. Thus, we expect the spectrum peak oscillation around some positive
magnitude.

This work was supported by JSPS Research for Future Programs in Area of Atomic-Scale Surface and Interface Dynamics, JSPS-RFTF96R16201, and the Ministry of Education, Science, Sports and Culture, Japan.

References

1. H. Watanabe, K. Kato, T. Uda, K. Fujita, M. Ichikawa, T. Kawamura, and K. Terakura, Phys. Rev. Lett. **80** (1998) 345.
2. T. Yasuda, D.E. Aspnes, D.R. Lee, C.H. Bjorkman, and G. Lucovsky, J. Vac. Sci. Technol. A**12** (1994) 1152.
3. H. Kageshima and K. Shiraishi, Phys. Rev. Lett. **81** (1998) 5936.
4. P. Vogl, H.P. Hjalmarson, and J.D. Dow, J. Phys. Chem. Solids **44** (1983) 365.
5. E.P. O'Reilly and J. Robertson, Phys. Rev. B**27** (1983) 3780.
6. T. Nakayama and M. Murayama, Jpn. J. Appl. Phys. **38** (1999) 3497.
7. N. Tit and M.W.C. Dharma-Wardana, J. Appl. Phys. **86** (1999) 387.
8. M. Murayama, K. Shiraishi, and T. Nakayama, Jpn. J. Appl. Phys. **37** (1998) 4109.
9. T. Nakayama, Phys. Stat. Sol., b**202** (1997) 741.

Negatively charged excitons in semiconductor quantum wells: effects of longitudinal electric and magnetic fields

Luis C.O. Dacal, Maria José S.P. Brasil, José A. Brum

DFMC-IFGW, Universidade Estadual de Campinas, C.P. 6165, 13083-970 Campinas-SP, Brazil

Abstract The effects of longitudinal electric and magnetic fields on the binding energy of the X^- in $Ga_{1-x}Al_xAs/GaAs$ quantum wells are studied variationally using the configuration interaction picture with a non-orthogonal basis set.

1 Introduction

Optical transitions in intrinsic semiconductor quantum wells (QWs) are dominated by the excitonic state formed by the binding of an electron and a hole through the Coulomb interaction. In the presence of a low-density electron gas, the dipole interaction between the neutral exciton and an extra electron may give rise to a bound negative complex, the trion (X^-), a negative hydrogen ion analogous. Since the trion is a charged complex, its response to external electric and magnetic fields is different than that of the neutral exciton. This complex was first observed in II-VI heterostructures[1] and has recently been studied, theoretically[2] and experimentally[3] in III-V systems.

In this work we present variational calculations of the trion binding energy in $Ga_{1-x}Al_xAs/GaAs$ QWs in the presence of electric and magnetic fields applied along the growth direction.

2 Model

The calculations are performed in the framework of the envelope function and effective mass approximations with in-plane parabolic dispersions for the electrons and the hole. The Hamiltonian is written in the centre-of-mass (CM) and relative coordinates for the in-plane motion.

The states in the growth direction (z) are represented by the QW solutions including the electric field, when it is present. The relative coordinates part of the wavefunction is represented by a two relative particles configuration space built from a single particle basis set formed by non-orthogonal gaussian-like functions. The X^- ground state is obtained diagonalizing the Hamiltonian matrix. The states are labelled by their total orbital angular momentum along the growth direction, L, and the total spin state. In the presence of a longitudinal magnetic field, the choice of the single particle basis is more difficult. For the neutral exciton, gaussian-like in-plane basis and single QW state is sufficient to describe the Coulomb interaction up to 10 T. However, the second electron is only weakly bound to the X^-. Consequently, for this complex a magnetic field of the order of 1 T is already relatively large, suggesting a description in terms of single particle Landau levels. To overcome this difficulty, it

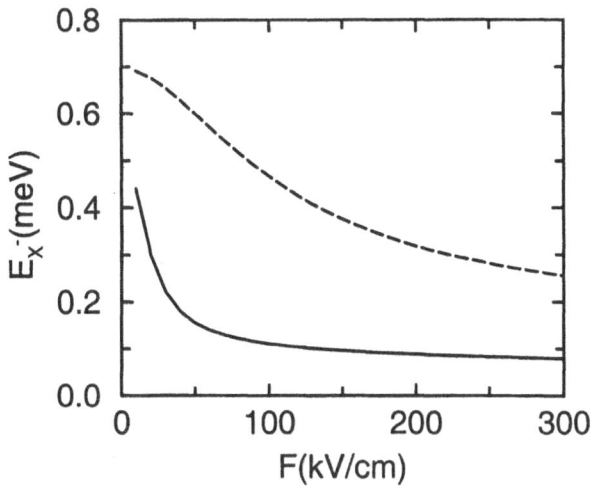

Fig. 1 X^- binding energy as a function of a longitudinal electric field for a quantum well width of 200 Å (full line) and 100 Å (dashed line).

is necessary to include several states in the basis, as it is discussed below.

3 Results and Discussions

In the absence of external fields, our results confirm the existence of only one bound state, a singlet with L=0, in agreement with previous works[3].

In Figure 1 we show the X^- binding energy as a function of the longitudinal electric field for two QW widths. The inclusion in the basis of more than one QW state for the hole enhances the X^- binding energy up to 10 % for a 300 Å QW width. The electric field pushes the two electrons and the hole towards opposite QW interfaces resulting in a decreasing of the trion binding energy. This effect saturates as we increase the electric field due to the presence of the interfaces and the trion remains bound even at fields of the order of 300 kV/cm. A similar effect is also observed on the neutral exciton[4]. These results explains the observation of X_- emission lines in one-side modulation doped samples, when an effective longitudinal electric field is present in the QW as a consequence of the charge transfer from the donors.

The case of a longitudinal magnetic field is particularly complex. Differently from the neutral exciton, it is no more possible to separate the CM and the relative motions. However, it is possible to express the CM

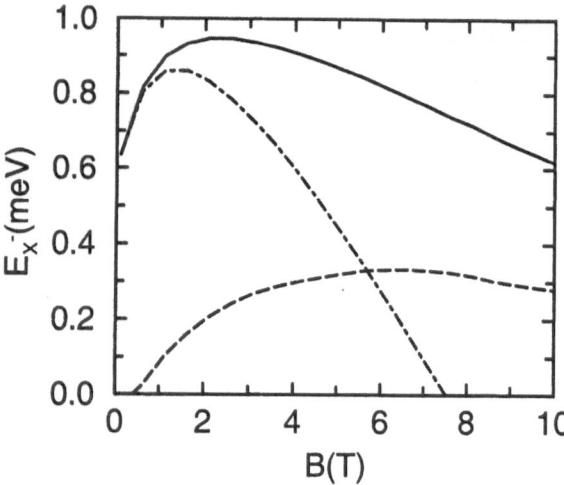

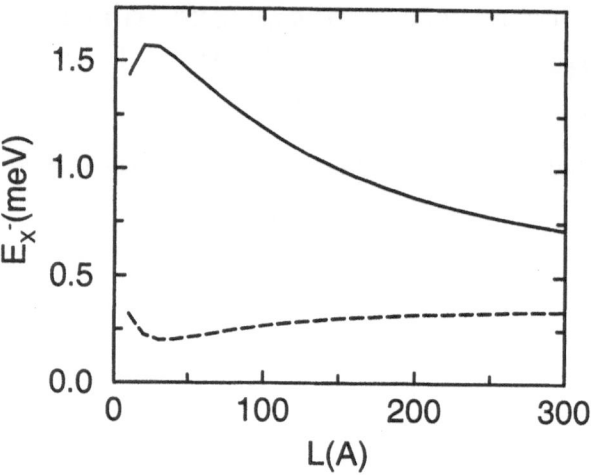

Fig. 2 Negative trion binding energy as a function of a longitudinal magnetic field for a quantum well width of 200 Å . L=0 singlet state (full line) and L=-1 triplet state (dashed line). We also show the case of L=0 singlet state when only one center-of-mass Landau level is included in the basis (dot-dashed line)

motion in terms of CM Landau levels with an appropriate unitary transformation[5]. These levels are coupled to the trion internal motion.

In Figure 2, we show the X$^-$ binding energy dependence with the magnetic field. The binding energy of the singlet (L=0) state initially increases up to $\sim$ 2 T, where this trend is inverted. This is a consequence of the interplay between the effect of the magnetic field diminishing the size of the electronic orbitals but also giving rise to a stronger Coulomb repulsion. It is important to observe that the inclusion of more than one CM Landal level changes considerably the magnetic field dependence. When the basis is limited to the first CM Landau level (dot-dashed line), the singlet trion becomes unbound at $\sim$ 7 T. The inclusion of more than one CM Landau level (two levels for the full line which showed to be sufficient for the convergence of the X$^-$ binding energy) increases significantly the trion binding energy. We do not observe the unbinding of the trion for the magnetic fields considered here, however the binding energy decreases with the magnetic field for fields higher than 2 T. This behavior is observed for QW widths up to 300 Å . Previous results[2] showed a different behavior for QWs of the order of 300 Å . On their results the trion binding energy always increases with the magnetic field. This discrepancy may be due to a limitation of our basis which includes only one electron QW state. This limit is being presently investigated. The triplet state L=-1, becomes a bound state for magnetic fields as low as 1 T. We do not observe energy crossing between the singlet and triplet states up to 10 T. The inclusion of more than one CM Landau levels in the basis is also fundamental to this effect. When we only consider the first CM Landau

Fig. 3 Negative trion binding energy as a function of the quantum well width for a fixed longitudinal magnetic field (5 T). L=0 singlet state (full line) and L=-1 triplet state (dashed line).

level, no triplet state is bound up to the magnetic fields considered here.

In Figure 3, the X$^-$ binding energy is shown as a function of the QW width for a fixed longitudinal magnetic field of 5 T. The different behaviors for the singlet (L=0) and the triplet (L=-1) states reflect the opposite spatial symmetry of their wave-functions. For the singlet state, the Coulomb attraction between the electrons and the hole is the most important interaction while for the triplet state the repulsion between the electrons dominates the interactions among the particles.

In conclusion, we have presented variational calculations for the X$^-$ binding energy in Ga$_{1-x}$Al$_x$As/GaAs QWs in the presence of longitudinal electric and magnetic fields. In the absence of external magnetic fields, only the singlet (L=0) state is bound and it remains so for electric fields up to 300 kV/cm. We have showed that with finite magnetic fields, the coupling between relative and CM motions is important and more than one CM Landau level has to be taken into account. No energy crossing between the singlet and triplet states is observed up to 10 T.

Acknowledgements: This work was supported by FAPESP (Brazil), CNPq (Brazil) and FAEP (Brazil). One of us (JAB) acknowledges G.A. Narvaez for helpful discussions.

References

1. K. Kheng et al., Phys. Rev. Lett. **71**, (1993) 1752.
2. D. Whittaker and A. Shields, Phys. Rev. B 56, (1997) 15185.
3. S. Glasberg et al., Phys. Rev. B 59, (1999) R10425.
4. J.A. Brum and G. Bastard, Phys Rev. B 31, (1985) 3893.
5. P. Schmelcher and L.S. Cederbaum, Phys. Rev. A **43**, (1991) 287 and B. Stébé and A. Moradi, Phys. Rev. B **61**, (2000) 2888.

The influence of intraband transitions on the resonant and extended electronic states in GaAs/AlGaAs asymmetric quantum wells as a function of their confinement

M. Levy[1], R. Beserman[1], R. Kapon[2], A. Sa'ar[2], V. Thierry-Mieg[3], R. Planel[3]

[1] Solid State Institute and Physics Department, Technion, Israel Institute of Technology, Haifa 32000, Israel. e-mail: sslevy@tx.technion.ac.il.

[2] Department of Applied Physics, The Fredi and Nadine Hermann School of Applied Science, The Hebrew University of Jerusalem, Jerusalem 91904, Israel.

[3] Laboratoire de Microstructures et Microelectronique - CNRS, 196 Avenue H. Ravera, BP107, 92225 Bagneux, France.

Abstract Electronic Bragg mirrors are used to confine carriers at energy levels above the barrier height in asymmetric coupled quantum wells. An electric field inside the quantum structure is created by transferring carriers from a wide quantum well into a narrow one. We used modulated photoluminescence and Raman spectroscopy to resolve the stark shifts of the bound and continuum levels as a function of the generated local field. Our results indicate that as a result of the photoinduced electric field, the shifts of the above the barrier levels are linked to their degree of localization and are much stronger than those of the bound states inside the well.

In this work we study the localization of the different classes of states inside the QW and above the barrier by using the recently developed method of locally modulated resonant Raman Scattering and modulated photoluminescence [?],[?]. The local electric field in the structure was obtained by using asymmetric coupled QW (ACQW) structure to transfer carriers from a wide QW (WQW) into a narrow QW (NQW) through intermediate barrier (Fig. 1). This charge transfer was created by the 10.6 μm line of a CO_2 laser, which induced intersubband electronic transition (ISBT) from the ground state of the WQW E_1 into the ground state of the NQW E_2 via an electronic state common to the asymmetric coupled QWs (ACQW)E_3. In fig. 1, we schematically show the

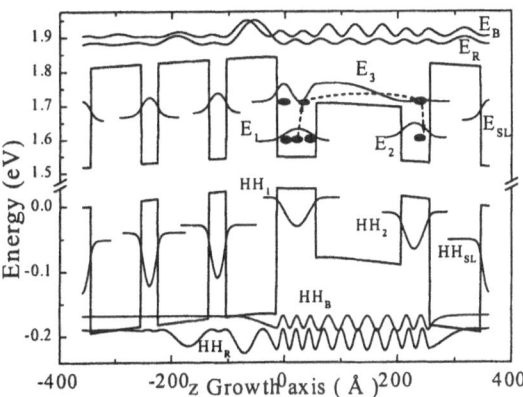

Fig. 1 Conduction and valence energy states of the Bragg confined structure, calculated by solving self-consistently the Ben Daniel-Duke and the Poison equations.

ISBT process. These electronic transitions give rise to a local electric field across the ACQW region. Hence local modulation of the photoluminescence signal inside the QW states, as well as local modulation of the resonant Raman scattering signal from the above the barrier localized states are observed. We measured the energy shift of the Bragg state, while other classes of states showed smaller shifts. The size of the shift is linked to their degree of localization in the ACQW region. We were able to distinguish between the different states by their amount of localization and to define several classes of states. The Bragg confining structure is composed of an ACQW which is grown between two $GaAs/Al_{0.34}Ga_{0.66}As$ Bragg mirrors. The ACQW is composed of a 7 nm GaAs WQW, a 15 nm $Al_{0.2}Ga_{0.8}As$ intermediate barrier and a 5 nm GaAs NQW. Each Bragg mirror consists of 4 periods of 3 nm-GaAs QW and 9 nm-$Al_{0.34}Ga_{0.66}As$ barrier. In order to populate the ground electronic state E_1 the sample was modulation doped with electron concentration of 2×10^{11} cm^{-2} creating a two dimensional electron gas in the WQW region (fig. 1).

Fig. 2(I)-(III) shows the PL and the photoluminescence excitation (PLE) spectra from the bound states inside the well and from the continuum states. The PL spectrum (Fig. 2(I)) shows three peaks E_1:HH_1, E_2:HH_2 and E_{SL}:HH_{SL} at 1.57 eV, 1.605 eV, 1.705 eV respectively. The PLE spectrum from above the barrier energy monitored at the WQW energy is shown in fig. 2(II). Three continuum transitions are resolved at 2.052 eV, 2.07 eV and 2.082 eV which are identified as the Bragg transition E_B:HH_B , the reflector transition E_R:HH_R and the E_B:HH_R transition. We assign the E_B:HH_B transition and E_B:HH_R to a class of highly localized energy states, while E_R:HH_R is an extended transition mostly localized in the reflector region. The high-energy part of the PL (fig. 2(III) near 2.07 eV, is assigned to the transition E_R:HH_R and it is seen also in the PLE spectrum. The PL signal from this state is an indication of the localization of this level in the reflector region. The electronic transitions E_B:HH_B and to a lesser extent the transition E_B:HH_R are more localized in the ACQW region than the E_R:HH_R one. As a result car-

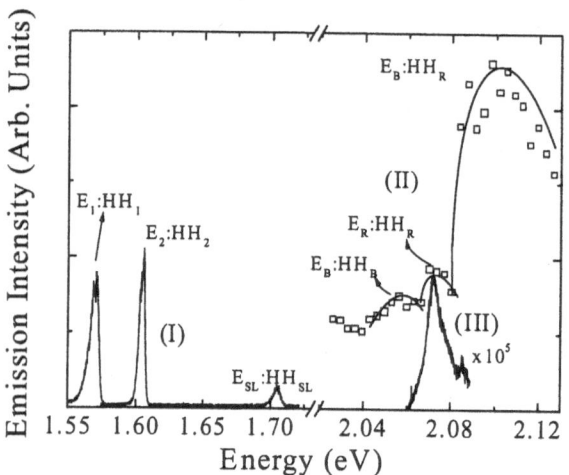

Fig. 2 PL and PLE of bound and above the barrier energy levels in the Bragg confined structure. (I)- PL of the three lowest bound subbands. (II)- PLE of above the barrier energy states monitored at the WQW energy transition. (III)- PL of above the barrier states in the reflector region.

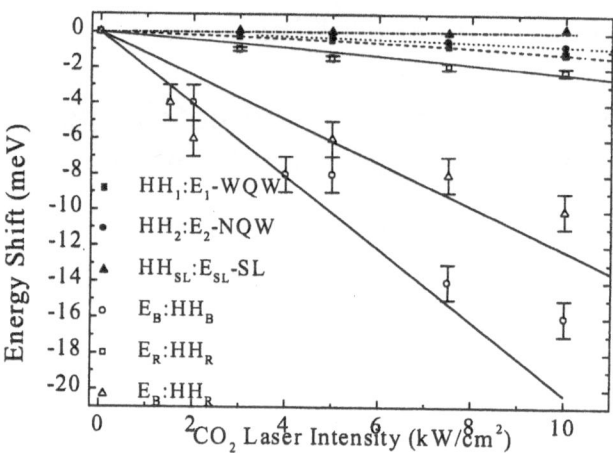

Fig. 3 Energy shifts of the bound and continuum energy levels under various infrared excitation intensities: solid squares - E_1:HH_1, solid circles- E_2:HH_2, solid triangles- E_{SL}:HH_{SL}, open squares-E_R:HH_R, open circles- E_B:HH_B, open triangles- E_B:HH_R

riers in the Bragg confined energy levels E_B and HH_B decay more rapidly to lower subbands and therefore do not recombine radiatively at this energy. On the other hand, carriers with wave functions that are more localized in the reflector region have a longer life time due to smaller overlap with the lower subbands and contribute significantly to the PL signal. Fig. 3 shows the red shift of the three bound transitions E_1:HH_1, E_2:HH_2, E_{SL}:HH_{SL} as a function of the infrared excitation intensity. When the IR excitation power intensity increases i.e. when the generated dc electric field increases, the red shifts of these energy transitions increase. Furthermore, the magnitude of the shifts depends on the QW size, the WQW shows a larger shift than the NQW. On the other hand, the transition E_{SL}:HH_{SL} takes place are outside the ACQW region and therefore does not show any shift when the localized electric field is applied as expected. When the local field is generated, electrons in the WQW and NQW accumulate in the left side of the wells while the holes accumulate in the right side of the wells, therefore they produce an electric dipole which red shifts the energy levels of these bound states. This electric dipole depends on the QW well size and therefore the energy levels are shifted by an amount, which is proportional to the well width. Three above the barrier transitions are resolved: E_B:HH_B, E_R:HH_R and E_B:HH_R. The red shifts of these levels under the infrared excitation are also reported in fig. 3. First we note that all above the barrier energy levels show stronger shifts than the bound states when subjected to the same photoinduced electric field. Second the transitions E_B:HH_B and E_B:HH_R show a stronger shift than the transition E_R:HH_R. These features are related to the degree of localization of these states and can be explained using the Stark effect model. Energy levels above the barrier that are confined in the ACQW region, can produce a much

larger electric dipole than the bound states, because the charge separation between electrons and holes which is of the order of the ACQW size (270 Å) is larger than the WQW which is the maximum charge separation of the bound states. The calculated average separation between electron and heavy holes of the Bragg confined level is 60 Å compared to a 5-6 Å for the transition E_1:HH_1 and even less for E_2:HH_2. This difference between the dipole moments of the bound states and the Bragg confined continuum state, leads to the stronger shift of the Bragg confined level as seen in fig. 3. Using perturbation theory we calculated the shifts of these continuum levels. The calculated results are shown as full lines in fig. 3 and are in a good agreement with the experimental data. The same argument explains the smaller shifts of the continuum transitions E_B:HH_R which has a smaller dipole moment of 30 Å. These two transitions E_B:HH_B and E_B:HH_R form a class of highly localized transitions in the ACQW region which are strongly influenced by the photoinduced electric field localized in the ACQW. On the other hand the transition E_R:HH_R has a much smaller shift under infrared excitation than the other two transitions E_B:HH_B and E_B:HH_R. The electrons and holes of this above the barrier transition are mostly localized in the reflector region, hence the joint density of electrons and holes in the ACQW region is much smaller than that of E_B:HH_B or E_B:HH_R, resulting in a much smaller electric dipole. As a result this state is less affected by the localized field and therefore shows smaller shifts. This confirms that this state belongs to a different class of continuum states which are mainly confined in the reflector region.

References

1. M. Bendayan, R. Kapon, R. Beserman, A. Sa'ar, R. Planel, Phys. Rev. B, **56** (1997), 9239.
2. M. Levy, R. Kapon, A. Sa'ar, R. Beserman, V. Thierry-Mieg, R. Planel, Physica E, **7** (2000), 245.

Proc. 25th Int. Conf. Phys. Semicond., Osaka 2000 (Eds. N. Miura and T. Ando)

467

Zeeman mapping of probability densities in square quantum wells using ultranarrow probes

G. Prechtl[1], W. Heiss[1], A. Bonanni[1], E. Janik[2], S. Mackowski[2], G. Karczewski[2], W. Jantsch[1]

[1] Institut für Halbleiter- und Festkörperphysik, Johannes Kepler Universität Linz, Altenbergerstrasse 69, 4040 Linz, Austria e-mail: gerhard.prechtl@jk.uni-linz.ac.at
[2] Institute of Physics, Polish Academy of Sciences, 02-668, Warsaw, Poland

Abstract: The exciton Zeeman splitting in quantum wells containing a single, ultranarrow magnetic layer is studied depending on the layer position. For various interband transitions we show that the dependence of the exciton Zeeman splitting on the position of the magnetic layer directly maps the probability density of free holes in the growth direction.

In this work, we demonstrate a novel method to probe experimentally the probability density of carriers confined in semiconductor quantum structures. We study by polarization dependent photoluminescence excitation (PLE) measurements the Zeeman splitting in quantum wells containing a single, ultranarrow magnetic layer as function of the layer position. In particular, a system consisting of a MnTe layer of a ¼ monolayer (ML) thickness embedded at varying positions in nonmagnetic CdTe/CdMgTe quantum wells is investigated.

The d-electrons of the Mn^{2+} ions in the magnetic layers with submonolayer thickness interact via spin-spin exchange with s-like conduction-band electrons and p-like valence-band holes . This sp-d exchange interaction results in a drastic increase of the Zeeman splitting which, because of the strongly localized nature of this interaction, sensitively depends on the position of the MnTe submonolayer in the quantum well.

The sp-d interaction is described by the exchange Hamiltonian

$$H_{ex} = \sum_{\vec{R}_i} J^{sp-d}(\vec{r} - \vec{R}_i)\vec{S}_i\vec{\sigma} \tag{1},$$

where $\vec{S}_i$ and $\vec{\sigma}$ are the spin operators of the Mn^{2+} ions and of the band electron, J^{sp-d} is the electron (hole) - ion s(p)-d exchange coupling constant, and $\vec{r}$ and $\vec{R}_i$ are the coordinates of the band electron and of the magnetic ions, respectively. Due to the localized character of the 3d electrons of the Mn ions, the function J is strongly peaked around $\vec{R}_i$ and vanishes quickly away from this point. Therefore, $J^{sp-d}(\vec{r} - \vec{R}_i)$ can be approximated by a delta function $J^{sp-d}(\vec{r} - \vec{R}_i) = A\Omega_0\delta(\vec{r} - \vec{R}_i)$, where A is

the exchange constant and Ω_0 the volume of the unit cell. The electron wave function can be written as:

$$\Psi(\vec{r};\vec{S}_1,...,\vec{S}_N) = \Psi(\vec{r};\{\vec{S}_i\}) = \psi_e(\vec{r};\{\vec{S}_i\})\Phi(\{\vec{S}_i\}) \tag{2}.$$

The symbol $\{\vec{S}_i\}$ denotes the set of quantum numbers required to label the states accessible to the system of local moments. The exchange term leads to an increased spin splitting of the conduction band, given by :

$$\Delta E_e = \sum_i \langle\Phi|S_i^z|\Phi\rangle A\Omega_0|\psi_e(R_i)|^2 \tag{3}.$$

Since in our structures the electrons are confined in growth direction, z, one can write the wave function as: $\psi_e(r = R_i) = \varphi_e(X_i, Y_i)\xi_e(Z_i)$. Moreover, all the magnetic ions are distributed within a single submonolayer, so that $Z_i = Z_{Mn}$ for all magnetic ions. Therefore, ΔE_e is proportional to $|\xi_e(Z_{Mn})|^2$ and the z dependence of the PD can be explored by measuring ΔE_e as function of Z_{Mn}. We note that in Eq. (3) the macroscopic average of S^z vanishes in the absence of an external magnetic field. Expressions analogous to Eq. (3) can be obtained for the splitting of the $\sigma=1/2$ light holes $(\Delta E_{lh}(Z_{Mn}))$ and of the $\sigma=3/2$ heavy holes $(\Delta E_{hh}(Z_{Mn}))$.

In particular we investigate a series of five single quantum-well structures (S1 to S5). The wells consist of 20 MLs of CdTe between 200 nm thick $Cd_{0.75}Mg_{0.25}Te$ barriers. Each quantum well contains a narrow, magnetic MnTe barrier with a Mn amount equivalent to 1/4 ML coverage. In sample S1 the MnTe layer is inserted after the third ML of CdTe, while in S2 to S5 the magnetic probes are embedded after the seventh, tenth, thirteenth, and seventeenth ML of CdTe, respectively.

The magneto-optical measurements were performed in Faraday-configuration where optical transitions are allowed for circularly polarized light with positive or negative helicity /1/. We performed polarization dependent photoluminescence excitation (PLE) experiments in magnetic fields up to 4 T. At a temperature of 1.7 K, the PLE spectrum of each sample consists of a Gaussian shaped line due to the e_1-hh_1 transition with a full width at half maximum of about 5 meV. Furthermore, we observe at higher energies a faint shoulder due to the e_1-hh_2 and a clear peak caused by the e_1-lh_1 transition.

We obtain similar PLE spectra for all samples S1-S5, the optical transition energies, however, continuously shift to higher energies when the MnTe barrier is moved from one side of the quantum well towards its center. With applied magnetic field, all optical transitions split into two distinct lines due to the Zeeman splitting.

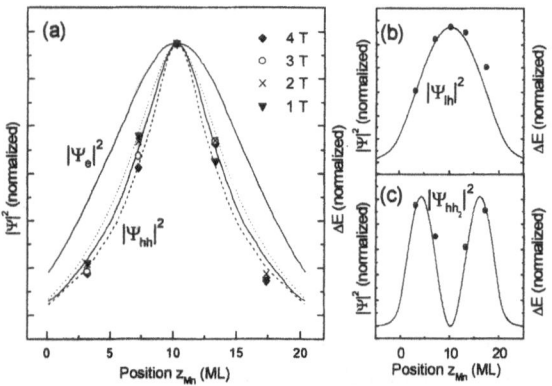

Fig. 1: (a) Probability density (PD) of the heavy hole ground state calculated for various valence band offsets (0.25 (dashed), 0.33 (solid) and 0.45 (dotted)) and of the electron ground state (valence band offset 0.33). The symbols display the experimentally observed interband Zeeman splitting of the e_1-hh_1 transition. (b) PD of the light hole ground state compared with the Zeeman splitting of the e_1-lh_1 transition. (c) as (b) but for the first excited heavy hole state hh_1.

Fig. 1(a) shows the dependence of the normalized Zeeman splitting $\Delta E(Z_{Mn})$ of the $e_1 - hh_1$ transition on the position of the MnTe layer in the quantum well. This dependence is compared with the electron and hole PDs, calculated for the ground state of our quantum wells perturbed at the position Z_{Mn} by a quarter ML of MnTe. For this calculation we used the following parameters [1,2]:

$$E_g(Cd_{1-x}Mn_xTe) = (1.606 + 1.654x)eV \qquad (1),$$
$$E_g(Cd_{1-x}Mg_xTe) = (1.606 + 1.592x)eV \qquad (2),$$

$m_e^* = 0.096m_0$, $m_{hh}^* = 0.63m_0$ [3,4] (m_0 is the free electron mass), and we took three values for the valence band offset, 0.25, 0.33 and 0.45, in the range given in

the literature [5]. Independent of the chosen valence band offset, we find an excellent agreement between the Z_{Mn} dependence of the experimental Zeeman splitting and the PD of the heavy hole ground state (Fig. 1(a)), for all applied fields. In contrast, the z dependence of the electron PD clearly deviates from the experimental $\Delta E(Z_{Mn})$ confirming that the interband Zeeman splitting is dominated by the heavy-hole spin splitting. This agrees with the situation in bulk $Cd_{1-x}Mn_xTe$, where the heavy hole splitting is four times larger than that of the electrons /1/. Therefore, by exploring $\Delta E(Z_{Mn})$ we map out the z dependence of the PD mainly of holes in contrast to the previous work, where wave functions of electrons have been probed /6,7,8/.

We also show for the e_1-lh_1 as well as for the e_1-hh_2 interband transition that the dependence of the Zeeman splitting on the position of the magnetic layer maps the probability density of free *holes* in the growth direction as demonstrated in Fig. 1(b) and Fig. 1(c), respectively.

Thus, the incorporation of ultranarrow magnetic layers allows direct mapping of the probability density distribution of *holes* confined in a quantum structure, with a vertical resolution of approximately one monolayer, by measuring the interband Zeeman splitting. Since the ultrathin magnetic probes are hardly influencing the probability densities under investigation, this technique can be applied not only to square quantum wells but also to more complicated kinds of multi-heterostructures.

This work has been supported by the FWF, Austria, and in Poland by the grant PBZ28.11 from the State Committee for Scientific Research.

References

1. J.K. Furdyna, J. Appl. Phys. **64**, R29 (1988)
2. D. Tönnies, G. Bacher, A. Forchel, A. Waag, T. Litz, G. Landwehr, Jpn. J. Appl. Phys. **33** L247 (1994)
3. K.K. Kanazawa, F.C. Brown, Phys. Rev. **135** A 1757 (1963)
4. R. Wojtal, A. Golnik, J.A. Gaj, Phys. Status Solidi (b) **96**, 241 (1979)
5. T. Lebihen, E. Deleporte, C. Delalande, Phys. Rev. B **55**, 1724 (1997)
6. P. H. Beton, J. Wang, N. Mori, L. Eaves, P. C. Main, T. J. Foster, M. Henini, Phys. Rev. Lett. **75**, 1996 (1995)
7. J. Y. Marzin, J. M. Gérard, Phys. Rev. Lett. **62**, 2172 (1989)
8. G. Salis, B. Graf, K. Ensslin, K. Campman, K. Maranowski, A. C. Gossard, Phys. Rev. Lett. **79**, 5106 (1997)

Tight-binding description of GaAs/AlAs quantum wells and superlattices: Gap transitions and intervalley couplings

R.B. Capaz[1], J.G. Menchero[1], T.G. Dargam[2], Belita Koiller[1] *

[1] Instituto de Física, Universidade Federal do Rio de Janeiro, Caixa Postal 68528, 21945-970 Rio de Janeiro RJ, Brazil
e-mail: capaz@if.ufrj.br
[2] Instituto de Física, Universidade Federal Fluminense, 24210-340, Niterói RJ, Brazil

Abstract We present two studies on the electronic properties of GaAs/AlAs heterostructures based on the tight-binding formalism. The first study addresses the problem of the real and reciprocal-space symmetry of the band-edge states for a quantum well near the critical width in which the system undergoes a direct- to indirect-gap transition. Results are analyzed for different tight-binding parametrizations. In the second study, a systematic investigation of the intervalley couplings in GaAs/AlAs superlattices is performed. We calculate the effect of interface alloying on the couplings, considering both even and odd monolayer numbers for GaAs and AlAs. Remarkably, all intervalley couplings are independent of the degree of interface alloying when averaged over the four possible even/odd configurations.

1 Introduction

In the tight-binding (TB) formalism, sometimes important differences arise between models containing only first-nearest-neighbor (1nn) interactions and those which also include second-nearest-neighbor (2nn) interactions. A well known example occurs in zincblende semiconductors, in which the 1nn treatment gives a dispersionless band structure between the X and W points [1]. This makes 1nn parametrizations unsuitable for describing any property or quantity that depends on the effective masses at those points. The 2nn models, in contrast, do not exhibit this anomalous behavior. Another example of a system in which a 2nn treatment is essential is the case of intervalley coupling at superlattice interfaces [2].

We present below a summary of the main results from two studies on GaAs/AlAs heterostructures recently performed within the TB formalism: In Sec. 2 we consider the band edge states behavior near the direct- to indirect-gap transition [3], while Sec. 3 deals with the role of interface imperfections on intervalley coupling [4].

2 Tight-binding description of the band-edge states in GaAs/AlAs quantum wells

We consider the first electron state, $|e_1 >$, and the first hole state, $|h_1 >$, in GaAs/AlAs quantum wells (QW's). We model the QW by a superlattice with very thick AlAs spacer layers. We compare TB results obtained from parametrizations proposed by Vogl *et al.*[5] and Boykin *et al.*[6] for 1nn, and Boykin [7] for 2nn, all of

* This work was partially supported by CNPq, PRONEX-MCT, FAPERJ and FUJB.

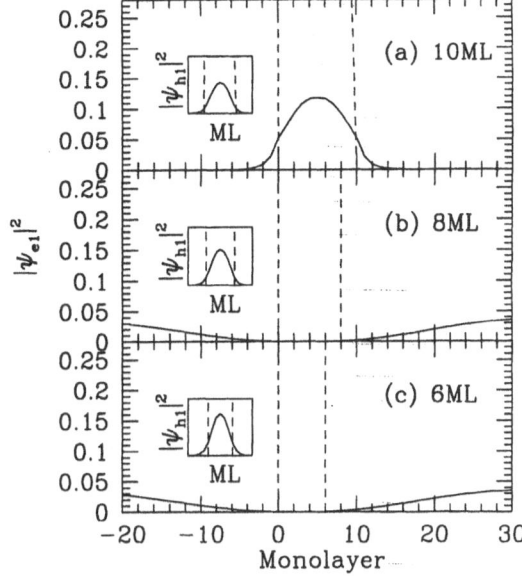

Fig. 1 Tight-binding envelope function squared for the first electron state in a 60 ML GaAs/AlAs supercell using B2nn parametrization, with the results for the first hole state shown in the inset. The vertical dashed lines indicate the GaAs/AlAs interfaces. (a) For 10 ML of GaAs, the wavefunction is localized in the well. For well widths of (b) 8 ML and (c) 6 ML confinement expels the electron from the well.

which adopt an $sp^3 s^*$ basis set. We refer to them as V1nn, B1nn and B2nn, respectively.

The spacial distribution of the states along the heterostructure growth direction z is conveniently depicted through a TB analogue of the usual envelope function [8]. In Fig. 1 we plot the TB envelope function squared for state $|e_1 >$ within the B2nn parametrization, obtained for a supercell of size $N_x = N_y = 2$, and $N_z = 30$, where N_i gives the number of conventional fcc unit cells along the cartesian direction i. For a well width of 10 ML (and above), $|e_1 >$ is localized within the GaAs region. The valence state $|h_1 >$, which is plotted in the inset, is also confined to the well region. We calculated the dipole moment of the band-edge transition, and found it to be strongly optically active. Therefore, QW's of this width are typical type-I materials. For the narrower widths, however, confinement effects expel the electron from the well. The hole state $|h_1 >$ remains localized in the well, and has essentially the same character as the 10 ML case. The narrower wells, therefore, exhibit typical type-II behavior. These results, including the critical transition width, are in agreement with those obtained from the empirical pseudopotential approach [9]. Considering the two-dimensional Brillouin zone for the QW, the point-group symmetry of both $|h_1>$ and $|e_1>$ states was identified to be $\bar{\Gamma}$ ($\mathbf{k} = 0$) in all cases [3].

The TB envelope distributions obtained from 1nn parametrizations are in agreement with the above results for the $|h_1>$ state at all well widths, as well as for the $|e_1>$ state in the type-I regime. However, for narrow wells, the $|e_1>$ state changes its point-group symmetry to $\bar{M}$ (V1nn) or $\bar{X}$ (B1nn). Moreover, for V1nn, it remains localized *within* the well even in the indirect-gap regime. We conclude that 1nn parametrizations are reliable for the description of the direct-gap (type-I) regime of QW's, and also for estimating the critical geometry at which the direct- to indirect-gap transition occurs. However, the description of the indirect-gap (type-II) regime requires a parametrization that is reliable in describing the X-point in the bulk materials, such as B2nn.

3 The role of interface imperfections on intervalley coupling in GaAs/AlAs superlattices

We consider a $(GaAs)_N(AlAs)_M$ superlattice. Within the TB B2nn scheme, the coupling potential between bulk states $|\Psi_{n\mathbf{k}}\rangle$ and $|\Psi_{n'\mathbf{k}'}\rangle$ is given by

$$V_{n'\mathbf{k}',n\mathbf{k}} = \frac{1}{N_c} \sum_{\mathbf{t}\mathbf{t}'} \sum_{mm'} a^*_{n'\mathbf{k}'}(m\prime) a_{n\mathbf{k}}(m) \exp(i\mathbf{k}\cdot\mathbf{t})$$
$$\times \exp(-i\mathbf{k}'\cdot\mathbf{t}')\langle\phi_{m'}(\mathbf{t}')|\mathcal{H}|\phi_m(\mathbf{t})\rangle , \quad (1)$$

where $\mathcal{H}$ is the superlattice Hamiltonian, and $a_{n\mathbf{k}}(m)$, $a_{n'\mathbf{k}'}(m\prime)$ are the TB expansion coefficients for the bulk eigenstates [4].

While the symmetry properties and selection rules of the various couplings [2] is interesting from a theoretical viewpoint, they are of little direct interest experimentally. In order to model the effects of interface alloying, we consider large supercells containing 128 As atoms at each interface. We maintain one interface of the supercell perfect, but for the second interface we assign Ga (Al) atoms with a probability ϵ to the first monolayer on the AlAs (GaAs) side of the interface. In other words, $\epsilon = 0$ results in a perfect interface, but $\epsilon = 1.0$ actually corresponds to interchanging the Ga and Al monolayers at the interface.

In Fig. 2(a) we plot the $\Gamma - X_{1z}$ coupling, both for (even,even) and (odd,odd) configurations. For the (even,even) configuration and $\epsilon = 0$, we recover previous results for the perfect interface [2]. As ϵ increases, interface alloying diminishes the $\Gamma - X_{1z}$ coupling, eventually destroying it completely at $\epsilon = 1.0$. For the (odd,odd) configuration we obtain zero coupling at $\epsilon = 0$, as required by symmetry. Amazingly, though, as the interface alloying ϵ increases, the (odd,odd) coupling rises by *exactly the same amount* as the (even,even) coupling falls.

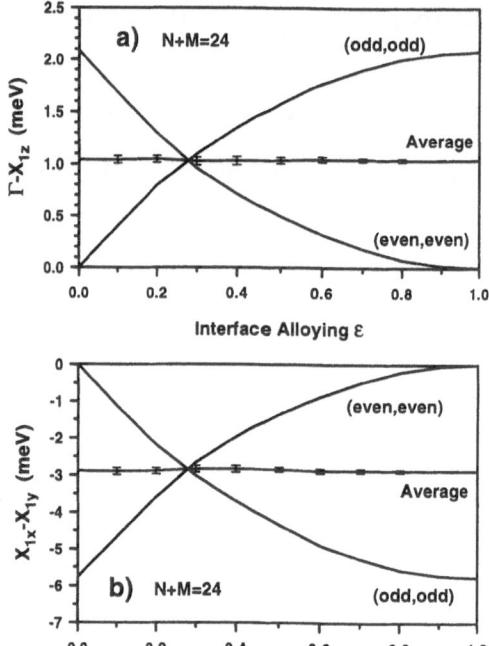

Fig. 2 Effect of interface alloying on X_1 coupling strengths for a superlattice period of 24 ML. (a) $\Gamma - X_{1z}$ coupling for (even,even) and (odd,odd) configurations. The average between the two is constant, within the small statistical uncertainty (error bars shown). (b) $X_{1x} - X_{1y}$ coupling strength. The characteristic behavior is the same as for $\Gamma - X_{1z}$.

Our results show that in the presence of steps the intervalley coupling is completely independent of the degree of interface alloying. In Fig.2(b) we plot the corresponding quantities for the $X_{1x} - X_{1y}$ coupling, and it is clear that the characteristic behavior is the same as for the $\Gamma - X_{1z}$ coupling.

References

1. D.J. Chadi and M.L. Cohen, Phys. Stat. Sol. (b) **68**, (1975) 405.
2. Y-T Lu and L.J. Sham, Phys. Rev. B **40**, (1989) 5567.
3. J.G. Menchero, T.G. Dargam, and B. Koiller, Phys. Rev. B**61**, (2000) 13021.
4. J.G. Menchero, B. Koiller, and R.B. Capaz, Phys. Rev. Lett. **83**, (1999) 2034.
5. P. Vogl, H. P. Hjalmarson, and J. D. Dow, J. Phys. Chem. Solids **44**, (1983) 365.
6. T. B. Boykin, G. Klimeck, R. C. Bowen, and R. Lake, Phys. Rev. B **56**, (1997) 4102.
7. T. B. Boykin, Phys. Rev. B **56**, 9613 (1997) 9613.
8. T. G. Dargam and B. Koiller, Solid State Commun. **105**, (1998) 211.
9. A. Franceschetti and A. Zunger, Phys. Rev. B **52**, (1995) 14664.

Proc. 25th Int. Conf. Phys. Semicond., Osaka 2000 (Eds. N. Miura and T. Ando)

471

Large quantum confinement effect of conduction electrons in ZnSe/BeTe type II heterostructures

R. Akimoto, Y. Kinpara, K. Akita

Electrotechnical Laboratory, 1-1-4 Umezono Tsukuba Ibaraki 305-8568, Japan e-mail: akimoto@etl.go.jp

Abstract The shortening of intersubband transition (ISBT) wavelength in ZnSe/BeTe type II superlattice (SL) featuring the large conduction band offset of $\sim$2.3 eV was examined aiming at potential applications at 1.55 μm optical communication wavelength. The electron subband structures as well as the above barrier states formed in ZnSe layers were investigated by photoluminescence (PL) and PL excitation (PLE) spectroscopy as a function of ZnSe layer width (4 to 50 ML) and fixed 14 ML-thick BeTe layer. Further, the ISBT from the first to the second electron subbands in ZnSe layer was observed directly by photo-induced infrared absorption measurements. We achieved the shortening of the ISBT wavelength down to 1.55μm for the SL structure with the 4$\sim$5 ML-thick ZnSe layer.

1 Introduction

ZnSe/BeTe heterostructures are newly developing material combinations with a type II band alignment[1]. Conduction electrons are localized in ZnSe layers due to very large confinement potential energy of $\sim$2.3 eV, while holes are localized in BeTe layers due to the valence band offset of $\sim$0.8 eV. Such deep potential energy especially for conduction electrons makes it possible to realize a large separation between the bound states. The large separation energy has advantages for potential applications of infrared optical devices utilizing intersubband transitions (ISBT). In the ZnSe/BeTe heterostructures, the conduction band offset is so large that ISBT energy between the first (E_1) and the second bound states (E_2) can be tuned up to near-infrared spectral region around 1.55 μm which is important wavelength for optical communications. Here, we will report on the large quantum confinement effect of conduction electrons in this novel ZnSe/BeTe heterostructures. The subband structures formed in ZnSe layers in ZnSe/BeTe super-lattices (SL) have been investigated by photoluminescence (PL) and PL excitation (PLE) spectroscopy as a function of ZnSe layer width. Further, the ISBT between E_1 and E_2 states in ZnSe layer has been observed directly by photo-induced near-infrared absorption measurement. We have achieved the shortening of the ISBT wavelength down to 1.55μm, which shows that ZnSe/BeTe heterostructure is promising for potential ISBT applications at the 1.55 μm optical communication wavelength.

2 Experimental procedure

ZnSe/BeTe SL is grown on GaAs (001) substrate by molecular beam epitaxy method. GaAs homoepitaxial layer with $\sim$2500Å-thickness is prepared in a separate III-V growth chamber to make the atomically flat As-

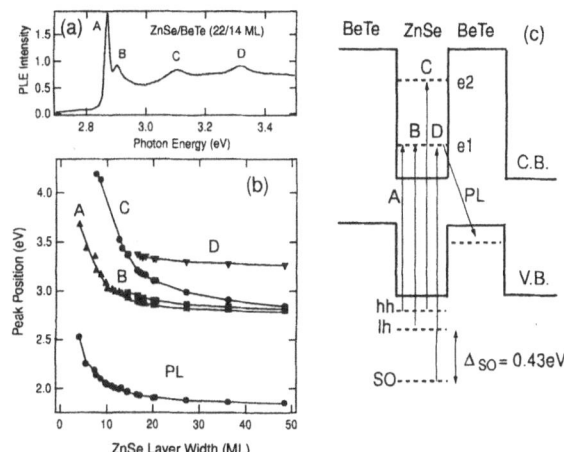

Fig. 1 (a)PLE spectrum for ZnSe/BeTe (22/14 ML) SL. (b) Peak position of PLE structures as a function of ZnSe layer width. (c) Optical transition corresponding to PLE structures.

terminated surface, before growing the ZnSe/BeTe SL. To examine the quantum confinement effect for conduction electrons localized in ZnSe layer, a thickness of the ZnSe layer is changed from 4 ML to 53 ML for each samples, while that of the BeTe layer is fixed at 14 ML.

For PL and PLE measurements at 4.2 K, a monochromated Xe arc lamp is used as an excitation source. For the photo-induced ISBT absorption measurements at 4.6 K, we employ a pump-probe method. The 375 nm laser light from second harmonic of mode-locked Ti: sapphire laser pumps the interband transition in ZnSe layer and populates electrons at the E_1 state, while a halogen lamp is used for probing the ISBT absorption at near-infrared spectral region. The probe light passing through a sample is dispersed by monochrometer. The pump laser beam is modulated by an optical chopper at 190 Hz, while a modulated transmission $-\Delta T$ of probe beam is taken by a phase-sensitive lock-in amplifier. The photo-induced absorption is expressed as $-\Delta T/T$, where T is the transmission without pump beam. The samples are prepared as a waveguide by polishing the two parallel facets at an angle of 40°[2] to prevent a infrared probe light from reflecting totally at the interface between GaAs(n=3.3) and ZnSe/BeTe epi-layer(n=$\sim$2.3).

3 Results and discussion

The samples exhibits a photoluminescence line due to a radiative recombination between electrons in a ZnSe

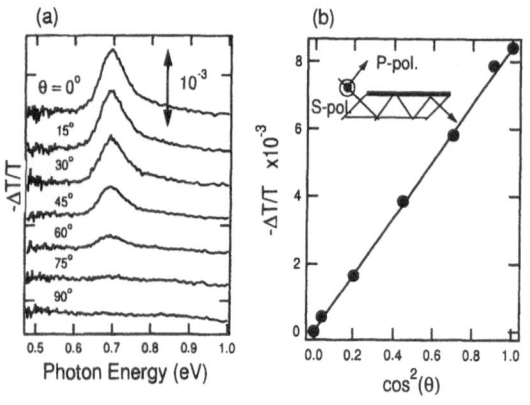

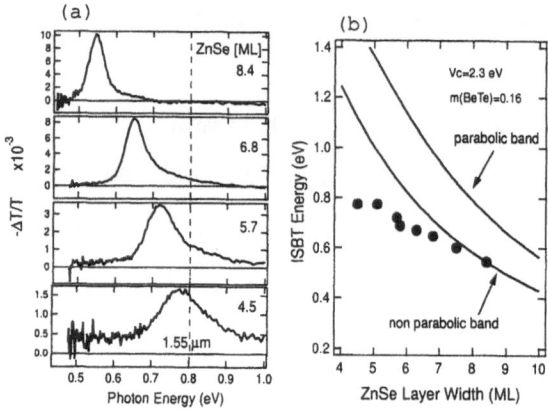

Fig. 2 (a)Photo-induced ISBT absorption spectra for ZnSe/BeTe(5.8/14 ML) SL with various polarization angle of incident probe light. P- and S-polarization correspond to $\theta=0°$ and 90°, respectively.(b)Absorption intesity as a function of $\cos^2(\theta)$.

Fig. 3 (a) Photo-induced ISBT absorption spectra for ZnSe/BeTe SLs with different ZnSe layer width. (b) ISBT energy as a function of ZnSe layer width and theoretical calculations.

layer and holes in a BeTe layer adjacent to the ZnSe layer. The subband structures for conduction electrons are investigated by PLE. We observed several fine structures in PLE spectra. PLE structures for ZnSe/BeTe (22/14 ML) SL is shown by A to D in fig.1(a). Fig.1(b) shows the peak position of the PLE structures as a function of ZnSe layer width. The PLE structures can be explained by optical transitions between higher order confinement states of electrons and hole states in ZnSe layers as indicated by fig.1(c).

The PLE structures A and C are focused, since these are optical transitions from a heavy hole state to E_1 and E_2 electron subbands in the ZnSe layer, respectively. From these data, we can estimate the separation energy between electron subbands. The optical transition energy are calculated by Kronig-Penny (KP) model, including non parabolic band effect by Kane three band model[3], a shift of band gap due to the compressive strain in ZnSe layer and the above barrier states for heavy hole. The calculation taking into account the above barrier states for heavy hole (hh1 and hh2) can fit the experimental data reasonably, while the calculation without it under-estimates the transition energy. This means that a simple energy difference between PLE structures A and C is not the electron ISBT energy. We estimate that the electron ISBT energy between E_1 and E_2 reaches 0.8 eV (1.5μm) at the ZnSe width of ~6ML.

The results of a photo-induced near-infrared absorption is shown in fig.2(a) for ZnSe/BeTe (5.8/14 ML) SL with the pump laser intensity of 0.67 W/cm^2 (averaged in time). We observed clear absorption signal located at ~0.7 eV. As the angle of linear polarization of the incident probe light changes from p- ($\theta=0°$) to s-polarization ($\theta=90°$), the signal intensity decreases gradually and vanishes completely at $\theta=90°$. The detailed analysis shows that the peak absorption intensity increases linearly as a function of $\cos^2(\theta)$ as shown in fig.2(b), suggesting the corresponding optical transition

occurs by a electric field component of the probe light normal to the epi-layer surface. Consequently, the dipole transition moment of this photo-induced absorption is directed normal to the grown epi-layers, which is the direct evidence that the photo-induced absorption is caused by electron ISBT in ZnSe layer or hole ISBT in BeTe layer. We can distinguish which contributes to the photo-induced absorption by changing ZnSe layer width with fixed BeTe layer width.

The photo-induced ISBT absorption spectra for several samples with different ZnSe layer width are shown in fig.3(a). The peak position of the ISBT absorption band shifts to higher energy for a sample with narrower ZnSe layer, indicating the ISBT absorption is caused by conduction electron in ZnSe layer. Further, the ISBT wavelength for the sample with 4.5 ML-thick ZnSe layer reaches 1.6 μm covering 1.55 μm sufficiently due to its rather broad band width of ~250 nm. In fig.3(b), ZnSe layer width dependence of the ISBT energy is plotted (dots). The results obtained by the KP model calculation (solid lines) assuming an optical transition between the first and the second subbands, with and without including the effect of a non-parabolic band are also shown. Obviously non-parabolic band effect is significant in the present SL structures having narrow ZnSe layers. Even in the calculation including the effect of non parabolic band by Kane model, the agreement with the experimental data becomes worse in the case of the ISBT energy > 0.6 eV, indicating more accurate band structure should be taken into account in the calculation.

References

1. A. Waag, F. Fischer, H. J. Lugauer, Th. Liz, J. Laubender, U. Lunz, U. Zehnder, W. Ossau, T. Gerhardt, M. Möller, and G. Landweher, J.Appl.Phys. **80**, (1996) 792.
2. M.Goppert, R.Becker, S.Petillon, M.Grun, C.Maier, A.Dinger, and C.Kligshirn, Physica E **7**, (2000) 89.
3. G. Bastard, *wave mechanics applied to semiconductor heterostructures* (Halsted press, 1988) p.89.

A study of the GaAs/partially ordered GaInP interface

T. Kobayashi[1], K. Inoue[1], A. D. Prins[1]*, K. Uchida[2], J. Nakahara[3]

[1] Department of Electrical and Electronics Engineering, Kobe University, 1-1 Rokkodai, Nada-ku, Kobe 657-8501, Japan   e-mail: kobayasi@eedept.kobe-u.ac.jp
[2] Department of Electronic Engineering, University of Electro-Communications, Chofu 182-8585, Japan
[3] Division of Physics, Graduate School of Science, Hokkaido University, Sapporo 060-0810, Japan

Abstract GaAs/partially ordered $Ga_{0.5}In_{0.5}P$ single quantum well structures grown by metalorganic vapor phase epitaxy were studied using time-resolved and continuous-wave photoluminescence at pressures up to ~5 GPa. Quantum well emission from GaAs/GaInP structures can be masked by anomalous bands at 1.35~1.46 eV. The introduction of two thin (~2 nm) GaP layers between the GaAs and GaInP restores this emission. The true nature of the 1.46 eV peak can only be determined at low excitation intensities and provides evidence of the spatially indirect transitions in a type II band alignment.

1 Introduction

When GaAs/$Ga_{0.5}In_{0.5}P$ quantum well (QW) structures are grown by metalorganic vapor phase epitaxy (MOVPE), emission from the GaAs QW is often masked by anomalous bands at around 1.46 eV [1-5]. This is closely related to the occurrence of partial CuPt-type ordering in GaInP. The emission from the GaAs QW is only restored if thin GaP layers are grown between the GaAs and GaInP [1,3-5]. Previous theoretical arguments [6] have suggested a type-II band alignment between GaAs and GaInP as the GaInP layer becomes more ordered. In such an alignment the spatially indirect transitions, GaInP(Γ_C)-GaAs (Γ_V), at the interface can occur at a smaller energy than the GaAs band gap (~1.51 eV). There are four possible combinations of this thin layer. We report the results of a study on GaAs/partially ordered GaInP QWs using time-resolved photoluminescence (TRPL) and continuous-wave photoluminescence (CWPL) at high pressures up to ~5 GPa.

2 Experiment

Four types of GaAs/GaInP single QW structures were grown on GaAs substrates by MOVPE under the same conditions [3]. Sample No. 1 consists of a 100-Å-thick GaAs well and 1100 Å partially ordered GaInP barriers. Sample No. 2 has a 14-Å-thick GaP layer applied between the lower GaInP and GaAs layers only, while sample No. 3 has a 28 Å GaP layer applied between the upper GaInP and GaAs layers. Sample No. 4 contains GaP layers sandwiching the GaAs well. The TRPL and CWPL measurements at pressures up to about 5 GPa were carried out with a diamond anvil cell. Luminescence was excited at 77 and 11 K by either the pulsed or CW 488 nm line of an Ar^+ laser.

---

* On leave from Queen Mary & Westfield College, University of London, UK

3 Results and Discussion

In CWPL measurements at atmospheric pressure (P=0) and at 11 K, sample No. 4 shows three main emission bands. The band at ~1.53 eV is from the GaAs QW, whose energy is slightly higher than the GaAs band-edge transition at ~1.51 eV. The impurity-related peak at 1.49 eV is also observed. However, the PL spectra of samples No. 1, No. 2 and No. 3 show no QW-related emissions in the energy range expected for type-I heterostructures, but are dominated by a peak at 1.42~1.46 eV whose energy is found to show a strong blueshift with the excitation intensity. In all samples the weak emissions from GaInP barriers are also observed at ~1.92 eV. This indicates that the GaInP layers used in this study have a smaller band gap than that of the disordered GaInP (~1.98 eV) and are partially ordered [3,4,7]. The presence of the GaAs/ partially ordered GaInP interface with no thin GaP layer seems to be closely related to the appearance of the ~1.46 eV emission.

Fig. 1 shows the pressure dependence of the CWPL peak energies of samples No. 2 and No. 3, measured at 77 K and at lower excitation intensities below 4 mW. Results for the GaAs QW emission at 1.53 eV in sample No. 4 measured at 10 mW are also shown for comparison. This QW emission peak behaves as expected with increasing pressure showing a peak shifting linearly at a rate of ~105 meV/GPa. At pressures above 3 GPa its intensity decreases and a new emission peak is detected at the low-energy side of the main PL peak which shifts toward lower energies for pressures up to ~5 GPa. This is attributed to the indirect recombinations, GaInP(X_C)-GaAs(Γ_V), both in real space and in momentum space [8].

For sample No. 2, the ~1.46 emission band strongly depends on the excitation intensity, unlike the QW emission at ~1.53 eV in sample No. 4. At lower pressures, its peak shows a sublinear shift toward higher energies. Its shift is rather similar to those obtained for the emissions from partially ordered GaInP layers, which again show the sublinear shift toward higher energies up to about 3.8 to 4.0 GPa [7]. Beyond ~3 GPa the emission band at 0.24 mW excitation almost disappears while the PL peak measured at 4 mW exhibits a shift toward lower energies for pressures up to ~5 GPa, with nearly same rate as X_C in GaInP. Above ~3 GPa another PL peak can be seen at the high-energy side of the main PL spectrum and shifts toward higher energies up

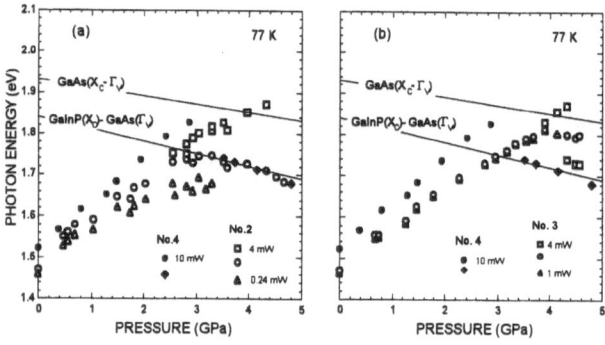

Fig. 1. PL peak energies of samples No. 2 and No. 3 as a function of pressure. Results for sample No. 4 are also shown for comparison. Solid lines represent results for descending X_c states in GaAs and disordered GaInP at 25 K [8].

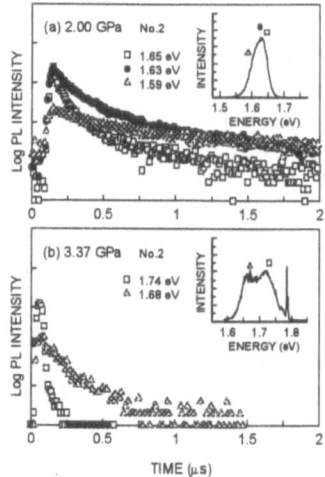

Fig. 2. Typical PL-decay profiles of sample No. 2. Pressures are (a) 2.0 and (b) 3.4 GPa.

to 4.4 GPa, nearly following the GaAs band-edge.

For sample No. 3 the ~1.46 eV emission shows a rather linear shift toward higher energies, and a weaker dependence of peak energy on the excitation intensity is obtained. At pressures above ~3.8 GPa the main PL peak shows a weak shift to lower energies, which is the typical character for the X state related transition after the Γ-X crossover in GaAs. In addition, two additional peaks can also be observed, one is the GaAs related 1.49 eV band the other the indirect recombinations, GaInP(X_C)-GaAs(Γ_V).

Fig. 2 shows the typical PL-decay profiles of sample No. 2 at 77 K and at pressures, 2.0 and 3.4 GPa. Each integrated PL spectrum is also shown in the inset. At 2.0 GPa, the main PL emission is dominated by a slower decay component and can be observed even at 5 μs. In addition the PL-decay profiles change remarkably for small changes in detection energy. This indicates the presence of, at least, two adjacent overlapping emissions of different nature, one with a shorter decay time of about 25 ns the other with a long decay time of 200~400 ns. Together with the prolonged nonexponential decay curves, a gradual redshift of the peak with delay time is

observed, indicating spatially indirect recombinations.

At 3.4 GPa, the PL-decay properties change drastically. PL emission band disappears nearly after 1.5 μs. At this pressure the descending X_C states in GaInP can affect such PL-decay profiles. In the inset two emission peaks can be observed. One at 1.67 eV is due to the ~1.46 eV emission with a longer decay time, while the other at 1.70 eV might be attributed to the emission from the X states in GaInP.

The CWPL spectra in samples No. 1 and No.3 at normal pressure appear to be similar to those in No. 2 but show a tendency closer to that of the GaAs layer at high pressures. The PL features observed for sample No. 2 and discussed earlier imply that the presence of ordered GaInP layers plays an important role in the radiative recombination at ~1.46 eV. Assuming a type-II band alignment between GaAs and partially ordered GaInP the 1.46 eV peak can be attributed to the spatially indirect transitions, GaInP(Γ_C)-GaAs (Γ_V), at the interface with no thin GaP layer, and hence would have a pressure dependence similar to that of partially ordered GaInP layer [4]. By using low excitation intensities, below those of the earlier study [4], we have confirmed that this is clearly the case in agreement with the theoretical arguments of [6].

This work was partly supported by the Grant-in-Aid for Scientific Research from the Ministry of Education, Science, Sports and Culture, Japan.

References

1. F.E.G. Guimarães, B. Elsner, R. Westphalen, B. Spangenberg, H. J. Geelen, P. Balk, and K. Heime, J. Cryst. Growth **124**, (1992) 199.

2. Q. Liu, S. Derksen, A. Lindner, F. Scheffer, W. Prost, and F. J. Tegude, J. Appl. Phys. **77**, (1995) 1154.

3. K. Uchida, T. Arai, and K. Matsumoto, J. Appl. Phys. **81**, (1997) 771.

4. S. H. Kwok, P.Y. Yu, K. Uchida, and T. Arai, J. Appl. Phys. **82**, (1997) 3630.

5. T. Kobayashi, A. Matsui, T. Ohmae, K. Uchida, and J. Nakahara, phys. stat. sol. (b) **211**, (1999) 247.

6. S. Froyen, A. Zunger, and A. Mascarenhas, Appl. Phys. Lett. **68**, (1996) 2852.

7. T. Kobayashi and R. S. Deol, Appl. Phys. Lett. **58**, (1991) 1289.

8. J. Chen, J. R. Seites, I. L. Spain, M. J. Hafich, and G.Y. Robinson, Appl. Phys. Lett. **58**, (1991) 744.

Analytical and Transfer matrix solutions of the $k.p$ Hamiltonian for the X-band in AlAs confined systems

Laura E. Bremme and Philip C. Klipstein

Clarendon Laboratory, Department of Physics, University of Oxford, Parks Road, Oxford, OX1 3PU, U.K.
Fax: +44-1865-272400; Email: l.bremme1@physics.ox.ac.uk

Abstract We present *analytical* and *transfer matrix* solutions of the two-band Hamiltonian for AlAs X-states. We offer a general yet relatively simple and flexible way to deal with all types of potential profile. We calculate energy spectra and wavefunctions for single quantum wells (QW) and discuss the implications of our results on Γ-X_Z mixing. We also determine the Landau level fan for AlAs systems with the magnetic field perpendicular to the camel's back axis.

1 Introduction

In the past much experimental and theoretical work was done to investigate the AlAs *bulk* dispersion around the X-point, which takes the form of a "camel's back". The energies of states in electrically or magnetically *confined* AlAs systems have, however, never been properly investigated. Usually an effective mass of $\sim 1.1m_0$ is used. Such an approach cannot be satisfactory since confinement energies in type II AlAs QWs and magnetic Landau level energies are each comparable with the camel's back depth [1]. There is therefore a great need to develop a more realistic treatment. We present here both *analytic* and *transfer matrix* solutions of the *two-band* Hamiltonian for AlAs X-states. We take into account the $k.p$-interaction [2], which has largely been ignored in QW physics, but which proves to be crucial. We present solutions for two confinement potentials along the axis of the camel's back: the finite square potential of a single QW and the parabolic potential due to a magnetic field. We also consider the implications of our results on Γ-X_Z interface band mixing.

2 Solutions of $k.p$ Hamiltonian

Restricting ourselves to the X_1-X_3 manifold, the confined states have wavefunctions $\Psi = \phi_1 u_1 + \phi_3 u_3$ where u_1 and u_3 are the X_1 and X_3 crystal periodic functions, and ϕ_1 and ϕ_3 the respective envelope functions [3]. The $k.p$ Hamiltonian is: $H = \underline{I}\left[-\frac{\hbar^2}{2m_z'}\frac{d^2}{dz^2} + V \right] - \underline{\sigma_z}\left[\frac{\Delta}{2}\right] + i\underline{\sigma_y}R\left[\frac{d}{dz}\right]$, where $R = -\left(\hbar^2/m_0\right)\langle u_1 \mid d/dz \mid u_3\rangle$, Δ is the X_1-X_3 energy splitting, $V(z) - \Delta(z)/2$ is the X_1 potential profile, and $\underline{\sigma_i}$ are Pauli matrices.

2.1 Analytical solution

We rewrite the Hamiltonian as

$$\underline{\phi}' = \frac{d}{dz}\begin{bmatrix}\phi_1(z)\\ \phi_3(z)\\ \Phi_1(z)\\ \Phi_3(z)\end{bmatrix} = \begin{bmatrix} 0 & 0 & 1 & 0\\ 0 & 0 & 0 & 1\\ f(E) & 0 & 0 & g(E)\\ 0 & s(E) & -g(E) & 0\end{bmatrix}\begin{bmatrix}\phi_1(z)\\ \phi_3(z)\\ \Phi_1(z)\\ \Phi_3(z)\end{bmatrix} = \underline{A}\underline{\phi}$$

where $\Phi_i = d\phi_i/dz$. By substituting $\underline{\phi} = \underline{P}\underline{\psi}$, where $\underline{P}$ is a matrix of the eigenvectors of $\underline{A}$, we obtain $\underline{\psi}' = \underline{J}\underline{\psi}$ in which $\underline{J} = \underline{P}^{-1}\underline{A}\underline{P}$ is diagonal with non-zero elements $J_{ii} = \lambda_i$, the eigenvalues of $\underline{A}$. We thus deduce the wavefunctions:

$$\underline{\phi} = \underline{P}\underline{\psi}, \text{ where } \psi_i(z) = \psi_i(0)\cdot\exp\left(-\int_0^z \lambda_i(z)dz\right).$$

Boundary conditions lead to a characteristic equation for the confined energies in a single QW: det[8×8-matrix]=0.

2.2 Transfer matrix solution

Assuming a slowly varying potential profile, the material can be divided into thin layers with constant potential. The solution is then a linear combination of plane waves.

$$\begin{bmatrix}\phi_1\\ \phi_3\end{bmatrix} = \begin{bmatrix} Ae^{ikz} + Be^{-ikz}\\ \beta\left(Ae^{ikz} - Be^{-ikz}\right)\end{bmatrix}$$

Inserting this into the Schrödinger equation, gives two values of $k \rightarrow k_+(E)$ and $k_-(E)$, and similarly two values of $\beta \rightarrow \beta_+(E)$ and $\beta_-(E)$. The general solution is thus:

$$\begin{bmatrix}\Psi_1\\ \Psi_3\end{bmatrix} = \begin{bmatrix} A_+ e^{ik_+z} + B_+ e^{-ik_+z} + A_- e^{ik_-z} + B_- e^{-ik_-z}\\ \beta_+\left(A_+ e^{ik_+z} - B_+ e^{-ik_+z}\right) + \beta_-\left(A_- e^{ik_-z} - B_- e^{-ik_-z}\right)\end{bmatrix}$$

Hence, a transfer matrix, $\underline{M_i}$, can be found between boundaries at z_i and z_{i+1}, such that $\Psi_{i+1} = M_i\Psi_i$, where

$$\underline{\Psi_i} = \begin{pmatrix}\Psi_1\\ \Psi_3\\ \Xi_1\\ \Xi_3\end{pmatrix}_{z_i} \text{ and } \begin{pmatrix}\Xi_1\\ \Xi_3\end{pmatrix} = \begin{bmatrix} \frac{1}{m_z'}\frac{d}{dz} & -\frac{2R}{\hbar^2}\\ \frac{2R}{\hbar^2} & \frac{1}{m_z'}\frac{d}{dz}\end{bmatrix}\begin{pmatrix}\Psi_1\\ \Psi_3\end{pmatrix}$$

The vector, $\underline{\Psi_i}$, is conserved across each boundary. Multiplying transfer matrices for the whole potential profile leads to a characteristic equation for the confined energies and corresponding wavefunctions. The transfer matrix approach is a general, yet relatively simple, precise formulation, being able to deal with any potential profile.

3 Results

We now present solutions for two examples: X_Z states confined in an AlAs single QW and x-confined X_X magnetic Landau levels ($B\|z$). For the $k.p$-parameters, values of $R = 3.4\ eV\text{Å}$ and $m_z' \sim 0.16m_0$ were used [1].

3.1 Quantum well results

Fig.1a shows the first six confined energies in an AlAs single QW as a function of well width, calculated with the widely used one-band model with an electron effective mass of 1.1m_0. The two-band method of

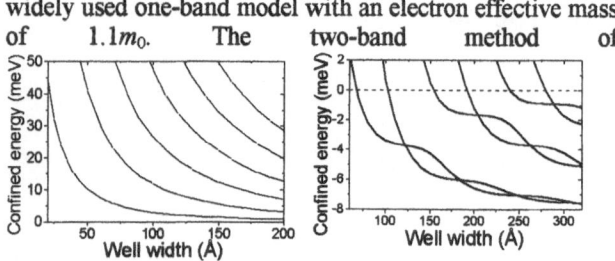

Fig. 1: Confined energies of QW, a) the one-band model ($m^* = 1.1m_0$) and b) the two-band model

calculating confined energies, where the **k.p**-interaction has been taken into account, shows a completely different picture (fig.1b). It becomes obvious that the simple effective mass approach is not an adequate description of the system. The energy levels of even and odd parity states cross periodically with varying well width. The periodicity appears to be ~65Å = π/k_0, where k_0 is the shift of the camel's back minimum from the X-point in bulk AlAs. This alternation of paired up energy levels breaks down at low well widths, for energies above the bulk X-point energy (dotted line). The *wavefunctions* also show an unusual behaviour in the two-band model. The energy alternation results in a parity alternation for the ground state. The X_1 and X_3 components of the wavefunction confined by the

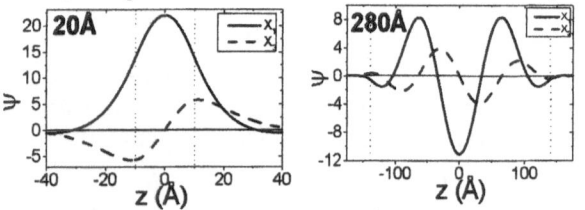

Fig. 2: Ground state wavefunctions for 20Å and 280Å QW, showing the components with X_1 and X_3 symmetry

X_1 profile have opposite parities and *both* have significant amplitude. Fig.2 compares both contributions for the ground state wavefunction in a 20Å and a 280 Å single QW. The larger the well width, the more nodes there are in the wavefunction.

3.2 Magnetic field results

In this section we present energy spectra for magnetic fields normal to the camel's back axis. The results are appropriate to Landau levels at the conduction band edge in bulk or in heterostructures with

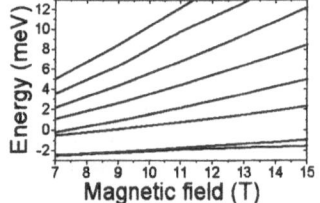

Fig. 3: Landau level fan for **B** normal to the camel's back axis

AlAs layers thicker than ~50Å [4] and magnetic fields parallel to the growth direction. The magnetic field creates a one dimensional *parabolic potential* at right angles to itself. Our results are consistent with previous work done on GaP and Te [5,6], which used an infinite set of harmonic oscillator functions, truncated to make a calculation possible. Using our transfer matrix treatment no such

approximation is needed. In fig.3, we calculate the Landau level fan for magnetic fields up to 15T. For low magnetic fields the Landau levels are doubly degenerate, and split, when the energies become comparable to the camel's back depth. The splitting is related to the two minima in the camel's back and is akin to the spitting between "bonding" and "antibonding" levels in coupled wells.

3.3 Γ-X_Z mixing

Recently it has been demonstrated that interfaces can have a strong influence on the energies and wavefunctions of nearly degenerate states with different symmetries [7,8,9]. The most notable example is Γ-X_Z mixing in GaAs/AlAs superlattices

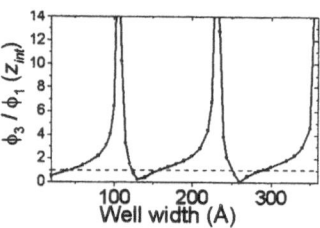

Fig. 4: Ratio of X_3 to X_1 interfacial amplitudes for QW ground state.

(SLs). Interface band mixing depends sensitively on the wavefunction amplitude at the interfaces. Fig.4 shows the ratio of the X_3 and X_1 components of the ground state wavefunction as a function of the QW width. Both components go to zero in turn for specific widths. Notably, the minority X_3 component can have a much larger interface amplitude than the majority X_1 component. This is significant, because the X_3 mixing potential is known to be much larger than the X_1 mixing potential [9], so in principle, there should be large variations in the Γ-X_Z mixing strength with QW thickness.

4 Conclusions

We have presented two methods for calculating X-state confinement energies, taking into account the camel's back dispersion: an analytical method appropriate to finite QWs, and a more general transfer matrix method appropriate to slowly varying potentials. We have shown that there is little justification for the effective mass treatment of X_Z-states in GaAs/AlAs SLs or of Landau levels in fields normal to the camel's back axis. We have found that the X_3-admixture with X_1-states is substantial and have discussed the implications on Γ-X_Z mixing.

1. L. E. Bremme, H. Im, H. Choi, G. W. Smith, R. Grey, P. C. Klipstein, *proc. 25th Int. Conf. Phys. Semicond.*, Osaka, (2000)
2. G. Bastard, "Wave Mechanics Applied to Semicond. Het.", Les Editions de Physique, p.35 (1988)
3. Y. Fu, M. Willander, E. L. Ivchenko, A. A. Kiselev, *Phys. Rev. B* **47**, (1993) 13498
4. For thinner layers the ground state changes from $X_{X,Y}$ to X_Z-see J. M. Smith *et al.*, *Phys. Rev. B* **57** (1998) 1740
5. H. Kamimura, K. Nakao, T. Doi, *proc. 10th Int. Conf. Phys. Semicond.*, Cambridge (1970)
6. N. Miura, G. Kido, M. Suekane, S. Chikazumi, *J. Phys. Soc. Japan* **52**, (1983) 2838
7. H. Im, P. C. Klipstein, R. Grey, G. Hill, *Phys. Rev. Lett.* **18**, (1999) 3693
8. T. Reker, H. Im, H. Choi, L. E. Bremme, P. C. Klipstein, R. Grey, G. Hill, *proc. 25th Int. Conf. Phys. Semicond.*, Osaka (2000)
9. P. C. Klipstein, in: *Semiconductors and Semimetals*, **55**, Chap. 2, Eds. T. Suski and W. Paul, Academic Press, New York 1998

Relation between the in-plane polarization anisotropy of the optical properties and the microscopic atomic configuration in $(Ga_{0.5}In_{0.5}As)/(InP)$ superlattices

Rita Magri and Stefano Ossicini

Dipartimento di Fisica, Universitá di Modena e Reggio Emilia, 41100 Modena, Italy and Istituto Nazionale per la Fisica della Materia
e-mail: magri@unimo.it

Abstract By first-principles calculations of the dielectric tensor elements we address the issue of the giant polarization anisotropy of the optical absorption experimentally observed in the $(Ga_{0.5}In_{0.5}As)/(InP)$ superlattices. We consider different atomic configurations of these superlattices and study how the polarization anisotropy depends on the strain not only at the interfaces but also in the bulk alloy.

1 Introduction

A strong in-plane polarization anisotropy (PA) of the optical properties has been observed in III-V semiconductors grown along the (001) direction that do not share a common atom, such as the type I (InGa)As/InP system[1]. The anisotropy is caused by the existence of subsequent *inequivalent* interfaces that leads to a reduction of the crystal field symmetry from the D_{2d} point group of common atom (001) superlattices (superlattices whose constituents share a common anion or a common cation) to a C_{2v} point group. Consequently, the crystal potential is not anymore symmetric with respect to a layer midpoint for all the roto-inversion operations of the D_{2d} point group changing z into -z. These symmetry effects and, thus, the PA, cannot be described by the standard implementation of the $\mathbf{k} \cdot \mathbf{p}$ theory. Krebs and Voisin[1] and Ivchenko et al.[2] have introduced phenomenological approaches that introduce a further coupling potential localized at the interfaces into the framework of the $\mathbf{k} \cdot \mathbf{p}$ theory. However, in both cases the magnitude of these interface potentials depends on heuristic parameters whose value is obtained by fitting the experimental data. Since the PA is due to the inequality of the interfaces we investigate here how the anisotropy depends on the different atomic configurations of the no-common atom $(Ga_{0.5}In_{0.5}As)_n/(InP)_n$ superlattices. In a zero spin-orbit approach, the in plane PA, that is the difference between the optical absorption corresponding to a polarization direction along the [110] axis and that corresponding to a polarizationd direction along the [-110] axis, is measured by the energy splitting between the two highest one-dimensional hole states[3]. Indeed, this crystal field splitting is directly related to the intensity of the C_{2v} symmetry term of the total crystal potential, that, in turn, depends on how much the two interfaces differ.

2 Theoretical approach

We have used two approximations to model the pseudobinary $Ga_{0.5}In_{0.5}As$ alloy: (i) the simple virtual crystal approximation, VCA, where a virtual cation M = $(Ga_{0.5}In_{0.5})$ is simulated by averaging the In and Ga pseudopotentials, (ii) the true ternary "special quasirandom structure", SQS-4, which is the ordered structure with only 4 atoms per unit cell that better mimic the first few radial correlation functions of a perfectly random structure[4]. For the VCA approximation we consider superlattices $(Ga_{0.5}In_{0.5}As)_n/(InP)_n$ with periods $n=6$ and $n=12$, while for the SQS-4 approximation only the $n=6$ case has been considered. For each layer material we consider always an even number of monolayers. Thus, all the superlattices have two inequivalent interfaces: one (Ga,In)-P and one In-As interface per unit cell.

We calculate directly the dielectric tensor elements in the dipole approximation:

$$\epsilon_{i,j}(\omega) = \frac{8\pi^2 e^2}{\omega^2 m^2 V} \sum_{v,c} \sum_{k} (P_{v,c}(k))^*_i (P_{v,c}(k))_j$$
$$\times \, \delta(E_c(k) - E_v(k) - \hbar\omega), \tag{1}$$

where $\overrightarrow{P}_{v,c}(k) = <v,k|\overrightarrow{P}|c,k>$ is the matrix element of the optical transition between the valence state $|v,k>$ and the conduction state $|c,k>$, $i,j = x,y,z$, $E_c(k)$ and $E_v(k)$ are the conduction and valence energy levels, and V is the unit cell volume. The in-plane polarization dependent imaginary part of the dielectric function in terms of the dielectric tensor elements is given by:

$$\epsilon_2(\omega, \theta) = \epsilon_{xx}(\omega)$$
$$+ [\epsilon_{yy}(\omega) - \epsilon_{xx}(\omega)] sin^2\theta + \epsilon_{xy}(\omega) sin2\theta, \tag{2}$$

where we indicate by θ the angle between the x axis and the polarization vector direction. In the superlattices grown along the [001] direction with no-common atom, the observed anisotropy between the absorption corresponding to polarization $\theta = \frac{\pi}{4}$ ([110] direction) and to polarization $\theta = -\frac{\pi}{4}$ ([-110] direction) is proporzional to the off-diagonal xy element of the dielectric tensor, which is different from zero in the case of the C_{2v} symmetry:

$$\Delta\epsilon(\omega) = \epsilon_2\left(\omega, \frac{\pi}{4}\right) - \epsilon_2\left(\omega, -\frac{\pi}{4}\right) = 2(\epsilon_{xy}(\omega)). \tag{3}$$

The superlattice wavefunctions and the level energies entering Eq. (1) have been calculated using the self-consistent density functional theory, within the local density approximation (DFT-LDA) and non local norm-conserving pseudopotentials[5]. To study the role of interface strain and atomic relaxations on the optical PA we consider separately the cases of unrelaxed systems with all atomic positions at the ideal sites of the zinc-blende lattice and fully relaxed systems.

Table 1 Calculated dielectric tensor elements for the unrelaxed and relaxed superlattices. Transition energies are in eV. The tensor is symmetric and for all the structures $\epsilon_{yy} = \epsilon_{xx}$ and $\epsilon_{yz} = \epsilon_{xz}$

Transition	Energy	ϵ_{xx}	ϵ_{zz}	ϵ_{xy}	ϵ_{xz}
\multicolumn{6}{c}{$(MAs)_{12}/(InP)_{12}$ unrelaxed}
V1-E1	0.627	4.074	0.000	-4.074	0.000
V2-E1	0.629	4.086	0.000	+4.086	0.000
V3-E1	0.664	0.000	7.110	0.000	0.000
\multicolumn{6}{c}{$(MAs)_{12}/(InP)_{12}$ relaxed}
V1-E1	0.592	4.102	0.000	-4.102	0.000
V2-E1	0.606	4.109	0.000	+4.109	0.000
V3-E1	0.678	0.000	6.967	0.000	0.000
\multicolumn{6}{c}{$(SQS\text{-}4)_6/(InP)_6$ unrelaxed}
V1-E1	0.623	4.206	0.000	-4.206	0.000
V2-E1	0.635	4.268	0.007	+4.268	0.175
V3-E1	0.654	0.004	7.984	+0.004	-0.176
\multicolumn{6}{c}{$(SQS\text{-}4)_6/(InP)_6$ relaxed}
V1-E1	0.574	4.281	0.000	-4.281	0.000
V2-E1	0.625	3.957	0.015	+3.957	0.246
V3-E1	0.704	0.007	6.879	0.007	-0.211

3 Results

The anisotropy effects are mainly observed over a 70 meV range[1], which corresponds to the optical transitions between the three upper hole states V1, V2, and V3, localized into the (InGa)As layers and the first electron state E1, which is also localized in the (InGa)As wells. In Table 1 we give the calculated dielectric tensor elements for the unrelaxed and relaxed $(MAs)_{12}/(InP)_{12}$ and $(SQS\text{-}4)_6/(InP)_6$ superlattices, corresponding to the optical transitions between these states at the Γ point, together with the corresponding transition energies. Since $\epsilon_{xy}(\omega_1) \sim -\epsilon_{xy}(\omega_2)$ and $|\epsilon_{xy}| = \epsilon_{yy} = \epsilon_{xx}$, in a zero spin-orbit calculation, the polarization direction enhances and quenches selectively the optical transitions associated to the C_{2v} crystal field splitted states. Because of this sort of selection rule the PA of these single transitions is therefore 100%. In Fig. 1 we plot the contribution to $\epsilon_2(\omega)$ of these transitions at Γ for the two polarization [110] and [-110] directions. We see that the atomic relaxations at the interfaces in $(MAs)_{12}/(InP)_{12}$ increases the splitting between the V1 and V2 highest hole states from 1.5 meV to 13.9 meV. Also the atomic relaxations within the bulk alloy in the $(SQS\text{-}4)_6/(InP)_6$ superlattice lead to a huge splitting between the same transitions, 51.0 meV. The experimental results[1] show two distinct ab-

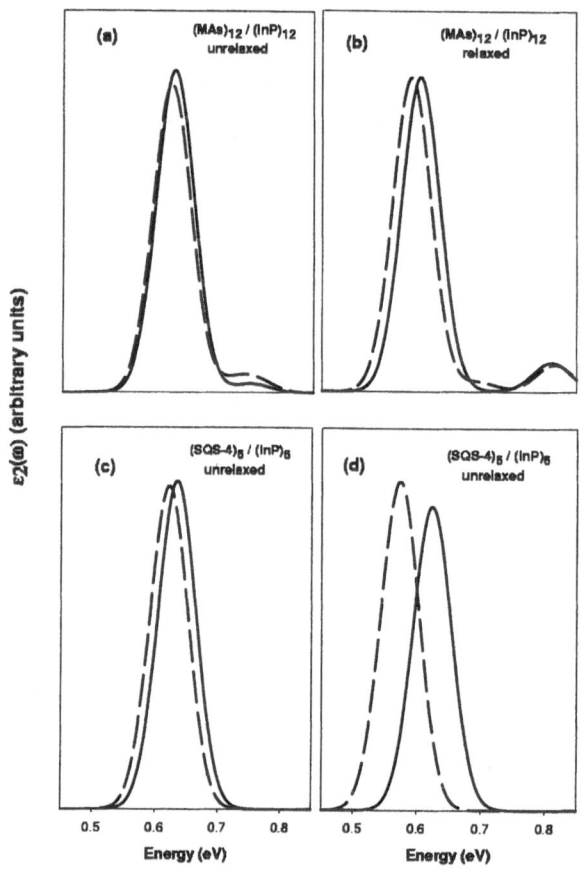

Fig. 1 Imaginary part of the dielectric function $\epsilon_2(\omega)$ for polarizations along [110] (solid curve) and along [-110] (dashed curve). Only the contribution of the interband transitions at the Γ point have been considered. The transitions have been dressed by a Gaussian broadening of 30 meV

sorption features split by about 40.0 meV. Considering that the real situation is somewhat in between the two limits of zero bulk strain of $(MAs)_n/(InP)_n$ and the large bulk strain of $(SQS\text{-}4)_n/(InP)_n$ (ordered structures have larger relaxation patterns than random structures), our values for the splitting at VBM compare well with the experiment.

Thus, we demonstrate here that the atomic relaxation processes, which differentiate the bond lengths at the interfaces and within the alloy layer, provide an important contribution to the intensity of the potential term of C_{2v} symmetry which is responsible for the in-plane PA.

References

1. O. Krebs and P. Voisin, Phys. Rev. Lett. **77**, (1996) 1829.
2. E. Ivchenko, A. Kaminski, and U. Rossler, Phys. Rev. B **54**, (1996) 5852. Jersey 1951).
3. R. Magri and S. Ossicini, Phys.Rev. B **58**, (1998) R1742.
4. A. Zunger, S.-H. Wei, L.G. Ferreira, and J.E. Bernard, Phys. Rev. Lett. **65**, (1990) 353.
5. G. Kerker, J. Phys. C **13**, (1980) L189.

Comparison of empirical pseudopotential and k · p calculations in p-doped strained layer SiGe quantum wells

Z. Ikonić, R. W. Kelsall, P. Harrison

Institute of Microwaves and Photonics, School of Electronic and Electrical Engineering, University of Leeds, Leeds LS2 9JT, United Kingdom, e-mail: z.ikonic@ee.leeds.ac.uk

Abstract The predictions of quantized state energies, as calculated by the empirical pseudopotential approach and by the 6×6 **k · p** model are compared for the case of SiGe based quantum wells, grown in the [001] direction. Generally, the agreement is very good for energies up to few hundred meV away from the valence band top.

1 Introduction

Intersubband transitions in p-doped strained-layer SiGe based quantum wells have recently become increasingly interesting, due to their possible use in intersubband quantum cascade lasers operating in the mid- to far-infrared wavelength range [1]. The commonly used approach to find the subband structure in the valence band (v.b.) is the 6×6 or (in case of intermediate or narrow gap semiconductors) the 8×8 **k · p** model. A recent study of the accuracy of the **k · p** model, as compared to the results of the more sophisticated empirical pseudopotential method (EPM) [2] has indicated that its accuracy in p-type AlGaAs based quantum wells is good for energies not far (≤ 200 meV) from the v.b. top. Here we investigate the accuracy of the 6×6 **k · p** model in strained SiGe based structures.

2 Details of calculation

The 6×6 **k · p** model was implemented by using the Fourier expansion of wave functions [3] and Burt-Foreman boundary conditions [4]. The structure is thus implicitly taken to be periodic, i.e. a superlattice (SL). In alloy layers the Luttinger γ parameters and the deformation potentials are taken as weighted averages of those for Si and Ge. The EPM was implemented as a supercell calculation, with the spin-orbit coupling included. Consistent with the model of homogeneous strain in individual layers, as in the **k · p** model, the atomic coordinates are taken as strain-scaled ideal ones. The atomic potentials are described by appropriate formfunctions $V(q)$. The virtual crystal approximation is used for alloys.

In normal **k · p** calculations one uses the material parameters extracted from experiments. Such parameters feature implicitly in the EPM, but their values are not usually in close agreement with the experimental values. For the purpose of comparison it is essential that the two calculations do actually use, explicitly or implicitly, the same set of parameters (the γ parameters, deformation potentials, and the average v.b. offset ΔE_v), and it is preferable that these are reasonably close to the ac-

Table 1 Parameters for Si and Ge implicit in the EPM calculation, also used in the **k · p** calculation for comparison, and parameters used in a regular **k · p** calculation.

	EPM		**k · p** [6]	
	Si	Ge	Si	Ge
γ_1	4.34	9.12	4.22	13.4
γ_2	0.37	2.76	0.39	4.25
γ_3	1.30	3.78	1.44	5.69
b [eV]	−2.39	−2.75	−2.10	−2.86
a_v [eV]	2.46	1.24	2.46	1.24

cepted experimental values. The γ parameters implicit to the EPM are extracted from dispersion of the HH, LH and SO bulk bands along the [001] or [111] directions. However, the problem is overdetermined (there are more equations than γ parameters) [2]. The precision margin of the results is a few percent, which one should bear in mind when comparing the EPM and **k · p** results. The deformation potential b is found from the HH-LH splitting after applying strain (i.e. alloy substrate) to Si or Ge, while a_v and ΔE_v are found from the differences of the v.b. tops in the unstrained and strained cases (the $Ge_{0.5}Si_{0.5}$ substrate was used). Extraction of these parameters also suffers from overdeterminancy. The form-functions from [5] are found to be the best in this respect, even with a lower cutoff energy of 4.5 Ry, the parameters showing good overall agreement with values used in normal **k · p** calculations, Table 1. The value of ΔE_v implicit in the EPM calculation is 0.438 eV, somewhat lower than 0.58 eV [7], but still good enough for the present purpose.

3 Results and discussion

In Fig. 1 the subband energies (measured downwards from the HH v.b. top in the well) in Ge_nSi_n SL ($n = 4-16$) on $Ge_{0.5}Si_{0.5}$ substrate, are given. Clearly, the degree of agreement is quite good for energies below ∼400 meV, even for the shortest-period structure Ge_4Si_4, but considerably decreases at higher energies.

The in-plane dispersion of subbands in Ge_4Si_4 is given in Fig. 2. At low energies the agreement is very good, slowly degrading as $k_{||}$ increases. Even for high-energy states, the *dispersions* agree reasonably well; the discrepancy is primarily due to the quantized part of energy.

Similar results are obtained in other structures, with alloy wells and/or barriers. That the agreement between

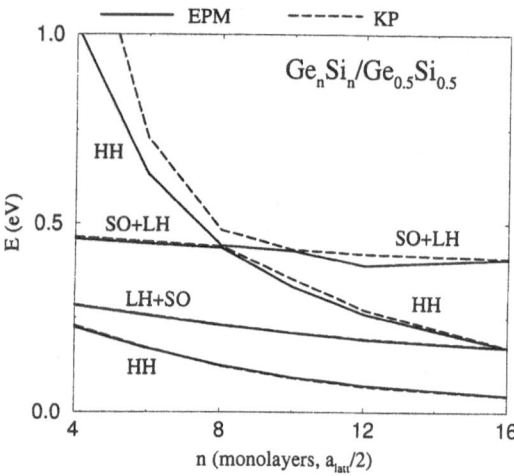

Fig. 1 Subband energies at $k_{\parallel} = 0$, calculated by EPM (solid lines) and $\mathbf{k} \cdot \mathbf{p}$ method (dashed lines) for Ge_nSi_n SL on $Ge_{0.5}Si_{0.5}$ substrate.

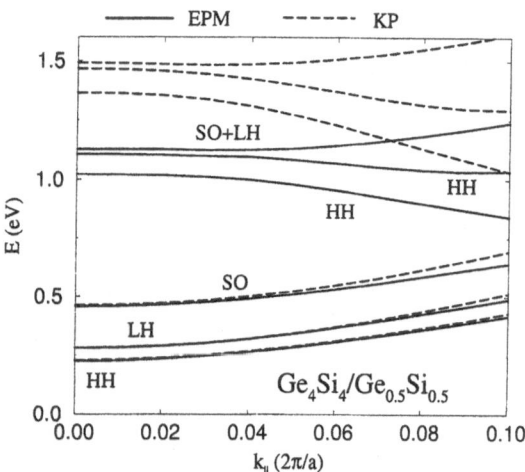

Fig. 2 The in-plane dispersion of subbands, with $k_{\parallel}$ along the <10> direction, calculated by EPM (solid) and $\mathbf{k} \cdot \mathbf{p}$ method (dashed lines) for Ge_4Si_4 SL on $Ge_{0.5}Si_{0.5}$ substrate.

the $\mathbf{k} \cdot \mathbf{p}$ and EPM becomes worse as $k_{\parallel}$ or the quantized state energy (i.e. the equivalent perpendicular wave vector in the well, k_z) increase, is in accordance with physical expectations. This is mostly due to the fact that the overlap of off-Γ bulk states and states at Γ decreases as $k_{\parallel}$ or k_z increase (an insufficiency of the $\mathbf{k} \cdot \mathbf{p}$ model even for bulk material [2]). An apparently larger discrepancy may occur in very wide wells; e.g. in the right-hand portion of Fig. 3. With such a dense energy spectrum there is a large interaction of states which amplifies the dicrepancy between the $\mathbf{k} \cdot \mathbf{p}$ and EPM results. This does not mean that the $\mathbf{k} \cdot \mathbf{p}$ method is unreliable, however. Similar discrepancies would occur when using different sets of material parameters that have appeared in the literature. It is unlikely that any method could provide good agreement with experiment in such cases.

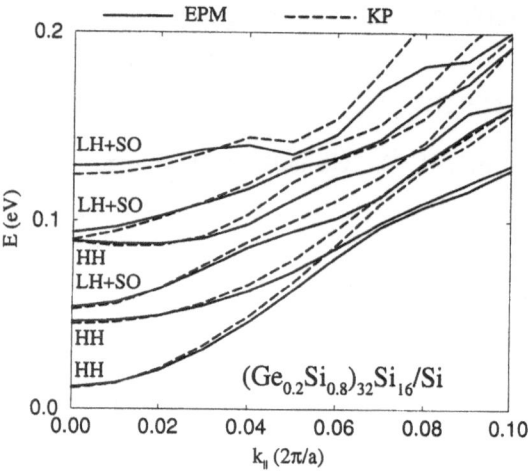

Fig. 3 Same as in Fig. 2, but for $(Ge_{0.2}Si_{0.8})_{32}Si_{16}$ SL on Si substrate.

We have also checked whether the agreement could be improved by introducing energy dependent γ parameters, in analogy to the energy-dependent effective mass which improves the accuracy of the effective mass method in the conduction band. Non-constant γ's improved the accuracy of the $\mathbf{k} \cdot \mathbf{p}$ model, but in an unstrained system and for $k_{\parallel} = 0$ [8]. However, this approach did not prove to be useful in strained SiGe system (once again due to the problem of strain-split induced overdeterminacy): whilst the accuracy was improved at high energies, it was degraded at low energies.

4 Conclusion

By performing the 6×6 $\mathbf{k} \cdot \mathbf{p}$ and EPM calculations for strained SiGe structures we found that the agreement between the two is very good for energies up to a few hundred meV, and for the range of in-plane wave vectors handled by $\mathbf{k} \cdot \mathbf{p}$ in most of practical calculations.

Acknowledgement. This work is supported by DARPA/ USAF contract No. F19628-99-C-0074. The authors thank W. Batty (Leeds) and R.A. Soref (Handscom AFB) for useful discussions.

References

1. R.A. Soref, L. Friedman, and G. Sun, Superlatt. Microstr. **23**, (1998) 427.
2. D.M. Wood, and A. Zunger, Phys. Rev. **B53**, (1996) 7949.
3. D. Gershoni, C.H. Henry, and G.A. Barraf, IEEE J. Quantum Electron. **29**, (1993) 2433.
4. B.A. Foreman, Phys. Rev. **B48**, (1993) 4964.
5. P. Friedel, M.S.Hybertsen, and M. Schlüter, Phys. Rev. **B39**, (1989) 7974.
6. A. Kahan, M. Chi, and L. Friedman, J. Appl. Phys. **75**, (1994) 8012.
7. C.G. Van de Walle, and R.M. Martin, Phys. Rev. **B34**, (1986) 5621.
8. F. Long, W.E. Hagston, P. Harrison, and T. Stirner, J. Appl. Phys. **82**, (1997) 3414.

Plasmons in laterally density modulated 2D electron gas in shallow etched single-heterostructures

Tomoya Tagawa and Shin-ichi Katayama

School of Materials Science, Japan Advanced Institute of Science and Technology, 1-1 Asahidai, Tatsunokuchi, 923-1292, Japan
e-mail: tagawa@jaist.ac.jp

Abstract Plasmons in two-dimensional electron gas (2DEG) have been explored theoretically in the long-wavelength limit. The lateral density modulation of 2DEG is calculated self-consistently by modeling the shallow etched single heterostructures. We discuss a possibility of opening of gaps at the zone center accompanied by a closing of gaps at the zone edge in the plasmon bands.

1 Introduction

Recently the resonant light scattering experiments have revealed the collective plasmon modes and single-particle excitations in laterally modulated 2DEG.[1-4] The theoretical works by Ager and Hughes [5] have pointed out the important effects of periodic gate screening on plasmon dispersions assuming density profile. More quantitative analysis including intersubband excitations were given by the authors in Ref.[6,7] based on a density response theory. Among the above experiments, Schdelbeck et al.[3] have reported on interesting data in n-type AlGaAs/GaAs single heterostructures. As long as we observe their data carefully, it seems that there appears an opening of gaps at the zone center accompanied by a closing of gaps at the zone edge in the plasmon bands. This behavior could not be understood by simple analysis assuming the same period of modulation with that of the metal gate. We explore theoretically the plasmon bands in a model to provide a possible explanation of such observation.

2 Electronic States and Plasmons

2.1 Lateral Modulation of 2D Electron Density

In the experiments of Ref.[2,3], the periodic metallic gate was fabricated on the rectangularly shallow etched surface as is shown in Fig. 1(a). The period is $a=a_1+a_2$ where a_1 and a_2 denote the width of the upper and lower flat portion in the gate. Those are located at distance d_1

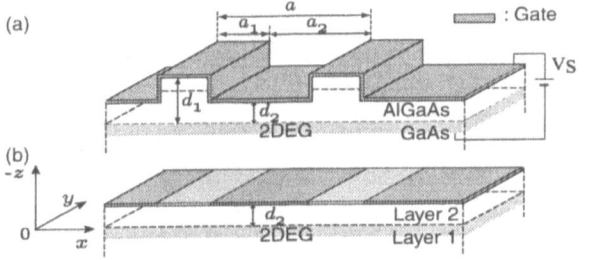

Fig. 1 (a)a shallow etched AlGaAs/GaAs single heterostructures.(b) two-layer model structures.

and d_2 from 2DEG. It might be difficult to estimate the electrostatic potential at 2DEG due to the rectangular metal gate including the doped ionized impurities in Al-GaAs layer.[8] We adopt here a two-layer model with a single periodic gate located at $-d_2$ in Fig. 1(b). The gate potential is assumed to be

$$V_{Gate}(x,z) = [V_{Gd}F(x) + V_{Gm}(1 - F(x))]\,\delta(z + d_2), \quad (1)$$

with

$$F(x) = \sum_n \theta\left(\left(\frac{a_1}{2}\right)^2 - (x - na)^2\right). \quad (2)$$

In this model, the effects from the upper flat portion in Fig. 1(a) is included by the following way. By assuming $V_{Gd} = rV_{Gm}$, we observe the influence on 2DEG potential at z=0 produced by the change of d_1. When $d_1 = d_2$, r=1, and as $d_1 \rightarrow \infty$, r goes to 0. Let the dielectric constants of layer 1, 2 and metallic gate be κ_1, κ_2 and κ_m. By making the Fourier series expansion along the x-axis, the coupled Shrödinger-Poisson equations with effective mass m^* become the following matrix form,

$$\sum_{\ell'=-\infty}^{\infty} [\frac{\hbar^2}{2m^*}k_{\ell x}^2\delta_{\ell\ell'} + V_{\ell-\ell'}]c_{\ell'}^{(n)}(k_x) = \varepsilon_n(k_x)c_\ell^{(n)}(k_x), \quad (3)$$

where $k_{\ell x} = k_x + K_\ell$ with $K_\ell = 2\pi\ell/a$, and

$$V_\ell = \int_{-d}^{\infty} \zeta_0(z')V_\ell(z')\zeta_0(z')dz', \quad (4)$$

ζ_0 being the wavefunction of the ground-state subband. According to Wulf [9], the Fourier component of the self-consistent potential is composed both of the Hartree term and the gate potential as

$$V_\ell(z) = I_\ell(z) + e^{-K_\ell z}\left(\frac{\epsilon}{a_\ell^+(d_2)}V_{G\ell} + \frac{a_\ell^-(d_2)}{a_\ell^+(d_2)}I_\ell(0)\right), \quad (5)$$

where $\epsilon = \kappa_2/\kappa_1$, $a_\ell^\pm(d) = \sinh(K_\ell d) \pm \epsilon\cosh(K_\ell d)$,

$$I_\ell(z) = \frac{2\pi e^2}{\kappa_1 K_\ell}\int_0^{\infty} n_{0\ell}(z')e^{-K_\ell|z-z'|}dz', \quad (6)$$

and

$$V_{G\ell} = (r - 1)V_{Gm}\frac{1}{\pi\ell}\sin\frac{\pi\ell a_1}{a}. \quad (7)$$

Ignoring the second term in parentheses of Eq.(5) and set $\epsilon = 1$, we carry out the calculation of electronic states by diagonalizing the matrix of Eq.(3).

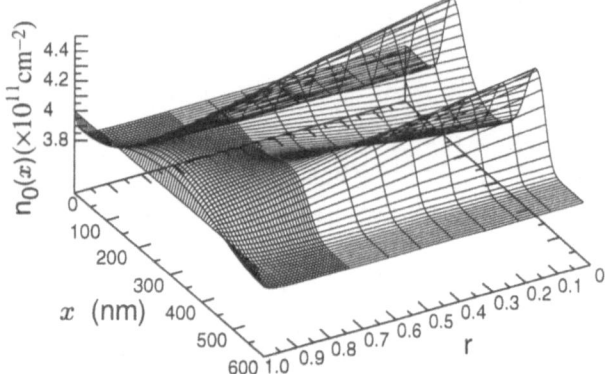

Fig. 2 Self-consistent lateral density distribution $n_0(x)$ of the 2D electron with $d_2=35$ nm as a function of x and r.

Figure 2 shows the calculated lateral density profile $n_0(x)$ for $d_2 = 35$ nm as a function of x and r. We choose $a = 576$ nm, $a_1 = 190$ nm, $N_s = 4 \times 10^{11} \text{cm}^{-2}$ and $V_{Gm} = 0.1$ V. We can see a dramatic growing of two peaks in $n_0(x)$ as r decreases. Such structure takes place for $r \leq 0.83$. We emphasize that the appearance of two peaks is associated with the formation of the self-consistent potential, and it depends crucially on the value of d_2/a. When $d_2/a \geq 0.1$ the density exhibits a sinusoidal modulation with a period of gate even at $r = 0$. Since $d_2/a = 0.06$, a period of $n_0(x)$ for $r = 0.7$ is about half of a as is seen from Fig. 3.

2.2 Plasmon Dispersion Relation

The dispersion relations of the plasmon band in the long-wavelength limit are calculated using $n_0(x)$ in the integro-differential equation for the scalar potential $\phi = \phi(q_y, x)$ associated with plasmons [10],

$$\omega^2 \phi(q_y, x) = -\frac{4\pi e^2}{m^* \kappa_1} \int_{-\infty}^{\infty} \mathrm{L}(q_y|x - x'|)\Delta n(q_y, x')dx', \quad (8)$$

with

$$\Delta n(q_y, x) = \frac{\partial n_0(x)}{\partial x}\frac{\partial \phi}{\partial x} + n_0(x)\left(\frac{\partial^2 \phi}{\partial x^2} - q_y^2\phi\right). \quad (9)$$

Further we take into account a screening by the periodic metallic gate on plasmon dispersions. Since $|\kappa_m| \gg |\kappa_1|$ and $\epsilon=1$, we approximate the Kernel for the Poisson equation as,

$$\mathrm{L}(q_y|x|) = \frac{1}{2\pi}\left(K_0(q_y|x|) - K_0(q_y[x^2 + 4d_2^2]^{1/2})\right). \quad (10)$$

where the second term expresses the reduction of energy by the screening, K_0 being the Bessel function. In Fig. 3(a) and (b), we plotted the dispersion relations of plasmon bands in the limit of $q_y \simeq 0$ for $d_2=15$ nm, $V_{Gm}= 0.4$ V and $d_2=65$ nm, $V_{Gm}= 0.3$ V, respectively. We have assumed that $r=0.6$, $N_s=6 \times 10^{11}$ cm^{-2}, and other parameters are the same with ones in Fig.2. If we assume $r = \exp(-2\pi D/a)$ with D=d_1-d_2, this gives d_1=65 nm and d_1=112 nm for Fig. 3(a) and (b), respectively. The solid and dashed lines indicate the dispersion relation in the presence and the absence of screening. It should be noted that there appears a opening of gaps at $q_x = 0$

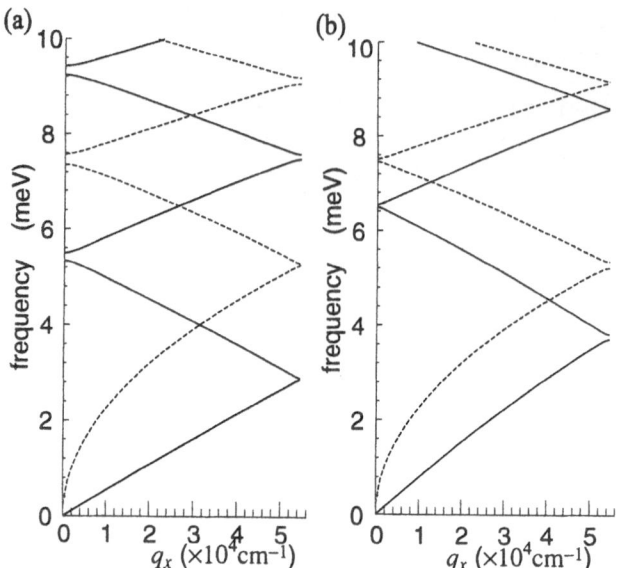

Fig. 3 Dispersion relation of plasmon bands in the reduced zone scheme.(a) $d_2=15$ nm and $V_{Gm}=0.4$ V. (b) $d_2=65$ nm and $V_{Gm} = 0.3$ V.

together with a closing at $q_x = \pi/a$ in Fig. 3(a). This is in contrast to the dispersion relation calculated by $d_2/a \sim 0.1$ in Fig. 3(b) where we can see the opening of gap at zone edge.

3 Conclusions

The present self-consistent calculation has demonstrated the appearance of dramatic density modulation by changing the values of d_2/a and a_1/a_2. Using the modulated density, we have pointed out a possible modification of plasmon bands in the long-wavelength limit. Though gaps of plasmon bands evaluated by using a probable electron density are small compared with observed ones, the features are consistent with experimental results.[3]

References

1. A.R. Goñi, A. Pinczuk, J.S. Weiner, J.M. Calleja,B.S. Dennis, L.N. Pfeiffer,and K.W. West, Phys.Rev.Lett.,**67**, (1991) 3298.
2. R. Strenz, V. Roßkopf, F. Hirler, G. Abstreiter, G. Böhm. G.Tränkle and G. Weimann, Semicond. Sci. Technol.**9**, (1994) 399.
3. G. Schdelbeck, R. Strenz, G. Abstreiter, G. Böhm. and G. Weimann, Solid State Commun. **93**, (1995) 569.
4. D. Richards, H. Hüsken, D. Bangert, H.P. Hughes, D.A. Ritchie, A.C. Churchill, M.P. Grimshaw, and G.A. Jones, Solid-State Electronics **40**, (1996) 203.
5. C.D. Ager and H. Hughes, Solid State Commun.**83**, (1992) 627, and references therein.
6. P.W. Park, A.H. MacDonald, and W.L. Schaich, Phys. Rev.B **46**, (1992) 12635. and references therein.
7. F.A. Reboredo and C.R. Proetto, Phys. Rev.B **50**, (1994) 15174.
8. J.H. Davies and I.A. Larkin, Phys. Rev.B **49**, (1994) 4800.
9. U. Wulf, Phys. Rev.B **35**, (1987) 9754.
10. F. Perez, B. Jusserrrand, C. Dahl, M. Filoche, Phys. Rev.B **54**, (1996) 11098.

Electronic properties of quantum wire superlattices elaborated by the 'Atomic Saw' method

F. Michelini[1], L. Ressier[1], G. Fishman[2], E. Vanelle[1], F. Laruelle[3], J. P. Peyrade[1]

[1] Laboratoire de Physique de la Matière Condensée, UMR-5830, INSA, Département de Physique, 135 avenue de Rangueil, 31077 Toulouse Cedex, France

[2] Institut d'Electronique Fondamentale, bât. 220, Université Paris Sud, 91405 Orsay Cedex, France

[3] Laboratoire de Microstructures et de Microélectronique, UPR 20, CNET, BP 107, avenue Henri Ravera, 92225 Bagneux Cedex, France

Abstract   We calculated numerically the one-dimensional band structure of GaAs/Ga$_{0.80}$Al$_{0.20}$As quantum wire superlattices, using the Luttinger formalism. These periodic heterostructures are elaborated from quantum well systems, by an original technique, the 'Atomic Saw' method. The optical properties of the superlattices have been investigated by Photoluminescence Excitation and linear polarization measurements. The observed 1D-confinement effects are fairly reproduced by the calculations.

1 Introduction

The calculation of the band structure in semiconductor heterostructures remains a crucial point in the understanding of their band-edge optical properties. We propose an general method of numerical calculations which takes into account the realistic orientation, shape, material composition and periodicity of the heterostructures. This method was applied to quantum wire superlattices elaborated by the 'Atomic Saw'.

2 Experimental investigations

The 'Atomic Saw' method takes advantages of the atomic scale deformation mechanisms in crystals to generate quantum wire structures from quantum well systems [1]. The initial heterostructure contains three GaAs/Ga$_{0.80}$ Al$_{0.20}$As quantum wells of widths equal respectively to 1.73 nm, 2.60 nm and 4.46 nm. Figure 1 schematically represents one of the obtained superlattices. The principle of the elaboration leads to coupled quantum wires of parallelogram section (according to the substrate active slipping planes, $\theta = 0.616$ rad). Previous statistical anal-

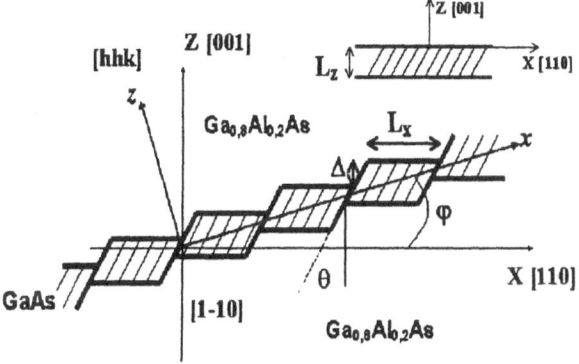

Fig. 1 Geometry of the quantum wire superlattices.

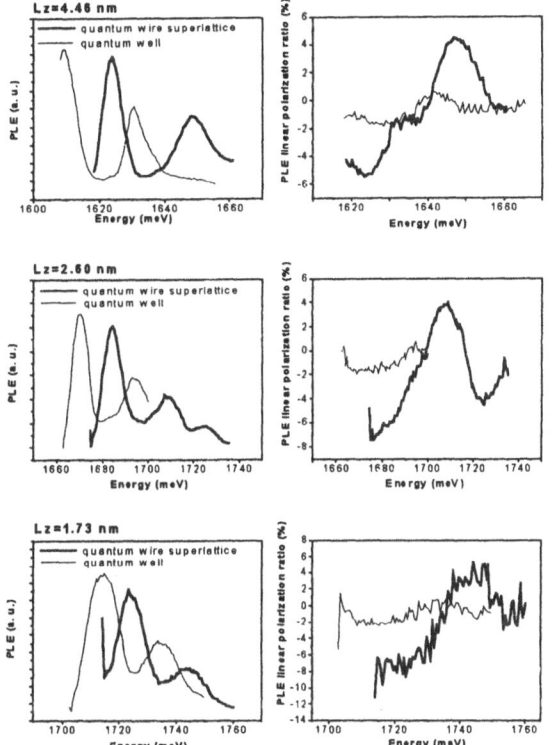

Fig. 2 PLE and linear polarization ratio defined as $P = \dfrac{I_X - I_Y}{I_X + I_Y}$, for Lz=4.46 nm, 2.60 nm, 1.73 nm.

ysis from Atomic Force Microscopy and Transmission Electron Microscopy observations gave a mean width Lx of 18 nm with a standard deviation of 9 nm. The mean shift Δ between the wires is 2.4 nm with a standard deviation of 0.6 nm [2]. Photoluminescence excitation (PLE) and polarization of photoluminescence excitation measurements were performed at 2 K (see Fig. 2). Each PLE spectrum reveals two well defined peaks corresponding to the optical transitions between the first level of electrons and heavy (light) holes $(e_1 - hh_1)$ $((e_1 - lh_1))$. The PLE peaks of the quantum wire superlattices are blue shifted with regards to the quantum well references. The spectra sketched in Fig. 2 exhibit an anisotropy of the polarization between 5 and 10 % arising after the structuration. The first optical transition is polarized parallel to the

Table 1 Experimental shift energies obtained for the three quantum wire superlattices.

Lz	4.46 nm	2.60 nm	1.73 nm
$\Delta(e_1 - hh_1)$ (meV)	14.6	14.0	9.8
$\Delta(e_1 - lh_1)$ (meV)	18.0	14.6	9.9
Δ(Stokes shift) (meV)	+2.3	1.4	-2.5

Table 2 Calculated shift energies obtained for the three quantum wire superlattices.

Lz	4.46 nm	2.60 nm	1.73 nm
$\Delta(e_1 - hh_1)$ (meV)	9.8	8.7	6.8
$\Delta(e_1 - lh_1)$ (meV)	10.2	9.7	6.6

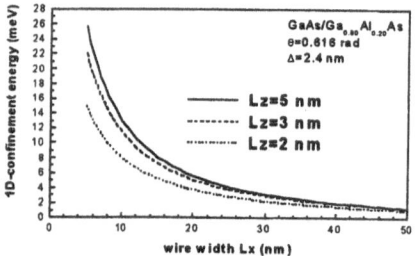

Fig. 3 Calculated 1D-confinement energy in the conduction band for three values of Lz upon the quantum wire width Lx.

wires while the second one is polarized perpendicular. Table 1 gives the shift energy values $\Delta(e_1 - h(l)h_1)$ of the first and second PLE peaks which correspond to the total 1D-confinement energy, and the variation of the Stokes shift, Δ(Stokes shift), deduced from PL measurements not presented here. All of these energy variations are defined by the quasi 1D-values minus the 2D-ones. These results show that the 1D-confinement energy generated by the structuration increases with the wire thickness Lz. The elaboration process induces a weak variation of the Stokes shift compared to $\Delta(e_1 - h(l)h_1)$, which assures of the good quality of the superlattices.

3 Theoretical approach

The principle of the band structure calculation has been presented in previous works on isolated heterostructures [3–5]. In the present case, the calculation must as well take into account the periodicity and the particular geometry of these heterostructures. We thus need to formulate the Luttinger Hamiltonian for a z spin quantization axis parallel to a $[hhk]$ direction [4]. The basis envelope functions are now :

$$|nm\rangle = \frac{1}{\sqrt{A}} \frac{1}{\sqrt{D}} \exp(ik_y y) \sin(\frac{n\pi}{2A} z) \exp\left(i\left(m\frac{2\pi}{D} + k_x\right)x\right) \tag{1}$$

A is a variationnal parameter and D the superlattice period. Table 2 lists the calculated shift energies for both $(e_1 - hh_1)$ and $(e_1 - lh_1)$ transition. As expected, these energies increase with the starting quantum well thickness Lz. Figure 3 and Figure 4 accurate this behavior upon the wire width Lx and the shift Δ between them.

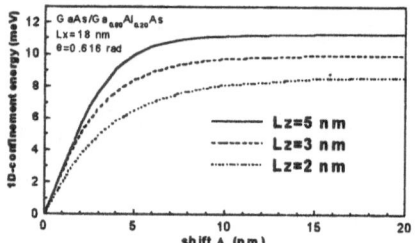

Fig. 4 Calculated 1D-confinement energy in the conduction band for three values of Lz upon the shift Δ between the wires.

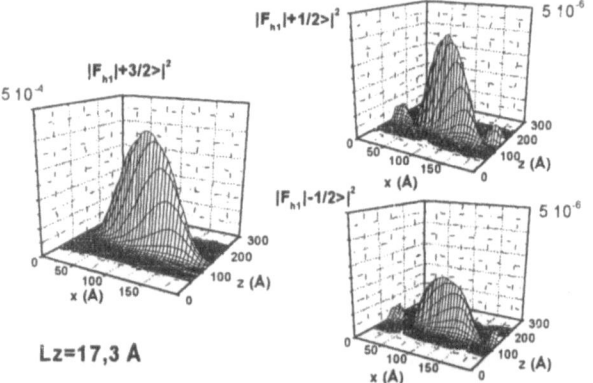

Fig. 5 Envelope functions of the valence band ground state for Lz=1.73 nm. The $\left|\pm\frac{1}{2}\right\rangle$ components are about one hundred times smaller than the $\left|\frac{3}{2}\right\rangle$ one.

Table 3 Theoretical values of the linear polarization ratio.

Lz	4.46 nm	2.60 nm	1.73 nm
$P[e_1 - hh_1]$ (meV)	-7	-7	-6
$P[e_1 - lh_1]$ (meV)	+19	+18	+2

The valence band mixing is illustrated in Fig. 5. An anisotropy of the polarization is thus expected and confirmed by the calculations presented in Table 3.

4 Discussion

The optical behavior of the superlattices are fairly reproduced by the numerical results, despite all the calculated 1D-confinement energies remain smaller than the corresponding experimental value. Considering Lx as a fitting parameter, a good agreement between the experimental and the numerical results is obtained for Lx=14 nm which is included in the standard deviation of Lx experimentally obtained.

References

1. J. P. Peyrade *et al.*, Appl. Phys. Lett. **60**, 2481 (1992).
2. L. Ressier, C. Vieu, and J. P. Peyrade, J. Appl. Phys. **81**, 35 (1997).
3. G. A. Baraff and D. Gershoni, Phys. Rev. B **43**, 4011 (1990).
4. G. Fishman, Phys. Rev. B **52**, 11132 (1995).
5. D. Brinkmann *et al.*, Phys. Rev. B **54**, 1872 (1996).

Cyclotron resonance of holes and shallow acceptor photoconductivity in strained MQW Ge/GeSi heterostructures in strong magnetic fields

V.Ya. Aleshkin, I.V. Erofeeva, V.I. Gavrilenko, O.A. Kuznetsov, M.D. Moldavskaya, V.L. Vaks, D.B. Veksler

Institute for Physics of Microstructures of Russian Academy of Sciences, GSP-105, Nizhny Novgorod, 603600, Russia
e-mail: gavr@ipm.sci-nnov.ru

Abstract Cyclotron resonance of photoexcited two-dimensional holes in undoped strained Ge/GeSi multiple-quantum-well heterostructures has been investigated in quantizing magnetic fields $h\nu \gg k_B T$ ($T = 4.2$ K, $\nu = 350$-700 GHz). Calculations of hole Landau levels have been performed and the observed "quantum deformation" of the CR line has been shown to result from the optical transitions from the different nonequidistant pairs of Landau levels. Shallow acceptor states as well as impurity photoconductivity spectra in strong magnetic fields have been calculated using the technique of acceptor envelope function expansion in the basis of the found Landau level wavefunctions.

1 Introduction

Two dimensional (2D) holes in strained SiGe-based heterostructures have been found to be sensitive to the built-in deformation and to the quantum confinement. The deformation splits light and heavy hole subbands thus decreasing the hole mass while the confinement results in the mixing of light and heavy hole states. Earlier in undoped $Ge/Ge_{1-x}Si_x(111)$ multiple-quantum-well (MQW) heterostructures the cyclotron resonance (CR) of photoexcited 2D holes of low mass ($m_c = 0.07m_0$) and high mobility ($\mu \approx 10^5$ cm²/V·s) in strain Ge layers in "semiclassical" case $h\nu \approx k_B T$ ($\nu \approx 130$ GHz, $T = 4.2$ K) have been observed [1,2]. In the first study of "quantum" CR in strained Ge/GeSi(111) system [3] selectively doped heterostructures were investigated using Fourier-transform spectrometer (cf. [4]). The

present paper is devoted to the CR investigation in quantizing magnetic fields in undoped samples using submillimeter backward wave tube oscillators. The results are discussed on the base of the carried out Landau level calculations in strained quantum wells (QWs).

2 Experimental

MQW $Ge/Ge_{0.88}Si_{0.12}$ heterostructure (#306, d_{Ge}=200A, d_{GeSi} = 260A, N_{QW}=162) was grown by CVD technique on Ge(111) substrate, the whole width of the structure being well in excess of the critical value thus providing the stress relaxation between the substrate and the heterostructure and biaxial elastic deformation of Ge layers of $2.1 \cdot 10^{-3}$. CR absorption spectra were studied in Faraday geometry at $T = 4.2$ K in the frequency range $\nu = 350$-700 GHz at the magnetic field sweeping. The sample was illuminated by GaAs light emitting diode ($\lambda \approx 0.9$ μm) that was triggered at $f = 1$ kHz, i.e. the spectra were measured at the modulation of photoexcitation. D.c. electric field was applied to the sample via strip ohmic contacts deposited on its surface.

Typical observed CR spectra are shown in Fig.1. The lowest curves represent the spectra at zero electric fields while the upper ones are obtained at some d.c. voltages applied to the sample. With the frequency increase from 350 to 700 GHz the CR absorption line proved to split into 3 lines. To distinguish these spectral features each curve was resolved into three or four Lorentians indicated in Fig.1 by

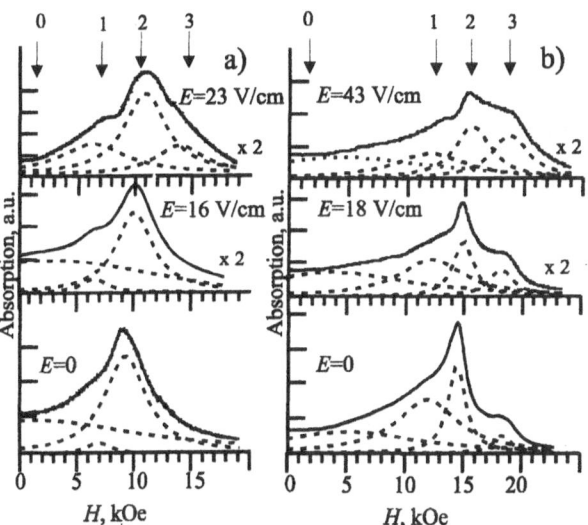

Fig. 1 CR spectra of photoexcited holes in Ge/GeSi sample #306 at $\nu = 370$GHz (a) and $\nu = 600$ GHz (b) measured at various d.c. electric fields. Each experimental curve (solid line) is resolved into three or four Lorentians (dashed lines).

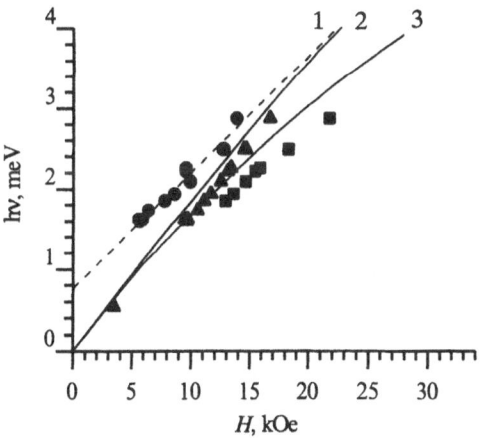

Fig. 2 Spectral positions of the absorption lines 1 (●), 2 (▲), 3 (■) (see Fig.1) versus the magnetic field. The point at $H = 3.5$ kOe corresponds to CR in "semiclassical" case [4]. Dashed line shows the extrapolated spectral position of the line1. Solid lines represent the calculated energies of the allowed optical transitions $0s_1 \to 1s$(2) and $3a \to 4a$ (3), respectively (see Fig.3).

arrows. The broad line 0 seems to result from nonresonant absorption in the lowest (nearest to the substrate) quantum wells where the hole mobility is low because of structure defects. By extrapolating the dependences of the lines frequency on the magnetic field to the zero field (Fig.2) two lines (2 and 3) were shown to result from CR transitions while the line 1 (whose spectral position extrapolates to non-zero energy at $H = 0$) seems to originate from optical transitions between excited states of shallow impurities (which become populated under the photoexcitation). In contrast to CR in "semiclassical" case [4] application of d.c. electric field (i.e. the increase of the hole effective temperature) does not lead to the CR line shift but to the change of relative amplitudes of the low-frequency CR line 3 ($m_c = 0.08m_0$) if compared with high-frequency one 2 ($m_c = 0.07m_0$). This implies that the low-frequency line results from CR transition from the higher Landau level.

3 Calculations and comparison with the experiment

The above interpretation is supported by the results of the calculation of hole Landau levels in rectangular QW in the strained Ge/GeSi heterostructure. Calculations were performed using 4x4 **k·p** Luttinger Hamiltonian in axial approximation, i.e. neglecting its off-diagonal elements proportional to ($\gamma_2 - \gamma_3$). The axial symmetry results in the conservation of the total angular momentum projection on the magnetic field direction M_j and the parity of the wave function with respect to the reflection in the plane that goes through the QW center perpendicular to the magnetic field direction. Thus each state could be classified by eigenvalue n= M_j+3/2 (n=0,1,2,...), and should be either symmetric (**s**) or antisymmetric (**a**) with respect to the above plane. This notation is used in Fig.3 where the fan chart of calculated lower Landau levels is plotted. The index in notation indicates the subband from which the given Landau level originates. In these calculations the actual axial asymmetry of the Hamiltonian was taken into account by the second order perturbation theory: this gives 3 to 10 percent correction in a Landau level position (increase of the cyclotron mass). In Faraday geometry dipole transitions are allowed between two states of the same parity if $\Delta n = \pm 1$. As it is seen from Fig.3 the lowest Landau levels are weakly interacting and their energies depend linearly on the magnetic fields up to 30 kOe. At $T = 4.2$ K the most of photoexcited holes populate the lowest Landau level $0s_1$. The allowed CR transition $0s_1 \rightarrow 1s_1$ corresponds to the cyclotron mass $m_c = 0.064m_0$ (solid line 2 in Fig.2) that is a little bit less than the observed mass for the line 2 of $0.07m_0$. Similarly, the calculated mass for the allowed transition from the next Landau level $3a_1 \rightarrow 4a_1$ $m_c = 0.073m_0$ (solid line 3 in Fig.2) is less than the observed one for the line 3 ($0.08m_0$). It is clearly seen in Fig.1 that the relative amplitude of the line 3 (with respect to the line 2) increases with d.c. electric field. The hole heating by the electric field results in the populating of the overlying Landau level $3a_1$ at the expense of devastating of the lowest one $0s_1$. The 10% discrepancy between the calculated and the observed mass values (see Fig.2) proba-

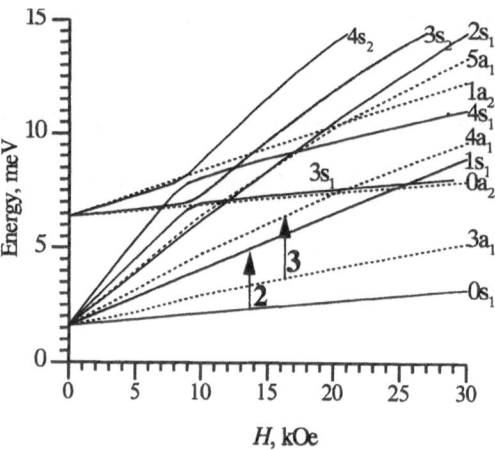

Fig. 3 Calculated Landau level energies in Ge QW (sample #306) versus the magnetic fields. CR transitions indicated by arrows 2, 3 correspond to the lines 2, 3 in Fig.1,2.

bly points out that the Luttinger parameters for holes in QWs could be different from those for bulk material.

The found eigenenergies and the wavefunctions of holes in QWs were used to calculated shallow acceptor states in the Ge/GeSi heterostructure in strong magnetic fields. The acceptor envelope function was expanded in the basis of the above Landau level wavefunctions. The results obtained allowed to interpret the observed behavior of the shallow acceptor photoconductivity spectra in the magnetic fields reported in [5].

Acknowledgements

The research described in this publication was made possible in part by RFBR (Grant #00-02-16568), Russian Scientific Programs "Physics of Solid State Nanostructures" (#99-1128), "Physics of Microwave" (#4.5)", "Fundamental Spectroscopy" (#8/02.08), "Leading Scientific Schools" (#00-15-96618), "Integration" (##540,541), "The Universities of Russia" (#015.01.01.94).and Research Grant for Young Scientists from RECSPM of N.Novgorod State University (#6).

References

1. V.I. Gavrilenko, I.N. Kozlov, O.A. Kuznetsov, M.D. Moldavskaya, V.V. Nikonorov, L.K. Orlov, and A.L. Chernov, JETP Lett. **59**, (1994) 348.
2. V.Ya. Aleshkin, N.A. Bekin, I.V. Erofeeva, V.I. Gavrilenko, Z.F. Krasil'nik, O.A. Kuznetsov, M.D.Moldavskaya, V.V Nikonorov, and L.V. Paramonov, *Proc. of 23th Int. Conf. Phys. Semicond.* edited by M.Scheffler and R.Zimmerman (World Scientific, Singapore, 1996) p.1907.
3. V.Ya. Aleshkin, N.A. Bekin, I.V. Erofeeva, V.I. Gavrilenko, M. Helm, Z.F. Krasil'nik, O.A. Kuznetsov, M.D.Moldavskaya, V. Nikonorov, and M.V. Yakunin, *Proc.Int. Symp. "Nanostructures: Physics and Technology,* (St.Petersburg, Russia, 1997) p.137.
4. C.M. Engelhardt, D. Tobben, M.Aschauer et al., Solid State Electron. **37**, (1994) 949.
5. V.I. Gavrilenko, I.V. Erofeeva, A.L. Korotkov, Z.F. Krasil'nik, O.A. Kuznetsov, M.D. Moldavskaya, V.V. Nikonorov, and L.V. Paramonov, JETP Lett. **65**, (1997) 209.

Breakdown of rotational symmetry at semiconductor interfaces

O. Krebs[1], S.Cortez[1], P. Voisin[1]

Laboratoire de Physique de la Matière Condendensée de l'Ecole Normale Supérieure, 24 rue Lhomond, F75005 Paris, France

Abstract The breakdown of roto-inversion symmetry at semiconductor hetero-interfaces leads to a coupling of the X and Y valence bands, giving rise to the recently discovered optical anisotropy of no-common atom heterostructures. This property is investigated within a pseudo-potential microscopic model, which reveals a direct dependence of the $X - Y$ (or heavy-light hole) mixing on the valence band offset.

1 Introduction

The cubic point group symmetry T_d of the zinc-blende lattice is reduced to C_{2v} at an abrupt interface: it looses not only the translational invariance along the z-axis, but also an element of rotational symmetry, namely the invariance by the fourfold roto-inversion around the [001] direction. The arrangement of chemical bonds in the vicinity of an interface anion, shown in Fig. 1, illustrates this symmetry breakdown which was until recently neglected in the classical envelope function theory (EFT). The main theoretical consequence of this effect consists in a zone center mixing of the X and Y valence bands, or equivalently of the heavy- and light-hole bands, when the spin-orbit interaction is taken into account [1,2]. This has been confirmed by recent experimental observations of a strong optical anisotropy between [110] and [$\bar{1}$10] polarization directions, for no-common atom heterostructures [3] like (InGa)As-InP or BeTe-ZnSe [4,5] quantum wells (QWs). By using effective parameters which quantify the symmetry breakdown at each hetero-interface, this optical anisotropy can be very efficiently predicted within the usual EFT, as well as its electric field dependence or Pockels effect [6]. Here, we present theoretical developments obtained in the framework of pseudopotential microscopic calculations, which give support to

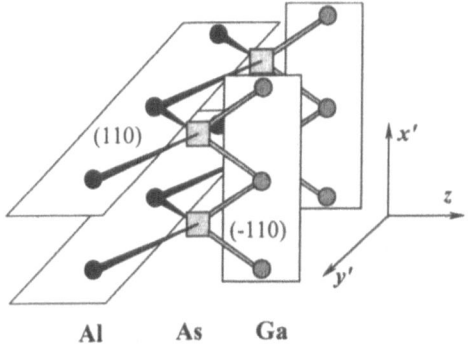

Fig. 1 Scheme of the geometrical arrangement of chemical bonds at an AlAs-GaAs interface.

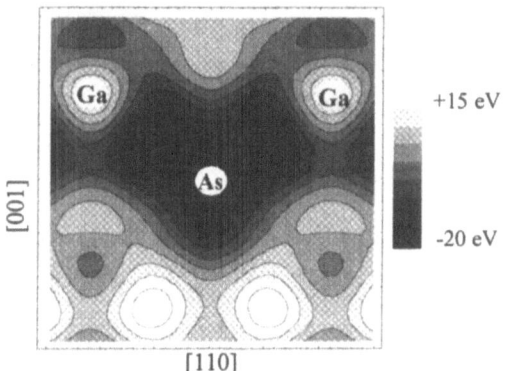

Fig. 2 Microscopic empirical pseudo-potential of bulk GaAs in the plane ($\bar{1}$10).

our original idea of localization of Bloch functions in every "half-monolayer". In particular, it confirms our original and recently questioned conclusion [7,8], that in the case of common atom QWs like GaAs-(AlGa)As, the $X - Y$ mixing term is governed by the valence band offset.

2 Bloch function localization and $X' - Y'$ splitting

We have used a simple pseudo-potential approach, ignoring the spin-orbit interaction. The microscopic potential of a bulk semiconductor is expanded on a plane wave basis limited to the first three harmonics, which corresponds to a kinetic energy cut-off of 60 eV. This accuracy is quite sufficient to correctly describe the potential wells associated with the chemical bonds. Using pseudopotential form factors given in Ref. [9], we show in Figs. 2-3 a few aspects of the microscopic potential and wavefunctions in GaAs. As can be seen in Fig. 2, the potential in the plane ($\bar{1}$10) which contains the so-called "forward" bonds of Fig. 1 (with respect to the growth axis z) has two deep wells nearly at the center of the Ga-As bonds. Typical figures are a well width of $L_b = 2$ Å and a well depth of 20 eV. The isodensity surface of the $X' = \frac{1}{\sqrt{2}}(X + Y)$ Bloch function displayed in Fig. 3 shows that this wavefunction is strongly localized, here in the forward part of the unit cell: 70% of the total X' electronic density is contained in the displayed surfaces. Similar localization occurs for the Y' Bloch function in the "backward" region corresponding to the bonds in the (110) plane. Let us note that, from the consideration of the features of the potential wells, this localization property must be very robust against perturbations of the order of a fraction of eV, like the spin-orbit interaction

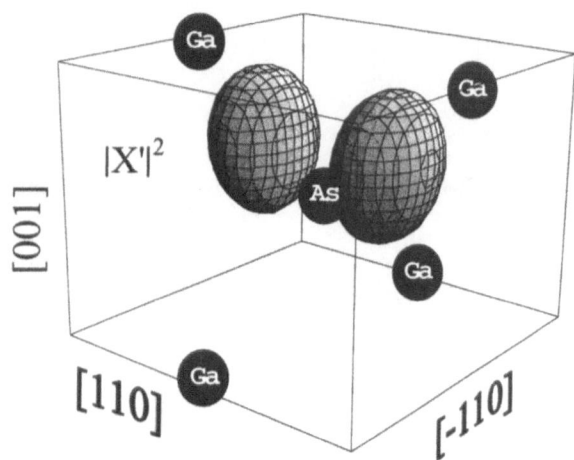

[001] [110] [-110]

$|X'|^2$

Ga Ga As Ga Ga

Fig. 3 Isodensity surface of the X' valence Bloch function at the zone center in bulk GaAs, showing the localization in forward half-cells.

or a valence band offset. Now, if we consider a GaAs-AlAs interface unit cell, it is obvious that the X' and Y' experience different potential wells, due to different form factors as well as to the valence band offset ΔE_v, so that a splitting appears between these states, corresponding to the above mentioned $X-Y$ mixing. Moreover, using the localization property of the Bloch functions, we obtain at the first order that :

$$\Delta_{X'Y'} = \langle X' \mid V(\mathbf{r}) \mid X' \rangle - \langle Y' \mid V(\mathbf{r}) \mid Y' \rangle = \Delta E_v \quad (1)$$

where the integration is performed over the interface unit cell and $V(\mathbf{r})$ is the difference of microscopic potential between the AlAs and GaAs regions. We have checked that this result lies within 10% of the exact numerical calculation of the splitting $\Delta_{X'Y'}$.

3 Heavy-Light hole mixing in QWs

In real heterostructures, we have to take into account the role of the envelope functions which describe the confined states as well as the spin-orbit interaction. Since the latter is diagonal in the Γ_8 representation, the previous $X-Y$ mixing transforms just into a direct heavy-light hole mixing denoted Δ_{HL}^{pq} between the p and q corresponding subbands. The envelope functions f_H^p, f_L^q act as a weighting factor at the positions where the symmetry is reduced, i.e. at every hetero-interface. By carefully decomposing the matrix element on successive monolayers [10], we obtain from (1) the following general expression:

$$\Delta_{HL}^{pq} = i \frac{a}{8\sqrt{3}} \sum_{z_j} (-1)^j \Delta E_v(z_j) f_H^p(z_j) f_L^q(z_j) \quad (2)$$

where the summation runs over all the atomic planes z_j (numbered in the growth direction z), and a is the lattice parameter. The factor $(-1)^j$ illustrates the competition between the successive anion or cation centered interfaces, which contribute with a different sign since the associated chemical bonds are rotated by $\frac{\pi}{2}$. The generalized $\Delta E_v(z_j)$ offset describes the change of materials

across the atomic plane z_j. It is equal to the valence band offset wherever the latter is properly defined on a microscopic scale like at common-atom interfaces, but has to be considered as an adjustable parameter wherever the potential of a given half-monolayer is undefined, e.g. due to large localized strains as it generally occurs at the no-common atom interfaces [4]. Nevertheless, in the usual situation of AlGaAs based heterostructures, equations (1) and (2) prevail and yield for a single quantum well:

$$\Delta_{HL}^{pq} = i \frac{a}{8\sqrt{3}} \Delta E_v \left[f_H^p(z_l) f_L^q(z_l) - f_H^p(z_r) f_L^q(z_r) \right] \quad (3)$$

In such a perfect QW, the above expression shows that the mixing term is non-zero only for the states of opposite parity, due to the compensation between the left (z_l) and right (z_r) interfaces. This result agrees with the D_{2d} symmetry of such a structure [1]. Nevertheless, this coupling gives rise to optical anisotropy when an electric field is applied along z: this is the quantum confined Pockels effect, first observed by Kwok [11] and that a standard $\mathbf{k} \cdot \mathbf{p}$ calculation missing the interface-induced mixing failed to explain [12].

4 Conclusion

We have discussed the physics of the symmetry breakdown at semiconductor interfaces on the basis of pseudopotential microscopic calculations. The present results give support to our previous empirical theory by proving the strong localization of Bloch functions and the dependence on the valence band offset of the heavy-light hole mixing.

References

1. O. Krebs and P. Voisin, Phys. Rev. Lett. **77**, (1996) 1829.
2. E.L. Ivchenko, A. Kaminski and U. Rössler, Phys. Rev. B **54**, (1996) 5852.
3. In this case, the symmetry breakdown of the two opposite QW interfaces does not compensate.
4. O. Krebs, D. Rondi, J. L. Gentner, L. Goldstein, and P. Voisin, Phys. Rev. Lett. **80**, (1998) 5770.
5. A. V. Platonov, V. P. Kochereshko, E. L. Ivchenko, G. V. Mikhailov D. R. Yakovlev, M. Keim, W. Ossau, A. Waag, and G. Landwehr, Phys. Rev. Lett. **80**, (1999) 3546.
6. E. L. Ivchenko, A. A. Toropov, and P. Voisin, Phys. Solid. State **40**, (1998) 1748.
7. B. Foreman, Phys. Rev. Lett. **82**, (1999) 1339.
8. O. Krebs and P. Voisin, Phys. Rev. Lett. **82**, (1999) 1340.
9. E. Caruthers, and P. J. Lin-Chung, Phys. Rev. B **17**, (1978) 2705.
10. A preprint LaTeX file giving the detailed derivation can be found at the internet address : http://www.lpmc.ens.fr/~krebs/Symmetry-Breakdown.zip.
11. S. H. Kwok, H. T. Grahn, K. Ploog, and R. Merlin, Phys. Rev. Lett. **69**, (1992) 973.
12. B. F. Zhu and Y. C. Chang, Phys. Rev. B **50**, (1994) 11932.

Proc. 25th Int. Conf. Phys. Semicond., Osaka 2000 (Eds. N. Miura and T. Ando)

489

Binding energy and internal magnetic field of exciton magnetic polarons in a single semimagnetic quantum dot

G. Bacher[1], A.A. Maksimov[1,2], A. McDonald[1], M.K. Welsch[1], H. Schömig[1], V.D. Kulakovskii[2], A. Forchel[1], Ch. Becker[3], G. Landwehr[3], L.W. Molenkamp[3]

[1] Technische Physik, Universität Würzburg, Am Hubland, 97074 Würzburg, Germany
[2] Institute of Solid State Physics, Russian Academy of Science, 142432 Chernogolovka, Russia
[3] Experimentelle Physik III, Universität Würzburg, Am Hubland, 97074 Würzburg, Germany

Abstract The formation of quasi-zero dimensional exciton magnetic polarons in semimagnetic single quantum dots (SQDs) is studied by means of photoluminescence (PL) spectroscopy. Comparing the temperature dependent PL energy shift of semimagnetic SQDs with their non-magnetic counterparts a low temperature polaron binding of 10.5 meV and an internal magnetic field of 3.5 T can be extracted. From the internal field, a dot diameter of about 6 nm is estimated.

In diluted magnetic semiconductors (DMS), the strong sp-d exchange interaction between the charge carriers and the magnetic Mn^{2+} ions results in a formation of exciton magnetic polarons (EMP) [1,2], i.e. small regions of the crystal with strongly correlated spins of localized carriers and magnetic ions. As compared to bulk or quantum wells, the EMP formation should be even more pronounced in quantum dots due to the three dimensional confinement of the charge carriers [3]. We have studied the formation of quasi-0D EMP in DMS SQDs using temperature dependent PL spectroscopy. The comparision of DMS SQDs with a non-magnetic reference sample allows an accurate estimate of the EMP binding energy and the internal exchange field between the charge carriers and the Mn^{2+} spins.

The structures under investigation consist of three monolayers of $Cd_{0.93}Mn_{0.07}Te$ (or $Cd_{0.93}Mg_{0.07}Te$, respectively) embedded in nonmagnetic $Cd_{0.6}Mg_{0.4}Te$ barriers grown by molecular beam epitaxy. So-called "natural" QDs are expected to form due to spatial variations of size and composition [4]. To achieve the high spatial resolution required for measuring SQDs, Al masks with apertures down to 150 nm in diameter have been prepared by lithographically. The samples were excited by the 514.5 nm line of an Ar^+-laser with a typical excitation density of 2 W/cm^2 in order to avoid a heating of the spin subsystem. The PL signal was detected with a liquid nitrogen-cooled CCD camera. Magnetoluminescence measurements have been performed using a split-coil system allowing magnetic fields up to 8T in Faraday geometry. The overall spectral resolution for the experiments was about 0.4 meV.

For apertures with large diameters of 5 μm or above, the PL spectra of both samples show an inhomogeneous broadening of 50 meV and 48 meV, respectively. A reduction of the aperture diameter results in the observation of individual lines, which can be attributed to the

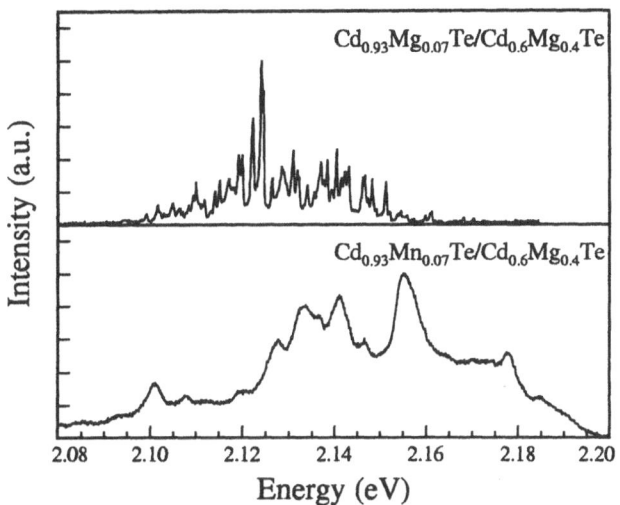

Fig. 1 Spatially resolved PL spectra for non-magnetic (top) and semi-magnetic (bottom) QDs. The diameter of the aperture was 350 nm (top) and 250 nm (bottom), respectively.

emission of SQDs. This is shown in figure 1, where spatially resolved PL spectra of both samples are compared. The most obvious difference between the two materials is the fact, that the PL linewidth of the semimagnetic SQDs is in the range of a few meV, exceeding the resolution limited PL linewidth of the non-magnetic reference by more than one order of magnitude. This can most likely be attributed to the dynamics of EMP formation and the occurence of magnetic fluctuations which may cause a linewidth broadening even in SQDs, as the dot is being probed repeatedly [5].

The study of SQDs allows to access the Zeeman shift and therefore the effective g-factor of the exciton. In figure 2, the PL energy shift of the σ^+ component of the Zeeman doublet is plotted versus magnetic field. The magnetic field induced energy shift of the non-magnetic SQD can be described by a linear relation, resulting in an effective g-factor of g = 2.33. No pronounced variation of the g-factor was found for different SQDs. The energy shift of the DMS SQD is strongly enhanced due to the giant Zeeman effect and a total shift of about 5 meV is obtained for 8 T. It is important to note here, that in case of EMP formation, a quantitative description of the PL shift in DMS SQDs should take into account both,

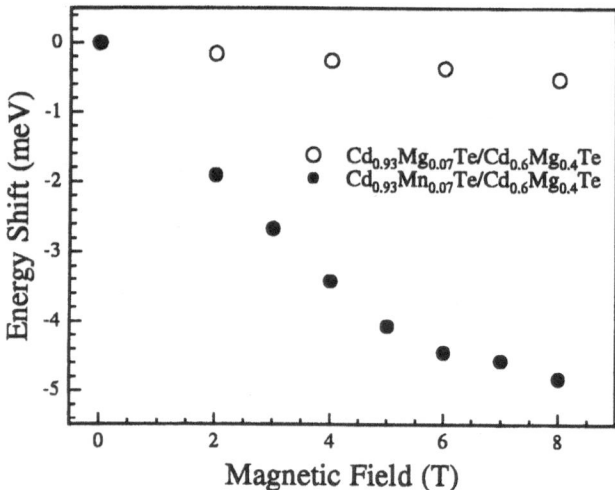

Fig. 2 PL energy shift for a non-magnetic and a semimagnetic SQD versus magnetic field.

the external magnetic field and the internal magnetic field in the EMP [5,6].

In order to proof the formation of quasi-0D EMP and to extract the polaron parameters, we have performed temperature dependent PL measurements. In figure 3, the PL energy is plotted versus temperature for both, the non-magnetic and the DMS SQD. While for the non-magnetic dot, a red shift of the PL energy is observed for increasing temperature due to the temperature dependent bandgap, a clear blue shift up to about 15 K is obtained in case of the DMS SQD, followed by a red shift, if the temperature is increased further. In striking contrast to semimagnetic quantum wells, where both, the magnetic and the electronic localization change with temperature, the temperature dependent PL shift of the semimagnetic SQD directly probes the magnetic localization. This allows a direct measurement of the temperature dependent binding energy of a quasi-0D polaron by comparing the non-magnetic and the semimagnetic SQD. As demonstrated in the inset of the figure, the experimental data follow quite nicely a Brillouin function. By fitting the data, we are able to exctract the low temperature polaron binding energy ($E_{MP} = 10.5$ meV) and the internal magnetic exchange field ($B_{MP} = 3.5$ T), which is enhanced by more than a factor of 4 as compared to bulk samples [6].

In $Cd_{1-x}Mn_x$Te the exchange interaction constant β for holes is four times larger than for electrons (α) and therefore the holes play the main role in the interaction of excitons with Mn ions [7]. The hole induced exchange field is proportional to the squared wavefunction $|\Psi(\mathbf{r})|^2$ and the relation between B_{MP} and the localization volume of the hole, V, can be written as [6]:

$$B_{MP} \approx \frac{\gamma}{3\mu g_{Mn}} \beta J \cdot \frac{1}{V}, \qquad (1)$$

where $J = 3/2$ is the hole spin, μ the Bohr magneton, and $g_{Mn} = 2$ the g-factor of the Mn^{2+} ion. The parameter γ takes into account that only a part of the

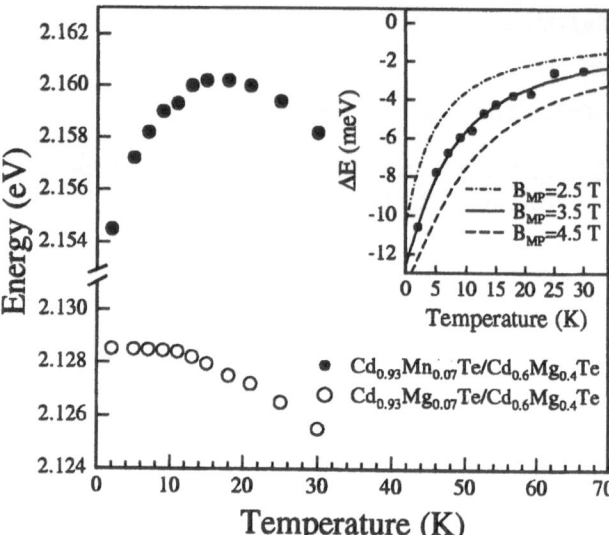

Fig. 3 PL energy for a non-magnetic and a semi-magnetic SQD versus temperature. The inset shows the experimental data (symbols) and the theory (lines) of the temperature dependent EMP binding energy

hole wavefunction is inside the DMS QD. For the investigated $Cd_{1-x}Mn_x$Te QDs, where the extension of the QD in growth z-direction is much smaller than in lateral direction, a value of $\gamma = 0.5 \pm 0.05$ was estimated. With $B_{MP} = 3.5$ T, an exciton localization volume $V \approx 540 N_0^{-1}$ is obtained and taking the nominal width of the $Cd_{0.93}Mn_{0.07}$Te layer (10 Å) as a first approximation of the QD height, this corresponds to an in-plane EMP diameter of about 6 nm.

In conclusion we have demonstrated the formation of quasi zero-dimensional exciton magnetic polarons in a DMS SQD. The temperature dependent magnetic localization can be well described by a Brillouin function and allows an estimate of the binding energy and the internal magnetic field of the EMP and therefore the diameter of the DMS SQD.

We gratefully acknowledge the expert technical assistance of M. Emmerling and the financial support of the Deutsche Forschungsgemeinschaft (SFB 410).

References

1. J.K. Furdyna, J. Appl. Phys. **64**, (1988), R29.
2. G. Mackh, W. Ossau, D.R. Yakovlev, A. Waag, G. Landwehr R. Hellmann, E.O. Gbel, Phys. Rev. B **49**, (1994), 10248.
3. A.K. Bhattacharjee, C. Benoit à la Guillaume, Phys. Rev. B **55**, (1997), 10613.
4. D. Gammon, E. S. Snow, B.V. Shanabrook, D.S. Katzer, D. Park, Science, **273**, (1996), 87.
5. A.A. Maksimov, G. Bacher, A. McDonald, V.D. Kulakovskii, A. Forchel, Ch. Becker, G. Landwehr, L.W. Molenkamp, Phys. Rev. B **62**, (2000), R7767.
6. K.V. Kavokin, I.A. Merkulov, D.R. Yakovlev, W. Ossau, G. Landwehr, Phys. Rev. B **60**, (1999), 16499.
7. J.A. Gaj, R. Planel, G. Fishman, Solid State Commun. **29**, (1979), 435.

Spin transport of excitons in asymmetric double quantum wells of $Cd_{1-x}Mn_xTe$

K. Kayanuma, E. Shirado, M.C. Debnath, I. Souma, S. Permogorov and Y. Oka

Research Institute for Scientific Measurements, Tohoku University, Katahira 2-1-1, Aoba-ku, Sendai 980-8577, Japan
CREST, Japan Science and Technology Corporation, Kawaguchi 332-0012, Japan
e-mail: oka@rism.tohoku.ac.jp

Abstract   Tunneling and transport of spin-polarized excitons were studied in the double quantum well system composed of diluted magnetic and nonmagnetic semiconductors. The time resolved exciton luminescence in magnetic field showed that the spin-polarized excitons diffuse from the magnetic quantum well to the non-magnetic quantum wells by conserving their spins. The tunneling time and the spin relaxation time of the excitons were determined from the analysis of circularly polarized exciton luminescence.

1 Introduction

The exchange interaction between carriers and localized magnetic ions in diluted magnetic semiconductors (DMSs) induces marked magnet-optical effects [1]. Double quantum well (DQW) structures have been intensively studied to clarify the tunneling processes of carriers and their spins [2-4]. The DQWs are also of practical importance for devices such as resonant tunneling diodes. In this paper, we study the tunneling and transport processes of the exciton spins in DQWs composed of the CdTe and $Cd_{1-x}Mn_xTe$ wells.

$CdTe/Cd_{1-y}Mg_yTe/Cd_{1-x}Mn_xTe$ DQW systems were grown by molecular beam epitaxy on GaAs (100) substrates [5]. The DQW structure consists of a narrow CdTe well (the well width: $L_{NW} = 20$ Å) and a wide $Cd_{0.90}Mn_{0.10}Te$ well ($L_{WW} = 200$ Å), which are separated by the $Cd_{0.65}Mg_{0.35}Te$ barrier layer ($L_B = 20$-300 Å). In the DQWs without the barrier ($L_B = 0$ Å) the widths of NW and WW were 380 and 1500 Å, respectively. Transient photoluminescence (PL) from the DQWs was measured by a spectrometer and streak camera system. The samples were excited by a frequency-doubled light (390 nm) of a mode locked titanium sapphire laser. External magnetic fields up to 7 T were applied in the Faraday configuration.

2 Results and Discussion

Circularly polarized PL spectra of the DQWs with L_B of 30 Å at B = 0.3 T are shown in Fig. 1. Two PL peaks are observed at 1.71 and 1.77 eV. The lower energy peak is the excitonic luminescence confined in

the NW of CdTe, while the higher energy peak at 1.77 eV is the excitonic luminescence from the WW of $Cd_{0.90}Mn_{0.10}Te$. For the samples with smaller barrier widths of 20-60 Å, the WW exciton luminescence intensity decreases. The decrease of the WW intensity indicates the increased probability of the carrier tunneling from WW to NW in the DQW with thin barrier.

At 0.3 T the PL of the WW is fully polarized to σ^+ component due to the large Zeeman splitting (6 meV) of the $Cd_{0.90}Mn_{0.10}Te$ WW. On the other hand the PL from the CdTe NW also shows the circular polarization degree of 8.6 %. In the DQWs with $L_B = 120$ Å, the circular polarization degree of the NW at 0.3 T was less than 0.5%. Therefore Zeeman-split levels of the excitons at 0.3 T in the isolated CdTe NW are fully thermalized. By comparing with the case of the isolated NW, the observed circular polarization degree (8.6 %) in the NW exciton PL in Fig. 1 is due to the population of the spin polarized excitons supplied from the DMS WW by tunneling.

Time-resolved PL shows that the WW emission in the DQWs with $L_B = 20$ and 30 Å decays with the short decay times of 30 and 100 ps, respectively. On the other hand, the PL from the NW shows the rise of the PL intensity with the rise time equal to the decay time of the WW. The appearance of the PL rise time in the NW for thin barrier samples indicates that both of electrons and holes created in the WW tunnel through

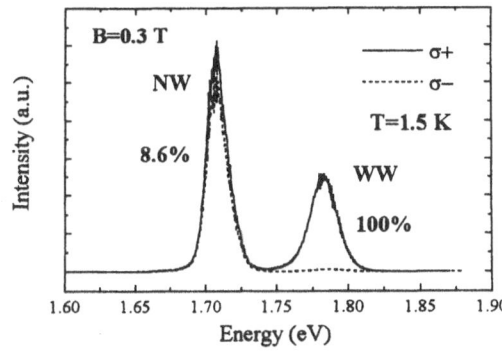

Fig. 1 Circularly polarized PL spectra of WW and NW in the DQWs of L_B=30 Å at B=0.3 T

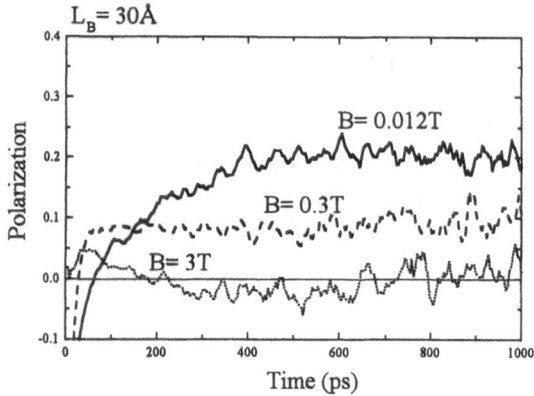

Fig. 2 Time variation of the circular polarization degree of the NW exciton PL in the DQWs of $L_B = 30$ Å.

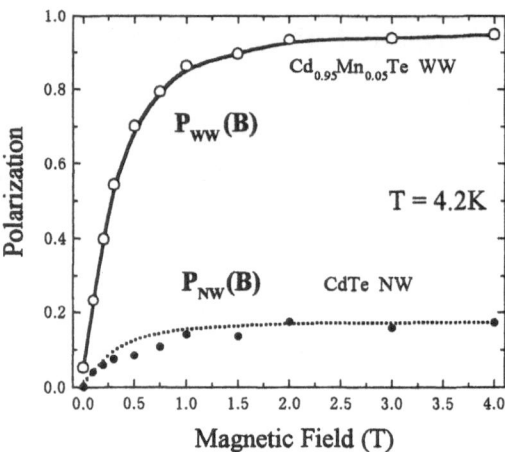

Fig. 3 Circular polarization degree of the exciton PL in the WW and NW in the DQWs with $L_B = 0$ Å.

the barrier and contribute to the radiative recombination in the NW. The observed time dependence of the exciton PL in the DQWs suggests the electron and hole tunneling with the same tunneling time.

Therefore we discuss our experimental results in the thin barrier DQWs by the exciton tunneling model [6]. Calculation of the exciton density in the WW and NW of the DQWs by changing the tunneling time shows the appearance of the rise time of the NW exciton density for short tunneling times. Thus the rise of the NW emission is due to the tunneling of the excitons from the WW. By comparing the experimental time evolution of the exciton PL in WW and NW with the calculation, we deduced the exciton tunneling time in the DQWs of $L_B = 30$ Å to be 100 ps.

Figure 2 shows the time variation of the circular polarization degree of the NW exciton PL in the DQWs with $L_B = 30$ Å. At very small magnetic field of B = 0.012 T, the polarization degree of the NW exciton PL gradually increases and reaches to 20 % at time after 400 ps. The gradual increase of the polarization degree of the NW is fitted by the calculation with the tunneling time of 200 ps, which mostly agrees with the tunneling time of 100 ps determined from the transient PL characteristics. At the field of 0.3 T, the degree of polarization decreases to 10% and at 3 T the degree of polarization becomes to be almost zero. The results are interpreted by the suppression of the exciton tunneling from the WW to the NW by the magnetic field. In the high magnetic field, the WW excitons have more localized character and cannot tunnel to the NW.

Figure 3 shows the circular polarization degree of the WW and NW exciton PL, $P_{WW}(B)$ and $P_{NW}(B)$ as a function of the magnetic field strength. $P_{WW}(B)$

increases with increasing the field and reaches to 95 % at 4 T. On the other hand $P_{NW}(B)$ also shows the increase and reaches to 18 %. In the case of an isolated CdTe NW at 4 T we cannot observe such high polarization degree. Therefore the circularly polarized exciton PL from the CdTe NW in Fig. 3 is due to the exciton spin injection from the WW to the NW. The exciton injection is caused by the diffusion of the polarized exciton spins to the NW. The spin relaxation time of the excitons in the WW is deduced as 30-55 ps at 0-4 T.

In summary, the DQWs with thin barriers of 30 Å showed the exciton tunneling from the WW to the NW with the tunneling time of 100-200 ps. In the DQWs without the barrier the polarized exciton spins in the WW are transferred to the lower lying NW and polarize the NW exciton PL by 18 %. Therefore the injected exciton spins become 18 % of the unpolarized excitons which are directly created by the optical excitation in the NW.

References

1. Y. Oka, J.X. Shen, K. Takabayashi, N. takahashi, H. Mitsu, I. Souma, R. Pittini, J. Lumines. **83-84** (1999) 83.

2. R. Fiederling, M. Keim, G. Reuscher, W. Ossau, G. Schmidt, A. Waag and L.W. Molenkamp, Nature **402** (1999) 787.

3. M. Oestreich, J. Huebner, D. Haegele, P.J. Klar, W. Heimbrodt, W.W. Ruehle, D.E. Ashenford and B. Lunn, Appl. Phys. Lett. **74** (1999) 1251.

4. K. Heike, W. Heimbrodt, W.W. Ruhle, Th. Pier, H.-E. Gumlich, D.E. Ashenford, B. Lunn, Solid State Commun. **93** (1995) 257.

5. M.C. Debnath, J.X. Shen, E. Shirado, I. Souma, T. Sato, R. Pittini and Y. Oka, J. Appl. Phys. **87** (2000) 6457.

6. N. Saito and Y. Kayanuma, Phys. Rev. **B51** (1995) 5453.

Raman scattering in MnTe/ZnTe and ZnSe/ZnMnSe multilayer Semiconductor superlattices

K. Ozawa[1], Y. Tanaka[1*], K. Iio[1], N. Nakajima[2], Y. Nabetani[3], T. Matsumoto[3]

[1] Dept. of Physics, Graduate School of Science and Engineering, Tokyo Inst. of Tech. 2-12-1 Oh-okayama, Meguto-ku, Tokyo 152-8551, Japan *e-mail:* `iio@lee.phys.titech.ac.jp`

[2] Dept. of Electronic information system, Faculty of Science and Engineering, Hirosaki University 1 Bunkyo-machi Hirosaki-shi, Aomori 036-8224, Japan

[3] Dept. of Electric Engineering, Faculty of Engineering, Yamanashi University 4-3-11 Takeda Kofu-shi, Yamanashi 400-8511, Japan

Abstract Raman scattering and magnetic susceptibility measurements were performed on MnTe/ZnTe and ZnSe/ZeMnSe semiconductor superlattices with magnetic spins. A basetone or overtones of the ZnTe-LO located within the band of photoluminescence (PL) spectra are found to be enhanced exclusively owing to the outgoing resonance in the MnTe/ZnTe system, besides observing phonon properties peculiar to the respective systems. A MnTe/ZnTe specimen, $[(\mathrm{MnTe})_{9.7}/(\mathrm{ZnTe})_{11}]_{600}$, remains to be paramagnetic down to 4K, though the interaction between Mn^{2+} spins is deduced to be antiferromagnetic.

1 Introduction

II-VI wide-band-gap semiconductor superlattices arouse interest because of the potential applicability of their unique optical and electronic properties. In particular, antiferromagnetic substances MnTe and MnSe can be ingredients of diluted magnetic semiconductors in alloying with II-VI compounds. One of the group of present authors, Nabetani and Matsumoto, have successfully synthesized MnTe/ZnTe and ZnSe/ZnMnSe multilayer semiconductor superlattices by using the Molecular-Beam Epitaxy (MBE) method on the GaAs(100)-substrates through buffer layers of ZnTe and ZnSe, respectively.

Motivated by interests in elementary excitations in those magnetic/nonmagnetic superlattices, the present authors have performed to study Raman scattering in both systems together by using a triple polychromator microscopic Raman systems installed with a CCD detector (Jobin Yvon T64000 systems). Detailed Raman spectra were measured at the liquid Helium temperatures. To gain information on magnetic properties of these superlattices, the magnetization as a function of the temperature have also been examined with a SQUID magnetometer.

2 Samples

MnTe and ZnTe (ZnSe and ZnMnSe) in superlattices were alternately grown on the buffer layer ZnTe (ZnSe). The superlattice structures of [MnTe/ZnTe] concerned here is described as $[(\mathrm{MnTe})_m/(\mathrm{ZnTe})_n]_x$, where m, n

represent numbers of monolayer and x does the periods of superlattice. Those of [ZnSe/ZnMnSe] are done as well.

We have studied $[(\mathrm{MnTe})_m/(\mathrm{ZnTe})_n]_x$ with (m,n,x) $=(8,2,120),(7,3,120),(7,5,120),(1.5,15,600),(9.7,11,600)$ and $[(\mathrm{ZnSe})_m/(\mathrm{ZnMnSe})_n]_x$ with $(m,n,x)=(7,6,60),(13,6,60),(21,6,60),(29,6,60)$. Quality and stacking structures of these superlattices have been examined with X-ray diffraction. Since they are sufficiently thick, strain relaxation in multilayers was confirmed by RHEED patterns to be established against the GaAs(100)-substrates [1,2].

3 Results and Discussion

3.1 Raman spectra

The Raman spectra of these systems were measured in the backscattering configuration with respect to their (001) surfaces by using the 488.0nm (2.54eV) or 514.5nm (2.41eV) lines (typical power 50mW) of an Ar^+ ion laser.

For the laser wavelength of 514.5nm, the spectra were found to show an interesting temperature dependence as seen in Fig.1. It seems that strong and broad photoluminescence (PL) spectrum is tied up with the phonon Raman spectra. A basetone ($206\mathrm{cm}^{-1}$) or overtones ($411\mathrm{cm}^{-1}$ and $618\mathrm{cm}^{-1}$) of the ZnTe-LO being located within the band of photoluminescence (PL) spectra, which is dependent of temperature, are enhanced extremely. This PL

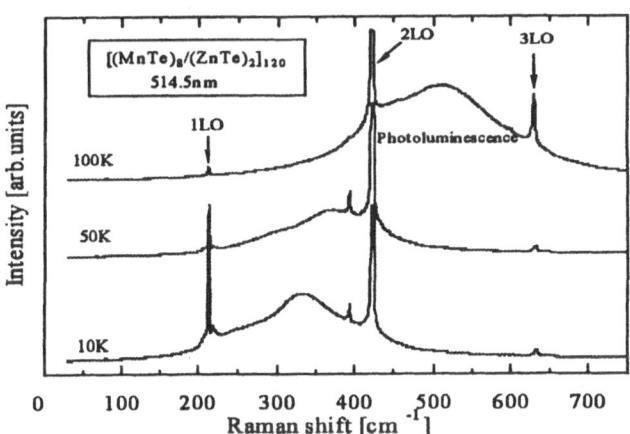

Fig. 1 The temperature dependence of the Raman scattering for $[(\mathrm{MnTe})_8/(\mathrm{ZnTe})_2]_{120}$ from 10K to 100K. The nLO peaks of ZnTe exhibit the outstanding enhancement in riding on the photoluminescence spectra of ZnTe layers.

* *Present address:* 2nd Examination Dept. Japanese Patent Office. *E-mail:* TYPA3009@jpo-miti.go.jp

spectrum can be produced by the excitons of the ZnTe layer whose band-gap energy is known to be 2.38eV, since the laser energy of the 514.5nm wavelength corresponds to 2.41eV and the Stokes Raman shift excited by this laser is less than 2.41eV. The PL peak estimated from the Raman spectra at 10K is reasonably located around 2.37eV. The strong enhancement of ZnTe-nLO is attiributed to an outgoing resonance in these systems.

Figure 2 shows the detailed polarization property of Raman spectra in [MnTe/ZnTe] measured at 4K in a region from 110cm^{-1} to 350 cm^{-1} at the excitation wavelength of 488.0nm.

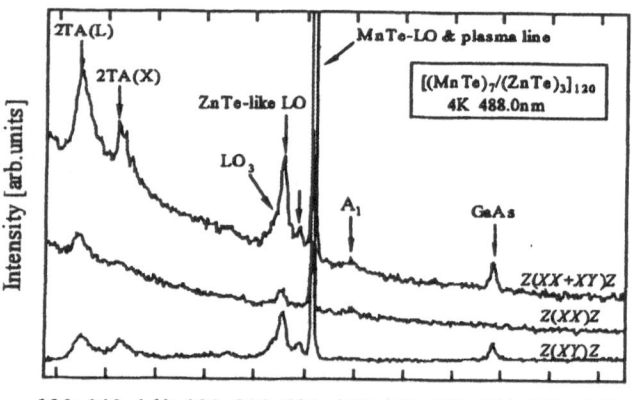

Fig. 2 The Raman spectra and the polarization property of $[(MnTe)_7/(ZnTe)_3]_{120}$ superlattice obtained for the Ar$^+$ laser wavelength 488.0nm

Unfortunately, the inevitable laser plasma line (associated with the 488nm line) lapped over the peak of the MnTe-like LO phonon located at 220 cm^{-1}. But the difference between the polarized $z(xx)z$ and depolarized $z(xy)z$ spectra of ZnTe-like LO can be found successfully. The peak in the $z(xx)z$ is located at 210 cm^{-1}, while that in the $z(xy)z$ is found at 211 cm^{-1}, in addition, a small peak at 207cm^{-1}. These Raman peaks are explained as a result of the confined optical mode of ZnTe layers, because the optical confined modes labeled by odd (B$_2$ symmetry) or even (A$_1$ symmetry). The two-phonon modes of acoustic phonon were observed only in the $z(xx)z$ configuration. Therefore these Raman shifts observed are recognized in terms of the conventional assignment for the bulk and superlattice phonon modes.

Figure 3 shows the Raman spectra of the [ZnSe/ZnMnSe] superlattices measured at the same condition. Apparently ZnTe-lile LO phonon and ZnMnSe-like LO phonon modes including the polarization properties of them are observed. Other samples with different numbers of monolayer also display the similar profile, too.

3.2 Magnetic susceptibility

Recently, spin-dependent semiconductor superlattices have attracted great interest from a view point of "spin technology" via exchange interactions between the localized magnetic ion spins and band carriers.

Zinc-blend MnTe (ZB-MnTe) deposited in thick lay-

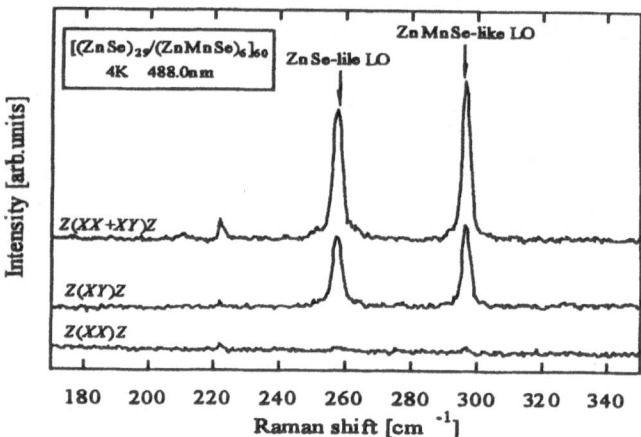

Fig. 3 The Raman spectra and the polarization property of $[(ZnSe)_{29}/(ZnMnSe)_6]_{60}$.

ers possesses a long-range antiferromagnetic order with spin frustration. The Néel temperature of the bulk ZB-MnTe have been reported to be nearly 60K [3]. So it is interesting to know magnetic properties of $[(MnTe)_{9.7}/(ZnTe)_{11}]_{600}$ superlattice, which can be a two-dimensional magnetic system similar to the Zinc-blend MnTe. The magnetization as a function of the temperature was measured under the field of 2 Tesla applied parallel to the surface of (001).

The data is shown in Fig.4 after correcting the inherent diamagnetism, which is due to the GaAs substrate and the ZnTe buffer layer. The reciprocal susceptibility having a negative Weiss temperature indicates that the interaction between Mn^{2+} spins in the superlattice are antiferromagnetic. However Mn^{2+} spins remains to be in a paramagnetic state at low temperatures, not necessarily being ordered.

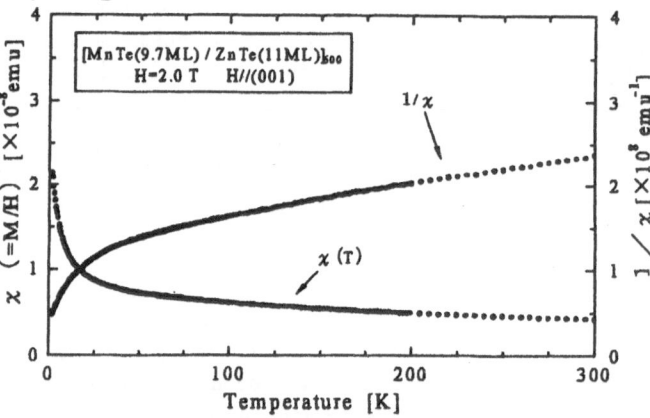

Fig. 4 The magnetic susceptibility of $[(MnTe)_{9.7}/(ZnTe)_{11}]_{600}$ as a function of the temperature.

References

1. I. Ishibe, Y. Nabetani, T. Kato and T. Matsumoto, J.Crystal Growth, in press.
2. T. Matsumoto, Y. Nabetani, T. Kato, M. Yasui, I. Ishibe and T. Suzuki, *International Symposium on Nanoscale Magnetism and Transport* ISNMT2000(Sendai), Abstracts p. 180.
3. K. Ando, K. Takahashi and T. Okuda, J. Magn. Magn. Mater. **104-107** (1992) 993.

Exciton-trion coupling in modulation doped quantum well structures

V.P. Kochereshko[1], G.V. Astakhov[1], R.A. Suris[1], D.R.Yakovlev[1,2], W.Ossau[2], J.Nurnberger[2], W.Faschinger[2], G.Landwehr[2].

[1] A.F.Ioffe Physico-Technical Institute RAS, 194021, St.Petersburg, Russia
[2] Physikalisches Institut der Universitat Wurzburg, 97074, Wurzburg, Germany

Abstract Modifications of exciton and trion reflectivity spectra with increasing 2D electron concentration have been studied in ZnSe/ZnMgSSe modulation doped quantum well structures. An idea of exciton - trion states mixing has been introduced from an analysis of the experimental results. A theory of such mixing has been developed.

Charged exciton complexes (trions) have been predicted in 1958 [1] and experimentally found in 1993 [2]. These states were observed in II-VI and III-V quantum well structures with 2DEG of low density (see review paper [3]). Till now excitons and trions were considered as independent excitations. Apparently the exciton-trion mixing via electromagnetic field is negligibly small because of small value of the coupling constant (ratio of exciton radiative and nonradiative dampings). However, the exciton trion mixing can be very effective. This coupling comes from a multiple exciton electron scattering, in which an exciton can trap an electron transforming into trion and a trion can loose an electron transforming into an exciton.

ZnSe/Zn$_{0.89}$Mg$_{0.11}$S$_{0.18}$Se$_{0.82}$ single quantum well structures have been investigated. Typical structures contain an 80Å single QW with modulation doping in the barriers at a 100Å distance from the QW. The electron concentration in the structures was varied from 10^9 cm^{-2} to 5×10^{11} cm^{-2}. The structure design allows us to control the electron concentration keeping all other parameters constant (i.e. QW width, barrier height, background impurity concentration, etc.). To determine the electron concentrations we used a method suggested in Ref.[4].

Fig.1 shows reflectivity spectra taken from 80Å single QW ZnSe/ZnMgSSe with various electron concentrations. In the spectrum of intentionally undoped structure (n_e=5×10^9 cm^{-2}) a pronounced exciton reflectivity line (X) can be found. A trion reflectivity line (X$^-$) appears in doped samples to low energy side from the exciton one when electron concentration increases. The amplitude of the X$^-$ line increases and the amplitude of the X line decreases with increasing electron concentration. The width of the X line increases with increasing electron concentration and at concentrations above 1.3×10^{12} cm^{-2} this line disappears from the spectra.

Using a method described in detail in Ref.[5] we extract X and X$^-$ parameters from reflectivity spectra and plot them as a function of electron concentration. Fig.2 shows the concentration dependencies of radiative damping (oscillator strength), nonradiative damp-

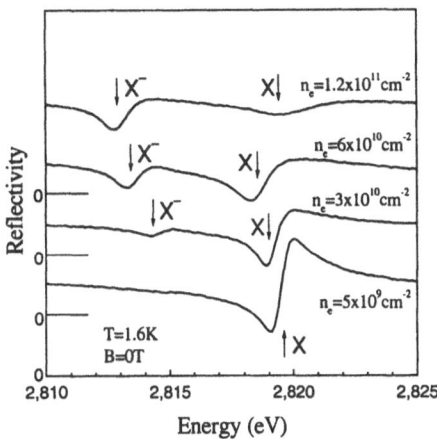

Fig. 1 Reflectivity spectra taken from ZnSe/Zn$_{0.89}$Mg$_{0.11}$S$_{0.18}$Se$_{0.82}$ modulation-doped QW structures as a function of 2DEG density.

ing (linewidth), and energy distance between X and X$^-$ lines. From the dependencies in Fig.2b one can found that the X linewidth (i.e. nonradiative damping) increases superlinearly with 2DEG density increase but the X$^-$ nonradiative damping keeps a constant level. The increase of the exciton damping is very natural because the increasing electron density induced an increasing inelastic exciton electron scattering with pair excitations: electron above Fermi level and hole below Fermi level. The independence of the X$^-$ linewidth (nonradiative damping) on the electron concentration is very strange because the trion is a charged particle and its scattering with 2D electrons should be very effective. We will discuss possible reasons for such behavior elsewhere.

Fig.2a shows X and X$^-$ radiative damping which is the oscillator strength as a function of the 2D electron density. One can see that the X$^-$ oscillator strength increases linearly for electron concentrations up to 5×10^{10} cm^{-2}. At higher concentrations the trion oscillator strength saturates. The linear dependence of the X$^-$ oscillator strength on the 2D electron density allows us to determine the effective trion radius. Let we introduce the trion oscillator strength per one electron as (Γ_0^T/N_e), and the exciton oscillator strength per unit cell as (Γ_0^X/N). Here N_e is the number of electrons in the QW, N is the number of unit cells per unit area. Following the approach developed in [6], [7] we obtain the relation

$$\left(\Gamma_0^T/N_e A_T\right) = \left(\Gamma_0^X/Na\right) \tag{1}$$

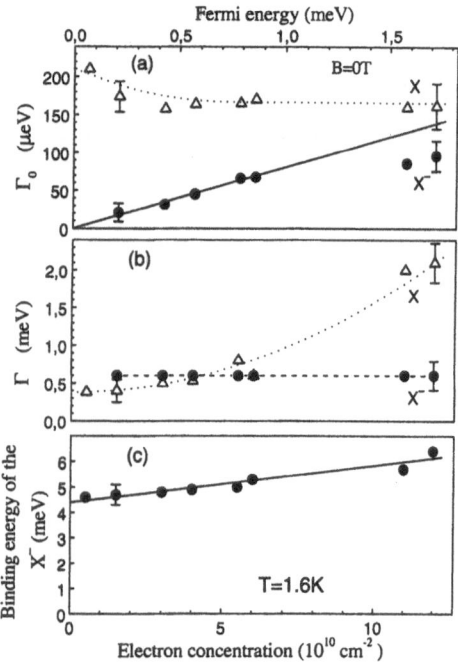

Fig. 2 Electron concentration dependence of exciton (triangles) and trion (circles) radiative damping - a), nonradiative damping - b) and the exciton - trion lines splitting - c).

Here $A_T = \pi a_T^2$ is the area of the trion with a radius a_T, a is the area of the unit cell, $Na = A$ is the unit area and $N_e/A = n_e$ is the two dimensional electron concentration. Using in Eq.(1) the experimentally determined parameters Γ_0^X ($n_e = 0$) $= 210 \mu eV$ and $\Gamma_0^T/n_e = 1.16 \times 10^{-9}$ $\mu eV cm^2$ we evaluate the radius of the trion as $a_T = 132 \text{Å}$. For comparison, the in-plane radius of 1s- and 2s exciton states in the studied QWs are 24Å and 120Å, respectively.

The X oscillator strength, in contrary to the X^- one, decreases with the electron concentration increase so that the total oscillator strength remains constant (see Fig.2a). Such redistribution of the oscillator strength between exciton and trion states led us to the conclusion of a possible mixing between these states.

Another evidence of such mixing one can find from the concentration dependence of the exciton - trion absorption line separation, which considered frequently as the trion binding energy. Fig.2c shows this separation as a function of the 2D electron density. The separation increases linearly with the electron concentration as $\Delta E = 4.4 \, meV + E_F$. The similar result has been reported recently for CdTe/CdZnTe QWs [8].

Three particles, two electrons and one hole could form a bound state (trion) with the binding energy E_b. Energy of the optical transition to the trion state is $E_{ex} - E_b$, where E_{ex} is the exciton energy. Additionally three particles can form an unbound state: exciton plus one free electron. Energy of the optical transition for an electron on the Fermi level is $E_{ex} - E_F$. Hence, if the Fermi energy is so high that $E_F = E_b$, these two states will be in exact resonance with each other. In these conditions

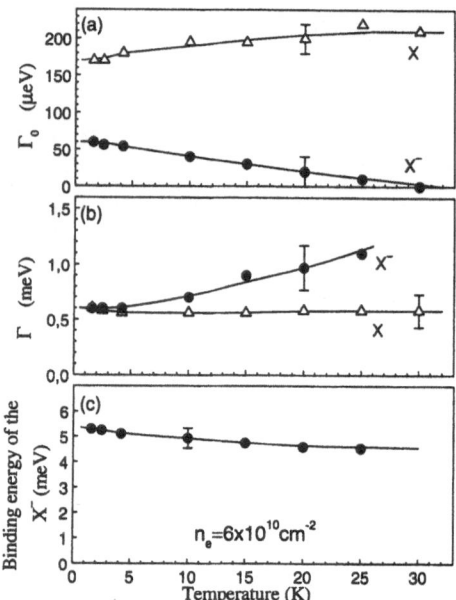

Fig. 3 Temperature dependence of the exciton (triangles) and trion (circles) parameters (see Fig.2).

we can not distinguish these states which can lead to their mixing by a small perturbation. Mixing of quantum states by any perturbation leads to their splitting and redistribution of the oscillator strength between them. This behavior is what we just observed in the experiment.

The concept of the exciton trion mixing can be confirmed by temperature dependence of the reflectivity spectra. Fig.3 display the temperature modifications of the exciton and trion parameters for the sample with electron concentration $n_e = 6 \times 10^{10}$ cm^{-2}. One can see from the dependence of exciton and trion oscillator strength (radiative damping) in Fig.3a that the exciton oscillator strength increases with the decreasing trion oscillator strength due to temperature dissociation. The total oscillator strength remains constant. Remarkable result has been found for the temperature modification of the exciton-trion separation. This splitting goes down when the trion state disappears with temperature, which is in complete agreement with the concept of the exciton-trion mixing.

Acknowledgements: This work was supported in part by RFFI-DFG grant No. Os98/6 and 00-02-04020.

References

1. M.A.Lampert, Phys. Rev. Lett.**1**, (1958) 450.
2. K. Kheng, et.al., Phys. Rev. Lett. **71**, (1993) 1752.
3. R.T. Cox, et.al., Acta Physica Polonica A **94**,(1998) 99.
4. G.V. Astakhov, et.al, Proc. Nanostructures: Physics and Technology, 7th International Symposium, St.Petersburg, Russia, June 14-18, 1999, pp.352-354.
5. E.L. Ivchenko and G.E. Pikus, *Superlattices and Other Heterostructures* (Springer-Verlag, Berlin 1997). Author, *Book title* (Publisher, place year) page numbers
6. E.I. Rashba, Sov. Phys. Semicond **8**, (1975) 807.
7. J. Feldmann, et.al., Phys. Rev. Lett. **59**, (1987) 2337.
8. V. Huard, et.al., Phys. Rev. Letters **84**, (2000) 187.

Rapid recombination process of free trions

D. Sanvitto[1,2], **R.A. Hogg**[1], **A.J. Shields**[1], **D.M. Whittaker**[1], **M.Y.Simmons**[2], **D.A. Ritchie**[2], **M. Pepper**[1,2]

[1] Toshiba Research Europe Ltd, Cambridge Research Laboratory, 260 Cambridge Science Park, Cambridge, CB4 0WE, UK.
[2] Cavendish Laboratory, University of Cambridge, Madingley Road, Cambridge CB3 0HE, UK.

Abstract In this paper we study the formation and recombination processes of free negative trions (X^-) in remotely doped quantum wells (QWs). We show that under non-resonant excitation, the neutral exciton (X) and X^- populations are not in thermodynamic equilibrium and the recombination dynamics are dominated by $X+e^- \longrightarrow X^-$ scattering when both species are present in the luminescence spectrum. The decay time of the thermalised X^- population is shown to be 1 order of magnitude faster than the decay time of free X. This difference is explained with a model which takes into account the three particle nature of the free X^-. We also discuss the effect of temperature, QW width, and homogeneous linewidth on the decay of free X^-.

1 Introduction

Negative trions (X^-), the analog of the H^-, are a bound state comprising two electrons and a hole. This species of exciton has recently been observed in semiconductor quantum wells (QWs)[1–4]. In some cases, the charged excitons are localised in the plane of the quantum well, resulting in a recombination behaviour similar to that of the donor bound exciton[3,5]. In our samples a Stokes shift is observed between absorption and luminescence, indicating some degree of localization, only in the case of thin QWs (~100 Å) at low temperatures (below 7 K). Conversely, no Stokes shift is observed for a 300 Å QW at temperatures as low as 2 K. These very high quality samples allow us to study the dynamics of free neutral and charged excitons under a wide range of excess electron densities.

2 Experimental details

The samples studied are remotely doped GaAs-Al$_{0.33}$Ga$_{0.67}$As QWs topped with 600 Å undoped Al$_{0.33}$Ga$_{0.67}$As and 2000 Å Si doped (10^{17}cm^{-3}) Al$_{0.33}$Ga$_{0.67}$As which were grown by molecular beam epitaxy on GaAs substrates[2]. A Schottky NiCr gate evaporated on top of the sample surface and an Ohmic contact with the well, allow the excess electron density to be varied in the QW. This results in the possibility of controlling the dominance of X or X^- in the luminescence spectra. The decay time was measured using time resolved PL with an excitation density sufficient to create ~ 2×10^7cm^2 e-h pairs per pulse. The temporal resolution was ~ 8 ps[6].

3 Results and discussion

By applying a bias between the gate and the Ohmic contact in the well we can tune the dominance of X or X^- in the photoluminescence (PL) spectra. Figure 1 (a) shows the decay time of the X and X^- luminescence as a func-

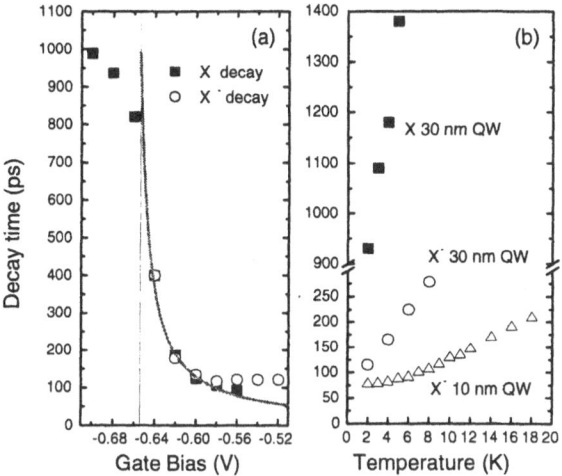

Fig. 1 (a) Decay time for X and X^- as a function of gate bias (electron density). The gray curve is the functional form of τ_X (see text). (b) Decay time of X and X^- in the 300 Å QW and X^- in a 100 Å QW as a function of temperature.

tion of applied front gate bias (and hence carrier density) for a 300 Å sample at 2 K. We observe a single exponential decay for both X and X^- at all carrier densities. At very high negative biases the QW is depleted and X dominates the spectra, giving a decay time of about 980 ps. As the bias is reduced the excess carrier density in the QW is increased and the trions appear in the time integrated PL. The appearance of X^- corresponds to a reduction of the exciton lifetime. At biases < -0.62 V where both X and X^- are observed, the decay times of the two species are identical. At higher biases (e.g. -0.52 V) we observe a faster decay time for the neutral exciton than the charged exciton. As a consequence of this behaviour the formation of X^- must be related to the depopulation of the excitons. As only a single exponential decay is observed for X^- we deduce that the charged exciton is formed from the neutral exciton population and is not formed directly from the e-h plasma. A model which allows the formation of X^- from the radiative X energy level gives:

$$\tau_X = \frac{1}{\Gamma_X + \Gamma_s} \tag{1}$$

$$\tau_{X^-} = \begin{cases} \frac{1}{\Gamma_X+\Gamma_s}, & \text{when } \Gamma_s < \Gamma_{X^-} - \Gamma_X \\ \frac{1}{\Gamma_{X^-}}, & \text{when } \Gamma_s > \Gamma_{X^-} - \Gamma_X \end{cases} \tag{2}$$

Where τ_X and τ_{X^-} are the measured X and X^- decay time, and Γ_X is the intrinsic exciton decay rate, Γ_s the $X + e^- \longrightarrow X^-$ scattering rate and Γ_{X^-} the X^- intrinsic decay rate which is ~ 0.01 ps^{-1}, almost 1 order of magnitude bigger than Γ_X. The gray line in fig. 1 (a) gives the functional form of τ_X. As we can see when the trion has reached its intrinsic decay time the X and X^- decay times are no longer identical as the X decay is dominated by scattering into X^-. In order to study the recombination process we tuned the bias to carrier densities where only one species of exciton is present in the PL spectra. Henceforth when referring to the X or X^- decay time we refer to the intrinsic decay of a thermalised population obtained at -0.7 V and -0.52 V gate voltage respectively. Previous work has shown that the exciton and trion populations thermalize to the lattice temperature [6].

Excitons in QWs can relax the momentum conservation in the direction normal to the plane of the QW, leading to a finite oscillator strength. However, the only radiative states for excitons are those which satisfy $\mathbf{k} < \frac{E_{photon}}{\hbar}$. In the case of X^- radiative decay releases an electron in addition to a photon. In this process the photon is no longer required to conserve the same trion-momentum, the mismatch may given up to the electron which remains after recombination. The absence of a lightcone and so the enlargement of the optical linewidth is the principal reason for the rapid radiative decay observed for charged excitons. Following the equation developed by Citrin and Andreani [7,8] for neutral excitons we substitute the light cone with a finite oscillator strength in $\mathbf{k}_{x,y}$[9,10], we obtain an equation for the X^- lifetime (τ_X^-):

$$\tau_{X^-} \propto \frac{1}{f_{//}^{X^-}} \left(k_B T + C\right) \frac{\hbar \Gamma_h}{k_B T} \left(\frac{1}{1 - e^{-\hbar \Gamma_h / k_B T}}\right) \quad (3)$$

Were $f_{//}^{X^-}$ is the trion oscillator strenght, k_B the Boltzmann constant, Γ_h the X^- linewith, T the temperature and C a constant.

Figure 1 (b) plots the decay time of X and X^- in a 300 Å QW and X^- in a 100 Å QW as a function of the lattice temperature. For free particles a linear dependence in temperature is predicted. We therefore conclude that for the 300 Å QW sample the charged exciton is free. The effect of temperature is to reduce the fraction of excitons at $\mathbf{k}$=0. For the exciton an increase in decay time with increasing temperature is therefore observed. The weaker dependence of the X^- decay on temperature is due to the wider range of $\mathbf{k}_{//}$ states which may couple to the photon. In the case of the 100 Å QW sample, the effects of localization due to QW roughness gives rise to an insensitivity of the decay time to temperature below 7 K. Above this temperature the linear trend is again observed,as expected for free trions. Comparing X^- in a 100 Å QW with the same species in a 300 Å QW we

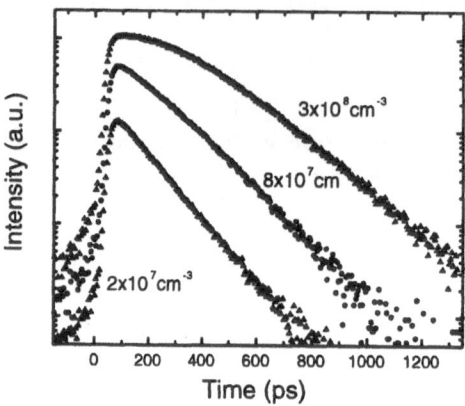

Fig. 2 X^- decay profile as a function of trion density n_{X^-}.

can determine the effect of well width on the oscillator strength. From our experiment,we find a factor of 2.6 difference. We note that the charged exciton oscillator strength is considerably more sensitive to well width than the neutral exciton which shows a factor of ~ 1.3 difference between a 100 and 300 Å QW[12].

The homogeneous linewidth (Γ_h) is a week function of temperature. As a result, the most effective way to vary this parameter is to increase the density of excitons[11]. We varied the excitation laser power in order to change the X^- density (n_{X^-}) and so Γ_h. The effect of varying Γ_h is presented in figure 2. Here the decay profile of X^- in a 300 Å QW at different laser powers is plotted. The decay obtained with a photo-injected density of n_{X^-}=2x10^{7}cm^{-3} shows a single decay time of 110 ps. Raising the laser power to give $n_{X^-} \sim$ 8x10^{7}cm^{-3} and 3x10^{8}cm^{-3} we observe a non exponential decay due to a time dependence of the X^- density and Γ_h during the decay process. At long times after excitation the low density decay time is recovered. This is the first observation of the effect of a finite linewidth on the trion decay. This Γ_h dependence is further evidence that X^- is a free particle.

References

1. K. Kheng *et al.*, Phys. Rev. Lett. **71**, (1993) 1752.
2. A. J. Shields *et al.*, Phys. Rev. B **51**, (1995) 18049.
3. G. Finkelstein *et al.*, Phys. Rev. Lett. **74**, (1995) 976.
4. A. Ron *et al.*, Solid State Commun. **97**, (1996) 741.
5. H. Okamura *et al.*, Phys. Rev. B **58**, (1998) R15985.
6. R. A. Hogg *et al.*, Physica B **272**, (1999) 412.
7. D. S. Citrin, Phys. Rev. B **47**, (1993) 3832.
8. L. C. Andreani *et al.*, Sol. Stat. Comm. **77**, (1991) 641.
9. B. Stébé *et al.*, Phys. Rev. B **58**, (1998) 9926.
10. A. Esser *et al.*, Proceedings OECS VI (1999).
11. R. Eccleston *et al.*, Phys. Rev. B **45**, (1998) 11403.
12. A. R. K. Willcox *et al.*, Super. and Micro. **16** (1994) 59.

Spin dependent exciton-exciton interaction in hot and cold 2D exciton gases controlled by an electric field

G. Aichmayr[1], L. Viña[1], S. P. Kennedy[2], R. T. Phillips[2], E. E. Mendez[3]

[1] Departamento de Física de Materiales C-IV, Universidad Autónoma de Madrid, Cantoblanco, E-28049 Madrid, Spain e-mail: guenther.aichmayr@uam.es

[2] Cavendish Laboratory, University of Cambridge, Cambridge CB30H3, U. K.

[3] Department of Physics and Astronomy, SUNY at Stony Brook, N. Y. 11794-3800, USA

Abstract We demonstrate the tunability of the exciton-exciton (X-X) interaction in a spin-polarised 2D exciton gas by applying an electric field ($\mathcal{E}$). The splitting between the two spin components (± 1) is observed in resonantly and non-resonantly created gases and it amounts up to 5 meV, which is compareable to the single exciton binding energy (8 meV). The splitting can be tuned continuosly to zero by increasing $\mathcal{E}$.

The spin dependence of X-X interactions was first studied by Damen *et al.* [1]. They observed an energy splitting between the +1 and -1 exciton spin components of a spin polarised exciton gas, which amounts to ~ 2 meV a few ps after pulsed excitation. Later on the origin of this splitting was explained in terms of two counteracting spin-dependent X-X interactions, one increasing the exciton binding energy E_{2D} and the other one decreasing it [2,3]. The former is the exchange interaction (I_{EC}) between electrons (and holes) of different excitons. It *increases* E_{2D} in proportion to the spin-polarized exciton density: $\Delta E_{2D}^{\pm} \propto n_X^{\pm} I_{EC}$, where $n_X^{\pm}$ refers to the spin +1 and -1 exciton populations, respectively. The latter is the vertex correction (I_{VC}) to the Coulomb interaction between electrons and holes of different excitons. This correction effectively *reduces* the electron-hole attraction due to the occupation of final states in the electron-hole scattering processes and lowers E_{2D} proportionally to $n_X^{\pm} I_{VC}$. For quantum wells at flat-band conditions, the repulsive interaction (I_{VC}) always outweighs the attractive one (I_{EC}) leading to a negative splitting, i.e. the minority excitons, in our case the -1 excitons, lie below the majority +1 excitons. The model predicts that the repulsive (attractive) interaction, I_{VC} (I_{EC}), becomes weaker (stronger) when the wavefunctions of the electron and hole forming the exciton are distorted by, for example, an electric field. This would lead to a decrease of the splitting and eventually to its sign reversal. A negative splitting, i.e. the majority excitons lie energetically below the minority ones, would lead to a transition to a ferromagnetic phase: the +1 exciton state would be energetically more favourable and the spin polarisation would maintain a finite value for long times, rather than decay with the characteristic exciton spin flip time ($\sim$100 ps). It has been argued that this ferromagnetic phase could favour Bose condensation of excitons [2].

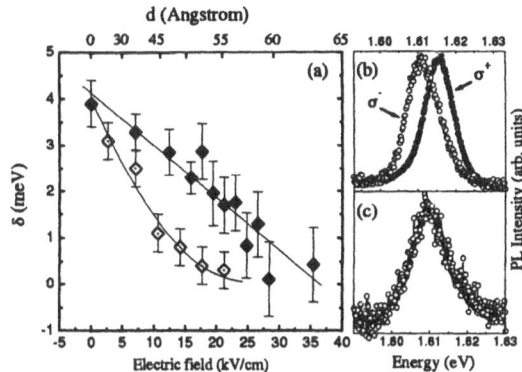

Fig. 1 (a) Energy splitting between the +1 and -1 spin components in a hot (full symbols) and cold (hollow symbols) exciton gas 10 ps after excitation. Upper x-axis is electron hole separation. (b) Spectra at 10 ps and $\mathcal{E}$=0. (c) Normalised spectra at 10 ps and $\mathcal{E}$=20 kV/cm.

We studied the change of energy splitting as a function of an electric field applied to a coupled double quantum well (CDQW). The field separates spatially the electrons and holes and confines them in the opposite wells of the DQW. Furthermore, not only the splitting but also the emission energies of the two spin components are investigated in order to obtain insight in the spin-independent direct coulomb interaction, i.e. the screening.

The time resolved photoluminescence was measured in a up-conversion set-up with 2 ps time resolution and in a two-colour set-up with 150 fs resolution, respectively. The CDQW sample was kept at 8 K. One period of the CDQW consisted of two 50 Å GaAs wells with a 20 Å $Al_{0.3}Ga_{0.7}As$ barrier. For details of the sample structure see Ref. [4]. We excite the CDQW structure with circularly σ^+ polarised laser pulses above the exciton resonance (hot gas) and at the resonance (cold gas), respectively. The emission energies are measured 10 ps after excitation.

Fig. 1a depicts the spin splitting for both cases. The full symbols represent the splitting in a hot exciton gas as a function of field. Although the excitons are surrounded by an electron hole plasma and the exciton distribution itself is not in thermal equilibrium, a clear energy splitting of up to 4 meV ($\sim E_{2d}/2$) is observed for

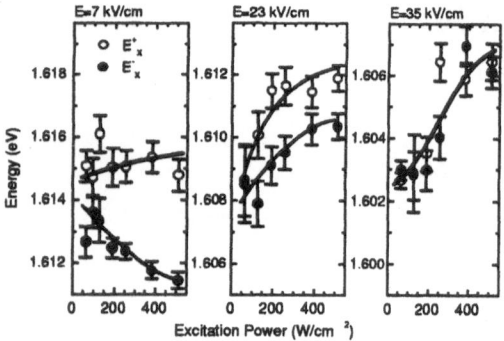

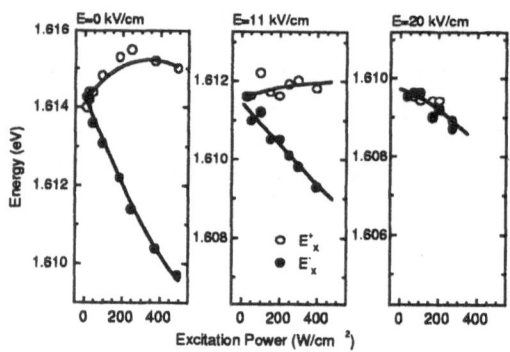

Fig. 2 Density dependence at non-resonant excitation, for three differnt field strength.

Fig. 3 Density dependence at resonant excitation, for three electric field strength.

a polarisation of 50%. On increasing the field the splitting decreases linearly and reaches zero at $\mathcal{E}=35$ kV/cm. The mean distance between electron and hole as a function of $\mathcal{E}$ is calculated from the wave functions (upper x-axis in Fig.1a). $\mathcal{E}=35$ kV/cm leads to a distance $d \simeq 60$ Å, which is a factor of 2.5 higher than the theoretical value from Ref.[2]. A further increase of the field does not lead to an inversion of the splitting and the ferromagnetic phase transition is not observed. At fields higher than 35 kV/cm the excitons are ionised and escape from the CDQW structure into the continuum. Nevertheless, a field induced alteration of the vertex and exchange corrections is clearly observed. At $\mathcal{E}=0$ $I_{VC} > I_{EC}$, while for $\mathcal{E}>0$ I_{VC} (I_{EC}) can be considerably reduced (enhanced) in order to yield a zero splitting, where $I_{VC} = I_{EC}$.

In a resonant experiment the exciton transition is excited directly creating a cold exciton gas, i.e. $K \simeq 0$, avoiding the electron hole plasma which screens interactions due to its charge. The open symbols in Fig.1a show the splittings obtained. In this case the polarisation is 95%. At zero field the splitting is about 4 meV for this density. In order to maintain the same initial exciton density on increasing the field, the excitation power has to be increased because of the reduced wave function overlap. Now the energy splitting is decaying exponentially rather than linearly and reaches 0 already at 20 kV/cm, which is closer to the theoretical value (see also the spectra in Fig.1b and c). A furhter increase of the field maintains the zero splitting. Its sign reversal, however, is again not observed. In the resonant experiment the carrier ionisation is not relevant at the available fields. Nevertheless, the injection of excitons at constant density at these fields becomes more difficult because of the reduced absorption. New structure geometries and/or lower temperatures will be employed in the future to explore the negative splitting.

The energy difference of the two spin components gives insight into the spin dependent X-X interaction. If we turn to the absolute emission energies we can also deduce effects of the direct, i.e. spin independent, X-X

interactions. In the non-resonant experiment we see a distinct blue shift of the energies of both components as a function of exciton density when the field is increased (see Fig.2). At low field the majority +1 excitons do not shift with density whereas the minority ones red-shift. Under resonant conditions the low field case is very similar, but on increasing the electric field the energies of the spin components behave differently (see Fig. 3). Now the +1 excitons maintain their energy or even decrease it at high fields and the -1 excitons always undergo a red-shift. Unfortunately, available models only deal with differences in binding energies of the spin components but not with absolute energies. Since the difference between the two cases is the presence of hot excitons and an electron-hole plasma in the first, and only cold excitons in the latter one, our experiments provide all the relevant information to develop a complete theory which should take exciton temperature and screening effects into account.

This work was partially supported by the European Union through the Training and Mobility of Researchers Ultrafast Quantum Optoelectronics Network, the austrian BMWF, the spanish DGICYT (PB96-0085), the CAM (07N/0026/1998) the Fundacion Ramon Areces, the Spain-US Joint Commission and the U.S. Army Research Office.

References

1. T. C. Damen *et al.*, Phys. Rev. Lett. **67**, 3432 (1991).
2. J. Fernández-Rossier and C. Tejedor, Phys. Rev. Lett. **78**, 4809 (1997).
3. L. Viña *et al.*, Phys. Rev. B **54**, R8317 (1996).
4. G. Aichmayr *et al.*, Phys. Rev. Lett. **83**, 2433 (1999).

From excitons to Fermi edge singularity

G. Yusa, H. Shtrikman, and I. Bar-Joseph

Department of Condensed Matter Physics, The Weizmann Institute of Science, Rehovot 76100, Israel

Abstract We study the evolution of the absorption spectrum of a modulation doped GaAs/AlGaAs semiconductor quantum well with decreasing the carrier density. We find that at some critical electron density there is a sharp change in the lineshape and the transitions energies of the exciton peaks. We show that this critical density marks an abrupt transition from a simple excitonic behavior to a Fermi edge singularity.

1 Introduction

The absorption spectrum of a semiconductor in the presence of a Fermi sea of electrons has been a subject of theoretical and experimental research for more than three decades [1-3]. In recent years there have been intensive studies of the behavior of the system at low electron densities. It was found that in this regime the absorption spectrum is characterized by the appearance of a bound state, the charged excitons, X^- [4-6], at some binding energy E_B below the neutral exciton, X. A natural question is how does the low density spectrum, of the X^- and X bound states, transform into the high density spectrum, characterized by a Fermi edge singularity (FES) [7-9].

2 Experimental

We investigate the evolution of the absorption spectrum of a single GaAs/AlGaAs quantum well with increasing electron density, from 1.5×10^{10} to 1.5×10^{11} cm^{-2}. We tune the electron density continuously by applying a voltage to a semi-transparent gate relative to an ohmic contact, and measure the absorption spectrum using photocurrent spectroscopy.

3 Results

Figure 1 shows a set of photocurrent spectra taken for various gate voltages. It is seen that as the density is reduced two broad peaks are formed and acquire an asymmetric singular lineshape (Figs. 1a-c). At further lower densities the singularity increases very fast with decreasing density, and eventually becomes a symmetric resonance, the heavy-hole exciton (Figs. 1d-e). We label the two peaks as ω_1 and ω_2. [10-13]

Figure 2 shows the density dependence of the energies of the energy difference $\omega_2-\omega_1$. It is evident that there exists a critical density value, $N_c=2.5 \times 10^{10}$ cm^{-2}, below which $\omega_2-\omega_1$ is constant and equals 1.2 meV. Only above this critical density this energy difference

grows, first rapidly and then more gradually. The dashed line describes the relation $\omega_2-\omega_1= E_B+E_F$, where $E_B=1.2$ meV is the charged exciton binding energy. It is evident that the measured points in Fig. 2b are below this curve at low density, and as the density gets higher the data points approach the dashed line.

A close examination of Fig. 1 reveals that at the critical density N_c not only the energy of the ω_2 peak

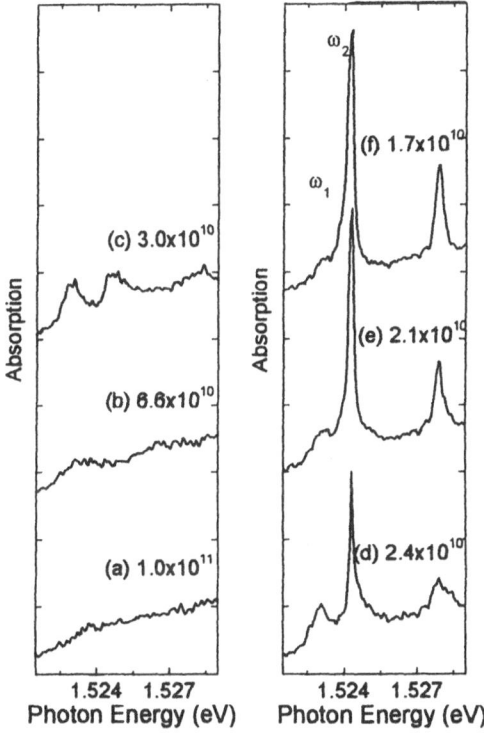

Fig. 1. A series of the absorption spectra at several gate voltages. Each spectrum is labeled by the corresponding electron density. The spectra are displayed after subtraction of the background signal (see text)

shifts but also its lineshape undergoes a drastic change: it can be seen that is symmetric below N_c (Figs. 1e-f) and is a singular asymmetric line above it (Fig. 1c-d). The critical density appears to be distinguishing between two regimes: a high density regime, in which the exciton energy and lineshape is sensitive to the presence of a

Fermi sea, and a low density regime where the exciton behavior is of a hydrogen-like object. The question which arises is what defines the density scale for the problem. One possible candidate is the X^- area, S_{X^-}. We note that when the density is lower than $1/S_{X^-}$ the X^-

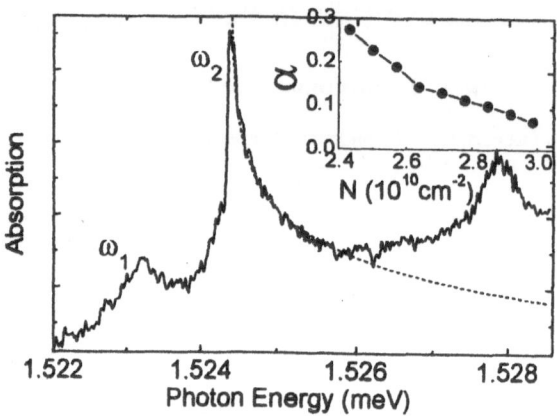

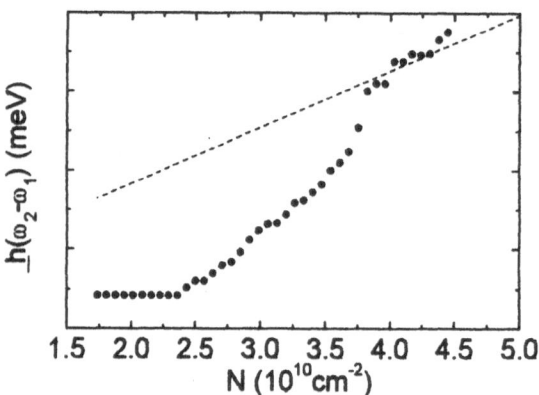

Fig. 2 The energy of difference $\omega_2-\omega_1$ as a function of electron density. The dashed line is E_B+E_F.

Fig. 3 A series of the absorption spectra at several gate voltages. Each spectrum is labeled by the corresponding electron density. The spectra are displayed after subtraction of the background signal (see text).

bound state is formed by the photo-excited electron and one of electrons of the 2DEG. As a result, this area is depleted of electrons, and there is no cost of E_F energy to pay in ionizing this state. Hence, the second peak is formed at an energy E_B above the first one and not at E_B+E_F. This argument also explains the change of the exciton lineshape around N_c. Only when a Fermi sea of electrons is present within this area and shake-up processes can occur [14], a high energy tail is formed. Indeed, the observed critical density agrees with the estimated size of X^-, which has a radius of ~ 30 nm. Another physical quantity, which defines a density scale, is the temperature. We note that at the critical density $E_F \approx 2.5$ kT, and consequently there are available states below E_F, which can participate in the $X-X^-$ absorption. This could explain why $\omega_2-\omega_1$ is independent of density below N_c. We find it difficult, however, to apply this argument to explain the changes in lineshape.

We notice that the evolution of the singularity above N_c is surprisingly fast. Figure 3 describes a fit of a power law singularity $A(\omega)=(\omega-\omega_0)^{-\alpha}$ to the high energy tail of ω_2 at $N=2.7 \times 10^{10}$ cm^{-2}. The inset shows the dependence of α on N. It is seen that α decreases by substantially (from 0.3 to 0.05) in a very narrow range, $\Delta N=8 \times 10^9$ cm^{-2}. The exponent α is related to the phase shift of the electrons at the Fermi surface, when scattering off the valence hole potential. In that sense it measures the efficiency of the Fermi sea electrons in screening that potential: the smaller α is, the better is the screening [15]. The fast change of α therefore represents the change of the screening properties of the 2DEG in the low density regime.

References

1. G. D. Mahan, Phys. Rev. Lett. **18**, (1967) 448.
2. M. Combescot and N. Noziers, J. Physique **32**, (1971) 913.
3. For a review see G. D. Mahan, *Many Particle Physics* (Plenum, New York, 1981).
4. K. Kheng *et al.*, Phys. Rev. Lett. **71**, (1993) 1752. G. Finkelstein *et al.*, Phys. Rev. Lett. **74**, (1995) 976.
5. A. J. Shields *et al.*, Phys. Rev. **B 52**, (1995) 7841.
6. V. Huard *et al.*, Phys. Rev. Lett. **84**, (2000) 187.
7. M.S. Skolnick *et al.*, Phys. Rev. Lett. **58**, (1987) 2130.
8. G. Livescu *et al.*, IEEE J. Quant. Electron. **QE-24**, (1988) 1677.
9. S. A. Brown *et al.*, Phys. Rev. **B 54**, (1996) R11082.
10. P. H,awrylak, Phys. Rev. **B 44**, (1991) 3821.
11. The notion of two thresholds already appears in Ref. 2.
12. X. Xia *et al.*, Phys. Rev. **B 45**, (1992) 1341.
13. J. Brum and P. Hawrylak, Comments on Cond. Matt. Phys. **18** (1997) 135.
14. G. Finkelstein *et al.*, Phys. Rev. **B 53**, (1996) 12593.
15. A. L. Efros *et al.*, Phys. Rev. **B 47**, (1993) 2233.

Magnetic traps for excitons in GaAs/Al$_x$Ga$_{1-x}$As quantum wells

J. A. K. Freire[1]*, F. M. Peeters[1], A. Matulis[2], V. N. Freire[3], G. A. Farias[3]

[1] Departement Natuurkunde, Universiteit Antwerpen, Universiteitsplein 1, B-2610 Antwerpen, Belgium
[2] Semiconductor Physics Institute, Goštauto 11, 2600 Vilnius, Lithuania
[3] Departamento de Física, Universidade Federal do Ceará, Caixa Postal 6030, Campus do Pici, 60455-760 Fortaleza, Ceará, Brazil

Abstract The trapping of excitons in a GaAs/Al$_x$Ga$_{1-x}$As quantum well due to a nonhomogeneous magnetic field arising from a ferromagnetic disk deposited on top of the well is studied. The dependence of the exciton confinement energy on the quantum well width, spin quantum number, and the applied magnetic field intensity is analysed. The results show that the spin contribution can be responsible for a hundred times increase of the exciton confinement energy as compared to the spinless situation. Furthermore, we find that the exciton confinement is a sensitive function of the interplay between the uniform applied magnetic field and the quantum well width.

1 Introduction

Whereas several methods were proposed for the trapping of charged particles and neutral atoms (mostly based on the use of magnetic fields) [1], only few have been successfully used to trap excitons [2,3]. One mechanism is the quantum-confined stark effect created by strong spatially varying electric fields [2]. The exciton confinement using harmonic-potential traps created by inhomogeneous strain and applied electric fields have also been reported [3]. Another possibility is the confinement of excitons by nonhomogeneous magnetic fields, which was recently proposed by Freire *et al.* [4,5]. They showed that excitons can be trapped in magnetic field inhomogeneities, with a confinement degree that depends strongly on their orbital and spin Zeeman energy.

In this work, we study the trapping of excitons in a GaAs/Al$_{0.3}$Ga$_{0.7}$As quantum well due to a nonhomogeneous magnetic field. The magnetic field profile is the one resulting from an uniform magnetised ferromagnetic disk which is placed above the quantum well in the presence of a background magnetic field.

2 Theoretical framework and results

The total exciton Hamiltonian is divided into a center-of-mass motion, a relative motion, a spin, a perturbative, and a well confinement contribution [4,5]. They give rise to an effective potential and an effective mass which are used to describe the motion of the exciton center-of-mass. The main approximations we have considered were: (i) the expansion of the vector potential in the center-of-mass and relative motion Hamiltonian up

to the second order in the relative motion coordinates [4]; (ii) a variational approach assuming that the difference between the 2D and 3D Coulomb interaction can be made very small by choosing a suitable variational parameter [6]; (iii) the expansion to zero order of the magnetic field in the spin Hamiltonian, where we assumed that there is no coupling between the spin contribution and those of the exciton center-of-mass and relative motion [5]. The resultant Schrödinger-like equation for the exciton center-of-mass motion is:

$$\left\{ -\frac{\hbar^2}{2R}\frac{d}{dR}\frac{R}{M^{eff}(R)}\frac{d}{dR} + V^{eff}(R) - E \right\}\psi(R) = 0, \quad (1)$$

with the following effective mass and effective confinement potential:

$$M^{eff}(R)/M = \left[1 - \frac{e^2\mu}{\hbar^2 Mc^2}\alpha^{n_r}_{m_r}\gamma^{-4}B_z(R)^2 \right]^{-1}, \quad (2)$$

and

$$V^{eff}(R) = \frac{\hbar^2}{2}\frac{1}{M^{eff}(R)}\frac{m_R^2}{R^2} + \frac{e^2}{8\mu c^2}\beta^{n_r}_{m_r}\gamma^{-2}B_z(R)^2 + \frac{e\hbar}{2\mu c}\xi m_r B_z(R) \pm \frac{1}{2}\mu_B g_{exc}B_z(R). \quad (3)$$

In the above equations, R is the radial coordinate of the center-of-mass motion, $M = (m_e^* + m_h^*)$ ($\mu = m_e^* m_h^*/M$) is the exciton mass (reduced mass), $\xi = (m_h^* - m_e^*)/M$, $\alpha^{n_r}_{m_r}$ and $\beta^{n_r}_{m_r}$ (in units of $a_B^{*\,4}$ and $a_B^{*\,2}$, respectively, where a_B^* is the effective Bohr radius) are constants related to the quantum numbers of the exciton relative motion (n_r, m_r) [4], $\mu_B = e\hbar/2m_{e,\parallel}^* c$ is the Bohr magneton, and γ is the variational parameter. The exciton spin orientation is controlled by the total spin quantum number ($m_z = S_{e,i} + J_{h,i} = \pm 1, \pm 2$), and the exciton g factor ($g_{exc} = g_{e,z} + g_{h,z}$) is related to the z component of the electron ($g_{e,z}$) and heavy hole ($g_{h,z}$) g factor [7].

The nonhomogeneous magnetic field profile used in this work, which can experimentally be created by depositing a ferromagnetic disk on top of a semiconductor quantum well with a uniform magnetic field (B_a) applied perpendicular to the plane of the disk [see inset in Fig. 1(a)], is shown in Fig. 1(a) for several values of the distance d of the magnetic disk to the quantum well. The

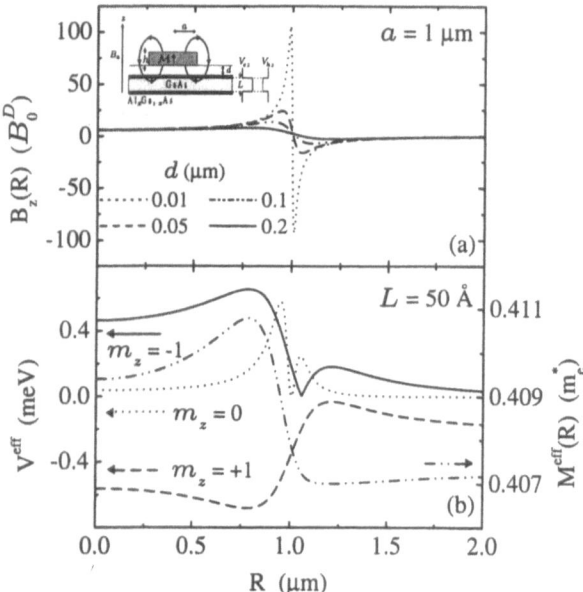

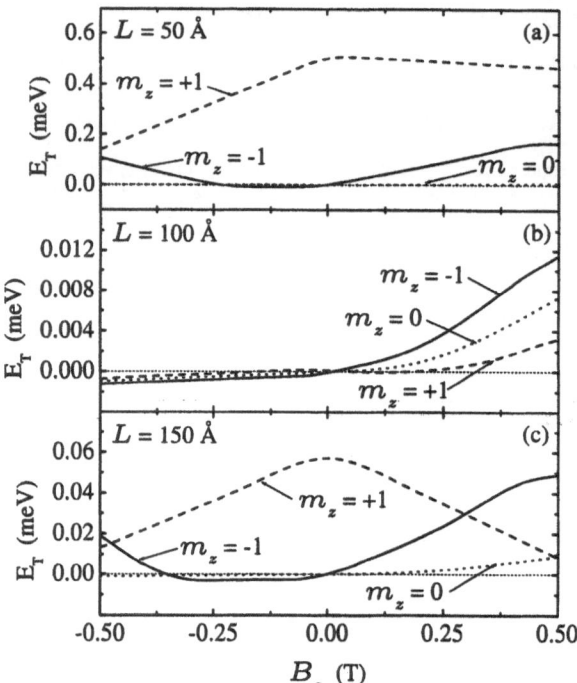

Fig. 1 (a) The magnetic field profile as a function of the radial coordinate R, for different values of d, for a disk magnetization $B_0^D = 0.174$ T, and applied field $B_a = 0$. In the inset, the side-view of the experimental setup. Here a (h) is the disk radius (thickness), and $\mathcal{M}$ the magnetization. (b) Effective potential and respective effective mass as a function of R, for the exciton ground state, for quantum well width $L = 50$ Å, $B_a = 0$ T, $B_0^D = 0.174$ T, $a = 1$ μm, and $d = 0.2$ μm.

Fig. 2 Trapping energy, E_T, for the exciton ground state, as a function of the applied field B_a, for different spin quantum numbers, for quantum well width (a) $L = 50$ Å, (b) $L = 100$ Å, and (c) $L = 150$ Å, for strength of the disk magnetization $B_0^D = 0.174$ T, for $a = 1$ μm, and $d = 0.2$ μm.

corresponding effective magnetic potential and effective mass [see Eqs. (3) and (2), respectively] for $d = 0.2$ μm is shown in Fig. 1(b).

The Schrödinger equation with the effective mass and effective potential, as depicted in Fig. 1(b), was solved numerically for the calculation of the trapping energy of the exciton ground state (E_T), defined as the difference between the energy of an excitonic state in the uniform applied field B_a and the exciton energy in the corresponding nonhomogeneous magnetic field. Its dependence on the applied field B_a is shown in Fig. 2, for quantum well widths (a) $L = 50$ Å, (b) $L = 100$ Å, and (c) $L = 150$ Å. In this figure, we use a disk magnetization strength B_0^D of 0.174 T, and a disk radius (distance to the quantum well) $a = 1$ μm ($d = 0.2$ μm). Notice that due to the strong dependence of the exciton g factor with the quantum well width [7], the energies related to the 100 Å quantum well are in some cases two orders of magnitude smaller that the energies of the 50 Å and 150 Å well widths. Also notice that the exciton is always trapped for positive applied fields, but this is not true for the negative field situation. In the latter case, for strong negative B_a the exciton is trapped only in the $m_z = +1$ state, but this situation is inverted for smaller negative fields, where now only the $m_z = -1$ spin state is trapped. Furthermore, there is a crossing in the trapping energies of the $m_z = \pm 1$ polarized states. This is due to the competition between the applied field and

the spin interaction, which can change the exciton localization [compare solid and dashed curves in Fig. 1(b)].

3 Acknowledgment

This work was partially supported by CNPq, IMEC, and IUAP. J. A. K. Freire was supported by the Brazilian Ministry of Culture and Education (MEC-CAPES), and F. M. Peeters was supported by the FWO-Vl.

References

1. For a recent review, see e.g., C. E. Wieman, D. E. Pritchard, and D. J. Wineland, Rev. Mod. Phys. **71**, (1999) 253.

2. A. O. Govorov and W. Hansen, Phys. Rev. B **58**, (1998) 12980; D. A. B. Miller, D. S. Chemla, T. C. Damen, A. C. Gossard, W. Wiegmann, T.H. Wood, and C. A. Burrus, Phys. Rev. B **32**, (1985) 1043.

3. V. Negoita, D. W. Snoke, and K. Eberl, Phys. Rev. B **60**, (1999) 2661.

4. J. A. K. Freire, A. Matulis, F. M. Peeters, V. N. Freire, and G. A. Farias, Phys. Rev. B **61**, (2000) 2895.

5. J. A. K. Freire, A. Matulis, F. M. Peeters, V. N. Freire, and G. A. Farias, Phys. Rev. B **62**, (2000), in press.

6. B.-H. Wei, Y. Liu, S.-W. Gu, and K.-W. Yu, Phys. Rev. B **46**, (1992) 4269.

7. E. Blackwood, M. J. Snelling, R. T. Harley, S. R. Andrews, and C. T. B. Foxon, Phys. Rev. B **50**, (1994) 14246.

Excess carrier effects upon the excitonic absorption thresholds of remotely doped GaAs/AlGaAs quantum wells.

R. Kaur[1,2], A. J. Shields[1], R.A. Hogg[1], J. L. Osborne[1,2], M. Y. Simmons[2], D. A. Ritchie[2], and M. Pepper[1,2]

[1] Toshiba Research Europe Limited, 260 Science Park, Milton Road, Cambridge CB4 OWE, U.K e-mail: rk232@cam.ac.uk
[2] Cavendish Laboratory, University of Cambridge, Madingley Road, Cambridge CB4 OHE, U.K

Abstract We report the evolution of the absorption thresholds seen in narrow GaAs/Al$_{0.33}$Ga$_{0.67}$As quantum wells with a continuously variable excess electron density using photoluminescence excitation, and electroreflectance spectroscopies. The doubly occupied state (X$^-$), evolves into the well known Fermi edge singularity at high electron densities. The evolution of the singly and unoccupied states with increasing electron density is discussed.

1 Introduction

Combescot and Nozieres (CN)[1] predicted the exisitence of two absorption thresholds in the x-ray spectra of metals due to the absorption into states where the hole is bound to an electron or unbound. Hawrylak[2], and later Brown et al.[3] applied the CN theory to a two-dimensional electron gas (2DEGs) confined in semiconductor quantum wells (QWs), and suggested there should be three thresholds, where the third threshold is due to the exisitence of a doubly occupied bound state, i.e. two electrons bound to the hole. Recently photoluminsecence (PL) spectra in GaAs/AlGaAs QWs containing intermediate carrier densities demonstrate that another type of exciton is bound at very low carrier densites of the order 10^{10}cm^{-2}[4–6]. These are the so-called charged excitons where a single excess carrier binds with an exciton to form a trion. The binding energy of the second electron is typically 1 to 2 meV in GaAs QWs. In this paper we report on the PL excitation (PLE) and electroreflectance (ER) spectra of high quality remotely doped narrow GaAs/Al$_{0.33}$Ga$_{0.67}$As QWs.

2 Experimental Details

We studied several remotely doped GaAs/Al$_{0.33}$Ga$_{0.67}$As single QWs with widths of 50, 100 and 300 Å, grown by MBE on (100) oriented GaAs substrate. Although the samples showed qualitatively similar behaviour we concentrate here on the results taken on the 100 Å QW. The layer structure and experimental details are described in detail in R. Kaur et al.[7].

3 Results and Discussion

Figure 1 plots PL and PLE spectra recorded on the 100Å QW sample. The PL spectra are plotted on a logarithmic scale, for different gate biases. At the most positive gate biases, for which the density is largest, the emission from the well consists of a broad band due to recombination of photoexcited holes with the 2DEG. This

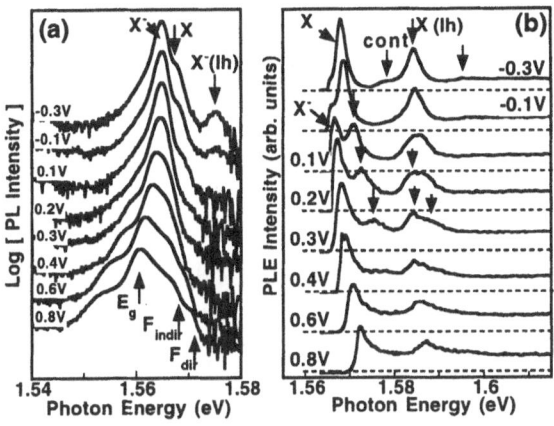

Fig. 1 . a) PL and b) PLE spectra obtained from the 100Å remotely doped GaAs/Al$_{0.33}$Ga$_{0.67}$As QW at different applied gate biases. The sample temperature was 2K.

lineshape is similar to that observed by Kuechler et al for 2DEGs in relatively narrow quantum wells.[8] The peak on the lower energy side (labelled E_g) we assign to recombination of electrons and holes across the bandgap at $k=0$, while the high energy side of this band displays two shoulders due to recombination of electrons at the Fermi energy. The higher energy shoulder (labelled F_{dir}) is assigned to k-vector conserving recombination of electrons and holes at the Fermi wavevector, while the lower energy shoulder (labelled F_{indir}) is due to recombination of electrons at the Fermi wavevector and holes at $k=0$. We suppose the latter, non-k-conserving process can occur due to interface roughness scattering. The feature to lower energy than the peak is ascribed to a shake-up process where optical recombination is accompanied by excitation of a 2DEG electron.

Reducing the gate bias causes the PL band to narrow due to the reduction in the Fermi energy, and the peak shifts to higher energies. At 0.1V a sharp peak ascribed to X$^-$ where two electrons in the bound state recombine with a hole[4–6]. At more negative gate biases a second peak emerges due to the neutral exciton (X). The separation of X and X$^-$ peaks yields a second electron binding energy of 2.1meV for 100Å QW.

Fig. 1b plots the PLE spectra measured on a 100Å QW under similar conditions to the PL. At the lowest gate

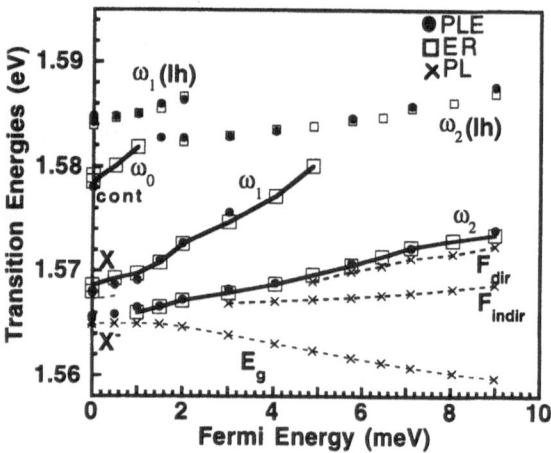

Fig. 2 Electron density dependence of the transition energies observed in the PL (crosses), PLE (solid circles), and ER (open squares) spectra.

bias (-0.3V) the PLE lineshape displays the well known form for undoped QWs. Fig. 1b demonstrates how the absorption spectrum of a QW evolves with electron density between the low and high density extremes. Notice that a second peak develops below X with increasing electron density. The splitting of this peak from X agrees closely with the energy separation of X and X^- in the PL. With further increasing electron density both the X and X^- peaks shift to higher energy. The X peak decreases in intensity with increasing carrier density and becomes difficult to resolve at gate biases above +0.3 V, for which the electron density is 8×10^{10} cm^{-2}. On the other hand, X^- evolves smoothly into the Fermi energy singularity seen at high carrier density. Notice that a X^- feature can also be seen to develop below the light hole X peak. This is more clearly resolved in the ER spectra which are plotted in [7].

Figure 2 plots the electron density dependence of the PL, PLE. and ER peak energies measured on the 100Å QW. The electron density is deduced from the separation of the PL peak and the non k-vector conserving Fermi energy transition. Notice that the PL peaks shifts to lower energy with increasing electron density. This is due to a combination of the Stark effect due to the electric field induced by the charge in the well and bandgap renormalisation. [9,4] The two shoulders on the PL band due to the Fermi edge show an increasing separation from the bandgap, due to the increase in the Fermi energy. Notice that the lowest energy peak of the PLE at intermediate and high carrier densities tracks the k-conserving Fermi energy transition in the PL, as expected for the Fermi energy transition. This again emphasises that at low carrier densities the Fermi energy transition evolves into the negatively charged exciton.

Our experimental results show the exisitence of all three thresholds (ω_2, ω_1, and ω_0) in the absorption spectra of a 2DEG. These three thresholds correspond to final states in which the hole is bound to two, one or zero electrons, respectively. In the low density limit, these states evolve to X^-, X and the continuum edge (unbound exciton), respectively. The ω_1 and ω_0 thresholds were calculated to be strongly suppressed when interactions were treated in a self-consistent Hartree calculation.[3] Consistent with this we see a weakening of the ω_1 threshold at densities around 10^{11} cm^{-2}, while the ω_0 threshold disappears at very low carrier densities. Notice too that a pronounced broadening appears on the high energy side of the lowest energy threshold of the PLE. This is the Auger tail predicted by Combescot and Nozieres[1] due to absorption assisted by a shake-up process where a 2DEG electron is promoted to an unoccupied state above the Fermi energy.

The ω_1 transition, which evolves from the neutral exciton at low carrier densities, shifts much more rapidly to higher energy with increasing density than the ω_2 transition. This is also true of the ω_0 transition, although the transition anticrosses with the ω_1(lh) ω_2(lh) transitions. The shift of ω_1, and ω_0 are not consistent with that predicted by Hawrylak. The shift of the ω_1 transition with increasing Fermi energy is assumed to be due to a direct excitation of an electron from the valence band to conduction band at twice the Fermi wavevector, in addition to the scattering of an electron from the bound state to the Fermi Edge. A similar non-k conserving transition also explains the shift of ω_0 with Fermi energy.

4 Summary

We have observed three thresholds in the absorption spectra of GaAs/Al$_x$Ga$_{1-x}$As QWs at finite electron densities. The thresholds all increase in energy with electron density, and show broadening effects at high electron density due to the shake up process of the Fermi electrons.

References

1. M. Combescot, and N. Nozieres, Journal de Physique, 32, 913 (1971).
2. P. Hawrylak, Phys. Rev. B, 44, 3821 (1991).
3. S.A. Brown et al, Phys. Rev. B 54, R11082 (1996).
4. A. J. Shields, M. Pepper, D. A. Ritchie, M. Y. Simmons, and G. A. C. Jones, SL and Microstruct. 15, 355 (1994).
5. A. J. Shields, M. Pepper, D. A. Ritchie, M. Y. Simmons, and G. A. C. Jones, Phys. Rev. B 52, 7841 (1995).
6. G. Finkelstein, H. Shtrikman, and I. Bar-Joseph, Phys. Rev. Lett. 74, 976 (1995).
7. R. Kaur, A. J. Shields, J. L. Osborne, M. Y. Simmons, D. A. Ritchie, and M. Pepper, Phys. Stat. Sol. 178, 465 (2000).
8. R. Kuechler et al, Semicond. Sci. and Technol. 8, 88 (1993).
9. Delalande et al, Phys. Rev. B 34, 2482 (1987).

Observation of heavy-hole minibands in ultra-short period superlattices

M. Eckardt[1], W. Geisselbrecht[1], S. Malzer[1], G.H. Döhler[1], K. Maranowski[2] and A.C. Gossard[2]

[1] Institut für Technische Physik I, Universität Erlangen-Nürnberg, Germany
[2] Materials Department, University of California, Santa Barbara, CA, USA

Abstract In this contribution we report on a rather exotic scenario where miniband formation in short-period superlattices (period d = 2 to 3nm) has unambigously been observed for heavy holes, for the first time, as we believe. In a study of electro-absorption in parabolic quantum wells (PQWs) we have observed a clear, but unexpected, structure at photon energies and fields which apparently are too high for transitions between *confined* valence and conduction band (harmonic oscillator) states to be possible. We have been able to solve this puzzle by taking into account the superlattice nature of the digital alloy (DA)-technique [1], used for the growth of the parabolic quantum wells and considering the resulting heavy-hole miniband structure.

1 Introduction

A simple nearly-free-electron estimate of miniband energies in short period superlattices yields the energy $E = \frac{\hbar^2}{2m^*}\left(\frac{\pi}{d}\right)^2$ for the center of the first miniband gap. For the GaAs conduction band (effective mass $m^* = 0.067m_o$) and periods d in the range of 2 to 3 nm (corresponding to about 7 to 11 monolayers only) the first minigap is expected to be centered at about 0.55 to 1.25 eV above the conduction band edge. These large values (which even exceed typical barrier heights), in combination with complications due to the realistic conduction band structure (Γ-X-valley mixing of well and barrier material, e.g.), are probably the reasons why miniband formation in such ultra-short period superlattices - to the best of our knowledge - has never been observed.

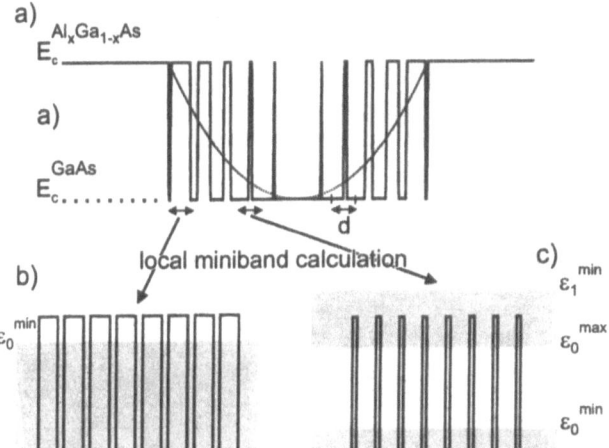

Fig. 1 (a) Schematic conduction band-edge profile of a PQW grown by the "digital-alloy" technique. (b) and (c) Typical local miniband structure in the region of wide and narrow barriers, respectively.

2 Theoretic Modeling

Parabolic quantum wells are usually grown by the "digital alloy"-technique [1], i.e. by a suitable variation of the barrier/well ratio in a short-period $GaAs/Al_xGa_{1-x}As$ superlattice with constant period d and Al concentration x in the barriers along the growth (z-) direction (Fig. 1(a)). In order to comprehend the digital effect on the electronic structure it is instructive to consider a z-dependent miniband structure. Based on the local barrier/well-width ratio a z-dependent miniband structure can be determined. In addition to the usually (exclusively) considered parabolically z-dependent lowest miniband edge, z-dependent miniband widths and miniband gaps are found (Fig. 1(b) and (c)).

For the theoretical calculation of our electro-absorption spectra the Schroedinger equation was solved numerically within the envelope function approximation for the realistic band edge profiles of the DA-structure with the respective applied external electrical fields. The broadening of the electronic states was accounted for by using a microscopic model of monolayer fluctuations [5].

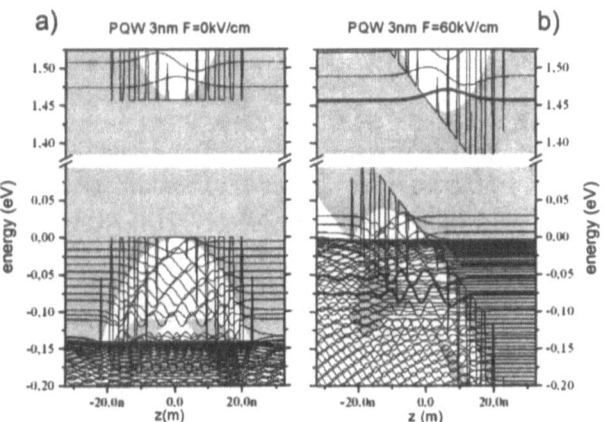

Fig. 2 (a) Local miniband structure calculated for a 44nm broad digital-alloy-PQW with superlattice period d = 3nm and x=0.03..0.33 at zero electric field. The boundary between shaded and white region indicates the effective parabolic potential which corresponds to the local upper heavy-hole miniband edge ϵ_{hh}. The lower shaded region indicates the local heavy-hole miniband gap. (b) same miniband structure, tilted by an external field. The (confined) Wannier-Stark states of the uppermost heavy-hole miniband and the conduction band state contributing to the unexpected structure are shown by heavy lines. (The discrete structure of the heavy hole continuum states is an artefact of the finite size of the model used in the calculations).

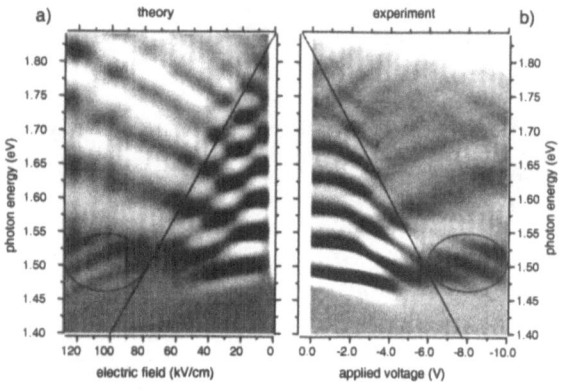

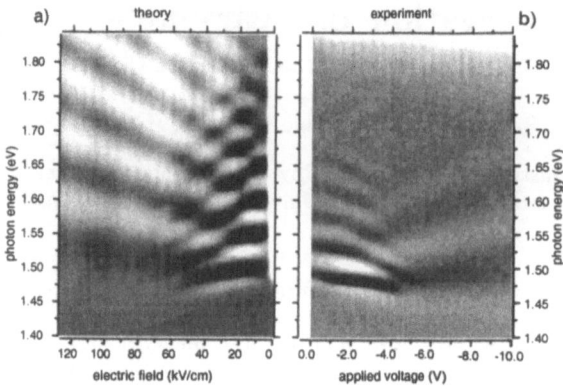

Fig. 3 Gray scale plot of electroabsorption spectra of a digital alloy PQW with superlattice period d = 3nm. (a) Theory (b) Experiment. The dashed line defines the range for which confined initial and final states for interband absorption exist in the parabolic quantum well, tilted by the electric field. The unexpected structure in the range of normally un-bound heavy-hole states, due to heavy-hole miniband formation, is encircled.

3 Digital Effect

At zero field the first 8 heavy hole harmonic oscillator wave functions are hardly affected by the finite miniband width as can be seen from Fig. 2a). When an external field is applied, the whole real-space band structure becomes tilted. At the fields where the unexpected structure in the electro-absorption is experimentally observed, strongly confined Wannier-Stark states of the first heavy hole miniband are, in fact, found (Fig. 2b). (Note that *continuum* hole states would be expected at these energies if there was no confinement by the miniband gap!). These states exhibit their largest overlaps with the two lowest (confined, harmonic oscillator) conduction subband states of the parabolic quantum well (which is practically unaffected by the miniband formation).

4 Experimental and Theoretical Results

Fig. 3 shows a comparison between the experimental and theoretical results for a digital alloy PQW with superlattice period d=3nm. As can be seen the theory perfectly reproduces all the features found in the experiment [2][3], including the unexpected structure due to the heavy-hole miniband formation. Besides the perfect agreement between theory and experiment the cause of the digital effect was further proven by electro-absorption measurements of a *analog* PQW. In contrast to the digital PQW an analog PQW is grown by a variation of the Al-concentration during the growth resulting in a smooth potential profile. This variation is obtained by changing the temperature of the Al-cell in a rather sophisticated way during the growth [4]. The electro-absorption spectra of this PQW do not exhibit the digital effect although a Franz-Keldysh-oscillation is found at the correspond-

ing position in the spectrum (Fig. 4b). Moreover, calculations for the corresponding analog PQW confirm that the "digital alloy" structure no longer shows up in the theoretical results (Fig. 4a). Finally, calculations and experiments for digital alloy PQWs with different superlattice period, even for d = 2 nm, confirm the expected variation of the "digital alloy" structure.

5 Conclusion

This observation is quite remarkable as it demonstrates that the superlattice in a digital alloy PQW is still perfect enough for the formation of minibands, although the period consists only of 7 to 11 monolayers. It should be noted also that our results are a unique effect of the digital alloy PQWs in high electric fields allowing for transitions from valence band minibands to harmonic oscillator states in the conduction band. Similar results for uniform superlattices of the same period exhibiting the familiar Wannier-Stark fans are *not* observable, because of the extremely wide conduction minibands associated with such short superlattice periods. Our interpretation in terms of heavy-hole miniband formation is corroborated by the excellent agreement between experiment and theory. In particular, the theoretically expected strong dependence of unexpected structure in the electro-absorption spectra on the superlattice period is quantitatively reproduced by the experiments and the structure is missing in PQWs grown by the „analog alloy" technique.

Fig. 4 Plot of electroabsorption spectra of an analog PQW. (a) Theory (b) Experiment. The digital effect is not observed in this "perfectly grown" PQW.

References

1. R.C. Miller, A.C. Gossard, D.A. Kleinman, O. Munteanu, Phys.Rev.B **29**, 3740 (1984)
2. W. Geisselbrecht et.al, Proceedings of the 24th International Conference on the Physics of Semiconductors, World Scientific, Singapore, (1999)
3. W. Geisselbrecht et.al, Superlattices and Microstructures **23**, 93, (1998)
4. W. Geisselbrecht, Doctoral Thesis, Erlangen 1999
5. M. Eckardt, Diploma Thesis, Erlangen 1999 (unpublished)

CdTe quantum wells as ideal system for the study of negatively charged excitons: A spectral and temporal analysis.

V. Ciulin[1], P. Kossacki[1,2], M. Kutrowski[3], A. Esser[4], S. Haacke[1], J.-D. Ganière[1], T. Wojtowitcz[3], B. Deveaud[1].

[1] Physics department, Swiss Federal Institute of Lausanne-EPFL, 1015 Lausanne, Switzerland.
  e-mail: ciulin@dpmail.epfl.ch
[2] Institute of Experimental Physics, Warsaw University, Hoza 69, Warsaw, Poland.
[3] Institute of Physics, Polish Academy of Sciences, Al. Lotników 32/46, Warsaw, Poland.
[4] Institut für Physik, Humboldt-Universität zu Berlin, AG Halbleitertheorie, Germany.

Abstract The negatively charged exciton lineshape and lifetime are measured simultaneously as a function of temperature in CdTe quantum wells at low electron concentrations $n_e < 2 \times 10^{10}$ cm^{-2}. We find that above 7 K both features are well described by considering delocalized and thermalized charged excitons.

1 Introduction

Negatively charged excitons (X^-), the semiconductor analogues of H$^-$, are now commonly observed in spectra of n-doped QWs [1,2]. For CdTe QWs, they appear 2-3 meV below the neutral exciton (X) spectral line. This relatively large X^- dissociation energy, compared to GaAs QWs, allows here the study of X^- photoluminescence (PL) over a significant temperature range where X^- thermal dissociation and localization effects are of minor importance. We present simultaneous measurements using picosecond time-resolved spectroscopy of both the lineshapes and lifetimes of X^- for quite low electron concentrations $n_e < 2 \times 10^{10}$ cm^{-2} in a modulation doped QW. Resonant excitation with X^- was used to enhance the signal and to allow for a precise measure of the lifetimes [3]. Our results can be summarized as follow: we observe an asymmetry on the X^- spectral line with a temperature dependent low-energy tail. In the time domain, the X^- lifetime is observed to increase linearly with temperature above 7 K. These experimental findings are well reproduced when considering that X^- with high k-vectors are allowed to recombine radiatively [4,5] and that they are thermalized.

2 Experimental observations

The sample is a one side graded modulation doped CdTe/-Cd$_{1-x}$Mg$_x$Te heterostructure containing a single 80 Å QW [6]. The inset in Fig. 1 shows cw PL and reflectivity spectra for an estimated electron density of $n_e < 2 \times 10^{10}$ cm^{-2}. Both X and X^- are seen in the reflectivity spectra thus confirming their intrinsic origin. Furthermore, they were unambiguously identified as they follow the appropriate polarisation selection rules in magnetic field [1].

The sample was excited by a 2 ps pulsed laser of 80 MHz repetition rate tuned in resonance with the X^- en-

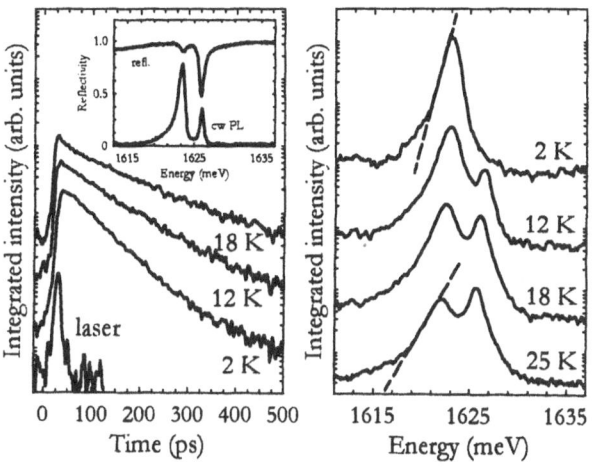

Fig. 1 Spectra for $n_e < 2 \times 10^{10}$ cm^{-2}: inset: cw PL and reflectivity spectra. Main panel: for resonant excitation with X^- and for several temperatures (a) Time-resolved PL decays integrated over X^- emission (b) Time-integrated spectra over about 800 ps starting some 10 ps after the initial laser excitation.

ergy. The linearly cross-polarized PL was collected by a streak camera which yields two dimensional images containing both temporal and spectral informations. Fig. 1a and 1b show, for an electron concentration $n_e < 2 \times 10^{10}$ cm^{-2}, the extracted time-resolved PL decays of X^- and the corresponding spectra for several temperatures. The time-integrated signal has no contribution from the exciting laser as it was excluded from the integration. X^- PL is seen to exhibit an exponential low-energy tail whose importance increases with temperature. To quantify this effect, the X^- lineshape was fitted by a low-energy decaying exponential convoluted with a Gaussian broadening function which reproduces well the low-energy part of the line. The exponential decay constants E$_{\text{tail}}$ from the best fits are shown in Fig. 2a.

The PL decays in Fig. 1a show a fast rise time and a mono-exponential decay. This latter is generally interpreted as a direct measurement of the radiative lifetime of X^- [7-9] when non-radiative processes and thermal dissociation of X^- can be neglected. Here we cannot neglect thermal dissociation of X^- at high temperatures

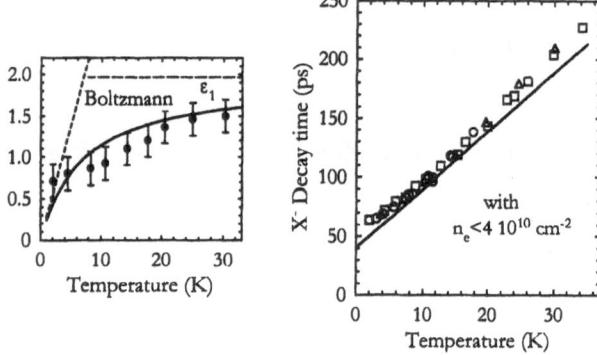

Fig. 2 E_{tail} and X^- decay times as a function of temperature, full lines give the model predictions (a) Fitted values of the spectral tail decay constants E_{tail} at $n_e < 2 \times 10^{10}\ cm^{-2}$, dash lines give the linear Boltzmann and the flat ϵ_1 contributions. (b) Decay time constants of the X^- time-resolved PL for $n_e < 2 \times 10^{10}\ cm^{-2}$ (open circles) and for two other electron concentrations with $n_e < 4 \times 10^{10}\ cm^{-2}$.

as the electron concentration is low. Indeed, in Fig. 1b, X is seen to dominate the PL spectra at 25 K. To decrease this effect, we have measured the X^- PL decays for slightly higher electron concentrations. The extracted mono-exponential decay constants for the lowest and two higher electron concentrations are presented in Fig. 2b. These X^- PL decays times are seen to be independent of the electron concentration (for $n_e < 2 \times 10^{10}\ cm^{-2}$ this is true up to about 20 K) and thus give a direct measure of X^- radiative lifetime. This latter is seen to increase linearly with temperature from about 7 K.

3 Model and discussion

The recombination of X^- is described by the optical matrix element of the transition $M(k)$ [4,5]. As the electron takes away the X^- center of mass momentum k (when neglecting the photon momentum), X^- with large k-vectors can recombine radiatively. We assume a Boltzmann thermal distribution for the X^- population. This is justified by the low densities of the photoexcited X^-. A fast thermalization is indicated by the fast X^- PL rise times and the following mono-exponential decays. The intensity of the PL can then be written as [5]:

$$I_{X^-}(\hbar\omega) \propto |M(k)|^2 \exp\left(-\frac{\epsilon}{k_B T}\frac{m_e}{M_X}\right)\theta(\epsilon)\,, \qquad (1)$$

where the photon energy is given via $\epsilon = E_{X^-}(k = 0) - \hbar\omega = \hbar^2 k^2 M_X/(2m_e M_{X^-})$ and θ is the Heaviside step function. $|M(k)|^2$ can be approximated by $\exp(-\epsilon/\epsilon_1)$ with $\epsilon_1 = 1.95\ meV$ being a material parameter calculated for our CdTe QW [5,10]. The spectral decay constant can be extracted from Eq. 1 and is given by $1/E_{tail} = (1/k_B T)(m_e/M_X) + 1/\epsilon_1$. These E_{tail} values are plotted as a full line in Fig. 2a and the two independent contributions arising from the Boltzmann distribution and the optical matrix element are shown as dashed lines. The model compares well with the experimental values specially when considering the many other possi-

ble contributions to low energy tails, e. g. arising from localization effects.

The radiative decay time is inversely proportional to the average of the emission probabilities of all populated k-states given in Eq. 1. The analytical analysis gives, A being a material dependent prefactor [5,10]:

$$\tau_{X^-} = A\left(k_B T + \epsilon_1 \frac{m_e}{M_X} + \frac{3}{5}\frac{M_{X^-}}{M_X}E_0\right)\,. \qquad (2)$$

Here we have also considered the finite photon momentum (light-cone effect), which gives $E_0 = 0.08\ meV$. The radiative lifetime increases linearly with temperature as the maximum recombination probability is around $k=0$ and the population of these states decreases with temperature. The result of this calculation is plotted as a full line in Fig. 2b and reproduces well the experimental decay times within the measuring error which increases with temperature from 3-10 ps. The main deviation between experiment and model occurs below 7 K and is interpreted as being due to localization effects which disappear when raising the temperature. As for the PL decay mono-exponential dependence, it is interesting to note that within the model, all the X^- k-states are able to decay radiatively. Thus, when thermalizing X^- does not scatter out of the radiative zone the way X does. For X, this phenomena is in part responsible for its non mono-exponential PL decay.

4 Conclusion

In conclusion, we have measured the optical properties of X^- PL as a function of temperatures in the limit of low electron densities in CdTe QWs. We find that our observations can be understood by considering the particular way in which X^- couples to light.

References

1. K. Kheng R. T. Cox, M. Y. d'Aubigné, F. Bassani, K. Saminadayar, S. Tatarenko, Phys. Rev. Lett. **71**, (1993) 1752.
2. A. Manassen, E. Cohen, A. Ron, E. Linder, L.N. Pfeiffer, Phys. Rev. B **54**,(1996) 10609.
3. B. Deveaud, F. Clérot, N. Roy, K. Satzke, B. Sermage, D. S Katzer, Phys. Rev. Lett. **67**,(1991) 2355.
4. B. Stébé, E. Feddi, A. Ainane, F. Dujardin, Phys. Rev. B **58**, (1998) 9926.
5. A. Esser, E. Runge, R. Zimmermann, W. Langbein, Phys. Rev. B **15**, (2000).
6. T. Wojtowicz, M. Kutrowski, G. Karczewski, J. Kossut, Appl. Phys. Lett. **73**, (1998) 1379.
7. G. Finkelstein, V. Umanski, I Bar-Joseph, V. Ciulin, S. Haacke, J. D. Ganiere, B. Deveaud, Phys. Rev. B **58**, (1998) 12637.
8. E. Vanelle, M. Paillard, X. Marie, T. Amand, P. Gilliot, D. Brinkmann, R. Lévy, J. Cibert, S. Tatarenko, Phys. Rev. B **62**, (2000) 2696.
9. R. A. Hogg, D. Sanvitto, A. J. Shields, M. Y. Simmons, D. A. Ritchie, M. Pepper, Physica B **272**, (2000) 412.
10. V. Ciulin, P. Kossacki, M. Kutrowski, A. Esser, S. Haacke, J. D. Ganiere, T. Wojtowitcz, B. Deveaud, sumitted to Phys. Rev. B.

Proc. 25th Int. Conf. Phys. Semicond., Osaka 2000 (Eds. N. Miura and T. Ando)

Collapse of the excitonic states at $r_s = 8$ in high quality GaAs/AlGaAs single quantum wells.

S. I. Gubarev[1], I. V. Kukushkin[1,2], S. V. Tovstonog[1], M. Yu. Akimov[1], L. V. Kulik[1,2] J. Smet[2], K. v. Klitzing[2], W. Wegscheider[3]

[1] Institute of Solid State Physics, RAS, Chernogolovka, 142432 Russia
[2] Max-Planck-Institut für Festkörperforschung, Heisenbergstr. 1, 70569 Stuttgart, Germany
[3] Walter Schottky Institute, Technische Universität München, Am Coulombwall, D-85748 Garching, Germany

Abstract The collapse of the excitonic states due to a very low carrier density of 2D-channel has been studied in GaAs/AlGaAs single quantum well (SQW) as a function of carrier density for electron and hole channels. The collapse of exciton states in high quality GaAs/AlGaAs SQW have been observed at surprisingly low electron density $n_e = 5 \cdot 10^9$ cm^{-2} which corresponds to dimensionelness parameter $r_s = 8$. This value is in a dramatic contrast with a previous finding as well as with the value $r_s \approx 2$ found in the 3D electron system.

1 Introduction

The screening of Coulomb interaction between electron and hole by mobile free carriers leads to a collapse of exciton states at certain carrier concentration when formation of excitonic states become impossible. In bulk material the carrier density at which this transition occurs corresponds to the average interparticle distances $r_s \approx 2$ expressed in units of the exciton Bohr radius [1]. The simple theoretical estimations indicate that screening of excitons by ideal 2D carriers is less effective than that in 3D system and collapse of exciton states occurs at $r_s \approx 1.9$ which corresponds to electron density of order 10^{11} cm^{-2} in GaAs/AlGaAs structures [2].

In this work we report the results of the exciton collapse study in high quality GaAsAlAs/AlGaAs SQWs detected by both the reflectivity and luminescence spectra measured simultaneously on the structures with electron and with hole conductivity channel. In contrast to the previous finding [3] we have found that in high quality AlGaAs structures with mobilities larger than $\mu_e \approx 5 \cdot 10^6$ the collapse of exciton states occurs at much lower electron density $n_e \approx 5 \cdot 10^9$ cm^{-2}, which corresponds r_s parameter as much as 8.

2 Experimental results and discussion

The reflectivity and luminescence spectra, as well as dimensional magnetoplasma resonance (DMR) have been measured simultaneously by the use of three fiber optical scheme for several high quality GaAs/AlGaAs SQWs of p- and n-types with the well width in the range from 200 Å to 300 Å.

The concentration of free carriers in the 2D-channel was smoothly varied by photodepletion effect, which allows for some samples even change the type of carri-

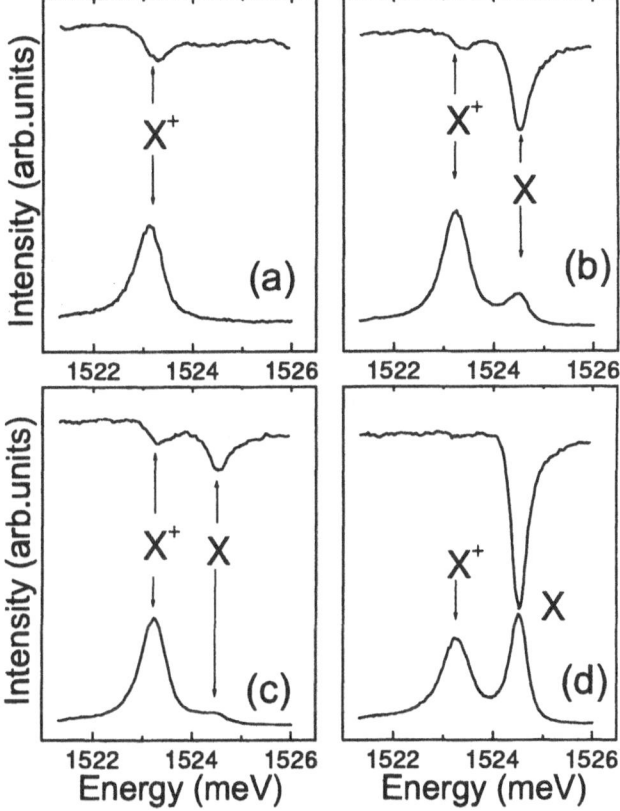

Fig. 1 Transformation of luminescence and reflectivity spectra of SQW with decrease of 2D carrier density in slightly doped SQW. Arrows indicate the positions of exciton X and localized trion X^+.

ers and pass through the zero concentration. The mw-ODDMR method in the frequencies range from 10 GHz to 40 GHz enables us a very accurate measurements of 2D-concentrations both for electron and hole channels down to extremely low carrier densities (as low as 10^9 cm^{-2}), at which the collapse of excitons occurs. The linewidth of the magnetoplasma resonance provides also the measurement of the carriers scattering time τ, which probes the sample quality. The collapse of excitons has been detected as a very sharp disappearance of the excitonic line both from reflectivity and luminescence spectra at the well defined threshold carrier density (see Fig. 1). At this density the exciton line, which dominates

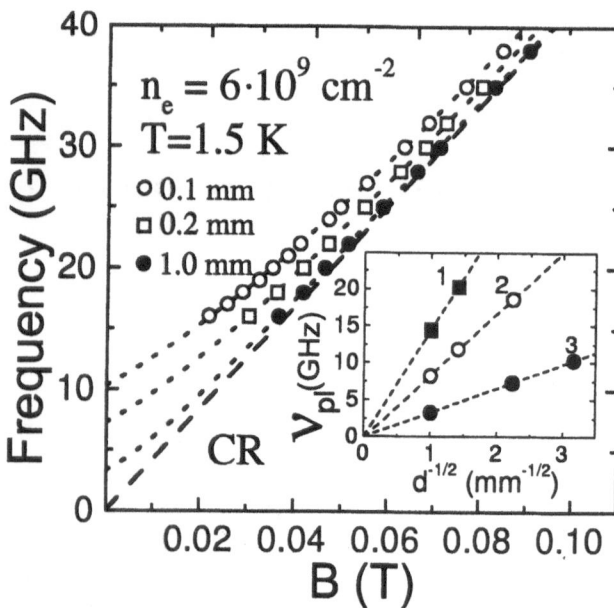

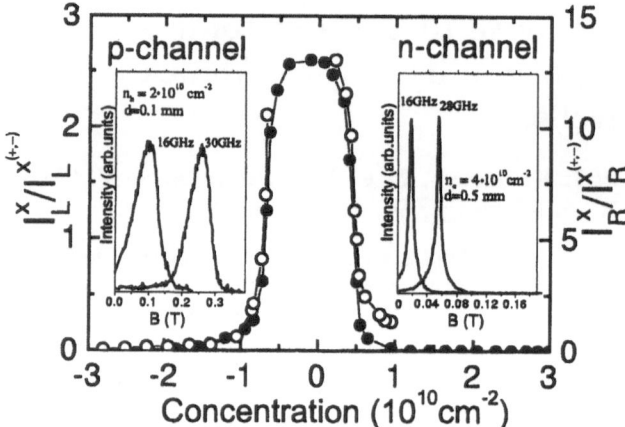

Fig. 3 The ratio of free exciton to bound trion intensities as function of 2D electron (hole) concentration measured from luminescence (solid circles) and reflectivity spectra (open circles). The inserts show profiles of DMR in the case of n channel (electron DMR) and in the case of p-channel (hole DMR).

Fig. 2 The field dependence of DMR resonance measured on mesa with different diameters d at the same 2D electron density. The insert shows the dependence of the plasma frequencies ν_{pl} on the mesa diameter d for 3 different 2D electron concentrations 1 -$n_e = 3 \cdot 10^{11}$ cm^{-2}, 2 -$n_e = 4 \cdot 10^{10}$ cm^{-2}, 3 - $n_e = 6 \cdot 10^{9}$ cm^{-2}.

in the both luminescence and reflectivity spectra at low concentrations, disappears from spectra and instead the line of negatively (positively) charged trion X^- (X^+) occurs which transferred to the line of 2DEG or 2DHG with increase of concentration of 2D carriers.

Both magnetooptical and magnetotransport methods fail to measure carrier concentration at low carrier densities less than $3 \cdot 10^{10}$ cm^{-2} To obtain 2D carrier density at extremely low concentrations we have used the technique of dimensional magnetic resonance (DMR). The coupling between cyclotron and plasma modes in confined 2D channel causes the frequency of the magneto-plasma resonance depends on both the size of the structure and carrier concentration and for disk shaped mesa with diameter d the DMR frequencies can be written as [4, 5]

$$\omega_{DMR} = \omega_{CR}/2 \pm \sqrt{\omega_p^2 + (\omega_{CR}/2)^2} \qquad (1)$$

where $\omega_{CR} = (eH)/(m^*c)$ is usual CR frequency and $\omega_p^2 = (3\pi^2 n_s e^2)/(4m^*\epsilon_{eff}d)$ is plasma frequency of 2DEG with density n_s in mesa with diameter d.

The frequencies of DMR upper branch for mesastructures of different diameters are displayed on Fig. 2. The typical resonance profiles for the n- and p-type of channel are shown on the Fig.3. The fitting by expression (1) gives the plasma frequency and allows an accurate measurement of electron density. For the best structures the halfwidth of resonance profile was measured as small as 0.03 mT, which corresponds to the electron scattering time τ_e as long as 210 ps. At low concentration the scat-

tering time decreases but the ODDMR study is possible for electron densities as low as 10^9 cm^{-2}.

From threshold behavior of exciton line intensity we have found that collapse of exciton states occurs at electron density $n_e \approx 5 \cdot 10^9$ cm^{-2}, which corresponds to $r_s = 8$. The difference between 3D and 2D screening experiments is connected with the fact that in 3D case impossible to realize the screening of exciton state by a charged electron (hole) gas. In all experiments the screening of Coulomb interaction in 3D systems was studied when both types of charges persist in the system. In contrast to that in 2D systems the screening by a charged electron gas is possible because of charge separation in the space.

In conclusion the collapse of the free exciton states have been observed at surprisingly low concentration of 2DEG and 2DHG by study the luminescence and reflectivity spectra of high quality SQWs as function of carrier density. The method of ODDMR enables an accurate measurements of 2D threshold density at which the exciton collapse occurs. The critical concentration at which collapse of exciton occurs depends strongly on the sample quality and decreases for high quality samples.

3 Acknowledgements

The authors gratefully acknowledge the financial support by Russian Fund for Fundamental Research and INTAS grant N99-1146.

References

1. I. V. Kukushkin et al., Zh. Eksp. Teor. Fiz **84** 1145 (1984).
2. G. E. W. Bauer, Phys. Rev. B **45**, 9153 (1992).
3. G. Finkelstein, H. Strikman, and I. Bar-Josef, Phys. Rev. Lett. **74**, 976 (1995).
4. B. A. Allen, Jr., H. L. Stormer, and J. C. M.Hwang, Phys. Rev. B **28**, 4875, (1983);
5. B. M. Ashkinadze et al., Phys. Rev. Lett. **83**, 812 (1999).

Proc. 25th Int. Conf. Phys. Semicond., Osaka 2000 (Eds. N. Miura and T. Ando)

Phase sensitive detection of transmitted femtosecond pulses in GaAs quantum wells

R. Kawahara, T. Kuroda, Y. Mitsumori, and F. Minami

Department of Physics, Tokyo Institute of Technology,
Meguro-ku, Tokyo 152-8551, Japan

Abstract The phase characteristics of transmitted optical pulses were investigated in the energy- and the temporal domain using the spectrally resolved cross-correlation technique. The temporal evolution of fs pulses passing through GaAs/AlGaAs quantum wells indicates a fairly long tail due to the coherent emission from the exciton. The phase discontinuity was observed in the time-domain data. The phase of the coherent emission signal is found to differ by 180° from that of the incident light field.

1 Introduction

Transmission of ultrashort light pulses allows one to investigate coherent properties of semiconductors in a straightforward way. Up to now, several pulse propagation experiments have been reported, studying polariton effects [1], quantum beats [2], exciton dephasing [3] and so on. In those experiments, however, the temporal and/or spectral profiles of transmitted pulses have been measured, and phase information of the light field was lost. Since the resonance effect between light and matter is expected to distort the phase of the light pulse, the phase characterization of the transmitted pulse should provide valuable information on coherent properties of semiconductors.

In the present paper, we report the observation of the phase distortion of transmitted pulses in the excitonic resonance region. The experiments were based on the frequency-resolved optical gating (FROG) which was recently developed to characterize phase property of the laser pulses [4]. In this method, a nonlinear auto-correlation signal is spectrally resolved, and from this data both of the phase and the amplitude of the light field are reconstructed. We applied this technique to a spectrally resolved second harmonic cross-correlation (SHG-crossFROG). We determined the phase of the weak transmitted pulse as a function of energy and time with the SHG-crossFROG method.

2 Experiment

The sample used in this work was GaAs/AlGaAs multiple quantum wells (MQW's) consisting of 20 periods of 8-nm-wide GaAs wells alternating with 10-nm-thick $Al_{0.3}Ga_{0.7}As$ barriers. It shows a heavy-hole (hh) exciton line peaked at ~1.57 eV, with a linewidth of ~3.6 meV. For excitation we used a fs mode-locked Ti-Sapphire laser. The wavelength was tuned to cover the hh-exciton line.

In the experiment, we first measured the spectrally resolved auto-correlation signals of the incident pulse. With use of the FROG algorithm, namely, iterative Fourier transform algorithm with generalized projection [4], we determined the phase of the incident pulse, together with the am-

plitude, as a function of energy and time. Then we measured the spectrally resolved cross-correlation between the transmitted pulse passing through the sample and the incident pulse, which had been characterized in advance. The phase characteristics of the weak transmitted pulse were reconstructed using a modified FROG algorithm developed in [5]. For the correlation measurements, we utilized a 300-μm thick KDP crystal. All experiments were performed at the temperature of 5K.

3 Results and Discussion

The intensity of the incident pulse and its phase are shown in Fig. 1(a) as a function of energy. These are also represented as a function of time in Fig.1 (b). It is found that the present laser pulse is not transform-limited but linearly chirped. The phase is observed to change gradually both in spectral domain and in time domain. The magnitude of chirping is around 100 degree.

Figure 2(a) illustrates the spectrum and the phase of the transmitted pulse in resonance with the hh-exciton. The anomalous spectrum, compared with that in Fig. 1(a), should arise from the excitonic absorption peaked at ~ 1.572 eV. Furthermore, the phase distortion of ~ 60° is clearly observed at the resonance region. Since the phase distortion is considered to reflect the dispersion of refractive index at the resonance region, we calculated an expected phase change based on the Lorentzian oscillator model. In this treatment, the phase anomaly $\Delta\phi(\omega)$ is given by,

$$\Delta\phi(\omega) = \frac{2\pi}{\lambda} d \times \Delta n(\omega), \qquad (1)$$

$$\Delta n(\omega) = \text{Re}\left[\frac{f_0}{(\omega - \omega_0) - i\gamma/2}\right], \qquad (2)$$

where λ and d represents the wavelength and sample thickness, respectively. The calculated phase profile is indicated by the dotted line, together with the calculated transmission spectrum using the same parameters. Both of the calculated curves are found to agree well with the experimental data. This fact shows that the light absorption appearing in the fs resonant transmitted pulse originates from a coherent process.

The temporal shape of the transmitted pulse and its phase are shown in Fig. 2(b). The temporal shape indicates a long tail, which is not found in the incident pulse shape (Fig. 1(b)). This tail should arise from the free induction decay of the excitonic polarization. Furthermore, the dip-profile is

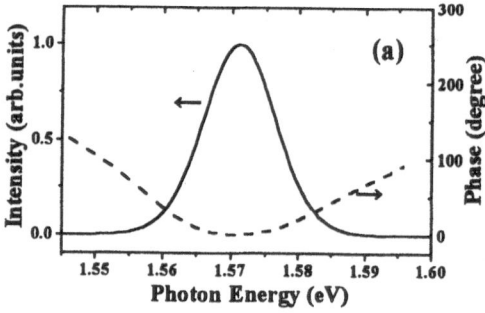

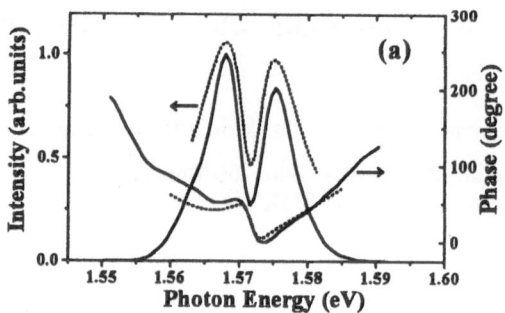

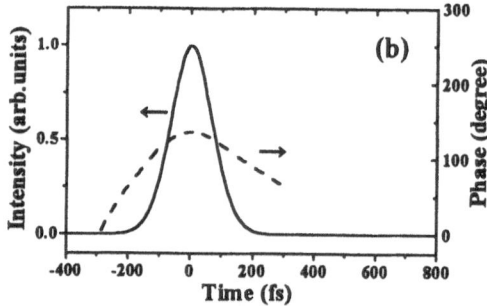

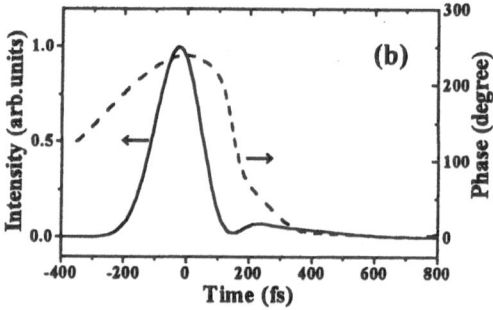

Figure 1. Change of the phase of the incident pulse together with the intensity in spectral domain (a), and time domain (b). The solid line refers the intensity, and the dashed line to the phase.

Figure 2. Change of the phase of the transmitted pulse together with the intensity in spectral domain (a), and time domain (b). The solid line refers the intensity, and the dashed line to the phase. The dotted line in (b) represents the calculated curves based on the Lorentzian model. The parameters used are, $f_0=1.5\times10^{12}$, $\omega_0=2.39\times10^{15}$ $[s^{-1}]$, $\gamma=4.6\times10^{12}$ $[s^{-1}]$.

apparently observed at ~ 140 fs. The appearance of the dip profile, known as the coherent dip, has been also observed in [1], and was interpreted by the destructive interference effect between the input light field and the coherent emission signal. Here, we can determine the phase of the transmitted pulse, and actually found the phase change appearing at this time delay, as shown in Fig. 2(b). The magnitude of the phase change is almost 180°. The result shows that the phase of the coherently emitted light is 180° out of phase in comparison with that of the incident light field, as has been theoretically expected in [6]

4 Conclusion

With use of the SHG-crossFROG technique, we determined the intensity and the phase of the transmitted pulse passing through MQW's. In the energy domain, we have observed the phase distortion around the exciton resonance. The spectral phase distortion is in agreement with the theoretical calculation based on a simple Lorenzian model. Furthermore, we have observed the coherent dip and the temporal phase discontinuities of ~ 180°. Such phase anomalies reflect the phase difference of the electrical field between the input pulse and the coherent emission from the excitonic polarization.

Acknowledgment

This work was supported by a Grand-in-Aid for Scientific Research.

References

1. T. Mishina, and Y. Masumoto, Phys. Rev. Lett. **71**, (1993) 2785.

2. E. O. Göbel, K. Leo, T. C. Damen, J. Shah, S. Schmitt-Rink, W. Schäler, J. F. Müller, and K. Köhler, Phys. Rev. Lett. **64**, (1990) 1801.

3. L. Schultheis, A. Honold, J. Kuhl, and K. Köhler, Phys. Rev. B **34**, (1986) 9027.

4. R. Trebino, K. W. DeLong, D. N. Fittinghoff, J. N. Sweetser, M. A. Krumbëgel, and B. A. Richman, Rev. Sci. Insttrum. **68**, (1997) 3277, and references therein.

5. S. Linden, H. Giessen, and J. Kuhl, Phys. Stat. Sol. (b) **206**, (1998) 119.

6. A. Laubereau, and W. Kaiser, Rev. Mod. Phys. **50**, (1978) 607.

Oscillatory behavior of the Γ-X coupling with AlAs thickness in type II GaAs/AlAs heterostructures

C. Gourdon[1], D. Martins[1], V. Voliotis[1], P. Lavallard[1], E. L. Ivchenko[2]

[1] Groupe de Physique des Solides, Universités Paris 6 et 7, CNRS UMR 75-88
Tour 23 - 2, Place Jussieu - 75251 Paris cedex 05 - France
[2] A. F. Ioffe Physico-Technical Institute, Russian Academy of Sciences
194021 St Petersburg, Russia

Abstract We have succeeded to demonstrate experimentally the predicted oscillatory behavior of the Γ-X coupling with AlAs layer thickness in GaAs/AlAs/GaAs type II heterostructures. We have developed a model describing the Γ-X coupling for excitons localized in regions with layer thickness fluctuations. The amplitude of variation of the experimental radiative time $\tau_{\Gamma X}$ is well reproduced within this model.

1 Introduction

An oscillating dependence of the interband radiative time upon the AlAs layer thickness in a symmetrical heterostructure follows from the analysis of parity of the electron and hole lowest-subband states at $k_x = k_y = 0$. The point-group symmetry of the structure is D_{2d} with the origin of symmetry operations located at any atom in the central atomic plane of the AlAs layer which is an As plane if the number M of AlAs monomolecular layers (ML) is even and an Al plane if M is odd. Whatever the value of M, the lowest hole subband states $|hh1, \pm 3/2\rangle$ transform according to the Γ_6 spinor representation of the D_{2d} group. In contrast, the electron states $|e1, \pm 1/2\rangle$ correspond either to the Γ_6 or to the Γ_7 representation depending on the parity of M. Indeed, they contain an X-like contribution with the bulk Bloch function of symmetry X_1 and an even envelope function $\psi_X(z)$. It is known that, under the reflection rotation S_4 that changes z into $-z$, the Bloch function $|X_1\rangle$ is invariant if the S_4 operation is centered at an anion (As) and changes sign if the origin of the S_4 operation is taken at a cation. Thus, if M is odd the functions $|e1, \pm 1/2\rangle$ transform according to the Γ_7 spinor representation and their Γ-like admixtures has an odd envelope function $\psi_\Gamma(z)$. As a result, for odd M, the overlap integral $\int \psi_\Gamma(z)\psi_{hh1}(z)dz$ vanishes and the interband optical transition $hh1 \longrightarrow e1$ is forbidden. This selection rule also applies if the $X1$-$X3$ mixing is taken into account [1].

In order to show the dependence of the radiative properties on the parity of AlAs MLs we have investigated the photoluminescence (PL) of GaAs/AlAs/GaAs type II pseudodirect double quantum wells (DQW) with GaAs and AlAs thickness in the range 3-7 MLs. The DQW is grown with a layer thickness gradient along the sample. Additional quantum wells are grown in the structure to provide accurate calibration of GaAs and

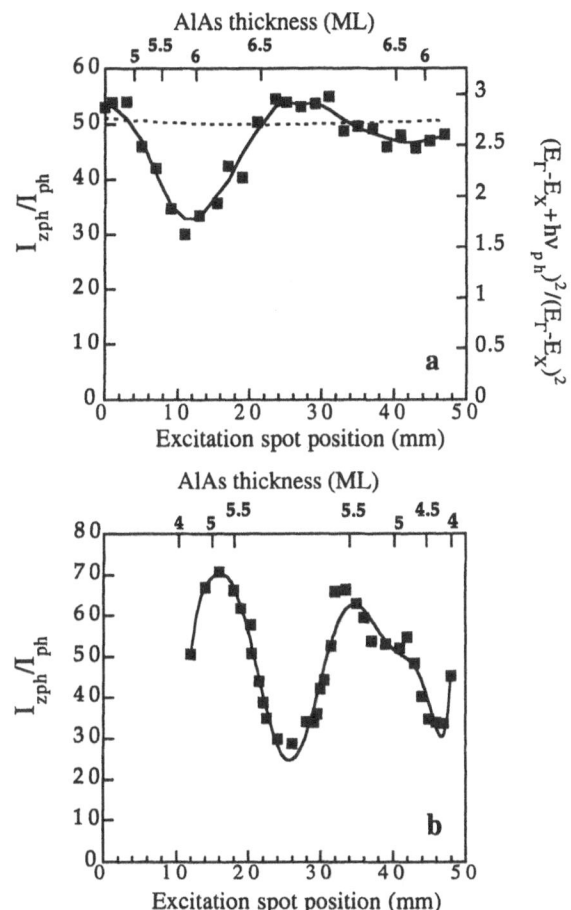

Fig. 1 (a) I_{zph}/I_{ph} (squares) and $(E_\Gamma - E_X + h\nu_{ph})^2/(E_\Gamma - E_X)^2$ (dotted line) as a function of position on sample A1. (b) I_{zph}/I_{ph} for sample C1. The solid lines are a guide for the eyes.

AlAs thicknesses by optical spectroscopy (OS) and transmission electron microscopy (TEM) [2,3].

2 Experimental results and model

We have studied the ratio I_{zph}/I_{ph} of the intensities of the zero-phonon (ZPH) line and its LO$_{AlAs}$ phonon (PH) replica. It is displayed in Fig. 1 for two samples. It shows a non-monotonous variation with AlAs thickness. In a simple approximation I_{zph}/I_{ph} is equal to the ratio of the radiative transition rate $w_{\Gamma X}$ due to the Γ-X coupling and the radiative rate w_{ph} due to the phonon-assisted transition. In second order perturbation theory, ignoring the AlAs ML dependence of $w_{\Gamma X}$, we have

$$\frac{w_{\Gamma X}}{w_{ph}} \propto \left| \frac{E_\Gamma - E_X + h\nu_{ph}}{E_\Gamma - E_X} \right|^2 , \qquad (1)$$

where E_Γ (E_X) are the energies of the $hh1 \longrightarrow e\Gamma1$-like($eX1$-like) transitions and $h\nu_{ph}$ the phonon energy. This ratio is plotted for sample A1 in Fig. 1. The experimental results show a markedly different behavior from this ratio. This is a clear evidence of the AlAs ML dependence of the Γ-X coupling. For sample C1 with optimized growth conditions and large AlAs gradient, the amplitude of variation of I_{zph}/I_{ph} is more than a factor of 2 between odd and even M and the oscillatory behavior with AlAs thickness is clearly observed. Localization of excitons in sites with an effective AlAs larger by about one ML than the average thickness determined by OS and TEM explains well the dependence of the radiative recombination rate on the parity of M.

To obtain more quantitative results, time-resolved PL was studied. The decay of the ZPH line and the PH replica is non-exponential. This is attributed to a distribution of $w_{\Gamma X}$ values due to localization of excitons in sites with a distribution of AlAs effective thicknesses. The analysis of the decay in those PL lines gives the Γ-X average radiative time $\tau_{\Gamma X} = (w_{\Gamma X})^{-1}_{av}$ and the radiative time $\tau_{ph} = (w_{ph})^{-1}$. The dependence of τ_{ph} on AlAs thickness is in very good agreement with the one calculated using second order perturbation theory. The dependence of $\tau_{\Gamma X}$ is in good accordance with the ratio I_{zph}/I_{ph} and clearly differs from that of τ_{ph}.

To obtain a quantitative description of the behavior of $\tau_{\Gamma X}$ we have developed a phenomenological model of the Γ-X coupling for excitons localized by layer thickness fluctuations. In the generalized envelope-function approximation, the coupling of Γ_1 and X_1 states is described by the δ-functional mixing potential $V_{\Gamma X}$ which acts on the envelope functions ψ_Γ, ψ_X and has the form [1]:

$$V_{\Gamma X}(z) = a_0 V \sum_m t_{\Gamma X} \eta(z_m) \delta(z - z_m) . \qquad (2)$$

Here the interfaces are taken as As planes at positions $z_m = m a_0/2$, a_0 is the lattice constant and m is an integer, $V = \hbar^2/(2m_0 a_0^2)$, m_0 is the free electron mass, $t_{\Gamma X}$ is a dimensionless coupling parameter, $\eta(z_m) = (-1)^m$ is the phase factor of the Bloch X zone-edge wavefunction. We have extended Eq. 2 to account for fluctuating or non-abrupt interfaces. A non-abrupt GaAs/AlAs interface is modelled by considering one ML of $Ga_{1-\xi}Al_\xi As$ alloy between pure GaAs and pure AlAs and defining an equivalent interface coordinate as $z_m = z_n + (1-\xi)a_0/2$ where z_n is the position of the As plane after the last pure Ga plane. From $\xi = 1$ to $\xi = 0$ the interface moves from $z_m = z_n$ to $z_m = z_{n+1} = z_n + (a_0/2)$, and the value of $t_{\Gamma X}(z_m) = t_{\Gamma X}\eta(z_m)$ jumps from $+t_{\Gamma X}$ to $-t_{\Gamma X}$, with arbitrary choice $\eta(z_n) = 1$. For any z_m between z_n and z_{n+1} we define a function $t_{\Gamma X}(z_m)$ which enters

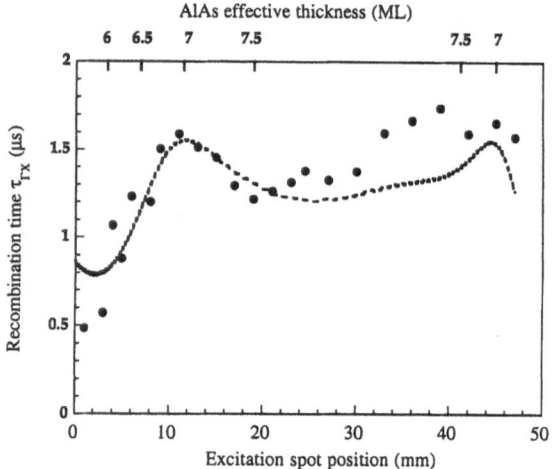

Fig. 2 Sample A1: radiative time due to the Γ-X coupling $\tau_{\Gamma X}$ (experimental: dots, calculated: dotted line).

Eq. 2 instead of $t_{\Gamma X}\eta(z_m)$ and continuously varies between the above limits. The properties of this function allow to expand it as $t_{\Gamma X}\sum_{p=0}^\infty C_p \cos \pi(2p+1)z_m$ with $\sum C_p = 1$. In order to simplify the calculations we set $t_{\Gamma X}(z_m) = t_{\Gamma X} \cos \pi z_m$. For each y-position along the sample the Γ-X coupling matrix element for a single localized exciton is calculated as:

$$M_{\Gamma X}(y) = t_{\Gamma X} a_0 V \{\psi_\Gamma^L(y)\psi_X^L(y) \cos \pi z_L(y) + \\ \psi_\Gamma^R(y)\psi_X^R(y) \cos \pi z_R(y)\} , \qquad (3)$$

with z_L ($z_R = z_L + e$) the position of the left (right) interface of the AlAs layer of thickness e. The radiative decay time is:

$$\tau_{\Gamma X}^{calc} \propto \frac{|E_\Gamma - E_X|^2}{|M_{\Gamma X}|^2_{av} |\langle \psi_\Gamma \mid \psi_{hh_1}\rangle|^2} . \qquad (4)$$

$|M_{\Gamma X}|^2_{av}$ is calculated as an average over a distribution of the position of interfaces around their mean value under the laser excitation spot. $\tau_{\Gamma X}^{calc}$ is shown in Fig. 2 as a function of the effective AlAs thickness for localized excitons. The best agreement between calculated and experimental results is obtained for partial correlation of the position of the left and right interfaces. This is achieved by using uncorrelated Gaussian distribution functions of z_L and e with full width at half maximum of one ML. The dimensionless coupling parameter $\tau_{\Gamma X}$ is found in the range 0.23-0.33. The value of the coupling parameter is of the right order of magnitude compared to the theoretically calculated one[4].

References

1. I.L. Aleiner and E. L. Ivchenko, Semiconductors **27**, (1993) 330 [Fiz. Tekh. Poluprovodn. **27**, (1993) 594].
2. D. Martins, C. Gourdon, P. Lavallard, and R. Planel, Solid State Commun. **114**, (2000) 389.
3. C. Gourdon , D. Martins , P. Lavallard, E. L. Ivchenko, Yun-Lin Zheng, R. Planel, to be published.
4. T. Ando, Phys. Rev. B **47**, 9621 (1993) 9621.

Blue Stark shift in composite quantum wells

A. Yu. Silov*, B. Aneeshkumar, M. R. Leys, H. Vonk, and J. H. Wolter

COBRA Inter-University Research Institute, Eindhoven University of Technology, P. O. Box 513, 5600 MB Eindhoven, The Netherlands

Abstract We have observed a substantial blue shift of the lowest energy transition in a composite double quantum well made from materials with a type-II band line-up. The photocurrent measurements demonstrate a linear Stark shift due to the separate confinement in real space for electrons and holes. The charge separation is up to 45 Å in the strain balanced $InAs_{0.42}P_{0.58}/Ga_{0.67}In_{0.33}As$ samples. The low temperature photoluminescence reveals that an electric field of only 30 kV/cm produces the blue shift of more than 25 meV. All our measurements agree with the calculations within the framework of Bir-Pikus strain Hamiltonian.

1 Introduction

In recent years there has been a great deal of interest in structures with separate confinement of electrons and holes in real space. They are especially used to realize a two dimensional exciton condensate [1] and for exploring the blue shifting nature of the absorption edge [2]. Here, for the first time, we demonstrate a large blue shift by using composite quantum well structures with separate confinement of electrons and holes. This has been accomplished by growing the composite well region out of materials with reverse band offset ratio with respect to the barrier material [3].

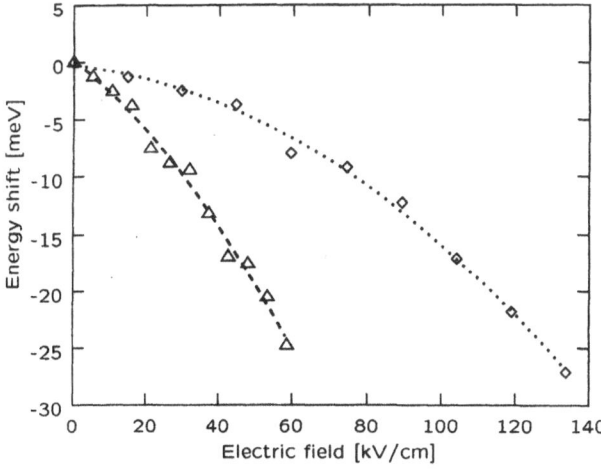

Fig. 1 Stark shift in the ground state: a) $InAs_{0.42}P_{0.58}$ under 1.3% compression (triangles); b) $InAs_{0.35}P_{0.65}$ under 1.1% compression (diamonds).

2 Samples

To realize the separate confinement we made use of $InAs_yP_{1-y}$ and $Ga_xIn_{1-x}As$ layers surrounded by InP barriers, where y was either 0.35 or 0.42, and x = 0.47 or 0.67, respectively. At the $Ga_xIn_{1-x}As$ / InP interface

---
* *E-mail:* A.Silov@phys.tue.nl

the conduction band to valence band offset ratio is assumed to be 40:60 and at the $InAs_yP_{1-y}$ / InP interface it is taken to be 70:30. The photocreated electrons were thus confined within the $InAs_yP_{1-y}$ while the holes are confined within the $Ga_xIn_{1-x}As$.

Both quantum wells are approximately 40 Å thick and are separated by a 10 Å thick InP barrier. All the structures were grown by Chemical Beam Epitaxy on n-type (100) InP substrates misorientated 0.5° towards the (111)B [4]. The quantum well region was grown in the intrinsic region of n-i-p structure.

3 Results and Discussion

3.1 Photocurrent measurements

Figure 1 shows the results of photocurrent measurements. They were performed at 100 K under reverse bias varying up to 13.5 V. Tunable infrared laser light was used as an excitation source. The photocurrent measurements on samples with a 1.3% compression on the $InAs_{0.42}P_{0.58}$ side of the composite quantum well and containing the lattice matched $Ga_{0.47}In_{0.53}As$ show linear Stark shift in electric field up to 15 kV/cm. The red shift is due to the fact that the applied electric field pulls the photocreated electrons and holes further apart.

Stark shift of the lowest energy transition as a function of the applied field is shown in Fig. 1. Here the line is the theoretical fit to the experimental data. The initial linear Stark shift is a clear indication of the separated confinement for electrons and holes in real space. The photocurrent measurements on sample with 1.1% compression on the $InAs_{0.35}P_{0.65}$ show that the slope of the linear red shift of the absorption edge is smaller. From this slope we can calculate the electron-hole separation. For the $InAs_{0.42}P_{0.58}$ sample, the electron-hole separation is 40Å, whereas for the sample with 1.1% compression it is decreased by 15%.

3.2 Photoluminescence

Using photoluminescence we also investigated the strain balanced samples. The photoluminescence measurements were done at 5 K. The wavelength of 532 nm from a Nd:YAG laser was used as the excitation source. An unfocused beam was used and the power density was about 400 mW/cm². The photoluminescence spectra at 5 K are shown in Fig. 2.

Two pairs of samples were investigated: one with lattice matched $Ga_{0.47}In_{0.53}As$ and the other with $Ga_{0.67}In_{0.33}As$ under tension. In each pair one is the inverted structure of the other, so that the built-in electric field is reversed. The $InAs_yP_{1-y}$ has the same compo-

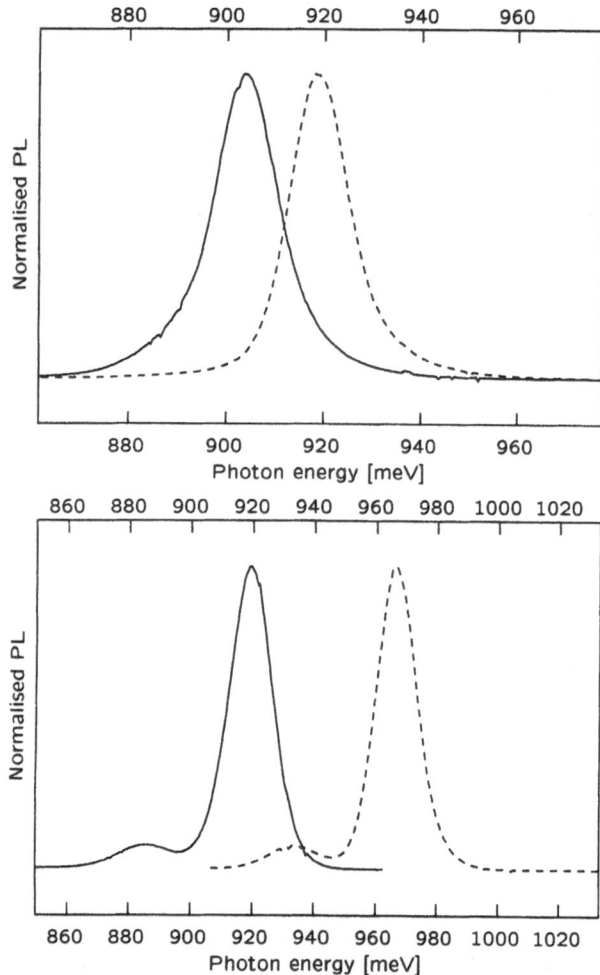

blue shift of 25 meV. All our measurements agree with the calculations in the framework of Bir-Pikus strain Hamiltonian in the Γ_8 representation. We exemplify one of these computations in Fig. 3.

In conclusion, we achieved large blue Stark shift in the composite quantum well structures using separate confinement of electrons and holes in real space. Due to the high asymmetry in the confinement, these structures can be used to achieve both red and blue shift of the absorption edge.

4 Acknowledgements

This work is a part of the research program of the "Stichting voor Fundamenteel Onderzoek der Materie (FOM)", which is financially supported by the "Nederlandse Organisatie voor Wetenschappelijk Onderzoek (NWO)". B. A. is also grateful for financial support from NUFFIC through a project for strengthening the International School of Photonics, CUSAT.

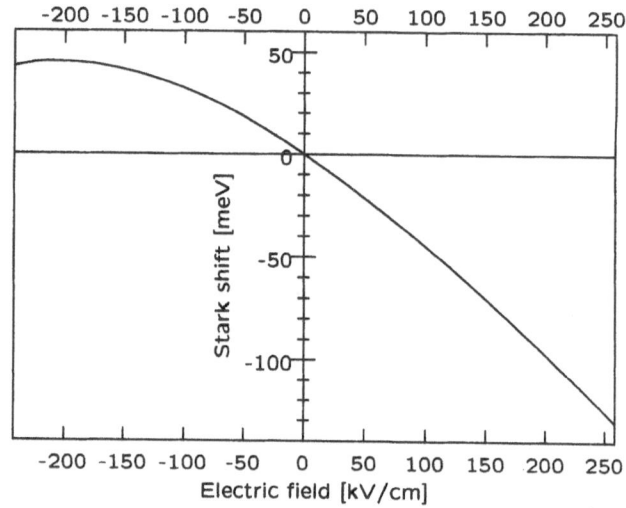

Fig. 2 Low-temperature photoluminescence spectra.
Top panel: a) $InAs_{0.42}P_{0.58}/InP/Ga_{0.47}In_{0.53}As$ (solid line) and b) $Ga_{0.47}In_{0.53}As/InP/InAs_{0.42}P_{0.58}$ (dashed line). The spectral shift between the two peaks is 15 meV.
Bottom panel: c) $InAs_{0.42}P_{0.58}/InP/Ga_{0.67}In_{0.33}As$ (solid line) and d) $Ga_{0.67}In_{0.33}As/InP/InAs_{0.42}P_{0.58}$ (dashed line). The shift is 50 meV.

Fig. 3 Calculation of the Stark shift. Both blue and red Stark shifts of the ground state are available: an example for $x = 0.47$ and $y = 0.42$.

sition in all structures with y=0.42. For the pair with $Ga_{0.47}In_{0.53}As$ composition, the shift between the photoluminescence peaks is 15 meV, while for the pair with the $Ga_{0.67}In_{0.33}As$ well the shift between the photoluminescence peaks is 50 meV. In $Ga_{0.67}In_{0.33}As$ well the ground state for the valence band becomes light hole in character. The strain increases the spatial separation between the electrons and holes, and the calculated separation is 45 Å.

There is no coupling between the electronic states in the valence band for the lattice matched $Ga_{0.47}In_{0.53}As$, but there exists a residual coupling in the conduction band. By putting more tension on the $Ga_xIn_{1-x}As$ side of the composite well, this residual coupling can be greatly reduced. Thus the corresponding photoluminescence peak shifts towards the high-energy side, compared to the lattice matched case. Additionally, it is found that an electric field of about 30 kV/cm can produce a large

References

1. L. V. Butov and A. I. Filin, Phys. Rev.B, **58**, (1998), 1980.
2. P. N. Stavrinou, S. K. Haywood and G. Parry, Appl. Phys. Lett., **64**, (1994), 1251.
3. A. Yu. Silov, M. R. Leys, H. Vonk and J. H. Wolter, *Proceedings of the ICPS-24, Jerusalem* (World Scientific, Singapore 1999) On CD: Section IV, Heterostructures and superlattices, Subsection D, Interband transitions, Excitons, No-10.
4. C. A. Verschuren, M. R. Leys, R. T. H. Rongen, H. Vonk, J. H. Wolter, J. Cryst. Growth, **200**, (1999), 19.

Spectroscopic Studies of the Interaction of an Electron Gas with Photoexcited Electron-hole Pairs in Modulation-doped GaAs/AlGaAs Quantum Wells

T. Yeo[1], H.A. Nickel[1], G. Comanescu[1], H.D. Cheong[1], M. Furis[1], A. Petrou[1], B.D. McCombe[1]

Department of Physics, SUNY at Buffalo, Buffalo, NY 14260, USA

Abstract The Optically detected resonance (ODR) spectroscopy has been used to study internal transitions of neutral (X) and negatively charged (X^-) excitons and to explore the effects of many-electron interactions in undoped and modulation-doped GaAs/AlGaAs quantum-well structures. Studies as a function of excess electron density in the wells reveal blue-shifted X^--like transitions in the magnetic field region corresping to Landau level filling factors $\nu < 2$. The blue shifted transitions are attributed to the collective response of the few-hole/many-electron system, and demonstrate the many-body collective nature of the optical response.

There have been numerous studies of the band-edge photoluminescence (PL) and/or reflectivity/absorption features of GaAs/AlGaAs quantum-well structures as a function of electron density and/or magnetic field. The PL evolves from bandgap-renormalized, band-to-band recombination to exciton-like recombination (initially negatively charged excitons, X^- and finally neutral excitons (X)) as the electron density is decreased. For high electron density samples with symmetric wells a similar evolutions takes place with increasing magnetic field; for filling factors $\nu < 2$, an X^--like line appears, and for $\nu < 1$ a line that appears to be X is evidenced [1]-[2].

We have combined magneto-photoluminescence (PL) and optically detected resonance (ODR) spectroscopies to explore the nature of the correlated electron gas and its response to photoinjected electron-hole pairs in nominally undoped and modulation-doped GaAs/AlGaAs quantum-well structures. We have used internal transitions of X^- and the effects of many-electrons on these spectroscopic signatures to probe the many-body state in conjunction with standard magneto-PL measurements. The experiments were carried out on four GaAs/AlGaAs multiple-quantum-well samples grown by molecular beam epitaxy. Sample 1 and 2 are undoped with the same nominal well-width (20nm) and $Al_xGa_{1-x}As$ barrier composition (x=0.3). Samples 3 and 4 have 24nm wells and are modulation doped in the central third of the barrier at 8×10^{10} and $2.8 \times 10^{11} cm^{-2}$, respectively. The far infrared(FIR) absorption resonances in a magnetic field were studied by the ODR. The samples are photo-excited by the 6328 Å line of He-Ne laser via an optical fiber. The intensity change of a particular band-edge luminescence feature was monitored while FIR laser light was shone to the sample as a function of magnetic field (details of the experimental techniques can be found in [3]).

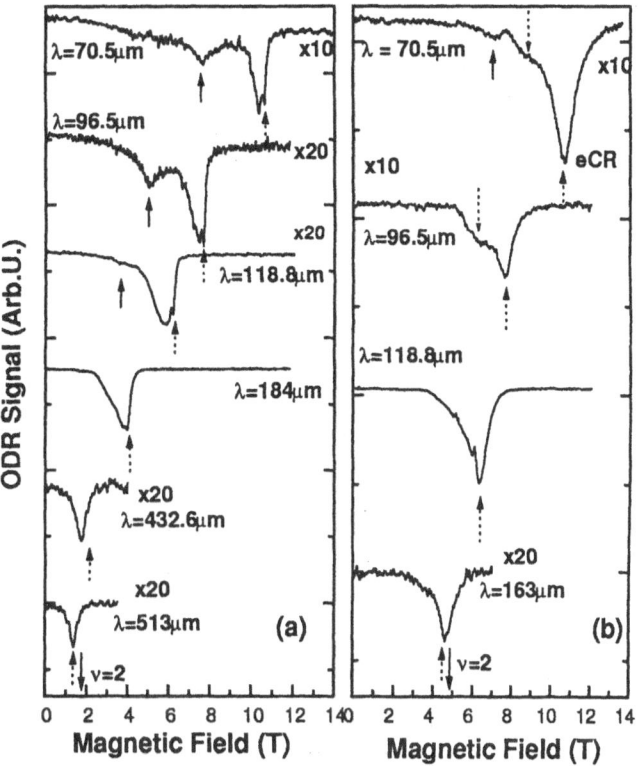

Fig. 1 ODR results for sample 3 (a) and 4 (b) at several FIR laser wavelengths. Electron cyclotron resonance (eCR) is indicated by the upward dotted arrows, and a feature that has beenidentified in lower density samples as a band of internal transitions of X^- (singlet ionizing transitions) is indicated by the upward solid arrows. The broad asymmetry of eCR to lower field is due to a band of ionizing triplet-like transition which is indicated by down dotted arrow for sample 4. The field corresponding to filling factor $\nu=2$ is indicated for each sample.

The magneto-PL spectra of sample 1 and 2 show fomation of X and X^- phase all magnetic field studied, however, sample 3 and 4 show bandgap renormalized Landau-level to Landau-level transitions at high filling factor (ν) and a discontinuity in the slope of the PL energy vs. field for the N = 0 Landau transition at $\nu = 2$. For $\nu < 2$ the dominant PL is depressed below the extrapolated N = 0 Landau level transition and has an excitonic field dependence.

The ODR spectra from sample 3 (Fig. 1 (a)) are similar to those from a nominally undoped QW sample 1 and

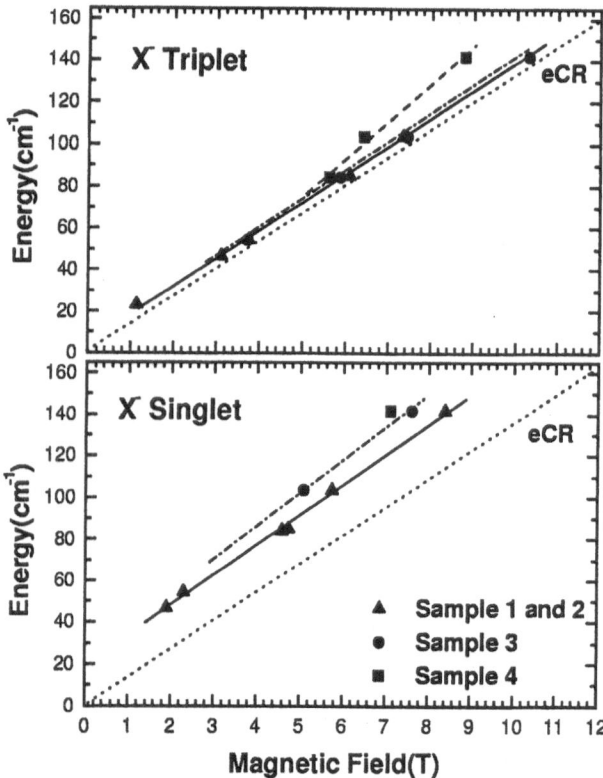

Fig. 2 Summary plot of the observed X^- singlet (lower panel) and triplet (upper panel) transitions for all four samples. The dotted line indicates the position of electron CR. The other lines are guidess to the eye, indicating the blue-shift in the transitions as the carrier density is increased.

2. For laser wavelengths shorter than 184µm an ODR band just below the sharp electron cyclotron resonance (eCR) is observable. We attribute the observed features to bands of ionizing internal transitions of X^- triplet-like ($X^- \rightarrow X^* + e^0$, electron in 0th Landau level, X in excited state X^*). The another observable feature is the ODR band of ionizing internal transition of X^- singlet-like which is indicated by up solid arrows. The resonant fields for both singlet-like and triplet-like X^- features occur above the field corresponding to $\nu = 2$ (1.6T). As the FIR wavelength increases (lower field for resonance), the triplet feature remains strong, but the singlet feature weakens and is not observable at 184µm (it should occur at a field for which $\nu > 2$). For longer wavelengths (432.6µm, 513µm) the only observable feature is eCR. For the heavily doped sample (Fig. 1 (b)) the behavior is rather different. At the highest photon energy, where the singlet band should lie just above the field corresponding to $\nu = 2$ for this sample, there is a weak feature that appears to be a blue-shifted, singlet-like band, and a band that lies just below eCR (the triplet-like band, indicated by dotted down arrow). At 96.5mm there is no singlet-like feature observable, but a rather broad band that begins slightly below eCR and extends to lower field; the spectrum at 118.8µm behaves similarly. At 163µm eCR line shape becomes rather symmetric and the broad band

below eCR is suppressed significantly. Fig. 2 shows summary plot of the observed X^- singlet (lower panel) and triplet (upper panel) transitions for all four samples. It is clearly seen that the both transitions are blue-shifted (lower resonant magnetic field) as the electron density increases. For FIR laser 70.5 µm, the X^- single transition occurs at magnetic field B = 8.4T, 7.6T and 7.1T for sample 1 and 2, sample 3 and sample 4, respectively.

The blue shift is the signature of a many-electron phenomenon whose origin is exchange lowering of the many electron ground state relative to the excited state due to electron-electron exchange interaction [4]. This phenomenon has been observed in negatively charged donor ion (D^-) with excess electrons [5]. Systematic studies of the dependence of the maxima of the observed bands on electron density and filling factor have shown: (1) there is no singlet or triplet-like bands appearing at fields corresponding to filling factors $\nu > 2$; and (2), when observed, the bands exhibit a systematic blue shift relative to the corresponding bands in nominally undoped samples. From this behavior we conclude that magneto-PL feature observed at fields above that correponding to $\nu = 2$ is indeed related to X^-, but represents a collective response of the system.

Work supported in part by NSF DMR 972265.

References

1. G. Finkelstein et al., Phys. Rev. Lett **74**, (1996) 976.
2. D. Gekhman et al., Phys. Rev. B **54**, (1996) 10320.
3. H. Nickel et al., Phys. Rev. B **62**, (2000) 2773.
4. P. Hawrylak, Phys. Rev. Lett **72**, (1994) 2943.
5. J.-P. Cheng et al., Phys. Rev. Lett **70**, (1993) 489.

Resonant Rayleigh scattering of exciton-polaritons in multiple quantum well structures

A. Kavokin[1], G. Malpuech[1], W. Langbein[2] and J.M. Hvam[3]

[1]LASMEA - Université Blaise Pascal Clermont II - CNRS 24 avenue des Landais - 63177 AUBIERE CEDEX- France
[2] Lehrstuhl für Experimentelle Physik EIIb, Universität Dortmund,Otto-Hahn Str.4, 44227 Dortmund, Germany
[3] Research Center COM, Technical University of Denmark, Bldg. 349, 2800 Kgs. Lyngby, Denmark

A theoretical concept of resonant Rayleigh scattering (RRS) of *exciton-polaritons* in multiple quantum wells (QWs) based on an analytical solution of Maxwell equations for light incident on an irregular grating of quantum dots is presented. The optical coupling between excitons in different QWs is shown to strongly affect the RRS dynamics, giving rise to characteristic temporal oscillations on a pico-second scale. Experimental data on the RRS from multiple QWs show the predicted strong temporal oscillations at small scattering angles, which are well explained by the presented theory.

This communication is aimed to show that the resonant Rayleigh scattering (RRS) spectra of multiple quantum wells (MQWs) are a result of exciton-polariton scattering involving all the quantum wells (QWs). In particular, the RRS signal is sensitive to the period of the MQW structure and the number of QWs. Experimental data on single QW and MQWs are presented. Strong temporal oscillations of the RRS have been observed for small scattering angles in a very good agreement with the presented theory.

Recent experiments on the RRS from different MQW structures have demonstrated a complex non-monotonic decay sometimes showing oscillations with a period of a few pico-seconds[1-3]. This oscillations have been associated in Ref. 2 to the interference between light-waves scattered by different localized exciton states in a QW plane. Experiments on single QWs[3] do not show such features. The observed discrepancies between the RRS from SQWs and MQWs point out the need for an exciton-polariton theory, which requires a revision of the physical concept of the RRS in QWs, namely introducing the concept of scattering of coherent exciton-polariton states. In our very recent paper[4] the optical coupling between different QWs within a MQW structure has been shown to result in the characteristic oscillations in the RRS spectra of MQWs. In Ref. 4 we have used the scattering function of a SQW obtained neglecting the polariton effect[5]. Here we develop further the model of Ref. 4, fully taking into account the polariton (retardation) effects within each individual QW.

We consider a QW containing N localized exciton states as an ensemble of N quantum dots and apply a Green function method of resolution of Maxwell equations for a light wave incident on a grating of quantum dots[6]. Unlike Ref. 6, in our case the grating is irregular, and the exciton resonance frequency ω_n changes from one dot to another thus yielding the exciton inhomogeneous broadening. This irregularity in distribution of dots allows the RRS, while a regular grating would only allow reflection and diffraction. Neglecting the deviations in a shape of exciton wave

function localized in different dots, we are able to resolve analytically the Maxwell equations and the obtain the amplitude of light scattered with a wave-vector $\vec{k}_s$ as

$$d(\vec{k}_s) = \sum_{n=1}^{N} \frac{i\Gamma_0 e^{i(\vec{k}-\vec{k}_s)\vec{R}_n}}{\omega_n - \omega - i(\gamma + \Gamma_0)} \quad , \quad (1)$$

where $\vec{k}$ is the wave-vector of incident light, $\vec{R}_n$ is the radius-vector of n-th quantum dot, γ is the exciton homogeneous broadening,

$$\Gamma_0 = \frac{1}{6}\omega_{LT}k^3 a_B^3 \left[\int d\vec{r}\,\Phi(\vec{r})\cos(\vec{k}_s\vec{r})\right]^2 , \quad (2)$$

where ω_{LT} and a_B are the exciton longitudinal-transverse splitting and Bohr radius in the bulk material, $\Phi(\vec{r})$ is the exciton wave-function taken with equal electron and hole co-ordinates. In the calculation presented here we have taken N=512x512, the average distance between dots of $0.1\ \mu m$, the exciton radiative life-time in a single dot

$$\tau = \frac{1}{2\Gamma_0} = 400 \text{ ps},$$ the distribution of ω_n to be given by a

Gaussian function with an inhomogeneous broadening parameter of Δ=0.08 meV (MQW), Δ=0.09 meV (SQW), γ = 6 μeV.

The algorithm that allowed us to calculate the RRS spectrum of a MQW structure taking into account all the acts of reflection, absorption and transmission (but only a single scattering act for each light-harmonics) is described in detail in Ref. 4. It is based on the transfer matrix method and scattering state technique. We do calculations for different scattering directions (within a range of about 3 degrees centered on the scattering angle corresponding to a given experimental configuration). Then we average the scattered intensity within this 3-degree range in order to take into account the speckle averaging performed experimentally.

Note, that secondary scattering (i.e. scattering of the same light wave by two different QWs within the structure) can be taken into account in our procedure (that requires, however, quite heavy angle integration).

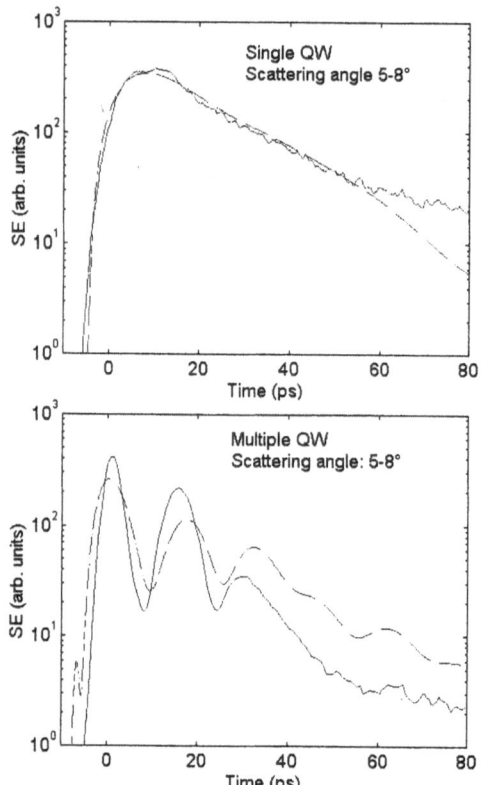

Figure 1. Experimental (solid) and theoretical (dashed) RRS spectra from single and multiple QW structures under small scattering angles.

The secondary scattering may be very important for the large scattering angles, that will be a subject of our further publication.

Let us compare the results of the present theory with experimental data on the RRS from SQW and MQWs. In the following we discuss experiments performed on GaAs/Al$_{0.3}$Ga$_{0.7}$As QWs grown by molecular-beam epitaxy without growth interruption. The MQW structure consists of 10 16 nm-thick wells and have barrier thicknesses of 15nm. The samples have been placed in a helium cryostat at 5 K temperature. The fundamental hh1-e1 1s exciton resonance is excited by optical pulses from a mode-locked Ti:sapphire laser of about 500 fs Fourier-limited pulse duration. The secondary emission (SE) is passed through an analyzer parallel to the linear excitation polarization (TM-polarization), a monochromator, and is detected time and angular resolved by a synchroscan streak camera with a time resolution of 3 ps. For details of our experimental technique see Refs. 3,4. Using the speckle analysis technique[3], we deduce from the temporally and directionally resolved emission intensity $I(t, \vec{k}_s)$ the average emission intensity $\overline{I(t)}$ where the average is taken over the scattering directions $\vec{k}_s$ at fixed time.

In Fig.1 $\overline{I(t)}$ for a 16nm MQW and a 15nm SQW sample are shown. Since both samples have nearly equal thicknesses, the inhomogeneous broadening of the individual wells due to the interface-roughness is comparable (just slightly more in the SQW structure having a thinner QW). The SE line-width (FWHM) for large scattering angles is 170 µeV for the SQW and 340µeV for the MQW. The non-radiative homogeneous broadening γ is about 6 µeV in both samples, as deduced from the decay of the SE coherence for long times. Clearly, the MQW shows strong modulations of the RRS within the first 40 ps, which are not observed for the SQW. The theoretical calculation shown by dashed lines in figure 1 is in a very good agreement with these data.

In conclusion, RRS from multiple quantum well structures are governed by the dynamics of the exciton-polaritons, which are extended mixed exciton-photon states occupying the entire system. The present semi-classical model takes into account the exciton-polariton effect in MQWs, and describes correctly the experimental RRS dynamics from GaAs MQWs for small scattering angles, which are dominated by the exciton-polariton dynamics.

Acknowledgment
This work has been supported in part by the European Commission in the framework of Contract "CLERMONT" No. HPRN-CT-1999-00132.

References
1.   H. Wang et al, Phys. Rev. Lett. **74,** (1995) 3065; S. Haacke et al, Phys. Rev. Lett. **78,** (1997), 2228; M. Woerner and J. Shah, Phys. Rev. Lett. **81,** (1998), 4208; N. Garro et al, Phys Rev B **60,** (1999), 4497.
2.   V. Savona, S. Haacke, and B. Deveaud, Phys. Rev. Lett. **84,** (2000), 183.
3.   W. Langbein, J.M. Hvam, R. Zimmermann, Phys. Rev. Lett. **82,** (1999), 1040.
4.   G. Malpuech, A. Kavokin, W. Langbein, J.M.Hvam, Phys. Rev. Lett, **85** , (2000), 650.
5.   V. Savona and R. Zimmermann, Phys. Rev B **60,** (1999), 4928.
6.   E.L.Ivchenko, A.V.Kavokin, Sov. Phys .Solid State, **34,** (1992), 1815.

Speckle-correlation spectroscopy on localized spin-split exciton states in quantum wells

R. Zimmermann[1], W. Langbein[2], E. Runge[1], J.M. Hvam[3]

[1] Institut für Physik der Humboldt-Universität zu Berlin, 10117 Berlin, Germany; e-mail: zim@physik.hu-berlin.de
[2] Lehrstuhl für Experimentelle Physik EIIb, Universität Dortmund, 44227 Dortmund, Germany
[3] Research Center COM, The Technical University of Denmark, 2800 Kgs. Lyngby, Denmark

Abstract The resonant secondary emission of excitons in a GaAs quantum well is investigated with special emphasis on spin-related properties. The measured co- and cross-polarized emission show a rich dynamical behaviour, which is analyzed using the speckle technique. The coherence decays at an equal rate in all channels, evidencing polarization beating (no dephasing). The long-time saturation value of the linear polarization degree as well as the nonzero speckle correlation reveal a preferential orientation of the excitonic eigenstates along [110]. This finding can be related to elongated interface islands.

The radiative (spin-singlet) exciton states in GaAs quantum wells (QWs) form a nearly degenerate doublet which is split into two (linearly polarized) states via the long-range exchange interaction,

$$\Delta_\alpha e^{2i\theta_\alpha} \sim \int d^2K \, |\psi_{\alpha K}|^2 (K_x + iK_y)^2 / K \, . \quad (1)$$

The wave function of the exciton center-of-mass (COM) motion in momentum space is denoted by $\psi_{\alpha K}$. Obviously, only *anisotropic* COM states exhibit a nonzero splitting. Spatially resolved photoluminescence spectra show indeed a rich variety of these spin-split states [1]. In the present work, we concentrate on ensemble properties which can be deduced from the wide-focus time-resolved secondary emission in different polarization channels. As shown below, information on both the splitting magnitude Δ_α and the wave function orientation θ_α can be extracted.

Spectra are taken at 5 K from a multiple GaAs QW (4 nm well width) with substantial inhomogeneous broadening (FWHM 8 meV), and the standard polarization degree (both linear and circular) is deduced. Further, the recently introduced speckle analysis [2] is extended to look for the correlation of speckles in different polarization channels (symbols in Fig. 1). The findings are: (i) The coherence degree decays slowly at an equal rate in all channels, proving that the rapid polarization transfer is not related to dephasing but simply polarization beating. (ii) The circular polarization degree shows a sign reversal before approaching zero at times larger than 100 ps. (iii) The long-time saturation value of the linear polarization degree depends on the polarization direction with respect to the sample orientation, being at maximum when exciting along [110]. (iv) A positive correlation between speckle features in the co- and cross-polarized spectra is seen.

For the coherent part of the secondary emission after short-pulse excitation, an expansion into COM eigenstates is used,

$$I^{in,out}(t) = I_0 \sum_{\alpha,\beta} M_\alpha^{in} M_\alpha^{out*} M_\beta^{in*} M_\beta^{out} e^{i(\epsilon_\alpha - \epsilon_\beta)t} \, . \quad (2)$$

The eigenenergies include the spin splitting, $\epsilon_\alpha = \epsilon_{\alpha 0} \pm \Delta_\alpha/2$, and the matrix elements contain the information on polarization direction,

$$M_{\alpha,+}^x = +\cos(\theta_\alpha)\psi_{\alpha 0} \; ; \quad M_{\alpha,-}^x = -i\sin(\theta_\alpha)\psi_{\alpha 0} \; ; \quad (3)$$
$$M_{\alpha,+}^y = +\sin(\theta_\alpha)\psi_{\alpha 0} \; ; \quad M_{\alpha,-}^y = +i\cos(\theta_\alpha)\psi_{\alpha 0} \, .$$

Note the underlying selection rules of heavy-hole to conduction band transitions. At times $\sigma t > 1$ (σ - inhomogeneous width of $\epsilon_{\alpha,0}$), the evolution is given exclusively by temporal interference between the spin-split pairs. After an ensemble average, the degreee of circular polarization $\mathcal{P}^{circ} = (I^{\sigma+\sigma+} - I^{\sigma+\sigma-})/(I^{\sigma+\sigma+} + I^{\sigma+\sigma-})$ can be reduced to

$$\mathcal{P}^{circ}(t) = \frac{\sum_\alpha \psi_{\alpha 0}^4 \cos(\Delta_\alpha t)}{\sum_\alpha \psi_{\alpha 0}^4} \stackrel{def}{=} \langle \cos(\Delta_\alpha t) \rangle \, . \quad (4)$$

At times large compared to the average spin splitting, the polarization vanishes asymptotically, but exhibits a sign reversal before (see Fig. 1). This is a clear indication of polarization beating, in contrast to spin dephasing via scattering. A similar analysis for linear polarization reveals an interesting dependence on the absolute choice of polarization direction (ϕ) with respect to the crystallographic axes (say [110]),

$$\mathcal{P}_\phi^{lin}(t) = \langle \cos^2(2\theta_\alpha - 2\phi) \rangle + \langle \sin^2(2\theta_\alpha - 2\phi)) \cos(\Delta_\alpha t) \rangle$$

For a random orientation of all wave functions, no dependence on ϕ is to be expected. However, the data, in particular their long-time limit

$$\mathcal{P}_\phi^{lin}(t \to \infty) = \frac{1}{2}(1 + \cos(4\phi)\, C_4) \, , \quad (5)$$

give clear evidence for a *preferential* orientation of the wave functions along [110]. This is quantified by a nonzero average $C_4 = \langle \cos(4\theta_\alpha) \rangle$. Fitting the long-time behaviour of the data in Fig. 1 gave a value of $C_4 = 0.37$.

Signatures of wave function orientation are expected in the speckle analysis as well. For the correlated appearance of speckles in the $\sigma+$ and $\sigma-$ channels, we define the mixed coherence degree as

$$C_{\text{mix-circ}}^2 = \frac{\overline{I^{\sigma+\sigma+} \cdot I^{\sigma+\sigma-}}}{\overline{I^{\sigma+\sigma+}} \cdot \overline{I^{\sigma+\sigma-}}} - 1 \, . \quad (6)$$

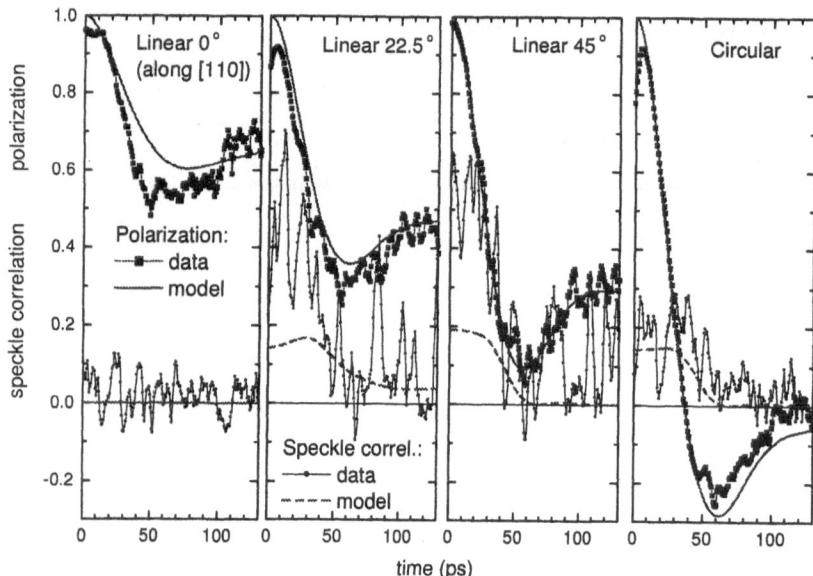

Fig. 1 Polarization degree and speckle correlation as a function of time for linear and circular polarization. Experimental data (symbols) are compared with the anisotropic model ($\eta/\xi = 0.5$), using simulation results for the distributions.

After a lengthy algebra, we get

$$C^2_{\text{mix-circ}}(t) = \tag{7}$$
$$\frac{\langle\cos(2\theta_\alpha)\sin(\Delta_\alpha t)\rangle^2 + \langle\sin(2\theta_\alpha)\sin(\Delta_\alpha t)\rangle^2}{1 - \langle\cos(\Delta_\alpha t)\rangle^2} \ .$$

A nonzero result is directly related to preferential orientation, and clearly seen in Fig. 1. Corresponding expression have been derived for linear polarization.

For obtaining the different quantities derived above, we need to calculate disorder eigenstates in an anisotropic potential landscape, which was done adding an analysis of spin splitting to standard simulations [3]. To overcome statistical fluctuations, a huge ensemble averaging is needed. For a detailed understanding, we have therefore applied a much easier model based on the seminal work by Wilkinson et al. [4] where the probability distribution of minima (value and shape) in a disordered landscape is considered. This method has been used before in

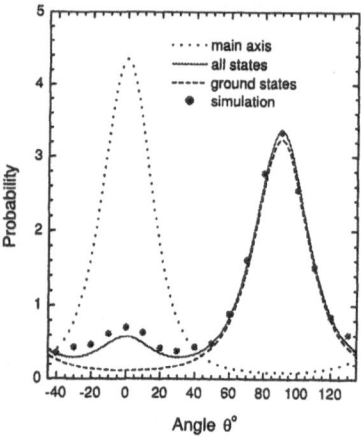

Fig. 2 Main axis distribution (dotted) and angle distribution of spin splitting at $E = -3\,\text{meV}$. Simulation data (symbols) refer to $\eta/\xi = 0.5$. In the minimum model (curves), $\eta/\xi = 0.6$ has been used to give the same value of $C_4 = 0.37$.

the present context of exciton spin splitting [5, 6]. We improve here in using an *anisotropic* Gauss-correlated potential characterized by two different correlation lengths (ξ and η). The potential-potential correlation reads

$$\langle V(x,y)\,V(0,0)\rangle = \sigma^2 \exp\left(-x^2/2\xi^2 - y^2/2\eta^2\right) \ . \tag{8}$$

In a first step, the probability distribution for energy, curvature and main axis direction of local minima is generated. In the resulting harmonic oscillator problem, it is important to include not only the local ground states but excited states as well. As a consequence, the angle distribution exhibits two peaks, since ground state and related excited states can have *orthogonal* directions in K space (Fig. 2).

The spectra, and in particular C_4, can be reasonably well fitted using a landscape correlation anisotropy of $\eta/\xi = 0.5$ and an averaged spin splitting of $\langle\Delta_\alpha\rangle = 31\,\mu\text{eV}$. Taking the potential distribution width ($\sigma = 3\,\text{meV}$) from the experiment, the splitting is consistent with correlation lengths of $\xi = 30\,\text{nm}$ and $\eta = 15\,\text{nm}$.

In conclusion, we have shown that polarization degree and speckle correlation can be successfully used to analyze the spin splitting of exciton states. The extracted information on exciton wave function anisotropy is related to the underlying disorder landscape. We attribute the preferential orientation of the minima along [110] to elongated island formation on the interfaces during growth.

References

1. D. Gammon et al., Phys. Rev. Lett. **76**, (1996) 1205.
2. W. Langbein, J. M. Hvam, and R. Zimmermann, Phys. Rev. Lett. **82**, (1999) 1040.
3. R. Zimmermann and E. Runge, J. Lumin. **60-61**, (1994) 320.
4. Fang Yang, M. Wilkinson et al., Phys. Rev. Lett. **70**, (1993) 323; erratum Phys. Rev. Lett. **72**, (1994) 1945.
5. H. Nickolaus et al. Phys. Rev. Lett. **81**, (1998) 2586.
6. M. Z. Maialle, Phys. Rev. B **61**, (2000) 10 877.

Optical properties of trions in semiconductor nanostructures

Axel Esser, Erich Runge, Roland Zimmermann

Institut für Physik der Humboldt-Universität zu Berlin, 10117 Berlin, Germany, e-mail: `axel.esser@physik.hu-berlin.de`

Abstract Using density-matrix theory we obtain an expression for the absorption coefficient of trions, which in combination with a numerical solution of the three-particle Schrödinger equation for electron and hole trions is used to model photoluminescence lineshapes, the radiative lifetime and the influence of an electrical field (quantum confined Stark effect). The triplet state of positively charged trions in GaAs quantum wells is found to be bound at zero magnetic field and we discuss their optical properties. In particular, it is shown that optical transitions are forbidden at in-plane momentum $k = 0$.

In the presence of an excess electron or hole density in semiconductor quantum wells (QW), generated either by modulation doping or by optical excitation, experimental absorption or photoluminescence (PL) spectra show temperature and density-dependent contributions of both negatively (X^-) and positively (X^+) charged excitons (trions) [1,2]. Within density-matrix theory the amplitude of the X^- trion transitions are related to the four-operator amplitude ($c^\dagger$-electron, $d^\dagger$-hole creation operator) [3]

$$T_\mathbf{q}(\mathbf{k}_1\,\sigma_1, \mathbf{k}_2\,\sigma_2) = \langle c^\dagger_{\mathbf{k}_1\sigma_1} c^\dagger_{\mathbf{k}_2\sigma_2} d^\dagger_{\mathbf{q}-\mathbf{k}_1-\mathbf{k}_2-\sigma_1} c_{\mathbf{q}\sigma_2}\rangle , \quad (1)$$

which couples directly to the interband polarization. The singlet and triplet trion states are given by $\sigma_1 = -\sigma_2$ and $\sigma_1 = \sigma_2$, respectively. Exploiting a generalized dynamical truncation scheme, we solve the equation of motion for $T_\mathbf{q}(\mathbf{k}_1\,\sigma_1, \mathbf{k}_2\,\sigma_2)$ in linear order of density and optical field $(\chi^{(1)})$ by a resolvent expansion in trion eigenstates $\Psi^{nl}_T(\boldsymbol{\rho}, \boldsymbol{\xi})$ with corresponding energies E^{nl}_T. We use the relative coordinates $\boldsymbol{\rho} = \mathbf{r}_{e1} - \mathbf{r}_{e2}$ and $\boldsymbol{\xi} = \mathbf{r}_{e1} + \mathbf{r}_{e2} - 2\mathbf{r}_h$. Then, the susceptibility is obtained, and we find for the absorption coefficient of trions

$$\mathrm{Im}\chi^T(\omega) = \chi_0 \sum_{l=s,t}(1 + \delta_{l,t})\sum_{nk} f_e(k)\,|M^l_n(\mathbf{k})|^2$$

$$\times\delta\left(E^{nl}_T - \frac{\hbar^2 M_X}{2m_e M_T}k^2 - \hbar\omega\right), \quad (2)$$

with a material-dependent prefactor χ_0, exciton mass $M_X = m_e + m_h$, and trion mass $M_T = 2m_e + m_h$. Note that the summation over spin states l in Eq. (2) yields an additional factor of 2 for triplet contributions. The energy conservation involves the electron-recoil energy $\epsilon = \hbar^2 M_X k^2 / 2m_e M_T$ for transitions at in-plane momentum k. These transitions are given by the optical matrix element

$$M^l_n(\mathbf{k}) = \int d\boldsymbol{\rho}\,\psi^{nl}_T(\boldsymbol{\rho}, \boldsymbol{\rho})\exp\left(-i\mathbf{k}\cdot\boldsymbol{\rho}M_X/M_T\right). \quad (3)$$

We obtain the trion wave function $\Psi^{nl}_T(\boldsymbol{\rho}, \boldsymbol{\xi})$ and the corresponding trion binding energies $B_T = E_X - E_T$ from

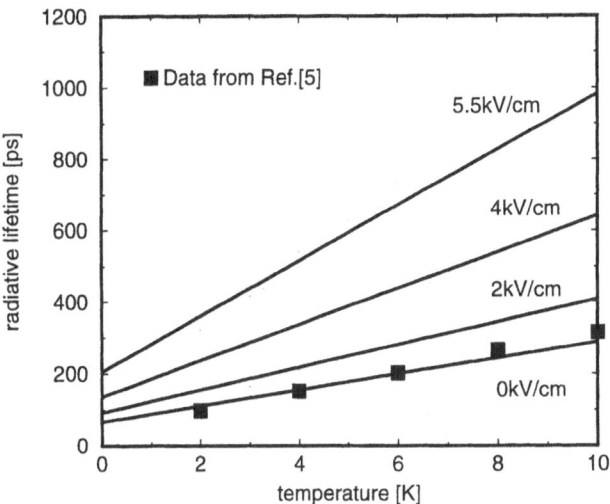

Fig. 1 Radiative lifetime of electron trions in the presence of an electric field in growth direction for a 30 nm GaAs QW.

the numerical solution of the three-particle Schrödinger equation by taking into account image-charge effects and different band offsets for the electron and the heavy hole. The effective Coulomb interaction is given by the overlap of the lowest electron and hole-sublevels. Having the trion binding energy and the wave function at our disposal, the optical properties can be studied. A specific feature is the inverse Boltzmann lineshape of the trion PL. According to the optical matrix element, the PL intensity of the singlet state shows approximately an exponential low-energy tail $I \propto \exp[-(1/\epsilon_1 + m_e/k_B T M_X)\epsilon]$ due to occupied transitions at $k > 0$, which can radiatively decay, but with a larger lifetime. In particular, we have shown that both a thermalized electron-trion and hole-trion distribution shows a linear increase of lifetime with temperature [4] according to

$$\tau_T = \frac{3M_X \tau_{0,T}}{4M_T E_0}\left(k_B T + \epsilon_1\frac{m_e}{M_X} + \frac{3}{5}\frac{M_T}{M_X}E_0\right), \quad (4)$$

which takes into account the finite photon momentum by $E_0 = \hbar^2 q_0^2 / 2M_X$ (q_0 is the light cone momentum of the QW). The prefactor $\tau_{0,T}$ is given by specific sample parameters, see [4] for details. Recently, this linear behaviour has been found independently in GaAs [5] and CdTe [6] QW's, and compares favorably with Eq. (4), see Fig. 1 and Ref. [6].

Due to an electric field along the growth direction, which is inherently present in modulation doped QW's or can be controlled by changing back and front gates as done in Ref. [7], the single-particle electron and hole

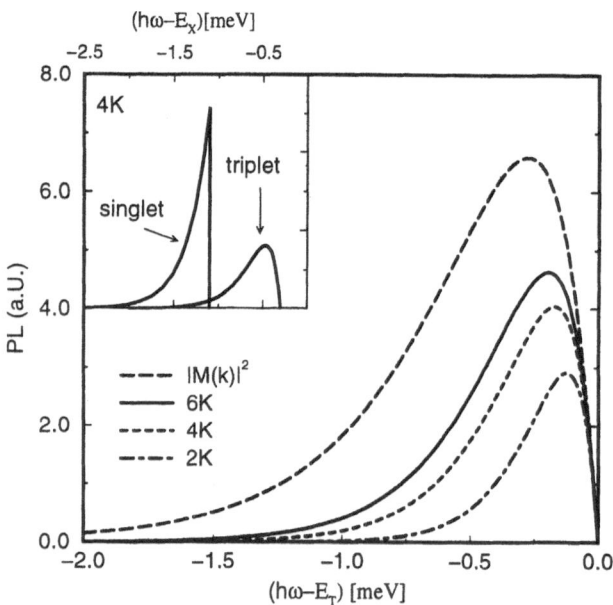

Fig. 2 Photoluminescence lineshape of the hole trion triplet state as function of temperature. Inset: Singlet and triplet-trion PL lineshape at 4 K for a 20 nm GaAs QW without additional broadening. An equal density of both states has been assumed.

sublevel states are polarized within the QW. The effective interparticle Coulomb interaction is then modified due to the variation of wave function overlap. In particular, the repulsive interaction for equal particles (e-e and h-h) is increased and the attractive interaction (e-h) is reduced. This gives rise to the well-known quantum confined Stark effect of excitons. Trions are even more influenced by the electric field as can be seen by the strong reduction of the trion binding energy [3,7]. More distinct for time-resolved experiments is the behaviour of the radiative lifetime, which is shown in Fig. (1) for a 30 nm GaAs QW. Both the temperature slope and the zero field offset are significantly increased.

The triplet states of trions are typically observed by applying a magnetic field [8] and are identified by polarization-dependent spectroscopy. However, based on magnetical translation symmetry, Dzybenko *et al.* [9] have shown that triplet states are optically inactive. This statement has lead to some discussion on their appearance in experiments, the possible role of localization, which breaks the magnetical translation symmetry, and the contribution of higher (bright) triplet states [10].

In contrast to the X^- triplet state, which is unbound at zero magnetic field, our numerically obtained binding energy of the triplet X^+ state is 0.3 meV for a 20 nm GaAs QW. This surprising result, contrary to the general believe of the unbound triplet state at zero field, is related to the reduction of exchange effects between two heavy particles (exchange is completely absent in the limit $m_h \rightarrow \infty$, and therefore singlet and triplet energy are the same). The experimental observation of

this triplet state is clearly hindered by the small binding energy, which is of the order of typical inhomogeneous exciton line widths in GaAs. However, there is also a peculiarity for the optical transitions. The optical property of this triplet X^+ state at zero magnetic field is again related to the optical matrix element $M(\mathbf{k})$, see Eq. (3). In particular, the antisymmetricity of the trion wavefunction $\psi_T^{nt}(\boldsymbol{\rho}, \boldsymbol{\xi}) = -\psi_T^{nt}(-\boldsymbol{\rho}, \boldsymbol{\xi})$ implies

$$M_n^t(k = 0) = 0. \tag{5}$$

Thus, for triplet states at the Brillouin center (or equivalent at temperature $T = 0$) the optical transition is forbidden. Only for higher k states (at finite temperatures) we find a nonvanishing matrix element and again a low-energy tail. See Fig. 2 for the temperature dependence and the full spectrum, which is obtained after summation of two degenerate triplet states (they would be Zeeman splitted by a magnetic field). To summarize, there are three limitations for the experimental observation of X^+ triplet states in PL without magnetic field: (i) their relatively small binding energy, (ii) optical transitions are dipole-allowed only for momentum $k > 0$, and (iii) the smaller oscillator strength in comparison with the singlet state. We therefore suggest to apply photoreflectance as in Ref. [8], where indeed a small feature below the exciton is visible.

References

1. A. J. Shields, M. Pepper, D. A. Ritchie, . Y. Simmons, and G. A. C. Jones, Phys. Rev. B **51**, 18 049 (1995).
2. S. Glasberg, G. Finkelstein, H. Shtrikman, and I. Bar-Joseph, Phys. Rev. B **59**, R10 425 (1999).
3. A. Esser, E. Runge, R. Zimmermann, and W. Langbein, Phys. Status Solidi B **221**, 281 (2000).
4. A. Esser, E. Runge, R. Zimmermann, and W. Langbein, Phys. Rev. B **15**, (2000).
5. R. A. Hogg, D. Sanvitto, A. J. Shields, M. Y. Simmons, D. A. Ritchie, and M. Pepper, Physica B **272**, 412 (1999).
6. V. Ciulin, P. Kossacki, S. Haacke, J. Garniere, B. Deveaud, A. Esser, M. Kutrowski, T. Wojtowicz, submitted.
7. A. J. Shields, F. M. Bolton, M. Y. Simmons, M. Pepper, and D. A. Ritchie, Phys. Rev. B **55**, 1970 (1997).
8. A. J. Shields, J. L. Osborne, M. Y. Simmons, M. Pepper, and D. A. Ritchie, Phys. Rev. B **52**, R5523 (1995).
9. A. B. Dzyubenko and A. Yu. Sivachenko, Phys. Rev. Lett. **84**, 4429 (2000).
10. A. Wojs, J. J. Quinn, and P. Hawrylak, Phys. Rev. B **62**, 4630 (2000).

Internal transitions of charged magneto excitons in II–VI quantum well heterostructures

C.J. Meining[1,2], M. Furis[1], H.A. Nickel[1], D.R. Yakovlev[2,3], W. Ossau[2], A. Petrou[1], B.D. McCombe[1]

[1] Department of Physics, SUNY at Buffalo, Buffalo, NY 14260, USA
[2] Physikalisches Institut der Universität Würzburg, 97074 Würzburg, Germany
[3] A.F. Ioffe Physico-Technical Institute RAS, 194021 St. Petersburg, Russia

Abstract   We report the observation of internal triplet transitions of negatively charged magneto excitons (X⁻) in doped and undoped CdTe/CdMgTe quantum wells via optically detected resonance (ODR) experiments. The transitions appear as a shoulder on the low field side of electron cyclotron resonance for filling factors $\nu < 2$ similar to previous observations in GaAs/AlGaAs quantum well heterostructures. In various ZnSe based structures only non-resonant absorption, which oscillates with filling factor, was observed.

In the past few years optically detected resonance (ODR) spectroscopy has been used extensively to investigate the electronic states of free charge carriers and their complexes (neutral (X) and charged (X⁻ and X⁺ magnetoexcitons), as well as bound states of neutral and charged donor impurities [1]. This technique combines far infrared (FIR) and near infrared methods by measuring the small (typically 0.1 %–10 %) intensity changes of a particular band-edge luminescence feature induced by resonant absorption of FIR radiation [1]–[3].

We report here FIR-ODR experiments on II–VI quantum wells (QW's). We have observed electron cyclotron resonance (e-CR) corresponding to an effective mass $m^* = 0.105\,m_0$, and what we believe to be internal transitions of X⁻ for two different CdTe/ Cd₀.₇Mg₀.₃Te QW's. Sample 1 consists of 10 80 Å wide CdTe wells separated by 280 Å Cd₀.₇Mg₀.₃Te barriers and is modulation doped (nominal two-dimensional electron gas (2DEG) density 3 to $4 \times 10^{11}\,\mathrm{cm^{-2}}$). The excess electron density of sample 2, which is nominally undoped, can be tuned via optical pumping above a 1000 Å thick superlattice (SL) (30 Å/30 Å) miniband, separated by 200 Å Cd₀.₇Mg₀.₃Te barriers from the 80 Å CdTe well [4]. In sample 3, a 100 Å wide ZnSe/Zn₀.₈₀Be₀.₀₈Mg₀.₁₂Se modulation doped ($n_e = 4.5 \times 10^{11}\,\mathrm{cm^{-2}}$) single QW, we observe only non-resonant absorption. All samples have been grown by molecular-beam epitaxy on (100) GaAs substrates.

The inset in Fig. 1 displays examples of ODR spectra for sample 1 at various FIR laser wavelengths. For filling factors greater than $\nu = 2$ (i.e. $B \leq 7.5\,\mathrm{T}$) a broad but symmetric CR is observed that can be approximated reasonably well by a Lorentzian line shape. However, below filling factor $\nu = 2$ ($\lambda_{FIR} = 117.7\,\mu\mathrm{m}$, $96.51\,\mu\mathrm{m}$) a strong asymmetry is observed in the ODR data which we attribute to a band of ionizing X^- triplet transitions as previously observed for III–V based structures [3]. The emergence of the X^- triplet transitions only for $\nu < 2$ is in accordance with the evolution of the photo-

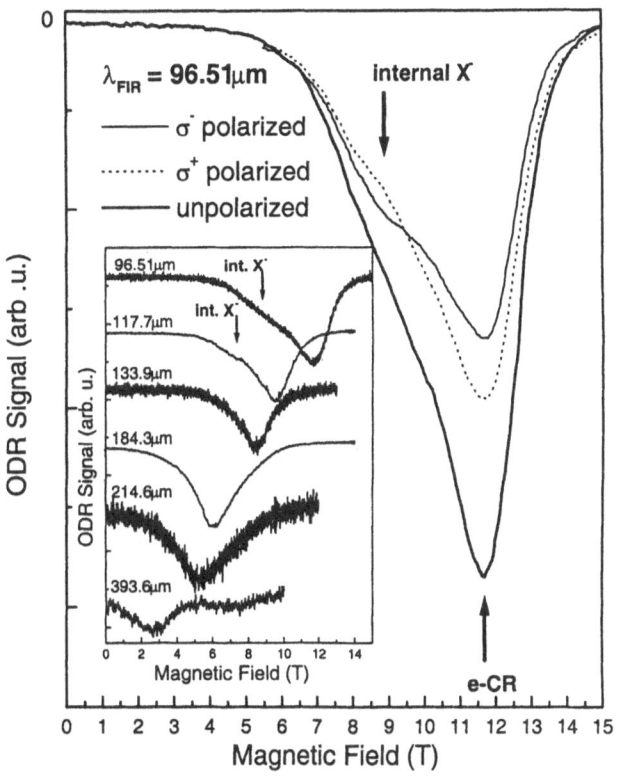

Fig. 1 ODR data for 80 Å CdTe/Cd₀.₇Mg₀.₃Te modulation doped QW's (sample 1): The low field feature (indicated by arrows) leads to asymmetry of the CR and occurs only for $\nu < 2$ (inset). FIR radiation is modulating the σ^- polarized PL component considerably stronger ($T = 4.2\,\mathrm{K}$).

luminescence (PL) spectrum from band-gap renormalized Landau-level-like behavior to exciton-like recombination with increasing magnetic field [5]. Figure 1 also shows a polarization study for $\lambda_{FIR} = 96.51\,\mu\mathrm{m}$ where the asymmetry is particularly strong. FIR Modulation of the polarized PL signal is considerably enhanced on the low field side of CR for σ^- polarized PL compared to the ODR signal for σ^+ polarized PL, and a clear shoulder is seen for σ^-. We believe this is due to a polarization dependence of the ODR mechanism for the two principal bands of ionizing internal X^- transitions [3]. PL modulation by FIR radiation of the band corresponding to the process $X^- \rightarrow X^* + e^0$ (electron in 0th Landau level, X in excited state X^*), is believed to be enhanced for one polarization direction, since $\nu < 2$ the final electron in this process is spin-polarized due to complete filling of

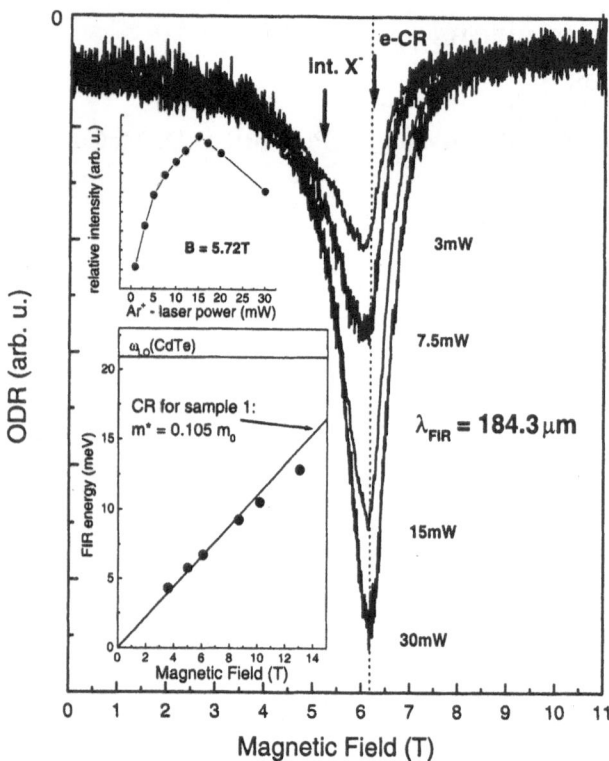

Fig. 2 ODR data for 80 Å CdTe/Cd$_{0.7}$Mg$_{0.3}$Te QW's (sample 2) at various Ar$^+$ laser powers for $\lambda_{FIR} = 184.3\,\mu$m. The Ar$^+$ laser power is given next to each trace ($P_{FIR} = 40$ mW). Upper inset: relative PL peak intensities $I(X^-)/I(X)$ vs. Ar$^+$ laser power. Lower inset: position of CR as a function of FIR photon energy. The dotted line is a guide to the eye ($T = 4.2$ K).

the lowest Landau level spin state. However, this mechanism is not fully understood yet.

The ODR data for sample 2 exhibit a feature on the low field side of CR for all FIR photon energies. This nominally undoped sample shows excitonic luminescence for all magnetic fields investigated. However, at high FIR photon energies e-CR is shifted to higher fields than expected from an effective mass $m^* = 0.105\,m_0$, as seen in sample 1 (see lower inset of Fig. 2). This deviation is attributed to the resonant magneto polaron effect, which is large in CdTe (Fröhlich coupling constant $\alpha = 0.3$). It is not seen in sample 1 due to Pauli Principle effects and screening in this high density sample. The low field deviation of CR for low FIR photon energies is likely to be due to long-range lateral potential fluctuations which lead to localization effects on the CR [7]. This is also responsible for the slight shift of CR to lower fields for low Ar$^+$ laser powers (see dotted line in Fig. 2).

Figure 2 also shows a power study of the ODR signal for $\lambda_{FIR} = 184.3\,\mu$m. The low field asymmetry gains strength relative to CR as the Ar$^+$ laser power is reduced below 30 mW. At 3 mW a clear shoulder is observed (indicated by an arrow). This can be explained in terms of the internal X$^-$ transitions by examining the relative intensities of the X and X$^-$ features in PL. The upper inset

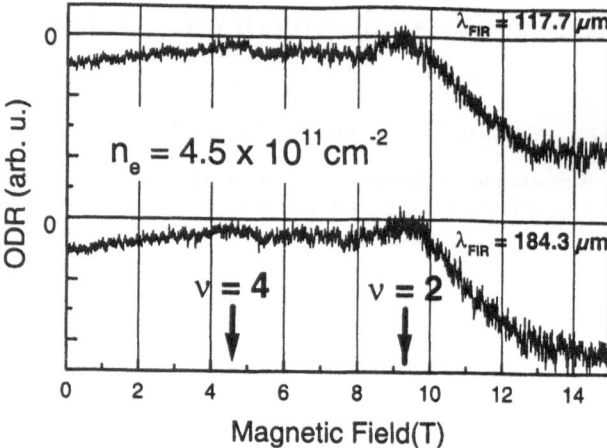

Fig. 3 ODR data for a 100 Å wide ZnSe/Zn$_{0.80}$Be$_{0.08}$Mg$_{0.12}$Se modulation doped QW ($T = 4.2$ K). With $m^* = 0.15\,m_0$ (bulk ZnSe) CR occurs at $B = 13.6$ T for $\lambda_{FIR} = 117.7\,\mu$m and at $B = 8.7$ T for $\lambda_{FIR} = 184.3\,\mu$m at $B = 8.7$ T.

of Fig. 2 shows the relative peak intensities $I(X^-)/I(X)$ of the PL signal for $B = 5.72$ T. The X$^-$ PL intensity increases up to about 15 mW relative to the X intensity and then decreases for higher Ar$^+$ laser powers, probably due to the dissociation of X$^-$ by carrier heating.

The positions of the X$^-$ internal transitions of the doped sample 1 are shifted to lower fields (i.e. higher energies) relative to those of the undoped sample 2. This blue-shift has also been observed recently for the internal X$^-$ transitions in GaAs based QW's [6] and has been extensively studied in the case of the negatively charged donor D$^-$ [8]. For the X$^-$ case it is attributed to excitation of collective magnetoplasma modes bound to the mobile hole.

ODR studies of sample 3 have revealed only non-resonant ODR signals (independent of FIR frequency), which oscillate with magnetic field, with minima at $\nu = 2$ and 4, and which become large for filling factors $\nu < 2$. No ODR signal is observed at the e-CR resonant field (see Fig. 3). This unusual behavior appears to be related to non-resonant heating of the 2DEG which becomes very small when the Fermi level lies in the cyclotron gap. This seems to indicate a lack of an effective heating mechanism at these fields since the PL is not affected by the FIR radiation. This phenomenon is to be investigated further.

Work supported in part by NSF DMR 9722625 and by Deutsche Forschungsgemeinschaft (grant OS98/6).

References

1. J. Kono et al., Phys. Rev. B **52**, (1995) R8654.
2. M.S. Salib et al., Phys. Rev. Lett. **77**, (1996) 1135.
3. A.B. Dzyubenko et al., Physica E **6**, (2000) 156.
4. D.R. Yakovlev et al., Phys. Rev. Lett. **79**, (1997) 3974.
5. D. Gekhtman et al., Phys. Rev. B **54**, (1996) 10320.
6. H.A. Nickel et al., to be published
7. J.P. Cheng et al., Phys. Rev. Lett. **64**, (1990), 3171
8. J.P. Cheng et al., Phys. Rev. Lett. **70**, (1993) 489.

Proc. 25th Int. Conf. Phys. Semicond., Osaka 2000 (Eds. N. Miura and T. Ando)

529

Many-body corrections in n-type 2D telluride structures

V. Huard[1,3] *, R.T. Cox[1,3], K. Saminadayar[1], C. Bourgognon[2], S. Tatarenko[2] and L. Besombes[1]

[1] CEA-Grenoble, Département de Recherche Fondamentale sur la Matière Condensée / SP2M, 17 avenue des Martyrs, 38054 Grenoble Cedex 9, France.
[2] Laboratoire de Spectrométrie Physique, Université J.Fourier, B.P. 87, 38402 Saint Martin d'Hères Cedex, France.
[3] Grenoble High Magnetic Field Laboratory, 38042 Grenoble Cedex 9, France

Abstract Many-body effects seen in optical spectra of CdTe quantum wells containing a 2D electron system are described.

1 Introduction

Electron-electron interactions have been shown to be important in 2D electron systems (2DES) by transport measurements, in the regime of both the integer and the fractional quantum Hall effect. These techniques give only an insight into the electronic states in the vicinity of the Fermi level whereas optical study of a 2DES allows probing of electron states below (emission) as well as above (absorption) the Fermi energy.

Our experiments were carried out on single, 100 Å thick CdTe and CdMnTe quantum wells (QWs) between CdZnMgTe barriers, grown by MBE, strained on CdZnTe substrates (12 % Zn), and modulation doped with Indium or Aluminium donors either symetrically or on one side only. The spacer thickness between doping planes and well edges varied from 100 to 300 Å. The corresponding electron densities n_e were between $3.7\ 10^{11}$ cm^{-2} and 10^{12} cm^{-2}. The PL spectra were taken at T=2K under Ar+ laser excitation (50 μW/cm^2), in magnetic fields up to 20T normal to the 2DEG plane. The specificity of our samples is that they are grown on transparent substrates, which allows us to obtain transmission spectra directly. Also, the strain-induced heavy-light hole splitting in the QW implies that, in our samples, light hole absorption transitions are split off to high energy and only heavy hole transitions are seen in emission. Nevertheless, there is still a small coupling between heavy- and light-holes. This coupling is small enough to keep the band structure parabolic close to the Brillouin zone center but enhances the heavy-hole effective mass. Numerical calculations, following Fishman [1], yield an effective mass in the plane for heavy holes of $m^*_{hh} = 0.25$ m_o for our 100 Å-wide QWs.

2 Density dependence in zero field

The simplest "many-body" effect is a three body effect, the existence of the trion resonance. For infinitesimal n_e, the photohole only interacts with just two electrons with antiparallel spins. With increasing n_e, the exciton peak is seen to move off rapidly to high energy as the energy separation between trion and exciton resonances increases linearly with the electron density [2]. The exciton

* E-mail : vhuard@yahoo.com

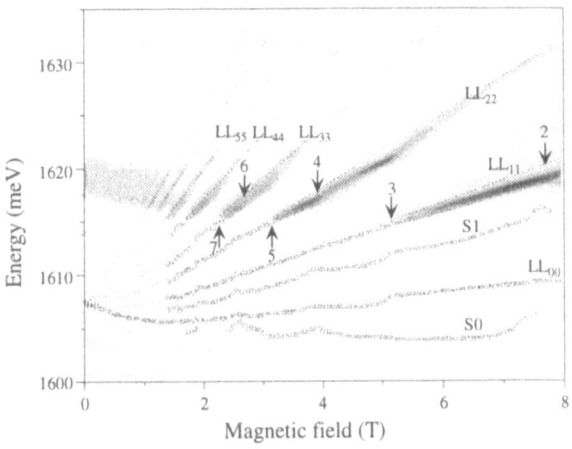

Fig. 1 Absorption intensities (in gray scale) and energies of emission peaks (symbols) against magnetic field for a CdMnTe QW (0.2% Mn) with n_e=3.7 10^{11} cm^{-2}.

peak then broadens and eventually disappears, while the trion absorption peak evolves to the asymmetric Fermi Edge Singularity (FES) shape characteristic of a photohole interacting with many electrons. All this is very well explained by the many-body theory of P. Hawrylak [3] in which a bound state is created where the photohole binds two electrons with antiparallel spins.

3 Magnetic field dependence

Spin-dependent interactions are also very significant in interband magneto-optical spectra of CdTe quantum wells. Absorption energies plotted against magnetic field B show the usual inter-Landau level, diagonal transitions LL$_{nn}$, between electron and hole Landau levels with the same quantum number n ($\Delta n = 0$), see figure 1. In emission spectra, some peaks correspond to these diagonal transitions, but other peaks (S$_0$, S$_1$) do not correspond to any transitions observed in absorption.

We attribute the satellite lines to off-diagonal transitions between Landau levels with different quantum numbers ($\Delta n > 0$), breaking the usual selection rules $\Delta n = 0$ [4]. The selection rules are strictly valid only for first order processes. Lyo [5] showed that the combination of Landau level mixing and low temperature can allow the observation of off-diagonal transitions. The admixture of electron (hole) Landau levels through an impurity- or disorder-electron (hole) matrix element allows off-diagonal transitions. Nevertheless, they should

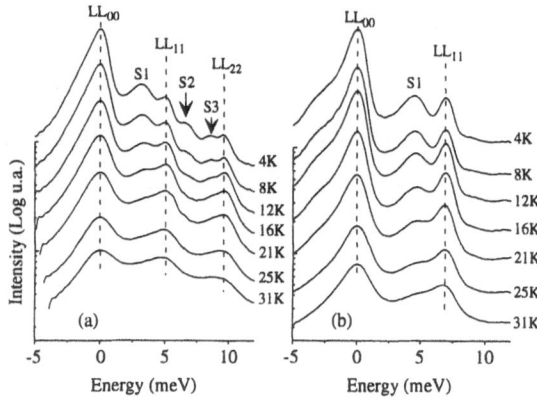

Fig. 2 Evolution of emission spectra as a function of temperature in polarization $\sigma+$ for a CdTe QW with $n_e = 4.7\,10^{11}$ cm^{-2} for two filling factors : (a) $\nu = 5.2$, (b) $\nu = 4.5$

have much lower intensity than standard diagonal transitions related to first-order processes. But, under low-temperature and low laser power conditions, it becomes more favorable to observe off-diagonal transitions as the hole population does not follow a Boltzmann distribution but concentrates on higher Landau levels, especially n=0. On the contrary, at higher temperatures, the hole population follows a Boltzmann distribution and standard diagonal transitions are expected to become dominant. At T = 2 K, we observe the coexistence of standard transitions LL$_{nn}$ and satellite lines. When the temperature is increased, the satellite lines lose intensity till finally only diagonal transitions remain, see figure 2.

4 Many-body corrections

Without considering many-body corrections, the energy difference between two successive peaks corresponding to diagonal transitions should be the sum of electron and hole cyclotron energies $\hbar\omega_{ce} + \hbar\omega_{ch}$. Figure 3 (a) displays the energy difference between LL$_{00}$ and LL$_{11}$ transitions for two CdTe QW with $n_e = 1.08\,10^{12}$ cm^{-2} (open circles) and $n_e = 4.7\,10^{11}$ cm^{-2} (black squares). This experimental energy difference is seen to be equal to the expected sum of electron and hole cyclotron energies (full line) for filling factors greater than 4, when the two Landau levels (n = 0 and 1) are filled. We take $m_{hh}^* = 0.25\,m_o$ and $m_e^* = 0.105\,m_o$ for the electron effective mass (value deduced from the field dependence of a shake-up line).

Figure 3 (b) displays the energy difference between the LL$_{00}$ diagonal and the LL$_{10}$ off-diagonal transitions for the two CdTe QWs previously described. The experimental energy difference deviates from the expected electron cyclotron energy even for $\nu > 4$. These two transitions are characterized by the same initial state before recombination (a hole in the n=0 valence band Landau level), but different final states : formation of a hole in the conduction band n=0 and n=1 Landau levels respectively. The energy difference of these two transitions is equal to the electron cyclotron energy plus the difference of the Hartree-Fock energies $\Sigma_{HF}(0)$ and $\Sigma_{HF}(1)$ re-

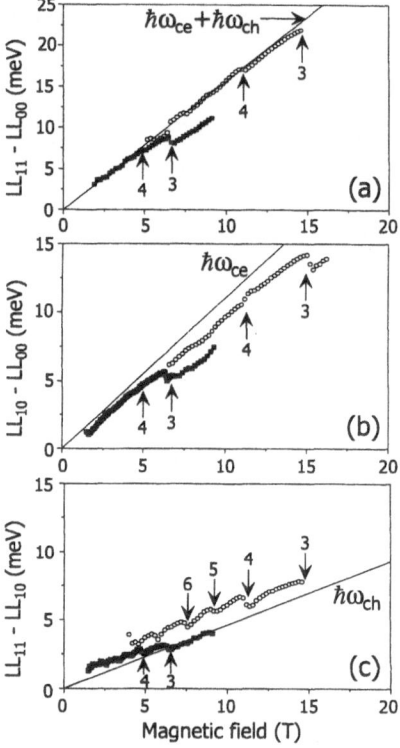

Fig. 3 Energy differences between : (a) successive diagonal transitions LL$_{nn}$ and LL$_{(n+1)(n+1)}$, successive diagonal and off-diagonal transitions (b) LL$_{nn}$ and LL$_{(n+1)(n)}$, (c) LL$_{nn}$ and LL$_{(n)(n-1)}$.

sulting from the interaction of the final state conduction band hole with electronic excitations (electron-electron corrections)[6].

Similarly, Figure 3 (c) displays the energy difference between the LL$_{11}$ diagonal and the LL$_{10}$ off-diagonal transitions. The experimental difference is greater than the expected hole cyclotron energy and presents oscillations as a function of filling factor. These oscillations are related to the existence of excitonic states as initial states before the recombination (excitonic corrections).

In conclusion, we have shown the coexistence of both diagonal and off-diagonal transitions in CdTe quantum wells at low temperatures. Measurement of the energy differences between the two types of transitions provides a new way of studying many-body corrections : it allows us to extract the electron-electron corrections and the electron-valence hole corrections separately, for comparison with theoretical models.

References

1. G. Fishman, Phys. Rev. B **52**, 11132 (1995).
2. V. Huard *et al.*, Phys. Rev. Lett. **84** 187 (2000).
3. P. Hawrylak, Phys. Rev. B **44**, 3821 (1991).
4. S.K. Lyo *et al.*, Phys. Rev. Lett. **61**, 2265 (1988).
5. S.K. Lyo, Phys. Rev. B **40**, 8418 (1989).
6. P. Hawrylak and M. Potemski, Phys. Rev. B **56**, 12386 (1997).

Possibility of excitonic polymers in type-II superlattices

Takuma Tsuchiya

Japan Advanced Institute of Science and Technology (JAIST)
1-1 Asahidai, Tatsunokuchi, Ishikawa 923-1292, Japan

Abstract The ground states of polyexcitons in GaAs/AlAs type-II superlattices were investigated by means of the diffusion Monte Carlo method. It was found that the excitons make bound states in which they form a line along the growth direction with the binding energy above 3 meV. The possibility of polymers of excitons was discussed.

1 Introduction

Polyexcitons, i.e. the bound state of excitons more than two, are possible, when one-particle states are degenerated. In Si and Ge, the conduction band has valley degeneracy and polyexcitons were predicted theoretically [1, 2] and were observed for Si and SiGe. [3–5]

$(GaAs)_n(AlAs)_n$ superlattices become type-II when the layer thickness does not exceed 13 monolayers (ML) or 37 Å, and electrons and holes are confined in AlAs and GaAs layers respectively. [6] In this structure, confinement for electrons and holes is strong, because of the heavy effective mass of electrons along the growth direction and the high barrier height for holes. As a result, the one-particle states in well layers are degenerated. Because of this degeneracy, there is a possibility that many excitons form polymers of excitons, in which excitons are stand in a line along the growth direction. Biexcitons (X_2) in this structure were observed by Nakayama and co-workers [7] and were investigated theoretically. [8]

In order to clarify the possibility of the excitonic polymer state, we investigate in this study the ground state of tri- (X_3) and tetra-exciton (X_4). We employ the diffusion Monte Carlo method, which gives us exact ground state energy within a small statistical error and particle configurations approximately.

2 Method of Calculation

The Hamiltonian for the present problem consists of the sum of the kinetic energies of electrons and holes in the effective mass approximation, the confinement potentials of each particle and the Coulomb potentials between particles. Using the diffusion Monte Carlo method, we can obtain the exact ground state energy, E, within a statistical error. [9] Average of inter-particle distances can also be obtained approximately. From the ground state energies, we can obtain binding energies defined by $B_{X_n} = (E_{X_{n-1}} + E_X) - E_{X_n}$ and $B_X = (E_e + E_h) - E_X$, where 'e' and 'h' denote an electron and a hole.

For many Fermion systems, the fixed-node approximation is indispensable, to take into account the Pauli principle. [9] In the present problem, however, this is not necessary, as long as the number of identical particles in a layer does not exceed $2g_v$, where g_v is the valley degeneracy. It is because the confinement of individual particles in well layers is strong.

3 Numerical Results

Parameters used in the present calculations are summarized in Table 1. Though the band structure of valence bands is complicated, the effective mass for holes is assumed to be isotropic and close to the heavy hole mass of the bulk. The effective mass of electrons is anisotropic, because the confined ground state of electrons comes from anisotropic X_z-states of the bulk AlAs. [6] The valley degeneracy g_v is 2 for electrons and 1 for holes. In the present calculations, the thickness d of the GaAs and the AlAs layers is assumed to be the same.

The calculated binding energies for X_2, X_3, and X_4 in $(GaAs)_n/(AlAs)_n$ superlattices are shown in Fig. 1. For $n < 8$, the confinement of particles in the wells is weak and the approximation of the present calculation is invalid. The result that all the binding energies are positive indicates that the polyexcitons are stable in the present structures. The binding energies for X_3 and X_4 are larger than those of X_2, but the difference is not large.

In order to understand the particle configuration, we show the inter-particle distance along the growth direction, Δz, for X_3 in Fig. 2. In this figure, the distance between an outermost electron (e1) and the other electrons (e2, and e3) and holes (h1, h2, and h3) are shown. The particles are numbered in order of the distance. All the distances almost equal to integral multiple of the layer thickness and Δz_{e1e2} and Δz_{e1e3} and Δz_{e1h1} and Δz_{e1h2} are close to one another. The resulting particle configuration of X_3 is shown in Fig. 3. The configuration for X_4, obtained by the same procedure, is also shown.

4 Discussions

The present results demonstrate that the polyexciton states are stable in type-II superlattices and the binding energy is almost independent of the number of excitons. These results can be explained as follows, if we assume that each exciton forms a dipole and the cou-

Table 1 Parameters used in the present calculations.

$m^*_{e,xy}$ (m_0)	$m^*_{e,z}$ (m_0)	m^*_h (m_0)	ΔV_e (meV)	ΔV_h (meV)	ϵ
0.19	1.1	0.5	280	530	11.3

m_0: electron rest mass

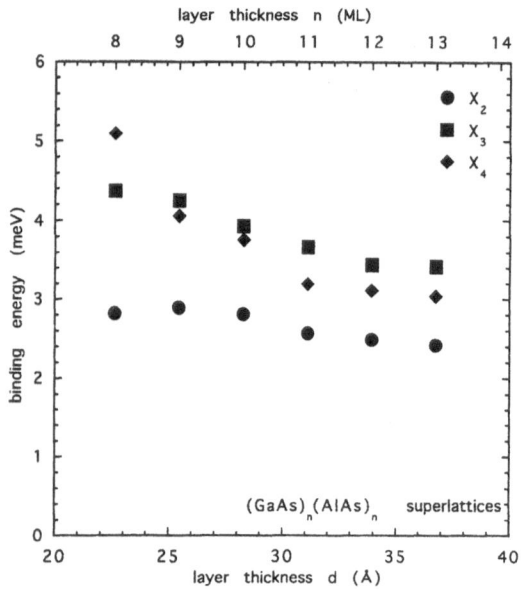

Fig. 1 Binding energies for X_2 (solid circles), X_3 (solid squares), and X_4 (solid diamonds) in $(GaAs)_n(AlAs)_n$ type-II superlattices.

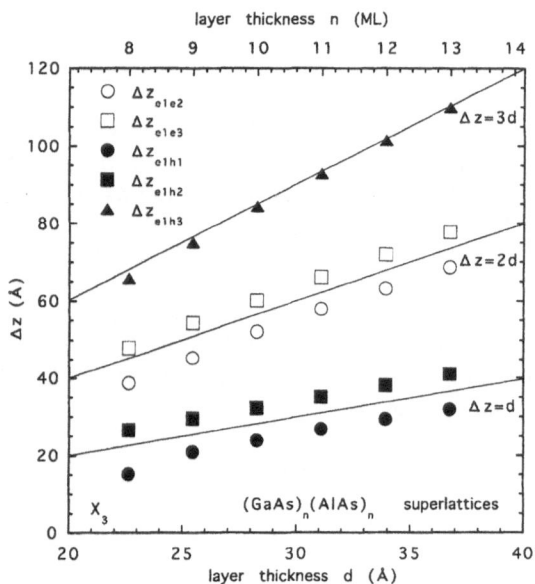

Fig. 2 Distance along the growth direction, Δz, between an outermost electron (e1) and the other electrons, e1 and e2, and holes, h1, h2, and h3, for X_3 in $(GaAs)_n(AlAs)_n$ type-II superlattices.

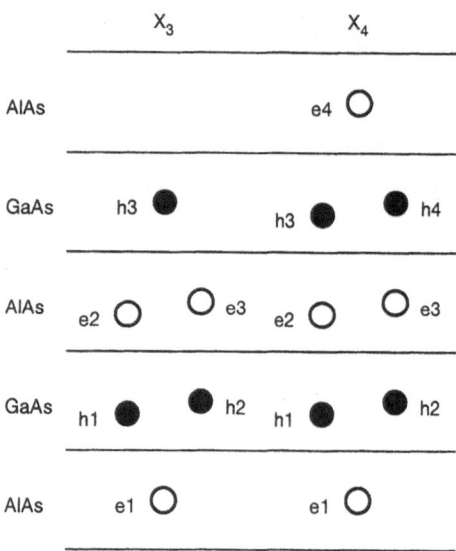

Fig. 3 Schematic illustration for inter-layer configurations of electrons (open circles) and holes (solid circles) of X_3 and X_4 in $(GaAs)_n(AlAs)_n$ type-II superlattices.

uration of particles is locally determined. Consequently, the excitons are expected to form an excitonic polymer along the growth direction.

5 Summary

We have investigated polyexcitons in GaAs/AlAs type-II superlattices, using the diffusion Monte Carlo method. It was found that these states are stable with binding energies comparable to biexcitons. The particle configuration that excitons stand in a line along the growth direction indicates the possibility of the excitonic polymer state in type-II superlattices.

Acknowledgments

This work is partially supported by the Grant-in-Aid for Scientific Research (C) of the Ministry of Education, Science, Sports, and Culture, Japan.

References

1. J.S. Wang, and C. Kittel, Phys. Lett. **42A** (1972) 189.
2. A.C. Cancio and Y.-C. Chang, Phys. Rev. B **42** (1990) 11317.
3. A.G. Steele, W.G. McMullan, and M.L.W. Thewalt, Phys. Rev. Lett. **59** (1987) 2899.
4. M.L.W. Thewalt, V.A. Karasyuk, D.A. Harrison, and D.A. Huber, *Proceedings of the 23rd International Conference on the Physics of Semiconductors* (World Sicentific, Singapore 1996) 341.
5. K. Shum, P.M. Mooney, and J.O. Chu, Phys. Rev. B **60** (1999) 5786.
6. See for example, M. Nakayama, I. Tanaka, I. Kimura, and H. Nishimura, Jpn. J. Appl. Phys. **29** (1990) 41.
7. M. Nakayama, K. Suyama and H. Nishimura, Phys. Rev. B **51** (1995) 7870; Nuovo Cimento **17D** (1995) 1629.
8. T. Tsuchiya, J. Lumin. **87-89** (2000) 509.
9. See for example, B.L. Hammond, W.A. Lester, Jr., and P.J. Reynolds, *Monte Carlo methods in ab initio quantum chemistry* (World Scientific, Singapore 1994).

pling between excitons is dipole-dipole coupling. Actually, excitons, e1h1, e2h2, e3h3, and e4h4 of X_4 in Fig. 3, for example, form dipoles. Because dipole potential decreases rapidly like r^{-3} as the distance r from the dipole increases, the coupling with distant excitons can be ignored. Therefore, the binding energy is dominated by coupling with the nearest excitons and is almost independent of the number of excitons. Further, the config-

Proc. 25th Int. Conf. Phys. Semicond., Osaka 2000 (Eds. N. Miura and T. Ando)

533

Localised Exciton Transitions in High-Quality GaAs/AlGaAs Quantum Wells

A. G. Steffan[1], A. García-Cristóbal[2], R. T. Phillips[1], D. A. Ritchie[1]

[1] Cavendish Laboratory, Madingley Road, Cambridge CB3 0HE, United Kingdom, e-mail: ags26@cus.cam.ac.uk

[2] Departament de Física Aplicada, Universitat de València, E-46100 Burjassot, Spain

Abstract We investigate the origin of optical interband transitions in high-quality GaAs/AlGaAs quantum wells. Using a novel microphotoluminescence technique based on the excitation and collection through a standard single-mode fibre, we are able to detect very narrow emission lines. Furthermore a model describing the localisation at quantum well interfaces is developed, which gives more insight into the nature of the states involved in optical emission.

1 Introduction

Optical and electronic properties of semiconductor heterostructures grown by molecular-beam epitaxy are influenced by their microscopic interface structures. A detailed understanding of the heterointerface is therefore very important to take full advantage of modern growth techniques.

A common method of characterising these interfaces is photoluminescence spectroscopy (PL). Previous studies [1] have demonstrated that quantum wells with interfaces planarised by growth interruption show a monolayer splitting in broad-area experiments and sharp optical transitions if small areas are examined, characteristic of laterally-confined excitonic states.

We have now performed microphotoluminescence measurements in high-quality GaAs/AlGaAs quantum wells grown without growth interruption, which show very narrow emission lines (FWHM $\approx 20\mu eV$), demonstrating that lateral potential fluctuations on a much smaller length-scale (in the range of the exciton radius) can give rise to exciton localisation in quantum wells. To gain a better understanding of the nature of the states involved, we have extended the interface model of H. Castella [2] et al. and applied it to our μ-PL experiments.

2 The Experimental Apparatus

The spatial resolution of this novel μ-PL setup is based on the excitation (633nm) and collection through a standard single-mode fibre with core diameter $4\mu m$ and numerical aperture 0.12. The sample is attached to the cleaved fibre with an adhesive cured under UV light (which allows accurate alignment of the fibre on the sample before fixing) and it is then immersed in liquid helium. The spectral resolution of the measurements was limited to $20\mu eV$ by the spectrometer.

The sample studied was grown at 650^oC without growth interruption or change in growth conditions to produce high-quality QWs with narrow photoluminescence linewidth. It consists of a GaAs buffer grown on a (100) substrate followed by 12 GaAs quantum wells with vary-

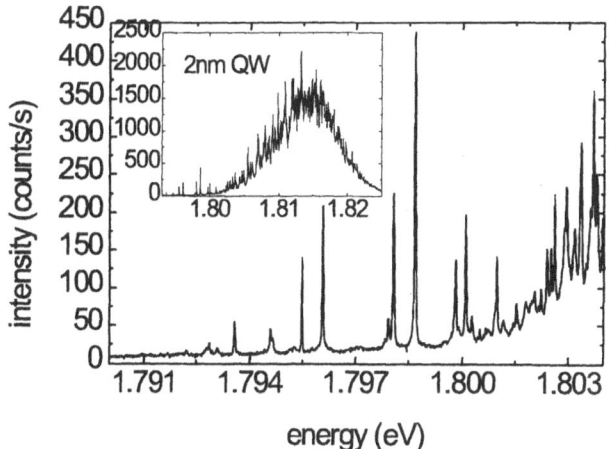

Fig. 1 Low energy tail of the μ-PL spectrum, the inset shows the whole QW peak.

ing width from 2.8 to 30nm, separated by 15nm $Al_{0.3}Ga_{0.7}As$. The last barrier is followed by a 50nm GaAs capping layer.

Previous measurements of the wider quantum wells have verified the high quality of the sample, with a FWHM linewidth of the 30nm QW of $165\mu eV$.

3 Results

Broad-area photoluminescence spectra of narrow QWs show no monolayer splitting, but in μ-PL measurements very narrow optical transitions (FWHM $\approx 20\mu eV$) show up (see Fig. 1). These emission lines come from well-confined states, and are very sharp because of the suppression of scattering mechanisms. The potential fluctuations in these wells are expected to be on a much smaller length scale than that arising in growth interruption - extending to the range of the exciton radius: clearly these fluctuations can also give rise to exciton localisation. In addition to the narrow emission lines a superimposed 'background' is visible. This background arises from more delocalised states, which coexist with the more localised states in energy. In the background, the emission lines merge together due to increased scattering among the closely-spaced states (see also Sec. 4). These sharp transitions can be observed in quantum wells up to 8nm in width and even wider QWs show a definite structure of localised transitions, unresolved because of the reduced lateral confinement in wider quantum wells. This suggests that all states in even "good" quantum wells are subject to lateral confinement.

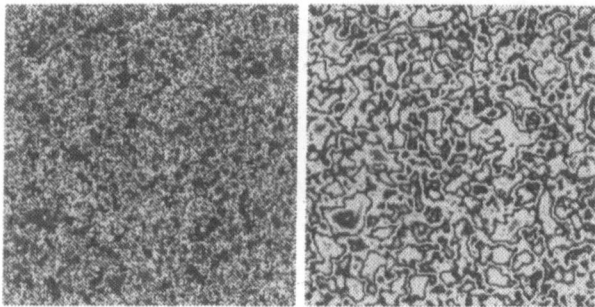

Fig. 2 Interface (left) and potential (right) distribution calculated with $DEN_{-1ML} = DEN_{+1ML} = 0.3$, $\xi = 2.5nm$ and dimensions $WIN_{X,Y} = 200nm$, $DIM_{X,Y} = 256$.

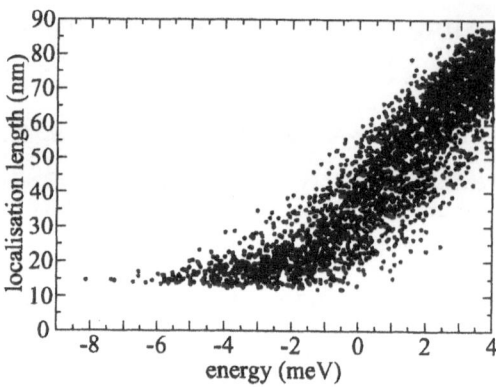

Fig. 4 Corresponding localisation length defined via the participation ratio.

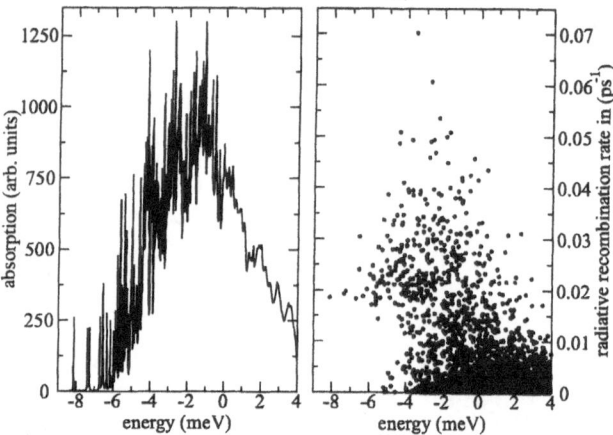

Fig. 3 Modeled absorption spectrum (left) and calculated radiative recombination rates (right).

total outscattering rate out of the state (radiative recombination + acoustic phonon scattering rate) (final state approximation). The radiative recombination rates illustrates the fundamental difference between the low energy tail states ($E \leq -5meV$) (Lifshitz states) with a narrow matrix element- and therefore radiative-rate-distribution, and the higher-lying states (Anderson localised states) which show a much broader (Porter-Thomas) distribution. The localisation length distribution (Fig. 4) demonstrates the same coexistence of localised & more delocalised states, with a minimal localisation length ζ_0 of around the exciton radius ($R_{EX}=10nm$) and an effective mobility edge at zero energy.

4 Calculation

To gain a better understanding of the μ-PL spectra in narrow quantum wells, a model of the interface was developed which includes potential variations introduced by interface-related disorder due to monolayer well-width fluctuations and interface roughness.

4.1 The interface model

Two level-cuts in a Gaussian Random Field (GRF) [2] define the regions of -1, 0 and +1 monolayer, relative to the nominal width of the QW (here $3nm$) with the densities DEN_{+1ML}, DEN_{-1ML} and the correlation length ξ as parameter. The interface potential distribution is then calculated by convoluting the well-width distribution with the squared relative electron and heavy-hole wavefunction, assumed to be a exponential function ($-10meV \leq P \leq 10meV$) (see Fig. 2).

4.2 Solving the problem

With this potential the center-of-mass Schrödinger equation for the 1s exciton state is calculated in real space [3] using the PARPACK package. Figure 3 and 4 show spectra resulting from the addition of 20 realisations, corresponding to a total area of $0.8\mu m^2$. The absorption spectrum shows the same behavior as seen in μ-PL spectra of high-quality QWs; very sharp transition lines on top a broad 'background' of more delocalised states that coexist in energy with a linewidth determined by the

5 Conclusion

Localisation by interface roughness in high quality quantum well samples has already been inferred from measurements of the Stokes shift of the quantum well luminescence peak from the absorption peak. Our measurements now show this explicitly. The fact that sharp optical transitions are visible in wider QWs suggests that we should consider all states even in "good" quantum wells to be subject to lateral confinement.

The qualitative agreement between the experiment and the model allows a further investigation of the states involved, which demonstrates fundamental differences between the strongly localised, low energy, tail states (Lifshitz states) and higher lying, more delocalised states (Anderson-localised states). [1]

References

1. D. Gammon, E. S. Snow, B. V. Shanabrook, D. S. Katzer, and D. Park, Phys. Rev. Lett. **76**, (1996) 3005
2. H. Castella and J. W. Wilkins, Phys. Rev. B **58**, (1998) 16186
3. E. Runge and R. Zimmermann, Advances in Solid State Physics **38**, (1998) 251

[1] This work has been supported by the EPSRC, the EU TMR programme "Ultrafast Quantum Optoelectronics" and by Renishaw plc. Our calculations were performed on the IRIX64 at the Cambridge High Performance Computing Facility.

Field-induced optical anisotropy in semiconductor superlattices : the Wannier-Pockels effect

S. Cortez[1], O. Krebs[1], J.C. Harmand[2], J.L. Gentner[3] and P. Voisin[1]

[1]Laboratoire de Physique de la Matière Condensée de l'Ecole Normale Supérieure
24 rue Lhomond, F75005 Paris, France
[2]Laboratoire de Concepts et Dispositifs Photoniques,
196 Av. H. Ravera, F92220 Bagneux, France
[3]Opto+, route de Nozay, F91460 Marcoussis, France

Abstract: We present measurements and 14 bands **k.p** calculations of the optical anisotropy of semiconductor superlattices under an applied electric field, or superlattice Pockels effect. In the common anion system InGaAs/InAlAs, the effect is governed by the bulk inversion asymmetry, while in the no common atom systems InGaAs/InP, the anisotropy is dominated by the interface symmetry breakdown.

The nature of the eigenstates in a short period superlattice (SL) [1] with an applied electric field is fairly different from the case of a single quantum well. Under an applied electric field, the eigenstates become localized Wannier functions centered on a given position with wings in the surrounding wells. This enables the observation of optical transitions which are direct or indirect in real space. Both the well-known Stark and Wannier-Stark effects are isotropic electro-optical effects and even functions of the applied electric field. Conversely, by analyzing the optical polarization dependence of the absorption spectra, one can observe a linear, anisotropic electro-optical effect associated with the breakdown of roto-inversion symmetry [2]. The native D_{2d} symmetry of a common atom SL like InGaAs/InAlAs is reduced to C_{2v} by an external electric field, yielding an in-plane optical anisotropy. There are two contributions to this "Wannier-Pockels" effect: the bulk inversion asymmetry and the interface induced heavy-light hole mixing [2]. Conversely, "no-common atom" systems like InGaAs-InP have a native C2v symmetry and optical anisotropy associated with the existence of specific interface bonds.

Two samples are studied here. Sample1 is a 20 period InGaAs(55Å)/ InAlAs(11Å) SL, lattice matched to InP. In this case, the interface-induced heavy-light hole mixing has opposite contributions on each side of the wells. Under an applied field of 60 kV/cm, we observe (Fig.1) a 2% polarization anisotropy on the indirect transition, while the direct transition shows an almost negligible anisotropy. Sample2 is a 20 period InGaAs (30Å) / InP (30Å) SL. As seen in Fig. 2, under a field of 110 kV/cm, the indirect (direct) transition has a 7.5% (2.5%) polarization.

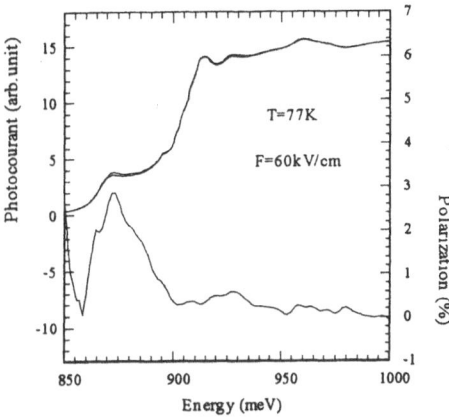

Fig1: Absorption of light polarized along [110] and [-110] and in-plane anisotropy of Sample1 (lower trace)

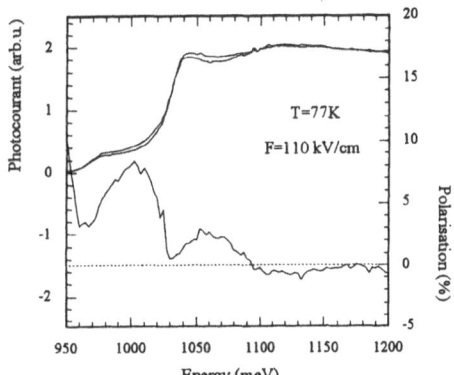

Fig2: Absorption and in-plane anisotropy of Sample2.

Our calculations are based on Pfeffer and Zawadzki 14 band **k**.**p** hamiltonian [3], including extra heavy-light hole mixing introduced by the interface C_{2v} symmetry [2]. A 15 period supercell is used, with 60Å thick barriers on each side, in order to reduce boundary effects when a saw-tooth periodic external electric field is applied. The hamiltonian is expanded and fully diagonalized in a plane wave basis. Eigenstates are calculated only at zero transverse k_t vector in order to reduce the computational effort. A first approximation of the optical spectra is then obtained by assuming a parabolic band dispersion with respect to k_t, and considering the optical matrix elements as independant of k_t. In order to account properly for the resonant bulk Pockels effect, we have included all the momentum matrix elements in the calculation of the **A**.**p** matrix elements. In this framework, the Γ_8, Γ_7 conduction components of the conduction states are optically coupled to the valence band by the momentum matrix element Q. (see ref [3] for notations).

A simulation of sample1 under an applied field of 60kV/cm is presented below. Realistic Kane parameters for both InGaAs and InAlAs are E_{P0}=21eV, E_{P1}=2eV and E_{Q1}=15eV. The remote bands effects are adjusted to fit the experimental effective masses [4,5]. Over 150 valence states and 60 conduction states have been used to obtain the optical spectra. A good quantitative agreement is found with respect to the experiment, although the calculated value is about 30% smaller then the experimental one. In this sample, the interface induced heavy light hole mixing yields only a small contribution to the optical anisotropy (typically 0.2%), compared to the bulk induced anisotropy (1.3% on the indirect transition).

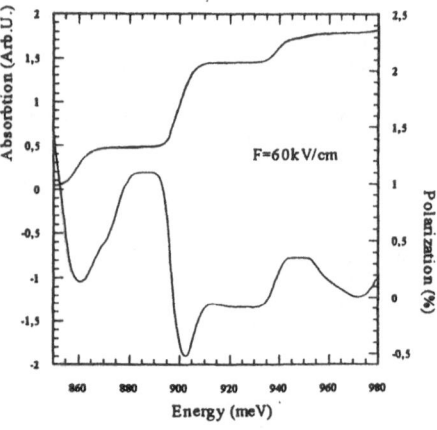

Fig3: Numerical simulation of sample1. The absorption and in-plane anisotropy spectra are plotted with a 16 meV hard shift to account for finite temperature and excitonic effects. A gaussian broadening of 3 meV has been applied.

The much larger anisotropy of Sample2 can only be fitted by a large heavy-light hole interface mixing characteristic of its specific no common atom interface bonds. [2] In that case, the localized heavy hole state is strongly coupled not only to light hole states in the same well, but also to light holes lying essentially in the surrounding wells, yielding a large anisotropy of the indirect transitions.

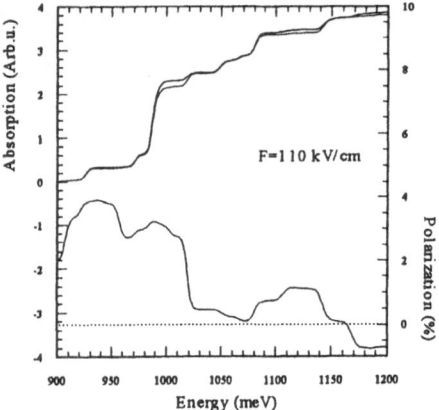

Fig4: Numerical simulation of sample2.

The 'non common atom' interface coupling parameters used are fitted to the electric field dependence of the optical anisotropy of the direct transition between F=50kV/cm and F=190kV/cm. The calculated value of the indirect transition optical anisotropy remains 30% smaller than experimental measurements.

In conclusion, we have studied both experimentally and theoretically the optical anisotropy of short period semiconductor superlattices. The numerical simulation reveals that in the system (InGaAs/InAlAs) the observed anisotropy is a direct consequence of the bulk inversion asymmetry. This effect is partly responsible for the large enhancement of the resonant transmitive electrooptic sampling (TEOS) induced by Bloch oscillations under coherent excitation of semiconductor superlattices. [6]

Aknowledgements The LPMC-ENS is "Unité Associée au CNRS (UMR 8551) et aux Universités Paris 6 et Paris 7". This work is supported in part by INTAS.

References,
1 J.Bleuse, G.Bastard and P.Voisin, Phys.Rev.Lett 60,220 (1988)
2 O.Krebs, P.Voisin Phys.Rev.Letters 77, 1829 (1996) see also:S.Cortez, O.Krebs and P.Voisin, Journal of Vac. Sci. and Tech. B, 18, 2232 (2000)
3 Pfeffer and Zawadzki, Phys. Rev B 53,12813 (1996)
4 Landolt-Bornstein, (ed. Springer-Verlag Berlin)
5 S.Adachi, Physical properties of III-V semiconductor compounds. (ed. Wiley-Interscience Publication)
6 Dekorsky et al., Phys.Rev B 50, 8106 (1994) see also Phys.Rev B 61, R10563 (2000)

Origin of optical-phonon Raman spectra in multiple quantum wells

Bang-fen Zhu[1], Shang Fen Ren[2]

[1] Center for Advanced Study, Tsinghua University, Beijing 100 084, P. R. China
[2] Department of Physics, Illinois State University, Normal IL 61790, USA

Abstract A new model, Fröhlich enhanced density of optical phonon modes, is proposed to clarify the origin of the resonant scattering lineshape in GaAs/AlAs multiple quantum wells, which has recently been assigned to the minima due to anticrossing of interface modes with the odd-order confined modes.

1 Introduction

The optical phonon modes and their Raman scattering spectra in $GaAs/AlAs$ multiple quantum wells (MQW) as the prototype have been extensively studied since last decade. [1,2]

There are two types of optical phonons in GaAs/AlAs MQW, the confined modes and the dispersive interface modes (IF). According to the symmetry analysis and microscopic theory, the even-order confined modes are allowed in the polarized backscattering configuration and the odd-order confined modes are observable for depolarized configuration; on the other hand, the IF modes evolved in backscattering are associated with zero parallel-wavevector q_{xy}, which are forbidden for Fröhlich scattering.

This theory has obtained wide supports both experimentally and theoretically [1,3]. Recently, Shields et al. has challenged this theory, [4,5], demonstrating that, because of the imperfections, the resonant Raman scattering peaks in MQW are not due to the even-order confined modes (except for $n = 2$), but result from the dips produced by anticrossings of the IF bands with the odd-order confined modes. By a model calculation they have successfully explained the resonant Raman spectra in MQW of identical wellwidth but different barrier thickness.

However, since all the optical phonon modes may contribute to Raman scattering to some extent if the wavevector conservation law is broken, the prerequisite of the "dip"-assignment depends critically on the predominance of the IF modes with non-zero wavenumbers over other modes in Fröhlich scattering. Unfortunately, generally speaking, this is not true. Particularly, for MQW with barrier thicker than well, the potential associated with the upper IF-bands in the well region is mainly antisymmetric, [6] which is a much weaker channel for intra-subband scattering compared to the more symmetric even-order modes.

There is, of course, another possibility, i.e. the minimum of the phonon density of states(DOS), the anticrossing regions, and the spectrum dips lie at identical frequencies; however, as shown by our microscopic calculation (Fig.1), the anticrossing region does not correlate to the minimum of the DOS, some even is very close to the maximum of DOS.

In this presentation we intend to clarify the origin of the "peaks" in the out-going resonant Raman spectra by a special model, the "DOS enhanced by Fröhlich-interaction" model.

2 Model

The optical-phonon modes in MQW $(GaAs)_{16}(AlAs)_m$ $(m = 8, 16, 30)$ are calculated by the microscopic dipole superlattice model proposed by Huang and Zhu[7,8]. This model is parallel to usual effective mass model for electronic states in MQW. To work out the vibrational modes of superlattice A/B, the relative motion of the oppositely charged ions in the unit cell of the bulk material is simulated as a charged oscillator with an intrinsic frequency ω_0, and $\Delta\omega_0^2 = \omega_0^2(B) - \omega_0^2(A)$ is thought as the phonon barrier added in B-region. We expand the phonon mode in superlattice with wavevector $\mathbf{q}$, mode-

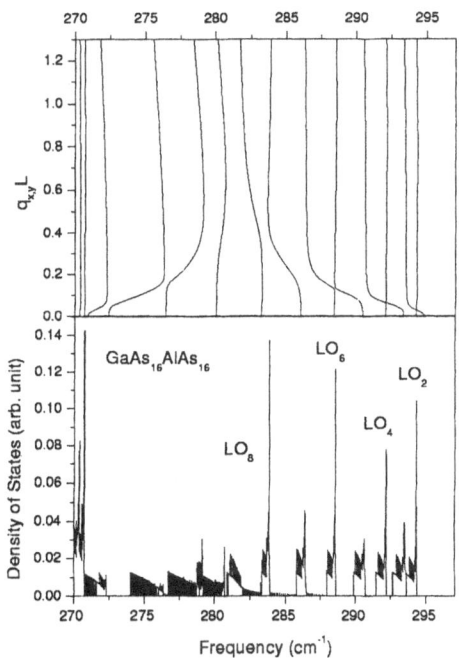

Fig. 1 The optical phonon dispersion for fixed q_z and the corresponding phonon density of states.

order index n and polarization i, $\mathbf{u}_n(\mathbf{q}, i)$, in terms of bulk phonon waves, $\mathbf{u}(\mathbf{q}_S, j)$,

$$\mathbf{u}_n(\mathbf{q}, i) = \sum_{S,j} a_{n,S,i} \mathbf{u}(\mathbf{q}_S, j). \tag{1}$$

Owing to the superlattice periodicity of L, this "perturbation" will couple together the bulk A-modes with different $\mathbf{q}_S$. Here supperlattice wavevector $\mathbf{q}$ is restricted to the minizone, $\mathbf{q}_S = \mathbf{q} + S2\pi/L\hat{z}$ (S is integer), and j denotes the polarization index. The calculated phonon dispersion curves based on this model are very close to that with other microscopic models.

In the "DOS enhanced by Fröhlich-interaction" model, two effects are taken into account. One is the DOS effect resulted from the relaxation of the conservation law of quasi-momentum owing to various imperfections in MQW. Another is the matrix element effect of the electron-phonon Fröhlich interaction. Unlike amorphous materials, phonon wave vector is still defined and meaningful in MQW, and the matrix element of Fröhlich interaction in ideal superlattices should play some role. Thus it is assumed that the out-going resonance Raman intensity can be expressed as

$$I(\omega) = \frac{1 + f(\omega)}{\omega} \sum_{n,\mathbf{q},i}^{q_{xy,max}} C_{n,\mathbf{q},i} \delta(\omega - \omega_{n,\mathbf{q},i}). \tag{2}$$

The weighting factor, $C_{n,\mathbf{q},i}$, for all modes is taken to be unity, except for the mode which is of approximate even-parity and its expansion coefficient for the $\mathbf{u}(\mathbf{q}_0, LO)$-bulk component (cf. Eq. 1) is beyond a certain parameter C,

$$|a_{n,S=0,LO}| \geq C. \tag{3}$$

The weighting factor for such a mode is then taken as

$$C_{n,\mathbf{q},LO} = \left(\frac{a_{n,S=0,LO}}{C|q|}\right)^2. \tag{4}$$

The reason why the weighting factor is expressed as above is simply due to the fact that the Fröhlich interaction is proportional to the wavelength of the optical phonon mode. Although the Fröhlich Raman scattering is forbidden in bulk materials because of the cancellation between the electron's and the hole's contributions in the dipole approximation, it becomes allowed in MQW under the resonance condition, because of the heavy- and light hole mixing in the MQW and the different penetration into the barrier for the electron and the heavy hole.[2]

3 Results and Discussion

The numerical results based on such model as shown in Fig. 2 have well reproduced the main features of the outgoing resonant Raman scattering experiments in MQW samples of identical wellwidth but different barrier thickness and agree with the numerical results by Shields et al. Regarding the fact that the first three calculated peaks have nearly the same Raman shifts in three MQW of $(GaAs)_{16}(AlAs)_m$ ($m = 8, 16, 30$) which deviate little from the calculated frequencies of DOS peaks, it is clear

that they should be assigned to the $LO_2 LO_4$, and LO_6 peaks. On the other hand, the peak around $280cm^{-1}$, which does not correlate to a sharp DOS peak and is absent in $(GaAs)_{16}(AlAs)_8$, results from the enhanced Fröhlich interaction of the lower IF branch in sample with thicker barrier. The numerical results above are not sensitive to chosen parameters, such as C, $q_{xy,max}$, the broadening factor etc.

We'd like also to point out that, the $4th$ peak in our calculated Raman spectra corresponds in frequency to the anticrossing region in the dispersion curve of optical phonons. This further confirms that the Raman spectra in GaAs MQW do not come from the anticrossing dips.

In conclusion, a model of "DOS enhanced by Fröhlich-interaction" has been proposed, demonstrating that the combination of the matrix element effect and the density of states of the optical phonons are responsible for the structures in resonant Raman spectra.

References

1. B. Jusserand and M. Cardona *Light Scattering in Solids V*, ed. M. Cardona (Springer-Verlag, Berlin 1989).
2. Kun Huang, Bang-fen Zhu, and Hui Tang, Phys. Rev. B **41**,1990 5825; Bang-fen Zhu, Kun Huang, and Hui Tang, Phys. Rev. B **40**,1989 6299.
3. A.K. Sood, M. Mendendez, M. Cardona, and K. Ploog, Phys. Rev. Lett.**54**,1985 2111.
4. A.J. Shields, M. Cardona, and K. Eberl Phys. Rev. Lett.**72**,1994 412.
5. A.J. Shields, M.P. Chamberlain, M. Cardona, and K. Eberl, Phys. Rev. B **51**1995, 17728.
6. A.K. Sood, M. Mendendez, M. Cardona, and K. Ploog, Phys. Rev. Lett. **54**1985 2115.
7. Kun Huang, and Bang-fen Zhu, Phys. Rev. B **38**,1988 2183; **38**,1988 13377.
8. Bang-fen Zhu, Phys. Rev. B **38**,1988 7694.

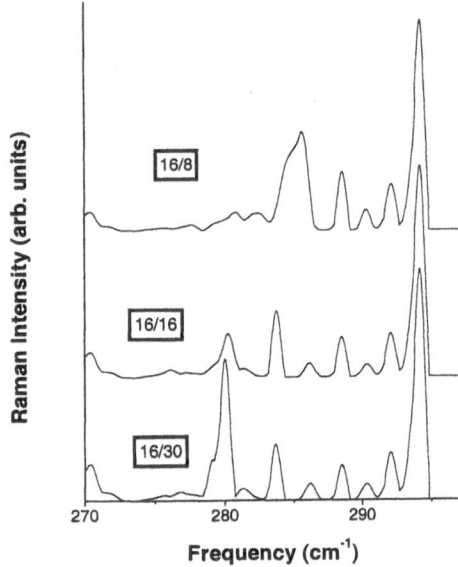

Fig. 2 The Raman intensity calculated by the model for three MQW of $(GaAs)_{16}(AlAs)_m$ ($m = 8, 16, 30$))

Proc. 25th Int. Conf. Phys. Semicond., Osaka 2000 (Eds. N. Miura and T. Ando)

539

Temperature dependence of the band gap in (InGa)(AsN)/GaAs single quantum wells

A. Polimeni[1], M. Capizzi[1], M. Geddo[2], M. Fischer[3], M. Reinhardt[3], A. Forchel[3]

[1] INFM-Dip. di Fisica, Università "La Sapienza", Piazzale A. Moro 2, I-00185 Roma, Italy e-mail: polimeni@roma1.infn.it
[2] INFM-UdR Pavia, via Bassi 6, I-27100 Pavia and Dip. di Fisica Università di Parma, Viale delle Scienze 7a, I-43100 Parma, Italy
[3] Universität Würzburg, Technische Physik, Am Hubland 97074 Würzburg, Germany

Abstract The electronic properties of $In_xGa_{1-x}As_{1-y}N_y$/GaAs single quantum wells (QW's) have been investigated by photoluminescence and photoreflectance spectroscopy. The thermal red-shift of the ground state recombination energy between low ($T=10$ K) and room temperature in QW's containing N is smaller than that found in N-free samples. An anticrossing between states of the conduction band edge and a N-induced localized level resonant with the conduction band accounts for this finding. An estimate of the strength of the band interaction as a function of N concentration is obtained in good agreement with two-band model predictions.

1 Introduction

(InGa)(AsN)/GaAs heterostructures are a feasible alternative to the (InGa)(AsP)/InP system for accessing the transmission windows of silica fibres [1]. This is possible thanks to the incorporation of a small amount of nitrogen in the (InGa)As matrix. In fact, a dramatic red-shift of the alloy band-gap is commonly observed – up to 100 meV per percent of N [2]. The physical origin of such band-gap shrinkage is still a matter of debate. Pressure-dependent photoreflectance measurements have been phenomenologically interpreted by assuming that N gives rise to a level, E_N, localized in real-space and degenerate with the (InGa)As conduction band (CB) [2]. The repulsive interaction between the CB extended states and E_N leads to the observed band gap shrinkage. In a second model, pseudopotential supercell calculations in $GaAs_{1-y}N_y$ show that nitrogen causes a mixing of Γ and L-states of the conduction band, which increases with N content [3]. Γ-L intermixing effects have been observed by ballistic electron emission spectroscopy [4] and Raman scattering experiments [5] in Ga(AsN) epilayers.

We investigated optically several $In_xGa_{1-x}As_{1-y}N_y$/GaAs single quantum wells (QW's) differing for the indium (x=0.25-0.38) and nitrogen (y=0-0.053) concentration. The comparison of the thermal red-shift, RS, of the ground state transition energy on going from $T=10$ K to room temperature in samples with and without nitrogen indicates a strong influence of nitrogen on the thermal properties of the $In_xGa_{1-x}As_{1-y}N_y$ QW's, namely, a decrease of RS with increasing y. In a band anticrossing framework [2], we fully account for the thermal red-shift observed in all investigated samples and estimate the interaction energy, V_{MN}, between the (InGa)As CB edge and the N localized level for different y's. The dependence of V_{MN} on y agrees well with sp^3s^* tight binding predictions [6].

2 Experiment and results

Ten $In_xGa_{1-x}As_{1-y}N_y$/GaAs single quantum wells grown by solid source molecular beam epitaxy have been grown. The (InGa)(AsN) well thickness, L, ranges between 6.0 and 8.2 nm. Photoluminescence (PL) was excited by the 515 nm line of an Ar^+ laser and detected by a Ge detector. Photoreflectance (PR) measurements were performed with a 100 W halogen lamp as a probe source. The excitation source was provided by a He-Ne laser (power 2 mW).

Figure 1 shows a comparison between room temperature PL and PR spectra of a sample having x=0.32, y=0.033, and L=6.0 nm. The PL band is due to heavy-hole free exciton recombination in the well. In PR, two transitions are observed and attributed to heavy- (HH) and light- (LH) hole excitons. The HH recombination energies as measured by PR and PL are offset as a consequence of the Stokes shift between absorption and emission measurements. As T is decreased, the temperature dependence of the PL line shape shows a progressive carrier trapping from extended to localized states.

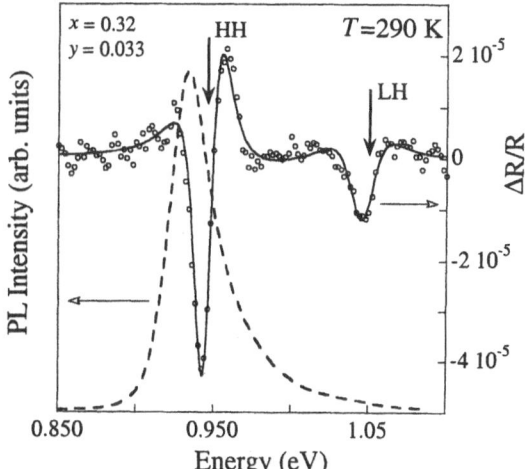

Fig. 1 Room temperature photoluminescence (dashed line) and photoreflectance (open circles) spectra of a QW having x=0.32, y=0.033, and L=6.0 nm. The continuous line is a simulation of the PR spectrum. HH and LH are heavy- and light-hole exciton transitions, respectively. Vertical arrows indicate the transition energies as determined by the simulation to the data.

Apart from localization effects, nitrogen can influence the rate at which the band gap shrinks when the temperature increases. In fact, in the interaction model mentioned above [2], N gives rise to a localized level whose energy E_N should not vary much with T, as observed in the case of deep centres [7]. In turn, the lowest electronic level resulting from

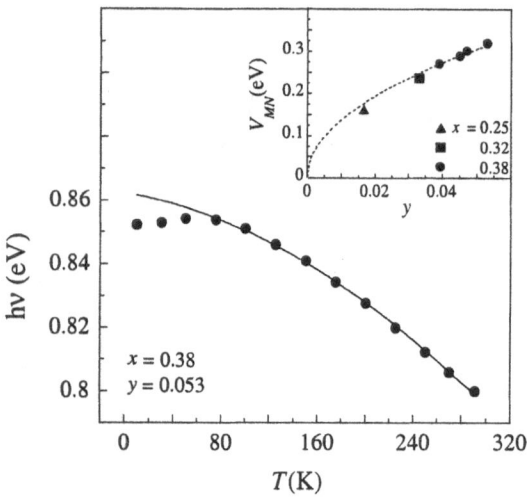

Fig. 2 PL peak energy $h\nu$ (full dots) vs. temperature for a sample with $x=0.38$, $y=0.053$, $L=8.2$ nm. The continuous line is the best fit of the interaction model to the data, as discussed in the text. Inset: The values of V_{MN} as obtained by a fit similar to that shown in the figure are shown as a function of N concentration. Dashed line is a fit of a square root law to the data.

the interaction between E_N and the CB edge "bears" part of the T-insensitivity of E_N. We propose that the temperature dependence of the resulting exciton recombination energy can be written as

$$h\nu_{exc}(T) = 1/2 \{ E_N + E_M(T) - [(E_N - E_M(T))^2 + 4 V_{MN}^2]^{1/2} \}, \quad (1)$$

where E_M is the energy of the (InGa)(AsN) *unperturbed* conduction subband, and V_{MN} describes the strength of the band mixing between E_N and the (InGa)As CB edge. For $V_{MN} = 0$ (i. e., $y=0$) Eq. (1) gives the temperature dependence of the recombination energy of N-free QW's. For increasing V_{MN} (or equivalently y), the term in square brackets tends to lower the ground state energy as well as to reduce the thermal shift of $h\nu_{exc}$. Equation (1) can be applied to the experimental data provided that the dependences of E_N, E_M, and V_{MN} on the nitrogen and indium content of the well are known. Unfortunately, these quantities (E_M and V_{MN}, at least) are not directly accessible by experiments. In Ref. 6, the functional dependences of E_N, E_M, and V_{MN} on nitrogen concentration have been derived by an sp^{3}s* tight-binding Hamiltonian model for the GaAs$_{1-y}$N$_y$/GaAs system. In that work, it is found that $E_M(y) = E_c^0 - \alpha y$, $E_N(y) = E_N^0 - \gamma y$, and $V_{MN}(y) = -\beta\sqrt{y}$. In our analysis, we set $E_c^0(T)$ equal to the T dependence of the free exciton recombination energy experimentally determined in the QW's *without* nitrogen. Furthermore, we set $\alpha=1.55$ eV, $E_N^0=1.65$ eV (as in Ref. 6), and $\gamma=5.0$ eV in order to take into account the expected lower energy position of the N level in the (InGa)As matrix with respect to the GaAs matrix. Figure 2 shows a best fit of Eq. (1) (continuous line) to the temperature dependence (full dots) of the PL peak position of a QW having $x=0.38$, $y=0.053$, and $L=8.2$ nm. The deviation of the theoretical data

from the experimental points observed for $T<100$ K is due to an increasing contribution of localized excitons to the low temperature PL spectra. A quite satisfactory agreement between the experimental data and the model is obtained for $V_{MN}=0.336$ eV. This quantity has been used as a fit parameter throughout the analysis performed on the samples considered in this work (leaving all the other parameters unchanged). In the inset of Fig. 2, the values of V_{MN} so obtained have been plotted as a function of y. Since the interaction energy is expected to be proportional to the square root of the nitrogen concentration, a square root law has been fitted to the data (dashed line in the inset of Fig. 2; $V_{MN}(y) = \beta\sqrt{y}$). The value of β (=1.36 eV) is comparable to that (1.5 eV) calculated for Ga(AsN) [6]. This provides an experimental support to theoretical predictions as well as a quantitative test for extending the sp^{3}s* tight-binding model of Ref. 6 to (InGa)(AsN) alloys.

3 Conclusions

We studied the influence of temperature on the PL and PR properties of (InGa)(AsN) QW's. We found that localized and free carriers dominate the radiative recombination processes at low and high T, respectively. A quantitative analysis of the thermal red-shift of the recombination energy provides an estimate of the interaction energy, V_{MN}, between the states of the bottom of the unperturbed conduction band of the (InGa)(AsN) lattice and a localized level induced by nitrogen. The estimated dependence of V_{MN} on y is in good agreement with the predictions of a recent two-band model.

References

1. M. Kondow, K. Uomi, A. Niwa, T. Kitatani, S. Watahiki, and Y. Yazawa, Jpn. J. Appl. Phys. **35**, (1996) 1273.

2. W. Shan, W. Walukiewicz, J. W. Ager III, E. E. Haller, J. F. Geisz, D. J. Friedman, J. M. Olson, and S. R. Kurtz, Phys. Rev. Lett. **82**, (1999) 1221.

3. T. Mattila, S. H. Wei, and A. Zunger, Phys. Rev. B **60**, (1999) R11245.

4. M. Kozhenikov, V. Narayanamurti, C. V. Reddy, H. P. Xin, C. W. Tu, A. Mascarenhas, and Y. Zhang, Phys. Rev. B **61**, (2000) R7861.

5. H. M. Cheong, Y. Zhang, A. Mascarenhas, and J. F. Geisz, Phys. Rev. B **61**, (2000) 13687.

6. A. Lindsay, and E. P. O'Reilly, Solid State Commun. **112**, (1999) 443.

7. P. W. Yu, Phys. Rev. B **42**, (1990) 11889.

Charge transfer of carriers by interband photoexcitation in asymmetric GaAs/AlGaAs coupled quantum wells

M. Levy[1], Yu. L. Khait[1], R. Beserman[1], A. Sa'ar[2], V. Thierry-Mieg[3], R. Planel[3]

[1] Solid State Institute and Physics Department, Technion, Israel Institute of Technology, Haifa 32000, Israel. e-mail: sslevy@tx.technion.ac.il.
[2] Department of Applied Physics, The Fredi and Nadine Hermann School of Applied Science, The Hebrew University of Jerusalem, Jerusalem 91904, Israel.
[3] Laboratoire de Microstructures et Microelectronique - CNRS, 196 Avenue H. Ravera, BP107, 92225 Bagneux, France.

Abstract The influence of interband photoexcitation intensity on the photoluminescence line shape was investigated in GaAs/AlGaAs modulation doped asymmetric structure composed of a wide quantum well weakly coupled to a narrow quantum well. The emission spectra show a broadening and a narrowing of the emission line widths from the wide quantum well and from the narrow quantum well respectively, with increasing laser power. In addition we observe a fast increase of the emission intensity from the narrow quantum well with respect to that from the wide quantum well. These processes are shown to be the result of carrier transfer into the narrow quantum well. We propose a charge transfer model to explain semi-quantitatively our results.

Asymmetric coupled quantum wells (ACQW) have recently gained much interest due to their potential applications such as mid-infrared (IR) photodetectors. These asymmetric ACQW are composed of a wide and a narrow GaAs quantum wells (QW's), separated by AlGaAs barrier which prevents the easy tunneling of free carriers between the QW's. The introduction of Si dopant results in the accumulation of a two dimensional electron gas (2DEG) in the wide QW (WQW), part of the 2DEG can be transfer to the narrow QW (NQW) by IR excitation, creating a photoinduced electric field. The asymmetry of this structure makes it a superior device for the study of non-linear effects such as the optical non-linear rectification [1], and for the study of the influence of IR photoinduced local electric fields on the localization of energy levels [2],[3]. The IR radiation in these structures is absorbed by 2DEG often introduced by modulation doping. The knowledge of the physical fundamental properties such as the 2DEG density, the built-in electric field and the band bending are crucial for structural engineering.

In this paper we study the influence of the laser excitation power on the photoluminescence (PL) line shape from ACQW structure. From the dependence of the PL line shape parameters such as the line width and integrated intensity we try to gain insight into the microscopic mechanisms that dominate the charge transfer processes in the ACQW. The charge transfer model explains well the behavior of the PL line shape as a function of laser excitation power. The sample grown

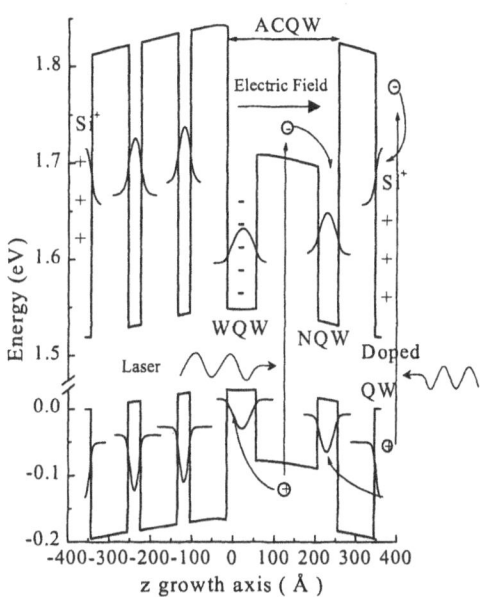

Fig. 1 Description of the structure, with the GaAs WQW (7 nm) and NQW (5 nm) separated by 15 nm Al0.2Ga0.8As intermediate barrier. Also shown is the doped 3 nm QW separated by 9 nm Al0.35Ga0.65As barrier. The arrows show schematically the photoexciting laser and the motion of photoexcited electron-hole pairs in the presence of the built in electric field.

on semi-insulating GaAs substrate contains 25 periods of ACQW's (fig. 1).

Each ACQW is composed of a WQW (7 nm) which contains the 2DEG and a NQW (5 nm). These two weakly coupled QW's are separated by a 15 nm $Al_{0.2}Ga_{0.8}As$ intermediate barrier. The whole ACQW structure is enclosed between a Bragg reflectors to increase the confinement of the electronic wave function in the ACQW. The sample is modulated doped with Si atoms to the level of 2×10^{11} cm^{-2} and the doping is inserted in a 3 nm QW separated by 9 nm $Al_{0.35}Ga_{0.65}As$ from the NQW (fig. 1). The Si dopands release their electrons into the WQW, which has the lowest electronic level so that it contains the whole 2DEG at low temperatures while the NQW is initially empty.

Under laser excitation, electron-hole pairs are created, which recombine radiatively in the WQW and in

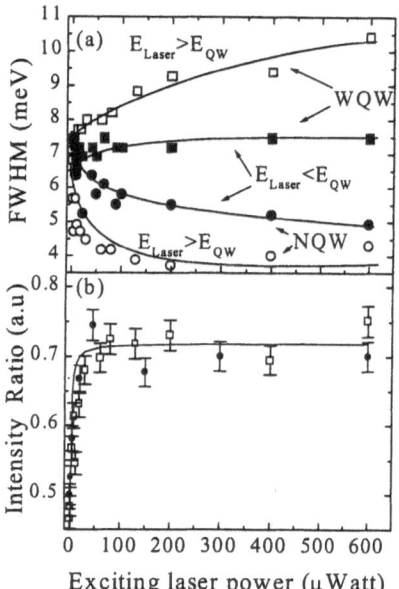

Fig. 2 (a)-The FWHM of the PL line from the WQW and NQW as a function of laser power. The open squares and circles are from the high-energy laser (1.825 eV), the solid squares and circles are measured by lower energy laser (1.62 eV). The solid lines are a guide to the eye. (b)-The Intensity ratio between the NQW and WQW as a function of laser power , the symbols are the same as in (a).

the NQW. Hence the PL spectrum is composed of two peaks. These two peaks behave differently as a function of the interband laser excitation power. We observe a broadening of the PL line shape from the WQW which contain the 2DEG and a narrowing of the PL line from the initially empty NQW. These changes in the FWHM as a function of the exciting laser intensity are summarized in fig. 2. Fig. 2(a) shows the line-width as a function of the exciting laser power when the laser photon energy was tuned to 1.825 eV above the band gap of the doped QW (open squares and circles). The figure shows that with increasing laser power the PL line-width from the WQW (open squares) is broadened by 50 7 meV to 11 meV) while the PL line-width associated with the NQW (open circles) becomes narrower by the same amount (from 7 meV to 3.5 meV). The full circles and squares show the change in the line width when the laser energy is below the band gap of the doped QW. at this lower photon energy this behaviour is less pronounced. The integrated emission ratio between the NQW and the WQW at these two photon energies are shown in fig. 2b (open squares). This ratio increases initially fast with the laser excitation and then it saturates. This enhancement of the ratio is due to a more intense emission from the NQW than from the WQW as the exciting laser power increases.

We explain this increase of the PL intensity and the narrowing of the PL line width from the NQW by suggesting that the NQW has an additional source of electron-hole recombination. Due to the modulation doping of

the sample, the WQW contains a 2DEG i.e. a negative charged sheet. On the other hand the ionized Si impurities in the 3 nm doped QW create a positive charged sheet. During the high-energy photon excitation (1.825 eV), electron-hole pairs are generated in the WQW, in the NQW and in the 3 nm Si doped QW (which is located 9 nm away from the NQW, see fig 1). Part of the photogenerated electrons in the WQW are repelled by the electric force toward the NQW contributing to an excess of electrons in the NQW, while the holes are captured and contribute to a radiative recombination with the 2DEG in the WQW. In the 3 nm doped QW which contains ionized Si+ impurities, part of the photoexcited electrons are captured by these impurities leaving free photoexcited holes. These holes are repelled by the positively ions and can tunnel efficiently to the NQW which is close to this doped QW, but they do not tunnel as efficiently to the WQW which is further away. Therefore the NQW contains additional electron-hole pairs which originates from the repelled electrons from the WQW and holes from the 3 nm Si doped QW which tend to recombine from the edge of the fermi levels in the conduction and valence bands. These higher-energy electron-hole recombinations increases the luminescence intensity from the NQW with respect to that of the WQW (Fig. 2(b)) and they are also responsible for the narrowing of the PL line of the NQW as can be predicted by the self consistent Born approximation formula for the line width [4].

$$\Gamma_{\alpha,k} = \sum_{k'} \frac{U_q^{\alpha,\alpha} \Gamma_{\alpha,k'}}{((\varepsilon_{\alpha,k} - \varepsilon_{\alpha,k'})^2 + \Gamma_{\alpha,k'}^2)} \qquad (1)$$

Here $\alpha=c$ ($\alpha=v$) defines the conduction (valence) electrons, where k and k' are the initial and final wave vectors. ε is the renormalized kinetic energy measured from the band edges and U is the scattering energy of the electrons and holes from the ionized-impurity [4]. Eq. 1 shows that the PL line width is relatively large for contributions from a low-energy thermally occupied electrons and holes and small for contributions arising from relatively high-energy electrons and holes. Hence we explain the narrowing of the NQW PL line width with increasing photoexcitation power by the increased contribution of electron-hole pairs with a higher energy recombining from the edge of the Fermi sea. On the other hand the emission from the WQW is dominated by the low energy thermally occupied holes. These low energy carriers recombining with the bottom electrons of the Fermi sea give rise to a broad PL line width.

References

1. E. Rosencher, Ph. Bois, B. Vinter and D. Kaplan, Appl. Phys. Lett. **56**, (1990) 1822.
2. M. Bendayan, R. Kapon, R. Beserman, A. Sa'ar, R. Planel, Phys. Rev. B, **56** (1997), 9239.
3. M. Levy, R. Kapon, A. Sa'ar, R. Beserman, V. Thierry-Mieg, R. Planel, Physica E, **7** (2000), 245.
4. S. K. Lyo and E. D. Jones, Phys. Rev. B, **38** (1988), 4113.

Optical Properties of InGaAsN / GaAs QWs for long-wavelength lasers on GaAs

H. Riechert, A. Yu. Egorov[1*], Gh. Dumitras[2] and B. Borchert

Infineon Technologies AG, Corporate Research Photonics, D – 81 730 München, Germany
email: henning.riechert@infineon.com
[1]A.F. Ioffe Physico-Technical Institute, St. Petersburg 194021, Russian Federation
[2]Technical University of Munich, Physikdepartment E16, D – 85 748 Garching, Germany
[*]guest scientist with Infineon Corporate Research

Abstract   We report on a range of optical properties of InGaAsN / GaAs QWs, focussing on the change of bandgap with N-content and In-content. Our structures are grown by MBE using an RF plasma source. We present evidence that during growth of InGaAsN, In-rich clusters are formed, and that thermal annealing at 700°C leads to a much more uniform composition. In addition to the well-known change of bandgap of GaAsN with N-content, we present data for InGaAsN lattice-matched to GaAs. From the PL of InGaAsN/GaAs QWs we observe that the change of bandgap for fixed N-content decreases with In-content. This can well be described in the model of Shan. et.al., using an interaction parameter of 375 meV for x_N=1.7%.

1 Introduction

InGaAsN has attracted much attention as a novel material, which allows to extend the emission wavelength for GaAs-based structures to 1.3 μm and possibly beyond. We have recently demonstrated high quality 1.3 μm edge-emitting lasers using InGaAsN/GaAs QWs as a gain medium, whith a performance comparable to that of conventional InGaAsP lasers, but with the advantage of significantly enhanced high temperature characteristics [1]. While work on devices using this material is progressing rapidly, basic studies on its optical properties are still scarce. The strong bowing (i.e. decrease) of the band gap of GaAs with the addition of small amounts of nitrogen has been explained by the interaction of a narrow band of quasilocalized N-states with the extended conduction band states [2]. The same authors have predicted a reduced bowing of the bandgap for InGaAsN with increasing In-content. In this paper we present the first report on this behavior in QW structures similar to those used in InGaAsN lasers.

2 Experimental

We have grown InGaAsN / GaAs QWs by molecular beam epitaxy using an RF plasma source to generate reactive nitrogen. By this approach, the nitrogen incorporation is unity and can be easily controlled.

Furthermore, unlike in MOCVD, there is very little -if any- interaction between the incorporation of In and N, such that the composition of quaternary material can be well controlled. This is important since X-ray diffraction cannot distinguish between different compositions leading to the same strain.

3 Results and discussion

We analyze photoluminescence of 6.2 nm wide InGaAsN / GaAs QWs with three fixed N-contents (1.7, 2.4 and 2.8%) with the In-contents varying between 0 and 37%. The uncertainty with respect to the real N-content is estimated to be around ± 0.1%. For reference, several of these structures were grown without N also. All structures were subjected to thermal annealing at 700°C to obtain highest optical quality.

3.1 Evidence for clustering in QWs

We find that unusually low growth temperatures are needed to obtain most intense and narrowest PL. Still, it is observed that the PL width of InGaAsN is significantly larger than for InGaAs, and that the PL will blue-shift under thermal anneal. From our data, we can explain these facts by assuming that InGaAsN (both for thick, lattice-matched layers or strained QWs) tends to form clusters, which we assume to be In-rich. Evidence are the appearance of sharper features in absorption after annealing and an untypical temperature shift of PL in unannealed samples [3].

3.2 Change of bandgap with In-content

Fig.1a shows the positions of the ground state transitions for all QW samples at 300K, grouped by N-contents. Fig. 1b shows only the differences between transition energies for samples with and without N. As can be seen, for the N-content of 1.7%N, the lowering of the transition energy ("bowing") decreases from 215 meV for GaAsN to 175 meV for In $_{0.37}$ Ga $_{0.63}$ AsN. For higher N-contents, this effect is even larger. We have modeled this behavior using the theory given in [2]. For calculation of the valence band structure we have used a 4x4 kp Hamiltonian approach and included the net strain reduction by the nitrogen. Open circles in fig. 1a denote the calculated transition energies. Excellent agreement is obtained if the interaction parameter V_{MN} is chosen to be 375 meV for $x_N = 1.7\%$. Likewise we can

describe the intensity of the luminescence by calculating transition matrix elements. As observed experimentally (see fig. 2), there is a strong increase of the luminescence for increased In-content, which we attribute to an increasing confinement of holes due to an increased valence band offset.

References

1. B. Borchert, A.Yu. Egorov, S. Illek, M. Komainda and H. Riechert, Electron.Lett. 35, (1999) 2204-2206.

2. W. Shan, W. Walukiewicz, J.W. Ager III, E.E. Haller, J.F. Geisz, D.J. Friedman, J.M. Olson and S.R. Kurtz, Phys. Rev. Lett., 82, (1999) 1221-1224.

3. A.Yu. Egorov, D. Bernklau, A. Rucki, M. Schuster, A. Kaschner, A. Hoffmann, G. Dumitras and H. Riechert, to be published in proceedings of MBE-XI, Beijing 2000

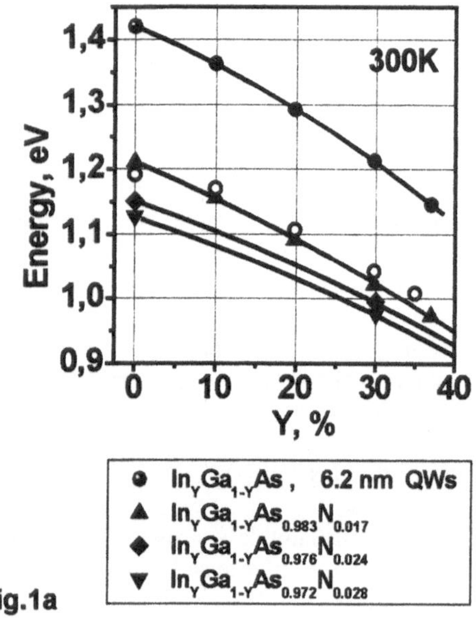

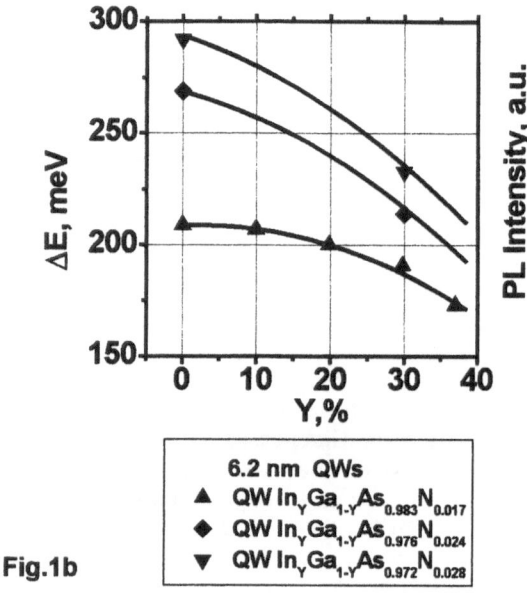

Fig.1a

Fig.1b

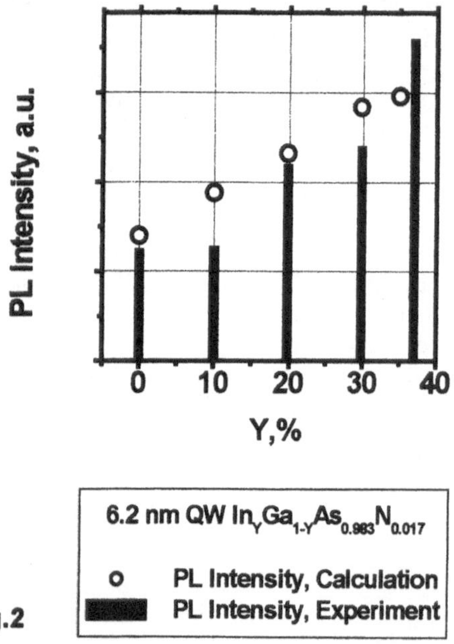

Fig. 1a
Ground state transition energy of 6,2 nm thick ternary and quaternary QWs with GaAs barriers
    full symbols – experiment
    open symbols – simulation for $In_xGa_{1-x}As_{0.983}N_{0.017}$ QWs

Fig. 1b
Difference in transition energy between ternary and quaternary QWs due to nitrogen incorporation

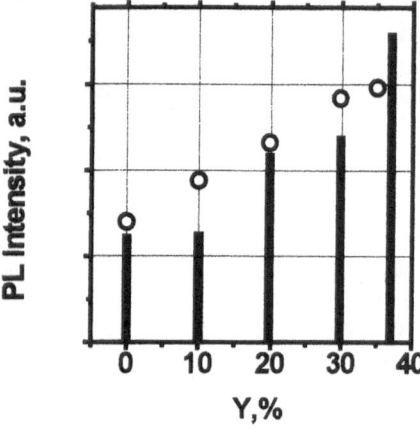

Fig.2

Fig. 2
Intensity of photoluminescence of $In_xGa_{1-x}As_{0.981}N_{0.019}$ QWs with GaAs barriers at room temperature
    circles - simulation
    bars   - experiment

Effective mass theory for a magneto-exciton in the type II superlattices or confinement structures and the origin of anomalous photoluminescence

K. Kouzu[1]*, M. Nishimura[1] and Hiroshi Kamimura[1]

Institute of Physics, Graduate School of Science, Science University of Tokyo, 1-3 Kagurazaka, Shinjuku-ku, Tokyo 162-8601, Japan e-mail: `kamimura@klabsun.yy.kagu.sut.ac.jp`

1 Introduction

Magneto-photoluminescence (Magneto-PL) in type II superlattices such as $(GaP)_n(AlP)_n$ with $n = 4$ to 9 [1] and also in type II confinement structures consisting of neighboring GaP and AlP quantum wells [2] have showed a redshift of PL peak energy and simultaneously an anomalous reduction of PL intensity when an applied magnetic field along the superlattice direction (Faraday configuration) increases up to 40 T. Then Kobayashi et al. [3] proposed that a bound exciton trapped in a quantum-dot like defect at the interface is responsible for the above unusual phenomena. In this paper, in order to clarify the origin of anomalous PL quantitatively, we first develop a general effective mass theory for a bound exciton trapped in a quantum-dot like defect at the interface in the type II systems by choosing GaP/AlP superlattices as an object of study, where an electron and a hole in a type II system lie in differnt layers. Then we calculate the binding energy of a bound exiton as a function of magnetic fields. Finally we calculate the PL intensity as a function of magnetic fields, and we show that the PL intensity decreases sharply when a magnetic field increases, consistent with experimental results.

2 Effective mass theory for a bound exciton at the interface of type II superlattices

By adopting the effective mass approximation for treating the excitonic interaction H_{ex} in a strong magnetic field, the effective Hamiltonian for a bound exciton in the presence of a strong magnetic field along the superlattice direction (the z axis) is given below,

$$H_1 = H_0 + H_{ex} = (H_e + H_h) - \frac{e^2}{4\pi\epsilon|\boldsymbol{r}_e - \boldsymbol{r}_h|} \quad , \quad (1)$$

$$H_e = \frac{p_{ex}^2 + p_{ey}^2}{2m_{e\perp}} + \frac{p_{ex}^2}{2m_e} + \frac{1}{2}\hbar\omega_{ec}L_{ez}$$
$$+ \frac{1}{8}m_{e\perp}\omega_{ec}^2(x_e^2 + y_e^2) + V(x_e, y_e) + V_c(z_e) \quad , \quad (2)$$

$$H_h = \frac{p_{hx}^2 + p_{hy}^2}{2m_{h\perp}} + \frac{p_{hx}^2}{2m_h} + \frac{1}{2}\hbar\omega_{hc}L_{hz}$$
$$+ \frac{1}{8}m_{h\perp}\omega_{hc}^2(x_h^2 + y_h^2) + V_v(z_h) \quad , \quad (3)$$

where $m_{e\perp}$ and $m_{h\perp}$ are the effective masses of an electron and a hole along AlP and GaP layers, respectively, and m_e and m_h the effective masses of a conduction electron in bulk AlP and a hole in bulk GaP, respectively. V_c and V_v in Eq.(3) are the confined potentials for an electron ($\boldsymbol{r}_e$) in AlP layer and that for a hole ($\boldsymbol{r}_h$) in the neighboring GaP layer, respectively, and $\omega_{ec} = eB/m_{e\perp}$ and $\omega_{hc} = eB/m_{h\perp}$. Further $V(x_e, y_e)$ represents the defect potential for a electron due to a quantum-dot like defect at the interface, where a conduction electron with a lighter effective mass becomes localized around the defect whose size L is comparable to the de Broglie wavelength of the conduction electron. When $V(x_e, y_e)$ vanishes, the above Hamiltonian corresponds to a free exciton in the presence of a magnetic field $\boldsymbol{B}$ for the type II systems.

The effective mass equation for a bound exciton is given below,

$$H_1 f(z_{e0}, z_{h0})\Psi(\boldsymbol{r}_e, \boldsymbol{r}_h) = E_1 f(z_{e0}, z_{h0})\Psi(\boldsymbol{r}_e, \boldsymbol{r}_h) \quad , \quad (4)$$

where z_{e0} and z_{h0} represent the centers in the charge densities of an electron in a AlP layer and of a hole in the neighboring GaP layer, respectively. By integrating H_1 by the envelope function $\Psi(\boldsymbol{r}_e, \boldsymbol{r}_h)$, we obtain the effective mass equation for $f(z_{e0}, z_{h0})$ along the z direction;

$$\left(-\frac{\hbar^2}{2m_{e\parallel}}\frac{\partial^2}{\partial z_{e0}^2} - \frac{\hbar^2}{2m_{h\parallel}}\frac{\partial^2}{\partial z_{h0}^2} + W_{ex}\right)f(z_{e0}, z_{h0})$$
$$= Ef(z_{e0}, z_{h0}) \quad . \quad (5)$$

* *Present address:HOYA CORPORATION Vision Care Company, Production Itsukaichi Laboratory, 1-1 Kowada, Akiruno-shi, Tokyo 190-0151 Japan*

Here we can show that the excitonic interaction W_{ex} can be expressed in the following analytical form;

$$W_{ex} = \int\int (\Psi^*(\boldsymbol{r}_e, \boldsymbol{r}_h) H_{ex} \Psi(\boldsymbol{r}_e, \boldsymbol{r}_h))$$

$$= \frac{e^2}{2(2\pi)^{3/2}\epsilon} \int_{-1}^{1} \frac{1}{\sqrt{g(t)}} \exp\left\{ -\frac{(z_{e0} - z_{h0})^2 t^2}{2g(t)} \right\} dt$$

$$= -const. + \lambda r^2 \quad , \tag{6}$$

where $r = z_{e0} - z_{h0}$.

From this result the excitonic interaction for the envelope function $f(z_{e0}, z_{h0})$ is expressed in the form of parabolic potential. Thus a solution of Eq. (5) is expressed as $f(r)$ and given by a form of

$$f(r) = \frac{1}{\sqrt{(2\pi)^{1/2}\gamma}} \exp[-r^2/4\gamma^2] \quad . \tag{7}$$

As variational functions for $\Psi(\boldsymbol{r}_e, \boldsymbol{r}_h)$ and $f(z_{e0}, z_{h0})$, we adopt Eq. (??) for $f(r)$ and the product form of $\phi_e(\boldsymbol{r}_e)\phi_h(\boldsymbol{r}_h)$ for $\Psi(\boldsymbol{r}_e, \boldsymbol{r}_h)$, where $\phi_e(\boldsymbol{r}_e)$ and $\phi_h(\boldsymbol{r}_h)$ are the envelope functions of an electron and a hole for H_e and H_h in Eq. (3), respectively.

3 The calculated results for the binding energy and the PL intensity

By using $\Psi(\boldsymbol{r}_e, \boldsymbol{r}_h) = \phi_e(\boldsymbol{r}_e)\phi_h(\boldsymbol{r}_h)$ and $f(r)$ in Eq. (??), we calculate the expectation value of H_1 in Eq. (4); $E_1(B) = \langle f\Psi|H_1|f\Psi\rangle$. Then, by adopting Gaussian functions for $\phi_e(\boldsymbol{r}_e)$ and $\phi_h(\boldsymbol{r}_h)$ following Kobayashi *et al.* [3], we minimize E_1 by the variational principle. The binding energy of a bound exciton defined as $E_b(B) = E_0(B) - E_1(B)$ is calculated as a function of B, where $E_0(B)$ is the binding energy of a bound exciton in the absence of a magnetic field. The calculated binding energies of a bound exciton in $(GaP)_n(AlP)_n$ short period superlattice with $n = 4$ is shown in Fig. ??, as a function of a magnetic filed B for various values of L, the dimension of the quantum-dot like defect. Then, using the obtained wavefunctions we have calculated the PL intensity following Kobayashi *et al.* [3]. The calculated results are shown as a function of B in Fig. ??. It is seen from Figs. ?? and ?? that the calculated results explain successfully the observed redshift of PL peak energy and also the anomalous decrease of PL intensity. From these results we conclude that the interplay between a quantum-dot like defect at the interface and the excitonic interaction is essentially important in causing the anomalous behaviors in PL of the type II systems in a strong magnetic field.

This work is partically supported by Hoso-Bunka Foundation.

References

1. K. Uchida *et al.*, *Phys. Rev.* B **53** (1996) 4809.
2. N. Usami *et al.*, *Phys. Rev.* B **60** (1999) 1879.
3. Y. Kobayashi, K. Kouzu and H. Kamimura, Solid State Commun. **109** (1999) 583.

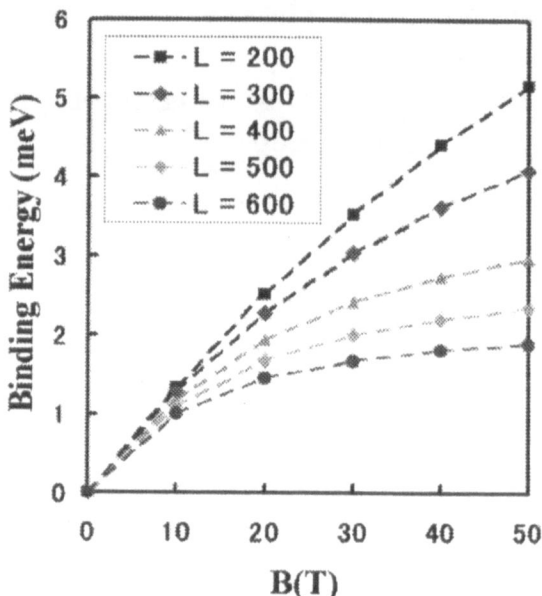

Fig. 1 The binding energy of a bound exciton in $(GaP)_4(AlP)_4$ as a function of a magnetic field for various size of the defect, L.

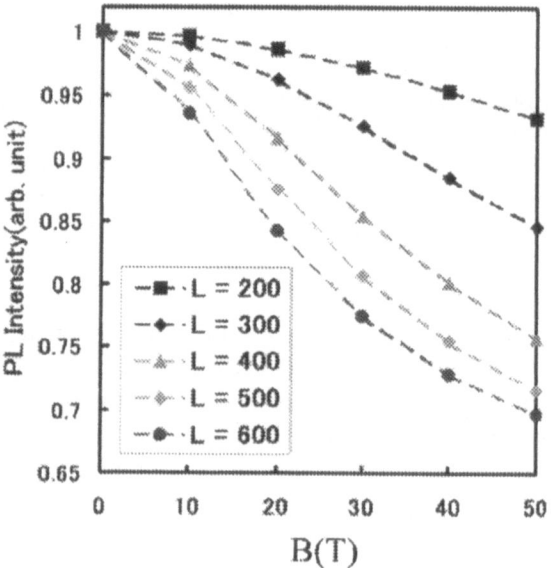

Fig. 2 The PL intensity from a bound exciton state in $(GaP)_4(AlP)_4$ as a function of a magnetic field for various values of L.

Modulation spectroscopy of critical points in excitonic energy structure of thick GaAs quantum wells

V. G. Davydov[1,2*], Yu. K. Dolgikh[2], Yu. P. Efimov[3], S. A. Eliseev[3], A. V. Fedorov[2], I. Ya. Gerlovin[2], I. V. Ignatiev[1,3], I. E. Kozin[1,3], V. V. Petrov[3], V. V. Ovsyankin[2], I. A. Yugova[3], H.-W. Ren[1,4**], S. Sugou[4], and Y. Masumoto[1,5]

[1] Single Quantum Dot Project, ERATO, JST, 5-9-9 Tokodai, Tsukuba-shi, Ibaraki 300-2635, Japan

[2] S. I. Vavilov State Optical Institute, Birzhevaya lin. 12, St.-Petersburg, 199034, Russia

[3] Institute of Physics, St.-Petersburg State University, St.-Petersburg, 198904, Russia

[4] Opto-Electronics Research Lab., NEC Corp., Tsukuba, Ibaraki 305-8501, Japan

[5] Institute of Physics, University of Tsukuba, Tsukuba 305-8571, Japan

Abstract An oscillations were observed in a wide spectral range in modulated photo- and electroreflection spectra of heterostructures with the thick GaAs quantum wells. It is proved experimentally and theoretically that the observed phenomenon manifests the quantum size effect of the exciton in the GaAs quantum wells.

1 Introduction

An epitaxial layer of GaAs with a thickness of more than 10-20 manometers generally is considered as a bulk material, basing on the fact that the energy of the lowest excitonic state in thick layer almost coincides with the bulk exciton energy. However, in high quality epitaxial layers at the low temperature carriers may have a free path larger than the layer thickness. Thus, even the remote hetero-interfaces plays an important rôle[1]. As a result, the distance between interfaces affects the whole energy structure of the excitonic levels.

In this paper we report about the quantum size effect in epitaxial layers with the thickness up to 150 nm. We have seen a large number of the quantum confined excitonic states in the photo-reflectance (PR) and electro-reflectance (ER) spectra of different samples with the GaAs quantum wells (QW's) of various thickness.

2 Experimental

The studied heterostructures were grown by molecular beam epitaxy on the n^+ GaAs (sample QDP1779) or semi-insulating GaAs substrates (samples e187 and e188). The QW's in e187 and e188 with the thickness of 50 and 100 nm, respectively, were embedded between the $Al_{0.3}Ga_{0.7}As$ barrier layers. The QW in QDP1779 with the thickness of 150 nm was embedded between thin AlAs layer and the AlAs/GaAs superlattice.

The PR and ER spectra were measured using a continuous wave tunable Ti:sapphire or dye laser as a source of a probe light. Another laser with wavelength 532 or 800 nm was used as a pump source in the PR experiments. The amplitude modulation of the pump and

* e-mail: val@sqdp.trc-net.co.jp

** *Present address:* Space Vacuum Epitaxy Center, University of Houston, Houston, TX77204-5507, USA.

probe beams at different frequencies (1 MHz and 2 kHz, respectively) and a double lock-in detection of the reflected probe signal modulated at the differential frequency allowed us to avoid noises from the scattered pump light and to detect fractional reflection changes as low as 10^{-7}. For the ER measurements, we applied to the sample a small ac voltage at frequency of about 3 MHz by means of the semitransparent gold electrode deposited on the sample surface. A constant electric bias was applied along with the modulation. All experiments were done at the sample temperature 2 or 5 K.

3 Results and Discussion

PR and ER spectra of the sample QDP1779 are shown in Fig. 1. One can see the large number of quasi-regular oscillations which start from the resonance energy of the

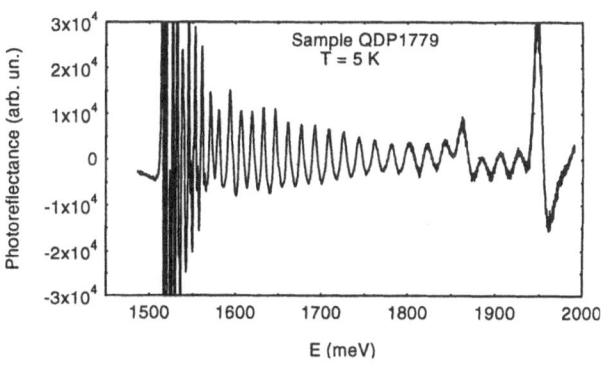

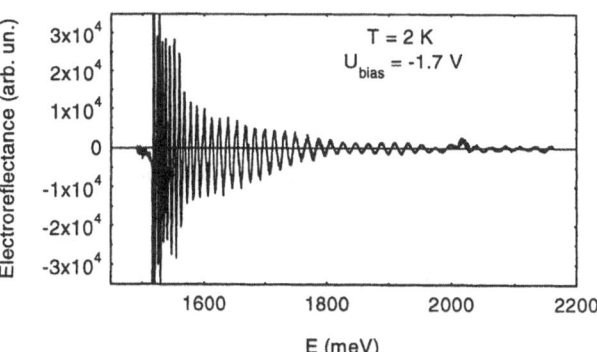

Fig. 1 PR (top) and ER (bottom) spectra of the sample QDP1779 with 150 nm GaAs well.

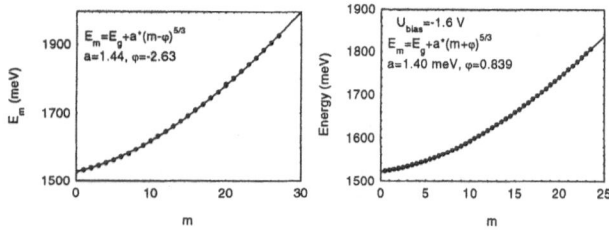

Fig. 2 Energy positions of the maxima (left) and extrema (right) of the PR and ER spectra, respectively, of the sample QDP1779, phenomenologically fitted by the 5/3 power law.

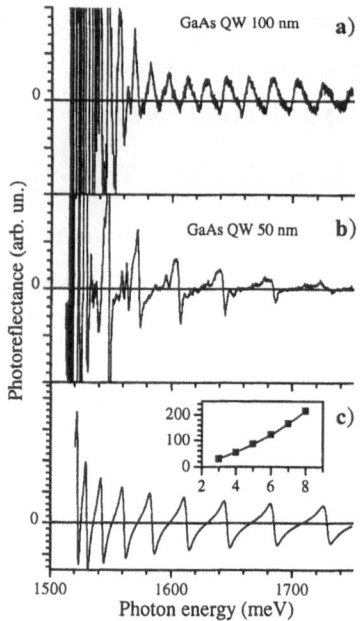

Fig. 3 PR spectra of the samples e188 (a) and e187 (b) and the theoretical simulation of the PR spectrum (c) as described in the text. The energy positions of right zeros in the spectrum (b) and their fit by the equation $\hbar\omega_{0,n} = E_g + 3.6n^{1.966}$ [meV] are shown in the inset.

GaAs bulk exciton and continue up to the end of the studied spectral region. Energy positions E_m of maxima or/and minima of these oscillations versus their number m are shown in Fig. 2. They can be well fitted by the phenomenological power law $E_m = E_g + a(m - \varphi)^{5/3}$, where E_g coincides with the GaAs bandgap and φ takes into account an uncertainty of the maxima numbering.

PR spectrum of the sample does not depend strongly on the wavelength of the pump beam. The only necessary condition is that the pump photon energy should be larger than the GaAs bandgap. ER spectrum also does not strongly depend on the applied bias in the range of negative voltages up to about -2 V. Outside this range, a considerable electric current starts to flow through the sample and oscillations vanish.

This oscillations can be caused by a Franz-Keldysh effect in built-in electric field[2] which was indeed found in the studied sample as shown in our early paper[3]. However, Franz-Keldysh oscillations should lead to the $E_m - E_g \sim m^{2/3}$ law in the homogeneous field. The field inhomogeneity can change this law only to the linear.

Another possibility is that the PR and ER spectra reflect the ladder of quantum confined excitonic states with large quantum numbers. In this case, the energy dependence of maxima in the spectra shows the energy dispersion for carriers in the bulk material, which is almost quadratic in a small energy region and approaches linear for higher energies. Our fit by 5/3 power law is a good approximation of these dependencies.

To check this assumption we have grown and studied the samples e187 and e188 with QW's of different (lower) thickness than in the discussed above sample QDP1779. PR spectra of these samples are presented in Fig. 3. As seen, the energy distance between the adjacent maxima depends inversely on the QW thickness.

We developed a simple model to describe our data in the effective mass approximation. The dielectric function of quantum well $\varepsilon = \varepsilon_1 + i\varepsilon_2$ has been calculated with regarded to the width of optical transitions, where

$$\varepsilon_1 = \varepsilon_\infty + \frac{2\mu |d_{CV}|^2}{\hbar^2 L} \sum_n \ln\left(\frac{\tilde{E}^2}{\Delta_n^2 + \gamma_n^2}\right), \quad (1)$$

$$\varepsilon_2 = \frac{4\mu |d_{CV}|^2}{\hbar^2 L} \sum_n \left(\frac{\pi}{2} + \arctan\left(\frac{\Delta_n}{\gamma_n}\right)\right), \quad (2)$$

ε_∞ is the background dielectric constant, μ is the reduced mass of electron and hole, L is the well thickness, d_{CV} is the dipole moment of the band-to-band transition, $\tilde{E}$ is the energy parameter taking into account the finite widths of valence and conduction band, n and γ_n are the quantum number and width of optical transition, $\Delta_n = \hbar\omega - E_g - \frac{\hbar^2\pi^2 n^2}{2\mu L^2}$, $\hbar\omega$ is the photon energy.

The modulation spectroscopy[2] measures the derivative of the dielectric function on modulated parameter. In the PR, the pump light produces hot carriers which accelerate the exciton relaxation rate[4]. In this case the modulated parameter is the line width γ_n and the reflectivity change is determined by the imaginary part (2) of dielectric constant: $\Delta R \sim \frac{d\varepsilon_2}{d\gamma} = -\frac{4\mu |d_{CV}|^2}{\hbar^2 L} \sum_n \frac{\Delta_n}{\Delta_n^2 + \gamma_n^2}$. This function oscillates with respect to $\hbar\omega$ (Fig. 3c). The right zeros of ΔR are determined by the conditions, $\Delta_n = 0$, i.e. $\hbar\omega_{0,n} = E_g + \frac{\hbar^2\pi^2 n^2}{2\mu L^2}$.

In the case of the ER, the electric field modulates[1] the energies Δ_n and the change in the reflectivity is determined by the real part (2) of dielectric constant.

4 Conclusion

In conclusion, a big number of quantum confined excitonic states were observed in the PR and ER spectra of the thick QW's. The observed phenomenon opens up the wide possibilities for precise study of the physical properties of bulk GaAs material.

References

1. S. Fafard, E. Fortin and A.P. Roth, *Phys. Rev. B*, **45**, 13769 (1992).
2. H. Shen and M. Dutta, *J. Apll. Phys.*, **78**, 1251 (1995).
3. V. Davydov *et al.*, *Apll. Phys. Lett.*, **74**, 3002 (1999).
4. I. V. Ignatiev *et al.*, *Phys. Rev. B*, **61**, 15633 (2000).

---

[1] Several mechanisms can cause this modulation. The Stark effect is an obvious example, but others are also possible.

Exciton-phonon coupling in wide bandgap II-VI quantum wells

B. Urbaszek[1], A. Balocchi[1], C. Morhain[1], C. Bradford[1], X. Tang[1], C.M. Townsley[2], C.B. O'Donnell[1], S.A. Telfer[1], K.A. Prior[1], B.C. Cavenett[1], R.J. Nicholas[2]

[1] Department of Physics, Heriot-Watt University, Edinburgh, EH14 4AS (U.K.)
[2] Clarendon Laboratory, Department of Physics, University of Oxford, Parks Road, Oxford OX1 3PU (U.K.)

Abstract Magnetic field and temperature dependent measurements are used to determine exciton binding energies and to study the exciton-LO phonon scattering processes in high quality ZnSe and ZnS quantum wells. An exciton binding energy as large as 43.9meV was measured in a MgS/ZnSe quantum well, for which the scattering with LO phonons is partially suppressed. For ZnMgS/ZnS quantum wells values for the exciton binding energies have been deduced which also indicate the possibility of suppressing exciton-LO phonon scattering.

1 Introduction

The stabilization of excitons at room temperature by the suppression of the exciton-LO phonon scattering is a key factor for high performance optoelectronic devices. However, the strong interaction between the excitons and LO phonons usually results in the dissociation of the excitons at high temperatures, reducing device efficiencies. This should be prevented in systems where all the excitations of the exciton have an energy larger than that of the LO phonon, namely for $E^X(1s-2s) > h\nu_{LO}$ due to the absence of final states for the scattering process [1]. This relation is expected to hold in quantum wires and dots but in these cases exciton stabilization has not yet been investigated in detail due to the difficulty in growing samples with a uniform distribution in size and shape. On the other hand, quantum wells (QWs) exhibit high uniformity leading to narrow spectra, but the enhancement of the exciton binding energy is usually not sufficiently large to satisfy the above condition. These problems can, however, be overcome using MgS/ZnSe QWs, where large Rydberg values are expected due to a confinement potential of 2eV. Also ZnMgS/ZnS QWs are investigated, where a ZnS bulk exciton binding energy of 36meV compared to an LO phonon energy of 44meV makes suppression of exciton ionization in quantum well structures achievable.

2 MgS/ZnSe/MgS quantum wells

Multiple quantum well structures containing 10 wells were grown by Molecular Beam Epitaxy (MBE) on GaAs (001) substrates by a novel method [2]. The data presented in this paper were measured by optical transmission after selectively removing the substrate by wet etching. In ZnSe, the bulk exciton binding energy is 20meV and the LO phonon energy is 32meV. Since the difference in bandgap between ZnSe (2.823eV) and MgS (>4.8eV) is in the order of 2eV, the condition $E^X(1s-2s) > 32$

meV is easily achievable in narrow QWs. Exciton binding energies were measured directly for two samples by magneto-transmission, a technique already used to study the ZnSe/ZnCdSe system [3]. In a magnetic field parallel to the growth direction, higher optical transitions are detectable, as shown in figure 1, where the measured energies of the exciton absorption peaks are plotted against the magnetic field strength. Since unpolarized light was used, the expected Zeeman splittings of the order of 0.5meV [3] were not observed. Zero field values have been extrapolated using the model of Engbring and Zimmermann [4]. For the 5.0nm and the 10.0nm wide quantum well samples the 2s heavy hole exciton state emerges at higher fields and the values obtained for the zero field $E^X_{HH}(1s-2s)$ are 22.3meV for the 10.0nm well and 37.1meV for the 5.0nm well. The condition $E^X(1s-2s) > h\nu_{LO}$ is thus fulfilled for the 5nm well, where the exciton binding energy is 43.9meV and corresponds to a remarkable enlargement by a factor of 2.2 compared with the bulk ZnSe value, caused by the large difference in bandgaps.

To verify directly the effect of the large exciton binding energy on the stabilization of the exciton at higher temperature the line widths of the heavy hole transitions were measured. For individual exciton states the broadening can be written as

$$\Gamma = \Gamma_0 + \Gamma_{ac}T + \Gamma_{LO}\left[exp\left(h\nu_{LO}/k_BT\right)-1\right]^{-1} \quad (1)$$

where Γ_0 represents the inhomogeneous broadening, Γ_{ac}

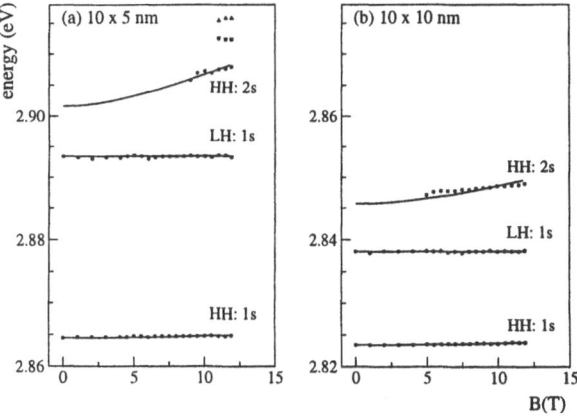

Fig. 1 Energy of exciton transition vs. magnetic field as measured by transmission for 5.0nm (a) and 10.0nm (b) multiple QWs of ZnSe in MgS. Solid lines. fit of the data [4]

is the contribution due to the scattering with acoustic phonons and Γ_{LO} is the contribution due to the scattering with LO phonons. The values obtained by fitting the experimental data with (1), shown in figure 2, are summarized in table 1. For the fitting procedure values

Table 1 *ZnSe* : Parameters for the heavy hole excitons obtained by fitting the magnetic field and temperature dependencies

L_w (nm)	E_{HH}^X(1s-2s) (meV)	E_{HH}^X(1s) (meV)	Γ_0 (meV)	Γ_{LO} (meV)
5.0	37.1	43.9	(11.0)	(31.8)
10.0	22.3	28.3	6.2	59.0

of $3.0\mu eVK^{-1}$ for the 10.0nm sample and $5.0\mu eVK^{-1}$ for the 5.0nm sample were assumed for Γ_{ac}, in accordance with the work of Blewett et al. [5]. For the 10.0nm wide quantum well the value for Γ_{LO} of 59.0meV is very close to that of 60meV reported for bulk ZnSe [6]. For the 5.0nm wide well the fit obtained with (1) is clearly not satisfactory, as two different regimes are observed. For temperatures up to 150 K the FWHM of the transition remains unchanged, whereas for higher temperatures its dependence is similar to that of the 10.0nm wide well. Although an interpretation involving scattering with phonons from the MgS barrier gives a better fit, this explanation is unlikely as (i) the two investigated structures are similar and so should be the contributions of LO phonons from the well and the barrier and (ii) the corresponding Γ_{Lo} of 340meV appears to be unrealistic. Instead we think that the two regimes correspond to a transition from the $E^X(1s-2s) > h\nu_{LO}$ regime, where the exciton-LO phonon scattering is suppressed, to the $E^X(1s-2s) < h\nu_{LO}$ regime, where exciton ionization occurs. The changeover is due to inhomogeneous broadening of the 1s exciton level and happens when delocalization, possibly through scattering with acoustic phonons, becomes effective. Although $E^X(1s-2s) > h\nu_{LO}$ holds for individual states, the line widths of which are defined by their spatial extensions, this condition is not met by exciton levels broadened by around 10meV after delocalization in the narrow QW sample.

3 ZnMgS/ZnS/ZnMgS quantum wells

A set of ZnS QWs with varied width has been investigated. In these structures the QWs are under tensile strain in ZnMgS barriers grown lattice matched on GaP by MBE. Strong improvements of the growth conditions allow the observation of light and heavy hole exciton transitions in reflectivity spectra with line width as narrow as 5meV. Higher excited states were not unambiguously identified in magneto-reflectivity spectra in pulsed fields up to 53 Tesla. Values for exciton binding energies were derived [7] from the measured diamagnetic shifts of heavy and light hole exciton ground states, which are summarized in table 2. The values for $E^X(1s-2s)$ were obtained from the 1s values by applying an analytical formula given by Mathieu et al [8]. These first results for this material indicate that $E^X(1s-2s) > h\nu_{LO}$ is achievable for heavy and light hole excitons for QWs narrower than 4nm.

Table 2 *ZnS* : Heavy and light hole exciton binding energies obtained by fitting the magnetic field dependence, note $h\nu_{LO} = 44meV$

L_w (nm)	E_{HH}^X(1s-2s) (meV)	E_{HH}^X(1s) (meV)	E_{LH}^X(1s-2s) (meV)	E_{LH}^X(1s) (meV)
3.5	-	-	47.4	57.8
4.0	48.0	58.5	34.1	43.6
5.0	38.2	48.0	31.7	41.0
10.0	28.6	37.7	24.2	33.0

4 Conclusions

We have measured the exciton binding energies for two ZnSe samples with different well widths. The exciton-LO phonon scattering process is suppressed for the 5.0nm quantum well for temperatures up to 150K. For higher temperatures the scattering is enabled due to the inhomogeneous broadening of the exciton levels. In ZnS large exciton binding energies for QWs narrower than 4nm indicate the possibility of suppressing ionization due to LO phonon scattering in this material.

The authors would like to thank Dr. I.Galbraith for helpful discussions.

References

1. S. Nojima, Phys. Rev. B **46** (1992) 2302.
2. C. Bradford et al, Appl. Phys. Lett. **76** (2000) 3929.
3. J. Puls et al, Phys. Rev. B **54** (1996) 4974.
4. J. Engbring, R. Zimmermann, phys. stat. sol. (b) **173** (1992) 733.
5. I.J. Blewett et al, Phys. Rev. B **59** (1999) 9756.
6. N.T. Pelekanos et al, Phys. Rev. B **45** (1992) 6037.
7. P.A. Shields et al, to be published.
8. H. Mathieu et al, Phys. Rev. B **46** (1992) 4092.

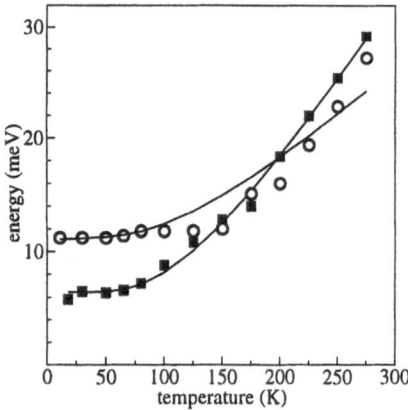

Fig. 2 FWHM for the heavy hole transitions vs. temperature as measured by transmission for 5.0nm (open circles) and 10.0nm (squares) multiple QWs of ZnSe in MgS. Solid lines. fit of the data with (1)

Magnetooptical Evidence of Many-body Effects in Spin-polarized 2D Electron Gas

C. Testelin[1], A. Lemaître[1], C. Rigaux[1], T. Wojtowicz[2], G. Karczewski[2] and F. Teran[3]

[1] Groupe de Physique des Solides, CNRS, Universités Paris 6 et 7, 2 place Jussieu, 75251 Paris, France,
[2] Institute of Physics, Polish Academy of Sciences, Al. Lotników, Warszawa, Poland
[3] Grenoble High Magnetic Field Laboratory, MPI/FKF and CNRS, 25 avenue des Martyrs, 38042 Grenoble Cedex 9, France

Abstract　Modulation doped CdMnTe/CdMgTe quantum wells are studied at 1.7 K by magnetoabsorption near the $HH_1 - E_1$ transition. In the investigated range of electron concentration, a full spin down polarization of the electron gas is found above $H = 1.5$ T. A simplified theoretical model including exchange energy and electron-hole interaction provides a coherent description of the interband magnetooptical spectra.

We report magnetoabsorption studies in (n-type) modulation doped quantum wells (MDQW) made of a diluted magnetic semiconductor (DMS) $Cd_{1-x}Mn_xTe$ surrounded by non magnetic barriers $Cd_{1-y}Mg_yTe$. Within the investigated electron concentration ($n_e \simeq 2\,10^{11}$ cm^{-2}), the electron-hole Coulomb interaction is considerably reduced and the magnetoabsorption spectra near the $HH_1 - E_1$ fundamental edge are dominated by the inter-Landau level transitions. The large electron spin splitting in the DMS quantum well is used to polarize the 2D electron gas at relatively low field. When only one spin component ($n_\downarrow$) of nth electron Landau level is populated, one may expect strong evidence of electron exchange interactions in magnetooptical spectra.

$Cd_{1-x}Mn_xTe/Cd_{1-y}Mg_yTe$ were grown by molecular beam epitaxy on (100)-oriented GaAs substrates. The structures, containing a 100 Å thick single $Cd_{1-x}Mn_xTe$ quantum well (QW) were modulation-doped with iodine in the barrier at a distance of 400 Å from the QW[1]. The electron concentration $n_e = 2.3\,10^{11}$cm^{-2} in the QW was determined from the period of the Shubnikov-de Haas oscillations observed in transport measurements [2].

Magnetoabsorption experiments are carried out at 1.7 K near the $HH_1 - E_1$ transition, in Faraday configuration ($\sigma^\pm$ polarization) in magnetic fields up to 14 T, applied along the growth axis. Magnetotransmission spectra are reported in Fig.1 for a structure of $x = 0.018$ and $y = 0.147$. The zero field transmission spectrum consists of an asymmetrical line (A) at 1639.5 meV and a very weak feature (B) at 1649.5 meV (Fig. 1). At non-zero magnetic field, a large Zeeman effect is observed: the line (A) splits into a σ^- polarized high energy component, of a nearly constant intensity, with a Zeeman shift nearly saturated at $H \sim 5$ T, and a σ^+ polarized low energy component which weakens with increasing field and disappears at $H \sim 1.5$ T. In contrast, the feature (B), hardly visible at zero field, gains intensity and evolves into a set of L_n lines visible in the σ^+ polariza-

tion. With increasing magnetic field the energies of L_n lines shift towards higher energies, as shown in Fig.2.

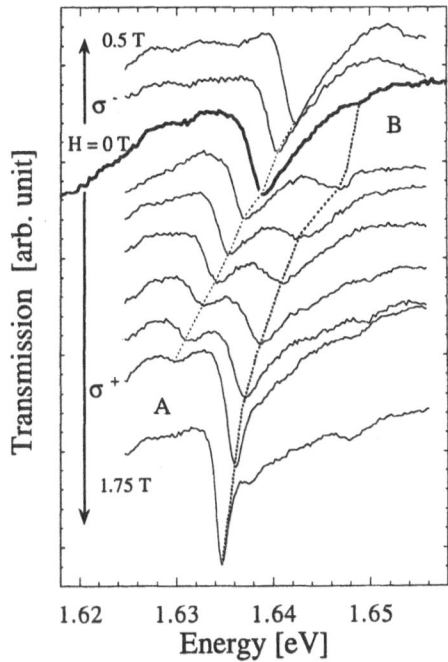

Fig. 1 Evolution of the transmission spectrum at low field in Faraday configuration with $\sigma^\pm$ polarization. The magnetic field increases with the step of 0.25 T between curves

We assign the lines L_n to interband transitions ($n \to n$) between the HH_1 heavy hole states with angular momentum $j_h = 3/2$ and the E_1 electron states with spin component $j_e = -1/2$ along the magnetic field. Every absorption line L_n appears at a discrete magnetic field, H_n, when the $n_\downarrow$ electron Landau level reaches the Fermi level. At $H > H_n$, emptying states become available for optical excitation. The H_n values correspond then to the integer values of the electron filling factor $\nu = n + 1$. The relation $H_n = hcn_e/e\nu$ yields an electron density $n_e = 2.25\,10^{11}$ cm^{-2} in excellent agreement with the results obtained from Shubnikov-de Haas measurements. This corresponds to a zero field electron Fermi energy, $E_f = 5.2$ meV. The observation of the continuous sequence of integer filling factor $\nu = 1,..,6$, in the only σ^+ polarization, between 14 and 1.5 T, indicates the full spin down polarization of the electron gas above $H = 1.5T$.

In the region $H < 1.5$ T where both spin states $j_e = \pm\frac{1}{2}$ are populated, negatively charged excitons (trions) can be formed. We assign the line (A) observed below

1.5 T to the X^- trion transition[3,4]. Since the spin down electron states are always populated, the high energy trion state, formed of $(j_h = -\frac{3}{2}, j_e = \frac{1}{2})$ photocreated exciton and $(j_e = -\frac{1}{2})$ electron states, is observed in σ^- polarization at any magnetic field. In contrast, the spin up states become empty at about $H = 1.5$ T and the low energy component of line (A), i.e., the $(j_h = \frac{3}{2}, j_e = -\frac{1}{2})(j_e = \frac{1}{2})$ trion state vanishes with increasing field, as shown in Fig. 2 in the σ^+ polarization.

As shown in different studies[3,5], the energy of the lowest excitation (X^-) is almost independent of the carrier density. One may thus estimate the Mn concentration from the free exciton energy (E_X) expected for an undoped QW of identical characteristics, taking $E_{b1} = 2.1$ meV[3] for the X^- binding energy. The energy value $E_X = 1641.6$ meV corresponds to a QW of Mn composition $x = 0.018$.

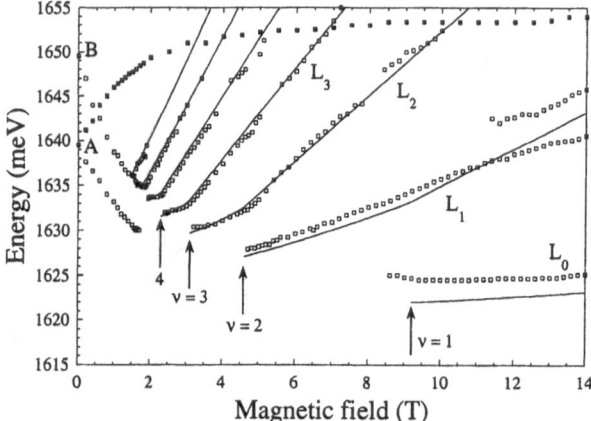

Fig. 2 Magnetic field dependence of the transition energies. Symbols: experiments (open symbols) σ^+, (closed symbols) σ^-. Solid lines: theoretical fit obtained for parameters $x = 0.018$, $E_a^* = 1644.5$ meV, $\hbar\omega^*/H = 1.5$ meV/T and $R_0 = 2.5$ meV. The integer filling factors are indicated by arrows. The anticrossing around $H = 12$ T (line L_1) is not discussed in the text

We analyze the field dependence $E_n(H)$ of the lines Ln_n by using a simplified model in the high field approximation : we assume that the Coulomb interaction between photocreated electron and hole issued from E_1 and HH_1 landau levels is weaker than the reduced cyclotron energy $\hbar\omega^*$. We determine the binding energy of the e-h pair associated to each inter-Landau level transition by the first order of the perturbation theory. We also include the exchange energy of conduction electrons due to the 2D electron gas. The energies E_n are:

$$E_n(H) = E_a^* + (n + \frac{1}{2})\hbar\omega^* + E^{sp-d}(H)$$
$$+ E_n^{exch}(H) - E_n^b(H), \qquad (1)$$

where $E_n^{exch}(H)$ and $E_n^b(H)$ are the exchange energy experienced by an electron in the $(n \downarrow)$ Landau level and the electron-hole binding energy, respectively. $\hbar\omega^* = \hbar eH/\mu c$ is the reduced cyclotron energy $(1/\mu = 1/m_e + 1/m_h$; m_e and m_h are the in-plane effective masses of

electron and hole, respectively). E_a is the $HH_1 \rightarrow E_1$ energy gap, after subtracting of the zero field exchange energy, already included in the fourth term of Eq. 1. The $E^{sp-d}(H)$ term describes the spin dependent contribution caused by Mn-carrier exchange interaction. This term is calculated for the Mn composition $x = 0.018$ using the expression of S_z vs x and H, given in ref.[6]. The electron-hole Coulomb, for σ spin electron, is described by the relation [7]:

$$E_{n\sigma}^0(H) = R_0 \frac{a_0}{\ell_c} \sqrt{2\pi} C_{nn}(1 - \tau_{n\sigma}) \qquad (2)$$

where R_0 and a_0 are effective parameters defined by $R_0 = \mu e^4/2\kappa^2\hbar^2$ and $a_0 = \hbar^2\kappa/\mu e^2$ (κ is the static dielectric constant). $\tau_{n\sigma}$ is the filling factor of conduction Landau level n of spin σ. The $\tau_{n\sigma}$ dependence of the binding energy included in eq. 2 accounts for the phase-space filling effect as explained in ref[2]. The coefficients $C_{nn'}$ are given by:

$$C_{nn'} = \frac{2}{\sqrt{\pi}} \int_0^{+\infty} dx e^{-x^2} L_n^0(x^2) L_{n'}^0(x^2) \qquad (3)$$

$L_n^m(x)$ are Laguerre polynomials. The exchange energy is given by [8,9]:

$$E_{n,\sigma}^{exch}(H) = -R_0 \frac{a_0}{\ell_c} \sqrt{2\pi} \sum_{n'} \tau_{n'\sigma} C_{n,n'} \qquad (4)$$

We consider only electrons with spin down component as the electron gas is polarized ($H \geq 1.5$ T). The parameter $\hbar\omega^*/H = 1.5 \pm 0.1$ meV/T is estimated from the periodicity of the L_n lines at high field, when $E^{sp-d}(H)$ is saturated (the two last terms of eq. 1 do not significantly affect the lines periodicity). Energies E_n of eq. 1 are fitted to the experimental lines L_n, taking R_0 and E_a as free parameters. The best theoretical fit, between 1.5 T and 14 T, is achieved for $R_0 = 2.5$ meV and $E_a = 1644.5 \pm 1$ meV (see Fig. 2). The validity of the basis assumption ($E_n^0 < \hbar\omega^*$) is confirmed by the parameters deduced from the fit. The present study emphasizes the strong influence of electron exchange energies on the electron Landau levels.

References

1. G. Karczewski, J. Jaroszyński, A. Barcz, M. Kutrowski, T. Wojtowicz, J. Kossut, J. Crystal Growth **184/185**, 814 (1998).
2. A. Lemaître et al., Phys. Rev. B **62**, 5089 (2000).
3. V.Huard, R.T.Cox, K.Saminadayar, A. Arnoult and S.Tatarenko, Phys. Rev. Lett. **84**, 187 (2000).
4. K. Kheng et al., Phys. Rev. Lett. **71**, 1752 (1993).
5. T. Wojtowicz et al., Phys. Rev. B **59**, R10437 (1999).
6. W. Grieshaber et al., Phys. Rev. B **53**, 4891 (1996).
7. A. H. MacDonald and D. S. Ritchie, Phys. Rev. B **33**, 8336 (1986).
8. T. Ando and Y. Uemura, J. Phys. Soc. Japan **37**, 1044 (1974).
9. A. H. MacDonald, H. C. A. Oji, and K. L. Liu, Phys. Rev. B **34**, 2681 (1986).

Forbidden interband transitions and the indirect valence band in strained GaInAs/InP quantum wells

A. Dörnen[1,2], J. Shao[1], V. Härle[1], F. Scholz[1]

[1] 4. Physikalisches Institut, Universität Stuttgart, Pfaffenwaldring 57, D-70550 Stuttgart, Germany
[2] e-mail: a.doernen@physik.uni-stuttgart.de

Abstract Magneto-optical absorption spectroscopy was carried out on tensile and compressively strained $Ga_x In_{1-x} As$ quantum wells lattice-matched to the InP barrier. The observation of P and D excited states allows to determine effective masses of electrons and holes separately. The results evidence an indirect valence band maximum approximately $0.2\,k_{max}$ from the Γ point for samples with a gallium fraction of $x = 0.6$.

1 Introduction

The effect of strain on the band structure of quantum wells is of great importance for the design of electronic and opto-electronic devices and has therefore induced thorough investigations. Recently, in tensile strained GaInAs quantum wells (QW) an unexpectedly long minority carrier lifetime has been observed and assigned to an indirect optical transitions due to an indirect valence band (VB) structure [1,2]. Aim of this paper is to evidence this by magneto-absorption (MA) spectroscopy, which has not been addressed in prior work [3].

We describe the excitonic lines by the theory of a quasi-two-dimensional exciton [4]. A confining potential with the shape of a finite rectangular box has been considered. The equation not only describes S states ($m = 0$, m: magnetic quantum number). Also included are P and D states due to the orbital momentum of the exciton ($m = \pm 1$ and ± 2). For states of $m \neq 0$ a term in the Hamiltonian is involved which is proportional to $(1/m_e^* - 1/m_h^*)$. Therefore these states de-

pend on effective masses of electrons (m_e^*) and holes (m_h^*) separately, while S states depend on the reduced effective exciton mass μ^* only.

2 Experiments and Discussion

Multi-QW samples of $Ga_x In_{1-x} As$/InP with 10 QWs were grown by low-pressure metal-organic vapor-phase epitaxy, the gallium fraction x was set to 0.3, 0.4, 0.47, 0.5, 0.6, and 0.7. The width of the well and of the barrier was chosen to 10 and 20 nm, respectively. Further details are given in Ref. [5].

Magneto-optical absorption experiments were carried out in the Faraday configuration of a superconducting magnet up to 6.8 T at a temperature of 1.8 K. A Fourier spectrometer equipped with a GaInAs p-i-n diode was used to record the spectra at a resolution of 4 cm^{-1} (0.5 meV).

Fig. 1 shows a series of absorption spectra at selected magnetic fields obtained from a sample with a gallium fraction of $x = 0.3$. The spectra are dominated by transitions into S states. These transitions are expected to be pronounced according to the strong overlap of the corresponding envelope functions [6]. Also observable are two types of forbidden transitions: (i) Transitions between subbands of different sub-band indexes n_e and n_h like the lines $1e - 2hh$ and $1e - 3hh$, which are observable already at $B = 0$. As benefit from these transitions the effective masses of electrons and holes along the quantization direction z of the confining well potential

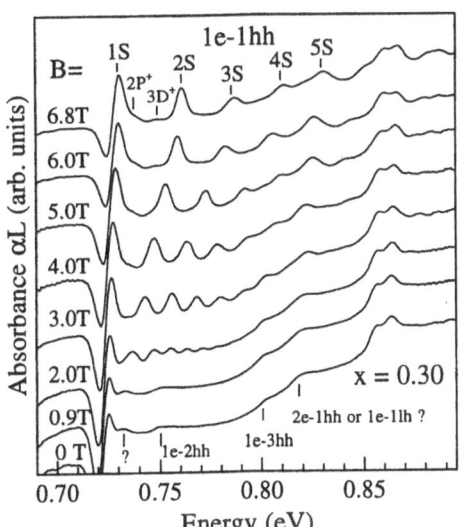

Fig. 1 Magneto-optical absorption spectrum of a $Ga_{0.3}In_{0.7}As$/InP (compressively strained) QW: S-, P- and D-type exited states.

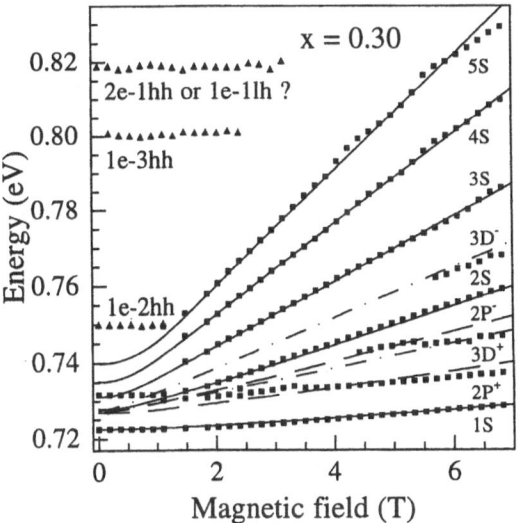

Fig. 2 Calculated allowed (full lines) and forbidden (broken lines) transitions to the experimental line positions (dots) of Fig. 1.

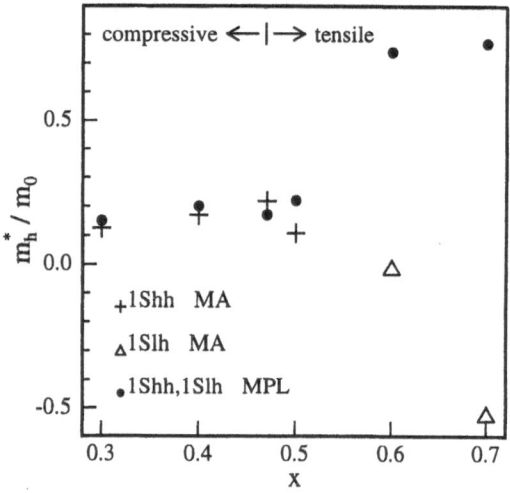

Fig. 3 Effective masses of holes, MPL (dot) and MA: for $x \leq 0.5$ m_{hh}^* (cross) and for $x > 0.5$ m_{lh}^* (triangle).

$(m_{e,z}^*$ and $m_{h,z}^*)$ can be determined. (The depth of the potential was calculated from the *model solid* theory) (ii) At magnetic fields above 5 Tesla transitions into P and D states occur. As outlined above, the diamagnetic shift of such lines allows to determine m_e^* and m_h^* separately from the same spectrum. In the fan-chart plot of Fig. 2 the experimental line positions are plotted together with the modeling.

Both types of forbidden transitions mentioned above cannot be observed in lattice matched $Ga_{0.47}In_{0.53}As/InP$ samples. In tensile strained samples with $x = 0.5$, 0.6 and 0.7 the full width at half maximum is significantly enhanced compared to the compressively strained samples. Thus, forbidden transitions of type (i) are observable but cannot be identified unambiguously. Probably for the same reason the observation of transitions into P-and D states fails, too.

Forbidden transitions of case (i) are possible since the overlap integral between the corresponding envelope functions is not orthogonal with respect to the subband index. Such transitions have been observed before in the GaInAs/ GaAs QW systems [7]. Additionally, in strained systems piezoelectric fields could also induce admixture between the envelope functions, which enhance the oscillator strength further. Transitions into P and D states may also be allowed by electric fields induced by the space charge layer due to the InP substrate, as argued in Ref. [8].

In Fig. 3 results from MA spectra are compared with data obtained from the magneto photoluminescence (MPL) of n-type modulation-doped GaInAs single QWs [9] (in MPL m_e^* and m_h^* are obtained from the Landau states of electrons and holes). For samples with $x \leq 0.5$, the heavy hole VB subband $1hh$ is highest in energy and thus m_h^* obtained in MPL should be compared with m_{hh}^* obtained from MA. As can be seen in Fig. 3 both data sets agree well. On the other hand for the tensile strained samples $(x > 0.5)$ the light hole VB subband $1lh$ is highest in energy, thus m_h^* results from MPL have to be compared to data of m_{lh}^* reported here. As shown in Fig. 3, MA and MPL yield completely different values.

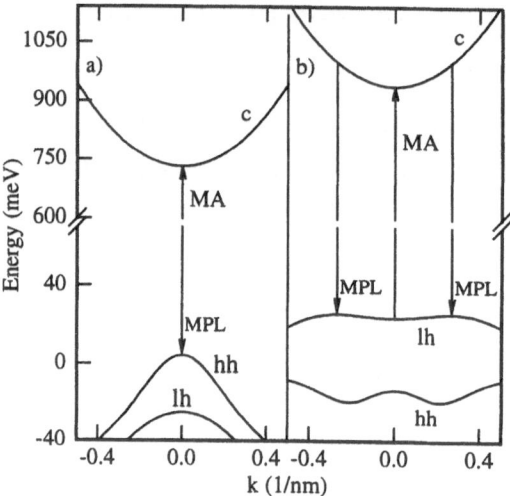

Fig. 4 Optical transitions in the (a) compressively and (b) tensile strained quantum well, respectively.

The results of Fig. 3 are understandable on a the basis of a **kp** theory (6-Band Kohn-Luttinger Hamiltonian) for the valence subbands, as shown in Fig. 4. Due to the interaction between the hh and lh subbands in tensile strained QWs the absolute maximum of energy is not at the Γ point. An additional maximum shows up [Fig. 4(b)]. In MPL therefore, m_h^* refers to the indirect VB maximum. Due to the n-type modulation doping in these samples, electrons with a wide range of k are available from the Fermi sea, which allows direct optical transitions in MPL. On the other hand in MA reported here, the optical transitions occur at the Γ point, which also is the optical threshold. Contrary, in compressively strained samples, transitions in MA as well as in MPL occur at the Γ point [Fig. 4(a)] and thus provide the same results for m_h^*.

The magneto-optical measurements therefore yield complementary information for a sample $Ga_{0.6}In_{0.4}As/InP$ at a well width of 10 nm: (i) in absorption $m_{lh}^*(\Gamma)$ and (ii) in luminescence $m_{lh}^*(k_1)$, with k_1 at ≈ 0.2 nm^{-1}.

We are grateful to the Deutsche Forschungsgemeinschaft for supporting our work.

References

1. P. Michler, A. Hangleiter, A. Moritz, G. Fuchs, V. Härle, and F. Scholz, Phys. Rev. B **48**, 11991 (1993).
2. V. Härle, H. Bolay, E. Lux, P. Michler, A. Moritz, T. Forner, A. Hangleiter, and F. Scholz, J. Appl. Phys. **75**, 5067 (1994).
3. M. Sugawara, N. Okazaki, T. Fujii, and S. Yamazaki, Phys. Rev. B **48**, 8102 (1993).
4. G. Duggan, Phys. Rev. B **37**, 2759 (1988).
5. J. Shao, D. Haase, A. Dörnen, V. Härle, and F. Scholz, J. Appl. Phys. **87**, 4303 (2000).
6. C. Weisbuch and B. Vinter, *Quantum semiconductor structures* (Academic, Boston, 1991).
7. K. J. Moore, G. Duggan, K. Woodbridge, and C. Roberts, Phys. Rev. B **41**, 1095 (1990).
8. M. Volk, S. Lutgen, T. Marschner, W. Stolz, E. O. Göbel, P. C. M. Christianen, and J. C. Maan, Phys. Rev. B **52**, 11096 (1995).
9. G. Reyher, Diploma Thesis, Universität Stuttgart, 1994, unpublished.

Proc. 25th Int. Conf. Phys. Semicond., Osaka 2000 (Eds. N. Miura and T. Ando)

Magnetic Field and Pressure Effects of Photoluminescence in a GaP/AlP Short-Period Superlattice

K. Uchida[1], N. Miura[1], F. Issiki[2], Y. Shiraki[2]

[1] Institute for Solid State Physics, University of Tokyo, 5-1-5 Kashiwanoha, Kashiwa, Chiba 277-8581, Japan
[2] Research Center for Advanced Science and Technology, University of Tokyo, 4-6-1 Komaba, Meguro-ku, Tokyo 153-8904, Japan

Abstract We report measurements of photoluminescence (PL) from a GaP/AlP short-period superlattice consisting of two indirect band-gap materials in high magnetic fields and at high pressure. The X_z-X_{xy} crossover accompanied by a strikingly large PL enhancement takes place under uniaxial pressure perpendicular to the heterointerfaces. In addition, the magneto-PL spectra under uniaxial pressure exhibit intensity decrease and spectral red-shifts. The results of the anomalous field behaviour are discussed on the basis of the carrier localization and the dynamical change of exciton radiative decay process with confinement potential fluctuation in a magnetic field.

1 Introduction

One of the ways to apply indirect band-gap materials for optoelectronic devices is the formation of superlattice (SL) structures, which is predicted to realize a *pseudo-direct* (direct only in the *k*-space) band-gap by Brillouin zone folding effect along the growth axis. Actually, there has recently been much interest in strong photoluminescence (PL) observed from GaP/AlP short-period SLs, where the possibility of a *pseudo-direct* band-gap transition has been an intriguing problem [1]. However, there has been a controversy concerning whether the conduction band minima are the folded-X_z or the X_{xy} valleys [2-4]; this is a crucial point for the *pseudo-direct* transition in the sense that the *k*-vector is conserved in the band-gap transition.

In this paper, we describe results of PL spectroscopy of $(GaP)_3/(AlP)_3$ SL in high magnetic fields and under high pressure. We observed for the first time a strikingly large PL enhancement under uniaxial pressure. In addition, the magneto-PL spectra under uniaxial pressure reveal anomalous features in a manner similar to that observed in previous work [5].

2 Experiment

The sample we have investigated was grown by gas source molecular beam epitaxy on a GaP (100) substrate; it consists of 50 periods of GaP/AlP layers and a 150 Å GaP capping layer [6].

PL measurements were performed at 4.2 K in magnetic fields (up to 42.6 T) under uniaxial pressure (up to 0.165 GPa)

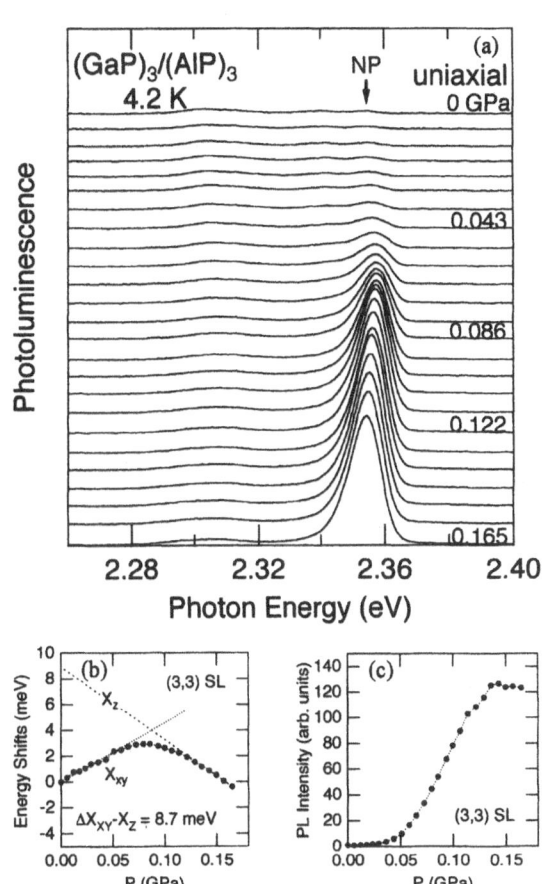

Fig. 1 (a) PL spectra for GaP/AlP (3,3) SL under uniaxial pressure up to 0.165 GPa at 4.2 K. (b) Energy shifts and (c) integrated intensity as a function of pressure.

both applied perpendicular to the hetero-interfaces. Pulsed magnetic fields were generated by a pulsed magnet. Uniaxial pressures were obtained by a new type of clamp cell specifically made for use in a pulsed high magnetic field [7]. The PL spectra were taken using a system combining a single grating spectrometer and a liquid-nitrogen-cooled CCD. The 351 nm line of an Ar-ion laser was used for the excitation.

The dependence of PL on uniaxial pressure for $(GaP)_3/$ $(AlP)_3$ SL at 4.2 K is shown in Fig. 1. At atmospheric pressure, the *no-phonon* line is centered at 2.355 eV, phonon replicas with comparable strength to the *no-phonon* line are observed on the lower energy side in the spectrum. With increasing uniaxial pressure up to about 0.05 GPa, the *no-phonon* line shifts linearly to higher energy. As the pressure is increased further, the peak position begins to turn to the lower energy side, the negative linear dependence is observed above 0.1 GPa. The energy shifts of phonon replicas are quite similar to those of the *no-phonon* line. Additionally, it was found that a strikingly large PL enhancement (more than 120 times) was observed in the *no-phonon* line accompanied by the turn-around of the energy shifts. To the contrary, under hydrostatic pressure, the PL peak was observed to shift linearly to the lower energy side with increasing pressure up to 0.4 GPa, in good agreement with calculation [3].

In the presence of a magnetic field at atmospheric pressure, the *no-phonon* line shifted slightly to lower energy and the intensity showed almost no change with increasing magnetic field up to 43 T applied perpendicular to the heterointerfaces. However, the magneto-PL spectra under uniaxial pressure demonstrated that the intensity decreased remarkably and the peak positions shifted largely to lower energy with asymmetric line-narrowing, as can be seen in Fig. 2. By contrast, when in-plane magnetic field is applied, no such anomalies were observed. Such features are quite similar to those of (n,n) SLs with n=4-7 in another work [8].

3 Discussion and conclusion

The dependence of PL on uniaxial pressure clearly shows that the conduction band minimum of $(GaP)_3/(AlP)_3$ SL at atmospheric pressure is the X_{xy} state, and the X_z state is located about 8.7 meV higher than the X_{xy} state. The X_{xy}-X_z crossover accompanied by large PL enhancement takes place at uniaxial pressure of about 0.1 GPa. This suggests that $(GaP)_3/(AlP)_3$ SL is indirect in both real and k- spaces at atmospheric pressure, and the *pseudo-direct* transition is realized under uniaxial pressure.

A large decrease in the strength and a spectral red-shift of the magneto-PL originating from the X_z-Γ transition is attributable to the dynamical change of the exciton radiative decay process due to carrier localization induced by the magnetic field. At zero field, electrons and holes are captured by local potential minima caused by the interface fluctuations. With an external magnetic field applied perpendicular to the heterointerfaces, the spatial overlap of the electron and hole wavefunctions decreases due to the shrinkage of in-plane wavefunctions, and a striking reduction of PL intensity takes place. Futhermore, if the exciton recombination lifetime

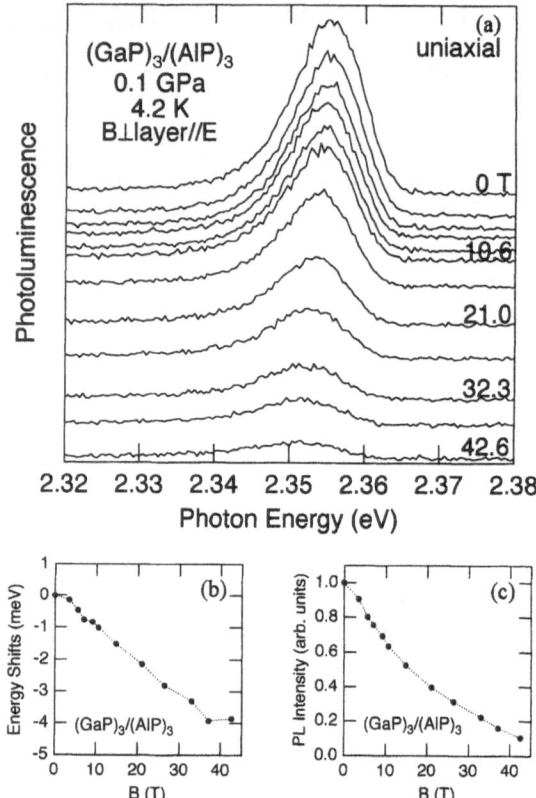

Fig. 2 (a) Magneto-PL spectra for GaP/AlP (3,3) SL in pulsed high magnetic fields up to 42.6 T and under uniaxial pressure of 0.1 GPa perpendicular to the hetero-interfaces at 4.2 K. (b) Energy shifts and (c) integrated intensity as a function of magnetic field.

becoms longer in a magnetic field, the exciton is expected to decay radiatively at the lower-energy side of potential fluctuations, which give rise to the red-shifts with asymmetric line-narrowing. The anomalies of magneto-PL suggest that exciton localization plays a key role in the PL.

References

1. U. Gnutzmann and K. Clausecker, Appl. Phys. **3**, 9 (1974).
2. M. Kumagai, T. Takagahara and E. Hanamura, Phys. Rev. B **37**, 898 (1988).
3. Y. Kobayashi and H. Kamimura, Solid State Commun. **98**, 957 (1996).
4. G. Shibata, T. Nakayama and H. Kamimura, Jpn. J. Appl. Phys. **33**, 6121 (1994).
5. K. Uchida, N. Miura, J. Kitamura, and H. Kukimoto, Phys. Rev. B **53**, 4809 (1996).
6. F. Issiki, S. Fukatsu and Y. Shiraki, Appl. Phys. Lett. **67**, 1048 (1995).
7. K. Uchida, N. Miura, T. Sugita, F. Issiki, N. Usami and Y. Shiraki, Physica B **249-251**, 909 (1998).
8. K. Uchida, N. Miura, F. Issiki and Y. Shiraki, Physica B (in press).

Optical properties of interacting excitons in quantum wells

S. de-Leon and B. Laikhtman

Racah Institute of Physics, Hebrew University, Jerusalem, Israel

Abstract   The gas of interacting excitons in quantum wells is studied. We obtain the Hamiltonian of this gas by the projection of the electron-hole plasma Hamiltonian to exciton states and an expansion in a small density. The mean field approximation of the exciton Hamiltonian gives the blue shift and spin splitting of the exciton luminescence lines. We succeeded to explain some recent experimental results which have not been explained so far.

1 Introduction

Optical properties of dense exciton gas in quantum wells had been studied intensively both experimentally and theoretically in the past decade. Recent experiments on density dependent blue shift of the exciton luminescence line,[1–3] its spontaneous spin splitting[7–9] and the decay of this splitting[8,10,11] have stressed the importance of exciton-exciton interaction of two-dimensional (2D) excitons in quantum wells.[4–6].

Theoretical investigation of interacting excitons encounters significant difficulties even when the exciton density is small and the interaction between excitons could be considered with the help of perturbation theory. Indeed, let us consider two electron - hole pairs (e1,h1) and (e2,h2) bound in excitons. One could expect that the Coulomb interaction within pairs (i.e., the interaction between e1 and h1 and the interaction between e2 and h2) has to be taken into account exactly because it provides bound states, while the Coulomb interaction between particles belonging to different excitons (e.g., the interaction between e1 and h2 and the interaction between e2 and h1) can be considered as a perturbation. However, due to electron-electron and hole-hole exchange it is impossible to say if bound pairs are really (e1,h1) and (e2,h2) or (e1,h2) and (e2,h1). For this reason it is not clear what part of the Coulomb interaction is the interaction between excitons and can be considered as a perturbation.

Attempts to overcome these difficulties were reduced so far to formal approaches, e.g., the introduction of exciton creation and annihilation operators as linear combinations of the products of electron and hole operators. The coefficients of the linear combinations were found with the help of a variational principle.[12,13] In our work we amend the method of Stolz et al.[14] explicitly separating exciton states from all possible states of the electron-hole gas. Then instead of a variational principle which accuracy is difficult to control we make use of the small density to obtain a tractable expression for the Hamiltonian of weakly interacting exciton gas.

2 Exciton Hamiltonian

The condition of a small concentration is necessary to consider electron-hole gas as gas of excitons. If it is not satisfied excitons overlap, the interaction between the particles belonging to different excitons becomes comparable with the interaction within excitons and excitons cannot be defined. Another necessary condition is low temperature compared to the exciton ionization energy.

To derive the exciton Hamiltonian we first write down the Hamiltonian of the electron-hole gas in the basis of functions

$$\Phi_{\{\nu\}}(\mathbf{r}_{e1},\sigma_{e1};\mathbf{r}_{h1},\sigma_{h1};\dots,\mathbf{r}_{eN},\sigma_{eN};\mathbf{r}_{hN},\sigma_{hN}) = \frac{1}{N!} \times$$

$$\sum(-1)^P \prod_{j=1}^{N} \Psi_{\nu_j}(\mathbf{r}_{ej_1},\sigma_{ej_1};\mathbf{r}_{hj_2},\sigma_{hj_2})\zeta_e(z_{ej})\zeta_h(z_{hj}), (1)$$

where $\mathbf{r}_{ej}, z_{ej}, \sigma_{ej}$ and $\mathbf{r}_{hj}, z_{hj}, \sigma_{hj}$ are electron and hole coordinates and spin variable respectively, $\zeta_{e,h}(z_{e,h})$ are the function describing the electron and hole quantization in the well, $\nu = (\mathbf{K}, \alpha, s)$ is the set of quantum numbers characterizing a state of one exciton,

$$\Psi_{\mathbf{K}\alpha,s}(\mathbf{r}_e,\sigma_e;\mathbf{r}_h,\sigma_h) = \frac{g_s(\sigma)}{\sqrt{S}} e^{i\mathbf{K}\mathbf{R}}\phi_\alpha(|\mathbf{r}_e-\mathbf{r}_h|), \quad (2)$$

is the wave function of one exciton, $\mathbf{R}$ is the exciton center of mass coordinate, S is the area of the sample, α is the quantum number characterizing the exciton internal state, $\sigma = \sigma_e + \sigma_h$, and $s = \pm 1, \pm 2$ is the exciton spin. We consider heavy hole excitons and use the term "spin" for the spin component perpendicular to the quantum well plane. The summation in Eq.(1) is carried out over all possible transposition of the particles and P is the parity of the transposition.

The equation for the function $\Phi_{\{\nu\}}$ where all quantum numbers $\alpha = 0$, i.e., correspond to the ground state exciton can be reduced to the form of Schrödinger equation. This equation defines the exciton Hamiltonian. To study optical properties only the matrix elements of the Hamiltonian diagonal with respect to exciton occupation numbers $n_{\mathbf{K}s}$ are necessary. They can be written as

$$E_N = \sum_{K,s} E_K n_{\mathbf{K}s} + \frac{V_b}{2S}N^2 + \frac{V_x}{4S}\sum_s (N_s - N_{-s})^2$$

$$+ \frac{U_d}{2S}\sum_s N_s^2, \quad (3)$$

where $N_s = \sum_{\mathbf{K}} n_{\mathbf{K}s}$, $N = \sum_s N_s$ is the total number of excitons, E_K is the exciton kinetic energy, U_d is the direct exciton interaction, V_x is the exchange interaction, and $V_b = U_d + V_x$. The coefficients U_d and

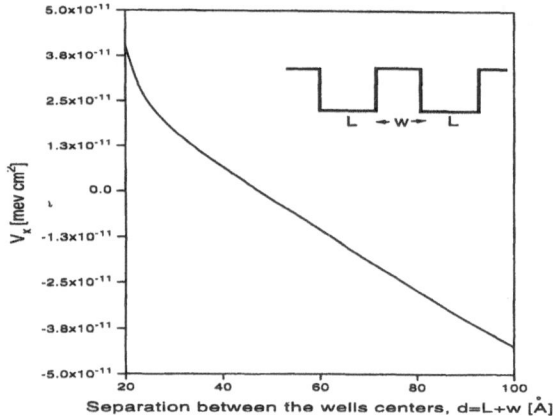

Fig. 1 The dependence of the exchange coefficient, V_x, on the separation between the centers of the wells, $d = w + L$, where $w = 10$ Å and L is changing. V_x is positive for a small enough separation between the wells and is negative for a large separation.

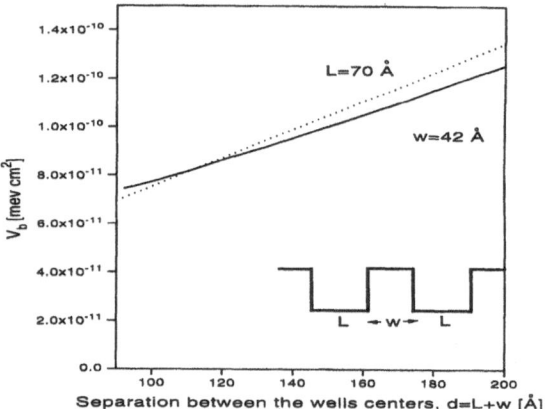

Fig. 2 The dependence of the coefficient, V_b, on the separation between the wells, $d = w + L$. The continuous line is for fixed barrier width, $w = 42$ Å, and changing wells width. The dashed line is for fixed wells widths, $L = 70$ Å, and changing the barrier width.

V_x are expressed in integrals of the bare exciton wave function and interaction potential between the particles. They are very sensitive to geometrical parameters of the heterostructure. Their behavior for a double well structure where electrons are confined in one well that holes are confined in the other is given in Fig.1 and Fig.2.

3 Comparison with experiments

The luminescence lines correspond to the energies of the exciton levels with different spins

$$\epsilon_s = -\epsilon_b + V_b n + V_x(n_s - n_{-s}) + U_d n_s , \qquad (4)$$

where ϵ_b is the exciton binding energy, $n_s = N_s/S$ and $n = N/S$ are exciton concentrations. Comparing of this expression with experimental results it is important to keep in mind that the direct interaction U_d is negligible when electrons and holes are in the same well and monotonically grows with their separation. At a large separation this is just dipole-dipole repulsion. The ex-

change coefficient V_x monotonically decreases with the separation between electrons and holes.

We compare our results with the experiment of Aichmayr et al. who observed a nontrivial behavior of spin split luminescence line of polarized excitons.[9] Excitons were optically excited in a double well structure where the separation between electrons and holes was changed with the help of gate electric field. The spin splitting between minority and majority excitons decayed with time apparently due to spin relaxation because the luminescence decay time was much longer. According to Eq.(4) the line shift with respect to its position after the relaxation is $\Delta\epsilon_s = V_x(n_s - n_{-s}) + U_d n_s$. At low gate voltage the separation between electrons and holes is small, the direct interaction is negligible and the shifts of the majority (+1) and minority (-1) lines are $\Delta\epsilon_{\pm 1} = \pm V_x(n_{+1} - n_{-1})$. With increase of the voltage the separation between electrons and holes grows leading to growth of U_d and decrease of V_x. At some intermediate voltage $V_x = U_d n_{-1}/(n_{+1} - n_{-1})$ and then $\Delta\epsilon_{+1} = U_d(n_{+1} + n_{-1})$ while $\Delta\epsilon_{-1} = 0$. At high voltage V_x is negative, and if $V_x \approx -U_d/2$ then $\Delta\epsilon_{\pm 1} = U_d(n_{+1} + n_{-1})/2$. This behavior completely corresponds to Fig.2 of Ref.[9].

In a double structure, in a weak magnetic field, Negoita et al.[15] observed a red shift of the exciton luminescence line which grew with the gate voltage that separated electrons and holes. The most striking result is that the red shift reaches values of 10 or 20 meV (depending on the gate voltage) at magnetic field around 1 T. The explanation that we suggest is based on a very narrow laser line that resonantly pumped the excitons. A weak magnetic field splits the exciton lines with different polarizations which leads to the reduction of pumping and a relative red shift. An increase of the external electric field increases the separation between electrons and holes leading to larger values of the coefficients V_b and U_d and a larger red shift, as it is observed in the experiment.

References

1. T. Amand et al., Phys. Lett. A **193**, (1994) 105.
2. L. V. Butov et al., Phy. Rev. B **60**, (1999) 8753.
3. V. Negoita et al., Phys. Rev. B **61**, (2000) 2779.
4. S. Schmitt-Rink et al., Phys. Rev. B **32**, (1985) 6601.
5. J. Fernández-Rossier et al., Phys. Rev. B **54**, (1996) 11582.
6. C. Ciuti et al., Phys. Rev. B **58**, (1998) 7926.
7. T. C. Damen et al., Phys. Rev. Lett. **67**, (1991) 3432.
8. P. Le. Jeune et al., Phys. Rev. B **58**, (1998) 4853.
9. G. Aichmayr et al., Phys. Rev. Lett. **83**, (1999) 2433.
10. A. Vinattieri et al., Phys. Rev. B **50**, (1994) 10868.
11. T. Amand et al., Phys. Rev. B **55**, (1997) 9880.
12. E. Hanamura, J. Phys. Soc. Japan, **29**, (1970) 50.
13. H. Haug and S. Schmitt-Rink, Prog. Quant. Electr. **9**, (1984) 3.
14. H. Stolz et al., Phys. Stat. Sol. (b) **105**, (1981) 585.
15. V. Negoita et al, Solid State Commun. **113**, (2000) 437.

Recombination Processes of GaNAs/GaAs structures: Effect of Rapid Thermal Annealing

I. A. Buyanova[1], W. M. Chen[1], G. Pozina[1], P. N. Hai[1], N. Q. Thinh[1], H. P. Xin[2], and C. W. Tu[2]

[1]Department of Physics and Measurement Technology, Linköping University, S-581 83 Linköping, SWEDEN
[2]Department of Electrical and Computer Engineering,University of California, La Jolla, CA 92093-0407, USA

Abstract The mechanisms for the improved properties of of undoped GaNAs/GaAs structures after the rapid thermal annealing (RTA) are investigated. Two effects are suggested to account for the observed dramatic improvement: (i) a decrease in potential fluctuations of the GaN_xAs_{1-x} band edge, deduced from the photoluminescence (PL), PL excitation and time-resolved measurements; and (ii) a significant reduction in concentration of competing non-radiative defects, revealed by the optically detected magnetic resonance studies.

1 Introduction

The remarkable, fundamental properties of the Ga(In)NAs alloys, such as the huge bowing in band gap energy, in combination with the possibility to obtain a desired lattice constant of the alloy by optimizing the N content offer promises for major improvements in the performance of optoelectronic and photonic devices based on III-V materials. Unfortunately, an increase in N composition, needed to achieve the desired band-gap energy, has been found to cause severe degradation of the material quality. According to recent studies [1.3] the alloy quality can be substantially improved by using post-growth treatments, such as oven annealing or rapid thermal annealing (RTA). The mechanism of the improvement, however, is currently not well understood.

In this work we carry out a systematic investigation of the RTA effect on the optical properties of the GaNAs/GaAs structures, aiming at determination of the mechanism responsible for the RTA-induced improvements in the material quality.

2 Samples and Methods

All the investigated structures were undoped and were grown by gas source MBE with N composition x ranging from 1.2 % up to 4.5 %. Both 7 periods $GaAs/GaN_xAs_{1-x}$ (70Å/200Å) multiple quantum well (MQW) structures and 1100 Å thick GaN_xAs_{1-x} epilayers were studied. Samples were grown at both low temperature (LT) of 420 °C and high temperature (HT) of 580 °C. RTA was done for 10 s with halogen lamps in a flowing N_2 ambient at 850 °C.

PL was excited either by the 514 nm line of an Ar^+ ion laser or by a tunable Ti:sapphire solid state laser. Time resolved PL measurements were performed using femtosecond pulses from a solid state Ti:sapphire laser, with an excitation wavelength of 800 nm. A time resolution of detection system was better than 200 ps. Optically detected magnetic resonance (ODMR) experiments were done at 4 K in a modified 9.22 GHz ESR spectrometer.

3 Experimental Results and Discussion

3.1 Radiative Recombination

Effect of RTA on the PL spectra of the investigated structures is demonstrated in Fig. 1, taking as an example the MQW structure with N content of 1.1%. In addition to a strong RTA-induced increase in PL efficiency (up to 20 times for the samples grown at 420 °C), which is observed for all investigated samples, a substantial decrease in the PL linewidth is observed.

According to our previous studies [4], the near-band gap emission in the GaNAs alloy at low temperatures is caused by the recombination of localized excitons (LE). Mechanism for the PL emission does not change after RTA since all characteristic PL properties remain qualitatively identical after annealing. However, the value of the localization potential, which reflects the degree of disorder in the GaNAs alloy, is strongly reduced after RTA based on the following experimental findings:

(i) a strong narrowing of the PL linewidth after RTA, since the low energy slope of the LE PL spectra is determined [5] by the energy distribution of the localized states within potential fluctuations; (ii) a decrease in the Stokes shift between the energies of the PL and PLE maxima (Fig.1); (iii) results of time-resolved measurement to be discussed hereafter.

Fig.2 shows results of the PL transient measurements for the as-grown and RTA-treated MQW structure. The PL decay is single exponential and PL decay time decreases with increasing emission energy, as typically observed for

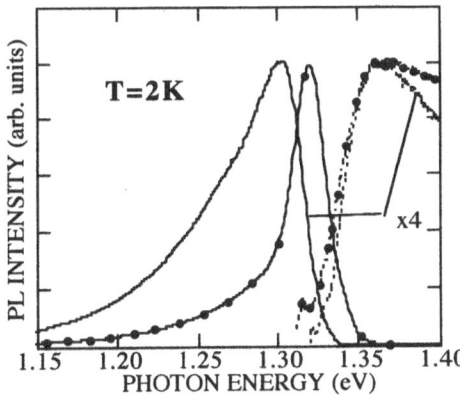

Fig. 1. Effect of RTA on the PL spectra of the HT-grown $GaAs/GaN_{0.012}As_{0.988}$MQW structure (c). The solid and dashed curves represent PL and PLE spectra, respectively. The spectra of the as-grown samples are shown by the plain lines, whereas the curves with dots are related to the RTA-treated samples. The spectra are normalized by their peak intensity for easy comparison

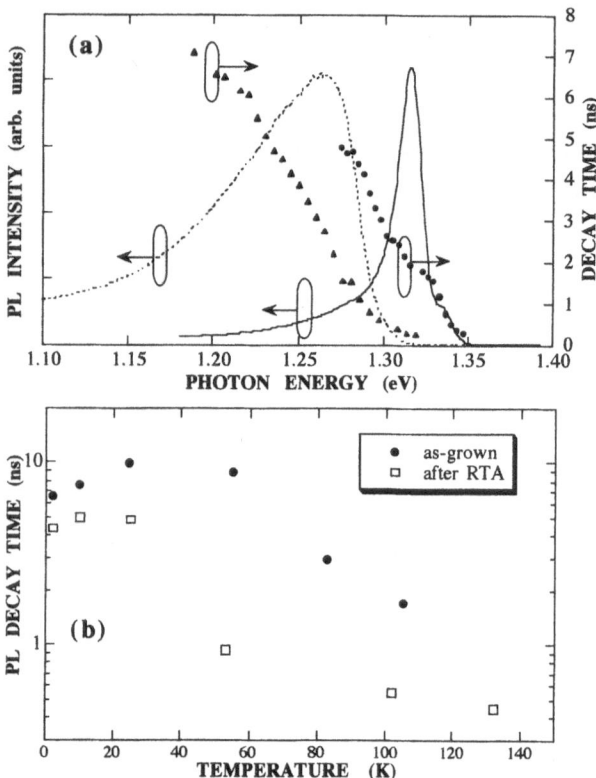

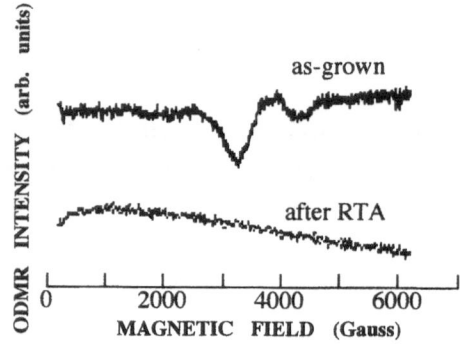

Fig.3. ODMR spectra detected via GaNAs-related emissions from the LT-grown GaAs/GaN$_{0.028}$As$_{0.972}$ MQW structure

Fig.2. (a) Spectral dependence of the PL decay time detected at 2K from the LT-grown GaAs/GaN$_{0.012}$As$_{0.988}$MQW structure before (triangles) and after (dots) RTA, respectively. Corresponding PL spectra for as-grown and RTA-treated structures are shown by the dashed and solid lines, respectively. (b) Temperature dependencies of the PL decay detected before (dots) and after RTA (rectangles).

the LE emission [4, 5]. The PL decay time is mainly radiative for the strongly localized excitons (i.e. at low emission energies) whereas exciton activation into the delocalized states can be monitored as the shortening of the PL decay.

Temperature dependence of the PL decay time is shown in Fig.2b. For the as-grown samples a complete activation of the LE excitons into the delocalized states requires temperatures higher than 100K, evident from the long (> 1 ns) decay time, observed at 100K. On the contrary, in the annealed samples a strong reduction of the PL decay time for all PL energies occurs below 60K. Since thermal energy required to activate excitons from the localized states is determined by the degree of localization, the transient data provide an additional piece of evidence for RTA-induced reduction in the potential fluctuations of the GaNAs band edge.

3.2 Nonradiative recombination

The narrowing in the energy distribution of the localized states after RTA can partly account for the increase in the PL efficiency at low temperatures- Fig.1. However, it can not explain the strong RTA-induced increase in the PL intensity observed at room temperatures

(RT), where the PL emission is due to free carrier recombination, neither the about 2 times increase in the RT PL decay time. The latter is usually determined by the nonradiative (NR) recombination. Thus, the RT PL data indicate a reduction in concentration of the NR channels after RTA. To gain insights on the nature of the NR defects in the studied structures we have performed the ODMR measurements.

In Fig.3 we show the ODMR spectra from the LT-grown GaN$_{0.033}$As$_{0.967}$/GaAs MQW structure before and after RTA. At least two negative ODMR signals are clearly detected in the as-grown samples. The same negative ODMR signals were detected via both GaNAs-related emissions, i. e. the near band gap PL and the deep defect-related PL band [4]. This fact implies [6] that defects giving rise to the ODMR signals are not involved in the PL process, but rather related to competing NR channels. The defects should be primarily introduced during the LT growth, since the ODMR signals are strongest in the LT-grown structures. Most importantly, their concentration can be drastically reduced by RTA – see Fig.3. The observed correlation between an increase in the PL intensity and the annealing out of the NR defects monitored via ODMR suggests that they may represent the dominant NR channel.

4. Conclusions

In conclusion, two effects are suggested to account for the improvement in the quality of the GaN$_x$As$_{1-x}$/GaAs structures after RTA: (i) a decrease in the potential fluctuations of the GaN$_x$As$_{1-x}$ band edge, deduced from the PL spectroscopies; and (ii) a significant reduction in the concentration of competing NR defects, revealed by the ODMR studies.

References
1. E. V. K. Rao et al, Appl. Phys. Lett. **72**, (1998) 1409.
2. S. Francoeur, G. Sivaraman, Y. Qiu, S. Nikishin, and H. Temkin, Appl. Phys. Lett. **72**, 1857 (1998).
3. L. H. Li, Z. Pan, W. Zhang, Y. W. Lin, Z. Q. Zhou, and R. H. Wu, J. Appl. Phys. **87**, 245 (2000).
4. I. A. Buyanova et al, Appl. Phys. Lett. **75**, (1999) 501.
5. M. Queslati et al, Phys. Rev. B **32**, (1985) 8220.
6. W. M. Chen and B. Monemar, Appl. Phys. A **53**, (1991) 130.

Dressed excitons and shallow impurities in low-dimensional semiconductor systems within a renormalized effective-mass approach

H. S. Brandi[1], A. Latgé[2], and L. E. Oliveira[3]

[1] Instituto de Física, Universidade Federal do Rio de Janeiro, Rio de Janeiro - RJ, 21945-970, Brazil
[2] Instituto de Física, Universidade Federal Fluminense, Niterói - RJ, 24210-340, Brazil
[3] Instituto de Física, Unicamp, CP 6165, Campinas - São Paulo, 13083-970, Brazil

Abstract A study of the effects of a laser field on the energy spectra of low-dimensional semiconductor systems is presented with the laser-semiconductor interaction being incorporated through a renormalization of the semiconductor gap and of the conduction/valence effective masses. We use a three-band Kane model and work within an extended dressed-atom approach. As an application of the present renormalized approach, we study the laser effects on exciton and on-center donor impurity states of GaAs-(Ga,Al)As quantum wells. We found that the exciton-peak weigth and blue laser-induced shifts in the exciton and donor peak energies may be quite significant.

1 Introduction

Recently, a simple scheme based in the inclusion of the effect of the laser interaction[1] with a two-band model semiconductor through the renormalization or dressing of the electron and hole effective masses has been proposed by Brandi et al[2]. This scheme may be very useful to treat situations in which the laser is tuned far below any resonances provided the effective mass approximation constitutes an appropriate physical description, as it is the case of a calculation of energy levels of shallow impurities under magnetic fields and evaluation of the optical Stark shifts in excitonic states, or other related phenomena in semiconductor heterostructures under laser fields. It has been shown[2] that, under certain laser intensities and detuning parameters, the effects of band dressing on the donor and exciton binding energies are quite important, suggesting that the contribution due to more realistic band-dressing calculations may be as important as those obtained by many-body techniques[3]. In order to better understand the importance of band-structure effects, we consider here a model semiconductor within a three-band Kane model in which spin-orbit coupling is included[4].

2 Results and discussion

A theoretical description of the effects of a laser field on the band structure of a Kane model semiconductor is performed within the framework of the k.p approximation. We consider the set of states formed by the lowest conduction band (Γ_6), the highest light- and heavy-hole valence bands ($\Gamma_8^{lh,hh}$), and the splitted spin-orbit band (Γ_7). In what follows we essentially extend the dressed-atom approach[1] to a Kane model semiconductor. Details of the calculation will be presented elsewhere. The

effect of a monochromatic homogeneous laser field on the semiconductor band structure may be obtained from the Hamiltonian[2]

$$H = H_o + \hbar\omega a^+ a + \frac{|e|}{m_o c} A_\omega \hat{p}.\hat{\epsilon}(a^+ + a), \tag{1}$$

where H_o is the diagonal matrix obtained from the Kane model, and $a^+(a)$ is the creation (annihilation) photon operator associated with the laser mode of frequency ω and polarization $\hat{\epsilon}$. The vacuum field amplitude is given by A_ω which is related to the classical amplitude of the vector potential A_o by $A_o = 2(N_o)^{1/2} A_\omega$ for $N_o >> 1$, where N_o is the number of photons in the field. Using as a basis the states obtained from the diagonalization of the Kane matrix[4], the dressed-band Hamiltonian matrix may be diagonalized, and one obtains the laser-dressed electronic bands and effective masses to order k^2. Essentially, laser effects correspond to a renormalization or dressing of the semiconductor gap and conduction/valence effective masses.

The dependence of the laser dressed conduction-band effective mass on the laser intensity is shown in Fig. 1(a), for a GaAs bulk semiconductor of gap ϵ_o, taking into account a detuning of 0.05 ϵ_o. Results for both the two-band[2] and Kane models, in the range of laser intensities considered, are qualitatively the same, differing approximately by 20% for $I/I_o \approx 10^{-4}$, where[2] $I_o \approx 5x10^7 MW/cm^2$. This indicates that laser dressing does not strongly depend on the modelling of the band structure, and therefore the present renormalized dressed effective-mass approach may be used to provide an adequate indication of the laser effects on any semiconductor heterostructure for which the effective-mass approximation is a good physical description.

As an application of the present renormalized approach, we study the laser effects on the on-center donor-impurity states of a 100 Å $GaAs - Ga_{0.7}Al_{0.3}As$ quantum well (QW). The importance of the laser dressing on the impurity levels may be inferred via a comparison with the corresponding effects of an externally applied magnetic field. In that sense, we consider the effects of a laser on the 1s-like shallow donor state under the presence of an applied magnetic field perpendicular to the QW interfaces[2,5]. Results for the magnetic-field dependence of the donor peak energy (transitions from the 1s-like donor to the first valence mini-band) are dis-

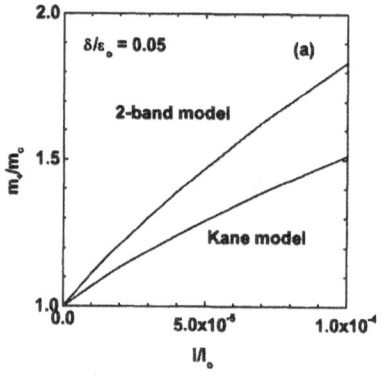

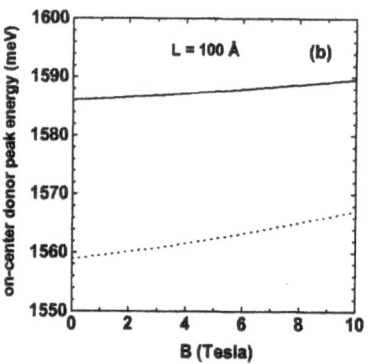

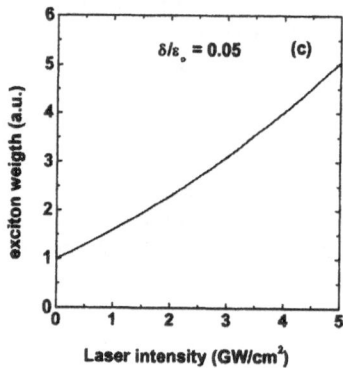

Fig. 1 Laser-dressed (a) effective mass for the conduction GaAs band as a function of the laser intensity; (b) on-center donor peak energy for a GaAs-(Ga,Al)As QW - dotted lines for the undressed results; (c) weigth of the exciton line as a function of the pump-laser intensity.

played in Fig. 1(b) for $I/I_o = 10^{-4}$ and a detuning parameter $\delta = 0.05\ \epsilon_o$. One notices that the laser-dressed shift in the donor peak energy is of the same order of magnitude of the QW confined donor binding energy, and much stronger than the corresponding shift induced by the applied magnetic field. The strong blue shift on the donor peak energy could be easily observable, although, to our knowledge, there are no experimental measurements concerning laser effects on impurity states

in doped GaAs-GaAlAs QWs. Calculations[6] were also performed for laser-dressed exciton states in a 100 Å $GaAs-Ga_{0.7}Al_{0.3}As$ QW for $\delta = 0.05\ \epsilon_o$. The blue laser-induced shift in the exciton peak energies is quite significant, and results qualitatively agree with femtosecond measurements by Mysyrowicz et al[7]. The absorption of a probe beam in the presence of pump photons has been theoretically investigated by adopting a dressed-exciton Greens function formalism and performing the exact summation of the corresponding renormalized ladder diagrams. Calculated results for the weigth of the exciton line in a 100 Å $GaAs-Ga_{0.7}Al_{0.3}As$ QW and for $\delta = 0.05\ \epsilon_o$ are presented in Fig. 1(c), showing a significant increase with the pump-laser intensity, as expected.

This work shows that, far below any resonances, one-body effects of band dressing on donor and exciton binding energies may be quite important and of the same order of magnitude as those obtained from many-body techniques[3]. Therefore, band-structure effects should be taken into account if one is interested in phenomena for which many-body corrections are important. The present model allows the study of complex laser-induced effects in semiconductor heterostructures, with a variety of possible applications in optoelectronic devices based on phenomena such as optical absorption, resonant tunneling, and transport properties in general.

Acknowledgements. The authors would like to thank the CNPq, FAPERJ, FAPESP, and FAEP-UNICAMP Agencies for partial financial support.

References

1. C. Cohen-Tannoudji, J. Dupont-Roc, and G. Grynberg, *Processus d'interaction entre photons et atomes* (Editions du CNRS, Paris 1988).
2. H. S. Brandi, A. Latgé, and L. E. Oliveira, Sol. State. Commun. **107**, (1998) 31; H. S. Brandi and G. Jalbert, Sol. State Commun. **113**, (2000) 207.
3. M. Combescot, Phys. Reports **221**, (1992) 167.
4. G. Bastard, "Wave mechanics applied to semiconductor heterostructures" (Les Editions de Physique, Les Ulis 1988).
5. A. Latgé, N. Porras-Montenegro, M. de Dios-Leyva, and L. E. Oliveira, Phys. Rev. B**53**, (1996) 10160.
6. A. Matos-Abiague, L. E. Oliveira, and M. de Dios-Leyva, Phys. Rev. B **58**, (1998) 4072.
7. A. Mysyrowicz, D. Hulin, A. Antonetti, A. Migus, W. T. Masselink, and H. Morkoç, Phys. Rev. Lett. **56**, (1986) 2748.

Superlinear photoluminescence in GaAs/(GaAl)As heterojunctions

Jinxi Shen*, R. Pittini, Y. Oka

RISM, Tohoku University, Katahira, Sendai 980-8577, Japan
CREST, Japan Science and Technology Corporation, Kawaguchi 332-0012, Japan

Abstract   We present two effects arising from the superlinearity of the photoluminescence in GaAs/(GaAl)As heterojunctions: (*a*) a significant enhancement of the photoluminescence intensity observed in the pump-probe experiments and (*b*) a strong anti-Stokes photoluminescence. In the latter experiment, an intense emission of the free excitons was observed when the heterojucntion was excited with laser pulses tuned below the free exciton energy. While in the pump-probe experiment, the luminescence intensity detected exciting the sample with the probe pulse increases linearly with the intensity of the pump pulse, even though the time delay between the two pulses is far beyond the exciton lifetime. These two phenomena are interpreted by the same mechanism, i.e. the bimolecular formation of excitons with the vertical transport of the photo-excited carriers driven under the interface electric field.

1. Introduction

Very much understanding has been reached in the recent years on the photoluminescence (PL) of heterojunctions [1-5]. However, the discussion about the origin of the PL in heterojunctions still remains complex. The interpretation incorporating the phonon interaction [1] and the vertical carrier transport [2, 3] attracts much attention although evidences are given that for some sample structures the emission comes directly from the two-dimensional electron gas [4, 5]. In this brief report, we present two important superlinear properties, i.e. the anti-Stokes PL and the pump-probe PL detected in heterojunctions. These results emphasize the importance of the vertical transport of the carriers and the bimolecular exciton formation as the fundamental processes for the luminescence in heterojucntions. It determines the characteristic properties of PL in herojunctions.

2. Experimental results and discussion

Experiments have been performed in n-type heterojunctions with a two-dimensional electron gas at a density of 4×10^{11}/cm^2. The sample was immersed in liquid helium to prevent the local heating by laser excitation. Tunable laser pulses generated by a mode locked titanium sapphire laser and an optical parametric amplifier were narrowed down energetically by a monochromator and directed on to the sample in a cryostat. For the pump-probe experiment, a beam splitter with 50%:50% ratio split the light with one beam passing through a delay stage. Luminescence singles were collected by a ¼-meter spectrometer followed by a streak camera. The temporal resolution was 5 ps and the spectral resolution was 1 Å

2.1 The anti-Stokes PL

The transient PL image with resonant excitation is shown in Fig. 1. The FX recombination is always observable when one tunes the excitation pulse across the free exciton (FX) energies. When one excites below the FX, a very strong anti-Stokes PL is observed (Fig. 1.c). In this case, the anti-Stokes PL intensity of the FX peak is reaches 10% of the intensity detected when exciting above the FX energy. In addition, the build-up time of the exciton emissions is strongly energy dependent. In fact, the rise time of the PL intensity is as long as 1 ns for both Fig. 1 (a) and (c) when exciting above and below the FX. But the rise time becomes much shorter when the excitation energy is tuned exactly on the FX peak as shown in Fig. 1 (b).

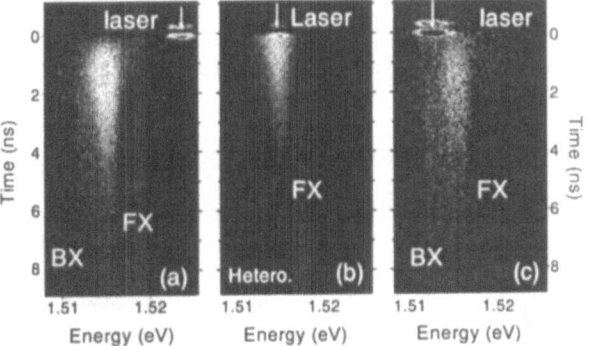

Fig. 1 Transient PL images taken under resonant excitation with excitation energy (a) above, (b) at, and (c) below the FX peak.

Considering the energy conservation, the anti-Stokes PL is attributed to the bimolecular exciton formation combined with the carrier vertical drift, as shown in Fig. 2. The free carriers are generated from impurities when the excitation energy is smaller than the FX recombination energy. When electrons and holes are driven together by the self-built interface electric field, they recombine and emit FXs with energies larger than the photon energy of the excitation. The self-built potential variation across the heterointerface provides the driving force for the carriers to move and the additional energy for the anti-Stokes recombination. Similar anti-Stokes PL is hard to be observed in simple GaAs, not only because of the small cross section of the interband impurity absorption but also the difficulty of the spatial movement in absence of a spatial potential variation. The photo-excited carriers in simple GaAs finally being trapped by the impurities locally and no FX transitions can be expected.

The rise time for the building up of the FX recombination intensity is related to the carrier drifting from their donors (acceptors) to the place of recombination. Similar recombinations happen for the carriers excited above the bandgap. In this case, the photo-excited electron-hole pairs are dissolved by the interface electric field and free carriers are available for the bimolecular exciton formations. Whereas exciting the heterojunction exactly on the FX energy, a large amount of FXs are excited directly in the flat band region. The FXs are generated geminately within the time of the photon travelling in the medium with the depth of the absorption length. As a result, The rising time is short as observed experimentally in Fig. 1 (b).

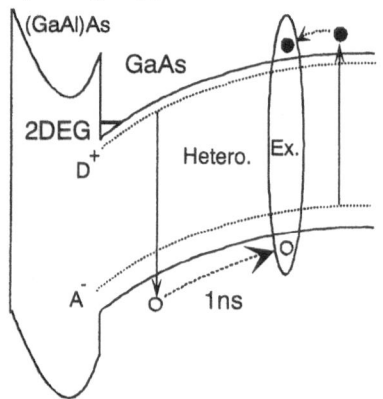

Fig. 2 Potential variation of a n-type heterojunction across the heterointerface and the mechanism of anti-Stokes PL by exciting below the FX.

2.2 The pump-Probe PL

The probability of bimolecular exciton formation is proportional to the product of the electron and hole densities. Therefore, the PL intensity attributed by the bimolecular exciton formation has a square dependence on the excitation power. However, an experimental non-linear power dependence is not necessarily related to the bimolecular exciton formation. For example, the non-linear power dependence detected before the onset of stimulated emission might be well related to the competition between the efficiency of the radiative and non-radiative recombinations. Therefore, a pump-probe PL experiment with adjustable time delay is a useful tool for the conclusive assignement of the PL signal to bimolecular exciton formation ruling out high order field effects by monitoring the recombination dynamics.

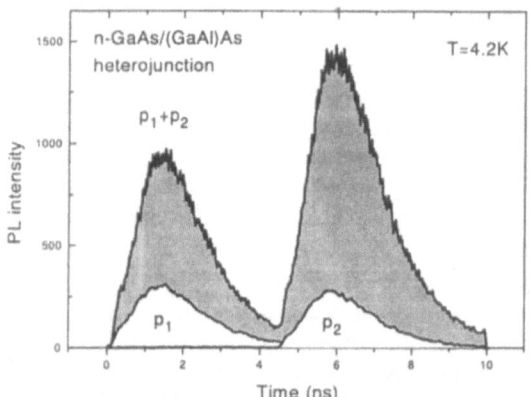

Fig. 3 Transient PL intensity of the pump-probe PL on n-type heterojunction.

Experimentally, we have focused two laser pulses to a sample area. The time delay used in Fig. 3 is 4.4 ns. The repetition period of the laser pulses is 12 ns. In Fig. 3, the two pulses have the same intensity. The two lower intensity curves filled with white areas are the PL intensities detected when the sample was excited by each single pulse. The curve with stronger PL intensity and filled gray area was detected when the sample was excited with both pulses. Therefore, the gray area equals to the PL intensity enhancement. It can be concluded from Fig. 3 that both, the rise time and the decay times are the same in the single pulse and in the two-pulses experiments. Most remarkable is the PL enhancement of the second pulse detected when the PL intensity of the first pulse already decayed. As a result, only the free carrier density, instead of the photon field, contributes to the superlinerity observed in Fig. 3. Therefore, the exciton formation process in heterojunctions is unambiguously bimolecular. The PL intensity of the P_1 pulse also increases due to the pulse repetition of the laser. The maximum enlargement of the transient PL intensity after the pulse P_2 is larger than the maximum PL enhancement achieved after the pulse P_1. This is attributed to the decay of the free carriers available for the bimolecular formation. In addition to the temporal properties in shown Fig. 3, the PL intensity enhancement is found to be proportional to the excitation intensity of the pump pulse.

3. Conclusion

We have presented two experimental results about the superlinearity of the PL in heterojuctions. The anti-Stokes PL and the enlargement of the PL intensity in the pump-probe PL clearly reveal the nature of the bimolecular exciton formation and the carrier vertical transport in heterojunctions. The experimental results show that the history of the carrier movement before the exciton formation determines the dynamic properties of the exciton recombinations, and so the fine structure of the spectra of heterojunctions versus those of GaAs, even though the excitons finally recombining are subjected to the same spatial condition like the simple GaAs.

*Present address:
E-Tek Dynamics Inc.; 1865 Lundy Ave.; San Jose, CA 95131; USA
References:
1. B. M. Ashkinadze, E. Linder, V. Umansky, to be appeared in Phys. Rev. B, October 2000.
2. J. X. Shen, Y. Oka, C. Y. Hu, W. Ossau, G. Landwehr, K. –J Friedland, R. Hey, K. Ploog, G. Weiman, Phys. Rev. **B 59** (1999) 8093.
3. J. X. Shen, R. Pittini, Y. Oka, E. Kurtz, Phys. Rev. **B 61** (2000) 2765.
4. Kazunori Aoki, Physica **B 272** (1999) 146.
5. K. Fuji, M. Saitoh, H. Nakata, T. Ohyama, Physica **B 272** (1999) 454.

The exciton dead layer revisited

M. Combescot[1], **R. Combescot**[2], **B. Roulet**[1]

[1] Groupe de Physique des Solides des Universités Paris VI and Paris VII,CNRS, Tour 23, 2 place Jussieu, 75231 Paris Cedex, France

[2] Laboratoire de Physique Statistique, Ecole Normale Supérieure,24 rue Lhomond, 75005 Paris , France

Abstract We show that, at fixed Bohr radius, the dead layer for an exciton confined in a semiconductor quantum well must increase with the exciton total mass, *in contradiction* with previous results. A reliable determination of this dead layer imposes a very accurate determination of the exciton energy in the wide well limit. We propose a new procedure to obtain this energy based on an adiabatic approach to the exciton motion associated to a novel calculation of the energy of an electron bound to an impurity in a quantum well, which relies on the determination of the *exact* envelope function. This envelope function turns out to be quite different from the free electron one used by everyone up to now.

Close to a surface, the bulk exciton has to distort in order to keep its electron (e) and its hole (h) inside the semiconductor. As this distorsion induces an energy increase one usually thinks that the exciton tries to avoid it by "staying away" from the surfaces. This leads to write the energy of an exciton in a large width quantum well as the energy of a bulk exciton $(-R_X = -\hbar^2/2\mu a_X^2)$ plus the energy of its center of mass M confined in a well somewhat narrower than the true well width L :

$$E_X = -R_X + \frac{\hbar^2}{2M}\frac{\pi^2}{(L-2d)^2} \qquad (1)$$

This equation is quite questionable as it states that the large L limit of the exciton energy should start as :

$$E_X \simeq R_X[-1 + \pi^2\frac{\mu}{M}(\frac{a_X}{L})^2 + 4\pi^2\frac{\mu}{M}\frac{d}{a_X}(\frac{a_X}{L})^3 + ...](2)$$

without linear term and with a quadratic term which is *exactly* the kinetic energy of the center of mass confined in the well L. It is a priori far from obvious that the change of the e-h relative motion, induced by the distorsion does add a contribution to the exciton binding energy $-R_X$ in $(a_X/L)^3$ only and not in (a_X/L) or $(a_X/L)^2$. By noting that the e-h coordinates are confined in a square $-L/2 < (z_e, z_h) < L/2$ when written in terms of the e-h positions (z_e, z_h), but in a parallelogram $-L/2 < (Z + \alpha_h z, Z - \alpha_e z) < L/2$ with $\alpha_{e,h} = m_{e,h}/M$ when written in terms of the center of mass and relative motion positions (Z, z) (see Fig.1), and by finding appropriate lower and upper bounds for this (Z, z) confinement, it is however possible to *prove* [1] that the two first terms of Eq.(2) are indeed exact (the energy being a monotonic function of the confinement).

The dead layer d being linked to the $(a_X/L)^3$ prefactor of the exciton energy expansion, it is clear that a very accurate procedure is necessary to obtain it in a reliable way. Before outlining this procedure let us briefly comment on previous results. From a very naive argument

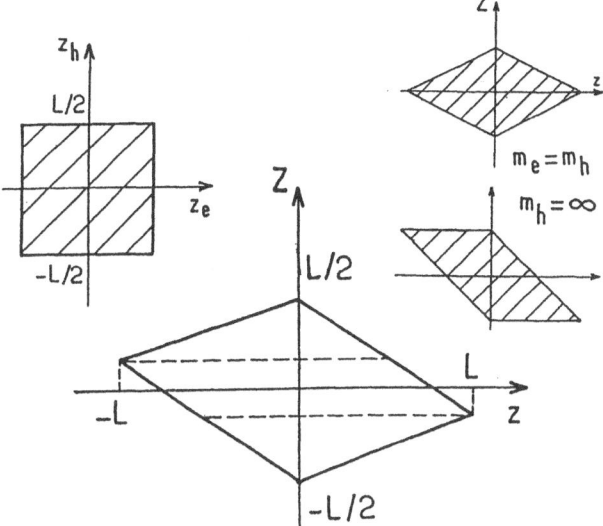

Fig. 1 e-h confinement in a quantum well of width L, shown in e-h coordinates (z_e, z_h) or in center of mass and relative coordinates (Z, z).

which sees the exciton as a rigid ball with e,h staying at a_X apart, the dead layer should decrease when μ/M increases: indeed, for $m_e = m_h$, the center of mass is in the middle of the e and h while for $m_h = \infty$, it is on the hole; so that in order to have the (e,h) inside the material, a rigid ball exciton should have its center of mass at a_X from the surface when $m_h = \infty$, while it should be at $a_X/2$ only when the two masses are equal. On the opposite, Del Sole and collaborators [2,3], using a much more elaborate approach, find a dead layer $(1/P \simeq a_X\sqrt{\mu/M}$ in their notations) which increases with μ/M.

As μ/M is always small ($\mu/M = 0$ for $m_h = \infty$ and $1/4$ only for $m_e = m_h$), an adiabatic procedure is quite appropriate to calculate the exciton energy. For a center of mass fixed at Z, the problem of the exciton relative motion is similar to the one of an electron bound to an impurity located at Z in a quantum well of width $L(Z)$ with $L(Z)$ shrinking to 0 when Z approaches $\pm L/2$ (see Fig. 1). If one uses for the bound electron motion Bastard's [4] trial function $f_\alpha(z)\exp(-|\mathbf{r}-\mathbf{r}_i|/\alpha a_X)$ with an envelope function $f_\alpha(z)$ taken as the free electron one $\cos(\pi z/L(Z))$, one approaches the large L limiting energy $-R_X$ within corrective terms of the order of $(a_X/L)^2$, instead of $\exp(-2L/a_X)$ as it should be. These terms have here a dramatic effect since they produce a spurious $(a_X/L)^2$ contribution to E_X which affects the second term of Eq.(2), i.e. the center of mass energy.

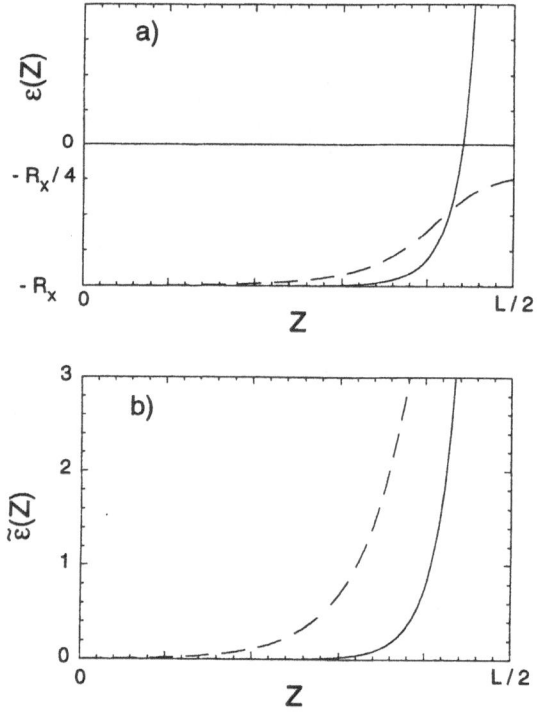

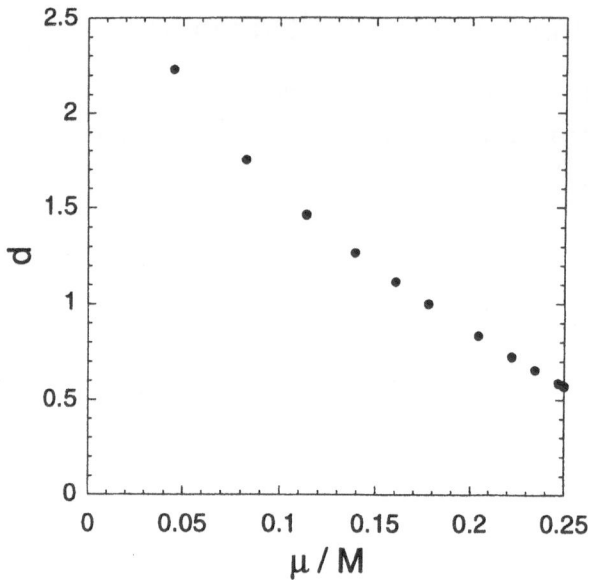

Fig. 3 Dead layer in Bohr radius i.e. for fixed μ.

Fig. 2 Potential $\epsilon(Z)$ and effective potential $\tilde{\epsilon}(Z)$ for the center of mass motion in the well (solid line $m_e = m_h$, dashed line $m_h = \infty$ for a) and $m_h = 10 m_e$ for b)).

This forced us to reconsider the problem of a bound electron in a quantum well. From a formal functional derivative procedure, we [5] have derived the differential equation satisfied by the $f_\alpha(z)$ which minimize the energy for a given α. An analytical solution exists in terms of degenerate hypergeometric functions. It is of importance to mention that this exact envelope function, which strongly depends on the impurity position, is very different from the usual free electron one. The obtained bound electron energy, which now reaches its large L value within exponentially small corrective terms, is up to 10% lower than the one obtained with the usual free electron envelope function.

The energy of an electron bound to an impurity located at Z in a well of width $L(Z)$, as calculated with this exact envelope function, gives the exciton relative motion energy $\epsilon(Z)$ for a center of mass at Z.

i) if $M = \infty$, $L(Z) = L$ is always large (see Fig. 1). For Z far from $\pm L/2$, the solution is a 3D hydrogenoid 1S state with an energy $-R_X$ while for $Z \simeq \pm L/2$, the impurity is on the wall and the solution is half a 3D 2P$_z$ state with an energy $-R_X/4$ so that $\epsilon(Z)$ varies from $-R_X$ to $-R_X/4$ when Z goes from 0 to $\pm L/2$ (see dashed curve of Fig. 2a)

ii) if M is finite, $L(Z)$ stays large and $\epsilon(Z)$ close to $-R_X$ except for $Z \simeq \pm L/2$, where the extension $L(Z)$

shrinks dramatically to 0 (see Fig. 1). In this limit, one has a 2D bound state for the (x, y) motion, with an energy $-4R_X$, and a strongly confined motion for the z direction, so that $\epsilon(Z)$ diverges when $Z \to \pm L/2$. Although not obvious at first, $\epsilon(Z)$ is found to monotonically increase from $-R_X$ to ∞ when Z goes from 0 to $\pm L/2$, the $-4R_X$ contribution for $\pm L/2$ making however the curves close to $Z = 0$ flater for finite M than for $M = \infty$ (see Fig. 2a).

This $\epsilon(Z)$ energy serves as a potential for the center of mass Schrödinger equation:

$$[-\frac{\hbar^2}{2M}\frac{\partial^2}{\partial Z^2} + \epsilon(Z)]\psi(Z) = E_X \psi(Z) \qquad (3)$$

In order to physically understand the μ/M dependence of the dead layer, we must however note that the effective potential which controls the Z motion is not $\epsilon(Z)$, but $R_X \tilde{\epsilon}(Z) = [\epsilon(Z) - \epsilon(0)]M/\mu$: while the $\epsilon(Z)$ curves cross for various μ/M, the $\tilde{\epsilon}(Z)$ curves do not, the smaller the value of μ/M, the narrower the effective potential $\tilde{\epsilon}(Z)$ (see Fig. 2b). It becomes then clear that this effective potential, which tends to keep the exciton farther from the walls if the center of mass is heavier, must produce an exciton dead layer which decreases when μ/M increases. The result of our new calculation indeed confirms [5] this behaviour (see Fig. 3).

References

1. M. Combescot, R. Combescot and B. Roulet (submitted)
2. A. D'Andrea and R. Del Sole, Phys. Rev. **25** (1982) 3714; Phys. Rev. **41** (1990) 1413.
3. N. Tomassini, A. D'Andrea, R. Del Sole, H. Tuffizo-Ulmer and R. T. Cox, Phys. Rev. **51** (1995) 5005.
4. G. Bastard, Phys. Rev. **24** (1981) 4714.
5. M. Combescot, R. Combescot and B. Roulet (submitted)

Picosecond dynamics of polarized resonance photoluminescence in the GaAs/AlGaAs- superlattices.

I. Ya. Gerlovin[1]*, Yu. K. Dolgikh[1], S. A. Eliseev[1], V. V. Ovsyankin[1], Yu. P. Efimov[2], I. V. Ignatiev[2,3], I. E. Kozin[2,3], V. V. Petrov[2], Y. Masumoto[3]

[1] Vavilov State Optical Institute,199034, Birzhevaya 14, St.-Petersburg, Russia
[2] Institute of Physics, St.-Petersburg State University, Russia
[3] Masumoto Single Quantum Dot project, ERATO, JST, Japan

Abstract Dynamics of the resonance excitonic photoluminescence excited in the GaAs/AlGaAs superlattices by a linearly or circularly polarized light is studied experimentally at 5 K. It is found that the depolarization of the photoluminescence is related to a reversible dephasing caused by statistical distribution of frequencies of the right- and left-hand polarized excitonic transitions.

1 Introduction

Polarization dynamics of excitonic radiation of the GaAs/AlAs quantum wells and superlattices is controlled by processes of the energy and phase relaxations between the fine excitonic states [1]. A study of the polarization dynamics makes it possible to determine the spin relaxation rates of such systems. However the experimental data obtained to date [2,3] are in rather poor agreement with the theory considered mainly the energy relaxation of excitonic spin [4].

In this paper, we report the study of the HH-exciton resonance photoluminescence (PL) kinetics excited in the GaAs/AlGaAs superlattices by short pulses of either linearly or circularly polarized light. Analysis of the experimental data has allowed us to make certain conclusions about mechanisms of both energy and phase relaxation of the excitonic magnetic momenta in the structures under study.

2 Experimental.

We studied a heterostructure comprised of 50 periods of the superlattice with the thickness of the GaAs and $Ga_{0.56}Al_{0.44}As$ layers being, respectively, 30 and 38 Å. The sample was excited by a tunable Ti:sapphire laser with a pulse duration of 5 ps. The PL of the sample was selected by a 0.25-m double subtractive dispersion monochromator. The PL kinetics was detected using a streak-camera. The pulsed exciton density in the sample did not exceed 10^9 cm^{-2}. All the experiments were performed at 5 K. We have studied the resonant PL kinetics under excitation by the linearly and circularly polarized light within the excitonic line. The experiments were performed at normal incidence of the exciting beam. The PL emission was collected in a narrow cone ($\approx 15°$) around the normal.

Figure 1 shows kinetics of the resonant PL, measured at various polarizations of the excitation and PL detection. We have found that kinetics of nonpolarized emission can be well approximated by an exponential function with the time constant $\tau_{PL} = 58$ ps, both for linear and circular polarization of the excitation. At the same time, the PL kinetics in co-

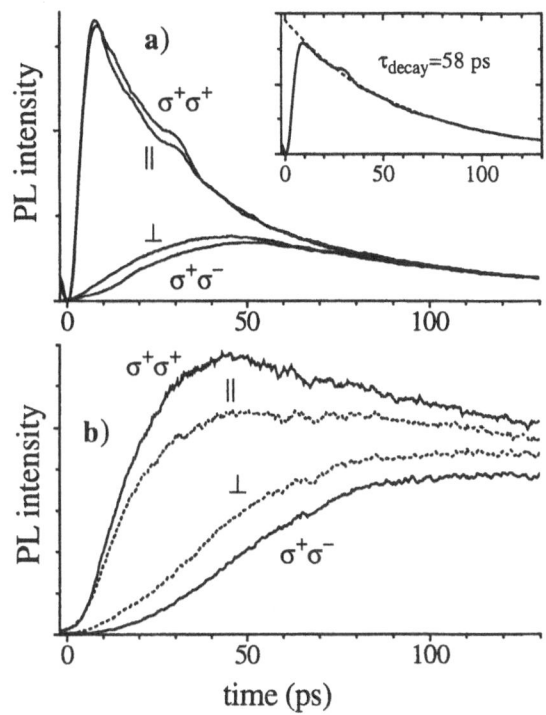

Fig. 1 PL kinetics at spectral position of the laser (a) and in spectral point red shifted by $\Delta E = 1.5$ meV (b) in different polarizations. Scattered laser light is subtracted. The co- and cross-polarized kinetics are marked by $\parallel$ and $\perp$ for linear polarizations and by $\sigma^+\sigma^+$ and $\sigma^+\sigma^-$ for circular polarizations, respectively. Inset: Decay of the nonpolarized PL measured at the laser line spectral position. Dashed curve is the exponential fit.

and cross-polarizations is strongly different. As seen from Fig.1, the PL detected in co-polarizations builds up with no delay, whereas the PL pulses in cross-polarizations show gently sloping front edges. Note that in the linear polarization the luminescence builds up faster than in the circular one.

We have studied time dependence of the degree of the PL polarization, $\rho = (I_1 - I_2)/(I_1 + I_2)$, where I_1 and I_2 are, respectively, the PL intensities in co- and cross-polarizations. The degree of polarization exhibits an exponential like decay which is slower for circular polarization We have performed the polarization measurements also upon detuning from the resonance with the exciting light (Fig. 1b). It has been found that the detuning is not accompanied by instant depolarization of the PL and even does not affect essentially the polarization decay time. Such a high stability of polarization char-

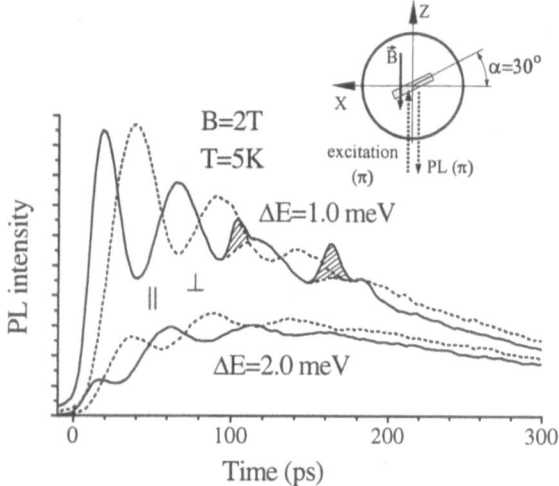

Fig. 2 PL kinetics in magnetic field $B = 2$ T at the spectral points with the Stokes shifts $\Delta E = 1.0$ meV (two upper curves) and $\Delta E = 2.0$ meV. Geometry of the experiment is shown in the inset. The kinetics shown by solid and dashed lines is measured in the linear co- and cross-polarizations, respectively. The hatched peaks are due to scattered laser light.

acteristics of the emission in the phonon-assisted process of diffusion looks rather surprising. PL kinetics of the sample in magnetic field reveals an oscillation like behavior as shown in Fig. 2. Analysis shows that the oscillations are caused by the quantum beats (QB's) of Zeeman components of the excitonic state. It is important that decay time of the QB's which is of about 100 ps considerably exceeds that of the polarization measured in absence of the magnetic field.

3 Analysis of the results

In the GaAs-based quantum-confined structures, the lowest states of free excitons are formed by electrons with the spin projection upon the quantization axis (z-axis) $M_z = \pm 1/2$ and by heavy holes with the spin projections $M_z = \pm 3/2$. As a result, four excitonic states are formed with two of them ($M_z = \pm 1$) being optically active and the two others ($M_z = \pm 2$) being not coupled with light [1]. Polarization dynamics of such a system should be determined by method of its excitation and by mechanisms of relaxation of these states. If the optically active doublet ($M_z = \pm 1$) is not split, the only depolarizing mechanism is the relaxation mediated by the "dark" states ($M_z = \pm 2$). As shown in [1], such a process should result in a pronounced nonexponentiality in decay of the nonpolarized emission. The absence of such a nonexponentiality, noted above, allows us to rule out the relaxation into the "dark" states. The depolarization we observed can be attributed only to splitting of the states ($M_z = \pm 1$).

Dynamics of depolarization via the spin relaxation essentially depends on the states which are initially excited by light. If polarization of the excitation corresponds to polarization of one of the oscillators and is orthogonal to that of another one, the only mechanism of depolarization of the emission is the energy relaxation between the states. Otherwise, when the light excites a coherent superposition of the both

states, the depolarization is controlled not only by the energy relaxation but also by dephasing capable of destroying the coherency. So in the latter case, the rate of depolarization should be higher. Depending on perturbation causing the fine splitting, the split states ban be either the linear oscillators aligned along the anisotropy axes or the right- and left-hand circular oscillators. Since in the structure under study the decay time of circular polarization (≈ 60 ps) proved to be larger than that of linear polarization (40 ps), we can conclude that the splitted states of the HH-exciton in this system are the states of circular oscillators with $M_z = +1$ and $M_z = -1$. So the perturbation that induces the splitting is oriented along the growth axis of the structure and therefore this perturbation has a nature similar to magnetic field. The second important conclusion which follows from essential difference between the depolarization rates is that the dephasing rate of the exciton magnetic moment is relatively high, at least is not lower than the energy relaxation rate.

Both conclusions are fairly nontrivial. First, it is commonly supposed that the optically active states of HH-excitons in the GaAs/AlAs-based heterostructures are not split [1,2]. Second, as the experiments in magnetic field show (see Fig. 2), excitonic spins in the studied structure are characterized by a high phase stability. To explain the obtained results, we have to suppose that the perturbation which splits the optically active states is of magnetic nature and random one both in magnitude and in sign. As a result, statistical distribution of the splittings exceeds their mean value that prevents from their direct observation. At the same time, this distribution destroys coherence of the superposition states, i.e. it serves as a source of the reversible phase relaxation. It can be supposed that the random perturbation is caused by exchange interaction with spins of photo-excited carriers localized at defects of the structure. Further studies are needed for more detailed information about the nature of this perturbation.

4 Conclusion.

The experimental study of polarization dynamics of the resonant PL in the GaAs/AlGaAs superlattice has allowed us to establish that only optically active states of the HH-exciton fine structure participate in relaxation leading to the PL depolarization. We propose that the splitting of the optically active excitonic state in the studied structure is of magnetic nature and is most likely to be related to random magnetic fields produced by spins of the carriers localized on defects.

The work was supported by the Russian Foundation for Basic Research.

References

1. E. L. Ivchenko, Pure and Appl. Chem. **67**, 463 (1995).
2. C. Gourdon, P. Lavallard, Phys. rev. B **46**, 4644 (1992).
3. P. Le Jeune, X. Marie, T. Amand, F. Romstad, F. Perez, J. Barrau, M. Brouseau, Phys. Rev. B **58**, 4853 (1998).
4. M. Z. Maialle, E. A. de Andrada e Silva, L. J. Sham, Phys. Rev B **47**, 15776 (1993).

Proc. 25th Int. Conf. Phys. Semicond., Osaka 2000 (Eds. N. Miura and T. Ando)

569

Magneto-photoreflectance of the above barrier state transitions in GaAs/Al$_{0.3}$Ga$_{0.7}$As double quantum wells

G. Sęk[1], M. Nowaczyk[1], L. Bryja[1], K. Ryczko[1], J. Misiewicz[1], M. Bayer[2], J. Koeth[2], A. Forchel[2]

[1] Institute of Physics, Wrocław University of Technology, Wybrzeże Wyspiańskiego 27, 50-370 Wrocław, Poland
  e-mail: grzes@if.pwr.wroc.pl
[2] Insitute of Physics, University of Würzburg, Am Hubland, D-97074 Würzburg, Germany

Abstract We report on the low temperature photoreflectance (PR) investigation of the resonant state related transitions in GaAs/Al$_{0.3}$Ga$_{0.7}$As double quantum wells (DQW) separated by thin AlAs barriers. The energy levels of confined and resonant states in the structures have been calculated using envelope function and transfer-matrix formalisms, respectively. The evolution of the PR spectra in the magnetic field up to 6 T has been discussed and the diamagnetic shifts has been derived and compared with their values for confined state excitonic transitions.

1 Introduction

There are several reports on the investigations of confined state excitonic transitions in double quantum wells of different material systems [1-5]. However, there is very little experimental work reported concerning the resonant state transitions in DQW's [6], and there is, to our knowledge, no work concerning the magnetic field dependence and the exciton diamagnetic shifts of such transitions. We present for the first time, a contactless electromodulation technique called photoreflectance, as a useful tool for the investigation of the magnetic field dependence of the resonant state excitonic transitions in GaAs/Al$_{0.3}$Ga$_{0.7}$As DQW structure.

2 Experimental

An undoped, symmetric GaAs/Al$_{0.3}$Ga$_{0.7}$As/ Al$_{0.35}$Ga$_{0.65}$As double quantum well structure separated by 1 monolayer (ML) AlAs barrier was grown by solid source molecular beam epitaxy along [001] direction on semi-insulating substrate. For the two quantum wells a width of 7.5 nm was chosen. The DQW was cladded between two 150 nm thick Al$_{0.3}$Ga$_{0.7}$As and Al$_{0.65}$Ga$_{0.65}$As layers. The scheme of the conduction band edge for this structure is shown in the inset of Fig. 1.

The PR measurements were performed at 2 K and in magnetic field up to 6 T using the setup similar to that described in ref. [7]. The power density of the pump beam of 488 nm line of an Ar$^+$ laser was reduced to about 1 mW/cm^2 to eliminate the photoluminescence background and to avoid photovoltaic effects and heating of the sample.

3 Results and discussion

Figure 1 shows PR spectra for the DQW sample with 1 ML AlAs barrier for 0 and 6 T. The spectra can be divided into several parts. The low energy part, below the Al$_{0.3}$Ga$_{0.7}$As barrier feature at 1.94 eV, consisting of several very sharp features is related to an excitonic transition between the heavy hole confined symmetric (*s*) and antisymmetric (*a*) states and the electron resonant states above the energy of the conduction band edge of the higher Al$_{0.35}$Ga$_{0.65}$As barrier. Very strong oscillatory features observed above 1.94 eV can be attributed to the transitions between very dense ladder of electron and hole states in the very broad

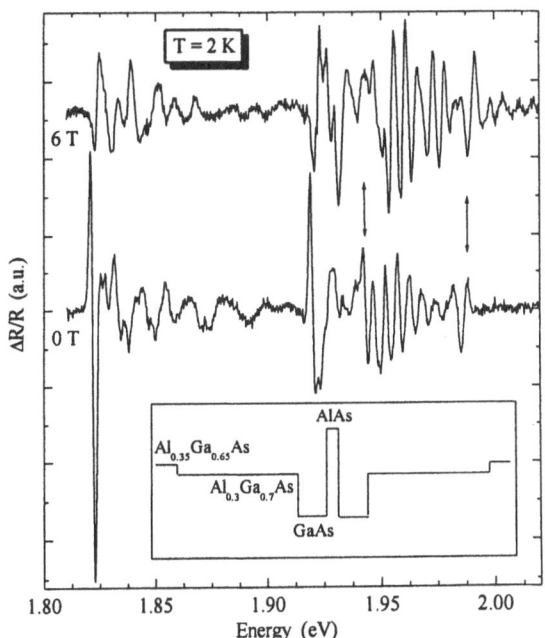

Fig. 1 PR spectra for DQW structure with 1 ML separating AlAs barrier. The inset shows a scheme of the conduction band edge. The transition energies of the AlGaAs barriers are marked by arrows.

and shallow potential step between the edges of both Al$_{0.3}$Ga$_{0.7}$As and Al$_{0.35}$Ga$_{0.65}$As barriers. We do not distinguish any particular transitions in this region, because our calculations show that the energetic distance between these states is lower than 1 meV and a quasi continuum of states is formed. Therefore the transitions may be not resolved even at 2 K. The possibility that these oscillations are related to the Franz-Keldysh effect has been also ruled out since the period of oscillations does not decrease like in Franz-Keldysh oscillations. The strong features at 1.99 eV originates from the Al$_{0.35}$Ga$_{0.65}$As barrier. Some extra features appear above this transition in higher magnetic fields. We connect them with the transitions between Landau levels in the barrier.

The energies of the observed transitions have been obtained from the fitting procedure according to the first and third derivative Lorentzian lineshape for the excitons in DQW and in bulk-like barriers, respectively [8]. The transitions have been assigned using the envelope function and transfer matrix formalisms. The material parameters have been taken after ref. [9].

The magnetic field dependence of the transition energies is presented in Fig. 2. Only the energies of

the strongest and well resolved resonant state related features are plotted. As it can be seen, all the transitions shift nonlinearly to higher energies with increasing magnetic field. The dependence has been described by a parabola reflecting a typical tendency of transition energies of excitons in weak magnetic fields. The values of the diamagnetic shift energies have been determined. For the case of the lowest resonant states transition (H1s - R1) the diamagnetic shift is about 2 meV and then increases for higher energy states, up to 11 meV in the case of H1a – R8 transition. This is significantly larger shift than for the ground confined state exciton in such a DQW, which shifts only about 1 meV in 6 T [4].

4 Conclusions

We have measured the low temperature magneto-photoreflectance on GaAs/Al$_{0.3}$Ga$_{0.7}$As double quantum wells separated by 1 ML thick AlAs barrier. We have observed sharp and strong excitonic transitions involving resonant states in the energy region above 1.8 eV. We have obtained the magnetic field dependence of the energy of these transitions and derived the values of the diamagnetic shifts, which we have compared with previously published for confined state transitions.

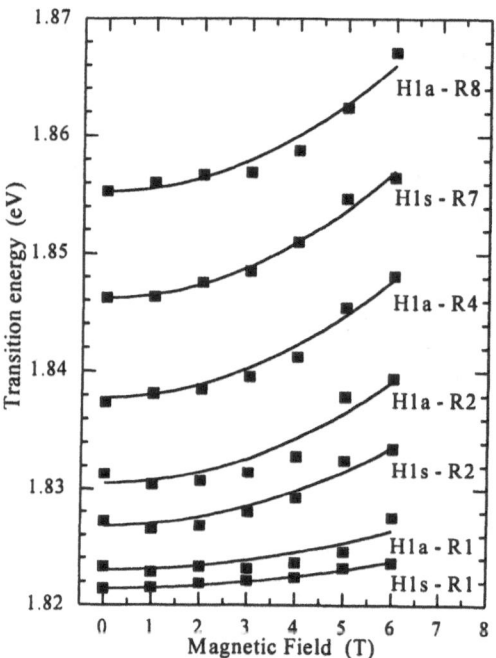

Fig. 2 Magnetic field dependence of the resonant state transition energies. The transitions are labelled according to the notation where H$ms(a)$ means the m.-th confined symmetric (antisymmetric) heavy hole state and Rn the n-th electron resonant state.

References

1. T. Westgaard, Q. X. Zhao, B. O. Fimland, K. Johannessen, L. Johnsen, Phys. Rev. B **45**, (1992) 1784.
2. Q. X. Zhao, B. Monemar, P. O. Holtz, M. Willander, B. O. Fimland, K. Johannessen, Phys. Rev. B **50**, (1994) 4476.
3. L. V. Butov, A. Zrenner, G. Abstreiter, A. V. Petinova, K. Eberl, Phys. Rev. B **52**, (1995) 12153.
4. M. Bayer, V. B. Timofeev. F. Faller, T. Gutbrod, A. Forchel, Phys. Rev. B **54**, (1996) 8799.
5. T. Wang, M. Bayer, A. Forchel, Phys. Rev. B **58** (1998) R10183.
6. G. Sęk, K. Ryczko, M. Ciorga, L. Bryja, M. Kubisa, J. Misiewicz, M. Bayer, J. Koeth, A. Forchel, *Proc. of the NATO Workshop on Optical Properties of Semiconductor Nanostructures*, edited by M. Sadowski et al., (Kluwer Academic Publishers, The Nederlands 2000) p. 91.
7. J. Misiewicz, K. Jezierski, P. Sitarek, P. Markiewicz, R. Korbutowicz, M. Panek, B. Ściana, M. Tłaczała, Adv. Mater. Opt. Electron. **5** (1995) 321.
8. F. H. Pollak, *Handbook on Semiconductors*, edited by M. Balkanski, (North Holland, Amsterdam, 1994) p. 527.
9. S. Adachi, Phys. Rev. B **58**, (1985) R1.

Proc. 25th Int. Conf. Phys. Semicond., Osaka 2000 (Eds. N. Miura and T. Ando)

571

Indirect transitions between barrier (X) electrons and two-dimensional hole gas in mixed type I - type II quantum wells

R. Guliamov[1], E. Lifshitz[1], E. Cohen[1], Arza Ron[1], and L. N. Pfeiffer[2]

[1] Solid State Institute, Technion-Israel Institute of Technology, Haifa 32000, Israel
[2] Bell Laboratories, Lucent Technologies, Murray Hill, New Jersey 07974, USA

Abstract We studied the photoluminescence (PL) spectrum resulting of the indirect recombination of barrier electrons and the two-dimensional hole gas (2DHG) that are excited in a structure of mixed type I - type II GaAs/AlAs quantum wells. The indirect transitions consist of a no - phonon band and momentum conserving (zone - edge) phonon sidebands. All these bands are blue shifted with increasing photoexcitation intensity. This shift is well explained by the energy shift of the lowest X subband in the electrostatic potential that is generated by the separate 2DEG and 2DHG charges.

Spatially separated 2DEG and 2DHG can be photoexcited in undoped, mixed type I - type II GaAs/AlAs quantum wells (MTQW) [1–3]. The 2DEG that forms in the wide GaAs QW's of this structure shows spectroscopic (and dynamic) properties that are similar to those observed in n-type modulation-doped QW's. On the other hand, we show in this study that the 2DHG that forms in the narrow GaAs QW's, recombines radiatively with electrons that are confined to the AlAs barrier (X conduction band) and have a very low density. The resulting 2DHG photoluminescence (PL) spectrum consists of indirect transitions that are very weak compared to the direct transitions of the 2DEG in the wide wells. Some of these PL bands are phonon sidebands (PSB) that involve momentum conserving phonons, similar to those observed in the indirect exciton spectrum of type-II GaAs/AlAs QW structures [4,5].

We utilize these indirect X electron - 2DHG transitions in order to study the dependence of the X_z and X_{xy} barrier electronic states on the 2DEG and 2DHG density, as it is varied by the photoexcitation intensity. We find that the observed blue shift of these transitions and their relative intensities are well explained by the conduction band bending that is caused by the electrostatic potential of the 2DEG and 2DHG layer charges.

The PL spectrum of the MTQW structure has bands in three ranges, centered around 1.52 eV, 1.77 eV and 1.92 eV. These transitions correspond to direct wide well (WW), indirect barrier - narrow well (NW) and direct NW radiative recombination processes, respectively. Due to the rapid $e_{1N} - X - e_{1W}$ electron transfer, the direct NW transitions are about 10^4 times weaker than those of the direct WW transitions. The intensity of the barrier-NW indirect transitions is about 10^2 times weaker than that of the direct WW transitions.

2.6nm/10nm/20nm GaAs/AlAs/GaAs MTQW

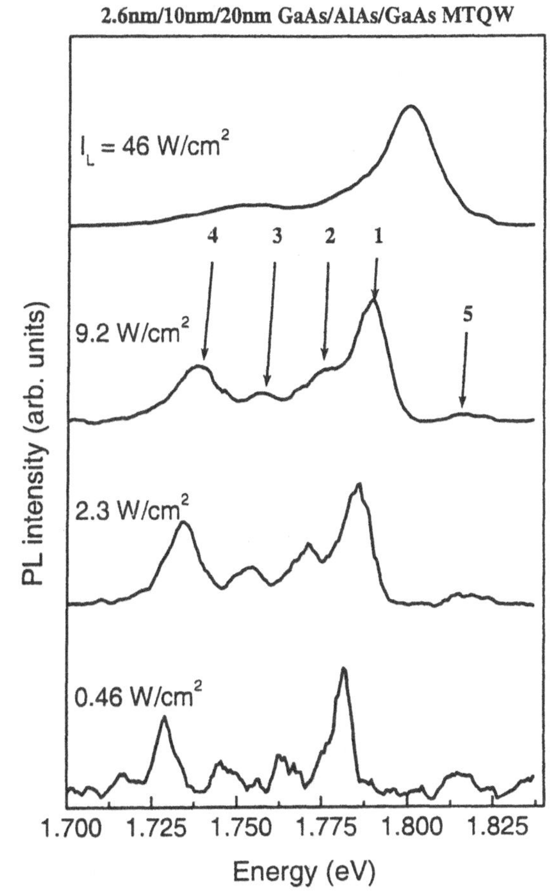

Fig. 1 PL spectra of the 2DHG in the narrow QW's that recombines with the X electrons in the barrier. The bands marked 1 and 5 are the zero-phonon indirect transitions and the other three are phonon sidebands.

The WW PL spectra exhibit an asymmetric band, which is red-shifted and broadened with increasing of excitation intensity (I_L). This band is associated with radiative recombination of the 2DEG with holes in the WW (2DEG - hh_{1W} recombination). Its spectral shape was fitted using the model given in [6]. From these fittings we extracted the WW - 2DEG density (n_e) and the renormalized band-gap energy (E_g). These n_e values are used in calculating the electrostatic potential that leads to the barrier-NW recombination shift (see below).

Figure 1 shows the barrier-NW PL spectra that are photoexcited above the NW bandgap ($E_L = 2.41$eV) with intensities $0.4 < I_L < 50W/cm^2$. They consist of five bands, which are blue-shifted and change their relative intensities with increasing I_L. Bands 1-4 have a similar blue shift (a maximum of $\sim 20meV$), wile band 5 is only slightly blue-shifted ($\sim 3meV$), followed by a red shift. We compared these spectra with those obtained from a type II QW structure having a GaAs well and AlAs barrier widths of $L_W = 26\text{Å}$ and $L_B = 102\text{Å}$, respectively. The spectra are similar, but that of the type II structure is independent of the excitation intensity. Following the analysis of the PL bands in type II QW's [4,5], band 1 is identified as the $hh_{1N} - X_{xy}$ no-phonon (ZP), indirect recombination band. The energy of bands 2, 3 and 4 appears at 13 meV, 32 meV and 49 meV below that of the ZP band, respectively. These values, are in close agreement with the transverse acoustic (TA), longitudinal acoustic (LA) and longitudinal optic (LO) phonon energies in AlAs. Thus, bands 2-4 are identified as the phonon sidebands of the $hh_{1N} - X_{xy}$ transition. The energy of band 5 is about 20 meV higher than that of band 1 and it is virtually independent of I_L. It was identified as the $hh_{1N} - X_z$ indirect recombination band of a type-II QW, since the X_z state lies above the X_{xy} state in AlAs barriers having a width $L_B > 60\text{Å}$ [7].

The PL bands of a type-II QW are due to the recombination of indirect excitons formed of X electrons and hh_1 holes. Under cw photoexcitation, only excitons are formed in this QW, and therefore, the PL band energies do not depend on I_L. Since a 2DEG and a 2DHG are photoexcited in the MTQW structure, we attribute the PL bands blue shift to band bending that is caused by the inter-well electrostatic potential. This potential is due to the charges of the separate 2DEG and 2DHG.

We calculated the electrostatic potential variation along the growth direction (z-axis), $\Delta V(z)$. For the calculation of the electrostatic potential we used an elementary, one-dimensional well-barrier sequence along the z-axis, that contains one NW, its two barriers and two WW's. In order to maintain charge neutrality we start this sequence from the middle of the WW and ending in the middle of an adjacent WW. Obviously, this sequence is repeated along the z-axis in order to describe the potential of the entire MTQW structure. An example of the band bending due to the variation of electrostatic potential, calculated for $n_e = 5.1\times10^{11}cm^{-2}$, is shown in Fig. 2a. The energy shifts of the X_{xy} and X_z subbands as a function of I_L (relative to the $I_L = 0$ value) are calculated using the triangular potential of the barrier, and are shown in solid lines in Fig. 2b. The experimental shifts of bands 1-5 are shown by the symbols. Evidently, there is a good agreement between the electrostatic model and the observed experimental shifts for bands 1-4. The blue shift of band 5 in MTQW is smaller than that of bands 1-4, as expected, since $m^*(X_{xy}) < m^*(X_z)$. This further supports the PL band assignment. As Fig.1 shows, there is an enhancement of the $hh_{1N} - X_{xy}$ ZP transition with respect to its PSB's

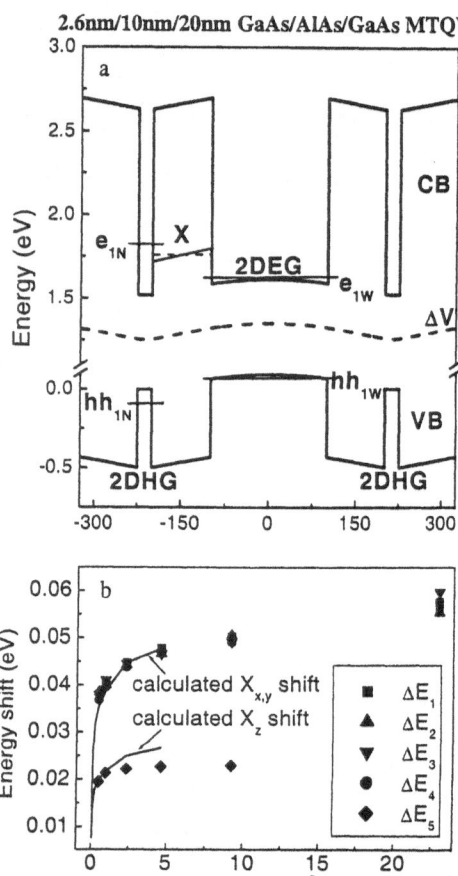

2.6nm/10nm/20nm GaAs/AlAs/GaAs MTQW

Fig. 2 a)The band bending caused by the 2DEG and 2DHG charges is calculated for a density $5 \times 10^{11} cm^{-2}$. The calculated electrostatic potential curve $\Delta V(z)$ is shown in the middle of the figure by the dashed line. The arrow indicates the recombination of a barrier-confined electron with the 2DHG in the narrow well. b) Measured $X_{x,y}$ and X_z level shifts (symbols) as a function of photoexcitation intensity. The solid lines are calculated using a simple electrostatic model for the level shifts.

as I_L increases. This maybe due to an admixing of the X_{xy} state with the X_z state that increases with increasing electric field strength. The X_z state overlaps more with hh_{1N} state than does the X_{xy} state.

Acknowledgment.

The research was done in the Barbara and Norman Seiden Center for Advanced Opto-Electronics Research and was supported by the Fund for the Promotion of Research at the Technion.

References

1. I. Galbraith et al., Phys. Rev. **B45**, (1992) 13499.
2. J. Feldman et al., Solid State Commun. **83**, (1992) 245.
3. A. Manassen et al., Phys. Rev. **B54**, (1996) 10609.
4. P. Dawson et al., Semicond. Sci. Technol. **5**, (1990) 54.
5. M. Maaref et al., Phys. Stat. Sol. **170**, (1992) 637.
6. S. Munnix et al., Superlatt.& Microstruct. **6**, (1989) 369.
7. H. W. van Kesteren et al, Phys. Rev. **B39**, (1989) 13426.

Electronic states of interface Al–2p core excitons in GaAs/AlAs/GaAs heterostructures

Koichi Inoue[1], Youichi Ishiwata[2], Shik Shin[2]

[1] Institute of Science and Industrial Research, Osaka University, 8-1 Mihogaoka, Ibaraki-shi, Osaka 567-0047, Japan.
[2] Institute for Solid State Physics, University of Tokyo, 5-1-5 Kashiwanoha, Kashiwa-shi, Chiba 277-8581, Japan.

Abstract The electronic states of Al–2p core excitons in the GaAs/AlAs/GaAs heterostructures are calculated by a finite element method in an effective mass approximation. Localized exciton states at the interface due to the valley mixing are found below the X-point core exciton energy.

1 Introduction

Type-II quantum well structures, where the lowest conduction state and the highest valence state are separated in real space, have recently attracted considerable attention owing their unique electrical and optical properties[1]. Even in Type-I quantum well structures, core-excited states of barrier materials have similar features: core holes remain in the barrier while the lowest energy state of the excited electrons is located outside the barrier. In such systems, electronic states at the hetero-interfaces play important roles. Recently, Al–2p core excitons in GaAs/AlAs/GaAs heterostructures have been studied experimentally, and distinguished structures have been found at the low energy side of the bulk core exciton peak in the reflectance spectra[2]. In this paper, to understand those structures, the electronic states of core excitons in the GaAs/AlAs/GaAs heterostructures are calculated in an effective mass approximation.

2 Calculation

Figure 1 schematically shows the sample structure grown on the GaAs (001) surfaces and the relevant conduction band energies. Cylindrical coordinates, ρ, φ, and z are adopted around the core hole, which is located at $(\rho, z) = (0, z_h)$. In this paper we consider only three conduction valleys, Γ_1, X_1, and X_3 in [001] direction, since they will be mixed at the interface. The three-valley effective

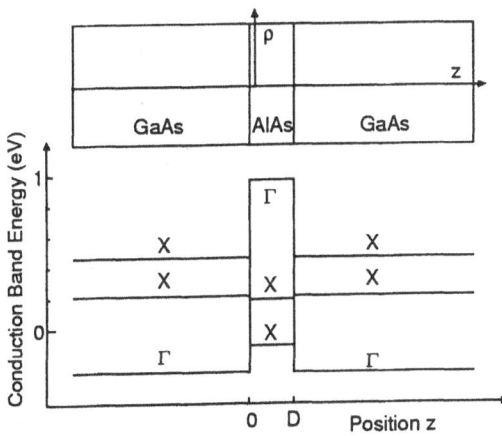

Fig. 1 Sample structure and conduction bands.

mass equation consist of envelope functions $\phi_\Gamma(\rho, z)$ at Γ, $\phi_u(\rho, z)$ at X_1, and $\phi_v(\rho, z)$ at X_3 points is given by

$$\left[E_\Gamma + V - \frac{\hbar^2}{2m_0} \nabla_\Gamma^2 \right] \phi_\Gamma = E\phi_\Gamma, \tag{1}$$

and

$$\begin{bmatrix} E_u + V - \frac{\hbar^2}{2m_0}\nabla_u^2 & i\frac{\hbar^2 P}{2m_0 a}\frac{d}{i dz} \\ -i\frac{\hbar^2 P}{2m_0 a}\frac{d}{i dz} & E_v + V - \frac{\hbar^2}{2m_0}\nabla_v^2 \end{bmatrix} \begin{bmatrix} \phi_u \\ \phi_v \end{bmatrix} = E \begin{bmatrix} \phi_u \\ \phi_v \end{bmatrix}, \tag{2}$$

where E_Γ, E_u, and E_v are the band energies, m_0 is the free electron mass, a is the lattice constant and P is the momentum matrix element between ϕ_u and ϕ_v. We restrict our attention only to cylindrically symmetric states. Thus,

$$\nabla_\zeta^2 = \frac{1}{m_\zeta^\rho \rho} \cdot \frac{d}{d\rho}\left(\rho \frac{d}{d\rho} \right) + \frac{1}{m_\zeta} \cdot \frac{d^2}{dz^2}, \tag{3}$$

where we express the valleys Γ, u, and v as the suffix ζ. Effective masses m_ζ^ρ and m_ζ are measured in the unit of m_0. Potential V is written as follows.

$$V(\rho, z) = -\frac{e^2}{\varepsilon\sqrt{(\rho^2 + (z - z_h)^2)}} + V_i \\ + (f_{x,\zeta} - f_c)\delta(\rho)\delta(z - z_h), \tag{4}$$

where e is the electron charge, ε is the dielectric constant, V_i is the image-charge potential, and the last term is due to the electron-hole exchange interaction (EX)[3], $f_{x,\zeta}$, and the central cell correction (CC) of the Coulomb interaction, f_c, both of which is assumed to be δ-functions, since core hole is strongly localized. The valley mixing at the interface is treated by the 6×6 interface matrix T developed by T.Ando [4], which is defined by

$$\begin{bmatrix} \phi_{\zeta,A} \\ \frac{1}{m_A} \cdot \frac{d}{dz}\phi_{\zeta,A} \end{bmatrix} = T(\text{AlAs} \leftarrow \text{GaAs}) \begin{bmatrix} \phi_{\zeta,G} \\ \frac{1}{m_G} \cdot \frac{d}{dz}\phi_{\zeta,G} \end{bmatrix}, \tag{5}$$

where suffix A and G means in AlAs and in GaAs, respectively. The matrix proposed in Ref. [4] is,

$$T = \begin{bmatrix} 1.24 & 0 & 0 & 0 & 0 & 0 \\ 0 & 0.81 & 0 & 0 & -0.83p & 0 \\ 0 & 0 & 0.97 & 0 & 0 & 0 \\ 0 & 0 & 0 & 1.03 & 0.26 & 0 \\ 0 & 0 & 0 & 0 & 0.83 & 0 \\ -1.24p & 0 & -0.4 & 0 & 0 & 1.2 \end{bmatrix}, \tag{6}$$

with the mixing parameter p. The parameters are summarized in Table 1. We solve above equations by a finite element method in a rectangular shape with a monolayer thickness in the z direction. Adjustable parameters

Table 1 Parameter values taken from Refs.[4]–[6].

	$m_\Gamma[4]$	$m_u[4]$	$m_v[4]$	$m_\Gamma^\rho[4]$	$m_u^\rho[6]$	$m_v^\rho[6]$
GaAs	0.068	1.35	1.57	0.068	0.25	0.25
AlAs	0.222	1.00	1.00	0.222	0.19	0.19

	$\epsilon[6]$	$P[4]$	$E_\Gamma[5]$	$E_u[5]$	$E_v[5]$	$p[4]$
GaAs	12.85	1.4	1.544	2.073	2.293	0.5
AlAs	10.06	2.7	3.095	2.300	2.119	0.5

are $f_{x,\zeta}$ and f_c. The thickness of AlAs is 7 monolayers (1.98nm), and core hole is located at the position in 1 through 7 monolayer depth in AlAs. Spectral intensities are obtained by the sum of the results for different core hole positions.

3 Results and Discussion

Some considerations about EX and CC will be necessary before we discuss the results. The CC is caused at the very vicinity of core hole mainly by following two effects: (1) insufficient dielectric screening enhance the Coulomb attractive potential, and (2) the effective mass approaches to the bare electron mass. The adjustable parameter f_c is relating to the first effect. To take account of the second effect in the finite element calculation, we use the free electron mass instead of effective masses at the cell including the core hole. As for EX, $f_{x,\zeta}$ is expected to be valley dependent since EX is proportional to the overlap between the core hole (a δ-function) and Wannier functions[3]. Thus s–like portions of Wannier functions at Al site are important, since they have large amplitudes near the Al inner core. As a rough estimation by an empirical LCAO calculation, it will be proposed that $f_{x,\Gamma} : f_{x,u} : f_{x,v} = 1 : 0 : 0.5$, approximately. In addition, it should mentioned that this term makes off-diagonal element in the form $\sqrt{f_{x,\Gamma} f_{x,v}}$ between ϕ_Γ and ϕ_v in Eqs. (1) and (2). As shown in the inset of Fig. 2, the spin-orbit partner, j=3/2 and j=1/2, have nearly equal intensities to each other. This suggests that the EX energy is near to one third of the spin-orbit splitting energy[3] about 0.45 eV. Although we cannot determine the precise parameters f_c and $f_{x,\zeta}$ at present, we use following values as the first trial: $f_{x,\Gamma} = 0.8$, $f_{x,u} = 0.0$, $f_{x,v} = 0.4$, and $f_c = 0.09$. The consistent EX energy is obtained by these values. We take the image-charge effect into account, but it is a minor effect.

Since the core exciton transitions occurs from Al-2p states to s– and d–like states at Al sites, the probability will approximately be proportional to $I = |\phi_\Gamma|^2 + 0.5|\phi_v|^2$ at the core hole position, if d-states contribution is negligible. In Fig. 2 (b) the calculated intensities I are shown as vertical bars. The spectra can be divide into three regions practically. The most intense state at 74.8 eV is due to the (001) X valley (v) core exciton component, whose energy is adjusted to the experimental peak position. It is shifted toward the higher energy from the calculated bulk core exciton position indicated

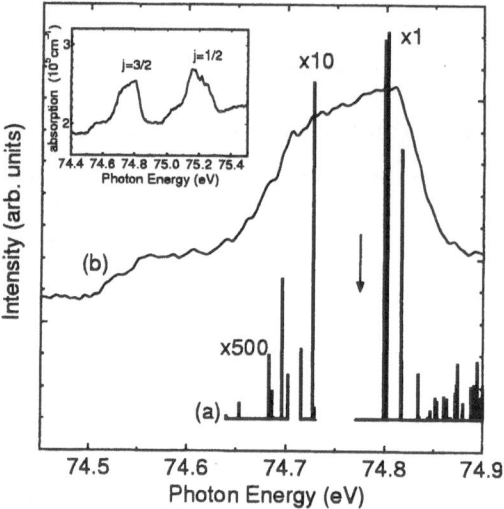

Fig. 2 Comparison between (a) calculated overlap spectra and (b) experimental absorption spectra of GaAs/AlAs(1.9nm thick)/GaAs heterostructure. In the inset the wide-range experimental spectrum is shown.

by an arrow, owing to the quantum confinement into the thin layer. The core excitons due to other (100) and (010) X valleys, which are not considered in the present model, are expected between the arrow and the intense peak because of their heavy masses toward the confinement direction. In Fig. 2 two peaks appear at 74.71 and 74.73 eV. They are localized exciton states due to the valley mixing at the interface. The splitting is due to the different core hole position: the state at 74.71 eV is due to the exciton whose core hole is located at the 2nd monolayer depth, while the core hole of the 74.73 eV state is at the 1st monolayer depth [7]. Weak transitions below 74.7eV are mainly due to the transitions from Al atoms at the 1st monolayer depth to the GaAs Γ states which are proximating a little into AlAs layer.

4 Conclusion

By the three valley effective mass calculation of the core exciton states in GaAs/AlAs/GaAs, it is concluded that localized exciton states due to the valley mixing at the interface appear below the X-point core exciton energy.

References

1. G.N.Carneiro and G.Weber, Phys. Rev. **B58**, (1998) 7829.
2. Y. Ishiwata, Thesis for master degree, The University of Tokyo (1999).
3. Y.Onodera and Y.Toyozawa, J. Phys. Soc. Jpn. **22**, (1967) 833.
4. T.Ando, Phys. Rev. **B47**, (1993) 9621.
5. T.Ando and H.Akera, Phys. Rev. **B40**, (1989) 11619.
6. L.Pavesi and M.Guzzi, J. Appl. Phys. **75**, (1994) 4779.
7. We take the As plane as the interface. The Al atoms at 1 monolayer depth from the interface means at a quarter of lattice constant.

Enhancement of Fermi energy optical emission induced by the band structure in strained-layer InGaAs/InP quantum wells

H. A. P. Tudury[1], F. Iikawa[1], E. Ribeiro[1,2], J. A. Brum[1], W. Carvalho Jr.[2], A. A. Bernussi[2], A. L. Gobbi[2]

[1] Instituto de Física "Gleb Wataghin", Universidade Estadual de Campinas, 13083-970 Campinas - SP, Brazil
[2] Laboratório de Optoeletrônica, Fundação CPqD/LNLS, 13088-061 Campinas - SP, Brazil

1 Introduction

We studied the electronic band structure of strained-layer $In_{1-x}Ga_xAs$/InP quantum wells using optical techniques. There are interesting features in the band structure of this system which are produced by the spatial quantum confinement along the growth direction, and by the built-in biaxial strain. These two effects induce a strong valence band mixing that may lead to an indirect gap for certain Ga concentrations, x, of the InGaAs alloy (quantum well). Recently, the indirect band structure has been investigated experimentally [1] and theoretically [2–4] in this system. Time-resolved photoluminescence results obtained in Ref. [1] show evidence of the indirect band structure in undoped samples. The crossover direct-to-indirect gap occurs for x around 0.55: direct gap is found for $x < 0.55$ and indirect for $x > 0.55$. The presence of an electron gas would suppress the excitonic effects and for this situation only theoretical predictions [2–4] are available. With this in mind, in the present work we investigated n-type modulation-doped InGaAs/InP quantum wells in order to directly assess the formation of an indirect gap in doped samples.

2 Experimental results and discussion

The set of samples studied in this work consist of a series of n-type InGaAs/InP modulation-doped single quantum wells. The samples were grown by low-pressure metal-organic chemical vapor deposition on an n-type doped InP substrate. The Ga concentration varies from 0.47 to 0.60 and the quantum well width is kept fixed (6 nm). We used an Ar laser [an halogen lamp] as light source for photoluminescence (PL) [electrore-flectance (ER)] measurements. Both experiments were performed at 2 K in a liquid-helium immersion cryostat. For ER measurements we deposited a semi-transparent gold film on the surface of the sample in order to apply the required electric fields.

Figure 1 shows the photoluminescence spectra of five samples with different Ga concentration. The peak position of the emission line shifts to higher energies as the Ga concentration of the sample increases. This shift is a combination of two different mecanisms. The dominant contribution comes from the increase of the gap energy of $In_{1-x}Ga_xAs$ alloy with x. The built-in tensile strain also increases with x, however it acts in the op-

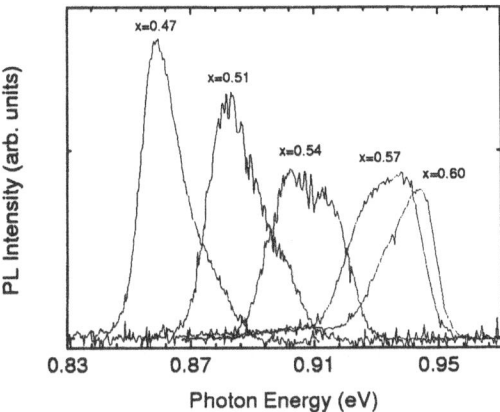

Fig. 1 Photoluminescence spectra of $In_{1-x}Ga_xAs$/InP modulation doped quantum wells

posite direction, lowering the band gap. Although strain has little effect on the net energy shift it also changes the relative separation among the bands, which leads to strong modifications on the valence band dispersion.

The optical transitions of n-type modulation-doped quantum wells are dominated by the photocreated hole distribution, since the electron states are filled up to the Fermi energy. The holes occupy different wave-vector k, parallel to the layer plane, around the top of the valence band. Based on a single particle model, the optical transitions of direct gap quantum wells occur around $k = 0$. The recombination should be centered around E_0 (gap energy) but broadened to higher energy region (electrons are available from k_0 to k_F), forming an asymmetric emission band. The PL line shapes of samples with $x = 0.47$ and 0.51 (see Fig. 1), which are expected to be direct gap quantum wells, are consistent with this description. The PL spectrum of sample with $x = 0.51$ is broader than the one with $x = 0.47$ because the hole effective mass is larger, which allows these carriers to be distributed over a broader range of k.

The PL spectra of indirect-gap quantum wells with $x = 0.57$ and 0.60, shown in Fig. 1, present the peak position at the high energy region of the PL spectra with a tail at the low-energy side. This is an opposite behavior to that observed in direct-gap samples ($x = 0.47$ and 0.51). The slightly indirect valence band dispersion allows the distribution of the photocreated holes for large values of k around k_0, where k_0 is the wavevector at the top of the valence band in an indirect gap quantum well. The hole occupation decreases when k decreases to zero. This means that the optical emission around $k \sim 0$ is weaker than those for $k \sim k_0$. Since the Fermi wavevector k_F of our samples is smaller than k_0, the PL intensity is stronger for $k \sim k_F$. The emission intensity must decrease with decreasing values of k. In this case, the PL band must present a peak around the Fermi energy and a tail at the low-energy region, as shown in Fig. 1 for samples with $x = 0.57$ and 0.60.

For the sample with intermediate Ga concentration, $x = 0.54$, the valence band is expected to have an almost flat dispersion. In this case, the photocreated holes are distributed over a large range of k, allowing optical emission for k varying from 0 to k_F. This results in a square-like PL line shape, as shown in Fig. 1.

The ER results reinforce the interpretation used to explain the PL data shown in Fig. 1. The ER spectrum is related to the optical absorption, which has a unique final state in our modulation-doped quantum wells: the Fermi energy. On the other hand, the PL emission occurs around either the gap energy or the Fermi energy, depending on the valence band structure. The experimental results obtained by ER and PL measurements as a function of electron density (which depends on the DC voltage applied to the samples) are shown in Fig. 2. In our samples, the electron gas density decreases sensitively for DC voltages lower than +80 mV. For voltages above this value the electron density decreases slightly and after that remains practically constant. We concentrate the discussion for DC voltages below +80 mV.

The ER transition energies follow the Fermi energy variation. All the samples exhibit practically the same energy range of variation for DC voltages below +80 mV. However, the same does not occur for the PL data. For example, in the sample with $x = 0.51$ (direct gap) PL peak energy decreases when DC voltage increases from −200 to +80 mV, while for the other samples (indirect gap) we observe an increase of the PL peak position with increasing electron density. The decreasing of PL peak position with DC bias for the sample with $x = 0.51$ is attributed to the electrostatic potential variation and the gap renormalization effects. For indirect-gap quantum wells the PL peak position is related to the Fermi energy. Thus, it follows the Fermi energy variation, which is dominant when compared to the electrostatic potential variation and the gap renormalization effects. These results show a clear evidence of the PL line shape dependence on the valence band structure. The ER data do not present any significant variation among the investigated samples, because the final state for absorption is the Fermi energy. An intriguing result obtained in our data is the large value of the Stokes shift energy (difference between absorption and emission energies at minimum carrier concentration) for indirect gap samples (as large as 30 meV for the sample with $x = 0.60$). A pos-

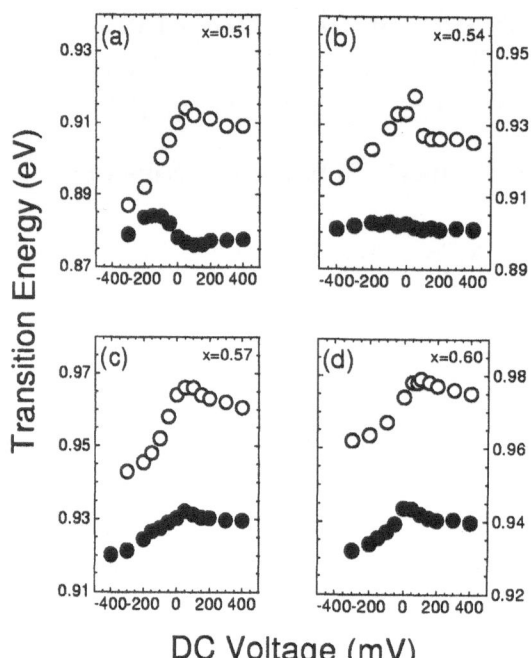

Fig. 2 Electroreflectance (open circles) and photoluminescence (solid circles) data of four samples as a function of DC voltage.

sible explanation is the effect of carrier localization due to an increase of the interface roughness as the built-in strain increases with x. However, additional experiments will be necessary for a conclusive explanation on this additional contribution.

3 Conclusions

We studied n-type $In_{1-x}Ga_xAs/InP$ modulation-doped single quantum wells by photoluminescence and electroreflectance techniques. The results of PL and ER are consistent with the single particle model and the band structure theory, attesting the formation of an indirect gap for samples with larger values of x, which also presented the large values of the Stokes shift. An increase of carrier localization effects with x might be a possible origin for these observations.

Acknowledgements

This work was supported by Fundação de Amparo à Pesquisa do Estado de São Paulo, CNPq and FAEP.

References

1. P. Michler, A. Hangleiter, A. Moritz, G. Fuchs, V. Härle, and F. Scholz, Phys. Rev. B **48**, (1993) 11991.
2. C. Y.-P. Chao and S.L. Chuang, Phys. Rev. B **46**, (1992) 4110.
3. M. Sugawara, N. Okazaki, T. Fujii, and S. Yamazaki, Phys. Rev. B **48**, (1993) 8102.
4. A. L. C. Triques and J. A. Brum, *Proceedings of the 22nd International Conference on the Physics of Semiconductors*, ed. by D. J. Lockwood (World Scientific, 1995), p. 1328.

Band offset determination and excitons in SiGe/Si(001) quantum wells

H. H. Cheng[1], S. T. Yen[2], R. J. Nicholas[3]

[1]Center for Condensed Matter Sciences, Taiwan University, 1, Roosevelt Road, Section 4, Taipei, Taiwan

[2]Department of Electrical Engineering, National Dong Hwa University, 1, Section 2, Da Hseuh Road, Shou Feng, Hualien, Taiwan

[3]Department of Physics, University of Oxford, Parks Road, Oxford, OX1, 3PU, United Kingdom

Abstract

We report both experimental and theoretical studies on $Si_{1-x}Ge_x/Si$ multi quantum wells. A self-consistent calculation is employed to model the excitonic transition. It shows that, in the large conduction band offset region the Δ_2-hh exciton is the lowest transition while in the small offset region the Δ_4-hh exciton is the lower. From an analysis of the data, a type II conduction band offset ratio of $30\text{Å}\pm3\%$ is concluded.

1 Introduction

Silicon germanium/silicon heterostructures have attracted great attention in recent years for its potential application in the Si-based optoelectronic devices [1,2]. Intensive research has been made to study the fundamental physical parameters of this system, such as effective masses, mobility and in particularly the band alignment.

In previous studies of the band alignment of SiGe/Si system, both type I and type II structures have been suggested by analyzing photoluminescence (PL) energy from quantum wells [3,4]. In this paper, we present both PL measurements and a theoretical treatment to study the band alignment of this system using Multi-Quantum Well (MQW) structures. A self-consistent excitonic model taking into account the Coulomb interaction between electron and hole is used to analyze the PL energy. From the analysis of the ground state transition, it is concluded that there is a relatively large type II conduction band offset ratio of $30\pm3\%$ [5,6].

2 Experiment

The samples consist of a thick buffer layer (2000 Å) of Si, followed by a ten period SiGe/Si multi-quantum well. Luminescence was excited by optic fiber using a diode laser at intensity levels of around 10mW/cm^2. The emission was collected by a fiber bundle and dispersed by using a 0.5 meter spectrometer and a North Coast germanium detector.

The measurements were performed at 4.2K. A summary of the sample structure is listed in Table I.

Table I. Summary of sample structure and the transition energies of the no-phonon (NP), transfer acoustic (TA), and transverse optical (TO) lines.

Sample	Ge (%)	w/b Å	NP meV	TA meV	TO MeV
N1	16.3	46/93	1028.5	1010.8	969.8
N2	23	52/183	961.6		902
N3	28	75/85	918.4		861.7

A typical is shown in Fig. 1. Several PL lines can be seen in a wide energy range between 850 and 1200 meV. These features come from two groups of lines originating from the bulk Si and the $Si_{1-x}Ge_x$ quantum wells. On the high energy side of the spectra, a family of Si-related transitions are observed at peak energies of 1155.9, 1136, and 1098.9 meV corresponding to the no-phonon (NP), transverse acoustic (TA), and transverse optical (TO) mode phonon replicas. In the low energy regime, several strong QW features are also observed as marked by the solid arrows in the Fig. 1. These lines are assigned as the NP, TA, and TO transitions [7]. The transition energies are summarized in Table 1.

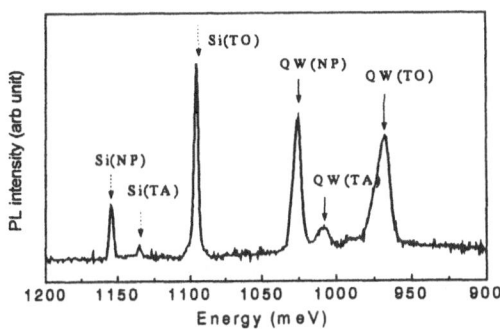

Fig. 1. A typical PL spectrum showing the transition originating from Si and SiGe.

3 Discussion

To quantitatively interpret the interband optical transition energies of the excitons, a self-consistent calculation of the excitonic energy level taking into account of the Coulomb interaction is performed. The details of this method can be found elsewhere [6]. Here, we describe the characteristic of the model. Figure 2. shows that, without the self-consistent procedure, the electron wavefunction is distributed symmetrically in the region of the Si layer, as indicated by the dashed line. Including the Coulomb attraction, the wavefunctions fe are modified and are pulled toward the Si/SiGe interface as plotted by the solid line. A similar behavior is also found for the hh state. As a consequence, the energy levels of both electrons and holes are shifted, hence the optical transition.

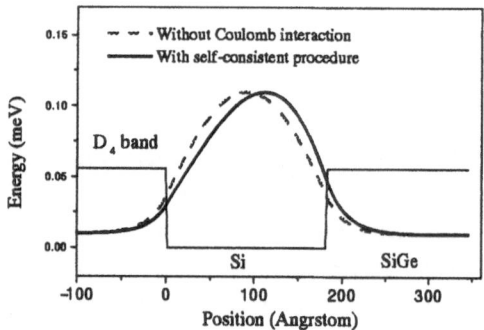

Fig. 2. Calculated carrier wave functions for a system of SiGe/Si quantum well, with and without the influence of Coulomb attraction.

We now discuss the interband transition energies. For a type II structure, the band offset is determined by the band gap of the two materials. The conduction and valence band offset are $[E_g(Si)-E_g(SiGe)]\times Qc$ and $[E_g(Si)-E_g(SiGe)]\times(1+Qc)$. $E_g(Si)$ and $E_g(SiGe)$ are the band gap of Si and SiGe. Qc is the conduction band offset ratio. Knowing the potential profile, the interband transition energy is given by $E=E_g(SiGe)-[E_g(Si)-E_g(SiGe)]\times Qc+E_e+E_h-E_b$. E_e and E_h are the modified confinement energy od electron and hole. E_b is the binding energy. To model the Δ_2-hh transition an energy of Ξe_T is added to obtained the potential of the Δ_2 band [8]. (This is the energy splitting of Δ_2 and Δ_4 band) The calculated transitions for the Δ_2-hh and Δ_4-hh excitons are plotted in Fig. 3. This shows that both transitions are strongly dependent on the band offset ratio and decrease with increasing Qc reflecting a type II band alignment. In the large

conduction band offset region the Δ_2-hh exciton is the lowest transition while in the small offset region the Δ_4-hh exciton is the lower. This is attributed to the enhanced binding energy due to the small band offset for the Δ_4 valleys. Considering only the transition energies it require a values of Qc 0.30±3% to fit the data.

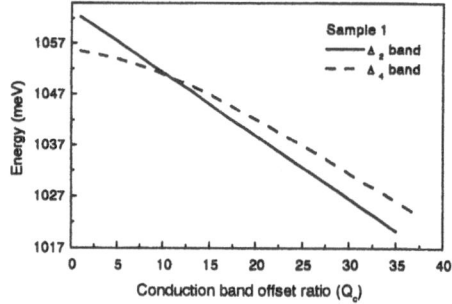

Fig. 3. Calculated transition energy for the Δ_2-hh (solid line) and Δ_4-hh (dashed line)

In conclusion we can say that there is growing evidence for the assignment of a type II band alignment in SiGe/Si quantum wells. By taking into account the Coulomb binding, a conduction band offset ratio of 30±3% is concluded.

Acknowledgements

This work is supported by National Science Council (Taiwan).

Reference

1. J. C. Bean, Proceedings of the IEEE, Vol. 80. No. 4. 571 (1992).

2. R. A. Soref, Proceedings of the IEEE, Vol. 81. No. 12. 1687 (1993).

3. D. C. Houghton, G. C. Aers, N. L. Rowell, K. Brunner, W. Winter, and K. Eberl, Phys. Rev. Lett **78**, 866 (1997).

4. M. L. W. Thewalt, D. A. Harrison, C. F. Reinhart, and J. A. Wolk, and H. Lafontaine, Phys. Rev. Lett **79**, 269 (1997).

5. H. H. Cheng, S. T. Yen, R. J. Nicholas, Phys. Rev. B **62**, 4638 (2000).

6 C. Penn, F. Schäffler, G. Bauer, and S. Glutsch, Phys. Rev. B **39**, 13314 (1999).

7. K. Brunner, K. Eberl, and W. Winter, Phys. Rev. Lett **76**, 303 (1996).

8. R. People, Phys. Rev. B **32**, 1405 (1985).

Few-Cycle THz Spectroscopy of Semiconductor Quantum Structures

K. Unterrainer[1], R. Bratschitsch[1], T. Müller[1], R. Kersting[1]*, J.N. Heyman[2], G. Strasser[1]

[1] Institute for Solid State Electronics, Technical University Vienna, A-1040 Vienna, Austria

[2] Macalester College, Department of Physics&Astronomy, St. Paul, MN55105, USA

Abstract Optically excited plasma oscillations in n-doped GaAs epilayers emit intense THz pulses. In optically excited parabolic quantum wells we observe coherent THz emission from intersubband plasmons. Using a THz-pump and THz-probe technique we observe the response of the intersubband polarization in semiconductor quantum structures. THz cross-correlation measurements of modulation doped semiconductor quantum structures allow to determine the absorption, the dispersion, and the dephasing times of the quantized electrons.

email: karl.unterrainer@tuwien.ac.at

Introduction

Following recent advances in femtosecond laser technology, several groups showed that ultrafast photoexcitation of semiconductors and semiconductor heterostructures can be used to generate electromagnetic radiation at THz frequencies. Most of these works were performed using photoconductive antennas or transmission lines [1]. THz generation from semiconductor surfaces or by optical rectification was introduced by Zhang [2]. Roskos et al. achieved for the first time THz emission from coherently oscillating electrons in a double-well potential [3]. Charge oscillations due to light and heavy hole excitons in a quantum well [4] and THz emission from Bloch oscillations [5] were of similar concept.

An alternative concept is coherent plasma oscillations of charge carriers. Here, the charge carriers are bound by their coulomb potential and the plasma frequency depends on the carrier density. Since an oscillating current density leads to the emission of electromagnetic waves, plasma oscillations should emit THz radiation. THz emission from coherent two-dimensional (2D) plasmons has been observed in time-resolved experiments in the accumulation layer of a GaAs heterostructure [6]. We have shown that THz pulses are emitted from coherent three-dimensional (3D) plasmon oscillations in n-doped GaAs epilayers [7].

We have used these few-cycle THz pulses to study intersubband transitions in quantum structures. The cross-correlation technique allows simultaneous measurement of both amplitude and phase.

We have used the results of the cross correlation spectroscopy to design an improved THz emitter based on intersubband plasmon oscillations in parabolic quantum wells.

Experiment

All experiments presented here are performed using a mode-locked Ti:Sapphire laser emitting 100 fs pulses at 800 nm (1.55 eV) with a pulse energy of 13 nJ. The pulses are transmitted through a Michelson interferometer and focused onto the emitter sample to spot sizes between 100 and 500 μm. The experimental setup is purged with nitrogen to prevent absorption by atmospheric humidity. Time resolution is achieved by focusing two delayed laser pulses on the sample. In this correlation technique the THz signal emitted from the sample is detected with a time integrating 4.2 K bolometer as a function of the delay time between the two exciting pulses.

In our cross-correlation measurements the sample is placed in one arm of an asymmetric THz interferometer. The Ti-Sapphire laser pulses are used to generate sample and analysis THz beams. The sample beam passes through the sample, and is superimposed with the analysis beam at a germanium beamsplitter. The detector measures the intensity i.e. the product of the two THz fields as a function of the delay between them. Both the absorption and dispersion of the sample can be extracted from cross-correlation measurements of the sample and of a reference.

THz emission from n-doped GaAs epilayers

For the plasmon emission experiments, n-doped GaAs epilayers were used. The doping concentrations were $1.9 \cdot 10^{15}$ cm^{-3}, $1.7 \cdot 10^{16}$ cm^{-3}, and $1.1 \cdot 10^{17}$ cm^{-3}, respectively.

---

* now: Renssealer Polytechnic Institute, Troy, NY12180.

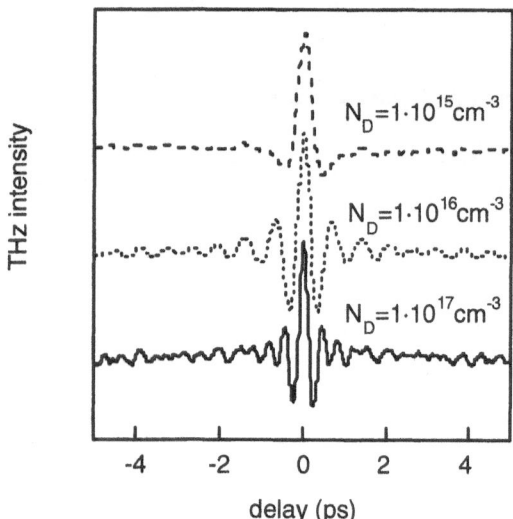

Fig. 1 THz auto correlation signal from three GaAs samples with different doping concentration.

The thickness of the epilayers was a few μm. The plasma oscillation of the extrinsic electrons is caused by the ultrafast dynamics of the surface field which follows the femtosecond laser excitation. The extrinsic electrons are confined between the undoped substrate of the structures and the surface depletion region. When photocarriers are excited by the femtosecond laser pulse in the surface field region they first perform an ultrafast ballistic motion and later a drift motion, both screening the surface field. Additionally, the difference of the diffusion currents of photogenerated electrons and holes builds up a Dember field [8]. In our structures the cold electrons of the epitaxial layer respond to the single-sided field change which initiates their plasma oscillation.

Time-resolved THz emission data from these samples are shown in Fig.1. Multiple oscillations are clearly visible from the THz auto correlation signal at higher doping concentrations. The oscillation period depends on the doping concentration and is well explained by the plasma frequency ω_p of the extrinsic carriers $\omega_P = \sqrt{ne^2/\varepsilon\varepsilon_0 m*}$, where n is the doping concentration. The emission intensity reaches a maximum for a doping concentration of about $1\cdot10^{16}$ cm^{-3}. Fig. 2 shows the Fourier transformation spectra of the auto correlation data. For all doping concentrations distinctive amplitude spectra are observed. From the observation that the frequencies do not depend on the density of photo

excited carries we deduce that the emission results exclusively from the coherent plasma excitation of the extrinsic electrons in the n-doped GaAs layer /8/.

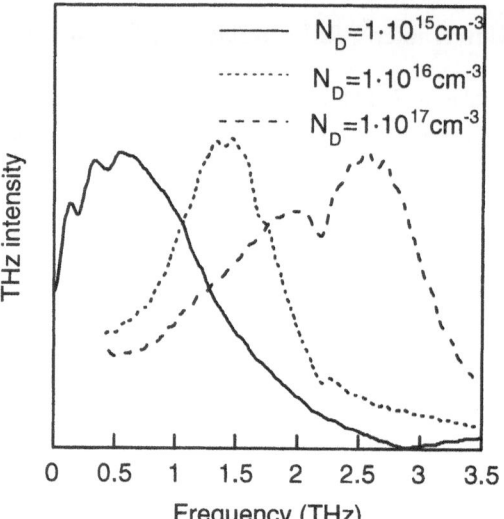

Fig. 2 Frequency spectra of the few-cycle THz pulses obtained from the Fourier transformation of the auto correlation data.

THz time domain spectroscopy of quantum structures

We investigated quantum well structures with intersubband transitions in the range of 1-3 THz. We are able to determine phase and amplitude of an intersubband transition [9]. Here we present the study of a parabolic quantum well which consists of a single period of a modulation-doped parabolically graded AlGaAs/GaAs structure. According to Kohn´s theorem the electrons in parabolically confined potentials will interact with light only at the bare harmonic oscillator frequency $\omega_{HO} = \sqrt{\dfrac{8\Delta}{L^2 m*}}$ (Δ is the depth of the parabolic potential, L the width, and m* the effective mass) independent of the carrier concentration in the well [10]. The parabolic well was designed to be 2000Å wide and to have a resonance frequency ω_{HO} of about 1.8 THz . The structure was modulation doped to give an electron concentration in the well of about $3\cdot10^{11}$ cm^{-2}. A metal grating (8 μm wide Au stripes, separated by 8 μm spacers) was deposited on the surface of the structure to allow normal incidence coupling to the intersubband transition.

Fig. 3 shows the time resolved THz pulse after passing through the sample. At short times (t < 1.5 ps), the signal shows mainly the exciting THz pulse which has a center frequency of 1.5 THz. At longer time delays (t > 1.5ps) the dielectric response of the electrons in the parabolic well becomes visible. The harmonic oscillator frequency of this well is calculated to be at 1.8 THz which corresponds very well with the observed oscillation period.

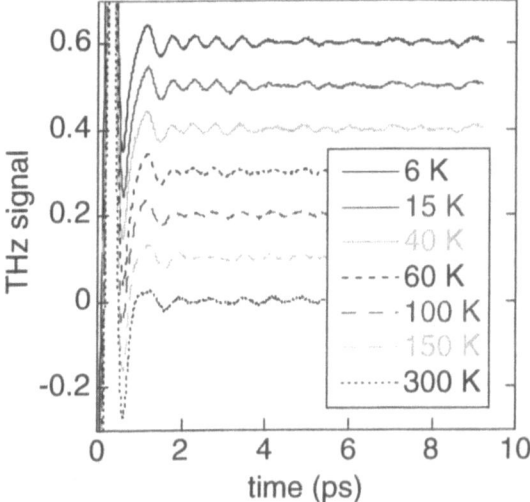

Fig. 3 Transmitted THz Cross correlation signal of the parabolic quantum well for different temperatures.

The temperature dependence shows a strong increase of the damping above 40 K. The same temperature dependence is found for the intersubband relaxation in THz electroluminescence measurements [11].

THz emission from parabolic wells

The potential of a doped parabolic quantum well (PQW) is special because it only allows for coupling of long wavelength radiation (as compared to the width of the well) to the center of mass coordinate of the electron system. Furthermore, the parabolic potential and hence the emission line is completely unaffected by an applied electric field. These unique features together with the fact that PQWs are relatively stable at elevated temperatures make them a promising candidate for use as a robust THz emitter. Unfortunately, optically pumped THz emission due to quantum beats cannot be expected in symmetric potentials like PQWs. However, THz radiation has been detected from *doped bulk* GaAs where the excitation mechanism is the ultrafast screening of the surface depletion field by a femtosecond laser pulse. Similar to doped bulk GaAs, the carriers inside a parabolic quantum well form a nearly homogeneous electron layer given by the curvature of the parabola.

In our experiments we use *modulation-doped* GaAs/AlGaAs PQWs which are grown with the digital alloy technique. Their widths L range from 1200 to 2000 Å. The carrier sheet densities n_s for the modulation-doped PQWs of 1.7×10^{11} - 5×10^{11} cm^{-2} (under illumination) are determined by Hall measurements. In contrast to the absorption measurements, no metallic grating on top of the sample is required to couple to the THz radiation.

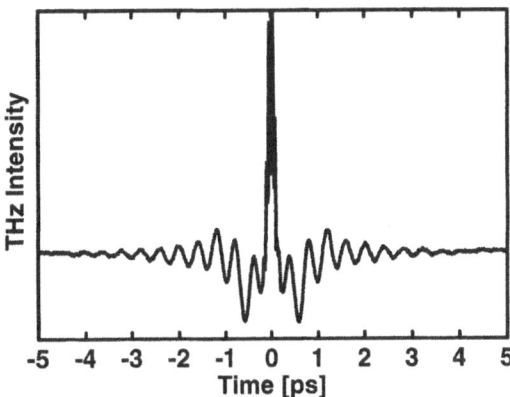

Fig. 4 THz autocorrelation signal of the 1400 Å PQW excited by 780 nm laser pulses (T = 5 K).

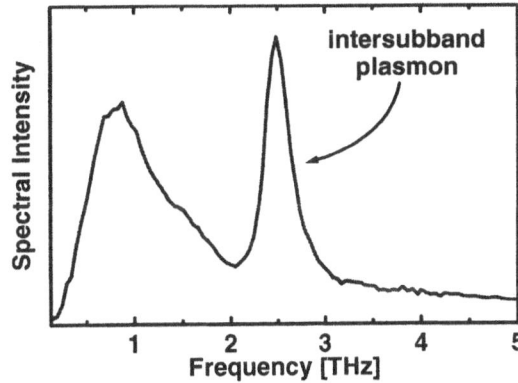

Fig. 5 Fourier transform of THz autocorrelation signal of the 1400 Å PQW.

The samples are mounted on the cold finger of a He continuous flow cryostat where a sample temperature of approximately 5 K can be obtained.

Figure 4 shows the THz AC trace of a modulation-doped PQW (L = 1400 Å, n_s=5x10^{11} cm^{-2}) excited by 780 nm laser pulses. The density of the optically generated carriers is kept well below the carrier density due to the modulation doping inside the PQW. The AC signal consists of a peak at 0 ps with superimposed NIR interference fringes followed by distinct oscillations that are visible for up to 4 ps. The spectrum of the emitted THz radiation (Fig. 5) consists of two components; a broad background peaking around 0.8 THz and a narrow line (FWHM: 0.3 THz) with a center frequency of 2.55 THz. The origin of the broadband component is due to THz generation at the surface of the sample. The narrowband emission results from the oscillation of the electrons inside the PQW. The observed frequency corresponds to the intersubband plasmon of the PQW [12]. The expected frequency of the intersubband plasmon for the 1400 Å wide PQW is 2.2 THz. FTIR absorption, THz-Time Domain Spectroscopy, and electroluminescence measurements [11] show nearly the same resonance frequency (2.2 and 2.37 THz) as observed in the femtosecond emission experiments.

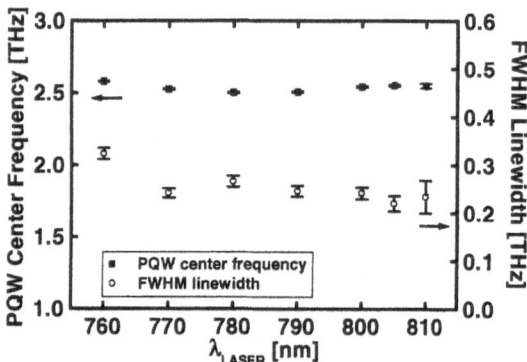

Fig. 6. Center frequency (filled squares) and full width at half maximum (FWHM) linewidth (hollow squares) of the intersubband emission depending on the excitation wavelength.

The excitation mechanism for the intersubband plasmon is due to screening of the surface depletion field by the electron-hole pairs injected by the ultrafast laser pulse. In this way the electrons inside the quantum well experience a kick and begin to oscillate with their eigenfrequency

This is supported by the fact that the oscillation can be excited over a large wavelength range of the femtosecond pulses (Fig. 6). The onset of intersubband plasmon emission appears at 815 nm near the 819 nm GaAs bandgap (T = 5 K) and can be observed to up to 760 nm. This large range implies that the excitation mechanism is clearly a non-resonant phenomenon, in contrast to the THz quantum beat experiments. This is also confirmed by the fact that we observe no THz radiation from an identical but undoped PQW.

The width of the intersubband emission is nearly unaffected when increasing the excitation wavelength. However, it shows a linear dependence on the number of optically excited carriers. These

optically injected carriers add to those from the modulation doping and cause a faster dephasing of the intersubband plasmon oscillation.

Conclusion

Few-cycle THz radiation is emitted from coherent 3-D plasmons in n-doped epitaxial GaAs layers. The generation mechanism is ultrafast screening of the surface field by optically generated carriers. We have performed time-domain measurements of intersubband charge oscillations in quantum well systems. Our all-THz measurements quantitatively determine both the absorption and dispersion in the samples. This can be used to determine the resonance frequencies, oscillator strengths and dephasing rates of quantized transitions in semiconductor nanostructures. In addition, we have demonstrated optically driven THz emission from intersubband plasmons in modulation-doped parabolic quantum wells. The combination of the designability of the transition frequency and the narrowband emission make modulation-doped PQWs attractive and easy-to-use THz emitters.

The authors acknowledge support by the Austrian Science Foundation (Start Y47) and the EU-TMR "INTERACT" and J.N.H. acknowledges support by the Petroleum Research Foundation and the National Science Foundation (DMR-007622).

References

1. D.H. Auston, K.P. Cheung, R.P. Smith, Appl. Phys. Lett. 45, 284 (1984); Ch. Fattinger, D. Grischkowsky, Appl. Phys. Lett. 54, 212 (1987).
2. X.-C. Zhang et al., Appl. Phys. Lett. 56, 866 (1990); B.B. Hu et al., Appl. Phys. Lett. 56, 506 (1990).
3. H. Roskos et al., Phys. Rev. Lett. 68, 2216 (1992).
4. P.C. M. Planken et al., Phys. Rev. Lett. 69, 3800 (1992).
5. C. Waschke et al., Phys. Rev. Lett. 70, 3319 (1993).
6. N. Sekine et al., Appl. Phys. Lett. 74, 1006 (1999).
7. R. Kersting et al., Phys. Rev. Lett. 79, 3038 (1997).
8. R. Kersting et al., Phys. Rev. B58, 4553 (1998).
9. J.N. Heyman et al., R. Kersting, K. Unterrainer, Appl. Phys. Lett. 72, 644 (1998).
10. L. Brey, N.F. Johnson, B.I. Halperin, Phys.Rev. B40, 10647 (1989).
11. J. Ulrich, R. Zobl, K. Unterrainer, G. Strasser, E. Gornik, K.D. Maranowski, A.C. Gossard, Appl. Phys. Lett. 74, 3158 (1999).
12. R. Bratschitsch, T. Müller, R. Kersting, G. Strasser, K. Unterrainer, Appl. Phys. Lett. 76, 3501 (2000).

Extreme Mid-Infrared Nonlinear Optics in Semiconductors

J. Kono[1], A. H. Chin[2*], and O. G. Calderon[2**]

[1] Department of Electrical and Computer Engineering, Rice University, Houston, Texas 77005, U.S.A.
[2] W.W. Hansen Experimental Physics Laboratory, Stanford University, Stanford, California 94305, U.S.A.

Abstract   We have observed multiple-order nonlinear optical phenomena induced by intense mid-infrared (MIR) pulses in semiconductors. We observed multiple (up to ±3 MIR photons) sidebands, multiple (up to seventh) MIR harmonics, and significant broadening of harmonic spectra. These extreme MIR nonlinear optical phenomena are primarily due to self-phase modulation that provides additional bandwidth and alleviates phase mismatch. These phenomena are suggestive of non-perturbative behavior.

1 Introduction

The strong desire to provide increased bandwidth has led to increasing interest in the development of terahertz (THz) electronics, i.e., the area between electronics and photonics where currently a technology gap exists. Because nonlinear optical phenomena (especially with strong fields) will likely play a role in future THz devices, mid-infrared (MIR) and far-infrared (FIR) nonlinear optics may serve as a basis for future device development. While many interesting nonlinear optical phenomena, such as high-order harmonic generation [1] and continuum generation [2,3] have been extensively studied using ultrafast laser sources in the visible or near-infrared (NIR), relatively few studies of nonlinear optical phenomena have been made using intense light in the MIR [4–6] or FIR [7,8].

Here, we examine non-degenerate sideband generation and harmonic generation using intense MIR pulses in semiconductors under conditions where extreme behavior is observed due to the strong fields at long wavelengths used. Under these conditions, we observed the generation of sidebands involving the interaction of up to 3 MIR photons with a NIR photon and the generation of MIR harmonics up to the seventh harmonic. In addition, we observed significant spectral broadening of MIR harmonics in semiconductors. The appearance of these extreme nonlinear optical phenomena demonstrates the influence of self-phase modulation on these nonlinear optical processes. The generation of these phenomena is aided by the low dispersion in many semiconductors in the MIR range, and is indicative of non-perturbative behavior.

---

* Present address: Lawrence Livermore National Laboratory, Information Science and Technology Division, Livermore, California 94550, U.S.A.

** Present address: Departamento Optica, Universidad Complutense de Madrid, Ciudad Universitaria s/n, 28040 Madrid, Spain.

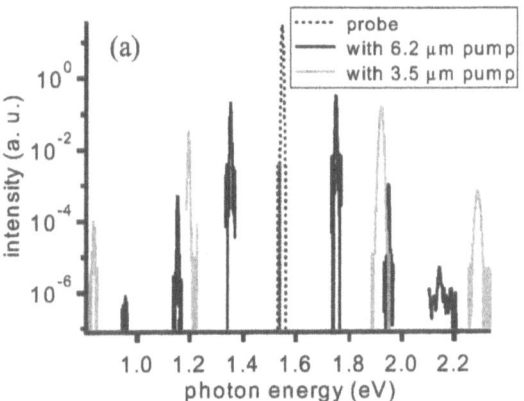

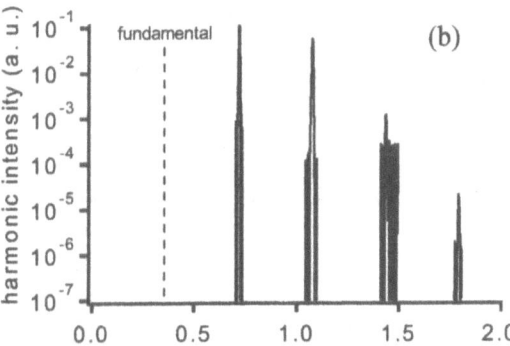

Fig. 1 (a) Optical sidebands in ZnSe (3 mm thick). The sidebands generated by 3.5 μm at ~2 × 10^10 W/cm² (gray) and 6.2 μm at ~ 3 × 10⁹ W/cm² (black) MIR pump are shown. (b) MIR harmonics in ZnS (2 mm thick) using 3.5 μm, ~1 ps, ~2 × 10^10 W/cm² MIR pulses.

2 Experimental Details

The source of intense MIR pulses used for these extreme nonlinear optics studies is an optical parametric amplifier (OPA) pumped by a Ti:Sapphire based regenerative amplifier. The system produced pulses at a 1 kHz repetition rate with either ~1 ps or ~200 fs pulse duration, and (with difference frequency mixing of the signal and idler) wavelengths from 3 to 10 μm. For these studies, we used 3.5 or 6.2 μm MIR pulses, i.e., wavelengths in regions where atmospheric absorption is negligible. We used two methods to study extreme nonlinear optics in semiconductors: 1) measuring the NIR spectrum after mixing of intense NIR and MIR beams in a semiconductor and 2) measuring the spectrum of an intense MIR beam after transmission through a semiconductor.

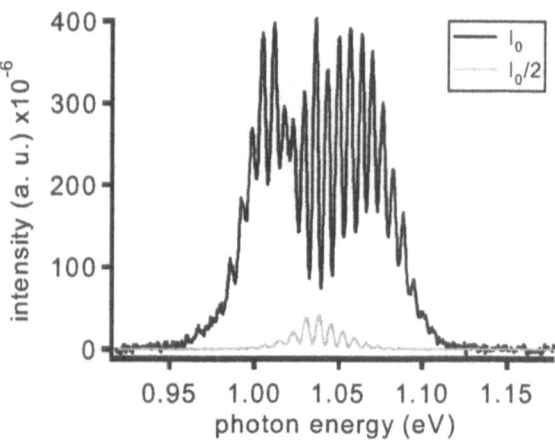

Fig. 2 Continuum generation in the third harmonic of 3.5 μm, $\sim$200 fs MIR in 350 μm thick GaAs(100), using $I_0 \sim 10^{11}$ W/cm^2 (black) and $I_0/2$ (gray).

3 Results and Discussion

Using the first method, we observed multiple optical sideband generation. As shown in Fig. 1a, the NIR probe spectrum (1.55 eV) is significantly modified to possess multiple sidebands that are spaced by the MIR photon energy. We observed not only ±1 sidebands but also higher-order wave mixing. Previous studies of sideband generation relied on resonant enhancement of the nonlinear susceptibility [7,8]. Here, we observed multiple *non-resonant* sidebands. We also observed multiple optical sidebands in ZnS and ZnTe, demonstrating the lack of system specificity due to the non-resonant nature of the observed multiple sideband generation.

Using the second method, we observed multiple MIR harmonics. Figure 1(b) shows the multiple harmonics observed in ZnS (2 mm thick). While third and higher harmonics have been observed from solid surfaces [9], and MIR harmonics have been observed in gases [5] and liquids [6], we observed multiple (*up to fifth*) MIR harmonics in bulk semiconductors. With higher MIR intensity ($\sim$10^{11} W/cm^2) using shorter ($\sim$200 fs) pulses at $\sim$3.5 μm, we also observed an interesting broadening of the bandwidth of MIR harmonics. Shown in Fig. 2 is the third harmonic in a GaAs(100) crystal (350 μm thick). Here, the broadened spectrum is reminiscent of self-phase modulation of the fundamental of an intense laser pulse in a gas or solid, but *the broadening is present in a harmonic of the fundamental instead*. In addition, we observed *significant spectral modulation* in the third harmonic in GaAs.

Under the excitation conditions used, significant self-phase modulation (SPM) is expected to occur. The additional bandwidth in the fundamental due to SPM alleviates any thickness dependence due to phase mismatch by providing a distribution of wavevectors. This, in combination with the low dispersion that exists in between phonon absorption and interband absorption in semiconductors, allows multiple sideband generation and multiple harmonics to be more easily observable using MIR rather than optical frequencies. The overall spectral width in Fig. 2 is thus due to spectral broadening of the fundamental MIR beam being mapped onto the harmonic. We attribute the significant spectral modulation to pulse splitting in the fundamental [3], which translates into pulse splitting in the harmonic. Interference between the time separated pulses produces the periodic spectral modulation. To qualitatively verify this, numerical calculations described above were also performed for the higher intensity case, assuming that the fundamental pulse consists of two temporally separated pulses [10,11]. Assuming fundamental pulses ($\sim$ 200 fs) separated by $\sim$ 400 fs with one pulse 1.5 times larger than the other, but with the overall energy in the pulse preserved, the simulation qualitatively matches the data. This suggests that pulse splitting in addition to self-phase modulation of the fundamental plays a role in the harmonic generation process in GaAs under these extreme conditions.

4 Summary

We have observed extreme MIR nonlinear optical phenomena in semiconductors. We observed multiple sidebands, multiple harmonics below the band edge of semiconductors and significant spectral broadening in harmonics using MIR pulses where significant self-phase modulation and pulse splitting of the fundamental is expected. The observed phenomena are suggestive of nonperturbative phenomena, and may play a role in future THz electronics.

5 Acknowledgements

We thank Prof. H.A. Schwettman for his support and Dr. T. Kimura for his help with the setup of the laboratory. We acknowledge support from NSF Grant DMR-9970962, ONR N00014-94-1024, the Japan Science and Technology Corporation PRESTO Program, and the NEDO International Joint Research Grant Program. OGC is supported by Becas Complutense "del Amo".

References

1. A. Rundquist et al., Science **280**, 1412 (1998).
2. A. Brodeur and S. L. Chin, Phys. Rev. Lett. **80**, 4406 (1998).
3. J. K. Ranka, R. W. Schirmer, and A. L. Gaeta, Phys. Rev. Lett. **77**, 3783 (1996).
4. A. H. Chin, J. M. Bakker, and J. Kono, Phys. Rev. Lett. **85**, 3293 (2000).
5. B. Sheehy et al., Phys. Rev. Lett. **83**, 5270 (1999).
6. R. Zurl and H. Graener, Appl. Phys. B **66**, 213 (1998).
7. J. Kono et al., Phys. Rev. Lett. **79**, 1758 (1997).
8. C. Phillips et al., Appl. Phys. Lett. **75**, 2728 (1999).
9. Y.-S. Lee, M. H. Anderson, and M. C. Downer, Opt. Lett. **22**, 973 (1997).
10. A. H. Chin, O. G. Calderon, and J. Kono, submitted to Phys. Rev. Lett.
11. O. G. Calderon, A. H. Chin, and J. Kono, submitted to Phys. Rev. A.

Proc. 25th Int. Conf. Phys. Semicond., Osaka 2000 (Eds. N. Miura and T. Ando)

Fine structure of the amplified spontaneous emission of ZnSe laser structures

R. Heinecke, U. Neukirch, P. Michler, J. Gutowski

Institute of Solid State Physics - Semiconductor Optics, University of Bremen, P.O. Box 33 04 40, D-28334 Bremen, Germany

Abstract   We present spectra of the edge emission of a series of high-quality separate-confinement-heterostructure lasers for stripe-excitation with ultrashort laser pulses in resonatorless geometry. Spectrally separated biexcitonic and electron-hole-plasma emission is observed.

1 Introduction

While there has been considerable progress in the theoretical understanding of emission processes in wide band gap semiconductors exhibiting strong Coulomb correlations, the nature of the lasing mechanism is still under debate. Different models such as emission out of a highly correlated dense electron hole plasma [1],[2] and biexcitonic lasing [3],[4],[5] have been used for the explanation of experimental results. Thus, further experiments with an eye to finer details are needed. Optical excitation with ultrashort laser pulses at a high repetition rate is ideally suited to obtain emission spectra with a high signal-to-noise ratio. Short pulse excitation above the bandgap of the waveguide creates hot charge carriers which undergo different relaxation processes to lower energy states and into the quantum well. Time-integrated spectra show the energetic fingerprints of these relaxation processes. Thus, they provide complementary information to quasi-steady-state measurements of laser emission.

2 Experimental Details

We have investigated three ZnSe-based separate-confinement-heterostructure samples at cryogenic temperatures using stripe excitation in resonatorless geometry: one 7 nm-thick ZnSe quantum well (QW) with a ternary $ZnS_{0.06}Se_{0.94}$ waveguide and two 6 nm and 10.5 nm ZnSe QWs embedded in quaternary $Zn_{0.9}Mg_{0.1}S_{0.16}Se_{0.84}$ waveguides. The high quality of the structures is reflected by the PL linewidth of the 1S heavy-hole exciton (X_{hh}) of 0.7 meV and a Stokes shift of less than 0.2 meV for the sample with ternary waveguide while an X_{hh} linewidth of 3 meV and a Stokes shift of 1.9 meV have been determined for the samples with quaternary waveguide. The samples were excited above the bandgap of the waveguide with 120 fs pulses from a frequency-doubled Ti:sapphire laser running at 82 MHz repetition rate. This way, high excitation densities up to $3 * 10^{12} cm^{-2}$ and a signal-to-noise ratio better than 10^4 were achieved. The excitation intensity was continuously varied by a sequence of two linear-wedge neutral-density filters over

five orders of magnitude. The geometry of the excitation focus was a stripe of 10 μm width and variable length. The emitted amplified spontaneous emission (ASE) from the cleaved sample edge was detected spectrally resolved by a CCD camera. Series of emission spectra at constant stripe length were taken for decreasing excitation density, the stripe length being varied between series.

3 Results

In all three samples, the edge emission spectra exhibit a peak at the X_{hh} energy whose spectral position is independent of the excitation density. Roughly one biexciton binding energy (6 meV, determined by PLE) below this excitonic peak, the two quaternary samples show a strong TE-polarized emission which undergoes a red-shift of about 3 meV over the whole density range. The intensity of this biexcitonic peak increases stronger with excitation density as does the intensity of the excitonic peak. In addition, one- and two-LO phonon replicas of the excitonic peak are observed.

For the sample with ternary waveguide, the situation is much more complex 1. At low densities, the spectra are dominated by a peak at 2.8035 eV, slightly more than one biexciton binding energy (4.8 meV) below the excitonic peak at 2.810 eV. For increasing density, however, a shoulder at 2.800 eV evolves into a new peak, rising exponentially and eventually dominating the spectra while the biexcitonic signature is still observable in our time integrating detection. At a density of $3 * 10^{12} cm^{-2}$ in the waveguide the density within the active layer can definitely be assumed to lie above the 2D-Mott density. Therefore, we ascribe the ASE maximum at highest densities to an electron-hole plasma.

Further, a series of ASE spectra was taken for a short stripe of only 15 μm and increasing density (not shown). Here, the biexcitonic peak dominates at all densities. At 2.800 eV, a distinct second maximum occurs at low density and remains clearly resolved for increasing density but never dominates. The energetic positions of the ASE maxima relative to the exciton peak are summarized in the table below 3 for different stripe lengths and densities. As indicated, electron-hole-plasma emission dominates only at high excitation densities and a long stripe.

The emission observed at the sample with ternary waveguide has been resolved as two distinct peaks. The ASE spectra of the samples with quaternary waveguide, which have a higher PL linewidth show a broad emission maximum. However, the energetic position of the maximum relative to the exciton energy in ASE is for all three

---

* *The authors would like to thank Prof. Hommel, University of Bremen, and his epitaxy group for sample preparation.*

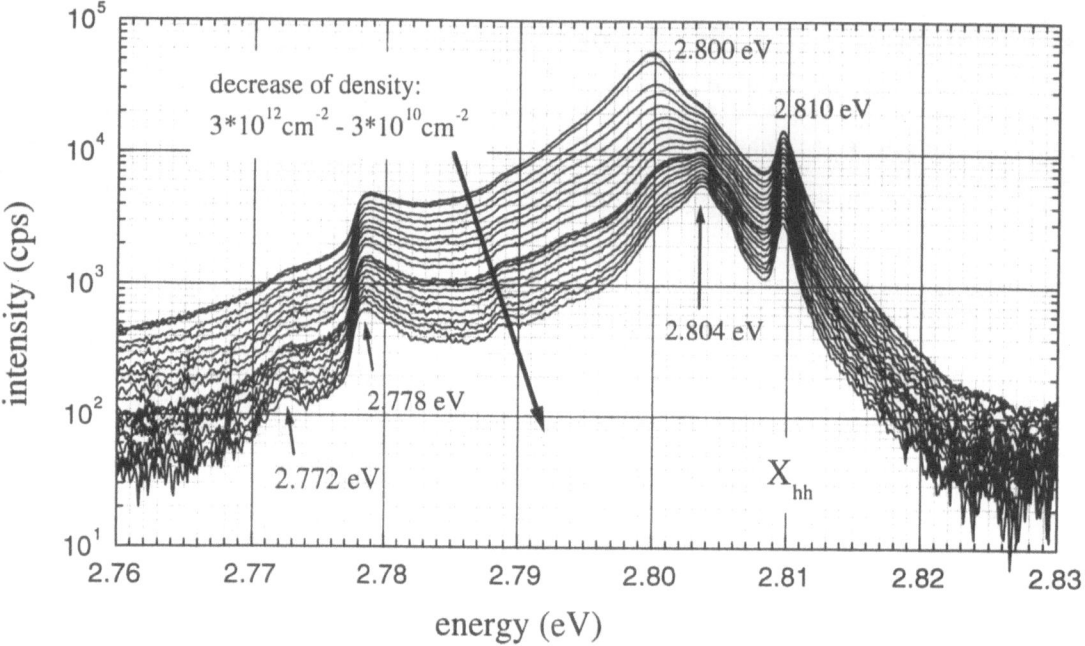

Fig. 1 Density-dependent edge emission from a ZnSe heterostructure laser with ternary waveguide for excitation with a 360 μm stripe. Exciton and biexciton peak are resolved together with their one-LO-phonon replicas. The structure at 2.800 eV evolves into a distinct peak for increasing density.

height	long stripe	short stripe
high density	-10 meV	-6 meV
low density	-6 meV	-6 meV

samples -6 meV for low excatation density and -10 meV for high excitation density. The shift of the ASE emission maximum with increasing density is therefore ascribed to a change from the biexcitonic peak to the electron-hole-plasma. As the spectra for short stripe excitation of the sample with ternary waveguide show, both emissions occur simultaneously.

4 Conclusion

Edge emission spectra of the ZnSe heterostucture laser samples confirm the relevance of biexcitonic processes in nearly homogeneous systems without localization. However, for the sample with ternary waveguide, we have been able to observe additional structures both within and clearly below the energy region of the biexciton. Specifically, a superlinear increase in intensity is observed at 2.8 eV that evolves independently of the still resolved biexcitonic features dominating the low excitation density spectra. Our results clearly show, that in high quality samples the biexcitonic emission at low densities is not completely replaced by plasma emission at high den-

sities, at least in a time integrated measurement. Rather, both emissions exhibits distinctly different spectral positions. It has still to be cleared whether these both emissions occur during different time windows (evolution of carrier densities) or stem from different locations in the samples or from different parts of the stripe which most probably does not provide perfectly homogeneous carrier densities.

References

1. L. Calgagnile et al., Phys. Rev. B **55** (1997) 13413
2. P. Michler et al., Phys. Rev. B **58** (1998) 2055
3. F. Kreller et al., Phys. Rev. Lett. **75** (1995) 2420
4. V. Kozlov et al., Phys. Rev. B **93** (1996) 10873
5. O. Homburg et al., Phys. Rev. B **60** (1999) 5743

Biexcitonic signatures in femtosecond pulse propagation

J. Meinertz, I. Gösling, U. Neukirch, J. Gutowski

Institut für Festkörperphysik, Universität Bremen, Postfach 33 04 40, 28334 Bremen, Germany

Abstract We have studied the nonlinear transmission of 120 fs probe pulses through a high-quality ZnSe/Zn(S,Se) sample excited by strong pump pulses in the vicinity of the exciton-biexciton transition. Pump and probe pulses were either co- or counter-circularly polarized. At a probe delay of 2 ps we observe induced absorption at the exciton-biexciton resonance in the spectrally resolved probe transmission for the counter-circular configuration. This is accompanied by characteristic signatures in the time resolved transmission.

In recent years, the nonlinear propagation of ultra-short laser pulses in the spectral vicinity of excitonic resonances has attracted much interest [1]. However, work has mostly been devoted to an *external* modification of the light-exciton coupling by using specially prepared geometries like microcavities [2] or multiple-quantum-well Bragg or anti-Bragg structures [3]. In this contribution, we investigate how pulse propagation is affected by the biexcitonic nonlinearity (BNL). This is appealing from a fundamental point of view for at least two reasons. First, the BNL is a dominant *intrinsic* optical nonlinearity in wide-gap semiconductors. Second, this nonlinearity can effectively be switched on and off experimentally by making use of selection rules.

We have performed pump-and-probe measurements with 120 fs pulses from a frequency-doubled Ti:sapphire laser system on a high-quality ZnSe/Zn(S,Se) sample. The central energy of the pulses was tuned to 2.807 eV, half-way between the first center-of-mass (COM) quantized heavy-hole (hh) excitonic polariton hh^1 at 2.809 eV and the exciton-biexciton resonance (EBR) 4.5 meV below. Pump and probe pulses were either co-circularly ($\sigma^+\sigma^+$) or counter-circularly($\sigma^-\sigma^+$) polarized. Switching between the two polarization configurations was done with Pockels cells, so that no manual adjustments of polarizers was neccessary. Behind the sample, the probe pulse was detected either spectrally or time resolved. The latter was achieved by up-converting the probe pulse with variable-delay 80 fs infrared pulses in a β-barium borate nonlinear optical crystal and detecting at the sum-frequency with a photomultiplier. The sample, pseudomorphically grown by molecular-beam-epitaxy on GaAs substrates, consists of a binary 23 nm ZnSe central layer embedded in ternary 350 nm Zn(S,Se) barriers. No indication of inhomogeneous broadening was seen in four-wave-mixing experiments which have yielded a biexciton binding energy of 4.5 meV. During the experiments, the sample was kept at 2 K inside a helium bath cryostat.

Fig. 1 shows probe transients and transmission spectra (inset) for pump off (linear case) and for a pump fluence of $2.2 * 10^{12}$ cm^{-2} with the pump preceding the probe by 2 ps. The linear transmission is dominated by a series of sharp absorption lines which can be attributed to COM quantized heavy- and light-hole (lh) excitonic polaritons inside the central layer [4]. The second of these, hh^2, is strongly subdued due to Fabry-Perot effects in the whole heterostructure (central and barrier layers). It is of no significance for the following discussion. Also, there is some slight absorption from donor-bound excitons around 2.804 eV. The probe spectrum has appreciable overlap with polariton resonances (PRs) hh^1 and hh^3. Consequently, the linear probe transient shows strong quantum beats (QBs) with a period of 0.5 ps, corresponding to the energetic separation of 7.9 meV between these two PRs. In the nonlinear transmission spectrum for the $\sigma^-\sigma^+$ polarization configuration, induced absorption is observed one biexciton binding energy below the first PR hh^1. The absence of this spectral feature for $\sigma^+\sigma^+$ excitation confirms the biexcitonic origin of this absorption which is thus identified as arising from the EBR. For both polarization combinations, the PRs show bleaching and broadening of roughly the same magnitude. This carries over to the nonlinear probe transients which, at first glance, look very similar. Both exhibit excitation-induced dephasing accompanied by a strong reduction of the amplitude of the QBs on the trailing part – these two effects are just the signatures in the time regime of the broadening and bleaching observed in the frequency regime. In the $\sigma^-\sigma^+$configuration, however, there are now three resonances seen in the probe spectrum: hh^1, hh^3, and the EBR. Thus, the corresponding probe transient should display a superposition of polariton and exciton-biexciton QBs. This reasoning is supported by subtle differences between the modulations on the trailing parts of both transients. While the transient in $\sigma^+\sigma^+$ configuration shows a modulation closely resembling the QBs on the linear transient albeit with a strongly reduced amplitude, every second QB on the other transient ($\sigma^-\sigma^+$) is stronger. The period of 0.9 ps defined by the three most prominent QB maxima, marked by arrows in the figure, corresponds directly to the biexciton binding energy. Meanwhile, traces of the original QBs are still seen on the transient.

In Fig. 2, the data for a pump fluence of $5.9*10^{12}$ cm^{-2} is presented. Absorption by the EBR has increased for counter-circular excitation while the PRs have undergone further bleaching and broadening. For co-circular excitation this is accompanied by a blue-shift of the hh^1

PR that can be ascribed to phase-space-filling. At this
density, the $\sigma^+\sigma^+$-transient shows an almost monoexponential decay. On the $\sigma^-\sigma^+$-transient, two QBs are just
resolved which are 0.9 ps apart. Again, this corresponds
to the biexciton binding energy.

At a probe delay of 2 ps, there is still some coherent
polarization left from the pump pulse. In order to avoid
these coherent contributions to the pump-and-probe signal, we have repeated the measurements at 5 ps and 10
ps probe delay. At a probe delay of 10 ps the difference
between the two polarization configurations in the spectra and the transients begins to blur, and induced absorption at the EBR in the nonlinear transmission spec-
tra is observed also in $\sigma^+\sigma^+$ configuration. This points to
the existence of an efficient spin-flip mechanism causing
equilibration between the two optically active excitonic
spin species on a picosecond time scale.

We have attempted a quantitative analysis of the observed features. Obviously, we need to take into account
spatial dispersion of excitons and the confinement, since
these two phenomena lead to the PRs within spectrum.
Also, the biexcitonic nonlinearity must be included at
least phenomenologically. To this end, we have extended
a previous approach based on an oscillator model with
spatial dispersion [4]. Proceeding as in Ref. [5], effects
like bleaching, broadening and energy shifts of PRs can
be reproduced. The BNL has been accounted for by including the EBR as an additional oscillator for counter-
circular polarizations, since a σ^--polarized pump pulse
effectively adds the EBR to the linear susceptibility χ^1
seen by a σ^+ polarized probe pulse by exciting an ensemble of spin -1 excitons.

The calculated transients qualitatively reproduce the
main features observed on the transients. For instance,
the position of beat maxima, excitation-induced dephasing, and the reduction of the amplitude of QBs with
increasing pump density can be simulated within the
model. In general, the correspondence between calculated and measured transients is better for the $\sigma^+\sigma^+$
configuration, i.e. for forbidden biexcitons. For the opposite configuration, model and experiment disagree significantly in two important points. First, the intensity
of the trailing part of the $\sigma^-\sigma^+$-transient is overesti-
mated. As can be seen in Figs. 1 and 2, the intensity of
the $\sigma^-\sigma^+$-transient always less than that of the $\sigma^+\sigma^+$-
transient. The calculated transients, however, have com-
parable intensity. Second, the amplitude of the exciton-
biexciton QBs is underestimated. Obviously, the BNL
calls for a more sophisticated treatment, e.g. on the level
of modern many-body semiconductor theory.

In conclusion, exciton-biexciton quantum beats have
been observed on real-time transients of 120 fs probe
pulses in a pump-and-probe experiment. At the highest pump fluences, these dominate the transients for
counter-circular polarization of the pump and probe pulses,
i.e. for allowed biexcitons, whereas for co-circular polarization, the probe transients show a monoexponential

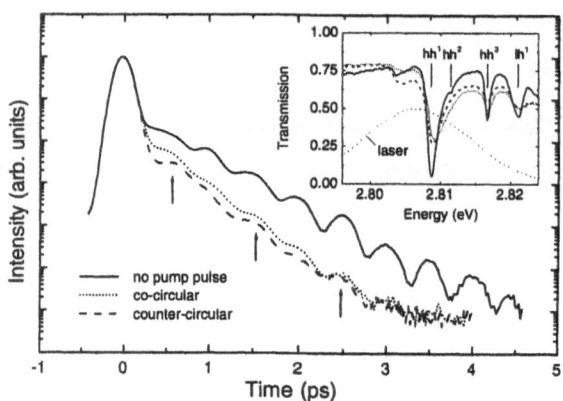

Fig. 1 Time-resolved probe pulse at 2 ps delay for a
pump-fluence (photons per pulse / irradiated area) of 2.2 $*$
10^{12} cm^{-2}.

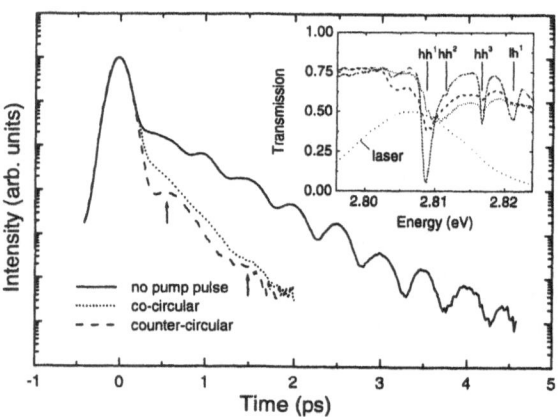

Fig. 2 Time-resolved probe pulse at 2 ps delay for a pump-
fluence of $5.9 * 10^{12}$ cm^{-2}.

decay. At medium excitation levels, $\sigma^-\sigma^+$-transients exhibit a superposition of polariton and exciton-biexciton
quantum beats. Qualitative explanation of the experimental data has been possible with an oscillator model
that accounts for spatial dispersion of excitons, confinement, and pump-induced optical nonlinearities. However, a more advanced theoretical approach is still needed
for a quantitative analysis. With our work, we hope to
stimulate efforts in this direction.

References

1. see, e.g., H. Giessen et al., Superlattices and Microstructures **26** (1999) 103, and references therein.
2. M. Kira et al., Solid State Comm. **102** (1997) 703.
3. M. Hübner et. al., Phys. Rev. Lett. **76** (1996) 4199.
4. U. Neukirch et al., phys. stat. sol. **196** (1996) 473.
5. U. Neukirch and K. Wundke, Phys. Rev. B **55** (1997)
   15408.

Femtosecond response times of large optical nonlinearity in low-temperature grown GaAs/AlAs multiple quantum wells

T. Okuno[1], Y. Masumoto[1], Y. Sakuma[2], M. Ito[2], H. Okamoto[2]

[1] Institute of Physics, University of Tsukuba, Tsukuba, Ibaraki 305-8571, Japan e-mail: okuno@sakura.cc.tsukuba.ac.jp
[2] Department of Materials Technology, Faculty of Engineering, Chiba University, 1-33 Yayoi-cho, Inage-ku, Chiba 263-8522, Japan

Abstract We have investigated optical nonlinearity in low-temperature (LT) molecular-beam-epitaxy grown GaAs/AlAs multiple quantum wells (MQWs). The LT MQW is shown to have larger nonlinearity than LT bulk GaAs. The response time of the LT MQW becomes less than 1 ps, which is about half that of LT bulk GaAs and ∼1/400 of the standard-temperature-grown MQW. These results demonstrate a clear advantage of the LT MQW having large optical nonlinearity as well as fast response time. In 310 °C-grown Be-doped MQW, even faster response time (0.3 ps) is obtained.

1 Introduction

Low-temperature (LT) molecular-beam-epitaxy (MBE) grown III-V semiconductors have attracted much attention due to the fast recovery time of their optical nonlinearity after interband excitation [1]. Excess As included by LT-growth is thought to bring fast carrier trapping in LT GaAs. Besides the recovery time, the strength of the nonlinearity is another important parameter of a material for applications such as all-optical switches. Room temperature excitons in multiple quantum wells (MQW) are known to show larger nonlinearity than bulk GaAs. There are only a few reports on LT MQW, and none of them except ref. [2] on LT InGaAs/GaAs MQWs describe both the response time and the strength of the nonlinearity. In addition, no paper compares the MQW and the bulk film. In this paper, we report fast response times as well as large nonlinearity in undoped [3] and beryllium-doped LT GaAs/AlAs MQWs, comparing with bulk LT GaAs.

2 Low-temperature grown MQW

Bulk-GaAs epilayers (700 nm thick) and GaAs(7 nm) / AlAs(7 nm) MQW (100 periods) were grown at various temperatures T_g between 700 (standard temperature) and 310 °C (LT) on (100) GaAs substrate by the standard solid state MBE. The strength of optical nonlinearity was evaluated by the saturation density I_s. This was determined from excitation density dependence of the absorption coefficient α, and smaller I_s means larger nonlinearity. The response time τ was measured by a time-resolved pump-probe method. These measurements were done at room temperature by using 200-fs pulses with wavelength around exciton peaks (∼825 nm) for MQWs or band gap (∼865 nm) for bulk GaAs.

Figure 1 shows absorption spectra of 700, 360, and 310 °C-grown GaAs/AlAs MQWs. Heavy-hole and light-

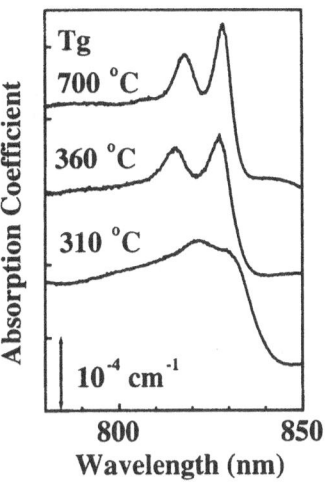

Fig. 1 Absorption spectra of GaAs/AlAs MQWs.

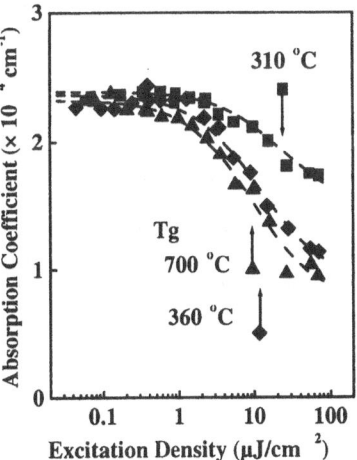

Fig. 2 Excitation density dependence of the optical absorption coefficient in GaAs/AlAs MQWs.

hole exciton peaks are observed in the 700 and 360 °C-grown MQWs. Even in the 310 °C-grown MQW, exciton peaks can be noticed. These facts indicate high quality of our LT MQWs. Figure 2 shows excitation-density (I) dependence of α in the MQWs. The results are semiempirically fitted by the expression $\alpha = \alpha_1/(1+I/I_s)+\alpha_2$,

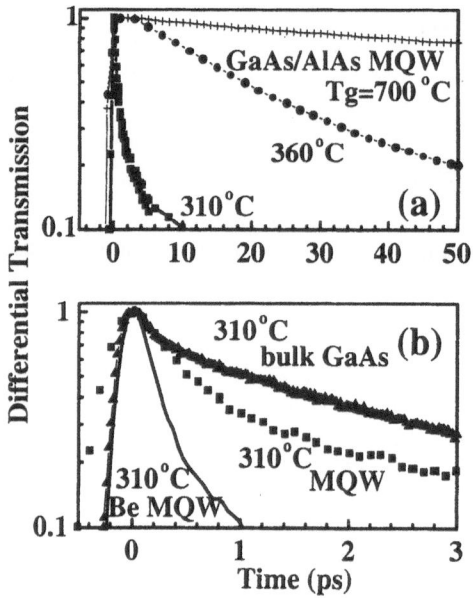

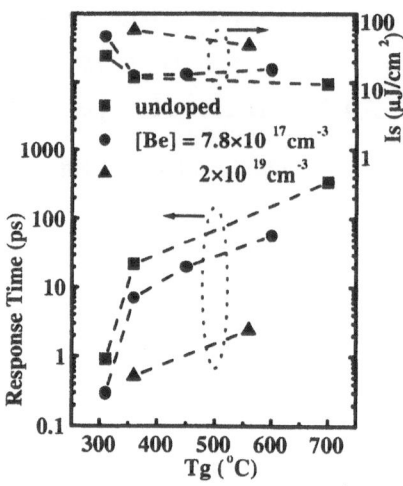

Fig. 4 Response time and saturation density of GaAs/AlAs MQWs plotted against growth temperature.

Fig. 3 Normalized temporal changes of differential transmission.

where α_1 and α_2 are fitting parameters. The obtained I_s (denoted by the arrows in the figure) are 9-24 μJ/cm^2. We also measured I_s in bulk GaAs grown in the same condition. The obtained values were 45-65 μJ/cm^2, and they are larger than those in the MQWs. This demonstrates excitonic enhancement of optical nonlinearity in the LT MQWs as well as in the standard-temperature grown MQW [3].

Figure 3 shows normalized temporal changes of differential transmission. We define the response time τ by the 1/e decay time. In Fig. 3(a), τ decreases with the decrease of T_g (340 ps for 700 °C, 27 ps for 360 °C, and 0.9 ps for 310 °C). In Fig. 3(b), it is observed that τ of the 310 °C-grown MQW (0.9 ps) is faster than that of the 310 °C-grown bulk GaAs (2 ps). These results show the larger optical nonlinearity having faster response time of the LT MQW than those of the standard-temperature grown MQW and LT bulk GaAs.

It is reported that in LT GaAs/AlAs MQW, the antisite As diffuses from AlAs barrier layers to GaAs well layers and As precipitates accumulate in GaAs layers [4]. Thus in the LT MQW, increased density of excess As in GaAs layers or at heterointerfaces is thought to lead to reduction in τ.

3 Effect of Be doping

Be-doping is reported to reduce response time in LT In-GaAs/InAlAs MQW [5] and in LT bulk GaAs [6]. We compared τ and I_s in Be-doped LT GaAs/AlAs MQW with those in undoped MQW. During MQW growth, Be was uniformly doped (7.8×10^{17} or 2×10^{19} cm^{-3}). In Fig. 3(b), τ of Be-doped (7.8×10^{17} cm^{-3}) 310 °C-

grown MQW is shown to be 0.3 ps, which is faster than that of the undoped 310 °C-grown MQW (0.9 ps). The saturation density I_s was 47 μ J/cm^2. This value is a little larger than that of the undoped 310 °C-grown MQW (24 μJ/cm^2), but was smaller than that of Be-doped 310 °C-grown bulk GaAs (90 μJ/cm^2). From these results, Be-doping in LT GaAs/AlAs MQW is effective for reducing τ, and the excitonic enhancement of the optical nonlinearity is still observed in the Be-doped LT MQW when compared with Be-doped LT bulk GaAs.

Figure 4 summarizes τ and I_s in undoped and Be-doped MQWs obtained in this research. We find that Be-doping decreases τ while not increasing I_s so much.

4 Conclusion

In the undoped MQWs, τ of the LT MQW is $\sim$1/400 of the standaed-temperature grown MQW, while the strength of the nonlinearity is almost the same within a factr of 3. This result demonstrates the advantage of the LT MQW having large optical nonlinearity as well as fast response time. In the 310 °C-grown Be-doped MQW, τ as fast as 0.3 ps is obtained.

References

1. S. Gupta, J.F. Whitaker, and G.A. Mourou, IEEE J. Quantum Electron. **28** (1992) 2464.
2. U. Keller, Appl. Phys. B **58** (1994) 347.
3. T. Okuno, Y. Masumoto, M. Ito, and H. Okamoto, Appl. Phys. Lett. **77** (2000) 58.
4. B. Lita, S. Ghaisas, R.S. Goldman, and M.R. Melloch, Appl. Phys. Lett. **75** (1999) 4082.
5. R. Takahashi, Y. Kawamura, T. Kagawa, and H. Iwamura, Appl. Phys. Lett. **65** (1994) 1790.
6. M. Haiml, U. Siegner, F. Morier-Genoud, U. Keller, M. Luysberg, P. Specht, and E.R. Weber, Appl. Phys. Lett. **74** (1999) 1269.

Proc. 25th Int. Conf. Phys. Semicond., Osaka 2000 (Eds. N. Miura and T. Ando)

591

Asymmetric short-period GaAs/AlAs superlattices for observation of low-threshold laser effect

V. G. Litovchenko[1], D. V. Korbutyak[1], S. G. Krylyuk[1], A. I. Bercha[1], H. T. Grahn[2], K. H. Ploog[2]

[1] Institute of Semiconductor Physics, National Academy of Sciences of Ukraine, Prospect Nauki 45, 03028 Kiev, Ukraine, e-mail: lvg@div9.semicond.kiev.ua

[2] Paul-Drude-Institut für Festkörperelektronik, Hausvogteiplatz 5–7, 10117 Berlin, Germany

Abstract Asymmetric short-period GaAs/AlAs superlattices, for which the well thickness is at least a factor of two larger than the barrier thickness, exhibit a direct band structure, while their symmetric counterparts are indirect. Therefore, asymmetric superlattices are characterized by an enhanced intensity of the luminescence and, thus, may be used as light-emitting devices. This conjecture is supported by the observation of stimulated emission at $T = 80$ K for a GaAs/AlAs superlattice with 6 monolayers well and 3 monolayers barrier width.

1 Introduction

Semiconductor quantum structures with carrier confinement, in particular superlattices (SLs), are considered to be very promising candidates for stimulated emission of light and, hence, lasing at very low threshold currents. The threshold current was predicted to decrease with a decrease of the active layer thickness [1,2]. However, for symmetric GaAs/AlAs SLs with equal well and barrier thickness, a reduction of the layer thickness below 12 monolayers, i. e., a SL period of 6.8 nm, results in an indirect band structure [3,4]. The X-minimum of AlAs in the conduction band becomes lower than the respective Γ-state in GaAs. In this case, the electron-hole transitions become indirect in both, real and momentum space, so that the observation of stimulated emission becomes impossible. Since indirect-gap semiconductors have inferior light emitting properties than their direct-gap counterparts, short-period GaAs/AlAs SLs are not very suitable for light-emitting devices, although their energy gap falls in the red part of the visible spectral region.

In our previous work, we showed that this drawback may be overcome by using asymmetric SLs with different well and barrier thickness [5]. If the ratio of well to barrier thickness becomes two or larger, short-period GaAs/AlAs SLs with any well thickness below 3.4 nm will always exhibit a direct energy gap, resulting in a significant enhancement of the photoluminescence (PL) intensity. Here, we report the observation of stimulated emission in a short-period asymmetric GaAs/AlAs SL with 6 monolayers well and 3 monolayers barrier width (1 monolayer corresponds to 0.283 nm).

2 Experimental

Two $(GaAs)_n/(AlAs)_m$ SLs with $n = 6$ monolayers and $m = 3$ and 4 monolayers were grown by molecular beam epitaxy on (001)-GaAs substrates. We will label these SLs n/m. The SL period and the composition have been confirmed by x-ray diffractometry. The difference between the nominal and actual layer thicknesses was about 1%. The PL experiments at low excitation levels were performed at 5 K using an Ar$^+$-laser (516 nm) for excitation. To observe stimulated emission, the second harmonic of a Nd:YAG-laser (532 nm) was used. The laser beam was focused onto the edge of the sample on a narrow strip with 1000 μm length and about 50 μm width. The spontaneous PL signal I_{spont} was detected from the surface of the sample, while the stimulated PL signal I_{stim} was detected from a cleaved edge [6].

The optical gain spectra were determined by measuring I_{stim} and I_{spont}. Using the optical gain coefficient g and the length of the excitation strip l, their product $g \cdot l$ can be derived from the following equation

$$\frac{I_{stim}}{I_{spont}} = \frac{exp(g \cdot l) - 1}{g \cdot l} , \qquad (1)$$

where I_{stim} and I_{spont} are taken at the same wavelength.

3 Results and discussion

In Fig. 1, the PL spectra recorded at low temperatures (5 K) of the 6/3 and 6/4 SLs are compared. This figure clearly demonstrates the difference in the energy band structure of the two samples, which is caused by the different barrier width. The PL spectrum of the 6/4 SL is typical for quasi-direct SLs and consists of an intensive zero-phonon line due to recombination of the $X_z - \Gamma$ excitons and two phonon satellites on the low-energy side of the zero-phonon line. However, only one PL line is observed for the 6/3 SL. This difference demonstrates that this sample exhibits a direct energy gap, in which photoexcited electrons and holes are confined within the GaAs layers. Therefore, recombination occurs via the respective Γ-levels in the conduction and valence bands.

Stimulated emission from the 6/3 SL starts to appear at 80 K for an optical pumping power of $P_{ex} = 4$ MW cm^{-2}. With increasing laser power, it becomes more prominent. PL spectra of the spontaneous and stimulated emission signal for $P_{ex} = 8$ MW cm^{-2} are shown in Fig. 2. The red shift of the maximum of the stimulated PL with respect to the spontaneous PL is explained by a large increase of the absorption coefficient in the spectral range of the spontaneous PL maximum. In contrast, no

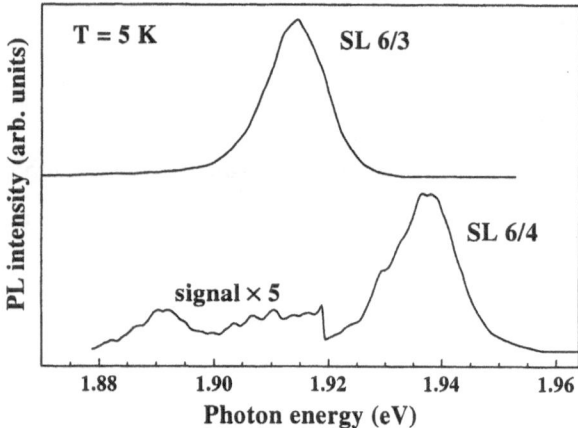

Fig. 1 PL spectra of the 6/4 and 6/3 SLs at 5 K and $P_{ex} = 50$ mW cm^{-2}. The spectra are normalized to their respective maximum.

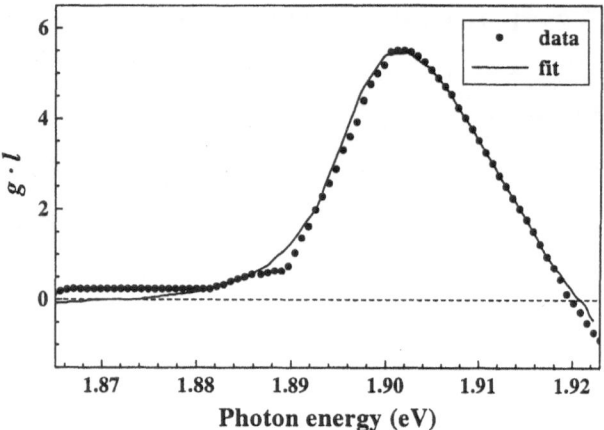

Fig. 3 Optical gain spectra of the 6/3 SL at 80 K calculated by Eq. 1. The solid line is a fit as described in [7–9].

stimulated PL signal was detected from the 6/4 SL at similar experimental conditions because of the indirect character of the energy gap in this sample.

In Fig. 3, the gain spectrum obtained according to Eq. 1 is plotted together with a fit. The fitting curve was calculated using the approach described in Refs. [7–9]. Because of the very large separation of the first and second minibands in the conduction and valence bands for this system, only the first electron and first hole miniband were taken into account in the calculations. The value of g was found to exhibit a super-linear behavior for increasing excitation power. For the highest value of $P_{ex} = 8$ MW cm^{-2}, the gain coefficient became as high as 50 cm^{-1}.

4 Conclusions

Stimulated PL has been observed at 80 K for a $(GaAs)_6/(AlAs)_3$ SL. The appearance of stimulated PL in such a short-period SL became possible because of the asymmetric thickness of well and barrier layer. For a ratio of well to barrier thickness of at least two, a transition from

a quasi-direct to a direct band structure occurs. Thus, when asymmetric GaAs/AlAs SLs have a direct energy gap, they can exhibit superior emission properties even for very small well widths. This observation opens new perspectives for the utilization of ultra-short-period SLs in light-emitting devices operating in the red region of the visible spectrum, in particular for the development of low-threshold lasers.

Acknowledgements We would like to thank A. Žukauskas and G. Tamulaitis (IMSAR, Vilnius, Lithuania) for their help in performing the experiments on stimulated emission and Yu. Kryuchenko for helpful discussions. This work was supported in part by the Fundamental Research Foundation of the Ministry for Education and Science of Ukraine.

References

1. H. C. Casey, Jr., J. Appl. Phys. **49**, (1978) 3684.
2. N. Holonyak, Jr., R. M. Kolbas, R. D. Dupuis, and P. D. Dapkus, IEEE J. Quantum Electron. **QE-16**, (1980) 170.
3. Yan-Ten Lu and L. J. Sham, Phys. Rev. B **40**, (1989) 5567.
4. R. Cingolani, L. Baldassarre, M. Ferrara, M. Lugarà, and K. Ploog, Phys. Rev. B **40**, (1989) 6101.
5. S. Krylyuk, D. V. Korbutyak, V. G. Litovchenko, R. Hey, H. T. Grahn, and K. H. Ploog, Appl. Phys. Lett. **74**, (1999) 2596.
6. R. Baltrameyunas and E. Kuokstis, Sov. Phys. — Collections **22**, (1982) 73.
7. E. Zielinski, H. Schweizer, S. Hausser, R. Stuber, M. H. Pilkhun, and G. Weimann, IEEE J. Quantum Electron. **QE-23**, (1987) 969.
8. D. V. Korbutyak, Yu. V. Kryuchenko, V. G. Litovchenko, R. Baltrameyunas, E. Gerazimas, and E. Kuokshtis, Sov. Phys. — JETP **69**, (1989) 757.
9. R. Cingolani, K. Ploog, A. Cingolani, C. Moro, and M. Ferrara, Phys. Rev. B **42**, (1990) 2893.

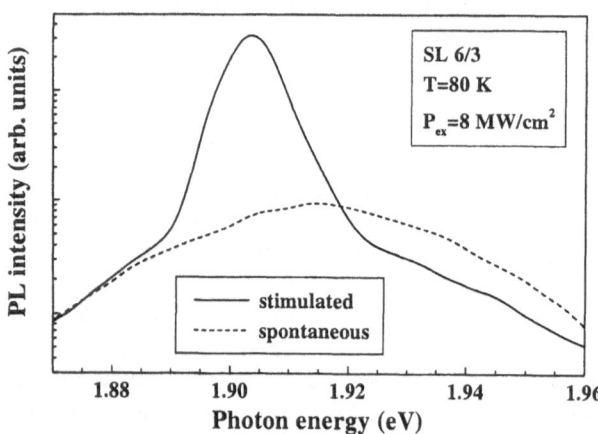

Fig. 2 Spontaneous (dotted line) and stimulated (solid line) PL spectra of the 6/3 SL at 80 K and excitation power of 8 MW cm^{-2}.

Nonlocality induced size-enhancement of excitonic nonlinear response in high quality samples

H. Ishihara[1], K. Cho[1], K. Akiyama[2], N. Tomita[2], Y. Nomura[2], T. Isu[2]

[1] Department of Physical Science, Graduate School of Engineering Science, Osaka University, Toyonaka, Osaka 560-8531, Japan e-mail: ishi@mp.es.osaka-u.ac.jp

[2] Mitsubishi Electric Corporation, 1-1, Tsukaguchi Honmachi 8-Chome, Amagasaki, Hyogo 661-8661, Japan

Abstract We investigate, both experimentally and theoretically, the size dependence of the degenerate four wave mixing (DFWM) of GaAs thin films confining the center-of-mass motion of excitons. As a result, we confirm that the large DFWM signal observed for a particular thickness is caused by the enhancement of the internal field with a non-dipole type spatial pattern, as predicted by the nonlocal response theory.

1 Introduction

The spatial dispersion is one of the remarkable features of semiconductors. This nature requires nonlocal description of the optical response, which is essential for mesoscopic systems. In contrast to a considerable number of studies of this problem in the linear response regime[1], little attention has been given to the problem how the nonlocality affects the nonlinear response. However, this problem is highly attractive because it is possible that the additional degrees of freedom for the spatial variation of radiation field brings about the nonlinear effects beyond the scheme of electric dipole transitions in the long wavelength approximation (LWA). Actually, the large nonlinear response due to the resonant enhancement of the internal field with a nano-scale spatial pattern has theoretically predicted[2], and the similar effect has also been reported for GaAs thin films[3]. In this contribution, we establish, both theoretically and experimentally, this new type of enhancement mechanism of the nonlinear response, investigating the thickness dependence of the degenerate four-wave mixing (DFWM) of GaAs thin films.

2 Experiment

We prepared double-hetero-structures of (high quality) GaAs sandwiched by $Al_{0.3}Ga_{0.7}As$ layers on a GaAs (100) substrate by molecular beam epitaxy. To evaluate the quality of the samples, we analyzed the linear reflectance spectra measured in near Brewster angle. Fig.1 (a) shows the spectrum of the 110-nm-thick layer, where we can see the several peak structures. Those structures are attributed to the quantized center-of-mass (CM) motion of excitons confined in the layer. The calculation with the nonlocal theory well reproduces these structures. (The theory will be explained in the following section.) With this analysis, we found the non-radiative damping constant of exciton considerably small, i. e., Γ

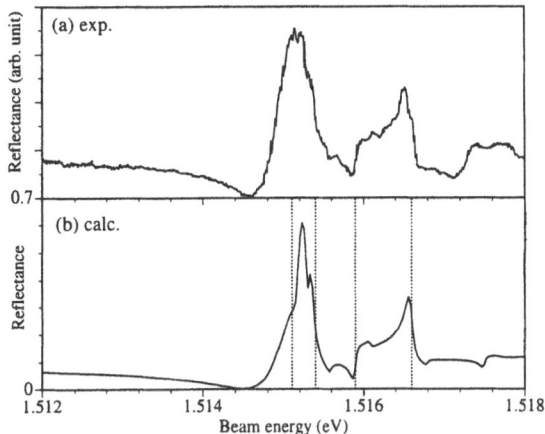

Fig. 1 The reflectance spectrum for the 110-nm-thick film. The vertical dotted lines in (b) indicate the quantized excitonic levels determined in the theory.

= 0.03 meV, which means the extremely good quality of the sample.

For the DFWM measurements, we detected backward signals in two-beam self-diffraction configuration. The pulse width was 2 ps and its intensity was set at 2.5 kW/cm^2 which is in the region such that the detected DFWM signal intensity shows a cubic dependence on the excitation intensity. The sample was kept at 5K during measurements. Fig. 2 shows the measured DFWM intensities for the several values of thickness. The signal intensity for around 110 nm is strongly enhanced, which is 25 times larger than that of the bulk (1-μm-thick) sample[3].

3 Theoretical Model and Calculated Results

For the analysis of the above results, we use the model of the CM confinement of excitons. Although the Bohr radius of exciton is relatively large for this material, it is known that this model is well applicable for the considered thickness region[4]. The relative motion of excitons is reflected in the parameter of the bare exciton - radiation coupling (e.g. LT-splitting) in the present model. Since the analysis of the linear response shows no contribution of the light hole excitons in the observed energy region, we consider the contributions from the heavy hole excitons alone. The eigenfunctions and the eigenenergies of quantized CM levels can be written

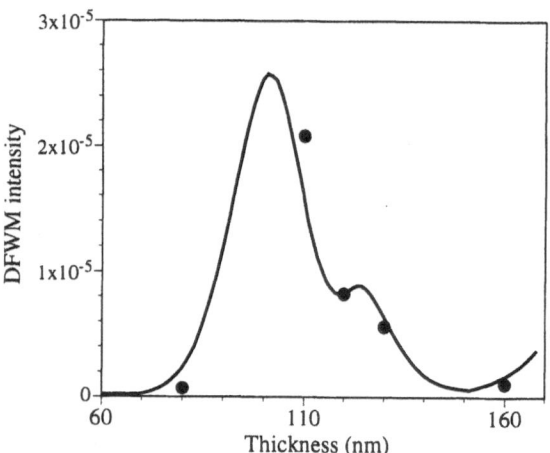

Fig. 2 The size dependence of the peak value of DFWM signal. Closed circles are measured values in arbitrary unit. Solid line shows calculated values normalized by the incident probe intensity.

as $\Phi_{K_{||},K_n}(\mathbf{R}) = \sqrt{1/S} \exp i K_{||} \mathbf{R}_{||} \sqrt{2/d} \sin K_n Z$ and $E_{K_{||},n} = \hbar\omega_T + \hbar^2(K_{||}^2 + K_n^2)/2M$, respectively, where S is the area of the film surface, d is the film thickness, $\mathbf{R}_{||}$ and Z are the coordinates parallel and perpendicular to the film surface, respectively, $\mathbf{K}_{||}$ and K_n are the corresponding wavevectors, $\hbar\omega_T$ is the exciton energy at the bottom of band and M is the total mass. The allowed values of K_n are $n\pi/d(n = 1, 2, 3, ...)$. The linear response analysis in the previous section has been done with this model, applying the Additional Boundary Condition (ABC)-theory to the p-polarization incidence[5].

For the third order nonlinear response, we also consider the CM motion of the biexciton states. Assuming the normal (Z-direction) incidence, we calculate the third order polarization with the above bases, and obtain the DFWM signal by means of the nonlocal theory for the nonlinear response[6].

The solid line in Fig. 2 shows the calculated intensity of the DFWM signal, where the peak value in the spectrum is indicated for each thickness. The result well agrees with the measured thickness dependence.

4 Discussions

The key to understand the above signal enhancement is the behavior of the internal field. We define the quantities $F_{K_n} = \int_0^d \sqrt{2/d} \sin(K_n Z) E(Z) dZ$, where $E(Z)$ is the electric field. The nonlocal third order polarization can be expressed with this quantities, and the expression of the mainly contributing part is

$$\sum_{n,m} \sin(K_n Z) A(K_n, K_m, \omega) F_{K_n}^{pump} F_{K_m}^{pump} F_{K_m}^{probe*}, \quad (1)$$

where the factor $A(K_n, K_m, \omega)$ includes the resonance poles of transitions between the ground and one-exciton states. (By the examination of the individual terms, we found that the signals are mainly from the one-exciton resonance.) The meaning of $\{F_{K_n}\}$ is the amplitude of each component of the internal field when we expand

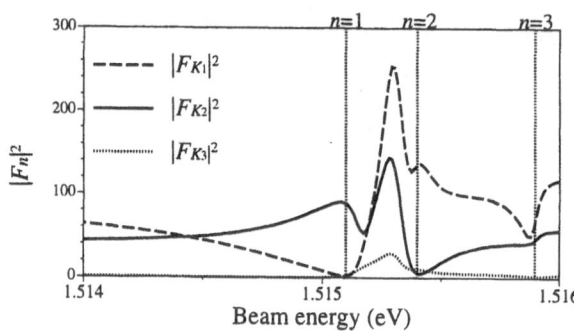

Fig. 3 Energy dependence of $| F_{K_n}^i |^2$ for d =110 nm. The vertical dotted lines indicate the quantized excitonic levels

$E(Z)$ with the bases of the function $\sqrt{2/d} \sin(K_n Z)$. Thus, F_{K_2}, for example, means the component with the quadrupole type spatial structure (having one node in the surface normal direction). From eq.(1), we understand that the transition to the n-th quantized level is strengthened if F_{K_n} has a large value near the resonance pole of the n-th level. Evaluating the individual terms in (1), we found that the large nonlinear signal mainly comes from the enhancement of F_{K_2}. Fig.3 explains this situation, showing the energy dependence of $| F_{K_n} |^2$ for d=110 nm. We can see that $| F_{K_2} |^2$ has a sharp peak very near the n=2 exciton state, which is prominent particularly for this thickness. This means that the transition to this level is strengthened by the internal field. On the other hand, the separation between the peak of $| F_{K_1} |^2$ and the lowest one-exciton level is comparatively large. Because of this off-resonance, the enhancement of $| F_{K_1} |^2$ is not very effective for the present signal enhancement.

It should be remarked that the spatial pattern of F_{K_2} component has one node in the surface normal direction, and the relevant transition is forbidden if the internal field is uniform in the sample. Therefore, we can say that the present enhancement effect is based on a new type of mechanism beyond the LWA as predicted in [2], which can be elucidated only within the nonlocal framework.

Acknowledgments

The theoretical part of this work was supported in part by Grants-in-Aid for COE Research (10CE2004) of the Ministry of Education, Science, Sports and Culture of Japan. The experimental part was performed under the management of FESTA supported by NEDO.

References

1. For example, the articles in "Excitons in Confined Systems", ed. by R. Del Sole, A. D'Andrea, and A. Lapiccirella, (Springer Proc. Phys. vol.25, 1988).
2. H. Ishihara and K. Cho: Phys. Rev. B**53**, (1996) 15823
3. K. Akiyama, N. Tomita, Y. Nomura and T. Isu: Appl. Phys. Lett. **75**, (1999) 475
4. K. Cho, A. D'Andrea, R. Del Sole and H. Ishihara: J. Phys. Soc. Jpn **59** (1990) 1853
5. K. Cho: J. Phys. Soc. Jpn. **55** (1986) 4113 .
6. H. Ishihara and K. Cho: Phys. Rev. B**48** (1993) 7960.

Recombination mechanism of anti-Stokes photoluminescence in partially ordered GaInP-GaAs heterostructure

S. J. Xu[1], Q. Li[1], H. Wang[1*], M. H. Xie[1], S. Y. Tong[1], J.-R. Dong[2]

[1] Department of Physics, The University of Hong Kong, Pokfulam Road, Hong Kong, China. e-mail: sjxu@hkucc.hku.hk
[2] Institute of Materials Research and Engineering, National University of Singapore, Singapore 119260, Singapore.

Abstract   In this article, we focus on recombination mechanism of anti-Stokes photoluminescence (ASPL) at partially ordered $GaInP_2$-GaAs interface. We find that the spectral features, i.e., energy position, lineshape, linewidth, and thermal activation energy, of the ASPL are totally different from that of the normal excitonic PL. Our results unambiguously demonstrate that the ASPL has different recombination channels from the normal PL. Our results also confirm that how to prevent the up-converted carriers captured by the interface localized states from thermally going back to the low band gap semiconductor is essential for the ASPL.

1 Introduction

The efficient ASPL at partially ordered $GaInP_2$-GaAs interface was first observed by Driessen [1] and later confirmed by other groups [2–5]. The most studies focus on the energy gain mechanism of the ASPL, that is, how the up-converted carriers get energy from the thermal source. For the $GaInP_2$-GaAs heterostructures, two types of energy gain mechanisms, namely, the cold Auger [1] and the two-step two photon absorption [2–4], were proposed. An easily neglected but essentially important question is whether the ASPL has the same recombination channels as the normal PL. Here, we are trying to find the answer to the question.

2 Experiments

The $GaInP_2$ films used in this study were grown on GaAs (001) substrates by metalorganic vapor phase epitaxy. The thickness of the film is 1 μm. In the PL measurements, the sample was mounted on the cold finger of a Janis closed cycle cryostat with varying temperature from 3.5 K to 300 K. The emission signal from the sample was dispersed by a Spex 750M monochromator and was detected with a Peltier-cooled Hamamatsu R928 photomultiplier. Standard lock-in amplification technique was employed to separate the emission signal from the background noise.

3 Results and Discussions

The left figure in Fig.1 shows the measured near band-edge transitions in GaAs substrate at 3.5 K under excitation of two laser lines with different wavelengths but equal powers. The right figure shows the normal PL

* On leave from National Laboratory for Superlattices and Microstructures, Institute of Semiconductors, Chinese Academy of Sciences, Beijing 100083, China

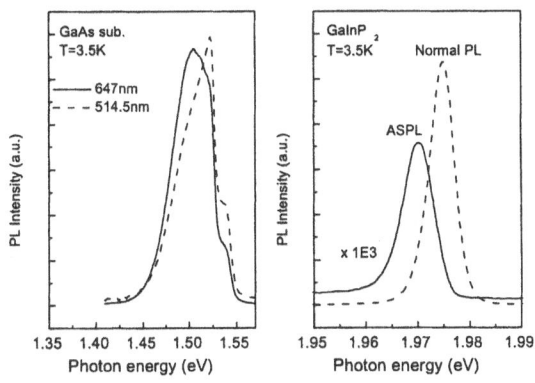

Fig. 1 The left figure: the 3.5 K PL spectra from GaAs excited by the 647 nm (1.916 eV) line and the 514.5 nm (2.41 eV) line. The right figure: the anti-Stokes PL and normal PL from the $GaInP_2$ epilayer. Note that for both the two laser lines (647 nm and 514.5 nm) the power of the laser beams are kept the same.

(dash line) and the ASPL (solid line) of $GaInP_2$ layer at 3.5 K. The ordering parameter η of the sample investigated here is calculated to be about 0.25 with the formula in the literature [6]. The linewidth of the normal excitonic PL shown in the right figure in Fig. 1 is 5.7 meV. These values are consistent with those recently reported by Zhang et al.[7] The linewidth of the ASPL line is 8.0 meV, which is wider than that of the normal excitonic PL line. The more interesting phenomenon is that the ASPL line has a ~ 5 meV red-shift with respect to the normal PL line. Furthermore, the ASPL line and the normal excitonic line possess different lineshape. The normal PL line exhibits a typical Lorentzian symmetric lineshape while the ASPL line has an asymmetric lineshape. The intensity of the ASPL is weaker by 3 orders of magnitude than the normal excitonic line. The former three differences between the ASPL and the normal PL definitely indicate the recombination mechanism difference between the two types of luminescence processes. In other words, for two types of luminescence processes, their recombination paths are different.

The blueshift of the ASPL peak with increasing temperature is a typical behavior of localized states related radiative recombination. The emission intensity of the ASPL at low-energy side decreases with increasing tem-

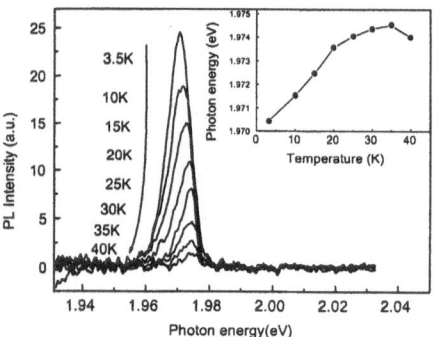

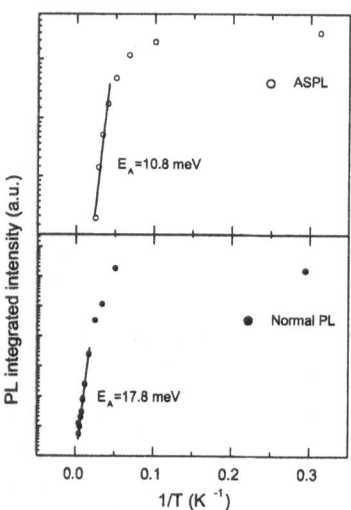

Fig. 2 The anti-Stokes PL recorded at different temperatures. The inset figure shows temperature dependence of the peak position of the ASPL.

Fig. 3 Integrated intensity of the ASPL and the normal PL versus 1/T.

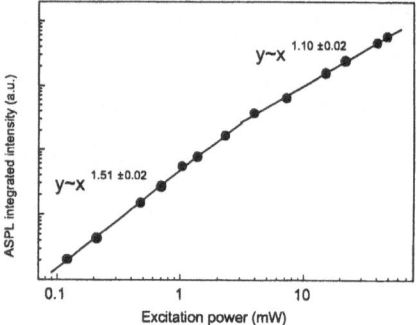

Fig. 4 The integrated intensity of ASPL versus excitation power. The temperature is 3.5K.

thermal activation energies are 10.8 meV and 17.8 meV for the ASPL line and the normal PL line, respectively.

The integrated intensity of the ASPL depends on the excitation intensity in a power function (Fig.4), that is, $I \propto P^a$. The value of a is 1.51 ± 0.02 at low excitation intensity and 1.10 ± 0.02 at high excitation intensity. However, a definite conclusion on the energy gain mechanism is still not given only according to the excitation intensity dependence of the ASPL, because the ASPL depends not only on the energy gain mechanism but also on the radiative recombination mechanism. As a localized states related PL, the ASPL also may exhibit the special excitation-power dependence like that shown in Fig.4 due to the limited localized states.

4 Conclusion

The efficient anti-Stokes PL is observed at the partially ordered $GaInP_2$-GaAs interface at low temperatures. We definitely demonstrate that the ASPL and normal PL have different recombination channels. It is found that the up-converted carriers recombine radiatively via the interface localized states giving the anti-Stokes PL. How to prevent the up-converted carriers from thermal escaping from the localized states is a key factor to obtain the ASPL, in particular, at higher temperatures.

References

1. F. A. J. M. Driessen, Appl. Phys. Lett. **67**, (1995) 2183.
2. Z. P. Su, K. L. Teo, P. Y. Yu, and K. Uchida, Solid State Communi. **99**, (1996) 933.
3. J. Zeman, G. Martinez, P. Y. Yu, and K. Uchida, Phys. Rev. B **55**, (1997) R13428.
4. K. Yamashita, T. Kita, and T. Nishino, J. Appl. Phys. **84**, (1998) 359.
5. W. Heimbrodt, M. Happ, and F. Henneberger, Phys. Rev. B **60**, (1999) R16326.
6. P. Ernst, C. Geng, F. Scholz, H. Schweizer, Y. Zhang, and A. Mascarenhas, Appl. Phys. Lett. **67**, (1995) 2347.
7. Y. Zhang, A. Mascarenhas, S. Smith, J. F. Geisz, J. M. Olson, and M. Hanna, Phys. Rev. B **61**, (2000) 9910.

perature more rapidly than that of the ASPL at high-energy side so that the ASPL peak becomes more asymmetric and narrowing. This can be explained as a result of the carriers thermal transfer from the localized states at lower energy to the localized states at higher energy. The final thermal escape of the carriers captured by the localized states at higher energy results in quenching of the ASPL.

The intensity of the ASPL peak strongly depends on temperature and almost completely quenches at ~ 50 K. Moreover, the thermal quenching of the ASPL line is accompanied with an about 4 meV blueshift of the peak position. The normal PL line remains observable up to room temperature. The logarithmic integrated-intensity of the two lines versus 1/T is plotted in Fig.3. The fitted

Proc. 25th Int. Conf. Phys. Semicond., Osaka 2000 (Eds. N. Miura and T. Ando)

597

Spontaneous Raman scattering in GaP-AlGaP heterostructure waveguides

T. Saito[1], K. Suto[1,2], M. Kawasaki[1], T. Kimura[2], A. Watanabe[2], and J. Nishizawa[2,3]

[1] Department of Material Science and Engineering, Tohoku University, Aramaki, Aoba02, Sendai 980-8579, Japan
  E-mail: sutoken@argon.material.tohoku.ac.jp
[2] Telecommunication Advancements Organization, Nagamati-koegi 19, Aoba, Sendai, 980, Japan
[3] Semiconductor Research Institute, Kawauti, Aoba, Sendai, 980-0862, Japan

Abstract   Spontaneous Raman scattering in micron order heterostructure GaP waveguides were investigated for [100], [110], and [11-2] directional waveguides. The backward Raman scattering shows the same result as expected from GaP bulk crystal. However, the forward Raman scattering shows the different modes from the backward scattering. This unexpected mode in the forward scattering is considered as the rotation of phonon wavevector, because of micron order cross section dimensions of GaP waveguides.

1 Introduction

Recently, stimulated Raman scattering in waveguides with micron order cross section has been used for fiber Raman amplifiers, semiconductor Raman lasers and amplifiers.   [1,2] They have been used or hoped for key components of future dense optical communications.

Semiconductor Raman lasers and amplifiers use the Raman effect in micron order waveguides, however, spontaneous and stimulated Raman scattering in optical waveguides have been little reported [3] The analysis of the Raman effect in waveguide will lead to improvement of the Raman scattering efficiency.

In this paper, we study the spontaneous Raman scattering in GaP waveguides surrounded by AlGaP cladding formed along different crystal directions.

2 Experiment

The GaP waveguides were fabricated with liquid phase epitaxial method, lithography, and reactive ion etching. The substrates were (001) surface for [100] and [110], (111)$_A$ for [11-2] directional waveguides. Anti-reflection coatings are evaporated on both facets of a waveguide to reduce the resonance effect in waveguide. The details are described in other paper. [4]

A Ti-Sapphire laser was used for the pump light with 100~200 mW power and a wavelength of 824 nm. Backward scattering mode means that the pump light and the scattering light propagate in the opposite directions, and forward scattering mode means that the two lights propagate in the same direction. The scattering lights are detected by GaAs photo-multiplier through a double-monochromator.

3 Result and discussion

The dimensions for each waveguide are shown in Table 1, together with the allowed Raman scattering modes for bulk crystals, [5] which come from the Raman selection rule. [6]

Figure 1(a) shows the Raman scattering in a [100] directional waveguide. For backward scattering, the result of Raman scattering indicates the same result as bulk GaP crystals, which are determined by the Raman selection rule. High intensity and sharp LO phonon mode is observed at 402 cm^{-1}. On the contrary, for forward scattering, the result shows a broad TO phonon mode at 365 cm^{-1}. This difference is also observed for [110] and [11-2] directional waveguides. Figure 1-(b) and -(c) shows the result for [110] and [11-2] directional waveguides. Although, the result of backward scatterings for [110] and [11-2] waveguides are just the expected results from the apparent Raman selection rule for a bulk crystal, the result of forward scattering shows the unexpected results, as well as for the [100] waveguide. This unexpected phonon modes in forward scatterings is considered as follows.

In GaP (refractive index is 3.16), the wavevector of Raman interacting LO phonons are estimated to be $\lambda_{LO} \approx 0.13$ μm for backward scattering and $\lambda_{LO} \approx 7.67$ μm for forward scattering, respectively. We have defined a factor η_z as follows,

$$\eta_z = \frac{q_z^2}{q_x^2 + q_y^2 + q_z^2} \approx \frac{\lambda_{LO}^{-2}}{d^{-2} + w^{-2} + \lambda_{LO}^{-2}} \quad (1)$$

where $q_{x,y,z}$ are the phonon wavevector components in x, y, z directions. Here, z is parallel to the waveguide direction and x is parallel to the substrate surface direction. The thickness d and width w are the cross-sectional dimensions of a waveguide in x and y directions, respectively. When we consider the case for a [100] waveguide, which has width $w = 3.8$ μm,

Direction	l (mm)	t (μm)	w (μm)	Coupling mode
[100]	6.0	1.1	3.8	LO
[110]	6.0	1.0	2.4	TO
[11-2]	5.2	2.0	3.4	TO

Table 1 Parameters of wach waveguide

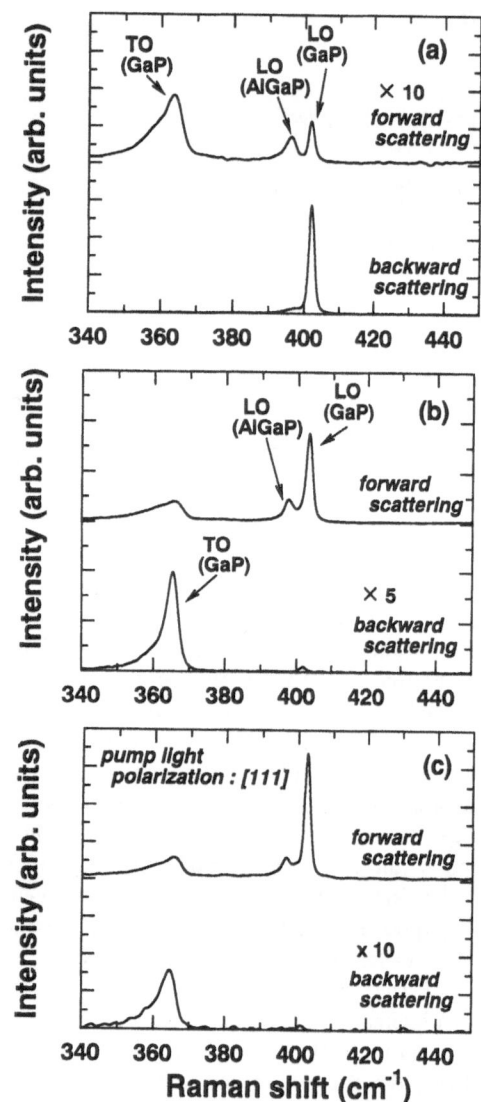

Fig.1 Raman scattering results for (a) [100], (b) [110],

(c) [11-2] directional waveguides.

The other waveguides [110] and [11-2] for forward scattering are also explained by the above equation and the same discussion. And also, the theoretical calculations of LO phonon and TO phonon intensity ratio, which comes from the above equation, are in good agreement with the intensity ratios from experiments.

4 Conclusion

Raman scatterings in micron order GaP waveguides were investigated both for forward and backward scattering modes for the first time. We found that the unexpected modes were observed for forward scattering modes in all [100], [110] and [11-2] directional waveguides. It was found that this result came from a rotation of wavevector of phonons depending on a size of micron order waveguides. It indicates that we can control the interacting mode with LO and TO phonons by changing the dimensions and directions of waveguides.

References

J. Nishizawa, Denshi Kagaku **14**, (1963)

2 T. Saito, K. Suto, T. Kimura, A. Watanabe, and J. Nishizawa, J. Appl. Phys. **87**, (2000) 3399

3. T. Saito, K. Suto, T. Kimura, A. Watanabe, and J. Nishizawa, J. Luminescence **87-89**, (2000) 883

4. K. Suto, T. Kimura, and J. Nishizawa, International J. Infrared and Millimeter Waves **16**, (1995) 691

5. C. H. Henry and J. J. Hopfield, Phy. Rev. Letter **15**, (1965) 964

6. R. Loudon, adv. Phys. **13**, (1964) 423.

thickness $d = 1.1$ μm, and length $l = 6.0$ mm,

$d < w < \lambda_{LO}$ is satisfied for forward scattering. So that $\eta_z \approx 1.0$ is calculated for backward scattering, but, $\eta_z \approx 0.02$ and $\eta_x \approx 0.9$ for forward scattering. Therefore, the wavevector of LO phonons changes almost parallel to x direction. It should be reminded that TO phonon frequency appears when ξ is perpendicular to the wavevector q. This fact means that, ξ parallel to the waveguide direction changes its frequency from the LO to the TO mode. The experimental result shown in Fig.1(a) reveals a main TO phonon spectral line, together with residual low intensity LO phonon lines.

Generation Mechanisms of coherent phonons in Quantum wells Revealed

K.J. Yee[1], Y.S. Lim[2], D.S. Kim[1]

[1] Department of Physics, Seoul National University, Seoul 151-742, Korea e-mail: denny@phya.snu.ac.kr
[2] Department of Applied Physics, Konkuk University, Chungju, Chungbook 380-701, Korea

Abstract   Generation mechanisms of coherent LO phonons in GaAs quantum wells are revealed. With the excitation energy near the exciton resonance, the dominant generation mechanism in quasi-2D quantum well appears to be the impulsive Raman scattering. This is in stark contrast with bulk GaAs, where the generation mechanism is known to be the screening of the surface field by photo-generated carriers.

1 Introduction

With the invention of ultrashort pulse lasers, generating and detecting coherent lattice vibrations in various materials became possible [1–4]. Coherent phonons are generated by exciting materials with ultrashort pulse laser whose pulse width is smaller than the period of the generated phonon oscillations.

One intriguing problem in this area is the generation mechanism and dephasing processes of coherent phonons. There are several generation mechanisms suggested, to explain the symmetry of the excited modes and the polarization consideration of exciting photons. In general, impulsive stimulated Raman scattering (ISRS) is known as the generation mechanism in transparent materials. Other generation mechanisms were proposed in opaque materials. The DECP (displacive excitation of coherent phonons) model in semimetals such as Bi or Te was suggested to explain the selective excitation of the A_1 symmetry mode [2]. In bulk GaAs, which is a polar material, coherent LO phonons are known to be generated by the ultrafast screening of the surface space-charge fields [5]. Although the generation mechanism in bulk GaAs is different from the DECP model, the two mechanisms are both displacive excitation of coherent phonons.

Up to now, there have been intensive studies on coherent LO phonons in bulk GaAs, and generation mechanism is more or less agreed upon. However, only a few studies were done about the coherent LO phonon oscillations in GaAs/AlGaAs quantum wells compared with bulk GaAs, and little study was done especially for the generation mechanism. In quantum well structures, the screening process along the growth direction is not so easy as in bulk GaAs due to the confinement effect of carriers in each well region [6]. In this paper, we study the generation mechanism of coherent LO phonons in quasi-two dimensional GaAs/AlGaAs quantum well structures. We find that the dominant generation mechanism with the excitation energy near the excitonic resonance is the ISRS mechanism for a MQW structure in contrast to the case of bulk GaAs. We find that the amplitude and phase of coherent LO phonons excited in multiple quantum wells (MQW's) near the excitonic resonance is consistent with the selection rules of the ISRS mechanism.

2 Experimental

Using femtosecond laser pulses, we performed the reflective electro-optic sampling technique (REOS) on bulk GaAs and a GaAs/Al$_{0.36}$Ga$_{0.64}$As MQW at the sample temperature of 12 K. REOS is a pump-probe technique with a polarization sensitive analysis of the probe pulse based on the electro-optic effect (Pockels effect). Changes of the probe polarization induced by electric fields parallel to the growth direction of GaAs materials can thus be obtained with high sensitivity. A Kerr-lens mode-locked ti:sapphire laser was used to generate transformed-limited pulses of about 50 fs. The center wavelength was tuned slightly above the lowest exciton resonance. All samples were grown on (001)-GaAs substrates by MBE machines, and the multiple quantum well (MQW) sample has 30 period of 15 nm well width and 5 nm barrier.

With the polarization of probe beam fixed parallel to one of principal axes ([100], [010]), the pump beam polarization was rotated using a half wave plate to characterize the polarization dependence of the coherent LO phonon oscillations.

3 Results and Discussion

Figure 1 shows the coherent phonon oscillations at different pump-beam polarizations in our MQW sample (Fig. 1(a)) and a bulk GaAs (Fig. 1(b)), which were obtained after subtracting low frequency in the REOS signal. Here the 'minimum' polarization in GaAs guantum well corresponds to the angle of minimum phonon oscillation. For GaAs quantum wells, drastic changes with pump-beam polarization in both the amplitude and the phase of the oscillations are observed. In addition, noticeable asymmetry in the oscillation amplitude is observed between the [110] and the [1$\bar{1}$0] directions. These behaviors show stark contrasts with the results for bulk GaAs where little polarization dependencies are observed. The angle dependence for the GaAs quantum well of the Fourier transformed intensity shows the behaviors more clearly as is shown in Fig. 2. While the overall angle dependence is sinusoidal, there is almost a factor of 2 difference between the two peaks at the [110] and the [1$\bar{1}$0] directions.

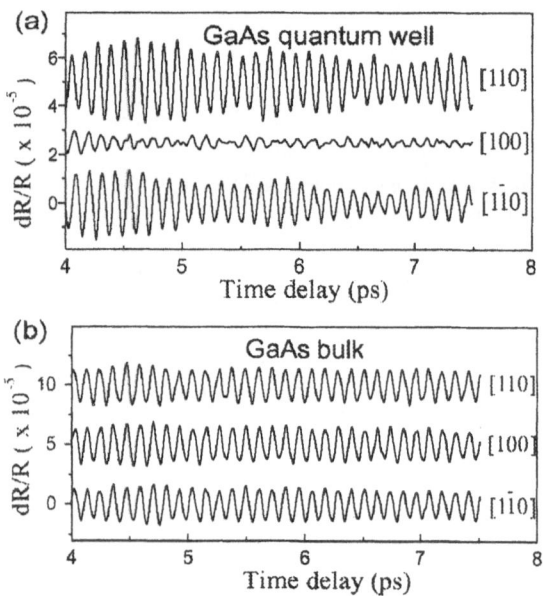

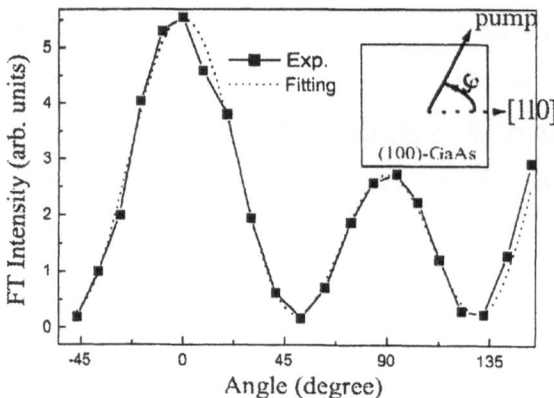

Fig. 1 The REOS signals in time domain obtained for a GaAs/AlGaAs MQW(a), and a bulk GaAs(b) at pump beam polarizations of [110], [100], [1$\bar{1}$0] and at 'minimum' at which the phonon oscillation is minimum.

All of our observations can be understood if we introduce both the allowed and forbidden Raman scattering tensors as well as their interference as the generation mechanisms of coherent phonons [7]. The "allowed" deformation potential Raman tensor is given as follows:

$$R_A = \begin{pmatrix} 0 & a \\ a & 0 \end{pmatrix} \qquad (1)$$

whereas the "forbidden", Fröhlich Raman tensor reads:

$$R_F = \begin{pmatrix} b & 0 \\ 0 & b \end{pmatrix} \qquad (2)$$

Where a is the polarizability for the allowed-Raman interaction, and b is the polarizability for the Fröhlich interaction. Noting that the two photons involved in the process of the impulsive stimulated Raman scattering (ISRS) have the same polarization, the angle dependence of the coherent phonon oscillation intensity is given as follows:

$$\frac{dS}{d\Omega} \propto |b + a \times \cos(2\varphi)|^2. \qquad (3)$$

This equation fits Fig. 2 rather well, as shown in dotted lines with $b/a = 0.25$. From the intensities at the two peaks, it is revealed that the two Raman amplitudes a and b interfere constructively in the [110] direction and destructively in the [1$\bar{1}$0] direction. Thus, we have shown that with the excitation energy near the exciton resonance, the the coherent phonon generation in the GaAs MQW with L_w=15 nm, and L_b=5 nm is dominated by the ISRS mechanism, in sharp contrast to that screening mechanism is more dominant for bulk GaAs.

Fig. 2 FFT intensity of phonon oscillations for the GaAs quantum well as a function of the pump polarization angle φ which is defined in inset, the dotted line is the fitting to the data.

4 Acknowledgement

This work was supported by MOST (the National Research Laboratory Program, Nanostructure Technology Project) and KOSEF (the Center for Strongly Correlated Materials Research).

References

1. G. C. Cho, W. Kütt, and H. Kurz, Phys. Rev. Lett. **65**, (1990) 764.
2. T. K. Cheng, S. D. Brorson, A. S. Kazeroonian, J. S. Moodera, G. Dresselhaus, M. S. Dresselhaus, and E. P. Ippen, Appl. Phya. Lett. **57**, (1990) 1004.
3. T. Dekorsy, G.C. Cho, and H. Kurz, *Light Scattering in Solids VIII*, edited by M. Cardona and G. Guintherodt (Springer, Berlin, 2000).
4. R. Merlin, Solid State Commun. **102**, (1997) 207.
5. T. Pfeifer, T. Dekorsy, W. Kütt, and H, Kurz, Appl. Phys. A: Solids Surf. **55**, (1992) 482.
6. K. J. Yee, D. S. Yee, D. S. Kim, T. Dekorsy, G. C. Cho, And Y. S. Lim, Phys. Rev. B **60**, (1999) R8513.
7. Jose Menendez and Manuel Cardona, Phys. Rev. Lett. **51**, (1983) 1297.

Probing and controlling spin-relaxation in GaAs quantum wells

Y. Ohno, R. Terauchi, T. Adachi, F. Matsukura, and H. Ohno

Laboratory for Electronic Intelligent Systems, Research Institute of Electrical Communication, Tohoku University, 2-1-1 Katahira, Aoba-ku, Sendai 980-8577, Japan

Abstract We investigated electron spin relaxation mechanisms in GaAs / AlGaAs quantum well (QWs) grown on (100) and (110)-oriented substrates with or without n-doping in order to control electron spin relaxation time τ_s in this system. In GaAs/AlGaAs (110) QWs, a predominant spin scattering mechanism [D'yakonov-Perel' (DP) mechanism] for conventional (100) QWs at higher temperatures, is substantially suppressed and τ_s in undoped (110) QWs was found to reach nanosecond order even at room temperature. The mechanism responsible for the spin relaxation in (110) QWs was examined by studying the quantized energy, electron doping and its mobility, and temperature dependences of τ_s. The results suggest that in the absence of DP interaction, electron-hole exchange interaction plays a significant role on the electron spin relaxation in a wide temperature range.

1 Introduction

Electron spin dynamics in semiconductor nanostructures has attracted much attention from both viewpoints of physics and applications, and understanding of the spin relaxation mechanisms is of great importance for the practical use of spins as information processing, transfer and storage. A number of time-resolved optical experiments have been carried out in order to elucidate the spin relaxation processes in bulk semiconductors and their heterostructures. For instance, extremely long ($\sim$ 100 ns) electron spin relaxation time τ_s in bulk n-type GaAs at 5 K[1], and $\tau_s \sim$ 1.5 ns in n-type ZnCdSe/ZnSe quantum wells (QWs) at 275 K[2] have been reported. The mechanism which limits τ_s in n-GaAs at low temperatures ($\sim$ 5 K) is considered to be due to the electron-hole exchange, since τ_s becomes longer when electrons and holes are spatially separated by applying external electric fields[3]. In the case of undoped and n-doped GaAs / AlGaAs (100) QWs, typical τ_s at high temperatures ($\sim$room temperature (RT)) is at most several tens of picoseconds. Recent experimental studies on the electron quantization energy (E_1) and mobility μ dependence of τ_s at RT offered that the D'yakonov-Perel' (DP) interaction is the most effective spin relaxation mechanism in undoped and n-doped GaAs/AlGaAs (100) QWs at RT[4,5], while the importance of the excitonic electron-hole exchange interaction has been stressed[6].

In the DP mechanism, the lack of inversion symmetry of zinc-blende structure results in spin precession due to spin-splitting of the conduction band via spin-orbit coupling, leading to spin relaxation[7]. In two dimensional (2D) systems, this process is expected to have strong dependence on the growth axis, although in most ex-

periments QWs formed on a (100) plane have been investigated. Here we report our study on τ_s in undoped and n-doped GaAs / AlGaAs QWs grown on (100) and on (110) substrates. We first show that the temperature dependence of τ_s in n-doped (100) QWs above 30 K is in agreement with the DP theory[5]. Then we show that τ_s in undoped (110) QWs reaches $\sim$ 2 ns at RT, about 30 times longer than that of similarly prepared (100) QWs[8]. This long τ_s is explained in terms of anisotropic dependence of the DP mechanism on the growth axis: the contribution from the DP interaction is expected to be suppressed for electron spins polarized perpendicular to the (110) plane. To elucidate the remaining spin relaxation mechanism in (110) QWs, we examined the dependence of τ_s on characteristic parameters and compared it with the results of (100) QWs. It is found that 2D electrons in modulation-doped (110) QWs have quite long τ_s in a wide temperature range.

2 Experimental

For a systematic study, we prepared several GaAs / AlGaAs QW structures grown on semi-insulating (100) and (110)-just GaAs substrates by molecular beam epitaxy. All the samples consist of 60 periods of GaAs QWs separated by 10-12 nm-thick $Al_{0.4}Ga_{0.6}As$ barriers. Undoped samples with different well widths L_W were used to study the electron quantized energy E_1 dependence of τ_s, while doped QW samples (with different doping concentration) and modulation-doped ones were prepared in order to investigate the effect of doping as well as the mobility μ dependence of τ_s.

We employed a degenerate circularly-polarized pump-probe transmission technique to measure τ_s, using an ultrashort optical pulses ($\sim$ 110 fs duration) generated by a mode-locked Ti:Al$_2$O$_3$ laser and tuned at heavy hole exciton resonance. τ_s was calculated from $\tau_s = 2\tau$, where τ is the decay of the polarization P ($P = P_0 \exp(-t/\tau)$). The pump power I_{pump} was $\sim$ 2 mW, except for the study on I_{pump} dependence of τ_s. The temperature (T) dependence of τ_s was obtained in a wide range of 5 $\sim$ 300 K with the sample set in a liquid He cryostat.

3 Results and Discussions

3.1 GaAs/AlGaAs (100) QWs

We first examine the spin relaxation mechanism in n-doped GaAs / AlGaAs QWs grown on (100) substrates. Figure 1 shows the temperature dependence of τ_s in n-doped (100) QWs with L_W = 7.5 nm and the doping

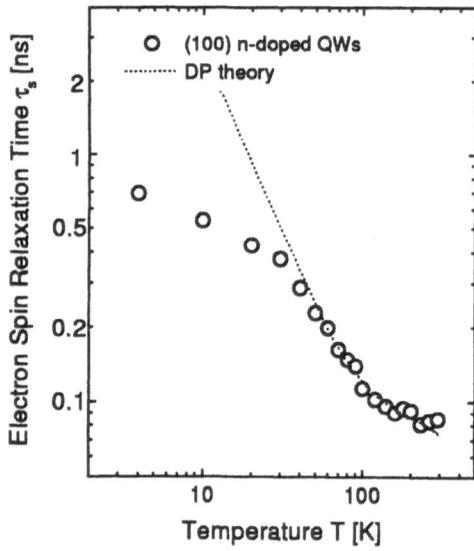

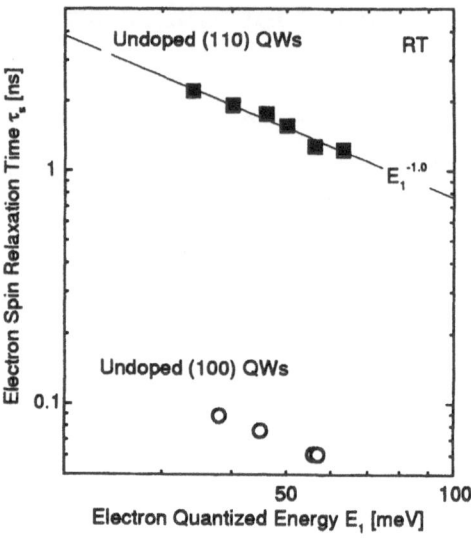

Fig. 1 Electron spin relaxation time τ_s of n-doped (100) QWs ($L_W = 7.5$ nm, $n = 4\times10^{10}$ cm^{-2} in each QW) is plotted as a function of the temperature T. A dotted line is the calculated result of τ_s based on the DP theory.

Fig. 2 Plots of spin relaxation time τ_s vs. electron quantized energy E_1 in (110) QWs are indicated by solid squares, and in (100) QWs by open circles, respectively.

density $n = 4 \times 10^{10}$ cm^{-2} per QW. As T decreases, τ_s increases down to $T \sim 30$ K. Below 30 K, on the other hand, τ_s tends to saturate. The dotted line in Fig. 1 indicates the calculated τ_s from the DP theory[7] which is fitted to the experimental data by multiplying a prefactor of ~ 10 with the calculated values, as described below. The DP theory predicts $\tau_s \propto E_1^{-2}T^1\tau_p^{-1}$, where τ_p is the electron momentum relaxation time. In the calculation, we assumed that $\tau_p = (m^*/e)\mu$, where m^* is the effective mass of electrons, and the T dependence of τ_p was taken into account by using μ at each T evaluated by Hall measurements. When multiplied by a factor of 10, the calculated results reproduce the experimental data in the range of T from RT down to ~ 30 K[9]. Although it is not fully understood at the moment, the constant factor of 10 might have come from a prefactor for τ_p which depends on the momentum relaxation mechanism[7]. When $T < 30$ K, the T dependence of τ_s deviates from the DP prediction, indicating that other spin relaxation mechanisms become dominant. In undoped (100) QWs, the spin relaxation in this temperature regime is considered to be due to the excitonic electron-hole exchange interaction[10,11].

3.2 GaAs/AlGaAs (110) QWs

Next, we examine the dependence of electron spin relaxation rate on the growth axis of QWs by comparing τ_s of (100) and (110) QWs. Figure 2 shows τ_s of undoped (110) QWs (solid squares) and (100) QWs (open circles) with different L_W measured at RT. The param-

eter E_1 is determined from the energy of the heavy-hole exciton absorption. A solid line in Fig. 2 indicates the least square fit for the data of (110) QWs which results in $\tau_s \propto E_1^{-1.0}$. The difference between the (100) case and the (110) case is dramatic; τ_s of (110) QWs with $L_W = 7.5$ nm ($E_1 \sim 47$ meV) reaches ~ 2 ns, whereas with the same E_1, τ_s was ~ 70 ps for (100) QWs. The 30-fold increase of τ_s can be attributed to the fact that the DP mechanism dominant in GaAs/AlGaAs (100) QWs is quite anisotropic and can be substantially suppressed when the growth axis and the spin polarization axis are taken along [110] direction[7].

In order to investigate the mechanism responsible for the slow spin relaxation in (110) QWs, we measured τ_s of undoped (110) QWs ($L_W = 7.5$ nm) at $T = 5 \sim 200$ K. In Fig. 3, τ_s is plotted as a function of T by solid squares. Contrary to the (100) case, τ_s now decreases with decrease of T as $\tau_s \propto T^{0.6}$ in high temperature regime ($T > 20$ K), while it saturates below $T < 20$ K. ¿From this temperature dependence, we can rule out the possibility of the DP mechanism being the dominant spin relaxation mechanism above 20 K: the DP mechanism is no longer responsible in (110) QWs.

Now we consider other possible spin relaxation mechanisms: the band mixing effect (Elliott-Yafet (EY) mechanism[12]) and the electron-hole exchange interaction (Bir-Aronov-Pikus (BAP) mechanism[13]). The E_1 dependence of τ_s of (110) QWs (Fig. 2) can be explained by both mechanisms. The EY theory in two dimension predicts $\tau_s \propto E_1^{-1}$[14]. The spin scattering due to the electron-hole exchange interaction has been shown to be

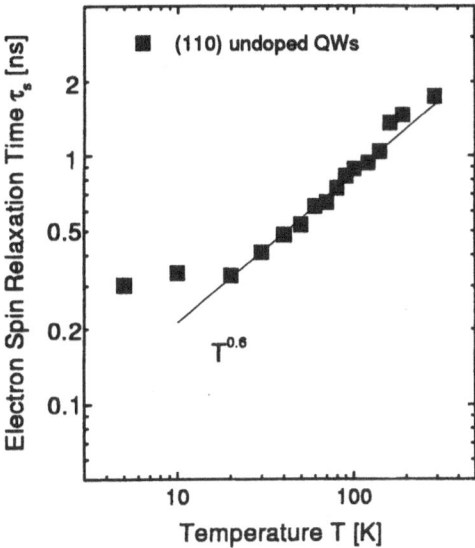

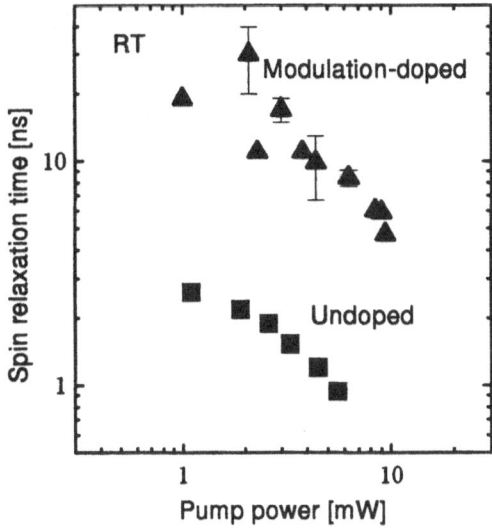

Fig. 4 The pump power dependence of the electron spin relaxation time τ_s of n-modulation doped and undoped GaAs/AlGaAs(110) QWs measured at RT.

Fig. 3 Electron spin relaxation time τ_s of undoped (110) QWs (solid squares) are plotted as a function of T. A solid line ($\tau_s \propto T^{0.6}$) is a guide for eyes.

enhanced in narrower QWs[6]. However, neither can account for the increase of τ_s with T (positive T dependence of τ_s). When $T < 20$ K, τ_s saturates at ~ 300 ps. Here, it is most probable that τ_s in undoped (110) QWs is limited by the same mechanism as that of the (100) case, i.e. by the excitonic exchange interaction[6, 10]: note that the results of (100) QWs shown in Fig. 1 are about the same order as that of Fig. 3 below 20 K. Since the DP interaction is absent, the exchange interaction may remain predominant in (110) QWs even above 20 K. Although the positive T dependence of τ_s cannot be explained within the framework of any simple exciton spin relaxation model, the electron-hole exchange interaction seems to be important, since the faster thermal ionization of excitons as T increases may result in weaker electron-hole exchange interaction and longer τ_s[2].

Next, we investigate the I_{pump} dependence of τ_s of undoped and n-modulation doped GaAs / AlGaAs (110) QWs[15]. We measured τ_s's of an undoped (110) QW sample ($L_W = 7.5$ nm) and a modulation-doped (110) QW sample ($L_W \sim 8$ nm, electron density $n \sim 1 \times 10^{11}$ cm^{-2} per each QW) at RT, and plotted them in Fig. 4 by solid squares and triangles, respectively. In the case of (110) QWs, it is found that τ_s decreases with the increase of I_{pump}, while τ_s does not depend on I_{pump} in (100) QWs (not shown). Moreover, doping of electrons is found to be quite effective for increasing τ_s, similar to the results reported in Refs.[1] and [2]: even at RT, τ_s of n-modulation doped (110) QWs exceeds 10 ns under weak excitation regime. Assuming that the electron-hole interaction is effective in (110) QWs at RT, these

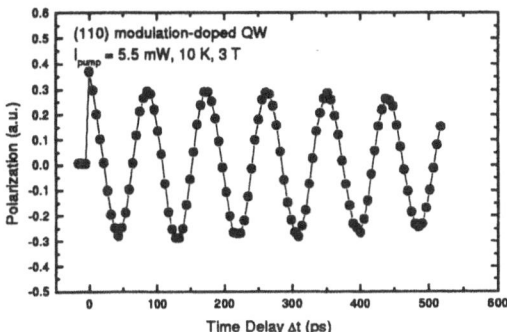

Fig. 5 Temporal evolution of the polarization of electrons in n-modulation doped (110) QWs in an in-plane magnetic field $B = 3$ T.

observations can be understood qualitatively: as I_{pump} increases, the electron-hole interaction is expected to be enhanced and thus τ_s becomes short since photo-excited hole density increases with I_{pump}. By doping electrons in QWs, on the other hand, the interaction might become weak due to the screening effect and/or shorter recombination lifetime of holes.

The temperature dependence of τ_s of n-modulation doped (110) QWs is also studied[15]. We found that, contrary to the case of undoped QWs, τ_s of n-modulation doped (110) QWs is almost independent of T: in the whole range of $T = 10 \sim 300$ K, τ_s remains at several nanoseconds. Figure 5 shows temporal evolution of the polarization when an in-plane magnetic field $B = 3$ T is applied. The oscillation of the polarization due to the precession is clearly observed: the transverse spin lifetime T_2^* is $\sim$ 6 ns at 10 K. Such a long-lived electron spin coherence in (110) QWs has turned out to be effective in controlling the nuclear polarization[16].

4 Conclusion

In conclusion, we investigated the spin relaxation mechanism in GaAs/AlGaAs (100) and (110) QWs. We demonstrated that the DP theory can explain the temperature and growth axis dependence of τ_s in GaAs/AlGaAs QWs. The temperature dependence $\tau_s \propto T^{0.6}$ observed in undoped (110) QWs at $T > 20$ K cannot be explained by any simple theoretical model, but electron-hole interaction is most probable to govern the electron spin relaxation in (110) QWs.

Acknowledgement

This work is partly supported by "Research for the Future" Program from the Japan Society for the Promotion of Science, and Grant-in Aid for Scientific Research (No. 09244103 and No. 11650002) from the Ministry of Education, Science, Sports, and Culture, Japan.

References

1. J.M. Kikkawa and D.D. Awschalom, Phys. Rev. Lett. **80**, (1998) 4313.
2. J.M. Kikkawa, et al., Science **277**, (1997) 1284.
3. D. Hagele et al., Appl. Phys. Lett. **73**, (1998) 1580; J.M. Kikkawa and D.D. Awschalom, Nature **397**, (1999) 139.
4. A. Tackeuchi et al., Appl. Phys. Lett. **68**, (1996) 797; R.S. Britton et al., Appl. Phys. Lett. **73**, (1998) 2140.
5. R. Terauchi et al., Jpn. J. Appl. Phys. **38**, (1999) 2549.
6. M.Z. Maialle et al., Phys. Rev. B **47**, (1993) 15776.
7. M.I. D'yakonov and V.Yu. Kachorovskii, Sov. Phys. Semicond. **20**, (1986) 110.
8. Y. Ohno et al., Phys. Rev. Lett. **83**, (1999) 4196.
9. Y. Ohno et al., Physica E, **6**, (2000) 817.
10. S. Adachi et al., J. Lumin., **72-74**, (1997) 307.
11. L. Munoz et al., Phys. Rev. B, **51**, (1995) 4247.
12. R.J. Elliott, Phys. Rev. **96**, (1954) 266.
13. G. Fishman and G. Lampel, Phys. Rev. B **16**, (1977) 820.
14. A. Tackeuchi et al., Jpn. J. Appl. Phys. **38**, (1999) 4680.
15. T. Adachi et al., to be published in Physica E.
16. G. Salis et al., cond-mat/0010139.

Speckle-averaged resonant Rayleigh scattering from quantum well excitons

G. R. Hayes[1], **B. Deveaud**[1], **V. Savona**[2], **S. Haacke**[3]

[1] Department of Physics, Institute of Micro- and Opto-electronics, Swiss Federal Institute of Technology, Lausanne-EPFL, Switzerland e-mail: `hayes@dpmail.epfl.ch`
[2] Institut für Physik der Humboldt-Universität zu Berlin, Hausvogteiplatz, 5-7, D-10117, Berlin, Germany
[3] Institute of Condensed Matter Physics, Lausanne University, CH-1015, Switzerland

Abstract The ultrafast resonant Rayleigh scattering from quantum well excitons has been measured using Fourier transform spectral interferometry. The experiment simultaneously measures a statistically significant number of speckles with one hundred femtosecond temporal resolution. We show that theories based on ensemble-averaging can be compared directly with the experimental resonant Rayleigh scattering dynamics.

The secondary radiation (SR) emitted after ultrafast resonant excitation of quantum well excitons is due to two sources. On an early timescale (<10 ps: the exact number is sample dependent) resonant Rayleigh scattering (RRS) is known to be the dominant mechanism. On a longer timescale incoherent photoluminescence (PL) provides an important contribution. Despite the experimental advances the ability to measure the time-resolved RRS (TR-RRS) unimpeded by PL, on a femtosecond timescale, has remained elusive [1-4]. In this article we present a novel technique, that of multiple-speckle spectral interferometry. This overcomes the limitations inherent in the other approaches. As a corollary we answer a highly controversial question namely, what is the correct theoretical treatment of resonant Rayleigh scattering?

Our experimental approach is that of Fourier transform spectral interferomety [5]. This field-correlation measurement is an excellent technique as it is only sensitive to the coherent RRS. The sample under investigation consists of twenty five GaAs quantum wells with well widths of 18 nm separated by 7 nm $Al_{0.4}Ga_{0.6}As$ barriers. It has an inhomogeneous linewidth of 0.6 meV and was held at 4K unless otherwise stated. We resonantly excited the sample at the lowest heavy-hole exciton transition using 120 fs transform-limited pulses from a mode-locked Ti:Sapphire laser. The spot size was 160 μm and therefore we calculate that the size of each speckle is about 2.2 mrad. The SR in the near-normal direction was collimated and overlapped with a reference pulse from the laser which arrived several picoseconds before the surface-scattered light. The signal was dispersed by a 1 m focal length spectrometer and detected by a CCD detector. The innovation in our experiment is to expand the reference beam using lenses. The enables us to simultaneously analyse a statistically significant number of speckles. Essentially the vertical axis of the CCD represents an angular spread in the SE of 63 mrad while the horizontal axis records the SE's spectral content. Addi-

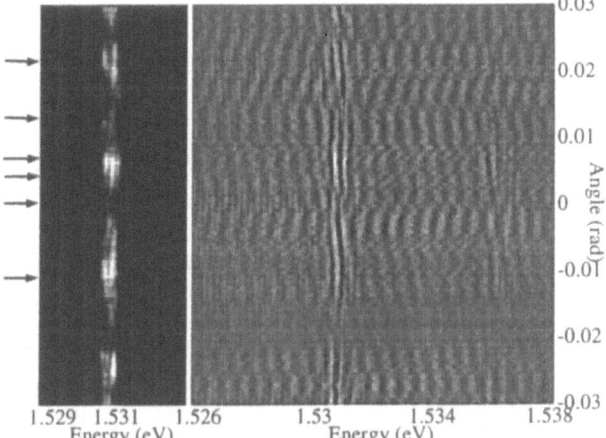

Fig. 1 (a) Time-integrated SR at the heavy hole exciton after resonant excitation. The excitation density is 1.4×10^8 excitons/cm^2. (b) Interferogram of the SR on an expanded energy scale. The arrows to the left of figure 1(a) indicate the angles at which the spectra plotted in figure 2 are taken.

tionally our setup is actively stabilised to less than $\lambda/20$ enabling measurements to be taken for several minutes with no noticeable loss of fringe contrast.

We show in figure 1(a) the time-integrated SR at an excitation density of 1.4×10^8 excitons/cm^2 and in (b) the corresponding spectral interferogram. The interferogram is obtained by first recording the spectrum resulting from the interference of the SR and the reference beam and then subtracting spectra of the SR with the reference beam blocked and the reference beam with the SR blocked. The spectra exhibit a complex and fine structure that varies from speckle to speckle. This manifests itself in the time domain as a speckle-dependent temporal response. Note that this experiment differs from a conventional PL setup in that it does not collect the light emitted over a large solid angle. In figure 1(b) fringes due to interference between the surface-scattered light from the sample and the reference beam are clearly observable below and above the heavy-hole exciton. Finer more pronounced fringes, implying a larger time-difference between the reference beam and the SR, can also be seen at the heavy-hole (HH) exciton energy. A weaker set of fringes is also apparent at the light-hole (LH) energy (1.5346 eV) The phase variations that occur across the interferogram are due to slight differences between the

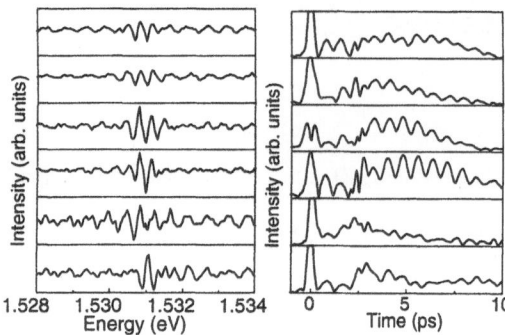

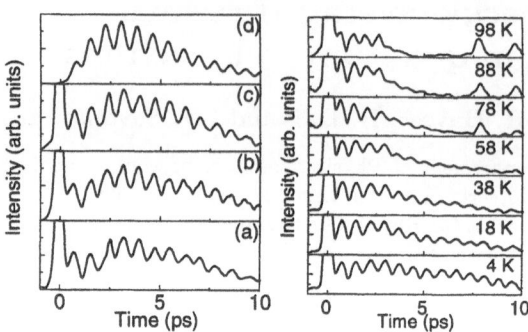

Fig. 2 (a) Interferograms measured at specific angles w.r.t. the normal. (b) Speckle-dependent TR-RRS calculated by Fourier-transforming the data plotted in figure 2(a).

Fig. 3 (Left) (a)Experimental TR-RRS obtained from figure 1(a). (b) TR-RRS measured from a different dset of speckles. (c) TR-RRS measured from a different point on the sample. (d) Theoretical TR-RRS signal. (Right) TR-RRS as a function of temperature. Additional peaks observed at higher temperatures are due to stray reflections.

optical path length of different parts of the reference beam and/or surface-scattered light.

We obtain the temporal dynamics of the coherent RRS by Fourier-transforming the interferogram one row at a time. In figure 2(a) we present some of the interferograms and in figure 2(b) the corresponding time-resolved dynamics measured at the angles indicated by the arrows to the left of figure 1(a). Note that the graphs are not normalised and represent the relative intensities of the various speckles. Figure 2(b) shows that there are pronounced differences in the temporal dynamics of individual speckles. This is due to the statistical nature of speckles. One speckle is the result of a stochastic summation of the complex electric fields emitted by the excitons. Therefore, and this is the main point of the article, the study of a single-speckle does not enable one to reveal the underlying properties of the disorder such as the correlation length or the disorder potential. This is in direct contradiction to the conclusions of Birkedal and Shah [3]. Instead a statistical analysis of a large number of speckles is required. In figure 3 we present the results of averaging over the interferogram (a speckle-average). As figure 3 shows, we obtain almost identical dynamics when we either (i) select a different set of angles by displacing the collimating lens horizontally or (ii) measure a different point on the sample. These results show that in our setup one interferogram averages over a statistically significant number of speckles. In ref. [6] it was shown theoretically that a directional average is equivalent to performing an ensemble-average and therefore experiments that measure an adequate number of speckles can be modelled correctly using the theories that incorporate ensemble-averaging. This is the situation in most experiments that employ large-angle collection. We demonstrate the validity of the ensemble-average approach by comparing, in figure 3, the speckle-averaged RRS signal with a quantum theory of resonant Rayleigh scattering. Full details of the calculation are provided elsewhere [7]. Again, our results directly contradict the claims of Birkedal and Shah [3].

Our experimental technique offers the possibility not only of obtaining the ratio of RRS to PL but, in conjunction with upconversion experiments, of extracting the full temporal dynamics of the incoherent PL. By performing both experiments under identical experimental conditions one can, in principle, subtract the coherent RRS from the total upconversion signal to isolate the time-resolved incoherent PL.

The intensity (not presented) and temperature dependence (figure 3(right)) of the SR provides us with further confirmation of the of the static disorder-related origin of the SR. In particular up to an exciton density of 1.9×10^9 excitons/cm^2 the SR intensity (measured at the peak of the emission) is linear with the exciton density as expected for RRS. Additionally the shape of the TR-RRS is density independent over this density range. We can fit the results at higher temperatures from the curve measured at 4 K multiplied by an exponentially decaying term to take into account dephasing processes.

In conclusion we have measured, for the first time, multiple-speckle ultrafast RRS. We stress once again that only a statistical analysis or averaging of many speckles provides meaningful information.

References

1. S. Haacke, R. A. Taylor, R. Zimmermann, I. Bar-Joseph and B. Deveaud, Phys. Rev. Lett. **78**, 2228, (1997).
2. M. Gurioli, F. Bogani, S. Ceccherini, and M. Colocci, Phys. Rev. Lett. **78**, 3205 (1997).
3. D. Birkedal and J. Shah, Phys. Rev. Lett. **81**, 2372 (1998).
4. W. Langbein, J M. Hvam and R. Zimmermann, Phys. Rev. Lett. **82**, 1040 (1999).
5. L. Lepetit, G. Cheriaux and M. Joffre, J. Opt. Soc. Am. B **12**, 2467 (1995).
6. V. Savona and. R. Zimmermann, Phys. Rev. B **60**, 4928 (1999).
7. G. R. Hayes, B. Deveaud, V. Savona and S. Haacke, Phys. Rev. B (to be published, Sep15 2000).

Exciton and spin transport by surface acoustic waves in GaAs quantum wells

T. Sogawa[1,2], P. V. Santos[1], S. K. Zhang[1], S. Eshlaghi[3], A. D. Wieck[3], K. H. Ploog[1]

[1] Paul-Drude-Institut für Festkörperelektronik, Hausvogteiplatz 5–7, 10117 Berlin, Germany
[2] NTT Basic Research Laboratories, 3-1, Morinosato-Wakamiya, Atsugi, Kanagawa 243-0198, Japan
[3] Lehrstuhl für Angewandte Festkörperphysik, Ruhr-Universität Bochum, 44780 Bochum, Germany

Abstract Spin transport in GaAs quantum wells induced by surface acoustic waves (SAWs) is investigated by spin-resolved micro-photoluminescence spectroscopy. We demonstrated that spins can be transported by the travelling piezoelectric field of a SAW over lengths up to 3 μm.

1 Introduction

Spin manipulation of free carriers in semiconductors is essential for the realization of 'spintronic' devices such as spin transistors and quantum-computing components. Spin transport by a DC electric field in n-type GaAs has been studied by Kikkawa and Awschalom [1]. Ohno *et al.* recently demonstrated hole-spin injection from a ferromagnetic p-type GaMnAs layer into an intrinsic InGaAs quantum well (QW) [2]. The moving piezoelectric field of a surface acoustic wave (SAW) provides an efficient way of transporting photogenerated electron-hole (e-h) pairs over macroscopic distances in piezoelectric materials [3,4]. Transport takes place through ionization of e-h pairs by the SAW electric field and subsequent sweep of electrons and holes in the maxima and minima, respectively, of the moving piezoelectric potential. An interesting question associated with the SAW-induced transport mechanism is whether the carriers retain their spin polarization during the motion. In this study, we employ spin-resolved micro photoluminescence (μ-PL) spectroscopy to investigate the simultaneous transport of carriers and of spin by SAWs in GaAs QWs.

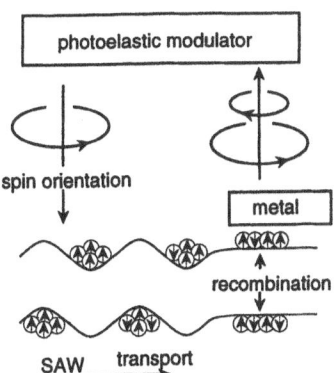

Fig. 1 Setup for spin transport measurements. Spin-polarized carriers (small circles) are generated by circularly polarized light from a photoelastic modulator (long ↓ arrow). The net spin after transport by a SAW is detected by analysing the circular PL polarization (long ↑ arrow).

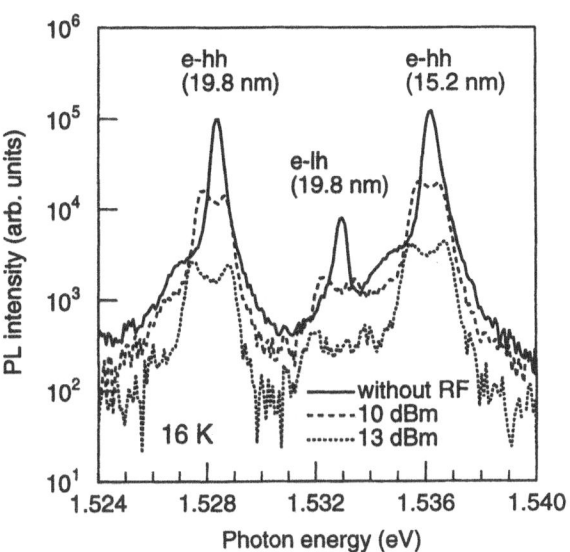

Fig. 2 PL spectra under SAWs with different amplitudes.

2 Experiment and discussion

The GaAs multiple QWs with thicknesses of 19.8 nm (QW1) and 15.2 nm (QW2) sandwiched by (AlAs/GaAs) superlattice barrier layers were grown by MBE as described in Ref. [5]. The QWs are placed about 300 nm below the surface. SAWs, with a wavelength of 5.6 μm and a frequency of 520 MHz, were generated by split-finger interdigital transducers oriented along the [110] direction. The optical measurements were performed at 16 K using a special μ-PL setup with spatially separated photoexcitation and detection spots, each with a radius of approx. 2 μm. Spin sensitivity was achieved by including a photoelastic modulator in the μ-PL setup to measure the degree of circular polarization of the PL after excitation with circularly polarized light from a cw Ti-sapphire laser (750 nm), as illustrated in Fig. 1. From these measurements, we determined the steady-state net-spin density, ($n_{up} - n_{down}$), and the degree of spin polarization, where n_{up} (n_{down}) denotes the density of electrons with up (down)-spin.

Figure 2 shows PL spectra under SAWs of various amplitudes (indicated by the nominal RF power applied to the transducer). The PL peaks around 1.528 eV, 1.533 eV, and 1.536 eV correspond, respectively, to the electron (e)-heavy hole (hh) and e-light hole (lh) transitions in QW1, and to the e-hh in QW2. The PL lines de-

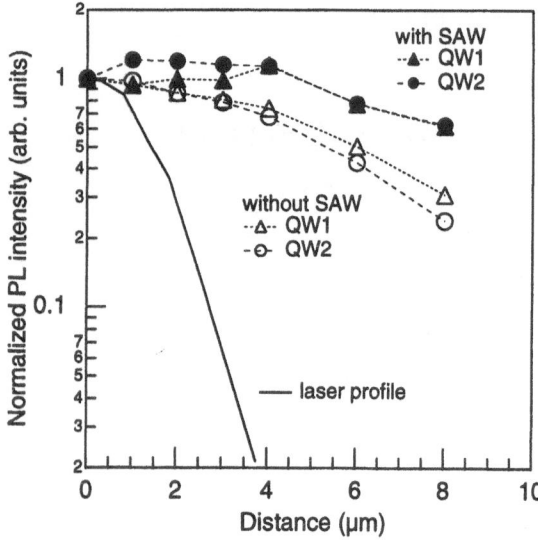

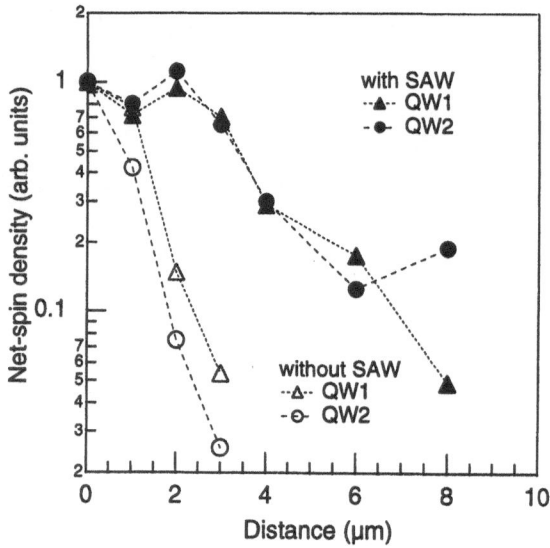

Fig. 3 PL intensity profiles in the absence and under the influence of a SAW. The solid line displays the laser reflection profile.

Fig. 4 Net-spin density profiles in the absence and under a SAW.

crease in intensity and split under a SAW. In order to distinguish between the effects of the strain and of the piezoelectric fields in the μ-PL spectra, measurements were also performed under a semitransparent metal film. The metal film screens the longitudinal component of the SAW piezoelectric field [3]. While under the metal film the spectra (not shown here) reveal the same splitting of the peaks as illustrated in Fig. 2, the integrated PL intensity is not affected by the SAW. The screening of the lateral piezoelectric field below the metal film thus enhances the recombination of the e-h pairs. The splitting of the PL peaks under a strong SAW is attributed to the band gap modulation induced by the SAW strain.

Exciton transport by a SAW is illustrated in Fig. 3. This figure displays the integrated PL intensity of the e-hh line for QW1 (triangles) and of QW2 (circles) in the absence (empty symbols) and under the influence of a SAW of 12 dBm (filled symbols) as a function of the separation between the generation and detection spots. In these measurements, the detection position was kept at the edge of the semitransparent metal film, while the photoexcitation position was varied (cf. Fig. 1). For comparison, the solid line shows the reflection profile of the excitation laser. In the absence of a SAW, the PL profiles correspond to Gaussian-like exciton diffusion profiles with a diffusion length of approx. 7 μm. Under a SAW, the excitons are transported over larger distances by the travelling SAW piezoelectric field.

The corresponding profiles for the normalized net-spin density, $(n_{up} - n_{down})$, are illustrated in Fig. 4. The measurements were performed under conditions similar to those in Fig. 3, except that a SAW excitation power of 10 dBm (instead of 12 dBm) was used. The spin diffusion profile in the absence of a SAW shows diffusion

lengths comparable to the half width of the laser reflection profile (cf. Fig. 3). In contrast, spin transport over considerably larger distances is observed under a SAW. The characteristic spin transport length obtained from a single exponential fit of the data is of about 3 μm. The enhanced spin transport length is attributed to the suppression of exciton spin relaxation induced by the spatial separation of electrons and holes. The separation reduces the exciton exchange interaction and thus the spin-flip probability [6].

3 Conclusion

Investigations of microscopic exciton and spin transport induced by SAWs in GaAs QWs demonstrate that spins can be transported over lengths of 3 μm, which are sufficiently long for applications in interconnections between spintronic device components.

Acknowledgement:
We thank the Deutsche Forschungsgemeinschaft, DFG, for partial financial support from projects SA598/2-1, GRK384, and SFB491.

References

1. J. M. Kikkawa and D.D. Awschalom, Nature **397,** (1999) 139.
2. Y. Ohno, D. K. Young, B. Beschoten, F. Matsukura, H. Ohno, and D.D. Awschalom, Nature **402,** (1999) 790.
3. C. Rocke, S. Zimmermann, A. Wixforth, J. P. Kotthaus, G. Böhm, and G. Weimann, Phys. Rev. Lett. **78,** (1997) 4099.
4. P. V. Santos, M. Ramsteiner, and F. Jungnickel, Appl. Phys. Lett. **72,** (1998) 2099.
5. S. Eshlaghi, C. Meier, D. Suter, D. Reuter, and A.D.Wieck, J. Appl. Phys. **86,** (1999) 6605.
6. A. Vinattieri, Jagdeep Shah, T. C. Damen, K. W. Goossen, L. N. Pfeiffer, M. N. Maialle, and L. J. Sham, Appl. Phys. Lett. **63,** (1993) 3164.

Coherent vs. Incoherent Emission in Quantum Wells studied by Polarisation- and Time-Resolved Spectroscopy

G. Aichmayr[1], L. Viña[1], S. P. Kennedy[2], R. T. Phillips[2], K. Ploog[3]

[1] Departamento de Física de Materiales C-IV, Universidad Autónoma de Madrid, Cantoblanco, E-28049 Madrid, Spain e-mail: guenther.aichmayr@uam.es

[2] Cavendish Laboratory, University of Cambridge, Cambridge CB30H3, U. K.

[3] PDI für Festkörperphysik, Hausvoigteiplatz 5-7, D-10117 Berlin, Germany

Abstract We present a new method to discriminate between coherent and incoherent light emission from a two dimensional exciton gas. It is based on the resonant creation of spin-polarized excitons by means of ultrashort light pulses and the subsequent analysis of the polarized emission using dynamic rate equations. Introducing two dephasing mechanisms, radiative and nonradiative, we found that increasing the excitation density leaves the latter unaffected while it enhances considerably the radiative one.

Due to the advances in femtosecond spectroscopy during the last few years, the very early stages of light emission dynamics in semiconductor heterostructures became an object of intense study. To investigate the processes in these structures at times <10 ps several techniques have been used: time resolved secondary emission [1], four wave mixing [2], speckle resolved spectrospcopy [3] and spectral interferometry [4]. We present here an alternative method: light emission from the two spin components of a circularly polarised exciton gas are detected separately with a time resolution of 150 fs. The spin dynamics is explained by an existing rate equation model [5], which was extended by including a coherent exciton population and exciton-exciton (XX) scattering effects.

The measurements were performed on a 50 period, multi quantum well (MQW) consisting of 77 Å GaAs wells separated by 72 Å wide AlAs barriers. The sample

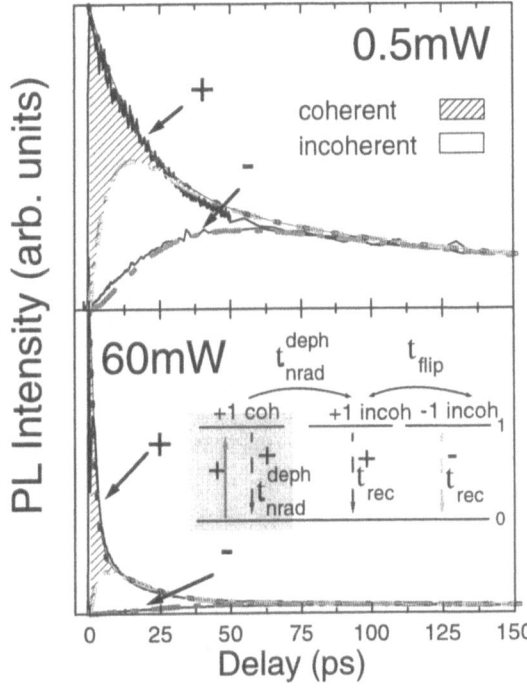

Fig. 2 Fits to the experimental data for 0.5 mW (a) and 60 mW (b) excitation power. The inset shows a simplified scheme of the rate equation model.

was held at 8 K. The excitation with a σ^+ circularly polarised pulse was done in resonance with the exciton energy with a two-colour up-conversion set-up, detecting σ^+ (from spin +1 excitons, X^{+1}) and σ^- (from spin -1 excitons, X^{-1}) emission separately.

After resonant σ^+ excitation, the majority X^{+1} component decays from its maximum while the minority X^{-1} component becomes populated by spin-flip processes reaching its maximum value after ~100 ps. The striking effect, shown in Fig. 1a, is that on increasing the excitation density, the initial decay of the X^{+1} becomes steeper, while the X^{-1} population is rising at about the same rate and its maximum population is decreasing. Exponential fits of the slow decay of σ^+ and σ^-, and on the rise of σ^- gives the following results: a long emission decay time $\tau_{dec}^{(\pm 1)} = 245 \pm 15$ ps independent on excitation density; rise time of σ^-, $\tau_{rise}^{(-1)} = 23$ ps at 0.5 mW, which

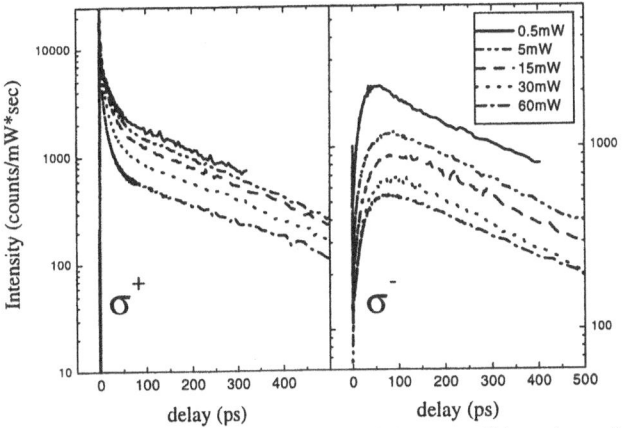

Fig. 1 Emission dynamics for +1 (a) and -1 (b) excitons for several excitation densities. Intensities are normalised to the excitation power.

becomes constant $\tau_{rise}^{(-1)} = 42 \pm 4$ ps at higher excitation densities. Finally a separate fit of the initial fast decay of the σ^+ emission yields a value of $\tau_{fdec}^{(+1)} = 18$ ps at 0.5 mW, which decreases to $\tau_{fdec}^{(+1)} \simeq 2$ ps at 60 mW. In resonant Rayleigh scattering experiments the behavior of $\tau_{fdec}^{(+1)}$ is generally explained by an enhanced dephasing rate due to XX interaction, whereas in spin-dynamic measurements this fast initial decay is attributed to the hole spin-flip and momentum scattering to optically inactive high-K states. Vinattieri *et al.* [5] presented a model which explains the fast initial decay in terms of this hole flip and momentum scattering. They could obtain good agreement with experiment at low densities having a time resolution of ~ 5 ps. However, it was not possible to reproduce our observations with this model. Adjusting the fast decay leads to an overestimation of the σ^- emission. Since in the model of Ref. [5] the initial coherent emission or resonant Rayleigh scattering is neglected we added a new decay path to the model. Assuming that the created X^{+1} exciton population is initially in phase with the laser pulse and coherent, we introduce two dephasing mechanisms: a radiative dephasing characterizing the coherent emission and a nonradiative one responsible for the transfer of excitons from coherent to incoherent $X^{\pm1}$ populations, which then gives rise to PL and further spin-flip processes. The fast decay of the σ^+ emission is characterized by the radiative dephasing time τ_{rad}^{deph} and the rise of the incoherent σ^+ emission by the nonradiative time τ_{nrad}^{deph}. The basic scheme of the model is drawn in the inset of Fig. 2. Exciton spin-flip (with τ_{flip}) allows a transfer between the incoherent X^{+1} and X^{-1} components. A detailed description of the incoherent dynamics is given in Ref. [5]. Additionally, since a density-dendent energy splitting between X^{+1} and X^{-1} exists, a Boltzman factor was added to the exciton spin-flip [6,7]. The mechanism responsible for direct flip from ±1 to ±2 states due to inter-excitonic exchange was also included [8,9]. As in any fit procedure with many parameters, some of these can be interdependent, making difficult the separation of the influence of different mechanisms. However, the dominant role of the coherent part in the total exciton population and its dynamic is not influenced by that fact. In all the fits we fixed the hole flip time to 1 ns [10] and the electron flip time to 300 ps [5]. The Boltzman factor for the exciton flip was taken from measured values [7] and the inter-excitonic exchange scattering rates from Ref. [9]. The exciton recombination time, τ_{rec}, was 120 ps, which differs a factor of 2 from the one that was fitted directly to the emission by exponential decay ($\tau_{dec}^{(\pm1)} = 245 \pm 15$ ps). This is not surprising since the population of the optically dark $X^{\pm2}$ states contributes to the long decay of the emission signal.

In Figs. 2a and b the resulting fits for two excitation densities are shown. The assumption of an initial coherent population and the two dephasing mechanisms is cru-

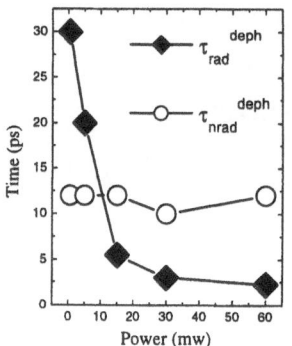

Fig. 3 Times for emission, τ_{rad}^{deph}, and dephasing, τ_{nrad}^{deph}, as obtained from the fit with the rate equation model.

cial to obtain a good agreement with the measurements. Now the initial rise of σ^- and the overall X^{-1} population is fitted satisfactorily. The hatched area in Fig. 2 between the total σ^+ and the fit for the incoherent σ^+ emission, represents the part of emitted coherent light in the total emission. On increasing the excitation by two orders of magnitude, the initial decay is considerably faster whereas the time to populate the incoherent X^{+1} component, τ_{nrad}^{deph}, is practically unchanged. The dephasing times obtained with the extended model for several excitation densities are shown in Fig. 3. The radiative decay time, τ_{rad}^{deph}, is decreasing very quickly between 5 and 15 mW and only reduces by a factor 2 from 15 to 60 mW, whereas the non-radiative time, $\tau_{nrad}^{deph} = 11 \pm 1$ ps, is constant. Dispite this clear behaviour further experiments are needed, and are under way, to confirm the nature of the initial emission. In principle, emission in angular directions which were not detected, non-radiative processes or the different dynamics of localised and extended excitons could contribute to the initial, density dependent fast decay.

This work was partially supported by the European Union through the Training and Mobility of Researchers Ultrafast Quantum Optoelectronics Network, the austrian BMWF, the spanish DGICYT (PB96-0085), the CAM (07N/0026/1998) and the Fundacion Ramon Areces.

References

1. N. Garro *et al.*, Phys. Rev. B **60**, 4497 (1999).
2. W. Langbein *et al.*, Phys. Stat. Sol. **178**, 13 (2000).
3. W. Langbein *et al.*, Phys. Rev. Lett. **82**, 1040 (1999).
4. S. Haake *et al.*, Phys. Rev. B **61**, R5109 (2000).
5. A. Vinattieri *et al.*, Phys. Rev. B **50**, 10868 (1994).
6. T. C. Damen *et al.*, Phys. Rev. Lett. **67**, 3432 (1991).
7. G. Aichmayr *et al.*, Phys. Rev. Lett. **83**, 2433 (1999)
8. M. Z. Maialle *et al.*, Phys. Rev. B **47**, 15776 (1993).
9. C. Ciuti *et al.*, Phys. Rev. B **58**, 7926 (1998).
10. R. Ferreira *et al.*, Phys. Rev. B **43**, 9687 (1991).

Ultrafast dynamics of holes in $Si_{1-x}Ge_x$/Si multiple quantum wells

Robert A. Kaindl[1], Matthias Wurm[1], Klaus Reimann[1], Michael Woerner[1], Thomas Elsaesser[1], Christian Miesner[2], Karl Brunner[2], and Gerhard Abstreiter[2]

[1] Max-Born-Institut für Nichtlineare Optik und Kurzzeitspektroskopie, Max-Born-Straße 2A, 12489 Berlin, Germany
   e-mail: reimann@mbi-berlin.de

[2] Walter-Schottky-Institut, Technische Universität München, Am Coulombwall, 85748 Garching, Germany

Abstract Ultrafast hole dynamics in $Si_{0.5}Ge_{0.5}$/Si multiple quantum wells is studied with resonant excitation of the HH1→ HH2 intersubband transition. Spectrally resolved pump-probe experiments with 150-fs mid-infrared pulses yield an intersubband relaxation time of 250 ± 100 fs, which is dominated by optical deformation potential scattering.

1 Introduction

Optical intersubband excitations in semiconductor nanostructures provide direct insight into the ultrafast dynamics of coherent optical polarizations and nonequilibrium carriers and play a key role for novel devices such as quantum cascade lasers. So far, most experiments have concentrated on intersubband excitations of quasi-two-dimensional electron plasmas [1]. In contrast, intersubband transitions of quasi-two-dimensional holes, which are highly relevant for device applications [2], have been much less explored. Here, we present the first time-resolved study of hole plasmas in quantum wells using direct intersubband excitation and probing of heavy holes. We demonstrate intersubband scattering of holes in p-type modulation-doped $Si_{0.5}Ge_{0.5}$/Si multiple quantum wells on a subpicosecond time scale. Emission of optical phonons via the deformation potential represents the dominant mechanism for intersubband relaxation.

2 Experiment

We investigate high-quality p-type $Si_{1-x}Ge_x$/Si multiple quantum wells grown by solid-source molecular beam epitaxy [3]. The structure consists of ten periods of modulation-doped ($n = 1.2 \times 10^{12}$ cm^{-2}) $Si_{1-x}Ge_x$ quantum wells (4.4 nm wide, $x = 50\%$) separated by 18-nm Si barriers. Polarized absorption spectra reveal an intersubband line due to HH1 → HH2 transitions, centered at 167 meV [(see Fig. 1(a)]. Mid-infrared pump and probe pulses of 150 fs duration are derived from the output of a 1-kHz optical parametric amplifier by difference frequency mixing in GaSe [4]. Broad pulse spectra allow to probe across the whole intersubband line by spectral selection after the sample.

3 Results and discussion

To determine the effects of the two-dimensional confinement and of the $Si_{0.5}Ge_{0.5}$/Si built-in strain we have performed an eight-band $\mathbf{k} \cdot \mathbf{p}$ band structure calculation (see Fig. 1) [3, 5]. The calculated intersubband absorption agrees well with the measured spectrum [(see Fig. 1(a)]. One should note that the HH2 – HH1 transition matrix element is strongly $\mathbf{k}$-dependent. Therefore only holes in a limited region of $\mathbf{k}$ space

are accessible in optical experiments. The mixed light-hole–split-off band LHSO1, which lies between the HH2 and HH1 subbands, is not seen in optical absorption for p-polarized light. As will be seen below, this band is very important for intersubband relaxation.

In Fig. 2 we show $\Delta T/T_0$ (T_0 is the transmission through the sample without pump pulse and ΔT the change in transmission at a certain delay after the pump pulse) as a fuction of probe photon energy for different delays between pump and probe. For these spectra the pump intensity was chosen to excite 30% of the holes initially present in the HH1 subband into the HH2 subband.

If the HH2 – HH1 transition matrix element were constant in $\mathbf{k}$ space, the number of holes in the HH2 subband as a function of time could simply be determined by spectrally integrating $\Delta T/T_0$. However, the matrix element is *not* constant [see Fig. 1(c)]. Therefore the time-resolved transmission changes are determined not only by the population difference between HH1 and HH2 subbands, but also by the hole distribution in $\mathbf{k}$ space.

Since the specific heat of a Fermi gas is very low at low temperatures, already a very small number of holes scattered back after excitation from the HH2 subband into the HH1 subband suffices to considerably raise the HH1 temperature. Therefore this rise is extremely fast. An increase of the carrier temperature leads to a depopulation of states at small $|\mathbf{k}|$ and to a population of states at large $|\mathbf{k}|$. Because of the different dispersions of HH1 and HH2 bands, this translates into a negative $\Delta T/T_0$ (induced absorption) at the low-energy side of the intersubband absorption line and into a positive $\Delta T/T_0$ (bleaching) at the high-energy side. The population difference between HH1 and HH2, on the other hand, can only give rise to a positive $\Delta T/T_0$. Therefore we have determined the HH1 temperature by integrating the area of negative $\Delta T/T_0$ and finding a theoretical spectrum with the same area. These spectra are shown by solid lines in Fig. 2, together with the temperatures obtained. One can see that the experimental data (dots) agree with the solid lines for delays $t \geq 2$ ps. Thus for these times no holes are present in the HH2 subband and the signal is caused solely by the heating of the HH1 distribution. At earlier times, we subtract a spectrum of a hot HH1 distribution with the same area of induced absorption from the data. The spectrally integrated area of this difference spectrum, which is shown in Fig. 3, is proportional to the number of holes in the HH2 subband. A single-exponential fit yields an intersubband relaxation time of 250 ± 100 fs.

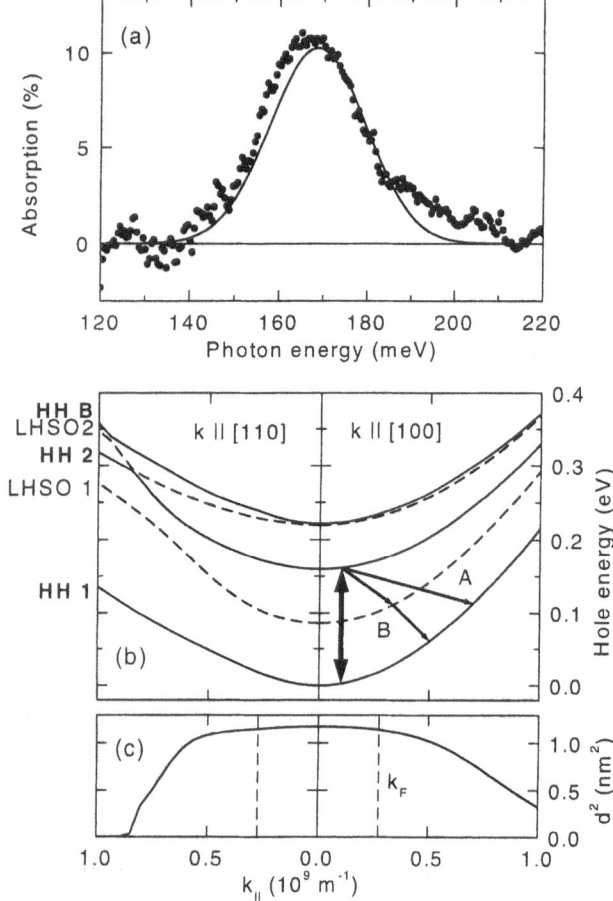

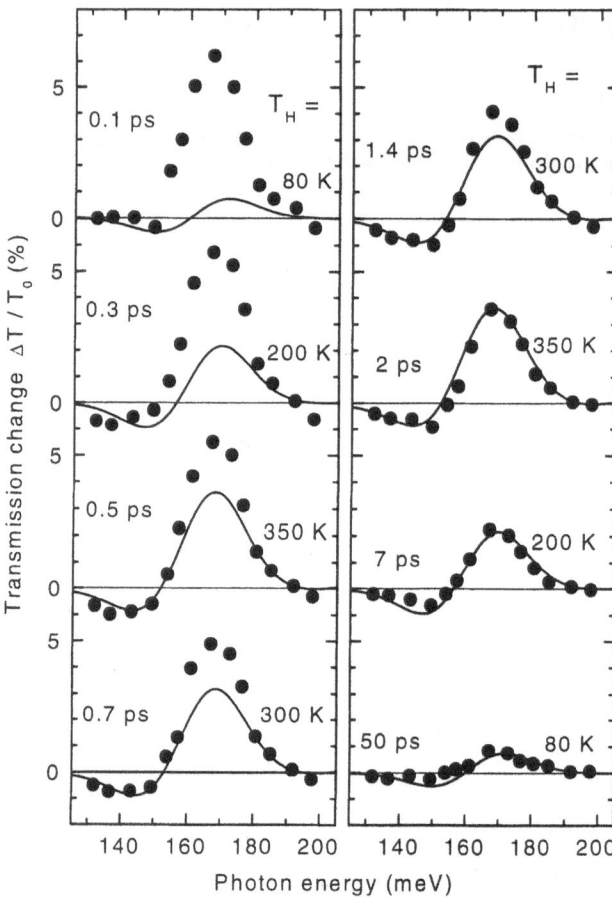

Fig. 1 (a) Measured stationary intersubband absorption at a temperature of 15 K (dots), compared with the calculated spectrum (solid line). (b) Calculated valence band dispersion for the first heavy-hole [HH1, HH2, and HHB (located in the barrier)] and mixed light-hole–split-off (LHSO1 and LHSO2) subbands. (c) Square of the dipole matrix element for the HH1 → HH2 transition. The Fermi wavevector k_F is indicated.

The dominant scattering process for holes, in contrast to Γ-valley electrons, is optical deformation potential scattering. With the eigenvectors from the bandstructure calculation and a deformation potential constant of 36 eV [6,7], we calculate an intersubband relaxation time of 225 fs, which agrees very well with the experimental value. It is interesting to note that the cascaded process [HH2 → LHSO1 → HH1, arrows B in Fig. 1(b)] is about three times faster than the direct process (HH2 → HH1, arrow A). This result has important consequences for the design of quantum cascade devices [2]. It shows that long HH2 lifetimes can be achieved if scattering to the LHSO1 state is not possible, e.g., if the energy difference between HH2 and LHSO1 is less than one optical phonon energy.

Fig. 2 Time- and spectrally-resolved transmission changes after exciting ≈ 30% of the holes present by doping from the HH1 to the HH2 subband (dots). The solid lines are spectra calculated for the HH1 temperatures T_H indicated (see text).

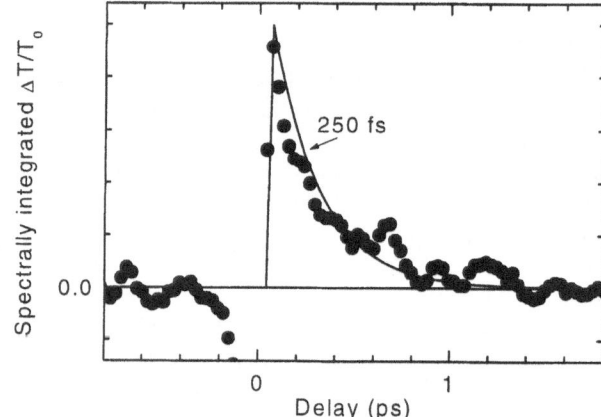

Fig. 3 Area of the difference spectrum obtained as described in the text. From an exponential fit (solid line) we derive a decay time of ≈ 250 fs.

References

1. T. Elsaesser and M. Woerner, Phys. Rep. **321**, (1999) 253.
2. L. Friedman, R. A. Soref, and G. Sun, J. Appl. Phys. **83**, (1998) 3480.
3. T. Fromherz, E. Koppensteiner, M. Helm, G. Bauer, J. F. Nutzel, and G. Abstreiter, Phys. Rev. B **50**, (1994) 15073.
4. R. A. Kaindl, M. Wurm, K. Reimann, P. Hamm, A. M. Weiner, and M. Woerner, J. Opt. Soc. Am. B, in print.
5. D. Gershoni, C. H. Henry, and G. A. Baraff, IEEE J. Quantum Electron. **29**, (1993) 2433.
6. W. Poetz and P. Vogl, Phys. Rev. B **24**, (1981) 2025.
7. M. Woerner and T. Elsaesser, Phys. Rev. B **51**, (1995) 17490.

Mechanism of Positively Charged Exciton Spin Relaxation in CdTe and CdMnTe Quantum Wells

E. Vanelle [1], **P. Kossacki** [2], **J. Cibert** [3], **T. Amand** [1], **P. Renucci** [1], **X. Marie** [1] and **S. Tatarenko** [3]

[1] Laboratoire de Physique de la Matière Condensée, INSA-CNRS, 135 av de Rangueil, 31077 Toulouse Cedex, France
[2] Institute of Experimental Physics, Warsaw University, ul.Hoza 69, 00-681 Warsaw, Poland
[3] Laboratoire de Spectrométrie Physique, CNRS – Université Joseph Fourier Grenoble, BP87, 38402 Saint Martin d'Hères, France

Abstract   We present a study of the charged excitons spin relaxation dynamics. We find that depending on the helicity of the exciting beam the spin relaxation time increases or decreases in the presence of a magnetic field. This behavior is discussed in the frame of former theories developed by Bir-Aronov-Pikus, Dyakonov-Perel, and Elliott-Yafet.

1   Introduction

With the recent growing interest in spin-electronics [1] in semiconductors, it is highly desirable to obtain a clear understanding of the mechanisms governing spin relaxation in heterostructures. These mechanisms have been studied in bulk material [2] and more recently in heterostructures [3] for free electrons (holes) and for excitons. The spin relaxation mechanisms involved within charged excitons may be different.

2   Experiments

Two 8 nm width quantum wells (QW's) have been studied: a p-doped ($\sim 7\ 10^{10}$ cm^{-2}) CdMgZnTe/CdTe [4], and a p-doped ($\sim 3\ 10^{11}$ cm^{-2}) CdMgZnTe/Cd$_{1-x}$M$_x$nTe QW (x=0.2%). The CdTe QW photoluminescence (PL) spectra exhibits two lines corresponding to the recombination of neutral excitons (X) and, 2.7 meV below, to the positively charged excitons (X$^+$) [4]. On the other hand, the CdMnTe QW PL spectra is dominated by the X$^+$ line. Time-resolved PL experiments have been carried out at 1.7 K under a longitudinal magnetic field **B** (Faraday configuration). The samples are excited by 1.4 ps pulses resonant with the X$^+$ transition (no neutral excitons are generated [4]). The time resolution is 1.5 ps.

We measured the X$^+$ PL polarization decay as a function of B, for σ$^+$ and σ$^-$ exciting beam. As the X$^+$ are formed with two holes of opposite spin, the polarization of the X$^+$ PL reflects the electronic spin state only.

The inset in figure 1-b describes the simplified band structure of the CdTe QW under an applied magnetic field. The X$^+$ PL intensity decreases (increases) while B increases for σ$^+$ (σ$^-$) exciting beam [5]. Indeed a σ$^-$ (σ$^+$)exciting beam creates an electron with a spin Sz=+1/2 (-1/2) and a hole with total angular-momentum projection J_z = -3/2 (+3/2); in order to create an X$^+$ a hole with J_z = +3/2 (-3/2) has to be captured. Due to the Zeeman splitting of the valence band (VB) the density of J_z = +3/2 hole is higher than the J_z = -3/2 hole density

and thus the oscillator strength of σ$^-$ polarized charged exciton is higher than the σ$^+$ polarized one.

Figure 1-a displays the results for the CdTe QW for B=4 T and 0. We observed a spin relaxation time of 65 ps at zero field, decreasing down to 25 ps at 4 T for σ$^+$ excitation and increasing up to 115 ps for σ$^-$ excitation. Figure 1-b displays the evolution of the polarization decay time versus B for both exciting helicities.

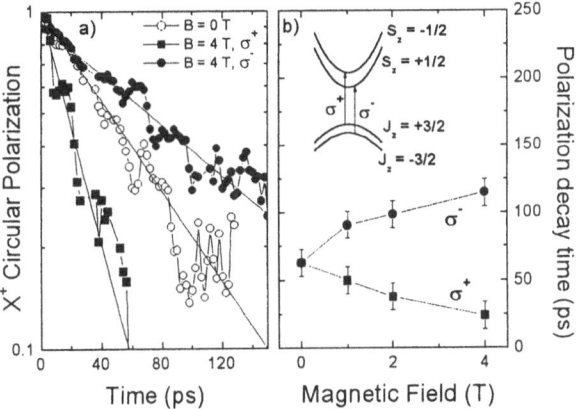

Fig. 1 Results for the CdTe QW: a) Time evolution of the X$^+$ circular polarization after resonant σ$^+$and σ$^-$ excitation with B = 4 T (full squares and full circles respectively) and after resonant excitation with B = 0 (open circles). b) X$^+$ polarization decay time as a function of B for σ$^+$ and σ$^-$ excitation (full squares and full circles respectively). The inset shows the conduction and valence band structure of the sample under an applied magnetic field.

For the CdMnTe QW we observe a spin relaxation time of 35 ps at zero field, increasing up to 150 ps at 1 T for σ$^-$ exciting beam (figure 2-a). Figure 2-b displays the evolution of the depolarization time versus B. For σ$^+$ exciting beam the PL intensity was below our experimental sensitivity due to the decrease of the X$^+$ oscillator strength.

A key point is that the introduction of Mn in the QW has a different effect in the VB and in the CB. In the VB, the hole spins tend to align parallel to an applied field **B** in both cases (but the effect is enhanced by the giant Zeeman effect due to the Mn spins in the CdMnTe QW). In the CB, the effective Lande g-factor has an opposite sign in CdTe and CdMnTe QWs (see the inset of figure 2-b).

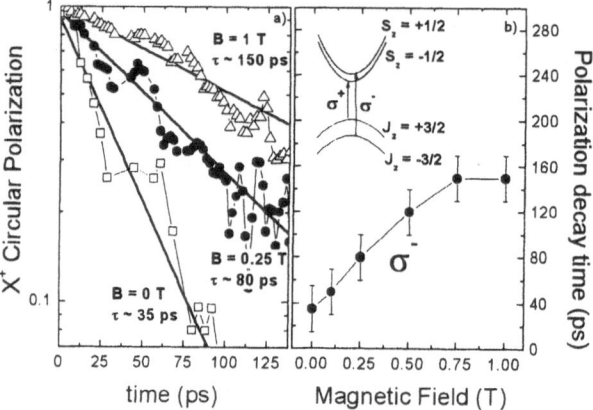

Fig. 2 Results for the CdMnTe QW: a) Time evolution of the X^+ circular polarization after resonant σ^- excitation with $B = 0$, 0.25 and 1 T (full squares, full circles and open triangles respectively). b) X^+ polarization decay time as a function of B for σ^- excitation (full circles). The inset shows the CB and VB structure of the sample under an applied magnetic field.

3   Discussion

Three mechanisms are generally invoked to account for the electronic spin relaxation rate: the Bir-Aronov-Pikus (BAP), Dyakonov-Perel (DP), and Elliott-Yafet mechanisms. The former can be ruled out for the present study as it is known to be efficient only at high temperature. The DP mechanism is based on the absence of inversion center in the Zinc-Blend structure. The degeneracy of the CB is thus lifted for $k_{//} \neq 0$. The spin splitting of the CB states is equivalent to the presence in the crystal of a transverse effective magnetic field with magnitude and orientation dependent on $k_{//}$. This field couples the $S_z = +1/2$ and $S_z = -1/2$ states. In the usual case where $\Omega \tau_p << 1$ (Ω is the Larmor frequency in this effective magnetic field and τ_p is the average elastic scattering time), the spin relaxation occurs as a result of random small rotations. Applying **B** increases the electronic spin relaxation time regardless of the initial S_z value. Indeed, due to the CB Zeeman splitting, the coupling between the $S_z = +1/2$ and $S_z = -1/2$ states (due to the effective magnetic field) is reduced when B increases. Therefore the relaxation time should increase and thus the DP mechanism must be ruled out.

The BAP mechanism is based on the short-range exchange interaction between electrons and holes: an electron [$k_{//} = k_e$, $S_z = 1/2$ (-1/2)] is scattered to the state [$k_{//} = k_e + q$, $S_z = -1/2$ (1/2)] while a hole [$k_{//} = k_h$, $J_z = -3/2$ (3/2)] is scattered to the state ($k_{//} = k_h - q$, $J_z = +3/2$ (-3/2)], with energy conservation. This description is valid for free electrons and free holes while we are dealing with localized X^+ [4]. Nevertheless let us examine

what is the effect of **B**. Note that for both samples the hole gas is degenerate.

First of all let's consider a cold electron, i.e. with a kinetic energy $\sim k_B T$. With $B = 0$ the electron spin flip occurs with an energy loss of $\sim k_B T$ so that the hole gains this energy and is scattered to an empty hole state (the reverse process is less probable due to space filling). The holes which can be involved in such processes are thus the ones located around the Fermi surface. With $B \neq 0$ let's consider the CdTe QW case first. A cold electron with $S_z = -1/2$ (σ^+ incident beam) can loose energy up to ΔEc (the CB Zeeman splitting) and thus the density of holes which can participate to the spin flip increases roughly like ΔEc. The relaxation time should thus decrease in the CdTe QW with σ^+ incident beam and increase in σ^-, and, considering the inversed CB structure, it should decrease in the CdMnTe QW with σ^- incident beam. Therefore this description cannot account for the experimental data.

Second, let's consider the case of hot electron, i.e. with a kinetic energy larger than the hole Fermi energy. Then, regardless of the initial S_z value, all holes can participate to the spin flip process and the relaxation time is then governed by the total density of $J_z = +3/2$ and $J_z = -3/2$ holes. The behavior is the same than for cold electrons in the case of the CdTe QW, but is opposite for the CdMnTe one: the relaxation time is expected to increase with B for σ^- incident beam which is the experimental behavior.

The BAP mechanism is the only one which can explain the **B** dependence and the polarization dependence of the spin relaxation time. Nevertheless we have analyzed it considering free electrons while we are dealing with bound electrons. Further studies are thus requested to describe in more details the present experiments but the fact that the electrons are bound may be an important point : their electronic wave functions are a combination of free electron wave functions with different $k_{//}$. The preliminary calculations we have performed (see also ref. 6) shows that the spin flip time of hot electrons is considerably reduced compared to that of cold electrons. The spin flip time of a bound electron could be then governed by the large wave vectors components of its wave function.

References

1. R. Fiederling *et al.* Nature **402**, 787 (1999)
2. G.L. Bir *et al.* JETP **42**, 705 (1976) ; M.I. Dyakonov and V.I. Perel, Soviet Phys. - Semicond. **10**, 208 (1976)
3. M.D. Martin *et al.* Phys. Stat. Sol. (b) **215**, 229 (1999)
4. Vanelle *et al.*, Phys. Rev B **62**, 2696 (2000)
5. P. Kossaki *et al.*, Phys. ReV. B **60**, 16018 (1999)
6. M.Z. Maialle and M.H. Degani, Phys. Rev. B **55**, 13771 (1997)

Microscopic simulation of hot-carrier intersubband relaxation in quantum-cascade lasers

R. C. Iotti[123], **F. Rossi**[12]

[1] Istituto Nazionale per la Fisica della Materia (INFM)
[2] Dipartimento di Fisica, Politecnico di Torino, Corso Duca degli Abruzzi 24, I-10129 Torino, Italy
[3] Scuola Normale Superiore, Piazza dei Cavalieri 7, I-56126 Pisa, Italy

Abstract   In this work we present a microscopic analysis of the carrier dynamics in intersubband light-emitting devices. In particular, a non conventional Monte Carlo simulation scheme has been developed in order to provide access to dynamical details without employing phenomenological parameters.

1 Introduction

Since their appearance [1], quantum cascade lasers (QCLs) have been subject of a rapid experimental development in emission wavelength, lasing threshold, output power and operating temperature. This has stimulated considerable theoretical interest, mainly motivated by the desire to improve the device performances by optimising the semiconductor-heterostructure design. Various theoretical models have been proposed to describe gain spectra and characteristics of QCLs. However, these approaches generally resort on macroscopic rate-equations to describe the carrier dynamics, whose application does not have a true justification [2].

The main difficulty a quantitative theory of QCL response has to face is, in fact, the inadequacy of the usual approach which treats intersubband lasers by analogy with a n-level atomic system. Rather than within a multi-level system, the electron dynamics in QCLs occurs within a multi-subband structure, and the existence of transverse degrees of freedom, i.e., the in-plane dynamics, should be properly taken into account. The most commonly used approach to describe carrier dynamics within a kinetic or Boltzmann-like formulation is the Monte Carlo (MC) method: It allows to take into account intra- as well as inter-subband scattering processes on equal footing and to include, at a kinetic level, a large variety of scattering mechanisms. In this work we present the first microscopic analysis of electron dynamics in unipolar light-emitting devices, which is based on a suitable non-conventional MC simulation scheme.

2 Monte Carlo simulation scheme

QCLs are complex structures, whose core is made up of repeated stages of active regions, sandwiched between electron-injecting and collecting regions [1]. As the name suggests, electron cascade along the quantized-level energy staircase of this system of coupled quantum wells. A proper tayloring of optical matrix elements and relaxation times is crucial to achieve a population-inversion condition. This can be done by designing the intersubband energy separations to benefit of an energy-selective channel by which electrons can dissipate their energy: optical-phonon-emission cascade.

In this work, we study the hot-carrier dynamics within the full core region (injector+active region+collector) of QCL devices. To this end we employ a fully 3D description of energy relaxation based on microscopic intra- as well as inter-subband scattering mechanisms. The three-dimensional single-particle electron states forming the basis set of our simulation scheme are obtained within the standard envelope function approximation. Electron dynamics is treated within an ensemble multisubband Monte Carlo simulation scheme. The time evolution of the carrier distribution function $f_{\mathbf{k}\nu}$ in the various subbands ($\mathbf{k}$=electron in-plane wavevector, ν=label of the confined level) is governed by the following Boltzmann-like equation:

$$\frac{d}{dt}f_{\mathbf{k}\nu} = \sum_{\mathbf{k}'\nu'} [\; f_{\mathbf{k}'\nu'}P_{\mathbf{k}'\nu',\mathbf{k}\nu}(1-f_{\mathbf{k}\nu}) \\ - f_{\mathbf{k}\nu}P_{\mathbf{k}\nu,\mathbf{k}'\nu'}(1-f_{\mathbf{k}'\nu'})] \qquad (1)$$

where $P_{\mathbf{k}\nu,\mathbf{k}'\nu'}$ is the scattering probability for a process connecting state $\mathbf{k}\nu$ to state $\mathbf{k}'\nu'$.

To describe on a microscopic level vertical transport processes through the localised states of the whole core region, we study the carrier dynamics in the presence of both carrier-carrier and carrier-optical phonon interaction [3], within a Frölich description. As an approach to the intercarrier scattering problem in multi-quantum-well systems, we have adopted the Born approximation. The effect of multisubband screening of the bare scattering potential has been taken into account within the static and long wavelength random-phase approximation [4].

The crucial point in our simulation scheme is to properly "close the circuit" without phenomenological parameters. This is achieved by imposing suitable boundary conditions which allow to simulate just one stage of the structure: Whenever an electron is scattered out of the simulated region, it is injected back into the main level-sequence. In the present Monte Carlo approach the current density across the whole structure is then an output of the simulation and the current/voltage characteristics of the semiconductor heterostructure is fully calculated within such a microscopic description.

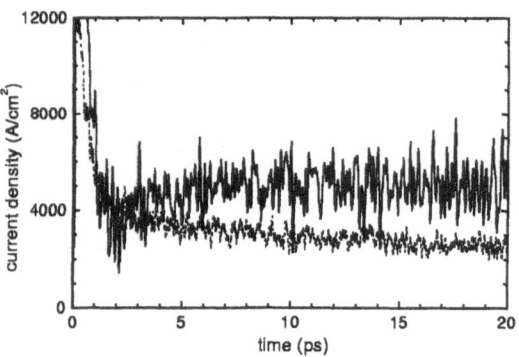

Fig. 1 Simulated current density as a function of time for the core region of the QCL device of [5], in presence (solid line) and absence (dashed line) of electron-electron interaction.

3 Application to a case of interest

The proposed simulation scheme has been applied to finite multiple quantum wells forming the active region of state-of-the-art QCLs. In particular, we have investigated the GaAs/(Al,Ga)As-based diagonal-configuration device of Ref. [5].

The time evolution of the carrier population of the various subbands has been studied both in the presence and absence of intercarrier interaction, in order to focus on the relative weigth of the carrier-carrier and carrier-phonon competing energy-relaxation channels. The carrier distribution function evolves from a phonon replica scenario to a thermalised distribution on varying the electron density: for low carrier densities ($< 10^{11}$ cm^{-2}) intercarrier scattering is weak and electrons relax their energy via a cascade of successive optical-phonon emissions, for high carrier densities intercarrier scattering is more effective in setting up a heated Maxwellian distribution. Our analysis shows that the operation of the QCL structure of Ref. [5] (sheet-density $3.9 \cdot 10^{11}$ cm^{-2} per period) is dominated by hot-electron effects and presents a pronounced non-equilibrium carrier distribution. At stationary conditions, the desired regime of inversion of population between the highest and the intermediate level of the active region is achieved.

Figure 1 shows the simulated current density across the core region of the QCL device with (solid curve) and without (dashed curve) the inclusion of intercarrier interaction. Simulations have been performed at the threshold operating parameters reported in [5]. The current density in the presence of both electron-phonon and electron-electron scattering mechanisms is about 5000 A/cm^2, in good agreement with experimental findings [5]. These results, together with a detailed analysis of the dynamical evolution of the carrier distribution, demonstrate that the electron-phonon interaction alone is not able to couple the injector subbands to the active region ones: While carrier-phonon relaxation is sufficient for a realistic description of the electron dynamics in

the bare active region [6] carrier-carrier scattering has to be included for a proper analysis of charge transport in the full core region. Moreover, the operation of the structure can be reproduced on the basis of incoherent intercarrier–assisted tunneling processes between injector/collector and active region, and thus within a purely semiclassical transport theory.

4 Summary and conclusions

Beside their fundamental interest, hot-electron relaxation processes play a crucial role in determining the operation efficiency of many semiconductor-based quantum devices. A typical example is represented by quantum-cascade lasers. In this work, we have presented the first global simulation of carrier dynamics in these intersubband light-emitting devices. Our fully three-dimensional approach allows to study in a purely microscopic way — without resorting to phenomenological parameters— the current-voltage characteristics of state-of-the-art unipolar nanostructures. Based on the proposed theoretical scheme, we are able to answer the controversial question: *is charge transport in Quantum-Cascade Lasers mainly coherent or incoherent?* Our analysis shows that a proper inclusion of carrier-phonon as well as carrier-carrier scattering within a semiclassical framework gives good agreement with experimental results.

Discussions with F. Beltram, S. Barbieri and C. Sirtori are gratefully acknowledged. This work has been partially supported by the European Commission through the Brite Euram UNISEL project and the TMR Network "Ultrafast Quantum Optoelectronics".

References

1. J. Faist *et al.*, Science **264**, (1994) 553.
2. P. Harrison, Appl. Phys. Lett. **75**, (1999) 2800.
3. T. Kuhn and F. rossi, Phys. Rev. Lett. **69**, (1992) 977.
4. S. M. Goodnick and P. Lugli, Phys. Rev. B **37**, (1998) 2578.
5. C. Sirtori *et al.*, Appl. Phys. Lett. **73**, (1998) 3486.
6. R. C. Iotti and F. Rossi, Appl. Phys. Lett. **76**, (2000) 2265.

Reduction of Coulomb Scattering in a GaAs/AlGaAs Mesoscopic 2DEG Disk

N. Suzumura, M. Yamaguchi, N. Sawaki

Department of Electronics, Nagoya University, Chikusa-ku, Nagoya 464-8603, Japan e-mail: n-suzumu@nuee.nagoya-u.ac.jp

Abstract The carrier relaxation phenomena in GaAs/Al GaAs 2DEG quantum wells are investigated via femtosecond reflectance change. The LO phonon emission time of 0.4 ps was obtained in a quantum well, while it was one order of magnitude longer in a disk shape sample and depended on the geometry. Possible origin of the modification of the Coulomb scattering in a disk of mesoscopic size is discussed.

1 Introduction

The carrier-carrier scattering has a crucial effect on the intra- and inter-subband carrier relaxation in quantum wells, which determines the ultra-fast dynamics of non-equilibrium carriers[1]. One of the role is the screening of scattering centers as ionized impurities and polar optical phonons[2]. Since the phenomenon is due to the modification of the charge density distribution to neutralize the charged particle or polarization, if the size of the device is smaller or comparable to the screening length, then the manner of screening might be modified as compared to that in a large device.

The other role of the carrier-carrier scattering is the randomization of the momentum and/or the energy exchange. Numerical results suggested that the scattering of electrons with ionized impurities as well as longitudinal phonons in semiconductors is modified by the presence of carrier-carrier scattering[3,4]. The critical density is such that the mean carrier density is equal to the screening length. Recent experimental results supported the suggestion and showed that the energy relaxation rate due to emission of phonons is reduced at low carrier densities[5,6]. The studies so far has been limited to samples with large area. If the size of the sample is comparable to the screening length or the mean distance between carriers, the manner of carrier-carrier scattering and hence the decay properties might be modified. In this report, we will demonstrate that the energy relaxation rate is much reduced in a 2DEG disk of mesoscopic size, and it depends on the geometry. The possible origin is discussed in relation to the modification of the Coulomb scattering rate in the disk.

2 Sample Structure and Experimental Method

An asymmetric double quantum well which has two quantum wells of different widths (5 and 10 nm) separated by a 4 nm $Al_{0.3}Ga_{0.7}As$ barrier layer was grown on a semi-insulating GaAs by MBE. The GaAs top layer and $Al_{0.3}Ga_{0.7}As$ cladding layer is n-type doped to get 2DEG in the quantum wells. Three kinds of disk were made by EB lithography so that the 2DEG is confined within the disk; (a) A circle of which diameter is 240 nm, (b) A

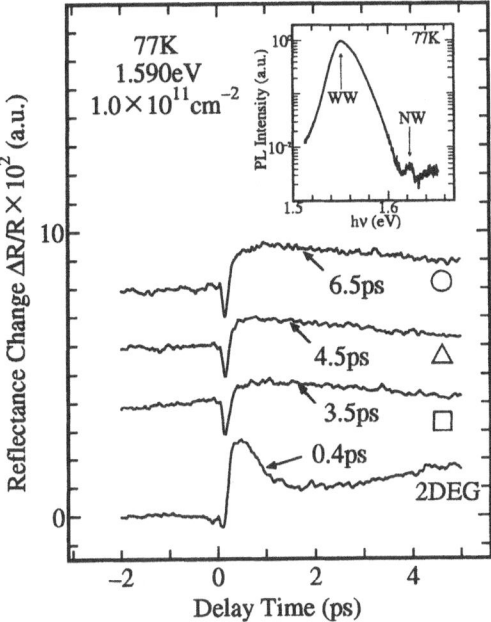

Fig. 1 Typical traces of the reflectivity change $\Delta R/R$ in various sample as a function of the delay time measured at $h\nu$=1.590eV. The inset shows a typical PL spectra at 77K.

triangle of which side is 325 nm, and (c) A square of 215 nm × 215 nm. The sizes are comparable to the carrier mean free path of 300-500 nm at 77 K.

The reflectance change at 77 K was measured as a function of time and energy using a mode locked Ti-sapphire laser (peak wavelength 785 nm, the auto correlation pulse width at the sample surface 80 fs). The pulse train was split into a pump and a probe beam, which were linearly polarized perpendicular to each other. The electrons excited in the quantum wells have two dimensional kinetic energy of 0 - 120 meV, and the decay processes was detected by measuring the reflectance change of the probe pulse[7,8]. The electron density excited was $10^{10}-10^{12}cm^{-2}$, where the carrier-carrier scattering would have a role in the decay process[9].

3 Reflectance change and Energy Relaxation in Patterned Structures

Figure 1 shows typical traces of the reflectance change $\Delta R/R$, at $h\nu$=1.590 eV in various samples as a function of time. The photoluminescence (PL) spectrum exhibits two peaks. The dominant peak at 1.550 eV is from the wide well and the subsidiary peak at 1.624 eV is from the narrow well. The apparent LO phonon energy is $\hbar\omega^{*}_{LO} =$

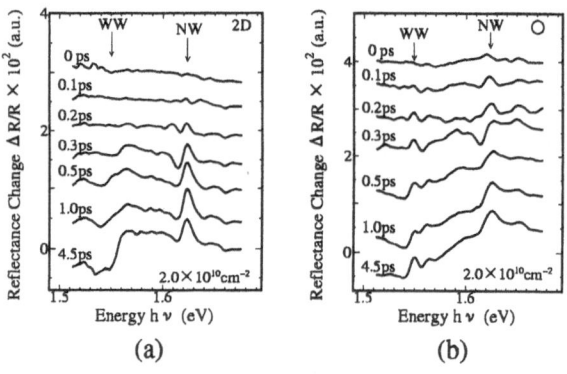

Fig. 2 The PR spectra for the as-grown 2DEG sample (a) and circular disk (b) are shown for various delay time.

$\hbar\omega_{LO}(1 + m_e^*/m_h^*) \sim 40\text{meV}$, so the characteristic time obtained at 1.590 eV represents the LO phonon emission time in the wide well. The fast decay on the order of 0.4 ps shown in a 2DEG sample disappeared in the disk samples, and we have longer time constants which have been achieved by assuming exponential decay with two time constants (rise and fall basically).

The decay time constant of 0.4 ps obtained for the 2DEG is on the same order but longer than that of LO phonon emission time in a bulk material of 0.24 ps [5]. It is notable that the decay time constants in disk samples are one order of magnitude longer. We investigated the decay time at various excitation intensity and found similar results. These results show that the buildup of the carrier density at the bottom of the quantum well, or the decay of the photo-excited carriers at high energies to the bottom, has been suppressed. To see this in more detail, the reflectance spectra were investigated as a function of time, which represent the change of the carrier density[7]. Figure 2 shows the results (The pump pulse is at t = 0). In Fig.2(a), the build up of excess carriers at $h\nu$=1.560 - 1.600 eV is obvious at t=0.3 ps, which persists for long time. This implies that we have achieved the increase of carriers in the wide quantum well at t=0.3 ps. Since the tunneling time from the narrow well is on the order of several pico-seconds[8], the build up of carriers is mainly due to the intra-band relaxation. In case of disk sample, on the other hand, Figure 2(b) shows that the build up of carriers in the wide quantum well is obvious at energies higher than 1.590 eV. Considering the apparent LO phonon emission energy of 40 meV, the shoulder at $h\nu$=1.590 eV shown in Fig. 2(b) might represent the bottle neck of LO phonon emission process. The weak increase of carriers at 1.500 - 1.590 eV indicates that the relaxation process is strongly suppressed in the sample.

As the origin of the suppression of LO phonon emission rate in semiconductors, various mechanisms have been proposed. The screening effect, Auger effect and the hot phonon were found to dominate the phenomena[10]. In any case, the suppression is a strong function

of the carrier density. We have tested the excitation intensity dependence of the decay time, but we could not find such an anomalous change within the density range studied. Therefore, we might think of another mechanism. One possibility is the strong modification of the carrier-carrier scattering in a disk, i.e, the modification of the screening, the modification of the Auger effect, or the modification of the scattering cross-section. The last two mechanisms are more or less due to the modification of the Coulomb scattering between carriers[6]. In Figure 1, the modification depends on the geometry of the sample. Therefore, we might take into account of the shape of the potential wall at the boundary of the disk.

On the basis of semi-classical model, the scattering cross section for the Rutherford scattering was investigated in terms of the collision parameter b. If the collision occurs after the reflection of carrier at the potential wall, the collision parameter b is modified by the trajectory change. We tested numerically how the collision parameter is modified by the geometry as well as the size, and found that the average value of b is a strong function of the geometry and size. We found that the scattering cross section is reduced in a circle as compared to square, in agreement with the observation. Further studies with Monte Calro method are desirable to clarify it.

4 Summary

The reflectance change in a GaAs/AlGaAs 2DEG disk has been studied using femto-second pump and probe method. In a conventional 2DEG sample, the energy relaxation time of 0.4 ps was obtained which is attributed to the net emission time of an LO phonon in a quantum well. In the disks, the time constants were found to be increased one order of magnitude. More over it depended on the geometry. It was attributed to the modification of the Coulomb scattering by the potential wall (geometry) of the disks.

Acknowledgments

This work has been partly supported by the Grant in Aid from the Ministry of Education, Science , Sports and Culture. Valuable and critical discussions by Prof. P. Vogl of Walter Schottky Institute, TU-Munich, is acknowledged. Use of the facilities of Nagoya University Venture Business Laboratory is acknowledged.

References

1. J. F. Ryan, *Hot Electrons in Semiconductors (ed) N. Balkan* (Oxford Univ.Press, 1998) Chap.8.
2. T. Ando *et al.*, Rev. Modern Phys. **54** (1982) 437.
3. J. Yamashita, Prog. Theor. Phys. **24** (1960) 357.
4. T. P. McLean *et al.*, J. Phys. Chem. Solids **16** (1960) 220.
5. M. Betz *et al.*, Phys. Rev. B **60** (1999) 11265.
6. F. X. Camescasse *et al.*, Phys. Rev. Lett. **77** (1996) 5429.
7. W. Fischler *et al.*, Appl. Phys. Lett. **68** (1996) 2778.
8. N. Suzumura *et al.*, Physica B **272** (1999) 378.
9. M. Hartig *et al.*, Phys. Rev. B **60** (1999) 1500.
10. K. Leo *et al.*, Phys. Rev. B **38** (1988) 1947.

Collective plasma response of interacting electrons localized in disordered GaAs/AlGaAs superlattices

Yu.A.Pusep[1], W.Fortunato[1], P.P.González-Borrero[2], A.I.Toropov[3], and J.C.Galzerani[1]

[1] Departamento de Fisica, Universidade Federal de São Carlos, CP 676, 13565-905 São Carlos, Brazil
[2] Instituto de Fisica de São Carlos, Universidade de São Paulo, 13560-970 São Carlos, SP, Brazil
[3] Institute of Semiconductor Physics, 630090 Novosibirsk, Russia

Abstract The collective plasmon-like excitations were studied by Raman scattering in intentionally disordered GaAs/AlGaAs superlattices with various strengths of disorder and different electron densities. We found that in the presence of disorder, the collective excitations tend to form coherently oscillating clusters with a finite extent. It was shown that both the temperature and disorder distroy them.

The problem of interacting electrons in disordered potential attracted much attention soon after the importance of Coulomb electron-electron interaction in the metal-to-insulator transition was realized. The interplay between disorder and interaction influences the high frequency (optical) response of electrons. As a matter of fact, the electron-electron interaction plays a determinative role in the collective plasma oscillations. Therefore, it is clear that the reaction of the disordered electron system to a field of electromagnetic radiation provides a staightforward way to probe the effects of interaction and disorder.

It has been shown in [1,2], that in the presence of disorder, the localized electrons find resonance conditions adapting their phases and forming the coherently oscillating spatial clusters.

We already showed that Raman scattering serves as a tool to probe both the spatial (the coherence length) and the energy (the resonance frequency and the damping) characteristics of the collective excitations in disordered semiconductors, where the Raman selection rules are relaxed resulting in a specific asymmetry of the Raman lines associated with the collective excitations [3,4].

In this paper we study the collective reaction of electrons subjected to the random potential of the semiconductor superlattices, where both the localization and the strength of the electron-electron interaction can be controlled by the growth. The disorder was introduced by a controlled random variation of the well thicknesses, while the interaction was changed by the different doping.

The strength of disorder was characterized by the disorder parameter $\delta = \Delta/W$, where Δ is the full width at half maximum of a Gauss distribution of the energy of the noninteracting electrons calculated in the isolated quantum well and W is the width of the nominal miniband in the absence of disorder.

Some selected Raman spectra of the GaAs/AlGaAs superlattices with a fixed disorder and different electron concentrations and with a fixed electron concentration

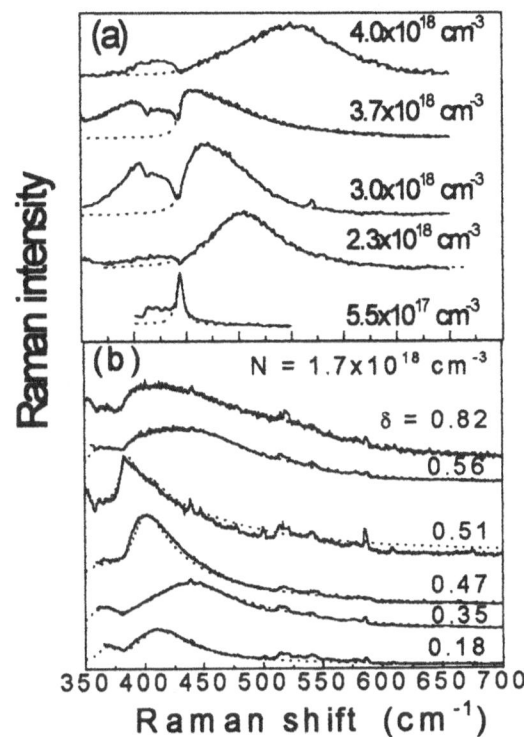

Fig. 1 Raman spectra of the AlAs-like collective coupled excitations measured at $T=10$ K in the GaAs/AlGaAs superlattices with different electron densities and a fixed disorder $\delta = 0.18$ (a) and with different disorders and a fixed electron density (b).

and varied disorder are demonstrated in the frequency range of the AlAs like optical phonons in Fig.1. The dotted lines present the calculated Raman intensities. The values of the parameters determining the shapes of the relevant Raman lines are collected in Fig.2.

With the increase of the doping level the plasmon-like line shifts to the high frequencies and acquires a strong asymmetry. Meanwhile, the broadening Γ rises due to the increase of the disorder induced by the random impurity potential. However, a strong line narrowing accompanied by a pronounced red shift was found beginning with the electron concentration $N = 2.5 \times 10^{18}$ cm^{-3}.

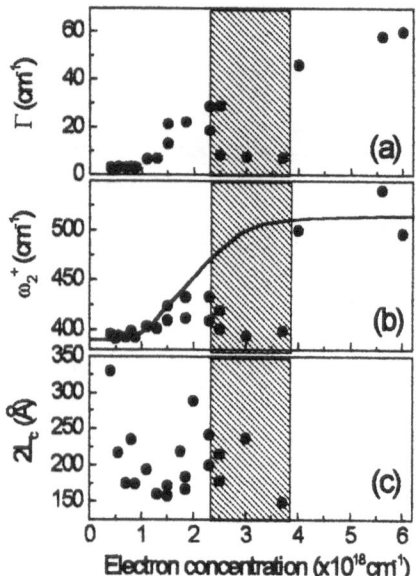

Fig. 2 Dependencies of the damping constant (Γ), the frequency (ω_2^+) and the coherence length (L_c) of the AlAs-like collective plasmon-LO phonon excitations on the electron density measured at $T=10K$ in the $(GaAs)_{17}(AlAs)_2$ superlattices with a fixed disorder ($\delta = 0.18$).

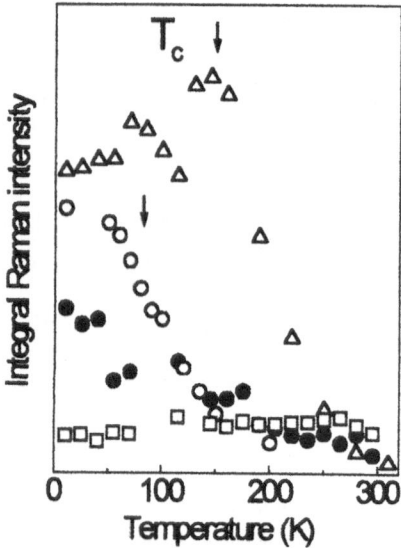

Fig. 3 Temperature dependencies of the integral intensities of the Raman lines associated with the collective plasmon-LO phonon excitations in the GaAs/AlGaAs intentionally disordered superlattices with different doping (N) and disorder (δ): $N = 7.0 \times 10^{17}$ cm^{-3}, $\delta = 18$ (triangles), $N=2.5 \times 10^{18}$, $\delta = 0.18$ (opened circles), $N = 1.7 \times 10^{18}$, $\delta = 0.47$ (closed circles). The integral intensity of the LO(GaAs) Raman line is given as a reference (squares).

According to the theory presented in [1,2], the coherent collective response of interacting localized electrons is responsible for the observed spectral narrowing. At low electron densities, the long distances between electrons do not allow an effective interaction between them. By increasing the electron densities, the cooperative behavior of the interacting electrons becomes dominating, resulting in the formation of a strongly correlated electron plasma in the presence of disorder, when the dynamical many-particle interaction almost completely screens out the effects of the random potential fluctuations. As a consequence, the broadenings Γ acquire the values very close to those found in the low-doped samples.

With a further increase of the doping, an abrupt increase of both the line broadening and the spectral line shift was observed at $N = 4.0 \times 10^{18}$ cm^{-3}. At such high electron concentrations the impurity disorder breaks down the correlation effects.

The dependence of the frequency of the coupled excitations on the electron concentration calculated for the case when the line shift is determined by the occupation of the miniband of the perfectly ordered superlattice is shown in Fig.2 by the solid line.

The effect of the line narrowing was found also with the increase of the disorder around $\delta = 0.5$.

The effects of the temperature on the collective excitations in the disordered superlattices are shown in Fig.3 where the dependencies of the integral Raman intensities are plotted. The integral intensity has a meaning of the number of collective states contributing to the Raman process. Therefore, the decrease of the integral intensity is associated with the destruction of the coherently oscillating clusters, which takes place at the critical temperature T_c.

Thus, we can state that qualitative agreement with the theory[1,2] was found; however, abrupt alterations of both the damping and the line shift, which were not theoretically predicted, were indeed observed in the experiment. These abrupt changes probably evidence phase transitions occoring in the electron plasma in the presence of disorder, caused by the formation of a strongly correlated plasma state in the form of coherently polarized dynamic clusters. The formation of such dynamic clusters is clearly observed in the temperature behavior of the integral intensities of the Raman lines associated with the coupled collective modes.

Acknowledgments: the financial supports from FAPESP and CNPq are gratefully acknowledged.

References

1. C.Metzner and G.H.Döhler, Phys. Rev B 60, (1999) 11005.
2. C.Metzner and G.H.Döhler, Physica E, 7, (2000) 718.
3. Yu.A.Pusep, M.T.O.Silva, J.C.Galzerani, N.T.Moshegov and P.Basmaji, Phys.Rev. **B58**, 10683 (1998)
4. Yu.A.Pusep, M.T.O.Silva, N.T.Moshegov and J.C.Galzerani, Phys. Rev. B 61, (2000) 4441.

Ultrafast relaxation dynamics of photoexcited carriers in an $In_{0.53}Ga_{0.47}As/InP$ multiple-quantum well

Y. Hamanaka[1], A. Nakamura[1], K. Tanase[1], R. Ohga[2], Y. Nonogaki[2], Y. Fujiwara[2], Y. Takeda[2]

[1] Center for Integrated Research in Science and Engineering and Department of Crystalline Materials Science, Nagoya University, Chikusa-ku, Nagoya 464-8603, Japan e-mail: hamanaka@cirse.nagoya-u.ac.jp
[2] Department of Materials Science and Engineering, Nagoya University, Chikusa-ku, Nagoya 464-8603, Japan

Abstract We heve investigated relaxation dynamics of photoexcited carriers and formation dynamics of excitons in a 10 nm-$In_{0.53}Ga_{0.47}As$/InP multiple quantum well at 77 K by femtosecond pump and probe spectroscopy. The observed spectral change of the exciton absorption band induced by the pump pulse is composed of broadening and bleaching components. The spectral broadening is attributed to scattering between excitons created by the probe pulse and free carriers generated by the pump pulse, and the bleaching is mainly due to the phase space filling. The time evolution of the absorption change shows that formation of excitons occurs on a time scale of ~ 20 ps after the intraband relaxation of photoexcited carriers.

1 Introduction

An understanding of the dynamics of carrier energy loss processes in InGaAs bulk and quantum well semiconductors with a band gap corresponding to the wavelength of optical communications is of importance in the design of optoelectronic devices as well as fundamental interest [1], [2]. Thermalization and cooling of carriers and exciton formation have been extensively investigated for quantum well (QW) structures of GaAs [3], [4], ZnCdSe [5] and CdSe [6]. A recent study of $In_{0.47}Ga_{0.53}As/In_{0.47}Al_{0.53}As$ heterostructures has shown carrier cooling dependent on the well width due to reduction of the electron-LO phonon coupling [7]. In QWs one can investigate hole relaxation between light-hole and heavy-hole bands which are split due to the quantum confinement effect.

In this paper, we report on relaxation dynamics of photoexcited carriers and formation dynamics of excitons in an $In_{0.53}Ga_{0.47}As$/InP multiple-QW with a well-width of 10 nm at 77 K in the time region of 100 fs to 1ns. We have found the intraband relaxation within ~ 1 ps followed by the exciton formation on a time scale of ~ 20 ps.

2 Experimental

A multiple-quantum well (MQW) structure was stacked on an InP substrate by using organo-metallic vapor phase epitaxy (OMVPE) separated by a 30 nm-thick InGaAs buffer layer. We used a MQW sample which was composed of 125 periods of 10 nm-wide $In_{0.53}Ga_{0.47}As$ well layers and 50 nm-wide InP barrier layers. Transient absorption spectra were measured using a femtosecond optical parametric amplifier based on a Ti:sapphire laser

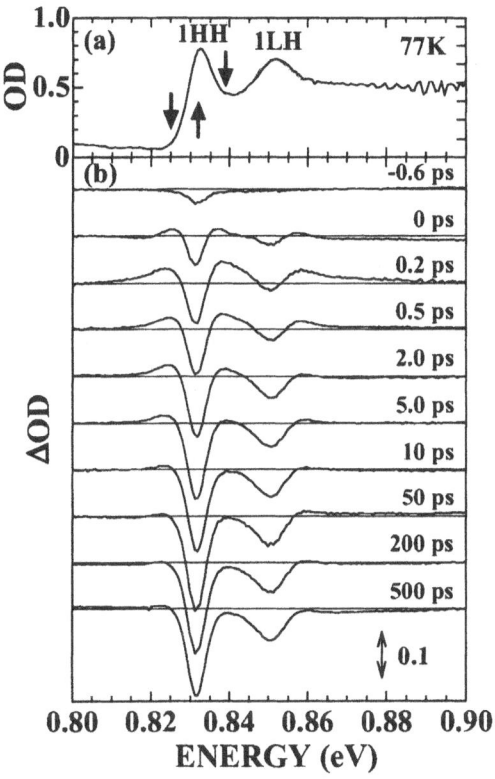

Fig. 1 (a) Linear absorption spectrum around the fundamental absorption edge at 77K. (b) Differential absorption spectra for various delay times.

amplifier operating at 1 kHz and a polychromator equipped with an InGaAs photodiode array. The pulse duration and photon energy of the pump pulse were 120 fs and 1.1 eV (below the band gap of InP), respectively. The probe pulse with a bandwidth in the range of 0.75 - 1.10 eV was generated by use of a water cell. Excitation density of electron-hole pairs was 4×10^9 cm^{-2}.

3 Results and Discussion

Figure 1 (a) shows the absorption spectrum around the fundamental absorption edge in $In_{0.53}Ga_{0.47}As$/InP at 77 K. We clearly see absorption peaks due to the $n = 1$ heavy-hole (1HH) and light-hole (1LH) excitons at 0.832 and 0.852 eV, respectively, and the energy splitting between them is 20 meV. This energy is smaller than energies of InGaAs LO phonons which consist of GaAs-like (34 meV) and InAs-like modes (29 meV). The differential absorption spectra are shown in Fig. 1 (b) for several delay times (t_d) between the pump and probe pulses. At $t_d = 0.2$ ps, we observe absorption bleaching at the 1HH and 1LH exciton peaks and absorption increase at the both sides of the peak. The absorption increase indicates that the 1HH and 1LH exciton bands are broadened by generation of carriers with pump pulse excitation. The bleaching at the peak arises from both broadening and reduction of the exciton oscillator strength due to a phase-space filling effect [5]. As the delay time proceeds, the absorption increase (0.825 and 0.839 eV) reduces, while the bleaching (0.832 eV) increases. At $t_d = 10$ ps, the absorption increase becomes very small, and the bleaching reaches a maximum.

Time evolutions of the differential absorption ΔOD were measured at 0.825, 0.839 eV (downward arrows) and 0.832 eV (upward arrow) within the 1HH exciton band. As shown in Fig. 2, ΔOD (> 0) at 0.825 and 0.839 eV exhibits a maximum around 1 ps, which yields a ~ 1 ps rise time of the additional broadening. Since the additional broadening is attributed to the increase in scattering rate between excitons created by the probe pulse and free carriers generated by the pump pulse, the rise time is ascribed to the intraband relaxation time of photo-generated carriers as has been discussed in Ref. 8 and 9. At the moment of excitation, the cross section of carrier-exciton scattering is small, because the wavevector of free carriers with a large kinetic energy is larger than the inverse of exciton Bohr radius. As the delay time proceeds, the cross section increases as a result of decrease in kinetic energy of free carriers via intraband relaxation. Consequently, ΔOD increases reflecting a relaxation of electrons and holes into the conduction and valence band edges with emission of LO phonons. The decay behaviors of ΔOD (0.825 and 0.839 eV) indicate the decrease in the scattering cross section. As the cross section of the exciton-exciton scattering is smaller than that of carrier-exciton scattering [10], the decay of ΔOD can be attributed to formation of excitons from electrons and holes around the band edge [9].

The exciton formation dynamics is also reflected on the time evolution of the absorption bleaching (0.832 eV). The bleaching exhibits a two-step rise behavior: ΔOD at the first step (< 1 ps) mainly arises from the spectral broadening, and at the second step the bleaching due to reduction of exciton oscillator strength increases. As the change in the oscillator strength is proportional to the exciton density, the observed rise time of ~ 20 ps is attributed to the exciton formation time. This rise time corresponds well to the decay time of ΔOD measured at 0.825 and 0.839 eV. Finally, we point out that the observed formation time is shorter than that for a bulk InGaAs film (~ 30 ps [2]), and longer than that for ZnCdSe/ZnSSe QWs (~ 5 ps [9]). The different for-

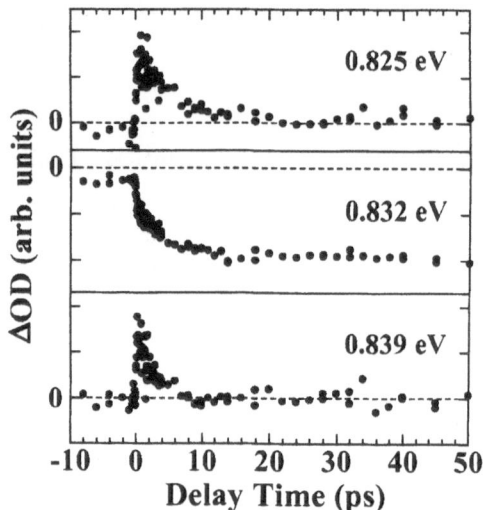

Fig. 2 Time evolutions of differential absorption at several photon energies at 77K.

mation times suggests the difference in electron-phonon coupling strength.

4 Summary

The intraband relaxation and exciton formation dynamics have been investigated for 10 nm-$In_{0.53}Ga_{0.47}As$/InP MQW at 77K. From the results of femtosecond pump and probe measurements, we have found that hot carriers relax to the band bottom within ~ 1 ps and excitons are formed on a time scale of 20 ps.

References

1. K. Kash and J. Shah, Appl. Phys. Lett. **45**, (1984) 401.
2. Y. Hamanaka, D. Nishiwaki, Y. Nonogaki, Y. Fujiwara, Y. Takeda and A. Nakamura, Physica B **272**, (1999) 391; J. Lumin. **83-84**, (1999) 49.
3. W. H. Knox, C. Hirlimann, D. A. B. Miller, J. Shah, D. S. Chemla and C. V. Shank Phys. Rev. Lett. **56**, (1986) 1191.
4. T. C. Damen, J. Shah, D. Y. Oberli, D. S. Chemla, J. E. Cunningham and J. M. Kuo Phys. Rev. B **42**, (1990) 7434.
5. T. Tokizaki, H. Sakai and A. Nakamura Phys. Rev. B **55**, (1997) 15776.
6. F. Yang, G. R. Hayes, R. T. Phillips and K. P. O'Donnell Phys. Rev. B **53**, (1996) R1697.
7. S. Bolton, G. Sucha and D. Chemla Phys. Rev. B **58**, (1998) 16326.
8. Y. Feng and H. N. Spector IEEE J. Quantum Electron. **24**, (1988) 1659.
9. T. Tokizaki, H. Sakai, G. Kogano and A. Nakamura Jpn. J. Appl. Phys. **38**, (1999) 3562.
10. A. Honold, L. Schultheis, J. Kuhl and C. W. Tu Phys. Rev. B **40**, (1989) 6442.

Spin relaxation of negatively charged excitons in CdTe-based quantum wells.

P. Kossacki[1,2], **V. Ciulin**[2], **M. Kutrowski**[3], **J-D. Ganière**[2], **T. Wojtowicz**[3], **B. Deveaud**[2]

[1] Institute of Experimental Physics, Warsaw University, Hoża 69, 00-681 Warsaw, Poland, E-mail: Piotr.Kossacki@fuw.edu.pl
[2] Physics Department, IMO, Swiss Federal Institute of Technology, Lausanne CH-1015, Switzerland
[3] Institute of Physics, Polish Academy of Sciences, Al.Lotników 32/46, Warsaw, Poland

Abstract We present a study of the spin relaxation of negatively charged excitons in CdTe-based quantum wells. The influence of the electron gas concentration and the excitation energy is discussed. We find that the relaxation time decreases dramatically when increasing the excitation energy. This time is shown to be a common function of the energy distance from the X^- photoluminescence independently on the carrier concentration.

1 Introduction

Phenomena related to spin relaxation in quantum wells became recently a subject of intense research. This interest has been stimulated by the idea of spin electronics [1]. So far most of the work was focused on the systems without preexisting carriers or containing high concentration of degenerate carrier gas [2,3]. Little work was devoted to the intermediate concentrations for which the optical properties are dominated by the charged exciton (trion) transition [4,5]. Such a system is particularly simple for a spin flip study. The trion state is a singlet and spin flip process is governed by the flip of an electron (for X^+) or a hole (for X^-). Moreover the absence of dark states makes polarisation analysis much simpler than in the case of neutral exciton.

2 Samples

The samples used in our work were one side modulation doped CdMgTe/CdTe heterostructures containing a single 80Å quantum well. The thickness of the doping layer was graded along the sample [6]. Therefore the concentration of electrons in the quantum well can be varied by selecting the spot on the sample. As a second way of control of the electron gas concentration we used above-barrier illumination, which increases carrier concentration due to transfer of electrons from the surface states. Such an arrangement assured a wide range of accessible concentrations (up to about $1.6 \times 10^{11} \text{cm}^{-2}$). Carrier concentration was monitored by the measurement of reflectivity spectra. At the lowest concentration only the neutral exciton line was observed. When increasing above-barrier illumination (increasing electron concentration) the line related to charged exciton appeared and became dominant in the higher concentration range. At the highest concentration, the line of charged exciton transition was transformed to the broad structure characteristic for Fermi edge singularity. As the electron concentration was increased the distance between X and X^- transitions increased from about 2.5 meV up to almost 6 meV [7,8]. At higher concentrations also the shift between absorption and photoluminescence was observed (Moss-Burstein shift). The identification of the neutral and

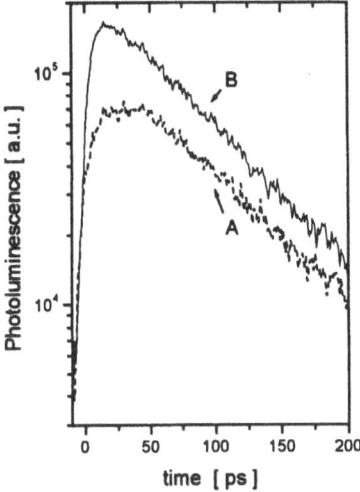

Fig 1. X^- photoluminescence versus time excited in σ^- polarisation (at time = 0) and detected in σ^+ polarisation. Lines A and B represent different energies of excitation as marked on Fig. 2. Measurement was done in helium bath at temperature 2K on quantum well containing about $4 \times 10^{10} \text{cm}^{-2}$ electrons.

charged exciton transitions was confirmed by measurement in magnetic field. The observed lines followed the appropriate selection rules known for X and X^- [4].

3 Experiment and discussion

Charged exciton spin relaxation was measured directly as the rise time of the photoluminescence excited by circularly polarised light of opposite helicity. The PL was excited by a mode-locked titanium sapphire laser providing 2 ps pulses with a repetition rate of 80MHz. The signal was collected by a streak camera with a final resolution better than 4 ps. Any significant heating of the carrier gas was avoided by the use of very low excitations (less than 50pJ/cm²). The excitation with energy lower than the energy of the neutral exciton transition assured a precise selection of the states participating in spin relaxation. In particular, no neutral exciton states were created. Therefore the whole spin relaxation process was determined by the spin flip of the hole in the singlet state of the charged exciton. The typical temporal profile of the PL signal is presented on Fig 1. It was analyzed in frames of simple two level rate equation model with two parameters (rise time τ_R and decay time τ_{dec}). The decay time was found to be always about 65 ps what is in good agreement with values reported previously for radiative recombination of X^- in

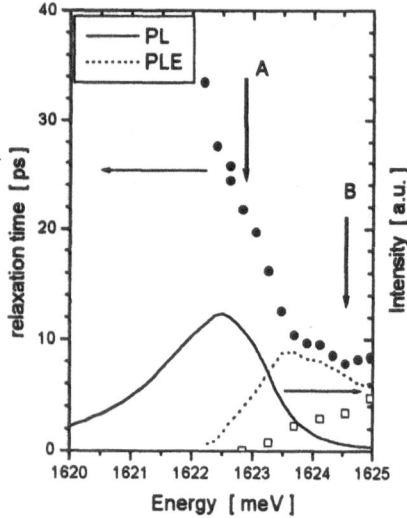

Fig. 2. X⁻ spin flip time versus energy of excitation (circles) measured on sample with intermediate electron concentration about 4×10^{10} cm^{-2}. The solid line represents time integrated photoluminescence spectra, dashed line represents PLE spectra. Vertical arrows A, B denote energies of time-resolved measurements shown on Fig. 1. Open squares denote photoluminescence rise time measured with crossed linear polarizations of excitation and detection

CdTe based quantum wells [9]. The relaxation time was obtained from τ_R and τ_{dec} using relation: $1/\tau_{SF} = 1/\tau_R - 1/\tau_{dec}$.

The example result of the excitation energy scan is presented on Fig. 2. We found that the relaxation time is about 35 ps, when the excitation coincided with the energy of the photoluminescence peak. When varying the excitation energy from the PL peak to the absorption maximum, a decrease of the relaxation time from 35 ps to 8 ps was observed. This means that charged excitons possessing a higher kinetic energy exhibit faster spin relaxation. One should note, however that the lower limit of relaxation time must be taken with precaution. The lowest measured value is limited by the energy relaxation within the spin subband. This time was estimated experimentally by the measurement of the PL rise time with crossed linear polarizations of excitation and detection. Indeed we found that the rise time is increasing with increasing the excitation energy and approaches the value measured in crossed circular polarizations (~5 ps). This shows the natural limit of our experiment, however longer relaxation times are reliable and can be interpreted as a spin flip time of the heavy hole forming singlet trion state. The higher limit 35 ps represents a spin relaxation of the hole in cold charged exciton in the state which participate in radiative transition. The excitation in higher energy leads to the formation of states possessing a nonzero kinetic energy and before radiative recombination, energy relaxation takes place. The spin flip time versus excess energy is presented on Fig 3 for three different concentrations of electron gas. It is interesting to note that the observed times almost do not depend on the

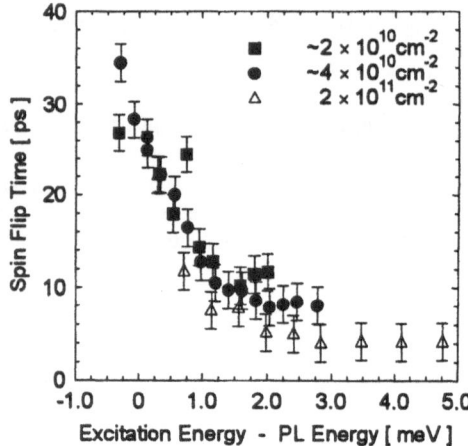

Fig. 3. X⁻ spin flip time versus energy difference between excitation and photoluminescence.

concentration over whole range used in experiment (from 2×10^{10} cm^{-2} to about 2×10^{11} cm^{-2}). Experimentally observed variation of the spin flip time might be interpreted in terms of well known mechanisms of spin relaxation such as Bir-Aronov-Pikus and Dyakonov-Perel. Both of them lead to a reduction of the hot carrier spin flip time in comparison to cold one (see also [5] on this conference). However the weak sensitivity on the electron concentration suggests negligible importance of relaxation processes based on the scattering on free carries. It allows to exclude some of the spin relaxation channels. Preliminary calculations show that it might be necessary to take into account the complex stucture of the valence band with the light-heavy hole mixing for higher wave vectors.

4 Conclusions

In conclusion we have measured the spin relaxation time of negatively charged excitons in CdTe-based quantum wells. We find that the relaxation time decreases dramatically when increasing the excitation energy. No significant influence of the electron concentration on the measured X⁻ spin flip time was observed (in the range from 2×10^{10} cm^{-2} to about 2×10^{11} cm^{-2}).

Acknowledgement

This work has been partially supported by Polish Committee for Scientific Research grant 2 P03B 045 19 and Swiss FNRS contract 2100-049538.96/1.

References

1. R. Fiederling et al., Nature 402, (1999) 787.
2. L. Vina, J. Phys. Cond. Matt. 11 (1999) 5929 and ref. therein
3. A. Vinattieri et al., Phys. Rev. B 50 (1994) 10868.
4. K. Kheng et al., Phys. Rev. Lett. 71, (1993) 1752.
5. E. Vanelle et al., Phys. Rev. B 62, (2000) 2696, and this conference
6. T. Wojtowicz et al., Appl. Phys. Lett. 73, (1998) 1379.
7. P. Kossacki et al., Phys. Rev. B 60, (1999) 16018.
8. V. Huard et al., Phys. Rev. Lett. 84, (2000) 187.
9. V. Ciulin et al., to be published and this conference.

Kinetics of a low density system of indirect excitons in double quantum wells

V.I. Yudson *

[1] Center for Frontier Science, Chiba University, 1-33 Yayoi-cho, Inage-ku, Chiba 263-8522, Japan
[2] Institute of Spectroscopy, Russian Academy of Sciences, Troitsk, Moscow reg. 142090, Russia

Abstract We study conditions for the formation of a spatially inhomogeneous state with puddles of a dense phase in a system of indirect excitons in an optically pumped double quantum well. It is shown that in a low excitation regime the dipole-dipole repulsion of indirect excitons modifies the balance between the rate of exciton collection by a puddle of a dense ("liquid") state and the recombination rate of the puddle. This leads to a new limitation on the puddle size and hence on the formation of an inhomogeneous state.

Recent experiments [1-4] with optically excited spatially separated electrons e and holes h in semiconductor double quantum wells (DQWs) have revealed rather complex behaviour of the system, e.g. temporal fluctuations of the photoluminescence [1-3]. Although some of these effects are claimed [1] to be connected with a coherent collective phase of an idealized system of spatially separated e and h [5], there has been proposed also another scenario [4] based on the formation of a e-h liquid state. Here we discuss this possibility for experimental conditions [4] with paying attention to effects of the repulsive dipole-dipole interaction of indirect excitons.

The paper [4] presents evidences for the formation of an e-h liquid state of density $n_l \sim 8.8 \times 10^{10} cm^2$ in 8nm GaAs - 5 nm AlGaAs - 8nm GaAs structure. However, for the optical excitation rate $P = 10$ mW/cm^2 and indirect exciton decay time estimated as $\tau = 2 \times 10^{-8}$, the maximal average exciton density n is not higher than $n = P\tau \sim 2 \times 10^8 cm^{-2}$. This means that the system state is not homogeneous but contains droplets (puddles) of a dense e-h phase. Because of the repulsion of indirect excitons with static dipole moments ed (d is the interlayer distance), the transition into a liquid state is not possible [5] for $d >> a$ (a is the exciton size), but is not excluded for small $d \sim a$ (a variational estimate [7] gives: $d < 1.1a$), that corresponds to the structure [4]. But the above estimate hold only for quasi-equilibrium systems. As is known since the 70s [8,9], the presence of generation-recombination processes in e-h systems leads to the limitation on the e-h droplet size R: a given drop tends to grow or to shrink until the rate of the exciton collection by the boundary of a two-dimensional (2D) puddle

$$G \sim \pi Rnu \qquad (1)$$

(u is the exciton mean velocity) balances both the exciton evaporation rate and the puddle recombination rate

* The work was supported in parts by the grant from the Ministry of Education of Japan, and by the program "Nanostructures" of the Russian Ministry of Science.

$\pi R^2 n_l/\tau$. This would give an estimate:

$$R < R_{max} \sim u\tau n/n_l \sim 2\mu m. \qquad (2)$$

But this estimate, valid for the "ordinary" excitons, ignores dipole- dipole interaction

$$V_{xx}(r) = e^2 d^2/\epsilon r^3 \qquad (3)$$

of indirect excitons. This interaction leads to the potential barrier

$$V(r) = 2e^2 d^2 n_l/\epsilon r \qquad (4)$$

for an exciton entering a e-h puddle. Eq.(2) is valid for

$$\max\{d, a, n_l^{-1/2}\} = r_{min} < r < R. \qquad (5)$$

For parameters [4] the barrier height $V_0 = V(r_{min})$ is estimated as

$$V_0 \sim 2e^2 d^2 n_l^{3/2}/\epsilon \sim 10K. \qquad (6)$$

At temperature $T = 2.5K$ this leads to a significant suppression of the exciton collection rate (1) by a factor

$$F_{th} = \exp[-V_o/T] \sim 0.02. \qquad (7)$$

A quantum-mechanical tunneling process is characterized by the factor

$$F_{tun} \sim \exp[-2\int_{r_{min}}^{r_{max}} dr \sqrt{2m[V(r) - E)]}], \qquad (8)$$

where r_{max} is the lowest quantity from the puddle size R and the classical turning point

$$r_2 = 2e^2 d^2 n_l/\epsilon E \qquad (9)$$

for the potential (4). The quantity F_{tun} is estimated as

$$F_{tun} \sim \exp[-\pi k_E r_2] \quad \text{for} \quad r_2 < R ; \qquad (10)$$

$$F_{tun} \sim \exp[-4k_E \sqrt{R r_2}] \quad \text{for} \quad R < r_2, \qquad (11)$$

where $k_E = \sqrt{2mE}$ is the wavevector of an exiton of energy E (with the typical $E \sim T$). The quantity F_{tun} turns out to be smaller than that F_{th} for the thermally activated penetration. With taking into account the factor F_{th} we arrive at the following new estimate for the maximal puddle size:

$$R < R_{max} \sim 40nm. \qquad (12)$$

This estimate shows that e-h puddles of a macroscopic size are hardly possible in the low excitation regime of [4], if these puddles are formed due to the "intrinsic" exciton condensation. Instead of this scenario of the nucleation in a spatially homogeneous system, there may be another scenario where e-h puddles are formed due to random potential variations. This random potential may stem from variations of the layer thickness, as well as from the alloy nature of the AlGaAs barrier. An estimate for the electron confinement energy variation δE, caused by the variation $\delta L = 0.4$nm of the QW thickness $L = 8$nm, gives $\delta E \sim 2.5$nm, which is comparable with the assumed binding energy of the e-h liquid [4]. In contrast to intrinsic e-h droplets, puddles formed due to potential variations do not need to be of the circular shape. Therefore, even configurations like "an e-h lake with entering e-h rivers" are possible. For such configurations excitons may be efficiently collected from the area of size $u\tau$. This would lead to a more moderate limitation on the lake size:

$$R_{max} \sim u\tau \sqrt{F_{th} n/n_l} \sim 3.5 \mu m. \tag{13}$$

One should have in mind that the above expression gives only the upper limit for R, while an actual puddle size depends on statistical properies of the disorder.

The formation of a mesoscopic e-h puddle would also be possible at the bottom of a sufficiently wide and deep potential well. However, such "regular" wells could hardly be formed by random potential variations.

In conclusion, we have shown that the interaction of indirect excitons forms a barrier for the exciton entering an e-h puddle. The barrier suppresses the exciton collection rate. In the presence of generation-recombination processes, this results in a strong limitation on the maximal puddle size and makes questionable the formation of e-h liquid puddles in ordered systems for low excitation regimes. However, we argue that the formation of mesoscopic e-h puddles may still take place in disordered double quantum wells.

References

1. L.V. Butov, et al., Phys. Rev. Lett. **73** (1994) 304.
2. V.V. Krivolapchuk, et al. Solid State Commun. **111** (1999) 49.
3. V. Negoita, D.W.Snoke and K.Eberl, Solid State Commun. **113** (2000) 437; Phys.Rev.B **61** (2000) 2779; V.Negoita, et al., Opt.Lett. **25** (2000) 572.
4. V.B. Timofeev, et al., Phys.Rev.B **61** (2000) 8420.
5. Yu.E. Lozovik and V.I. Yudson, JETP Letters **22** (1975) 274; Sov.Phys.JETP **44** (1976) 389.
6. Yu.E.Losovik and O.L.Berman, Zh. Eksp. Theor. Fiz. **111** (1997) 1879.
7. C.D.Jeffries and L.V.Keldysh, eds., *Electron-Hole Droplets in Semiconductors* (North-Holland, Amsterdam 1983).
8. R.M.Westervelt, in ref.[7], p.187.

Photoexcitation-energy-dependent capture dynamics of excitons in electronically isolated GaAs quantum wells

K. Fujiwara[1], H. T. Grahn[2], L. Schrottke[2], and K. H. Ploog[2]

[1] Kyushu Institute of Technology, Tobata, Kitakyushu 804-8550, Japan, e-mail: fujiwara@ele.kyutech.ac.jp

[2] Paul-Drude-Institut für Festkörperelektronik, Hausvogteiplatz 5–7, 10117 Berlin, Germany

Abstract The exciton capture and relaxation dynamics of a GaAs triple quantum well with different well thicknesses separated by thick $Al_{0.17}Ga_{0.83}As$ barriers has been investigated under various photoexcitation energy conditions. The three excitonic emission bands show clear quantitative differences in the exciton capture efficiency as well as in the emission dynamics, when the wells are excited close to the barrier band edge. The origin of these differences is probably related to the energy of the second subband of the thickest quantum well, which is located just below the barrier band edge.

1 Introduction

In a system consisting of quantum wells (QWs) with different thicknesses separated by sufficiently thick barriers, confined electron states in each well are considered to be independent. Therefore, the relevant photoluminescence (PL) properties of each well should exhibit an independent characteristics without any correlation between the QWs. However, in recent studies of electronically isolated QWs, unusual PL correlations [1,2] as well as Stokes and anti-Stokes transfer [3] between the QW excitonic bands have been reported, suggesting that the different wells are by no means independent with respect to excitonic radiative recombination as well as carrier capture and relaxation processes.

In this paper, we have investigated the exciton capture and relaxation dynamics in a GaAs triple QW system with different well thicknesses separated by thick barriers of $Al_{0.17}Ga_{0.83}As$ using steady-state and time-resolved PL experiments. We find that the excitonic PL dynamics varies between the different QWs and strongly depends on the energy state, where the excitons are photogenerated.

2 Experimental

A triple QW sample was grown on a GaAs (100) substrate by molecular beam epitaxy. Nominal widths of the undoped GaAs QWs are $L_Z = 7.8$ nm (QW1), 5.5 nm (QW2) and 3.5 nm (QW3) starting from the substrate side, which are electronically isolated by 36 nm thick $Al_{0.17}Ga_{0.83}As$ barriers. These heterostructures were embedded in a pair of 72-nm $Al_{0.17}Ga_{0.83}As$ layers and further confined using $Al_{0.3}Ga_{0.7}As$ barriers. Steady-state PL spectra for this sample were measured at 4 K in a He cryostat using a halogen lamp as a weak excitation light source and a monochromator and computer controlled photon counting system for detection. Spectrally

and temporally resolved PL transients were measured at (15–20) K in a closed-cycle He cryostat using a pyridine-2 dye laser with 10 ps pulses for excitation and a streak camera system for detection.

3 Results and discussion

Figure 1 shows a cw PL spectrum of the sample at an excitation wavelength of 696 nm and three PL excitation (PLE) spectra measured for QW1, QW2 and QW3. Note that the PL spectra of QW2 and QW3 occurring at higher energies exhibit higher intensities than the low-

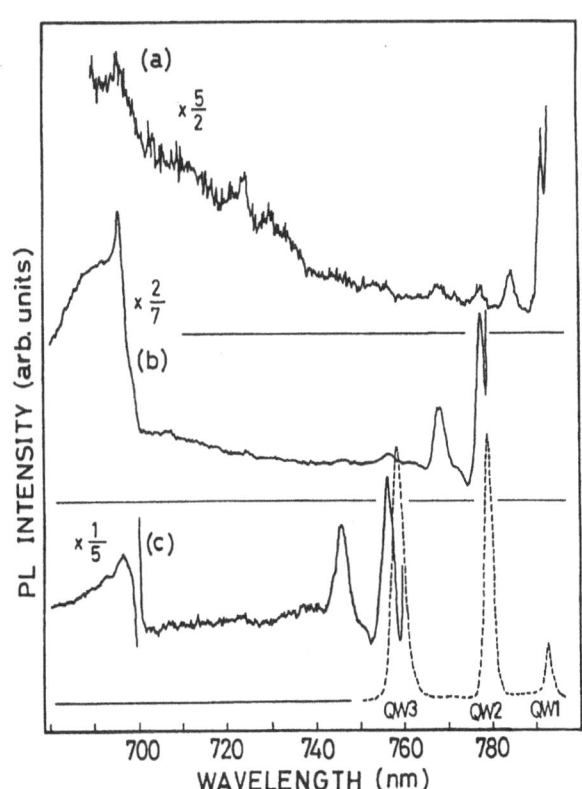

Fig. 1 PL (dashed curve) and PLE (solid curves) spectra of the triple QW sample at 4 K. The excitation wavelength for the PL spectra is 696 nm. The detection wavelengths for the PLE spectra are set to (a) 794.7 nm (QW1), (b) 780.5 nm (QW2) and (c) 761.0 nm (QW3). The spectra are shifted vertically for clarity.

est energy peak of QW1. All three lines are red-shifted from their corresponding exciton resonance, which are clearly observed in the PLE spectra. When comparing the three PLE spectra, clear correlations are observed between the three wells. For example, the PL intensity of QW1 is enhanced due to the excitonic resonances of the other two wells. These PLE correlations are seen both for the heavy-hole and light-hole exciton resonances. The weaker PLE peak appearing around 725 nm in Fig. 1(a) is due to the resonant excitation of electrons and holes in the second (n=2) subbands of QW1 [1]. The electron level is located just below the band edge of the barrier.

For even shorter wavelengths (about 700 nm) in Fig. 1, when the excitons are resonantly generated at the barrier excitonic state, the PL intensity of QW1 is much less enhanced than for the other two QWs. However, for QW2 and QW3, a distinct enhancement of the PL signal is observed at the barrier resonant excitation. This implies that excitons are efficiently captured by QW2 and QW3, for which the n=2 states are located near or above the barrier band edge.

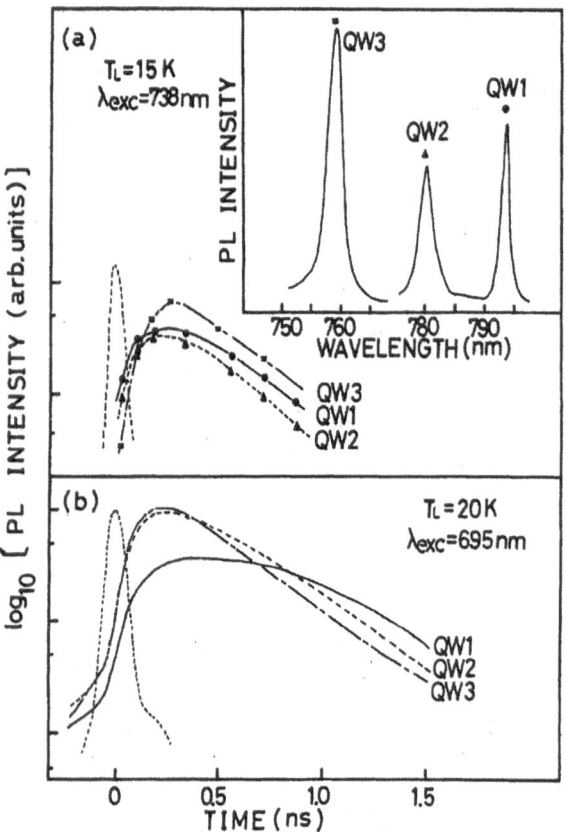

Fig. 2 Semi-logarithmic plots of the PL intensity as a function of time for the three excitonic emission bands measured at excitation wavelengths of (a) 738 nm and (b) 695 nm. The inset in (a) shows the corresponding time-integrated PL spectrum excited at 738 nm.

In order to study the dynamical behavior of exciton capture from the barrier to each well, the PL transients were measured at barrier resonant and off-resonant photoexcitation conditions. Figure 2(a) shows the PL intensity of the three QWs on a logarithmic scale as a function of time for excitation at 738 nm. In this case, all three wells are directly excited above the heavy-hole exciton resonance energies by more than 40 meV. The PL rise times are relatively fast for all three wells, since the fast energy relaxation processes assisted by phonon emission are permitted within each well. Although some differences in the PL intensities exist between the wells as shown in the inset of Fig. 2(a), the PL intensity of QW1 is even larger than the one of QW2 in contrast to the barrier resonant excitation. The wavelength-integrated PL transients for the three excitonic emission bands at barrier resonant excitation (695 nm) are displayed in Fig. 2(b). As expected, the PL rise time of QW1 is now significantly slower than the rise times of QW2 and QW3. This observation indicates that the exciton capture processes are strongly influenced by the energy position of the n=2 excited state relative to the barrier band edge. Therefore, under this excitation condition, an efficient resonant electron capture [4] in QW1 is not possible. We attribute the dramatically different PL dynamics observed for the three excitonic emission bands to the different energy position of the n=2 states relative to the barrier band edge. These excitation energy dependent PL dynamics directly indicate that exciton capture processes are not uniform across electronically isolated QW excitonic bands.

4 Conclusions

The PL properties of a quantum system consisting of three electronically isolated GaAs QWs with different well thicknesses have been investigated by low temperature cw and time-resolved PL experiments. It is found that the distribution of the PL intensity over the three excitonic emission bands and their dynamics strongly depend on the energy state, where the excitons are photogenerated. The photoexcitation energy dependent evolution of the carrier capture and radiative recombination processes indicate that the PL dynamics is significantly influenced by the energy positions of the n=2 states relative to the barrier band edge.

Acknowledgements   This work was supported in part by the Grand-in-Aid for Scientific Research (No.11650020) from the Ministry of Education, Science, Sports and Culture of Japan.

References

1. L. Schrottke, H. T. Grahn, and K. Fujiwara, Phys. Rev. B **56**, (1997) 13321.
2. K. Fujiwara *et al.*, Inst. Phys. Conf. Ser. No. **166** (IOP, Bristol, 2000) p. 103.
3. A. Tomita, J. Shah, and R. S. Knox, Phys. Rev. B **53**, (1996) 10793.
4. A. Fujiwara *et al.*, Phys. Rev. B **51**, (1995) 2291.

Magnetic polaron dynamics at high excitation densities in Cd₁₋ₓMnₓTe/ZnTe multiple quantum wells

R. Pittini[1,2], **J.X. Shen**[1,2], **M. Takahashi**[1,2], **Y. Oka**[1,2]

[1] RISM, Tohoku University, Katahira, Sendai 980-8577, Japan

[2] CREST, Japan Science and Technology Corporation, Kawaguchi 332-0012, Japan
  e-mail: pittini@rism.tohoku.ac.jp

Abstract   We carried out polarized optical pump–probe experiments in $Cd_{1-x}Mn_xTe/ZnTe$ multiple quantum wells under high excitation to study the dynamics of the excitons and the exciton magnetic polarons in these materials. The high pump excitation was tuned in the continuum of the quantum wells and saturated the excitons there. Under this condition, the (otherwise paramagnetic) quantum wells showed a long-range magnetic order (induced by the strong pump excitation) persisting over 300 ps after the pump excitation. New excitons (generated by the probe beam) were absorbed directly in the spin–polarized Mn^{2+} environment, yielding an enhancement of the density of states at 10-60 meV below the energy of the lowest excitonic state. Furthermore, we found that, at high excitation densities, the excitons couple antiferromagnetically to the long-range 'background' magnetization of the Mn spins.

1. Introduction

Diluted magnetic semiconductors (DMSs) continue to attract a wide interest in the physics of ultrafast phenomena [1]. The sp-d exchange interaction enhances the effective g-factor of the band electrons and holes and strongly influences the dynamics of the carrier and exciton spins interacting with the Mn^{2+} spins. In addition, in DMSs quantum wells (QWs), quantum wires, quantum dots, the exciton wavefunctions are squeezed and confined to the lower dimensionality of the nanostructure. The quantum confinement further increases the radial overlap of the sp-d orbitals and therefore enhances the sp-d exchange interaction. The strong magnetic coupling between the carriers and the Mn ions affects the dynamical transport and the lifetime of the carriers in DMSs nanostructures. Therefore, the Mn concentration and an external magnetic field, both influence the electronic performance of a DMSs based nanostructure device. Furthermore, the ionic spins of a DMS material order ferromagnetically at sufficiently high carrier densities [2]. Such a ferromagnetic order can be obtained by doping carriers [3], or by optical pumping (this work). The latter method offers the possibility of a direct control of the excitonic spins by means of the optical polarization. In particular, it permits to verify whether the assumption made in Ref. 2 of a ferromagnetic interaction between excitons and ionic spins (low-density limit) is still valid at the moderately high carrier densities required for the onset of long-range order.

2. Experimental details

$Cd_{1-x}Mn_xTe/ZnTe$ multiple quantum wells (MQWs) were grown by hot wall epitaxy on GaAs (001) substrates.

Transient pump-probe absorption experiments were performed using an optical parametric amplifier (OPA) seeded and pumped by 120 fs short pulses from a titanium–sapphire laser [2]. The output of the OPA was used as pump beam for the pump-probe experiments, with a pulse energy varying between 1 and 8000 $\mu J/cm^2$ on the sample. The photon energy of the pump pulse was tuned above the energy gap of the DMS well material but still below the energy gap of the barrier material. To probe the time variation of the optical absorption after the pump excitation, we used a white spectrum generated with a sapphire platelet. The polarization of the pump and probe pulses could be varied from linear to circular by inserting phase shifters. The optical experiments were performed between 1.5 and 300 K, in magnetic fields up to 7 T. To measure the optical transmission of the DMSs nanostructure samples, the GaAs substrate had to be etched away because it strongly absorbs the light at the energies of the excitons of the DMSs nanostructures.

3. Results and discussion

The optical pump-probe signal strongly depends on the exciton density, which is in turn controlled by the intensity of the pump pulse. In this work, we took care to saturate the excitons in the quantum wells [4,5], in order to observe the transient absorption signal of the Mn polarization and trying to induce a long-range magnetic order.

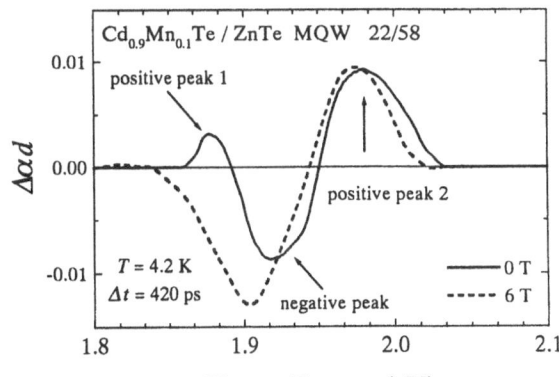

Fig. 1: Lineshape of the pump-probe spectrum of $Cd_{0.9}Mn_{0.1}Te/ZnTe$ MQWs (22 Å /58 Å) at 420 ps after the pump excitation.

In Fig. 1, we plot the transient absorption spectrum taken at 420 ps after the pump excitation in $Cd_{0.9}Mn_{0.1}Te/ZnTe$ MQWs with 100 repetitions of 22 Å wide quantum wells alternated to 58 Å wide barriers. In the pump-probe spectra, we observed three signals, a deep negative signal and two positive transient absorptions. The negative pump-probe signal arises from the bleaching of the exciton absorption in the wells and is proportional to the carrier and exciton density. The intensity of the negative pump-probe signal drops exponentially with a decay time of 409 ps, a little faster compared to the exciton recombination time of 505 ps observed for the PL intensity [6]. The two positive peaks are clearly recognized in Fig. 1. Positive pump-probe signals are very peculiar because they indicate that new states are created by the pump pulse in the quantum wells. Similar spectra were measured for comparable samples. The intensity of positive peak 1 was found to be proportional to the Mn concentration and disappears when a magnetic field of 6 T is applied parallel to the sample growth direction (Fig. 1). We remark that when a magnetic field of 6 T is applied, the pump-probe spectra measured by linearly polarized light (Fig. 1) shift only by 10 meV, i.e. much less than the Zeeman splitting observed for the excitonic absorption with circularly polarized light. Furthermore, with increasing time delay, the energy of 'positive peak 1' shifts exponentially towards the exciton energy [4,6]. This positive transient absorption is assigned to the signal of the exciton magnetic polaron and is detected only after a significant number of excitons have already decayed after polarizing the Mn spin through the sp-d exchange interaction. This peak appears below the QW exciton energy because energy is gained through the alignment of the Mn spins. Therefore, the energy difference between the negative pump-probe peak and the low-energy positive pump-probe peak (Fig. 1) is a direct measure of the Mn spin polarization in the sample [4,6]. Instead, 'positive peak 2' is attributed to the enhancement of the density of states appearing at the high-energy side of the band edge as a consequence of the Coulomb screening and becoming important for a high density of the carriers created in the wells by the pump pulse [6].

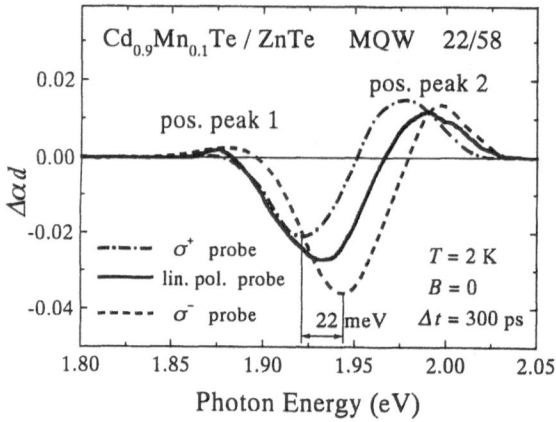

Fig. 2: Lineshape of the pump-probe spectrum of $Cd_{0.9}Mn_{0.1}Te/ZnTe$ MQWs (22 Å /58 Å) measured at 300 ps after the σ^+ polarized pump excitation for several probe polarizations.

The spectrum shown in Fig. 1 was measured with linearly polarized pump and probe pulses. This allowed to measure the mean pump-probe signal. But to study the exciton magnetic polaron dynamics it is necessary to use circularly polarized

pump and probe beams. In Fig. 2, we display the pump-probe spectra measured for the same sample with a time delay of 300 ps with circularly polarized probe pulses in zero field. The pump pulse with a fluence of 0.6 mJ/cm² was σ^+ polarized. The spectrum taken with a linearly polarized probe pulse is also shown in Fig. 2 and interpolates well between the two spectra taken with σ^+ and σ^- polarized probe. A small deviation is observed around 1.91 eV, where a higher density of uncoupled electron-hole plasma is observed in the experiment of the linearly polarized probe, in which the pump beam was particularly strong (1.0 mJ/ cm²) [5]. Several interesting features are observed in Fig. 2. The energy position of the negative pump-probe signal is shifted for the two circular polarizations. This indicates that the strong pump pulse induces a macroscopic magnetization in the sample (otherwise paramagnetic before the pump excitation) persisting over 300 ps after the pump excitation. Comparing the energy splitting of 22 meV detected between the bleaching peaks of the σ^+ and σ^- polarizations (Fig. 2) with the field dependent energy shift of the excitonic absorption, we estimate a molecular field of 0.6 T induced in the sample by the pump pulse and persisting after 300 ps. The bleaching of the σ^- polarization is stronger than the bleaching of the σ^+ polarized probe beam. This transient circular dichroism has been observed systematically and indicates that at 300 ps the excitonic density is still high enough to turn the RKKY exchange interaction between the excitons and the Mn^{2+} spins antiferromagnetic. A confirmation is found in the fact that the 'positive peak 1' is strong only for the σ^- polarization and is not observed for the σ^+ polarization. This means that the polaron-like states have, at high exciton densities, an antiferromagnetic coupling between the excitons and the Mn^{2+} spins. These experimental results are well reproducible, even though the amplitude of 'positive peak 1' is small. Finally, 'positive peak 2' is observed for both polarizations, but it is broader and stronger for the σ^+ polarization. This indicates that a polaronic signal (corresponding to the positive peak 1 observed for the σ^- polarization) mixes in the 'positive peak 2' for the σ^+ polarization.

In summary, we induced a long-range ferromagnetic order of the Mn^{2+} spins with an intense optical excitation. Furthermore, we found that, at high excitation densities, the excitons couple antiferromagnetically to the Mn spins.

References:
1. D.D. Awschalom, N. Samarth, J. Magn. Magn. Mat. **200**, 130 (1999)
2. T. Dietl, A. Haury, Y. Merle d'Aubigné, Phys. Rev. B **55**, 3347 (1997)
3. A. Haury, A. Wasiela, A. Arnoult, J. Cibert, S. Tatarenko, T. Dietl, Y. Merle d'Aubigné, Phys. Rev. Lett. **79** (1997) 511
4. R. Pittini, H. Mitsu, M. Takahashi, J.X. Shen and Y. Oka, J. Appl. Phys. **85**, 5938 (1999)
5. R. Pittini, J.X. Shen, M. Takahashi, Y. Oka, J. Cryst. Growth **214/215**, 801 (2000)
6. R. Pittini, M. Takahashi, J.X. Shen and Y. Oka, J. Lumin. **87-89**, 393 (2000)

Excitonic vs free-carrier spin-relaxation in III-V quantum wells

A.Malinowski[1], P.A.Marsden[1], R.S.Britton[1], K.Puech[2], A.C.Tropper[1], R.T.Harley[1]

[1] Department of Physics and Astronomy, University of Southampton, Southampton, United Kingdom.
[2] Optoelectronics Research Centre, University of Southampton Southampton, United Kingdom.

Abstract Spin and population dynamics of photoexcited carriers in GaAs/AlGaAs and InGaAs/InP quantum wells is investigated from 5K to 300K, revealing transition between excitonic and free carrier spin-relaxation at intermediate temperatures. Localisation is found to prolong spin relaxation of excitons by two orders of magnitude.

Spin dynamics of photoexcited carriers in semiconductor quantum structures shows remarkable and sometimes unexpected behaviour as well as having potential device applications. At low temperatures spin-relaxation in intrinsic GaAs quantum wells was found [1], in agreement with theory for itinerant excitons [2], to be dominated by processes involving simultaneous electron and hole spin-flips on 10ps timescale, much faster than individual electron or hole spin-flips. The mechanism involves an effective k-dependent magnetic field, originating from exchange coupling, in which the exciton spin precession is inhibited by momentum scattering to give spin-relaxation rate proportional to scattering time. At room temperature excitons become unstable (dissociation time ~ 400fs [3]) and spin relaxation is determined by free-carrier processes; hole relaxation is sub-ps so that free electron spin-relaxation determines the overall relaxation [4] [5]. Electron spin-relaxation is via the D'yakonov-Perel (DP) mechanism [6] in which the conduction band spin-splitting now acts as a k-dependent effective field which induces spin precession inhibited by electron momentum scattering. Here we report investigations of the intermediate temperature regime showing evolution from excitonic to free-carrier spin-relaxation in (100)-oriented, MBE-grown multi-single-quantum well GaAs/Al$_{0.35}$Ga$_{0.65}$As and In$_{0.47}$Ga$_{0.53}$As/InP samples. We also show, from measurements on two different GaAs/AlGaAs wafers, A and B, grown respectively without and with an interruption of 125s at each interface to control their morphology [7], that localisation of excitons at low temperatures in sample B dramatically extends their spin lifetime.

Pump-probe differential transmission or reflection methods based on a cw modelocked Ti:sapphire laser or synchronously pumped optical parametric oscillator (OPO) were used to study spin relaxation with ps resolution at the n=1 heavy-hole exciton resonance [5]. The pump pulses were circularly polarised, to generate spin polarised carriers, and the probe was linearly polarised. Pump-induced changes in probe reflection or transmission ('sum' signal) and probe polarisation rotation ('difference' signal) were recorded using a system of balanced

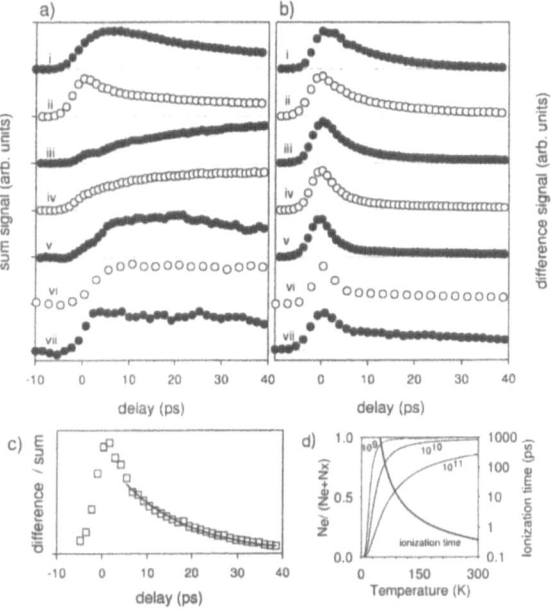

Fig. 1 Normalised plots of a) "sum" and b) "difference" signals from 10nm GaAs quantum well in sample A. i: 5.5K, ii: 40K, iii: 55K, iv: 70K, v: 85K, vi: 130K, *vii* : 200K. c) Pure spin decay at 5.5K giving spin relaxation time 20ps. d) Calculated exciton ionisation time and equilibrium free electron concentration for GaAs/AlGaAs quantum well ref [8] for excitation densities 10^9, 10^{10} and 10^{11}cm^{-2}

photodetectors and lock-in detection. The 'sum' indicates total excited population dynamics whereas the 'difference' is determined by imbalance of excited spin populations. Pump excitation density was ~ 5×10^{10}cm^{-2} with probe ~ 10^{-2} × pump.

Fig. 1 shows representative sum (a) and difference (b) signals for a 10nm well in GaAs/AlGaAs sample A, grown without interruption and having minimal localisation at low temperatures [7]. Up to 40K the data are essentially the same as previous results for excitons [1], indicating rapid excitonic spin relaxation; Fig. 1c shows extracted pure exciton spin decay at 5.5K. Between 55K and 85K new behaviour is observed in which the sum signal shows a gradual rise after the pump which we assign to exciton dissociation with time constant ~ 22ps at 55K falling to ~ 10ps at 70K. Above 85K the rise is too abrupt to resolve. Corresponding to this exciton dissociation the difference signal develops a slowly decaying component, significantly slower than the excitonic relaxation, which can be assigned to free electron spin-

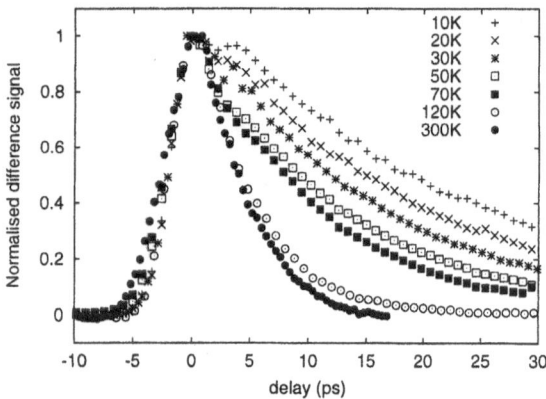

Fig. 2 Normalised "difference" signals from an 8nm In-GaAs/InP quantum well. The dip appearing in low temperature data at 2.5ps delay is an experimental artifact.

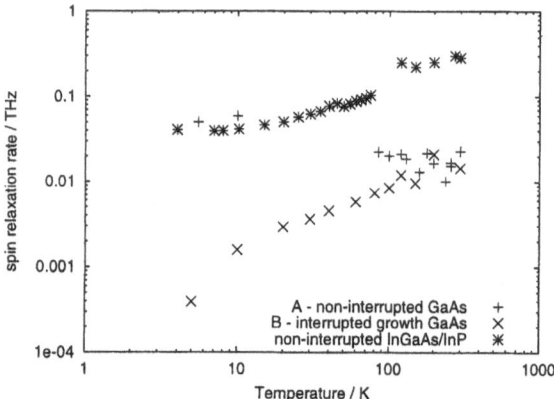

Fig. 3 Measured spin-relaxation rates for 10nm GaAs quantum wells and 8nm InGaAs quantum wells. Below 50K spin-relaxation is an excitonic process whereas above $\sim 100K$ spin-relaxation is of free electrons.

relaxation. Fig 1d shows calculations [8] [5] of relative concentrations of free carriers and excitons at thermal equilibrium and exciton ionisation times, which agree well with this interpretation considering the extremely rapid temperature dependence of the calculation.

Fig. 2 shows difference data for the InGaAs/InP sample which indicate pure spin decay because the sum signals show only weak decay over this timescale. Exciton dissociation should develop over a similar temperature range to that for GaAs/AlGaAs with a transition from exciton to free carrier spin relaxation between 50K and 100K. In this case, a slow electron spin relaxation component does not occur above 70K. This is consistent with the dramatically faster electron spin relaxation in the (InGa)As/InP material system in which well and barrier do not have common atoms, resulting in strong enhancement of the effective k-dependent field driving the spin reorientation [9]. In the low temperature regime where excitonic spin relaxation dominates, the rates for the two material systems are very similar. As shown in fig. 3, the measured spin relaxation rates at room temperature for InGaAs/InP are 10× those for GaAs/AlGaAs

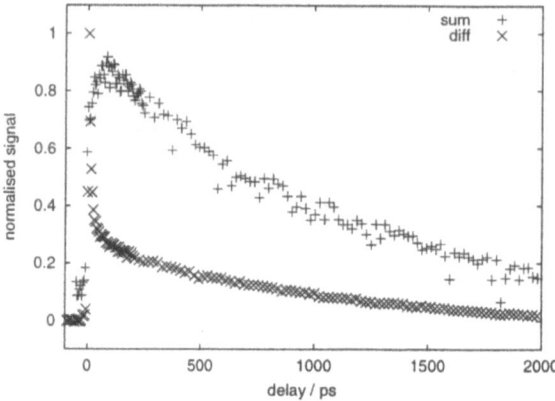

Fig. 4 Normalised "sum" and "difference" signals for 10nm quantum well in sample B at 5K

whereas at low temperatures the rates for InGaAs/InP and GaAs/AlGaAs sample A are similar. The weak temperature dependence for electron spin-relaxation is consistent with dominance of DP mechanism and expected mobility of electrons in these samples [5].

Fig. 3 also shows the spin relaxation rate for GaAs/-AlGaAs sample B in which carriers are strongly localised by atomically flat interface regions [7]. At room temperature the rates for samples A and B are similar due to dominance of electron-phonon momentum scattering. In B, the spin relaxation rate falls rapidly with temperature and is $\sim 10^{-2}$ that of the other samples at low temperatures. This difference is consistent with exciton localisation which will switch off the rapid relaxation mechanism of itinerant excitons. Study of the low temperature sum and difference signals for sample B (fig. 4) confirms this interpretation. Over the first 30ps following the pump the sum signal continues to rise, corresponding to migration of initially itinerant excitons to localisation sites. During this time the difference signal decays rapidly at a rate characteristic of spin-relaxation of itinerant excitons as seen in sample A. Subsequently the difference decay becomes extremely slow corresponding to localised excitons.

References

1. A.Vinattieri et al, Phys. Rev. B. **50**, (1994) 10868.
2. M.Z.Maialle et al, Phys. Rev. B. **47**, (1993) 15776.
3. W.H.Knox et al, Phys. Rev. Lett. **54**, (1985) 1306
4. R.S.Britton et al, Appl. Phys. Lett. **73**, (1998) 2140
5. A.Malinowski et al, To appear in Phys. Rev. B **62**, Nov 15 2000
6. M.I.D.D'Yakonov and V.I.Perel Sov. Phys, Solid State **13** (1972) 3023
   M.I.D.D'Yakonov and V.Yu. Kacharovski, Sov. Phys. Semicond. **20**, (1986) 110
7. N.Garro et al, Phys. Rev. B **55**, (1997) 13752
8. D.S.Chemla et al, IEEE J. Quantum Electron. **20**, (1984) 265
9. A.Tackeuchi et al, Appl. Phys. Lett. **70** (1997) 1131
   T.Geuttler et al, Phys. Rev. B **58**, (1998) R10179

1S - 2S - Continuum Coupling and Dephasing in 5 nm GaAs Quantum Wells

A.A. Busch[1], J.M. Watson[3], P. Paddon[3], Z. Wasilewski[2], Jeff F.Young[1]

[1] Department of Physics and Astronomy, University of British Columbia, Vancouver, BC, Canada
[2] Institute for Microstructural Sciences, National Research Council, Ottawa, Canada
[3] Now at Nortel Networks, Ottawa, Canada

Abstract   Absolute linear optical absorption and two-pulse degenerate four wave mixing (DFWM) are used to study nominally undoped, 5 nm wide, GaAs multiple quantum well samples. A common set of power law functions are used in the susceptibilities of both first and third order numerical models to fit the two sets of data. The DFWM simulations include the actual excitation pulse shape. These simulations are used to deduce power law exponents, exciton energies, and dephasing rates for the exciton (1S and 2S), biexciton, and continuum-edge states at 5 K.

1 Introduction

The interaction of light with semiconductor materials is known to be strongly influenced by Coulomb effects. This is especially true in systems of reduced dimensionality, where the absorption profile is dominated by the Coulomb attraction of the electrons and holes over at least several meV around the band edge. Despite an enormous amount of work, however, a quantitative understanding of these effects has been elusive. We previously reported a systematic quantitative study of the linear absorption spectra of a series of 5 nm GaAs quantum well samples with doping density ranges from 0-20×10^{10} cm^{-2} [1]. We subsequently studied the same samples using two-pulse degenerate four wave mixing (DFWM).

In this paper, we describe a limited set of these DFWM results. In particular, we report the exciton and biexciton parameters we deduce by forcably fitting the linear absorption and DFWM data from the undoped sample using a common set of material parameters. The sample was removed from its substrate and Van der Waals bonded to a glass substrate [1]. Linear absorption experiments were conducted in transmission using a broad-band white light source. The DFWM experiments reported here consist of time- and spectrally-integrated data as a function of the delay between the two ($\sim$100 fs) excitation pulses.

Linear and nonlinear models typically use Lorentzian functions to describe the resonant response of individual transitions that may be excited by a given applied field. However, we have chosen to parameterize each resonance with a more general function,

$$\frac{\mu_j}{(\omega - \omega_j - i\Gamma_j)^{\alpha_j}} , \qquad (1)$$

both because this is the natural response function that describes the Fermi-edge singularity observed in our doped samples [1,2], and because it conveniently allows us to represent the continuum in our undoped sample with a single response function. Clearly α_j should be $\sim$1 for bound exciton transitions in undoped amples and $\alpha_j \to 0$ for the continuum.

The DFWM simulations include the actual excitation pulse shape and inhomogeneous broadening. Linear and nonlinear simulations are used to deduce power law exponents, exciton energies, and dephasing rates for the exciton (1S and 2S), biexciton, and continuum-edge states in the quantum wells.

2 Results and Comparison

The linear absorption of the heavy-hole peak and continuum edge in the undoped GaAs multiple quantum well sample is shown in Figure 1, along with a fit to the data. The DFWM signals are shown in Figure 2 for both co-linear (CLP) and co-circular (CCP) input beam polarizations, at a single pulse detuning (shown in figure 1). As expected [3], the CCP data show only the single excitonic (1S & 2S) and continuum states, while the CLP data show, in addition, signals due to the biexciton levels.

The requirement that both the linear and nonlinear data be simultaneously fit helps reduce the range over which the parameters can be allowed to vary. From the linear absorption, information about the 2S state and continuum energies are difficult to extract accurately due to inhomogeneous broadening. For the same reason, the linear spectra are not as sensitive to the dephasing rates. In contrast, the four-wave mix-

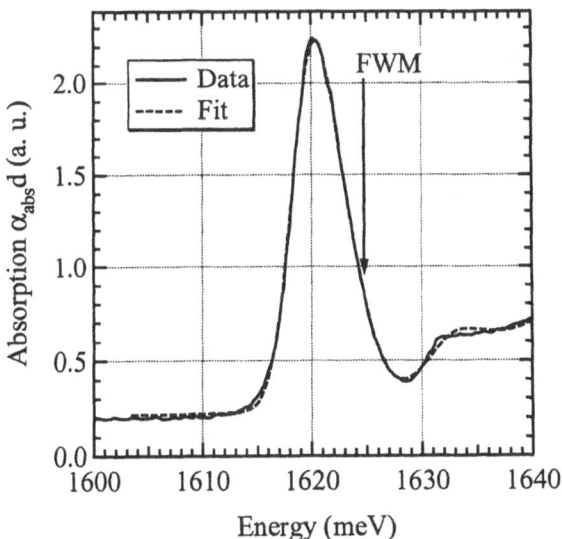

Fig. 1 Linear absorption of heavy hole exciton peak and continuum edge in undoped 5 nm multiple quantum wells at 5 K. The arrow indicates the center frequency (1624.8 meV) of the pulsed excitation resulting in the DFWM data shown in figure 2.

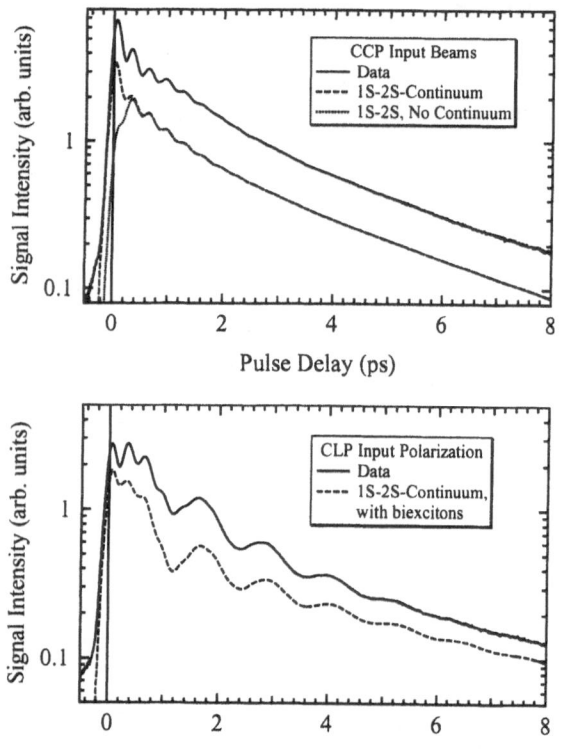

Fig. 2 DFWM data and simulations for co-circularly polarized (CCP) and co-linearly polarized (CLP) input laser beams. Data and simulations are shifted for clarity. The laser pulse central frequency was 1625.8 meV, with an intensity full-width half maximum of 134 fs.

ing curves are quite sensitive to the 2S parameters through the period, amplitude and decay rate of the 1S-2S beats. The DFWM simulations are also responsive to changes in the values of the continuum oscillator strength and power law lineshape parameter, α_j, used to describe the continuum.

We obtain good fits to both sets of data for different laser detunings and polarizations, but only if we (1) assume that there is a double-gaussian shaped inhomogeneous broadening profile, with a separation close to that expected for a half-monolayer fluctuation in well width, and (2) include all four of the triply resonant terms that contribute to $\chi^{(3)}$. The system thus consists of two 1S-2S-continuum systems not coherently coupled to each other. In addition, the CLP simulations also contain biexciton and two-free-exciton states coupled to the 1S states [3].

From the CCP fits, the 1S-2S separation is found to be 12.75 meV and the 1S-continuum edge separation is 16.4 meV. The lower 1S energy level is at 1619.8 meV, the separation of the two 1S-2S-continuum systems is 2.82 meV, and the inhomogeneous broadening is 4.2 and 4.7 meV for the upper and lower systems, respectively. The 2S dephasing rate is 1.77 ps^{-1}, while the longest 1S dephasing rate is 0.066 ps^{-1}. The ratio of the dipole moment strengths, μ_j/μ_{1S}, was found to be 0.48 and 0.73 for the 2S and continuum states, respectively.

The lower graph of Figure 2 shows the CLP DFWM signal. In addition to the 1S-2S-continuum levels described above, this contains a longer period beat signal due to biexcitons. Fixing parameters from the linear absorption and the CCP data for the 1S, 2S, and continuum states, the biexction parameters can be extracted from the CLP data. The binding energy is 3.49 meV, while the dephasing rate of the biexciton state is 0.44 ps^{-1}.

The 1S and biexciton dephasing rates increase as the laser pulse energy is increased, while the 2S dephasing rate remains approximately constant. As a function of temperature, the 1S, 2S, and biexciton dephasing rates all show an approximate T^2 dependence, although the same data can be fit by a more complicated function ($\exp\{AT^\alpha\}$ for T<15 K and $\exp\{-E_a/k_BT\}$ for T>10 K)[4]. All dephasing rates were found to be insensitive to changes of input beam intensity over 1.5 orders of magnitude, corresponding to changes of DFWM signal intensity of 2 orders of magnitude.

While the power law exponent for the 1S & 2S states is close to 1, the coefficient that best describes the shape of the continuum edge was found to be $\alpha = 0.062$. The effect of the continuum on the DFWM simulations is shown in Figure 2. With $\alpha = 0.06$, the continuum portion of the signal is centered around zero pulse delay, decaying 2-3 orders of magnitude within approximately three pulse widths (0.45 ps). The DFWM signal is quite insensitive to the continuum dephasing rate over a wide range. For these fits, a continuum dephasing rate of 2.5 ps^{-1} was used [5]. A power law response function with $\alpha \sim 0.06$ therefore provides a simple means of describing the continuum contribution to coherent dynamics in this system.

3 Conclusions

Absolute linear absorption and two-pulse degenerate four wave mixing (DFWM) were used to study undoped GaAs multiple quantum well samples. A common set of power law functions were used to fit experimental data with first and third order numerical models. The models adequately describe the experimental data, including the continuum response, and were used to deduce power law exponent, exciton energies, and dephasing rates for the exciton (1S and 2S), biexciton, and continuum-edge states at 5 K.

References

1. S.A. Brown et. al., Phys. Rev. B **54**, (1996) 11 082.
2. M. Combescot & P. Nozières, J. de Physique **32** (1971) 913.
3. G. Bongiovanni et. al., Semicond. Sci. Tech. **12**, (1997) 300.
4. M.D. Webb et. al., Phys. Rev. B **43**, (1991) 12 658.
5. S. Glutsch et. al., Phys. Rev. B **52**, (1995) 4941.

Electric-field-dependent carrier dynamics in an (Al,Ga)As/GaAs double-quantum-well superlattice

L. Schrottke, R. Hey, H. T. Grahn

Paul-Drude-Institut für Festkörperelektronik, Hausvogteiplatz 5-7, 10117 Berlin, Germany, e-mail: `lutz@pdi-berlin.de`

Abstract We investigate the electronic subband population in a double-quantum-well superlattice by time-resolved photoluminescence (PL) spectroscopy as a function of an applied electric field perpendicular to the layers. Using a rate equation model, the field-dependent electron density in the two wells and the transfer coefficients for holes between adjacent wells can be derived from the time-resolved PL intensities. The electron population of the narrower well exhibits a pronounced maximum as a function of the electric field.

1 Introduction

Since the invention of the quantum cascade laser [1], the interest in a quantitative evaluation of the electronic subband population in semiconductor heterostructures has considerably increased. Photoluminescence (PL) spectroscopy, which is a bipolar method as both the electron *and* hole concentrations determine the intensity, has been proven to be a powerful tool for the investigation of such unipolar systems, when only one hole subband is involved in the interband transitions [2]. In the more complex quantum cascade structures, so far the electron distribution has only been determined qualitatively [3]. However, for a precise quantitative discussion of the electric-field-dependent electronic subband population, the dynamics of electrons *and* holes have to be considered. The carrier concentrations are determined by the transport properties as well as by the recombination term, which couples the hole with the electron concentration.

Recently, we have investigated the relation between the cw-PL spectra and the subband population in a double-quantum-well (DQW) superlattice [4]. In such superlattices, each period consists of two QWs and two barriers with different thicknesses. An applied electric field can lead to an enhanced occupation in the QWs following the thinner barrier. The alternating well widths allow to identify the QWs by their PL energy. The electronic subband population due to current injection can be determined by photoluminscence (PL) spectroscopy, if the number of photoexcited carriers is small compared to the number of injected electrons. In Ref. [4], a simple model was shown to provide an upper limit for the required excitation intensity. However, any inhomogeneity in the hole distribution was completely neglected.

In order to achieve at least a homogeneous hole distribution at a certain time t, e. g. at $t = 0$, both wells have to be photoexcited with fs-pulses. For excitation intensities small compared to the electrically injected electron

density, the temporal behavior of the PL spectra not only allows to determine their ratio, which is derived from the PL intensity ratio at time $t = 0$, but also the values of the electron population in each QW.

2 Model for the hole dynamics

In this model, the electron concentration is assumed to be time independent, since it is mainly determined by the electrical injection of electrons through the contacts. Therefore, the dynamics of the photoexcited electrons can be neglected. However, the holes, which are only excited by the fs-pulses, exhibit a temporal behavior. The rate equations for the hole densities in QW1 and QW2 can be written in the following form

$$\frac{dp_1(t)}{dt} = t_{21}\, p_2(t) - (r_1 n_1 + t_{12})\, p_1(t)\,, \tag{1}$$

$$\frac{dp_2(t)}{dt} = t_{12}\, p_1(t) - (r_2 n_2 + t_{21})\, p_2(t)\,, \tag{2}$$

where $p_i(t)$ denotes the hole sheet density in the corresponding QW. The quantities t_{12} and t_{21} refer to the transfer coefficients per unit time of the holes between adjacent wells, i. e., the fraction of holes transferred from QW1 to QW2 and vice versa. n_1 and n_2 denote the constant sheet densities of the electrons, while r_1 and r_2 refer to the respective recombination coefficients. Within this simple model, the spatial distribution of carriers is assumed to be laterally homogeneous and the same in each period of the superlattice. Therefore, we have to consider only a single period (QW1 and QW2). This system of differential equations with constant coefficients has two eigenfrequencies $\omega_{1,2} = -\alpha \pm \beta$, where

$$2\alpha = r_1 n_1 + t_{12} + r_2 n_2 + t_{21}\,, \tag{3}$$

$$2\beta = \sqrt{(2\Delta)^2 + 4 t_{12} t_{21}}\,, \tag{4}$$

$$2\Delta = r_1 n_1 + t_{12} - (r_2 n_2 + t_{21})\,. \tag{5}$$

For a homogeneous excitation, the initial conditions are $p_1(0) = p_2(0) = p_0$ so that the analytic solution of the system of differential equations in Eqs. (1) and (2) is given by

$$p_1(t) = p_0 e^{-\alpha t}\left[\cosh(\beta t) + \frac{t_{21} - \Delta}{\beta}\,\sinh(\beta t)\right]\,, \tag{6}$$

$$p_2(t) = p_0 e^{-\alpha t}\left[\cosh(\beta t) + \frac{t_{12} + \Delta}{\beta}\,\sinh(\beta t)\right]\,. \tag{7}$$

The PL intensity $I_i(t)$ of QW1 and QW2 is proportional to $r_i n_i\, p_i(t)$ so that the ratio $\rho(t)$ of the PL intensities of the two wells can be written as

$$\rho(t) = \frac{r_1 n_1 p_1(t)}{r_2 n_2 p_2(t)} = \frac{r_1 n_1}{r_2 n_2} \frac{\beta + (t_{21} - \Delta)\ \tanh(\beta t)}{\beta + (t_{12} + \Delta)\ \tanh(\beta t)}\ . \quad (8)$$

In the two limits $t = 0$ and $t \to \infty$, we obtain

$$\rho_0 = \rho(0) = \frac{r_1 n_1}{r_2 n_2}\ , \quad \rho_\infty = \rho(\infty) = \rho_0\ \frac{\beta + t_{21} - \Delta}{\beta + t_{12} + \Delta}\ . \quad (9)$$

Solving these two equations for $r_1 n_1$ and $r_2 n_2$ using the expression for ω_1 leads to

$$r_1 n_1 = -\omega_1\ \frac{\rho_\infty + \rho_0}{1 + \rho_\infty}\ , \quad r_2 n_2 = -\frac{\omega_1}{\rho_0}\ \frac{\rho_\infty + \rho_0}{1 + \rho_\infty}\ . \quad (10)$$

The quantities t_{12} and t_{21} can be calculated from Eq. (9) applying the definition of Δ as well as the eigenfrequencies $\omega_1 + \omega_2 = -2\alpha$ and $\omega_1 - \omega_2 = 2\beta$. We obtain

$$t_{12} = \frac{\rho_0}{1 + \rho_\infty}\ \omega_1 - \frac{\rho_0}{\rho_0 + \rho_\infty}\ \omega_2\ , \quad (11)$$

$$t_{21} = \frac{\rho_\infty}{\rho_0(1 + \rho_\infty)}\ \omega_1 - \frac{\rho_\infty}{\rho_0 + \rho_\infty}\ \omega_2\ . \quad (12)$$

In order to be able to determine ω_i for $i = 1$ or 2 from the experimental data, we have to use an auxiliary function $\theta(t, \omega)$, which is defined as

$$\theta(t, \omega) = \frac{\frac{d}{dt}\left[e^{-\omega t}\ I_1(t)\right]}{\frac{d}{dt}\left[e^{-\omega t}\ I_2(t)\right]}\ . \quad (13)$$

This auxiliary function has the property that it becomes independent of time, i. e., constant, for $\omega = \omega_i$.

3 Experimental results

The investigated sample consists of a DQW superlattice with 4- and 5-nm GaAs wells and 10- and 14-nm $Al_{0.3}Ga_{0.7}As$ barriers embedded in an n-i-n structure. QW1 corresponds to the 4-nm well. The excitation was carried out with a fs-Ti:sapphire laser, the PL transients were detected using a streak camera with a time resolution of several ps. Figure 1 shows the measured ratio ρ of the PL intensities of the two QWs as a function of the applied voltage for different times between $t = 0$ and $t = 1$ ns. While $\rho(t)$ is approximately constant at

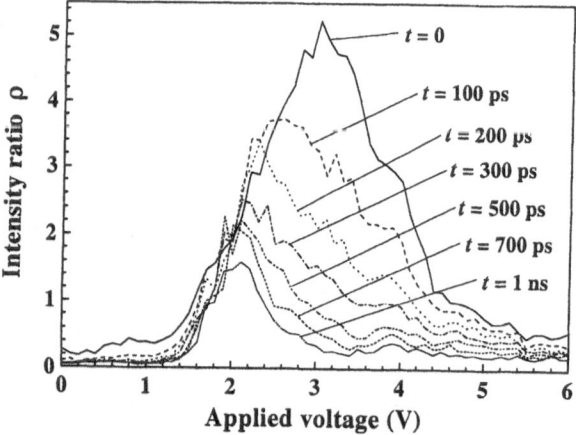

Fig. 1 Measured intensity ratio $\rho = I_1/I_2$ as a function of the applied voltage for different times as indicated.

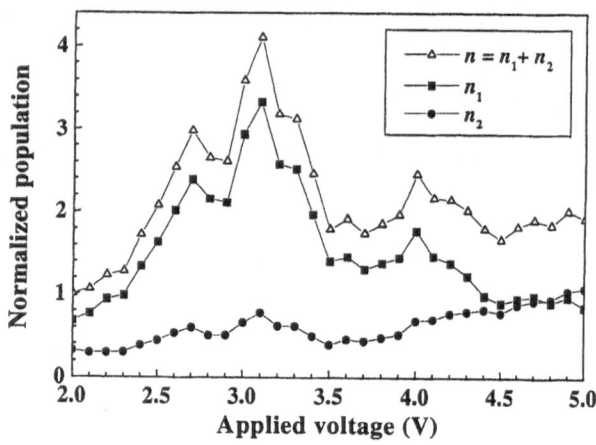

Fig. 2 Electronic subband population of QW1 and QW2 as well as $n = n_1 + n_2$ normalized to $n(2\ V)$ as derived from Eqs. (10) and (13).

2 V, it exhibits a dramatic decrease with time at 3 V. In Fig. 2, the electronic subband population in the narrower well as derived from Eqs. (10) and (13) exhibits a maximum at 3 V, which corresponds to a faster radiative decay. However, the population in the wider well increases only slightly with increasing voltage. The enhancement of the electron population in QW1 between 2 and 3 V is correlated with a very strong increase of the current. The decrease of the electron population of QW1 beyond 3 V may be a result of an increasing escape of electrons into the continuum so that the population in the narrower well begins to decrease. Another possibility to explain the decrease at higher fields may be an oversimplification of the hole dynamics. The values for the transfer coefficients t_{12} and t_{21} can also be evaluated using Eqs. (11) and (12). As expected for the field direction used for Fig. 1, t_{12} is larger than t_{21}. The decrease of n_1 at 3 V is correlated with an enhancement of t_{12}.

4 Conclusions

We have shown that the electronic subband population in a DQW superlattice can be determined from time-resolved PL spectroscopy, if the dynamics of the holes is modelled appropriately. Furthermore, the transfer coefficients for hole transport between adjacent wells can also be obtained. Future developments should take into account the finite length of the superlattice structure as well as consider p-i-p structures.

References

1. J. Faist, F. Capasso, D. L. Sivco, C. Sirtori, A. L. Hutchinson, and A. Y. Cho, Science **264**, (1994) 553.
2. Y. B. Li, J. W. Cockburn, J. P. Duck, M. J. Birkett, M. S. Skolnick, I. A. Larkin, M. Hopkinson, R. Grey, and G. Hill, Phys. Rev. B **57**, (1998) 6290.
3. L. R. Wilson, P. T. Keighley, J. W. Cockburn, J. P. Duck, M. S. Skolnick, J. C. Clark, G. Hill, M. Moran, and R. Grey, Appl. Phys. Lett. **75**, (1999) 2079.
4. L. Schrottke, R. Hey, and H. T. Grahn, Phys. Rev. B **60**, (1999) 16635.

Direct and indirect radiative recombination in strongly excited ZnSe/BeTe superlattices

A.A. Maksimov[1], S.V. Zaitsev[1], I.I. Tartakovskii[1], V.D. Kulakovskii[1], N.A. Gippius[2], D.R. Yakovlev[3], W. Ossau[3], G. Reuscher[3], A. Waag[3], and G. Landwehr[3]

[1] Institute of Solid State Physics, RAS, 142432 Chernogolovka, Moscow Region, Russia
[2] Institute of General Physics, Russian Academy of Sciences, 117942 Moscow, Russia
[3] Physikalisches Institut der Universität Würzburg, Am Hubland, 97074 Würzburg, Germany

Abstract A strong spectral shift (up to 0.5 eV) of the recombination band corresponding to the indirect transition from the ZnSe conduction band to the BeTe valence band in ZnSe/BeTe type-II superlattices and the radiative recombination time decrease by more than two orders of magnitude at high carrier density was found. For the recombination band corresponding to the spatially direct optical transition in the ZnSe layer a nonmonotonic dependence of the integral photoluminescence (PL) intensity on the excitation pumping level was observed. Numerical calculations of recombination times for the indirect optical transitions and nonmonotonic behavior of hole relaxation time from the ZnSe layer to the BeTe layer on carrier concentration are in a good quantitative agreement with experimental results.

1 Introduction

ZnSe/BeTe superlattices are semiconductor heterostructures with a type-II band alignment and large band offsets, leading to confining potentials of > 2.0 eV for electrons in ZnSe layers and ≈ 0.9 eV for holes in BeTe layers. The photoluminescence (PL) spectrum of a weakly-excited ZnSe/BeTe superlattice at temperatures $T = 4$–300 K shows two bands: a band with a maximum near 2.7–2.8 eV (optical transition direct in space, corresponding to radiative recombination of photoexcited electrons and holes in the ZnSe layer) and a wide band with a long-wavelength limit close to 2.0 eV, corresponding to an interband transition indirect in real space of photoexcited electrons in the ZnSe to the BeTe valence band [1, 2](Fig.1, dotted curves). The relative intensities of these two bands depend strongly on the thickness of ZnSe layer and the density of optical excitation P.

Deep potential wells for electrons and holes make it possible to realize under high levels of photoexcitation the presence of spatially separated layers of electrons and holes in ZnSe and BeTe, respectively, with extremely high carrier density exceeding 10^{13} cm^{-2}. The electric fields induced by such dense electron-hole layers in turn give rise to a strong bending of the conduction and valence bands. This leads to a large shift of the electron and hole energy levels, as well as to a strong change in the wavefunctions $\Psi_e(z), \Psi_h(z)$ and the spatial distributions of electron $n_e(z)$ and hole $n_h(z)$ densities in growth direction z [1].

We report on an investigation of a spectral reorganization in the regions of spatially indirect and direct

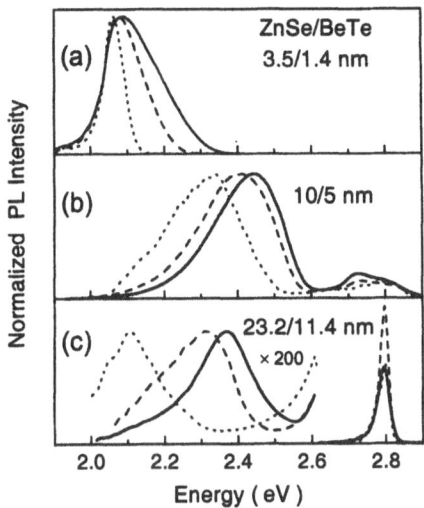

Fig. 1 PL spectra of ZnSe/BeTe superlattices at different levels of N$_2$-laser excitation (solid - 300, dashed - 150 and dotted line - 40 kW/cm^2), $T = 5$K.

transitions and nonlinear effects of PL in strongly excited type-II ZnSe/BeTe superlattices.

For the PL excitation a N$_2$-laser with pulse duration of 10 ns was used. The laser light is absorbed in the ZnSe layers only, as the BeTe direct band gap exceeds 4 eV.

2 Indirect transitions

Fig. 1 displays, that with the increase of the laser excitation a blue shift of the wide spectral band in the energy range 2 - 2.6 eV corresponding to radiative recombination of photoexcited electrons in the ZnSe layer and holes in the BeTe layer is observed.

We performed numerical calculations and by solving a self-consistent problem we determined the positions of the energy levels and wavefunctions of the electrons in the ZnSe layer and holes in the BeTe layer with different carrier densities, taking into account the bending of the conduction and valence bands. The experimentally observed spectral blue shifts demonstrate good agreement with the computational results (for details see [1,2]).

The measurements of the PL kinetics in a ZnSe/BeTe superlattice of the ZnSe/BeTe superlattice [3] at different spectral intervals, which correspond to different carrier concentrations, allow to estimate the recombi-

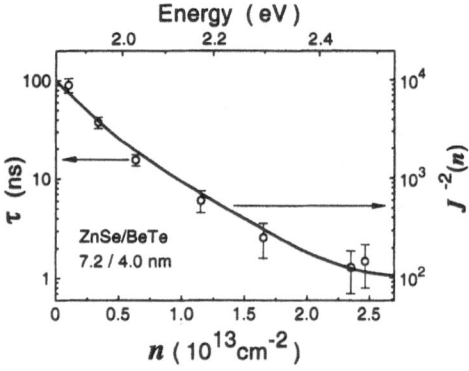

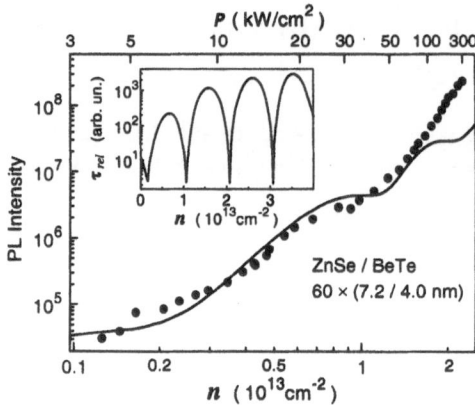

Fig. 2 The PL recombination times τ measured at different spectral positions at $T = 5K$ (open circles) and the dependence of the inverse square of the overlap integral $J^{-2}(n)$ (solid line).

Fig. 3 Measured at $T = 300K$ (points) and calculated (solid line) dependences of spatially direct optical transition PL intensities on carrier density n and excitation power P (the top x-axis), respectively. The *inset* shows the calculated dependence $\tau_{rel}(n)$.

nation times τ presented by points in Fig. 2. The values of τ vary by more than two orders of magnitude from $\tau \approx 100$ ns at very weak levels of optical pumping ($n \lesssim 10^{10} \text{cm}^{-2}$) down to $\tau \lesssim 1$ ns at very high excitation levels $n \gtrsim 2 \times 10^{13} \text{cm}^{-2}$. These experimental results are in good agreement with the calculated wavefunctions overlapping integral J for the lowest electron and hole energy levels:

$$J = \int \Psi_{e1}(z) \times \Psi_{h1}(z) \cdot dz, \qquad \tau \propto J^{-2}. \qquad (1)$$

Thus, as the carrier density increases, the Coulomb attraction of the electrons and holes results in strong carrier localization near the boundary of the interface, substantially increasing the overlap of their wavefunctions. This leads to a decrease of the radiative recombination time on the blue side of the spectral band corresponding to the spatially indirect optical transitions.

3 Direct transitions

The integral intensity of the spatially direct optical transition, corresponding to radiative recombination of photoexcited electrons and holes in the ZnSe layer depends drastically on the thickness of ZnSe layer (see Fig. 1). This dependence can be explained by different hole relaxation times τ_{rel} from the ZnSe layer to the BeTe layer. In the case of rather thin (3.5 and 10 nm, Fig. 1 a and b, respectively) ZnSe layers in the investigated superlattices τ_{rel} is much shorter than the radiative τ_r and nonradiative τ_{nr} times of the spatially direct transitions, and the PL intensity of the direct transitions is proportional to $\tau_{rel} \times P$. On the contrary, in the case of rather thick (23.2 nm) ZnSe layer $\tau_{rel} \gg \tau_r$, and the PL intensity of the direct transition depends only on the density of optical excitation P and exceeds considerably the PL intensity of the indirect transition (see Fig. 1c).

We stress here the astonishing experimental fact of direct transition PL intensity nonmonotonic dependence on the photocarrier concentration (see symbols in Fig. 3). The calculations show a very complicated, nonmono-

tonic behaviour for the dependence $\tau_{rel}(n)$ due to changes in resonance conditions between the lowest above barrier hole state in the ZnSe layer and the hole energy levels in the BeTe layer with increasing of band bending. The results of calculations for the sample with 7.2 nm ZnSe layers are presented in the inset of Fig. 3. A value proportional to n/τ_{rel} integrated over all 60 ZnSe layers with different concentrations n is shown by the solid line in Fig. 3. The analysis of the spectral shape of the corresponding spatially indirect transitions and computational results allow to reconstruct the relation between P and the carrier density n. There is a reasonable agreement between experimental and calculated results of the direct PL intensity changes by more than two orders of magnitude. The discrepancy at high excitation levels may be due to a saturation of the nonradiative processes at considerably increased hole concentration in the ZnSe layers. That in turn would cause the additional superlinear increase of the spatially direct PL intensity, which has not been taken into account in the calculations.

The work was supported by NATO grant PST.CLG 975344, INTAS grant 9915, RFBR grant 98-02-16651, and Deutsche Forschungsgemeinschaft (SFB410).

References

1. S. V. Zaitsev, V. D. Kulakovskii, A. A. Maksimov, D. A. Pronin, I. I. Tartakovskii, N. A. Gippius, M. Th. Litz, F. Fisher, A. Waag, D. R. Yakovlev, W. Ossau, and G. Landwehr, JETP Lett. **66**, (1997) 376.
2. A.A. Maksimov, S.V. Zaitsev, I.I. Tartakovskii, V.D. Kulakovskii, D.R. Yakovlev, W. Ossau, M. Keim, G. Reuscher, A. Waag, and G. Landwehr, Appl. Phys. Lett. **75**, (1999) 1231.
3. A. Waag, Th. Litz, F. Fisher, H.-J. Lugauer, T. Baron, K. Schüll, U. Zehnder, T. Gerhard, U. Lunz, M. Keim, G. Reuscher, and G. Landwehr, Festkörperprobleme / Advances in Solid State Physics **37**, ed. by R. Helbig (Vieweg, Braunschweig 1997), p.43.

Radiative recombination and spin relaxation of excitons in $Cd_{1-x}Mn_xTe$ quantum wells

M. C. Debnath [1,2*], J. X. Shen[1,2], I. Souma [1,2], R. Pittini [1,2], Y. Oka [1,2]

[1] Research Institute for Scientific Measurements, Tohoku University, Katahira 2-1-1, Aoba-ku, Sendai 980-8577, Japan
[2] CREST, Japan Science and Technology Corporation, Kawaguchi 332-0012, Japan

Abstract The quantum confinement of carriers and excitons spin have been investigated in $Cd_{1-x}Mn_xTe/Cd_{1-y}Mg_yTe$ single quantum wells and spin superlattices by time-resolved photoluminescence spectroscopy. The magnetic field induced of a spin superlattice structure has realized where carrier spins are spatially separated in $Cd_{1-x}Mn_xTe$ and $Cd_{1-y}Mg_yTe$ layers. Magnetic field dependencies of the formation of exciton magnetic polaron are attributed as a spin dependent confinement effect of excitons in single quantum well and spin superlattice structures.

1 Introduction

The study of quantum confinement of carriers or excitons spin in semiconductor heterostructures is a current research interest both in experimental and theory. In diluted magnetic semiconductors (DMSs), it is possible to magnetically tuning the band alignment due to the large spin splitting of the DMS band and made it possible to investigate the spin-dependent carrier confinement effect in magnetic field. The DMS quantum wells (QWs) and superlattices (SLs) of $Cd_{1-x}Mn_xTe/Cd_{1-y}Mg_yTe$ is an ideal structure to study the carrier confinement and spin-relaxation phenomena [1]. We have realized a spin SLs (SSLs) in this material where spin-up and spin-down states for the electrons and holes are spatially separated by the application of an external magnetic field [2]. The formation of an exciton magnetic polaron (EMP) in SSLs and single QWs (SQWs) will be discussed in magnetic field where confinement potential is controlled by the externally applied field.

2 Experimental

$Cd_{1-x}Mn_xTe/Cd_{1-y}Mg_yTe$ heterostructures were grown by molecular beam epitaxy on (100)-oriented GaAs substrate with $x = 0.05$, $y = 0.25$ for SQWs and $x = y = 0.05$ for SSLs to achieve a flat band offset in zero field. The experimental results on SQWs and SSLs (25 periods) with well widths (L_w) 150 Å and barrier widths (L_b) 230 Å will be presented here. Time-resolved photoluminescence (PL) measurements were performed with a femtosecond mode-locked titanium sapphire laser, frequency doubled for the excitation and a streak camera with a time resolution of 5 ps for detection. The incident light was circularly polarized as either σ^+ (left) or σ^- (right). The magnetic spin splitting of the exciton was

*Corresponding author: e-mail: debnath@rism.tohoku.ac.jp

studied in the Faraday configuration in the center of a superconducting magnet allowing fields up to 7 T.

3 Results and discussion

In Fig. 1(a), we show the magnetic field dependence of the PL spectra detected with circular polarizations σ^- and σ^+ in $Cd_{0.95}Mn_{0.05}Te/Cd_{0.95}Mg_{0.05}Te$ superlattices with $L_w = 150$ Å and $L_b = 230$ Å. In zero magnetic field, the excitonic transitions involve the n = 1 states of the electrons and heavy holes. This is a superposition of two different recombinations - one from the $Cd_{0.95}Mn_{0.05}Te$ layer and the other from the $Cd_{0.95}Mg_{0.05}Te$ layer. This feature becomes clear when a magnetic field above 0.5 T is applied. In a magnetic field and with σ^- polarization (here shown for 3 T and 5 T), we observe a broad band with two PL peaks. The peak on the higher energy side around 1.68 eV (here peak becomes less clear due to smoothing of the spectra) remains unchanged in the magnetic field. We attribute this peak to the transitions occurring in the non-DMS CdMgTe layer. On the other hand, the lower energy side shows a red shift in the magnetic field. We attribute this to the DMS CdMnTe layer. Detecting with σ^+, only the lower energy peak is observed in the PL spectra and shows a large red shift with increasing magnetic field. The comparison of the PL spectra detected with σ^- and σ^+ polarization in 5 T (Fig. 1 (a)) clearly indicates the different origin of the transitions.

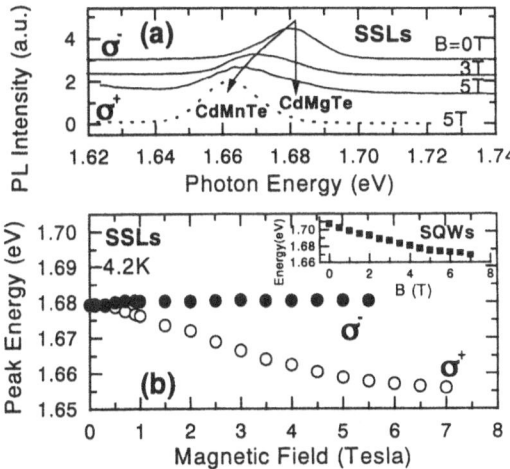

Fig. 1 (a) PL spectra taken in $Cd_{0.95}Mn_{0.05}Te/Cd_{0.95}Mg_{0.05}Te$ SSLs with σ^- and σ^+ polarization. (b) Transition energies of the Zeeman splitting in SSLs and SQWs (inset).

In Fig. 1 (b), we show the field dependence of the PL peak position detected with these two polarizations. We found that in σ^+ polarization the PL peaks show a negligible shift until 0.5 T and then exhibit a pronounced asymmetry of the splitting, which demonstrates the SSLs character as formerly was observed in ZnSe/Zn$_{1-x}$Mn$_x$Se multilayers [3]. The large red shift observed with increasing magnetic field is a characteristic of the DMS material as we observed in SQWs is shown in the inset of Fig.1 (b). With σ^- polarization, the peak energy saturates with increasing magnetic field which is a typical behavior of a non-DMS material. It is difficult to evaluate the PL peak energy taken with σ^- polarization above 5 T due to the overlap of the contributions of the DMS and non-DMS layers. From our analysis, we conclude that in the spectra taken with σ^+ polarization both spin-down electrons and heavy holes ($m_j = -1/2$, $m_j = -3/2$) are localized in the Cd$_{1-x}$Mn$_x$Te layer. Instead, with σ^- polarization both spin-up electrons and heavy holes ($m_j = +1/2$, $m_j = +3/2$) are confined inside the Cd$_{1-y}$Mg$_y$Te layer. Fig. 1 thus shows that the structure under investigation has formed a SSLs in an external magnetic field where the spin-down electrons/heavy holes and spin-up electrons/heavy holes are spatially separated in the Cd$_{1-x}$Mn$_x$Te and Cd$_{1-y}$Mg$_y$Te layers, respectively.

We have studied the EMP formation in SQWs and SSLs by following the magnetic field dependence of the exciton energy relaxation [4]. The energy relaxation in Cd$_{0.95}$Mn$_{0.05}$Te/Cd$_{0.75}$Mg$_{0.25}$Te SQWs with $L_w = 150$ Å are shown in the inset of Fig. 2 (a). The energy relaxation due to the EMP formation depends strongly on the magnetic fields. In zero magnetic field, the localized magnetic moments are polarized by the spin of the excitons. When an external field is applied stronger than the internal polarizing field of the polaron (here 3.5 T), all magnetic moments are aligned to the external field and the EMP energy is reduced. Therefore, the suppression of the energy relaxation in fields characterizes the reduction of the EMP binding energy. The difference of the energy relaxation between 0 and 3.5 T is 4.2 meV. We assign it to the EMP binding energy suppressed by the external field as shown in Fig. 2(a). On the other hand, no EMP formation is detected below the magnetic field in 0.5 T of our SSLs structure (Fig. 2(b)) where excitons are confined in the non-magnetic Cd$_{0.95}$Mg$_{0.05}$Te layer but the application of a magnetic field in 0.5 T, Cd$_{0.95}$Mn$_{0.05}$Te band edge splitted which leads to the formation of EMP and reaches the maximum value of EMP binding energy 3.5 meV in 2.5 T. The EMP binding energy then decreases with increasing the magnetic field as a similar pattern we observed in SQWs. Surprisingly, we found that the EMP formation still exists in 7 T which could not be detected in SQWs. In SSLs, the EMP formation shows a spin

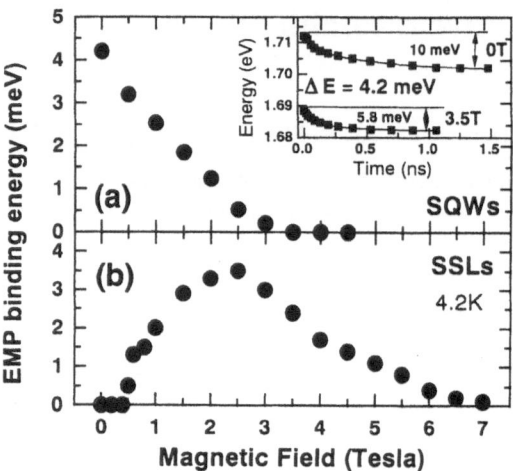

Fig. 2 The EMP binding energy as a function of magnetic field in Cd$_{0.95}$Mn$_{0.05}$Te/Cd$_{0.75}$Mg$_{0.25}$Te SQWs (a) and Cd$_{0.95}$Mn$_{0.05}$Te/Cd$_{0.95}$Mg$_{0.05}$Te SSLs (b) with $L_w = 150$ Å and $L_b = 230$ Å. Inset shows the energy relaxation in 0 and 3.5 T of SQWs.

dependent confinement effect in magnetic field. In the present SSLs, a magnetic field of 0.5 T localizes the lower lying state of the excitons. The localization energy is found to increase with increasing external field. Therefore, the EMP formation is enhanced by the application of a magnetic field and reaches the maximum EMP binding energy in 2.5 T. Increasing the magnetic field further increases the 'background' polarization of the Mn ions and therefore inhibits the EMP formation. The results, therefore, clearly shows that the Cd$_{0.95}$Mn$_{0.05}$Te/Cd$_{0.95}$Mg$_{0.05}$Te system behaves as a true SSLs and provides sufficient confinement potential in a magnetic field to form the EMP state where spin-down electrons and heavy holes are spatially separated in the Cd$_{0.95}$Mn$_{0.05}$Te well layer.

In conclusion, we have realized a SSLs in Cd$_{0.95}$Mn$_{0.05}$Te/Cd$_{0.95}$Mg$_{0.05}$Te heterostructure. The formation of the SSLs has been confirmed by the Zeeman splitting of the excitons and the EMP states in magnetic field. Due to the large spin splitting of the Cd$_{0.95}$Mn$_{0.05}$Te band edges, the potential can be controlled by the external field and the EMP formation is enhanced by the magnetic field due to the increased localization of the lower lying excitonic state. The EMP binding energy in SSLs is large in 2.5 T and then shows a gradual reduction with increasing the magnetic field which is, in fact, a similar character of SQWs.

References

1. L. J. Sham, J.Phys.: Condens. Matter, **5** (1993) A51.
2. M. von. Ortenberg, Phys. Rev. Lett., **49** (1982) 1041.
3. N. Dai, H. Luo, F. C. Zhang, N. Samarth, M. Dobrowolska, and J. K. Furdyna, Phys. Rev. Lett., **67** (1991) 3824.
4. M. C. Debnath, J. X. Shen, I. Souma, E. Shirado, T. Sato, R. Pittini, and Y. Oka, J. Appl. Phys., **87** (2000) 6457 and J. Cryst. Growth, **214/215** (2000) 797.

Nonequilibrium electrons in double quantum well structures :

Boltzmann equation approach

S. Khan-ngern and I. A. Larkin

Department of Physics & Astronomy, University of Sheffield, Sheffield S3 7RH U.K.

Abstract Kinetics of electron scattering in double quantum well heterostructures has been investigated. The electron distribution functions are calculated for the whole range of τ_0/τ_{ee} ratios using a model kinetic equation involving terms describing the e-e and electron-LO-phonon scattering, as well as electron escape and electron generation processes; where τ_0 and τ_{ee} refer to the electron-LO-phonon and electron-electron (e-e) scattering times, respectively. The resulting calculations obtain a comprehensive description of the lasing process in the intersubband lasers. Exact shapes of nonequilibrium distributions are essential for the correct physical understanding of the temperature behaviour of the intersubband lasers. The main conclusion of this work is that the strong nonuniform of the spectral density of gain causes the nonparabolicity to be harmless even at the small τ_0/τ_{ee} regime of operation.

1 Introduction

Studies of mid- and long-wavelength infrared (IR) lasers based on electronic intersubband transitions within the quantum wells (QWs) in semiconductor low dimensional heterostructures have attracted a great amount of interest since the first demonstration of a so-called Quantum Cascade Laser (QCL) was reported by Faist et al. [1]. To create lasing efficiency, in general, it requires specially designed structures providing sufficient global population inversion beween the two subbands involved.

In our earlier works [2,3] it has been shown that to achieve inverted population in a modified GaAs-AlGaAs double-quantum-well (DQW) structure, we should ensure efficient drain of carriers from E_1 subband. This happens when the device has appropriate design parameters providing a good electron confinement at E_2 subband and short electron lifetime at the E_1 subband. [2,3]. Recently, it has been shown by Faist et al. [4] that for lasing in the intersubband lasers, global population inversion is not a necessary condition but that band nonparabolicities combined with the thermal electron distribution in the lasing unit can make laser action possible. This idea has been studied theoretically by Gelmont et al. [5,6]. The calculations of spectral density of gain $g(\Omega)$ are described as a functional on electron distribution functions $f_1(\varepsilon_1)$ and $f_2(\varepsilon_2)$, corresponding to the occupation probabilities of kinetic energy states ε_1 and ε_2 in subbands E_1 and E_2, respectively. Generally, the distribution functions are nonthermal and their actual shapes affect strongly on the spectral density of gain.

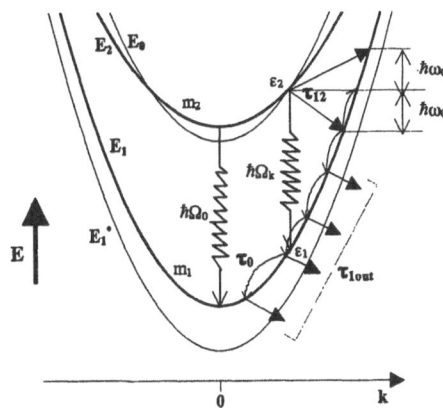

Fig. 1 Kinetics of electron scattering in a DQW structures presenting radiative intersubband transitions in the QW1, and also shown the nonradiative inter- and intrasubband transitions by emission or absorption of LO phonons.

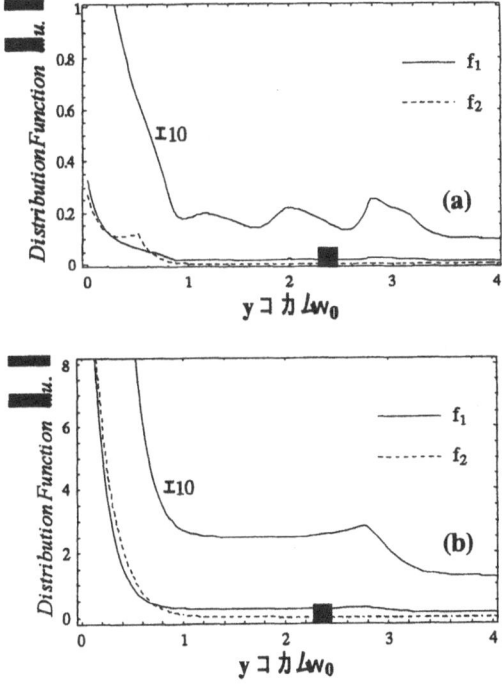

Fig. 2 Solutions of the kinetic equation under the monochromatic pumping $P(y) = \delta(y - \frac{1}{2})$ into the upper subband E_2 with equal subband population: $n_2 = n_1$; assuming the following parameters: $\tau_0 = 0.1$ ps, $\tau_{12} = \tau_1 = 1$ ps, the subband separation energy $\hbar\Omega_0 = 155$ meV, and $m_2 = 1.2m_1$, at temperature $T = 77$ K for different values of η : **(a)** $\eta = 1/20$ and **(b)** $\eta = 2$.

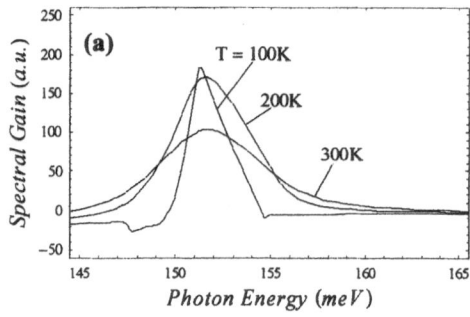

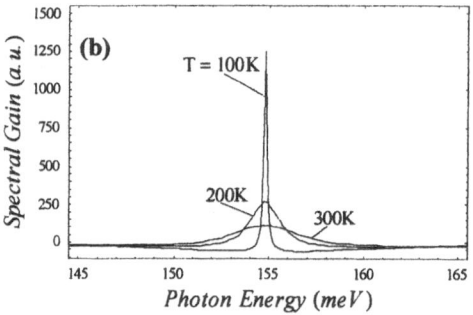

Fig. 3 Gain spectra in the low-concentration limit at different temperatures ranging from 100 to 300 K under the monochromatic pumping $P(y) = \delta(y - \frac{1}{2})$ into the upper subband E_2 with equal subband population: $n_2 = n_1$. **(a)** Assumed the following parameters: $\tau_0 = 0.1$ ps, $\tau_{12} = \tau_1 = 1$ ps, $\tau_{ee} = 2$ ps, subband separation energy $\hbar\Omega_0 = 155$ meV, and $m_2 = 1.2 m_1$. **(b)** Same spectra calculated in the parabolic model, $m_2 = m_1$.

2 Results and discussion

Kinetics of electron scattering as shown in Fig. 1 can be described by a model equation [7,8]

$$\frac{\partial f_i(\varepsilon_i;t)}{\partial t} = S_{LO}(\varepsilon_i) + C_{ee}(\varepsilon_i) + R_i(\varepsilon_i) + G_i(\varepsilon_i) \qquad (1)$$

where $f_i(\varepsilon_i)$ is the electron energy distribution function, corresponding to the occupation probability of kinetic energy states ε_i in the subband E_i; here E_i denotes the total energy of electron. In fact, the dispersion relations $\varepsilon_i(k)$ in the two subbands $i = 1, 2$ are different and nonparabolic. However, all the effects of interest here are simplified by regarding the subbands themselves as parabolic, but characterized by different effective masses m_1 and m_2 [5,6]. The term $S_{LO}(\varepsilon_i)$ is the electron scattering by LO phonons in subband E_i, $C_{ee}(\varepsilon_i)$ describes e-e scattering processes, $R_i(\varepsilon_i)$ stands for electron escape from subband E_i, and $G_i(\varepsilon_i)$ is the electron generation in the subband E_i. For further details see in Ref.[6-9].

We have solved analytically the model kinetic equation, eq.(1) for the subband distribution functions $f_1(y_1)$ and $f_2(y)$, where $y = \varepsilon/\hbar\omega_0 \equiv \varepsilon_2/\hbar\omega_0$ and $y_1 = \varepsilon_1/\hbar\omega_0 = (m_2/m_1)y$ are dimensionless energy variables. Calculated subband distribution functions for different values of $\eta = \tau_0/\tau_{ee}$ determining the concentration of electrons are

shown in Fig. 2. With increasing the values of η; i.e. increasing the concentrations of electrons, the subband distribution functions in the upper subband become close to a Maxwellian while for the lower subband they are always strongly nonequilibrium and deviate far from Maxwellian. The detailed shapes of these distribution functions are essential for the correct physical understanding of temperature dependence of the intersubband lasers. However, they have little contribution to the calculations of spectral density of gain $g(\Omega)$.

Gain spectra calculated in a low-concentration limit, $n_{2D} \ll 10^{11}$ cm^{-2} and so far $\eta = \tau_0/\tau_{ee} \ll 1$, at different temperatures are shown in Fig. 3. For further details of calculations see in Refs. [5] and [6]. The resulting calculations show that the gain is not strongly dependent upon temperature in the nonparabolic model. In contrast, it has a strong temperature dependence in the parabolic model. This implies that it is possible to achieve lasing action at high temperature due to the nonparabolicity effect.

3 Conclusions

In our approach, the distribution functions are investigated for the whole range of τ_0/τ_{ee} ratios using a model kinetic equation involving terms describing the e-e and electron-LO-phonon scattering, as well as electron escape and electron generation processes. We have solved analytically the kinetic equation for the subband distribution functions and obtain a comprehensive description of the lasing process in the intersubband lasers. The main conclusion of this work is that the strong nonuniform of the spectral density of gain causes the nonparabolicity to be harmless even at the small τ_0/τ_{ee} regime of operation.

Acknowledgments

We are grateful to L. R. Wilson and J. W. Cockburn for useful discussions. In addition, we would like to thank Prof. M. S. Skolnick for his various supports.

References

1. J. Faist *et al.*, Science **264**, (1994) 553.
2. Y. B. Li *et al.*, Phys. Rev. B **57**, (1998) 6290.
3. S. Khan-ngern and I. A. Larkin, Phys. Lett. A **266**, (2000) 209.
4. J. Faist *et al.*, Phys. Rev. Lett. **76**, (1996) 411.
5. B. Gelmont, V. B. Gorfinkel, and S. Luryi, Appl. Phys. Lett. **68**, (1996) 2171.
6. V. B. Gorfinkel , S. Luryi, and B. Gelmont, IEEE J. Quantum Electron. **32**, (1996) 1995.
7. S. E. Esipov, and Y. B. Levinson, Sov. Phys. JETP **63**, (1986) 191.
8. S. E. Esipov, and Y. B. Levinson, Adv. Phys. **36**, (1987) 331.
9. J. H. Smet, G. Fonstad, and Q. Hu, J. Appl. Phys. **79**, (1996) 9305.

Direct study of spin relaxation processes in 2D ZnCdSe/ZnSe systems through time resolved measurements of polarized exciton emission.

S. Permogorov[1,2], Y. Oka[1], R. Pittini[1], J.X. Shen[1], K. Kayanuma[1], A. Reznitsky[2], L. Tenishev[2], S. Verbin[3]

[1]RISM, Tohoku University, Katahira, Sendai 980-8577, Japan and CREST, Japan Science and Technology Corporation, Kawaguchi, 332-0012, Japan
[2]A.F.Ioffe Physical-Technical Institute, St Petersburg 194021, Russia
[3]St.Petersburg State University, St.Petersburg 198840, Russia

ABSTRACT Using femtosecond pulse excitation and circular polarized streak camera detection we have studied the spin relaxation of localized excitons in *ZnCdSe/ZnSe* superlattices in external magnetic fields. Spin relaxation times strongly decrease with magnetic field indicating phonon assisted mechanism of spin relaxation. Transfer of polarization during the exciton energy relaxation has been clearly observed.

1   Introduction

Study of photoluminescence with high time resolution in external magnetic fields with polarized detection provides extensive and most direct information on dynamical properties of excitons. The classical examples are the studies of magneto-circular polarization (MCPL), optical orientation and quantum beats. As a result, the information on the values of g-factors and the energy and spin relaxation times can be obtained. In this paper we present the results of MCPL study of decay kinetics of the exciton photoluminescence (PL) in low dimensional *ZnCdSe/ZnSe* superlattices..

2   Experimental

We have used the samples grown by ALE as superlattices with submonolayer insertions of *CdSe* in *ZnSe* matrices. The structural studies has shown [1] that *Cd* is distributed over 3-4 lattice periods, forming the insertions of *ZnCdSe* solid solution intermediate between Quantum Wells and Quantum Dots. The measurements of polarized PL kinetics at nonpolarized excitation were performed at T=1.6 K in external magnetic fields up to 8 Tesla in Faraday configuration with overall time resolution of 20 ps. Experimental results were obtained as two-dimensional energy-time streak-images. The analysis of these images allows to determine both integrated and resolved spectral or temporal PL profiles.

3   Results and Discussion

Typical time integrated PL spectra for two circular polarizations are shown in Fig.1. The main PL band is due to recombination of excitons localized in the tails of the density of states [2] by fluctuations of composition in *ZnCdSe* QW`s. The band is inhomogeneously broadened and shows a strong dispersion of emission decay times

ranging from several picoseconds at high energy wing up to ~ 300 ps at the low energy wing. This dispersion mostly reflects the probability of exciton tunnelling relaxation to the states with lower energy.

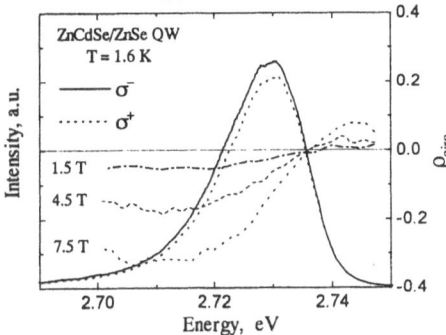

Fig. 1 Time-integrated PL spectra for two circular polarizations (solid and dotted curves) in magnetic field *B*=4T. Dashed curves show the spectral dependence of circular polarization degree of PL at different fields.

The same figure also shows the spectral dependence of circular polarization degree $\rho_{circ}=(I_+ - I_-)/(I_+ + I_-)$ for several values of magnetic fields. The polarization of PL arises due to the thermal redistribution of population between the sublevels of optically active exciton state split by magnetic field.

Spectral dependence of ρ_{circ} can be qualitatively understood as a result of competition between the spin relaxation and population decay of two magnetic sublevels. At high energies the spin relaxation is not completed during the exciton life-time, and values of polarization degree are small. In the region of low energy wing of PL band the spin relaxation is more complete and the absolute values of polarization degree are higher. A reversal of polarization sign at the high energy wing of PL band is a specific property of exciton relaxation across the inhomogeneously broadened tail of the density of localized states [3]. Due to the sharp increase of the density of states around the mobility edge the density of states for lower spin sublevel is much higher than that for the upper state at the same spectral position. As a result, the lower states are depopulated more efficiently, which lowers the corresponding emission intensity and brings up the reversal of polarization sign.

In order to obtain the quantitative information on the exciton spin relaxation times we have analysed decay of intensity for two polarized PL components as shown on Fig.2. At the same figure the decay kinetics of polarization degree ρ_{circ} is presented. It can be seen that at small times the absolute value of polarization degree grows almost linearly with the time delay.

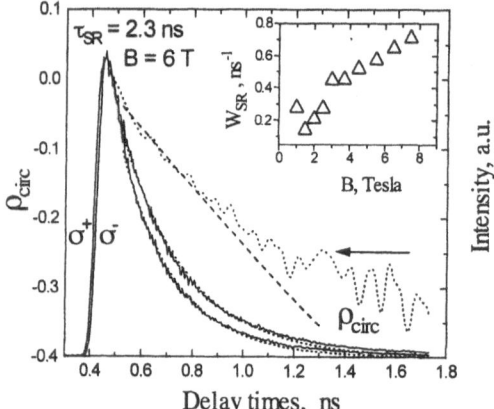

Fig.2. Decay curves of spectrally integrated PL intensity. Dotted line shows the time-behaviour of ρ_{circ}. Dashed line presents ρ_{circ} calculated with two-level model using exponential fits of intensity decays. The insert shows the field dependence of spin relaxation rate $W_{SR} = 1/\tau_{SR}$

This result is in agreement with a simple two-level model assuming equal initial population of two spin sublevels by pumping pulse. For the upper level the decay of population is governed by the time of radiative and nonradiative transitions τ_0 and the time of spin relaxation to the lower level τ_{SR}. For lower level the decay kinetics is given by the same total life time τ_0 and the time of thermally activated transitions to upper level $\alpha\tau_{SR}$, where $\alpha = \exp(\Delta/kT)$. For such a model the polarization degree at small times is given by the expression:

$$\rho_{circ}(t) = (t/\tau_{SR})(1 - \alpha^{-1})$$

At $t = 0$ the polarization should be zero and increases in absolute value proportionally to the delay time. Using this simple model we have determined the value of spin relaxation time at $B = 6$ T as $\tau_{SR} = 2.3$ ns. The insert to Fig.2 shows the dependence of the spin relaxation rate $W_{SR} = 1/\tau_{SR}$ on the strength of magnetic field B. It can be seen that W_{SR} increases almost linearly with B. Such a dependence suggests the spin relaxation mechanism assisted with simultaneous emission of acoustic phonons. The increase of magnetic field increases the splitting of exciton spin sublevels which allows the acoustic phonons with larger density of states to participate in the spin relaxation processes.

We have found that simple two level system does not

allow to study the dependence of spin relaxation times on the exciton localization energy. Fig. 3 shows the spectrally resolved decay curves of circularly polarized emission for the states in the region of the maximum of PL band (a) and at the low energy wing of PL band (b). It can be seen, that the decay kinetics for more deeply localized states is strongly nonexponential with a detectable rise-time. It can be concluded that the population of these states takes place also through the energy relaxation of localized excitons with higher energies. Nonzero PL polarization at $t=0$ indicates the polarization of excitons at the initial stages of relaxation with subsequent polarization transfer.

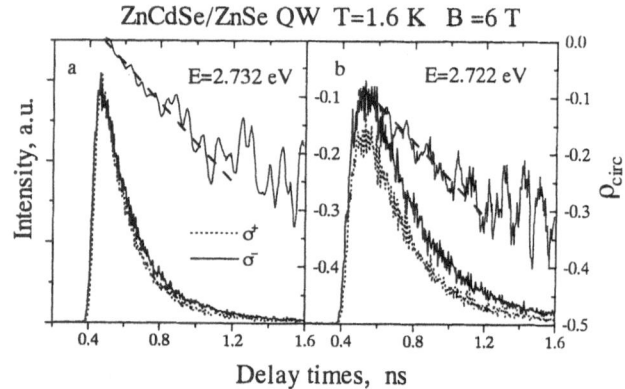

ZnCdSe/ZnSe QW T=1.6 K B =6 T

Fig.3. Spectrally resolved decay kinetics of PL intensity.

4 Conclusions

We have studied the temporal evolution of exciton PL in low-dimensional *ZnCdSe/ZnSe* superlattices. Study of PL time decay in polarized light allows to determine the times of exciton spin relaxation. Spin relaxation times strongly decrease with the increase of magnetic field indicating the phonon assisted mechanism of spin relaxation. The detailed analysis of the dependence of spin relaxation times on exciton localization energy is not possible due to the complicated character of exciton population kinetics for deep localized states. The transfer of polarization in the processes of exciton energy relaxation has been clearly observed.

ACKNOWLEDGEMENTS. This work was supported by CREST and Japanese Ministry of Education, Science and Culture, as well as by Russian RFBR Foundation and Russian Ministry of Science and Technology.

References

1. R.N.Kyutt, A.A.Toropov, S.V.Sorokin, T.V.Shubina, S.V.Ivanov, M.Karlsteen, M.Willander, Appl.Phys.Lett., **75**, (1999) 373.

2. S.A.Permogorov, A.N.Reznitsky, L.N.Tenishev, A.V.Kornievsky, S.Yu.Verbin, S.V.Ivanov, S.V.Sorokin, W.von der Osten, H.Stolz, M.Jutte, Phys. Solid State, **40**, (1998) 743.

3. A.N.Reznitsky, S.Yu.Verbin, S.A.Permogorov, A.G.Tsekun, A.Yu. Kaminskii, Phys. Solid State, **37** (1998) 1164.

Mixed neutral and negatively charged microcavity polaritons

R. Rapaport[1], E. Cohen[1], Arza. Ron[1], E. Linder[1], R. de Picciotto[2], R. Harel[2], L. N. Pfeiffer[2]

[1] Solid State Institute and Physics Department, Technion, Israel Institute of Technology, Haifa 32000, Israel.
[2] Bell Laboratories, Lucent Technologies, 600 Mountain Avenue, Murray Hill, New Jersey 07974.

Abstract Cavity polaritons (CP) are studied in a structure of a single GaAs/AlAs quantum well (QW) that contains a dilute 2-dimensional electron gas (2DEG), and is embedded in a λ-wide microcavity (MC). The 2DEG density is varied by photoexcitation, and for $n_e \leq 5 \times 10^{10} cm^{-2}$, negatively charged CP's are formed. They arise from the strong coupling of the MC-photon with the (e1:hh1)1S (X) and (e1:lh1)1S neutral excitons, and the negatively charged (e1:hh1)1S+e exciton (X-). Negatively charged CP's have several properties that are distinct from those of the neutral CP's: a. The X- coupling strength to the MC-photon increases as $\sqrt{n_e}$. b. CP's have a non-vanishing electric charge due to that of the bare X- and this should lead to their transport in an applied electric field. c. The bare X- is fermion-like while the bare X is boson-like and their admixture by the coupling to the MC-photon leads to CP's with energy dependent commutation relations. The experimental reflection and photoluminescence spectra spectra were analyzed using the coupled oscillators model. From the fitted spectra, the dependence of the exciton oscillator strengths on n_e is extracted. It is observed that the X- coupling strength increases with increasing n_e and there is an oscillator strength transfer from X to X-. The CP's effective charge and mass are calculated as a function of the MC-photon energy (namely, of the detuning energy). The calculated e/m can be ~ 300 times larger than that of a free electron in the bare GaAs QW. This suggests that the CP mobility should be very high, and that very fast electrical transport is expected. Indications for this are found in initial experiments of the reflection spectral variation under an electric field applied in the QW plane.

1 Introduction

The most frequently studied system that shows strong exciton - confined photon coupling is that of a quantum well (QW) embedded in a semiconductor microcavity (MC) [1-3]. In a GaAs/AlGaAs QW/MC structure, the strong X and X_{lh} excitons coupling to the MC mode results in the formation of neutral cavity polaritons (CP). (The notation used is: X = (e1:hh1)1S and X_{lh} = (e1:lh1)1S). These CP's show a Rabi splitting that is a measure of the coupling strength, and have a well defined in-plane dispersion. The dispersion curves have a large curvature, resulting in a very small polariton effective mass [1,3], that can be viewed as the reduced mass of the exciton translational mass ($M_X = m_e + m_h \simeq 0.2 m_e$ in GaAs) and a very small MC-photon mass ($m_{ph} \simeq 10^{-5} m_e$) [1]. Recently, we observed that when the GaAs QW contains a low density two dimensional electron gas (2DEG) and it is embedded in a MC, there is a strong interaction between the MC mode and both the neutral excitons (X

and X_{lh}) and the negatively charged exciton X- = (X+e). This results in the formation of charged CP's [4]. Similar observations were made for a CdTe QW that contains a two dimensional hole gas (2DHG) and is embedded in a MC, where a strong interaction was observed between the MC mode and the positively charged exciton (X+) [5]. The charged CP's have some distinct properties, in addition to those exhibited by the neutral CP's: a. The MC photon - X- coupling strength increases as $\sqrt{n_e}$. This is analogous to the coupling strength dependence of the confined photon - atom on the density of free atoms in a metallic cavity [6]. b. The charged CP's have a non-vanishing electric charge that is due to the bare X- charge. Moreover, the (model calculated) polariton charge and mass are continuously tunable with the MC photon energy, resulting in a very high charge to mass ratio. c. The bare X and X- exciton wavefunctions are admixed by their interaction with the MC mode. Since the bare X- is fermion-like while the bare X is boson-like, the creation and annihilation operators of the mixed cavity polaritons have energy dependent commutation relations. In this paper we describe spectroscopic experiments and results of model calculations that show some of the CP's properties.

2 Experiments

The MC structure under study was grown by molecular beam epitaxy on a (001)-oriented GaAs substrate. It consists of an $Al_{0.1}Ga_{0.9}As$ λ-cavity with a cladding of $AlAs/Al_{0.1}Ga_{0.9}As$ distributed Bragg reflectors (DBR), having 15/25 periods on the top and bottom sides, respectively. Embedded in the MC center is a mixed type I - type II QW's structure (MTQW) consisting of a single 102A/200A/102A AlAs/GaAs/AlAs wide QW, cladded on both sides by 26A/102A GaAs/AlAs narrow QW's. It was shown [7] that under laser photoexcitation with an energy below the narrow QW bandgap energy, $E_{L1} < Eg(e1-hh1)_N$, excitons are generated only in the wide QW. However, for a laser energy $E_{L2} > Eg(e1-hh1)_N$ (with intensity I_{L2}), a 2DEG is generated in the wide QW (and a 2DHG is generated in the narrow QW). The 2DEG density is linearly proportional to I_{L2}, and is found to vary, for our structure, approximately as $n_e(I_{L2})$ (cm^{-2}) $\sim 6 \times 10^9 I_{L2} (mW/cm^{-2})$, where I_{L2} is the intensity estimated to reach the MTQW [8]. Reflection spectra were taken with the samples at T=5K, using a Halogen-Tungsten filament lamp as a probe beam, and under excitation

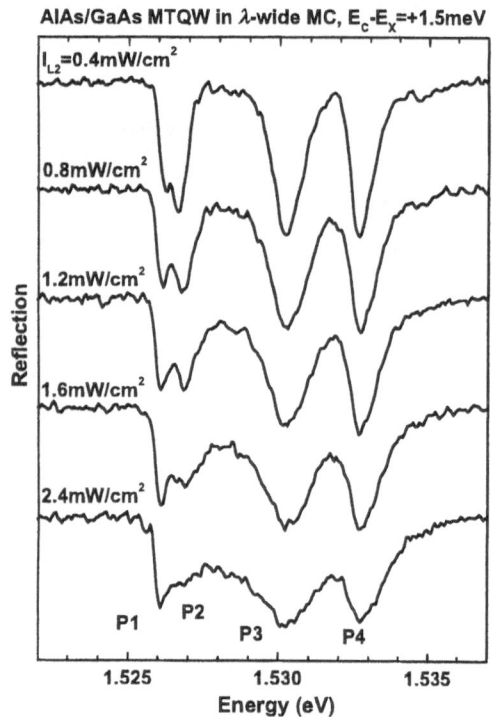

Fig. 1 Reflection spectra taken under excitation at E_{L2}=1.96eV, various I_{L2}, and monitored at a single spot on the sample, corresponding to a detuning energy, E_C-E_X $\cong$ +1.5meV. The labels of the resulting CP's are shown at the bottom of the figure.

with a He-Ne laser (E_{L2}=1.96eV) in the I_{L2} range corresponding to $0 < n_e < 5 \times 10^{10} cm^{-2}$.

The spectra were monitored at different illumination spots on the sample surface and this resulted in different MC mode energies. For the spectroscopy with an in-plane electric field, electrical contacts were deposited on the cleaved edges of narrow stripes made of the wafer. Fig. 1 presents reflection spectra observed at an illumination spot on the sample surface, corresponding to a MC mode energy of $E_C = E_X + 1.5$meV, with increasing I_{L2}. The four reflection dips correspond to CP's formed by the coupled X- , X, and X_{lh} exciton modes and MC-mode. It is noted that all the lines are significantly broadened with increasing I_{L2}, and this implies that the presence of the 2DEG increases the polariton dephasing rate via scattering with free electrons. The energy of the CP branches as a function of E_C, extracted from the lines of the reflection spectra observed with I_{L2}=1.2mW/cm^2, is shown in Fig. 2. The highest energy branch arises from the (e1:hh1)2S exciton and it will not be discussed in this paper. Fig. 3 shows reflection spectra observed with E_C value as in Fig.1, but with increasing electric field strength. The additional CP branch that is associated with X- disappears and the lines become narrower as the field intensifies.

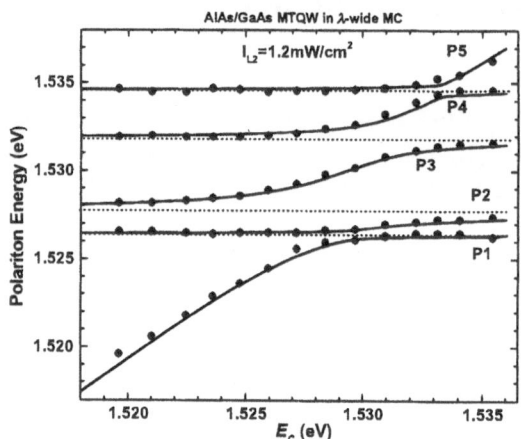

Fig. 2 Polariton energies as a function of the bare MC-mode energy, E_C, extracted from reflection spectra taken with $I_{L2} \simeq 1.2mW/cm^2$ (full circles). The solid lines are the calculated CP energies and the dotted lines mark the bare exciton energies.

3 Analysis

We first obtain the dependence on I_{L2} (namely, on n_e) of the X and X- coupling strengths to the MC-mode (V_X and V_{X-}, respectively). In order to do this, we diagonalized the coupled excitons - MC mode Hamiltonian for various E_C and I_{L2} values, and fit the CP energies as obtained from the experimental reflection spectra. The solid lines in Figs. 2 present the results of such a fit. We note that in fitting the reflection spectra in the case of I_{L2}=0, X- is excluded. In Fig. 4a we plot $(2V_{X-})^2$ as a function of I_{L2}. A clear linear dependence is found, and, due to the linear dependence of n_e on I_{L2}, this means that $V_{X-} \propto \sqrt{n_e}$. As we pointed out previously [4], this is in analogy to what is observed for atoms in a metallic cavity, where the collective coupling strength of N identical atoms with the confined cavity photon is proportional to $\sqrt{N}$ [6]. In contrast, the coupling strength of N neutral excitons is independent of N and is determined by the properties of the QW. In this sense, charged excitons are a better analogue to atoms in a cavity than neutral excitons. Note that since f_{X-}, the X- oscillator strength, is proportional to V_{X-}^2, then $f_{X-} \propto n_e$. In Fig. 4b we present the fitted $(2V_X)^2$ vs. $(2V_{X-})^2$ for various I_{L2} values. A decrease in $(2V_X)^2$ is observed as $(2V_{X-})^2$ increases. This implies that there is an oscillator strength transfer from X to X- as n_e increases. Previous studies on modulation p-doped CdTe QW containing a mixed X - X+ phase, reported an oscillator strength transfer from X to X+ as the density of holes increases [5]. Following it, we fit the data of Fig. 4b with a linear dependence:$V_X^2 = V_{X0}^2 - \alpha V_{X-}^2$. This can be interpeted as follows: α determines the change of the total oscillator strength ($f_T = f_X + f_{X-}$) with n_e. $\alpha > 1$ means that f_T decreases when n_e increases, $\alpha < 1$ means that it increases and $\alpha = 1$ means a fixed f_T.

The fitting parameters have the values:$\alpha = 0.9 \pm 0.4$ and $V_{X0}^2 = 4.1 \pm 0.25meV^2$. Due to the relatively large

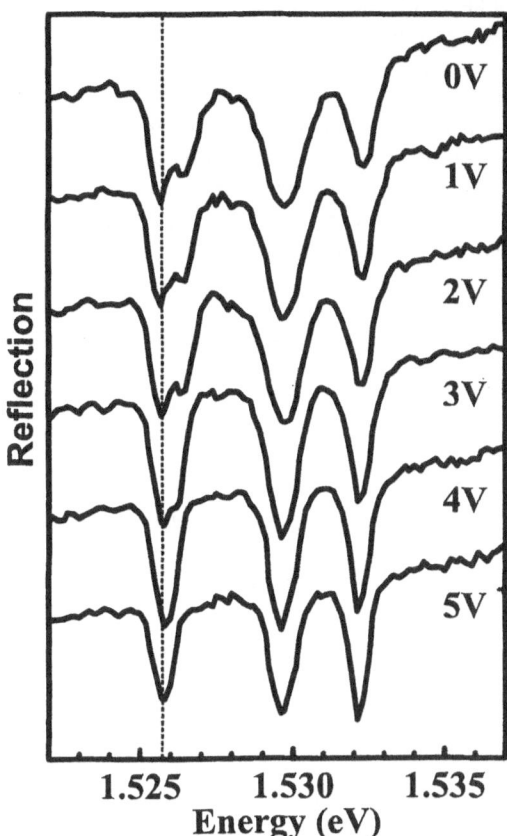

Fig. 3 Reflection spectra taken at the same detuning and photoexcitation energies as in Fig.1, with increasing applied voltage (increasing in-plane electric field strength).

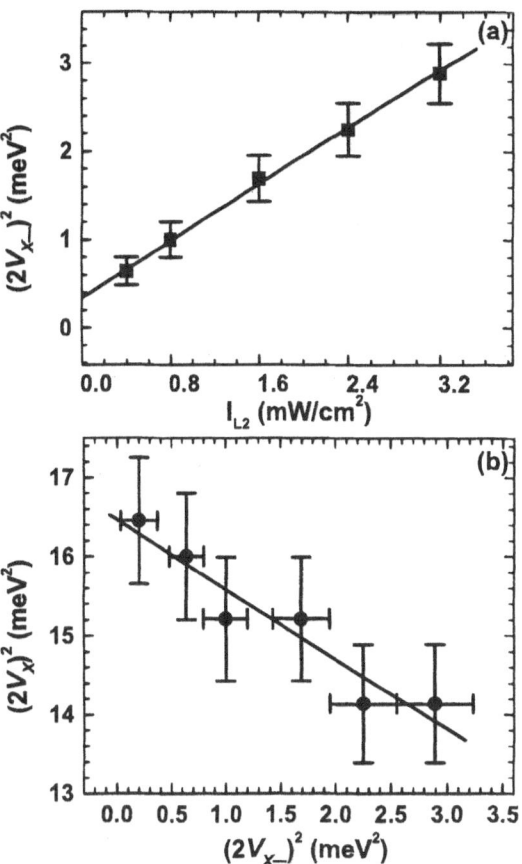

Fig. 4 a. $(2V_{X-})^2$ as a function of I_{L2} (full squares). b. $(2V_X)^2$ as a function of $(2V_{X-})^2$ (full circles). The solid lines in both a and b are linear fits to the experimental data.

error bars of $2V_X^2$, we cannot make a definite conclusion whether α is smaller, equal or larger than unity. Intuitively, assuming that all the electrons in the 2DEG have an equal probability to form X- under photoexcitation, the possible number of equivalent X- excitations per unit area is determined by n_e. In an analogous description to the case of two-level atoms in a cavity, if all these X- excitations are coherent, the single-excitation coupling strength is enhanced by a factor of $\sqrt{n_e}$ [4]. The total oscillator strength of the QW should be a constant that is proportional to the total QW density of states (DOS), which in turn is proportional to the density of unit cells. If we ignore the small reduction in the available DOS for free electrons due to the presence of the low density 2DEG (which amounts to neglecting phase space filling effects) and the effect of screening of the excitons (which amounts to neglecting oscillator strength transfer from bound to free states), we expect to have $V_X^2 = V_{X0}^2 - V_{X-}^2$. This is quite consistent with the experimentally extracted value of $\alpha \approx 1$. A more formal description is given in [9,10].

CP's composed of charged excitons are expected to carry a net electric charge, that is equal to the electron charge times the relative fraction of X- in the polariton

wavefunction. This fraction is obtained from the diagonalization of the coupled oscillators Hamiltonian, and it strongly depends on E_C and on n_e. Similarly, the effective mass at K=0 is calculated from the CP dispersion relations. Using these values, the effective charge to mass ratio, $(e/m)_{eff,j}$, is calculated for each CP branch (j = P1-P4), as a function of E_C . Here we used the following parameters: $m_X = 0.177m_e$, $m_{X-} = 0.244m_e$, $m_{X_{lh}} = 0.267m_e$ and $m_C = n_{GaAs}^2 E_C/c^2$, where n_{GaAs} is the GaAs index of refraction. The results are shown in Fig.5.

Near resonance (i.e. $E_C \simeq E_X$), the $(e/m)_{eff,j}$ ratio of the three lowest energy CP branches is orders of magnitude larger than the free electron e_0/m_0. For the P1 branch, $(e/m)_{eff,j}$ reaches a value of $\sim 3300e_0/m_0$ (~ 220 times larger than the e/m of the conduction electrons in GaAs). This should be manifest in high collision rates of CP's with free electrons, as we observe in the reflection linewidth dependence on n_e. Another possible consequence of the high $(e/m)_{eff}$, is an expected high in-plane mobility of charged CP's. The reflection spectra observed under an applied electric field (Fig. 3) indicate that P2 disappears as the field strength increases. The remaining three reflection lines become narrower with-

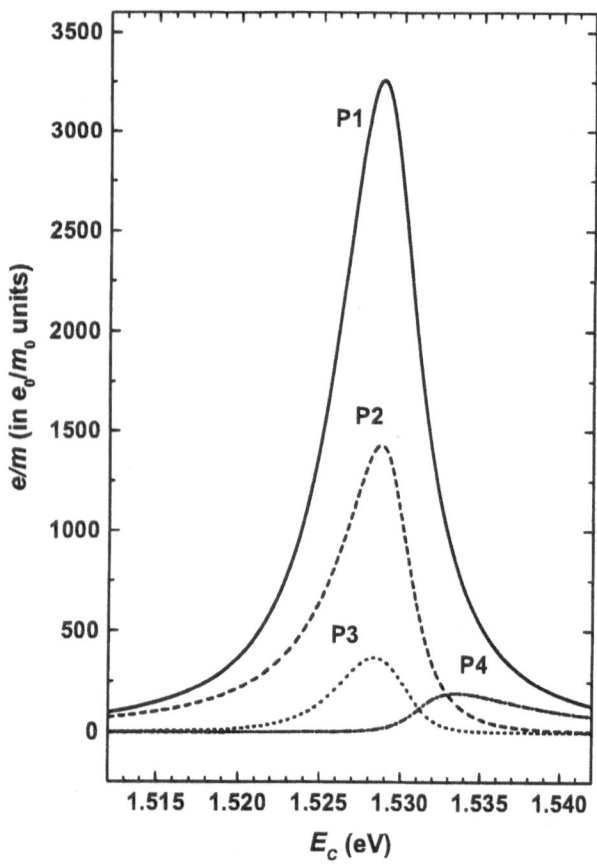

Fig. 5 The calculated charge to mass ratio, $(e/m)_{eff,j}$, of the P_j, j=1-4 CP branches, in units of the free electron charge to mass ratio, e_0/m_0.

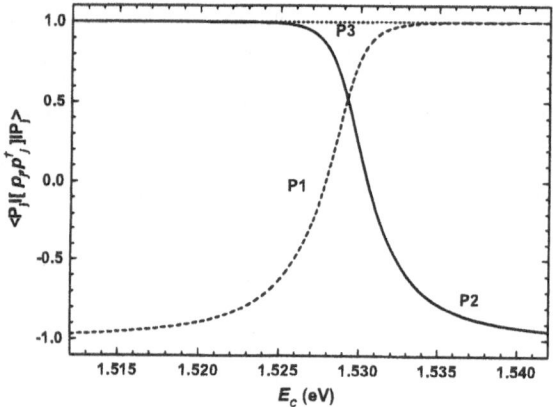

Fig. 6 The calculated expectation values of the commutation relations of the P_j, j=1-3 CP branches.

out a noticeable change in energy. At this level of understanding, we can only state that a strong enough electric field results in charged polariton dissociation, but we cannot conclude whether CP transport occurs.

For low densities, X and X_{lh} are boson-like quasi-particles, due to their integral spin. X- has a half integral spin, carried by its additional electron, and is thus

fermion-like. Since the X and X- are close in energy (relative to their coupling strength with the MC-photon), the resulting polaritons share both the X and the X- wavefunctions together with the MC-photon. The charged cavity polaritons can then be viewed as quasi-particles that bear the properties of their bare constituents. Consequently, in the simplest model, the commutation relations of their creation and annihilation operators will be determined by those of the bare constituents, weighted by their relative fraction in the mixed modes. Fig. 6 shows these expectation values calculated for P1, P2 and P3. P3 has a very small fraction of X-, and thus its commutation relation expectation value is always close to +1. Those of P1 and P2 vary continuously from -1 to +1.

4 Summary

We presented a spectroscopic study of charged CP's by measuring and analyzing the reflection spectra of a MC structure that has an embedded QW with continuously varied 2DEG densities. The analysis of the CP's energy is based on the coupled oscillators model. It shows that the oscillator strength of the charged exciton increases linearly with n_e and that there is an oscillator strength transfer form the neutral to the charged excitons. The calculations yield the relative fractions of each exciton and of the MC mode in the CP wavefunctions. From them, the effective charge to mass ratio is calculated for each CP branch, as a function of the MC mode energy. Preliminary experiments of an in-plane applied electric field on the reflection spectra indicate that charged CP's could be electrically transported in the QW plane. Aknowledgements: The work at the Technion was done at the Barbara and Norman Seiden Center for Advanced Opto-Electronics Research. It was supported by the United States - Israel Binational Sience Foundation (BSF), Jerusalem, Israel.

References

1. M. S. Skolnick, T. A. Fisher and D. M. Whittaker, Semicond. Sci. Technol. 13, 645 (1998).
2. G. Khitrova, H. M. Gibbs, F. Jahnke, M. Kira, S. W. Koch, Rev. Mod. Phys. 71, 1591 (1999).
3. V. Savona, in Confined Photon System, Fundamentals and Applicatons, Edited by H. Benisty, Springer Lecture Notes in Physics(Berlin,1999), p.173.
4. R. Rapaport, R. Harel, E. Cohen, Arza Ron, E. Linder, L.N. Pfeiffer, Phys. Rev. Lett. 84, 1607 (2000).
5. T. Brunhes, R. Andre, A. Arnoult, J. Cibert and A. Wasiela, Phys. Rev. B. 60, 11568 (1999).
6. J. M. Raimond and S. Haroche, in "Confined Electrons and Photons", E. Burstein and C. Weisbuch, Editors, Plenum, NY, 1995. p.383.
7. I. Galbraith et al., Phys. Rev. B. 45, 13499 (1992).
8. A. Manassen et al., Phys. Rev. B. 54, 10609 (1996).
9. R. Rapaport, R. Harel, E. Cohen, Arza Ron, E. Linder, and L.N. Pfeiffer,Phys. Stat. Sol. (a) 178, 481 (2000).
10. R. Rapaport, E. Cohen, Arza Ron, E. Linder, and L.N. Pfeiffer, Submitted for publication in Phys. Rev. B.

Confined optical modes in photonic molecules and crystals

M. Bayer[1], A. Forchel[1], T.L. Reinecke[2], P.A. Knipp[2]

[1] Physikalisches Institut, Universiẗ Würzburg, Am Hubland, D-97074 Würzburg, Germany e-mail: mbayer@physik.uni-wuerzburg.de
[2] Naval Research Laboratory, Washington DC, 20375, USA

Abstract The optical modes in microresonators with a three dimensional confinement of light (photonic dots) have been studied spectroscopically, from which detailed information about their energies and field distributions has been obtained. Due to the confinement their mode spectrum is dominated by sharp resonances. Photonic molecules can be assembled by connecting together several photonic dots. The electromagnetic field distributions in two-dot molecules bear strong resemblences to the bonding and anti-bonding orbitals in diatomic molecules. By increasing the number of coupled resonators in a linear chain the transition from a quasi-atomic system to a photonic crystal is obtained.

1 Introduction

The control of light-matter interaction has been one of the central topics of semiconductor physics [1]. One of these interactions is the spontaneous emission, the kinetics of which is determined by Fermi's golden rule: The emission rate for a transition of energy ω between an initial state $| I \rangle$ and the final states $| F \rangle$ is given by

$$\frac{1}{\tau} = \frac{4\pi}{\hbar} \sum_{F} | \langle F | \mathbf{d} \cdot \boldsymbol{\varepsilon}(\mathbf{r}) | I \rangle |^2 \cdot \rho(\omega). \qquad (1)$$

Here $\mathbf{d} = -e\mathbf{r}$ is the electric dipole operator and $\boldsymbol{\varepsilon}(\mathbf{r})$ is the vacuum electric field distribution. $\rho(\omega)$ is the density of the optical modes into which the electronic transitions can radiate. From this equation different ways of manipulating the transition rate become obvious:

1. The first possibility is the engineering of the electronic states in the optically active medium, e.g. by a reduction of its dimensionality which has obtained considerable attention during the last decades.
2. The second possibility is the engineering of the optical modes by a deliberate manipulation of the vacuum field distribution or by controlling the density of optical modes in the solid state system.

For modifying the optical modes two basic concepts, that of microresonators [2–5] and that of photonic crystals [6–8] have been developed. The microresonator concept relies on placing the optically active medium between highly reflecting mirrors, among which the electromagnetic field becomes confined. The photonic crystal concept, on the other hand, relies on a periodic modulation of the dielectric function with a lattice constant comparable to the wavelength of light. From such a modulation a photonic band structure can emerge: In the bands optical modes can propagate, whereas no modes exist in the

Fig. 1 Scanning electron micrograph of a photonic dot with a diameter of 3 μm.

band gaps. Here we want to present results, which represent in some sense an unification of these two different concepts.

2 Optical modes in photonic dots

Starting point for the fabrication of photonic dots were λ-cavities consisting of a GaAs layer with a height of $\sim$ 250 nm sandwiched between two high reflectance GaAs/AlAs mirrors. In the center of the GaAs layer a 7 nm wide $In_{0.14}Ga_{0.86}As$ quantum well was placed as optically active medium. The planar cavities were laterally patterned using lithography [9,10]. Figure 1 shows a scanning electron micrograph of a cylindrical micropillar with a diameter of 3 μm. The cavity was etched through the top reflector and the GaAs layer, whereas the bottom reflector essentially remained unpatterned except of a few mirror stacks. Due to the resulting discontinuity of the refractive index in lateral direction, the electromagnetic field is confined three-dimensionally in the resonator, as can be shown in optical studies: [9–11] For an unpatterned cavity the photoluminescence spectrum (not shown) is dominated by the fundamental resonator mode with an energy given by the separation of the two mirrors. The spectrum changes strongly for a patterned cavity. In this case emission from several discrete optical modes is observed. [9–11]. This is an indication that light is three dimensionally confined in the resonators.

Two characteristic features are to be noted for decreasing lateral resonator size: (1) The energies of the

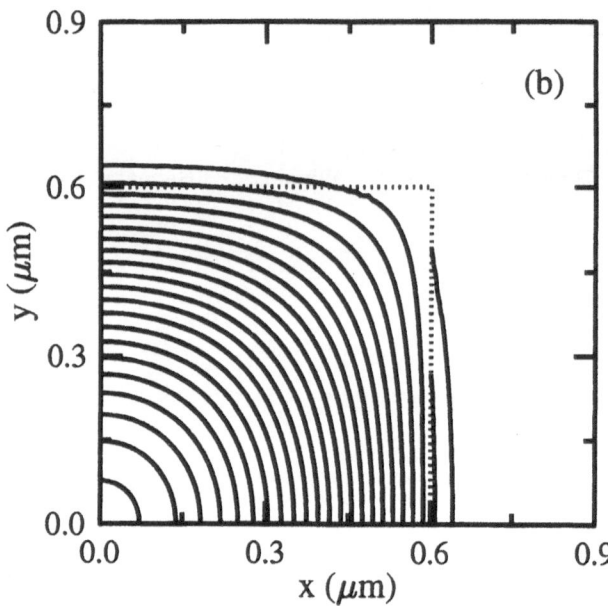

Fig. 2 Contour plot of the in-plane electric field distribution in a photonic dot with a lateral size of 1.2 μm.

optical modes increase, (2) the splittings between the optical modes increase as well. These size dependences are strongly reminiscent to those of the interband transition energies in electronic quantum dots. Therefore we call these structures 'photonic dots', to distinguish them from planar resonators. The energies of the optical modes in these square shaped cavities are given by the simple form

$$E = \sqrt{E_0^2 + \hbar^2 c^2 \left(q_x^2 + q_y^2\right)}, \qquad (2)$$

where E_0 is the light confinement energy between the Bragg mirrors. The lateral wave numbers $q_i, i = x, y$ can be approximated by standing wave conditions:

$$q_i = \frac{\pi}{L}(n_i + 1), \qquad n_i = 0, 1, \dots \qquad (3)$$

Each optical mode thus can be characterized by a set of two eigenvalues (n_x, n_y). The n_i give the number of nodes in the electric field distribution along the corresponding direction.

To obtain further insight into the optical mode confinement we have also calculated the electric field distributions in the photonic dots. [12,13] Figure 2 show a contour plot of the in-plane field of the (0,0) mode in a 1.2 μm large dot. Only one quarter of the dot, whose boundaries are indicated by the dotted lines, is shown in the panel. Indeed, for the 1.2 μm wide structure the electric field is well confined inside the dot. This structure size corresponds roughly to the minimum resonator size studied in the present experiments. Only when going to small resonators with a lateral size comparable to the light wavelength in the material, the electric field leaks significantly out of the resonator.

Fig. 3 Scanning electron micrograph of a photonic molecule. The size of each dot is 3μm by 3 μm. The dimensions of the channel are 1 μm length and 1 μm width [14].

3 Optical modes in photonic molecules

In a next step, we have coupled two photonic dots by a narrow channel. The strength of the coupling between the dots can be controlled by the dimensions of the channel, its width and its length. The optical modes in these structures have been studied as function of the coupling. From the coupling one expects a behavior similar to the splitting of the electronic levels in diatomic molecules, for which the interaction leads to the formation of bonding and anti-bonding molecule orbitals. Figure 3 shows a scanning electron micrograph of a molecule with a dot size of 3μm by 3μm. The length of the channel was 1 μm, as was the width of the channel. [14].

A series of coupled resonator structures has been fabricated in which the length of the interconnecting channel was fixed to 1 μm and its width was varied. The emission energies of the lowest lying optical modes are plotted against the channel width in Figure 4 [14]. Zero channel width corresponds to the case of separated dots for which there are two optical modes in the energy range of interest here. For non-zero channel width coupling between the dots is obtained and the ground mode (0,0) splits into two modes. The coupling increases with increasing channel width and so does the energy splitting. The higher lying mode is two-fold degenerate (neglecting polarization), because the (1,0) and (0,1) field distributions can be transformed into one another by a 90 deg. rotation in the cavity plane. Therefore one expects a splitting into four modes in the molecules, as is confirmed by the experimental data in Figure 4.

The analogy with diatomic molecules can be also seen from the calculated field distributions in Figure 5, which shows contour plots of the six lowest modes in the resonator plane [14]. The lowest lying molecule modes (a) and (b), which originate from the splitting of the (0,0) mode in a photonic dot, correspond to bonding and anti-bonding σ-orbitals formed from s-atom states. The next two higher lying modes (c) and (d) correspond to σ-like

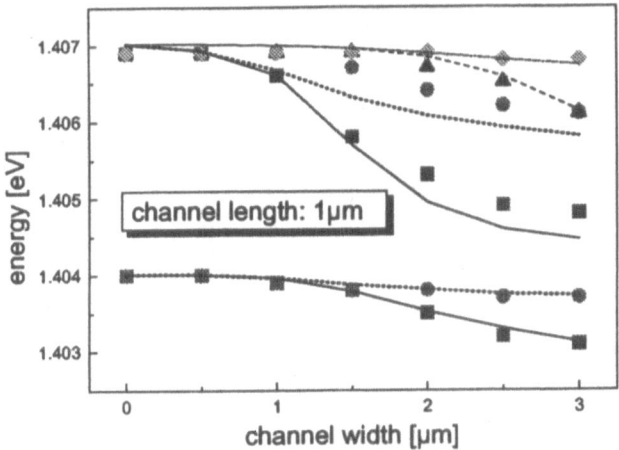

Fig. 4 Dependence of the optical mode energies in photonic molecules on the channel width. The channel length was 1 μm. Symbols give experimental data, lines the results of calculations [14].

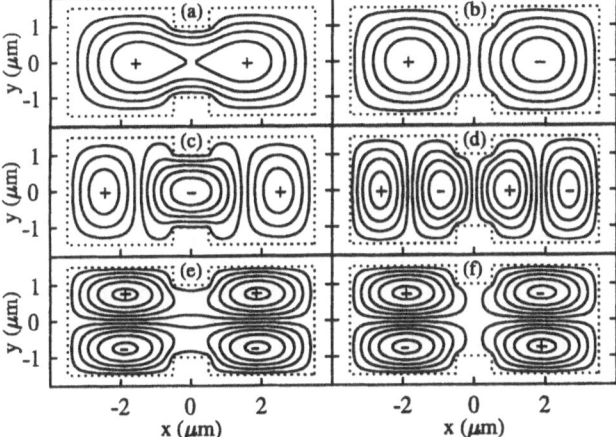

Fig. 5 In-plane electric field distributions of the six lowest lying optical modes in photonic molecules. The channel length is 1 μm, its width is 2 μm [14].

molecular orbitals formed from p-atom modes which are oriented along the molecule axis. Finally the (e) and the (f) modes are analogous to π-orbitals formed from p-modes, now oriented perpendicular to the molecule axis.

4 Photonic crystals

4.1 Evolution of a crystal band structure

Increasing the number of coupled dots, i.e. forming chains of coupled photonic dots, should lead to the formation of photonic bands, in which the optical modes show dispersion along the chains. Therefore these structures permit the study of the transition from a quasi-atomic to a molecular system and further to a crystal. They also give insight into band structure formation occuring in other systems as e.g. for the electronic bands in solids. Unlike the electronic case, in this approach we are able to construct a crystal-like system by adding individual building blocks one by one.

We have fabricated such chain structures consisting of a variable number of photonic dots. The optical modes

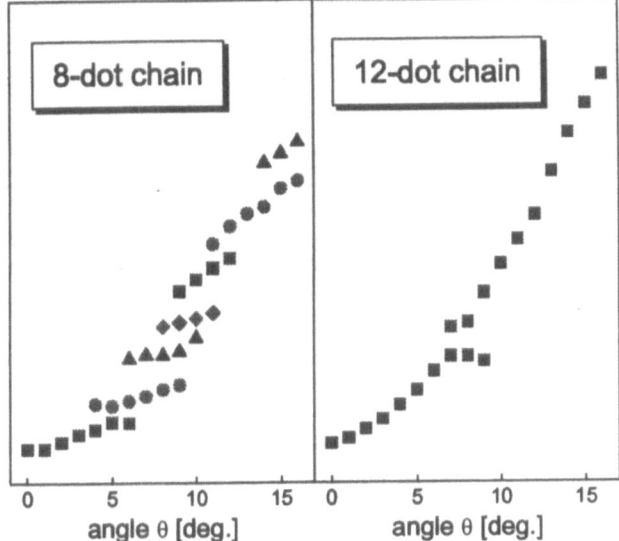

Fig. 6 Angle dependence of the optical mode energies in photonic chains formed from varying numbers of dots. The angle of detection was varied along the chains [15].

in these structures can be studied by angle-resolved spectroscopy. In such experiments only the emission along a certain spatial direction is registered by using an aperture and the direction of detection is varied, here along the chain. This direction can be characterized by the polar angle ϑ relative to the cavity normal. Each angle ϑ corresponds to a wavenumber $k = (E/\hbar c)\sin\vartheta$ along the chain. Figure 6 shows the optical mode energies versus the angle of detection for 8-dot and 12-dot chains [15]. The lengths and widths of the channels were 1 μm and 2.5 μm. With increasing chain length the energy separation between the optical modes decreases due to the reduced confinement along the chain. The energies in 2-dot and 4-dot chains (not shown) do not depend on ϑ. In this case the electromagnetic fields are three-dimensionally confined as in single cavities. For 8 coupled dots (left panel) a 'shift' of the observed mode to higher energies with increasing ϑ. Finally, for 12 coupled dots (right panel) a discrete optical mode spectrum can no longer be resolved. For this case the modes have collapsed to form a band resulting in dispersive behavior along the chain. Neglecting the weak refractive index modulation along the chain, the mode energies are approximately those of a photonic wire (equal widths of dots and channels) [16]

$$E = \left(E_0^2 + E_y^2 + \hbar^2 c^2 k_x^2\right)^{1/2} . \tag{4}$$

E_y is the optical confinement energy perpendicular to the chain. No apparent changes are observed when the number of coupled photonic dots is increased further.

Nevertheless, since the dielectric function ε is modulated periodically along the chain, also the observation of band gaps is expected. Indeed, for the 12-dot chain the optical mode does not shift smoothly to higher energies but a discontinuity in its dispersion relation is observed for detection angles around 7 deg. This detection angle corresponds to the wave number of the first Brillouin zone edge. The discontinuity of 0.2 meV is, however, small as can be expected from the weak modulation of ε in this case.

4.2 Crystal band structure

To enlarge the band gaps, a stronger modulation of the dielectric function is required. This was obtained experimentally by reducing the width of the channel to 1.5 μm. Figure 7 shows the energies of the photon modes (different symbols) versus the wave vector k_x along a chain consisting of 50 coupled photonic dots [15]. For comparison the ground photon mode dispersion in a 3μm wide wire without modulation also is shown (dashed-dotted line). Three discontinuities are clearly observed in the dispersion for the crystal photon. At these features there are strong deviations from the dispersion of a wire of constant width resulting in gaps in the frequency spectrum of $\sim$ 1.3, 0.8 and 2.1 meV. The energy gaps occur at the boundaries of the Brillouin zones , which are located at $k_{BZ} = n \cdot \pi/P$, with the lattice constant P and the zone number $n = \pm 1, 2, \ldots$. The dotted vertical lines in Figure 7 indicate the zone boundaries. Band gaps occur at the boundaries of the first, the second and the fourth zone. At the boundary of the third zone, surprisingly, no energy gap is found in the experiments.

The solid lines show the results of calculations for the photonic dispersions using the boundary element method [12,13]. We observe a good agreement with experiment for all parts of the photonic band structure. In particular, for the third zone boundary the calculated gap is 0.035meV, which is too small to be resolved in the experiment. This small band gap is characteristic for the photonic crystal under study: The band gap is determined by the Fourier component of the lattice potential at the wave number at which the gap occurs. For the present structure with a ratio of dot to channel length of 3 to 1 we find that the Fourier component at the third zone boundary is small resulting in a small band gap. Recent studies of photonic crystals in which the ratio was varied confirm that the third band gap opens up when the ratio is different from 3 to 1.

5 Summary

The optical modes in confined photon systems have been studied spectroscopically and have been compared to the results of numerical calculations. These studies furnish a quantitative understanding of the confined modes.

6 Acknowledgements

The experimental support by Timm Baars, Gregor Dasbach, Thomas Gutbrod, Günter Guttroff, Andreas Kuther, Andrew McDonald and Frank Weidner during the course of this work is gratefully acknowledged. We also thank the State of Bavaria and the Deutsche Forschungsgemeinschaft for the financial support of this work.

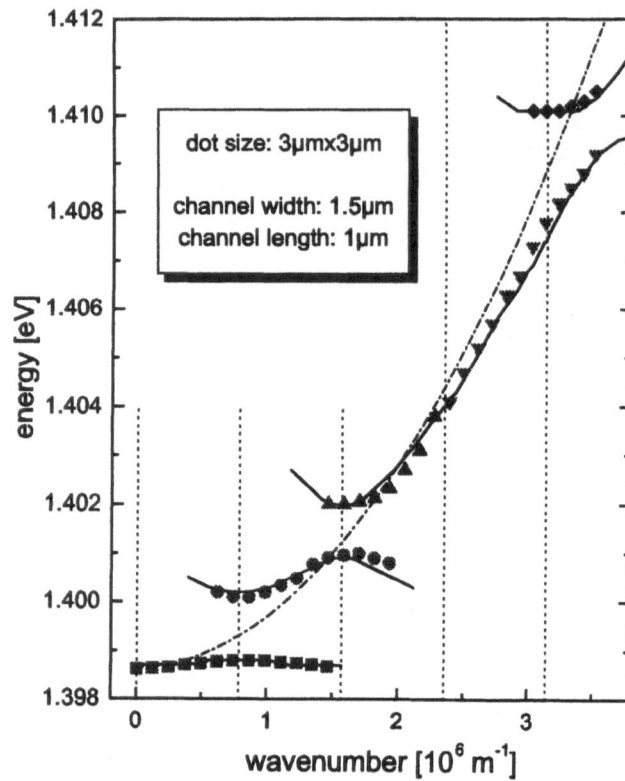

Fig. 7 Dispersion relation of the optical modes in a photonic crystal fabricated from 50 coupled photonic dots. Symbols give the experimental data, solid lines the results of the calculations. The dashed-dotted line shows the dispersion of the optical mode in a photonic wire. The dotted lines indicate the boundaries of the Brillouin zones [15].

References

1. see, for example, *Confined Electrons and Photons*, ed. by E. Burstein and C. Weisbuch, NATO ASI Series, Physics Vol. **340**, Plenum Press, New York, 1995.
2. H. Yokoyama et al., Appl. Phys. Lett. **57**, (1990) 2814.
3. G. Björk et al., Phys. Rev. A **44**, (1991) 669.
4. C. Weisbuch et al., Phys. Rev. Lett. **69**, (1992) 3314.
5. R. Houdré et al., Phys. Rev. Lett. **73**, (1994) 2043.
6. E. Yablonovitch, Phys. Rev. Lett. **58**, (1987) 2059.
7. S. John, Phys. Rev. Lett. **58**, (1987) 2486.
8. See, for example, J.D. Joannopoulos, P.R. Villeneuve, S. Fan, *Photonic Crystals, Molding the Flow of Light*, (Princeton University Press, Princeton, NJ, 1995), and J.D. Joannopoulos, P.R. Villeneuve, S. Fan, Nature **386**, (1997) 143.
9. J.M. Gérard et al., Appl. Phys. Lett. **69**, (1996) 449.
10. J.P. Reithmaier et al., Phys. Rev. Lett. **78**, (1997) 378; M. Röhner et al., Appl. Phys. Lett. **71**, (1997) 488.
11. T. Gutbrod et al., Phys. Rev. B **59**, (1999) 2223; *ibid.* **57**, (1998) 9950.
12. P.A. Knipp and T.L. Reinecke, Phys. Rev. B **54**, (1996) 1880.
13. P.A. Knipp and T.L. Reinecke, Physica **E2**, (1998) 920.
14. M. Bayer et al., Phys. Rev. Lett. **81**, (1998) 2582.
15. M. Bayer et al., Phys. Rev. Lett. **83**, (1999) 5374.
16. A. Kuther et al., Phys. Rev. B **58**, 15744 (1998).

Non-linear spin polarization dynamics in semiconductor microcavities

P. Renucci[1], **X. Marie**[1], **T. Amand**[1], **M. Paillard**[1], **P. Senellart**[2], **J. Bloch**[2]

[1] Laboratoire de Physique de la Matière Condensée CNRS- INSA, 135 avenue de Rangueil, F-31077 Toulouse, France
renucci@insa-tlse.fr
[2] Laboratoire de Microstructures et Microélectronique, 196, avenue H. Ravera, F-92220 Bagneux, France

Abstract We have studied the spin dynamics of resonantly excited lower branch polariton states in a semiconductor quantum microcavity. At low excitation intensity, the cavity detuning dependence of both the circular and linear polarization dynamics is evidenced. At higher excitation intensity, the spin-dependent non-linear emission is studied.

1 Introduction

Microcavity polariton dynamics is presently the subject of intense investigations. Strong non linearities in the emission [1-3] as well as in the reflectivity behaviour [4] reveal the role of polariton mutual scattering. However, much less studies have been devoted up to now to polariton spin dynamics [5]. We report here on the spin dynamics in semiconductor microcavities in the strong coupling regime.

2. Experiments

Two λ-microcavities have been investigated, one with a single 8 nm wide $Ga_{0.95}In_{0.05}As$ quantum well with a Rabi splitting $\Omega_R \approx 3.7$ meV (MC1, presented here), and one with four 12 nm wide $Ga_{0.86}In_{0.14}As$ quantum wells with $\Omega_R \approx 6$ meV (MC2). The experimental results are similar in both samples. At resonance (δ=0) the MC1 lower branch linewidth amounts to 0.65 meV (limited by the sample wedge).

The samples are excited by 1.5 ps pulses generated by a mode-locked Ti-doped sapphire laser. The time-resolved photoluminescence is then recorded by up-converting the luminescence signal in a $LiIO_3$ non-linear crystal with the picosecond pulses generated by an optical parametric oscillator synchronously pumped by the same Ti:Sa laser [6]. The time-resolution is limited by the laser pulse-width (~ 1.5 ps) and the spectral resolution is about 3 meV. The linear and the circular polarization degrees of the luminescence are defined as $P_{Lin}=(I^X-I^Y)/(I^X+I^Y)$ and $P_C = (I^+-I^-)/(I^++I^-)$ respectively. Here I^X (I^Y) and I^+ (I^-) denote respectively the X (Y) linearly polarized and the right (left) circularly polarized luminescence components. The incident angle is 8°, corresponding to an initial in plane wave vector $k_{//} \approx 10^4$ cm^{-1}; the detection is normal to the surface, with a small acceptance angle (about 10^{-3} steradian). In the following, the excitation and detection energies are resonant with

the lower polariton branch. The experiments have been performed at T=12 K.

3. Experimental results and discussions

Figure 1 displays the time-resolved linear polarization of the MC1 microcavity emission following a linearly polarized excitation pulse for different cavity detunings. For the positive detuning δ=4 meV (i.e. when the exciton character of the polariton is dominant), the linear polarization decay is governed by the exciton transverse spin relaxation time $T_{s2}^{exc.}$. When the exciton weight in the polariton state decreases (i.e. for smaller δ), we observe in figure 1 that the polariton spin relaxation time increases. For large negative detuning (e.g. δ=-6 meV), the spin relaxation time is quenched [owing to the short cavity lifetime (~2.5 ps), the spin polarization cannot be measured for t>15 ps].

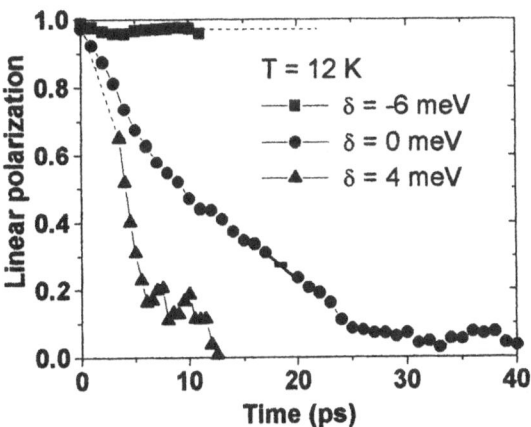

Fig. 1 Linear polarization dynamics of the lower branch microcavity emission for different cavity detunings.

The interpretation is that the polariton linear polarization decay reflects simply the transverse spin relaxation time of its exciton component : $1/T_{s2}^{pol.} = |C_e|^2 / T_{s2}^{exc.}$, where $|C_e|^2$ is the weight of the exciton component of the polariton state.

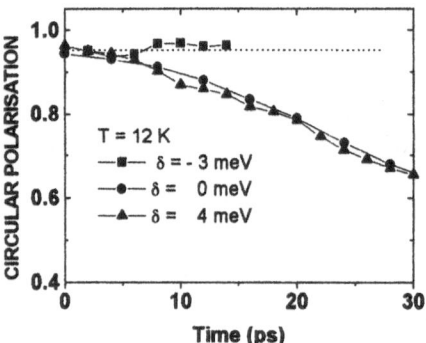

Fig.2 Circular polarization dynamics for three detuning values

The circular polarization of the polariton emission also depends on the cavity detuning δ. Figure 2 displays the circular polarization dynamics following a circularly polarized excitation for three detuning values. We observe a spin relaxation quenching for the negative detuning ($\delta=-3$ meV) but no measurable dependence of the spin relaxation time for $\delta \geq 0$. The circular polarization decay time corresponds to the longitudinal spin relaxation time ($T_{s1} \sim 70$ ps for $\delta = 0$). This spin relaxation is thus measured during the long total intensity decay time which corresponds to the return from the reservoir towards the $k_{//} \approx 0$ detected polariton states. The circular polarization dynamics reflects thus mainly the exciton spin relaxation within the reservoir. This explains the similar depolarization dynamics measured in figure 2 for $\delta=0$ or $\delta=4$ meV. In contrast, for $\delta<0$, the scattering to the reservoir is quenched. The detuning dependence is stronger in the case of linear polarization due to its fast decay ($T_{s2} \sim 15$ ps for $\delta = 0$, fig.1), which occurs before polaritons populating the reservoir are scattered back into the $k_{//} \approx 0$ detected states.

We have investigated the density dependence of the linear and circular polarization dynamics in the regime where the polariton emission is linear with excitation. As the density increases, the linear polarization decay becomes faster, which we attribute to mutual exchange interaction, which occurs through the exciton component of the polariton, in analogy with the corresponding known processes for excitons in bare quantum wells (QW) [7]. In contrast, the circular polarization decay time is not modified when the density increases since the circular depolarization mechanism is based on the electron-hole exchange within the polariton, in analogy again with bare excitons in QW [8].

For higher excitation density, the polariton emission is strongly super-linear. Figure 3 displays the time dependence of the circular polarization dynamics under circularly polarized excitation for two different excitation powers for a detuning $\delta=0$ meV. The small

incident photon density per pulse in these resonant excitation conditions ensures that the microcavity is still in the strong coupling regime. For instance, we estimate that the 150 μW experiment corresponds to a photogenerated polariton density of about 5×10^9 cm^{-2}, which is much smaller than the exciton saturation density. At this excitation, the circular polarization initially saturates at a value of about 100%, with a plateau which can last up to 20 ps (depending of the initial photogenerated density). The inset displays the luminescence intensity over power as a function of the excitation power. The co-polarized emission intensity displays a strong non-linear behaviour, similar to the previously observed one in cw experiments [1-3], while the emission intensity of opposite helicity keeps linear.

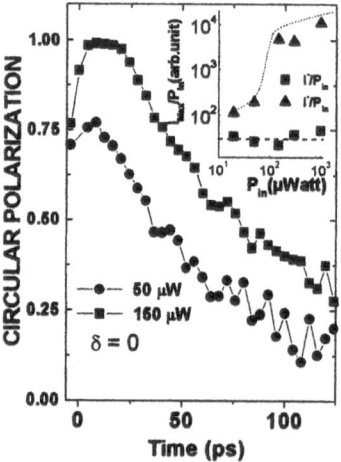

Fig.3 Time resolved circular polarization of the microcavity lower branch emission following a circularly polarized optical pulse for two excitation powers. Inset : Lower branch peak emission intensity over power for the copolarized (I^+/P_{in}) and counter polarized (I^-/P_{in}) components, as a function of the excitation power P_{in}

We attribute this effect to self seeded stimulated scattering of polaritons copolarized with the excitation light down to $k_{//} \approx 0$, while the counter polarized polaritons are still in a non- stimulated scattering regime [1,4]. The plateau in Fig. 3 is the result of the self seeded stimulated process occuring for I^+ and not for I^- which keeps linear with incident power.

1.   P. Senellart and J. Bloch, Phys. Rev. Lett. **82**, 1233 (1999)
2.   Le Si Dang *et al.* Phys. Rev. Lett. **81**, 3920 (1998)
3   A. I. Tartakovskii *et al.* Phys. Rev. B **60**, R11293 (1999)
4.   P. G. Savvidis *et al.* Phys. Rev. Lett. **84**, 1547 (2000)
5.   M. D. Martin *et al.* Phys. Stat. Sol. (a) **178**, 539(1999)
6.   X. Marie *et al.* Phys. Rev. B **59**, R2494 (1999)
7.   P. Le Jeune *et al.* Phys. Rev. B **58**, 4853 (1998)
8.   M. Z. Maialle *et al.* Phys. Rev. B **47**, 15776 (1993)

Optically pumped quantum dot lasers using high-Q microdisk cavities

P. Michler[1*], **A. Kiraz**[1], **C. Becher**[1], **Lidong Zhang**[1], **E. Hu**[1], **A. Imamoglu**[1], **W. V. Schoenfeld**[2], **P. M. Petroff**[2]

[1] Department of Electrical and Computer Engineering, University of California Santa Barbara, CA 93106, USA, e-mail: pmichler@physik.uni-bremen.de
[2] Materials Department,University of California Santa Barbara, CA 93106, USA

Abstract We report experimental observation of optically-pumped continuous-wave lasing from self-assembled quantum dots (QDs) embedded in high-quality factor (Q) microdisk laser structures. For large diameter (4.5 μm) microdisks containing InAs QDs, the emission spectra show simultaneous lasing on 1 to 5 well separated modes (Q = 10000 - 17000) in the wavelength range between 920 and 990 nm. For small diameter (2 μm) microdisks containing (In,Ga)As QDs, lasing occurs even when the estimated average number of QDs inside the cavity-mode volume is unity.

1 Introduction

Semiconductor quantum dot (QD) microdisk structures combine small mode volume, high-Q values, and zero dimensional electron density of states. These attributes offer the potential for low power, high speed optoelectronic devices. Recently, enhanced spontaneous emission efficiency [1] and optically pumped lasing [2,3] from InAs QDs embedded in GaAs microdisks have been observed. In this paper, we report power-dependent PL-measurements on microdisk structures.

2 Experimental

Our samples were grown by molecular beam epitaxy (MBE) on a semi-insulating GaAs substrate. The microdisks consist of a disk and a pedestal area. One or two layers of (In,Ga)As self-assembled quantum dots (QD) were grown at the center of the 250 nm disk layer. The QD density of each array is $10^{11} \mathrm{cm}^{-2}$. Microdisks with diameters ranging from 1.5 to 6 microns have been fabricated. Details of the microdisk structure and the processing can be found in Ref. [3]. The samples are mounted in a He gas flow cryostat and cooled to 6 K. Optical pumping is performed with a continuous-wave Ti-sapphire laser operating at 760 nm, generating free electron-hole pairs in the GaAs layer.

3 Results and Discussion

Figure 1 displays a typical PL-spectrum for a 4.5 μm diameter disk under low excitation conditions (30 W/cm^2) in the range between 925 and 985 nm. Several sharp peaks, superimposed on a weak background signal are observed. The background corresponds to free QD PL, whereas the sharp peaks correspond to emission from QDs which couple to microdisk modes. For comparison,

---

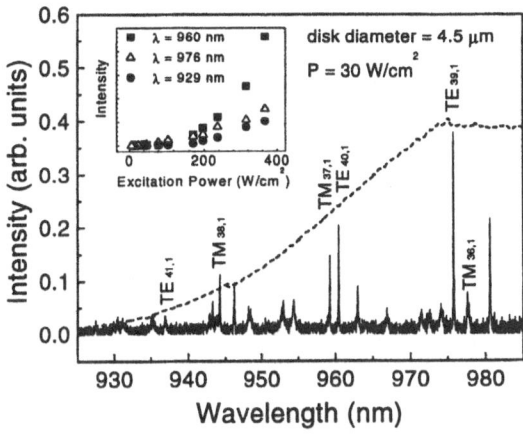

Fig. 1 PL spectrum of a 4.5 μm diameter microdisk in the range from 925 - 985 nm (solid line). For comparison the PL spectrum of the QDs of an unprocessed part of the wafer is also shown under similar excitation conditions (dashed line). Inset: Laser emission intensity versus incident pump power of a 4.5 μm diameter microdisk for three high-Q modes (960, 976, and 929 nm).

Fig. 1 also shows the QD emission (dashed line, FWHM = 66 meV) of an unprocessed part of the wafer under equivalent conditions. The modes of the microdisks can be approximated by so-called whispering gallery modes (WGMs) [4]. Our theoretical estimate of the WGMs with radial mode order n = 1 and azimutal mode number m are also given in Fig. 1. The Q-values of the modes are typically above 10000 and for some modes they are limited by the resolution of our detection system (Q $\leq$ 17000).

As the pump power increases, the peak intensity of the modes at 976, 960, 946, 929, and 927 nm increase drastically. The inset of Figure 1 displays the intensity as a function of pump power for the 929, 960 and 976 nm modes. The observed nonlinear dependence suggests the onset of laser action. The threshold pump power densities for the 5 modes are in the range between 70 - 180 W/cm^2. We could not observe if there is a linewidth reduction above transparency, as the measurement is limited by the monochromator resolution ($\sim$ 0.05 nm).

Figure 2 (a) shows the emitted intensity of the 909 nm mode of a 2 μm diameter disk as a function of the

---

* *Present address:* University of Bremen, Institut für Festkörperphysik, P.O. Box 330440, 28334 Bremen, Germany

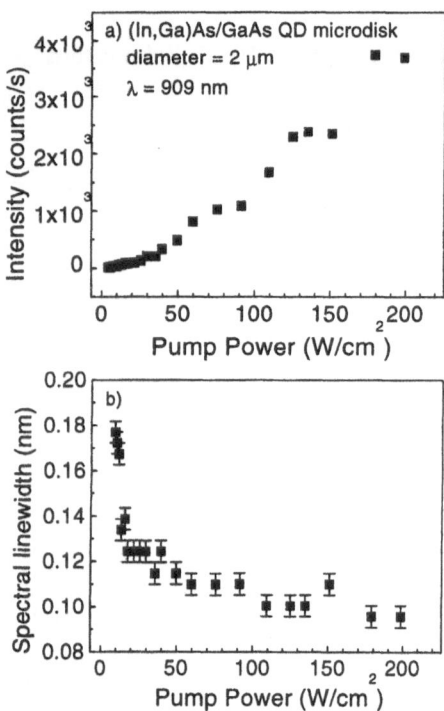

Fig. 2 Laser emission intensity (a) and spectral linewidth (b) versus incident pump power for a 2 μm diameter microdisk.

incident pump power density. A threshold behavior is also clearly seen for this mode. When the pump power density exceeds 20 - 30 W/cm^2, the emitted intensity increases much more rapidly with the pump power density. In that case, we were able to study the pump power dependence of the spectral linewidth of the mode which is plotted in Fig. 2 (b). Starting at linewidth of $\sim$ 0.18 nm at 10 W/cm^2, the linewidth decreases to $\sim$ 0.12 nm at threshold power. This indicates that Q is limited by QD absorption processes at low pump powers [5]. Above threshold (20 - 30 W/cm^2), we observe a further decrease of the linewidth until a value of $\sim$ 0.095 nm is reached at 200 W/cm^2. The linewidth decrease above threshold density corresponds to an increase of temporal coherence. This narrowing is in agreement with recent observation on comparable structures [2].

The photo-excited carriers are localized in QDs with vastly different transition energies, resulting in an inhomogeneously broadened gain spectrum. All QDs whose homogeneously broadened linewidth (due to dephasing) overlap with a cavity mode, contribute to lasing of that mode [6]. Modes which couple to spatially isolated dots with transition-energy separations larger than the homogeneous broadening can start lasing simultaneously if the optical gain is above the lasing threshold.

It is important to note that high resolution PL spectra on unprocessed, low density QD samples (not shown here) show that higher excited state transitions contribute to the QD emission for excitation densities above 50 W/cm^2 and the single exciton linewidth is below 80 μeV (resolution limited) up to 400 W/cm^2. Using these results, we can estimate the average number of QDs which contribute to lasing of one mode.

The 4.5 μm microdisk contains two layers of InAs QDs with a total QD density of $\sim 2 \times 10^{11}$cm^{-2}. Using the WGM mode area in the active disk plane [7], the linewidth of a typical mode at threshold of $\sim$ 0.13 meV (Q = 10000), and the broad distribution of QD PL ($\sim$ 66 meV), we find the average number of QDs that couple with their ground state transition to a WGM mode to be $\sim$ 8. We point out however that this average number in practice corresponds to a larger number of QDs that couple weakly to the WGM, either due to partial spectral or spatial overlap. Moreover, transitions from QDs with higher excited states might contribute to the emission at the estimated threshold power densities between 80 - 170 W/cm^2.

The 2 μm microdisk contains a single layer of InGaAs QDs with a total QD density of $\sim 1 \times 10^{11}$ cm^{-2} with peak emission at $\sim$ 910 nm. The threshold of the lasing mode shown in Fig 2 is estimated to be $\sim$ 20 W/cm^2. The linewidth of this mode at threshold $\sim$ 0.16 meV (Q = 8000) indicates that the average number of QDs that couple to the lasing mode is unity and contributions from higher excited states are negligible at this excitation power density. This result indicates that it should be possible to realize a microdisk laser that contains a single QD.

In conclusion, we have fabricated record high-Q ($\geq$ 17000) InAs/GaAs and InGaAs/GaAs QD microdisk structures. We have demonstrated optically-pumped cw-lasing from QD-based microdisk structures on 1 to 5 well separated modes in the wavelength range between 900 and 990 nm. For small diameter (2 μm) microdisks, lasing occurs even when the estimated average number of QDs inside the cavity-mode volume is unity.

References

1. J. M. Gérard, and B. Gayral, Journal of Lightwave Technol. **17**, (1999) 2089.
2. H. Cao, J. Y. Xu, W. H. Xiang, Y. Ma, S.-H. Chang, S. T. Ho, and G. S. Solomon, Appl. Phys. Lett. **76**, (2000) 3519.
3. P. Michler, A. Kiraz, Lidong Zhang, C. Becher, E. Hu, and A. Imamoglu, Appl. Phys. Lett. **77**, (2000) 184.
4. R. E. Slusher, A. F. J. Levi, U. Mohideen, S. L. McCall, S. J. Pearton, and R. A. Logan, Appl. Phys. Lett. **63**, (1993) 1310.
5. B.Gayral, J. M. Gérard, A. Lemaître, C.Dupois, L. Manin, and J. L. Pelouard, Appl. Phys. Lett. , (1999) 1908.
6. M. Sugawara, K. Mukai, and Y. Nakata, Appl. Phys. Lett. **74**, (1999) 1561.
7. M. K. CHIN, D. Y. CHU, AND S.-T. HO, J. Appl. Phys. **75**, (1994) 3302.

Proc. 25th Int. Conf. Phys. Semicond., Osaka 2000 (Eds. N. Miura and T. Ando)

Incoherent amplification phenomena in semiconductor microcavities

G. Dasbach *, T. Baars, M. Bayer, A. Larionov, and A. Forchel

Technische Physik, Universität Würzburg, Am Hubland, 97074 Würzburg, Germany e-mail: gdasbach@cip.physik.uni-wuerzburg.de

Abstract The non-linear optical properties of microcavities have been studied in pump and probe experiments. Coherent as well as incoherent amplification phenomena can be resolved. The gain at long delay times can be explained by polariton-polariton scattering of excitations in the predominantly excitonic part of the lower polariton branch.

1 Introduction

The coupling between excitons and photons in semiconductor micorcavities leads to the formation of so called cavity polaritons.[1] Like polaritons known from bulk, the dispersion relation of cavity polaritons exhibits pronounced anticrossing behaviour in the regime of strong exciton photon coupling. This leads to significant modifications of the scattering dynamics of polaritons as compared to excitons.

The investigation of scattering processes is a central topic of current research on these systems.[2–5] Recent theoretical and experimental studies focus on coherent amplification phenomena related to stimulated scattering of polaritons, that take advantage of the unique properties of the dispersion relation of polaritons. In the pump-and-probe experiments presented here, we investigate the properties of amplification phenomena with lifetimes exceeding the dephasing times of polaritons by more than one order of magnitude. It is demonstrated that the incoherent gain shows the same dependence on the relative polarization of the two beams, as the coherent one. Furthermore we explore the role of inter and intra branch scattering by selectively exciting polaritons in the lower (LPB) or upper (UPB) polariton branch.

2 Experimental observations

The microcavity under investigation contains a 7 nm wide $In_xGa_{1-x}As$ (x=0.14) quantum well (QW) at the antinode of a λ GaAs cavity. The sample shows a Rabi splitting of 3.8 meV at resonance ($\Delta = 0$), while the linewidths are in the 1 meV range. A spectrally narrow ($\approx 1.5meV$ FWHM) pump pulse generates polaritons in the LPB or UPB respectively. By changing the angle of the pump beam relative to the sample normal, the inplane wave vector of the polaritons induced by the pump pulse can be controlled. For the experiments presented here the pump angle (Θ) was fixed at 8°. The system is probe by spectrally wide probe pulses ($\sim 30meV$

---

* *Present address:* Technische Physik, Universität Würzburg, Am Hubland, 97074 Würzburg, Germany

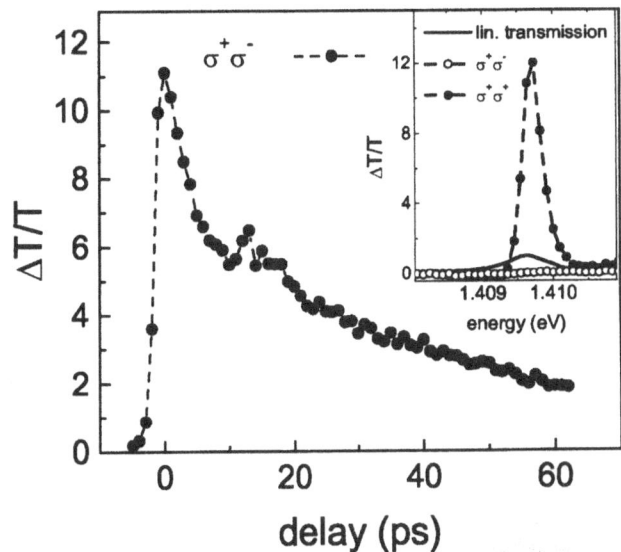

Fig. 1 Spectrally integrated gain as function of delay for zero detuning, when the LPB is pumped. The solid dots represent the gain for co-circular excitation. The inset shows the differential transmission signal ($\Delta T/T_0$) in the spectral region of the LPB as function of energy for zero detuning. The solid (open) dots show the differential transmission spectrum when pump and probe beam are co-circularly (anti-circularly) polarized. These spectra are recorded at zero delay between pump and probe. The solid trace shows the linear transmission of the cavity.

FWHM). All experiments were performed in transmission geometry at a sample temperature of $T = 2K$.

The inset of Fig. 1 shows transmission spectra in the spectral region of the LPB at zero delay for various excitation conditions. The open dots mark the differential transmission when both beams have opposite circular polarizations. We observe no non-linear signal for this configuration ($\Delta T \approx 0$). When switching to co-circular excitation conditions, i.e. both beams have identical circular polarizations, we find a pronounced amplification that exceeds the linear signal (solid trace) by more than one order of magnitude. No gain can be observed at the spectral position of the UPB. In Fig. 1 we analyze the time evolution of the spectrally integrated non-linear signal. For co-circular excitation we find a rapid rise of the signal reaching a maximum at zero delay. The decay exhibits a more complex behaviour. In the first regime ($\tau < 10ps$) we observe a fast decay ($\tau_{decay} \approx 5ps$). The second regime ($\tau > 20ps$) is characterized by a much slower decay ($\tau_{decay} \approx 45ps$).

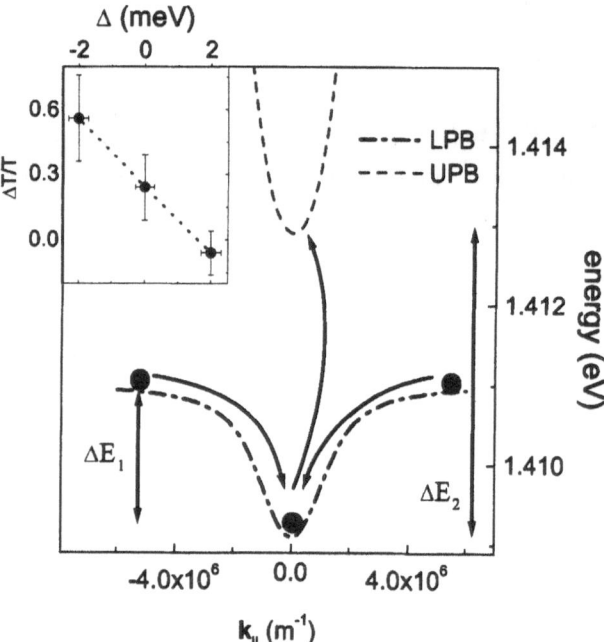

Fig. 2 Polariton dispersion relations for zero detuning. The suggested scattering mechanism is indicated by arrows. The positions of the solid dots indicate the initial states. The inset shows the gain (for pumping the UPB) as a function of cavity detuning. The dashed line is a guide to the eye.

3 Discussion

This observation leads to the assumption, that two different mechanisms cause the amplification. The first can be identified as the 'angle resonant stimulated polariton amplifier' described in by Savvidis *et al.*[2], which is due to coherent wavemixing [3]between the polariton population induced by the pump beam at a certain $k_\parallel$ and the probe polaritons at $k_\parallel = 0$. This model cannot explain lifetimes of the amplification exceeding the dephasing times of polaritons. The long decay times observed here indicate, that the amplification is related to incoherent polaritons located in the non radiative states of the LPB. In order to contribute to the detected gain they have to undergo scattering processes into the $k_\parallel \approx 0$ states of the LPB. Phonon assisted scattering cannot account for the pronounced polarization selection rules observed here. The polarization dependence indicates that long range exciton-exciton exchange interaction plays a key role for the amplification process in the incoherent regime. Fig. 2 shows polariton-polariton scattering channels that fulfill momentum and energy conservation and lead to an increase of the number of polaritons at $k_\parallel = 0$. In the three polariton process indicated by the arrows one probe polariton at $k_\parallel = 0$ and two polaritons of equal but opposite inplane momentum are involved. The excitations at large $k_\parallel$ get scattered into $k_\parallel = 0$-states of the LPB (triggered by the probe population), fulfilling momentum conservation. In order to satisfy energy conservation one probe polariton gets scattered into the UPB. This mechanism requires polaritons with neg-

ative, as well as positive in plane momentum therefore it is obvious that the polaritons injected into the LPB by the pump at a fixed $k_\parallel{}'$ need to get redistributed in momentum space before they can contribute to the amplification process. The suggested amplification process should show a strong dependence on the cavity detuning, in particular for positive detunings, when the energy difference between the bottom of the LPB and the UPB (ΔE_2) exceeds twice the energy difference between the bottom and the excitonic part of the LPB (ΔE_1), this process should be suppressed. The inset of Fig. 2 displays the observed detuning dependence, which agrees well with the expected behaviour.

When increasing the energy of the pump beam, to resonantly excite polaritons in the UPB, gain can be observed as well. Under these conditions the amplification is comparable to the linear transmission signal ($\Delta T/T \approx 1$). At the same time the decay constant is increased by about one order of magnitude $\tau_{decay} \approx 0.5ns$. These observations can be explained consistently within the framework of the model discussed above. When pumping the upper branch, fast relaxation into predominantly exitonic states of the LPB occurs and polaritons are spread out into a wide region of momentum space.[5] Therefore the probability of finding two polaritons with equal but opposite momentum is decreased. Consequently the intensity of the amplification is much lower and the gain remains for longer times. This three polariton scattering process can explain the observed amplification at longer delay times ($\tau \gg \tau_{coh}$), but does not describe the probe induced mechanism that triggers the process. Considering coherent wavemixing can help to find a suitable explanation. Even though the pump-polaritons loose their coherence after a few ps, the probe-polaritons form a coherent state. Wavemixing might occur with all pump-polaritons with matching phases, causing the amplification.

4 Acknowledgements

This work has been supported by the Deutsche Forschungsgemeinschaft and the state of Bavaria. We gratefully acknowledge C. Ciuti for the enlightening discussions and F. Schäfer as well as J.P. Reithmaier for the growth of the sample.

References

1. C. Weisbuch, M. Nishioka, A. Ishikawa, and Y. Arakawa, Phys. Rev. Lett. **69**, (1992) 3314 .
2. P. G. Savvidis, J. J. Baumberg, R. M. Stevenson, M. S. Skolnick, D. M. Whittaker, and J. S. Roberts, Phys. Rev. Lett. **84**, (2000) 1547.
3. C. Ciuti , P. Schwendimann, B. Deveaud, and A. Quattropani, Phys. Rev. B **62**, (2000) R4825.
4. T. Baars, M. Bayer, A. Forchel, F. Schfer, and J. P. Reithmaier , Phys. Rev. B. **61**, (2000) R2409.
5. J. J. Baumberg, A. Armitage, M. S. Skolnick, and J. S. Roberts , Phys. Rev. Lett. **81**, 661 (1998).

Selectively *in situ* probing of self-assembled InGaAs quantum dots in a planar GaAs microcavity by angle-resolved detection of Photoluminescence spectrum

J.H. Chen, J.H. Zhao, F.H. Yang, P.H. Tan, J.D. Zhang, Y.P. Zeng†, J.Q. You, and H.Z. Zheng

National Laboratory for Superlattices and Microstructures, and †Material Science Center, Institute of Semiconductor, Chinese Academy of Sciences, P. O. Box 912, Beijing 100083, P. R. China

Abstract   In the present work we propose a simple and convenient means to selectively *in situ* detect the PL spectra of the different sets of SADs, each set of which has a distinct PL transition energy, by using SADs microcavity structure and angle-resolved PL technique in combination.

There has been great interest in the observation of quantum optics[1] in a semiconductor microcavity, much like the early study of atoms in a microcavity[2]. Many physical phenomena newly discovered in such systems essentially result from manipulating both electronic and photonic wave functions. To this end, zero-dimensional quantum dots (QDs) in a semiconductor microcavity naturally become the very interesting system, since it should easily exhibit the strong coupling between QD's exciton and photon cavity mode due to its δ-function-like energy spectrum in close analogy with that of the atom. As a result, the growth of self-assembled quantum dots (SADs) inside the planar microcavity has become a common practice for obtaining such system. However, due to the discreteness of SADs in size and shape their photoluminescence (PL) spectrum usually appears to be much broadened, and of a Gaussian-type line-shape. That makes it difficult to *in situ* probe optical properties of a particular set of SADs, which have a unique optical transition energy. In the present work we propose a simple and convenient means to selectively *in situ* detect the PL spectra of the different sets of SADs, each set of which have a distinct PL transition energy, by using SADs microcavity structure and angle-resolved PL technique[3] in combination.

The structure of SADs microcavity with $In_{0.5}Ga_{0.5}As$ SADs as the active medium was grown on (100)-oriented, undoped GaAs substrate by molecular beam epitaxy (MBE) in an EPIGEN-II system. A nominally $3\lambda/2$-thick cavity spacer layer was sandwiched between the bottom and top $\lambda/4$ GaAs/AlAs distributed Bragg Reflector (DBR) mirrors, which consist of 19 and 15 pairs of GaAs/AlAs layer respectively. Two groups of three stacked layers of $In_{0.5}Ga_{0.5}As$ SADs (each was formed from 5 monolayers of $In_{0.5}Ga_{0.5}As$) were embedded at two antinodes of the planar cavity. All the layers were grown at 600°C except that the growth temperature for SADs was lowered to 500°C. A wedged cavity, as desirable in experiments for the sake of tuning was achieved by stopping the substrate rotation during the   MBE

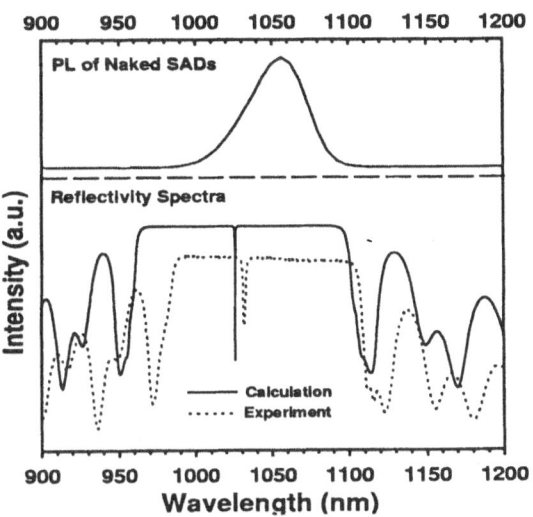

Fig. 1 Calculated and measured reflectivity spectra at 40°. For the former $n=3.543$ is used. The inset gives PL spectrum of naked $In_{0.5}Ga_{0.5}As$/GaAs SADs at 60K.

growth of the spacer layer, while two DBR mirrors were all grown under the rotation conditions. In the measurement of PL spectra the sample was mounted into either a closed He gas recycling cryostat with temperatures variable from 10K to 300K, or a liquid-nitrogen cryostat. The 488.0 nm line of Ar^+ laser was used for the excitation of PL. For the angle-resolved measurement, PL emission is collected via an optical fiber set at the angle θ and 20cm apart from the sample. The fiber set could be slided along a semicircular track. The reflectivity spectra were performed by shining the illumination of an unpolarized white light on the sample, and detecting the reflected light with a liquid nitrogen-cooled InGaAs photodiode.

To check the overall quality of the grown samples, the inset to Fig.1 gives a typical PL spectrum of naked $In_{0.5}Ga_{0.5}As$/GaAs SADs measured at 60K from the reference sample. It displays a Gaussian-type line-shape, peaking at the wavelength of 1060 nm with a full width at half maximum (FWHM) of 50 nm. The PL spectrum of SADs microcavity sample, collected by large numerical aperture (NA) lens exhibits a similar behavior (not displayed here). Also shown in Fig.1 is the calculated

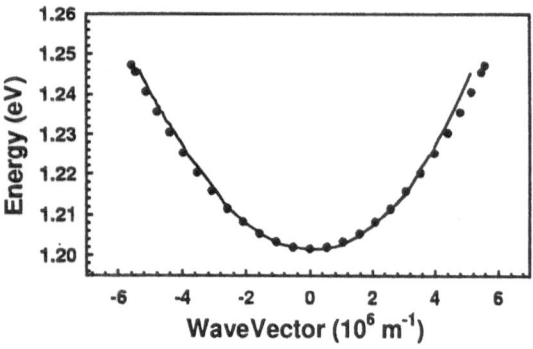

Fig. 2 Calculated(-) and measured($\bullet$) PL dispersions as a function of $k_{//}$.

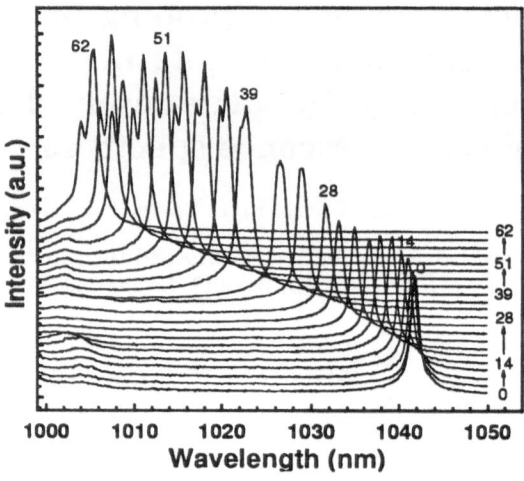

Fig. 3 A set of angle-resolved PL spectra of SADs microcavity at 77K.

and measured reflectivity spectra of the SADs microcavity sample at the angle of 40°. The former is that of an "empty" microcavity (i.e. in the absence of the absorbing SADs medium), calculated by using transfermatrix reflectivity (TMR) method[4], giving finesse factors from 5000 to 2500 as angle varies from 0° to 60°. As seen clearly, the close similarity in the overall features of the measured spectrum in comparison to the calculated one implies the perfectness of the SADs microcavity. The obtained FWHM in the presence of the SADs absorbing medium is as narrow as 1~1.5nm, and the finesse factor has a value of ~1000.

The proposal for selectively *in situ* detecting the PL of a distinct set of SADs comes from the well-known fact that the photon mode in an "empty" microcavity has an in-plane dispersion, described by the simple relations $E_{ph} = (c\hbar/n)[(m\pi/l_z)^2 + k_{//}^2]^{1/2}, m = 1, 2, \cdots$ and $k_{//} = k\sin\theta$. where, m is the index of the normal cavity mode, k is the wave number in free space, and θ is the detection angle at which PL is measured. c, n, l_z are the light speed, refractive index of the cavity, and the cavity length in the grown direction, respectively. A more serious calculation on the cavity resonance as a function of the angle θ is also performed by employing TMR method. The resultant dispersion relation is plotted in Fig.2. With increasing θ angle from 0° to 60° the cavity mode is continuously blue-shifted by about 40nm. At a particular angle θ, obviously, only the PL of SADs, the emission energy of which resonantly matches $E_{ph}(\theta)$, can be detected. Changing the detection angle amounts to probing another set of SADs with a different PL energy. To testify the above idea, the angle-resolved PL spectra of $In_{0.5}Ga_{0.5}As$ SADs microcavity were measured in an angle range from 0° to 62° at 77K, as shown in Fig.3.

Two main features of the data are worthy of notes. Firstly, in contrast to very broadened PL spectrum of the naked SADs, a set of extremely narrow PL peaks with a FWHM of only 1.5~2.5 nm show up at the different angles. Secondly, with increasing the detection angle θ from 0° to 62°, a continuous blue-shift from 1042nm to 1006nm is observed in accordance with the calculation.

This appear more evident in Fig.2, where the measured and calculated $E(k_{//})$ dispersions are in good agreement, if the refractive index of $n= 3.413$ is in use. We believe that by using the angle-resolved detection of PL from a SADs planar microcavity, one is able, for the first time, to selectively *in situ* probe a certain set of SADs with an unique PL frequency. That will bring us a lot of opportunities for *in situ* exploring the novel properties of QDs. The further work in this direction is under progress.

A doublet also features the angle-resolved PL spectra as the angle goes beyond 40°. With micro-scanning the angle through the central angle of each doublet, it undergoes an intensity-exchange process, much like an anticrossing behavior. Its physical origin is presumably attributed to the strong coupling between the cavity mode and zero-dimensional exciton. The details will be published elsewhere.

The work in National Laboratory for Supperlattices and Microstructures, Institute of Semiconductors, is supported by State Council of Science and Technology, Chinese Academy of Sciences and National Natural Science Foundation under the contract No.19734002

References

1. C. Weisbuch, R. Houdre, R. Stanley, Phys. Scripta **T66**, (1996) 121
2. P. R. Berman, *Cavity Quantum Electrodynamics* (Academic Press, New York, 1994)
3. R. Houdre, C. Weisbuch, R. P. Stanley *et. al*, Phys. Rev. Lett. **73**, (1994) 2043.
4. M. Born and E. Wolf, *Principles of Optics* (Pergamon, Oxford, 1986).

Spin quantum beats of exciton-polariton in semiconductor microcavities

P. Renucci , X. Marie, T. Amand, M. Paillard, E. Vanelle

Laboratoire de Physique de la Matière Condensée CNRS- INSA, 135 avenue de Rangueil, F-31077 Toulouse, France
renucci@insa-tlse.fr

Abstract We have observed for the first time polariton spin quantum beats under transverse magnetic field, and circular polarization excitation (Voigt configuration). From the dependence of the emission polarization oscillation period on the cavity detuning, we deduce that : (a) the effective electron Landé g-factor corresponds to the product of the bare electron one by the exciton weight within the polariton; (b) the electron and hole spins are uncorrelated.

1. Introduction

The problem of Coulomb correlations in semiconductor quantum structures is presently the subject of wide theoretical and experimental investigations. In particular, the question of correlations between electron and hole in bound or unbound pair states has been recently debated in the context of quantum well structures [1]. With this respect, electron and hole spin dynamics give some decisive informations : time resolved optical pumping experiments with circular excitation light performed under transverse magnetic field allow to trace out the electron and hole spin correlations through the analysis of the emission intensity and polarization oscillations [2]. In the context of microcavities, where quantum correlations have been recently observed through intraband coherences [3], it is of particular interest to investigate the electron-hole spin correlations in the strong coupling regime.

2. Experiments

We present here experiments performed under transverse magnetic field (Voigt configuration) at low temperature (1.7 K). Spin dynamics is monitored through time and polarization resolved polariton secondary emission using the two colour up-conversion scheme [4]. The time-resolution is limited by the laser pulse-width (~ 1.5 ps) and the spectral resolution is about 3meV. The excitation power is weak, in order to ensure that we stay within the strong coupling regime. The excitation angle is 8°, corresponding to an initial in plane wave vector $k_{//} \approx 10^4$ cm^{-1}; the detection is normal to the surface, with a small acceptance angle (about 10^{-3} steradian). The circular polarization degree of the emission is defined as $P_C = (\Gamma^+ - \Gamma^-)/(\Gamma^+ + \Gamma^-)$, where Γ^+ (Γ^-) denote respectively the right (left) circularly polarized luminescence components.

Two λ-microcavities have been investigated, one with a single 8 nm wide $Ga_{0.95}\,In_{0.05}\,As$ quantum well with a Rabi splitting $\Omega_R \approx 3.7$ meV (MC1), and one with four 12 nm wide $Ga_{0.86}\,In_{0.14}\,As$ quantum wells with $\Omega_R \approx 6$ meV (MC2). As previously shown in cw-reflectivity studies under longitudinal magnetic field [5], the coupling between the magnetic field and the $\left|J_z = \pm 1\right\rangle$ polaritons (here J_z is the polariton angular momentum projection) occurs through their exciton component, using the appropriate electron and hole Landé factors $g_{e(h)}$.

3. Experimental results and discussions

Figure 1 displays the polariton circular polarization oscillations under transverse magnetic field of the MC1 microcavity emission for the cavity detuning $\delta = 0$. Here

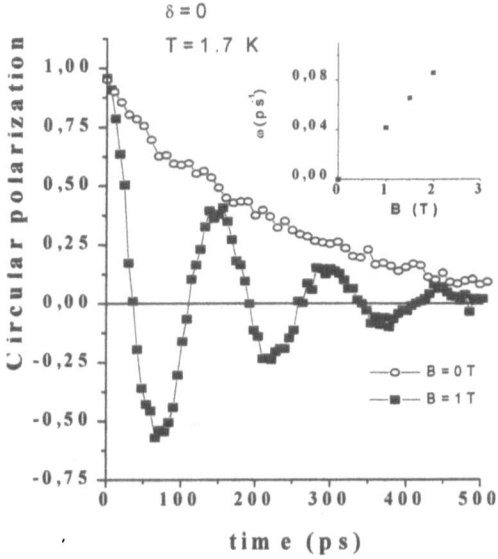

Fig. 1 Polariton circular polarization oscillations under transverse magnetic field, and *non resonant* right circularly polarized excitation. The polarization decay without magnetic field is also shown. Inset: the pulsation ω dependence on the magnetic field.

the excitation is tuned to the InGaAs QW gap (non resonant case). Note that the oscillations are symmetrical with respect to the time axis, and that their envelope coincides with the polariton circular polarization decay without magnetic field under righ or left (not shown) circularly polarized excitation. The beat pulsation ω depends linearly on the magnetic field intensity.

The figure 2 displays the polarization oscillations in the same conditions but now the excitation is resonant with the lower polariton branch. The beat frequency is the same as in the previous case, and the same dependence on magnetic field is measured. However, the modulation amplitude is reduced, as the circular polarization remains positive at any time delay.

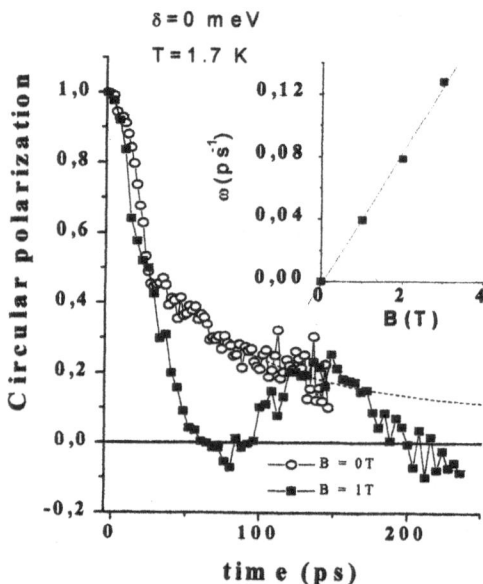

Fig. 2 Polariton circular polarization oscillations under transverse magnetic field, and *resonant* excitation conditions. The polarization decay without magnetic field is also shown. Inset: the pulsation ω dependence on the magnetic field.

We have measured the dependence of the electron g-factor on the cavity detuning δ. The figure 3 displays, for instance, the electron effective Landé factor g_e as a function of δ for the MC2 cavity. It increases monotonously with the detuning, and approaches the one of bare quantum well exciton for strong positive detuning ($\delta/\Omega_R>1$). We have checked that the measured values are the same under resonant or non resonant excitation conditions.

We interpret the previous results in the following way. In non resonant excitation conditions, electron-hole pairs scatter quickly to the reservoir of exciton states, before relaxing to the $k_{//} \approx 0$ lower branch states. The hole spin flip time is short, as is the case for excitons in bare quantum wells under non resonant excitation, so that the electron and the hole spins are uncorrelated [2]. The polarization oscillations are thus the manifestation of the electron spin Larmor precession within the polariton. As the hole spin is relaxed, the modulation amplitude of the beat is maximum [2]. The g-factor dependence on detuning prooves

unambiguously that *we are in the strong coupling regime* ; the measured value corresponds to the bare electron g-factor times the exciton weight in the polariton, and averaged over the lower polariton branch.

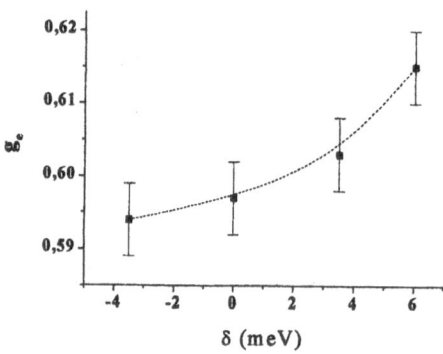

Fig. 3 The electron effective Landé factor g_e as a function of the cavity detuning δ (MC2).

This averaging explains the rather small detuning dependence measured in figure 3.

In resonant excitation condition, the measured periods of the polarization beats are the same, but their modulation amplitude is strongly reduced, in contrast to the previous case, the polarization taking now only positive values. This denotes that the hole spin is now stable [2]. However, the correlation between electron and hole spins is lost here, since the beat period is unchanged with respect to the non-resonant case. If the electron-hole spins were correlated, the beat period should not depend linearly on the magnetic field owing to the electron-hole exchange interaction. We suggest that this spin correlation loss is due to efficient collision scattering to the reservoir states [6]. Subsequent mutual exchange occurs, an efficient process due to the mixing of the $|J_z = \pm1\rangle$ and $|J_z = \pm2\rangle$ states induced by the magnetic field in the reservoir [7].

Acknowledgements : We thank V. Thierry-Mieg for the sample growth and P. Senellart and J. Bloch for fruitful discussions.

1. M. Ostreich *et al.*, Phys. Stat. Sol. (a) **178**, 27 (2000)
2. T. Amand *et al.*, Phys. Rev. Lett. **78**, 1355 (1997);
   M. D'Yakonov *et al.*, Phys. Rev. B **56**, 10412 (1997)
3. Y.-S. Lee *et al.*, Phys. Rev. Lett. **83**, 5338 (1999)
4. X. Marie *et al.* , Phys. Rev. Lett. **79**, 3222 (1997)
5. A. Armitage *et al.*, Phys. Rev. B **55**, 16395 (1997)
6. Savvidis *et al.*, Phys. Rev. Lett. **84**, 1547 (2000)
7. T. Amand *et al.*, Phys. Rev. B **55**, 9880 (1997)

Observation of enhanced spontaneous emission coupling factor in blue InGaN microcavities

S. Kako, T. Someya, Y. Arakawa

Reserch Center for Advanced Science and Technology, University of Tokyo, 4-6-1 Komaba, Meguro-ku, Tokyo 153-8904, Japan
e-mail: kako@iis.u-tokyo.ac.jp

Abstract   We demonstrated for the first time the enhancement of spontaneous emission coupling factor β in nitride-based vertical microcavity surface emitting laser at room temperature. From input-output measurements and analysis of the rate equations, the β of the lasing mode is estimated to be 1.6×10^{-2} by assuming the internal quantum efficiency η of 10% and the transparency carrier concentration of $1.0 \times 10^{19} cm^{-3}$. The estimated β can be well accounted for by a simple theoretical model.

1 Introduction

Controlling of the spontaneous emission process has attracted great attentions since the proposal of Purcell factor in 1946 [1]. Most of the experimental works on semiconductor microcavities have dealt with GaAs-based or ZnSe-based semiconductors so far.

Recent progress in growth technology of GaN-based materials makes it possible to fabricate a high quality InGaN multiple quantum wells (MQWs) and a highly reflective nitride-based distributed Bragg reflector (DBR)[2]. Recently, we have achieved the lasing action at room temperature in a blue 2.5λ vertical cavity surface emitting laser (VCSEL) with a semiconductor/dielectric hybrid cavity under optical pumping[3]. Very recently Y. - K. Song and co-worker also reported the lasing action of their InGaN MQWs VCSEL with a dielectric mirrors [4].

Our 2.5λ VCSEL can be considered as one-dimensional InGaN microcavity laser. Therefore we expected that the microcavity effects should be observed in our InGaN microcavity laser structures. To the best of our knowledge, there is, however, no report of GaN-based microcavity lasers about enhancement of spontaneous emission coupling factor β. In this paper, we examined the enhancement of the β which is a measure of the spontaneous emission coupling with the laser mode.

2 Experiment

The vertical microcavity surface emitting laser structure was grown on a sapphire substrate by atmospheric-pressure MOCVD [3]. A 2.5λ GaN microcavity containing InGaN MQWs as the active region is sandwiched between a nitride DBR and an oxide DBR. The reflectivities of the nitride and oxide DBRs were 98% and 99.5%, respectively. Details of the growth techniques can be found elsewhere. [2,3]

In order to estimate a cavity quality factor Q, we measured spontaneous emission under continuous-wave (CW) photo-excitation. Spontaneous emission spectra at room temperature are shown in Fig.1. The excitation

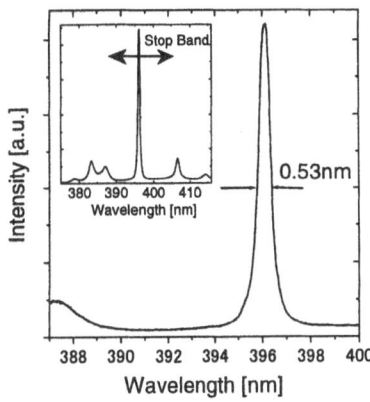

Fig. 1 Spontaneous emission spectra under CW photo-excitation at room temperature. Inset shows the same spectrum without the scale of the wavelength.

source was a CW He-Cd laser with a peak wavelength of 325 nm. The spontaneous emission was analyzed by a monochromator and a CCD camera system with a resolution of 0.03 nm. As can be seen in Fig.1, a cavity resonance mode at 396.1 nm is clearly observed. The other peaks are due to modulation produced by the transmission characteristics of the nitride DBR. The linewidth of the cavity resonance mode is about 0.54 nm. Therefore the cavity Q of the resonance mode is about 740. This value is in a good agreement with a theoretical prediction which is calculated by the reflectivities of the nitride and oxide DBRs, and assumption that an absorption coefficient of GaN at 396.1 nm is about $10^2 cm^{-1}$.

We have been investigating the lasing actions of the InGaN vertical microcavity surface emitting laser under a femto-second pulsed excitation. As an excitation source we used a frequency-doubled Ti:sapphire femto-second laser with a peak wavelength of 382 nm, a pulse width of about 300 fs, and a pulse repetition rate of 100 kHz. The excitation beam was focused into the <5 μm diameter spot on the side of the nitride DBR, and the emission collected from the side of the oxide DBR was analyzed by a monochromator and a cooled CCD camera system with a resolution of 0.03 nm. Input-output characteristics on logarithmic scales are shown in Fig.2. The jump around a threshold is proportional to the inverse of the β on logarithmic scales. Therefore we can estimate the beta from this experimental result.

In order to estimate the β of the laser mode, we analyzed the rate equations taking into account a non-radiative process and a transparency condition[5–7]. Because both the number of carriers and the number of

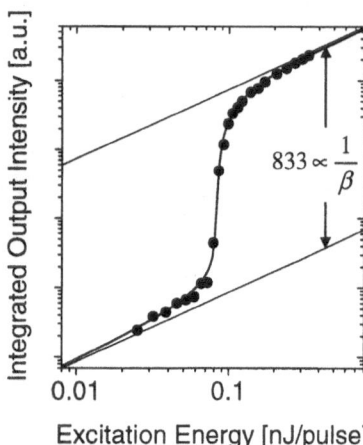

Fig. 2 Input-output characteristics on logarithmic scales. Solid symbols show the experimental data, and the curve shows the theoretical fitting line. The lines indicate liner dependence.

photons in the rate equations depend on the lateral size of the laser mode, we estimated this value to be $3\mu m$ from the experimental far-field image in the case of the lasing. In addition, we calculated the effective cavity length of 7.5λ which results from the penetration of the DBRs for the estimation of the effective optical volume.

In order to take into account a non-radiative process and a transparency condition, we need to make necessary estimations in a quantitatively meaningful way. The threshold of our microcavity laser was 0.08 nJ/pulse, which corresponds to $2.7 \times 10^{19} cm^{-3}$. A transparency carrier concentration of the active material of microcavity lasers should be the same order to the carrier concentration at the threshold. Therefore we assumed the transparency carrier concentration of InGaN MQWs to be $1.0 \times 10^{19} cm^{-3}$ in our microcavity laser. Time-resolved photoluminescence (TRPL) measurements and their temperature dependence of InGaN MQWs have been performed in order to estimate a non-radiative recombination process at room temperature. According to those measurements the internal quantum efficiency of InGaN MQWs is about 10%. Now we can estimate a radiative recombination rate and a non-radiative recombination rate from the carrier lifetime of 200 ps which was determined by TRPL measurements. The theoretical fittings were performed by assuming the internal quantum efficiency $\eta = \tau_r/\tau_{PL}$ to be 10%, where τ_r is the radiative recombination lifetime and τ_{PL} is the carrier lifetime, and the transparency carrier concentration to be $1.0 \times 10^{19} cm^{-3}$. By solving the rate equations under pulse excitation and fitting the results to the experimental data, we estimated the β to be 1.6×10^{-2}. This value is about three orders-of-magnitude improvement compared with the value of $\beta \approx 10^{-5}$ for a typical edge emitting lasers.

3 Discussion

In order to take account of a non-radiative process and a transparency condition, we carefully chose these values.

Here the discussion will focus on the influence of the internal quantum efficiency and the transparency carrier concentration on the β estimated by the theoretical fitting. If we neglect both a non-radiative process and a transparency condition, we obtain the β of 1.2×10^{-3}. According to the analysis of the rate equations, this value of β of 1.2×10^{-3} is the minimum value of the β which can describe the experimental data. The effect of both a non-radiative process and a transparency condition only contributes to increasing the β.

On the other hand, if the β is small enough, the β can be expressed by the Purcell factor F_p which is given by [1]

$$F_p = \frac{3}{4\pi^2} \frac{Q}{V_c/(\frac{\lambda}{n})^3} \qquad (1)$$

where Q is the cavity quality factor, λ is the wavelength of laser mode, V_c is the effective optical volume of the laser mode (see the section2), and n is the refractive index. Using Eq. (1), β is given by $F_p/(1 + F_p)$ [6]. Therefore we can obtain the β of 2.5×10^{-2}. Considering the simplicity of the models, the agreement between the β estimated from the experiments using the rate equations and the β obtained by Eq.1 is good.

4 Conclusion

In conclusion, we achieved a high quality factor of 740 in blue InGaN vertical microcavity surface emitting laser. We demonstrated for the first time the enhancement of the β in nitride-based vertical microcavity surface emitting lasers. By solving the rate equations under pulse excitation and fitting the results to the experimental data, we estimated the β to be 1.6×10^{-2}. The estimated β can be well accounted for by a simple theoretical model.

Acknowledgments

This work was supported by the Grant-in-Aid for COE Research from the Ministry of Education, Science, Sports, and Culture (#12CE2004" Control of Electrons by Quantum Dot Structures and Its Application to Advanced Electronics"), and the Hoso-Bunka Foundation.

References

1. E. M. Purcell, Phys. Rev. **69**, (1946) 681.
2. T. Someya, and Y. Arakawa, Appl. Phys. Lett. **73**, (1998) 3653.
3. T. Someya, R. Werner, R. Forchel, M. Catalano, R. Chingolani and Y. Arakawa, Sience **285**, (1999) 1905.
4. Y. -K. Song, H. Zhou, M. Diagne, A. V. Nurmikko, R. P. Schneider, Jr. ,C. P. Kuo, M. R. Krames, R. S. Kern, C. Cater-Coman, and F. A. Kish, Appl. Phys. Lett. **76**, (2000) 1662.
5. H. Yokoyama, and S. D. Brorson, J. Appl. Phys. **66**, (1989) 4801.
6. Y. Yamamoto, S. Machida, and G. Björk, Phys. Rev. A **44**, (1991) 657.
7. G. Björk, and Y. Yamamoto, IEEE J. Quantum Electron. **27**, (1991) 2386.

Ballistic transport of exciton-polaritons in a graded quantum microcavity

B. Sermage[1], G. Malpuech[2], A. Kavokin[2], and V. Thierry-Mieg[3]

[1] France TELECOM, Centre National d'Etudes des Télécommunications, BP 107, 92225
[2] LASMEA - Université Blaise Pascal Clermont II - CNRS 24 avenue des Landais - 63177 AUBIERE CEDEX- France
[3] Laboratoire de Microstructures et de Microélectronique, CNRS, BP 107, 92225 Bagneux, France

We report an experimental evidence of the acceleration of exciton-polaritons propagating in-plane of a semiconductor microcavity in the regime of the ballistic transport. The acceleration is achieved due to the gradient of the thickness of the cavity and results from the reciprocal space filtering effect of the cavity. The wave-packet propagating in the cavity plane is constantly filtered by the cavity eigen-mode that shifts towards larger wave-vectors as the cavity width increases. The theory based on the extended scattering state technique reproduces the polariton transport in a graded cavity with a good accuracy.

Ballistic propagation of mixed exciton-photon states, i.e. exciton-polaritons[1] in a plane of a semiconductor microcavity[2] is a striking manifestation of the photon-induced transport of electronic excitations in solids. Enhanced efficiency of the light-matter coupling and a macroscopic coherence length of the exciton-polaritons in high-quality microcavities allows one to expect pronounced ballistic transport phenomena which are of extreme interest for applications in ultra-fast opto-electronics. First indirect and controversial studies of the in-plane polariton transport in microcavities have been undertaken since the end of 1980s [3,4], while a direct evidence for such a phenomenon has been obtained quite recently with use of spatially and time-resolved optical spectroscopy[5]. The free-path length of tens to hundreds microns and the typical group velocities of the polaritons of the order of $10^6 - 10^7$ m/s have been measured.

Is there any way to alter the velocity of polariton drift in the cavities? This question has a double importance – for opto-electronic applications like optical transistors, and from the fundamental point of view. How far goes the analogy between electrons and photon-like neutral quasi-particles (exciton-polaritons)? Whether the latter ones can be accelerated and which factor plays role of the electric field for them? The present communication addresses these questions and gives an experimental evidence of the acceleration of the coherent polariton drift by the gradient of the thickness of the microcavity.

The polaritons propagating in the plane of the cavity represent the wave-packets containing harmonics characterized by a continuous spectrum of k_x. The weight $g(k_x)$ of each harmonic in the wave packet initially is given by the form of the incident pulse $f(x)$:

$$g(k_x) = \frac{1}{2\pi} \int_{-\infty}^{\infty} dx f(x) \exp(-ik_x x) \qquad (1)$$

As the wave packet starts to propagate in the plane its shape in the reciprocal space is modified due to the filtering effect of the cavity mode. Namely, harmonics characterized by k_x different from the k_x of the cavity mode at the given frequency and given coordinate are being suppressed (reflected), while those resonant with k_x of the cavity mode propagate freely. Thus, the cavity works as a spatially distributed filter in the reciprocal space. The shape of the propagating wave-packet experiences continuous changes while moving along the sample, so that the role of harmonics characterized by large values of k_x increases, and the group velocity of the entire wave-packet increases.

Below we present the experimental evidence of this phenomenon by means of time- and spatially-resolved spectroscopy.

We excite resonantly the lower branch of the exciton-photon coupled modes with 1.5 ps long pulses coming from a TiSa mode-locked laser. The sample maintained at low temperature (9 K) represents a λ GaAs microcavity with a 8 nm thick $In_{0.05}Ga_{0.95}As$ quantum well located at the antinode of the electromagnetic field. The Bragg reflectors are made of 26 and 22 pairs of $AlAs/Al_{0.1}Ga_{0.9}As$ quarter-wavelength layers, giving a calculated photon lifetime of 14 ps. The vacuum field Rabi splitting is $\Omega_R = 3.6$ meV. The sample has a thickness gradient along the x-axis, so that the bare cavity mode energy derivative over x obtained from the spatially resolved luminescence spectra is:

$$\partial E_c / \partial x = 0.135 \text{ eV/cm.} \qquad (2)$$

The exciton resonance energy in the quantum well is $E_{ex} = 1.4765$ eV. The laser beam is focused on the sample at normal incidence with a 80 mm focal length lens so that the diameter of the spot on the sample is about 35 μm. The averaged excitation power is maintained under 100 nW which corresponds to about $2.5 \; 10^7$ excitons/cm^2. The light coming from the sample is collected by the same lens and focalized by a second lens on the entrance slit of a monochromator and time dispersed by a synchroscan streak camera. It is very essential to note that we excite at normal incidence, i.e. the average in-plane wave vector of the incident pulse is zero.

The experimental procedure is as follows: In the focus mode (i.e. using the streak camera without temporal scanning), we mask the reflected beam and we put the grating at first order. We adjust the laser wavelength so that the secondary emission (Rayleigh scattering plus photoluminescence) is maximum (resonance condition). Then we suppress the mask on the reflected beam, we turn the grating at zero order, we open the entrance slit of the monochromator to maximum and we use the streak camera in the synchroscan mode. We then observe the displacement of the emission source as a function of time.

The traces obtained for different detunings $\delta = E_c - E_{ex}$ are shown on Fig. 1(a). At very short times we see a very intensive spot which is due to the reflection by the free surface of the sample and then a tail which is caused by the reflection by the microcavity itself. The intensity of the reflected signal decreases with a decay time which corresponds to the photon lifetime in the cavity. This tail has a velocity which is nearly zero at the early times and then increases with time achieving a value about $2 \cdot 10^6$ m/s.

a)

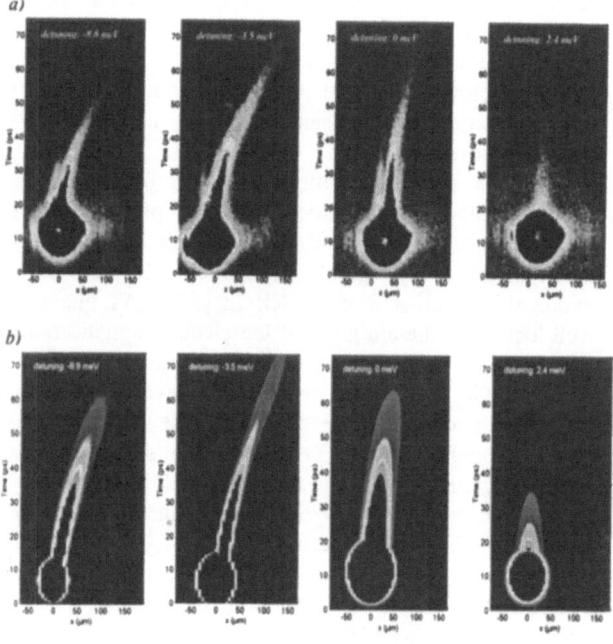

b)

Figure 1. . Movement of the reflection source for four different values of the detuning δ between the cavity mode energy and the exciton energy: experiment (a) and theory (b).

In the case of negative or zero detuning, the polaritons propagate as a unique beam. This shows that polaritons are very little scattered and that they keep their coherence in agreement with the fact the emission is in the specular direction. The coherence time is only limited by their

lifetime in this case. For the positive detuning, the intensity of the polariton beam is much smaller that means that the most part of polaritons is scattered towards the reservoir.

We model these experiments using the scattering state technique described in Ref. 6. We chose the incident pulse in the form

$$f(\omega, x) = j(\omega) \exp\left(-\left(\frac{x}{\Delta}\right)^2\right), \qquad (3)$$

where $j(\omega)$ is the spectral function of the incident pulse taken as a Gaussian function in our case. We have taken $\Delta = 30\mu$ m. The time-resolved x-dependent reflection coefficient is given by

$$R(x,t) = \frac{1}{2\pi} \int_{-\infty}^{+\infty} j(\omega) \exp(-i\omega t) \left(\int_{-\infty}^{+\infty} g(k_x) r(x,\omega,k_x) \exp(ik_x x) dk_x \right) d\omega \qquad (4)$$

with $r(x,\omega,k_x)$ being a reflection coefficient of the structure calculated at the given coordinate x for the light characterized by the frequency ω and the in-plane wave-vector k_x. We have calculated it taking into account the exciton inhomogeneous broadening as described in Ref. 7. The function g is given by Eqs. (1,3).

Figure 1(b) shows the calculated time- and spatially resolved intensity of the reflected light for the experimental configuration of the figure 1(a). One can see a fairly good agreement between the simulation and the experimental data. This is an additional evidence that the observed acceleration effect is purely due to the k-space filtering of the coherent ballistic polariton drift, and does not involve any substantial diffusion of the polaritons.

Acknowledgment
This work has been supported in part by the European Commission in the framework of Contract "CLERMONT" No. HPRN-CT-1999-00132.

References
[1]   V.M. Agranovich and V.L. Ginzburg, Crystal Optics with Spatial Dispersion and Excitons (Springer, Berlin, 1984).
[2] C. Weisbuch, et al., Phys. Rev. Lett. **69**, (1992) 3314; for review see M.S. Skolnick, T.A. Fisher, ,and D.M. Whittaker, Semicond. Sci. Technol. **13**, (1998) 645.
[3] K. Ogawa, T. Katsuyama, and H. Nakamura, Appl. Phys. Lett. **53**, (1988), 1077, Phys. Rev. Lett. **64**, (1990), 796.
[4] K. Oimatsu, T Iida, H. Nishimura, K. Ogawa and T. Katsuyama, J. Lumin. **48&49**, (1991), 713.
[5] T. Freixanet et al., Phys. Rev. B **61**, (2000), 7233.
[6] A. Kavokin et al, Phys. Rev. B **60**, (1999), 15554.
[7] L.C. Andreani et al. Phys Rev. B **57**, (1998) 4670.

Ultrafast polarization switching in a CdTe microcavity

M. D. Martín[1], H. Davies[1,2], L. Viña[1], and R. André[3]

[1] Dpt. Física de Materiales, Universidad Autónoma de Madrid, E-28049 Madrid, Spain
[2] Clarendon Laboratory, Dpt. of Physics, University of Oxford, Parks Road, OX1 3PU, UK
[3] Lab. Spectrométrie Physique (CNRS), Univ. Joseph Fourier 1, F-38402 Grenoble, France

Abstract We have studied the spin polarization dynamics of II-VI microcavity polaritons. Under non-linear emission conditions, we have observed a very fast and efficient reversal of the polarization of the emission. This effect is accompanied by a splitting of the σ^+ and σ^- emission energies. These two effects become evident in the photon-like polariton branch, but the excitons and the cavity are still strongly coupled.

Introduction

The polarization of the light emitted by semiconductor heterostructures has been profusely studied in the last decade [1]. The development of fast spin-based opto-electronic devices has rekindled the interest in the manipulation of the spin in semiconductors and a new field known as spintronics has arisen. In microcavities, new aspects in the spin properties are expected because of the mixed radiation-matter character of the fundamental excitations of the system, namely cavity-polaritons [2]. Strong anomalies in the polarization of the photon-like polariton branch emission of III-V microcavities have been reported [3-5]. In this paper, we will give a detailed description of the time-resolved spin polarization in a II-VI microcavity as a function of the exciton-cavity detuning and the excitation density.

Experimental details

The sample consisted of 90 Å CdTe quantum wells (QWs) placed in the antinodes of a $\lambda/2$ $Cd_{0.40}Mg_{0.60}Te$ microcavity. The cavity is wedge-shaped and allows tuning the cavity resonance to the transition in the QWs. A Rabi splitting of 10.3 meV was found from cw reflectivity measurements. The sample was kept constant at 5 K. A constant difference of 90 meV was kept between the excitation energy and the photon-like branch. The photoluminescence (PL), after excitation with σ^+ polarized pulses, was polarization- and time-resolved in a conventional up-conversion spectrometer with a time resolution of ~ 2 ps. We will refer to the PL's polarization degree, defined as $\wp=(I^+-I^-)/(I^++I^-)$ -where $I^{+/-}$ denotes the $\sigma^{+/-}$ emission intensity-, simply as polarization.

Experimental results and discussion

We have studied the recombination dynamics of microcavity polaritons as a function of excitation density and exciton-cavity detuning ($\delta = E_C - E_X$). We have found that the PL shows two peaks only at very short times (<50 ps). We have observed a Rabi splitting of 9.5 meV, slightly smaller than that extracted from cw reflectivity

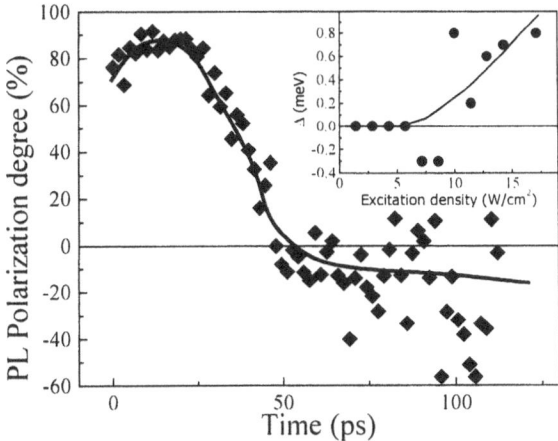

Fig. 1 Time evolution of the polarization degree of the PL of the UPB for $\delta>0$. Inset: σ^+/σ^- energy splitting as a function of excitation density for $\delta>0$. The line is a guide to the eye.

measurements. The time evolution of the PL of the upper/lower polariton branch (UPB/LPB) depends on the detuning. The characteristic decay times are: $\tau_d(LPB)$~375/175 ps, $\tau_d(UPB)$~15/100 ps for $\delta>0/\delta\sim0$, respectively. In the case of $\delta<0$, the decay time of the PL of both branches is approximately equal and amounts to τ_d~150 ps. An increase of the excitation density has several important consequences on the emission dynamics. The most important one is the exponential growth of the integrated intensity with increasing excitation power, for any detuning. A similar dependence has been reported recently [4] and it is characteristic of the final state population stimulated scattering processes in a gas of bosons. A second effect is a small blue shift (<2 meV) of the emission energies with increasing excitation density. It has been demonstrated that the strong coupling regime persists in spite of the appearance of this small blue shift [6]. We will now concentrate on the spin properties of microcavity polaritons for three different detunings and under non-linear emission conditions.

Positive detuning

The time evolution of the polarization of the UPB for an excitation density of ~10 W/cm^2 is presented in Fig. 1. The PL is co-polarized with the excitation, the maximum is attained at ~20 ps and it is followed by a fast decay, in contrast with the behavior of excitons in bare

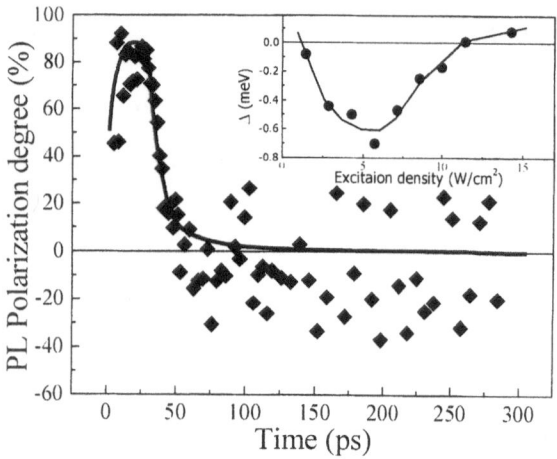

Fig. 2 Time evolution of the polarization degree of the PL of the LPB for δ~0. Inset: σ⁺/σ⁻ energy splitting as a function of excitation density for δ~0. The line is a guide to the eye.

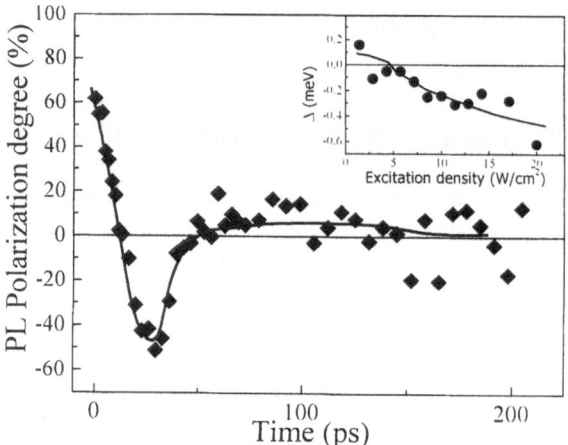

Fig. 3 Time evolution of the polarization degree of the LPB PL for δ<0. Inset: σ⁺/σ⁻ energy splitting as a function of excitation density for δ<0. The line is a guide to the eye.

QWs [7], which do not show any initial rise. An increase of the excitation density leads to a splitting ($\Delta = E^- - E^+ \sim 1$ meV) between the σ⁺ and the σ⁻ polarized components of the PL of the UPB at short times (inset of Fig. 1, @ 20ps). The initial pulse populates mainly the +1 spin level at K>0. The observed splitting, which still needs theoretical clarification, is responsible for the initial increase of the polarization under conditions of final-state stimulated scattering. A positive splitting means that the +1 spin state is the lower state and if the relaxation towards K~0 states is faster than the spin relaxation process then the population of the +1 spin state will be favored, giving rise to an increase of ℘. The fast decay of ℘ is due to the stimulated recombination of +1 spin states [5].

Zero detuning

The PL of the LPB is co-polarized with the excitation light and a similar behavior of the polarization has been reported recently [3-5]. The maximum polarization is attained in ~25 ps and is followed by a very fast decay, as shown in Fig. 2 for an excitation density of ~6 W/cm². The splitting Δ at ~13 ps is depicted in the inset. For small densities Δ is negative, and increases with increasing excitation density up to ~5 W/cm². At larger excitation densities, it decreases and reaches positive values. At δ~0, the emission spectra show a complex behavior, which is still under investigation.

Negative detuning

The behavior observed in this case is markedly different to those reported in the two previous sections: excitation with σ⁺ polarized light yields to a σ⁻ polarized PL (Fig. 3, ~14 W/cm²). It can be seen that the initial polarization of ~60% switches very quickly to negative values (-60%) in ~30 ps. As depicted in the inset of Fig. 3 for 20 ps delay, the spin splitting is negative for δ<0. This means that the -1 spin level (E^+) is the energetically

lowest one. As in the previous sections, the non-resonant excitation pulse mainly populates the +1 spin state of K>0. However, for δ<0, the final state stimulated scattering populates the -1 spin level (E^-) at K~0, leading to a larger σ⁻ emission intensity and thus to a negative polarization. Although the origin, and even more, the sign reversal of this spin splitting, is still to be understood it accounts for the observed polarization reversal.

Conclusions

We have presented the polarization-resolved emission dynamics of a II-VI microcavity under non-linear emission conditions. We have observed a large positive polarization for positive exciton-cavity detuning and a large negative polarization for negative detuning. The origin of these large polarizations is a splitting observed between the σ⁺ and the σ⁻ components of the PL. This spin splitting is positive for δ>0 and negative for δ<0, revealing the reversal of the spin alignment with changing the detuning. The origin of this spin splitting is still under study.

Acknowledgments

This work has been partially supported by the EU (TMR-Ultrafast Quantum Optoelectronics Network), the Spanish DGICYT (PB96-0085), the CAM (07N/0026/1998) and the Fundación Ramón Areces.

References

1. T. C. Damen et al., Phys. Rev. Lett. 67. (1991) 3432:
A. Vinattieri et al., Appl. Phys. Lett. 63. (1993) 3164
2. C. Weisbuch et al., Phys. Rev. Lett. 69. (1992) 3314.
3. A. I. Tartakowskii et al., Phys. Rev. B 60. (1999) R11293.
4. P. G. Savvidis et al., Phys. Rev. Lett. 84. (2000) 1547.
5. M. D. Martín et al., Solid State Commun. (2000) submitted.
6. L. S. Dang et al., Phys. Rev. Lett. 81. (1998) 3920.
7. L. Muñoz et al. Phys. Rev. B 51. (1995) 4247.

Cavity QED of quantum dots embedded in dielectric microspheres

Hailin Wang[1], **Xudong Fan**[1], **Mark Lonergan**[2]

[1] Department of Physics, University of Oregon, Eugene, OR 97403, USA
[2] Department of Chemistry, University of Oregon, Eugene, OR 97403, USA

Abstract Embedding core/shell CdSe/ZnS nanocrystals in a polystyrene sphere, we have demonstrated enhanced spontaneous emission when the photoluminescence is resonant with a whispering gallery mode. The manifestation of these cavity QED effects depends sensitively on nanocrystal sizes, providing a unique and sensitive probe to the underlying radiative dynamics.

Spontaneous emission reflects fundamental dynamical interactions between matter and vacuum and can be controlled by using optical cavities with dimensions of the optical wavelength. Tailoring spontaneous emission of atoms in a microcavity has led to the experimental demonstration of single-atom lasers, strong coupling between an atom and a cavity mode, and more recently quantum logic gates. Recent advances in fabricating semiconductor nanostructures and microcavities have also made these cavity QED studies accessible to semiconductors, especially nanostructures such as quantum dots (QDs) that feature discrete atomic-like transitions. Controlling spontaneous emission of collective optical excitations in these systems can lead to a better understanding of the underlying optical interactions and can also lead to a rich variety of applications such as high-efficiency light emitters and scalable quantum logic gates.

In this paper, we report observation of enhanced spontaneous emission from core/shell CdSe/ZnS nanocrystals embedded in a dielectric microsphere. We show that the manifestation of these cavity QED effects depends sensitively on the size of the QDs, providing unique and valuable information on the underlying optical transitions in these QDs.

The core/shell CdSe/ZnS nanocrystals used in our studies were fabricated by using a high temperature organometallic synthesis. Three groups of nanocrystals with average core radius R of 2 nm, 2.7 nm, and 4.5 nm (referred to as NC1, NC2, and NC3, respectively) were used. For cavity QED studies, polystyrene spheres with a diameter of 15 μm were doped with CdSe/ZnS nanocrystals within the interior surface of the sphere. An inset in Fig. 1 shows the optical image of a typical nanocrystal-doped polystyrene sphere.

Dielectric spheres are versatile optical cavities. In these spheres, whispering gallery modes (WGMs) form along the sphere surface via total internal reflection. Lowest order WGMs propagate along an equatorial ring of the sphere surface. Figure 1 shows the photoluminescence (PL) spectrum obtained at 10 K from NC2 em-

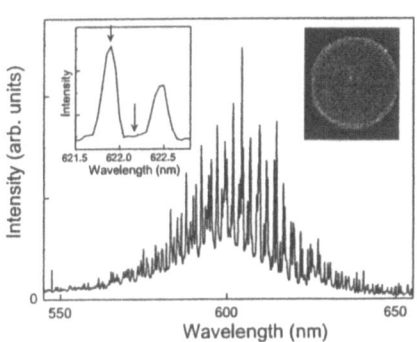

Fig. 1 PL spectrum from nanocrystals with R = 2.7 nm embedded in the interior surface of a polystyrene sphere. The bright ring in the right inset is due to PL from embedded nanocrystals. Arrows in the left inset indicate the spectral positions used for the time-resolved PL in Fig. 2.

bedded in a polystyrene sphere. The linewidth of the narrowest WGMs, as shown in the inset, is 0.2 nm, corresponding to a Q-factor of 3000.

Figure 2 shows time-resolved PL of nanocrystals embedded in polystyrene spheres. In each sub-figure, the lower curve is obtained at a wavelength resonant with a WGM while the upper curve is obtained at a wavelength near but is off-resonant ($\sim$ 3 Å away) with the given WGM. An example of the relevant wavelength positions is shown in the inset in Fig. 1. A spectral bandwidth of 1 Å is used for these studies in order to separate the resonant and off-resonant contributions. As shown in Fig. 2, pronounced cavity-induced modification in the time-resolved PL can be observed for NC2 and NC3 but not for NC1.

Modification in the spontaneous emission rate due to the presence of a resonant cavity can be characterized by the Purcell factor given by $F_p = \Gamma_c/\Gamma_0$ where Γ_c and Γ_0 are the spontaneous emission rate into the cavity mode and in free space, respectively [1]. For the polystyrene spheres used in the above study, $F_p = 0.3$ is expected if we assume that the nanocrystals are resonant with the cavity mode and are also at the maximum of the vacuum electric field. Taking into consideration the spatial and spectral distributions of QDs involved in a measurement, we estimate the actual effective Purcell factor to be of order 0.2. Note that the Purcell factor is independent of the size of the QDs embedded in the microsphere.

For simple systems where decay of the relevant excited state population is entirely radiative in origin, F_p

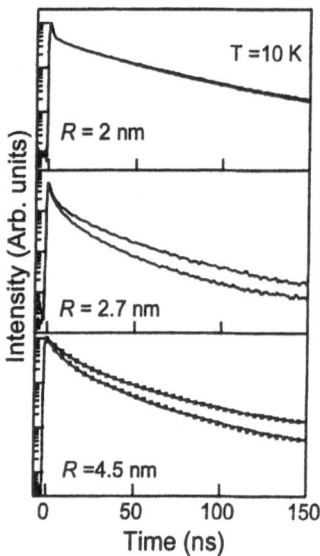

Fig. 2 Time-resolved PL from nanocrystals embedded in a polystyrene sphere at spectral positions resonant (the lower curve in each figure) and off-resonant (the upper curve in each figure) with given WGMs. For each figure, the amplitude is normalized to the same peak intensity. Results of numerical fits to PL from nanocrystals with R = 4.5 nm are shown as squares and are discussed in the text.

determines the relative enhancement in the excited state decay rate due to a resonant cavity. For more complicated systems such as QDs, the manifestation of enhanced spontaneous emission, however, also depends on relative contributions from other inherent nonradiative decay processes. If we assume a radiative decay rate Γ_r and nonradiative decay rate Γ_{nonr} for the given excited state, the relative change in the total decay rate, $\Gamma_t = \Gamma_{nonr} + \Gamma_r$, is given by

$$\varepsilon = F_p/(1 + \Gamma_{nonr}/\Gamma_r), \tag{1}$$

since only Γ_r is modified by the presence of a resonant cavity. For relativley small F_p, clear signature of cavity-induced changes in the total decay rate can be observed only when Γ_r is greater than or at least comparable to Γ_{nonr}. In this regard, enhanced spontaneous emission provides a unique spectroscopic probe for the quantum yield of the underlying individual transition, information that is not accessible to typical time-resolved PL. Note that for many cavity QED phenomena and especially for applications such as quantum logic gates, it is crucial that optical transitions involved feature near unity quantum yield since nonradiative decay processes cannot be directly manipulated or controlled by using an optical microcavity.

Time-resolved PL for colloidal QDs features multiple decay components, corresponding to contributions from various optical transitions in the QDs. We attribute

the decay component that exhibits pronounced cavity QED effects to optical emissions from the lowest dipole-allowed excitonic states since excitonic states at higher energies are expected to be characterized by rapid nonradiative relaxation to lower excited states. Note that within the effective mass approximation (EMA), the lowest excitonic states in CdSe nanocrystals are dipole forbidden [2]. Relaxation from the lowest dipole-allowed excitonic states to these dark states involves spin relaxation and is expected to be extremely slow [3].

The strong size dependence in the manifestation of the cavity QED effects reflects the underlying size dependence in the quantum yield of the lowest dipole-allowed excitonic states in CdSe nanocrystals. Earlier calculations based on EMA have shown that the oscillator strength of these states increases with nanocrystal size [2]. When the nanocrystal size increases from less than 1.5 nm to more than 3 nm in radius, radiative life time $1/\Gamma_r$ decreases from of order 100 ns to of order 10 ns. The ratio Γ_{nonr}/Γ_r is thus expected to increase with decreasing nanocrystal size. Hence, smaller nanocrystals should feature a smaller ε as well as a smaller relative contribution from the lowest dipole-allowed excitonic states to the overall PL, leading to negligible CQED effects, as observed for NC1.

For a quantitative analysis, we have focused on results obtained from NC3 and have used two exponential components to fit the PL in the first 150 ns. The decay time (in free space) obtained for the fast component that exhibits cavity QED effects is $1/\Gamma_t = 8.7$ ns. Decay time for the slow component is of order 30 ns. The numerical analysis also yields $\varepsilon = 0.2$, approaching the theoretically expected F_p and indicating that Γ_r is considerably greater than Γ_{nonr} since otherwise ε would be too small to have a significant effect, as shown by Eq. 1. This is further supported by the total decay rate Γ_t obtained since theoretically one expects $1/\Gamma_r \sim 10$ ns for very large nanocrystals.

In summary, we have demonstrated enhanced spontaneous emission from colloidal QDs coupling to WGMs of a dielectric sphere and have shown strong size dependence in the manifestation of cavity QED effects. At low temperature and for very large nanocrystals (R ~ 4.5 nm), our results further indicate that decay of the lowest dipole-allowed excitonic states is primarily radiative in origin. These remarkable properties reflect the very high quality of core/shell nanocrystals and open the door to a variety of applications of QDs as artificial atoms in quantum optics, including, for example, entanglement of QDs.

References

1. E. M. Purcell, Phy. Rev. **69**, 681 (1946).
2. Al. L. Efros et al., Phys. Rev. B **54**, 4843 (1996).
3. A. Gupta et al., Phy. Rev. B **59**, R10421 (1999).

Polarization of magnetopolaritons in a semiconductor microcavity

M. D. Martín[1], S. Burgas[1], M. Alonso[1], L. Viña[1], F. J. Terán[2], M. Potemski[2], and E. E. Mendez[3]

[1] Dpt. Física de Materiales, Universidad Autónoma de Madrid, E-28049 Madrid, Spain
[2] GHMFL, MPI/FKF & CNRS, 25 Av. Des Martyrs, F-38402 Grenoble, France
[3] Dpt. of Physics and Astronomy, SUNY at Stony Brook, N. Y. 11794-3800, USA

Abstract We have studied, by means of low temperature polarization-resolved photoluminescence, the influence of an external magnetic field on the spin properties of microcavity polaritons. Both, Faraday and Voigt configurations have been used. Our results reveal the robustness of the spin alignment in semiconductor microcavities.

Introduction

The optical properties of semiconductor quantum well (QW) microcavities have been the subject of an intense effort in the last decade. These systems offer the possibility of controlling the strength of the radiation-matter interaction since both, excitons and photons, are spatially confined inside them. Although exciton polaritons were postulated in the 60's [1], they were not experimentally observed until 1992 [2]. After the pioneering work of Weisbuch et al., polariton dispersion [3], saturation [4] and recombination dynamics [5] have been reported. More recently, there have been several reports on the optical properties of microcavity polaritons under an external magnetic field [6-8]. In most of these works, the magnetic field was applied in the Faraday configuration ($\vec{B} \parallel \vec{k}$) and the Zeeman shift of the photoluminescence (PL) of magnetopolaritons was studied. In the paper presented here, we concentrate on the spin polarization properties of magnetopolaritons considering both, Faraday and Voigt configurations. The analysis of the degree of polarization of the PL, defined as $\wp = (I^+ - I^-)/(I^+ + I^-)$, where $I^{+/-}$ denotes the PL emitted with $+1/-1$ helicity, reveals the robustness of the spin in III-V semiconductor microcavities, in agreement with recent reports [9, 10].

Experimental details

The sample consisted of 3 GaAs coupled QWs of 42Å, separated by 17Å $Al_{0.25}Ga_{0.75}As$ barriers, that were placed in the antinode positions of a $3\lambda/2$ $Al_{0.25}Ga_{0.75}As$ microcavity. The microcavity was sandwiched between the top/bottom Bragg reflectors made of 20.5/24 pairs of alternating $AlAs/Al_{0.35}Ga_{0.65}As$ layers. The sample was mounted in a Helium bath cryostat and inserted in the core of a superconducting magnet. Magnetic fields up to 14 Tesla were applied. For the optical excitation of the sample, in the first reflectivity minimum after the stopband of the mirrors (1.71 eV), the output of a tunable Ti:Sapphire laser was focused in an optical fiber. A linear polarizer and a $\lambda/4$ plate were placed at the end of this fiber. The PL emitted by the sample in both configurations, was analyzed into its σ^+ and σ^- components by means of a second pair of linear polarizer and $\lambda/4$ plate. A second optical fiber was used to collect the light. The experiments were carried focusing the excitation spot on a point of the sample characterized by a slight positive exciton-cavity detuning.

Experimental results and discussion

Voigt configuration

The PL spectra characteristic of the Voigt configuration are displayed in Figure 1, for zero (Fig. 1a) and 14 Tesla (Fig. 1b). Three lines can be observed in the spectra, which are labeled as A, X and C. The lines A and X have an excitonic character since their energies and intensities increase with increasing magnetic field. On the contrary, the high-energy side line (C), remains

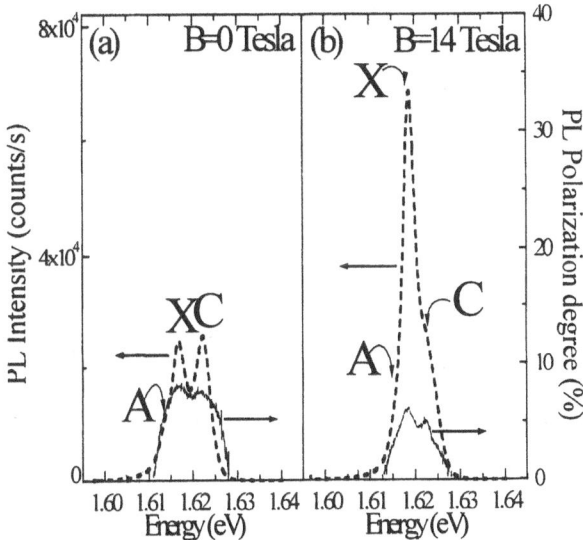

Fig. 1 Microcavity PL (dashed line) and polarization spectra (solid line) at (a) zero Tesla and (b) 14 Tesla. The magnetic field was applied in the Voigt configuration.

almost unaffected by the increase of the field, revealing its photonic character. Additionally, this assignment is confirmed by its energy shift when scanning the excitation spot across the sample. The energy of the line X corresponds to the exciton recombination in the QWs. This line shifts ~2 meV with increasing magnetic field up to 14 Tesla. The field induced enhancement of the exciton oscillator strength leads to an increase of its intensity by a factor ~3. The line A has also an excitonic character, as inferred by its magnetic field dependence (similar to that of line X), and it is rigidly shifted by ~5 meV with respect to line X, independently of the field strength. This line A is tentatively attributed to the excitons confined in a thicker QW, due to a monolayer fluctuation.

The degree of polarization is relatively small (<10%), due to the non-resonant excitation. The Hanle effect is responsible of the field-induced reduction of the polarization degree of the PL. One can extract from an analysis of the decrease of the polarization with field, a spin-relaxation time. In our sample, the reduction is very small, what renders very difficult to extract accurate values for this time. However, our experiments reveal the stiffness of the spin orientation in microcavities, in agreement with recent reports [9, 10].

Faraday configuration
Characteristic PL spectra for the Faraday configuration are depicted in Figure 2. There are some evident differences between these spectra and those presented in Figure 1, which are related with the different geometry of the collection of light by the fiber inside the magnet. The three lines are more clearly resolved in the Faraday configuration. Line C does not shift with increasing magnetic field, giving a further experimental evidence of its photonic character. Lines A and X shift with increasing magnetic field due to the Zeeman effect. The energy shift amounts to ~4 meV for both lines. The enhancement of the exciton oscillator strength due to the shrinkage of the wave function into the QW leads to an

increase in the emission intensity by a factor ~4. Although line A shifts ~4 meV with increasing magnetic field, it remains almost uncoupled of the cavity because their energy difference is too large.

The magnetic field enhances the polarization of the PL, making the σ^+ emission intensity much larger than that of the σ^-. The effect is observed for the three lines but it is noticeably larger for line A, whose polarization degree increases from <5% for zero Tesla to ~50% at 14 Tesla. This increase for the two polaritonic lines, X and C, is considerably smaller than that of line A: it changes from ~5% to ~35% and from ~5% to ~25% for lines X and C, respectively.

Conclusions
We have studied the influence of an external magnetic field on the optical properties of cavity polaritons. The PL spectrum has revealed the normal mode splitting between the exciton-like and the cavity-like polariton branches, and a third peak, attributed to excitons confined inside a thicker QW, due to a monolayer fluctuation.

The magnetic field in the Voigt configuration has only a little influence on the PL spectrum: a diamagnetic shift of ~2 meV is observed in the two excitonic peaks. The decrease of the polarization degree of the PL is so small that no accurate value for the spin relaxation time can be extracted. However, this fact confirms the stiffness of the spin orientation in microcavities.

In the Faraday configuration, the diamagnetic shift amounts to ~4 meV. The exciton is tuned into resonance with the cavity. The polarization degree of the PL increases considerably with increasing magnetic field, reaching values as high as 50%.

Acknowledgments
This work has been partially supported by the Spanish DGICYT (PB96-0085), the CAM (07N/0026/1998), the Fundación Ramón Areces, the Spain-US Joint Commission, the US Army Research Office and the "Picasso" Collaboration Program.

References
1. J. J. Hopfield. Phys. Rev. B **112**. (1958) 1555.
2. C. Weisbuch et al., Phys. Rev. Lett. **69**. (1992) 3314.
3. R. Houdré et al., Phys. Rev. Lett. **73**. (1994) 2043.
4. R. Houdré et al., Phys. Rev. B **52**. (1995) 7810.
5. B. Sermage et al., Phys. Rev. B **53**. (1996) 16516.
6. J. Tignon et al., Phys. Rev. Lett. **74**. (1995) 3967.
7. M. Opher-Lipson et al., Phys. Rev. B **59**. (1999) 10261.
8. S. N. Walck et al., Phys. Rev. B **60**. (1999) 10695.
9. M. D. Martín et al., Solid State Commun. (2000) submitted.
10. A. I. Tartakowskii et al., Phys. Rev. B **60**. (1999) R11293.

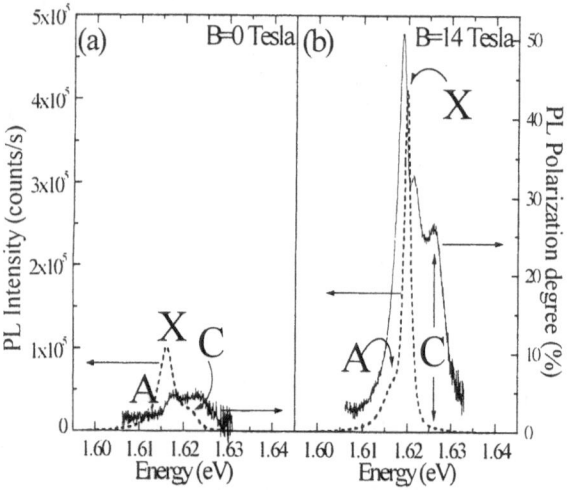

Fig. 2 Microcavity PL (dashed line) and polarization spectra (solid line) at (a) zero Tesla and (b) 14 Tesla. The magnetic field was applied in the Faraday configuration.

Rabi splitting enhancement in planar semiconductor microcavities.

A.D'Andrea and L.Pilozzi

Istituto di Metodologie Avanzate Inorganiche, CNR, c.p.10- I-00016 Monterotondo Scalo (Roma), Italy

Abstract Recently, strong enhancement of the Rabi energy is pointed out in a 3/2 λ planar cavity, embodying a dielectric grating with selected physical parameter values. Moreover, Rabi energy enhancement is obtained also in a planar laser cavity based on super-radiant emission of a multi-quantum wells at λ/2 separation. These two interesting effects are computed by model calculation in the semiclassical framework, and in the effective mass approximation. The possibility of their interplay in order to have strong enhancement of the Rabi energy is briefly discussed.

1 Introduction

The development of high-quality resonators obtained by the modern epitaxial growth techniques allows significant modifications of the photon density of states in optical systems [1]. Recently, strong intensity redistribution among different propagation modes of the Wannier exciton oscillator strength in a planar laser cavity with embedded a dielectric grating was pointed out [2,3]. Moreover, Bragg periodicity is also at basis of super-radiant emission in multi-quantum well (MQW) superstructures [4].

It is well known that the radiation-matter interaction at valence-conduction band edge is represented by the longitudinal-transverse splitting energy that is rather small in GaAs material (about 0.5meV), but this interaction energy increases of about one order of magnitude for quantum wells in a planar laser cavity (the Rabi energy is about 5meV). Moreover, a further enhancement of Rabi energy in a planar laser cavity can be obtained by using two different effects, namely: i)- the resonant diffractive phenomena in a dielectric grating, and ii)- the super-radiant effect in a MQWs Bragg structure.

The main aim of the present paper is to study these two basic phenomena in a semiconductor vertical microcavity in order to enhance the radiation matter-interaction value. Finally, the possibility of the interplay between these two different phenomena in order to obtain a very large Rabi splitting energy is briefly discussed.

2 Results and discussion

2a Dielectric grating in a planar cavity

The optical response of an InGaAs/GaAlAs/GaAs(001) planar cavity model, embodying the spatial dispersion of Wannier exciton in a quantum well and grating periodicity in the lateral direction, is computed in a semiclassical framework by a rigorous numerical solution based on polaritonic approach [3]. Let us consider a dielectric grating and a quantum well at the antinodes of the electric field in a 3/2 λ cavity. The physical parameter values for the quantum well, and for the dielectric grating are: the well thickness L_W=8nm, Kane's energy E_K=23eV, exciton total mass M=0.3m_o, non-radiative homogeneous

broadening Γ_{NR}=0.01eV, cavity refractive index $n_c = \sqrt{\varepsilon_b}$ =3.6, and 1S exciton energy E_{1S}=1.3eV, the grating thickness Lz=132.4nm, dielectric contrast $\Delta\varepsilon=\varepsilon_b-\varepsilon_0$=4.55, periodicity d=288nm and factor filling L_x/d=1/2. The calculation proceed as in Ref.3. In Fig.1 the S normal polarization reflectivity of the cavity with embedded the dielectric grating with the reciprocal lattice vector G=2π/d, very close to the characteristic light wave vector in the cavity (G=0.92k_c), is reported.

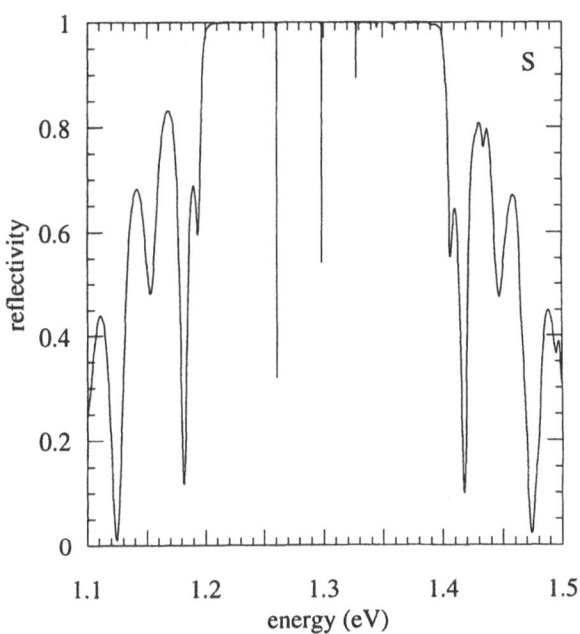

Fig.1 Stop band of the 3/2 λ cavity with embedded the dielectric grating.

The cavity dip at $\hbar\omega_c$ = 1.3eV is clearly shown, while two additional dips are also present due to the diffracted electric field components folded by the grating in the stop band (the so called "leaky modes" of the Bragg mirror).

In fact, in Fig.2 are shown the energy positions of these three dips as a function of the grating periodicity, and we observe that by increasing the periodicity value of the grating, the first traveling diffractive wave (k_x = ±2π / d) comes closer to the normal wave (k_x = 0) and dips move in the stop band from the high energy side towards to the lower energy.

This is in agreement with the dispersion curve of the leaky modes that are localized between the lowest energy cavity dispersion curve and the light dispersion in the bulk.

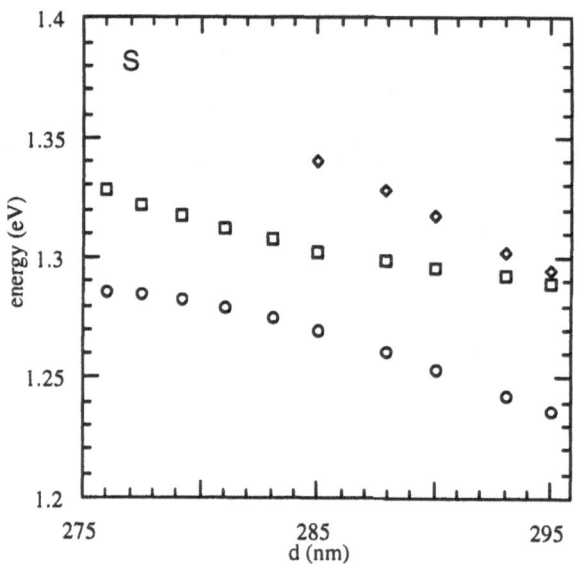

Fig.2. Reflectivity dip energies as a function of grating periodicity

The energy position of the absorbance peak, as a function of detuning, is shown in Fig.3. Notice that the Rabi splitting energy is 15 meV, while its value for the same cavity, but with an equivalent slab replacing the dielectric grating, is about 6.5meV.

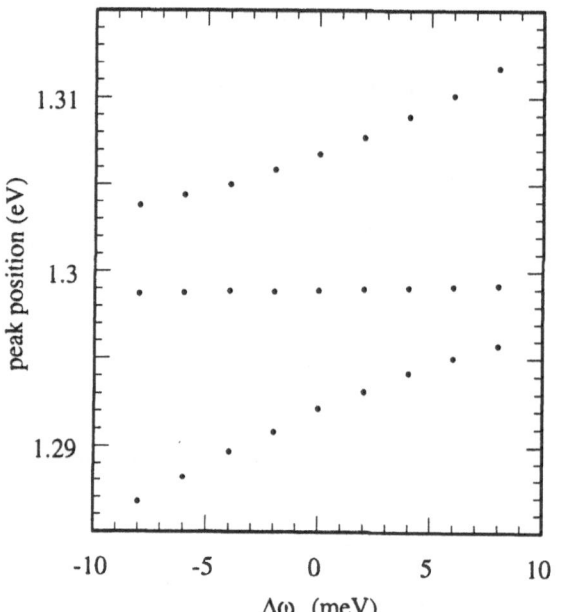

Fig.3 Energy peak positions as a function of detuning.

The diffractive properties are crucially dependent from the coherence length of the electric field, strongly affected by disorder. Our model refined embodying gaussian inhomogeneous broadening in the exciton polarization decreases dramatically the central peak intensity due to the bare exciton level.

The full width at half maximum of the absorbance peaks as a function of detuning is shown in Fig.4. Notice that the FWHM does not cross at zero detuning as expected in symmetric cavity, because of the large asymmetry of the *absorbance* lineshape in presence of a dielectric grating.

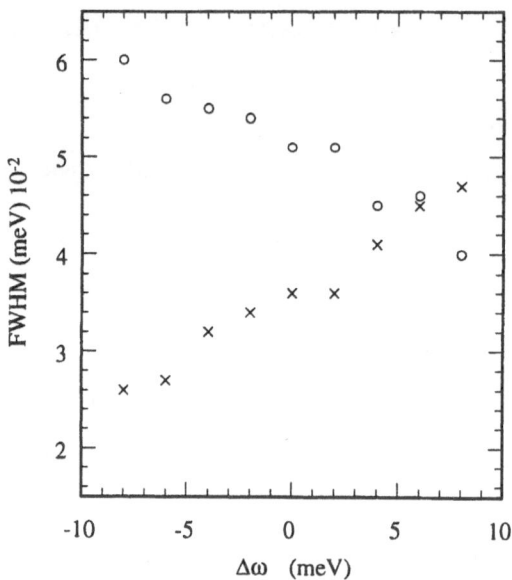

Fig.4 FWHM as a function of detuning.

In fact, two broad peaks are present in the absorbance spectrum due to two traveling waves ($G=0$, $2\pi/d$) that sample different points in the exciton dispersion curve (exciton-polaritons), the first peak is at 1.3eV and the second shifted of about 4meV in the high energy side of the spectrum.

Finally, the Rabi energy can be enhanced for the same cavity increasing the dielectric contrast of the grating (it can be as large as $\hbar\Omega = 23$meV for $\Delta\varepsilon=12$). This points out the resonant nature of the interplay between normal and first diffracted wave confined in the cavity.

2b MQWs Bragg structure in a planar cavity

Let us consider a $3/2\,\lambda$ cavity with two quantum wells at the antinodes of the electric field. The well located at the site towards Bragg/substrate interface shows a Rabi energy $\hbar\Omega = 7.7$meV, in close agreement with that obtained for the equivalent slab computation; a slight greater value is obtained for well in the site towards vacuum/Bragg interface (9.8meV), while an enhancement (14.4 meV) is computed for two wells located at the electric field antinodes. Notice, that this Rabi enhancement is connected with resonance scattering of the normal electric wave ($k_x=0$) in the cavity, while in the former case it is essentially due to the resonance of the first diffracted wave, and therefore, the relative intensity of these two effects in general should be in competition.

In conclusion, the merging of these two effects requires an accurate tailoring of the system. Calculation in a 2λ cavity with dielectric grating and multi-quantum well Bragg system are in progress.

References:
[1] C.Weisbuch, M.Nishioka, A.Ishikawa and Y.Arakawa, Phys.Rev.Lett.v.**69** (1992), 3314
[2] A.V.Kavokin, M.A.Kaliteevski and M. R. Vladimirova, Phys.Rev.B **54** (1996) 1490
[3] L.Pilozzi and A.D'Andrea, Phys.Rev.B v.**61** (2000) 4771
[4] E.L.Ivchenko, M.A.Kaliteevski, A.V.Kavokin and A.I.Nesvizhskii, Opt.Soc.of Am.v.**13** (1996) 1061

Proc. 25th Int. Conf. Phys. Semicond., Osaka 2000 (Eds. N. Miura and T. Ando)

675

Luminescence properties of CdS quantum dots embedded in monolithic II-VI microcavity

T. Tawara, H. Yoshida, H. Kumano, S. Tanaka, I. Suemune

Research Institute for Electronic Science, Hokkaido University, Kita-12, Nishi-6, Kita-ku, Sapporo 060-0812, Japan
e-mail: tawara@es.hokudai.ac.jp

Abstract   Optical properties of self-organized CdS quantum dots (QDs) embedded in monolithic ZnSe/MgS-superlattice microcavity were examined. In the photoluminescence (PL) of CdS QDs, the energy relaxation process was dominated by the photoexcited holes in the ZnSe barrier layers which is in resonance with the ZnSe LO phonon energies. Moreover this energy relaxation process formed the dispersion branches independent of the cavity mode. In the PL measurements of the on/off-resonant excitation, it was observed that the enhancement of the spontaneous emission occurs by the selected excitation of the CdS QDs which is in the resonance with the cavity mode and the LO phonon-assisted relaxations.

1 Introduction

Wide band gap II-VI semiconductor microcavities are of great interest in regard to the strong light-matter interactions since these materials have large exciton binding energies. Moreover self-organized quantum dot (QD) structures with discrete density of electronic states are expected to enhance this excitonic property. From these points of view, the optical properties of the QDs embedded in the microcavity are quite interesting issues.

Up to now we have demonstrated the growth of monolithic II-VI semiconductor microcavities in the blue-green region with DBRs based on ZnSe/MgS superlattice (SL) for the first time [1]. These DBRs with only 5-periods have showed the maximum reflectivity of 92% at the wavelength of 510 nm. We have also reported separately the structural and optical properties of II-VI semiconductor QDs [2, 3]. In this work, the luminescence properties of CdS QDs embedded in the ZnSe/MgS-SL microcavity were studied. Especially it is focused on the LO-phonon assisted resonant excitation of the CdS QDs in-resonance and off-resonance with the cavity mode.

2 Experimental Procedure

Substrates used were GaAs (001) and samples were grown by MOVPE. It is well known that Mg-based compounds grow more easily with the rocksalt structure, which is more stable than the zincblende structure. Therefore, ZnSe/MgS SLs were introduced to stabilize the zincblende structure. The growth temperature was set to 300°C for the ZnSe (λ/4 and cavity layers) and CdS QDs layers, and 400°C for the SL layers to obtain smoother surfaces.

The sample structure is shown in Fig. 1. The DBRs were grown with the alternate ZnSe and ZnSe/MgS-SL λ/4

wavelength layers. One SL layer is composed of 12.5-periods of ZnSe(9.5Å)/MgS(32.0Å). The maximum reflectivity of these DBRs is measured to be about 0.92 with 5-periods. The microcavity structure has 4.5(bottom)/4(top)-periods DBRs. The light emitting layers studied here are five-stacked CdS QDs with 2.0 mono-layers (MLs) thickness embedded in an anti-node position of the ZnSe λ cavity. This λ cavity is formed in wedge shape with the energy change of the cavity mode of about 360 meV/cm. Photoluminescence (PL) spectra of these samples excited using a multi-line Ar$^+$ laser were measured.

3 Results & Discussion

Figure 2 shows the relation of the cavity resonance frequency and the PL peak energy of the CdS QDs embedded in ZnSe/MgS-SL microcavity measured at 20 K. The excitation energy in the PL measurements is 2.5407 eV. The solid circles and squares are PL main peak and subpeak energies, respectively. The open circles show the cavity mode energies measured by reflection spectra. In the PL measurements, PL peaks showed discontinuous change as shown in the inset of Fig. 2 for the change of the cavity mode. The magnitude of this discontinuous shift is about 30 meV and is very close to

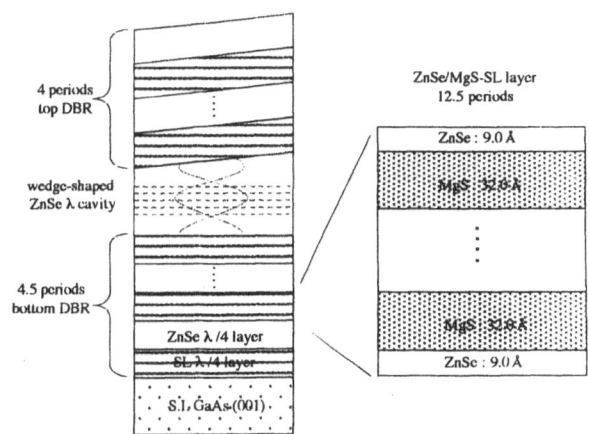

Fig. 1 The schematic diagrams of the ZnSe/MgS-SL microcavity structure with CdS QDs used in this study. The microcavity structure has 4.5-periods (bottom) and 4-periods (top) DBRs, and wedge-shaped ZnSe λ cavity. One SL layer is composed by 12.5-periods of ZnSe/MgS. Five stacked CdS QDs with 2.0 MLs thickness are embedded in an anti- node position of ZnSe λ cavity.

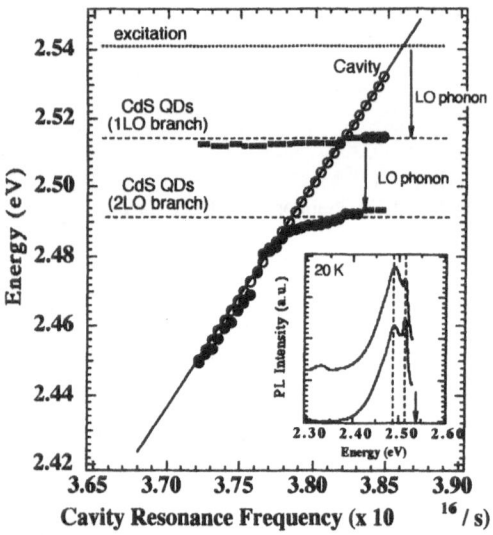

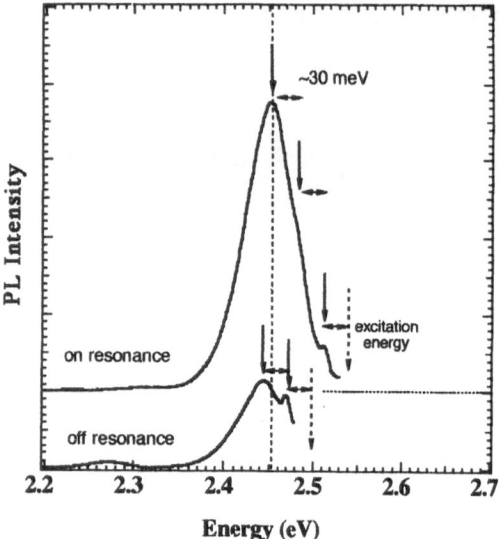

Fig. 2 The relation of the cavity resonance frequency and the PL peak energy of CdS QDs embedded in ZnSe/MgS-SL microcavity measured at 20 K. The excitation energy of PL measurements is 2.5407 eV. The solid circles and squares are PL main peak and subpeak energies, respectively. The open circles show the cavity mode energies measured by reflection spectra.

Fig. 3 Excitation energy dependence of PL spectra of 2.0 MLs CdS QDs in the microcavity structure. Excitation energies are 2.4507 eV (upper curve) and 2.4997 eV (lower curve). Dashed and solid arrows show the excitation energy and the subpeak positions, respectively. Dashed line shows the cavity mode position of 505 nm.

the LO phonon energy of 31 meV in the ZnSe barrier layer, which is much different from 37 meV in the CdS QDs. Taking into account the type-II band lineup and selective excitation of the CdS QDs, the amount of this shift shows that the energy relaxation processes are dominated by the photoexcited holes in the ZnSe barrier layers which is in resonance with the ZnSe LO phonon energies [4]. Therefore the formation of the two branches which are not the cavity mode in the energy dispersion as shown in Fig. 2 is originated from 1 and 2LO phonon relaxations. The PL emission energies of the CdS QDs were switched to the cavity-like mode below the 2LO phonon energy.

It is well known that the exciton-polariton has the anti-crossing behavior in the energy dispersion with Rabi splitting. In this dispersion as shown in Fig. 2, the only lower polariton branch was observed due to the recombination occur in lower energy states in PL measurements. Estimated Rabi splitting in this case is very low compared with one of about 15 meV in the previous reports for the II-VI exciton-polariton [5, 6]. This will be caused by the low cavity finesse (about 40) [1] and the broad luminescence linewidth of the CdS QDs originated by the dot size distribution.

Figure 3 shows the PL spectra of the on/off-resonant excitation between the cavity mode and the LO phonon energies in the cavity-like emission. The PL spectra of the CdS QDs in the microcavity excited with different photon energies were measured. When the cavity mode is excited in resonance with the multiple LO phonon energies, the PL intensity was

increased about 4 times compared with that of the off-resonant excitation. These results suggest that the enhancement of the spontaneous emission occurs by the selected excitation of the CdS QDs with the resonance between the cavity mode and the LO phonon-assisted relaxations.

4. Conclusion

Optical properties of the CdS QDs embedded in ZnSe/MgS-SL microcavity were examined. In the luminescence of CdS QDs in the microcavity, it was observed that the series of subpeaks separated by about 30 meV are superimposed on the luminescence spectra. This energy separation is caused by the LO phonon-assisted relaxation in the ZnSe barrier layer. Moreover this energy relaxation process formed the dispersion branches independent of the cavity mode. In the PL measurements of the on/off-resonant excitation, it was observed that the enhancement of the spontaneous emission occurs by the selected excitation of the CdS QDs with the resonance between the cavity mode and the LO phonon-assisted relaxations.

References

1. T. Tawara et al., J. Crystal Growth (to be published).
2. I. Suemune et al., Appl. Phys. Lett. **71** (1997) 3886.
3. T. Tawara, et al., Appl. Phys. Lett. **75** (1999) 235.
4. S. Farfad, et al., Phys. Rev. **B52** (1995) 5753.
5. P. Kelkar et al., Phys. Rev. **B52** (1995) R5491.
6. Le Si Dang et al. Phys. Rev. Lett. **81** (1998) 3920.

Strongly detuned IV-VI microcavity and microdisk resonances: mode splitting and lasing

T. Schwarzl[1], W. Heiss[1], G. Springholz[1], S. Gianordoli[2], G. Strasser[2], M. Aigle[3], H. Pascher[3]

[1] Institut für Halbleiterphysik, Johannes Kepler Universität Linz, Altenbergerstr. 69, A-4040 Linz, Austria.

[2] Institut für Festkörperelektronik, Technische Universität Wien, A-1040 Wien, Austria

[3] Experimentalphysik I, Universität Bayreuth, Universitätsstr. 30, D-95447 Bayreuth, Germany

Abstract PbTe/EuTe microcavities and microdisks supporting strongly detuned resonances were fabricated by molecular beam epitaxy and reactive ion milling. For detuned cavity modes, we find a pronounced angle dependent relative polarization splitting of up to 1 % and vertical laser emission at 4.8 μm, for sample temperatures between 35 and 85 K. Furthermore, lateral confinement effects in circular microdisks with diameters below 20 μm are demonstrated.

Microcavities have attracted immense interest in recent years due to their unique physical properties as well as their high potential for device applications.

The very short cavities with lengths comparable to an optical wavelength require high reflectivity mirrors, which are realized by Bragg interference mirrors. Such Bragg mirrors exhibit a stop band with high reflectivity around a certain target wavelength. The width of this stop band is determined by the refractive index contrast between the mirror materials. We have recently demonstrated PbTe/EuTe microcavities with Bragg mirrors having a very high refractive index contrast [1], which is more than four times higher than that of III-V [2] and II-VI [3] mirrors. In these cavities, the mirror stop band width is large enough to sustain Fabry–Perot resonances, which are highly detuned with respect to the stop band center wavelength. The range for this detuning is about six times larger than for III-V Bragg cavities.

In this work, we investigate the angle dependent polarization splitting between the TE and TM modes of detuned resonances (DRs) as well as mid infrared vertical lasing [4, 5] from a DR. Furthermore, a splitting and a blueshift of resonances in laterally structured planar cavities into microdisks are shown.

The microcavity samples were grown by molecular beam epitaxy (MBE) on (111) oriented BaF_2 substrates. They consist of two high reflectivity $Pb_{0.95}Eu_{0.05}Te/EuTe$ Bragg mirrors with only three layer pairs and with (2λ) and (4λ) $Pb_{0.95}Eu_{0.05}Te$ cavities in between. The samples intended for laser emission have inserted several 20 nm thick PbTe quantum wells in the cavity layer as active material. Circular microdisks with various diameters were formed by lateral structuring of the planar cavities by reactive ion milling. The optical characterization of the microcavities was performed with polarization dependent FTIR transmission measurements. For investigation of the microdisks, an IR microscope mounted on the FTIR spectrometer was used allowing

spatially resolved reflectivity measurements with a resolution of 8 μm. The laser samples were optically pumped with a pulsed Nd:Yag laser [5].

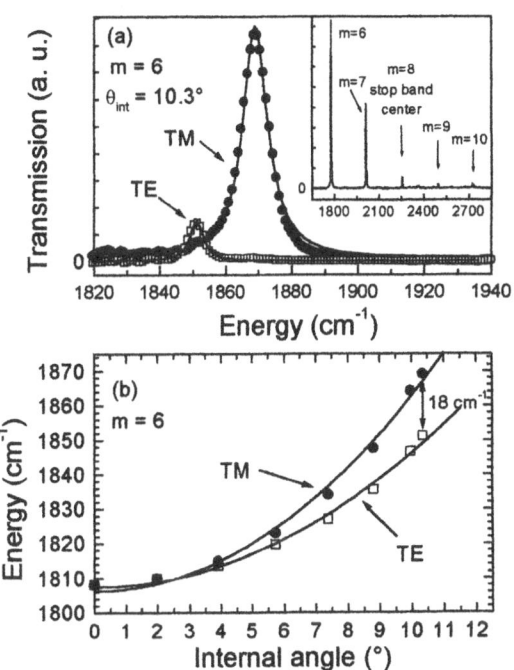

Fig. 1 (a) FTIR transmission of the m=6 resonance of a 4λ microcavity for TE and TM polarization at an internal angle of 10.3°. Inset: FTIR transmission spectrum of the stop band region of the microcavity.
(b) Angle dependent dispersion of the m=6 mode for TE and TM polarization.

Due to the polarization dependence of the reflection and transmission coefficients described by the the Fresnel equations an angle dependent splitting between the TE and TM modes of detuned cavity resonances is expected. This splitting gets larger with the detuning of the resonance energy from the stop band center, as was theoretically described in Ref. [6]. Our IV-VI microcavity with a length of 4λ supports five cavity resonances, as shown in the inset of Fig. 1a by the transmission spectrum of the cavity in the stop band region. The energy of the most detuned resonances with

678   T. Schwarzl et al.

order m=6 and m=10 is about 450 cm⁻¹ off from the mirror center. Fig 1a shows the polarization splitting of the m=6 mode at 1860 cm⁻¹. At an internal angle of only 10.3° (64° external) it amounts 18 cm⁻¹ yielding a relative splitting of 1 %. In comparison, in GaAs/AlAs cavities the tunability of the resonances is limited by the small stop band widths resulting in a polarization splitting of only 0.1 % at an external angle of 60° [7]. For the TE mode we observe, in addition, a considerable larger finesse than for the TM mode, which appears much higher and broader in the transmission spectrum. The difference of the finesse is due to a lower reflectivity for the TM polarization as predicted by the Fresnel formulas. The angular dispersion of the polarization modes of the m=6 resonance is shown in Fig 1b. Both dispersions are fitted with the same equation [7] by using different values for the effective refractive index (3.55 and 4.25 for TM and TE mode, respectively).

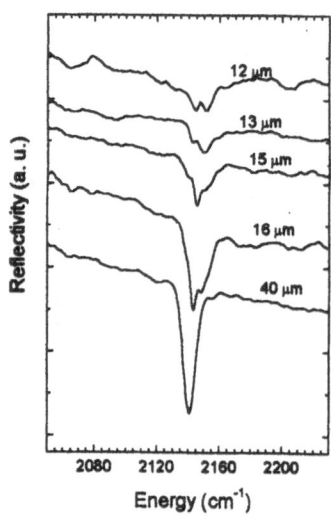

Fig. 2 Reflectivity spectra of circular microdisks with various diameters showing the m=5 resonance dip.

The DRs have a lower quality factor than the central resonances, due to a about 3.5 % lower Bragg mirror reflectivity at the edges of the mirror stop band. Nevertheless, optically pumped lasing has been observed also at DRs, which is in particular important for IV-VI semiconductor lasers exhibiting a strong temperature dependence of the energy band gap. This enables vertical single mode laser operation over a large temperature range via mode hopping between central mode and DRs. This is observed indeed in the (2λ) laser cavity exhibiting three resonances. The sample shows narrow forward directed stimulated emission. At low temperatures around 4 K, emission is observed at the central m=4 cavity mode at 5.87 µm, whereas at 35 K the emission switches to the m=5 DR at 4.82 µm. Above

85 K the lasing quenches, because the band gap energy exceeds the energy of the DR [5].

The lower quality of the DRs can be improved by laterally structuring of the planar microcavities to obtain three dimensionally confined photonic states [8]. In Fig. 2, the reflectivity of one DR (m=5) of circular microdisks with different diameters is shown. The disks were structured from a 2λ planar microcavity. For disk with diameters below 20 µm, the resonance dip shows a pronounced splitting into narrower modes. The distance between the disk modes gets larger with decreasing diameter, as is expected from lateral confinement. Furthermore, the resonances shift to higher energies with decreasing diameter, also indicating lateral confinement. The narrow microdisk modes are attributed to radial-like modes as well as high order whispering-gallery-like modes [9].

In conclusion, we have demonstrated microcavities with strongly detuned resonances with respect to the mirror stop band center. These resonances exhibited a pronounced angle dependent polarization splitting and enable vertical emitting lasing despite low cavity finesse. Furthermore, circular microdisks showing lateral confinement effects were fabricated by laterally structuring planar microcavities.

References

1. T. Schwarzl, W. Heiß, and G. Springholz, Appl. Phys. Lett. **75**, (1999) 1246.
2. R. P. Stanley, R. Houdre, U. Oesterle, M. Gailhanou, and M. Ilegems, Appl. Phys. Lett. **65**, (1994) 1883.
3. E. Hadji, J. Bleuse, N. Magnea, J. L. Pautrat, Appl. Phys. Lett. **68**, (1996) 2480.
4. T. Schwarzl, W. Heiß, G. Springholz, M. Aigle, and H. Pascher, Electron. Lett. **36**, (2000) 322.
5. G. Springholz, T. Schwarzl, M. Aigle, H. Pascher, and W. Heiss, Appl. Phys. Lett. **76**, (2000) 1807.
6. G. Panzarini, L. C. Andreani, A. Armitage, D. Baxter, M. S. Skolnick, V. N. Astratov, J. S. Roberts, A. V. Kavokin, M. R. Vladimirova, and M. A. Kaliteevski, Phys. Rev. **B 59**, (1999) 5082.
7. D. Baxter, M. S. Skolnick, A. Armitage, V. N. Astratov, D. M. Whittaker, T. A. Fisher, J. S. Roberts, D. J. Mowbray, M. A. Kaliteevski, Phys. Rev. **B 56**, (1997) R10032.
8. J. P. Reithmaier, M. Röhner, H. Zull, F. Schäfer, A. Forchel, P. A. Knipp, and T. L. Reinecke, Phys. Rev. Lett. **78**, (1997) 378.
9. D. Labilloy, H. Benisty, C. Weisbuch, T. F. Krauss, C. J. M. Smith, R. Houdre, and U. Oesterle, Appl. Phys. Lett. **73**, (1998) 1314.

Acknowledgements

We thank W. Krepper und O. Fuchs for technical assistance. This work is supported by the FWF P 13330 – TPH and by the GME, Vienna.

Bottleneck and resonant enhancement in free electron - cavity polaritons scattering in GaAs/AlGaAs microcavities

R. Rapaport[1], **A. Qarry**[1], **E. Cohen**[1], **Arza. Ron**[1], **E. Linder**[1], **L. N. Pfeiffer**[2]

[1] Solid State Institute and Physics Department, Technion, Israel Institute of Technology, Haifa 32000, Israel.
[2] Bell Laboratories, Lucent Technologies, Murray Hill, New Jersey 07974.

Abstract We report on an experimental study of the free electron scattering effects on the cavity polaritons (CP's) energy and linewidth in a GaAs/AlGaAs microcavity (MC), with an embedded quantum well (QW) that contains a variable density two dimensional electron gas (2DEG). Reflection spectra were measured (at T=80K) as a function of the MC mode energy, in the spectral range of the (e1:hh1)1S and the (e1:lh1)1S excitons, and with a 2DEG density $n_e < 5 \times 10^{11} cm^{-2}$. The main observations are: a. Over a large detuning energy range the linewidth of the lowest CP branch does not vary while those of the higher CP's increase strongly with increasing n_e. It indicates that the lowest CP branch bottleneck is not suppressed even by the efficient electron - CP scattering. b. When the MC mode energy is close to resonance with each of the excitons, a strong increase in the modes linewidths is seen, indicating a resonant enhancement in the free electrons - CP's scattering c. In the $n_e \sim 10^{11} cm^{-2}$ range, the linewidths of all CP branches increase and the exciton - MC confined photon system transforms from the strong to the weak coupling regime. This is attributed to an increased CP dephasing (resulting from electron - CP scattering) and not from screening and phase space filling effects.

Recently, the scattering dynamics of microcavity polaritons (CP) were intensively studied in order to identify polaritonic effects in the scattering process (as opposed to bare exciton effects) [1]. The main effects are: a. A bottleneck region in the lowest CP dispersion curve for either quasi-elastic polariton - polariton scattering [2] or acoustic phonon - polariton scattering [3]. These scattering processes were found to have a small effect on the CP linewidths. b. Stimulated scattering to the lowest CP that were studied mainly by time-resolved reflection or emission, usually under high excitation intensities [4]. It was reported that in bare QW's, the exciton - acoustic phonon and exciton - exciton scattering rates are much slower than that of the exciton - electron [5,6]. We thus utilize this efficient scattering in order to investigate the free electron effects on the CP linewidth. The present study was done at an ambient temperature of T=80K, so that in the limit of low 2DEG densities ($n_e < 5 \times 10^{11} cm^{-2}$), this gas is nondegenerate. The CP's studied here are formed of the X = (e1:hh1)1S and X_{lh} = (e1:lh1)1S excitons of a single GaAs/AlAs QW (160A or 200A wide) embedded in a λ-wide GaAs/GaAlAs MC. The 2DEG is photogenerated in the QW by utilizing the staggered subband alignment of a mixed type I - type II GaAs/AlAs QW structure (MTQW)[7]. The photoexcitation energy is E_{L2}=1.96eV

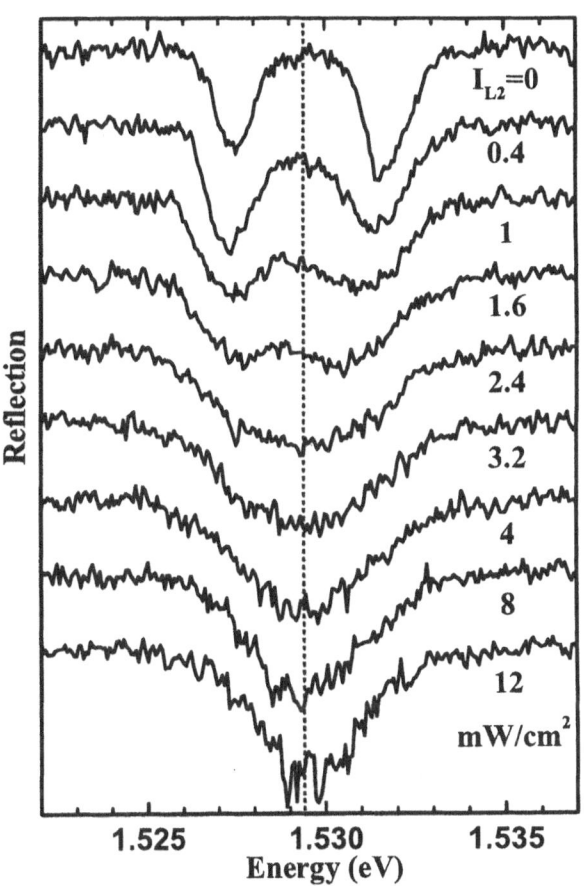

Fig. 1 Reflection spectra taken under excitation at E_{L2}=1.96eV, various I_{L2}, and monitored at a single spot on the sample (160A QW), corresponding to a detuning energy, δ=E_C-E_X=0.

and its intensity (I_{L2}) is varied in a range corresponding to a 2DEG density of $10^9 < n_e < 5 \times 10^{11} cm^{-2}$. Fig. 1 shows the reflection spectra observed for exact resonance between the MC mode and the X exciton: $\delta = E_C - E_X = 0$. For I_{L2}=0 (without photoexcited electrons), the two Rabi split lines have a linewidth of $\Gamma \sim 1meV$, a width typical of this QW / MC structure at T=80K. As I_{L2} increases, the two lines broaden and their splitting diminishes. Fig.2 shows the Γ dependence on I_{L2} of the two Rabi split lines, measured for $\delta = 0$ and -4.3 meV. The MC mode linewidth is measured well below the resonance and its value ($\Gamma_{MC} \simeq 2meV$) is

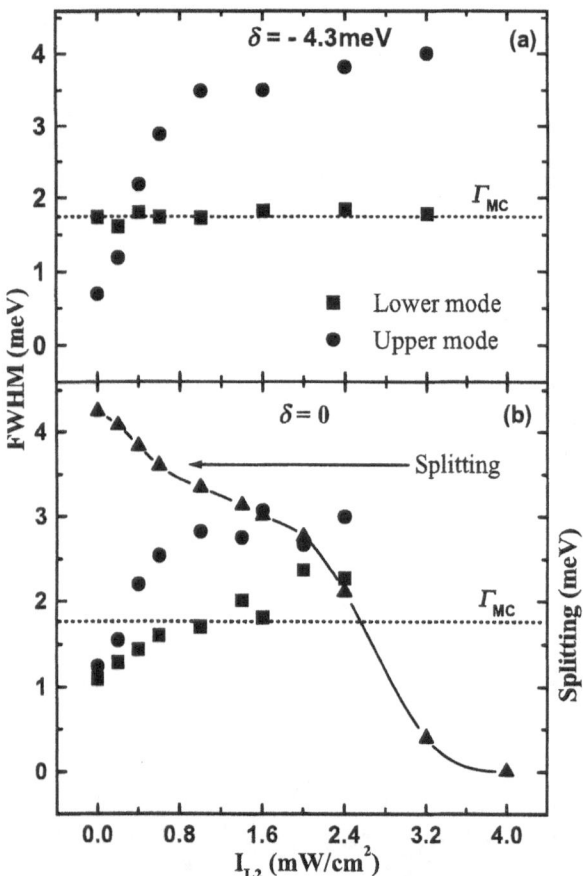

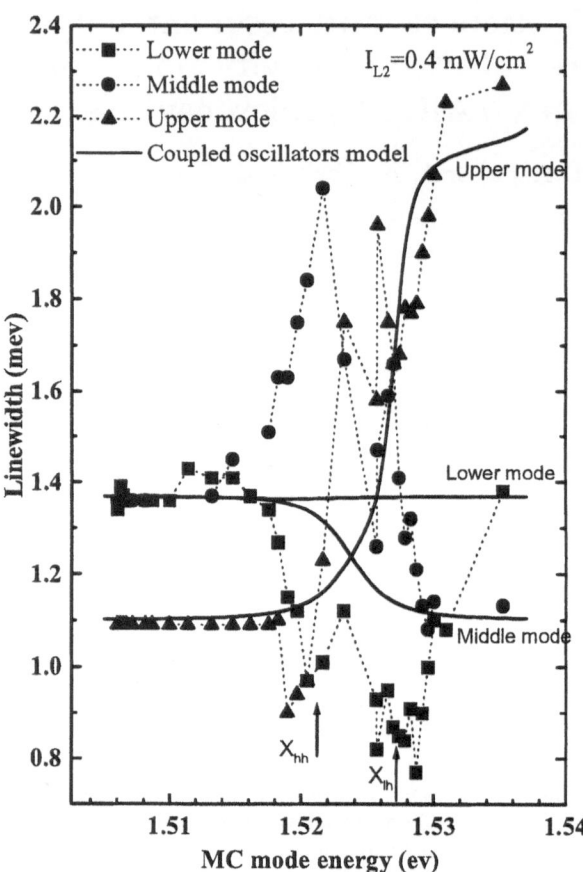

Fig. 2 Polariton mode linewidths as a function of I_{L2}, monitored at two different position on the sample (160A QW) corresponding to a. $\delta = -4.3meV$ and b. $\delta = 0$. Also shown in b. is the energy splitting between the corresponding polariton modes. The dotted lines in both a. and b. mark the bare MC-mode linewidth.

marked in the figure. Also shown is the Rabi splitting dependence on I_{L2} for the $\delta = 0$ case. Fig. 3 shows the linewidth dependence on the MC mode energy for a fixed density of the 2DEG, and for the three CP branches.

The data of Figs. 1 and 2 clearly indicate that electron - CP scattering is very efficient. The fact that a Rabi splitting is observed even for large linewidths means that the excitons are still bound when they are strongly scattered. The lowest CP branch broadens little with increased I_{L2} , leading to the conclusion that this branch has a bottleneck that is not suppressed even by the efficient electron - CP scattering. In order to interpret the data of Fig.3, we calculated the linewidth dependence on the MC mode energy for the three CP's using the coupled oscillators model[8] and assuming an n_e-dependent bare X and X_{lh} linewidths. The results are shown in Fig. 3 by the solid lines. Evidently, only the upper CP linewidth is well fitted while the other two branches are not. In particular, the calculation cannot reproduce the lower mode linewidth decrease up to $\delta \simeq +8meV$. This is another indication for the lowest cavity polari-

Fig. 3 Polariton linewidths as a function of the MC-mode energy (200A QW), measured for $I_{L2} = 0.4mW/cm^2$. The dotted lines are guides to the eye and the solid line are a fit to a coupled oscillators model. The arrows mark the energies of the bare excitons

ton bottleneck. The experimental linewidths in Fig. 3 show a clear increase at narrow MC-mode energy ranges around resonances with each of the excitons (X and X_{lh}). This result is also not reproducible by the simple coupled oscillators model, and it probably depends on the detailed mechanism of the free electrons - CP's scattering. **Aknowledgements:** The work at the Technion was done at the Barbara and Norman Seiden Center for Advanced Opto-Electronics Research. It was supported by the United States - Israel Binational Sience Foundation (BSF), Jerusalem, Israel.

References

1. M. S. Skolnick, T. A. Fisher and D. M. Whittaker, Semicond. Sci. Technol. 13, 645 (1998).
2. F. Tassone and Y. Yamamoto, Phys. Rev. B59, 10830 (1999).
3. G.Cassabois et al, Phys. Rev. B61, 1696 (2000).
4. R. Huang et al, Phys. Rev. B61, 7854 (2000).
5. J. Kuhl et al, Festkorperproblelme 29, 157 (1989).
6. A. Honold et al, Phys. Rev. B40, 6442 (1989).
7. I. Galbraith et al., Phys. Rev. B. 45, 13499 (1992).
8. R. Rapaport, et al., Phys. Rev. Lett. 84, 1607 (2000).

Biexcitons or bipolaritons in a semiconductor microcavity?

P. Borri[1], W. Langbein[1], U. Woggon[1], J. R. Jensen[2], J. M. Hvam[2]

[1] Experimentelle Physik IIb, Universität Dortmund, Otto-Hahn Str.4, D-44227 Dortmund, Germany
[2] Research Center COM, The Technical University of Denmark, Bldg. 349, DK-2800 Kgs. Lyngby, Denmark

Abstract   A well–resolved polariton to biexciton transition is measured in a high–quality quantum–well microcavity, using a pump-probe experiment. Even if the Rabi splitting exceeds the biexciton binding energy in the bare quantum well, the biexciton binding is found not to be significantly affected by the strong exciton-photon coupling. Moreover, the formation of bipolaritons is not resolved. These results agree with predictions in literature based on the in-plane wavevector dispersion in the microcavity.

The optical properties of semiconductor microcavities in the strong exciton-photon coupling regime have attracted increasing interest in the last decade. Recently, the presence of biexcitons in quantum well (QW) microcavities has been addressed both theoretically and experimentally. Theoretically, the bound state character of the biexciton (i.e. binding energy and radius) is predicted to be not significantly affected by the strong exciton-photon coupling, even if the Rabi splitting exceeds the biexciton binding energy in the bare quantum well[1,2]. This is explained when taking into account the in-plane wavevector dispersion in the microcavity. In fact, the exciton-photon coupling strongly modifies only a very small region of the in-plane wavevectors while, due to the exciton-exciton relative motion in the biexciton, the constituent excitons in the biexciton extend over a region of large wavevectors. Moreover, the lowest bipolariton bound state, formed by two lower polaritons (LP) with small in-plane wavevectors, is predicted to have a binding energy orders of magnitude smaller than the biexciton binding energy, due to the small effective mass of the lower polaritons. Thus, bipolaritons are not expected to be resolved even in very high quality microcavities, due to the unavoidable radiative damping of the LP.

Experimentally, biexcitonic effects, in the form of an attractive interaction of cross-circular polarized excitons, were found to be important to explain degenerate four-wave mixing in a GaAs microcavity, first by Gonokami et al.[3] in 1997. Further experimental work was performed by Fan et al.[4] using polarization dependent pump-probe spectroscopy. However, in both of these works the microcavity polariton resonances had large broadening. Thus, a well–resolved biexciton transition was not observed. Very recently, spectrally resolved coherent features associated with a polariton-biexciton transitions in the pump-probe response at negative delays of a nonmonolithic ZnSe microcavity have been reported[5]. However in the investigated structure the

biexciton binding energy exceeded the Rabi splitting, and thus this system was not suited to study the effect of a strong exciton-photon coupling on the biexciton binding. It has also been shown very recently that when creating a high exciton density, the exciton–biexciton transition can couple strongly to the cavity photons and form a new Rabi splitting in a quantum well microcavity, taking over the oscillator strength of the exciton-photon coupling[6]. This high density regime is also not suited to study the effect of a strong exciton-photon coupling on the biexciton binding. Thus, the mentioned predictions have never been verified until now.

In this work, we have measured a well–resolved polariton to biexciton transition using a pump-probe experiment on a high quality single quantum well microcavity. The sample consists of a 25 nm wide GaAs quantum well in the center of a wedged λ-cavity with Bragg reflectors of AlAs/AlGaAs. Details of the sample can be found in[7]. In a bare quantum well nominally identical to the one in the microcavity, we measured a heavy-hole (HH) biexciton binding energy of 1.1 meV, and a heavy-hole light-hole (LH) splitting of few meV[8]. Thus, both the HH and the LH excitons couple strongly to the cavity photons and three polariton resonances are observed in the microcavity (see Fig. 1), with a HH Rabi splitting of 3.6 meV and a LH Rabi splitting of 2.2 meV. The HH Rabi splitting well exceeds the HH biexciton binding energy. In the experiment, 100 fs probe pulses, and 500 fs pump pulses resonant only to the LP, hit the sample, held at T=5K, at small incident angles, and the probe reflection is spectrally analyzed. Details of the set-up are described in[9]. In Fig. 1, the reflectivity spectrum measured for different polarization configurations of pump and probe photons is shown at zero pump-probe delay. The reflectivity spectrum without pump pulses is also shown for comparison (lower curve). The cavity mode is in resonance with the HH exciton (zero detuning). The pump pulse leads to a well distinguished absorption resonance slightly below the middle polariton that disappears for co-circular polarizations (see arrow in figure). The formation of heavy-hole biexcitons in a bare QW is known to follow these polarization selection rules, namely is absent for co-circular excitation[8]. Actually, for co-circular polarizations an induced absorption slightly below the upper polariton is also observed. We suggest that it corresponds to a heavy-hole light-hole mixed biexciton, which is allowed for co-circular polarizations and is separated from the HH biexciton by

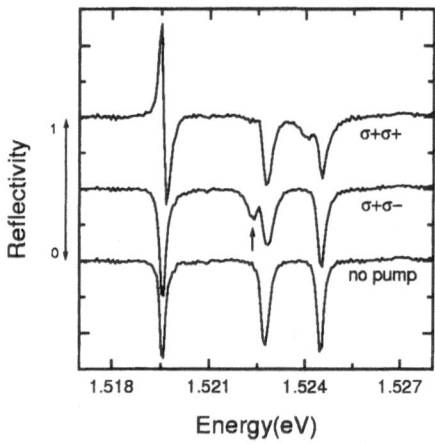

Fig. 1 Reflectivity spectra for co-circular and cross-circular pump and probe photons at zero pump-probe delay. The cavity mode is in resonance with the bare HH exciton (zero detuning). The pump fluence is $0.12\mu J/cm^2$. The lower curve is the reflectivity spectrum without the pump pulse, shown for reference.

the HH-LH splitting. Moreover, a strong contribution is observed at the LP, suppressed for cross-circular polarizations, and thus due to scattered light from the pump photons and/or to coherent nonlinearities such as four-wave mixing.

That the observed induced absorption shown in Fig. 1 corresponds to the lower polariton–biexciton (LP-XX) transition is further confirmed by changing the detuning between the cavity mode and the HH exciton. When shifting the cavity mode above the HH exciton (positive detuning) the LP transition shifts towards higher energies[7]. Since the biexciton energy is not significantly affected by the coupling to the cavity photons and therefore by the detuning, the LP-XX transition should shift towards lower energies. This is shown in Fig. 2 for three different detunings. The LP-XX transition is indicated by the arrows, and clearly shifts towards lower energies while the LP shifts towards higher energies. This also results in a reduced visibility of the LP-XX resonance while it shifts far from the cavity resonance. The expected shift of the LP-XX transition versus the detuning is given by: $E_{LP-XX} = 2E_X - E_{BXX} - E_{LP}$ with E_{LP}, E_X, E_{LP-XX} being the lower polariton, bare exciton, and LP-biexciton transition energies, respectively, and E_{BXX} the biexciton binding energy. Thus, the sum of E_{LP-XX} and E_{LP} is expected to be independent of the detuning. In the inset of Fig. 2 the difference between twice the bare HH exciton energy and the measured sum of LP and LP-XX transition energies is shown versus the detuning. Within the experimental uncertainty, a constant biexciton binding energy of 0.86 meV is obtained. This value is slightly smaller than the 1.1 meV measured in the comparable bare QW. We attribute the reduction to an increased dielectric screening of the Coulomb attraction, as a consequence of the higher refractive index of the QW barrier in the microcavity structure compared

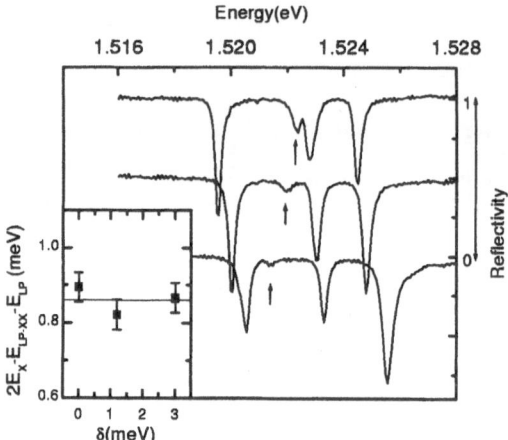

Fig. 2 Reflectivity spectra at zero pump-probe delay for different detuning between the cavity mode and the HH exciton. The pump fluence is $0.12\mu J/cm^2$. The inset shows the difference between twice the HH bare exciton energy and the measured sum of the LP and LP-XX transition energies. The line is a guide for the eyes. The detuning δ is defined as the energy difference between the cavity mode and the bare HH exciton.

to the bare QW[7,8]. Additionally, built-in electric fields would also reduce the biexciton binding. The presence of excess free carriers, that could lead to such electric fields, was in fact observed in the investigated structure[10]. A constant value of E_{BXX}, independent of the detuning, shows the negligible influence of the exciton-photon coupling to the biexciton binding.

We should mention that when changing the relative pump-probe delay time, the observed LP-XX transition has the dynamics ruled by the LP lifetime, given by the photonic lifetime in the cavity times the photonic content in the LP. In fact, when the LP density created by the pump is fully decayed into external photons, the LP-XX induced transition should disappear. Also, the time–integrated LP-XX induced absorption is linear in the pump intensity, as expected from an induced absorption proportional to the LP density[9].

References

1. G. C. L. Rocca, F. Bassani, and V. M. Agranovich, J. Opt. Soc. Am. B **15**, (1998) 652.
2. C. Sieh *et al.*, Eur. Phys. J. B **11**, (1999) 407.
3. M. Kuwata-Gonokami *et al.*, Phys. Rev. Lett. **79**, (1997) 1341.
4. X. Fan, H. Wang, H. Q. Hou, and B. E. Hammons, Phys. Rev. B **57**, (1998) R9451.
5. U. Neukirch *et al.*, Phys. Rev. Lett. **84**, (2000) 2215.
6. M. Saba *et al.*, Phys. Rev. Lett. **85**, (2000) 385.
7. J. R. Jensen, P. Borri, W. Langbein, and J. M. Hvam, Appl. Phys. Lett. **76**, (2000) 3262.
8. W. Langbein and J. M. Hvam, Phys. Rev. B **61**, (2000) 1692.
9. P. Borri *et al.*, Phys. Rev. B **62**, (2000) in press.
10. P. Borri, J. R. Jensen, W. Langbein, and J. M. Hvam, Phys. Rev. B **61**, (2000) R13 377.

Theory of Microcavity Resonant Rayleigh Scattering

D. M. Whittaker

Toshiba Research Europe Limited, 260 Cambridge Science Park, Cambridge, CB4 0WE, United Kingdom.

Abstract It is shown that resonant Rayleigh scattering from a microcavity containing a disordered quantum well can be understood in terms of a simple model where the cavity acts as frequency dependent filter for the scattering signal from the quantum well. The accuracy of this model is demonstrated by comparison with a full numerical treatment of the scattering process.

1 Introduction

The role of disorder in determining reflectivity line-widths in semiconductor microcavities has been widely debated over the last few years.[1–4] However, resonant Rayleigh scattering (RRS)[5] is a much more direct probe of disorder, since it involves a change in the photon wave-vector, and so occurs only in systems lacking translational invariance. Recently, RRS measurements on microcavities have been reported[6]; this paper provides a theoretical description of the observed results, which shows that the microcavity acts as a filter for the RRS from the quantum well within it.[7,8]

2 Microcavity Resonant Rayleigh Scattering

The model of microcavity reflectivity line-widths which is now established[3,4] includes a full treatment of the interaction between the quantum well exciton and the disorder, but completely neglects any mixing between photons with different wave-vectors. This model works because the extremely small polariton density of states inhibits such scattering - the effect termed motional narrowing[1]. Such a treatment is clearly inadequate for RRS, where some photon scattering is required to cause the change in wave-vector. However, the same motional narrowing argument suggests that the RRS signal will be dominated by processes involving only a single change in photon wave-vector. This picture leads to a semi-analytic model of microcavity-RRS, in which the microcavity acts as a filter for the underlying quantum well RRS, simply multiplying it by a smooth function determined by the detunings in the incident and scattering directions.

3 The Filter Model

The filter model of microcavity-RRS described in Sec.(2) can be expressed formally in terms of the Green's functions for the exciton and photon. Fig.(1) shows the type of single scattering processes which are included in the filter model, and an example of the photon multiple-scattering terms which are not included.

The diagrams of Fig.(1) are readily summed to give the diagonal[3] and off-diagonal microcavity polariton

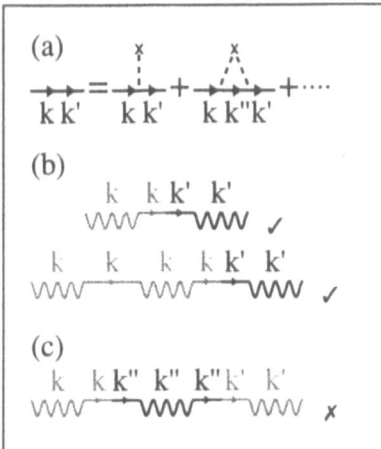

Fig. 1 (a) The full exciton Green's function, $G_e(k, k'; \omega)$, including disorder scattering to all orders. (b) Typical scattering processes included in the filter model. (c) A photon multiple-scattering terms which is *not* included.

Green's functions

$$G_p(k, \omega) = \frac{1}{\hbar\omega + i\hbar\gamma - \delta_k - (\hbar\Omega/2)^2\, G_e(k; \omega)} \tag{1}$$

and

$$G_p(k, k'; \omega) = G_p(k; \omega)\, G_e(k, k'; \omega)\, G_p(k'; \omega). \tag{2}$$

Here γ is the photon homogeneous width, $\hbar\Omega$ the vacuum Rabi splitting energy and δ_k the detuning for wave vector k.

The amplitude of the RRS signal at energy $\hbar\omega$ is proportional to $G(k, k'; \omega)$ where k, k' are the in-plane wave vectors corresponding to the excitation and detection directions. Hence the microcavity-RRS spectrum, $I_p(k, k'; \omega)$ is given by

$$I_p(k, k'; \omega) = |G_p(k; \omega)|^2\, I_e(k, k'; \omega)\, |G_p(k'; \omega)|^2 \tag{3}$$

where $I_e(k, k'; \omega) \propto |G_e(k, k'; \omega)|^2$ is the quantum well RRS spectrum.

Eq.(3) shows that the microcavity-RRS spectrum is determined by multiplying the quantum well spectrum by the functions $|G_p(k; \omega)|^2$, $|G_p(k'; \omega)|^2$ describing the filtering by the cavity for the excitation and detection directions. It is appropriate to describe this as a filtering effect because, unlike the quantum well RRS spectrum, they are smooth functions, independent of the microscopic details of a particular instance of the disorder potential. In fact, $|G_p(k; \omega)|^2$ is closely related to the microcavity absorption spectrum, which is proportional to $\mathrm{Im}\{G_p(k;\omega)\}$.

The filter model makes a simple prediction about the geometry under which the microcavity-RRS signal will be strongest. To obtain a good signal, it is necessary that the overlap between the filter functions in the excitation and detection directions is large. For a given photon energy, this is achieved by chosing angles which correspond to the polariton features in both directions. Although this can be done such that different branches are accessed in the two directions, the simplest way is to chose the same angle to the surface normal for excitation and detection, but different azimuthal angles. The filter function depends only on $|k|$, so this makes $|G_p(k;\omega)|^2$ and $|G_p(k';\omega)|^2$ identical, maximising their overlap. This leads to the prediction of a 'Rayleigh-ring' – a strong Rayleigh signal coming out in a ring at the same angle to the surface normal as the exciting beam.

4 Comparison with Numerical Simulations

Although the motional narrowing argument for minimal scattering which lead to the filter model is highly plausible, the accuracy of the approximation can best be demonstrated by comparison with an exact numerical simulation of the microcavity-RRS process. The simulation uses the method introduced in Ref.[3] to calculate reflectivity spectra for microcavities. Taking a particular instance of the disorder potential, the evolution of an initial plane wave photon state, of wave vector k, is followed by solving the time-dependent Schrödinger equation for the coupled exciton-photon system. The time dependent the Rayleigh amplitude, $G_p(k,k';t)$, is obtained by calculating the overlap of this state with a plane wave photon of wave vector k'. $G_p(k,k';t)$ is then transformed into the frequency domain by means of a discrete Fourier transform.

Fig.(2) shows a typical RRS spectrum calculated in this way, and compared with the predictions of the filter theory, obtained by simulating the RRS for the quantum well on its own, and multiplying by the filter functions $|G_p(k;\omega)|^2$ and $|G_p(k';\omega)|^2$. The two results are clearly very similar, with the same speckle pattern apparent in each. This comparison with the exact calculation demonstrates the accuracy of the filter model in describing the microcavity-RRS process. There are small discrepancies between the two spectra which may indicate contributions from higher order scattering processes. However, it is difficult to rule out the alternative explanation that they are merely the result of numerical inaccuracies.

5 Comparison with Experiments

The ideal experimental test of the filter model would be to compare the speckle pattern for the microcavity and the bare quantum well, just as in the numerical comparison of Sec.(4). This is probably not possible for a normally fabricated microcavity, though it may be feasible

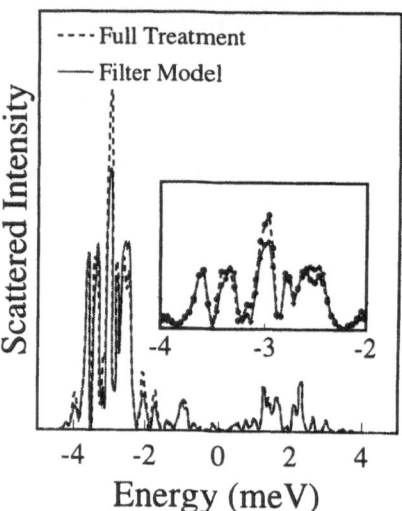

Fig. 2 Numerical microcavity-RRS spectrum for scattering on the Rayleigh ring. The calculation demonstrates the good agreement between the filter model (solid line) and the full treatment (dashed line).

using the non-monolithic structures recently described by Neukirch *et al.*[9].

The most relevant experiments which have been reported are those of Freixanet *et al.*[6], who performed angular dependent microcavity-RRS measurements. Their results are in good qualitative agreement with the predictions of the filter model, in that they observe a clear Rayleigh-ring. More quantitatively, their expression for the width of this ring is exactly what is implied by the filter theory: if the peak in the filter function has width $\hbar\Gamma$, then the change in k required to lose the overlap of the excitation and detection functions is just $\Delta k = \hbar\Gamma/(dE/dk)$, as in Eq.(1) of Ref.[6].

References

1. D. M. Whittaker, P. Kinsler, T. A. Fisher, M. S. Skolnick, A. Armitage, A. M. Afshar, M. D. Sturge and J. S. Roberts, Phys. Rev. Lett. **77**, (1996) 4792-4796.
2. V. Savona, C. Piermarocchi, A. Quattropani, F. Tassone and P. Schwendimann, Phys. Rev. Lett. **78**, (1997) 4470-4473.
3. D. M. Whittaker, Phys. Rev. Lett. **80**, (1998) 4791-4794.
4. C. Ell, J. Prineas, T. R. Nelson, Jr., S. Park, H. M. Gibbs, G. Khitrova, S. W. Koch and R. Houdré, Phys. Rev. Lett. **80**, (1998) 4795-4798.
5. J. Hegarty, M. D. Sturge, C. Weisbuch, A. C. Gossard and W. Wiegmann, Phys. Rev. Lett. **49**, (1982) 930-933.
6. T. Freixanet, B. Sermage, J. Bloch, J. Y. Marzin and R. Planel, Phys. Rev. B **60**, (1999) R8509-8512.
7. D. M. Whittaker, Phys. Rev. B **61**, (2000) R2433-2435.
8. A. V. Shchegrov, J. Bloch, D. Birkedal and J. Shah, Phys. Rev. Lett. **84**, (2000) 3478-3471.
9. U. Neukirch, S. R. Bolton, N. A. Fromer, L. J. Sham and D. S. Chemla, Phys. Rev. Lett. **84**, (2000) 2215-2218.

Spatially extended cavity polaritons arising from weakly confined excitons

H. Ishihara and J. Kishimoto

Department of Physical Science, Graduate School of Engineering Science, Osaka University, Toyonaka, Osaka 560-8531, Japan
e-mail: ishi@mp.es.osaka-u.ac.jp

Abstract The coupling between the cavity field and weakly confined excitons is theoretically studied, where the degree of freedom of the center-of-mass (CM) motion of excitons, which is confined in a thin film, is explicitly considered. With the calculation of linear response, we show that the exciton-cavity coupling can be controlled by manipulating the matching of the spatial patterns of the radiation field and the CM wavefunctions.

1 Introduction

Control of the radiation-matter coupling is currently of major interest. A variety of coupling schemes have been realized through the fabrication of various types of cavities (photonic crystals) and the matter nano-structures [1]. Particularly, the coupled systems of the cavity and the two-dimensional excitons in quantum wells have been extensively studied[2]. On the other hand, little attention has been given to the degree of freedom of the spatial variation of excitonic wavefunctions along the direction of the film thickness. One of the few examples is a study of the "bulk microcavities"[3] in which the whole cavity is filled with an excitonic active material. In this case, one can obtain a large coupling due to the spatially extended overlap between the radiation field and the wavefunctions of the center-of-mass (CM) motion of excitons. If we manipulate the matching of the spatial patterns of the radiation field and the CM wavefunctions, varying the thickness of the cavity and the excitonic active layer arbitrarily, it would further extend the degree of freedom to control the radiation-exciton coupling. However, a systematic study of how the quantized CM levels of excitons interact with the cavity field, depending on the thickness of the excitonic active layer, has so far been lacking. The purpose of this contribution is to perform this study by means of the nonlocal response theory[4].

2 Model and calculation

We consider a thin film with the thickness (d) lying at the center of a vertical cavity, which consists of a pair of distributed Bragg reflectors (DBR's). We explicitly treat the CM motion of excitons confined in the film, and treat the relative motion of excitons as in the bulk material. Restricting ourselves to the normal (Z-direction) incidence, we consider only the $k_\parallel = 0$ subspace for both excitons and electromagnetic field, where $k_\parallel$ is the component of wave vector parallel to the film surface. For this model, the eigenenergies and eigenfunctions of excitonic system in the film are $E_n = \hbar\omega_T + \hbar^2 k_n^2/2M$ and $\phi_n(Z) = \sqrt{2/d}\,\sin(k_n Z)$, where $\hbar\omega_T$ is the exci-

ton energy at the bottom of band and M is the total mass of the exciton. The allowed values of k_n are $n\pi/d$ ($n = 1,2...$). The nonlocal linear susceptibility $\chi(Z, Z')$ can be calculated with those bases in the following form, namely

$$\chi(Z, Z') = \sum_n \bar{\chi}_n(\omega)\rho_n^*(Z)\rho_n(Z') \qquad (1)$$

$$\bar{\chi}_n(\omega) = \frac{\varepsilon_b \Delta_{LT}/(4\pi)}{E_n - \hbar\omega - i\Gamma}, \qquad (2)$$

where ε_b is the background dielectric constant, Δ_{LT} is the LT splitting energy, and $\rho_n(Z)$ is the density of the transition dipole moment associated with the state $\phi_n(Z)$, and Γ is the phenomenologically introduced damping constant. The self-consistent field in the film can be obtained as a solution of the Maxwell equation including $\chi(Z, Z')$ by means of the ABC-free theory by Cho[4]. The solution includes the two arbitrary amplitudes. We can fix them with the Maxwell's boundary conditions.

3 Results and Discussion

As a model of the strong coupling of the radiation and excitons, we consider a CdSe thin film in the cavity. The DBR's can be constructed from CdS and $CdS_{0.36}Se_{0.64}$. The exciton parameters in CdSe are; $\hbar\omega_T = 1.8227$ eV, $\Delta_{LT} = 1.86$ meV, the lattice constant $a_0 = 7.02$ Å, $\Gamma = 0.1$ meV, $M = 2.63$. The background dielectric constants for $CdSe(\varepsilon_{CdSe})$, $CdS(\varepsilon_{CdS})$ and $CdS_{0.36}Se_{0.64}(\varepsilon_{CdSSe})$ are 7.02, 5.24 and 6.35, respectively. In the following calculations, we suppose DBR's consist of 20 pairs of CdS and CdSSe layers with $\lambda/4$-thickness, where λ is the wavelength of the light in each material.

In Fig. 1, we show the transmittance spectra for the several values of d. The cavity resonance is always tuned to the lowest CM level for each d. There appear several peaks of upper polaritons corresponding to the quantized CM levels. In the case of the $\lambda/2$-cavity, only the CM levels with the even parity with respect to the center of the film couple with the cavity field. The arrows indicating peaks show the orders in which they appear when d increases. Seeing these spectra, we notice that the main peak in the upper polariton region transfers to the mode associated with the higher CM level one by one with the increase of the thickness, while the splitting between this main peak and that of the lower polariton mode increases.

This behavior can be seen more clearly in Fig.2 (a) where the peak positions in the transmittance spectra

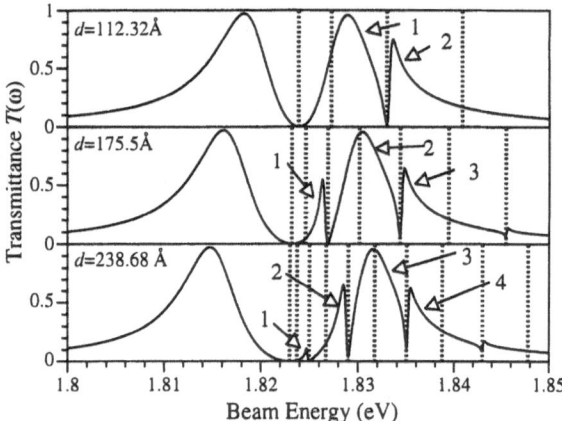

Fig. 1 The transmittance spectra for the several values of d. The vertical dotted lines indicate the quantized excitonic levels. The arrows indicating peaks show the orders in which they appear.

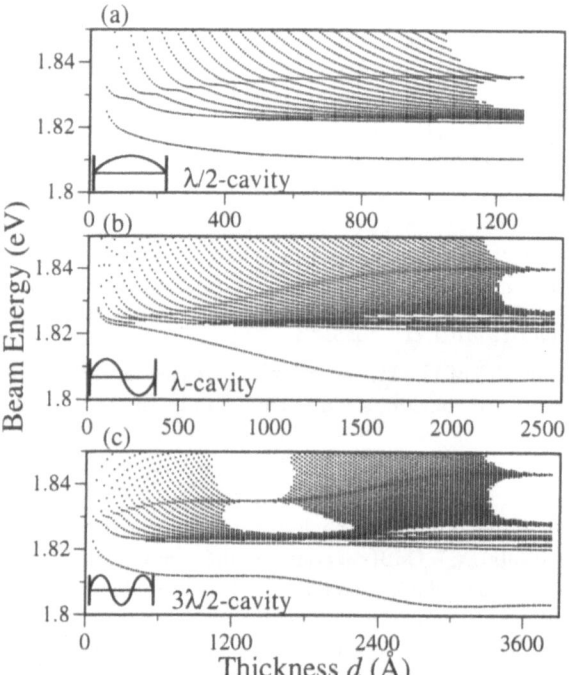

Fig. 2 Peak positions in transmittance spectra for each thickness. Insets indicate the spatial structures of the cavity field.

$T(\omega)$ are indicated for each d. We can see that the coupling strength between the cavity and excitons is enhanced in a certain thickness region, and that the mode transfers between the neighboring modes in the upper polaritons appear as the anti-crossing behavior. Fig. 2 (b) and (c) show the same quantities for the λ-cavity and the $3\lambda/2$-cavity, respectively. In these cases, only the excitonic levels with the odd parity couple with the cavity field in the λ-cavity, whereas only the levels with the even parity couple with the cavity field in the $3\lambda/2$-cavity. In both cases, the mode splitting is enhanced with the increase of d as in Fig 2(a). However, we should note that it does not always increase linearly with the thick-

ness, but shows a step like behavior. For example, in the case of $3\lambda/2$-cavity, the enhancement is saturated around $d = 900\mathring{A}$ and around $d = 3000\mathring{A}$ again.

It is well known that the oscillator strength of weakly confined excitons shows a size linear enhancement due to the coherent extension of the CM motion of excitons[5] as long as the sample size is much smaller than λ. The above enhancement of coupling is a manifestation of essentially the same physical effect. In the present case, however, the relationship between the spatial patterns of the radiation field and CM wavefunctions plays an important role, which is clearly reflected in the peculiar thickness dependence of the mode splitting. For example, when the film surfaces approach the anti-node positions of the radiation field while d increases, the overlap of the radiation and CM wavefunctions increases and the splitting is enhanced, whereas when the film surfaces approach the node positions of the radiation field, the enhancement is saturated.

Another characteristic effect of this system is that the cavity-exciton system increases their coupling strength with the increase of d while changing the coupling partner among the CM levels one by one. This means that the coupling between the different CM levels with the same parity are also induced remarkably by the confined cavity field, which is not remarkable for the bare film.

In this way, we have shown that the CM confined excitons in the cavity provide us a variety of coupling behavior, which means that we can extend the degree of freedom to control the radiation-matter coupling considering the spatial variations of both the radiation and matter wavefunctions. Although the present study has been performed within the linear response regime, the study of the nonlinear response of this system would also be an interesting subject in the future.

Acknowledgments

The authors are grateful to Prof. K. Cho for fruitful discussions and support. They also thank Dr. H. Ajiki for useful discussions. This work was supported in part by Grants-in-Aid for COE Research (10CE2004) of the Ministry of Education, Science, Sports and Culture of Japan.

References

1. For example, the articles in *Quantum Optics of Confined Systems* (NATO ASI Series), edited by M. Ducloy and D. Bloch (Kluwer Academic Publisher, 1996)

2. For example, N. Ochi, T. Shiotani, M. Yamanishi, Y. Honda and I. Suenume: Appl. Phys. Lett. **58** (1991) 2735, C. Weisbuch, M. Nshioka, A. Ishikawa, and Y. Arakawa: Phys. Rev. Lett. **69**, (1992) 3314, Y. Yamamoto and R. E. Slusher: Physics Today **46** (1993) 66.

3. A. Tredicucci, Y. Chen, V. Pellegrini, M. Börger, L. Sorba, F. Beltram and F. Bassani, Phys. Rev. Lett. **75**, (1995) 3906.

4. K. Cho: J. Phys. Soc. Jpn. **55** (1986) 4113 .

5. E. Hanamura: Phys. Rev. **B37** (1988) 1273 .

Stimulation of polariton emission in a homogeneously broadened semiconductor microcavity

V. Mizeikis[1]*, J. Erland[2], J. R. Jensen[2], N. A. Mortensen[3] J. M. Hvam[2]

[1] Institute for Materials Research and Applied Science, Vilnius University, Saulėtekio al. 9, build. III, LT-2040 Vilnius, Lithuania, e-mail: `vygantas.mizeikis@ff.vu.lt`
[2] Research Center COM, Technical University of Denmark, DK-2800 Kongens Lyngby, Denmark
[3] Mikroelektronik Centret, Technical University of Denmark, DK-2800 Kongens Lyngby, Denmark

Abstract We investigate transient light emission from the lower polariton branch in a homogeneously broadened GaAs microcavity excited by femtosecond laser pulses under angle-resonant conditions. The measured emission dynamics reveals stimulated emission due to the bosonic enhancement by the final-state occupation.

1 Introduction

Bosonic stimulation of cavity polariton (CP) emission by the final state occupation has attracted much interest recently [1]. So far the experiments were mostly performed in inhomogeneously broadened microcavities (MC), where polariton in-plane localization by quantum well (QW) disorder may reduce the number of interacting initial and final states. Therefore it is interesting to examine the less investigated polariton emission from homogeneously broadened MCs, where the localization is minimized and bosonic stimulation may be more pronounced.

2 Sample and experimental details

The sample investigated consists of a 25 nm GaAs QW embedded into a wedged GaAs/AlAs λ cavity (bare line width $\delta_{MC} = 260\ \mu eV$) [2] at the anti-node. At a temperature T= 4 K used for the emission measurements, bare QW exciton line is predominantly homogeneously broadened ($\delta_{QW} = 120\ \mu eV$). Because of the anti-crossing between the cavity photon dispersion and the quantum well heavy-hole (HH) and light-hole (LH) exciton dispersions, cavity polaritons (CP) having bosonic nature and strongly modified in-plane dispersions (denoted as upper (UP) middle (MP) and lower (LP) polariton branches) with a large Rabi splitting and ultra narrow polariton line widths $< 200\ \mu eV$, result. In this work we focus on the LP state. Fig. 1 shows the measured and calculated [3] LP in-plane dispersion E(k), and contributions of the HH, LH exciton, and cavity mode (CM) to the LP state at zero detuning $\delta = 0$ between the exciton and cavity resonances. It can be seen that the LH exciton contribution is negligible under the conditions used for the transient emission measurements.

---

* *Present address:* Satellite Venture Business Laboratory, Tokushima University, 2-1 Minamijyosanjima, Tokushima 770-8506, Japan

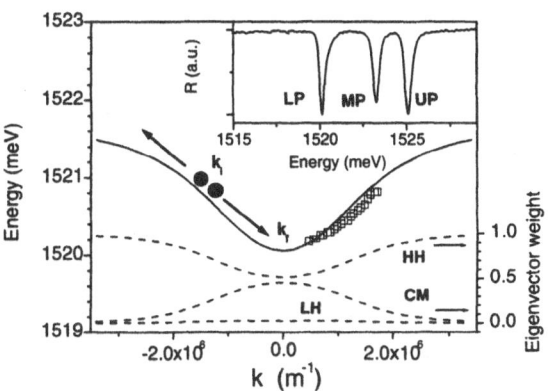

Fig. 1 Calculated (solid line) and measured (squares) LP in-plane dispersion. Dashed lines show the calculated eigenvector components corresponding to the HH, LH, and CM contributions to the LP state. Polariton-polariton scattering at k_i is shown schematically by arrows. Inset: the MC reflection at normal incidence with LP, MP, and UP dips.

In the transient emission experiments different values of the in-plane k component are accessed by varying the light incidence or detection angle. LP population ($N \approx 10^9$ cm^{-3}) is excited in the initial states $k_i \approx 1.5 \cdot 10^{-6}$ m^{-1} by spectrally shaped ($\tau_p \approx 1$ ps) pump pulses from a Ti:Saphire laser, and emission from the final states around $k_f \approx 0$ is monitored with a streak camera (temporal resolution 4 ps). The k_i value is optimized for the angle-resonant polariton-polariton scattering [4]. To further reveal the stimulation, the final states are additionally populated ($N \approx 10^8$ cm^{-3}) by variably delayed probe pulses.

3 Results and discussion

Fig. 2 shows transient emission after the excitation by the pump pulse for various excitation levels N_i. At the lowest excitation the emission is almost instantaneous, and decays exponentially with the time constant of 100 ps characteristic for spontaneous decay. At higher excitations the emission exhibits a distinct peak, delayed by about 15 ps, and then decays rapidly with time constant of less than 20 ps, approaching the subsequent slow spontaneous decay. We have found that the total time-

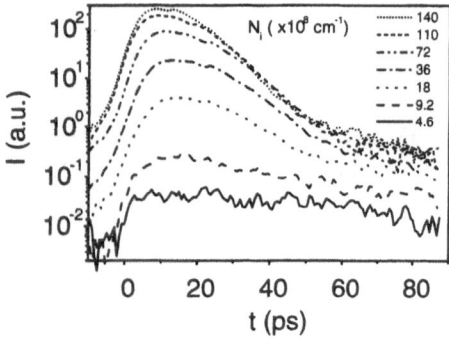

Fig. 2 Evolution of the time-resolved emission from MC with increasing excitation density.

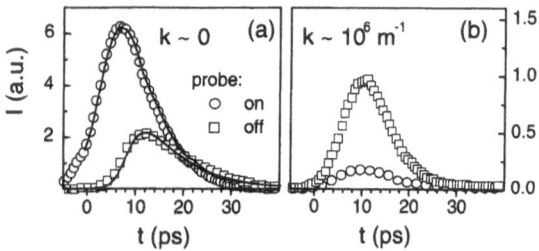

Fig. 3 LP emission after the pump and delayed ($\Delta t = 3$ ps) probe pulse excitation (circles) compared to the emission without probe (squares) at $k \approx 0$ (a), the same at $k \approx 10^6$ m^{-1} (b). Solid lines in (a) are fits using Eqs. 1

integrated emission intensity $I \sim N_i^2$. Below we will argue that the intense, delayed and peaked contribution to the emission arises due to i) angle-resonant exciton-exciton scattering within the LP branch (sketched by arrows in Fig. 1) which effectively populates the final states, ii) stimulation via the bosonic final state occupation.

In addition to this intrinsic mechanism, a population at $k_f \approx 0$ can be seeded by external injection of polaritons by the delayed probe pulse. Fig. 3 (a) shows comparison between the LP emissions with and without probe. It is obvious that the seeded emission peaks in time shortly after the probe arrival, and is more intense.

To understand the measured emission dynamics, we modeled it using the coupled rate equations which account for the processes (i) and (ii) together with polariton scattering with acoustic phonons [5]:

$$\frac{dN_f}{dt} = -\frac{N_f}{\tau_f} + AN_i(1 + N_f) + \frac{1}{2}BN_i^2(1 + N_f), \quad (1)$$

$$\frac{dN_i}{dt} = -\frac{N_i}{\tau_i} - AN_i(1 + N_f) - BN_i^2(1 + N_f),$$

where N_f and N_i are the polariton populations (per k-state per unit area), τ_f and τ_i are the polariton life times, A (B) is the coupling coefficient due to the polariton-phonon (polariton-polariton) scattering. Note the pres-

ence of $(1 + N_{i,f})$ type terms representing the bosonic nature of polaritons. The polariton generation terms have been omitted here. The simulation results in Fig. 3 (a) match perfectly the experimental data, using the measured lifetimes $\tau_f = 7$ ps, $\tau_i = 100$ ps, and $B = 2.4 \cdot 10^{-9}$ cm^{4}s^{-1} [5], and assuming $A \approx 0$ (insignificant phonon scatering). We stress here that inclusion of the third terms in Eqs. (1) which represent polariton-polariton scattering and bosonic enhancement, is crucial for obtaining the correct fits in Fig. 3 (a), and reproducing the $I \sim N_i^2$ dependence.

These findings show that the polariton-polariton scattering to $k_f \approx 0$ together with the bosonic stimulation cause the emission of the pronounced delayed pulse from the MC. Emission efficiency and timing can be controlled by probe, seeding additional polaritons into the $k_f \approx 0$ states. Although in this case wave-mixing between different coherent polariton polarizations can not be excluded, it can not explain the non-seeded data (Fig. 2) for the following reasons: 1) We only inject a coherent polariton population in the initial states with zero laser-pulse overlap with the final states 2) The measured line width of the polaritons limits the influence of coherent wave-mixing to a time-scale of 10 ps, much shorter than the transients shown in Fig. 2 [6]. Returning to the experiments with seeding, the data in Fig. 3 demonstrates that the enhancement of the emission at $k_f \approx 0$ is accompanied by its suppression at intermediate $0 < k < k_i$, which become populated due to slower scattering mechanisms (like the previously neglected phonon scattering from k_i) which destroy the coherence. Thus, the suppressed emission indicates the removal of polaritons from the $k \neq 0$ states, giving strong indication that in the seeding experiments population effects also play important role.

Concerning the role of the homogeneous broadening, we have recently compared between the LP emission in this sample, and in another one with inhomogeneously broadened QW [7]. The differences found therein support the assumption about stronger polariton-polariton interactions in the homogeneously broadened sample.

In conclusion, we have found that the dominant contribution to the LP emission from homogeneously broadened GaAs microcavities under the angle-resonant excitation comes from polariton-polariton scattering and bosonic final state occupation effects.

This work was supported by the Danish Natural Sciences Research Council. V.M. acknowledges financial support from Danish Rectors Conference and Research Center COM.

References

1. P. Senellart *et al.*, Phys. Rev. Lett. **82**, (1999) 1233.
2. J.R. Jensen *et al.*, Appl. Phys. Lett. **76**, (2000) 3262.
3. J. Erland *et al.*, phys. stat. sol. (b) **221** (2000) 115.
4. P.G. Savvidis *et al.*, Phys. Rev. Lett. **84**, (2000) 1547.
5. F. Tassone *et al.*, Phys. Rev. B **59**, (1999) 10830.
6. Y.-S. Lee *et al.*, Phys. Rev. Lett. **83**, (1999) 5338.
7. J. Erland *et al.*, to be published.

Bloch oscillations of light in laterally confined Bragg mirrors and multiple coupled microcavities

G. Malpuech[1], A. Kavokin[1], A. Di Carlo[2], P. Lugli[2] and G. Panzarini[3]

[1]LASMEA - Université Blaise Pascal Clermont II - CNRS 24 avenue des Landais - 63177 AUBIERE CEDEX- France
[2]INFM, Dipartimento di Ingegneria Elettronica, Università di Roma II "Tor Vergata", via Tor Vergata s.n.c., 00133, Roma, Italy
[3]INFM, Dipartimento di Fisica "A. Volta", Università di Pavia, via Bassi, I-27100, Pavia, Italy

Photons, as electrons, may experience a confinement within inclined allowed optical bands of photonic structures, that results in a characteristic Wannier-Stark ladder and photonic Bloch oscillations PBO analogous to electronic Bloch oscillations in crystals subjected to an electric field. In the photonic case the role of electric field is played by a gradient of size of the elementary cell of the optical crystal. We considered two systems promising for observation of PBO, namely, the laterally confined Bragg mirrors and planar multiple-microcavity structures with cavity size varying along the growth axis. Applying the scattering state technique we have modelled the PBO in a time-domain.

Bloch oscillations (BO), i.e. oscillations of an electron with a Brillouin zone induced by an electric field[1] are very hard to observe experimentally. Actually, the dephasing of electrons because of their collisions and interaction with acoustic phonons destroys the oscillations in conventional crystals[2]. The idea of this work is to show that BO equally exist for photons if they propagate through quasi-periodical dielectric structures. The advantage of photonic systems for observation of BO is in the minor role of the dephasing processes. The disadvantage is that the electric field has no effect on photons, so that to incline the photonic bands one has to change gradually the elementary cell of the photonic crystal. Recently, we have proposed a model structure (a Bragg mirror embedded in a planar wave-guide) where photonic Bloch oscillations (PBO) can be observed[3]. In this paper we present a theory of PBO in two new types of photonic structures, namely, pillar Bragg mirrors and planar coupled multiple microcavities.

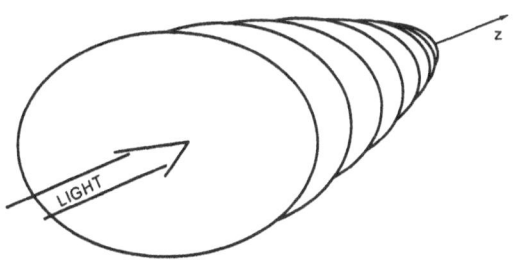

Figure 1 shows schematically the proposed structure of a pillar microcavity made of porous silicon layers with two different degrees of porosity. (having

refractive indices $n_a = 1.27$ and $n_b = 2.25$). The corresponding layer thicknesses are $L_a = 380nm$ and

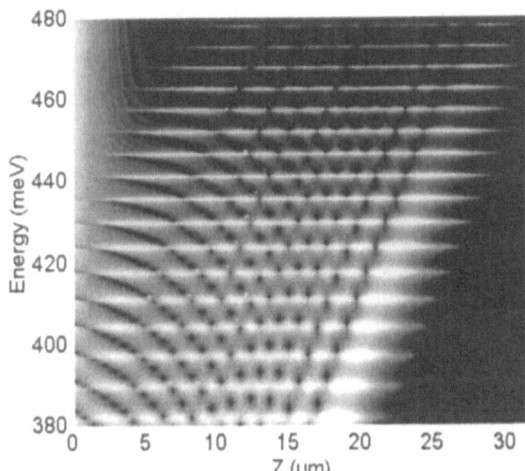

Figure 2. Photonic Wannier-Stark ladder in the structure under study.

$L_b = 300nm$, respectively. The structure contained 46 periods. Variation of the lateral dimension $L(z)$ of the structure can be achieved by chemical etching. Figure 2 shows the frequency and co-ordinate dependent intensity of the electric field in this structure illuminated by a delta-pulse of light. One can clearly see a Wannier-Stark ladder. Figure 3 shows the time- and co-ordinate resolved intensity of light propagating in this structure illuminated by a 0.3 ps-long pulse of light. One can clearly see the PBO with a period of about 0.7 ps.

Figure 4 shows schematically the model multiple-microcavity structure that contained 18 microcavities separated by Bragg mirrors containing 2.5 pairs of dielectric layer each one. The thicknesses of all

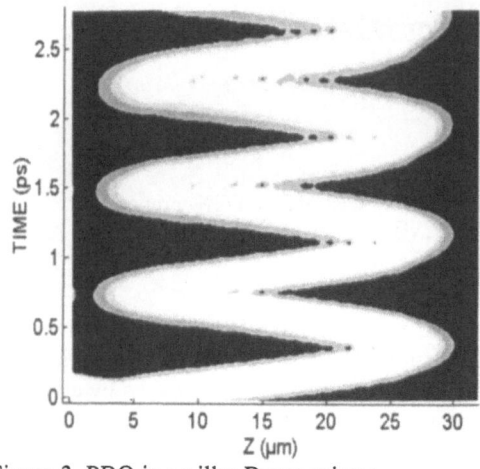

Figure 3. PBO in a pillar Bragg mirror.

layers increase gradually from the beginning till the end of the structure. All the structure parameters as well as the details of our calculation technique will be given elsewhere[4] Figure 5 shows the calculated photonic band structure of the system (in the growth direction). Allowed

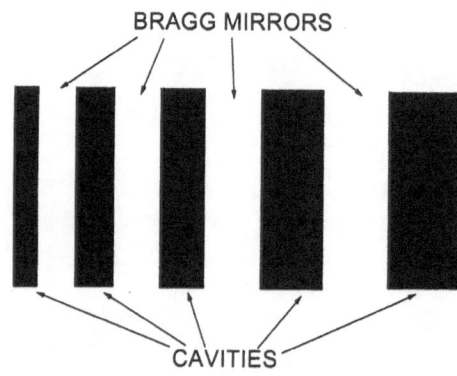

Figure 4. Scheme of a multiple coupled microcavity structure with gradually changing layer thicknesses.

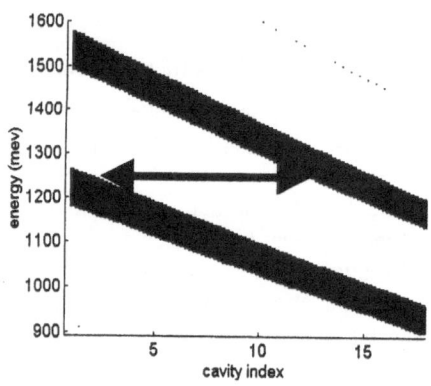

Figure 5. Photonic band structure of the model multiple microcavity. Stop-bands are shown in black.

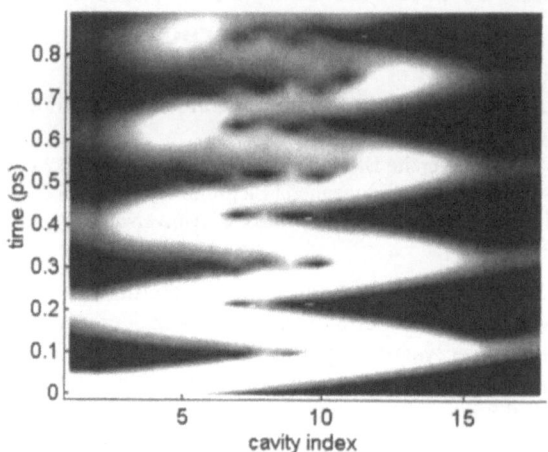

Figure 6. PBO in a multiple coupled microcavity structure.

bands are shown by white colour. One can see the spatial region where the photonic confinement can be achieved. Figure 6 shows the time- and co-ordinate dependent intensity of the electric field in this structure illuminated by a 0.3 ps pulse centred on the region marked by an arrow in Figure 5. One can see very well pronounced PBO. The period of oscillations is 0.2 ps in this particular case. It can be strongly altered by changing a number of microcavities within the structure[4]. In conclusion, PBO can be observed in various multilayer semiconductor and dielectric structures.

Acknowledgments

This work has been supported by the European Commission in the framework of Contract "CLERMONT" No. HPRN-CT-1999-00132.

References

1.   G. H. Wannier, Phys. Rev., (1955), 1227; Phys. Rev., 1835 (1956); Phys. Rev., (1960), 432; Rev. Mod. Phys. , (1962),645.
2.   J. Zak, Phys. Rev. Lett., (1968) 1477; Phys. Rev., (1969), 1366.
3.   A. Kavokin, G. Malpuech, A. Di Carlo, P. Lugli, F. Rossi, Phys. Rev. B **61**, (2000), 4413.
4.   G. Malpuech et al., unpublished.

Enhanced Light Transmission Through Nano-Structured Surfaces

S. Meinlschmidt[1], R. Windisch[2], A. Knobloch[1], P. Kiesel[1], P. Heremans[2], G. H. Döhler[1]

[1] Universität Erlangen-Nürnberg, Institut für Technische Physik I, Erwin-Rommel-Str. 1,D-91058 Erlangen, Germany
[2] IMEC, Kapeldreef 75, B-3001 Leuven, Belgium

Abstract We study the transmission properties of textured surfaces as employed in the fabrication of high-efficiency surface-textured light-emitting diodes. Surface texturing is found to enable the transmission of a significant fraction of light which is incident on the surface under an angle larger than the critical angle of total internal reflection.

1 Introduction

The efficiency of conventional light-emitting diodes (LED's) is limited by total internal reflection of light at the semiconductor–air interface. For GaAs/AlGaAs LED's, the angle of total internal reflection is 16°, resulting in the extraction of only 2% of the internally generated light per out-coupling surface. This problem can be solved by a combination of surface texturing and the application of a rear reflector on a thin-film LED [1]. Using this concept of the so-called non-resonant cavity (NRC)-LED, external quantum efficiencies up to 40% have been demonstrated [2].

The light extraction in these NRC-LED's is generally believed to be due to internal scattering of the internally reflected light [1]. This results in a change in the propagation angle, giving the light multiple chances to escape from the semiconductor material. In order to investigate this mechanism in more detail, we have studied the scattering and transmission properties of the textured surface.

For the investigation of the internal scattering, the angle dependence of the internally scattered light has been measured using a photodetector integrated onto the sample. This enables the measurement of light scattered during internal reflection into angles larger than the critical angle of total internal reflection [3]. Using this technique, we have found that the light that is scattered during internal reflection at a textured semiconductor-air interface is almost perfectly randomized [4]. In this paper, we investigate the influence of surface scattering on the transmission properties of the textured surface.

2 Surface texturing

The surface is textured using a technology often called "natural lithography" [5]. A monolayer of randomly positioned polystyrene spheres serves as a mask for dry etching. In order to obtain a more homogeneous distribution of the spheres on the surface, the sphere size can be reduced in oxygen plasma before etching. The dry etch process produces randomly positioned identical cylindrical pillars on the surface. For the application on the surface of LED's emitting at 870 nm, the optimum original

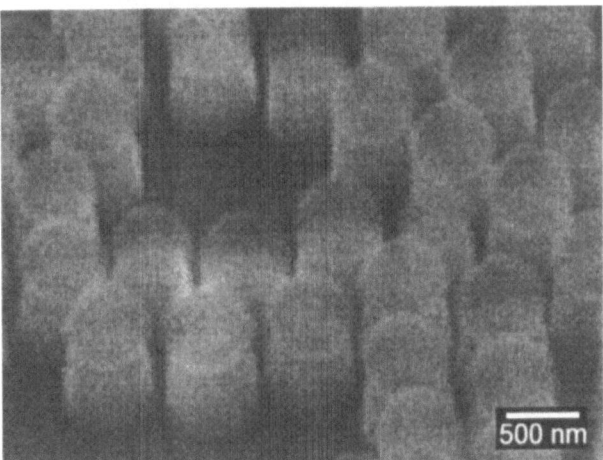

Fig. 1 S.E.M. photograph of a textured surface.

sphere size is 430 nm and the optimum etch depth is 180 nm [2]. The resulting surface is shown in Fig. 1.

3 Experimental results

3.1 Normal incidence

For the measurement of the light transmission through a textured surface, the surface is illuminated with an optical fiber imaged onto the bottom side of the sample. The angle dependence of the light transmitted through the textured top surface is measured. For normal incidence, the surface texturing results in a reduction of the total light transmission to 39%, compared to 68% for an untextured sample without anti-reflection coating.

Approximately half of the transmitted light is transmitted specularly, i.e. according to Snell's law. The illumination occurred with a numerical aperture of 0.34 (corresponding to an insertion angle of 20), and hence the specularly transmitted light is measured in the range between -20 and 20. The other half of the transmitted light is scattered during transmission, resulting an an approximately Lambertian emission from the surface.

3.2 Incidence under arbitrary angles

For the measurement of the light transmission through the textured surface for other incident angles than 0, one edge of the sample is bevelled in order to enable the controlled insertion of light under the desired angle. As long as the incident angle is smaller than the angle of total internal reflection, the result is similar to that for normal incidence. The total transmission is reduced, and a large fraction of the light is transmitted specularly.

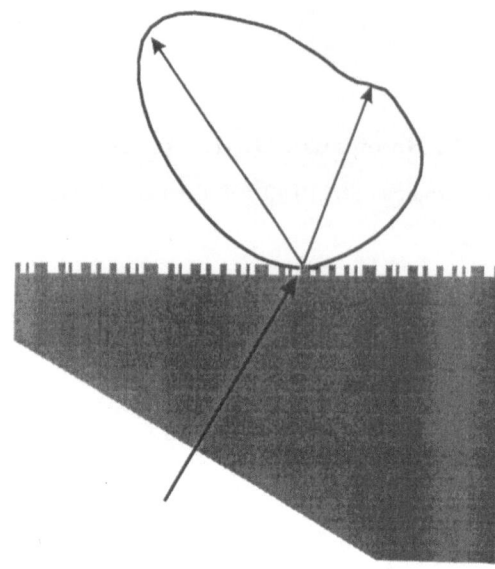

Fig. 2 Angular distribution of the transmitted light for an incident angle of 30°. The figure also shows the shape of the bevelled sample schematically. The two arrows in forward direction represent the theoretical diffraction maxima for the present surface.

For incident angles larger than the angle of total internal reflection, the transmission of an untextured sample is zero. For a textured surface with optimum texturing parameters, a significant fraction of the incident light is transmitted. In Fig. 2, the directionality of the transmitted light is shown for an incident angle of 30°.

The integral transmission through flat and textured surfaces is measured using a large-area photodetector. For angles below the critical angle for total internal reflection, surface texturing reduces the transmission. For larger incident angles, however, a significant fraction of light is transmitted though a textured surface. For example, at 30° incidence almost 10% of the light is transmitted.

For larger incident angles, the transmission is smaller. The angular distribution of the transmitted light varies strongly with both the incident angle and the wavelength. The observed variations are qualitatively identical to the behaviour of diffraction patterns. Figure 3 shows the observed total transmission as a function of the angle of incidence for both flat and textured surfaces.

4 Theoretical model

The angular distribution of the transmitted light can be described as a superposition of the electromagnetic waves originating from all points of the surface (pillars and flat regions in between). This superposition is a product of a factor describing the diffraction on a single cylindrical pillar (form factor), and a factor considering the position of the pillars on the surface (structure factor). For the latter, the positions of the pillars have been extracted from an S.E.M. photograph.

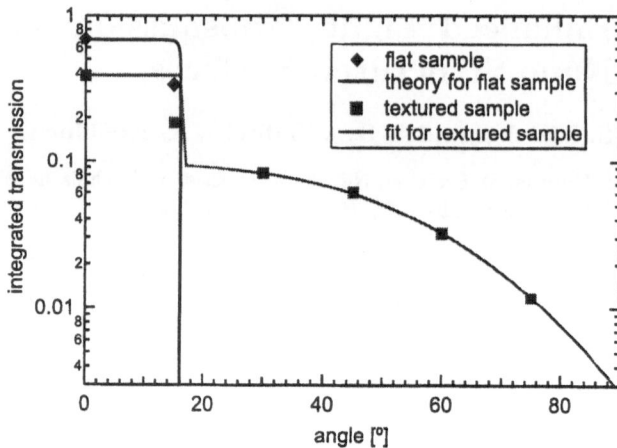

Fig. 3 Total transmission as a function of the angle of incidence.

Due to the short-range periodicity present in the distribution of the pillars on the surface, the calculated angular dependence of the transmission shows broadened diffraction peaks, which correspond to the direction of maximum transmission in the measurement. In Fig. 2, the direction of the calculated diffraction peaks is indicated by the two thin arrows.

5 Conclusions

The transmission properties of a semiconductor surface are heavily altered by surface texturing. For incident angles smaller than the critical angle for total internal reflection, the transmission probability is decreased by surface texturing. On the other hand, in the case of larger incident angles the transmission becomes possible, while the light is totally internally reflected in the case of a flat surface.

For isotropic incidence of light onto the semiconductor–air interface, as present in an LED, the total transmission through the textured surface is almost doubled, as compared to a flat surface. Together with the light scattering during internal reflection, the change in the transmission properties due to surface texturing enhances the light output from surface-textured LED's.

References

1. I. Schnitzer, E. Yablonovitch, C. Caneau, T. J. Gmitter, and A. Scherer, Appl. Phys. Lett. **63**, (1993) 2174
2. R. Windisch, B. Dutta, M. Kuijk, A. Knobloch, S. Meinlschmidt, S. Schoberth, P. Kiesel, G. Borghs, G. H. Döhler, and P. Heremans, IEEE Trans. Electron Dev. **47**, (2000) 1492
3. R. Windisch, S. Schoberth, S. Meinlschmidt, P. Kiesel, A. Knobloch, P. Heremans, B. Dutta, G. Borghs, G.H. Döhler, J. Opt. A: Pure Appl. Opt. **1**, (1999) 512
4. R. Windisch, M. Kuijk, B. Dutta, A. Knobloch, P. Kiesel, G. H. Döhler, G. Borghs, P. Heremans, Proc. SPIE **3938**, (2000) 70
5. H. W. Deckman and J. H. Dunsmuir, Appl. Phys. Lett. **41**, (1982) 377

Optically pumped lasing at 1.3 μm of GaInNAs-based VCSEL structures

M. Hetterich[1], M.D. Dawson[1], A.Yu. Egorov[2]*, H. Riechert[2]

[1] Institute of Photonics, University of Strathclyde, Glasgow G4 0NW, Scotland, United Kingdom
e-mail: `bc39@ph70.rz.uni-karlsruhe.de`
[2] Infineon Technologies, Corporate Research CPR 7, D-81730 Munich, Germany

Abstract We investigate optically pumped lasing of GaInNAs-based VCSELs emitting at 1.3 μm. The samples measured were full device structures including doping. Conventional p-i-n-type VCSELs showed high internal threshold excitation densities of around 10 MW/cm^2. Improvements, especially the introduction of a tunnel junction reduced the threshold to about 128 kW/cm^2 and should after further optimization enable us to realize an electrically pumped 1.3 μm VCSEL in the near future.

1 Introduction

GaInNAs/GaAs is a novel semiconductor material system promising for the realization of high-performance laser diodes emitting at the 1.3 μm optical fiber window. In comparison to the more common GaInAsP system it possesses a larger conduction band offset, which should result in a stronger electron confinement and thus in an improved high-temperature device performance. Rececently, we were able to confirm this large conduction band offset using photoluminescence excitation measurements. We also found direct evidence for an increased effective electron mass, an observation prospectively of benefit for laser diodes [1].

An additional advantage of GaInNAs is the fact, that it can be grown coherently on GaAs which enables the use of high refractive index contrast AlGaAs-based distributed Bragg reflectors (DBRs) in vertical-cavity surface-emitting lasers (VCSELs) for telecom wavelengths. However, while progress in the development of edge-emitting lasers at 1.3 μm by many groups is impressive (see e.g. [2] for some results obtained by our collaboration) the first electrically pumped VCSEL with continuous wave (cw) operation at 1294 nm and a maximum single mode output power of 60 μW at room temperature could only be demonstrated very recently [3]. There are also only a few reports on optical pumping of long-wavelength GaInNAs-VCSELs so far [4,5]. In this contribution we describe some of our results obtained from such experiments. The samples investigated were full device structures. Especially they included doping, which is important, because effects like losses due to free carrier absorption etc. can be expected to have a significant influence on e.g. the laser threshold in the final electrically pumped device.

2 Sample design and experiment

The VCSEL samples were grown on GaAs by solid-source molecular-beam epitaxy using an RF-coupled plasma source for nitrogen. Two different designs were investigated: The first was an ordinary p-i-n structure with a 27-period Si-doped AlAs/GaAs bottom DBR, an undoped λ-cavity and a 26-period Be-doped Al$_{0.80}$Ga$_{0.20}$As/GaAs top DBR. The active region contained 3 Ga$_{0.63}$In$_{0.37}$N$_{0.019}$As$_{0.981}$ quantum-wells (6.5 nm) in the electric field maximum of the cavity mode, embedded in quaternary Ga$_{0.95}$In$_{0.05}$N$_{0.017}$As$_{0.983}$ barrier layers. A similar design was recently used successfully for the realization of edge-emitting lasers [2]. In the second type of VCSEL structures investigated we mainly replaced the p-doping in the top DBR by a pn tunnel junction, an approach also suggested independently in [3]. The specific sample discussed in our paper consisted of a 32-period AlAs/GaAs bottom mirror, a $\frac{3}{2}\lambda$ cavity with 2 GaInNAs quantum-wells (6.5 nm) in each field maximum and a 27-period AlGaAs/GaAs top DBR.

For the optical pumping experiments described below we used a Q-switched Nd:YAG laser as excitation source which produced 7.7 ns pulses at $\lambda = 1064$ nm with a repetition rate of 2 kHz and a maximum pulse energy of about 17 μJ. The laser beam was attenuated by a fixed and a variable neutral-density filter and then focussed on the as-grown VCSEL structure using a 20$\times$ long working distance infrared microscope objective in a 45° configuration. The latter lead to an elliptical excitation spot with a measured diameter of 8.5 μm in the vertical and 12 μm in the horizontal direction, i.e. the spot size was diffraction-limited. The generated VCSEL emission perpendicular to the sample surface was dispersed by a 46 cm monochromator equipped with a 600 g/mm grating and detected by a thermoelectrically cooled Si/InGaAs detector using a lock-in amplifier.

3 Results and discussion

The first samples investigated were VCSELs with a conventional p-i-n structure. Reflectivity and temperature-dependent photoluminescence measurements (using a cw Nd:YAG laser as excitation source) confirmed, that the cavity resonance was in the center of the DBR stop-band at the design wavelength (1.3 μm) and well matched to the quantum-well emission. Despite that, some of the first samples did not show any lasing even for incident

---

* On leave from Ioffe Institute, St. Petersburg, Russia

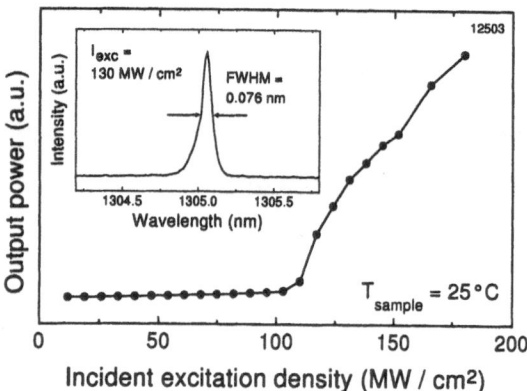

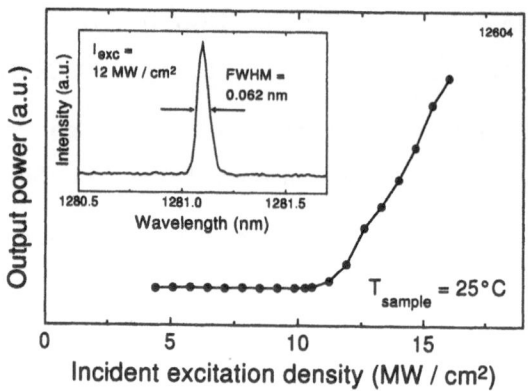

Fig. 1 Spectrally integrated output power versus incident excitation density for a *p-i-n*-type VCSEL with three quantum-wells and quaternary barriers in the active region. The inset shows the emission spectrum obtained at 130 MW/cm² excitation density.

Fig. 2 Spectrally integrated output power versus incident excitation density for a VCSEL with tunnel junction and four quantum-wells in a GaAs cavity. The inset shows the lasing spectrum obtained at 12 MW/cm² excitation density.

excitation densities as high as 2.5 GW/cm². The results of the best *p-i-n*-type VCSEL so far are shown in Fig. 1. For this sample lasing at a wavelength of 1305.1 nm could be obtained, but still with a very high threshold pump power density of 108 MW/cm². Due to the fact, that only the active region was pumped, the corresponding internal excitation density was much lower but difficult to determine exactly, mainly because the absorption coefficient of the GaInNAs quantum-wells and the quaternary barriers are not known accurately enough. Part of the pump light is also lost in the doped top DBR. Taking reflection into account, we estimate the internal threshold excitation density to be about 10 MW/cm², which would correspond to a threshold current density of around 9000 kA/cm². This value is clearly far too big for an electrically pumped device and suggests that high losses are present in the sample. Due to the excellent performance of edge-emitting devices [2], we concluded these losses to be at least partly the result of problems in the VCSEL design, not only insufficient material quality. In particular, we suspected free carrier absorption (inter- and intra-band) in the *p*-doped DBR to play an important role. A tunnel junction was therefore introduced to minimize these effects.

Fig. 2 shows the results of optical pumping experiments obtained for such a sample. Lasing at a wavelength of 1281.1 nm was observed for this VCSEL. For low pump intensities the linewidth was limited by the monochromator resolution to 0.062 nm (full width at half maximum), while for higher excitation levels the emission broadened and became more structured, probably due to the poor lateral mode confinement in the as-grown samples used. The latter also caused some modulation in the output versus pump power characteristics for higher pump powers, which was frequently observed for the samples investigated (see e.g. Fig. 1).

Compared to the VCSELs with *p-i-n* structure, the external threshold excitation density decreased by one order of magnitude to 11 MW/cm², only 230 kW/cm² of which were actually absorbed in the active region. This would correspond to a threshold current density of about 200 kA/cm² in an electrically pumped device. In previous publications values about two orders of magnitude lower were reported. However, these VCSELs had more quantum-wells in a longer cavity [5] or operated at shorter wavelength [4]. Additionally, our samples also include doping, which introduces significant losses due to e.g. free carrier absorption, especially for long-wavelength VCSELs. Indeed, lower thresholds were observed for samples lasing at shorter wavelength. Although the samples presented here show only spontaneous emission for electrical injection so far, we therefore believe, that a further optimization of the VCSEL design (e.g. the doping profile) and material quality should enable us to realize an electrically pumped VCSEL emitting at 1.3 μm in the near future.

Acknowledgements This work was supported by the EU under BriteEuram BRPR-CT98-0721 ("OPTIVAN").

References

1. M. Hetterich, M.D. Dawson, A.Yu. Egorov, D. Bernklau, and H. Riechert, Appl. Phys. Lett. **76**, (2000) 1030.
2. B. Borchert, A.Yu. Egorov, S. Illek, M. Komainda, and H. Riechert, Electron. Lett. **35**, (1999) 2204, and references therein.
3. K.D. Choquette, J.F. Klem, A.J. Fischer, O. Blum, A.A. Allerman, I.J. Fritz, S.R. Kurtz, W.G. Breiland, R. Sieg, K.M. Geib, J.W. Scott, and R.L. Naone, Electron. Lett. **36**, (2000) 1388.
4. M.C. Larson, M. Kondow, T. Kitatani, K. Tamura, Y. Yazawa, and M. Okai, IEEE Photon. Technol. Lett. **9**, (1997) 1549.
5. C. Ellmers, F. Höhnsdorf, J. Koch, C. Agert, S. Leu, D. Karaiskaj, M. Hofmann, W. Stolz, and W.W. Rühle, Appl. Phys. Lett. **74**, (1999) 2271.

The interaction between exciton states near the quantum well bandgap and confined photons in GaAs/AlAs microcavities

A. Qarry[1], R. Rapaport[1], E. Cohen[1], Arza. Ron[1], E. Linder[1], L. N. Pfeiffer[2]

[1] Solid State Institute and Physics Department, Technion, Israel Institute of Technology, Haifa 32000, Israel.
[2] Bell Laboratories, Lucent Technologies, 600 Mountain Avenue, Murray Hill, New Jersey 07974.

Abstract The strong resonant interaction between microcavity (MC) confined photons and quantum well (QW) interband excitations near the onset of the exciton continuum is investigated in λ - wide, planar GaAs/AlGaAs MC's, having a single GaAs/AlAs QW (200A wide). The spectra show sharp lines of the cavity polaritons that are formed of the (1e:hh1)1S, (1e:lh1)1S, (e1:hh1)2S and (e1:lh1)2S excitons, the interband transition between the N=1 electron and hole Landau states, as well as of unidentified magnetoexcitons with energies around that of the (e1:hh1)2S. All the near bandgap cavity polaritons are observed only when the magnetoexcitons, of which they are formed, are tuned into resonance with the MC confined photons, and their intensity is observed to increase with increasing B. The measured (1e:hh1)1S and (e1:hh1)2S diamagnetic shift and Rabi splitting are used in order to estimate the magnetoexcitons wavefunction in-plane radius decrease with increasing B (in the low field range).

The strong resonant interaction between microcavity (MC) confined photons and quantum well (QW) interband excitations, is clearly manifest in the optical spectra by the vacuum Rabi splitting and by the cavity polariton dispersion[1,2]. Since the lowest energy QW excitons, (e1:hh1)1S and (e1:lh1)1S, have large oscillator strengths[3], and their linewidths are usually smaller than the Rabi splitting, the strong coupling limit is easily achieved for them and therefore, the spectroscopic and dynamic properties of these excitons have been thoroughly investigated[2]. The effect of a magnetic field that is applied perpendicularly to the QW plane on the cavity polaritons that are formed of Landau transitions is mainly in inducing a transition from a weakly coupled magnetoexciton-MC mode system to a strongly coupled one[2,4].

Our main objective is to study the energy and oscillator strength dependence on B of the QW magnetoexcitons near the e1 - hh1 bandgap by utilizing their resonant interaction with the MC mode. This is done by studying the low temperature, circularly polarized reflection spectra. These spectra provide a very accurate determination of the cavity polaritons energy dependence on B due to the sharpness of the lines and to the increase of each magnetoexciton oscillator strength with B. This results in a tunability of the coupling strength with the MC mode. The peak energies of the reflection lines (at B=0) as a function of the MC mode energy (Ec) are shown in Fig.1. The highest energy branch results from the (e1:hh1)2S exciton that is separated by

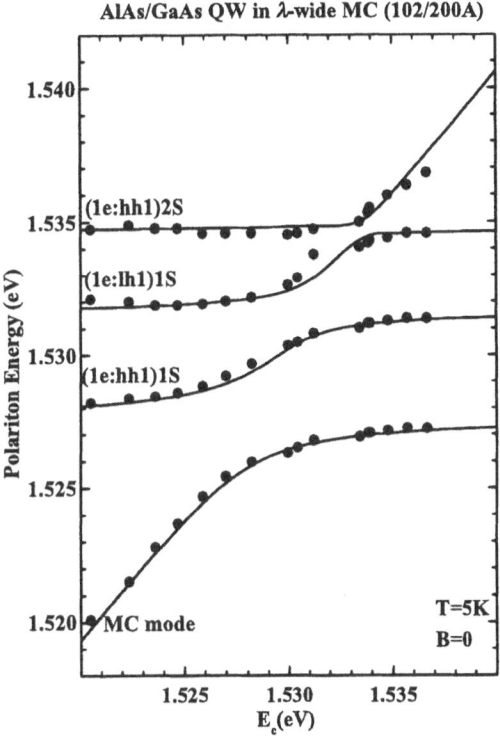

AlAs/GaAs QW in λ-wide MC (102/200A)

Fig. 1 The magnetoexcitons energy dependence on magnetic field (open and full circles). The slopes of the lineary extrapulated parts are indicated. Solid lines indicates the electron-heavy hole magnetoexciton transitions and dashed lines the electron light hole transitions.

7.0meV from the (e1:hh1)1S exciton (very close to the calculated value of 6.5meV[6]). A narrow linewidth (0.46 meV) is measured for the Rabi split cavity polaritons arising from the (e1:hh1)2S exciton. It indicates that only the n=2 exciton contributes since it has the largest oscillator strength. The energies shown in Fig.1 are fitted using the coupled oscillators model[1] in order to obtain the exciton - MC photon interaction strengths. Under B, many reflection lines appear due to the magnetoexcitons oscillator strength increase. The fan diagram shown in Fig.2 is the B-dependent energy of all the magnetoexcitons that are obtained from the reflection spectra.

The fan diagram is examined regarding the relative strength of the magnetic and Coulomb interactions. It is known that the (e1:hh1)1S and (e1:lh1)1S magnetoexcitons arise mainly from the electron and hole N=0 Landau states[5], and are most strongly affected by the e-h Coulomb interaction. For the near bandgap transi-

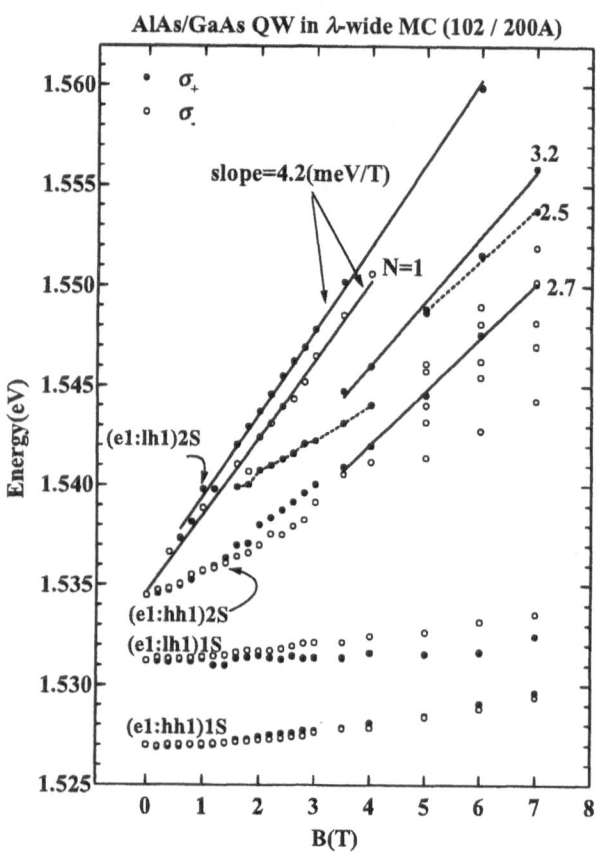

AlAs/GaAs QW in λ-wide MC (102 / 200A)

Fig. 2 The magnetoexcitons energy dependence on magnetic field (open and full circles). The slopes of the lineary extrapolated parts are indicated. Solid lines indicate the electron-heavy hole magnetoexciton transitions and dashed lines the electron light hole transitions.

$$\Delta E(B) = \frac{E_{\sigma+}(B) - E_{\sigma-}(B)}{2} + (E_{\sigma+}(B) - E(0)) \quad (1)$$

The $\Delta E(B)$ values are best fitted with a function of the form: $AB^2 + CB^4$. This fit holds for the (e1:hh1)1S and (e1:lh1)1S magnetoexcitons in the entire $0 \leq B \leq 7T$ range. For the (e1:hh1)2S magnetoexciton it holds only for $B < 4T$, while for higher magnetic fields it becomes linear in B. The slope of this line is smaller than that of the N=1 magnetoexciton, and it can be interpreted by the weak effect of the e-hh Coulomb interaction on the (e1:hh1)2S exciton, that results in the formation of its wavefunction by admixing the n=2 hydrogenic state with the N=1 Landau state[5].

The fitting parameter C is negative for the (e1:hh1)1S and (e1:hh1)2S magnetoexciton, meaning that the in-plane exciton radius decreases with increasing B. This is consistent with the observed increase in the (e1:hh1)2S Rabi splitting[2].

Since the magnetoexciton oscillator strength is related to the Rabi splitting, its increase indicates a reduction of the wavefunction in-plane dimensions with increasing B.

Finally, there are several weak reflection lines with energies between the cavity polaritons that are formed of the (e1:hh1)2S and the N=1 magnetoexcitons. Their energy is given by the points that, for $B > 4T$ in Fig.2, can be grouped into several straight lines with extrapolated slopes that are indicated in the figure. We propose to attribute them to magnetoexcitons with high orbital momentum components that are admixed with the s-type states by the magnetic interaction. These transitions are observed only when they are tuned into exact resonance with the MC mode and because their oscillator strength is greatly enhanced with increasing B.

Acknowledgments: The work at Technion was done in the Barbara and Norman Seiden Center for Advanced Optoelectronics and was supported by the U.S.- Israel Binational Science Foundation, Jerusalem, Israel.

tions we indicated the linear extrapolation slopes of the B - dependent magnetoexcitons energy. The experimental points of the highest energy magnetoexciton form a straight line with a slope of 4.2meV/T, which is exactly the value expected for the N=1 Landau transition. We thus conclude that the Coulomb interaction energy is very small for the N=1 magnetoexciton, even for low B values. The (e1:hh1)2S exciton was observed in the reflection spectra at B=0. The (e1:lh1)2S magnetoexciton cannot be observed at B=0 since it appears within the (e1-hh1) continuum. However, under an applied magnetic field, its oscillator strength increases and it emerges since its interaction with the MC mode transforms from weak to strong. It is extrapolated in Fig.2 to an energy of 1.540eV for B=0. The energy difference between the (e1:lh1)1S and (e1:lh1)2S magnetoexcitons that is measured here is 8.0meV, as compared to the calculated value of 7.3meV[6]. The (e1:hh1)1S, (e1:lh1)1S and (e1:hh1)2S magnetoexcitons energy dependence on B with respect to their energy at B=0 is measured here with great accuracy in the low field range. These energies are obtained by computing:

References

1. V. Savona, in Confined Photon System, Fundamentals and Applicatons, Edited by H. Benisty, Springer Lecture Notes in Physics(Berlin,1999), p.173.

2. M. S. Scolnick, T. A. Fisher and D. M. Whittaker, Semicon. Sci. Technol. **13**, 645, (1998).

3. L. C. Andreani, in Confined Electrons and Photons, Edited by C. Weisbuch and E. Burstein, (Plenum,Boston,1995), p.57.

4. J. Tignon, R. Ferreira, J. Wainstain, C. Delalande, P. Voisin, M. Voos, R. Houdre, U. Oesterle and R. P. Stanley Phys. Rev. B, **36**, 4068, (1997).

5. M. Sugawara, N. Okazaki, T. Fujii and S. Yamazaki Phys. Rev. B, **48**, 8848, (1993).

6. L. C. Andreani and A. Pasquarello Phys. Rev. B, **42**, 8928, (1990).

Resonant Raman scattering in an InAs/GaAs monolayer structure

J. Maultzsch, S. Reich, A. R. Goñi, C. Thomsen

Institut für Festkörperphysik, Technische Universität Berlin, Hardenbergstr. 36, 10623 Berlin, Germany

Abstract　We studied the excitation-energy dependence for Raman scattering on the GaAs optical phonon modes in an InAs/GaAs heterostructure. The GaAs LO phonon scattering exhibits a resonance with the InAs optical transitions, the Raman profiles depending on the polarization of the incoming light. We interpret our results as due to incoming resonance with E1-HH1 and the E1-LH1 transition of the monolayer. In addition, we demonstrate that the scattering intensity is influenced by an inhomogeneous exciton linewidth due to fluctuations in the well width.

1 Introduction

The vibrational properties of low-dimensional semiconductor structures have been widely studied by Raman spectroscopy.[1] Resonant-Raman experiments yield information also about the electronic properties as the scattering intensity is enhanced if the incoming or outgoing photon energy matches an allowed electronic transition. We consider a monolayer structure from zincblende-type materials as characterized by the point group D_{2d}. The reduction of dimensionality leaves the symmetry of the electron state (E1) at the condcuction band minimum unchanged (A_1), whereas the degeneracy of the hole states at the Γ point is lifted. The heavy hole state has E symmetry and is twofold degenerate; the light hole state has B_2 symmetry.

We present resonant-Raman scattering experiments on the GaAs optical phonon modes in an InAs/GaAs monolayer structure for different scattering geometries. We also investigated the dependence of the Raman profile on the exciton coherence lifetime as previously described by Shields et al.[2]

2 Experiment

The sample consisted of a single InAs layer (effective thickness about 1.5 ML), embedded in bulklike GaAs with two AlGaAs cladding layers forming a waveguide. The InAs-monolayer fundamental transition is about 100 meV below the GaAs bandgap. The light and heavy-hole state are separated energetically by a few meV [3].

Raman experiments were performed at low temperatures (2-40 K) in backscattering geometry with the light propagating along the monolayer plane in $(\bar{1}10)$ direction and polarized parallel or perpendicular to the monolayer. The inset of Fig. 1 shows a schematic picture of the setup and the coordinate system used. Raman spectra were excited with a Ti-Sapphire laser, tuned from 1.41 eV to 1.435 eV. The scattered light was analyzed by a triple-grating spectrometer and detected with a charge-coupled-device (CCD) camera.

3 Results

In Fig. 1 we show Raman spectra recorded at 5 K with an excitation energy of 1.424 eV. The lowest trace shows a spec-

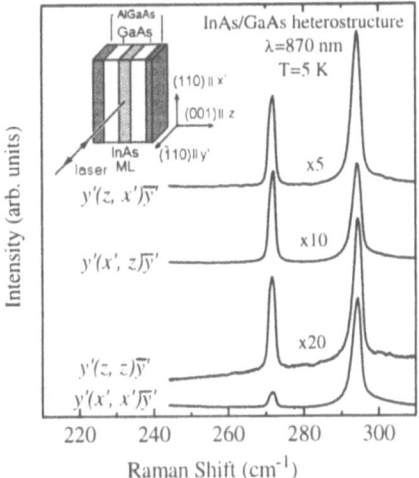

Fig. 1 Raman spectra recorded at T=5 K with an excitation energy of 1.424 eV. The scattering geometry is given in Porto's notation. The inset shows a sketch of the sample and the scattering geometry.

trum in $y'(x', x')\bar{y}'$ scattering geometry. We find the GaAs LO phonon (296 cm^{-1}), which for bulk GaAs is forbidden in this configuration, to be 10–20 times stronger than the allowed TO (273 cm^{-1}) mode. For $y'(z, z)\bar{y}'$ geometry and crossed polarizations the LO scattering is observed as well, although much weaker. Note that the Raman tensor of the LO phonon is not symmetric; the scattering intensity is different in the two crossed polarizations. By varying the excitation energy we found a resonance of the GaAs LO-phonon with the InAs optical transitions, while the TO scattering intensity was almost constant. Fig. 2 shows Raman profiles in $y'(x', x')\bar{y}'$ and $y'(z, z)\bar{y}'$ geometries, along with a photoluminescence (PL) and a photoluminescence excitation (PLE) spectrum. The maxima of the Raman profiles are at 1.424 eV for $y'(x', x')\bar{y}'$ and at 1.428 eV for $y'(z, z)\bar{y}'$ configuration; we observe the same energy shift of $\Delta = 4$ meV between the two cross-polarized spectra. The position of the Raman profile depends only on the polarization of the incoming light, i.e., 1.424 eV for (x', x') and (x', z) and 1.428 eV for (z, z) and (z, x') polarized light. The PLE peak in Fig. 2 at 1.439 eV corresponds to the HH1-exciton of the InAs monolayer. The photoluminescence maximum is shifted by about 20 meV to lower energy due to the relaxation of free carriers into localized, lower-energy states. Likewise, the Raman profiles are shifted to the *red* with respect to the PLE peak.

4 Discussion

As was shown by Shields et al. for a GaAs/AlAs multiple quantum well, the energetic separation between the Raman

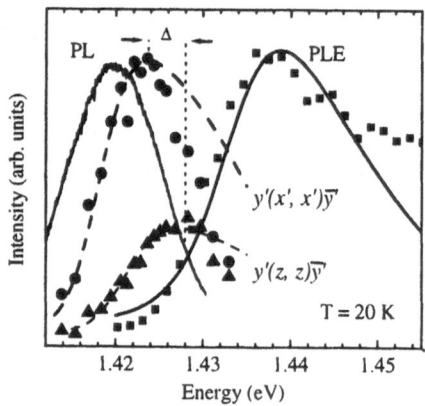

Fig. 2 Photoluminescence (PL), photoluminescence excitation (PLE) spectra, and Raman profiles of the 1.5 ML InAs/GaAs heterostructure. The PL and PLE spectra were recorded at 20 K with the light polarized parallel to the monolayer plane (x'). The Raman profiles show the scattering intensity of the GaAs LO-phonon mode normalized to the TO scattering intensity as a function of incoming photon energy. The solid and dashed lines were calculated with Eq. 2 and 3, respectively.

profile and the PLE peak can be attributed to an inhomogeneous broadening by fluctuations in the well width.[2] They assumed that an exciton at energy E_h with a homogeneous linewidth Γ_h can scatter with equal probability to all energetically lower, wider-well regions. Then $\Gamma_h(E_h)$ increases with E_h as the number of exciton states below E_h. Taking a Gaussian distribution $G(E_h)$ of homogeneously broadened exciton states centered at E_i and with a half-width Γ_i [2]

$$
\Gamma_h(E_h) = \int_{-\infty}^{E_h} G(E, E_i, \Gamma_i)\, dE
$$
$$
= \Gamma_0 + \frac{1}{2}\Gamma_1\left\{1 + \mathrm{erf}\left(\frac{E_h - E_i}{\sqrt{\ln 2}\,\Gamma_i}\right)\right\}, \qquad (1)
$$

where $\mathrm{erf}(x)$ is the error function, Γ_0 the low-energy limit of the homogeneous linewidth Γ_h, and Γ_1 the difference between the low and the high-energy limit. In the approximation that the PLE spectrum resembles the absorption, the PLE intensity is given by

$$
I_{PLE}(E) \propto \int_{-\infty}^{\infty} \frac{\Gamma_h \cdot G(E_h, E_i, \Gamma_i)}{\pi[(E_h - E)^2 + \Gamma_h^2]}\, dE_h. \qquad (2)
$$

The Raman scattering intensity for incoming resonance with the Gaussian-like distributed exciton states of energy E_h may be written as

$$
I_{RRS}(E) \propto \int_{-\infty}^{\infty} \frac{G(E_h, E_i, \Gamma_i)}{\pi[(E_h - E)^2 + \Gamma_h^2]}\, dE_h. \qquad (3)
$$

The PLE spectrum and the resonant Raman profile calculated with Eq. 2 and 3 are shown in Fig. 2. The parameters were $E_i = 1.444$ eV for the $y(x', x')\bar{y}$ Raman profile, $\Gamma_i = 14.5$ meV, $\Gamma_0 = 2 \cdot 10^{-6}$ eV, and $\Gamma_1 = 1 \cdot 10^{-2}$ eV. It is nicely seen, that the energy shift between the PLE and the Raman profile peak is reproduced well by this calculation. We find though, that the measured Raman profile is less asymmetric than the calculated one. A possible reason is that

the measured PLE spectrum resembles more the exciton continuum state than the ground state. The energetic difference between the Raman profiles and the PLE peak should actually be lowered by the exciton binding energy. A smaller energy difference in our calculations results in a less asymmetric Raman profile, as we find experimentally.

The energy shift $\Delta = 4$ meV for different polarizations of the incoming light can be explained by an incoming resonance with the two different optical transitions in the monolayer. Due to the symmetry of the electronic states x' polarized light interacts mainly with the E1-HH1 exciton, z polarized light with the E1-LH1 exciton of the monolayer [3]. Fröhlich-induced LO scattering then is allowed for $y'(x', x')\bar{y}'$ and $y'(z, z)\bar{y}'$ configuration in resonance with the E1-HH1 (lower excitation energy) and the E1-LH1 transition (higher excitation energy), respectively. In cross-polarized configuration mixing of the heavy and light-hole states can lead to Fröhlich-induced scattering involving defects [4]. Taking into account the energy separation between the heavy-hole and light-hole state, we calculated the $y'(z, z)\bar{y}'$ Raman profile in Fig.2 with the same parameters as given above, but with the center E_i of the distribution at $E_i = 1.448$ eV. This difference between the heavy- and the light-hole state (4 meV) obtained by fitting the Raman profiles agrees well with the value determined by luminescence experiments.[3]

5 Conclusion

In conclusion, we studied Raman scattering by the GaAs LO phonons in an InAs/GaAs monolayer structure in resonance with the InAs optical transitions. The Raman profiles were found to depend on the polarization of the incoming light due to incoming resonance with the E1-HH1 and the E1-LH1 transition. This is consistent with the symmetry of the electronic states in the monolayer predicted by group theory. We were able to reproduce well the measured Raman profiles and their peak energies together with PLE spectra by a calculation taking into account the influence of the exciton lifetime on the scattering intensity. Compared to the superlattice studied by Shields [2] our monolayer shows larger well-width fluctuations as evidenced by the larger Γ_1 which we obtained from our fits to the data.

6 Acknowledgements

We thank A. Fainstein for helpful discussions, F. Heinrichsdorff and D. Bimberg for fabrication of the sample. This work was supported by the Deutsche Forschungsgemeinschaft in the framework of Sfb 296.

References

1. B. Jusserand and M. Cardona, in *Light Scattering in Solids V*, ed. by M. Cardona and G. Güntherodt (Springer, Berlin 1989)
2. A. Shields, M. Cardona, R. Nötzel, and K. Ploog, Phys. Rev. B **46**, (1992) 10490.
3. A. R. Goñi, A. Cantarero, H. Scheel, S. Reich, C. Thomsen, P. V. Santos, F. Heinrichsdorff, and D. Bimberg, Solid State Comm.**116**, (2000) 121.
4. M. Cardona and C. Trallero-Giner, Phys. Rev. B **43**, (1991) 9959.

Light-exciton coupling in semiconductor microcavities of cylindrical and spherical symmetry

R.A. Abram[1], S. Brand[1], M.A. Kaliteevski[1], V.V. Nikolaev[2], M.V. Maximov[2], N.N. Ledentsov[2], C.M. Sotomayor Torres[3] A.V. Kavokin[4]

[1] Department of Physics, University of Durham, South Road, Durham, DH1 3LE, UK

[2] Ioffe Physicotechnical Institut of Russian Academy of Science, 26 Polytechnicheskaya, St. Petersburg, Russia

[3] Institute of Materials Science and Department of Electrical Engineering, University of Wuppertal, Gauss-Strasse 20, 42097 Wuppertal, Germany

[4] Université Blaise Pascal Clermont-Ferrand II, Complexe Scientifique des Cézeaux, 24 Avenue des Landais, 63177 Aubière, France

Abstract Exciton-light coupling in a cylindrical microcavity based on a cylindrical Bragg reflector and containing a quantum wire has been treated by using classical electromagnetism and a non-local dielectric response model for the exciton. It is shown that a converging cylindrical electromagnetic wave can be fully absorbed by a quantum wire exciton when the radiative and nonradiative damping of the exciton are equal. In addition, absorption spectra for a particular, simple cylindrical structure are calculated and are shown to exhibit a transition between the weak and strong coupling regimes. Approximate formulae for the frequencies of the TE and TM electromagnetic modes of a spherical microcavity formed by a spherical Bragg reflector have also been obtained, and provide the basis of a theory of exciton-light coupling in spherical structures.

1 Cylindrical structure

The structure that has been studied is an infinitely long circular cylindrical microcavity formed by a central cylinder with refractive index n_0, and radius ρ_0 of the order of the wavelength of light, surrounded by a cylindrical Bragg reflector [1] with layers of refractive indices n_1 and n_2. A relatively thin quantum wire of circular cross section is located at the centre of the structure. We consider the polariton states of the system when an electromagnetic mode of the cavity with no axial or azimuthal variation, and with its electric vector parallel to the cylindrical axis, is coupled to an exciton in the quantum wire. The equation defining the frequencies of the polariton states may be obtained once the frequency-dependent reflection coefficients of the quantum wire and Bragg mirror are known [2]. The latter has recently been derived by Kaliteevski et al. [1]. The reflection coefficient r_{QW} of the quantum wire can be obtained by solving Maxwell's equations with the inclusion of an excitonic contribution to the dielectric polarization, which can be described by a nonlocal dielectric susceptibility [2]. The result is

$$r_{QW} = 1 + \frac{2i\Gamma_0}{\omega_{ex} - \omega - i(\Gamma + \Gamma_0)} \qquad (1)$$

where Γ_0 and Γ are respectively the radiative and nonradiative damping factors and ω_{ex} is the renormalized exciton frequency. It is interesting to note that $|r_{QW}| = 1$

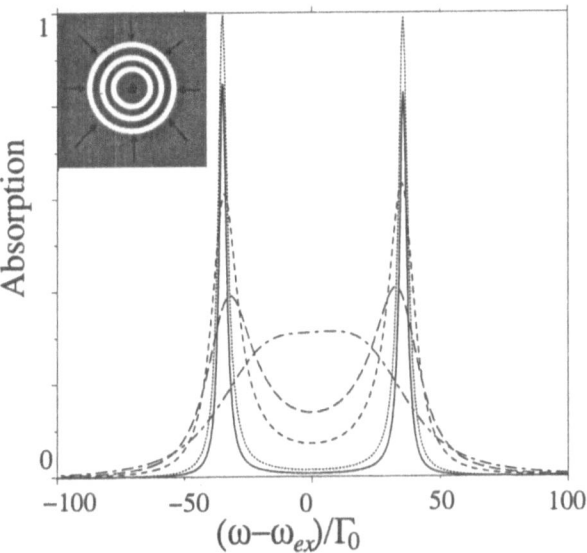

Fig. 1 Absorption spectrum of a cylindrical microcavity with a quantum wire. The refractive indices are 3.0 for the grey layers and 1.0 for the white layers. The reflector layer thicknesses are a quarter wavelength at frequency ω_{ex}. The core radius ρ_0 is chosen to tune an optical mode to the exciton frequency and is given by $\rho_0\omega_{ex}/2\pi c = 0.37015$. The radiative damping factor is $\Gamma_0 = 10^{-4}\omega_{ex}$. Solid line, $\Gamma = \Gamma_0$; dotted, $\Gamma = 2\Gamma_0$; short-dashed, $\Gamma = 10\Gamma_0$; long-dashed, $\Gamma = 20\Gamma_0$; dashed-dotted, $\Gamma = 50\Gamma_0$.

for all ω when $\Gamma = 0$ and $r_{QW} = 0$ when $\omega = \omega_0$ and $\Gamma = \Gamma_0$. This behaviour is different from that of an exciton in a quantum well [3], and can be attributed to the transition from incident converging wave to reflected diverging wave which occurs in the cylindrical structure under consideration here.

The polariton frequencies ω_{em} are given by [2]

$$\omega_{em} = 0.5(\omega_{0j} + \omega_{ex}) \pm 0.5\sqrt{(\omega_{0j} - \omega_{ex})^2 + 4\Delta^2} \qquad (2)$$

where ω_{0j} is the optical mode frequency closest to the exciton resonance, and Δ is the Rabi splitting given by

$$\Delta = 2\sqrt{(2\Gamma_0\omega_b c)/(bc + 2n_0\rho_0\omega_b)} \qquad (3)$$

in terms of the Bragg frequency of the reflector ω_b, and the quantity $b = \pi n_1 n_2/n_0(n_2 - n_1)$.

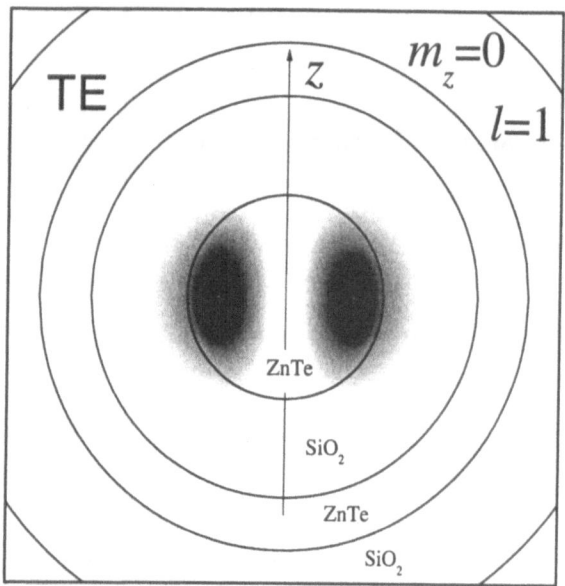

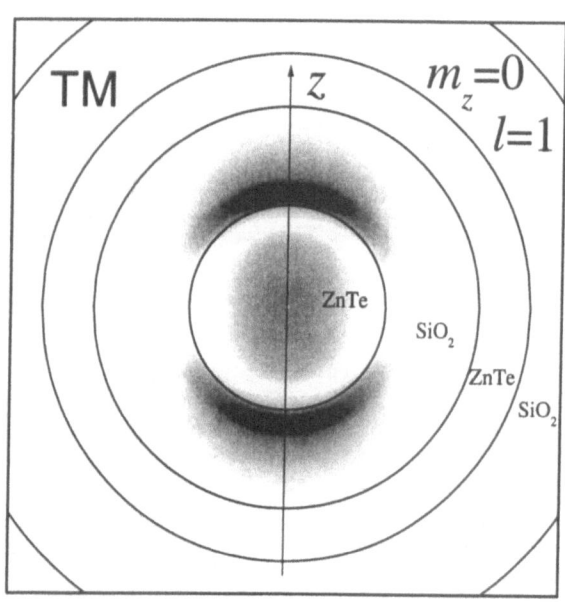

Fig. 2 The field profiles for the $l = 1$ and $m_z = 0$ TE mode (with $\omega = 1.1195\omega_b$) and TM mode (with $\omega = 0.8785\omega_b$) in a cavity of core radius obeying the relation $\rho_0\omega_b/2\pi c = 0.15$

Fig. 1 shows absorption spectra for a cylindrical microcavity with a quantum wire. The spectral postions of the absorption peaks correspond to the modes of the polariton system. If the exciton broadening is much smaller than the Rabi splitting Δ, the spectra exhibit two peaks with frequencies that are quite insensitive to the value of Γ, but widths and amplitudes that vary significantly. The existence of two peaks is evidence of the strong-coupling regime [4]. When $\Gamma \approx \Delta$, the weak-coupling

regime occurs and the two peaks merge into one, centred at the exciton frequency.

2 Spherical structure

Now consider a spherical microcavity with core of radius ρ_0 and refractive index n_0 surrounded by a spherical Bragg reflector with layers of refractives indices n_1 and n_2. In such a spherically symmetric structure, the electromagnetic field for $\omega \neq 0$ can be represented by a linear combination of the two independent polarizations (TE and TM) of spherical waves (multipole fields) [5][6] of different $l \geq 1$ and $-l \leq m_z \leq +l$. The frequencies of the eigenmodes of an empty cavity satisfy the equation

$$h_l^{(2)}(k\rho_0) = Rh_l^{(1)}(k\rho_0) \tag{4}$$

where $h_l^{(1)}$ and $h_l^{(2)}$ are spherical Bessel functions, $k = n_0\omega/c$, and R is the amplitude reflection coefficient of the Bragg mirror. For the TE modes, an approximate solution of eq. 4 is

$$\omega = \left\{ \frac{b + \pi(2N + l + 1)}{b + 2n_0\rho_0(\omega_b/c)} \right\} \omega_b \tag{5}$$

where N is an integer and the b and ω_b are as defined in sec. 1. The corresponding expression for the TM modes is

$$\omega = \left\{ \frac{b + \pi(2N + l)}{b + 2n_0\rho_0(\omega_b/c)} \right\} \omega_b \tag{6}$$

Fig. 2 illustrates the spatial dependence of the squared magnitude of the electric field for the $l = 1$ and $m_z = 0$ TE and TM modes of a particular cavity with a ZnTe core and a Bragg reflector composed of ZnTe ($n = 2.7$) and SiO$_2$ ($n = 1.45$) layers. For the TE mode, the electric field vanishes at the centre of the cavity, which is also the case for any value of l. In contrast, the electric field of the TM mode does not vanish at the centre, and for the mode shown is parallel to the z-axis. Therefore the TM mode can couple to excitons in a quantum dot placed at the centre of the structure, and the coupling can be analysed using the methods referred to in sec. 1 for the cylindrical case.

References

1. M.A. Kaliteevski, R.A. Abram, V.V. Nikolaev, and G.S. Sokolovski, J. Mod. Optics **46**, (1999) 875-890.
2. M.A. Kaliteevski, S. Brand, R.A. Abram, V.V. Nikolaev, M.V. Maximov, N.N. Ledentsov, C.M. Sotomayor Torres, and A.V. Kavokin, Phys. Rev. B **61**, (2000) 13791-13797.
3. L.C. Andreani, G. Panzarini, A.V. Kavokin, and M.R. Vladimirova, Phys. Rev. B **57**, (1998) 4670-4680.
4. C. Weisbuch, M. Hishioka, A. Ishikava, and Y. Arakawa, Phys. Rev. Lett. **69**, (1992) 3314-3317
5. W.K.H. Panofsky and M. Philips, *Classical Electricity and Magnetism* (Addison-Wesley, Reading Massachusetts, 1962)
6. J.D. Jackson, *Classical Electrodynamics* (Wiley, New York, 1999)

Near-field spectroscopy of delocalized excitons in single quantum wires

F. Intonti[1], V. Emiliani[1*], Ch. Lienau[1], T. Elsaesser[1], R. Nötzel[2], K. H. Ploog[2]

[1] Max-Born-Institut, Max-Born-Str. 2a, D-12489 Berlin, Germany, e-mail: intonti@mbi-berlin.de
[2] Paul-Drude-Institut für Festkörperelektronik, D-10117 Berlin, Germany

Quasi-one-dimensional (Q1D) nanostructures such as semiconductor quantum wires have recently attracted considerable interest because of, e.g., their predicted sharp density of states, enhanced carrier mobilities, or their potential for improved performance over quantum well lasers. A necessary condition for exploring and exploiting these quantum effects is to fabricate nanostructures of high-structural quality that avoid localization effects, i.e. contain Q1D excitons delocalized over mesoscopic length scales. Disorder effects counteract this exciton delocalization. In fact, spatially resolved PL studies on different QWR structures [1,2,3] did not show evidences of delocalized excitonic states, but suggest an optical behavior of QWR's arising from closely spaced localized excitons making these QWRs similar to a chain of quantum dots.

We report near field PL images of excitons confined in a single GaAs QWR. By combining high spatial and spectral resolution, we resolve the characteristic emission features of excitons in a complex quasi-one-dimensional disorder potential and demonstrate the coexistence of single 0D localized excitons and of excitons that are delocalized over a length of up to several microns along the QWR.

A novel coupled QWR-Dot nanostructure grown by molecular beam epitaxy of GaAs/(AlGa)As QW structures on patterned GaAs(311)A substrates [4] is studied by near-field photoluminescence (PL) spectroscopy. Near-field spectra are recorded at 10K in an illumination/collection geometry with a combined spatial/spectral resolution of 150 nm and 100 μeV, respectively. With near-field resolution, the broad QWR PL band that is observed in far-field spectra, breaks up into a set of sharp and intense emission peaks superimposed on a broad continuum (Fig. 1a). The near-field PL spectrum was recorded at a fixed position on the QWR.

Two-dimensional near-field images, recorded at the spectral position

of the sharp lines, (Fig. 1b) show that this kind of luminescence stems from regions resolution-limited in size. Thus, they are attributed to the emission of localized excitons out of local potential minima that arise from monolayer-height fluctuations of the QWR profile. When the detection energy is tuned to the low energy side of the broad continuum (out of the sharp resonances), the PL spatial distribution reveals the existence of regions with a larger average extension of 400-600 nm in extension (Fig. 1c). For detection energies in the high energy part of the emission band, a completely different situation appears. Here, the two-dimensional images (Fig. 1d) indicate an uniform PL distribution **delocalized** along the QWR axis on a length scale of more than 2 μm. An even larger extension is measured in the region where the QWRs are not interrupted by dots (not shown).

Delocalized PL components are observed for a broad range of detection energies. The corresponding spectral dependence is shown in Fig. 1(a) by the dotted curve. The shape of this spectrum is close to a Gaussian centered at 1.6740 eV with a width of 10 meV. These spectra are representative for different regions of the

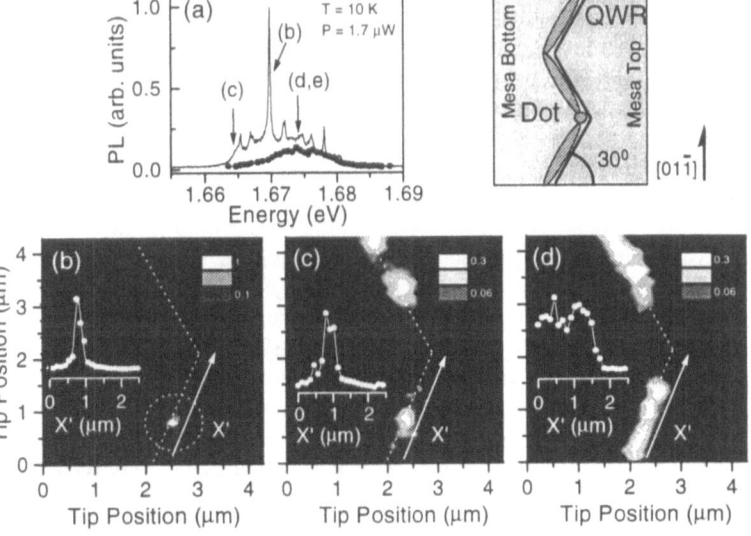

Fig 1. a) High spatial/spectral resolution near-field PL spectrum of the QWR at 10 K and schematic of the coupled QWR-dot structure. Two-dimensional near-field PL image recorded at (b) 1.6698 eV (localized exciton), (c) 1.6643 eV (weakly localized exciton) and at (d) 1.674 eV (delocalized exciton), respectively. The insets in (b) - (d) give cross-sections along the QWR axis X′.

sample. The delocalized component is present even at very low excitation densities where only a minor fraction of the available exciton states is populated and - thus - nonlinear effects of state and/or band filling can safely be neglected.

The occurrence of the three PL components clearly demonstrates that disorder induced local variations of the quasi-one-dimensional confinement potential of the QWR and the interplay between localized and delocalized states play a central role for the electronic properties of this type of QWR structure. We performed a theoretical study of the excitonic spectra in order to develop a quantitative understanding of the disordered QWR structure. In this model, we consider a QWR oriented along the x axis with a lateral confinement potential $V(y)$ of a width of 70 nm and a confinement energy of 55 meV, superimposed by a randomly generated Gaussian correlated potential of width $\sigma = 5$ meV and correlation length $L_c = 20$ nm (Fig. 2a). As in the experiments, we find three distinctly different types of 1s exciton states in the QWR: (i) excitons localized in a single potential minimum (Fig. 2b). These localized excitons couple strongly to light and give rise to sharp emission peaks in the near-field spectra. The axial density of these localized excitons is about $1/4L_c$, and they are distributed over an energy range of roughly 2σ. (ii) Several eigenstates, at low transition energies, with wavefunctions that extend over few local potential minina, showing extensions between 50 and 200 nm and a reduced coupling to light (Fig. 2c). (iii) A large density of about 10 states/(μm*meV) that are delocalized along the QWR with wavefunctions of highly complex shape and extensions between several 100 nm and several μm (Fig. 2d). In the optical spectra, these delocalized states give rise to a broad distribution of densely spaced weak resonances. If the linewidth of the individual resonances is broader than their energy separation, individual transitions are not resolved and broad structureless bands occur in the absorption and - for sufficient thermal population of emitting states - also in the PL spectrum. We attribute the spectrally broad and spatially delocalized luminescence in our near-field measurements (Fig. 1) to those delocalized excitons.

In Fig. 2 (e), the overall optical spectrum calculated from the model is summarized. We find a broad quasi-continuous contribution from delocalized states, superimposed by individual sharp resonances from localized excitons. Both contributions occur in a similar range of transition energies. The overall spectral width of the measured delocalized PL component (points in Fig. 2a) is well reproduced if we a assume a thermalized exciton gas at a temperature of $T_E \sim 30\text{-}40$ K, i.e. somewhat higher than the lattice temperature of nominally 10 K. The elevated exciton temperature is due to

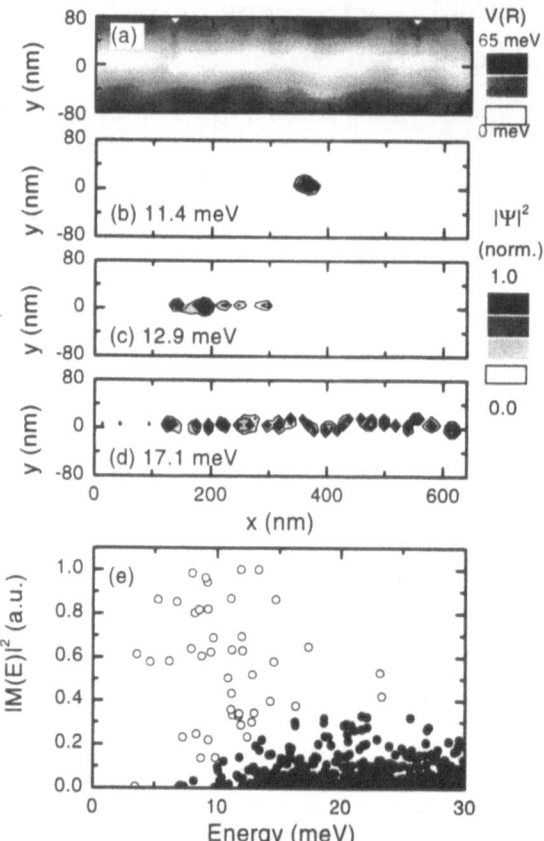

Fig. 2. Potential distribution $V(R=(x,y))$ of the exciton COM potential of a disorder QWR along (x) and perpendicular (y) to the QWR axis. Probability densities of representative (b) localized exciton wavefunction, $E_X=11.4$ meV, (c) weakly delocalized excitons, $E_X = 12.9$ meV, and (d) delocalized excitons, $E_X = 17.1$ meV. (e) Absorption spectrum calculated for a 2 μm long QWR. The different contributions from localized (open circles) and delocalized excitons (closed circles) are highlighted.

the nonresonant excitation with excess energies of several 10 meV.

In summary, highly spatially and spectrally resolved near-field PL spectroscopy allowed us to analyze in detail the electronic properties of a a single GaAs quantum wire. We experimentally and theoretically demonstrate the coexistence of localized and delocalized quasi-one-dimensional excitons and provide the basis for further, time-resolved studies of the carrier dynamics in such novel nanostructures.

References
1. J. Hasen et al., Nature **390**, 54 (1997).
2. J. Bellessa et al., Appl. Phys. Lett. **77**, 2481, (1997).
3. F. Vouilloz et al., Solid State Comm. **108**, 945 (1998).
4. J. Fricke et al., J. Appl. Phys. **85**, 3576 (1999).

Proc. 25th Int. Conf. Phys. Semicond., Osaka 2000 (Eds. N. Miura and T. Ando)

703

Scanning Near-Field Optical Spectroscopy of Buried Semiconductor Heterostructures

M. Hauert[1], R. Roshan[1], A.C. Maciel[1], J. Kim[1], J.F. Ryan[1], A. Schwarz[2], A. Kaluza[2], Th. Schäpers[2] and H. Lüth[2]

[1] Clarendon Laboratory, Department of Physics, University of Oxford, Oxford OX1 3PU, UK
[2] Institut für Schicht- und Ionentechnik, Forschungszentrum Jülich, 52425 Jülich, Germany

Abstract We present the first low-temperature scanning near-field optical microscopy cross-section images of v-groove quantum wire structures. Images with a resolution of ∼ 200 nm are obtained and spatial mapping of the photoluminescence provides direct identification of the spectral features.

1 Introduction

Scanning near-field optical microscopy (SNOM) has been developed as a technique for studying sub-wavelength sized objects. In this paper we describe its application to buried semiconductor hetrostructures, and in particular to v-groove quantum wire (QWR) structures. They are made by growing a quantum well (QW) on a (001) surface pre-patterned by (110) v-groove channels defined by (111) facets. Growth of the barrier and QW materials results in the formation of crescent-shaped QWRs at the bottom of the grooves, (111) sidewall quantum wells (SQWs), top quantum wells (TQW) on the flat (001) surfaces and a vertical quantum well (VQW) in the centre of the v-groove [1]. Photoluminescence (PL) has been used extensively to characterize v-groove QWRs, but the presence of the strongly interconnected quantum structures makes it difficult to interpret the spectra. In this paper we present submicron PL images obtained with a low-temperature near-field optical microscope, which allow one-to-one association of the spectral lines with the actual features of the structure.

2 Experimental details

The samples studied here contain vertically stacked GaAs/GaAlAs QWRs (for details of the sample growth see [2]). Experiments were performed at 10 K using a low-temperature scanning near-field optical microscope (LT-SNOM) in which the sample is attached to the cold finger of a continuous-flow liquid helium cryostat, while the SNOM scan head remains at room temperature [3]. PL images were acquired by photoexcitation at 514.5 nm through a nanometer-sized aperture in a tapered optical fiber tip while scanning it over the sample, the luminescence being collected in the far field.

3 Results

Fig. 1 presents PL spectra obtained from a sample containing an array of 30 verically stacked multiple QWR layers. The patterned substrate containing an array of v-grooves separated by 3 μm was first overgrown by a 50 nm Al_3Ga_7As buffer layer, followed by a 20-period, 2 nm GaAs/2 nm AlAs superlattice (SL), and finally a 250

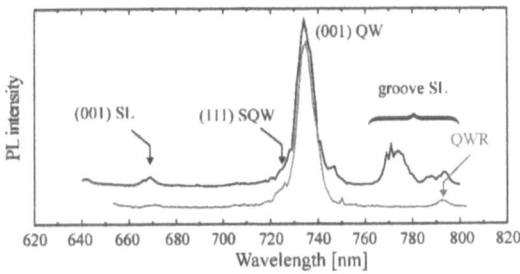

Fig. 1 Photoluminescence spectra of the stacked QWR array with the tip positioned close to the sample (001) surface (red curve) and further away from the surface (black curve).

nm Al_3Ga_7As layer. The QWRs were then produced by repeating the growth sequence: 3 nm GaAs, 100 nm Al_3Ga_7As. Due to the large amount of material deposited, the grooves fill and the surface becomes nearly planar. In order to probe the buried heterostructure the sample was cleaved perpendicular to the v-groove channels. When the SNOM tip is positioned close to the (001) surface, the PL spectrum contains two main features (red curve): a peak at 793 nm and a strong peak at 734 nm with a high energy shoulder. When the tip is positioned further below the (001) surface, additional features appear between 760 and 800 nm, and a peak at 670 nm (black curve).

The peak at 734 nm can be assigned to the 3 nm GaAs TQW, on the basis of theoretical estimates and

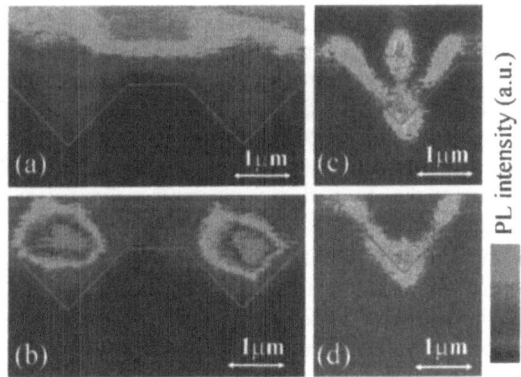

Fig. 2 Low temperature near-field optical microscope images with the detection wavelength set at 734 nm (a), 722 nm (b), 793 nm (c) and 788 nm (d), which correspond to the TQW, SQW, QWR + SL and SL respectively. The etched substrate is defined by the red line.

measurements of unpatterned regions of the sample. The corresponding PL mapping is shown in Fig. 2(a). It can be seen that the signal becomes much brighter when the tip approaches the (001) surface where only TQWs are present due to planarisation of the surface. Single quantum wells cannot be resolved, due mainly to carrier diffusion: even when the tip is positioned away from the QW, carriers photoexcited in the $Al_{.3}Ga_{.7}As$ barriers are captured into the wells and produce a strong signal at the TQW wavelength.

In Fig. 2(b), the luminescence is maximized when the tip lies on the groove channels, which is consistent with emission from the SQWs. Indeed, luminescence from the (111) SQWs is expected to be at higher energy than that of (001) QWs due to their reduced width.

The low energy region of the luminescence spectrum from v-groove structures arises from the quantum wires. However, the sample under consideration contains the added complication of a GaAs/AlAs superlattice which also gives rise to QWR states [4]. The SNOM images clearly resolve these features. The red curve in Fig. 1 is the luminescence from the QWR region, including QWR, TQW and SQW, whereas the black curve includes the SL luminescence: the peak at 670 nm corresponds to the luminescence from the SL (predominantly on the (001) surfaces) whereas the broad feature between 760 nm and 800 nm is due to the SL QWRs grown inside the v-channels. This is clearly indicated by Fig. 2(c) and (d), acquired at 793 nm and 788 nm respectively. The v-shaped luminescent region persists over a broad wavelength range, whereas luminescence from the QWRs is only apparent in Fig. 2(c). In this image, single QWRs cannot be resolved in the growth direction due to their small separation, but the luminescent region extends over $\sim 1\,\mu m$, and includes ~ 10 vertically stacked QWRs. The gap between the SL and QWR luminescence at the bottom of the grooves is due to the 250 nm AlGaAs layer grown on top of the SL.

The spatial resolution of the above experiment is given by the width of the luminescence intensity across the QWRs which is approximately 200 nm, in good agreement with the estimated probe aperture. Diffusion from the barrier into the QWRs, and carrier trapping from the SQWs thus seem to be negligible. Trapping from the SQWs in particular is expected to be inefficient in these structures due to a strong pinch-off region separating QWR and SQW [5].

Individual QWRs have been spatially resolved in a second sample in which the SL was omitted, and the QWR separation increased. The image presented in Fig. 3 shows wires that are resolution limited in the lateral direction. In the vertical (growth) direction however, the width of the PL is considerably larger, which is due to the diffusion of carriers from the VQW into the QWRs.

The PL intensity in the SNOM image is a convolution of the effective tip aperture and the wire width/ spatial distribution of carriers contributing to the QWR

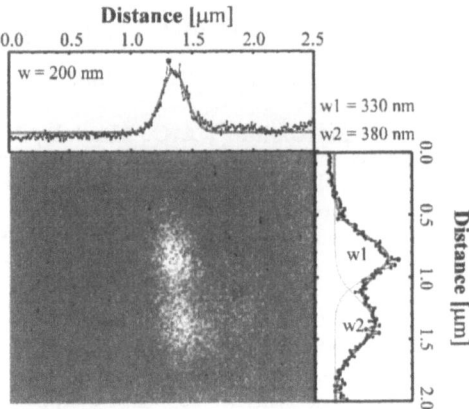

Fig. 3 2.5 $\mu m \times$ 2 μm scan with detection at the quantum wire wavelength. Gaussian fits of the experimental points and their respective widths w are indicated on the figure.

luminescence. The wire width in the vertical direction is about 10 nm and can be neglected; the tip aperture and the spatial distribution of carriers can be approximated by Gaussians of widths 200 nm and L_d respectively, where L_d is the diffusion length. From the Gaussian fits of the PL image, we deduce the diffusion length for the two QWRs: $L_{d1} = 260$ nm and $L_{d2} = 320$ nm. These values can be compared with the estimated diffusion length $L_d = \sqrt{D\tau}$, where D is the ambipolar diffusion coefficient and τ the carrier lifetime in the VQW. Using reasonable values of $D \simeq 10$ cm^2/sec and $\tau \simeq 100$ ps, we obtain $L_d \simeq 316$ nm, which is in good agreement with the values deduced from the SNOM image.

4 Conclusion

We have presented the first cross-sectional LT-SNOM images of v-groove quantum wire structures and have succeeded in studying single QWRs. We have obtained clear evidence of the important role of carrier capture from the vertical quantum wells, whereas carrier trapping from the barrier and the sidewalls quantum wells into the quantum wires is negligible at low temperatures.

5 Acknowledgements

This work was supported by the EPSRC (UK), and the EC through the ULTRAFAST network. We are grateful to Ch. Lienau (MBI, Berlin) for much useful help and advice on the LT-SNOM.

References

1. E. Kapon, D.M. Hwang and R. Bhat, Phys. Rev. Lett. **63**, (1989) 430
2. Th. Schäpers et al., Appl. Surf. Sci. **123/24**, (1997) 687
3. G. Behme, A. Richter, M. Süptitz and Ch. Lienau, Rev. Sci. Instrum. **68**, (1997) 3458
4. C. Kiener et al., Appl. Phys. Lett. **67**, (1995) 2851
5. R. Roshan et al., Proceedings of ICPS 25 (to be published).

Proc. 25th Int. Conf. Phys. Semicond., Osaka 2000 (Eds. N. Miura and T. Ando)

705

Hints for a Non-thermal Distribution of Excitons in CdSe/ZnSe quantum islands

G. von Freymann, E. Kurtz, C. Klingshirn, Th. Schimmel, M. Wegener

Institut für Angewandte Physik, Universität Karlsruhe (TH), 76128 Karlsruhe, Germany

Abstract Autocorrelation spectroscopy on the basis of thousands of individual near-field photoluminescence spectra of single ultrathin CdSe layers at low temperatures exhibits a strong positive correlation peak around 18 meV energy with a width of 5 meV. Using simulations and experiments as a function of temperature and laser intensity, we can exclude interpretations along the lines of biexcitons or phonon sidebands. We attribute this feature to the splitting of ground state and an excited state in individual quantum islands. This interpretation implies that the potential minima are rather uniform in size and that the distribution of excitons is non-thermal.

1 Introduction

Epitaxially grown, thin semiconductor layers often lead to a series of sharp photoluminescence lines in near-field or micro-photoluminescence experiments[1–6]. These individual spectra can exhibit many tens or even hundreds of individual lines and usually look like a random distribution at first sight. Autocorrelation spectroscopy allows to extract more information by focusing on statistically significant features in all spectra: calculation of the autocorrelation for each spectrum and subsequent averaging over all autocorrelations allows for global statements about the potential landscape. Correlations in the spectrum can arise from correlations in the potential or from a potential which is just noise. The first case would lead to positive correlations the latter to negative correlations, i.e. to the so called level repulsion[7,8].

Recent experiments with ultrathin CdSe/ZnSe samples have shown a strong positive correlation feature around 18 meV energy with a width of 5 meV[9,10]. In this article, we review the application of the method called autocorrelation spectroscopy on huge datasets and then briefly discuss the origin of the observed correlation feature.

2 Experiment

For the optical experiments we frequency-double about 1 W power at 800 nm wavelength from a continuous wave Ti:sapphire laser in a 2 mm thick BBO crystal to obtain about 3 μW power at 400 nm wavelength (3.10 eV photon energy). The excitation power can be continuously attenuated using a sequence of two polarizers. In the parallel polarized position, 200-300 nW are effectively coupled into an optical fiber. The light propagates towards a nanometer scale apex which is formed by a two-step selective etching process. Electron micrographs of identical fiber tips have been shown in [11]. The samples are mounted in a continuous-flow-cryostat and can be temperature stabilized in a range from 4.2 up to 310 K. The

photoluminescence is collected with the same fiber and sent into a 0.5 m grating spectrometer (1800 lines/mm grating), which is connected to a liquid nitrogen cooled, back-illuminated charge-coupled-device (CCD) camera. The spectral resolution of this system is 0.06 nm wavelength, which is equivalent to an energy resolution of 250 μeV in the spectral range investigated. We scan the uncoated fiber tip in a constant height of 100 nm above the surface, i.e. the feedback loop is inactive. Under these conditions, the measured spatial resolution is better than $\lambda/5$ of the photoluminescence wavelength (540 nm)[10]. The data sets shown in the following are always based on 1600 individual spectra, each with 0.5 s exposure time of the CCD-camera, taken in a 2 μm × 2 μm area with 50 nm (100 nm) separation between adjacent points. For the intensity-dependent measurements, only 400 spectra from the same area are collected with exposure times up to 10 s. We have carefully adjusted the exposure time for each measurement in order to maintain the same signal-to-noise ratio. The four samples A, B, C, and D are nominally ultrathin CdSe layers (2–3.6 ML) clad between ZnSe barriers. They are described in more detail in Ref. [9] and the references therein. The individual photoluminescence spectra of these samples look very different in terms of line density and width of the line distribution. Also, they are shifted in the overall spectral position $\hbar\omega_0$ (left hand side of Fig. 1). However, the correlation feature found by autocorrelation spectroscopy[9,10] is very similar for many different measurements on samples A, B, C, and D (solid lines in the right hand side of Fig. 1). Obviously, the correlations in the photoluminescence lines of these CdSe/ZnSe layers are generic for this particular material system.

3 Discussion

These correlations are consistent with spectra which consist of sets of pairs of lines. For each pair the energy spacing has to be around 18 meV. Can we learn more about the relative strength of the two lines in the pairs? For an interpretation along the lines of a phonon sideband or of biexcitonic emission we would like to know whether the second peak lies on the low or on the high energy side of the first peak. To obtain this information, we multiply the individual photoluminescence spectra with filter functions $f(\hbar\omega)$. Exponential functions, i.e. $f(\hbar\omega) \propto \exp(\hbar\omega/E_0)$, are especially simple to interpret. For the case of $E_0 < 0$ and for spectra which consist of pairs of lines with one main line and a weaker low-energy sideband, the sideband would be enhanced by a

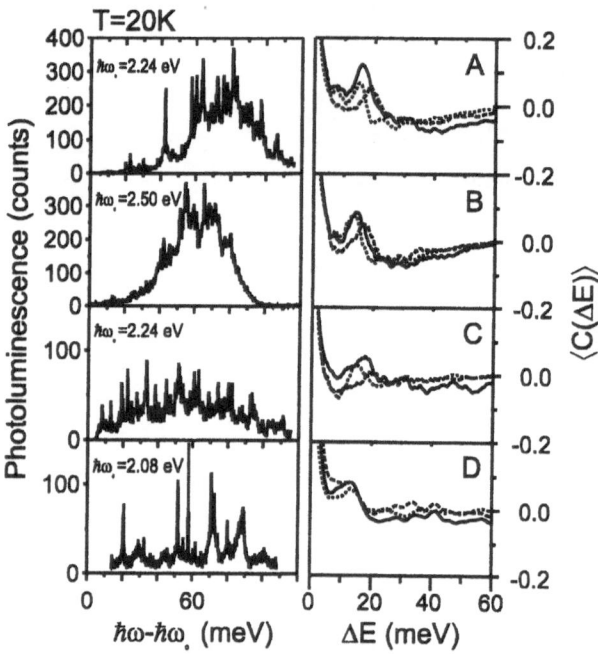

Fig. 1 One individual spectrum of each sample A, B, C, and D is shown on the left hand side. The averaged auto-correlations for these data are displayed on the right hand side: raw data (solid line), filtered data with $E_0 < 0$ (dashed line), filtered data with $E_0 > 0$ (dotted line). Filter function if $f(\hbar\omega) \propto \exp(\hbar\omega/E_0)$.

certain factor (independent on its absolute energetic position). Consequently, the correlation function $\langle C(\Delta E) \rangle$ would exhibit enhanced correlations. Similarly, a high-energy sideband would be suppressed for $E_0 < 0$ and vice versa for $E_0 > 0$. This idea has been verified explicitly by numerical simulation (not shown). Applying this procedure to the data shown in Fig. 1 on the left hand side, leads to the curves shown on the right hand side (full lines correspond to $E_0 = +18$ meV, dashed lines to $E_0 = -18$ meV). Obviously, the determined height of the correlation peak is roughly the same for $E_0 = \pm 18$ meV. This implies that we have pairs of lines with comparable strength – on the average. This can either mean that the two lines in the pairs have identical strength in each case or that they strongly fluctuate independently. By additional simulations we have found, that the latter scenario would lead to a correlation peak in $\langle C(\Delta E) \rangle$ which is definitely too weak to be consistent with the experiment. For lines of identical strength, our simulations give results which are consistent with the experiment.

Pairs of lines with comparable strength make an interpretation along the lines of a phonon sideband unlikely (the LO-phonon energy of bulk CdSe is 26 meV, that of ZnSe is 32 meV. The observed correlation energy is below 20 meV for each sample). Furthermore, phonon sidebands in this material system are not strong enough to explain the observed correlation feature ($I_{LO} = 0.04 I_X$, [12]). Biexcitonic effects could also lead to pairs of lines of comparable strength. A clear signature of biex-

citonic effects would be an intensity dependence. We have not found any intensity-dependence when attenuating the incident laser intensity by two orders of magnitude, i.e. from 200 nW down to 2 nW excitation power. However, it is interesting to note that the quantum dot biexciton binding energy of ≈ 20 meV observed in single dot experiments[12,13] is close to the numbers seen here. Islands of rather uniform size ('self-organized quantum dots') would lead to a rather uniform level spacing. Several researchers[3,14] have indeed reported islands in this material system with an areal density around 10^{11} cm^{-2} and with sizes of 5 to 10 nm. The internal motion of the electron and the hole within the exciton would lead to a further smearing-out, hence to a broadening, of the potential for the excitons[8] (for bulk CdSe, the exciton Bohr radius is 9 nm). The expected level spacing for such effective potential wells seems consistent with the correlation energy of 18 meV observed in our work[10] (the exciton mass of bulk CdSe is $1.3 \times m_0$, that of ZnSe $1.9 \times m_0$). Also, the number of lines in each spectrum is roughly consistent with an areal density of islands of 10^{11} cm^{-2}. If this interpretation concerning sets of pairs of lines from rather uniform islands is correct indeed, our observations would be evidence for a highly non-thermal distribution of excitons in these islands (due to the 'phonon bottle-neck').

4 Acknowledgements

This research has been supported by the DFG-SFB 195, the DFG-GK 284 and is performed within the Institut für Nanotechnologie der Universität Karlsruhe (TH). The research of M.W. is funded by the DFG Leibniz-Preis. We thank E. Runge, and R. Zimmermann for stimulating discussions.

References

1. H. F. Hess et al., Science **264**, 822 (1994)
2. A. Zrenner et al., Phys. Rev. Lett. **72**, 3382 (1994)
3. F. Flack et al., Phys. Rev. B **54**, 17312 (1996)
4. U. Jahn et al., Phys. Rev. B **56**, R 4387 (1997)
5. L. M. Robinson et al., Phys. Rev. Lett. **83**, 2797 (1999)
6. Q. Wu et al., Phys. Rev. Lett. **83**, 2652 (1999)
7. E. Runge et al., Phys. Stat. Sol. B **206**, 167 (1998)
8. E. Runge et al., Ann. Phys. **8**, 229 (1999)
9. G. von Freymann et al., Appl. Phys. Lett. **77**, 394 (2000)
10. G. von Freymann et al., Journal of Microscopy, in press (Jan/Feb 2001)
11. Ch. Adelmann et al., Appl. Phys. Lett. **74**, 179 (1999)
12. F. Gindele et al., Phys. Rev. B **60**, R 2157 (1999)
13. G. Bacher et al., Phys. Rev. Lett. **83**, 4417 (1999)
14. N. Peranio et al., Phys. Rev. B, submitted (2000)

Quantum optics and the observation of Electromagnetically Induced Transparency with QW subbands.

C. C. Phillips[a], E. Paspalakis[b], G. B. Serapiglia[a], C. Sirtori[c] and K. L.Vodopyanov[a]

[a] Experimental Solid State Group, Physics Department, Imperial College, London, SW7 2BZ, UK
[b] Optics Section, Physics Department, Imperial College, London, SW7 2BZ, UK
[c] LCR Thomson-CSF, Domaine de Corbeville,91404 Orsay Cedex, France
E.mail. chris.phillips@ic.ac.uk,

Abstract The phenomenon of electromagnetically-induced quantum coherence is demonstrated between 3 confined electron subband levels in a quantum well which are almost equally spaced in energy. Applying a strong coupling field, two-photon-resonant with the 1-3 intersubband transition, produces a pronounced narrow transparency feature in the 1-2 absorption line. This result can be understood in terms of all 3 states being simultaneously driven into "phase-locked" quantum coherence by a single coupling field. We describe the effect theoretically with a density matrix method and an adapted linear response theory.

1. Introductory Remarks.

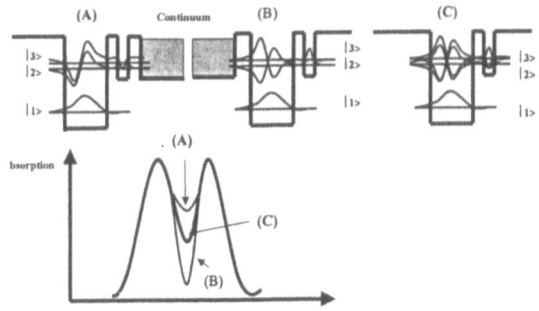

Fig.1. Fano effect. In (a) and (b) the quantum phases of states 2 and 3 are set by strong coupling to the same continuum, so that quantum interference increases (decreases) the absorption compared with the twin Lorentzian spectrum of case (C) where the states have no defined "coherence".

The calculation of an optical transition rate between two states ($|1>$ and $|2>$) of an electronic system normally only requires knowledge of the *amplitudes* of the state's wavefunctions and possibly the radiation field. This is because, although QM predicts many other scattering pathways involving other states of the system (e.g. $|3>$), these contributions are added coherently in the scattering rate calculation, and usually other non-optical scattering processes dephase these extra "coherences" making their time-averaged contribution tend to zero. An exception is the Fano effect (fig.1), recently demonstrated with QW subbands [1] where the QM phase relationship between e.g. states $|2>$ and $|3>$ is fixed by tunnel-coupling them sufficiently strongly

to the same continuum. Now the $|1>\rightarrow|2>$ and $|1>\rightarrow|2>\rightarrow|3>\rightarrow|2>$ pathways can interfere constructively (destructively) to increase (decrease) the absorption.

Quantum optical experiments typically use atoms in an intense "coupling" radiation field, ω_c to establish these coherences[2]. The coupling strength measured by the Rabi frequency, $\Omega_{Rabi} = \mu E/\hbar$, where E is the radiation field amplitude and μ the transition dipole. In Electromagnetically Induced Transparency (EIT) studies a three level system of well-defined parity is used, so there are two dipole-allowed transitions ($|1>\rightarrow|2>$ and $|2>\rightarrow|3>$) and a third ($|1>\rightarrow|3>$) dipole forbidden "Raman" transition. $\hbar\omega_c$ is tuned to ~ E_{23} (Autler Townes condition) to establish a "coherence" between $|2>$ and $|3>$. Absorption again becomes significantly affected by QM interference between $|1>\rightarrow|2>$ and $|1>\rightarrow|2>\rightarrow|3>\rightarrow|2>$) pathways whose probability amplitudes can, in the ideal case, destructively interfere to completely cancel the original absorption, measured with a weak probe beam, at ~E_{12}.

Practical EIT observations require transitions of well-defined energies (i.e. not bands), and need Ω_{Rabi} greater than the dephasing rates so that Rabi splittings can be resolved on a homogenously broadened transition. Previous studies have been only with in narrow-line atomic vapour systems [2,3], and defect levels in crystals [4] where the dephasing rates are low.

In contrast to other coherent effects in semiconductors (e.g. four wave mixing, photon echoes, exciton beating) the

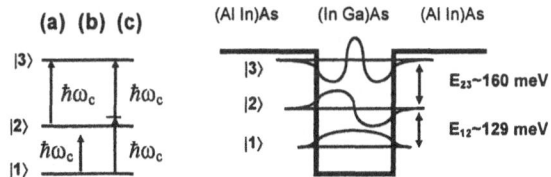

Fig.2 Schematic of Quantum well energy levels and coupling laser resonance conditions. a) "Autler-Townes" EIT , $\omega_c \sim E_{23}$, b) Strongly driven two level atom, $\omega_c \sim E_{12}$, c) "Phase-locked" coherence, $\omega_c \sim E_{13}/2$.

interference is essentially *between* the states of an *individual* oscillator, rather than between elements of ensemble of essentially 2-level oscillators, so it can in

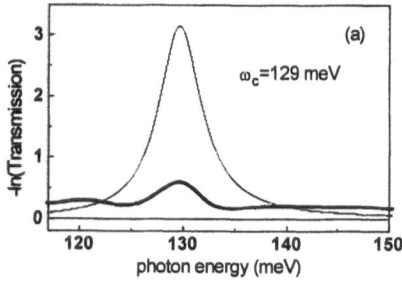

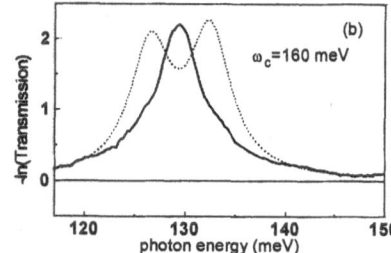

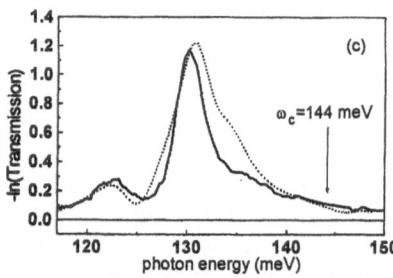

Figure 3. Measured (heavy solid lines) and calculated (dotted lines) absorption across the 1-2 resonance for different coupling laser frequencies (a) $\omega_c \sim E_{12}$, (b) $\omega_c \sim E_{23}$ (c) $\omega_c \sim E_{13}/2$. Fig 3a, light solid line, linear absorption spectrum.

principle be observed in a system containing only one oscillator

The coupling field significantly perturbs the electron Hamiltonian that determined the original "bare" electron states of a system, creating a coherent superposition of the original states, which strongly modifies macroscopic properties like photon absorption and spontaneous and stimulated emission. Once the "coupling" beam has created the necessary quantum coherences, both it and the probe beam (with $\hbar\omega_c \sim E_{12}$) can propagate without loss through a medium that would be otherwise highly opaque to the probe.

By allowing operation close to a transition resonance (where the material would otherwise absorb strongly), EIT enables the exploitation of highly efficient resonant non-linear optical processes [5]. Also, if e.g. the state |2> is populated incoherently, stimulated photon emission occurs at E_{12} without the corresponding absorption process being possible. This has led to predictions, and, in sodium vapour, recent demonstrations [6] of Lasing Without Inversion (LWI)

Although QW subbands have high dephasing rates ($\sim$100fsec^{-1}), compared with atoms they have the unique attraction that the transition energies, dipoles and even

symmetry properties can be engineered with considerable freedom and precision. Also, because of the light conduction band effective masses, QW intersubband dipole elements, and hence Rabi frequencies, can be large.

Here we study a novel three-level QW "ladder"-type system (fig.2), in which, for the first time, the states are close, compared with the transition linewidths, to being equally spaced in energy. In this case a single coupling field can simultaneously drive all three states into coherence with their quantum phases "locked" so as to produce an enhanced coherent transparency feature in the absorption spectrum. This allows us to observe EIT in a QW system and to demonstrate "phase-locked" quantum coherence, both for the first time.

2. Experimental Details.

The QW sample consists of forty symmetric 10 nm n-doped ($n_s = 6 \times 10^{11}$ cm^{-2}) In$_{0.47}$Ga$_{0.53}$As wells with 10nm Al$_{0.48}$In$_{0.52}$As barriers, lattice matched to an undoped InP substrate. Transition energies (matrix elements) were $E_{12} \sim 129$meV (2.34nm) and $E_{23} \sim 160$meV (2.64 nm). E_{12} was measured from the sample's linear absorption spectrum (fig.3a). The sample was polished into a 45° wedge to allow one double-pass of the beams through the QW's. Absolute transmission values, T, were obtained by ratioing spectra linearly polarised perpendicular and parallel to the QW normal since only the former couple to the ISB transitions. The interaction region was shorter than the wavelengths, rendering dispersion and propagation effects negligible. Independently tuneable ($\lambda \sim$ 6 - 12μm) synchronised coupling (ω_c) and probe (ω_p) $\sim$100psec laser pulses, were generated in separate Er^{3+}:Cr^{3+}:YSGG pumped Optical Parametric Generators [7]. The pulses were sufficiently narrow linewidth and long compared with the subband dephasing and relaxation times for the measurement to be effectively continuous-wave and monochromatic.

Effective coupling field intensities, of up to $\approx$ 2.6MW cm^{-2} were attained. All measurements were taken through a 300μm pinhole which was attached to the sample facet, and the probe was defocused to $\sim 10^{-3}$ of the coupling beam intensity with a 10° angular separation between the two. Linear absorption measurements (fig. 3a) gave a Lorentzian line centred at E_{12}=129meV with a homogenous linewidth, $\gamma_{12} \sim 5$meV at T=30 K sample.

Tuning the coupling energy to the lower transition, ($\hbar\omega_c = E_{12}$) simply saturates the|1> $\rightarrow$ |2> transition by exciting electrons into state |2> (fig. 3a).

Tuning ω_c the to E_{23} resonance, i.e. the classic "Autler-Townes" EIT condition, produced only small (save for the saturation caused by the partial populating of level |2>) modifications to the absorption spectrum (fig. 3b). In principle the ($\sim$4meV) Rabi frequency ($\sim$4meV) should have been enough to resolve Rabi splitting of the line here but in practice this is masked by spatial and temporal variations in the beam intensities.

The most striking result happens when the coupling beam frequency was tuned half-way between E_{12} and E_{23}, on the two-photon resonance for the $|1\rangle \rightarrow |3\rangle$ transition i.e. ($\hbar\omega_c \sim E_{13}/2$) (fig. 3c). A pronounced (67% reduction in absorption) dip suddenly appears in the absorption spectrum.

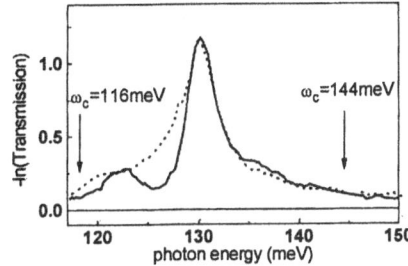

Figure 4. Comparison between measured absorption spectra with $\omega_c \sim E_{13}/2$ (heavy line) and with, $\omega_c = 116$ meV (light line) i.e at equal but opposite detunings from the E_{12} resonance.

At ~ 3.6 meV (FWHM) it is narrower than the original homogenous linewidth and it disappears if the coupling laser is detuned by an equal amount *below* E_{12} ($\hbar\omega_c = 116$ meV, fig.4). Its spectral narrowness and the non-obvious relationship between its position (126meV) and the coupling photon energy (144meV), both suggest an origin in quantum interference.

3. Theoretical Treatment.

The time evolution of the density matrix, using the rotating–wave and relaxation time approximations is

$$\dot\rho_{11} = \gamma_2\rho_{22} + i\Omega_{12}(\rho_{12} - \rho_{21})$$

$$\dot\rho_{22} = -\gamma_2\rho_{22} + \gamma_3\rho_{33} - i\Omega_{12}(\rho_{12} - \rho_{21}) + i\Omega_{23}(\rho_{23} - \rho_{32})$$

$$\dot\rho_{33} = -\gamma_3\rho_{33} - i\Omega_{23}(\rho_{23} - \rho_{32})$$

$$\dot\rho_{12} = \left(i\Delta_{21} - \frac{\gamma_{12}}{2}\right)\rho_{12} - i\Omega_{12}(\rho_{22} - \rho_{11}) + i\Omega_{23}\rho_{13}$$

$$\dot\rho_{13} = \left(i\Delta_{31} - \frac{\gamma_{13}}{2}\right)\rho_{13} - i\Omega_{12}\rho_{23} + i\Omega_{23}\rho_{12}$$

$$\dot\rho_{23} = \left(i\Delta_{32} - \frac{\gamma_{23}}{2}\right)\rho_{23} - i\Omega_{23}(\rho_{33} - \rho_{22}) - i\Omega_{12}\rho_{13}$$

where $\rho_{ij} = \rho_{ji}^*$, $\sum_{i=1}^{3}\rho_{ii} = 1$ for carrier conservation.and Ω_{ij} are the Rabi frequencies of the coupling laser field and the $|i\rangle \leftrightarrow |j\rangle$ transition. Note that, because both dipole transitions lie close to $\hbar\omega_c$ we allow for the simultaneous coupling of both to the coupling field, with important consequences. $\Delta_{21} = \hbar\omega_c - E_{12}$ and $\Delta_{32} = \hbar\omega_c - E_{23}$ are the

single photon detunings and $\Delta_{31} = \Delta_{21} + \Delta_{32}$ the two-photon detuning of the coupling laser frequency from the various electronic resonances.

The damping of the population states $|i\rangle$, γ_i are inversely proportional to the intersubband recombination times. The γ_{ij} ($i \neq j$) are the total coherence relaxation rates, given by $\gamma_{12} = (\gamma_2 + \gamma_{12}^{dph})$, $\gamma_{23} = (\gamma_2 + \gamma_3 + \gamma_{23}^{dph})$, $\gamma_{13} = (\gamma_3 + \gamma_{13}^{dph})$, where γ_{ij}^{dph} is the dephasing rate of the quantum coherence of the $|i\rangle \leftrightarrow |j\rangle$ transition. The γ_{ij}^{dph} here are determined by strong electron-electron, interface roughness and phonon scattering processes, and are a major obstacle to the observation of coherent effects such as EIT.

We solve these equations numerically in the steady state limit, using the method of ref [8] but adapted to coherently include absorption processes between both of the dipole allowed QW transitions.

The quantum regression theorem [8] is used to calculate the linear absorption spectrum as a function of ω_p. Measured intensities and transition dipoles give $\hbar\Omega_{12} \approx \hbar\Omega_{23} \approx$ 5meV. Dephasing rates, $\gamma_{12} = \gamma_{23} = \gamma_{13} = $ 5meV and $\gamma_2 \sim \gamma_3 = 1.3$ meV, are deduced from the $\tau = 1$ psec

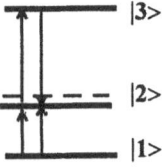

Figure 5. Schematic of the quantum interference at "two photon resonance". The dashed line represents the original "bare" energy of state 2, before being optically Stark shifted by the coupling laser.

intersubband relaxation time and other induced-absorption studies detailed in ref. [9].

With these parameters, when $\omega_c = E_{23}$ the theory predicts Autler-Townes splitting but, significantly, no coherent EIT. The theoretical spectrum is a superposition of 2 Lorentzian lines split by the Rabi effect, but there is no destructive interference at the line centre that would reduce the absorption below the split-Lorentzian lineshape.

In agreement with the experiment, theory predicts the strongest quantum interference effect when the coupling frequency is tuned to the "two-photon resonance", $\omega_c = E_{13}/2$ i.e. the "phase-locked" regime. This can be understood in terms of a four photon process (fig.5), where the absorption-emission path $1 \rightarrow 3 \rightarrow 2$ interferes destructively with the absorption path $1 \rightarrow 2$. At the same time the "coupling" field has Stark shifted the original "bare" $1 \rightarrow 2$ transition to the red and the $2 \rightarrow 3$ transition to the blue. This results in an overall red shift of the $1 \rightarrow 2$ transition to ~ 125meV, and a corresponding transparency feature at this energy. The coherence of all the states is strongly maintained by the "phase-locked" effect.

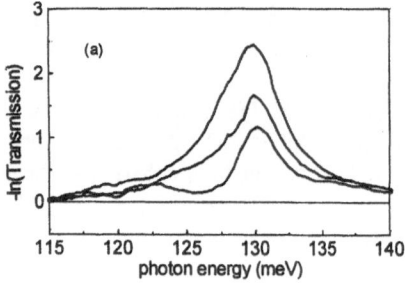

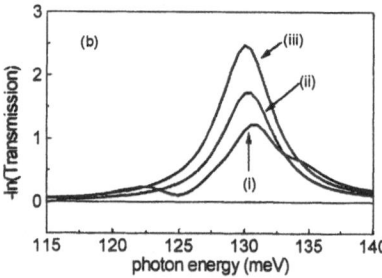

Figure 6. Experimental (a) and theoretical (b) coupling intensity dependence of the EIT feature for "two-photon resonance" case $\omega_c = E_{13}/2$ (i) Ω_{Rabi} =4.8meV (ii) Ω_{Rabi} =1.6 meV(iii) Ω_{Rabi} =0.48meV.

Also, at this ω_c, reducing the coupling Rabi frequency by as little as 20% destroys the interference, in good agreement with the intensity dependent measurements shown in fig. 6. Both of these facts suggest a strong dependence on two-photon coupling in the interpretation of the results.

The dominant inhomogenous broadening mechanism is conduction band non-parabolicity, [10] which contributes an inhomogenous linewidth component

$$\Delta_{ij}^{npb} = \hbar^2 k_F^2 / 2 M_{ij}$$
$$\text{where } M_{ij} = [m_i^{*-1} - m_j^{*-1}]^{-1},$$

amounting to ~4.9 meV in these samples. This is sufficiently small compared with the total transition linewidth that the "sole collective excitation" picture, as defined in ref.[11] and assumed in our three level treatment, should be valid under these measurement conditions.

4. Concluding Remarks.

In conclusion, we have observed a sharp interference feature associated with electromagnetically induced quantum interference in QW's, which is not related to saturation or spectral hole burning effects. We found that the induced transparency can be observed when the coupling field is two-photon-resonant with the E_{13} subband transition, this behaviour being fully consistent with a density matrix treatment which simultaneously includes coupling between all states in a three-level system [12]. The observations were made under similar experimental regimes as occur, or as could be engineered to occur if desired, in QC lasers at

threshold, and may have significant influence in their future design.

REFERENCES.

1. J. Faist et al., Nature (London) **390**, 589 (1997); J. Faist et al., Appl. Phys. Lett. **71**, 3477 (1997). H. Schmidt, K.L.Campman, A.C.Gossard and A. Imamoglu Appl. Phys. Lett. **70**, 3455 (1997).
2. For reviews, see S.E. Harris, Phys. Today **50**, No. 7, 36 (1997) and J. P. Marangos, J. Mod. Opt. **45**, 471 (1998).
3. K.-J. Boller, A. Imamoglu and S. E. Harris, Phys. Rev. Lett. **66**, 2593 (1990);J.E. Field, K.H. Hahn and S.E. Harris, Phys. Rev. Lett. **67**, 3062 (1991);M.M. Kash et al., Phys. Rev. Lett. **82**, 5229 (1999).
4. Y. Zhao et al., Phys. Rev. Lett. **79**, 641 (1997); B. S. Ham, P. R. Hemmer and M. S. Shahriar, Opt. Commun. **144**, 227 (1997); Phys. Rev. A **59**, R2583 (1999); K. Ichimura, K. Yamamoto and N. Gemma, Phys. Rev. A **58**, 4116 (1998).
5. M. Jain et al., Phys. Rev. Lett. **77**, 4326 (1996), G. Almogy , A.Yariv, J. of Nonlinear Optical Physics and Materials, **4**, 401 (1995).
6. G. Padmabandu et al., Phys. Rev. Lett. **76**, 2053 (1996).
7. K.L. Vodopyanov, V. Chazapis, Opt. Commun. **136**, 98 (1997); K.L. Vodopyanov, V. Chazapis, and C.C Phillips, Appl. Phys. Lett. **69**, 3405 (1996).
8. L. M. Narducci et al., Phys. Rev. A **42**, 1630 (1990).
9. Intersubband recombination times were measured using the absorption saturation method of K. L. Vodopyanov, et al. Semicond. Sci. and Technol. **12**, 708-714 (1997).
10. Conduction band non-parabolicity was treated as in C. Sirtori F.Capasso and J.Faist Phys.Rev.B **50**, 8663 (1994) using the materials parameters of ref.[9]. Relative to the well bottom, bare energy levels (in-plane effective masses) were $E_1 = 46$ meV (0.045 m_0); $E_2 = 176$ meV (0.053 m_0); $E_3 = 338$ meV (0.061 m_0).
11. D. E. Nikonov, A. Imamoglu, L.V.Butov and H. Schmidt. Phys. Rev. Lett. **79**, 4633 (1997); D. E. Nikonov, A. Imamoglu and M.O. Scully, Phys. Rev. B **59**, 12212 (1999).
12. G. B. Serapiglia, E. Paspalakis, C. Sirtori, K. L.Vodopyanov and C. C. Phillips." *Observation of Laser-Induced Quantum Coherence in a Semiconductor Quantum Well*." Phys. Rev. Letts, **84** (5), 1019-1022 (2000).

Towards quantum well hot hole lasers

P. Kinsler[*], W.Th. Wenckebach

Department of Applied Physics, Faculty of Applied Sciences, T.U. Delft,
Lorentzweg 1, 2628 CJ Delft, The Netherlands.

Abstract It should be possible to improve hot-hole laser performance by moving from bulk materials to a quantum well structure. The extra design parameters enable us to alter the band structure by changing the crystal orientation of the growth direction; to use the well width to shift the subband offsets, enabling the effect of the LO phonon scattering cut-off to be controlled; and to use modulation doping to ensure a high hole concentration to increase the gain without the dopants being present in the gain region. We present the first simulations of THz quantum well hot-hole lasers that can produce inversion and optical gain.

1 Introduction

Hot hole lasers [1,2] emit in the THz (far-infrared) with an unusually broad gain spectrum, allowing amplification and generation of laser pulses on a picosecond time scale. The THz band has important potential applications in (e.g.) medical imaging and office communications. Bulk hot-hole lasers have been realised in p-Ge, producing gains of $\sim 0.25\,\mathrm{cm}^{-1}$ around 4THz. Of the III-V materials, both GaAs and InSb are suitable for hot hole lasers; although their performance is not as good as in Ge [3]. Investigation of these is the most useful for industrial applications because of the existing ability to grow high quality III-V structures.

We present a discussion of likely modes of operation of quantum well hot-hole lasers in GaAs, together with predictions from Monte Carlo simulations for one design. The simulations use an infinite-well $k.p$ bandstructure, and include optical phonons, acoustic phonons, and piezoelectric phonons [4–6]. As yet they do not include ionised impurity scattering, but this is a small effect and should not affect the character of the results significantly.

2 Quantum Wells

Bulk hot-hole lasers can be described using the heavy and light hole valence bands: an electric field accelerates the heavy holes in a streaming motion to high energies $E > E_{LO}$; from where (ideally) they scatter into light hole cyclotron orbits formed by the magnetic field; and then emit a photon and return to low energy in the heavy hole band to repeat the cycle. In a quantum well each valence band breaks up into a set of subbands: heavy hole subbands HH1, HH2, HH3, ...; and similarly for light holes LH1, LH2, etc. The non parabolicity of the bulk bandstructure leads to a variety of possible quantum well bandstructures, depending on the orientation of the crystal axes in the well material. Figure 1 shows the lower subbands of two simple cases schematically. Note that the [101] well is not symmetric in x and y; and

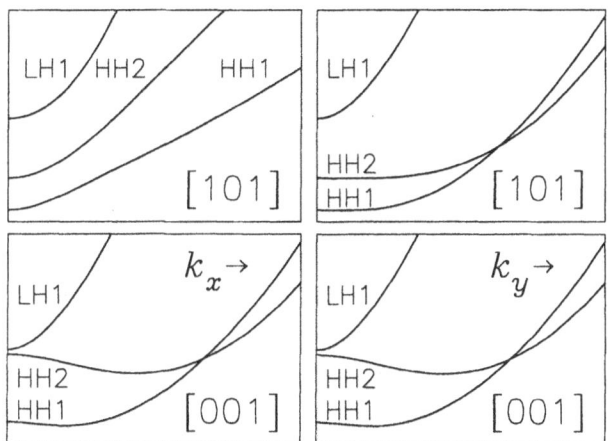

Fig. 1 Band structure of a quantum well hot-hole laser. Upper graphs: 100Å well with a [101] growth direction; lower graphs: a 100Å well with a [001] growth direction. The k_x variation is shown on the left-hand graphs, and k_y on the right. The vertical scale is 0–100meV, and the horizontal 0–$1\times10^9\mathrm{m}^{-1}$.

that for the [001] well the HH dispersions are no longer even approximately parabolic, with the HH2 having a noticeable local maxima at the origin.

The [101] well has two heavy-hole subbands (HH1, HH2) that need to be considered. For wells over about 50Å wide, the two lowest energy, and hence most heavily populated subbands will be HH1 and HH2. This means that in contrast to the bulk case we do not need a magnetic field to confine the light holes in cyclotron orbits. The LH1 is well above the HH2 for a 100Å well, and so has minimal effect on hot-hole laser operation. Along one direction (y) both HH subbands are very flattened at the base, but with a LH-like curvature for larger k_y; whereas along the other (x) HH2 sits between HH1 and LH1. This means that the heavy-hole distributions will be roughly rectangular in outline; and an increasing electric field will shift the distributions along its direction. In the x direction the two HH subbands (nearly) cross at a point about 15meV above the bottom of HH1, allowing for fast inter-subband scattering. Holes that get further up HH1 to an energy equal to E_{LO} above the bottom of HH2 will quickly emit an LO phonon and return to populate either HH2 or HH1. Note that HH2 is flat to larger k_y values than HH1, making it relatively easy to get inversion where the HH2 is flat but the HH1 is not.

Fig. 2(a) shows the difference in the distribution functions between HH2 and HH1, for an electric field of 250V/cm along the x direction. We see regions of inversion for $k_x \sim 0$ and $k_y \approx 5.00 \times 10^8\mathrm{m}^{-1}$. There is

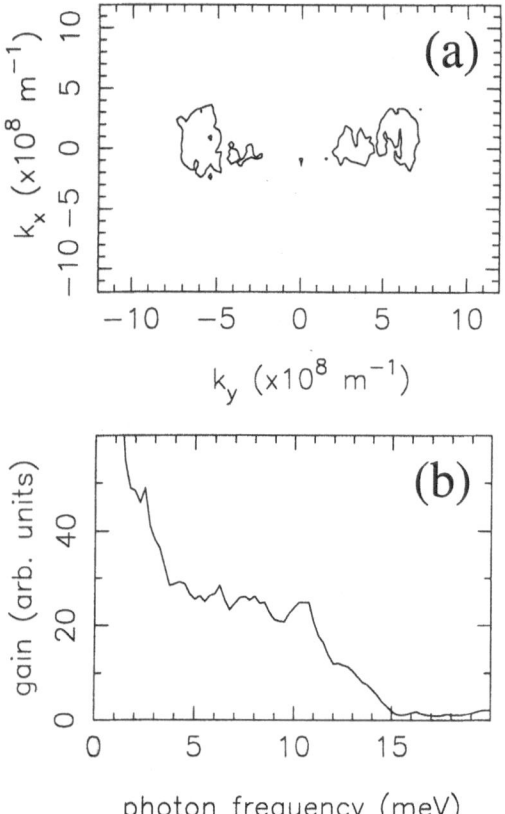

Fig. 2 [101] 100Å quantum well hot-hole laser, with electric field $F_x = 250$V/cm (a) Contour plot showing the region of inversion; (b) spectrum of the gain cross-section for y polarised light. Both graphs are affected by statistical noise from the simulations.

a small displacement in the direction of the field, and the amount of inversion decreases with increasing field strength. If the electric field is applied along the y direction, we see only one inversion peak, because the other is wiped out as the HH1 distribution is shifted underneath it; and a field applied along $x = y$ has a similar, but not so marked, effect. Figure 2(b) shows the optical gain due to the inversion obtained and shown in (a). We see optical gain occurring over the range of energies between the small splitting at the HH1-HH2 anti-crossing ($\sim$ 0meV) and the separation of the subband minima ($\approx$ 15meV).

The [001] well HH2 subband has a distinct local maxima at $k = 0$; and also LH1 is rather close to HH2, and should not be neglected. Our code does not yet allow for these more complicated dispersions, but we can see that a lasing cycle might be as follows: A HH1 of moderate positive k_x is accelerated by an electric field $+F_x$, where it might scatter into HH2 near the anti-crossing. Then it will scatter back onto the inverted part of HH2, either by acoustic phonon scattering or LO phonon emission. Here it will be accelerated *backwards* past $k = 0$ to the point of inflection ($-k_{LHi}$) where the HH2 effective mass is infinite; and subsequently emit a photon and drop down to HH1, from where it will eventually scatter to moderate positive k_x and will repeat the cycle. Note that HH1 also

has a local maxima, but it is smaller and has less effect; and its point of inflection where holes can collect is offset from the HH2 one, and so should not interfere with population inversion too much. In contrast to the [101] well, we expect the emission spectrum of this [001] well to be peaked and centered at $E_{HH2}(k_{LHi}) - E_{HH1}(k_{LHi})$, where the HH2's will accumulate.

Quantum well hot hole lasers also benefit from modulation doping. One significant source of unhelpful scattering is that due to holes scattering off either other holes or impurities. We can halve this contribution to the scattering simply by moving impurities used to add the holes to the device to outside the active region. Also, we can vary the well widths to vary the inter-subband separations to greater or less than E_{LO} to enhance or reduce optical phonon scattering as required; or even use the well's crystal orientation to adjust the densities of states (DOS) of the subbands. The [101] well has HH1, HH2 with flat $E(k)$ in the k_y direction, leading to an enhanced density of states, hence making those regions into relatively preferred destinations for scatterings. This differs markedly from the DOS effects in bulk material [3].

3 Conclusions

We have shown the potential for quantum well hot-hole lasers by an investigation of the possible bandstructure in combination with computer simulation. Although we have not yet explored the full parameter space of possible designs, our first attempt (the 100Å [101] quantum well) produced simulations which showed inversion over a range of electric field strengths and directions, with the optimum being for a 250V/cm field in the in-plane x direction. Further, the 100Å [001]well has a bandstructure which promises a good inversion. We aim to continue to test different designs and field combinations, including the addition of magnetic fields, in order to determine the most practical designs for experimental investigation.

Acknowledgements: This work is funded by the European Commission via the program for Training and Mobility of Researchers.

References

[*] Electronic mail: Dr.Paul.Kinsler@physics.org

1. Opt. Quantum Electron. Vol. **23**, Special Issue Far-infrared Semiconductor Lasers, (Eds. E. Gornik and A.A. Andronov, Chapman and Hall, London 1991).

2. J. N. Hovenier, A. V. Muravjov, S. G. Pavlov, V. N. Shastin, R. C. Strijbos, and W. Th. Wenckebach, Appl. Phys. Lett. **71**, 443-445 (1997).

3. P. Kinsler, W.Th. Wenckebach, submitted to J. Appl. Phys.

4. C. Moglestu, *Monte Carlo simulation of semiconductor devices*, (Chapman & Hall, London, 1993).

5. E.O. Kane, in *Semiconductors and Semimetals*, eds. R.K. Willardson and A.C. Beer, (Academic Press, New York, 1966), Vol. **1**, page 75.

6. B.K. Ridley *Quantum Processes in Semiconductors*, (Clarendon Press, Oxford, 1988).

Proc. 25th Int. Conf. Phys. Semicond., Osaka 2000 (Eds. N. Miura and T. Ando)

713

Character of electronic excitations in GaAs-AlGaAs quantum structures

E. Ulrichs, C. Steinebach, C. Schüller, Ch. Heyn, W. Hansen, and D. Heitmann

Institut für Angewandte Physik und Zentrum für Mikrostrukturforschung, Universität Hamburg, Jungiusstraße 11, 20355 Hamburg, Germany

Abstract We have performed Raman experiments on modulation-doped quantum dots. Besides charge-density and spin-density excitations we observe so called single-particle excitations (SPE). We elucidate the character of these SPE's by comparing mean field approaches with numerically exact diagonalization treatments of quantum dots with small electron numbers.

1 Introduction

The excitation spectrum of modulation-doped quantum dots containing a small number N of electrons can be calculated by numerically exact diagonalization methods. In such a picture, all excitations appear as many-body excitations of the N-electron system. Quantum dots with large numbers of electrons are usually treated within mean-field theories, like, for example, using first the Kohn-Sham local-density approximation (LDA) to calculate effective single-particle energies and then apply time-dependent LDA (TDLDA) to determine the dynamic excitations of the system. In both pictures some resonances can be classified as charge-density (CDE) or spin-density (SDE) excitations. CDE's can be observed with far-infrared absorption. The selection rules for Raman experiments are such that for parallel polarization of exciting and scattered radiation CDE's are observed, whereas in crossed polarization SDE's are measured.

2 Experiments and Discussion

Experimental Raman spectra on quantum dot arrays are shown in Fig. 1. These dots were prepared, starting from modulation-doped AlGaAs-GaAs quantum wells, by deep-mesa etching. The lithographic dot diameter was 240 nm and the electron number per dot N about 200. For an excitation energy $E_L = 1587$ meV two series of excitations are observed which can be classified according to the selection rules as SDE_i and CDE_i. The index i denotes the change of the lateral quantum numbers for the transitions which contribute predominantly to these excitations [1]. For decreasing excitation energy E_L, in particular when approaching the fundamental gap of the quantum well, one observes additional excitations which occur in both polarizations. Excitations obeying the same selection rules have also been observed in quantum wells and have been attributed to single-particle excitations [3]. What is the role of such SPE's in a quantum dot where, as said above, strictly speaking, all excitations are many-body excitations?

To get a deeper understanding of the character of these "SPE's" and the resonant scattering mechanism,

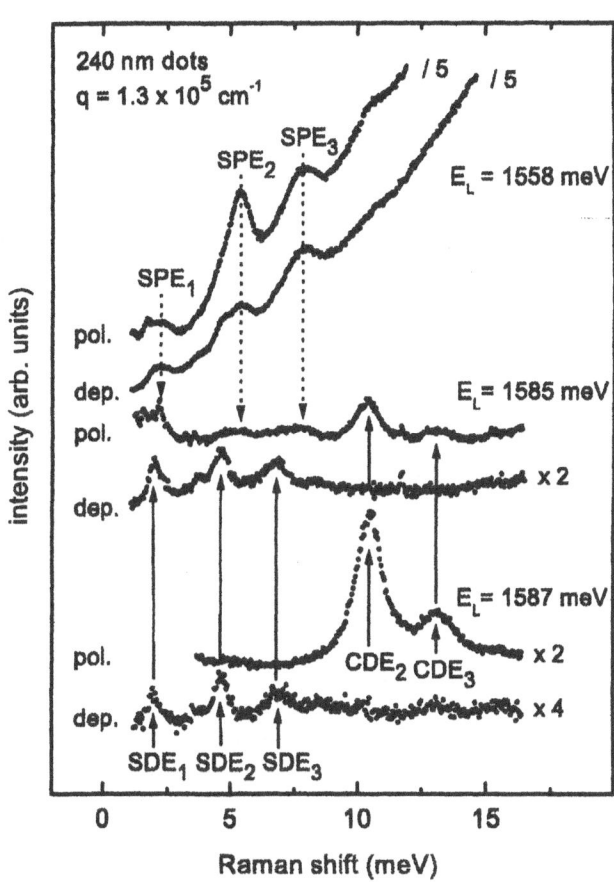

Fig. 1 Depolarized and polarized Raman spectra of quantum dots for different laser energies.

we have performed calculations of Raman spectra for 6-electron quantum dots using numerically exact diagonalization methods [2]. In such an exact treatment one starts with the 6-electron Hamiltonian

$$H = \sum_{i=1}^{6} -\frac{\hbar^2}{2m^*}\Delta_i + \sum_{i=1}^{6} V_{ext}(r) + \frac{1}{2}\sum_{i\neq j}\frac{e^2}{\epsilon|r_i - r_j|}. \quad (1)$$

Since the confinement arises from the remote donors we assume that the external potential $V_{ext}(r)$ has a parabolic shape. Transitions between the ground state and the excited many-body states are shown in Fig. 2 in the upper panel. One can identify the collective SDE and CDE as singlet-triplet and singlet-singlet transitions, respectively. Very surprisingly, there are two additional transitions in between these collective excitations, one, according to the polarization selection rules, with spin-density, the other with charge-density character.

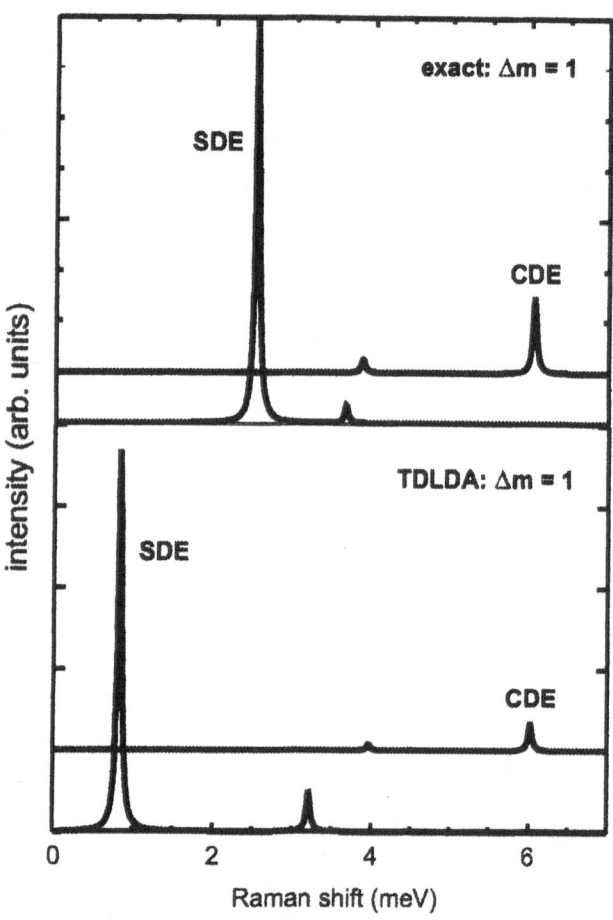

Fig. 2 Raman spectra for a 6-electron quantum dot calculated by exact numerical diagonalization (upper panel) and within TDLDA (lower panel). Polarized (upper spectra) and depolarized (lower spectra) are shifted against each other for clarity.

We have also performed a TDLDA calculation of the dynamic response starting from the self-consistent Kohn-Sham single-particle levels (lower panel of Fig. 2). On the contrary, here one starts by solving the effective one-electron Hamiltonian

$$H_{LDA} = -\frac{\hbar^2}{2m^*}\Delta + V_{ext}(r) + V_H(r) + V_{XC}(r) \qquad (2)$$

for the 6-electron problem. $V_H(r)$ and $V_{XC}(r)$ are the effective Hartree and LDA exchange-correlation potentials, respectively. We find that the additional spin-density and charge-density type of excitations of the exact many-body treatment are very close to some transitions between the Kohn-Sham single-particle levels. (Nevertheless, strictly speaking, show small collective shifts). From this we can conclude that even for only 6 electrons in the quantum dot, electronic excitations with energies close to the unrenormalized single-particle transitions occur. For 12 electrons per quantum dot, i.e., three completely filled shells, we found 4 corresponding transitions with energies close to the single-particle levels. We believe that from these excitations the experimentally observed SPE's in many-electron dots, as shown in Fig. 1, evolve.

An interesting point is, as demonstrated in Fig. 2, that the so-called SPE's have either CDE or SDE character. At the moment we cannot confirm this experimentally from the polarization selection rules since the differences in the excitation energies are too small. Here it would be interesting to perform other independent experiments. For example it would be interesting to tune single-particle-like transitions in quantum wells (or dots, if possible) through the Reststrahlen regime of the polar GaAs. Here one would expect that the CDE-like SPE would feel the interaction with LO and TO phonons, whereas the SDE-like SPE should not.

We acknowledge support from the German Science foundation through SFB 508 and GrK "Nanostructured Solids" and the NEDO program.

References

1. C. Steinebach, C. Schüller, and D. Heitmann, Phys. Rev. B **59**, (1999) 10240.
2. C. Steinebach, C. Schüller, and D. Heitmann, Phys. Rev. B **61**, (2000) 15600.
3. A. Pinczuk, S. Schmitt–Rink, G. Danan, J. P. Valladares, L. N. Pfeiffer, and K. W. West, Phys. Rev. Lett. **63**, (1989) 1633.

Correlation of vertical transport and infrared absorption in GaAs/AlGaAs superlattices

M. Helm[1] and G. Strasser[2]

[1] Institute of Ion Beam Physics and Materials Research, Forschungszentrum Rossendorf, P.O. Box 510119, D-01314 Dresden, Germany
[2] Institut für Festkörperelektronik, TU Vienna, A-1040 Wien, Austria

Abstract We have studied the interminiband absorption and the current-voltage characteristics on the same GaAs/AlGaAs superlattices. The Esaki-Tsu-type negative differential resistance is observed as well as the thermal (de)population of the minibands. The possibility of mapping the nonequilibrium electron distribution function through infrared absorption measurements under bias is discussed.

1 Introduction

Usual investigations of Bloch oscillations and Wannier-Stark ladders in superlattices are performed with optical spectroscopy (stationary or time resolved) or transport measurements. The former has revealed the Wannier-Stark energy levels [1] and the oscillatory electron motion [2], respectively, the latter shows the Esaki-Tsu type negative differential resistance (NDR) [3]. On the other hand, it has been shown that infrared interminiband spectroscopy can yield valuable information about the density of states [4] and the electron distribution function [5]. Recently we have combined transport and infrared absorption experiments on the same weakly coupled superlattice and found evidence for Wannier-Stark ladders in the continuum and next-nearest neighbor tunneling [6].

2 Samples

Our goal is the extension of this work to more strongly coupled superlattices. For this we have prepared two n-type GaAs/AlGaAs superlattices with the following parameters: well width 62 Å (65 Å), barrier width 36 Å (42 Å), and Al content in the barriers 34% (33%). These values were determined by X-ray diffraction and, partially, transmission electron microscopy. Each superlattice has 300 periods and is embedded between heavily doped GaAs contact layers. The resulting (calculated) widths, Δ_1, of the first miniband are 10 and 6 meV, respectively. This is narrow enough to avoid excessive current densities, but wide enough to ensure miniband transport through extended states.

3 Results

Fig. 1 shows temperature dependent transmission spectra of both samples, prepared in a multi-pass waveguide geometry. The absorption features, from low to high energy, are due to electrons at $k_z=\pi/d$,

due to the 1s-2p$_z$ impurity transition, and due to the states at $k_z=0$, respectively. The doping is such that at low temperature the first miniband is only partially filled, whereas at higher temperature also states at the zone edge are occupied, giving rise to the observed temperature dependence of the absorption. We would like to point that this temperature dependence is an unambiguous proof for the extended miniband character of the electronic states. This information could not be obtained from I-V measurements alone, where miniband transport cannot be distinguished from sequential tunneling through localized quantum-well states [7]. Also note that the miniband widths can be experimentally determined as indicated in the figure.

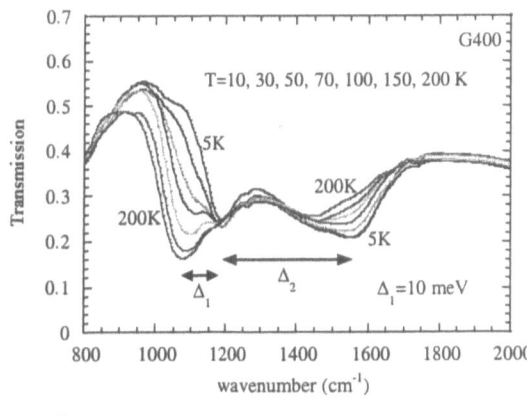

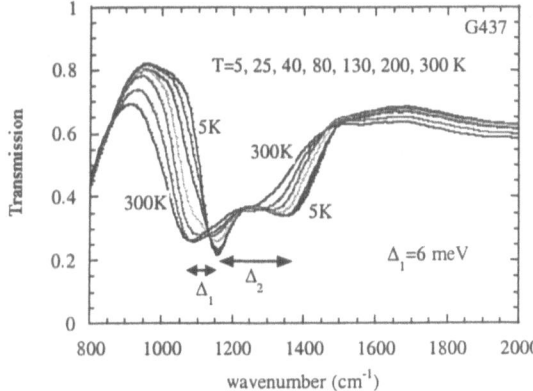

Fig. 1 Interminiband transmission spectra of both superlattices for different temperatures. The ratio of p- to s-polarization is plotted.

Both superlattices also exhibit a well developed NDR due to the negative curvature of the miniband dispersion, with peak current densities of 3500 A/cm^2 and 900 A/cm^2, respectively. The I-V characteristic of the 6 meV sample is shown in Fig. 2. It appears that at 80 K the NDR sets in, before the proper peak current is reached. This is supported by a polarity dependence of the I-V curve (not shown).

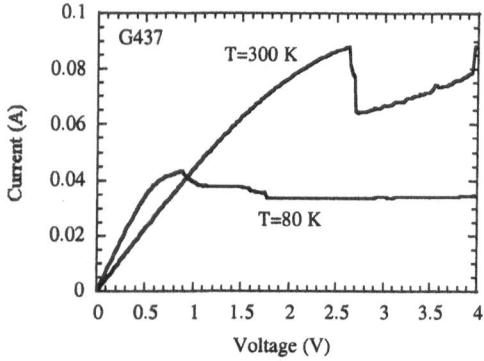

Fig. 2 Current-voltage characteristic of the superlattice with 6 meV miniband width, measured on a 100 μm square mesa.

4 Discussion

We have demonstrated previously that the inter-miniband absorption spectrum reflects the electron distribution function [5]. If the infrared measurements are performed under electric bias near the current peak and the NDR region, significant changes in the spectrum are expected. At low temperature, a vertical electric field will change the electron distribution function, giving rise to a change in the absorption spectrum. Inversely, the electron distribution function under Bloch oscillation conditions [8] can be deduced from the field-induced differential absorption spectrum. Measurements of the *far*-infrared/THz absorption under the same conditions should reveal the optical gain predicted below the Bloch frequency [9].

Acknowledgment: We are grateful to A. Mücklich for sample characterization by transmission electron microscopy.

References

1. E. E. Mendez, F. Agullo-Rueda, and J. M. Hong, Phys. Rev. Lett. **60**, (1988) 2426; P. Voisin et al., Phys. Rev. Lett. **61**, (1988) 1639.
2. K. Leo, Semicond. Sci. Technol. **13**, (1998) 249.
3. A. Sibille et al., Superlatt. Microstruct. **13**, (1993) 247.
4. M. Helm, W. Hilber, T. Fromherz, F. M. Peeters, K. Alavi, and R. N. Pathak, Phys. Rev. B **48**, (1993) 1601.
5. W. Hilber, M. Helm, K. Alavi, and R. N. Pathak, Appl. Phys. Lett. **69**, (1996) 2528.
6. M. Helm, W. Hilber, G. Strasser, R. DeMeester, F. M. Peeters, and A. Wacker, Phys. Rev. Lett. **82**, (1999) 3120.
7. A. Wacker and A.-P. Jauho, Phys. Rev. Lett. **80**, (1998) 369.
8. A. Wacker, A.-P- Jauho, S. Rott, A. Markus, P. Binder, and G. H. Döhler, Phys. Rev. Lett. **83**, (1999) 836.
9. S. J. Allen et al., Semicond. Sci. Technol. **7**, (1992) B1.

Influence of in-plane magnetic field on cyclotron resonance in double-layer two dimensional electron system

H. Aikawa[1], S. Takaoka[1], A. Kuriyama[1], K. Oto[1], K. Murase[1], S. Shimomura[2], S. Hiyamizu[2], T. Jungwirth[3], L. Smrčka[3]

[1] Graduate School of Science, Osaka University, 1-1 Machikaneyama, Toyonaka, Osaka 560-0043, Japan
   e-mail: aikawa@mmm.phys.sci.osaka-u.ac.jp
[2] Graduate School of Engineering Science, Osaka University, 1-3 Machikaneyama, Toyonaka, Osaka 560-8531, Japan
[3] Institute of Physics, Academy of Science of Czech Republic Cukrovarnická 10, 160 00, Praha 6, Czech Republic

Abstract   The in-plane magnetic field ($B_\parallel$) dependence of cyclotron effective mass (m_c) in a double-layer two dimensional electron system has been investigated by far-infrared cyclotron resonance spectroscopy. At a relatively small perpendicular magnetic field ($B_\perp = 5.0$ T), with increasing $B_\parallel$, m_c of the symmetric subband increases, while that of the antisymmetric subband decrease and vanishes above $B_\parallel = 4.0$ T. These behaviors can be qualitatively explained by the semi-classical calculation. For $B_\perp > 5.0$ T, we have observed the complicated $B_\parallel$ dependence of m_c that contains more than two kinds of m_c's.

1 Introduction

When an in-plane magnetic field is applied to a two dimensional electron system (2DES) in GaAs/AlGaAs heterostructure, the originally isotropic "Fermi loop" of 2DES is distorted to an *egg*-like shape, due to the asymmetric trianguler confinement potential [1]. We have investigated the distortion of the "Fermi loop" by the magnetic electron focusing effect [2], the Weiss oscillations [3], and the cyclotron resonance (CR) spectroscopy [4].

In a double-layer 2DES, the coupling of the electron wave function in each layer forms the symmetric subband (SS) and the antisymmetric subband (AS). The semi-classical calculation [5] predicts that the "Fermi loop" of the SS is distorted to a *peanut*-like shape and that of the AS is distorted to a *lens*-like shape by $B_\parallel$, as shown in Fig. 1(a). With increasing $B_\parallel$, m_c of the SS increases, on the other hand, that of the AS decreases and vanishes above a certain $B_\parallel$, since the "lens" of the AS becomes vacant. The distortions have been investigated mainly by the temperature damping of Shubnikov-de Haas (SdH) oscillations [6]. We have demonstrated the distortion of the "Fermi loop" by the CR spectroscopy [7].

The semi-classical expression of m_c related to the "Fermi loop" is

$$m_c = \frac{\hbar^2}{2\pi} \frac{dS_F}{dE}, \tag{1}$$

where S_F denotes the area surrounded by the "Fermi loop". We can directly obtain m_c from the CR measurement by using the formula of $m_c = eB_\perp/\omega_c$, where ω_c is the cyclotron frequency.

Electrons subjected to a confining potential $V(z)$ and a tilted magnetic field $\boldsymbol{B} = (0, B_\parallel, B_\perp)$ are described by the following Hamiltonian,

$$H = -\frac{\hbar^2}{2m^*}\frac{\partial^2}{\partial x^2} + \frac{e^2 B_\perp^2}{2m^*}x^2 - \frac{\hbar^2}{2m^*}\frac{\partial^2}{\partial z^2} + V(z)$$
$$+ \frac{e^2 B_\parallel^2}{2m^*}z^2 - \frac{e^2 B_\parallel B_\perp}{m^*}xz. \tag{2}$$

In Eq.(2), we choose the gauge $\boldsymbol{A} = (0, xB_\perp - zB_\parallel, 0)$, and use the transformation $x \rightarrow x - \hbar k_y/eB_\perp$ of the original x coordinate. The first four terms in Eq.(2) describe the quantization into Landau levels and electronic subbands. The fifth one results a positive diamagnetic shift which is proportional to $B_\parallel^2$, and the last one describes the coupling between the Landau levels of different subbands. In the semi-classical calculation where $B_\perp \rightarrow 0$, this term is neglected. However, $B_\perp$ is so strong in this study that this term shoud be taken into account to understand the experimental results.

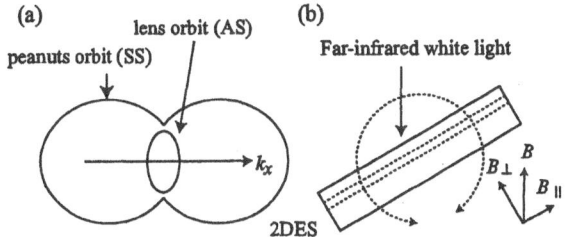

Fig. 1 (a) The sketch of each subband in k-space. The *peanut* shaped the symmetric subband (SS) and the *lens* shaped the antisymmetric subband (AS) in a double-layer 2DES. (b) The schematic top view of the exprimental configuration.

2 Experimental Procedure

The sample was grown by molecular beam epitaxy on a GaAs substrate. Unlike the usual double-layer 2DES, the GaAs well of width W = 40 nm is sandwiched between two modulation doped AlGaAs layers. Thus the double-layer 2DES is formed by Hartree potential instead of the AlGaAs barrier.

The sample is a 5×5 mm square with four ohmic contact placed at each corner. Only the SS is occupied in the dark condition, since the storng coupling of the electron wave function in each layer makes the symmetric-antisymmetric gap energy $\Delta_{\mathrm{SAS}} = 3.9$ meV larger than the Fermi energy E_F. The elecrton carrier density evaluated from SdH oscillation and the mobility are $N_{SS} = 1.7 \times 10^{11}$ cm^{-2} and $\mu = 1.0 \times 10^5$ cm^2/Vs at 2 K, respectively. After illminated by a GaAs light emitting diode, both the SS and AS are occupied with $N_{SS} = 2.4 \times 10^{11}$ cm^{-2} and $N_{AS} = 1.3 \times 10^{11}$ cm^{-2}, and the total mobility is $\mu = 8.0 \times 10^4$ cm^2/Vs.

The cyclotron resonance has been measured by a far-infrared Fourier spectrometer. We can adjust $B_\perp$ and $B_\parallel$ by rotating sample in the horizontal split pair of superconducting magnets. In this experiment, the magnetic field direction is parallel to the light pass direction [See, Fig. 1(b)].

3 Results and discussion

We have measured the $B_\parallel$ dependence of the CR at several fixed $B_\perp$'s. At $B_\perp = 5.0$ T in the dark condition, only one CR dip is observed as shown in Fig. 2(a), which agrees with SdH oscillation analysis [See, Fig. 2(b)]. With increasing $B_\parallel$, the corresponding m_c increases as shown in Fig. 3(a). On the other hand, the intensity of absorption rapidly decreases, which has not been observed in a single layer 2DES [4].

After illumination, both the SS and AS are occupied, which is confirmed by SdH double peaks as shown in Fig. 2(b). The CR dip splits into two dips; a small dip appears at a higher wave number side [See, Fig. 2(a)]. This dip shifts to higher wave number with $B_\parallel$, and it vanishes above $B_\parallel = 4.0$ T. In other words, the corresponding m_c decreases with $B_\parallel$ and vanishes as shown in Fig. 3(a), which agrees with the behavior of m_c of the AS predicted by the semi-classical calculation. The increase of m_c of the SS is not affected by changing the carrier density. The influence of $B_\parallel$ on m_c can be understood qualitatively by the the semi-classical calculation, except that the increasing rate of m_c is smaller than the value expected from the semi-classical calculation.

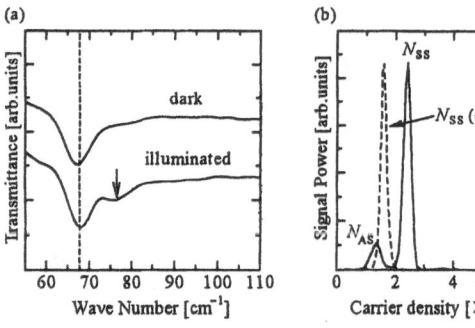

Fig. 2 (a) The far-infrared transmission spectra of double-layer 2DES in the dark and illuminated conditions at $B_\perp = 5.0$ T and $B_\parallel = 2.5$ T. The arrow indicates the dip of the AS. (b) The Fourier power spectra of SdH oscillation in the dark (dashed line) and illuminated (solid line) conditions.

At $B_\perp = 7.0$ T [See, Fig. 3(b)], no difference in the $B_\parallel$ dependence of m_c's is recognized by changing the carrier density. In the dark condition where only the SS

is occupied, three m_c's have been observed in $B_\parallel = 1.5 \sim 4.0$ T. The dominant m_c decreases with $B_\parallel$, which cannot be explained by the semi-classical calculation. We attribute the m_c's to the optical transition between Landau levels of different subbands.

In conclusion, the $B_\parallel$ dependence of m_c in a double-layer 2DES can be partially explained by the semi-classical calculation. To understand the $B_\parallel$ dependence of m_c in strong $B_\perp$ regime, the influence of $B_\perp$ to the coupling of Landau levels of different subbands should be taken into account.

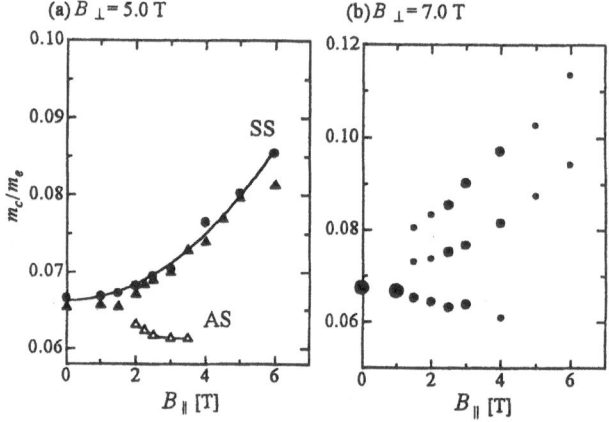

Fig. 3 (a) The $B_\parallel$ dependence of m_c/m_e at $B_\perp = 5.0$ T, where m_e is a bare electron mass. The solid lines are guides to the eye. In the dark condition (solid circles), only m_c of the SS is observed. After illumination, both m_c of the SS (solid triangles) and AS (open triangles) are observed. (b) The $B_\parallel$ dependence of m_c/m_e at $B_\perp = 7.0$ T. Three different m_c's are observed both in the dark and illuminated conditions. The sizes of solid circles indicate the intensity of absorption.

Acknowledgements

This work was supported by a Grand-in-Aid for COE Research (10CE2004) from Ministry of Education, Science and Culture (Japan).

References

1. L. Smrčka and T. Jungwirth, J. Phys. Condens. Matter **6** (1994) 55.
2. K. Ohtsuka, S. Takaoka, K. Oto, K. Murase, and K. Gamo, Physica B **249-251** (1998) 780.
3. K. Oto, T. Suzumura, S. Takaoka, K. Murase, M. Hayashi, and S. Yamada, *Proc. 25th Int. Phys. on Seniconductors* (2000, Osaka) to be published.
4. S. Takaoka, H. Aikawa, K. Oto, and K. Murase, *Proc. 14th Int. Conf. on High Magnetic Fields in Semiconductor Physics* (2000, Matsue) to be published.
5. L. Smrčka and T. Jungwirth, J. Phys. Condens. Matter **7** (1995) 3721.
6. M.A. Blonut, J. A. Simmons, and S. K. Lyo, Phys. Rev. B **57** (1998) 14822.
7. A.Kuriyama, S. Takaoka, K. Oto, K. Murase, S. Shimomura, S. Hiyamizu, M. Cukr, T. Jungwirth, and L. Smrčka, Solid State Commun. **111** (1999) 699.

Proc. 25th Int. Conf. Phys. Semicond., Osaka 2000 (Eds. N. Miura and T. Ando)

719

Circular photogalvanic effect in p−GaAs/AlGaAs MQW

S.D. Ganichev[1,2], **E.L. Ivchenko**[2], **H. Ketterl**[1], **L.E. Vorobjev**[3], **M. Bichler**[4], **W. Wegscheider**[1], **W. Prettl**[1]

[1] Institut für Experimentelle und Angewandte Physik, Universität Regensburg 93040 Regensburg, Germany e:-mail sergey.ganichev@physik.uni-regensburg.de

[2] A.F. Ioffe Physico-Technical Institute, Russian Academy of Sci. 194021 St. Petersburg, Russia

[3] St. Petersburg State Technical University, 195251 St. Petersburg, Russia

[4] Walter Schottky Institut, TU München, 85748, Garching, Germany

Abstract The circular photogalvanic effect (CPGE) has been observed in (001)- and (311)A-oriented p-GaAs/AlGaAs quantum wells at normal incidence of far-infrared radiation. It is shown that monopolar optical spin orientation of free carriers causes an electric current which reverses its direction upon changing from left to right circularly polarized radiation. As proposed, CPGE can be utilized to investigate separately spin polarization of electrons and holes.

1 Introduction

Spin polarization and spin relaxation in semiconductors are the subject of intensive studies of spin-polarized electron transport aimed at the development of spinotronic devices like a spin transistor [1,2]. So far, photo-induced spin polarization has been achieved by *interband* optical absorption of circularly polarized light with the photo-generation of spin-polarized electrons and holes [3–5].

Here we report on the first observation of the circular photogalvanic effect (CPGE) at *intersubband* transitions in quantum-well (QW) structures. Phenomenologically, the effect is a transfer of angular momenta of circularly polarized photons to the directed movement of free carriers, electrons or holes, and therefore depends on the symmetry of the medium. Microscopically, it arises due to optical spin orientation of holes in QWs and asymmetric spin-dependent scattering of spin-polarized carriers followed by an appearance of an electric current. The two states of light circular polarization $\sigma_\pm$ result in different spin orientations and, thus, in electric photocurrents of opposite directions. In contrast to the case of *interband* optical excitation, under *intersubband* transitions only one kind of carriers is involved leading to a monopolar spin orientation. The observed effect can be utilized to investigate spin polarization of free carriers.

2 Experimental set up and samples

The experiments have been carried out on (001)-MOCVD grown p-GaAs/AlGaAs multi QW structures with 400 undoped wells of 20 nm width separated by 10 nm wide doped barriers and on a p-GaAs/AlGaAs (311)A-MBE grown single QW of 15 nm width. Samples of 5×5 mm^2 size with hole densities in the QWs varying from $2 \cdot 10^{11}$ cm^{-2} to $2 \cdot 10^{12}$ cm^{-2} have been investigated. Two pairs of ohmic contacts (see inset Fig. 1) have been pre-

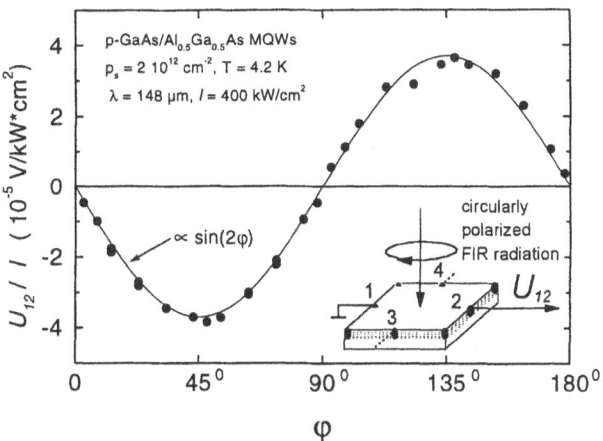

Fig. 1 Photogalvanic voltage signal U_{12} picked up across the contact pair 1-2 and normalized by the intensity I as a function of phase angle φ. The full line is fitted after Eq. 2 ($U_{12} \propto sin2\varphi$).

pared along [1$\bar{1}$0] and [110] for (001)-GaAs/AlGaAs and [0$\bar{1}$1] and [$\bar{2}$33] for (311)-GaAs/AlGaAs, respectively.

A high power far-infrared molecular laser pumped by a TEA-CO$_2$ laser has been used as a radiation source delivering 100 ns pulses with the intensity up to 1 MW/cm^2. NH$_3$ has been used as a FIR laser medium yielding strong linearly-polarized emissions at wavelengths $\lambda = 76\,\mu$m, $90\,\mu$m, and $148\,\mu$m [6,7]. The photon energies of the radiation correspond to transitions between heavy- and light- hole subbands of GaAs QWs. The linearly polarized laser light could be modified to a circularly polarized light by applying crystalline quartz $\lambda/4$–plates.

3 Experimental results and discussion

While irradiating the (001) GaAs QWs by normally incident circularly polarized radiation (see inset in Fig. 1) a fast photocurrent signal, U_{12}, has been observed in unbiased samples across one contact pair (contacts 1 & 2 in the inset). The photocurrent changes sign if the circular polarization is switched from σ_+ to σ_-. Measurements of U_{12} as a function of the angle φ between the optical axis of the $\lambda/4$-plate and the plane of polarization of the laser radiation, reveal that the photogalvanic current j is proportional to the degree of circular polarization $P_{circ} = \sin 2\varphi$ (Fig. 1). The magnitude and the

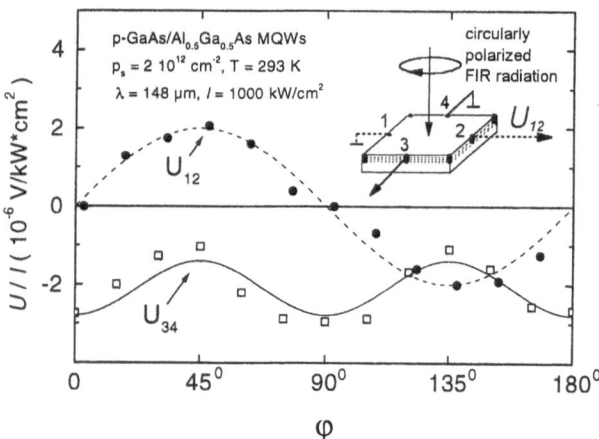

Fig. 2 Photogalvanic voltage signal U normalized by the intensity I for the two different contact pairs as a function of phase angle φ. Lines are fitted after Eq. 2 as $U_{12} \propto \sin 2\varphi$ and $U_{34} \propto (\chi_+ + \chi_- \cos^2 2\varphi)$, respectively.

sign of the CPGE are practically unchanged with variation of the angle of incidence in the range of -40° to +40°. The photogalvanic signal picked up at the other pair of contacts (contacts 3 & 4 in the inset of Fig. 2) does not change sign by switching of the circular polarization from σ_+ to σ_-. This is demonstrated in Fig. 2 where the voltage U_{34} is plotted as a function of the phase angle φ together with U_{12}. The signal U_{34} is periodic in φ with the period being one half of that of U_{12}. The photogalvanic current under study can be described by the following phenomenological expression

$$j_\lambda = \chi_{\lambda\mu\nu}(E_\mu E_\nu^* + E_\nu E_\mu^*)/2 + \gamma_{\lambda\mu}\, i(\mathbf{E} \times \mathbf{E}^*)_\mu\,, \qquad (1)$$

where $\mathbf{E}$ is the complex amplitude of the electric field of the electromagnetic wave and

$$i\,(\mathbf{E} \times \mathbf{E}^*) = P_{circ}\, E_0^2\, \frac{\mathbf{q}}{q}\,,$$

$E_0 = |\mathbf{E}|$, $\mathbf{q}$ is the light wavevector. In a bulk crystal, $\lambda = x, y, z$, while in a QW structure grown along the z direction, say $z \parallel [001]$, the index λ runs only over the in-plane axes $x \parallel [100], y \parallel [010]$ because the barriers prevent carriers from moving along the z axis and, definitely, $j_z = 0$. The photocurrent given by the tensor χ describes the so-called linear photogalvanic effect (LPGE) because it is usually observed under linearly polarized optical excitation. The circular photogalvanic effect (CPGE) described by the pseudotensor γ can be observed only under circularly polarized excitation.

In an (001)-grown QW structure with D_{2d} or C_{2v} point symmetry a generation of the circular photocurrent *under normal incidence* is forbidden. Thus in order to explain the experimental data the next possible subgroup C_s is needed, which contains only two elements: the identity and one mirror reflection. In this case under normal incidence and for light initially polarized along x' and transmitted through the $\lambda/4$ plate we have

$$j_{x'} = \gamma E_0^2 \sin 2\varphi = \gamma E_0^2 P_{circ}\,, \qquad (2)$$
$$j_{y'} = E_0^2(\chi_+ + \chi_- \cos^2 2\varphi) = E_0^2(\chi_2 - \chi_- P_{circ}^2)\,,$$

where $\chi_\pm = (\chi_{y'x'x'} \pm \chi_{y'y'y'})/2, \chi_2 \equiv \chi_{y'x'x'}$. Instead of x and y we use here the axes $x' \parallel [1\bar{1}0], y' \parallel [110]$. It follows then that a circular photocurrent is induced only along x', while the current induced along y' is a result of the linear photogalvanic effect described by the χ tensor components $\chi_{y'x'x'}$ and $\chi_{y'y'y'}$. by the projection of the circularly polarized radiation on x'. The CPGE described by γ can be related to an electric current in a system of free carriers with nonequilibrium spin-polarization [8]. The possible microscopic mechanism is the spin-dependent scattering of optically oriented holes.

Comparison of Eq. (2) and Figs. 1,2 demonstrates a good agreement between the theory and the experimental data. In order to prove that $j_{y'}$ is caused by the linear photogalvanic effect, we excited the sample by linearly polarized light and measured the voltage U_{34} as a function of the angle α between the plane of linear polarization and the axis x'. This signal is well described by $U_{34} \propto [\chi_+ + \chi_- \cos(2\alpha)]$ which follows from Eq. (1).

The low symmetry C_s gives an evidence for the in-plane anisotropy showing that in the investigated QWs (i) the [110] and [1$\bar{1}$0] crystallographic directions are different, and (ii) one of the reflection planes is removed as a symmetry element. This symmetry reduction is attributed to a tilt angle of 5° between the crystallographic [001] direction and the sample surface normal as has been verified by x-ray diffraction. In order to prove this assumption one sample grown on a (311)A-plane has been investigated. The CPGE has been observed with a magnitude comparable to the signal of the (001) sample. Since in the case of the (311)- sample only one QW has been grown in contrast to the (001) samples which consisted of 400 QWs, we conclude that the magnitude of the signal from a single (311)A-grown QW is several tens of times larger than that of tilted (001) samples.

To summarize, it is shown that the optical orientation of free carriers can be accompanied by their drift. The symmetry considerations show that the CPGE can be observed also in an ideal QW of the symmetry D_{2d} if the incident radiation is obliquely incident. The experiments presented here have been carried out in p-doped QWs but similar results are expected also for n-QWs.

Financial support by the DFG, NATO Linkage Grant and the RFFI is gratefully acknowledged.

References

1. S. Datta, and B. Das, Appl. Phys. Lett. **56**, (1990) 665.
2. G. Prinz, Phys. Today **48**, (1995) 58.
3. *Optical orientation*, F. Meier, and B.P. Zacharchenia, Eds. (Elsevier Science Publ., Amsterdam, 1984)
4. J.M. Kikkawa, and D.D. Awschalom, Nature **397**, (1999) 139.
5. J.M. Kikawa, and D.D. Awschalom, Phys. Rev. Lett. **80**, (1998) 4313.
6. S.D. Ganichev, W. Prettl, and I.N. Yassievich, review in Phys. Solid State **39**, (1997) 1703.
7. S.D. Ganichev, Physica B **273-274**, (1999) 737.
8. E.L. Ivchenko, Yu.B. Lyanda-Geller, and G.E. Pikus, Sov. Phys. JETP **73**, (1990) 550.

Cascading Effect in Type-II InAs/GaSb/AlSb Intersubband Light Emitter

K. Ohtani, H. Sakuma, and H. Ohno

Laboratory for Electronic Intelligent Systems, Research Institute of Electrical Communication, Tohoku University, 2-1-1 Katahira, Aoba-ku, Sendai 980-8577, Japan

Abstract Cascading effect in type-II InAs/GaSb/AlSb intersubband light emitting diodes have been demonstrated. Using the low group-V flux and control of interface bonds during molecular beam epitaxy, the samples with high structural and optical quality has been prepared. From current-light output characteristics, the slope efficiency is demonstrated to be proportional to the number of periods in agreement with theory, which indicates that injected carriers are recycled in the type-II InAs/GaSb/AlSb cascade structure.

1. Introduction

Using multi-stacking of active layers, cascading effect (carrier recycling) is important to realize high external quantum efficiency and high power operation in the wavelength region where photons have small energy. The successful implementation of the cascading effect was demonstrated in quantum cascade lasers (QCL) using intersubband transition in type-II GaInAs/AlInAs quantum well (QW) structures [1]. Recently high power continuous wave operation (5-8μm, 200mW/facet at 80K) in the region important for gas sensing and environment monitoring has been achieved [2, 3].

The intersubband light emitter based on type-II InAs/GaSb/AlSb quantum cascade structures has a great potential over the type-I material systems such as GaAs/AlGaAs and GaInAs/AlInAs structures. Type-II InAs/AlSb staggered structure has highly tunable intersubband transition energy due to the large conduction band offset energy (~1.35eV) compared with that of type-I systems. Type-II InAs/GaSb broken gap structure can be designed to block electrons injected into the excited state in InAs QW reducing the current path and can be used to extract electrons efficiently from the ground state by interband tunneling. Also since optical gain is larger than that of type-I systems because of the large dipole matrix element and small non-radiative relaxation rate, the laser threshold current is expected to be lower.

Although the cascading effect is essential ingredient in type-II InAs/GaSb/AlSb intersubband light emitter, it is difficult to demonstrate such effect because of the difficulty in obtaining run-to-run reproducibility. The previous experimental results about type-II interband cascade emitter show that due to the variance of the wafer-to-wafer structural

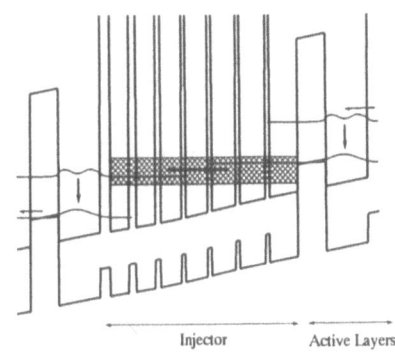

Fig. 1 Schematic band structure of type-II InAs/GaSb/AlSb intersubband light emitter

quality, the relationship between the number of active layers and the optical power does not agree with theory [4]. Here we have demonstrated the cascading effect in type-II InAs/GaSb/AlSb intersubband emitter grown by molecular beam epitaxy. In order to evaluate the effect, the samples with high structural and optical quality have been prepared by carefully optimizing growth condition.

2. Experimental

Intersubband light emitters were grown on undoped InAs (100) substrates by solid source molecular beam epitaxy. Two sets (sample A-B, C-D) of the samples containing different numbers of periods with the same active layer structure were prepared with careful control of the antimony flux and the interface bond configuration, which are essential for the high structural and optical quality. To minimize the lattice mismatch between the InAs substrate and the cascade structure, Al-As bond was adopted for both interfaces. Sample A-B have 5 and 10 periods while sample C-D have 10 and 20 periods. The injector structure consisted of modulation doped n-InAs/AlSb graded superlattices. The active layer for sample A-B consisted of an InAs/GaSb coupled QW, which was made of 7ML AlSb barrier, 30ML InAs quantum well (26ML for sample C-D) and 20ML GaSb quantum well. After the growth of the injector/active layer periods, an InAs contact layer was grown. The structures were processed into 400μm x 400μm mesa and a 10μm Cr/Au metal grating was fabricated to obtain intersubband emission in the vertical direction. The electroluminescence (EL) measurements were performed with a step-scan Fourier transform infrared (FT-IR) spectrometer.

3. Results and discussion

The structural quality was evaluated by double crystal X-ray diffraction (DCXRD) measurements. A number of sharp satellite peaks are observed for all the samples, showing that the cascade structure is of high structural quality and fully strained. The transmission electron micrograph taken on the 20-periods sample (sample D) indicates the absence of dislocations. Figure 2 shows the current-voltage characteristics of the two sets of the samples. The turn-on voltage (V_{on}), at which the current starts to flow by the alignment of injector potential, is given by,

$$V_{on}=N*(E_{12}-E_{qf})/e, \qquad (1)$$

where N is the number of periods, e is the unit of charge, E_{12} is the energy spacing between the first excited state and the ground state of InAs QW and E_{qf} is the quasi-Fermi level in the injector/active region. The turn-on voltage of sample B (sample D) is about two times larger than that of sample A (sample C), in accordance with the expectation that the turn-on voltage is proportional to the number of periods. The EL measurements shown in figure 3 revealed that both the emission peak energy and the full width at half-maximum (FWHM) are identical for the two samples with different periods, being at 253meV(255meV) and 11.8meV(11.2meV) for sample A (sample B), and being at 222meV(223meV) and 10.2meV(9.2meV) for sample C (sample D), respectively. This emission energy is in close agreement with the calculated energy between the ground state and the first excited state of InAs QW using a multi-band k p theory. Current-light output characteristics of all the samples were linear. The differential external quantum efficiency (η_{ex}) is given by,

$$\eta_{ex} =dP/dI=N*E_{12}*\tau_{NR}*\eta_{inj}*J/(e*\tau_{R}), \qquad (2)$$

where τ_{NR} is the non-radiative lifetime, τ_{R} is the radiative lifetime, η_{inj} is the injection efficiency and J is the injection current density. The slope of the current-light output characteristics for the two sets of the samples showed that the sample B (D) has 1.87 (1.73) times higher η_{ex} than the sample A (C), indicating that η_{ex} is proportional to the numbers of periods. This is a clear demonstration of injected carriers being recycled in the type-II InAs/GaSb/AlSb intersubband cascade structures.

4. Conclusion

We have demonstrated the cascading effect in intersubband emitting diode based on type-II InAs/GaSb/AlSb

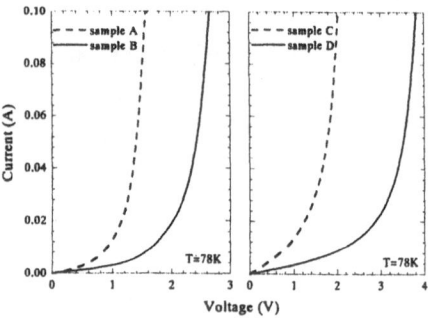

Fig. 2 Current-voltage characteristics of the samples

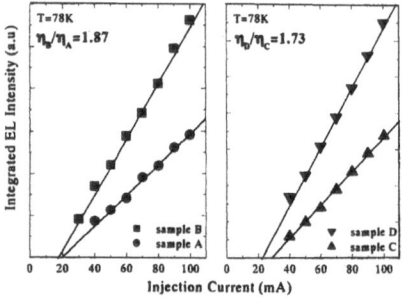

Fig. 3 Current-light output characteristics of the samples

quantum cascade structures for the first time. Many sharp satellite peaks of the DCXRD spectra and the same emission energy and FWHM of electroluminescence spectra indicate the successful preparation of the samples to evaluate such effect. Observed current-voltage and current-light output characteristics show that injected carriers are recycled in the type-II InAs/GaSb/AlSb intersubband quantum cascade structures.

References

1. J. Faist, F. Capasso, D. L. Sivco, C. Sirtori, A. L. Hutchinson, and A. Y. Cho, Science **264**, (1994) 553

2. J. Faist, A. Tredicucci, F. Capasso, C. Sirtori, D. L. Sivco, J. N. Baillargeon, and A.Y. Cho, IEEE J. Quantum Electron. **34**, (1998) 336.

3. C. Gmachl, A. Tredicucci, F. Capasso, A. L. Hutchinson, D. L. Sivco, J. N. Baillargeon, Appl. Phys. Lett. **72**, (1998) 3130.

4. E. Dupont, J. P. McCaffrey, H. C. Liu, C. H. Lin, M. Buchanan, and S. S. Pei, Appl. Phys. Lett. **72**, (1998) 1495.

Magneto-optical transitions involving a 2DEG confined in Cd(Mn)Te/CdMgTe quantum wells

F.J. Teran[1]*, M.L. Sadowski[1], M. Potemski[1], P.Kossacki[2] P. Hawrylak[3] and G. Karczewski[4]

[1] Grenoble High Magnetic Field Laboratory, MPI/FKF&CNRS, 38042 Grenoble, ˙e-mail: fjteran@labs.polycnrs-gre.fr
[2] Physics Department, IMO, Swiss Federal Institute of Technology, Lausanne, Switzerland
[3] Institute for Microstructural Sciences, NRC, Ottawa, K1A, OR6, Canada
[4] Institute of Physics, Polish Academy of Sciences, 02668 Warsaw

Abstract The main features of magneto-optical data which are characteristic for a large class of CdTe and CdMnTe modulation doped quantum well structures (MDQW) with 2D electron concentrations in the range of $3 - 7 \times 10^{11} cm^{-2}$ are discussed.

1 Introduction

In this paper we report the results of magneto-optical measurements performed on Cd(Mn)Te/CdMgTe 10 nm MDWQ's. Typical, low temperature luminescence (PL) and absorption type experiments in magnetic fields up to 23T are completed for our samples by transport, cyclotron resonance (CR) as well time-resolved and temperature dependent luminescence experiments.

Representative interband optical spectra measured at low temperatures in the absence of magnetic fields are shown in Fig.1. The broad PL spectrum is composed of a lower energy band (BB) and higher energy peak (EE) and shows a clear high energy cutoff which coincides with the onset of the absorption measured in the photoluminescence-excitation spectrum. The distance between the center of the BB band and the absorption onset can be identified with the Moss-Burnstein shift assuming $m_e=0.108m_0$ obtained from the CR experiments, and $n_e= 5.9 \times 10^{11} cm^{-2}$ determined from transport measurements, and finding $m_h \simeq 0.45m_0$ for the hole mass. The appearance of the EE peak is striking. Its amplitude depends strongly on the energy and intensity of laser excitation. The application of a magnetic field decomposes the PL spectrum into several lines.

2 Magnetic field dependence

Their energy positions versus magnetic field are shown in Figs. 2a and 2b for the CdMnTe/CdMgTe and CdTe/CdMgTe MDQWs, respectively. Both quantum wells have the same very similar electron concentrations. The diagrams are shown for spectra with preferential, σ^+. polarization for CdMnTe and σ^--polarization for the CdTe quantum well. Both samples show complex though similar diagrams[1], in which one can, however, distinguish two groups of transitions originating from the BB band, which resemble Landau-level fancharts . The EE PL-peak, even if well pronounced at zero field, becomes less visible at low magnetic field. Its energy position seems to be little influenced by the field but, surprisingly, this peak gains in amplitude in the vicinity of filling factor $\nu=2$, and dominates the PL spectrum at high fields. The most pronounced deviations from a linear field dependence are observed for two, lowest energy peaks L_0 and S_0. The irregularities in the positions of PL peaks correlate with the

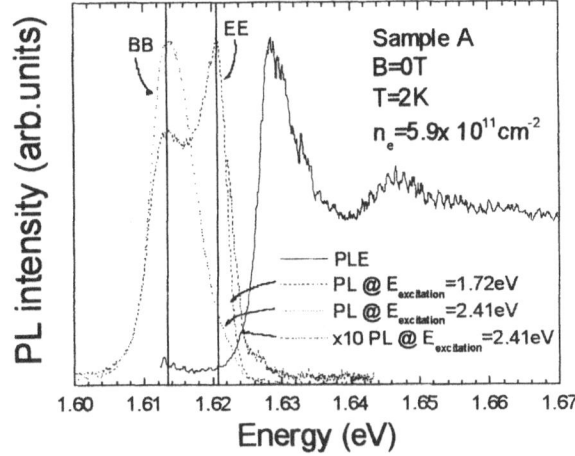

Fig. 1. PL (at two different excitation energies) and PLE spectra at T=2K and B=0T for a 10-nm $Cd_{0.995}Mn_{0.005}Te$/CdMgTe MDQW (sample A).

Landau level filling factor, as can be deduced from a comparison with transport data measured simultaneously with PL for the CdMnTe structure. These irregularities partially resemble the known effects observed previously in GaAs structures and are probably of many body origin[2]. In what follows we focus on higher energy transitions which show a more regular field behavior.

3 Optical transitions

The group of PL transitions denoted by L_N in Fig.2 can be identified with inter-band Landau level transitions which conserve the Landau level index ($\Delta n=0$). These transitions are observed in PL when the corresponding Landau levels are populated with electrons, but become progressively visible in absorption when Landau levels are emptied. Solid lines in Fig.2. show the dependence $E_{LN}=(N+1/2)\hbar\omega_r$ with the reduced mass of $m_r=0.0872$ for the CdMnTe sample. This again implies the hole mass of $m_h=0.5m_0$ when using the CR data for the electron mass.

Transitions from the second group, denoted by S_N, are not active in absorption and show a weaker field dependence. At first sight, the S_N transitions with $N>0$, could be recognized as "off diagonal" transitions[3] between the top valence band Landau level (preferentially occupied with holes at low temperatures) and the subsequent electronic Landau levels. Such a hypothesis implies, for example, the energy distance between L_1 and S_1 to be directly related to the hole mass, but the hole mass

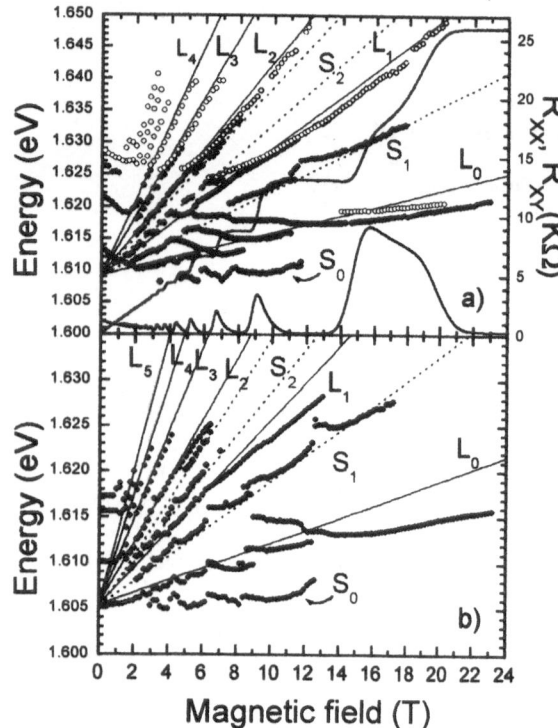

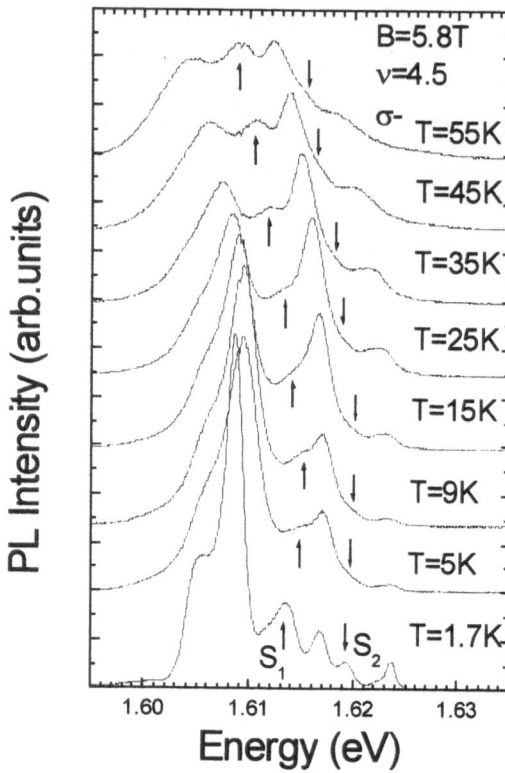

Fig.2. Energy positions of magneto-PL peaks (closed circles) and magneto-reflectance resonances (open circles) for: (a) sample A (σ^+ transitions) and (b) parent, nonmagnetic 10nm-wide CdTe/CdMgTe MDQW (sample B) (σ transitions). For sample A, the results of magneto-transport (R_{xx} and R_{xy}) are also shown. The 2D electron concentrations are $n_e = 5.9 \times 10^{11} \text{cm}^{-2}$ and $6.3 \times 10^{11} \text{cm}^{-2}$ for samples A and B, respectively.

value $m_h = 0.17 m_0$ determined in this way contradicts our previous estimations. One may think about different many body contributions to L_1 and S_1 transitions, but it is not clear why the above mentioned discrepancy should persist up to very high temperatures (Fig.3).

Fig.3. Temperature evolution of σ PL at B=5.8T for sample B.

Therefore, our investigations of Cd(Mn)Te MDQWS suggest that magneto-luminescence in these structures may partially reflect the effects of electron localization under conditions of quantum Hall effect.

4 Conclusion

We therefore conclude that sets of L_N and S_N transitions involve different electronic states. We suggest that electrons localized in the tails of Landau levels contribute to S_N transitions, whereas L_N luminescence involves delocalized electronic states (from the centers of Landau levels). We then expect, for example, that S_1 should involve electrons from the low energy tail of the first Landau level and holes from the top valence level if the temperature is sufficiently low. S_1 is now lowered with respect to L_1 by the sum of the cyclotron energy of the hole and an energy approximately equal to the half-width of the electronic Landau level. The latter component appears to be dominant, in agreement with estimations based on measured widths of quantum Hall plateaus. Localized electrons may recombine with any holes, i.e., also with those redistributed over different Landau levels at higher temperatures or at very short times after the laser excitation. This is the reason why S_1 is also observed under these extreme conditions as illustrated by our data shown in Figs 3 and 4.

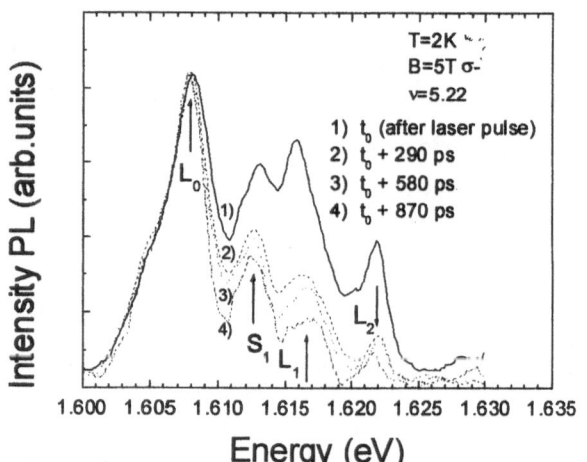

Fig.4 σ PL spectra integrated over the 290ps-wide time window measured at different delay times after the picosecond laser excitation for sample B, (t_0=50ps).

References

1. . F.J.Teran et al., Physica B **256-258** 577 (1998).
2. L.Gravier et al., Phys. Rev. Lett. **80** 3344 (1998)
3. S.K.Lyo et al., Phys. Rev. Lett. **61**, 2265 (1988).

Comparison of intersubband relaxation times in GaN/AlGaN and in InGaAs/AlGaAs quantum wells

T. Asano,[1] S. Yoshizawa,[1] S. Noda,[1] N. Iizuka,[2] K.Kaneko[2], N. Suzuki[2], and O. Wada[3]

[1]Department of Electronic Science and Engineering, Kyoto University, Yoshida-Honmachi, Sakyo-ku, 606-8501, Japan, email: tasano@kuee.kyoto-u.ac.jp
[2] Corporate Research and Development Center, Toshiba Corporation, 1, Komukai Toshiba-cho, Saiwai-ku, Kawasaki 212-8582, Japan
[3] FESTA Laboratories, The Femtosecond Technology Research Association, 5-5 Tokodai, Tsukuba 300-2635, Japan

Abstract The intersubband relaxation times in various quantum well structures are measured by femtosecond pump-probe technique. The dependence of the relaxation time on material/structure of the quantum wells is discussed. The relaxation time in GaN/AlN quantum wells is shown to be shorter than that in (In)GaAs/Al(Ga)As quantum wells by an order of magnitude. The importance of intrasubband relaxation process in short wavelength region is also shown.

1   Introduction

Intersubband transitions (ISB-Ts) in quantum well (QW) structures have been observed in various material systems such as GaAs/AlGaAs,[1] InGaAs/AlAs,[2-3] InGaAs/AlAsSb,[4] and GaN/AlGaN [5-6]. One of the most important properties of ISB-T is ultrashort energy relaxation time of excited carriers (picosecond ~ sub-picosecond) by means of LO phonon scattering. Although there are several reports on the ISB relaxation times that is measured with enough time resolution (~ 0.1 ps), it is not fully understood how the material and structure of the QW determine the ISB energy relaxation time. In this paper, we show the femtosecond pump-probe measurement results on the ISB carrier relaxation times in various QWs, and discuss the dependence of the relaxation time on the materials and structures.

2   Experimental

We prepared three different samples for relaxation time measurements: (a) 200 periods of $In_{0.3}Ga_{0.7}As(29Å)$ (:Si $1\times10^{19}cm^{-3}$) /AlAs(68Å) QWs [2] grown by MBE on a GaAs substrate, (b) 150 periods of GaAs (59Å) /$Al_{0.35}Ga_{0.65}As$ (150Å) (:Si $1.5\times10^{18}cm^{-3}$) QWs grown by MBE on a GaAs substrate (c) 200 periods of GaN (38Å) /$Al_{0.65}Ga_{0.35}N$ (18Å) (:Si $1\times10^{19}cm^{-3}$) QWs [6] grown by MOVPE on a sapphire substrate. The ISB absorption spectra (for TM polarization) of the samples are shown in Fig. 1. The ISB energy spacing is found to be 500meV (2.5µm) for sample (a), 172 meV (7.2µm) for sample (b), and 275meV (4.5µm) for sample (c) from the figure. The difference in the absorption magnitude is mainly determined by the difference in the transition dipole moment and the numbers of electrons effectively doped to the 1st conduction subband. Since the absorption in (a) is much smaller than that in (b) and (c), a 45°-prism-like configuration is used in the pump-probe measurements for the former while conventional Brewster angle configuration is used for the latter.

The ISB energy relaxation times were measured by one-color pump-probe method. The infrared ultrashort (< ~120 fs) optical pulses were generated by optical parametric amplification technique (for < 2.6µm) and also by differential frequency mixing technique (for 3.5~10 µm) from modelocked Ti-Sapphire laser pulses. The center

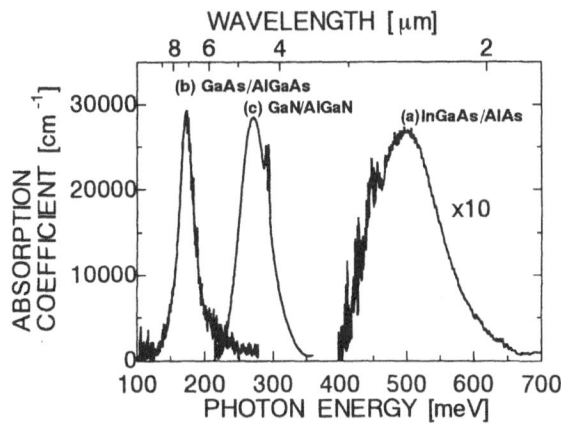

Fig. 1 Infrared absorption spectra of InGaAs/AlAs, GaAs/AlGaAs, and GaN/AlGaN quantum wells (TM polarization).

wavelength of the pulses was tuned to be resonant to the ISB absorption peak before the measurement of each sample.

3   Results and Discussions

The obtained pump-probe profiles are shown in Fig 2. The ISB relaxation times obtained for samples (a), (b) and (c) are ~ 2.7 ps, 0.6~0.7 ps, and ~ 0.15 ps, respectively. In Fig. 3, we plotted the measured intersubband relaxation times as a function of the intersubband energy together with the theoretical results. The sample (c) that consists of III-N materials has much shorter relaxation time than the samples (a) and (b) that consist of III-As materials. The difference can be explained from the viewpoint of the ionicity of the materials that mainly determines the electron-LO phonon scattering ratio.[7] Since the III-N materials have much lager ionicity than the III-As materials, the ISB electron scattering ratio in (c) is theoretically larger than those in (a) and (b) by an order of magnitude. It is shown in Fig. 3 that the calculated relaxation times almost reproduce the difference between the relaxation time in III-As materials and that in III-N materials. On the other hand, the difference in the relaxation time between (a) and (b) can be explained from the viewpoint of the ISB energy spacing. When the ISB energy becomes large, there are two factors which make ISB relaxation slower: (1) the ISB scattering time itself becomes longer since large momentum change is needed in the scattering process, and (2) intrasubband relaxation process begins to have additional effects [8] since the two subbands are not parallel. The results calculated by taking into account the both effects (dashed line) well reproduces the difference between (a) and (b) (Fig.3).

Therefore, we can expect from Fig.3 that the ISB relaxation time in GaN/AlGaN QWs may be on the order of sub-picosecond even if the wavelength is shortened to the communication wavelengths (1.3~1.55 μm).

4 Conclusion

We have measured and compared the ISB relaxation time in various QW structures by femtosecond pump-probe technique. The relaxation time in GaN/AlN QWs (~0.15ps) is an order of magnitude shorter than that in (In)GaAs/Al(Ga)As QWs (1~3ps), which is mainly due to the difference in the ionicity of the material. ISB-T in GaN/AlGaN is shown to be very promising for the application to ultrafast optical devices.

Acknowledgement

This work was supported in part by Grant-in-Aid for Scientific Research from the Ministry of Education, Science, Sports and Culture of Japan, the Femtosecond Technology Research Association (FESTA) which is supported by New Energy and Industrial Technology Development Organization (NEDO), and VBL Kyoto University.

References

1. L. C. West, and S. J. Eglash: Appl. Phys Lett. **46** (1985) 1156.
2. T. Asano, S. Noda, T. Abe, and A. Sasaki: Jpn. J. Appl. Phys. **35** (1996) 1285.
3. B. Shung, H. C. Chui, M. M. Fejer, and J. S. Harris, Jr.: Electronics Lett. **33** (1997) 818.
4. T. Mozume, H. Yoshida, A. Neogy, and M. Kudo, Jpn. J. Appl. Phys. **38** (1999) 1286.
5. C. Gmachl, H. M. Ng, and A. L. Cho: Appl. Phys. Lett. **77** (2000) 334.
6. N. Iizuka K. Kaneko, N. Suzuki, T. Asano, S. Noda, O. Wada: Appl. Phys Lett. **77** (2000) 648.
7. B. K. Ridley: Phys. Rev. **B39** (1989) 5282.
8. T. Asano, S. Noda, and K. Tomoda: Appl. Phys. Lett. **74** (1999) 1418.

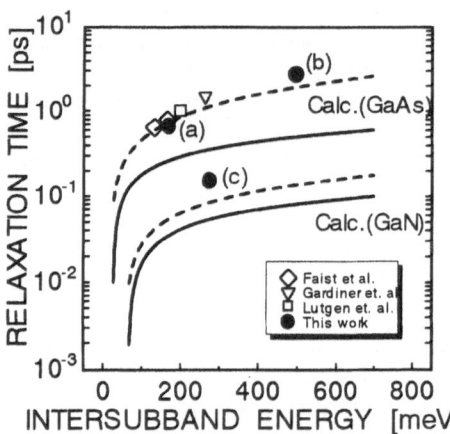

Fig. 3 Measured intersubband relaxation time plotted as a function of the intersubband energy. The data (a) ~ (c) in the figure correspond to the relaxation time measured for sample (a) ~(c) which is shown in Fig. 1. The solid and broken lines are the calculated relaxation time including and not including the intrasubband relaxation time, respectively. (The upper two lines: GaAs, the lower two lines: GaN.)

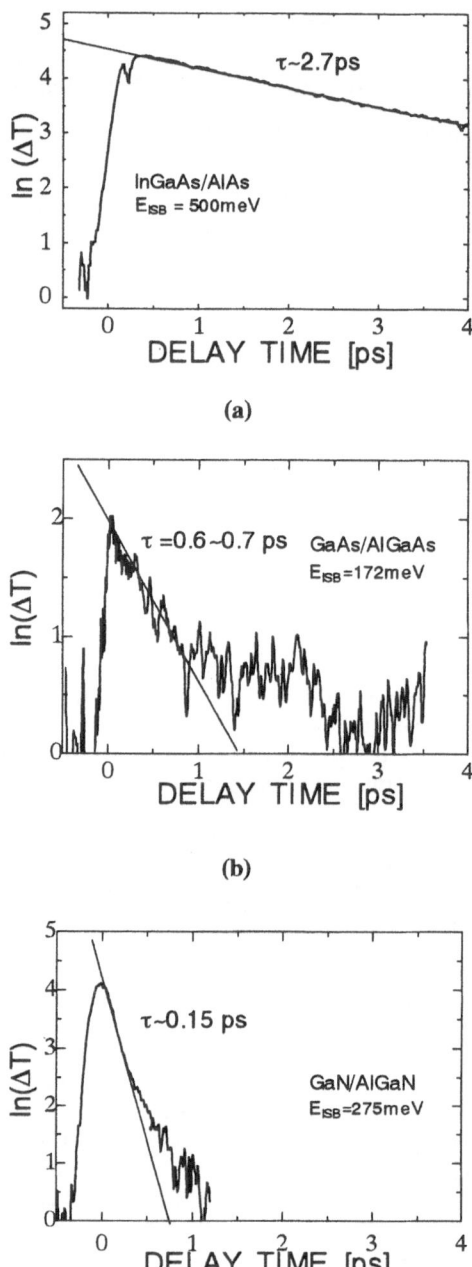

Fig. 2 Pump-probe profiles of transient absorption saturation of intersubband transition in (a) InGaAs(28Å)/AlAs quantum wells (b) GaAs(59Å)/AlGaAs quantum wells, and (c) GaN(40Å)/AlGaN quantum wells.

Electronic inelastic light scattering in a periodic δ-doping GaAs multiple quantum well structure

C. Kristukat[1], A.R. Goñi[1], S. Rutzinger[2], W. Wegscheider[2]*, G. Abstreiter[2], and C. Thomsen[1]

[1] TU Berlin, Institut für Festkörperphysik, PN 5-4, Hardenbergstr. 36, 10623 Berlin, Germany
[2] Walter Schottky Institut, TU München, Am Coulombwall 9, 85748, Garching, Germany

Abstract We have measured resonant inelastic light scattering by electronic intersubband excitations of a two-dimensional (2D) hole gas formed in the δ-doped layers of a so-called *nipi* structure. Light scattering spectra exhibit prominent features in the range from 12 to 20 meV, which are attributed to single-particle excitations between confined hole states in the deltalike p-doped regions. The dependence of their scattering intensity under resonance conditions have been studied as a function of polarization and temperature.

1 Introduction

The introduction of modulation doping in semiconductor technology has led to the fabrication of heterostructure devices with extremely high carrier mobilities. Growing a very thin layer with high impurity concentration, δ-like doping profiles can be achieved. In recent years, much effort has been devoted to the study of the electronic and optical properties of δ-doped systems [1–3] due to applications in modern high-mobility and high-speed devices. Inelastic light scattering is a powerful tool for the investigation of the spectrum of elementary excitations of low-dimensional electron and/or hole gases formed in doped semiconductor heterostructures [5]. Here we report on inelastic light scattering by intersubband electronic excitations of a two-dimensional hole gas (2DHG) formed in a δ-doped nipi structure. Spectra reveal a strong peak at about 12 meV, which exhibits marked collective character associated with excitations of the spin density. It also shows strong resonant enhancement of its scattering cross section when the laser energy matches that of an optical transition from an excited light-hole level, as indicated by the polarization selection rules. The resonant behavior of this peak was also studied as a function of temperature.

2 Experiment

The sample consists of an alternating n and p-doped (*nipi*) GaAs/AlGaAs multiple quantum well (MQW) structure grown by molecular beam epitaxy. The growth sequence starts with a 800 nm thick heavily p-doped GaAs layer followed by 31 periods of 2 nm GaAs and 28 nm Al$_{0.33}$Ga$_{0.67}$As. An alternating Si and Be δ-doping is incorporated in the center of the GaAs wells at a concentration of about $5 \cdot 10^{18}$ cm^{-3}. The structure is capped by another 800 nm thick p-doped GaAs layer. The use

* *Present address:*Institut für Angewandte und Experimentelle Physik, Universität Regensburg, 93040 Regensburg, Germany

of a MQW is to provide additional confinement for photoexcited minority carriers.

Resonant inelastic light scattering measurements were performed at low temperatures between 5 and 40 K in backscattering geometry from the cleaved edge of the sample. Hereafter, z is taken as the growth direction, thus, $x'(,)\bar{x}'$ being the scattering configuration with light linearly polarized either in y or z direction. Spectra were excited with a tuneable Ti:sapphire laser (700 nm - 850 nm) using a microscope to focus the light onto the sample with a spot diameter of about 2 μm. Power densities were kept below 150 W/cm^2. The scattered light was dispersed by a triple DilorXY spectrometer and detected with a LN$_2$ cooled charge-coupled device.

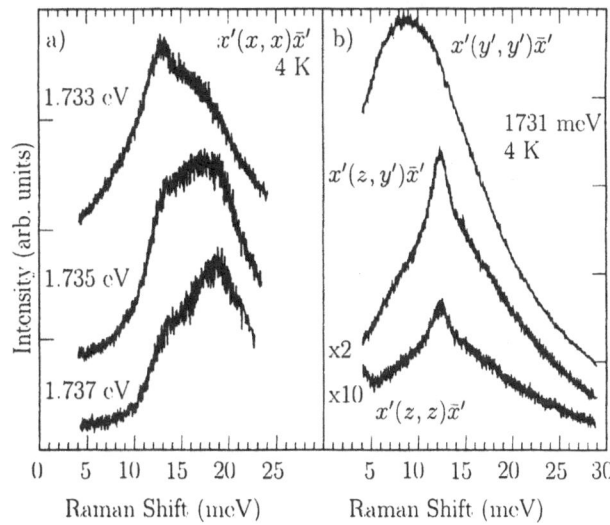

Fig. 1 (a) Light scattering spectra in $x'(z,z)\bar{x}'$ configuration from the cleaved edge of the sample for different excitation energies. **(b)** Spectra measured with different linear polarizations of the incident and scattered light.

3 Results and Discussion

Figure 1a shows polarized light scattering spectra measured at 4 K for different excitation energies. We find strong signals in the energy range between 12 and 19 meV, which are resonantly enhanced at an excitation energy of 1740 meV. They are apparent in spectra with both parallel and crossed linear polarization of incident and scattered light, thus, they can be identified as single-particle excitations (SPE). This SPE band exhibits at least two distinct peaks at 12 meV and 19 meV, which are observable in crossed polarization already at a very low exposure time of 1 s with 120 W/cm² excitation power density. This is an indication that these peaks possess a considerable contribution from collective excitations of the spin density. The linewidth of about 2 meV is most likely due to inhomogeneous broadening caused by fluctuations of the δ-doping potential. In contrast, the broad band between the two peaks is attributed to 2D intersubband pair excitations of the SPE continuum with wave vector $q > 0$, which are activated by disorder [4].

We interpret the features observed in light scattering spectra as due to intersubband excitations of a 2D hole gas (2DHG) formed in the p-doped regions. This assignment is based on the fact that for our nipi structure one expects an almost fully compensation between n and p doping [5], while the position of the Fermi level within the valence band is mainly determined by the uncompensated positive free charge of the two heavily p-doped GaAs cladding layers. Moreover, for the p-type δ-doped layers the repulsion of the minority carriers by the electrostatic potential is much less effective than for n-type ones [2], such that interband optical transitions are strongly suppressed in the n-doped regions. Hence, we assume that we are dealing with light scattering by excitations of the 2DHG. A further assignment of the excitations to particular hole eigenstates is not possible because of lack of knowledge about the Fermi energy. Nevertheless, by comparing with self-consistent calculations carried out for p-type δ-doped GaAs [2], we can infer that the peaks at 12 and 19 meV might correspond to intersubband transitions involving the first excited heavy and light-hole states.

Further information about electronic states can be obtained from the remarkable polarization dependence of the light scattering spectra. Figure 1b displays low-temperature spectra measured from the cleaved edge of the sample using different polarizations of the incident and scattered light. In $x'(y', y')\bar{x}'$ and $x'(y', z)\bar{x}'$ configuration the spectra are mostly the same and dominated by luminescence. Only in $x'(z, z)\bar{x}'$ and $x'(z, y')\bar{x}'$ polarization, i.e. with incident polarization parallel to the growth direction, the spectra show the features corresponding to intersubband excitations, being the signal in crossed polarization ten times stronger than for the parallel case. This gives clear evidence that the observed excitations pick up a large contribution from spin-flip pair transitions which are active in light scattering only for those resonances involving *light-hole* states. In contrast to heavy holes, the light-hole wavefunction has no definite spin component allowing for a spin flip in the inelastic light scattering process [5]. Since the spatial part of the light-hole wavefunction has predominantly p_z component, whereas heavy-hole states have only $p_{x,y}$ character, it is possible to select the valence-band state participating in the scattering process by setting the linear polarization of the incident photon either paral-

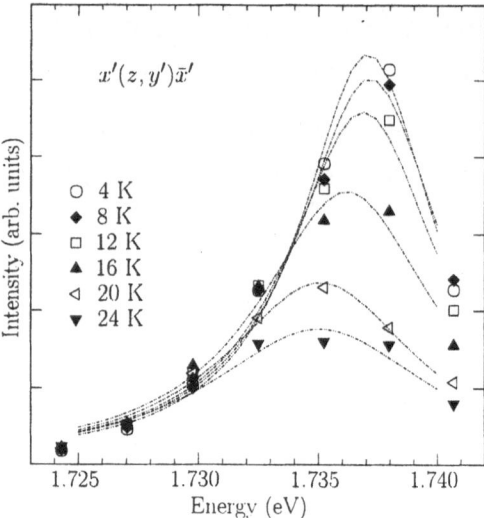

Fig. 2 Resonant profiles for light scattering by single-particle excitations of the 2DHG at different temperatures. Dotted lines represent fits to the data points using Lorentzian lineshapes.

lel (light holes) or perpendicular (heavy holes) to the growth direction [6].

Finally, we notice that the temperature dependence of the SPE feature at 12 meV exhibits an unexpected behavior for excitation energies near 1730 meV. The peak intensity increases up to 16 K before it begins to drop. A careful study of the photon-energy dependence of the resonance condition at different temperatures, however, demonstrates that this effect is simply due to the temperature renormalization of the energy of the optical transition giving rise to the resonance [6]. Figure 2 shows several resonance curves obtained in the temperature range between 4 and 25 K. With increasing temperature the maximum of the resonance curve at 1737 meV shifts to lower energies by about 2 meV while its width almost doubles pointing to a lifetime broadening of electron and hole states in the δ-doped layer. Placing the laser on the low-energy flank of the resonance curve results in a net increase of scattering efficiency for temperatures below 20 K.

References

1. J. Wagner, A. Hülsmann, G. Kaufel, and K. Köhler, Surf. Sci. **267**, (1992) 274.
2. D. Richards, J. Wagner, H. Schneider, G. Hendorfer, M. Maier, A. Fischer, and K. Ploog, Phys. Rev. B **47**, (1993) 9629.
3. C. A. Siqueira, W. N. Rodrigues, M. V. B. Moreira, A. G. de Oliveira, A. S. Chaves, and L. Ioriatti, Phys. Rev. B **59**, (1999) 5008.
4. A. Pinczuk, S. Schmitt-Rink, G. Danan, J. P. Valladares, L. N. Pfeiffer, and K. W. West, Phys. Rev. Lett. **63**, (1989) 1633.
5. G. Abstreiter, M. Cardona, and A. Pinczuk, *Light Scattering in Solids IV* (Springer, Berlin 1984), p. 5.
6. A. R. Goñi *et al.*, Solid State Commun. **116**, (2000) 121.

Proc. 25th Int. Conf. Phys. Semicond., Osaka 2000 (Eds. N. Miura and T. Ando)   729

Intersubband electroluminescence using X-Γ carrier injection in a GaAs/AlAs double-quantum-well superlattice

C. Domoto[1], N. Ohtani[1], K. Kuroyanagi[1], P. O. Vaccaro[1], T. Nishimura[1], H. Takeuchi[2], M. Nakayama[2]

[1] ATR Adaptive Communications Research Laboratories, 2-2 Hikaridai, Seika-cho, Soraku-gun, Kyoto 619-0288, Japan, e-mail: chiaki@acr.atr.co.jp

[2] Department of Applied Physics, Faculty of Engineering, Osaka City University, 3-3-138 Sugimoto, Sumiyoshi-ku, Osaka 558-0022, Japan

Abstract   We report the intersubband electroluminescence (EL) from a GaAs/AlAs superlattice consisting of asymmetric double quantum wells. Interband-photoluminescence properties provide confirmation that electron population at the second Γ (Γ2) subband in the GaAs layer results from the carrier injection into the Γ2 subband from the adjacent the lowest X (X1) subband in the AlAs layer, which is initiated by the X-Γ resonance. The energy of the EL, 190 meV, agrees with the energy spacing between the Γ2 and the lowest Γ (Γ1) subband in the GaAs layer.

1 Introduction

The intersubband emission in superlattices (SLs) and quantum wells (QWs) has been attracting much interest as a novel mid-infrared device: a quantum-cascade (QC) laser [1]. Current QC lasers, however, require complicated structures to achieve effective carrier injection into the higher-energy subbands. In this paper, we propose a simple carrier-injection mechanism for the intersubband emission.

In a GaAs/AlAs SL having a direct transition nature, the so-called type-I SL, the lowest energy electron and heavy-hole subbands (Γ1 and hh1) at Γ point are confined in the GaAs layer. In contrast, the AlAs layer, which is the barrier for the Γ subbands, is the quantum well for the X electron subbands. Recently, it has been reported that the optical and electronic properties of SLs are remarkably affected by the X subbands even in type-I SLs under applied bias voltages [2-5]. It should be noted that the photoluminescence (PL) due to the interband transition from the second Γ-electron subband (Γ2) to the hh1 subband is observable because of the carrier injection to the Γ2 subband in the GaAs layer from the first X subband (X1) in the adjacent AlAs layer.

2 Sample structures and experiment

In the present work, we have investigated intersubband electroluminescence (EL) in a GaAs/AlAs SL consisting of 20 periods of asymmetric double quantum wells (5.7-nm AlAs, 6.8-nm GaAs, 1.1-nm AlAs, and 4.5-nm GaAs layers). The samples are embedded in p-i-n (sample A) and n-n^--n (sample B) structures on a (001) n^+-GaAs substrate by molecular-beam epitaxy. For the sample B, the central 0.5-nm region in the 4.5-nm GaAs layer was doped with an n-type dopant of Si at ~2×10^{18} cm^{-3}. Optical measurements were performed at 20 K for interband PL and at 77 K for intersubband EL. Intersubband EL spectra in a mid-infrared region were observed with a Fourier-transform-infrared (FT-IR) spectrometer. In EL measurements, a pulsed bias voltage with the frequency of 1 kHz and the duty ratio of 50 % was applied, and the EL signal was detected with a lock-in amplifier.

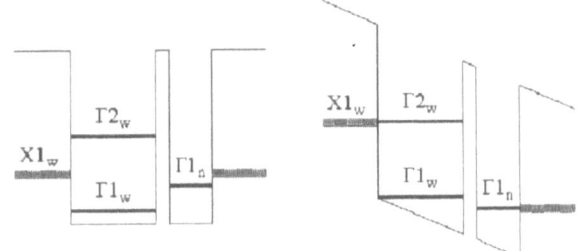

Fig. 1 Schematic energy-band diagrams of electron subbands under (a) the flat band condition and (b) the X1$_w$-Γ2$_w$ resonance condition, where the subscript w and n denote the wide and narrow layers in the asymmetric double quantum wells.

3 Results and discussion

Before discussing experimental results, we describe the expected mechanisms of the carrier injection and intersubband emission in the present sample. Figure 1 (a) and (b) show the schematic band diagrams of electron subbands under the flat band condition and the X1$_w$-Γ2$_w$ resonance one, respectively. The subscript w and n denote the wide and narrow layers in the asymmetric double quantum wells. In the framework of an effective-mass approximation, the calculated subband energies of Γ1$_w$, Γ2$_w$, Γ1$_n$, X1$_w$, hh1$_w$, and hh1$_n$ are 75, 266, 133, 185, 18, and 36 meV, respectively. From Fig. 1 (b) it is expected that the carrier injection to the Γ2$_w$ subband from the X1$_w$ one is initiated by the X1$_w$-Γ2$_w$ resonance. In addition, carriers of the Γ1$_w$ subband are quickly scattered to the Γ1$_n$ subband through the narrow AlAs layer, and then they are injected to the X1$_w$ subband in the next period. The scattering time from the Γ subband to the X one in GaAs/AlAs SLs is of the order of sub-ps [6]. Thus, the above cascade mechanism results in the intersubband emission from the Γ2$_w$ subband to the Γ1$_w$ one.

Figure 2 shows the interband PL spectra of sample A (p-i-n structure) as a function of reverse bias voltage, where the brightness is proportional to the logarithmic intensity. The white dotted lines indicate the calculated wavelengths of the Γ1$_w$-hh1$_w$, and Γ2$_w$-hh1$_w$ transitions under the flat band condition. We notice that two PL bands whose wavelengths are close to the calculated Γ2$_w$-hh1$_w$ one appear around the bias voltage of 6.5 V. The two PL bands are classified by the slope of the wavelength vs. the bias voltage. One PL band with a remarkable slope originates from the X1$_w$-hh1$_w$ transition because the indirect transition in real space produces an electrostatic

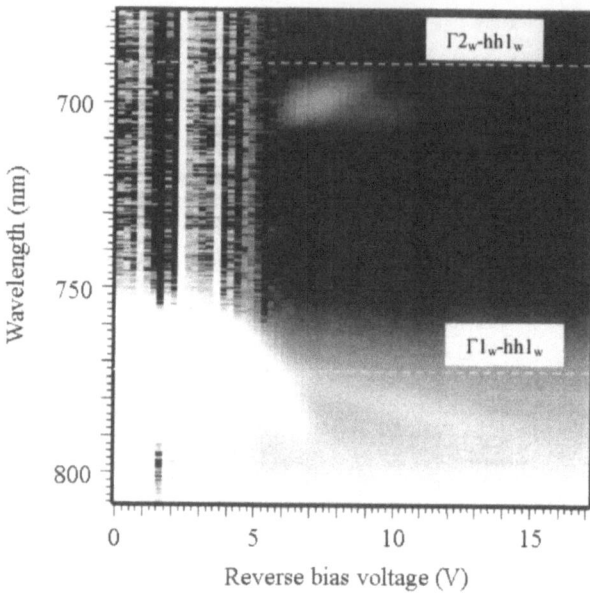

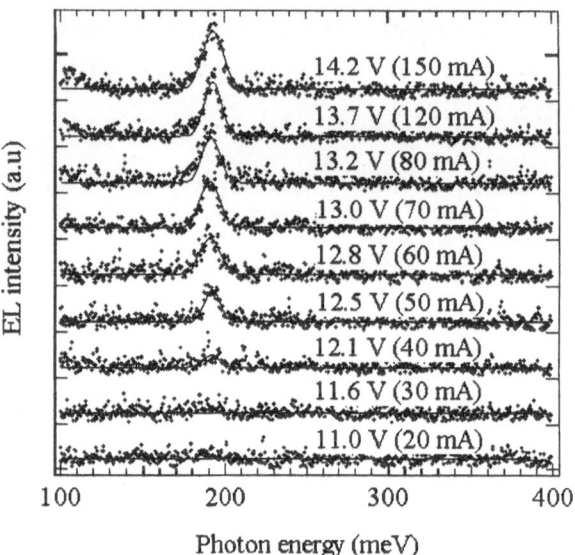

Fig. 2 Photoluminescence spectra as a function of the reverse bias voltage in sample A (*p-i-n* structure), where the brightness is proportional to the logarithmic PL intensity. The white dotted lines indicate the calculated wavelengths of the $\Gamma 1_w$-hh1$_w$ and $\Gamma 2_w$-hh1$_w$ transitions under the flat band condition.

Fig. 3 Electroluminescence spectra from the $\Gamma 2$-$\Gamma 1$ intersubband transition at different bias voltages in the sample B (*n-n⁻-n* structure).

potential proportional to the bias voltage. The other PL band is attributed to the $\Gamma 2_w$-hh1$_w$ transition. Although the $\Gamma 2_w$-hh1$_w$ transition is forbidden in principle, it is allowed because of the symmetry breaking of the envelope functions due to a quantum-confined-Stark effect. The bias voltage of ~6.5 V at which the PL intensity reaches a maximum corresponds to the X1$_w$-$\Gamma 2_w$ resonance condition as shown in Fig. 1(b), which is consistent with that estimated from the calculated subband energies. Thus, the appearance of the $\Gamma 2_w$- hh1$_w$ PL clearly indicates the carrier injection to the $\Gamma 2_w$ subband from the X1$_w$ one.

Figure 3 shows the mid-infrared EL spectra of sample B (*n-n⁻-n* structure) at the bias voltages from 11.0 and 14.2 V. The solid curves in Fig. 3 depict the mean spectra of the observed data points shown by dots, where scattered peaks are background noises in the FT-IR spectrometer. In Fig. 3, we can not observe any EL at the bias voltage below than 11.6 V. The EL starts at 12.1 V to be observed with the photon energy of 190 meV corresponding to the energy spacing between the $\Gamma 2_w$ and $\Gamma 1_w$ subbands. The appearance of the EL originates from the X1$_w$-$\Gamma 2_w$ resonance as shown in Fig. 1. The resonance voltage in sample B with the *n-n⁻-n* structure is much higher than that in sample A with the *p-i-n* structure. The reason for the difference of the resonance voltage seems to be considerable screening of the electric field in the *n-n⁻-n* structure. We have confirmed that the EL spectra are polarized perpendicular to the QW layer, which indicates that the EL is due to the intersubband carrier transition. We, therefore, consider that the intersubband emission results from the cascade mechanism based on the X1$_w$-$\Gamma 2_w$ carrier injection discussed with Fig. 1.

In conclusion, we have detected the mid-infrared EL originated from the intersubband transition in the GaAs/AlAs superlattice consisting of asymmetric double quantum wells. The polarization of the EL is perpendicular to the quantum well layer. The energy of the intersubband EL, 190 meV, has good agreement with the energy spacing between the $\Gamma 2_w$ and $\Gamma 1_w$ subbands. The electron population results from the carrier injection into the $\Gamma 2_w$ subband from the adjacent X1$_w$ one, which is initiated by the X1$_w$-$\Gamma 2_w$ resonance. This demonstrates that the X1$_w$-$\Gamma 2_w$ carrier-injection mechanism is useful for the simple designing of intersubband-emission devices.

Acknowledgements The authors would like to thank Tahito Aida, Norifumi Egami and Bokuji Komiyama for their encouragement throughout this work and Prof. Makoto Hosoda for his helpful discussions.

References
1. J. Faist, F. Capasso, D. L. Sivco, C. Sirtori, A. L. Hutchinson, and A.Y. Cho, Science, **264**, (1994) 553
2. M. Hosoda, N. Ohtani, H. Mimura, K. Tominaga, P. Davis, T. Watanabe, G. Tanaka, and K. Fujiwara, Phys. Rev. Lett., **75**, (1995) 4500
3. M. Hosoda, K. Tominaga, N. Ohtani, H. Mimura, and M. Nakayama, Appl. Phys. Lett., **70**, (1997) 1581
4. M. Hosoda, K. Tominaga, N. Ohtani, K. Kuroyanagi, N. Egami, H. Mimura, K. Kawashima, and K. Fujiwara, Appl. Phys. Lett., **71**, (1997) 2827
5. C. Domoto, N. Ohtani, K. Kuroyanagi, Pablo O. Vaccaro, and N. Egami, Jpn. J. Appl. Phys., **38** (4B), (1999) 2577
6. J. Feldman, J. Nunnenkamp, G. Peter, E. Gobel, J. Kuhl, K. Ploog, P. Dawson, and C. T. Foxon, Phys. Rev. B, **42**, (1990) 5809

Hot electron optical phenomena in GaAs/AlAs MQW structures in strong lateral electric field

L.E.Vorobjev[1], S.N.Danilov[1], I.E.Titkov[1], D.A.Firsov[1], V.A.Shalygin[1], A.E.Zhukov[2], A.R.Kovsh[2], V.M.Ustinov[2], V.Ya.Aleshkin[3], A.A.Andronov[3], E.V.Demidov[3], Z.F.Krasilnik[3]

[1] St. Petersburg State Technical University, St. Petersburg 195251, Russia, e-mail LVor@twonet.stu.neva.ru
[2] A.F.Ioffe Institute of Russian Academy of Science, St. Petersburg 194021, Russia
[3] Institute for Physics of Microstructures, Nizhny Novgorod 603600, Russia

Abstract Electrooptical phenomena due to intersubband electron transitions under heating electric field were investigated in GaAs/AlAs MQW structures. These structures are purposed for creation a mid infrared laser of new type. Experimental results on electron redistribution between size-quantization levels under electron heating were obtained up to electric field of 3500 V/cm.

1 Introduction

Recently a new intraband laser scheme was proposed [1]. It is based on hot electron phenomena in GaAs/AlAs MQW structures under lateral transport, namely, intervalley transfer and real space transfer of electrons. It should provide lasing in mid and far IR ranges. In the present paper an experimental study of electron redistribution between size-quantization levels under electron heating in such structures was performed, namely for the first time the electrooptical phenomena due to intersubband electron transitions were investigated in GaAs/AlAs MQW structures.

2 The samples

The structure under study was grown up by MBE method and consisted of 100 periods of 10 nm GaAs quantum wells and 2.5 nm AlAs barriers. The middle layer (5 nm width) of each quantum well was doped by Si ($N_D = 6 \cdot 10^{17}$ cm^{-3}). The considered MQW structure is such that the lowest electronic level of the system is the GaAs Γ-valley level, while the lowest level in the AlAs layer is the so-called X_Z-valley level. By X_Z we denote such X-valley which has the heaviest electron mass along the normal to heterointerface (Z axis). Calculated energies of the size-quantization levels are the following: $\Gamma 1 = 39$ meV, $\Gamma 2 = 148$ meV, $\Gamma 3 = 308$ meV, $\Gamma 4 = 500$ meV, $\Gamma 5 = 709$ meV; $X_z 1 = 195$ meV, $X_z 2 = 293$ meV, $X_z 3 = 436$ meV (zero energy corresponds to the Γ-valley bottom in GaAs).

3 Equilibrium absorption and birefringence

Experimental absorption spectra demonstrate strong absorption peak due to $\Gamma 1 - \Gamma 2$ transitions (Fig.1). This peak is observed at $h\nu = 117.4$ meV (at $T = 100$ K). Another peak near $h\nu \approx 150$ meV is probably connected with optical transitions $\Gamma 2 - \Gamma 3$. It has a very low intensity. The intensity of this peak increases with the temperature due to the increase of electron concentration in $\Gamma 2$ subband.

In accordance with selection rules, $\Gamma 1 - \Gamma 2$ optical transitions take place for light of p-polarization only. That is why there is a linear birefringence in the spectral range corresponding to these transitions, and refractive indexes for light waves of p- and s-polarization are not equal each other: $n_p \neq n_s$. We calculated the equilibrium spectra of birefringence magnitude ($n_p - n_s$) using Kramers–Kronig relationship and experimental spectra of optical absorption (α_p).

The calculated equilibrium birefringence spectra are shown in Fig.1.

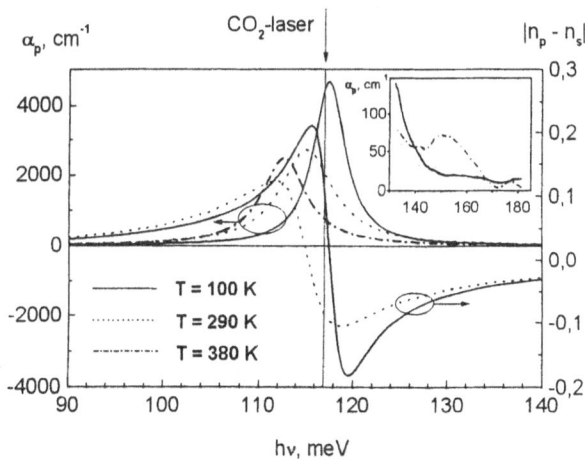

Fig. 1 Equilibrium absorption and birefringence spectra of GaAs/AlAs MQW structure at different temperatures.

4 Electrooptical phenomena

As it was predicted in [1], sufficiently strong lateral electric field can produce the electron population inversion between $\Gamma 2$ and $\Gamma 1$ levels, as well as between $X_Z 1$ and $\Gamma 1$ levels. Electrooptical phenomena are convenient tools to study experimentally the electron redistribution between these levels. We investigated the modulation of optical absorption and birefringence in strong lateral electric field using CO_2-laser. Its operating wavelength $\lambda = 10.6$ μm corresponds to the spectral range of $\Gamma 1 - \Gamma 2$ intersubband transitions. To increase the sensitivity we used multipass waveguide geometry for the measurements, so that the total length of the optical way through the MQW layers was 17.7 μm. The incident laser beam excited in the MQW layers the waves of both p- and s-polarization. We measured the amplitude modulation as well as the modulation of phase retardation for these waves. Experimental technique was the same as it was described in [2]. Electron heating was achieved with pulse lateral electric field ($\Delta t = 200$ ns).

The results of electrooptical experiments are presented in Fig.2. At a temperature of 100K there is a great decrease of optical absorption in electric field of 3500 V/cm. The magnitude of decrease is $\Delta\alpha_p = -1100$ cm^{-1}, i.e. 25% of the equilibrium value of the absorption coefficient α_p. At the same field the magnitude of birefringence $\Delta |n_p - n_s|$ is about 0.1.

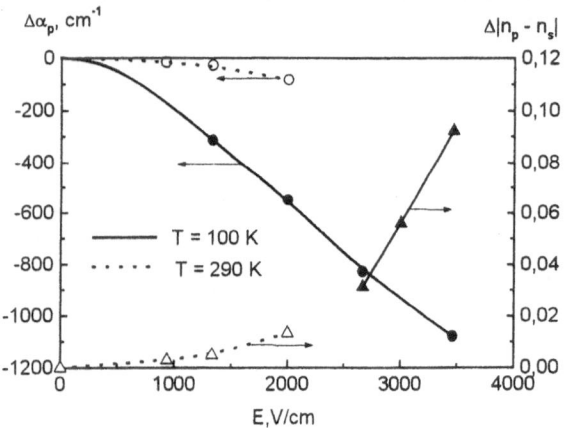

Fig. 2 Variation of the absorption coefficient and the birefringence magnitude with the lateral electric field.

There are two reasons of the modulation observed. In addition to the redistribution of electrons between the size-quantization levels, which we are interested in, we must also take into account the shift of the levels under electron heating due to exchange interaction

effect. The role of exchange interaction in intersubband absorption was investigated in [3]. It was shown that under electron heating the line of intersubband absorption undergoes a "red" shift due to exchange interaction effect.

As it is clear from Fig. 1, in the case of our experiment at $T = 100$ K and $\lambda = 10.6$ μm the line shift gives a great contribution to birefringence modulation, whereas redistribution of electrons between the size quantization levels gives a main contribution to absorption modulation. So we have extracted from our electrooptical measurements the value of spectral line shift and the rate of electron redistribution independently. At maximum electric field $E = 3500$ V/cm the magnitude of "red" shift is 0.7 meV, and relative decrease of difference $n_{\Gamma 1} - n_{\Gamma 2}$ is about 25%, where $n_{\Gamma 1}$ and $n_{\Gamma 2}$ are the electron concentrations in $\Gamma 1$ and $\Gamma 2$ subbands respectively. Similar value of "red" shift of absorption line can be estimated with the help of [3] and [4] taking into account relatively small electron mobility in our structure (1000 cm^2/V·s). The observed magnitude of $n_{\Gamma 1} - n_{\Gamma 2}$ is close to calculated one [1], so we can predict that population inversion between $\Gamma 1$ and $\Gamma 2$ subbands will appear at much higher fields: E > 8 kV/cm.

5 Conclusion

Optical absorption and birefringence in GaAs/AlAs MQW structures due to intersubband electron transitions were investigated both in equilibrium conditions and under electron heating by lateral electric field. Experimental results on electron redistribution between size-quantization levels were obtained up to field of 3500 V/cm. Relying on our experiments we concluded that population inversion between $\Gamma 1$ and $\Gamma 2$ subbands will appear at electric fields in excess of 8 kV/cm.

This work was supported by grants of INTAS, RFBR and Russian Programs "PhSSNS" and "Integration".

References

1. V.Ya.Aleshkin, A.A.Andronov, JETP Lett. **46**, (1998) 78.
2. L.E.Vorob'ev, I.E.Titkov, D.A.Firsov, V.A.Shalygin, et al., Semiconductors **32**, (1998) 757.
3. L.E.Vorobjev, I.I.Saydashev, D.A.Firsov, V.A.Shalygin, JETP Lett. **65**, (1997) 549.
4. E.Towe , L.E.Vorobjev, S.N.Danilov, Yu.V.Kochegarov, et al., Appl. Phys. Lett. **75**, (1999) 2930.

5 Heterostructures and Superlattices: Transport

Novel mid-infrared laser designs based on intraband and interband carrier transitions in quantum wells

L.E.Vorobjev[1], D.A.Firsov[1], G.G.Zegrya[2], V.L.Zerova[1], E.Towe[3], J.W.Cockburn[4], Z.F.Krasil'nik[5]

[1] St. Petersburg State Technical University, St. Petersburg 195251, Russia, e-mail LVor@twonet.stu.neva.ru
[2] A.F.Ioffe Institute of Russian Academy of Science, St. Petersburg 194021, Russia
[3] Laboratory for Optics and Quantum Electronics, University of Virginia, Charlottesville, Virginia 22903-2442
[4] Department of Physics, University of Sheffield, Sheffield S3 7RH, UK
[5] Institute for Physics of Microstructures, Nizhny Novgorod 603600, Russia

Abstract Two new types of mid infrared injection lasers containing quantum wells of asymmetrical funnel shape are proposed. Inversion of population arises between the third and the second levels of size quantization due to special potential profile of quantum well providing long electron lifetime on third excited level. Depopulation of the ground level is realized with intensive interband stimulated radiation generated in the same structure ("two-color laser", quantum wells are embedded into *i*-layer of diode laser structure) or due to resonant electron-hole Auger scattering in type II Sb-containing heterostructures ("Auger laser").

1 Introduction

There have been a big variety of interesting suggestions for realization of the mid-infrared (MIR, $\lambda > 4$ μm) quantum well (QW) lasers, based on intraband (intersubband) optical transitions. Only two of them have been so far experimentally realized. These are unipolar quantum cascade laser (UQCL) [1] and unipolar quantum fountain laser (UQFL) with intraband (intersubband) external optical pumping [2]. These new developments, having great scientific potential and importance, did not find yet any commercial applications. The UQCL requires complicated fabrication technology, while the UQFL needs a high level of MIR ($\lambda \approx 10$ μm) optical intraband pumping. These features pose a significant restraint to potential applications of these lasers.

In the present work, two new designs of the intraband QW lasers are proposed. In both cases population inversion (PI) is realized between third *e3* and second *e2* levels of size quantization in three-level QW system. Long relaxation lifetime on level *e3* exceeding the lifetime on level *e2* is obtained due to special shape of QW. Depopulation of the ground level *e1* is provided by different ways for "two-color laser" and "Auger laser" designs.

2 Principle of operation

2.1 Two-color laser

Arising inversion of population in two-color laser is clear from Fig. 1. We consider particular InGaAs/AlGaAs structure although the main conclusions are general and can be applied to other heterostructure materials. The structure parameters $L_1 = 4.6$ nm, $L_{NW} = 7.6$ nm, $L_2 = 13.8$ nm, were chosen for realization of the following conditions: the distance between two upper energy levels $E_{e3} - E_{e2}$ corresponds to MIR range; $E_{e3} - E_{e2} > \hbar\omega_0$ and $E_{e2} - E_{e1} > \hbar\omega_0$ ($\hbar\omega_0$ is the energy of optical phonon); the level *e3* is related to the wide portion of the QW, while the levels *e2* and *e1* are within the narrow portion of the QW. The obtained energy levels are: $E_{e1} = 52$ meV, $E_{e2} = 199$ meV, $E_{e3} = 318$ meV, $E_{hh1} = 12$ meV, $E_{hh2} = 45$ meV.

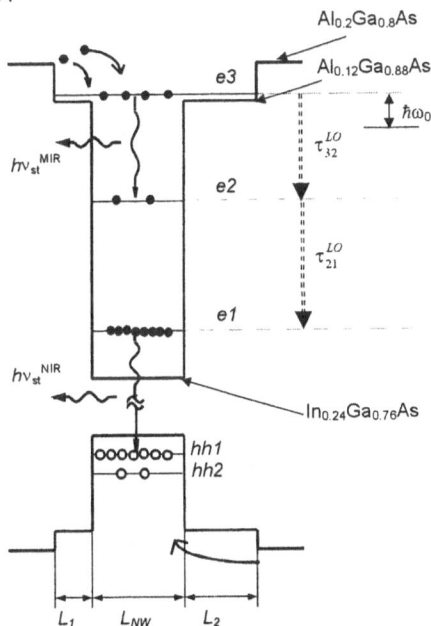

Fig. 1 Potential diagram and electron transitions in two-color laser.

Interaction with optical phonons determines the electron lifetime related to the intersubband transitions Electron lifetime on the level *e3* is relatively long because of a week overlap between the electron wave functions of *e3* and *e2*, *e1* levels. On the contrary, electron lifetime on level *e2* is small because of a large overlap of the electron wave functions of *e2* and *e1* levels.

Electrically or optically injected electrons remain on the level $e3$ for a long time τ_3, while the lifetime τ_2 on the level $e2$ is much shorter. Calculations give $\tau_{32}^{LO} = 5.4$ ps, $\tau_{21}^{LO} = 0.43$ ps, $\tau_{31}^{LO} = 19.6$ ps. This implies appearance of electron PI between the levels $e3$ and $e2$. If the optical gain exceeds the losses, the MIR lasing may begin. To prevent electron n_1 and hole p_1 accumulation at their respective ground levels $e1$ and $hh1$ (it may destroy the PI because of e-e and e-h interactions) stimulated interband ($e1 \rightarrow hh1$) near-infrared (NIR) emission is used. With this aim QW structure is embedded into i-layer of laser diode structure. Stimulated NIR emission efficiently removes electrons and holes from the levels $e1$ and $hh1$: after beginning of the NIR laser action the concentrations n_1 and p_1 become fixed and do not further increase when the injection current J exceeds threshold value J_{th}^{NIR}. To make simultaneous MIR and NIR lasing possible in the described structure, the appropriate combined optical waveguide for MIR and NIR radiation is constructed to accommodate both optical modes.

Using typical structure parameters and simple rate equations we calculated threshold current (or threshold intensity of interband optical pumping) for MIR stimulated emission: $J_{th}^{MIR} \cong 10$ kA/cm^2, $P_{vth}^{MIR} \cong 20$ kW/cm^2

2.2 Auger laser

Similar to the two-color laser the QW of asymmetrical funnel shape is used for creation of population inversion. Depopulation of the ground level $e1$ is provided here with resonant Auger recombination. To enhance the efficiency of Auger recombination it is convenient to use Sb-containing type II heterostructures (see Fig. 2). For the particular structure with the specific parameters $In_{0.65}Ga_{0.35}As_{0.45}Sb_{0.55}$ (narrow well), $Al_{0.1}Ga_{0.9}As_{0.05}Sb_{0.95}$ (wide well), $Al_{0.25}Ga_{0.75}As_{0.05}Sb_{0.95}$ (barrier), $L_1 = 3$ nm, $L_{NW} = 7$ nm, $L_2 = 18$ nm the following positions of energy levels were calculated: $E_{e1} = 102$ meV, $E_{e2} = 377$ meV, $E_{e3} = 536$ meV, $E_{e4} = 592$ meV, $E_{hh1} = -332$ meV, $E_{hh2} = -342$ meV, and the resonant Auger condition $E_{e3} - E_{e1} \cong E_{e1} - E_{hh1}$ is satisfied.

According to calculations the electron lifetime τ_A on level $e1$ is determined by probability of resonant Auger recombination:

$$\frac{1}{\tau_A} = 8\pi \frac{E_B}{\hbar} \left(\frac{E_{e1}^0}{k_B T} \right)^2 \frac{m_e}{m_h} \lambda_{E_g}^2 |\mathbf{Z}_{13}|^2 n_1 p_1 \cdot e^{-\frac{(E_{e3}-E_{e1})-(E_{e1}-E_{hh1})}{k_B T}}. \quad (1)$$

Here m_e and m_h are effective masses of electrons and holes; $\mathbf{Z}_{13}$ is dipole matrix element of electron transition from ground state $e1$ to excited state $e3$; $E_B = m_e e^4 / 2\hbar^2 \varepsilon_0$; $\lambda_{E_g} = \hbar / [2m_e (E_{e1} - E_{hh1})]^{1/2}$; n_1 and p_1 are surface concentrations of electrons and holes on levels $e1$ and $hh1$ ($n_1 \cong p_1$). Strong dependency of resonant Auger recombination probability upon electron concentration n_1 stabilize electron concentration at high injection levels as well as provides additional Auger pumping of excited level $e3$ (see Fig. 2). So, usually destructive for injection lasers Auger recombination is in our case significant positive factor.

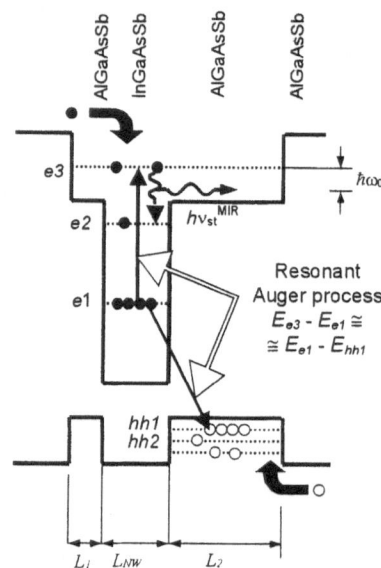

Fig. 2 Potential diagram and electron transitions in Auger laser.

Calculations of threshold current (or threshold intensity of interband optical pumping) for MIR stimulated emission: give the values: $J_{th}^{MIR} \cong 2.3$ kA/cm^2, $P_{vth}^{MIR} \cong 4.6$ kW/cm^2.

3 Laser design.

Electrical or optical interband pumping can be realized for the two-color and Auger lasers. It should be noted that relatively low values of threshold intensity of optical pumping allow using for pumping accessible semiconductor lasers. Simple MIR waveguide based on graded composition layers and p-i-n (electrical pumping) or undoped (optical pumping) structures can be easily developed.

This work is supported by INTAS, RFBR and Program PhSSNS of Russian Ministry of Science and Technology

References

1. J. Faist, F. Capasso, D.L. Sivco, C. Sirtori, A.L. Hutchinson, A.Y. Cho, Science **264**, (1994) 553.
2. O. Gauthier-Lafaye, P. Boucaud, F.H. Julien, S. Sauvage, S. Cabaret, J.-M. Lourtioz, V. Thierry-Mieg, R. Planel, Appl. Phys. Lett. **71** (1997) 3619.

Is there a true metallic state in two dimensions?

M.Y. Simmons[1], **A.R. Hamilton**[1], **M. Pepper**[2], **E.H. Linfield**[2], **P.D. Rose**[2] and **D.A. Ritchie**[2]

[1] School of Physics, University of New South Wales, Sydney 2052, Australia
[2] Cavendish Laboratory, Madingley Road, Cambridge CB3 0HE, United Kingdom

Abstract   We examine what has happened to the localising corrections to the conductivity due to both weak localisation and weak hole-hole interactions in high quality 2D GaAs hole systems that exhibit 2D metallic like behaviour. We demonstrate that these corrections are present in the so-called 'metallic' phase and are shown to increase as T→0. These results demonstrate that despite strong interactions ($r_s > 10$) 2D GaAs hole systems still behave like Fermi liquids. Furthermore we find that conventional disordered conduction with temperature dependent screening can account for many aspects of the so-called metallic state.

1 Introduction

The Quantum Hall effect is direct evidence for the existence of both localised and extended states in two-dimensional systems. It is still however unclear what happens to these extended states as B→0. In 1979 the one parameter scaling theory of localisation [1] showed that in the absence of a magnetic field the presence of any disorder in a non-interacting 2D system would localise all states at T=0. Whilst a 2D system might appear conducting at high temperatures, quantum corrections would take over as it was cooled to the absolute zero of temperature at which point it would become completely insulating. However in 1994 experimental evidence emerged suggesting the possible existence of a metallic state at B=0 in a strongly interacting, low disorder 2D system [2]. Since then metallic-like behaviour has been observed in other high quality 2D material systems including p-type SiGe and p-type GaAs [3–6].

In this paper we address the question of what has happened to the quantum corrections, originally predicted by scaling theory, in these high quality samples. We demonstrate that both weak localisation and electron-electron interaction effects are present in the metallic phase and give rise to localising corrections to the conductivity that increases as T→0. These localising corrections highlight that the system still behaves like a Fermi liquid, and that the drop in resistance with decreasing T is not quantum in origin [7]. Subtracting these corrections from the Drude conductivity reveals a linear decrease in conductivity with increasing temperature similar to that predicted for temperature dependent screening where $\Delta\sigma/\sigma \propto T$, for small T/T_F [8].

2 Experimental

Two different *in-situ* n+ back-gated modulation doped GaAs hole systems were used in this study. Sample A is a quantum well grown on a (311)A substrate with hole densities in the range $0 - 3.5\times10^{11}$ cm^{-2}, and a peak mobility of 2.5×10^5 cm^{2}V^{-1}s^{-1}. The critical carrier density at the transition is $\sim4.5\times10^{10}$ cm^{-2} ($r_s \sim$ 14, $m^* = 0.3m_e$). Sample B is a single GaAs/AlGaAs heterojunction [9], with a hole density in the range $0 - 10\times10^{10}$ cm^{-2}, a peak mobility of 1.1×10^6 cm^{2}V^{-1}s^{-1} and a critical density of $\sim1\times10^{10}$ cm^{-2} ($r_s \sim 25$).

Four terminal magnetoresistance measurements were performed at temperatures down to 150mK using low frequency ac lockin techniques and currents of 0.1-5 nA to avoid electron heating.

3 From insulating to metallic behaviour

Figure 1(a) shows the temperature dependence of the B=0 resistivity ρ of sample A plotted for six different carrier densities close to the transition, from $p_s = 3.5 - 5.6\times10^{10}$ cm^{-2}. Strongly localised behaviour is observed at the lowest carrier densities, with ρ taking the familiar form for variable range hopping:
$\rho(T) = \rho_{VRH} \exp[(T/T_{VRH})^{-m}]$, with m=1/2 far from the transition and m=1/3 close to the transition. A transition from insulating to metallic behaviour occurs as the carrier density is increased, with a critical density of $p_c \approx 4.5\times10^{10}$ cm^{-2} ($r_s = 12$) at the transition. Above this critical density the resistivity decreases exponentially as the temperature is reduced although the magnitude of the resistivity drop is much smaller than that found in silicon systems [2]. Previous studies of this sample have demonstrated that the ρ(T) traces can be scaled along the T-axis to collapse all the data onto one of two separate branches [5]. This ability to scale the data both in the strongly localised and metallic regimes has been taken as evidence for the existence of a quantum phase transition between insulating and metallic states in 2D systems [2].

The zero field conductivity, $\sigma(T)$ of a Fermi liquid with $r_s < 1$ is given by [10]:

$$\sigma(T) = \sigma_D(T) + \Delta\sigma_{WL}(T) + \Delta\sigma_I(T) \qquad (1)$$

where $\sigma_D(T)$ is the Drude conductivity, $\Delta\sigma_{WL}(T)$ and $\Delta\sigma_I(T)$ are the quantum corrections due to weak localisation and electron-electron (hole-hole) interactions respectively. If quantum corrections are present then we would expect to observe a weak increase in the resistivity at low temperatures near the transition. On the logarithmic scale of figure 1 the resistivity data closest to the transition ($p_s = 4.52\times10^{10}$ cm^{-2}) shows no evidence for any temperature dependence. Similar apparently temperature independent resistivity traces have been reported in silicon MOSFET samples [11]. Indeed such a temperature independent separatrix between metal-

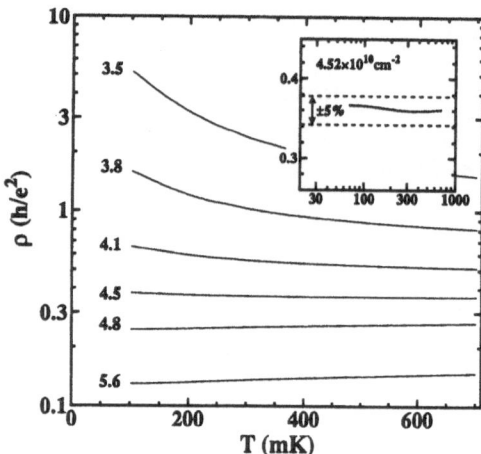

Fig. 1 Temperature dependence of the resistivity at densities from $p_s = 3.5 - 5.6 \times 10^{10}$ cm^{-2} for sample A. The inset shows the temperature dependence of the resistivity on a linear scale at $p_s = 4.52 \times 10^{10}$ cm^{-2} showing a small increase at temperatures below 300mK.

lic and insulating behaviour was taken as further evidence for a zero temperature quantum phase transition. However if we zoom in and look in detail at this curve (inset to Fig. 1), we can clearly see a weak (approximately 2%) increase in the resistivity as $T \to 0$.

A small logarithmic increase in the resistivity of similar magnitude $(1 - 5\%)$ with decreasing temperature has been observed in earlier studies of weakly interacting, disordered 2D systems $(r_s \sim 4)$ [12]. In these studies both weak localisation and weak electron-electron interactions caused a small correction to the Drude conductivity as $T \to 0$. It is difficult to unambiguously detect the presence (or absence) of quantum corrections solely from the $B = 0 \rho(T)$ data. To resolve this issue a more rigorous method is to study the temperature dependence of the magneto- and Hall resistivities.

4 Quantum corrections to the resistivity

4.1 Weak localisation

Figure 2 shows the temperature dependence of the $B=0$ resistivity (left hand panel) and magnetoresistance (right hand panel) at different densities on both sides of the "metal"-insulator transition. In Fig. 2(a) we are just on the insulating side of the transition. The left hand panel shows that $\rho(T)$ is essentially T-independent down to 300mK and then increases by 2.5% as the temperature is further reduced. The origins of this increase can be ascertained from the magnetoresistance shown in the right hand panel of Fig. 2(a). Below ~300mK a characteristic negative magnetoresistance peak develops as phase coherent weak localisation effects become important, mirroring the small increase in the resistivity at $B=0$.

Increasing the carrier density brings us into the metallic regime (Fig. 2(b)) where the exponential drop in the resistivity with decreasing temperature starts to become visible. The upturn in $\rho(T)$ marked by the arrow has

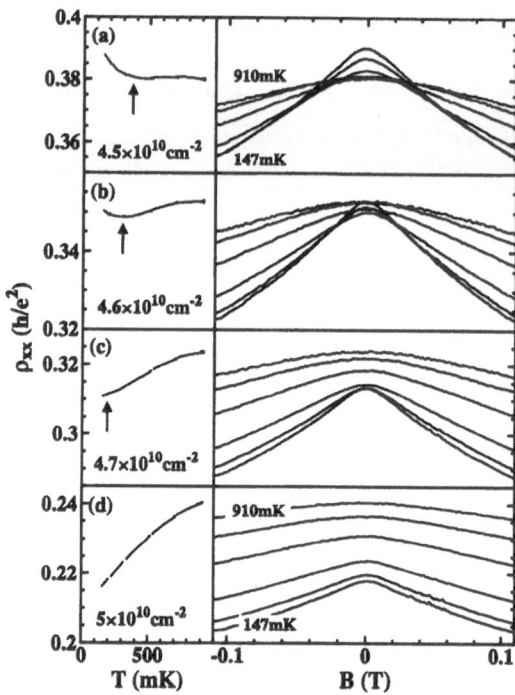

Fig. 2 (a-d) The left hand panels show the $B = 0$ resistivity as a function of temperature, illustrating the transition from insulating to metallic behaviour as the density increases. The right hand panels show the corresponding magnetoresistance traces for temperatures of 147, 200, 303, 510, 705 and 910mK for sample A.

moved to lower temperatures and the negative magnetoresistance in the right hand panel has become less pronounced. Further increasing the density (Fig. 2(c)) causes the metallic behaviour to become stronger, with the upturn in $\rho(T)$ moving to even lower temperatures, until at $p_s = 5 \times 10^{10}$ cm^{-2} the upturn is no longer visible within the accessible temperature measurement range. However, the magnetoresistance still exhibits remnants of the weak localisation temperature dependent peak at $B=0$. The weak localisation is therefore always present and is neither destroyed in the metallic regime, nor is it "swamped" by the exponential decrease in resistivity with decreasing temperature. Instead what can clearly be seen in the left hand panel of Fig. 2, is that the upturn in $\rho(T)$ due to weak localisation marked by the arrows, moves to lower T as the carrier density is increased. This is not surprising since as we move further into the metallic regime both the conductivity and therefore the mean free path $(l \propto \sigma/\sqrt{p_s})$ increase, such that the weak localisation corrections $\Delta\sigma_{WL} \propto l_\phi/l$ are only visible at lower temperatures (larger l_ϕ).

We can extract the phase relaxation time τ_ϕ by transposing the magnetoresistance data to magnetoconductance and fitting it to the Hikami formula for weak localisation [13]. Figure 3 shows the temperature dependence of the phase breaking rate, $1/\tau_\phi$ for three different densities on both sides and close to the "metal"-insulator

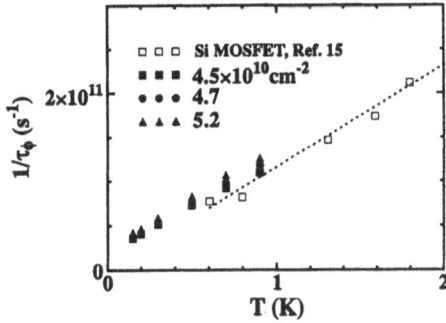

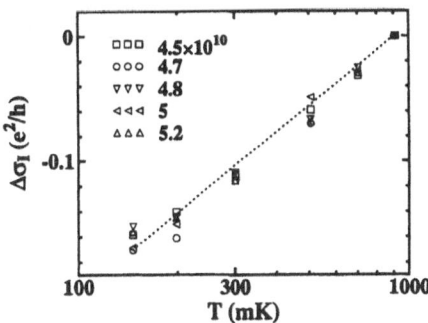

Fig. 3 Phase breaking rate versus temperature for densities close to the "metal"-insulator transition for sample A. Solid symbols are data obtained from this study; open symbols are data from Si MOSFETs, Ref. [15]

Fig. 4 The logarithmic correction to the conductivity in sample A from the hole-hole interactions at densities close to the transition.

transition. The phase breaking rate falls approximately linearly with decreasing temperature for all three traces in agreement with that predicted for disorder enhanced hole-hole scattering [14], where $1/\tau_\phi \sim 2k_B T/(\hbar k_F l)$. This phase breaking mechanism only depends on $k_F l$ and it is therefore particularly noteworthy that the phase breaking rates in these low density p-GaAs samples, with $2.5 < k_F l < 5$, are almost identical to those found in n-type silicon MOSFETs [15] with $k_F l \sim 1$, despite a factor of 20 difference in the carrier densities. This agreement with scattering limited electron lifetime suggests that the electron states are only mildly perturbed by the strong interactions and essentially remain Fermi liquid like.

Another important feature of these results is that there is little variation in τ_ϕ with density and in particular there is no dramatic change in τ_ϕ as we cross from insulating ($p_s = 4.5 \times 10^{10}$ cm^{-2}) to metallic behaviour ($p_s = 5.2 \times 10^{10}$ cm^{-2}). There is therefore no reflection of the exponential decrease of $\rho(T)$ with decreasing temperature in the phase breaking rate. This implies that whatever mechanism is causing metallic behaviour does not suppress weak localisation as originally believed and is further evidence that the system is behaving as a Fermi liquid. Since all models of the resistivity in the metallic phase [16,17] predict that the exponential drop saturates at low temperatures, our data shows that localisation effects will again take over as T→0.

4.2 Electron-electron interactions

Whilst our results indicate that weak localisation takes over as $T \to 0$ there remains the possibility that electron-electron interactions could stabilise a metallic state. To investigate this possibility we examine the low field Hall effect. Unlike weak localisation, interactions not only affect the $B = 0$ resistivity, but also cause a correction to the Hall resistance:

$$\frac{\Delta R_H}{R_H} = -2\frac{\Delta\sigma_I}{\sigma} \qquad (2)$$

From the temperature dependence of the Hall resistivity we extract the interaction correction to the zero field conductivity, $\Delta\sigma_I$ using equation (2). Figure 4 shows

a plot of the interaction correction for different carrier densities on both sides of the transition. All the data collapses onto a single line, clearly demonstrating a $\log(T)$ dependence of $\Delta\sigma_I$, which reduces the conductivity to zero as T→0.

It is perhaps surprising that results observed in, and derived from, weakly interacting systems apply to our system where interactions are strong and $r_s > 10$. Nevertheless, we find reasonable agreement between the magnitude of the logarithmic corrections due to interactions in our system and those predicted by Altshuler et al. [14] (within a factor of 2). As with the phase coherent effects this logarithmic correction due to hole-hole interactions is independent of whether we are in the insulating or metallic phase and is present despite the exponential drop in resistivity. This suggests that electron-electron interactions are not responsible for the 2D "metal"-insulator transition observed in high mobility systems.

5 Temperature dependent screening

We can subtract the corrections due to weak localisation and hole-hole interactions from the measured zero-field conductivity to extract the bare Drude conductivity $\sigma_D(T)$ using equation 1. These results are plotted in figure 5 against T/T_F at a carrier density of 4.7×10^{10} cm^{-2} - just on the metallic side of the transition. Immediately we can see that the resulting Drude conductivity shows a linear metallic temperature dependence in agreement with that expected for temperature dependent screening in the clean limit [8,19]. Similar results have been reported previously in high quality p-GaAs [5] and more recently in p-SiGe samples [20].

The universality of this result in p-GaAs is highlighted by comparing the results from sample A with that from a higher quality sample – sample B [21]. Figure 6(a) shows the temperature dependence of the $B=0$ resistivity at different carrier densities in the so-called metallic regime for sample A. In the metallic regime where $\rho \ll h/e^2$ the quantum corrections are negligible, and we can replot the data (Fig. 6(b)) to show the fractional change in the Drude conductivity $\Delta\sigma/\sigma$ against T/T_F. Here $\Delta\sigma$ is the change (decrease) in conductivity

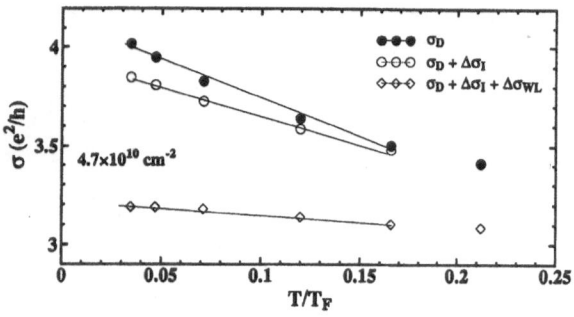

Fig. 5 Extraction of the Drude conductivity $\sigma_D(T)$ in sample A from the measured zero field conductivity at a density of 4.7×10^{10} cm^{-2}.

with increasing T, and T_F is the Fermi temperature. All the data for the different densities collapse onto a common curve, which is linear at low T/T_F, and saturates at higher T/T_F as the system becomes non-degenerate.

Figure 6(c) shows the T-dependent resistivity for a range of carrier densities from $1.6 - 3.5 \times 10^{10}$ cm^{-2} for sample B. This sample shows much stronger metallic-like behaviour than sample A, with up to a factor of two decrease in resistivity as the temperature is reduced from 700mK to 30mK. The corresponding fractional change in conductivity is plotted in Figure 6(d), where the data again collapse onto a common trace. In both cases $\Delta\sigma/\sigma$ is linear in T/T_F at low temperatures ($T/T_F < 0.2$), and flattens off at higher temperatures.

The magnitude of the temperature dependent screening correction to the conductivity $\Delta\sigma/\sigma$ is larger for the higher quality sample B than for sample A. This is consistent with the theory of Ref. [8], which predicts that the effect of temperature dependent screening is more significant at low densities. It is only possible to reach such low densities in very high quality samples such as sample B.

6 Conclusions

We show that both weak localisation and electron-electron interaction effects are present in the so-called metallic phase and give rise to localising corrections to the conductivity that increases as T→0. These localising corrections highlight that the system still behaves like a Fermi liquid, and that the drop in resistance with decreasing T is not quantum in origin. In the absence of these corrections saturating at low (inaccessible) temperatures the results argue against the existence of a true metallic phase even when interactions are strong ($r_s = 10$).

Furthermore we show that temperature dependent screening can account for many aspects of the metallic like drop in the resistivity observed in two different low disordered GaAs hole systems. The fractional change in the conductivity $\Delta\sigma/\sigma$ depends only on T/T_F such that the so-called metallic state can be described by conventional disordered conduction.

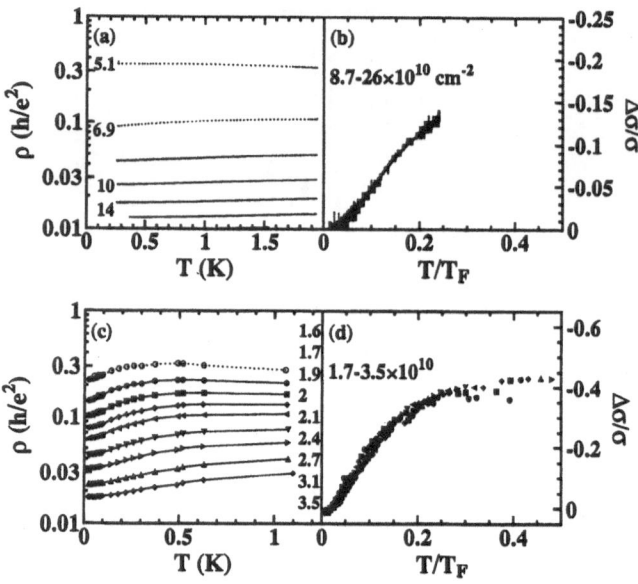

Fig. 6 Comparison of the metallic behaviour in samples A and B. (a) Temperature dependence of the $B = 0$ resistivity at carrier densities in the range $5.1 - 14 \times 10^{10}$ cm^{-2} for sample A. (b) Corresponding fractional change in the conductivity $\Delta\sigma/\sigma$ as a function of the dimensionless temperature T/T_F. (c,d) Equivalent data for sample B for $p_s = 1.6 - 3.5 \times 10^{10}$ cm^{-2}

References

1. E. Abrahams *et al.*, Phys. Rev. Lett. **42**, (1979) 690; L.P. Gor'kov *et al.*, JETP Lett. **30** (1979) 229.
2. S.V. Kravchenko *et al.*, Phys. Rev. B **50** (1994) 8039; Phys. Rev. B **51** (1995) 7038.
3. P.T. Coleridge *et al.*, Phys. Rev. B **56** (1997) R12764.
4. M.Y. Simmons *et al.*, Physica B **249-251** (1998) 705.
5. M.Y. Simmons *et al.*, Phys. Rev. Lett. **80** (1998) 1292.
6. Y. Hanein *et al.*, Phys. Rev. Lett. **80** (1998) 1288.
7. M.Y. Simmons *et al.* Phys. Rev. Lett. **84** (2000) 2489.
8. A. Gold and V.T. Dolgopolov, Phys. Rev. B **33**, (1986) 1076.
9. M. Y. Simmons *et al.*, Appl. Phys. Lett. **70** (1997) 2750.
10. B.L. Altshuler and A.G. Aranov, Electron-electron interactions in Disordered Conductors, (Elsevier Science, Holland 1985).
11. S.V. Kravchenko *et al.*, Phys. Rev. Lett. **84** (2000) 2909.
12. M.J. Uren *et al.*, J. Phys. C**13** (1980) L985; D.J. Bishop *et al.*, Phys. Rev. Lett **44**, (1980) 1153.
13. S. Hikami *et al.*, Prog. Theor. Phys. **63** (1980) 707.
14. B.L. Altshuler *et al.*, J. Phys. C **15** (1982) 7367.
15. R.A. Davies and M. Pepper, J. Phys. C **16**, (1983) L353.
16. V.M. Pudalov *et al.*, JETP Lett. **66** (1997) 175.
17. A.P. Mills *et al.*, Phys. Rev. Lett. **83** (1999) 2805.
18. M.J. Uren *et al.*, J. Phys. C **14** (1981) 5737.
19. S. Das Sarma and E.H. Hwang, Phys. Rev. Lett **83** (1999) 164.
20. V. Senz *et al.*, cond-mat/0004312.
21. A.R. Hamilton *et al.*, cond-mat/0003295.

Self-induced Shapiro effect in semiconductor superlattices

F. Löser[1], M. M. Dignam[2,3], Yu. A. Kosevich[2,4], K. Köhler[5], K. Leo[1]

[1] Institut für Angewandte Photophysik, Technische Universität Dresden, Germany, e-mail: leo@iapp.de
[2] Max-Planck-Institut für Physik komplexer Systeme, 01187 Dresden, Germany
[3] Physics Department, Queen's University, Kingston, ON, Canada K7L 3N6
[4] Moscow State University of Technology "STANKIN", 101472 Moscow, Russia.
[5] Fraunhofer-Institut für Angewandte Festkörperphysik, 79108 Freiburg, Germany

Abstract We observe that the periodic motion of Bloch-oscillating carriers in semiconductor superlattices is accompanied by a quasi-DC current: The photo-injected electron hole pairs interact with their self-induced oscillating field mediated by the nonequilibrium electron-hole plasma. This novel quantum effect is a coherent anolog of the Shapiro effect in Josephson junctions. The direction of the motion can be controlled by the laser excitation.

Among the macroscopic quantum effects known in physics, the Josephson effect in weakly coupled superconductors and Bloch oscillations (BO) in semiconductor superlattices (SLs) [1] are both related with the generation of an AC effect by application of a DC field to the system. In case of the Josephson effect, the phase between the wave functions varies linear in time; in case of BO, the k-vector of an electron in a periodic potential subject to a static electric field F varies linear in time. Due to the periodicity of the band structure, the electron performs a periodic spatial motion with the Bloch frequency: $\omega_B = edF/\hbar$ and a semiclassical spatial amplitude $Z_B^s = \Delta/(2eF)$, where d is the lattice constant and Δ is the width of the band. The frequency space analog of BO is the Wannier-Stark ladder (WSL) eigenstates with resonances at $E_p = E_0 + p\hbar\omega_B$, $p = 0, \pm1, \pm2, ...$, and BO are simply quantum beats between the different levels. Both the WSL [2,3] and BO [4–10] in time domain were recently observed in SL by using optical techniques. Both Josephson effect and BO show new phenomena if DC and AC voltages are applied: The current across Josephson junction shows the so-called Shapiro steps [11] which appear at voltages corresponding to multiples of the photon energy of the AC field. A similar effect for the *incoherent* (phase-independent) current in a semiconductor SL has been predicted by Ignatov et al. [12] and observed by Unterrainer et al. [13].

Here, we report about a novel *coherent* (phase-dependent) Shapiro effect by directly measuring the carrier motion using optical techniques [14]. The effect is caused by the AC depolarization field created by the Bloch-oscillating carriers themselves in a SL in an external DC field. The direction of the quasi-DC motion of the carriers can be externally controlled by changing the central frequency of the exciting optical pulse.

We use a 67/17 Å GaAs/Al$_{0.3}$Ga$_{0.7}$As SL with a combined miniband width of $\Delta \approx 40\,meV$ held at $10\,K$

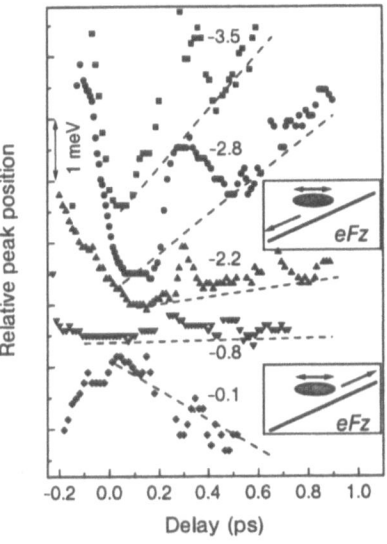

Fig. 1 Shift of the $p = -1$ heavy-hole WSL ladder transition as a function of the delay time. The parameter for the different curves is the spectral position of the laser pulse, $\hbar\omega_c$, relative to the experimentally observed $p = 0$ transition, E_0^{EX}, given in units of the WSL splitting $\hbar\omega_B$: $(\hbar\omega_c - E_0^{EX})/\hbar\omega_B$. The carrier density is $2.9 \times 10^9\,cm^{-2}$ per well and the electric field 15 kV/cm.

in an applied field of $F = 15\,kV/cm$. This corresponds to a semiclassical amplitude $Z_B^{sc} = 1.6d$ and a BO period of 315 fs. The wave packets are optically excited by a mode-locked Ti-Sapphire laser with a spectral width (FWHM) of $21\,meV \triangleq 1.7\hbar\omega_B$, centered at an energy $\hbar\omega_c$ and their motion detected via transient two-beam four-wave mixing. The electric field related to the dipole field of the Bloch oscillating carrier ensemble modulates, during a BO cycle, the energies of the WSL transitions causing peak shifts [7,15]. Figure 1 shows the peak shift of the heavy-hole $p = -1$ excitonic WSL transition as a function of the delay time. The data clearly show a linear shift in time of the peak corresponding to a linear motion of the carrier ensemble, superimposed by the well-known BO of the carriers [7]. In the following, we show that the observed linear shifts are due to a *quantum (phase-dependent) coherent motion* .

First, we have checked the dependence of the excitonic wave packet (EWP) motion on the laser excitation.

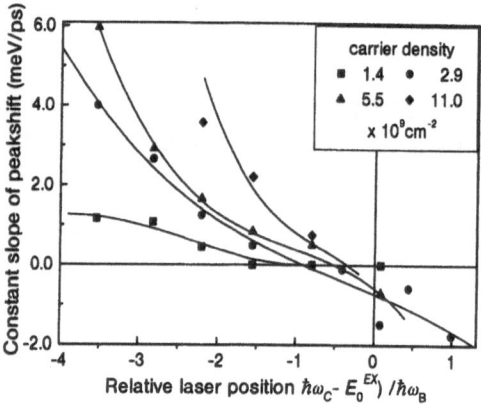

Fig. 2 Slope of the linear (in time) carrier motion as a function of the laser excitation energy.

Optical excitation energy well below or above the center of the WSL creates an EWP that performs BO with an amplitude Z_B approaching the semi-classical amplitude Z_B^{sc}. Excitation close to the WSL center creates an EWP which performs breathing-mode oscillations, as has been shown theoretically [16] and experimentally [8,9]. In our case, the amplitude minimum is at $\hbar\omega_c = E_0^{EX} - 0.8\hbar\omega_B$, with E_0^{EX} as the energy of the $p = 0$ heavy-hole exciton WSL state [8,9]. The dependence of the BO amplitude on excitation confirms that the linear shift of the carrier ensemble is caused by the coupling of the DC-field-induced BO with the self-induced AC field: The linear shift disappears when the excitation creates a breathing mode (Fig. 1). For excitation below the position of minimum amplitude, the linear shift of the $p = -1$ WSL peak is *upward in energy* corresponding to an electron motion *downward in the WSL*. For excitation above the amplitude minimum position, the linear shift of the $p = -1$ WSL peak is *downward in energy*, which corresponds to the electron motion *upward in the WSL*. Figure 2 plots the slope of the linear motion vs. laser spectral position confirming that for $\hbar\omega_c - E_0^{EX} > -0.8\hbar\omega_B$ the slope becomes negative, i.e., the direction of the coherent quasi-DC carrier motion is inverted compared to incoherent carrier drift motion. Second, we have checked the density dependence of the coherent current: Usual drift transport velocity is independent of the carrier density N to first order, yielding a depolarization field that increases linearly with the density. The data of Fig. 3 show that the field shift is increasing roughly quadratically with density, which implies that the coherent motion velocity is proportional to the carrier density.

We have developed a model based on the dynamically-controlled truncation calculation of the intraband polarization P_{intra} arising from excitons with the inclusion of a screened depolarization field [17–19]. The self-generated AC field is the depolarization field of the excitons along the SL growth direction and is given by $E_{AC}(\omega) = -4\pi P_{intra}(\omega)/\epsilon(\omega)$, where $\epsilon(\omega)$ is the di-

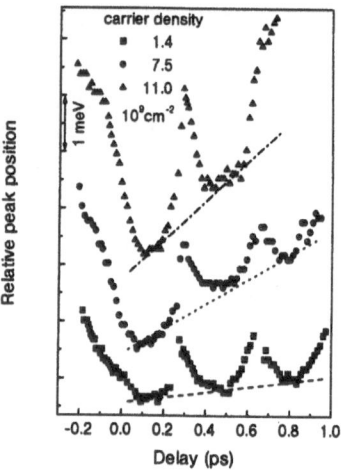

Fig. 3 Wannier-Stark ladder peak shift vs. delay time for different carrier densities, taken at a laser energy $2.2 \times \hbar\omega_B$ below the WSL center E_0^{EX}.

electric constant due to the *incoherent*, non-equilibrium, electron-hole plasma (EHP). A full model which contains the the EHP is not available at present, thus a standard model of a free-carrier plasma is employed. For excitation below the WSL center, the plasma effective mass is positive and the electron delivers work to the plasma. For excitation above the center, the plasma has negative mass and delivers energy to the excitons. It can be shown that the observed upward motion of the electrons proves the existence of *gain at THz frequencies arising from the non-equilibrium incoherent EHP* [14].

We thank DFG, VW-Stiftung, and NSERC for support.

References

1. L. Esaki and R. Tsu, IBM J. Res. Dev. **61**, (1970) 61.
2. E. E. Mendez et al., Phys. Rev. Lett. **60**, (1988) 2426.
3. P. Voisin et al., Phys. Rev. Lett. **61**, (1988) 1639.
4. J. Feldmann et al., Phys. Rev. B **46**,(1992) 7252.
5. K. Leo et al., Solid State Comm. **84**, (1992) 943.
6. C. Waschke et al., Phys. Rev. Lett. **70**, (1993) 3319.
7. V. G. Lyssenko et al., Phys. Rev. Lett. **79**, (1997) 301.
8. M. Sudzius et al., Phys. Rev. B **57**, (1998) R12693.
9. F. Löser et al. Phys. Rev. B **61**, (2000) R13373 .
10. T. Dekorsy et al., Phys. Rev. Lett. **85**, (2000) 1080.
11. S. Shapiro, Phys. Rev. Lett. **11**, (1963) 80.
12. A.A. Ignatov, K.F. Renk, and E.P. Dodin, Phys. Rev. Lett. **70**, (1993) 1996.
13. K. Unterrainer et al., Phys. Rev. Lett. **76**, (1996) 2973.
14. F. Löser et al., submitted to Phys. Rev. Lett.
15. K. Leo, Semicond. Sci. Technol. **13**, (1998) 249 .
16. M. Dignam, J.E. Sipe, and J. Shah, Phys. Rev. B **49**, (1994) 10502.
17. J. M. Lachaine et al., Phys. Rev. B **62**, (2000) R4829.
18. M. M. Dignam, Phys. Rev. B **59**, (1999) 5770 .
19. Alternatively, the experiments can be described by a semiclassical model (Yu. A. Kosevich, Ann. Phys. (Leipzig) **8**, (1999) 145).

Effects of a parallel magnetic field on the novel metallic behavior in two dimensions

K. Eng[1,2], X. G. Feng[1], Dragana Popović[1], S. Washburn[2]

[1] National High Magnetic Field Laboratory, Florida State University, Tallahassee, FL 32310, USA
[2] Department of Physics and Astronomy, University of North Carolina at Chapel Hill, Chapel Hill, NC 27599, USA

Abstract Magnetoconductance (MC) in a parallel magnetic field B has been measured in a two-dimensional electron system in Si, in the regime where the conductivity decreases as $\sigma(n_s, T, B = 0) = \sigma(n_s, T = 0) + A(n_s)T^2$ (n_s – carrier density) to a *non-zero* value as temperature $T \to 0$. Very near the $B = 0$ metal-insulator transition, there is a large initial drop in σ with increasing B, followed by a much weaker $\sigma(B)$. At higher n_s, the initial drop of MC is less pronounced.

1 Introduction

We have recently reported [1] the observation of a novel two-dimensional (2D) metallic behavior in Si metal-oxide-semiconductor field-effect transistors (MOSFETs). In this regime, the conductivity decreases as $\sigma(n_s, T) = \sigma(n_s, T = 0) + A(n_s)T^2$ (n_s – carrier density) to a *non-zero* value as temperature $T \to 0$. This simple $\sigma(T)$ spans two decades in T ($0.020 \le T < 2$ K). Several samples have been studied in detail and they all exhibit qualitatively the same behavior. Fig. 1 shows some typical results obtained in the lowest T range. The substrate bias $V_{sub} = +1$ V was used to maximize the range of T where this novel metallic behavior is observed [1].

In 2D, the existence of a metal with $d\sigma/dT > 0$ is very surprising as it contradicts any theoretical description available to date. All the other related research efforts in recent years have considered only the metallic behavior with $d\sigma/dT < 0$. However, some of the tremendous amount of work that was done on Si MOSFETs in earlier decades has been largely ignored. In particular, all the samples from previous studies could be divided into two groups: one group exhibited the behavior consistent with that seen in high-mobility Si MOSFETs [2], but the other displayed various "anomalies" [3] that were never understood. Our samples are representative of the latter, a *broad* class of Si MOSFETs historically known as "nonideal" samples [3]. We have established [1] the precise form of $\sigma(T)$ in these devices. The data strongly suggest [1] the existence of a metallic phase at $T = 0$ and of a novel, *continuous* metal-insulator transition (MIT) in 2D. New analysis of some of the early data [4] has established [5] that the same $\sigma(T)$ is exhibited by a variety of both n-type and p-type Si inversion layers, in both circular and linear geometries, with different channel lengths, substrate dopings, and oxide thickness.

2 Results and Discussion

Our samples are standard Si MOSFETs of Corbino geometry (channel length = 0.4 mm, mean circumference = 8 mm), and a peak mobility $\sim 1\text{m}^2/\text{Vs}$ at 4.2 K. Other sample details have been given elsewhere [6,1]. Conductance was measured as a function of gate voltage V_g using a low-noise current preamplifier and a low-noise analog lock-in at ~ 17 Hz. The excitation voltage was kept low enough to avoid electron heating. Measurements were carried out in either a He^3 cryostat or a $He^3 - He^4$ dilution refrigerator. $\sigma(T, n_s)$ was measured in fields of up to 18 T and for $0.020 \le T \le 4$ K.

While a detailed study of $\sigma(T, n_s, B)$ will be presented elsewhere [7], here we show some typical magnetoconductance data obtained at $T = 0.25$ K for several n_s above the zero-field n_c (Fig. 2). Very near the $B = 0$ MIT, a large initial drop of MC is observed with increasing B, followed by a weaker dependence at higher fields. At higher n_s, the initial drop of MC appears to be less pronounced although σ continues to decrease with B at all temperatures (Fig. 3). In fact, a closer inspection of the data (Fig. 3 inset) reveals that, in the high field regime (here $B > 5$ T), σ decreases *exponentially* with B. Such a strong $\sigma(B)$ in the high field regime has been, so far, observed only in the very dilute 2D hole system in GaAs [8] with the "conventional" metallic conduction ($d\sigma/dT < 0$). The effect has been attributed [9] to the coupling of the parallel field to the orbital motion arising from the finite 2D layer thickness. In Si-based devices, on the other hand, a saturation in conductivity has been observed [10,11] at high fields. Therefore, the behavior of our samples at high B differs from that of other Si MOSFETs. For our experimental conditions, the thickness of the 2D inversion layer is less than 5 nm [12]

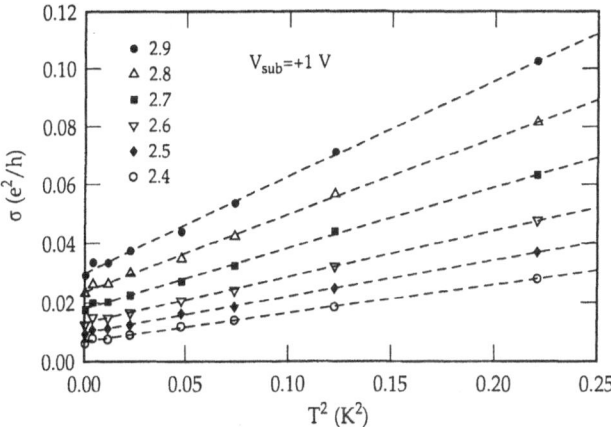

Fig. 1 Sample 19: $\sigma(T)$ in zero magnetic field plotted *vs.* T^2 for different $n_s(10^{11}\text{cm}^{-2})$ given on the graph. The dashed lines are fits. The data are shown for $0.020 \le T < 0.5$ K.

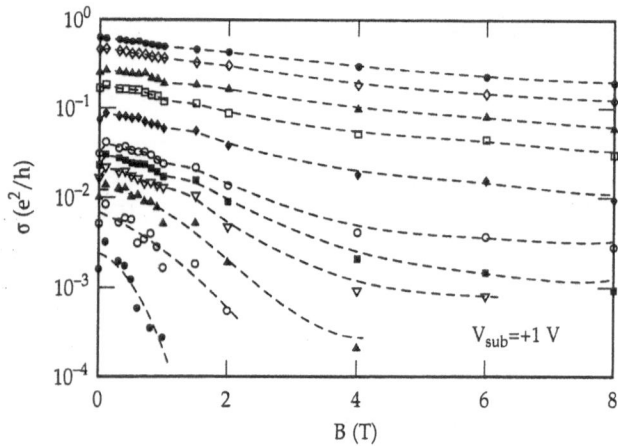

Fig. 2 Sample 9: σ vs. B applied parallel to the 2D plane at $T = 0.25$ K for $n_s(10^{11}\mathrm{cm}^{-2}) = 3.0, 2.8, 2.5, 2.3, 2.0$ and 1.7 to 1.2 in steps of 0.1 (top to bottom). At $B = 0$, $n_c = 0.95 \times 10^{11}\mathrm{cm}^{-2}$. Dashed lines are guides to the eye.

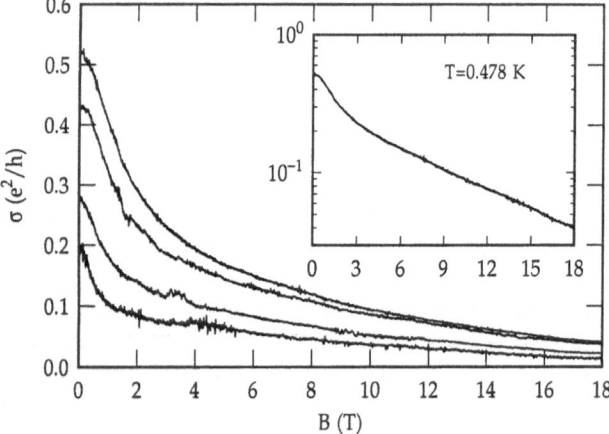

Fig. 3 Sample 19: σ vs. parallel magnetic field B for $n_s = 4.0 \times 10^{11}\mathrm{cm}^{-2}$ and $T = 0.478, 0.403, 0.269, 0.152$ K (top to bottom.); $V_{sub} = +1$ V. Inset: $\sigma(B)$ at $T = 0.478$ K on a semi-log scale. $\sigma(B)$ is exponential at the highest fields.

(the magnetic length varies from 12 nm to 6 nm for $5 < B < 18$ T), i. e. comparable to that in other Si MOSFETs [10]. Thus it seems unlikely that the difference in the high field response can be explained by the orbital effects [9].

In the low field regime, all devices seem to exhibit similar behavior. It has been argued [11,13] that the magnetic field above which the conductivity "saturates" is the field required to fully polarize the spins of free 2D carriers. Other experiments, on the other hand, seem to support [14] an alternative possibility in which the parallel field polarizes spins in an "external" band of localized states, causing a change in scattering. So far, however, there has been no experimental attempt to identify the origin of this "external" band and to vary its population in samples with $d\sigma/dT < 0$. In our samples, on the other hand, we have established [1] that the T^2 form of $\sigma(T)$ in the metallic regime is related to the population of localized states in the tail of the upper 2D subband. The number of occupied "upper tail" states was

varied systematically by applying V_{sub}. It is intriguing that the same effects might be responsible for the behavior of magnetoconductance in these two, apparently very different, conduction regimes (with $d\sigma/dT < 0$ and $d\sigma/dT > 0$, respectively).

While all experiments seem to suggest that spin degrees of freedom are playing an important role in dilute 2D systems, further work is clearly needed in order to reach a detailed microscopic understanding of the 2D metallic phase. In particular, "non-ideal" Si MOSFETs are proving to be an invaluable resource.

This work was supported by the National High Magnetic Field Laboratory (NHMFL) through NSF Cooperative Agreement DMR-9527035, an NHMFL In-House Research Program grant, and NSF Grants DMR-9796339 and DMR-0071668.

References

1. X. G. Feng, D. Popović, S. Washburn, and V. Dobrosavljević, submitted to Phys. Rev. Lett.; X. G. Feng, D. Popović, and S. Washburn, Phys. Rev. Lett. **83**, (1999) 368.

2. S. V. Kravchenko, et al., Phys. Rev. B **50**, (1994) 8039; S. V. Kravchenko, et al., Phys. Rev. B **51**, (1995) 7038.

3. T. Ando, A. B. Fowler, and F. Stern, Rev. Mod. Phys. **54**, (1982) 437, and references therein.

4. M. E. Sjöstrand and P. J. Stiles, Solid State Commun. **16**, (1975) 903; M. E. Sjöstrand, T. Cole, and P. J. Stiles, Surf. Sci. **58**, (1976) 72.

5. K. Walther, D. Popović, and P. J. Stiles (unpublished).

6. D. Popović, A. B. Fowler, and S. Washburn, Phys. Rev. Lett. **79**, (1997) 1543.

7. K. Eng, X. G. Feng, D. Popović, and S. Washburn (to be published).

8. J. Yoon, et al., Phys. Rev. Lett. **84**, (2000) 4421.

9. S. Das Sarma and E. H. Hwang, Phys. Rev. Lett. **84**, (2000) 5596.

10. D. Simonian, et al., Phys. Rev. Lett. **79**, (1997) 2304; V. M. Pudalov, et al., JETP Lett. **65**, (1997) 932; S. V. Kravchenko, et al., Phys. Rev. B **58**, (1998) 3553; D. Simonian, et al., Physica B **256-258**, (1998) 607; K. M. Mertes, et al., Phys. Rev. B **60**, (1999) R5093.

11. T. Okamoto, et al., Phys. Rev. Lett. **82**, (1999) 3875.

12. U. Kunze, Phys. Rev. B **35**, (1987) 9168.

13. S. A. Vitkalov, et al., Phys. Rev. Lett. **85**, (2000) 2164.

14. V. M. Pudalov, et al., preprint cond-mat/0004206 (2000).

Optics with Ballistic Electrons: Anti-Reflection Coatings for GaAs-AlGaAs Superlattices

C. Pacher[1], G. Strasser[1], E. Gornik[1], F. Elsholz[2], A. Wacker[2], and E. Schöll[2]

[1] Institut für Festkörperelektronik und Mikrostrukturzentrum, TU Wien, A-1040 Wien, Austria
[2] Institute for Theoretical Physics, Technical University of Berlin, D-10623 Berlin, Germany

Abstract We try to find an analogy between ballistic electron transport and classical optics: the influence of a quantum mechanical anti-reflection coating for ballistic electron transport through short period superlattices is investigated. Furthermore transmission resonances at the five eigenstates of the second miniband are experimentally observed for the normal superlattice, whereas the transmission is smooth for the superlattice with anti-reflection coating.

1 Introduction

In the past many experiments have been performed studying the wave nature of electrons in lateral transport experiments in two dimensional electron gas systems using GaAs/AlGaAs heterostructures [1]. On the other hand an energy spectroscopy technique has given evidence for coherent transport in vertical direction transport [2]. We try to adopt the principle of an optical anti-reflection coating to ballistic transport through superlattice minibands.

2 Theoretical Investigations

We calculate the transmission for ballistic electrons through superlattices embedded between one additional barrier on each side with a barrier width smaller than those of the superlattice barriers. Using the transfer-matrix method in envelope function approximation including non-parabolicity we studied the transmission through a GaAs/Al$_{0.3}$Ga$_{0.7}$As-superlattice with five periods (barrier width 25 Å, well width 65 Å) while varying the width of the additional barrier and the distance between it and the superlattice. As a measure for the transmission we integrate the transmission $T(E)$ over the width of the first superlattice miniband: $T_1 = \int T(E) \mathrm{d}E$. We found that the maximum of the transmission T_1 is achieved when the barrier width equals half the width of the barriers constituting the superlattice (12.5Å) in a distance equal to the well width of the superlattice (65Å). The calculations of T(E) for both samples show a very strong enhancement of the integrated transmission for the first miniband (by a factor of 3.1) and a significant enhancement for the second miniband (+79 %) due to the additional barriers (fig.1).

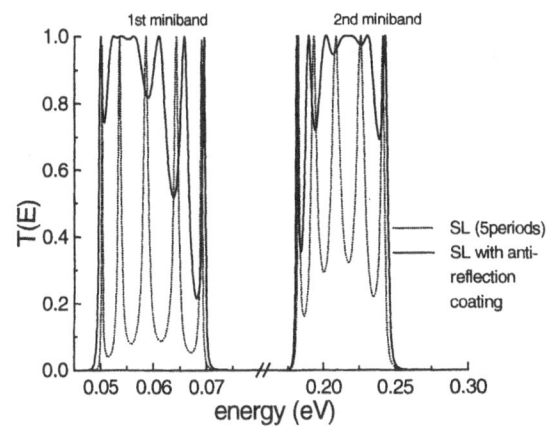

Fig. 1 Calculated transmission of the superlattice and the superlattice with anti-reflection coating

3 Experiments

In a three-terminal device [3,4] hot electrons are injected through a tunnel barrier. These hot electrons are tunable in their energy by means of an appliedvoltage V$_{EB}$ between the emitter contact (E) and the base contact (B) behind the barrier. After traversing a drift region the hot electrons hit the superlattice. In a third contact, the collector (C), the

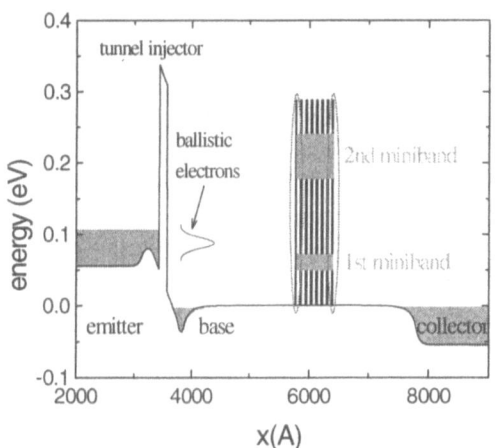

Fig. 2 Bandstructure of the three-terminal device for the superlattice sample with anti-reflection coating

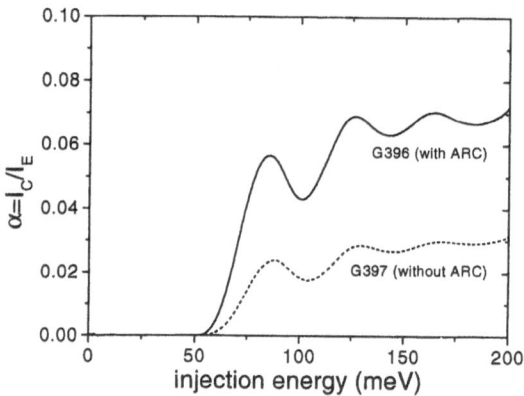

Fig. 3 Experimentally determined transfer ratios for the superlattice with and without anti-reflection coating The multiple resonances have their origin in ballistic electrons which emitted one or more LO phonons in the drift region in front of the superlattice.

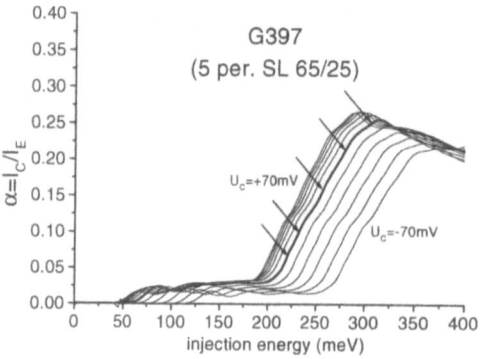

Fig. 4 Transmission for superlattice with five periods without anti-reflection coating under different bias conditions, the thick line marks zero bias. The arrows mark the positions of the five eigenstates in the second miniband; they correspond to maxima in the transmission. The energy offset in the peak position compared to the calculation (fig. 1) is due to series resistance in the base layer.

electrons, which are transmitted through the superlattice, are detected as collector current (fig.2). From the ratio $\alpha = I_C/I_E$ of the measured currents at T=4.2K an energy-resolved spectrum T(E) of the analyzed superlattice is obtained. We fabricated two 3TDs similar to Ref. [2], which were identical apart from two additional barrier(12.5Å)-well(65Å) pairs, surrounding the superlattice (GaAs/Al$_{0.3}$Ga$_{0.7}$As, five periods, barrier 25Å, well 65Å) in one sample. Recently it was shown that ballistic transport dominates in our structures [2]. This implies that the majority of the electrons, which enter the superlattice "feel" the anti-reflection layer at the other side of the superlattice. From the measured data we get an increase by a factor of 2.4 compared to 3.1 from the convolution of the calculated transmission with an Gaussian injector distribution of 25 meV FWHM, whereas the position of the first miniband is the same in both samples (fig.3). The transmission in the second miniband is also enhanced as predicted by our calculations. In the region of the second miniband we observe a series of peaks for the sample without anti-reflection coating (fig.4) due to transport through eigenstates, which are split by 15meV (fig. 1). This behavior cannot be seen in the sample with anti-reflection coating, which is in good agreement with the calculated transmission that shows very broad and overlapping peaks. Applying a positive or negative voltage to the collector of the sample the influence of an electric field on the electron transport can be studied. Consistent with Ref. [2] the transmission of the first miniband

does not depend on the direction of the applied electric field since the transmission time is much shorter than the scattering time of the dominant interface roughness scattering. Since the width of the second miniband exceeds the energy of an LO phonon (36meV), LO phonon scattering becomes the dominant scattering mechanism leading to an asymmetric behavior of the transmission of the second miniband as can be seen in fig. 4. This effect can be observed in both samples, but is weaker in the sample with anti-reflection coating, although the superlattice is about 33% longer.

References

1. A. Yacobi, U. Sivan, C.P. Umbach, and J.M. Hong, Phys. Rev. Lett. **66**, 1938 (1991); A. Yacobi, M. Heiblum, V. Umansky, H. Shtrikman, and D. Mahalu, Phys. Rev. Lett. **73**, 3149 (1994); E. Buks, R. Schuster, M. Heiblum, D. Mahalu, V. Umansky, Physica B, **249-251**, 295 (1998); R. Schuster, E. Buks, M. Heiblum, D. Mahalu, V. Umansky, H. Shtrikman, Nature, **385**, 417 (1997); E. Buks; R. Schuster; M. Heiblum; D. Mahalu; V. Umansky, Nature **391**, 871 (1998).
2. C. Rauch, G. Strasser, K. Unterrainer, W. Boxleitner, and E. Gornik, Phys. Rev. Lett. **81**, 3495 (1998).
3. M. Heiblum, M. I. Nathan, D. C. Thomas, and C. M. Knoedler, Phys. Rev. Lett. **55**, 2200 (1985).
4. C. Rauch, G. Strasser, K. Unterrainer, E. Gornik, B. Brill, Appl. Phys. Lett. **70**, 649 (1997).

Proc. 25th Int. Conf. Phys. Semicond., Osaka 2000 (Eds. N. Miura and T. Ando)

745

Chaotic Quantum Transport in Superlattices

T.M. Fromhold, A.A. Krokhin[*], A.E. Belyaev, C.R. Tench, S. Bujkiewicz[†], P.B. Wilkinson, F.W. Sheard, L. Eaves, and M. Henini

School of Physics and Astronomy, University of Nottingham, Nottingham NG7 2RD, UK

Abstract We calculate the current-voltage characteristics of a three-terminal GaAs/AlAs superlattice in a high magnetic field. When the magnetic field is tilted relative to the superlattice axis, electrons confined to a single miniband follow chaotic semiclassical trajectories. The onset of chaos extends the electron orbits thereby increasing the current, which is modulated by resonances between the cyclotron and Bloch frequencies.

1 Introduction

Our recent theoretical work [1] has introduced semiconductor superlattices (SLs) with a tilted magnetic field as a new type of quantum chaotic system that is accessible to experiment. In contrast to previous structures that have been used to study quantum chaos, the classical Hamiltonian for electron motion in the SLs has an intrinsically quantum-mechanical origin. For low electric fields, SLs have well-defined minibands. The energy-wavevector dispersion relations define an effective Hamiltonian that determines semiclassical orbits for electrons confined to a single miniband. Our calculations for this system have shown that tilting an applied magnetic field B at an angle θ to the SL axis induces a transition from stable to chaotic classical motion. The onset of chaos strongly delocalizes both the classical orbits and the corresponding quantum wavefunctions [1]. But the effect of this delocalization on measurable transport properties has not yet been investigated.

2. Current-voltage characteristics

In this paper we calculate current-voltage $I(V)$ characteristics for the three-terminal SL structure [2] shown schematically in Fig. 1, using the classical Kubo formula

$$I(V) \propto \frac{1}{\tau} \sum_s \int_0^\infty \exp(-t/\tau) v_x^s(t) dt, \qquad (1)$$

where $v_x(t)$ describes the time evolution of the electron velocity along the SL (x-) axis (determined by numerical solution of the semiclassical equation of electron motion through the miniband [1]), the electron scattering time $\tau = 1$ ps is obtained from experiment [2], and the summation is over all trajectories consistent with the electron injection energy.

Fig. 2(a) shows $I(V)$ curves calculated for electrons which

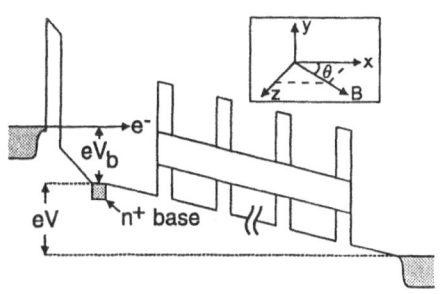

Fig. 1: Schematic conduction band diagram of a three-terminal SL showing n⁺ emitter, base, and collector contacts (shaded), and lowest SL miniband. Inset: orientation of tilted magnetic field relative to the x-direction, parallel to SL axis.

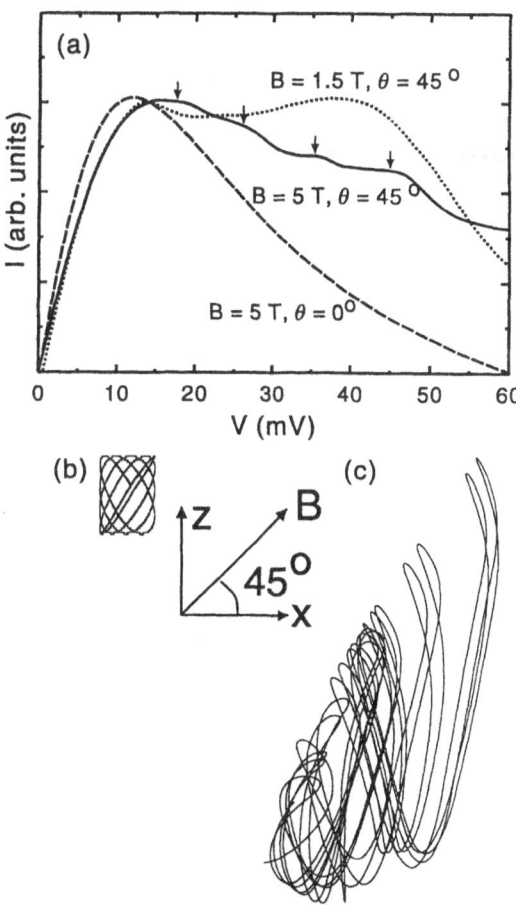

Fig. 2: $I(V)$ curves (a) calculated for $B = 5$ T, $\theta = 0°$ (dashed curve), $B = 1.5$ T, $\theta = 45°$ (dotted curve), $B = 5$ T, $\theta = 45°$ (solid curve). Arrows indicate resonant maxima. Classical orbits in x-z plane (axes inset) for $V = 52$ mV, $B = 5$ T and (b) $\theta = 0°$, (c) $\theta = 45°$.

---

[*] On sabbatical from Universidad Autonoma de Puebla, Mexico

[†] *Present address:* Institute of Physics, Wroclaw University of Technology, Wybrzeze Wyspianskiego, 50-370 Wroclaw, Poland

are injected at the bottom of the lowest (8 meV wide) miniband in a SL whose unit cell consists of a 9.3 nm wide GaAs quantum well and a 1.3 nm wide AlAs barrier. Similar results are obtained for a wide range of injection energies.

When $\theta = 0°$, the electrons perform Bloch oscillations, of angular frequency ω_B, along the SL axis. The $I(V)$ plot (dashed curve in Fig. 2(a)) peaks at the voltage V_B for which $\omega_B = 1/\tau$ [3]. The solid (dotted) traces in Fig. 2(a) show $I(V)$ curves calculated for $\theta = 45°$ and $B = 5$ T (1.5 T). For $V \lesssim V_B$, both of these curves are similar to that for $\theta = 0°$. This is because for low V, the electron motion remains approximately separable even when $\theta \neq 0°$. The electron orbits are similar to the Bloch oscillations found for $\theta = 0°$, but are inclined at an angle θ to the SL axis. Because these orbits are qualitatively similar to the Bloch oscillations, they produce current values comparable to those at $\theta = 0°$.

For $V \gtrsim V_B$, the current is much higher in the tilted field geometry. This can be explained by considering the regular and chaotic orbits (Figs. 2(b) and (c)) which the electrons perform for $\theta = 0°$ and 45° respectively. When $\theta = 0°$, the energy associated with motion along the x-direction is conserved. This limits the amplitude of the Bloch oscillations and confines the electrons to a small region of the x-axis (Fig. 2(b)). Consequently, the electrons cannot travel far along the SL before they scatter, and so the current is low. By contrast, a tilted magnetic field transfers momentum between the x- and z-directions, and so produces chaotic trajectories which

extend much further along the SL axis (Fig. 2(c)). This chaos-induced delocalization increases the distance that the electrons travel before scattering, and thus raises the current.

3. Oscillatory structure in $I(V)$ curves for $\theta \neq 0°$

The $I(V)$ curves for $\theta = 45°$ contain oscillatory structure (dominant peaks in 5 T trace marked by arrows in Fig. 2(a)), which is emphasized in the derivative plots shown in Fig 3. These fluctuations originate from classical resonances between the Bloch frequency and the cyclotron frequency $\omega_C \cos \theta$ corresponding to the x-component of $\boldsymbol{B}$. For V, B, and θ values which satisfy the resonance condition $l\omega_C \cos \theta = n\omega_B$ ($l, n = 1, 2, 3, \ldots$), most electron orbits remain stable and localized even when $\boldsymbol{B}$ is tilted. To illustrate this, the inset in Fig. 3(b) shows a Poincaré section (slice through the classical phase space) calculated for the $\omega_C \cos \theta = \omega_B$ resonance at $B = 5$ T, $\theta = 45°$. The scattered points show the z-co-ordinate and corresponding velocity component at each turning point along x. The phase space is almost completely filled by stable islands. But small changes in V, B, or θ, destroy the resonance and produce a predominantly chaotic phase space. Consequently, the electron orbits change repeatedly between regular and chaotic as V is increased. This modulates the orbital width along the SL axis, and thereby produces fluctuations in $I(V)$. The resonance condition gives rise to fluctuations with two distinct periodicities. The oscillatory structure within the box in Fig. 3(a) is periodic in $1/V$ (Fig. 3(b)). But at higher V, the dominant fluctuations (arrows in Fig. 2(a)) are periodic in V.

The repeated transition between regular and predominantly chaotic dynamics is a unique feature of SLs which occurs because the characteristic frequencies of the motion $\omega_C \cos \theta$ and ω_B are *both* independent of energy. Because the resonances between $\omega_C \cos \theta$ and ω_B occur at the *same* V, B, and θ values for all energies, the associated current fluctuations should be observable in real SL structures in which electrons enter the miniband with a narrow range of energies [2]. We note that $I(V)$ curves measured for three-terminal SLs when $B = 0$ T [2] are in excellent quantitative agreement with the Esaki-Tsu formula obtained by evaluating the integral in eq. (1) analytically. Our $I(V)$ curves in a tilted magnetic field are also calculated from eq. (1), and should therefore give accurate predictions for real devices. Experiments in this regime could therefore probe chaotic quantum transport through energy bands.

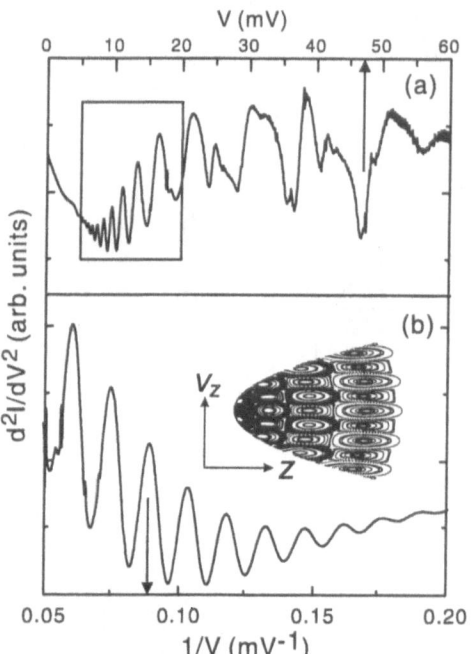

Fig. 3: (a) d^2I/dV^2 versus V curve calculated for $B = 5$ T and $\theta = 45°$ (derivative of solid curve in Fig. 2), showing oscillatory structure. (b) shows curve within box in (a) plotted as a function of $1/V$. Inset in (b): Poincaré section showing co-ordinates (z, v_z) whenever $v_x = 0$, at $B = 1.5$ T and $\omega_B = \omega_c \cos \theta$.

References

1. S. Bujkiewicz, C.R. Tench, T.M. Fromhold, M.J. Carter, F.W. Sheard, and L. Eaves, Physica E **6**, (2000) 306.
2. C. Rauch, G. Strasser, K. Unterrainer, W. Boxleitner, E. Gornik, and A. Wacker, Phys. Rev. Lett. **81**, (1998) 3495.
3. L. Esaki and R. Tsu, IBM J. Res. Dev. **14**, (1970) 61.

Proc. 25th Int. Conf. Phys. Semicond., Osaka 2000 (Eds. N. Miura and T. Ando)

747

The magnetoresistance of a two-dimensional electron gas in the presence of a spatially random magnetic field

A. W. Rushforth[1], B. L. Gallagher[1], P. C. Main[1], C. H. Marrows[2], B. J. Hickey[2], E. D. Dahlberg[3], A. C. Neumann[1], M. Henini[1]

[1] School of Physics and Astronomy, University of Nottingham, University Park, Nottingham, NG7 2RD, UK.

[2] Department of Physics and Astronomy, University of Leeds, Leeds, LS2 9JT, UK.

[3] Department of Physics and Astronomy, University of Minnesota, 116 Church St. S.E. Minneapolis, MN 55455, USA.

Abstract We have studied the magnetoresistance of a near-surface two-dimensional electron gas (2DEG) in the presence of a random magnetic field produced by CoPd multilayers deposited onto the surface of the heterostructure. The presence of the random magnetic field is confirmed by magnetic force microscopy (MFM) measurements in the presence of an external magnetic field. We compare our magnetoresistance measurements with a theoretical model and we obtain a good quantitative agreement by using a scattering time close to the quantum scattering time instead of the momentum relaxation time.

1 Introduction

The properties of a two-dimensional electron gas (2DEG) in the presence of a spatially random magnetic field has attracted great theoretical interest and has relevance for the study of composite fermions [1,2]. Recently there have been several experimental realisations of a random field at the site of a 2DEG [3–7].

We present here a novel method for producing a large-amplitude random magnetic field with a correlation length much smaller than the electron mean free path. We use Magnetic Force Microscopy (MFM), in the presence of an external magnetic field, to obtain detailed information about the spatial variation of the random magnetic field. This allows us to compare our experimental results with the theory of Hedegard and Smith [1].

2 Experiment

Our sample consists of a GaAs/Al$_x$Ga$_{1-x}$As heterostructure containing a 2DEG 35nm below the surface. At 4.2K the electron density, $n = 4.2 \times 10^{15}$ m^{-2} and the mobility, $\mu = 24$ m^{2}V^{-1}s^{-1} corresponding to a mean free path, $\lambda = 2.5$ μm. The device geometry is a modified Hall bar arrangement which we have described elsewhere [8]. We used dc magnetron sputtering to deposit a 1mm square of CoPd multilayers over the active region of the device. The multilayers consist of 200 bilayers with a Co layer thickness of 6 Å and a Pd layer thickness of 19 Å. The multilayers were deposited in an argon atmosphere at a pressure of 3 mtorr.

CoPd multilayers have perpendicular magnetic anisotropy, and at certain parts of the hysteresis loop, the magnetic domains break up into a random configuration. This produces a random magnetic field at the plane of the 2DEG. Figure 1 shows MFM images of the CoPd multilayers at zero external field and with the application of an external field of up to 0.3 T applied perpendicular to the plane of the multilay-

ers. The white areas of the image represent domains pointing in the direction of the applied field and the dark areas represent domains pointing in the opposite direction. As the external field is applied, the domains pointing in the direction of the applied field grow at the expense of domains pointing in the opposite direction. The width of the domains is typically ∼ 500 nm which will produce a random field with a correlation length much smaller than the electron mean free path.

Figure 2a (solid lines) shows the magnetoresistance of the 2DEG in the presence of an external magnetic field applied perpendicular to the plane of the 2DEG. The magnetic field is swept through a hysteresis loop from ± 1 T. Figure 2b shows the corresponding magnetic hysteresis loop obtained by VSM at 4.2 K. The random domain configuration is present between points A and B on the down sweep, and between points A' and B' on the up sweep. We see a large magnetoresistance as the random field is switched on at points A and A' which decreases as the domain pattern evolves. At points B and B' the multilayers become saturated and the magnetoresistance returns to zero as the random field is switched off. The dashed lines show the magnetoresistance when the multilayers are swept through a minor hysteresis loop from ± 0.35 T.

3 Comparison with theory

Hedegard and Smith [1] propose a model in which they solve the Boltzmann equation by including the random magnetic field in the driving force term and they include scattering in the relaxation time approximation. They obtain an equation for the magnetoresistance which requires, as input parameters, the Fermi velocity, v$_F$, the momentum relaxation time, τ, and the correlation function of the random magnetic field. We are able to use MFM images to calculate the correlation function of the random field at the 2DEG as a function of the external magnetic field [9]. We obtain a good quantitative fit between the model and our experimental data for $\tau = 1.8 \times 10^{-12}$s (Figure 3a). The relevant experimental curve for comparison is the dashed line, since the external field was varied between ± 0.3 T during the MFM measurements. For our sample the momentum relaxation time, $\tau_t = 9 \times 10^{-12}$s and the quantum scattering time, $\tau_s = 1 \times 10^{-12}$s. Therefore, the relevant scattering time to put into the model is closer to the quantum scattering time, τ_s than the momentum relaxation time, τ_t. This indicates that small angle scattering is playing a significant role in determining the transport properties of the system. To test this fur-

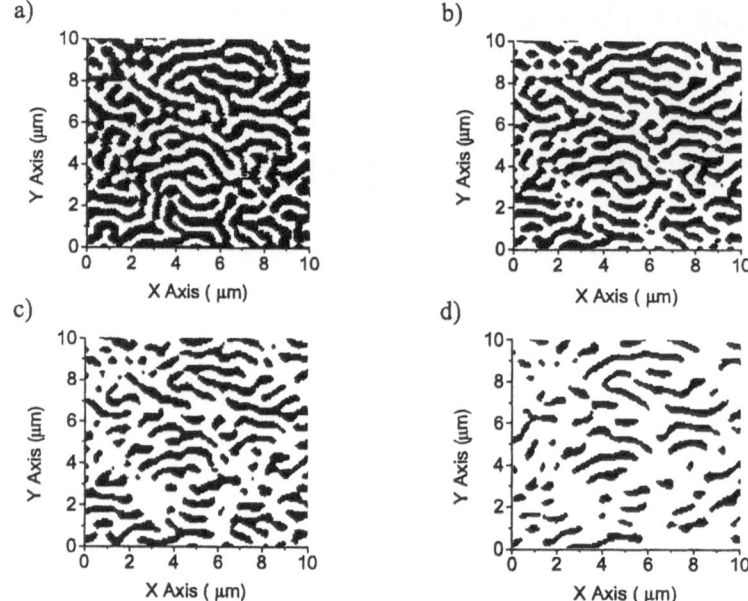

Fig. 1 MFM images of CoPd multilayers in (a) no applied field, and an external field of (b) 0.1 T, (c) 0.2 T and (d) 0.3 T applied perpendicular to the plane of the multilayers.

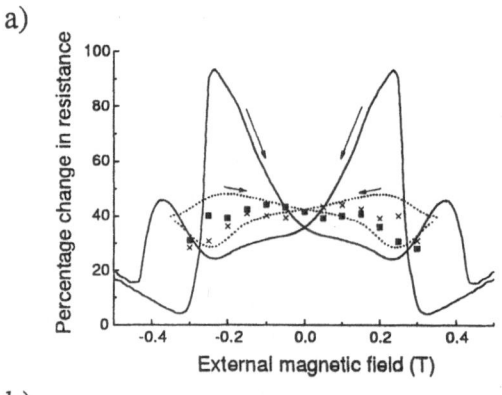

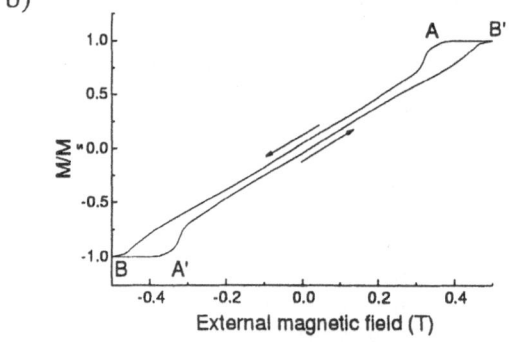

Fig. 2 (a) Magnetoresistance of the 2DEG as a function of the external magnetic field applied perpendicular to the plane of the 2DEG. Solid curves are for the field sweeping from ± 1.0 T. Dashed curves are for the field sweeping from ± 0.35 T. Dots (crosses) represent the theoretical fit for the field sweeping up (down) from ± 0.35 T. (b) Magnetic hysteresis loop of the multilayers with the field applied perpendicular. The magnetisation is normalised to the saturation magnetisation.

ther, we are currently performing Monte-Carlo simulations to calculate the magnetoresistance of the device by calculating electron trajectories in a random field with the presence of scattering.

4 Summary

We have used CoPd multilayers as a novel method for producing a large amplitude, small correlation length random magnetic field at the level of a 2DEG. We have used MFM images to deduce the correlation length of the random field. This has allowed us to compare the measured magnetoresistance with a theoretical model. We find a good quantitative agreement when a scattering time close to the quantum scattering time is used in the model.

5 Acknowledgements

We gratefully acknowledge funding by the EPSRC (UK) and the ESPRIT SPIDER programme (EU).

References

1. Per Hedegard and Anders Smith, Phys. Rev. B **51**, (1995) 10896.
2. AD Mirlin, J Wilke, F Evers, DG Polyakov and P Wölfle, PRL **83**, (1999) 2801.
3. FB Mancoff et al., Phy. Rev. B **51**, (1995) 13269.
4. A Smith et al., Phys. Rev. B **50**, (1994) 14726.
5. PD Ye et al., in: Scheffer and Zimmerman (Eds.), Proceedings of the 23rd International Conference on the Physics of Semiconductors, World Scientific, Singapore, P. 1537.
6. AK Geim, SJ Bending, IV Grigorieva, Phys. Rev. Lett. **69**, (1992) 2252.
7. GM Gusev et al., Phys. Rev. B **59**, (1999) 5711.
8. AW Rushforth et al., Physica E **6**, (2000) 751
9. The details of this calculation will be published elsewhere.

Giant negative magnetoresistance in two-dimensional antidot arrays: ballistic orbital effect and ballistic weak localization

T. Osada[1], H. Nakamura[1], Y. Shiraki[2]

[1] Institute for Solid State Physics, University of Tokyo, 5-1-5 Kashiwanoha, Kashiwa, Chiba 277-8581, Japan
 e-mail: osada@issp.u-tokyo.ac.jp
[2] Research Center for Advanced Science and Technology, University of Tokyo, 4-6-1 Komaba, Meguro-ku, Tokyo 153-8904, Japan

Abstract   The negative magnetoresistance (MR) in two-dimensional dense antidot arrays has been studied. In order to enhance negative MR, we employed very dense triangular lattices of anisotropic antidots, and realized large negative MR without any oscillatory structures. The negative MR survived even above 100K, and depended on the current direction. So that, the main part of negative MR originates from classical orbital effect. At low temperatures and low magnetic fields, the negative MR have another temperature-sensitive component due to ballistic weak localization. It seems to show no logarithmic behavior.

1 Introduction

Two-dimensional (2D) antidot array is one of the typical ballistic systems. The antidot size and distance are smaller than the mean free path (elastic scattering length). When they are much larger than electron wave length, we can regard an electron as a classical billiards ball. Since the system size is much larger than the phase coherence length (inelastic scattering length), however, the electron transport over the whole system must be treated as bulk transport.

Antidot arrays have shown rich magnetotransport phenomena, such as commensurability oscillations [1], Aharonov-Bohm (AB)-type oscillations [2], and ballistic Altshuler-Aronov-Spivak (AAS) oscillations. The commensurability oscillation is a classical orbital effect which occurs when the cyclotron orbit can encircle one or several antidots without any scattering. The AB-type oscillation relates to the quantization around an antidot. The ballistic AAS oscillation is caused by the interference of electron waves.

In addition to these oscillatory effects, it has often been observed that the background magnetoresistance (MR) shows negative field dependence. As the origin of negative MR in antidot arrays, both classical and quantum mechanisms can be considered. Gusev et al. experimentally showed that disorder enhances negative MR in antidot arrays. Since the negative MR had no temperature dependence, they concluded that it is not a quantum effect but a classical one, and ascribed its origin to the increase of mean free path [4]. Hennig et al. carried out classical calculations of MR for antidot arrays with potential saddle points at antidot vacancies, and led negative MR numerically as a classical orbital effect [5]. On the other hand, Nihey et al. mentioned that negative MR is caused by the ballistic weak localization due to quantum interference effect [3]. Recently, Yavtushenko et al. have studied the ballistic weak localization in antidot arrays in detail [6].

In this paper, we demonstrate the large negative MR by choosing appropriate antidot shape and arrangement. Using these antidot arrays, we study the negative MR in antidot arrays experimentally. Two different origins, classical orbital effect and ballistic weak localization, are discussed.

2 Experiments

In order to emphasize the negative MR in 2D antidot arrays, we tried to interrupt electron orbits contributing to the commensurability oscillations, and to enhance the back scattering at zero field. So, we employed very dense array of antidots with anisotropic shapes, the "brick-type" antidot array and its modified version, "bow-tie-shaped" antidot array. The carrier density and mobility were $n=3.0 \times 10^{11}$ cm^{-2} and $\mu=2.3 \times 10^5$cm^2/Vs at $T=4.2$K, respectively. The mean free path is estimated as $l=2.1\mu$m at $T=4.2$K. The size of the "brick-type" and "bow-tie" antidots are 0.4μm $\times$ 1.6μm and 0.6μm $\times$ 1.6μm, respectively. The channel width W between antidots is about 0.4μm. As shown in Fig. 1, we succeeded to realize the clear negative MR reaching 80% at 1.0K without any commensurability (or AB/AAS) oscillations in the "bow-tie-shaped" antidot array. As indicated by an arrow, negative slope finishes at $B=0.43$T, above which the classical cyclotron ornit becomes smaller than channel width W. It is remarkable that negative MR survives even above 100K.

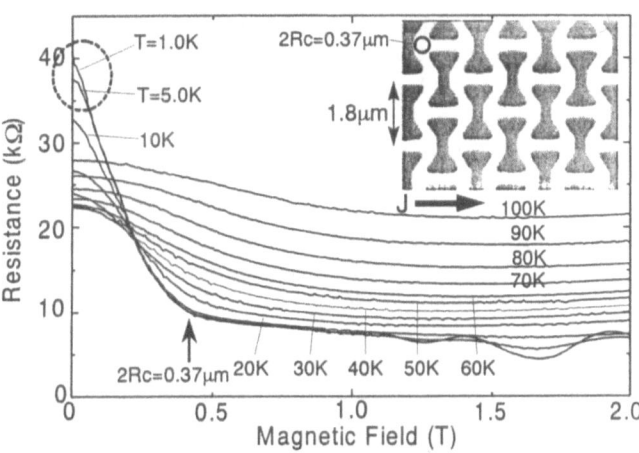

Fig. 1 Magnetoresistance of the "bow-tie-shaped" antidot array at various temperatures. Inset shows the SEM image of the sample.

Although the "brick-type" antidot array shows broader negative MR than the "bow-tie" antidot array, we could study all elements of resistivity tensor in this sample. As shown in Fig. 2(a), the negative MR has remarkable dependence on the current direction. It appears only when the current flow is perpendicular to the long axis of a rectangular antidot. No clear negative MR was observed when the current is parallel to the long axis of an antidot.

3 Discussion

The experimental results strongly suggest that negative MR originates mainly from classical orbital effect. To check if classical orbital effect could cause negative MR in the present case, we carried out numerical calculations of MR for the "brick-type" antidot array. The conductivity can be evaluated by the following Kubo type formula:

$$\sigma_{ij} \approx \int dE(-\partial f/\partial E)\int_0^\infty \langle v_i(t)v_j(0)\rangle e(-t/\tau)dt , \quad (1)$$

where $f(E)$ is the Fermi distribution function, $v_i(t)$ is velocity of an electron moving along an orbit, and τ is bulk scattering time. The ensemble average is taken over the initial states of all possible orbits. Each orbit was calculated assuming hard-wall potential. The calculation, which considers only classical orbital effect, qualitatively reproduced experimental results as shown in Fig. 2(b). This result supports the fact that most features of the negative MR are explained as classical orbital effect.

Next, we see the quantum contribution to the negative MR. As seen in Fig. 1, MR seems to lose its temperature dependence at low temperatures. In fact, the traces at 1K and 5K almost coincide except low field region (B<0.1T). This is reasonable because the classical effect is expected to have only weak temperature dependence. In the low field region, however, MR still has temperature dependence even at low temperatures. This temperature dependent part is considered to have a quantum origin.

In the ballistic antidot arrays, electrons are not scattered by random impurities but reflected by antidot walls. The electron waves reflected by antidot walls interfere with each other, and form localized states in the same way as conventional Anderson localization in diffusive metals. This "ballistic weak localization" state is also destroyed by external magnetic fields, showing negative MR.

Ballistic weak localization has been studied mainly in single ballistic cavity [7-9]. It has been known that the chaotic nature of the cavity is important for their magneto-transport features.

As mentioned above, the negative MR has two components, the classical part and the localization part. For detailed studies, it is important to separate them. The classical part hass no temperature depencence at low temperatures except low magnetic fields. So, assuming that the localization part vanishes (but the classical part is still constant) at higher temperature (T~5K), we extracted the localization part by subtracting the higher temperature conductivity from measured one.

The extracted localization part of conductivity (positive magnetoconductivity) did not show clear logarithmic field dependence expected from the scaling theory for the 2D

weak localization. In addition, the zero-field conductivity did not show clear logarithmic temperature dependence. These facts mean failure of the 2D scaling. This suggests that the localization occurs very locally in the very "dense" antidot arrays used in the present study. Elctrons are strongly localized in a cavity surrounded by antidots, and the system is regarded as connected ballistic cavities. In this situation, the localization length is in the same order of the antidot dimension, so the system is not in the "weak" localization region showing logarithmic behaviors.

This work was financially supported by the Toray Science Foundation and Grant–in–Aid for Pioneer Research from the Japan Society for the Promotion of Science.

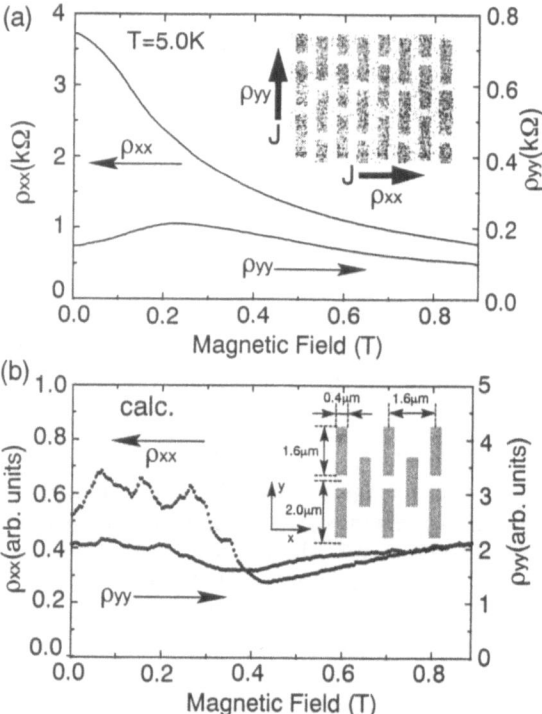

Fig. 2 Anisotropy of magnetoresistance in the "brick-type" antidot array. (a) experiment and (b) calculation.

References

1. D. Weiss, M. L. Loukes, A. Menschig, P. Grambow, K. von Klitzing, and G. Weimann, Phys. Rev. Lett. **66** (1991) 2790.
2. D. Weiss, K. Richter, A. Menschig, R. Bermann, H. Schweizer, K. von Klitzing, and G. Weimann, Phys. Rev. Lett. **70** (1993) 4118.
3. F. Nihey, S. W. Hwang, and K. Nakamura, Phys. Rev. **B51** (1995) 4649.
4. G. M. Gusev, P. Basmaji, Z. D. Kvon, L. V. Litvin, Yu. V. Nastaushev, and A. I. Tropov, Surf. Sci. **305** (1994) 443.
5. R. Hennig, P. Rotter, M. Suhrke, and U. Rossler, Physica **B249-251** (1998) 321.
6. O. Yevtushenko, G. Lutjering, D. Weiss, and K. Richter, Phys. Rev. Lett. **84** (2000) 542.
7. A. M. Chang, H. U. Baranger, L. N. Pfeiffer, and K. W. West, Phys. Rev. Lett. **73** (1994) 8857.
8. H. U. Baranger, R. A. Jalabert, and A. D. Stone, Phys. Rev. Lett. **70** (1993) 3876.
9. Y. Takane and K. Nakamura, J. Phys. Soc. Jpn. **66** (1997) 2977.

Correlation of optical and transport properties
of (AlGa)As/GaAs heterostructures

L. Gottwaldt[1,2], F.J. Ahlers[2], E.O. Göbel[2], G. Hein[2], S. Nau[1], K. Pierz[2], W. Stolz[1]

[1] Materials Science Center and Department of Physics, Philipps-University, D-35032 Marburg, Germany
[2] Physikalisch-Technische Bundesanstalt, D-38116 Braunschweig, Germany, e-mail: lars.gottwaldt@ptb.de

Abstract We have investigated the influence of disorder in (AlGa)As/GaAs heterostructures on both the optical and the electronic properties. We show that the optimization of samples for optics and transport, respectively, may require different growth parameters. We further report for our samples a pronounced anisotropy of the electron mobility with regard to the crystal axis even for exactly orientated substrates. By means of AFM experiments this anisotropy can be definitely correlated with the anisotropy of topological islands formed at the interface.

1 Introduction

Molecular beam epitaxy (MBE) as well as metalorganic vapor phase epitaxy (MOVPE) are the major tools for the fabrication of complex semiconductor heterostructures for device applications. The quality of the semiconductor samples depends on numerous growth parameters, like the pressure during the growth, misorientation of the substrate, growth temperature and growth interruption. In a complex layer system the interior interfaces, which generally differ from the surface, are of particular importance. The disorder of interior interfaces of a quantum well for example influences both the optical and the electronic properties directly.

We report systematic studies of the correlation of these disorder mediated effects in optics and transport for (AlGa)As/GaAs quantum well structures. In addition, atomic force microscopy (AFM) in combination with selective etching is applied for imaging the real topology of the interior interfaces.

2 Sample structure

The measurements were performed on $Al_{0.3}Ga_{0.7}As$/GaAs quantum well (QW) structures, with a single 10 nm thick QW. Using a modulation doped structure with a semi-transparent top Schottky gate allows variation of the electron concentration between zero and typically $4 \cdot 10^{11}$ cm^{-2}. We have investigated three inverted (modulation doping below the QW) and one normally doped (doping layer above the QW) structure grown by MBE at a growth temperature of 630°C. The inverted sample labelled I1 was grown without any interruption, I2 with 30 s after growth of the GaAs QW and I3 with 30 s growth interruption before the growth of the QW.

3 Correlation of the results

3.1 Correlation of optical and transport data

The used sample structure allowed optical experiments (photoluminescence (PL) and photoluminescence excita-

tion spectroscopy (PLE)) in the excitonic regime, i.e. at very low electron density, as well as magnetotransport measurements (high electron density) to be performed on one and the same sample by varying the gate voltage.

The relevant quantities which we have considered for the correlation are the low temperature exciton luminescence linewidth and Stokes shift in PLE spectra on the one hand and electron mobility as obtained from temperature dependent Hall and Shubnikov de Haas experiments as well as the width of the quantum Hall plateaus on the other hand.

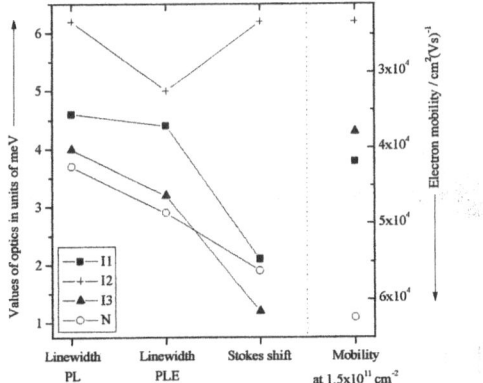

Fig. 1 A comparison of the relevant quantities in optics (on the left) and transport (right) for the investigated samples; quality increases from top to bottom.

According to the common quality criteria, from the three inverted structure samples I3 exhibits the best optical properties while I2 is worst (Fig. 1). Obviously growth interruption at the different QW interfaces affects the optical quality differently: While 30 s growth interruption at the (AlGa)As layer improves them, the same interruption at the GaAs interface deteriorates them.

By comparison the results for the normally doped structure show slightly better values for the linewidth then the best inverted structure, with the Stokes shift being larger and comparable to that of sample I1. However, with regard to the linewidth data it should be considered that in the inverted samples an additional broadening due to the quantum confined Stark effect may occur.

The data for the electron mobility measured at 40 mK of the four different samples are also depicted in Fig. 1 for a fixed carrier density of $1.5 \cdot 10^{11}$ cm^{-2}. Obviously these data reveal a different quality ranking as the optical spectroscopy experiments. Sample I1 without any growth interruption exhibits the highest mobility of the inverted structures. Furthermore a pronounced decrease of the mobility at low temperatures is detected, which indicates that ionized impurity scattering in addition to interface roughness scattering reduces the mobility of the samples, especially of the samples grown with growth interruption.

The fact that sample N (normally doped) exhibits a higher mobility at all temperatures is caused by the different relevant interior interface. Due to the band bending the electron mobility of the normally doped structure is mainly influenced by the GaAs/(AlGa)As interface, whereas in the inverted structures electrons are confined at the (AlGa)As/GaAs interface with islands on a smaller lateral scale (proved by AFM images).

With regard to the width of the quantum Hall plateaus the classification of the samples becomes even more complicated since the width depends not only on interface properties (localized states) and electron mobility but also on carrier density. If we consider the width of the plateaus normalized to the carrier density as a relevant figure we find for all the different inverted samples a maximum width for a mobility of about 30000 $cm^2(Vs)^{-1}$, independent on details of the growth procedure.

3.2 Anisotropic electron mobility

The electron mobility measured at 40 mK exhibits a pronounced anisotropy with regard to the crystal axis even for exactly oriented samples   while optical spectra naturally provide averaged information only. This is depicted in Fig. 3 for the samples I2 and I3, respectively. The mobility measured in the [0-11] direction is always larger than in the perpendicular [011] direction [1].

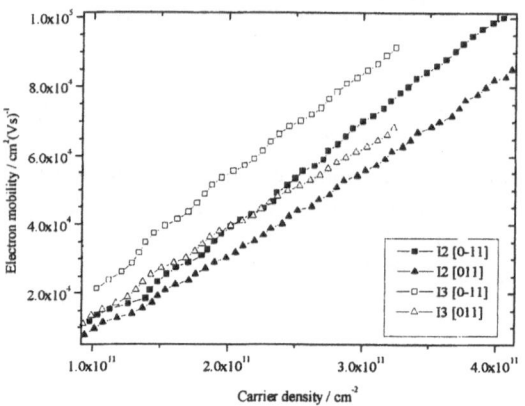

Fig. 2 Mobility measured at 40 mK with respect to the crystal axis vs. carrier density. Growth was interrupted for 30 s, once after the GaAs QW (I2) and once after the (AlGa)As layer (I3).

We attribute this difference to the contribution of interface roughness scattering which is revealed in the AFM experiments. We used a combination of highly selective etching and subsequent AFM studies for the structural investigation of the interior QW interfaces [2]. The interfaces exhibit an anisotropy of topological islands elon-

gated in the [0-11] direction. The anisotropic islands have their fundamental nature in the zinc blende crystal structure of (AlGa)As and GaAs, which lacks inversion symmetry [3]. These AFM data clearly demonstrate that the contribution of interface scattering is related to the scale of the topological roughness, becoming more important as the characteristic length scale decreases.

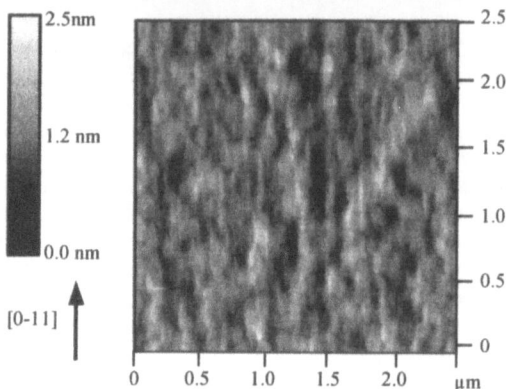

Fig. 3 AFM image of an interior GaAs interface, exposed by selective etching. Growth was interrupted for 30 s before overgrowth.

Due to the pronounced fluctuation of the island heights by several atomic layer thicknesses interface roughness scattering is the dominant scattering mechanism in our QW at higher carrier densities.

The value of the anisotropy increases with increasing carrier densities, because, at low densities, the dominant scattering mechanism is Coulomb scattering, which is essentially isotropic. Thus interface roughness scattering becomes more important as the density increases due to free carrier screening of the charged scattering centers[3]. Consequently, in the density regime where interface roughness scattering dominates, measurements of the anisotropy of the electron mobility may provide important conclusions on the topology of the relevant interior interfaces.

4 Summary

Our results prove that for our sample structures and the chosen growth conditions in the MBE, optimization of the samples for optics and transport, respectively, requires different growth parameters. The measured anisotropy in the electron mobility can be correlated with the anisotropy of topological islands formed at the interface as revealed by AFM experiments.

Acknowledgments: This research has been supported by the DFG within the framework of the SFB 383

References

1. T. Saku, Y. Horikoshi, Y. Tokura, Jpn. J. Appl. Phys. Vol 35, 34 (1996)
2. G. Benartz, et al., J. Appl. Phys. 86, 6757 (1999)
3. Y. Markus, U. Meirav, H. Shtikman, B. Laikhtman, Semicond. Sci. Technol.9, 1297 (1994)

A method of determining potential barrier heights at semiconductor heterointerfaces

Gil-Ho Kim[1,2], H.-S. Sim[3], M. Y. Simmons[4], D. A. Ritchie[1], C.-T. Liang[5], A. C. Churchill[1] and W. S. Han[2]

[1] Cavendish Laboratory, Madingley Road, Cambridge CB3 OHE, United Kingdom
[2] Telecommunication Basic Research Laboratory, ETRI, Yusong P.O. Box 106, Taejon 305-600, Korea
[3] Department of Physics, Korea Advanced Institute of Science and Technology, Taejon 305-701, Korea
[4] School of Physics, University of New South Wales, Sydney 2052, Australia
[5] Department of Physics, National Taiwan University, Taipei 106, Taiwan

Abstract   We report low–field magnetoresistance measurements of a two–dimensional electron gas formed in a GaAs quantum well, in which half a monolayer of AlAs has been inserted into the centre of the well. A large anisotropy is observed in both the mobility and the low field magnetoresistance in the orthogonal $[\bar{1}10]$ and $[110]$ directions. We describe a method of using the anisotropic low field magnetoresistance to calculate the magnitude of the effective potential of the AlAs sub–monolayer at the GaAs/AlGaAs heterointerface.

The ability to control and measure the electronic potential at a semiconductor heterostructure interface is important for understanding systems such as lateral superlattices [1], one–dimensional quantum wires, and single quantum wells as well as a basic understanding of transport at heterointerfaces. An interesting route for the fabrication of one–[2] and zero–[3] dimensional nanostructures is by natural formation during the growth procedure. Whilst there are a few studies of the physics of such systems there is a notable dearth in the literature of detailed magnetotransport measurements of naturally formed submonolayer potential heterostructures. In this paper we present detailed magnetotransport studies of a two–dimensional electron gas (2DEG) in a GaAs quantum well in which half a monolayer of AlAs has been inserted into the centre of the well. Using a full surface Schottky gate it is possible to control the carrier density of the 2DEG in the well and observe the effect of the AlAs potential on the magnetoresistance in both the $[110]$ and $[\bar{1}10]$ directions. A strong positive magnetoresistance is observed at low–field in the $[110]$ direction, with a corresponding magnetic breakdown at a critical field B_c. From the theory of magnetic breakdown we can calculate the effective potential height of the AlAs submonolayer as a function of carrier density. Our results demonstrate a method using electrical transport measurements to determine the potential barrier height formed at a heterointerface.

A n-AlGaAs/GaAs heterojunction grown by molecular beam epitaxy on an undoped GaAs (001) substrate deliberately mis–oriented by 0.09 degrees. The structure consists of a 0.6 μm thick undoped GaAs buffer layer, followed by a 500 Å undoped $Al_{0.33}Ga_{0.67}As$ barrier, a 200 Å undoped GaAs quantum well, a 400 Å undoped $Al_{0.33}Ga_{0.67}As$ spacer layer, a 400 Å Si–doped

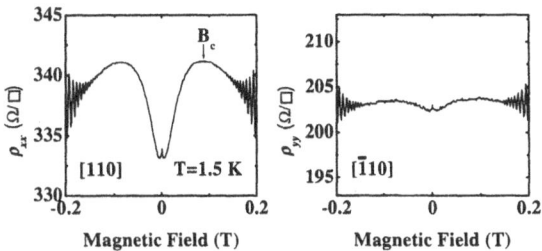

Fig. 1 Low–field magnetoresistivity traces at a carrier density of 2.3×10^{11} cm^{-2} in the (a) $[110]$ and (b) $[\bar{1}10]$ directions at a carrier density of 2.6×10^{11} cm^{-2} in both orthogonal directions. All traces are at 1.5 K.

$(1 \times 10^{18}$ cm$^{-3})$ $Al_{0.33}Ga_{0.67}As$ layer, and finally a 170 Å GaAs capping layer. Using a GaAs substrate with an intentional misorientation from the (001) plane, it is possible to form a periodic potential of AlAs islands along the $[110]$ surface using migration enhanced epitaxy.[4] Knowing the misorientation angle and the monolayer thickness, it is possible to calculate the period of the effective corrugated potential using $d = a\cot\alpha \approx 180$ nm. The devices were processed into an orthogonal Hall bar structure, where the current flows in either the $[110]$ and $[\bar{1}10]$ direction. A Schottky barrier gate was constructed onto the top of the structure using NiCr/Au which allowed control of the carrier density n_s In the 2DEG channel.

Figs 1(a) and 1(b) show the low–field magnetoresistance of the 2DEG for the $[110]$ and $[\bar{1}10]$ directions, respectively. The measurements were taken at 1.5 K at a carrier density of 2.34×10^{11} cm^{-2} ($V_g = 0$) estimated from the slope of the Hall resistance for both directions. The zero–field resistance in the $[110]$ direction is about one–and–a–half times higher than that in the $[\bar{1}10]$ direction leading to a marked anisotropy in the mobility in the two orthogonal directions (with a peak mobility of 1.2×10^5 cm^2/Vs). A positive magnetoresistance is observed in both directions but which is much more pronounced in the $[110]$ direction. The occurence of a strong magnetoresistance in the $[110]$ direction is discussed in more detail below. At higher fields ($B > 0.15$ T) Shubnikov–de Haas oscillations corresponding to 2D transport are observed.

Figure 2(a) shows the mobility μ versus the carrier density n_s at 1.5 K in the $[110]$ and $[\bar{1}10]$ directions for

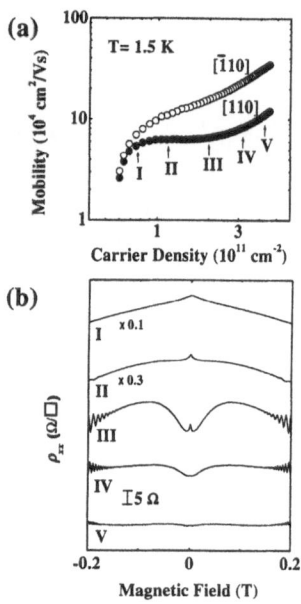

(a)

(b)

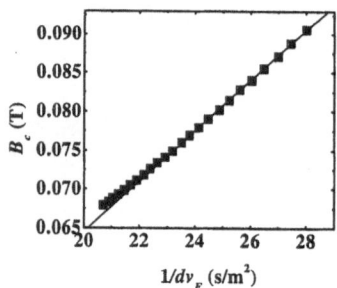

Fig. 3 B_c as a function of $1/dv_F$.

Fig. 2 (a) The mobility along the [110] and [$\bar{1}$10] directions as a function of carrier density n_s at T=1.5 K. Arrows with Roman numerals indicate the points at which the magnetoresistance data in figure 3 (b) were taken. (b) Magnetoresistance in the [110] direction are measured at gate voltages -0.20 (top), -0.15, -0.05, 0.10, and 0.25 V (bottom), respectively.

sample A. At low densities (0.7×10^{11} cm^{-2}) the mobility is independent of direction as the electrons are mainly scattered by isotropic impurity scattering. At higher densities anisotropic interface scattering due to the quasi–periodic 1D potential dominates and the mobility becomes anisotropic. In order to provide further insight as to what is happening at the different carrier densities (marked I–V) we present the corresponding magnetoresistance traces along the [110] direction in Fig. 2(b).

At low carrier densities ($< 0.68 \times 10^{11}$ cm^{-2}) where the mobility is isotropic, a large negative magnetoresistance is observed [see Fig. 2(b) at $V_g = -0.20$ V]. As the carrier density is increased interface roughness scattering starts to dominate and a large positive magnetoresistance develops, with a maxima at $B = B_c$ (see III). In order to understand the occurence of a large positive magnetoresistance observed in the [110] direction for the intermediate field range (0.02 to 0.2 Tesla) we turn to an earlier study of 1D surface superlattices by Beton *et al.*[5,6] They observed a positive magnetoresistance in a periodically modulated 2DEG when the direction of the current flow is orthogonal to the superlattice axis. The magnetoresistance shows a maxima at a critical field B_c, where B_c is given by:

$$B_c = 2\pi V_{eff}/dv_F, \tag{1}$$

where V_{eff} is the amplitude of the effective potential, d is the period, and v_F is Fermi velocity. The positive magnetoresistance arises from electron orbits "streaming" along the 1D superlattice potential. When the force

due to the magnetic field is greater than that due to the electric field ($B > B_c$) a form of magnetic breakdown occurs and the streaming orbits are suppressed. This is a purely classical effect which ignores the effect of tunneling through the AlAs potential barriers. Increasing the carrier density even further (traces IV and V) causes the magnitude of the positive magnetoresistance to decrease. This is because the Fermi energy increases with increasing carrier density, so the relative height of the fixed potential due to the periodic 1D AlAs superlattice decreases.

Using Eq. (1), the effective 1D periodic potential energy of the AlAs submonolayer, V_{eff} can be determined from B_c and n_s. The experimentally determined B_c is plotted in Fig. 3 as a function of calculated value $1/(dv_F)$, where Fermi velocity v_F is obtained from n_s. From the linear fit we obtain the effective potential V_{eff} of the AlAs submonolayer to be ≈ 0.512 meV. It can be seen that the effective potential V_{eff} is almost constant over all this range of densities but at low electron densities, V_{eff} is observed to increase and deviate from 0.512 meV. This increase in V_{eff} at low densities can be understood as a consequence of the electron drift in the [110] direction.

In summary, we have incorporated half a monolayer of AlAs into a GaAs quantum well to induce an imperfectly periodic 1D potential in a 2DEG. A strong low–field positive magnetoresistance is observed in the [110] direction, with a corresponding critical magnetic breakdown field. From the theory of magnetic breakdown we can calculate the effective potential of the AlAs submonolayer as a function of carrier density and found that the potential is fixed at 0.512 meV. The results demonstrate a method of inducing a fixed periodic potential at a heterostructure interface and using electrical transport measurements to determine the effective height of potential.

This work was funded by the UK EPSRC. G.H.K. is grateful for financial support from the Korean MIC.

References

1. C.-T. Liang *et al.*, Phys. Rev. B **49**, (1994) R8518.
2. Y. Nakamuta *et al.*, Appl. Phys. Lett. **69**, (1996) 4093.
3. G. H. Kim *et al.*, Appl. Phys. Lett. **73**, (1998) 2468.
4. M. Tanaka *et al.*, Appl. Phys. Lett. **54**, (1989) 1326.
5. P. H. Beton *et al.*, Phys. Rev. B **42**, (1990) 9229.
6. P. H. Beton *et al.*, Phys. Rev. B **43**, (1991) 9980.

Conductance quantization in an array of ballistic constrictions

S. de Haan[1], **A. Lorke**[2], **J. P. Kotthaus**[1], **W. Wegscheider**[3], **M. Bichler**[4]

[1] Sektion Physik and Center for NanoScience, LMU München, Geschwister-Scholl-Platz 1, 80539 München, Germany e-mail: stephan.dehaan@physik.uni-muenchen.de

[2] Laboratorium für Festkörperphysik, Gerhard-Mercator-Universität, Lotharstr. 1-21, ME 245, 47048 Duisburg, Germany

[3] Institut für Angewandte und Experimentelle Physik, Universität Regensburg, 93040 Regensburg, Germany

[4] Walter Schottky Inst., TU München, Am Coulombwall, 85748 München, Germany

Abstract   We have investigated the transport properties of a two-dimensional array of ballistic constrictions and found that the conductivity as a function of gate voltage increases in approximately equally spaced steps. Surprisingly, the height of the observed steps is not an integer multiple of $2e^2/h$, but roughly one third of this fundamental unit. Numerical simulations strongly suggest that this increase in resistance is due to coupling between parallel channels in our superlattice.

1 Introduction

Since advances in nanofabrication have made it possible to access the ballistic transport regime, many interesting results associated with electronic transport through quantum point contacts (QPC's, i. e. short and narrow constrictions that are much smaller than the electron mean free path), have been obtained. Single [1] and double QPC's both in serial [2] and in parallel [3] configuration have been investigated. In most cases, the conductance of these systems is found to be quantized in units of $2e^2/h$.

Here, we report on transport experiments on a large two-dimensional array of such ballistic constrictions and show that we have realized a *material with a quantized conductance*.

2 Fabrication

The device was fabricated from a GaAs/AlGaAs-heterostructure with a shallow two-dimensional electron gas with typical carrier density of 6×10^{11} cm^{-2} and mobility of 8×10^5cm^2/Vs. High-resolution electron beam lithography and wet chemical etching techniques have been used to define a lateral superlattice of 55×96 quantum point contacts, i.e. parallel rows of narrow constrictions that are linked every 700 nm (see right inset of Fig. 2). A thin NiCr-gate, covering the entire structure allows us to vary the electron density and thus the number of occupied one-dimensional subbands in the constrictions.

3 Experiment

Fig. 1 shows the gate voltage dependence of the longitudinal conductivity σ_{xx}[1] at two different temperatures. As the measurements have been carried out in different cooling cycles, the two curves are slightly offset. Sweeping the gate voltage at sufficiently low temperature (250

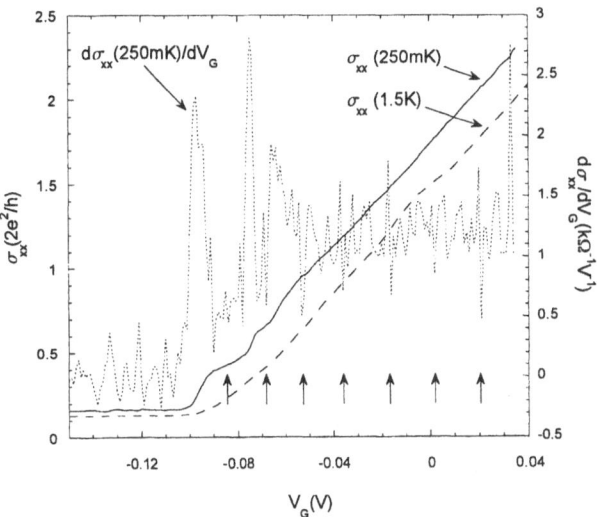

Fig. 1 Longitudinal conductivity σ_{xx} at 250 mK (solid line), 1.5 K (dashed line) and derivative $d\sigma_{xx}(250\text{mK})/dV_G \propto dI/dV_G$ (dotted line) vs. gate voltage of a lateral superlattice of quantum point contacts. Minima in the differential transconductance, corresponding to plateaus in the conductivity, are indicated by arrows.

mK), both in two-terminal (shown here) and in four-terminal measurements σ_{xx} reveals a step-like structure of equally spaced plateaus. Up to seven oscillations can be identified in the differential transconductance dI/dV_G (dotted line in Fig. 1). As the ocurrence of these steps is qualitatively independent of the actual geometry, our device can be regarded as the realization of a material with a quantized conductance.

The height of the observed conductivity steps is $\approx 0.3 \frac{2e^2}{h}$, and thus not an integer multiple of the fundamental unit of conduction. Previous results show that the conductance of parallel QPC's is practically additive [3], i.e. $G_{total} = (\frac{2e^2}{h}) \sum_i n_i$, where n_i is the number of modes propagating in contact i. The resistance of two series point contacts however is not simply additive [2]. In this case, the conductance can vary between $G_i/2$ and G_i, if G_i is the conductance of an individual contact. Thus, our experimental data is neither consistent with the simple model of a mesh of Ohmic resistors, each of the value $\frac{h}{2ne^2}$, which would lead to a conductivity quantized in units of $2e^2/h$, nor with the picture of coherent

---

[1] $\sigma_{xx} = \frac{I}{V}\frac{96}{55}$ in the present case

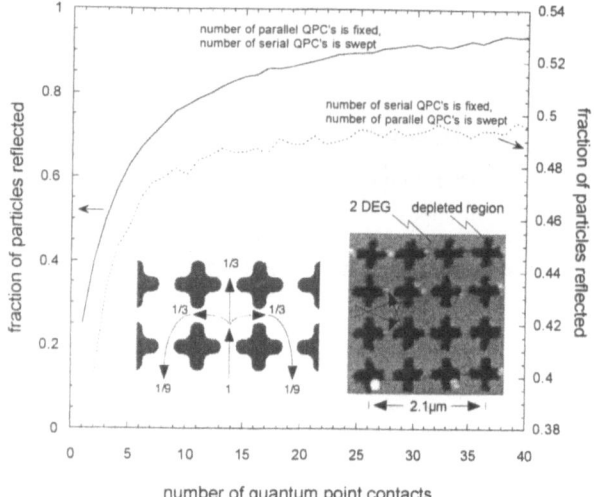

number of quantum point contacts

Fig. 2 Fraction of particles reflected in the numerical simulation when sweeping the number of series (solid line, left scale) and the number of parallel (dotted line, right scale) point contacts. Right inset: AFM micrograph of a section of the fabricated superlattice. Arrows represent electron trajectories that are backscattered by the concave boundaries. Left inset: Illustration of the model used in the simulation.

transport through a series of quantum point contacts, which would result in even larger conductance steps.

4 Numerical simulation

As schematically indicated in the right inset of Fig. 2, a mechanism that could explain our observations is backscattering from the boundaries of the coupled quantum dots formed between the constrictions. Due to the fact that parallel rows of point contacts are linked every 700 nm, coherent backscattering of electrons should become possible in this system (see left inset of Fig. 2).

In order to find out, how the characteristics of an individual point contact are modified in a large array, we employ a simple model, simulating electronic transport in our structure. As a first approximation, we assume that an electron entering a dot will leave it through either of the "exits" with equal probability (see left inset of Fig. 2). Thus, coupling of the rows is introduced by a finite probability for leaving the dot to the sides. If we now inject particles from one side of the structure, we can trace their trajectories through the lattice and determine the fraction R of those that do not manage to get to the other side. Fig. 2 shows the results of this Landauer-Büttiker-type description for two different situations. When the number n_p of parallel QPC's is kept at a fixed value (in this case $n_p = 2$) and the number n_s of series QPC's is swept, we obtain the solid curve. Keeping n_s constant (here $n_s = 2$ again) and sweeping n_p, we obtain the dotted curve. Obviously, the n_s-dependence of R is much more distinct than the dependence on n_p. However, the important result is that a growth of the lattice in either of the two directions leads to an increase of R.

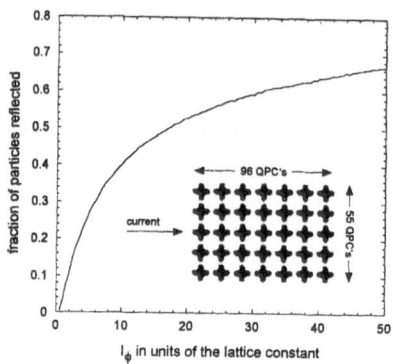

Fig. 3 Fraction of particles reflected in the numerical simulation when sweeping the phase coherence length. Inset: Sketch of the superlattice geometry used for this calculation.

To get a more realistic model of the device, in a second step we introduce a finite phase coherence length l_ϕ. The number of electrons that, after having travelled a distance l through the lattice, can still be reflected in a coherent way decays like e^{-l/l_ϕ}. Fig. 3 shows the result for the special geometry of our sample, i.e. an array of 55×96 quantum point contacts. It can be seen that the fraction of coherently reflected particles strongly depends on the phase coherence length. However, even at small values of l_ϕ the reduction of transmission, compared to an array of totally uncoupled QPC's, is maintained.

Of course, this simple model is not suitable for making quantitative statements. However, it illustrates that in a superlattice of our size the coupling of adjacent contacts in connection with a sufficiently large phase coherence length can significantly reduce electronic transport through the system. It therefore offers an explanation for the fact that the height of the experimentally observed conductivity plateaus is smaller than the fundamental unit. Magnetotransport experiments on the same structure in the classsical regime ($T = 4.2\,\mathrm{K}$) show that the concave shape of the dot boundaries strongly influences the transport properties of the device [4]. The conductance measurements discussed here suggest a similar mechanism.

5 Acknowledgments

It is a pleasure to thank S. Ulloa for stimulating discussions. Financial support by BMBF through grant 01BM914 is gratefully acknowledged.

References

1. B. J. van Wees *et al.*, Phys. Rev. Lett. **60**, (1988) 848; D. A. Wharam *et al.*, J. Phys. C **21**, (1988) L209.
2. D. A. Wharam *et al.*, J. Phys. C **21**, (1988) L887; C. W. J. Beenakker and H. van Houten, Phys. Rev. B **39**, (1989) 10445.
3. C. G. Smith *et al.*, J. Phys. Condens. Matter **1**, (1989) 6763; E. Castaño and G. Kirczenow, Phys. Rev. B **41**, (1990) 5055; M. E. Sherwin *et al.*, Appl. Phys. Lett. **65**, (1994) 2326.
4. S. de Haan *et al.*, Phys. Rev. B **60**, (1999) 8845.

Transport in asymmetric two-dimensional lateral surface superlattices

S. Chowdhury[1], A. R. Long[1], J. H. Davies[2], D. E. Grant[1], E. Skuras[1], C. J. Emeleus[1]

[1] Department of Physics and Astronomy, Glasgow University, Glasgow, G12 8QQ, U. K.
[2] Department of Electronics and Electrical Engineering, Glasgow University, Glasgow, G12 8QQ, U. K.

Abstract   We have studied square two-dimensional lateral surface superlattices of period 100 nm formed by shallow etching of holes into a GaAs/AlGaAs heterostructure. In some samples, the symmetry was broken using piezoelectric modulation. The commensurability oscillations in the magnetoresistance characteristics of our samples were fully in accord with a recent semi-classical model of guiding center drift in two-dimensional superlattices.

1 Introduction

The transport of electrons in a two-dimensional electron gas (2DEG) subject to a periodic potential and a perpendicular magnetic field has been of theoretical and experimental interest for many years. Experimental magnetoresistance studies of the problem in lateral surface superlattices (LSSLs) show oscillatory structures known as commensurability oscillations (COs).[1] These are generally much weaker if the modulation is two-dimensional (2D) [2] than if the modulation varies only in 1D. This was initially attributed [2] to suppression of the conductivity by the predicted sub-structure in the Landau bands. Recently we have predicted [3] and experimentally confirmed [4] that asymmetry is the key to the observation of large COs in 2D LSSLs and the COs are weak, irrespective of the magnitude of the potential, if the potential landscape is symmetric. In this paper we present some further confirmatory data.

According to our theory [3], in symmetric square 2D LSSLs in which the two main axial Fourier components of the potential are equal in amplitude, all the trajectories of the guiding center motion are closed and any COs should be very weak if not totally absent. In asymmetric 2D LSSLs however, in which the two axial Fourier components are unequal in amplitude, besides closed trajectories there are open guiding center trajectories that always run along the direction of the weaker potential. Thus transport along the direction of the larger potential component will show very large COs, reminiscent of 1D LSSLs, while there should be no COs at all for transport along the direction of the weaker potential. In presence of a dominant diagonal Fourier component, drift will occur along its contours and the COs will have the period of the diagonal Fourier component which is $a/\sqrt{2}$ for a square lattice of period a. These predictions are at variance with those of a previous semi-classical theory of 2D transport [5] which suggests that each Fourier component should contribute independently to the magnetoresistance and give rise to its own series of COs.

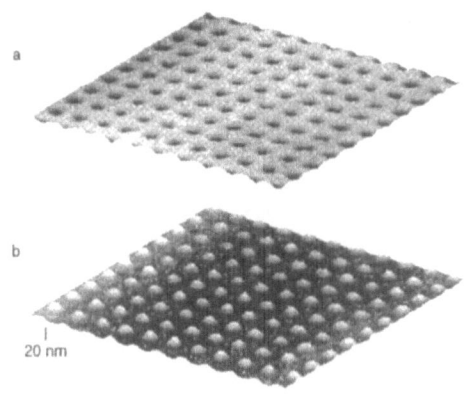

Fig. 1 (a) AFM image of 100 nm square lattice of shallow etched holes after removal of resist and before gating. (b) Inverted image showing depth and uniformity of etching.

2 Experimental Procedures

We have studied 100 nm square 2D LSSLs realised by shallow etching a pattern of holes of 50 nm diameter some 20 nm into the surface of Hall-bars fabricated on a GaAs/AlGaAs heterostructure containing a high mobility 2DEG (Fig. 1). After the etching we removed the etch mask, and completed the fabrication by depositing a continuous thick metal gate over the patterned area. Such a system is essentially symmetric between the principal axes of the pattern, which were aligned along and across the Hall-bars. Modulation is brought about because the surface states are periodically brought closer to the 2DEG, the so called *surface effect*. [6] The surface potential and hence the modulation from this source can be reduced by forward biasing the overlying gate. Asymmetry was realised in certain samples by introducing a 6 nm thick strained layer of $In_{0.2}Ga_{0.8}As$ 10 nm below the surface, which was also patterned in the etching process. Besides the surface effect, this produces a periodic *stress modulation* that breaks the symmetry of the perturbation via the anisotropic piezoelectic effect. [6] On a (100) GaAs wafer, the piezoelectric effect is of equal magnitude but opposite sign in the $[01\bar{1}]$ and $[011]$ directions while there is no piezoelectic effect along $[010]$ and $[001]$. The modulation due to surface effect has the same sign in all directions. In the $[01\bar{1}]$ direction (for a period of 100 nm) the two contributions add to each other, while along $[011]$ they partially cancel each other [7]. In this work we prepared Hall-bars aligned with all four of the important directions, the crystal axes and the cleavage directions in between. We call symmetric samples prepared on the plain heterostructure *unstressed*, and those on material containing the InGaAs layer which are naturally asymmetric we call *stressed*.

In our original studies [4], we reported measurements made on samples containing pillars on the surface and etched regions in between. We expect the current samples containing a superlattice of holes to behave similarly, as both sources of modulation will be reversed in sign.

3 Results and Discussion

Symmetric unstressed 2D LSSLs were found to show very weak COs in all directions as expected from our theory [3]. Asymmetric 2D LSSLs showed very large COs of 100 nm period in $[01\bar{1}]$ Hall-bars (see Fig. 2), while no COs were observed in the [011] direction at any gate bias. Because we expect significant modulation in both principal directions in these samples, this result is in striking agreement with our theory [3]. Results for transport in the [010] and [001] directions, still with 100 nm period along the transport direction, are shown in Fig. 3, for positive gate bias. This bias reduces the axial surface potential components and leaves the diagonal Fourier components, which result from the stress modulation and are unaffected by biasing the gate, to dominate. One of these, along $[01\bar{1}]$, will be slightly larger than the other, along [011], due to a residual surface effect, and hence the guiding centre drift will be orthogonal to $[01\bar{1}]$. The effective period of the superlattice in these diagonal directions is $100/\sqrt{2}$ nm, and the diagonal drift will generate COs for transport in both the crystal axis directions with this effective period. As Fig. 3 shows we get very similar magnetoresistance traces for Hall-bars in these two directions and the effective period of the COs is indeed $100/\sqrt{2}$ nm period, again in striking agreement with our theory [3]. There is no evidence for 100 nm period COs in this data.

These phenomena were all seen for the 100 nm square superlattices of pillars [4], but are clearer for these hole samples. We believe that this reflects the higher quality of the etch patterns in the hole samples.

4 Conclusions

Agreement between our theory and experiment demonstrates that we have for the first time achieved clear understanding of the importance of asymmetry for electron transport in 2D LSSLs. Our observations suggest that different Fourier components do not contribute independently to the COs, in agreement with our theory [3], but contradicting the previous model [5] based on perturbation theory.

References

1. D. Weiss, K. von Klitzing, K. Ploog, and G. Weimann, Europhys. Lett. **8**, (1989) 179

2. R. R. Gerhardts, D. Weiss, and U. Wulf, Phys. Rev. B **43**, (1991) 5192

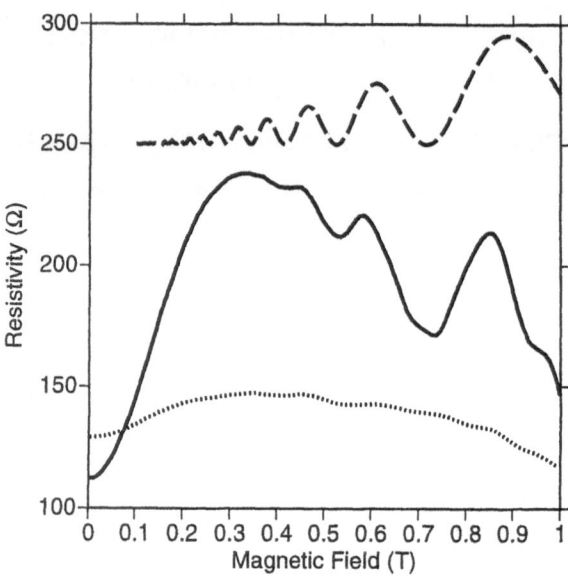

Fig. 2 Magnetoresistivity data for $01\bar{1}$ (full line) and [011] (dotted line) measured without gate bias at 4 K. The dashed line is calculated for COs of 100 nm period at the experimental carrier density.

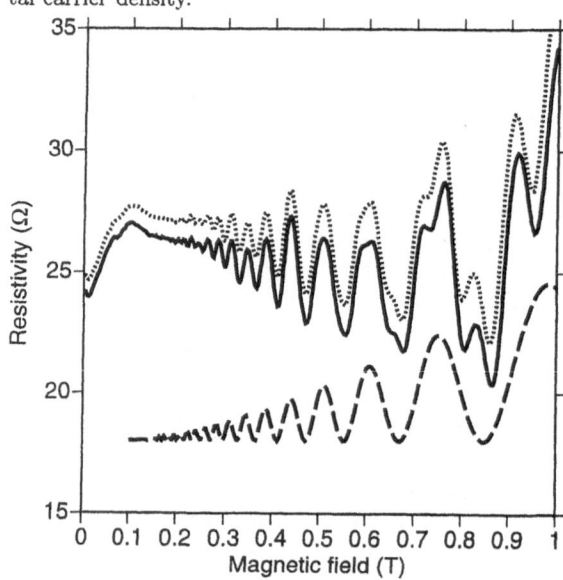

Fig. 3 Magnetoresistivity data for [010] (full line) and [001] (dotted line) measured with +0.3 V gate bias at 4 K. The dashed line is calculated for COs of $100/\sqrt{2}$ nm period at the experimental carrier density.

3. D. E. Grant, A. R. Long and J. H. Davies, Phys. Rev. B **61**, (2000) 13127

4. S. Chowdhury, C. J. Emeleus, B. Milton, E. Skuras, A. R. Long, J. H. Davies, G. Pennelli and C. R. Stanley, Phys. Rev. B **62**, (2000) R4821

5. R. R. Gerhardts, Phys. Rev. B **45**, (1992) 3449

6. C. J. Emeleus, B. Milton, A. R. Long, J.H.Davies, D. Petticrew, and M. C. Holland, Appl. Phys. Lett. **73**, (1998) 1412

7. B. Milton, C. J. Emeleus, K. Lister, J. H. Davies and A. R. Long Physica E, **6**, (2000) 555

Suppression of miniband transport in magnetic field: role of injection process

A.A. Krokhin *, A.E. Belyaev, T.M. Fromhold, C.R. Tench, H.M. Murphy, S. Bujkiewicz, P.B. Wilkinson, F.W. Sheard, L. Eaves, M. Henini

School of Physics and Astronomy, University of Nottingham, Nottingham NG7 2RD, UK

Abstract We present experimental and theoretical results of the current-voltage characteristics of undoped GaAs-AlGaAs superlattices (SL) with strong magnetic field applied parallel to the SL axis. The key result is that the magnetic field strongly suppresses the current flow. We propose a mechanism for suppression which is related to the rate of injection of electrons into the SL region and determines the concentration of carriers in the miniband. This mechanism shows a good agreement with our experimental data.

1 Introduction

Semiconductor superlattices (SLs) exhibit a strong nonlinearity in the current-voltage characterstcs, $I(V)$. A magnetic field $\mathbf{B}$ applied to the SL leads to specific singularities on the $I(V)$ curve which are the manifestations of different resonances effects, e.g., cyclotron-Stark-phonon resonance[1] and magnetophonon resonance[2]. Apart from sharp resonance effects, the magnetic field also gives rise to a smooth positive magnetoresistance. So far this influence of the *parallel* magnetic field has received much less attention. In undoped SLs the suppression of conductivity by magnetic field is manifested in a wide range of B from 1T to 20T[3]. Although rather extensive numerical Monte-Carlo simulations of electron dynamics in the miniband have been developed[4], the physical reasons of the effect of suppression are still unclear, specially in the region of low bias voltages.

Here we attribute the effect of suppression to the decreasing of the probability of the electron tunneling from the emitter contact into the SL miniband and show that the effect has a pure statistical nature since parallel magnetic field has no direct force effect on longitudinal dynamics.

2 Experimental results

The sample used in our experiments was grown by MBE on an N-type GaAs substrate. It contains 19 SL periods and has a total length $L \approx 270$ nm. A unit cell of the SL is formed from an AlGaAs barrier of width 2.08 nm and a GaAs well of width 9.72 nm. At zero bias, the bottom of the first miniband is at an energy $E_1 \approx 12$ meV *above* the Fermi level of the emitter contact. Thus there is a potential barrier of height E_1 which, prevents thermal excitation of an electron from the emitter into the SL. Therefore at $T = 0$K and zero voltage the de-

* *Present address:* Instituto de Física, Universidad Autónoma de Puebla, Apdo. Post. J-48, Puebla, 72570, México

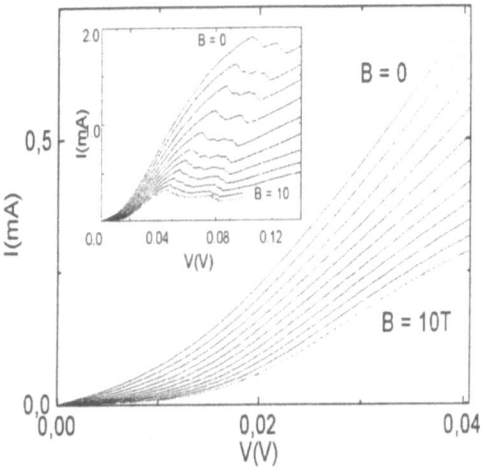

Fig. 1 $I(V)$ characteristics measured at $T = 4.2$K and $B = 1, 2, \ldots, 10$T. Insert shows the same curves over a wider range of voltages.

vice possesses no intrinsic conductance. Since our sample is undoped, the current which flows when a finite bias voltage is applied is due exclusively to *non-ohmic effects*. This is seen from the experimental $I(V)$ curves shown in Fig. 1. The effect of current suppression induced by the magnetic field can be seen already for $B = 1$T and becomes more pronounced at stronger fields.

3 Mechanisms of electron injection

The classical probability of surmounting the barrier E_1 is $\sim exp(-E_1/kT) \approx 5 \times 10^{-15}$ at $T = 4.2$K, which is too small to produce any measurable current. In this case we need to take into account an alternative mechanism of electron injection which is a quantum-mechanical tunneling. Under an applied voltage the energy bands are tilted, which gives rise to a finite probability of tunneling between the states in the emitter contact and those in the SL miniband.

The total injection current produced by the occupied emitter states is given by

$$j(F) = e \int_{-\infty}^{\infty} v_x D(\varepsilon) f(\varepsilon) \rho(\varepsilon) \, d\varepsilon , \qquad (1)$$

where v_x is the electron velocity, $f(\varepsilon)$ and $\rho(\varepsilon) \propto 1/v_x$ is the distribution function and the density of states of electrons in the emitter, associated with motion along the x-axis. The integrand in Eq. (1) is a product of two

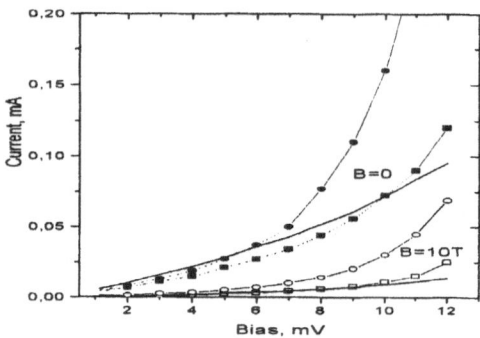

Fig. 2 $I(V)$ curves in the Boltzmann regime for $B = 0$ (3 upper curves) and for $B = 10T$ 3 lower curves. Solid curves without simbols show experimental data. Squares (circles) indicate theoretical results including (neglecting) the space-charge effects.

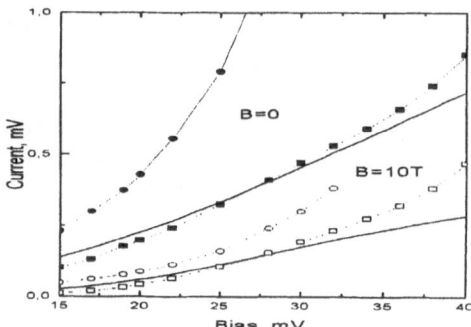

Fig. 3 The same as Fig. 2 but for the Fermi regime

sharp functions, $f(\varepsilon)$ and $D(\varepsilon)$. The principal contribution to the integral comes from a narrow interval near the optimal tunneling energy $\varepsilon_0(F)$. This energy satisfies a transcendental equation which has simple solutions in two limiting cases of weak and strong electric fields.

Boltzmann regime, $F \ll F_c = 2kT\sqrt{2mE_1}/e\hbar$. In this case of weak fields the injected electrons come from the states in the emitter which lie higher than the Fermi level, i.e. they come from the Boltzmann tail of the Fermi distribution. The electric current through the contact is given by asymptotic formula,

$$j_B(F) = \frac{e^2}{h^2} F \sqrt{2\pi mkT} \exp\left[-\frac{E_1}{kT} + \frac{(e\hbar F)^2}{24m(kT)^3}\right] . \quad (2)$$

Fermi regime, $F \gg F_c$. For the case of strong fields the tunneling occurs from the states which lie just below the Fermi level and the current is given by,

$$j_F(F) = \frac{e^2}{h^2} \sqrt{\frac{\pi mk^2T^2}{E_1}} \sqrt{\frac{F}{F_c}} \exp\left(-\frac{2E_1}{3kT}\frac{F_c}{F}\right) . \quad (3)$$

Having two asymptotics (2) and (3) we can estimate the height of the barrier E_1 from our experimental data by plotting $\ln(I/V)$ vs V. According to Eqs. (2) and (3) the curvature of this plot changes from concave (in Boltzmann regime) to convex (in Fermi regime) at the $V_c = F_c L$. In the experiment the transition between the two regimes occurs (for $B = 0$) at $V_c \approx 13$ meV ($F_c = 0.5$ kV/cm). Evaluating E_1 from the formula for F_c we obtained that $E_1 \approx 3$ mEv. This is much lower the E_1 value of 12 meV which follows from the band structure calculations. We attribute this difference to the edge effects in the band structure of a finite SL.

The parallel magnetic field affects the distribution function $f(\varepsilon)$ which now contains summation over Landau levels instead of integration over transverse momenta p_y and p_z. The origin of the suppression is due to the inrease of the *minimal* transverse energy by $\hbar\omega_c/2$, where ω_c is cyclotron frequency. Since the magnetic field does not change the total energy, the longitudinal energy is decreased and this gives rise to the decrease of the tun-

neling rate $D(\varepsilon)$. Our calculations show that the suppression is almost independent on the bias voltage (for $V < 40$ meV)and can be described by a B-dependent factor,

$$j(F, B) = j(F)q(B) , \quad q(B) = \frac{1}{2}\tanh\left(\frac{\pi\nu}{\omega_c}\right) , \quad (4)$$

where $\nu \simeq 7kT$ is the broadening of the Landau states due to collisions with defects and phonons.

To calculate $I(V)$ of the device we need to consider a self-consistent problem of electron injection and electron propagation within the miniband. The latter requires a solution of the Poisson equation which gives the distribution of the electric field $F(x)$ along the finite sample. This distribution is inhomogeneous and depends on the local concentration of the electrons which together with conservation of flux determines the total current through the sample. The results of solution of the self-consistent set of equation are shown in Fig. 2 (Fig. 3) for Boltzmann (Fermi) limit. When space-charge effects are included, we obtain good agreement between the theory and experiment for both $B = 0$ and $10T$. Similar correspondence is obtained for all $B < 10T$; this cofirms the proposed mechanism of magneto-suppression. In Fig. 3 the calculated $I(V)$ curves deviate strongly from the experimental data for $V > 35$ mV. This is because the transmittance of the potential barrier between the emitter and the SL miniband becomes close to unity, and so the electrons are able to enter the miniband directly. In this region the conductivity of the SL is explained satisfactory by the velocity quenching model[3] and confirmed by Monte Carlo simulations[4].

References

1. L. Canali, M. Lazzarino, L. Sobra, and F. Beltram, Phys. Rev. Lett. **76**, (1996) 3618; P. Kleinert and V.V. Bryksin, Phys. Rev. B **56**, (1997) 15827.
2. H. Noguchi, H. Sakaki, T. Takamasu, and N. Miura, Phys. Rev. B **45**, (1992) 12148; W.M. Shu and X.L. Lei, Phys. Rev. B **50**, (1994) 17378.
3. H.M. Murphy, et al., Microelectronic Eng. **47** (1995) 65; L. Eaves, et al., Physica B **272**, (1999) 190.
4. N. Mori, C. Hamaguchi, L. Eaves, and P.C. Main, unpublished.

Quantitative Evaluation of Electron-Electron Scattering Rate in Two-Dimensional Electron Gas by Magnetic Lateral Superlattice

Mayumi Kato[1], Akira Endo[1] Makoto Sakairi[1], Masahiro Hara[1], Shingo Katsumoto[1,2], Yasuhiro Iye[1,2]

[1] Institute for Solid State Physics, University of Tokyo, Kashiwanoha, Kashiwa, Chiba 277-8581 Japan

[2] CREST, Japan Science and Technology Corporation, Mejiro, Toshima-ku, Tokyo 171-0031 Japan

Abstract Application of spatially periodic magnetic field gives rise to an excess resistivity in a two-dimensional electron gas(2DEG), which is expressed as $\Delta\rho(T) = AT^2 + C$ at temperatures below about 15K. This system offers an unique opportunity to quantitatively evaluate the electron-electron umklapp scattering rate in 2DEG.

1 Introduction

Electron-electron scattering process plays such an important role in transport as establishing thermal equilibrium within the electron system. It does not, however, directly contribute to resistivity in a system having continuous translational symmetry, because the total momentum of colliding electrons is conserved. For the momentum relaxation to occur, the participation of crystal lattice, *i.e.* umklapp process, is necessary. In this work, we investigate the electron-electron umklapp scattering rate in two-dimensional electron gas (2DEG) at the GaAs/AlGaAs heterointerface.

Electron-electron scattering process is not visible in the resistivity of a plain GaAs/AlGaAs 2DEG, because the electron density is such that the Fermi wavevector is much smaller than the reciprocal lattice vector of the GaAs crystal. In order to switch on the umklapp scattering process, we introduce an artificial magnetic lateral superlattice (MLSL), *i.e.* periodical modulation of magnetic field[1,2]. Similar attempts have been made by using electrostatic potential modulation[3] and magnetic field modulation[4]. The use of MLSL for lifting the translational invariance possesses two distinct points of advantage over a similar experiment using a lateral electrostatic superlattice (ELSL). Firstly, the madulation amplitude of MLSL can be changed without affecting the density of 2DEG, which cannot be done so easily for ELSL. Secondly, the modulation profile of MLSL can be evaluated by straightforward magnetostatic calculation, while the corresponding calculation for ELSL has to take account of rather complicated screening effects.

2 Experimental

The samples are fabricated from a GaAs/AlGaAs single heterojunction wafer with electron density $n_e \approx 3 \times 10^{15} \mathrm{m}^{-2}$ and mobility $\mu \approx 60\mathrm{m}^2/\mathrm{Vs}$ at 4.2K. The heterointerface resides 75nm below the surface. A standard Hall bar is formed by wet chemical etching with the current path oriented along the $\langle 100 \rangle$ direction. An array of ferromagnetic metal stripes (Ni or Co) is fabricated on

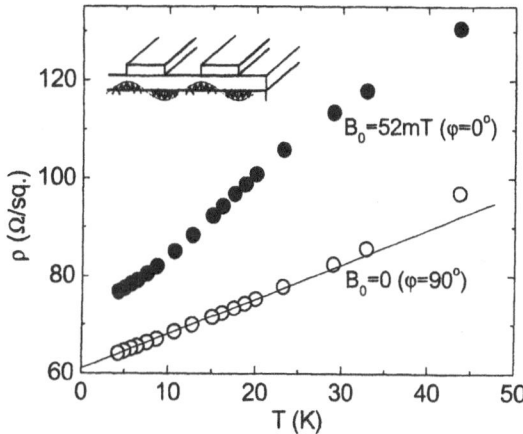

Fig. 1 The inset shows a sketch of the MLSL structure. The main panel shows the temperature dependence of the resistivity with and without the magnetic field modulation.

the surface electron beam lithography and vacuum deposition and lift-off. The period of the array is $a = 500\mathrm{nm}$ and the the width and thickness of the stripes are 250nm and 60nm, respectively. The choice of the $\langle 100 \rangle$ crystalline orientation is to minimize a strain-induced electrostatic potential modulation[5]. The measurements are made in a cross-coil magnet system consisting of a 6T Helmholtz coil for horizontal magnetic field and a 1T solenoid for vertical field. Temperature is regulated to $\pm 2\mathrm{m}K$ in a variable temperature insert by closed-loop feedback control of heater power. A rotating stage sample holder is used for coarse adjustment of the field orientation with respect to the 2DEG plane. (The fine adjustment is done by the cross-coil system.)

3 Results and Discussion

Periodically modulated magnetic field of the form $B_z(x) = B_0 \cos(Kx)$, $(K = 2\pi/a)$ is produced by the fringing field of the MLSL magnetized to saturation by a sufficiently strong magnetic field (typically 5T) applied parallel to the 2DEG plane (see the inset of Fig. 1). The magnetoresistance as a function of a perpendicular magnetic field shows commensurability oscillation (magnetic Weiss oscillation). The analysis of the commensurability oscillation gives an estimate of the magnetic field modulation amplitude. In the case of a sample with Co stripes, the

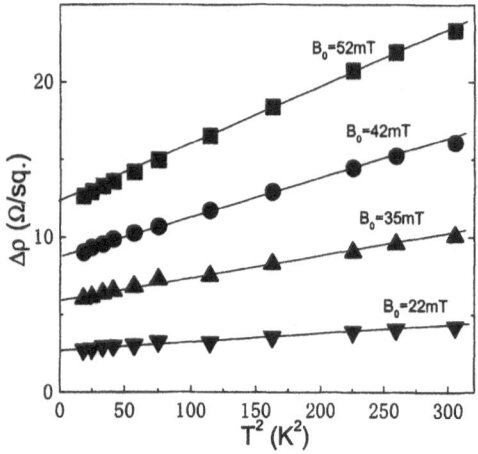

Fig. 2 The excess resistivity of 2DEG for different values of the magnetic field modulation amplitude B_0.

maximum modulation amplitude (obtained when the in-plane field is perpendicular to the stripes) is found to be $B_0 = 52$mT. The modulation amplitude can be varied from the maximum value to zero by changing the direction of the in-plane field.

In what follows, we focus on the case in which the mean value of the perpendicular field is zero. Figure 1 shows the temperature dependence of the resisivity of the 2DEG in a 52mT magnetic field modulation and in the absence of modulation. The T-linear behavior of the latter is due to acoustic phonon scattering. The resistivity increase $\Delta\rho(T)$ caused by the MLSL is plotted in Fig. 2 as a function of T^2 for different values of the modulation amplitude B_0. The excess resistivity is expressed as $\Delta\rho = AT^2 + C$. The temperature quadratic term is attributed to electron-electron scattering.

In order to corroborate this, we have made similar measurements as a function of the bias current, while keeping the sample at the lowest temperature. The null result for the $B_0 = 0$ case rules out Joule heating effect as the origin of the J-dependence. The electron temperature was estimated from the analysis of the Shubnikov-de Haas oscillation amplitude. The solid circles in Fig. 3 shows the result, which is in good agreement with the temperature dependence (open triangles) obtained by varying the lattice temperature.

A similar experiment on a sample with the MLSL oriented perpendicular to the current direction yields a null result. When the MLSL is oriented at an angle 45 degree with respect to the current direction, a "transverse resistivity" with the same magnitude as the excess longitudinal resistivity appears. Such anisotropic with respect to the relative orientation of the modulation wavevector and the current direction confirms that the excess resistivity is caused by the umklapp process in the MLSL.

The present system is unique in that the all the relevant parameters are precisely known, so that quanti-

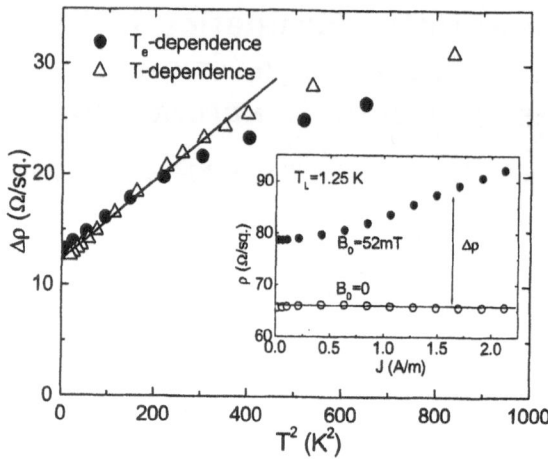

Fig. 3 The inset shows the resistivity as a function of the bias current density. The dependence of the excess resistivity on the electron temperature T_e in comparison with the T-dependence for $B_0 = 52$mT.

tative comparison with theoretical calculation can be carried out without adjustable parameters. Such theories are currently developed. Sasaki and Fukuyama[6] treat the effect of MLSL and interelectron Coulomb interaction in a perturbation scheme. The change in conductivity at $T = 0$ is calculated as, $\Delta\sigma(T = 0) = -(e^2/h)(\omega_0\tau)^2[(2k_F)^2 - K^2]^{1/2}/2K$, which gives good agreement with the experimental data. In the perturbation scheme, the effect is proportional to $\omega_0^2 \propto B_0^2$. Uryu and Ando[7], on the other hand, emphasize the importance of minigap formation, and treat the problem in the Boltzmann tranport formalism. They find that the coefficient A of the T^2-term diminishes with decreasing ratio of the modulation wavenumber and the Fermi wavenumber, and its dependence on B_0 changes from sublinear to superlinear.

4 Conclusion

We have shown that the MLSL offers a unique tool for exploration of the electron-electron scattering process in 2DEG. Quantitative comparison of the experimentally obtained scattering rate with the recently developed theoretical calculations are now under way.

We thank T. Sasaki, H. Fukuyama, S.Uryu and T. Ando for helpful discussions.

References

1. M. Kato, A. Endo, S. Katsumoto and Y. Iye, Phys. Rev. **B58** (1998) 4876.
2. M. Kato, A. Endo, S. Katsumoto and Y. Iye, J. Phys. Soc. Jpn. **68** (1999) 1492; **68** (1999) 2870.
3. A. Messica *et al.*, Phys. Rev. Lett. **78** (1997) 705.
4. A. Nogaret *et al.*, Phys. Rev. **B55** (1997) 16037.
5. E. Skuras *et al.*, Appl. Phys. Lett. 70 (1997) 871.
6. T. Sasaki and H. Fukuyama, submitted to J. Phys. Soc. Jpn.
7. S. Uryu and T. Ando, this Proceedings.

Shubnikov-de Haas effect with several filled subbands

U. Ekenberg

Theoretical Physics, Royal Institute of Technology, SE-10044 Stockholm, Sweden

Abstract　There is recent evidence that the common interpretation of Shubnikov-de Haas oscillations does not hold for a rather complex system like a two-dimensional hole gas with a spontaneous spin splitting caused by inversion asymmetry. We compare subband and Landau level calculations to analyse when the common interpretation starts to break down, starting from simple electron systems and introducing complications one at a time. For several hole systems we show that incorrect conclusions can be drawn such that cases with only one filled subband may appear to show spin splitting. For parabolic electron bands the Shubnikov-de Haas spectra for single and double occupancy are qualitatively different but the conclusions about individual subband populations can depend on the broadening and the range of B-values considered. It is proposed that a more careful analysis of localized and delocalized states is important.

1 Introduction

The standard method to deduce subband populations in two-dimensional systems is to examine the periodicities in $1/B$ of the Shubnikov-de Haas (SdH) oscillations, i.e. the longitudinal magnetoresistance ρ_{xx}. For a single parabolic band this is well established. Sometimes several periodicities are seen in the SdH spectra, which is usually interpreted as evidence that several subbands are filled. From these periodicities the populations of the individual subbands are often deduced.

For a two-dimensional hole gas in an asymmetric potential, there is recent evidence that this standard interpretation does not hold.[1, 2] Hole subbands are strongly nonparabolic because of the interaction between heavy and light hole subbands and the combination of spin-orbit coupling and inversion asymmetry gives rise to spin splittings which are of the same order of magnitude as the subband separations.

It is the purpose of this paper to evaluate when subband populations deduced from the standard interpretation of SdH oscillations start to deviate from the actual subband populations by comparing results from subband calculations (for $B = 0$) and Landau level calculations (for $B \neq 0$). For this reason we start from the simple case with simple parabolic electron subbands and add one or several of the following complications: multiple subband occupation, Zeeman splitting, asymmetric potential (leading to spontaneous spin splitting) and hole coupling (leading to strong nonparabolicity).

The basic reason for the discrepancies is that SdH oscillations mirror the Landau levels but are often interpreted in terms of the subband structure at $B = 0$. In the limit $B \to 0$ the conclusions drawn from Lan-

dau levels should in principle be consistent with those from subbands. However, SdH oscillations are not resolved experimentally at very low magnetic fields, so the conclusions are typically drawn for magnetic fields ≥ 1 T, although for very good samples the SdH oscillations can be seen for even lower B. Furthermore, calculations become very demanding numerically for very small magnetic fields where a large number of Landau levels is filled. We here concentrate on the case with intermediate magnetic fields, typically 1-3 T.

2 Method

The calculations are performed within the envelope function approximation. For holes the Luttinger-Kohn matrix Hamiltonian is used. When appropriate the subband calculations are self-consistent while the Landau levels are calculated in the obtained fixed potential. The density of states is computed assuming the same Gaussian broadening for all the Landau levels. The broadening parameter Γ (standard deviation) is taken as an adjustable parameter to see how the results are influenced by the sample quality. The density of states at the Fermi level is taken as a measure of ρ_{xx}. Although the exact B dependence of ρ_{xx} depends on other factors it is a reasonable approximation to assume that the periodicities, which is of our interest here, mirror the periodicities in the density of states.

3 Results

We first consider the case with two populated parabolic subbands separated by 10 meV. The parameters are typical for GaAs/AlGaAs quantum wells. The total carrier concentration is $N_s = 5 \cdot 10^{11}$ cm^{-2} and the carriers are distributed between the subbands in the ratio 73:27. The slope in a diagram of index of maxima and minima in the SdH spectra vs. $1/B$ changes gradually in the range 1-7 T. In the range 1.1 - 1.4 T the slope corresponds to the population of the subband with the higher population and for higher B the slope starts to correspond to N_s. However, we find that the concluded populations can depend on the broadening parameter Γ and the range of B-values considered. For example, if Γ is reduced new minima become resolved and the apparent slope changes if the new extrema are included.

If we include the Zeeman splitting using the bare g-factor 0.44 the SdH spectra are hardly changed. It is known that exchange effects can enhance the effective g-factor considerably. However, even with a ten-fold enhancement of the chosen g-factor a noticable spin splitting is typically seen at magnetic fields above ~ 5 T and the populations deduced at low magnetic fields are unaffected.

In the valence band the Landau level separations are much smaller because of the larger heavy hole mass and the nonlinearity of their B dependence, with stems from the repulsion between the heavy hole and light hole subbands. To see clear SdH oscillations around 1 T we need to use an anomalously small Γ value ~ 0.05 meV. The fact that such oscillations have been seen experimentally can be explained in terms of the localization of the states in the tails of the Landau levels, that we have not considered here. Since only the delocalized states in the middle of the Landau levels contribute to the current, localization acts as an effective narrowing of the Landau levels.

Interesting effects can occur when we compare the results with Zeeman splitting, which is quite strong in hole systems, with those where the Zeeman coupling (given by the parameter κ) has been turned off. For the actual value the SdH can appear regular but when κ is set to zero an apparent spin-splitting of a SdH peak results.

We have considered an asymmetric $In_{0.25}Ga_{0.75}As$ quantum well with p-doping in one barrier only and compared to a similar quantum well with symmetric doping. The parameters are the same as in our previous work [1], where we found that the Landau level fans and the spin splittings at the Fermi level were very similar down to about 1 T in the two cases although we have a spin splitting in the asymmetric well which results in a 10% population difference between the two spin subbands.

We have now also simulated the Shubnikov-de Haas spectra and displayed the results in Fig. 1 for the symmetric and the asymmetric quantum well. Only below 1.2 T significant differences are seen. Near 1.5 T one of the curves shows a tendency to a splitting, but this occurs for the symmetric well where it is not expected.

In Fig. 2 we have made the standard plot of the index of the DOS extrema vs. $1/B$. Also here the results for the two quantum wells are very similar down to 1.2 T. It is remarkable that even for the symmetric well there appear to be two distinctly different slopes, which normally would be interpreted as evidence for the existence of two filled spin subbands. The slopes correspond to populations of $7.9 \cdot 10^{11}$ and $3.6 \cdot 10^{11}$ cm^{-2} while the actual total carrier density is $3.0 \cdot 10^{11}$ cm^{-2}.

4 Discussion and conclusions

We have examined a number of two-dimensional systems and simulated Shubnikov-de Haas oscillations. For electrons in a parabolic band the conclusions about subband populations drawn from these spectra agree fairly well with the actual subband populations at $B = 0$, but the conclusions can be sensitive to the choice of broadening parameter and magnetic field region where the periodicities of the SdH oscillations are determined. For hole subbands conclusions from SdH spectra must be taken with

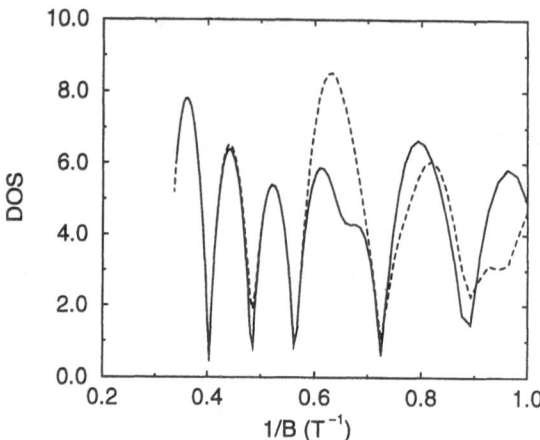

Fig. 1 Density of states at the Fermi level vs. $1/B$. Solid lines: symmetric quantum well, dashed lines: asymmetric quantum well

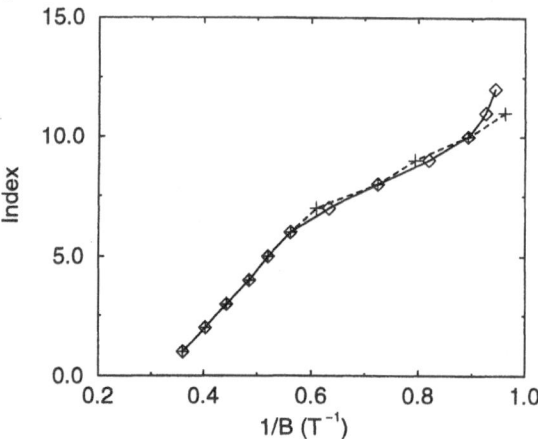

Fig. 2 Index of extrema in the density of states vs. $1/B$. Solid lines and diamonds: symmetric quantum well, dashed lines and plus signs: asymmetric quantum well

considerable caution. We have shown cases where the Zeeman coupling has been turned off but still appears to show spin splitting compared to the case with Zeeman coupling. Even in the case of one spin-degenerate subband the standard plot of extremum index vs. $1/B$ can show two straight lines with different slopes. Spectra for an asymmetric quantum well and an otherwise similar symmetric quantum without spontaneous spin splitting only display a difference below about 1 T.

It is proposed that the localized states in the Landau level tails improve the possibilities to interpret SdH spectra. We have shown a number of cases where the standard interpretation of multiple SdH oscillations fails but in spite of a large number of calculations a clear picture of when it holds has not yet emerged. Further theoretical and experimental studies are desirable.

References

1. O. Mauritz and U. Ekenberg, Phys. Rev. B **60**, (1999) R8505.
2. R. Winkler, S.J. Papadakis, E.P. De Poortere and M. Shayegan, Phys. Rev. Lett. **84**, (2000) 713.

On the origin of beat patterns in the quantum magneto-resistance of gated InAs/GaSb and InAs/AlSb quantum wells

A.C.H. Rowe[1], R.S. Ferguson[2], R.A. Stradling[1]

[1] Experimental Solid State Physics Group, Imperial College of Science Technology and Medicine, Prince Consort Road, London SW7 2BZ United Kingdom e-mail: a.c.rowe@ic.ac.uk

[2] The Department of Electrical and Electronic Engineering, Imperial College of Science Technology and Medicine, Exhibition Road, London SW7 2BZ, United Kingdom

Abstract This paper details measurements of the quantum magneto-resistance (QMR) in InAs/GaSb wells over a range of temperatures (0.4 K < T < 31 K). Beating at high carrier concentrations is a result of the mixing of the first confined subband Shubnikov-de Haas (SdH) series and a magneto-intersubband series. Beating due to two narrowly spaced SdH series is observed at the lowest carrier concentrations, but it remains unclear whether this is a result of the Rashba interaction. No such beating is observed in the InAs/AlSb sample over the largest concentration range yet reported.

1 Introduction

In polar semiconductors lacking a centre of inversion symmetry there is an in-built crystal electric field. As a consequence of the spin-orbit interaction, this creates a fine energy splitting of the conduction band states denoted ΔE_{BIA}. In heterostructures of these materials, there is an additional energy splitting due to the structural asymmetry, ΔE_{SIA}. Lommer *et al.* first pointed out that the latter in the form of the Rashba effect, should be the dominant form of splitting in narrow-gap heterostructures (NGH) [1]. The beat patterns observed in early tilted field QMR measurements were assigned to two narrowly split SdH series resulting from Rashba-type spin splitting of the first confined subband [2]. It is now common practise to *assume* that Rashba-type spin splitting is the cause of beat patterns observed in *perpendicular field* QMR measurements in NGHs. The beat patterns are then used to estimate values for ΔE_{SIA} and α_0, the spin-orbit coefficient.

This work shows that beat patterns can arise from another source unrelated to any form of zero field spin splitting (ZFSS).

2 Results and Discussion

Low magnetic field gated QMR measurements are presented for two samples—sample A, a 20 nm InAs/GaSb quantum well (QW) and sample B, a 15 nm InAs/AlSb QW. Both wafers were electrically characterised at 4 K before gate processing. Sample A has a concentration and mobility of 0.9×10^{12} cm^{-2} and 130,000 cm^2/Vs respectively. The equivalent values for sample B are 0.5×10^{12} cm^{-2} and 250,000 cm^2/Vs. Approximately 100 nm of high quality gate oxide (SiO$_2$) is then deposited on each wafer using a photolytically assisted CVD process.

Experiments on sample A over a range of temperatures (0.4 K < T < 31 K) demonstrate that the measured beat patterns are a result of mixing of the first subband SdH and a MIS series.

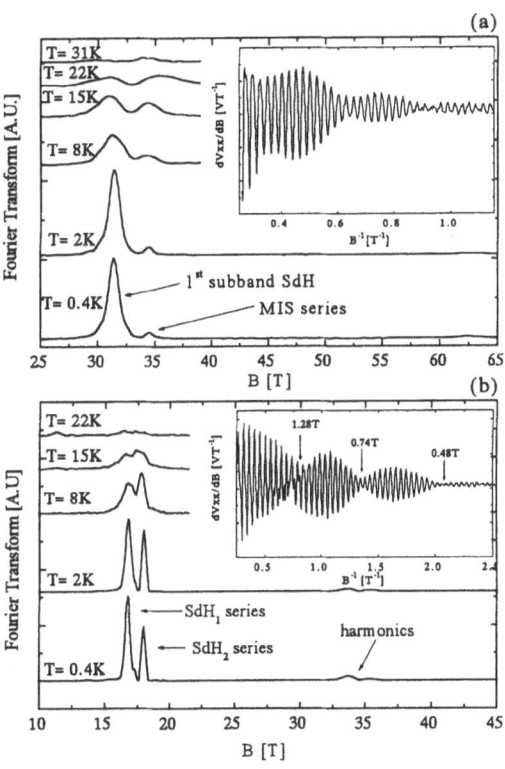

Fig. 1 Fourier transforms of QMR data in sample A over a range of temperatures. (a) $V_g = +6$ V. (b) $V_g = -7$ V. Insets show QMR derivatives at 0.4 K.

MIS scattering is a quasi-elastic process thought to be mediated by either defect or acoustic phonon interactions and can only occur when the second subband becomes populated [3] i.e. when $E_F > E_2$ where E_F is the Fermi energy and E_2 is the second subband energy. A maximum in the MIS scattering occurs each time the Landau levels of the first and second subbands cross. Consequently the MIS scattering rate varies periodically in B^{-1} with frequency

$$B_{f,MIS} = \frac{E_{12} m^*}{\eta e},$$ (1)

where E_{12} is the intersubband energy spacing, m^* is the energy dependent effective mass and e is the electron charge. This produces oscillations in the transverse resistivity (ρ_{xx}) with the same frequency.

Importantly, MIS oscillations persist even in the presence of significant thermal excitation whereas SdH oscillation amplitudes are strongly temperature dependent. At low fields (i.e. in the absence of harmonic content), SdH amplitudes drop with increasing temperature according to $X/\sinh(X)$ where $X = 2\pi^2 k_B T / \hbar \omega_c$ and $\omega_c = eB/m^*$ is the cyclotron frequency. It is this difference that allows MIS and SdH series to be distinguished in measurements of the QMR at different temperatures.

Although the amplitude of the MIS peak in the Fourier spectrum is lower than that of the SdH peak (see Fig. 1(a)), a significant beat effect still occurs at low fields (see Fig. 1(a) [inset]).

Fig. 1(a) shows a sequence of Fourier transforms of QMR oscillations measured in sample A over a range of temperatures at a particular gate voltage, $V_g = +6V$. The lower frequency of the two varies according to that expected for SdH, while the higher frequency mode has an amplitude only weakly temperature dependent (see Fig. 2). The latter mode is therefore assigned to a MIS series.

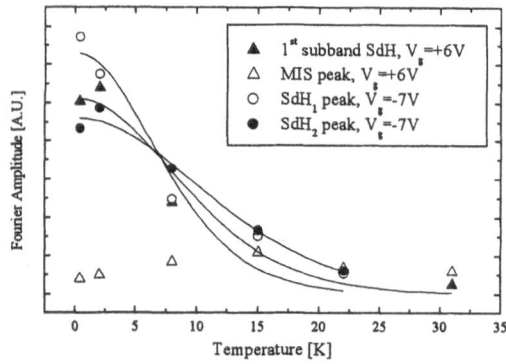

Fig. 2 Peak power versus temperature for modes shown in Fig. 1. SdH series decay according to $X/\sinh(X)$ [solid lines].

It is important to note the absence of the second subband SdH series whose amplitude is usually expected to be weaker than that for the first subband [4]. The relative amplitude can be decreased further still if $\mu_2 < \mu_1$, where μ_i is the mobility of the ith subband. Therefore, an absence of the second subband SdH is not sufficient evidence that $E_F < E_2$, from which it would follow that the observed beat patterns must be a result of two narrowly split SdH series. Assumptions that the second subband remains unoccupied based on perpendicular field QMR measurements can therefore lead to an erroneous identification of ZFSS. *Some identifications of Rashba-type ZFSS that have appeared in the literature may need to be re-examined.*

Evidence that $E_F > E_2$ is found in the variation of the low field mobility μ, with gate bias. When the second subband becomes occupied, the additional (intersubband) scattering channel acts to reduce the overall mobility. As this subband fills, μ reduces. The gate bias at which $E_F \sim E_2$ can be estimated, and is marked in Fig. 3.

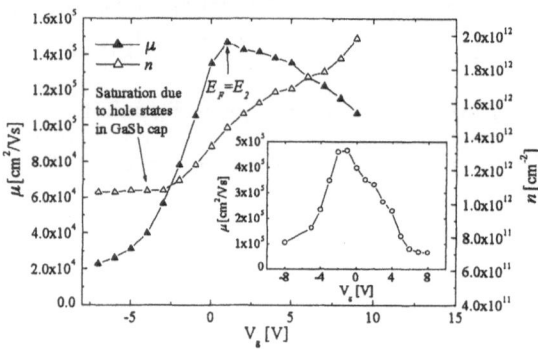

Fig. 3 Low field variation of n and μ versus V_g for sample A and sample B [inset]. The onset of a downwards trend in μ marks the point where $E_F \sim E_2$.

Over most of the concentration range in sample A, MIS related beating is absent. A beating of different origin appears only at the lowest carrier concentrations ($n \sim 0.82 \times 10^{12}$ cm^{-2}). It is clear from the similar temperature dependence of the amplitude of the two Fourier components shown in Figs. 1(b) and 2, that this beating is a result of two narrowly split SdH series. If the appearance of the second series is assigned to ZFSS, an upper limit for ΔE_{SIA} of 4.8meV can be estimated using the series method of Ref. 2. However, E_F is close to the GaSb valence band on both sides of the well as evidenced in the saturation of the gate effect for $V_g < -4V$—the result of the population of near surface hole states. Thus valence band effects such as electron-hole hybridisation which result in a minigap of order ~1 meV cannot be ruled out. Furthermore, the saturation in the gate effect may well correspond to an approximately symmetric well profile which would imply minimal Rashba-type splitting. It remains unclear therefore whether the observed beating at low concentrations is due to the Rashba effect.

Since spin-orbit effects should be large in InAs based heterostructures, this lack of unambiguous ZFSS is counter to expectations. This is highlighted in sample B where there is an absence of any beating at all over the largest first subband concentration range yet reported (0.48×10^{12} cm^{-2} to 2.4×10^{12} cm^{-2}), but in agreement with results over a more limited concentration range [5]. Beating observed at the highest concentrations (not shown) is associated with a rapid drop in μ (see inset Fig. 3) and is therefore likely to be MIS related.

References

1. G. Lommer, F. Malcher, U. Rössler, Phys. Rev. Lett. **60**, (1988), 728
2. B. Das, S. Datta, R. Reifenberger, Phys. Rev. B **41**, (1990), 8278
3. M.E. Raikh and T.V. Shahbazyan, Phys. Rev. B **49**, (1994), 5531
4. R. Kubo, S.J. Miyake and N. Hashitsume, in *Solid State Physics* (eds. Seitz and Turnbull), (1965), 270
5. S. Brosig, K. Ensslin, R.J. Warburton, C. Nguyen, B. Brar, M. Thomas and H. Kroemer, Phys. Rev. B **57**, (1999) R13989

Rashba spin splitting in 2D electron and hole systems: Implications for the metal-insulator transition

R. Winkler

Institut für Technische Physik III, Universität Erlangen-Nürnberg, Staudtstr. 7, D-91058 Erlangen, Germany e-mail: Roland.Winkler@physik.uni-erlangen.de

Abstract According to the familiar Rashba model, the inversion asymmetry induced spin splitting in two-dimensional (2D) electron systems increases linearly with in-plane wave vector $k_\parallel$. For 2D heavy hole systems, on the other hand, the spin splitting is of third order in $k_\parallel$ so that spin splitting becomes negligible in the limit of small 2D hole densities. We discuss consequences of this behavior in the context of recent arguments on the origin of the metal-insulator transition observed in 2D systems.

At zero magnetic field B spin splitting in quasi two-dimensional (2D) semiconductor quantum wells (QW's) can be a consequence of the bulk inversion asymmetry (BIA) of the underlying crystal (e.g. a zinc blende structure) and of the structure inversion asymmetry (SIA) of the confinement potential [1]. Here we want to focus on the SIA spin splitting which is usually the dominant part of $B = 0$ spin splitting in 2D systems [2]. In 2D hole systems spin splitting differs remarkably from the more familiar situation in 2D electron systems [3]. We discuss consequences of this behavior in the context of recent arguments on the origin of the metal-insulator transition observed in 2D systems.

To lowest order in the in-plane wave vector $k_\parallel$ SIA spin splitting in 2D electron systems is given by the so-called Rashba model [4]

$$H_{6c}^{SO} = \alpha\, \mathbf{k} \times \mathbf{E} \cdot \boldsymbol{\sigma}. \tag{1}$$

Here $\boldsymbol{\sigma} = (\sigma_x, \sigma_y, \sigma_z)$ denotes the Pauli spin matrices, α is a material-specific prefactor [1], and $\mathbf{E}$ is an effective electric field that breaks the inversion symmetry [5]. Assuming $\mathbf{E} = (0, 0, E_z)$ we obtain for the spin splitting of the subband dispersion to lowest order in $\mathbf{k}_\parallel = (k_x, k_y, 0)$

$$\mathcal{E}_{6c}^{SO}(\mathbf{k}_\parallel) = \pm\langle\alpha E_z\rangle k_\parallel, \tag{2}$$

i.e., SIA spin splitting increases linearly as a function of $k_\parallel$. For small $k_\parallel$ Eq. (2) thus becomes the dominant term in the energy dispersion $\mathcal{E}_\pm(\mathbf{k}_\parallel)$, so that SIA spin splitting of electron states is most important for small 2D densities. In particular, we get a divergent van Hove singularity of the density-of-states (DOS) at the bottom of the subband [6] which is characteristic for a k linear spin splitting. As an example, we show in Fig. 1 our self-consistently calculated [6] results for an MOS inversion layer on InSb obtained by means of an 8 × 8 Hamiltonian. For small $k_\parallel$ the spin splitting increases linearly as a function of $k_\parallel$, in agreement with Eq. (2). Due to nonparabolicity the spin splitting for larger $k_\parallel$ converges toward a constant [6].

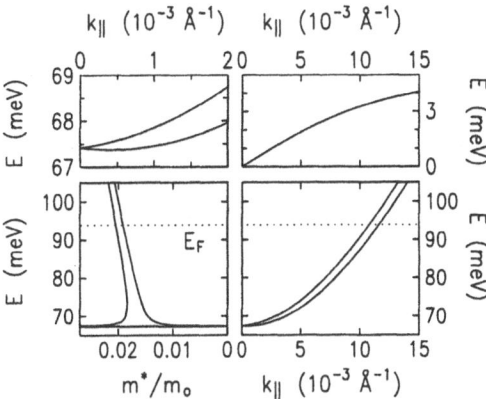

Fig. 1 Self-consistently calculated subband dispersion $\mathcal{E}_\pm(k_\parallel)$ (lower right), DOS effective mass m^*/m_0 (lower left), spin splitting $\mathcal{E}_+(k_\parallel) - \mathcal{E}_-(k_\parallel)$ (upper right) and subband dispersion $\mathcal{E}_\pm(k_\parallel)$ in the vicinity of $k_\parallel = 0$ (upper left) for an MOS inversion layer on InSb with $N_s = 2 \times 10^{11}$ cm^{-2} and $|N_A - N_D| = 2 \times 10^{16}$ cm^{-2}. The dotted line indicates the Fermi energy E_F.

The Rashba model (1) can be derived by purely group-theoretical means. The electron states in the lowest conduction band are s like (orbital angular momentum $l = 0$). With spin-orbit (SO) interaction we have total angular momentum $j = 1/2$. Both $\mathbf{k}$ and $\mathbf{E}$ are polar vectors and $\mathbf{k} \times \mathbf{E}$ is an axial vector (transforming according to the irreducible representation Γ_4 of T_d) [1]. Likewise, $\boldsymbol{\sigma}$ is an axial vector. The dot product (1) of $\mathbf{k} \times \mathbf{E}$ and $\boldsymbol{\sigma}$ therefore transforms according to the identity representation Γ_1, in accordance with the theory of invariants [7]. In the Γ_6^c conduction band the scalar triple product (1) is the only term of first order in $\mathbf{k}$ and $\mathbf{E}$ that is compatible with the symmetry of the band.

Now we want to compare the Rashba model (1) with the SIA spin splitting of hole states. The topmost valence band is p like ($l = 1$). With SO interaction we have $j = 3/2$ for the HH/LH states (Γ_8^v) and $j = 1/2$ for the SO states (Γ_7^v). For the Γ_8^v valence band there are two sets of matrices which transform like an axial vector, namely $\mathbf{J} = (J_x, J_y, J_z)$ and $\boldsymbol{\mathcal{J}} = (J_x^3, J_y^3, J_z^3)$ (Ref. [8]). Here J_x, J_y and J_z are the angular momentum matrices for $j = 3/2$. Thus we get [3]

$$H_{8v}^{SO} = \beta_1\, \mathbf{k} \times \mathbf{E} \cdot \mathbf{J} + \beta_2\, \mathbf{k} \times \mathbf{E} \cdot \boldsymbol{\mathcal{J}}. \tag{3}$$

Similar to the Rashba model the first term has axial symmetry with the symmetry axis being the direction of the electric field $\mathbf{E}$. The second term is anisotropic, i.e., it depends on both the crystallographic orientation of $\mathbf{E}$ and $\mathbf{k}$. Using $\mathbf{k} \cdot \mathbf{p}$ theory we find that the prefactor β_2 is always much smaller than β_1, i.e., the dominant term in Eq. (3) is the first term. This is due to the fact that the $\mathbf{k} \cdot \mathbf{p}$ coupling between Γ_8^v and Γ_6^c is isotropic, so that it contributes to β_1 but not to β_2. The prefactor β_2 stems from $\mathbf{k} \cdot \mathbf{p}$ coupling to more remote bands such as the higher conduction bands Γ_8^c and Γ_7^c.

The first term in Eq. (3) couples the two LH states ($j_z = \pm 1/2$) and the HH states ($j_z = \pm 3/2$) to the LH states. But there is no k linear splitting of the HH states proportional to β_1. The second term in Eq. (3) contains a k linear coupling of the HH states. Neglecting in Eq. (3) the small terms proportional to β_2 we obtain for the HH spin splitting to lowest order in $\mathbf{k}_\parallel$

$$\mathcal{E}_{\mathrm{HH}}^{\mathrm{SO}}(k_\parallel) \propto \pm \langle \beta_1 E_z \rangle k_\parallel^3, \tag{4a}$$

whereas for the LH states we have to lowest order in $\mathbf{k}_\parallel$

$$\mathcal{E}_{\mathrm{LH}}^{\mathrm{SO}}(k_\parallel) \propto \pm \langle \beta_1 E_z \rangle k_\parallel. \tag{4b}$$

Thus we have a qualitative difference between the spin splitting of electron and LH states which is proportional to $k_\parallel$ and the splitting of HH states which essentially is proportional to $k_\parallel^3$. The former is most important in the low-density regime whereas the latter becomes negligible for small densities. Note that for 2D hole systems the first subband is usually HH like so that for low densities the SIA spin splitting is given by Eq. (4a). It is crucial that, basically, we have

$$\alpha, \beta_1, \beta_2 \propto \Delta_0 \tag{5}$$

with Δ_0 the SO gap between the bulk valence bands Γ_8^v and Γ_7^v, i.e., we have no SIA spin splitting for $\Delta_0 = 0$.

As an example, we show in Fig. 2 our results for the topmost HH subband of a GaAs/Al$_{0.5}$Ga$_{0.5}$As heterostructure obtained by means of a 14×14 Hamiltonian (Γ_8^c, Γ_7^c, Γ_6^c, Γ_8^v, and Γ_7^v). It fully took into account both SIA and BIA. The weakly divergent van Hove singularity of the DOS effective mass at the subband edge indicates that the k linear splitting is rather small. Basically, the spin splitting in Fig. 2 is proportional to $k_\parallel^3$. For comparison, in the upper, right part of Fig. 2 the dotted line shows the spin splitting of the first LH subband. For small $k_\parallel$ the splitting is linear in $k_\parallel$, but for larger $k_\parallel$ due to both HH-LH mixing and nonparabolicity it is dominated by terms of higher order in $k_\parallel$.

Recently, spin splitting in 2D systems has gained renewed interest because of an argument by Pudalov [9] which relates the metal-insulator transition (MIT) in low-density 2D systems with the SIA spin splitting. Based on the Rashba model [4] it was argued that the SIA spin splitting "results in a drastic change of the internal properties of the system even without allowing for the Coulomb interaction" [10]. In the low-density regime required for the MIT, however, this argument holds only for electron and LH states. As noted above, spin splitting in low-density HH systems is rather small. The MIT has been observed also in pure HH systems in, e.g., Si/SiGe

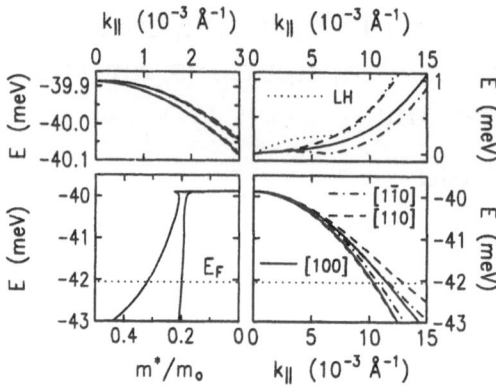

Fig. 2 Self-consistently calculated anisotropic dispersion $\mathcal{E}_\pm(\mathbf{k}_\parallel)$ (lower right), DOS effective mass m^*/m_0 (lower left), spin splitting $\mathcal{E}_+(\mathbf{k}_\parallel) - \mathcal{E}_-(\mathbf{k}_\parallel)$ (upper right) and dispersion $\mathcal{E}_\pm(\mathbf{k}_\parallel)$ in the vicinity of $k_\parallel = 0$ (upper left) of the topmost HH subband of a [001] grown GaAs/Al$_{0.5}$Ga$_{0.5}$As heterostructure with $N_s = 2 \times 10^{11}$ cm^{-2} and $|N_A - N_D| = 2 \times 10^{16}$ cm^{-2}. Different line styles correspond to different directions of the in-plane wave vector $\mathbf{k}_\parallel$ as indicated. In the lower, right figure, the dotted line indicates the Fermi energy E_F. In the upper, right part, the dotted line shows the spin splitting of the first LH subband for $\mathbf{k}_\parallel \parallel$ [100].

QW's [11,12] for which already the bulk SO interaction is very small. Therefore it appears unlikely that here the broken inversion symmetry of the confining potential is responsible for the MIT. We note that spin-orbit coupling in 2D electron systems is enhanced due to many-body effects [13]. It can be expected that similar effects are also relevant for the $B = 0$ spin splitting of 2D hole systems, though this will not affect our general conclusions.

References

1. U. Rössler, F. Malcher, and G. Lommer, in *High Magnetic Fields in Semiconductor Physics II*, Vol. 87 of *Solid-State Sciences*, edited by G. Landwehr (Springer, Berlin, 1989), p. 376.
2. P. Pfeffer and W. Zawadzki, Phys. Rev. B **59**, R5312 (1999).
3. R. Winkler, Phys. Rev. B **62**, 4245 (2000).
4. Y. A. Bychkov and E. I. Rashba, J. Phys. C: Solid State Phys. **17**, 6039 (1984).
5. R. Lassnig, Phys. Rev. B **31**, 8076 (1985).
6. R. Winkler and U. Rössler, Phys. Rev. B **48**, 8918 (1993).
7. G. L. Bir and G. E. Pikus, *Symmetry and Strain-Induced Effects in Semiconductors* (Wiley, New York, 1974).
8. J. M. Luttinger, Phys. Rev. **102**, 1030 (1956).
9. V. M. Pudalov, Pis'ma Zh. Eksp. Teor. Fiz. **66**, 168 (1997), [JETP Lett. **66**, 175 (1997)].
10. M. A. Skvortsov, Pis'ma Zh. Eksp. Teor. Fiz. **67**, 118 (1998), [JETP Lett. **67**, 133 (1998)].
11. J. Lam, M. D'Iorio, D. Brown, and H. Lafontaine, Phys. Rev. B **56**, R12741 (1997).
12. P. T. Coleridge, R. L. Williams, Y. Feng, and P. Zawadzki, Phys. Rev. B **56**, R12764 (1997).
13. G.-H. Chen and M. E. Raikh, Phys. Rev. B **60**, 4826 (1999).

Weak localization in periodically modulated magnetic field

Y. Blum[1], A. Tsukernik[1], A. Palevski[1] , T. A. Shutenko[2], B. L. Altshuler[2,3], I. L. Aleiner[4], A. Rudra[5], E. Kapon[5]

[1] School of Physics and Astronomy, Tel Aviv University, Tel Aviv 69978, Israel
[2] Physics Department, Princeton University, Princeton, NJ 08544
[3] NEC Research Institute, 4 Independence Way, Princeton, NJ 08540
[4] Department of Physics and Astronomy, State University of New York, Stony Brook, NY 11794
[5] Physics Department, Swiss Federal Institute of Technology, 1015 Lausanne, Switzerland

Abstract We study weak localization in $GaAs/Al_xGa_{1-x}As$ V-grooved heterostructures in periodically modulated magnetic field. We find theoretical expression for striped magnetic field magnetoconductance that agrees with our experimental data.

Weak localization (WL) correction to the classical conductivity of disordered metals was extensively studied both theoretically and experimentally [1]. However these investigations were mostly restricted to uniform magnetic fields.

It is well known[2] that WL effects in strictly 2D systems are affected only by the component of magnetic field which is perpendicular to the 2D plane. *Uniform* magnetic field applied to the 2D V-grooved film as shown on Fig. 1, results in the effective *modulation* of the normal magnetic field component, because the direction of the normal to 2DEG is not uniform. This modulation depends on the orientation of magnetic field relative to

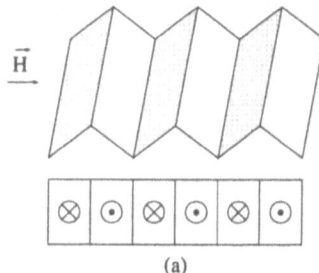

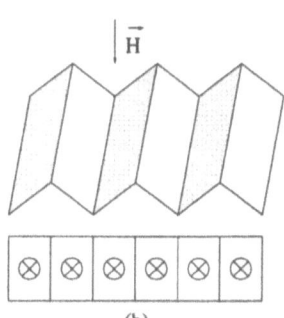

Fig. 1 Different orientations of magnetic field to the V-grooves substrate and magnetic field component normal to the 2DEG: (a) parallel to the substrate; (b) perpendicular to the substrate.

the film. Namely, the normal component of the magnetic field turns out to be modulated when the external field is *parallel* to the substrate as shown on Fig. 1a. Contrarily, if the external field is *perpendicular* to the substrate (Fig. 1b), the normal component of the magnetic field is uniform. We measured magnetoconductance of 2DEG for both magnetic field orientations and extended the WL theory to the case of the striped magnetic field. The theory and the experimental data are in a good agreement.

The 2DEG was obtained at $GaAs/Al_xGa_{1-x}As$ structure grown by low-pressure organometallic chemical-vapor deposition on top of an undoped (100)-GaAs V-groove substrate patterned with 0.5-μm-pitch,[011]-orientated V-grooves. We estimate the angle between the sidewalls of the V-groove to be approximately 90^0 based on SEM analysis. Multiple probe sample of rectangular shape was patterned using standard photolithography and measured at low temperatures using four probe AC lock-in amplifier techniques. Based on complimentary resistivity and Hall effect measurements, we estimate the low temperature mobility of the sample $\mu = 85,000 cm^2 V^{-1} s^{-1}$ and the electron density $n = 3.4 \times 10^{11} cm^{-2}$. The measurements were done at sufficiently low currents $0.1 \mu A$ to avoid electronic heating.

Table 1 Relevant sample's parameters.

$T[K]$	$\tau_\varphi [10^{-11} s]$	$\tau_{so} [10^{-11} s]$	$D[cm^2 s^{-1}]$	$d[\mu m]$
0.3	8.5 ± 1.6	10.5 ± 0.3	1000	0.25
1.3	2.7 ± 1.3	10.5 ± 0.3	1000	0.25
4.2	0.60 ± 0.03	10.5 ± 0.3	1000	0.25

The magnetoconductance data was taken at T =0.3K, 1.3K, 4.2K. The data for a uniform magnetic field orientation of the sample at 1.3K is shown on Fig.2 by circles. It is a standard positive magnetoconductance with low spin orbit scattering rate in a low uniform magnetic field which is well described by familiar WL theory [2, 1,4]. The two fitting parameters used for the theoretical curve (dashed line), namely dephasing time, τ_φ, and time of spin-orbital relaxation τ_{so} are listed in the Table 1. The rest of parameters entering the theoretical expression are calculated based on the resistivity and density

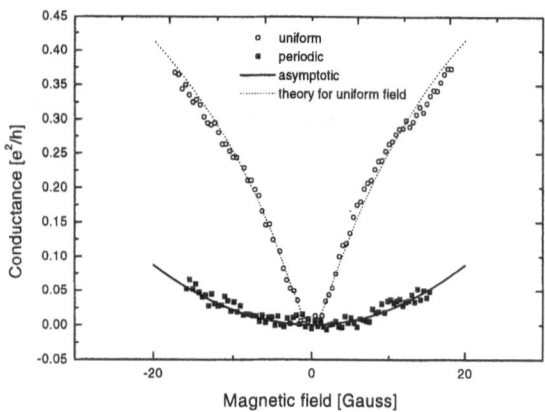

Fig. 2 Experimental data for magnetoconductance in the uniform and striped magnetic fields at $T = 1.3$K, theoretical curve for uniform magnetic field magnetoconductance and theoretical asymptotics of for the weak striped magnetic field magnetoconductance.

measurements. The parameters of the sample are presented in Table 1.

The data at $T = 1.3$K for modulated magnetic field (i.e., for the parallel) orientation is presented on Fig.2 by squares. The effect is much weaker and the suppression of WL correction to the conductivity for this orientation occurs on a much larger scale of magnetic fields.

Below we discuss the calculation of the effect of a striped magnetic field on the conductivity [3]. The results of this calculation, for the set of parameters as in Table 1 at $T = 1.3$K, are shown on Fig.2 by solid line.

We start our with the expression for the spatially averaged weak localization correction [2,4]

$$\sigma_{WL} = -\frac{2e^2 D}{\pi \hbar} \int d\mathbf{r} \ C(\mathbf{r}, \mathbf{r}) \qquad (1)$$

where D stands for the diffusion constant, and $C(\mathbf{r}_1, \mathbf{r}_2)$ is the familiar Cooperon propagator.

If this magnetic field is weak, one can expand Cooperon up to the second order in the vector potential and substitute the result into the Eq. (1). Subtracting the WL correction in the absence of magnetic field one obtains the magnetoconductance in the weak striped magnetic field in the form

$$\delta\sigma\left(\frac{d}{L_H}, \frac{L_\varphi}{d}\right) = \frac{e^2}{2\pi^2\hbar}\left(\frac{d}{L_H}\right)^4 f\left(\frac{L_\varphi}{d}\right) \qquad (2)$$

where L_H is the a single electron magnetic length, L_φ is the phase coherence length and $f(y)$ is a dimensionless function [3] shown on Fig. 3.

In our experiments at parallel magnetic field the magnetoconductance was quadratic in H in the whole interval of field variation. It means that one can use the low field asymptotics. However the Eq.(2) is not applicable directly, since it does not take spin-orbit scattering into account. For weak magnetic field one can use Eq.(2) and present magnetoconductance with the spin-orbit correc-

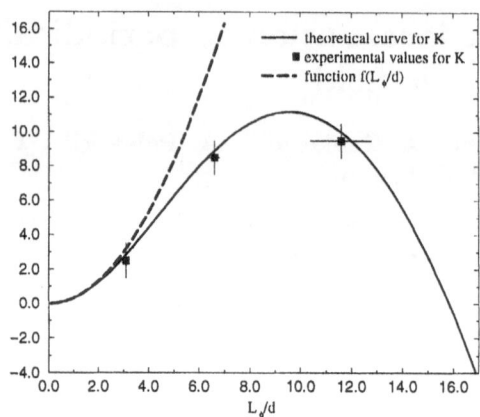

Fig. 3 The theoretical dependence of magnetoconductance on the ratio of stripe width to the dephasing length, experinmetal points of this dependence for 0.3K, 1.3K and 4.2K. and the plot of function $f(L_\varphi/d)$.

tions [1] in the form

$$\delta\sigma_{so}\left(\frac{d}{L_H}, \frac{L_\varphi}{d}, \frac{L_\varphi^*}{d}\right) = \frac{e^2}{2\pi^2\hbar}\left(\frac{d}{L_H}\right)^4 \times$$

$$\times \left[\frac{3}{2}f\left(\frac{L_\varphi^*}{d}\right) - \frac{1}{2}f\left(\frac{L_\varphi}{d}\right)\right] \qquad (3)$$

where

$$L_\varphi^* = \sqrt{L_\varphi^2 + \frac{3}{4}D\tau_{so}} \qquad (4)$$

The striped magnetic field magnetoconductance,Eq.(3), may be expressed as

$$\delta\sigma_{so}\left(\frac{d}{L_H}, \frac{L_\varphi}{d}, \frac{L_\varphi^*}{d}\right) = \frac{e^2}{2\pi^2\hbar}\left(\frac{d}{L_H}\right)^4 K \qquad (5)$$

The dependence of K in the Eq.(5), given by Eq.(3), on L_φ/d is shown on Fig. 3 by solid line, for values of $D\tau_{so}$ taken from Table 1. The values of the coefficient K obtained from fitting of experimental striped magnetic field magnetoconductance data to the Eq.(5) at temperatures $T = 0.3$K, 1.3K and 4.2K are shown on Fig. 3 by points with error bars. The experimental points fit the theoretical curve surprisingly well. This implies that the Eq.(3) is a proper description of the WL magnetoconductance for striped magnetic field geometry.

In conclusion, we observed experimentally magnetoresistance in periodically modulated magnetic field. We proposed the theoretical description of this effect that seems to agree well with the experiment and predicts the behavior of the conductance at stronger fields.

References

1. G. Bergmann, Phys. Reports **107**, 1 (1984).
2. B. L. Altshuler, D. Khmel'nitzkii, A. I. Larkin, P. A. Lee, Phys. Rev B **22**, 5142 (1980).
3. Detailed calculation will be published.
4. B. L. Altshuler and A. G. Aronov, in *Electron-Electron Interactions in Disordered Systems*, A. L. Efros and M. Pollak eds., (North-Holland, Amsterdam, 1985).

Proc. 25th Int. Conf. Phys. Semicond., Osaka 2000 (Eds. N. Miura and T. Ando)

771

Intrinsic mobility limits in polarization induced two-dimensional electron gases.

D. Jena*, Y. P. Smorchkova, C. R. Elsass, A. C. Gossard, U. K. Mishra

Department of Eletrical and Computer Engineering, University of California, Santa Barbara, CA, 93106, USA.
e-mail: djena@indy.ece.ucsb.edu

Abstract Unusually large spontaneous and piezoelectric polarization make the III-V nitrides a very different material system from the widely studied III-V arsenides. The 2DEG in such heterostructures can be entirely polarization induced, without the need of modulation doping. We show that alloy scattering and interface roughness scattering mechanisms in the nitrides are coupled with the strong polarity of the material, and leads to a severe scattering mechanism.

1 Introduction

III-V nitride semiconductors, due to their large polarization fields, provide a new method of creating two-dimensional electron gases (2DEG) at heterojunctions. The sheet densties of such 2DEGs are much higher than the sheet densities possible in AlGaAs/GaAs heterostructures, where confined electrons are supplied by modulation doping ,[1],[2].

III-V nitrides have wide bandgaps, leading to large band discontinuities at heterointerfaces. In addition, the large spontaneous and piezoelectric polarization leads to the confinement of high 2DEG sheet densities confined at heterointerfaces even without intentional modulation doping [3]. The source of the electrons is still unclear, though recent work strongly indicates existence of donor-like surface states $1.42eV$ [4] below the conduction band edge at the surface.

1.1 Transport and 2DEG mobility

The novel effects of polarization and the large number of charged dislocations are the new features in the nitride heterostructures. Scattering of conduction electrons by charged dislocations in bulk GaN was studied by Look et. al.[5], and shown to considerably affect the electron mobility. Scattering by charged dislocations in AlGaN/GaN 2DEGs was studied theoretically and expermiental data compared with experimetal results by D. Jena et. al [6]. Published work on transport in AlGaN/GaN 2DEGs [7], [8] do not take dislocations into account. In addition, a scattering mechanism is shown to originate from the coupling of alloy disorder with the difference in polarization between AlN and GaN in the AlGaN barrier. We call it 'dipole scattering' due to nature of the charge origin. We incorporate these novel scattering mechanisms to find the total electron mobility. By using a more accurate model of the defect nature,

* *Present address:* Department of Electrical and Computer Engineering, University of California, Santa Barbara, CA 93106

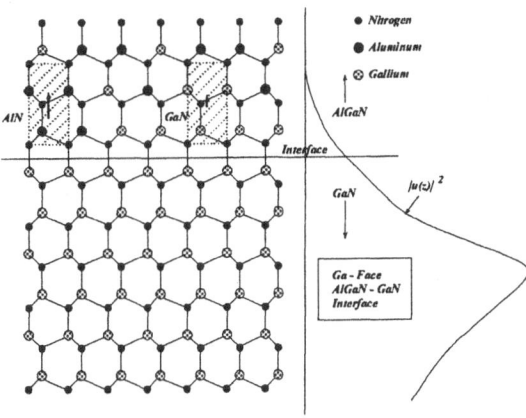

Fig. 1 Interface at AlGaN/GaN heterojunction, showing the dipole moments in AlN and GaN unit cells, with the squared wavefunction for the bound state.

we find that we can explain electron mobilities we measure in our high quality MBE grown samples, and it's dependence on the sheet density n_s and the temperature T.

In this short report we describe qualitatively the new dipole scattering mechanism and how the incorporation of dislocations and dipole scattering gives an accurate way to explain electron mobility. Figure [1] shows the interface of a AlGaN/GaN heterostructure. The barrier is made of layers of Al(Ga) atoms alternating with N atoms. The classical theory of polarization treats the macroscopic polarization field as a contribution of dipole moments from each unit cell. The dipole moment is related to the macroscopic polarization by $\mathbf{p}_{dipole}/\Omega = \mathbf{P}$ (Ω is the volume of a unit cell). The dipole moment is thus given by the macroscopic polarization as $\mathbf{p_{dipole}} = \Omega \cdot (\mathbf{P_{sp}} + \mathbf{P_{pz}})$, where $\mathbf{P_{sp}}, \mathbf{P_{pz}}$ are the spontaneous and piezoelectric polarizations respectively. There is a large difference in spontaneous and piezoelectric polarizations between AlN and GaN, and thus the alloy disorder is coupled with polarization disorder. An AlN unit cell and a GaN unit cell are shown in a $Al_x Ga_{1-x} N$ barrier. The dipole moments are indicated, and can be calculated using the piezoelectric and spontaneous polarizations of AlN and GaN [9],[1].

We treat the disorder in the virtual crytal approximation by assuming a perfect crystal barrier with the average dipole moment in every unit cell, and then calcu-

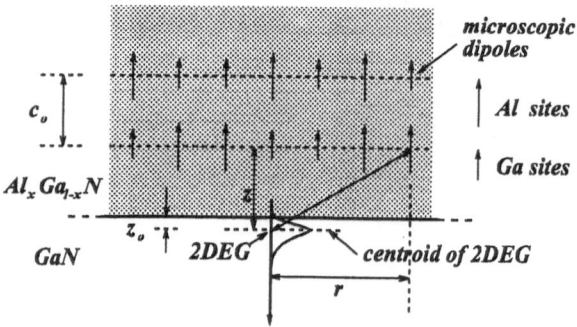

Fig. 2 Schematic diagram of barrier shown embedded with dipoles of different magnitudes, whose fluctuation leads to scattering.

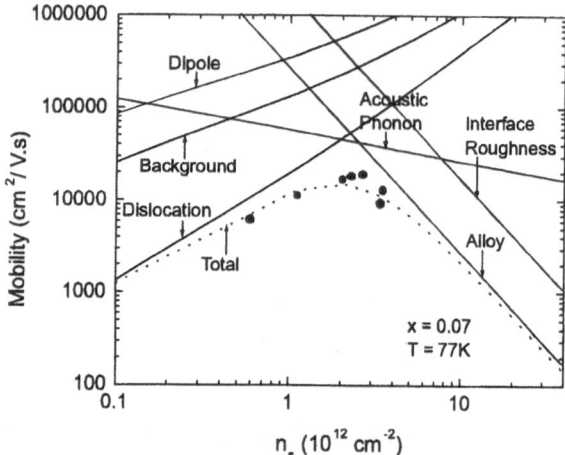

Fig. 3 Theoretical prediction of totla mobility (dotted line) showing the fit to experimentally measured mobilities (filled circles).

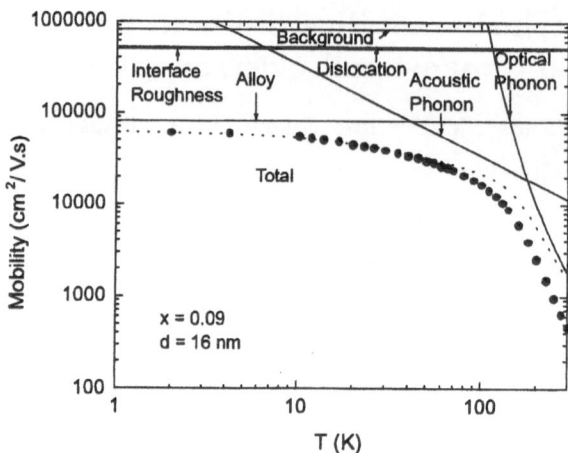

Fig. 4 Theoretically calculated mobility explaining the measured temperature dependent electron mobility (filled circles) of the highest mobility sample reported in literature.

loy scattering being the dominant scattering mechanism for this sample, the fit between theory and experiment shown in Figure [4]. The deviation of theory from measured mobility at $T > 100K$ is because of our assumption of a constant 2DEG density, which does not hold for this temperature range. Further details will presented in a later work.

References

1. O. Ambacher *et al*, J. Appl. Phys. **87**, (2000) 334.
2. T. Ando, A. B. Fowler, and F. Stern, Rev. Mod. Phys. **54**, (1982) 437.
3. M. A. Khan *et al*, Appl. Phys. Lett. **75**, (1999) 2806.
4. J. P. Ibbetson *et al*, Appl. Phys. Lett. **77**, (2000) 250.
5. D. C. Look and J. R. Sizelove, Phys. Rev. Lett. **82**, (1999) 1237.
6. D. Jena *et al*, Appl. Phys. Lett. **76**, (2000) 1707.
7. L. Hsu, and W. Walukiewicz, Phys. Rev. B. **56**, (1998) 1520.
8. Y. Zhang and J. Singh, J. Appl. Phys. **85**, (1999) 587.
9. F. Bernardini *et. al*, Phys. Rev. B. **56**, (1997) R10 024.
10. D. Jena *et al*, J. Appl. Phys. **88**, (2000) (To be published).
11. I. P. Smorchkova *et al*, J. Appl. Phys. **86**, (1999) 4520.

late the fluctuation at individual AlN and GaN unit cells. This is shown schematically in Figure [2]. We evaluate the dipole scattering rates for a two-dimensional electron gas using Fermi's golden rule for transition rates, and find the dipole scattering limited elecron mobility. Refer to [10] for details of the procedure.

We include all scattering rates to calculate the final expected electron mobility. The results for a sample of alloy composition $x = 0.07$ and temperature $T = 77K$ is shown in Figure [3]. The mobility is calculated as a fuction of the 2DEG sheet density, which is varied by changing the barrier thickness, keeping x constant. The fit is able to explain the maximum in mobility, and shows surprisingly that for low sheet densities, dislocations are the mobility limiting mechanism. It can be attributed to the weak screening at low n_s. In addition, dipole scattering is not negligible, though not the mobility limiting mechanism for this case.

We calculate the 2DEG mobility for a sample grown here in UCSB with $x = 0.09$ and $d = 16nm$ barrier, which showed the highest low temperature mobility reported in literature to date [11]. The theory identifies al-

Combined S- and Z-shaped current bistability induced by charging of quantum dots

A.E. Belyaev[1,3], L. Eaves[1], S.A. Vitusevich[2,3], P.C. Main[1], M. Henini[1], A. Förster[2], N. Klein[2], and S.V. Danylyuk[3] *

[1] School of Physics and Astronomy, University of Nottingham, NG7 2RD Nottingham, UK
[2] Forschungszentrum-Jülich, D-52425 Jülich, Germany
[3] Institute of Semiconductor Physics, NASU, 03028 Kiev, Ukraine

Abstract S- and Z-shaped current bistability are observed in a p-i-n diode comprising a layer of quantum dots (QD) in the intrinsic region. The phenomena are analyzed in terms of QD charging effect. A model presented in the work shows that switching between the two current states within the first as well as the second bistable regions is controlled by the charges stored in the QD.

1 Introduction

A double barrier resonant tunneling structure (DBRTS) which is embedded into the intrinsic region of an n-i-n diode represents by itself well-known system that shows N- (or in some cases Z-) shaped Current-Voltage Characteristics (CVC). Asymmetric design of barriers provides charge accumulation inside the quantum well (QW) and a hysteresis loop is observed when voltage is swept up and down [1],[2]. A giant bistability has been observed if charge accumulation occurs in triangular well, that is formed in the front of emitter barrier [3]. A DBRTS which is embedded into the intrinsic region of a p-i-n diode can be a source for S-shaped CVC and, consequently, shows inverted current bistability due to different rates for electron and hole injection in and escape out the QW [4]. In this communication, we report a new type of DBRTS that exhibits both S- and Z-shaped current bistability. We realize such a system in GaAs p-i-n diode incorporated two 3-nm AlAs barriers and 6.2-nm GaAs QW in the undoped intrinsic (i) region. Undoped spacer layers of 100 nm (60 nm) separated the barriers from $2 * 10^{16} cm^{-3}$ n-doped ($2 * 10^{18} cm^{-3}$ p-doped) contact layers. A 1.8 ML plane of InAs is embedded in the middle of QW to form a plane of quantum dots (QD). A schematic diagram of the heterostructure is shown in Fig.1. To carry out the two-terminal electrical measurements described herein, samples were processed into circular mesas of 100-400 μm diameter.

2 Experimental results and discussion

To investigate mechanisms responsible for current bistability we analyze current-voltage (I-V) and capacitance-voltage (C-V) characteristics simultaneously. The I-V curves are taken on an HP4145B Semiconductor Parameter Analyzer configured to input a voltage sweep while

* *A.E.B. acknowledges the Royal Society (U.K.) and DAAD (Germany) for financial support.S.A.V. acknowledges the Office of Naval Research for Grant N 00014-00-1-4054*

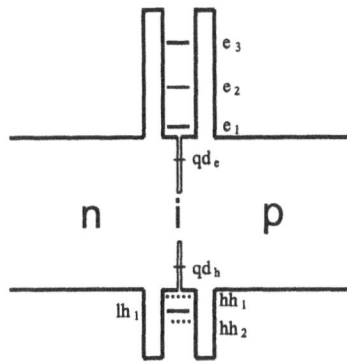

Fig. 1 Schematic band diagram of the structure (neglecting space charge effects).

measuring current. The C-V characteristics were measured with a LRC meter (HP 4275A), treating the device as a resistance and a capacitance in parallel. Figure 2 is a helium temperature I-V curve displayed by a 100 μm-diam structure taken by sweeping the voltage from 0 up to 2.5 V and then back down to 0 V, while measuring the current through the device. The arrows in the figure show predicted positions for resonances through electron and hole quasibound levels in the GaAs QW. A pronounced hysteresis is displayed in the figure around the voltage of 2.3 V with the upper current line corresponding to the sweep from low to high bias. The observation of the hysteresis in the dc I-V curve is indicative of a bistability in device current for a given applied bias. The negative differential resistance and hysteresis displayed in Fig.2 result from resonant tunneling of electrons through the second quasibound level in the GaAs quantum well. Precise measurements reveal also another region of double-valued current behavior at voltage bias between 1.2 V and 1.4 V. The feature becomes more visible with decreasing temperature. Upper inset in Fig.2 represents low bias portion of the I-V curve measured at 30 mK. In contrast, the lower current line corresponds to the sweep from low to high bias. The S-shaped bistable behavior observed in this case is intimately related to the presence of the QD states and controlled by their net charge. To prove this statement we have to examine what processes contribute to the current flow. At low temperature and bias far away from flat band condition the main contribution to the current arises from hole and electron injection into the i-region and their sequential recombination there. If the quasi- Fermi level in

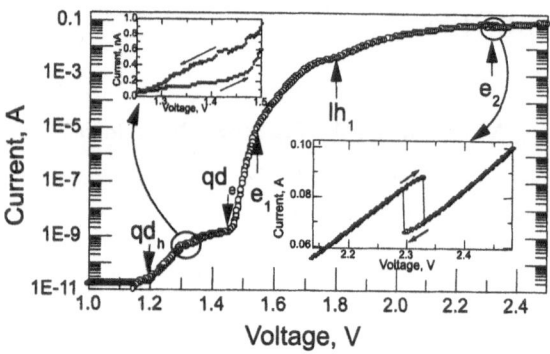

Fig. 2 Current-voltage characteristics of the DBRTS. Solid (open) circles correspond to sweeping V up (down). Labeled arrows correspond to predicted resonances. qd_h (qd_e) correspond to voltage at which QD hole (electron) states begin to charge (discharge) under sweeping up (down). The first hysteresis loop, which exhibits inverted behaviour, is shown in upper inset. Enlarged picture of the second resonance is shown in lower inset.

the positively biased side coincides with lowest QD hole state the hole tunneling becomes possible. The arrow labeled as qd_h in Fig.2 shows the bias at which the QD hole states begin to fill when voltage sweeps up while the qd_e mark fixes the bias at which the QD electron states start to discharge for the voltage sweeping down. The double-valued current behavior can be understood by considering different voltage distribution through the structure at the bias sweeps up and down. Indeed, the electric field across i-region is uniform until a hole capturing onto the QD states becomes possible. Because of low recombination probability the current remains low and positive charges accumulates in the QW. As a result, the electric field on the electron emitter side is smaller than one on the hole emitter side. Further increase of bias leads to alignment of the quasi-Fermi level in the negatively biased side with the lowest QD electron state. Electrons captured onto QD states will neutralize the positive charge leading to the recovering of uniform electric field distribution. The current also increases due to enhanced e-h recombination. Thus, at the same bias two electric field configurations and, consequently, two current states should exist depending on the direction of sweeping V. It should be noted that magnitude of the effect depends on a number of parameters - tunneling and emission rates for both electrons and holes as well as e-h recombination rate - and can be varied by structure design or by external factors (for instance, temperature). Temperature has no influence on the tunneling process but changes considerably the thermionic emission rates that causes the changes of conditions for bistability. The fact is confirmed by C-V measurements. Figure 3 shows temperature dependent C-V characteristics measured at $f = 10\ kHz$. It is seen that increasing temperature leads to the appearance of capacitance switching between two charging states of the device.

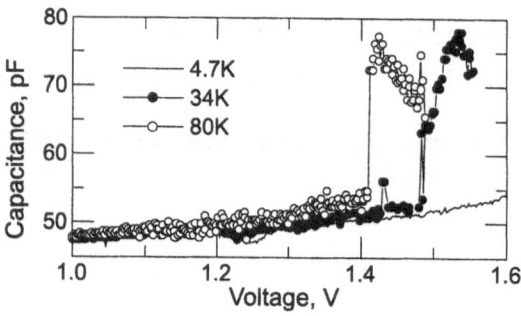

Fig. 3 Temperature dependence of C-V characteristics measured at $f = 10\ kHz$.

The charging state of the QD also plays an important role in the formation of the second hysteresis loop. Indeed, for applied voltages large than about 2 V the holes injected from p-doped contact layer may pass directly over the top of the second emitter barrier. This process reduces the number of holes entering the QW and also the filling of the QD hole states. At the same time electrons should be captured onto the QD electron states resulting in the net negative charge accumulated in the QW. Thus, two electric field configurations are possible depending on the sweeping direction while switching from high current state to low one occurs at higher voltage than the reverse switching.

3 Conclusion

We have investigated a bipolar transport in p-i-n double barrier diode with a layer of self-organized InAs quantum dots in the intrinsic region. Two kinds of current bistability are observed in the structure. The first hysteresis loop appears in the I-V curve at bias below flat-band conditions. The hysteresis exhibits inverted behavior as compared to the second unstable region that is revealed at bias above flat-band regime. We analyze the observed phenomena in terms of QD charging effect. Sequential charging of QD hole and electron states disturbs uniform potential distribution differently depending on the voltage sweeping direction. Also, the presence of the InAs layer causes a modification of the electronic levels of the QW, induces a redistribution of the electron charge in the device, and leads to additional changes in the electrostatic potential profile. Our model shows that switching between the two current states within the first as well as the second bistable regions is controlled by the charges stored in the QD.

References

1. A. Zaslavsky, V.J. Goldman, D.C. Tsui, J.E. Cunningham. Appl. Phys. Lett., **53**, (1988) 1408.
2. E. S. Alves, L. Eaves, M. Henini, et al. Electronics Lett. **24** (18), (1988) 1190.
3. A.E. Belyaev, S.A. Vitusevich, B.A. Glavin, et al. Inorganic Materials, **33**, (1997) 116.
4. O. Kuhn, J. Genoe, D.K. Maude, et al. Physica E, **2**, (1998) 483.

Interaction of Mu with Spin Current in GaAs/GaAsP/Si

E. Torikai[1], Y. Ikedo[1], A. Ihori[1], K. Shimomura[2], K. Nagamine[2,3], T. Saka[4] and T. Kato[5]

[1] Faculty of Engineering, Yamanashi University, Takeda 4, Kofu   400-8511, Japan
  e-mail: et@es.yamanashi.ac.jp
[2] Meson Science Laboratory, High Energy Accelerator Research Organization (KEK), Oho 1, Tsukuba 305-0801, Japan
[3] Muon Science Laboratory, The Institute for Physical and Chemical Research (RIKEN), Hirosawa, Wako, 351-0198, Japan
[4] Department of Applied Electronics Eng., Daido Institute of Technology, Daido-cho, Minami-ku, Nagoya 457-8531, Japan
[5] New Materials Research Laboratory, Daido Steel Co. Ltd., Daido-cho 2-30, Minami-ku, Nagoya 457-8545, Japan

Abstract   We have studied a feasibility of probing the spin current in semiconductors by using spin-dependent scattering of conduction electrons at polarized muonium (Mu) centers and also a possibility of spin injection into Si wafer from the strained GaAs epilayers in which the spin-polarized conduction electrons were excited by circularly polarized laser illumination. The Mu spin depolarization showed the significant dependence on the laser wavelength in GaAs/GaAsP/Si and GaAs/Si systems but not in the Si, giving a clear evidence of interaction of Mu with conduction electrons. A slight difference in the triplet-singlet population of Mu depending on the polarization direction of the spin current is also at the resonant wavelength of the strained GaAs layer

Spin injection into semiconductors attracts much attention with expectation to develop a new field of electronics controllable by spin [1,2]. Despite the increasing importance of the spin-dependent transportation properties, there is so far few direct microscopic observation of spin-dependent scattering of the carriers involved. Torikai et al. have proposed a new mechanism to probe the polarized electron flow (spin current) by using the exchange scattering of conduction electrons at a spin-polarized muonium (Mu) center in the triplet state [3]. If conduction electrons have spin σ parallel to spin s of a bound electron of Mu, the scattering cross section would be smaller than that in the anti-parallel case due to the Pauli exclusion principle. Thus the population change of Mu going from the triplet state to singlet will be greater in the anti-parallel spin current case than that in the opposite case. In the presence of photo-induced polarized electrons, the Mu spin depolarization in Si covered by a strained GaAs overlayer showed a significant dependence on the wavelength of the laser light as well as a slight difference in the triplet-singlet population of Mu depending on the polarization direction of the spin current. Their results suggest the spin injection into the conduction band of the Si substrate across the double interfaces of the GaAs/GaAsP/Si structure.

For the further understanding of such a mechanism, we have conducted a series of Mu spin relaxation (MuSR) experiments in Si (100) wafers covered by strained GaAs overlayers in the presence of photo-induced spin-polarized electrons. In a thin GaAs epilayer which is strained by a lattice mismatch with a substrate or buffer layers in-between, the degeneracy of the top of the valence band is removed and hence energy levels of light holes (l.h.) with magnetic quantum number $|m_j|=1/2$ are separated from those of heavy holes (h.h.) with $|m_j|=3/2$. The energy splitting between h.h. and l.h. attained up to 35 meV in the GaAs/GaAsP/GaAs system at room temperature[4]. When the GaAs epilayer is illuminated by circularly polarized light with resonant wavelength, conduction electrons are excited to one of the pure states of $|1/2, 1/2\rangle$ or $|1/2, -1/2\rangle$ selectively. Nakanishi et al have observed the electron spin-polarization up to 86 % when they were emitted from the GaAs epilayer with surface treatment of the negative electron affinity [5]. The relaxation time of the polarized holes in the valence band was the order of nanosecond [6]. On the other hand, the spin polarization of conduction electrons in a light element such as Si, if it is possible to transport, is expected to survive some orders of magnitude longer than in GaAs. Attempting a spin injection into Si wafers from a strained GaAs epilayer, we have chosen the sample structures of (a) p-GaAs/GaAs$_{0.83}$P$_{0.17}$/n-Si and (b) p-GaAs/n-Si. In the latter sample, the polarization of the conduction electrons will not as high as in the sample (a) because of rather poor morphology compared with the sample (a), but it might have a greater chance to survive because there is only one interface between GaAs layer and Si substrate.

The GaAs epilayers were grown by the metal organic vapor-phase epitaxy method on Si (100) substrates with a GaAsP buffer layer in sample (a) and directly in sample (b). The thickness of each layers are (a) GaAs 0.14 μm, GaAsP 2 μm and (b) GaAs 1 μm. The Si substrates with 25x25 mm^2 in area and 290 μm in thickness were cut from a same wafer together with sample (c) Si for a reference. On both surfaces of the samples, the gold electrodes were evaporated to apply the bias voltage perpendicular to the surface.

A series of MuSR experiments has been carried out at the pulsed surface muon facility in KEK. The pulsed Ti-saphire laser, with 30 ns in pulse width and 150 mJ/pulse in intensity was delivered in coincidence with the arrival of muons. Muons were introduced with a direction incident from the Si side and laser light illuminated on the strained GaAs surface. The diameters of both the laser and the muon beams were about 20 mm at the sample surfaces. It is noted that in all samples with the different structures the MuSR signals are mainly from those stopped in Si because muons stopped in thin epilayers are negligibly small. The temperatures of samples were 16 K without laser illumination and 43(2) K with illumination.

The time evolutions of the muon spin depolarization were observed under the laser illumination of wavelength between 750 and 860 nm and without illumination in the longitudinal magnetic field of 3 kG. Typical examples of them in the GaAs/GaAsP/Si, GaAs/Si and in the Si are shown in Fig. 1 (a), (b), and (c) respectively. It is seen that the aspects of depolarization spectra depend significantly on the laser wavelength in (a) and (b) but not in (c).

The wavelength dependence in the former case can be attributed to the effect of conduction electrons excited in the GaAs/GaAsP overlayers and transported to the Si substrate. Since the overlayers are transparent for light greater than 830 nm in wavelength, the depolarization in longer wavelength was mainly due to direct illumination effect of Si substrate which is independent on the laser wavelength as is seen in Fig. 1(c). On the other hand, the laser light of 750 nm was absorbed completely by the GaAs/GaAsP layers, generating photo-excited conduction electrons. In the intermediate range, a part of the incident laser light was absorbed exciting the conduction electrons and the rest was incident directly to Si. Therefor a fraction of electrons and photons injected to Si substrates was changed from almost zero to hundred percent. MuSR spectra in the GaAs/Si sample showed the significant dependence on the wavelength. The detailed comparison will be given elsewhere. These results give clear evidence that the interaction of Mu with conduction electrons was observed separately from that with photons for the first time.

In order to study the spin dependence of the interaction of Mu with conduction electrons, we have compared the difference of the MuSR spectra obtained by switching the right and left polarization of the laser light, which in turn influence the spin polarization of the conduction electrons. In GaAs/GaAsP/Si, the polarization asymmetry of muon was greater when the laser polarization vector was parallel than in the anti-parallel case in the whole time range of MuSR spectra observed at 66 K with laser wavelength of 815 nm both in the magnetic field of 30 G and 3 kG. The difference of asymmetries was 0.4 (2)% beyond the statistic error of the measurements in both fields. Such spin dependence was not detected at 860 nm. In case of Si and GaAs/Si systems, we did not find a difference of the asymmetry depending on the spin direction beyond the statistic error in the conditions so far investigated. Although we need to confirm the present results by further experiments with higher statistics, these results imply the possibility that highly polarized conduction electrons that were optically induced in the strained GaAs and injected into the Si substrate across the interfaces of GaAs/GaAsP and GaAsP/Si. The possible origin of the spin dependence of Mu spin relaxation was discussed in ref. [3].

This work was partially supported by a Grant-in-aid for Scientific Research from the Ministry of Education, Science, Sports and Culture of Japan.

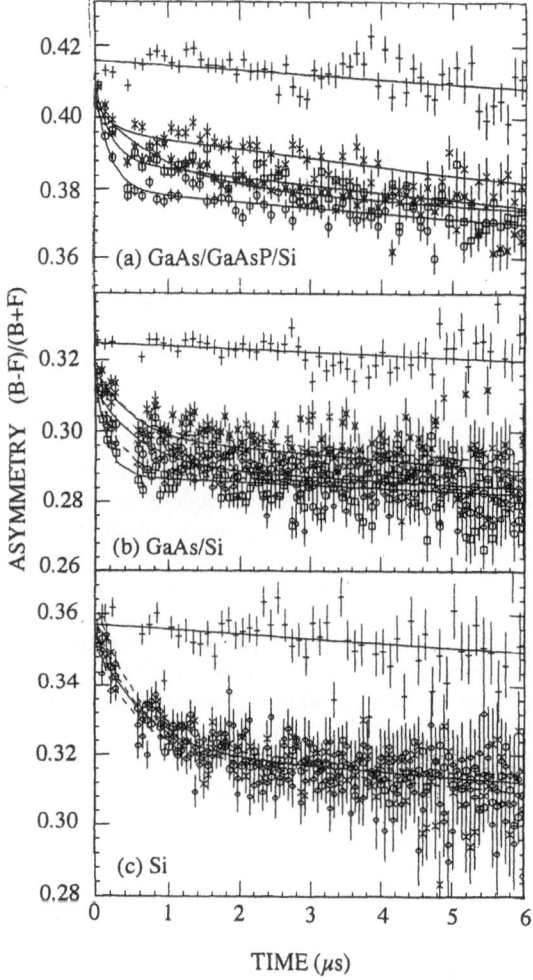

Fig.1 The time evolution of the muon spin depolarization in (a) GaAs/GaAs/Si , (b) GaAs/Si and (c) Si, in the laser illumination with the wavelength between 750 nm (x) and 810 nm (circles) and without illumination (+) in longitudinal magnetic field of 3 kG. Temperatures are 43 K with illumination and 16 K without illumination.

References
1. P. R. Hammer et al., Phys. Rev. Lett. **83**, 203 (19999)
2. J. M. Kikkawa and D. D. Awschalom, Nature **397**, 139 (1999)
3. E. Torikai et al., Physica B **289-290**, 558 (2000).
4. T. Saka et al. Jour. Crystal Growth **124**. 346 (1992) and H.Aoyagi et al., Phys. Lett. **A167** (1992).
5. T. Nakanishi et al., Phys. Lett. **A158**, 345 (1991).
6. H. Horinaka et al., Jpn. J. Appl. Phys. **34**. 179 (1995).

Proc. 25th Int. Conf. Phys. Semicond., Osaka 2000 (Eds. N. Miura and T. Ando)

777

Metallic behaviour and temperature dependent screening in p-SiGe

V. Senz[1], T. Ihn[1], T. Heinzel[1], K. Ensslin[1], G. Dehlinger[2], U. Gennser[2], D. Gruetzmacher[2]

[1] Solid State Physics Laboratory, ETH Zurich, ETH Hoenggerberg, CH-8093 Zurich, Switzerland
[2] Paul-Scherrer Institute, Switzerland

Abstract The temperature dependent resistivity of p-SiGe is discussed in the temperature range from 90 K to 180 mK. We find that the metallic behaviour at the lowest temperatures and at high densities can be well described by temperature dependent screening.

The so-called metal-insulator transition (MIT) in two-dimensional electron and hole gases (2DHGs) at zero magnetic field has been under intensive discussion in recent years [1]. A similar phenomenology has been found in many material systems, among them the p-SiGe 2DHGs [2–4]. In this paper we present data of the temperature dependent resistivity in p-SiGe and show that the metallic behaviour at low temperatures and high densities can be understood as a result of temperature dependent screening [5].

In our MBE-grown samples the 2DHG resides in a 20 nm $Si_{0.85}Ge_{0.15}$ quantum well sandwiched between two undoped Si layers. Remote Boron doping was introduced at a distance of 15 nm above the quantum well and a Ti/Al Schottky gate allowed tuning the hole density [3]. Transport measurements on conventional Hall bars were performed at temperatures between 180 mK and 90 K using standard four-terminal DC- and AC techniques.

Figure 1 shows the temperature dependent resistivity between 1.7 K and 90 K for carrier densities between $2.0 - 5.1 \times 10^{11} cm^{-2}$ across the MIT. Above 55 K the Hall-density (not shown) depends strongly on temperature and the 2DHG is not the only conducting channel in the sample. This leads to the strong resistivity increase as the temperature is lowered down to about 55 K.

Lowering the temperature further to about 25 K freezes out phonons and the resistivity decreases for all densities. This behaviour agrees with the linear temperature dependence of acoustic deformation potential scattering for which a larger temperature coefficient of the resistivity is expected for lower carrier densities [6].

Below 25 K one enters the Bloch-Grüneisen regime where phonon scattering looses its importance compared to other scattering mechanisms like in Si-MOSFETs. Scattering off charged interface impurities has been reported to be the dominant scattering mechanism in p-SiGe [7]. Other mechanisms like alloy disorder scattering, surface roughness scattering and remote impurity scattering may also contribute to the observed resistivity but their importance may be discussed elsewhere. For the purpose of this paper consideration of the dominant scattering mechanism is sufficient.

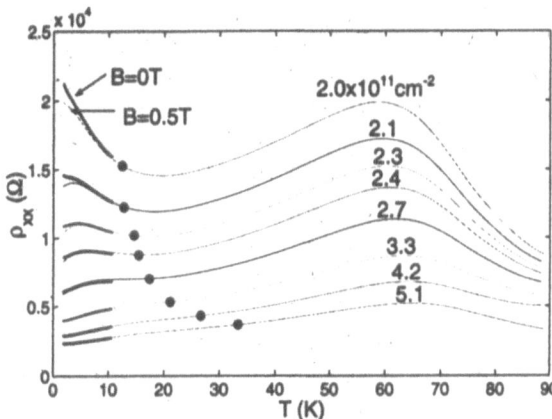

Fig. 1 Temperature dependent resistivity from 1.7 K to 90 K. Details in the text.

At low enough temperatures the 2DHG undergoes the transition from a non-degenrate to a degenerate system. As a measure for this transition we mark the temperature $kT_c = E_F/2$ with filled dots on each curve in Fig. 1. Below this temperature the curves look similar to the ones presented for p-GaAs/AlGasAs systems in Ref. [8] and for Si-MOSFETs in Ref. [9]: at intermediate densities $(2.1 - 2.4 \times 10^{11} cm^{-2})$ when the temperature is lowered the resistivity tends to increase initially, but the increase is overpowered at lower temperatures by a strong resistivity drop such that a resistivity maximum occurs. At higher densities the maximum disappears gradually and a purely metallic temperature dependence is observed. At lower densities the resistivity increases with decreasing temperature indicating insulating behaviour.

The behaviour at intermediate and at high densities has been interpreted for the p-GaAs/AlGaAs system in terms of temperature dependent screening and the quantum-classical crossover associated with the transition between the degenerate and the non-degenerate regime [10]. According to this theory, temperature dependent screening leads to the strong resistivity increase with increasing temperature and the crossover to the classical regime tends to invert the slope of the resistivity.

As an alternative interpretation we mention another scenario which includes effects of strong localisation and carrier freeze out. This interpretation is stimulated by the observation that the region with $d\rho/dT < 0$ between 5 K and 10 K develops gradually as the carrier

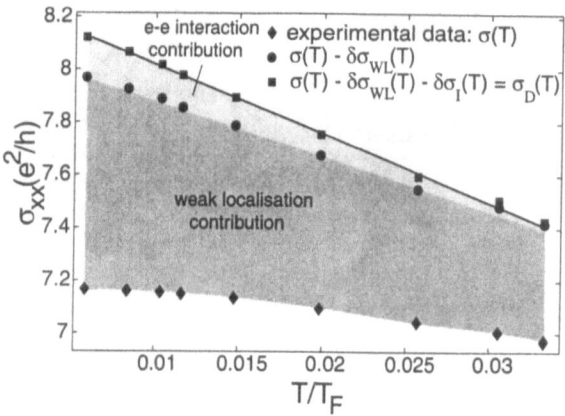

Fig. 2 Contributions to the metallic conductivity

density is reduced. It is conceivable that the freeze-out of a small fraction of holes in the strongest potential minima could drive the remaining 2DHG above the percolation threshhold through the greatly improved (non-linear) screening which would in turn lead to the subsequent $d\rho/dT > 0$ region below about 5 K. Such an interpretation is supported by recent local compressibility measurements on p-GaAs/AlGaAs structures [11]. Both interpretations are essentially based on screening effects and it may well be that both are correct in a single sample, each in its own distinct density range.

In the recent controversy about the MIT interaction effects have been widely discussed in the literature [12] and even the validity of the Fermi-liquid description was doubted [13]. Within this paper we will stay within the Fermi-liquid description of the 2DHG in spite of typical values of the interaction parameter $r_s \approx 8$. Interactions will be considered inasmuch as they are contained in screening effects and in the so-called interaction corrections to the conductivity [14].

The above interpretations do not contain the effects of coherent backscattering occurring at the lowest temperatures in our samples. In order to demonstrate the importance of these effects, Fig. 1 shows the resistivities below 10 K measured at zero magnetic field (thick lines) where coherent backscattering is important and at 0.5 T (thin lines) where coherent backscattering is suppressed.

Above, the general behaviour of the temperature dependent resistivity has been discussed and relevant effects have been named. In the following we concentrate on a detailed analysis of the resistivity in the metallic regime at high densities. Our analysis of the data includes classical and quantum contributions to the resistivity. In this spirit we have analyzed our data according to

$$\sigma(T) = \sigma_D(T) + \delta\sigma_{WL}(T) + \delta\sigma_I(T),$$

where $\sigma_D(T)$ denotes the classical contributions to the conductivity including temperature dependent screening, $\delta\sigma_{WL}(T)$ is the weak localisation correction to the conductivity and $\delta\sigma_I(T)$ are interaction corrections.

The weak localisation contribution to the conductivity $\delta\sigma_{WL}(T)$ has been extracted from the magnetoresistance at low magnetic fields [5]. The data can be fitted by the standard theory and the phase coherence rate depends linearly on temperature [4].

Interaction corrections to the conductivity $\delta\sigma_I(T)$ can be experimentally obtained from a detailed analysis of the temperature dependence of the Hall slope at low magnetic fields [5]. It is found that these corrections lead to an additional insulating contribution to the temperature dependence of the conductivity.

In p-SiGe with its dominant large angle interface impurity scattering $\sigma_D(T)$ can be fitted by [15]

$$\sigma_D(T) = \sigma_D(0)(1 - C \cdot T/T_F).$$

Figure 2 shows the result of the above procedure which is in detail described in Ref. [5] for a density of 2.6×10^{11} cm^{-2}. From the measured curve we first subtract the weak localisation correction and the interaction correction. The remaining part of the conductivity is linear in T and gives the experimental value $C = 3.1$ which is in good agreement with the theoretically predicted value of 2.8.

In conclusion, we have discussed relevant scattering mechanisms in p-SiGe samples in which the MIT was observed. We find that the metallic behaviour of the conductivity can be well described by the theory of temperature dependent screening. This finding argues against the existence of a true MIT in our p-SiGe system.

References

1. For references to the extensive literature see S.V. Kravchenko et al., Phys Rev B **59**, (1999) R12740; S.V. Kravchenko, Braz. J. Phys. **29**, (1999) 623.
2. P.T. Coleridge et al, Phys. Rev. B **56**, (1997) R12764.
3. V. Senz et al, Ann. Phys. (Leipzig) **8**, (1999) SI 237.
4. V. Senz et al, Phys. Rev. B **61**, (2000) R5082.
5. V. Senz et al, cond-mat/0004312v2, 27 April 2000.
6. T. Ando, A.B. Fowler, F. Stern, Rev. Mod. Phys. **54**, (1982) 437.
7. C.J. Emeleus et al, J. Appl. Phys. **73**, (1993) 3852; T.E. Whall, Appl. Surf. Sci. **102**, (1996) 221.
8. A.P. Mills, Jr., A.P. Ramirez, L.N. Pfeiffer, K.W. West, Phys. Rev. Lett. **83**, (1999) 2805.
9. A. Prinz, V.M. Pudalov, G. Brunthaler, G. Bauer, in: Proc. Int. Meeting SIMD-99, Mauri, (1999), edited by K. Hess. To be published in *Superlattices and Microstructures* (2000).
10. S. Das Sarma, E.H. Hwang, Phys. Rev. B **61**, (2000), R7838.
11. S. Ilani, A. Yacoby, D. Mahalu, H. Shtrikman, Phys. Rev. Lett. **84**, (2000) 3133.
12. T. Vojta, F. Epperlein, M. Schreiber, Phys. Rev. Lett. **81**, (1998) 4212; J.S. Thakur, D. Neilson, Phys. Rev. B **59**, (1999) R5280.
13. V. Dobrosavljevic, E. Abrahams, E. Miranda, S. Chakravarty, Phys. Rev. Lett. **79**, (1997) 455.
14. B.L. Altshuler, A.G. Aronov, *Electron-electron interactions in disordered conductors* (Elsevier Science Publishers B.V., 1985), editors A.L. Efros and M. Pollak.
15. A. Gold et al, Phys. Rev. B **33** (1986) 1076.

Proc. 25th Int. Conf. Phys. Semicond., Osaka 2000 (Eds. N. Miura and T. Ando)

779

Anisotropic magneto-transport properties of 70 nm-period lateral surface superlattices in high magnetic fields

Masashi Akabori*, Junichi Motohisa and Takashi Fukui

Research Center for Interface Quantum Electronics, Hokkaido University, North 13, West 8, Kita-ku, Sapporo 060-8628, Japan

*e-mail: akabori@rciqe.hokudai.ac.jp

Abstract We study magneto-transport properties of a 70 nm-period lateral surface superlattice (LSSL) utilizing self-organized GaAs multiatomic steps formed on GaAs (001) vicinal substrates. We measure four-terminal magneto-resistance (MR) and Hall resistance (HR) for current flow perpendicular and parallel to the multiatomic steps. Unique MR characteristics such as positive MR, amplitude modulation, phase inversion are clearly observed in high magnetic fields in spite of less anisotropy in HR. The results are clearly explained by the predicted theories based on the spectra of Landau subbands.

1 Introduction

Transport properties of lateral surface superlattices (LSSLs) have been extensively studied because of their unique electronic states [1-2]. However, as most of the work focused on LSSLs with period of around 200 nm, their unique properties like commensurability oscillations [2] was only observed at very small magnetic field less than 1 T where the cyclotron radius is comparable to the period.

On the other hand, we have reported novel LSSLs that are formed on GaAs multiatomic steps [3-5]. They are promising because the period of LSSLs is determined by the multiatomic steps and is around 70 nm, which is shorter than that of LSSLs fabricated by lithography techniques. In additon, it is possible to achieve large potential modulation if the multiatomic steps can directly be embedded in the channel of two-dimensional electron gas (2DEG). Therefore, they are expected to offer new regime for electron transport properties including magneto-transport.

In this paper, we report on magneto-transport properties of a novel LSSL utilizing self-organized GaAs multiatomic steps with the period of 70 nm and describe their novel features above 1 T.

2 Experiment

Figure 1(a) shows layer structure of LSSLs. We grew an n-$Al_{0.3}Ga_{0.7}As/In_{0.2}Ga_{0.8}As/GaAs$ selectively doped double heterostructure on a GaAs vicinal substrate by metal-organic vapor phase epitaxy (MOVPE) [5]. As the thickness of InGaAs channel layer is periodically modulated as a result of step-bunching during GaAs growth, 2DEG accumulated in InGaAs layer is subject to a strong lateral periodic modulation. Figure 1(b) shows a surface morphology of 4 nm-thick $In_{0.2}Ga_{0.8}As$ layer on GaAs multiatomic steps. The misorientation angle was 4° toward [$\bar{1}10$] direction.

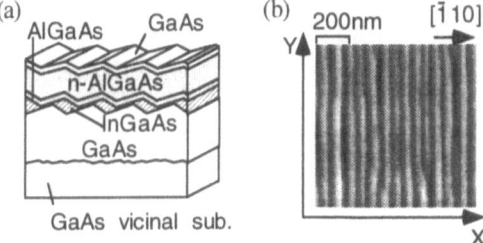

Fig.1 (a) Schematic illustration of layer structure of the LSSL utilizing multiatomic steps. (b) Surface morphology of InGaAs layer on GaAs multiatomic steps.

Coherent and straight step array was observed and the average period is 63 nm. The image indicates that the channel has coherent corrugation.

After the growth, two types of Hall-bar with a gate electrode were fabricated whose channels were defined in either perpendicular ($\perp$) or parallel (//) to the multiatomic steps. By using the Hall-bars, four-terminal magnetoresistance (MR) R_{XX} ($\perp$) and R_{YY} (//), and Hall resistance (HR) R_{XY} ($\perp$) and R_{YX} (//) were measured at low temperature.

3 Results and discussions

Figure 2 shows MR (R_{XX} and R_{YY}) and HR (R_{XY} and R_{YX}) characteristics of the LSSL measured at $T = 1.6$ K for electron concentration $N_S = 6.7 \times 10^{11}$ cm^{-2}. The observed HR in R_{XY} and R_{YX} showed isotropic behavior, and the positions of quantum HR plateaus were nearly the same resulting from the same N_S. On the other hand, strong anisotropy was observed in MR. For $B <$

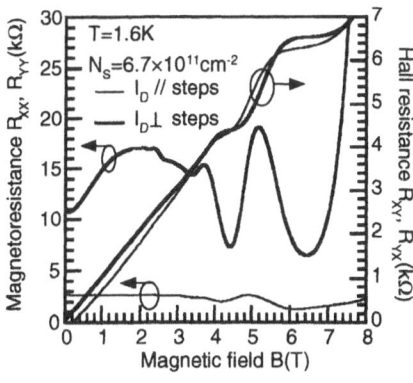

Fig.2 MR and HR characteristics of the LSSL for different current directions.

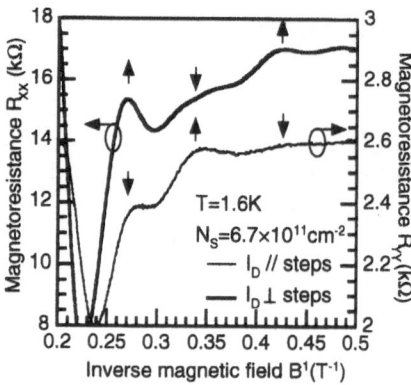

Fig. 3 MR characteristics as a function of B^{-1}.

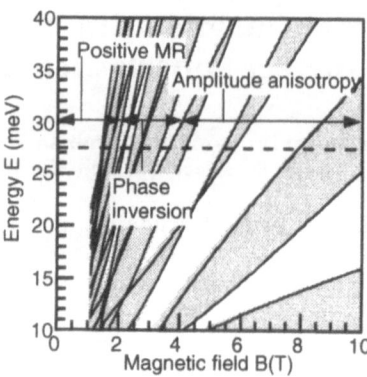

Fig. 4 Landau subband spectrum as a function of B. Dotted line represents the Fermi energy for $N_S = 6.7 \times 10^{11}$ cm^{-2}.

2 T, large positive MR was observed in R_{XX}. This positive MR originates from the enhancement of electron drift in the strong periodic potential [6], as discussed in our previous report [5]. For $B < 4$ T, the amplitude of Shubnikov-de Haas (SdH) oscillations for R_{XX} became larger as magnetic field became higher, while the enhancement for R_{YY} was less pronounced in spite of lower zero field resistance in R_{YY}. The MR ratio at the peak around $B = 5$ T compared to zero field resistance is 1.74 for R_{XX} and 0.96 for R_{YY}, respectively. The result indicates that the anisotropy of MR amplitude does not originate from the difference of mobility between current directions, because high mobility samples usually show large SdH oscillations.

Furthermore, we confirmed MR amplitude modulation in both R_{XX} and R_{YY} for 2 T < B < 4 T. In Fig. 3, MR is plotted as a function of inverse magnetic field B^{-1}. We can see that the phase of modulation is inverted between R_{XX} and R_{YY}. That is, when the peak of R_{XX} is enhanced (upward arrow), that for R_{YY} is suppressed (downward arrow), and vice versa. Such phase-inversion behavior of MR oscillaitons was also observed in previous LSSLs with the period around 200 nm at weak magnetic field less than 1 T [2].

In order to clarify such anisotropy of MR characteristics, we calculated Landau subband spectrum in the LSSL. For the calculation, we used two methods, that is, numerical solution of Schrödinger equation and first-order perturbation theory [2, 7]. We assumed that the potential modulation was given by the following equation;

$$V(x) = V_0 \sin\left(\frac{2\pi x}{L}\right) \qquad (1)$$

The potential amplitude was estimated to be $V_0 = 7$ meV by the peak position of positive MR in R_{XX} [5]. The calculated spectrum as a function of magnetic field B is shown in Fig. 4. In the figure, the width of Landau subbands is periodically modulated. In particular, the subband width is widely expanded and the gaps between neighboring subbands become

narrower or extinguished under $B < 4$ T because of large amplitude and short periodicity of the potential. Therefore, strong modulation of conductivity σ_{YY} along the potential modulation of LSSL is expected, and it is consistent with the experimental results. In addition, the expansion of Landau subbands corresponds to modulation of the density of states (DOS). As a result, the large anisotropy in MR amplitude at higher magnetic fields and their phase inversion, which have been predicted previously [7], could be observed in the LSSL. We think that these characteristics in 70 nm-period LSSL are the manifest of commensurability oscillations, i.e. Weiss oscillations [2] observed in LSSLs with larger periods.

4. Summary
We measured magnetoresistance (MR) and Hall resistance (HR) characteristics of a novel 70 nm-period lateral surface superlattice (LSSL) on GaAs multiatomic steps. We found strong anisotropy of MR for current directions as observed in positive MR, amplitude modulation and their phase inversion. These unique characteristics can be explained by calculated Landau subbands and predicted theories.

References
1. H. Sakaki, K. Wagatsuma, J. Hamasaki, and S. Saito, Thin Solid Films **36** (1976) 497.
2. R. R. Gerhardts, D. Weiss, and K. v. Klitzing, Phys. Rev. Lett. **62** (1989) 1173.
3. M. Akabori, J. Motohisa, T. Irisawa, S. Hara, J. Ishizaki, and T. Fukui, Jpn. J. Appl. Phys., **36** (1997) 1966.
4. M. Akabori, J. Motohisa, and T. Fukui, J. Cryst. Growth **195** (1998) 579.
5. M. Akabori, J. Motohisa, and T. Fukui, Physica E **7** (2000) 766.
6. P. H. Beton, E. S. Alves, P. C. Main, L. Eaves, M. W. Dellow, M. Henini, O. H. Hughes, S. P. Beaumont, and C. D. W. Wilkinson, Phys. Rev. B **42** (1990) 9229.
7. C. Zhang and R. R. Gerhardts, Phys. Rev. B **41** (1990) 12850.

Thermopower of a two-dimensional antidot lattice

A.G.Pogosov, M.V.Budantsev, A.E.Plotnikov, A.K.Bakarov, A.I.Toropov

Institute of Semiconductor Physics. SBRAS, 630090 Novosibirsk, Russia

Abstract Off-diagonal component of the thermopower of a two-dimensional antidot lattice is experimentally studied. It was found that commensurability oscillations of thermopower are much more pronounced than that of magnetoresistance, which originates from the drastic difference between the effective mobilities of quasielectrons (above the Fermi surface) and quasiholes (below the Fermi surface) near geometrical resonances. Comparable analysis of the thermopower and magnetoresistance behavior has shown the crucial role of smoothness of the potential defining the antidots. In such a potential the effective diameter of antidots is different for electrons with different energies, which leads to anomalously strong dependence of the electron mobility on energy.

Up to date the study of the transport properties of high mobility two-dimensional electron gas (2DEG) in the presence of a periodic lattice of circular artificial scatters (antidots) has been largely restricted to the investigation of magnetoresistance while another kinetic coefficient — the thermopower — remains practically uninvestigated. Quasiclassical phenomena in antidot lattices related to geometrical resonances are usually described through the conductivity tensor σ, whereas the relevant experimental quantity is the resistivity tensor ρ [1]. Thermopower being related to the energy derivative of conductivity provides additional information about transport phenomena in these systems.

In the present work we report the first experimental investigation of thermopower in a two-dimensional lattice of antidots. The lattice was fabricated from a 2DEG in a GaAs/AlGaAs heterojunction with electron density $Ns=(2-5)\cdot10^{11}cm^{-2}$ and mobility $(5-7)\cdot10^{5}cm^{2}/V\,s$ by means of electron lithography and subsequent plasma etching. The lattice with the period $d=0.9$ μm and the antidot diameter $a=0.2-0.3$ μm covers the Hall bar with the dimensions $L{\times}W=9{\times}6$ μm^2.

For thermal voltage measurements, two opposing long leads (12 μm) connected to the unstructured part of the Hall device (contacts 3 and 6, see the inset of Fig. 1) were used to apply heating current at a frequency f. The voltage drop across the region of the device with antidots was measured at frequency $2f$ as a function of magnetic field normal to the 2DEG plane. Joule heating due to the applied ac current locally raises the temperature of the electron gas with respect to that of the lattice. At sufficiently low frequencies the temperature of the electron system oscillates with frequency $2f$, and the voltage drop measured at this harmonic is proportional to the thermopower of the antidot lattices. In the present work the off-diagonal component of the thermopower tensor (Nernst-Ettingshausen effect) is investigated. Thus a thermal voltage V_{47} between contacts 4 and 7 was measured as a function of magnetic field. The same Hall device was also used for complementary measurements of the magnetoresistance. The measurements were carried out at liquid helium temperature in magnetic field up to 1 T.

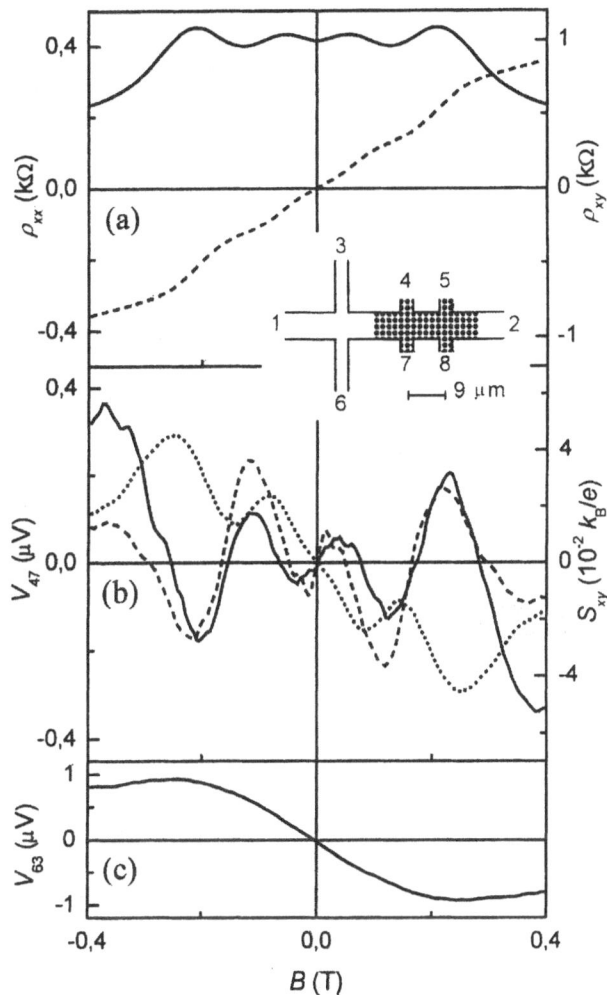

Fig. 1 (a) Longitudinal ρ_{xx} (solid line) and Hall ρ_{xy} (dashed line) magnetoresistivities. (b) Experimentally measured thermal voltage V_{47} (solid line), thermopower S_{xy} (dashed line) obtained from $\rho_{ij}(B)$ curves for two states of sample with slightly different N_s and thermopower of 'hard wall billiard' (dotted line), obtained from expression (5). (c) Thermovoltage of unstructured 2 DEG V_{63}. Insert schematically shows the layout of the experimental samples. Dark areas are the regions of the sample covered by antidots.

Experimental magnetoresistance curves ρ_{xx} and ρ_{xy} are shown in Fig. 1a. In a region of weak magnetic fields ($|\mathrm{B}|{<}0.1$ T) one can observe the increase of ρ_{xx} with magnetic field and a slight quenching of ρ_{xy}. These effects are known as originating from the channeling electron trajectories [2]. In a stronger magnetic field B≈0.22 T, when the cyclotron diameter of the electron orbit becomes comparable to the antidot array period $2R_c=d$, a peak of ρ_{xx} and a plateau of ρ_{xy} arise. These features are explained in the frames of the dynamic chaos theory by the existence of the islands of a stable motion in a phase space, which give essential contribution to the kinetic coefficients [3,4].

A thermal voltage V_{47}, measured from the part of the sample, covered by the antidots, is plotted in Fig. 1b. It exhibits oscillations that correlate with the features of ρ. This allows us to relate these oscillations to the features of the classical electron dynamics in the antidot lattice, which are responsible for the oscillations of ρ. It should be noted that these oscillations are much more pronounced than that of the magnetoresistance. The most significant anomalies are observed near zero magnetic field and at the commensurability condition $2R_c=d$, where $V_{47}(B)$ dependence even change the sign. For the comparison the thermovoltage from the nonmodulated part of the 2DEG (V_{63}) was also measured using contacts 4 and 7 to pass the heating current (Fig. 1c). One can see that it has non-oscillating behaviour. Plotting V_{47} and V_{63} a special care of the sign of the signal was taken, so one can notice, that the signal of the thermal voltage of antidot lattice in the regions of the anomalous transport has inverted sign in comparison to that measured from non-modulated 2DEG. Earlier the inversion of the sign in kinetic coefficients was observed in antidot lattices as so called negative Hall effect, which was explained by predominant reflection of channeling trajectories, colliding with antidots to the 'wrong' direction, countering the Laurenz force. In our case the Hall resistance is just slightly quenched, while thermal voltage changes the sign both for $|B|<0.1$ T and at $2R_c=d$.

Another way to obtain the thermopower is to use the Mott-like relations [5]:

$$S = \frac{\pi^2 k_B^2 T}{3e} \rho \frac{d(\rho^{-1})}{dE}, \qquad (1)$$

where k_B is the Boltzmann constant and e is the electron charge. Using experimental $\rho(B)$ dependencies for two states of the system with slightly different electron densities N_{S1} and $N_{S2}\approx 1.1 N_{S2}$, obtained by illumination with an infrared LED, enabled us to estimate $d\rho/dE \approx (\rho(N_{S2})-\rho(N_{S1}))/(N_{S2}-N_{S1})\nu$, where $\nu=dN_S/dE$ is the density of states in a 2DEG. Function $S_{xy}(B)$, calculated from Eq. (1) is presented in Fig. 1b (dashed curve). One can see, that it fits rather well the experimentally measured thermal voltage V_{47}, which confirms both the correctness of the proposed method of the thermopower measurement and the applicability of the Mott-like relations for the thermopower of the billiard with antidot lattice. Some difference between $V_{47}(B)$ and $S_{xy}(B)$ curves can be explained by the appearance of $\nabla_y T \neq 0$ with the increase of the magnetic field (Righi-Leduc effect) that can be a source of additional thermal voltage.

Now let us discuss the probable physical mechanisms responsible for the thermopower in the antidot lattice. Consider the resistance of the lattice in a simple model of hard wall potentiall billiard. Using the Kubo formula for the resistance [4], one can obtain the following expression for this case:

$$\rho_{ij} = \frac{F_{ij}(R_c/d, a/d, l/d)}{e^2 \nu d v_F}, \qquad (2)$$

Here F_{ij} is a dimensionless function, i and j denote x and y axes, l is a mean free path in the non-modulated 2DEG and v_F is the Fermi velocity. To make this model more realistic, suppose that a is dependent on the Fermi energy, and moreover $da/dE_F<0$ since the increase of E_F (i.e. N_s) leads to the decrease of the depletion regions. Thus the derivative of ρ_{ij} can be written in the following way

$$\frac{d\rho_{ij}}{dE} = \frac{\partial \rho_{ij}}{\partial R_c}\frac{\partial R_c}{\partial E} + \frac{\partial \rho_{ij}}{\partial a}\frac{\partial a}{\partial E} + \frac{\partial \rho_{ij}}{\partial l}\frac{\partial l}{\partial E} + \frac{\partial \rho_{ij}}{\partial v_F}\frac{\partial v_F}{\partial E} \qquad (3)$$

Neglecting the third term in Eq. 3 (we found that for the 2DEG, used for the experimental structures, $l(N_s)$ is almost constant) and using Eqs. (1–3) one can obtain

$$S_{xy} = S_{xy}^h + S_{xy}^a, \qquad (4)$$

where

$$S_{xy}^h = \frac{\pi^2 k_B^2 T}{3e(\rho_{xx}^2 + \rho_{xy}^2)}\left(\rho_{xx}\frac{d\rho_{xy}}{dB} - \rho_{xy}\frac{d\rho_{xx}}{dB}\right)\frac{B}{2E_F} \qquad (5)$$

is the contribution to the thermopower due to the change of the cyclotron radius, and

$$S_{xy}^a = -\frac{\pi^2 k_B^2 T}{3e(\rho_{xx}^2 + \rho_{xy}^2)}\left(\rho_{xx}\frac{d\rho_{xy}}{da} - \rho_{xy}\frac{d\rho_{xx}}{da}\right)\frac{da}{dE_F} \qquad (6)$$

is the contribution due to the change of the antidot radius. Function $S_{xy}^h(B)$ calculated from Eq. (5) using the experimental magnetoresistance curves is plotted in Fig. 1b (dotted line). The evident difference between $S_{xy}(B)$ and $S_{xy}^h(B)$ dependencies allows us to conclude that the contribution of S_{xy}^a to the thermopower dominates the total signal.

Acknowledgments

This work was supported by INTAS through the grant 99-1661.

References

1. E.M. Baskin, A.G.Pogosov, M.V. Entin, JETP **83**,(1996), 1135.
2. S.Luthi, T.Vancura, K.Ensslin, Rschuster, G.Bohm, W.Klein, Phys.Rev.**B.68** (1992), 1367.
3. E.M.Baskin; G.M.Gusev, Z.D.Kvon, A.G.Pogosov, M.V.Entin, JETP Let.**55** (1992), 678
4. R.Fleischmann, .T.Geisel, R. Ketzmerick, Phys.Rev.Let. **68** (1992), 1367.
5. L.Smreka and P.Streda, J.Pys.**C 10** (1977), 2153

Magnetotransport in two-dimensional lateral superlattices in strongly coupled electron-hole gases

B. Kardynał, R. J. Nicholas, J. Rehman, K. Takashina, and N. J. Mason

Department of Physics, Clarendon Laboratory, Oxford University, Parks Road, Oxford OX1 3PU, UK.

Abstract We have studied magnetotransport in a coupled electron-hole system in an InAs/GaSb heterostructure with two-dimensional periodic potential modulations of different strengths. Strong oscillations in low magnetic field longitudinal magnetoresistance have been observed. The amplitude of these oscillations is a function of the magnetic field applied parallel to the two-dimensional system, which modifies the coupling between the electrons and holes.

1 Introduction

Coupled electron-hole systems can be easily formed in InAs/GaSb heterostructures, in which electrons are transferred from the valence band of GaSb into the conduction band of InAs, forming a two-dimensional hole gas (2DHG) in the GaSb and a two-dimensional electron gas (2DEG) in the InAs without the need for doping. It has been shown that tunnelling between these closely separated layers leads to hybridisation of the electron and hole energy levels, opening an energy gap at the crossing point between the electron and hole dispersion curves [1] [2] [3] [4]. In this paper, we study the dynamics of the electrons in this coupled system in the presence of a two-dimensional periodic potential.

In a single 2DEG, the introduction of an antidot array potential leads to oscillations in low field magnetoresistance [5] [6], which are associated with electrons being trapped around a group of antidots in the classical limit (when the period of the potential modulation is much larger that the Fermi wavelength) [5]. To a first approximation, this is equivalent to the cyclotron radius being equal to the radius of the circle enclosing this group of antidots. However it is best modelled by considering the chaotic motion of the electrons, which move ballistically as they are scattered from the antidot potentials [7].

2 Experimental results

2.1 Sample preparation

Samples were prepared on InAs/GaSb double heterostructures, which consisted of a 30 nm thick InAs well embedded in a thick layer of GaSb and located 90 nm below the surface of the wafer. Two different samples were used for this study. The first sample (A) had electron (hole) concentrations of $7.1 \cdot 10^{15}(6.0 \cdot 10^{15})$ m^{-2} and a mobility of 20.1 (0.9) m^{2}V^{-1}s^{-1} The second sample (B) had electron (hole) concentrations of $5.8 \cdot 10^{15}(3.8 \cdot 10^{15})$ m^{-2} and a mobility of 16.2 (0.8) m^2/(Vs). After defining Hall bars, a potential modulation was introduced in the electron-hole gases by etching arrays of holes with a 650

nm period using either reactive ion etching (sample A) or wet chemical etching (sample B).

Since the carriers in this system originate from direct transfer between GaSb and InAs, the potential modulation resulting from etching depends strongly on the etch depth. If the etch is stopped within the GaSb but deep enough for the hole gas to be within the surface band bending region, the hole concentration in the etched regions decreases, the electron concentration increases and an antidot and dot array potentials can be obtained for holes and electrons, respectively (Fig. 1a). This is the case in sample B. Only when the etch penetrates through the InAs well do antidot array potentials form in both systems, which is the case for sample A and is shown in Fig. 1b.

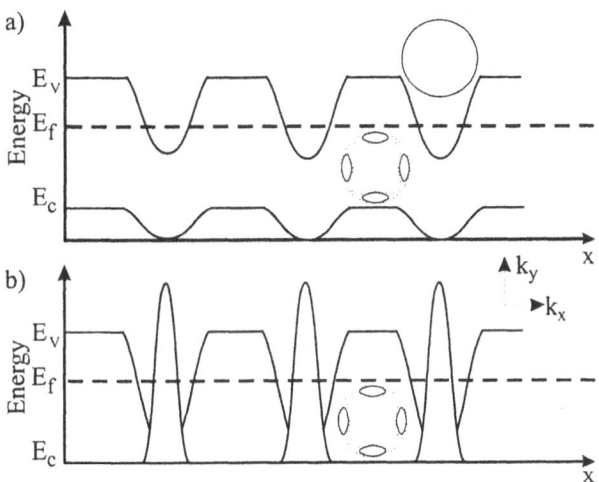

Fig. 1 Schematic potential profile showing conduction band of InAs and valence band of GaSb if etching is (a) terminated in the GaSb cap layer and (b) penetrates through the InAs layer (b). Also shown are schematic pictures of the Fermi surfaces in different regions of the potential landscape.

2.2 Results of magnetotransport measurements

The results of measurements of low field longitudinal magnetoresistance in both samples (A and B) are shown in Fig. 2. The main figure shows the oscillatory part of the longitudinal magnetoresistance obtained by subtracting a parabolic background from the data shown in the inset to the figure. The measured pattern of oscillations is similar in both samples, with the largest peak at about 0.4T and another sharp peak at about 0.18T. Also shown in the figure are the expected positions of the peaks originating from the pinned orbits of electrons in a single 2DEG that has an electron concentration N_e. These positions are given by the expression $2\hbar\sqrt{2\pi N_e}/eB = f(n)a$, where $f(n)$ is a function of the number of antidots enclosed by an electron's orbit, n ($=1, 2, 4...$) and a is the period of the antidot array. A discrepancy between the experimental and model peak positions is commonly observed in single 2DEGs [5] and is not unexpected since the commensurability condition only approximately explains the oscillations in antidot array potentials [7].

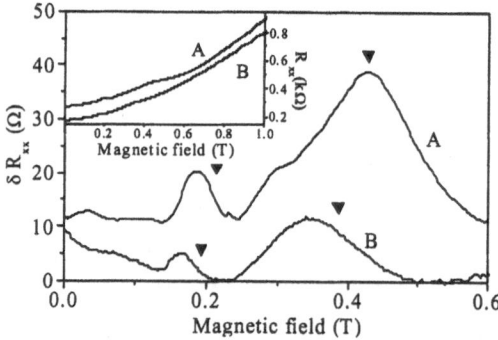

Fig. 2 Oscillatory part of magnetoresistance for samples A and B, with arrows marking the expected positions of the peaks for a single 2DEG. The total magnetoresistance is shown in the inset.

The Fermi surface of the coupled electron-hole system is altered by the hybridisation of electron and hole states. When the carrier concentrations are similar, it consists of a few separate pockets in k-space as shown in Fig. (1). Each closed orbit in k-space can contribute to oscillations in the magnetoresistance. However, at the fields at which the oscillations occur, there is a high probability of magnetic breakdown between separate parts of the Fermi surface [8]. As the mobility of electrons is much higher than that of holes in these samples, only the oscillations associated with the closed orbits of the electrons can be distinguished in the magnetoresistance measurements (Fig. 2).

In order to establish the role of coupling in the observed magnetoresistance, the samples were also measured with a magnetic field applied parallel to the plane of the electron-hole system. This has the effect of shifting the electron and hole Fermi surfaces in respect to each other in k-space and removing the electron-hole hybridisation [1] [2]. The oscillatory part of the magnetoresistance measured in the presence of an in-plane field for both samples is shown in Fig. 3.

In sample A, for which the antidot array potential is defined in the electron gas, the oscillations are always present in the magnetoresistance even if the coupling between the electrons and holes is removed. Their amplitude, however, falls at intermediate value, when the coupling between the electron and hole layers can cause the magnetic breakdown to be inhibited.

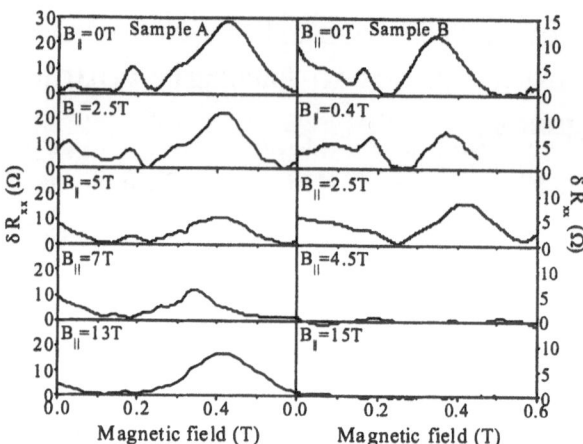

Fig. 3 Oscillatory part of magnetoresistance measured for the in-plane magnetic fields marked.

The situation is different for sample B, in which the antidot array potential is only defined in the hole gas. In this case, oscillations associated with the pinned orbits of electrons are observed in the magnetoresistance only when there is significant coupling between the electrons and holes. At about 4.5T, the coupling becomes weaker and electrons are subject only to the dot array potential defined by etching (Fig. 1b); oscillations in the magnetoresistance disappear.

3 Conclusion

We have measured the low field magnetoresistance of an electron-hole system with a two-dimensional periodic potential. We have observed oscillations associated with the electron gas, which are similar to those observed in antidot arrays in single two-dimensional electron gases. Coupling between the electrons and holes is responsible for a variation in the amplitude of oscillations when an in-plane magnetic field is applied to the samples. Interestingly the oscillations are observed even if the antidot potential is only defined in the two-dimensional hole gas, with a dot array potential present in the electron gas. Strong coupling between the electrons and holes is a necessary condition for the oscillations to occur in this case.

References

1. M. Lakrimi *et al.*, Phys. Rev. Lett. **79**, (1997) 3034-3037.
2. M. J. Yang *et al.*, Phys. Rev. Lett. **78,** (1997) 4613-4616.
3. D. M. Symons *et al.*, Phys. Rev. B **58**, (1998) 7292-7299.
4. Yu. Vasilyev *et al.*, Phys. Rev. B **60**, (1999) 10636-10639.
5. D. Weiss *et al.*, Phys. Rev. Lett. **66**, (1991) 2790-2793.
6. C. Albrecht *et al.*, Phys. Rev. Lett **83**, (1999) 2234-2237.
7. R. Fleishmann *et al.*, Phys. Rev. Lett. **68**, (1992) 1367-1370.
8. J. Hu and A. H. MacDonald, Phys. Rev. B **46**, (1992) 12554-12559.

Semiclassical origin of the 2D metallic state in high mobility Si-MOS and Si/SiGe structures

G. Brunthaler[1], A. Prinz[1], G. Pillwein[1], G. Bauer[1], K. Brunner[2], G. Abstreiter[2], T. Dietl[3], V.M. Pudalov[4]

[1] Institut für Halbleiterphysik, Johannes Kepler Universität, A-4040 Linz, Austria
[2] Walter Schottky Institut, TU München, Am Coulombwall, D-85748 Garching, Germany
[3] Institute of Physics, Polish Academy of Sciences, PL-02668 Warszawa, Poland
[4] P. N. Lebedev Physics Institute of the Russian Academy of Sciences, Moscow 117924, Russia

Abstract   From the temperature dependence of the phase coherence time $\tau\varphi$ an upper limit for single-electron quantum interference effects is extracted. It follows that the "metallic" state in Si-MOS is caused by semiclassical effects and not quantum coherent ones. The comparison with Si/SiGe reveals that disorder is not a determining parameter for the resistivity drop.

1 Introduction

The strong resistivity drop towards low temperatures T in two dimensions (2D) has been observed in several material systems (see [1] and references therein). This apparent "metallic" state has attracted a great deal of interest as it seems to contradict the one-parameter scaling theory [2]. A central question is now whether the "metallic" state is induced by quantum coherent effects or if it is of semiclassical nature.

We show in this work that most of the strong resistivity drop in Si-MOS is in the semiclassical regime, i.e. that quantum interference effects do not contribute to the conductivity. This is achieved by a comparison of the phase coherence time τ_φ and the momentum relaxation time τ. When $\tau_\varphi < \tau$, no single-electron quantum interference effects exist, as the coherence time is too short to allow electrons a coherent return to their origin. The temperature dependence of τ_φ is extracted from the magnetoresistivity $\rho(B)$ due to weak localization. We also determine the threshold for quantum interference corrections to the conductivity due to electron-electron (e-e) interaction effects from the equality $k_b T = \hbar/\tau$.

In the second part of the paper, we compare the "metallic" properties of high-mobility Si-MOS with high-mobility Si/SiGe:P structures and find that disorder $k_F\ell$ is not the driving parameter for the resistivity drop. We find that screening and scattering effects may explain the different behavior of the two material systems.

2 Experiment

Our investigations were performed on two Si-MOS samples (Si-15 and Si43 with peak mobilities of $\mu = 31,000$ and $20,000$ cm^2/Vs, respectively) and on one Si/SiGe:P structure (R510C, $\mu = 180,000$ cm^2/Vs). The latter contains a 15 nm Si cannel, a 14 nm Si$_{0.7}$Ge$_{0.3}$ spacer and a 11.5 nm Si$_{0.7}$Ge$_{0.3}$:P doping layer. Resistivity and Hall measurements were performed in four-terminal ac-technique at 17 Hz.

3 Results and conclusions

The weak localization effect was investigated by $\rho(B)$ measurements over a wide density and temperature range in Si-MOS samples Si-15 and Si-43. The peak in $\rho(B)$ was fitted by the usual expression for single-electron coherent backscattering [3]. The T dependence of τ_φ and τ was obtained for several densities from 1.9 to 15×10^{11} cm^{-2}. At $T = 300$ mK, τ_φ exceeds τ by up to two orders of magnitude. But since τ_φ strongly decays with increasing T, the extrapolated τ_φ and the evaluated τ cross each other at higher temperature. The temperature T_q at which $\tau_\varphi = \tau$ is an upper limit for single-electron quantum interference effects. The dependence $\tau_\varphi(T)$ can be fitted by a T^{-p} law with p between 1.1 and 1.7.

The density and temperature dependence of the temperature limit T_q identifies the region of the strong drop in $\rho(T)$ where no single-electron quantum interference effects take place. For 1.9×10^{11} cm$^{-2} < n < 3 \times 10^{11}$ cm^{-2}, the limit T_q decreases from about 6 to 2 K with increasing density n and is thus in the region of the strong $\rho(T)$ drop. But for $n > 3 \times 10^{11}$ cm^{-2}, T_q is completely below the strong $\rho(T)$ drop (for details see [4]), which is thus not related to single-electron quantum interference effects.

The equation $\hbar/\tau = k_B T_{ee}$ defines a temperature T_{ee}, which gives an upper limit for the occurrence of quantum interference corrections to the conductivity induced by e-e interaction [5]. Again the behavior of T_{ee} does not coincide at all with the density and temperature dependence of $\rho(T)$ (see [4]). Thus, the "metallic" state can not be caused by quantum interference effects due to e-e interaction either. Only close to the "metal-insulator" transition at the critical concentration n_c quantum interference effects will occur, but it is not clear whether they dominate the behavior there or not.

In order to get further information about the so called "metallic" state, we compare the behavior for two different material systems. Figure 1 shows $\rho(T)$ at several different densities for Si-MOS and Si/SiGe. We find that the drop in $\rho(T)$ is much stronger in Si-MOS than in Si/SiGe at the same resistivity values. As resistivity is equivalent to disorder $k_F\ell$, this shows that the latter is not a single driving parameter for the "metallic" conduction.

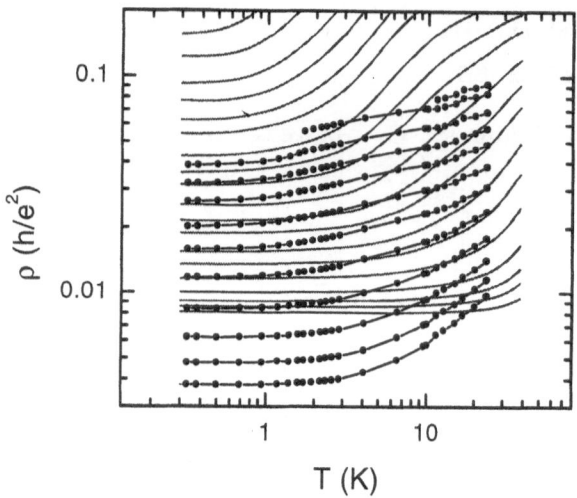

Fig. 1 Resistivity versus temperature for Si-MOS Si-43 (full curves, $1.34 < n < 35 \times 10^{11}$ cm^{-2}) and Si/SiGe R510C (curves with symbols, $2.2 < n < 4.6 \times 10^{11}$ cm^{-2}).

The electronic structure of the two systems is quite similar, both systems are n-type and the electrons move in a Si-layer with two-fold valley degeneracy. The phonon density is also comparable in the two systems. What are the differences? The dielectric constant are different due to the influence of the oxide layer. Also the interface between the 2D channel and the barrier material is different, it is Si/SiO$_2$ for the one and Si/SiGe for the other. Further the electron mobilities behave quite different.

The Si-MOS samples Si-15 and Si-43 have peak mobilities of 31,000 and 20,000 cm^2/Vs, respectively, whereas the Si/SiGe structure has a maximum mobility of 180,000 cm^2/Vs. In both samples μ decreases strongly towards lower density. From the analysis of the Shubnikov-de Haas effect, we get the quantum life time τ_q, which is related to single scattering events, whereas the momentum relaxation time τ may consist of many small angle scattering events. Our analysis gives a ratio τ/τ_q of 15 for Si/SiGe, whereas it is a small number in Si-MOS. This shows that in Si-MOS the scattering centers are in direct vicinity of the electron layer whereas in Si/SiGe remote impurity scattering takes place. The high value for τ/τ_q in Si/SiGe is in very good agreement with calculations for such structures [7].

For p-Si/SiGe structures with a small "metallic" effect of about 10%, it was recently shown that the drop in $\rho(T)$ can be explained by temperature dependent screening effects of impurity scattering [8]. For the T dependence of the screening, analytical expressions have been given for Si-MOS structures [9]. In addition, numerical calculations for Si-MOS have revealed that changes in $\rho(T)$ by up to an order of magnitude may take place [10]. On the other hand, for remote impurity doping in GaAs/AlGaAs structures, the T dependence of screening should be much smaller [11]. This is in agreement with the much smaller $\rho(T)$ drop in Si/SiGe with dominant remote impurity scattering. In addition, the observed

strong resistivity changes at $k_B T < E_F/4$ are consistent with the continuing temperature dependence in the strong degeneracy limit. Screening is thus expected to give an important contribution to the strong $\rho(T)$ drop.

Further, the different material interfaces in the two systems may lead to different scattering effects. At both interfaces, the surface roughness scattering will contribute, but mainly at high density. For lower n, charged hole traps in the oxide layer of Si-MOS seem to be able to explain the strong temperature and magnetic field dependence [6], as the filling of these states and thus the efficiency of scattering depends strongly on the Fermi energy. The same type of traps is not expected to be present in the Si/SiGe system.

In conclusion, we have shown that in the "metallic" state the resistivity drop occurs mainly for such temperatures and densities which correspond to the range of semiclassical physics and quantum interference effect do not contribute. The scattering due to charged trap states in the Si/SiO$_2$ interface and temperature dependent screening effects are consistent with the strong resistivity drop into the "metallic" state.

4 Acknowledgments

We thank B. L. Altshuler and A. Gold for stimulating discussions. The work was supported by the Austrian Science Fund (FWF) Project P13439, INTAS (99-1070), RFBR, NATO (PST.CLG.976208), NSF (DMR-0077825), Programs 'Physics of solid state nanostructures', 'Statistical physics' and 'Integration'.

References

1. V. Senz, T. Heinzel, T. Ihn, K. Ensslin, G. Dehlinger, D. Grützmacher, U. Gennser, Phys. Rev. B **61**, (2000) R5082.
2. E. Abrahams, P. W. Anderson, D. C. Licciardello, and T. V. Ramakrishnan, Phys. Rev. Lett. **42**, (1979) 673.
3. B.L. Altshuler, D.E. Khmelnitskii, A.I. Larkin, and P.A. Lee, Phys. Rev. B **22**, (1980) 5142; H. Fukuyama, Surf. Sci. **113**, (1982) 489.
4. G. Brunthaler, A. Prinz, G. Bauer, and V. M. Pudalov, cond-mat/0007230.
5. B.L. Altshuler and A.G. Aronov, *Electron-Electron Interaction in Disordered Systems*, edited by A.L. Efros and M. Pollak, (North Holland, Amsterdam 1985) p. 1. H. Fukuyama, ibid.
6. B. L. Altshuler and D. L. Maslov, Phys. Rev. Lett. **82**, (1999) 145.
7. F. Stern and S.E. Laux, Appl. Phys. Lett. **61**, (1992) 1110.
8. V. Senz, T. Ihn, T. Heinzel, K. Ensslin, G. Dehlinger, D. Grützmacher, U. Gennser, cond-mat/0004312.
9. A. Gold and V. T. Dolgopolov, Phys. Rev. B **33**, (1986) 1076.
10. S. Das Sarma and E.H. Hwang, Phys. Rev. Lett. **83**, (1999) 164.
11. A. Gold, Phys. Rev. B **41**, (1990) 8537.

Zero-bias conductance anomaly in GaAs/AlGaAs modulation doped field-effect transistors

S. Skaberna[1], U. Kunze[1*], D. Reuter[2], A.D. Wieck[2]

[1] Werkstoffe der Elektrotechnik, Ruhr-Universität Bochum, D-44780 Bochum, Germany (E-mail: kunze@lwe.ruhr-uni-bochum.de)
[2] Angewandte Festkörperphysik, Ruhr-Universität Bochum, D-44780 Bochum, Germany

Abstract　A minimum is observed in the low-temperature differential conductance of a modulation-doped field-effect transistor around zero drain voltage. The anomaly disappears at about 30 K and is sensitive to magnetic fields of only few 10 mT. We assume that a quantum correction of the conductance arises from Fermi velocity mismatch in the two-dimensional electron gas at the boundary below and outside the gate electrode. This correction is suppressed when at nonzero drain voltage the phase coherence of the electron waves is destroyed by carrier heating.

1 Introduction

The zero-bias anomaly (ZBA) is a common phenomenon occurring in any kind of a tunneling experiment, which appears as a peak or dip in the differential conductance characteristic in a range of a few mV around zero applied bias voltage. Its origin is, e.g., the interaction of tunneling carriers with phonons or paramagnetic impurities, a super-conducting energy gap in the electrode, or the Coulomb blockade mechanism [1]. Apart from this "classical" effect a ZBA has also been observed in semiconductor/super-conductor weak links due to Andreev reflection [2,3], in the quantized conductance regime of ballistic constrictions fabricated from a high-mobility two-dimensional electron gas (2DEG) [4], and in the single-electron transport via a lithographically defined quantum dot due to Kondo resonances [5]. In this work we report on a ZBA observed in the drain conductance characteristic of a GaAs/AlGaAs modulation doped field-effect transistor (MODFET).

2 Experiments

The MODFETs were fabricated on GaAs/AlGaAs hetero-structures grown by molecular-beam epitaxy with high (sample 1) and low (sample 2) electron mobility and with the 2DEG separated by 40 nm and 15 nm from the surface, respectively. The as-grown data at $T = 4.2$ K in the dark were $\mu = 1\times10^6$ cm^2/Vs, $n = 4.4\times10^{11}$ cm^{-2} (sample 1), and $\mu = 1\times10^5$ cm^2/Vs, $n = 2.7\times10^{11}$ cm^{-2} (sample 2). The transistor geometry was defined by wet-chemical mesa etching. Channel dimensions are 30 μm length and 10 μm width, and a central part of 8 μm was constricted to a width ranging from 1–10 μm. A 16 μm long active channel was defined by the Au gate electrode. Additionally, source and drain contacts of 300×300 μm^2 area were formed by alloying an evaporated AuNiGe layer. The drain current-voltage (I-U_D) characteristic and its first derivative with respect to U_D, the differential conductance g_D, were taken by modulating the dc drain voltage U_D with an ac signal of 0.1 mV rms at 470 Hz frequency. Unless otherwise noted all measurements were performed at $T = 4.2$ K.

*Corresponding author

3 Results and discussion

The I-U_D characteristic at fixed gate voltage U_G shows a distinct deviation from Ohm's law due to carrier heating by the electric field (Figure 1). The corresponding conductance characteristic (Fig. 2) reveals a ZBA minimum in the range of $U_D \approx \pm7$ mV which is imposed upon a bell-shaped background. The amplitude of the conductance dip amounts to about 10–15% of the maximum conductance, which is 3–8 mS according to the channel width. In a wide range of U_G the ZBA amplitude is nearly constant, only at U_G close to the threshold voltage the dip vanishes. As shown in Fig. 3 the differential conductance of the low-mobility MODFETs exhibits the same conductance minimum of roughly equal magnitude.

While further cooling below $T = 4.2$ K does not change

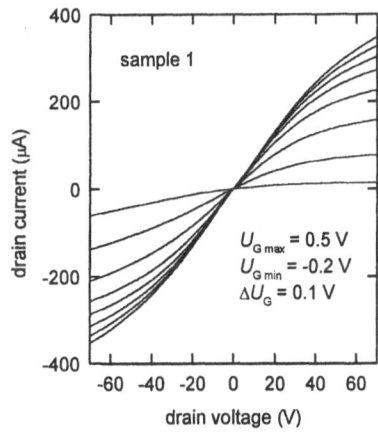

Fig. 1 Current-voltage characteristic at different gate voltage U_G of a 3 μm wide high-mobility channel at $T = 4.2$ K.

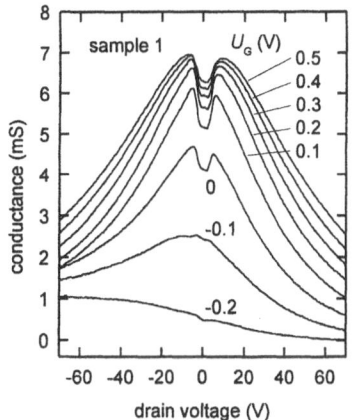

Fig. 2 Differential conductance characteristic of the same MODFET as used in Fig. 1 at $T = 4.2$ K.

the ZBA in both types of samples the anomaly gradually disappears at increasing temperature up to 30 K (Fig. 4). It should be noted that Figs. 3 and 4 were taken in different cooling cycles, and a slight shift of the threshold voltage led to the increase of the conductance in Fig. 4 compared with Fig. 3. Finally we tested the influence of a weak magnetic field on the ZBA. As shown in Fig. 5 the conductance minimum gradually vanishes as a field is raised to $B = 100$ mT, while simultaneously a maximum develops around zero bias. This maximum persists at least up to 500 mT, where the increasing magnetic field reduces the background signal due the classical magnetoresistance.

A simple explanation of the ZBA refers to a nonlinear current-voltage characteristic of the alloyed source/drain contacts and a substantial contact resistance. However, in this case the width of the conductance dip should depend on the channel resistance, which is controlled by the gate voltage. We interpret the ZBA as a nonclassical conductance correction that is suppressed when the long-range coherence of the electron waves is destroyed by an increased scattering rate, either due to phonon scattering at elevated temperatures or by heating the carriers in an electric field. One possible mechanism is coherent backscattering, called weak localization (WL) [6]. WL is

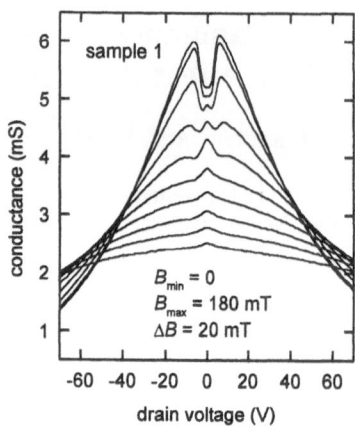

Fig. 5 Conductance characteristic of the MODFET used in Fig. 1 at $U_G = 0.2$ V at $T = 4.2$ K under different magnetic field B.

known to reduce the conductance by typically a few percent of the fundamental conductance in units of e^2/h. In our experiment the correction is more than one order of magnitude larger, therefore it is highly unlikely that WL is the origin of the ZBA. An alternative explanation is based on the Fermi wave vector mismatch between the 2DEG below and outside the gate region. Different electron densities in the gate and the series channel region lead to a reduced transmission probability at the boundaries between the regions, which causes an additional resistance [7]. This zero-bias conductance minimum should vanish at equal electron densities in the adjacent regions. From the geometry of the MODFET and the sheet conductance of the as-grown 2DEG we can estimate that the electron density is low in the series region and that equal densities are achieved at about $U_G = -0.15$ V. Unfortunately, this regime close to the threshold voltage has not been studied in detail so far. Further studies including magnetotransport are required to identify the condition of equal densities. Future work is also needed in order to understand the influence of the magnetic field on the sign and shape of the ZBA.

Acknowledgement. This work has been supported by the Bundesministerium für Bildung und Forschung (BMBF) under Grant No. 01BM920.

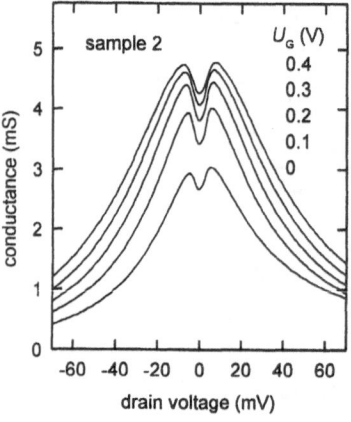

Fig. 3 Differential conductance of a 3 μm wide low-mobility channel at $T = 4.2$ K with gate voltage U_G as parameter.

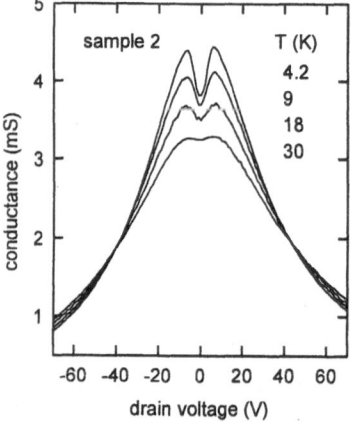

Fig. 4 The same MODFET as used in Fig. 3 at $U_G = 0.2$ V with temperature as parameter.

References

1. E.L. Wolf, *Principles of Electron Tunneling Spectroscopy* (Oxford University Press, New York 1985).
2. H. Takayanagi and T. Akazaki, Jpn. J. Appl. Phys. **34** (1995) 6977.
3. K.-M. H. Lenssen, M.R. Leys, and J.H. Wolter, Phys. Rev. B **58** (1998) 4888.
4. L. Martín-Moreno, J.T. Nicholls, N.K. Patel, and M. Pepper, J. Phys.: Condens. Matter **4** (1992) 1323.
5. D. Goldhaber-Gordon, H. Shtrikman, D. Mahalu, D. Abusch-Magder, U. Meirav, and M. A. Kastner, Nature **391** (1998) 156.
6. H. van Houten, B.J. van Wees, and C.W. Beenakker, in *Springer Series in Solid-State Sciences* Vol. 83, eds. H. Heinrich, G. Bauer, and F. Kuchar (1988), pp. 198-207.
7. N.A. Mortensen, K. Flensberg, and A.-P. Jauho, Phys. Rev. B **59** (1999) 10176.

Magnetotransport and Capacitance Investigations of Strongly Coupled but Spatially Separated 2D Electron and Hole Systems

M. Pohlt[1], M. Lynass[1], W. Dietsche[1], K. v. Klitzing[1], K. Eberl[1], R. Mühle[2]

[1] Max-Planck-Institut für Festkörperforschung, Heisenbergstr. 1, Stuttgart 70569, Germany
[2] Swiss Federal Institute of Technology Zürich, ETH Zentrum, Zürich 8092, Switzerland

Abstract   We report on the preparation of a two-dimensional electron gas (2DEG) coupled to a two-dimensional hole gas (2DHG), separated by a barrier of only 150 Å. We show how we fabricate the samples by MBE and focused ion implantation. We also studied the thermodynamical properties of the interacting electron hole system by capacitance measurements with and without magnetic fields at low temperatures.

1 Introduction

Due to the attractive coulomb interaction of the charges in different layers coupled 2DEG/2DHG systems have been of great interest to theorists. They predicted the formation of spatially indirect excitons with long lifetimes due to the suppression of tunneling by a high quality barrier. At low temperatures Bose Einstein condensation of these excitons is expected[1][2][3].

Therefore the study of spatially indirect electron hole systems is of great interest. Nevertheless there are not yet many experimental investigations due to the difficulty in preparing samples with separate contacts to the extremely closely spaced 2DEG/2DHG systems. So far two attempts to fabricate such structures have been reported[4][5], both were based on nonequilibrium methods.

2 Sample Fabrication

Our method consists also on the application of a voltage to a pn Junction that has an AlGaAs barrier between the electron and the hole region. The structure of the sample is shown in fig. 1. A forward voltage between the electron and the hole contacts will push the electrons and holes towards the barrier on their respective sides and form there the 2DEG and 2DHG. The most difficult step in the preparation is the formation of p-type contacts which have to extend *underneath* the barrier without deteriorating its quality. To achieve these contacts to the hole layer it is necessary to form patterned buried p-conducting layers as indicated in fig 1. In our samples these layers were formed by in situ focussed ion implantation.

First of all a GaAs buffer is grown on a semiinsulating substrate. The 1000 Å thick n-type backgate follows and then 3000 Å intrinsic GaAs and the degenerate C doped layer (1000 Å which will act as the contacts to the 2DHG).

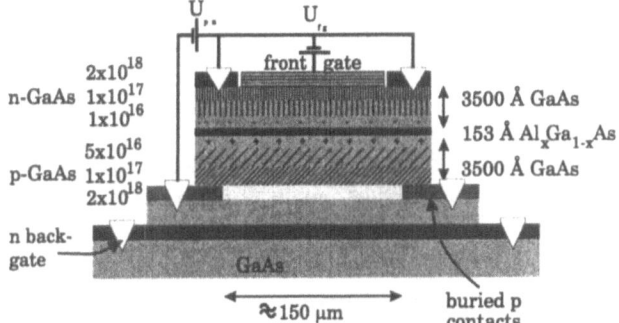

Fig. 1 Sample Structure. The ohmics are made of Au/Ge/Ni for the n side and Au/Zn/Au for the p side

At this point the growth is interrupted and the sample is transferred through a UHV tube to the focused ion beam implantation chamber. To avoid contamination during the transfer we use a thin As_2 cap on the wafer through which the ions are implanted. The As_2 cap and the contaminations are removed just before overgrowth. To build *individual* contacts Si ions are maskless implanted to overcompensate the p-type conductivity in certain areas in the C-doped layer and isolate the individual contacts from each other. This method was necessary because a simple Be implantation would cause a low quality barrier due to its strong segregation towards the surface during the overgrowth.

After returning to the MBE the patterned p contacts are overgrown by a GaAs layer with weak p type doping to ensure good ohmic contact between the patterned region and the hole channel. The 153 Å to 200 Å AlGaAs barrier is grown on top of this followed by the weakly Si doped GaAs layer that connects to the degenerate n-type GaAs top layer. The weakly doped layers are provided with doping gradient towards the 2D systems to improve mobility without loosing ohmic contact. The doping profiles are controlled with MBE precision and therefore no dopants will reach the barrier and cause leaks.

Contacts to the three-dimensional degenerate layers are provided by evaporated Au/Ge/Ni for n and Au/Zn/Au for p contacts. The front gate on top is evaporated into a hole in the degenerate n-layer with the same photomask, thus isolating the individual n-contacts from each other.

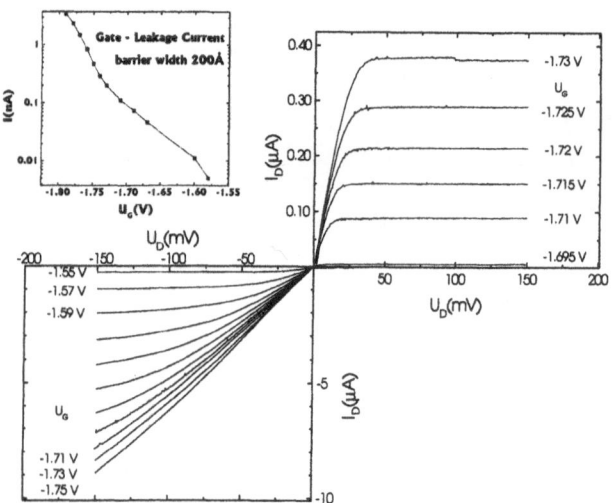

Fig. 2 Output Characteristic of the electron-hole system (electrons are on the right and holes on the left hand side). T = 4.2 K; Top left inset: barrier leakage current versus interlayer voltage at 4.2 K.

3 Device Characterisation

In our new kind of electron-hole transistor we accumulate the two-dimensional systems at a threshold voltage of about 1.55 V (cf fig 2). Depending on the particular doping profile either the holes or the electrons appear first so that the intrinsic electron and hole densities in general differ. With the aid of the front gate the electron density can be changed independently.

By optimising the MBE growth parameters and with the aid of the protective As_2 cap during the transfer we achieved a very high quality barrier. The leakage current of the electron hole system is only 10 pA at the threshold voltage of charge accumulation and there is a wide useful range of interlayer voltages accessible, where barrier leakage remains below 1 nA.

An elegant way to establish equal densities of electrons and holes also in the low density regime is the use of a SISFET. The front gate is removed in this type of device and instead the 2DEG is contacted with one big contact from the top. In this structure there is always only one threshold voltage at which both the electrons and the holes accumulate. For the capacitance measurements presented below we have used this device.

4 Capacitance study

The capacitance of the 2DEG/2DHG system with a barrier of 200 Å is shown in fig 3 for various voltages between electron- and hole-system as a function of the magnetic field. From the comparison with the resistance curves (not shown here) of the hole system it is evident that the minima in the capacitance correspond to minima in the thermodynamic density of states (TDOS) of the electron system.

The interpretation of the minima of the TDOS as caused by the Landau levels of the electron system gives values of the density that agree well with the absolute

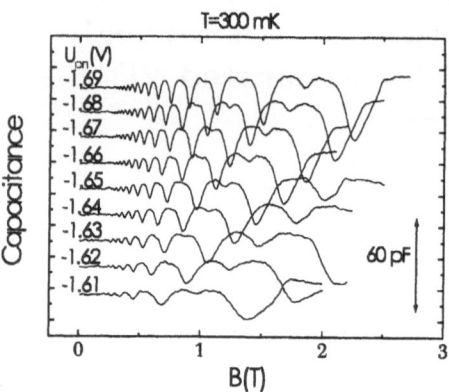

Fig. 3 Capacitance between the electron and the hole plate as a function of the magnetic field for various interlayer voltages. n_s increases linearly from $7 \cdot 10^{10} cm^{-2}$ at -1.61 V to $2.2 \cdot 10^{11} cm^{-2}$ at -1.69V.

magnitude of the capacitance. The positions of the minima in C vs the magnetic field give straight lines. Deviations from linearity that would hint at exciton binding are not found at these relatively high densities. The achieved sensitivity of our measurement system is not sufficient to study the very low density regime, where the oscillations are very small due to the relatively large width of the landau levels.

5 Conclusion

We have shown how it is possible to fabricate closely coupled 2DEG/2DHG systems with the combination of MBE and FIB. The samples that were obtained had separate highly reliable ohmic contacts to both layers and barrier thicknesses of only 153 Å. We have also shown a study of the thermodynamic properties of the double layer system performed by capacitance measurements. At higher density the system shows dominant 2DEG character. At present improvements are being made to study the low density regime where interaction effects are expected to dominate the transport properties.

6 Acknowledgments

This work was supported by the BMBF. M. P. wants to acknowledge useful discussions with J. G. S. Lok and U. Sivan.

References

1. S. I. Shevchenko, Sov. J. Low. Temp. Phys. **2**, (1976) 251.
2. Yu. E. Lozovik, and V. I. Yudson, Sov. Phys. JETP **44**, (1976) 389.
3. G. Vignale, and A. H. MacDonald, Phys. Rev. Lett. **76**, (1996) 2786.
4. U. Sivan, P. M. Solomon, and H. Shtrikman Phys. Rev. Lett. **68**, (1992) 1196.
5. B. E. Kane, J. P. Eisenstein, W. Wegscheider, L. N. Pfeiffer, and K. W. West, Appl. Phys. Lett. **65**, (1994) 3266.

First principles study of spin-electronics: Zero-field spin-splitting in superlattices

J. A. Majewski[1], P. Vogl[1]*, P. Lugli[2]

[1]Walter Schottky Institute, Technical University Munich, Am Coulombwall 3, D-85748 Garching, Germany email: majewski@wsi.tum.de
[2]Department of Electronic Engineering, University of Rome "Tor Vergata", I-00133 Rome, Italy

Abstract We present first-principles calculations of the fundamental coupling mechanisms that give rise to spin-splittings of the electronic energy bands in semiconductors at zero magnetic field. We show that these effects are induced by asymmetric chemical bonds in heterostructure interface layers but are neither caused nor influenced by macroscopic electric fields such as charge depletion or piezoelectric fields, in contrast to widely accepted notions. The k-linear Rashba coupling is found to be negligible for GaAs/AlAs heterostructures and superlattices.

1 Introduction

Zero-field spin-splittings of electronic band edge states near III-V heterostructure interfaces are an important ingredient for the realization of spin-transistors [1,2] and have been observed mostly near interfaces [3,4]. It is well established by now (see [5] for an excellent review) that there are two mechanisms that give rise to splittings of band edge Kramer pair states that are linear in the Bloch wave vector, namely the Dresselhaus effect or bulk inversion asymmetry (BIA) [6] and the Rashba effect [7,8] or structure inversion asymmetry (SIA) [5,6,8,9]. However, the physical origin of the latter, i.e. its magnitude and the relevance and role of macroscopic electric fields and strain has remained controversial [5]. No quantitative calculations of BIA plus SIA effects beyond semi-empirical k·p-theories have been published so far.

2 Results and discussion

We have studied the spin splitting of conduction and valence bands by performing first-principles local density functional calculations of pseudomorphic AlAs/s-GaAs and InP/s-AlSb superlattices with [001] and [111] growth orientation. The latter orientation gives rise to large built-in piezoelectric fields. These allow us to study the effect of macroscopic fields on the spin-splittings. Within the first-principles total-energy pseudopotential method [10], we have generated fully relativistic, separable pseudopotentials [11] and determined the ground state energy by minimizing the electronic and ionic degrees of freedom simultaneously. The band structure of superlattices has been calculated with a preconditioned conjugate gradient algorithm that allowed us to include up to 25000 plane waves.

2.1 Effective Hamiltonian

The zero-field spin splitting of a nondegenerate (Kramers doublet) band edge state can be characterized by an effective spin Hamiltonian $H = \sigma \cdot \mathbf{B}_{\text{eff}}(\mathbf{k}_\parallel)$ [6] where σ is the Pauli matrix vector and $\mathbf{k}_\parallel$ is the lateral wave vector. The magnitude and the direction of the effective field $\mathbf{B}_{\text{eff}}$ is determined by symmetry and depends on $\mathbf{k}_\parallel$ and the considered band state but we omit the latter label for brevity. The spin splitting of a Kramers pair near Γ is then given by $\Delta E_{\text{spin}}(\mathbf{k}_\parallel) = 2|\mathbf{B}_{\text{eff}}(\mathbf{k}_\parallel)|$ and contains a term linearly proportional to $\mathbf{k}_\parallel$. Generally, the effective magnetic field term can be written as $\mathbf{B}_{\text{eff}} = \mathbf{B}_R + \mathbf{B}_B$, where the Rashba or SIA term equals $\mathbf{B}_R = \alpha_R\, \mathbf{k} \times \mathbf{n}$ ($\mathbf{n}$ is the unit vector along the growth direction) and the bulk Dresselhaus or BIA term (see e.g. [5,6]) equals $\mathbf{B}_B = \alpha_B(-k_x, k_y, 0)$ and $\mathbf{B}_B = \alpha_B(k_y - k_z, k_z - k_x, k_x - k_y)$ for the [001] and [111] growth directions, respectively. For the [111] growth direction, $\mathbf{B}_B$ and $\mathbf{B}_R$ are parallel to each other and one obtains $\Delta E_{\text{spin}}(\mathbf{k}_\parallel) = 2\,|\mathbf{k}_\parallel\,(\alpha_R + \alpha_B)|$ independently of the direction of $\mathbf{k}_\parallel$. Thus, α_R and α_B cannot be determined separately in an ab-initio calculation., For the [001] growth direction, on the other hand, the spin splitting is $\Delta E_{\text{spin}} = 2\,|\mathbf{k}_\parallel|\,[\alpha_B{}^2 + \alpha_R{}^2 - 2\alpha_B\alpha_R\sin(2\theta)]^{1/2}$ and depends on the angle θ between $\mathbf{k}_\parallel$ and [100] which allows a separate determination of α_R and α_B.

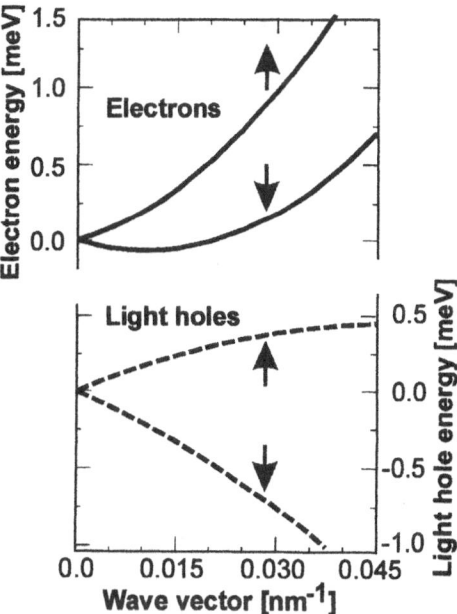

Fig. 1 Calculated dispersion relations for lowest conduction and light hole bands near the Γ point for a [111] $(\text{AlAs})_3(\text{GaAs})_3$ superlattice as a function of the lateral wave vector.

* Present address: Nagoya University, Venture Business Laboratory, 464-8603 Nagoya, Japan

2.2 AlAs/GaAs superlattices

We first consider strained layer short-period AlAs/s-GaAs superlattices and depict in Fig.1 the 2 lowest conduction and top light hole bands very close to the Γ-point. The two lowest conduction bands in both [001] and [111] superlattices have predominantly Γ-character with some mixing of X and L states, respectively. In contrast to the [001] superlattice, the lower symmetry in the biaxially strained [111] superlattice induces a macroscopic field that we find to be equal to 0.5 MV/cm (see Fig. 2).

In the [001] (AlAs)$_3$(GaAs)$_3$ superlattice, our analysis of the linear-k spin splittings yields $\alpha_B = 0.034$ eVÅ and $\alpha_R = -0.002$ eVÅ for the lowest conduction band. These values change only weakly for longer superlattice periods; we find $\alpha_B = 0.029$ eVÅ and $\alpha_R = -0.001$ eVÅ in an (AlAs)$_6$(GaAs)$_6$ superlattice. This extremely small Rashba term originates in the interface-induced microscopic structural asymmetry.

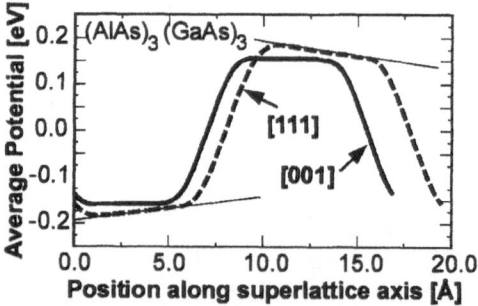

Fig. 2 The macroscopic average of the electrostatic potential in (AlAs)$_3$(GaAs)$_3$ [001] and [111] superlattices along the superlattice axis. The thin lines guide the eye to show the macroscopic electric field. The potential change across the interface corresponds to the dipole contribution to the valence band offset. Note that the atomic distances differ for [111] and [001].

In the [111] (AlAs)$_3$(GaAs)$_3$ superlattice, on the other hand, the sum $\alpha_R+\alpha_B = 0.13$ eVÅ is much larger for the lowest conduction band. This large difference is caused by pronounced band folding effects in the short-period [111] superlattice and originates in the mixed Γ and Λ character of the lowest conduction band. Indeed, in bulk GaAs, we find $\alpha_R+\alpha_B = 0.20, 0.15,$ and 0.135 eV Å for $\mathbf{k} = 1/3L,$ $2/3L,$ and the L-point, respectively. The values in AlAs are slightly smaller. The macroscopic electric field can be altered by varying the lateral lattice constant and has only a small effect on this result, consistent with the detailed results given below. In the absence of band folding - such as in a single GaAs/AlAs heterostructure - we thus find the Rashba-coupling to be negligible, irrespective of the presence or absence of any macroscopic electric field.

2.3 InP/AlSb superlattices

To further investigate the role of the interface asymmetry versus macroscopic field, we have performed calculations for lattice matched [001] and [111] InP/s-AlSb superlattices. Contrary to the AlAs/GaAs system, the bulk constitu-

ents share no common elements in this case which, by symmetry, leads to a macroscopic electric field even in the [001] case. Quantitatively, this macroscopic field is about 2 MV/cm for the unrelaxed (InP)$_3$(AlSb)$_3$ superlattice but becomes nearly zero once the atoms are allowed to relax so as to minimize the total crystal energy. For the lowest conduction band (which has dominantly InP-type Γ-character), we find $\alpha_B = 0.11$ and $\alpha_R = 0.015$ eVÅ for the unrelaxed case with the high electric field. In the relaxed case, where the field is almost zero, $\alpha_R = 0.10$ eVÅ becomes larger whereas α_B remains unchanged. This result clearly shows that the microscopic arrangement of atoms at the interface, rather than the macroscopic field, determines the magnitude of Rashba effect.

In the [111] (InP)$_3$(AlSb)$_3$ superlattice, the piezoelectricity of the strained AlSb leads to a strong macroscopic electric field also in the relaxed superlattice. Nevertheless, we find $\alpha_B + \alpha_R = 0.10$ eVÅ for the lowest conduction band which implies a smaller value for the linear-k spin splittings than for the [001] superlattice.

2.4 Rashba effect in strained bulk

Finally, we have calculated the strain dependence of the Rashba coupling in bulk AlSb in order to analyze the strain and interface induced linear-k splitting separately. For the first conduction band in tetragonally strained bulk AlSb, we find a very small $\alpha_R^{001} = (1/2)C_4' \mid e_{xx} - e_{zz} \mid$ with the constant $C_4' = 0.077$ eVÅ. In [111]-strained AlSb, we predict a value of $\alpha_R = C_3 e_{xy}/2$, with $C_3 = 6.3$ eVÅ. This result depends sensitively on the (self-consistently calculated) internal strain parameter. In fact, C_3 decreases by 30% if the internal strain parameter is set to zero. This additionally confirms the Rashba effect to be controlled by microscopic bonding asymmetries rather than by macroscopic fields.

References

1. S. Datta and B. Das, Appl. Phys. Lett. **56**, (1990) 665.
2. V. Moroz and C. H. Barnes, Phys. Rev. B **60**, (1999) 14272 and references therein.
3. T. Schäpers, G. Engels, J. Lange, T. Klocke, M. Hollfelder, and H. Lüth, J. Appl. Phys. **83**, (1998) 4324 and references therein.
4. P. Ramvall, B. Kowalski, and P. Omling, Phys. Rev. B **55**, (1997) 7160.
5. P. Pfeffer, Phys. Rev. B **59**, (1999) 15902 and references therein.
6. E. A. de Andrada e Silva, G. C. La Rocca and F. Bassani, Phys. Rev. B **55**, (1997) 16 293.
7. Bychkov and E. I. Rashba, J. Phys. C **17**, (1984) 6039.
8. F. G. Pikus and G. E. Pikus, Phys. Rev. B **51**, (1995) 16918.
9. L. Wissinger, U. Rössler, R. Winkler, B. Jusserand, and D. Richards, Phys. Rev. B **58**, (1998) 15 375.
10. W. E. Pickett, Computer Phys. Rep. **9**, (1989) 115.
11. N. Troullier and J. Martins, Phys. Rev. B **43**, (1991) 1993; L. Kleinman and D. M. Bylander, Phys. Rev. Lett. **48**, (1982) 1425.

A comparative study of 'metallic' and 'insulating' behaviour of the two dimensional electron gas on (100) and vicinal surfaces of Si MOSFET

S. H. Roshko[1], S. S. Safonov[1], A. K. Savchenko[1], A. G. Pogosov[1,2], Z. D. Kvon[2]

[1] School of Physics, University of Exeter, Stocker Road, Exeter, EX4 4QL, UK, e-mail: S.H.Roshko@ex.ac.uk
[2] Institute of Semiconductor Physics, Siberian Branch Russian Academy of Science, 630090, Novosibirsk, Russia

Abstract We have performed a detailed analysis of the temperature dependence of the 2DEG resistance of Si MOS-FETs for two different orientations of the Si surface: conventional (100), where unexpected metal-to-insulator transition (MIT) has been earlier observed, and a vicinal surface which is tilted by a small angle from the (100) surface. We have seen a low-temperature crossover from $dR/dT < 0$ ('insulator') to $dR/dT > 0$ ('metal') in the vicinal sample when the electron concentration is varied, and attribute it to the existence of a narrow impurity band at the interface. The presence of the impurity band has been directly detected by a hysteresis in the resistance with varied gate voltage. With increasing T, both structures exhibit high temperature crossover which we describe by temperature dependent impurity scattering near the transition from the degenerate to non-degenerate state.

1 Introduction

The possibility of a metal-insulator transition in two-dimensional systems has attracted wide attention in the last few years. This resulted from observations of an anomalous crossover from a negative to a positive dR/dT in 2DEG on (100) Si, when the concentration of electrons is increased [1–3]. The 'metallic' character is in obvious contradiction with the 2D scaling theory of electron localisation for non-interacting electrons [4]. Several models have appeared where the transition to unusual 'metallic' behaviour was predicted as a result of non-critical crossover for conventional, though non-trivial, electron transport. Two of these models are based on electron scattering by impurities: i) Altshuler and Maslov suggest that the crossover in Si MOSFETs can be attributed to an impurity band (IB) in the oxide [5], ii) Das Sarma and Hwang argue that the 'metal' originates from the temperature dependent screening [6].

In this work we have performed resistance measurements, over a broad temperature range from 50 mK to 70 K, in a 2DEG on the surface of Si which is different from the (100) plane. We expected that the difference in the surface and impurity states at the interface would affect the manifestation of $R(T)$. Indeed, we have observed *two* crossovers (at low and high temperatures) and provided the first experimental evidence that in these structures the crossovers are due to impurity scattering [5,6].

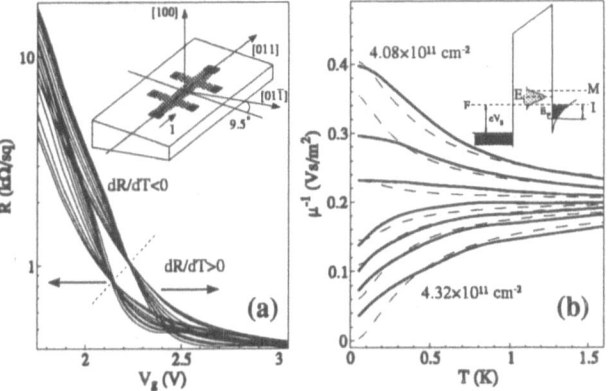

Fig. 1 (a). Resistance as a function of gate voltage for a vicinal sample at $T = 50 - 920$ mK. Two arrows show different direction of V_g sweeps. Insert: Schematic diagram of a vicinal MOSFET. (b). Temperature dependence of the mobility near the crossover at different concentration (from top to bottom $V_g = 2.1 - 2.24$ V). Dashed lines show the fit by Eq. (1) with the width of the IB, $W \simeq 0.08$ meV. Inset: A diagram of the impurity band in the oxide giving rise to the 'metallic' and 'insulating' behaviour of the 2DEG

2 Samples

The vicinal samples are n-channel MOSFETs fabricated on a surface which is tilted from the (100), with a peak mobility of 2×10^4 cm^2/Vs at $T = 4.2$ K, Fig. 1a, inset. We have also measured 'normal' samples grown on the (100) Si which are made by the same technology as the vicinal samples, and have maximum mobility around 1.5×10^4 cm^2/Vs. The oxide thickness in both types of structure is 120 nm.

3 Results and Discussion

Fig. 1a shows the resistance as a function of V_g for a vicinal sample. A change in the sign of dR/dT is clearly seen near $R_{sq} \sim 1$ kOhm $\sim 0.04 \times h/e^2$, with 'metallic' behaviour at larger V_g. When the gate voltage is slowly swept in the two opposite directions, two distinct groups of curves are obtained. The hysteresis disappears above 4 K and is most pronounced in the crossover region. The Shubnikov-de Haas measurements have been performed in two different resistance states, $R^{(1)}$ and $R^{(2)}$, obtained when a particular V_g was approached from the two opposite points outside the crossover region. They have shown that electron concentrations are equal with accuracy of

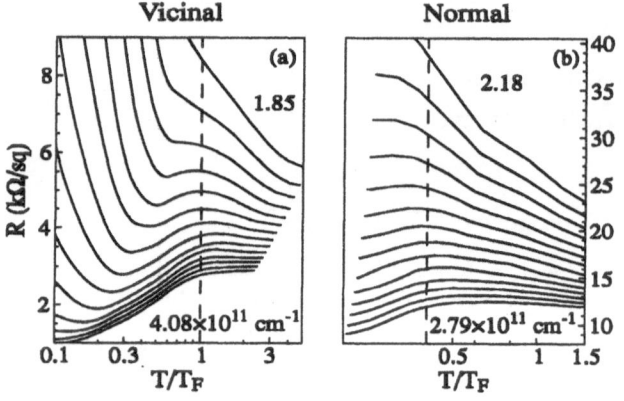

Fig. 2 Resistance as a function of T/T_F for vicinal and normal samples in the range of the high temperature crossover. The variation of T_F with concentration corresponds to the shift of the hump shown by the dashed lines.

a few percent and that the resistance difference is caused by the difference in the electron mobility.

We suggest that the crossover in the resistance ·is due to the effect on the mobility of a *narrow impurity band* at the interface Si/Si oxide. Its character is similar to that in [5]: it scatters electrons when it is positively charged, and, with increasing the Fermi level, the number N_+ of ionised impurities in the IB decreases which reduces scattering of the 2DEG. The origin of the hysteresis in Fig. 1a is a slow (at low temperatures) electron exchange between the 2DEG and the IB separated by a barrier. When the 2DEG concentration is decreased, some of electrons remain trapped in the IB, and this gives rise to a smaller value of $R^{(2)}$ compared with $R^{(1)}$. The sign of dR/dT depends on the position of the Fermi level in the IB. When the Fermi level F is above and close to the IB, the IB becomes more positively charged with increasing T and the resistance of the 2DEG increases: $dR/dT > 0$. The crossover in the sign of dR/dT occurs, when the Fermi level is close to the centre of the IB, and the 'insulating' $R(T)$ corresponds to the Fermi level in its lower part, Fig. 1b, inset. Then with increasing T the positive charge of the IB decreases, resulting in reduced scattering: $dR/dT > 0$.

The crossover points in Fig. 1a correspond to the half filled IB. Assuming the IB to be at the level E_i, the variation of the mobility for different F is given by

$$\mu^{-1}(T) \propto N_+ \propto 1 \Big/ \left(1 + \exp \frac{F - E_i}{k_B T} \right). \qquad (1)$$

Fig. 1b shows the temperature dependence of the resistance $R^{(2)}$ near the crossover at $n_c \simeq 4.18 \times 10^{11} \ cm^{-2}$, in the range $T = 0.05 - 4$ K, presented as μ^{-1} to illustrate the simple model of IB scattering.

At higher temperatures, another crossover is seen at $R_{sq}^c \simeq 3$ kOhm, Fig. 2a. Near this transition, there is a non-monotonic $R(T)$ with a gradual change from $dR/dT > 0$ to $dR/dT < 0$ with increasing T. We note that in the temperature range $T > 4$ K the system experiences a transition from degenerate to nondegener-

ate state. The variation of T_F with concentration corresponds to the shift of the hump in Fig. 2a. The metallic behaviour at $T < T_F$ is explained in [6] by the temperature dependence of the screening function which produces a linear rise in the resistance with temperature. All main features of the model [7,6] for the temperature dependent ionised impurity scattering are seen in this crossover: linear drop of the resistance at $T \ll T_F$, a shift of the position of the hump in $R(T)$ with T_F, and the existence of a small, positive magnetoresistance when $T \sim T_F$. The latter decreases with either increasing concentration or decreasing temperature, i.e. when the system is driven towards the degenerate state, which is in agreement with the classical behaviour of degenerate semiconductors.

We have also performed a comparative study of a normal sample (100) Si, Fig. 2b. At temperatures below 1 K, the normal sample shows similar behaviour to that in the vicinal samples around $R \sim 1$ kOhm. At the same time, no hysteresis has been observed in this case, which does not allow us to link the low-temperature crossover in this sample directly to an IB. At high temperatures the sample shows a crossover in $R(T)$ but at a higher critical resistance, $R^c \sim 15$ kOhm, and lower concentration, $n_c \simeq 2.7 \times 10^{11} \ cm^{-2}$ than the vicinal sample. The position of the resistance hump near the transition is shifted from $T_c = T_F \sim 20$ K to $T_c = 0.4 T_F \sim 5$ K in the normal sample. This supports the applicability of the model [6] for the high temperature transition in both samples.

4 Conclusion

We have observed several unusual features of the crossover from 'metallic' to 'insulating' behaviour of the 2DEG on a vicinal Si surface and have been able to explain them by classical electron conduction with temperature dependent impurity scattering.

5 Acknowledgments

We are grateful to B. L. Altshuler and D. L. Maslov for stimulating discussions, EPSRC and ORS award fund for financial support.

References

1. S. V. Kravchenko *et al.*, Phys. Rev. B **50**, 8039 (1994); S. V. Kravchenko *et al.*, Phys. Rev. Lett. **77**, 4938 (1996); D. Simonian *et al.*, Phys. Rev. Lett. **79**, 2304 (1997).
2. S. V. Kravchenko and T. M. Klapwijk, Phys. Rev. Lett. **84**, 2909 (2000).
3. D. Popovic, A. B. Fowler, and S. Washburn, Phys. Rev. Lett. **79**, 1543 (1997);
4. E. Abrahams, P. W. Anderson, D. C. Licciardello, and T. V. Ramakrishnan, Phys. Rev. Lett. **42**, 673 (1979).
5. B. L. Altshuler and D. L. Maslov, Phys. Rev. Lett. **82**, 145 (1999).
6. S. Das Sarma and E. H. Hwang, Phys. Rev. Lett. **83**, 164 (1999).
7. A. Gold and V. T. Dolgopolov, Phys. Rev. B, **33**, 1076 (1986).

The role of surface-localized states in the in-plane transport properties of superlattices

A.B.Henriques

Instituto de Física, Universidade de São Paulo, Caixa Postal 66318, São Paulo 05315-970, Brazil

Abstract The role played by the surface-localized states (Tamm states) in the in-plane transport properties of doped superlattices was investigated theoretically. The scattering rate of electrons in the Tamm states is smaller than the same rate for electrons occupying extended miniband states, but at each scattering event they are on average deflected by a larger angle than electrons from the miniband. As a result of this all electrons present similar transport mobilities. The relative contribution of the surface-localized states to the total in-plane conductivity is approximately equal to $1/(N_w - 1)$, where N_w is the number of wells in the finite superlattice.

1 Introduction

Surface-localized states in solids were first described theoretically by I.Tamm[1], and they are often denominated "Tamm states". Superlattices can be described as one-dimensional analogs of real solids in which the atoms are substituted by interacting quantum wells, and may contain Tamm states at their internal surfaces. In undoped superlattices surface-localized states arise if structural disorder is present [2], or if the superlattice is submitted to an external electrical field [3]. Recently, it was shown that in doped superlattices Tamm states below the Fermi energy arise even in unperturbed ideal structures [4]. The Tamm localization in this case is a consequence of a strong electrical field present near the borders of the superlattice. Charge carriers in a Tamm state are localized within an epitaxial layer, but along the epilayer they remain mobile and can contribute to the conductivity. In the present report we investigate theoretically the scattering of electrons in a Tamm state and establish their contribution to the in-plane conductivity of the sample.

2 Results

Fig. 1 shows the threshold energies of the electronic subbands, obtained from a self-consistent solution of Schrödinger and Poisson equations, as a function of the number of wells, in a structure with strongly interacting quantum wells. The structure considered constisted of N_w In$_{0.53}$Ga$_{0.47}$As quantum well layers, each of thickness $L_w = 50$Å, separated by $N_w - 1$ InP barrier layers (thickness $L_B = 40$Å), which where doped with Si in their center (areal concentration $N_d = 3.5 \times 10^{12}$ cm^{-2}). The silicon atoms where assumed to be distributed in a Gaussian of full width at half maximum, Δ_{Si}, and supposed to be fully ionized. The whole structure was imbedded within thick layers of undoped InP. The origin of the energy scale was fixed at the Fermi level. The $N_w - 2$ lowest energy states are associated with wave functions which extend along the whole structure,

and give rise to a quasi-continuous energy miniband (see Fig.1) when N_w increases. The energy width of the fundamental mininiband is approximately 27 meV. The next two states have their wave functions completely localized in the wells at the ends of the superlattice, which characterizes them as Tamm states. Because of the symmetry of our structure, the two Tamm states, associated with both surfaces of the slab, become degenerate when N_w is large. The Fermi energy for this Tamm state is roughly half of the same value for the fundamental electronic miniband.

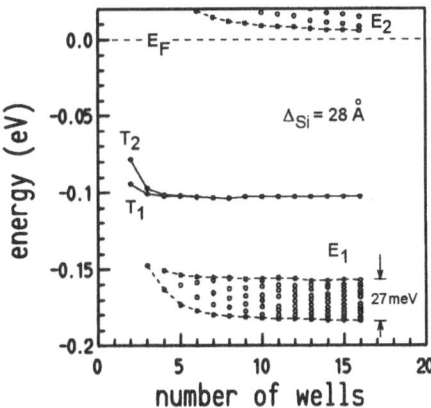

Fig. 1 Energy threshold of subbands as a function of N_w. E_1 and E_2 label the sets of levels which originate the first and second minibands, respectively.

The quantum lifetime, τ_Q, of the electrons in individual subbands was calculated in the random phase approximation, which is known to produce reliable results [5]. The main electron scattering mechanism was assumed to be the their Coulomb interaction with the ionized Si atoms. The calculations where done in the lines of Ref.[6], but in the present case the effective mass dependence on position was also taken into account. Fig. 2 shows the dependence of the calculated width of the energy levels, $\Delta E = \hbar/2\tau_Q$, as a function of the number of wells. All levels below the Fermi energy were included in the calculation. The dielectric matrix did not include contributions from states above the Fermi level, which are very small [5]. The curves labeled T_1 and T_2 represent ΔE for the states localized in the outer wells. Since the computing time increases roughly with the 4th power of the number of subbands taken into account, it is convenient to keep this value as low as possible. Fig. 2 shows that ΔE already saturates at $N_w \sim 5$, so in subsequent calculations we fixed $N_w = 5$. The Tamm states are described by longer quantum lifetimes than the states from the lowest energy miniband. This is due to their greater localization in the ternary layer than electrons from the extended states, hence a weaker interaction with charged donors situated in the InP layers. For the Tamm and miniband states to be isolated from each other, they must be separated by an energy greater than the added halfwidth of the two levels; calculations show that this is guaranteed if $N_d > 2.0 \times 10^{12} cm^{-2}$. When Δ_{Si} increases, ΔE also increases for all subbands, by approximately 10% when Δ_{Si} is incremented by 25Å, due to the shortening of the average electron-impurity distances when Δ_{Si} increases.

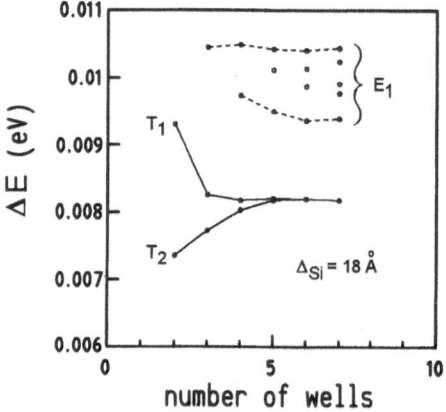

Fig. 2 Halfwidth of the energy levels as a function of N_w.

The theoretical result shown in Fig.2 can be compared with experimental estimates, reported in Ref. [4], obtained from Shubnikov-de Haas (SdH) measurements on a sample with the same parameters i.e. $N_d = 3.5 \times 10^{12}$ cm^{-2}, 16 well layers of thickness 50Å and separated by 15 barrier layers of thickness 40Å. The cyclotron mass of carriers at the Fermi level, m_c, was determined from the temperature dependence of the SdH component, whereas the quantum mobility, μ_Q, was obtained from the field dependence of the SdH component on the magnetic field intensity, at a fixed temperature. The halfwidth of the energy levels, given by $\Delta E = e\hbar/2\,m_c\,\mu_Q$, is estimated to be $\Delta E_T = 0.0098$ eV for the Tamm state, whereas the miniband is characterized by an average value of $\Delta E_1 = 0.013$ eV. This compares with the theoretical values of $\Delta E_T = 0.008$ eV and $\Delta E_1 = 0.010$ eV, respectively, as Fig. 2 shows. The theoretical values are larger than the experimental ones, which is an indication that additional scattering mechanisms - notably scattering by interface defects [7] - are operating.

The conductivity from each electronic subband was obtained by solving Boltzmann equation as described in Ref.[6], but now also taking into account the dependence of the effective mass on position. The calculated transport mobilities of electrons from the miniband and from the Tamm states are found to be nearly the same. This is because Tamm electrons, although scattered less often

than electrons from the miniband states (see above), are on average scattered by a larger angle, given that their wave vector is smaller, and that the Coulomb interaction is long range. The net result is a nearly constant transport mobility for all electrons at the Fermi level.

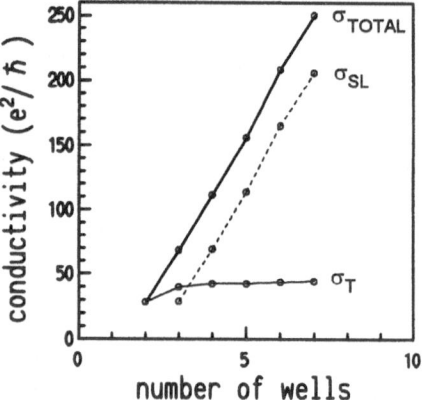

Fig. 3 Conductivity components as a function of the number of wells in the finite superlattice. σ_T and σ_{SL} are the contributions to the conductivity stemming from electrons in Tamm states and superlattice miniband, respectively. σ_{TOTAL} is the total in-plane conductivity.

The calculated conductivity along the epitaxial layers is depicted in Fig.3. The contribution from the surface localized state saturates, whereas the contribution from the extended superlattice states increases approximately linearly when N_w increases. Fig.3 shows that the saturation value for conductivity contribution from Tamm electrons is roughly equal to the same contribution from each additional superlattice period. This is because, although there are two Tamm states, their Fermi energy is only about half of the Fermi energy for the electronic miniband. Since only $N_w - 2$ periods contribute to the formation of the superlattice miniband, the relative contribution from Tamm electrons will be, to a good approximation, equal to $1/(N_w - 1)$ of the overall in-plane conductivity.

3 Acknowledgements

We acknowledge support by CNPq, Grant No. 306335/88, and FAPESP, Grants No. 99/10359-7 and No. 00/07399-6.

References

1. I.Tamm, Phys. Z. Sovjetunion **1**, (1932) 733.
2. H.Ohno, E.E.Mendez, J.A.Brum, J.M.Hong, F.Agulló-Rueda, L.L.Chang, and L.Esaki, Phys. Rev. Lett. **64**, (1990) 2555.
3. Roger H. Yu, Phys. Rev. B **47**, (1993) 1379; Roger H. Yu, Phys. Rev. B **47**, (1993) 15692.
4. A.B.Henriques, Phys. Rev. B **61**, (2000) R13369.
5. Guo-Qiang Hai, Nelson Studart, and François M. Peeters, Phys. Rev. B **52**, (1995) 8363.
6. A.B.Henriques, Phys. Rev. B **53**, (1996) 16365.
7. G.Bastard, J.A.Brum, and R.Ferreira, Solid State Physics **44**, (1991) 229.

High-Mobility Heterostructure as a New Kind of Chaotic Billiards

L.D. Shvartsman

The Racah Institute of Physics, The Hebrew University of Jerusalem. Jerusalem, 91904, Israel

Abstract. We show that p-heterostructures of cubic semiconductors and double-well structure of InAs-GaSb are good model systems for new kind of chaotic billiards. For these structures, the chaotic behavior results from non-trivial dispersion of particles even for very simple circular geometry of billiards.

1 Introduction

Chaotic dynamics of the particles in various real and model systems are widely studied. This problem has a fundamental importance for understanding of the origin of statistical laws. The most popular billiards support the chaotic behavior for a "normal" particle, i.e. when the dispersion of the particle is parabolic and isotropic. $\mathcal{E}(p) = \dfrac{\vec{p}^2}{2m}$ The chaotic dynamics in these cases results from the non-trivial geometry of external potential. For example, this geometry may be multi-connected, as for Sinai, Cassini billiards, etc [1,2]. In [3,4] the stochastic dynamics was studied for heterostructures in magnetic field. In this paper we show that some systems based on high-mobility heterostructures are excellent candidates for new kind of chaotic billiards. Here the bizarre geometry of external potential becomes unnecessary because the dispersion of the particle may be rather complicated.

2. Conjugated Billiards

The analysis of an arbitrary billiard based on the elastic reflection of the particle can be trivially reduced to the solution of the Hamilton equations:

$$\frac{\partial H(\vec{r}, \vec{p})}{\partial \vec{r}} = -\frac{d\vec{p}}{dt}; \frac{\partial H(\vec{r}, \vec{p})}{\partial \vec{p}} = \frac{d\vec{r}}{dt};$$

For the particles with trivial parabolic dispersion the energy conservation requires that the angle of incidence is equal to the reflection angle. For the non-trivial dispersion law this statement is obviously not valid anymore, and the trajectories of the particle have to be considered simultaneously in p-space and in r-space. In the Fig.1 we consider this methodology for the simplest case of circular billiard and normal particle. Two following equations describe the boundaries of conjugated billiards in r- and p-space:

$$U(x, y) = \text{Const.} = E \quad (1)$$

$$\mathcal{E}(p) = E = \text{Const.} \quad (2)$$

The particle trajectories in both real and momentum space are the set of straight lines in between the boundary variables $\vec{r_i}$ and $\vec{p_i}$. All the consecutive values of boundary variables have to satisfy (1) for r_i and the (2) for p_i. Besides, the ballistic flight in a real space is defined by the vector $(\frac{\partial \mathcal{E}}{\partial \vec{p}})_{\vec{p}=\vec{p_0}}$, while in p-space the direction of the flight is given by the vector $(-\frac{dU}{d\vec{r}})_{\vec{r}=\vec{r_i}}$ (see Fig.1). Naturally, each succeeding value of boundary variables is entirely defined by the previous one following (3) and (4):

$$\vec{r_i} = \vec{r_{i-1}} + (\frac{\partial \mathcal{E}}{\partial \vec{p}})_{\vec{p}=\vec{p_{i-1}}} t_i \quad (3)$$

$$\vec{p_i} = \vec{p_{i-1}} + (\frac{\partial U}{\partial \vec{r}})_{\vec{r}=\vec{r_{i-1}}} t_i \quad (4)$$

There are two kinds of trajectories: closed ones and opened ones. For the closed trajectories, the sets of boundary variables contain final number of elements, and for opened trajectories this number is infinite. Nevertheless, for the both cases in the Fig.1 there is a clear one- to-one correspondence: $\vec{r_i} \leftrightarrow \vec{p_i}$. Both opened and closed trajectories for this simple case certainly do not exhibit the chaotic behavior because correlation function:

$$G(\tau) = \int (\vec{r}(t+\tau)\vec{r}(t) + \vec{p}(t+\tau)\vec{p}(t))dt \text{ does not}$$

decay properly. The same fact may be reflected by the absence of clear stochastic domains as islands, scars, etc., when mapping the trajectory in the plane of boundary variables. The necessary condition for the chaotic behavior includes both infinite sets for boundary variables and absence of one- to-one correspondence between them. It may be realized by the change of the geometry of p-space isoenergetic contours. Next section contains the brief review of the topology of the typical dispersion laws of p-type heterostructure.

3. Isoenergetic Curves for Quantum Hetero-structures

We have considered circular billiards for such structures as: 1. p-heterostructure of GaAs and other cubic semiconductors [5-7] and 2. double-well structure of InAs-GaSb [8]. The energy spectrum of the carriers in p-type heterostructure is known to be the system of quantum-sized subbands arising from the originally degenerated in p=0 light hole and heavy hole bands. The bands have Γ_8 symmetry. We have based our calculation on 4x4 Kohn-Luthinger Hamiltonian with proper chosen boundary conditions [5-7]. The shape of the isoenergetic curve depends on the energy. It can be multi-valley, it can consist of two contours, and finally it can be one contour shaped as a corrugated sphere having a square symmetry. The last case is shown in the Fig.2b.

4. Results and Discussion

For these structures, the chaotic behavior of particles is supported even for simplest circular geometry of billiard. Figures 2a,b show that both r- and p-trajectories may be very complicated for anisotropic single-connected p-contour of ground subband of Ge quantum well. Closed trajectories still exist: e.g. the square one close to denoted as 1-4 in Fig.2a, provided the billiards walls at p.1 are hit by the particle with momentum $p_y=0$. For the actual trajectory shown in Fig.2a,b the momentum is different, and the particle leaves the vicinity of the closed trajectory and in future forgets of its existence. To prove the chaotic behavior, we simulated the map of trajectories in the plane of boundary variables. The calculations have been performed according to the methodology described above (1-4), and the map clearly indicates the presence of chaotic islands. The chaotic behavior is also proven by trajectory mapping for InAs-GaSb double wells, when the dispersion of the particle is multi-connected (the p-contour alike Fig. 2a but accomplished with the circular opening centered at p=0). This billiard is similar to the Sinai billiard with switched variables $p \leftrightarrow r$, and it exhibits the similar mapping.

References

1. G. M. Zaslavsky, *Chaos in Dynamic Systems,* Harwood, New York (1984).
2. Ya. Sinai, Yu. Bunimovich, Commun. Math. Phys. **78**, 247 (1980).
3. E. M. Baskin, A.G.Pogosov, M.V.Entin, JETP, **83**, 1135, 1996.
4. E. M. Baskin, G.M.Gusev, Z.D.Kvon, A.G.Pogosov, M.V.Entin, JETP Lett., **55**, 678, 1992.
5. L.D.Shvartsman, Solid State Communications, 46, 787,1983.
6. Gilad Schechter, L.D.Shvartsman, Phys.Rev.,**B58**, 3941, 1998.
7. A.V.Chaplik, L.D.Shvartsman,,Poverhnost. No. 2, 73, 1981. [English transl.: Sov.Phys.-Surface].
8. Smadar de-Leon, L.D.Shvartsman, B. Laikhtman, Phys.Rev., **B61,**1891,1999.

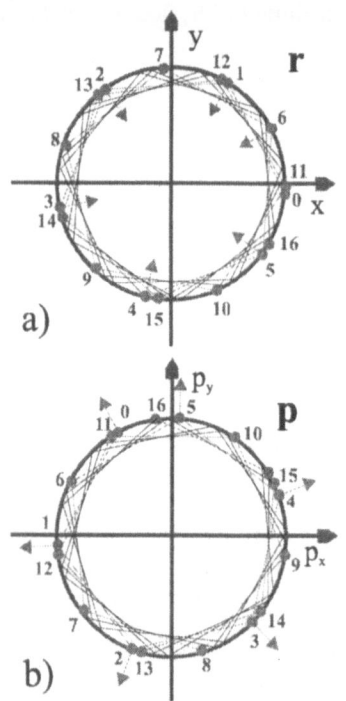

Fig. 1

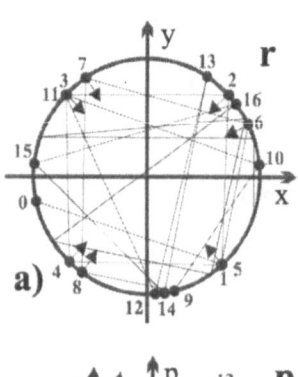

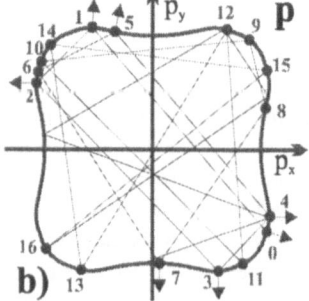

Fig. 2

Gate-Controlled Very Large Spontaneous Spin-Splittings in Normal $In_{0.75}Ga_{0.25}As$ / $In_{0.75}Al_{0.25}As$ Heterojunctions Grown on GaAs Substrates

S. Yamada*, Y. Sato, S. Gozu and T. Kikutani

School of Materials Science, JAIST-Hokuriku, 1-1, Asahidai, Tatsunokuchi, Ishikawa, 923-1292 Japan

*shooji@jaist.ac.jp

Abstract In normal $In_{0.75}Ga_{0.25}As$ / $In_{0.75}Al_{0.25}As$ heterojunction, spin-orbit coupling parameters of ~30 $(x10^{-12}eVm)$ as well as their tunability by the front-gate are confirmed. In the spin-FET with NiFe electrodes, we observed gate-dependent two terminal resistance with a strong oscillatory component suggesting a spin-presession.

1 Introduction

Spontaneous or zero-filed spin splittings (ZFSSs) have so far been reported in various heterojunctions such as HgTe, InAs / (Al)GaSb, InGaAs / InAlAs, InGaAs / InP etc. However, experiments as well as theories are still controversy about the Rashba term itself and the estimated (gated) splittings still remain relatively small (~5 meV). To manifest the pure origin of ZFSS and to design optimum material / structure for spin-FETs, it is desirable to create new (narrow gap) and high quality heterojunctions with a larger ZFSS. $In_{0.75}Ga_{0.25}As$ / $In_{0.75}Al_{0.25}As$

modulation-doped heterojunctions grown via InAlAs buffer on GaAs substrates could be an alternative candidate, since they have a record low-temperature two-dimensional electron gas (2DEG) mobility of < $5x10^5$ $cm^2/Vsec$ for a sheet electron density, $n=1x10^{12}/cm^2$ [1]. In this paper, we report detailed study of the ZFSS of the heterojunction, which describes a first finding of a very large gate-controlled ZFSS (~10 meV) in normal-type modulation-doped heterojunction.

2 Experiments and Results

ZFSSs are estimated from low-temperature magnetoresistance beat analysis in van der Pauw (5 x 5 mm^2) and gated-Hall bar (40 x 500 μ m^2) samples. Maximum spin-orbit coupling constant of $\alpha_{zero}s$ ~30 $(x10^{-12}eVm)$ were obtained and the initially gradual and succesively rapid decrease of α_{zero} as the gate voltage (V_g) decreased was

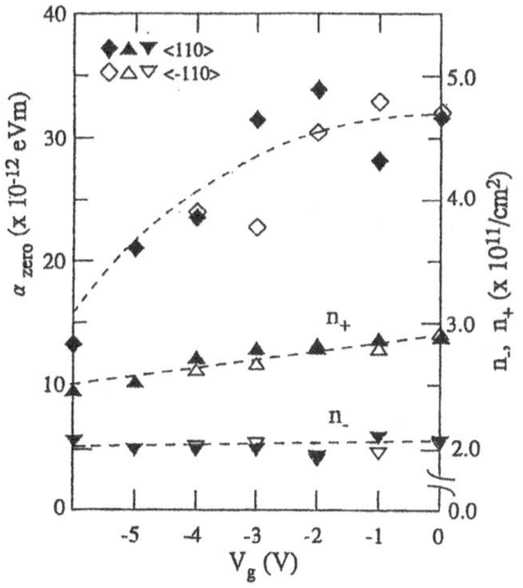

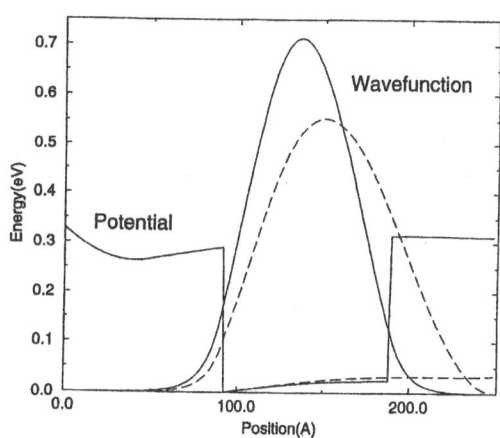

Fig. 1 Gate voltage (V_g) dependencies of $\alpha_{zero}s$ and sheet electron densities of the splitted bands n(-) and n(+) in gated-Hall bar samples. <110> and <-110> means the longer-side directions of the samples, suggesting no in-plane anisotropy of α_{zero}.

Fig. 2 Examples of self-consistent calculation for 30 (dashed) and 10 (solid) nm well heterojunctions. In 30 nm case, the confining potential for the 2DEG is mostly triangular giving a strong penetration asymmetry due to the lack of right-side barrier.

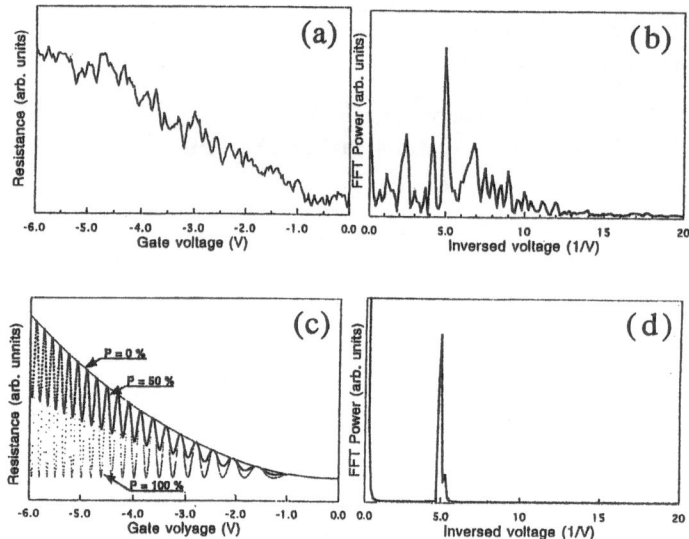

Fig. 3 (a) obtained two-terminal resistance (R_{sd}) as a function of Vg in our NiFe electrodes FET, (b) FFT result for the trace in (a), (c) calculated R_{sd} assuming the feature in Fig. 1 and various spin-polarizations from the NiFe source, and (d) corresponding FFT result. Source-drain distance is 8 μ m and the gate length is 3 μ m.

confirmed at 1.5 K (Fig. 1). This dependence is unique and reasonable when the normal heterostructure is used. Main origin of such large α_{zero}s seems to be a relatively thick (30 nm) $In_{0.75}Ga_{0.25}As$ well in our samples, which has been impossible due to strain especially in pseudmorphic heterojunctions on InP substrates . In fact, there appeared no beat oscillations in a 10 nm well sample. This suggest that, in the 30 nm case, a stronger asymmetry between the wave function penetrations into the both barriers exists, as supported by the calculation (Fig. 2) and hence not the well-field dependent but the interace contribution in the Rashba term should play a major role in our case. Even when this is the case, V_g dependent change of α_{zero}s can be explained by the change of the penetration asymmetry due to the applied V_g . Also in theory, possible dominance of this contribution have recently been pointed out [2].

Figure 3 demonstrates the results of fundamental spin-FET operation in the device made using this heterojunction and NiFe source-drain electrodes at 1.5 K. The distance between the electrodes is 8 μ m and the length and width of the gate are 3 and 10 μ m, respectively. Figure 3(a) is the two-terminal resistance, R_{sd}, as a function of V_g observed under the in-plane magnetic field of 6000 (gs) applied parallel to the current. Simultaneously, we confirmed the magneto-resistance hysteresis between

-200 and +200 Gs resemble to that reported by Hammer et al [3] in the similar sample. Corresponding Fourier analysis result is shown in Fig. 3(b), where the main peak is found at 5 (1/V). In Fig. 3(c), the R_{sd} calculated by assuming the V_g dependency of α_{zero} shown in Fig. 1 is shown, the oscillation of which is due to spin-presession by the spin-orbit interaction. Fig. 3(d) is a Fourier analysis result for the calculated oscillation. As can be seen, the result exihibits a main peak at 5 (1/V), which almost conincides with that observed in the experiment (Fig. 3(b)). This result strongly supports a possible realization of fundamental spin-FET operation in our sample.

3 Summary

In normal $In_{0.75}Ga_{0.25}As$ / $In_{0.75}Al_{0.25}As$ heterojunction, we have confirmed spin-orbit coupling parameters of up to 30 (x10^{-12}eVm) as well as their tunability against the front-gate voltage. In the spin-FET composed of this heterojunction and NiFe source-drain electrodes, we observed gate-dependent two terminal resistance which has a strong frequency component as that expected assuming spin-presession due to the spin-orbit coupling.

References
1. S.Gozu et al, Jpn.J.Appl.Phys.,37,L1501 (1998).
2. P.Pfeffer et al, Phys.Rev.B59,R5312 (1998).
3. P.R.Hammer et al, Phys.Rev.Lett.,83,203 (1999).

Field domains in semiconductor superlattices: Dynamic scenarios of multistable switching

A. Amann[1], A. Wacker[1], L. L. Bonilla[2], E. Schöll[1,3]

[1] Institut für Theoretische Physik, Technische Universität Berlin, Hardenbergstr. 36, 10623 Berlin, Germany
[2] Escuela Politécnica Superior, Universidad Carlos III, Avenida de la Universidad 20, 28911 Leganés, Spain
[3] Center of Nonlinear and Complex Systems, Duke University, Box 90305, Durham, NC 27708-0305, USA

Abstract The dynamical switching behavior between different branches of the current–voltage characteristic in semiconductor superlattices and the current response during a voltage sweep over several branches are analyzed. Based on a microscopic model we simulate the dynamics of the charge distribution in the superlattice. The complex behavior observed in recent experiments is explained by different mechanisms for the domain boundary formation.

1 Introduction

Weakly coupled semiconductor superlattices exhibit a multistable current–voltage characteristic with many branches associated with static field domains. Recent experiments [1, 2] on the current response of superlattices at non-stationary external voltage bias have revealed various scenarios which are not well understood.

Based on a microscopic model for sequential tunneling in weakly coupled superlattices [3–5], we find that different mechanisms for the domain formation are effective. Our simulations consistently explain the dynamic switching behavior both for application of voltage steps $V_{\rm step}$ at individual points of the current–voltage characteristic, and for fast sweeps over the whole characteristic. For small positive values $V_{\rm step}$ the charge distribution located in the domain boundary moves upstream *against* the direction of the charge transport to form the new boundary. The current relaxes monotonically to its new stationary value on the up-sweep branch. For larger positive $V_{\rm step}$ a dipole wave provides the mechanism for the domain relocation, resulting in a non-monotonic current response. Hereby the multistable system is switched to a different operating point located on the down-sweep branch, although the voltage step is in up-sweep direction. For negative values of $V_{\rm step}$ the charge distributions always moves in the direction of the field. Without fitting parameters our model reproduces quantitatively the experimental results.

We also explain the frequency dependence of the current–voltage characteristic of superlattices studied in [1]. The experimentally observed softening of the current–voltage characteristic and the overall increase of the current is reproduced by considering up-sweeps which are fast in comparison with the domain relocation time. For even faster up-sweeps we observe a transition to a dipole domain formation, leading to an N-shaped current—voltage characteristic for which there is also experimental evidence in Ref. 1.

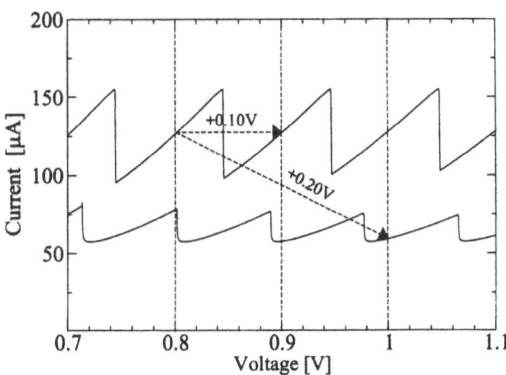

Fig. 1 Simulated current–voltage characteristic of a GaAs/ AlAs superlattice (upper trace: up-sweep; lower trace: down-sweep). The arrows indicate the switching processes discussed in Sec. 2.

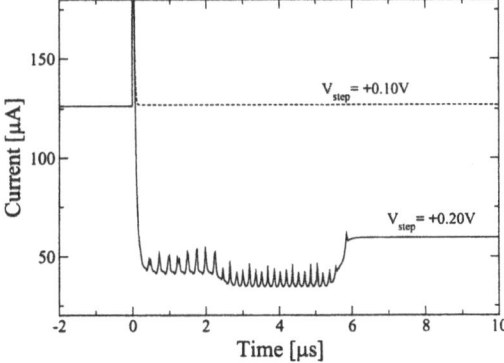

Fig. 2 Current response vs. time. At $t = 0$ a voltage step $V_{\rm step}$ is applied on top of an initial voltage $V_i = 0.8$V.

2 Dynamic switching

Within the model for sequential tunneling [5], we simulated an $N = 40$ period superlattice with barrier width $b = 4$nm, well width $w = 9$nm, 2D doping density $N_A^{2D} = 1.5 \times 10^{11}$cm^{-2}, scattering width $\Gamma = 10$meV and cross section 14000μm^2 at 5K numerically . The parameters are chosen to be similar to the superlattice in [2]. The boundaries are modeled by Ohmic contacts with conductivity $\sigma = 0.01(\Omega$m$)^{-1}$, additionally an external resistor of $R = 100\Omega$ was assumed.

The static current–voltage characteristic in Fig. 1 exhibits the typical sawtooth pattern of stationary domains in superlattices. The current response vs. time for two different voltage steps is shown in Fig. 2. While for $V_{\rm step} = 0.1$V the current relaxes after a sharp initial peak to a final value on the up-sweep branch, for $V_{\rm step} = 0.2$V

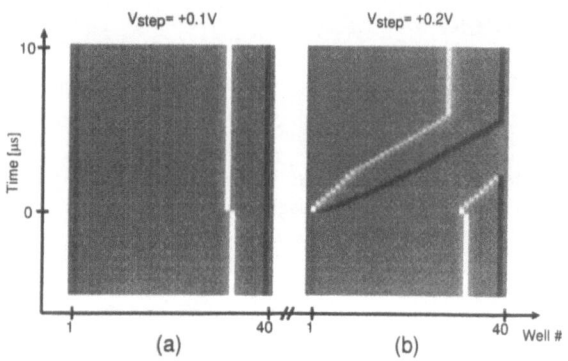

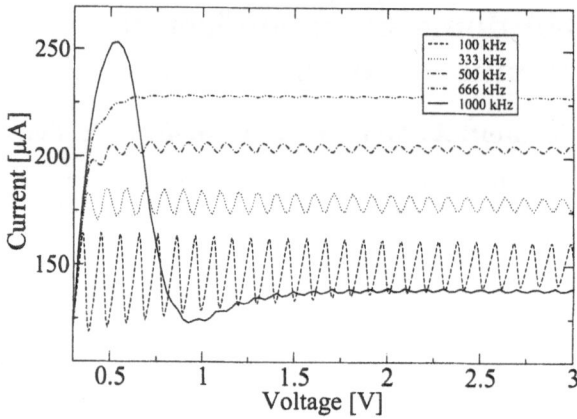

Fig. 3 Evolution of the electron densities in the quantum wells during the switching process for two different switching voltages V_{step}. High and low electron densities are indicated by white and black, respectively. Well 1 is located at the emitter, well 40 at the collector. At $t = 0$ a voltage step V_{step} is applied on top of an initial voltage $V_i = 0.8V$.

Fig. 4 Current–voltage characteristic during a triangular voltage sweep at various frequencies. The sweep starts at $V_i = 0.3V$ and ends at $V_f = 3.0V$.

the current response is more complicated. After a peak at time $t = 0$ the current drops to a value below the one corresponding to the static current–voltage characteristic. For $0 < t < 6\mu s$, the current exhibits small oscillations with an amplitude of about $10\mu A$ at a frequency of $15MHz$. At $t > 6\mu s$ the current reaches a stationary value which is now on the down-sweep branch of the static current–voltage characteristic.

The reason for this different behavior is revealed by the dynamic evolution of the carrier density distribution, Fig. 3. We observe that two different mechanisms for the domain formation are effective. In Fig. 3 (a) the new domain boundary is formed by a shift of the electron accumulation front by one superlattice period towards the emitter (cathode), against the direction of the charge transport. In Fig. 3 (b) a dipole domain consisting of an accumulation and a depletion front is formed at the emitter. While the accumulation front of the dipole relocates as the new domain boundary, the old domain boundary and the depletion front traverse the superlattice to the collector where they vanish. Since during this relocation process the new domain boundary moves towards the collector, this corresponds to a voltage down-sweep under quasi-static conditions. Accordingly, the final current corresponds to a value on the down-sweep branch, although the voltage is increasing throughout the switching process.

3 Fast sweeping

In recent experiments [1], the current response of weakly coupled superlattices was studied under triangular voltage sweeps over various branches. It was found that the characteristic saw-tooth pattern of the stationary current–voltage characteristic softens and the overall current increases with increasing sweep frequency. For higher frequency the stationary pattern disappears.

In Fig. 4 this behavior is numerically reproduced for the superlattice discussed in Sec. 2. For frequencies up to about 100kHz the current vs. voltage essentially fol-

lows the static characteristic. For higher frequencies we observe a softening of the sawtooth pattern, which is due to the finite relaxation time for a transition to the next branch. At the same time the overall current rises. The reason for this behavior is that the velocity of the electron accumulation is connected with the current through the superlattice [6]. By sweeping the external voltage, we impose a constraint on the velocity of the accumulation front, by which the corresponding current is determined. For even higher frequency, the static pattern completely disappears. At about 1000kHz the accumulation front is too slow to follow the sweep. Instead, a dipole domain, as discussed in Sec. 2 for switching at higher voltages, now appears. The current decreases dramatically as also observed during the dipole switching mechanism of Fig. 3.

4 Conclusion

We have shown that two different domain formation mechanisms are effective in superlattices with static field domains. If the increase of the external voltage bias V_{step} is small, the usual well-to-well hopping of the domain boundary occurs. For a larger V_{step}, a dipole wave traversing the superlattice from the emitter provides the mechanism for the domain boundary relocation. This mechanism can also be triggered by a fast sweep of the external voltage over several branches.

References

1. Y. Shimada and K. Hirakawa, Jpn. J. Appl. Phys. **30**, (1997) 1944.
2. K. J. Luo, H. T. Grahn, and K. H. Ploog, Phys. Rev. B **57**, (1998) 6838.
3. F. Prengel, A. Wacker, and E. Schöll, Phys. Rev. B **50**, (1994) 1705, ibid **52**, (1995) 11518.
4. L. L. Bonilla, J. Galán, J. A. Cuesta, F. C. Martínez, and J. M. Molera, Phys. Rev. B **50**, (1994) 8644.
5. A. Wacker, in *Theory of Transport Properties of Semiconductor Nanostructures*, edited by E. Schöll (Chapman and Hall, London, 1998), Chap. 10.
6. A. Carpio, L. L. Bonilla, A. Wacker, and E. Schöll, Phys. Rev. E **61**, (2000) 4866.

Proc. 25th Int. Conf. Phys. Semicond., Osaka 2000 (Eds. N. Miura and T. Ando)

Angular dependent magnetoresistance oscillations in semiconductor superlattice with incoherent interlayer coupling

M. Kuraguchi[1], E. Ohmichi[1], T. Osada[1], Y. Shiraki[2]

[1] Institute for Solid State Physics, University of Tokyo, 5-1-5 Kashiwanoha, Kashiwa, Chiba 277-8581, Japan
[2] Research Center for Advanced Science and Technology, University of Tokyo, 4-6-1 Komaba, Meguro-ku, Tokyo 153-8904, Japan

Abstract We have studied the angular dependence of the interlayer magnetoresistance in semiconductor superlattices whose interlayer coupling is systematically changed. We found that when the system loses its interlayer coherence, the angular dependence of the background magnetoresistance drastically changes, but the angular dependent magnetoresistance oscillations (AMRO) still survive. In addition, we found that AMRO appear even in the quantum limit where only the lowest Landau level is occupied and the conventional AMRO cannot be expected. We also discuss the orbital confinement into a single well under magnetic fields parallel to the layers.

1 Introduction

Angular dependent magnetoresistance oscillations (AMRO) have been discovered in various quasi-two-dimensional conductors such as organic conductors, semiconductor superlattices and layered oxides. AMRO have been explained by semiclassical theories as the topological effect of the warped cylindrical Fermi surface, supposing that the interlayer coupling is coherent. Recently, McKenzie et al. proposed a new theory [1]. According to their theory, AMRO appear whether or not the interlayer coupling is coherent, while the peak effect at $\theta = \pm 90°$ requires the interlayer coherence. To elucidate the validity of their theory, semiconductor superlattices are thought to be the best subjects, because the simple Kronig-Penny model describes well the band structures and the interlayer coupling strength can be controlled by designing the superlattice structures. In AlGaAs/GaAs superlattices, Yagi et al. firstly reported appearance of AMRO [2] and Kawamura et al. found the peak effects [3]. In this paper, we report the angular dependence of the interlayer magnetoresistnce of semiconductor superlattices whose interlayer coupling is intentionally controlled.

2 Experiment

Superlattice samples were grown by molecular beam epitaxy on conducting substrates. To measure the interlayer transport, we patterned the samples into post-type devices. Table 1 shows the sample parameters. The first miniband width $4t_c$ was estimated using the Kronig-Penny model, and the carrier density n and the in-plane scattering time τ were obtained from Shubnikov-de Haas oscillations. The interlayer coupling was changed by the barrier height and width. In sample #2, $4t_c$ is smaller than $\hbar/\tau$, so that the interlayer transport is expected to be incoherent. On the other hand, the interlayer coupling of sample #1 is coherent because $4t_c > \hbar/\tau$.

3 Results and discussion

Fig. 1(a) shows the typical angular dependence of the interlayer magnetoresistance in sample #1 with coherent interlayer coupling. The measurements were performed at rather high temperature in order to avoid the super-position of Shubnikov-de Haas oscillations, following the previous works [2,3]. A peak structure appeared at the field orientation parallel to the layers ($\theta = \pm 90°$). In addition, a broad peak which seemed to be a peak of AMRO was observed at about $\theta = \pm 40°$ for each field strength. Its position almost satisfies the following Yamaji condition for the conventional AMRO [4],

$$ck_F \tan\theta = \pi\left(i - 1/4\right), \tag{1}$$

where c, k_F and i are the spacing between neighboring layers, the Fermi wave number in a layer, and an integer, respectively. On the other hand, Fig. 1(b) shows the angular dependence of the magnetoresistance in sample #2 with incoherent interlayer coupling. In this incoherent sample, no peak effect was observed up to 13 T while clear AMRO was visible satisfing the Yamaji condition for each field strength, in accordance with McKenzie's prediction.

In this way, we observed AMRO in both coherent and incoherent samples. Therefore, the present result seems to support McKenzie's theory. However, all AMRO appearing in Fig. 1 cannot be identified to the conventional AMRO. At the AMRO peaks observed in higher fields, the normal field components are large enough for the

Table 1 Sample parameters. $4t_c$, $\hbar/\tau$ and n are the first miniband width, the rate of the in-plane scattering and the carrier density per layer, respectively.

Sample	Well Width (Å)	Barrier Width (Å)	Barrier Height (meV)	n ($\times 10^{11} cm^{-2}$/layer)	$4t_c$ (meV)		$\hbar/\tau$ (meV)	interlayer coupling
#1	188	38	153	1.7	2.14	>	0.98	coherent
#2	150	150	85	6.0	0.120	<	1.03	incoherent

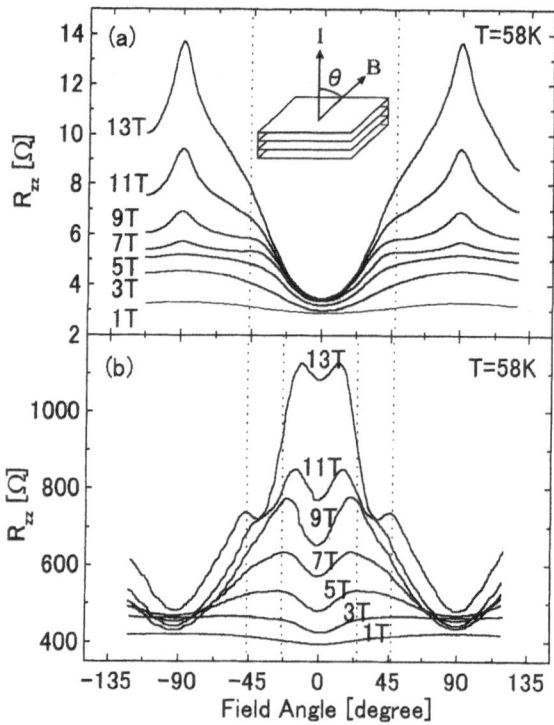

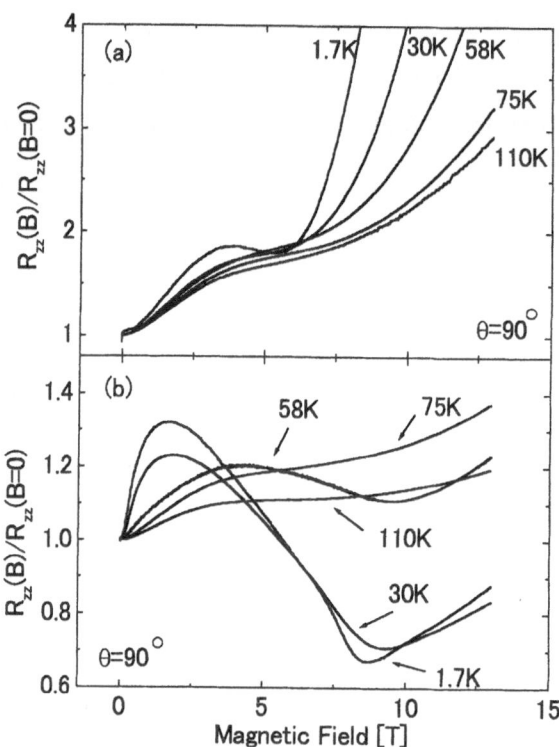

Fig. 1 (a) Angular dependence of the interlayer magnetoresistance of sample #1. (b) Interlayer magnetoresistance of sample #2 at the same temperature. Vertical dotted lines indicate calculated Yamaji's angles.

Fig. 2 Magnetic field dependence of the interlayer magnetoresistance of sample #1 (a) and sample #2 (b) at $\theta = 90°$.

system to enter the quantum limit. In the quantum limit, only the lowest Landau level, whose width does not oscillate with field rotation is occupied, so that we cannot expect the conventional AMRO. This indicates that the high field AMRO have different origins from the conventional ones. To explain the novel type AMRO in the quantum limit, we might have to take into account the finite layer thickness.

Next, we switch to the angular dependence of the background magnetoresistance. The background magnetoresistance of sample #2 exhibits an anomalous behavior as shown in Fig. 1(b). In the low magnetic field region, the magnetoresistance shows a usual behavior like sample #1: it takes the maximum and minimum values at $\theta = \pm 90°$ and $\theta = 0°$, respectively. However, the angular dependence of the background resistance in the high magnetic field region is opposite to the one in the low field region. This feature can not be explained by the semiclassical theory, since the Lorentz force should work most effectively at $\theta = \pm 90°$. It seems that the normal component of the magnetic field dominantly contributes to the interlayer magnetoresistance in this sample.

Fig. 2(a) and (b) show the magnetic field dependence of the interlayer resistance of sample #1 and sample #2 at $\theta = 90°$, respectively. We found a resistance dip at about 6 T and 9 T in sample #1 and #2, respectively. The position of this dip structure almost coincides with the field where the cyclotron orbit becomes the same

size as the well width (6.1 T for sample #1 and 10 T for sample #2). Especially, in the sample #2, the interlayer resistance showed very large decrease around this dip. This decrease causes the anomalous angular dependence of background magnetoresistance in the incoherent system.

Finally, we have to note the effect of temperature. The measurements were carried out at rather high temperatures to kill the Shubnikov-de Haas oscillations. As a result, the temperature became too high to resolve the interlayer band structures ($T > 4t_c$). Nevertheless, AMRO survive in the systems with both coherent and incoherent interlayer coupling and the peak structure also survives in the coherent system. Therefore, the observed angular effects seem to be insensitive to temperature, in other words, the thermal distribution. This may reflect the fact that these effects originate from the modification of properties of single electronic state, such as the averaged velocity or the tunneling probability.

The authors greatly acknowledge valuable discussions with Mr. M. Kawamura. This work was financially supported by the Torey Science Foundation and Grant-in-Aid for Scientific Research from the Japan Society for the Promotion of Science.

References

1. R. H. McKenzie et al., Phys. Rev. Lett. **81**, (1998) 4492.
2. R. Yagi et al., J. Phys. Soc. Jpn. **60**, (1991) 3784.
3. M. Kawamura, et al., Physica B **249-251**, (1998) 882.
4. K. Yamaji, J. Phys. Soc. Jpn. **58**, (1989) 1520.

Optical and transport properties of modulation doped InAs/GaAs superlattices

V.A. Kulbachinskii[1], R.A. Lunin[1], V.G. Kytin[1], A.V. Golikov[1], V.A. Rogozin[1], V.G. Mokerov[2], Yu.V. Fedorov[2], A.V. Hook[2]

[1] Low Temperature Physics Department, Moscow State University, 119899, Moscow, Russia
[2] Institute of Radioengineering and Electronics, RAS, Moscow, Russia

Abstract The photoluminescence, magnetoresistance and Hall effect have been investigated in modulation doped InAs/GaAs superlattices as a function of InAs layer thickness Q in the range $0.33 \leq Q \leq 2.7$ monolayer (ML). The investigation of photoluminescence spectra has shown that when InAs layer thickness achieves 2.7 ML the quantum dots are formed. The dependence of the anisotropy of resistivity on thickness of InAs layers has peculiarities at integral numbers of monolayers.

1 Introduction

In recent years considerable attention has been focussed on the research of the self-organized ensembles of quantum dots, that is the quasi-zero-dimensional objects (with dimensions 5-20 nm) shaped during heteroepitaxial growth in case of misalignment of lattice parameters with a substrate [1-3]. While the mechanism of a nucleation and formation of quantum dots and optical properties of undoped structures with quantum dots have been widely studied, the optical and electrical properties of doped structures at initial stage of quantum dot formation have not been investigated enough. In the present work the investigation of a photoluminescence at the temperature 77 K, magnetoresistance and Hall effect in modulation doped InAs/GaAs superlattices with very thin InAs layers in magnetic fields up to 8 T in the temperature range 0.4-4.2 K was conducted.

2 Samples

The structures were grown by MBE at 490^0C on semi-insulating (001) GaAs substrates. The samples consisted of 1μm thick GaAs buffer layer, superlattice InAs/GaAs (with different periods and total thickness 14 nm), a 10 nm thick spacer layer $Al_{0.2}Ga_{0.8}As$, a 35 nm thick Si-doped layer $Al_{0.2}Ga_{0.8}As$ and a 6 nm thick GaAs cap layer. In investigated structures the effective thickness of InAs layers was changed from 0.33 ML up to 2.7 ML. The thickness of GaAs layers in superlattice was also proportionally changed to keep the mean composition of the superlattice equivalent to solid solution $In_{0.16}Ga_{0.84}As$.

3 Results and discussion

The photoluminescence spectra of samples with thickness less than 2 monolayers showed a low-energy peak with a maximum at photon energy 1.356–1.375 eV and a high-energy peak with a maximum at photon energy 1.406–1.434 eV (Fig. 1). These two peaks correspond to optical transitions from the two electronic subbands to hole subband. However, when the effective thickness of InAs layers equal to or exceeds 2.7 monolayers, the photoluminescence spectrum essentially transforms. A new broad and intensive photoluminescence peak with a maximum at 1.265 eV appears in long-wavelength region of the photoluminescence spectrum. As shown in Ref. [1], such change of photoluminescence spectrum may be ascribed to the phase transition from two-dimensional layer-by-layer growth to the formation of a vertically stacked quantum dots. The second feature of the photoluminescence spectra is not monotonous dependence of the intensity I_{PL} on the InAs layer thickness Q. Maximal intensity was observed for Q=0.33 ML.

A negative magnetoresistance was observed at low temperatures in low magnetic fields B<0.1T in all samples. In high magnetic fields the Shubnikov - de Haas effect and quantum Hall effect was observed (Figs. 2,3). The Hall electron concentration is equal to $(2-15)*10^{11}$ cm^{-2} in different samples at temperature 4.2 K and coincides with concentration determined from the Shubnikov-de Haas effect. The maximum value of electron mobility 9400 cm^2/Vs is observed in a sample with InAs layer thickness 0.33 ML. When the thickness of InAs layers increases the mobility falls down to approximately 3000 cm^2/Vs at temperature 4.2 K. The relatively high mobility in sample with minimum submonolayer thickness Q=0.33 ML of InAs may be explained by a small fluctuation of elastic strains and most perfect crystal lattice. The maximal intensity of photoluminescence also was observed in this sample.

In all samples the anisotropy of conductivity is

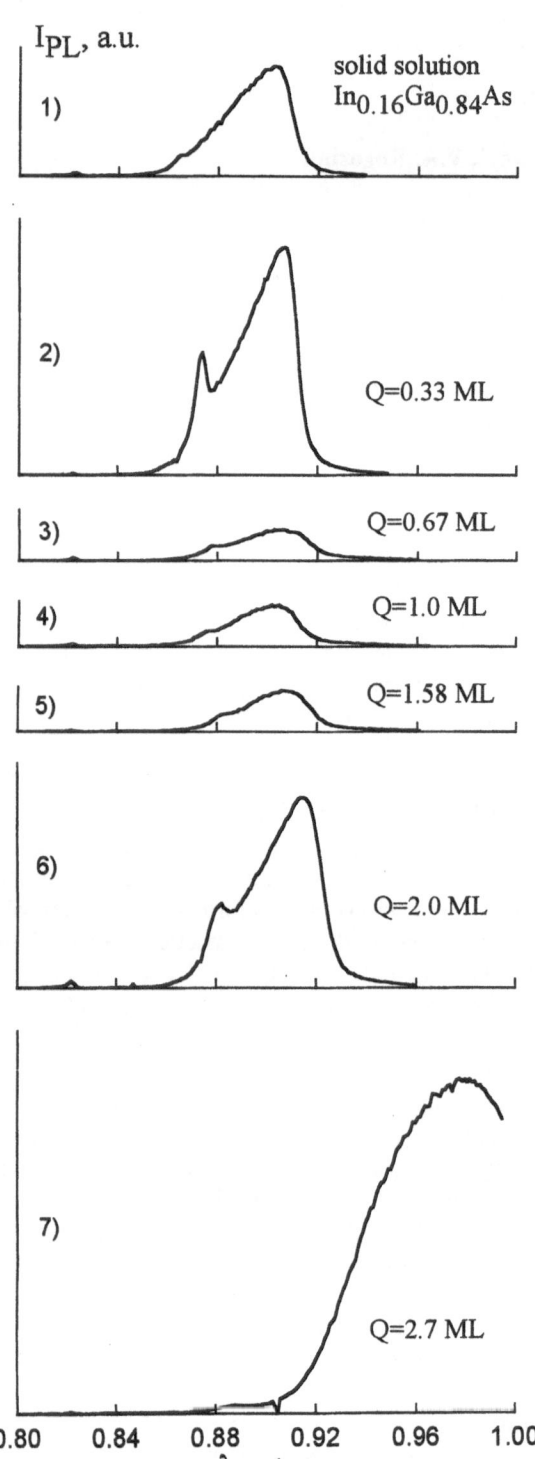

Fig. 1 Photoluminescence spectra of modulation doped InAs/GaAs superlattices with InAs layer thickness Q from 0.33 up to 2.7 monolayer and the mean composition $In_{0.16}Ga_{0.84}As$ sample.

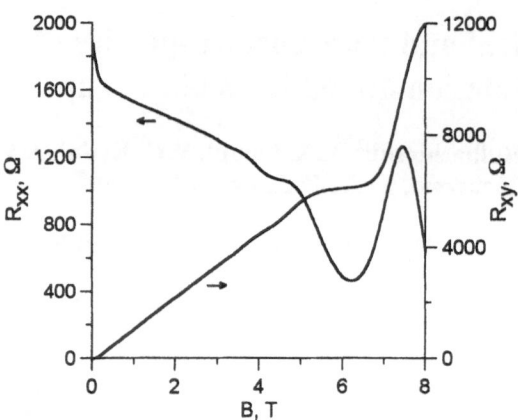

Fig. 2 Transverse magnetoresistance R_{xx} and Hall resistance R_{xy} for sample with InAs layer thickness Q=1.58 ML at T=0.4 K.

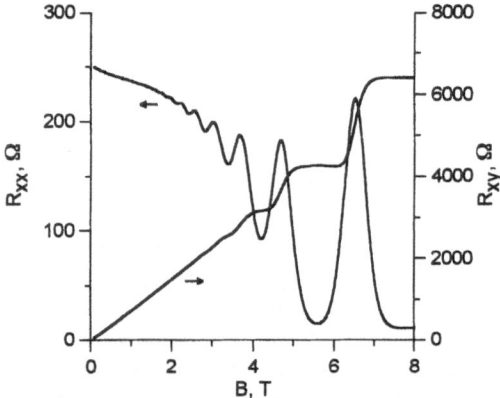

Fig. 3 Transverse magnetoresistance R_{xx} and Hall resistance R_{xy} for sample with solid solution $In_{0.16}Ga_{0.84}As$ T=0.4 K.

observed. The anisotropic electron mobility may correlate to the asymmetric dislocation distribution [4]. The dependence of anisotropy on thickness of InAs layers has peculiarities at integral numbers of monolayers.

The work was supported by the RFBR (Grant N 00-02-17493).

References
1. N.N. Ledentsov, V.M. Ustinov, V.A. Shchukin *et al.*, Fiz. Tekh. Poluprovodn. **32**, (1998) 385 (Semiconductors **32**, (1998) 343).
2. D. Bimberg, V.A. Shchukin, N.N. Ledentsov *et al.*, Appl. Surf. Sci. **130-132**, (1998) 713.
3. V.A. Kulbachinskii, V.G. Kytin, R.A. Lunin *et al.*, Fiz. Tekh. Poluprovodn. **33**, (1999) 316 (Semiconductors **33**, (1999) 318).
4. T. Schweizer, K. Kohler, W. Rothemund et al., Appl. Appl. Phys. Lett. **59**, (1991) 2736.

Weak localization effects in a wide parabolic quantum well

N. M. Sotomayor[1], G. M. Gusev[1], J. R. Leite[1], N. T. Moshegov[2], A. I. Toropov[2]

[1] Instituto de Física da Universidade de São Paulo, CP 66318, CEP 05315-970, São Paulo, SP, Brasil, e-mail: sotomayo@macbeth.if.usp.br

[2] Institute of Semiconductor Physics, Novosibirsk, Russia

Abstract   We measured weak localization magnetoresistance peak in a wide parabolic quantum well with 5 occupied subbands. We find almost linear line shape of the negative magnetoresistance in perpendicular magnetic field, and quadratic dependence in parallel magnetic field. We assume that this unusual line shape for the weak localization peak for different experimental geometries is related to the area distribution function of closed path, which strongly depends on the specific type of scatterers. In perpendicular B magnetoresistance is related to the closed electron trajectories, which are formed due to the in-plane large angle scattering in the presence of a smooth disordered potential. In parallel B the electron motion perpendicular to the layer should be taken into account. In this geometry the diffuse surface scattering leads to the enhanced probability of the small areas enclosed by the electron trajectories.

1 Introduction

Recently the interference of the electron waves scattered along closed trajectories in opposite directions, which leads to the quantum localization, called weak localization, has attracted much attention in electronic billiard systems [1]. Magnetic field introduces the phase difference between clockwise and counterclockwise trajectories, destroys the interference and leads to negative magnetoresistance. In electron billiards a striking difference in the shape of the weak localization magnetoresistance peak centered at $B = 0$ for chaotic and nonchaotic scattering has been found [1]. In the case of the nonchaotic cavities, a linear dependence with magnetic field in contrast to Lorentzian behavior in chaotic cavities has been observed. As was shown in [2] and [3], quantum corrections to the conductivity can be represented as a sum of contributions of closed paths with N collisions which has area S. It has been assumed that the difference in the shape of the negative magnetoresistance in chaotic and nonchaotic cavities was originated from the difference in the areas distribution enclosed by electron trajectories. These differences can occur not only in electron billiard systems, but also in system with specific type of scattering. It can be result from some correlation in distribution of scatterers, scattering anisotropy, or long range potential fluctuations in two-dimensional system.

2 Results and discussions

The samples studied were parabolic quantum wells grown by molecular beam epitaxy with electron mobility $\mu = 65 \times 10^3 \text{cm}^2/\text{Vs}$ and density $n_s = 3.9 \times 10^{11} \text{cm}^{-2}$. It cor-

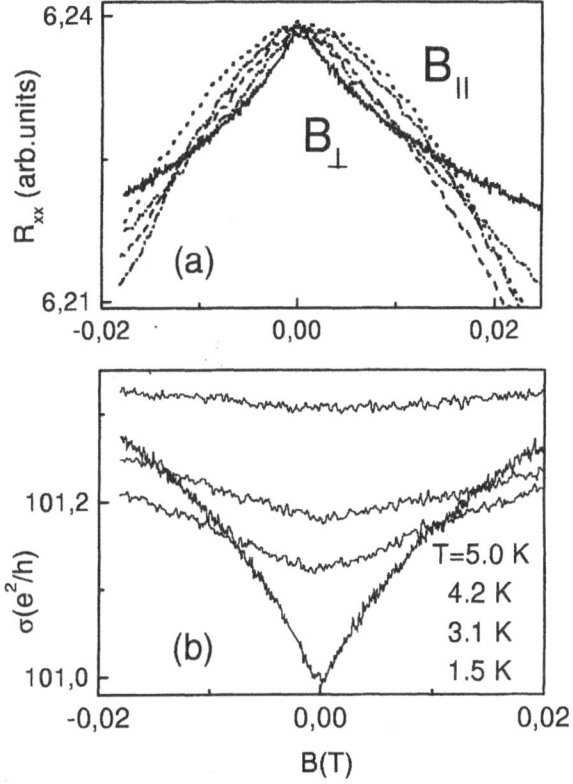

Fig. 1 (a) The magnetoresistance for wide parabolic quantum well for different tilt angles. Solid line $\Theta = 90^o$, dotted line 0°. (b) The temperature evolution of the magnetoconductance for the wide parabolic well in perpendicular magnetic field.

responds to the classical width of the quasi-three-dimensional electron gas $W_e = 1900 A$. Self-consistent calculations of the energy levels in well demonstrate 5 occupied subbands. The magnetic field dependence of the conductivity for different angles Θ between the field and the normal of the quantum well plane is shown in Fig. 1a. Fig. 1b shows the temperature dependence of the magnetoconductance traces, which supports that magnetoconductance has a quantum origin, because it is rapidly suppressed by the temperature. We see linear line shape for the weak localization magnetoresistance peak in perpendicular magnetic field, which gradually transforms to almost quadratic B dependence in parallel magnetic field. We are not able to fit experimental trace with theoretical curves described by two–dimensional theory for isotropic scattering by randomly distributed scatteres with a short–range potential[4]. We assume that this deviation may result from the different contribution of the short range and long range scatterers in the in–plane well motion and motion in direction perpendicular to the well. For analysis of negative magnetoresistance due to weak localization suppression we have to consider new approach, which based on the information about the statistics of closed paths during the electron scattering. This method was developed for ballistic electron billiards [2] [3], however, as was shown recently, may be used for diffusive systems too [5]. In this case the magnetoconductance is determined by the probability density $W(S)dS$ of return after N collisions along a trajectory, which encloses the area dS. Thus, the area distribution function $W(S)$ contains the all information about behaviour of the negative magnetoresistance due to quantum interference effects. We calculate this area distribution function of closed paths for randomly distributed scattering centers with potential $V(x,y) = V_i exp[-(x_i - x)^2/\xi^2]exp[-(y_i - y)^2/\xi^2]$, where V_i, x_i and y_i are varied with the use of a random number generator. 2D system is represented as a lattice M×M, with total number of scatterers about 10^4. Figure 2 shows the magnetoconductance calculated for the impurity scattering with short and long range potentials. The inset shows the dependence of the probability to form closed path W on the area S for both type of the potential. We see the deviation from the S^{-1} dependence of the area distribution function of closed path S, which holds in the short range scatterers regime in two–dimensional systems, the probability for small areas decreases, and for large areas increases. Such deviation can be explained by the suppression of the small angle scattering events in long range potential. Such suppression also exist in nonchaotic electron billiard systems and leads to the quasi–linear shape of the weak localization magnetoresistance peak [1]. Importance of the large angle scattering in modulation doping samples, such as HEMT transistors and quantum wells, has been found from the measurements of the damping of the Shubnikov de Haas oscillations. Thus, the excess of the closed paths with large area in comparison with short range scatterers regime leads to increase of the low field part of the magnetoconductance, as we see in Fig.2. Now let us consider the situation, when magnetic field is parallel to the quantum well plane. In this case electrons are scattered by the surface. Weak localization in thing films with diffuse surface scattering in the limits of small and large magnetic field has been studied analytically in [6] and numerically in [7]. We perform numerical calculation in the diffuse scattering case. Fig.2 shows the results of the simulations. We see, that now the probability to form small area is larger, than for isotropic short range scat-

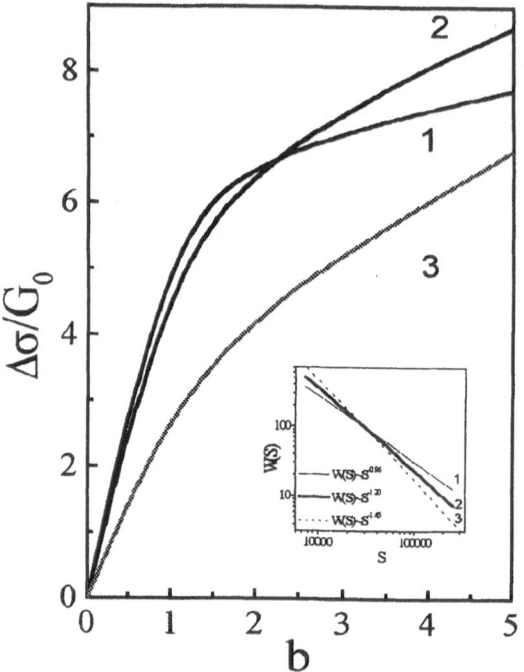

Fig. 2 Magnetoconductance in the presence of the different type of scatterers. 1- short range scatterers, 2- long range scatterers, 3- diffuse surface scatterers, b=B/B$_0$, B$_0$=hc/(2el^2), l is the mean free path. Inset- Area distribuition function of closed paths $W(s)$.

terers case. It results to decrease of the low field part of the magnetoconductance.

Acknowledgments

This work was supported by CNPq and FAPESP

References

1. A. M. Chang, H. U. Baranger, L. N. Pfeifer, K. W. West, Phys. Rev. Lett., **73**, (1994) 15.
2. H. U. Baranger, R. A. Jalabert, A. D. Stone, Phys. Rev. Lett., **70**, (1993) 3876.
3. G. M. Gusev, Z. D. Kvon, A. G. Pogosov, P. Basmaji, JETP, **83**, (1996), 375.
4. P. A. Lee, T. V. Ramakrishnan, Rev. Mod. Phys., **57**, (1985), 287.
5. G. M. Minkov, A. V. Germanenko, V. A. Larionova, S. A. Negashev, I. V. Gornyi, Phys. Rev. B, **61**, (2000), 13164.
6. V.K. Dugaev and D.E. Khmelnitskii, Sov.Phys. JETP, 59, 1038 (1984).
7. C.W.J. Beenakker and H. van Houton, Phys. Rev.B, 38, 3232 (1987).

Proc. 25th Int. Conf. Phys. Semicond., Osaka 2000 (Eds. N. Miura and T. Ando)

809

Magnetotransport properties of multisubband semiconductor structures

N.S. Averkiev[1], L.E. Golub[1,2], S.A. Tarasenko[1], M. Willander[2]

[1] A.F. Ioffe Physico-Technical Institute, Russian Academy of Sciences, 194021 St. Petersburg, Russia
e-mail: nsab@les.ioffe.rssi.ru
[2] Physical Electronics and Photonics, Department of Physics, Chalmers University of Technology and University of Göteborg,
SE-412 96 Göteborg, Sweden

Abstract Magnetotransport in semiconductor structures with several occupied subbands is investigated theoretically. The conductivity tensor is calculated in the regimes of weak localization. classical magnetic field and Shubnikov-de Haas oscillations. The effect of intersubband scattering on magnetoresistance is clearly demonstrated.

1 Introduction

A magnetic field is well known to effect on kinetic phenomena in semiconductor structures if the magnetic length, l_B. is comparable with a characteristic length, determining kinetic properties of a semiconductor. When several size-quantized subbands are occupied in a semiconductor structure, each of the parameters mentioned above depends on a subband number. When l_B becomes consequently comparable with each of the characteristic lengths. the peculiarities appear in magnetotransport effects. Moreover. intersubband scattering may give new characteristic times and lengths. In this report. we have studied magnetoresistance in the regimes of weak localization. classical magnetotransport and Shubnikov-de Haas oscillations in a multisubband system. The case of two occupied subbands was investigated in detail. For simplicity, spin effects are omitted.

2 Weak localization

The weak localization phenomenon consists in the interference of waves propagating along the same path in opposite directions. With increasing of magnetic field, l_B is first much larger than the mean free path, l. The motion along the paths of corresponding size has diffusion character, and the anomalous magnetoresistance takes place in fields so that l_B is equal to the length of dephasing. Intersubband scattering leads to the non-additive contributions to the conductivity from different subbands and to the dependence of magnetoresistance on the intersubband scattering times and on the dephasing times in each subband. The results are published in Refs. [1–3].

With further increasing of a magnetic field. weak localization is destroyed at the paths with the size of order of l. This weak localization regime is called non-diffusion. Interlevel scattering leads to the magnetoconductivity correction determined by relations between the total relaxation times in the subbands. τ_j. and the time of intersubband scattering. $\tau_{jj'}$.

Fig. 1 presents the dependence of weak localization correction to conductivity on magnetic field for a sys-

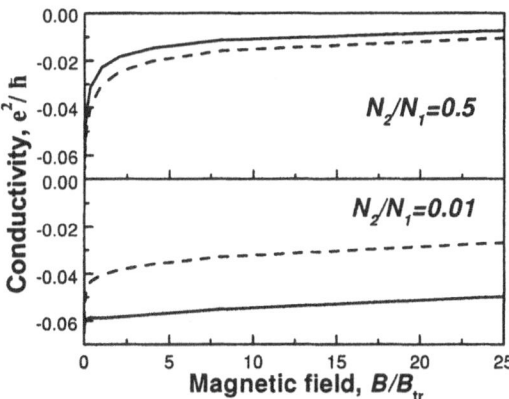

Fig. 1 Anomalous magnetoresistance at different subband occupations. $\tau_1 = \tau_2$, solid curves: $\tau_{12}/\tau_1 = 2$. dashed: $\tau_{12}/\tau_1 = 10$.

tem with two size-quantized levels at intensive (solid curves) and weak (dashed) intersubband scattering. N_j are the subband occupations and $B_{\mathrm{tr}} = \hbar/el_1^2$. where l_1 is the mean free path for the 1st subband. One can see the intersubband transition influences essentially on the anomalous magnetoresistance: the weak localization corrections at $N_2/N_1 = 0.01$ differ by the factor of 1.5.

3 Classical magnetoresistance

In stronger magnetic fields. when $l_B \sim \sqrt{l\lambda} \ll l$ ($\omega_c\tau_j \lesssim 1$). a classical motion of electrons on cyclotron orbits takes place. Here λ is the electron wavelength and ω_c is the cyclotron frequency. The calculations of the conductivity tensor in a classical magnetic field were performed with the method of a kinetic equation. We took into account both intra- and intersubband scattering. Doing so. we obtained the system of coupled kinetic equations for distribution functions in subbands under magnetic and electric fields. Then calculating the current density. we have got the conductivity tensor. For two occupied subbands. it has the form:

$$\sigma_{xx} = \frac{e^2}{m}\left[(N_1 A_2 + N_2 A_1 + 2C\sqrt{N_1 N_2})\right. \tag{1}$$
$$\cdot(A_1 A_2 - C^2 - \omega_c^2) + (N_1 + N_2)(A_1 + A_2)\,\omega_c^2\right]$$
$$/[(A_1 A_2 - C^2 - \omega_c^2)^2 + (A_1 + A_2)^2\omega_c^2].$$

$$\sigma_{xy} = -\frac{e^2}{m}\,\omega_c[(N_1 A_2 + N_2 A_1 + 2C\sqrt{N_1 N_2}) \tag{2}$$
$$\cdot(A_1 + A_2) - (N_1 + N_2)(A_1 A_2 - C^2 - \omega_c^2)]$$
$$/[(A_1 A_2 - C^2 - \omega_c^2)^2 + (A_1 + A_2)^2\omega_c^2],$$

$$A_j = 1/\tau_j - \oint \frac{d\varphi}{2\pi} W_{jj}(\varphi) \cos\varphi , \qquad (3)$$

$$C = \oint \frac{d\varphi}{2\pi} W_{12}(\varphi) \cos\varphi ,$$

where $W_{jj'}(\varphi)$ are the probabilities of intrasubband ($j = j'$) or intersubband ($j \neq j'$) scattering by an angle φ at the Fermi surface.

It is seen that one cannot present the conductivity tensor components as a sum of two terms proportional to the carrier concentrations in subbands. Due to non-zero first Fourier harmonic of the intersubband scattering probability, C, there is the contribution, proportional to $\sqrt{N_1 N_2}$. This term may be small at, for example, short-range scattering. In this case the size-quantized subbands contribute quasi-independently to the conductivity. However, in general case, the analysis of experimental data has to be performed with help of the equations (1), (2).

4 Shubnikov-de Haas oscillations

It is known that the Shubnikov-de Haas effect, displaying in the conductivity oscillations under quantized magnetic field, is caused by consequent crossing of the Fermi level by Landau ones. The period of oscillations is found proportional to the carrier concentration what allows to use this phenomenon for its determination. Temperature increasing leads to broadening of carrier distribution and, therefore, to suppression of the conductivity oscillations.

In 2D systems, small Shubnikov-de Haas oscillations are observed in such fields that $\omega_c \tau_j \sim 1$. The main and the combine frequencies appear in the first and second order in the parameter $\exp(-\pi/\omega_c \tau_j)$ respectively.

To calculate the conductivity tensor, we used the diagrammatic method. The Green function of a quasi-two-dimensional electron gas at finite temperature was calculated in the Matsubara technique. The self-energy parts were found from the system of coupled Dyson equations.

At zero temperature, the main contribution appears in the first order in $\exp(-\pi/\omega_c \tau_j)$ terms [4]. The terms of higher orders are small and therefore are not observable. However, with increasing of temperature, the amplitudes of the main harmonics decrease exponentially and the non-damping with temperature oscillations of higher orders may manifest themselves. In the second order in $\exp(-\pi/\omega_c \tau_j)$, the former have the following form for two occupied subbands:

$$\sigma_{\alpha\beta}^{(nd)} = S_{\alpha\beta} \cos\left(2\pi \frac{\Delta}{\hbar \omega_c}\right) \exp\left[-\frac{\pi}{\omega_c}\left(\frac{1}{\tau_1} + \frac{1}{\tau_2}\right)\right] , \quad (4)$$

where Δ is the energy gap between subbands.

At further increasing of temperature, these oscillations dominate. The expressions for smoothly magnetic field-dependent coefficients $S_{\alpha\beta}$ and for the thermally-damping parts of the conductivity tensor are presented in Ref. [5].

In Fig. 2 the resistance is presented as a function of magnetic field. Calculating we assumed that the carrier

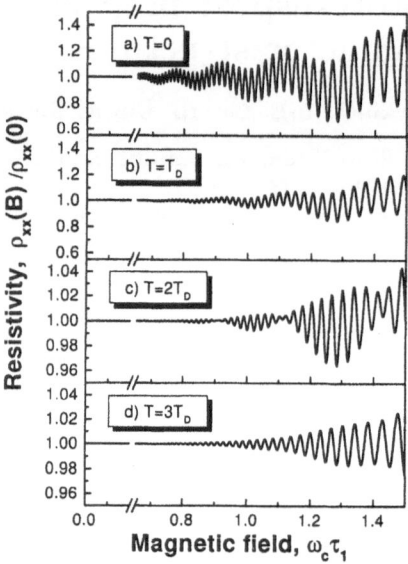

Fig. 2 The temperature behavior of the Shubnikov-de Haas oscillations. $E_F \tau_1/\hbar = 50$, $(E_F - \Delta)\tau_2/\hbar = 5$, $\tau_1 = \tau_2$.

concentration in the excited subband is relatively small: $(E_F - \Delta)/E_F = 0.1$, where E_F is the Fermi energy, and intersubband scattering is intensive: $\tau_1/\tau_{12} = 0.5$.

At $T = 0$, the oscillations from both subbands are clear. The low-frequency harmonic exists due to intensive intersubband scattering [4]. In Fig. 2c the case of $T = 2T_D$ is presented, where $T_D = \hbar/2\pi k_B \tau_1$ is the Dingle temperature. The amplitudes of the main and differential harmonics are comparable, and the differential frequency is close to the main one due to small filling of the excited subband. Therefore the beats are observable. At $T = 3 T_D$, the main harmonic damps out. The oscilations are described only by the non-damping terms (4).

5 Conclusion

We have shown that intersubband scattering has strong influence on magnetotransport in multisubband systems. The conductivity tensor has been calculated in a wide range of magnetic fields.

Acknowledgments

This work was supported by the RFFI, projects 00-02-17011 and 00-02-16894, and the Russian State Programme "Physics of Solid State Nanostructures".

References

1. N.S. Averkiev, L.E. Golub and G.E. Pikus, Solid State Commun. **107**, (1998) 757.
2. N.S. Averkiev, L.E. Golub and G.E. Pikus, *Proc. of 24th Int'l. Conf. Phys. Semicond.*, (Jerusalem, 1998) CD.
3. N.S. Averkiev, L.E. Golub and G.E. Pikus, Semiconductors **32**, (1998) 1087.
4. N.S. Averkiev, L.E. Golub and S.A. Tarasenko, JETP, **90**, (2000) 360.
5. N.S. Averkiev, S.A. Tarasenko, L.E. Golub and M. Willander, to be published.

Effect of buffer layer thickness on improvement of modulation doped CdTe/CdMgTe heterostructures grown on GaAs substrate

D. Wasik[1], M. Baj[1], L. Dmowski[2], J. Siwiec-Matuszyk[1], J. Przybytek[1], E. Janik[3], T. Wojtowicz[3], G. Karczewski[3]

[1] Institute of Experimental Physics, Warsaw University, Hoza 69, 00-681 Warsaw, Poland, e-mail: daw@fuw.edu.pl
[2] High Pressure Research Center "Unipress", PAS, Sokolowska 29/37, 01-142 Warsaw, Poland
[3] Institute of Physics, Polish Academy of Sciences, Al. Lotnikow 32/46, 02-668 Warsaw, Poland

Abstract We have shown that the increase of the buffer layer thickness in low-dimensional CdTe/$Cd_{1-x}Mg_x$Te structures leads to significant improvement of the sample quality and stability. In such hetrostructures high quality 2DEG in the quantum well is formed without any illumination and without parallel conduction. Moreover, the process of pressure-induced degradation is much less essential than for samples having thin buffer layer.

1 Introduction

The potential hybridisation of II-VI semiconducting materials with III-V optoelectronics has led to significant efforts in optimising the epitaxial growth of II-VI layers on GaAs substrates. In the case of modulation doped CdTe/$Cd_{1-x}Mg_x$Te quantum well structures, although a real progress in the quality has been achieved, there still remain important problems, which so far have not been fully resolved. The first problem concerns parallel conduction bypassing the conductance of the quantum well. This parasitic conductance channel could be cut out by the decrease of dopant concentration. However, it results in a strong increase of the resistance of the samples at low temperatures, which makes electrical measurements possible by use of illumination only [1]. The second problem refers to stability of the heterostructures under external hydrostatic pressure. Our previous results have shown that application of pressure leads to creation of huge number of structural defects in the vicinity of the interface between the II-VI structure and the substrate which, propagating in the samples, permanently damage them [2].

The idea of the present work was to search for the improvement of the structure quality and stability by introducing thick CdTe/$Cd_{1-x}Mg_x$Te buffer layer, which separates the QW and highly defected interface region.

2 Samples

In our studies we used two types of modulation doped CdTe/$Cd_{1-x}Mg_x$Te samples grown by molecular beam epitaxy method. In the first case hetrostructures were grown on (100) GaAs/CdTe hybrid substrates, which were fabricated in separate MBE process by covering GaAs wafer first by 1 nm of ZnTe and then by about 3 μm of undoped CdTe. In the second case samples were grown as previously on (100) GaAs substrate covered by 1 nm of ZnTe but then by about 6 μm of undoped $Cd_{1-x}Mg_x$Te buffer layer. For all the samples the CdTe/$Cd_{1-x}Mg_x$Te structures consisted of a single quantum well (10 nm,

CdTe) embedded in $Cd_{1-x}Mg_x$Te barriers. Magnesium content in the barrier x was equal to 0.20 and 0.13. The quantum well was separated from the iodine doped $Cd_{1-x}Mg_x$Te layer by the undoped spacer.

Measurements of the resistance tensor were performed in magnetic field up to 7.5 T at temperature 4.2 K and hydrostatic pressure up to 0.9 GPa.

3 Results and Discussion

All the samples with thin buffer layer reveal either (1) very large resistance (greater than 20 MΩ) at low temperatures or (2) significant contribution of parallel conductance if the resistance is much lower ($\sim$ 100 kΩ). The case (1) corresponds to heterostructures with dopant concentration of 10^{17} cm^{-3} and with Mg composition in the barrier x=0.20. We have checked (by means of capacity experiments) that high resistance is due to the absence of free carriers in the samples but not due to the resistance of electric contacts. Illumination during about 5 minutes causes persistent generation of free electrons in the QW. Spectrum of persistent photocurrent reveals that this process is via valence band -> conduction band transitions in the barrier (E_{photon}>2.1 eV) or via release of electrons from deep traps (for E_{photon}>1.3 eV). The second type of thin buffer layer samples (2) is related to heterostructures with greater dopant concentration (10^{18} cm^{-3}) and shallower quantum well (Mg composition x=0.13). Analysis of the conductance tensor versus magnetic field shows that contribution of electron transport in parallel channel is comparable or even greater than 2D conductance in the QW, which causes for example an absence of plateaux in the Hall ρ_{xy} resistance (for more details see [3]). To explain the presence of this phenomenon in the CdTe/$Cd_{1-x}Mg_x$Te structures we performed theoretical calculations of the conduction band energy profiles adopting numerical procedure used for InP/InGaAs hetrostructures [4] assuming: no surface states, presence of donors in the doping layer only and the position of iodine deep donor level as in the $Cd_{1-y}Mn_y$Te [5]. Calculations show that for the depth of the quantum well corresponding to Mg content x<0.20 a potential valley is formed spontaneously outside the QW in the region doped with iodine donors (see Fig. 1). This leads to redistribution of free electrons between the QW and the valley and thus an additional conductance channel in the barrier is formed.

Experimental data obtained for thick buffer layer samples show substantial improvement in their electrical properties with respect to those described above. All the samples (both with Mg content, x equal to 0.13 and 0.20) reveal formation of high quality 2DEG in the quantum well without any illumination and without parallel conductance which is evidenced by well resolved quantum oscillations in ρ_{xx} resistance and plateaux in Hall ρ_{xy} resistance. 2D sheet carrier densities of about $3 \div 5 \times 10^{11}$ cm^{-2} extracted from Shubnikov - de Haas period are in good agreement with the values determined from low-field Hall measurements. The values of ρ_{xy} related to plateaux match well with theoretical predictions for filling factors $2 \leq v \leq 6$. Electron mobility achieves value of 0.9×10^{5} cm^{2}/Vs.

For all investigated samples (both with thin and thick buffer layer) application of external hydrostatic pressure leads to irreversible increase of the ρ_{xx} resistance which after releasing pressure remains as measured at high pressure. This effect is caused by persistent strong decrease of the mobility, resulting from formation of structural defects in the vicinity of the interface between the GaAs substrate and II-VI material due to mismatch of the compressibility [2]. For thin buffer layer samples this process causes elimination of electron transport in the quantum well. After releasing pressure the samples reveal 3D conductance only (SdH oscillations disappear while Hall resistance ρ_{xy} has in magnetic field a shape of straight line). For thick buffer layer samples the effect of hydrostatic pressure is less essential due to reduction of structural defects propagation into quantum structure (see Fig. 2). After pressure experiments the Hall plateaux become less pronounced but the presence of 2DEG is evident. Values of mobility (≈ 1000 cm^{2}/Vs) are significantly lower than before application of pressure but they are still almost two orders greater than Hall mobilities observed in samples having thin buffer layers.

4 Conclusion

Our experimental data show that the increase of the buffer layer thickness leads to essential improvement in electrical properties of the CdTe/Cd$_{1-x}$Mg$_x$Te structures via reduction of structural defect propagation into the region of quantum structure. This enables formation of 2DEG in the QW without any illumination and without parallel conductance as well as an increase of sample stability under pressure.

Acknowledgements The authors are very much indebted to Leszek Baj for the preparation of photolithographic masks. This work was partially supported by the State Committee for Scientific Research (Republic of Poland) under Grant 2P03B05215.

References

1. G. Karczewski, J. Jaroszynski, M. Kutrowski, A. Barcz, T. Wojtowicz, and J. Kossut, J. Cryst. Growth **184/185**, (1998) 814.
2. D. Wasik, J. Mikucki, J. Siwiec, M. Baj, J. Gronkowski, J. Jasinski, and G. Karczewski, *Proc. of 37th EHPRG Meeting* "High Pressure Research" - in press.
3. D. Wasik, M. Baj, L. Dmowski, J. Siwiec-Matuszyk, E. Janik, T. Wojtowicz, and G. Karczewski, *Proc. of 9th HPSP*, Sapporo 2000, to be published in physica status solidi.
4. J. Mikucki, M. Baj, D. Wasik, W. Walukiewicz, W.G. Bi, and C.W. Tu, Phys. Rev. **B61**, (2000) 7199.
5. D. Wasik, K. Kudyk, M. Baj, J. Jaroszynski, G. Karczewski, T. Wojtowicz, A. Barcz, and J. Kossut, Phys. Rev. **B59**, (1999) 12917.

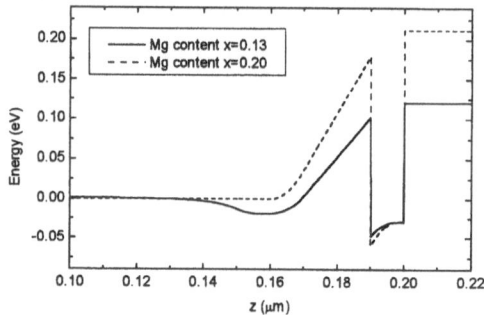

Fig. 1 Potential profiles of CdTe/Cd$_{1-x}$Mg$_x$Te structures. Zero of the energy is put at the Fermi level.

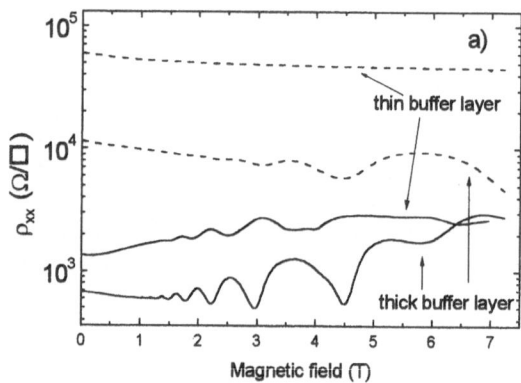

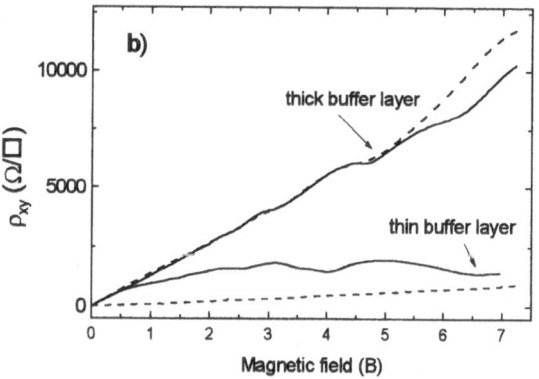

Fig. 2 Shubnikov - de Haas ρ_{xx} a) and Hall ρ_{xy} b) resistivities measured at 4.2 K and ambient pressure before (solid lines) and after (dashed lines) application of pressure of p=0.9 GPa

Domain formation in a one-dimensional superlattice with phonon scattering

L.G.Mourokh[1,2], A.Yu.Smirnov[2,3], N.J.M.Horing[2], V.I.Gavrilenko[4]

[1] Department of Physics, Brooklyn College of the City University of New York, Brooklyn, NY 11210-2889
[2] Department of Physics and Engineering Physics, Stevens Institute of Technology, Hoboken, NJ 07030
[3] D-Wave Systems Inc., Vancouver, B.C., Canada
[4] Institute for Physics of Microstructures RAS, 603600 Nizhny Novgorod, Russia

Abstract   We have analyzed the dynamics of the temporal formation of high electric field domains in a one-dimensional superlattice miniband subject to inelastic phonon scattering. Our formulation is based on the derivation of equations of motion for generating functions of electron momentum fluctuations, which take account of electron scattering from acoustic phonons, facilitating the determination of explicit analytical expressions for both the electron drift velocity and the diffusion coefficient as functions of the applied bias voltage. While our considerations here are limited to a one-dimensional superlattice, they can also be applied to transverse degrees of freedom for a three-dimensional superlattice. Our results provide a clear determination of the domain formation growth rate and its dependence on temperature and carrier concentration as well as bias field.

The physics of semiconductor structures exhibiting various current and voltage instabilities has been a focal point of intense research interest for device purposes. These structures have proven to be very useful in the generation, amplification, switching, and processing of microwave signals. In structures having an N-shape negative differential conductivity (NDC), certain fluctuations can grow into a steady nonuniform spatial field distribution that leads to the formation of moving or static layer-like field inhomogeneties (domains) [1]. In this regard, fluctuation phenomena and system stability have been of special interest. The influence of fluctuations on domain formation is twofold. First of all, small spatial fluctuations of charge lead to the creation of domains under conditions of negative differential conductivity in the current-voltage characteristics. On the other hand, the fluctuation phenomenon of carrier diffusion works against domain development. Special conduction properties of importance in this context may be found in superlattices having several distinct transport mechanisms, including miniband transport, resonant tunneling, and so called $\Gamma \to X$ transport, which give rise to complex current-voltage characteristics with several NDC regions.

This paper is concerned with the microscopic determination of the growth rate of electric field domains in a superlattice biased in the NDC regime. To this end we provide fully quantum mechanical analytic forms for the drift velocity and the diffusion coefficient. Considering a simple one-dimensional model of a miniband (within the tight-binding approximation), including electron scattering from one-dimensional acoustic phonons, we show that, for electric field strength within the NDC region at low carrier concentration and/or at high temperature the diffusion processes dominate and domain formation is impossible; whereas high carrier concentration and/or low temperature lead to domain formation. For a small field fluctuation $\delta E(x,t)=E(x,t)-E_0$ from a constant, uniform steady field, E_0, the solution of the one-dimensional equation for total current flow (including diffusion) jointly with the Poisson equation takes the form of a growing or decaying wave [1]:

$$\delta E(x,t) \sim e^{\lambda t} \exp\left\{ i\frac{2\pi}{L}\left(x - V_d t\right)\right\}, \qquad (1)$$

where L is the length of the superlattice, V is the steady state electron drift velocity, which depends on the bias electric field E, and $V_d=V(E_0)$. The growth (decay) rate λ is determined by

$$\lambda = -\left(\frac{N_0 e[dV/dE]_{E_0}}{\varepsilon} + D\left(\frac{2\pi}{L}\right)^2 \right), \qquad (2)$$

where N_0 is the fixed negative uniform background charge density, e is the electron charge, D is the diffusion coefficient and ε is the dielectric permittivity. It is clear from Eq.(2) that when E_0 falls within the NDC region and $[dV/dE]_{E0} < 0$, small fluctuations of the electric field tend to grow, but diffusion works against this process.

The tight-binding Hamiltonian is given by

$$\hat{H} = \frac{\Delta}{2}\left[1 - \cos\left(\frac{\hat{p}d}{\hbar}\right)\right] - \frac{1}{\sqrt{L}}\sum_k \hat{Q}_k(t)e^{ik\hat{x}} - eEx + H_{ph}. \qquad (3)$$

Here, $\hat{x}$ and $\hat{p}$ are the Heisenberg operators for electron position and momentum, respectively, Δ is the miniband width, d is the superlattice period, $\hat{Q}_k(t)$ is a spatial harmonic of the phonon displacement variable, and $\hat{H}_{ph}$ is the free bath Hamiltonian. The Heisenberg equations of motion for the operator functions $\hat{f}_n(t) = \exp\{in\hat{p}(t)d/\hbar\}$ (where n is an integer) yield equations for the momentum generating functions $F_n(t) = \langle \hat{f}_n(t) \rangle$ and the fluctuating parts $\tilde{f}_n(t) = \hat{f}_n(t) - \hat{I}F_n(t)$ ($\hat{I}$ is the unit operator) as

$$\dot{F}_n(t) + \sum_{m=-\infty}^{m=\infty} G(n,m)F_{m+n}(t) = in\Omega_B F_n(t), \qquad (4)$$

and

$$\dot{\tilde{f}}_n(t) + (\gamma_n - in\Omega_B)\tilde{f}_n = \xi_n. \qquad (5)$$

Microscopic expressions for the damping rate γ_n, the fluctuation source ξ_n and the "collision term" $G(n,m)$ are given in Refs.[2,3]. Here, $\Omega_B = eE_0 d/\hbar$ is the Bloch oscillation frequency. The resulting expressions for the drift velocity and diffusion coefficient are also provided in Refs. [2,3]. Employing these relations in Eq.(2), we obtain an exact expression for the domain growth (decay) rate. Of special interest are the temperature and concentration dependences of this rate and the conditions for domain growth ($\lambda > 0$). The temperature dependence of the domain growth rate is presented in Fig.1 for Bloch frequency $\Omega_B = 1.2 \times 10^{12} \text{s}^{-1}$, concentration $N_0 = 3.88 \times 10^{15} \text{cm}^{-3}$ and GaAs-based SL parameters (with period of superlattice $d=57\text{nm}$ and miniband width $\Delta=100\text{meV}$). One can see that, for these parameters, domain formation is possible only for temperatures below 150K.

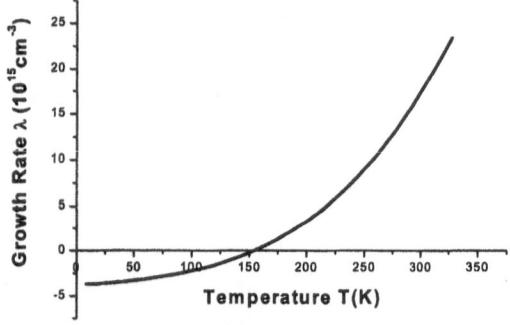

Fig. 1 Growth rate λ as a function of temperature for a GaAs-based superlattice.

We have also evaluated the critical charge density (minimal concentration necessary for domain formation) and its temperature dependence solving the equation

$$\lambda(N_0) = 0 \qquad (6)$$

The results are shown in Figure 2. As one may expect, the critical charge density increases with increasing temperature, but the changes in absolute value are rather small up to room temperature.

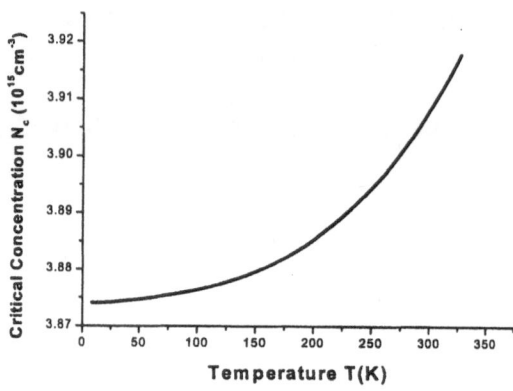

Fig. 2 Critical concentration as a function of temperature for a GaAs-based superlattice.

References

1. M.P.Shaw, V.V.Mitin, E Scholl, and H.L.Grubin, *The Physics of Instabilities in Solid State Electron Devices* (Plenum Press, New York 1992).

2. A.Yu.Smirnov and L.G.Mourokh, Phys.Lett.A **231**, (1997) 429.

3. L.G.Mourokh, A.Yu.Smirnov and N.J.M.Horing, Phys.Lett.A **269**, (2000) 175.

Asymmetric carrier diffusion and phonon-wind-driven transport in an InGaAs-InP quantum well

A.F.G Monte*, S.W. da Silva, P.C. Morais, J.M.R. Cruz, and A.S. Chaves

Universidade de Brasília, Instituto de Física, Núcleo de Física Aplicada, CP 04453, 70919, Brasília-DF, Brazil email: adamo@fis.unb.br

Abstract The diffusion of electron-hole plasma in an intrinsic InGaAs-InP single quantum well was investigated by measuring the photoluminescence (PL) intensity profile around the illuminated area. We find an asymmetric profile on the carrier distribution in a quantum well grown on a tilted substrate. The asymmetry is temperature dependent and also depending on the phonon flux in the laser spot.

1 Introduction

The investigation of the transport properties of quantum well (QW) systems has been achieved through microphotoluminescence techniques [1,2]. These experiments also show that the regime of diffusion is identified with a Gaussian profile. However, more recently the effects of the substrate misorientation on the carrier transport properties in InGaAs thin layers have showed that the profile of the carrier diffusion is not always Gaussian, and can be asymmetric if the QW is grown on tilted substrates direction [3]. The diffusion in the QW with the normal oriented along the [001] direction is clearly distinct from the diffusion in a QW 2° tilted towards the direction [111]. In the first case, the diffusion is axially symmetric, whereas in the second one it is asymmetric, showing a tail along the [110] direction. The PL image of the tilted quantum wells also shows that the diffusion can not be described by a Gaussian. Indeed, the diffusion of the carriers in an asymmetric medium is anomalous and can be described by a Lévy distribution [4]. In this study, we report experiments on diffusion of photogenerated electron-hole plasmas in a tilted semiconductor QW which directly demonstrate the temperature dependence of the asymmetry in the diffusion. In particular, we focus our attention in the high optical excitation and low temperature conditions. Under such conditions, a phonon-wind driving mechanism may have influence on the photocarrier distribution [5].

2 Experiment

The sample was grown by vapor levitation epitaxy which has been described in detail elsewhere [6]. The sample used is grown on a Sn-doped InP substrate oriented 2° off the [001] towards the [111], and the growth time of the InGaAs layer was 10s which corresponds approximately to 4 monolayers [7].

The sample is mounted in a temperature-controlled optical cryostat and is optically excited using an Ar^+-ion laser tuned at 514 nm, which provided energy excitation above the InP band gap. The laser beam was focused on the sample surface down to a spot of 5 μm in diameter, thus allowing lateral measurements with very good resolution. Photoexcitation creates electron-hole pairs, which are captured by the low energy InGaAs. We used band-pass filters or a 0.5m spectrometer to select the E_1-HH_1 transition from the InGaAs QW. Lateral diffusion is observed by monitoring the luminescence region around the excitation spot. The luminescence is collected from the sample surface, through the cryostat window, and imaged at the plane of a scanning pinhole (25 μm wide) attached to step motors and coupled to a liquid Ni-cooled Ge detector, as described elsewhere [5]. The flexibility of this technique allows optical transport measurement to be performed with spatial resolution of about 1.9 μm in systems where ohmic contact could prove to be difficult, such as in nanostructures, since we can directly measure the excitation region. By generating carriers in the spot region and measuring the total luminescence power $I(x)$, the spatial distribution of carriers are quantified with a one-dimensional line scan through the luminescence image.

3 Results and Discussions

A spatial profile of the carrier distribution, which is obtained from the square root of the spectrally integrated PL intensity, is plotted in Fig. 1 with a power intensity of 5 mW. The quantity Δ is the full width at half maximum (FWHM) of the spatial carrier distribution. The carriers are mainly confined in the InGaAs layer. The luminescence intensity I_{PL} is proportional to the square of the carrier concentration for radiative electron-hole pair recombination. Therefore, the observed PL distribution represents the profile of the carrier density (n), considering that I_{PL} is proportional to the square of the carrier density, $I_{PL} \propto n^2$. Besides, important information can be obtained by varying the lattice temperature. So, the profile of the carrier distribution in the InGaAs ultrathin layer (4 monolayers) is shown as a function of the temperature. At 60K the distribution looks quite

---

* *Present address:* Department of Physics, University of Sheffield S3 7RH, UK

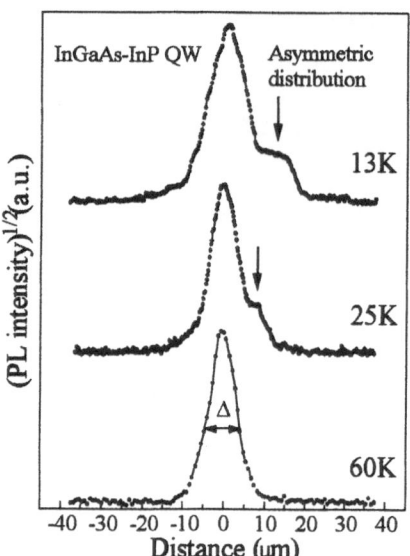

Fig. 1 Profile of the carrier distribution as a function of the temperature in a thin layer of InGaAs (4 monolayers) grown on top of an InP substrate oriented 2° off from the [001] towards the [111].

symmetric (See Fig. 1). However, as the temperature is decreased the carrier profile is composed of two structures; one symmetric with respect to the laser spot and another asymmetric with the peak displaced from the diffusion center (spot of the laser). We have already observed that the diffusion in a QW 2° tilted towards the direction [111] is distinct from the diffusion in a QW normally oriented, as reported in [3]. In this Ref., we have assumed and proved that both distributions are not Gaussian and they should correspond to a true Lévy distribution. Besides, in the tilted QW the Lévy distribution is asymmetric. The appearance of two regimes of diffusion has been studied in terms of a fractal diffusion described by a generalized Lévy distribution [4].

The asymmetric distribution (shown by the arrows in Fig. 1) becomes separated from the main distribution at low temperature. At 25 K, a shoulder on the right-hand side of the main distribution begins to appear. With decreasing temperature this shoulder moves to the right. This seems to be the same phenomenon that was reported in Ref. [3], though in the low temperature regime.

In Fig. 1, the width of the carrier expansion is larger at 13 K, therefore the assumption of a phonon-wind is introduced in the low temperature range (T < 15 K), once this mechanism will push the carriers away from the optical excitation region, thus spreading up the plasma expansion [5]. On the other side, the reduction in Δ up to 15 K, when T increases, is consistent with the picture of a sharp decrease of the diffusion coefficient as a result of the thermal damping of the phonon-wind-driven force.

The rugosity at the interfaces which limit the QW is assumed to be the main cause for the fractal-like diffusion, as discussed elsewhere [3]. The interfaces of the QW are terraces separated by orthogonal steps of height equal to one or more atomic monolayers. Therefore, the influence of the interface morphology on the carrier diffusion would explains the asymmetric diffusion in the tilted QW. Carrier diffusion may be limited by kinks and bends at the QW interfaces which create potential fluctuations that affect the transport. These effects are certainly expected to be temperature dependent, as small potential barriers can be surmounted by thermal activation [8]. Thus, the study of the temperature dependence of the asymmetric carrier expansion is also very important. Once the temperature decreases the asymmetry increased, showing that the carrier diffusion normal to the steps is thermally activated. On these grounds, the asymmetry decreases at high temperatures due to the escape of the carriers out of the potential barrier to the surroundings in the QW.

In summary, we investigated the transport properties of electron-hole pairs in an $In_{0.53}Ga_{0.47}As$-InP QW by spatially-resolved PL measurements. It is revealed that the carrier diffusion normal to the steps is asymmetric and highly temperature dependent. At low temperatures the influence of the phonon-wind should be take into account to understand the spatial shift of the asymmetric distribution.

Acknowledgments

The authors would like to thank for financial support from the TWAS and the Brazilian agencies CNPq, FAP-DF, and CAPES.

References

1. Y. Toda, S. Shinomori, T. Arakawa, and Y. Arakawa, Physica E **2**, (2000) 987.

2. T. Sogawa, H. Ando, and S. Ando, Physica E **7**, (2000) 1020.

3. A.F.G. Monte, S.W. da Silva, J.M.R. Cruz, P.C. Morais, and A.S. Chaves, Phys. Lett. A **268**, (2000) 430.

4. A.S. Chaves, Phys. Lett. A **239**, (1998) 13.

5. A.F.G. Monte, S.W. da Silva, J.M.R. Cruz, P.C. Morais, and A.S. Chaves, Phys. Rev. B **62**, (2000).

6. H.M. Cox, S.G. Hummel, and V.G. Keramidas, J. Cryst. Growth **79**, (1986) 900.

7. J.M. Worlock, F.M. Peeters, H.M. Cox, and P.C. Morais, Phys. Rev. B **44**, (1991) 8923.

8. Y. Tang, D.H. Rich, A.M. Moy, and K.Y. Cheng, Appl. Phys. Lett. **72**, (1998) 55.

Nonzero Hall resistance in a spatially fluctuating magnetic field with zero mean

A.A. Bykov[1], G.M. Gusev[2], J.R. Leite[2], A.K. Bakarov[1], N.T. Moshegov[1], D.K. Maude[3], M. Casse[3,4], J.C. Portal[3,4,5]

[1] Institute of Semiconductor Physics, Novosibirsk, Russia,
[2] Instituto de Física da Universidade de São Paulo, SP, Brazil, e-mail: :gusev@macbeth.if.usp.br,
[3] GHMF, MPI-FKF/CNRS, BP 166, F-38042, Grenoble, Cedex 9, France
[4] INSA-Toulouse, 31077, Cedex 4, France,
[5] Institut Universitaire de France

Abstract We study Hall effect of a nonplanar stripe-shaped two-dimensional electron gas. Electrons in such system experience sign alternating magnetic field with zero average, when uniform field is applied parallel to the substrate. Nonzero Hall resistance is observed in strong magnetic field, when electron transport is characterized by the propagation of the chiral snake-like trajectories. The nonzero Hall conductance arises from the incomplete statistical averaging of the number of snake states channeling in positive and negative directions.

1 Introduction

In periodical and random magnetic field with field distribution which is symmetric about zero mean value $< B >=0$, electron motion consists of the percolation of the snake-like trajectories along B=0 lines [1]. Naively, it can be assumed that the random magnetic field with zero mean leads to zero Hall conductance. However, we demonstrate some possibility to obtain nonzero Hall current for $< B >=0$. Consider periodical magnetic field in X direction, when the average cyclotron radius R_c is smaller then half of the magnetic field periodicity d. In this regime transport is determined by the snake trajectories, which are channeling along Y direction. Snake states running in the positive and negative Y directions are spatially separated. The asymmetry in the propagation of the snake-like trajectories could lead to the nonzero Hall voltage. One of the example of such field could be sawtooth magnetic field. Snake-like trajectories travel in very different field gradient, and, therefore, have a different velocities even for $< B >=0$. When the electric current flows in X direction, the net current in Y direction in such structures is unbalanced, and nonzero Hall voltage appears. In the case of the sinusoidal magnetic field the asymmetry in the propagation of the snake-like orbits can arise from the difference in the electron density for the region with positive and negative field gradient. Snake states running in the positive and negative Y directions have a different number of channels, which also leads to nonzero Hall conductance.

2 Results and discussions

We study Hall effect of a nonplanar stripe-shaped two-dimensional electron gas (2DEG) with stripes oriented perpendicular to the current in the presence of the in-plane external magnetic field B_{ext}. Since 2DEG is sen-

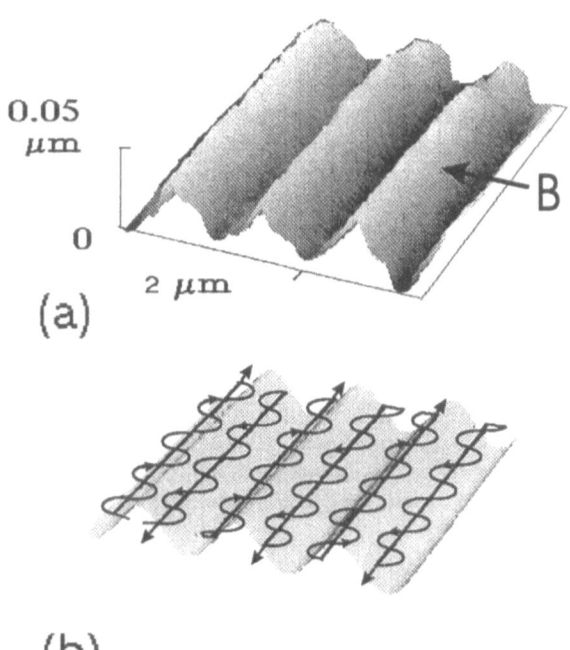

Fig. 1 (a) Atomic force microscope image of the sample surface. (b) Schematic view of the periodical effective magnetic field produced in corrugated 2DEG by the external uniform magnetic field applied in XY plane and the classical picture of the electron trajectories subjected by this field.

sitive only to the normal component of B_{ext}, electrons move in sign alternating magnetic field with zero mean $< B >=0$ and nonzero variance, which is proportional to B_{ext}.

Samples were fabricated employing overgrowth of GaAs and $Al_xGa_{1-x}As$ materials by molecular beam epitaxy on the prepatterned (100) GaAs substrate [2] The sample surface consists of the periodical array of the ridges. The average periodicity of the ridges d is 1 μm, and the corrugation height h is 300 $\AA$. Fig.1 a shows the atomic force image of the corrugated surface. If the applied magnetic field is parallel to the substrate, the effective magnetic field becomes essentially nonuniform and sign alternating as shown in fig.1b. For magnetic field oriented

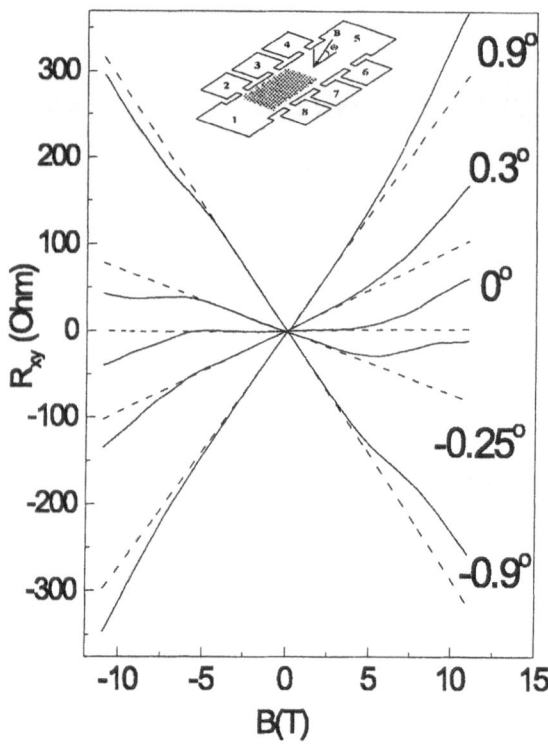

Fig. 2 Hall resistance as a function of B for different angles between the applied magnetic field and substrate plane. Dashes - resistance measured between voltage probes on the planar region. Insert - schematic view of the sample and experiment geometry.

in the x direction, the normal component of B can be expressed as B_N (x,y,z) $= (\overline{B}\ \overline{\nabla}f)/|\ \overline{\nabla}f\ | \sim$ (df/dx)×B_{ext} where gradient $\overline{\nabla}f$(df/dx, df/dy, df/dz) is defined for the surface f(x,y,z)=0. If the height of the stripes h is smaller than periodicity, therefore, df/dx $\sim$ h/d $\ll$1, and we have $B_N \approx \pm$ (h/d)×B_{ext}. When the amplitude of the magnetic field fluctuations due to the sample corrugation is large enough to bind electrons between maxima and minima of the field and create snake-like trajectories (fig 1b).

Fig.2 shows the results of the Hall effect measurement, when the applied magnetic field is exactly parallel to the substrate with in-plane component perpendicular to the ridges and slightly tilted away from the substrate plane. Zero Hall resistance is observed at low B_{ext}, as expected for $< B >$=0. However, at $B_{ext} >$5 T a nonzero Hall voltage is found. Hall voltage changes the sign for reverse B-polarity, which justify that the probes are aligned, and there is no mixing of the diagonal resistance component. For small tilt angles Hall voltage of the nonplanar 2DEG is still deviate from the linear B-dependence. Considering realistic profile of the surface, we find that the average cyclotron radius approaches $R_c \sim$ 0.5-0.6 μm $\sim$ d/2, and snake states ap-

pear in B>5 T. As was shown in [1] the snake states form a bundle of the width given by

$$W = \Lambda(2k_F\Lambda)^{1/2} \qquad (1)$$

where $\Lambda= (h/e\nabla B)^{1/3}$, ∇B is gradient of the magnetic field. In accordance with Landau-Buttiker formalism the conductance of the ballistically propagating modes is given by

$$\sigma = (2e^2/h)(k_F W/\pi) \qquad (2)$$

where k_F is electron Fermi vector. Substituting equation (1) in (2) we find

$$\sigma = (2^{3/2}e^2/\pi h)(k_F\Lambda)^{3/2} \sim \nabla B^{-1/2} \qquad (3)$$

In our case ∇ B$\sim B_{ext}$, and the conductance due to the snake states is proportional to $B_{ext}^{-1/2}$. If ∇B_+ for states running in positive Y direction is larger than ∇B_- for trajectories running in negative Y directions

$$\sigma_+ - \sigma_- \sim \nabla B_+^{-1/2} - \nabla B_-^{-1/2} \neq 0$$

It is worth noting here, that we also measure diagonal resistance in the presence of the in-plane external magnetic field. When the in-plane component is perpendicular to the stripes, resistance across the ridges increases from 160 Ohm at B=0 T to 22000 Ohm at B=11 T, a ratio 140. For the quantitative comparison of the experimental results with the theoretical model we use the magnetic field gradient obtained from the realistic profile of the surface, shown in fig. 1a, which is approximately equal to 4G/A at B_{ext}=10 T. From this we obtain the value of the band width W=0.6 μm, and number of the snake states k_FW/π=30. The magnetoresistance of the sample with multiple barriers in series is equal to the number of sidewalls M=200 (twice the number of barriers) multiplies the magnitude of the resistance in one barrier. It gives the value of the resistance at B=10 T R_{xx} =83000 Ohm, 4 times larger, than we observed in experiment. Therefore, we assume that dissipation occurs near the some of the corners of the barriers region, and we have intermediate case between really diffusive and drift regime [3]. The nonzero Hall conductance arises from the incomplete statistical averaging of the number of snake states channeling in positive and negative directions. Total number of the channeling states is N$\approx$ 30×200=6000. From fig.2 we see that the Hall resistance at B=10 T approaches 50 Ohm. Therefore we can obtain that the average difference between the snake states channeling in positive and negative directions is equal to ΔN$\approx NR_{xy}/R_{xx} \simeq 4$.

Acknowledgments

The work is supported by CNPq, FAPESP, USP-COFECUB and RFFI N 00-02-17896 .

References

1. J.E. Muller, Phys.Rev.Lett., 68, 385 (1992).
2. G.M. Gusev, J.R. Leite, A.A. Bykov, N.T. Moshegov, V.M. Kudryashov, A.I. Toropov, Yu.V. Nastaushev, Phys.Rev.B 59, 5711 (1999).
3. I.S.Ibrahim, V. A Schweigert, F.M.Peeters, Phys.Rev.B 56, 7508 (1997).

Proc. 25th Int. Conf. Phys. Semicond., Osaka 2000 (Eds. N. Miura and T. Ando)   819

Optical detection of ballistically injected electrons in III/V heterostructures

M. Kemerink[1], K. Sauthoff[2], P.M. Koenraad[1], J.W. Gerritsen[3], H. van Kempen[3] and J.H. Wolter[1]

[1]COBRA Inter-University Research Institute, Eindhoven University of Technology, P.O. Box 513, 5600 MB Eindhoven, The Netherlands email: m.kemerink@phys.tue.nl.
[2]4. Physikalisches Institut, Universität Göttingen, Bunsenstraße 13, D-37073 Göttingen, Germany
[3]Research Institute for Materials, University of Nijmegen, Toernooiveld, 6525 ED Nijmegen, The Netherlands

Abstract. We present a novel spectroscopic technique which is based on the ballistic injection of minority carriers from the tip of an STM into a semiconductor heterostructure. By analyzing the resulting electro-luminescence spectrum as a function of tip-sample bias, both the injection barrier height and the carrier relaxation rate Γ_s after injection can be determined. From current dependent measurements we find that carrier trapping by impurities causes a significant non-radiative recombination channel at room temperature.

1 Introduction

Since the pioneering work of Bell and Kaiser [1,2] it is well known that a scanning-tunneling microscope (STM) can be used to inject carriers ballistically into semiconductor heterostructures. The main strength of this technique, ballistic electron emission spectroscopy (BEES), is that it combines the spatial resolution of an STM with sensitivity to the height of a buried injection barrier. However, since the electrical detection scheme only tells which fraction of the injected carriers passes the barrier ballistically, the path followed after injection is not determinable by this measuring method. However, if some of the layers that contain the majority carriers are optically active, injected minority carriers can recombine radiatively and the ballistically injected current can be detected optically. In this paper we will demonstrate the 'optical detection of ballistic electrons' (ODBE). We will show that, from the bias dependence of the peaks in the optical spectrum, not only the injection barrier height can be extracted, but also the carrier scattering rate in the deeper layers of the structure. We estimate that the detection limit of our setup sets the minimal current that has to be injected over the first AlGaAs barrier at 1 fA. In comparison, the detection limit in a BEES experiment is about 20 fA.

2 Experimental

The sample that was used in the ODBE experiments is shown in Fig. 1. The barrier material is $Al_{0.25}Ga_{0.75}As$, the p-type ($p=6*10^{15}$ m^{-2}) quantum well (QW) consists of $In_{0.1}Ga_{0.9}As$ and the superlattice (SL) consists of alternating GaAs and AlAs layers. Both barrier layers contain a Be delta-doping layer with a nominal density of $1*10^{12}$ cm^{-2}, all other layers are nominally undoped. Prior to mounting the samples are Sulfur-passivated. After passivation, the samples were mounted in the head of a home-built low-temperature STM with optical access [3].

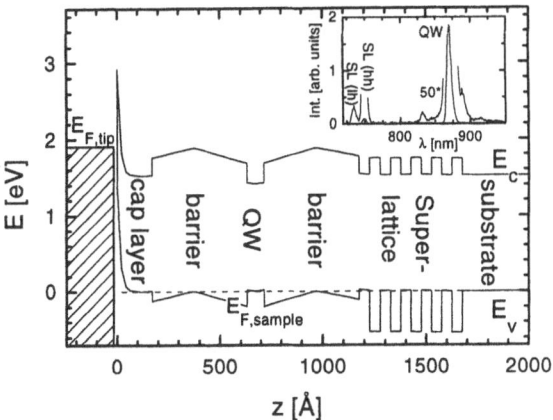

Fig. 1 Self-consistent band diagram of the sample used. T=4.2 K, U_t=1.91 V. Inset: ODBE spectrum obtained at 4.2 K with I_t=10 nA and U_t=3.5 V.

3 Results and discussion

A low-temperature ODBE-spectrum is displayed in the inset of Fig. 1. From the integrated intensity we estimate the number of photons created in the QW per electron emitted by the STM tip to be $7 \cdot 10^{-4}$ at $U_t = 3.5$ V.

Fig. 2 shows the current dependence of the QW

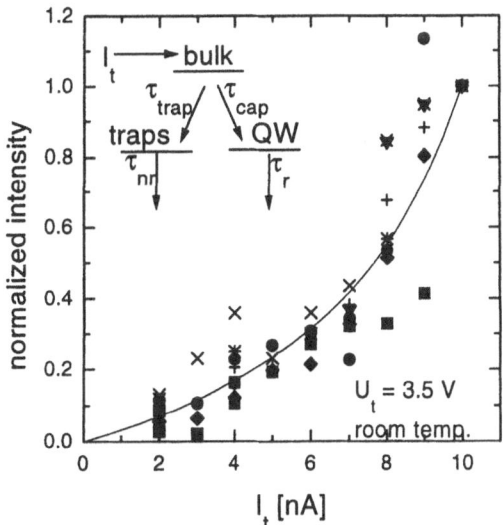

Fig. 2 QW ODBE intensity vs. current at room temperature. The solid line is generated with the rate-equation model illustrated in the inset. The time constants τ_{trap}, τ_{cap}, τ_{nr} and τ_r are the trapping time, the capture time, the non-radiative recombination time and the radiative recombination time, respectively.

STM-induced luminescence (STL) peak at room temperature. We interpret the apparent non-linearity in terms of saturation of a non-radiative recombination channel, which is illustrated in the inset. For low current densities the large majority of the injected electrons is trapped in non-radiative recombination centers. At high enough currents the traps saturate and electrons can reach the quantum well where they can recombine radiatively. We tentatively assign the unionized Beryllium acceptors in the first doping layer as the non-radiative traps. The line in Fig. 2 is a fit to the data based on this model which describes the experimental data very well. At low temperatures (77/4.2 K) the non-linearity is fully absent over three orders of magnitude, i.e. for all currents giving observable signal. Since the electron binding energy to a Be acceptor is about 30 meV, thermal excitation of trapped electrons becomes negligible at these temperatures. This strongly decreases the current needed to saturate the traps.

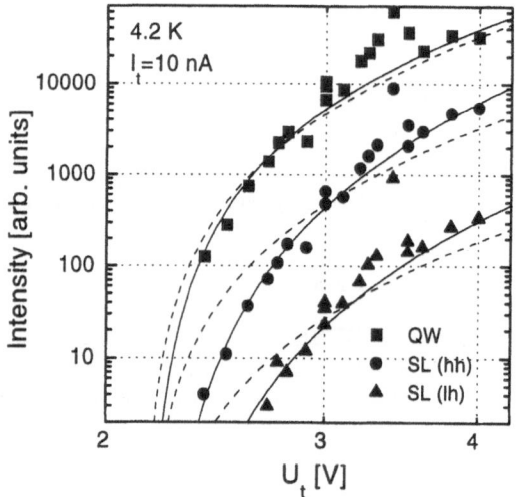

Fig. 3 Bias dependence of QW and SL peaks at $I_t = 10$ nA. The solid (dashed) lines are fits to the full (conventional BEEM) model, see text.

The dependence of the electro-luminescence intensity on the applied bias is shown in Fig. 3. To interpret the data, we made an extension to the conventional BEES model, which yields a $U_t^{2.5}$-dependence of the intensity on bias, to account for the scattering of the injected carriers. Prior to recombination in the QW or SL an electron has, after transmission over the first AlGaAs barrier, to relax to the bottom of the conduction band. However, electrons that reach the bottom of the conduction band before the first doping δ-layer or after the second δ-layer are driven away from the QW region by the built-in electric fields. For recombination in the SL, the electron has to travel ballistically past the second doping layer. By assuming that the injected electrons, with group velocity v, are scattered with rate Γ_s we find for the number of unscattered (ballistic) electrons as a function of depth:

$$\frac{dn_b(E,\mathbf{k})}{dz} = \frac{dn_b(E,\mathbf{k})}{dt}\frac{dt}{dz} = -\frac{n_b\Gamma_s(E,\mathbf{k})}{v(E,\mathbf{k})} \quad (1)$$

To solve this differential equation we assume parabolic bands, with effective mass m^*, and a scattering rate that is independent of energy. We assume that at the used injection energies the dominant scattering mechanism is LO-phonon scattering. Since LO-phonon scattering is isotropic, the direction of the electron is fully randomized after a single scattering event. Therefore it will, on average, reach the bottom of the conduction band at the position where the first scattering event took place. It should be stressed that this means that the electron path to the layer in which the electron recombines is purely ballistic. Under the above conditions we find for the number of electrons that reaches the QW:

$$L \propto I_t \cdot \left(\exp\left[-\frac{\sqrt{m^*/2} \cdot \Gamma_s \cdot z_{Be,l}}{\sqrt{e(U_t - U_0)}} \right] - \right.$$
$$\left. \exp\left[-\frac{\sqrt{m^*/2} \cdot \Gamma_s \cdot z_{Be,r}}{\sqrt{e(U_t - U_0)}} \right] \right) \cdot \left(e(U_t - U_0)\right)^{2.5} \quad (2)$$

where $z_{Be,l}$, $z_{Be,r}$ are the z values of the left and right Be doping layers, and U_0 the height of the first barrier and I_t the tunneling current. A similar formula can be derived for the STL signal of the superlattice. The last term accounts for the voltage dependence due to k-conservation and quantum-mechanical reflection at the first barrier, which is well-known from BEES theory [4]. The solid lines in Fig. 3 are fits to (2) with $U_0 = 2.1$ V and $\Gamma_s = 5 \cdot 10^{13}$ s^{-1}, the dashed lines are fits to the last term of (2) only, which is the usual BEES formula. From Fig. 3 it is clear that carrier relaxation needs to be part of any proper description of ODBE experiments. The value found for U_0 is in reasonable agreement with the value obtained from the self-consistent calculation shown in Fig. 1, 1.9 V. From the fitted scattering rate Γ_s we find for electrons with an excess energy of 0.1 eV a ballistic mean free path of 15 nm. From similar measurements at 77 K, we find $U_0 = 1.8$ V and $\Gamma_s = 8 \cdot 10^{13}$ s^{-1}, which yields a ballistic mean free path of 9 nm at 0.1 eV. Of course, the mean free path increases with increasing excess energy (bias).

Acknowledgements. The research of dr. M. Kemerink has been made possible by a fellowship of the Royal Netherlands Academy of Arts and Sciences. One of the authors (K.S.) likes to acknowledge financial support by the Deutsche Forschungsgemeinschaft (SFB 345).

References

[1] W.J. Kaiser and L.D. Bell, Phys. Rev. lett. **60**, 1406 (1988).
[2] L.D. Bell and W.J. Kaiser, Phys. Rev. lett. **61**, 2368 (1988).
[3] M. Kemerink, J. W. Gerritsen, J. G. H. Hermsen, P. M. Koenraad, H. van Kempen, and J. H. Wolter, unpublished.
[4] M. Prietsch and R. Ludeke, Phys. Rev. Lett. **66**, 2511 (1991).

Electron tunneling time between double quantusm dot

A. Tackeuchi[1], Y. Nakata[2], T. Kuroda[1], K. Mase[1], N. Yokoyama[2]

[1] Department of Applied Physics, Waseda University, 3-4-1 Okubo, Shinjuku-ku, Tokyo 169-8555, Japan
e-mail:atacke@mn.waseda.ac.jp
[2] Fujitsu Laboratories Ltd., 10-1 Morinosato-Wakamiya, Atsugi 243-0197, Japan

Abstract We have directly measured carrier tunneling times between vertically aligned double quantum dots (QDs) using time-resolved photoluminescence measurement. The vertically aligned double QD structure consists of the $In_{0.9}Al_{0.1}As$ QDs layer, GaAs barrier layer and InAs QDs layer. The tunneling times were measured for the three different barrier thicknesses. The dependence of the tunneling time on the barrier thickness is in agreement with the Wentzel-Kramers-Brillouin approximation. The nonresonant tunneling rate between QDs is found to be suppressed to one-tenth of the tunneling rate between quantum wells.

We have directly measured carrier-tunneling times between vertically aligned double quantum dots (QDs). The tunneling time between the QDs will determine the operation time of future devices using QD structures. The vertically aligned double QDs consists of the $In_{0.9}Al_{0.1}As$ Stranski-Krastanow (SK) mode QDs, GaAs barrier and InAs SK mode QDs. In this structure, the carriers are expected to transfer from the ground state of the InAlAs QDs to that of the InAs QDs by energetically non-resonant tunneling. The vertically aligned QDs is expected to give the barrier thickness dependence of the tunneling time [1,2]. Three samples of vertically aligned double QDs with different GaAs barrier thicknesses, $L_B = $ 6 nm, 8 nm, and 10 nm, are grown. Figure 1 shows a cross-sectional TEM picture of a sample with GaAs barrier thickness of 8 nm. The dot density is 1×10^{11} cm^{-2}. By cross-sectional TEM pictures, we confirmed those thicknesses and that upper InAs QDs are vertically aligned to the position of lower InAlAs QDs. Xie et al. reported that the pairing probability of the stacked QDs is maintained at a level of 0.9 for the GaAs thickness not more than 46 MLs [3]. For reference, two samples of $In_{0.9}Al_{0.1}As$ SK QDs and InAs SK QDs are also separately grown.

Figure 2a shows the CW photoluminescence (PL) spectra of the vertically aligned QDs pumped by a Ti:sapphire laser (770 nm). The reference samples of $In_{0.9}Al_{0.1}As$ QDs and InAs QDs have PL peaks at 910 nm (FWHM 70 nm) and 1040 nm (FWHM 90 nm), respectively. These peak wavelengths are nearly same as those of the vertically aligned double QDs. Since SK-dots have large inhomogeneous broadening of about 100 meV, we designed the energy difference between the PL peaks of InAlAs QDs and InAs QDs to be larger than the inhomogeneous broadening by using Al composition of 0.1 for the InAlAs QDs. The relative intensity of the

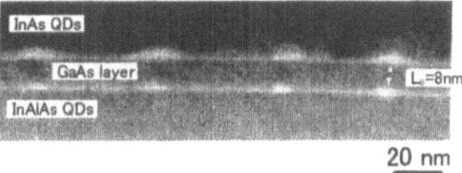

Fig. 1 Cross-sectional TEM picture of vertically aligned QDs with GaAs barrier thickness of 8 nm.

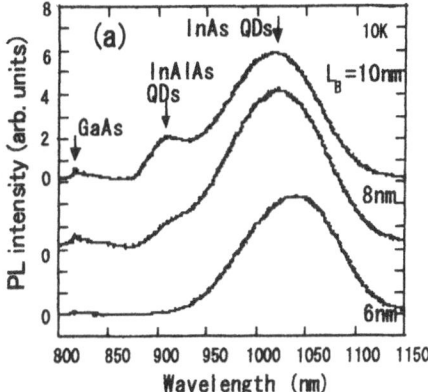

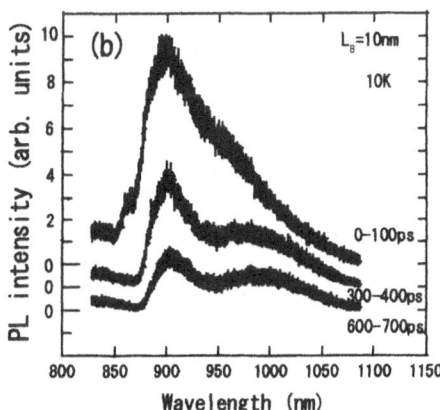

Fig. 2 (a) CW photoluminescence spectra of vertically aligned double quantum dots measured by a near infrared photomultiplier (Hamamatsu Photonics R5509-72) at 10 K. Intensity of the InAlAs PL peak decreases with barrier thickness due to carrier tunneling. (b) Time resolved spectra of vertically aligned double quantum dots of L_B=10 nm at 300 ps time interval. Each spectrum is time-integrated during 100 ps. The sensitivity of the streak camera (Hamamatsu Photonics C4334-04) at the wavelength of 1000nm is about half that of 900 nm.

910 nm peak is smaller than that of the 1020 nm PL. This feature can be explained by the carrier tunneling from InAlAs QDs to InAs QDs. The PL decay was measured using a streak camera (Hamamatsu Photonics C4334-04) with a time resolution of 15 ps. Figure 2b depicts the time resolved spectra of the sample with L_B = 10 nm. Although the CW PL peak of InAlAs QDs in Fig. 2a looks to be mostly buried in the PL of InAs, we can clearly separate the PL decay of InAlAs QDs from the time resolved spectra. Figure 3 shows the decay time of the PL peak at 910 nm as a function of barrier thickness. The tunneling times, τ_t, are directly evaluated using $\tau_{PL} = \tau_t \tau_r / (\tau_t + \tau_r)$ which is derived from a simple rate equation. The recombination lifetime, τ_r, is measured to be 1.1 ns for a reference InAlAs QDs sample. The least square fitting of the dependence of the measured tunneling time on the barrier thickness gives τ_t [ps] = 8.42exp (0.464L_B [nm]). This is in agreement with the Wentzel-Kramers-Brillouin approximation, τ_t = C exp (4L_B (2m*/h^2 (V-E))$^{1/2}$), where C is a constant, m* is the effective mass in the barrier layer, V is the band discontinuity of the conduction band, and E is the quantum level energy.

We fitted the tunneling times as a function of L_B (m*(V-E))$^{1/2}$ to compare the obtained tunneling times with those of quantum wells [4-7]. The energy difference V-E is estimated to be 70 meV based on the PL energies and the infrared photo current measurement for InAs QDs [8]. The tunneling time between the coupled quantum wells (QWs) has been studied since the late 1980s using time-resolved optical measurement. Those measurements clarified that the nonresonant tunneling between the QWs follows the Wentzel-Kramers-Brillouin (WKB) approximation and is assisted by longitudinal optical (LO) phonon scattering [9]. The comparison shows that the tunneling rate between QDs is one order of magnitude slower than those between QWs. In the nonresonant tunneling between QDs, the LO phonon scattering rate is expected to be suppressed unless the

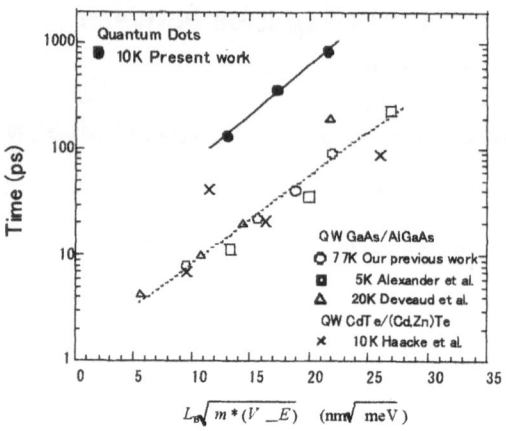

Fig. 4 Tunneling times of the vertically aligned double quantum dots (closed circles) and the tunneling times of the coupled quantum wells. The solid line is fitted to the present QD data by the least squares method. The dashed line is fitted to our previous results for coupled quantum wells.

energy difference between the two levels is equal to the LO phonon energy. The difference in the tunneling rate indicates that the LO phonon scattering is suppressed in QDs. Our experimental result supports the existence of a phonon relaxation bottleneck in QDs [10]. To evaluate the precise difference of LO phonon scattering rate, we need more information about wave functions and scattering processes in the QDs.

References

1. R. Heitz, I. Mukhametzhanov, P. Chen and A. Madhukar, Phys. Rev. **B 58** (1998) R10151.
2. A. Tackeuchi, T. Kuroda, K. Mase, Y. Nakata and N. Yokoyama, Phys. Rev. B. **62** (2000) 1568.
3. Q. Xie, A. Madhukar, P. Chen and N. P. Kobayashi, Phys. Rev. Letters **75** (1995) 2542.
4. A. Tackeuchi, Y. Sugiyama, T. Inata and S. Muto, Jpn. J. Appl. Phys. **31** (1992) 3823., A. Tackeuchi et al., Appl. Phys. Lett. **58** (1991) 1670.
5. B. Deveaud, F. Clerot, A. Chomette, A. Regreny, R. Ferreira and G. Bastard, Europhys. Lett. **11** (1990) 367.
6. M. G. W. Alexander, M. Nido, W. W. Ruehle and K. Koehler, Superlattices and Microstructures **9** (1991) 83.
7. S. Haacke, N. T. Pelekanos, H. Mariette, M. Zigone, A. P. Heberle and W. W. Ruehle, Phys. Rev. B. **47** (1993) 16643.
8. N. Horiguchi, T. Futatsugi, Y. Nakata, N. Yokoyama, T. Mankad and P. M. Petroff, Jpn. J. Appl. Phys. **38** (1999) 2559.
9. S. Muto, T. Inata, A. Tackeuchi, Y. Sugiyama and T. Fujii, Appl. Phys. Lett. **58** (1991) 2393.
10. H. Benisty, C. M. Sotomayor-Torres and C. Weisbuch, Phys. Rev. B. **44** (1991) 10945.

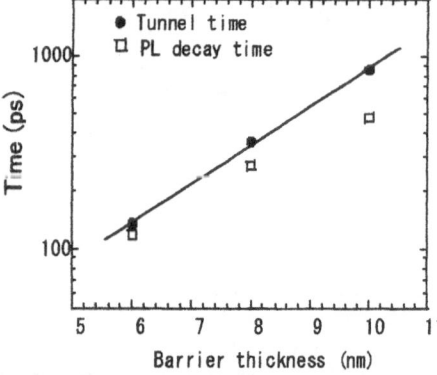

Fig. 3 Decay times of photoluminescence peak of InAlAs QDs (open square) and tunneling times (closed circles) deduced from the PL decay times. The solid line is fitted to tunneling times by the least squares method.

Direct Measurement of the AlAs X-band Fermi Surface

Hyunsik Im[1+], L E Bremme[1], P C Klipstein[1], A V Kornilov[2], R Grey[3], G Hill[3]

[1] Clarendon Laboratory, Department of Physics, University of Oxford, Parks Road, Oxford, OX1 3PU, UK
  E- mail: p.klipstein@physics.ox.ac.uk and im@nano.iis.u-tokyo.ac.jp
[2] Lebedev Physical Institute, Moscow, 117924, Russia
[3] EPSRC III-V Facility, Dept of E. & E. Eng., University of Sheffield, Mappin Street, Sheffield SI 3JD, U.K.

Abstract We report the first direct measurement of the shape of the AlAs Fermi surface, using magneto-resonant tunneling at high hydrostatic pressure. The six lobes of the Fermi surface are almost spherical, contrary to the widely accepted model of strongly ellipsoidal lobes. We also demonstrate that when a large electric field is applied along the [001]-confinement direction, the in-plane constant energy surfaces in an AlAs quantum well (QW) rotate from the two <100> directions to a single <110> direction, due to X_X-X_Y interface band mixing.

1 Introduction

The electron Fermi surface of Si or AlAs has almost become a trademark of traditional semiconductor physics. The six constant energy surfaces in the <100> directions, near the X-point faces of the Brillouin zone, are ellipsoidal and often labelled X_X, X_Y and X_Z. There has been much experimental and theoretical work to try to establish the exact shape of the Fermi surface in AlAs. Major to minor axis ratios between 2.4 and 7 have been reported [1,2]. In the present work we report the first *direct* measurement of this ratio, using 2D→2D magneto-tunneling between $X_{X,Y}$ states in AlAs QWs grown in the z-direction: $X_{XY}(1)$→$X_{XY}(m)$, with m=1, 2 and 3. We show that for Fermi wave-vectors close to $0.02 \times 2\pi/a_0$ (a_0 is the cubic lattice parameter) the ratio is close to unity, demonstrating that the AlAs Fermi surface is in fact almost spherical. These experimental results make possible a complete re-evaluation of the presently accepted $k \cdot p$ parameters and the camel's back in AlAs, as reported by Bremme *et al.* [3]. We also show that the direction of the major axis of the Fermi surface at low electric fields can rotate by 45° to point along [110] at large electric fields applied parallel to z. We explain this behaviour by X_X-X_Y interface band mixing, predicted more than ten years ago but only recently observed [4]. We show for the first time how the constant energy surface at large electric fields can rotate by 90° when the field direction is reversed.

2 Results and analysis

We have studied pressure-induced 2D→2D tunneling between confined X_{XY} states in 60 or 70Å wide AlAs *wells* separated by a 40Å wide GaAs *barrier*. Hydrostatic pressure was generated by a miniature (22mm diameter) cell that could be rotated inside a 15T superconducting magnet. The tunneling mechanism and further experimental details are documented in ref. [5]. The electronic dispersion of the X_{XY} states can be investigated by applying an in-plane magnetic field which causes a change in electron wave-vector, $\Delta k_{//} = eB_{//}\Delta z/\hbar$, where Δz is the distance traveled by electrons when tunneling between X wells, leading to a shift in the resonance bias. This shift as function of in-plane field angle allows mapping of the constant energy surface in the collector well.

2.1 Direct measurement of the X-band Fermi surface

Fig. 1 shows the m=1 resonance of the 70-40-70 sample, in zero magnetic field (dashed) and as function of a 15T-field rotated in the xy-plane. The bias separation between the two resonance peaks in forward and reverse bias, is plotted as a function of field angle. This characteristic dumbbell shape, with a ratio of major to minor axes of 1.23, corresponds to a Fermi surface at the AlAs 'camel's back' which is nearly spherical. A similar ratio was also observed for the 60-40-60 sample. The twofold symmetry can be attributed to the existence of a uniaxial stress component in the pressure cell, which splits the X_X and X_Y minima, causing only the lowest pair of minima in the emitter to be occupied [6].

The 'camel's back' dispersion for bulk AlAs, is described by the eigenvalue, $E(k)$, of the Hamiltonian:

$$H(k) = \underline{I}(\hbar^2 k^2 / 2m_z') - \sigma_z[\Delta/2] - i\sigma_y Rk$$

where σ_y and σ_z are Pauli spin matrices, Δ is the X_1-X_3 energy splitting, and $R = -(\hbar^2/m_0)<u_1|d/dz|u_3>$ is the $k \cdot p$-interaction in which u_1 and u_3 are the X_1 and X_3 crystal periodic functions. Currently accepted values for the $k \cdot p$-parameters are R =1 eVÅ and m_z' = $1.56m_0$ [7]. However, these can neither explain the longitudinal to transverse mass ratio determined by Faraday rotation [8], if the transverse mass is $0.284m_0$ [9], nor the experimental results on X_X-X_Y mixing [4]. This suggests

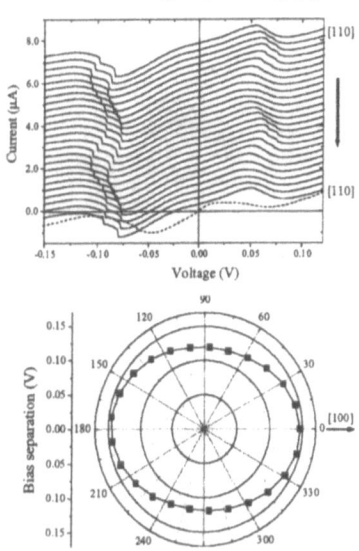

Fig. 1: $X_{XY}(1)$→$X_{XY}(1)$ at 4.2K and 9 kbar for the 70-40-70 sample and its bias shift as a function of magnetic field angle.

that the currently accepted values of the **k.p**-parameters need to be re-evaluated. The direct measurement of the Fermi surface in AlAs now provides the missing information for a re-evaluation of the camel's back. However, it cannot be determined solely from the shape of the Fermi surface presented here. A new determination of R and m'_z and hence the new shape of the camel's back are presented by Bremme et al. [3], taking also into account a whole range of other experimental data: magneto-tunneling between $X_{X,Y}$ Landau levels, cyclotron resonance and PL data. They find $R = 3.4eV\text{Å}$ and $m'_z = 0.16m_0$ at ambient pressure.

2.2 Effect of X_X-X_Y mixing on the Fermi surface

In comparison with the in-plane dispersion for the lower resonance, the resonances $X_{XY}(2)$ and $X_{XY}(3)$ show a substantially modified Fermi surface. Fig 2 shows the results for the $m=3$ resonance of the 60-40-60 sample at 4.2K in forward and reverse bias at ambient pressure. We plot the angle dependence of the bias shift with respect to its position at zero magnetic field. The constant energy surface is elliptical, but its principal axes are along [110] and [$\bar{1}$10]. Similar behaviour was also observed at 10kbar (see fig. 3). The change of the constant energy surface from elongation along the <100> directions for $m=1$ to an elliptical surface oriented along a single <110> direction for $m=3$ is due to X_X-X_Y interface band mixing [4]. The mixing depends sensitively on an *electric field* applied across the sample and increases with the confinement number, m. It can modify the in-plane dispersion of the collector well substantially, so that for $m=1$, when the mixing is weak, we observe the bulk camel's back dispersion whereas for $m=3$, when the mixing is strong, we observe a new dispersion with a minimum at the X-point, leading to a 45°-rotation of the constant energy surfaces. Fig. 3 shows also the angle dependence of the $m=2$ resonance, in comparison with $m=3$. Interestingly, it can be observed that the shape is almost circular and that the major axis is oriented *between* [100] and [110].

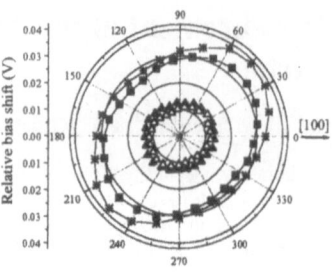

Fig. 3: Bias shift of the $X_{XY}(1) \rightarrow X_{XY}(m)$ peak at 15T for $m=2$ (▲forward, 10k*bar*; • reverse, 10k*bar*) and $m=3$ (■forward, 10k*bar*; * forward, 1bar).

Another interesting feature for $m=3$ (see fig. 2) is the observation that reversing the bias may rotate the orientation of the major axis by 90°. In an ideal sample, the Ga-As bonds at adjacent interfaces of an AlAs QW lie in the [110] and [$\bar{1}$10] planes respectively. An electric field applied in the z-direction, breaks the symmetry of the system and causes the [110] and [$\bar{1}$10] effective masses to be different. On reversing the electric field, this will lead to an interchange of the effective masses in the two <110> directions and hence to a 90° rotation. The effect is quite general and Reker et al. [10] have observed analogous mass anisotropy for Γ electrons.

3 Conclusions

In conclusion, we report the first direct mapping of the Fermi surface at the conduction band minimum in AlAs. The Fermi surface is found to be much more spherical than previously supposed. We have demonstrated in the same AlAs QW that X_X-X_Y mixing can lead to a rotation of the axis of the constant energy surface from the <100> directions at small applied bias to a single <110> direction at large bias, and have studied the mixing strength as a function of electric field. In addition, the Fermi surface is also seen to rotate between [110] and [$\bar{1}$10], when the electric field direction is reversed.

+ Presently at the I.I.S., University of Tokyo
1. M Z Huang, W Y Ching, *J. Phys. Chem. Sol.*, **46**,977 (1985)
2. A A Kopylov, *Solid State Commun.* **56**, 1 (1985)
3. L E Bremme, H Im, H. Choi, R Grey and P C Klipstein, *proc. of this conference*, Osaka (2000)
4. H Im, P C Klipstein, R Grey and G Hill, *Phys. Rev. Lett.* **83**, 3693 (1999)
5. P C Klipstein, Chapter 2 of "High Pressure Semiconductor Physics II, Series: "Semiconductors and Semimetals", **55**, 45, eds. T. Suski and W. Paul (Academic Press, 1998)
6. H Im, L E Bremme, P C Klipstein, A V Kornilov, R Grey and G Hill, *to be published*
7. O Madelung, Semiconductors-Basic Data (2nd revised edition), Springer-Verlag (Berlin, Heidelberg, New York, 1996), p76
8. B Rheinländer, H Neumann, P Fischer and G Kuhn, *Phys. Stat. Sol. (b)* **49**, K167 (1972)
9. H Im, P C Klipstein, R Grey and G Hill, *Phys. Rev. B*, accepted for publication (2000)
10. T Reker, H Im, L E Bremme, H Choi, Y Chung, R Grey, G Hill and P C Klipstein, *proc. of this conference*, Osaka (2000)

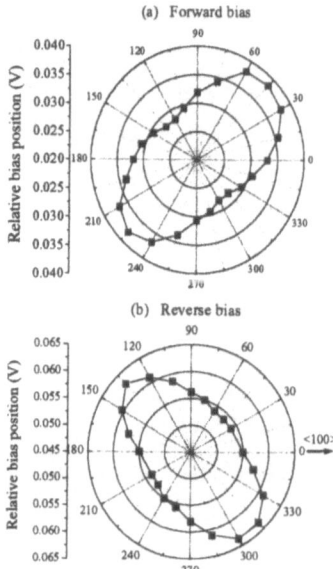

Fig. 2: Bias shift of the $X_{XY}(1) \rightarrow X_{XY}(3)$ peak at 15T and 1*bar*

Miniband transport and Stark-Cyclotron-Resonance in InAs/GaSb superlattices

V. J. Hales, R. J. Nicholas and N. J. Mason.

Clarendon Lab, Department of Physics, University of Oxford, Parks Road, Oxford, OX1 3PU, U.K.
Fax: +44-1865-272400; Email: v.hales1@physics.ox.ac.uk

Abstract Electron transport in InAs/GaSb superlattices under intense parallel electric and magnetic fields is investigated. At low electric fields we observe a conduction threshold, which exhibits magnetic field and temperature dependence. At higher biases the superlattice exhibits a large magneto-resistance and in the quantum limit we observe conduction peaks, which satisfy the Stark-Cyclotron-Resonance condition. Eight band **k.p** calculations show that the miniband width in this system is strongly **k** dependent.

1 Introduction

Since the pioneering work of Esaki and Tsu [1], extensive research on electron transport in semiconductor superlattices has been published. Vertical transport in the high electric field limit has demonstrated a negative differential conductance, NDC beyond a critical electric field, F_c [2,3]. The NDC stems from the field-induced localisation of the miniband states which split into a Wannier-Stark ladder, a series of subbands separated in energy by eFL, where F is the applied electric field and L is the superlattice period. The application of an additional intense magnetic field parallel to the growth direction splits the states into Wannier-Stark-Landau (WSL) states separated in energy by $n\hbar\omega_c$, where n is the landau level index, $\omega_c = eB/m^*$ is the cyclotron frequency, m^* is the effective mass and B is the magnetic field. In general conduction proceeds by inelastic hopping between WSL states, but for particular ratios of electric and magnetic field, elastic scattering can occur. This condition is called Stark-Cyclotron-Resonance (SCR) [3,4] where

$$eFL = n\hbar\omega_c \qquad (1)$$

We have studied vertical transport in InAs/GaSb short period superlattices in intense electric and magnetic fields. Previous vertical transport studies in this system have been confined to weakly coupled single layers [5].

2 Results

We have studied undoped 100 period InAs/GaSb superlattices grown on InAs substrates by MOVPE. Well and barrier widths of 50 Å and 80 Å respectively, yield a ground state miniband width of 15meV and inter-miniband tunnelling is suppressed by the large minigap of 131meV. Although nominally undoped a sheet carrier density of 1×10^{11} cm^{-2} was extracted from in-plane transport measurements on a sample grown simultaneously on an insulating substrate. Pulsed and dc $I(V)$ measurements were performed on 300μm mesas from 0T to 14.5T at 4.2K.

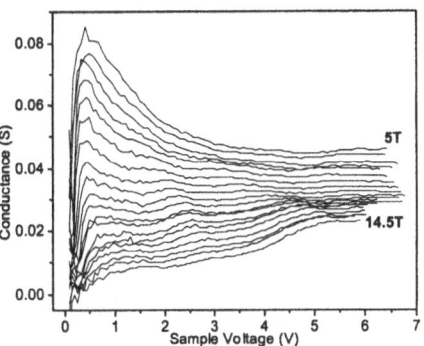

Fig. 1. Pulsed $G(V)$ curves from 5T to 14.5T at 4.2K.

2.1 I(V) measurements at 4.2K

In Fig. 1 pulsed $G(V)$ characteristics shown from 5T to 14.5T exhibit three sets of peaks, which we attribute to SCR. The positions of these peaks have been plotted against magnetic field in Fig. 2. Also plotted is a fit to the peaks using Eq. (1) with an effective mass of $0.054m_e$, which was obtained from eight-band **k.p** calculations. The peaks are

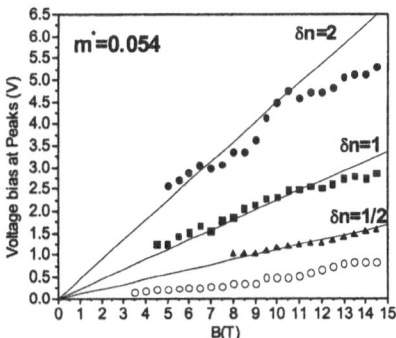

Fig. 2. Voltages at conductance peaks versus magnetic field. Solid points are experimental data and solid lines are fits from eight-band **k.p** calculations. Open circles are the conduction threshold peaks from Fig. 1.

assigned as $\delta n = 1$, 2 and ½, corresponding to resonant tunnelling between Landau levels in adjacent wells ($\delta n = 1$ and 2) and next nearest neighbour wells ($\delta n = 1/2$).

The magnitude of the peaks is much larger than previously observed in single carrier GaAs/AlAs systems (10% when compared with 1% in ref [3]). A second striking characteristic, which is peculiar to this data, is the large magneto-resistance. Canali et al [3] observed little change in the $I(V)$ curves of GaAs/AlGaAs superlattices with the application of 8T. The magneto-resistance is particularly large at bias voltages below 1V, as can be seen in Fig.3, a $\log_{10}$ plot of dc current against magnetic field. The magneto-resistance is very clearly demonstrated and varies by a factor of 10^3 to 10^2 with the bias value. The low bias plots also exhibit strong Shubnikov-de-Haas oscillations, which yield a sheet carrier density of $1 \times 10^{11} \text{cm}^{-2}$, in agreement with the value extracted from in-plane transport measurements.

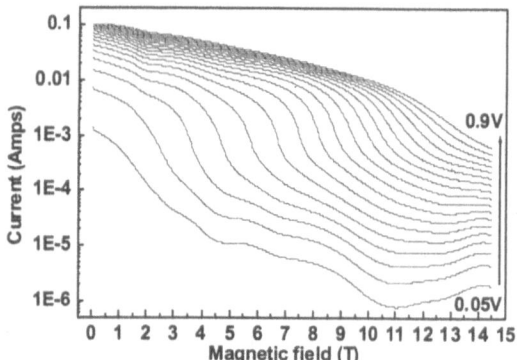

Fig. 3. $\log_{10}$ plot of current against magnetic field for biases from 0.05V to 0.9V at 4.2K

There is a strongly field dependent conduction threshold, which manifests itself as a sharp fall in the current, at around 1mA in Fig. 3 The field dependence of the current drop has also been plotted for each bias voltage, in Fig. 2. The voltage turn-on is well below the condition for Stark localisation of carriers, and hence electrons would be expected to scatter elastically through the miniband. However, the very large low bias magneto-resistance confirms that this is not the case.

2.2 Temperature Dependence

The temperature dependence of the conduction threshold has also been investigated. Pulsed $I(V)$ measurements taken at 15T for a range of temperatures, 1.8K, 2.4K, 60K and 200K are displayed in Fig. 4. The first observation is that the overall shape of the $I(V)$ curve for the high bias data and the magnitude of the SCR resonance peaks remain unchanged with temperature. This confirms that impurity and layer fluctuations are the dominant scattering mechanisms and that quasi-elastic acoustic phonon scattering is negligible.

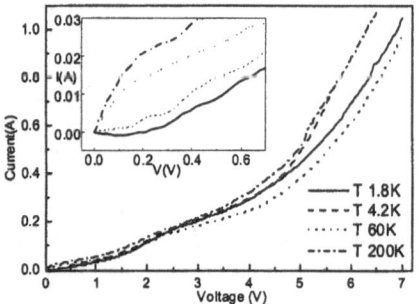

Fig. 4. $I(V)$ curves at 15T, at 1.8K, 2.4K, 60K and 200K.

The inset in Fig. 4 shows the behaviour of the low voltage conduction threshold with temperature. The threshold is clearly strongly temperature dependent. By 60K the activated turn on has completely disappeared.

3 Discussion

The observation of enhanced SCR resonances and a large magneto-resistance in InAs/GaSb superlattices can be explained by the unusual band alignment of the system. In the broken gap system tunnelling occurs due to interband mixing and hence is strongly wave vector dependent. When a strong electric or magnetic field is applied parallel to the superlattice the effective band gap is increased and consequently the in-plane momentum k decreases. Eight band $k.p$ calculations for this system have shown that the miniband width decreases rapidly with increasing energy and magnetic field [6]. For example, the width of the n = 3 landau level is less than one third of the width of the n = 0 landau level at 10T. Similar calculations for GaAs/AlAs superlattices showed no appreciable change in the miniband properties with k. The enhancement of the resonance peaks is also attributed to the miniband dependence on k, where tunnelling preferentially occurs through a sequential process using the wider lower lying minibands.

The threshold to conduction exists in a region where carriers are in principle non-localised. However the temperature activation of the threshold suggests that some disorder such as well width fluctuations is causing a degree of localisation. This behaviour will be the subject of further investigations.

4 Conclusions

We present the first vertical transport measurements in InAs/GaSb superlattices. SCR resonances involving up to two landau level index changes are observed which agree with an eight-band $k.p$ fit. The existence of a large magneto-resistance is explained by the strong k dependence of the miniband properties, which causes a suppression of conduction with increasing magnetic field and energy. The enhancement of the resonance peaks can also be explained by the miniband dependence on k.

5 Acknowledgments

This work is supported by the United Kingdom Engineering and Physical Sciences Research Council (EPSRC), and Nortel Networks plc.

1. L. Esaki, R. Tsu, *IBM Res. and Dev.* **61** (1970) 14
2. L. Eaves et al., *Physica B.* **272** (1990) 190
3. L. Canali et al, *Phys. Rev. Lett.* **76**, (1996) 3618
4. J. Lui et al., *Semicond. Sci. Tech.* **12** (1997) 142214
5. P. Klipstein, Semiconductors and Semimetals, ed. R.K. Willardson and A.C. Beer (Academic press, 1998) Vol **55**
6. V. J. Hales et al., *proc. 25th Int. Conf. Phys. Semicond.*, Osaka, September 2000

Effective mass anisotropy of Γ electrons in GaAs/AlGaAs quantum wells due to interface band mixing

T. Reker[1], H. Im[1*], H. Choi[1], L. E. Bremme[1], Y. C. Chung[1], R. Grey[2], G. Hill[2], P. C. Klipstein[1]

[1] Clarendon Laboratory, Dept. of Physics, Univ. of Oxford, Parks Road, Oxford OX1 3PU, U.K.
[2] III-V Central Facility, Dept. of Electrical and Electronic Eng., Univ. of Sheffield, Mappin Street, Sheffield S1 3JD, U.K.

Abstract We have studied the resonant tunneling of Γ-electrons in a GaAs/Al$_{1-x}$Ga$_x$As double barrier structure (DBS) as a function of the angle of an in-plane magnetic field. We observe a clear anisotropy of the in-plane effective mass for the two $\langle 110 \rangle$ directions. We explain the behaviour in terms of interface band mixing in which both Γ_1-Γ_{15z} and Γ_{15x}-Γ_{15y} mixing are involved.

1 Introduction

Interface Band Mixing can play a subtle but highly significant role in the properties of semiconductor heterostructures. Important examples include Γ -X$_Z$ and X$_X$ - X$_Y$ mixing in the conduction band and Γ_{15x} - Γ_{15y} mixing in the valence band. Recently it has been found that such mixing can lead to a strong anisotropy of both electrical and optical properties [1][2][3]. In this work we present a new type of anisotropy for GaAs/AlGaAs quantum wells, in which the Γ conduction band effective mass is different for the two in-plane $\langle 110 \rangle$ directions, when an electric field is applied perpendicular to the well.

The origin of the anisotropy may be understood on a qualitative level by reference to Fig. 1, where the bonding arrangement in a GaAs/ AlAs quantum well is shown. The Al-As bonds originating from an As site at each interface lie in the two orthogonal $\langle 110 \rangle$ planes. If both interfaces are perfect, the symmetry cannot distinguish one of the $\langle 110 \rangle$ directions over the other. However, when an electric field is applied perpendicular to the interfaces, the symmetry is broken since one bond orientation may now be preferred. Differences in the roughness of the two interfaces could also lead to a symmetry breaking.

In this paper we also outline the steps of a quantum mechanical treatment for the observed anisotropy of the Γ electron effective mass. We show that interface band mixing both between Γ_1 and Γ_{15z} and between Γ_{15x} and Γ_{15y} is necessary to account for the behaviour. This contrasts with the optical case in which Γ_{15x} − Γ_{15y} mixing alone can explain the observed polarisation anisotropy[4].

2 Experimental

2.1 Samples

The DBS was grown by MBE on a [001] n$^+$-doped GaAs substrate. The sample had Ga$_{0.94}$Al$_{0.06}$As emitter and

* *Present address: I.I.S., Univ. of Tokyo, 7-22-1 Roppongi, Minato-Ku, Tokyo, Japan*

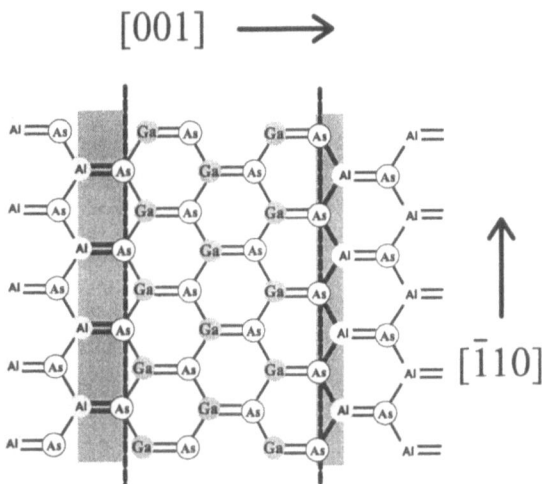

Fig. 1 An AlAs/GaAs quantum well. The AlAs bonds at opposite interfaces lie in the [110] and [$\bar{1}$10] planes respectively.

collector layers doped with $n = 1 \cdot 10^{17} cm^{-3}$ and undoped Ga$_{0.94}$Al$_{0.06}$As spacers 50Å wide. These were separated from the undoped GaAs quantum well (120Å) by undoped Ga$_{0.72}$Al$_{0.28}$As barriers (80Å). Two-terminal current- and conductance-voltage characteristics were measured on circular mesas of 20 μm diameter.

2.2 Results

For an in-plane magnetic field, B_x, an electron traveling over a distance Δz in the growth direction, from the emitter to the quantum well of the DBS, undergoes a change of in-plane momentum due to the field of $\Delta k_{\parallel} = -\frac{eB_x\Delta z}{\hbar}$ [5]. This effect may be used to map out the sub-band dispersion along a fixed direction in the well, by varying the field. Fig.2(right) shows the shift in bias position caused by the field for the $\Gamma(3)$ resonance in forward bias, reproducing the expected quadratic behaviour. A smaller quadratic shift was also observed for the $\Gamma(2)$ resonance, with a total shift of 0.15V at 9T.

Shown in Fig. 2(left) are current-voltage traces in forward bias for the $\Gamma(2)$ resonance as a function of the angle of an in-plane magnetic field of 9T . The bias voltage attains it maximum or minimum values when the field passes through the two orthogonal $\langle 110 \rangle$ directions. This shows clearly that the effective mass of the $\Gamma(2)$ sub-band exhibits a small two-fold anisotropy with its extreme values along the $\langle 110 \rangle$ directions.

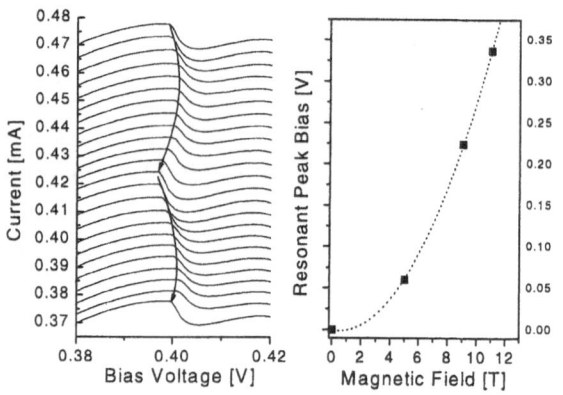

Fig. 2 Left: Experimental $I - V$ traces at $B = 9T$ for the $\Gamma(2)$ resonance as a function of the in-plane magnetic field direction, at an increment of $15°$ starting in a $\langle 110 \rangle$ direction, (traces are offset for clarity). Right: In-plane dispersion for the $\Gamma(3)$ resonance for a fixed magnetic field direction (points) and parabolic fit to the data (dashed line).

3 Discussion

It can be shown from simple $k.p$ theory [6] that the wave functions for the confined states in the quantum well take the form

$$\Gamma(m) = \{\phi_m(z)|\Gamma_1\rangle + [\Xi_m(z)k_x + i\Theta_m(z)k_y]|X\rangle$$
$$+ [\Xi_m(z)k_y - i\Theta_m(z)k_x]|Y\rangle + \Psi_m(z)|Z\rangle\}e^{ik_\| \cdot \rho} \quad (1)$$

where m is the sub-band index, $|\Gamma_1\rangle = |iS\rangle$ is a crystal periodic function (CPF) with s-orbital symmetry, $|X\rangle, |Y\rangle, |Z\rangle$ are the basis from which the Γ_{15} conduction or valence band CPFs are constructed, ρ and $k_\| = ik_x + jk_y$ are in-plane positon- and wave-vectors, and ϕ_m, Ξ_m, Θ_m, and Ψ_m are slowly varying envelope functions (on the length scale of the cubic lattice parameter, a_0). Ψ_m has the opposite parity to ϕ_m, Ξ_m and Θ_m due to the form of the $k.p$ interaction.

Following the treatment in [7], the tetrahedral bonding at an interface, centred at z_i, with two differing cation species on either side (Fig. 1) leads to a microscopic "interface potential" V_{INT} with the property

$$V_{INT}(x', y', z - z_i) = V_{INT}(y', x', z_i - z) \quad (2)$$

where x' and y' denote the [110] and $[\bar{1}10]$ directions, respectively. This is the potential which when added to the average for GaAs and AlAs yields the true microscopic potential of the crystal. It reflects the change in crystal potential when going from AlAs to GaAs (or vice versa) and acts over a region L, where $L \geq \frac{a_0}{2}$ but is smaller than a characteristic length over which the Γ_1 and Γ_{15} envelope functions change appreciably. From Eq. (2) it may easly be shown that

$$\langle X|V_{INT}|X\rangle = \langle Y|V_{INT}|Y\rangle = \langle Z|V_{INT}|Z\rangle = 0$$
$$\langle \Gamma_1|V_{INT}|\Gamma_1\rangle = 0 \quad (3)$$

$$\langle X|V_{INT}|Y\rangle = \langle Y|V_{INT}|X\rangle = \alpha$$
$$\langle \Gamma_1|V_{INT}|Z\rangle = -i\beta \quad (4)$$

Restricting ourselves to the $\Gamma(1)$, $\Gamma(2)$ basis, the above relations lead to mixing of the form

$$V_{12}(k_x, k_y) = \langle \Gamma(1)|V_{INT}|\Gamma(2)\rangle$$
$$= A_{12}k_xk_y + iB_{12}(k_x^2 - k_y^2) + C_{12} \quad (5)$$

Here, A_{12} and B_{12} are real *even* functions of the electric field, E, and depend on the mixing potential α and on the amplitudes of the envelope functions at all of the interfaces: $\Xi_m(z_i), \Theta_m(z_i)$ for $m = 1, 2$ and all z_i. Similarly, C_{12} is a real *odd* function of E, and depends on β and on $\phi_m(z_i), \Psi_m(z_i)$ for $m = 1, 2$ and all z_i. In a non zero electric field, inspection of Eq. (5) shows that the mixing potential is different for the two $\langle 110 \rangle$ directions, when $k_x = k_y$ and $k_x = -k_y$ respectively. Therefore, the dispersion of a given sub-band will be different in each of these directions. Furthermore, on reversing the electric field, the sign of C_{12} changes, thereby rotating the constant energy surface of each sub-band by $90°$, consistent with the earlier symmetry breaking argument based on Fig. 1. The current argument may easily be generalised to a larger set of quantum well basis states [6]. Eq. (5) therefore explains nicely the mass anisotropy observed in experiment.

Finally, we note that a similar treatment to that presented here, but involving only mixing between the Γ_{15x} and Γ_{15y} valence states, can explain the optical anisotropy observed in Fig. 2 of Ref. [3]. A value $\alpha \sim 0.5eV\text{\AA}$, comparable in size with the predictions of Foreman [8], leads to good agreement with experiment[6][9].

4 Conclusion

We have presented experimental evidence for the anisotropy of the in-plane effective mass of Γ states in a GaAs/AlGaAs quantum well subjected to a perpendicular electric field. The effect is explained by Interface Band Mixing involving two contributions: Γ_1 mixing with Γ_{15z} and Γ_{15x} mixing with Γ_{15y}.

5 Acknowledgments

This work was funded by the E.P.S.R.C. (UK). T.R. and L.E.B. acknowledge support from the Friedrich Flick Förderungsstiftung and the Rhodes Trust, respectively.

References

1. H. Im, P. C. Klipstein, R. Grey, and G. Hill, Phys.Rev.Lett. **83**, (1999) 3093
2. O. Krebs, D. Rondi, J.L. Genter, L. Goldstein, and P. Voisin, Phys.Rev.Lett. **80**, (1998) 5770
3. S.H. Kwok, H.T. Grahn, K. Ploog, and R. Merlin, Phys. Rev. Lett. **69**, (1992) 973
4. O. Krebs and P. Voisin, Phys. Rev. Lett. **77**, (1996) 1829
5. R.K. Hayden et al., Phys. Rev. Lett. **66**, (1991) 1749
6. T. Reker, H. Im, L.E. Bremme, Y.C. Chung, P.C. Klipstein, R. Grey, G. Hill, to be published
7. P.C. Klipstein, Proc. ICPS24, 1998, CDROM 1265.pdf
8. B.A. Foreman, Phys. Rev. Lett. **81**, (1998) 425
9. B. A. Foreman, Phys Rev Lett. **82**, (1999)1339

Vertical transport and interband luminescence in type II InAs / GaSb / InAs heterostructures

M.Roberts, N.J. Mason, S.G. Lyapin, Y.C. Chung and P.C. Klipstein

Clarendon Laboratory, University of Oxford, Parks Road, Oxford OX1 3PU, UK. email : m.roberts1@physics.ox.ac.uk

Abstract The resonant conduction and electro-luminescence (EL) at 1bar $< P <$ 11kbar in type II InAs/GaSb/InAs double heterojunctions with a wide range of GaSb thickness are presented alongside self consistent band profiles and hole dispersions. Our results show that coherent electron–light-hole tunneling, as previously proposed by others, is not the cause of the negative differential resistance (NDR). In contrast, an inelastic interaction between electron and heavy hole sub-bands gives a more plausible explanation. In addition, interband EL is observed beyond NDR. Our self consistent model shows that it is due to Zener tunneling, which must therefore also contribute to the background current.

1 Introduction

InAs/GaSb single heterojunctions (SHETs) with 'InSb like' or 'GaAs like' interface bonding exhibit a current resonance followed by a region of NDR. Evidence presented previously [1] showed that the principle NDR for 'InSb like' interfaces was consistent with a quasi-elastic electron–heavy-hole interaction, whereas 'GaAs like' interfaces appeared to involve inelastic processes. In both cases there was an abnormally high background current of unknown origin beyond the NDR. In contrast, Yu *et al.* [2] only observed NDR in InAs/GaSb/InAs double heterojunctions (DHETs) for GaSb > 60Å. Basing their analysis on flat-band calculations, they suggested that the principal NDR is due to resonant electron–light-hole tunneling and that the heavy-hole contribution gives only weak secondary features [3].

In this work, we present vertical transport and EL results on MOVPE grown DHETs with 'GaAs like' interface bonding at pressures up to P = 11kbar. Self-consistent calculations of the band profiles, which appear to be dramatically different to those found in the literature, were carried out to compare with our results. We argue that this discrepancy has its origins in the significant 2D hole concentration that accumulates in the central GaSb layer. Our results provide strong evidence in favour of an *inelastic* electron–heavy-hole interaction.

We also explain the origin of interband mid-infra-red emission which plays a crucial role in the background current observed at higher bias.

2 Sample details and experimental method

n-InAs/p-GaSb/n-InAs nominally undoped DHETs were grown on n-InAs substrates with a GaSb layer thickness 0-1200Å followed by a 3000Å InAs cap. The wafers were processed into 100-150μm mesa structures and metalised with thermally evaporated, non annealed Cr/Au pads. 3-20μsec pulsed *I-V* and EL were measured at 25 - 300K.

3 Results

All the results presented have been corrected for parasitic

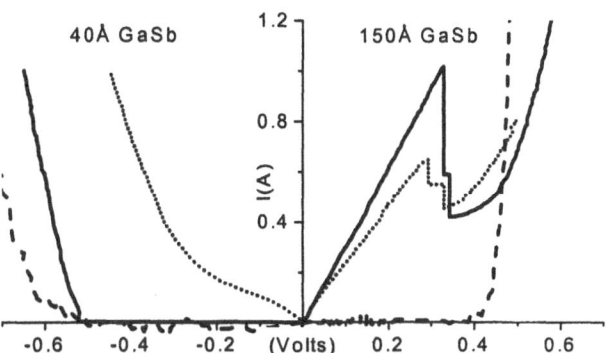

Fig. 1 30K *I-V* (solid), 300K *I-V* (dotted) and 30K *EL-V* (dashed) for 40Å and 150Å DHETs at 1 bar. All the characteristics are nearly symmetric about the origin and only half of each is shown.

resistances in the circuit. Maximum and minimum values of this resistance were deduced from the conditions that (i) the NDR region have a negative slope and (ii) the post NDR region have a positive slope at all bias. This gives a resonance peak bias at 1 bar of $328 \pm 25mV$ and $306 \pm 165mV$ respectively, for the 150Å and 60Å GaSb DHETs presented here (Fig. 1(right) and Fig. 2). The NDR vanishes at 77K for GaSb thicknesses below 60Å, but a weak feature returns at higher temperatures (Fig. 1(left)). Also shown in Fig. 1 is the EL vs. bias. The EL threshold is at the valley bias for the 150Å DHET, and the conduction onset in the 40Å DHET. The EL spectrum corresponds to the PL of bulk InAs, (peak at 410meV or 3.02μm).

The resonant current is suppressed under hydro-static pressure, as shown in Fig 2 for a 60Å DHET. A weak NDR feature is still discernible at 11 kbar. The parasitic resistance correction in Fig. 2 was taken to be independent of pressure.

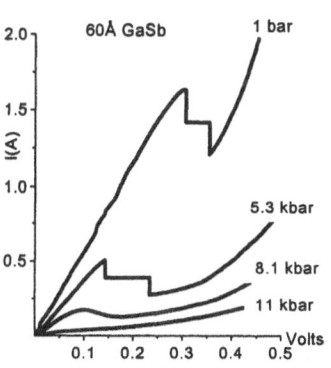

Fig. 2 *I-V* of 60Å GaSb DHET at 77K vs. hydrostatic pressure.

4 Band Profile Calculations

Self consistent calculations on DHET structures in the literature [4] ignore any confinement effects in the emitter, and use the 3D Thomas-Fermi approximation to relate the Fermi level to the band edges. Charge build-up in the GaSb is also ignored, resulting in potential profiles that deviate only slightly from the flat band approximation. In this work, the emitter is treated as a 2D well for electrons, consistent

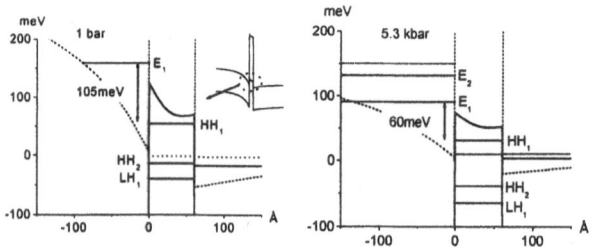

Fig. 3 60Å GaSb DHET band profiles at the NDR peak bias. 1 bar at 306mV, 5.3 kbar at 142mV. The emitter 2D electron concentrations are 2.15×10^{12} cm^{-2} and 7.8×10^{11} cm^{-2} respectively.

with the true band bending expected for a type II interface. We also point out that the rapid leakage of GaSb valence electrons into the InAs conduction band at the +ve electrode will result in a significant build-up of holes in the GaSb. As a result, the solutions for all of the biased band profiles presented here have the hole quasi Fermi level pinned to the electron quasi Fermi level in the +ve electrode.

5 Discussion

The band profile given in Fig. 3 for the 60Å DHET, biased to its current peak at 1 bar, shows that the lowest electron sub-band is far from being resonant with *any* of the hole sub-bands. The electron and hole ground states are separated by ~ 105meV, suggesting that an inelastic process is responsible for the NDR. This is in agreement with the analysis of NDR in 'GaAs like' SHETs reported previously [1].

Hydrostatic pressure increases the bulk band gaps and decreases the band overlap at ~10 meV/kbar [1]. This causes the peak current and bias to shift towards the origin. The decrease in peak current is adequately explained by the reduction in the emitter electron concentration. What determines the peak bias is not entirely clear, but analysis of the band profiles for all of the pressures studied shows that the electrons and holes are always uncrossed by 55 − 105meV at the peak bias, again consistent with an inelastic conduction mechanism.

Hydrostatic pressure pushes the light-hole sub-band below the region of band overlap, as shown in Fig 3. The persistence of clear NDR provides conclusive proof that the resonance peak does not correspond to electrons tunneling through the barrier with the energy of a light-hole sub-band.

Elastic tunneling can still give rise to NDR in the absence of resonant hole sub-bands, as observed in InAs/GaAsSb/InAs single barrier structures, due to the bias dependence of the transmission coefficient [5]. However, the tunnel current calculated with the 2 band model of ref. [2] is orders of magnitude higher than that observed in thin DHETs at low temperature. For example, the 40Å GaSb DHET of Fig.1 passes a current of only 14 A cm^{-2} at 300mV, which is much less than the ~ 400kA cm^{-2} predicted by the 2 band model. Moreover, the reappearance of a weak resonant feature at 300K is very strong

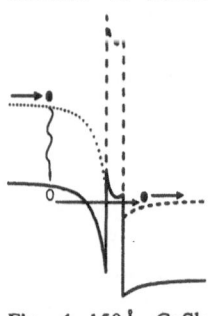

Fig. 4 150Å GaSb DHET at 518mV, showing Zener tunneling and EL.

evidence for the *inelastic* nature of the NDR. We conclude that single barrier elastic tunneling is also an unlikely cause of the NDR in these structures.

The 3μm luminescence observed in all samples is explained by the onset of Zener tunneling. Once the electron quasi Fermi levels are separated by more than the InAs bandgap, valence electrons in the emitter can tunnel into the conduction band of the +ve contact. This is consistent with the band profile shown in Fig. 4 and the observation of bulk InAs luminescence from the emitter.

Hole dispersions were calculated using a 4 band *k.p* model to give an insight into the thickness and temperature dependence of the resonant transport. It turns out that only the HH$_1$ sub-band is occupied at low temperatures in all the cases studied, due to the large density of states offered by the flattening of the HH$_1$ dispersion. The HH$_1$ states in this flattened region of *k*-space have a substantial light-hole component, which we believe may be a key element in the inelastic conduction process. For all GaSb thicknesses above 60Å, HH$_1$ states with a substantial light-hole component are occupied at low temperatures. However, this is not the case for the thinner samples, where the holes need to gain thermal energy to access the states with a large light-hole component. This is shown in Fig. 5, where the hole Fermi level crosses the flattened region of HH$_1$ for the 80Å GaSb DHET, but lies in the nearly parabolic region in the 40Å GaSb DHET. It could partly explain the dramatic temperature dependence of the thinner samples, and the relative insensitivity to temperature of the thicker samples.

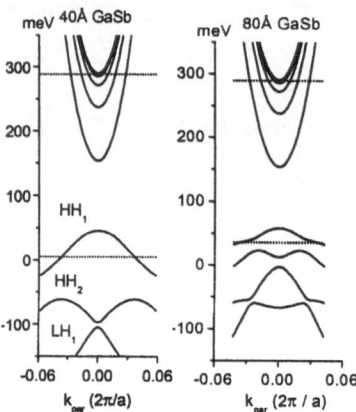

Fig. 5 Dispersion curves for 40Å and 80Å GaSb DHETs at ~280mV bias. Quasi electron and hole Fermi levels are shown by dotted lines.

4 Conclusion

The background current and interband luminescence in InAs/GaSb/InAs DHETs is explained by Zener tunneling. The resonant conduction and NDR in these structures at lower biases cannot be explained adequately in terms of a coherent electron–light-hole resonance. The evidence suggests that the resonant conduction is inelastic and that it may be strongly dependent on the occupation of heavy-hole states with a significant light-hole component.

3 References

1. U.M.Khan-Cheema, P.C.Klipstein, N.J. Mason *et al.*, Proc. **ICPS23** (1996) 2271.
2. Yu et al., Appl. Phys. Lett. **57**, (1990) 2675.
3. Liu et al., Phys. Rev. B. **55**, (1997) 7073
4. Ting et al, Phys. Rev. B. **45**, (1992) 3583
5. Soderstrom et al, Appl. Phys. Lett. **53**, (1989) 1348

Resonant Tunneling of Holes under in-plane uniaxial stress

Y. C. Chung[1], T. Reker[1], L. E. Bremme[1], R. Grey[2], and P. C. Klipstein[1]

[1] Clarendon Lab., Dept. of Physics, Univ. of Oxford, Parks Road, Oxford OX1 3PU, U.K. email: y.chung1@physics.ox.ac.uk
[2] III-V Facility, Dept. of Elec. Eng., Univ. of Sheffield, Mappin Street, Sheffield SI 3JD

Abstract It has long been recognised that the resonant tunneling of holes through *p*-type double-barrier structures (DBSs) is considerably more complicated than that of electrons in *n*-type devices, due to **k.p** mixing between heavy-hole (HH) and light-hole (LH) states. In this work we use an in-plane uniaxial stress to tune continuously the mixing between the HH and LH states. We also solve the four band **k.p** Hamiltonian (including stress terms) and achieve good agreement between experimental and calculated results. Our results show clearly that the flow of holes through *off zone centre* states dominates the resonant tunneling current in *p*-type structures.

1 Introduction

Resonant tunneling of electrons was first investigated by Tsu and Esaki [1] in GaAs/Ga$_{1-x}$Al$_x$As superlattices. Since then, an extensive amount of work has been done on electron tunneling in DBSs, both theoretically and experimentally, e.g. refs [2-4]. Resonant tunneling of holes was first investigated by Mendez el. al.[5]. They found that it was not possible to explain their results without considering the mixing between light hole (LH) and heavy hole (HH) states. It was quickly recognised that a multi-band model must be employed to consider the tunneling properly, e.g. refs [6-8]. In such models, the momentum parallel to the layers, $k_\parallel$, needs to be considered since it affects the mixing between LH and HH states and this can change the transmission probability dramatically. Many attempts have been made to prove experimentally that the current through off-zone centre states dominates the *I-V* characteristics in such devices. In all cases, the basic idea is to compare the shapes and peak positions of experimental

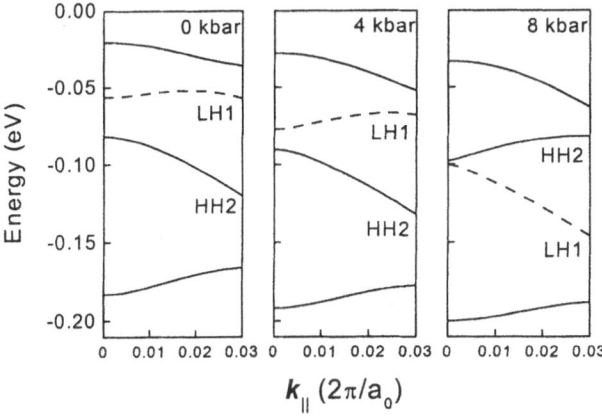

Fig. 1 Sub-band dispersions of a 60 Å GaAs well, grown along [001] and subjected to uniaxial stress along [100].

and calculated *I-V* curves. However, such an approach is prone to be very unreliable since either a one monolayer difference in the well or barrier thickness, or a slight change in the doping concentration in the emitter layer, can lead to a huge change in the shape of the predicted *I-V* characteristic or its peak positions. Also, the accumulation and depletion regions, charge build up in the well, and any inelastic tunneling processes are very difficult to model. In this work, an in-plane uniaxial stress is used to tune the mixing between HH and LH states in a DBS. To show that the off-zone centre current is dominant, we compare experimental results not with the full *I-V* trace but rather with the calculated dispersion relation at various uniaxial stress values.

2 Uniaxial stress and current calculation

To calculate the sub-band dispersions and the current, the four band **k.p** Hamiltonian (including uniaxial stress terms) was solved by a finite element approach [8], which is known to be numerically stable. The transmitted current density through the DBS is:

$$J = \frac{2 \times 4}{(2\pi)^3} \int_{k_\parallel=0}^{k_F} \int_{k_z=0}^{k_F} \int_{\theta=0}^{\pi/2} T(k_\parallel, \theta, k_z) \qquad (1)$$
$$\cdot v(k_\parallel, \theta, k_z) \cdot f_E \cdot (1 - f_C) k_\parallel dk_\parallel dk_z d\theta$$

where v is the emitter group velocity in the z direction, and f_E and f_C are the hole occupation factors in the emitter and the collector, respectively. The transmission coefficient, T, was calculated from the probability flux, J_p, in the emitter and collector, where it can be shown that

$$J_p = \frac{i\hbar}{2} \sum_{r=1}^{4} \left(\phi_r^* \frac{\partial \phi_r}{\partial z} - \frac{\partial \phi_r^*}{\partial z} \phi_r \right) / m_r$$
$$+ \frac{i}{\hbar} \left[S^* \left(\phi_1 \phi_2^* - \phi_3 \phi_4^* \right) - S \left(\phi_1^* \phi_2 - \phi_3^* \phi_4 \right) \right] \qquad (2)$$

in which $\phi_1,\ldots,\phi_4$ are the envelope functions of the four spin components $^3/_2$, $^1/_2$, $-^1/_2$, and $-^3/_2$ respectively, $m_0/m_r = (\gamma_1 \pm 2\gamma_2)$ where the positive sign is for $r=2$ and $r=3$, $S = -\sqrt{3}(\hbar^2/m_0)\gamma_3 (k_x - ik_y)$, and γ_i are the GaAs Luttinger parameters.

An in-plane uniaxial stress applied along [100] to a quantum well grown along [001] induces considerable changes to the sub-band structure. Fig. 1 shows the calculated dispersions of the sub-bands for a 60Å GaAs well at different stress values. The LH1 and HH2 states cross at the zone centre just below 8 *kbar*. On the other hand, the LH1 and HH2 states at $k_\parallel \sim 0.01$ to $0.02 \times 2\pi/a_0$ (a_0 is the cubic lattice parameter) do not cross and their energies decrease at about the same rate.

3 Experiment, results, and discussion

A GaAs/AlAs DBS was grown by MBE along [001] with 51Å undoped AlAs barriers, a 60Å undoped GaAs well, and 51Å undoped GaAs spacers. The surrounding GaAs layers were Be-doped with $p=5\times10^{17}\text{cm}^{-3}$. Circular mesas of 200µm diameter were fabricated on 3.2mm × 12mm rectangular specimens of the 0.4mm thick GaAs wafer. Uniaxial stress was applied along [100] by pressurising the shorter sides of the specimens in a specially designed pressure cell.

Figure 2(c) shows how the LH1 and HH2 current peaks are shifted by the applied uniaxial stress. From the dispersion of the sub-bands in the well (Fig. 1), it can easily be seen that for a current flowing through the zone centre the peak positions will behave as in Fig. 2 (a), with the LH1 peak crossing the HH2 peak at around 7 *kbar*. Surprisingly, the experimental results show no evidence of such a crossing. However, the experimental results agree rather well with Fig. 2 (b), which shows the confinement energies of the sub-bands at $k_\parallel = 0.16 \times 2\pi/a_0$, a wavevector comparable with typical Fermi wavevectors in the emitter. This clearly demonstrates that the current through *off zone centre* states is dominant. It has long been thought that, even for the LH resonance, most of the current arises due to

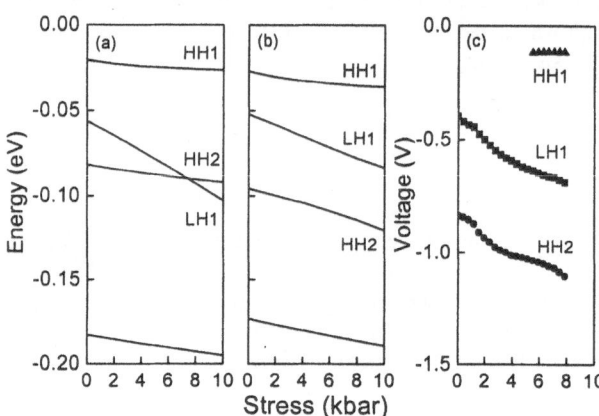

Fig. 2 Calculated confinement energies for a 60 Å wide GaAs well as a function of uniaxial stress, at (a) $k_\parallel = 0$ and (b) $k_\parallel = 0.16 \times 2\pi/a_0$. (c) Measured positions of the HH1, LH1 and HH2 current peaks.

incoming HH states, since $k_\parallel$ values are much larger for HH compared to LH states and the transmission coefficient, T, increases with $k_\parallel$ by several orders of magnitude [6]. Our results provide the first definitive experimental proof of this principle.

Figure 3 compares the experimental and calculated I-V results for various uniaxial stresses. The emitter and collector are treated as flat bands and no inelastic tunneling or charge build up in the well is considered in the

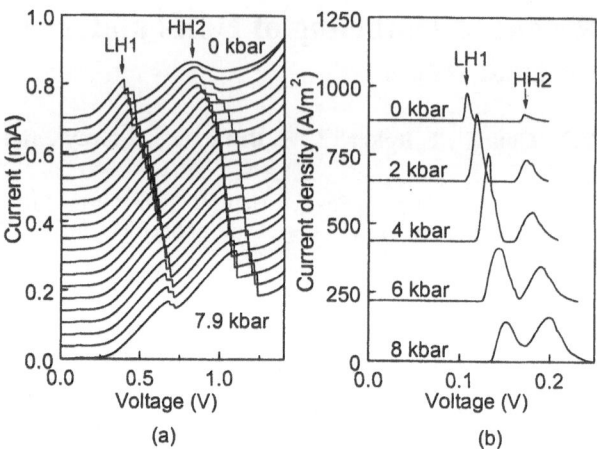

Fig. 3 Experimental and calculated I-V traces.

calculation. As can be seen from Fig. 3(a) the resonant component of the current measured for the LH1 peak goes through a maximum at around 3.5 *kbar*, while that for HH2 increases monotonically with stress. Qualitatively, this matches the calculated current in Fig. 3(b) quite well. Also, the calculated peak positions in Fig. 3(b) follow the same trend as in Figs. 3(a) and 2(b).

4 Conclusions

The dependence of hole resonant tunneling on uniaxial stress has been demonstrated. Stress allows the crucial mixing between LH and HH states to be tuned continuously. Our results provide the first clear experimental confirmation that the current through a hole tunneling device is dominated by *off zone centre* tunneling.

5 Acknowledgements

This work was funded by the E.P.S.R.C. (UK). T.R. and L.E.B. acknowledge support from the Friedrich Flick Förderungsstiftung and the Rhodes Trust, respectively.

References

1. R. Tsu, and L. Esaki, Appl. Phys. Lett. **22**, (1973) 562
2. H. Ohnishi, T. Inata, S. Muto, N. Yokoyama, and A. Shibatomi, Appl. Phys. Lett. **49**, (1986) 1248
3. P. J. Turley, C. R. Wallis, and S. W. Teitsworth, Phys. Rev. B **47**, (1993) 12 640
4. G. Klimeck, R. Lake, R. C. Bowen, W. R. Frensley, and T. S. Moise, Appl. Phys. Lett. **67**, (1995) 2539
5. E. E. Mendez, W. I. Wang, B. Ricco, and L. Esaki, Appl. Phys. Lett. **47**, (1985) 415
6. R. Wessel, and M. Altarelli, Phys. Rev. B **39**, (1989) 12 802
7. C. Y. Chao, and S. L. Chuang, Phys. Rev. B **43**, (1991) 7027
8. Y. X. Liu, D. Z. Ting, and T. C. McGill, Phys. Rev. B, **54** (1996) 5675

Resonant tunneling through zero-dimensional impurity states: Effects of a finite temperature

P. König, U. Zeitler, J. Könemann, T. Schmidt, R. J. Haug

Institut für Festkörperphysik, Universität Hannover, Appelstraße 2, 30167 Hannover, Germany
e-mail: zeitler@nano.uni-hannover.de

Abstract We have performed temperature dependent tunneling experiments through a single impurity in an asymmetric vertical double barrier tunneling structure. In particular in the charging direction we observe at zero magnetic field a clear shift in the onset voltage of the resonant tunneling current through the impurity. With a magnetic field applied the shift starts to disappear. The experimental observations are explained in terms of resonant tunneling through a spin degenerate impurity level.

Resonant tunneling experiments through a single impurity [1] mainly reflect the energetic position of an impurity level and the Fermi-Dirac distribution in the emitter. In order to refine the concept of such tunneling processes, in particular as far as the temperature dependence is concerned, we investigated an asymmetric double-barrier resonant-tunneling device (DBRTD) consisting of a 10 nm GaAs quantum well (QW) and two 5 nm and 8 nm wide $Al_{0.3}Ga_{0.7}As$ barriers. The structure is sandwiched between highly doped GaAs electrodes (Si-doped with $n_{Si} = 4 \times 10^{17}$ cm^{-3}) separated from the barriers by a 7 nm thick nominally undoped GaAs spacer. From this wafer we processed a vertical tunneling diode with a mesa diameter of 2 μm. When applying a bias voltage V between the top and the bottom electrode the sample displays the normal behavior of a resonant tunneling diode with pronounced current peaks at $V_r^+ = 82$ mV and $V_r^- = -183$ mV. For lower voltages additional small current steps are observed. We assign them to tunneling through zero-dimensional impurity levels originating from donor atoms unintentionally present in the GaAs QW [1,2]. In this work we concentrate on the first current step due to resonant tunneling through the energetically lowest lying impurity state. We will show that the spin degeneracy of this ground state and the Coulomb blockade causes a shift of the step position as a function of temperature [3].

For a theoretical description of the effects observed we regard the situation as sketched in Fig. 1. A zero-dimensional impurity with a spin-degenerate ground state ϵ is situated inside the central well of the DBRTD. It is coupled via two tunneling barriers (characterized by tunneling rates $\Gamma_L < \Gamma_R$) to three-dimensional reservoirs with chemical potentials μ_L and μ_R. Applying a finite transport voltage V to the structure induces a difference in the chemical potentials, $eV = \mu_L - \mu_R$. A resonant current through the impurity shows up as a current step at a voltage $V_+ > 0$ (or $V_- < 0$, respectively) when μ_L (or μ_R, respectively) equals ϵ [3]. Two possible spin

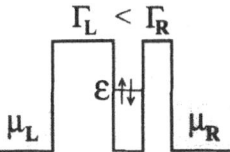

Fig. 1 Energy-level scheme showing the two-fold spin-degenerate ground state of an impurity at energy ϵ coupled to electrodes with chemical potentials μ_L and μ_R. The tunneling rates through the two barriers are Γ_L and Γ_R.

states can be occupied by an electron during a tunneling event. However, Coulomb blockade prohibits simultaneous occupancy of both states at the same time. With this condition the tunneling current I through the impurity can be calculated as [4]:

$$I = 2e\,\Gamma_L\Gamma_R \frac{f_L(\epsilon) - f_R(\epsilon)}{\Gamma_L + \Gamma_R + \Gamma_L f_L(\epsilon) + \Gamma_R f_R(\epsilon)} \qquad (1)$$

where $f_L(\epsilon)$ and $f_R(\epsilon)$ are the Fermi-Dirac distributions in the left and right reservoirs.

For a finite bias $|eV| \gg kT$ we have $f_R \ll 1$ for $V > 0$ (and $f_L \ll 1$ for $V < 0$) and the current as a function of an applied bias V simplifies to:

$$I = 2e\Gamma_L\Gamma_R \frac{f_L(\alpha_L e(V_+ - V))}{\Gamma_L + \Gamma_R + \Gamma_L f_L(\alpha_L e(V_+ - V))} \quad (V > 0) \tag{2a}$$

$$I = -2e\Gamma_L\Gamma_R \frac{f_R(\alpha_R e(V - V_-))}{\Gamma_L + \Gamma_R + \Gamma_R f_R(\alpha_R e(V - V_-))} \quad (V < 0) \tag{2b}$$

The prefactors α_L and α_R account for the fact that only a part of the bias voltage drops between the emitter reservoir and the impurity, see also below.

An experimental example for the I-V characteristic of our sample is shown in Fig. 2. Clear current steps due to resonant tunneling through the energetically lowest lying impurity level are observed for both bias directions in the two top figures. For positive bias (right panel) the electrons tunnel into the impurity through the thicker barrier Γ_L and leave it through the thinner one Γ_R. The impurity is mostly empty and both degenerate states of ϵ are available for tunneling (non-charging direction). For negative bias the tunneling current is limited by the small tunneling Γ_L for electrons leaving the impurity. As a consequence, one of the two states of ϵ is mostly occupied and Coulomb blockade suppresses a simultaneous tunneling event through the other spin state (charging direction). Therefore, the resonant current is smaller than in the non-charging direction. This current

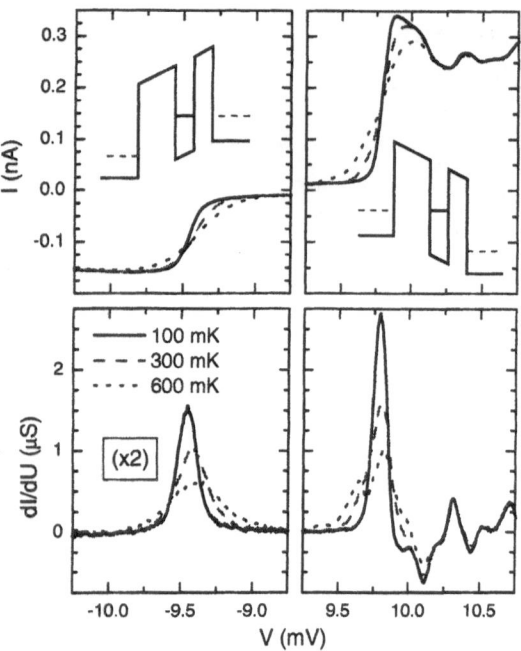

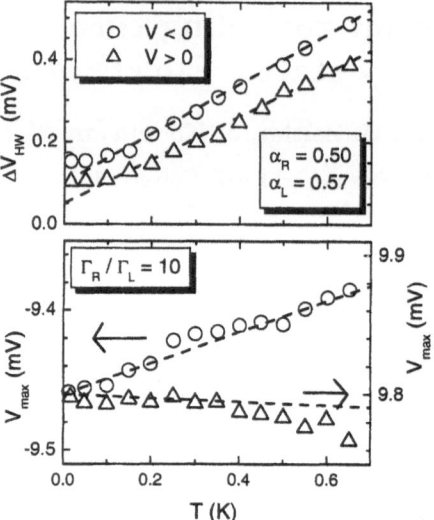

Fig. 3 Top: Temperature dependence of the half-width ΔV_{HW} of the conductance peak for the two bias directions. The lines show fits for lever factors.
Bottom: Experimental temperature dependent shift of the conductance-peak position for both bias directions (symbols) compared to the theoretical behavior using $\Gamma_R/\Gamma_L = 10$ (dashed lines).

Fig. 2 Current voltage characteristics (top) and differential conductance (bottom) of a vertical asymmetric DBRTD for different temperatures. The steps are due to resonant tunneling through a single impurity. The left panels represent the charging direction, the right panels show the non-charging direction at positive bias voltages. The insets sketch the energy-level schemes of the structure at the step voltages V_- and V_+ for the two bias directions.

suppression follows from Eqs. (2a) and (2b) which predict for $\Gamma_R \gg \Gamma_L$ and $T = 0$ current steps $\Delta I^+ = 2e\Gamma_L$ in the non-charging direction and $\Delta I^- = -e\Gamma_L$ in the charging direction.

Effects of temperature can be visualized more clearly when plotting the differential conductance $G = dI/dV$ as a function of bias voltage. The current steps show up as peaks in G. Fitting its half width ΔV_{HW} as a function of temperature to Eqs. (2a) and (2b) allows to determine the lever factors α_L and α_R, see Fig. 3, top panels. For both bias directions we find $\alpha \approx 0.5$ indicating that the impurity is approximately situated half way between the two reservoirs.

Eqs. (2a) and (2b) also predict that the maximal conductance at finite temperature is not always observed at the voltage where the chemical potential in the source (i.e. μ_L for positive bias and μ_R for negative bias) equals ϵ. In particular in the charging direction the step voltage shifts to a lower absolute value when T is increased. This is indeed observed experimentally and shown in the left bottom curve of Fig. 3. The solid line shows a theoretical fit as expected from Eq. (2b) and yields $\Gamma_R/\Gamma_L = 10$ in reasonable agreement with the expected tunneling rates through the two barriers. For the non-charging direction the shift observed is negligible and consistent with $\Gamma_R/\Gamma_L = 10$.

When applying a magnetic field the Zeeman effect lifts the spin-degeneracy of ϵ. As a consequence the cur-

rent step in the SET direction splits-into two steps corresponding to tunneling of electrons with a different spin orientation [5,6]. From the magnetic field dependence of this split we extract a Landé factor $g^* = -0.14$ for the impurity ground state [6]. In the charging direction Coulomb blockade still prohibits simultaneous tunneling through both energy levels and only one current step is observed. However, since the degeneracy is lifted the shift of the step voltage is no more observed as long as $3.5k_BT > g^*\mu_B B$. For $3.5k_BT < g^*\mu_B B$ the energy splitting of the ground state ϵ is no more resolved and ϵ can be again regarded as a virtually degenerate level. Finally it is worthwhile remarking that the effects described in this paper also influence the spin splitting of the current steps in the SET direction and a direct determination of g^* is only possible if $3.5k_BT < g^*\mu_B B$.

In conclusion we have shown that the actual voltage position of a resonant current step through a zero-dimensional state is not only given by the energetic position of this state but can also be strongly influenced by temperature.

References

1. M. A. Reed et al., Phys. Rev. Lett. **60** (1988) 535.
2. T. Schmidt et al., Europhys. Lett. **36** (1996) 61.
3. Here and in the following we define the step voltages V_+ and V_- as the voltages where the current steps exhibit the largest slope.
4. H. Schoeller, Transport Theory of Interacting Quantum Dots, in Mesoscopic Electron Transport, NATO ASI Series E - Vol. 345, pp. 291-330, (Kluwer, Amsterdam, 1997).
5. M. R. Deshpande et al., Phys. Rev. Lett. **76** (1996) 1328.
6. P. König, T. Schmidt, and R. J. Haug, cond-mat/0004165.

Temperature Dependence of Resonant Tunneling Characteristics in a p-type GaAs/AlAs Double-Barrier Structure

M. Ono, N. Nishioka, M. Morifuji, and C. Hamaguchi

Department of Electronic Engineering, Osaka University, Suita City, Osaka 565-0871, Japan

Abstract   We report temperature dependence of resonant tunneling characteristics of holes in a GaAs/AlAs double-barrier structure. The measured I-V curves show distinct peaks associated with resonant tunneling through quasi-bound light-hole, heavy-hole, and spin-orbit split-off states in the quantum-well region. From measurements at various temperatures, we have found that each peak shows different behavior with rising temperature. Complex behavior of peak shifts suggests effect from valence band dispersion.

1 Introduction

Since the first observation of resonant tunneling of holes in 1985 [1], it has shown various interesting behavior mainly due to complexity of the valence band structure. [2]-[4] For example, resonant peaks of holes measured in magnetic fields show shifts reflecting in-plane dispersion curves of the valence states in quantum well region. Inter-band tunneling effect has been also pointed out, since a few bands can be involved in a tunneling process. [5] However, the resonant tunneling of holes has not attracted so much attention as that of electrons. In this paper, we investigate thermal behavior of resonant tunneling phenomena in a p-type GaAs/AlAs double-barrier structure where holes are majority carriers.

2 Measurements

Table 1 shows the layer structure of the p-type resonant tunneling diode used in this study. The 5.1 nm undoped-GaAs quantum well is cladded by 4.2 nm undoped-AlAs barriers. This double-barrier structure is put between p-doped reservoir regions with 5.1 nm GaAs buffer layers in between. The wafer grown in an MBE was fabricated

Table 1 Layer structure of p-type resonant tunneling device.

Material	Doping	Layer thickness
GaAs	$p=2\times10^{18}cm^{-3}$	600 nm
GaAs	$p=1\times10^{18}cm^{-3}$	100 nm
GaAs	$p=5\times10^{17}cm^{-3}$	100 nm
GaAs	undoped	5.1 nm
AlAs	undoped	4.2 nm
GaAs	undoped	5.1 nm
AlAs	undoped	4.2 nm
GaAs	undoped	5.1 nm
GaAs	$p=5\times10^{17}cm^{-3}$	100 nm
GaAs	$p=1\times10^{18}cm^{-3}$	100 nm
GaAs	$p=2\times10^{18}cm^{-3}$	2000 nm

n+ GaAs substrate

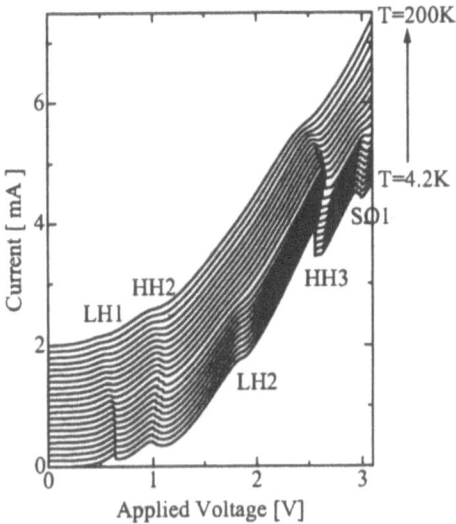

Fig. 1 I-V curves measured from the resonant tunneling diode with negative applied bias at temperatures ranging from 4.2 K to 200 K. LH, HH, and SO stands for the resonance signal with the light-hole, heavy hole, and spin-orbit split-off states, respectively.

by photolithography and chemical wet etching to define device area. Deposition of AuZn alloy at the upper surface, lifting off, and sintering at 350 °C for 1 min. in N_2 gas ambient were sequentially carried out in order to form ohmic contacts. AuGe alloy was deposited on the backside. The size of mesa structure is $20 \times 20 \ \mu m^2$.

The current-voltage (I-V) characteristics measured at temperatures from 4.2 K to 200 K are shown in Figure 1. Five peaks are clearly observed in the I-V curves. These peaks were assigned to the resonance with light hole (LH), heavy hole (HH), and spin-orbit split-off (SO) states in the quantum well region as shown in Fig. 1, where the number 1, 2, and 3 denote the index of the quasi-bound states in the quantum well region. This assignment was done regarding Ref. [3] where resonant tunneling phenomena in the similar structure was investigated. We note that almost similar I-V characteristics were obtained with both positive and negative applied bias. From this figure, we see that these peaks show shifts as temperature increases, along with distinct broadening at higher temperatures.

Fig. 2 shows the peak voltage plotted as functions of the temperature. Enlarged figures for LH1, HH2, and

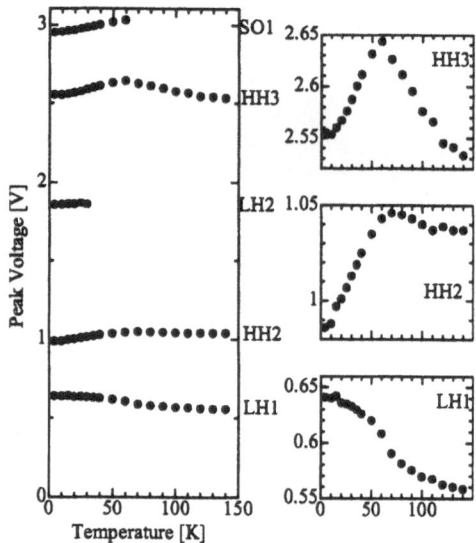

Fig. 2 Peak voltages of resonant tunneling plotted as functions of temperature (left hand side). Right hand side figures are enlarged ones for LH1, HH2, and HH3 resonant peaks.

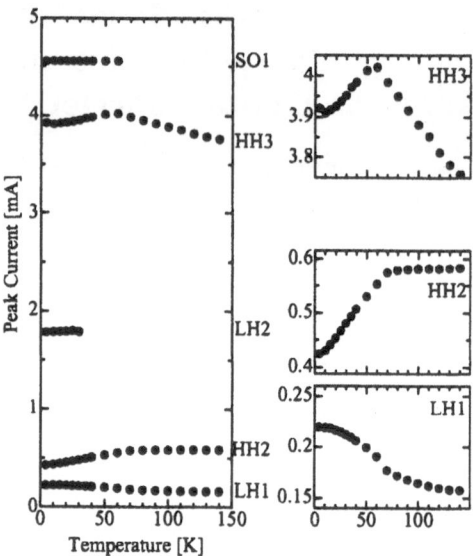

Fig. 3 Peak current of resonant tunneling plotted as functions of temperature (left hand side). Right hand side figures are enlarged ones for LH1, HH2, and HH3 resonant peaks.

HH3 peaks are also shown in the right hand side of this figure. These figures indicate that the shift of peak voltage is different for each resonant level. The LH1 peak shifts to lower voltage with rising temperature. On the other hand, the HH2 peak shifts to higher with rising temperature at low temperature region (from 4 K to 60 K), whereas at higher temperatures, this peak is almost independent of temperature. Like the HH2 peak, HH3 peak shows shift to higher voltage at temperatures lower than 60K, however, this peak shows lower shift at higher temperatures.

Fig. 3 shows change of peak current of the resonant peaks plotted as functions of temperature. The peak current also shows complex behavior as temperature increases. It can be seen that behavior of peak current is similar to that of peak voltage.

3 Discussion

It is known that tunneling characteristics is affected directly by change of carrier distribution. Thermal change of carrier density and the Fermi energy in the hole reservoir region can also affect tunneling characteristics. Furthermore, effect from phonon scattering may be a cause of the peak change. These effects are known to give rise to broadening of resonant peaks. The observed peak broadening at higher temperatures can be explained by these effects. However, none of such mechanisms can not solely explain the observed complex shifts which depends on the resonant levels .

The observed shift may be explained by considering the valence band dispersion curves in to consideration. It has been known that resonant peaks of holes show shift in magnetic fields perpendicular to the current. [4] Like the peak shifts in this study, the peak shift in per-

pendicular magnetic fields are different for different resonant levels. This phenomenon is explained as follows: in a perpendicular magnetic field, a hole can have an in-plane momentum, and thus resonant peaks show shifts corresponding to the in-plane valence dispersion curves. Since in-plane dispersion curves are different for each level, peak shifts can be different for each level.

It is, however, still unclear how the dispersion curves affect the thermal behavior of tunneling current. The observed complex behavior of resonant peaks may be attributed to combined effects of a few mechanisms. The voltage shift at lower temperatures suggests that dispersion curves in the valence band play an important roll. On the other hand, the peak shift in the higher temperature region is accompanied with broadening of peaks, which suggests effect of thermal change of carrier distribution.

References

1. E. E. Mendez, W. I. Wang, B. Ricco, and L. Esaki, Appl. Phys. Lett. **47** (1985) 415.
2. U. Gennser, V. P. Kesan, D. A. Syphers, T. P. Smith III, S. S. Iyer, and E. S. Yang, Phys. Rev. Lett. **67**, (1991) 3828.
3. R. K. Hayden, L. Eaves, M. Henini, D. K. Maude, J. C. Portal, and G. Hill, Appl. Phys. Lett. **60**, (1992) 1474.
4. R. K. Hayden, D. K. Maude, L. Eaves, E. C. Valadares, M. Henini, F. W. Sheard, O. H. Hughes, J. C. Portal, and L. Cury, Phys. Rev. Lett. **66**, (1991) 1749.
5. M. Morifuji and C. Hamaguchi, Phys. Rev. B **52**, (1995) 14131.

Proc. 25th Int. Conf. Phys. Semicond., Osaka 2000 (Eds. N. Miura and T. Ando)

Negative differential resistance and current self-oscillation in doped GaAs/AlAs superlattices

J.N. Wang [a],[*], C.Y. Li [a], X.R. Wang [a], B.Q. Sun[b], Y.Q. Wang [a], W.K. Ge [a], D.S. Jiang [b], and Y.P. Zeng [b]

[a] Physics Department, Hong Kong University of Science and Technology, Clear Water Bay, Kowloon, Hong Kong, China

[b] Institute of Semiconductors, Chinese Academy of Sciences 100083, Beijing, China

Abstract We observed the transverse magnetic field induced transition from static to dynamic electric field domain formation in doped GaAs/AlAs superlattices. A dynamic dc voltage band was found emerging from each sawtooth-like branch of the current-voltage characteristics in the transition process. Within each dynamic dc voltage band at a fixed magnetic field the current self-oscillation frequency was increased while the average current was decreased. These results were explained by a general analysis of stability of the sequential tunneling current in superlattices.

1. Introduction

In last few years, self-sustained temporal current oscillations, or so-called current self-oscillation, under a fixed dc bias corresponding to dynamic electric-field domain (EFD) formation have been observed in weakly coupled semiconductor superlattices (SL's) [1-5]. It has been shown that current self-oscillation only occurs within a certain range of carrier concentration present in these SL's. However, above this carrier concentration range the current-voltage characteristics of these SL's exhibit discontinuities on the sequential resonance tunneling plateau and show a series of sawtooth-like current branches corresponding to static EFD formation [6,7]. Here, we report that applying an external magnetic field parallel to the SL layers (so-called transverse field) can also produce the transition between static and dynamic EFD formation. During the transition process, dynamic dc voltage band, within which temporal current self-oscillation was observed, was found emerging from each sawtooth-like branch in the transition process. These results were explained by a general analysis of the stability of the sequential tunneling current in superlattices.

2. Results and discussion

The doped GaAs/AlAs SLs studied in this work were grown by molecular beam epitaxy. The SL consists of 10-30 periods of 14nm GaAs well and 4nm AlAs barrier. The SL is sandwiched between two n^+-GaAs layers. The central 10nm of each GaAs well was doped with Si ($n=2\times10^{17}cm^{-3}$). The sample was fabricated into $0.2\times0.2mm^2$ mesas. The current-voltage, I(U), characteristics of the samples was measured with

Corresponding Author, E-mail:phjwang@ust.hk

magnetic field ranging from 0 to 14T and sample temperature at 1.6K by a HP4155A semiconductor parameter analyzer. The current self-oscillations were recorded by a HP54600A digital oscilloscope.

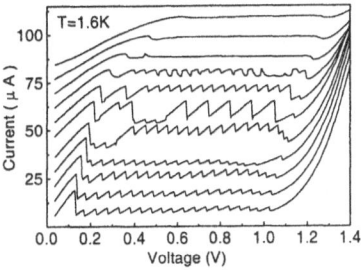

Figure 1. I(U) curves measured at 1.6K with increasing B showing the transition from static to dynamic EFD formation for the first current plateau. Curves are offset for clarity and from bottom to top B=0T-10T with 1T steps.

A typical I(U) curve of the SL samples studied at low temperature showed three sequential resonance tunneling plateaus with a series of sawtooth-like current branches corresponding to static EFD formation. In this study, we focus on the first plateau. A typical transition process from static to dynamic EFD formation caused by increasing transverse magnetic field is shown in Figure 1. When B < 6T, the I(U) of the SL exhibits a series of sawtooth-like current branches on the plateau corresponding to static EFD formation. At B > 8T, the sawtooth-like branches disappear, a flat current plateau is observed in the I(U), and the current self-oscillations are observed under dc biases across the whole plateau indicating the dynamic EFD formation. During this transition process and at intermediate magnetic fields, as is more clearly shown in Figure 2 which is an enlargement of Figure 1, a dynamic dc voltage band (indicated by open squares in Fig.2) emerges from each sawtooth-like current branch at the beginning of transition process (see I(U) curve at B=6.7T in Fig.2), within this band current self-oscillations or dynamic EFD formation are observed. These dynamic dc voltage bands expand towards higher voltage side within each sawtooth-like current branch, squeeze out the static regions (indicated by solid squares in Fig.2) in which static EFD's are formed, and join up together to turn the whole plateau into dynamic EFD formation as the B increases (see I(U) curves at B>7.1T).

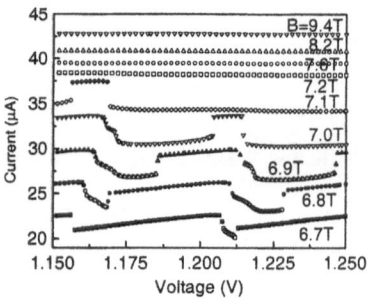

Figure 2. Enlargements of Fig.1 show the development of dynamic voltage bands as a function of B. Open symbols indicate dynamic regions while solid symbols static regions. Curves are offset for clarity.

These results can be understood in terms of the general analysis of instabilities of the sequential tunneling current in SL given by Wang et al. [8]. They show that so long as the absolute magnitude of the NDR of one barrier does not exceed the sum of the positive differential resistance (PDR) of the remaining barriers, static EFD's can form. If this condition is not met, the EFD's are unstable and current self-oscillation occurs.

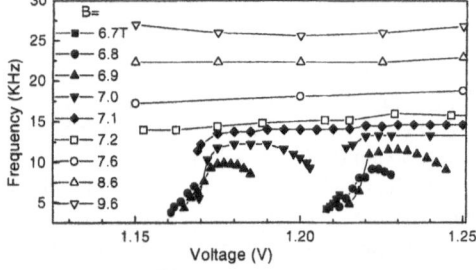

Figure 3. Voltage dependence of the current self-oscillation frequency in the dynamic dc voltage bands shown in Fig.2 at various B indicated.

An applied transverse magnetic field causes redistribution of the tunneling electron momentum and energy. In order to conserve the momentum and energy in the tunneling process, the resonant peak voltage shifts to a higher value, the peak current decreases, and the width of the resonance peak increases with increasing B. As a result, the magnitude of the NDR increases with increasing B and produces the transition from static to dynamic EFD formation [9]. In this general analysis it is also predicted that the current self-oscillations, associated with the dynamic EFD formation, occur initially within a certain dc voltage range (or voltage band) on each sawtooth like current branch. Therefore, a series of dynamic voltage band is formed across the plateau. The dynamic voltage bandwidth is increased with increasing the magnitude of the NDR [10].

Before the dynamic dc voltage bands join up together, the average current within the dynamic dc voltage band is bias dependent and slightly decreases with increasing B, but after the dynamic voltage bands join up together, the average current is almost bias Running title

independent and increases with increasing B (see Fig.2). Figure 3 shows the voltage dependence of the current self-oscillation frequency at various B for the dynamic dc voltage bands shown in Fig.2. In comparison to the I(U) curves in Fig.2, we find that the frequency is voltage dependent and inversely proportional to the average current at a fixed B before the dynamic dc voltage bands join up, as it is more clearly shown in Figure 4. However, after the dynamic dc voltage bands join up together, the frequency is both voltage and the average current independent at a fix B. If we eliminate B induced bias increase for higher B, the frequency is increased with increasing B. Some of above results can be understood in term of motion of the system along a limit cycle [11]. Some are not clear.

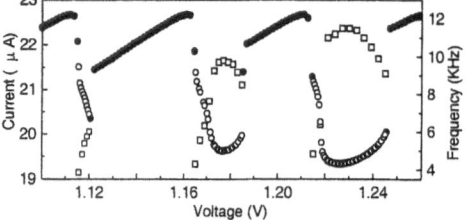

Figure 4. The voltage dependence of the average current (circle) and the current oscillation frequency (open square) at B=6.9T. Solid and open squares show the static and dynamic regions in I(U), respectively.

Acknowledgement: the work is supported by the Research Grant Council, Hong Kong, China.

Reference:
[1] R.Merlin, S.H.Kwok, T.B.Norris, H.T.Grahn, K.Ploog, L.L.Bonilla, J.Galan, J.A.Cuesta, F.C.Martinez, and J.Moleara, in Proc. 22nd Int.Conf. on the Physics of Semiconductors, editor by D.J.Lockwood (Singapore, World Scientific 1995) vol.2 1039
[2] H.T.Grahn, J.Kastrup, K.Ploog, L.L.Bonilla, J.Galan, M. Kindelan, and M.Moscoso, J.J..Appl.Phys.34, 4526 (1995)
[3] Baoquan Sun, Desheng Jiang, and Xiaojie Wang, Semicon. Sci.Technol. 12, 401 (1997)
[4] J.Kastrup, R.Hey, K.H.Ploog, H.T.Grahn, L.L.Bonilla, M.Kindelan, M.Moscoso, A.Wacker, and J.Galan, Phys. Rev. B55, 2476 (1997)
[5] N.Ohtani, N.Egami, H.T.Grahn, K.Ploog, and L.L.Bonilla, Phys. Rev. B 58, R7528 (1998)
[6] L.L.Chang, L.Esaki and R.Tsu, Appl. Phys. Lett. 24, 593 (1974); L.Esaki and L.L.Chang, Phys. Rev. Lett. 8, 496 (1974)
[7] H.T.Grahn, R.J.Hang, W.Muller, and K.Ploog, Phys. Rev. Lett. 67, 1618 (1991)
[8] X.R.Wang and Q.Niu, Phys. Rev. B 59 R12755 (1999)
[9] J.N.Wang, B.Q.Sun, X.R.Wang, and H.L.Wang Solid State Com. 112 p371(1999)
[10] J.N.Wang, B.Q.Sun, X.R.Wang, Y.Q.Wang, W.K.Ge, and H.L.Wang, Appl. Phys. Lett. 75 p2620 (1999)
[11] X.R.Wang, J.N.Wang, B.Q.Sun, and D.S.Jiang, Phys. Rev. B61 p7261 (2000)

Photocurrent self-oscillations in weakly coupled, type-II GaAs/AlAs superlattices embedded in p-i-n and n-i-n diodes

N. Ohtani[1], M. Rogozia[2], C. Domoto[1], T. Nishimura[1], H. T. Grahn[2]

[1] ATR Adaptive Communications Research Laboratories, 2-2 Hikaridai, Seika-cho, Soraku-gun, Kyoto 619-0288, Japan
e-mail: ohtani@acr.atr.co.jp
[2] Paul-Drude-Institut für Festkörperelektronik, Hausvogteiplatz 5–7, 10117 Berlin, Germany

Abstract We have performed a systematic investigation of the carrier density and temperature dependence of photocurrent self-oscillations in undoped GaAs/AlAs superlattices (SLs) embedded in p-i-n and n-i-n diodes. The SLs have almost identical well and barrier thicknesses, but their oscillating frequencies are considerably different. Two possibilities are discussed to explain the differences in the observed frequencies.

1 Introduction

In the last several years, self-sustained current oscillations have been observed in weakly coupled superlattices (SLs) originating from unstable electric-field domain formation [1,2]. The electric field distribution in SLs is divided into a high and low electric-field region by a monopole formed at the domain boundary. The oscillating frequency is determined by the recycling motion of the monopole. Self-sustained photocurrent (PC) oscillations have been observed in undoped, photoexcited SLs embedded in p-i-n [3,4] and n-i-n diodes [5,6]. However, the influence of the difference in the doping polarity of the cap layer on the oscillating frequency has not been clarified yet. We have conducted a systematic investigation of the carrier density and temperature dependence of PC self-oscillations observed in undoped GaAs/AlAs SLs embedded in p-i-n and n-i-n diodes.

2 Sample structures and experiment

2.1 Sample structures

Two samples were grown on n-GaAs substrates, one with a p-type cap layer, the other one with an n-type cap layer. Thus, one sample represents a p-i-n diode, while the other correspond to an n-i-n diode. The intrinsic regions of both samples contain a GaAs/AlAs SL with 75 periods, whose layer thicknesses are quite similar. For the p-i-n (n-i-n) sample, the GaAs layer thickness amounts to 3.1 nm (3.4 nm), while the AlAs layer thickness is 5.1 nm in both samples. These SLs are sandwiched between $Al_{0.4}Ga_{0.6}As$ cladding layers. Since they are type-II SLs, the carrier transport is dominated by X-electrons so that the GaAs layer corresponds to the barrier layer.

2.2 Experimental setup

The samples are structured into mesas with 120 μm diameter and supplied with Ohmic contacts covering only a small portion of the full mesa area. The diodes are mounted in a He-flow cryostat. A solid-state laser with a wavelength of 680 nm is focused onto the top of the mesa in order to photoexcite carriers in the undoped SL region. The carrier density can be varied by changing the laser intensity. The frequency spectra of the PC oscillations are detected with an Advantest R3361A spectrum analyzer.

3 Results and Discussion

We have measured PC oscillations in both samples for different laser intensities at different temperatures. Figure 1 shows the frequency spectra of PC self-oscillations for a laser intensity of 14 mW and a temperature of 20 K. These oscillations are known to originate from unstable electric-field domains, for which a low-field domain ($X_1 \rightarrow X_1$ resonant tunneling) and a high-field domain ($X_1 \rightarrow X_2$ resonant tunneling) coexist [5]. In the p-i-n sample, the frequency of the PC oscillations is around 10 MHz, while for the n-i-n sample it is below 2 MHz. The oscillation frequency in the p-i-n sample appears to be a factor of about 7 larger than in n-i-n sample. Similar differences in the detected frequencies are found for other temperatures and laser intensities. For example, as shown in Fig. 2 for a laser intensity of 20 mW at 120 K, the oscillation frequency is above 70 MHz in the p-i-n sample, while for the n-i-n sample it is around 40 MHz. However, the ratio of the oscillation frequencies at higher temperatures (120–150 K) is only about 2, which is considerably smaller than the ratio at 20 K. In addition, the n-i-n sample exhibits some static and chaotic regions as function of the applied voltage.

There are two possibilities to explain the higher frequency in the p-i-n sample. According to our calculations using a rather simple semiclassical model [7], the frequency in the p-i-n sample is expected to be larger by a factor of 7 at low temperatures, because the thickness of the GaAs barrier layer of the p-i-n sample is thinner by one monolayer. This factor is roughly equal to the observed ratio of the frequencies shown in Fig. 1. However, in the high-temperature regime, the observed frequencies in both samples become higher, and the ratio between them is decreased. The increase of the frequencies may be related to a thermal activation of electrons, which reduces the effective barrier height.

Another possible reason for the higher frequency in the p-i-n sample is related to different potential distri-

N. Ohtani, et al.

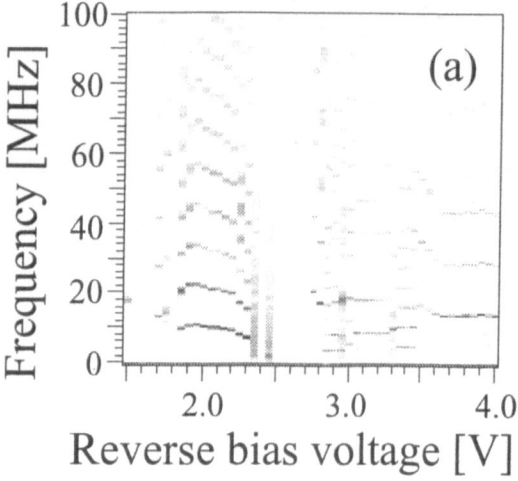

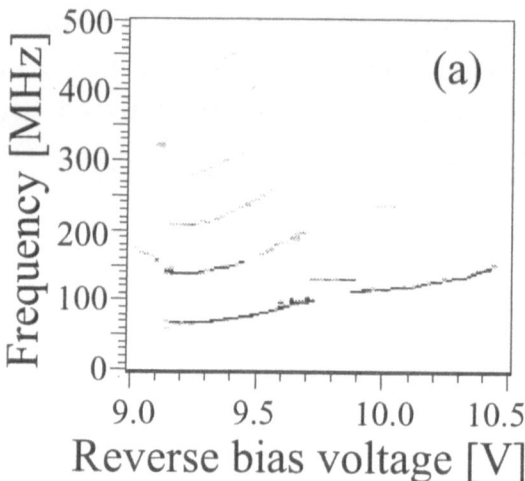

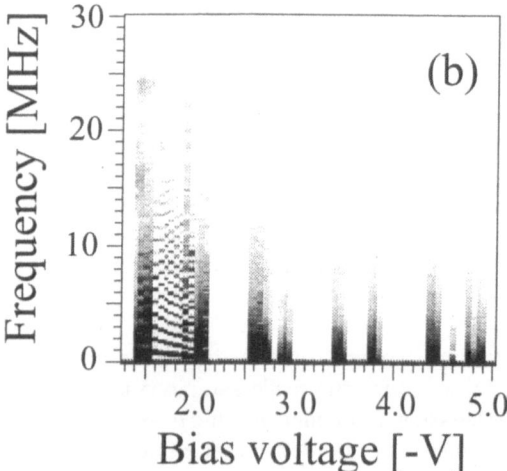

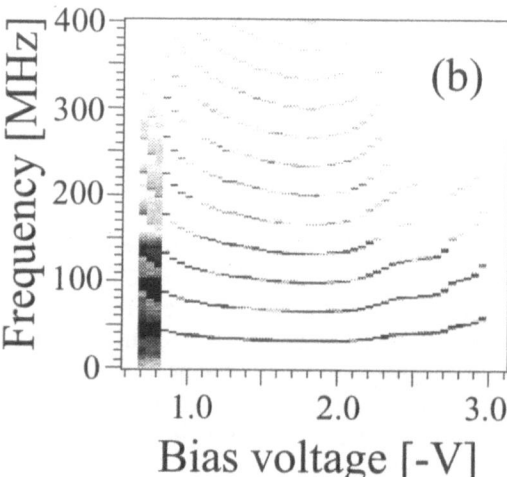

Fig. 1 Frequency spectra of self-sustained PC oscillations vs bias voltage for a laser intensity of 14 mW at 20 K in the (a) *p-i-n* sample and (b) the *n-i-n* sample. The darker areas correspond to larger amplitudes.

Fig. 2 Frequency spectra of self-sustained PC oscillations vs bias voltage for a laser intensity of 20 mW at 120 K in the (a) *p-i-n* sample and (b) *n-i-n* sample. The darker areas correspond to larger amplitudes.

butions in the two samples. In order to maintain a constant Fermi level, the potential profile near the cladding layers in the *n-i-n* sample has to be bent. This bending potential prevents photoexcited holes from escaping from the SL region. This delayed hole sweep-out from the SL could result in a reduction of the oscillation frequency due to the heavier effective mass of the holes with respect to the electrons. However, in the *p-i-n* sample, holes can directly escape from the SL. We therefore expect that the oscillation frequency in the *p-i-n* sample is affected by the hole sweep-out. In order to distinguish between the two possibilities, it is necessary to investigate two identical SL structures in the two different configurations.

Acknowledgements The authors would like to thank Bokuji Komiyama and Tahito Aida for their encouragement throughout this work. One of us (N. O.) would like to thank Paul-Drude-Institut für Festkörperelektronik for its hospitality.

References

1. H. T. Grahn, J. Kastrup, K. H. Ploog, L. L. Bonilla, J. Galán, M. Kindelan, and M. Moscoso, Jpn. J. Appl. Phys. **34**, (1995) 4526.
2. J. Kastrup, R. Hey, K. H. Ploog, H. T. Grahn, L. L. Bonilla, M. Kindelan, M. Moscoso, A. Wacker, and J. Galán, Phys. Rev. B **55**, (1997) 2476.
3. N. Ohtani, N. Egami, H. T. Grahn, and K. H. Ploog, Phys. Rev. B **61**, (2000) R5097.
4. N. Ohtani, N. Egami, H. T. Grahn, K. H. Ploog, and L. L. Bonilla, Phys. Rev. B **58**, (1998) R7528.
5. H. Mimura, M. Hosoda, N. Ohtani, K. Tominaga, K. Fujita, T. Watanabe, H. T. Grahn, and K. Fujiwara, Phys. Rev. B **54**, (1996) R2323.
6. N. Ohtani, N. Egami, K. Fujiwara, and H. T. Grahn, Solid State Electron. **42**, (1998) 1509.
7. M. Rogozia and H. T. Grahn, submitted to Appl. Phys. A.

Zener-phonon resonances in the quantum transport of multiband semiconductor superlattices

P. Kleinert*

Paul-Drude-Institut für Festkörperelektronik, Hausvogteiplatz 5–7, 10117 Berlin, Germany, e-mail: kl@pdi-berlin.de

Abstract The influence of electron-phonon scattering on intersubband tunneling in two-band semiconductor superlattices is treated within a quantum-kinetic approach. Intracollisional field effects are taken into account. In addition to the known electro-phonon and intersubband resonances, combined Zener-phonon resonances are predicted to occur in the current density. These nonanalytic current resonances are most pronounced, when the widths of the upper and lower miniband differ remarkably.

1 Introduction

Recently, there have been intensive investigations of the influence of external electric fields on transport properties of semiconductor superlattices (SLs). When an electric field E is applied perpendicular to the layers of a SL with a period d, different transport regimes are commonly identified. At low fields, the current increases linearly with field. In the adjacent region, when the carriers approach the minizone boundary (where they are Bragg diffracted), the current is expected to decrease with increasing field. Under these circumstances, the energy levels of the Wannier-Stark (WS) ladder can be resolved ($\hbar\Omega = eEd > \hbar/\tau_{eff}$, τ_{eff} is an effective scattering time), and the carriers become increasingly localized in space. The related negative differential conductivity (NDC) is a manifestation of the phenomenon that electrons accelerated perpendicular to the SL layers might probe the negative-effective-mass region of the miniband. The main problem in observing NDC is the large current density, which make the field in the SL nonuniform and cause the formation of electric-field domains. At sufficiently high electric fields ($\Omega\tau_{eff} > 1$), the miniband splits into a WS ladder of localized states, and only scattering induced hopping transitions lead to a current. In this electric field region, the appearance of so-called electro-phonon resonances was predicted to occur [1–3], giving rise to a nonmonotonic current-voltage (I-V) dependence. These resonant-type current anomalies are due to intra-collisional field effects (ICFEs) [1] and have no analogy in a semiclassical picture. At still higher electric field strengths, intersubband tunneling (also called Zener tunneling) leads to additional peculiarities in the current density. However, also in this electric field region, inelastic scattering plays an important role determining the linewidth of the tunneling res-

onance and its position. The latter situation is realized by optical-phonon assisted tunneling. It is the aim of this paper to study the influence of scattering on intersubband tunneling in two-band SLs by taking into account ICFEs in a quantum transport approach.

2 The model

Within the density-matrix approach, we exploit the kinetic equations for the subband distribution functions $f_\nu^\nu(\mathbf{k})$ to express the current density perpendicular to the SL layers by [4]

$$
\begin{aligned}
j_z = -\frac{n}{E}\sum_{\mathbf{k},\mathbf{k}'}\Big\{ & \varepsilon_1(k_z)f_1^1(\mathbf{k}')W_{11}^{11}(\mathbf{k}',\mathbf{k}) \\
& +\varepsilon_2(k_z)f_2^2(\mathbf{k}')W_{22}^{22}(\mathbf{k}',\mathbf{k}) +\varepsilon_1(k_z)f_2^2(\mathbf{k}')W_{21}^{21}(\mathbf{k}',\mathbf{k}) \\
& +\varepsilon_2(k_z)f_1^1(\mathbf{k}')W_{12}^{12}(\mathbf{k}',\mathbf{k})\Big\},
\end{aligned}
\tag{1}
$$

where n denotes the electron density. This equation is only valid for homogeneous electric fields, where an explicit spatial dependence of the distribution functions disappears. The dispersion relations of the two-band SL for the first ($\nu = 1$) and second ($\nu = 2$) minibands are given by

$$
\varepsilon_1(\mathbf{k}) = \varepsilon(\mathbf{k}_\perp) + \frac{\Delta_1}{2}(1 - \cos(k_z d)),
\tag{2}
$$

$$
\varepsilon_2(\mathbf{k}) = \varepsilon(\mathbf{k}_\perp) + \varepsilon_g + \Delta_1 + \frac{\Delta_2}{2}(1 + \cos(k_z d)),
\tag{3}
$$

where Δ_1 (Δ_2) are the miniband widths and ε_g the gap energy at zero electric field. We assume equal effective masses for the lateral electron motion in the lower and upper miniband. In Eq. (1), $W_{\nu\nu'}^{\nu\nu'}(\mathbf{k}',\mathbf{k})$ denotes the scattering probability, which depends on the electric field. In the limit of strong electric fields ($\Omega\tau_{eff} \gg 1$), we exploit the Stark ladder representation and retain only the main Fourier component $[\sum_{k_z} f_\nu^\nu(\mathbf{k})]$ of the distribution functions on the right-hand side of Eq. (1) (cf. [4]). These lateral, non-equilibrium distribution functions have to be calculated from coupled kinetic equations. However, in the sequential tunneling regime, one can avoid this additional complication by assuming Boltzmann-type distribution functions of the form

$$
f_\nu^\nu(\varepsilon(\mathbf{k}_\perp)) = \frac{2\pi\hbar^2}{mk_BT}F_\nu\exp\left(-\frac{\varepsilon(\mathbf{k}_\perp)}{k_BT}\right),
\tag{4}
$$

where F_ν ($\nu = 1, 2$) denotes the occupation of the minibands. For simplicity, these quantities are calculated in

* *Acknowledgements:* The author acknowledges partial financial support by the Deutsches Zentrum für Luft- und Raumfahrt.

the relaxation-time approximation. Following the calculation outlined in Ref. [4], we obtain

$$F_2 = A/(2A + \tau/\tau_{21}) \quad \text{and} \quad F_1 = 1 - F_2 \qquad (5)$$

with

$$A = 2Q_{12}^2 (\Omega\tau)^2 \sum_{l=-\infty}^{\infty} \frac{J_l^2((\Delta_1 + \Delta_2)/2\hbar\Omega)}{(l\Omega\tau - \omega_g\tau)^2 + 1}. \qquad (6)$$

Q_{12} is the dipole matrix element, J_l the Bessel function, and $\hbar\omega_g = \varepsilon_g + (\Delta_1 + \Delta_2)/2$ the effective gap. Carrier recombination is described by τ_{21}, and the width of the tunneling resonance is determined by τ. At the tunneling resonance $\Omega = \omega_g$, a sizeable redistribution of carriers may occur. However, according to Eqs. (5) and (6), a global population inversion is not possible in the considered two-band model.

The current density of the two-band SL is calculated from Eq. (1) by considering Eqs. (2-6) and the expressions for the electric field dependent scattering probabilities published in Ref. [4].

3 Numerical results and discussion

The current density has been calculated by taking into account inelastic scattering on polar-optical phonons. In the context of our current interest, where we intend to provide a better understanding of how inelastic scattering influences intersubband tunneling, we do not need to consider the details of the electron-phonon interaction. Therefore, the intra- and inter-miniband matrix elements have been replaced by only one coupling parameter γ, which enters the constant current density j_0. Figure 1 shows numerical results for the relative current density as a function of the electric field. Our approach is only valid in the region of high electric fields, when $\Omega\tau_{eff} > 1$ is satisfied. To reproduce the linearly increasing part of the I-V characteristics at low electric fields, we have to give up our restriction to the lowest order Fourier component of the distribution function and must consider an increasing number of Stark levels in the Stark ladder approximation. However, the most interesting carrier transitions take place at high electric fields, where only few WS levels are involved. Three different kinds of current resonances can be identified. First, there are electro-phonon resonances (denoted by vertical dash-dotted lines) at $l\Omega = \omega_0$ ($l = 1, 2, 3, \ldots$), which are strongly temperature dependent and most pronounced at low temperatures. To our knowledge, such resonant-type current anomalies were observed in narrow-band semiconductors [5], but not in semiconductor SLs. A second group of current resonances, indicated by vertical dashed lines, appears at $l\Omega = \omega_g$. These Zener-intersubband resonances are associated with the field dependent redistribution of the carrier density. Therefore, both intra- and intersubband current components contribute to these resonances via the distribution function. Finally, there are new Zener-phonon resonances, which are predicted to appear at $l\Omega = \omega_g \pm \omega_0$ (vertical solid lines). These peculiarities in the I-V charac-

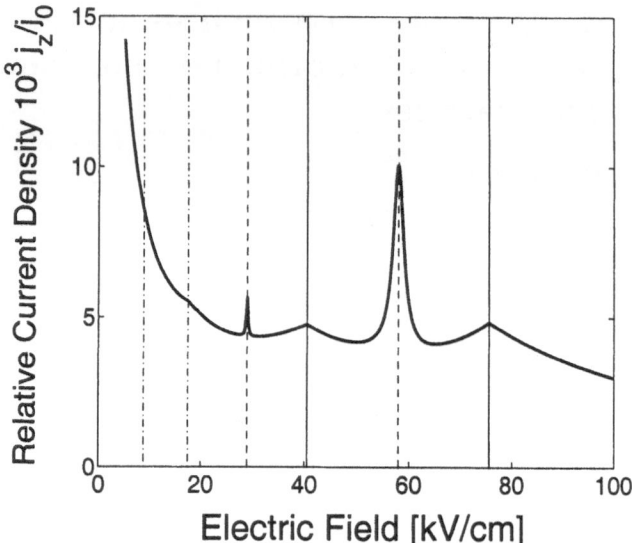

Fig. 1 Electric field dependence of the relative current density for $\varepsilon_g = 100$ meV, $\Delta_1 = 2$ meV, $\Delta_2 = 30$ meV, and $T = 300$ K. The carrier occupation has been calculated from Eqs. (5) and (6) using the scattering times $\tau = 1$ ps and $\tau_{12} = 1$ ps. The constant reference current density is $j_0 = enm^*\omega_0 \,|\, \gamma\,|^2 /\hbar^3$, where γ denotes the electron-phonon coupling constant and n the carrier density. The SL period is given by $d = 20$ nm.

teristics are quite similar to electro-phonon resonances. Both current anomalies exhibit a nonanalytic field dependence, which becomes evident at low temperatures. In the limit of vanishing temperature ($T \to 0$), these resonances are composed of sharp edges, since the dispersion of polar-optical phonons is neglected. The origin of these current anomalies are ICFEs, which were included in our approach. However, there is also a striking discrepancy between electro-phonon and Zener-phonon resonances. As an intersubband effect, Zener-phonon resonances depend on the properties of both minibands and are most pronounced, when the widths of the lower and upper minibands differ remarkably. In contrast, electro-phonon resonances belong to a single miniband and do not probe intersubband properties. Therefore, the conditions for an experimental verification of Zener-phonon and electro-phonon resonances are quite different. We expect that our theoretical work stimulates experimental studies of these novel quantum-mechanical current resonances, which are due to ICFEs.

References

1. V. V. Bryksin and P. Kleinert, J. Phys. Condens. Matter **9**, (1997) 7403.
2. S. Rott, P. Binder, N. Linder, and G. H. Döhler, Physica E **2**, (1998) 511.
3. A. Wacker et al., Phys. Rev. Lett. **83**, (1999) 836.
4. V. V. Bryksin, V. C. Woloschin, and A. W. Rajtzev, Sov. Phys. Solid State **22** (1980) 1796. [Fiz. tverd. Tela **22**, (1980) 3076].
5. D. May and A. Vecht, J. Phys. C **8**, (1975) L505.

Correlation between a Remote Electron and a Two-Dimensional Electron Gas in Resonant Tunneling Devices

Hatsuhiro Kato[1], François M. Peeters[2]

[1] Faculty of Engineering, Yamanashi University 4-3-11 Takeda, Kofu, Yamanashi 400-8511, Japan
[2] Departement Natuurkunde, Universiteit Antwerpen (UIA), Universiteitsplein 1, 2610 Antwerpen, Belgium

Abstract We investigated theoretically the Coulomb correlation between a tunneling electron and a two-dimensional electron gas (2DEG) formed at the emitter of a double barrier resonant tunneling device and compare it with experimental data. The novelty of the present approach is that we describe the tunneling or remote electron as a polaron, whose Coulomb interaction (i.e. Coulomb correlations) with the 2DEG is viewed as due to excitation of virtual "phonons". This study is also relevant for bilayer systems in which one of the 2DEG's is very dilute such that a one-electron approximation is valid in this layer.

1 Introduction

The electronic Coulomb interaction in a doped semiconductor leads to exchange and correlation contributions to the electron energy similar to those in metals[1]. Recently resonant tunneling revealed this Coulomb interaction as a shift in the bias position of the tunneling peak. A quantum resonance device was used which consisted of a double quantum well containing a single two-dimensional electron gas (2DEG) as injector and a lateral magnetic field is applied to the system [2]. Lok *et al.* [2] pointed out that this energy shift is due to an acoustic-phonon-like excitation which appears at the vacancy created by the tunneling electron. We propose a model to explain this vacancy as a polarization of the 2DEG due to the tunneling electron.

In Fig. 1(a) a schematic diagram of the system is shown. The solid circle represents the tunneling electron which is a distance d away from the 2DEG, whose schematic density profile is shown in Fig. 1(b). The tunneling electron is to be regarded as a remote electron, which interacts through the Coulomb interaction with the 2DEG and forms a *remote polaron* whose dressing (i.e. polarization) appears in the 2DEG. If the 2DEG is

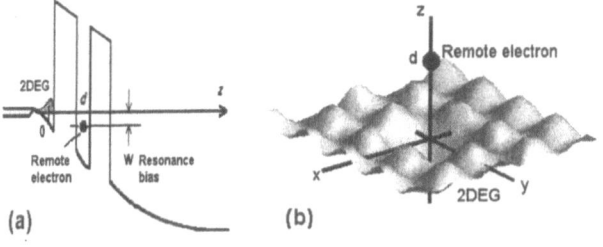

Fig. 1 Band diagram of resonant tunneling device (a), and schematic profile of the 2DEG density and the remote electron configuration (b).

in a crystalline state, this polarization is equivalent to the displacement or a vacancy of the lattice electron [3]. In fact, the electrons form a Wigner crystal (WC) when a strong magnetic field is applied perpendicular to the system. This was confirmed experimentally by Andrei *et al.* [4] in a GaAs heterojunction where usually the electrons behaves as a quantum gas in the absence of a magnetic field.

This study is also relevant for bilayer systems in which one of the 2DEG's is very dilute such that a one electron approximation is valid in this layer. Such a bilayer system is realized in a high quality double quantum-well or in a wide single quantum-well.

2 Tunneling Elecron as a Remote Polaron

2.1 Formalism

The Hamiltoniam of a remote electron is essentially equivalent to that of a polaron, which is given by

$$H = \frac{1}{2m^*}\left(\mathbf{p} + \frac{e}{c}\mathbf{A}(\mathbf{r})\right)^2 + H_{\text{ph}} + H_{\text{int}}. \quad (1)$$

Here, $\mathbf{r}$ and $\mathbf{p}$ are the two-dimensional position and momentum of the remote electron, e the elementary charge, c the velocity of light, $\mathbf{A}(\mathbf{r})$ the vector potential in the symmetric gauge, m^* the effective mass of the remote electron, H_{ph} the "phonon" Hamiltonian and H_{int} the interaction Hamiltoninan[5,6]. If the 2DEG is in a non-crystalline state, m^* is the effective mass of the host material, but if the 2DEG forms a WC, the electron mass is enhanced by the periodic potential of the lattice electrons. This mass enhancement causes a shrinkage of the polaron radius and affects the resonant bias.

The "phonons" are : 1) the Wigner phonons when the 2DEG is crystallized into a WC, or 2) plasmons when the 2DEG is in the non-crystalline state. If the electrons are crystallized into a triangular lattice in the absence of a magnetic field, there are longitudinal and transverse modes. These modes are hybridized into two new modes when a magnetic field is applied. Consequently, the phonon Hamiltonian H_{ph} is given by

$$H_{\text{ph}} = \sum_{j\mathbf{k}} \hbar\omega_{j\mathbf{k}}\left(a_{j\mathbf{k}}^\dagger a_{j\mathbf{k}} + \frac{1}{2}\right). \quad (2)$$

Here, $\omega_{j\mathbf{k}}$ is the "phonon" frequency with mode index j and wave vector $\mathbf{k}$, $a_{j\mathbf{k}}$ and $a_{j\mathbf{k}}^\dagger$ are annihilation and creation operators. If the 2DEG is in the non-crystalline state, no transverse mode is present and the "phonons" are essentially plasmons, which are labeled only by their wave number $\mathbf{k}$. The explicit expression of the phonon frequency was derived using a single mode assumption in conjunction with the f-sum rule [5].

The interaction Hamiltonian, H_{int}, between the remote electron and the 2DEG is

$$H_{\text{int}} = \sum_{j\mathbf{k}} \frac{V_{j\mathbf{k}}}{\sqrt{\Omega}} \exp(i\mathbf{k} \cdot \mathbf{r})(a_{j\mathbf{k}} + a^{\dagger}_{j-\mathbf{k}}), \qquad (3)$$

with the system area Ω. The coupling strength $V_{j\mathbf{k}}$ is $\mathbf{k}$- and d-dependent and given by

$$V_{\mathbf{k}} = \frac{2\pi e^2 n_{\text{s}}^{1/2}}{\epsilon} \ell_{j\mathbf{k}} \exp(-kd), \qquad (4)$$

with n_{s} the density of the 2DEG and ϵ the dielectric constant of the host material. The detail expression for $\ell_{j\mathbf{k}}$ was derived in the previous papers[5,6].

We apply a variational scheme which is based on the path-integral formalism of the polaron problem and use the free electron as a trial state [7]. The resulting expression for the free energy, F, becomes up to second order in $V_{j\mathbf{k}}$,

$$F = \frac{1}{\beta} \log[2\sinh(\hbar\beta\omega_{\text{c}}/2)]$$
$$-\frac{1}{2} \int_0^{\beta} d\tau \sum_{j\mathbf{k}} \frac{|V_{j\mathbf{k}}|^2}{\Omega} \exp[-\mathbf{k}^2\ell_B^2 D(\tau)]G(\tau), \qquad (5)$$

with

$$D(\tau) = \frac{\sinh(\tau\hbar\omega_{\text{c}}/2)\sinh[(\beta - \tau)\hbar\omega_{\text{c}}/2]}{\sinh(\beta\hbar\omega_{\text{c}}/2)}, \qquad (6)$$

$$G(\tau) = \frac{\cosh[(\beta/2 - \tau)\hbar\omega_{j\mathbf{k}}]}{\sinh(\beta\hbar\omega_{j\mathbf{k}}/2)}. \qquad (7)$$

Here, $\omega_{\text{c}} = eB/m^*c$, $\ell_B = (\hbar/m^*\omega_{\text{c}})^{1/2}$, and $\beta = 1/k_BT$ with k_B the Boltzmann constant and tempetature T. In the zero temperature limit ($\beta \to \infty$) the free energy (5) reduces to the ground state energy

$$E = \frac{\hbar\omega_{\text{c}}}{2} - \int_0^{\infty} dt \sum_{j\mathbf{k}} \frac{|V_{j\mathbf{k}}|^2}{\Omega}$$
$$\times \exp\left[-\frac{\mathbf{k}^2}{2}\ell_B^2(1 - e^{-t\omega_{\text{c}}}) - t\omega_{j\mathbf{k}}\right]. \qquad (8)$$

2.2 Energy Shift of the Resonant Peak

When an electron tunnels from a 2DEG at $z = 0$ to the quantum well at $z = d$, this electron loses the energy $W(d,B) = E(d,B) - E(0,B)$. Because of energy conservation the energy loss $W(d,B)$ affects the lateral resonance bias (see Fig. 1(a)). When a magnetic field changes from zero to B, the resonance bias changes due to the energy shift $\Delta(d,B) = W(d,B) - W(d,0)$.

In Fig. 2, $\Delta(d,B)$ is plotted as a function of the external magnetic field B for the tunneling distance $d = 15$nm and 2DEG density $n_{\text{s}} = 3 \times 10^{11}\text{cm}^{-2}$. The solid circles represent the experimental data of Lok *et al.* [2], and the solid curves are our theoretical results when the "phonons" are Wigner phonons (Wigner polaron)

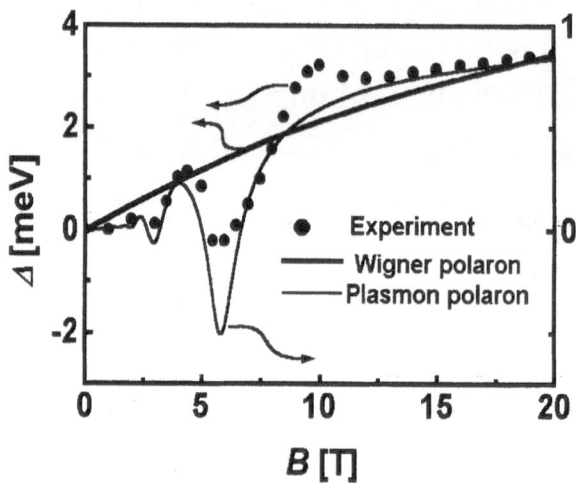

Fig. 2 The B dependence of Δ for the experimental data (solid circle), the Wigner polaron (thick curve) and the plasmon polaron (thin curve). The vertical axis for the plasmon polaron result is on the right hand side.

or plasmons (plasmon polaron). Although the theoretical data for the Wigner polaron does not account for the oscillation in Δ with B, its absolute value agrees rather well with the experiment data. On the other hand, the theory of the plasmon polaron is able to account for those oscillations in Δ but it is about a factor of 4 smaller than the experimental one. The oscillations are due to a variation of the filling factor of the Landau levels which affects the polarization of the 2DEG.

3 Conclusion

We formulated the Coulomb correlation between a remote electron and a 2DEG as a *remote polaron* problem and explained the experimentally measured magnetic field dependence of the energy shift in a resonant tunneling device.

References

1. A. W. Overhauser, Phys. Rev. B **3** (1971) 1888.
2. J.G.S. Lok, A.K. Geim, J.C. Maan, L. Eaves, A. Nogaret, P.C. Main and M. Henini, Phys. Rev. B **56** (1997) 1053.
3. A formalism based on this point of view is also possible and will be published elsewhere.
4. E. Y. Andrei, D. Deville, D. C. Glattli, F. I. B. Williams, E. Paris and B. Etienne, Phys. Rev. Lett. **60** (1988) 2765.
5. H. Kato, F. M. Peeters and S. E. Ulloa , Europhys. Lett. **40** (1997) 551; H. Kato and F. M. Peeters, Europhys. Lett. **45** (1999) 235.
6. H. Kato and F. M. Peeters, Phys. Rev. B **59** (1999) 14342.
7. R. P. Feynman and A. R. Hibbs, *Quantum mechanics and path integrals* (MacGraw-Hill, New York, 1965).

Proc. 25th Int. Conf. Phys. Semicond., Osaka 2000 (Eds. N. Miura and T. Ando)

Spin effects in InAs quantum dots: Tunneling experiments in tilted magnetic fields

J. M. Meyer[1], I. Hapke-Wurst[1], U. Zeitler[1], R. J. Haug[1], H. Frahm[2], A. G. M. Jansen[3], K. Pierz[4]

[1] Institut für Festkörperphysik, Universität Hannover, Appelstraße 2, 30167 Hannover, Germany
   e-mail: zeitler@nano.uni-hannover.de
[2] Institut für Theoretische Physik, Universität Hannover, Appelstraße 2, 30167 Hannover, Germany
[3] Grenoble High Magnetic Field Laboratory, MPIF-CNRS, B.P. 166, 38042 Grenoble Cedex 09, France
[4] Physikalisch-Technische Bundesanstalt Braunschweig, Bundesallee 100, 38116 Braunschweig, Germany

Abstract The Landé factor g^* of individual InAs quantum dots is measured experimentally by means of magneto-tunneling experiments. With the magnetic field applied parallel to the growth direction z we find $g^*_\parallel = 0.7...0.9$ for different dots investigated. When B is tilted away from z by an angle ϑ an increase of g^* following a phenomenological behavior $g^*(\vartheta) = \sqrt{(g^*_\parallel \cos \vartheta)^2 + (g^*_\perp \sin \vartheta)^2}$ is observed. In high magnetic fields where only the lowest Landau level in the three-dimensional emitter is occupied a strong enhancement of the resonant current is observed. These results are discussed in terms of a field-induced Fermi-edge singularity.

The quantized energy levels of InAs quantum dots (QDs) can be accessed efficiently with resonant tunneling experiments [1,2]. In particular it is possible to measure directly the g-factor of such structures [3–5]. In this paper we will show experimentally that the g-factor of InAs QDs strongly depends on the orientation of the magnetic field and the results will be modeled by a g-factor tensor with two independent components.

Our samples consist of InAs quantum dots embedded in an AlAs barrier and sandwiched between two highly doped GaAs electrodes. A detailed description of the sample structure can be found in [2]. Macroscopic AuGeNi contacts with a typical diameter of 50 μm were annealed into the top electrode and vertical tunneling diodes with the same diameter were processed using wet-chemical etching.

Applying a voltage between the top and the bottom electrode we measure the typical I-V characteristics of a single-barrier tunneling device [2]. Distinct current steps superimposed on the coarse I-V curve are assigned to resonant tunneling through individual InAs quantum dots [1,2]. Such a typical step for $T = 0.5$ K is shown in Fig. 1 for $B = 0$ T and $B = 9$ T applied parallel to the growth direction. As can be seen in the figure, the single step observed at zero magnetic field splits into two steps which we assign to the tunneling through the spin-split ground state of the dot. The splitting ΔV is linear in magnetic field and given by $\Delta V = g^*_\parallel \mu_B B / \alpha e$. Here $\alpha = 0.3$ the is energy-to-voltage conversion factor given by the ratio of the energy separation emitter-dot to the total voltage drop. For numerous dots investigated we find a Landé factor $g^*_\parallel$ in the range of 0.7 ... 0.9.

From the temperature dependence of the step heights in high magnetic fields we conclude that the low voltage

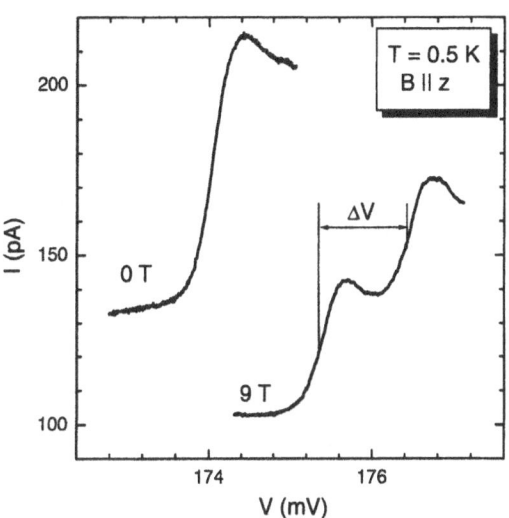

Fig. 1 Typical current step due to resonant tunneling through an InAs quantum dot. In a magnetic field the spin splitting of the ground state in the dot is resolved as two distinct sub-steps.

step is related to the tunneling of the minority-spin electrons from the emitter [4]. Since the g-factor in GaAs is negative this results in a positive Landé factor in the InAs QDs. Our measured values of g^* are consistent with other experiments [3]. The strong deviation from the InAs bulk value ($g^* = -14.8$) is explained in terms of size quantization effects, strain and possible other effects such as alloying of AlAs into the InAs dot and leakage of the electronic wave function into the AlAs barriers and the GaAs electrodes.

When the magnetic field B is tilted away from the growth direction z the splitting of the current steps gradually increases. As a function of the tilt angle ϑ between B and z the Landé factor deduced from this splitting follows a phenomenological behavior

$$g^*(\vartheta) = \sqrt{(g^*_\parallel \cos \vartheta)^2 + (g^*_\perp \sin \vartheta)^2}. \tag{1}$$

An example for $g^*(\vartheta)$ is shown in Fig. 2: I-V characteristics were measured at $B = 20$ T and $T = 0.5$ K for different tilt angles between the magnetic field and the growth direction and the Landé factor was extracted from the voltage split ΔV.

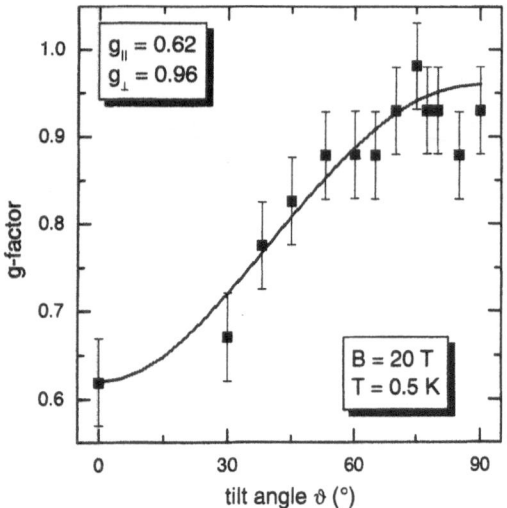

Fig. 2 Angular dependence of the experimentally measured g-factor at $B = 20$ T compared to the phenomenological model.

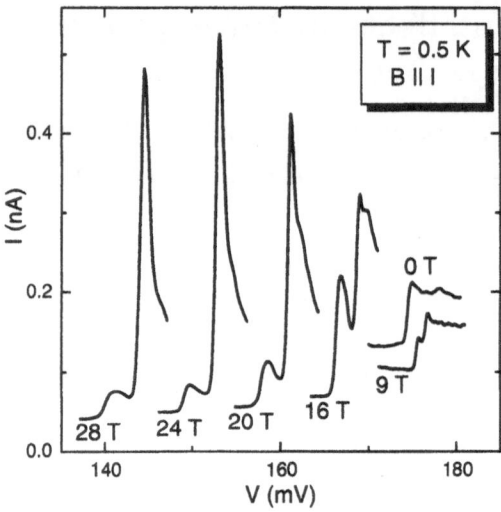

Fig. 3 Evolution of the spin-split current peaks in strong magnetic fields.

The behavior observed can be understood by modeling an InAs quantum dot by a flat disc with a height $h \approx 3$ nm and a diameter $d \approx 15$ nm [2]. Then the g-factor tensor with generally nine independent real components [6] reduces to a diagonal tensor with two independent components $g_{\perp}^* = g_{xx} = g_{yy}$ and $g_{\parallel}^* = g_{zz}$ describing precisely the angular dependence observed.

Our experimental results clearly prove the anisotropic nature of the spin in an InAs quantum dot. Due to size quantization effects one would expect a maximum g-factor when the magnetic field is applied in the direction of the strongest confinement, i.e. the growth direction z. Such quantum-confinement effects have indeed been observed in quantum wells [7] and quantum wires [8]. However, in our case we observe the minimum value of g^* along the strongest confinement. Therefore, other mechanisms such as spatially dependent strain, AlAs alloying into the InAs quantum dots and leakage of the electronic wave functions into the AlAs barriers are dominating in the observed g-factor anisotropy.

Apart from the spin splitting of the two current steps another spin effect can be obseved in very high magnetic fields (up to 28 T) where a dramatic enhancement of the spin-split current steps is observed, see Fig. 3. In particular electrons carrying the majority spin in the emitter display an current enhancement of more than an order of magnitude in high magnetic fields. These spectacular experimental observations can be explained in terms of a field-induced Fermi-edge singularity [4].

The singularity is due to the Coulomb interaction between the local potential in the dot and the Fermi edge in the emitter. In a strong quantizing magnetic field applied parallel to the current direction all the electrons in the emitter are confined in the lowest Landau level. Therefore, the emitter can be regarded as a one-dimensional (1D) channel with a momentum perpendicular to the

boundary. The strong variation of the Fermi momentum for these 1D channels leads to the singularities observed, for more details see [4].

In conclusion we have determined the anisotropic nature of the Landé factor in InAs QDs by means of resonant tunneling experiments in tilted magnetic field. We find an angular variation of g^* which can be described by two independent tensor components $g_{\parallel}^*$ and $g_{\perp}^*$. In high magnetic fields a huge enhancement of the spin-split current steps is observed which we assign to a field-induced Fermi-edge singularity.

References

1. I. E. Itskevich, T. Ihn, A. Thornton, M. Henini, T. J. Foster, P. Moriarty, A. Nogaret, P. H. Beton, L. Eaves, and P. C. Main, Phys. Rev. B **54** (1996) 16401;
   T. Suzuki, K. Nomoto, K. Taira, and I. Hase, Jpn. J. Appl. Phys. **36** (1997) 1917;
   M. Narihiro, G. Yusa, Y. Nakamura, T. Noda, and H. Sakaki, Appl. Phys. Lett. **70** (1997) 105.
2. I. Hapke-Wurst, U. Zeitler, H. W. Schumacher, R. J. Haug, K. Pierz, and F. J. Ahlers, Semicond. Sci. Technol. **14** (1999) L41.
3. A. S. G. Thornton, T. Ihn, P. C. Main, L. Eaves, and M. Henini, Appl. Phys. Lett. **73** (1998) 354.
4. I. Hapke-Wurst, U. Zeitler, H. Frahm, A. G. M. Jansen, R. J. Haug, and K. Pierz, accepted for publication in Phys. Rev. B (cond-mat/0003400).
5. J. M. Meyer, I. Hapke-Wurst, U. Zeitler, R. J. Haug, and K. Pierz, *Proc. of Int'l. Conf. Quantum Dots 2000, edited by A. Zrenner (to be published in physica status solidi)*.
6. A. A. Kiselev, E. L. Ivchenko, and U. Rössler, Phys. Rev. B **58** (1998) 16353.
7. P. Le Jeune, D. Robart, X. Marie, T. Amand, M. Brousseau, J. Barrau, V. Kalevich, and D. Rodichev, Semicond. Sci. Technol. **12** (1997) 380.
8. M. Oestreich, A. P. Heberle, W.W. Rühle, R. Nötzel, and K. Ploog, Europhys. Lett. **31** (1995) 399.

Carrier transport affected by hole-subband resonances in a strained GaAs/InAlAs superlattice

M. Hosoda[1], K. Kuroyanagi[2], N. Ohtani[2], T. Aida[2]

[1] Department of Applied Physics, Osaka City University, Sugimoto, Sumiyoshi-ku, Osaka-city 558-8585, Japan e-mail: hosoda@a-phys.osaka-cu.ac.jp

[2] ATR Adaptive Communications Research Laboratories, Seika-cho, Soraku-gun, Kyoto 619-0288, Japan

Abstract A switch in hole transport, from the heavy hole (hh) to light hole (lh) tunneling path, is found in a GaAs/InAlAs strained superlattice under an electric field. This phenomenon is caused by a hh-lh transfer, as demonstrated by photoluminescence and photocurrent response. Under a high electric field, most holes flow through the lh path.

1 Introduction

In the study of carrier transport in superlattices (SLs), investigations have mainly focused on electron transport in the conduction band. When SLs are irradiated by light, this generates holes, which may also contribute to the current output. Hole transport, however, has not been clearly observed until now. More specifically, no clear evidence has emerged for photocurrent impulse responses. This is because the photocurrent temporal responses generated by holes are weaker than the same responses generated by electrons, due to the slow tunneling time of holes, as determined by their heavier effective masses. However, when a certain resonant tunneling mechanism exists between hole states, especially when heavy hole (hh) to light hole (lh) resonant tunneling occurs, a rather obvious hole transport by lh's can be observed. Using photoluminescence (PL) measurement, we recently found that hole subband resonances can be clearly observed in a strained GaAs/InAlAs SL. [1] In this SL, we proved hole occupation in higher energy lh states, using a PL emission from the lh Stark-ladder states. However, the carrier transport has yet to be studied. In this report, we demonstrate that the hole transport effected by the hole subband resonances in the above SL can influence photocurrent impulse responses sufficiently.

2 Experiments

The sample was a p-i-n heterostructure diode grown on a (100)-oriented n^+-GaAs substrate. The intrinsic region consisted of a 60-period GaAs (3.7 nm)/$In_{0.3}Al_{0.7}As$ (0.85 nm) SL, sandwiched by undoped 15−20 nm $Al_{0.4}Ga_{0.6}As$ cladding layers. The lattice mismatch between the GaAs and $In_{0.3}Al_{0.7}As$ was 2.1%. A focused Ti-Sapphire laser (0.6-ps pulse width, 720-nm fundamental wave, 430-nm SHG) was irradiated on the p-cap layer of the 50×50 μm^2 sample in a cryostat. The photocurrent response was measured with a sampling oscilloscope. The PL was measured with a monochromator

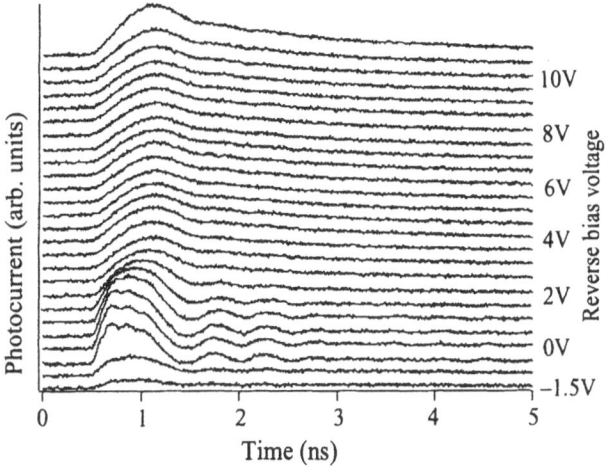

Fig. 1 Photocurrent impulse responses under various reverse bias voltages. Optical excitation intensity was 0.75 mW.

plus streak-camera under a 633-nm HeNe laser excitation. All measurements were done under various reverse bias voltages at 20K. In the following figures, minus voltage means forward bias, and the flatband of the sample was −1.5 V.

3 Results and Discussion

The sample showed normal photocurrent spectra, as observed in ordinary short-period SLs, indicating good sample quality. Using the photocurrent spectra, we determined the ground state energy levels of lh (lh1), hh (hh1), electron (e1) in quantum well (QW), and the resonance voltages between the hole states. These values agreed well with our calculation regarding compressive strain into the $In_{0.3}Al_{0.7}As$ barrier. The above evaluation proves that only the barrier suffers the strain, because the barrier is sufficiently thinner than the QW. Comparing hole state in this GaAs/$In_{0.3}Al_{0.7}As$ sample SL to GaAs/AlAs SLs, more efficient hole resonance and tunneling are expected due to lower barrier hight and lighter effective mass of holes.

The photocurrent response reflects the above effect. Figure 1 shows the photocurrent response under 720-nm wavelength excitation. It shows a slow decay tail at greater bias voltages than 1 V, where hh1(0)-lh1(+2) resonance occurs. Figure 2 illustrates the above condition. The photocurrent peak maximizes its delay at 3 V, where

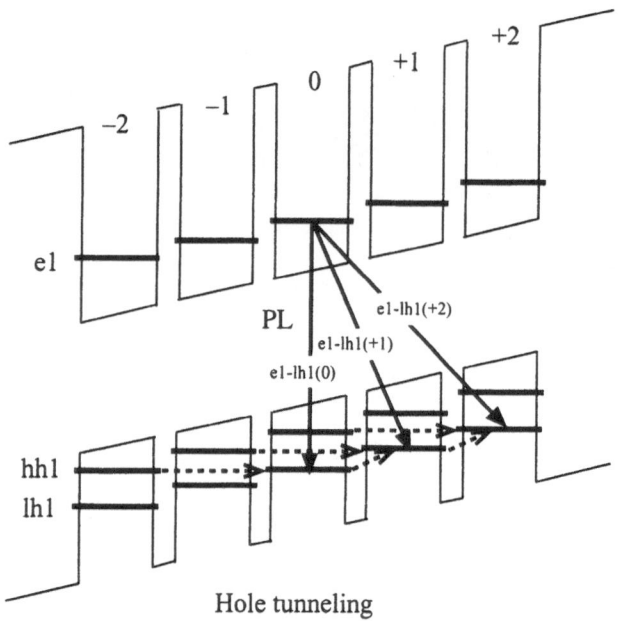

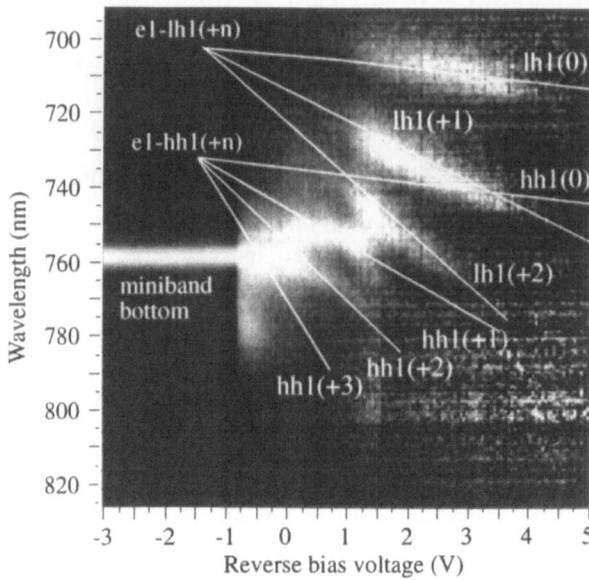

Fig. 3 Normalized PL spectra versus reverse bias voltage under 0.5-mW HeNe laser excitation. The brightness corresponds to the PL intensity. PL detection gain below −1 V is switched to low, due to strong PL emission from the miniband bottom of e1. Various resonances of the n-th Stark-ladder states of holes (lh1(+n) and hh1(+n)) can be observed.

Fig. 2 Schematic figure of hole transport and hole Stark-ladder PL emission.

hh1(0)-lh1(+1) resonance occurs. From this voltage, the slow decay tail further evolves. On the other hand, optical excitation under a wavelength of 430 nm showed almost normal photocurrent responses with slight decay tail. The 430-nm irradiation created e-h pairs near the p-cap region due to a very short penetration depth. Thus electrons were the main carriers running through the SL under the reverse bias voltage. In contrast, under the 720-nm irradiation, e-h pairs were created deep in the SL. Therefore, holes also ran through the SL. When the hole transport was enhanced by the lh occupation via the hh-lh resonant tunneling, the lh transport could be detected to have a faster tunneling time due to the light effective masses, as the fast drift velocity generated a sufficient photocurrent peak intensity.

The hole resonance was also supported by the PL measurement. Figure 3 shows the PL spectra. The lh Stark-ladder PLs arise after the hh1(0)-lh1(+2) resonance at 1 V, as PL emission from hh1 vanishes simultaneously. This indicates strong hole occupation in the lh1 state caused by hole injection from the hh1 state. Since our calculation showed a wide spread of lh wave function through about two to three SL periods in this low-barrier and lighter lh effective-mass system, hole injection via the hh1(0)-lh1(+2) resonance is possible. After the hh1(0)-lh1(+1) resonance at 3 V, most of the holes occupy the lh1 state, and are swept out very quickly. Our calculated lh1-lh1 non-resonant sequential tunneling time was approximately 40 fs. Therefore, lh1 Stark-ladder PL becomes quenched.

The voltage of the slow decay tail generation corresponded to the resonance voltage and hole occupation of the lh state supplied from the hh state, which was sup-

ported by our PL measurements. The voltages and the PL spectral behaviors also agreed well with our calculated hole resonance voltages and PL intensity behaviors as a function of the bias voltage, through a calculation of the overlap integral between the electron and hole wave functions. [1]

Consequently, when the hole transport was enhanced by the lh occupation via the hh-lh resonant tunneling, the lh transport had a faster tunneling time, because the fast drift velocity generated a sufficient photocurrent peak intensity. This process generates observable influence in the photocurrent response, such as the slow decay tail shown in Fig. 1. In the case of the very slow hh transport, e.g. GaAs/AlAs SL systems, it causes a weak peak current intensity with a long-lasting current pulse width for the impulse response, and accordingly, is not able to be observed due to a disturbance of the noise level in equipment.

4 Conclusion

Hole transport in a barrier-compressive strained GaAs/InGaAs superlattice was studied. PL measurements confirm slow decay tail generation in the photocurrent response, due to lh1-lh1 tunneling transport after hh1-lh1 resonant tunneling.

References

1. K. Kuroyanagi, N. Ohtani, N. Egami, K. Tominaga, M. Hosoda, and M. Nakayama, Physica B **272,** (1999) 198-201.

Disorder-enhanced tunneling transport through doping barriers

R. Elpelt, O. Wolst, H. Willenberg, S. Malzer*, G.H. Döhler

Institut für Technische Physik I, Universität Erlangen-Nürnberg, 91058 Erlangen, Germany
*electronic address: malzer@physik.uni-erlangen.de

Abstract We report on the effects of doping induced disorder on the transport through δ-doped triangular barriers. Calculations based on the assumption of a homogeneously doped background (jellium model) yield a smooth current-field characteristic in the case of 3D-3D transport. Taking into account the discrete subband structure in a 2D-2D transport model, distinct peaks are expected in the current signal. Neither of them is found in the experiments where the smooth current-voltage characteristic exceeds the theoretical curves by orders of magnitude. Taking into account the random doping potential which results in a local lowering of the transport barrier, however, both models agree reasonably well with the experimental findings.

1 Introduction

The random distribution of impurity atoms in doped semiconductors leads to strong potential fluctuations, even in bulk crystals, where these fluctuations are strongly screened by the free carriers. The potential fluctuations are particularly pronounced in systems like n-i-p-i doping superlattices [1], where the fixed ionized-impurity charge in the n- and p-layers is only partially compensated by free carriers ("filling factor" < 1). These disorder effects have been studied in detail for the case of luminescence in n-i-p-i doping superlattices. At low excitation densities the luminescence is dominated by the recombination of electrons occupying electronic states in the local *minima* of the fluctuating conduction band edge with holes located at the *maxima* of the fluctuating valence band edge, both of them representing regions of locally high dopand densities ("clusters"). Excellent agreement between experiment and theory has been obtained by using a Monte Carlo method for calculating the potential fluctuations of a random impurity distribution in the (δ-doped) n- and p-layers and using the wave functions for the partly screened potential [2].

2 Vertical transport

The influence of disorder is completely different, if vertical transport through doping-induced barriers is considered. In the following we consider only the case of an n-type δ-doped n-i-p-i structure (defined by excess doping density in the donor layers). For the transport between neighboring n-layers, not the maximum peaks of the "potential fluctuation mountains" associated with "acceptor clusters", but rather a few saddle points, corresponding to "acceptor voids" (which actually cover only a minor fraction of the total area) turn out to be responsible for the transport. This is true, independently

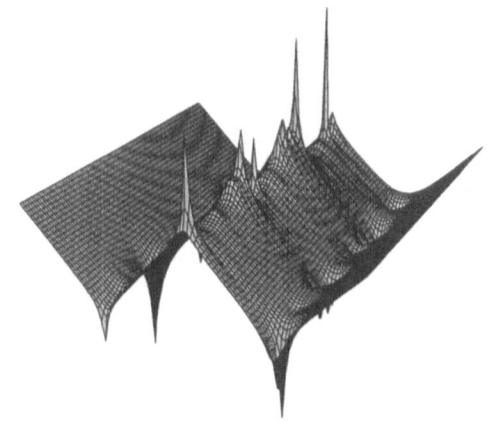

Fig. 1 Threedimensional plot of the impurity potential mountains v(x, y_0, z) of a δ-doped p-layer with two neighboring δ-doped n-layers for a fixed value of the y-coordinate.

on whether *tunneling through* the layers, or thermally activated transport *over* the barrier is dominant. We have performed both realistic calculations and measurements of the transport under the influence of an electric field in growth direction. As all the acceptors in the p-layers are ionized, the lack of screening by holes yields particularly pronounced potential fluctuations. In fact, we find excellent agreement between theory and experiment over the full temperature range from liquid helium to room temperature (see Fig. 3), if the random potential is taken into account, whereas there is dramatic disagreement if a uniform acceptor charge is assumed (cp. Fig. 2).

In the jellium case the two different transport theories that have been investigated, a simple 3D-3D model and the more sophisticated 2D-2D model, do not even show a similar *qualitative* behaviour. The 3D-3D model assumes ballistic electrons running against the barriers with an energy dependent attempt frequency. If the perpendicular component of the electron energy is below the barrier height the possibility of tunneling is determined by a WKB transmission coefficient. Thermally activated transport sets in, if the energy exceeds the barrier, where the transmission coefficient is simply set equal to 1. Both transport mechanisms exhibit a smooth dependence on the electron energy. Contrary to this, the 2D-2D model starts from subband states on either side of the barrier. While the description of thermally activated hopping is kept the same as for the 3D-3D model, the tunneling mechanism is determined within 2nd order perturbation theory. In the dominant 2D-2D process electrons cross the barrier by a combined process: resonant tunneling

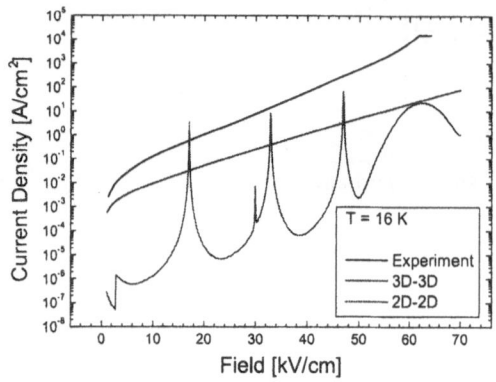

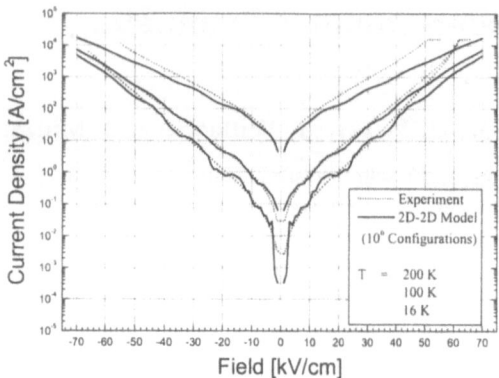

Fig. 2 Comparison of calculated and measured current density vs. field curves with disorder effects neglected (impurity charges assumed to be uniformly distributed). Both theoretical models, the semiclassical 3D-3D model as well as the sophisticated 2D-2D- model yield much too low currents as the carriers cannot take advantage of the saddle points of the realistic potential model. The characteristic obtained from the 2D-2D- model exhibits a pronounced subband structure due to elastic tunneling between subbands of neighboring n-layers and due to the corresponding inelastic phonon-assisted interlayer transitions.

through the barrier and subsequent relaxation via (optical) phonons. Note that tunneling is an intrinsically oscillatory process and there would be no transport without a relaxation to a final state. The resulting current-voltage characteristic shows distinct peaks whenever two neighboring subbands are aligned by an appropriate external field. Thereby, the width of the peak reflects the strength of relaxation behind the barrier. In Fig. 3 the 3D-3D model and the 2D-2D model are plotted together with the experimental curve. No resonances are found experimentally and the current density is by orders of magnitude higher than expected.

The disagreement results from the fact that the uniform potential barrier is by orders of magnitude less transparent for tunneling and requires a much higher thermal activation energy compared to the saddle points in the random potential mountains. Also, as expected, the subband structure turns out to be completely obscured by the potential fluctuations, both in theory and experiment [3]. This is due to the lack of correlations between initial and final states in neighboring n-layers (Note that this situation is in strong contrast to the case of intralayer-intersubband transitions in n-i-p-i superlattices, where subbands have clearly been observed in electronic Raman scattering experiments [4]).

3 Conclusion

In conclusion, we have shown that doping induced disorder in δ-doped barriers leads to enhanced tunneling and thermal transport due to local saddle points in the barrier. While cluster of acceptor atoms are responsible for the luminescence signal, transport is strongly

Fig. 3 Comparison of calculated (2D-2D) and measured current density vs. electric field curves with disorder taken into account for three temperatures. Note that there is good agreement in the tunneling regime (T=16K), the intermediate and the purely thermally activated transport (T=300K).

dominated by voids of acceptor atoms. In addition, the naively expected resonant fine structure in the current-voltage characteristic is smeared out by the random potential fluctuations.

We have also tested our theoretical approach by a comparison with experimental results obtained from linearly graded AlGaAs-barrier samples, which do not exhibit potential fluctuations. Quantitative agreement of the current-voltage characteristics with theory is obtained also in this case. In this case, however, as expected, also the subband structure can be observed in the transport measurements.

4 Acknowledgements

This work has been support by the Deutsche Forschungsgemeinschaft DFG under contract Do356/16-2.

References

1. G.H. Döhler, IEEE Journal of Quantum Electronics **22**, 1682 - 1695 (1986)
2. C. Metzner, K. Schrüfer, U. Wieser, M. Luber, M. Kneissl and G.H. Döhler, Phys. Rev. B **51**, 5106-5115 (1995)
3. R. Elpelt, *Untersuchung des schichtsenkrechten Transports und der optischen Eigenschaften von δ-dotierten Galliumarsenid-übergittern* in *Physik mikrostrukturierter Halbleiter* (S. Malzer, T. Marek, und P. Kiesel, eds.), Erlangen 2000
4. G.H. Döhler, H. Künzel, D. Olego, K. Ploog, P.Ruden, H.J. Stolz, and G. Abstreiter, Phys. Rev. Lett. **47**, 864 (1981)

Response time of the double-barrier heterostructures with resonant tunneling

M. N. Feiginov

Institute of Radioengineering and Electronics RAS, 11 Mokhovaya St., Moscow 103907, Russia e-mail: misha@mail.cplire.ru

Abstract We have shown that, first, the response time (τ_{resp}) of the double-barrier heterostructures (DBHS) can be much smaller as well as much larger than the quasibound-state lifetime in the quantum well (τ_{dwell}). Second, the real part of the DBHS conductance can be negative and large at frequencies higher than the reciprocal τ_{dwell} in the DBHSs with heavily doped collector. The displacement current and Coulomb interaction of the electrons in the quantum well with emitter and collector are responsible for the effects. A simple analytical expression for the impedance of DBHS has been derived, it is in fairly good agreement with experimental data. An equivalent circuit is proposed.

Introduction. The double-barrier heterostructures (DBHS) attract a lot of attention since, first, their N-shaped I-V curve has negative differential conductance (NDC) region; second, they are one of the fastest operating devices nowadays. A huge number of publications are devoted to the optimization of DBHSs to increase their operating frequencies. The following parameters are widely believed to be determining the high-frequency (HF) operation of DBHSs: peak to valley ratio (PVR), one assumes that the greater PVR or $[-G^0]$ (G^0 is the static differential conductance), the greater is $[-G^\infty]$ (G^∞ is the real part of the HF differential conductance); the barrier are made as thin as possible, that should decrease the electron tunnel dwell time (τ_{dwell}) in the quantum well (QW) between the barriers; the undoped spacer layers are inserted on the collector side of DBHS to decrease DBHS capacitance. The analytical theory developed in the present paper shows that G^∞ is substantiated by different parameters: generally speaking, G^0 has nothing to do with G^∞; τ_{dwell} has effect on the value of G^∞ rather than the frequency range where $G^\infty < 0$; the increase of the spacer layer thickness on the collector side of DBHS leads to sharp decrease of $[-G^\infty]$ in the NDC region and G^∞ becomes positive in the DBHSs with the typical thickness of the spacer layer. The physical reasons of the effects are the displacement currents in DBHSs and the variation of the tunnel transparency of the collector barrier with bias. The latter effect is generally supposed to be not essential.

The linear response of DBHS has been considered analytically in the present work in the sequential-tunneling approximation. The problem has been treated self-consistently. The details of derivation can be found in[1,2].

Response time. It has been shown that the response time (τ_{resp}) of DBHS is smaller and much smaller than τ_{dwell} in the positive differential conductance region of the I-V curve. By τ_{resp} we call the tunnel relaxation

time of the charge fluctuations in the QW. The physical reason of that could be explained as follows. Let us consider, for example, a kind of alpha-decay problem: let us put N electrons in a QW, then the time required for the number of electrons is decreased by 1 electron is τ_{dwell}/N. If N is large, then the required time can be much smaller than τ_{dwell}. The example shows that the characteristic time constant in a tunnel problem can be much smaller than τ_{dwell}. The similar situation is realized in DBHS. Let us decrease the number of electrons in QW by 1. Then the bottom of the QW lowers down due to the Coulomb interaction of the electrons in QW with emitter and collector. As a result the number of the empty states in the QW available for tunneling of electrons from emitter is increased by $(1 + \beta)$ rather than 1, where $\beta = \delta U_w/\delta E_{fw} = e^2\rho_{2D}/C$ is the ratio of the variation of the energy of the bottom of the 2D-subband in QW (U_w) and that of the Fermi energy in QW (E_{fw}), $\rho_{2D} = m^*/\pi\hbar^2$ is the 2D density of states in QW, $C = \epsilon(l + d)/4\pi ld$ is the QW capacitance, the effective emitter-well distance (d) is more than the emitter-barrier thickness by the Thomas-Fermi screening length and the half width of the QW; l is the similar well-collector distance, that includes also the thickness of the depletion region. The typical value of β is $\approx 5 \div 10$. Hence, the time required for a new electron to tunnel into QW from emitter is $(1 + \beta)$ times less than the electron dwell time in the QW due to the tunneling to emitter. That is, τ_{resp} can be much smaller than τ_{dwell}.

In NDC region an opposite situation is realized. The tunnel transparency of the emitter barrier significantly decreases, when the bottom of the QW lowers down – the resonant condition is violated. That strongly slow down the charge relaxation processes: τ_{resp} is always large than τ_{dwell} in the region and $\tau_{resp} \to \infty$ when $G^0 \to -\infty$.

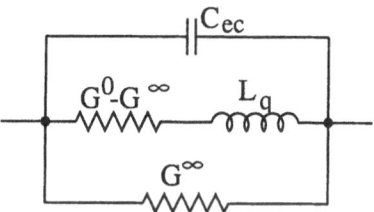

Fig. 1 The equivalent circuit of DBHS, here $L_q \equiv \tau_{resp}/(G^0 - G^\infty)$ is an inductance. Note that in the original paper [4] the inductance was supposed to be equal to $L_q = \tau_{dwell}/G^0$

Conductance of DBHS. The following equations has been derived for the linear conductance of DBHS:

$$G(\omega) \equiv \frac{\delta E_{fc}}{e\delta J} = i\omega C_{ec} + G^\infty + \frac{G^0 - G^\infty}{1 + i\omega\tau_{resp}}, \quad (1)$$

where E_{fc} is the collector Fermi-energy, emitter is supposed to be grounded, i.e. $\delta E_{fe} = 0$; J is the total DBHS current that includes the displacement one also, $C_{ec} = \epsilon/4\pi(l + d)$ is the emitter-collector capacitance. G^∞ and G^0 have the form:

$$G^\infty = \frac{d}{l+d}C_{wc}\left[\frac{1}{\tau_{resp}} - \frac{1}{\tau_{dwell}}\right] + \frac{l-d}{l+d}e^2 N_{2D}\nu_c', \quad (2)$$

$$G^0 = \nu_c C_{wc}\left(1 - \tau_{resp}/\tau_{dwell}\right). \quad (3)$$

where $C_{wc} = \epsilon/4\pi l$ is the well-collector capacitance, N_{2D} is the electron concentration in QW, $\nu_c(V_{wc})$ is the tunnel escape rate of electrons from QW via collector barrier; $V_{wc} \equiv U_w - U_c$, U_c is the conduction band bottom energy in collector. The equation for the DBHS conductance (1) contains just 4 parameters, with 2 of them are well known: C_{ec} is known with a good accuracy, as a rule, and G^0 is also known, if the static I-V curve is measured. The DBHS conductance (1) can be represented in the form of the equivalent circuit shown in Fig.1.

At low frequencies ($\omega\tau_{resp} \ll 1$) the DBHS conductance can be approximated by the RC-circuit:

$$G \approx i\omega\tilde{C} + G^0, \quad \tilde{C} = C_{ec} + \tau_{resp}\left[G^\infty - G^0\right]. \quad (4)$$

As it follows from (4) (see [1,2]), the low-frequency capacitance $\tilde{C}$ can be more or less then the geometrical one (C_{ec}). It should be noted that $\tilde{C} \to \infty$ when $G^0 \to -\infty$.

At high frequencies one can approximate DBHS conductance by an other RC-circuit:

$$G \approx i\omega C_{ec} + G^\infty. \quad (5)$$

The approximation is valid even when $\omega\tau_{dwell} \gg 1$. Note that G^0, generally speaking, has nothing to do with G^∞. The displacement currents are responsible for the HF operation of DBHS. The currents contribute to the imaginary part of the conductance only in the case of the simplest RC-circuit. But in the case of a more complicated systems like DBHS, which has 3 space-charge regions (emitter, collector and QW), or more complicated circuits (e.g., two RC-circuits connected in series) the displacement currents have significant effect on the real part of the HF conductance.

In the NDC region $\tau_{resp} > \tau_{dwell}$ [see (3)] and the first term in (2) is negative. The last term in (2) is positive if $l > d$. Nowadays the DBHSs for HF applications are made with the spacer layers on the collector side of the structure (i.e. $l \gg d$). The assessments show [2] that in the kind of structures the last term in (2) is approximately the same (and more) as the first one. As a result $G^\infty > 0$ in the structures in the NDC region (see, e.g., Fig.2, [3]) and variation of the transparency of the collector barrier with bias (ν_c') is responsible for the effect.

To achieve negative G^∞ one should exclude spacer layers on the collector side of DBHS and heavily dope collector, i.e. l should be of the order of d. In the DBHSs

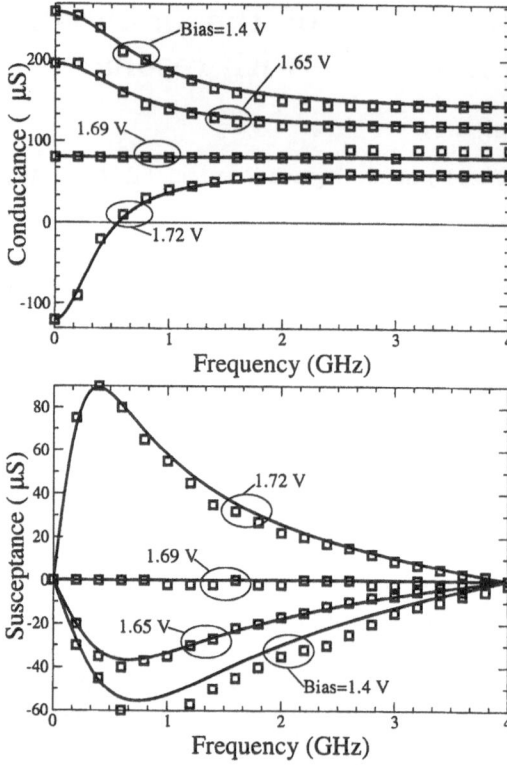

Fig. 2 The experimental points are taken from [3]; the continuous lines are calculated with the help of Eq.(1). $\mathrm{Re}[G(\omega)]$ and $\mathrm{Im}[G(\omega)] - \omega C_{ec}$ are plotted in the upper and lower figures, respectively. The necessary parameters for the theoretical calculation were obtained as follows: G^0 and G^∞ are taken from the experimentally measured $\mathrm{Re}[G(\omega)]$; $\mathrm{Re}[G(\omega)] = (G^0 + G^\infty)/2$ at the frequency $\omega = 1/\tau_{resp}$, that gives possibility to get τ_{resp} from the measured $\mathrm{Re}[G(\omega)]$.

with large NDC – such that the condition ($-G^0 \gg \nu_c C_{wc}$) is satisfied (the condition is met in many experimental works, e.g., in [3]) – $\tau_{resp} \gg \tau_{dwell}$, see (3). In the structures the last term in (2) is negligibly small, and $G^\infty \approx -[d/(l+d)]C_{wc}/\tau_{dwell}$. Note that G^∞ grows as $\propto 1/l(l+d)$ with decrease of l. Thus, we see that in DBHSs with heavily doped collector the HF conductance should be negative and large at frequencies much higher than $1/\tau_{dwell}$. Namely that kind of structures can be used as an active element for HF applications.

The comparison of the frequency dependence of the DBHS conductance (1) and the experimental data is shown in Fig.2. The theory and experiment are in excellent agreement with each other.

Acknowledgements. The work was supported by Programs FTNS (99-1124) and PAS (3.1.99); INTAS (97-11475) and RFBR (99-02-17592).

References

1. M. N. Feiginov, Appl. Phys. Lett. **76**, (2000) 2904.
2. M. N. Feiginov, *to appear in* Nanotechnology **11**, (2000).
3. J. P. Mattia, A. L. McWhorter, R. J. Aggarwal, F. Rana, E. R. Brown, and P. Maki, J. Appl. Phys. **84**, (1998) 1140.
4. E. R. Brown, C. D. Parker, and T. C. L. G. Sollner, Appl. Phys. Lett. **54**, (1989) 934.

Current self-oscillations with discrete frequencies in weakly coupled semiconductor superlattices

M. Rogozia, H. T. Grahn, R. Hey

Paul-Drude-Institut für Festkörperelektronik, Hausvogteiplatz 5–7, 10117 Berlin, Germany, e-mail: `rogozia@pdi-berlin.de`

Abstract The frequency of current self-oscillations in weakly coupled superlattices usually varies continuously with increasing applied voltage within a single plateau of the I-V characteristics. However, a more detailed investigation reveals the existence of discrete frequencies, which are locked to particular values. The locking frequency is an integer multiple of a fundamental frequency f_0, which is determined by the length of the used coaxial cables.

1 Introduction

In moderately doped, weakly coupled semiconductor superlattices (SLs), current self-oscillations have been observed with frequencies ranging from values below 1 MHz up to 10 GHz [1]. The frequency is mainly determined by the resonant coupling between adjacent wells. The oscillation regime is usually accompanied by a structureless current plateau observed in the time-averaged I-V characteristics. Within a single sample, the frequency typically increases continuously with increasing applied voltage for a particular current plateau of the I-V characteristics [2].

We show that the voltage dependence of the frequency of the current self-oscillations exhibits steps. The observed frequencies are almost integer multiples of a fundamental frequency f_0, which depends on the length of the used coaxial cables.

2 Experimental

Three different GaAs/AlAs SLs with 40 periods each have been investigated. The samples are grown by molecular beam epitaxy on n^+-GaAs. In sample A, one period consists of a GaAs well with a thickness $d_W = 20.0$ nm and an AlAs barrier with a thickness $d_B = 2.0$ nm, while in sample B (C) these widths are $d_W = 10.1$ nm (9.0 nm) and $d_B = 1.4$ nm (1.5 nm). The central region of the GaAs wells is Si-doped in the 10^{16} cm^{-3} range. The n-type SLs are sandwiched between n^+-contacts. The complete structures are supplied with Ohmic contacts and etched into mesas of 16, 35, and 70 μm diameter.

The mesas are mounted onto a sapphire plate with gold stripes evaporated on it. The sample is either put directly on a sample holder without any extra cables or put into a He-flow cryostat equipped with two high frequency coaxial cables each of about 0.26 m length. The DC voltage (Keithley SMU236) is applied through the back contact via the inductive part of a bias-tee (Picosecond pulse Labs model 5808), whereas the top contact is grounded. The power spectra are recorded with a

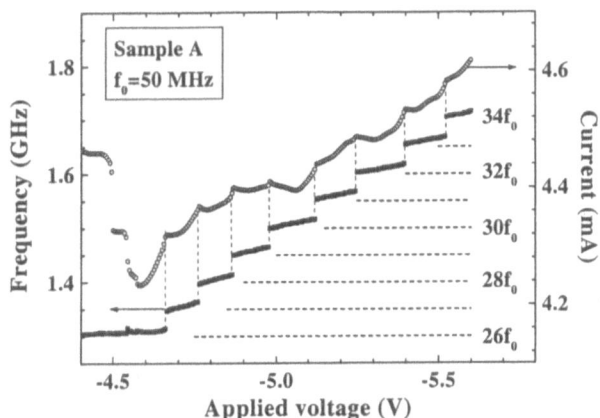

Fig. 1 Current and frequency versus applied voltage in the third plateau the I-V characteristics of sample A recorded at 296 K. The bias-tee is directly connected to the sample holder without using the cryostat. The cable length from the bias-tee to the spectrum analyzer amounts to 2 m. The dashed lines indicate integer multiples of $f_0 = 50$ MHz.

spectrum analyzer (Advantest R3361A), which is either connected through the capacitor of the same bias-tee as above or put in series with the sample. Outside the cryostat, we used coaxial cables of 20 GHz band width.

3 Results

All samples exhibit at room temperature current self-oscillations with frequencies between 0.8 and 3 GHz. Figure 1 shows the dominant oscillation frequency and the current as a function of the applied voltage in the third plateau of the I-V characteristics of sample A at room temperature. The bias-tee is directly connected to the sample holder without using the cryostat. The frequency increases discontinuously with a step separation of 50 MHz. The jump in frequency is correlated with cusps in the current-voltage characteristics. The frequencies of each branch at maximum power are integer multiples of a fundamental frequency f_0 of 50 MHz as indicated in the figure.

The fundamental frequency changes, when the sample is put into the cryostat. Figure 2 shows the power spectra as a function of applied voltage in sample A at room temperature with the sample in the cryostat using otherwise the same setup as for Fig. 1. The fundamental frequency is now determined by the cable length from the sample in the cryostat to the bias-tee, which is directly connected to one of the sockets outside the

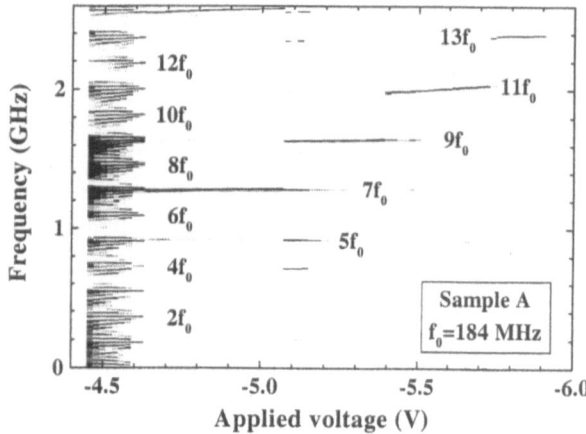

Fig. 2 Frequency versus applied voltage in the third plateau of the I-V characteristics of sample A at 296 K with $f_0 = 184$ MHz. The sample is mounted in the cryostat with the bias-tee being directly connected to the sockets outside the cryostat. Darker areas correspond to larger power.

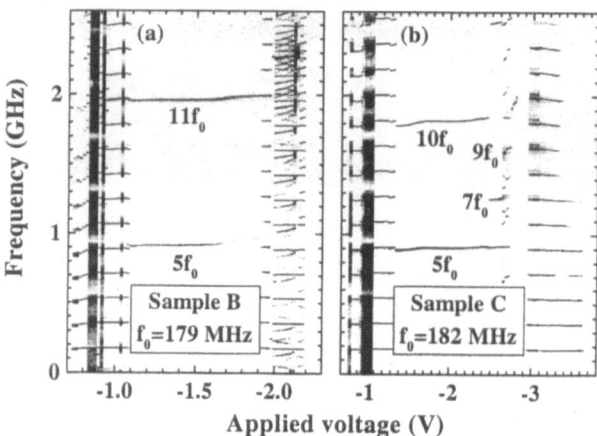

Fig. 3 Frequency versus applied voltage in the first plateau of the I-V characteristics of (a) sample B and (b) sample C at 296 K. We used the same setup as for Fig. 2. The fundamental frequencies f_0 in (a) and (b) have about the same value as in Fig. 2.

cryostat. Since the cable length is now much shorter, the fundamental frequency increases to $f_0 = 184$ MHz. However, in addition to the locking observed in Fig. 1, there is additional fine structure, which looks very similar to the quasi-periodic oscillations in AC driven SLs [3]. The cable acts as a resonator, which supplies an AC modulation of the DC voltage.

The same principle behavior is found in samples B and C. The corresponding power spectra are shown in Fig. 3(a) and (b), respectively, using exactly the same setup as for Fig. 2. The fundamental frequencies are 179 and 182 MHz in agreement with the 184 MHz in Fig. 2. The difference between these frequencies is within the experimental error bars. Note that the locking in the configuration of Figs. 2 and 3 usually occurs in odd integer multiples of the fundamental frequency.

We also investigated the temperature dependence of this behavior in sample B between 10 and 350 K. The fundamental locking frequency hardly changes with temperature supporting the above explanation. With decreasing temperature, the frequencies remain a multiple of f_0, but the integer varies with temperature. Furthermore, the fundamental frequency f_0 or the separation between two adjacent frequencies in the power spectra Δf does not depend on the mesa size.

4 Discussion

The fundamental frequency f_0 in the frequency-locked state is clearly related to the cable length used between the sample and the spectrum analyzer, when the sample is mounted on a sample holder outside the cryostat. By using different cable lengths, we were able to show that $1/f_0$ varies linearly with the cable lengths, i. e.

$$\frac{1}{f_0} = \frac{2(L_{cable} + L_0)}{v_{meas}}, \tag{1}$$

where $L_0 = 0.25$ m denotes an offset due to additional cable lengths inside the bias-tee and the spectrum ana-

lyzer. Note the factor of two for the cable length, since the inverse round-trip time through the cables determines f_0. The value of the velocity of electromagnetic waves derived from this measurement ($v_{meas} = 2.30 \times 10^8$ m/s) agreed with an independent measurement of the velocity in the cables alone ($v_{cable} = 2.32 \times 10^8$ m/s). When the sample is put into the cryostat, the length of the cables from the sample to the sockets of the cryostat determines f_0. The larger value of f_0 is in agreement with the shorter lengths of the cables.

In strongly coupled GaAs/AlAs SLs at oscillation frequencies of about 25 and 70 GHz, frequency jumps of $\Delta f = 3.2$ and 4.9 GHz have been reported by Schomburg *et al.* [4,5]. Since Δf does not seem to vary for the two different samples, it is possible that their observation can also be explained by the round-trip time effect described above. Using our signal velocity v_{cable}, a frequency of 4 GHz corresponds to a cable length of about 3 cm, which is probably the effective cable length between the sample and their frequency divider directly attached to the sample.

5 Summary

We have observed frequency locking of the current self-oscillations in weakly coupled GaAs/AlAs SLs without any AC driving voltage. The fundamental frequency f_0 is determined by the inverse round-trip time of the coaxial cables.

References

1. M. Rogozia, R. Hey, H. Kostial, and H. T. Grahn, *Proceed. 26th Int. Symp. on Compound Semiconductors*, edited by K. H. Ploog, G. Tränkle, and G. Weimann (IOP, Bristol, 2000) p. 147.
2. J. Kastrup *et al.*, Phys. Rev. B **55**, (1997) 2476.
3. K. J. Luo *et al.*, Phys. Rev. Lett. **81**, (1998) 1290.
4. E. Schomburg *et al.*, Appl. Phys. Lett. **72**, (1998) 1498.
5. E. Schomburg *et al.*, Physics E **2**, (1998) 295.

Observation of the scattered electrons in the resonant tunneling regime using a three-terminal quantum-well heterostructure

Gyungock Kim, Dong Wan Roh, Seung Won Paek, Kwang Man Koh, Kwang E. .Pyun, and Chong Hoon Kim

Electronics and Telecommunications Research Institute, Yusong P. O. Box 106, Taejon, Korea 305-600.

Abstract The electron resonant tunneling is probed using a three-terminal Ga(Al)As/GaAs(001) tunneling heterostructure. The measured base current turns out to be the direct measurement of the scattered electrons in the resonant tunneling regime. The onset of a sudden sharp increase in the base current occurs at the resonant peak voltage, revealing the 'turn on (activation)' of the major scattering of electrons. The observed base current near resonant voltage is the time-averaged phenomena caused by the dynamic electron tunneling and scattering mechanism. The experimental results indicate that the dynamical balance between the scattering effect and the tunneling time (dwelling time) of electrons at a resonant peak voltage may limit the resonant peak current density of a quantum-well heterostructure intrinsically.

The topic of electron resonant tunneling in quantum-well heterostructures, known as the fastest electron transport phenomenon, has received a great deal of attention for the study of the physics and also due to the applicability to high-speed electrical and optical devices.[1-5] Although extensive experimental and theoretical efforts have been devoted to the understanding of the electron tunneling phenomena through quantum-well confined states, many questions remain to be answered, such as, the electron tunneling and scattering mechanism of a quantum well in the resonant tunneling regime, electron tunneling time, resonant tunneling peak current which is much lower than the theoretically predicted values, etc..[2]

In the present paper, we investigate the electron scattering effect in the resonant tunneling regime, using a Ga(Al)As/AlAs (001) three-terminal resonant tunneling heterostructure (RTS). The three-terminal RTS consists of an emitter, an emitter barrier, a base, a quantum-well structure and a collector. The quantum-well structure embedded between the base layer and the collector layer includes a triple-barrier quantum-well tunneling heterostructure (TBS) and wide spacer layers on both sides of the TBS. The electrons are injected toward the TBS, either from the emitter or from the collector. With this structual configuration, a thin base terminal is used to probe the scattered electrons, which are injected from the emitter. The structure was grown by molecular beam epitaxy (MBE) on (100) n$^+$ GaAs substrate. The 100 nm-thick undoped GaAs spacer layer is formed on the n$^+$-GaAs collector. The TBS on the GaAs spacer layer

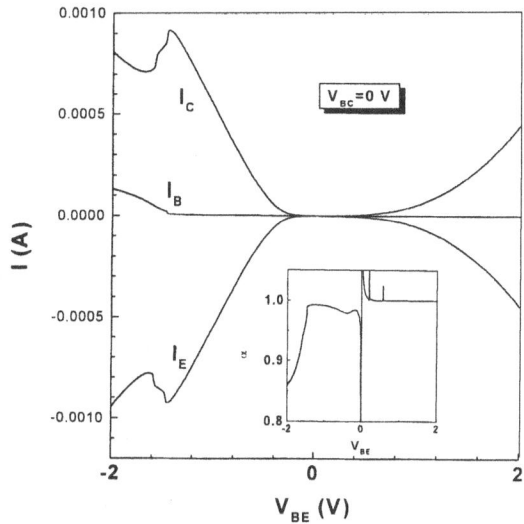

Fig.1 The measured tunneling current-voltage (I-V) characteristics, and the electron transfer ratio ($\alpha=|I_C/I_E|$) of the three-terminal Ga(Al)As/GaAs(001) tunneling heterostructure in the base-common configuration. A sudden sharp increase in the base current occurs at the resonant peak voltage. Here subscripts, E, B, and C indicate emitter, base and collector, respectively.

comprises of an 1.6 nm AlAs bottom barrier, a 4.5 nm undoped GaAs quantum well, a 2.9 nm AlAs middle barrier, an undoped 5.8 nm GaAs quantum well, and an undoped 1.6 nm AlAs top barrier.[5] The 100 nm-thick undoped GaAs spacer is formed between the top barrier of the TBS and the n$^+$-GaAs base layer. The 30 nm-thick GaAs base layer is doped with Si to 1×10^{18} cm^{-3}. The undoped 6 nm Al$_{0.4}$Ga$_{0.6}$As emitter-barrier layer, which prevents the equilibrium electrons in the base from entering the emitter, is sandwiched between a 5 nm-thick undoped GaAs spacer and a GaAs emitter layer. The double-mesa structures were fabricated using the standard technique of photolithography, dry and wet mesa etching, and lift-off processes.

Fig.1 shows the measured tunneling current-voltage (I-V) characteristics, and the electron transfer ratio ($\alpha=|I_C/I_E|$) of the three-terminal Ga(Al)As /GaAs(001) tunneling heterostructure in the base-common configuration. Figure shows the electron transfer ratio from the collector to the emitter, α (= the transfered current/ the injected current) is ~1 with positive V_{BE} (base-emitter bias). This implies that with positive V_{BE}, negligible electron scattering occurs in the electron transit from the collector (or TBS) to the emitter. Also, very low base current is observed up to the resonant voltage, indicating low electron scattering effect, since $I_C + I_E + I_B = 0$. The onset of a sudden sharp increase in the base current occurs at the resonant peak voltage. This sudden increase of base current at the resonant peak voltage reveals the 'turn on' of the major scattering of electrons. The observed base current is the static observation (time-averaged phenomena) of the dynamic electron tunneling and scattering mechanism.

Fig. 2 shows the scatter ratio ($\beta=1-\alpha$) of electrons, and the derivative of the scatter ratio with respect to the electron injection voltage, V_{BE}. The derivative of the scatter ratio ($\beta=1-\alpha$, where $\alpha=|I_C/I_E|$) of electrons with respect to the electron injection voltage has a sharp peak localized at the resonant voltage, which is related to the scattering cross section of electrons at the resonant voltage as a function of energy. From $\Delta E \, \tau \sim \hbar /2\pi$, the estimated characteristic time, τ, of the scattering is around a few tens of femtoseconds, which is within the coherent regime. The experimental result indicates that the electron scattering suddenly turned on in the resonant peak voltage may be the Pauli exclusion-type restriction, which is activated on the fraction of the injected electrons to the quantum-well at the resonant peak voltage, where the tunneled electrons inside the quantum-well have already exhausted all the possible 2D quantum-well states, due to their finite dwelling time (tunneling time). The dynamical balance between the scattering effect and the tunneling time of electrons near a resonant peak voltage may determine the resonant peak current density through a quantum-well heterostructure, and this may be the major intrinsic reason for limiting a resonant peak current density to a lower value than the theoretically predicted ones.

We have investigated the electron scattering effect in the resonant tunneling regime, using a Ga(Al)As/AlAs (001) three-terminal resonant tunneling heterostructure (RTS), and presented the static observation of the dynamic electron tunneling and scattering effect. The experimental results indicate that the dynamical balance between the tunneling time (dwelling time) of electrons and the Pauli exclusion-type scattering effect at a resonant peak voltage may determine the resonant peak current density of a quantum-well heterostructure, and result in limiting the resonant peak current density of the structure intrinsically.

References

1.   R. Tsu and L. Esaki, Appl. Phys. Lett, 22, 562 (1973); L. L. Chang, L. Esaki and R. Tsu, Appl. Phys. Lett, 24, 593 (1974);

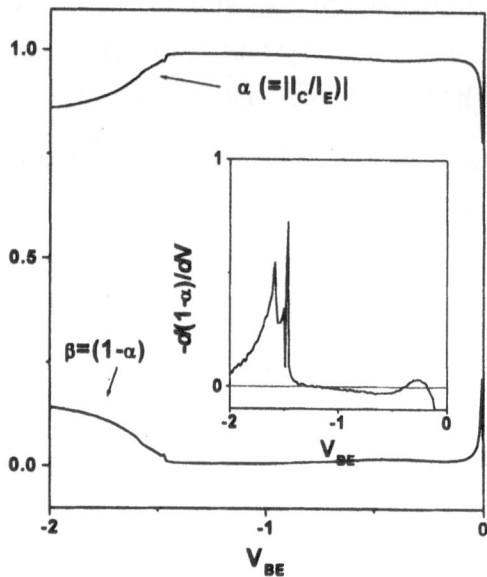

Fig. 2 The scatter ratio ($\beta=1-\alpha$) of electrons, and the derivative of the scatter ratio with respect to the electron injection voltage, V_{BE} (base-emitter bias), which has a sharp peak localized near resonant voltage.

S. Luryi, Appl. Phys. Lett., 47, 490,1985; V. Goldman and D. Tsui, Phys.Rev. Lett., 58, 1256 (1987); J. Young, B. Wood, G. Aers, R. Devine, H. Liu, D.

2.   Landheer, M. Buchanan, A. Spring Thorpe and P. Mandeville, Phys. Rev. Lett., 60, 2085 (1988); H. Toshimura, J. Schulman, and H. Sakaki, Phys. Rev. Lett., 64, 2422, 1990; M. Tsuchiya, T. Matsusue, and H. Sakaki, Phys. Rev. Lett., 59, 2356, 1987; M. Jackson, M. Johnson, D. Chow, T. McGill, and C. Nieh, Appl. Phys. Lett.,454, 552,1989, V. Goldman, D. Tsui, and J. Cunningham, Phys. Rev. B, 35, 9387, 1987.

3.   G. Kim, Y. Choi, P. Park, H. Chu, E. Lee and G. Arnold, Phys. Rev. B, 50, 7582 (1994); G. Kim, H. Suh and E. Lee, Phys. Rev. B, 52, 2632 (1995).

4.   G. Kim, D. Roh, and S. Paek, J. Appl. Phys., 81, pp7070-7072, 1997; G. Kim, D. Roh, S. Paek, and E. Lee, J. Appl. Phys., 82, pp3368-3373, 1997;

5.   E. Koenig, B. Jogai, M. Paulus, C. Huang, and C. Bozada, J. Appl. Phys., 68, pp3425-3430, 1997.

Nature of the localized phase in a two-dimensional electron system

I.E. Itskevich[1]*, R.J.A. Hill[1], S.T. Stoddart[1], H.M. Murphy[1], A.S.G. Thornton[1], P.C. Main[1], L. Eaves[1], M. Henini[1], D.K. Maude[2], J.-C.Portal[2]

[1] School of Physics and Astronomy, University of Nottingham, University Park, Nottingham NG7 2RD, United Kingdom
[2] GHMFL MPI-CNRS, 38042 Grenoble, France, and INSA-CNRS, 31077 Toulouse, France

Abstract The tunnel current through InAs self-assembled quantum dots embedded in a barrier of a tunnelling diode was used as a local probe of the localized phase of a two-dimensional electron system (2DES) in the accumulation layer of the diode. By means of high pressure, we were able to tune the 2DES concentration n corresponding to a particular resonant peak. We have found direct evidence that at low n, the localized electrons form relatively large, high-density clusters.

The nature of the localized phase in a low-density two-dimensional electron system (2DES) has been a matter of debate for years. Experimental results have been interpreted both in terms of electron clusters [1,2] and localization of single electrons [3]. However, only indirect information can be obtained from most of the measurements; to study directly the nature of the *localized* electrons, a *local* probe of the 2DES is required.

We have studied resonant tunnelling through localized electron levels of InAs self-assembled quantum dots (SAQD) [4] embedded in the barrier of a tunnelling diode. This provides an effective *local* probe of the 2DES that is formed in the accumulation layer of the diode under applied bias V [5,6]. The 2DES density n can be tuned over a wide range by varying V, but each tunnelling peak is observed in a current-voltage $I(V)$ characteristic of a diode at particular V and hence n. However, an additional degree of flexibility can be achieved by using high pressure P. The effect of P is to shift down the electron levels in the SAQD with respect to the conduction-band states in GaAs [7], so that the peaks are observed at lower V and n. Therefore, P can be used to tune n corresponding to a particular peak.

Details of the sample and measurement techniques have been described earlier [5,6]. Hydrostatic P was applied in a liquid clamp cell and measured using a calibrated InSb gauge. The conduction-band profile of the sample under applied V is shown schematically in Fig. 1. A layer of InAs SAQD was embedded in the centre of a 10 nm AlAs barrier. A series of sharp resonant peaks is observed in $I(V)$ at low T, each peak due to tunnelling via a discrete electron level of an individual dot [5]. Resonant tunnelling through a state at energy E_{QD} occurs if, roughly, E_{QD}/e equals the voltage drop V_1 from the 2DES to the dot layer. Note that V_1 is a small fraction

* *Present address:* School of Engineering, University of Hull, Cottingham Road, Hull HU6 7RX, United Kingdom; e-mail: I.Itskevich@eng.hull.ac.uk

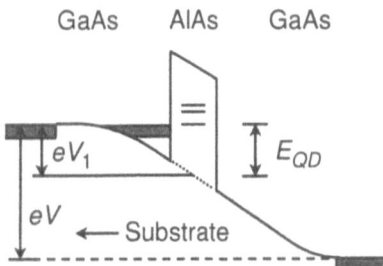

Fig. 1 Schematic conduction-band diagram of the tunnelling device under applied bias V, with the 2DES accumulated on the emitter side of the barrier. The SAQD levels in the barrier are shown schematically.

of the external voltage V; $f = (dV_1/dV)^{-1}$ defines the electrostatic leverage factor. Note also that if the 2DES is separated into localized clusters at low n, then V_1 in the vicinity of the cluster is, to a first approximation, proportional to the local concentration of electrons n_{loc}, while V is roughly proportional to the average concentration n_{av}. In a cluster, n_{loc} is obviously less sensitive to bias than n_{av}, and to achieve n_{loc} and V_1 necessary for the resonance with the same SAQD state shifted by P, a greater change in V is required. Therefore, the pressure coefficient for the peak bias for a peak due to tunnelling from a cluster should be higher than that for a peak due to tunnelling from the homogeneous 2DES.

A representative set of $I(V)$ characteristics at various P is shown in Fig. 2; a smoothly varying background current has been subtracted from each $I(V)$ for clarity. Also, the $I(V)$ are offset proportionally to P: this allows us to plot the peak voltage as a function of P on the same diagram. We first discuss the $I(V)$ at $P < 4.3$ kbar. With increasing P, the peak shifts gradually to lower biases at the rate of ≈ 30 mV/kbar; the vast majority of peaks observed in the $I(V)$ of our devices exhibit the same behaviour. We will call these peaks "slowly-moving". In most cases, their intensity falls with decreasing peak bias until, at $V = 50 - 120$ mV, the peaks vanish. This is consistent with the bias required for the formation of an accumulation layer in the diode. We estimate f to be 12 ± 1 in our sample. Therefore, the coefficient for the resonant level is about 2.5 meV/kbar lower than that for the Γ-valley in the conduction-band of bulk GaAs, which is 10.7 meV/kbar [8]. This observation provides unambiguous confirmation that the single-electron tunnelling occurs through Γ-valley-related resonant states of individual SAQD [5]. The observed is consistent with our

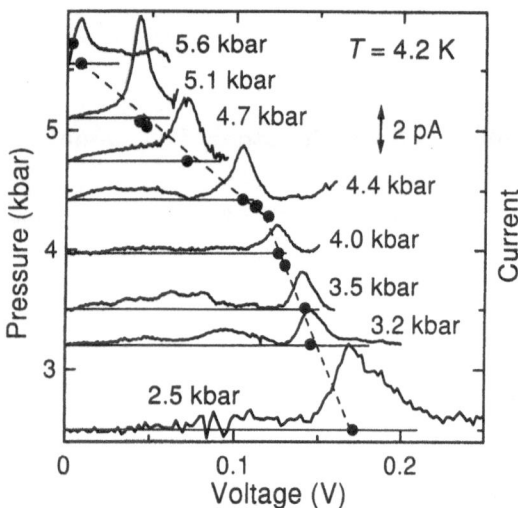

Fig. 2 $I(V)$ curves with subtracted background, right axis, and peak positions as a function of P, left axis. For each curve, the offset is proportional to P. Around 4.3 kbar a slowly-moving peak transforms to a fast-moving peak.

earlier data obtained from high-pressure photoluminescence [7].

However, there are also other patterns of peak behaviour. In particular, we have observed peaks that shift much faster with P, at about 100 mV/kbar; we call them "fast-moving" peaks. Many of them persist up to P high enough to shift the peaks to very low V. Eventually, these peaks may even give finite conductance at zero V. Furthermore, there is evidence that some peaks may change their character, i.e., change from a slowly-moving to a fast-moving peak with increasing P. An example of such a transformation is shown in Fig. 2. The transformation occurs at about $4.2 - 4.3$ kbar. The peak can be observed at very low V and, at $P = 5.5$ kbar, it gives finite conductance at zero V.

We suggest that the fast-moving peaks may be due to tunnelling from isolated electron clusters rather than from a homogeneous 2DES. This assumption is based, first, on the higher pressure coefficient which can be expected for the peaks due to tunnelling from clusters, as discussed above. Also, only the fast-moving peaks persist to very low V, while the slowly-moving peaks quench around $50 - 100$ mV. Finally, only *localized* electron clusters can persist to zero bias; these are necessary to account for the finite zero-bias conductance.

To verify our assumption, we measured n_{loc}. It was obtained from the magneto-oscillations of the voltage at which the onset of the resonant peak occurs [6]. The value of n_{av} at various V was obtained from magnetocapacitance measurements at zero P [9]; we do not expect it to change under P. Figure 3 shows n_{loc} measured for both slowly- and fast-moving peaks. While for a slowly-moving peak (above 120 mV) the values of n_{loc} and n_{av} agree within experimental error, for a fast-moving peak (below 120 mV), n_{loc} is clearly higher than n_{av}. This is the fingerprint of a localized electron cluster.

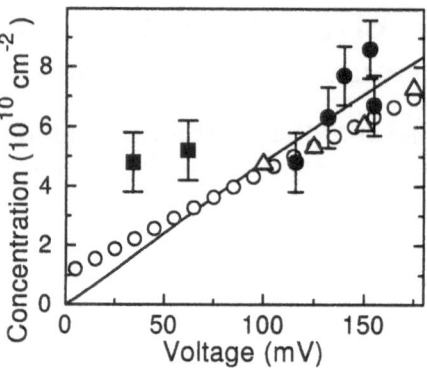

Fig. 3 The electron concentration in the 2DES as a function of V. Open symbols: the *average* values obtained at zero P from magneto-capacitance measurements (triangles) and from the integration of the $C(V)$ [9] (circles). Solid line: the prediction of a simple electrostatic model. Solid symbols: the *local* values measured as described in the text for the slowly-(circles) and fast-moving (squares) peaks at different P.

Our results indicate that the localized phase of the 2DES consists of relatively large electron clusters rather than localized single electrons. The observed magneto-oscillations of the peak onset voltage require the formation of Landau levels. The clusters must contain a number of electrons at least a few times higher than the filling factor. This provides an estimate of a lower limit for the cluster size of $100 - 200$ nm. The effective local mobility in a cluster should be large enough for the Landau levels to be formed at B as small as 0.5 T. Our observations also suggest that the observed transformation of the peak from slowly- to fast-moving, as shown in Fig. 2, may be evidence for a localization transition in the 2DES, from a homogeneous gas to localized clusters.

To conclude, by means of high pressure, we have obtained unambiguous evidence for resonant tunnelling through Γ-valley-related electron states in self-assembled InAs quantum dots. We have used the tunnel current through a SAQD as a local probe of the electron density of a 2DES and found that, at low densities, the electrons form localized clusters which are relatively large and have a relatively high effective local mobility.

Acknowledgements Supported by EPSRC (UK) and GNTP "Nanostructures", grant 97-1068 (Russia). L.E. and I.E.I. are grateful to the EPSRC and to the Royal Society, respectively.

References

1. R. C. Ashoori et al., Phys. Rev. B **48**, 4616 (1993).
2. I. V. Kukushkin et al., Phys. Rev. B **53**, R13260 (1996).
3. G. Eytan et al., Phys. Rev. Lett. **81**, 1666 (1998).
4. D. Bimberg, M. Grundmann, and N. N. Ledentsov, *Quantum Dot Heterostructures* (Wiley, New York, 1998).
5. I. E. Itskevich et al., Phys. Rev. B **54**, 16401 (1996).
6. P. C. Main et al., Phys. Rev. Lett. **84**, 729 (2000).
7. I. E. Itskevich et al., Appl. Phys. Lett. **70**, 505 (1997); I. E. Itskevich et al., Phys. Rev. B **58**, 4250 (1998).
8. D. J. Wolford and J. A. Bradley, Solid State Commun. **53**, 1069 (1985).
9. S. T. Stoddart et al., Phys. Rev. B, to be published.

Evidence for Screening Breakdown near the Metal-to-Insulator Transition in two Dimensions

W. Jantsch[1], Z. Wilamowski[2], N. Sandersfeld[1] and F. Schäffler[1]

[1]Institut für Halbleiterphysik, Johannes Kepler Universität, A-4040 Linz, Austria
  Fax: +43 732 2468 9696; e-mail: wolfgang.jantsch@jk.uni-linz.ac.at

[2]Institute of Physics, Polish Academy of Sciences, Al Lotnikow 32/46, PL 02-668 Warsaw, Poland

Abstract We evaluate the magnetic susceptibility of a high mobility 2d electron gas in a modulation doped SiGe/Si/SiGe quantum well structure from spin resonance (ESR) experiments. With decreasing carrier density the magnetic susceptibility decreases and tends to zero at the critical concentration for the metal to insulator transition (MIT). Assuming weak interaction, this finding can be attributed to a breakdown in screening.

The experimentally inferred existence of a metallic phase of two-dimensional (2D) carriers [1,2] appears to contradict scaling theory [3] which poses one of the unsolved problems in solid state physics. In this contribution we investigate the recently discovered ESR of high mobility electrons in a modulation doped Si quantum well [4-7]. We find a very narrow resonance whose integral absorption (which is proportional to the magnetic susceptibility, χ_m) extrapolates to zero as we approach the critical density for the MIT – the ESR is seen only in the metallic phase and χ_m apparently constitutes an order parameter for the MIT.

The normalized mean distance of the electrons, $r_s = 1/(a_B \sqrt{\pi n_e})$, describes also the ratio of Coulomb interaction and kinetic energy (here a_B is the Bohr radius and n_e the carrier density). In our experiment r_s always exceeds one and therefore a simpleminded evaluation of χ_m in terms of the Pauli susceptibility a priori is not justified. Nevertheless, we find an a posteriori justification for this approach by evaluating the screening wave vector allowing us to calculate the electron mobility: a comparison to experimental data shows good agreement. Within this approach we show that the screening efficiency vanishes in the insulating phase, i.e., the metallic phase is stabilized by screening.

We investigate samples grown by molecular beam epitaxy. A pseudomorphic Si quantum well between $Si_{0.75}Ge_{0.25}$ barriers was deposited on a strain-relaxed, compositionally graded $Si_{0.75}Ge_{0.25}$ buffer layer on a Si(001) substrate. The quantum well is 12 nm thick, and the top barrier is modulation doped with an undoped spacer of 12 nm thickness. Sample 1 (S1) has a 250 nm wide doping layer with an Sb concentration of

$7 \cdot 10^{17}$ cm^{-3}. Because of the natural surface depletion layer, n_e in this sample is less than $1 \cdot 10^{10}$ cm^{-2} after cooling in darkness.

Upon illumination with photon energies above the Si band gap, the sample shows persistent photoconductivity which is used to adjust n_e up to a saturation value of $n_s = 3 \cdot 10^{11}$ cm^{-2} by increasing the illumination dose. S2 is prepared from the same wafer, but has a self-aligned palladium Schottky gate that covers almost the entire 4x8 mm^2 large specimen area. The highly doped S3 has a very high mobility of 300,000 cm^2/Vs at 300 mK. The ESR experiments are done with a standard X-band spectrometer (Bruker 200 SRC) and a TE102 cavity.

Results for the integral ESR absorption are given in Fig. 1, normalized to the saturation value that is common for all three samples. The carrier density was determined for S1 from the cyclotron resonance (CR) which appears also as a broad resonance in the ESR spectra. For S2 and S3 the carrier density was obtained from Shubnikov-de-Haas measurements. The data in Fig. 1 show an onset at a density of $7 \cdot 10^{10}$ cm^{-2}, which

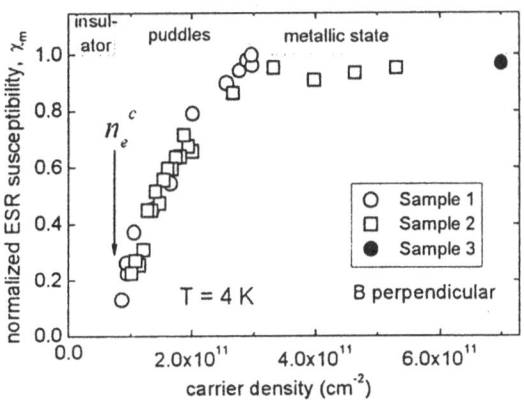

Fig. 1: Integral CESR absorption normalized to its extrapolated saturation value. This quantity is proportional also to the density of states at the Fermi level and inversely proportional to the screening length. Unity corresponds to a screening length of 8 Å.

coincides with the critical carrier density n_e^c for the MIT in Si MOSFET's. [8]

For an interpretation of the data in Fig. 1 we tentatively assume behavior according to Pauli susceptibility. The latter is proportional to the density of states at the Fermi level, $D(E_F)$. Within this admittedly oversimplified model, the data in Fig. 1 indicate that $D(E)$ does not have the sharp band edge of an ideal 2D electron gas but it approaches the constant value for large carrier density. At the band edge, there is a continuous decay indicative of the formation of tail states. These tail states are a consequence of potential fluctuations, δV. In principle δV may originate from fluctuations in the alloy composition, the well width, impurities, strain and in the density of donors in the doping layer. In view of the high mobility of these samples the latter appears to be the most important mechanism.

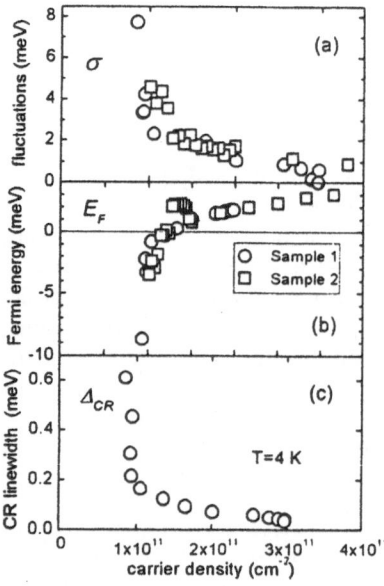

Fig. 2: (a) Fluctuation amplitude, (b) Fermi energy, E_F, and. (c) Cyclotron resonance line width, Δ_{CR}, as a function of carrier concentration.

From the data of Fig. 1 we can evaluate the fluctuation amplitude δV assuming some distribution of V. For Gaussian broadening, we can describe our two experimental quantities, χ_m and n_e, in terms of integrals containing δV and E_F as parameters. The latter two quantities can be evaluated by a simultaneous fit. Results are given in Fig. 2. The results show that δV diverges as the Fermi level approaches the band edge. Simultaneously the CR line width also tends to diverge. All this happens as n_e approaches the critical value for the MIT, n_e^c.

Figs. 2a,b also indicate that δV at high density is smaller than E_F, but for lower concentrations this is not the case anymore. Closer to the MIT apparently the 2D electron system breaks up into puddles. The puddle size can be estimated here from the ESR line width assuming hyperfine interaction of 2D electrons localized to a puddle with Si^{29} isotopes therein. As a result we obtain a typical puddle size of 1 μm.

For a weakly interacting electron system, the Thomas Fermi screening vector is also proportional to $D(E_F)$. Therefore the data of Fig. 1 also represent q_{TF} within this approximation. We see that if we approach n_e^c, also the screening efficiency goes to zero. The driving mechanism obviously is the following vicious cycle: as the Fermi level goes down, the density of states decreases which in turn decreases the screening efficiency and thus the potential fluctuations increase which lowers the density of states at the Fermi level and so on - δV diverges because of positive feedback.

Finally, since the connection between the CR line width and the macroscopic mobility is not obvious here, and we want to test the applicability of our weak interaction approach, we calculate the mobility using the results for q_{TF} of Fig. 1 and compare it to measured values [9]. The calculated values for $\mu(n_e)$ agree within a factor of two with the experimental data. Even more striking is the fact that the Thomas-Fermi mobility perfectly reproduces the experimentally found $n_e^{2.34}$ power law down to about $1.5 \cdot 10^{11}$ cm^{-2}.

Acknowledgements:

Work supported by the *FWF*, *GMe* and *OeAD*, Vienna, and by *KBN* grant 8 T 11B 003 15 in Poland

References:.

1. S.V. Kravchenko and T.M. Klapwijk, Phys. Rev Lett. **84**, 2909 (2000)

2. J. Yoon, C.C. Li, D. Shahar, D.C. Tsui, and M. Shayegan, Phys. Rev. Lett. **82**, 1744 (1999)

3. E. Abrahams *et al.*, Phys. Rev. Lett. **42**, 673 (1979) and Ann. Phys. **8**, 539 (1999)

4. N. Nestle, G. Denninger, M. Vidal, C. Weinzierl, Phys. Rev. **B56**, R4359 (1997)

5. W. Jantsch, Z. Wilamowski, N. Sandersfeld and F. Schäffler, Phys. stat. sol. (b) **210**, 643 (1998)

6. Z. Wilamowski, W. Jantsch, N. Sandersfeld, and F. Schäffler, Ann. Phys. (Leipzig) **8**, 507 (1999)

7. C.F.O. Graeff, M.S. Brandt, M. Stutzmann, M. Holzmann, G. Abstreiter, F. Schäffler, Phys. Rev. **B59**, 13242 (1999)

8. S.V. Kravchenko, W.E. Mason, G.E. Bowker, J.E. Furneaux, V.M. Pudalov, M. D'Iorio, Phys. Rev. B **51**, 7038 (1995)

9. D. Többen: PhD thesis approved by the Fakultät für Physik der Technischen Universität München, 1995, and private communication.

Coulomb Interaction and Density of States in Amorphous Si_xGe_{1-x} Films

K. Nakada, K. Nara, N. Aoki, Y. Ochiai

Department of Materials Technology, Chiba University, 1-33 Yayoi-cho, Inage-ku, Chiba 263-8522, Japan

Abstract We have measured ESR, optical absorption and the low temperature resistance in amorphous Si/Ge alloy films and studied their dependences on the ratio of the alloy composition in Si_xGe_{1-x}. The density states and the coulomb gap have been determined. And an effective band-width of the electron states can be estimated by the result of ESR.

1 Introduction

Recently, the interlayer coupling between super-conducting layers has been studied in metal-semiconductor multi-layer films, so as to discuss the origin of superconductivity in low dimensional super-conducting materials [1]. Frequently, the vertical transport of the layer shows Mott type variable range hopping (VRH) between metallic layers, which become super-conducting at low temperatures. In multi-layer systems of amorphous Si/Nb films, the magneto-resistance is affected by super-conducting fluctuations, and the difference between quantum tunneling and hopping conduction appears in the weak localization correction to inelastic scattering processes of the normal transport. The coupling effect is strongly related to the transport parameters of the insulating layer of amorphous Si [2]. And also it is noted that a characteristic magneto-resistance, due to weak localization is largely different between Si and Ge as in insulating amorphous layer[3]. Therefore, a precise analysis of the insulating layer of amorphous Si_xGe_{1-x} is important in order to investigate the superconductivity via electron correlation in such layered systems. In this paper, we have studied on density of states and the Coulomb gap (CG) in amorphous Si_xGe_{1-x} films [4] in order to clarify on relation between hopping transport and the alloy ratio.

2 Experiments

The films were fabricated by using electron beam evaporation of Si, and Ge in a high vacuum chamber. The thickness of the Si_xGe_{1-x} film sample is 2000 A and the size is 5 x 1 mm^2. The sample, known as a Hall bar type, has two current and four voltage leads. The alloy ratio, x, was controlled with gradually changing composition of Si and Ge, each other. In order to determine the real ratio values, x, the optical gap of the film was measured from their absorption spectra and the ESR line width was measured at room temperature. The two-terminal resistance below room temperature was measured down to liq. N_2 temperature. Optical absorption was measured by a conventional visible/near-ultraviolet photometer, JASCO/UVIDEC-50. ESR measurement of the films has been performed at room temperature by using JEOL/FE1XG. The ESR spin concentration were determined for various alloy ratios of the film using a calibrated standard sample

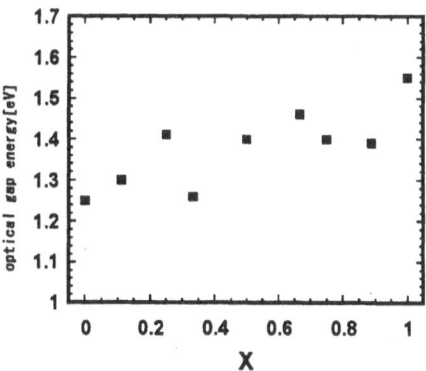

Fig. 1 The alloy ratio dependence of the optical gap energy in Si_xGe_{1-x}

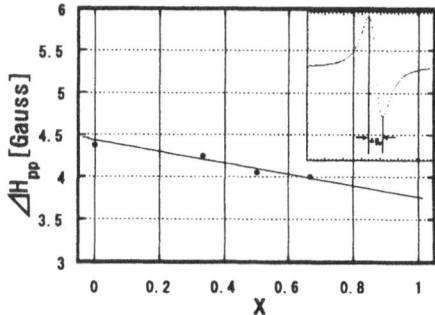

Fig. 2 The ESR line width in Si_xGe_{1-x}

3 Results and discussions

The alloy ratio dependence of the optical gap energies are shown in Fig.1. The gap of amorphous Si is larger than that of amorphous Ge and is

width has a small linear dependence on the ratio as shown in Fig. 2. Also the ESR spin density of unpaired electron in the film was determined by comparison with a standard sample. We summarize in Fig.3 the ESR spin density and the density of states estimated from previous transport measurements [5]. It shows that neither has a significant dependence on the ratio x. However, from a crossover between Mott type VRH and CG type hopping located at high and low temperature regions, respectively, a systematic change of CG depending on x can be observed in Si_xGe_{1-x}[5]. The CG of amorphous Si is about twice of that of amorphous Ge film as shown in Fig.4. It means that for Ge rich film, the CG hoping remains at slightly higher temperatures than that in Si rich case. By comparing the density of states with the ESR spin density, it is found that the effective bandwidth is almost half of the optical gap. The CG energy clearly depends on the ratio and is about 10^{-2} of the optical gap. This shows that electron-electron interaction is significant at low temperatures. On the other hands, as for their localization lengths, there is no clear dependence on the ratio value x, while such lengths are clearly dependent on temperature. It indicates that above difference in the weak localization between amorphous Si and Ge can be discussed with their coulomb interaction based on the change of CG value. It is found that structural disorder in amorphous semiconductors is important. There exists also a clear difference on electron correlation between Si and Ge. Both weak localization and superconductivity must be related to coulomb interaction. Therefore, the superconductivity of the multi-layer film should be strongly affected by the insulating and/or dielectric properties in amorphous semiconductor between metallic layers.

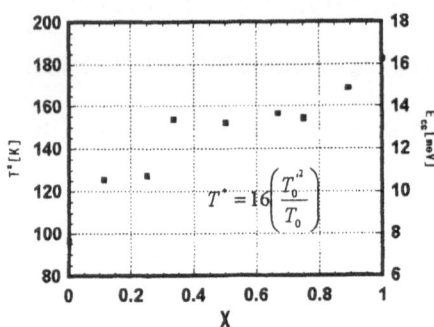

Fig. 4 Coulomb gap energy in Si_xGe_{1-x}

4 Conclusions

Two kinds of transport regime, Mott and ES type hopping conductions, have been observed in the low temperature transport in amorphous Si_xGe_{1-x} and the crossover temperature between both transports has been determined as a function of the ratio of the alloy composition. Therefore, at low temperatures, the difference of coulomb interaction is remarkably affected to the quantum transports. On the other hands, the localization length and the density of states are almost independent on the ratio of the alloy composition of amorphous Si_xGe_{1-x}. It is expected that those latter parameters must be an intrinsic in amorphous tetrahedral semiconductors.

Acknowledgements; The authors are indebted to Prof. Micheal Pollak for his important suggestions and valuable comments in our discussion for electron-electron interactions and also for his interesting discussions in 8th International Conference on Hopping and Related Phenomena at Murcia, Spain, on 7-10, Sept. 1999.

References

1. T. Kitatani, J.P. Bird and Y. Ochiai, Surf. Sci. **267** (1992) 583.
2. Y. Ochiai, J.P. Bird and T. Kitatani, J.Non-Cryst.Sol. **198-200** (1996) 804.
3. K. Mima and Y. Ochiai, Proc. of 4th Int. Symp. on Advanced Physical Fields, Tsukuba, **139** (1999) 139.
4. A.L. Efros and M. Pollak, Modern Problems in Condensed Matter Science, **10**, *Electron-Electron Interactions in Disordered Systems*, North-Holland Physics Publishing a division of Elsevier Science Publishers B.V., Amsterrdam 1985.
5. N. Aoki, K. Nara and Y. Ochiai, phys. stat. sol. (b) **218** (2000) 5.

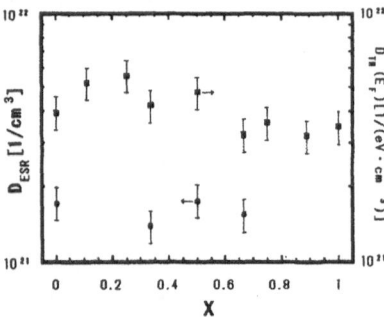

Fig. 3 The ESR spin density and the density of states in Si_xGe_{1-x}

Studies of localization in an interacting two-electron system [*]

J. Talamantes[1], M. Pollak[2], I. Varga[3] [**]

[1] California State University, Bakersfield, CA 93311, USA. e-mail: jtalamantes@csub.edu
[2] University of California, Riverside, CA 92651, USA.
[3] Philipps-Universität Marburg, Germany.

Abstract   We investigate the effect of electron-electron interactions on the localization properties of a two-electron, two-dimensional random system. The model includes the random energy, direct and exchange Coulomb interaction energies, and off-diagonal energies due to quantum tunneling. The systems studied consist of sites on a square lattice and random energies comparable to the Coulomb interaction energy. We use several criteria for studying the degree of localization: level spacing statistics, inverse participation ratios and entropy functions.

1 Introduction

The problem of the interplay between disorder, interaction and quantum tunneling, and their combined effect on electronic localization is not new – the possibility that quantum correlation due to the electron-electron interaction (EEI) may act to delocalize the electrons was proposed twenty years ago [1], but a firm answer has not been achieved yet. It is now fairly well established [2] that the EEI can, under certain circumstances, delocalize the electrons. Here, we report on various ways in which one might hope to quantify electron delocalization in an interacting system.

A tool used frequently to assess localization has been the distribution $p(s)$ of nearest-level spacings s. In the absence of interactions, it has been shown [3] that $p(s)$ shifts from Poisson to Wigner as the system goes from being strongly localized to delocalized. For *interacting* systems, such correspondence has never been proven. Still, level spacing statistics has been used often to study localization also with interactions [2].

Another quantity used commonly to study localization in non-interacting systems is the inverse participation number P^{-1} [4], which measures over how many sites the (one-particle) wave function spreads. Here, we allow for interactions, and use two similar quantities: *(i)* a configuration-space inverse participation number R which measures over how many configurations the many-particle wave function spreads; and *(ii)* a real-space inverse participation number P^{-1} which measures over how many sites the wave function extends. In the limits of strong localization and of complete delocalization it is easy to see the connection between P^{-1} and

R: the strong-localization limit, $R = 1$, implies the presence of a single configuration, which in turn implies that each particle is localized on a single site. In the complete delocalization limit the wave function extends uniformly over all configurations, implying that the particles are uniformly spread over the system in real space. Furthermore, we study two entropy functions: one each in configuration and in real space. They are similar in spirit to the two inverse participation numbers in that they measure the spread of the eigenfunctions over localized states.

2 The model

We consider two spinless electrons, and express the Hamiltonian by

$$H = \sum_{\alpha=1}^{2} (T_\alpha + V_\alpha + \varepsilon_\alpha) + \frac{e^2}{\kappa r_{12}}, \qquad (1)$$

where α labels the electrons; T, V, and ε are the operators corresponding to the kinetic energy, the interaction of the electrons with the cores, and the random potential, respectively; e is the electronic charge; κ is the dielectric constant; r_{12} is the electron-electron (e-e) distance. We choose the energies of sites i from a box distribution $-W/2 < \epsilon_i < W/2$, with W equal to the nearest-neighbor (n-n) Coulomb energy e_C. We vary the site concentration by changing the n-n distance r_{nn} as a parameter. We write H in the representation of the two-electron wave functions ϕ_{ij}, which are constructed by antisymmetrizing products of local one-electron orbitals (centered on i and j) φ for electrons 1 and 2. We take $\varphi \sim \exp(-r/a_B)$, with r the electron-core distance and a_B the microscopic (Bohr) radius corresponding to those orbitals.

Analytic solutions for $\langle \phi_{ij}|H|\phi_{kl}\rangle$ were derived and computed for processes $\{ij\} \to \{kl\}$ which involved *(i)* no electron transfers (*i.e.* diagonal matrix elements); or *(ii)* a one-electron n-n transfer; or *(iii)* a next n-n transfer; or *(iv)* a next-to-next n-n transfer; or *(v)* two simultaneous n-n transfers. All other off-diagonal matrix elements were set to zero. We diagonalize H – the result is the set of eigenenergies E_I, and the corresponding states Φ_I in which the ϕ_{ij} come in with amplitudes $A_{I,ij}$: $\Phi_I = \sum_{\{ij\}} A_{I,ij}\phi_{ij}$.

In this work we take W to always equal e_C; thus, $W = e_C \sim 1/r_{nn}$. Since r_{nn} is a parameter here, W changes accordingly. We do this because then the effect

[*] This work was supported by the National Science Foundation Grant DMR-9803686

[**] *Permanent address:* Technical University, Budapest, Hungary

of interaction is most important, and also because this condition prevails in many experiments, for example, in the vast experimental literature on impurity conduction at moderate compensation [5].

3 Procedure

We set up "samples" on the computer with n_s $(= L \times L)$ sites arranged on a lattice, and diagonalize eq. (1) for the parameters L and r_{nn}. For definiteness, we take a_B and κ to be 10Å and 3 respectively, which yields and effective mass $m^* = 0.16m$, with m the electron mass – these values seem appropriate for $2D$ systems. Cyclic boundary conditions are used. In what follows, we use a_B as the unit of distance. In every case the number of samples was sufficient to obtain at least 1.5×10^4 levels for each pair $\{n_s, r_{nn}\}$.

As one measure of localization, we compute $p(s)$. (We dropped up to 100 states from the band edges.) As in [6], a parameter $\eta = (\text{var}(s) - 0.273)/(1.0 - 0.273)$ is computed as a measure of how close $p(s)$ is to a Poisson $(\eta = 1)$ or Wigner $(\eta = 0)$ distribution. As an alternate measure of localization we compute the configuration-space inverse participation number: $R_I = \sum_{\{ij\}} |A_{I,ij}|^4$, and a real-space counterpart $P_I^{-1} = \sum_i p_{i,I}^2$, where $p_{i,I}$ is the electron density on site i in eigenstate I. Finally, we investigate the behavior of the entropy function (in configuration space) $S_{C,I} \equiv - \sum_{\{ij\}} A_{I,ij}^2 \ln A_{I,ij}^2$, and a real-space match: $S_{R,I} \equiv - \sum_i p_{i,I} \ln p_{i,I}$.

4 Results and Conclusions

We first present the basic result regarding the effect of the EEI on localization. Fig. 1 shows a comparison of $\eta(r_{nn})$ between the interacting and non-interacting systems [i.e. eq. (1) without the last term]. It is very clear that this criterion shows EEI to enhance delocalization – while with EEI delocalization occurs at $r_{nn} \sim 10$, without EEI delocalization is somewhere at $r_{nn} < 7$. We note that there is no clear small-size scaling behavior, suggesting a crossover $(r_{nn} \sim 9\text{-}11)$, rather than a critical transition.

We also investigated the behavior of R_I, P_I^{-1}, $S_{C,I}$ and $S_{R,I}$ for $L = 7$. Our preliminary observation is that R_I and P_I^{-1} roughly track each other; the same is true for $S_{R,I}$ and $S_{C,I}$. We emphasize that such tracking is not automatic, e.g. the behavior in level spacing statistics was shown in [2] to indicate a localization-delocalization transition at a different concentration than does the wave function in real space.

It was argued in [2] that (unlike in the non-interacting case) for interacting systems delocalization in *real space* requires a somewhat larger overlap for n-n sites than does the transition to Wigner statistics. This is in keeping with previous work [7], where it was observed that the wave functions are not compact in real space without EEI, but space-filling with the interactions, i.e. while the EEI may make the wave function extend over more sites, it does not similarly increase its *spatial* extent.

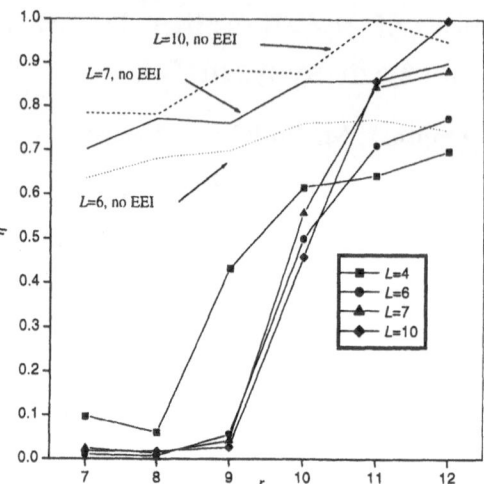

Fig. 1 Distribution $p(s)$ of nearest level spacing. The lines without symbols correspond to the non-interacting system. The lines with symbols correspond to the full Hamiltonian.

It is clear that $p(s)$ is Poisson well inside the localized regime, and Wigner well into the delocalized regime; however, the crossover in level statistics and its relationship to a localization-delocalization transition in *real space* is not as well established as in the non-interacting problem, and requires further study. Whereas it is clear that deep localization in configuration space implies deep localization in real space, and the same holds true for well-delocalized systems, the actual transition need not occur simultaneously in the two domains.

References

1. M. Pollak and M. L. Knotek, Journal of Non-Cryst. Sol. **32**, (1979) 141; M. Pollak, Phil. Mag. B **42**, (1980) 781.
2. see, e.g., J. Talamantes and M. Pollak, cond-mat/0009033 *preprint* (2000); to appear in Phys. Rev. B, and references therein.
3. see, e.g. B. Shklovskii, B. Shapiro, B. R. Sears, P. Lambrianides and H. B. Shore, Phys. Rev. B **47**, (1993) 11487.
4. D. J. Thouless, Physics Reports **13**, (1974) 93.
5. see, e.g., N. F. Mott, Adv. Phys. **16**, (1967) 49.
6. E. Cuevas, Phys. Rev Lett. **83**, (1999) 140.
7. J. Talamantes and M. Pollak in *Hopping and Related Phenomena*, edited by O. Millo and Z. Ovadyahu (Racah Institute of Physics, Jerusalem 1995) p. 56; J. Talamantes and M. Pollak, *Physics of Semiconductors*, edited by D. J. Lockwood (World Scientific Publishing, Singapore 1995) p. 97.

Bridging the gap with cleaved edge overgrowth superlattices

on minigaps, magnetic breakdown and quantum interference in artificial bandstructures

R.A. Deutschmann[1], W. Wegscheider[1,2], C. Albrecht[3], J.H. Smet[3], M. Rother[1], M. Bichler[1], G. Abstreiter[1]

[1] Walter Schottky Institut, Technische Universität München, 85748 Garching, Germany
[2] Universität Regensburg, Naturwissenschaftliche Fakultät II, 93040 Regensburg, Germany
[3] Max-Planck-Institut für Festkörperforschung, 70569 Stuttgart, Germany

Abstract The purpose of this paper is twofold. First we introduce a system that gives independent control over the strength and period of an atomically precise one-dimensional potential modulation imposed on a two-dimensional electron system. Second, along with low temperature magnetotransport experiments on such a system, we employ a semiclassical framework to explain all observed magnetoresistance oscillation frequencies. In particular two distinct quantum interference processes are recognized. a) self-interference along closed orbits, partly rendered possible by magnetic breakdown b) mutual interference between open electron orbits. The notion of the latter mechanism bridges the gap between the known commensurability oscillations and the artificial bandstructure. We suggest a close relation also to quantum-interference effects observed in magnesium [1] and organic conductors [2].

1 Introduction

Superlattices (SLs) have long been a fascinating system to study physical effects, such as Bloch oscillations, otherwise unobtainable with conventional crystals [3]. Furthermore the creation of man made artificial crystals gives control over lattice periods and coupling strengths, such that the resulting artificial bandstructures can be engineered for the specific purpose. A particular example are two dimensional electron systems (2DESs) in a periodic potential, which serve as a testbed to study bandstructure effects by means of magnetotransport experiments, since oscillations in the magnetoresistance provide immediate information on the area encircled by closed electron orbits at the Fermi energy E_F [4]. Electrons are confined in one direction to a quantum well, and a lateral periodic potential modulation is additionally imposed from the surface of the sample with for example lithographically defined top gates. As in the conventional vertical SL-geometry, an artificial bandstructure derives from the reduced width of the Brillouin zone and zone folding. As the magnetic field is raised and if E_F is located within a higher miniband, not only the zero field closed Fermi contours, but also closed trajectories composed of Fermi contour sections belonging to other occupied minibands are traced out by virtue of a tunnelling process referred to as magnetic breakdown [5]. All these closed orbits, that encircle different areas, leave sig-

natures in the magnetoresistance and should be instrumental in uncovering the detailed bandstructure. Contrary to metallic systems, E_F can be tuned over a wide range in these semiconductor based SLs, so that it should in principle be possible to map out the entire energy dispersion.

Surprisingly, to date, such experimental evidence for an artificial bandstructure is sparse. Only very recently, using two dimensional modulation, unambiguous proof of two different closed electron orbits was achieved [6]. Although different one dimensionally modulated systems have been investigated in the past [7], similar data for this textbook case has not been reported. This lack of evidence may be related to the inherent inadequacy of lateral modulation schemes in producing concurrently a high quality 2DES and a sufficiently short period and large amplitude modulation to guarantee the occupation of only few, well-isolated minibands.

2 Sample design

We use a new concept to fabricate lateral SLs based on the cleaved edge overgrowth technique [8], that overcomes the limitations of previous geometries by periodically modulating the material composition *directly adjacent* to the 2DES. In this way, both the period and the modulation strength can be tailored with unprecedented precision by MBE growth. The sample structure is shown in Fig. 1. In a first MBE-step, an undoped SL

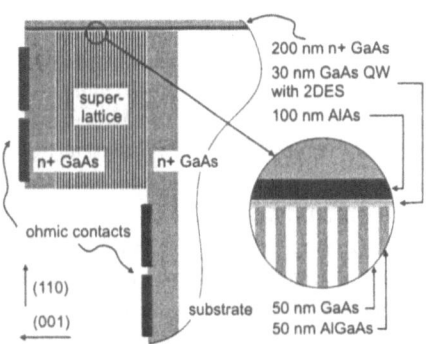

Fig. 1 Sample cross section. The modulated 2DES is located under the gate in the overgrown quantum well.

with lattice constant $d = 100$ nm of 30 periods of 50 nm GaAs and 50 nm $Al_{0.32}Ga_{0.68}As$ is grown between two n^+-GaAs contacts, that act as source and drain. In a second MBE-step, the sample is cleaved *in situ* and immediately thereafter overgrown by a 30 nm undoped GaAs quantum well (CEO qw), a 100 nm AlAs barrier and an n^+-GaAs gate contact. Similar samples with 15 nm period are discussed in Ref. [9]. By applying a positive gate voltage with respect to source and drain a 2DES is induced at the GaAs/AlAs heterointerface.

3 Electron density distribution

Since our method of generating a potential and density modulation on a 2DES is quite different from conventional methods, we have conducted self consistent Poisson/Schroedinger calculations to understand the influence of the parameters gate voltage, CEO qw thickness and period length. One particular example of the electron density distribution is shown in Fig. 2. The extent

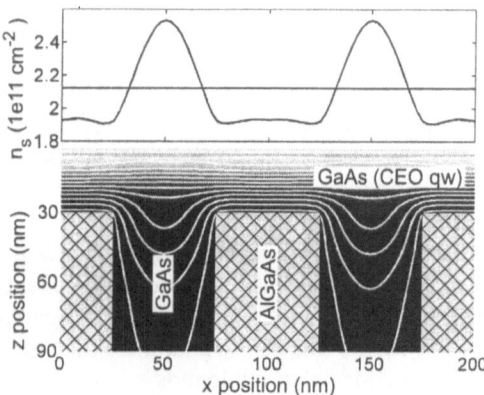

Fig. 2 Electron distribution for a CEO qw thickness of 30 nm and mean 2D density of $n_s = 2.1 \times 10^{11}/cm^2$.

of the electron wavefunctions in z direction is modulated by the SL, resulting in a modulated 2D electron density in x direction as the 3D bulk density is integrated over the z coordinate. Surprisingly the local 3D bulk electron density at the T-junctions between CEO qw and SL qw is reduced. We define an effective potential modulation amplitude V_{eff} by weighting the local potential amplitude with the local electron density. For a 30 nm CEO qw V_{eff} lies between 0.6 meV and 1.4 meV, peaked at $n_s = 2.6 \times 10^{11}/cm^2$. V_{eff} decreases exponentially as the CEO qw thickness is increased.

4 Magnetoresistance measurements

We have obtained magnetotransport data, with the magnetic field oriented perpendicular to the 2DES (in z direction), at 0.3 K with lock-in techniques. When sweeping the magnetic field, measured magnetoresistance oscillations were periodic in inverse magnetic field, but different oscillation frequencies could be distinguished depending on the field strength. For example at an electron density of $n_s = 1.92 \times 10^{11}/cm^2$ for $B < 0.2T$ the period was equal to 0.367/T, whereas for $B > 0.2T$

a period of 0.252/T dominated. In the latter regime an additional beating pattern with a longer period of 0.693/T appeared in the envelope. In order to obtain all frequency components in dependence of the electron density, we recorded the magnetoresistance from $B = 0T$ to 1.4T while systematically varying the electron density between $n_s = 0.45 \times 10^{11}cm^{-2}$ and $n_s = 4.6 \times 10^{11}cm^{-2}$. Each curve was then Fourier transformed with respect to the inverse field. The result is summarized in Fig. 3. The amplitudes of the frequency components are plotted on a logarithmic gray scale and the ordinate is scaled to units of electron density. Large and small amplitudes appear as bright and dark, respectively.

5 Semiclassical model

To analyze the transport data, the contours of constant energy E_F of the modulated 2DES are considered, that are obtained with the calculated bandstructure in the SL-direction, while keeping in mind the free electron motion parallel to the equipotential lines. Hereafter, minibands are assigned an index n that runs from 0 for the energetically lowest lying miniband to N for the last, partially filled miniband. The minibands with index n and $n + 1$ are separated by a minigap at $k_x = \pm\pi/d$ or $k_x = 0$. For the density range covered in this experiment three to six minibands are occupied. In general, Fermi contour N is closed, whereas all other contours $(0, \ldots, N - 1)$ describe open electron trajectories. The electrons trace these contours in a direction fixed by the sign of the magnetic field. A transition from contour n to its neighbour $n + 1$ entails quantum mechanical tunnelling across the minigap, a process referred to as magnetic breakdown [5]. The tunnelling probability depends on B, E_F and the gap size. It vanishes at $B = 0$ and increases exponentially with B. At the lowest B-values electrons can only execute the closed contour N. As B is raised though, other closed orbits, composed of segments of contour N as well as segments of open contours with lower index, become possible by virtue of magnetic breakdown. The product of the corresponding tunneling probabilities determines the orbit's overall probability [10]. We call A_F^n the area enclosed by an electron where n is the lowest index contour segment along the closed orbit. In the case of large fields the electron will tunnel across all encountered gaps and circle an orbit of area A_F^0. The area A_F^n can be calculated to a very good approximation in the limit $V_0 \to 0$:

$$A_F^n = 2 k_F^2 \left(\arccos n \frac{k_0}{k_F} - n \frac{k_0}{k_F} \sqrt{1 - (n \frac{k_0}{k_F})^2} \right), \quad (1)$$

where k_F is the Fermi wavenumber and $k_0 = \pi/d$. According to Onsager [4], electrons that orbit around an arbitrarily shaped Fermi surface A_F give rise to 1/B-periodic oscillations in the magnetoresistance with a frequency Δ^{-1},

$$\Delta^{-1} = \frac{\hbar}{2\pi e} A_F. \quad (2)$$

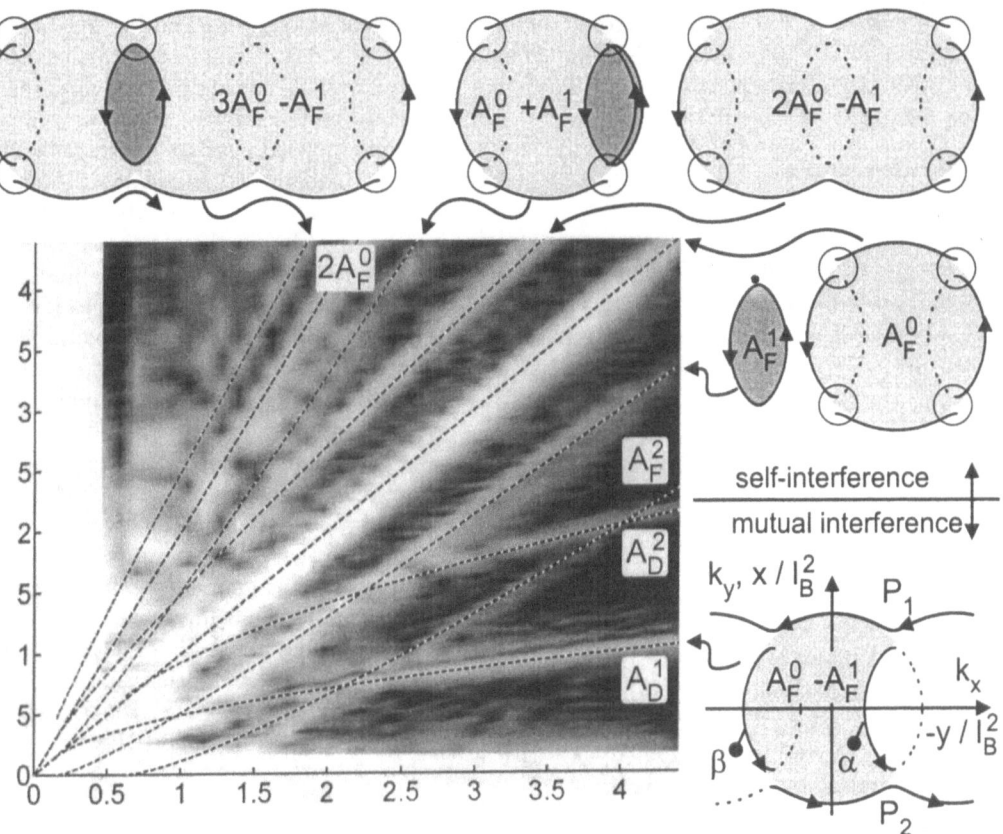

Fig. 3 Gray scale plot of the Fourier transformed magnetoresistance data. The density dependent peaks are directly related to different electron trajectories in the artificial bandstructure, which are also shown.

Our density dependent study in Fig. 3 of the frequency components contained in the magnetoresistance now enables us to identify the maxima marked A_F^0, A_F^1 and A_F^2 as caused by electrons performing closed orbits around the respective areas. The dashed lines in this figure are calculated according to Eq. 1 and using the Onsager relation. For simplicity the orbits displayed here have been calculated for Fermi energy in the second miniband. The tunneling junctions are indicated by thin circles. It is beyond the scope of this paper to analyze the probability of the different orbits depending on the magnetic field, but in general at low fields orbits associated with A_F^n and large n dominate (small areas, magnetic breakdown unlikely), while at large fields the orbit associated with A_F^0 prevails (large area, tunneling probability unity at all junctions). This orbit in fact brings about the commonly known Shubnikov-de Haas oscillations. The orbits with area A_F^n are the most obvious closed trajectories, however more complicated closed paths with these simple surfaces as constituents are in fact also resolved in our experiment and are shown in Fig. 3 (for example $A_F^0 + A_F^1$, $2A_F^0 - A_F^1$ and $3A_F^0 - A_F^1$).

Up to this point, the discussed oscillations were a direct consequence of the constructive self-interference of the wavefunction along *closed* orbits and the subsequent quantization in a magnetic field. This mechanism leaves unexplained our observation of the frequency com-

ponents determined by the surfaces $A_D^1 = A_F^0 - A_F^1$ and $A_D^2 = A_F^0 - A_F^2$ in Fig. 3, since an electron circling along the closed boundary of this surface would violate the chirality imposed by the B-field along part of the perimeter. We assert that oscillations with such frequencies originate from the $1/B$-periodic modulation of the backscattering probability due to quantum-mechanical interference between two open trajectories with common start and end points, as illustrated in the lower right part of Fig. 3 for surface A_D^1 ("mutual interference"). Electrons travelling in negative k_x direction from point P_1 follow either path α or β, depending on whether they do or do not tunnel at this starting point, and rejoin at point P_2. Constructive interference of the coherent superposition of both paths maximizes the backscattering probability and consequently the conductivity σ_{yy} approaches a minimum, which implies a minimum in the longitudinal resistivity ρ_{xx} as well. In the case of destructive interference, the electron will effectively proceed along the open Fermi contour and thus σ_{yy} and ρ_{xx} reach their maximum value. This qualitatively different interference phenomenon reminds of an Aharonov-Bohm interferometer with the important disparity that in the case at hand the area in real space enclosed by the interfering paths scales with B^{-2}, since real space orbits have the same shape apart from a $\pi/2$-rotation as their counterparts in reciprocal space but are scaled with the square of the magnetic length, $l_B^2 = \hbar/(eB)$. As a result, one anticipates a $1/B$-periodic rather than a B-periodic phase difference between the interfering trajectories. The phase accumulated along path $i = \alpha$ or β is given by

$$\Phi_i = \frac{1}{\hbar} \int_i (\hbar \mathbf{k} - e\mathbf{A}) \, d\mathbf{r} + \frac{\pi}{2} C_i, \qquad (3)$$

where $-e$ is the electron charge, $\mathbf{A}$ the vector potential, and C_i a constant. As a result, the B-dependent contribution to the phase difference $\Phi_D = \Phi_\beta - \Phi_\alpha$ is equal to $l_B^2 A_D^{\beta-\alpha}$, where $A_D^{\beta-\alpha}$ is the area in reciprocal space bounded by a pair of paths α and β. The area $A_D^{\beta-\alpha}$ is nothing but the difference between surfaces A_F^i and A_F^j, that have common borders except for path α and β. The alternation frequency of the backscattering probability is then obtained from the condition $l_B^2 A_D^{\beta-\alpha} = 2\pi m$ ($m = 1, 2, \ldots$), i.e. from Eq. 2 when substituting $A_D^{\beta-\alpha}$ for A_F. Now it is clear that this mutual quantum interference effect explains the remaining peaks in Fig. 3 marked A_D^1 and A_D^2.

From Eq. 1, it follows that $4k_F\pi/d$ is a very good approximation of A_D^1 for the density range covered in the experiment of Fig. 3 and is exact in the high density limit. Strikingly, the resulting periodicity is identical to the one of the well-known commensurability oscillations (COs) [11], that up to now in the literature have not been derived directly from the bandstructure. While in our picture quantum mechanics is essential (calculation of the bandstructure, quantum interference), the COs in ρ_{xx} can already be understood classically [12]. Whenever the lattice period d is commensurate with the free electron cyclotron radius R_c in accord with $2R_c = (m - 1/4)d$, ρ_{xx} reaches a minimum. In this expression, the term equal to 1/4 fixes the absolute position on the magnetic field axis. In our derivation so far we were only able to calculate the *frequency* of the magnetooscillations induced by the mutual interference effect, but not the absolute position of the resistance extrema. The reason is that we do not know the constant C_i in Eq. 3. We remark that for a closed orbit as in the Shubnikov-de Haas case this constant represents the Maslov index, and has a value of -2.

6 Conclusions

The fabrication of a device with few occupied minibands enabled the unambiguous demonstration of the artificial bandstructure through the observation of multiple closed orbits and $1/B$-periodically enhanced backscattering related to quantum interference of open orbits. We think that based on the latter mechanism a quantitative theory for also the amplitude and position, and not only the frequency, of the COs could be developed.

Although the observation of COs of course has been possible earlier using conventional surface lateral superlattices, additionally we were able to resolve a second type of CO (feature A_D^2), plus two circular shaped surfaces (A_F^1 and A_F^2) besides the fundamental Shubnikov-de Haas surface A_F^0. This became possible by engineering a close to ideal lateral superlattice, which excels in many ways over standard designs due to the unparalleled combination of attractive features: virtually no limitations on modulation period and strength, monolayer precision of the superlattice, exceptional density tunability and high quality due to the absence of modulation doping. With this design the stage is set for finding further exciting physics at even smaller modulation periods.

We think that this work bridges the gap between quantum interference phenomena observed in semiconductor SLs, in metals [1] and organic conductors [2]. These systems have in common a bandstructure with narrow gaps, allowing for different electron trajectories at finite magnetic field. The large body of theoretical work developed for semiconductor SLs may favorably be applied also to the latter two systems. In particular at high fields even in these systems the lattice potential may be regarded as a pertubation to the free electron gas in magnetic field.

We would like to thank P. Vogl, R. R. Gerhardts and M. Brack for helpful discussions. This work has been supported by the DFG via SFB 348 and the BMBF under contracts 01BM912 and 01BM919/5.

References

1. R. W. Stark and C. B. Friedberg, Phys. Rev. Lett. **26**, 556 (1971).
2. T. Sasaki and H. Sato and N. Toyota, Solid State Comm. **76**, 507 (1990).
3. L. Esaki and R. Tsu, IBM J. Res. Dev. **14**, 61 (1970).
4. L. Onsager, Phil. Mag. **43**, 1006 (1952).
5. M. H. Cohen and L. M. Falicov, Phys. Rev. Lett. **7**, 231 (1961); E. I. Blount, Phys. Rev. **126** 1636 (1962).
6. C. Albrecht et al., Phys. Rev. Lett. **83**, 2234 (1999).
7. T. Cole, A. A. Lakhani, and P. J. Stiles, Phys. Rev. Lett. **38**, 722 (1977); T. G. Matheson and R. J. Higgins, Phys. Rev. B **25**, 2633 (1982); L. J. Sham et al., Phys. Rev. Lett. **40**, 472 (1978); T. Evelbauer et al., Z. Phys. B **64**, 69 (1986); Y. Nakamura, T. Inoshita, and H. Sakaki, Physica E **2**, 944 (1998).
8. H. L. Stormer et al., Appl. Phys. Lett. **58**, 726 (1991); Y. Ohno et al., Phys. Rev. B **52**, R11619 (1995).
9. R. A. Deutschmann et al., Physica E **6**, 561 (2000); R. A. Deutschmann et al., Physica E **7**, 294 (2000).
10. A. B. Pippard, Proc. Roy. Soc. A **270**, 1 (1962); L. M. Falicov and H. Stachowiak, Phys. Rev. **147**, 505 (1966).
11. D. Weiss et al., Europhys. Lett. **8**, 179 (1989); R. R. Gerhardts, D. Weiss, and K. v. Klitzing, Phys. Rev. Lett. **62**, 1173 (1989); R. W. Winkler et al., *ibid.*, 1177 (1989); C. Zhang and R. R. Gerhardts, Phys. Rev. B **41**, 12850 (1990).
12. C. W. J. Beenakker, Phys. Rev. Lett. **62**, 2020 (1989)

Electron Wires Driven by a Surface Acoustic Wave and Nonlinear Acoustoelectric Interactions in Quantum Wells

A. O. Govorov[1], A. V. Kalameitsev[1], V. M. Kovalev[1], H.-J. Kutschera[2], M. Streibl[2], M. Rotter[2], and A. Wixforth[2]

[1] Institute of Semiconductor Physics, Russian Academy of Sciences, Siberian Branch, Lavrent'eva Av. 13, 630090 Novosibirsk, Russia; E-mail: Govor@isp.nsc.ru

[2] Sektion Physik der Ludwig-Maximilians-Universität and Center for NanoScience, Geschwister-Scholl-Platz 1, D-80539 München, Germany

Abstract   We study both theoretically and experimentally the nonlinear interaction of intense surface acoustic waves (SAW's) with electron and electron-hole plasmas in quantum wells. The experiments performed on hybrid semiconductor-piezoelectric structures exhibit strongly nonlinear acousto-electric effects due to the formation of moving electron wires. To describe the nonlinear phenomena, we develop a coupled-amplitude method for a two-dimensional system in the strongly nonlinear regime of interaction. Theory and experiment are found to be in a good agreement. Also, we show theoretically that the nonlinear interaction of a SAW with a photo-generated electron-hole plasma qualitatively differs from the acousto-electric interaction in the case of an unipolar electron system. For low temperatures, we consider the regime when the intense SAW forms moving quantum wires and develop a theory of the quantum SAW attenuation.

1 Introduction

The interaction between SAW's and mobile carriers in quantum wells is a powerful method to study dynamic properties of two-dimensional (2D) systems. The SAW can trap carriers and induce acoustic charge transport (ACT) [1]. Also, the SAW-method was applied to study the quantum Hall effects, electron transport through a quantum-point contact, and commensurability effects in a 2D system [2]. However, most of those experiments have been done in the regime of small signals and linear interaction. The linear interaction in 2D systems and in nanostructures was theoretically studied in a number of papers [3,4]. Here we study the transition from the linear regime of acousto-electric interaction to the limit of strongly nonlinear effects in a 2D electron plasma and in a system with photogenerated carriers [5–8].

2 Nonlinear acousto-electric interaction in an electron system

The room-temperature experiments with SAW's were performed on the hybrid semiconductor-$LiNbO_3$ structures fabricated by the epitaxial lift-off technique [9]. These structures contain a semiconductor quantum well tightly bonded to the $LiNbO_3$ host crystal and a top metallic gate [5]. Due to the strong piezoelectricity of a hybrid structure a SAW can break up a formerly 2D electron plasma into moving wires. The effect of the wire formation was clearly seen in the ACT experiments from

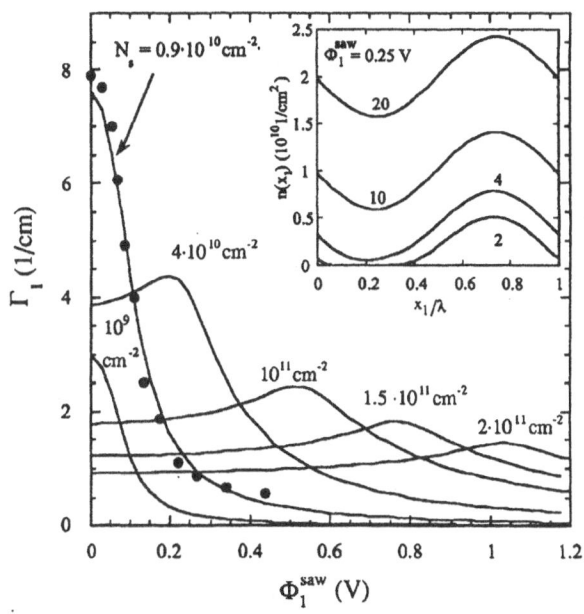

Fig. 1 The calculated absorption coefficient Γ_1 as a function of the SAW piezopotential amplitude Φ_1^{saw} for various electron densities N_s. The dots show the experimentally measured absorption coefficient at the top gate voltage $-7.5\ V$. In the inset we plot the calculated local carrier density n as a function of the in-plane coordinate x_1 for different N_s. The numbers attached to the plots correspond to N_s in units of $10^{10}\ cm^{-2}$.

the saturation of the transport velocity [5]. Also, the nonlinear acousto-electric interaction results in an increase of the SAW velocity c_s and in a strong modification of the SAW attenuation. To describe the experimental results we develop a coupled-amplitude theory assuming $K_{eff}^2 \ll 1$, where K_{eff}^2 is the effective electromechanical coupling coefficient of the hybrid structure. In a short sample, for the SAW-velocity shift δc_s and the absorption coefficient Γ_1 we obtain [6]

$$\frac{\delta c_s}{c_s} = \frac{< j_e \Phi_{SAW} >}{2 I_{SAW}}, \qquad \Gamma_1 = \frac{< j_e E_{SAW} >}{I_{SAW}}. \tag{1}$$

where $< ... >$ means averaging, j_e is the 2D current induced by the SAW, and E_{SAW} and Φ_{SAW} are the electric field and the potential induced by the SAW, respectively. I_{SAW} is the SAW intensity.

The electron current was numerically calculated from the nonlinear hydrodynamic equations. We use the following parameters. The electron mobility at room temperature is 5000 cm^2/Vs. The SAW wave length $\lambda = 33 \ \mu m$ and $c_s = 3.8 * 10^5 \ cm/s$. Using our theoretical results we can explain experimental observations. For the case of the nonlinear SAW absorption coefficient we find a good quantitative agreement between theory and experiment (Fig. 1). At low electron densities the SAW absorption coefficient decreases with increasing sound intensity, whereas at high electron density the absorption coefficient is not a monotonous function of the sound intensity (Fig. 1). This behavior is explained in terms of the nonlinear dynamic screening and the formation of wires. In the limit $I_{SAW} \to \infty$, $\Gamma_1 \propto 1/I_{SAW}$ and $\delta c_s \propto 1/\sqrt{I_{SAW}}$ [6,10].

3 Photogenerated electron-hole plasma

To model the acousto-electric effects in an electron-hole plasma of a quantum well we include into the 2D hydrodynamic equations the generation and nonlinear recombination terms. We see from our results that with increasing SAW intensity the plasma turns into electron and hole wires and the average carrier density N_e increases due to the spatial separation of electrons and holes. In this regime the sound dissipation increases with increasing the SAW-intensity [8]. At the same time, the SAW-velocity decreases [8] (Fig. 2). In the limit $I_{SAW} \to \infty$, the SAW dissipation $Q \propto \sqrt{I_{SAW}}$ and $\delta c_s \to -K_{eff}^2/2$. This is in contrast to the case of an electron plasma, where the sound dissipation saturates and c_s increases at high SAW-intensity [6,10]. Experiments were performed on InGaAs quantum wells for the case of a spatially-inhomogeneous plasma generated by a laser beam [7], so far.

4 Driven quantum wires

At a high SAW intensity electrons in the moving wires can be quantized at low temperature [11]. Our estimations show that the quantization energy under the realistic conditions can be about 2 meV. Using the motion equation for the electron density matrix we calculate the impurity-assisted SAW absorption in the regime of moving quantum wires. A quantum mechanism of the nonlinear acousto-electric interaction arises from the formation of 1D subbands. The SAW absorption is an oscillating function of the acoustic intensity due to the density of states in wires. Furthermore, we find that the SAW absorption does not vanish even in the strictly 1D limit. This is due to electron scattering by impurities in the direction perpendicular to the sound wave vector.

5 Acknowledgements

We would like to thank J. P. Kotthaus and A. V. Chaplik for very helpful discussions. We gratefully acknowledge

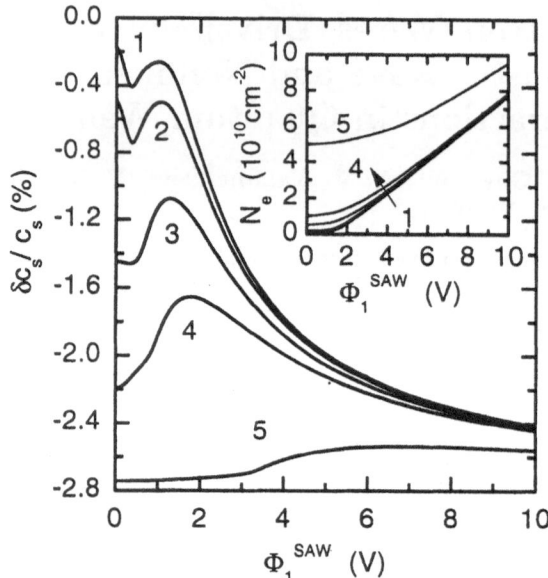

Fig. 2 The calculated shift of the SAW-velocity due to an electron-hole plasma as a function of the SAW potential amplitude Φ_1^{saw} for various optical excitation powers. The numbers 1-5 correspond to the photogenerated 2D carrier densities in the absence of a SAW $n_{s0} = 10^9$, $2*10^9$, $5*10^9$, 10^{10}, and $5*10^{10} \ cm^{-2}$, respectively. $\lambda = 60 \ \mu m$; the electron mobility $\mu_e = 2000 \ cm^2/Vs$ and the hole mobility $\mu_h = \mu_e/6$. The insert shows the average density $N_e(\Phi_1^{SAW})$ for various n_{s0}. $K_{eff}^2 = 0.056$.

financial support by the Volkswagen-Foundation and by the Russian Foundation for Basic Research.

References

1. M. J. Hoskins, H. Morko, and B. J. Hunsinger, Appl. Phys. Lett. **41**, (1982) 332; W. J. Tanski et al., *ibid* **52**, (1988) 18.
2. A. Wixforth et al., Phys. Rev. **B 40**, (1989) 7874; R. L. Willett et al., Phys. Rev. Lett. **71**, (1993) 3846; V. I. Talyanskii et al., Phys. Rev. **B 56**, (1997) 15180; J. M. Shilton et al., Phys. Rev. **B 51**, (1995) 14770.
3. K. A. Ingebrigtsen, J. Appl. Phys. **41**, (1970) 454; A. V. Chaplik, Sov. Tech. Phys. Lett. **10**, (1984) 584.
4. V. L. Gurevich, V. B. Pevzner, and G. J. Iafrate, Phys. Rev. Lett. **77**, (1996) 3881; Y. Levinson et al., Phys. Rev. **B 58**, (1998) 7113; G. R. Aizin, G. Gumbs, and M. Pepper, Phys. Rev. **B 58**, (1998) 10 589; C. Eckl, Yu. A. Kosevich, and A. P. Mayer, Phys. Rev. **B 61** (2000), in press.
5. M. Rotter et al., Phys. Rev. Lett. **82**, (1999) 2171; M. Rotter et al., Appl. Phys. Lett. **75**, (1999) 965.
6. A. O. Govorov et al., Phys. Rev. **B 62**, (2000) 2659.
7. M. Streibl et al., Appl. Phys. Lett. **75**, (1999) 4139.
8. A. V. Kalameitsev, A. O. Govorov, H.-J. Kutschera, and A. Wixforth, JETP. Lett. **72**, (2000), in press.
9. E. Yablonovich et al., Appl. Phys. Lett. **56**, (1990) 2419; M. Rotter et al., *ibid* **70**, (1997) 2097.
10. V. L. Gurevich and B. D. Laikhtman, Sov. Phys. JETP **19**, (1964) 407; Yu. V. Gulyaev, Sov. Phys. - Solid State, **12**, (1970) 328.
11. L. V. Keldysh, Fiz. Tverd. Tela **4**, (1962) 1015 [Sov. Phys. - Solid State]; V. V. Popov and A. V. Chaplik, Zh. Eksp. Teor. Fiz. **73**, (1977) 1009 [Sov. Phys. JETP].

Generation and detection of picosecond acoustic phonon pulses in a double quantum well structure

I. Ishii[1], O. Matsuda[1], T. Fukui[2], J. J. Baumberg[3], O. B. Wright[1]

[1] Department of Applied Physics, Faculty of Engineering, Hokkaido University, Sapporo 060-8628, Japan
[2] Research Center for Interface Quantum Electronics, Hokkaido University, Sapporo 060-8628, Japan
[3] University of Southampton, Department of Physics and Astronomy, Southampton, UK

Abstract   Coherent acoustic phonon pulses are excited and detected in a GaAs/Al$_{0.3}$Ga$_{0.7}$As double quantum well structure using femtosecond optical pulses. The wavelength of the infrared pump light is chosen to selectively generate picosecond acoustic phonon pulses in two buried quantum wells. The wavelength of the probe light pulses is chosen to allow phonon pulse detection in the near-surface region of the sample. Using an interferometric detection technique both reflectivity and phase variations of the probe beam are monitored, and the phonon pulses generated in the well regions are clearly observed in both these variations. An excellent agreement is obtained when the data is compared to a model accounting for the transient surface displacement combined with the spatially distributed variations in refractive index.

The excitation and detection of acoustic phonon pulses in solids using the ultrafast optical pump-probe technique has been the subject of much interest in recent years. The method not only has practical application to the non-contact and nondestructive evaluation of thin films, particularly in the sub-micron thickness region[1], but can also be used to probe a variety of basic ultrafast phenomena in metals and semiconductors such as ultrafast photoexcited carrier relaxation, ultrafast carrier diffusion, the strength of the electron-phonon coupling, and high frequency acoustic phonon propagation in the 10 GHz–1 THz region [2,3].

Following on from work on single quantum well structures[4], we investigate in this paper picosecond coherent acoustic phonon generation and propagation in a GaAs/Al$_{0.3}$Ga$_{0.7}$As double quantum well structure using a picosecond pump–probe technique. Infrared pump light pulses are tuned for preferential absorption in the two buried quantum wells, while visible probe light pulses allow detection of the arrival of these pulses at the sample surface.

The GaAs/Al$_{0.3}$Ga$_{0.7}$As double quantum well structure is prepared by MOVPE on a GaAs (100) substrate. The details of the dimensions and energy band gaps at room temperature [5] are shown in Fig. 1. In order to characterize the transition energies for the quantum well structure, photoluminescence measurements were carried out at 300 K using an excitation photon energy of 1.61 eV. The spectra in Fig. 2 show two emission peaks at 1.47 eV and 1.49 eV, corresponding to the two wells.

The acoustic phonon pulses are generated by infrared normally incident pump light pulses ($h\nu=1.49$ eV) from

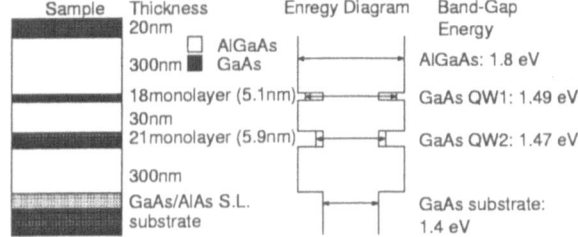

Fig. 1 Left: structure of the GaAs/Al$_{0.3}$Ga$_{0.7}$As double quantum well structure. Right: simplified schematic energy band gap diagram of the structure at room temperature.

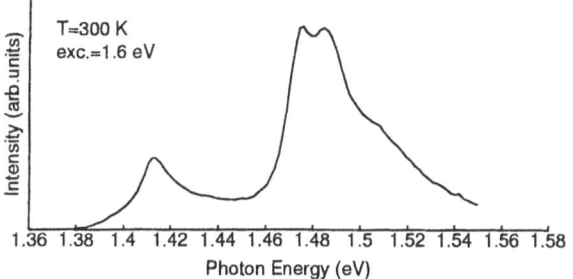

Fig. 2 Photoluminescence spectra of the quantum well structure at 300 K. The peaks due to two quantum wells with different well widths are clearly distinguished at 1.47 eV and 1.49 eV.

a mode-locked Ti-sapphire laser. From the measured photoluminesence spectra this photon energy is expected to be able to efficiently excite both quantum wells. The optical pump pulse duration is ~700 fs, the incident fluence is ~0.04 mJ/cm^2 (spot size ~30 μm in diameter), and the laser repetition rate is 82 MHz. Delayed frequency-doubled blue probe light pulses ($h\nu=2.99$ eV), focused at normal incidence to the same 30 μm spot as the pump pulses, mainly sample the top 20 nm GaAs layer of the structure. (The optical absorption lengths of GaAs and Al$_{0.3}$Ga$_{0.7}$As for the probe light are 17 nm and 32nm, respectively[6].) We estimate the excited carrier density in the quantum wells to be ~10^8 cm^{-2}. For probe light detection, we use a modified Sagnac interferometer which is sensitive to the transient changes in the amplitude and phase of the reflected probe light[7].

The solid lines in Fig. 3 show typical results of the pump-probe measurement. Reflectivity and phase changes of the probe light are plotted as a function of

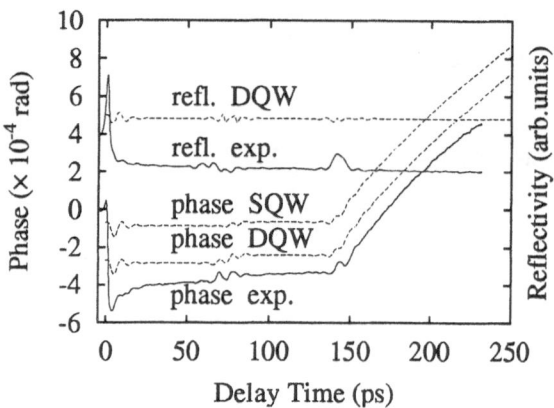

Fig. 3 Solid line : Reflectivity and phase change of the probe light as a function of delay time. Dashed line : Theoretical curve of reflectivity and phase change for single and double quantum well (SQW and DQW).

delay the time between pump and probe pulses. In both the phase and reflectivity signals, we obtain small features around 70~80 ps corresponding to the arrival of the phonon pulses from each well at the surface. In contrast to the reflectivity signal, the phase signal shows a step in signal level after this time and a large increase after 140 ps.

In order to interpret these results quantitatively the localized strain generation in the quantum wells and in the substrate should be considered. The incident pump photons excite electron-hole pairs, and these relax to the band edge or form excitons within a few picoseconds. Owing to the relatively long lifetime of the band edge carriers or excitons (several hundred ps or more) and the significant coupling between the carrier concentration and the strain in GaAs, the deformation potential mechanism is dominant for acoustic phonon generation in the ~100 GHz frequency region [2,3].

We have incorporated this mechanism in a simulation program to calculate the generation, propagation, and detection of the acoustic phonons in the multilayer structure. The details will be given in a future publication, but the essence of the calculation is outlined here. In the generation phase of the process, the absorption of the pump light is calculated taking the multiple reflection of the light into consideration. We assume here that the induced stress is proportional to the absorbed photon number or carrier density. The thermoelastic stress is estimated to be ~3 times smaller, and is neglected. In the propagation phase, the strain wave is set up by the photoinduced stress, and propagates in the sample according to the wave equation. The sound velocity and density of each layer is incorporated in the program to handle the reflection and transmission of the sound wave at each interface and the surface. The sound velocities used here are 4.73 nm/ps and 4.95 nm/ps for GaAs and $Al_{0.3}Ga_{0.7}As$ layers, respectively[5]. Almost all the acoustic energy is transmitted by the interface between

GaAs and $Al_{0.3}Ga_{0.7}As$. In this way, the strain distribution is calculated as a function of space and time. In the detection phase, the amplitude and phase of the reflected probe light is calculated as the solution to the problem of the light reflection by the multilayer structure taking account of the photoelastic effect and the displacement of the surface and interfaces. The photoelastic constants of the materials and the effective energy absorption of the quantum wells and superlattice were treated as fitting parameters. Any pulse broadening due to the ultrasonic attenuation is ignored.

As shown in Fig. 3, the agreement between the experimental curve for the phase and the calculation is excellent in the region of the delay times > 50 ps. (For delay times < 50 ps other effects such as carrier relaxation and transient temperature changes not included in the model are important.) A reasonable correspondence is also found for the reflectance. In particular the model successfully accounts for the shape and symmetry of the echoes in the phase signal, and the step in this signal indicates that the phonon pulses are compressional and unipolar. On the same plot is shown the theoretical curves for the case in which one well in turn is artificially removed from the structure and replaced by $Al_{0.3}Ga_{0.7}As$. This suggests that we have indeed resolved the signals from both wells. Moreover, the sudden rise in signal at 140 ps can be explained as a broad coherent phonon pulse arising from the GaAs substlate.

In conclusion we have demonstrated spatially selective coherent acoustic phonon generation in a $GaAs/Al_{0.3}Ga_{0.7}As$ double quantum well structure with tuned femtosecond pump light pulses. By choosing quantum wells tuned to wavelengths further separated than in the present study it should be possible achieve both generation and detection within the buried region of the sample, and thus to fabricate ultrahigh frequency optically–addressed ultrasonic transducers in this way.

This work is partially supported by Grant-in-Aid for Scientific Research from the Ministry of Education, Science, and Culture (Japan) and the Murata Science Foundation.

References

1. C. Thomsen, H. T. Grahn, H. J. Maris, and J. Tauc, Phys. Rev. B **34** (1986) 4129.
2. N. V. Chigarev, D. Yu. Paraschuk, X. Y. Pan, and V. E. Gusev, Phys. Rev. B **61**, (2000) 23.
3. O. B. Wright and V. E. Gusev, Appl. Phys. Lett **66**, (1995) 1190.
4. J. J. Baumberg, D. A. Williams, and K. Köhler, Phys. Rev. Lett. **78**, (1997) 3358.
5. S. Adachi, J. Appl. Phys. **58**, (1985) R1.
6. *Properties of aluminium gallium arsenide*, ed. S. Adachi (INSPEC, London, UK, 1993) 87.
7. D. H. Hurley and O. B. Wright, Opt. Lett. **24**, (1999) 1305.

Amplification and generation of high-frequency coherent acoustic phonons under the drift of 2D-electrons

M. A. Stroscio[1], S. M. Komirenko[2], K. W. Kim[2], A. A. Demidenko[3], V. A. Kochelap[3]

[1] U.S. ARO, Research Triangle Park, NC 27709-2211
[2] Department of Electrical and Computer Engineering, North Carolina State University, Raleigh, North Carolina 27695-7911
[3] Institute for Semiconductor Physics, National Academy of Sciences of Ukraine, Kiev, 252650, Ukraine

Abstract   We analyze the Cerenkov emission of high-frequency confined acoustic phonons by drifting electrons in a quantum well. We find that the electron drift can cause strong phonon amplification (generation). For the example of a $Si/SiGe/Si$ device it is shown that the amplification coefficients of the order of hundreds of cm^{-1} can be achieved in the sub-THz frequency range.

1 Introduction

High-frequency lattice vibrations with a high degree of spatial and temporal coherence have been observed for a number of semiconductor materials and heterostructures. These include Si, Ge, GaAs as well as SiGe and AlGaAs superlattices. Typically, high-frequency coherent phonons are excited optically by ultrafast laser pulses. The development of electrical methods of coherent phonon generation is an important problem.

An electric current flowing though a semiconductor can produce high-frequency coherent acoustic phonons. If the current is due to free electron motion in an electric field, phonon amplification (generation) can be achieved via the Cerenkov effect if the electron drift velocity exceeds the velocity of sound. This effect is well-known for bulk samples. High drift velocities and large densities of electrons are necessary for practical use of the Cerenkov effect. Advanced technology of semiconductor heterostructures opens new possibilities to employ this effect for high-frequency phonon generation. Indeed, such phenomena as high electron mobility at large electron density and phonon confinement in a quantum well (QW) can greatly facilitate achieving phonon amplification and generation by electron drift. In this report, we analyze the generation of high-frequency confined acoustic phonons under the electron drift in a QW layer.

2 Calculation of amplification of confined phonons

Consider a symmetric heterostructure $B/A/B$ with electrons confined in the layer A of thickness $2d$. Assuming isotropic elastic properties for both semiconductors A and B one can introduce the longitudinal, V_{LA} and V_{LB}, and transverse, V_{TA} and V_{TB}, sound velocities. If $V_{TA} < V_{TB}, V_{LA}, V_{LB}$, then localization of acoustic waves near the embedded layer A will occur [1]. These localized waves propagate along the layer and decay outside it. Let the waves propagate along x-coordinate and

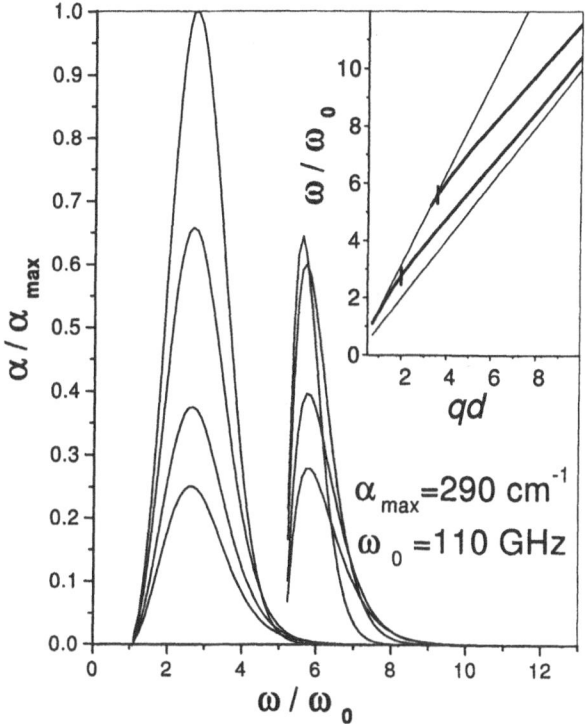

Fig. 1 The amplification coefficient versus frequency for two lowest SSV phonon branches in the $Si/SiGe/Si$-heterostructure. T= 50, 100, 200 and 300 K. Increasing of T leads to decreasing of α at maximum. The values of ω and q which correspond to the maxima of α at T=50 K are indicated on the dispersion relations shown in the Insert.

z- is direction perpendicular to the layers. There are two classes of the localized waves: the shear-horizontal (SH) waves with the displacement vector $\mathbf{u} = (0, u_y, 0)$, and the shear-vertical (SV) waves with $\mathbf{u} = (u_x, 0, u_z)$. Dispersion relations for each class of waves are represented by a set of branches $\omega = \omega_\nu(q)$, with ω and q being the wave frequency and wave vector, ν is an integer. For a given ν, localization of the waves depends on q. Let $\mathbf{u}_{\nu,q}(x, z, t) = \mathbf{w}_{\nu,q}(z)e^{(iqx - i\omega t)}$ be solutions of the elastic equations describing the localized waves. Solutions with different "quantum numbers" $\{\nu, q\}$ are orthogonal. We normalize the solutions by imposing the condition that the $\{\nu, q\}$ wave has an elastic energy equal to $\hbar\omega_{\nu,q}$. The set of such solutions (modes) allows one to quantize the lattice vibrations, introduce *confined phonons* and analyze processes of absorption and emission of the phonons.

For the interaction of a confined mode with electrons we assume that a) only the lowest two-dimensional electron subband is populated and b) presence of higher subbands can be ignored. Then the electron wavefunctions have the form $\Psi_{\mathbf{k}}(\mathbf{r}, z) = e^{i\mathbf{k}\mathbf{r}}\chi(z)$, where $\mathbf{k}$ is the two-dimensional electron wavevector. We suppose that electrons interact with phonons via the deformation potential (DP); thus, the energy of this interaction is $H = b\,div\,\mathbf{u}$, where b is the DP constant.

For the phonon number of the mode $\{\nu, q\}$ we can introduce the kinetic equation:

$$\frac{dN_{\nu,q}}{dt} = \gamma_{\nu,q}^{(+)}(1 + N_{\nu,q}) - \gamma_{\nu,q}^{(-)}N_{\nu,q} - \beta_{\nu,q}N_{\nu,q}, \qquad (1)$$

where $\gamma_{\nu,q}^{(\pm)}$ are parameters which determine evolution of the phonon number, $N_{\nu,q}$, in time due to the interaction with electrons. The parameter $\beta_{\nu,q}$ describes phonon losses. They can include phonon scattering or phonon absorption due to non-electronic mechanisms, phonon decay due to anharmonicity of the lattice, etc. In Eq. (1) the terms which correspond to stimulated processes can be represented by $\left(\gamma_{\nu,q}^{(+)} - \gamma_{\nu,q}^{(-)}\right)N_{\nu,q} \equiv \gamma_{\nu,q}N_{\nu,q}$, with the phonon increment (decrement) equal to

$$\gamma_{\nu,q} = \frac{m^*}{\pi\hbar^3 q}|M(q)|^2\left(\mathcal{I}^{(+)}(q) - \mathcal{I}^{(-)}(q)\right), \qquad (2)$$

where $|M|^2 = |M_{DP}|^2/(\kappa^{(el)})^2$ and M_{DP} being the matrix element

$$M_{DP}(q) \equiv b\left(\int_{-\infty}^{\infty}div(\mathbf{w}_{\nu,q})\chi_1^2(z)dz\right).$$

We take into account the effect of electron screening of the DP by introducing the electron permittivity: $\kappa^{(el)}(q) = 1 + 2\pi e^2 d\mathcal{A}(q)\mathcal{B}(qd)/\kappa$; $\mathcal{A}(q)$ is the polarization operator of two-dimensional electrons, κ is the dielectric constant and the factor $\mathcal{B}(s)$ is

$$\mathcal{B}(s) = \frac{1}{s}\int_{-\infty}^{\infty}\int_{-\infty}^{\infty}d\zeta d\zeta'\chi^2(\zeta d)\chi^2(\zeta'd)e^{-s|\zeta-\zeta'|}.$$

We introduce also the population factor calculated via the electron distribution function $F(\mathbf{k}) = F[k_x, k_y]$:

$$\mathcal{I}^{(+)}(q) = \int_{-\infty}^{\infty}dk_y F\left[sign(q)\frac{m^*\omega_{\nu,q}}{\hbar|q|} \perp \frac{1}{2}q, k_y\right], \qquad (3)$$

m^* is the effective mass.

Depending on the shape of the electron distribution function, $F[k_x, k_y]$, the value $\gamma_{\nu,q}$ can be either positive, or negative. If the phonon increment caused by the electron-phonon interaction is positive and, in addition, it exceeds phonon losses, $\gamma_{\nu,q} > \beta_{\nu,q}$, the population of corresponding mode(s) should increase in time, i.e., we obtain the effect of phonon generation.

One can introduce the amplification (absorption) coefficient for the confined acoustic modes which describes the rate of increase in the acoustic wave intensity per

unit length: $\alpha_{\nu,q} = \gamma_{\nu,q}/V_g$, where V_g is the group velocity of the wave. The signs of $\gamma_{\nu,q}$ and $\alpha_{\nu,q}$ are determined by the factor $(\mathcal{I}^{(+)} - \mathcal{I}^{(-)})$, which is to be calculated from the distribution function.

We suppose that the electrons drift in an applied electric field along the QW layer. Under the realistic assumption of strong electron-electron scattering, the distribution function can be thought as the shifted Fermi distribution: $F[k_x, k_y] = F_F\left[k_x - \frac{m^*}{\hbar}V_{dr}, k_y\right]$, where $F_F(\mathbf{k})$ is the Fermi function, V_{dr} and T are the electron drift velocity and temperature, respectively. From Eq. (2) for phonons propagating along the electron flux ($q > 0$), we immediately find that $\gamma_{\nu,q}, \alpha_{\nu,q} > 0$ if the electron drift velocity exceeds the confined phonon phase velocity: $V_{dr} > \omega_{\nu,q}/|q|$. This criterion is the well-known condition of the Cerenkov generation effect.

3 Discussion

It is easy to see that the deformation-potential interaction couples only SV phonons with electrons. We define the symmetric shear-vertical (SSV) modes as those with $w_x(z) = w_x(-z), w_z(z) = -w_z(-z)$ and the antisymmetric ones with $w_x(z) = -w_x(-z), w_z(z) = w_z(-z)$. For a symmetric QW, the electrons are coupled with the SSV phonons.

We have analyzed of the amplification coefficient for different heterostructures. For $III - V$ and SiGe heterostructures, the acoustic mismatch is typically small and the lowest mode is an antisymmetric SV mode. Consequently, all SSV modes have finite frequency onsets. This determines two important features: a low-frequency cut-off of the amplification and a nonmonotonous dependence of the matrix element $M(q(\omega))$. The population factor, in turn, limits phonon amplification at high frequencies. As a result, amplification band [2] for each SSV phonon branch is relatively narrow. Two typical amplification bands are illustrated in Fig. 1. These results are obtained for a p-doped $Si/Si_{0.5}Ge_{0.5}/Si$ structure. We set $d = 5\,nm$, the hole density is taken as $10^{12}cm^{-2}$ and the drift velocity is $V_{dr} = 2.5\,V_{TA}$ with $V_{TA} = 3.4\,10^5 cm/s$ for the $SiGe$ layer. One can see that amplification coefficient of the order of tens to hundreds cm^{-1} can be achieved for confined modes in the sub-THz frequency range.

Our results suggest that a simple electrical method for generation of high-frequency coherent phonons can be developed on the basis of the Cerenkov effect.

The work was supported by ARO and AFOSR-MURI grants.

References

1. V. V. Mitin, V. A. Kochelap and M. A. Stroscio, *Quantum Heterostructures* (Cambridge University Press, New York 1999).
2. S. M. Komirenko, K. W. Kim, A. A. Demidenko, V. A. Kochelap, M. A. Stroscio, Appl. Phys. Lett. **76**, (2000) 1869.

Observation of a two-dimensional plasmon in a metallic monolayer on silicon surface

Tadaaki Nagao[1,2,3], Torsten Hildebrandt[4], Martin Henzler[4], and Shuji Hasegawa[2,3]

[1]Department of Physics, Graduate School of Science, The University of Tokyo, 7-3-1 Hongo, Bunkyo-ku, Tokyo 113-0033, Japan, Fax +81-3-5841-4209, E-mail: nagao@phys.s.u-tokyo.ac.jp

[2]Precursory Research for Embryonic Science and Technology (PRESTO) and [3]Core Research for Evolutional Science and Technology (CREST), Japan Science and Technology Corporation (JST), 4-1-8 Honcho, Kawaguchi, Saitama 332-0012, Japan

[4]Institut für Festkörperphysik der Universität Hannover, Appelstraße 2, D-30167, Hannover, Germany

Abstract   By use of electron-energy-loss spectrometer with high momentum resolution, we have measured a collective electronic excitation localized at the topmost surface layer on the Si(111)-√3 × √3-Ag surface. The obtained energy dispersion was found to be isotropic with respect to the azimuthal orientation and agreed very well with the plasmon dispersion calculated from 2D nearly-free-electron theory using random phase (RPA) approximation.

1. Introduction

Two-dimensional (2D) plasmon (or sheet plasmon) constitutes an interesting research subject since the charge density distribution of which is restricted in the 2D space and thus shows very different electrodynamical properties compared with those of the 3D-type plasmons (Fig. 1). For example, bulk and surface contributions are always inseparable in the properties of the ***surface plasmons***. On the contrary, properties of the ***2D plasmons*** are mainly determined by those of the 2D electron systems (2DESs) themselves, especially in the cases of 2DESs supported in (semi-)insulating bulk media. We can take advantage of this characteristic for utilizing the 2D plasmon as a sensitive noncontact probe of electronic properties of the 2D electron system (2DES) in the atomically thin region, such as metallic monolayer supported on less conductive dielectric media.

By using electron-energy-loss spectroscopy (EELS), we can determine the plasmon energy as a function of wave number $q_\parallel$ with excellent accessible energy range covering from the far-infrared to ultraviolet regimes [1]. The measurement of the energy dispersion curve combined with the theoretical analysis provides us the atomistic information of static and dynamical properties of the 2DES in a great detail.

In the present paper, we report on the direct measurement of the energy dispersion of a 2D plasmon found at the topmost surface layer on the

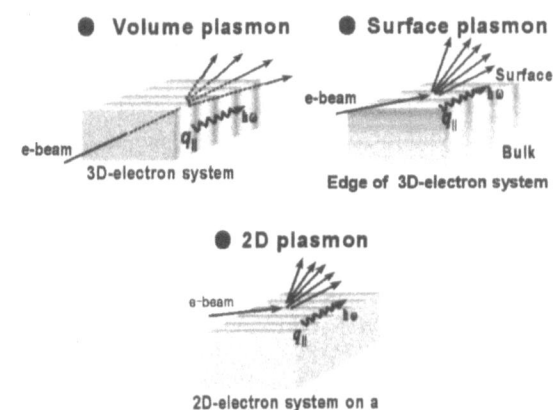

Fig. 1. Charge density wave distribution for the 3D and 2D plasmons.

Si(111)-√3 × √3-Ag surface. We have clarified that the quantum statistics and nonlocal effects manifest themselves in the large wave numver region $q_\parallel \approx k_F$, where k_F is the Fermi wave number of the 2DES. We have also found good agreement with the calculated dispersion from 2D nearly-free-electron theory [2], and clarified the quantitative agreement with the static electronic properties previously measured by photoelectron spectroscopy.

2. Experimental

The experiment was performed in ultrahigh vacuum with 6×10^{-11} Torr base pressure. The √3×√3-Ag surface was prepared by depositing precisely one monolayer (1 ML) of Ag onto a clean Si(111)-7×7 surface held at around 800 K while keeping the pressure below 1×10^{-9} Torr. Angle resolved electron-energy-loss spectra was recorded by ELS-LEED (energy-loss-spectroscopy with low-energy electron diffraction) which has an energy resolution in meV regime and momentum resolution in 10^{-3} Å^{-1} regime [3].

3. Results and discussion

The Si(111)-√3 × √3-Ag surface is one of the most promising prototype system for investigating 2D plasmons in atomically thin system [4,5]. There is an electron pocket centered at the Γ point of the surface Brillouin zone of a nearly parabolic band, and constitutes a nearly-free 2DES at the topmost surface layer. The carrier density in this electron pocket is $N_{2D} = (1.6 \pm 0.3) \times 10^{13}$ /cm^2 and the effective mass $m^* = (0.29 \pm 0.05)m_e$ (m_e is the free- electron mass), estimated from photoelectron spectroscopy (PES) measurements [5].

The plasmon energy dispersion is plotted as a function of wave vector $q_{\parallel}$ in Fig. 2. As clearly seen, there is no dependence of the energy dispersion on the electron primary energy E_p (electron probing depth) and the azimuthal orientation. These fact imply that the observed loss has a strong 2D nature. Also, vanishing excitation energy at small $q_{\parallel}$ agrees with the expected behavior of 2D plasmon [2,6].

Using the "nonlocal" response theory with random phase approximation (RPA), we analyzed the plasmon in a 2DES on a semi-infinite dielectric substrate. For convenience, we show an approximation up to the second-order term in $q_{\parallel}$,

$$\omega_{2D}(q_{\parallel}) = [\, 4\pi N_{2D} e^2 m^{*-1} (1+\varepsilon_{Si})^{-1} q_{\parallel} \\ + 6N_{2D} \hbar^2 \pi (2m^*)^{-2} q_{\parallel}^2 + O(q_{\parallel}^3) \,]^{1/2}.$$

Here, $\omega_{2D}(q_{\parallel})$ is the 2D plasma frequency, N_{2D} is the areal density of electrons in the 2DES, $\varepsilon_{Si} = 11.5$ is the dielectric constant of the Si, m^* is the electron's effective mass, and e is the elementary electric charge.

The first term is identical to the $\sqrt{q_{\parallel}}$ term from the classical local response theory [6]. Second- and higher-order terms originate from nonlocal effects and reflect the quantum statistics of the 2DES; for example, the second term is rewritten as $(3/4)v_F^2 q_{\parallel}^2$ with the Fermi velocity v_F of the 2DES and exhibits the degeneracy of the electron system.

The solid curve in Fig.2 is the best fit to the nonanalitic full RPA dispersion. The electron density and the electron effective mass are determined to be $N_{2D} = 1.9 \times 10^{13}$ /cm^2 and $m^* = 0.30 \ m_e$, respectively which agreed very well with the values form the PES measurements given above. The overall fit is excellent, and compared with the "local" $\sqrt{q_{\parallel}}$ dispersion calculated using the same N_{2D} and m^* values shown by the dashed curve, improvement in the fit at larger $q_{\parallel}$ value is clearly seen. This indicates the importance of including nonlocal effects and quantum statistics in analyzing the energy dispersion of the 2D plasmons. Moreover, fitting over a wide $q_{\parallel}$ range with the "nonlocal" full RPA dispersion yields the values of N_{2D} and m^* simultaneously, while in the "local" $\sqrt{q_{\parallel}}$ case, only their ratio is obtained

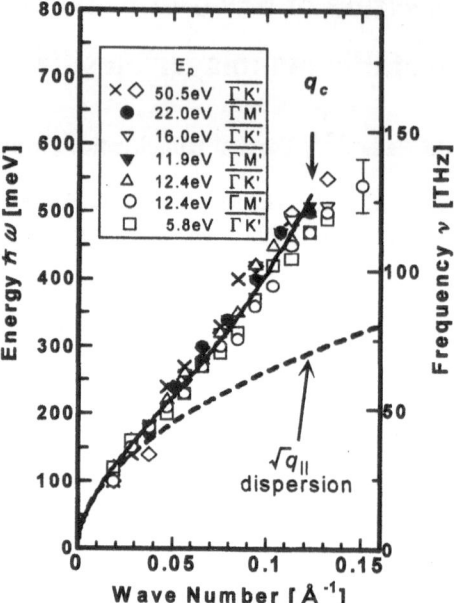

Fig. 2. Plasmon dispersion plot measured with $q_{\parallel}$ scanning along ΓK′ and ΓM′ directions at 90 K.

4. Summary

We have measured the dispersion of a 2D plasmon localized at the topmost atomic layer of the Si(111)-√3 × √3-Ag surface. We have found that the energy dispersion is isotropic with respect to the azimuthal orientation and shows no dependence on the electron probing depth. We have clarified that the observed energy dispersion does not agree with $\sqrt{q_{\parallel}}$ dispersion by the classical local response theory, but agrees very well with the full RPA dispersion, especially at large $q_{\parallel}$ values.

Acknowledgments

T. N. acknowledges Professors T. Inaoka, C. Oshima, A. Liebsch, and H. Ishida for valuable and helpful discussions.

Rerences

1. H. Ibach and D. L. Mills, *Electron Energy Loss Spectroscopy and Surface Vibrations*. (Academic Press, New York, 1982).
2. F. Stern, Phys. Rev. Lett. **18**, 546 (1967).
3. H. Claus, A. Buessenschuett, and M. Henzler, Rev. Sci. Instrum. **63**, 2195 (1992).
4. T. Takahashi, and S. Nakatani, Surf. Sci. **282**, 17 (1993) and references therein.
5. Y. Nakajima, *et al.*, Phys. Rev. B **56**, 6782 (1997) and references therein.
6. R. H. Ritchie, Phys. Rev. **106**, 874 (1957).

First- and Second-Order Raman Spectroscopy of $^{70}Ge_n/^{76}Ge_n$ Isotope Superlattices

K. Morita[*1], **K. M. Itoh**[*1, *2], **M. Nakajima**[*3], **H. Harima**[*3], **K. Mizoguchi**[*4], **Y. Shiraki**[*5], and **E. E. Haller**[*6].

[*1]Dept. Applied Physics and Physico-Informatics, Keio University and [*2]PRESTO-JST, Yokohama, 223-8522 Japan
[*3] Dept. Applied Physics, Osaka University, Osaka 565-0871, Japan
[*4]Dept. Applied Physics, Osaka City University, Osaka, 558-8585 Japan
[*5]RCAST, Univ. of Tokyo, Tokyo 153-8904 Japan
[*6]UC Berkeley and Lawrence Berkeley National Labs, Berkeley, CA 94720 USA

Abstract

First- and second-order Raman spectra have been obtained from novel superlattice (SL) structures, $^{70}Ge_n/^{76}Ge_n$ isotope superlattices (SLs) (n=4, 8, 12, 20 and 32). LO phonon frequencies obtained from first-order Raman spectra were compared with calculations based on the planar bond-charge (PBC) model. A number of first-order phonon peaks that have not been predicted by the PBC model, have appeared in the spectra. The second-order Raman spectra suggest that the phonons are SL-like in the n≤20 samples while they become bulk-like in the n=32 sample.

1   Introduction

$^{70}Ge_n/^{76}Ge_n$ isotope SLs are new type SLs, in which ^{70}Ge and ^{76}Ge stable isotope layers have been stacked periodically with the thickness n in units of atomic layers. While the electronic structure of the isotope Ge SLs is practically the same as that of bulk Ge, the phonon dispersion is zone-folded and change dramatically with respect to that of bulk Ge. The properties of the isotope Ge SLs are significantly different from those of conventional SLs, e. g., GaAs/GaAlAs and Si/SiGe, in which both the electronic and vibrational properties are affected in the direction perpendicular to the plane. Consequently, isotope SLs allow for a detailed and systematic investigation of the phonons in low dimensional semiconductor structures. Cardona and his colleagues have pioneered the Raman investigations involving isotope SLs [1,2]. Fuchs, et al., have calculated the zone-folded LO phonon frequencies and their Raman intensities in $^{70}Ge_4/^{76}Ge_4$ isotope SLs [1] in the framework of the planar bond-charge (PBC) model [3]. Spitzer, et al., have investigated the first-order Raman spectra in a series of $^{70}Ge_n/^{74}Ge_n$ isotope SLs [2]. The good agreement between calculation and the experimental results clearly shows the validity of the PBC model and the high quality of the $^{70}Ge_n/^{74}Ge_n$ isotope SLs.

Compared to $^{70}Ge_n/^{74}Ge_n$ SLs that have been used in previous studies [2, 4], the $^{70}Ge_n/^{76}Ge_n$ SLs employed in the present study have larger mass differences between the two constituents (^{70}Ge and ^{76}Ge instead of ^{70}Ge and ^{74}Ge), i.e., they allow for a closer look at zone-folded phonons. The second-order Raman spectroscopy is also interesting since the first-order Raman spectroscopy probes the phonons at the zone-center while the second-order Raman effect reflects the two-phonon joint density of the states over the entire Brillouin zone [5]. The observation of the density of the states allows us to determine whether phonons are SL-like or bulk-like. Therefore, second-order Raman spectroscopy has been employed to estimate the layer thickness at which phonon characteristics changes from SL-like to bulk-like in our Ge isotope SLs.

2   Experiments

We have grown five $^{70}Ge_n/^{76}Ge_n$ isotope SLs with n=4, 8, 12, 20, and 32 by the solid-source molecular beam epitaxy (MBE) technique. A detailed description of our growth method is given elsewhere [4]. The ^{70}Ge and ^{76}Ge solid sources were isotopically enriched to 96.3% and 85.1%, respectively. The SLs were grown on top of a 200Å ^{70}Ge buffer layers in the direction <100>. The temperature of the substrate during the growth was

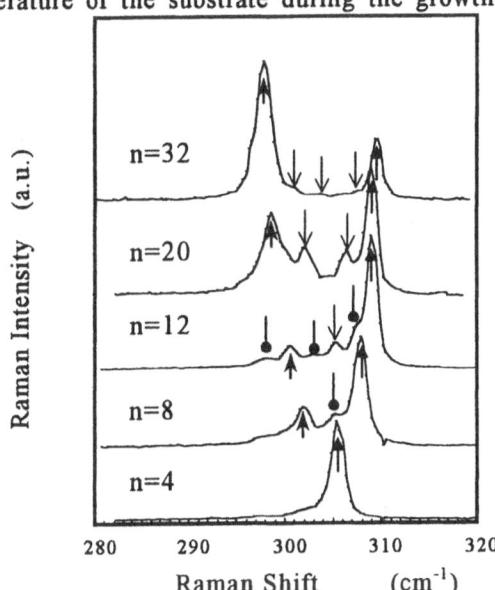

Fig.1 First-order Raman spectra of n=4, 8, 12, 20 and 32 $^{70}Ge_n/^{76}Ge_n$ isotope SLs. Positions of Raman active modes, $LO_1(^{70}Ge)$ (↑) and $LO_1(^{76}Ge)$ (▲), $LO_3(^{70}Ge)$, $LO_5(^{70}Ge)$, $LO_7(^{70}Ge)$ (↓) and of newly discovered peaks (●) are indicated.

maintained at 350 ℃. The first and second-order Raman spectra of the isotope SLs were recorded with a spectral resolution of 1cm⁻¹ in back scattering geometry at T = 10K using the Ar^+ 488 nm-line.

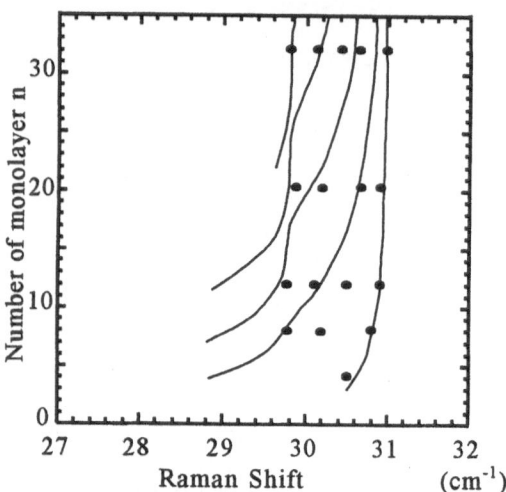

Fig.2 Direct comparison of LO phonon frequencies obtain by the first-order Raman experiments (filled circles) and calculations (solid curves). The PBC model has been employed for the phonon calculation.

3 Result and discussion

3.1 Fist-order Raman spectra of Ge isotope SLs

Our first-order Raman spectra of the n=4, 8, 12, 20 and 32 $^{70}Ge_n/^{76}Ge_n$ isotope SLs are shown in Fig.1. Fig. 2 shows the comparison between experimentally obtained LO phonon frequencies and the results of the PBC calculation. In Fig.1, the Raman active confined LO phonons that are expected for each isotope SL by the PBC model are observed. Two phonon peaks, $LO_1(^{70}Ge)$ and $LO_1(^{76}Ge)$, appearing at the edges of the spectra and correspond to modes strongly confined in the ^{70}Ge and ^{76}Ge layers, respectively. The peaks, $LO_3(^{70}Ge)$, $LO_5(^{70}Ge)$, and $LO_7(^{70}Ge)$, correspond to Raman active phonons predicted by the PBC model. Other unlabeled peaks indicated by (↓) are newly discovered ones that are not predicted by the PBC model. The polarization dependent measurements show clearly that these newly discovered peaks are not associated with the zone-folded phonons that are predicted to be *Raman inactive*. Further investigations are being performed in order to clarify the origin of these unexpected peaks.

3.2 Second-order Raman spectra of Ge isotope SLs

The second-order Raman features appearing between 530 and 630cm⁻¹ of the Ge isotope SLs, ^{70}Ge bulk, ^{76}Ge bulk, and $^{70}Ge/^{76}Ge$ hetero-structure are shown in Fig.3. Ge isotope SLs of n=8, 12, and 20 show broad peaks that are characteristic of SLs. The sharp features that are characteristics of bulk samples disappear since the number of Van Hove singularities

in the density of the states is increased due to zone-folding. As the results, they are smeared out to show one broad peak in each spectrum. On the other hand, the second-order spectrum of n=32 SL shows sharp features that are due clearly to phonons in ^{70}Ge

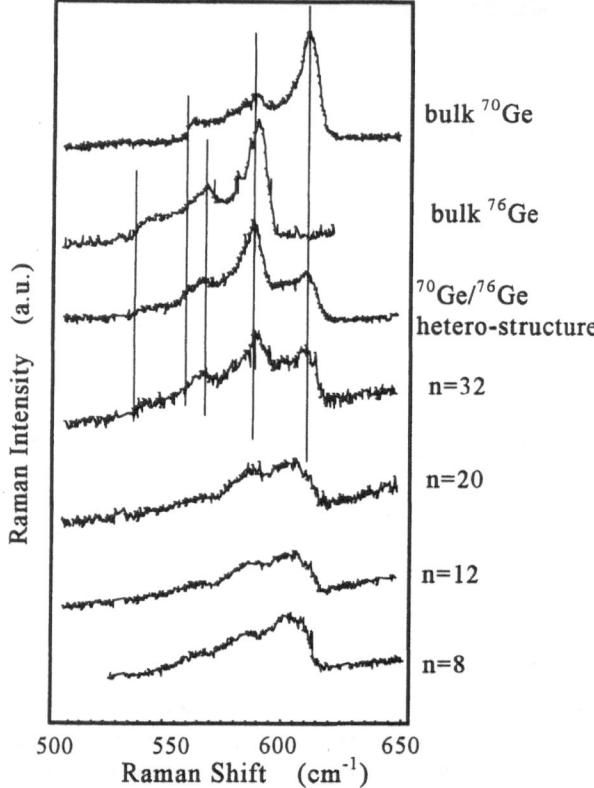

Fig.3 Comparison of second-ordered Raman spectra between the n=8, 12, 20, and 32 $^{70}Ge_n/^{76}Ge_n$ isotope SLs, ^{70}Ge bulk sample, ^{76}Ge bulk sample, and $^{70}Ge/^{76}Ge$ hetero-structure.

and ^{76}Ge bulk crystals. At n=32 the thickness of the each isotope layer has become large enough to act like a bulk material.

4 Conclusion

First- and second-order Raman spectra have been obtained from $^{70}Ge_n/^{76}Ge_n$ isotope SLs. The first–order Raman spectra show a number of phonon peaks that have been predicted by the PBC model. A numbers of unpredicted peaks have been discovered. The second-order Raman spectra suggest that the phonons are SL-like in the n≤20 samples while they become bulk-like in the n=32 sample.

Reference

1. H. D. Fuchs, *et al.*, Superlattices Microstruct. **13**, 447 (1993).
2. J. Spitzer, *et al.*, Phys. Rev. Lett. **72**, 1565 (1994).
3. P. Molinàs-Mata, *et al.*, Phys. Rev. B **43**, 9799 (1991).
4. K. Morita, *et al.*, Thin Solid Films, **369**, 405 (2000).
5. B. A. Weinstein, *et al.*, Phys. Rev. B. **46**, 2545 (1973).

Proc. 25th Int. Conf. Phys. Semicond., Osaka 2000 (Eds. N. Miura and T. Ando)

879

Raman scattering in Ge quantum dot superlattices

A.Milekhin[1,2], N.Stepina[2], A.Yakimov[2], A.Nikiforov[2], S.Schulze[1], T.Kampen[1], and D.R.T.Zahn[1]

[1]Institut für Physik, Technische Universität Chemnitz, D-09107 Chemnitz, Germany e-mail:
a.milekhin@physik.tu-chemnitz.de
[2]Institute of Semiconductor Physics, 630090, Novosibirsk, Russia

Abstract Self-organised Ge quantum dot superlattices are investigated using macro- and micro-Raman experimental setups.

The analysis of frequency positions of both transverse optical (TO) and longitudinal optical (LO) Ge and Ge-Si phonons observed allows the strain and atomic intermixing of Ge and Si atoms in Ge QDs to be determined independently. It was found that samples with almost fully strained Ge QDs (strain is about 4%) demonstrate an abrupt interface while samples with partially relaxed QDs show an atomic intermixing (up to x=0.09) in the interface region which is very likely to be responsible for the strain relaxation.

In the acoustic spectral range a number of peaks (up to 15th order) assigned to Raman scattering by folded longitudinal acoustic (LA) phonons superimposed on a intense and broad band of continuous emission due to the breakdown of crystal-momentum conservation is observed. The folded LA phonons in the Ge dot superlattice are well described by the elastic continuum model. Fitting of the frequency positions of folded doublets allows the actual period of the structures to be determined.

1 Introduction

Self-assembled Ge quantum dot superlattices attract much attention because of their unique electronic and optical properties making them perspective for the design of optoelectronic devices compatible with silicon technology. It is well established that the growth of Ge on (001) Si progresses according to the following steps. For small coverage up to 6Å Ge grows in a layer-by-layer mode. At higher thickness of Ge (up to 14Å) small regularly shaped dislocation-free islands on a wetting layer of about 6Å are formed which are called "hut clusters" or quantum dots. At the same time the understanding of phenomena in such structures still requires significant theoretical and experimental effort.

In this paper we report on Raman studies of self-organised Ge quantum dot superlattices performed using both macro- and micro-Raman setups.

2 Experimental details

Samples were grown by MBE utilizing Stranski-Krastanov growth mode on (001)-oriented Si substrates. The set of samples under investigation consists of Ge and Si layers with nominal thicknesses 6-14Å and 15-300Å, respectively. Number of periods of the structures was 10.

The dot base size determined using high-resolution transmission electron microscopy (HRTEM) and scanning tunnelling microscopy (STM) is approximately 15 nm and a QD height is 1.5 nm.

The macro-Raman scattering experiments were performed using the 514.5nm line of an Ar^+ laser, that is in the vicinity of the E_1 resonance in bulk Ge. The scattered light was analysed in a backscattering geometry using a Dilor XY triple monochromator. Using the micro-Raman setup, light of the same wavelength with a spot size of 1μm was focused on a cleaved (110)-oriented sample edge.

3 Results and discussion

Due to the biaxial strain the zone-centre optical phonons of the Ge QD superlattices split into two modes with wave vector parallel and perpendicular to the [001] direction (LO phonon and double degenerated TO phonons, respectively). According to the selection rules for Raman scattering in planar Ge/Si superlattices the odd confined longitudinal optical phonons can be seen in z(xy)-z scattering geometry while only LA phonons are observable in z(xx)-z geometry. The transverse optical phonons in the superlattices can be observed in y'(x'z)-y' geometry. Here and furtheron x, y, z, x', y' refer to directions parallel to [100], [010], [001], [1-10], [110], respectively. For Raman scattering within Ge QD superlattices, selection rules are expected to be lifted due to the lowering of the symmetry. Therefore, the prohibited vibrational modes are expected to be observable in the Ge QD superlattices.

Experimental Raman spectra of sample 14/300 (here the numbers indicate the nominal thickness of Ge and Si layers in Å, respectively) measured in different scattering configurations are shown in Fig. 1. As expected the strong peaks are observed in the Raman spectra at 315 cm^{-1} and 415 cm^{-1}, denoted as LO and L which correspond to LO phonon in Ge QDs and longitudinal Ge-Si vibrational mode, respectively. Two additional features at 308 cm^{-1} and 404 cm^{-1} related to TO phonon in Ge QDs, and transversal Ge-Si vibrational mode (shown as TO and T) appear in the Raman spectra taken in the in-plane scattering geometries. Observation of TO and both TO and LO phonons in the "prohibited" z(xx)-z and y'(zz)-y' scattering geometries, respectively, manifests the weakening of Raman selection rules in Ge QD

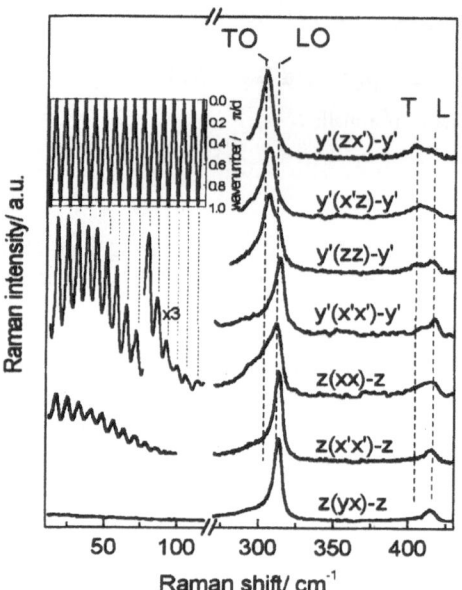

Fig. 1 Raman spectra of sample 14/300 measured in different scattering configurations. The inset shows the calculated dispersion of LA phonons.

superlattices. The L-T splitting observed is explained on the basis of *ab initio* microscopic calculation by the different character and the spatial localisation of the atomic clusters contributing to the L and T vibrations [1]. The strain-induced shifts of the LO and TO phonon frequency positions for ideally strained planar superlattices were estimated as $17 cm^{-1}$ and $12 cm^{-1}$, respectively [2]. The frequency positions of LO and TO Ge phonons determined from the experiment are somewhat below the calculated value.

With decreasing thickness of Si layers the frequency positions of the Ge and Ge-Si optical phonons are shifted towards lower frequencies with respect to their values for ideally strained superlattices. The influence of confinement effect is not significant in the samples. For a nominal thickness of the Ge layers of 14Å this shift should be about $2 cm^{-1}$ because of a rather flat dispersion of Ge optical phonons near the centre of the Brillouin zone. Taking into account the frequency shift of Ge optical phonons the compressive strain of Ge bonds in the structures with a thickness equal or higher than 45Å was estimated to be 4%. For the lower Si layer thickness the strain in the structures is partially relaxed in the interface region and corresponds to 2.8%.

The atom intermixing (alloying) at the Si/Ge interface is considered as the only reason which may contribute to the shift of the phonon frequency position and causes a partial strain relaxation. The frequency position of the Ge-Si vibrational mode allows the atomic intermixing to be evaluated. Assuming that a graded interface-involving two atomic layer is formed

and taking the masses of atoms at the Si-Ge and Ge-Si interfaces as $m_{Si-Ge}=(1-x)\cdot m_{Ge}+x\cdot m_{Si}$ and $m_{Ge-Si}=(1-x)\cdot m_{Si}+x\cdot m_{Ge}$ [3] the intermixing was estimated as x=0.09 in samples with thicknesses of Si layers below 45Å and and less than x=0.04 for Si layer thicknesses equal or exceeding 45Å.

The low frequency region of the Raman spectrum measured in z(xx)-z geometry is presented in Fig.1. A broad and intense band located around $40 cm^{-1}$ appears. The origin of this continuous emission is the scattering and recombination of electron-hole pairs created in the same photon absorption process due to partial breakdown of crystal-momentum conservation [4]. In addition, superimposed on this band, a number of periodic oscillations up to 15th order with decreasing intensities attributed to scattering by folded longitudinal acoustic phonons is clearly observed. The doublets of the folded phonons were not resolved in the measurements because of small splitting value (of about $1 cm^{-1}$) at the scattering wave vector used. The inset to the Fig.1 shows the acoustic phonon dispersion curve calculated using the Rytov model [5]. Parameters used in the calculation were taken from [6]. The agreement with the Rytov model is excellent and no adjusting of parameters is required. The actual period of structures obtained from the calculation is 379 Å that exceeds the nominal thickness by about 20%. The validity of the obtained value is confirmed by HRTEM.

4 Summary

Raman scattering by acoustical and optical phonons in self-organized Ge dot superlattices was investigated using macro- and micro-Raman setup. The weakening of selection rules for Ge QD superlattices was observed. Strain and atomic intermixing in the structures were estimated from the analysis of Ge and Ge-Si phonon features. The folded LA phonons and low frequency continuous emission were observed in the acoustic spectral range. The frequency positions of the folded LA phonons are in good agreement with those calculated using the elastic continuum model.

References

1. S. de Gironcoli, E.Molinari, R.Schrorer, and G.Abstreiter, Phys.Rev.B **48**, (1993) 8959.
2. F.Cerdeira, C.J.Buchenauer, F.H.Pollak, and M.Cardona, Phys.Rev.B **5**, (1972) 580.
3. M.W.C.Dharma-wardana, G.C.Aers, D.J.Lockwood, an J.-M.Baribeau, Phys.Rev.B41 (1990) 5319.
4. A.Mlayah, A.Sayari, R.Grac, A.Zwick, R.Carles, M.A.Maaref, R.Planel, Phys.Rev.B **56**, (1997) 1486.
5. S.M.Rytov, Akust. Zh. **2**, (1956) 71.
6. A.Milekhin, N.Stepina, A.Yakimov, A.Nikiforov, S.Schulze, D.R.T.Zahn, Eur. Phys. J. B **16**, (2000) 355.

Plasmon-phonon coupled mode excitation in a hot 2DEG observed by a phonon pulse technique

J.K. Wigmore, H.A. Al Jawhari, A.G. Kozorezov, and M. Sahraoui-Tahar

Physics Department. Lancaster University. Lancaster. LA1 4YB United Kingdom

Abstract   By studying the dependence of phonon emission from GaAs/AlGaAs 2DEGs on sheet density we obtained evidence that plasmon-phonon coupled mode excitation is the dominant mechanism for energy loss in the intermediate range of carrier temperatures.

The 2DEG energy loss process in GaAs/AlGaAs heterostructures at helium temperatures has previously been studied by several groups using, for example, optical fluorescence [1] and electrical transport [2]. At the lowest hot carrier temperatures, below 10K, emission of acoustic phonons is the primary mechanism, whilst at high temperatures, above 60K longitudinal optic (LO) phonon excitation is dominant. However details in the intermediate temperature range are still unresolved. First there is disagreement over the power level at which a different mechanism takes over, the so-called crossover. In addition the observed magnitude of the input power at crossover would imply a value of the deformation potential considerably larger than is acceptable. Other problems are identified by the phonon pulse technique, which is valuable in providing a direct probe of the excitations emitted, rather than of the heated electrons [3,4].

In our experiments the 2DEGs studied were formed at GaAs/AlGaAs heterojunctions fabricated as HEMT (high electron mobility transistor) structures, so that the sheet density, N_s, could be varied via gate depletion. A 2DEG was heated by applying a voltage pulse lasting a few nanoseconds to the drain of the HEMT, and the emitted phonon flux was measured by an aluminium superconducting bolometer on the back face of the substrate wafer [4]. The spatial resolution of the phonon flux measurement was a few degrees and the nanosecond time resolution allowed separation of phonons of different polarisation. As in earlier studies [3,4] only transverse acoustic (TA) phonons were observed to propagate along a (100) GaAs cubic axis. If deformation potential coupling had been responsible longitudinal acoustic (LA) phonons should have been at least as strongly coupled as TA to the hot electrons.

The primary measurement was of the dependence on the input power per electron, P_e, of the emitted phonon flux in the given direction. A change of slope was observed at P_{c1} which in earlier work was identified, erroneously we now believe, as the crossover from energy loss of the 2DEG by acoustic phonons to that by optic phonons; a typical data set is shown in Fig.1. The change of slope implied that the fraction of total

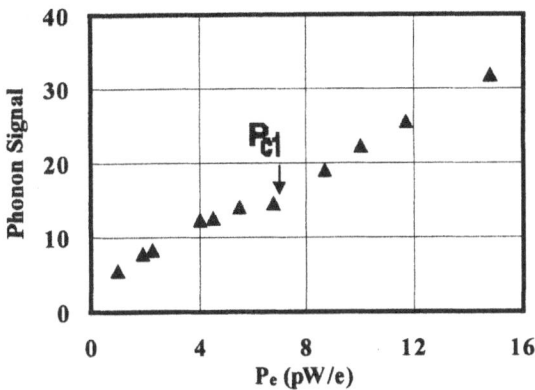

Fig.1 Magnitude of TA phonon flux as a function of input power per electron, exhibiting a crossover at 6.8 pW/e.

emitted energy appearing as acoustic phonons travelling in the direction of the bolometer depended on the type of excitation emitted by the hot electrons. Above the crossover the variation was linear, implying that the characteristics of the emitted phonons, such as frequency and anisotropy, did not vary with power. This is consistent with excitation of a single LO mode within a narrow frequency range. Below the crossover the TA flux varied sub-linearly with power, implying a changing phonon distribution reaching the bolometer.

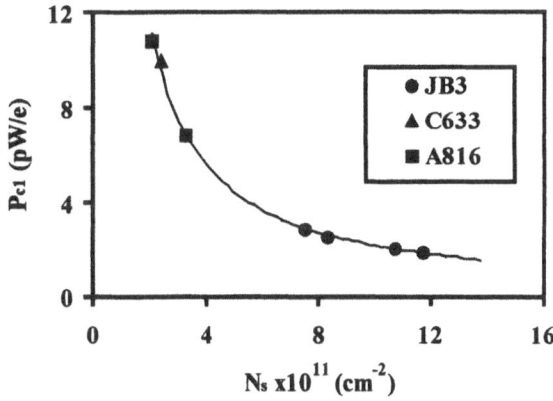

Fig. 2 Dependence of the crossover on 2DEG sheet density for HEMT samples JB3(Fujitsu), C633 (Cambridge) and A816 (Glasgow).

The effect of varying the sheet density by applying a bias to the HEMT gates is shown in Fig. 2 for three totally different samples. The value of Pc1 varied by a factor of 5 over a similar density range, and the points lay on a smooth curve regardless of sample origin. This behaviour could not be explained by a model of the crossover being due to a switch from acoustic to optic phonon emission as the dominant energy loss process. Instead we propose that there exists an intermediate range of power over which energy loss takes place via excitation and decay of plasmon-phonon coupled modes.

The theory of these modes is well known. Wu Xiaoguang et al [5] gave the complete dispersion curves for sheet densities of 1 and $10 \times 10^{11} cm^{-2}$, spanning the range of our experiments. Two-dimensional plasmons and LO phonons couple to give two propagating modes, the so-called upper (phonon-like) and lower (plasmon-like) branches. The former has the frequency Ω_{LO} of the LO phonon at $k=0$, increasing with increasing wavevector to the edge of the Landau damping regime. In contrast the plasmon-like mode extends to zero frequency at $k=0$, and hence can be excited at low carrier temperatures. However the requirements of energy and momentum conservation give rise to a threshold frequency, Ω_-, dependent on sheet density, below which the mode cannot be excited through emission by a single hot electron.

We propose that the observed crossover should be interpreted as the switch from plasmon-like coupled mode to LO phonon emission. In the vicinity of the crossover the total energy loss can be written as the sum of these components as follows

$$P_e = \left(\frac{\hbar\Omega_-}{\tau_-}\right)\exp\left(-\frac{\hbar\Omega_-}{kT}\right) + \left(\frac{\hbar\Omega_{LO}}{\tau_+}\right)\exp\left(-\frac{\hbar\Omega_{LO}}{kT}\right)$$
(1)

Strictly the first of these terms should be an integral over all frequencies above Ω_- but we assume the largest population is at the threshold. The quantities τ_+ and τ_- are the electron lifetimes against energy loss via the corresponding modes. The first of these has been measured in a number of different experiments, yielding values in the range 100 -500 fs. We will assume a typical value of 300fs; it is then possible to estimate τ_- for each sheet density from the measured value of Pc1 (Table I). The values obtained close to 1ps imply that energy loss via the plasmon-like coupled mode is a very efficient process. This mechanism has not previously been considered because of the assumption that the coupled mode would decay

back into single electron excitations, with no net energy loss from the 2DEG [6].

$N_s \times 10^{11} cm^{-2}$	2	4	6	8	10
$\tau_- \times 10^{-12} s$	2.5	1.1	0.8	0.7	0.7

Table I   Values of τ_- determined from Pc1 for different values of sheet density.

However it has been now been demonstrated experimentally that the decay of a coupled mode in GaAs takes place primarily via emission of a pair of phonons [7]. Kozorezov et al [8] showed theoretically that this interaction occurs via the second order dipole moment and lattice anharmonicity, both of which are known for GaAs. At low temperatures other processes such as excitation of electron-hole pairs and collision damping are significamtly weaker than phonon emission. Although this work was for 3D modes, the principle is valid also for the 2D situation.

In this model the phonons detected by the bolometer below the crossover result from anharmonic down-conversion of the decay products. of the lower (plasmon-like) coupled mode This process should give rise primarily to TA phonons, in agreement with experiment. Further evidence for the coupled mode model is provided by the observation at higher input power levels of a second crossover, Pc2, which we attribute to the transition from LO phonon emission to excitation of the upper (phonon-like) coupled mode.

We are grateful to J.E.F. Frost (Cambridge), C.R. Stanley (Glasgow) and the Fujitsu Company, Japan, for samples, and to EPSRC for supporting the work.

References

1.   J. Shah, A. Pinczuk. A.C . Gossard and W. Wiegmann Phys. Rev. Lett. **54** (1985) 2045
2.   K. Hirakawa and H. Sakaki Appl. Phys. Lett. **49** (1986) 889
3.   P. Hawker, A.Kent, O.H. Hughes and L.J. Challis Semicond. Sci . Technol. 7 (1992) B29
4.   J.K. Wigmore, M. Erol, M. Sahraoui-Tahar, C.D.W. Wilkinson, J.H. Davies and C. Stanley Semicond. Sci. Technol. **8** (1993) B22
5.   Wu Xiaoguang, F.M. Peeters and J.T. Devreese Phys. Rev. **B32** (1985) 6982
6.   S. Das Sarma, J.K. Jain and R. Jalabert Phys. Rev. **B37** (1988) 4560
7   M. Giltrow, A. Kozorezov, M. Sahraoui-Tahar, J.K. Wigmore, H. Davies, C. Stanley, and C.D.W. Wilkinson Phys. Rev. Lett. **75** (1995) 1827
8.   A. Kozorezov, J.K. Wigmore and M. Giltrow J. Phys: Condens. Matter **9** (1997) 4863

Peculiarities of phonons in strained short-period GaN/AlN superlattices: A *first principles* study

J.-M. Wagner and F. Bechstedt

Institut für Festkörpertheorie und Theoretische Optik, Friedrich-Schiller-Universität Jena; IFTO, Max-Wien-Platz 1, D-07743 Jena, Germany

Abstract The structural and vibrational properties of strained short-period GaN/AlN superlattices are investigated by first-principles density-functional calculations. The stability of the ordered structures is determined. The phonon frequencies show interesting features in dependence on polytype and layering. The angular dependence of the IR-active modes is dicussed.

1 Introduction

Compared to the well-studied case of a GaAs/AlAs superlattice (SL), the physical situation is remarkably different for a GaN/AlN SL: first, the layers are mutually strained due to different lattice constants, second, there is a partial overlap of the optical phonon dispersions of the bulk materials, and, third, the common anion is the lightest atom. To clarify the basic vibrational properties of strained short-period nitride SLs and to sustain the use of Raman and IR spectroscopy for material characterization we investigate the consequences of these conditions for the phonon modes by first-principles calculations of structural and lattice-dynamical properties. For comparison we consider material layers of both the cubic and hexagonal polytypes.

2 Computational Method

Within the framework of the density-functional theory, the local-density approximation and a plane-wave pseudopotential method, the effect of the Ga-3d electrons is also taken into account [1]. The vibrational properties (including the phonon eigendisplacements) are determined using the density-functional perturbation theory [2]. For a given primitive cell of the SL, the elastic deformation of the layers is determined by total energy minimizations. The internal degrees of freedom of the hexagonal SLs are taken into account by the calculation of Hellmann-Feynman forces. To determine the stability of the SLs with various arrangements of the layers, their total energies are compared with the corresponding values of strained and unstrained bulk compounds [3].

3 Results and Discussion

3.1 Structure and Stability

We restrict ourselves to the case of homogeneously strained layers with an elastically relaxed in-plane lattice constant which follows from the elastic constants. Effects of varying biaxial strain will be considered elsewhere [4]. The thicknesses of the two material layers have been found to follow the elastic properties of the bulk constituents as given by the elastic constants. The stability of the short-period SLs is found to depend on the polytype and the number of bilayers: whereas for cubic structures one has an energy gain compared to bulk layers (strained or unstrained), for hexagonal ones only the 1×1 SL is stable. Also for larger numbers of bilayers, SLs based on both polytypes are unstable for the considered strain situations, but with energy differences smaller than the thermal energy $k_B T$ at room temperature. Nevertheless, we expect a stabilization of the SL formation due to kinetic effects.

3.2 Phonons

Because of the different lattice constants of GaN and AlN the strain in the SL layers is compressive for GaN and tensile for AlN. It was found experimentally and theoretically [5,6] that the bulk phonon frequencies increase with increasing compressive strain. The only exception is the low-frequency E_2 mode of the wurtzite structure whose frequency decreases. The same holds for zone-boundary TA phonons of the cubic polytype, similar to the case of hydrostatic pressure. Figure 1 shows the resulting spectra for Γ-point phonons of a hexagonal and a cubic 3×3 SL in dependence on the propagation direction. In contrast to the case of GaAs/AlAs SLs, we obtain separate spectral regions for all LO and all TO modes, respectively, and a mixing in each of them with respect to the two materials.

Only the uppermost LO phonons are confined AlN modes. The LO modes that are located in the range of the strain-enhanced branch overlap show propagating behavior, and some of them exhibit displacements of the nitrogen atoms only, in agreement with the picture of folded phonons [corresponding, e. g., to cubic LO(X) modes]. The TO phonons are confined to one material layer, with an energetical separation between AlN and GaN modes and an interface mode reaching into the gap. One GaN-derived TO mode changes its character and shows an angular dispersion, which will be discussed below. Because of the opposite strain behavior of the transversal acoustic phonons, their branch overlap is reduced and TA modes confined to the AlN layers occur. The uppermost LA phonon is dominated by AlN vibrations, but here the damping of the GaN vibrations is less pronounced than for the confined TA mode.

In Fig. 1, the modes are further be classified by their dynamical dipole moment $\mathbf{p}_j = \sum_{\kappa=1}^{N} \mathbf{Z}_\kappa^B \cdot \mathbf{u}_\kappa^{(j)}$ (with N the number of atoms in the primitive cell, $\mathbf{Z}_\kappa^B$ the tensor of the Born effective charge of the κth atom, and $\mathbf{u}_\kappa^{(j)}$ its amplitude of vibration for the considered mode j), which determines the strength of their IR activity. In the cubic SL, for some modes $\mathbf{p}$ vanishes exactly due to the symmetric displacement of the atoms. The corresponding modes in the hexagonal SL possess a nonvanishing $\mathbf{p}$ because of the lower symmetry.

The angular dependence of the Γ-point phonon modes is a manifestation of the axial symmetry of the SL structure and of the influence of the long-range electric field, therefore it

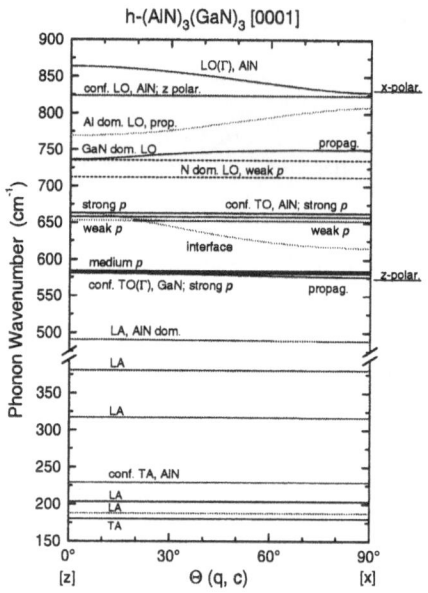
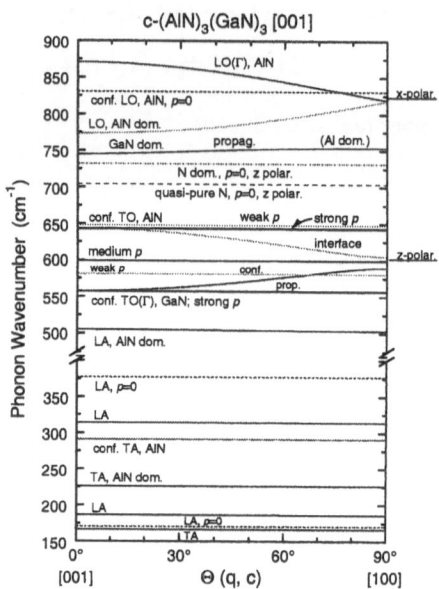

Fig. 1 Angular dispersion of zone-center phonon modes of a hexagonal (left panel) and a cubic-based (right panel) 3×3 GaN/AlN superlattice. Θ denotes the angle between the phonon propagation direction **q** and the c axis. The mode type is given for $\Theta = 0$. The lowest-energy TA modes are not shown. Where significant, the dominant (dom.) material character is indicated. For the dynamical polarization **p** see the text.

deserves our particular attention. All transversal phonons in Fig. 1 split into a dispersionless transversal and an angular-dependent mode of mixed polarization. For the TA modes, this splitting is negligible except for 1×1 SLs. In addition, also LO-derived modes show such an angular dispersion, except for those with vanishing **p**. The GaN-TO mode that exhibits angular dispersion changes its character from confined to propagating for increasing angle Θ. Its displacement pattern is similar to a bulk TO(Γ) mode, and it is strongly IR active (large **p**). The frequency value resulting for complete in-plane propagation is an intrinsic one, i. e., it is an eigenfrequency of the crystal obtained by neglecting the macroscopic field, and it is z polarized (cf. Fig. 1). In contrast, the $\Theta = 90°$ value for the interface mode in the TO gap is determined by the field in the layer plane. In the hexagonal SL, it is similar to an A_1 mode. In the cubic SL the limiting intrinsic value has a considerably higher frequency, causing a larger dispersion including mode (anti)crossing (similarly to the upper end of the interface mode in the TO gap), and the characteristic properties of strong polarization and z displacements are found at a different branch for $\Theta = 90°$.

Among the LO modes, there is a difference between the uppermost and all the others: while the latter stay mostly z polarized, the former one changes from z to x polarization (cf. Fig. 1), and it is completely governed by the accompanying electric field. For all other LO modes, the limiting value for in-plane propagation is an intrinsic frequency. Therefore, if the field would be screened out or the coupling to the field would be suppressed (i. e., $\varepsilon_\infty \to \infty$ or $\mathbf{Z}^B \to 0$), all LO modes would remain constant at the intrinsic frequency except for the uppermost one which would vanish and show up again as an additional TO mode, taking one of the two intrinsic limiting values of the above-described TO(Γ)-like mode. Also this mode possesses a large **p**, and its displacement re-

sembles the bulk LO(Γ) pattern, therefore this assignment in Fig. 1. Altogether, this behavior shows a similarity to bulk material with respect to the lifting of the LO-TO degeneracy.

4 Summary

In conclusion, we found that some general properties of the phonons in nitride SLs can be derived from the known bulk dispersions, taking into account the frequency shift caused by strain. However, there are some peculiarities which can only be determined by a microscopic study, taking into account the full symmetry and influence of the long-range electric field. The dynamical dipole moment vanishes exactly for some longitudinal modes only in the cubic structure. The most strongly IR active LO and TO mode show a bulk-like behavior concerning the LO-TO splitting, making the SL similar to an ordered bulk crystal.

Acknowledgments

This work has been supported by the Deutsche Forschungsgemeinschaft (contract Be 1346/8-4). The calculations were performed on the Cray T90 of the NIC in Jülich.

References

1. K. Karch, F. Bechstedt, and T. Pletl, Phys. Rev. B **56** (1997), 3560.
2. P. Giannozzi, S. de Gironcoli, P. Pavone, and S. Baroni, Phys. Rev. B **43** (1991), 7231.
3. D. M. Wood, S.-H. Wei, and A. Zunger, Phys. Rev. Lett. **58** (1987), 1123.
4. J.-M. Wagner, J. Gleize, and F. Bechstedt, to appear in *Proc. of Int'l. Workshop Nitride Semicond.*, Jap. J. Appl. Phys. (2000).
5. J.-M. Wagner and F. Bechstedt, Phys. Status Solidi B **216** (1999), 793.
6. V. Yu. Davydov, N. S. Averkiev, I. N. Goncharuk, D. K. Nelson, I. P. Nikitina, A. S. Polkovnikov, A. N. Smirnov, M. A. Jacobson, and O. K. Semchinova, J. Appl. Phys. **82** (1997), 5097.

Study of Ambipolar Diffusion and Drift of Spatially Separated Charge Carriers

M. Beck, M. Vitzethum, D. Streb, P. Kiesel, C. Metzner, S. Malzer, G.H. Döhler

Institut für Technische Physik I, Universität Erlangen-Nürnberg, D-91058 Erlangen, Germany

Abstract The direct observation of the ambipolar transport of photogenerated carriers in bulk semiconductors is strongly limited by both, short recombination lifetimes and slow ambipolar diffusion. The spatial separation of electrons and holes into the doping layers of a pin-diode or a n-i-p-i superlattice leads not only to strongly enhanced recombination lifetimes. Also, the diffusion transport process along the doping layers is no longer driven by the random thermal motion of the carriers, but by internal fields originating from the reduced screening. We have investigated this transport process in pin-diodes, using an all-optical pump-&-probe experiment. The density dependence of the diffusion coefficient has been investigated by depleting the n-layer with a pn-voltage in a stationary experiment. Moreover, the ambipolar drift of separated electrons and holes in a lateral electric field has been observed using a time-resolved technique. The experimental results are in quantitative agreement with numerical simulations based on a theory assuming Ohmic transport and ambipolarity.

1 Introduction: Ambipolar Transport of Separated Electrons and Holes

The ambipolar diffusion constant of electrons and holes separated by space charge fields in pin-diodes or in n-i-p-i structures was shown to exceed the respective value in bulk semiconductors by several orders of magnitude about ten years ago [1]. The reason for this strong enhancement of the diffusion process is the reduced attraction between carriers of opposite charge, whereas the repulsion between carriers of the same type is unaltered. Furthermore, the spatial separation of the photogenerated carriers leads to much larger carrier lifetimes compared to bulk semiconductors ($\tau^{\mathrm{nipi}} \approx$ ms...s [2], $\tau^{\mathrm{bulk}} \approx$ ns). Both these effects result in a strongly increased diffusion length ($\Lambda^{\mathrm{nipi}} \approx$ cm...m, $\Lambda^{\mathrm{bulk}} \approx \mu$m), thus enabling a direct observation of ambipolar transport on a macroscopic length scale and on an easily accessible time scale.

Although the spatial separation of electrons and holes alters the transport process significantly, it can still be described by a single diffusion equation [1] for both types of excess carriers (ambipolar diffusion). Allowing for an externally applied electric field E_{ext} along the doping layers, the theory predicts the same ambipolar *drift* behavior as in bulk material [4], in contrast to the strongly enhanced *diffusion* process. For the stripe-shaped ($L_x \gg L_y$; cf. Fig. 1) structures used in our experiments, a one-dimensional description is justified. The transport equation for the (two-dimensional) excess carrier distribution

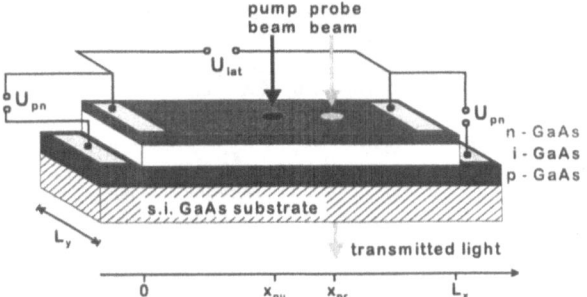

Fig. 1 Schematic view of the sample and illustration of the pump-&-probe experiment

$\rho(x, t)$ then reads

$$\frac{\partial \rho}{\partial t} = \frac{\partial}{\partial x}\left[\left(\underbrace{\frac{\mu_n \mu_p (n+p)}{\mu_n n + \mu_p p} \frac{k_B T}{e}}_{D_{\mathrm{ambi}}^{\mathrm{bulk}}} + \underbrace{\frac{\mu_n n \mu_p p}{\mu_n n + \mu_p p} \frac{e}{C_{\mathrm{pn}}}}_{D_{\mathrm{ambi}}^{\mathrm{pin}}}\right)\frac{\partial \rho}{\partial x}\right]$$

$$+ \underbrace{\frac{\mu_n \mu_p (p-n)}{\mu_n n + \mu_p p}}_{\mu_{\mathrm{ambi}}} E^{\mathrm{ext}} \frac{\partial \rho}{\partial x} . \qquad (1)$$

We simulate the ambipolar transport process including realistic boundary conditions into the numerical integration of Eq. (1).

In this contribution, we report on a new all-optical method to study ambipolar transport and on its application for the investigation of two interesting aspects of ambipolar transport. First, we have investigated the density dependence of the ambipolar *diffusion* coefficient by accumulating carriers within the sample using stationary excitation. Second, the influence of an externally applied electric field along the doping layers, i.e. the ambipolar *drift* has been examined.

2 Experimental Technique

We have set up a pump-&-probe experiment making use of the electro-absorption (Franz-Keldysh-Effect) in the intrinsic (i-) layer of a pin-diode. Electron hole pairs excited in the i-layer of the diode by a pump laser ($\hbar\omega > E_{\mathrm{gap}}$, cw or pulsed) at x_{pu} (Fig. 1) are separated into the doping layers by the internal electric field normal to the layers, $F_\perp^0$, within a few ps. The separated excess carriers now diffuse (and drift) laterally within the layers and, in turn, induce changes of the field $F_\perp^0$ depending on x (and on t in the case of pulsed excitation). Thus, the transmission of a weak probe laser beam ($\hbar\omega < E_{\mathrm{gap}}$, cw or pulsed) at x_{pr} changes due to the Franz-Keldysh-Effect.

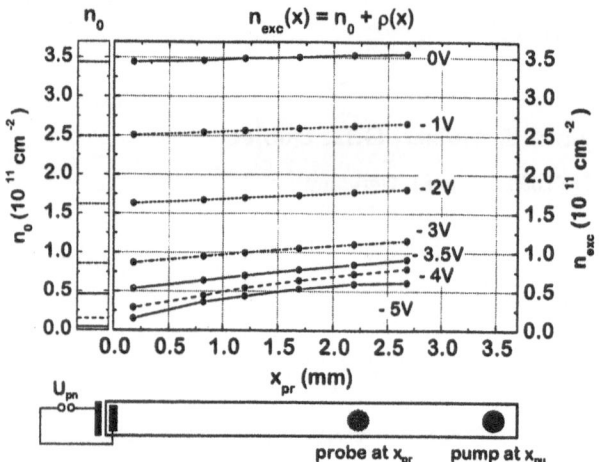

Fig. 2 Upper part: steady state electron density $n_{\text{exc}}(x) = n_0 + \rho(x)$ vs. x_{pr} for various values of the reverse bias U_{pn}, determined from the observed transmission changes for excitation at x_{pu}, close to the right end of the sample. The corresponding dark electron densities $n_0(U_{\text{pn}})$ are indicated on the left. Lower part: illustration of the experiment.

Continuous excitation creates a stationary excess carrier distribution. The stationary solution of Eq. (1) can therefore be determined directly from the observed absorption changes. Temporally and spatially resolved measurements are possible by adjusting the temporal delay and the spatial distance on the sample between the two laser pulses.

3 Stationary Experiment: Investigation of the Large Signal Case

Applying a reverse bias U_{pn} to the diode, the dark carrier concentrations n_0 and p_0 can be reduced until the (thin and weakly doped top-) n-layer depletes at the threshold voltage $U_{\text{th}} \approx -4.2$ V. The density dependent coefficients $D_{\text{ambi}} = D_{\text{ambi}}^{\text{bulk}} + D_{\text{ambi}}^{\text{pin}}$ and μ_{ambi} (cf. Eq. (1)) can thus be influenced by U_{pn}. If the excess density ρ is much smaller than the dark carrier concentrations n_0 and p_0 (small signal case), D_{ambi} and μ_{ambi} become constants and Eq. (1) can be solved analytically.

Under stationary excitation, carriers excited locally at $x = x_{\text{pu}}$ in a sample with contacts only at one end of the stripe (cf. Fig. 2) diffuse towards the ends. Whereas they can recombine at $x = 0$, they accumulate between x_0 and L_x. Between the contacts and the excitation spot ($0 < x < x_{\text{pu}}$), the small signal solution of the steady state transport equation is given by

$$n(x) = -\frac{G}{D_{\text{ambi}}L_y}x + n_0 . \qquad (2)$$

Since the generation rate G can be obtained from the measured photocurrent, D_{ambi} can be deduced from the slope of the curves shown in Fig. 2. In the small signal case ($U_{\text{pn}} = 0, -1, -2, -3$ V), we obtain diffusion constants in the range of 10^4 cm^2/s. With increasing reverse bias, however ($U_{\text{pn}} = -3.5, -4, -5$ V), the n layer becomes more and more depleted such that the small

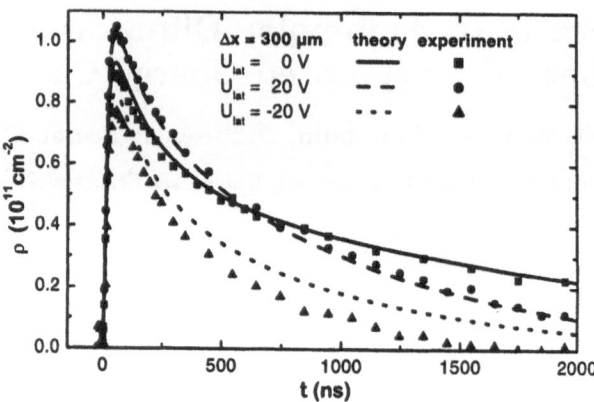

Fig. 3 Measured and calculated excess carrier concentration vs. time after excitation for $x_{\text{pr}} = x_{\text{pu}} + 300$ μm without ambipolar drift (solid line / squares) and drifting from $x_{\text{pu}} = L_x/2$ towards or away from the probe position x_{pr}.

signal condition no longer holds (large signal case) and the curves deviate from straight lines.

4 Time Resolved Experiment: Investigation of Ambipolar Drift

In the small signal case, the solution of the ambipolar transport equation (Eq. (1)) for temporally and spatially δ-shaped excitation in an infinite sample, the excess carrier concentration is given by

$$\rho(x, t > 0) = \frac{N}{\sqrt{4\pi t D_{\text{ambi}}}} \exp\left(-\frac{(x - v_{\text{drift}}t)^2}{4D_{\text{ambi}}t}\right), \qquad (3)$$

In the large signal case or a finite sample however, the density dependence of the coefficients $D_{\text{ambi}} = D_{\text{ambi}}^{\text{bulk}} + D_{\text{ambi}}^{\text{pin}}$ and $v_{\text{drift}} = \mu_{\text{ambi}}E^{\text{ext}}$ only allows for a numerical solution.

We investigate the ambipolar drift using excitation conditions close to the small signal case. Fig. 3 shows the time dependence of the measured and the calculated excess carrier density at $x_{\text{pr}} = x_{\text{pu}} + 300$ μm ($x_{\text{pu}} = L_x/2$) for different lateral voltages U_{lat} (cf. Fig. 1). The distribution of excess carriers drifts from x_{pu} towards x_{pr} for $U_{\text{lat}} = +20$ V and away from x_{pr} for $U_{\text{lat}} = -20$ V.

We use Eq. (3) to estimate v_{drift} from the measured and calculated densities at two different x_{pr}. A thorough analysis of the data yields $v_{\text{drift}} \approx 0.9$ μm/ns for both experiment and theory. Slight deviations from the small signal case results in a slightly time-dependent drift velocity in the deviation from the theoretical small signal drift velocity $v_{\text{drift}} \approx 1.4$ μm/ns obtained by inserting n_0 and p_0 into the term $\mu_{\text{ambi}}E^{\text{ext}}$ in Eq.(1).

More details about the described experiments can be found in [3,4].

References

1. K. H. Gulden et al.,Phys. Rev. Lett. **66,** (1991) 373.
2. G. H. Döhler, Phys. Stat. Solidi **52,** (1972) 79.
3. D. Streb et al., Superlatt. and Microstr. **25,** (1999) 21.
4. D. Streb, *Ambipolarer Transport in Halbleiter-Schichtstrukturen,* Eds.: T. Marek, S. Malzer, P. Kiesel (Universität Erlangen-Nürnberg, 2000), ISBN 3-932392-25-6

Localization of carriers and Bloch oscillations in quantum dot superlattices in dc electric field

R.A. Suris[1], I.A. Dmitriev[2]

[1] A.F.Ioffe Physical-Technical Institute, 194021, Polytechnicheskaya 26, St.Petersburg, Russia
[2] St.Petersburg State Technical University, 194021, Polytechnicheskaya 29, Russia

Abstract   Analysis of localization of electrons in ideal 2D and 3D Quantum Dot Superlattices (QDSL) in homogeneous dc electric field and Bloch oscillations in such structures is presented. In our consideration we suppose that electric field and tunneling matrix elements between dots are small enough to describe QDSL in single miniband approach.

The spectrum and wave functions of dc biased QDSL

The spectrum of ideal QDSL looks unusual owing to its very strong dependence on field orientation.

For the rational directions of field (when ratio of field projections on basic vectors of QDSL is rational) chains of dots (or planes in 3D QDSL) with the same electric potential appear in directions, normal to the field (Fig. 1). Neglecting of tunneling of carriers between quantum dots in these chains (planes) leads to degenerate localized states. Carrier tunneling removes the degeneracy and creates transverse minibands with widths exponentially depending on field orientation with respect to the QDSL axes: the larger is the distance between quantum dots in the chain (plane) the less is the miniband width (Fig. 2).

$$E_N(k_\perp) = 2\Delta_\perp \cos(k_\perp d_\perp) - NeFd_\parallel \qquad (1)$$

We underline the nontrivial fact that only transverse tunneling matrix elements can remove the degeneracy though they can be much smaller then matrix elements between adjacent quantum dots.

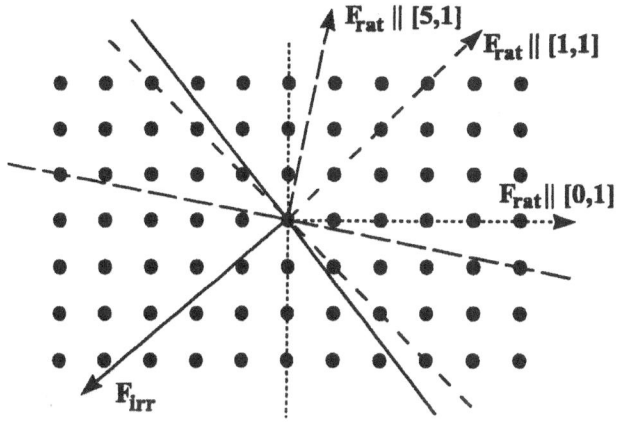

Fig. 1 Rational and irrational orientations of field for 2D QDSL.

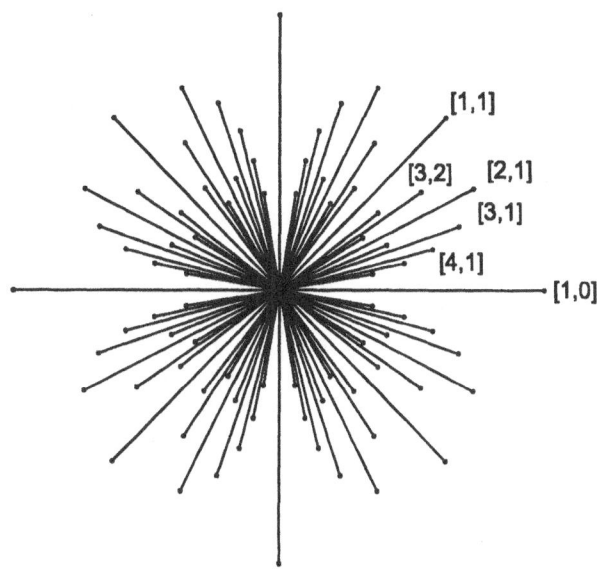

Fig. 2 The transverse miniband width dependence on rational field orientation (in logarithmic scale).

For irrational field orientations QDs are not in resonance (Fig. 1), spectrum is pure discrete and forms 2D (3D) Stark ladder:

$$E_{N1,N2} = N_1 eF\mathbf{a}_1 + N_2 eF\mathbf{a}_2, \qquad (2)$$

where a_1, a_2 are basic QDSL vectors (2D case).

Near rational field orientations the abrupt modifications of localizaton area occur (from localised in transverse directions states to delocalizated ones and vice versa).The smaller is the transverse resonance tunneling matrix element, corresponding to a rational field orientation, the sharper is this modification.

We see that for ideal QDSL in quantizing electric field an infinitely small variation of field orientation leads to the qualitative modification of spectrum and wave functions. So it's time to say a few words about changes that will come into play in real structures.

First of all, the real structures are finite, so number of rational orientations is limited.

The second is a technological dispersion of QD sizes and distances between QD, which for infinite 2D QDSL must lead to localization in transverse (one-dimensional) minibands. It's shown [1] that for diagonal disorder in a chain electron localizes on the length $l_{loc} = d \, 4\Delta^2 / < \delta\varepsilon^2 >$, where d is chain period, $< \delta\varepsilon^2 >$ is mean square dispersion of quantization levels. We can conclude that for given rational field orientation in real 2D QDSL may be realized one of three situations:

for strong disorder $l_{loc} < d$ and transverse tunneling matrix elements do not play role in formation of spectrum and wave functions (this case corresponds to full localization in isolated transverse chain);

for mediate disorder transverse localization length is Anderson localization length, and here it is interplay between resonance tunneling and localization, and,

at least, for weak disorder localization length is whole transverse QDSL size. Only in the last case we can expect the density of states close to unperturbed one and use the ideal QDSL description.

Furthermore, in our consideration we ignored Coulomb interaction, which for high carrier concentrations also must lead to localization of carriers.

We see that in real 2D QDSL delocalization (in the meaning we noticed above) may occur only for such few rational field orientations, which provide the widest transverse bandwidth (Fig. 2). In 3D QDSL the direction of field can be normal to whole planes of QDs. In this case delocalization conditions are not so strict.

Bloch oscillations

In this part we describe the Bloch oscillations, arising in QDSL after abrupt switching on the electric field. For such initial conditions semiclassical description appears to be correct for the whole fields range. For given QDSL miniband spectrum $E(\mathbf{k}) = \sum 2\Delta_{\mathbf{R}} \cos(\mathbf{kR})$, $\mathbf{R} = \sum N_i \mathbf{a}_i$, we arrive at the expression for current oscillations:

$$\mathbf{j} = \sum_R \frac{2en_e\Delta_R \mathbf{R}}{\hbar} < \cos \mathbf{kR} > \sin(\Omega_R t), \qquad (3)$$

where $\Omega_R = e\mathbf{FR}/\hbar$ are Stark frequinces of QDSL.

As tunneling matrix elements exponentially decrease with distance between quantum dots, oscillations perform at two (or three in 3D QDSR) main frequencies, which can be independently tuned by applied field value and orientation [2]. Other harmonics frequencies are various combinations of the main ones. Amplitudes of these harmonics are exponentially smaller. Oscillations are periodic for rational field orientations, when ratio of frequencies is rational number, and quasiperiodic for irrational orientations of field.

The damping of Bloch oscillations

In quantum well superlattices observation of Bloch oscillations meet many obstacles due to fast dephasing processes. The wide transverse spectrum of quantum well provides fast optical and acoustic phonon and elastic scattering of electrons for any field value.

In QDSL we have a possibility to vary transverse miniband width by means of appropriate choice of field orientation and thus eliminate elastic and optical phonon scattering. For that transverse miniband width should

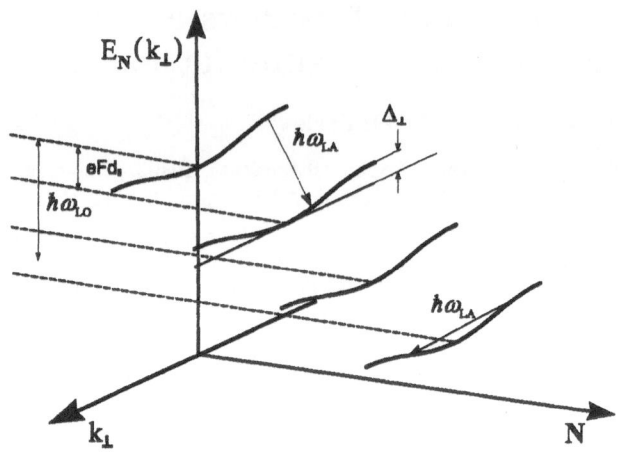

Fig. 3 Scattering processes in QDSL.

be smaller than optical phonon energy and distance between adjacent steps of Stark ladder, optical phonon frequency should not be close to an integer number of Stark frequences (Fig. 3).

Furthermore, the acoustic phonon scattering also can be significantly suppressed. One can easily make sure that for very narrow transverse minibands intraminiband acoustic phonon scattering is forbidden due to energy and transverse momentum conservation laws.

Wavelength of phonons, participating in interminiband acoustic phonon scattering in quantizing electric field, can be much smaller than QD size. It's obvious that interaction of such short-wave phonons with electrons should be very weak.

At least, when distance between Stark ladder steps exceeds optical phonon frequency, all one-phonon processes become forbidden and damping of Bloch oscillations is possible only due to multi-phonon processes.

We should notice that even in absence of scattering in real QD structures the dispersion of Stark frequencies will lead to damping of oscillations. So the dispersion should be much smaller than Stark frequency for acceptable field values.

The work was supported by Russian Foundation for Basic Research (Grants No 99-02-16796 and 00-15-96812) and by Program "Physics of Solid State Nanostructures" (Project No 97-1035)

References

1. A.P. Dmitriev, Zh.Eksp.Teor.Fiz.,**95**, (1989)234
2. R.A. Suris,*Prospects for Quantum Dot Applications in Electronics and Optoelectronics, in Future Trends in Microelectronics, Reflections on the Road to Nanotechnology*, NATO ASI Series, Series E: Applied Sciences, edited by S. Luryi, J. Xu an A. Zaslavski (Dordrecht/ Boston/ London, Kluwer Academic Publishers, 1996), **323**, p. 197

A novel quantum transport mechanism in biased quantum-box superlattices under terahertz irradiation

P. Kleinert[1], V. V. Bryksin[2]*

[1] Paul-Drude-Institut für Festkörperelektronik, Hausvogteiplatz 5-7, 10117 Berlin, Germany, e-mail: kl@pdi-berlin.de
[2] Physical Technical Institute, Politekhnicheskaya 26, 194021 St. Pertersburg, Russia

Abstract   The influence of strong terahertz irradiation on nonlinear quantum transport in semiconductor superlattices is studied within the double-time Kadanoff-Baym Keldysh non-equilibrium Greens function theory. By going beyond the Kadanoff-Baym ansatz, we predict the existence of a new photon-induced phononless current contribution, which has no counterpart in former density-matrix approaches. This current density is intimately related to the double-time character of the correlation functions and reveals the hopping nature of the transport in narrow miniband superlattices.

1 Introduction

Nonlinear carrier dynamics in semiconductor superlattices (SLs) driven by strong dc and ac electric fields has been the subject of intense experimental and theoretical studies. At sufficiently high dc electric field strengths, the miniband of extended states splits into a Wannier-Stark ladder. Consequently, the energy spectrum of a one-dimensional transport model or a SL under a quantizing magnetic field becomes completely discrete. In these molecular like systems, scattering giving rise to lifetime broadening plays a fundamental role. Without any collisional broadening, the current-voltage characteristics becomes completely singular with δ-like peaks at resonance positions. The non-perturbative treatment of scattering is indispensable, because physical quantities depend non-analytically on the coupling constant of the respective scattering process. In addition, the quantum-kinetic properties of carriers and its energy spectrum depend on each other in a self-consistent manner. This requires the consideration of the double-time nature of correlation functions beyond the Kadanoff-Baym (KB) ansatz. We identify a specific phononless current contribution, which appears only when these quantum correlations are explicitly taken into account. Most former transport studies relied on the KB ansatz and are therefore not able to identify this novel photon-assisted current, which is characterized by a deep relationship between the carrier spectrum and its kinetic properties. Our study addresses a topic of fundamental interest, which refers to the interplay between the quantum localization of eigenstates induced by strong fields and their scattering mediated decay.

* *Acknowledgements*: The authors acknowledge partial financial support by the Deutsches Zentrum für Luft- und Raumfahrt.

2 Description of the approach

We treat nonlinear carrier transport in the lowest narrow miniband of a SL driven by strong homogeneous dc and ac electric fields applied along the SL axis. It is assumed that the carriers remain essentially within the lowest miniband and the electron gas is non-degenerate in the limit of low carrier density. The non-equilibrium Kadanoff-Baym-Keldysh Greens function (GF) approach [1] is used to calculate the double-time GFs $G^{\lessgtr}$ from the Dyson equation written down in the wave-number representation. Crucial for drastic simplifications of the Dyson equation is a strict consideration of the following symmetry properties of the correlation functions

$$G^{\lessgtr}(\mathbf{k}, t \mid \mathbf{k}', t') = G^{\lessgtr}(\mathbf{k} \mid t, t')\delta(\mathbf{k}' - \mathbf{k} - \mathbf{A}(t', t)), \quad (1)$$

where $\mathbf{A}(t)$ is the vector potential of the external fields $\mathbf{E}(t) = \mathbf{E}_{dc} + \mathbf{E}_{ac} \cos \omega_{ac} t$ $(d\mathbf{A}(t)/dt = e\mathbf{E}(t)/\hbar)$ and $\mathbf{A}(t', t) = \mathbf{A}(t') - \mathbf{A}(t)$. The symmetry between the GFs defined on the upper and lower time branch becomes particularly transparent for functions, which are introduced by

$$\widetilde{G}^{\lessgtr}(\mathbf{k} \mid t, t') = G^{\lessgtr}(\mathbf{k} - \frac{1}{2}\mathbf{A}(t', t) \mid t, t'). \quad (2)$$

These GFs satisfy the symmetry relation

$$\widetilde{G}^{\lessgtr}(\mathbf{k} \mid t, t')^* = -\widetilde{G}^{\lessgtr}(\mathbf{k} \mid t', t). \quad (3)$$

This relation is most important in our approach and has been exploited in our previous study of stationary quantum transport [2]. Both $\widetilde{G}^>$, which is closely related to the density of states, and $\widetilde{G}^<$, which leads to the carrier distribution function, are calculated from a coupled set of self-consistent equations. This approach generalizes former treatments based on the density matrix [3] or the Boltzmann equation [4], where the distribution function $f(\mathbf{k} \mid T)$ has been determined as a function of the center-of-mass time T by considering its periodicity $(f(\mathbf{k} \mid T + 2\pi/\omega_{ac}) = f(\mathbf{k} \mid T))$. However, the microscopic treatment of collisional broadening requires the introduction of an additional time variable, namely the difference time t, which is associated with the finite duration of all scattering events. To account for the self-consistent interplay between the carrier kinetics and its spectrum, one has to solve an additional integral equation for $f(\mathbf{k} \mid T, t)$ with respect to the t dependence, which appears just beyond the KB ansatz. $f(\mathbf{k} \mid T, t)$ or its Fourier transform $f_m(\mathbf{k} \mid \omega)$ relates the GFs $\widetilde{G}^>$ and $\widetilde{G}^<$ to each other in a self-consistent manner without returning to the quasi-equilibrium situation, when the t dependence of the distribution function is dropped. Starting from the Dyson equation and the self-consistent Born approximation for elastic scattering, analytical solutions for $\widetilde{G}^>$ and $\widetilde{G}^<$ were derived for narrow-band SLs subject to strong electric dc and ac fields. From this solution, we calculated a photon-stimulated, phononless current contribution, which dominates when inelastic scattering is suppressed. This novel quantum-mechanical transport mechanism is intimately related to the double-time nature of the correlation functions and has no counterpart in the density-matrix theory as long as this approach relies implicitly on the KB ansatz.

3 Numerical results and discussion

We focus on a photon-induced, phononless steady-state current contribution, which appears only due to the scattering mediated finite width of initial and final Wannier-Stark states. The collisional broadening due to elastic scattering must be treated beyond perturbation theory. Numerical results are obtained for the sequential tunneling regime, when a characteristic thermalization time is much shorter than the hopping time.

Figure 1 shows the relative current density as a function of the dc electric field. When the ac field vanishes (dash-dotted line), this photon-induced stationary current contribution becomes non-zero only below the field strength $E_{dc} = 4\hbar\sqrt{U}/ed$ (indicated by an arrow in Fig. 1), where our high-field approach is no longer applicable (U is the strength of the elastic impurity scattering and d the SL period). This is in line with the expectation that the phononless current disappears, when the ac field is switched off. This holds true independently of the dc electric field strength. As a consequence, the photon-induced, phononless current $j_z(E_{dc}, E_{ac})$ cannot be expressed by the Tien-Gordon type formula [5] of the form

$$j_z(E_{dc}, E_{ac}) = \sum_{k=-\infty}^{\infty} J_k^2\left(\frac{eE_{ac}d}{\hbar\omega_{ac}}\right) j_z^{(dc)}(eE_{dc}d + k\hbar\omega_{ac}),$$

(4)

where $j_z^{(dc)}$ is the current density at $E_{ac} = 0$ and J_k the Bessel function. Indeed, if $E_{ac} = 0$, we obtain $j_z(E_{dc}, 0) = j_z^{(dc)}(eE_{dc}d)$ for all E_{dc} so that according to Eq. (4) also the photon-induced current at $E_{ac} \neq 0$ would vanish. This stresses the fact that we identified a new photon-induced transport mechanism, because almost all former approaches relied on Eq. (4).

In Fig. 1, the solid (dashed) line shows the photon-induced relative current density, which is obtained at $E_{ac} = 10$ kV/cm when $\omega_{ac}/\omega_0 = 0.1$ (0.2) (ω_0 is the frequency of polar-optical phonons, which serve only as a frequency scale). Photon-assisted current resonances occur at $E_{ac} = k\hbar\omega_{ac}/ed$ (with k being any integer) indicated by thin vertical lines. These sharp features in the current-voltage characteristics are separated by real current gaps, which disappear only when inelastic scattering

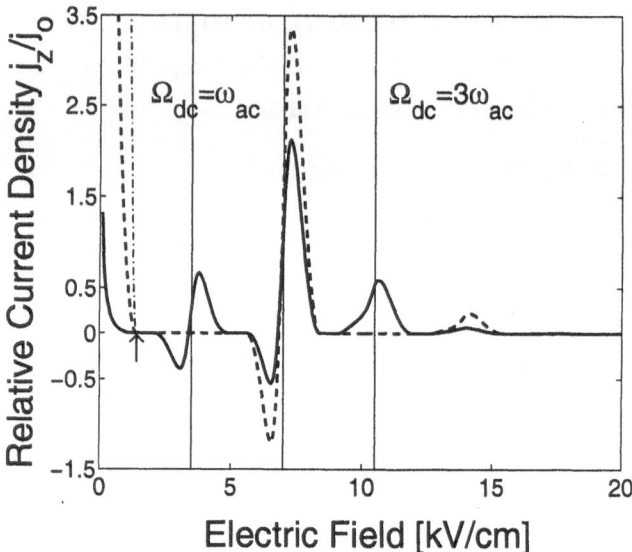

Fig. 1 Photon-induced, phononless stationary current density (measured in units of a field-independent reference current density j_0 given by $j_0 = en_s\Delta^2/16\hbar^2\omega_0$, where n_s is the sheet density, Δ the miniband width, and ω_0 the frequency of polar-optical phonons, which we chose as our frequency scale) as a function of the dc electric field for $\omega_{ac}/\omega_0 = 0.1$ (solid line) and $\omega_{ac}/\omega_0 = 0.2$ (dashed line). The field strength of the ac field is $E_{ac} = 10$ kV/cm for the solid and dashed lines and $E_{ac} = 0$ for the dash-dotted line (only shown up to the arrow, because otherwise the current disappears), respectively. Lifetime broadening is included via elastic impurity scattering, the strength of which is $\sqrt{U}/\omega_0 = 0.01$. The broadening of the Bloch resonance is considered by a phenomenological scattering time of $\tau_{ac} = 1$ ps, and the temperature is 4 K.

gives rise to some background current. The existence of current gaps reveals the hopping nature of the carrier transport. Furthermore, Fig. 1 shows that the current resonances exhibit some internal structure. These features depend sensitively on the parameter τ_{ac}, which describes the damping of current resonances induced by the radiation field. At some dc electric field strengths, the current becomes negative due to dominating photon absorption, which provides the energy for the upward carrier motion in the Wannier-Stark ladder.

From a technical point of view, our approach can cope with both the quasi-classical transport and the quite different hopping transport regime, which becomes relevant when the states are strongly localized by the external fields.

References

1. L. V. Keldysh, Zh. Eksp. Teor. Fiz. **47**, (1964) 1515.
2. V. V. Bryksin and P. Kleinert, J. Phys. A **33**, (2000) 233.
3. V. V. Bryksin and P. Kleinert, Phys. Rev. B **59**, (1999) 8152.
4. O. M. Yevtushenko, Phys. Rev. B **54**, (1996) 2578.
5. P. K. Tien and J. P. Gordon, Phys. Rev. **129**, (1963) 647.

On the spectra of field and current oscillations due to laser irradiation in superlattices

Yu.A. Romanov, Ju.Yu. Romanova

Institute for Physics of Microstructures RAS, N.Novgorod, 603600, Russia

Abstract The nonlinear oscillations of field and current in semiconductor superlattices excited by a terahertz laser radiation are studied within a self-consistent multifrequency internal field approach. It is shown that the commonly used approximation of a single frequency internal field is inadequate, especially for superlattices with a high electron concentration.

In study of the current harmonics generation in semiconductor superlattices (SL) under a terahertz laser radiation it is generally assumed that the field inside them is specified as a single-frequency one [1]. This approach is good for SL with a low concentration of electrons ($\omega_0 << \omega_1$, ω_0 is the plasma frequency, ω_1 the frequency of the external field). With the high electron concentration SL the nonlinear current excited by the "original" harmonic field induces rather strong fields at multiple frequencies, which in their turn largely change all of the current harmonics. The factors still more important for their effect on the spectrum of oscillations in a superlattice may be the dissipative and parametric instabilities, which give rise to generation of field harmonics with frequencies that are both multiple to and fractional and incomparable with the frequency of an external source. In this case it is fairly often that a static field appears spontaneously in a SL. The collective impact of these two factors may result in a considerable broadening and enrichment of the spectra from the current excited in SL and the radiation emerging from the SL. In particular, spontaneous generation of a static field in an SL triggers further generation of the even harmonics of radiation and new hysteretic effect. This calls for a necessity of investigating the behavior of SL electron plasma in a self-consistent multifrequency field.

The electrodynamic system is similar to the one proposed in [1], in which the laser radiation was coupled to the superlattice by a broad band bow-tie antenna. The self-consistent set of equations include the Boltzmann kinetic equation within a relaxation time approximation for collisions (τ),the the continuity equation for the total current and the condition that the static current is zero.

If all harmonics of the field inside the SL, except for the fundamental one and the static field, are negligibly small, i.e. $E(t) = E_c + E_1 cos(\omega_1 t + \varphi)$, then they connect with external field $E_0 cos(\omega_1 t)$, by equations set:

$$V_0 exp(i\varphi) = g_1 - 2w \sum_{n=-\infty}^{\infty} \frac{J_n(g_1)J_{n+1}(g_1)}{1 + i(\Omega_c + n\omega_1)\tau},$$

$$\sum_{n=\infty}^{\infty} \frac{\Omega_c + n\omega_1}{1 + i(\Omega_c + n\omega_1)^2\tau^2} J_n^2(g_1) = 0, \qquad (1)$$

where $w = (\omega_0/\omega_1)^2$, $\Omega_c = eE_c d/\hbar$ is the Stark frequency, d is the SL period, $g_1 = eE_1 d/\hbar\omega_1$, $J_n(g_1)$ is the n-th order Bessel function, V_0 dimensionless external field amplitude, φ is the phase shift of the fundamental harmonic of the SL field relative to the external field. Fig.1 shows dependences $g_1(V_0)$ at $\omega_1\tau = 10$ and $E_c = 0$ for different electron concentrations. The hysteretic character of this dependence (which isn't usefull at high w) leads to multivaluedness of radiation spectra.

Fig.2 offers the voltage and current spectra for three steady states arising in SL with $\omega_1\tau = 10$, $w = 2.5$ for the same value of $V_0 = 4.5$, but at different initial conditions. The appearance of these three states is unrelated to the multivaluedness of the fundamental harmonic of the internal field, described by Eq.(1) (see Fig.1). (It is nearly the same in all three states). They appear by generation of the third subharmonic (the parametric resonance -a)) and by spontaneous generation of a static field (dissipative instability - b),c)),which enriches the oscillation spectrum with even harmonics. The last largely reduce the third harmonic amplitude in the current oscillations spectrum, which has to be taken into account in experimental investigations.

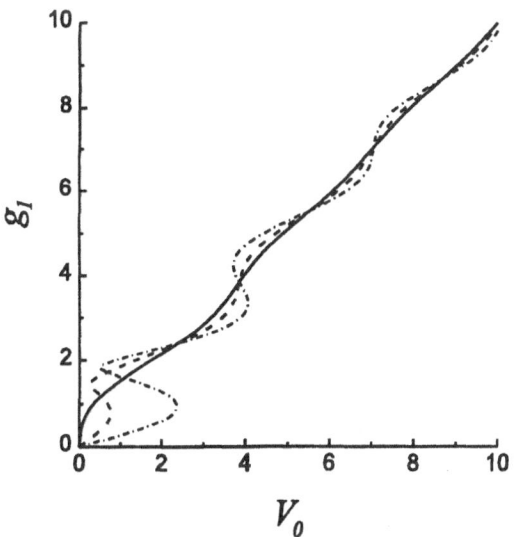

Fig. 1 Dependences of the amplitude of the fundamental harmonic of field in a SL from external field amplitude at $\omega_1\tau = 10$ for different values of w. $w = 1$(solid), $w = 2.5$ (dush), $w = 5$ (dush-dot).

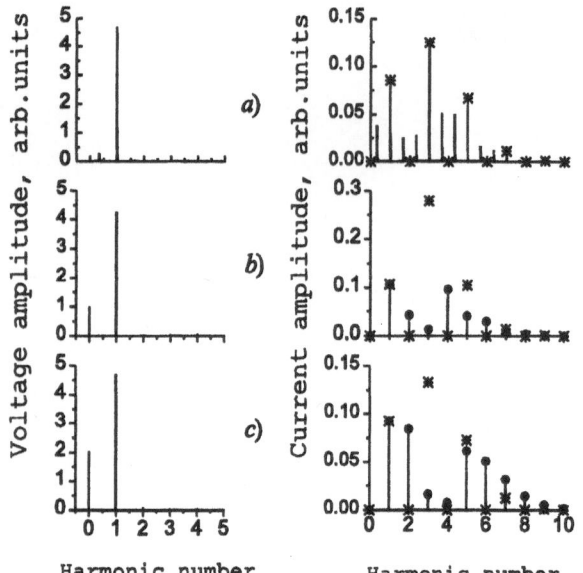

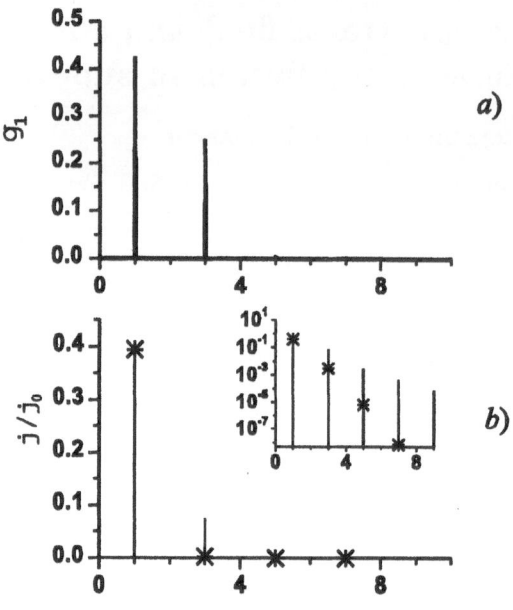

Fig. 2 The voltage and current spectra for three steady states in SL with $\omega_1\tau = 10$, $w = 2.5$ for $V_0 = 4.5$, a- nonlinear oscillation with triple period and without static field; b and c - nonlinear oscillations with the static field $g_c = -0.99$ and $g_c = -2.025$ respectively. Solid line is exact solution, stars - results of rough calculations (single-frequency field in SL) using formula (1), and circle is the same including the static field.

Fig. 3 The voltage (a) and current (b) spectra in SL with high electron concentration. $\omega_1\tau = 10$, $w = 10$, $V_0 = 3.5$. The insert shows current spectrum in logarithmic scale.

Another example showing why it is necessary to account for the frequency multiplicity of the internal field is given in Fig.3. High electron concentrations in SL (for typical SL parameters: $d = 100\text{Å}$, $\Delta = 0.018eV$ at room temperature it is $6 \cdot 10^{15}cm^{-3}$) caused parametric generation of a large-amplitude third harmonic of field, which determines much higher amplitude values of the upper current harmonics than is the case in the single-frequency limit. This strong parametric relation between the harmonics in this oscillations also causes in a decrease of the fundamental harmonic amplitude in the SL field. Analysis of the phase relationships of g_ω and $g_{3\omega}$ shows that their contributions in the higher harmonics currents are not in sinchronism. Due to the frequency multiplicity of the internal field one may, therefore, expect both expansion and narrowing in the spectra of the nonlinear current in SL.

This work was supported by the Interdisciplinary Science and Technology Program of Russia "Physics of Solid State Nanostructures" (Grant N 99-1129).

References

1. A.W.Ghosh, M.C.Wanke, S.J.Allen, J.W.Wilkins, Appl. Phys.Lett. **74**, (1999) 2164.

Springer Proceedings in Physics

1 *Fluctuations and Sensitivity in Nonequilibrium Systems*
Editors: W. Horsthemke and D. K. Kondepudi

2 *EXAFS and Near Edge Structure III*
Editors: K. O. Hodgson, B. Hedman, and J. E. Penner-Hahn

3 *Nonlinear Phenomena in Physics*
Editor: F. Claro

4 *Time-Resolved Vibrational Spectroscopy*
Editors: A. Laubereau and M. Stockburger

5 *Physics of Finely Divided Matter*
Editors: N. Boccara and M. Daoud

6 *Aerogels* Editor: J. Fricke

7 *Nonlinear Optics: Materials and Devices*
Editors: C. Flytzanis and J. L. Oudar

8 *Optical Bistability III*
Editors: H. M. Gibbs, P. Mandel, N. Peyghambarian, and S. D. Smith

9 *Ion Formation from Organic Solids (IFOS III)*
Editor: A. Benninghoven

10 *Atomic Transport and Defects in Metals by Neutron Scattering*
Editors: C. Janot, W. Petry, D. Richter, and T. Springer

11 *Biophysical Effects of Steady Magnetic Fields*
Editors: G. Maret, I. Kiepenheuer, and N. Boccara

12 *Quantum Optics IV*
Editors: J. D. Harvey and D. F. Walls

13 *The Physics and Fabrication of Microstructures and Microdevices*
Editors: M. J. Kelly and C. Weisbuch

14 *Magnetic Properties of Low-Dimensional Systems*
Editors: L. M. Falicov and J. L. Morán-López

15 *Gas Flow and Chemical Lasers*
Editor: S. Rosenwaks

16 *Photons and Continuum States of Atoms and Molecules*
Editors: N. K. Rahman, C. Guidotti, and M. Allegrini

17 *Quantum Aspects of Molecular Motions in Solids*
Editors: A. Heidemann, A. Magerl, M. Prager, D. Richter, and T. Springer

18 *Electro-optic and Photorefractive Materials*
Editor: P. Günter

19 *Lasers and Synergetics*
Editors: R. Graham and A. Wunderlin

20 *Primary Processes in Photobiology*
Editor: T. Kobayashi

21 *Physics of Amphiphilic Layers*
Editors: J. Meunier, D. Langevin, and N. Boccara

22 *Semiconductor Interfaces: Formation and Properties*
Editors: G. Le Lay, J. Derrien, and N. Boccara

23 *Magnetic Excitations and Fluctuations II*
Editors: U. Balucani, S. W. Lovesey, M. G. Rasetti, and V. Tognetti

24 *Recent Topics in Theoretical Physics*
Editor: H. Takayama

25 *Excitons in Confined Systems*
Editors: R. Del Sole, A. D'Andrea, and A. Lapiccirella

26 *The Elementary Structure of Matter*
Editors: J.-M. Richard, E. Aslanides, and N. Boccara

27 *Competing Interactions and Microstructures: Statics and Dynamics*
Editors: R. LeSar, A. Bishop, and R. Heffner

28 *Anderson Localization*
Editors: T. Ando and H. Fukuyama

29 *Polymer Motion in Dense Systems*
Editors: D. Richter and T. Springer

30 *Short-Wavelength Lasers and Their Applications*
Editor: C. Yamanaka

31 *Quantum String Theory*
Editors: N. Kawamoto and T. Kugo

32 *Universalities in Condensed-Matter*
Editors: R. Jullien, L. Peliti, R. Rammal, and N. Boccara

33 *Computer Simulation Studies in Condensed-Matter Physics: Recent Developments*
Editors: D. P. Landau, K. K. Mon, and H.-B. Schüttler

34 *Amorphous and Crystalline Silicon Carbide and Related Materials*
Editors: G. L. Harris and C. Y.-W. Yang

35 *Polycrystalline Semiconductors: Grain Boundaries and Interfaces*
Editors: H. J. Möller, H. P. Strunk, and J. H. Werner

36 *Nonlinear Optics of Organics and Semiconductors*
Editor: T. Kobayashi

37 *Dynamics of Disordered Materials*
Editors: D. Richter, A. J. Dianoux, W. Petry, and J. Teixeira

38 *Electroluminescence*
Editors: S. Shionoya and H. Kobayashi

39 *Disorder and Nonlinearity*
Editors: A. R. Bishop, D. K. Campbell, and S. Pnevmatikos

40 *Static and Dynamic Properties of Liquids*
Editors: M. Davidović and A. K. Soper

41 *Quantum Optics V*
Editors: J. D. Harvey and D. F. Walls

42 *Molecular Basis of Polymer Networks*
Editors: A. Baumgärtner and C. E. Picot

43 *Amorphous and Crystalline Silicon Carbide II: Recent Developments*
Editors: M. M. Rahman, C. Y.-W. Yang, and G. L. Harris

44 *Optical Fiber Sensors*
Editors: H. J. Arditty, J. P. Dakin, and R. Th. Kersten

45 *Computer Simulation Studies in Condensed-Matter Physics II: New Directions*
Editors: D. P. Landau, K. K. Mon, and H.-B. Schüttler

46 *Cellular Automata and Modeling of Complex Physical Systems*
Editors: P. Manneville, N. Boccara, G. Y. Vichniac, and R. Bidaux

47 *Number Theory and Physics*
Editors: J.-M. Luck, P. Moussa, and M. Waldschmidt